WIRELESS COMMUNICATIONS

WIRELESS COMMUNICATIONS

From Fundamentals to Beyond 5G

Third Edition

Andreas F. Molisch
Fellow, IEEE
University of Southern California, USA

Contents

Preface to the Third, Expanded and Completely Revised, Edition
From the Fundamentals to Beyond 5G

Wireless communications is one of the most important modern technologies and is interwoven with all aspects of our daily lives. When we wake up, we check social media, email, and news on our smartphones. Before getting up, we adjust the room temperature through a Bluetooth-connected thermostat. After we leave the house and activate the Wi-Fi security cameras, we order a rideshare on a phone app that recognizes our location through GPS and are driven to a factory where manufacturing robots are connected and controlled via 5G. And that is only the start of the day....

It is thus no wonder that wireless infrastructure, user devices, and networks are among the largest and most critical industries in most countries. As the demands for wireless services constantly increase, so do the requirements for new products and for engineers that can develop these products and bring them to market. Such engineers need a deep understanding of both the fundamentals that govern the behavior of wireless systems, the current standardized systems implementations, and more recent research developments that will influence the next generation of products.

The goal of this book is to help students, researchers, and practicing engineers to acquire, refresh, or update this knowledge. It is designed to lead them from the *fundamental principles and building blocks*, such as digital modulation, fading, and reuse of spectrum, to *more advanced technologies that underly modern wireless systems*, such as multicarrier and multiantenna transmission, to a description of the *standardized systems dominating 5G cellular, Wi-Fi, and short-range communications*, to the cutting-edge research that *will form the basis for beyond-5G systems*. In brief, the book aims to lead the reader **from the fundamentals to beyond 5G.**

Why This New Edition

The changing landscape of wireless systems over the past decade influenced both my research and my teaching. Thus, when - following the positive reception of the first two editions – the publisher asked me to develop a third edition, I decided to not only considerably expand the covered material, but also to completely revise and re-organize the whole book.

One of the key challenges in studying wireless communications is the amazing breadth of topics that interact, and the need for an understanding of *all* aspects that impact the performance of systems. This has always been true, but over the past years, the joint optimization of different "layers", from hardware to applications, has become even more important, thus, the breadth of knowledge a wireless engineer needs is constantly increasing. Consequently, the new edition covers topics that were consciously omitted in the earlier editions, such as transceiver hardware, scheduling, and localization.

Another challenge is that both practical wireless systems and the science on which they are based are constantly changing. Since 2010, when the second edition of this book appeared, wireless communications has undergone a number of revolutionary changes. The first was caused by the emergence of smartphones as the digital device of choice for the majority of people, replacing computers. A second revolution is the emergence of an all-pervasive Internet-of-Things, in which many billions of sensors, actuators, and computation devices are connected wirelessly. All this has led to a major shift of wireless communications from voice-centric to data-centric communication systems and considerably diversified the types of traffic, and the requirements in terms of latency, energy efficiency, and data rate. To satisfy all these new requirements, novel scientific ideas and technologies have been developed. Numerous new sections have been added to describe these developments. Furthermore, the data-centric nature of modern wireless communications is now reflected throughout the book.

Finally, the relevant wireless standards have changed profoundly in the past decade. The world of cellular has simplified: it is dominated by the 4G standard LTE and the 5G standard NR. This allowed to eliminate, or move to appendices, the description of several older standards, while giving more space for an in-depth description of LTE/NR. On the other hand, standards for IoT and low-power applications, in particular Bluetooth and Zigbee, have gained importance, and are now described in a new chapter.

Increasing the page size and page count, moving some material (including the end-of-chapter examples and the list of acronyms and symbols) to a website, and removing outdated material, allowed to add some 350 pages of completely new material. Some of the key additions of the book are:

- Extensions in scope:
  - Hardware and its impact on system design
  - Scheduling
  - Localization

- New scientific developments
  - Massive MIMO and distributed MIMO
  - NOMA
  - Interference alignment
  - Millimeter-wave and THz communications
  - OTFS and generalized multicarrier techniques
  - Spatial modulation, intelligent reflective surfaces, and OAM
  - Stochastic-geometry analysis of cellular networks
- New standards
  - 5G (3GPP New Radio) standard
  - Wi-Fi 5 (802.11ac) and Wi-Fi 6 (802.11ax) standards
  - A preview of Beyond 5G standards
  - Bluetooth, Bluetooth Low Energy, and Zigbee standards

Given the length to which the current edition already runs, it is inevitable that not all relevant topics could be covered, with the choice often reflective of how well such topic could be integrated into the overall framework – for example, satellite communications, while important, would require too much background on the principles and constraints of satellites to make an inclusion practical. Still, I strove to select the topics in this book to give the reader a well-rounded view of wireless communications systems and prepare them for working in research and product development of this vibrant area.

Synopsis

The book is divided into six parts. Part 1 provides a high-level overview of wireless communications. Chapter 1 first gives a taxonomy of different wireless services, and then describes the requirements for data rate, range, and energy consumption, for different applications, and different types of systems. Chapter 2 describes the basic challenges of wireless communications, like multipath propagation and limited spectrum resources. Chapter 3 provides a high-level overview of wireless systems that serves to put the various chapters in the remainder of the book into their proper context.

Part 2 is dedicated to antennas and propagation channels, i.e., the medium over which communications happens in wireless systems. A key fact about these channels is the existence of multiple propagation paths over which the signal can propagate from transmitter to receiver. Chapter 4 describes the fundamental propagation processes such as reflection and diffraction, which form the basis for deterministic descriptions of multipath channels. Chapters 5 and 6 then consider statistical descriptions, such as amplitude fading statistics (like Rayleigh fading), and delay dispersion. Chapter 7 gives specific models for propagation channels in different environments, covering path loss as well as wideband and directional models. Chapter 8 discusses antennas, which translate between electric signals in the transceivers, and the electromagnetic waves propagating over the channel. The part finishes with a discussion of channel sounding, i.e., techniques for measuring impulse responses and other characteristics of the channel.

Part 3 describes the techniques for communicating data between one transmitter and one receiver – in other words, the basics of conveying data on a single link. We start in Chapter 10 with a description of different modulation formats, their abstract representations (such as signal space), and characteristics (such as spectral efficiency), as well as a discussion of various popular modulation formats. Chapter 11 then describes demodulation techniques, starting with a discussion of optimum transceivers and then progressing to methods for computing error probabilities in different types of channels. Since typical error probabilities in fading channels are unacceptably high, the next two chapters discuss countermeasures. Diversity (discussed in Chapter 12) is used to avoid high errors in fading dips and increase the probability that information is successfully received. Alternatively, or in combination with it, error correction coding can to be used. Chapter 13 provides a survey of coding, starting with the fundamental performance limits, and then describing both classical (block codes, convolutional codes) and modern (turbo codes, LDPC codes, polar codes) techniques. The next two Chapters deal with techniques for communicating in delay-dispersive channels. Chapter 14 discusses equalizers, which are used to avoid intersymbol interference caused by delay dispersion in "standard" modulation techniques. Chapter 15 then discusses OFDM, a modulation format that is especially suited for operation in delay-dispersive channels, and which has become the dominant modulation format in both current cellular and Wi-Fi communications, as well as other related techniques that are under consideration for beyond 5G systems. An important way to reduce interference and improve data rate is based on the use of multiple antenna elements at transmitter and receiver. These so-called MIMO (or SIMO or MISO) techniques have become one of the most important and intensely researched topics in wireless over the past 20 years, and Chapter 16 provides an extensive description. Part 3 wraps up with a description of wireless transceiver hardware, and the impact that it has on systems performance.

Part 4 is dedicated to multiuser communications, reflecting the fact that in almost all circumstances, multiple wireless users want to operate in the same area. This problem can be broken into two parts: (i) multiple users want to talk to one central node (e.g., a base station), and (ii) multiple central nodes are in a given area, each talking with their own set of users. A variety of methods for the first problem are discussed in Chapters 18 and 19. Chapter 18 concentrates on the methods where each user gets exclusive use of a part of the available spectrum for part of the time; Chapter 19 deals with so-called spread-spectrum techniques where the signals from multiple users are on the air at the same time but encoded (spread) in a way that allows separation of the different signals. Another important question is *which specific* time-frequency resources to allocate to each user, a problem known as scheduling. While this question was not very relevant in second-generation voice-based systems, it is very consequential in modern data systems; it is thus treated in Chapter 20. Next, Chapter 21 considers the question of how multiple base stations can co-exist in a geographical area, leading to the cellular principle. Not only classical frequency reuse, but also modern techniques that improve spectral efficiency beyond this approach, are discussed. Chapter 22 describes how multiple antenna elements at the BS are used for multiuser communications, both in the context of a single cell (multiuser MIMO, massive MIMO) and multicell communications (Cooperative MultiPoint, distributed massive MIMO). Chapters 18–22 all deal with infrastructure-based systems, such as cellular or Wi-Fi systems, where (typically mobile) user equipment communicates with fixed infrastructure nodes. Chapter 23 finally treats ad-hoc networks, where nodes that all have equal capabilities (and are often all mobile) communicate with each other, without need for fixed infrastructure. Such systems are popular, e.g., for sensor networks and for emergency communications where infrastructure has been destroyed by natural disasters. The different nature of communicating gives rise to a number of new challenges whose solutions are discussed in this chapter.

Part 5 deals with special aspects of wireless systems that are either related to specific applications, or encompass advanced techniques. Chapters 24 and 25 deal with the encoding and transmission of speech and video, respectively. Since brute-force sampling and quantization of these types of information would lead to requirement for very high data rates, smart compression and encoding schemes have been developed over the past decades, which are surveyed here. Chapter 26 considers the adaptive use of spectrum, where a particular frequency band is not reserved for a particular operator/system, but rather multiple systems can co-exist, by being "considerate neighbors." Given the high value of spectrum, such cognitive radio techniques have obvious potential for more economical communications. Chapter 27 considers relaying and cooperative communications, where particular nodes help to forward information from the transmitter to the receiver; this can serve, e.g., to improve the coverage range of an infrastructure node. When multiple such helper nodes are available, network coding can help to improve the efficiency of transmission to multiple receivers. Chapter 28 then discusses advanced techniques for handling interference, such as multiuser detection, NOMA, and interference alignment. Finally, Chapter 29 discusses localization techniques – while not strictly a wireless communications approach, localization through GPS or other methods is built into many cellphones and other wireless devices, and location information supports various aspects of wireless communications systems. We thus survey both the fundamentals of wireless localization and tracking, as well as specific systems including GPS.

The last chapters, in Part 6, then finally describe wireless standards. Standardization of wireless systems has played an enormous role in their proliferation, allowing to combine the advantages of competition with those of economies of scale. After a brief introduction into the motivations and principle of standardization (Chapter 30), which also includes (as appendices) extensive description of still-used second- and third-generation systems GSM and WCDMA, we describe the dominant cellular standards LTE (Chapter 31) and 5G (Chapter 32), the standard for wireless local area networks Wi-Fi – 802.11 (Chapter 33), and two important standards for IoT applications, Bluetooth and Zigbee (Chapter 34). An outlook of possible topics and developments in Beyond 5G standards in Chapter 35 concludes this book.

A companion website (www.wiley.com/go/molisch) contains some material that is as useful, but which would have made the printed version of the book overly bulky. Most importantly, this includes the end-of-chapter exercises, the list of symbols, and the list of acronyms. Further, it includes the appendices to the various chapters.

How to Use This Book for Courses or Self-Study

The book is designed as a textbook for undergraduate and graduate courses, for self-study, as well as a reference book for researchers and engineers. Readers are assumed to have an understanding of elementary communication theory, like modulation/demodulation as well as of basic aspects of electromagnetic theory, though a brief review of these fields is given at the beginning of the corresponding chapters of the book. The material tries to get readers to a stage where they can read more advanced monographs, and even research papers; for all those readers who want to dig deeper, "further reading" sections at the end of the chapters list some key references. The text includes both mathematical formulations and intuitive explanations. I firmly believe that such a dual approach leads to the deepest understanding of the material. **To help the readers who first want to build a fundamental understanding, the book denotes (with a star ∗) the parts that cover advanced material and can be skipped without losing context with the later (unstarred) sections**. However, this does *not* mean that these parts are unimportant - on the contrary, they are often current and research-active topics. It is also noteworthy that Parts V and VI by definition cover advanced materials.

In addition to being a textbook, the text is also intended to serve as a reference for researchers and practitioners. For this reason, I have tried to make it easier to read isolated chapters. All acronyms are explained the first time they occur in each chapter (not just at their first occurrence in the book) and are further listed and explained on the companion website; a list of symbols (see the companion website) explains the meaning of symbols used in the equations. Also, frequent cross-references should help for this purpose.

Suggestions for Courses

While there is clearly too much material to be covered in a single course, instructors can select material for one-semester (or two-semester) courses, and adapt the selection to the level of the class and target audience. The book contains worked-out examples in the main text and a large number of homework exercises at the end of the chapters. Solutions to these exercises, as well as presentation slides, are available to instructors on the companion website of this book.

A typical undergraduate/early graduate course might cover

- Introduction/ types of systems (Sections 1.2–1.3) and basic problems of wireless (Chapter 2)
- Link budgets and system overview (Chapter 3)
- Basic propagation processes (Sections 4.1, 4.2.1, 4.3.1, 4.4.1)
- Statistical channel models (Sections 5.1–5.5, 5.6.1, 5.8.1), wideband channels (6.1, 6.2, 6.5.1, 6.5.2), and channel models (7.1.1–7.1.3, 7.2.1–7.2.3)
- Antennas (8.1.1, 8.1.2, 8.2.1, 8.3.1, 8.3.3, 8.3.6, 8.4.1, 8.4.2)
- Modulation (10.1, 10.2, 10.3.1–10.3.4) and demodulation (11.1, 11.2.1)
- Diversity (12.1, 12.2, 12.4.3)

- Channel coding (13.1.1–13.1.4, 13.2.1–13.2.3, 13.3.1, 13.3.2,13.10.1, 13.10.3)
- Equalizer (14.1, 14.2.1–14.2.2, 14.4)
- OFDM (15.1–15.4, 15.9)
- Multiple-antenna systems (16.1.1–16.1.5, 16.2.1-16.2.3, 16.2.9.1–16.2.9.1.2, 22.4.1, 22.5.1)
- Multiple access (18.1–18.3, 18.5, 19.1)
- Cellular principles (21.1, 21.2.1, 21.4.1)
- Wireless standards/5G (Chapter 30 and select sections from Chapter 31/32).

For advanced graduate courses, there is an almost infinite set of combinations that can be put together. Some example "advanced topics" courses are:

- *Wireless propagation channels and wireless transceiver hardware*: Chapters 4–9, 17
- *Multiantenna systems*: Sections 6.7, 7.3–7.7, 8.4, 9.6, Chapters 12, 16, 22, 31.3.5–31.3.6, 32.3.5–32.3.7, 33.3.3, 33.5.3–33.5.5
- *Network planning and spectrum management*: Chapters 20, 21, 23, 26, 27.1–27.4, Chapters 30–32
- *5G:* Sections 15.5–15.8, 13.5, 13.7, 13.8, 16.1.5–16.1.7, 16.2.5–16.2.10, Chapters 22, 31, 32

As mentioned, an extensive set of slides (~1000 pages) and a solution manual for the end-of-chapter exercises are available to instructors.

Acknowledgments for the Third Edition

Many people have helped in the process of writing this new edition. Special thanks to Antony Vetro, who updated the chapter on video coding (Chapter 25) he had written for the second edition. Several experts generously donated their time to review the new and/or revised material in this book: Katsuyuki Haneda (Chapter 7), Buon Kiong (Vincent) Lau (Chapter 8), Emanuele Viterbo (Chapter 13), Bashar Tahir (Sections 13.6–13.10, 18.2, and Chapter 28). Francois Rottenberg (Chapter 15), Ismail Guvenc (Sec. 15.10), Ronny Hadani (Sec. 15.12), Angel Lozano (Chapter 16), Hossein Hashemi (Chapter 17), Henrik Sjoland (Chapter 17), Lin Cai (Chapter 18), Li-Chun Wang (Chapter 20), Jang-Won Lee (Chapter 20), Harpreet Dhillon (Chapter 21), Erik Larsson (Chapter 22), Lee Swindlehurst (Chapter 22), Bhaskar Krishnamachari (Chapter 23), Frank Fitzek (Section 27.5), Robert Schober (Sec. 28.1-28.4), Syed Jafar (Section 28.5), Mike Buehrer (Chapter 29), Andrea Conti (Chapter 29), Ed Tiedeman (Chapter 30) Aris Papasakellariou (Chapter 31), Erik Dahlman (Chapter 32), Wanshi Chen (Chapter 32), Robert Stacey (Chapter 33), and two anonymous reviewers (Chapter 34). To all of them my profound gratitude. I would also like to reiterate my thanks to the reviewers of the first and second edition (see the acknowledgments for those editions); material reviewed by them is contained throughout the book. Of course, any remaining errors are mine. I also thank the students from my classes at USC, as well as readers and students from all over the world, who provided suggestions for corrections and improvements.

The development of the current edition was guided by Juliet Booker with great expertise and dedication. It has been truly a joy working with her, and she saved the day on many occasions. I would like to thank her and her team for their help, advice, and patience. The book was expertly copy-edited and typeset by Mahalakshmi Pitchai, Thanigaivel Thirumal. I thank Aagam Shah for translating many of the new drawings from my handwritten scribbles into nice digital drawings, Ruiyi Shen for formatting of the references and Sayantan Adhikary for creating solutions for many of the new end-of-chapter exercises.

No book is written in a vacuum, and certainly, this book would not have been written without the support of many people. The generosity of Andrew and Erna Viterbi has funded the chair I hold at USC. The support and encouragement of Yannis Yortsos (Dean of the engineering school), and Richard Lcahy, Sandeep Gupta, and Sandy Sawchuk (current and former chairs of the electrical engineering department) at USC, as well as governmental funding agencies and industry sponsors, have enabled my research and educational activities during the years I have been writing on this edition. My views of wireless systems have been strongly impacted by my research collaborations with colleagues at USC and all over the world, as well as my PostDocs and graduate students. To all of them goes my gratitude. Most of all, I want to thank Betty for her support, patience, encouragement, and humor during the long process of writing this book.

Errata Corrections

On the topic of errors: despite great efforts, multiple proofreadings, and the invaluable input of the reviewers, it is almost unavoidable that a tome like this contains misprints and errors. Those that are discovered will be posted on my website, https://wides.usc.edu/students.html#textbooks. If you discover errors, please email them to Andreas.Molisch@ieee.org.

Disclaimer

Information contained in this work has been obtained by the author from sources believed to be reliable. However, neither the author nor the publisher guarantees the accuracy or completeness of any information published herein, and to the fullest extent of the law, neither author nor publisher shall be responsible for any errors, omissions, or damages arising out of use of this information, nor do they assume any liability for any injury and/or damage to persons or property as a matter of products liability, negligence or otherwise, or from any use or operation of any methods, products, instructions, or ideas contained in the material herein. Practitioners and researchers must always rely on their own experience and knowledge in evaluating and using any information, methods, compounds, or experiments described herein. In using such information or methods they should be mindful of their own safety and the safety of others, including parties for whom they have a professional responsibility. This work is published with the understanding that the author and the publisher are supplying information but are not attempting to render engineering of other professional services. If such services are required, the assistance of an appropriate professional should be sought.

Preface and Acknowledgements to the Second Edition

Since the first edition of this book appeared in 2005, wireless communications research and technology continued its inexorable progress. This fact, together with the positive response to the first edition, motivated a second edition that would cover the topics that have emerged in the last years. Thus, the present edition aims to bring the book again in line with the breadth of topics relevant to state-of-the-art wireless communications engineering.

There are more than 150 pages of new material, covering

- cognitive radio (new Chapter 21);
- cooperative communications, relays, and ad hoc networks (new Chapter 22);
- video coding (new Chapter 23);
- 3GPP Long-Term Evolution (new Chapter 27);
- WiMAX (new Chapter 28).

There are furthermore significant extensions and additions on the following:

- MIMO (in Chapter 20), in particular a new section on multi-user MIMO (Section 20.3).
- IEEE 802.11n (high-throughput Wi-Fi) in Section 29.3.
- Coding (bit-interleaved coded modulation) in Section 14.5.
- Introduction to information theory in Sections 14.1, 14.9, and Appendix 17.
- Channel models: updates of standardized channel models in Appendix 7.
- A number of minor modifications and reformulations, partly based on feedback from instructors and readers of the book.

These extensions are important for students (as well as researchers) to learn "up-to-date" skills, Most of the additional material might be best suited for a graduate course on advanced wireless concepts and techniques. However, the material on LTE (or WiMAX) is also well suited as an example for standardized systems in a more elementary course (replacing, e.g., discussions of GSM or WCDMA systems).

As for the first edition, presentation slides and a solutions manual are available *for instructors that adopt the textbook for their course*. This material can also be obtained from the publisher or from a new website, wides.usc.edu/teaching/textbook. This site will also contain important resources for all readers of the book, including an "errata list," updates, additional references, and similar material.

The writing of the new material was a major endeavor and was greatly helped by the support of Sandy Sawchuk, Chair of the Department of Electrical Engineering at the University of Southern California. Particular thanks to Anthony Vetro, who wrote the new chapter on videocoding (Chapter 23). I am also grateful to the experts that kindly agreed to review the new material, namely Honggang Zhang and Natasha Devroye (Chapter 21), Gerhard Kramer, Mike Neely, Bhaskar Krishnamachari (Chapter 22), Erik Dahlman (Chapter 27), Yang-Seok Choi Hujun Jin, and V. Shashidar (Chapter 28), Guiseppe Caire (new material in Chapters 14 and 17), Robert Heath and Claude Oestges (new material of Chapter 20), and Eldad Perahia (Section 29.3). As always, responsibility for any residual errors lies with me.

I also thank the students from my classes at USC, as well as readers and students from all over the world, who provided suggestions for corrections and improvements. Exercises for the new chapters were created by Junyang Shen, Hao Fang, and Christian Mehlfuehrer. Thanks to Neelesh B. Mehta for providing me with several figures on LTE.

As for the first edition, Mark Hammond from J. Wiley acted as acquisition editor; Sarah Tilley was the production editor. Special thanks to Dhanya Ramesh of Laserwords for her expert typesetting.

Preface and Acknowledgements to the First Edition

When, in 1994, I wrote the very first draft of this book in the form of lecture notes for a wireless course, the preface started by justifying the need for giving such a course at all. I explained at length why it is important that communications engineers understand wireless systems, especially digital cellular systems. Now, more than 10 years later, such a justification seems slightly quaint and outdated. Wireless industry has become the fastest growing sector of the telecommunications industry, and there is hardly anybody in the world who is not a user of some form of wireless technology. From the ubiquitous cellphones, to wireless LANs, to wireless sensors that are proliferating – we are surrounded by wireless communications devices.

One of the key challenges in studying wireless communications is the amazing breadth of topics that impacts this field. Traditionally, communications engineers have concentrated on, for example, digital modulation and coding theory, while the world of antennas and propagation studies was completely separate – "and never the twain shall meet." However, such an approach does not work for wireless communications. We need an understanding of *all* aspects that impact the performance of systems and make the whole system work. This book is an attempt to provide such an overview, concentrating as it does on the physical layer of wireless communications.

Another challenge is that not only practical wireless systems, but also the science on which they are based is constantly changing. It is often claimed that while wireless systems rapidly change, the scientific basis of wireless communications stays the same, and thus engineers can rely on knowledge acquired at a given time to get them through many cycles of system changes, with just minor adjustments to their skill sets. This thought is comforting – and unfortunately false. For example, 10 years ago, topics like multiple-antenna systems, OFDM, turbo codes and LDPC codes, and multiuser detection, were mostly academic curiosities, and would at best be treated in PhD-level courses; today, they dominate not only mainstream research and system development, but represent vital, basic knowledge for students and practicing engineers. I hope that, by treating both new aspects as well as more "classical" topics, my book will give today's students and researchers knowledge and tools that will prove useful for them in the future.

The book is written for advanced undergraduate and graduate students, as well as for practicing engineers and researchers. Readers are assumed to have an understanding of elementary communication theory, like modulation/demodulation as well as of basic aspects of electromagnetic theory, though a brief review of these fields is given at the beginning of the corresponding chapters of the book. The core material of this book tries to get students to a stage where they can read more advanced monographs, and even research papers; for all those readers who want to dig deeper, the majority of chapters include a "Further Reading" section that cites the most important references. The text includes both mathematical formulations and intuitive explanations. I firmly believe that such a dual approach leads to the deepest understanding of the material. In addition to being a textbook, the text is also intended to serve as a reference tool for researchers and practitioners. For this reason, I have tried to make it easier to read isolated chapters. All acronyms are explained the first time they occur in each chapter (not just at their first occurrence in the book); a list of symbols (see p. xlvii) explains the meaning of symbols used in the equations. Also, frequent cross-references should help for this purpose.

Synopsis

The book is divided into five parts. The first part, the introduction, gives a high-level overview of wireless communications. Chapter 1 first gives a taxonomy of different wireless services, and then describes the requirements for data rate, range, energy consumption, etc., that the various applications impose. This chapter also contains a brief history, and a discussion of the economic and social aspects of wireless communications. Chapter 2 describes the basic challenges of wireless communications, like multipath propagation and limited spectrum resources. Chapter 3 then discusses how noise and interference limit the capabilities of wireless systems, and how link budgets can serve as simple system-planning tools that give a first idea about the achievable range and performance.

The second part describes the various aspects of wireless propagation channels and antennas. As the propagation channel is the medium over which communication happens, understanding it is vital to understanding the remainder of the book. Chapter 4 describes the basic propagation processes: free space propagation, reflection, transmission, diffraction, diffuse scattering, and waveguiding. We find that the signal can get from the transmitter to the receiver via many different propagation paths that involve one or more of these processes, giving rise to many multipath components. It is often convenient to give a statistical description of the effects of multipath propagation. Chapter 5 gives a statistical formulation for narrowband systems, explaining both small-scale (Rayleigh) and large-scale fading. Chapter 6 then discusses formulations for wideband systems and systems that can distinguish the directions of multipath components at the transmitter and receiver. Chapter 7 then gives specific models for propagation channels in different environments, covering path loss as well as wideband and directional models. Since all realistic channel models have to be based on (or confirmed by) measurements, Chapter 8 summarizes techniques that are used for measuring channel impulse responses.

Finally, Chapter 9 briefly discusses antennas for wireless applications, especially with respect to different restrictions at base stations and mobile stations.

The third part of the book deals with the structure and theory of wireless transceivers. After a short summary of the components of a RF transceiver in Chapter 10, Chapter 11 then describes the different modulation formats that are used for wireless applications. The discussion not only includes mathematical formulations and signal space representations, but also an assessment of their advantages and disadvantages for various purposes. The performance of all these modems in flat-fading as well as frequency-selective channels is then the topic of Chapter 12. One critical observation we make here is the fact that fading leads to a drastic increase in error probability, and that increasing the transmit power is not a suitable way of improving performance. This motivates the next two Chapters, which deal with diversity and channel coding, respectively. We find that both these measures are very effective in reducing error probabilities in a fading channel. The coding chapter also includes a discussion of near-Shannon-limit-achieving codes (turbo codes and low-density parity check codes), which have gained great popularity in recent years. Since voice communication is still the most important application for cellphones and similar devices, Chapter 15 discusses the various ways of digitizing speech, and compressing information so that it can be transmitted over wireless channels in an efficient way. Chapter 16 finally discusses equalizers, which can be used to reduce the detrimental effect of frequency selectivity of wideband wireless channels. All the chapters in this part deal with a single link – i.e., the link between one transmitter and one receiver.

The fourth part then takes into account our desire to operate a number of wireless links simultaneously in a given area. This so-called *multiple-access* problem has a number of different solutions. Chapter 17 discusses frequency domain multiple access (FDMA) and time domain multiple access (TDMA), as well as packet radio, which has gained increasing importance for data transmission. This chapter also discusses the cellular principle, and the concept of frequency reuse that forms the basis not only for cellular, but also many other high-capacity wireless systems. Chapter 18 then describes spread spectrum techniques, in particular CDMA, where different users can be distinguished by different spreading sequences. This chapter also discusses multiuser detection, a very advanced receiver scheme that can greatly decrease the impact of multiple-access interference. Another topic of Part IV is "advanced transceiver techniques." Chapter 19 describes OFDM (orthogonal frequency domain multiplexing), which is a modulation method that can sustain very high data rates in channels with large delay spread. Chapter 20 finally discusses multiple-antenna techniques: "smart antennas," typically placed at the base station, are multiple-antenna elements with sophisticated signal processing that can (among other benefits) reduce interference and thus increase the capacity of cellular systems. MIMO (multiple-input-multiple-output) systems go one step further, allowing the transmission of parallel data streams from multiple-antenna elements at the transmitter, which are then received and demodulated by multiple-antenna elements at the receiver. These systems achieve a dramatic capacity increase even for a single link.

The last part of the book describes standardized wireless systems. Standardization is critical so that devices from different manufacturers can work together, and also systems can work seamlessly across national borders. The book describes the most successful cellular wireless standards – namely, GSM (Global System for Mobile communications), IS-95 and its advanced form CDMA 2000, as well as Wideband CDMA (also known as UMTS) in Chapters 21, 22, and 23, respectively. Furthermore, Chapter 24 describes the most important standard for wireless LANs – namely, IEEE 802.11.

A companion website (www.wiley.com/go/molisch) contains some material that I deemed as useful, but which would have made the printed version of the book overly bulky. In particular, the appendices to the various chapters, as well as supplementary material on the DECT (Digital Enhanced Cordless Telecommunications) system, the most important cordless phone standard, can be found there.

Suggestions for Courses

The book contains more material than can be presented in a single-semester course, and spans the gamut from very elementary to quite advanced topics. This gives the instructor the freedom to tailor teaching to the level and the interests of students. The book contains worked examples in the main text, and a large number of homework exercises at the end of the book. Solutions to these exercises, as well as presentation slides, are available to instructors on the companion website of this book.

A few examples for possible courses include:

Introductory course:

- Introduction (Chapters 1–3).
- Basic channel aspects (Sections 4.1–4.3, 5.1–5.4, 6.1, 6.2, 7.1–7.3):
  - elementary signal processing (Chapters 10, 11, and Sections 12.1, 12.2.1, 12.3.1, 13.1, 13.2, 13.4, 14.1–14.3, 16.1–16.2);
  - multiple access and system design (Chapters 17, 22 and Sections 18.2, 18.3, 21.1-21.7).
- Wireless propagation:
  - introduction (Chapter 2);
  - basic propagation effects (Chapter 4);
  - statistical channel description (Chapters 5 and 6);
  - channel modeling and measurement (Chapters 7 and 8);
  - antennas (Chapter 9).
  This course can also be combined with more basic material on electromagnetic theory and antennas.

- Advanced topics in wireless communications:
  - introduction and refresher: should be chosen by the instructor according to audience;
  - CDMA and multiuser detection (Sections 18.2, 18.3, 18.4);
  - OFDM (Chapter 19);
  - ultrawideband communications (Sections 6.6, 18.5);
  - multiantenna systems (Sections 6.7, 7.4, 8.5, 13.5, 13.6, and Chapter 20);
  - advanced coding (Sections 14.5, 14.6).
- Current wireless systems:
  - TDMA-based cellular systems (Chapter 21);
  - CDMA-based cellular systems (Chapters 22 and 23);
  - cordless systems (supplementary material on companion website);
  - wireless LANs (Chapter 24); and
  - selected material from previous chapters for the underlying theory, according to the knowledge of the audience.

Acknowledgments to the First Edition

This book is the outgrowth of many years of teaching and research in the field of wireless communications. During that time, I worked at two universities (Technical University Vienna, Austria and Lund University, Sweden) and three industrial research labs (FTW Research Center for Telecommunications Vienna, Austria; AT&T (Bell) Laboratories–Research, Middletown, NJ, U.S.A.; and Mitsubishi Electric Research Labs., Cambridge, MA, U.S.A.), and cooperated with my colleagues there, as well as with numerous researchers at other institutions in Europe, the U.S.A., and Japan. All of them had an influence on how I see wireless communications, and thus, by extension, on this book. To all of them, I owe a debt of gratitude. First and foremost, I want to thank Ernst Bonek, the pioneer and doyen of wireless communications in Austria, who initiated this project, and with whom I had countless discussions on technical as well as didactic aspects of this book and the lecture notes that preceded it (these lecture notes served for a course that we gave jointly at TU Vienna). Without his advice and encouragement, this book would never have seen the light of day. I also want to thank my colleagues and students at TU Vienna, particularly Paulina Erätuuli, Josef Fuhl, Alexander Kuchar, Juha Laurila, Gottfried Magerl, Markus Mayer, Thomas Neubauer, Heinz Novak, Berhard P. Oehry, Mario Paier, Helmut Rauscha, Alexander Schneider, Gerhard Schultes, and Martin Steinbauer, for their help. At Lund University, my colleagues and students also greatly contributed to this book: Peter Almers, Ove Edfors, Fredrik Floren, Anders Johanson, Johan Karedal, Vincent Lau, Andre Stranne, Fredrik Tufvesson, and Shurjeel Wyne. They contributed not only stimulating suggestions on how to present the material but also figures and examples; in particular, most of the exercises and solutions were created by them and Section 19.5 is based on the ideas of Ove Edfors. A special thanks to Gernot Kubin from Graz University of Technology, who contributed Chapter 15 on speech coding. My colleagues and managers at FTW, AT&T, and MERL – namely, Markus Kommenda, Christoph Mecklenbraueker, Helmut Hofstetter, Jack Winters, Len Cimini, Moe Win, Martin Clark, Yang-Seok Choi, Justin Chuang, Jin Zhang, Kent Wittenburg, Richard Waters, Neelesh Mehta, Phil Orlik, Zafer Sahinoglu, Daqin Gu (who greatly contributed to Chapter 24), Giovanni Vanucci, Jonathan Yedidia, Yves-Paul Nakache, and Hongyuan Zhang, also greatly influenced this book. Special thanks and appreciation to Larry Greenstein, who (in addition to the many instances of help and advice) took an active interest in this book and provided invaluable suggestions.

A special thanks also to the reviewers of this book. The manuscript was critically read by anonymous experts selected by the publisher, as well as several of my friends and colleagues at various research institutions: John B. Anderson (Chapters 11–13), Anders Derneryd (Chapter 9), Larry Greenstein (Chapters 1–3, 7, 17–19), Steve Howard (Chapter 22), Thomas Kaiser (Chapter 20), Achilles Kogantis (Chapter 23), Gerhard Kristensson (Chapter 4), Thomas Kuerner (Chapter 21), Gerald Matz (Chapters 5–6), Neelesh B. Mehta (Chapter 20), Bob O'Hara (Chapter 24), Phil Orlik (Section 17.4), John Proakis (Chapter 16), Said Tatesh (Chapter 23), Reiner Thomae (Chapter 8), Chintha Tellambura (Chapters 11–13), Giorgia Vitetta (Chapter 16), Jonathan Yedidia (Chapter 14). To all of them goes my deepest appreciation. Of course, the responsibility for any possible remaining errors rests with me.

Mark Hammond as publisher, Sarah Hinton as project editor, and Olivia Underhill as assistant editor, all from John Wiley & Sons, Ltd, guided the writing of the book with expert advice and considerable patience. Manuela Heigl and Katalin Stibli performed many typing and drawing tasks with great care and cheerful enthusiasm. Originator expertly typeset the manuscript.

List of Abbreviations

It clarifies the abbreviations and acronyms in Chapters 1–30 and 35, as well as "standard use" acronyms in Chapters 31–34.

Acronyms that are specific to the standards considered in Chapters 31 (LTE), 32 (NR), 33 (Wi-Fi), and 34 (Bluetooth and Zigbee), as well as Appendix 30 (GSM, WCDMA, DECT) are given in dedicated glossaries at the end of those chapters.

List of Symbols

A list of symbols is available as PDF, Symb.pdf, at wiley.com/go/molisch.

This list gives a brief overview of the use of variables in the text. Due to the large number of quantities occurring, the same letter might be used in different chapters for different quantities. Those variables that are used only locally (in a single section), and explained directly at their occurrence, are not mentioned here.

About the Companion Website

This book has an Instructor Companion Website and a Student Companion Website:

www.wiley.com/go/molisch/wireless3e

Supplementary resources for this new edition include:

- Companion Site (Open Access)
  - List of Acronyms
  - List of Symbols
  - Exercises (by chapter)
  - Appendices (by chapter)
  - Selected color figures
  - Table corresponding updates between 2e and 3e
  - Errata and Updates
- Instructor Companion Website (Password Access)
  - Solutions Manual
  - Presentation Slides

Part I
Introduction

In the first part of this book, we give an introduction to the basic applications of wireless communications, as well as the technical problems inherent in this communication paradigm. After a brief history of wireless, Chapter 1 describes the different types of wireless services and works out their fundamental differences. The subsequent Section 1.3 looks at the same problem from a different angle: what data rates, ranges, etc., occur in practical systems, and especially, what combination of performance measures are demanded (e.g., what data rates need to be transmitted over short distances; what data rates are required over long distances?). Chapter 2 then describes the technical challenges of communicating without wires, putting special emphasis on fading and co-channel interference. Chapter 3 describes the most elementary problems of designing a wireless system, such as setting up a link budget in either a noise-limited or an interference-limited system. It also provides a high-level overview of wireless systems, so that the reader can place the subsequent chapters in the overall picture.

After studying this part of the book, the reader should have an overview of different types of wireless services, and understand the technical challenges involved in each of them. The solutions to those challenges are described in the later parts of this book.

Wireless Communications: From Fundamentals to Beyond 5G, Third Edition. Andreas F. Molisch.
© 2023 John Wiley & Sons Ltd. Published 2023 by John Wiley & Sons Ltd.
Companion website: www.wiley.com/go/molisch/wireless3e

1

Applications and Requirements of Wireless Services

Wireless communications is one of the big engineering success stories of the last 40 years – not only from a scientific point of view, where the progress has been phenomenal, but also in terms of market size and impact on society. Companies that were completely unknown 40, or even 10, years ago are now household names all over the world, due to their wireless products, and in several countries, the wireless industry is dominating the whole economy. Working habits, and even more generally the ways we all communicate and live our lives, have been changed by the possibility of being connected "anywhere, anytime."

For a long time, wireless communications had been associated with cellular *telephony*, as this was the biggest market segment and has had the highest impact on everyday lives. The ability to talk to a *person*, instead of calling a *location* (and hoping that the desired person was there) constituted a paradigm shift in communications. After about 2005, emphasis shifted to wireless *data connectivity* – in other words, smartphones and Wi-Fi-connected laptops. The ability to connect to the internet anytime, anywhere, has led to a significant change in working habits and mobility of workers – answering emails in a coffee shop, or sitting on a train, has become an everyday occurrence. Furthermore, personal communications, including use of social media, can now happen everywhere. But besides these widely publicized cases, a large number of less obvious applications have been developed, and are starting to change our lives. Wireless sensor networks monitor factories, wireless links replace the cables between computers and keyboards, and wireless positioning systems monitor the location of trucks that have goods identified by wireless Radio Frequency (RF) tags. This variety of new applications causes the technical challenges for the wireless engineers to become bigger with each day. This book aims to give an overview of the solution methods for current as well as future challenges.

Quite generally, there are two paths to developing new technical solutions: engineering driven and market driven. In the first case, the engineers come up with a brilliant scientific idea – without having an immediate application in mind. As time progresses, the market finds applications enabled by this idea.[1] In the other approach, the market demands a specific product and the engineers try to develop a technical solution that fulfills this demand. In this chapter, we describe these market demands. We start out with a brief history of wireless communications, in order to convey a feeling of how the science, as well as the market, has developed in the past 100 years. Then follows a description of the types of services that constitute the majority of the wireless market today. Each of these services makes specific demands in terms of data rate, range, number of users, energy consumption, mobility, and so on. We discuss all these aspects in Section 1.3. We wrap up this section with a description of the interaction between the engineering of wireless devices and the behavioral changes induced by them in society.

1.1 History

1.1.1 How It All Started

When looking at the history of communications, we find that wireless communications is actually the oldest form – shouts and jungle drums did not require any wires or cables to function. Even the oldest "electromagnetic" (optical) communications are wireless: smoke signals are based on the propagation of optical signals along a line-of-sight connection. However, wireless communications as we know it started only with the work of Maxwell and Hertz, who laid the basis for our understanding of the transmission of electromagnetic waves. It was not long after their groundbreaking work that Tesla demonstrated the transmission of information via these waves – in essence, the first wireless communications system. In 1898, Marconi made his well-publicized demonstration of wireless communications from a boat to the Isle of Wight in the English Channel. It is noteworthy that while Tesla was the first to succeed in this important endeavor, Marconi had the better public relations, and is widely cited as the inventor of wireless communications, receiving a Nobel prize in 1909.[2]

---

[1] The second chapter gives a summary of the main technical challenges in wireless communications – i.e., the basis for the engineering-driven solutions. Chapters 3–29 discuss the technical details of these challenges and the scientific basis, while Chapters 30–35 expound specific systems that have been developed in recent years.

[2] Marconi's patents were actually overturned in the 1940s.

Wireless Communications: From Fundamentals to Beyond 5G, Third Edition. Andreas F. Molisch.
© 2023 John Wiley & Sons Ltd. Published 2023 by John Wiley & Sons Ltd.
Companion website: www.wiley.com/go/molisch/wireless3e

In the subsequent years, the use of radio (and later television) became widespread throughout the world. While in everyday language we usually do not think of radio or TV as "wireless communications," they certainly are, in a scientific sense, information transmission from one place to another by means of electromagnetic waves. They can even constitute "mobile communications," as evidenced by car radios. A lot of basic research – especially concerning wireless propagation channels – was done for entertainment broadcasting. By the late 1930s, a wide network of wireless information transmission – though unidirectional – was in place.

1.1.2 The First Systems

At the same time, the need for bidirectional mobile communications emerged. Police departments and the military had obvious applications for such two-way communications and were the first to use wireless systems with closed user groups. Military applications drove a lot of the research during, and shortly after, the Second World War. This was also the time when much of the theoretical foundations for communications in general were laid. Claude Shannon's groundbreaking work *A Mathematical Theory of Communication* [Shannon 1948] appeared during that time and established the possibility of error-free transmission under restrictions for the data rate and the Signal-to-Noise Ratio (SNR). Some of the suggestions in that work, like the use of optimum power allocation in frequency-selective channels, have only recently been introduced into wireless systems.

The 1940s and 1950s saw several important developments: the use of *Citizens' Band* (CB) radios became widespread, establishing a new way of communicating between cars on the road. Communicating with these systems was useful for transferring vital traffic information and related aspects within the closed community of the drivers owning such devices, but it lacked an interface to the public telephone system, and the range was limited to some 100 km, depending on the power of the (mobile) transmitters. In 1946, the first mobile telephone system was installed in the U.S.A. (St. Louis). This system did have an interface to the *Public Switched Telephone Network* (PSTN), the landline phone system, though this interface was not automated, but rather consisted of human telephone operators. However, with a total of six speech channels for the whole city, the system soon met its limits. This motivated investigations of how the number of users could be increased, even though the allocated spectrum would remain limited. Researchers at AT&T's Bell Labs found the answer: the cellular principle, where the geographical area is divided into cells; different cells might use the same frequencies. To this day, this principle forms the basis for the majority of wireless communications.

Despite the theoretical breakthrough, cellular telephony did not experience significant growth during the 1960s. However, there were exciting developments on a different front: in 1957, the Soviet Union launched the first satellite (*Sputnik*) and the U.S.A. soon followed. This development fostered research in the new area of satellite communications.[3] Many basic questions had to be solved, including the effects of propagation through the atmosphere, the impact of solar storms, the design of solar panels and other long-lasting energy sources for the satellites, and so on. To this day, satellite communications is an important area of wireless communications (though not one that we address specifically in this book). Nowadays, the most widespread applications lie in satellite TV transmission, wireless speech/data transmission between relay stations on different continents, and satellite phones.

1.1.3 Analog Cellular Systems

The 1970s saw a revived interest in cellular communications. In scientific research, these years saw the formulation of models for path loss, Doppler spectra, fading statistics, and other quantities that determine the performance of analog telephone systems. A highlight of that work was Jakes' book *Microwave Mobile Radio* that summed up the state of the art in this area [Jakes 1974]. The 1960s and 1970s also saw a lot of basic research that was originally intended for landline communications but later also proved to be instrumental for wireless communications. For example, the basics of adaptive equalizers, as well as multi-carrier communications, were developed during that time.

For the practical use of wireless telephony, the progress in device miniaturization made the vision of "portable" devices more realistic. Companies like Motorola and AT&T vied for leadership in this area and made vital contributions. Nippon Telephone and Telegraph (NTT) established a commercial cellphone system in Tokyo in 1979. However, it was a Swedish company that built up the first system with large coverage and automated switching: up to that point, Ericsson AB had been mostly known for telephone switches while radio communications was of limited interest to them. However, it was just that expertise in switching technology and the (for that time, daring) decision to use digital switching technology that allowed them to combine different cells in a large area into a single network, and establish the *Nordic Mobile Telephone* (NMT) system [Meurling and Jeans 1994]. Note that while the switching technology was digital, the radio transmission technology was still analog, and the systems became therefore known as *analog* systems (also called first-generation, 1G, systems). Subsequently, other countries developed their own analog phone standards. The system in the U.S.A., e.g., was called *Advanced Mobile Phone System* (AMPS) and achieved considerably uptake.

An investigation of NMT also established an interesting method for estimating market size: business consultants equated the possible number of mobile phone users with the number of Mercedes 600 cars (the top-of-the-line luxury car at that time) in Sweden. Obviously, mobile telephony could never become a mass market, could it? Similar thoughts must have occurred to the management of the inventor of cellular telephony, AT&T. Upon advice from a consulting company, they decided that mobile telephony could never attract a significant number of participants and stopped business activities in cellular communications.[4]

---

[3] Satellite communications – specifically by geostationary satellites – had already been suggested by science fiction writer Arthur C. Clark in the 1940s.
[4] These activities were restarted in the early 1990s, when the folly of the original decision became clear. AT&T then paid more than 10 billion dollars to acquire McCaw, which it renamed "AT&T Wireless."

The analog systems paved the way for the wireless revolution. During the 1980s, they grew at a frenetic pace and reached market penetrations of up to 10% in Europe. At the beginning of the 1980s, most phones were "portable," but definitely not handheld. In most languages, they were just called "carphones," because the battery and transmitter were stored in the trunk of the car and were too heavy to be carried around. But at the end of the 1980s, handheld phones with good speech quality and quite acceptable battery lifetime abounded. The quality had become so good that in some markets digital phones had difficulty establishing themselves – there just did not seem to be a need for further improvements.

1.1.4 GSM and the Worldwide Cellular Revolution

Even though the public did not see a need for changing from analog to digital, the network operators knew better. Analog phones have a bad spectral efficiency (we will see why in Chapters 3 and 21), and due to the rapid growth of the cellular market, operators had a high interest in making room for more customers. Also, research in communications had started its inexorable turn to digital communications and that included digital wireless communications as well. In the late 1970s and the 1980s, research into spectrally efficient modulation formats, the impact of channel distortions, and temporal variations on digital signals, as well as multiple access schemes and much more, were explored in research labs throughout the world. It thus became clear to the cognoscenti that the real-world systems would soon follow the research.

Consequently, the *European Telecommunications Standards Institute* (ETSI) group started the development of a digital cellular standard that would become mandatory throughout Europe and was later adopted in most parts of the world: *Global System for Mobile communications* (GSM). The system was developed throughout the 1980s; deployment started in the early 1990s and user acceptance was swift. Due to additional features, better speech quality, and the possibility for secure communications, GSM-based services overtook analog services typically within 2 years of their introduction. In the U.S.A., the change to digital systems was somewhat slower, but by the end of the 1990s, this country also was overwhelmingly digital.

Digital phones turned cellular communications, which was already on the road to success, into a blockbuster. By the year 2000, market penetration in Western Europe and Japan had exceeded 50%, and though the U.S.A. showed a somewhat delayed development, growth rates were spectacular as well.

The development of wireless systems also made clear the necessity of standards. Devices can only communicate if they are compatible, and each receiver can "understand" each transmitter – i.e., if they follow the same standard. But how should these standards be set? Different countries developed different approaches. The approach in the U.S.A. is "hands-off": allow a wide variety of standards and let the market establish the winner (or several winners). When frequencies for digital cellular communications were auctioned off in the 1990s, the buyers of the spectrum licenses could choose the system standard they would use. For this reason, three different standards were originally used in the U.S.A. A similar approach was used by Japan, where two different systems fought for the market of Second Generation (2G) cellular systems. In both Japan and the U.S.A., the networks based on different standards work in the *same* geographical regions, allowing consumers to choose between different technical standards.

The situation was different in Europe. When digital communications were introduced, usually only one operator *per country* (typically, the incumbent public telephone operators) existed. If each of these operators would adopt a different standard, the result would be high market fragmentation (i.e., a small market for each standard), without the benefit of competition between operators. Furthermore, roaming from country to country, which for obvious geographical regions is much more frequent in Europe than in the U.S.A. or Japan, would be impossible. It was thus logical to establish a single common standard for all of Europe. This decision proved to be beneficial for wireless communications in general, as it provided the economy of scales that decreased cost and thus increased the popularity of the new services.

1.1.5 New Wireless Systems and the Burst of the Bubble

Though cellular communications defined the picture of wireless communications in the general population, a whole range of new services was introduced in the 1990s. Cordless telephones started to replace the "normal" telephones in many homes. The first versions of these phones used analog technology; however, also for this application, digital technology proved to be superior. Among other aspects, the possibility of listening in to analog conversations, and the possibility for neighbors to "highjack" an analog cordless Base Station (BS) and make calls at other people's expense, led to a shift to digital communications. While cordless phones never achieved the spectacular market size of cellphones, they constituted a solid market, until later on the success of cellphones led to a general demise of "home phones."

Another market that seemed to have great promise in the 1990s was *fixed wireless access* and *Wireless Local Loop* (WLL) – in other words, replacing the copper lines to the homes of the users by wireless links but without the specific benefit of mobility. A number of technical solutions were developed, but all of them ultimately failed. The reasons were as much economical and political as they were technical. The original motivation for WLL was to give access to customers for alternative providers of phone services, bypassing the copper lines that belonged to the incumbents. However, regulators throughout the world ruled in the mid-1990s that the incumbents *have* to lease their lines to the alternative providers, often at favorable prices. This eliminated much of the economic basis for WLL. Interestingly, at the time of this writing (2022), fixed wireless access is re-considered as a scheme to provide broadband data access at competitive prices, as we will discuss below.

The biggest treasure thus seemed to lie in a further development of cellular systems, establishing the *Third Generation* (3G) (after the analog systems and 2G systems like GSM) [Bi et al. 2001]. 2G systems were essentially pure voice transmission systems (though some simple data services, like the *Short Message Service* (SMS) were included as well). The new 3G systems were to provide data transmission at rates comparable with the ill-fated *Integrated Services Digital Network* (ISDN) (144 kbit/s), and even up to 2 Mbit/s, at speeds of up to 500 km/h. After long deliberations, two standards were established: one designed by the *Third Generation Partnership Project* (3GPP) (supported by Europe, Japan, and some American companies) and one by 3GPP2 (supported by another group of mainly American companies). The new standards also required a new spectrum allocation in most of the world, and the selling of this spectrum became a bonanza for the treasuries of several countries.

The development of 3GPP, and the earlier introduction of the IS-95 CDMA (*Code Division Multiple Access*) system in the U.S.A., sparked a lot of research into CDMA and other spread spectrum techniques (see Chapter 19) for wireless communications; by the end of the decade, multi-carrier techniques (Chapter 15) had also gained a strong footing in the research community. Multi-user detection – i.e., the fact that the effect of interference can be greatly mitigated by exploiting its structure (see Chapter 28) – was another area that many researchers concentrated on, particularly in the early 1990s. Finally, the field of multi-antenna systems (Chapters 16 and 22) saw an enormous growth since 1995, and for some time accounted for almost half of all published research in the area of the physical layer design of wireless communications.

The spectrum sales for 3G cellular systems and the Initial Public Offerings (IPOs) of some wireless start-up companies represented the peak of the "telecom bubble" of the 1990s. In 2000/2001, the market collapsed with a spectacular crash. Developments on many of the new wireless systems were stopped as their proponents went bankrupt, while the deployment of other systems, including 3G cellular systems, was slowed down considerably. Most worrisome, many companies slowed or completely stopped research, and the general economic malaise led to decreased funding of academic research as well.

1.1.6 Wireless Revival

Between 2003 and 2008, several developments led to a renewed interest in wireless communications. The first one was a continued growth of 2G and 2.5G cellular communications, stimulated by new markets as well as new applications. To give just one example, in 2008, China had more than 500 million cellphone users – even before the first 3G networks became operative. Worldwide, about 3.5 billion cellphones were in use in 2008.

Furthermore, 3G networks had become widely available and popular – especially in Japan, Europe, and the U.S.A. (in 2008, overall cellphone market penetration of cellphones in Western Europe was more than 100% and was approaching 90% in the U.S.A.). Data transmission speeds comparable to cable (5 Mbit/s) were available. This development, in turn, spurred the proliferation of devices that not only allow voice calls but also Internet browsing and even reception of streaming audio and video.

A second important development was the unexpected success of wireless computer networks (wireless Local Area Networks [LANs]). Devices following the Institute of Electrical and Electronics Engineers (IEEE) 802.11 standard (Chapter 33) enabled computers to be used in a way that is almost as versatile and mobile as cellphones. The standardization process had already started in the mid-1990s, but it took several versions, and the impact of intense competition from manufacturers, to turn this into a mass product. By the late 2000s, wireless access points were widely available not only at homes and offices but also at airports, coffee shops, and similar locations. As a consequence, many people who depended on laptops and Internet connections to do their work now got more freedom to choose when and where to work.

1.1.7 The Smartphone Revolution and the Internet of Thing

A turning point in the evolution of wireless communications was the introduction of *smartphones*, i.e., devices that not only enable constant connectivity to the internet but also provide a number of other applications (calendar, camera, etc.). The first such devices were already introduced in the early 2000s, but it was really the introduction of the *iPhone* that led to widespread adoption, due to an emphasis on fashionable design and simple user interfaces. The subsequent introduction of the Android operating system, available for free to a multitude of device manufacturers, further significantly increased the market by providing a larger variety of devices, at various price points.

Among the most popular applications on smartphones are video streaming, video telephony, and use of social media. These applications led to an enormous increase in data consumption (typically a doubling every year) as well as demand for higher data rate. This demand could largely be satisfied by the introduction of the fourth-generation cellular communications standard, Long-Term Evolution (LTE). For it, Multiple Input Multiple Output-Orthogonal Frequency Division Multiplexing (MIMO-OFDM), see Chapters 15 and 16, is the modulation method of choice, which has spurred research in this area.

Even though LTE was able to satisfy the demand for high data rates for a while, the remorseless increase in demand is expected to overload the existing networks within the next few years (i.e., the early 2020s). For this reason, the fifth-generation cellular standard, 3GPP NR (New Radio) is being developed. A first version was approved in 2018, and deployment started in the same year. One of the main developments is the use of millimeter-wave spectrum to drastically increase the data rate; this has led to a revival of research in this field (much research was already done in the 1990s, though it did not lead to commercial systems). At first, mm-wave-based fixed wireless access systems were deployed and provided data rates comparable to cabled connections with considerably smaller deployment costs (no need to dig up the streets to lay the cables).

A parallel development has been the improvement of wireless LANs (Wi-Fi). While the standard in 2008 was able to provide raw data rates of around 300 Mbit/s, the most recent incarnations provide more than 9 Gbit/s. This great increase was achieved through the enhanced use of MIMO, as well as increase of the employed bandwidth. Wi-Fi is used not only for linking a computer to the internet via a cable modem but also enables communication of many devices within a home to each other.

Yet another area of development is the *Internet of Things (IoT),* and other machine-to-machine communications. Wireless sensors and actuators offer new possibilities of monitoring and controlling factories and even homes from remote sites, ranging from automated thermostats to remote reading of power meters. In the context of autonomous cars, establishing and maintaining communications links between cars is critically important. Finally, there are also many military and surveillance applications. Recent developments envision a "tactile internet" that allows, e.g., for remote surgery, piano lessons, and other applications with tactile feedback. Many of these applications require very low latency, giving rise to new fundamental research, because much of the classical communication theory is based on the assumption of large blocks of data and associated absence of latency constraints. Much of the next phase of 5G cellular development will be dedicated to achieving such low latency.

The IoT and sensor networks have also spurred a wave of research into ad hoc, peer-to-peer, and mesh networks. Such networks do not use a dedicated infrastructure. If the distance between source and destination is too large, other nodes of the network help in forwarding the information to the final destination. Since the structure of those networks is significantly different from the traditional cellular networks, a lot of new research is required.

Beyond the technologies and applications envisioned by 5G, research has already started for wireless systems of the next generation, generally called Beyond 5G (B5G) or 6G. These new technologies, which are surveyed in Chapter 35, will lead to an even deeper interweaving of wireless technology into our daily lives, and also enable completely new applications.

Summarizing, the current developments are based on three tendencies: (i) a much broader range of products, (ii) data transmission with a higher rate for already existing products, and (iii) higher user densities. These trends determine the directions of research in the field and provide a motivation for many of the more recent scientific developments.

1.2 Types of Services

1.2.1 Cellular Telephony

Cellular communications is the economically most important form of wireless communications. Figure 1.1 shows a block diagram of a simple cellular system. A mobile user (with a device called the *User Equipment* [UE]), is communicating with a BS that has a good radio connection with that UE. The BSs are connected to a mobile switching center or similar, which is connected to the internet and/ or the public telephone system.

Cellular communications is characterized by the following important properties:

1. The information flow is bidirectional. A user can transmit and receive information at the same time.
2. The user can be anywhere within a (nationwide or international) network. Neither the user nor the calling party needs to know the user's location or the BS the UE has connection to; it is the network that has to take the user location into account.
3. The location of a user can change significantly during a call, and it can move from the coverage range of one BS to another.
4. A call or data packet can originate from either the network or the user. In other words, a cellular customer can be called or can initiate a call.
5. A call or data packet is generally intended only for a single user; other users of the network should not be able to listen in.

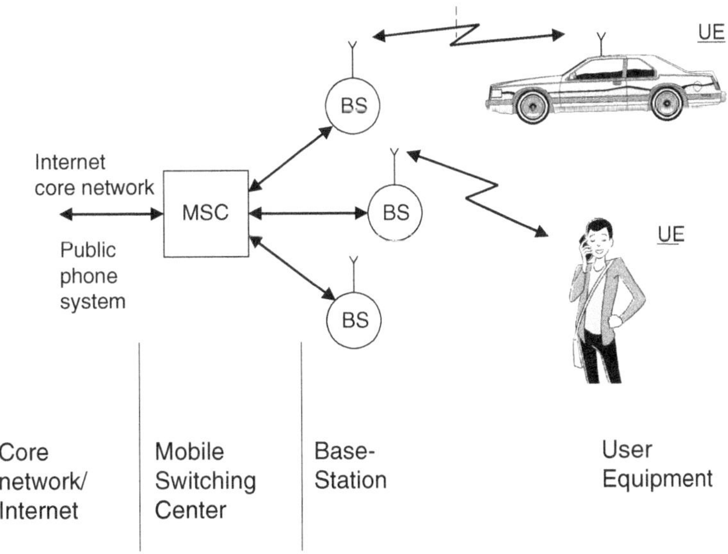

Figure 1.1 Principle of a cellular system. *In this figure*: MSC, Mobile Switching Center; BS, Base Station; UE, User Equipment. Color version available at wiley.com/go/molisch/wireless3e.

Since each user wants to transmit or receive different information, the number of active users in a network is limited. The available bandwidth must be shared between the different users; this can be done via different "multiple access" schemes (see also Chapters 18 and 19).

In order to increase the number of possible users, the *cellular principle* is used: the area served by a network provider is divided into a number of subareas called *cells*. Within each cell, different users have to share the available bandwidth: let us consider in the following the case that each user employs a different carrier frequency. Even users in neighboring cells have to use different frequencies, in order to avoid interfering with each other. However, for cells that are sufficiently far apart, the same frequencies can be used, because the signals get weaker with increasing distance from their transmitter. Thus, within one country, there can be hundreds or thousands of cells that are using the same frequencies (Chapter 21).

Another important aspect of cellular telephony is the unlimited mobility. The user can be anywhere within the coverage area of the *network* (i.e., the user is not limited to a specific cell), in order to be able to communicate. Also, they can move from one cell to the other during one call. The cellular network interfaces with the internet and the PSTN as well as with other wireless systems.

As mentioned in our brief wireless history, cellphones started to become popular in the 1980s and are now a dominant form of communications. Due to this reason, this book often draws its examples from cellular telephony, even though the general principles are applicable to other wireless systems as well. Chapters 31 and 32 give a detailed description of the most popular cellular systems.

1.2.2   Satellite and UAV Cellular Communications

Satellite cellular communications mostly have the same operating principles as land-based cellular communications, with satellites operating as BSs. However, there are some key differences.

The distance between the "BS" (i.e., the satellite) and the UE is *much* larger: for geostationary satellites, that distance is 36,000 km; for *Low Earth Orbit* (LEO) satellites, it is several hundred kilometers. Consequently, the transmit powers need to be larger, high-gain antennas need to be used on the satellite (and in many cases also on the UE), and communications from within buildings is almost impossible.

Another important difference from the land-based cellular system lies in the cell size: due to the large distance between the satellite and the Earth, it is impossible to have cells with diameters less than tens of km even with LEO satellites; for geostationary satellites, the cell areas are even larger. This large cell size is the biggest advantage as well as the biggest drawback of the satellite systems. On the positive side, it makes it easy to have good coverage even of large, sparsely populated areas. On the other hand, the area spectral efficiency is low, which means that (given the limited spectrum assigned to this service) fewer people can communicate at the same time.

The costs of setting up a "BS" – i.e., a satellite – are much higher than for a land-based system. Not only is the launching of a communications satellite very expensive but it is also necessary to build up an appropriate infrastructure of ground stations for linking the satellites to the internet.

As a consequence of all these issues, the business case for satellite communications systems has long been quite different: it is based on supplying a small number of users with vital communications at a much higher price. Emergency workers and journalists in disaster and war areas, ship-based communications, and workers on offshore oil drilling platforms are typical users for such systems. The INMARSAT system is the leading provider for such communications. In the late 1990s, the IRIDIUM project attempted to provide lower priced satellite communications services by means of some 60 LEO satellites but ended in bankruptcy.

However, recent advances in reusable launch rockets have changed the economics. For example, the StarLink system of SpaceX will in its full version consist of tens of thousands of LEO satellites, which also provide relatively low latency. Data rates in the tens or even hundreds of Mbit/s have been demonstrated in proof-of-principle tests.

A related topic is the use of Unmanned Aerial Vehicles (UAVs), commonly also known as drones, as flying BSs.[5] They can then connect to fixed infrastructure either directly, by relaying via satellites, or by relaying via other UAVs.

UAVs have the advantage that they can be easily repositioned and thus adapt to the demand of coverage. Such adaptability can range from a "personalized" BS that flies with a driving car, to temporary BSs that are placed at an event venue to supply coverage to tens of thousands of people. Furthermore, the cell size of a UAV-BS can be easily adapted by changing the height above ground. The main drawback is the limited flight time, and limited weight and form factor that a UAV can carry.

1.2.3   Trunking Radio

Trunking radio systems are an important variant of cellular phones, where there is no connection between the wireless system and other networks; therefore, it allows communication between members of closed user groups. Obvious applications include police departments, fire departments, taxis, and similar services. The closed user group allows implementation of several technical innovations that are not possible (or more difficult) in normal cellular systems:

1. *Group calls*: a communication can be sent to several users simultaneously, or several users can set up a conference call between multiple users of the system.

---

[5] Another type of flying BSs that has attracted attention are high-altitude platforms such as balloons; however, the largest trials in this field have been stopped by 2021 as uneconomical.

2. *Call priorities*: a normal cellular system operates on a "first-come, first-serve" basis. Once a call is established, it cannot be interrupted.[6] This is reasonable for cellphone systems, where the network operator cannot ascertain the importance or urgency of a call. However, for the trunk radio system of, e.g., a fire department, this is not an acceptable procedure. Notifications of emergencies have to go through to the affected parties, even if that means interrupting an existing, lower priority call. A trunking radio system thus has to enable the prioritization of calls and has to allow dropping a low-priority call in favor of a high-priority one.

3. *Relay networks*: the range of the network can be extended by using each UE as a relay station for other UEs. Thus, a UE that is out of the coverage region of the BS might send its information to another UE that is within the coverage region, and that UE will forward the message to the BS; the system can even use multiple relays to finally reach the BS. Such an approach increases the effective coverage area and the reliability of the network.

A number of trunking systems are nowadays implemented as "virtual" trunking systems within a regular cellular network. Thus, BSs are part of the regular cellular networks, while they can establish group calls that through proper encryption are kept separate from all other users. The UEs of such systems are typically capable of direct communication to enable functionality also in the absence of cellular infrastructure.

1.2.4 Wireless LANs and Cordless Telephony

Wireless LANs establish a wireless link between a UE (computer, smartphone, home appliance) and an Access Point (which in this book we also call BS, to unify notation) that provides connection to the internet. The main difference from a cellular system is that the UE is associated with and can communicate with, only a single BS (see Figure 1.2). There is thus no *mobile switching center*; rather, the BS is directly connected to the internet. This has several important consequences:

1. The BS does not need to have any network functionality. When a transmission is coming in from the internet, there is no need to find out the location of the UE. Similarly, there is no need to provide for handover between different BSs.

2. There is no central system. A user typically has one BS for their apartment or business under control but no influence on any other BSs. For that reason, there is no need for (and no possibility for) frequency planning or other coordination between different BSs.

3. The fact that both the BS and the UE are under the control of the user also implies a different pricing structure: there are no network operators that can charge fees for connections from the UE to the BS; rather, the only occurring fees are the fees from the BS into the internet.

In many other respects, a WLAN is similar to the cellular network: it allows mobility *within* the cell area; the information flow is bidirectional; transmissions can originate from either the internet or the UE, and there have to be provisions such that calls cannot be intercepted or listened to by unauthorized users and no unauthorized calls can be made. If different users are connected to one BS, they have to share the spectral resources. Furthermore, the WLAN can also establish connectivity between two UEs that are within its reach without requiring a connection to the outside internet.

Wireless LANs have also evolved into wireless *Enterprise LANs* (see Figure 1.3). Such a network contains several BSs that are connected to each other, and possibly a central control station. Such a system has essentially the same functionality as a cellular system; it is only the size of the coverage area that distinguishes such a full functionality wireless Enterprise LAN from a cellular network.

WLAN devices can, in principle, connect to any BS that uses the same standard. However, the owner of the BS can restrict the access – e.g., by appropriate security settings. A number of communities and companies are operating WLAN networks that are either open to the public, or open to customers of those companies, and can cover cities or whole states (though with spotty coverage). Such systems are even more similar to cellular networks.

The original purpose of WLANs was to connect computers (desktops or laptops) to the internet; sensors, household appliances, etc., which show even lower mobility, were also added. However, later, the additional purposes arose of connecting smartphones to the internet. This changed the mobility requirements from nomadic setups (where UEs stayed in one location for a long time) to slow mobility, where UEs might be moved around by people carrying them, or changing position while sitting.

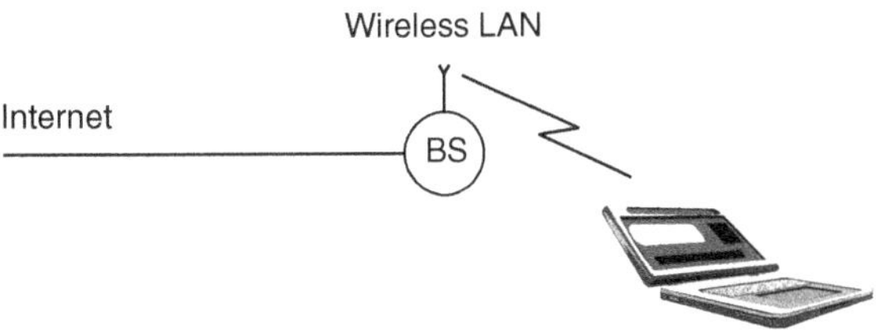

Figure 1.2   Principle of a simple WLAN. Color version available at wiley.com/go/molisch/wireless3e.

---

[6] Except for interrupts due to technical problems, like the user moving outside the coverage region.

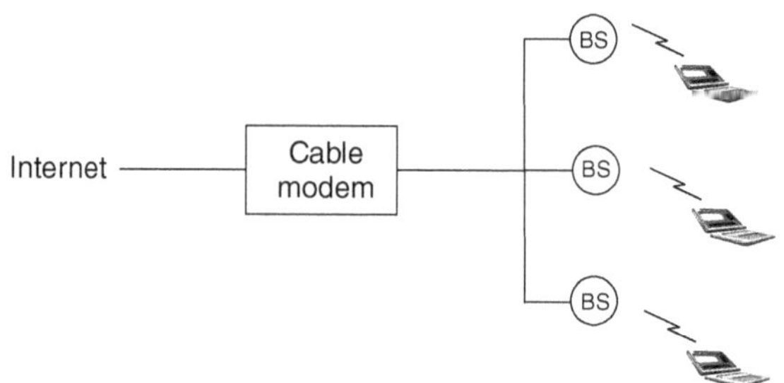

Figure 1.3 Principle of an enterprise WLAN. Color version available at wiley.com/go/molisch/wireless3e.

WLANs also bear a strong resemblance to cordless phones, which establish a wireless link between a handset and a BS that connects into the PTSN. Larger systems, called *Private Automatic Branch eXchange (PABX)* are similar in functionality to enterprise WLANs. Cordless phones were very popular in the 1990s but are in less widespread use now, since many households have given up landline phone connections in favor of cellular phones.

1.2.5 Personal Area Networks

When the coverage area becomes even smaller than that of WLANs, we speak of *Personal Area Networks* (PANs). Such networks are mostly intended for simple "cable replacement" duties. For example, devices following the *Bluetooth* standard allow to connect a hands-free headset to a phone without requiring a cable; in that case, the distance between the two devices is less than a meter. In such applications, data rates are fairly low (<1 Mbit/s). Recently, wireless communications between components in an entertainment system, between computer and peripheral devices (printer, mouse), and similar applications have gained importance, and a number of standards for PANs have been developed by the IEEE 802.15 group. For these applications, data rates in excess of 100 Mbit/s are used.

Networks for even smaller distances are called *Body Area Networks* (BANs), which enable communications between devices located on various parts of a user's body. Such BANs play an increasingly important role in the monitoring of patients' health and of medical devices (e.g., pacemakers).

PANs and BANs can either have a network structure similar to a cellular approach or they can be ad hoc networks as discussed in Section 1.2.8. The main characteristics are:

- Communication is bi-directional.
- The BS (if it exists) tends to send different information to the different devices in the network, thus necessitating the sharing of the available spectral resources.
- There is no significant mobility, and devices do not change the BS they are associated with.
- There is no coordination between different BSs.

1.2.6 Fixed Wireless Access

Fixed wireless access systems can also be considered as a variant of WLANs, essentially replacing a dedicated cable connection between the user and the internet access point. The main differences from a WLAN system are

(i) There is no mobility of the UEs.
(ii) The BS almost always serves multiple users.
(iii) The distances bridged by fixed wireless access devices are much larger (between 100 m and several tens of kilometers) than those bridged by WLANs.
(iv) Different BSs in an FWA system are usually under the control of the same operator, so that frequency planning and similar coordination between different users are possible.

The purpose of fixed wireless access lies in providing users with internet connections without having to lay cables from a central switching office to the office or apartment the user is in. Considering the high cost of labor for the cable-laying operations, this can be an economical approach.

It is noteworthy that FWA systems (mainly for phone connections) were attempted in the 1990s, and were not successful there due to economical reasons (see Section 1.1.5). However, there is a revived interest at the time of this writing to use FWA to providing high-speed data connectivity via FWA. This case might have different economics, in particular when operated in regions where incumbent internet providers have a (quasi-) monopoly and charge high prices, or for providing fast access to underserved areas where laying of cables is uneconomical, such as spread-out rural communities.

1.2.7 Broadcast

In broadcast radio/TV, information is transmitted to different, possibly mobile, users (see Figure 1.4). Four properties differentiate broadcast radio from, e.g., cellular telephony:

1. The information is only sent in one direction. It is only the broadcast station that sends information to the radio or TV receivers; the listeners (or viewers) do not transmit any information back to the broadcast station.
2. The transmitted information is the same for all users.
3. The information is transmitted continuously.
4. In many cases, multiple transmitters send the same information. This is especially true, e.g., in Europe, where national broadcast networks cover a whole country and broadcast the same program in every part of that country.[7]

The above properties led to many simplifications in the design of broadcast radio networks. The transmitter does not need to have any knowledge or consideration about the UEs. There is no requirement to provide for duplex channels (i.e., for bringing information from the receiver to the transmitter). The number of possible users of the service does not influence the transmitter structure either – irrespective of whether there are millions of users, or just a single one, the transmitter sends out the same information. This is an important difference from cellular systems, where the available spectral resources have to be shared between the different users that want to get different information.

The above description has been mainly true for traditional broadcast TV and radio. Satellite TV and radio differ in the fact that often the transmissions are intended only for a subset of all possible users (pay-TV or pay-per-view customers), and therefore, encryption of the content is required in order to prevent unauthorized viewing. Note, however, that this "privacy" problem is different from regular cellphones: for pay-TV, the content should be accessible to all members of the authorized user group ("multi-cast"), while for cellphones, each call should be accessible only for the single person it is intended for ("unicast") and not to all customers of a network provider.

Despite their undisputed economic importance, broadcast networks are not at the center of interest for this book – space restrictions prevent a more detailed discussion. Still, it is useful to keep in mind that they are a specific case of wireless information transmission.

1.2.8 Ad Hoc Networks

Up to now, we have dealt with "infrastructure-based" wireless communications, where certain components (BSs, TV transmitters, etc.) are intended by design to be in a fixed location, to exercise control over the network and interface with other networks. The size of the networks may differ (from LANs covering just one apartment to cellular networks covering whole countries), but the central principle of distinguishing between "infrastructure" and "UE" is common to them all. There is, however, an alternative in which there is only one type of equipment, and those devices, all of which may be mobile, organize themselves into a network according to their location and according to necessity. Such networks are called *ad hoc networks* (see Figure 1.5). There can still be "controllers" in an ad hoc network, but the choice of which devices act as controllers is done opportunistically whenever a network is formed. There are also ad hoc networks without any hierarchy. While the actual transmission of the data (i.e., physical layer communication) is almost identical to that of the infrastructure-based networks, the medium access and the networking functionalities are very different.

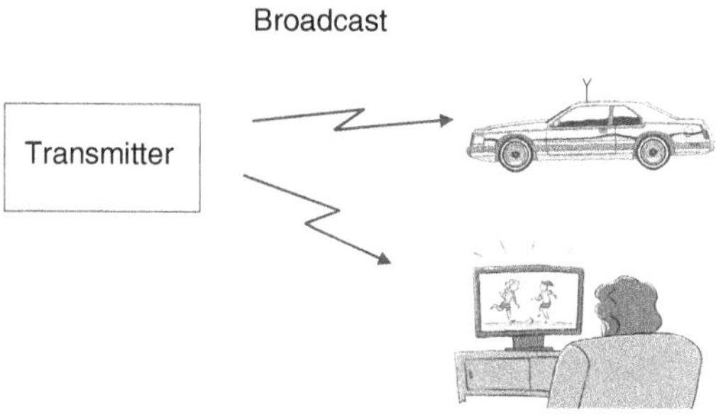

Figure 1.4 Principle of broadcast transmission.

---

[7] The situation is slightly different in the U.S.A., where a "local station" usually covers only a single metropolitan area, often with a single transmitter.

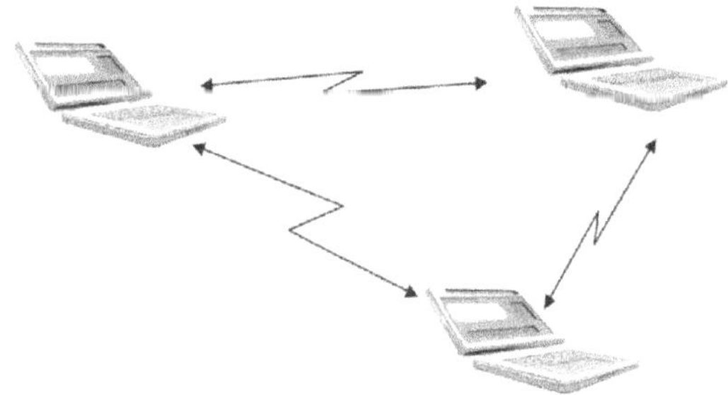

Figure 1.5 Principle of an ad hoc network.

The advantages of ad hoc networks lie in their low costs (because no infrastructure is required) and high flexibility. The drawbacks include reduced efficiency, smaller communication range, and restrictions on the number of devices that can be included in a network. Ad hoc networks also play a role in emergency communications (when infrastructure was destroyed, e.g., by an earthquake) as well as military communications.

1.2.9 Internet of Things and Sensor Networks

IoT and sensor networks are different types of categories compared to the other systems mentioned above. They do not provide a different network structure, but rather a different set of applications. The underlying network structure can be either a cellular/FWA-like structure or an ad hoc network. For applications in which sensors just send measurement data to a central BS, it is possible that only unidirectional communication takes place. While most sensors are static, and thus no mobility and handovers need to be considered, other applications (such as sensors on cars or trains) can show high dynamics. Also, latency requirements can vary widely, from 1 ms (for sensors/actuators that are part of the "tactile internet") to many minutes for sensor measuring air temperature.

1.3 Requirements for the Services

A key to understanding wireless design is to realize that different applications have different requirements in terms of data rate, range, mobility, energy consumption, and so on. It is *not* necessary to design a system that can sustain gigabit per second data rates over a 100 km range when the user is moving at 500 km/h. We stress this fact because there is a tendency among engineers to design a system that "does everything but wash the dishes"; while appealing from a scientific point of view, such systems tend to have a high price and low efficiency. In the following, we list the range of requirements encountered in system design and enumerate which requirements occur in which applications.

1.3.1 Data Rate

Data rates for wireless services span the gamut from a few bits per second to several gigabit per second, depending on the application:

- *Sensor networks and IoT* usually require data rates from a few bits per second to about 1 kbit/s. Typically, a sensor measures some critical parameter, like temperature, speed, etc., and transmits the current value (which corresponds to just a few bits) at intervals that can range from milliseconds to several hours. Higher data rates are often required for the central nodes of sensor networks that collect the information from a large number of sensors and forward it for further processing. In that case, data rates of up to 10 Mbit/s can be required. These "central nodes" show more similarity to WLANs or fixed wireless access.
- *Speech communications* usually require between 5 and 64 kbit/s depending on the required quality and the amount of compression. For cellular systems, which require higher spectral efficiency, source data rates between 5 and 10 kbit/s are standard. For cordless systems, less elaborate compression and therefore higher data rates (32 or 64 kbit/s) are used.
- *Elementary data services* require between 10 and 100 kbit/s. One category of these services uses the display of a standard (nonsmart) cellphone to provide Internet-like information. Since the displays are smaller, the required data rates are often smaller than for conventional Internet applications. Another type of data service provides a wireless mobile connection to laptop computers. In this case, speeds that are at least comparable with dial-up (around 50 kbit/s) are demanded by most users, though elementary services with 10 kbit/s (exploiting the same type of communications channels foreseen for speech) are sometimes used as well. Elementary data services are almost completely replaced by high-speed data services in most countries, but still play an important role in some parts of the world.
- *Communications between computer peripherals and similar devices*: for the replacement of cables that link computer peripherals, like mouse and keyboard, to the computer (or similarly for cellphones), wireless links with data rates around 1 Mbit/s are used. The functionality of these links is similar to the previously popular infrared links but usually provides higher reliability.

- *Video streaming and conferencing*: data rates for video streaming range from 200 kbit/s to 20 Mbit/s, depending on the required quality. Considering that "raw" HDTV requires 2 Gbit/s, strong data compression is required. For video conferencing, data rates > 2 Mbit/s are required, because compression in real-time is not as efficient.
- *Web browsing and related data services*: While web browsing typically does not require high average data rates (users take a while to read a file before they look at the next), high *peak* data rates are essential. Furthermore, use of social media often requires high data rates. Data rates between 1 and 10 Mbit/s are typical.
- PANs refers mostly to the range of a wireless network (up to 10 m) but often also has the connotation of high data rates, mostly for linking the components of consumer entertainment systems (streaming video from computer or game console to a TV or Virtual Reality [VR] headset), high-speed computer connections (wireless Universal Serial Bus [USB]), or gaming. The required rates can range from 100 Mbit/s to 10 Gbit/s. However, requirements for data rates are constantly increasing; for example, to provide nausea-free virtual reality with 8k resolution, data rates in excess of 100 Gbit/s are required; see also Chapter 35.

1.3.2 Range and Number of Users

Another distinction among the different networks is the range and the number of users that they serve. By "range," we mean here the distance between one transmitter and receiver. The coverage area of a *system* can be made almost independent of the range, by just combining a larger number of BSs into one big network.

- *BANs* cover the communication between different devices attached to one body – e.g., from a cellphone in a hip holster to a headset attached to the ear. The range is thus on the order of 1 m. BANs are often subsumed into PANs.
- *PANs* include networks that achieve distances of up to or about 10 m, covering the "personal space" of one user. Examples are networks linking components of computers and home entertainment systems. Due to the small range, the number of devices within a PAN is small, and all are associated with a single "owner." Also, the number of overlapping PANs (i.e., sharing the same space or room) is small – usually less than five. That makes cell planning and multiple access much simpler.
- *WLANs*, as well as cordless telephones, cover still larger ranges of up to 100 m. The number of users is usually limited to about 10, though with the increased use of various connected home devices, this number is increasing fast. For number of users in the hundreds (e.g., at conferences or meetings), the efficiency of the system decreases.
- *Cellular systems* have a range that is larger than, e.g., the range of WLANs. Microcells typically cover cells with 300 m radius, while macrocells can have a radius of up to 30 km. Depending on the available bandwidth and the multiple access scheme, the number of *simultaneously active* users in a cell is usually between 5 and 500, while the number of potential users can be in the tens of thousands. If the system is providing high-speed data services, the number of simultaneously active users usually shrinks.
- *Fixed wireless access services* cover a range that is similar to that of cellphones – namely, between 100 m and several tens of kilometers. Also, the number of users is of a similar order as for cellular systems.
- *Satellite systems* provide even larger cell sizes, often covering whole countries and even continents. Cell size depends critically on the orbit of the satellite: geostationary satellites provide larger cell sizes (1000 km radius) than LEOs.

Figure 1.6 gives a graphical representation of the link between data rate and range. Obviously, higher data rates are easier to achieve if the required range is smaller. One exception is fixed wireless access, which demands a high data rate at rather large distances.

1.3.3 Mobility

Wireless systems also differ in the amount of mobility that they have to allow for the users. The ability to move around while communicating is one of the main charms of wireless communication for the user. Still, within that requirement of mobility, different gradations exist:

- *Fixed devices* are placed only once, and after that time communicate with their BS, or with each other, always from the same location. The main motivation for using wireless transmission techniques for such devices lies in avoiding the laying of cables. Even though the devices are not mobile, the propagation channel they transmit over can change with time: both due to people/cars passing by and due to changes in the environment (moving foliage, rearranging of machinery, furniture, etc.). Fixed wireless access is a typical case in point.
- *Nomadic devices*: nomadic devices are placed at a certain location for a limited duration of time (minutes to hours) and then moved to a different location. This means that during one "drop" (placing of the device), the device is similar to a fixed device. However, from one drop to the next, the environment can change radically. Laptops are typical examples: people do not operate their laptops while walking around but place them on a desk to work with them. Minutes or hours later, they might bring them to a different location and operate them there.
- *Low mobility*: many communications devices are operated at pedestrian speeds. Cellphones operated by walking human users are typical examples. The effect of the low mobility is a channel that changes rather slowly, and – in a system with multiple BSs – handover from one cell to another is a rare event.

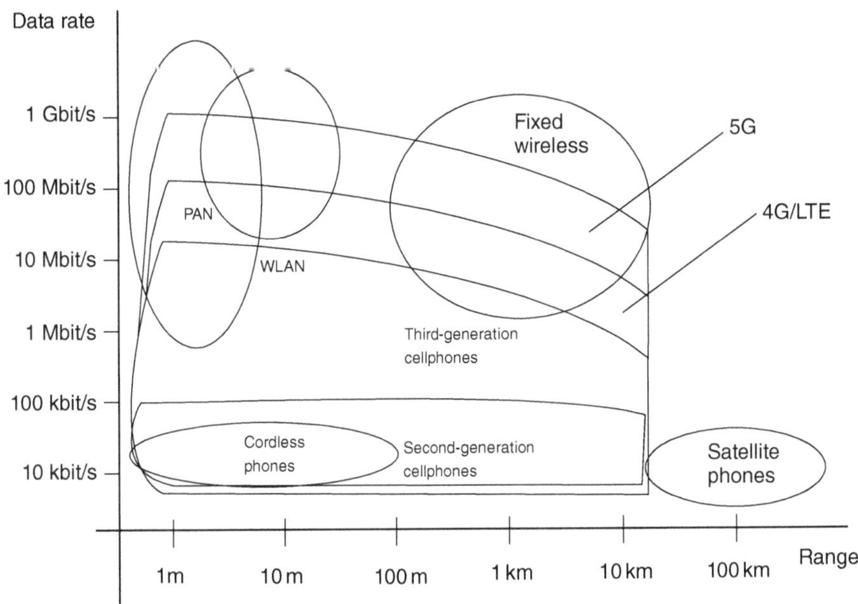

Figure 1.6 Data rate versus range for various applications.

- *High mobility* usually describes speed ranges from about 30 to 150 km/h. Cellphones operated by people in moving cars are one typical example.
- *Extremely high mobility* is represented by high-speed trains and planes, which cover speeds between 300 and 1000 km/h. These speeds pose unique challenges both for the design of the physical layer (Doppler shift, see Chapter 5) and for the handover between cells.

Figure 1.7 shows the relationship between mobility and data rate.

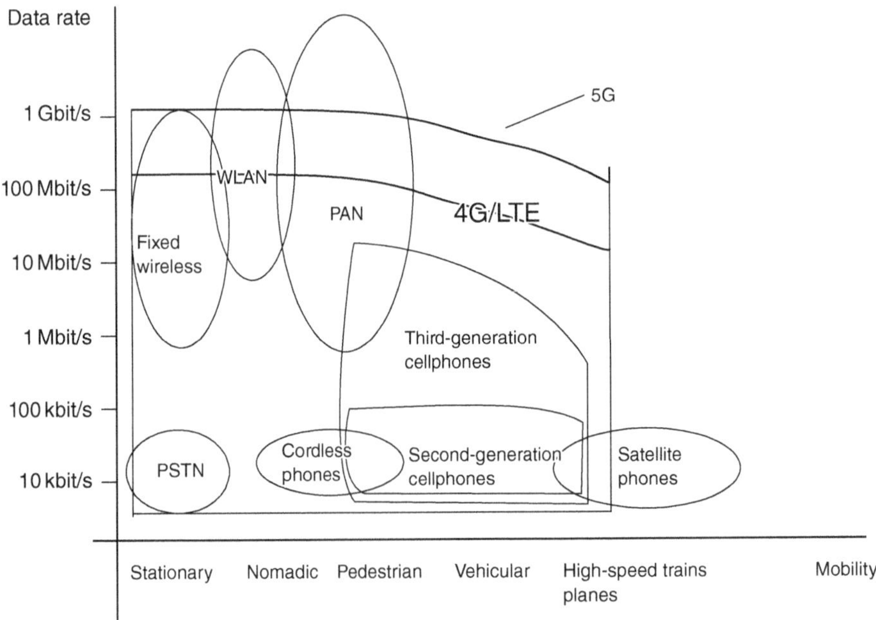

Figure 1.7 Data rate versus mobility for various applications.

1.3.4 Energy Consumption

Energy consumption is a critical aspect for wireless devices. Most wireless devices use (one-way or rechargeable) batteries, as they should be free of *any* wires – both the ones used for communication and the ones providing the power supply.

- *Rechargeable batteries*: nomadic and mobile devices, like laptops, cellphones, and cordless phones, are usually operated with rechargeable batteries. Standby times as well as operating times are one of the determining factors for customer satisfaction. Energy consumption is determined on one hand by the distance over which the data have to be transmitted (remember that a minimum SNR has to be maintained), and on the other hand, by the amount of data that are to be transmitted (the SNR is proportional to the energy per bit). The energy density of batteries has increased slowly over the past 100 years, so that the main improvements in terms of operating and standby time stem from reduced energy consumption of the devices. For cellphones (used for talking only), talk times of more than 6 hours and standby times of more than 48 hours are considered minimum requirements. For laptops, power consumption is not mainly determined by the wireless transmitter, but rather by other factors like hard drive usage and processor speed. For smartphones, the energy consumption of the processor and of the wireless connection is of the same order, and both have to be considered for maximizing battery lifetime. The minimum requirement is operation for a full day (16 hours), so that nightly recharging is sufficient. This holds for both cabled and wireless recharging of cellphones, since the latter case it requires placing phone and charging station in very close proximity.
- *One-way batteries*: sensor network nodes often use one-way batteries, which offer higher energy density at lower prices. Furthermore, changing the battery is often not an option; rather, the sensor including the battery and the wireless transceiver is often discarded after the battery has run out. It is obvious that in this case energy-efficient operation is even more important than for devices with rechargeable batteries.
- *Energy harvesting*: A possibility to avoid change of batteries is the use of "energy harvesting," where RF energy transmitted by surrounding devices is used to power the batteries. A related approach is remote powering, where electromagnetic energy is transmitted over larger distances with the explicit purpose of powering a device (sometimes in conjunction with information transmission). Since the received power levels are typically small, either of those approaches can only recharge batteries of devices like sensors that have very low energy consumption.
- *Power mains*: BSs and other fixed devices can be connected to the power mains. Therefore, energy efficiency is not a major concern for them. It is thus desirable, if possible, to shift as much functionality (and thus energy consumption) from the UE to the BS. However, even for BSs, power consumption is an important design factor. Power accounts for more than 50% of the operating expenses of network operators, so that reducing power consumption has clear economic benefits and also helps to achieve "green" (environmental protection) goals.

User requirements concerning batteries are also important sales issues, especially in the market for cellular handsets: The weight of a UE is determined mostly by the battery. Weight and size of a handset are critical sales issues. It was in the mid-1980s that cellphones were commonly called "carphones," because the UE could only be transported in the trunk of a car and was powered by the car battery. By the end of the 1980s, the weight of the battery had decreased to about 2 kg, so that it could be carried by the user in a backpack. By the year 2000, the battery weight had decreased to about 200 g. Part of this improvement stemmed from more efficient battery technology, but to a large part, it is caused by the decrease of the power consumption of the handsets. The trend was somewhat reversed with the introduction of the smartphone, since high-resolution screens, as well as the signal processing for mobile games, etc., have high power consumption. Battery size is typically designed to enable 16 hours of continuous operation. This is in contrast to voice-only phones that might work up to a week on a single battery charge.

These "commercial" aspects determine the maximum size (and thus energy content) of the battery, and consequently, the admissible energy consumption of the phone during standby and talk operation.

1.3.5 Use of Spectrum

Spectrum can be assigned on an exclusive basis, or on a shared basis. That determines to a large degree the multiple access scheme and the interference resistance that the system has to provide:

- *Spectrum dedicated to service and operator*: in this case, a certain part of the electromagnetic spectrum is assigned, on an exclusive basis, to a service provider. A prime case in point is cellular service, where the network operators buy or lease the spectrum on an exclusive basis (often for a very high price). Due to this arrangement, the operator has control over the spectrum and can plan the use of different parts of this spectrum in different geographical regions, in order to minimize interference.
- *Spectrum allowing multiple operators*:
  - *Spectrum dedicated to a service*: in this case, the spectrum can be used only for a certain service (e.g., car-to-car communications) but is not assigned to a specific operator. Rather, users can set up qualified equipment without a license. Such an approach does not require (or allow) interference planning. Rather, the system must be designed in such a way that it avoids interfering with other users in the same region. Since the only interference *can* come from equipment of the same type, coordination between different devices is relatively simple. Limits on transmit power (identical for all users) are a key component of this

approach – without them, each user would just increase the transmit power to drown out interferers, leading essentially to an "arms race" between users.

o *Free spectrum*: is assigned for different services as well as for different operators. The ISM band at 2.45 GHz is the best known example – it is allowed to operate microwave ovens, Wi-Fi LANs, and Bluetooth wireless links, among others, in this band. Also for this case, each user has to adhere to strict emission limits, in order not to interfere too much with other systems and users. However, coordination between users (in order to minimize interference) becomes almost impossible – different systems cannot exchange coordination messages with each other, and often even have problems determining the exact characteristics (bandwidth, duty cycle) of the interferers.

After 2000, two further approaches have been promulgated:

- *Ultra-Wide Bandwidth Systems* (UWB) spread their information over a very large bandwidth, while at the same time keeping a very low-power spectral density. Therefore, the transmit band can include frequency bands that have already been assigned to other services, without creating significant interference. UWB is discussed in more detail in Section 26.8.
- *Adaptive spectral usage*: another approach relies on first determining the current spectrum usage at a certain location and then employing unused parts of the spectrum. This approach, also known as *cognitive radio*, is described in detail in Chapter 26.

1.3.6   Direction of Transmission

Not all wireless services need to convey information in both directions.

- *Simplex systems* send the information only in one direction – e.g., broadcast systems and pagers.
- *Semi-duplex systems* can transmit information in both directions. However, only one direction is allowed at any time. Walkie-talkies, which require the user to push a button in order to talk, are a typical example. Note that one user must signify (e.g., by using the word "over") that they have finished their transmission; then the other user knows that now they can transmit.
- *Full-duplex systems* allow simultaneous transmission in both directions – e.g., cellphones and cordless phones.
- *Asymmetric duplex systems*: for data transmission, we often find that the required data rate in one direction (usually the downlink) is higher than in the other direction. However, even in this case, full-duplex capability is maintained.

1.3.7   Service Quality

The requirements for service quality also differ vastly for different wireless services. The first main indicator for service quality is *speech quality* for speech services (and similar for video) and *file transfer speed* for data services. Speech quality is usually measured by the *Mean Opinion Score* (MOS). It represents the average of a large number of (subjective) human judgments (on a scale from 1 to 5) about the quality of received speech (see also Chapter 24). Similar measures are established for video quality, though establishing a consistent standard is more difficult in this case. The speed of data transmission is simply measured in bit/s – obviously, a higher speed is better.

An even more important factor is the availability of a service. For cellphones and other speech services, the *service quality* is often computed as the complement of "fraction of blocked calls[8] plus 10 times the fraction of dropped calls." This formula takes into account that the dropping of an active call is more annoying to the user than the inability to make a call at all. Similar requirements occur for video streaming, where blurry pictures (low video quality) are considered as slightly annoying, while complete halting of the playback (rebuffering) is considered a major inconvenience by the users.

For emergency services and military applications, service quality is better measured as the complement of "fraction of blocked calls plus fraction of dropped calls." In emergency situations, the inability to make a call is as annoying as the situation of having a call interrupted. Also, the systems must be planned in a much more robust way, as service qualities better than 99% are required. "Ultra-reliable systems," which are required, e.g., for factory automation systems, require service availability in excess of 99.999%.

A related aspect is the *admissible delay (latency)* of the communication. The most stringent requirements occur in the context of sensor/actuator communications like those related to the tactile internet. In this context, where fast control loops are involved, latencies on the order of 1 ms are required. Similar requirements occur for virtual reality and augmented reality, in particular in the context of gaming. For voice communications, the delay between the time when one person speaks and the other hears the message must not be larger than about 100 ms. For streaming video and music, delays can be larger, as buffering of the streams (up to several tens of seconds) is widely used. In both voice and streaming video communications, it is important that the data transmitted first are also the ones made available to the receiving user first. For data files, the acceptable delays can be usually larger and the sequence in which

---

[8] Here, "blocked calls" encompasses all failed call attempts, including those that are caused by insufficient signal strength, as well as insufficient network capacity.

the data arrive at the receiver is not critical (e.g., when downloading email from a server, it is not important whether the first or the seventh of the emails is the first to arrive). A number of systems establish different Quality of Service (QoS) classes and prioritize data transmission accordingly.

1.4 Economic and Social Aspects

1.4.1 Economic Requirements for Building Wireless Communications Systems

The design of wireless systems not only aims to optimize performance for specific applications but also to do that at a reasonable cost. As economic factors impact the design, scientists and engineers have to have at least a basic understanding of the constraints imposed by marketing and sales divisions. Some of the guidelines for the design of wireless *devices* are as follows:

- Move as much functionality as possible from the (more expensive) analog components to digital circuitry. The costs for digital circuits decrease much faster with time than those of analog components.
- For mass-market applications, try to integrate as many components onto one chip as possible. Most systems strive to use only two chips; one for analog RF circuitry and one for digital (baseband) processing. Further integration into a single chip (system on a chip) is desirable. Exceptions are niche market products, which typically try to use general-purpose processors, Application-Specific Integrated Circuits (ASICs), or off-the-shelf components, as the number of sold units does not justify the cost of designing more highly integrated chips.
- As human labor is very expensive, any circuit that requires human intervention (e.g., tuning of RF elements) is to be avoided. Again, this aspect is more important for mass-market products.
- In order to increase the efficiency of the development process and production, the same chips should be used in as many systems as possible.

When it comes to the design of wireless *systems and services*, we have to distinguish between two different categories:

- Systems where the mobility is of value by itself – e.g., in cellular telephony. Such services can charge a premium to the customer – i.e., be more expensive than equivalent, wired systems. Smartphones are a case in point: the per-gigabyte price has been higher than that of wired internet connections in the past and is expected to remain so. Despite this fact, the services might compete (and ultimately edge out) traditional wired services if the price difference is not too large, as happened in the case of cellular voice telephony, which largely supplanted landlines.
- Services where wireless access is only intended as a cheap cable replacement, without enabling additional features – e.g., fixed wireless access. Such systems have to be especially cost-conscious, as the buildup of the infrastructure has to remain cheaper than the laying of new wired connections, or buying access to existing ones.

1.4.2 The Market for Wireless Communications

Cellphones are a highly dynamic market that has grown tremendously. Still, different countries show different market penetrations. Some of the factors influencing this penetration are:

- *Price of the offered services*: the price of the services is in turn influenced by the amount of competition, the willingness of the operators to accept losses in order to gain greater market penetration, and the external costs of the operators (especially, the cost of spectrum licenses). However, the price of the services is not always the decisive factor for market penetration.
- *Price of the UEs*: the UEs are usually subsidized by the operators and are either free or sold at a nominal price, if the consumer agrees to a long-term contract. Exceptions are "prepaid" services, where a user buys a certain number of minutes of service usage (in that case, the handsets are sold to the consumer at full cost); at the other end of the market, high-end devices usually require a significant co-payment by the consumer.
- *Attractiveness of the offered services*: in many markets, the price of the services offered by different network operators is almost identical. Operators try to distinguish themselves by different features, like better coverage, free music streaming services, etc. The offering of these improved features also helps to increase the market size in general, as it allows customers to find services tailored to their needs.
- *General economic situation*: obviously, a good general economic situation allows the general population to spend more money on such "nonessential" things as mobile communications services. In countries where a very large percentage of the income goes to basic necessities like food and housing, the market for cellular telephony is obviously more limited. Still, wireless internet connectivity is nowadays often counted among the "basic necessities."
- *Existing telecom infrastructure*: in countries or areas with a bad existing wired telecom infrastructure, cellular telephony, and other wireless services can be the only way of communicating. This can enable high market penetration.

- *Predisposition of the population*: there are several social factors that can increase the cellular market: (i) people have a positive attitude to new technology; (ii) people consider communication as an essential part of their lives; and (iii) high mobility of the population, with people being absent from their offices or homes for a significant part of the day.

Wireless communications has become such a huge market that most of the companies in this business are not even known to most of the consumers. Consumers tend to know *network operators* and *handset manufacturers*. However, other industries abound:

- *Infrastructure manufacturers*: providing infrastructure (BSs, switches, etc.) to the network operators. At the time of this writing, there are four major manufacturers for cellular infrastructure, while larger numbers exist for WLANs and other wireless infrastructure.
- *Component manufacturers*: most handset manufacturers buy chips, batteries, antennas, etc., from external suppliers. This trend was accelerated by the fact that many manufacturers and system integrators spun off their semiconductor divisions. There are even handset companies that do not manufacture anything but are just design and marketing operations with outsourced production.
- *Software suppliers*: software and applications are becoming an increasingly important part of the market. Apps for smartphones, ranging from games to word processors to social media apps, constitute a market worth several tens of billions of dollars.
- *Systems integrators*: WLANs and sensor networks need to be integrated either into larger networks or combined with other hardware (e.g., sensor networks have to be integrated into a factory automation system). This offers new fields of business for OEMs and system integrators.

1.4.3 Behavioral Impact

Engineering does not happen in a vacuum – the demands of people change what the engineers develop and the products of their labor influences how people behave. Cellphones have enabled us to communicate anytime – something that most people think of as desirable. But we have to be aware that it changes our lifestyle. In former times, one did not call a person, but rather a location. That meant a rather clean separation between professional or personal life. Due to the cellphone, anybody can be reached at any time – somebody from work can call in the evening; a private acquaintance can call in the middle of a meeting; in other words, professional and private lives get intermingled. On the positive side, this also allows new and more convenient forms of working and increased flexibility.

Another important behavioral impact is the development of (or lack of) *cellphone etiquette*. Most people tend to agree that hearing a cellphone ring during a concert is exasperating – and still, there is a significant number of people who are not willing to turn their phones to the "silent" mode during such occasions. There also seems to be an innate reluctance in humans to just ignore a ringing cellphone. People are willing to interrupt whatever they are doing in order to answer a ringing phone. Caller identification, automatic callback features, etc., are solutions that engineers can provide to alleviate these problems. A related problem is checking one's smartphone while being in the company of, or even having conversations with, others, and the general reduction of attention span due to distractions from smartphones. "Screentime limits" and other schedulers for limiting distractions, which are enforced by software, can be one remedy.

On a more serious note, wireless devices, and especially cellphones, can be a matter of life and death. Being able to call for help in the middle of the wilderness after a mountaineering accident is definitely a lifesaving feature. Location devices for victims of avalanches have a similar beneficial role. On the downside, drivers who are distracted by their cellphone conversations or text messages constitute a serious hazard on the roads. Studies at Virginia Technological University showed that talking on a cellphone while driving – even when using hands-free equipment – constitutes a hazard similar to being drunk. Text messaging while driving is even worse! It is estimated that in the U.S.A. alone, more than 10,000 traffic fatalities per year are caused by distracted drivers, with the majority of the distractions related to cellphone use.

While the author of this book hopes that the readers will remember many of the technical facts presented throughout the text, the by far most important message to remember from this book is: **Do not text or phone while driving !!!** The problem is, again, not purely a technical one, as a multitude of solutions (including the "off" button) have already been developed to solve this issue. Rather, it is a matter of behavioral changes by the users and the question of what the engineers can do to further these changes.

For updates and errata for this chapter, see https://wides.usc.edu/students.html#textbooks

Exercises

See Sec. 36.1 of Exercises.pdf at wiley.com/go/molisch/wireless3e

2

Technical Challenges of Wireless Communications

In the previous chapter, we described the requirements for wireless communications systems stemming from the applications and user demands. In this chapter, we give a high-level description of the technical challenges to wireless communications systems. Most notably, they are as follows:

- Broadcast effect, i.e., the fact that a signal sent from a Transmitter (TX) inherently can reach multiple Receivers (RX).
- Multi-path propagation: i.e., the fact that a transmit signal can reach the RX via different paths (e.g., reflections from different houses or mountains);
- Spectrum limitations;
- Energy limitations;
- User mobility.

This sets the stage for the rest of the book, where these challenges, as well as remedies, are discussed in more detail.

As a first step, it is useful to investigate the differences between wired and wireless communications. Let us first repeat some important properties of wired and wireless systems, as summarized in Table 2.1.

2.1 Broadcast Effect

A first fundamental difference between wired and wireless systems is that waves sent out by the TX go into all directions and can thus reach many RXs simultaneously. This effect, which is called the *broadcast effect*, can be an advantage or a drawback. It allows to supply multiple RXs with the same information, without requiring any additional energy or spectral resources to reach them. This is the basic principle underlying broadcast TV and radio, such that the number of people who can listen to a transmission is essentially unlimited.

On the other hand, the broadcast effect is also the fundamental reason for interference in wireless systems. The information intended for a particular RX (and only it) cannot be constrained so that it reaches only the intended target but will also be incident onto other RXs, where it acts as interference. Signals intended for different users thus usually have to be distinguished by other means, e.g., by sending them in different parts of the spectrum. The resulting mechanisms for *multiple access* will be discussed in more detail in Part IV of this book.

2.2 Multi-path Propagation

For wireless communications, the transmission medium is the radio channel between TX and RX. The signal can get from the TX to the RX via a number of different propagation paths. In some cases, a Line of Sight (LOS) connection might exist between TX and RX. Furthermore, the signal can get from the TX to the RX by being reflected at or diffracted by different *Interacting Objects* (IOs) in the environment: houses, mountains (for outdoor environments), windows, walls, etc. The number of these possible propagation paths is typically very large. As shown in Figure 2.1, each of the paths has a distinct amplitude, delay (runtime of the signal), direction of departure from the TX, and direction of arrival; most importantly, the components have different *phase shifts* with respect to each other. In the following, we discuss some implications of the multi-path propagation for system design.

Wireless Communications: From Fundamentals to Beyond 5G, Third Edition. Andreas F. Molisch.
© 2023 John Wiley & Sons Ltd. Published 2023 by John Wiley & Sons Ltd.
Companion website: www.wiley.com/go/molisch/wireless3e

Table 2.1 Wired and wireless communications

Wired communications	Wireless communications
The communication takes place over a more or less stable medium like copper wires or optical fibers. The properties of the medium are well defined and time-invariant	Due to user mobility as well as multi-path propagation, the transmission medium varies strongly with time
Increasing the transmission capacity can be achieved by using a different frequency on an existing cable, and/or by stringing new cables	Increasing the transmit capacity must be achieved by more sophisticated transceiver concepts and smaller cell sizes (in cellular systems), as the amount of available spectrum is limited
The range over which communications can be performed without repeater stations is mostly limited by attenuation by the medium (and thus noise); for optical fibers, the distortion of transmitted pulses can also limit the speed of data transmission	The range that can be covered is limited both by the transmission medium (attenuation, fading, and signal distortion) and by the requirements of spectral efficiency (cell size)
Interference and crosstalk from other users either do not happen or the properties of the interference are stationary (for a given set of scheduled users)	Interference and crosstalk from other users are inherent in the principle of cellular communications. Due to the mobility of the users, they also are time-variant, even for a given set of active users
The delay in the transmission process is also constant, determined by the length of the cable and the group delay of possible repeater amplifiers	The delay of the transmission depends partly on the distance between BS and UE and is thus time-variant
The *Bit Error Rate* (BER) decreases strongly (approximately exponentially) with increasing *Signal-to-Noise Ratio* (SNR). This means that a relatively small increase in transmit power can greatly decrease the error rate	For simple systems, the average BER decreases only slowly (linearly) with increasing average SNR. Increasing the transmit power usually does not lead to a significant reduction in BER. However, more sophisticated signal processing helps
Due to the well-behaved transmission medium, the quality of wired transmission is generally high	Due to the difficult medium, transmission quality is generally low unless special measures are used
Jamming and interception of dedicated links with wired transmission is almost impossible without consent by the network operator[a]	Jamming a wireless link is straightforward, unless special measures are taken. Interception of the on-air signal is possible. Encryption is, therefore, necessary to prevent unauthorized use of the information
Establishing a link is *location* based. In other words, a link is established from one outlet to another, independent of which *person* is connected to the outlet	Establishing a connection is based on the (mobile) equipment, usually associated with a specific person. The connection is not associated with a fixed location
Power is either provided through the communications network itself (e.g., for traditional landline telephones), or from traditional power mains (e.g., fax). In neither case is energy consumption a major concern for the designer of the device	UEs use rechargeable or one-way batteries. Energy efficiency is thus a major concern

[a]Note, though, that interception of wired Internet communication is simple due to the design of this particular communication protocol, making encryption vital.

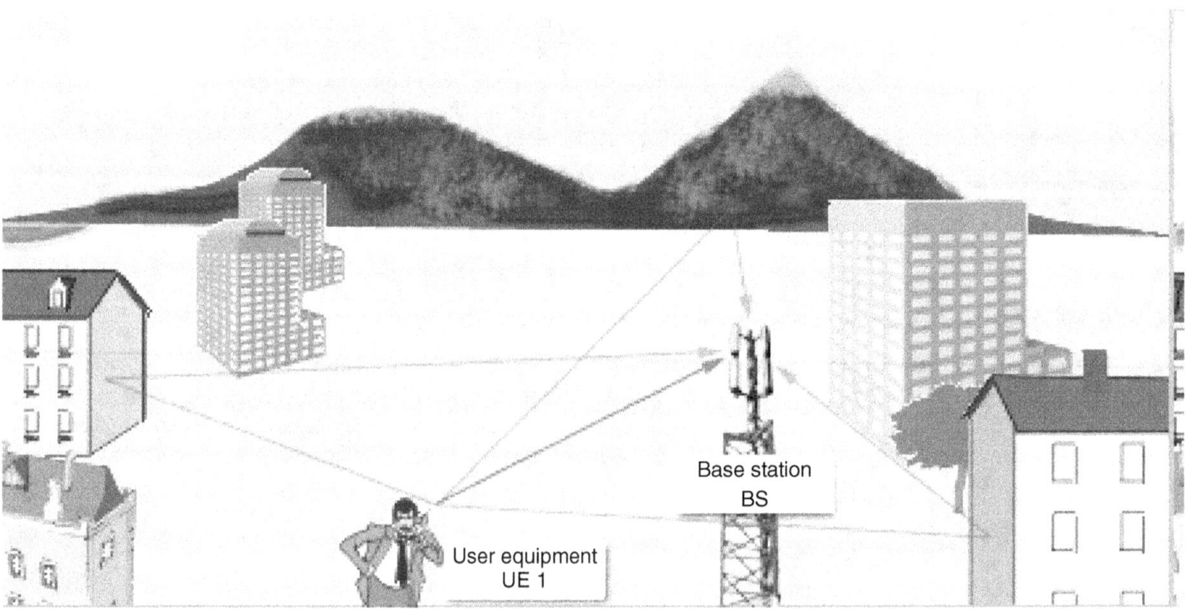

Figure 2.1 Multi-path propagation. Color version available at wiley.com/go/molisch/wireless3e.

2.2.1 Fading

A simple RX cannot distinguish between the different *Multi-Path Components* (MPCs); it just adds them up, so that they interfere with each other. The interference between them can be constructive or destructive, depending on the phases of the MPCs, (see Figure 2.2). The phases, in turn, depend mostly on the run length of the MPC, and thus on the position of the User Equipment (UE) and the IOs. For this reason, the interference, and thus the amplitude of the total signal, changes with time if either TX, RX, or IOs is moving. This effect – namely, the changing of the total signal amplitude due to interference of the different MPCs – is called *small-scale fading*. In other words, even a small movement can result in a large change in signal amplitude. A similar effect is known to all owners of car radios – moving the car by less than 1m (e.g., in stop-and-go traffic) can greatly affect the quality of the received signal. For cellphones, it can often be sufficient to move one step in order to improve signal quality, since at 2 GHz carrier frequency, a movement by less than 10 cm can already create a change from constructive to destructive interference and vice versa.

As an additional effect, the amplitude of each separate MPC changes with time (or with location). Obstacles can lead to a shadowing of one or several MPCs. Imagine, e.g., the UE in Figure 2.3 that at first (at position A) has LOS to the Base Station (BS). As the UE moves behind the high-rise building (at position B), the amplitude of the component that propagates along the direct connection (LOS) between BS and UE greatly decreases. This is due to the fact that the UE is now in the radio shadow of the high-rise building, and any wave going through or around that building is greatly attenuated – an effect called *shadowing*. Of course, shadowing can occur not only for a LOS component but also for *any* MPC. Note also that obstacles do not throw "sharp" shadows: the transition

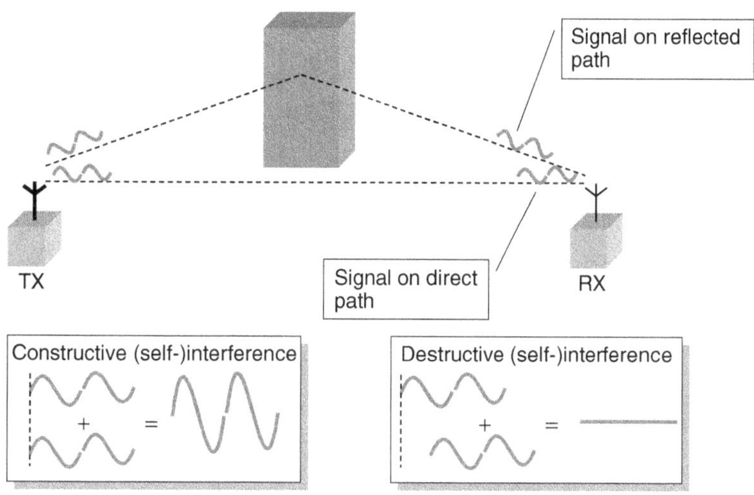

Figure 2.2 Principle of small-scale fading.

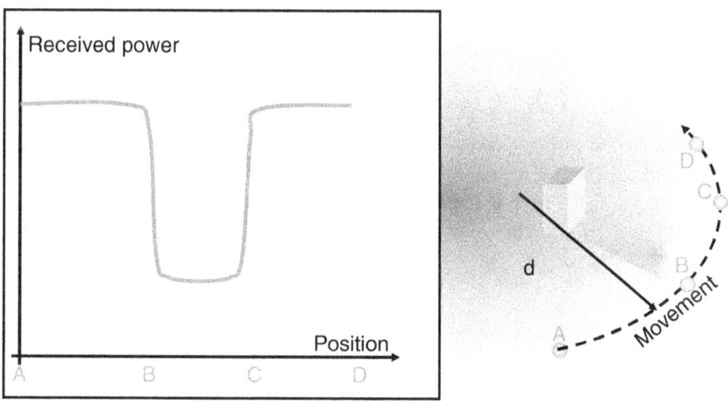

Figure 2.3 The principle of shadowing.

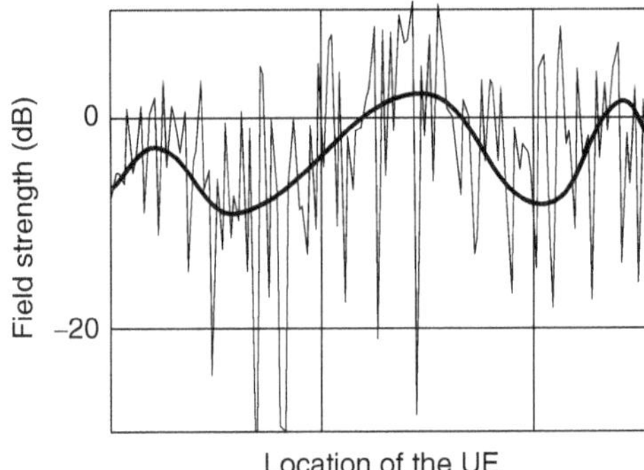

Figure 2.4 Typical example of fading. The thin line is the (normalized) instantaneous field strength; the thick line is the average over a 1 m distance.

from the "light" (i.e., LOS) zone to the "dark" (shadowed) zone is gradual.[1] The UE has to move over large distances (from a few meters up to several hundreds of meters) to move from the light to the dark zone. For this reason, shadowing gives rise to *large-scale fading*.

Large-scale and small-scale fading overlap, so that the received signal amplitude can look like the one depicted in Figure 2.4. Obviously, the transmission quality is low at the times (or places) with low signal amplitude. This can lead to bad speech quality (for voice telephony), high Bit Error Rate (BER), and low data rate (for data transmission), and – if the quality is too low for an extended period of time – to termination of the connection.

It is well known from conventional digital communications that for nonfading communications links, the BER decreases approximately exponentially with increasing Signal-to-Noise Ratio (SNR) if no special measures are taken. However, in a fading channel, the SNR is not constant; rather, the probability that the link is in a *fading dip* (i.e., location with low SNR) dominates the behavior of the BER. For this reason, the average BER decreases only *linearly* with increasing average SNR. Consequently, improving the BER often cannot be achieved by simply increasing the transmit power. Rather more sophisticated transmission and reception schemes have to be used. Most of Part III of this book is devoted to such techniques.

Due to fading, it is almost impossible to exactly predict the received signal amplitude at specific locations. For many aspects of system development and deployment, it is considered sufficient to predict the mean amplitude and the *statistics* of fluctuations around that mean. Completely deterministic predictions of the signal amplitude – e.g., by solving approximations to Maxwell's equations[2] in a given environment – are very sensitive to even small uncertainties in the locations and properties of the IOs. More details on fading can be found in Chapters 4–7.

2.2.2   *Intersymbol Interference*

The runtimes for different MPCs are different. We have already mentioned above that this can lead to different phases of MPCs, which lead to interference in narrowband systems. In a system with large bandwidth, and thus good resolution in the time domain,[3] the major consequence is signal dispersion: in other words, the impulse response of the channel is not a single delta pulse but rather a sequence of pulses (corresponding to different MPCs), each of which has a distinct arrival time in addition to having a different amplitude and phase (see Figure 2.5). This signal dispersion leads to InterSymbol Interference (ISI) at the RX. MPCs with long runtimes, carrying information from bit k, and MPCs with short runtimes, carrying contributions from bit $k + 1$ arrive at the RX at the same time and interfere with each other (see Figure 2.6). Assuming that no special measures[4] are taken, this ISI leads to errors that cannot be eliminated by simply increasing the transmit power, and are therefore often called *irreducible errors*.

ISI is essentially determined by the ratio between symbol duration and the duration of the impulse response of the channel. This implies that ISI is not only more important for higher data rates but also for multiple access methods that lead to an increase in transmitted *peak* data rate (e.g., time division multiple access, see Chapter 18). Finally, it is also noteworthy that ISI can even play a role when the duration of the impulse response is *shorter* (but not *much* shorter) than bit duration (see Chapter 11).

---

[1] This is due to (i) diffraction effects, as also explained in more detail in Chapter 4 and (ii) the fact that secondary radiation sources like houses are spatially extended (compare how a long fluorescent tube never throws a sharp shadow).

[2] The most popular of these deterministic prediction tools are "ray tracing" and "ray launching," which are discussed in Sec. 4.7.

[3] Strictly speaking, we refer here to resolution in the delay domain. An explanation for the difference between the time domain and delay domain is given in Chapter 6.

[4] Special measures include equalizers (Chapter 14), Rake receivers (Chapter 19), and Orthogonal Frequency Division Multiplexing (OFDM) (Chapter 15).

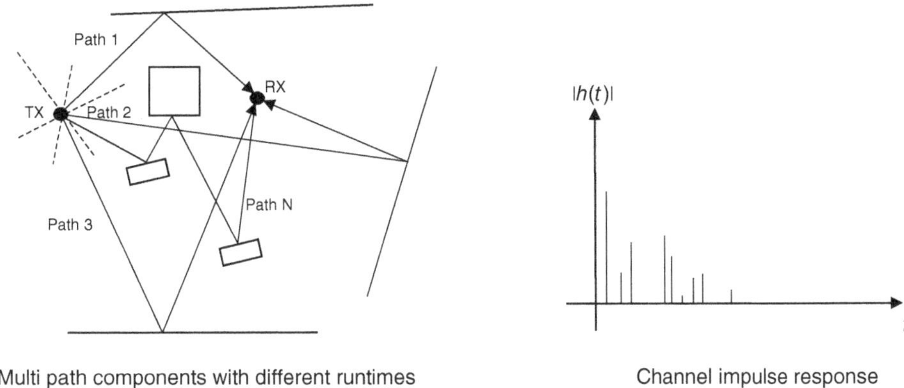

Figure 2.5   Multi-path propagation and resulting impulse response.

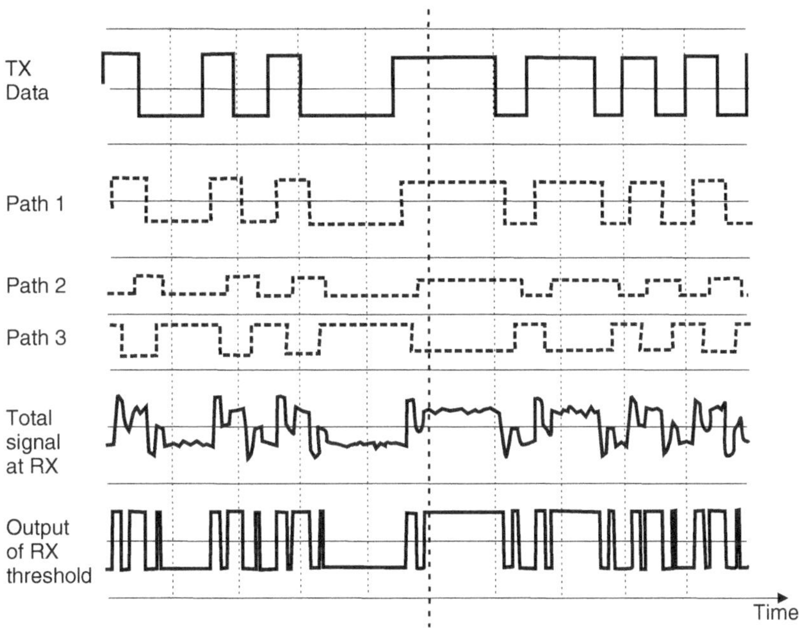

Figure 2.6   Intersymbol interference.

2.3   Spectrum Limitations

The spectrum available for wireless communications services is limited, and its permissible usage is determined by international agreements. For this reason, the spectrum has to be used in a highly efficient manner. Two approaches are used: regulated spectrum, where a single network operator has control over the usage of the spectrum, and unregulated spectrum, where each user can transmit without additional control, as long as they comply with certain restrictions on the emission power and bandwidth. In the following, we first review the frequency ranges assigned to different communications services. We then discuss the basic principle of frequency reuse for both regulated and unregulated access.

2.3.1   Assigned Frequencies

The frequency assignment for different wireless services is regulated by the *International Telecommunications Union* (ITU), a specialized agency of the United Nations. In its tri-annual conferences (*World Radio Conferences*), it establishes worldwide guidelines for the usage of spectrum in different regions and countries. Further regulations are issued by the frequency regulators of individual countries, including the *Federal Communications Commission* (FCC) in the U.S.A., for more details see Sec. 30.2.4. While the exact frequency assignments differ, similar services tend to use the same frequency ranges all over the world (the following is a rough outline; many exceptions and additional services exist):

- Below 100 MHz: at these frequencies, we find Citizens' Band (CB) radio, pagers, aeronautical and maritime services.
- 100–700 MHz: these frequencies are mainly used for broadcast (radio and TV) applications, plus some maritime frequencies (around 150 MHz) and public safety bands (450–520 MHz).
- 400–500 MHz: a number of cellular and trunking radio systems make use of this band. It is mostly systems that need good coverage but show low user density.
- 700–1000 MHz: cellular systems use this band. Also, some emergency communications systems make use of this band, and an Industrial, Scientific, and Medical (ISM) band is located around 915 MHz in some regions of the world.
- 1200–1600 MHz: these bands contain, *inter alia*, the frequencies at which GPS transmits its signal (1228 and 1575 MHz). Since the received signal is both weak and especially important for many applications, strict guard bands around it are used.
- 1.7–2.2 GHz: this is the main frequency band for cellular communications.
- 2.4–2.5 GHz: the most important ISM band. Wireless Local Area Networks (WLANs) and wireless Personal Area Networks (PANs) operate in this band; they share it with many other devices, including microwave ovens.
- 3.3–3.8 GHz: the so-called "mid-band," which has been assigned to cellular communications around 2020 (different parts of this spectrum are used in different parts of the world).
- 4.8–7.1 GHz: most WLANs (and unlicensed cellular operation) can be found (depending on the region) between 4.8 and 5.8 GHz. The frequency range around 5.8 GHz has been assigned to car-to-car communications. Finally, the frequency band from 6 to 7.1 GHz was assigned to WLANs in America and parts of that band in Europe and Australia.
- 6–8 and 10–12 GHz: these bands are mainly used for satellite communications, though note the use of 6–7 GHz for WLAN as described above.
- 11–15 GHz: in this range, we can find the most popular satellite TV services, which use 14.0–14.5 GHz for the uplink, and 11–12.5 GHz for the downlink.
- 24–100 GHz: this range contains (in the U.S.A.) almost 4 GHz of licensed spectrum for cellular applications (28 GHz band and 37–40 GHz), as well as 14 GHz of unlicensed spectrum (the 57–71 GHz band); other countries are in the process to establish similar rules. Furthermore, car radar systems have long operated in the 24 GHz band but are moving to 77 GHz.
- 100 GHz. There is also increasing interest in the frequency ranges above 100 GHz, in particular for ultra-high-speed, short-distance communication as well as radar-like applications and sensing. The frequency allocations are currently in flux.

More details about the exact frequencies for specific services can be obtained from the national frequency regulators, as well as from the ITU. The detailed specifications are highly complex; the summary table of the FCC, for example, runs over 120 pages.

Different frequency ranges are optimum for different applications. Low carrier frequencies usually propagate more easily (see also Chapter 4), so that a single BS can cover a large area. On the other hand, absolute bandwidths are smaller and also the frequency reuse is not as efficient as it is at higher frequencies.[5] For this reason, low-frequency bands are best for services that require good coverage but have a small required aggregate rate of information. A typical case in point is television, since only a single information stream is sent to *all* users (thus the bandwidth does not scale with the number of users). For cellular systems, low carrier frequencies are ideal for covering large regions with low user density such as rural areas, or for providing outdoor-to-indoor coverage. For cellular systems with high user densities, as well as for WLANs, higher carrier frequencies are usually more desirable.[6]

The amount of spectrum assigned to the different services does not always follow technical necessities but rather historical developments. For example, for many years, the amount of precious low-frequency spectrum assigned to TV stations was much higher than would be justified by technical requirements. Using appropriate frequency planning and different transmission techniques (including simulcast), a considerable part of the spectrum below 1 GHz could be freed up for alternative usage – a process that took place in the North America and Europe around 2008/2009. The process took a long time because broadcast stations tended to fight such a development, as it required modifications in their TXs. As these stations have a considerable influence on public opinion as well as lobbying power, frequency regulators were hesitant to enforce appropriate rule changes.

It is also noticeable that the financial terms on which spectrum is assigned to different services differ vastly – from country to country, from service to service, and even depending on the time at which the spectrum is assigned. Obviously, spectrum is assigned to public safety services (police, fire department, and military) without monetary compensation. Even television stations usually get the spectrum assigned for free. In the 1980s, spectrum for cellular telephony was often assigned for a rather small fee, in order to encourage the development of this then-new service. Since the mid- and late-1990s, spectrum auctions are generally seen as a welcome method to increase a country's revenues; for example, the price paid in 2008 for a US-wide license for 15 MHz of "beachfront"

---

[5] As we will see in the next subsection, frequency reuse requires that a signal is attenuated strongly outside the cell it is assigned to. However, low carrier frequency results in good propagation so that the signal can remain strong far outside its assigned cell.

[6] However, the carrier frequency should not become *too* high: at extremely high frequencies, it becomes difficult to cover even small areas.

spectrum (in the 700 MHz band) was on the order of US$5 billion, while the auction of the 3.5 GHz midband in 2020 provided US$80 billion to the government coffers. Other countries chose to assign spectrum based on a "beauty contest," where the applicant had to guarantee a certain service quality, coverage, etc., in order to obtain a license. Unregulated services, like WLANs, are assigned spectrum without fees.

2.3.2   Frequency Reuse in Regulated Spectrum

Since spectrum is limited, the *same* spectrum has to be used for *different* wireless connections in *different* locations. To simplify the discussion, let us consider in the following a cellular system where different connections (different users) are distinguished by the frequency channel (band around a certain carrier frequency) that they employ. If an area is served by a single BS, then the available spectrum can be divided into N frequency channels that can serve N users simultaneously. If more than N users are to be served, multiple BSs are required, and frequency channels have to be reused in different locations.

For this purpose, we divide the area (a region, a country, or a whole continent) into a number of *cells*; we also divide the available frequency channels into several groups. The channel groups are now assigned to the cells. The important thing is that channel groups can be used in multiple cells. The only requirement is that cells that use the same frequency group do not interfere with each other *significantly*.[7] It is fairly obvious that the same carrier frequency can be used for different connections in, say, Rome and Stockholm, at the same time. The large distance between the two cities makes sure that a signal from the UE in Stockholm does not reach the BS in Rome and can therefore not cause any interference at all. But in order to achieve high efficiency, frequencies must actually be reused much more often – typically, several times within each city. Consequently, *intercell interference* (also known as *co-channel interference*) becomes a dominant factor that limits transmission quality. More details on co-channel interference can be found in Part IV.

Area spectral efficiency describes the effectiveness of reuse – i.e., the traffic density that can be achieved per unit bandwidth and unit area. It is therefore given in units of Erlang/(Hz m^2) for voice traffic and bit/(s Hz m^2) for data. Since the area covered by a network provider, as well as the bandwidth that it can use, are fixed, increasing the area spectral efficiency is the only way to increase the number of customers that can be served, and thus revenue. Methods for increasing the area spectral efficiency are thus at the center of wireless communications research.

Since a network operator buys a license for a spectrum, it can use that spectrum according to its own planning – i.e., network planning can make sure that the users in different cells do not interfere with each other significantly. The network operator is allowed to use as much transmit power as it desires; it can also dictate limits on the emission power of the UEs of different users.[8] The operator can also be sure that the only interference in the network is created by its own network and users.

2.3.3   Frequency Reuse in Unregulated Spectrum

In contrast to regulated spectrum, several services use frequency bands that are available to the general public. For example, some WLANs operate in the 2.45 GHz band, which has been assigned to "ISM" services. Anybody is allowed to transmit in these bands, as long as they (i) limit the emission power to a prescribed value, (ii) follow certain rules for the signal shape and bandwidth, and (iii) use the band according to the (rather broadly defined) purposes stipulated by the frequency regulators.

As a consequence, a WLAN RX can be faced with a large amount of interference. This interference can either stem from other WLAN TXs or from microwave ovens, cordless phones, and other devices that operate in the ISM band. For this reason, a WLAN link must have the capability to deal with interference. That can be achieved by selecting a frequency band within the ISM band at which there is little interference, by using spread spectrum techniques (see Chapter 19), or some other appropriate technique.

There are also cases where the spectrum is assigned to a specific service (e.g., vehicle-to-vehicle communications) but not to a specific operator. In that case, RXs might still have to deal with strong interference, but the structure of this interference is known. This allows the use of special interference mitigation techniques.

Unregulated spectrum often employs *cognitive radio* techniques (see Chapter 26, Sec. 31.6.2, 32.6.2), where a device senses which part of the spectrum is currently unused in the location of interest, and dynamically moves the transmit frequency accordingly.

2.4   Limited Energy

Truly wireless communications requires not only that the information is sent over the air (not via cables) but also that the UE is powered by one-way or rechargeable batteries; otherwise, a UE would be tied to the "wire" of the power supply. Batteries in turn impose restrictions on the power consumption of the devices. The requirement for small energy consumption results in several technical imperatives:

- The power amplifiers in the TX have to have high efficiency. As power amplifiers account for a considerable fraction of the power consumption in many devices, mainly amplifiers with an efficiency above 50% should be used in UEs. Such amplifiers – specifically,

---

[7] The threshold for *significant* interference (i.e., the admissible signal-to-interference ratio) is determined by modulation and reception schemes, as well as by propagation conditions.

[8] There are some exceptions to that rule – e.g., emission limits dictated by health concerns, as well as limits imposed by the standard of the system used by the operator (e.g., LTE).

class-C or class-F amplifiers – are highly nonlinear.[9] As a consequence, wireless communications tend to use modulation formats that are insensitive to nonlinear distortions, see also Chapter 17.

- Signal processing must be done in an energy-saving manner. This implies that the digital logic should be implemented using power-saving semiconductor technology like Complementary Metal Oxide Semiconductor (CMOS), while the faster but more power-hungry approaches like Emitter Coupled Logic (ECL) do not seem suitable for UEs. This restriction has important consequences for the algorithms that can be used for interference suppression, combating of ISI, etc.
- The RX (especially at the BS) needs to have high sensitivity. Already the second-generation Global System for Mobile Communications (GSM) was specified so that even a received signal power of -100 dBm leads to an acceptable transmission quality. Such an RX is several orders of magnitude more sensitive than a TV RX. If the GSM standard had defined -80 dBm instead, then the transmit power would have to be higher by a factor of 100 in order to achieve the same coverage. This in turn would mean that – for identical talktime – the battery would have to be 100 times as large – i.e., 20 kg instead of the typical 200 g. But the high requirements on RX sensitivity have important consequences for the construction of the RX (low-noise amplifiers, sophisticated signal processing to fully exploit the received signal) as well as for network planning.
- Maximum transmit power should be used only when required. In other words, transmit power should be adapted to the channel state, which in turn depends on the distance between TX and RX (*power control*). If the UE is close to the BS, and thus the channel has only a small attenuation, transmit power should be kept low.
- For cellular phones, and even more so for sensor networks, an energy-efficient "standby" or "sleep" mode has to be defined.

Several of the mentioned requirements are contradictory. For example, the requirement to build an RX with high sensitivity (and thus, sophisticated signal processing) is in contrast to the requirement of having energy-saving (and thus slow) signal processing. Engineering tradeoffs are thus called for.

2.5 User Mobility

Mobility is an inherent feature of most wireless systems and has important consequences for system design. Fading was already discussed in Section 2.2.1. A second important effect is particular to mobile users in cellular systems: the system has to know at any time which cell a user is in:[10]

- If there is an incoming call for a certain UE (user), the network has to know in which cell the user is located. The first requirement is that a UE emits a signal at regular intervals, informing nearby BSs that it is "in the neighborhood." Two databanks then employ this information: the *Home Location Register* (HLR) and the *Visitor Location Register* (VLR). The HLR is a central database that keeps track of the location a user is currently at; the VLR is a database associated with a certain BS that notes all the users who are currently within the coverage area of this specific BS. Consider user A, who is registered in San Francisco, but is currently located in Los Angeles. It informs the nearest BS (in Los Angeles) that it is now within its coverage area; the BS enters that information into its VLR. At the same time, the information is forwarded to the central HLR (located, e.g., in New York). If now somebody calls user A, an enquiry is sent to the HLR to find out the current location of the user. After receiving the answer, the call is rerouted to Los Angeles. For the Los Angeles BS, user A is just a "regular" user, whose data are all stored in the VLR.
- If a UE moves across a cell boundary, a different BS becomes the *serving BS*; in other words, the UE is *handed over* from one BS to another. Such a handover has to be performed without interrupting the call; as a matter of fact, it should not be noticeable at all to the user. This requires complicated signaling.

For updates and errata for this chapter, see https://wides.usc.edu/students.html#textbooks

Exercises

See Sec. 36.2 of Exercises.pdf at wiley.com/go/molisch/wireless3e

---

[9] Linear amplifiers, like class A, class B, or class AB, have efficiencies of less than 30%.

[10] This effect is not relevant, e.g., for simple WLAN systems: either a user is within the coverage region of the (one and only) BS, or they are not.

3

Wireless System Design Overview

This chapter explains the principles of wireless transmissions and the planning of wireless systems with one or multiple users. This serves to provide a first "bird's eye" view of such systems and helps to put the later chapters into the correct context. By necessity, this chapter makes a number of simplifying assumptions and looks at special cases; in order to not clutter the text with footnotes and caveats, these will not be explicitly mentioned every time. Rather, the subsequent chapters (to which we will reference) will contain the detailed formulations.

3.1 Noise-limited Systems and Link Budgets

3.1.1 Link Budget Assumptions

A rudimentary analysis of a wireless communication system can be done based on Figure 3.1. The system uses digital modulation of a high-frequency carrier, such that the signal of interest is up-converted to a passband and sent over the air to the Receiver (RX), where it is downconverted to baseband, and the transmitted signal recovered. We consider a situation where only a single TX (assumed to be a Base Station (BS)) transmits, and a User Equipment (UE) receives; thus, the performance of the system is determined only by the strength of the (useful) signal and the noise. A simple analysis of such a system can be based on the following assumptions:

1. A Transmitter (TX) sends out a wireless signal. The transmit signal is characterized by the carrier frequency f_c on which it is sent, the bandwidth B that it occupies, and the power of the transmission P_{TX}.
2. The signal is attenuated during the propagation from TX to RX. The inverse attenuation, i.e., the ratio of the received power P_{RX} to P_{TX} consists of several factors: (i) the path gain (inverse path loss), a function that depends on distance between TX and RX, the carrier frequency f_c, and the environment, (ii) the *antenna (power) gain*, which describes how much the power is enhanced by focusing the transmission into the "right" direction (and similarly, by collecting signal energy from the "right" direction at the RX), (iii) attenuation of lossy components at TX (note that the effect of lossy components at the RX are subsumed into the RX noise figure, see below).
3. The attenuation of the power is not a monotonous (smooth) function of the distance but also shows fading, as discussed in Section 2.2.1. To account for these fluctuations and thus guarantee a high probability that a certain P_{RX} is available, a safety margin, called *fading margin*, *FM*, has to be incorporated in the system planning.
4. In order to satisfy a minimum required quality of transmission, e.g., in terms of bit error rate, the signal demodulated at the RX has to have a minimum operating *signal to noise ratio*, $\gamma_{th,op}$.

As the UE moves further away from the BS, the received signal power decreases, and at a certain distance, the Signal-to-Noise Ratio (SNR) does not achieve the required threshold for reliable communications anymore. Therefore, the range of the system is noise-limited; equivalently we can call it signal-power limited. Depending on the interpretation, it is too much noise or too little signal power that leads to a bad link quality.

3.1.2 Noise Power, Path Gain, and Antenna Gain

Thermal Noise

The following discussion provides more details about the components of a link budget. First, consider the noise power. The noise that disturbs the signal is mainly *thermal noise*, which is usually assumed to have a noise power spectral density

$$N_0 = k_B T_e \tag{3.1}$$

Wireless Communications: From Fundamentals to Beyond 5G, Third Edition. Andreas F. Molisch.
© 2023 John Wiley & Sons Ltd. Published 2023 by John Wiley & Sons Ltd.
Companion website: www.wiley.com/go/molisch/wireless3e

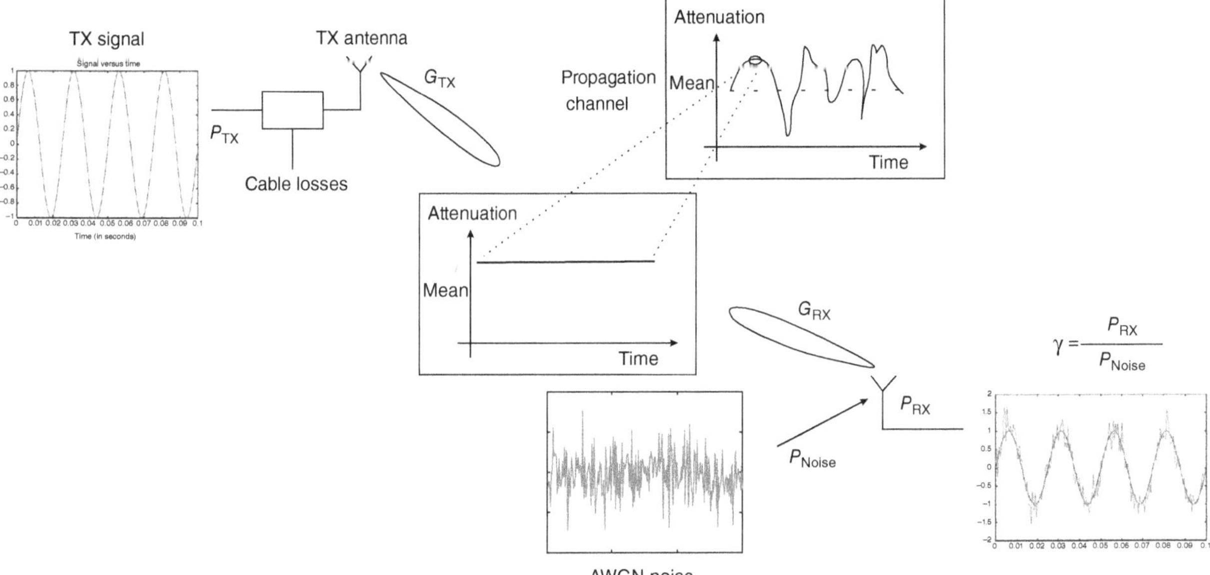

Figure 3.1 Block diagram of a wireless system with flat fading. Note that the channel changes slowly, so that at a given time (where signal and noise vary) it is effectively constant. Color version available at wiley.com/go/molisch/wireless3e.

where k_{B} is Boltzmann's constant, $k_{\text{B}} = 1.38 \cdot 10^{-23}$ Joules/Kelvin, and the environmental temperature is 300 K; this is independent of the considered frequency, i.e., the noise is spectrally white (called this way in analogy to white light). The noise power is then

$$P_{\text{n}} = N_0 B \tag{3.2}$$

where B is the RX bandwidth (in units of Hz). It is common to write Eq. (3.1) using logarithmic units:

$$N_0 = -174 \, \text{dBm/Hz}. \tag{3.3}$$

This means that the noise power contained in 1 Hz bandwidth is −174 dBm. The noise power contained in a bandwidth B is

$$-174 + 10 \log_{10}(B) \, \text{dBm}. \tag{3.4}$$

The logarithm of the bandwidth B, specifically $10\log_{10}(B)$, has the units dBHz. Additional noise is created by RX components, which are characterized by the noise figure, F_{dB}. The "effective" noise power is then $-174 + 10\log_{10}(B) + F_{\text{dB}}$ dBm, or $F k_{\text{B}} T_{\text{e}} B$ on a linear scale.

The amplitude of the thermal noise is Gaussian distributed, a fact that can be explained by the central limit theorem.

Other Noise Sources

In addition to the thermal noise, also manmade noise needs to be taken into account.

- *Spurious emissions*: Many electrical appliances, as well as radio TXs designed for other frequency bands, have spurious emissions over a large bandwidth that includes the frequency range in which wireless communications systems operate. For urban outdoor environments, car ignitions and other impulse sources are especially significant sources of noise. In contrast to thermal noise, the noise created by impulse sources decreases with frequency. At typical cellular frequencies, the noise enhancement by manmade noise is on the order of a few dB, though it does show significant variations with the type of environment. Note that frequency regulators in most countries impose limits on "spurious" or "out-of-band" emissions for all electrical devices. In contrast to thermal noise, man-made noise does not necessarily have a Gaussian amplitude distribution. However, as a matter of convenience, most system-planning tools, as well as theoretical designs, assume Gaussianity anyway.
- *Other intentional emission sources*: Several wireless communications systems operate in unlicensed bands. In these bands, everybody is allowed to operate (emit electromagnetic radiation) as long as certain restrictions with respect to transmit power, etc. are fulfilled. The most important of these bands is the 2.45-GHz Industrial, Scientific, and Medical (ISM) band. The amount of interference in these bands can be considerable but is often not predictable. This type of interference is also commonly treated like noise.

For communications operating in licensed bands, spurious emissions are the only source of man-made noise. It lies in the nature of the license (for which the license holder usually has paid) that no other intentional emitters are allowed to operate in this band.
It must finally be noted that spurious and intentional emissions are not worsened by the noise figure of the RX.

Path Gain

We next turn to the computation of the received power. If the signal propagates in free space, then the relationship between P_{TX} and P_{RX} can be written as

$$P_{RX} = P_{TX} G_{RX} G_{TX} \left(\frac{\lambda}{4\pi d} \right)^2 \tag{3.5}$$

where G_{RX} and G_{TX} are the antenna gains (discussed below) of the TX and RX antennas, respectively, and λ is the wavelength, c_0/f_c. A physical interpretation of this so-called "Friis' law" will be given in Chapter 4. However, for earth-bound wireless systems, this law is only valid up to distances $d < d_{\text{break}}$; beyond that point, the power is proportional to $d^{-\alpha_{\text{LBP}}}$, where α_{LBP} is a constant that typically lies between 3.5 and 4.5. For $d > d_{\text{break}}$ the received power is thus

$$P_{RX}(d) = P_{RX}(d_{\text{break}}) \left(\frac{d}{d_{\text{break}}} \right)^{-\alpha_{\text{LBP}}} \quad \text{for } d > d_{\text{break}}. \tag{3.6}$$

This is shown in Figure 3.2. More extensive discussions of pathloss models can be found in Chapter 7.

Fading Margin

Wireless systems suffer from temporal and spatial variations of the transmission channel (*fading*), as discussed in Section 2.2.1. In other words, even if the distance is fixed, the received power can change significantly with time, and/or with the movement of the UE over very small distances (keeping d effectively constant). The power computed from (3.5 and 3.6) is only a *mean* value. If it is used as the basis for the link budget, then the transmission quality will be above the threshold only in approximately 50 % of the times and locations.[1] This is completely unacceptable reliability. Therefore, we have to add a *fading margin*, which makes sure that the minimum receive power is exceeded in at least 90 % of all cases (or some other reliability value defined by the system designer), see Figure 3.3. The value of the fading margin depends on the amplitude statistics of the fading and will be discussed in more detail in Chapter 5.

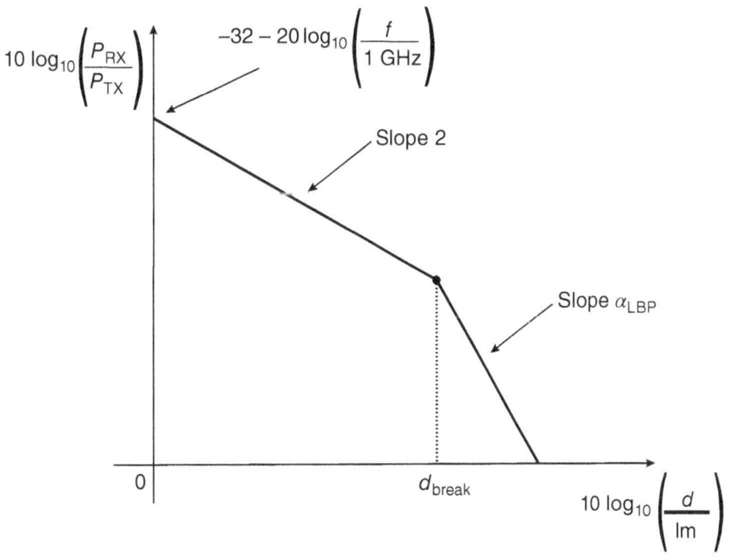

Figure 3.2 Breakpoint model for the pathloss.

---

[1] For the purpose of this chapter, we neglect the difference between *mean* and *median* power.

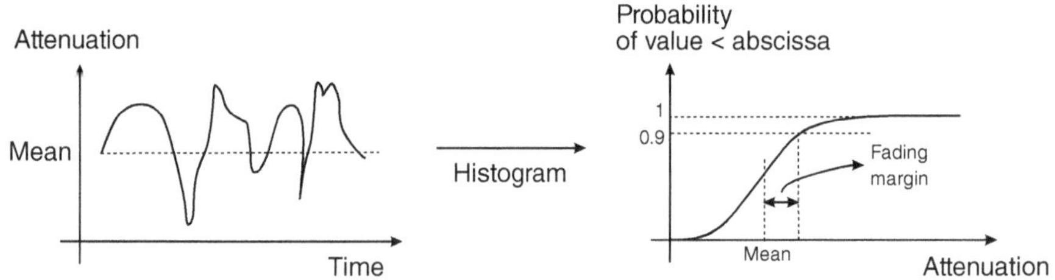

Figure 3.3 Fading margin to guarantee a certain outage probability.

Antenna Gain

The term

$$\left(\frac{\lambda}{4\pi d}\right)^{2}$$

is often called the isotropic free space path gain. It is the path gain in free space that is encountered when isotropic antennas are used at TX and RX. An isotropic antenna transmits a signal equally into all directions. There are, however, antennas that focus energy into certain directions – for example, the ground station for a satellite uplink, which clearly focuses the transmitted energy toward the receiving satellite. The signal power is thus enhanced in this particular direction; this enhancement is often described by the *antenna gain* (more precisely, the *peak antenna power gain*). The product $P_{\mathrm{TX}}G_{\mathrm{TX}}$ is often called the *Equivalent Isotropically Radiated Power, EIRP*.

When the antenna has losses, then the gain is reduced; this is particularly relevant for the UE, since the absorption of signal energy in the nearfield by the user leads to such losses; see also Chapter 8. While most link budgets assume an isotropic lossless UE antenna, which thus has a gain of 0 dB, a negative gain is more realistic and will be used in some of the examples below.

3.1.3 Link Budget

After these preliminaries, we are now ready to compute link budgets. A link budget is the clearest and most intuitive way of computing the required TX power [Sklar 2001, Chapter 5]. In this link budget, we write all equations that connect the TX power to the received SNR in a tabular form. It has to be noted, however, that the link budget gives only an approximation (often a worst-case estimate) for the total SNR, because some interactions between different effects are not taken into account.

As most factors influencing the SNR enter in a multiplicative way, it is convenient to write all equations in a logarithmic form, specifically, in dB (decibel)

$$P_{\mathrm{dB}} = 10\log_{10}(P_{\mathrm{linear}}). \tag{3.7}$$

A table with the conversion of elementary factors into dB is given below. Since most numbers can be written as products of these factors, conversions can be easily made without a calculator. For example, 160 on a linear scale is $8 \cdot 2 \cdot 10 = 9 + 3 + 10 = 22$ dB. Further values can be obtained from other logarithmic computation rules. For example, 7 is the square root of $49 \simeq 50 = 100/2 = 17$ dB. Since the square root on a linear scale corresponds to a division by 2 on a logarithmic scale, 7 corresponds to 8.5 dB.

Multiplicative factor	Additive term in dB
2	3
3	5
5	7
8	9
10	10

Link budgets can be created both for the *Downlink, DL* (the link from the BS to the UE), or the *uplink, UL*, (UE to BS). The pathloss is the same for uplink and downlink and that antenna gains are the same no matter whether the antenna is transmitting or receiving – thus, $G_{\mathrm{BS}} = G_{\mathrm{RX,\,UL}} = G_{\mathrm{TX,\,DL}}$, and similarly for G_{UE}. However, in many other respects, the link budgets for uplink and downlink differ:

- The transmit power for the downlink, i.e., the power emitted by the BS, is much higher than the transmit power for the uplink. This is motivated by the fact that a BS is connected to the power mains, while the UE has to use battery power to transmit. Furthermore, safety limits place an upper limit on UE TX power, typically at a few hundred mW.

- The noise figures of BS and UE are typically quite different. As UEs have to be produced in quantity, it is desirable to use low-cost components, which typically have higher noise figures.

For the pathloss, a handy number to keep in mind is that the pathloss at 1 m distance, at 1 GHz carrier frequency, is 32 dB. From this, the pathloss at 1 m distance at any arbitrary frequency can be computed from the fact that the pathloss goes with f_c^2, so that

$$2 \cdot 10 \log_{10}\left(\frac{f_c}{1\ \text{GHz}}\right)$$

needs to be added to this number. Similarly, when considering distances larger than 1 m (but smaller than d_{break}), we simply add

$$2 \cdot 10 \log_{10}\left(\frac{d}{1\ \text{m}}\right)$$

and when considering distances larger than d_{break}, we add first the pathloss from 1 m to the breakpoint,

$$2 \cdot 10 \log_{10}\left(\frac{d_{\text{break}}}{1\ \text{m}}\right)$$

and then additionally the loss from the breakpoint to the distance of interest

$$\alpha_{\text{LBP}} \cdot 10 \log_{10}\left(\frac{d}{d_{\text{break}}}\right).$$

All of these factors are easy to compute in one's head with the dB calculation guidelines given above.

Example 3.1 Link Budget *Consider the downlink of a GSM system: GSM is an old (second-generation) cellular system, that is still in widespread use in parts of the world, see Appendix 30.A. The carrier frequency is 950 MHz, the RX sensitivity is (according to the GSM specifications) −102 dBm. The output power of the TX amplifier is 30 W. The antenna gain of the TX antenna is 10 dB, the losses in connectors, combiners, etc., are 5 dB. The fading margin is 12 dB, the breakpoint d_{break} is assumed to be at a distance 1 m. What distance can be covered?*

TX side			
TX power	P_t	30 W	45 dBm
Antenna gain	G_t	10	10 dB
Losses (combiner, connector,...)	L_f		−5 dB
EIRP (Equivalent Isotropically Radiated Power)			50 dBm
RX side			
RX sensitivity	P_{min}		−102 dBm
Fading margin	FM		12 dB
Minimum RX power (median)			−90 dBm
Admissible pathloss (difference EIRP and min. RX power)			140 dB
Pathloss at $d_{\text{break}} = 1$ m	$[\lambda/(4\pi d)^2]$		32 dB
Pathloss beyond breakpoint	prop. $d^{-\alpha_{\text{LBP}}}$		108 dB

Depending on the pathloss exponent, which for non-line-of-sight is typically

$$\alpha_{\text{LBP}} = 3.5...4.5.$$

we obtain the coverage distance

$$d_{\text{cov}} = 10^{108/(10\alpha_{\text{LBP}})}\ \text{m}. \tag{3.8}$$

If, for example, $\alpha_{\text{LBP}} = 3.5$, then the coverage distance is 1,200 m.

This example was particularly easy, because the RX sensitivity was prescribed by the system specifications. If it is not available, the computations at the RX become more complicated.

Example 3.2 Link Budget   *Consider a 5G NR cellular system (see Chapter 32) at 2,500 MHz carrier frequency, that is affected by thermal noise only (temperature of the environment $T_e = 300\ K$). Antenna gains at TX and RX sides are 10 dB and -2 dB, respectively. Losses in cables, combiners, etc. at the TX are 2 dB. The noise figure of the RX is 7 dB, the bandwidth of the signal is 20 MHz. The required operating SNR is 3 dB, the desired range of coverage is 0.5 km. The breakpoint is at 10 m distance; beyond that point, the pathloss exponent is 4, and the fading margin is 9 dB. What is the minimum TX power?*

Due to the formulation of this problem, we can work our way backward from the RX to the TX.

Noise spectral density $k_B T_e$:	-174 dBm/Hz
Bandwidth:	73 dBHz
Thermal noise power at the RX	-101 dBm
RX excess noise	7 dB
Required SNR	3 dB
Required RX power	-91 dBm
Pathloss from 10 m to 0.5 km distance $(4 \cdot 10 \cdot \log_{10}(50))$	68 dB
Pathloss from TX to breakpoint at 10 m $(\lambda/(4\pi d))^2 = 32 + 20 + 8$	60 dB
Antenna gain at the UE G_{RX} (2 dB loss):	$-(-2)$ dB
Fading margin	9 dB
Required EIRP:	48 dBm
TX antenna gain G_{TX} (10 dB gain):	-10 dB
Losses in cables, combiners, et. at TX L_f	2 dB
Required TX power (amplifier output)	40 dBm

The required TX power is thus 40 dBm, or 10 W. The link budget is also represented in Figure 3.4.

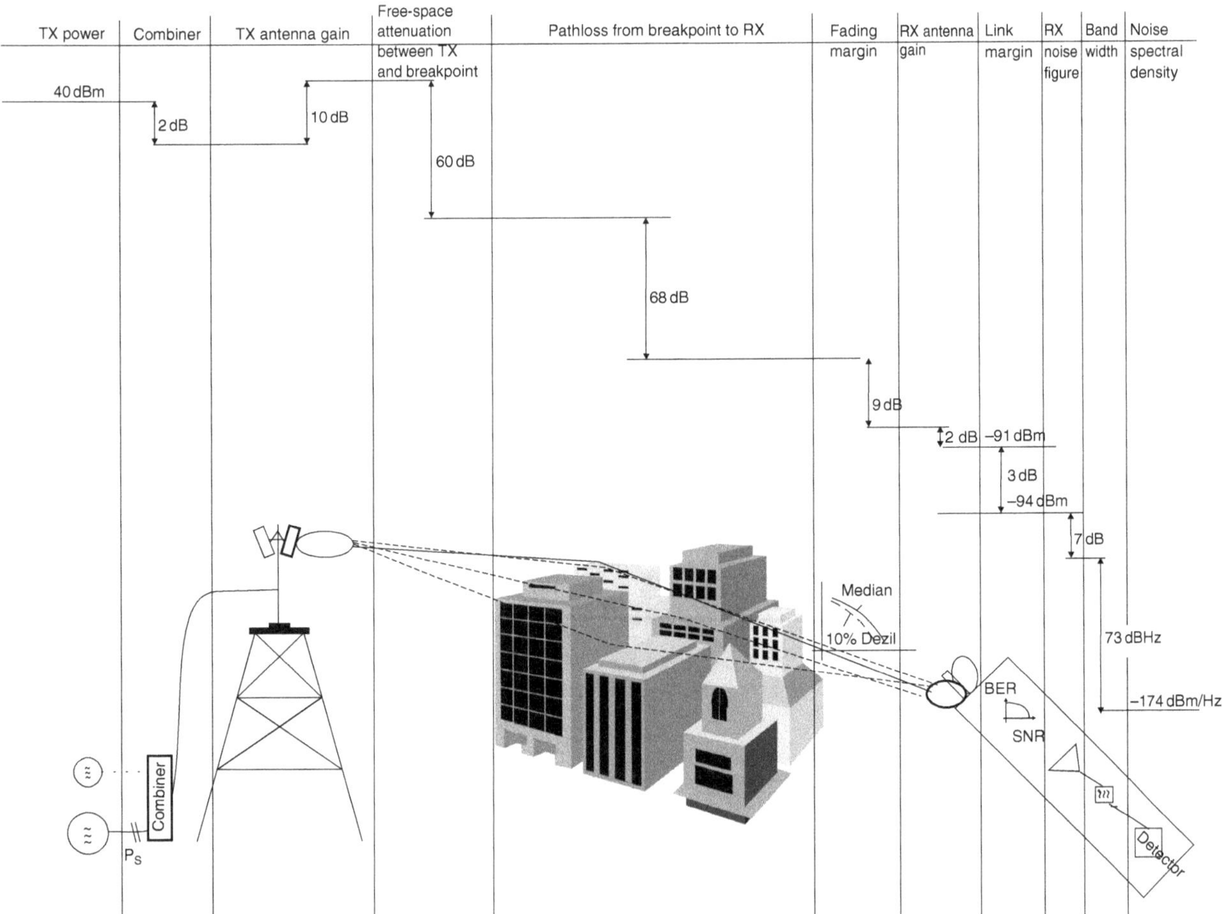

Figure 3.4   Link budget of example 3.2. Color version available at wiley.com/go/molisch/wireless3e.

3.1.4 More on Channels and Antennas

Propagation Channels

For the link budget, we subsumed the effect of the wireless propagation channel into two quantities: pathloss and fading margin. This description is appropriate if the performance is mainly determined by the signal power and noise power, and signal distortions are not really relevant. It furthermore assumes a simple interaction between the propagation channel and the antennas, namely that the antennas simply add a certain gain to the power budget.

In many situations, we need a more detailed description of the propagation channel. An important such description is tied to the picture of multi-path propagation as described in Section 2.2.1. We see that signals propagate from the TX to the RX via different paths; the associated electromagnetic waves are the Multi-Path Components (MPCs). If the transmitted signal is a pure sinusoid, say $\cos(2\pi f_c t)$, then the received signal stemming from the ℓth MPC is $|a_\ell|\cos(2\pi f_c t + \varphi_\ell)$. Thus, the MPC can be described by the amplitude and phase, which often are subsumed into the complex amplitude $a_\ell = |a_\ell|e^{j\varphi_\ell}$. Furthermore, the MPC is also characterized by a delay τ_ℓ (i.e., how long it takes a pulse emitted at the TX to reach the RX when propagating along the path of this MPC). Finally, the MPC is characterized by its direction of departure Ω into which it leaves the TX, and the direction of arrival Ψ from which it arrives at the RX.

The total received signal can be interpreted as the sum of the contributions of the different MPCs, weighted by the complex antenna gain, as described in more detail in the next subsection. The resulting description is very general, and from it follow various models that explain the fundamental challenges and opportunities for communication system design for wireless channels.

Antennas

The peak antenna power gain only describes the signal power enhancement into a single direction. A more general description of the antenna characteristics is the antenna (power gain) *pattern* $G_{\mathrm{TX}}(\Omega)$, which describes the enhancement or attenuation of the signal into any direction (spatial angle coordinate) Ω. Since an antenna is a passive element, the total power, integrated over all directions, cannot be increased by the antenna; it can only focus the available energy. Thus, what is gained in one direction must be lost in other directions, so that $(1/4\pi)\int G(\Omega)d\Omega = 1$. Consequently, a high-gain antenna is only useful if the direction of the maximum gain coincides with the direction of the MPCs carrying the main power. If the MPCs are widely spread out, an effective high gain cannot be realized, because some MPCs get enhanced while others get attenuated. In the remainder of this chapter, we assume that this effect does not occur, but rather that all important MPCs experience the maximum antenna gain.

Even more generally, an antenna can be interpreted as weighting the MPC amplitudes with the antenna gains in the directions of the amplitudes. This weighting affects not only the amplitude but also the phase of the signal. Thus, imagine an MPC propagation from the TX location to the RX location; when this MPC emanates from the TX in the direction Ω and arrives at the RX from the direction Ψ, then its contribution to the total received signal is (see also Figure 3.5)

$$\widetilde{G}_{\mathrm{TX}}(\Omega)\widetilde{G}_{\mathrm{RX}}(\Psi)\,|\,a\,|\,e^{j\varphi}. \tag{3.9}$$

The quantity $\widetilde{G}_{\mathrm{TX}}(\Omega)$ is the (complex) amplitude antenna pattern. It follows immediately that $G(\Omega) = \left|\widetilde{G}(\Omega)\right|^2$. The impact of the antenna phase on the total signal will play an important role when we discuss ray tracing (Section 4.7), as well as for phased array antennas. More extensive descriptions of all these aspects will be provided in Chapter 8.

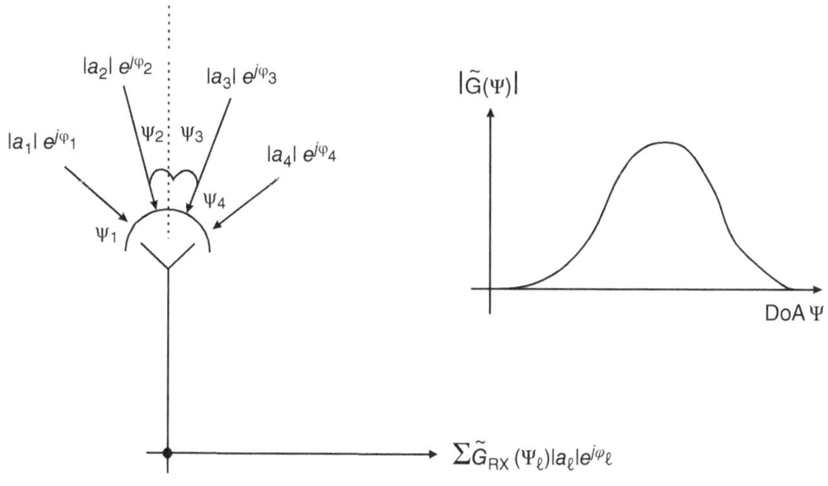

Figure 3.5 Weighted addition of MPCs at the antenna. $\widetilde{G}_{\mathrm{TX}}(\Omega) = 1$ is assumed.

3.2 Digital Modulation and Receiver Signal Processing

Up to now, we have characterized TX and RX purely by power and bandwidth.[2] We now detail our view of a wireless system by considering how these quantities relate to the actual system performance. We proceed in two steps: we first consider a system where the propagation channel does not introduce any signal distortions and relate power and bandwidth to data rate and error probability. We then consider what special processing must be done at TX and RX to obtain good performance when the channel does distort the signal.

3.2.1 Systems Without Signal Distortion

The signal that is sent over the air is an analog waveform. Yet the information we want to convey is digital, i.e., bits. The mapping from the bits to the waveforms is called *modulation*, the reverse process at the RX is *demodulation*. The best-known modulation format is binary modulation, where, e.g., a pulse with amplitude $+1$ is transmitted to represent a logical 1, and a pulse with amplitude -1 represents logical 0. Such a format is not always very spectrally efficient, so that multi-level modulation is often preferred: there exist multiple (e.g., 4 or 16) waveforms, and the TX first maps groups of bits onto one symbol (e.g., the bit combination 10 is mapped onto symbol 3), and then the waveform corresponding to this digital symbol is transmitted. Since wireless communications deals with bandpass signals (i.e., occupying a narrow bandwidth around a carrier f_c), the symbols can be distinguished by the amplitude and the phase of the signal that is being transmitted, which can be represented as a complex number. Modulation formats that distinguish their symbols by both amplitude and phase are often called complex modulation, for example see Figure 3.6. For more detailed and precise descriptions of modulation formats, see Chapter 10.

If the channel introduces no signal distortion, then the waveform that arrives at the RX is an attenuated, phase shifted, and noisy version of the TX waveform. From this signal, the RX now needs to reconstruct the symbols and thus the bits that were transmitted. Consider first the case that all the symbols are independent of each other. Then the RX can perform a symbol-by-symbol decoding. The main detrimental effect is the noise, and the RX has, for each symbol, a certain probability of "misinterpreting" the received waveform and making a wrong conclusion about which symbol (and thus which group of bits) was sent. The resulting error probability is called the *uncoded bit error probability*, or *Bit Error Rate* (*BER*). Different modulation formats have different BERs, see Chapter 11 for details.

An elementary TX thus consists of a modulator that maps the bits onto the analog waveform in baseband and an upconverter that translates the signal to the carrier frequency. Similarly, an elementary RX consists of a downconverter, plus a demodulator that takes a waveform and processes it. The simplest possible output of the demodulator is a decision about which symbol the RX thinks was sent – this is known as a "hard decision," since there is a single bit as output for every transmitted bit. The RX can also output "soft decisions," which not only describe this decision but also the confidence that the RX has into this decision, see Figure 3.7. Such soft decisions form the basis for many more advanced RX structures. A block diagram of a system incorporating the soft decision, followed by a hard decision, is shown in Figure 3.8.

Figure 3.6 Mapping from bits to symbols to transmit waveform with an 8-symbol modulation alphabet. Color version available at wiley.com/go/molisch/wireless3e.

---

[2] Carrier frequency and noise figure will play a minor role in the discussions of this subsection, and thus are not explicitly mentioned.

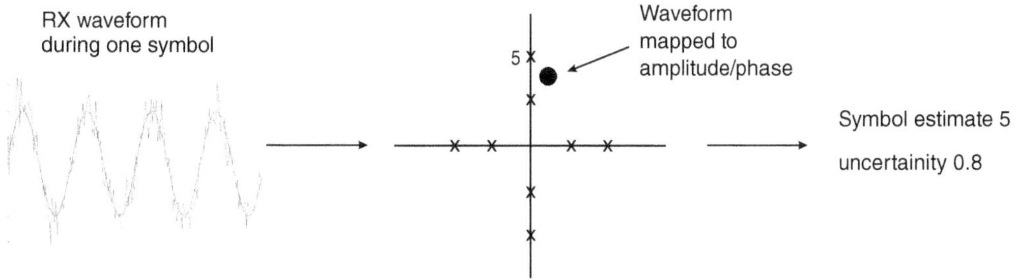

Figure 3.7   Soft detection of symbols in the presence of noise. Color version available at wiley.com/go/molisch/wireless3e.

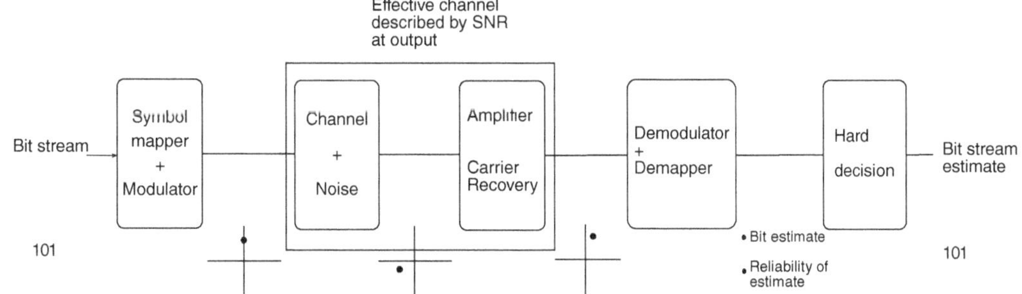

Figure 3.8   Block diagram of system with soft decision.

The raw BER obtained in the presence of noise is not acceptable for most applications. To reduce the error probability, redundancy must be introduced in the form of error correction codes. The simplest form one can imagine is repetition coding: the same bit is sent multiple times, and the RX performs a majority decision on which bit was sent. Even though this code is too crude for practical applications, some important intuitive insights can be gained from it: the number of coded bits that need to be transmitted is larger than the number of source bits; the larger the amount of redundancy, the greater the ratio of coded over source bits, and the stronger the protection that can be achieved against errors. The shape of the BER vs SNR curve changes, see Figure 3.9 – while for uncoded transmission, it is approximately exponential, with increasingly strong coding it becomes more similar to a step function – a block of bits can be transmitted without errors if the SNR is above a certain threshold (called the channel capacity), or has very high probability to be in error if the SNR is below the threshold. From a "macro" point of view, the SNR is a good measure for the performance of the system, since it implies both the uncoded BER, and the achievable transmission rate when coding is used. The design of the encoders and decoders is in itself an important and intricate aspect of wireless system design. Much work has been done over the past 70 years in this field, and Chapter 13 will describe the most important types of codes.

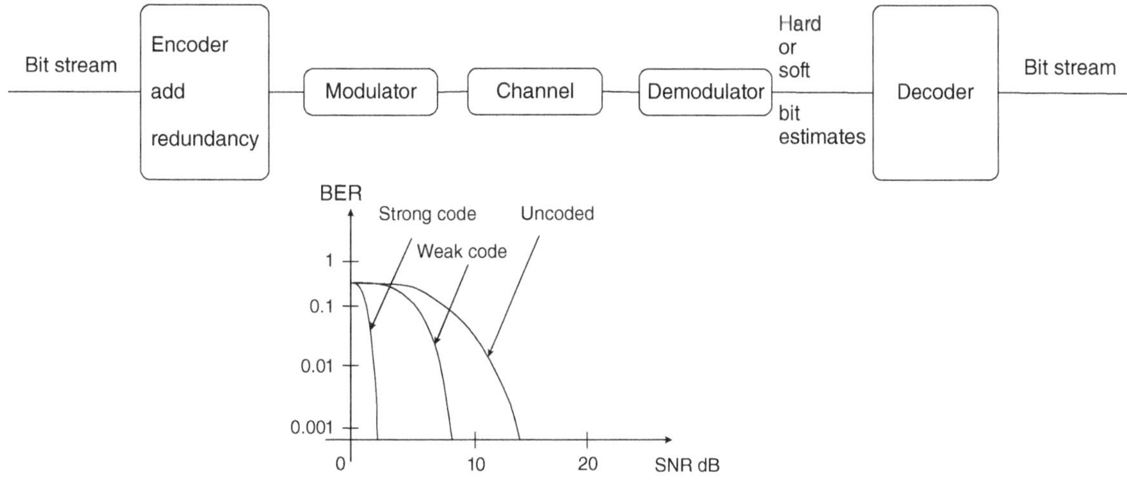

Figure 3.9   Block diagram of coded system, and BER versus SNR curves.

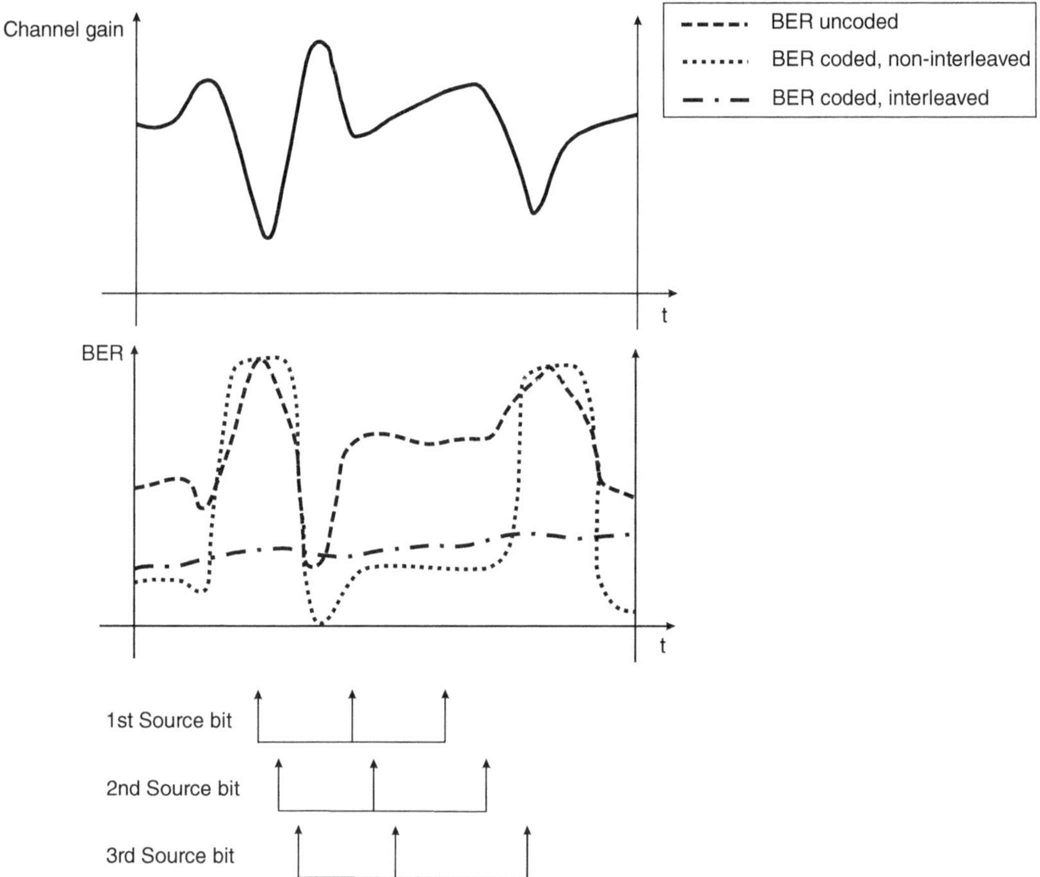

Figure 3.10 Channel gain and raw BER, noninterleaved coded BER, and interleaved coded BER in time-varying channel.

The description of modulation and coding above assumed (implicitly) that the channel stays constant over time. However, in reality, the channel is fading, i.e., changing its attenuation over time (see Chapter 2). This has a major impact on the performance analysis of digital modulation formats, as well as on the design of coders. Consider first uncoded systems: when the channel is in a "good" state (fading peaks), then the BER will be extremely low (due to the exponential decrease of the BER with the SNR). However, when the channel is in a bad state (fading dips), then the BER will be close to 0.5. The average BER is thus dominated by the cases where the channel is in a fading dip; the computation of the average BER will be discussed in Section 11.2. Similar considerations hold for coded systems: when the channel is in a fading peak, a packet will be transmitted without error, but in a fading dip, every bit is affected by a lot of noise, so that even the added redundancy cannot prevent the packet from being in error. In order to make the best use of the coding, the bits of a codeword have to be transmitted such that they experience different fading states (also known as different "diversity branches"), see Figure 3.10. Take again the example of a repetition code: the three repetitions of a bit can be transmitted at times that are so much spaced apart that the fading state changes in between; this is achieved by using, e.g., a suitable interleaver. There are also other forms of diversity that are discussed in Chapter 12. The interleaver, encoder/decoder, and modulator/demodulator need to be designed jointly, and need to take into account the properties of the propagation channel, such as the speed at which it changes between fading states. Most importantly, the decoder needs to take into account the fading state that the different bits in the code block encounter, since this provides information about the reliability of a received symbol (the above-discussed "soft information"). The combination of interleaver, modulator, and propagation channel can be seen as an "equivalent" channel which the coder/decoder combination "sees," and whose performance can be assessed, see Figure 3.11.

3.2.2 Systems with Signal Distortion

Up to now, we have included the attenuation introduced by the channel, and the fading that is created by the constructive/destructive interference of the MPCs. However, as we saw in Section 2.2.2, the multi-path propagation also leads to intersymbol interference, or – more generally – to delay dispersion that induces distortions of the signal on its way from the TX to the RX. We can describe the channel thus as a time-varying filter, which is characterized either by its (time-varying) impulse response, or a (time-varying) transfer function. The former interpretation is more natural for the use of so-called single-carrier systems – the type of systems described above, where symbols are transmitted one after the other from the TX, each represented by a short Quadrature Amplitude Modulation (QAM)-modulated pulse (duration is 1/symbol rate). It is clear then that the transmitted total waveform is distorted by the channel, and

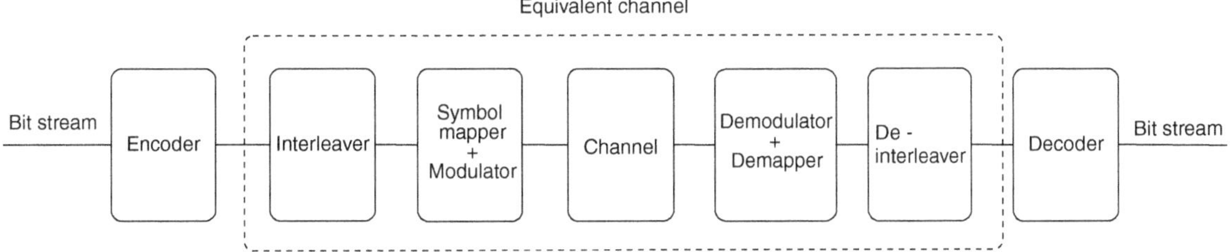

Figure 3.11 Block diagram of system with coder/decoder.

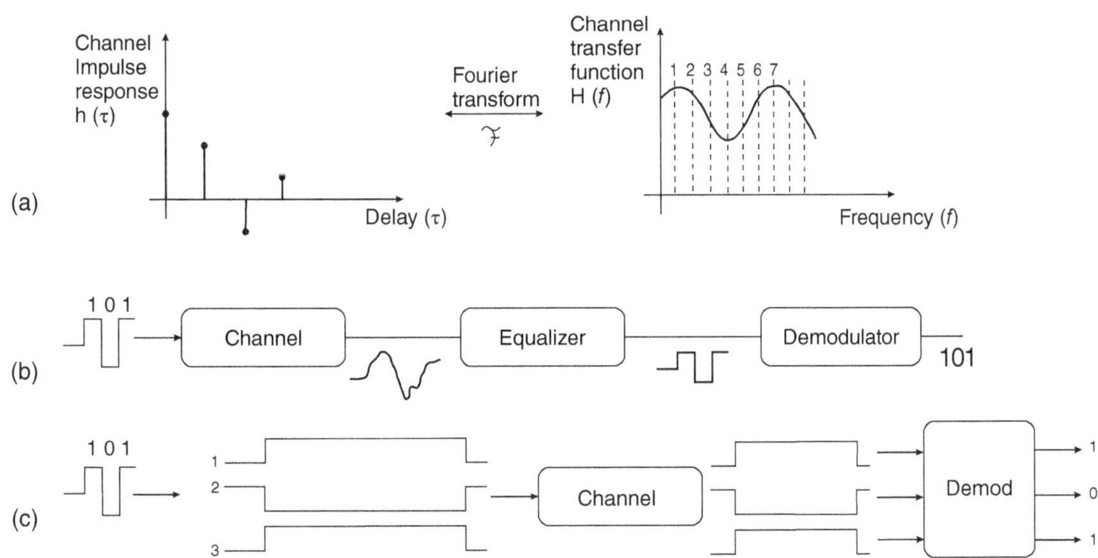

Figure 3.12 (a) Impulse response (left) and transfer function of a delay-dispersive channel. (b) Modulated signal with distortions, and equalizer to combat the effect. (c) OFDM splits up the data stream into parallel streams that are transmitted on different subcarriers.

due to intersymbol interference, the received waveform at any time is impacted by multiple transmitted symbols. Reversing this intersymbol interference and recovering the originally transmitted symbols is the task of the equalizer, described in Chapter 14, see Figure 3.12b for the basic principle. It usually consists of two parts: (i) a channel estimator that determines the channel state information CSI, and thus, what distortions are induced on the signal on its way, and (ii) the actual equalizer, which can be realized as an adaptive filter that essentially undoes the channel distortion.

Another popular method for handling the intersymbol interference is the use of multi-carrier modulation, in particular Orthogonal Frequency Division Multiplexing (OFDM). This approach is most easily interpreted in the frequency domain: the high-rate data stream is split into many low-rate streams, each of which is transmitted on a slightly different carrier frequency (called subcarrier), see Figure 3.12c. As a consequence, each of those low rate streams "sees" a different transfer function, but – because it is low-rate and therefore narrow bandwidth – does not experience any significant distortions. The RX can thus demodulate the signals on the different subcarriers separately, and in an easy way. This approach, and methods for implementing it efficiently, are discussed in Chapter 15.

It is obvious that in order to understand the signal distortions and design the systems to combat them, it is first necessary to characterize the corresponding characteristics of the channel. A complete description would be given by the complex amplitude and delay of all the MPCs, as well as how those parameters change over time. Such detailed knowledge might not always be necessary, but rather a condensed statistical description can be sufficient; more details about the formal wideband channel characterizations are discussed in Chapter 6, and popular models for their emulation in Chapter 7. The design of the equalizers (e.g., how many degrees of freedom the adaptive filter should have), or the OFDM system (e.g., into how many subbands the available system band is divided) depends on the anticipated behavior of the channel. Furthermore, the instantaneous state of the channel determines the settings of the equalizer (and equivalent quantities for OFDM).

While the delay dispersion of the channel seems like a negative effect (because it introduces intersymbol interference), it can actually be a benefit. This can be understood most easily in the frequency-domain (OFDM) interpretation: the different subcarriers are "diversity branches" that experience different fading; while one subcarrier might be in a fading dip, another can be in a fading peak. Thus, coding across subcarriers provides significant advantages in terms of error probability. Take again the example of the repetition code: when the different copies of a bit are on subcarriers that are in different fading states, then the probability that at least some of them arrive with "good quality" is much higher. Thus, the probability that the transmitted bits can be reconstructed from those "good" symbols is greatly

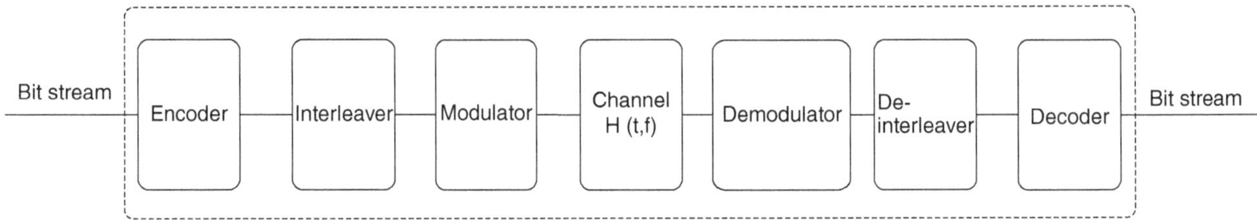

Figure 3.13 An ideal system can be described by its capacity, whose value depends only on the propagation channel. The capacity is an upper bound on the possible achievable data rate.

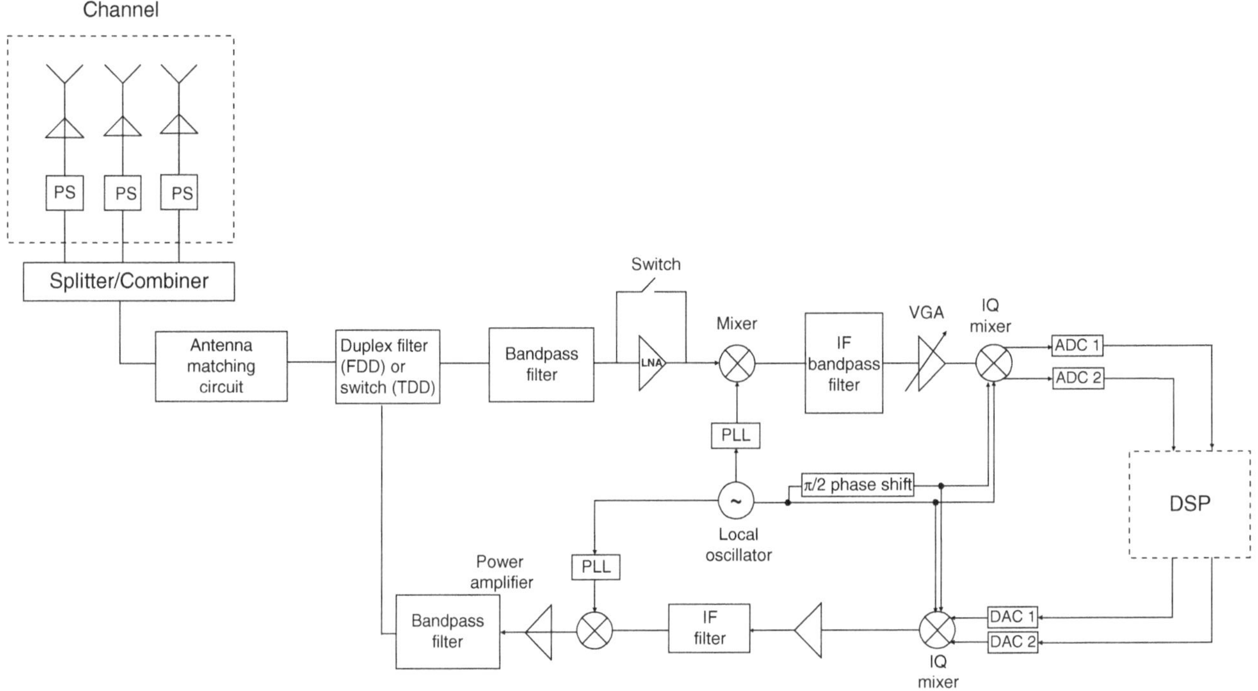

Figure 3.14 Block diagram of a wireless transceiver.

increased. Again, the way that bits are mapped onto the symbols on the different subcarriers needs to be carefully considered and designed, possibly using frequency interleaving. Similar to the case of interleaving in time, the decoder also needs to know the CSI on the different subcarriers, since the decoding process needs to take into account the reliability of the different received symbols.

For many system analysis considerations, we can assume that the encoder/decoder is ideal, so that the system can operate "at capacity." In that case, the design of the equalizer or OFDM, and the performance analysis, can be separated from the encoder/decoder design, and the performance analysis is based only on the former components, plus the channel properties. A block diagram of such a system can be found in Figure 3.13.

In addition to downconverter and demodulator, an RX also needs to contain many "auxiliary" components, from suitable amplifiers and filters to synchronization circuitry and channel estimators. These components can lead to additional signal distortions, many of which lead to considerable theoretical complications. For example, analog-to-digital converters are not ideal but have residual quantization errors that might limit the performance. Furthermore, upconverters/downconverters use local oscillators that are afflicted by phase noise, which can be a limiting factor in particular for cheap oscillators and/or operation at high frequencies. More extensive discussions of the hardware components will be found in Chapter 17. A block diagram showing some of the essential components is shown in Figure 3.14.

3.2.3 Systems with Multiple Antennas

A very important area of wireless communications, that actually has increased in relevance over the past 20 years, is systems with multiple antenna elements at the TX, at the RX, or at both. There are a variety of ways in which such multiple antenna elements can be exploited. The first one is as diversity branches, a situation most easily explained if the multiple antennas are at the RX. The signal from the TX to each of the RX antennas goes through a different channel, whose fading state will be different from those of the channels to the

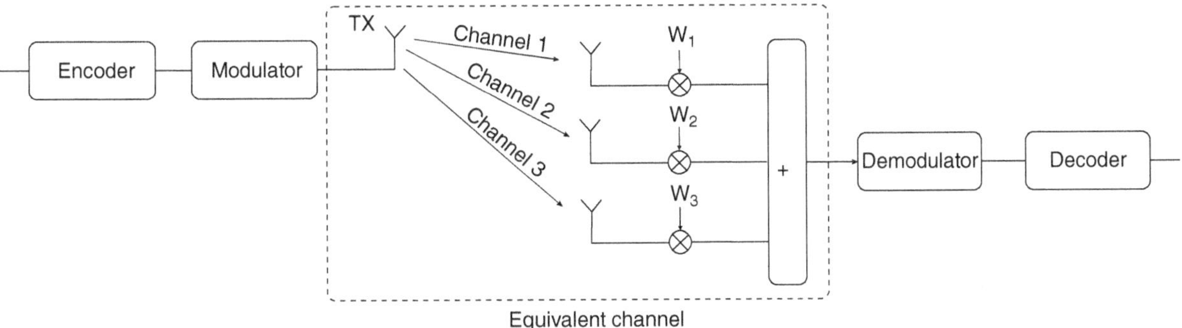

Figure 3.15 Receive antenna diversity. The signals at different antenna elements are weighted and added up. The combination of the multiple propagation channels and the combiner can be seen as an "effective" channel.

other antennas. The RX can then, in some form, combine (weight and add up) the signals from all the different antenna elements, see Figure 3.15. The channel from the TX to the output of this combiner can be seen as an "effective" channel, and with a suitable choice of the combiner weights, this effective channel is better than any of the constituent channels from TX to an RX antenna element. In particular, the probability of experiencing a deep fading dip can be dramatically reduced, which improves the overall performance of the system. The same method works (under some constraints) when the TX has multiple antenna elements. These antenna combiners can be designed independently of modems and codecs, though conversely, the changed statistics of the effective channel might impact the modem and codec design. In the link budget analysis of Section 3.1, the effect of such diversity combining is both an increase in average SNR (effectively, an increased antenna gain) and smaller channel fluctuations (less difference between residual fading peaks and dips), which reduces the required fading margin. Diversity antennas are discussed in Chapter 12 and Section 16.1.

A related approach is the use of beamforming. Combining signals from multiple antenna elements with certain linear weights can be seen as forming an antenna pattern whose main beam (and thus maximum gain) "looks" into a particular direction. A similar effect can be achieved at the TX. We can thus form high-gain antennas that can be adaptively steered such that the receive power is maximized. This is easiest to visualize in the case of a LOS connection between the TX and RX. The TX can then form a beam in the direction of the RX, and vice versa. The effect of such beamforming can thus be directly included in the link budget as an increased antenna gain. Beamforming is discussed in Section 16.1.

Either of those cases requires additional knowledge of the propagation channel. Note that different MPCs superpose (interfere) differently at the different antenna elements, and the way how they superpose depends on the directions of the MPCs. Thus, we need to know the directions of arrival (for multiple antenna elements at the RX) or directions of departure (for multiple antenna elements at the TX), in addition to the amplitudes and delays of the MPCs mentioned above. Again, more compact descriptions, possibly relying on statistical characterization, can be found, see Section 6.7. Alternatively, the joint description of the (nondirectional) channels from the TX to each RX antenna element, i.e., a multi-antenna channel vector, can be used to describe the channel. This can be measured directly (as happens most often in system operation), or it can be derived from a directional characterization of the MPCs (which is often used for performance simulations); the two formulations are equivalent. The statistics of those characterizations have an impact on the design of the antenna combiners or beamformers, as well as on the absolute performance.

Yet another way of using multiple antenna elements becomes possible when they are present at both TX and RX. In this case, the data stream at the TX can be split into multiple data streams that are transmitted in parallel, at the same time, and in the same frequency band, and is thus called spatial multiplexing. From the multiple signals at the RX antenna elements, the streams can be separated and detected. The effectiveness of that technique strongly depends on the directions of the MPCs and in particular the "spreading" of these directions – the wider the spread, the better this scheme works. The scheme can be set up in a way that each of the data streams has its own encoder and modulator, and at the RX there is a single stage that performs the stream separation, after which the demodulation and decoding is done separately for each stream, see Figure 3.16a. This makes the overall system design easier, as the stream separation can be designed separately from the other aspects of the transceivers. However, better performance can often be achieved by performing a joint encoding, and in particular, a joint detection of the different streams, as shown in Figure 3.16b. Details about spatial multiplexing can be found in Section 16.2.

3.3 Multi-user Systems

3.3.1 Separation of Users in a Cell

Section 3.2 described the models and designs for single-link wireless systems, i.e., systems that communicate between one BS and one UE. However, in most practical situations, one BS communicates, essentially simultaneously, with multiple UEs. Wireless system design thus has to establish how the different UEs are separated, both in the uplink and the downlink, by appropriate division of resources (time/frequency/space/code). We will call this the Multiple-Access (MA) design. It is important to ensure that the various links between the BS and the UEs do not interfere with each other; i.e., each link can be considered as interference-free. In other words, once the resources have been assigned, each link can be treated like in the single-link case discussed in Section 3.2.

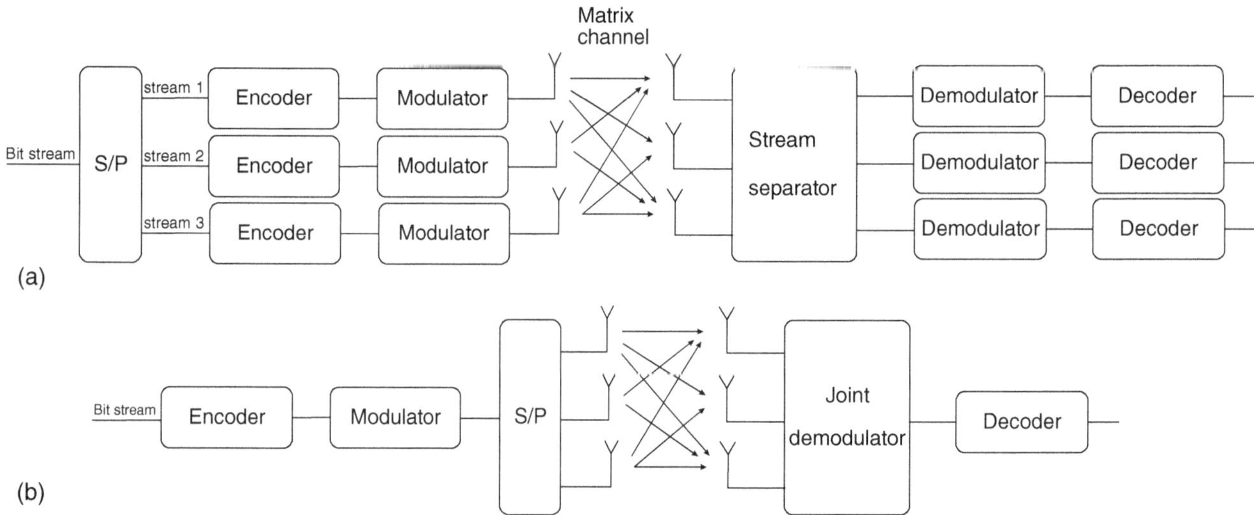

Figure 3.16 Spatial multiplexing with separate encoding of spatial streams (a) and joint encoding of spatial streams (b). S/P is serial/parallel conversion.

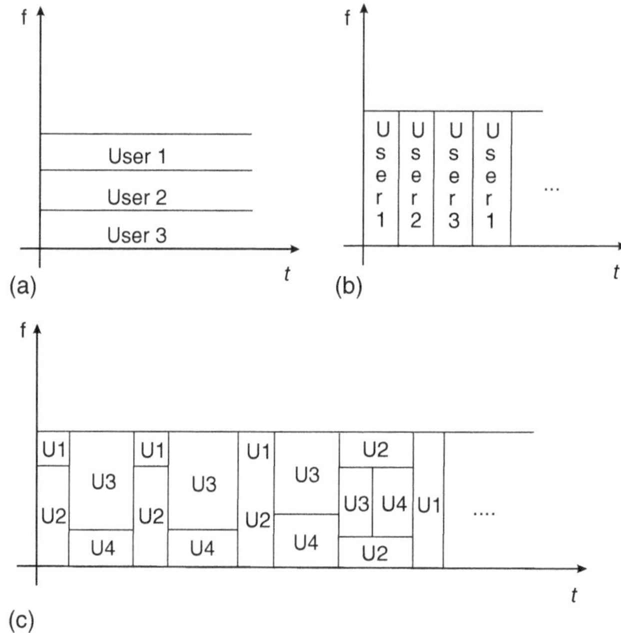

Figure 3.17 Various multiple access methods dividing the time–frequency plane: (a) FDMA, (b) TDMA, (c) OFDMA.

A conceptually simple type of MA format is the time/frequency MA. Consider a plot of the time–frequency plane, see Figure 3.17. The x-axis shows the different times at which a user might transmit, the y-axis the frequency bands (the union of the frequency bands for the transmission of the different users should be a subset of, or identical to, the overall bandwidth assigned to the service). Then different parts of this time–frequency plane can be assigned to different users. For example, each user can be given a subband for the duration of a call/data session [this is known as Frequency Division Multiple Access (FDMA)], Figure 3.17a, or different users transmit at different times [Time Division Multiple Access (TDMA)], Figure 3.17b, or different users partition the time–frequency plane in an almost-arbitrary fashion [Orthogonal Frequency Division Multiple Access (OFDMA)], Figure 3.17c. The assigned portions can be periodic over time or can be assigned by the system on-demand. Details about time/frequency MA are described in Chapter 18; an alternative MA format called CDMA is discussed in Chapter 19.

If the channel is neither time-nor frequency selective, then the computation of the capacity for each user is simple: it is the capacity that the user would have in a single-link system, multiplied with the fraction of the time–frequency resources that is assigned to this user. If the channel is time- and/or frequency selective, i.e., shows changes over time, or in the transfer function, then the capacity depends on which portions of the spectrum are assigned to which user – a process known as scheduling. Consider the situation in Figure 3.18a, and assume a pure TDMA system, and that the channels vary with time but are independent of

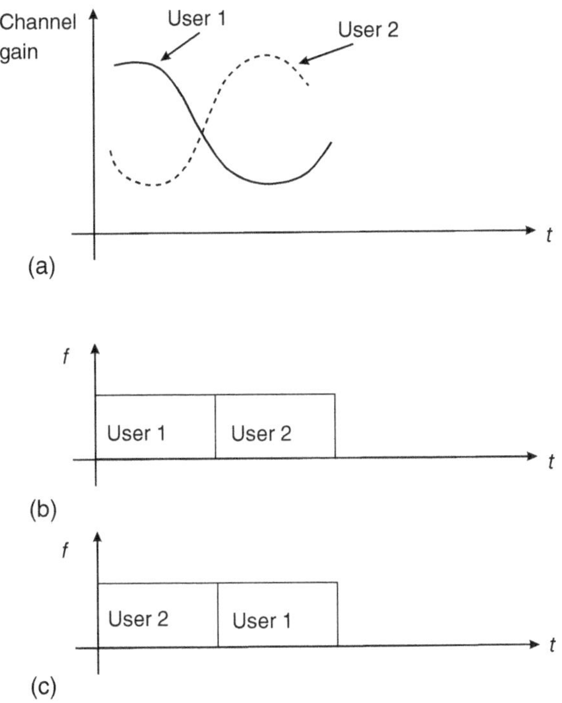

Figure 3.18 (a) Channel gains for users 1 and 2, (b) possible scheduling, (c) alternative possible scheduling.

frequency. There are two channels, from the BS to UE 1, and from BS to UE 2. Then we can assign timeslots such that scheduling occurs as shown in Figure 3.18b, or as shown in Figure 3.18c. It is obvious that the former is the better solution, and a good scheduler will assign the users accordingly. Similarly, in a frequency selective channel, subbands can be assigned to different UEs in a smart fashion. Note that this approach exploits the fact that channels show variations in time and frequency; yet to exploit it, the scheduler needs to know the instantaneous channel states. If that is not possible, it is preferable to use diversity to eliminate variations of the channels, and thus recover the situation described at the beginning of this paragraph. In addition to the optimization of the throughput, a scheduler also needs to consider aspects such as fairness and latency of the different users. All this will be discussed in more detail in Chapter 20.

The choice of the MA format interacts with the scheduler and the design of the modem. In particular, single-carrier modulation is used when at any given point in time a user is assigned a contiguous frequency band (this band can change from time to time). On the other hand, if OFDM is used as modulation, then multiple, separated, subbands can be assigned to a user at a time. This can be especially advantageous when these subbands are widely spaced so that they experience different fading states; in this case coding across those subbands becomes more effective (compare Sec. 3.2).

Another method of separating the signals from different UEs is space-division multiple access, also known as Multi-User MIMO, MU-MIMO. Consider first the UL. We already mentioned that an RX antenna with multiple antenna elements can form an adaptive beampattern, placing the main beam into a particular direction and – equally importantly – placing nulls in other directions (Figure 3.19). Furthermore, an RX with N antenna elements can form N such beams simultaneously. A BS can thus form, for each of N UEs, a beam that receives the signal from this UE with strong power, while suppressing the signals from the other users. Thus,

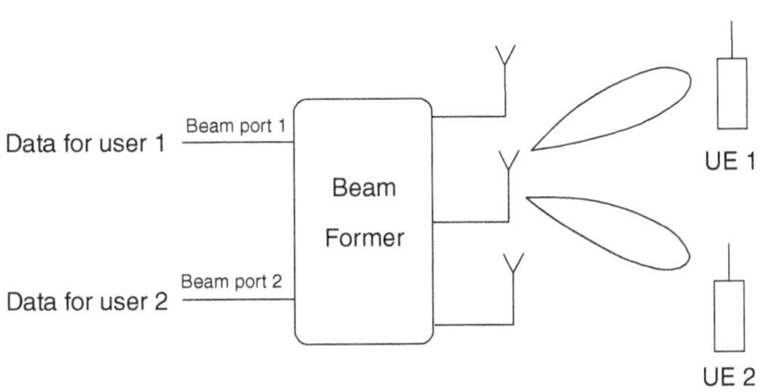

Figure 3.19 Principle of MU-MIMO.

the BS can separate the data streams from the different UEs, and then demodulate and decode them separately (notice the similarity of the description with spatial multiplexing described above – this is not a coincidence, because the two systems have strong similarities, as discussed in the introduction of Chapter 22). For the downlink, similar considerations hold – as long as the BS (which is now the TX) knows the channels to the different UEs, it can form beams that enhance the signal to the desired UE and minimize transmission in the direction of the other UEs. Thus, multiple UEs can communicate with the BS in the same time–frequency resources. The MU-MIMO can be combined with other MA formats, especially OFDMA. Then in each subband, and at each point in time, a different set of N users can communicate with the BS. Obviously, the scheduler and the beamformer have a strong interaction and cannot be separated from each other.

3.3.2　Interference-limited Systems

The Cellular Principle

We now come to the situation where different links do interfere with each other. The practically most important case where this occurs is in cellular systems, where the BS-UE links in different cells interfere with each other. Consider the situation in Figure 3.20, where we have multiple BSs, each of which is communicating with one UE (we assume here that we are looking at a specific time instant and assume that the BSs are single-antenna, so that no MU-MIMO can be used), located within its hexagonally shaped "cell." We consider first the downlink, and assume that all transmissions occur in the same frequency band. Then the transmission from BS 1 is the desired signal at UE 1. However, because it is transmitted by an omni-directional antenna, it also reaches UE 2, for which it is interference (UE 2 gets its desired signal from BS 2). If UE 2 is close to the cell boundary, then it experiences strong interference. It is common to treat the interference as Gaussian, which is an approximation that allows to treat the interference as equivalent noise. Consequently, the performance is determined not by the SNR, but rather the Signal-to-Interference-and-Noise Ratio (SINR).

The most common way to reduce this interference is to disallow the use of the same frequency band in adjacent cells. In other words, the available bandwidth is divided into subbands, and different subbands are used within one group (cluster) of cells, as shown in the example of Figure 3.21. As a consequence, a BS that operates on the same frequency band, and can thus create interference for UE 1, is much farther away. How far such an interfering BS has to be away can be determined from a link budget, which will be described next.

Link Budget for Interference-limited Systems

Consider the case that the interference is so strong that it completely dominates the performance, so that the noise can be neglected. Let a BS cover an area (cell) with radius R around the BS. Furthermore, there is an interfering BS at distance D from the "desired" BS, which operates at the same frequency, and with the same transmit power. How large does D have to be in order to guarantee satisfactory transmission quality 90% of the time, assuming (as a worst case) that the UE is at the cell boundary? The computations follow the link budget computations of Section 3.1.

One difference between interference and noise lies in the fact that interference suffers from fading, while the noise power is typically constant. For determination of the fading margin FM, we thus have to account for the fact that (i) the desired signal is weaker than its median value during 50% of the time, and (ii) the interfering signal is stronger than its median value 50% of the time.

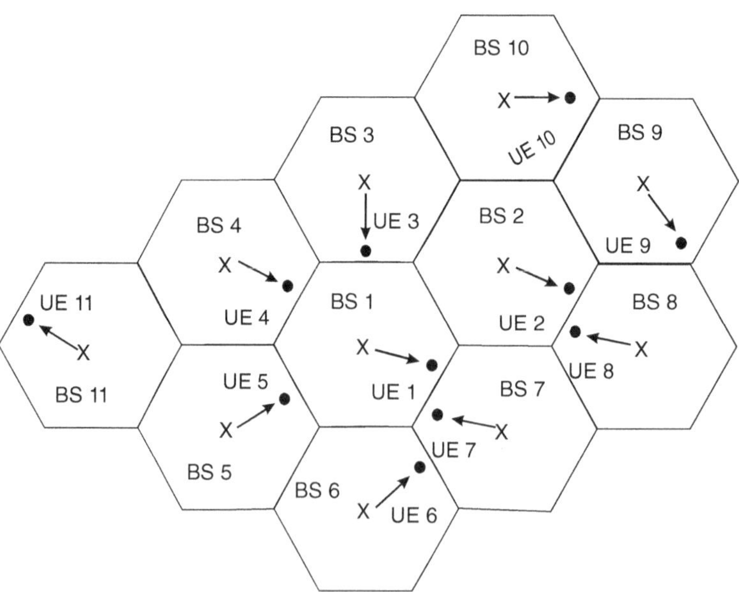

Figure 3.20　Division of area into cells.

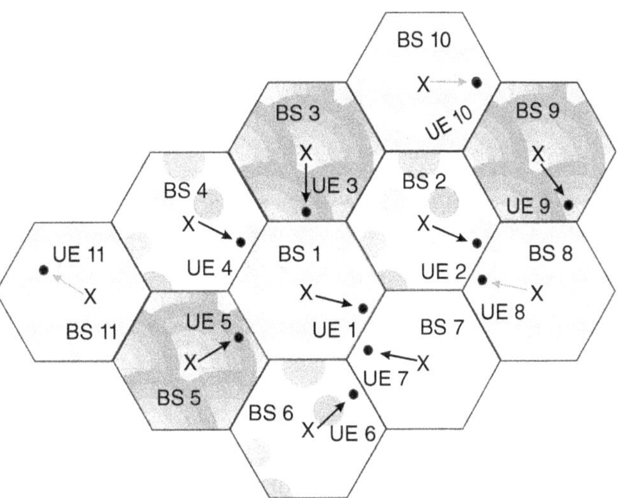

Figure 3.21 Base stations in different cells using different frequencies, indicated by shading. Color version available at wiley.com/go/molisch/wireless3e.

Mathematically speaking, the complementary cumulative distribution function of the Signal-to-Interference Ratio (SIR) is the probability that the ratio of two random variables is larger than a certain value in $x\%$ of all cases (where x is the percentage of locations in which transmission quality is satisfactory), see Chapter 5. As a first approximation, we can add the fading margin for the desired signal (i.e., the additional power we have to transmit to make sure that the desired signal level exceeds a certain value $x\%$ of the time, instead of 50%), and the fading margin of the interference (i.e., the power *reduction* to make sure that the interference exceeds a certain value only $(100 - x)\%$ of the time, instead of 50% of the time, see Figure 3.22), so that $FM = FM_{\mathrm{des}} + FM_{\mathrm{inf}}$. This results in an overestimation of the true fading margin. Therefore, if we use that value in system planning, we are on the safe side.

Link budgets of SIRs can sometimes be simpler than those for the SNR because certain properties of the BSs and UE that are identical for the desired and the interfering signal cancel out. For example, the TX power, and the attenuation of the signals in cables at the BSs do not play a role if they are the same for all BSs. Similarly, the antenna gain at the UE might cancel out – but it does not always: remember that the antenna gain depends on the direction of the signal. Thus, since the desired and interfering signal are incident onto the UE from different directions, different antenna gains might be relevant for it. Similarly, the desired and interfering signals will leave the BS antennas at different angles and thus might experience different gains. The noise figure at the UE is irrelevant because the

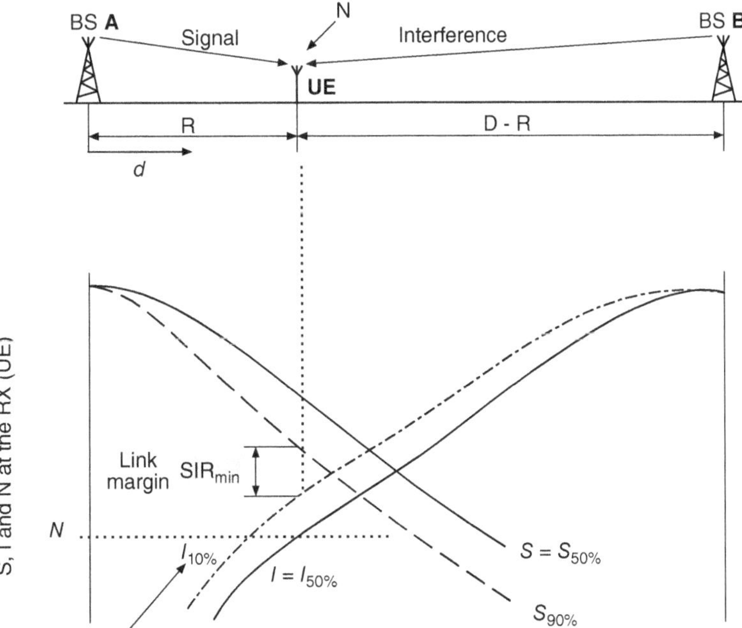

Figure 3.22 Relationship between cell radius and reuse distance. Solid lines: median values. Dashed lines: 90% – decile of the desired signal. Dash-dotted line: 10% – decile of the interfering signal.

system is assumed to be interference-limited. Consequently, assuming that $R > d_{\mathrm{break}}$, the ratio of the average power from the desired BS to interfering BS is in the simplest case

$$\frac{P_{\mathrm{des}}}{P_{\mathrm{inf}}} = \mathrm{SINR} = \left(\frac{R}{D-R}\right)^{-\alpha_{\mathrm{LBP}}}.$$

If it is now required that $\mathrm{SINR} = \gamma_{\mathrm{th,\,op}} + FM$, the required D/R can be easily computed. Note that the larger the ratio D/R is, the less often a frequency can be "reused," and thus we have to divide the available spectrum into narrower bands – which in turn means that less bandwidth is available for each cell. All of these aspects will be discussed in greater detail in Chapter 21.

Example 3.3 *Consider the downlink of a cellular system operating at 2 GHz, with 20 MHz bandwidth, and TX power of 40 dBm. The antenna gains at the BS are 12 dB for the desired signal and 9 dB for the interfering signal; the gains at the UE are 2 dB for both signals. All pathlosses follow a breakpoint law with a breakpoint at 20 m, and a pathloss coefficient of 4 beyond the breakpoint. The fading margin for both desired and interfering signal is 7 dB, and the required SIR is 5 dB. Assuming that the UE is 300 m from the desired BS, what is the required D between the BS?*

From the above input, the required difference (on a dB scale) of the pathlosses is

Minimum SIR	5 dB
Fading margin for desired signal	7 dB
Fading margin for interfering signal	7 dB
Antenna gain for interfering signal	9 dB
Antenna gain for desired signal	−12 dB
Required difference in path gain	16 dB

Since the UE is beyond the breakpoint distance both with respect to the desired and the interfering BSs, the distance ratio $40 \log_{10}(R/(D-R)) = 16$, corresponding to $R/(D-R) = 2.5$.

We finally note that the interference power from multiple interfering BSs add up. At the same time, when these interfering BSs are of somewhat similar strength, the fading of the various interference sources will tend to average out – in other words, the interference level becomes approximately constant, and the fading margin then is determined only by the fading of the desired signal. Exact calculations of the interference statistics for a finite number of interferers can become quite complicated; for details see Chapter 21.

Frequency Planning and Methods for Interference Reduction

The above description of interference is at the core of the frequency planning in cellular networks. It is strongly tied to the large-scale properties of the propagation channel, since the pathloss coefficient and the fading statistics (and thus the fading margin) impact the reuse distance. On the other hand, there is relatively little interaction with the design of the single-link transceivers – modulation, coding, diversity, etc., all enter through the aggregate parameters of $\gamma_{\mathrm{th,\,op}}$ and FM. However, there are other, more advanced methods that allow reduction of the reuse distance, be it through scheduling, and/or the use of multiple antenna elements at the BS, see Section 22.11.

Ad hoc Networks

While ad hoc networks have a different operating principle, the computation of the amount of interference, and the impact on the performance stays in principle the same. When a device receives a signal from a particular partner that is a distance R away, there must not be another transmission from an undesired device within a certain radius D that transmits as well. The challenges lie in the scheduling of different transmissions, and in the protocol of how the different devices agree on who is allowed to transmit at a particular time, since there is no central infrastructure that can dictate the scheduling. These aspects are treated in Chapter 23.

3.4 Summary

We see that wireless systems are highly complex systems with many components that interact with each other. In order to get an intuitive understanding and perform insightful performance analysis, simplified system models are required that idealize certain other blocks and components. In the majority of this book, we mostly follow this approach. Yet, in order to truly optimize systems, the interactions between all these components, which include cross-layer design, have to be taken into account. We will thus mention repeatedly which building blocks interact with others, and how their interaction impacts the overall system design. Furthermore, Part

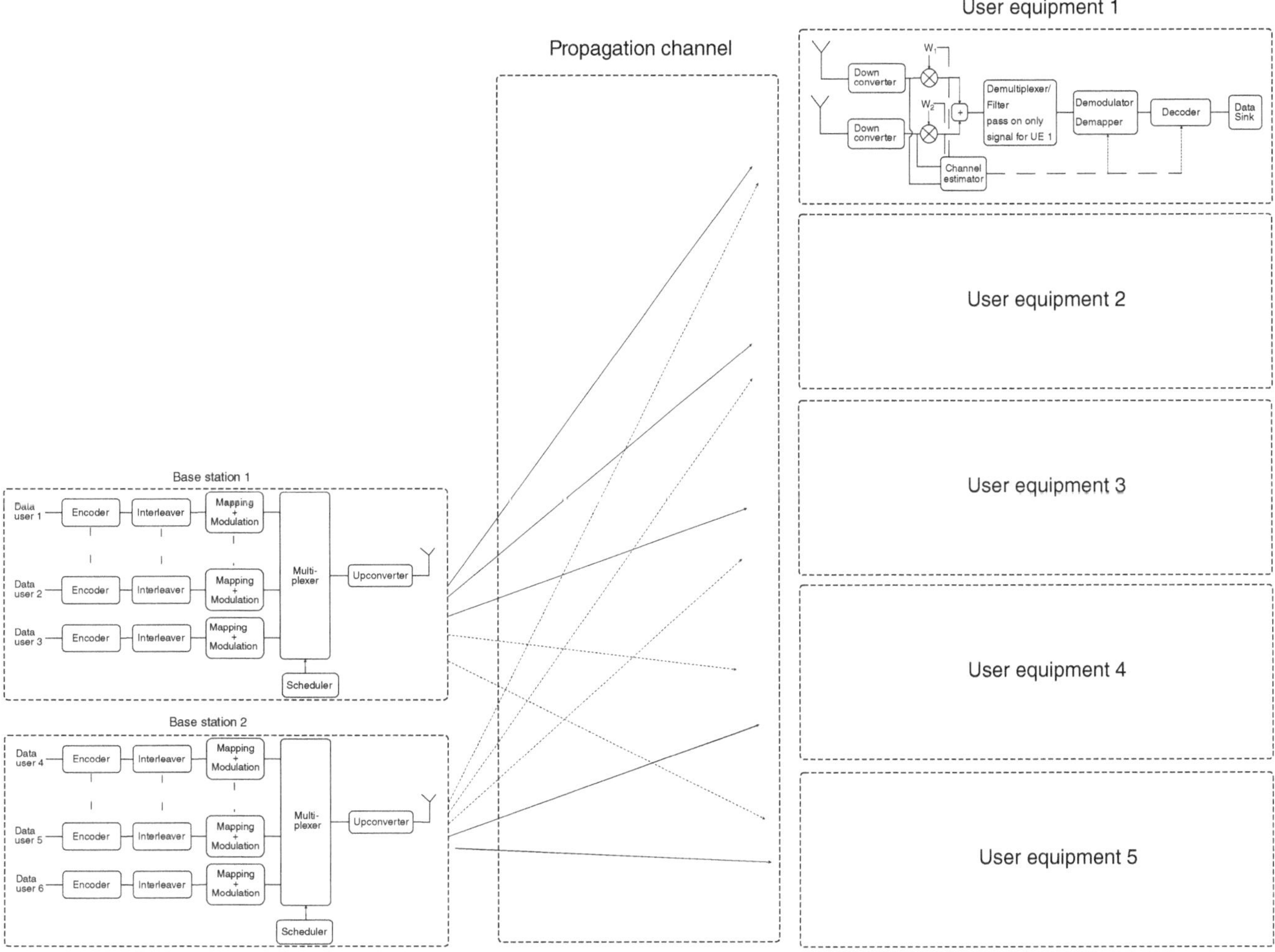

Figure 3.23 Full block diagram of a wireless system.

VI will discuss more integrated system design, and how this is realized in standardized systems. There we will see that in addition to the components discussed here, also a lot of signaling is required, partly providing control information on how the payload data transmission is to be executed, and partly as feedback; Parts II–V will not consider this control signaling in detail.

To wrap up this chapter, Figure 3.23 shows a detailed block diagram of a wireless system for the downlink.

For updates and errata for this chapter, see https://wides.usc.edu/students.html#textbooks

Exercises

See Sec. 36.3 of Exercises.pdf at wiley.com/go/molisch/wireless3e

Part II

Wireless Propagation Channels

Every wireless transmitter (TX) generates electrical signals (currents, guided waves) that are converted by the antennas into propagating waves. The propagation of those waves from the TX to the receiver (RX) (where the signals are converted back to currents by the receive antenna) is described by the wireless propagation channel. The channel is thus the medium linking the TX and RX.

The properties of the channel determine the ultimate performance limit any wireless communication system can achieve, and also determine how specific wireless systems behave. It is thus essential that we know and understand the channel and apply this knowledge to system design. Part II of this book, consisting of Chapters 4–9, is intended to provide such an understanding.

Wireless channels differ from wired channels by *multipath propagation* (see also Chapter 2), i.e., the existence of a multitude of propagation paths from TX to RX, and for each of them the signal can be reflected, diffracted, or scattered along its way. One way to understand the channel is to consider all those different *propagation phenomena*, and how they impact each *Multi-Path Component* (MPC). After describing the individual propagation phenomena in Sections 4.1–4.6, Section 4.7 will explain how to apply this knowledge to deterministic channel models and prediction (ray tracing).

The alternative is a more phenomenological view. We consider important channel parameters, like received power, and analyze their statistics. In other words, we do not care how the channel looks in a specific location, or how it is influenced by specific MPCs. Rather, we are interested in describing the *probability* that a channel parameter attains a certain value. The parameter of highest interest is the received power or field strength; it determines the behavior of narrowband systems in both noise- and interference-limited systems. We will find that the received power *on average* decreases with distance. The physical reasons for this decrease are described in Section 5.1; models are detailed in Section 7.1. However, there are variations around that mean that can be modeled stochastically. These stochastic variations are described in detail in Sections 5.2–5.8.

The interference of the different MPCs creates not only variations of the received power with time and/or place but also delay dispersion. Delay dispersion means that if we transmit a signal of duration T, the received signal has a *longer* duration T'. Naturally, this leads to *Inter Symbol Interference* (ISI). This effect is of critical importance to essentially all modern wireless systems, and we will discuss in Part III of this book a number of methods to mitigate the negative effects of delay dispersion, and even find ways to turn it into something helpful. Furthermore, *angular dispersion* is important in systems with multiple antennas, including LTE, 5G, and high-throughput Wi-Fi. These developments require the introduction of new parameters and description methods to quantify delay and angular dispersion, as described in Chapter 6. Chapter 7 provides typical values for those parameters, as well as complete models, in both outdoor and indoor environments.

Chapter 8 gives a summary of the operating principle, and the main characteristics, of antennas. Also, typical designs of antennas for base stations and user equipment, as well as adaptive antennas (antenna arrays) are discussed.

For both the understanding of the propagation phenomena, and for the stochastic description of the channel, we need channel measurements. Chapter 9 describes the measurement equipment, and how the output from that equipment needs to be processed. While measuring the received power is fairly straightforward, finding the delay dispersion and angular characteristics of the channel is much more involved.

Wireless Communications: From Fundamentals to Beyond 5G, Third Edition. Andreas F. Molisch.
© 2023 John Wiley & Sons Ltd. Published 2023 by John Wiley & Sons Ltd.
Companion website: www.wiley.com/go/molisch/wireless3e

4

Propagation Mechanisms

In this chapter, we describe the basic mechanisms that govern the propagation of electromagnetic waves, emphasizing those aspects that are relevant for wireless communications.

As discussed in Chapter 3, the transmit antenna is sending out electromagnetic waves (for the moment, we will just consider waves with a pure sinusoidal time variation; the effect of information modulated on those waves will be discussed later). At a sufficient distance,[1] the emitted field can be interpreted as consisting of a sum of plane waves, each weighted by the (complex) antenna pattern, that are propagating into different directions. These plane waves interact with the environment, or more precisely, with dielectric and conducting obstacles (*Interacting Objects* [IOs]) in the environment. Thus, not only the plane wave transmitted in the direction of the Line-of-Sight (LOS) between Transmitter (TX) and Receiver (RX), but also waves transmitted into other directions, can reach the RX. We thus have *multi-path propagation*, i.e., the transmitted signal can reach the RX via different paths, and the different plane waves reaching the RX are called the *Multi-Path Components* (MPCs). Figure 4.1 (replicating Figure 2.1 for the convenience of the reader) shows this principle. Note that the TX sends out many plane waves; however, only the few shown here reach the RX and are thus relevant for the link from BS to UE1. At the RX, the MPCs impinge on the antenna, and are weighted by the complex RX antenna pattern and added up.

To understand the multi-path propagation, we have to investigate the interaction of the MPCs with the environment. The simplest case is free space propagation – in other words, one TX and one RX antenna in free space. In a more realistic scenario, there is interaction with the IOs in the environment. If these IOs have a *smooth* surface, waves are specularly *reflected* and a part of the energy penetrates the IO (*transmission*). If the surfaces are *rough*, the waves are diffusely *scattered*.[2] Finally, waves can also be *diffracted* at the edges of the IOs. These effects will now be discussed one after the other.[3] We note that the discussion uses a ray approximation of the waves, which works well for the frequency ranges (>100 MHz) and environment types we are interested in. However, there are a number of phenomena (such as over-the-horizon propagation) that play a role in particular at lower frequencies, which we neglect here.

4.1 Free Space Attenuation

We start with the simplest possible scenario: a TX and an RX antenna in free space, and derive the received power as a function of distance.

Energy conservation dictates that the integral of the power density over any closed (aka Gaussian) surface surrounding the TX antenna must be equal to the transmitted power. If the closed surface is a sphere of radius d, centered at the TX antenna, and if the TX antenna radiates isotropically, then the power density on the surface is $P_{\mathrm{TX}}/(4\pi d^2)$. It is instructive to imagine the TX power as the coloring evenly distributed on the surface of a balloon (a sphere of radius d). As we pump up the balloon, i.e., increase d, the coloring will be spread out, thus becoming more and more transparent, though of course the total amount of color stays the same (Figure 4.2).

The RX antenna has an "effective area" denoted by A_{RX}. We can envision that all power impinging on that area is collected by the RX antenna. Then the received power is given by:

$$P_{\mathrm{RX}}(d) = P_{\mathrm{TX}} \frac{1}{4\pi d^2} A_{\mathrm{RX}}.$$

---

[1] In the so-called far field, see Eq. (4.4).

[2] Criteria for which a surface is smooth or rough will be given in Section 4.4.

[3] In the literature, the IOs are often called "scatterers," even if the interaction process is not scattering.

Wireless Communications: From Fundamentals to Beyond 5G, Third Edition. Andreas F. Molisch.
© 2023 John Wiley & Sons Ltd. Published 2023 by John Wiley & Sons Ltd.
Companion website: www.wiley.com/go/molisch/wireless3e

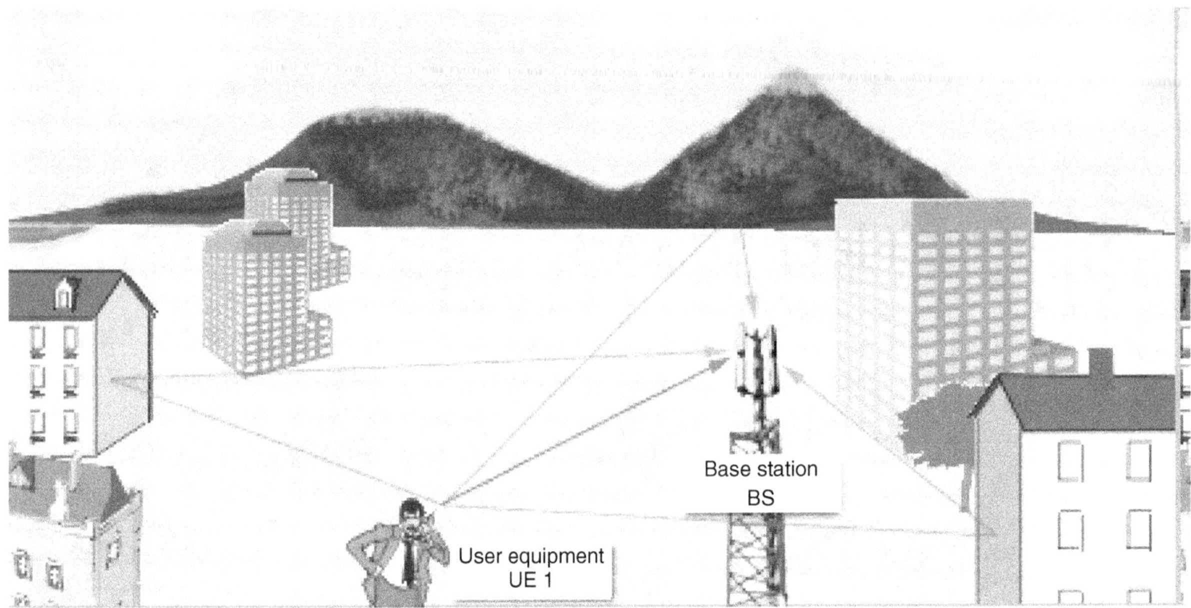

Figure 4.1 Multi-path propagation. Color version available at wiley.com/go/molisch/wireless3e.

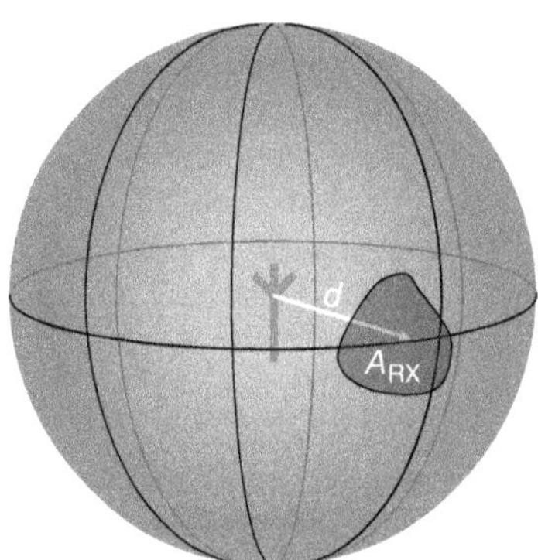

Figure 4.2 Received power area density at distance d, and collection of power by antenna area A_{RX}. Color version available at wiley.com/go/molisch/wireless3e.

If the TX antenna is not isotropic, then the energy density has to be multiplied with the (power) antenna gain, G_{TX}, in the direction of the RX antenna location.[4]

$$P_{\text{RX}}(d) = P_{\text{TX}} G_{\text{TX}} \frac{1}{4\pi d^2} A_{\text{RX}}. \tag{4.1}$$

The product of TX power and gain in the considered direction is also known as *Equivalent Isotropically Radiated Power* (EIRP), i.e., it is the power that an isotropic antenna would have to radiate such that the same energy density is seen in the considered direction.

The effective antenna area is proportional to the power that can be extracted from the antenna connectors for a given energy density. For a parabolic antenna, e.g., the effective antenna area is roughly the geometrical area of the surface. However, antennas with very small geometrical area – e.g., dipole antennas – can also have a considerable effective area, see Chapter 8.

It can be shown that there is a simple relationship between the effective area and the antenna gain:

$$G_{\text{RX}} = \frac{4\pi}{\lambda^2} A_{\text{RX}}. \tag{4.2}$$

Most noteworthy in this equation is the fact that – for a fixed antenna area – the antenna gain increases with increasing frequency (decreasing wavelength λ). This is also intuitive, as the directivity of an antenna is determined by its electrical size (size in units of wavelengths).

---

[4] If not stated otherwise, this book always defines the antenna gain as the gain over an isotropic radiator. For further discussions of antenna properties, see Chapter 8.

Substituting (4.2) into (4.1) gives the received power P_{RX} as a function of the distance d in free space, also known as *Friis' law*:

$$P_{RX}(d) = P_{TX}G_{TX}G_{RX}\left(\frac{\lambda}{4\pi d}\right)^2. \tag{4.3}$$

The factor $(\lambda/4\pi d)^2$ is also known as the *free space loss factor*.

Friis' law seems to indicate that the "attenuation" in free space increases with frequency. This is counterintuitive, as the energy is not lost, but rather redistributed over a sphere surface of area $4\pi d^2$. This mechanism has to be independent of the wavelength. This seeming contradiction is caused by the fact that we assume the *antenna gain* to be independent of the wavelength. On the other hand, if we assume that the *antenna area* of the RX antenna is independent of frequency, then the received power becomes independent of the frequency (see Eq. (4.1)). When both the TX and the RX antenna area are held constant, the received power actually *increases* with frequency.[5]

The question now naturally arises whether it is more meaningful to consider fixed gain or fixed antenna size. There is no unique answer to this question, but the following points should be considered:

- When comparing the same system (e.g., LTE) operating in different bands (e.g., 700 MHz and 2.1 GHz), it is often useful to assume constant gain, as such systems tend to use the same antenna *type* (e.g., $\lambda/2$ dipole or monopole), and not the same antenna *size*.
- When comparing systems with vastly different operating frequencies (e.g., Wi-Fi at 2.45 and 60 GHz), we usually do not have the constraint of using the same antenna type. Rather, the *form factor* (i.e., antenna size) needs to be considered – e.g., the antenna has to fit easily into a handset or onto an access point. This implies that a constant-area assumption is more meaningful.
- Finally, it has to be kept in mind that a very high gain is only helpful *if the gain is obtained in the right direction*, i.e., in the TX antenna has to have high gain in the direction of the RX and vice versa. For a directional link in free space (using, e.g., a parabolic or horn antenna), this is easy to achieve – the antenna can be oriented manually in the right direction. However, in *mobile* environments, the "right" direction is not known a priori. Thus, high gain can only be meaningfully obtained with *adaptive* systems, such as antenna arrays with adaptable weights (see Section 8.4) that allow to adjust the main transmit direction. While beneficial from the point of view of link budget, such antennas increase cost, and also require extra signaling to determine the right direction.

The validity of Friis' law is restricted to the far field of the antenna – i.e., the TX and RX antennas have to be at least one *Rayleigh distance* apart.[6] The Rayleigh distance (also known as the *Fraunhofer distance*) is defined as:

$$d_R = \frac{2L_a^2}{\lambda} \tag{4.4}$$

where L_a is the largest dimension of the antenna; furthermore, the far field requires $d \gg \lambda$ and $d \gg L_a$. We also note that the plane-wave approximation is only valid at distances beyond the Rayleigh distance.

Example 4.1 *Compute the Rayleigh distance of a square antenna with 20 dB gain.*

The gain is 100 on a linear scale. In that case, the effective area is approximately

$$A_{RX} = \frac{\lambda^2}{4\pi}G_{RX} = 8\lambda^2. \tag{4.5}$$

For a square-shaped antenna with $A_{RX} = L_a^2$, the Rayleigh distance is given by:

$$d_R = \frac{2 \cdot 8\lambda^2}{\lambda} = 16\lambda. \tag{4.6}$$

For setting up link budgets, it is often advantageous to write Friis' law on a logarithmic scale. Equation (4.3) then reads

$$P_{RX|dBm} = P_{TX|dBm} + G_{TX|dB} + G_{RX|dB} + 20\log\left(\frac{\lambda}{4\pi d}\right) \tag{4.7}$$

---

[5] An important corollary of this consideration is the assessment of millimeter-wave (and similarly for THz) radiation. It is often claimed that "mm-wave systems suffer from high pathloss." Such a statement is oversimplified, as it implies a comparison with constant antenna gain. With suitable (constant-area) antennas it is possible to obtain the same, or even lower, free-space pathloss than cm-waves.

[6] It will be obvious to the reader that Friis' law cannot hold for arbitrarily small d, since otherwise the RX power could be larger than the TX power, thus violating the first law of thermodynamics.

where $|_{dB}$ means "in units of dB" and log in this chapter always means $\log_{10}$. In order to better point out the distance dependence, it is advantageous to first compute the received power at 1 m distance:

$$P_{RX|dBm}\,(1\,m) = P_{TX|dBm} + G_{TX|dB} + G_{RX|dB} + 20\log\left(\frac{\lambda|_m}{4\pi \cdot 1}\right). \tag{4.8}$$

The last term on the r.h.s. of Eq. (4.8) is about -32 dB at 1 GHz and decreases by 6 dB for every doubling of the frequency. The actual received power at a distance d (in meters) is then:

$$\begin{aligned}
P_{RX|dBm}(d) &= P_{RX|dBm}(1\,m) - 20\log(d|_m) \\
&= P_{TX|dBm} + G_{TX|dB} + G_{RX|dB} - 32 - 20\log\left(\frac{f}{1\,GHz}\right) - 20\log\left(\frac{d}{1\,m}\right).
\end{aligned} \tag{4.9}$$

4.2 Reflection and Transmission

4.2.1 Snell's Law

Electromagnetic waves are often reflected at one or more IOs before arriving at the RX. The reflection coefficient of the IO, as well as the direction into which this reflection occurs, determines the power that arrives at the RX position. In this section, we deal with *specular reflections*. This type of reflection occurs when waves are incident onto smooth, large (compared with the wavelength) objects. A related mechanism is the *transmission* of waves – i.e., the penetration of waves into and through an IO. Transmission is especially important for wave propagation into, and inside, buildings: if the Base Station (BS) is either outside the building, or in a different room, then the waves have to penetrate a wall (dielectric layer) in order to get to the User Equipment (UE).

We first summarize the most important results:

1. A wave incident on the interface between air and a dielectric, possibly lossy, material is partially reflected. The angle of reflection is equal to the angle of incidence; the reflection coefficient depends on the angle as well as the polarization of the incident wave, and the material properties.
2. A wave incident onto a dielectric layer is partly reflected, and partly transmitted through the layer, where the direction of the wave that went through the layer (and emerges on the other side) is the same as the direction of incidence.

We now look at these aspects in more detail and provide mathematical expressions.

We first derive the reflection and transmission coefficients of a homogeneous plane wave incident onto a dielectric halfspace (most commonly, the ground). The dielectric material is characterized by its dielectric constant $\varepsilon = \varepsilon_0 \varepsilon_r$ (where ε_0 is the vacuum dielectric constant 8.854×10^{-12} Farad/m, and ε_r is the relative dielectric constant of the material) and conductivity σ_e. Furthermore, we assume that the material is isotropic and has a relative permeability $\mu_r = 1$.[7] The dielectric constant and conductivity can be merged into a single parameter, the *complex* dielectric constant:

$$\delta = \varepsilon_0 \delta_r = \varepsilon - j\frac{\sigma_e}{2\pi f_c}. \tag{4.10}$$

where f_c is the carrier frequency, and j is the imaginary unit. Though this definition is strictly valid only for a single frequency, it can actually be used for all *narrowband* systems, where the bandwidth is much smaller than the carrier frequency (typically a bandwidth $<0.1\,f_c$), as well as much smaller than the bandwidth over which the quantities σ_e and ε vary significantly.[8] Note that the reflection coefficients do change when the frequency changes drastically, e.g., from traditional cellular bands (2 GHz) to mm-wave bands. Tables of the dielectric characteristics of various common materials such as brick or concrete in different frequency bands can be found in the literature.

Let the plane wave be incident on the halfspace at an angle Θ_e, which is defined as the angle between the wave vector $\mathbf{k}$ and the unit vector that is orthogonal to the dielectric boundary. We have to distinguish between the *Transversal Magnetic* (TM) case, where the magnetic field component is parallel to the boundary between the two dielectrics, and the *Transversal Electric* (TE) case, where the electric field component is parallel (see Figure 4.3).

---

[7] This is approximately true for most materials relevant for the cellular and WLAN systems covered in this book.
[8] Note that this means "narrowband" in the Radio Frequency (RF) sense. We will later encounter a different definition for narrowband that is related to delay dispersion in wireless channels.

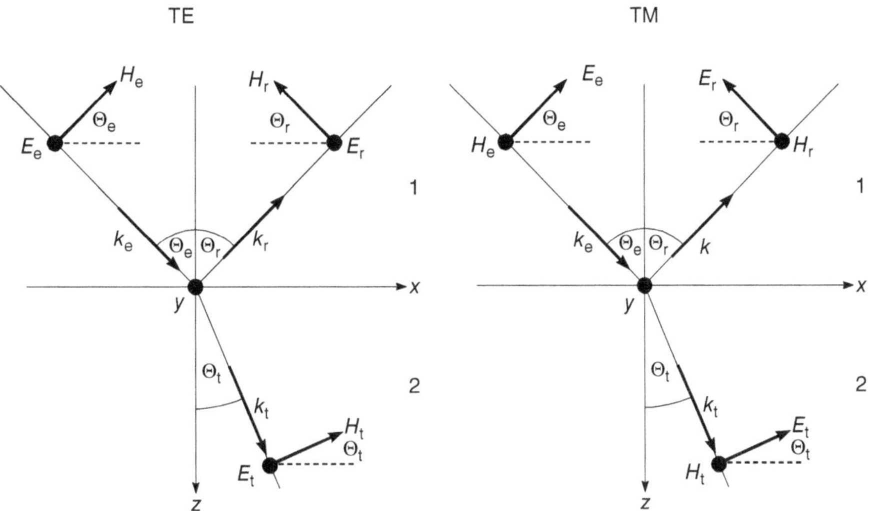

Figure 4.3 Reflection and transmission: for the TE case the E vectors point into the page, and for the TM case, the H vectors point out of the page.

The reflection and transmission coefficients can now be computed from postulating incident, reflected, and transmitted plane waves, and enforcing the continuity conditions at the boundary – see, e.g., [Ramo et al. 1967]. From these considerations, we obtain Snell's law: the angle of incidence is the same as the reflected angle:

$$\Theta_r = \Theta_e \tag{4.11}$$

and the angle of the transmitted wave is given by:

$$\frac{\sin \Theta_t}{\sin \Theta_e} = \frac{\sqrt{\delta_1}}{\sqrt{\delta_2}} \tag{4.12}$$

where subscripts 1 and 2 index the considered medium. Note that the angles become complex in lossy materials, giving rise to more complicated propagation [Orfanidi 2016].

The reflection and transmission coefficients are different for TE and for TM waves. For TM polarization:

$$\rho_{TM} = \frac{\sqrt{\delta_2} \cos \Theta_e - \sqrt{\delta_1} \cos (\Theta_t)}{\sqrt{\delta_2} \cos \Theta_e + \sqrt{\delta_1} \cos (\Theta_t)} \tag{4.13}$$

$$T_{TM} = \frac{2\sqrt{\delta_1} \cos (\Theta_e)}{\sqrt{\delta_2} \cos \Theta_e + \sqrt{\delta_1} \cos (\Theta_t)} \tag{4.14}$$

and for TE polarization:

$$\rho_{TE} = \frac{\sqrt{\delta_1} \cos (\Theta_e) - \sqrt{\delta_2} \cos (\Theta_t)}{\sqrt{\delta_1} \cos (\Theta_e) + \sqrt{\delta_2} \cos (\Theta_t)} \tag{4.15}$$

$$T_{TE} = \frac{2\sqrt{\delta_1} \cos (\Theta_e)}{\sqrt{\delta_1} \cos (\Theta_e) + \sqrt{\delta_2} \cos (\Theta_t)} \tag{4.16}$$

where $\rho_{TM} = E_r/E_e$, and $T_{TM} = E_t/E_e$ (and similarly for ρ_{TE} and T_{TE}). Note that the reflection coefficient has both an amplitude and a phase. Figure 4.4 shows both for a dielectric material with complex dielectric constant $\delta = 4 - 0.25j$. It is noteworthy that for both TE and TM waves, the reflection coefficient becomes -1 (magnitude 1, phase shift of 180°) at grazing incidence ($\Theta_e \rightarrow 90°$). This is the same reflection coefficient that would occur for reflection on an ideally conducting surface. We will see later that this has important consequences for the impact of ground-reflected waves in wireless systems.

In highly lossy materials, the transmitted wave is not a homogeneous plane wave, so that Snell's law must be interpreted with caution. However, these considerations are not practically important for cellular and WLAN systems, which are at the center of our attention.

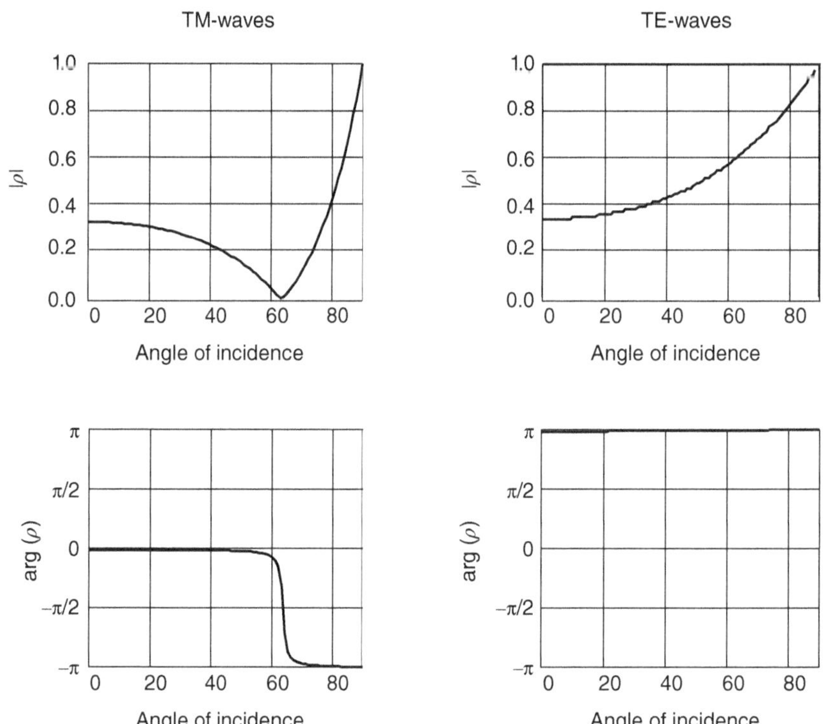

Figure 4.4 Reflection coefficient for a dielectric material with complex dielectricity constant $\delta = (4 - 0.25j)\varepsilon_0$.

*4.2.2 Reflection and Transmission for Layered Dielectric Structures

The previous section discussed the reflection and transmission in a dielectric halfspace. This is of interest, e.g., for ground reflections and reflections by terrain features, like mountains. A related problem is the problem of *transmission through* a dielectric layer. It occurs when a user inside a building is communicating with an outdoor BS, or in a picocell where the UE and the BS are in different rooms. In that case, we are interested in the attenuation and phase shift of a wave transmitted through a wall. Fortunately, the basic problem of dielectric layers is well known from other areas of electrical engineering – e.g., optical thin-film technology [Heavens 1965], and the results can be easily applied to wireless communications.

The simplest, and practically most important, case occurs when a dielectric layer is surrounded on both sides by air. The reflection and transmission coefficients can be determined by summation of the partial waves, resulting in a total transmission coefficient:

$$T = \frac{T_1 T_2 e^{-j\alpha_{\mathrm{diel}}}}{1 + \rho_1 \rho_2 e^{-2j\alpha_{\mathrm{diel}}}} \tag{4.17}$$

and a reflection coefficient:

$$\rho = \frac{\rho_1 + \rho_2 e^{-j2\alpha_{\mathrm{diel}}}}{1 + \rho_1 \rho_2 e^{-2j\alpha_{\mathrm{diel}}}} \tag{4.18}$$

where T_1 is the transmission coefficient of a wave from air into a dielectric halfspace (with the same dielectric properties as the considered layer) and T_2 is the transmission coefficient from dielectric into air; they can be computed from the results of Section 4.2.1. The quantity α_{diel} is the electrical length of the dielectric as seen by waves that are at an angle Θ_t with the layer:

$$\alpha_{\mathrm{diel}} = \frac{2\pi}{\lambda} \sqrt{\varepsilon_{r,2}} d_{\mathrm{layer}} \cos(\Theta_t) \tag{4.19}$$

where d_{layer} is the geometrical length of the layer. Note also that there is a waveguiding effect in lossy materials (see discussion in the previous section), so that the results of this section are not strictly applicable for dielectrics with losses.

In *multi-layer* structures, the problem becomes considerably more complicated [Heavens 1965]. However, in practice, even multi-layer structures are described by "effective" dielectric constants or reflection/transmission constants. These are measured directly for the composite structure. The alternative of measuring the dielectric properties for each layer separately and computing the resulting effective dielectric constant is prone to errors, as the measurement errors for the different layers add up.

Example 4.2 *Compute the effective ρ and T for a 50 cm-thick brick wall at 4 GHz carrier frequency for perpendicularly incident waves.*

Since $\Theta_e = 0$, Eqs. (4.11) and (4.12) imply that also $\Theta_r = \Theta_t = 0$. At $f = 4$ GHz ($\lambda = 7.5$ cm), brick has a relative permittivity ε_r of 4.44 [Rappaport 1996]; we neglect the conductivity. With the air having $\varepsilon_{air} = \varepsilon_1 = \varepsilon_0$, Eqs. (4.13)–(4.16) give the reflection and transmission coefficients for the surface between air and brick as:

$$\left. \begin{aligned} \rho_{1,\mathrm{TM}} &= \frac{\sqrt{\varepsilon_2} - \sqrt{\varepsilon_1}}{\sqrt{\varepsilon_2} + \sqrt{\varepsilon_1}} = 0.36 \\[4pt] \rho_{1,\mathrm{TE}} &= \frac{\sqrt{\varepsilon_1} - \sqrt{\varepsilon_2}}{\sqrt{\varepsilon_2} + \sqrt{\varepsilon_1}} = -0.36 \\[4pt] T_{1,\mathrm{TM}} &= \frac{2\sqrt{\varepsilon_1}}{\sqrt{\varepsilon_2} + \sqrt{\varepsilon_1}} = 0.64 \\[4pt] T_{1,\mathrm{TE}} &= \frac{2\sqrt{\varepsilon_1}}{\sqrt{\varepsilon_2} + \sqrt{\varepsilon_1}} = 0.64 \end{aligned} \right\} \tag{4.20}$$

and between brick and air as:

$$\left. \begin{aligned} \rho_{2,\mathrm{TM}} &= \rho_{1,\mathrm{TE}} \\ \rho_{2,\mathrm{TE}} &= \rho_{1,\mathrm{TM}} \\ T_{2,\mathrm{TM}} &= T_{1,\mathrm{TE}} \cdot \sqrt{\varepsilon_2/\varepsilon_1} = 1.36 \\ T_{2,\mathrm{TE}} &= T_{1,\mathrm{TM}} \cdot \sqrt{\varepsilon_2/\varepsilon_1} = 1.36 \end{aligned} \right\} \tag{4.21}$$

Note that the transmission coefficient can become larger than unity (e.g., $T_{2,\mathrm{TE}}$). This is not a violation of energy conservation: the transmission coefficient is defined as the ratio of the amplitudes of the transmitted and incident field. Energy conservation only dictates that the flux (energy) density of a reflected and transmitted field equals that of the incident field.

The electrical length of the wall, α_{diel}, is determined using Eq. (4.19) as:

$$\alpha_{\mathrm{diel}} = \frac{2\pi}{\lambda}\sqrt{\varepsilon_r}d = \frac{2\pi}{0.075}\sqrt{4.44} \cdot 0.5 = 88.26. \tag{4.22}$$

Finally, the total reflection and transmission coefficients can be determined using Eqs. (4.17) and (4.18), which gives:

$$\left. \begin{aligned} T_{\mathrm{TM}} &= \frac{T_{1,\mathrm{TM}}T_{2,\mathrm{TM}}e^{-j\alpha_{\mathrm{diel}}}}{1 + \rho_{1,\mathrm{TM}}\rho_{2,\mathrm{TM}}e^{-2j\alpha_{\mathrm{diel}}}} = \frac{0.64 \cdot 1.356 e^{-j88.26}}{1 - 0.36^2 e^{-j176.53}} = 0.90 - 0.36j \\[6pt] \rho_{\mathrm{TM}} &= \frac{\rho_{1,\mathrm{TM}} + \rho_{2,\mathrm{TM}}e^{-2j\alpha_{\mathrm{diel}}}}{1 + \rho_{1,\mathrm{TM}}\rho_{2,\mathrm{TM}}e^{-2j\alpha_{\mathrm{diel}}}} = \frac{0.36(1 - e^{-j176.53})}{1 - 0.36^2 e^{-j176.53}} = 0.086 + 0.22j \\[6pt] T_{\mathrm{TE}} &= \frac{T_{1,\mathrm{TE}}T_{2,\mathrm{TE}}e^{-j\alpha_{\mathrm{diel}}}}{1 + \rho_{1,\mathrm{TE}}\rho_{2,\mathrm{TE}}e^{-2j\alpha_{\mathrm{diel}}}} = \frac{T_{2,\mathrm{TM}}T_{1,\mathrm{TM}}e^{-j\alpha_{\mathrm{diel}}}}{1 + \rho_{2,\mathrm{TM}}\rho_{1,\mathrm{TM}}e^{-2j\alpha_{\mathrm{diel}}}} = T_{\mathrm{TM}} \\[6pt] \rho_{\mathrm{TE}} &= \frac{\rho_{1,\mathrm{TE}} + \rho_{2,\mathrm{TE}}e^{-2j\alpha_{\mathrm{diel}}}}{1 + \rho_{1,\mathrm{TE}}\rho_{2,\mathrm{TE}}e^{-2j\alpha_{\mathrm{diel}}}} = \frac{\rho_{2,\mathrm{TM}} + \rho_{1,\mathrm{TM}}e^{-2j\alpha_{\mathrm{diel}}}}{1 + \rho_{2,\mathrm{TM}}\rho_{1,\mathrm{TM}}e^{-2j\alpha_{\mathrm{diel}}}} = \frac{-\rho_{1,\mathrm{TM}} - \rho_{2,\mathrm{TM}}e^{-2j\alpha_{\mathrm{diel}}}}{1 + \rho_{2,\mathrm{TM}}\rho_{1,\mathrm{TM}}e^{-2j\alpha_{\mathrm{diel}}}} = -\rho_{\mathrm{TM}} \end{aligned} \right\}. \tag{4.23}$$

It is easily verified that in both cases $|\rho|^2 + |T|^2 = 1$.

In practice, the transmission through a wall depends both on the wall structure/thickness and the building material. The most common wall structures are (i) concrete and steel/concrete, (ii) brick/stone walls, and (iii) stud/post – plus – drywall. The attenuation created by those walls decreases in the sequence mentioned here. The absolute values also depend on the frequency. At the usual cellular frequencies (1–2 GHz), typical attenuations are on the order of 20, 10, and 3 dB, respectively. At mm-wave frequencies, steel concrete outer walls can have attenuations of up to 60 dB, while interior drywall attenuation remains on the order of 5–10 dB. Importantly, the attenuation of outer walls also depends on the windows. Standard single-pane windows have low attenuation (on the order of a few dB) even at mm-wave frequencies. Thus, such windows can be efficient pathways for radiation into/out of a building. However, energy-saving windows, which have a thin metal coating to reflect heat, show *much* higher attenuation – around 20 dB at cm-wave frequencies and some 40–60 dB at mm-wave frequencies. Figure 4.5 shows the attenuations of different structures as a function of frequency. Finally, we also note that the attenuation depends on the angle of incidence of the radiation, in line with the discussion above. Thus, waves that are incident perpendicularly on a wall are transmitted with less attenuation.

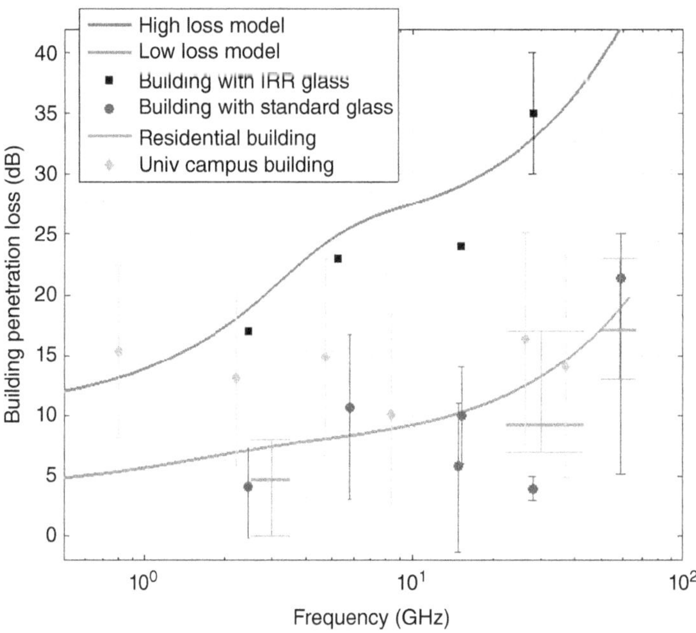

Figure 4.5 Building penetration as function of frequency. Color version available at wiley.com/go/molisch/wireless3e.

Reproduced with permission from [Haneda et al. 2016] © IEEE.

4.2.3 The d^{-4} Power Law

One of the "folk laws" of wireless communications says that the received signal power is inversely proportional to the *fourth* power of the distance between TX and RX. This law is often justified by computing the received power for the case that only a direct (LOS) wave, plus a ground-reflected wave, exists. For this specific case, the following equation is derived in Appendix 4.A (see www.wiley.com/go/molisch/wireless3e):

$$P_{RX}(d) \approx P_{TX} G_{TX} G_{RX} \left(\frac{h_{TX} h_{RX}}{d^2} \right)^2 \tag{4.24}$$

where h_{TX} and h_{RX} are the height of the TX and the RX antenna, respectively; it is valid for distances larger than:

$$d_{break'} \simeq \frac{4 h_{TX} h_{RX}}{\lambda}. \tag{4.25}$$

This equation, which replaces the standard Friis' law, implies that the received power becomes independent of frequency if constant-gain antennas are used. Furthermore, it follows from Eq. (4.24) that the received power increases with the square of the height of both BS and UE. For distances $d < d_{break'}$, Friis' law remains approximately valid, though significant variations occur even for small changes of d, due to the constructive or destructive interference of the two waves.

For the link budget, it is useful to rewrite the power law on a logarithmic scale. Assuming that the power decays as d^{-2} until a breakpoint d_{break},[9] and from there with $d^{-\alpha_{LBP}}$, then the received power is given by (see also Chapter 3):

$$P_{RX|dBm}(d) = P_{RX|dBm}(1m) - 20 \log \left(d_{break} \big|_m \right) - \alpha_{LBP} 10 \log \left(d/d_{break} \right). \tag{4.26}$$

Figure 4.6 shows the received power when there is a direct wave and a ground-reflected wave, both from the exact formulation (see Appendix 4.A) and from Eq. (4.24). We find that the transition between the attenuation coefficient $\alpha_{PL} = 2$ and $\alpha_{LBP} = 4$ is actually not a sharp breakpoint, but rather smooth. According to Eq. (4.25), the breakpoint is at $d = 90$ m; this seems to be approximated reasonably well in Figure 4.6.

The equations derived above (and in Appendix 4.A) are self-consistent, but it has to be emphasized that they are not a *universal* description of wireless channels. After all, propagation does not always happen over a flat Earth. Rather, multiple propagation paths

---

[9] Note that $d_{break'}$ and d_{break} have different definitions: $d_{break'}$ is defined via the phase difference between direct and ground-reflected wave, while d_{break} is defined via the intersection of the curve first to d^2 and d^4. d_{break} applies in general scenarios, while $d_{break'}$ is only applicable for the case of direct-plus-ground reflected wave. In the latter, special, case, the numerical values of d_{break} and $d_{break'}$ are similar.

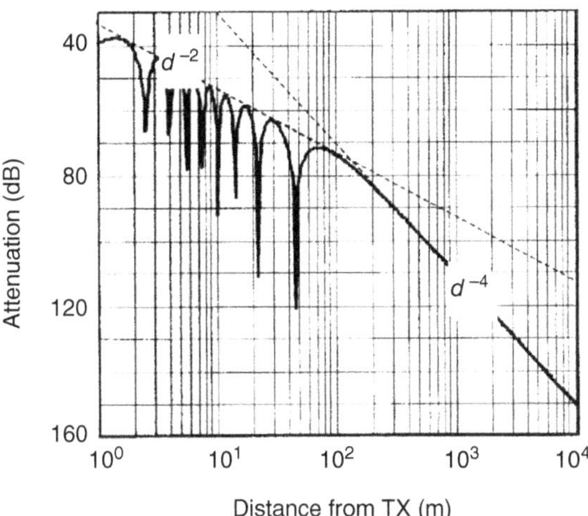

Figure 4.6 Propagation over an ideally reflecting ground. Height of BS. 5 m. Height of UE: 1.5 m. Frequency: 900 MHz.

are possible, LOS connections are often shadowed by IOs, etc. Thus the derived theory often does not agree with the measurement results in realistic channels in several respects:

1. $\alpha_{\mathrm{LBP}} = 4$ is not a universally valid decay exponent at larger distances. Rather, values between $1.5 < \alpha_{\mathrm{LBP}} < 5.5$ have been measured, and the actual value strongly depends on the surrounding environment. The value of $\alpha_{\mathrm{LBP}} = 4$ is, at best, a mean value of various environments (see Section 7.1 for typical values in different environments).
2. The transition between $\alpha = 2$ and $\alpha_{\mathrm{LBP}} = 4$ rarely occurs at $d_{\mathrm{break'}}$ predicted by (4.25).
3. Measurements also show that there is a second breakpoint, beyond which a pathloss coefficient >6 is valid. This effect is not predicted at all by the above model. For some situations, the effect can be explained by the radio horizon (i.e., the curvature of the Earth), which is not included in the model above [Parsons 1992].

4.3   Diffraction

All the equations derived up to now deal with infinitely extended IOs. However, real IOs like buildings, cars, etc., have a finite extent. A finite-sized object does not create sharp shadows (the way geometrical optics would have it), but rather there is diffraction due to the wave nature of electromagnetic radiation. Only in the limit of very small wavelength (large frequency) does geometrical optics become exact.

In the following, we first treat two canonical diffraction problems: diffraction of a homogeneous plane wave (i) by a knife edge or screen and (ii) by a wedge, and derive the diffraction coefficients that tell us how much power can be received in the shadow region behind an obstacle. Subsequently, we consider the effect of a concatenation of several screens.

4.3.1   Diffraction by a Single Screen or Wedge

The Diffraction Coefficient

The simplest diffraction problem is the diffraction of a homogeneous plane wave by a semi-infinite screen, as sketched in Figure 4.7. Diffraction can be understood from Huygen's principle that each point of a wavefront can be considered the source of a spherical wave. For a homogeneous plane wave, the superposition of these spherical waves results in another homogeneous plane wave, see transition from plane A' to B'. If, however, the screen eliminates parts of the point sources (and their associated spherical waves), the resulting wavefront is not plane anymore, see the transition from plane B' to C'. Constructive and destructive interferences occur in different directions.[10]

The electric field at any point in the far field ($x \gg \lambda$) can be expressed in a form that involves only a standard integral, the *Fresnel integral*. With the field that would be obtained if the screen were not present represented as $\exp(-jk_0x)$, the total field becomes [Vaughan and Andersen 2003]:

$$E_{\mathrm{total}} = \exp\left(-jk_0x\right)\left(\frac{1}{2} - \frac{\exp\left(j\pi/4\right)}{\sqrt{2}}F(\nu_{\mathrm{F}})\right) = \exp\left(-jk_0x\right)\widetilde{F}(\nu_{\mathrm{F}}) \qquad (4.27)$$

---

[10] For more accurate considerations, it is noteworthy that Huygen's principle is not exact. A derivation from Maxwell's equations, which also includes a discussion of the necessary assumptions, is given, e.g., in [Marcuse 1991].

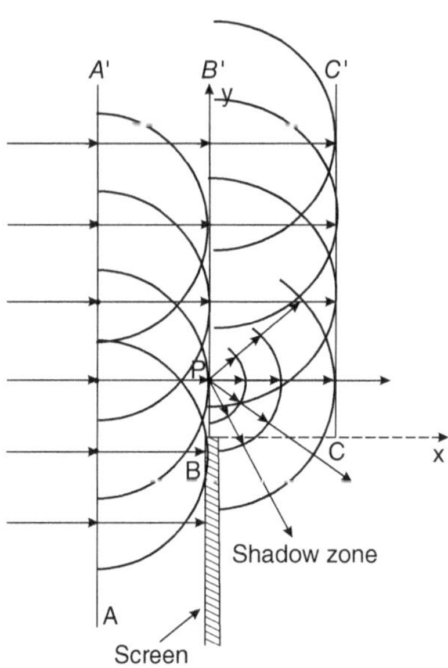

Figure 4.7 Huygen's principle.

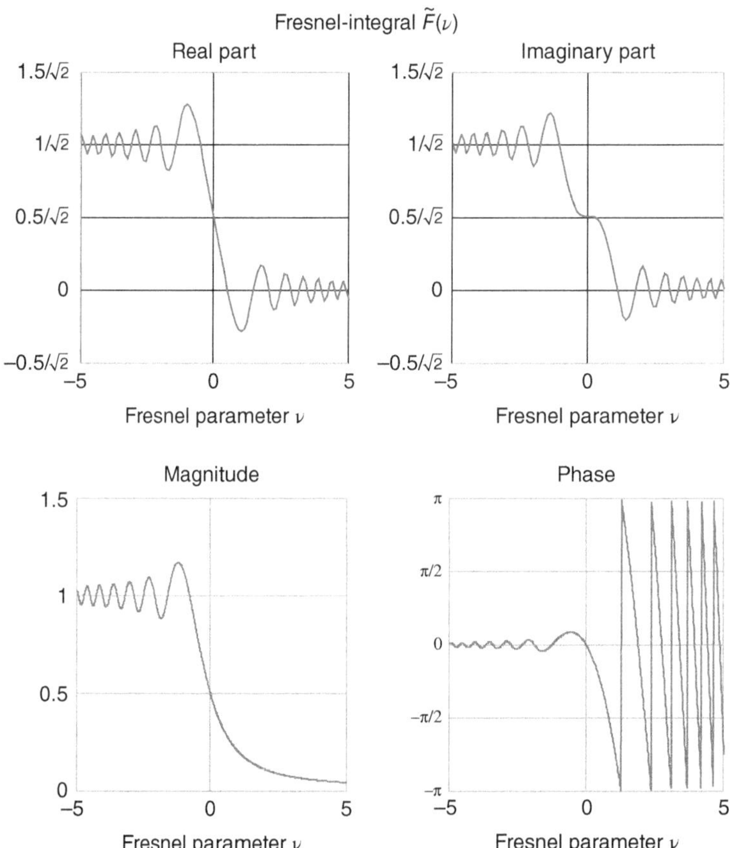

Figure 4.8 Fresnel integral. Color version available at wiley.com/go/molisch/wireless3e.

where the Fresnel parameter $\nu_F = -y\sqrt{2/(\lambda x)}$ and the Fresnel integral $F(\nu_F)$ is defined as:

$$F(\nu_F) = \int_0^{\nu_F} \exp\left(-j\pi\frac{t^2}{2}\right) dt. \tag{4.28}$$

Figure 4.8 plots this function. It is interesting that $\widetilde{F}(\nu_F)$ can become larger than unity for some values of ν_F. This implies that the received power at a specific location can actually be *increased* by the presence of screen. Huygen's principle again provides the

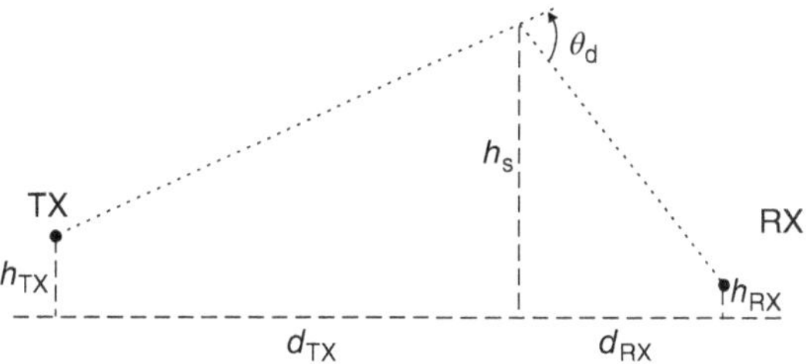

Figure 4.9 Geometry for the computation of the Fresnel parameters.

explanation: some spherical waves that would normally interfere destructively in a specific location are blocked off. However, note that the *total* energy (integrated over the whole wavefront) *cannot* be increased by the screen.

Another important observation is that the magnitude of the Fresnel parameter for a given point in the shadow zone decreases roughly with $\sqrt{f}$ (this is a loss *in addition* to the free-space pathloss, since we are considering here a plane wave that by definition does not have free-space pathloss). This implies that diffraction becomes less efficient as the frequency increases, or in other words, the shadows thrown by objects such as houses become "sharper" for higher frequencies. This again agrees with intuition and observation from daily life: long-wavelength signals can more easily "bend" around an obstacle, whereas short wavelength signals such as light or X-rays throw sharper shadows.

Consider now the more general geometry of Figure 4.9. The TX is at height h_{TX}, the RX at h_{RX}, and the screen extends from $-\infty$ to h_s. The diffraction angle θ_d is thus:

$$\theta_d = \arctan\left(\frac{h_s - h_{TX}}{d_{TX}}\right) + \arctan\left(\frac{h_s - h_{RX}}{d_{RX}}\right). \tag{4.29}$$

and the Fresnel parameter ν_F can be obtained from θ_d (for small θ_d) as:

$$\nu_F = \theta_d\sqrt{\frac{2d_{TX}d_{RX}}{\lambda(d_{TX} + d_{RX})}}. \tag{4.30}$$

The field strength can again be computed from Eq. (4.27), just using the Fresnel parameter from Eq. (4.30). Note that this equation requires that $d_{TX}, d_{RX} \gg h, \lambda$.

Note that the result given above is approximate in the sense that it neglects the polarization of the incident field. More accurate equations for both the TE and the TM case can be found in [Bowman 1987].

Example 4.3 *Consider diffraction by a screen with $d_{TX} = 200$ m, $d_{RX} = 50$ m, $h_{TX} = 20$ m, $h_{RX} = 1.5$ m, $h_s = 40$ m, at a center frequency of 900 MHz. Compute the diffraction coefficient.*

A center frequency of 900 MHz implies a wavelength $\lambda = 1/3$ m. Computing the diffraction angle θ_d from Eq. (4.29) gives:

$$\begin{aligned}
\theta_d &= \arctan\left(\frac{h_s - h_{TX}}{d_{TX}}\right) + \arctan\left(\frac{h_s - h_{RX}}{d_{RX}}\right) \\
&= \arctan\left(\frac{40 - 20}{200}\right) + \arctan\left(\frac{40 - 1.5}{50}\right) = 0.756 \text{ rad.}
\end{aligned} \tag{4.31}$$

Then, the Fresnel parameter is given by Eq. (4.30) as:

$$\nu_F = \theta_d\sqrt{\frac{2d_{TX}d_{RX}}{\lambda(d_{TX} + d_{RX})}} = 0.756\sqrt{\frac{2 \cdot 200 \cdot 50}{1/3 \cdot (200 + 50)}} = 11.71. \tag{4.32}$$

Evaluation of Eq. (4.28), with MATLAB or [Abramowitz and Stegun 1965], yields

$$F(11.71) = \int_0^{\nu_F} \exp\left(-j\pi\frac{t^2}{2}\right) dt \approx 0.527 - j0.505. \tag{4.33}$$

Finally, Eq. (4.27) gives the total received field as:

$$\begin{aligned}
E_{\text{total}} &= \exp\left(-jk_0 x\right)\left(\frac{1}{2} - \frac{\exp\left(j\pi/4\right)}{\sqrt{2}}F(11.71)\right) \\
&= \exp\left(-jk_0 x\right)\left(\frac{1}{2} - \frac{\exp\left(j\pi/4\right)}{\sqrt{2}}(0.527 - j0.505)\right) \\
&= (-0.016 - j0.011)\exp\left(-jk_0 x\right).
\end{aligned} \tag{4.34}$$

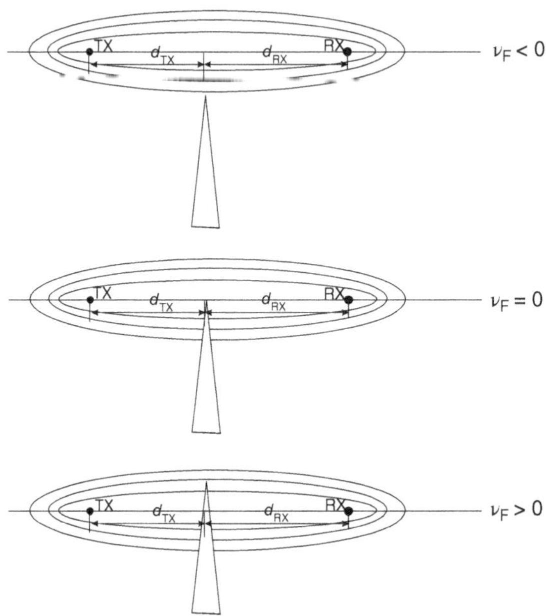

Figure 4.10   The principle of Fresnel ellipsoids.

Fresnel Zones

The impact of an obstacle can also be assessed qualitatively, and intuitively, by the concept of *Fresnel zones*. Figure 4.10 shows the basic principle. Draw an ellipsoid whose foci are the BS and the UE locations. According to the definition of an ellipsoid, all rays that are reflected at points on this ellipsoid have the same run length (equivalent to runtime). The eccentricity of the ellipsoid determines the extra run length compared with the LOS – i.e., the direct connection between the two foci. Ellipsoids where this extra distance is an integer multiple of $\lambda/2$ are called "Fresnel ellipsoids." Now extra run length also leads to an additional phase shift, so that the ellipsoids can be described by the phase shift that they cause. More specifically, the ith Fresnel ellipsoid is the one that results in a phase shift of $i \cdot \pi$. An object that does not overlap with the first Fresnel zone is usually considered to not lead to significant diffraction. It is thus, for example, possible to define a "line of sight" condition between a TX and an RX if the first Fresnel zone is free of obstacles.

Fresnel zones can also be used for explanation of the d^{-4} law. The propagation follows a free space law up to the distance where the first Fresnel ellipsoid touches the ground. At this distance, which is the breakpoint distance, the phase difference between the direct and the reflected ray becomes π. As we increase the distance further, the phase difference between direct and reflected ray decreases monotonically from π to 0. Due to the reflection coefficient -1, this means that the interference between the two rays changes from constructive at the breakpoint to more and more destructive (the more destructive the smaller the phase difference between the rays, i.e., the larger the d), leading to a faster decay of received power with distance than in free-space.

*Diffraction by a Wedge

The semi-infinite absorbing screen is a useful tool for the explanation of diffraction, since it is the simplest possible configuration. However, many obstacles especially in urban environments are much better represented by a wedge structure, as sketched in Figure 4.11. The problem of diffraction by a wedge has been treated for some 100 years and is still an area of active research. Depending on the boundary conditions, solutions can be derived that are either valid at arbitrary observation points or approximate solutions that are only valid in the far field (i.e., far away from the wedge). The latter solutions are usually much simpler, and will thus be the only ones considered here.

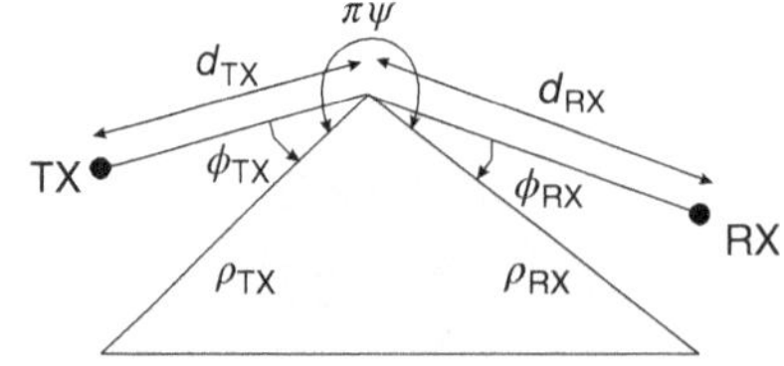

Figure 4.11   Geometry for wedge diffraction.

The part of the field that is created by diffraction can be written as the product of the incident field with a phase factor $\exp(-jk_0 d_{RX})$, a geometry factor $A(d_{TX}, d_{RX})$ that depends only on the distance of TX and RX from the wedge, and the diffraction coefficient $D(\phi_{TX}, \phi_{RX})$ that depends on the diffraction angles [Vaughan and Andersen 2003]:

$$E_{\text{diff}} = E_{\text{inc},0} D(\phi_{TX}, \phi_{RX}) A(d_{TX}, d_{RX}) \exp\left(-jk_0 d_{RX}\right). \tag{4.35}$$

The diffracted field has to be added to the field as computed by geometrical optics.[11]

The definition of the geometry parameters is shown in Figure 4.11. The geometry factor is given by:

$$A(d_{TX}, d_{RX}) = \sqrt{\frac{d_{TX}}{d_{RX}(d_{TX} + d_{RX})}}. \tag{4.36}$$

The diffraction coefficient, D, depends on the boundary conditions – namely, the reflection coefficients ρ_{TX} and ρ_{RX}. Explicit equations are given in Appendix 4.B.

*4.3.2 Diffraction by Multiple Screens

Diffraction by a single screen is a problem that has been widely studied, because it is amenable to closed-form mathematical treatment, and forms the basis for the treatment of more complex problems. However, in practice, we usually encounter situations where *multiple IOs* are located between TX and RX. Such a situation occurs, e.g., for propagation over the rooftops of an urban environment. As we see in Figure 4.12, such a situation can be well approximated by diffraction by multiple screens. Unfortunately, diffraction by multiple screens is an extremely challenging mathematical problem, and – except for a few special cases – no exact solutions are available. Still, a wealth of approximate methods has been proposed in the literature, of which we give an overview in the remainder of this section.

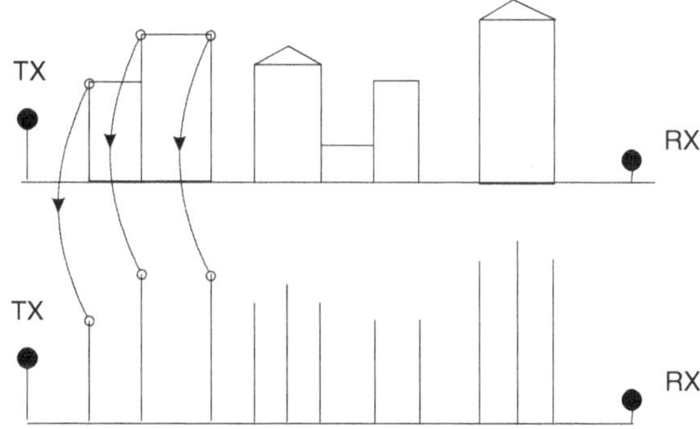

Figure 4.12 Approximation of multiple buildings by a series of screens.

Bullington's Method

Bullington's method replaces the multiple screens by a single, "equivalent" screen. This equivalent screen is derived in the following way: put a tangential straight line from the TX to the real obstacles, and select the steepest one (i.e., the one with the largest elevation angle), so that all obstacles either touch this tangent, or lie below it. Similarly, take the tangents from the RX to the obstacles, and select the steepest one. The equivalent screen is then determined by the intersection of the steepest TX tangent and the steepest RX tangent (see Figure 4.13). The field resulting from diffraction at this single screen can be computed according to Section 4.3.1.

The major attraction of Bullington's method is its simplicity. However, this simplicity also leads to considerable inaccuracies. Most of the physically existing screens do not impact the location of the equivalent screen. Even the highest obstacle might not have an impact. Consider Figure 4.13: if the highest obstacle lies between screens 01 and 02, it could lie below the tangential lines, and thus not influence the "equivalent" screen, even though it is higher than either screen 01 or screen 02. In reality, these high obstacles *do* have an effect on propagation loss and cause an additional attenuation. The Bullington method thus tends to give optimistic predictions of the received power.

---

[11] If the field incident on the wedge can be written as $E_{\text{inc},0} = E_0 \exp(-jk_0 d_{TX})/d_{TX}$, the above equations become completely symmetrical with respect to d_{TX} and d_{RX}.

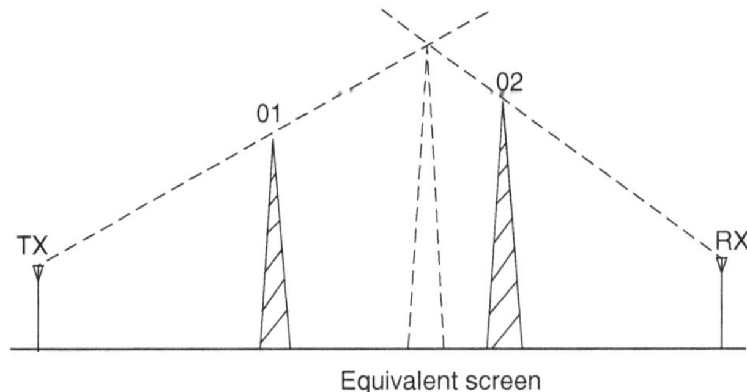

Figure 4.13 Equivalent screen after Bullington.
Reproduced with permission from [Parsons 1992] © J. Wiley and Sons, Ltd.

The Epstein–Petersen Method

The low accuracy of the Bullington method is due to the fact that only two obstacles determine the equivalent screen, and thus the total diffraction coefficient. This problem can be somewhat mitigated by the Epstein–Petersen method [Epstein and Peterson 1953]. This approach computes the diffraction losses for each screen separately. The attenuation of a specific screen is computed by putting a virtual "TX" and "RX" on the tips of the screens to the left and right of this considered screen (see Figure 4.14). The diffraction coefficient, and the attenuation, of this one screen can be easily computed from the principles of Section 4.3.1. Attenuations by the different screens are then added up (on a logarithmic scale). The method thus includes the effects of *all* screens.

Despite this more refined modeling, the method is still only approximate. It uses the diffraction attenuation (Eq. 4.27) that is based on the assumption that the RX is in the far field of the screen. If, however, two of the screens are close together, this assumption is violated, and significant errors can occur.

The inaccuracies caused by this "far-field assumption" can be reduced considerably by the *slope diffraction method*. In this approach, the field is expanded into a Taylor series. In addition to the zeroth-order term (far field), which enforces continuity of the electrical field at the screen, also the first-order term is taken into account, and used to enforce continuity of the first derivative of the field. This results in modified coefficients A and D, which are determined by recursion equations [Andersen 1997].

Deygout's Method

The philosophy of Deygout's method is similar to that of the Epstein–Petersen method, as it also adds up the attenuations caused by each screen [Deygout 1966]. However, the diffraction angles are defined in the Deygout method by a different algorithm:

- In the first step, determine the attenuation between TX and RX if only the ith screen is present (for all i).
- The screen that causes the largest attenuation is defined as the "main screen" – its index is defined as i_{ms}.
- Compute the attenuation between the TX and the tip of the main screen caused by the jth screen (with j running now from 1 to i_{ms}). The screen resulting in the largest attenuation is called the "subsidiary main screen." Similarly, compute the attenuation between the main screen and the RX, caused by the jth screen ($j > i_{ms} + 1$).
- Optionally, repeat that procedure to create "subsidiary subsidiary screens," etc.
- Add up the losses (in dB) from all considered screens.

The Deygout method works well if there is actually one dominant screen that creates most of the losses. Otherwise, it can create considerable errors.

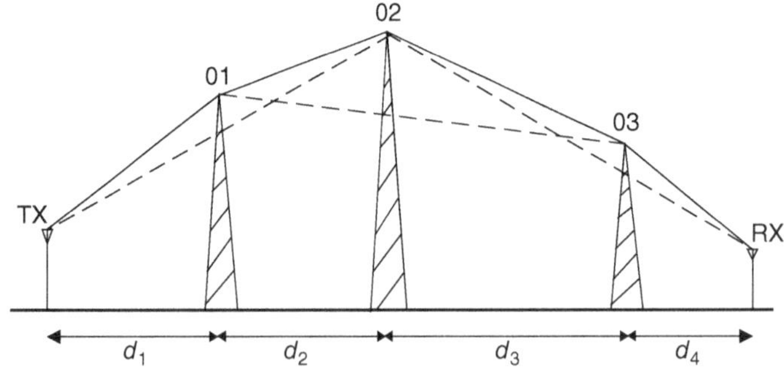

Figure 4.14 The Epstein–Petersen method.
Reproduced with permission from [Parsons 1992] © J. Wiley and Sons, Ltd.

Example 4.4   *There are three screens, 20 m apart from each other and 30, 40, and 25 m high. The first screen is 30 m from the TX, the last screen is 100 m from the RX. The TX is 1.5 m high, the RX is 30 m high. Compute the attenuation due to diffraction at 900 MHz by the Deygout method.*

The attenuation L caused by a certain screen is given as:

$$L = -20 \log \widetilde{F}(\nu_F) \tag{4.37}$$

where $\widetilde{F}(\nu_F)$ as defined in Eq. (4.27) is given by:

$$\widetilde{F}(\nu_F) = \frac{1}{2} - \frac{\exp(j\pi/4)}{\sqrt{2}} F(\nu_F). \tag{4.38}$$

First, determine the attenuation caused by screen 1. The diffraction angle θ_d, Fresnel parameter ν_F, and attenuation L are given by:

$$\left. \begin{aligned}
\theta_d &= \arctan\left(\frac{30-1.5}{30}\right) + \arctan\left(\frac{30-30}{140}\right) = 0.760 \,\text{rad} \\
\nu_F &= 0.760\sqrt{\frac{2\cdot 30\cdot 140}{1/3\cdot(30+140)}} = 9.25 \\
L_1 &= -20\log\left(\left|\frac{1}{2} - \frac{\exp(j\pi/4)}{\sqrt{2}}F(9.25)\right|\right) \\
&\approx -20\log\left(\left|\frac{1}{2} - \frac{\exp(j\pi/4)}{\sqrt{2}}\cdot(0.522-j0.527)\right|\right) = 32.28 \,\text{dB}
\end{aligned} \right\} \tag{4.39}$$

where the Fresnel integral $F(\nu_F)$ is numerically evaluated. Similarly, the attenuation caused by screens 2 and 3 is $L_2 = 33.59$ dB and $L_3 = 25.64$ dB, respectively. Hence, the main attenuation is caused by screen 2, which therefore becomes the "main screen."

Next, the attenuation from the TX to the tip of screen 2, as caused by screen 1, is determined. The diffraction angle θ_d, Fresnel parameter ν_F, and attenuation L become:

$$\left. \begin{aligned}
\theta_d &= \arctan\left(\frac{30-1.5}{30}\right) + \arctan\left(\frac{30-40}{20}\right) = 0.296 \,\text{rad} \\
\nu_F &= 0.269\sqrt{\frac{2\cdot 30\cdot 20}{1/3\cdot(30+20)}} = 2.51 \\
L_4 &= -20\log\left(\frac{1}{2} - \frac{\exp(-j\pi/4)}{\sqrt{2}}F(2.51)\right) \\
&\approx -20\log\left(\frac{1}{2} - \frac{\exp(-j\pi/4)}{\sqrt{2}}\cdot(0.446-j0.614)\right) = 21.01 \,\text{dB}
\end{aligned} \right\}. \tag{4.40}$$

Similarly, the attenuation from the tip of screen 2 to the RX, as caused by screen 3, is $L_5 = 0.17$ dB. The total attenuation caused by diffraction is then determined as the sum of all attenuations – i.e.:

$$L_{\text{total}} = L_2 + L_4 + L_5$$
$$= 33.59 + 21.01 + 0.17 = 54.77 \,\text{dB}.$$

Empirical Models

The *International Telecommunications Union* (ITU) proposed an extremely simple, semi-empirical model for diffraction losses (i.e., losses in addition to free space attenuation):

$$L_{\text{total}} = \sum_{i=1}^{N} L_i + 20 \log C_N \tag{4.41}$$

where L_i is the diffraction loss from each separate screen (in dB), and C_N is defined as:

$$C_N = \sqrt{\frac{P_a}{P_b}} \tag{4.42}$$

where

$$P_a = d_{p1} \prod_{i=1}^{N} d_{ni} \left(d_{p1} + \sum_{j=1}^{N} d_{nj} \right)$$
$$P_b = d_{p1} d_{nN} \prod_{i=1}^{N} \left(d_{pi} + d_{ni} \right)$$

(4.43)

Here, d_{pi} is the (geometrical) distance to the preceding screen tip, and d_{ni} is the distance to the following one. [Li et al. 1997] showed that this equation leads to large errors and proposed a modified definition:

$$C_N = \frac{P_a}{P_b}.$$

(4.44)

Comparison of the Different Methods

The only special case where an exact solution can be easily computed is the case when all screens have the same height and are at the same height as the TX and RX antennas. In that case, diffraction loss (on a linear scale!) is proportional to the number of screens, $1/(N_{screen} + 1)$. Let us now check whether the above approximation methods give this result (see Figure 4.15):

- The Bullington method is independent of the number of screens, and thus obviously gives a wrong functional dependence.
- The Epstein–Petersen method adds the attenuations *on a logarithmic scale* and thus leads to an exponential increase of the total attenuation on a linear scale.
- Similarly, the Deygout method and the ITU-R method predict an exponential increase of the total attenuation as the number of screens increases.
- The slope diffraction method (up to 15 screens) and the *modified* ITU method lead to a linear increase in total attenuation and thus predict the trend correctly.

However, note that the above comparison considers a specific, limiting case. For a small number of screens of different height, both the Deygout and the Epstein–Petersen method can be used successfully.

4.4 Scattering by Rough Surfaces

Scattering on rough surfaces (Figure 4.16) is a process that is very important for wireless communications. Scattering theory usually assumes roughness to be *random*. However, in wireless communications it is common to also describe *deterministic*, possibly periodic, structures (e.g., bookshelves or windowsills) as rough. For ray-tracing predictions (see Section 4.7), "roughness" thus describes all (physically present) objects that are not included in the used maps and building plans. The justifications for this approach are rather

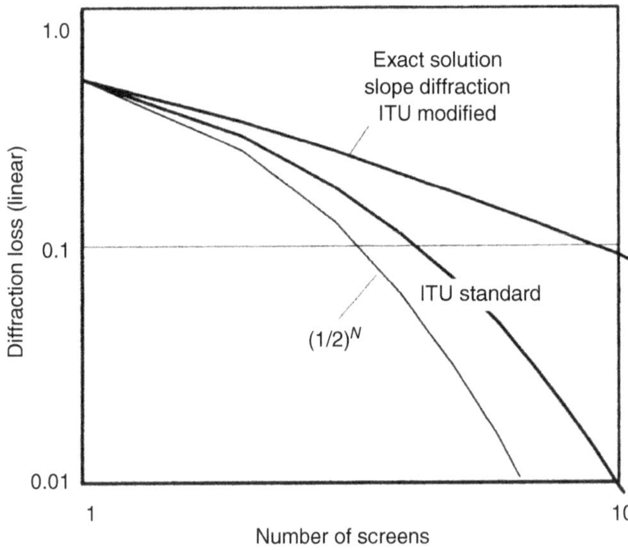

Figure 4.15 Comparison of different computation methods for multiple-screen diffraction.

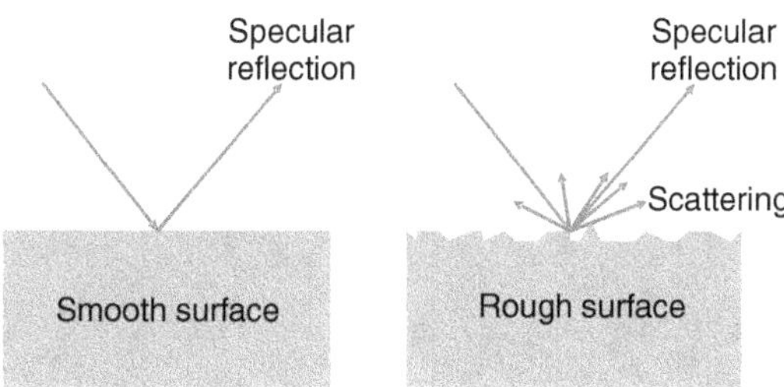

Figure 4.16 Scattering by a rough surface.

heuristic: (i) the errors made are smaller than some other error sources in ray-tracing predictions and (ii) there is no better alternative that achieves low computational complexity.

That being said, the remainder of this section will consider the mathematical treatment of genuinely rough surfaces. This area has been investigated extensively in the last 50 years, mostly due to its great importance in radar technology. Two main theories have evolved: the Kirchhoff theory and the perturbation theory.

4.4.1 The Kirchhoff Theory

The Kirchhoff theory is conceptually very simple and requires only a small amount of information – namely, the probability density function of surface amplitude (height). The theory assumes that height variations are so small that different *scattering points* on the surface do not influence each other – in other words, that one point of the surface does not "cast a shadow" onto other points of the surface. This assumption is actually not fulfilled very well in wireless communications.

Assuming that the above condition is actually fulfilled, surface roughness leads to a reduction in power of the specularly reflected ray, as radiation is also scattered in other directions (see r.h.s. of Figure 4.16). This power reduction can be described by an *effective* reflection coefficient ρ_{rough}. In the case of Gaussian height distribution, this reflection factor becomes:

$$\rho_{\mathrm{rough}} = \rho_{\mathrm{smooth}} \exp\left[-2(k_0 \sigma_{\mathrm{h}} \sin \psi)^2 \right] \tag{4.45}$$

where σ_{h} is the standard deviation of the height distribution, k_0 is the wavenumber $2\pi/\lambda$, and ψ is the angle of incidence (defined as the angle between the wave vector and the surface). The term $2k_0\sigma_{\mathrm{h}} \sin \psi$ is also known as Rayleigh roughness.[12] Note that for grazing incidence ($\psi \approx 0$), the effect of the roughness vanishes, and the reflection becomes specular again.

*4.4.2 Perturbation Theory

The perturbation theory generalizes the Kirchhoff theory, using not only the probability density function of the surface height but also its spatial correlation function. In other words, it takes into account the question "how fast does the height vary if we move a certain distance along the surface?" (see Figure 4.17).

Mathematically, the spatial correlation function $W(\Delta_{\mathrm{r}})$ is defined via:

$$\sigma_{\mathrm{h}}^2 W(\Delta_{\mathrm{r}}) = E_{\mathbf{r}}\{h(\mathbf{r})h(\mathbf{r} + \Delta_{\mathrm{r}})\} \tag{4.46}$$

where $\mathbf{r}$ and $\Delta_{\mathbf{r}}$ are (two-dimensional) location vectors, and $E_{\mathbf{r}}$ is expectation with respect to $\mathbf{r}$.[13] We need this information to find whether one point on the surface can "cast a shadow" onto another point of the surface. If extremely fast amplitude variations are allowed, shadowing situations are much more common. The above definition enforces spatial statistical stationarity – i.e., the correlation is independent of the absolute location $\mathbf{r}$. The correlation length L_{c} is defined as the distance so that $W(L_{\mathrm{c}}) = 0.5 \cdot W(0)$.[14]

---

[12] Some papers call $k_0\sigma_{\mathrm{h}} \sin \psi$ the Rayleigh roughness.

[13] We make here implicitly the assumption of "spatial ergodicity," i.e., that taking the expectation over the locations $\mathbf{r}$ is identical to taking the expectation over the statistical ensemble.

[14] A more extensive discussion of expectation values and correlation functions will be given in Chapter 6.

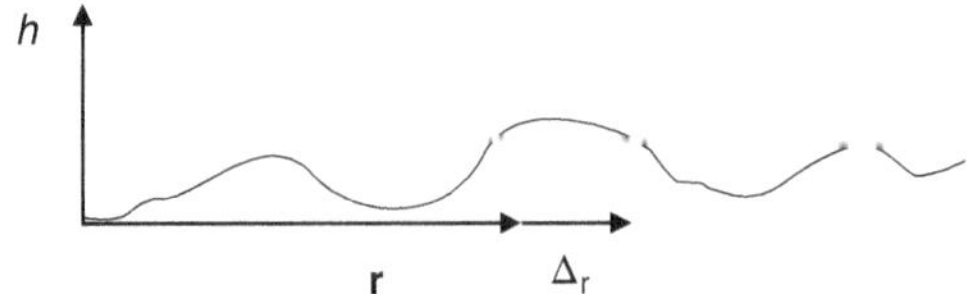

Figure 4.17 Geometry for perturbation theory of rough scattering.

The effect of surface roughness on the amplitude of a specularly reflected wave can be described by an "effective" (complex) dielectric constant δ_{eff}, which in turn gives rise to an "effective" reflection coefficient, as computed from Snell's law.[15] For vertical polarization, the δ_{eff} is given by [Vaughan and Andersen 2003]:

$$\frac{1}{\sqrt{\delta_{\text{r,eff}}}} = \begin{cases} \dfrac{1}{\sqrt{\varepsilon_\text{r}}} + j\dfrac{k_0\sigma_\text{h}^2 \sin\left(2\psi\right)}{2L_\text{c}} \displaystyle\int_0^\infty \dfrac{1}{x}\dfrac{d\hat{W}(x)}{dx}\,dx, & k_0L_\text{c} \ll 1 \\[3ex] \dfrac{1}{\sqrt{\varepsilon_\text{r}}} + (k_0\sigma_\text{h})^2(\sin\psi)^3, & k_0L_\text{c} \gg 1, \psi \gg \dfrac{1}{\sqrt{k_0L_\text{c}}} \\[3ex] \dfrac{1}{\sqrt{\varepsilon_\text{r}}} - \dfrac{\sigma_\text{h}^2}{2L_\text{c}^2}\dfrac{\sqrt{k_0L_\text{c}}}{\sqrt{2\pi}}\exp\left(j\pi/4\right)\displaystyle\int_0^\infty \dfrac{1}{x\sqrt{x}}\dfrac{d\hat{W}(x)}{dx}\,dx, & k_0L_\text{c} \gg 1, \psi \ll \dfrac{1}{\sqrt{k_0L_\text{c}}} \end{cases} \tag{4.47}$$

where $\hat{W}(x) = W(x/L_\text{c})$, while for horizontal polarization, it is

$$\frac{1}{\sqrt{\delta_{\text{r,eff}}}} = \begin{cases} \dfrac{1}{\sqrt{\varepsilon_\text{r}}} + j\dfrac{k_0\sigma_\text{h}^2}{2L_\text{c}} \displaystyle\int_0^\infty \dfrac{1}{x}\dfrac{d\hat{W}(x)}{dx}\,dx, & k_0L_\text{c} \ll 1 \\[3ex] \dfrac{1}{\sqrt{\varepsilon_\text{r}}} + (k_0\sigma_\text{h})^2 \sin\psi, & k_0L_\text{c} \gg 1, \psi \gg \dfrac{1}{\sqrt{k_0L_\text{c}}} \\[3ex] \dfrac{1}{\sqrt{\varepsilon_\text{r}}} - \dfrac{(k_0\sigma_\text{h})^2}{\sqrt{k_0L_\text{c}}}\dfrac{2}{\sqrt{2\pi}}\exp\left(-j\pi/4\right)\displaystyle\int_0^\infty \dfrac{1}{\sqrt{x}}\dfrac{d\hat{W}(x)}{dx}\,dx, & \sqrt{k_0L_\text{c}} \gg 1, \psi \ll \dfrac{1}{\sqrt{k_0L_\text{c}}} \end{cases} \tag{4.48}$$

Comparing these results to the Kirchhoff theory, we find that there is good agreement in the case $k_0L_\text{c} \gg 1$, $\psi \gg 1/\sqrt{k_0L_\text{c}}$. This agrees with our above discussion of the limits of the Kirchhoff theory: by assuming that the coherence length is long compared with the wavelength, there cannot be diffraction by a sudden "spike" in the surface. And by fulfilling $\psi \gg 1/\sqrt{k_0L_\text{c}}$, it is assured that a wave incident under angle ψ cannot cast a shadow onto other points of the surface.

*4.4.3   Directional Characteristics of the Diffuse Radiation

The above equations describe the strength of the specular reflection coefficient, and thus implicitly the amount of energy contained in the diffuse reflection (specular and diffuse energy has to equal the incident energy). For a complete description, we also need to know the directional characteristics of the diffuse reflection. The most popular models for this are the following:

- *Lambertian model:* here it is assumed that the signal is radiated equally into all directions. This implies that the power is proportional to $\cos(\gamma_\text{s})$, where γ_s is the angle between the surface normal and the wave vector of the scattered signal.
- *Single-lobe model:* this assumes that the scattered signal is concentrated in a beam (lobe) that is centered on the specular direction. The shape and width of the beam are further parameters of the model that need to be specified.
- *Double-lobe model:* this model is similar to the single-lobe model but assumes that in addition to the lobe centered on the specularly reflected component, there is also a lobe centered on the incident direction, i.e., some of the radiation is scattered back into the direction from which the wave came.

4.5   Waveguiding

Another important process is propagation in (dielectric) waveguides. This process models propagation in street canyons, corridors, and tunnels. The basic equations of dielectric waveguides are well established [Collin 1991, Marcuse 1991]. However, those waveguides occurring in wireless communications deviate from the idealized assumptions of theoretical work:

---

[15] We assume here that the material is not conductive, $\sigma_\text{e} = 0$. The imaginary contribution arises from surface roughness. Treatment for conductive materials follows the same basic formulation as the one described here.

- The materials are lossy.
- Street canyons (and most corridors) do not have continuous walls but are interrupted at more or less regular intervals by cross-streets. Furthermore, street canyons lack the "upper" wall of the waveguide.
- The surfaces are rough (window sills, etc.).
- The waveguides are not empty but filled with metallic (cars) and dielectric (pedestrian) IOs.

Propagation prediction can be done either by computing the waveguide modes or by a geometric optics approximation. If the waveguide cross-section as well as the IOs in it are much larger than the wavelength, the latter method gives good results [Klemenschits and Bonek 1994].

Conventional waveguide theory predicts a propagation loss that increases exponentially with distance. Some measurements in corridors observed a similar behavior. The majority of measurements, however, fitted a $d^{-\alpha_{\rm PL}}$ law, where $\alpha_{\rm PL}$ varies between 1.5 and 5. Note that a loss exponent smaller than 2 does not contradict energy conservation or any other laws of physics. The d^{-2} law in free space propagation just stems from the fact that energy is spread out over a larger surface as the distance is increased. If energy is perfectly guided, even d^0 becomes (theoretically) possible.

*4.6 Atmospheric Absorption

While propagation through a vacuum does not lead to any attenuation, propagation through air might. The air molecules, as well as water vapor and other particles, can lead to absorption and/or scattering. At frequencies below 40 GHz, the impact of this attenuation is mostly negligible for the distances covered by typical wireless systems.[16] Above 40 GHz, we observe from Figure 4.18 that there are certain frequency bands at which strong attenuation occurs, such as the 60 GHz band, where oxygen attenuation is quite pronounced. At these frequencies, communication is only possible over short distances (this can also be sometimes positive, since interference is strongly attenuated as well). Furthermore, the attenuation may depend strongly on the moisture content in the air. This can constitute a problem in the link budget planning, since rainstorms might lead to outages of links that work well under normal, dry, weather conditions.

In contrast to free-space attenuation, which follows a *power law*, atmospheric attenuation is *exponential*, i.e., the attenuation in dB simply follows as $A_{\rm dB} = \zeta d$, where ζ is the attenuation coefficient shown in Figure 4.18, and d is the distance in km.

4.7 Deterministic Channel Modeling

With the knowledge of the various propagation processes, we can now find a method to characterize the wireless propagation channel for a specific location of TX and RX, e.g., when analyzing a cellular deployment. We can interpret then the wireless channel as *deterministic* – once the location of TX, RX, and IOs is fixed, the characteristics of the channel can be obtained from the well-known laws of electromagnetics (a more detailed discussion about such deterministic versus stochastic channel descriptions can be found in Chapter 7).

Due to issues of computational complexity, it is common to use the *high-frequency* approximation (also known as *ray approximation*[17]) for the computation. In this approximation, electromagnetic waves are modeled as rays that follow the laws of geometrical optics (Snell's laws for reflection and transmission); further refinements allow inclusion of diffraction and diffuse scattering in an approximate way. This is essentially the picture we described at the beginning of this section – the MPCs are approximated as plane waves, and the interaction of those plane waves with the environment follows the equations and discussions of Sections 4.1–4.6.

The goal is then to find the characteristics of each MPC, namely

- amplitude, which can be computed from the sum (on a logarithmic scale) of the attenuations of the various interaction processes the MPC goes through on its way from the TX to the RX, as well as the phase shift created by those interactions
- delay (or equivalently, the pathlength the MPC covers)
- direction of departure from the TX
- direction of arrival at the RX
- polarization.

The characteristics of all of these MPCs constitute a complete description of the channel. It can be converted to the *double-directional impulse response*, from which a number of other interesting channel properties can be derived, see Section 6.7. For the purpose of this section, we restrict ourselves to the question of how to obtain the MPC characteristics enumerated above. In the following, we describe various implementations.

---

[16] Different considerations occur, e.g., in radio astronomy and satellite communications, where propagation through many kilometers of atmosphere occur, so that even small attenuations per unit length are relevant.

[17] In the literature, such methods are often generally known as *ray tracing*. However, the expression "ray tracing" is also used for one specific implementation method (described below). We will thus stick with the name "high-frequency approximation" for the general class of algorithms.

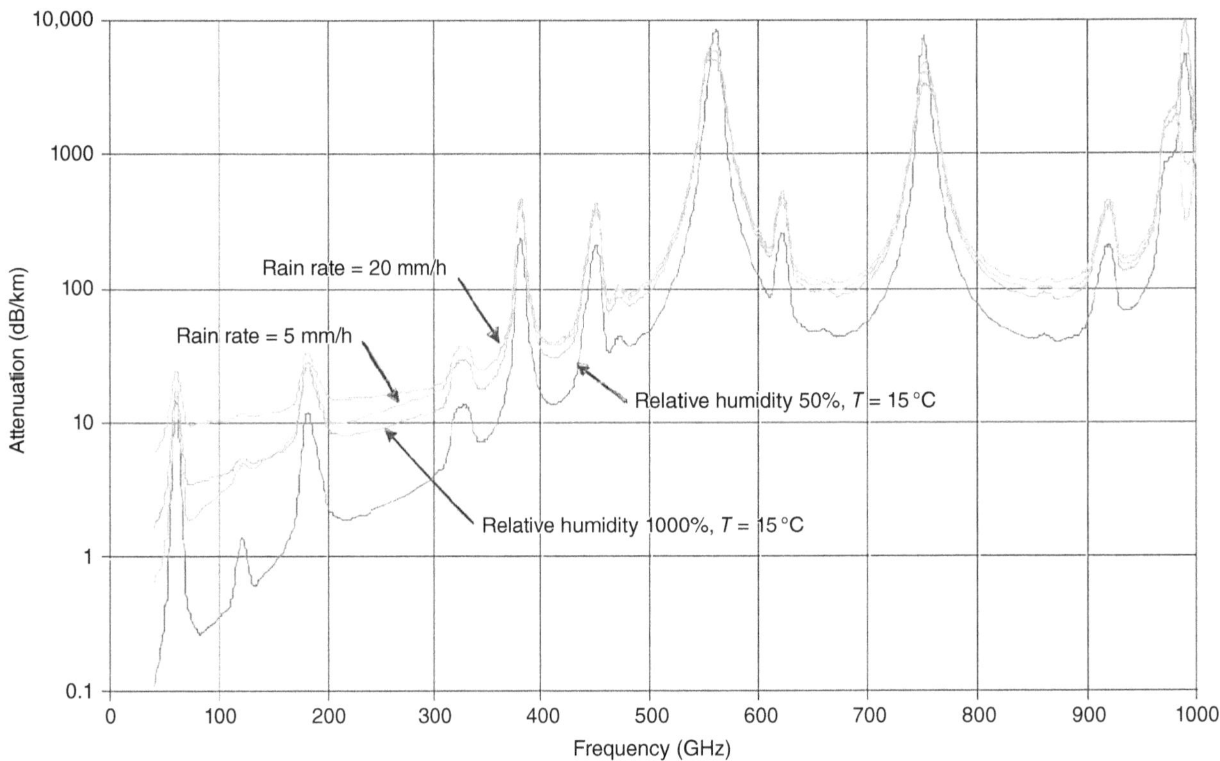

Figure 4.18 Atmospheric attenuation 40–1000 GHz. Color version available at wiley.com/go/molisch/wireless3e. Reproduced with permission from [McMillan 2005] © Springer Nature.

4.7.1 Ray Launching

In the *ray-launching* (aka shoot-and-bounce) approach, the transmit antenna sends out (launches) rays into different directions. Typically, the total spatial angle 4π is divided into N units of equal magnitude, and each ray is sent in the direction of the center of one such unit (i.e., uniform sampling of the spatial angle) (see Figure 4.19). The number of launched rays is a tradeoff between accuracy of the method and computation time.

The algorithm follows the propagation of each ray until it either hits the RX or becomes too weak to be significant (e.g., drops below the noise level). When following a ray, a number of effects have to be taken into account:

- *Free space attenuation*: as each ray represents a certain spatial angle, the energy *per unit area* decreases d^{-2} along the path of the ray. Atmospheric attenuation as described in Section 4.6 can be included as well if relevant (i.e., usually for mm-wave and THz frequencies).
- *Reflections* change the direction of a ray and cause an additional attenuation. Reflection coefficients can be computed from Snell's laws (see Section 4.2) depending on the angle of incidence and possibly the polarization of the incident ray.

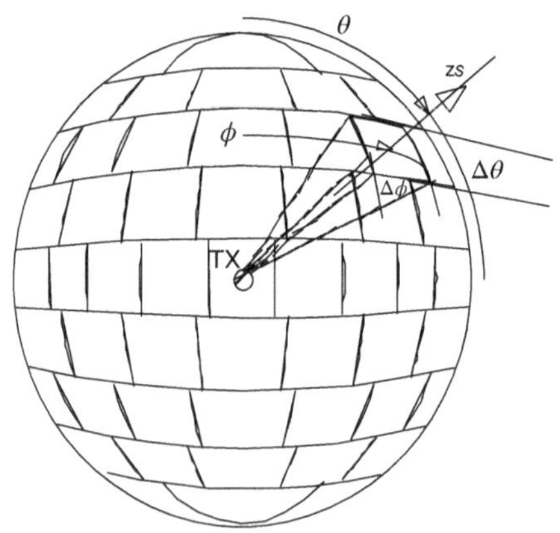

Figure 4.19 Principle of ray launching. Reproduced with permission from [Damosso and Correia 1999] © European Union.

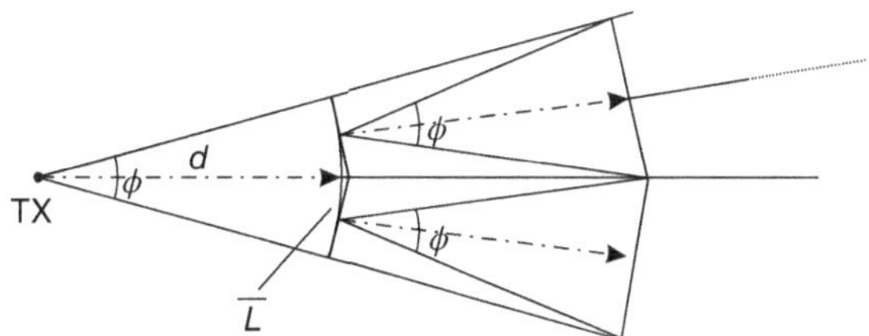

Figure 4.20 Principle of ray splitting, shown here in the two-dimensional plane.

- *Diffraction and diffuse scattering* are included in more advanced models. In those cases, a ray that is incident on an IO gives rise to several new rays. The amplitudes of diffracted rays are usually computed from the geometrical or uniform theory of diffraction, as discussed in Section 4.3, and the diffuse scattering from the equations in Section 4.4.

The *ray-splitting* algorithm is an important improvement in the accuracy of the method. The algorithm is based on the premise that the effective cross-section of the ray should never exceed a certain size (e.g., the size of a typical IO). Thus, if a ray has propagated too far from the TX, it is subdivided into two rays. Let us explain that principle in more detail using Figure 4.20. To simplify the discussion, we consider only the two-dimensional case. Each ray represents not only a certain angle but rather an angular *range* of width ϕ – corresponding to the angle between two launched rays. The intersection of such an angular range with a circle of radius d has a length of approximately ϕd (for the three-dimensional case, think "cross-section" instead of "length"). Thus, the farther we get away from the TX, the larger the length that is covered by the ray. In order to maintain high accuracy of the simulation, this length should not become too large. As soon as it reaches a length $\bar{L}$, the ray is split (thus reducing the length to $\bar{L}/2$). The resulting subrays (which again represent a whole angular range of width ϕ) then propagate until they reach a length $\bar{L}$, etc.

Ray launching gives the channel characteristics in the whole environment – i.e., for many different RX positions and a given TX position. In other words, once we have decided on a BS location, we can compute coverage, delay spread, and other channel characteristics in the whole envisioned cell area. Furthermore, there exist preprocessing schemes that allow the inclusion of multiple TX locations: the environment (the IOs) is subdivided into "tiles" (areas of finite size, typically the same size as the maximum effective area of a ray) and the interaction between all tiles is computed. Then, for each TX position, only the interaction between the TX and the tiles that can act as first IOs has to be computed [Hoppe et al. 2003].

4.7.2 Ray Tracing

Classical *ray tracing* determines all rays that can go from *one* TX location to *one* RX location. The method operates in two steps:

1. First, all rays that can transfer energy from the TX location to the RX location are determined. This is usually done by means of the image principle. Rays that can get to the RX via a reflection show the same behavior as rays from a virtual source that is located where an image of the original source (with respect to the reflecting surface) would be located (see Figure 4.21).
2. In a second step, attenuations (due to free space propagation and finite reflection coefficients) are computed, thus providing the parameters of all MPCs.

Ray tracing allows fast computation of single- and double-reflection processes and also does not require ray splitting. On the downside, effort increases exponentially with the order of reflections that are included in the simulation. Also, the inclusion of diffuse scattering and diffraction is nontrivial. Finally, the method is less efficient than ray launching for the computation of channel characteristics over a wide area.

*4.7.3 Geographical Databases

The foundation of all deterministic methods is the information about the geography and morphology of the environment. The accuracy of that information determines the achievable accuracy of any deterministic channel model.

For *indoor environments*, that information can usually be obtained from building plans, which nowadays are often available in digital form.

In rural areas, geographical databases are available with a resolution of 10–100 m. These databases are often created by means of satellite observations. In many countries, morphological information (land usage) is also available; however, obtaining this information in an automated and consistent way can be quite challenging.

In *urban areas*, digital databases use two different types of data: vector data and pixel data. For vector data, the actual location of building endpoints is stored. For pixel data, a regular grid of points is superimposed on the area, and for each pixel, it is stated whether it falls on "free space" (streets, parks, etc.) or is covered by a building. In both cases, building heights and materials might be included in the database. Of particular interest are public domain databases such as OpenStreetMaps.

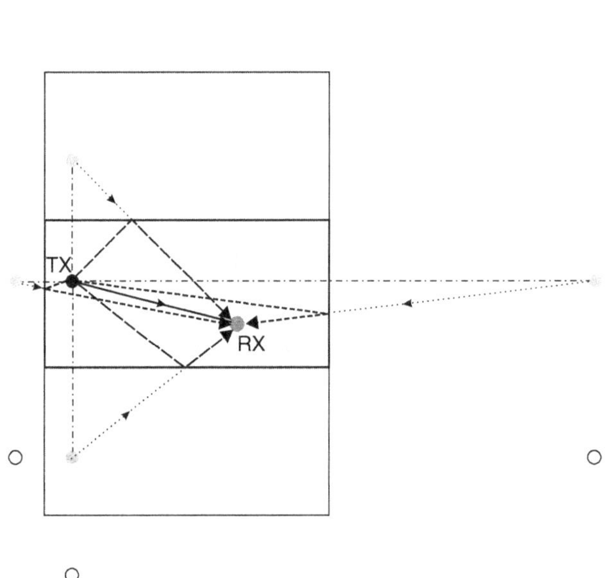

Figure 4.21 The image principle. Grey circles: virtual sources corresponding to a single reflection. Hollow circles: virtual sources corresponding to double reflections. Dotted lines: rays from the virtual sources to the RX. Dashed lines: actual reflections. Solid lines: line of sight.

An interesting recent development is the use of point clouds obtained from LIDAR scanning of an environment. The resolution of such point clouds can be on the order of centimeters, or even better. Such point clouds have been used especially in conjunction with ray tracing. This approach increases the runtime, but at the same time provides much better accuracy, in particular at high frequencies.

*4.7.4   Achievable Accuracy

In order to quantify the accuracy of ray tracers, we first have to distinguish between the "point-wise" and the "area-averaged" accuracy. Pointwise accuracy means the deviation of the channel characteristic, e.g., received power, obtained by ray-tracing from the measured value at that point. These values can often show large deviations: due to small-scale fading the received power changes over a few cm, and ray tracing is not capable of tracking such fast variations. It can provide *average* characteristics in an area (i.e., the power averaged over a 10 m× 10 m area), and even the *statistics* of the power in that area. For the average power, typical commercial ray tracers/launchers can reach 3–6 dB mean error, and 6–10 dB standard deviation of the error, when considering the received power at typical cellular frequencies (0.5–2.5 GHz). This level of fidelity makes ray launching, e.g., useful for cell planning of a network operator based on coverage. However, the accuracy in the prediction of the delay spread (compare Chapter 6) is considerably lower and even worse for angular spread; the reason is that these quantities are more sensitive to errors in the amplitude of weaker components, so that any inaccuracies of the underlying geographical database, or the electromagnetic computations, are amplified. For example, parked cars and traffic signs, which are not contained in most databases, have very little impact on the received signal power, but might drastically change the angular spread.

*4.7.5   Machine Learning for Coverage Prediction

While ray launching is an efficient coverage prediction method, the computational effort for large-scale simulations is still considerable. This is particularly true when many simulations have to be performed for different possible BS positions. For this reason, alternatives based on machine learning (ML) with supervised learning have drawn considerable attention. In this approach, a neural network is trained by providing both photographs (and/or GIS databases) of an environment, and pathloss data obtained from ray launching as the associated labels. An area of interest can be divided into pixels, so that both the geography and the ray-launching-simulated RX power are pictures. ML can now be used to establish mappings from the input parameters (geographical database) to the RX power. Since convolutional neural networks (CNN) are well suited for image processing, they have been used in such investigations, see Figure 4.22 for an example.

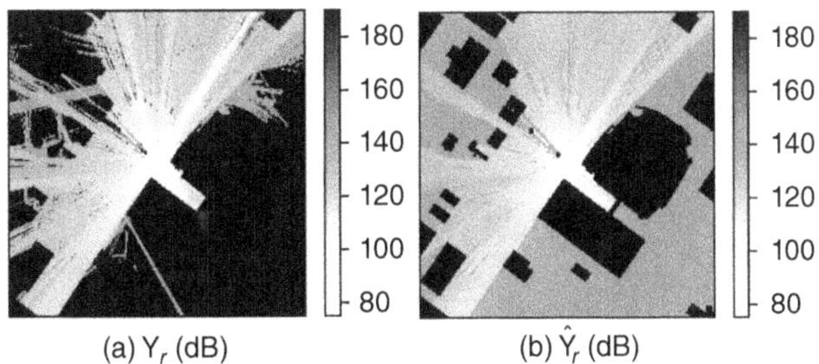

Figure 4.22   Comparison between ray tracing (a) and ML prediction (b) in a microcell in Houston. Color version available at wiley.com/go/molisch/wireless3e. Reproduced with permission from [Ratnam et al. 2020] © IEEE.

*4.8   Appendices

App. 4.A: Derivation of the d^{-4} Law

See App4.pdf at wiley.com/go/molisch/wireless3e

App. 4.B: Diffraction Coeffcients for Diffraction by a Wedge or Cylinder

See App4.pdf at wiley.com/go/molisch/wireless3e

Further Reading

Several books on wireless propagation processes give an overview of all the phenomena discussed in this chapter: [Bertoni 2000, Blaunstein 1999, Parsons 1992, Vaughan and Andersen 2003, Haslett 2008, Barclay 2002]; MATLAB programs for simulating them are described in [Fontan and Espineira 2008]. The basic propagation processes, especially reflection and transmission, are described in any number of classic textbooks on electromagnetics – e.g., [Ramo et al. 1967], which is a personal favorite – but also in many others. More involved results can be found in [Bowman 1987] and [Felsen and Marcuvitz 1973]. The description of transmission through multi-layer dielectric films can be read in [Heavens 1965]. The geometric theory of diffraction was originally invented by [Keller 1962], extended to the uniform theory of diffraction in [Kouyoumjian and Pathak 1974], and summarized in [McNamara et al. 1990]. The different theories for multiple knife edge diffraction are nicely summarized in [Parsons 1992]; however, continuous research is being done on that topic, and further solutions are being published [Bergljung 1994]. Scattering by rough surfaces, mostly inspired by radar problems, is described in [Bass and Fuks 1979], [de Santo and Brown 1986], [Ogilvie 1991]; an excellent summary is given in [Vaughan and Andersen 2003]. The theory of dielectric waveguides is described in [Collin 1991], [Marcuse 1991], and [Ramo et al. 1967].

Ray tracing originally comes from the field of computer graphics, and a number of books [Glassner 1989] are available from that perspective; its application to wireless is described, e.g., in [Valenzuela 1993]. Descriptions of the ray launching algorithm can be found in [Lawton and McGeehan 1994]. Ray splitting was introduced in [Kreuzgruber et al. 1993]. There is also a considerable number of commercial software programs for radio channel prediction that use ray tracing/launching. Point cloud simulations were pioneered by [Jarvelainen and Haneda 2014]. Machine Learning for pathloss prediction is discussed, e.g., in [Ratnam et al. 2020] and [Levie et al. 2021]. Methods that do not rely on the ray approximation include the Method of Moments, which is described in the classical book by [Harrington 1993]. The FEM is described in [Zienkiewicz and Taylor 2000], while the FDTD method is described in [Kunz and Luebbers 1993].

For updates and errata for this chapter, see https://wides.usc.edu/students.html#textbooks

Exercises

See Sec. 36.4 of Exercises.pdf at wiley.com/go/molisch/wireless3e

5

Statistical Description of the Wireless Channel

5.1 Introduction

The deterministic channel description discussed in Chapter 4 is very useful for the deployment planning of networks, since it provides information, e.g., how the placement of a Base Station (BS) impacts the received power at various locations within a cell. However, in many circumstances, it is too complicated, and not necessary, to describe all reflection, diffraction, and scattering processes that determine the different Multi-Path Components (MPCs). Rather, it is preferable to describe the *statistics* of the propagation channel, e.g., the *probability* that a channel parameter attains a certain value, without worrying at *exactly what location* this parameter is attained. This type of modeling is particularly important for the design and performance assessment of transmitters/receivers and the signaling methods that they employ; we will thus use this description extensively in the later parts of the book.

The most important parameter is channel gain, as it determines received power or field strength; it will be at the center of our interest in this chapter.[1]

In order to arrive at a better understanding of the channel, let us first look at a typical plot depicting received power as a function of distance (Figure 5.1). The first thing we notice is that received power can vary strongly – by 100 dB or more. We also see that variations happen on different spatial scales:

- On a very-short-distance scale (on the order of a wavelength), power fluctuates around a (local) mean value, as can be seen from the insert in Figure 5.1. Since these fluctuations happen on a scale that is comparable with one wavelength, they are called *small-scale fading*. The reason for these fluctuations is interference between different MPCs, as mentioned in Chapter 2. Fluctuations in field strength can be well described statistically – namely, by the (local) mean value of the power and the statistics of the fluctuations around this mean.

- Mean power, averaged over about 10 wavelengths, itself shows fluctuations. These fluctuations occur on a larger scale – typically a few hundred wavelengths. These variations can be seen most clearly when moving on a circle around the Transmitter (TX) (Figure 5.2). The reason for these variations is shadowing by large objects (Chapter 2) and is thus fundamentally different from the interference that causes small-scale fading. However, this *large-scale fading* can also be described by a mean and the statistics of fluctuations around this mean.

- The large-scale mean itself depends monotonically on the distance between TX and Receiver (RX). This effect is related to free space path loss or some variation thereof. This effect is usually described in a deterministic manner. A simple model was discussed in Section 4.2.3; more extensive discussions and more detailed models are relegated to Section 7.1.

This chapter only concentrates on the statistical variations of the channel gain; delay dispersion and other effects will be treated later. It is thus sufficient for this chapter to consider the impact of the channel onto an unmodulated (sinusoidal) carrier signal, though the considerations are also valid for *narrowband* systems (as defined in Section 6.1). Sections 5.2 and 5.3 explain the *two-path model* – it is the simplest model explaining small-scale fading effects. Sections 5.4 and 5.5 generalize these considerations to more realistic channel models and give statistics of the received field strength for various scenarios. The next two sections describe the temporal statistics of fading, including how fast the state of the channel changes, and the frequency and average duration of dips (instances of very low received power). Finally, the statistical variations due to shadowing effects are described in Section 5.8.

---

[1] Note that we talk of channel gain (and not attenuation) because it is directly proportional to receive power (receive power is transmit power times channel gain); of course, this gain is smaller than unity. In the following, we often assume unit transmit power and use "receive power" and "channel gain" interchangeably.

Wireless Communications: From Fundamentals to Beyond 5G, Third Edition. Andreas F. Molisch.
© 2023 John Wiley & Sons Ltd. Published 2023 by John Wiley & Sons Ltd.
Companion website: www.wiley.com/go/molisch/wireless3e

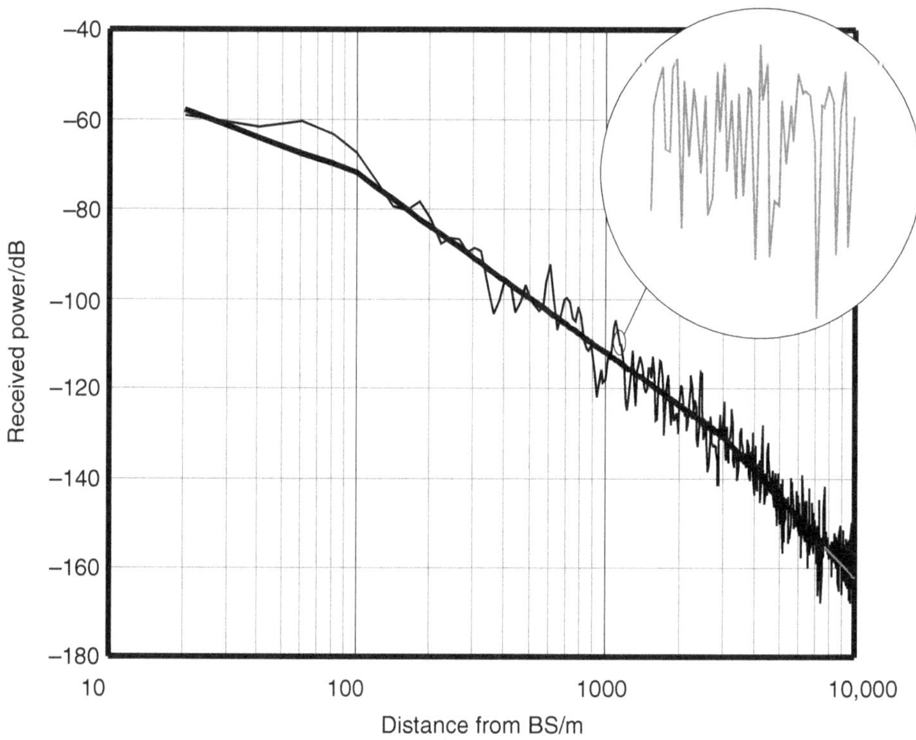

Figure 5.1 Received power as a function of distance from the TX.

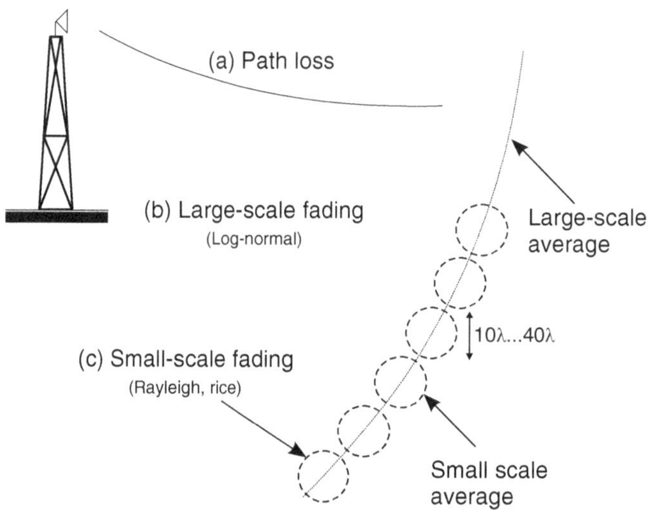

Figure 5.2 Types of received power variations.

5.2 The Time-Invariant Two-Path Model

As an introduction to the rather involved subject of multipath propagation and fading, we consider the simplest possible case – two MPCs whose properties do not change with time. This is called, in brief, the time-invariant two-path model. We transmit a sinusoidal waveform and determine the (complex) transfer function at the location of the RX.

First, consider a single MPC. Let the transmit signal be a sinusoidal wave:

$$E_{TX}(t) \propto \cos(2\pi f_c t). \tag{5.1}$$

Let the received signal be approximated as a homogeneous plane wave. If the runlength between TX and RX is d, the received signal can be described as:

$$E(t) = E_0 \cdot \cos(2\pi f_c t - k_0 d) \tag{5.2}$$

where k_0 is the wavenumber $2\pi/\lambda$, and E_0 is the amplitude at the RX. Using complex baseband notation,[2] this reads

$$E = E_0 \exp(-jk_0 d). \tag{5.3}$$

Note that the real part of the field in complex representation, $\mathrm{Re}\{E\}$ is equal to the instantaneous value of the field strength at time $t = 0$.

Now consider the case that the transmit signal gets to the RX via two different propagation paths, created by two different *Interacting Objects* (IOs), see Figure 5.3. These paths have different runtimes:

$$\tau_1 = d_1/c_0, \text{ and } \tau_2 = d_2/c_0. \tag{5.4}$$

The RX is assumed to be in the far field of the IOs, so that the two waves arriving at the RX position $\mathbf{r}$ are two homogeneous plane waves whose absolute amplitudes do not vary as a function of RX position (we vary RX positions only within an area that has less than about 10λ diameter). We assume furthermore that both waves are vertically polarized, and have amplitudes $E1$ and $E2$ at the reference position (origin of the coordinate system) $\mathbf{r} = 0$. We get the following expression for the superposition of two plane waves:

$$E(\mathbf{r}) = E1 \exp(-j\mathbf{k}_1\mathbf{r}) + E2 \exp(-j\mathbf{k}_2\mathbf{r}) \tag{5.5}$$

where $\mathbf{k}_1$ is the wave vector (i.e., has the absolute magnitude k_0, and is pointing into the direction of propagation) of wave 1. Note that due to our assumptions, the location vector $\mathbf{r}$ impacts the phase, but not the amplitude, of each of the two MPCs.

The upper part of Figure 5.4 depicts the real part of $E(\mathbf{r})$, and the lower part shows the magnitude, which is proportional to the square root of received power. We see locations of both constructive and destructive interference – i.e., location-dependent fading. There are fading dips where the phase differences due to the different runtimes are between 90° and 270°; the dips have their smallest amplitude at the locations of destructive interference, i.e., where the phase difference is 180°. At these points, the RX sees a signal whose total amplitude is the difference of the amplitudes of the constituting waves. If the amplitudes of the constituting waves are equal, destructive interference can be complete. In the points of constructive interference, the total amplitude is the sum of the amplitudes of the constituting waves. This phenomenon can be seen in the lower right part of Figure 5.4. Since fading dips are on the order of a wavelength apart (corresponding to 30 cm at a carrier frequency of 1 GHz), this fading is called *small-scale fading*.[3]

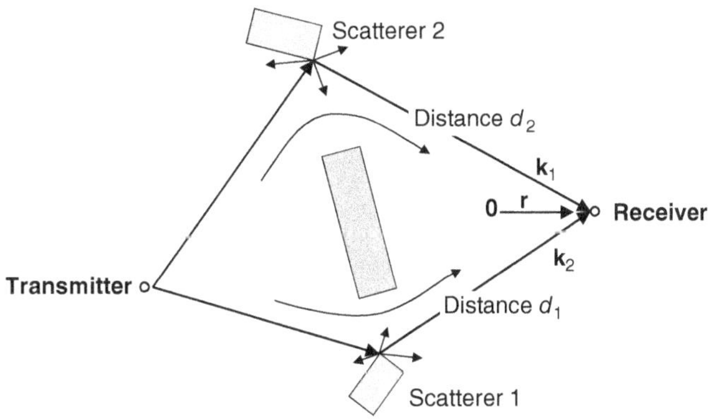

Figure 5.3 Geometry of the time-invariant two-path model.

---

[2] The bandpass signal – i.e., the physically existing signal – is related to the complex baseband (lowpass) representation as $s_{BP}(t) = \mathrm{Re}\{s_{LP}(t)\exp[j2\pi f_c t]\}$ (see also Section 10.1).

[3] It is also known as *short-term fading*, or *fast fading*. Unfortunately, the expression *fast* fading is often used for two completely different phenomena. On one hand, it is used as a synonym for small-scale fading, independent of the actual *temporal* scale of the fading (which depends on the movement speed of the TX, RX, and IOs). On the other hand, it is used to denote channel variations within the duration of a symbol length (in contrast to the "quasi-static" channel), which depends both on the speed of the TX/RX/IO, and the duration of the symbol. It is thus preferable to call the fading due to interference effects *small-scale fading*, as this is unambiguous; this is the notation we adhere to in this book.

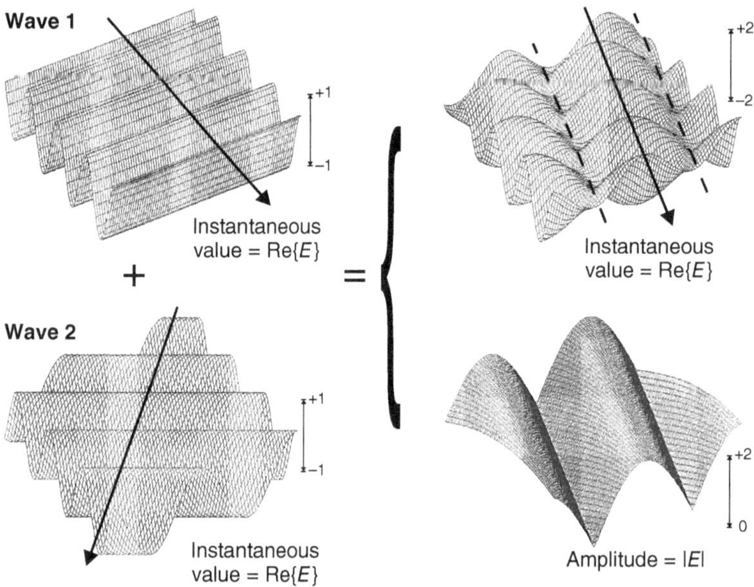

Figure 5.4 Interference of two planewaves with $E1 = E2 = 1$ and arg $(\mathbf{k}_1, \mathbf{k}_2) = 30°$.

5.3 The Time-Variant Two-Path Model

In general, the runtime (path length) difference between the different propagation paths changes with time. This change can be due to movements of the TX, the RX, the IOs, or any combination thereof; to simplify discussion, we henceforth assume only movement of the RX – this is the situation, e.g., in a cellular downlink, i.e., transmission from the BS to the mobile User Equipment (UE). The RX then "sees" a time-varying interference pattern; we can imagine that the RX moves through the "mountains and valleys" of the field strength plot. Spatially varying fading thus becomes time-varying fading.

The movement of the RX also leads to a shift of the received frequency, called the Doppler shift. In order to explain this phenomenon, let us first revert to the case of a *single* sinusoidal wave reaching the RX, and also revert to real passband notation. If the RX moves away from the TX with speed v, then the distance d between TX and RX increases with that speed. Thus:

$$
\begin{aligned}
E(t) &= E_0 \cdot \cos\left(2\pi f_{\mathrm{c}} t - k_0[d_0 + \mathrm{v}t]\right) \\
&= E_0 \cdot \cos\left(2\pi t \left[f_{\mathrm{c}} - \frac{\mathrm{v}}{\lambda} \right] - k_0 d_0 \right)
\end{aligned}
\tag{5.6}
$$

where d_0 is the distance at time $t = 0$. The frequency of the received oscillation is thus decreased by v/λ – in other words, the Doppler shift is given by:

$$
\nu = -\frac{\mathrm{v}}{\lambda} = -f_{\mathrm{c}} \cdot \frac{\mathrm{v}}{c_0}.
\tag{5.7}
$$

Note that the Doppler shift is negative when the TX and RX move away from each other – familiar from the fact that the siren of an emergency vehicle driving away from an observer sounds lower, while an approaching one sounds higher. Since the speed of the movement is always small compared with the speed of light, the Doppler shifts are relatively small.

In the above example, we had assumed that the direction of RX movement is aligned with the direction of wave propagation. If that is not the case, the Doppler shift is determined by the speed of movement *in the direction of wave propagation*, $\mathrm{v}\cos(\gamma)$ (see Figure 5.5). The Doppler shift is then:

$$
\nu = -\frac{\mathrm{v}}{\lambda} \cos(\gamma) = -f_{\mathrm{c}} \cdot \frac{\mathrm{v}}{c_0} \cos(\gamma) = -\nu_{\max} \cos(\gamma).
\tag{5.8}
$$

The maximum Doppler shift $\nu_{\max}$ typically lies between 1 Hz and 1 kHz. Note that in general, the relationship $\nu_{\max} = f_{\mathrm{c}} \cdot \mathrm{v}/c_0$ is based on several assumptions – e.g., static IOs, no reflections on moving objects, etc.

Since the Doppler shifts are so small, it seems natural to ask whether they have a significant influence on the radio link. To answer this question, we first of all have to distinguish between a Doppler *shift* and the Doppler *spread*. A pure frequency shift of the signal (if it were the same for all MPCs) would indeed have a negligible impact on system performance. The RX anyway estimates the carrier frequency from the received signal (this is called *carrier recovery*), and thus would simply estimate the Doppler-shifted carrier. However, this situation occurs only if there is only a single MPC.

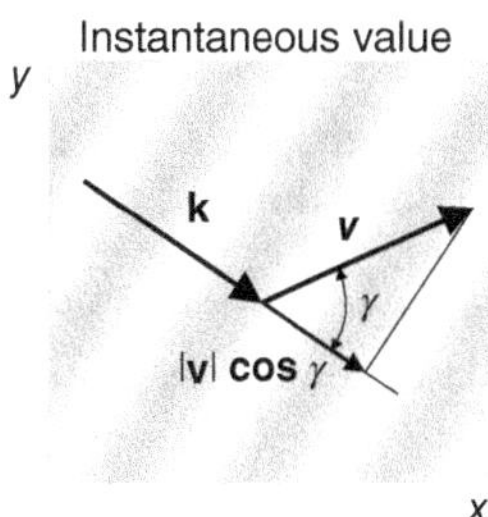

Figure 5.5 Projection of velocity vector |v| onto the direction of propagation **k**.

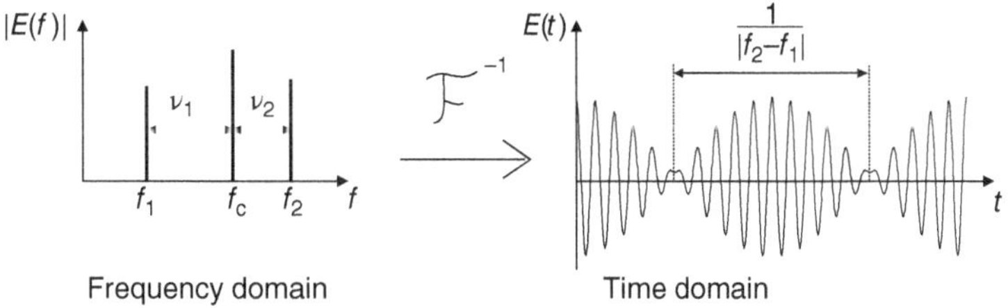

Figure 5.6 Superposition of two carriers with different frequencies (beating).

Different MPCs, on the other hand, which arrive from different directions, have *different* Doppler shifts, according to Eq. (5.8). This leads to a Doppler *spreading*, i.e., frequency dispersion. This effect has a much more significant impact on system performance.

The superposition of several Doppler-shifted waves creates the sequence of fading dips. Again, this can be demonstrated using the two-path model. As the RX moves, it receives two waves that are each Doppler shifted, but by different amounts. By Fourier transformation to the time domain, we get the well-known effect of *beating* of two oscillations with slightly different frequencies (see Figure 5.6). The frequency of this beating envelope is equal to the frequency difference between the two carriers – i.e., the difference of the two Doppler shifts. The RX thus sees a signal whose amplitude undergoes periodic variations – this is exactly the small-scale fading that is created by traversing the "mountains and valleys" of the field strength plot of Figure 5.4.

Summarizing, we can obtain the fading rate in the two-path model by two equivalent considerations:

1. We superimpose two incident waves, plot the resulting interference pattern (field strength "mountains and valleys"), and count the number of fading dips per second that an RX sees when moving through that pattern.
2. Alternatively, we can think of superimposing two signals with different Doppler shifts at the receive antenna, and determine the fading rate from the beat frequency – i.e., the difference of the Doppler shifts of the two waves.

Doppler spread is thus an important parameter of the channel, even though it is so small:

- Doppler spread is a measure for the rate of change of the channel, as we have discussed above.
- Furthermore, the superposition of many slightly Doppler-shifted signals leads to phase shifts of the total received signal that can impair the reception of angle-modulated signals (Chapters 10 and 11). These phase shifts lead to a random Frequency Modulation (FM) of the received signal (see Section 5.7), and are especially important for signals with low bit rates.
- In Orthogonal Frequency Division Multiplexing (OFDM) systems, where a signal consists of many parallel subcarriers that are closely spaced in frequency, Doppler spread can lead to inter-carrier interference, see Section 15.7. This is particularly relevant at high carrier frequencies, such as mm-wave systems.

5.4 Small-Scale Fading Without a Dominant Component

Following these basic considerations utilizing the two-path model, we now investigate a more general case of multipath propagation. We consider a radio channel with many IOs and a moving RX. Due to the large number of IOs, a deterministic description of the radio channel is not efficient anymore, which is why we take refuge in stochastic description methods. This stochastic description is essential for the whole field of wireless communications and is thus explained in considerable detail. We start out with a computer experiment in Section 5.4.1, followed by a more general mathematical derivation in Section 5.4.2.

5.4.1 A Computer Experiment

Consider the following simple computer experiment. The signals from several IOs are incident onto an RX that moves over a small area. The IOs are distributed approximately uniformly around the receiving area. They are also assumed to be sufficiently far away so that all received waves are homogeneous plane waves, and that movements of the RX within the considered area do not change the amplitudes of these waves. The different distances and strength of the interactions are taken into account by assigning a random phase and a random amplitude to each wave. We are then creating eight constituting waves E_ℓ with absolute amplitudes $|a_\ell|$, angle of incidence (with respect to the x-axis) ϕ_ℓ and phase $\varphi_{0,\ell}$:

	$|a_\ell|$	ϕ_ℓ	$\varphi_{0,\ell}$
$E_1(x, y) = 1.0 \exp[-jk_0(x \cos(169°) + y \sin(169°))] \exp(j311°)$	1.0	169°	311°
$E_2(x, y) = 0.8 \exp[-jk_0(x \cos(213°) + y \sin(213°))] \exp(j32°)$	0.8	213°	32°
$E_3(x, y) = 1.1 \exp[-jk_0(x \cos(87°) + y \sin(87°))] \exp(j161°)$	1.1	87°	161°
$E_4(x, y) = 1.3 \exp[-jk_0(x \cos(256°) + y \sin(256°))] \exp(j356°)$	1.3	256°	356°
$E_5(x, y) = 0.9 \exp[-jk_0(x \cos(17°) + y \sin(17°))] \exp(j191°)$	0.9	17°	191°
$E_6(x, y) = 0.5 \exp[-jk_0(x \cos(126°) + y \sin(126°))] \exp(j56°)$	0.5	126°	56°
$E_7(x, y) = 0.7 \exp[-jk_0(x \cos(343°) + y \sin(343°))] \exp(j268°)$	0.7	343°	268°
$E_8(x, y) = 0.9 \exp[-jk_0(x \cos(297°) + y \sin(297°))] \exp(j131°)$	0.9	297°	131°

We now superimpose the constituting waves, using the complex baseband notation again. The total complex field strength E thus results from the sum of the complex field strengths of the constituting waves. We can also interpret this as adding up complex random phasors. Figure 5.7 shows the instantaneous value of the total field strength at time $t = 0$ – i.e., Re$\{E\}$, in an area of size $5\lambda \cdot 5\lambda$.

Let us now consider the statistics of the field strengths occurring in that area. As can be seen from the histogram (Figure 5.8), the values Re$\{E\}$ follow, to a good approximation, a zero-mean Gaussian distribution. This is a consequence of the central limit theorem: when superimposing N statistically independent random variables, none of which is dominant, the associated *probability density function* (pdf) approaches a normal distribution for $N \to \infty$ (see Appendix 5.A for a more exact formulation of this statement). The conditions for the validity of the central limit theorem are fulfilled approximately: the eight constituting waves have random angles of incidence and phases, and none of the amplitudes a_ℓ is dominant. Figures 5.9 and 5.10 show that the imaginary part of the field strength, Im$\{E\}$, is also normally distributed.[4]

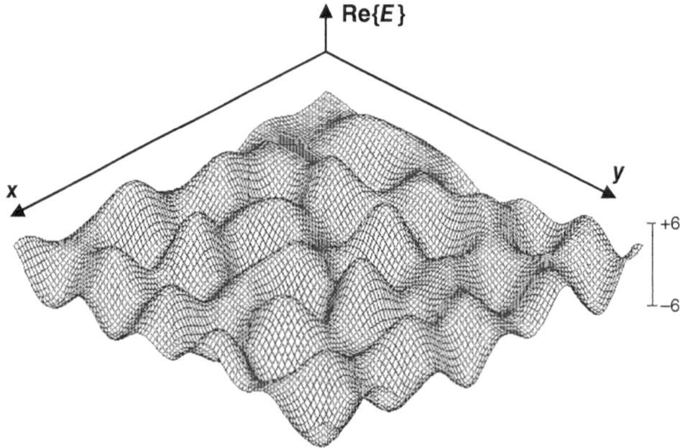

Figure 5.7 Instantaneous value of the field strength at time $t = 0$ – i.e., Re$\{E\}$. Superposition of the eight constituting waves in the area $0 < x < 5\lambda$, $0 < y < 5\lambda$.

---

[4] The imaginary part represents the instantaneous value of the field strength at a time that corresponds to a quarter period of the radio frequency oscillation, $\omega t = \pi/2$.

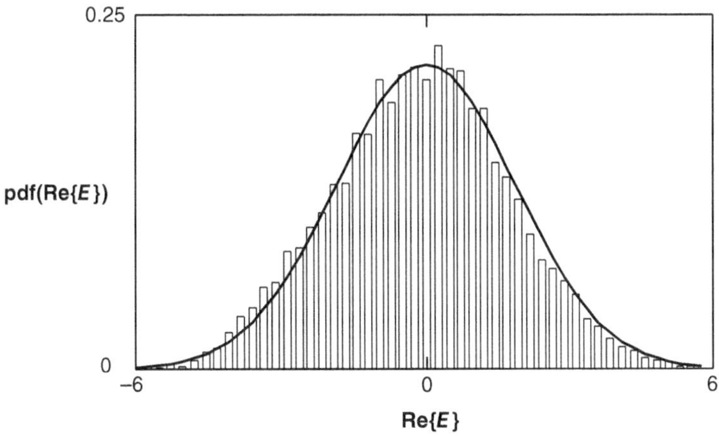

Figure 5.8   Histogram of the field strength of Figure 5.7. A Gaussian pdf is shown for comparison.

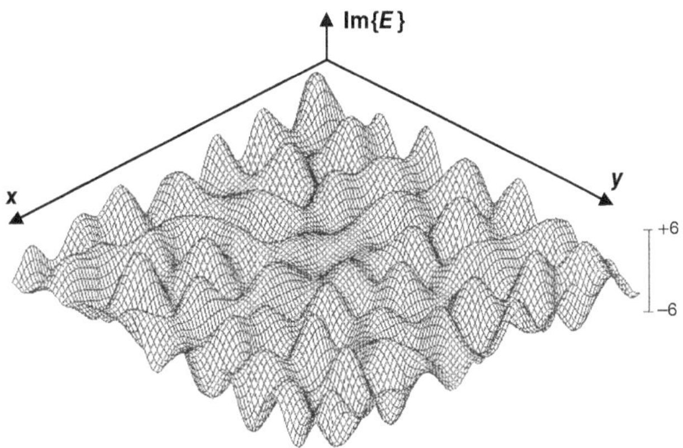

Figure 5.9   Imaginary part of E.

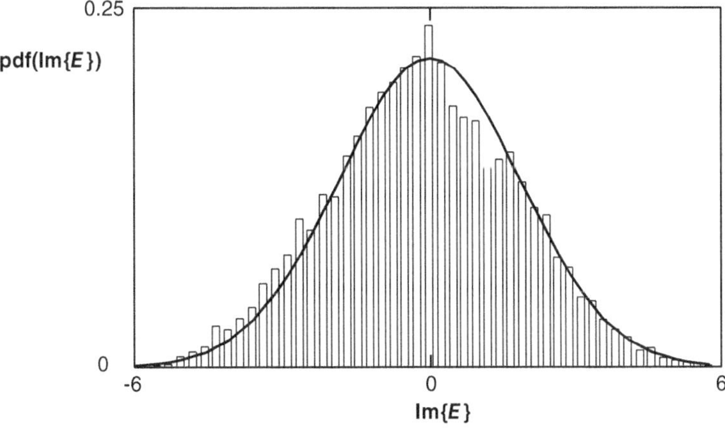

Figure 5.10   Histogram of the field strength of Figure 5.9. A Gaussian pdf is shown for comparison.

The behavior of most RXs is determined by the received (absolute) *amplitude (magnitude)*. We thus need to investigate the distribution of the envelope of the received signal, corresponding to the magnitude of the (complex) field strength phasor. Figure 5.11 shows the field strength seen by an RX that moves along the y-axis of Figures 5.7 or 5.8. The left diagram of Figure 5.12 shows the complex field strength phasor, and the right side the absolute amplitude of the received signal.

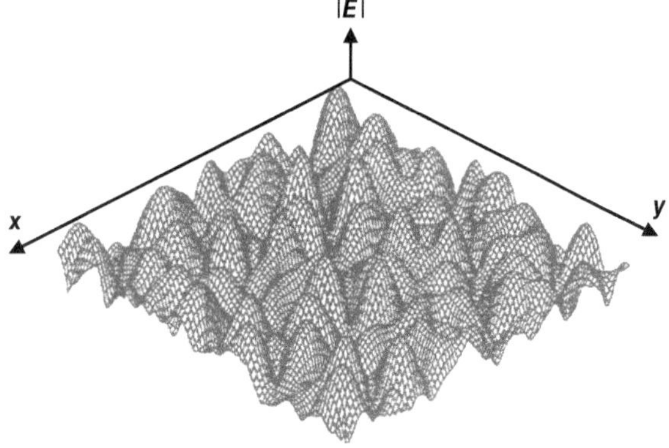

Figure 5.11   Amplitude of the field strength.

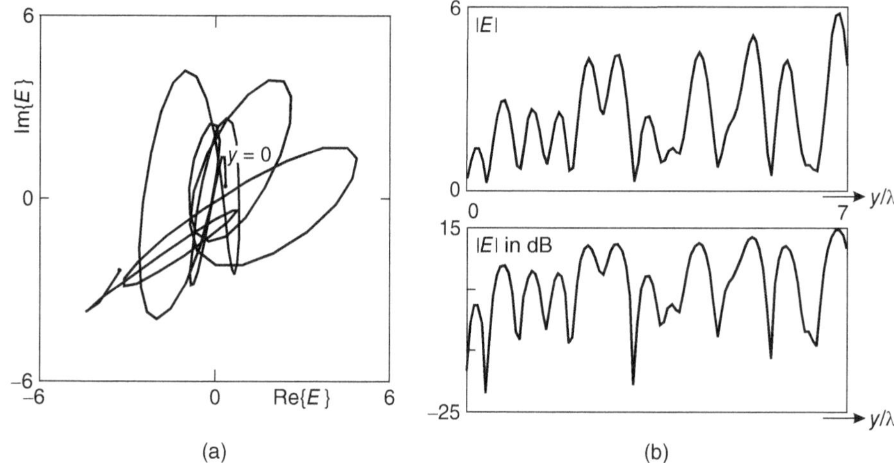

(a)   (b)

Figure 5.12   Complex phasor of the field strength (a) and received amplitude |E| (b) for a RX moving along the y-axis.

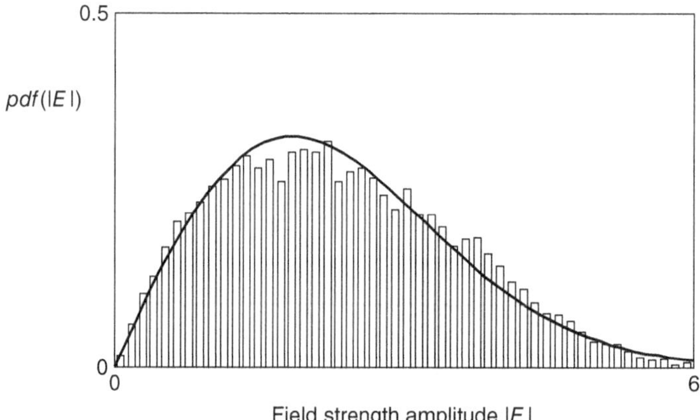

Figure 5.13   Pdf of the received amplitude.

Figures 5.11 and 5.13 show a three-dimensional representation of the amplitudes and the statistics of the amplitude over that area, respectively. Figure 5.13 also exhibits a plot of a Rayleigh pdf. A Rayleigh distribution describes the magnitude of a complex stochastic variable whose real and imaginary parts are independent and normally distributed; more details are given below. Figure 5.14 shows that the *phase* is approximately uniformly distributed.

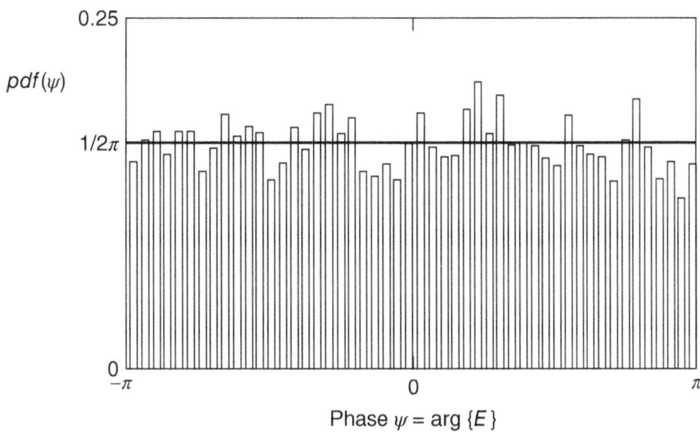

Figure 5.14 Pdf of the received phase.

5.4.2 Mathematical Derivation of the Statistics of Amplitude and Phase

After these pseudo-experimental considerations, we now turn to a more detailed, and more mathematically sound, derivation of Rayleigh distribution. Consider a scenario where N homogeneous plane waves (MPCs) have been created by reflection/scattering from different IOs. The IOs and the TX do not move, and the RX moves with a velocity v. The real part of the received field strength due to the ℓth MPC is thus $|a_\ell| \cos(\varphi_\ell)$, the imaginary part is $|a_\ell| \sin(\varphi_\ell)$.

As above, we assume that the absolute amplitudes of the MPCs do not change over the region of observation. The sum of the squared amplitudes is thus:

$$\sum_{\ell=1}^{N} |a_\ell|^2 = C_{\mathrm{P}} \tag{5.9}$$

where C_{P} is a constant. However, the phases φ_ℓ vary strongly and are thus approximated as random variables that are uniformly distributed in the range $[0, 2\pi]$.

More precisely, the phases at different times (or equivalently, locations) are interpreted as different realizations of uniformly distributed random variables. This can be justified by considering the Doppler shift for computation of the total field strength $E(t)$. If we look at an unmodulated carrier, we get (in real passband notation):

$$E(t) = \sum_{\ell=1}^{N} |a_i| \cos \left[2\pi f_{\mathrm{c}} t - 2\pi \nu_{\max} \cos(\gamma_\ell) t + \varphi_{0,\ell} \right] \tag{5.10}$$

where $\varphi_{0,\ell}$ is the phase of the ℓth MPC at time 0. Rewriting this in terms of in-phase and quadrature-phase components in real passband notation, we obtain

$$E_{\mathrm{BP}}(t) = I(t) \cdot \cos(2\pi f_{\mathrm{c}} t) - Q(t) \cdot \sin(2\pi f_{\mathrm{c}} t) \tag{5.11}$$

with

$$I(t) = \sum_{\ell=1}^{N} |a_\ell| \cos \left[- 2\pi \nu_{\max} \cos(\gamma_\ell) t + \varphi_{0,\ell} \right] \tag{5.12}$$

$$Q(t) = \sum_{\ell=1}^{N} |a_\ell| \sin \left[- 2\pi \nu_{\max} \cos(\gamma_\ell) t + \varphi_{0,\ell} \right]. \tag{5.13}$$

The arguments of the cos() and sin() can thus be interpreted as random variables that take on different values at different times (for sufficiently large time intervals, these realizations can be considered independent). The observations at different times then constitute the "ensemble" over which we take the statistics. Note that for the determination of the amplitude statistics in this section, we are not interested in the temporal/spatial correlation of the I and Q samples as the RX moves along a trajectory; this aspect will be discussed in Section 5.6.

Based on the above assumptions, both the in-phase and the quadrature-phase component are the sum of many random variables, none of which dominate (i.e., $|a_\ell| \ll C_p$). It follows from the central limit theorem that the pdf of such a sum is a normal (Gaussian) distribution, regardless of the exact pdf of the constituent amplitudes – i.e., we do not need knowledge of the a_ℓ or their distributions!!! (see Appendix 5.A for more details). A zero-mean Gaussian random variable has the pdf:

$$pdf_x(x) = \frac{1}{\sqrt{2\pi}\sigma} \exp\left(-\frac{x^2}{2\sigma^2}\right) \tag{5.14}$$

where σ^2 denotes the variance.

Starting with the statistics of the real and imaginary parts, Appendix 5.B derives the statistics of amplitude and phase of the received signal. The pdf is a product of a pdf for ψ – namely, a uniform distribution:

$$pdf_\psi(\psi) = \frac{1}{2\pi} \qquad -\pi < \psi \leq \pi \tag{5.15}$$

and a pdf for r – namely, a Rayleigh distribution:

$$pdf_r(r) = \frac{r}{\sigma^2} \cdot \exp\left[-\frac{r^2}{2\sigma^2}\right] \quad \text{for } r \geq 0. \tag{5.16}$$

For $r < 0$ the pdf is zero, as absolute amplitudes are by definition positive.

5.4.3 Properties of the Rayleigh Distribution

A Rayleigh distribution has the following properties, which are also shown in Figure 5.15:

$$\left.\begin{array}{rl} \text{Mean value} & \bar{r} = \sigma\sqrt{\dfrac{\pi}{2}} \\[2mm] \text{Mean square value} & \overline{r^2} = 2\sigma^2 \\[2mm] \text{Variance} & \overline{r^2} - (\bar{r})^2 = 2\sigma^2 - \sigma^2\dfrac{\pi}{2} = 0.429\sigma^2 \\[2mm] \text{Median value} & r_{50} = \sigma\sqrt{2\cdot ln\,2} = 1.18\sigma \\[2mm] \text{Location of maximum} & \max\{pdf(r)\} \text{ occurs at } r = \sigma \end{array}\right\} \tag{5.17}$$

where the bar denotes expected value (we deviate here from our usual notation $E\{\}$ in order to avoid confusion with the field strength E).

The *cumulative distribution function*, cdf (x), is defined as the probability that the realization of the random variable has a value smaller than or equal to x. The cdf is thus the integral of the pdf:

$$\text{cdf}(r) = \int_{-\infty}^{r} pdf(u)du. \tag{5.18}$$

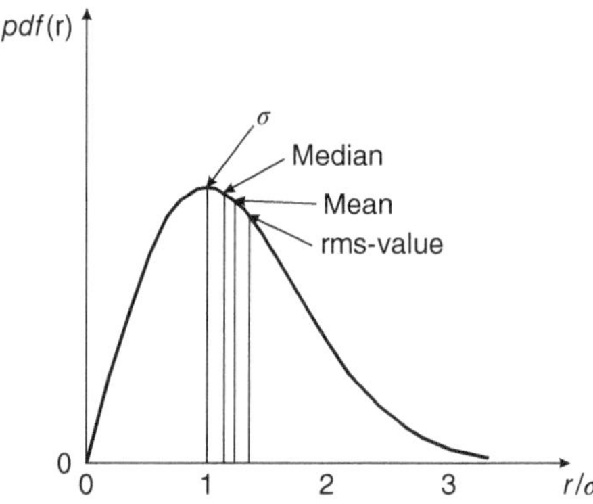

Figure 5.15 Pdf of a Rayleigh distribution.

Applying this equation to the Rayleigh pdf, we get

$$\text{cdf}(r) = 1 - \exp\left(-\frac{r^2}{2\sigma^2}\right). \tag{5.19}$$

For small values of r this can be approximated as:

$$\text{cdf}(r) \approx \frac{r^2}{2\sigma^2}. \tag{5.20}$$

By transformation of variables, we can easily see that the squared amplitude, and by extension the power, has an exponential distribution

$$pdf_{\text{P}}(P) = \frac{1}{\overline{P}} \exp\left(-P/\overline{P}\right)$$

where $\overline{P}$ is the mean power.[5] It is straightforward to check whether a measured ensemble of field strength values follows a Rayleigh distribution: the empirical cdf of the square (i.e., power) is plotted on the so-called Weibull paper (see Figure 5.16). The cdf of the exponential distribution is a straight line with slope one on it. For small values of P, an increase of P by 10 dB has to increase the value of the cdf by a factor of 10.

The Rayleigh distribution is widely used in wireless communications. This is due to several reasons:

- It is an *excellent approximation* in a large number of practical scenarios, as confirmed by a multitude of measurements. However, it is noteworthy that there *are* scenarios where it is not valid. These can occur, e.g., in *Line Of Sight* (LOS) scenarios, some indoor scenarios, and in (ultra) wideband scenarios.
- It describes a *worst-case scenario* in the sense that there is no dominant signal component, and thus there is a large number of fading dips. Such a worst-case assumption is useful for the design of robust systems.[6]
- It depends only on a *single parameter*, the mean received power – once this parameter is known, the complete signal statistics are known. It is easier, and less error-prone, to obtain this single parameter either from measurements or deterministic prediction methods than to obtain the multiple parameters of more involved channel models.
- *Mathematical convenience*: computations of error probabilities and other parameters can often be done in closed form when the field strength distribution is Rayleigh.

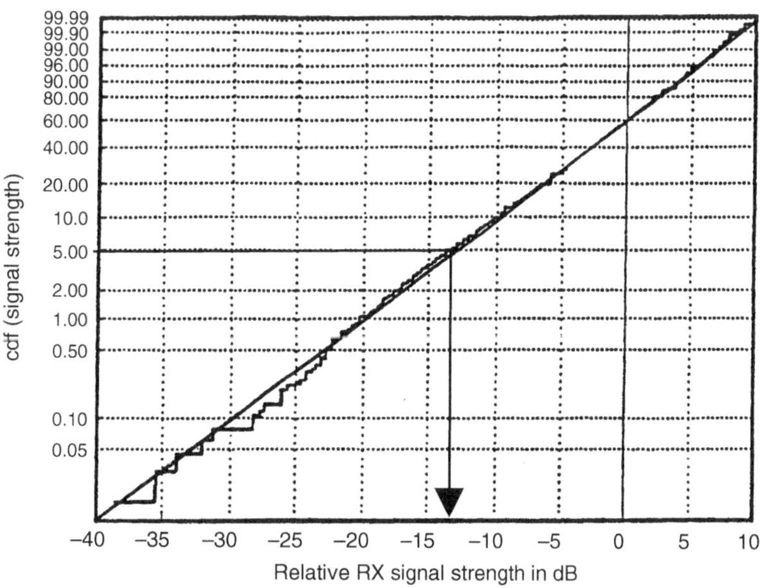

Figure 5.16 Measured cdf of the (normalized) receive power of an indoor nonline of sight scenario.
Reproduced with permission from [Gahleitner 1993] © R. Gahleitner.

---

[5] To simplify discussion, we do not distinguish in this chapter between the squared absolute amplitude and the power, though they are actually related by a proportionality constant.
[6] As we will see later on, there are some fading distributions that show a larger number of fading dips – e.g., Nakagami distributions with $m < 1$; furthermore, the large number of MPCs can also be an advantage for specific systems, see Section 16.2.

5.4.4 Fading Margin for Rayleigh-Fading

Knowledge of the fading statistics is extremely important for the design of wireless systems. We saw in Chapter 3 that for noise-limited systems the received power[7] determines the performance. As field strength and thus power is a random variable, even a large *mean* power does not guarantee successful communications at *all* times. Rather, the power exceeds a minimum value only in a certain *percentage* of situations. The task is therefore to answer the following question: "Given a minimum receive power required for successful communications, how large does the mean power have to be in order to ensure that communication fails in no more than $x\%$ of all situations?" In other words, how large does the *fading margin* have to be?

The cdf gives by definition the probability that a certain field strength level is not exceeded. In order to achieve an $x\%$ outage probability, it follows that:

$$x = cdf(P_{\min}) \approx \frac{P_{\min}}{\overline{P}} \tag{5.21}$$

where the r.h.s follows from Eq. (5.20). From this, we can immediately compute the required average power $\overline{P}$ as $\overline{P} = P_{\min}/x$.

Example 5.1 *For a signal with Rayleigh-distributed amplitude, what is the probability that the received signal power is at least 20, 6, 3 dB below the mean power. Compare the exact result and the result from the approximate formulation of Eq. (5.20).*

From a Rayleigh-distributed signal envelope P, the average power is by definition:

$$\overline{P} = 2\sigma^2. \tag{5.22}$$

A power level 20 dB below the mean power corresponds to $\frac{P_{\min}}{\overline{P}} = \frac{1}{100}$:

$$\Pr\{P < P_{\min}\} = 1 - \exp\left(-\frac{1}{100}\right) = 9.95 \times 10^{-3}. \tag{5.23}$$

Similarly, the exact results for 6 and 3 dB are 0.221 and 0.393, respectively.

The approximate formulation of Eq. (5.20) gives $\frac{P_{\min}}{\overline{P}} = 0.01, 0.25$, and 0.5, respectively. It is thus reasonably accurate for power levels 6 dB below the mean power but breaks down for higher values of $P_{\min}$.

For the interference-limited case, the situation is somewhat more complicated: not only does the desired signal fade but so do the interferers. For the computation of the statistics of the amplitude ratio of signal and interference, we note that both the desired signal and the interference are Rayleigh-fading; we thus need the pdf of the ratio of two random variables, each of which is Rayleigh-distributed:

$$pdf(r) = \frac{2\widetilde{\theta}r}{\left(\widetilde{\theta} + r^2\right)^2} \quad \text{for } r \geq 0 \tag{5.24}$$

where r now denotes the amplitude ratio and $\widetilde{\theta} = \sigma_1^2/\sigma_2^2$ is the ratio of mean signal power to mean interference power. The associated cdf is given by:

$$cdf(r) = 1 - \frac{\widetilde{\theta}}{\left(\widetilde{\theta} + r^2\right)} \quad \text{for } r \geq 0. \tag{5.25}$$

We can from this easily derive the pdf and cdf of the signal – to interference power ratio θ

$$pdf(\theta) = \frac{\widetilde{\theta}}{\left(\widetilde{\theta} + \theta\right)^2} \quad \text{for } \theta \geq 0$$

---

[7] Field strength and power can be mapped onto each other. The following considerations hold for both, with the mapping $r^2 \leftrightarrow P$ and $2\sigma^2 \leftrightarrow \overline{P}$.

and

$$cdf(\theta) = 1 - \frac{\widetilde{\theta}}{\left(\widetilde{\theta} + \theta\right)} \quad \text{for } \theta \geq 0.$$

This formulation is helpful for computation of the reuse distance (see Chapters 3 and 21).

5.5 Small-Scale Fading with a Dominant Component

5.5.1 A Computer Experiment

Fading statistics change when a dominant MPC – e.g., a LOS component or a dominant specular component – is present. We can gain some insights by repeating the computer experiment of Section 5.4.1, but now adding an additional wave with the (dominant) amplitude 5:

	$|a_9|$	ϕ_9	$\varphi_{0,9}$
$E_9(x, y) = 5.0 \exp[-jk_0(x \cos(0°) + y \sin(0°))]\exp(j0°)$	5.0	0°	0°

Figure 5.17 shows the real part of E, and the contribution from the dominant component is visible; while Figure 5.18 shows the absolute value. The histogram of the absolute value of the field strength is shown in Figure 5.19. It is clear that the probability of deep fades is much smaller than in the Rayleigh-fading case.

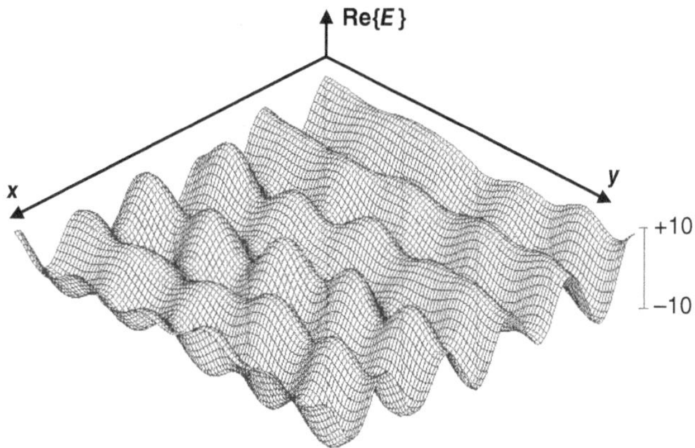

Figure 5.17　Re(E) – i.e., the instantaneous value at $t = 0$, in the presence of a dominant MPC.

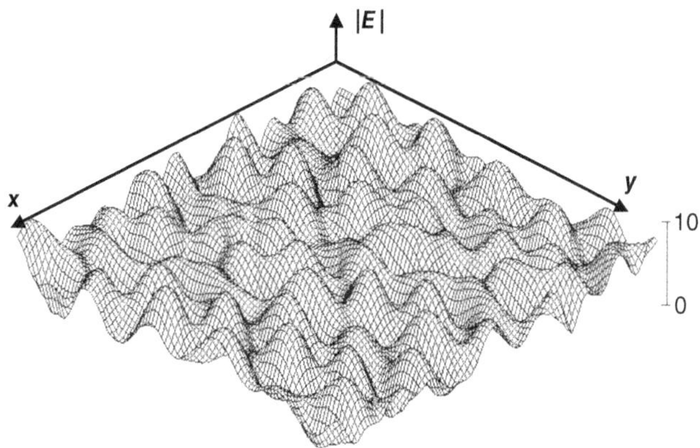

Figure 5.18　Magnitude of the electric field strength, |E|, in an example area, in the presence of a dominant MPC.

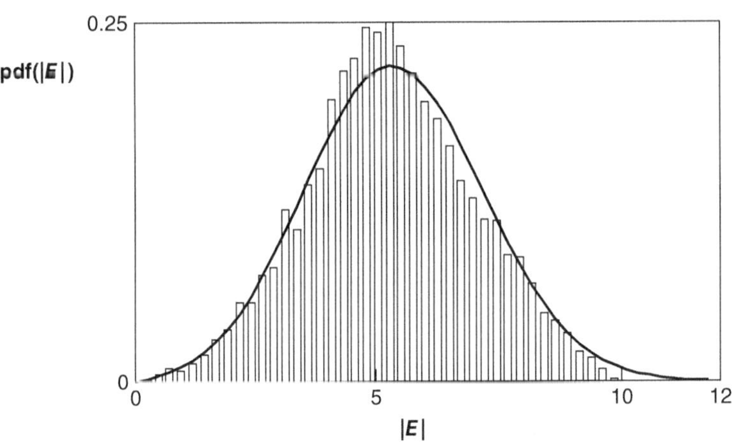

Figure 5.19   Histogram of the amplitudes in the presence of a dominant MPC.

5.5.2   *Derivation of the Amplitude and Phase Distribution*

The pdf of the amplitude can be computed in a way that is similar to our derivation of the Rayleigh distribution (Appendix 5.B). Without restriction of generality, we assume that the LOS component has zero phase, so that it is purely real. The real part thus has a nonzero-mean Gaussian distribution, while the imaginary part has a zero-mean Gaussian distribution. Performing the same type of variable transformation as in Appendix 5.B, we get the joint pdf of amplitude r and phase ψ [Rice 1947]:

$$pdf_{r,\psi}(r,\psi) = \frac{r}{2\pi\sigma^2}\exp\left(-\frac{r^2 + A^2 - 2rA\cos(\psi)}{2\sigma^2}\right) \quad \text{for } r \geq 0,\ -\pi < \psi \leq \pi \tag{5.26}$$

where A is the amplitude of the dominant component. In contrast to the Rayleigh case, this distribution is not separable. Rather, we have to integrate over the phases to get the amplitude pdf, and vice versa.

The pdf of the amplitude is given by the *Rice distribution* (solid line in Figure 5.19):

$$pdf_r(r) = \frac{r}{\sigma^2}\cdot\exp\left[-\frac{r^2 + A^2}{2\sigma^2}\right]\cdot I_0\left(\frac{rA}{\sigma^2}\right) \quad \text{for } r \geq 0. \tag{5.27}$$

$I_0(x)$ is the modified Bessel function of the first kind, zero order [Abramowitz and Stegun 1965]. The mean square value of a Rice-distributed random variable r is given by:

$$\overline{r^2} = 2\sigma^2 + A^2. \tag{5.28}$$

The ratio of the power in the LOS component to the power in the diffuse component, $A^2/(2\sigma^2)$, is called the Rice factor K_r.

Figure 5.20 shows the Rice distribution for three different values of the Rice factor. The stronger the LOS component, the rarer the occurrence of deep fades. For $K_r \to 0$, the Rice distribution becomes a Rayleigh distribution, while for large K_r it approximates a Gaussian distribution with mean value A.

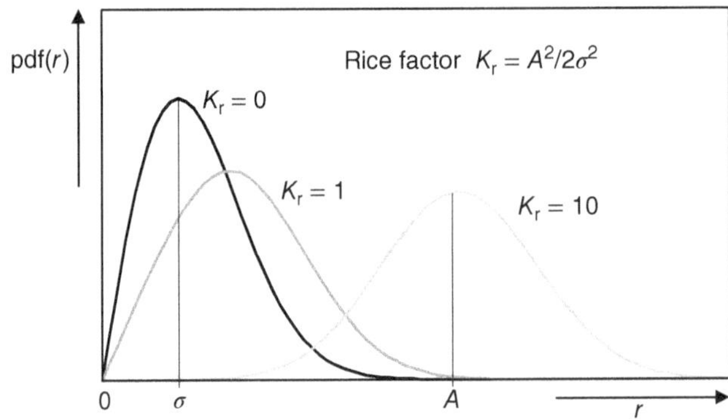

Figure 5.20   Rice distribution for three different values of K_r – i.e., the ratio between the power of the LOS component and the diffuse components.

Example 5.2 *Compute the fading margin for a Rice distribution with K_r = 0.3, 3, and 20 dB so that the outage probability is less than 5%.*

The outage probability can be expressed in terms of the *cdf* of the Rician power, or the Rician absolute field strength. In this example, we use the latter approach, so that:

$$P_{\text{out}} = cdf(r_{\text{min}}). \tag{5.29}$$

The cdf of a Rician field strength is given as:

$$cdf(r_{\text{min}}) = \int_0^{r_{\text{min}}} \frac{r}{\sigma^2} \cdot \exp\left[-\frac{r^2 + A^2}{2\sigma^2}\right] \cdot I_0\left(\frac{rA}{\sigma^2}\right) \ dr \quad \text{for } r_{\text{min}} \geq 0$$

$$= 1 - Q_M\left(\frac{A}{\sigma}, \frac{r_{\text{min}}}{\sigma}\right) \tag{5.30}$$

where $Q_M(a, b)$ is Marcum's Q-function (see also Chapter 11) given by:

$$Q_M(a, b) = e^{-(a^2 + b^2)/2} \sum_{n=0}^{\infty} \left(\frac{a}{b}\right)^n I_n(ab). \tag{5.31}$$

$I_n(\cdot)$ is the modified Bessel function of the first kind, order n. It is given by:

$$I_n(x) = \frac{1}{\pi} \int_0^{\pi} e^{x \cos\theta} \cos(n\theta) d\theta$$

where n is integer. The fading margin is given by:

$$\frac{\overline{r^2}}{r_{\text{min}}^2} = \frac{2\sigma^2(1 + K_r)}{r_{\text{min}}^2}. \tag{5.32}$$

The *cdf* of the squared amplitude is plotted in Figure 5.21. The required fading margins at different K_r can be found from that figure: they are 11.5, 9.7, and 1.1 dB, for Rice factors of 0.3, 3, and 20 dB, respectively.

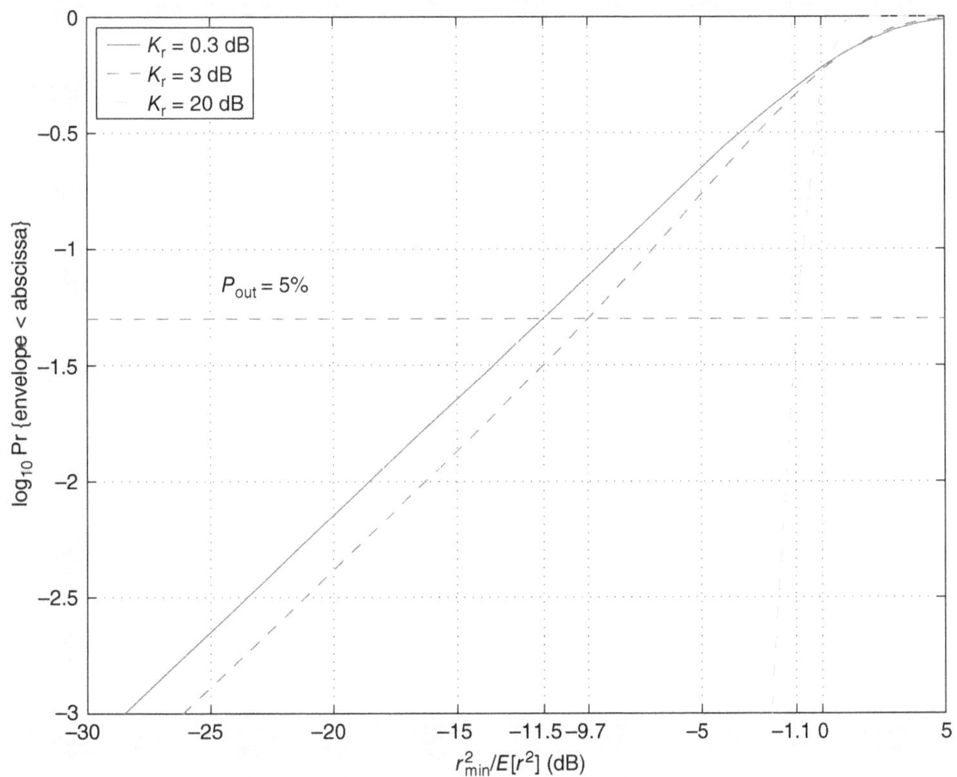

Figure 5.21 The cdf of the squared amplitude for Rice fading, $\sigma = 1$.

The presence of a dominant component also changes the *phase distribution*. This becomes intuitively clear by recalling that for a very strong dominant component, the phase of the total signal must be very close to the phase of the dominant component – in other words, the phase distribution converges to a delta function. For the general case (remember that we define the phase of the LOS component as $\psi = 0$), the pdf of the phase can be computed from the joint pdf of r and ψ and becomes:

$$\text{pdf}(\psi) = \frac{1 + \sqrt{\pi K_{\mathrm{r}}}\, e^{K_{\mathrm{r}} \cos^2(\psi)} \cos(\psi)\left(1 + \text{erf}\left[\sqrt{K_r} \cos(\psi)\right]\right)}{2\pi e^{K_{\mathrm{r}}}} \quad \text{for } -\pi < \psi \le \pi \tag{5.33}$$

where $erf(x)$ is the error function [Abramowitz and Stegun 1965]:

$$\text{erf}(x) = \left(2/\sqrt{\pi}\right) \int_0^x \exp\left(-t^2\right) dt.$$

Figure 5.22 shows the phase distribution for $\sigma = 1$ and different values of A.

The pdf of the power is given as

$$pdf_P(P) = \frac{1 + K_{\mathrm{r}}}{\overline{P}} \exp\left(-K_{\mathrm{r}} - \frac{(K_{\mathrm{r}} + 1)P}{\overline{P}}\right) I_0\left(2\sqrt{\frac{K_{\mathrm{r}}(K_{\mathrm{r}} + 1)P}{\overline{P}}}\right) \quad \text{for } P \ge 0. \tag{5.34}$$

From a historical perspective, it is interesting to note that all the work about Rice distributions was performed without the slightest regard for wireless channels. The classical paper [Rice 1947] considered the problem of a sinusoidal wave in additive white Gaussian noise. However, from a mathematical point of view, this is just the problem of a deterministic phasor (giving rise to a non-zero-mean) added to a zero-mean complex Gaussian distribution – exactly the same problem as in the field strength computation. Existing results thus just had to be reinterpreted by wireless engineers. This fact is so interesting because there are probably other wireless problems that can be solved by such "reinterpretation" methods.

*5.5.3 *Nakagami Distribution*

Another probability distribution for field strength that is in widespread use is the Nakagami m-distribution. The pdf is given as:

$$pdf_r(r) = \frac{2}{\Gamma(m)} \left(\frac{m}{\overline{P}}\right)^m r^{2m-1} \exp\left(-\frac{m}{\overline{P}} r^2\right) \tag{5.35}$$

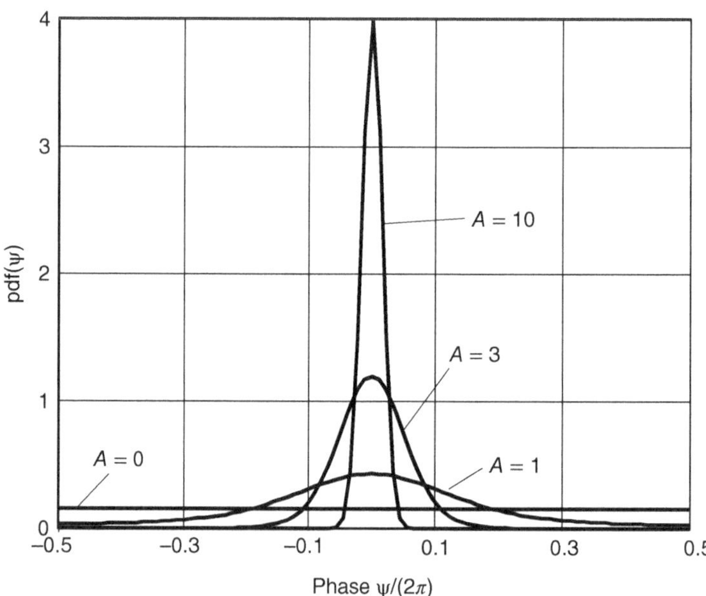

Figure 5.22 Pdf of the phase of a non-zero-mean complex Gaussian distribution, with $\sigma = 1$, $A = 0, 1, 3, 10$.

for $r \geq 0$ and $m \geq 1/2$; $\Gamma(m)$ is Euler's Gamma function, defined as

$$\Gamma(x) = \int_0^\infty t^{x-1} \exp(-t) dt$$

which reduces to $\Gamma(x) = (x-1)!$ for integer values of x. The parameter $\overline{P}$ is the mean square value $\overline{P} = \overline{r^2}$, and the parameter m is

$$m = \frac{\overline{P}^2}{\overline{\left(r^2 - \overline{P}\right)^2}}. \tag{5.36}$$

It is straightforward to extract these parameters from measured values. If the absolute field strength (amplitude) is Nakagami-fading, then the power follows a Gamma distribution:

$$pdf_P(P) = \frac{m}{\overline{P}\Gamma(m)} \left(\frac{mP}{\overline{P}}\right)^{m-1} \exp\left(-\frac{mP}{\overline{P}}\right) \qquad \text{for } P \geq 0. \tag{5.37}$$

Nakagami and Rice distribution have a quite similar shape, and one can be used to approximate the other. For $m > 1$ the m-factor can be computed from K_r by (see [Stueber 2017]):

$$m = \frac{(K_r + 1)^2}{(2K_r + 1)} \tag{5.38}$$

or conversely

$$K_r = \frac{\sqrt{m^2 - m}}{m - \sqrt{m^2 - m}}. \tag{5.39}$$

While Nakagami and Rice pdfs show good "general" agreement, they have different slopes close to $r = 0$. This in turn has an important impact on the achievable diversity order (see Chapter 12).

The main difference between the two pdfs is that the Rice distribution gives the *exact* distribution of the amplitude of a non-zero-mean complex Gaussian distribution – this implies the presence of one dominant component and a large number of nondominant components. The Nakagami distribution describes *in an approximate way* the amplitude distribution of a vector process where the central limit theorem is not necessarily valid (e.g., ultrawideband channels, Section 6.6). Further models for small-scale fading are mentioned in Section 7.1.8.

5.6 Doppler Spectra and Statistics of Temporal Channel Variations

5.6.1 Temporal Variations for Moving UE

Section 5.3 showed the physical interpretation of the frequency shift by movement – i.e., the Doppler effect. If the UE is moving, then different directions of the MPCs arriving at the UE give rise to different frequency shifts. This leads to a broadening of the received spectrum. The goal of this section is to derive this spectrum, assuming that the transmit signal is a sinusoidal signal (i.e., the narrowband case). The more general case of a wideband transmit signal is treated in Chapter 6.

Let us first repeat the expressions for Doppler shift when a wave only comes from a single direction. Let γ denote the angle between the velocity vector v of the UE and the direction of the wave at the location of the UE. As shown in Eq. (5.8), the Doppler effect leads to a shift of the received frequency f by the amount ν, so that the received frequency is given by:

$$f = f_c \left[1 - \frac{v}{c_0} \cos(\gamma)\right] = f_c + \nu \tag{5.40}$$

where $v = |\mathbf{v}|$. Obviously, the frequency shift depends on the direction of the wave, and must lie in the range $f_c - \nu_{max} \cdots f_c + \nu_{max}$, where $\nu_{max} = f_c \, v/c_0$.

If there are multiple MPCs, we need to know the distribution of power of the incident waves as a function of γ. As we are interested in the *statistical* distribution of the received signal, we consider the pdf of the received power; in a slight abuse of notation, we call it the pdf of the incident waves $pdf_\gamma(\gamma)$. It describes the power density coming from the angular range $[\gamma, \gamma + d\gamma]$, which depends both on the density of MPCs (how many MPCs per unit angle) and the powers of those MPCs.

The MPCs arriving at the RX are also weighted by the antenna pattern of the UE; therefore, the power of an MPC arriving in the direction γ has to be multiplied by the pattern $G(\gamma)$. The received power spectrum as a function of direction is thus:

$$S(\gamma) = \overline{\Omega}\left[pdf_\gamma(\gamma)G(\gamma)\right] \tag{5.41}$$

where $\overline{\Omega}$ is the mean power of the arriving field that would be received with an omni-antenna.

In a final step, we have to perform the variable transformation $\gamma \to \nu$. The Jacobian can be determined as:

$$\left|\frac{d\gamma}{d\nu}\right| = \left|\frac{1}{\frac{d\nu}{d\gamma}}\right| = \frac{1}{\left|\frac{\mathrm{v}}{c_0}f_c\sin(\gamma)\right|} = \frac{1}{\sqrt{\left(f_c\frac{\mathrm{v}}{c_0}\right)^2 - (f - f_c)^2}} = \frac{1}{\sqrt{\nu_{\max}^2 - \nu^2}} \quad \text{for} \ -\nu_{\max} \le \nu \le \nu_{\max}. \tag{5.42}$$

We furthermore have to take into account the fact that waves from the direction γ and $-\gamma$ lead to the same Doppler shift, and thus need not be distinguished for the purpose of deriving a Doppler spectrum. The Doppler spectrum thus becomes:

$$S_D(\nu) = \begin{cases} \overline{\Omega}\left[pdf_\gamma(\gamma)G(\gamma) + pdf_\gamma(-\gamma)G(-\gamma)\right]\dfrac{1}{\sqrt{\nu_{\max}^2 - \nu^2}} & \text{for} \ -\nu_{\max} \le \nu \le \nu_{\max} \\ 0 & \text{otherwise} \end{cases} \tag{5.43}$$

For further use, we also define

$$\Omega_n = (2\pi)^n \int_{-\nu_{\max}}^{\nu_{\max}} S_D(\nu)\nu^n d\nu \tag{5.44}$$

as the nth moment of the Doppler spectrum.

More specific equations can be obtained for specific angular distributions and antenna patterns. A very popular model for the angular spectrum at the UE is that the waves are incident uniformly from all azimuthal directions, and all arrive in the horizontal plane, so that:

$$pdf_\gamma(\gamma) = \frac{1}{2\pi} \quad \text{for} \ -\pi < \gamma \le \pi. \tag{5.45}$$

This situation corresponds to the case when there is no LOS connection, and a large number of IOs are distributed uniformly around the UE (compare Section 7.6.2). Assuming furthermore that the antenna is a vertical short dipole (Section 8.3.1), with an antenna pattern $G(\gamma) = 1.5$, the Doppler spectrum becomes

$$S_D(\nu) = \frac{1.5\overline{\Omega}}{\pi\sqrt{\nu_{\max}^2 - \nu^2}} \quad -\nu_{\max} \le \nu \le \nu_{\max}. \tag{5.46}$$

This spectrum, known as the *classical* or *Jakes* spectrum, is depicted in Figure 5.23. It has the characteristic "bathtub" shape – i.e., (integrable) singularities at the minimum and maximum Doppler frequencies $\nu = \pm\nu_{\max} = \pm f_c\mathrm{v}/c_0$. These singularities thus correspond to the direction of the movement of the UE, or its opposite. It is remarkable that a uniform azimuthal distribution can lead to a highly *non*uniform Doppler spectrum.

Naturally, *measured* Doppler spectra do not show singularities; even if the model and underlying assumptions would be strictly valid, it would require an infinite number of measurement samples to arrive at a singularity. Despite this fact, the classical Doppler spectrum is the most widely used model. Alternative models include the following:

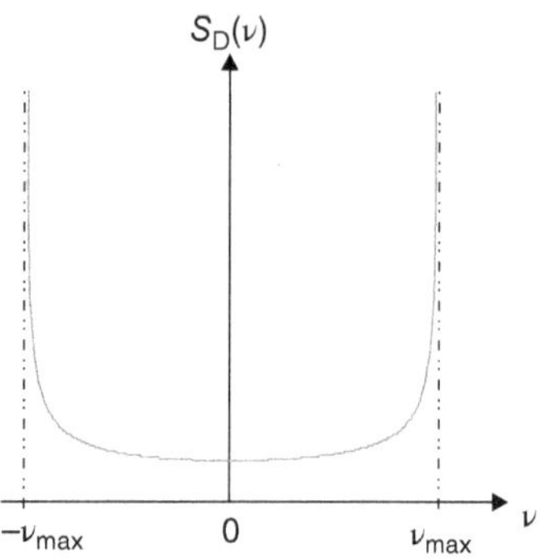

Figure 5.23 Jakes Doppler spectrum.

- Aulin's model [see Parsons 1992], which limits the amplitude of the Jakes spectrum at and near its singularities;
- Gaussian spectrum;
- uniform spectrum (this corresponds to the case when all waves are incident uniformly in all three dimensions, and the antenna has an isotropic pattern).

As already mentioned in Section 5.2, the Doppler spectrum has two important interpretations:

1. It describes *frequency dispersion*. For narrowband systems, as well as OFDM, such frequency dispersion can lead to transmission errors. This is discussed in more detail in Chapters 11 and 15. It has, however, no *direct* impact on most other wideband systems (like single-carrier Time Division Multiple Access (TDMA) or Code Division Multiple Access (CDMA) systems).
2. It is a measure for the *temporal variability* of the channel. As such, it is important for *all* systems.

The temporal dependence of fading is best described by the autocorrelation function of fading, which describes how similar the fading states at two times separated by Δt are. More precisely, the normalized correlation between the in-phase component at time t and the in-phase component at time $t + \Delta t$ can be shown to be, under the above assumptions:

$$\frac{\overline{I(t)I(t + \Delta t)}}{\overline{I(t)}^2} = J_0(2\pi\nu_{\max}\Delta t) \tag{5.47}$$

which is proportional to the inverse Fourier transform of the Doppler spectrum $S_D(\nu)$, while the correlation $\overline{I(t)Q(t + \Delta t)} = 0$ for all values of Δt. $J_0(x)$ is the Bessel function zero order first kind. The normalized covariance function of the envelope is thus approximately:

$$\frac{\overline{r(t)r(t + \Delta t)} - \overline{r(t)}^2}{\overline{r(t)^2} - \overline{r(t)}^2} = J_0^2(2\pi\nu_{\max}\Delta t). \tag{5.48}$$

The overbar denotes expectation over the ensemble of channel realizations, which here is assumed to be constituted by the channels at different times t (this in turn implies stationarity of the fading process, see Appendix 6.A for a more detailed discussion).

Example 5.3 *Assume that a UE is located in a fading dip. On average, what minimum distance should the UE move so that it is no longer influenced by this fading dip?*

The temporal correlation function when traveling at a speed v through a static environment is completely equivalent to the spatial autocorrelation function with (spatial) displacement vΔt. As a first step, we plot the envelope correlation function (Eq. 5.48) in Figure 5.24. If we now define "no longer influenced" as "envelope correlation coefficient equals 0.5," then we see that (on average) the RX has to move through 0.18 λ. If we want complete decorrelation from the fading dip, then moving through 0.38 λ is required.

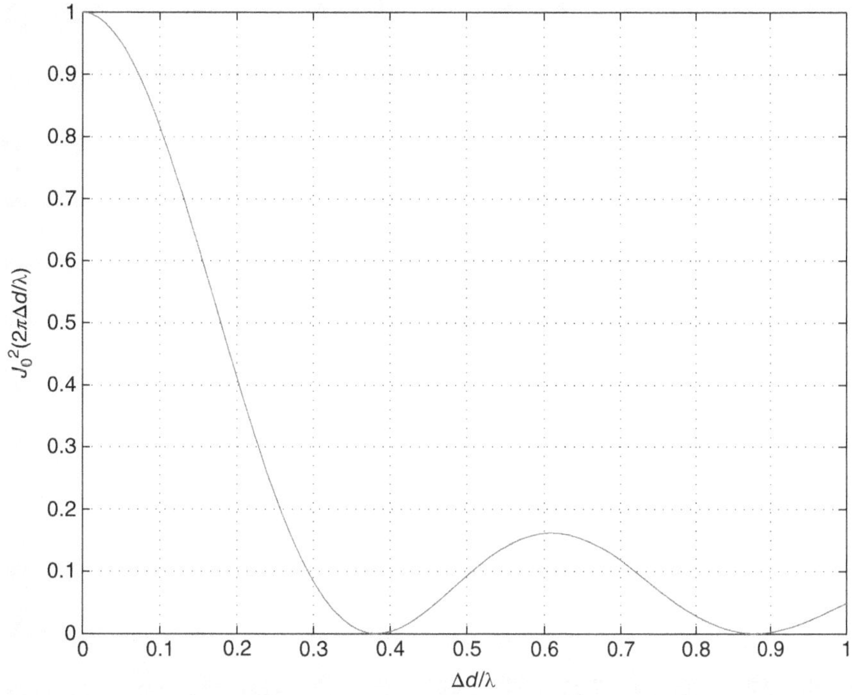

Figure 5.24 Amplitude correlation as a function of displacement of the RX.

5.6.2 Temporal Variations in Fixed and Nomadic Wireless Systems

In a fixed wireless system (i.e., a system where the UE never moves, see [Anderson 2003]), as well as nomadic systems, where the UE does not move over the time period of interest, temporal variations arise only from movement of the IOs.[8] The resulting fading leads to a different way of thinking about, and modeling of, the temporal channel changes. In a mobile link, a UE tracks the spatial fading (i.e., the fading "landscape" of Figure 5.11) as it moves. The spatial and temporal fading can thus be mapped onto each other, the scaling factor being the mobile speed v. In a fixed wireless link, the variation is due instead to the fact that some scattering objects are moving, a prime example being cars, or vegetation moving in the wind.

Measurements have shown that in many cases, the amplitude statistics of the fading (taken over the ensemble of values observed at different times) are Rician. However, the Rice factor now represents the ratio of the power in the *time-invariant* MPCs vs. the power in the *time-variant* MPCs. In other words, the temporal Rice factor has nothing to do with LOS (or other dominant) components: in a pure NLOS situation with many equally strong components, the temporal Rice factor can approach infinity if all MPCs are time invariant.

The Doppler spectrum of such a system consists of a delta pulse at $\nu = 0$, and a continuous spectrum that depends on the movement and location of the moving IOs. Various measurement campaigns have shown that the shape of this diffuse spectrum is approximately Gaussian.

5.6.3 Generation of Fading

The prevalent approach of simulating Rayleigh fading with Jakes Doppler spectrum is based on the discrete model

$$h(t) = \sum_{i=1}^{\widetilde{N}} C_i \exp\left[\, j2\pi t \nu_{\max} \cos\left(\gamma_i\right) + \varphi_i\right]$$

where one can set

$$C_i = \frac{1}{\sqrt{\widetilde{N}}}, \gamma_i = \frac{2\pi i + \theta_i}{\widetilde{N}}.$$

For the phases φ_i, the original model proposed by Jakes is to set them all to zero, $\varphi_i = 0$. However, it can be shown that this leads to a nonstationary model. Rather, random phases should be used, more precisely, both φ_i and θ_i should be independent uniformly distributed random variables [Xiao et al. 2006].

Another approach would place IOs geometrically and then simulate through simple ray tracing the total accumulating signal. This approach will be discussed in more detail in Section 7.6.

*5.7 Temporal Fading Characterization

5.7.1 Level Crossing Rate

The Doppler spectrum is a complete characterization of the temporal statistics of fading. However, it is often desirable to have a different formulation that allows more direct insights into system behavior. A quantity that allows immediate interpretation is the occurrence rate of fading dips – this occurrence rate is known as the *Level Crossing Rate* (LCR). Obviously, it depends on which level we are considering (i.e., how a fading dip is defined): falling below a level that is 30 dB below the mean happens more rarely than falling 3 dB below this mean. As the admissible depth of fading dips depends on the mean-field strength, as well as on the considered system, we want to derive the LCR for arbitrary levels (i.e., depth of fading dips).

Providing a mathematical formulation, the LCR is defined as the expected value of the rate at which the received field strength crosses a certain level r in the positive direction. This can also be written as:

$$N_{\mathrm{R}}(r) = \int_0^\infty \dot{r} \cdot pdf_{r,\dot{r}}(r, \dot{r})\, d\dot{r} \quad \text{for } r \geq 0 \tag{5.49}$$

where $\dot{r} = dr/dt$ is the temporal derivative, and $pdf_{r,\dot{r}}$ is the joint pdf of r and $\dot{r}$.

In Appendix 5.C, we derive the LCR as:

$$N_{\mathrm{R}}(r) = \sqrt{\frac{\Omega_2}{\pi\Omega_0}} \frac{r}{\sqrt{2\Omega_0}} \exp\left(-\frac{r^2}{2\Omega_0}\right) \qquad \text{for } r \geq 0. \tag{5.50}$$

---

[8] To simplify notation, we henceforth just speak of fixed wireless systems, even when this covers the nomadic case as well.

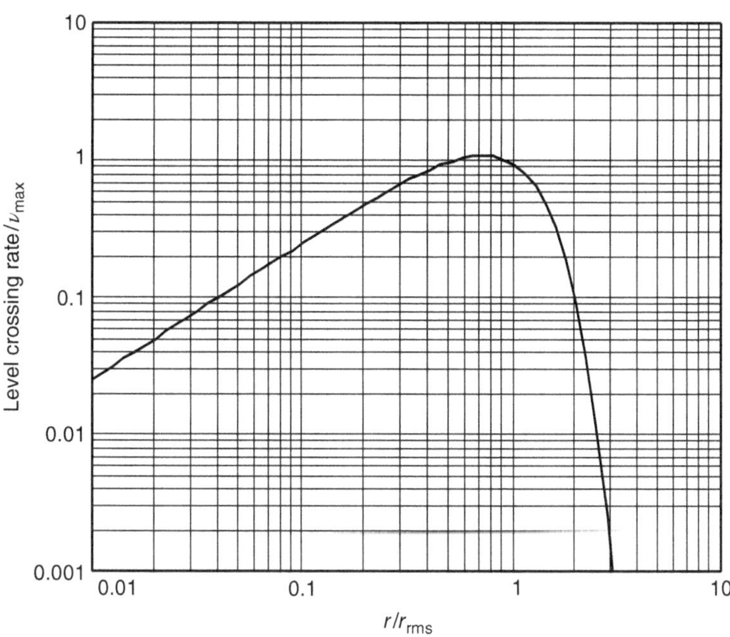

Figure 5.25 Level crossing rate normalized to the maximum Doppler frequency as a function of the normalized level r/r_{rms}, for a Rayleigh-fading amplitude and Jakes spectrum.

Note that $\sqrt{2\Omega_0}$ is the root-mean-square value of amplitude. Figure 5.25 shows the LCR for a Rayleigh-fading amplitude and Jakes Doppler spectrum.

5.7.2 Average Duration of Fades

Another parameter of interest is the *Average Duration of Fades* (ADF). In the previous sections, we have already derived the rate at which the field strength goes below the considered threshold (i.e., the LCR), and the total percentage of time the field strength is lower than this threshold (i.e., the cdf of the field strength). The ADFs can be simply computed as the quotient of these two quantities:

$$ADF(r) = \frac{cdf_{\mathrm{r}}(r)}{N_{\mathrm{R}}(r)} \qquad \text{for } r \geq 0. \tag{5.51}$$

A plot of the ADF is shown in Figure 5.26.

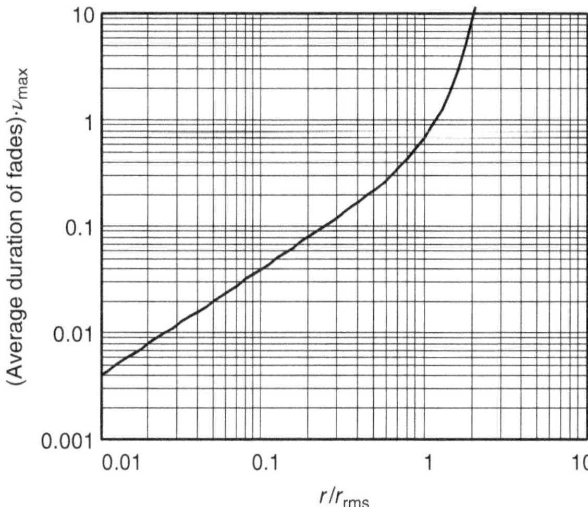

Figure 5.26 Average duration of fades normalized to the maximum Doppler frequency as a function of the normalized level r/r_{rms}, for a Rayleigh-fading amplitude and Jakes spectrum.

Example 5.4 *Assume a multipath environment where the received signal has a Rayleigh distribution and the Doppler spectrum has the classical bathtub (Jakes) shape. Compute the LCR and the ADF for a maximum Doppler frequency $\nu_{\max} = 50$ Hz, and amplitude thresholds:*

$$r_{\min} = \frac{\sqrt{2\Omega_0}}{10}, \frac{\sqrt{2\Omega_0}}{2}, \sqrt{2\Omega_0}.$$

The LCR is computed from Eq. (5.50). For a Jakes scenario the second moment of the Doppler spectrum is given as:

$$\Omega_2 = \frac{1}{2}\Omega_0(2\pi\nu_{\max})^2 \tag{5.52}$$

therefore,

$$N_{\mathrm{R}}(r_{\min}) = \sqrt{2\pi}\cdot\nu_{\max}\cdot\frac{r_{\min}}{\sqrt{2\Omega_0}}\exp\left(-\frac{r_{\min}^2}{2\Omega_0}\right). \tag{5.53}$$

The ADF is computed from Eq. (5.51) where:

$$cdf(r_{\min}) = 1 - \exp\left(-\left(\frac{r_{\min}}{\sqrt{2\Omega_0}}\right)^2\right). \tag{5.54}$$

These expressions are evaluated for different values of the threshold, $r_{\min}$. The results are tabulated in Table 5.1.

Table 5.1 Effect of threshold on ADF and LCR.

$r_{\min}$	$N_{\mathrm{R}}(r_{\min})$	$cdf(r_{\min})$	ADF (ms)
$\dfrac{\sqrt{2\Omega_0}}{10}$	12.4	0.01	0.8
$\dfrac{\sqrt{2\Omega_0}}{2}$	48.8	0.22	4.5
$\sqrt{2\Omega_0}$	46.1	0.63	13.7

5.7.3 Random Frequency Modulation

A random channel leads to a random phase shift of the received signal; in a time-variant channel, these phase shifts are time variant as well. By definition, a temporally varying phase shift is an FM. In this section, we compute this *random FM*.

The pdf of the instantaneous angular frequency $\dot{\psi}$ can be computed from the joint pdf $pdf_{r,\dot{r},\psi,\dot{\psi}}$ of Appendix 5.C. We now have to integrate over the variables r, $\dot{r}$, and ψ. This results in:

$$pdf_{\dot{\psi}}(\dot{\psi}) = \frac{1}{2}\sqrt{\frac{\Omega_0}{\Omega_2}}\left(1 + \frac{\Omega_0}{\Omega_2}\dot{\psi}^2\right)^{-3/2} \tag{5.55}$$

see Figure 5.27.

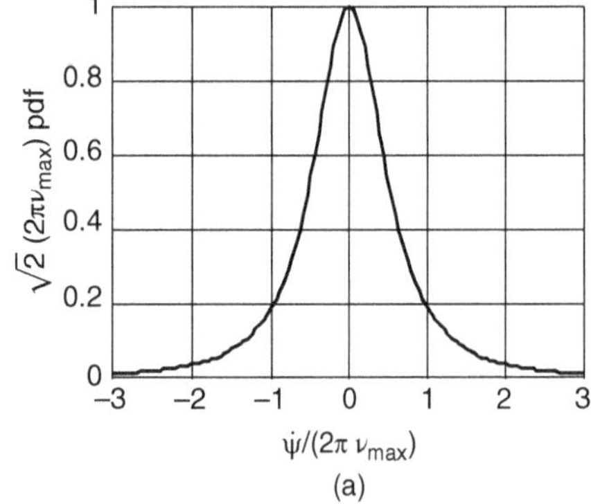
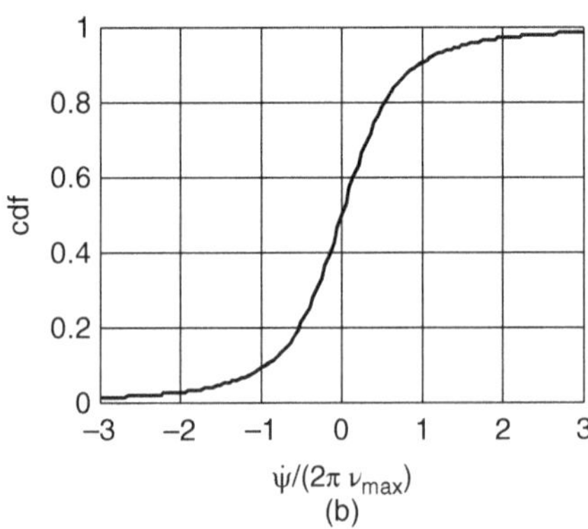

Figure 5.27 Normalized pdf (a) and cdf (b) of a random FM for Rayleigh fading and Jakes spectrum.

This is a "student's *t*-distribution" with two degrees of freedom [Mardia et al. 1979]. The cumulative distribution is given as:

$$cdf_{\dot{\psi}}(\dot{\psi}) = \frac{1}{2}\left[1 + \sqrt{\frac{\Omega_0}{\Omega_2}}\,\dot{\psi}\left(1 + \frac{\Omega_0}{\Omega_2}\dot{\psi}^2\right)^{-1/2}\right]. \tag{5.56}$$

It is remarkable that all instantaneous frequencies in a range $-\infty$ to ∞ can occur – they are not restricted to the range of possible Doppler frequencies!

Strong FM most likely occurs in fading dips: if the signal level is low, very strong relative changes can occur. This intuitively pleasing result can be shown mathematically by considering the pdf of the instantaneous angular frequency conditioned on the amplitude level r_0 [Jakes 1974]:

$$pdf(\dot{\psi}\mid r_0) = \frac{r_0}{\sqrt{2\pi\Omega_2}}\exp\left(-r_0^2\frac{\dot{\psi}^2}{2\Omega_2}\right). \tag{5.57}$$

This is a zero-mean Gaussian distribution with variance Ω_2/r_0^2. The variance is the larger the smaller the signal level is.

We thus see that fading dips can create errors in two ways: on one hand, the lower signal level leads to higher susceptibility to noise. On the other hand, they increase the probability of strong random FMs, which introduces errors in any system that conveys information by means of the *phase* of the transmitted signal. In addition, we find in Chapter 11 that fading dips are also related to inter-symbol interference.

5.8 Large-Scale Fading

5.8.1 Large-Scale Fading Distribution

Small-scale fading, created by the superposition of different MPCs, changes rapidly over the spatial scale of a few wavelengths. If the field strength is averaged over a small area (e.g., ten by ten wavelengths), we obtain the *Small Scale Averaged* (SSA) field strength.[9] In the previous sections, we treated the SSA field strength as a constant. However, as explained in the introduction, it varies when considered on a larger spatial scale, due to shadowing of the MPCs by IOs.

Many experimental investigations have shown that the SSA field strength F, *plotted on a logarithmic scale*, shows a Gaussian distribution around a mean μ. Such a distribution is known as lognormal, and its pdf is given by:

$$pdf_{\mathrm{F}}(F) = \frac{20/\ln(10)}{F\sigma_{\mathrm{F}}\sqrt{2\pi}}\cdot\exp\left[-\frac{\left(20\log_{10}(F)-\mu_{\mathrm{F,dB}}\right)^2}{2\cdot\sigma_{\mathrm{F}}^2}\right]\quad\text{for }F\geq 0 \tag{5.58}$$

where σ_{F} is the standard deviation of F, and $\mu_{F,\mathrm{dB}}$ is the mean of the values of F expressed in dB. Also, power is distributed lognormally. However, there is an important fine point: when fitting the logarithm of SSA *power* to a normal distribution, we find that the median value of this distribution, $\mu_{\mathrm{P,\ dB}}$, is related to the median of the field strength distribution $\mu_{F,\mathrm{dB}}$ as $\mu_{P,\mathrm{dB}} = \mu_{F,\mathrm{dB}} + 10\log_{10}(4/\pi)$ if the small-scale fading is Rayleigh. The pdf for the power thus reads

$$pdf_{\mathrm{P}}(P) = \frac{10/\ln(10)}{P\sigma_{\mathrm{P}}\sqrt{2\pi}}\cdot\exp\left[-\frac{\left(10\log_{10}(P)-\mu_{\mathrm{P,dB}}\right)^2}{2\cdot\sigma_{\mathrm{P}}^2}\right]\quad\text{for }P\geq 0 \tag{5.59}$$

where $\sigma_P = \sigma_F$. Typical values of σ_{F} are 4–10 dB.

The lognormal variations of F have commonly been attributed to shadowing effects. Consider the situation in Figure 5.28. If the UE moves, it changes the angle γ, and thus the diffraction parameter. If the distance between the edge and the UE is now large, the UE

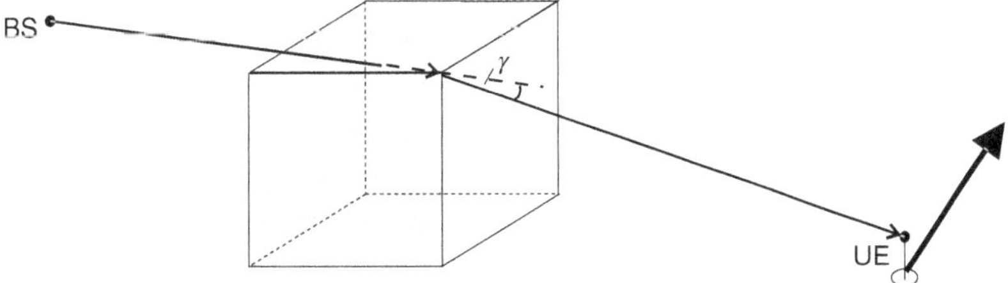

Figure 5.28 Shadowing by a building.

---

[9] Strictly speaking, this is only an approximation to the SSA field strength, as there can only be a finite number of statistically independent sample values within the finite area over which we average.

must move over a large distance (e.g., several tens of wavelengths) for the field strength to change noticeably. Note that this effect changes the absolute amplitude of the MPC, $|a|$, and has nothing to do with an interference effect. Consider now the situation where an MPC undergoes several of these or similar processes on its way from the TX and RX, each of which contributes a certain attenuation. The received field strength then depends on the *product* of the attenuations; on a logarithmic scale, these attenuations add up. We can thus model the effect as the sum of random variables *on a dB scale* – i.e., a lognormal distribution. However, the mechanism just described is not necessarily valid in all physical situations.[10]

In many applications, the sum of several lognormally distributed random variables is required. Such situations occur, e.g., in channel modeling where superposition of multiple MPC "clusters," each of which is lognormally fading, give the total receive power. They also arise in interference modeling, where the total interference is the sum of interference from multiple devices, each of which sends its signal through a channel with shadowing. It has been found that the sum of lognormally distributed variables is again *approximately* lognormally distributed. The mean and the variance of this new lognormal variables can be computed through so-called *moment matching*, where the mean and variance of the sum are related to mean and variance of the constituent random variables. The simplest – but not very accurate – such result is the Fenton–Wilkinson approximation, which states that (assuming equal variances σ_c^2 of the constituents)

$$\sigma_{\text{sum}}^2 = \ln\left[\left(\exp(\sigma_c^2) - 1 \right) \frac{\sum_i \exp(2\mu_i)}{\left(\sum_i \exp(\mu_i) \right)^2} + 1 \right]$$

$$\mu_{\text{sum}} = \ln\left[\sum_i \exp(\mu_i) \right] - \frac{\sigma_{\text{sum}}^2 - \sigma_c^2}{2}$$

where all variables are represented in the natural logarithm domain. More accurate approximations can be obtained from matching the moment-generating function at specific points, see "Further Reading." Also note that the product of lognormal random variables is exactly lognormally distributed.

We finally turn to the spatial correlation of the shadowing. The value of the shadowing changes only slowly with the location of the UE. Thus, realizations of the shadowing field strength (or power) that are measured at a distance Δx apart are correlated. The most common model is that of an exponential correlation, i.e., $E_x\{F(x)F(x + \Delta x)\} = \exp(-\Delta x/\bar{x})$, where the decorrelation distance $\bar{x}$ is typically on the order of 5–50 m, depending on the environment. More detailed models are described in Section 7.1.7.

*5.8.2   Joint Small-Scale and Large-Scale Fading

Consider now the statistics of the field strength based on samples taken from a large area (i.e., including both lognormal fading and MPC interference effects). The pdf of these samples is given by the so-called Suzuki distribution, which follows in a straightforward manner from the laws of conditional statistics. The mean value of the field strength, taken over a small area, is $\bar{r} = \sigma\sqrt{\pi/2}$, compare (5.17), and the distribution of the local value of the field strength conditioned on it is

$$pdf(r \mid \bar{r}) = \frac{\pi r}{2\bar{r}^2} \exp\left(-\frac{\pi r^2}{4\bar{r}^2} \right) \quad \text{for } r \geq 0. \tag{5.60}$$

The local mean (i.e., the expected value used above) is distributed according to lognormal statistics. Unconditioning results in the pdf of the field strength:

$$pdf_r(r) = \int_0^\infty \frac{\pi r}{2\bar{r}^2} \exp\left(-\frac{\pi r^2}{4\bar{r}^2} \right) \frac{20/\ln(10)}{\bar{r}\sigma_F\sqrt{2\pi}} \exp\left(-\frac{\left(20\log_{10}(\bar{r}) - \mu_{F,\text{dB}} \right)^2}{2\sigma_F^2} \right) d\bar{r} \quad \text{for } r \geq 0. \tag{5.61}$$

This pdf is also known as the Suzuki distribution, an example is shown in Figure 5.29. A similar function can be defined if the small-scale statistics are Rician or Nakagami. Note, however, that the relationship between mean of the amplitude and mean power of the small-scale fading is different when the small-scale statistics are not Rayleigh.

The pdf for the power reads

$$pdf_P(P) = \int_0^\infty \frac{1}{\Omega} \exp\left(-\frac{P}{\Omega} \right) \frac{10/\ln(10)}{\Omega\sigma_P\sqrt{2\pi}} \exp\left(-\frac{\left(10\log_{10}(\Omega) - \mu_{P,\text{dB}} \right)^2}{2\sigma_P^2} \right) d\Omega \quad \text{for } P \geq 0. \tag{5.62}$$

Since both large-scale and small-scale fading occurs in practical situations, the fading margin must account for the combination of the two effects (see also Chapter 3). One possibility is to just add up the fading margin for the Rayleigh distribution and the fading

---

[10] An alternative explanation, based on double-scattering processes, was proposed by [Andersen 2002]. But, independently of the mechanism that leads to its creation, many measurements have confirmed that the pdf of F is well-approximated by a lognormal function.

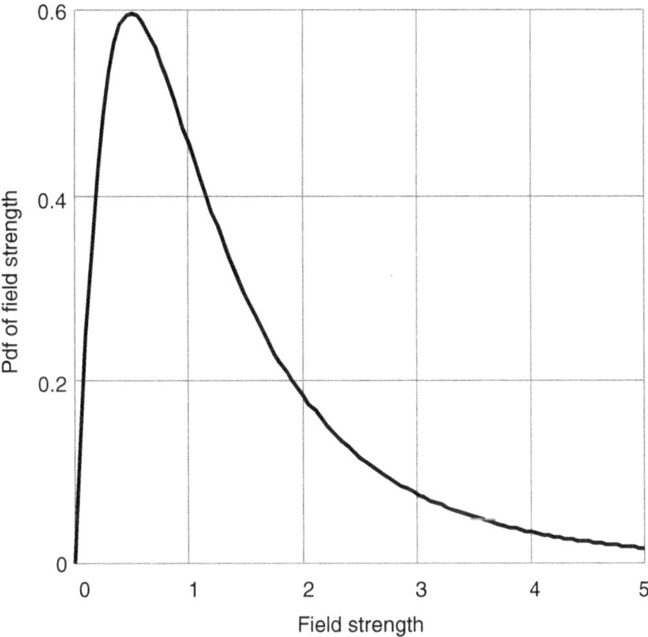

Figure 5.29 Suzuki distribution with $\sigma_F = 5.6\,\text{dB}$, $\mu_{F,\,\text{dB}} = 1.4\,\text{dB}$.

margin for a lognormal distribution. This method is commonly used because of its simplicity but overestimates the required fading margin. The more accurate method is based on the cdf of the Suzuki distribution, which can be obtained by integrating Eq. (5.61) from $-\infty$ to x. This then allows computation of the necessary mean field strength r_0 for a given admissible outage probability (e.g., 0.05), as shown in the following example.

Example 5.5 *Consider a channel with $\sigma_F = 6\,dB$ and $\mu_{F,\,dB} = 0$ dB. Compute the fading margin for a Suzuki distribution relative to the mean-dB value of the shadowing so that the outage probability is smaller than 5%. Also compute the fading margin for Rayleigh fading and shadow fading separately.*

For the Suzuki distribution:

$$\text{Pr}_{\text{out}} = cdf(r_{\min}) = \int_0^{r_{\min}} pdf_r(r)dr$$
$$= \int_0^{\infty} \left(1 - \exp\left(-\frac{\pi r_{\min}^2}{4\bar{r}^2}\right)\right) \frac{20/\ln(10)}{\bar{r}\sigma_F\sqrt{2\pi}} \exp\left(-\frac{\left(20\log_{10}\bar{r} - \mu_{F,\text{dB}}\right)^2}{2\sigma_F^2}\right) d\bar{r}. \tag{5.63}$$

Inserting $\sigma_F = 6$ dB and $\mu_{F,\text{dB}} = 0$, the *cdf* is expressed as:

$$cdf(r_{\min}) = \frac{20/\ln 10}{\sqrt{2\pi}\cdot 6} \int_0^{\infty} \frac{1}{\bar{r}} \exp\left(-\frac{\left(20\log_{10}r - 0\right)^2}{2\cdot 36}\right)\left(1 - \exp\left(-\frac{\pi}{4}\cdot\frac{r_{\min}^2}{\bar{r}^2}\right)\right) d\bar{r}. \tag{5.64}$$

The fading margin is defined as:

$$FM = \frac{\mu^2}{r_{\min}^2}. \tag{5.65}$$

The *cdf* plot is obtained by evaluating Eq. (5.65) for different values of $r_{\min}$, the result is shown in Figure 5.30. A fading margin of 15.5 dB is required to get an outage probability of 5%.

The outage probability for Rayleigh fading only is evaluated as:

$$\text{Pr}_{\text{out}} = cdf(r_{\min}) = 1 - \exp\left(-\frac{r_{\min}^2}{2\sigma^2}\right)$$
$$= 1 - \exp(-1/FM). \tag{5.66}$$

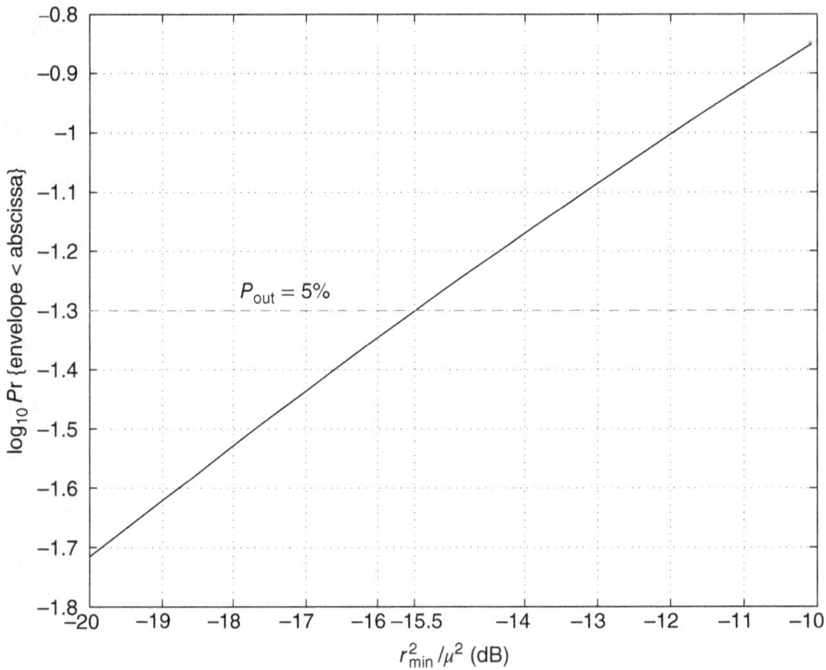

Figure 5.30 The Suzuki *cdf*, $\sigma_F = 6\,\mathrm{dB}$ and $\mu_{dB} = 0$.

After some manipulation, we get:

$$
\begin{aligned}
FM_{\mathrm{Rayleigh,dB}} &= -10\,\log_{10}(-\ln(1 - P_{\mathrm{out}})) \\
&= 12.9\,\mathrm{dB}.
\end{aligned}
$$
(5.67)

For shadow fading, the pdf of the field strength values in dB is a standard Gaussian distribution. The complementary cdf (i.e., unity minus the cdf) is thus given by a Q-function defined as:

$$
Q(a) = \frac{1}{\sqrt{2\pi}} \int_a^\infty \exp\left(-\frac{x^2}{2}\right) dx.
$$

The outage probability is given as:

$$
\mathrm{Pr}_{\mathrm{out}} = Q\left(\frac{M_{\mathrm{large\text{-}scale_{dB}}}}{\sigma_F}\right).
$$
(5.68)

Inserting $\sigma_F = 6\,\mathrm{dB}$ and $P_{\mathrm{out}} = 0.05$ into Eq. (5.68) we get:

$$
\begin{aligned}
FM_{\mathrm{large\text{-}scale_{dB}}} &= 6 \cdot Q^{-1}(0.05) \\
&= 9.9\ \mathrm{dB}.
\end{aligned}
$$
(5.69)

When computing the fading margin as the sum of the margin for Rayleigh fading and shadowing, we obtain

$$
FM_{\mathrm{dB}} = 12.9 + 9.9\,\mathrm{dB} = 22.8\,\mathrm{dB}.
$$
(5.70)

Compared with the fading margin obtained from the Suzuki distribution, this is a more conservative estimate.

It turns out that a Suzuki distribution can, for large variances, be approximated reasonably well by a lognormal distribution, where the parameters of approximating lognormal distribution are related to the parameters of the shadowing alone (i.e., without the Rayleigh fading) by (see [Stueber 2017])

$$
\mu_{\mathrm{P,approx,dB}} = \mu_{\mathrm{P,shadow,dB}} - 2.5
$$
(5.71)

$$
\sigma^2_{\mathrm{P,dB}} = \sigma^2_{\mathrm{P,shadow,dB}} + 31.
$$
(5.72)

*5.9 Appendices

App. 5.A: The Central Limit Theorem

See App5.pdf at wiley.com/go/molisch/wireless3e

App. 5.B: Derivation of the Rayleigh Distribution

See App5.pdf at wiley.com/go/molisch/wireless3e

App. 5.C: Derivation of the Level Crossing Rate

See App5.pdf at wiley.com/go/molisch/wireless3e

Further Reading

The mathematical basis for the derivations in this chapter is described in a number of standard textbooks on statistics and random processes, especially the classical book of [Papoulis 1991]. The statistical model for the amplitude distribution and the Doppler spectrum was first described in [Clarke 1968]. A comprehensive exposition of the statistics of the channel, derivation of the Doppler spectrum, LCR, and ADFs, for the Rayleigh case, can be found in [Jakes 1974]. Since Rayleigh fading is based on Gaussian fading of the I- and Q-component, the rich literature on Gaussian multivariate analysis is applicable [Muirhead 1982]. Derivation of the Nakagami distribution, and many of its statistical properties, can be found in [Nakagami 1960]; a physical interpretation is given in [Braun and Dersch 1991]; the Rice distribution is derived in [Rice 1947]. The Suzuki distribution is derived in [Suzuki 1977]. More details about the lognormal distribution can be found in [Stueber 2017], [Cardieri and Rappaport 2001]. [Mehta et al. 2007] present a matching of the moment-generating function that can be used to compute the sum of both uncorrelated and correlated lognormals, as well as sum of Suzuki variables. The combination of Nakagami small-scale fading and lognormal shadowing is treated in [Tjhung and Chai 1999]. The fading statistics, including ADF and LCR, of the Nakagami and Rice distributions are summarized in [Abdi et al. 2000]. Another important aspect of statistical channel descriptions is the generation of random variables according to a prescribed Doppler spectrum. An extensive description of this area can be found in [Paetzold 2002].

For updates and errata for this chapter, see https://wides.usc.edu/students.html#textbooks

Exercises

See Sec. 36.5 of Exercises.pdf at wiley.com/go/molisch/wireless3e

6

Wideband and Directional Channel Characterization

6.1 Introduction

In the previous chapter, we considered the effect of multi-path propagation on the received field strength and the temporal variations if the transmit signal is a pure sinusoid. These considerations are also valid for all systems where the bandwidth of the transmitted signal is "very small" (see below for a more precise definition). However, most current and future wireless systems use a large bandwidth, either because they are intended for high data rates or because of their multiple access scheme (see Chapters 14, 15, 18, and 19). We therefore have to describe variations of the channel over a large bandwidth as well. Furthermore, for characterization of multi-antenna systems (Chapters 12, 16, 22), we need to understand the variations of the channel as a function of space or direction. All of these description methods are the topic of the current chapter.

The impact of multi-path propagation in wideband systems can be interpreted in two different ways: (i) the transfer function of the channel varies over the bandwidth of interest (this is called the *frequency selectivity* of the channel) or (ii) the impulse response of the channel is not a delta function; in other words, the arriving signal has a longer duration than the transmitted signal (this is called *delay dispersion*). These two interpretations are equivalent, as can be shown by performing Fourier transformations between the delay domain and the frequency domain.

In this chapter, we first explain the basic concepts of wideband channels using again the simplest channel – namely, the two-path channel. We then formulate a general statistical description methods for wideband, time-variant channels (Section 6.3), and discuss its most common special form, the WSSUS – Wide Sense Stationary Uncorrelated Scatterer – model (Section 6.4). Since these description methods are rather complicated, condensed parameters are often used for a more compact description (Section 6.5). Section 6.6 considers the case when the channel is not only wide enough to show appreciable variations of the transfer function but even so wide that the bandwidth becomes comparable with the carrier frequency – this case is called "ultra wideband."

Systems operating in wideband channels have some important properties:

- They suffer from InterSymbol Interference (ISI). This can be most easily understood from the interpretation of delay dispersion. If we transmit a symbol of length T_S, the arriving signal corresponding to that symbol has a longer duration, and therefore interferes with the subsequent symbol. This effect was briefly discussed in Chapter 2, for more details see Section 11.3, which describes the effect of this ISI on the Bit Error Rate (BER) if no further measures are taken; Chapter 14 describes equalizer structures that can actively combat the detrimental effect of the ISI and even turn it into an advantage.
- They can reduce the detrimental effect of fading. This effect can be most easily understood in the frequency domain: even if some part of the transmit spectrum is strongly attenuated, there are other frequencies that do not suffer from attenuation. Appropriate coding and signal processing can exploit this fact, as explained in Section 13.10 and Chapters 14, 15, and 19.

The properties of the channel can vary not only depending on the frequency at which we consider it but also depending on the location. This latter effect is related to the directional properties of the channel – i.e., the directions from which the Multi-Path Components (MPCs) are incident. Section 6.7 discusses the stochastic description methods for these directional properties; they are especially important for antenna diversity (Chapter 12) and multi-element antennas (Chapters 16 and 22).

Wireless Communications: From Fundamentals to Beyond 5G, Third Edition. Andreas F. Molisch.
© 2023 John Wiley & Sons Ltd. Published 2023 by John Wiley & Sons Ltd.
Companion website: www.wiley.com/go/molisch/wireless3e

6.2 The Causes of Delay Dispersion

6.2.1 The Two-Path Model

Why does a channel exhibit delay dispersion – or, equivalently, why are there variations of the channel over a given frequency range? The simplest picture arises again from the two-path model, as introduced in Sec. 5.2. The transmit signal gets to the Receiver (RX) via two different propagation paths with different runtimes:

$$\tau_1 = d_1/c_0 \quad \text{and} \quad \tau_2 = d_2/c_0. \tag{6.1}$$

We assume now that runtimes and amplitudes do not change with time (this occurs when neither Transmitter (TX), RX, nor Interacting Objects (IOs) move. Consequently, the channel is linear and time invariant, and has an impulse response (in complex baseband representation, compare Section 10.1.3):

$$h(\tau) = a_1\delta(\tau - \tau_1) + a_2\delta(\tau - \tau_2) \tag{6.2}$$

where again the complex amplitude $a = |a| \exp(j\varphi)$. Clearly, such a channel exhibits delay dispersion; the support (duration) of the impulse response has a finite extent, namely $\tau_2 - \tau_1$.

A Fourier transformation of the impulse response gives the transfer function $H(j\omega)$:

$$H(f) = \int_{-\infty}^{\infty} h(\tau) \exp\left[-j2\pi f\tau\right]d\tau = a_1 \exp\left[-j2\pi f\tau_1\right] + a_2 \exp\left[-j2\pi f\tau_2\right]. \tag{6.3}$$

The magnitude of the transfer function is

$$|H(f)| = \sqrt{|a_1|^2 + |a_2|^2 + 2|a_1||a_2|\cos\left(2\pi f\cdot\Delta\tau - \Delta\varphi\right)} \tag{6.4}$$
$$\text{with } \Delta\tau = \tau_2 - \tau_1 \text{ and } \Delta\varphi = \varphi_2 - \varphi_1.$$

Figure 6.1 shows the transfer function for a typical case. We observe first that the transfer function depends on the frequency, so that we have *frequency-selective fading*. We also see that there are dips (*notches*) in the transfer function at the so-called *notch frequencies*. In the two-path model, the notch frequencies are those frequencies where the phase difference of the two arriving waves becomes 180°. The frequency difference between two adjacent notch frequencies is

$$\Delta f_{\text{Notch}} = \frac{1}{\Delta\tau}. \tag{6.5}$$

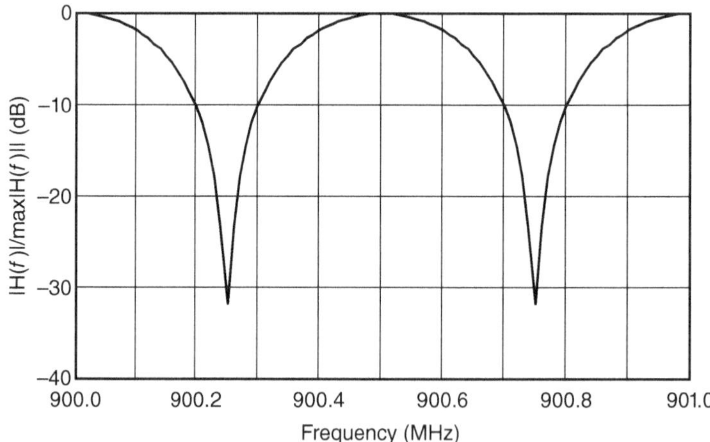

Figure 6.1 Normalized transfer function for $|a_1| = 1.0$, $|a_2| = 0.95$, $\Delta\varphi = 0$, $\tau_1 = 4\,\mu s$, $\tau_2 = 6\,\mu s$ at the 900-MHz carrier frequency.

The destructive interference between the two waves is the stronger the more similar the amplitudes of the two waves are.

Channels with fading dips distort not only the amplitude but also the phase of the signal. This can be best seen by considering the group delay, which is defined as the derivative of the phase of the channel transfer function $\phi_{\text{H}} = \arg(H(f))$:

$$\tau_{\text{Gr}} = -\frac{1}{2\pi}\frac{d\phi_{\text{H}}}{df}. \tag{6.6}$$

As can be seen in Figure 6.2, group delay can become very large in fading dips. As we will see later, this group delay can be related to ISI.

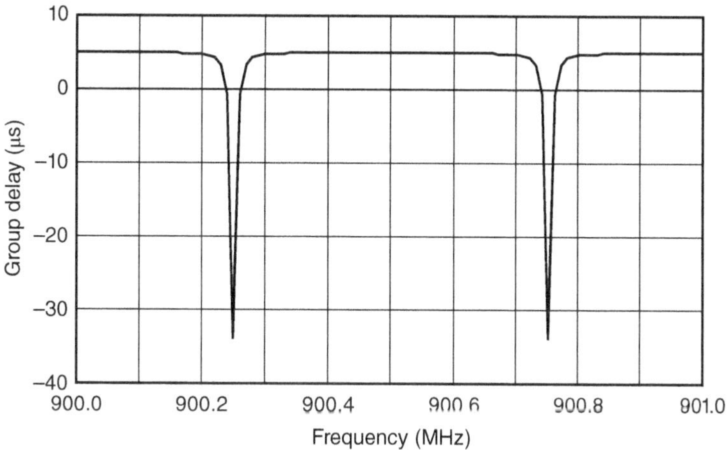

Figure 6.2   Group delay as a function of frequency (same parameters as in Figure 6.1).

6.2.2   The General Case

After the simple two-path model, we now progress to the more general case where IOs can be at any place in the plane. Again, the scenario is static, so that neither TX, RX, nor IOs move.

First, consider the situation in the frequency domain. At a given frequency the (complex) MPCs add up coherently, i.e., according to their phases. For the case that there are many MPCs and none of them are dominant, the transfer function can be interpreted as a realization of Rayleigh fading, adding up the different MPC amplitudes with random complex weights

$$H(f) = \sum_{\ell=1}^{N} | a_\ell | \, e^{j\varphi_\ell} e^{-j2\pi f \tau_\ell}. \tag{6.7}$$

From this equation, we can also see that as the frequency changes, the complex weights for the addition changes, so that different frequencies result in different complex transfer functions. Thus, the variations of the transfer function over frequency are mathematically similar to the variation of the transfer function of a narrowband signal over location (time): there, the different complex weights arise from the phase shifts tied to different Doppler shifts of the MPCs at the User Equipment (UE), while for frequency selectivity the phase shifts are tied to the different delays of the MPCs. This duality between Doppler shift and delay shift will recur in the discussions of this chapter. A phase shift of 2π occurs for the ℓth contribution when the frequency changes by $1/\tau_\ell$.

We now draw the ellipses that are defined by their focal points – TX and RX – and the eccentricity determining the runtime.[1] All rays that undergo a single interaction with an object on a specific ellipse arrive at the RX at the same time. Signals that interact with objects on different ellipses arrive at different times. Thus, the channels are *delay-dispersive* if the IOs in the environment are not all located on a single ellipse.

It is immediately obvious that in a realistic environment, IOs never lie exactly on a single ellipse. The next question is thus: How strict must this "single ellipse" condition be fulfilled so that the channel is still "effectively" nondispersive? The answer depends on the system bandwidth. An RX with bandwidth W cannot distinguish between echoes arriving at τ and $\tau + \Delta\tau$, if $\Delta\tau \ll 1/W$ (for many qualitative considerations, it is sufficient to consider the above condition with $\Delta\tau = 1/W$). Thus echoes that are reflected in the donut-shaped region corresponding to runtimes between τ and $(\tau + \Delta\tau)$ arrive at "effectively" the same time (see Figure 6.3).

A time-discrete approximation to the impulse response of a wideband channel can thus be obtained by dividing the impulse response into bins of width $\Delta\tau$ and then computing the sum of echoes within each bin. If enough nondominant IOs are in each donut-shaped region, then the MPCs falling into each delay bin fulfill the central limit theorem. In that case, the amplitude of each bin can be described statistically, and the probability density function (pdf) of this amplitude is Rayleigh. Thus, all the equations of Chapter 5 are still valid; but now they apply for the field strength *within* one delay bin. We furthermore define the minimum delay as the runtime of the direct path between the Base Station (BS) and the UE d/c_0 and we define the maximum delay as the runtime from the BS to the UE via the farthest "significant" IO – i.e., the farthest IO that gives a measurable contribution to the impulse response.[2] The *maximum excess delay* $\tau_{\max}$ is then defined as the difference between minimum and maximum delay.

---

[1] These ellipses are thus quite similar to the Fresnel ellipses described in Chapter 4. The difference is that in Chapter 4 we were interested in excess runtimes that introduce a phase shift of $i \bullet \pi$, while here we are interested in delays that are typically much larger.
[2] We see from this definition that the maximum delay is a quantity that is extremely difficult to measure, and depends on the measurement system.

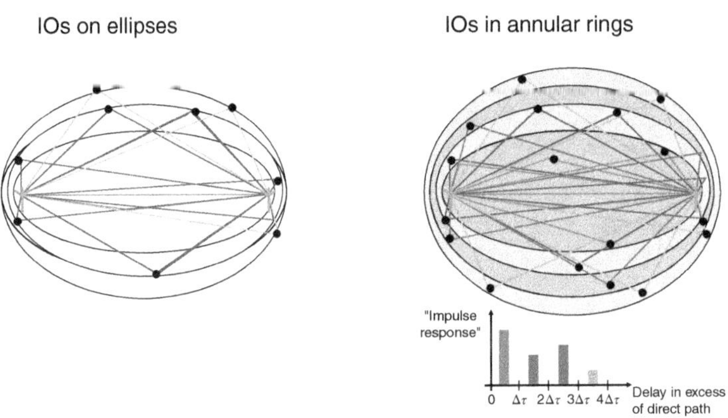

Figure 6.3 Scatterers located on the same ellipses lead to the same delays.

The above considerations also lead us to a mathematical formulation for *narrowband* and *wideband* from a time domain point of view: a system is narrowband if the inverse of the system bandwidth $1/W$ is much larger than the maximum excess delay $\tau_{\max}$. In that case, all echoes fall into a single delay bin, and the amplitude of this delay bin is $\alpha(t)$. A system is wideband in all other cases. In a wideband system, the *shape* and duration of the arriving signal are different from the shape of the transmitted signal; in a narrowband system, they stay the same.

If the impulse response has a finite extent in the delay domain, it follows from the theory of Fourier Transforms (FTs) that the transfer function $\mathcal{F}\{h(\tau)\} = H(f)$ is frequency dependent. Delay dispersion is thus equivalent to *frequency selectivity*. A frequency-selective channel cannot be described by a simple attenuation coefficient, but rather the details of the transfer function must be modeled. Note that any real channel is frequency selective if analyzed over a large enough bandwidth; in practice, the question is whether this is true over the bandwidth of the considered system. This is equivalent to comparing the maximum excess delay of the channel impulse response with the inverse system bandwidth. Figure 6.4 sketches these relationships, demonstrating the variations of wideband systems in the delay and frequency domain.

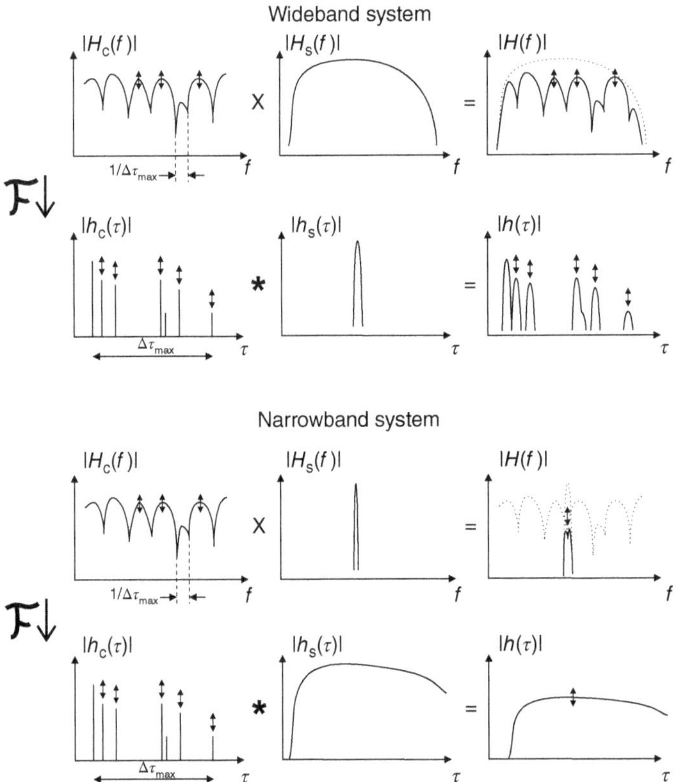

Figure 6.4 Narrowband and wideband systems. $H_{\rm C}(f)$, channel transfer function; $h_{\rm C}(\tau)$, channel impulse response. $h_{\rm s}(\tau)$ describes the "system impulse response," incorporating their "fundamental pulse shape" that we wish to transmit, as well as the effects of filters, etc.

We stress that the definition of a wideband wireless system is fundamentally different from the definition of "wideband" in the usage of Radio Frequency (RF) engineers. The RF definition of wideband implies that the system bandwidth becomes comparable with carrier frequency.[3] In wireless communications, on the other hand, we compare the properties of the *channel* with the properties of the system. It is thus possible that the same system is wideband in one channel, but narrowband for another.

6.3 System-Theoretic Description of Wireless Channels

As we have seen in the previous section, a wireless channel can be described by an impulse response; it thus can be interpreted as a linear filter. If the BS, UE, and IOs are all static, then the channel is time invariant, with an impulse response $h(\tau)$. In that case, the well-known theory of *Linear Time-Invariant* (LTI) systems [Oppenheim et al. 2015] is applicable. In general, however, wireless channels are time variant, with an impulse response $h(t, \tau)$ that changes with time; we have to distinguish between the absolute time t and the delay τ. Thus, the theory of the *Linear Time-Variant* (LTV) system must be used. This is not just a trivial extension of the LTI theory, but gives rise to considerable theoretical challenges and causes the breakdown of many intuitive concepts. Fortunately, most wireless channels can be classified as *slowly* time-variant systems, also known as *quasi-static*. In that case, many of the concepts of LTI systems can be retained with only minor modifications.

6.3.1 Characterization of Deterministic Linear Time-Variant Systems

As the impulse response of a time-variant system, $h(t, \tau)$, depends on two variables, τ and t, we can perform Fourier transformations with respect to either (or both) of them. This results in four different, but equivalent, representations. In this section, we investigate these representations, their advantages, and drawbacks.

From a system-theoretic point of view, it is most straightforward to write the relationship between the system input (transmit signal) $x(t)$ and the system output (received signal) $y(t)$ as:

$$y(t) = \int_{-\infty}^{\infty} x(\tau)K(t, \tau)d\tau \tag{6.8}$$

where $K(t, \tau)$ is the *kernel* of the integral equation, which can be related to the impulse response. For LTI systems, the well-known relationship $K(t, \tau) = h(t - \tau)$ holds. Generally, we define the time-variant impulse response as:

$$h(t, \tau) = K(t, t - \tau) \tag{6.9}$$

so that

$$y(t) = \int_{-\infty}^{\infty} x(t - \tau)h(t, \tau)d\tau. \tag{6.10}$$

An intuitive interpretation is possible if the impulse response changes only slowly with time – more exactly, the duration of the impulse response (and the signal) should be much shorter than the time over which the channel changes significantly. Then we can consider the behavior of the system at one time t like that of an LTI system. The variable t can thus be viewed as "absolute" time that parameterizes the impulse response, i.e., tells us which (out of a large ensemble) impulse response $h(\tau)$ is currently valid. Such a system is also called *quasi-static*.

Fourier transforming the impulse response with respect to the variable τ results in the *time-variant transfer function H* (t, f):

$$H(t, f) = \int_{-\infty}^{\infty} h(t, \tau) \exp\left(-j2\pi f\tau\right)d\tau. \tag{6.11}$$

The input–output relationship is given by:

$$y(t) = \int_{-\infty}^{\infty} X(f)H(t, f) \exp\left(j2\pi ft\right)df. \tag{6.12}$$

The interpretation is straightforward for the case of the quasi-static system – the spectrum of the input signal is multiplied by the spectrum of the "currently valid" transfer function, to give the spectrum of the output signal. If, however, the channel is quickly time varying, then Eq. (6.12) is a purely mathematical relationship. The spectrum of the output signal is given by a double integral

---

[3] In wireless communications, it has become common to denote systems whose bandwidth is larger than 20% of the carrier frequency as *Ultra Wide Bandwidth* (UWB) systems (see Section 6.6). This definition is similar to the RF definition of wideband.

$$Y\left(\tilde{f}\right) = \int_{-\infty}^{\infty}\int_{-\infty}^{\infty} X(f)H(t,f)\exp\left(j2\pi ft\right)\exp\left(-j2\pi\tilde{f}t\right)df\,dt \tag{6.13}$$

which in that case does *not* reduce to $Y(f) = H(f)X(f)$ [Matz and Hlawatsch 1998].

A Fourier transformation of the impulse response with respect to t results in a different representation – namely, the Doppler-variant impulse response, better known as *spreading function* $s(\nu, \tau)$:

$$s(\nu,\tau) = \int_{-\infty}^{\infty} h(t,\tau)\exp\left(-j2\pi\nu t\right)dt. \tag{6.14}$$

This function describes the spreading of the input signal in the delay and Doppler domains.

Finally, the function $s(\nu, \tau)$ can be transformed with respect to the variable τ, resulting in the *Doppler-variant transfer function* $B(\nu, f)$:

$$B(\nu,f) = \int_{-\infty}^{\infty} s(\nu,\tau)\exp\left(-j2\pi f\tau\right)d\tau. \tag{6.15}$$

A summary of the interrelations between the system functions is given in Figure 6.5. Figure 6.6 shows an example of a measured impulse response; Figure 6.7 shows the spreading function computed from it.

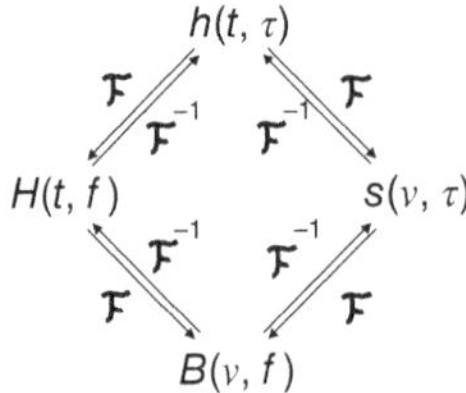

Figure 6.5 Interrelation between deterministic system functions.

6.3.2 Stochastic System Functions

We now return to the stochastic description of wireless channels. Interpreting them as time-variant stochastic systems, a complete description requires the multi-dimensional pdf of the impulse response – i.e., the joint pdf of the complex amplitudes at all possible values of delay and time. However, this is usually much too complicated in practice. Instead, we restrict our attention to a second-order description – namely, the *AutoCorrelation Function* (ACF).

Let us first repeat some facts about the ACFs of one-dimensional stochastic processes (i.e., processes that depend on a single parameter t). The ACF of a stochastic process y is defined as:

$$R_{yy}(t,t') = E\{y^*(t)y(t')\} \tag{6.16}$$

where the expectation is taken over the *ensemble of possible realizations* (for the definition of this ensemble see Appendix 6.A). The ACF describes the relationship between the second-order moments of the amplitude pdf of the signal y at different times. If the pdf is zero-mean Gaussian, then the second-order description contains all the required information. If the pdf is non-zero-mean Gaussian, the mean

$$\bar{y}(t) = E\{y(t)\} \tag{6.17}$$

together with the autocovariance function

$$\tilde{R}_{yy}(t,t') = E\{[y(t)-\bar{y}(t)]^*[y(t')-\bar{y}(t')]\} \tag{6.18}$$

constitutes a complete description. If the channel is non-Gaussian, then the first- and second-order statistics are not a complete description of the channel. In the following, we mainly concentrate on zero-mean Gaussian channels.

Let us now revert to the problem of giving a stochastic description of the channel. Inserting the input–output relationship into Eq. (6.16), we obtain the following expression for the ACF of the received signal:

$$R_{yy}(t,t') = E\left\{\int_{-\infty}^{\infty} x^*(t-\tau)h^*(t,\tau)d\tau \int_{-\infty}^{\infty} x(t'-\tau')h(t',\tau')d\tau'\right\}. \tag{6.19}$$

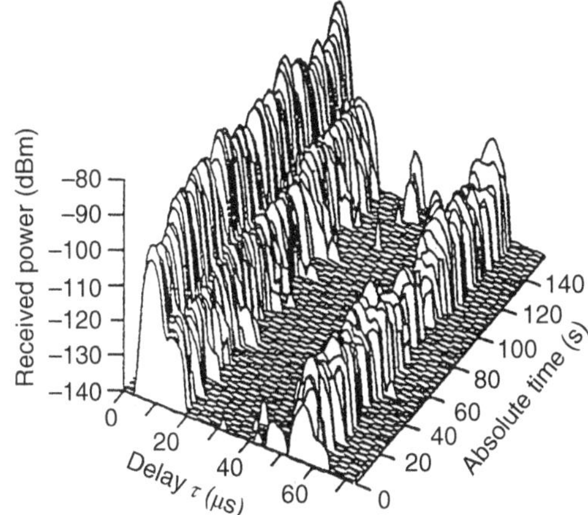

Figure 6.6 Squared magnitude of the impulse response $|h(t, \tau)|^2$ measured in hilly terrain near Darmstadt, Germany. Measurement duration 140 seconds; center frequency 900 MHz. τ denotes the excess delay.
Reproduced with permission from [Liebenow and Kuhlmann 1993] © U. Liebenow.

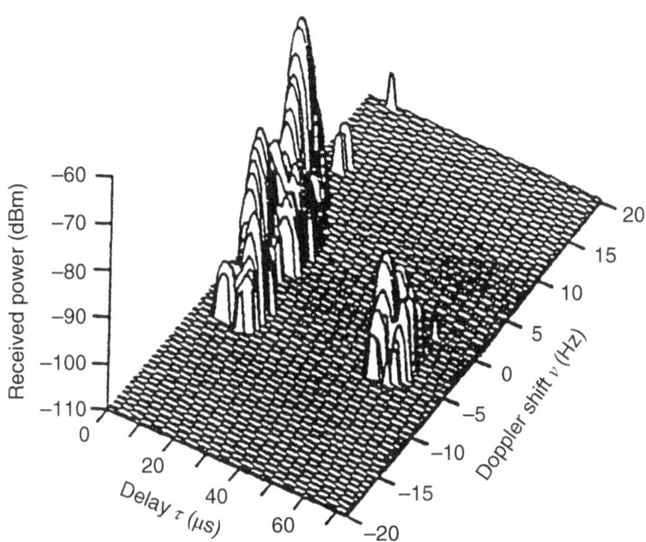

Figure 6.7 Squared magnitude of spreading function computed from the data of Figure 6.6.
Reproduced with permission from [Liebenow and Kuhlmann 1993] © U. Liebenow.

The system is linear, so that expectation can be interchanged with integration. Furthermore, the transmit signal can be interpreted as a stochastic process that is independent of the channel, so that expectations over the transmit signal and over the channel can be performed independently. Thus, the ACF of the received signal is given by:

$$R_{yy}(t, t') = \int_{-\infty}^{\infty} \int_{-\infty}^{\infty} E\{x^*(t-\tau)x(t'-\tau')\}E\{h^*(t,\tau)h(t',\tau')\}d\tau d\tau'$$

$$= \int_{-\infty}^{\infty} \int_{-\infty}^{\infty} R_{xx}(t-\tau, t'-\tau')R_{\mathrm{h}}(t,t',\tau,\tau')d\tau d\tau' \tag{6.20}$$

i.e., a combination of the ACF of the transmit signal and the ACF of the channel:

$$R_{\mathrm{h}}(t, t', \tau, \tau') = E\{h^*(t, \tau)h(t', \tau')\}. \tag{6.21}$$

Note that the ACF of the channel depends on four variables since the underlying stochastic process is two-dimensional.

We observe a formal similarity of the channel ACF to the impulse response of a deterministic channel: we can form stochastic system functions by Fourier transformations. In contrast to the deterministic case, we now have to perform a *double Fourier transformation*, with respect to the pair of variables t, t' and/or τ, τ'. From that, we obtain in an elementary way the relationships between the different formulations of the ACFs of input and output – e.g., $R_s(\nu, \nu', \tau, \tau') = E\{s^*(\nu, \tau)s(\nu', \tau')\} = \iint R_h(t, t', \tau, \tau') \, exp(+j2\pi\nu t) \, exp(-j2\pi\nu't')dt \, dt'$.

6.4 The WSSUS Model

The correlation functions depend on four variables and are thus a rather complicated form for the characterization of the channel. Further assumptions about the physics of the channel can lead to a simplification of the correlation function. The most frequently used assumptions are the so-called *Wide-Sense Stationary* (WSS) assumption and the *Uncorrelated Scatterers* (US) assumption. A model using both assumptions simultaneously is called a *WSSUS* model.

6.4.1 Wide-Sense Stationarity

The mathematical definition of *wide-sense stationarity* is that the ACF depends not on the two variables t, t' separately, but only on their difference $t - t'$. Consequently, *second-order amplitude statistics* do not change with time.[4] We can thus write

$$R_{\rm h}(t, t', \tau, \tau') = R_{\rm h}(t, t + \Delta t, \tau, \tau') = R_{\rm h}(\Delta t, \tau, \tau'). \tag{6.22}$$

Physically speaking, WSS means that the *statistical properties* of the channel do not change with time. This must not be confused with a static channel, where fading *realizations* do not change with time. For the simple case of a flat Rayleigh-fading channel, WSS means that the mean power and the ACF do not change with time, while the instantaneous amplitude can change.

According to the mathematical definition, WSS has to be fulfilled for any arbitrary time, t. In practice, this is not possible: as the UE moves over larger distances, the mean received power changes because of shadowing and variations in path loss. Rather, WSS is typically fulfilled over an area of about 10λ diameter (compare also Section 5.1). We can thus define quasi-stationarity over a stationarity time, i.e., a finite time interval (associated with a movement distance of the UE) over which statistics do not change noticeably.

WSS also implies that components with different Doppler shifts undergo uncorrelated fading. This can be shown by considering the Doppler-variant impulse response $s(\nu, \tau)$. Inserting Eq. (6.22) into the definition of $R_{\rm s}$, we get

$$R_{\rm s}(\nu, \nu', \tau, \tau') = \int_{-\infty}^{\infty} \int_{-\infty}^{\infty} R_{\rm h}(t, t + \Delta t, \tau, \tau') \exp[2\pi j(\nu t - \nu'(t + \Delta t))] dt d\Delta t \tag{6.23}$$

which can be rewritten as:

$$R_{\rm s}(\nu, \nu', \tau, \tau') = \int_{-\infty}^{\infty} \exp[2\pi jt(\nu - \nu')] dt \int_{-\infty}^{\infty} R_h(\Delta t, \tau, \tau') \exp[-2\pi j\nu' \Delta t] d\Delta t. \tag{6.24}$$

The first integral is an integral representation of the delta function $\delta(\nu - \nu')$. Thus, $R_{\rm s}$ can be factored as:

$$R_{\rm s}(\nu, \nu', \tau, \tau') = \tilde{\tilde{P}}_{\rm s}(\nu, \tau, \tau')\delta(\nu - \nu'). \tag{6.25}$$

This implies that contributions undergo uncorrelated fading if they have different Doppler shifts. The function $\tilde{\tilde{P}}_{\rm s}()$, which is implicitly defined by Eq. (6.25), is discussed below in more detail.

Analogously, we can write $R_{\rm B}$ as:

$$R_{\rm B}(\nu, \nu', f, f') = P_{\rm B}(\nu, f, f')\delta(\nu - \nu'). \tag{6.26}$$

6.4.2 Uncorrelated Scatterers

The US assumption is defined as "contributions with different delays are uncorrelated," which is written mathematically as:

$$R_{\rm h}(t, t', \tau, \tau') = P_{\rm h}(t, t', \tau)\delta(\tau - \tau') \tag{6.27}$$

or for $R_{\rm s}$ as:

$$R_{\rm s}(\nu, \nu', \tau, \tau') = \tilde{P}_{\rm s}(\nu, \nu', \tau)\delta(\tau - \tau'). \tag{6.28}$$

---

[4] Strict sense stationarity means that fading statistics of arbitrary order do not change with time. For Gaussian channels, WSS implies strict sense stationarity.

The US condition is fulfilled if the phase of an MPC does not contain any information about the phase of another MPC with a different delay. If scatterers are distributed randomly in space, phases change in an uncorrelated way even when the UE moves only a small distance.

For the transfer function, the US condition means that R_H does not depend on the absolute frequency, but only on the frequency difference[5]:

$$R_\mathrm{H}(t, t', f, f + \Delta f) = R_\mathrm{H}(t, t', \Delta f). \tag{6.29}$$

This implies that the frequency correlation function is independent of the carrier frequency, and only depends on the frequency difference Δf between the two considered frequencies. Again, it is obvious that this cannot hold for all frequencies – the correlation function at 100 MHz will be different from that at 100 GHz. We can thus define a *stationarity bandwidth* within which the US assumption is valid, while for larger changes in carrier frequency the assumption will not hold.

6.4.3 WSSUS Assumption

The US and WSS assumptions are duals: US defines contributions with different delays as uncorrelated, while WSS defines contributions with different Doppler shifts as uncorrelated. Alternatively, we can state that US means that R_H depends only on the frequency difference, while WSS means that R_H depends only on the time difference.

It is thus natural to combine these two definitions in the WSSUS condition, so that the ACF has to fulfill the following conditions:

$$R_\mathrm{h}(t, t + \Delta t, \tau, \tau') = P_\mathrm{h}(\Delta t, \tau)\delta(\tau - \tau') \tag{6.30}$$
$$R_\mathrm{H}(t, t + \Delta t, f, f + \Delta f) = R_\mathrm{H}(\Delta t, \Delta f) \tag{6.31}$$
$$R_\mathrm{s}(\nu, \nu', \tau, \tau') = P_\mathrm{s}(\nu, \tau)\delta(\nu - \nu')\delta(\tau - \tau') \tag{6.32}$$
$$R_\mathrm{B}(\nu, \nu', f, f + \Delta f) = P_\mathrm{B}(\nu, \Delta f)\delta(\nu - \nu'). \tag{6.33}$$

In contrast to the ACFs, which depend on four variables, the P-functions on the r.h.s. depend only on *two* variables. This greatly simplifies their formal description, parameterization, and application in further derivations. Because of their importance, they have been given distinct names. Following [Kattenbach 1997], we define

- $P_\mathrm{h}(\Delta t, \tau)$ as delay cross power spectral density;
- $R_\mathrm{H}(\Delta t, \Delta f)$ as time-frequency correlation function;
- $P_\mathrm{s}(\nu, \tau)$ as scattering function;
- $P_\mathrm{B}(\nu, \Delta f)$ as Doppler cross power spectral density.

The scattering function has special importance because it can be easily interpreted physically. If only single interactions occur, then each differential element of the scattering function corresponds to a physically existing IO. From the Doppler shift, we can determine the Direction Of Arrival (DoA); the delay determines the radii of the ellipse on which the scatterer lies.

The WSSUS assumption is very popular, but not always fulfilled in practice. Appendix 6.A gives a more detailed discussion of the assumptions and their validity.

6.4.4 Tapped Delay Line Models

A WSSUS channel can be represented as a tapped delay line, where the coefficients multiplying the output from each tap vary with time. The impulse response is then written as:

$$h(t, \tau) = \sum_{i=1}^{\widetilde{N}} c_i(t)\delta(\tau - \tau_i) \tag{6.34}$$

where $\widetilde{N}$ is now the number of taps, $c_i(t)$ are the time-dependent complex coefficients for the taps, and τ_i is the delay of the ith tap. For each tap, a Doppler spectrum determines the changes of the coefficients with time. This spectrum can be different for each tap, though many models assume the same spectrum for each tap, see also Chapter 7.

One interpretation of a tapped delay line is as a physical representation of the multi-path propagation in the channel. Each of the $\widetilde{N}$ components corresponds to one group of closely spaced MPCs: the model would be purely deterministic only if the arriving signals

---

[5] Proof is left as an exercise for the reader.

consisted of *completely resolvable* echoes from discrete IOs. However, in most practical cases, the resolution of the RX is not sufficient to resolve all MPCs. We thus write the impulse response as:

$$h(t, \tau) = \sum_{i=1}^{\widetilde{N}} \sum_{k(i)} a_{i,k(i)}(t)\delta(\tau - \tau_i) = \sum_{i=1}^{\widetilde{N}} c_i(t)\delta(\tau - \tau_i) \tag{6.35}$$

where the $k(i)$ index the MPCs in the ith delay bin centered around τ_i. Note that the second part of this equation makes sense in a band-limited system. In that case, each complex amplitude $c_i(t)$ represents the sum of several MPCs, which fades. WSSUS implies that all the taps are fading independently and that their average power does not depend on time.

Another interpretation of the tapped delay line is based on the sampling theorem. Any wireless system, and thus the channel we are interested in, is band limited. Therefore, the impulse response can be represented by a sampled version of the continuous impulse response $\widetilde{h}_{\mathrm{bl}}(t, \tau) = \sum c_i(t)\delta(\iota - \iota_i)$; similarly, the scattering function, correlation functions, etc., can be represented by their sampled versions. Commonly, the samples are spaced equidistantly, $\tau_i = i \cdot \Delta\widetilde{\tau}$, where the distance between the taps $\Delta\widetilde{\tau}$ is determined by the Nyquist theorem. The continuous version of the impulse response can be recovered by interpolation:

$$h_{\mathrm{bl}}(t, \tau) = \sum_i c_i(t) \, \mathrm{sinc}(\pi W(\tau - \tau_i)) \tag{6.36}$$

where W is the bandwidth and $\mathrm{sinc}(x) = \sin(x)/x$. Note that if the *physical* IOs fulfill the WSSUS condition, but are not equidistantly spaced, then the tap weights $c_i(t)$ are *not* necessarily WSSUS.

Many of the standard models for wireless channels (see Chapter 7) were developed with a specific system and thus a specific system bandwidth in mind. It is often necessary to adjust the tap locations to a different sampling grid for a discrete simulation: in other words, a discrete simulation requires a channel representation $h(t, \tau) = \sum \widetilde{c}_i(t)\delta(\tau - iT_s)$, but $\Delta\widetilde{\tau}/T_s$ is a noninteger. This can be done by using the interpolation formula – i.e., resampling $h_{\mathrm{bl}}(t, \tau)$ in Eq. (6.36) at the desired rate. Alternatively, we can describe the channel in the frequency domain and transform it back (with a discrete FT) with the desired tap spacing. Note, however, that a channel model based on measurements with a certain bandwidth W should not be extended to a larger bandwidth.

6.4.5   Interpretation and Limits of the WSSUS Assumption

The WSSUS assumption is in widespread use and also has important physical considerations. Remember that it states that the second-order fading statistics are independent of absolute time and of absolute frequency. Define furthermore the statistical ensemble as different realizations of the small-scale fading (see also the discussion in Appendix 6.A). This means first of all that – as a UE moves around – the signal power averaged over the ensemble does not change. This assumption is violated as soon as the realization of the shadowing changes significantly (note that shadowing by moving objects, such as trucks blocking certain MPCs, also leads to violation of WSS). Similarly, the US assumption requires that the small-scale-averaged RX power does not change over frequency; this assumption is violated when, e.g., the antenna gain or even just the antenna pattern changes significantly. But the constraints of WSSUS go further: they require, e.g., that the Power Delay Profile (PDP) does not change over time, which implies that the delays of the MPCs must not change significantly (i.e., more than a fraction of the resolvable delay bin) as the UE moves.

We can thus define a *stationarity region*, i.e., an area within which the WSS assumption is fulfilled. Note that this is fundamentally different from the *coherence region*, i.e., the area within which the impulse response is constant. Typically, the size of the coherence region is on the order of a wavelength (due to small-scale fading), while the stationarity region is determined by the shadowing correlation length, which is tens of meter or more.[6] Similarly, the *stationarity bandwidth* is the bandwidth within which the US assumption is fulfilled, and can be much larger than the coherence bandwidth we will discuss below. Ultrawideband systems with large relative bandwidth, as discussed in Section 6.6, generally have a bandwidth larger than the stationarity bandwidth and thus violate the US assumption.

The concept of stationarity region is important because many wireless transceiver algorithms need to know the second-order statistics, and thus have to estimate them. The stationarity region/time thus characterizes how often those second-order statistics have to be re-estimated.

6.5   Condensed Parameters

The correlation functions are a somewhat cumbersome way of describing wireless channels. Even when the WSSUS assumption is used, they are still *functions* of two variables. A preferable representation would be a function of one variable, or even better, just a single parameter. Obviously, such a representation implies a serious loss of information, but this is a sometimes acceptable price for a compact representation.

---

[6] Different criteria for stationarity, i.e., average power, or no change in the power delay profile, might result in different sizes of the stationarity region. Note that the distance over which the PDP changes significantly decreases with increasing bandwidth.

6.5.1 Integrals of the Correlation Functions

A straightforward way of getting from two variables to one is to integrate over one of them. Integrating the scattering function over the Doppler shift ν gives the *delay power spectral density* $P_h(\tau)$, more popularly known as the Power Delay Profile (PDP). The PDP contains information about how much power (from a transmitted delta pulse with unit energy) arrives at the RX with a delay between $(\tau, \tau + d\tau)$, irrespective of a possible Doppler shift. The PDP can be obtained from the complex impulse responses $h(t, \tau)$ as:

$$P_h(\tau) = \lim_{T \to \infty} \frac{1}{2T} \int_{-T}^{T} |h(t, \tau)|^2 dt \tag{6.37}$$

if ergodicity holds. Note that in practice the integral will not extend over infinite time, but rather over the time span during which quasi-stationarity is valid (see above).

Analogously, integrating the scattering function over τ results in the *Doppler power spectral density* $P_B(\nu)$.

The *frequency correlation function* can be obtained from the time-frequency correlation function by setting $\Delta t = 0$ – i.e., $R_H(\Delta f) = R_H(0, \Delta f)$. It is noteworthy that this frequency correlation function is the FT of the PDP. The *temporal correlation function* $R_H(\Delta t) = R_H(\Delta t, 0)$ is the inverse FT of the Doppler power spectral density.

6.5.2 Moments of the Power Delay Profile

The PDP is a *function*, but for obtaining a quick overview of measurement results, it is preferable to have each measurement campaign described by a single *parameter*. While there are a large number of possible parameters, normalized moments of the PDP are the most popular.

We start out by computing the zeroth-order moment – i.e., delay-integrated power:

$$P_m = \int_{-\infty}^{\infty} P_h(\tau) d\tau. \tag{6.38}$$

The normalized first-order moment, the *mean delay* is given by:

$$T_m = \frac{\int_{-\infty}^{\infty} P_h(\tau) \tau d\tau}{P_m}. \tag{6.39}$$

The square root of the normalized second-order central moment is known as *rms delay spread* and is defined as:

$$S_\tau = \sqrt{\frac{\int_{-\infty}^{\infty} P_h(\tau) \tau^2 d\tau}{P_m} - T_m^2}. \tag{6.40}$$

The rms delay spread has obtained a special stature among all parameters. It has been shown that under some specific circumstances, the bit error probability due to delay dispersion is proportional to the rms delay spread only (see Section 11.3), while the actual *shape* of the PDP does not have a significant influence. In that case, S_τ is all we need to know about the channel. It cannot be stressed enough, however, that this is true only under specific circumstances, and that in other cases the rms delay spread might not have a clear quantitative relationship with system behavior. It is also noteworthy that S_τ does not attain finite values for all physically reasonable signals. A channel with $P_h(\tau) \propto 1/(1 + \tau^2)$ for $\tau > 0$ is physically possible, and does not contradict energy conservation, but $\int_{-\infty}^{\infty} P_h(\tau) \tau^2 d\tau$ does not converge.

Example 6.1 *Compute the rms delay spread of a two-spike profile*

$$P_h(\tau) = \delta(\tau - 10\,\mu s) + 0.3\delta(\tau - 17\,\mu s).$$

The time-integrated power is given by Eq. (6.38):

$$P_m = \int_{-\infty}^{\infty} \left(\delta(\tau - 10^{-5}) + 0.3\delta(\tau - 1.7 \cdot 10^{-5}) \right) d\tau \tag{6.41}$$

$$= 1.30$$

and the mean delay is given by Eq. (6.39):

$$T_{\mathrm{m}} = \int_{-\infty}^{\infty} \left(\delta(\tau - 10^{-5}) + 0.3\delta(\tau - 1.7 \cdot 10^{-5}) \right) \tau d\tau / P_{\mathrm{m}}$$
$$= \left(10^{-5} + 0.3 \cdot 1.7 \cdot 10^{-5} \right) / 1.3 = 1.16 \cdot 10^{-5} \mathrm{s}. \tag{6.42}$$

Finally, the rms delay spread is computed according to Eq. (6.40):

$$S_{\tau} = \sqrt{ \frac{\int_{-\infty}^{\infty} \left(\delta(\tau - 10^{-5}) + 0.3\delta(\tau - 1.7 \cdot 10^{-5}) \right) \tau^2 d\tau}{P_{\mathrm{m}}} - T_{\mathrm{m}}^2 }$$
$$= \sqrt{ \frac{(10^{-5})^2 + 0.3(1.7 \cdot 10^{-5})^2}{1.3} - (1.16 \cdot 10^{-5})^2 } = 3\,\mu\mathrm{s}. \tag{6.43}$$

6.5.3 Moments of the Doppler Spectra

Moments of the Doppler spectra can be computed in complete analogy to the moments of the PDP. The integrated power is

$$P_{\mathrm{B,m}} = \int_{-\infty}^{\infty} P_{\mathrm{B}}(\nu) d\nu \tag{6.44}$$

where obviously $P_{\mathrm{B,m}} = P_{\mathrm{m}}$. The mean Doppler shift is

$$\nu_{\mathrm{m}} = \frac{\int_{-\infty}^{\infty} P_{\mathrm{B}}(\nu)\nu d\nu}{P_{\mathrm{B,m}}}. \tag{6.45}$$

The rms Doppler spread is

$$S_{\nu} = \sqrt{ \frac{\int_{-\infty}^{\infty} P_{\mathrm{B}}(\nu)\nu^2 d\nu}{P_{\mathrm{B,m}}} - \nu_{\mathrm{m}}^2 }. \tag{6.46}$$

It is noteworthy that the rms Doppler spread for a Jakes spectrum, extending from $[-\nu_{\max}, \nu_{\max}]$ is $\nu_{\max}/\sqrt{2}$.

6.5.4 Coherence Bandwidth and Coherence Time

In a frequency-selective channel, different frequency components fade differently. Obviously, the correlation between fading at two different frequencies is the smaller the more these two frequencies are apart. The *coherence bandwidth* B_{coh} defines the frequency difference that is required so that the correlation coefficient is smaller than a given threshold.

A mathematically exact definition starts out with the frequency correlation function $R_{\mathrm{H}}(0, \Delta f)$, assuming WSSUS (see Figure 6.8 for examples). The coherence bandwidth can then be defined as:

$$B_{\mathrm{coh}} = \frac{1}{2} \left[\underset{\Delta f > 0}{\arg\max} \left(\frac{|R_{\mathrm{H}}(0, \Delta f|}{R_{\mathrm{H}}(0,0)} = 0.5 \right) - \underset{\Delta f < 0}{\arg\min} \left(\frac{|R_{\mathrm{H}}(0, \Delta f)|}{R_{\mathrm{H}}(0,0)} = 0.5 \right) \right]. \tag{6.47}$$

Since $R_{\mathrm{H}}(0, \Delta f)$ is the FT of a real function (the PDP), it is Hermitian symmetric, and its absolute value is an even function of f. Thus we can simplify the above to

$$B_{\mathrm{coh}} = \underset{\Delta f > 0}{\arg\max} \left(\frac{R_{\mathrm{H}}(0, \Delta f)}{R_{\mathrm{H}}(0,0)} = 0.5 \right).$$

This is essentially the half-width half-maximum bandwidth of the correlation function. The somewhat complicated formulation stems from the fact that the correlation function need not decay *monotonically*. Rather, there can be local maxima that exceed the threshold. A precise definition thus uses the bandwidth that encompasses all parts of the correlation function exceeding the threshold.[7]

---

[7] An alternative definition would define the coherence bandwidth as the second central moment of the correlation function; this would circumvent all problems with local maxima. Unfortunately, this second moment becomes infinite in practically important cases.

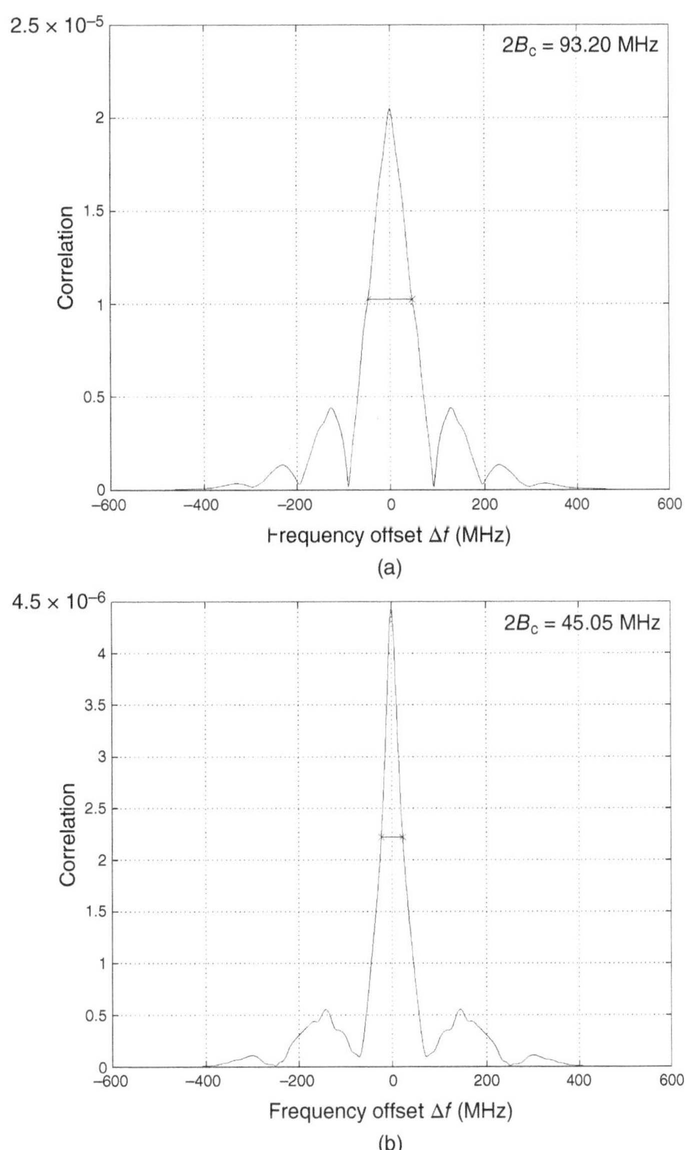

Figure 6.8 Typical frequency correlation function.
Reproduced with permission from [Kattenbach 1997] © Shaker Verlag.

The rms delay spread S_τ and the coherence bandwidth B_{coh} are obviously related: S_τ is derived from the PDP $P_h(\tau)$ while B_{coh} is obtained from the frequency correlation function, which is the FT of the PDP. Based on this insight, [Fleury 1996] derived an "uncertainty relationship":

$$B_{coh} \gtrsim \frac{1}{2\pi S_\tau}. \tag{6.48}$$

Equation (6.48) is an inequality and therefore *does not* offer the possibility to obtain one parameter from the other (though this is often done anyway). The question thus arises whether B_{coh} or S_τ better reflects the channel properties. An answer to that question can only be given for a specific system. For a Frequency Division Multiple Access (FDMA) or Time Division Multiple Access (TDMA) system without an equalizer, the rms delay spread is the quantity of interest, as it is related to the BER, though generally, it over-emphasizes long-delayed echoes. For Orthogonal Frequency Division Multiplexing (OFDM) systems (Chapter 15), where the information is transmitted on many parallel carriers, the coherence bandwidth is a better measure.

The temporal correlation function is a measure of how fast a channel changes. The definition of the coherence time T_{coh} is thus analogous to the coherence bandwidth; it also has an uncertain relationship with the rms Doppler spread.

Figure 6.9 summarizes the relationships between system functions, correlation functions, and special parameters. Ergodicity is assumed throughout this figure.

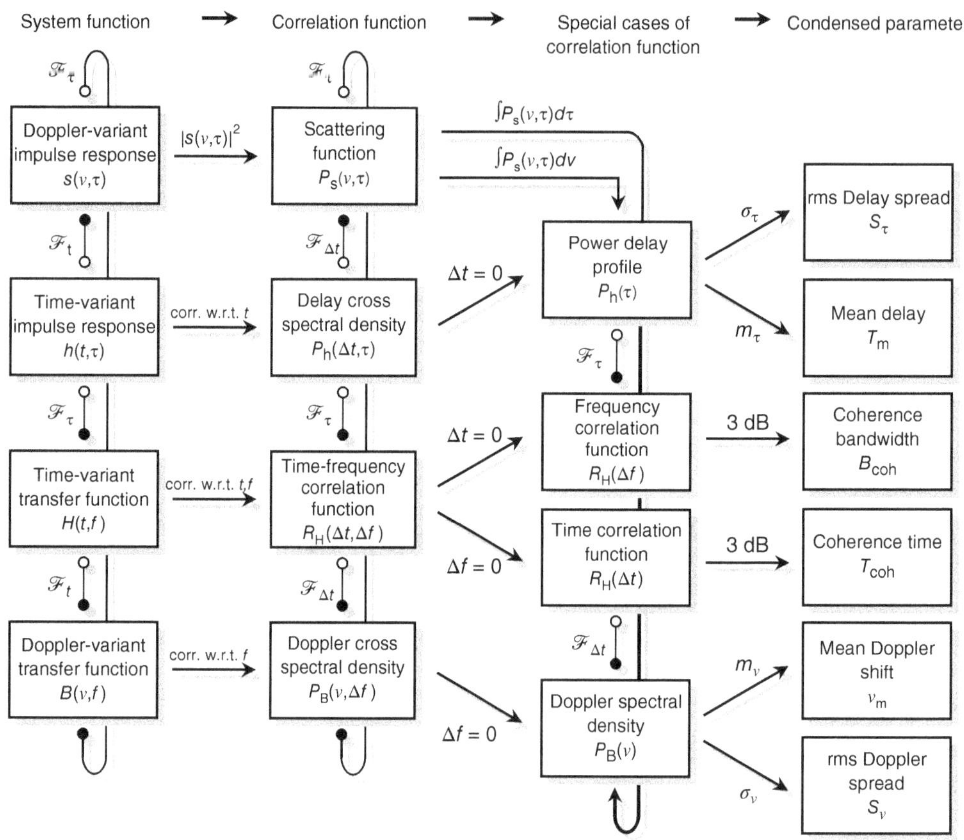

Figure 6.9 Relationships between system functions, correlation functions, and condensed parameters for ergodic channel impulse responses. The exact relationships for going from the first to the second column are given in Eqs. (6.30)–(6.33).
Reproduced with permission from [Kattenbach 1997] © Shaker Verlag.

*6.5.5 Window Parameters

Another useful set of parameters are the so-called *window parameters* [de Weck 1992], more precisely the *interference quotient* Q_T and the *delay window* W_Q. They are a measure for the percentage of energy of the average PDP arriving within a certain delay interval. In contrast to the delay spread and coherence bandwidth, the window parameters need to be defined in the context of specific systems.

The *interference quotient* Q_T is the ratio between the signal power arriving within a time window of duration T, relative to the power arriving outside that window. It characterizes the self-interference due to delay dispersion. If, e.g., a system has an equalizer that can process MPCs with a delay up to T, then every MCP within the window is "useful," while energy outside the window creates interference – these components carry information about bits that cannot be processed and thus act as independent interferers.[8] A similar statement holds for OFDM systems (Chapter 15), where signal echoes falling outside a system-specific guard interval (called the cyclic prefix) create interference.

A mathematical definition of the interference quotient is given as:

$$Q_T = \frac{\int_{t_0}^{t_0+T} P_h(\tau)d\tau}{P_m - \int_{t_0}^{t_0+T} P_h(\tau)d\tau}. \tag{6.49}$$

This quotient depends not only on the PDP and the duration T but also on the starting delay of the window t_0. This latter dependence is often eliminated by setting the starting delay to the minimum excess delay (i.e., the first MPC determines the start of the window) $t_0 = \tau_{min}$. Alternatively, the t_0 can be chosen to maximize Q_T:

$$Q_T = \max_{t_0} \left\{ \frac{\int_{t_0}^{t_0+T} P_h(\tau)d\tau}{P_m - \int_{t_0}^{t_0+T} P_h(\tau)d\tau} \right\}. \tag{6.50}$$

---

[8] The interpretation is only an approximate one. There is no sharp "jump" from "useful" to "interference" when the delay of an MPC exceeds the equalizer length. Rather, there is a smooth transition, similar to the effect of delay dispersion in unequalized systems, see Section 11.3.

This definition makes sense because an RX can often adapt equalizer timing to optimize performance.

A related parameter is the delay window W_Q (see Figure 6.10). This defines how long a window has to be so that the power within that window is a factor of Q larger than the power outside the window. The defining equations are the same as for the interference quotient. The difference is just that now T is considered as variable, and Q as fixed.

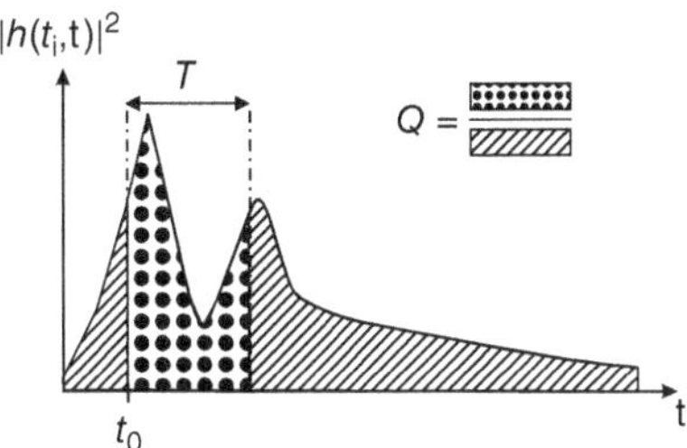

Figure 6.10 Definition of window parameters.
Reproduced with permission from [Molisch 2000] © Prentice Hall.

Example 6.2 *For an exponential PDP, $P_h(\tau) = \exp(-\tau/2\,\mu s)$, compute the delay window so that the interference quotient becomes 10 and 20 dB, respectively – i.e., so that 91% and 99% of the energy are contained in the window. Do the same of the two-spike profile $\delta(\tau - 10\,\mu s) + 0.3\delta(\tau - 17\,\mu s)$.*

Let the starting delay be equal to the minimum excess delay. For an exponential PDP, the interference quotient, given by Eq. (6.49), is

$$
Q_{\mathrm{T}} = \frac{\int_0^T e^{-\tau/2\cdot 10^{-6}}\,d\tau}{\int_0^\infty e^{-\tau/2\cdot 10^{-6}}\,d\tau - \int_0^T e^{-\tau/2\cdot 10^{-6}}\,d\tau}
$$

Solving for T yields

$$
T = 2\cdot 10^{-6}\ln\left(Q_{\mathrm{T}} + 1\right).
$$

and the 91% and 99% windows are thus $T_{91\%} = 4.8\,\mu s$ and $T_{99\%} = 9.2\,\mu s$, respectively. For the two-spike profile, the starting delay is $t_0 = 10^{-5}$ and the energy within the window is

$$
\int_{10^{-5}}^{T + 10^{-5}} \left(\delta\left(\tau - 10^{-5}\right) + 0.3\delta\left(\tau - 1.7\cdot 10^{-5}\right)\right)d\tau = \begin{cases} 1, & 0 < T < 7\,\mu s \\ 1.3, & T > 7\,\mu s \\ 0, & \text{otherwise} \end{cases}.
$$

Hence, the interference quotient is greater than 10 dB and/or 20 dB for $T > 7\,\mu s$.

*6.6 Ultra Wideband Channels

6.6.1 *UWB Signals with Large Relative Bandwidth*

The above models are wideband in the sense that they model the delay dispersion caused by multi-path propagation. However, they are still based on the following two assumptions.

1. The reflection, transmission, and diffraction coefficients of the IOs are constant over the considered bandwidth.
2. The relative bandwidth of the system (bandwidth divided by carrier frequency) is *much* smaller than unity.

Note that these conditions are met for the bandwidth of most currently used wireless systems. However, for certain applications, a technique called *Ultra Wide Band* (UWB) transmission (see also Section 26.8) has gained increased interest. UWB systems have a relative bandwidth of more than 20%. In that case, the different frequency components contained in the transmitted signal "see" different propagation environments. For example, the diffraction coefficient of a building corner is different at 100 MHz compared with 1 GHz; similarly, the reflection coefficients of walls and furniture can vary over the bandwidth of interest. A channel impulse response realization is then given by:

$$h(\tau) = \sum_{\ell=1}^{N} a_\ell \chi_\ell(\tau) * \delta(\tau - \tau_\ell) \tag{6.51}$$

where $\chi_\ell(\tau)$ denotes the distortion of the ℓth MPC by the frequency selectivity of IOs and $*$ denotes convolution. Expressions for these distortions are given, e.g., in [Qiu 2002]; one example for a distortion of a short pulse by diffraction by a screen is shown in Figure 6.11.

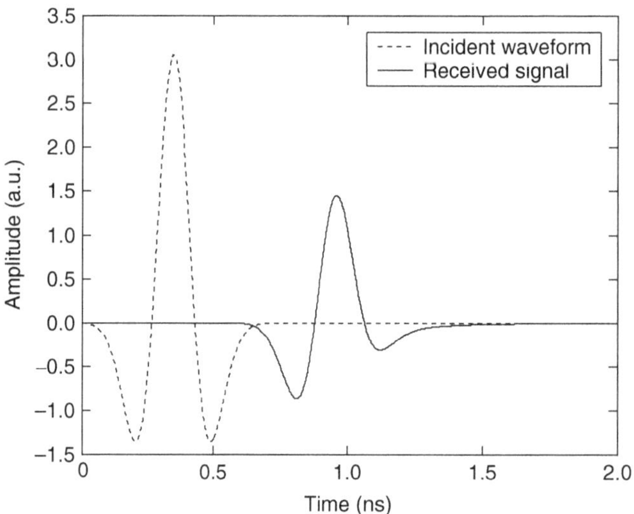

Figure 6.11 UWB pulse diffracted by a semi-infinite screen. Reproduced with permission from [Qiu 2002] © IEEE.

As mentioned above (and detailed in Chapter 4), propagation effects are fundamentally frequency dependent: path loss is a function of frequency if the antennas have constant gain; similarly, diffraction and reflection are frequency dependent. Thus, the higher frequency components of the transmitted signal are usually attenuated more strongly by the combination of antenna and channel. Also, this effect leads to a distortion of each individual MPC since *any* frequency dependence of the transfer function leads to delay dispersion; the amount of distortion can be different for each MPC, depending on the propagation processes it undergoes.

As a consequence of these factors, statistical channel models also change. First of all, the path gain has to be redefined according to $G_{\mathrm{pr}}(d,f) = E\left\{\int_{f-\Delta f/2}^{f+\Delta f/2} |H(f,d)|^2 df\right\}$, where the expectation is taken over both small-scale and large-scale fading and Δf is smaller than the stationarity bandwidth, i.e., sufficiently small so that all propagation effects stay constant within it. Furthermore, when representing the impulse response as a tapped delay line, the fading at the different taps becomes correlated: the MPC distortion makes each multi-path contribution influence several subsequent taps.

6.6.2 UWB Channels with Large Absolute Bandwidth

Another definition of UWB signals is that they have more than 500 MHz absolute bandwidth. Despite the high temporal resolution of UWB systems, there is still an appreciable probability that several MPCs fall into one resolvable delay bin, and add up there; in other words, there is fading even in UWB. The difference to conventional system lies mainly in the number of MPCs that fall into one bin. This number is influenced by the environment: the more objects are in the environments, the more MPCs can occur. For example, residential environments tend to have fewer MPCs than industrial environments. Furthermore, the delay of the considered bin plays a role: for larger excess delays, there are more feasible paths causing this particular delay. Thus, fading depth increases with increasing delay. Finally, the frequency band is important; the number of MPCs tends to be lower for very high frequency ranges (though even there the number of MPCs can be considerable, see also Chapter 7). Depending on these factors, a Rayleigh distribution of the amplitudes might or might not be suitable. Nakagami, Rice, or lognormal distributions have been suggested.

At a high absolute bandwidth, not every resolvable delay bin contains MPCs, so that delay bins containing MPCs are interspersed with "empty" delay bins, i.e., not containing any discrete (plane wave) MPCs. The resulting PDP is called "sparse." The bandwidth required for these phenomena to occur depends on the environment, the frequency range, and whether directional antennas are used that restrict the angular range from which MPCs can be received, see Section 6.7.

6.7 Directional Description

6.7.1 Basic Double-Directional Impulse Response

The wideband description we have given in the previous sections is not a completely general description of the channel:

- It models only the temporal properties of the channel. The directions of the MPCs do not enter the description. These directional properties are important for spatial diversity (Chapter 12) and multi-element antennas (Chapters 16 and 22).
- The impulse response actually does not describe only the propagation channel but also includes the effect of the antennas. It adds up the weighted MPCs, where the weighting depends on the specific antenna used (see Sec. 3.2). Thus, changing the antenna changes the impulse response, even though the true propagation channel remains unchanged.

These problems are eliminated by using the most fundamental deterministic description of the propagation channel, the *double-directional impulse response* [Steinbauer et al. 2001],[9] which consists of a sum of contributions from the MPCs:

$$h(t, \tau, \boldsymbol{\Omega}, \boldsymbol{\Psi}; \mathbf{r}_{\mathrm{TX}}, \mathbf{r}_{\mathrm{RX}}) = \sum_{\ell=1}^{N(\mathbf{r})} h_\ell(t, \tau, \boldsymbol{\Omega}, \boldsymbol{\Psi}; \mathbf{r}_{\mathrm{TX}}, \mathbf{r}_{\mathrm{RX}}). \tag{6.52}$$

The impulse response depends on the time t, the delay τ, the *Direction-of-Departure (DoD)* $\boldsymbol{\Omega}$, the *DoA* $\boldsymbol{\Psi}$, and the number of MPCs, $N(\mathbf{r})$, for the specific combination of the locations of TX $\mathbf{r}_{\mathrm{TX}}$ and RX $\mathbf{r}_{\mathrm{RX}}$. Note that $\boldsymbol{\Omega}$ and $\boldsymbol{\Psi}$ are spatial angles, describing both azimuth and elevation. The $h_\ell(t, \tau, \boldsymbol{\Omega}, \boldsymbol{\Psi}; \mathbf{r}_{\mathrm{TX}}, \mathbf{r}_{\mathrm{RX}})$ is the contribution of the ℓth MPC, modeled as

$$h_\ell(t, \tau, \boldsymbol{\Omega}, \boldsymbol{\Psi}; \mathbf{r}_{\mathrm{TX}}, \mathbf{r}_{\mathrm{RX}}) = |a_\ell| e^{j\varphi_\ell} \delta(\tau - \tau_\ell) \delta(\boldsymbol{\Omega} - \boldsymbol{\Omega}_\ell) \delta(\boldsymbol{\Psi} - \boldsymbol{\Psi}_\ell) \tag{6.53}$$

which describes each MPC as a homogeneous plane wave. Besides the absolute amplitude $|a|$ and the delay, also the DoA and DoD vary slowly with time and the TX and RX locations (over many wavelengths), while again, the phase φ varies quickly; note that we have not written down these dependencies explicitly in (6.53). The model above furthermore assumes that the strengths of the MPCs stay constant over the area of interest around $\mathbf{r}_{\mathrm{TX}}$, $\mathbf{r}_{\mathrm{RX}}$, see the discussion in Section 6.7.3.

Knowledge about the parameters of the MPCs can stem from (i) ray-tracing simulations, as discussed in Section 4.7, (ii) stochastic channel models, see Chapter 7, or (iii) measurements, see Chapter 9.

*6.7.2 Generalizations of the Double-Directional Impulse Response

The above description is not the most general one, though we will use it in the following frequently to explain points without an excess of notation. For a completely general formulation, the following items need to be additionally taken into account:

- *Diffuse multi-path:* while the "finite sum of discrete multi-path" model described above is very popular, it does not reflect the complete physical reality. Diffuse scattering, as well as wavefront curvature from nearby scatterers, can give rise to other components that (irrespective of their physical origin) are commonly called Diffuse Multi-path Components (DMC). The total impulse response is then

$$h(t, \tau, \boldsymbol{\Omega}, \boldsymbol{\Psi}; \mathbf{r}_{\mathrm{TX}}, \mathbf{r}_{\mathrm{RX}}) = \sum_{\ell=1}^{N(\mathbf{r})} h_\ell(t, \tau, \boldsymbol{\Omega}, \boldsymbol{\Psi}; \mathbf{r}_{\mathrm{TX}}, \mathbf{r}_{\mathrm{RX}}) + h_{\mathrm{DMC}}(t, \tau, \boldsymbol{\Omega}, \boldsymbol{\Psi}; \mathbf{r}_{\mathrm{TX}}, \mathbf{r}_{\mathrm{RX}}). \tag{6.54}$$

The DMC, due to its nature, is most efficiently not described by a sum of delta functions, but rather by a continuous version of the delay-angle-Doppler dispersion profile. The DMC is commonly interpreted as a random process, and its description is limited to the parameters of this random process. In the simplest form, this random process is assumed to be uniform in angle, and (one-sided) exponential in delay, though more advanced modes are possible, see Section 7.5.4.

- *Doppler shift:* in the case of movement, an additional phase shift term occurs for each of the MPCs, representing the Doppler shift of the component:

$$|a_\ell| e^{j\varphi_\ell} \delta(\tau - \tau_\ell) \delta(\boldsymbol{\Omega} - \boldsymbol{\Omega}_\ell) \delta(\boldsymbol{\Psi} - \boldsymbol{\Psi}_\ell) \exp(j2\pi\nu_\ell t) \tag{6.55}$$

---

[9] To be completely general, we also have to include a description of polarization as well as several other effects, see Sec. 6.7.2.

where ν_ℓ describes the Doppler shift of the ℓth MPC. It can be seen that when describing the spreading function instead of the time-variant impulse response, a completely symmetric formulation is achieved, i.e.,

$$s_\ell(\nu, \tau, \boldsymbol{\Omega}, \boldsymbol{\Psi}; \mathbf{r}_{\mathrm{TX}}, \mathbf{r}_{\mathrm{RX}}) = \mid a_\ell \mid e^{\,j\varphi_\ell}\delta(\tau - \tau_\ell)\delta(\boldsymbol{\Omega} - \boldsymbol{\Omega}_\ell)\delta(\boldsymbol{\Psi} - \boldsymbol{\Psi}_\ell)\delta(\nu - \nu_\ell). \tag{6.56}$$

This is the most general description that makes no statements about possible interrelationship between the Doppler and other parameters. For the (practically important) case that the Doppler shifts are created only by the movement of the UE, the Doppler shift has a strict mapping to the DoA: $\nu_\ell = (\mathrm{v}/c_0)\cos(\gamma_\ell)$, where γ_ℓ is the angle between the velocity vector $\mathbf{v}$ (with magnitude v) and the DoA of the ℓth MPC (implicitly assuming here that the UE is the RX).

- *Polarization:* the above description is for a single polarization direction. For a general description, two orthogonal polarizations (assuming we are in the far-field) should be described. In that case, the complex amplitude $a_\ell = \mid a_\ell \mid e^{\,j\varphi_\ell}$ becomes a matrix

$$\begin{pmatrix} a_\ell^{\mathrm{V,V}} & a_\ell^{\mathrm{V,H}} \\ a_\ell^{\mathrm{H,V}} & a_\ell^{\mathrm{H,H}} \end{pmatrix} \tag{6.57}$$

where superscripts V and H refer to vertical and horizontal polarization for concreteness, but any other set of orthogonal polarizations could be used as well. Thus, the impulse response becomes

$$\mathbf{h}_\ell(t, \tau, \boldsymbol{\Omega}, \boldsymbol{\Psi}; \mathbf{r}_{\mathrm{TX}}, \mathbf{r}_{\mathrm{RX}}) = \begin{pmatrix} a_\ell^{\mathrm{V,V}} & a_\ell^{\mathrm{V,H}} \\ a_\ell^{\mathrm{H,V}} & a_\ell^{\mathrm{H,H}} \end{pmatrix}\delta(\tau - \tau_\ell)\delta(\boldsymbol{\Omega} - \boldsymbol{\Omega}_\ell)\delta(\boldsymbol{\Psi} - \boldsymbol{\Psi}_\ell)\,exp\,(j2\pi\nu_\ell t). \tag{6.58}$$

We also note that all the above models make a number of implicit assumptions:

- *Narrowband assumption:* this typically means that the bandwidth is within 10% of the carrier frequency (thus, "narrowband" in the RF sense, meaning "not ultrawideband," as opposed to "narrowband" in the sense of Section 6.1). The assumption shows itself in the statement that the impulse response of each separate MPC is a delta impulse, or equivalently, that the transfer function of each MPC is frequency-flat. Furthermore, the narrowband assumption is required for the Doppler effect to lead to a Doppler *shift*, not a scaling. The narrowband assumption cannot be considered as a channel property alone but depends on the observation system, namely the bandwidth of transmission.
- *Far-field assumption:* this means that the distance between the TX and RX and the scattering objects is sufficiently large so that the propagation over each path can be represented as a plane wave *over the area of interest*, which typically will be the size of the antenna array aperture, and thus related to the observation system.
- Furthermore, we assume that the runtime of the signal over the size of the antenna or antenna array is smaller than the inverse system bandwidth, which is reflected in the assumption that τ_ℓ is not a function of the angles of incidence.

6.7.3 Stochastic Channel Description and Condensed Parameters

The stochastic description of directional channels is analogous to the nondirectional case. The ACF of the impulse response can be generalized to include the directional dependence so that it depends on six or eight variables. We can also introduce a "generalized WSSUS condition" so that contributions coming from different directions are fading independently. Note that the directions of the MPCs at the UE on one hand, and the Doppler spreading on the other hand, are linked, and thus ν and $\boldsymbol{\Psi}$ are not independent variables anymore (we assume in the following a downlink situation, so that $\boldsymbol{\Psi}$ are the directions at the UE).

Analogously to the nondirectional case, we can then define condensed descriptions of the wireless channel. We first define

$$E\{s^*(\nu, \tau, \boldsymbol{\Omega}, \boldsymbol{\Psi})s(\nu', \tau', \boldsymbol{\Omega}', \boldsymbol{\Psi}')\} = P_s(\nu, \tau, \boldsymbol{\Omega}, \boldsymbol{\Psi})\delta(\boldsymbol{\Omega} - \boldsymbol{\Omega}')\delta(\boldsymbol{\Psi} - \boldsymbol{\Psi}')\delta(\tau - \tau')\delta(\nu - \nu') \tag{6.59}$$

from which the *Double-Directional Delay Power Spectrum (DDDPS)* is derived as

$$DDDPS(\tau, \boldsymbol{\Omega}, \boldsymbol{\Psi}) = \int P_s(\nu, \tau, \boldsymbol{\Omega}, \boldsymbol{\Psi})d\nu. \tag{6.60}$$

The DDDPS is also often generally called the second-order statistics of the channel, since it describes the second-order (correlation) functions of this random process.

From this, we can establish the *Angular Delay Power Spectrum (ADPS)* as seen from the BS

$$ADPS_{\mathrm{BS}}(\tau, \boldsymbol{\Omega}) = \int DDDPS(\tau, \boldsymbol{\Omega}, \boldsymbol{\Psi})G_{\mathrm{UE}}(\boldsymbol{\Psi})d\boldsymbol{\Psi} \tag{6.61}$$

where G_{UE} is the antenna power pattern of the UE. Analogously, we can define the ADPS at the UE, $ADPS_{\mathrm{UE}}(\tau, \boldsymbol{\Psi})$. In many cases, the antenna pattern is simply assumed to be isotropic.

The double-directional *Angular Power Spectrum (APS)* is given by

$$APS_{\mathrm{DD}}(\mathbf{\Omega}, \mathbf{\Psi}) = \int DDDPS(\tau, \mathbf{\Omega}, \mathbf{\Psi})d\tau \tag{6.62}$$

from which the angular power spectra at BS and UE can be computed as

$$
\begin{aligned}
APS_{\mathrm{BS}}(\mathbf{\Omega}) &= \int ADPS_{\mathrm{BS}}(\tau, \mathbf{\Omega})d\tau = \int APS_{\mathrm{DD}}(\mathbf{\Omega}, \mathbf{\Psi})G_{\mathrm{UE}}(\mathbf{\Psi})d\mathbf{\Psi} \\
APS_{\mathrm{UE}}(\mathbf{\Psi}) &= \int ADPS_{\mathrm{UE}}(\tau, \mathbf{\Psi})d\tau = \int APS_{\mathrm{DD}}(\mathbf{\Omega}, \mathbf{\Psi})G_{\mathrm{BS}}(\mathbf{\Omega})d\mathbf{\Omega} .
\end{aligned}
\tag{6.63}
$$

Note also that an integration of the ADPS over the angles recovers the PDP. For simplicity of notation, we will consider in the following the APS at the BS, and drop the subscript BS.

The *azimuthal spread* is defined as the second central moment of the APS if all MPCs are incident in the horizontal plane, so that $\mathbf{\Omega} = \phi$. In many papers, it is defined in a form analogous to Eqs. (6.44)–(6.46), namely

$$S_\phi = \sqrt{\frac{\int APS(\phi)\phi^2 d\phi}{\int APS(\phi)d\phi} - \left(\frac{\int APS(\phi)\phi d\phi}{\int APS(\phi)d\phi}\right)^2}. \tag{6.64}$$

However, this definition is ambiguous because of the periodicity of the azimuthal angle: by this definition, $APS = \delta(\phi - \pi/10) + \delta(\phi - 19\pi/10)$ would have a different angular spread from $APS = \delta(\phi - 3\pi/10) + \delta(\phi - \pi/10)$, even though the two APSs differ just by a constant offset. A better definition is (see [Fleury 2000])

$$S_\phi = \sqrt{\frac{\int |\exp(j\phi) - \mu_\phi|^2 APS(\phi)d\phi}{\int APS(\phi)d\phi}} \tag{6.65}$$

with

$$\mu_\phi = \frac{\int \exp(j\phi)APS(\phi)d\phi}{\int APS(\phi)d\phi}. \tag{6.66}$$

In this definition, the angular spread is dimensionless and ranges between 0 and 1. For small values of the angular spread, the two definitions can sometimes lead to similar results.

Example 6.3 *Consider the APS defined as $APS = 1$ for $0° < \phi < 90°$ and $340° < \phi < 360°$, compute the angular spread according to the definition Eqs. (6.64) and (6.65), respectively.*

According to Eq. (6.64) we have

$$
\begin{aligned}
S_\phi &= \sqrt{\frac{\int_0^{\pi/2} \phi^2 d\phi + \int_{17\pi/9}^{2\pi} \phi^2 d\phi}{\int_0^{\pi/2} d\phi + \int_{17\pi/9}^{2\pi} d\phi} - \left(\frac{\int_0^{\pi/2} \phi d\phi + \int_{17\pi/9}^{2\pi} \phi d\phi}{\int_0^{\pi/2} d\phi + \int_{17\pi/9}^{2\pi} d\phi}\right)^2} \\
&= 2.09 \text{ rad} = 119.7°.
\end{aligned}
\tag{6.67}
$$

In contrast, Eqs. (6.65) and (6.66) yield

$$
\begin{aligned}
\mu_\phi &= \frac{18}{11\pi} \int_{-\pi/9}^{\pi/2} \exp(j\phi)d\phi = 0.7 + 0.49j \\
S_\phi &= \sqrt{\frac{18}{11\pi} \int_{-\pi/9}^{\pi/2} \left(\left(\cos(\phi) - \mathrm{Re}(\mu_\phi)\right)^2 + \left(\sin(\phi) - \mathrm{Im}(\mu_\phi)\right)^2\right)d\phi} \\
&= 0.521.
\end{aligned}
\tag{6.68}
$$

The values obtained from the two methods differ radically. We can easily see that the second value makes more sense: the APS extends continuously from $-20°$ to $90°$. The angular spread should thus be the same as for an APS that extends from 0 to 110°. Inserting this in Eq. (6.64), we obtain an angular spread of 0.56, i.e., a radical change. The value from Eq. (6.68) remains at 0.52, as it should. It is interesting that this value is close to $(\phi_{\max} - \phi_{\min})/(2\sqrt{3})$ – and remember that for a rectangular PDP, the relationship between rms delay spread and maximum excess delay is also $S_\tau = \tau_{\max}/(2\sqrt{3})$.

Similarly to the delay spread, also the angular spread is only a partial description of the angular dispersion. It has been shown that the correlation of signals at the elements of a uniform linear array depends mainly on the rms angular spread and not on the shape of the APS; however, this is valid only under some very specific assumptions.

We now come to the range of validity of the above descriptions. For the channel statistics to stay constant, the (generalized) WSSUS condition must be fulfilled, which is only true within a limited spatial range called the *region of stationarity*. Strictly speaking, we can define different regions of stationarity, depending on whether the absolute power,[10] the shape of the PDP, or the APS need to remain constant. Each of those channel characteristics can have different sizes of the region of stationarity; if a single number is required, the size of the smallest of these different regions should be used.

6.7.4 Connection to Conventional Impulse Response

The double-directional impulse response is valuable in a number of respects. Firstly, it provides a description of the propagation channel that is independent of all antenna effects. This provides a clear connection to the actual physics of propagation, and the parameters of the MPCs can be obtained from ray tracing or measurements.

Secondly, the conventional impulse response of the radio channel (including the antennas) can be recovered *for any antenna pattern* by weighting the DDIR with the antenna pattern, and then integrating over all angles

$$h(t, \tau) = \int \int h(\tau, \mathbf{\Omega}, \mathbf{\Psi}) \widetilde{G}_{\mathrm{TX}}(\mathbf{\Omega}) \widetilde{G}_{\mathrm{RX}}(\mathbf{\Psi}) d\mathbf{\Psi} d\mathbf{\Omega} \tag{6.69}$$

where $\widetilde{G}_{\mathrm{TX}}$ and $\widetilde{G}_{\mathrm{RX}}$ are the complex (amplitude) patterns of the TX and RX antenna elements, respectively. If required, a single-directional impulse response at TX and RX can be obtained by integrating the DDIR over $\mathbf{\Psi}$ or $\mathbf{\Omega}$, respectively. This principle is shown in Figure 6.12. A similar principle underlies the definition of the mean effective gain of an antenna, see Chapter 8.

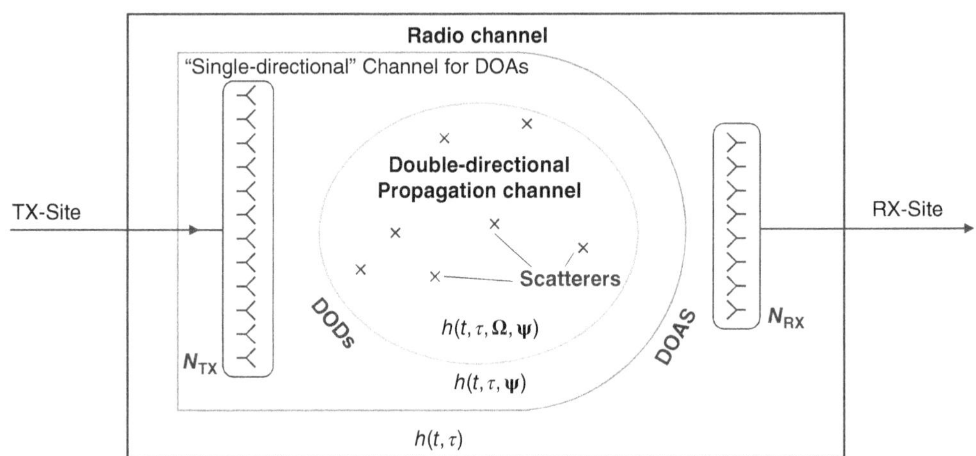

Figure 6.12 Double-directional impulse response and its relation to nondirectional impulse response. Reproduced with permission from [Steinbauer et al. 2001] © IEEE.

Thirdly, directional channel descriptions are especially valuable in the context of multiple-antenna systems. In that case, we are often interested in obtaining the joint impulse responses at the different antenna elements. The impulse response thus becomes a matrix if we have antenna arrays at both link ends, and a vector if there is an array at one link end. We denote the transmit and receive element coordinates as $\mathbf{r}_{\mathrm{TX}}^{(1)}, \mathbf{r}_{\mathrm{TX}}^{(2)}, \dots \mathbf{r}_{\mathrm{TX}}^{(N_{\mathrm{TX}})}$, and $\mathbf{r}_{\mathrm{RX}}^{(1)}, \mathbf{r}_{\mathrm{RX}}^{(2)}, \dots \mathbf{r}_{\mathrm{RX}}^{(N_{\mathrm{RX}})}$, respectively, so that the impulse response from the jth transmit to the ith receive element becomes

$$\begin{aligned}
h_{i,j} &= h\left(\mathbf{r}_{\mathrm{TX}}^{(j)}, \mathbf{r}_{\mathrm{RX}}^{(i)}\right) \\
&= \sum_{\ell} h_{\ell}\left(\tau, \mathbf{\Omega}_{\ell}, \mathbf{\Psi}_{\ell}; \mathbf{r}_{\mathrm{TX}}^{(1)}, \mathbf{r}_{\mathrm{RX}}^{(1)}\right) \widetilde{G}_{\mathrm{TX}, j}(\mathbf{\Omega}_{\ell}) \widetilde{G}_{\mathrm{RX}, i}(\mathbf{\Psi}_{\ell}) \\
&\quad \times \exp\left(j < \mathbf{k}(\mathbf{\Omega}_{\ell}), \left(\mathbf{r}_{\mathrm{TX}}^{(j)} - \mathbf{r}_{\mathrm{TX}}^{(1)}\right) >\right) \exp\left(-j < \mathbf{k}(\mathbf{\Psi}_{\ell}), \left(\mathbf{r}_{\mathrm{RX}}^{(i)} - \mathbf{r}_{\mathrm{RX}}^{(1)}\right) >\right).
\end{aligned} \tag{6.70}$$

---

[10] For the received power, the size of the region of stationarity is similar to the shadowing correlation length, though possible changes in the pathloss also factor in.

Here $\widetilde{G}_{\mathrm{TX},j}$ and $\widetilde{G}_{\mathrm{RX},i}$ are the complex (amplitude) patterns of the transmit and receive antenna elements, respectively when such element is placed at the reference location;[11] $\mathbf{k}$ is the unit wave vector in the direction of the ℓth DoD or DoA, and $<\cdot,\cdot>$ denotes the inner product. This formulation again makes the assumptions discussed in Section 6.7.2. It can be interpreted in that the contribution of the ℓth MPC endures a phase shift in the amount of $< \mathbf{k}(\mathbf{\Omega}_\ell), \left(\mathbf{r}_{\mathrm{TX}}^{(j)} - \mathbf{r}_{\mathrm{TX}}^{(1)}\right) >$ between the reference location $\mathbf{r}_{\mathrm{TX}}^{(1)}$ and the location of the jth antenna elements, compare also Section 8.4, and similar at the RX. If the arrays at TX and RX are uniform linear arrays, we can write Eq. (6.70) as

$$\mathbf{H} = \int\int h(\tau, \mathbf{\Omega}, \mathbf{\Psi})\widetilde{G}_{\mathrm{TX}}(\mathbf{\Omega})\widetilde{G}_{\mathrm{RX}}(\mathbf{\Psi})\mathbf{u}_{\mathrm{RX}}(\mathbf{\Psi})\mathbf{u}_{\mathrm{TX}}^{\dagger}(\mathbf{\Omega})\mathrm{d}\mathbf{\Psi}\mathrm{d}\mathbf{\Omega} \tag{6.71}$$

where we used the *steering vector*

$$\mathbf{u}_{\mathrm{TX}}(\mathbf{\Omega}) = \frac{1}{\sqrt{N_t}}\left[1, \exp\left(-j2\pi\frac{d_{\mathrm{a}}}{\lambda}\cos\left(\gamma(\mathbf{\Omega})\right)\right), \dots \exp\left(-j2\pi(N_t-1)\frac{d_{\mathrm{a}}}{\lambda}\cos\left(\gamma(\mathbf{\Omega})\right)\right)\right]^T$$

where $\cos(\gamma(\mathbf{\Omega}))$ is the directional cosine between the DoD and the array axis, and analogously defined $\mathbf{u}_{\mathrm{RX}}(\mathbf{\Psi})$.

We thus see that it is always possible to obtain the impulse response matrix from a double-directional impulse response, while the reverse operation is very difficult.

*6.7.5 Direction-Dependent Condensed Parameters

Condensed parameters such as pathloss and delay spread are generally defined from the nondirectional impulse response $h(t, \tau)$. However, we have seen that this $h(t, \tau)$ depends on the particular antenna pattern used at TX and RX. In many cases, it is desirable to avoid this ambiguity and thus define pathloss and delay spread for the specific case of omni-directional antenna patterns at TX and RX. Thus, for example, measurements of the impulse response can be done with omnidirectional antennas or should be done in such a way that the impulse response that *would* occur if the pattern *were* omni-directional can be synthesized from the measured results.

In other situations, it is interesting to find the impulse response and pathloss when the antenna pattern is strongly directional, such as for a horn antenna. It is then often assumed that the horn is oriented in the direction that gives the maximum power, and from the thus-measured impulse response, the parameters like pathloss and delay spread are computed. They are then called the "directional pathloss" and "directional delay spread," compare Section 7.1.1. Besides the orientation of the horn, they also depend on the specific shape of the antenna pattern of this horn. Such directional parameters are mostly employed for the characterization of mm-wave and THz propagation channels, since in that frequency range use of highly directional antennas is widespread.

*6.8 Appendices

App. 6A: Validity of WSSUS in Mobile Radio

See App6.pdf at wiley.com/go/molisch/wireless3e

App. 6B: Instantaneous Channel Parameters

See App6.pdf at wiley.com/go/molisch/wireless3e

Further Reading

The theory of linear time-variant systems is described in the classical paper of [Bello 1963]; further details are discussed in [Kozek 1997] and [Matz and Hlawatsch 1998]. The theory of WSSUS systems was established in [Bello 1963], and further investigated in [Hoeher 1992, Kattenbach 1997, Molnar et al. 1996, Paetzold 2002]; more considerations about the validity of WSSUS in wireless communications can be found in [Fleury 1990, Kattenbach 1997, Kozek 1997]. A method for characterizing non-WSSUS channels is described in [Matz 2003], and a detailed and extensive description of such channels is provided in [Hlawatsch and Matz 2011]. An overview of condensed parameters, including delay spread, is given in [Molisch and Steinbauer 1999]. Survey descriptions of UWB channels can be found in [Molisch 2009, Qiu 2004].

---

[11] i.e., the phase shift due to the displacement $r_{\mathrm{TX}}^{(j)} - r_{\mathrm{TX}}^{(1)}$ is taken into account separately, by the term in the third line of the equation.

Description methods for spatial channels are discussed, e.g., in [Durgin 2003], [Ertel et al. 1998], [Yu and Ottersten 2002]. Generalizations of the WSSUS approach to directional models are discussed in [Fleury 2000], [Kattenbach 2002]. The double-directional channel description was introduced in [Steinbauer et al. 2001]; it also forms the basis for most MIMO channel models, see Chapter 7. Polarization of wireless channels is discussed in [Shafi et al. 2006].

For updates and errata for this chapter, see https://wides.usc.edu/students.html#textbooks.

Exercises

Sec. 36.6 of Exercises.pdf at wiley.com/go/molisch/wireless3e

7

Channel Models

For the design, simulation, and planning of wireless systems, we need *models* for the propagation channels. In the previous chapters, we have discussed some basic properties of wireless channels, and how they can be described mathematically – amplitude-fading statistics, scattering function, delay spread, etc. In this chapter, we discuss in a more concrete way how these mathematical description methods can be converted into generic simulation models, and how to parameterize these models.

There are two main applications for channel models:

1. The designers of *wireless networks* are interested in analyzing the performance of a given system in a certain geographical region, so that they can optimize locations of Base Stations (BSs) and other network design parameters. For such applications, *location-specific channel models* that make good use of available geographical and morphological information are desirable. However, the models should be robust with respect to small errors in geographical databases. These models are of critical importance for network planning and deployment, since finding good BS locations, and fine-tuning the system parameters for optimum network performance, need to be done on a computer (instead of by trial-and-error in the actual system). Location-specific models are almost always obtained from the ray-tracing/launching simulations we discussed in Section 4.7.
2. The design, testing, and type approval of *wireless systems* needs models that reflect the properties that impact system performance. This is usually achieved by simplified channel models that describe the *statistics* of the impulse response in parametric form. The number of parameters is small, and parameters are *independent of specific locations*. Such models sometimes lead to insights due to closed-form relationships between channel parameters and system performance. Furthermore, they can easily be implemented by system designers for testing purposes, and the system performance can be assessed by simulation using these channel models. If the models are standardized, the simulation results from different people or companies are comparable.

This chapter will focus on the stochastic channel models, since location-specific models were already treated in Chapter 4. We will discuss both the generic way of modeling the channels and the specific parameterization of those models in various environments. Stochastic channel models can describe narrowband characteristics (Section 7.1), delay dispersion (Section 7.2), as well as directional dispersion (Section 7.3); the joint delay and directional dispersion are discussed in Section 7.4. We will in the following discuss the characteristic quantities and models in that sequence. We will also find that there are three basic approaches for modeling channels with dispersion, namely (i) generalized tapped delay lines (Section 7.5), (ii) geometry-based stochastic channel models (Section 7.6), and (iii) semi-deterministic models (Section 7.7).

All models need to be either based on measurements or verified by measurements (see Chapter 9). The models also involve a tradeoff between the model accuracy, number of available measurements, and model complexity (runtime). There is therefore not a single "best model," but rather different models may best fit different applications.

7.1 Narrowband Models

This section first provides definitions of the narrowband channel characteristics, putting into mathematical language the intuitive descriptions given in Section 5.1. It then proceeds to actual models for those narrowband characteristics, namely pathloss, shadowing, and small-scale fading.

7.1.1 Definitions for Pathloss and Shadowing

The channel power gain between a TX at location $\mathbf{r}_{\mathrm{TX}}$ and an RX at location $\mathbf{r}_{\mathrm{RX}}$ at time t and frequency f can be written as

$$|h(t,f,\mathbf{r}_{\mathrm{TX}},\mathbf{r}_{\mathrm{RX}})|^2 = \frac{P_{\mathrm{RX}}(t,f,\mathbf{r}_{\mathrm{RX}})}{P_{\mathrm{TX}}(t,f,\mathbf{r}_{\mathrm{TX}})}. \tag{7.1}$$

Wireless Communications: From Fundamentals to Beyond 5G, Third Edition. Andreas F. Molisch.
© 2023 John Wiley & Sons Ltd. Published 2023 by John Wiley & Sons Ltd.
Companion website: www.wiley.com/go/molisch/wireless3e

This channel gain is usually defined under the assumption that isotropic antennas are used (see the "directional pathloss" discussion for further details). As we discussed previously, changes to the path gain are due to small-scale fading, shadowing, and large-scale distance changes. Averaging over the small scale fading eliminates [assuming fulfillment of the Wide Sense Stationary Uncorrelated Scatterer (WSSUS) condition] the dependence on frequency and time, such that the Small-Scale Averaged (SSA) path gain becomes

$$PG(\mathbf{r}_{\mathrm{TX}}, \mathbf{r}_{\mathrm{RX}}) = \frac{1}{T_{\mathrm{stat}}} \frac{1}{B_{\mathrm{stat}}} \int\limits_{T_{\mathrm{stat}}} \int\limits_{B_{\mathrm{stat}}} |h(t, f, \mathbf{r}_{\mathrm{TX}}, \mathbf{r}_{\mathrm{RX}})|^2 df dt \tag{7.2}$$

where T_{stat} and T_{stat} are stationarity-time and -bandwidth, respectively. If the antennas have isotropic patterns, such that the N Multi-Path Components (MPCs), which come from different directions, are weighted by the same antenna gains, the path gain can be shown to be equal to the sum of the powers of the MPCs,

$$PG(\mathbf{r}_{\mathrm{TX}}, \mathbf{r}_{\mathrm{RX}}) = \sum_{\ell=1}^{N} |a_\ell(t)|^2. \tag{7.3}$$

It is common to model this path gain as a function of the Euclidean distance between TX and RX $d = \|\mathbf{r}_{\mathrm{TX}} - \mathbf{r}_{\mathrm{RX}}\|$, as we have already done in previous sections. Of course, not all $\mathbf{r}_{\mathrm{TX}}$, $\mathbf{r}_{\mathrm{RX}}$ pairs lead to exactly the same PG; we thus model the path gain as a deterministic function of d plus random variations around it. Expressed on a logarithmic scale, this becomes

$$PG_{\mathrm{dB}}(d) = \overline{PG}_{\mathrm{dB}}(d) + S_{\mathrm{dB}}(d) \tag{7.4}$$

where $\overline{PG}$ is the mean path gain (often in the literature simply called "path gain"), and S_{dB} is a random variable describing the deviation from the mean path gain, commonly called shadowing; it will be discussed in more detail in Sections 7.1.5 and 7.1.7. Inverting the sign of $PG_{\mathrm{dB}}(d)$ gives the pathloss, $PL_{\mathrm{dB}}(d)$.

$$PL_{\mathrm{dB}}(d) = -PG_{\mathrm{dB}}(d) \tag{7.5}$$

and we also set $\overline{PL}_{\mathrm{dB}}(d) = -\overline{PG}_{\mathrm{dB}}(d)$.[1]

In some situations, models based on the Euclidean distance d do not provide good results. Rather, it is the so-called Manhattan distance that is relevant. As shown in Figure 7.1, this is the distance measured along axes at right angle, which corresponds to the distance a wave guided down a street canyon, and diffracting into another canyon intersecting at right angle, has to cover. This distance measure is meaningful, e.g., for non-LOS (NLOS) situations in urban environments in which the BS is significantly below the rooftop level.

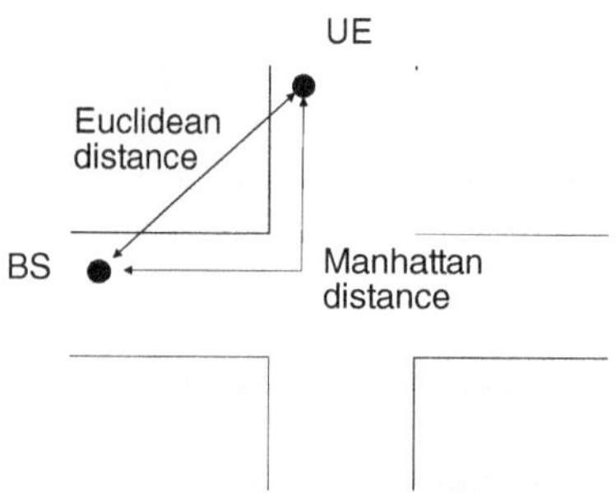

Figure 7.1 Euclidean and Manhattan distance.

*Directional Pathloss

The received power, as used in (1) depends, among other things, on the antenna patterns (compare (6.69)). If we do not use the assumption of isotropic antenna patterns, but rather assume a directional pattern that admits only MPCs from a certain angular range, e.g., a horn antenna, the pathloss becomes dependent on both the shape of the horn pattern and the antenna orientation. By taking the maximum over all antenna orientations at TX and RX, and subtracting (on a dB scale) the antenna gain, we obtain the *directional path gain*. Note that for the case of free-space propagation, "standard" (isotropic) path gain and directional path gain are identical, though the received power is higher by the antenna gain when using a directional antenna. Conversely, in an isotropic channel (directions of arrival and directions of departure uniformly distributed in angle), the received power is independent of the antenna pattern, while

---

[1] Consequently, we should define a shadowing loss that is the negative of the shadowing gain. However, since S is generally modeled as a zero-mean random variable, such a distinction is usually not explicitly done.

the directional path gain is lower when using a directional antenna, since only a small percentage of the MPCs is collected by the antenna.

The directional pathloss is used in particular for description of high-frequency channels. The subsequent discussions will be based on the conventional pathloss, though the many of the techniques described below can be applied to the directional pathloss as well.

7.1.2   The Power-Distance Law

A popular model for the average path gain is the power distance law, also called the Floating-Intercept (FI) or $\alpha-\beta$ model. It describes the average pathloss (on a linear scale) as a power law of the distance, $\overline{PL}(d) \propto \overline{PL}(d_0)(d/d_0)^{\alpha_{\rm PL}}$, or, on a logarithmic scale,

$$\overline{PL}_{\rm dB}(d) = \alpha_{\rm PL} \cdot 10 \log_{10}\left(\frac{d}{d_0}\right) + \beta_{\rm PL}. \tag{7.6}$$

Here, the slope $\alpha_{\rm PL}$ is also called the pathloss coefficient, and the value $\beta_{\rm PL}$ is the pathloss at a reference distance d_0. The reference distance is typically set either as $d_0 = 1$ m or as $d_0 = d_R$, i.e., the Rayleigh (Fraunhofer) distance. The parameters $\alpha_{\rm PL}$ and $\beta_{\rm PL}$ can be obtained from the best fit of a straight line to the set of measured pathloss values on a logarithmic scale, according to (7.6), see Figure 7.2 for an example.

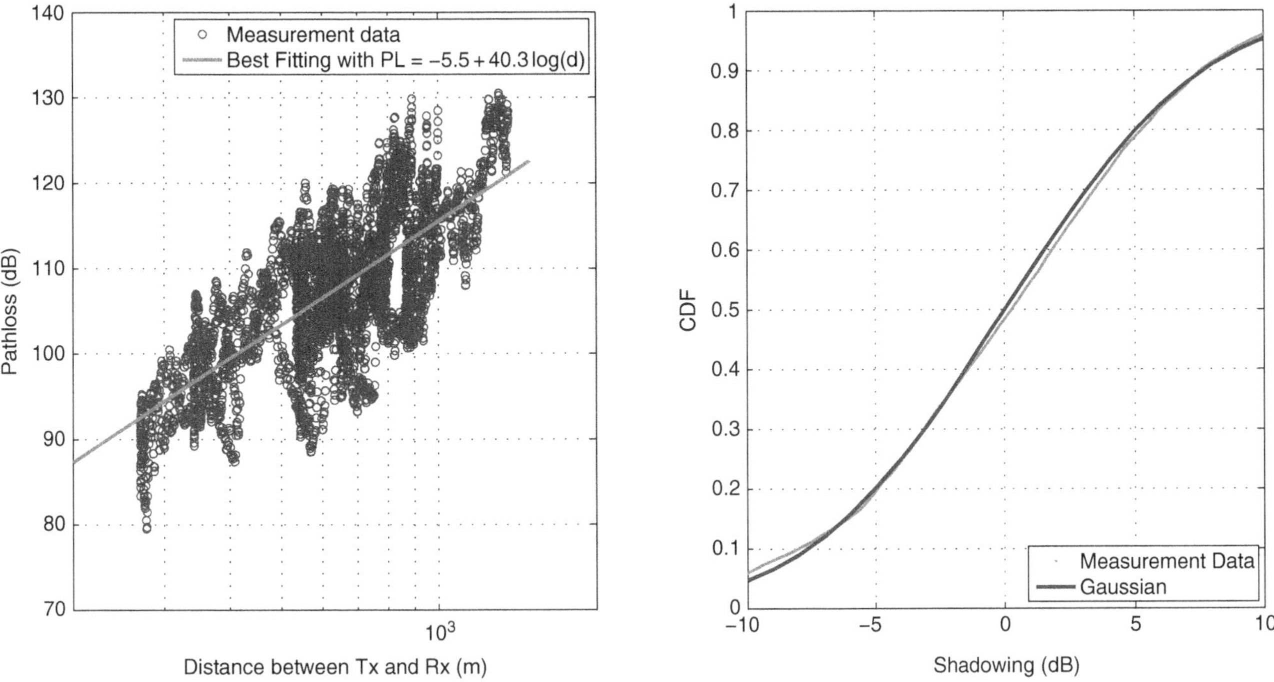

Figure 7.2   Example for fitting of pathloss and shadowing from measurement data in suburban environment. Color version available at wiley.com/go/molisch/wireless3e.
Reproduced with permission from [Biglieri et al. 2013] © Cambridge University Press.

A variant of the $\alpha-\beta$ model is the Close-In (CI) model, which sets $\beta_{\rm PL,CI} = PL_{\rm freespace}(d_0)$ equal to the free-space pathloss,

$$\overline{PL}_{\rm dB}(d) = \alpha_{\rm PL,CI} \cdot 10 \log_{10}\left(\frac{d}{d_0}\right) + \beta_{\rm PL,CI} \tag{7.7}$$

so that there is only a single parameter α to be fitted; this can make comparisons between different measurement campaigns easier. The $\alpha-\beta$ model has been widely used for more than 50 years, though some caveats are in order:

- The model is obtained from fitting to measurement data. Consequently, the model is only valid within the range of distances at which actual measurements are available to which the data are fitted. Ignoring this caveat can lead to unphysical results.
- Great care must be taken in the estimation process of the parameters $\alpha_{\rm PL}$ and $\beta_{\rm PL}$, in particular when locations at which measured data are available are nonuniformly distributed; see [Karttunen et al. 2016] for a detailed discussion. Furthermore, the model does

not incorporate any antenna gain; yet many of the measurements are performed with directional antennas. Care must again be taken in the extraction of parameters under those circumstances.

- The model inherently assumes stationarity, so that both pathloss and shadowing variance depends only on the distance but not the absolute location of TX and RX. It is thus generally used to represent, and be derived from, a wide range of different locations as well as a wide range of distances without being tied to a specific BS. These aspects, and some remedies, will be discussed in Section 7.1.5.

The pathloss coefficients in free space are 2, as we have seen in Section 4.1. In situations where there is both a LOS component and multi-path, the slope of the CI model, $\alpha_{\mathrm{PL,CI}}$ is somewhat *smaller* than 2 while in NLOS situations, it typically ranges from about 2 to 5. It is often useful to distinguish there between *obstructed LOS* (OLOS), where some smaller objects block the LOS component, and *heavy clutter*, also called *severe NLOS*, where many and large objects obstruct the LOS; $\alpha_{\mathrm{PL,CI}}$ in the former case are usually between 2 and 3, while in the latter case they range from 3 to 5. The α_{PL} in the general α–β model can vary over a larger range, because it is related to the β, as well as the range of distances the model is valid for. A theoretical justification of the α–β model, and a derivation of α_{PL} and β in a variety of typical scenarios is given in [Chizhik et al. 2021].

We now turn to typical values of (average) pathloss and shadowing observed in measurements. For indoor environments, such as office (including corridors, labs, etc.), as well as residential and factory halls, it was generally found that pathloss coefficients in LOS situations are in the range [1.2, 2.3], with lower values obtained in corridors, while higher values are obtained in factory halls. A similar range of parameters is also obtained for outdoor environments, with pathloss coefficients in street canyons around 1.5–1.8, while in more open LOS environments they are around 2.0. However, it must be noted that for larger distances, a breakpoint model may be suitable (compare Sections 4.2 and 7.1.3), with a pathloss coefficient around 4 beyond the breakpoint. Standard deviations of the shadowing are small, with 1–4 dB occurring both indoor and outdoor.

For NLOS, outdoor (microcellular and microcellular) environments show pathloss coefficients between 2.2 and 4. Note that the pathloss coefficient and other model parameters can be influenced by the height of the BS and User Equipment (UE), as well as geometry parameters of the environment, see Section 7.1.4. In indoor environments, pathloss coefficients are between 2 and 5 for NLOS depending on the severity of the clutter. The impact of the environment tends to be stronger at higher frequencies (e.g., mm-wave), with higher pathloss coefficients if the signal has to either pass through walls or suffers at least one diffraction. The shadowing standard deviation in NLOS is on the order of 4–8 dB at typical cellular/Wireless Local Area Network (WLAN) frequencies (<6 GHz) but increases for higher frequencies. At mm-wave frequencies, it also exhibits a pronounced dependence on distance and has been modeled as increasing from typically 5–10 dB at 30 m to more than 20 dB at 200 m.[2]

In most environments, the pathloss does not exhibit a strong frequency dependence beyond the f^2 dependence of the free-space pathloss, in both LOS and NLOS conditions. However, as mentioned above, the shadowing standard deviation for NLOS is significantly larger at higher frequencies, i.e., there is a higher probability that RX locations receive insufficient power for communications.

Specific values of the model parameters in various environments for standardized models are discussed in Section 7.1.4.

7.1.3 Breakpoint Models

The fit to measurements can sometimes be improved by a *dual-slope* or *breakpoint* (BP) model, where a different (usually larger) pathloss coefficient is used beyond a breakpoint distance d_{break}:

$$\overline{PL}(d) = \begin{cases} \overline{PL}_{\mathrm{SBP}}(d), & d \le d_{\mathrm{break}} \\ \overline{PL}_{\mathrm{LBP}}(d), & d > d_{\mathrm{break}} \end{cases} \tag{7.8}$$

where d_{break} is known as the breakpoint distance, $\overline{PL}_{\mathrm{SBP}}(d)$ is the pathloss $\overline{PL}(d)$ as defined in (7.7), or possibly (7.6) and

$$\overline{PL}_1(d) = \alpha_1 \cdot 10 \log_{10}\left(\frac{d}{d_{\mathrm{break}}}\right) + \beta_1. \tag{7.9}$$

As we have seen in Section 4.2.3, this type of model occurs in long-distance transmission with a LOS and a ground-reflected component. It is also useful when the pathloss exponent changes notably from OLOS to severe NLOS. Alternatively, the breakpoint model can characterize the transition from LOS to NLOS (this is the case we assumed in Chapter 3), though in this case, a probabilistic LOS model (see Section 7.1.6) may be preferable. Continuity of the pathloss at the breakpoint may or may not be enforced.

---

[2] However, as we will in Section 7.1.5, much of this effect actually stems from different streets having different pathloss coefficients.

7.1.4 Impact of Frequency, Transceiver Height, and Environment

The parameters of the log-distance law, $\alpha_{\rm PL}$ and $\beta_{\rm PL}$, depend on a number of other parameters, in particular frequency, height of the BS and UE, as well as the structure of the environment. In the following, we mention some popular models that describe this parametric dependence.

The Okumura–Hata Model

The Okumura–Hata model is based on the extensive measurements of [Okumura et al. 1968] in Japan and was brought into a form suitable for computer simulations by [Hata 1980]. The pathloss (in dB) is written as

$$\overline{PL}_{\rm dB}(d) = A + B\log_{10}(d) + C \tag{7.10}$$

where A, B, and C are factors that depend on frequency and antenna height. The factor A increases with the carrier frequency and decreases with increasing height of BS and UE. Also, the pathloss exponent (proportional to B) decreases with increasing height of the BS. Appendix 7.A describes the details of those correction factors. The model is only intended for large cells, with the BS being placed higher than the surrounding rooftops.

The COST 231-Walfisch-Ikegami Model

The Okumura Hata model is not suitable for either microcells or small macrocells, due to the restrictions on the distance between BS and UE and the antenna height. Based on [Walfisch and Bertoni 1988] and [Ikegami et al. 1984], the research and standardization group COST 231 suggested a model that is valid also for these cases [Damosso and Correira 1999].

In that model, the total pathloss consists of the free-space pathloss $PL_{\rm fs}$, the *multi-screen loss* $L_{\rm msd}$ along the propagation path, and the attenuation from the last roofedge to the UE, $L_{\rm rts}$ (*roof-top-to-street diffraction and scatter loss*). The free-space loss depends on the carrier frequency and the distance, while the rooftop-to-street diffraction loss depends on the frequency, the width of the street, and the height of the UE, as well as on the orientation of the street with respect to the connection line BS–UE. The multi-screen loss depends on the distance between buildings and the distance between BS and UE, as well as on the carrier frequency, the BS height, and the rooftop height, see Figure 7.3. Detailed equations are given in Appendix 7.B.

The model assumes a Manhattan street grid (streets intersecting at right angles), constant building height, and flat terrain. Furthermore, the model does not include the effect of propagation in the horizontal plane, i.e., waveguiding through street canyons, which can lead to an underestimation of the received power.

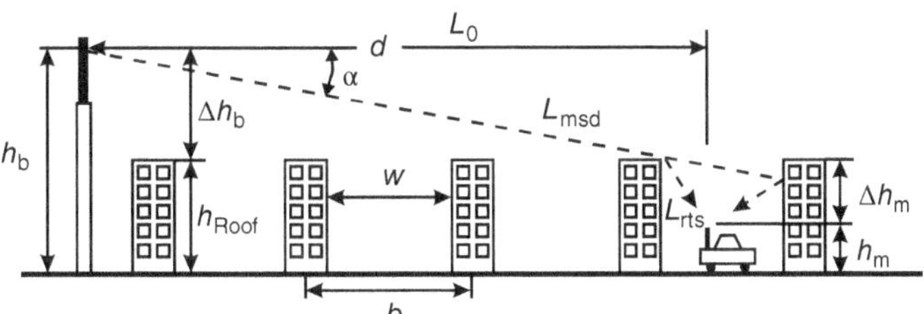

Figure 7.3 Parameters in the COST 231 Walfisch–Ikegami model.
Reproduced with permission from [Damosso and Correia 1999] © European Union.

*The Motley–Keenan Model

For indoor environments, the wall attenuation plays an important role. Based on this consideration, the Motley–Keenan model suggests that the pathloss (expressed in dB) can be written as [Motley and Keenan 1988]

$$\overline{PL} = \overline{PL}(d_0) + 10\alpha_{\rm PL}\log_{10}(d/d_0) + F_{\rm wall} + F_{\rm floor} \tag{7.11}$$

where $F_{\rm wall}$ is the sum of the attenuations by the walls that a MPC has to penetrate on its way from TX to RX; similarly, $F_{\rm floor}$ describes the summed-up attenuation of the floors that are located between the BS and the UE. Depending on the building material, the attenuation by one wall can lie between 1 and 20 dB in the 300 MHz–5 GHz range and can be much higher at higher frequencies.

The Motley–Keenan model is a site-specific model, in the sense that it requires knowledge of the location of BS and UE, and the building plan. It is, however, not very accurate, as it neglects the propagation paths that "go around" the walls. For example, propagation between two widely separated offices can occur either through many walls (quasi-LOS), or through a corridor (signal leaves the office, propagates down a corridor, and enters from there into the office of the receiver, RX). The latter type of propagation path can often be more efficient but is not taken into account by the Motely–Keenan model.

The 3GPP Model

A widely used model for system analysis is that of 3GPP. The pathloss model in it distinguishes between a number of different environments and uses a mixture of distance-power (using either Euclidean or Manhattan distance) and breakpoint models. For example, in the urban microcell, the pathloss for the LOS case is

$$\overline{PL}_{\text{UMi}-\text{LOS}} = \begin{cases} 32.4 + 21\log_{10}(d_{3\text{D}}) + 20\log_{10}(f_c) & 10\,\text{m} \le d_{2\text{D}} \le d_{\text{break}} \\ 32.4 + 40\log_{10}(d_{3\text{D}}) + 20\log_{10}(f_c) & \\ \quad -9.5\log_{10}\big((d_{\text{break}})^2 + (h_{\text{BS}} - h_{\text{UE}})^2\big) & d_{\text{break}} \le d_{2\text{D}} \le 5\,\text{km} \end{cases}$$

and for the NLOS case is given by

$$\overline{PL}_{\text{UMi}-\text{NLOS}} = 35.3\log_{10}(d_{3\text{D}}) + 22.4 + 21.3\log_{10}(f_c) - 0.3(h_{\text{UE}} - 1.5) \quad \text{for } 10\,\text{m} \le d_{2\text{D}} \le 5\,\text{km}$$

and lower-bounded to become never smaller than the LOS pathloss; f_c is in GHz. Note that the pathloss is different for LOS and NLOS situations, requiring also to assess the probability that LOS exists, see Section 7.1.6. Generally speaking, the pathloss coefficients are around 2 for LOS (depending on the environment between 1.8 and 2.2), and between 3 and 4 for NLOS environments (though more extreme values occur). The detailed equations and parameterizations are given in Appendix 7.D. The models claim validity for frequencies ranging from 0.5 to 100 GHz, but the validation by measurement has been done only for a subset of those frequencies, and even then is often based on a small number of measurements.

*7.1.5 Power Laws with Random Parameters

The α–β model (and its variants) are based on the assumption that spatial variations are stationary, i.e., that the mean and the variance of the pathloss fit equation depend only on the distance between TX and RX but not the absolute position. This is enforced by creating a large ensemble of UE positions that are parameterized only by their Euclidean distance to their BS, and further increasing the ensemble size by considering multiple BSs; the fit is then done to this large ensemble of points. As a consequence, the model can be used to represent a wide range of BS and UE locations.

However, this approach also has two drawbacks: (i) it obscures the dependence of the pathloss on the absolute position of the BS and UE, and (ii) it inevitably leads to a larger standard deviation – oftentimes *much* larger – than what would otherwise be witnessed for a UE moving on a trajectory in a single cell. Work over the past 20 years has shown experimentally that both the pathloss coefficient and the shadowing variance changes from cell to cell (for a cellular system) or house to house (for indoor systems), and that these quantities can be described as random variables over the ensemble of BSs (see "Further Reading").

Even more, we can find that the pathloss coefficient changes from street to street within a cell – an effect that is especially pronounced in urban microcells at higher frequencies, but which also exists at lower frequencies. An example is illustrated in Figure 7.4 in which waveguiding takes place in an urban-canyon environment. Different streets are signified by different colors/shades of grey. Thus, a UE moving along a trajectory within one street would only experience pathloss of that particular color. We can clearly observe that different streets have greatly differing pathloss coefficients: e.g., points on the street NLOS2 show a negligible slope, while the street NLOS6 corresponds to an almost vertical line in Figure 7.4; these effects can be explained physically in terms of waveguiding and diffraction. Thus, the difference in pathloss between two points clearly depends not only on the difference between the points but also

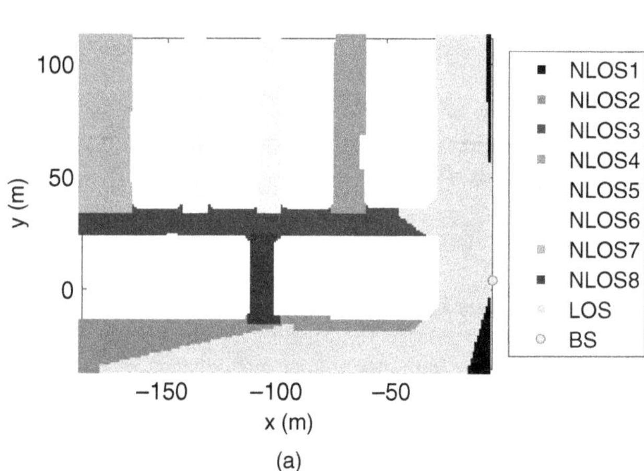

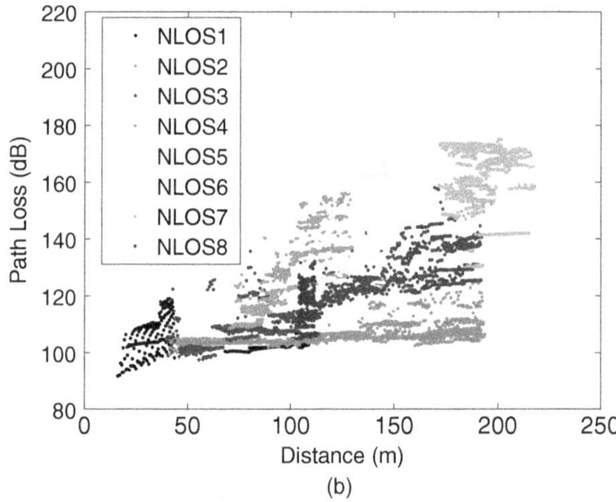

(a)   (b)

Figure 7.4 (a) Environment map with color coding of streets. (b) Pathloss as a function of link distance in NLOS, different greyshades indicate different streets. Reproduced with permission from [Karttunen et al. 2017] © IEEE. Color version available at wiley.com/go/molisch/wireless3e.

on the absolute position – most notably, which street the two points are in; this is in clear contradiction to the assumption of stationarity that underlies the usual pathloss models.

7.1.6   LOS Probability and Composite Models

We have seen above that in several models, such as the 3GPP model, the pathloss parameters are different in LOS and NLOS situations. The resulting equations can be interpreted as the pathloss conditioned on the LOS state, i.e., the presence/absence of a LOS between UE and BS. An assessment of the total pathloss as a function of distance thus needs to combine this information with the LOS probability. For a simulation, we thus first choose at random, and according to the LOS probability, whether LOS exists, and based on this choice pick the pathloss parameters. Equivalently, we can describe the pathloss pdf not as a lognormal distribution around $\overline{PG}(d)$, but rather as a bimodal distribution that has its peaks near $\overline{PG}^{\mathrm{LOS}}(d)$ and $\overline{PG}^{\mathrm{NLOS}}(d)$.

LOS probability is usually assumed to be unity in an area close to the BS, and then exponentially decaying outside this area, with the parameters and various refinements depending on the environment. The LOS probability models of 3GPP for different environments are shown in Table 7.1.

Table 7.1   LOS probability model of 3GPP.

Scenario	LOS probability (distance is in meters)
Rural macro	$\mathrm{Pr_{LOS}} = \begin{cases} 1 & , d_{\mathrm{2D\text{-}out}} \leq 10\mathrm{m} \\ \exp\left(-\dfrac{d_{\mathrm{2D\text{-}out}} - 10}{1000}\right) & , 10\mathrm{m} < d_{\mathrm{2D\text{-}out}} \end{cases}$
Urban micro – street canyon	$\mathrm{Pr_{LOS}} = \begin{cases} 1 & , d_{\mathrm{2D\text{-}out}} \leq 18\mathrm{m} \\ \dfrac{18}{d_{\mathrm{2D\text{-}out}}} + \exp\left(-\dfrac{d_{\mathrm{2D\text{-}out}}}{36}\right)\left(1 - \dfrac{18}{d_{\mathrm{2D\text{-}out}}}\right) & , 18\mathrm{m} < d_{\mathrm{2D\text{-}out}} \end{cases}$
Urban macro	$\mathrm{Pr_{LOS}} = \begin{cases} 1 & , d_{\mathrm{2D\text{-}out}} \leq 18\mathrm{m} \\ \left[\dfrac{18}{d_{\mathrm{2D\text{-}out}}} + \exp\left(-\dfrac{d_{\mathrm{2D\text{-}out}}}{63}\right)\left(1 - \dfrac{18}{d_{\mathrm{2D\text{-}out}}}\right)\right]\left(1 + C'(h_{\mathrm{UT}})\dfrac{5}{4}\left(\dfrac{d_{\mathrm{2D\text{-}out}}}{100}\right)^3 \exp\left(-\dfrac{d_{\mathrm{2D\text{-}out}}}{150}\right)\right) & , 18\mathrm{m} < d_{\mathrm{2D\text{-}out}} \end{cases}$ where $C'(h_{\mathrm{UT}}) = \begin{cases} 0 & , h_{\mathrm{UT}} \leq 13\mathrm{m} \\ \left(\frac{h_{\mathrm{UT}} - 13}{10}\right)^{1.5} & , 13\mathrm{m} < h_{\mathrm{UT}} \leq 23\mathrm{m} \end{cases}$
Indoor – mixed office	$\mathrm{Pr_{LOS}} = \begin{cases} 1 & , d_{\mathrm{2D\text{-}in}} \leq 1.2\mathrm{m} \\ \exp\left(-\dfrac{d_{\mathrm{2D\text{-}in}} - 1.2}{4.7}\right) & , 1.2\mathrm{m} < d_{\mathrm{2D\text{-}in}} < 6.5\mathrm{m} \\ \exp\left(-\dfrac{d_{\mathrm{2D\text{-}in}} - 6.5}{32.6}\right)\cdot 0.32 & , 6.5\mathrm{m} \leq d_{\mathrm{2D\text{-}in}} \end{cases}$
Indoor – open office	$\mathrm{Pr_{LOS}} = \begin{cases} 1 & , d_{\mathrm{2D\text{-}in}} \leq 5\mathrm{m} \\ \exp\left(-\dfrac{d_{\mathrm{2D\text{-}in}} - 5}{70.8}\right) & , 5\mathrm{m} < d_{\mathrm{2D\text{-}in}} \leq 49\mathrm{m} \\ \exp\left(-\dfrac{d_{\mathrm{2D\text{-}in}} - 49}{211.7}\right)\cdot 0.54 & , 49\mathrm{m} < d_{\mathrm{2D\text{-}in}} \end{cases}$
InF-SL InF-SH InF-DL	$\mathrm{Pr_{LOS,subsce}}(d_{\mathrm{2D}}) = \exp\left(-\dfrac{d_{\mathrm{2D}}}{k_{\mathrm{subsce}}}\right)$ where $k_{\mathrm{subsce}} = \begin{cases} -\dfrac{d_{\mathrm{clutter}}}{\ln(1-r)} & \text{for } \mathrm{InF-SL} \text{ and } \mathrm{InF-DL} \\ -\dfrac{d_{\mathrm{clutter}}}{\ln(1-r)} \bullet \dfrac{h_{\mathrm{BS}} - h_{\mathrm{UT}}}{h_c - h_{\mathrm{UT}}} & \text{for } \mathrm{InF-SH} \text{ and } \mathrm{InF-DH} \end{cases}$
InF-DH	The parameters d_{clutter}, r, and h_c are defined in Table 7.2-4 of [3GPP 38.901]
InF-HH	$\mathrm{Pr_{LOS}} = 1$

In most standardized models, the LOS state is defined as having an optical line of sight between TX and RX; consequently, the LOS probability is independent of frequency.[3] However, the impact of small IOs, such as trees, cars, or lantern masts, is usually not taken into account for the definition of LOS probability, but rather modeled separately via additional shadowing/blockage terms.

---

[3] An alternative definition for LOS probability tests whether the first Fresnel zone (compare Section 4.3) is obstructed by an object; this definition results in frequency-dependent LOS probability.

7.1.7 Shadowing

Shadowing is, according to (7.4), the deviation of the SSA-averaged path gain from its (distance-dependent) average. Thus, if a UE moves on a circle around the BS at a distance d_1, it will show variations around a constant value, namely $\overline{PG}(d_1)$. If the UE moves radially away from the BS, the UE experiences an average path gain that decreases monotonically with increasing distance and shadowing variations superimposed on it. These variations are usually modeled as lognormally distributed, as described in Section 5.8. Furthermore, shadowing at different locations may be correlated.

Let the BS location be fixed, and consider S_k and S_m as the shadowing at locations $\mathbf{r}_{\mathrm{UE},k}$ and locations $\mathbf{r}_{\mathrm{UE},m}$ that are separated by Δ_{km}. Then the normalized shadowing correlation function is

$$R_{\mathrm{shad}}(\Delta_{km}) = \frac{E\{S_k S_m\}}{\sqrt{E\{S_k^2\}E\{S_m^2\}}} \tag{7.12}$$

where the expectation in the numerator is taken over the ensemble defined by the different location pairs that fulfill this separation condition, and the expectations in the denominator are taken simply over all $\mathbf{r}_{\mathrm{UE},k}$ and $\mathbf{r}_{\mathrm{UE},m}$, respectively.

The most common model for this correlation function is the Gudmundson model

$$R_{\mathrm{shad}}(\Delta_{km}) = \exp\left[-\frac{\Delta_{km}}{d_{\mathrm{shad\text{-}cor}}}\right] \tag{7.13}$$

where $d_{\mathrm{shad\text{-}cor}}$ is the distance between two points at which the correlation has decayed to a value of $1/e$. It is typically on the order of 10 m (urban microcells) to 100 m (rural macrocells), though it does also exhibit some frequency dependence (smaller at higher frequencies). It is obvious that (7.12) models shadowing as stationary, as it only depends on the distance between locations. Thus, the ensemble over which the expectation is taken in (7.12) can be, e.g., different points $\mathbf{r}_{\mathrm{UE},k}$.

Another model can be based on the angle under which the BS sees the different UEs. Denoting the angle difference between the two observation points as γ_{km}, the correlation model is described as

$$R_{\mathrm{shad}}(\mathbf{r}_{\mathrm{UE},k}, \mathbf{r}_{\mathrm{UE},m}) = c_1 + c_2 \cos(\gamma_{km}), \qquad c_1 + c_2 \leq 1 \tag{7.14}$$

indicating that UEs that are placed along the same radial line from the BS are more strongly correlated, because they share at least part of the propagation environment. The ensemble in this case is different location pairs that have the same angle difference γ between them. More elaborate models that combine distance and angles have been suggested as well.

When the correlation from one BS to multiple UEs needs to be established, the correlations can be written into a matrix, or the shadowing can be seen as a random process, whose realizations at the different locations are determined by the correlation function. For example, lognormal shadowing at multiple locations can be determined by filtering independent normal samples with a filter whose impulse response is determined by the correlation function.

While in most situations the correlation of shadowing from a BS to multiple UEs is of interest, as outlined above, there are also situations where correlation of shadowing from one UE to multiple BSs is required. This becomes relevant, e.g., for the analysis of inter-cell interference, as well as for BS cooperation and distributed Multiple Input Multiple Output (MIMO) (compare Section 22.11). Fewer measurement results exist for those situations, and the correlation is often described by a constant correlation coefficient.

*More Details about Shadowing

The above description follows the "standard" treatment of shadowing in much of the literature. However, this treatment makes a number of simplifying assumptions that are not always fulfilled in practice. Thus, this subsection gives a more exact description. Shadowing is generally the deviation of the Small-Scale Averaged (SSA)-averaged path gain $PG(\mathbf{r}_{\mathrm{BS}}, \mathbf{r}_{\mathrm{UE},k})$ from its large-scale average $\overline{PG}(\mathbf{r}_{\mathrm{BS}}, \mathbf{r}_{\mathrm{UE},k}) = E_{\mathrm{lsf}}\{PG(\mathbf{r}_{\mathrm{BS}}, \mathbf{r}_{\mathrm{UE},k})\}$. In many cases, stationarity of shadowing is not fulfilled. Let the BS location be fixed, and consider the case where the average path gain depends only on the distance between BS and UE

$$PG(\mathbf{r}_{\mathrm{BS}}, \mathbf{r}_{\mathrm{UE},k}) = \overline{PG}(\mathbf{r}_{\mathrm{UE},k} \mid \mathbf{r}_{\mathrm{BS}}) + S_k \tag{7.15}$$

in other words, consider S_k to be the shadowing at $\mathbf{r}_{\mathrm{UE},k}$, the location of the kth UE. The normalized shadowing correlation function then depends on the locations of *both* the UEs, and not just their Euclidean distance:

$$R_{\mathrm{shad}}(\mathbf{r}_{\mathrm{UE},k}, \mathbf{r}_{\mathrm{UE},m}) = \frac{E_{\mathrm{lsf}}\{S_k S_m\}}{\sqrt{E_{\mathrm{lsf}}\{S_k^2\}E_{\mathrm{lsf}}\{S_m^2\}}}. \tag{7.16}$$

where we make use of the fact that shadowing is modeled to be zero-mean. *Only* in the case of stationarity does this become

$$R_{\text{shad}}(\mathbf{r}_{\text{UE},k}, \mathbf{r}_{\text{UE},m}) = R_{\text{shad}}(\|\mathbf{r}_{\text{UE},k} - \mathbf{r}_{\text{UE},m}\|) = R_{\text{shad}}(\Delta_{km}) \tag{7.17}$$

and only depend on the distance between locations, following (7.12).

However, in many situations, the variations seen when moving along a trajectory depend on the starting point, and/or the particular trajectory, even when the covered absolute distance is the same. In this case, stationarity is violated. A typical example would be two streets where one has uniform building height, and the other has strong variations of that height; obviously, in the latter, the shadowing variance can be expected to be much larger.

In the nonstationary case, definition of the ensemble of large-scale fading is more difficult, because we obviously cannot simply use different locations $\mathbf{r}_{\text{UE},k}$. The most common way is to use a set of points that are within a "reasonably close" distance from $\mathbf{r}_{\text{UE},k}$, though the exact threshold is not a settled matter. For example, in a street-by-street pathloss model Section 7.1.5, different points within one street could constitute such an ensemble.

In the nonstationary case, it is important to distinguish the shadowing variance that can be observed when moving along a trajectory in a cell from the variance of S in (7.4). The latter describes the deviations of the observations in a very large ensemble (including different locations within a cell and multiple cells) from its mean, and can thus be much larger than the shadowing variance on a trajectory. Using this deviation from the mean of a large ensemble can be well applied to so-called "drops" of the UEs, i.e., simulations of trajectories that are shorter than a stationarity distance. However, for simulations over longer trajectories, the "true" shadowing variance has to be used, with different trajectories having different variances.

7.1.8 Small-Scale Fading

The different types of small-scale fading have already been described at length in Chapter 5. As we saw, the most commonly used models are Rayleigh fading, Rice fading, and Nakagami fading, which will also be at the center of this section. For some applications, though, different distributions are required. A generalization of the Rice distribution is the so-called *two path with diffuse power* distribution; it is used when, as the name indicates, there are two strong paths (e.g., LOS and ground reflected), plus a number of small components; it has been used, e.g., for propagation over open sea. Furthermore, the Weibull distribution and the $\kappa - \mu$ distribution have been used; see the references in the "further reading" section for details.

*Distribution Extraction from Measurements

The various distributions are described by their parameters – in the case of a Rayleigh distribution that is the average power; in the case of Rice, the power and the Rice factor, and so on. Extraction of these parameters from measurement is done via various estimation algorithms, such as maximum-likelihood estimation or moment matching.

When analyzing measurement results, it is usually not obvious which of the many possible fading distributions is the best for the available results. The "goodness of fit" can be tested by the *Kolmogorov–Smirnov* (KS) test, i.e., it tests the hypothesis that the empirical samples come from the theoretical distribution (with the estimated optimum parameters). This is repeated for all candidate theoretical distributions; the distribution with the highest pass rate is then chosen to be the true one. Alternatively, the *Akaike Information Criterion* (AIC) can be used to decide between different distributions; this criterion also takes into account the number of free parameters in a candidate distribution, preferring simpler distribution, and thus aiming to avoid overfitting.

Modeling of the Small-Scale Fading

The parameters of the fading distributions can show large-scale variations, and this variation is often described as stochastic, i.e., the parameters are seen themselves as random variables with certain distributions. In the case of Rayleigh fading, the parameter is the SSA-power, which varies according to the shadowing, and is therefore (usually) distributed lognormally. For Rician fading, it is often assumed that the K-factor is lognormally distributed as well; importantly, a Rice factor can be nonzero for both LOS and NLOS situations, but the mean Rice factor is obviously much higher in the LOS case (typically on the order of $[10, 20]$ dB) compared to the NLOS case (typically $[-10, 3]$) dB. Statistical distributions of parameters for other small-scale fading parameters are less well established.

Another important factor of small-scale fading is the temporal autocorrelation function or, equivalently, the Doppler spectrum. In the case of Rayleigh fading, the Jakes spectrum (see Section 5.6) is the most widely used form, and implementations were discussed in Chapter 5. Other Doppler spectra have been derived for various special assumptions of IO distribution around the UE. It must be kept in mind that for the important case of fixed BS and IOs and moving UEs, the Doppler spectrum follows immediately from the angular spectrum of the MPCs at the UE (see Section 5.6). Thus, the directional channel models that are prevalent nowadays have no need for explicit Doppler (or autocorrelation function) models, since the angular spectrum at the UE is already modeled anyway.

7.2   Delay Dispersion Models

7.2.1   Tapped Delay Line

Delay dispersion is characterized by the channel impulse response or the Power Delay Profile (PDP). For theoretical derivations, continuous models of these quantities are popular; however, for computer simulations, discretized (sampled) models are used. Many such discretized approaches use some variant of a tapped delay line model, i.e.,

$$h(t,\tau) = \sum_{i=1}^{\widetilde{N}} c_i(t)\delta(\tau - \tau_i), \tag{7.18}$$

where the τ_i are often regularly spaced, i.e., $\tau_i = i \cdot \Delta\tau$ and the $c_i(t)$ are fading, because they describe the sum of many MPCs, sampled (usually) at the Nyquist rate, see Figure 7.5. The tap delays and the fading statistics (including the average power) of each tap characterize the impulse response within the constraints of the model. Those constraints are: (i) the impulse response does not reflect pathloss and shadowing, which have to be superimposed on $h(t, \tau)$, (ii) the number of the MPCs, and the average power carried by each tap, is constant over the considered time range (area), and (iii) the delays of the taps are constant over the simulation range. Conditions (ii) and (iii) imply that the simulation range is smaller than the "region of stationarity."

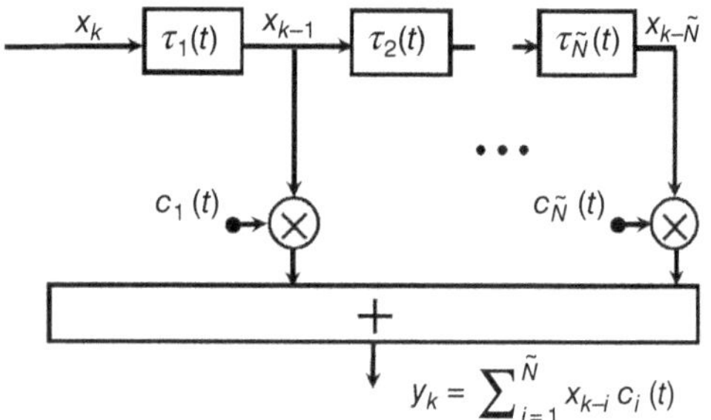

Figure 7.5   Tapped-delay line model. Color version available at wiley.com/go/molisch/wireless3e.
Reproduced with permission from [Rappaport et al. 2022] © Cambridge University Press.

    Tapped delay line models are usually derived from measurements with a specific bandwidth, and/or are intended for the emulation of systems with a particular bandwidth, which in turn determines the spacing between the tap delays τ_i. We note that using models for a bandwidth larger than the one they were designed for (i.e., for which measurements are available) is *not* admissible and can cause completely misleading simulation results.

    The tap small-scale fading statistics are commonly the same as the narrowband small-scale fading statistics discussed in Section 7.1.8: in the delay bin corresponding to a line-of-sight path, Rician fading statistics usually apply while in other bins, where a large number of similar-strength components fall into a delay bin, fading statistics are Rayleigh. As signal bandwidth increases and delay bin width correspondingly decreases, fewer MPCs fall into each bin, leading to changed statistics (e.g., Nakagami or Rice); this transition occurs typically for bandwidths of a few hundred MHz. In the limiting case of very large bandwidth, the MPCs become truly resolvable and represent individual MPCs, and thus may be nonfading over time and space.

    When time variations of the channel are to be included, also the temporal correlation function (or equivalently the Doppler spectrum) needs to be defined, just like in the narrowband case discussed in Section 7.1.8. This has to be done for every delay tap, though some models assume a separable scattering function (i.e., the Doppler spectrum is independent of the delay). Simulating such a (single) delay tap with a given Doppler spreading is formally the same as simulating a flat fading channel, and can be done using the same methods as for flat-fading simulation, e.g., sum of sinusoids.

    An especially simple case occurs when the number of taps is limited to $\widetilde{N} = 2$, and no LOS component is allowed. This is the simplest stochastic fading channel exhibiting delay dispersion,[4] and thus very popular for theoretical analysis. It is alternatively called *two-path channel, two-delay channel,* or *two-spike channel.*

    Another popular channel model consists of a purely deterministic LOS component, plus *one* fading tap ($\widetilde{N} = 1$) whose delay τ_0 can differ from τ_1. This model is widely used for satellite channels – in those channels, there is almost always a LOS connection, and the reflections from buildings near the RX give rise to a delayed fading component. The channel reduces to a flat-fading Rician channel when $\tau_0 = \tau_1$.

    Generalizations of the model can take into account changes in local area channel statistics. This includes changes in the average powers, and in the extreme case, the appearance and disappearance of paths as the UE move over larger distances. This effect can be implemented as a stochastic birth-death process, where at randomly chosen times a path appears, with the average generation rate

---

[4] Note that each of the taps of this channel exhibits Rayleigh fading; the channel is thus different from the two-path model used in Chapters 5 and 6 for purely didactic reasons.

being a model parameter (and similarly for the disappearance of paths). Furthermore, changes in the delays of the paths can be emulated. Alternative ways of implementing these effects will be discussed in Section 7.6.

7.2.2 Clustered Models

It has been observed in many measurements that the (continuous) PDP can be approximated by a one-sided exponential function

$$P_h(\tau) \propto P_{sc}(\tau) = \begin{cases} \exp\left(-\tau/S_\tau\right) & \tau \geq 0 \\ 0 & \text{otherwise} \end{cases}. \tag{7.19}$$

In a more general model (see also Section 7.4.2), the PDP is the sum of several delayed exponential functions, corresponding to multiple *clusters* of IOs

$$P_h(\tau) = \sum_m \frac{E_m^c}{S_{\tau,m}^c} P_{sc}\left(\tau - \tau_{0,m}^c\right) \tag{7.20}$$

where E_m^c, $\tau_{0,m}^c$, S_m^c are the energy, delay, and delay spread of the mth cluster, respectively. The sum of all cluster powers has to add up to the narrowband power described in Section 7.1.

7.2.3 The Saleh–Valenzuela Model

For systems with high bandwidth, MPCs can be resolved in the delay domain. In that case, it is advantageous to describe the PDP by the arrival times of the MPCs, plus an "envelope" function that describes the power of the MPCs as a function of delay. This again leads to a discrete (tapped delay line) channel model, though with different interpretations: the taps do not represent the (fading) sum of many MPCs sampled at the Nyquist rate, but rather actual, discrete MPCs that are not subject to small-scale fading.

In order to statistically model the arrival times of MPCs, a first-order approximation assumes that objects that cause reflections in an urban area are located randomly in space, giving rise to a *Poisson distribution* for the excess delays. However, measurements have shown that MPCs arrive in clusters, as discussed above. This fact is popularly modeled by the *Saleh–Valenzuela (SV) model*. It assumes that *within* each cluster, the MPCs are arriving according to a Poisson distribution and that the arrival times of the *clusters themselves* are Poisson distributed (but with a different interarrival time constant). Furthermore, the powers of the MPCs within a cluster decrease exponentially with the delay, and the powers of the clusters follow a (different) exponential distribution, see Figure 7.6.

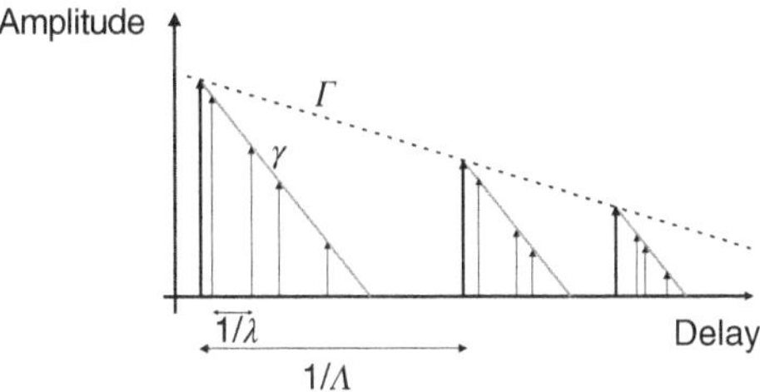

Figure 7.6 The Saleh–Valenzuela model.

Mathematically, the following discrete-time impulse response is used

$$h(\tau) = \sum_m \sum_k a_{k,m}(\tau)\delta(\tau - T_m - \tau_{k,m})$$

where $\{a_{k,m}\}$ are the MPC amplitudes,[5] and the distribution of cluster arrival time and the ray arrival time are given by

$$pdf(T_m|T_{m-1}) = \Lambda \exp\left[-\Lambda(T_m - T_{m-1})\right], m > 0$$
$$pdf\left(\tau_{k,m}|\tau_{(k-1),m}\right) = \lambda \exp\left[-\lambda\left(\tau_{k,m} - \tau_{(k-1),m}\right)\right], k > 0$$

---

[5] Note that the grouping of MPCs into clusters is fundamentally different from the grouping of MPCs into fading taps that form the basis of the tapped delay line model. For the clustering, we group together MPCs that behave in a similar way, but which may be separable by the RX. For the tapped delay line, we group together MPCs that are not resolvable by the RX. A tap and a cluster may coincide by accident, but they are fundamentally different quantities.

where T_l is the arrival time of the first path of the mth cluster; $\tau_{k,m}$ is the delay of the kth path within the mth cluster relative to the first path arrival time (by definition, $\tau_{0,m} = 0$); Λ is the cluster arrival rate; λ is the ray arrival rate, i.e., the arrival rate of paths within each cluster. All dependencies of the parameters on absolute time have been suppressed in the above equations.

The PDP within each cluster is

$$E\{|a_{k,m}|^2\} = \frac{E_m^c}{\gamma_m} \exp\left(-\tau_{k,m}/\gamma_m\right) \tag{7.21}$$

where E_m^c is the energy of the mth cluster, and γ_m is the intra-cluster decay time constant; in many situations, it is assumed to be independent of m. The cluster power decreases exponentially as well,

$$E_m^c \propto \exp\left(-T_m/\Gamma\right) \tag{7.22}$$

where Γ is the inter-cluster decay time constant.

*7.2.4 Standardized Models

A special case of the tapped delay-line model is the COST 207 model, which specifies the PDPs or tap weights and Doppler spectra for four typical environments. These PDPs were derived from numerous measurement campaigns in Europe. The model distinguishes between four different types of macrocellular environments, namely *Typical Urban* (TU), *Bad Urban* (BU), *Rural Area* (RA), and *Hilly Terrain* (HT). Depending on the environment, the PDP has a single-exponential decay, or it consists of two single-exponential functions (clusters) that are delayed with respect to each other. The second cluster corresponds to groups of far-away high-rise buildings or mountains that act as efficient IOs, and thus give rise to a group of delayed MPCs with considerable power. Details are given in Appendix 7C.

The COST 207 models are based on measurements with a rather low bandwidth and are applicable only for systems with 200 kHz bandwidth or less. For the simulation of third-generation cellular systems, which have a bandwidth of 5 MHz, the *International Telecommunications Union ITU* specified another set of models that accounts for the larger bandwidth. This model distinguishes between pedestrian, vehicular, and indoor environments. Additional tapped delay line models were also derived for indoor wireless LAN systems and Personal Area Networks; for an overview, see [Molisch and Tufvesson 2004]. However, most of these models have been superseded by the more general double-directional models discussed in Section 7.4.

7.2.5 Typical Values for Delay Dispersion

Delay dispersion is often described by means of the rms delay spread. While this parameter is sensitive to measurement errors and does not always describe how delay dispersion will impact the system performance, still, the vast majority of measurement campaigns available in the literature provide (only) this parameter to characterize an environment.

The following describes typical values of the delay spread for different environments. Note that the specifics of the surroundings play a role, and thus both smaller and larger values have been measured. Most results have been measured for below 6 GHz, though results for higher frequencies, where they are available, tend to be in a similar range. Remarkably, the use of directional antennas has a somewhat limited impact on the delay spread, with antenna beamwidth on the order of 20° reducing the delay spread by less than a factor of 2 (though this depends, of course, on the environment). For more details and references see [Molisch and Tufvesson 2004] for below 6 GHz, and [Molisch et al. 2021] for mm-wave.

- *Indoor residential buildings*: 5–10 ns are typical; but up to 30 ns have been measured
- *Indoor office environments* show typically delay spreads between 10 and 100 ns in NLOS but is significantly smaller in LOS, but even 300 ns have been measured. The room size has a clear influence on the delay spread. Also, building size and shape have an impact.
- *Factory halls, malls, and airport halls* have typical delay spreads that range from 50 to 200 ns, with up to 500 ns occurring in some cases.
- In *microcells*, delay spreads range from around 5–100 ns (for LOS situations) to 100–500 ns (for NLOS). For mm Wave frequencies, values between 10 and 60 ns were usually obtained, though at certain locations within the measurement area much larger values (several hundred ns) were found.
- *Tunnel and mines*: Empty tunnel delay spreads are on the order of 20 ns, while car-filled tunnels exhibit up to 100 ns.
- *Typical urban and suburban environments* show (for <6 GHz) delay spreads between 100 and 800 ns, although values up to 3 µs have also been observed. Delay spreads depend on the height of the BS (compared to the surrounding buildings), with macrocells (BS above the surrounding rooftops) resulting in higher delay spreads than microcells.
- *Bad urban and hilly terrain environments* show clear examples of multiple clusters that lead to much larger delay spreads. Delay spreads up to 18 µs, with cluster delays of up to 50 µs have been measured in macrocells in various European cities, while American cities show somewhat smaller values. Cluster delays of up to 100 µs occur in mountainous terrain.

The delay spread is a function of the BS–UE distance, increasing with distance approximately as d^ε, where $\varepsilon = 0.5$ in urban and suburban environments, and $\varepsilon = 1$ in mountainous regions. The delay spread also shows considerable large-scale variations. Several papers find that the delay spread has a lognormal distribution (i.e., $10\log_{10}(S_\tau/1s)$ has a Gaussian distribution) with a variance of typically 2–3 dB in suburban and urban environments. A comprehensive model containing all those effects was first proposed by [Greenstein et al. 1997].

It must be noted that generally, the delay spread does not decrease significantly with frequency when measuring in the same environment, the same TX/RX heights, and the same antennas. The often made assumption that delay spread decreases with increasing frequency seems to stem from measurements with equipment that had lower dynamic range at higher frequencies, or because comparisons were made between low-frequency channels measured with omni-directional antennas versus high-frequency channels with directional antennas.

7.3 Angular Dispersion

In this section we separately analyze the angular dispersion at the BS and the UE, without discussing the joint characteristics and/or the impact of the delay on these characteristics; those will be discussed further in Section 7.4.

7.3.1 Angular Dispersion at the BS

The most common model for the azimuthal Angular Power Spectrum (APS) at the BS is a Laplacian distribution [Pedersen et al. 1997], i.e., there is only a single cluster of MPCs, and shape of the APS is given as

$$APS(\phi) = \exp\left[-\sqrt{2}\frac{|\phi - \phi_0|}{S_\phi}\right], \tag{7.23}$$

where ϕ_0 is the mean azimuthal angle; for the LOS case, ϕ_0 coincides with the LOS direction.

Also for the angular dispersion, the rms (angular) spread (see Section 6.7) is a common measure. The following range of rms angular spreads can be considered typical (again, more extreme values might occur in specific surroundings):

- *Indoor office environments*: rms angular spreads between 10° and 30° for NLOS situations, and typically 5° and 15° for LOS at frequencies <6 GHz. For mm-wave frequencies, in NLOS situations, values on the order of 20°–70° have been measured.
- *Industrial environments*: rms angular spreads between 20° and 30° for NLOS situations.
- *Microcells*: rms angular spreads between 5° and 20° for LOS, and 10°–40° for NLOS have been observed for frequencies <6 GHz. For mm-wave, values on the order of 10° have been measured.
- *Typical urban and suburban environments*: measured rms angular spreads on the order of 3°–20° in dense urban environments, with macrocells being at the lower end of the range, and microcells on the upper end. In suburban environments, the angular spread is smaller, due to a frequent occurrence of LOS.
- *Bad urban and hilly terrain environments*: rms angular spreads of 20° or larger, due to the existence of multiple clusters.
- *Rural environments*: rms angular spreads between 1° and 5° have been observed.

In outdoor environments, the distribution of the angular spread over large areas has also been found to be lognormal and correlated with the delay spread. This permits the logarithms of the spreads to be treated as correlated Gaussian random variables. The dependence of the angular spread on the distance is still a matter of discussion.

For BS arrays that can resolve both azimuth and elevation, such as rectangular arrays, the elevation characteristics are of importance as well. The elevation spectrum at the BS is typically modeled as Laplacian

$$APS(\theta) = \exp\left[-\sqrt{2}\frac{|\theta - \theta_0|}{S_\theta}\right]. \tag{7.24}$$

The measured elevation spreads S_θ are typically around 3°–5° in macrocellular environments, while larger values, typically 5°–15° have been measured in microcells. As is common for angular spreads, values tend to be smaller in rural environments, and also in LOS situations. Elevation spreads in indoor environments are generally larger, because floor and ceiling reflections can have a significant impact. However, the values can depend significantly on the clutter on the floor, as well as the material and structure of the ceiling.

7.3.2 Angular Dispersion at the UE

For outdoor environments, it is commonly assumed that radiation is incident from all azimuthal directions onto the UE, because the UE is surrounded by "local IOs" (cars, people, houses, etc.). This model dates back to the 1970s. However, more recent studies indicate that the azimuthal spread can be considerably smaller, especially in street canyons. The azimuthal APS is then again approximated as Laplacian; cluster angular spreads on the order of 20° have been suggested. Furthermore, the angular distribution is a function of the delay. For UEs located in street canyons without LOS, MPCs/clusters with small delays are related to over-the-rooftop propagation, which may result in large azimuthal spreads of those components, and mean in the geometrical direction of the LOS. Later components are waveguided through the streets and thus confined to the smaller angular range and have a mean azimuth along the axis of the street. In Indoor environments with (quasi-) LOS, early components have a small angular spread, while components with larger delay have an almost uniform APS, and often can be modeled as multiple clusters stemming, e.g., from reflections on the different walls of the room.

For mm-wave frequencies, azimuthal spreads are often in the range $30°$–$70°$. While ray tracing/launching generally predicts the angular spreads at the BS well, it tends to significantly underestimate the angular spreads at the UE, because many scattering objects such as street signs, parked cars, etc., are not included in the geographical databases used for ray tracers.

For the elevation spectrum, extensive measurements in [Kalliola et al. 2002] showed that an (asymmetrical) double-exponential function provides, in outdoor situations, a good fit for the elevation power spectrum measured at the UEs, see Figure 7.7. Elevation spreads measured in that reference were typically less than $10°$–$15°$, though other references have also reported larger values. The elevation spreads may also depend on the delay, with early-arriving components (propagating over the rooftops and diffracted into the street in which the UE is located) showing larger elevation spread than later components that have been propagating mostly in the horizontal plane by waveguiding through the street canyons.

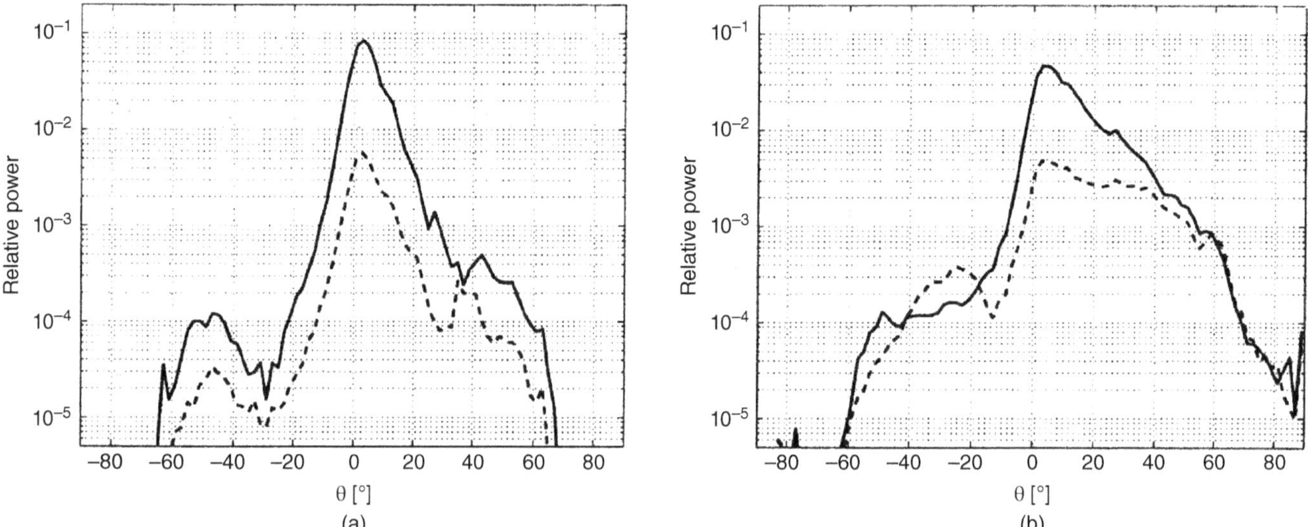

Figure 7.7 Elevation spectra for urban microcell (a) and macrocell (b) at the UE. θ here denotes co-elevation, i.e., $\theta = 0$ is horizontal. Reproduced with permission from [Kalliola et al. 2002] © IEEE.

7.4 Joint Dispersion Characteristics and Clustering

7.4.1 General Model Structure and Factorization

The previous two sections discussed the marginal distributions, PDP and APS, of the delay and angle dispersion at BS and UE. However, for most system simulations, we are interested in the joint characteristics; this is particularly true for MIMO systems (see also Chapters 16 and 22).

The simplest case is obtained if the Double-Directional Delay Power Spectrum, DDDPS, can be factored into three functions; in other words that the joint distribution is the product of the marginals (which we discussed in Sections 7.2 and 7.3)

$$DDDPS(\tau, \Omega, \Psi) = APS^{\mathrm{BS}}(\Omega)APS^{\mathrm{UE}}(\Psi)P_h(\tau). \tag{7.25}$$

This implies that the angular power spectrum at the BS is independent of the delay, as is the angular power spectrum at the UE. Furthermore, the angular power spectrum at the UE is independent of the direction into which the BS transmits, and vice versa.[6] Such a factorization greatly simplifies theoretical computations and also the parameterization of channel models. However, it does not always correspond to physical reality.

---

[6] Note that we here implicitly assume a downlink transmission, equating the BS with the TX. This is purely a matter of notational convenience.

7.4.2 Clustering

Measurement results show that MPCs typically arrive in clusters, i.e., groups of MPCs that have similar characteristics, while being significantly different from the characteristics of MPCs in other clusters. The clustering effect arises, e.g., because the MPCs are created by the interaction of the TX signal with objects located in a certain region, like a group of high-rise buildings or mountains, or from waves undergoing similar waveguiding processes in a corridor or street canyon. To be more precise, we need to distinguish between *MPC clusters* and *IO clusters*. The former is a group of MPCs that have "similar" characteristics, while the latter are groups of physical objects that – through interaction with incident waves – can give rise to MPC clusters.

Among MPC clusters, we furthermore distinguish two definitions. The most physically motivated is that a cluster is a group of MPCs that show similar large-scale behavior: if a UE moves over a large distance (much larger than a stationarity region), the variations of power, delay, Direction-of-Arrivals (DoAs), and Direction-of-Departures (DoDs) of the MPCs are highly correlated (the threshold for this correlation to be sufficient is an arbitrary parameter to be set for analysis). Clearly, these definitions can only be applied if large-scale measurements (or deterministic simulations) tracking the UE over large distances are available. Alternatively, we can define an MPC cluster as a group of MPCs with similar parameters (τ, Ω, Ψ), surrounded by "areas" (in the (τ, Ω, Ψ) space) that contain no MPCs. In the latter case, clustering can be done (i) by visual inspection, which is easy to do for small sets of measurement data but does not scale well to large measurement campaigns, and also may suffer from the fact that human inspection is by its nature subjective, or (ii) automated clustering algorithms, such as K-power means, or K-density; the reader is referred to [Huang et al. 2019b] for an extensive review of these techniques.

With either definition, the indices of the MPCs $h_i(\tau, \Omega)$, $i = 1, ..., N$ can be grouped into $M \leq N$ *disjoint* classes (or clusters) $C_1, ..., C_M$, where each class has $N_m \geq 1$ elements, and

$$\sum_{m=1}^{M} N_m = N, \tag{7.26}$$

so that the impulse response can be rewritten as

$$h(t, \tau, \Omega, \Psi; \mathbf{r}_{\text{TX}}, \mathbf{r}_{\text{RX}}) = \sum_{m=1}^{M} \sum_{k \in C_m} h_k(t, \tau, \Omega, \Psi; \mathbf{r}_{\text{TX}}, \mathbf{r}_{\text{RX}}). \tag{7.27}$$

The concept of clusters is useful because the parameters of a cluster change only over very large scales. To give an example, the PDP of a single cluster is always an exponential function. When the Average Power Delay Profile (APDP) consists of three clusters, we have a total of three exponentials. When the UE moves over large areas, the position of the exponentials relative to each other changes, and even the decay constants of the clusters may change, but the shape of the cluster PDP remains unchanged. In addition, it is often convenient and reasonably accurate to assume that each of the cluster DDDPS can be decomposed, into a product of PDP, APS at the BS, and APS at the UE, even when this decomposition does not hold for the *total* DDDPS:

$$DDDPS(\tau, \Omega, \Psi) = \sum_m P_m^{\text{c}} APS_m^{\text{c,BS}}(\Omega) APS_m^{\text{c,UE}}(\Psi) P_{h,m}^{\text{c}}(\tau) \tag{7.28}$$

where superscript ^{c} stands for "cluster," P_m^{c} is related to the power of the mth cluster, and m indexes the clusters. Obviously, this model reduces to Eq. (7.25) if only a single cluster exists.

*7.4.3 Physical Propagation Effects Impacting Clustering

The clustering process and in particular the relationship between the clusters can be best understood from analyzing the physical propagation processes.

Urban Environments

In urban outdoor scenarios, there are three main effects dominating the propagation from BS to UE:

1. *MPCs propagating over the rooftops*, and then being diffracted in the vertical plane toward the UE. The mean azimuths at the BS are mainly in the direction of the UE (even when no LOS exists), and the mean elevation is in the direction under which the BS "sees" the roof edges. The elevation at the UE is determined by the angle under which the UE "sees" the roof edge in the street canyon it is located in; these angles can be quite large (on the order of 60°, typically).

2. *MPCs waveguided in street canyons*: as discussed in Section 4.5, a street canyon essentially constitutes a "waveguide," even though this waveguide has a missing top, and sidewalls that have "holes" (cross-streets), and surfaces with nonideal reflection coefficients. In such a structure, the waves propagate essentially in the horizontal plane, leading to small elevation angles at BS and UE. The mean azimuth direction at the UE is determined by the orientation of the street canyon. At the BS of microcells, the mean azimuth

of such a canyon-guided cluster is also determined by the street orientation. In a macrocell, the dominant direction at the wave propagation is determined by the angle between the BS and the location of the canyon opening into which the waves couple.

3. *MPCs reflected at "far IOs"*: in a number of environments, high-rise structures can constitute efficient far IOs for MPCs. Such a structure would have LOS to both the BS and the UE, and a specularly reflecting surface. Thus, the MPCs suffer relatively little attenuation (proportional to $(d_{\text{BS-scatt}} + d_{\text{scatt-UE}})^2$). The elevation angle under which such a an IO is seen at the BS is usually positive but with a relatively small value.

An example of these mechanisms is shown in Figure 7.8.

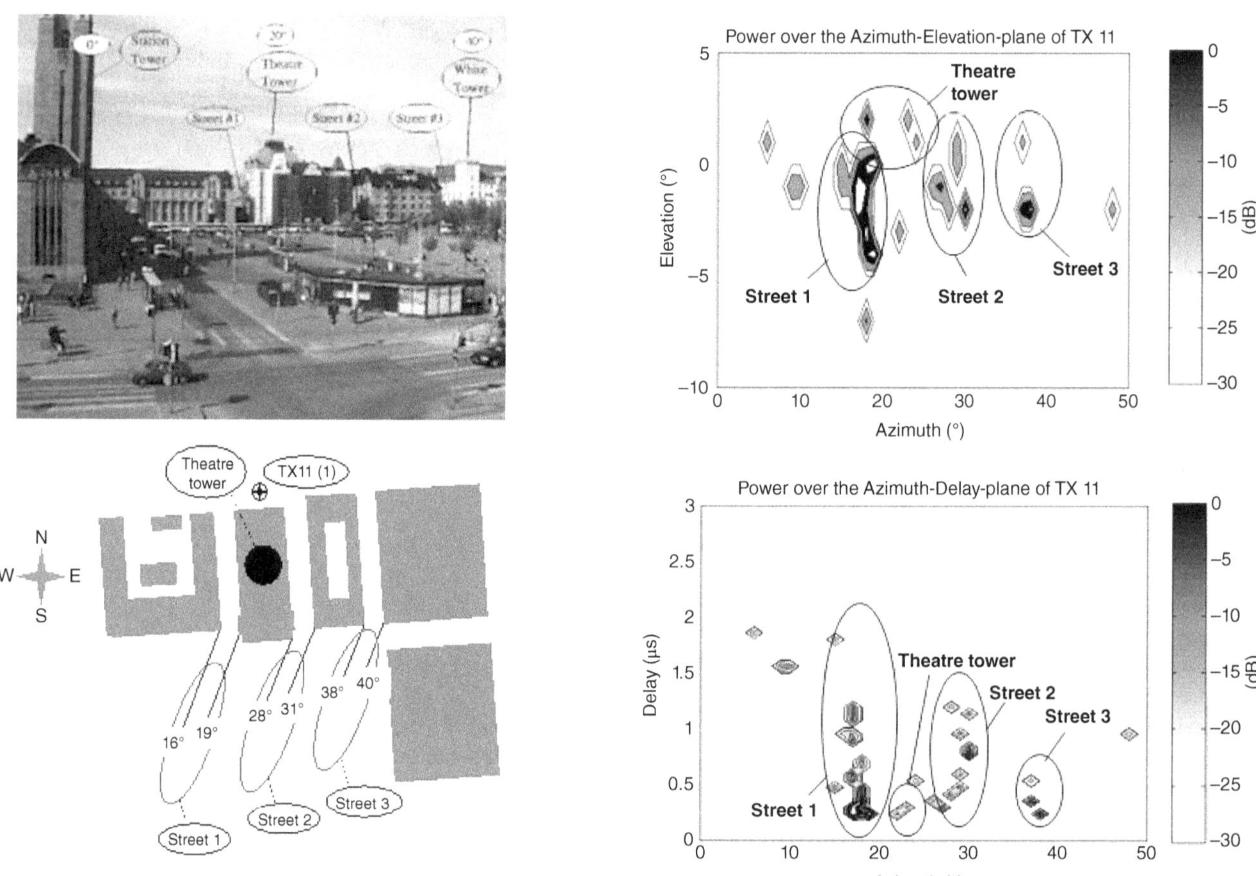

Figure 7.8 Photograph from BS location, and APS with propagation paths for the clusters. Theatre tower is a tall building acting as far IO. Color version available at wiley.com/go/molisch/wireless3e.
Reproduced with permission from [Toeltsch et al. 2002] © IEEE.

Outdoor-to-Indoor

In outdoor to indoor propagation two main factors determine the clustering structure:

- The attenuation by the building wall and windows. At low frequencies, and for thin walls (e.g., post-and-drywall construction), penetration loss by the building walls is small. In this case, the angular characteristics at the BS do not change significantly compared to the case when the UE is outdoor, while the characteristics at the UE are mainly determined by the angles of the MPCs that are incident on the building, as well as reflections from wall and ceiling of the room in which the UE is located. Steel-concrete walls have much higher attenuation, such that windows might be the major entry points of the MPCs into the building; if the windows have energy-saving glazing, then they have high attenuation as well, in particular at higher frequencies. The penetration loss might also depend significantly on the angle at which the MPCs are incident onto the building, with clusters at grazing angle attenuated much more significantly than those at normal incidence. All this can significantly change the clustering structure. This is shown, e.g., in Figure 7.9, which compares the MPCs for the situations that the UE is outdoors and indoors for a 28 GHz link.

- The height of the floor in which the UE is located impacts the *effectiveness* of the different processes, and thus their relative weights. This is most noticeable for the over-the-rooftop propagation, since the diffraction angle with respect to the rooftops changes as the UE height increases. In extreme cases, a UE on an upper floor can have LOS to the BS, even though the UE at street level does not. A special case occurs when a BS is located outside a high-rise building and has LOS to the different floors. In this case, we can expect that the mean elevation at the BS is approximately the LOS direction (plus some variations).

(a)

(b)

(c)

Figure 7.9 Bird's eye view of propagation scenario (a). Strengths and directions of mm-wave MPCs superimposed on photograph from UE location outside (b) and inside (c) a building. It can be seen that components that are strong outdoors, but incident on the building at a grazing angle, vanish or show very weak amplitude indoors, while components entering through the glass door show significant strength. Color version available at wiley.com/go/molisch/wireless3e.

Reproduced with permission from [Bas et al. 2019] © IEEE.

Indoor

If BS and UE are in the same room, then the directions of the clusters follow from the geometrical relationship of LOS angle and the angles under which single or multiple reflections depart from/arrive at the link ends. Reflections from different walls typically provide different clusters. In setups where many small offices are along a corridor, the strongest propagation path might be out the door of the office in which the BS is located, waveguided down the corridor, and through the door into the office with the UE; again the cluster angles follow from the geometrical relationships.

7.4.4 Cluster Parameter Correlation

The above considerations suggest, and numerous measurement campaigns have confirmed, that shadowing, delay dispersion, and angular dispersion are correlated with each other. More specifically, the cluster rms delay spread, rms angle spread at the BS, and shadowing, can be modeled as correlated lognormally distributed random variables according to

$$S_m = 10^{\sigma_{\mathrm{shf}} X_m/10},$$

(7.29)

$$S_{\varphi,m} = \mu_{s\varphi}10^{\sigma_{s\varphi}Y_m/10}, S_{\theta,m} = \mu_{s\theta}10^{\sigma_{s\theta}W_m/10}, \tag{7.30}$$

$$\bar{S}_{\tau,m} = \mu_{s\tau}\left(\frac{d}{d_{\text{ref}}}\right)^{\varepsilon}10^{\nu_{s\tau}Z_m/10}, \tag{7.31}$$

where X_m, Y_m, Z_m, W_m are random Gaussian variables with zero mean, unit variance, and a 4×4 cross-correlation matrix that can be parameterized from measurements. The correlations between different clusters are zero. The parameters σ_{shf}, $\sigma_{s\tau}$, $\sigma_{s\varphi}$, $\sigma_{s\theta}$ are standard deviations expressed in dB. The median azimuth spread is $\mu_{s\varphi}$ while $\mu_{s\tau}$ is the median delay spread at a reference distance d_{ref}. The delay spread has been observed to increase with distance as d^{ε}, as discussed above (some models set $\varepsilon = 0$ to simplify the model).

*7.4.5   Modeling Hierarchies

Radio propagation depends on topographical and electromagnetic features of the operating environment. To account for variations in these characteristics, stochastic and semi-stochastic models have to have different parameters (and possibly different structures) in different environments. To systematically accommodate a wide variety of environments and transmission situations, some models have adopted a hierarchical structure. For example, the COST 259/273/2100 models, as well as the 3GPP use the 3-level structure:

At the top level, a first distinction is made by the cell type, namely macro-, micro-, and picocells, which are defined as the BS antenna being above rooftop, below rooftop, or indoor, respectively. For each cell type, a number of *Radio Environments* (RE's) have been identified, which represent a whole class of multi-path conditions that give rise to similar radio channel characteristics. Such REs can be related to the surroundings in which the system operates, e.g., indoor office, factory hall, outdoor hotspot, etc. The features of an RE are defined by a number of *external parameters*, such as the frequency band, the average height of BS and UE antennas, their average distance from each other, and average building heights and separations. The bottom layer of the hierarchical structure of modeling constructs consists of the *propagation scenarios*, which are defined as random realizations of multi-path conditions.

The propagation conditions of each RE are described by a set of fixed parameters, as well as probability density functions for stochastic parameters, such as the mean and the standard deviation of the cluster delay spread. Since they characterize the propagation conditions of the entire radio environment, they are called *Global Parameters* (GPs). Propagation scenarios are controlled by the GPs of the associated RE. *Local Parameters* (LPs) are random realizations of parameters that describe channel conditions in a local area. For movements of the UE within a sufficiently small *local area*, not larger than some tens of wavelengths, the LPs determining the propagation scenario remain approximately constant. The simulation of the small-scale fading, whose characteristics are determined by the local parameters, then provides the different realizations of the impulse responses.

7.5   Generalized Tapped-Delay Line Models

7.5.1   Fundamental Considerations

The concept of tapped-delay line models can be extended to model the directional characteristics of the MPCs as well. Equation (7.18) can be rewritten as

$$h(t, \tau, \Omega, \Psi) = \sum_{i=1}^{\widetilde{N}} c_i(t)\delta(\tau - \tau_i)\delta(\Omega - \Omega_i)\delta(\Psi - \Psi_i) \tag{7.32}$$

or, in clustered form

$$h(t, \tau, \Omega, \Psi) = \sum_{m=1}^{M} \sum_{k \in C_m} c_{k,m}(t)\delta(\tau - \tau_{k,m})\delta(\Omega - \Omega_{k,m})\delta(\Psi - \Psi_{k,m}). \tag{7.33}$$

In this latter form, it is often assumed that the tap coefficients $c_{k,m}(t)$ are nonfading, or only slowly fading, while the small-scale fading is created by the superposition of taps within a cluster. The prescription of the inter-cluster and intra-cluster characteristics in both delay and angular domains then determines the model.

*7.5.2   The 3GPP Model

Among the many existing models of this type, the practically most important is the 3GPP channel model, which is used in the context of the standardization of cellular systems, compare Chapters 31 and 32. The model claims validity for a frequency 0.5–100 GHz and only a few parameters are modeled to be frequency dependent.

The model defines clusters C_m as a group of paths that have exactly the same delay but slightly different angles. The delays of the clusters are either given deterministically (this is used in the so-called "link-level model"), or are assigned at random, according to a given (parameterized) probability density function (this is used in what is called the "system level model"). Power of the cluster is assigned according to the path delay, with the average power (over shadowing) decreasing with increasing delay.

Turning now to the angular dispersion: the baseline of the angles is the LOS connection between the two link ends (such a connection can be drawn even for those cases that an actual physical LOS component does not exist); this establishes a geometrical relationship between UE location and angles at BS and UE.[7] Each cluster has a mean cluster angle that deviates from the baseline (equally likely to the right and the left from the LOS), which is created according to a specified probability density; and which increases with increasing delay. Note that the angle deviations at BS and UE are chosen independently, but they are still somewhat correlated in that large delays lead to large deviations from the baseline angle at *both* link ends. Last but not least, each cluster has itself an angular spread, which takes on a deterministic value, such as 5°. This is realized in that a cluster consists of 20 taps, which all have slightly different angles. Each of the taps has the same amplitude and random phases; their superposition thus provides not only an angular spread but also small-scale fading when either the UE moves, or different values of the random phases are chosen.

The 3GPP model also includes polarization and can thus be combined with dual-polarized antenna setups. Determination of the transfer functions between the various TX and RX antenna elements can be done according to (6.71).

The rms angular spreads and delay spreads in the different environments can be described as random variables that are lognormally distributed, so that their cdf is sufficiently characterized by their mean and standard deviation. These values are prescribed in the standard separately for LOS, NLOS and (in applicable cases) for Outdoor-to-Indoor (O2I) situations, see Appendix 7.D. For the four environments of Umi street canyon, UMa, RMa (up to 7 GHz), and InH, one can notice a frequency dependency of the angular spreads, and delay spread mean and standard deviations. However, apart from this, the ultimately adopted specifications show little dependence on the carrier frequency. Thus, while claiming to be valid up to 100 GHz, the model mainly has an experimental basis only for <6 GHz.

A stepwise procedure for the generation of impulse response is shown in Figure 7.10. First, one can choose a radio environment, e.g., rural macrocellular, urban macrocellular, urban microcellular, etc., as well as the associated radio system layout, antenna parameters (numbers of elements, antenna gain, and beamforming architecture). Then the pathloss is computed for the UEs in LOS and NLOS conditions, followed by computing a realization of the correlated large-scale propagation parameters. At this stage, the generation of small-scale parameters begins: compute the realization of the cluster random delays, DoA and DoD, and cluster powers. All of these parameters are random variables with distributions that are governed by the large-scale parameters mentioned above. The cluster realizations are then used to generate the impulse response: each cluster contains 20 paths with the same delay and (deterministic) angle deviation from the cluster center. Each of the 20 possible DoAs is associated randomly with one of the possible 20 AoDs, and the initial phase of the path corresponding to the DoA/DoD pair is selected at random as well. From this, and the knowledge of the complex antennas patterns, the fading coefficients at the different antenna elements can be computed (6.71), providing the normalized impulse response. Finally, the pathloss and shadowing are applied, providing the properly scaled impulse response.

This process is repeated for all the UEs via the principles of a "drop" where each drop initializes a different position and a set of LPs for a UE; this establishes the relationship of delays and angles between the links a BS has to multiple UEs. More details of the model and the implementation procedure can be found in Appendix 7.D.

7.5.3   MIMO Matrix Models

The previous sections have described models that include the directional information of the MPCs. An alternative concept that is popular in particular for theoretical analysis of multiple-antenna systems is to stochastically model the impulse response matrix (see Section 6.7) of a MIMO channel. In this case, the channel is characterized not only by the amplitude statistics of each matrix entry (which in NLOS is usually zero-mean complex Gaussian) but also by the *correlation* between those entries. The correlation matrix (for each tap) is defined by first "stacking" all the entries of the channel matrix in one vector $\mathbf{h}_{\text{stack}} = [H_{1,1}, H_{1,2}, ...H_{1,N_{\text{TX}}}, H_{2,1}, ...H_{N_{\text{RX}},N_{\text{TX}}}]^T$ and then computing the correlation matrix as $\mathbf{R} = E\{\mathbf{h}_{\text{stack}}\mathbf{h}_{\text{stack}}^\dagger\}$ where superscript † denotes Hermitian transpose. One popular simplified model assumes that the correlation matrix can be written as a Kronecker product of the correlation matrices at the transmitter (TX) and the RX, $\mathbf{R} = \mathbf{R}_{\text{TX}} \otimes \mathbf{R}_{\text{RX}}$, where $\mathbf{R}_{\text{TX}}$ and $\mathbf{R}_{\text{RX}}$ are the correlation matrices at the TX and the RX, respectively. This model implies that the correlation matrix at the RX is independent of the direction into which transmission occurs, and thus also independent of TX beamforming; this is equivalent to assuming that the DDDPS can be factored into independent APSs at the BS and at the UE. In that case, the channel transfer function matrix can be generated as

$$\mathbf{H} = \mathbf{R}_{\text{RX}}^{1/2}\mathbf{G}_{\text{G}}\mathbf{R}_{\text{TX}}^{1/2} \tag{7.34}$$

where $\mathbf{G}_{\text{G}}$ is a matrix with independent identically distributed (i.i.d.) zero-mean complex Gaussian entries. Note that the Frobenius norm $\|\mathbf{H}\|_{\text{F}}$ of (7.34) is not normalized anymore. A contribution from a LOS component is usually modeled separately in a deterministic way, where the phase shifts of the entries follow from simple geometrical considerations.

*7.5.4   Diffuse Multi-Path

Numerous measurements showed that the channel impulse response consists of several well-concentrated strong paths (specular MPCs) and a huge number of weak paths [Dense or Diffuse MPCs (DMC)]. Most of the multi-path models are based on a superposition of a finite number of specular MPCs. The accuracy of this approach can be controlled – within certain limits – via the number of propagation paths. However, in order to keep the number of model parameters reasonably small, it is often preferable to model a

---

[7] However, all other parameters are chosen in the delay/angle space, not in geometrical space. Unfortunately, 3GPP itself calls the model "geometry-based stochastic channel model," though it is different from the established usage of GSCMs, see Section 7.6.

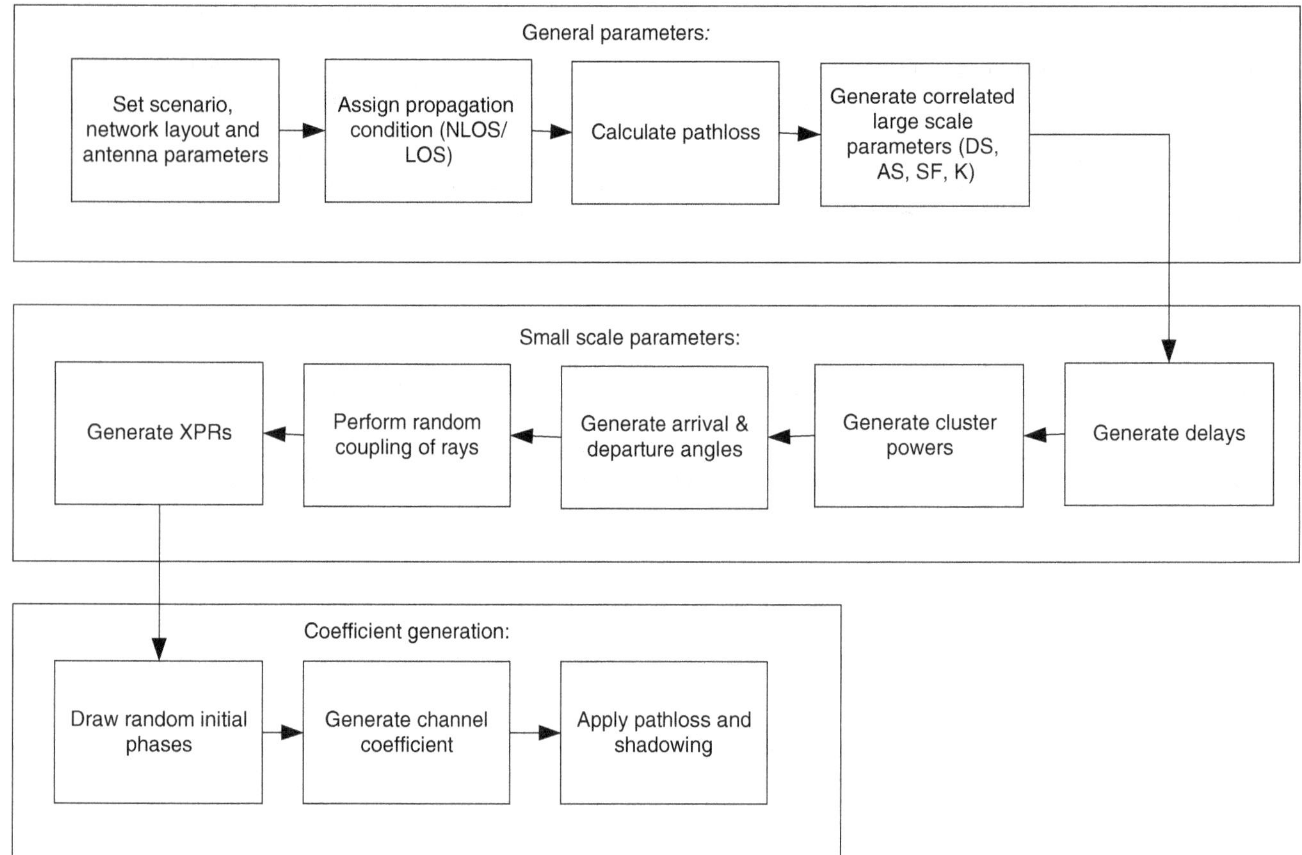

Figure 7.10 Standardized 3GPP / ITU-R channel impulse response generation procedure for propagation channel models in 4G and 5G systems. Color version available at wiley.com/go/molisch/wireless3e.

relatively low number of specular MPCs, and include separately the DMC. The diffuse components are usually described by their (continuous) DDDPS, which is often assumed to be a product of an APDP, and an APS at TX and RX.

The most common model for the DMC is that the APDP is a single-exponential decay that starts at the delay of the first observed MPC, and the APSs at TX and RX are uniform functions. However, measurements have shown that in many situations the APS emphasizes a particular range of directions; it is then often described by the vonMises distribution

$$pdf_x(x) = \frac{\exp\left[\kappa \cos\left(x - \mu\right)\right]}{2\pi I_0(\kappa)}$$

where μ is the direction of the maximum, and κ describes the angular concentration: the larger, the more concentrated the angular distribution is. In all of the above cases, the DMC is assumed by a single distribution that is independent of the MPCs. Alternatively, DMCs have also been modeled as a sum of diffuse clusters (with nonuniform APSs), associated with discrete MPC clusters.

*7.5.5 Polarization

Most channel models analyze only the propagation of vertical polarization, corresponding to the transmission and reception with vertically polarized antennas. However, there is increased interest in polarization diversity, i.e., antennas that are co-located but receive waves with different polarizations. In order to simulate such systems, models for the propagation of dual-polarized radiation are required.

The transmission from a vertically polarized antenna will undergo interactions that result in power being leaked into the horizontal polarization component before reaching the RX antenna (and vice versa). The amplitude coefficients for the MPCs thus have to be written as a polarimetric 2×2 matrix, so that the complex amplitude $\mathbf{a}_\ell$ becomes,

$$\mathbf{a}_\ell = \begin{pmatrix} a_\ell^{VV} & a_\ell^{VH} \\ a_\ell^{HV} & a_\ell^{HH} \end{pmatrix}$$

where V and H denote vertical and horizontal polarization, respectively. For example, a_ℓ^{VH} denotes the amplitude when the TX transmits on the horizontal polarization and the RX receives on the vertical.

The most common polarimetric channel model assumes that the entries in the matrix are statistically independent, complex Gaussian variables. The *mean* powers of the *VV* and the *HH* components are assumed to be identical; similarly, the *mean* powers of the *VH* and *HV* components are the same. The cross-polarization ratio, *XPD*, which is the ratio (expressed in dB) of the mean powers in *VV* and *VH*, is modeled as a Gaussian random variable. The mean and variance of the XPD can depend on the propagation environment, and even on the delay of the considered components. Typical values for the mean of the XPD lie between 0 and 12 dB (higher in LOS situations); for the variance, around 3–6 dB.

7.6 Geometry-Based Stochastic Channel Models

7.6.1 General Principle

In the *Geometry-Based Stochastic Channel Model*, GSCM, it is not the strength and direction of the MPCs that is modeled stochastically, but rather the location of IOs and the strength of the interaction processes (Figure 7.11). The directionally resolved impulse response is then obtained in two steps:

1. assign locations to the IOs, according to the probability density function of their position;
2. based on simplified ray tracing, determine the contributions of the MPCs to the double-directional impulse response. Each MPC (corresponding to one IO) has a unique DoA, DoD, delay, and phase shift.

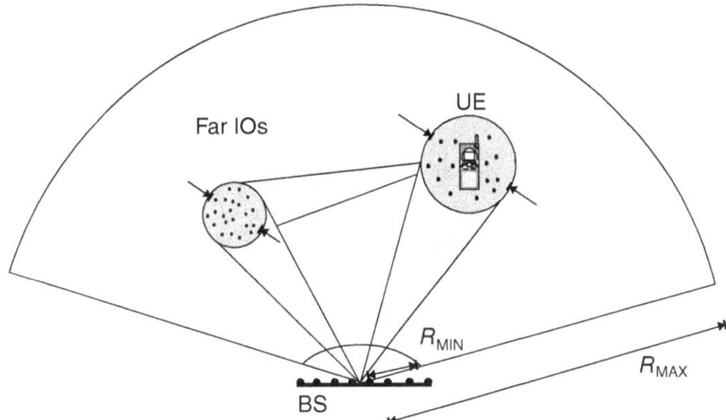

Figure 7.11 Principle of GSCM modeling, assuming single-interaction. BS assumed to have an antenna array.

Assuming furthermore that only single-bounce interactions can occur, the ray tracing becomes extremely simple: apart from the LoS, all paths consist of two subpaths connecting the IOs to the TX and RX, respectively. These subpaths characterize the DoD, DoA, and propagation time (which in turn determines the overall attenuation, usually according to a power law). The IO interaction itself can be taken into account via an additional random phase shift and/or attenuation. Note that in this case, DoA, DoD, and delay are not independent of each other, but rather every set of two determines the third one.

7.6.2 IO Distributions

The simplest model for the IO distribution is based on the assumption that all relevant IOs are close to the UE. This case occurs, e.g., in macrocells with regular buildings structures like suburban environments. In that case, the radiation from the UE interacts with IOs around the UE but can proceed without further interaction from those objects to the BS.[8] Different models exist for the distribution of the IOs around the UE:

- Some papers place all IOs on a circle around the UE (this model is also very suitable for theoretical analysis, and has been used since the 1970s, see [Lee 1973]).
- Other papers suggest a uniform distribution within a disk. When the UE moves, also the disk around the UE (but not the IOs) moves. Some IOs thus "fall out" of the IO disk, while new IOs enter, see Figure 7.12. This corresponds to the physical reality that IOs that are far away from the UE do not make significant contributions.

---

[8] By default, IOs are assumed to be at fixed locations, though moving IOs can be emulated when required.

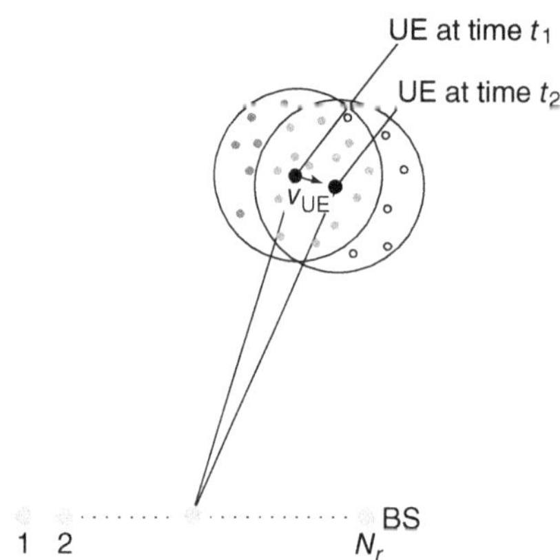

Figure 7.12 "Vanishing" and "appearing" of IOs when the UE moves. It is assumed that all the IOs are in a disk around the UE. Scatterers that are active only at time t_1 (t_2) are shown as black (empty) circles; scatterers that are active at both time instants are shown in grey. Reproduced with permission from [Fuhl et al. 1998] © IET.

- A one-sided Gaussian distribution $pdf(r) = \exp(-r^2/2\sigma^2)$, $r \geq 0$, has also been suggested, where r is the distance from the UE. Computing the PDP and the APS from this distribution gives results that are fairly similar to an exponential PDP and Laplacian APS.

As mentioned above, when a UE moves over larger distances, some IOs "vanish," while others "appear." Of course, the IOs do not really vanish, but rather they become irrelevant (because the UE cannot "see" them anymore, compare the visibility region in Section 7.6.4), and similarly for the "appearing" IOs. Consequently, a circle of IOs is always centered on the UE.

Generalizations include so-called *far IOs* (also known as far scatterers), which correspond to high-rise buildings (compare Section 7.4.3) or mountains. Such a far IO can be modeled either as a single specular reflector (corresponding, e.g., to a high-rise building with a smooth glass front) or a cluster of IOs. When the UE moves, far IOs tend to stay visible far longer than local IOs.

*7.6.3 Multi-Bounce Modeling

In the single-bounce scattering model, the position of an IO completely determines DoD, DoA, and delay. However, in many environments, multiple scattering processes occur. If the directional channel properties need to be reproduced only for *one* link end (i.e., multiple antennas only at the TX or RX), multiple-bounce scattering can be incorporated into a GSCM via the concept of *equivalent scatterers*, i.e., (virtual) single-bounce scatterers whose position is chosen such that they mimic multiple bounce contributions in terms of their delay and DoA. For a given delay and DoA, closed-form equations exist that allow the computation of the location of those equivalent scatterers [Molisch et al. 2003b].

In a MIMO system, the equivalent scatterer concept fails since the angular channel characteristics are reproduced correctly only for one link end (remember that for single-bounce models, DoA, DoD, and delay cannot be chosen independently). A possible remedy is the *twin-cluster* concept, see Figure 7.13. A twin cluster is composed of two representations of IO clusters, in particular the first cluster after the TX that the MPCs "experience" on their way, and the last cluster before the RX. Their locations and sizes are chosen such that the correct angular spectra are seen at the TX and RX. The multi-bounce propagation between the first and the last cluster is essentially a "black box" from a channel model point of view where the details of the propagation do not matter; only the aggregate delay and attenuation caused by the multi-bounce propagation plays a role. This excess delay is thus another variable describing the twin cluster set, usually modeled as a random variable following a measurement-based distribution, and similar for excess attenuation. The ratio of twin clusters to the total number of clusters is determined by a selection factor parameterized from measurements.

7.6.4 Modeling of Large-Scale Movement

Geometric channel models have advantages especially when movement is to be simulated. Whenever the UE moves, adjustments to the parameters of the MPCs are automatically made. Thus, the correct fading correlation results automatically from the movement; also the correlation between the DoAs at the UE and the Doppler is taken into account. Furthermore, any changes in the mean DoAs, DoDs, and delays due to large-scale movement of the UE are automatically included. This is in contrast to generalized tapped delay lines, where the tap locations are usually assumed to be fixed.

In order to model the appearance and disappearance of far clusters, the concept of *Visibility Region, VR* has found wide acceptance. For each cluster, we define certain physical regions in a coverage area so that if the UE is in such a region, the cluster is active, i.e., the MPCs belonging to that cluster contribute to the DDDPS, otherwise, they are not, see Figure 7.14. The visibility regions are placed at random in the cell area, with the pdf of the visibility region centers being a parameter of the model. For each cluster, a separate set of visibility regions must be generated. It is furthermore common to define a *transition function* that ensures that when a UE enters a visibility region, the MPC cluster does not activate all of a sudden (which would lead to a discontinuity in power) but gradually.

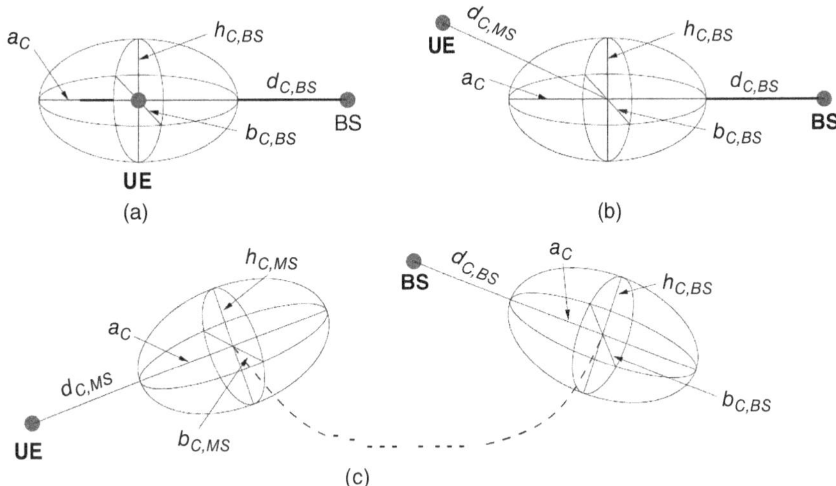

Figure 7.13   Three kinds of clusters: (a) local cluster, (b) far single cluster, (c) twin cluster. The acronym MS in the figure is UE in the notation of this book. Color version available at wiley.com/go/molisch/wireless3e.
Reproduced with permission from [Verdone and Zanella 2012] © Springer.

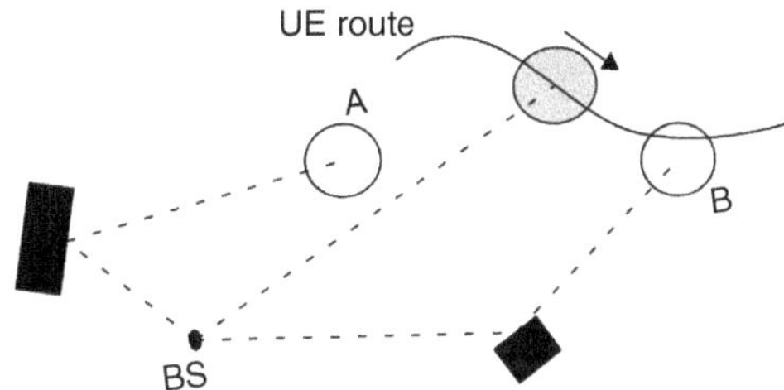

Figure 7.14   Illustration of the visibility region concept. Black squares: far IOs. Hollow circles: associated visibility region. Grey circle: Local IO cluster. MPCs from the far IOs are active when the UE is in the visibility region associated with that IO.
Reproduced with permission from [Asplund et al. 2006] © IEEE.

The appearance/disappearance of an actual LOS component can also be modeled by a visibility region. However, the cluster of MPCs corresponding to scattering around the UE is always present (though IOs in that cluster might appear/disappear, see above); the appearance and disappearance of clusters only occur for far clusters.

The environment (i.e., the clusters and the VRs) is generated independently of the UE position, so a single environment realization can be combined with a long trajectory of a UE (or several UEs). The movement of the UE in this simulation area causes the visibility of different clusters to change as the UE enters and leaves different VRs, resulting implicitly in nonstationary channel simulations. This also implies that the model structure and parametrization are independent of the UE speed, yet the speed ultimately enters into the time-variant channel impulse response: the higher the speed, the faster the UE moves in and out of visibility regions, decreasing the stationarity time of the channel.

With these mechanisms, GSCMs have a built-in *spatial consistency*, i.e., when a UE moves on a loop over a large area, the channel at the end point is the same as the channel at the starting point (since end point and starting points are the same). Also, changes of directions of the MPCs from different clusters are consistent with each other. Note that spatial consistency is very difficult to achieve with generalized tapped delay line models, which often use approximations, or revert to the mechanisms of GSCMs for this purpose.

*7.6.5   Multi-Link Models

While the correlation of channels from one BS to multiple UEs is well described by the models discussed above, the situation is more complicated for modeling the links from a single UE to multiple BSs. It cannot be assumed that the large-scale parameters such as rms delay spread, angular spread, and large-scale fading are independent just because the BSs are widely separated. Rather the fact that the vicinity of the UE is the same for the different links, tends to create some degree of correlation for these large-scale parameters such as shadowing, angular spread, etc. When the considered BSs are in the same direction compared to the movement of the UE, the large-scale parameters of the different links have a tendency to be positively correlated, but slightly negatively correlated when the BSs are located in different directions compared to the movement of the UE.

In GSCMs, correlation properties in multi-link scenarios can be modeled by the concept of *common clusters*. The basic idea behind common clusters is that links between a UE and different BSs (and vice versa) sometimes use the same clusters and hence they partially use the same propagation paths in the simulations, with the result of correlation of the link parameters. If the visibility regions from two BSs with respect to a cluster overlap, then a UE located in that visibility region experiences this cluster as a common cluster. Examples of common clusters are shown in Figure 7.15.

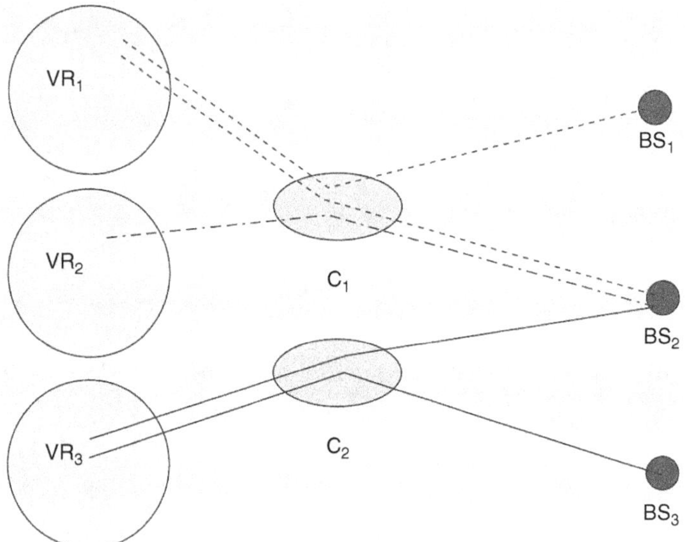

Figure 7.15 Examples of common clusters. The clusters C_1 and C_2 are associated to three BSs and three users in the visibility regions. The lines define the visibility of the different users to the BSs. Color version available at wiley.com/go/molisch/wireless3e. Reproduced with permission from [Liu et al. 2012] © IEEE.

*7.6.6 COST 259/273/2100 Models

The above principles have been implemented in a series of channel models that were standardized by the European COST initiative, see App. 7.F. The resulting model has gained wide acceptance because it is very realistic, incorporating a wealth of effects and their interplay, for a number of different environments. The first of these models, COST 259, introduced the basic model structure and principles such as the visibility region, with an emphasis on directional characteristics at one link end only. COST 273 introduced the twin-cluster principle and thus enabled better modeling of double-directional characteristics. The COST 2100 model further refined these ideas and provided parameterizations in a variety of different environments; it also introduced the concept of common clusters. In later COST actions, such as COST IC 1004 and COST IRACON, the COST 2100 model was also extended for massive MIMO channels by modeling nonstationarities over a large antenna array, and death–birth processes for individual MPCs.

7.7 Semi-Deterministic Models

While GSCMs place the IOs purely stochastically, there are many situations where it is desirable to create a deterministic geometry, e.g., in an indoor scenario to have four walls at a specific location. On the other hand, some of the fine structure, like the placement of chairs, tables, books, etc., should be modeled stochastically, both because it simplifies the model and associated ray tracing, and because it is not even possible to know the location of all these smaller objects exactly.

Based on this philosophy, semi-deterministic (also called quasi-deterministic) channel models describe a certain portion of the propagation channel, such as strong specularly reflected components, as deterministic and add certain random components to this deterministic model. These random components represent either spreading around the (deterministically computed) cluster center, or unpredictable events such as mobile IOs, or smaller objects that might scatter into different directions.

More specifically, the deterministic geometry is used to generate, via (simplified) ray tracing, strong MPCs that are created by inter-action with large fixed objects such as buildings. These "main MPCs," known as directed or *D-rays*, are thus quasi-deterministic in nature. They follow simple ray tracing or waveguiding that is either restricted to single reflections or incorporates multiple reflections. Each of these D-rays is associated with a cluster of MPCs whose directions and delays are spread around the main MPC. A number of weaker reflections from static objects in the environment known as *R-rays* are then introduced; in contrast to the D-rays, their para-meters are not obtained from simplified ray tracing but rather chosen randomly. Finally, flashing or *F-rays* are introduced as a result of interaction of the signal with mobile randomly located IOs such as moving cars. The modeling of mobility is fairly straightforward in semi-deterministic models, as the Doppler shifts and time variations follow from the deterministic superposition of the MPCs. Just like GSCMs, these models provide inherent spatial consistency. The principle of the model is outlined in Figures 7.16 and 7.17.

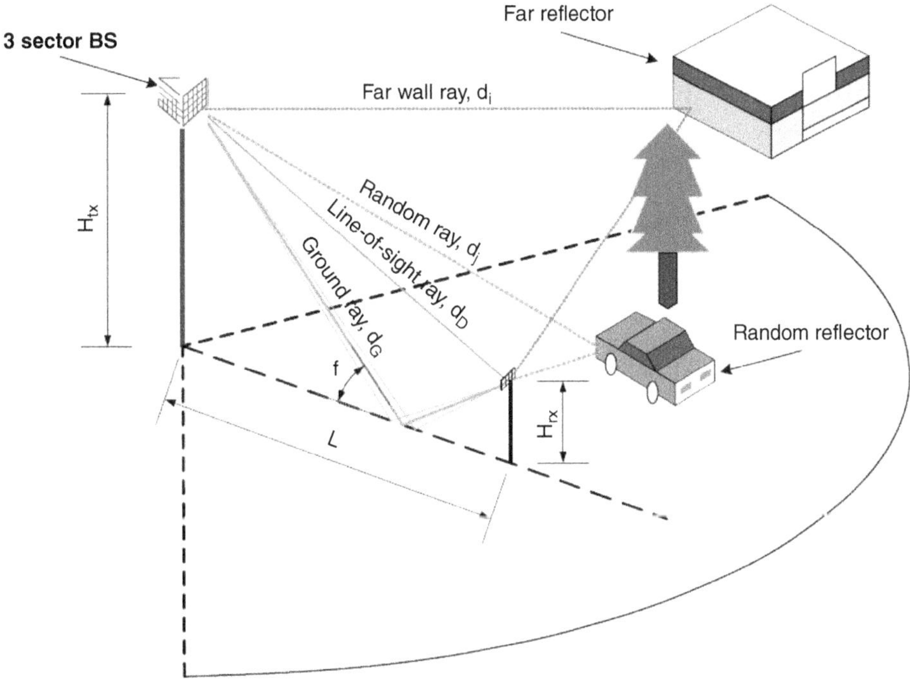

Figure 7.16 Components of a semi-deterministic channel model. Color version available at wiley.com/go/molisch/wireless3e. Reproduced with permission from [Weiler et al. 2016] © EuraSIP.

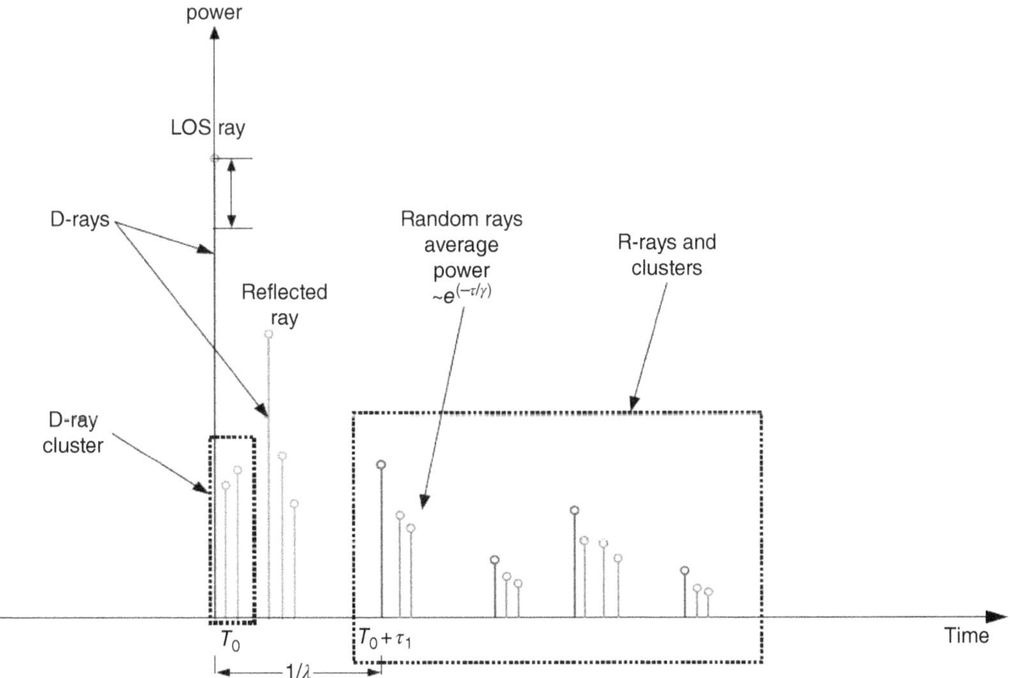

Figure 7.17 Structure of a semi-deterministic channel model. Color version available at wiley.com/go/molisch/wireless3e. Reproduced with permission from [Weiler et al. 2016] © EuraSIP.

The semi-deterministic channel model principle, first suggested in the early 2000s, was adopted in the mid-2010s by the EU projects METIS and MiWEBA and has also been adopted as an optional alternative to the tapped-delay line model for 3GPP standardization.

7.8 Blockage

Humans, cars, and similar objects can create time-varying blockages of MPCs, and thus significantly impact the channel characteristics. While these effects are not included in the channel models described above, their impact can be incorporated with some straightforward methods.

7.8.1 Static Human Blockage

Channels are usually modeled under the assumption that there are no objects in the immediate vicinity of the antennas. While this assumption is fulfilled well for BS antennas, it is violated in the case of a UE that is held by a human person. The presence of this person will both distort the antenna pattern and attenuate signals from a range of directions, leading to enhanced shadowing.

In terms of the pathloss, it is possible to account for this effect through defining and quantifying rotational shadowing. In other words, the received signal power varies depending on the orientation of the user in space. This shadowing is characterized by a mean attenuation, a variance, as well as a coherence angle (instead of the usual coherence distance).

A more detailed characterization can be done by interpreting the antenna and the human body as a "superantenna," whose pattern can then be combined with the double-directional description of the channel without the body, compare also Section 8.5.2.

7.8.2 Dynamic Blockage

Another type of blockage occurs when humans or cars that are in the far field of the UE block either the LOS or other MPCs. It is again most straightforward to investigate the impact of those losses in a double-directional channel description, since the geometry of the MPCs and the blockage objects allows to determine the interactions (penetration and diffraction) with some simple mathematical formulae. In principle, equations for diffractions around arbitrary objects can be used. For simplicity, it has become common to model any object by a rectangular screen, and use equations based on uniform theory of diffraction. The equations can be found in [Peter 2017] and in Appendix 7.D, since they are also used for the 3GPP channel model. In essence, an object can lead to significant blockage if it enters the first Fresnel zone of a ray.

The blockage effect is significantly more pronounced at higher frequencies. For example, while typical body blockage might result in an attenuation of 5–10 dB at <6 GHz, it can lead to 20–30 dB attenuation at 30–60 GHz, and even more at THz frequencies. Similar statements hold for blockages by cars and buses. This is due both due to the higher penetration loss through the blocker, and the smaller diffraction gain at higher frequencies.

*7.9 Special Models

7.9.1 D2D and V2V Models

Device-to-Device (D2D) radio channels have fundamentally different properties compared to those of conventional cellular channels. The main reason for this is that both the TX and the RX antennas are located at low heights, and hence there is more interaction with objects in the close neighborhood of the devices. The difference from traditional Device-to-Infrastructure (D2I) links is especially pronounced for outdoor links, where a BS would be high above ground (typically 10 m for microcell, and up to 100 m for macrocells), while all devices are at street level. Consequently, over-the-rooftop propagation is not a significant mechanism in D2D, and even street canyon propagation is more strongly affected by shadowing objects such as cars and trucks. The most frequently occurring application of outdoor D2D systems is Vehicle-to-Vehicle (V2V) communications, though other applications are also thinkable. In indoor situations, the difference between D2I and D2D propagation mechanisms is less pronounced, and the range of validity for many indoor channel models includes the D2D case.

Pathloss

A first impact of the different propagation conditions is the pathloss model. For outdoor situations, the pathloss coefficients are generally in the range 1.6–2 when the two devices communicating with each other are in the same street (in urban, suburban, or rural environments) or on the same highway. When the two devices are on orthogonal streets, models based on Manhattan distances plus a diffraction loss provide good agreement with measurements, e.g., [Mangel et al. 2011]:

$$PL(d_r, d_t, w_r, x_t, i_s) = C + i_s L_{SU} + \begin{cases} 10 \log_{10}\left(\left(\frac{d_t^{E_T}}{(x_t w_r)^{E_S}} \frac{4\pi d_r}{\lambda} \right)^{E_L} \right), & \text{if } d_r \leq d_b \\[2ex] 10 \log_{10}\left(\left(\frac{d_t^{E_T}}{(x_t w_r)^{E_S}} \frac{4\pi d_r^2}{\lambda d_b} \right)^{E_L} \right), & \text{if } d_r > d_b \end{cases} \tag{7.35}$$

where d_t and d_r denote the distance of the TX and RX to the intersection center, respectively, w_r is the width of the RX street, and x_t is the distance of the TX to the wall. In the model $C = 3.75$ dB is the so-called curve shift, $L_{SU} = 2.94$ dB is the sub-urban loss, $i_s = 0$ is the urban loss factor, $i_s = 1$ the sub-urban loss factor, $E_L = 2.69$ is the loss exponent, $E_S = 0.81$ is the street exponent, $E_T = 0.957$ is the TX distance exponent, and finally $d_b = 180$ m is the breakpoint distance.

Delay Dispersion

As far as the delay dispersion is concerned, the rms delay spread for V2V environments varies with the location, and – similar to the cellular case – can be modeled as a random variable (log normally distributed in most cases). Mean rms delay spreads are on the order of 100–200 ns for rural and suburban environments, and up to 400 ns for urban environments. High rms delay spreads are observed when the LOS component is blocked and there are large reflecting objects close to the road.

Temporal Variations

There are two distinct groups of D2D channels, depending on the dynamics of the nodes, i.e., if the devices themselves are moving or not. In the first case, devices at *both* link ends can move, sometimes very fast. In addition, scatterers and shadowing objects can also move. This is, e.g., the case for V2V channels, for which extensive research has shown that the channel statistics typically change quickly over time, and hence the conventional assumption about WSSUS (see Section 6.4) is only fulfilled for rather short time intervals, and moderate frequency intervals. Figure 7.18 shows an example of a PDP; it is obvious that the relative delay between the MPCs created by the LOS component and various specular MPCs changes significantly over time, thus limiting the duration over which the Wide-Sense Stationary (WSS) assumption is valid. Stationarity bandwidths of around 100 MHz and a stationarity time corresponding to a movement of 10 m are typical. Modeling of nonstationarities can be done most easily according to the GSCMs described in 7.6, though approaches for generalized tapped delay lines have been suggested as well, see Section 7.5.

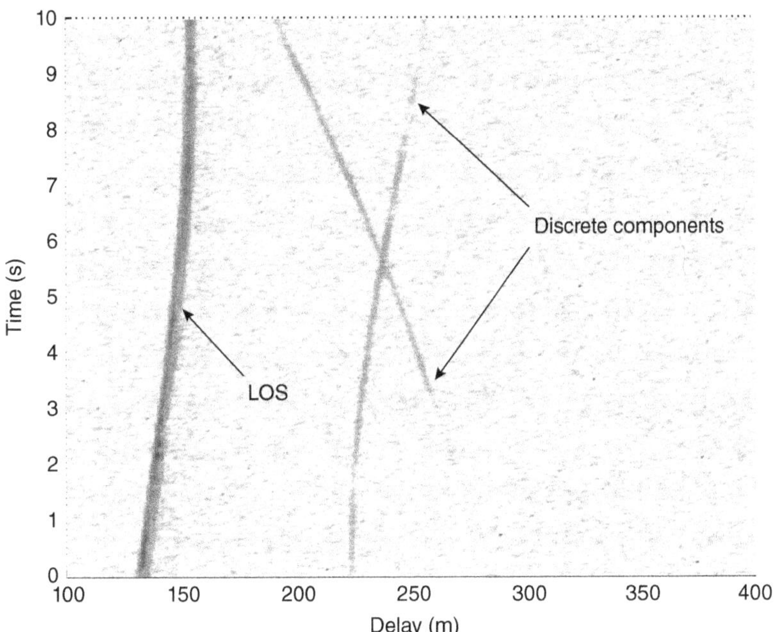

Figure 7.18   Example plot of a time-varying power delay profile of V2V communications on a highway. Delay is multiplied with speed of light. Color version available at wiley.com/go/molisch/wireless3e.
Reproduced with permission from [Karedal et al. 2009] © IEEE.

In static or nomadic scenarios, the two nodes do not change with respect to each other. Temporal variations occur due to objects moving in the vicinity, and can be mostly described by the methods of Section 7.8.2.

7.9.2   Drone-to-Ground Channels

Another type of channel in which mobility plays a major role describes the links between Unmanned Arial Vehicles (UAVs), also known as drones, and between UAVs and a ground station. In the former case, there is usually a strong LoS component, which might be affected by Doppler and possible wobble by the UAV, but – in particular for high-flying drones – a very limited amount of multi-path. The UAV-to-ground channels show a larger variety of channels. In this case, a LoS might or might not exist, and MPCs reflected from buildings and/or terrain features can be significant. One can also think of a UAV-to-ground channel as a channel with a highly elevated BS that is furthermore mobile.

In open terrain, the pathloss in LoS situations is often described well by the two-path model (direct plus ground reflection) discussed in Section 4.2, or – for larger distances – a curved earth model. In many cases, pathloss and shadowing variations are due to change of the antenna gains as the DoA and DoDs change as the UAV is progressing on its trajectory; also the change in polarization mismatch can impact the received power. The small-scale fading statistics usually follow Rician distributions, with K-factors in the 5 GHz range in urban areas of more than 25 dB, almost 30 dB in hilly/mountainous area, and even higher over the sea. Rms delay spreads are typically small, on the order of tens of ns, but in some cases values of several hundred ns have been identified.

When a drone is flying close to the rooftops in an urban area, so that LOS is intermittent, significant variations of the received power can be anticipated. Figure 7.19 shows the measured field strength at 3.5 GHz carrier frequency when the drone is flying at approximately rooftop height (inner color plot) and significantly above (outer color plot) for a closed trajectory in an urban area.

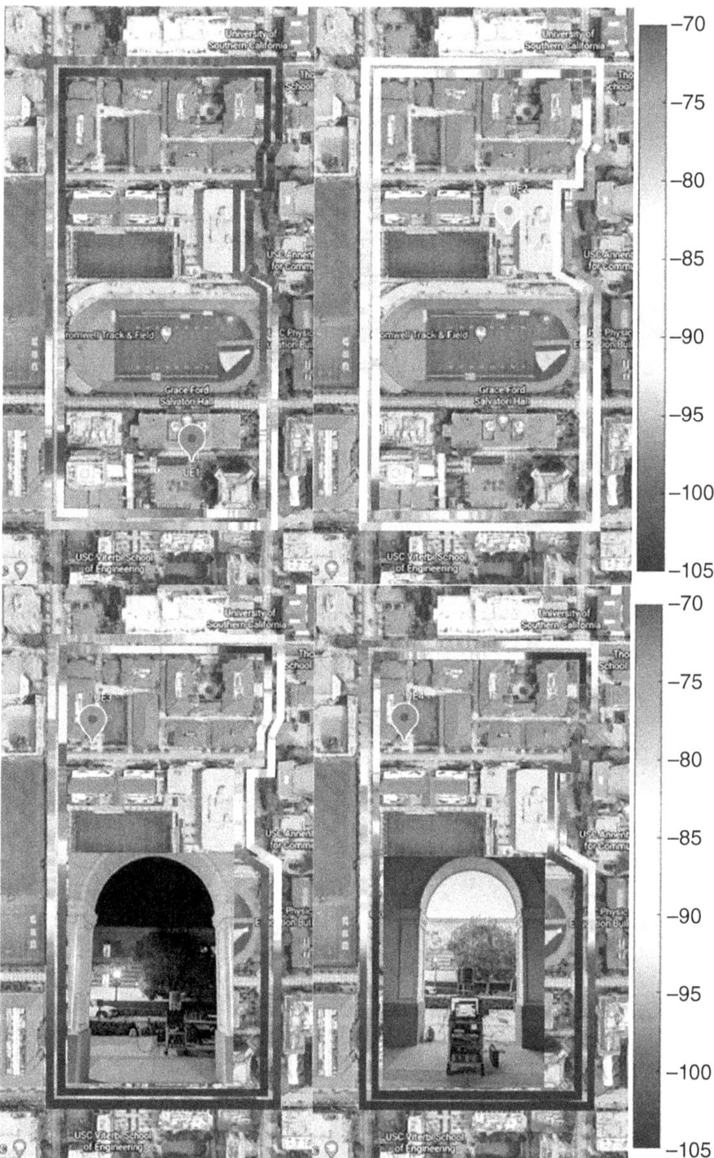

Figure 7.19 Channel measurement environment in an urban environment (university campus): colorbars show channel gains (in dB) between all AP locations (drone positions) along the trajectory at 35 m (inner loop)/70 m height (outer loop) and four different UE locations. Color version available at wiley.com/go/molisch/wireless3e.
Reproduced with permission from [Choi et al. 2021] © IEEE.

At the time of this writing, this is an area with a lot of active measurement campaigns, and significant change in the state of the art can be anticipated over the next years.

7.9.3 Mm-wave and THz Channels

The propagation conditions at mm-wave and THz frequencies (at the time of this writing, interest focuses on the 30–300 GHz band) can be quite different from the traditional bands for cellular and WiFi (0.7–7 GHz), often called cmWave band. This occurs because most of the propagation processes are frequency dependent: (i) pathloss may increase or decrease with frequency, depending on whether constant-gain or constant-area antennas are assumed (compare Section 4.1), (ii) effectiveness of diffraction decreases with frequency (compare Section 4.4), (iii) specular reflection coefficients decrease, and diffuse scattering increases with increasing frequency (compare Section 4.4), (iv) penetration loss into buildings, and through the human body, increases with frequency (Section 4.2), (v) attenuation during propagation through foliage increases, and (vi) atmospheric attenuation increases, though not necessarily monotonically, with frequency (Section 4.6). The pathloss coefficients of mm-wave and THz channels do not differ significantly from those of cm-wave channels. However, due to the propagation effects, the coverage probability of mm-wave and THz is lower than in the cm-wave band; i.e., there are more locations with outage even if the average signal strength may be quite good.

A wide range of measurements have been performed over the past 30 years, and models have been derived from them. However, even at the time of this writing, work is ongoing, and further insights and refinements in particular at higher frequency ranges can be expected. Up to now, the most widely explored indoor environment is the office (including corridors, labs, etc.). Outdoor deployment of mm-wave systems is typically associated with cell sizes on the order of 100–200 m, so that urban microcells are the most common scenario; it is thus also the environment in which most outdoor mm-wave channel measurements have been done. Existing results for the most popular parameters were mentioned in Sections 7.1–7.8, more results are in the references in "Further reading".

7.9.4 Body Area Network Channels

In wireless Body Area Networks (BANs), both the TX and the RX are on/in a human body. Such networks have gained great attention because of their medical applications (e.g., blood pressure sensor or insulin sensor communicating with a "hub" such as a cellphone worn on the body), as well as communication between smartwatch and hub, gaming, etc.

Due to the different location and mounting of at least one of the link ends, the propagation mechanisms, and the resulting channels, show significant differences to the traditional cellular or WLAN channels. The first and most obvious difference is the short distance, typically around 50 cm, and never exceeding 2 m. Secondly, both the TX and RX are on/in the body, and thus experiencing significant distortions of their antenna patterns. Methods that in cellular communications allow the separation of antenna and channel (see Section 8.5.2) are more difficult, or even impossible, in such channels. Thirdly, propagation might occur via surface waves propagating on the body surface or clothing – an effect not applicable in traditional wireless systems. Finally, the body shape and tissue composition of the person wearing the devices have a significant impact on the channel, determining, e.g., whether there is a LOS between TX and RX, or whether that LOS is blocked by tissue.

A number of measurement campaigns, as well as high-accuracy simulations (e.g., Finite Difference Time Domain), have been performed to investigate BAN network channels. A few key insights are

- The pathloss does not show a very pronounced dependence on the distance between TX and RX, since the distance variations are naturally limited. Rather, the relative positioning on the body is essential. For similar Euclidean distances, two devices on the front of the body (with LOS between them) might have 30 or 40 dB higher path gain than when one of the devices is on the front and the other on the back of the body. Furthermore, in particular for directive (patch) antennas, the matching of the orientations, and of the polarizations, is important.
- Variations of the pathloss due to body shape are significant, e.g., up to 15 dB difference have been measured in one campaign with 60 participants.
- Many investigations have been done for the situation where the user is in an open area or an anechoic chamber. However, for indoor environments, reflections on the surrounding walls might create more efficient pathways from front to back. This both increases the received power, and also the delay dispersion.

*7.10 Appendices

App. 7.A: The Okumura-Hata Model

See App7.pdf at wiley.com/go/molisch/wireless3e

App. 7.B: The COST 231-Walfisch-Ikegami Model

See App7.pdf at wiley.com/go/molisch/wireless3e

App. 7.C: The COST 207 GSM Model

See App7.pdf at wiley.com/go/molisch/wireless3e

App. 7.D: The 3GPP Spatial Channel Model

See App7.pdf at wiley.com/go/molisch/wireless3e

App. 7.E: The 802.15.4a UWB Channel Model

See App7.pdf at wiley.com/go/molisch/wireless3e

App. 7.F: The COST 259/273/2100 Channel Model

See App7.pdf at wiley.com/go/molisch/wireless3e

Further Reading

There is a rich literature on channel models. Besides the original papers already mentioned in the main text, [Andersen et al. 1995] and [Molisch and Tufvesson 2005] give an overview of different models. Definitions for pathloss and shadowing are reviewed in [Gentile 2022], while other parameters are discussed in [Molisch 2022]. A simple recursive pathloss model for urban microcells was proposed in [Berg 1995]. The modeling of pathloss coefficient and shadowing variance as random was first suggested by [Erceg et al. 1999], and for indoor in [Ghassemzadeh et al. 2004]; a street-by-street pathloss model for mm-wave frequencies is derived in [Karttunen et al. 2017]. Extensive measurements in the mm-wave range and a resulting model are described, e.g., in [Rappaport et al. 2013, 2015a,b].

The Okumura–Hata model is based on the extensive measurements of [Okumura et al. 1968] in Japan, and was brought into a form suitable for computer simulations by [Hata 1980]. Extensions exploiting the terrain profile are discussed in [Badsberg et al. 1995]. The COST 231–Walfisch–Ikegami model was developed by the research and standardization group COST 231 [Damosso and Correia 1999] based on the work of [Walfisch and Bertoni 1988] and [Ikegami et al. 1984].

A discussion of different shadowing correlation models and their physical reasonableness can be found in [Szyszkowicz et al. 2010]. The Two-path With Diffuse Power (TWDP) model is described in [Durgin et al. 2002]; the $\kappa - \mu$ distribution in [Yacoub 2007]; Weibull fading in [Sagias and Karagiannidis 2005]. The importance of diffuse multi-path and a first model was described in [Richter 2005]; an extensive review is given in [Jiang et al. 2022]. An extensive discussion of polarization can be found in [Shafi et al. 2006].

For the implementation of the tapped delay line model, we recommend [Paetzold 2002] and [Zheng and Xiao 2003]. Generalizations of the tapped delay line approach to the MIMO case were suggested by [Xu et al. 2002]. The Poisson approximations for the arrival times were developed by [Turin et al. 1972] and later improved and extended by [Suzuki 1977] and [Hashemi 1979]. The Saleh–Valenzuela model was first proposed in [Saleh and Valenzuela 1987]. Clustering algorithms are reviewed in [Huang et al. 2019a,b, 2022]. The correlation between shadowing and delay spread was first established by [Greenstein et al. 1997].

The fundamental approach for 3GPP-type models was developed in [Calcev et al. 2007], the most recent version (at the time of this writing) can be found in [3GPP TR 38.901 v16.0.1 2019]. Fundamental propagation effects linking cluster parameters are discussed in [Molisch and Tufvesson 2014]. The IEEE 802.11n model covers spatial models for indoor environments [Erceg et al. 2004]. The first statistical channel model for ultrawideband channels was proposed in [Cassioli et al. 2002]; the IEEE 802.15.4a channel model [Molisch et al. 2004; 2006b] describes a the standardized model for this case.

The GSCM was proposed in one form or the other in [Blanz and Jung 1998], [Fuhl et al. 1998], [Norklit and Andersen 1998], and [Petrus et al. 2002], see also [Liberti and Rappaport 1996]. More details about efficient implementations of a GSCM can be found in [Molisch et al. 2003], while a generalization to multiple-interaction processes is described in [Molisch 2004]. The Laplacian structure of the power azimuthal spectral was first suggested in [Pedersen et al. 1997], and though there has been some discussion about its validity, it is now in widespread use. The twin cluster approach was introduced by [Molisch and Hofstetter 2006] and [Hofstetter et al. 2006]; the concept of visibility regions by [Molisch et al. 2006a]. The COST 259, 273, and 2100 models are described, respectively in [Molisch et al. 2006a] and [Asplund et al. 2006], [Molisch and Hofstetter 2006], and [Liu et al. 2012] and [Haneda et al. 2012].

The description of the channels for MIMO systems by transfer function matrices stems from the classical work of [Winters 1987] and [Foschini and Gans 1998]. The Kronecker assumption was proposed in [Kermoal et al. 2002]; a more general model encompassing correlations between directions of arrival and directions of departure was introduced by [Weichselberger et al. 2006]; other generalized models were proposed by [Gesbert et al. 2002] and [Sayeed 2002]. Different types of MIMO models are surveyed in [Almers et al. 2007].

Semi-deterministic channel models were first suggested in [Molisch et al. 2002a] and [Kunisch and Pamp 2003], and later used by the METIS and MiWeBa European projects. V2V channel models are surveyed in [Mecklenbrauker et al. 2011]; railway channels in [Wang et al. 2015]; mm-wave channels in [Molisch et al. 2021] and [Rappaport et al. 2022] and the key differences been cm-wave and mm-wave channels described in [Shafi et al 2018]; THz channels are surveyed in [Han et al. 2022], and outdoor measurements at 140 GHz described in [Gomez-Ponce et al. 2022]. UAV channel models are surveyed in [Kahwaja et al. 2019] and [Yan et al. 2019]. For body area networks, an extensive measurement campaign as well as a review of other measurements can be found in [Sangodoyin and Molisch 2018].

For updates and errata for this chapter, see https://wides.usc.edu/students.html#textbooks

Exercises

Sec. 36.7 of Exercises.pdf at wiley.com/go/molisch/wireless3e

8

Antennas

8.1 Introduction and Brief Characterization

8.1.1 Antennas as Part of a Wireless System

Antennas convert electric currents or conducted waves, i.e., signals from an electronic circuit such as a modulator, to electromagnetic waves propagating in space, or vice versa. They are the interface between the wireless propagation channel and the Transmitter (TX) or Receiver (RX) circuits and thus impact wireless system behavior. This chapter discusses some important aspects of antennas for wireless systems, both at the *Base Station* (BS) and at the *User Equipment* (UE). We concentrate on the antenna aspects that are specific to practical wireless systems, while leaving the fundamental theory to the textbooks mentioned in the "Further Reading" section.

There are two main parameters describing antennas: (i) how efficiently they convert between currents/conducted waves and propagating waves (i.e., antenna efficiency) and (ii) how much they concentrate the radiation into specific directions (i.e., antenna directivity). Besides these quantities, a number of other parameters, such as bandwidth, polarization, etc., play a role. From a practical point of view, the antenna size and the manufacturing cost also need to be considered.

8.1.2 Why Do Antennas Radiate

It is well known that static electric charges give rise to a static electric field, and charges moving with a constant speed (i.e., DC) give rise to a static magnetic field. Propagating electromagnetic fields, on the other hand, are created by accelerating or decelerating charges. Thus, an antenna needs to be excited by a time-varying current, which will cause the electrons of the metallic antennas to move into a particular direction, then stop, and move back to the other direction. The resulting acceleration/deceleration will create a time-varying electromagnetic field that can detach from the antenna and propagate away, see Figure 8.1. Generally, the current should be resonant with the antenna structure, so that the dimensions of efficient antennas are often related to multiples of half a wavelength, $\lambda/2$.

Another interpretation can be given for antennas such as horn antennas, which are stimulated by guided propagating waves (e.g., from a rectangular hollow waveguide). The propagating electromagnetic field changes shape as it enters the horn openings, and finally turn into waves that propagate away from the horn opening, see Figure 8.2. In this interpretation, the antenna can also be seen as an impedance transformer. This is because the waveguide and free space present different impedances to the waves, and the antenna makes it possible for the guided waves to smoothly transition to an unguided medium with a different wave impedance.

*8.1.3 The Hertzian Dipole

Now, we consider the electromagnetic fields at a very simple antenna, the Hertzian (short) dipole. This is a thin piece of wire of length L_H (with $L_H \ll \lambda$), which we henceforth assume to be oriented along the z-axis. We furthermore assume that all signals and fields vary with time sinusoidally, i.e., $\exp(j\omega_c t)$, where $\omega_c = 2\pi f_c$, and we assume that the waves are propagating in vacuum (essentially the same as air), for which magnetic and dielectric constants are denoted as μ_0 and ε_0, respectively. In this case, the electromagnetic field in the far-field (distances larger than the Rayleigh distance, see Section 4.1 and also Section 8.1.4) can be written as

$$E_\theta(r,\theta) = \eta_0 H_\phi(r,\theta) = \frac{j\omega_c\mu_0 L_H I_0}{4\pi r} e^{-jk_0 r} \sin(\theta) \tag{8.1}$$

where $\eta_0 = \sqrt{\mu_0/\varepsilon_0}$ is the free-space wave impedance, (r, ϕ, θ) are the coordinates in a spherical coordinate system, and I_0 is the peak amplitude of the time-varying current. This means that there are only two field components, which are perpendicular to each other, and are perpendicular to the direction of the wave propagation. We will derive other key characteristics of the Hertzian dipole in Section 8.3.1 from this equation.

Wireless Communications: From Fundamentals to Beyond 5G, Third Edition. Andreas F. Molisch.
© 2023 John Wiley & Sons Ltd. Published 2023 by John Wiley & Sons Ltd.
Companion website: www.wiley.com/go/molisch/wireless3e

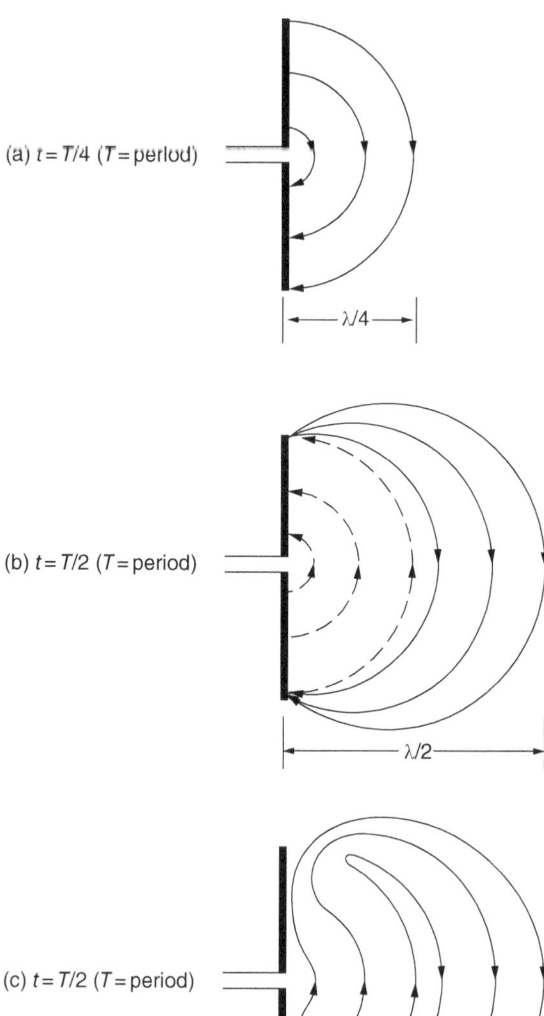

Figure 8.1 Evolution of electric field lines in a short dipole, and their detaching from the antenna.
Reproduced with permission from [Balanis 2005] © J. Wiley & Sons, Ltd.

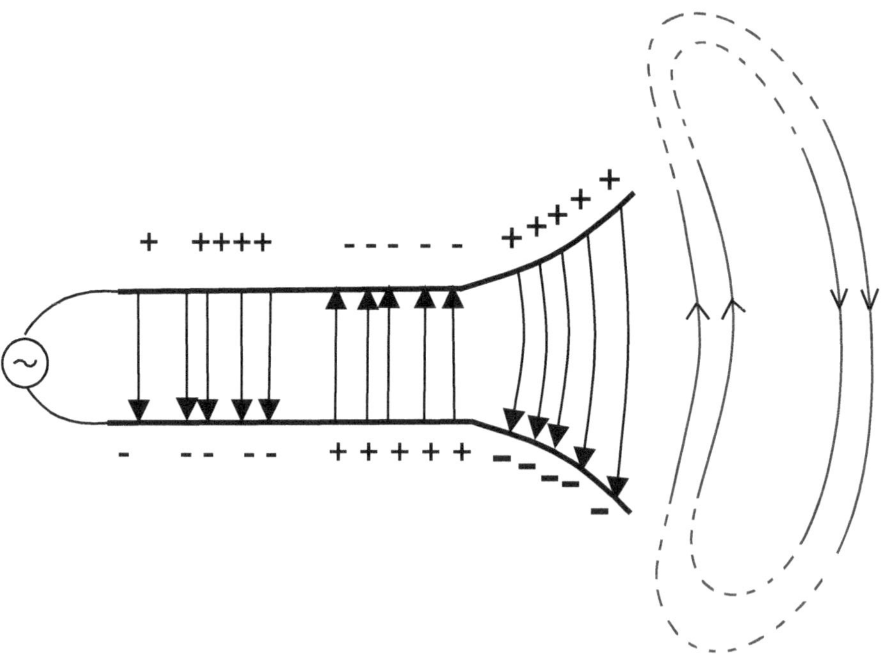

Figure 8.2 Evolution of electric field lines in a horn antenna, and their detaching from the antenna.

Example 8.1 *Derive the general field expressions for the Hertzian dipole.*

From classical electromagnetic theory (see references in "Further Reading"), we know that it is useful to define an auxiliary function, namely the retarded vector potential $\mathbf{A}$:

$$\mathbf{A} = \mu_0 \int_V \frac{\mathbf{I}(V')e^{-jk_0 r'} dV'}{4\pi r'} \tag{8.2}$$

where $\mathbf{I}$ is a vector describing the volume density of the flowing current (including the *direction* of flow) in a small volume element dV'; for a Hertzian dipole, the volume elements are the pieces of a thin wire, and the integration over V becomes an integration along the length of the wire. Furthermore, r' is the distance between dV' and the location at which we evaluate $\mathbf{A}$, and μ_0 is the vacuum magnetic constant; for later purposes, we define ε_0 as the vacuum dielectric constant. The electrical and magnetic fields, $\mathbf{E}$ and $\mathbf{B}$ can be computed from $\mathbf{A}$ as

$$\mathbf{E} = -\frac{j\omega_c}{k_0^2} \nabla(\nabla \cdot \mathbf{A}) - j\omega_c \mathbf{A} \tag{8.3}$$

$$\mathbf{B} = \nabla \times \mathbf{A} \tag{8.4}$$

where ∇ is the Nabla operator and k_0 is the wavenumber, i.e., $2\pi/\lambda_c$.

For a Hertzian dipole of length L_H, through which a uniform current $I_0 \exp(j\omega_c t)$ flows, the retarded potential only has a component in z direction, namely

$$A_z = \mu_0 \frac{L_\mathrm{H} I_0}{4\pi r} e^{-jk_0 r}. \tag{8.5}$$

In spherical coordinates, $A_r = A_z \cos(\theta)$, and $A_\theta = -A_z \sin(\theta)$. From this follows the electric and magnetic field components, using (8.2–8.3) and that (in vacuum), $\mathbf{H} = \mathbf{B}/\mu_0$:

$$H_\phi(r,\theta) = \frac{L_\mathrm{H} I_0}{4\pi} e^{-jk_0 r} \left(\frac{jk_0}{r} + \frac{1}{r^2} \right) \sin(\theta) \tag{8.6}$$

$$E_r(r,\theta) = \frac{L_\mathrm{H} I_0}{4\pi} e^{-jk_0 r} \left(\frac{2\eta_0}{r^2} + \frac{2}{j\omega_c \varepsilon_0 r^3} \right) \cos(\theta) \tag{8.7}$$

$$E_\theta(r,\theta) = \frac{L_\mathrm{H} I_0}{4\pi} e^{-jk_0 r} \left(\frac{j\omega_c \mu_0}{r} + \frac{\eta_0}{r^2} + \frac{1}{j\omega_c \varepsilon_0 r^3} \right) \sin(\theta). \tag{8.8}$$

We see that each of the fields has terms that decay with different order as r increases. At large distances, the electric field in radial direction, E_r is negligible compared to E_θ. By considering only the terms decreasing with $1/r$, i.e., the terms dominating at large distances, (8.1) follows.

Based on the equations of the Hertzian dipole, it is possible to compute the fields of more general wire antennas. Consider a thin wire of length L_a, which need not be short, with center at $z = 0$, oriented again along the z-axis, with a nonuniform current distribution $I(z)$. Each differential piece of wire of length dz creates a contribution to the overall field, which we obtain by integrating over the current distribution.

Example 8.2 *What is the electric field for a $\lambda/2$ dipole in the far field?*

Consider a dipole of length $\lambda/2$, where the current distribution is given as $I_0 \sin[k_0(-|z| + \lambda/4)]$. For the far-field, we can make the approximation that the different distances of the differential currents do not influence the amplitudes and that the phase shift between the contributions is approximated as $z\cos(\theta)$, i.e., the projection of the current (running along the direction of the z-axis) onto the line connecting the antenna center to the observation point. Thus

$$A_z(r,\theta) = \frac{\mu_0}{4\pi r} I_0 e^{-jk_0 r} \int_{-\lambda/4}^{\lambda/4} \sin\left[k_0\left(\frac{\lambda}{4} - |z'| \right) \right] e^{jk_0 z' \cos(\theta)} dz'. \tag{8.9}$$

Splitting the integration range into $[-\lambda/4, 0]$ and $[0, \lambda/4]$, and using the integral formula (which can be easily verified using the product rule)

$$\int e^{ax} \sin(bx + c) dx = \frac{e^{ax}}{a^2 + b^2} [a \sin(bx + c) - b \cos(bx + c)] \tag{8.10}$$

this becomes

$$A_z = \frac{\mu_0}{4\pi r} I_0 e^{-jk_0 r'} \frac{?}{k_0 \sin^2(\theta)} \left[\cos\left(k_0 \frac{\lambda}{4} \cos(\theta)\right) - \cos\left(k_0 \frac{\lambda}{4}\right) \right]. \tag{8.11}$$

Using again $A_\theta = -A_z \sin(\theta)$, the electric field becomes, with some simplifications

$$E_\theta = \frac{j\eta_0 I_0}{2\pi r} e^{-jk_0 r} \frac{\cos[(\pi/2)\cos(\theta)]}{\sin(\theta)}. \tag{8.12}$$

In many situations, it is helpful to approximate the current as piecewise-constant where the length of the pieces is L_H

$$I(z) = \sum_{m=0}^{M-1} I_m g_R(z - mL_H, L_H) \tag{8.13}$$

where $M = L_a/L_H$, and $g_R(z, \Delta)$ is a function that is 1 if $0 \leq z < \Delta$ and 0 otherwise (assuming here the antenna extends from 0 to L_a). For each of those current pieces, we can compute the induced field, where r now is the distance between the location of the considered current and the observation point. With this principle, it is possible to compute the field strength *for a given distribution of the current*. The equations linking the current strengths I_m and the E and H fields are then linear matrix equations. By imposing the correct electromagnetic boundary conditions, it then becomes possible to compute numerically the actual current distribution; this latter approach is called the *Method of Moments* and is used in many commercial antenna design programs.

Numerical calculations can be also performed to obtain the inherent resonant properties of an arbitrarily shaped metallic object, using a method called *Characteristic Mode Analysis* (CMA). This method yields the resonant frequencies, currents, and near/far-fields of the structural-dependent radiation modes (characteristic modes), allowing designers to implement appropriate feeds in a systematic manner to utilize these modes to fulfill different requirements. Such an approach is especially valuable where the antenna in question is not of the popular types described in Section 8.3, in which case design equations/rules are not available. Even for standard antenna types, insights on resonant properties from CMA enable the antenna to be adapted to requirements beyond standard design rules.

*8.1.4 Field Regions

From the above field equations, we can conclude that there are three main regions, according to the distance of the observation point from the antenna:

1. *Far-field*: this is the region we most commonly deal with; it is the regions where the radiated field can be described, in any direction, as a transverse plane wave, so that the field vectors are orthogonal to the direction of propagation. The functional dependence of the field on the angles (essentially, the radiation pattern, see Section 8.2.1) is independent of the distance from the source, and the amplitudes fall off as $1/r$. The (inner) boundary of the far-field is the Rayleigh distance (also known as *Fraunhofer distance*), which is commonly defined as (see also Section 4.1)

$$d_R = \frac{2L_a^2}{\lambda} \tag{8.14}$$

where L_a is the largest dimension of the antenna, and we assume $L_a > \lambda$.

2. *Radiative near-field*: in this region the radiative field dominates, but the radial field component is appreciable and the angular dependence of the field is a function of the distance from the antennas. The outer boundary of this region is the Rayleigh distance, while the inner boundary is at

$$d_{NF} = 0.62 \sqrt{\frac{L_a^3}{\lambda}}. \tag{8.15}$$

When dielectric or metallic objects are located in the radiative near-field, the antenna pattern can be distorted significantly. This plays an especially important role for handsets, see Section 8.5.2.

3. *Reactive near-field*: in this part of the near-field region, the reactive part of the field dominates. This region exists within distances $<d_{NF}$; for short dipoles, where this value might not be meaningful, $d_{NF} = \lambda/2\pi$ is commonly used. If a dielectric or metallic object is introduced into the near-field, energy can be re-directed towards the source. This is generally undesirable, though it can be sometimes exploited, e.g., for sensing purposes such as Radio Frequency Identification (RFID) (Section 29.8).

8.2 Characterization of Antennas

8.2.1 Antenna Pattern, Directivity, and Beamwidth

Antenna patterns describe the distribution of the energy in the far-field, as a function of observation angle: for the transmit case, the antenna pattern describes what fraction of the total radiated power is transmitted in what direction; for the receive case it is a measure of how efficiently it can receive power incident from a particular direction. From the reciprocity theorem of antennas, it follows that the antenna patterns are the same for the transmit and the receive case; we thus use transmit and receive case interchangeably in the following; any statement about the transmit case has a similar interpretation in the receive case and vice versa.

Let us now give mathematically more precise definitions, neglecting, for the moment, the effect of polarization (see Section 8.2.4 for characterization of polarization). Consider the far-field distribution on a sphere of unit radius around a transmitting antenna. Each point on the unit sphere is representative of a particular direction, characterized by the spatial angle $\Omega = (\phi, \theta)$. We can then first define the *antenna (power) directivity pattern* $D(\Omega)$ of an antenna (often simply called the *antenna pattern* or *radiation pattern*), as a measure for how much a transmit antenna concentrates the emitted radiation into a certain direction [Vaughan and Andersen 2003]:

$$D(\Omega) = \frac{\text{total power radiated per unit solid angle in a direction } \Omega}{\text{average power radiated per unit solid angle}}. \tag{8.16}$$

The directivity pattern is thus normalized

$$\frac{1}{4\pi} \int D(\Omega)d\Omega = 1. \tag{8.17}$$

From the directivity pattern, we can derive several important condensed parameters:

- *Directivity* $D_{\max}$ is defined as

$$D_{\max} = \max_{\Omega} [D(\Omega)]. \tag{8.18}$$

Thus, when the maximum of the pattern points in the desired direction, the directivity quantifies how much the transmit power is enhanced compared to the case of, e.g., an isotropic antenna – in this case the directivity is often expressed as dBi (which stands for dB compared to isotropic). For future reference, we denote the angle in which $D_{\max}$ occurs as $\Omega_{\max} = (\phi_{\max}, \theta_{\max})$.

- *Beamwidth* Δ_{BW}: We discuss first the azimuthal beamwidth, and thus consider the "cut" of the antenna pattern in the horizontal plane, $D(\phi, \pi/2)$, shown in Figure 8.3. The *Full-Width Half Maximum (FWHM)* beamwidth is the angle between directions ϕ_1 and ϕ_2, which are to the left and right of the maximum, and for which $D(\phi_{1, 2}) = 0.5D_{\max}$ holds; this quantity is often also called the 3 dB beamwidth. The definition of the elevation beamwidth is completely analogous, though the numerical values of the beamwidths in azimuth and elevation can be quite different, depending on the antenna design.

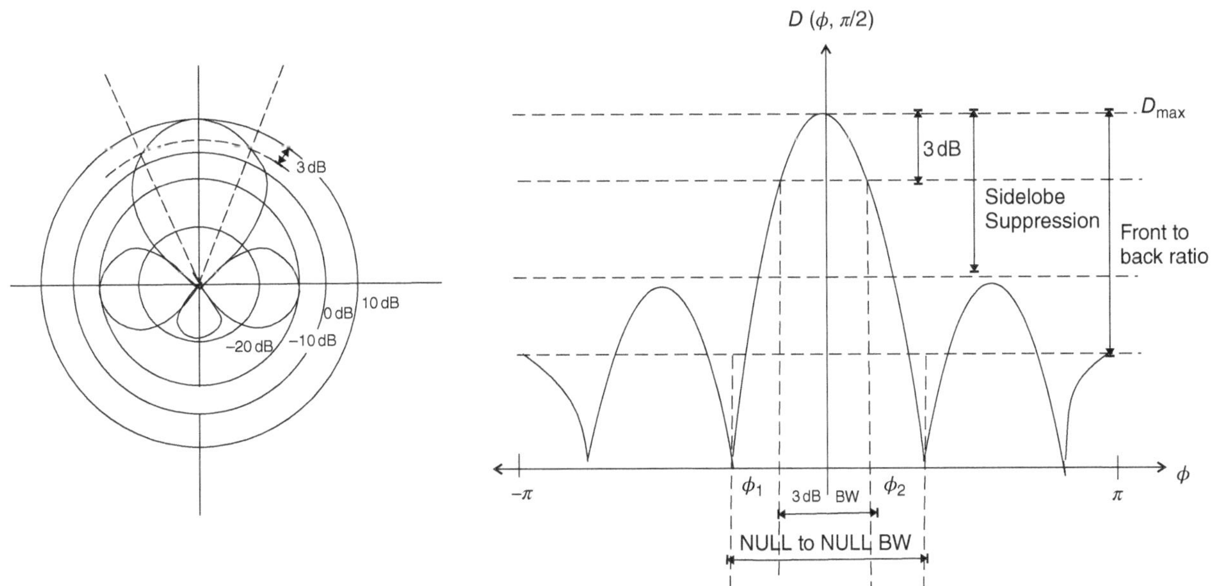

Figure 8.3 Sketch of a typical beampattern as represented in data sheets (left), and definition of quantities related to beampatterns (right).

Note that the beamwidth and the directivity are related to each other. A high directivity implies that $D(\Omega)$ must be small in most directions, so that the beamwidth is small. As a rough first approximation,

$$D_{\max} \approx \frac{4\pi}{\Delta_{\mathrm{BW},\phi}\Delta_{\mathrm{BW},\theta}}. \tag{8.19}$$

Note that the 3 dB beamwidth is not the only possible definition of the beamwidth; we can, e.g., define the 10 dB beamwidth via $D(\phi_{1,\,2}) = 0.1 D_{\max}$. Similarly, we can define the beamwidth between the two nulls of the antenna pattern that are closest to the maximum. The main beam is defined as the set of angles (and associated directivities) between those nulls.

Example 8.3 *Consider a directivity pattern $D(\Omega) = C(1 - (\phi/(0.1\pi))^2)\sin^2(\theta)$ for $\phi = [-0.1\pi \ldots 0.1\pi]$ and 0 otherwise. What are the 3 dB and null-to-null beamwidths in azimuth and elevation, and what is the maximum directivity?*

To compute the proportionality constant C, we use (8.17):

$$\begin{aligned}
4\pi &= \int_{-0.1\pi}^{0.1\pi} \int_0^\pi C\big(1 - (\phi/(0.1\pi))^2\big) \sin^2(\theta) \sin(\theta)\, d\theta\, d\phi \\
&= C\left[(2\cdot 0.1\pi)\left(1 - \left(\frac{(0.1\pi)^2}{3(0.1\pi)^2}\right)\right)\right]\left[-\cos(\theta) + \frac{\cos^3(\theta)}{3}\right]_0^\pi \\
&= C\frac{4}{3}0.1\pi\frac{4}{3} = 4\pi C\frac{4}{90} \Rightarrow C = 22.5.
\end{aligned}$$

The maximum of the pattern occurs at $\phi = 0$, $\theta = \pi/2$, and the maximum directivity is thus $D_{\max} = 22.5$ or approximately 13 dB. The 3 dB beamwidth in azimuth is $0.2\pi/\sqrt{2}$, and in elevation $\pi/2$. The approximate directivity as computed from (8.19) is then $D_{\max} = 18$, which is reasonably close to the exact result. Finally, the null-to-null beamwidth is 0.2π in azimuth and π in elevation.

- *Sidelobe suppression* is described by the ratio

$$\frac{\displaystyle\max_{\Omega \neq \mathrm{mainbeam}} [D(\Omega)]}{D_{\max}}. \tag{8.20}$$

- *Front-to-back ratio* has variously been defined as either $D_{\max}/D(-\Omega_{\max})$ or as the sidelobe suppression in the half-space that is centered around $-\Omega_{\max}$.

The directivity is also related to another useful quantity, the *antenna area* (also called *antenna aperture*). The antenna area is, essentially, the proportionality constant relating (for a receive antenna) the power density of the radiation S_{area} (in W/m^2) to the actually received power when the antenna is perpendicular to the direction of propagation (more precisely, when the normal of the antenna surface is collinear with the Poynting vector of the radiation field):

$$P_{\mathrm{RX}} = S_{\mathrm{area}} A_{\mathrm{RX}}. \tag{8.21}$$

For a quasi-planar antenna, such as a parabolic antenna, this antenna area is closely related to the geometrical area. However, for other antenna types, such as dipoles, the area can be divorced from a physical area (a line antenna does not really have a geometrical area). Rather, it needs to be viewed as an abstract quantity according to the above definition. Importantly, it can be shown that there is a simple relationship between effective area and antenna directivity

$$D_{\mathrm{RX}} = \frac{4\pi}{\lambda^2} A_{\mathrm{RX}}. \tag{8.22}$$

Most noteworthy in this equation is the fact that – for a fixed antenna area – the antenna directivity increases with frequency. This is also intuitive, as the directivity of an antenna is determined by its size in units of wavelengths. Thus, for a fixed geometrical size of an aperture, the directivity increases with frequency, as also discussed in Section 4.1.

We will see below that the *gain* of an antenna in a certain direction is related to the directivity – it is, essentially, the directivity multiplied with the antenna efficiency. An ideal antenna has efficiency $\eta = 1$, while real antennas have $\eta < 1$ due to mismatch and ohmic losses. These losses are generally independent of the direction of radiation, so that the only effect on the antenna pattern is a scaling by a constant factor (which, typically, is close to 1). For this reason, it has become common to treat "gain" and "directivity" almost interchangeably; we will similarly speak of "gains" in other chapters even when strictly speaking directivities are meant.

How can engineers obtain antenna patterns? The first approach is to measure the radiation characteristics. A reference (known) transmit antenna[1] sends out an electromagnetic field, and the antenna of interest is placed – with a particular orientation – as RX in the far-field of the transmit antenna, and the received power is recorded. From the distance between the transmit and receive antenna, and the known transmit characteristics, the antenna gain for the used orientation of the receive antenna can be determined. This measurement is then repeated for many different orientations of the receive antenna. Note that such measurements have to be performed in an anechoic chamber, a room in which no reflections from walls occur, and thus the pattern of a device can be measured without interference from such reflections. Alternatively, measurements of the near-field can be done in an anechoic chamber, and the far-field pattern is obtained from those results by a suitable transformation. A motivation for this approach is the fact that avoiding the need to place the antenna in the far-field allows the use of smaller anechoic chambers, which is important in particular for measurements at lower frequencies.[2] Finally, antennas can be simulated with suitable software (e.g., HFSS™, CST™, etc.). Such simulations are very helpful in the design phase and can – depending on whether *all* parts of the antenna, including feed connectors, etc., are included – give realistic results. Still, results ultimately have to be verified by measurements.

Antenna manufacturers often provide gain patterns in a datasheet.[3] Such information is useful, but a few caveats are in order. Firstly, many manufacturers use an idealized (simulated) antenna pattern instead of an actually measured one. Even if they provide measurements from a prototype, due to process variations in the manufacturing, any sample delivered to a customer can have a different antenna pattern than the prototype. Furthermore, data sheets usually do not provide the full antenna pattern $D(\Omega) = D(\phi, \theta)$ but only so-called "cuts," such as $D(\phi, \pi/2)$ and $D(0, \theta)$.[4] It must be stressed that only in a few rare case does $D(\phi, \theta) = D(\phi, \pi/2)D(0, \theta)$ hold. Last but not least, manufacturers generally provide only the pattern averaged over polarization, and do not include any phase information (see below for a discussion of those aspects).

When the antenna is operated in a random scattering environment, it becomes meaningful to investigate the *Mean Effective Gain (MEG)*, which is an average of the gain over different directions when the directions of the incident radiation are determined by a random environment [Andersen and Hansen 1977, Taga 1990]. The MEG is obviously considerably lower than $D_{\max}$, since the maximum of the pattern might not point into the direction from which the radiation is incident.

*8.2.2 Equivalent Circuits, Antenna Efficiency, and Matching

We now turn to the topic of equivalent circuits and matching for antennas; readers not familiar with matching can find a brief primer in the appendix of Chapter 17. From the point of view of the TX circuitry, the antenna is just one more circuit element, and as such can be described by an input impedance $Z_A = R_A + jX_A$. Let furthermore the antenna impedance be described by a series of two resistors R_{ohmic} and R_{rad}. Here, R_{ohmic} describes the ohmic losses in the antenna. R_{rad} describes the "dissipation" of energy that occurs because energy is radiated from the antenna. In other words, it is defined as the resistance of an equivalent network element so that the radiated power P_{rad} can be written as $0.5|I_0|^2 R_{\text{rad}}$ where I_0 is the magnitude of the current running through the antenna. Finally, the imaginary part of the input impedance, X_A, can be inductive or capacitive.

Let furthermore the whole TX before the antenna be represented as an equivalent source with voltage V, and source impedance Z_S. The most effective power delivery occurs when the source impedance and the antenna impedance are conjugate-matched, i.e., $R_S = R_{\text{ohmic}} + R_{\text{rad}}$, and $X_S = -X_A$. In this case, half of the power delivered by the generator is dissipated in the antenna, by radiation, and by ohmic losses; the ratio between those components is $R_{\text{rad}}/R_{\text{ohmic}}$. The radiation efficiency is then

$$\eta_{\text{rad}} = \frac{R_{\text{rad}}}{R_{\text{rad}} + R_{\text{ohmic}}}. \tag{8.23}$$

This formulation shows the importance of a reasonably high radiation resistance: for constant ohmic losses described by R_{ohmic}, the smaller the radiation resistance, the worse the antenna efficiency (and also, the more difficult the matching). This is one of the key reasons why electrically small antennas are generally undesirable in wireless systems from an efficiency point of view, though they are often used for other (e.g., cosmetic), reasons.

For a dipole with a *uniform current distribution*, the radiation resistance is

$$R_{\text{rad}}^{\text{uniform}} = 80\pi^2 (L_{\text{a}}/\lambda)^2. \tag{8.24}$$

For dipoles with a tapered current distribution (maximum at the feed, and linear decrease towards the end), the radiation resistance is $0.25 R_{\text{rad}}^{\text{uniform}}$.

---

[1] There are also alternative measurement methods that do not need calibrated reference antennas, see, e.g., [Evans 1990].
[2] Remember that for antennas whose size is proportional to the wavelength, $K\lambda$, the Rayleigh distance increases as $d_R = 2(K\lambda)^2/\lambda \propto \lambda$.
[3] Remember that due to reciprocity, the antenna pattern is the same for the transmit and the receive case.
[4] More generally, it is common to show the cuts of the E-plane (defined as the plane that contains the electric field vector and the maximum of the radiation pattern) and the H-plane (similarly defined for the magnetic field vector).

Example 8.4 *Compute the radiation resistance for a half-wave dipole with a sinusoidal current distribution and compare it to the above approximations.*

From Example 8.3, we know that

$$E_\theta(r,\theta) = \frac{j\eta_0 I_0}{2\pi r} e^{-jk_0 r} \frac{\cos\left[(\pi/2)\cos(\theta)\right]}{\sin(\theta)}. \tag{8.25}$$

The total transmitted power P_T is

$$
\begin{aligned}
P_T &= \int_0^{2\pi}\int_0^\pi \frac{|E_\theta(r=1,\theta)|^2}{2\eta_0}\sin(\theta)d\theta d\phi \\
&= \frac{\eta_0 I_0^2}{4\pi}\int_0^\pi \frac{\cos^2[(\pi/2)\cos(\theta)]}{\sin(\theta)}d\theta.
\end{aligned}
\tag{8.26}
$$

This integral needs to be evaluated numerically (or by using tabulations of special functions); the radiation resistance then follows as $R_{\rm rad} = 2P_T/I_0^2 = 73\ \Omega$. Compare this to 197 and 49 Ω for the uniform current and linear taper assumption, respectively. This demonstrates that the input impedance is considerably more sensitive than the radiation pattern to the assumption of the current distribution.

The loss resistance of a (thin) dipole can be computed as

$$R_{\rm loss} = \sqrt{\frac{\pi f_c \mu_0}{\sigma_{\rm cond}}}\frac{1}{2\pi a}\frac{L_a}{3} \tag{8.27}$$

where $\sigma_{\rm cond}$ is the conductivity of the metal the antenna is made of, and a is the radius of the dipole (assuming that the cross-section is circular). The loss resistance increases with $\sqrt{f_c}$, while the radiation resistance increases with f_c^2 (all these approximations are for electrically short antennas, as mentioned above). This demonstrates that the radiation efficiency increases (for fixed antenna length) with increasing frequency; similarly we can observe for a fixed frequency the increase of radiation efficiency with L_a.

Finally, the reactance for a dipole with tapered current can be approximated as

$$X_A = -\frac{120}{\pi L_a/\lambda}\left[\ln\left(\frac{L_a}{2a}\right) - 1\right]. \tag{8.28}$$

The approximation holds for short ($L_a < \lambda/2$) and thin ($L_a \gg a$), dipoles; progressively more complicated equivalent circuits consisting of multiple reactances and resistors exist for more general antennas. An example for the real and imaginary part of the input impedance can be found in Figure 8.4.

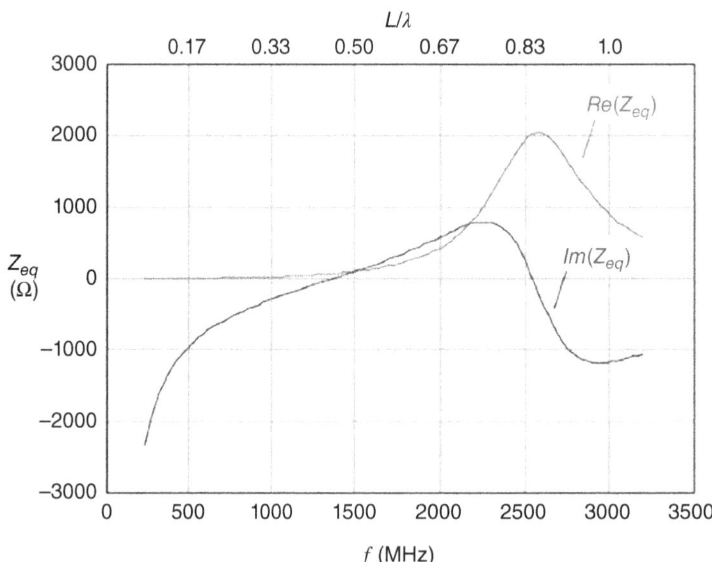

Figure 8.4 Input impedance for a thin dipole ($L_a/a = 100$), $L_a = 10$ cm. Color version available at wiley.com/go/molisch/wireless3e. Reproduced with permission from [Dobkin 2005] © D. Dobkin.

The equivalent circuit is valid for the receive case as well. As we can see from Figure 8.5, a receive antenna can be represented as a voltage source (or several voltage sources, representing desired signal, external noise, and internal noise), in series with two resistors (one representing the radiation resistance, and the other the conductive losses in the circuitry), plus a reactance. The question of available power, matching of impedances, etc., can then in principle be handled by the general techniques that will be discussed in Section 17.2.3.

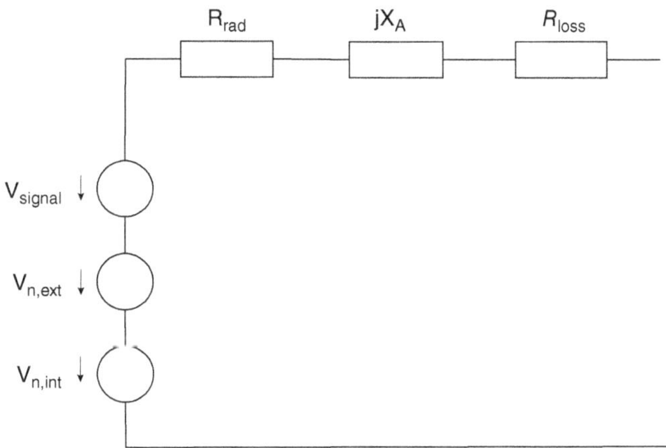

Figure 8.5 Equivalent circuit for a receive antenna.

In other words, the receive antenna is a source with internal impedance Z_A, while the RX circuitry is a load Z_L (to which as much power as possible should be delivered). The voltage applied to the antenna connectors is $V_{oc} = \mathbf{E}_{inc} \cdot \boldsymbol{\ell}_{eff}$, where $\mathbf{E}_{inc}$ is the incident electric field, and $\boldsymbol{\ell}_{eff}(\phi, \theta)$ is the electric length that will be explained in more detail in Section 8.2.4 (for the moment, just note that for a vertical dipole with tapered current distribution, $\boldsymbol{\ell}_{eff}$ is $(L_a/2)\mathbf{e}_z$, where $\mathbf{e}_z$ is the unit vector in the z-direction. The power that can be delivered to the load is then

$$\frac{1}{2} \operatorname{Re} \left\{ |\mathbf{E}_{inc} \cdot \boldsymbol{\ell}_{eff}|^2 \frac{Z_L}{|Z_A + Z_L|^2} \right\} \tag{8.29}$$

and the maximum is achieved again by conjugate matching, so that in the lossless case

$$P_{L,\,max} = \frac{|\mathbf{E}_{inc} \cdot \boldsymbol{\ell}_{eff}|^2}{8R_{rad}}. \tag{8.30}$$

This implies that at most half of the power "seen" by the antenna can actually be delivered to the RX circuitry; the other half is sometimes interpreted as being re-radiated by the antenna.[5]

Related to the antenna efficiency is the Q-factor (compare also Chapter 17). It can be defined as [Vaughan and Andersen 2003]

$$Q = 2\pi \frac{\text{energy stored}}{\text{energy dissipated per cycle}}. \tag{8.31}$$

The Q-factor can be related to the input impedance as

$$Q = \frac{f_c}{2R_A} \frac{\partial X_A}{\partial f}. \tag{8.32}$$

For an antenna contained in a sphere with diameter L_a, the Q factor is given as

$$Q \geq \frac{1}{(k_0 L_a/2)^3} + \frac{1}{k_0 L_a/2}. \tag{8.33}$$

Besides the ohmic losses, we also experience reduced efficiency because the antenna might be mismatched to the impedance of the cable/waveguide connected to it, leading to reflections. The efficiency can thus be written as

---

[5] The interpretation that half of the power is re-radiated is a controversial one and actively discussed in the electromagnetics community.

$$\eta = \eta_{\mathrm{rad}}\left(1 - |\Gamma|^2\right) \tag{8.34}$$

where Γ is the reflection coefficient. For antennas that are to operate only at a single frequency, perfect matching can be achieved. However, there are limits to how well an antenna can be matched over a larger band. The so-called Bode–Fano bound states if the antenna impedance can be modeled as a resistor and a capacitor in series

$$\int \frac{1}{(\omega)^2} \ln\left(\frac{1}{|\Gamma(\omega)|}\right) d\omega < \pi RC \tag{8.35}$$

where R and C are the resistance and capacitance, respectively (for additional bounds when other equivalent models apply, see Appendix 17.B).

Matching is most difficult in two cases:

- *Electrically small antennas*: such antennas have a large, negative (i.e., capacitive) reactance. While matching is possible, the matching is inherently narrowband (this follows immediately from the Fano bound and the equivalent circuit of small antennas). Wideband matching is possible by consciously reducing the transducer power gain, which can be achieved by introducing losses: e.g., the insertion of an attenuator.
- *Broadband antennas*: antennas with large bandwidth are difficult to match unless losses are accepted, as described above.

The existence of a large dielectric or conducting object (such as a human body) in the near-field of the antenna changes not only the radiation characteristics but also the impedance. Thus, handset antennas are often designed with adaptive matching networks that can adjust to the distortions of the antenna impedances. Alternatively (or in addition), the position of the antenna elements can be designed to reduce the probability for the presence of foreign objects in the near field.

A high efficiency is a key criterion for any wireless antenna. From the point of view of the transmit antenna, a high efficiency reduces the required power of the amplifier to achieve a given field strength at the RX. From the point of view of the receive antenna, the achievable SNR is directly proportional to the antenna efficiency. Antenna efficiency thus directly affects the battery lifetime of the UE, as well as the transmission quality of a link. It is difficult, however, to define an absolute goal for the efficiency. Ideally, $\eta = 1$ should be obtained over the whole bandwidth of interest. However, it must be noted that in recent years, antenna efficiency of UE antennas has *decreased*. It has been sacrificed mainly for cosmetic reasons, namely to decrease the size of the antennas. This does not pose a significant problem in places where systems are interference-limited anyway. However, in locations where the system is limited by noise, such reduced efficiency is problematic. It must also be mentioned that for transmit antennas, minimization of VSWR (or at least keeping VSWR below a certain threshold) is also vital, as high-power reflections from the TX antenna can damage electronic components.

8.2.3   Bandwidth

Antenna bandwidth is defined as the bandwidth over which the antenna characteristics (reflection coefficient, antenna gain, etc.) fulfill the specifications. However, the manufacturer often only specifies the bandwidth within which the reflection coefficient (or the S-parameter $|S_{11}|$, see Chapter 17) remains below a certain threshold, such as 10 dB. In many other situations, it is of interest over what bandwidth the antenna patterns stay (approximately) constant.

It is usually desirable to have a single antenna cover the whole system bandwidth. In particular, for Frequency Domain Duplexing (FDD, see Section 18.5), this system bandwidth must encompass both TX and RX bands. For example, the bandwidth of an LTE uplink channel might be only 5 MHz, while the separation between uplink and downlink bands is about 200 MHz, and the carrier frequency is around 1.8 GHz. The system bandwidth in this case is 200 MHz, and the relative antenna bandwidth in this case must be $200/1800 \approx 10\%$. As we have mentioned above, a large bandwidth implies that the matching circuits cannot be perfect anymore. This effect is the stronger, the smaller the physical dimensions of the antenna are.

Another interesting special case is dual- or multi-mode devices. For example, most 2G and 3G devices have to be able to operate at both the 700 and the 1700–2100 MHz carrier frequencies. Antennas for such devices can be constructed by requiring good performance in the whole 0.7–2.1 GHz band. Yet it is often sufficient to design antennas that have good performance in the required bands only; this goal is easier to fulfill if the different band center frequencies are integer multiples of each other. Still, the design of antennas that cover two or more bands with high efficiency is a very challenging task. The large number of bands in 4G and 5G (see Secs. 31.2.4 and 32.2.4) has made the task even more difficult. On the other hand, recent progress in the field of CMA provided new tools to solve such tasks. For example, an efficient multi-band antenna can be achieved by constructing a common antenna feed that can excite different resonant modes at different frequencies.

For BS antennas, it thus often is preferable from a technical point of view to use different antennas for different frequency bands. However, multi-band antennas are to be preferred for aesthetical reasons and reduced visual impact.

8.2.4   Polarization and Complex Antenna Patterns

In Section 8.1.3, we have discussed the electric field vector that is generated by a sinusoidal signal applied to antenna ports. The temporal evolution of the field vector orientation and magnitude, at a fixed location in space, is called the polarization of the wave.

Furthermore, the polarization of an antenna is, by definition, identical to the polarization of the electric field that it emits in the direction of maximum gain. The most common antenna polarizations are

1. Linear polarization including vertical polarization (so that the field vector is along the θ-direction) and horizontal polarization (such that the field vector is along the ϕ direction), see Figure 8.6a. In case the propagation direction is the x-axis, the electric field vector is in z-direction for vertical, and y-direction for horizontal polarization.
2. Circular polarization, where the field vector moves in a circle either in clockwise (right-hand polarization) or counter-clockwise (left-hand polarization) direction, see Figure 8.6b. A circular polarization can be interpreted as consisting of two orthogonal linear polarizations with equal magnitude that have a fixed phase relationship (namely, $\pm\pi/2$ phase shift) between them.

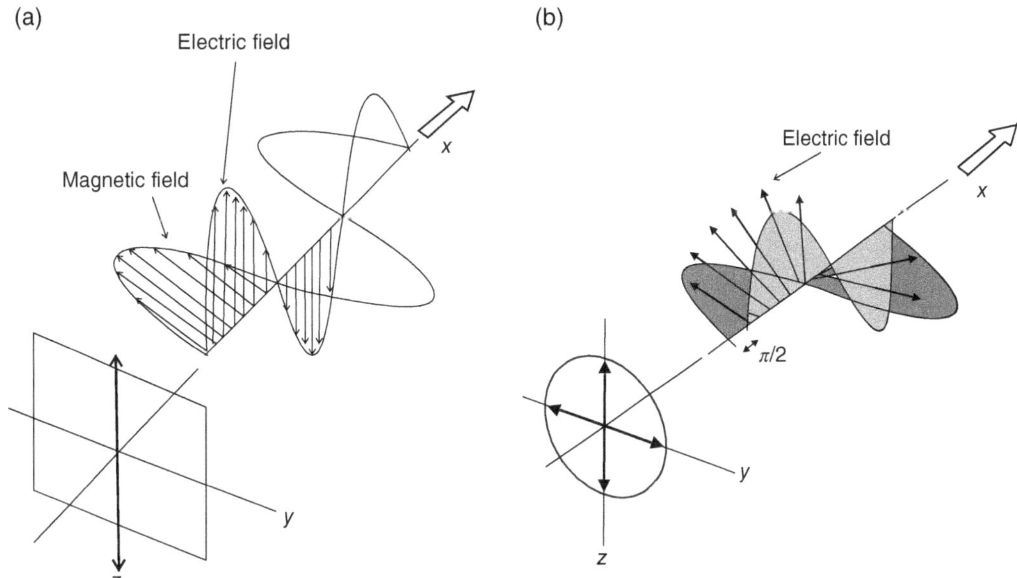

Figure 8.6 Polarization vectors: (a) Linear polarization. (b) Circular polarization. Color version available at wiley.com/go/molisch/wireless3e.

Polarization is a far-field characteristic, i.e., we consider the electric field vector at a sufficient distance from the TX antenna. Thus, the propagating wave is generally characterized by two field components (the component in the direction of propagation is negligible). Importantly, these two components are along the ϕ and the θ coordinate of a spherical coordinate system (they are not along x, y, and z-axis – only in the case that we consider propagation along, e.g., the x-axis does ϕ and θ become identical to y and z). It is commonly used notation (also used in this chapter) that the components along the ϕ and the θ coordinate are *always* called "horizontal" and "vertical," even if the propagation direction is not in the horizontal plane.

The total field can be described as consisting of these two field components. Importantly, the relative strength of these components can be different in different directions. Thus, each component can have its own antenna pattern, $D_\phi(\phi_0, \theta_0)$ and $D_\theta(\phi_0, \theta_0)$, respectively. The total antenna power pattern is thus

$$D(\phi_0, \theta_0) = \frac{D_\phi(\phi_0, \theta_0) + D_\theta(\phi_0, \theta_0)}{\frac{1}{4\pi} \int\int \left(D_\phi(\phi, \theta) + D_\theta(\phi, \theta)\right) \sin(\theta) d\theta d\phi}. \tag{8.36}$$

It is important to note that even though D_ϕ and D_θ can be interpreted as the separate antenna patterns for the two polarizations, the normalization is applied to them jointly; this ensures the correct representation of situations where one polarization is stronger than the other.

Interaction with objects such as walls in the environment can change the orientation and relative strength of those components, as we explored in more detail in Chapter 4.

The directivity, as described above, refers only to power radiated to/from a particular direction. In many situations, we also need to know the dependence of the phase on the direction of transmission (or reception), or – more generally – the *complex antenna pattern*. To obtain a more precise definition, we first introduce the concept of the *vector effective length* $\mathbf{l}_{\mathrm{eff}}$ of an antenna:

$$\mathbf{l}_{\mathrm{eff}}(\phi, \theta) = \ell_\phi(\phi, \theta)\mathbf{e}_\phi + \ell_\theta(\phi, \theta)\mathbf{e}_\theta \tag{8.37}$$

where $\mathbf{e}_\phi$ and $\mathbf{e}_\theta$ are unit vectors in the direction of ϕ and θ, respectively, see Figure 8.7. For a TX antenna, it relates the current between the antenna ports, I, to the created electric field in the far-field as

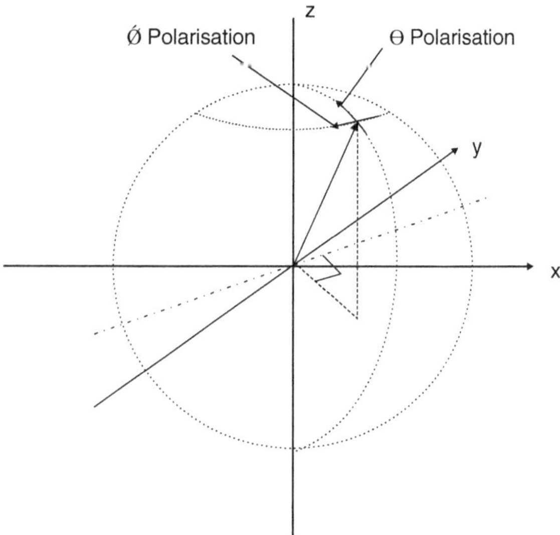

Figure 8.7   Definition of ϕ and θ polarization.

$$\mathbf{E}_{\mathrm{rad}} = -\frac{j\eta_0 k_0 I}{4\pi} \frac{e^{-jk_0 r}}{r} \boldsymbol{\ell}_{\mathrm{eff}}(\phi, \theta) \tag{8.38}$$

and is thus obviously strongly related to the vector potential $\mathbf{A}$. Conversely, we can say that for a RX antenna, if there is an incident field with field vector $\mathbf{E}_{\mathrm{inc}}$ then the voltage created at the antenna port (when left as open-circuit) is

$$V_{\mathrm{oc}} = \mathbf{E}_{\mathrm{inc}} \cdot \boldsymbol{\ell}_{\mathrm{eff}}(\phi, \theta) \ . \tag{8.39}$$

This shows that $\boldsymbol{\ell}_{\mathrm{eff}}(\phi, \theta)$ is an indication of the polarization of the antenna. Furthermore, we see that an antenna will most efficiently receive a signal if the electric field vector of the incident field is aligned with the direction of the vector effective length. For example, a vertical dipole has a $\boldsymbol{\ell}_{\mathrm{eff}}(\phi, \theta)$ that is oriented along the z-axis; thus it most efficiently receives waves that are vertically polarized as well. If there is an angle mismatch, i.e., the vector effective length and the field vector are at an angle $\alpha_{\mathrm{mismatch}}$, then the efficiency with which the power can be received is reduced by a factor $\cos^2(\alpha_{\mathrm{mismatch}})$.

The effective length now also helps to define the complex antenna pattern or complex directivity $\widetilde{\mathbf{D}}(\phi_0, \theta_0)$: in brief, the complex pattern is directly proportional to the effective length,

$$\widetilde{\mathbf{D}}(\phi_0, \theta_0) = C_{\mathrm{nd}} \boldsymbol{\ell}_{\mathrm{eff}}(\phi_0, \theta_0) \tag{8.40}$$

i.e., a vector consisting of the components $\widetilde{D}_\phi$ and $\widetilde{D}_\theta$ representing the two polarizations. Note that this is an "amplitude directivity." It is related to the "conventional" (power) directivity as

$$D_\phi(\phi_0, \theta_0) = \left|\widetilde{D}_\phi(\phi_0, \theta_0)\right|^2 \tag{8.41}$$

$$D_\theta(\phi_0, \theta_0) = \left|\widetilde{D}_\theta(\phi_0, \theta_0)\right|^2. \tag{8.42}$$

From the normalization condition

$$\frac{1}{4\pi} \int\int \left(D_\phi(\phi, \theta) + D_\theta(\phi, \theta)\right) \sin(\theta) d\theta d\phi = 1 \tag{8.43}$$

we can easily determine C_{nd}.

Antennas can often be intended to receive only one polarization. The *Cross-Polarization Discrimination (XPD)* of the antenna describes how well this goal is fulfilled. For example, a vertical dipole is intended to receive only vertically polarized radiation (assuming incidence in the horizontal plane). An XPD of, say, -20 dB implies that radiation incident with horizontal polarization is not suppressed completely, but rather attenuated by 20 dB compared to radiation of the same strength with vertical polarization. XPD of antennas plays a particularly important role in microwave directional links and satellite TV, where different information is transmitted in the ϕ and θ polarizations, and the system relies on the antenna polarization to suppress the undesired information. Thus, the polarizations of the radiation and the receive antenna should be carefully aligned.

For non-line-of-sight scenarios, requirements for a specific polarization of the antennas are typically not very stringent. Even if the transmit antenna sends mainly with a single polarization, the propagation through the wireless channel leads to a depolarization (see Chapter 4), so that the cross-polarization ratio at the RX is rarely higher than 10 dB. This fact is advantageous in many practical situations: as the orientation of UE antennas cannot be predicted (different users hold handsets in a different way), a low sensitivity of the antenna to the polarization of the incident radiation and/or a uniform polarization of the incident radiation, is advantageous.

The question arises now: can we transmit (or receive) both polarizations simultaneously? The answer is yes, and it can actually be achieved in two ways: (i) we can place two orthogonally polarized antennas close to each other. For example, a dipole antenna (for vertical polarization) and a loop antenna (for horizontal polarization) can be essentially co-located, as shown in Figure 8.8; (ii) we can use a dual-polarized antenna. A typical example for the latter type is a patch antenna (see Sec. 8.3.3), which can excite either polarization depending on the location of the antenna feed (i.e., where the cable carrying the signal is connected to the patch). Then, by placing two feeds on the patch, both polarizations can be excited simultaneously. Generally, the ability to transmit both polarizations with either a single, or two co-located antennas (implying a small form factor), is highly beneficial for antenna diversity and other multi-antenna techniques, see Chapters 12 and 16.

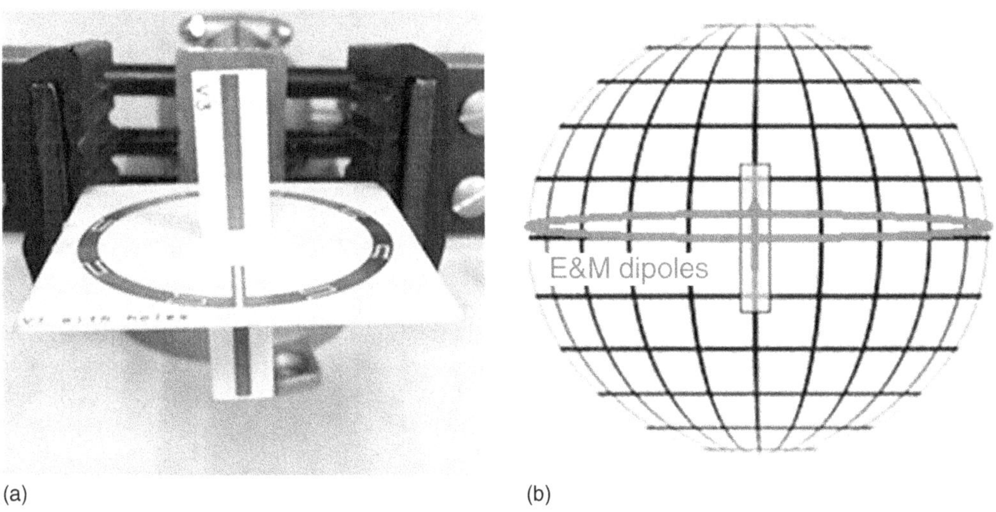

Figure 8.8 Electric dipole and loop antenna (magnetic dipole) for dual-polarized collocated antennas. Color version available at wiley.com/go/molisch/wireless3e.
Reproduced with permission from [Mirza et al. 2017] © Intech Open.

8.3 Popular Antenna Types

8.3.1 Monopole- and Dipole Antennas

Linear antennas are the "classical" antennas for UEs and have long determined the typical look of those devices. The most common ones are electric monopoles, located above a conducting plane (the casing), and dipoles. The antenna pattern of a short (Hertzian) dipole oriented along the z-axis, see Figure 8.9, is uniform in azimuth, and sine-shaped in elevation angle θ (compare (8.1)).

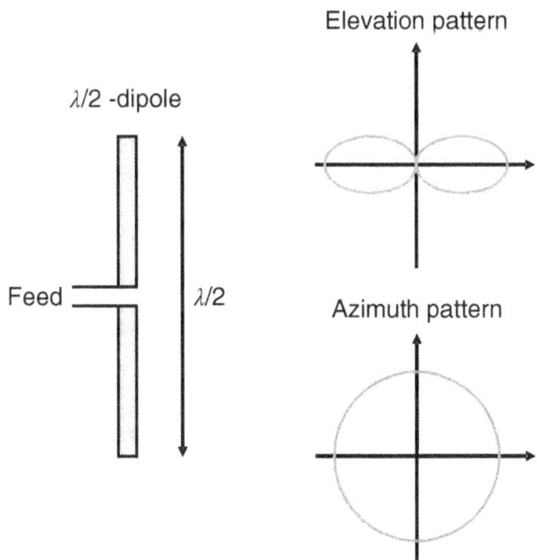

Figure 8.9 Shape and radiation pattern of $\lambda/2$ dipole antenna.

$$D(\varphi,\theta) \propto \sin^2(\theta) \tag{8.44}$$

with a maximum directivity

$$D_{\text{max}} = 1.5. \tag{8.45}$$

A $\lambda/2$ dipole has the following properties (compare (8.12)):

$$D(\varphi,\theta) \propto \left[\frac{\cos\left(\frac{\pi}{2}\cos(\theta)\right)}{\sin(\theta)} \right]^2 \tag{8.46}$$

and a maximum directivity

$$D_{\text{max}} = 1.64. \tag{8.47}$$

We can thus see that the patterns of the $\lambda/2$ dipole and the Hertzian dipole do not differ dramatically. However, the radiation resistance can be quite different, see Section 8.2.2.

From the image principle, it follows that the radiation pattern of a monopole located above a conducting plane is identical to that of a dipole antenna in the upper half-plane. Since the energy transmitted into the upper half-space $0 \leq \theta \leq \pi/2$ is twice that of the dipole, the maximum gain is twice, and the radiation resistance is half, that of the dipole. Note, however, that the image principle is valid only if the conducting plane extends infinitely – this is not the case for UE casings. We can thus anticipate considerable radiation also into the lower half-space even for monopole antennas mounted on UEs.

We also note that the feed of a dipole antenna is balanced (symmetric), while for the monopole the feed is unbalanced. Depending on the type of signal generated in the transmit circuit, namely differential or grounded, there needs to be a so-called balun between the antenna and the circuitry to convert between *bal*anced and *un*balanced signals. For example, on a coaxial cable, the signals on inner and outer conductor are not symmetrical. A signal from such a cable thus has to be "symmetrized" by a balun before it can be connected to a dipole.

The reduction of the radiation resistance characteristic for a monopole is often undesirable, since it makes matching more difficult, and leads to a reduction of the efficiency due to the ohmic losses. One way to increase the efficiency without increasing the physical size of the antenna is the use of *folded* dipoles. A folded dipole consists of a pair of half-wavelength wires that are joined at the nonfeed end [Vaughan and Andersen 2003]; these increase the input impedance.

Another variant of the dipole antenna is the bowtie antenna (also called biconical antenna), where the thickness (radius of the cross section) of the antenna increases with increasing distance from the feedpoint. The main advantage of this antenna is its increased bandwidth, made possible by the dual-triangle shape providing more resonances due to multiple current path lengths (relative to single path length of the wire version).

The biggest plus of monopole- and dipole antennas is that they can be produced easily and cheaply. Not only can "thin metallic rods" be produced easily, but it is also possible to print dipole antennas on dielectric substrate, see Figure 8.8 for an example. The relative bandwidth is sufficient for most applications in single-band systems. For UEs, the main disadvantage is the size relative to a typical cellphone casing. In the 900 MHz band, a $\lambda/4$ monopole is 8 cm long – sometimes longer than the UE itself (especially for compact flip phones, still widely used in parts of the world; smart phones may be considerably larger). Retractable antennas were popular in the 1990s but were easily damaged and unsightly.

For this reason, shorter and/or integrated antennas, which are less efficient, have become dominant. For example, a popular approach for hidden UE antenna design is to utilize the conducting plate in the phone (i.e., Printed Circuit Board [PCB]) as part of the radiation structure, instead of acting as a ground plane for the antenna element [Vainikainen et al. 2002]. This approach allows even low bands (e.g., 900 MHz) to obtain good bandwidth despite the small size of the hidden antenna element. In essence, the "antenna" plays the role of coupling the power into the ground plane to radiate indirectly, more than it is radiating by itself. The use of the ground plane for radiation has become more popular in UE antenna design in recent years as it is opportunistic in reusing the ground plane and the design tool of CMA has become more mature, as will be further discussed in Section 8.5.1.

*8.3.2 Helical Antennas

The geometry of helical antennas is outlined in Figure 8.10. A helical antenna can be seen as a combination of a loop antenna and a linear antenna, and thus has two modes of operation. The dimensions of the antenna determine which mode it is operating in. If the dimensions of the helix are much smaller than a wavelength, then the antenna is operating in the normal mode. It behaves similar to a linear antenna and has a pattern that is shaped mainly in the radial direction. Otherwise, it is operating in axial mode, and the antenna pattern is mainly shaped in the direction of the axis.

In general, a helical antenna has lower bandwidth and a smaller input impedance than a monopole antenna. In general, the polarization of the transmission is elliptical, though it can become circular if the ratio $2\lambda d/(\pi D)^2$ becomes unity [Balanis 2005]. The polarization becomes vertical if the helical antenna is arranged over a conducting plane, as the horizontal components of the actual antenna and its image cancel out. When the circumference of the helix is on the order of one wavelength, the antenna pattern has its maximum along the axis of the helix, and the polarization is almost circular. Helical antennas were popular in the 2000s, since

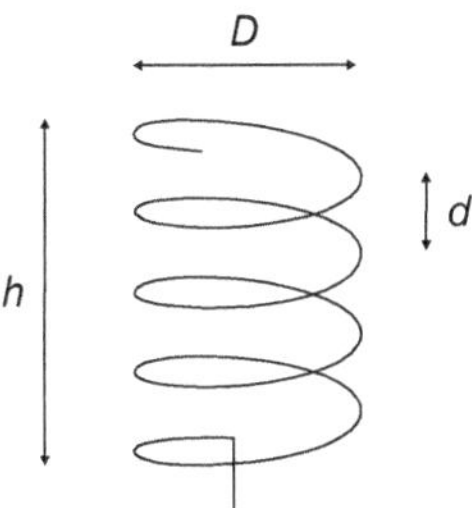

Figure 8.10 Geometry of a helical antenna. Color version available at wiley.com/go/molisch/wireless3e.

they are typically shorter than monopole antennas, but they still are external antennas that protrude from the UE casing and have thus not been adopted for smartphones.

8.3.3 Microstrip Antennas

A microstrip antenna (patch antenna) consists of a dielectric substrate, which is covered on one side by a layer of conducting material (ground plane), while on the other side there is a *patch* of conducting material. The configuration is outlined in Figure 8.11, see also [Fujimoto 1987, 2008].

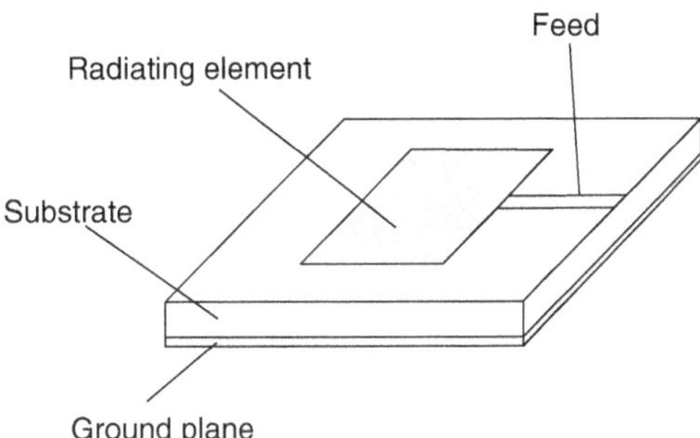

Figure 8.11 Geometry of a microstrip antenna.

The properties of a microstrip antenna are determined by the shape and dimension of the metallic patch, as well as by the dielectric properties of the used substrate. Essentially, the patch is a resonator whose dimensions have to be multiples of half the effective dielectric wavelength. Thus, a high dielectric constant of the substrate allows the construction of small antennas. The most commonly used patch shapes are rectangular, circular, and triangular patches. The patch is usually fed either by a coaxial cable or a microstrip line.

As mentioned above, the size and efficiency of the microstrip antenna are determined by the parameters of the dielectric substrate. A large dielectric constant ε_r reduces the size. The length of one side must be approximately

$$L_a = 0.5\lambda_{\text{eff}} \tag{8.48}$$

which can be expressed via the effective dielectric constant

$$\lambda_{\text{eff}} = \frac{\lambda_0}{\sqrt{\varepsilon_{\text{eff}}}} \quad \text{with} \quad \varepsilon_{\text{eff}} = \frac{\varepsilon_{\text{substrate}} + 1}{2}. \tag{8.49}$$

Unfortunately, a reduction of the physical size also leads to a smaller bandwidth, which is usually undesirable. For this reason, substrates used in practice usually have a very low ε_r – even air is used quite frequently. A further possibility for reducing the size of patch antennas is the use of short-circuited resonators, which reduces the required size of a resonator from $\lambda/2$ to $\lambda/4$, see Figure 8.12.

The bandwidth of microstrip antennas can be increased by various measures. The most straightforward one is an increase of the antenna volume, i.e., the use of thicker substrates with a lower ε_r. Alternatives are the use of matching circuits, and the use of parasitic elements.

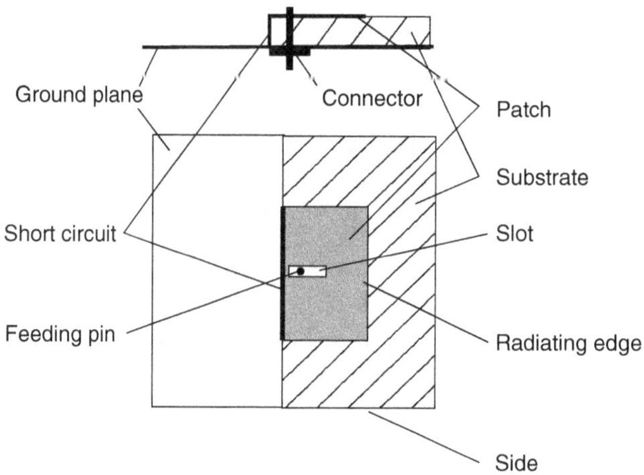

Figure 8.12 Short-circuited $\lambda/4$ patch antenna.

Microstrip antennas have several important advantages for wireless applications

- They are small and can be manufactured cheaply.
- The feedlines can be manufactured on the same substrate as the antenna.
- They can be integrated into the UE, without sticking out from the casing.
- For very high frequencies (i.e., small wavelength), they can even be integrated on a chip.

However, they also have serious weaknesses

- They have a low bandwidth (usually just a few percent of the carrier frequency).
- They have a low efficiency.

A variant of the microstrip antenna does not feed the patch directly, but rather via electromagnetic coupling. This case uses a configuration where the ground plane is sandwiched between two layers of dielectric material. On the top of the one material is the patch, while the feedline is at the bottom of the other dielectric layer. The coupling is effected through a slot (aperture) in the ground plane. These antennas are thus called *aperture-coupled patch antennas*, whose structure is depicted in Figure 8.13. This design has the advantage that the dielectric properties of the two layers can be chosen differently, depending on the requirements for the patch and the feedline. A better isolation of the radiation patch from the remainder of the TX/RX can be achieved, and the achievable bandwidths can be significantly better than for regular microstrip antennas.

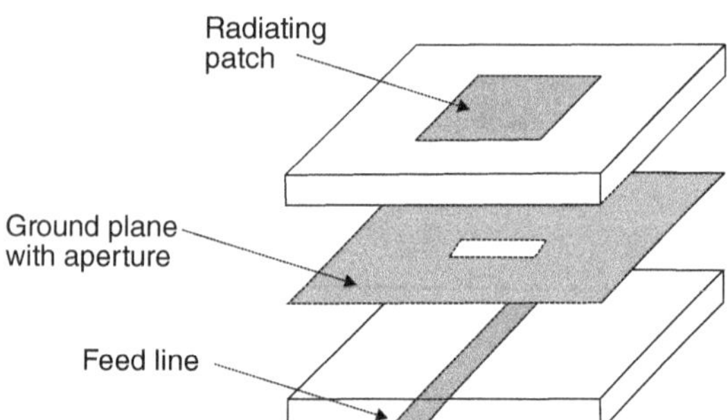

Figure 8.13 Aperture coupled patch antenna.

8.3.4 PIFA-Antennas

Some of the problems of the microstrips antennas can be alleviated by the PIFA (*Planar Inverted F Antenna*). The shape of the PIFA is similar to that of a $\lambda/4$ short-circuited microstrip antenna, see Figure 8.14. A planar, radiating element is located parallel to a ground plane. This element is short-circuited over a distance W. If W is chosen equal to the length of the edge L, then we obtain a short-circuited $\lambda/4$ microstrip antenna. If W is chosen smaller, then the resonance length increases, and the current distribution on the radiating element changes.

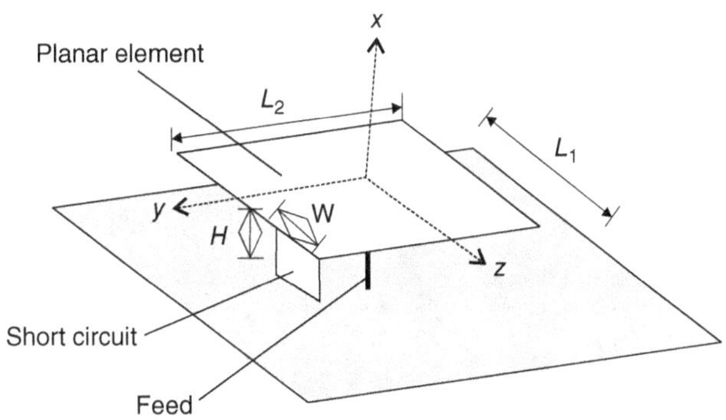

Figure 8.14 PIFA antenna.

8.3.5 Multi-Band Antennas

Modern cellular handsets are anticipated to be able to handle different frequencies for communications. As more and more frequency bands have been made available for communications, requirements for multiple band antennas have increased. For example, LTE uses frequency bands 600–900, 1700–2200, with NR additionally using 3.3–5.0 GHz, while Bluetooth and Wi-Fi use 2.4–2.5 GHz, and Wi-Fi in addition 5.2–7.1 GHz, and GPS operating at 1.45 GHz. The design of internal multi-band antennas is very complicated, and few rules for closed-form design are available. However, as mentioned in Section 8.2.3, CMA yields the resonant properties of the antenna structure over frequency, which can be used to guide the design of multi-band antennas. Figure 8.15 shows an example of a microstrip multi-band antenna for a model operating at 900, 1800, and 2450 GHz bands.

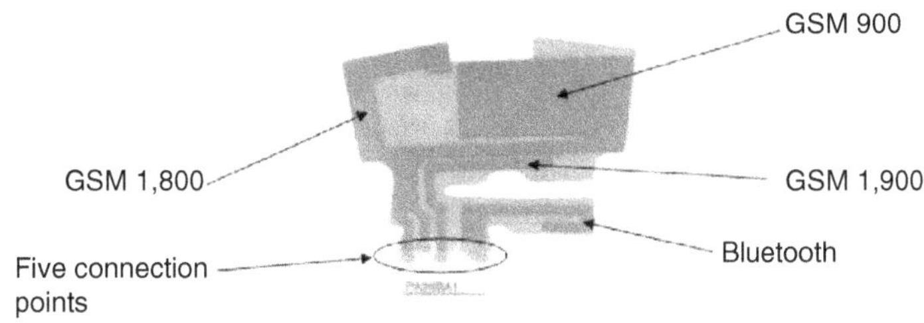

Figure 8.15 Integrated multi-band antenna.
Reproduced with permission from [Ying and Anderson 2003] © Z. Ying.

It must be noted that multi-band antennas are not an antenna type per se (like, e.g., helical antennas), but rather can be based on different standard antenna architectures. In general, many antenna types can be multi-band, not only at multiples of the fundamental resonant frequencies but also at other frequencies by suitable adjustments of the structure to create and feed the appropriate resonant paths.

8.3.6 Horn Antennas

In principle, any open waveguide is a radiating element, as the guided waves propagate into space when they hit the open end; however, the directivity of such a waveguide is low, and the sudden change in impedance leads to strong reflections. Much higher directivity can be achieved by having the waveguide flare in a pyramid (for rectangular waveguide) or conical (for cylindrical waveguide) shape. The achievable directivity depends mainly on the aperture, i.e., the cross section of the horn at the actual opening, though horns with very steep opening angle (short flange) tend to reduce this ideal directivity more than very long horns. Due to various parasitic effects, the maximum gain of horn antennas is limited to about 30 dB. Generally, horn antennas are mostly used at higher frequencies, especially at mm-wave and THz frequencies, both because of the form factor (as follows from their high directivity, horn antennas are generally much larger than a wavelength), and due to cost reasons. Pyramidal horns are the most commonly used horns, see Figure 8.16; they create linear polarization. For circular polarization, conical horns are used.

Horn antennas can operate over very large bandwidths (a factor of 1:10 for lower to upper-frequency limit is common). Further improvements in bandwidth can be achieved by corrugated horns, which have grooves transverse to the horn axis; those grooves not only provide better bandwidth but also smaller sidelobes.

A planar version of a horn antenna is the so-called Vivaldi antenna, see Figure 8.17. It consists of a slotline that is fed by a coaxial cable and ends in a tapered opening that is essentially a planar version of a horn. It can be printed on a circuit board, which greatly reduces the production cost. It also has a high bandwidth and is thus suitable for wideband and ultrawideband transceivers.

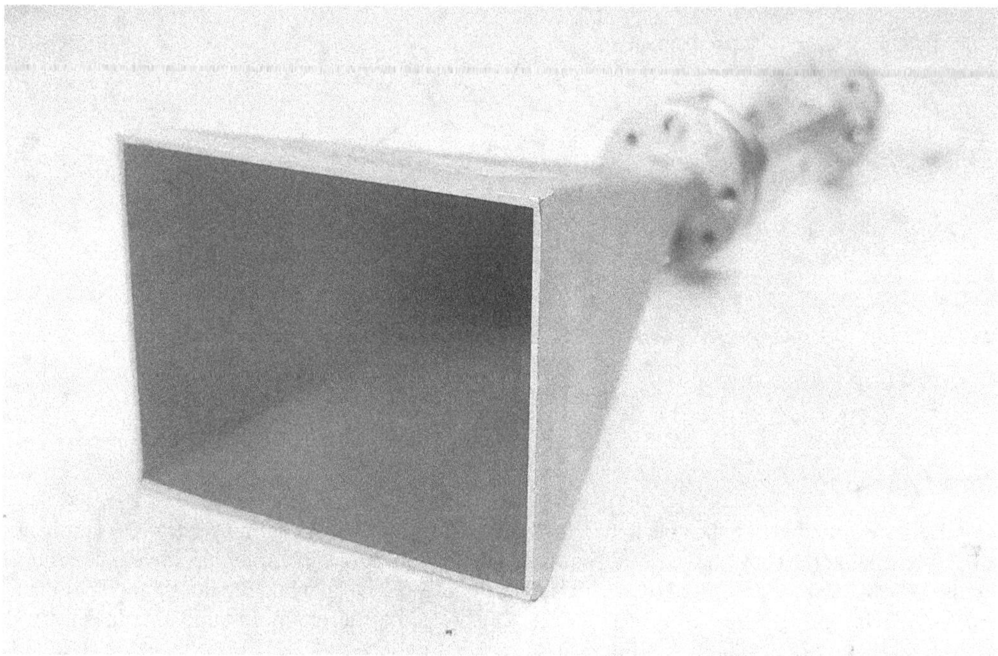

Figure 8.16 60 GHz standard gain horn antenna. Color version available at wiley.com/go/molisch/wireless3e.

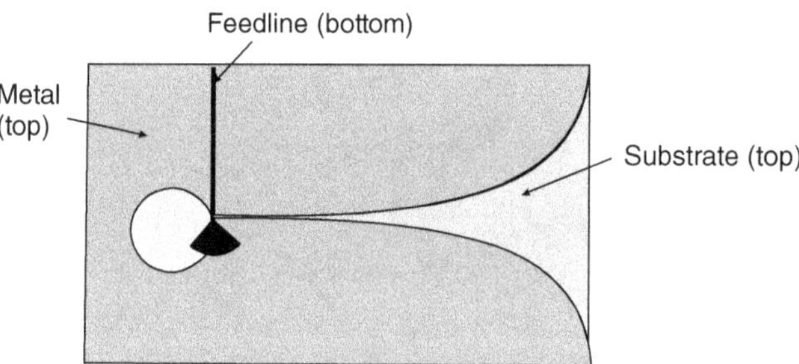

Figure 8.17 Typical Vivaldi antenna.

8.3.7 Reflector Antennas, Yagi-Uda Antennas, and Switched Parasitic Antennas

An effective way to increase the directivity of antennas is to combine a low-directivity radiating element with a passive reflector (and/or director). The most intuitive example of these is the parabolic antenna, widely known as the "satellite dish" for reception of TV signals. There, a radiating element is located in the focal point of a parabolic-shaped mirror, see Figure 8.18. The antenna pattern of the overall arrangement thus is essentially determined by the directivity of the dish, which is very high – remember that an infinitely large parabola redirects all rays incident along the axis to a sharp focal point. Thus, the limitation on the antenna directivity is the finite radius a_P of the parabolic mirror; it is given as

$$D_{\max} \approx \pi^2 \left(\frac{2a_P}{\lambda} \right)^2 \tag{8.50}$$

and the half-power beamwidth is approximately $(1.2\lambda/2a_P)$.

Another widely used type of antenna is the Yagi–Uda antenna. The basic outline is shown in Figure 8.18. Multiple dipoles are perpendicular to the antenna axis. One of the elements is actively fed, while the other elements are parasitic (radiation coupled). The one (or more) element behind the active element acts as reflector, whereas the other elements serve as the directors. The number of directors is typically much larger, in the range of 3–10. Overall directivity of 9–15 dB is typical. The length and spacing of the directors and reflectors need to be carefully optimized, and neither length nor spacing needs to be the same for the different elements. As a matter of fact, the bandwidth of the antenna can be increased significantly by letting the length of the reflectors vary, so that each is particularly effective at a certain frequency. Yagi–Uda antennas have long been popular as receive antennas for broadcast TV.

The Yagi–Uda antenna, as well as other parasitic antenna structures, can be made adaptive by using switches to change the effect of the parasitic elements. This change can be achieved by (i) short-circuit/leave open the connection of the parasitic element to the

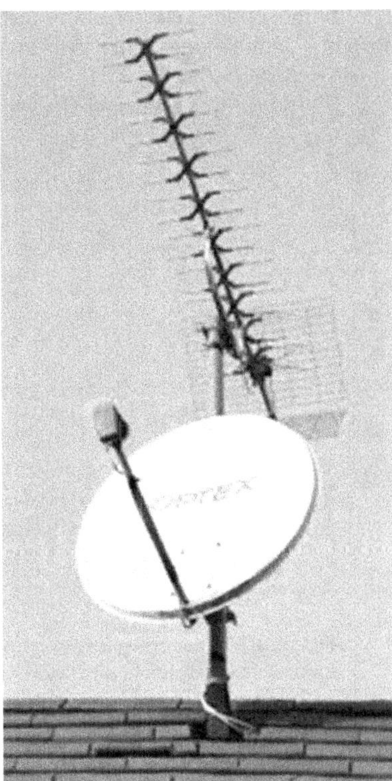

Figure 8.18 Picture of parabolic antenna (bottom) and Yagi–Uda antenna (top). Color version available at wiley.com/go/molisch/wireless3e. Unknown Author © Wikimedia Commons.

ground or other elements, or (ii) add (or not) reactances that increase or decrease the effective length of the parasitic elements. Such switched parasitic antenna structures have the capability of adapting the beam pattern, just as phased arrays (see below) do, but require considerably less hardware effort. On the other hand, there is less control over the exact shape of the pattern and in particular the depth of nulls than in phased arrays.

8.3.8 Lens Antennas

Lenses are a familiar tool in optics to focus radiation from a source with low directivity (e.g., radiation from a point source) into a (quasi-)parallel beam (high directivity), or vice versa. More mathematically speaking, a lens performs a Fourier transformation, providing a transformation between a spatial domain (i.e., location of a point source) and the beam domain (direction of the quasi-parallel beam). The propagation of millimeter-wave and THz radiation is quasi-optical, i.e., follows similar rules as lightwaves.[6] Thus, dielectric lenses can be used, e.g., in transmit antennas to improve the directivity of the transmitted radiation.

A lens antenna thus consists of a feed (possibly a point source, more commonly a directive antenna), and a piece of dielectric material. Both the shape of, and the distribution of the dielectric constant within, this (possibly nonhomogenous) material determine the focusing properties of the lens, or more generally, the achieved beam pattern. A simple example is shown in Figure 8.19, where the material is homogenous, and the shape of the lens determines the focusing, similar to a traditional optical telescope.

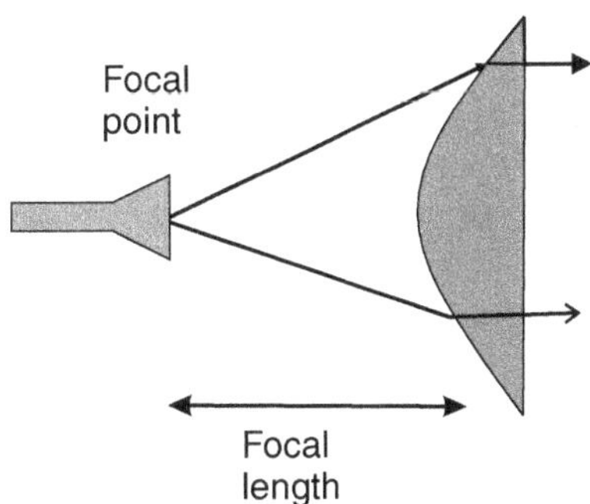

Figure 8.19 Traditional lens antenna.

---

[6] In principle, lens antennas can also be used at lower frequencies.

An important example of a lens antenna with nonhomogenous material is the Luneberg antenna. It is spherical in shape and has a dielectric constant that changes continuously from a high value in the center to a low value near the surface of the sphere (Figure 8.20); it is designed such that excitation with point sources in different locations leads to radiation into different directions. This makes it easy to implement antennas that send different signals into different beam directions (compare Chapter 22).

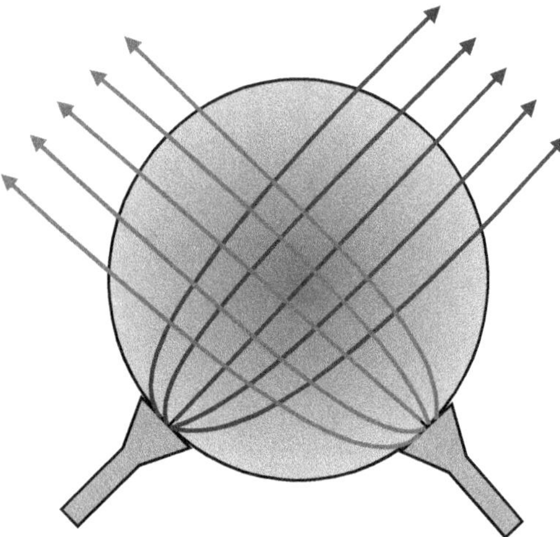

Figure 8.20 Luneberg lens antenna. Darker shading indicates higher dielectric constant. Color version available at wiley.com/go/molisch/wireless3e.

*8.3.9 On-Chip Antennas

Motivation

For antennas in the mm-wave bands,[7] it is common to have the antennas directly integrated on the Radio Frequency (RF) chip, or on the packaging of the chip. There are multiple reasons for such an arrangement:

- Since the size of antenna elements is proportional to the wavelength, it is actually possible to fit antennas, and even of antenna arrays (Section 8.4), on a chip. This is in contrast to typical antennas at cm-wave and lower frequencies, since those would not find enough space.
- Connection lines between signal generation and external antennas are more difficult to build at mm-waves. Most mm-wave connectors are hollow waveguides that have to be matched carefully to the outputs of the RF generators in order to avoid excessive reflections. Furthermore, microstrip lines have small geometrical dimensions at high frequencies (making them less robust to manufacturing tolerances), and are sensitive to interference. The difficulties of external connection lines become even more daunting for arrays, where many of those lines are required.

Despite these strong motivations, there are also significant drawbacks of integrated antennas.

- Most notable is the low antenna efficiency. As we will describe in more detail below, typical antenna efficiencies lie on the order of 10% or less (as compared to better than 90% for typical cm-wave antennas).
- It is difficult to include a radome on the chip/packaging.
- An integrated antenna is closely connected to a very complex dielectric/metallic structure, i.e., the chip. Such a structure in the near-field of the antenna distorts the antenna characteristics that would be obtained in "normal" antenna design, where the antenna is located on a homogeneous dielectric material (possibly with a metallic backplane). This in turn implies that the antenna has to be designed together with the chip, which not only is a complicated process, but which also requires the process to be repeated every time the chip changes even when the desired antenna characteristics might remain the same.

We note finally that for a number of specialized applications, discrete mm-wave antennas (in particular horn or parabolic antennas, as discussed above, remain popular. This is especially true for directional links, where no antenna adaptivity is required, and it is cheaper to build a single transceiver – with a single, highly efficient, and highly directional antenna – than to create an antenna array that can electronically enforce a particular beam direction. Such discrete horn antennas have been built to create directivities of up to 30 dB, while specialized variants such as Cassegrain antennas achieve up to 50 dB at each link end, and can thus sustain high SNR over large distances even for wideband systems.

---

[7] The same considerations apply to THz frequencies; for conciseness we just speak about mm-wave bands henceforth.

On-Chip Antenna Structures

The fundamental structures used for mm-wave antennas are the same that are most common for cm-wave antennas, namely dipoles, slot antennas, and patch antennas.

For either on-chip or in-package realizations, dipole antennas have to be planar dipoles. A dipole might be printed as a simple dipole shape, or a zig-zag or meander line, to increase the effective length of the radiating elements within given geometrical constraints.

Patch antennas (or, more precisely, microstrip patch antennas) are also well suited for integration on-chip or in-substrate. The efficiency of the patch increases significantly with the thickness of the substrate, i.e., separation of the patch from the ground plane.

Antenna efficiencies can be enhanced by placing a dielectric lens either on top, or the back of the antenna. Top-mounted lenses enhance the radiation into free space, i.e., the desired direction. Different types of lenses such as hemispherical, elliptical, or hyper-hemispherical have been suggested. Lenses that are mounted in the back of the antennas mainly serve to make the substrate look infinite. This decreases, e.g., the creation of surface guided modes, and thus enhances efficiency. The most common technology for realizing on-chip antennas is CMOS (Complementary Metal Oxide Semiconductor). It is by far the most cost-effective process, and the emergence of CMOS chips with transit frequencies in the hundreds of GHz has enabled the current interest in mm-wave communications. The substrate is formed by a layer of doped Si, which is approximately 300–700 μs thick. On top of it sits a layer of silicone dioxide (SiO_2) or similar material, in which many layers of metal traces are located, which are usually made of copper or aluminum; this overall structure is called the Back End of Line (BEOL) layer.

The main challenges for antenna efficiency arise from the properties of the substrate on which the antenna is located. Due to the high permittivity of the substrate, the energy created by the metal structures is mainly directed towards the substrate. For "normal" dielectric structures, in particular at mm-wave frequencies, this is not a problem: the metallic backplane at the bottom ensures that the energy ultimately is radiated into the desired direction. However, for on-chip antennas, the losses in the substrate, as well as the excitation of surface waves, which propagate into the "wrong" directions, create significant reduction of the radiation efficiency, which typically is less than 10%.

In-Package Antennas

An alternative to integrating the antennas in the semiconductor structure of the chip is the integration into the packaging of the chip. The chip is typically placed within a small air cavity in the packaging, connected to the rest of the package through thin bond wires. The in-package antenna can comprise multiple layers of antennas, baluns, feed lines, and hybrids. Connections between the different layers are provided by metallic vias. There are a number of different packaging processes that can be used, such as liquid crystal polymers, and teflon. They differ in their relative permittivity, which allows a tradeoff: a higher permittivity has the advantage that the antenna structures can be realized on a smaller area, but leads to a lower bandwidth.

In-package antennas are insulated from the lossy substrate that is the chip, and thus can provide significantly better efficiencies. Furthermore, the vertical dimensions of the packaging are less constrained than those of the chip. For these reasons, the efficiency of in-package antennas is significantly better than those of on-chip antennas – efficiencies better than 80% can be obtained.

8.4 Antenna Arrays

An antenna array consists of multiple antenna elements whose signals are controlled in such a way that they essentially create an electronically steerable antenna pattern. Some specific forms of antenna arrays (especially phased arrays, discussed below) have been used for many decades, but it is really only in the past 10–20 years that their full flexibility has been widely exploited in wireless systems. Today they constitute critical components of most modern cellular systems and Wi-Fi, both at the BS and the UE. In this section we will discuss some basics; further details in particular with respect to processing of the signals will be described in Chapters 9, 12, 16, and 22.

8.4.1 Data Model for Array

Let us establish a mathematical model for receive array and incoming signals (the transmit case works completely analogous). For simplicity, we analyze first the case of a *Uniform Linear Array* (ULA) consisting of N_r elements, where the signal $r_n(t)$ is detected at the nth element. To further simplify the discussion, we will assume for the moment that all waves are propagating in the horizontal plane and that the elements of the linear array are on the x-axis.

Consider now the case that a plane waves is incident on the array, with a direction of arrival ϕ, see Figure 8.21. The relationship between the incident signals s and the signal it creates at the first antenna element is simply:

$$r(t) = a_1 s(t - \tau_1) + n_1(t) \tag{8.51}$$

where τ_1 is the runtime between the source of the signal and the first antenna element, a_1 the (complex) amplitude of the channel from the TX antenna to the first receive antenna element, and $n_1(t)$ is the noise at the first antenna element. Consider now the second antenna element. Here, the received signal is

$$r_2(t) = a_2 s(t - \tau_2) + n_2(t). \tag{8.52}$$

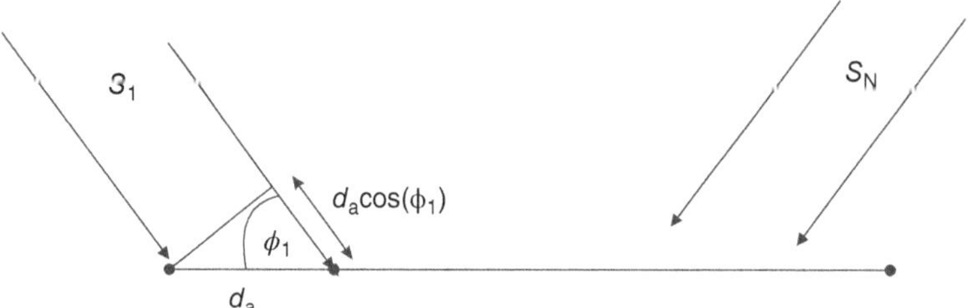

Figure 8.21 Plane waves incident at angle ϕ on a uniform linear array.

If the source is in the far-field, then $a_2 = a_1$, because the impact on the amplitude by the slightly longer distance the wave has to cover is negligible. The difference of the runtimes of a plane wave to elements 1 and 2 is

$$\tau_2 - \tau_1 = \frac{d_a}{c_0} \cos(\phi). \tag{8.53}$$

Let us furthermore assume that the signal is narrowband in the RF sense, i.e., the bandwidth of the signal is much smaller than the carrier frequency. Then the runtime difference can be converted into an equivalent phase shift $(\tau_2 - \tau_1)2\pi f_c$, so that

$$s(t - \tau_2) = s(t - \tau_1) \exp\left(-j(\tau_2 - \tau_1)2\pi f_c\right). \tag{8.54}$$

The relationship between r and s is thus

$$r_1(t) = \widetilde{s}(t) + n_1(t) \tag{8.55}$$
$$r_2(t) = \widetilde{s}(t) \exp\left(-jk_0 d_a \cos(\phi)\right) + n_2(t) \tag{8.56}$$

where $\widetilde{s}(t) = a_1 s(t - \tau_1)$, and $k_0 = 2\pi/\lambda$ is the wavenumber. For the next antenna element, we get similarly

$$r_3(t) = \widetilde{s}(t) \exp\left(-j2k_0 d_a \cos(\phi)\right) + n_3(t). \tag{8.57}$$

From this, we can conclude the general relationship between r and s

$$\mathbf{r}(t) = \mathbf{u}\widetilde{s}(t) + \mathbf{n}(t) \tag{8.58}$$

where $\mathbf{r}(t) = [r_1(t), r_2(t), ... r_{N_r}(t)]^T$, $\mathbf{n}(t) = [n_1(t), n_2(t), ... n_{N_r}(t)]^T$, and

$$\mathbf{u} = \begin{pmatrix} 1 \\ \exp\left(-jk_0 d_a \cos(\phi)\right) \\ \exp\left(-j2k_0 d_a \cos(\phi)\right) \\ \vdots \\ \exp\left(-j(N_r - 1)k_0 d_a \cos(\phi)\right) \end{pmatrix} \tag{8.59}$$

is the array *steering vector*. Its entries are the phase shifts between the antenna elements associated with a signal from direction ϕ. We will encounter it again below when discussing phased arrays, which aim to steer an antenna pattern into a specified direction.

We can generalize the above description for arbitrary antenna configurations, and remove the assumption that the waves have to be incident from the horizontal direction. We can define the location of each of the antenna elements by a location vector $\mathbf{l}_n$ that points from an (arbitrary) origin of the coordinate system to the center of the antenna element.[8] Each element is furthermore characterized by its complex element pattern $\widetilde{G}_n(\phi, \theta)$.[9] In some situations, such as a ULA, all element patterns can be identical. In some other cases, such as circular arrays, the element patterns can be rotated versions of a basic pattern, i.e., $\widetilde{G}_n(\phi, \theta) = \widetilde{G}_{\text{basic}}(\phi - n\phi_{\text{offset}}, \theta)$. However, in the most general case every antenna element can have a different pattern. Furthermore, we can concisely describe the propagation of a wave with direction (ϕ, θ) by its associated wave vector $\mathbf{k}_0 = k_0(\cos\phi)\sin(\theta), \sin(\phi)\sin(\theta), \cos(\theta))^T$. Consider now the phase shift that a signal from a particular direction creates at an antenna element. Firstly, it will be weighted by the element pattern in this direction. Secondly, it will experience a phase shift equal to the inner product of the wave vector with the location vector, $\langle \mathbf{k}_0, \mathbf{l}_n \rangle$. Thus, the general steering vector is

---

[8] It is common to pick the origin of the coordinate system either as the first element of the array, or as the center.
[9] Remember we are assuming gain and directivity to be equivalent.

$$\mathbf{u} = \begin{pmatrix} \widetilde{G}_1(\phi,\theta)\exp\left(-j\langle \mathbf{k}_0,\mathbf{l}_1\rangle\right) \\ \widetilde{G}_2(\phi,\theta)\exp\left(-j\langle \mathbf{k}_0,\mathbf{l}_2\rangle\right) \\ \widetilde{G}_3(\phi,\theta)\exp\left(-j\langle \mathbf{k}_0,\mathbf{l}_3\rangle\right) \\ \vdots \\ \widetilde{G}_{N_r}(\phi,\theta)\exp\left(-j\langle \mathbf{k}_0,\mathbf{l}_{N_r}\rangle\right) \end{pmatrix}. \tag{8.60}$$

In later chapters, we will also assume that the noise at the different antenna elements is independent and identically distributed, so that the correlation matrix of the noise is a diagonal matrix with entries σ_n^2 on the main diagonal. If the signals from the antenna elements are linearly combined with time-variant, complex antenna weights $w_n{}^*$, then the combiner output $y(t)$ can be written as

$$y(t) = \mathbf{w}^\dagger(t)\mathbf{r}(t) \tag{8.61}$$

where $\mathbf{w} = [w_1, w_2, ... w_{N_r}]^T$. Note that the received signals are multiplied with $w_n{}^*$, not w_n – this is purely a matter of notation convention, the reason for which will become clearer in later chapters.

8.4.2 Beam Pattern of Array Antennas

Based on this signal model, it now becomes easy to synthesize antenna patterns with desired characteristics. This shaping can either be done in a predetermined way (see Section 8.5.3) or adaptively (see Chapters 12, 16, and 22).

The complex gain due to the multiple antenna elements, often also called array factor, of a ULA follows from the above discussion as

$$M(\phi,\theta) = \sum_{n=0}^{N_r-1} w_n{}^*\exp\left[-jnk_0 d_a\cos(\gamma)\right] \tag{8.62}$$

where γ is the angle between the direction of interest (ϕ,θ) and the axis of the array.

For the case that all $|w_n| = 1$, and $\arg(w_n{}^*) = n\Delta$, the array factor becomes

$$M(\phi,\theta) = \frac{\sin\left[\frac{N_r}{2}\left(k_0 d_a\cos\gamma - \Delta\right)\right]}{\sin\left[\frac{1}{2}\left(k_0 d_a\cos\gamma - \Delta\right)\right]}. \tag{8.63}$$

The phase shift Δ determines the direction of the antenna main lobe. This principle forms the basis for phased array antennas [Hansen 1998], which are also widely used in radar. By imposing $\varphi_n = n\Delta$, the degrees of freedom have reduced to one – while the main lobe can be put into an arbitrary direction, the placement of the sidelobes and the nulls follows uniquely from that direction.

Example 8.5 *Consider beamsteering in the vertical direction. Let a BS antenna mounted at 51 m height, providing coverage for a cell with 1 km radius. The antenna consists of eight short dipoles, separated by $\lambda/2$. How much should the phase shift Δ be so that the maximum is pointed onto the cell edge at UE height (1 m)?*

In order to get constructive interference of the contributions from the different antennas in the angle θ_0, the Δ should fulfill

$$\Delta = k_0 d_a\cos\theta_0. \tag{8.64}$$

Obviously, if $\theta_0 = \pi/2$, then $\Delta = 0$, i.e., no phase shift is necessary. In our example, we aspire for a tilt angle

$$\theta_0 = \frac{\pi}{2} + \arctan\left(\frac{51-1}{1000}\right) = \frac{\pi}{2} + 0.05. \tag{8.65}$$

The phase shift thus has to be

$$\Delta = k_0 d_a\cos\left(\frac{\pi}{2} + 0.05\right) = -0.05\pi. \tag{8.66}$$

For a rectangular array in the xy plane, the array factor is simply the product of the array factors along the two axes, to wit

$$M(\phi,\theta) = \frac{\sin\left[\frac{N_{r,x}}{2}\left(k_0 d_{a,x}\sin\theta\cos\phi - \Delta_x\right)\right]}{\sin\left[\frac{1}{2}\left(k_0 d_{a,x}\sin\theta\cos\phi - \Delta_x\right)\right]}\frac{\sin\left[\frac{N_{r,y}}{2}\left(k_0 d_{a,y}\sin\theta\sin\phi - \Delta_y\right)\right]}{\sin\left[\frac{1}{2}\left(k_0 d_{a,y}\sin\theta\sin\phi - \Delta_y\right)\right]} \tag{8.67}$$

where $N_{r,x}$, $d_{a,x}$, and Δ_x are number of antenna elements, spacing, and phase shift of excitation for elements along the x-axis, and similarly for the elements along the y-axis.

Note that the above derivations assumed $|w_n| = 1$. This implies, in the transmit case, that the total signal power emitted increases linearly with the number of antenna elements. This can be a meaningful assumption if, e.g., the TX power is limited by the power handling capacity of the TX power amplifier, and each antenna element has its own power amplifier. Another meaningful normalization is $\Sigma|w_n|^2 = 1$; we denote the array factor with this specific normalization as $M'(\phi, \theta)$. This means that the total TX power is assumed to be independent of N_r, and improvements of the received power result only from a higher directivity. This normalization can be applied to antenna patterns; we thus should use $M'(\phi, \theta) = M(\phi, \theta)/\sqrt{N_r}$. It furthermore needs to be used whenever the transmit power is limited, e.g., because of limitations in the power supply of the TX. We also note that $\Sigma|w_n|^2 = 1$ allows the freedom to shift power from one antenna to another (as opposed to, e.g., assumed $|w_n| = 1, \forall\, n$, and thus increases the degrees of freedom for the design of the array factor.

The overall pattern of an antenna array can be written as the product of the pattern of a single element and the array factor $M'(\phi, \theta)$

$$\widetilde{G}_{\text{array}}(\phi, \theta) = \widetilde{G}_{\text{element}}(\phi, \theta)M'(\phi, \theta) \tag{8.68}$$

where we note that the array factor describes the impact of the array structure on the field (not the power), while the (power) antenna pattern is given as

$$G_{\text{array}}(\phi, \theta) = G_{\text{element}}(\phi, \theta)|M'(\phi, \theta)|^2. \tag{8.69}$$

In all of the numerical examples above, we have assumed that $d_a \leq \lambda/2$. This condition is necessary to avoid spatial aliasing, i.e., periodicities of the antenna pattern); Figure 8.22 shows an example of spatial aliasing when the antenna distance is chosen $d_a \leq 1.1\lambda$. We see that in this case the array pattern shows multiple peaks of equal magnitude. This is, of course, undesirable since it means, e.g., in the transmit case considerable energy is wasted, being sent into a direction in which the intended RX is not placed.

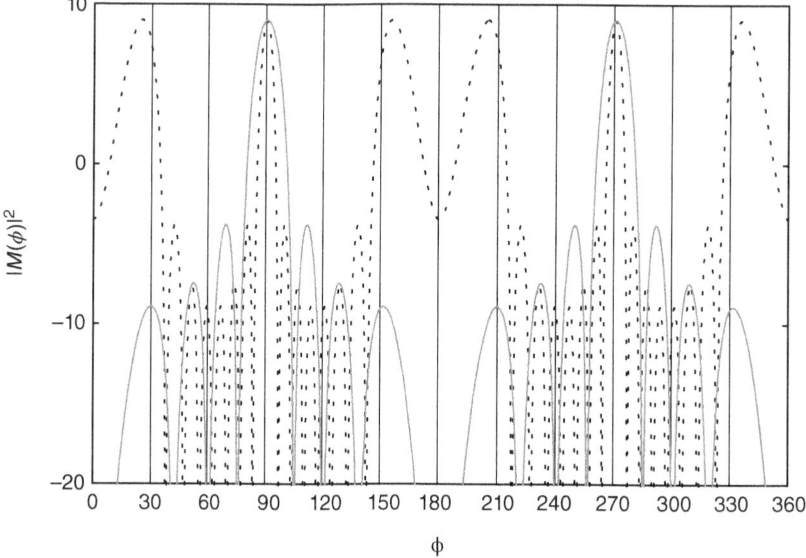

Figure 8.22 Array factor $M'(\phi)$ of a uniform linear array with $d = 0.5\lambda$ (solid) and $d = 1.1\lambda$ (dotted), and $N_r = 8$ antenna elements. Color version available at wiley.com/go/molisch/wireless3e.

However, there are some situations in which a spacing $d_a \geq \lambda/2$ is meaningful. This occurs when the element pattern has a limited beamwidth and good sidelobe suppression. Remember that the overall pattern of an antenna array is the product of the element pattern with the array pattern, Eq. (8.69). Thus, a spatial aliasing lobe of the array factor that is sufficiently suppressed by the element pattern does not show up in $G_{\text{array}}(\phi, \theta)$. For example, most patch antennas exhibit a "sufficient suppression" of the element patterns for all angles $>50°$ from broadside; in this case the antenna spacing can be increased to $\sim\lambda$, see Figure 8.22. This is important because the dimensions of patches are close to $\lambda/2$, so that placing their centers at distance of $\lambda/2$ would mean that their edges are very close together, increasing the mutual coupling (see below). Moreover, the larger element spacing also facilitates smaller beamwidth, which improves the spatial resolution of the array.

We finally comment on the limited bandwidth of arrays. In the derivation of the system model, we have stated that incidence at a certain angle results in a phase shift between the signals that are separated by distance d, where the amount of the phase shift is $2\pi fd \cos(\phi)/c_0$. However, the situation becomes more complicated in a wideband (in the RF sense) systems such that the frequencies at upper and lower band edge are significantly different.

Example 8.6 *Consider a ULA with element spacing of d_a. What is the phase at first and last antenna element when we consider upper and lower frequency $f_l = f_c - \delta/2$ and $f_u = f_c + \delta/2$?*

The phase difference between first and last antenna element at the two considered frequencies is

$$2\pi(f_c \pm \delta/2)(N_r - 1)d_a \cos(\gamma)/c_0. \tag{8.70}$$

Assuming furthermore $\lambda/2$ spacing of the antenna elements, the difference between these two values becomes

$$\frac{2\pi f_c}{c_0}\frac{\delta}{f_c}(N_r - 1)\frac{\lambda}{2}\cos(\gamma) = 2\pi\frac{(N_r - 1)}{2}\cos(\gamma)\frac{\delta}{f_c}. \tag{8.71}$$

The maximum deviation of the array pattern from its nominal value (the value at center frequency) thus depends on the relative bandwidth, as well as on the angle, and the number of antenna elements.

The above example shows that the steering vector changes its direction as the frequency changes; an effect known as *beam squinting*. Thus, for a wideband signal, different frequency components would need different phase shifts in order for their steering vectors to all point into the same direction, i.e., towards the desired angle ϕ. This can be particularly problematic if the phase shifters are implemented by analog circuitry, which provides the same shift for all frequencies. Beam squinting can be combatted by either implementing adjustable delay elements instead of phase shifters (though the former are more difficult to realize), or to create frequency-dependent phase shifts in baseband (in which case each antenna element needs a separate upconversion from baseband to RF).

*8.4.3 Mutual Coupling

Antennas that are placed in close proximity to each other influence each other's radiation characteristics. Consider the receive case first: an incident wave induces a voltage on antenna A. However, we know that even with perfect matching, a part of the radiation is scattered into space. A fraction of that scatterered radiation then arrives at antenna B, and induces there a voltage in addition to the one that is created by wave when it is directly incident on antenna B. This effect is known as *mutual coupling*. It leads both to a change of the antenna pattern, and the antenna impedance.

That sum of the direct, and the induced, voltage is obviously different from the one that would occur when the scattering were absent, and thus changes the array characteristics. The total effect on the received signal at a particular antenna has to take into account the scattering from all other antennas in an array; of course the impact decreases with the distance between the antenna elements. In an infinite array, the effect of the mutual coupling is the same for all antenna elements, because every element has the same number of neighbors at given distances, and thus "sees" the same amount of scattering from all the other antennas. For a finite array, the elements at the edge have different number of neighbors, and thus experience different distortions than the center elements. This makes the design of beam shapes more complicated, and is thus undesirable; it is thus common to have "dummy" elements, which are not connected to a RX chain, but still create mutual coupling, so that all active antenna elements see the same number of neighbors that create coupling. Finally, we note that mutual coupling exists also in the transmit case, as can be expected from the reciprocity theorem. The signal transmitted from antenna A is scattered by antenna B, and the resulting wave adds up with the directly transmitted wave (equivalently, one can say that the currents in antenna A induce currents in antenna B, which also radiate).

The mutual impedances between array elements, and thus the mutual coupling, can be computed in closed form, for at least some simple structures such as dipoles. However, these results depend on a number of simplifications and thus often deviate from reality. Detailed numerical computations, e.g., using the Finite Difference Time Domain (FDTD) method, or actual measurements, are a more reliable way of obtaining the results.

8.5 Special Aspects of Antennas for BS and UE

Antenna design for wireless communications is influenced by two factors: (i) performance considerations, and (ii) size and cost considerations. The latter aspect is especially important for antennas on UEs. Those antennas must show not only good electromagnetic performance, but must be small, mechanically robust, and easy to produce. Furthermore, the performance of those antennas is influenced by the casing on which they are mounted, and by the person operating the handset. For BS antennas, on the other hand, both size and cost are less restricted, and – at least for outdoor applications – the immediate surroundings of the antennas are less filled with obstacles.

8.5.1 Antenna Mounting on the UE

The antennas do not operate in empty space but are placed on top of, or inside, the casing, which can be considered to be part of the radiator. Furthermore, the antenna characteristics are influenced by the hand and the head of the user; this influence also depends on the mounting of the antenna on the UE. It is therefore important to investigate different options of placing the antenna, and to see how this placement influences the performance.

External antennas are usually linear and helical antennas. Such antennas are nowadays only used for specialty UEs, such as for emergency personnel, where communication range is critical. These antennas are usually placed on the upper, narrow side of the casing, i.e., they stick out from that part of the casing. This is done mainly for ergonomic reasons – if they were sticking out from the lower side, they would feel uncomfortable to the user, and the hand of the user would often cover the antenna, leading to additional attenuation. Such external antennas can also be used for devices not used by humans, i.e., machine-to-machine communication, where the placement of the antennas is dictated by the need to avoid mechanical damage.

For microstrip antennas and PIFAs, there are more options for placements. Those antennas are usually used as internal antennas, i.e., integrated into the casing, or enclosed within the casing. This greatly reduces the danger of mechanical damage. However, there is an increased probability that the user will place the hand over the antenna, which increases absorption of the electromagnetic energy, and thus worsens the link performance [Erätuuli and Bonek 1997]. Examples for the positioning can be found in Figure 8.23. Usually, multiple antennas need to be placed to avoid the covering by the hand, in particular for smartphones, which can be held in different ways, see below.

Figure 8.23 Placement an external antenna on the casing of an UE. Similar placements inside a UE are common for internal antennas.

For lower frequencies (i.e., below 1 GHz), the PCB of the UE can help to increase the bandwidth, as well as to create more antenna ports for Multiple Input Multiple Output (MIMO) operation (see Chapter 16). This is because the PCB presents additional "real estate" for the antenna currents to flow, and thus to exploit more resonant modes. It is even possible to opportunistically exploit additional modes created by a large display screen. In addition, suitable choices of the utilized modes can also lead to smaller effects of the user hand and head [Aliakbari and Lau 2021].

8.5.2 Effects of Objects in the Near-Field of the UE

A key factor influencing the radiation efficiency is the presence of dielectric and/or conducting material, namely the user, in the vicinity of a UE. This effect has to be taken into account, as it leads to strong distortions of the antenna pattern, and absorption of energy, and thus decreasing the gain.

One approach of incorporating the impact is an empirical change in the effective gain of the antenna. The ratio of the effective gain in the presence of the user, divided by the effective gain of the antenna in free space, is then used for all the computations. For the determination of the empirical effective gain it should be analyzed for a large number of users in order to eliminate specific properties of any one user (at which angle with respect to the head is the UE being held, is the user right-handed or left-handed, what are the dielectric properties of the hand holding the device, etc.). Figure 8.24 shows an exemplary cumulative distribution function for a patch antenna and a helical antenna. Note that the difference between the effective antenna gain and the theoretical values can exceed 10 dB. This has a large impact on network planning.

A more detailed picture of the distortion of the antenna pattern can also be obtained. We can consider the antenna and the human body as a "superantenna," whose characteristics (efficiency, radiation pattern) can be measured and characterized in the same manner as "regular" antennas, e.g., by measurement in an anechoic chamber, and can be employed in the same way as "standard" antenna patterns. Figure 8.25 shows example measurements of antenna patterns by a human head and body.

The impact of the human body, and the human hand in modern smartphones depends on the orientation of the phone, and the locations of the hand relative to the phone casing, depending on whether the phone is used in phone mode (held to the ear), browsing in portrait mode, or video-watching (or browsing) in landscape mode. This often necessitates the installation of multiple antennas, since there is no single antenna can be ensured to be free of being covered by the hand in all situations.

We furthermore note that the impact of the human hand, head, and body, increases with increasing frequency, so that these considerations are even more important, e.g., for mm-wave systems. Figure 8.26 shows the pattern distortions for different holding modes at mm-wave frequencies.

8.5.3 BS Antenna Types

The design requirements for BS antennas are different from those of UE antennas. The cost has a smaller impact, as BSs are generally much more expensive, so that relatively speaking the cost of antennas play less of a role. Also, the size restrictions are less stringent: for

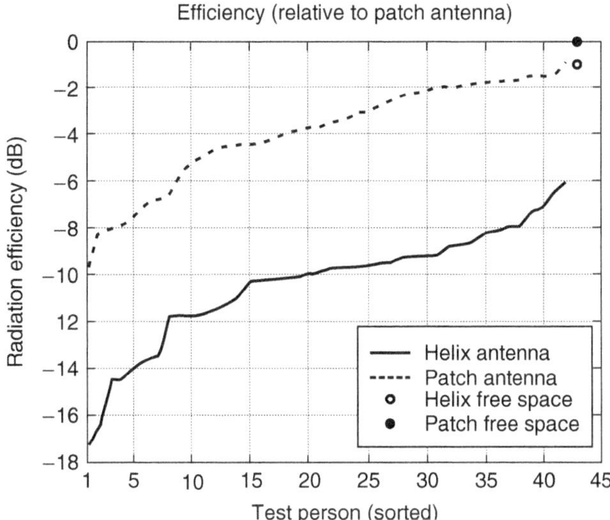

Figure 8.24 Efficiency of UE antennas close to a human body, relative to the efficiency of the antennas in free space. Different users and positions of the UE were used in the ensemble.
Reproduced with permission from [Pedersen et al. 1998] © IEEE.

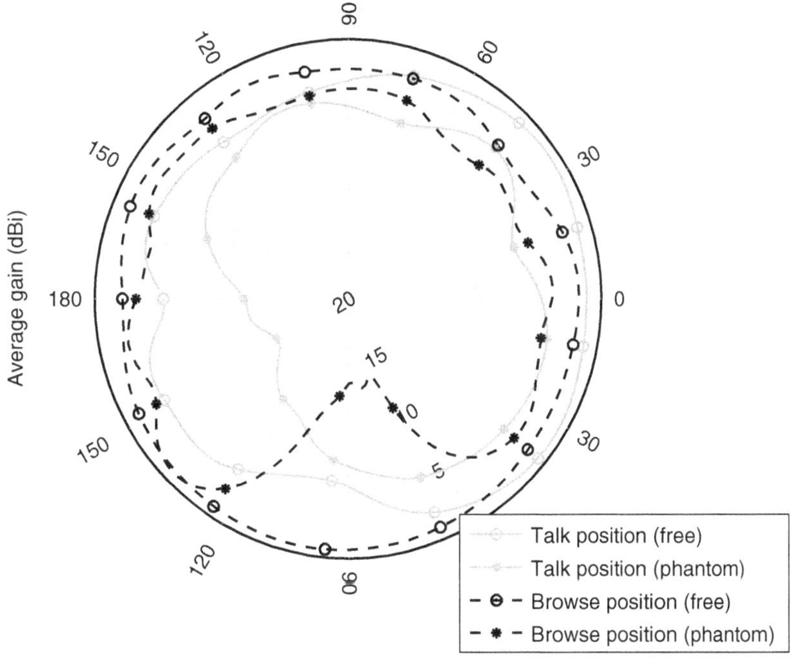

Figure 8.25 Pattern of antenna distortion by human body in talk and data-browsing position.
Reproduced with permission from [Harryson et al. 2010] © IEEE.

macrocells, it is only required that (i) the mechanical stress on the antenna mast, especially due to wind forces, must remain reasonable, and (ii) the "cosmetic" impact on the surroundings must be small. For micro- and picocells, the antennas need to be considerably smaller, as they are mounted on building surfaces, street lanterns, or on office walls. The desired antenna pattern is also quite different for BS antennas compared to UE antennas. As the physical placement and orientation of the BS antenna is known, the patterns should be shaped in such a way that no energy is wasted.

To elaborate: it is undesirable that a BS antenna radiates isotropically. Radiation emitted into the direction with large elevation angle (i.e., into the sky) is not only wasted, but actually increases the interference to other systems. The optimum antenna pattern is achieved when the received power is constant within the whole cell area, and vanishes outside. The problem is then to synthesize such an elevation pattern. This can be achieved approximately by means of array antennas, where the complex weights are chosen in such a way as to minimize the mean-square error of the pattern. An even simpler approach is to use an antenna array where all

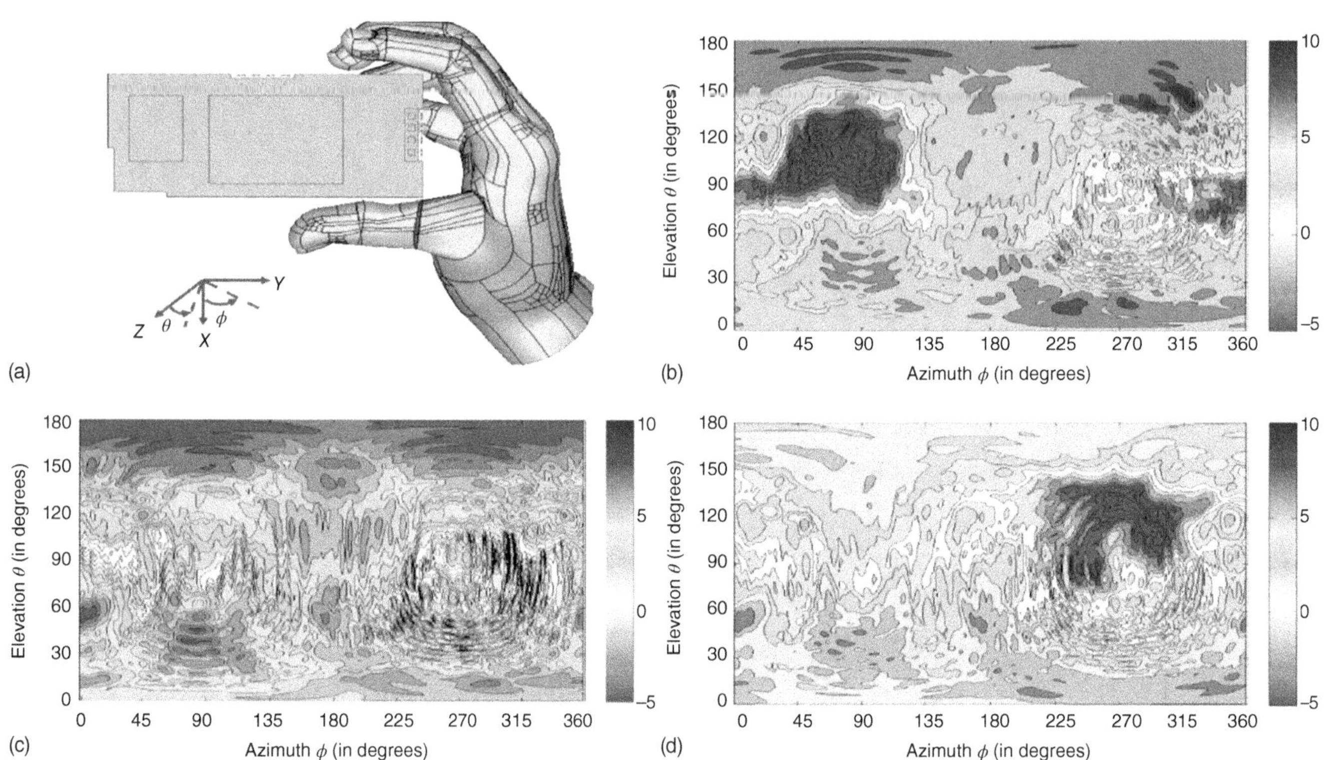

Figure 8.26 (a) A typical UE design with multiple subarrays (top and long edges) in Landscape mode. Received gain as a function of azimuth and elevation angles for the UE design at 28 GHz in (b) freespace mode, and with hand blocking in (c) landscape and (d) portrait modes. Color version available at wiley.com/go/molisch/wireless3e.
Reproduced with permission from [Raghavan et al. 2018] © IEEE.

elements have the same magnitude of the weight, so that the overall pattern is the fundamental (sinc) shape, see Section 8.4.2, but tilting the main lobe down such that the peak of the antenna pattern points to a location within the cell (usually near the cell edge). This downtilt can be achieved either mechanically (by tilting the whole antenna array) or electrically (by means of the phases).

The desired azimuthal antenna pattern is either (i) omnidirectional, i.e., uniform in $[0, 2\pi)$, or (ii) uniform within a sector, and zero outside. The angular range of a sector is usually 60° or 120°. For the omnidirectional antennas, which are mostly used in rural macrocells, linear antennas are used (or vertical arrays of linear antennas). For sector antennas, either linear antennas with appropriately shaped reflectors, or patch antennas (which have inherently nonuniform antenna patterns) can be used. Patch antennas are particularly popular for micro- and picocells, since they can fit flush to a wall. All considerations from Section 8.2 about bandwidth, efficiency, etc., remain valid; except for the size requirement, which is relaxed considerably. Another interesting aspect is the common use of polarization to provide additional antenna ports, particularly for MIMO operation. Typically, ±45° polarization are used (i.e., vertical and horizontal polarizations rotated by 45°) to ensure that the receive signals at the two antennas have - on average - balanced powers.

For 4G and particularly 5G systems, adaptive antennas (antenna arrays that can adapt the beampattern in the horizontal, and also possibly vertical dimension) have become common. These are usually realized as rectangular arrays (in the x–z plane) of microstrip antennas. Moreover, the emergence of massive MIMO systems in 5G require larger arrays to be designed. These can involve hundreds of elements, typically consisting of dual-polarized elements with directional patterns for sectorized cell coverage (e.g., patch antennas and dipoles with reflectors). Due to the element spacing being half-wavelength or more, these arrays can be very large physically for typical cellular frequencies of up to 6 GHz. This large size/weight and the resulting high wind load will in turn require larger and more expensive supporting structures. One recent idea is to create three antenna ports instead of two (dual-polarized) ports per basic element of the array [Chiu et al. 2020], which can then reduce overall size of array by a third.

*8.5.4   Effect of Objects in the BS Near-Field

The antenna pattern of BS antennas is usually defined for the case that the antenna is in free space, or above an ideally conducting plane. This is what is measured in an anechoic chamber, and also the easiest to compute. However, at its location of operation, a BS antenna might still have in its surroundings objects of finite extent and conductivity. The antenna mast, which is usually metallic, and thus highly conductive, can lead to distortions. Similarly, roofs made out of certain building materials, like reinforced concrete, can distort the antenna patterns, see Figure 8.27. The antenna patterns in such surroundings can then deviate significantly from the theoretical patterns.

Distortions of the vertical pattern are plotted in Figure 8.28. They arise from two effects: the fact that the (finite-extent) roof is much closer to the antenna than the ground, and that the dielectric properties of the roof are different from those of the ground.

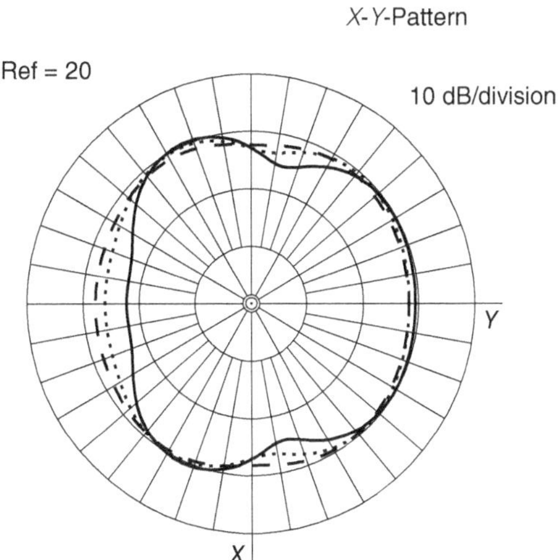

Figure 8.27 Antenna pattern of an omnidirectional antenna close to an antenna mast. Distance antenna – mast 30 cm. Diameter of the mast: very small (dashed), 5 cm (dotted), 10 cm (solid).

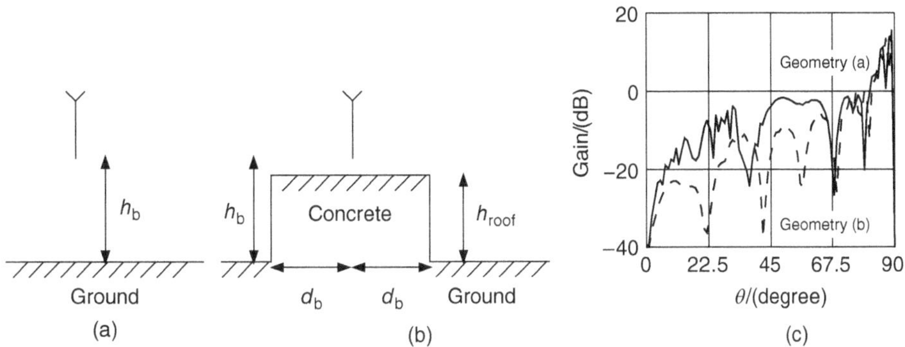

Figure 8.28 Distortion of the vertical antenna pattern. (a) Idealized geometry (b) real geometry, including the roof (c) distortions of the antenna patterns. Properties of all materials: relative dielectric constant $\varepsilon_r = 2$; Conductivity 0.01 S/m. $d_b = 20$ m.
Reproduced with permission from Molisch et al. [1995] © European Microwave Association.

Further Reading

For a general introduction to antenna theory, we just refer to the many excellent books on antenna theory, e.g., [Ramo et al. 1967], [Stutzman and Thiele 1997], [Balanis 2005]; [Kraus and Marhefka 2002], and the somewhat more advanced text of [Collin 1985].

For more details on antennas specifically for wireless communications, see [Godara 2001] and [Vaughan and Andersen 2003]. Antennas for UEs are discussed in [Hirasawa and Haneishi 1991] and [Fujimoto 2008]. Antenna design for BSs is surveyed in the monograph of [Chen and Luk 2009] and in the conference paper by [Beckman and Lindmark 2007]. A specific discussion of lens antennas can be found in [Fernandes et al. 2014]. The measurement of antenna characteristics is discussed in [Evans 1990]. A recent example for the use of PCB and screen for broadband antenna design is [Aliakbari et al. 2021].

Finally, phased array antennas are treated in detail in [Mailloux 1994] and [Hansen 1998]. Discussions on beamtilting can be found in [Manholm et al. 2003]. Antenna arrays for mmWave systems are surveyed in [Ghosh and Sen 2019]. The impact of the human head and body on antenna characteristics is discussed, e.g., in [Ogawa and Matsuyoshi 2001], [Kivekaes et al. 2004] and [Harryson et al. 2010].

For updates and errata for this chapter, see https://wides.usc.edu/students.html#textbooks

Exercises

See Sec. 36.8 of Exercises.pdf at wiley.com/go/molisch/wireless3e

9

Channel Sounding

9.1 Introduction

9.1.1 Requirements for Channel Sounding

The measurement of the impulse responses and related characteristics of wireless channels, better known as *channel sounding*, is a fundamental task for wireless communications engineering because any meaningful channel model must be based on, or verified by, measurement data. For stochastic channel models, parameter values have to be obtained from extensive measurement campaigns, while for deterministic models the quality of the prediction has to be checked by comparisons with measured data.

As the systems, and the required channel models, have become more complex, so did the tasks of channel sounding. Measurement devices in the 1960s only had to measure the received field strength. The transition to wideband systems necessitated the development of a new class of channel sounders that could measure impulse responses – i.e., delay dispersion. The focus on directional propagation properties that arose in the 1990s, caused by the interest in multi-antenna systems, also affects channel sounders. Suitable sounders now have to be able to measure double-directional impulse responses.

In addition to the changes in measured quantities, the *environments* in which the measurements are done are changing. Up to 1990, measurement campaigns were usually performed in macrocells. Since then, microcells, outdoor-to-indoor settings, and especially indoor propagation have become the focus of interest. Furthermore, the carrier frequency range of interest has changed – while early measurements were mostly in the 1–2 GHz range, now interest has expanded to both lower frequencies, as well as the 3–7 GHz range, and the 20–500 GHz range, i.e., the millimeter-wave (mm-wave) and Terahertz (THz), regimes.

In the following, we discuss the most important channel-sounding approaches. After a discussion of the basic requirements of wideband measurements, different types of sounders are described. This is followed by an outline of spatially resolved channel sounding techniques. A description of high-resolution parameter estimation techniques that are frequently used for evaluating channel sounder measurements is also given.

9.1.2 Generic Sounder Structure

The word *channel sounder* gives a graphic description of the functionality of such a measurement device. A Transmitter (TX) sends out a signal that excites – i.e., "sounds" – the channel. The output of the channel is observed ("listened to") by the Receiver (RX), and stored. From the knowledge of the transmit and the received signal, the time-variant impulse response or one of the other (deterministic) system functions is obtained.

Figure 9.1 shows a block diagram of the channel sounder that is conceptually simplest. The TX sends out a signal $s(t)$ that consists of periodically repeated pulses $p(t)$ that have a duration T each:

$$s(t) = \sum_{i=0}^{N_\mathrm{p}-1} p\left(t - iT_\mathrm{rep}\right) \tag{9.1}$$

where T_rep is the repetition interval of the transmitted pulses. One *measurement run* consists of N_p pulses that are transmitted at fixed intervals. The pulses are the convolution of a basis pulse $\widetilde{s}(t)$ created by a pulse generator and a transmit filter:

$$p(t) = \widetilde{s}(t) * g(t) \tag{9.2}$$

where $g(t)$ is the impulse response of the transmit filter. The waveform used for $p(t)$ depends on the type of sounder and can greatly differ for sounders working in the time domain, compared with sounders working in the frequency domain, as discussed in Sections 9.2 and 9.3.

Wireless Communications: From Fundamentals to Beyond 5G, Third Edition. Andreas F. Molisch.
© 2023 John Wiley & Sons Ltd. Published 2023 by John Wiley & Sons Ltd.
Companion website: www.wiley.com/go/molisch/wireless3e

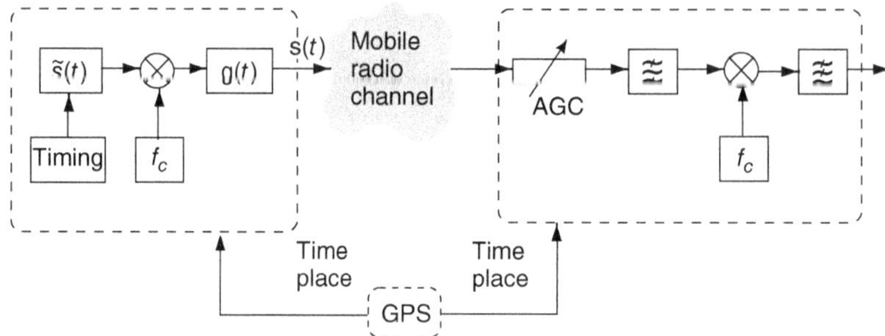

Figure 9.1 Principle of a channel sounder. Correct synchronization between TX and receiver is especially important. *In this figure*: AGC, Automatic Gain Control. $g(t)$ can be realized in baseband or RF (or a combination).

This block diagram is generic; the properties of the sounder are mostly determined by the choice of the sounding signal. In order to perform efficient measurements, the following requirements should be fulfilled by the sounding signal:

- *Large bandwidth*: the bandwidth W is inversely proportional to the shortest temporal changes in the sounding signal and thus determines the achievable *delay resolution*.
- *Large time-bandwidth product*: it is often advantageous if the sounding signal has a duration that is longer than the inverse of the bandwidth – i.e., a time-bandwidth product *TW* larger than unity. For many systems, the *instantaneous* transmit power (as opposed to *average* transmit power) is limited. In this case, a large *TW* allows the transmission of high *energy* in the sounding signal, and thus obtains a higher Signal-to-Noise Ratio (SNR) at the RX. Sounding schemes with large *TW* are related to spread spectrum systems (Chapter 19). At the RX, special signal processing (despreading) is required in order to exploit the benefits of large *TW*.
- *Signal duration*: the effective signal duration must be adapted to channel properties. On one hand, a long-sounding signal can give a large time-bandwidth product, which is beneficial (see above). On the other hand, the sounding signal should not be longer than the coherence time of the channel – i.e., the time during which the channel can be considered to be approximately constant. For practical reasons, the pulse repetition time $T_{\rm rep}$ should be larger than the duration of the constituent pulse $p(t)$ and the maximum excess delay of the channel.
- *Power spectral density*: the power spectral density of the sounding signal, $|P_{\rm TX}(j\omega)|^2$, should be uniform across the bandwidth of interest. This allows to have the same quality of the channel estimate at all frequencies. Due to efficiency considerations, little energy should be transmitted outside the bandwidth of interest.
- *Low crest factor*: signals with a low crest factor

$$C_{\rm crest} = \frac{\text{Peak amplitude}}{\text{rms amplitude}} = \frac{\max\{s(t)\}}{\sqrt{\overline{s^2(t)}}} \tag{9.3}$$

allow efficient use of the transmit power amplifier. A first estimate for the crest factor can be obtained from [Felhauer et al. 1993]:

$$1 < C_{\rm crest} \leq \sqrt{TW}. \tag{9.4}$$

- *Good correlation properties*: correlation-based channel sounders require signals whose Auto-Correlation Function (ACF) has a high *Peak to Off-Peak* (POP) ratio, and a zero mean. The latter property allows *unbiased estimates*. Correlation properties are critical for channel estimates that directly use the correlation function while it is less important for parameter-based estimation techniques (see Section 9.5).

The design of optimum, digitally synthesized sounding signals thus proceeds in the following steps:

1. Choose the duration of the sounding signal according to the channel coherence time and the required time-bandwidth product.
2. For a constant power spectral density, all frequency components need to have the same absolute value.
3. Now the only remaining free parameters are the phases of the frequency components. These can be adjusted to yield a low crest factor.

*9.1.3   Identifiability of Wireless Channels

The temporal variability of wireless channels has an impact on whether the channel can be identified (measured) in a unique way.

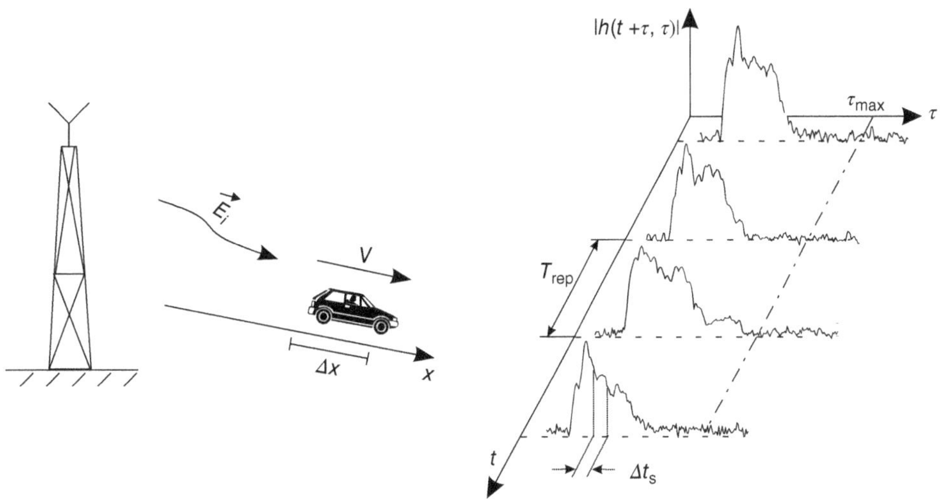

Figure 9.2 Time-variant impulse response of the channel and channel identifiability. A new snapshot can only be taken after the impulse response of the previous excitation has died down.

A band-limited *time-invariant* channel can always be identified by appropriate measurement methods, the only requirement being that the RX fulfills the Nyquist theorem [Proakis and Salehi 2005] in the delay domain – i.e., samples the received signal sufficiently fast.

In a *time-variant* system, the repetition period T_{rep} of the sounding pulse $p(t)$ is of fundamental importance. The channel response to any excitation pulse $p(t)$ can be seen as one "snapshot" (sample) of the channel (see Figure 9.2). In order to track changes in the channel, these snapshots need to be taken sufficiently often. Intuitively, T_{rep} must be smaller than the time over which the channel changes. This notion can be formalized by establishing a sampling theorem in the time domain. Just as there is a minimum sampling rate to identify a signal with a band-limited spectrum, so is there a minimum temporal sampling rate to identify a time-variant process with a band-limited Doppler spectrum. Thus, the temporal sampling frequency must be at least twice the maximum Doppler frequency ν_{max}:

$$f_{rep} \geq 2\nu_{max}. \tag{9.5}$$

Rewriting Eq. (9.5), and using the relationship between the movement speed of the User Equipment (UE) and the Doppler frequency $\nu_{max} = f_c v_{max}/c_0$ (see Eq. 5.8), the repetition frequency for the pulses can be written as

$$T_{rep} \leq \frac{c_0}{2f_c v_{max}}. \tag{9.6}$$

In that case,

$$\frac{v}{\Delta x_s} \geq 2\frac{v_{max}}{\lambda_c} \tag{9.7}$$

holds, so that the distance Δx_s between the locations at which the sounding has to take place is upper bounded as

$$\Delta x_s \leq \frac{v}{v_{max}}\frac{\lambda}{2} \leq \frac{\lambda}{2}. \tag{9.8}$$

Equation (9.8) thus tells us that for an aliasing-free measurement at least two snapshots per wavelength are required.

Strongly time-varying channels can be fundamentally unidentifiable because requirements for the design of sounding signals can become contradictory. On one hand, the repetition frequency T_{rep} has to be larger than the maximum excess delay of the channel τ_{max}; otherwise, the impulse responses from the different excitation pulses start to overlap. On the other hand, we have just shown that the repetition frequency has to fulfill $T_{rep} \leq 1/2\nu_{max}$. Thus, channels can be identified in an unambiguous way only if

$$2\tau_{max}\nu_{max} \leq 1. \tag{9.9}$$

A channel that fulfills these requirements is known as *underspread*. If Eq. (9.9) is not fulfilled, then the channel can only be identified by making specific assumptions – e.g., a certain parametric model. Fortunately, the overwhelming majority of wireless channels are underspread; in many cases, even $2\tau_{max}\nu_{max} \ll 1$ is fulfilled. We will assume that this is fulfilled in the following. This also implies that the channel is *slowly time variant* (see Chapter 6), so that $h(t, \tau)$ can be interpreted as the impulse response $h(\tau)$ that is valid at a certain (fixed) time instant t.[1]

Example 9.1 *A channel sounder is in a car that moves along a street at 36 km/h. It measures the channel impulse response at a carrier frequency of 2 GHz. At what intervals does it have to measure? What is the maximum excess delay the channel can have so as to still remain underspread?*

$$\left.\begin{aligned} v &= 36\,\text{km/h} = 10\,\text{m/s} \\ \lambda_c &= \frac{c_0}{f_c} = \frac{3\cdot10^8}{2\cdot10^9} = 0.15\,\text{m} \end{aligned}\right\}. \tag{9.10}$$

The channel must be sampled in the time domain at a rate that is, at minimum, twice the maximum Doppler shift. Using Eq. (9.5),

$$f_{\text{rep}} = 2\cdot v_{max} = 2\cdot\frac{v}{\lambda_c}. \tag{9.11}$$

The sampling interval T_{rep} is given as

$$T_{\text{rep}} = \frac{1}{f_{\text{rep}}} = \frac{\lambda_c}{2\cdot v} = 7.5\,\text{ms}. \tag{9.12}$$

At a mobile speed of 36 km/h, this corresponds to a channel snapshot taken every 75 mm. To calculate the maximum excess delay τ_{max}, we make use of the fact that the channel must be underspread in order to be identifiable. Hence from Eq. (9.9),

$$\left.\begin{aligned} 2\cdot\tau_{max}\cdot v_{max} &= 1 \\ \tau_{max} &= \frac{1}{2\cdot v_{max}} = T_{\text{rep}} = 7.5\,\text{ms} \end{aligned}\right\}. \tag{9.13}$$

This is orders of magnitude larger than the maximum excess delays that occur in typical wireless channels (see Section 7.2.5).

If τ_{max} is smaller, then the repetition period T_{rep} of the sounding pulse should be decreased; this would allow averaging of the snapshots and thus improvement of the SNR.

9.1.4 Influence on Measurement Data

When performing the measurements, we have to be aware of the fact that measured impulse responses carry undesired contributions as well. These are mainly:

- interference from other (independent) signal sources that also use the channel;
- additive white Gaussian noise.

Interference is created especially when measurements are done in an environment where other wireless systems are already active in the same frequency range. Wideband measurements in the 2-GHz range, e.g., become quite difficult, as these bands are heavily used by various systems. If the number of interferers is large, the resulting interference can usually be approximated as equivalent Gaussian noise. This equivalent noise raises the noise floor, and thus decreases the dynamic range. Channel sounding campaigns thus often have to choose a time/place where interference is minimized, and/or use sophisticated postprocessing, including outlier detection, to eliminate measurement results that are heavily impacted by interference.

9.2 Time-Domain Measurements

A time-domain measurement directly measures the (time-variant) impulse response. Assuming that the channel is slowly time variant, the measured impulse response is the convolution of the *true channel impulse response* with the *impulse response of the sounder*:

$$h_{\text{meas}}(t_i, \tau) = \tilde{p}(\tau) * h(t_i, \tau) \tag{9.14}$$

---

[1] Strictly speaking, use of the "slow time variance" concept also requires that the sounding signal has a duration that is much smaller than the coherence time of a channel. We will assume this in the following, though care has to be taken in particular for signals with large TW.

where the effective *sounder impulse response* $\widetilde{p}(\tau)$ is the convolution of the transmitted pulse shape and the RX filter impulse response:

$$\widetilde{p}(\tau) = p_{\mathrm{TX}}(\tau) * p_{\mathrm{RX}}(\tau) \tag{9.15}$$

if the channel and the transceiver are linear.[2] The sounder impulse response should be as close to an ideal delta (Dirac) function as possible.[3] This minimizes the impact of the measurement system on the results. If the impulse response of the sounder is not a delta function, it has to be eliminated from the measured impulse response by a deconvolution procedure, which leads to noise enhancement and other additional errors.

9.2.1 Impulse Sounder

This type of channel sounder, which is comparable with an impulse radar, sends out a sequence of short pulses $p_{\mathrm{TX}}(\tau)$. These pulses should be as short as possible, in order to achieve good spatial resolution but also contain as much energy as possible, in order to obtain a good SNR. Figure 9.3 shows a rough sketch of a transmit pulse, and the received signal after this pulse has propagated through the channel.

The receive filter is a bandpass filter – i.e., has a constant-magnitude spectrum in the frequency range of interest. Ideally, $p_{\mathrm{RX}}(\tau)$ should not have an impact, so that

$$\widetilde{p}(\tau) = p_{\mathrm{TX}}(\tau). \tag{9.16}$$

When comparing the sounding signal with the requirements of Section 9.1.2, we find that the signal has a small time-bandwidth product, a short signal duration, and a high crest factor. The requirement of high pulse energy and short duration for transmitted pulses implies that the pulses have a very high peak power. Amplifiers and other Radio Frequency (RF) components that are designed for such high peak powers are expensive, or show other grave disadvantages (e.g., nonlinearities). A further disadvantage of an impulse sounder is its low resistance to interference. As the sounder interprets the received signal directly as the impulse response of the channel, any interfering signal – e.g., from a cellphone active in the band of interest – is interpreted as part of the channel impulse response.

9.2.2 Correlative Sounders

The time-bandwidth product can be increased by using correlative channel sounders. Equations (9.14) and (9.15) show that it is not the transmit pulse shape alone that determines the impact of the measurement system on the observed impulse response. Rather, it is the convolution of $p_{\mathrm{TX}}(\tau)$ and $p_{\mathrm{RX}}(\tau)$. This offers additional degrees of freedom for designing transmit signals that result in high delay resolution but low crest factors.

The first step is to establish a general relationship between the desired $p_{\mathrm{TX}}(\tau)$ and $p_{\mathrm{RX}}(\tau)$. As is well known from digital communications theory, the SNR of the RX filter output is maximized if the receive filter is the *matched filter* with respect to the transmit waveform, compare Section 11.1.[4] Concatenation of the TX and RX filters then has an impulse response that is identical to the ACF of the TX filter:

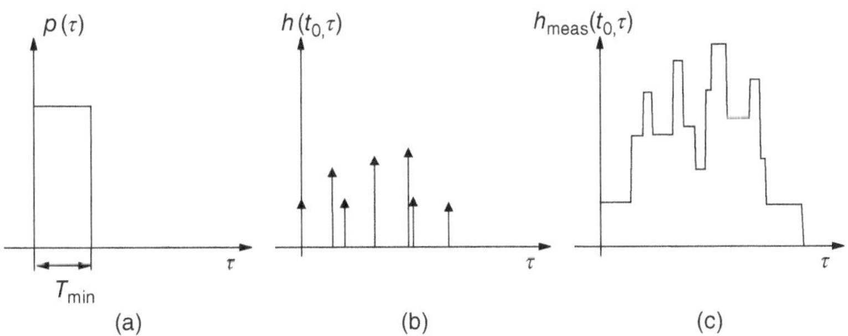

Figure 9.3 Principle of pulse-based measurements. (a) shows a (zoomed-in) sample of a (periodically repeated) transmit pulse. (b) shows the impulse response of the channel. (c) shows the output from the channel, as measured by the RX.

---

[2] Strictly speaking, a sounder impulse response is also time variant, due to second-order effects like temperature drift or aging of electronic components. However, it is more common to recalibrate the sounder, and consider it as time invariant until the next calibration.

[3] This corresponds to a spectrum that is flat over all frequencies. For practical purposes, it is sufficient that the spectrum is flat over the bandwidth of interest.

[4] Strictly speaking, the SNR at the output of the RX-matched filter is maximized if the receive filter is matched to the signal that is actually received at the RX antenna connector. However, that would require knowledge of the channel impulse response – the very quantity we are trying to measure. Thus, matching the filter to the transmit signal is the best we can do.

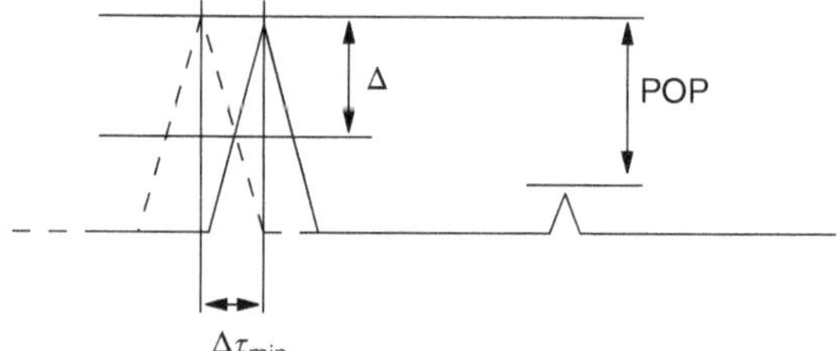

Figure 9.4 Definition of the *peak to off-peak ratio* and the delay resolution $\Delta\tau_{min}$·($\Delta = 6\,\mathrm{dB}$).

$$\widetilde{p}(\tau) = p_{\mathrm{TX}}(\tau) * p_{\mathrm{RX}}(\tau) = R_{p_{\mathrm{TX}}}(\tau). \tag{9.17}$$

The sounding pulses thus should have an ACF that is a good approximation of a delta function; in other words, a high autocorrelation peak $R_{p_{\mathrm{TX}}}(0)$, as well as low ACF sidelobes. The ratio between the height of the autocorrelation peak and the largest sidelobe is called the Peak-to-Off-Peak, *POP*, ratio and is an important quantity for characterization of correlative sounding signals. Figure 9.4 shows an example of an ACF, and the delay resolution that can be achieved with such a signal [de Weck 1992].

In practice, *Pseudo Noise* (PN) *sequences* or linearly frequency modulated signals (chirp signals) have become the prevalent sounding sequences. Maximum-length PN sequences (*m-sequences*), which can be created by means of a shift register with feedback, are especially popular. Such sequences are well known from Code Division Multiple Access (CDMA) systems and have been extensively studied both in the mathematical and the communications engineering literature (see Section 19.3.2 for more details). The ACF of an m-sequence with periodicity M_{c} has only a single peak of height M_{c}, and a POP of M_{c}. Following the CDMA literature, each of the M_{c} elements of such a sequence is called a *chip*.

For constant chip duration and increasing length of the m-sequence, the POP, as well as the time-bandwidth product, increases: signal duration increases linearly with M_{c}, while bandwidth, which is approximately the inverse of chip duration, stays constant. The increased time-bandwidth product improves immunity to noise and interference. More exactly, noise and interference are suppressed by a factor M_{c}. The reason for this is discussed in more detail in Section 19.2, as the principle is identical to that of direct-sequence CDMA.

The interpretation of measurements with correlative channel sounders in time-varying channels requires some extra care. The basic principle of correlative channel sounders is that $p_{\mathrm{TX}}(\tau) * h(t, \tau) * p_{\mathrm{RX}}(\tau)$ is identical to $[p_{\mathrm{TX}}(\tau) * p_{\mathrm{RX}}(\tau)] * h(t, \tau)$. In other words, we require that the channel at the beginning of the PN sequence is the same as the one at the end of the PN sequence. This is a good approximation for slowly time-variant channels. However, if this condition is not fulfilled, correction procedures need to be used [Matz et al. 2002].

9.3 Frequency Domain Analysis

The techniques described in the previous section directly estimate the impulse response of the channel in the time domain. Alternatively, we can try to directly estimate the transfer function – i.e., measure in the frequency domain. The fundamental relationship Eq. (9.1) still holds. However, the shape of the waveform $p(t)$ is now different. The main criterion for its design is that it has a power spectrum $|P(j\omega)|^2$ that is approximately constant in the bandwidth of interest, and that it allows interpretation of the measurement result directly in the frequency domain.

9.3.1 Frequency Stepping and Vector Network Analyzers

The most straightforward way of measuring in the frequency domain is by determining the transfer function at a number of frequency points, where each frequency is measured at a different time instant. Thus, the excitation signal is

$$p(t) = \sum_{k=0}^{N_{\mathrm{f}}-1} \mathrm{rect}[t - kT_{\mathrm{sc}}, T_{\mathrm{sc}}] \cos\left[j2\pi(f_{\mathrm{L}} + k\Delta_{\mathrm{f}})t\right]$$

where f_{L} is the lowest measured frequency, i.e., the starting point of the frequency stepping, rect[t,a] is a function that is 1 between t and t+a and 0 otherwise, Δ_{f} is the stepping width, and N_{f} the number of frequency points the sounder steps through. T_{sc} is the time that one particular frequency is being measured; it determines the bandwidth (called IF bandwidth in a Vector Network Analyzer [VNA]) that the signal occupies around each of the frequencies it steps through; it should be chosen at least as $1/T_{\mathrm{sc}}$. Increasing T_{sc} increases the measurement duration (the total duration is $N_{\mathrm{f}}T_{\mathrm{sc}}$) but allows to improve the SNR.

This measurement principle is used in VNAs. These devices are widely used for channel measurement campaigns, as they are calibrated precision devices, and available in many RF labs. A VNA generally measures the S-parameters of a *Device Under Test* (DUT); if this DUT is a wireless channel, the parameter S_{21} is the channel transfer function at the frequency that is used to excite the channel.

Measurements using a VNA are usually accurate and can be performed in a straightforward way. However, there are also important disadvantages:

- Such measurements are slow, so that repetition rates typically cannot exceed a few Hz (!). Since we require that the channel does not change significantly during one measurement, VNA measurements are limited to static environments.
- The TX and RX are often placed in the same casing. This puts an upper limit on the distance that TX and RX antennas can be spaced apart, since the signal from the VNA has to be transmitted through a cable to the locations of the antennas. Note that cable losses increase significantly with frequency. While standard cables might have more than 1 dB/m attenuation at 2.4 GHz, special cables with 0.05 dB/m are available. On the other hand, at 60 GHz, even expensive specialty cables can have an attenuation of 5 dB/m and more. To combat this problem, mm-wave measurements either need to transmit low-frequency sounding signals to the location of the antennas where the signals are upconverted to the mm-wave band, or use radio-over-fiber to send the signals to the antennas.
- While VNAs at lower frequencies (up to 8 GHz) are relatively low cost, VNAs for mm-wave frequencies are very expensive. At even higher frequencies, there are (at the time of this writing) no available commercial VNAs. Instead, frequency extender units are combined with existing lower-frequency VNAs.

From these restrictions, it follows that VNAs are mainly suitable for indoor measurements.

9.3.2 Chirping

A related method of frequency domain analysis is based on chirping. The transmit waveform is given as

$$p(t) = \exp\left[2\pi j\left(f_{\mathrm{L}}t + \mathrm{W}\frac{t^2}{2T_{\mathrm{chirp}}}\right)\right] \quad \text{for } 0 \leq t \leq T_{\mathrm{chirp}}. \tag{9.18}$$

Consequently, the instantaneous frequency is

$$f_{\mathrm{L}} + \mathrm{W}\frac{t}{T_{\mathrm{chirp}}} \tag{9.19}$$

and thus changes linearly with time, covering the whole range W of interest. The receive filter is again a matched filter. Intuitively, the chirp filter "sweeps" through the different frequencies, measuring different frequencies at different times, and can thus be seen as a continuous version of a frequency-stepping method. However, in contrast to frequency stepping, real-time operation *is* possible with chirping. Conversely, a chirp can also be seen as a spread-spectrum time-domain signaling with flat (over the bandwidth) power density spectrum. The de-chirping with a matched filter, therefore, recovers the impulse response.

9.3.3 Multi-Tone Sounding

Alternatively, we can sound the channel on different frequencies at the same time. The conceptually most simple way is to generate different, sinusoidal sounding signals with different weights q_k, phases φ_k, and frequencies and transmit them all from the TX antenna simultaneously:

$$p(t) = \sum_{k=0}^{N_{\mathrm{tones}}-1} q_k \, \exp\left[2\pi jt\left(f_{\mathrm{L}} + k\mathrm{W}/\left(N_{\mathrm{tones}}-1\right) + j\varphi_k\right] \quad \text{for } 0 \leq t \leq T_{\mathrm{ss}}. \tag{9.20}$$

Due to hardware costs, calibration issues, etc., analog generation of $p(t)$ using multiple oscillators to generate multiple frequencies is not practical. However, it is possible to generate $p(t)$ digitally, similar to the principles of Orthogonal Frequency Division Multiplexing (OFDM) described in Section 15.3, and then use just a single oscillator to upconvert the signal to the desired passband (and similarly at the RX).

A multi-tone waveform that is well suited for channel sounding has a low crest factor (ideally, a constant envelope), and constant power spectral density in the band of interest. A waveform that fulfills these requirements are Zadoff–Chu sequences, which are defined as

$$X^{(u_{\mathrm{ZC}})}(k) = e^{-j\pi u_{\mathrm{ZC}}k(k+1)/M_{\mathrm{ZC}}} \qquad 0 \leq k < M_{\mathrm{ZC}} \tag{9.21}$$

where M_{ZC} is the length of the sequence, and u_{ZC} is the index of the sequence. In line with the requirements, Zadoff–Chu sequences have the remarkable property that they have constant amplitude in the time domain (important for the power amplifiers) and the frequency domain (important for adherence to a spectral mask). Furthermore, the ACF of such a sequence is a delta function. We also immediately see the similarity to chirp waveforms – as a matter of fact, these sequences can be interpreted as discrete versions of a chirp.

*9.4 Modified Measurement Methods

9.4.1 Swept Time Delay Cross Correlator (STDCC)

The Swept Time Delay Cross Correlator (STDCC) is a modification of correlative channel sounders that aims to reduce the sampling rate at the RX. Normal correlative channel sounders require sampling at the Nyquist rate. In contrast, the STDCC samples at rate $T_{\rm rep}$, i.e., use just a single sample value for each m-sequence – namely, at the maximum of the ACF. The position of this maximum is changed for each repetition of the m-sequence, by shifting the time base of the RX with respect to the TX. Thus, $K_{\rm scal}$ transmissions of the m-sequence give the sampled values of a single impulse response $h(\tau_i)$, $i = 1,...,K_{\rm scal}$. The delay resolution is thus better, by a factor $K_{\rm scal}$, than the inverse sampling rate. This drastically reduces the sampling rate and the requirements for subsequent processing and storing of the impulse response. It also reduces the speed at which the data need to be streamed from the acquisition unit to the storage medium; in particular, in channel sounders for high bandwidth, the available speed for this bus can constitute a bottleneck.

In an STDCC, shifting of the maximum of the ACF for consecutive repetitions of the sounding signal is achieved by using different time bases in the TX and RX. In particular, the delayed time base of the RX correlator (compared with the TX sequence) is achieved by using a chipping frequency (inverse of the chip duration) that is smaller by Δf. This results in a slow relative shift of the TX and RX sequences. During each repetition of the sequence, the correlation maximum corresponds to a different delay. After a duration

$$T_{\rm slip} = \frac{1}{f_{\rm TX} - f_{\rm RX}} = \frac{1}{\Delta f} \tag{9.22}$$

the TX versus RX signals have shifted by one chip duration – i.e., the ACF maximum occurs at delay $\tau = 1/f_{\rm TX}$. This means that (for a static channel) the output from the sampler should be sampled with the so-called *slip rate*:

$$f_{\rm slip} = f_{\rm TX} - f_{\rm RX}. \tag{9.23}$$

The ratio:

$$K_{\rm scal} = \frac{f_{\rm TX}}{f_{\rm slip}} \gg 1 \tag{9.24}$$

is the *scaling factor* $K_{\rm scal}$ of the impulse response. The actual impulse response can be obtained from the measured sample values as

$$\hat{h}(t_i, k\Delta\tau) = C \cdot h_{\rm STDCC}(t_i, k\Delta\tau K_{\rm scal}) \tag{9.25}$$

where C is a proportionality constant.

Note that this measurement method has increased measurement duration, as the duration of each measurement is increased by the factor $K_{\rm scal}$. Remember that a channel is identifiable only if it is underspread – i.e., $2\nu_{\max}\tau_{\max} < 1$, which is usually fulfilled in wireless channels. For an STDCC, this requirement changes to $2K_{\rm scal}\,\nu_{\max}\tau_{\max} < 1$, which in time-variant channels requires careful choice of $K_{\rm scal}$.

Example 9.2 *Consider an STDCC that performs measurements in an environment with 500-Hz maximum Doppler frequency and maximum excess delay of 1 μs. The sounder can sample at most with 1 Msample/s. What is the maximum delay resolution (inverse bandwidth) that the sounder can achieve?*

In order for the channel to remain identifiable (underspread) when measured with an STDCC, the following condition has to hold:

$$2 \cdot K_{\rm scal} \cdot \tau_{\max} \cdot \nu_{\max} = 1$$
$$K_{\rm scal} = 1/(2 \cdot \tau_{\max} \cdot \nu_{\max}) = 1000.$$

The sounder can take one sample for each repetition of the sounding pulse – i.e., one sample per μs. Hence, the STDCC sounder can resolve Multi-Path Components (MPCs) separated by a delay of 1 μs/1000 = 1 ns.

9.4.2 Inverse Filtering

In some cases, it is advantageous to use a receive filter that optimizes the POP ratio but is *not* ideally matched to the transmit signal. At first glance, it sounds paradoxical to use such a filter, which results in a worse SNR. However, there can be good practical reasons for this approach. Small variations of the SNR are usually less important than the sidelobes of the ACF: while sidelobes can be eliminated by appropriate deconvolution procedures, they can give rise to additional errors. It is thus meaningful to optimize the receive filters with respect to the POP ratio, not with respect to the SNR.

Let us in particular consider *inverse filtering*, and compare it to matched filtering. For the matched filter, the receive filter transfer function is chosen as $P_{\rm TX}^*(f)$, so that the total filter transfer function $P_{\rm MF}(f)$ (concatenation of transmit and receive filter) is given as

$$P_{\mathrm{MF}}(f) = P_{\mathrm{TX}}(f) \cdot P_{\mathrm{TX}}^*(f). \tag{9.26}$$

For inverse filtering, the receive filter transfer function is chosen as $1/P_{\mathrm{TX}}(f)$ in the bandwidth of interest, so that the total transfer function is made as close to unity as possible:

$$P_{\mathrm{IF}}(f) = P_{\mathrm{TX}}(f) \cdot \frac{1}{P_{\mathrm{TX}}(f)} \approx 1. \tag{9.27}$$

The inverse filter is thus essentially a *zero-forcing equalizer* (see Section 14.2.1) for compensation of distortions by the transmit filter. It is important that the transmit spectrum P_{TX} does not have any nulls in the bandwidth of interest. The inverse filter leads to noise enhancement, and thus to a worse SNR than a matched filter. On the positive side, the inverse filter is *unbiased*, so that the estimation error is zero-mean.

9.4.3 Averaging

It is common to average over several, subsequently recorded, impulse responses or transfer functions. Assuming that the channel does not change during the whole measurement time and that the noise is statistically independent for the different measurements, then the averaging of M profiles results in an enhancement of the SNR by $10 \cdot \log_{10} M$ dB. However, note that the maximum measurable Doppler frequency decreases by a factor of M.

Averaging over different realizations of the channel is used to obtain, e.g., the Small Scale Averaged (SSA) power. Exact measurement of the SSA power would require infinitely many statistically independent samples. For a finite number of samples, as always occurs in practical measurements, a residual uncertainty remains.

Example 9.3 *A UE moves along a straight line and can measure a statistically independent sample of the impulse response every 15 cm. Measurements are taken over a distance of 1.5 m, so that shadowing can be considered constant over this distance. The goal is estimation of channel gain (attenuation), averaged over small-scale fading. Assume first that the measurements are only taken at a single frequency. What is the standard deviation of the estimate of channel gain? What is the probability that the estimator is more than 20% off?*

Since measurement is only taken at a single frequency, we assume a flat Rayleigh-fading channel with mean power $\overline{P}$. Each sample of the power of the impulse response is then an exponentially distributed random variable with mean power $\overline{P}$. A reasonable estimate of mean power is

$$\hat{\overline{P}} = \frac{1}{N_{\mathrm{samp}}} \sum_{k=1}^{N_{\mathrm{samp}}} P_k$$

where P_k is the kth sample of the power of the impulse response and N_{samp} is the number of samples. As a measure of the error, we choose the normalized standard deviation, $\sigma_{\hat{\overline{P}}}/\overline{P}$. Since the P_k are identically distributed and independent, we have

$$\frac{\sigma_{\hat{\overline{P}}}}{\overline{P}} = \frac{1}{\overline{P}} \sqrt{\mathrm{Var}\left(\frac{1}{N_{\mathrm{samp}}} \sum_{k=1}^{N_{\mathrm{samp}}} P_k\right)} = \frac{1}{\overline{P}} \sqrt{\frac{1}{N_{\mathrm{samp}}^2} N_{\mathrm{samp}} \overline{P}^2} = \frac{1}{\sqrt{N_{\mathrm{samp}}}}.$$

Using 11 samples, the relative standard deviation is 0.3. Approximating the probability density function (pdf) of the estimator to be Gaussian with mean $\overline{P}$ and variance $\overline{P}^2/N_{\mathrm{samp}}$, the probability that the estimate is more than 20% off is then

$$1 - \mathrm{Pr}\left(0.8\overline{P} < \hat{\overline{P}} < 1.2\overline{P}\right) = 2\,Q\left(0.2\sqrt{N_{\mathrm{samp}}}\right). \tag{9.28}$$

For $N = 11$ this probability becomes 0.5.

Example 9.4 *Consider now the case of wideband measurements, where measurements are done at ten independently fading frequencies. How do the results change?*

Again aiming to estimate the narrowband channel attenuation averaged over small-scale fading, usage of wideband measurements just implies that $N = 110$ measurements are now available. Modifying Eq. (9.28), we find that the probability for more than 20% error has decreased to 0.036.

9.4.4 Synchronization

The synchronization of TX and RX is a key problem for wireless channel sounding. It is required to establish synchronization in frequency and time at a TX and RX that can be separated by distances up to several kilometers. This task is made more difficult by the presence of multi-path propagation and time variations of the channel. Several different approaches are in use:

1. In indoor environments, *synchronization by cables* is possible. For distances up to about 10 m, coaxial cables are useful; for larger distances, fiber-optic cables are preferable. In either case, the synchronization signal is transmitted on a known and well-defined medium from the TX to the RX.
2. For many outdoor environments, the *Global Positioning System* (GPS) offers a way of establishing common time and frequency references. The reference signals required by channel sounders are an integral part of the signals that GPS satellites transmit. An additional benefit lies in the fact that the measurement location is automatically recorded as well. The disadvantage is that this method requires that both TX and RX have line-of-sight connection to GPS satellites; the latter condition is rarely fulfilled in micro-cellular and indoor scenarios. Furthermore, depending on the specific GPS RX, the accuracy of the 1 pulse-per-second GPS timing signal is affected by errors of up to 50 ns, introducing a timing jitter that may be problematic.
3. *Rubidium clocks* at the TX and RX are an alternative to GPS signals. They can be synchronized at the beginning of a measurement campaign; as they are extremely stable (relative drifts of 10^{-11} are typical), they retain synchronization for several hours.
4. *Measurements without synchronization*: it is possible to synchronize via the wireless link itself – i.e., the received signal self-triggers the recording at the RX by exceeding a certain threshold. The advantage of this technique is its simplicity. However, the drawbacks are twofold: (i) noise or interference can erroneously trigger the RX and (ii) it is not possible to determine absolute delays.

9.4.5 Calibration

The measured impulse response is determined by two factors: the actual propagation channel, and the impulse response of the measurement system. The latter can be decomposed further into the impulse response we expect based on the mathematical form of the excitation signal and the postprocessing (this may be a delta function in an ideal case), and the impulse response of the actual hardware, including filters, up/downconverters, etc. Any effect of the system impulse response needs to be eliminated from the measurement results in order to provide the "pure" channel characteristics. For this purpose, it is necessary to measure the system impulse response, which is done through a so-called *back-to-back* calibration. There, the sounder TX and RX are directly connected with a cable, and the overall system response is measured. Specifically, in the case of VNAs, a *SOLT calibration* (Short Open Loss Termination) is used. This calibration establishes the reference planes and measures the frequency response of the network analyzer.

The above-described calibration does not include the antennas. This is not a problem if the antennas are to be considered a part of the channel. If, however, antenna effects are to be eliminated, a separate calibration of the antennas has to be performed, and taken into account during evaluation of the measurements. If the antennas are completely isotropic, then the frequency dependence of the antennas can be determined (e.g., in an anechoic chamber), and calibrated out just like any other frequency-dependent component of the system. However, if the antenna is nonisotropic, then accounting for the for antenna pattern is possible only if the directions of the MPCs are known, or determined as part of the evaluation (see Section 9.5).

Also, possible time variations of the system response have to be considered. Due to temperature drift, mechanical influences (vibrations), and the time-varying presence of objects in the vicinity of the antennas that can distort the antenna pattern (see Section 8.5.2), the system response might change between the time where the calibration measurements are taking, and the time the channel measurements are executed. The resulting calibration errors lead to errors in the estimated channel impulse response.

9.5 Directionally Resolved Measurements

9.5.1 Data Acquisition

Directionally resolved channel measurements, and models based on those measurements, are important for the design and simulation of multi-antenna systems. Here, we first discuss how to make measurements that are directionally resolved at just the RX (to simplify notation, we assume only variation of the azimuth angle ϕ; generalization to including multiple elevation angles is straightforward). These concepts are then generalized to Multiple Input Multiple Output (MIMO) measurements (directionally resolved at both link ends) at the end of this section.

Fortunately, it is not necessary to devise directional channel sounders from scratch. Rather, a clever combination of existing devices can be used to allow directional measurements. We can distinguish two basic approaches: *measurements with directional antennas* and *array measurements*.

- *Measurements with directional antennas*: a highly directive antenna is installed at the RX. This antenna is then connected to the RX of a "regular" channel sounder. The output of the RX is thus the impulse response of the combination of the channel and the antenna pointing in a specific direction, i.e.,

$$h(t, \tau, \phi_i) = \int h(t, \tau, \phi)\widetilde{G}_{\mathrm{RX}}(\phi - \phi_i)d\phi \tag{9.29}$$

where ϕ_i is the direction in which the maximum of the receive antenna pattern is pointing. By stepping through different values of ϕ_i, we can obtain an approximation of the directionally resolved impulse response. One requirement for this measurement is that the channel stays constant during the *total* measurement duration, which encompasses the measurements of all the different ϕ_i. If the antenna has to be rotated mechanically in order to point to a new direction, the total measurement duration can be several seconds or even minutes. The better the directional resolution, the longer the measurement duration. Such measurements with mechanically rotating directional (especially horn) antennas are popular in particular at high frequencies, e.g., mm-wave. A much faster alternative is enabled by the use of phased array antennas, which allow an *electronic* switching of the beam direction, which can be done within microseconds or less. However, this requires the availability of phased arrays in the desired frequency band and the desired beamwidth.

- *Measurement with an antenna array*: a uniform linear antenna array (compare Section 8.4) consists of a number of antenna elements, each of which has low (or no) directivity, which are spaced apart at a distance d_a that is on the order of one wavelength. The impulse response is measured at all these antenna elements (quasi-) simultaneously. The resulting vector of impulse responses is either useful by itself (e.g., for the prediction of diversity performance) or the directional impulse response can be extracted from it by appropriate signal-processing techniques (*array processing*).

Measurement of the impulse response at the different antenna elements can be done by means of three different approaches (see Figure 9.5):

- *Real arrays*: in this case, one downconversion chain exists for each receive antenna element. Measurement of the impulse response thus truly occurs at all antenna elements simultaneously. The drawbacks include high costs, as well as the necessity to calibrate multiple downconversion chains.

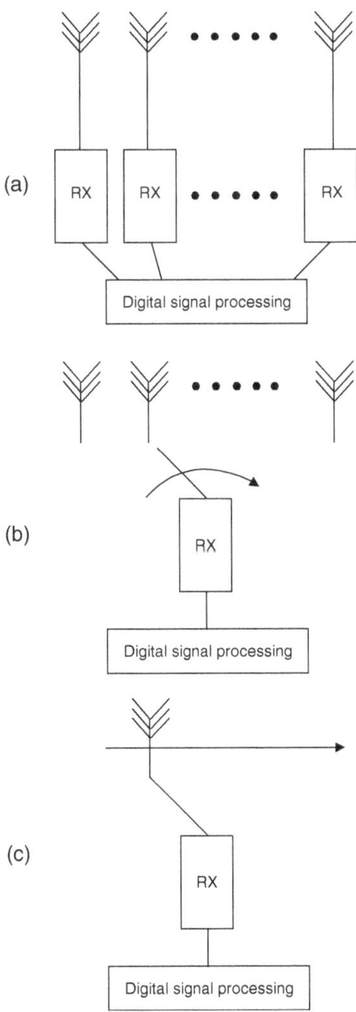

Figure 9.5 Types of arrays for channel sounding: real array (a), switched array (b), virtual array (c).
Reproduced with permission from [Molisch and Tufvesson 2005] © Hindawi.

- *Multiplexed arrays*: in this technique, multiple antenna elements, but only one downconversion chain, exist. The different antenna elements are connected to a downconversion chain via a fast RF switch [Thomae et al. 2000]. The RX thus first measures the impulse response at the first antenna element, then it connects the switch to the second element, measures its impulse response, and so on.
- *Virtual array*: in this technique, there is only a single antenna element, which is moved mechanically from one position to the next, measuring the impulse responses at the different antenna elements.

A basic assumption for evaluation is again that the environment does not change during the measurement procedure. "Virtual arrays" (which – due to the need of mechanically moving an antenna – require a few seconds or even minutes for one measurement run) can thus only be used in static environments.[5] This precludes scenarios where cars or moving persons are significant Interacting Objects (IOs), or where channel measurements are done with TX and/or RX moving during the measurements. In nonstatic environments, multiplexed arrays are usually the best compromise between measurement speed and hardware effort. A related aspect is the impact of frequency drift and loss of synchronization of the TX/RX. The longer a measurement lasts, the higher the impact of these impairments.

Comparing now rotating directional antennas and virtual arrays, we find that in both cases the resolution is limited by the aperture size (see Section 9.5.2). Relative advantages and drawbacks arise from practical considerations: rotating directional antennas can extract angular information even without phase coherence over the duration of the measurement – the power observed into the different directions is independent of possible relative phase drift of the local oscillators over time. However, at low frequencies, directive antennas are mechanically large and heavy, and may be difficult to move. Using an omni-directional antenna and moving it linearly with, e.g., a stepper motor can be preferable. Furthermore, at low frequencies, a cable connection synchronizing the local oscillators is often feasible. For these reasons, virtual arrays are commonly used at low frequencies, while rotating directional antennas are popular at high frequencies.

We now turn to the question of how to extract directional information from array measurements.

9.5.2 Beamforming

Having established how we obtain the measurement that is needed to determine the directions of arrival, we now proceed with the evaluation. We will for the moment assume a flat fading channel, i.e., consider only directional dispersion. Generalization to the delay-dispersive/frequency-selective case can be done, e.g., by applying the below-described procedures separately for each delay bin.

The simplest possible approach is conventional (Fourier) beamforming. In the case of rotating directional antennas, the signal recorded for the different antenna orientations is the *convolution* of the directional impulse response with the antenna pattern. The "Fourier processing" then consists simply of measuring the impulse responses with different horn orientations and accepting the fact that the resolution of the measurement is limited by the beamwidth of the directional antenna, which in turn is determined by the antenna aperture size – the smaller the aperture, the poorer the resolution. It is thus common to call the results from such processing "aperture-size limited." The limitation is similar to the delay domain, where simple Fourier processing can resolve the delay of an MPC only within the inverse bandwidth.

The situation is fairly similar when measuring with an antenna array, e.g., a Uniform Linear Array (ULA). A Fourier transform maps the locations of the antenna elements onto the directions of arrival ϕ_i. The phase resolution of the array is determined by the size of the array; the corresponding angular resolution is (for large N_r) $\lambda/(N_r d_a)$. The advantage of this method is its simple implementability (requiring only a Fast Fourier Transform (FFT)); the drawback is its coarse resolution.

More exactly, the angular spectrum $P_{\mathrm{BF}}(\phi)$ is given as

$$P_{\mathrm{BF}}(\phi) = \frac{\mathbf{u}^\dagger(\phi)\mathbf{R}_{\mathrm{rr}}\mathbf{u}(\phi)}{\mathbf{u}^\dagger(\phi)\mathbf{u}(\phi)} \tag{9.30}$$

where $\mathbf{R}_{\mathrm{rr}}$ is the correlation matrix of the incident signal and, for a ULA,

$$\mathbf{u}_{\mathrm{RX}}(\phi) = \begin{pmatrix} 1 \\ \exp\left(-jk_0 d_a \cos(\phi)\right) \\ \exp\left(-j2k_0 d_a \cos(\phi)\right) \\ \vdots \\ \exp\left(-j(N_r - 1)k_0 d_a \cos(\phi)\right) \end{pmatrix} \tag{9.31}$$

is the steering vector into direction ϕ (compare also Eqs. (8.59) and (8.60)). The beamformer $P_{\mathrm{BF}}(\phi)$ is also known as the *Bartlett* beamformer.

---

[5] Note that a mechanically rotating directional antenna can also be interpreted as a virtual array. Similarly, beamswitching with a phased array can be interpreted as a switched array approach.

*9.5.3 High-Resolution Algorithms – General Aspects

The problem of low resolution of Fourier techniques can be eliminated by means of so-called *high-resolution methods*. The resolution of these methods is not limited by the size of the antenna array but only by modeling errors and noise. This advantage is paid for by high computational complexity. Furthermore, there is often a limit on the *number* of MPCs whose directions can be estimated.

High-resolution methods include the following:

- Minimum Variance Method (MVM: Capon's beamformer): this method is a pure spectral search method, determining an angular spectrum such that for each considered direction the sum of the noise and the interference from other directions is minimized. It will be described in Section 9.5.4.
- ESPRIT: it determines the signal subspace, and extracts the directions of arrival in closed form. A description of this algorithm, which is mainly suitable for regular array structures such as ULAs, is given in Appendix 9.A.
- MUltiple SIgnal Classification (MUSIC): this algorithm also requires determination of the signal and noise subspaces, but then uses a spectral search to find the directions of arrival.
- Maximum-likelihood estimation of the parameters of incident waves: the problem of maximum-likelihood parameter extraction is its high computational complexity. These will be described in more detail in Section 9.5.5.

Some of the algorithms use certain assumptions about the array: the antenna patterns of all elements are identical, no *mutual coupling* between antenna elements, and the distance between all antenna elements is identical. For other algorithms, in particular Maximum-likelihood estimators, such conditions are not necessary, but it is a prerequisite to *accurately know* the complex antenna pattern of every array element. The measurement of these characteristics, i.e., the array calibration, has to be performed in an anechoic chamber, and is rather time-consuming, since the antenna pattern is on a dense angular grid (1–5 degree spacing).[6] More discussion of the use of the antenna pattern can be found in Section 9.5.5.

*9.5.4 Minimum Variance Method – Capon's Beamformer

The MVM is a simple subspace-based beamforming method that achieves high resolution by minimizing interference from other directions. The spectrum is easy to compute, namely,

$$P_{\mathrm{MVM}}(\phi) = \frac{1}{\mathbf{u}^\dagger(\phi)\mathbf{R}_{\mathrm{rr}}^{-1}\mathbf{u}(\phi)}. \tag{9.32}$$

However, it can be seen that for this – and many other high-resolution algorithms (especially subspace-based algorithms) – it is required that the correlation matrix does not become singular. Such singular $\mathbf{R}_{\mathrm{rr}}$ typically occurs if the sources of the waves from the different directions are correlated. For channel sounding, all signals typically come from the same source, so that they are completely correlated. In that case, subarray averaging ("spatial smoothing" or "forward–backward" averaging) has to be used to obtain the correct correlation matrix [Haardt and Nossek 1995]. The drawback of subarray averaging is that it decreases the effective size of the array.

Example 9.5 *Three independent signals with amplitudes 1, 0.8, and 0.2 are incident from directions 10°, 45°, and 72°, respectively. The noise level is such that the SNR of the first signal is 15 dB. Compute first the correlation matrix, and steering vectors for a five-element linear array with $\lambda/2$ spacing of the antenna elements. Then plot $P_{\mathrm{BF}}(\phi)$ and $P_{\mathrm{MVM}}(\phi)$.*

Let us first assume that the three signals arrive at the linear array with zero initial phase. Thus we have the following parameters for the three MPCs:

$$\begin{aligned}
a_1 &= 1 \cdot e^{j \cdot 0}, &\quad \phi_1 &= 10\pi/180 \text{ rad} \\
a_2 &= 0.8 \cdot e^{j \cdot 0}, &\quad \phi_2 &= 45\pi/180 \text{ rad} \\
a_3 &= 0.2 \cdot e^{j \cdot 0}, &\quad \phi_3 &= 72\pi/180 \text{ rad}.
\end{aligned} \tag{9.33}$$

We assume that measurement noise has a complex Gaussian distribution, and is spatially white. The common variance σ_{n}^2 of the noise samples is given as

$$\sigma_{\mathrm{n}}^2 = \frac{1}{10^{\frac{15}{10}}} = 0.032. \tag{9.34}$$

Take now the steering vectors for the five-element linear array with element spacing $d = \dfrac{\lambda}{2}$ and create a matrix such that each column is a steering vector corresponding to one MPC:

---

[6] While the description here concentrates on determination of azimuth, in reality the *joint* azimuth and elevation pattern has to be measured for every antenna element in the array – it is *not* sufficient to measure just the azimuth and elevation cuts separately.

$$
\mathbf{U} = \begin{bmatrix}
1 & 1 & 1 \\
\exp\left(-j\cdot\pi\cdot\cos\left(\phi_1\right)\right) & \exp\left(-j\cdot\pi\cdot\cos\left(\phi_2\right)\right) & \exp\left(-j\cdot\pi\cdot\cos\left(\phi_3\right)\right) \\
\exp\left(-j\cdot2\cdot\pi\cdot\cos\left(\phi_1\right)\right) & \exp\left(-j\cdot2\cdot\pi\cdot\cos\left(\phi_2\right)\right) & \exp\left(-j\cdot2\cdot\pi\cdot\cos\left(\phi_3\right)\right) \\
\exp\left(-j\cdot3\cdot\pi\cdot\cos\left(\phi_1\right)\right) & \exp\left(-j\cdot3\cdot\pi\cdot\cos\left(\phi_2\right)\right) & \exp\left(-j\cdot3\cdot\pi\cdot\cos\left(\phi_3\right)\right) \\
\exp\left(-j\cdot4\cdot\pi\cdot\cos\left(\phi_1\right)\right) & \exp\left(-j\cdot4\cdot\pi\cdot\cos\left(\phi_2\right)\right) & \exp\left(-j\cdot4\cdot\pi\cdot\cos\left(\phi_3\right)\right)
\end{bmatrix}. \tag{9.35}
$$

Inserting values for the directions of arrival gives

$$
\mathbf{U} = \begin{bmatrix}
1 & 1 & 1 \\
-0.9989 - 0.0477j & -0.6057 - 0.7957j & 0.5646 - 0.8253j \\
0.9954 + 0.0953j & -0.2663 + 0.9639j & -0.3624 - 0.9320j \\
-0.9898 - 0.1427j & 0.9282 - 0.3720j & -0.9739 - 0.2272j \\
0.9818 + 0.1898j & -0.8582 - 0.5133j & -0.7374 + 0.6755j
\end{bmatrix}. \tag{9.36}
$$

The three incident signals result in an observed array response given by

$$
\mathbf{r}(t) = \mathbf{U}\cdot\mathbf{s}(t) + \mathbf{n}(t). \tag{9.37}
$$

where $\mathbf{s}(t) = [1\quad 0.8\quad 0.2]^T$ and $\mathbf{n}(t)$ is the vector of additive noise samples. We evaluate the correlation matrix $\mathbf{R}_{rr} = E[\mathbf{r}(t)\mathbf{r}^\dagger(t)]$ for 10,000 realizations (where the columns of the steering vectors have different phases φ_i, corresponding to independent realizations[7]) resulting in:

$$
\mathbf{R}_{rr} = \begin{bmatrix}
1.7295 & -1.3776 + 0.5941j & 0.8098 - 0.6763j & -0.4378 + 0.3755j & 0.4118 + 0.1286j \\
-1.3776 - 0.5941j & 1.7244 & -1.3621 + 0.5957j & 0.7993 - 0.6642j & -0.4275 + 0.3666j \\
0.8098 + 0.6763j & -1.3621 - 0.5957j & 1.7080 & -1.3493 + 0.5850j & 0.7932 - 0.6599j \\
-0.4378 - 0.3755j & 0.7993 + 0.6642j & -1.3493 - 0.5850j & 1.6903 & -1.3439 + 0.5780j \\
0.4118 - 0.1286j & -0.4275 - 0.3666j & 0.7932 + 0.6599j & -1.3439 - 0.5780j & 1.6875
\end{bmatrix}. \tag{9.38}
$$

The angular spectrum for the conventional beamformer is given by Eq. (9.30), and it is plotted as the dashed line in Figure 9.6. We see that the conventional beamformer fails to resolve the three incident signals. The angular spectrum for the MVM (Capon's beamformer) is given by Eq. (9.32); its result is plotted as a solid line in Figure 9.6. Three peaks can be identified in the vicinity of the true angles of arrival 10°, 45°, and 72°.

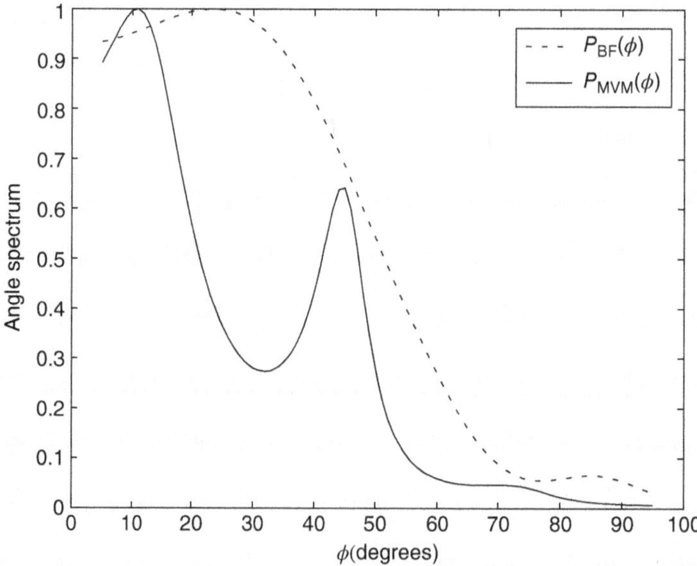

Figure 9.6 Comparison of conventional beamformer and the MVM angle spectrum for a five-element linear array. The true angles are 10°, 45°, and 72°.

[7] We *assume* here that we have different realizations available. As discussed above, many channel-sounding applications require additional measures like subarray averaging to obtain those realizations.

*9.5.5 High-Resolution Parameter Estimation – General Model

The principle of High-Resolution Parameter Estimation (HRPE) is to write the received signal in parametric form, i.e., as a sum of plane waves. Consider the case of SIMO, i.e., an antenna array with N_r antenna elements at the RX, and a single antenna at the TX. Based on the discussion in Section 6.7, the baseband representation of the directional channel in the frequency domain, can be expressed as a superposition of N planar waves:

$$\mathbf{h}(f_k) = \sum_{\ell=1}^{N} \mathbf{h}_\ell(f_k) = \sum_{\ell=1}^{N} a_\ell \widetilde{\mathbf{g}}_{\mathrm{RX}}(\phi_{\mathrm{RX},\ell}, \theta_{\mathrm{RX},\ell}, f_k) e^{-j2\pi f_k \tau_\ell} + \mathbf{n}(f_k) \in \mathbb{C}^{N_r \times 1}, \tag{9.39}$$

where f_k is the index for a frequency point of interest, $\widetilde{\mathbf{g}}_{\mathrm{R}}(\phi_{\mathrm{R},\ell}, \theta_{\mathrm{R},\ell})$ is the calibrated Rx antenna array pattern vector that takes the displacement of the antenna elements into account, such that the ith entry of this vector is

$$\left[\widetilde{\mathbf{g}}_{\mathrm{RX}}(\phi_{\mathrm{RX},\ell}, \theta_{\mathrm{R},\ell})\right]_i = \widetilde{G}_i(\phi_{\mathrm{RX},\ell}, \theta_{\mathrm{RX},\ell}) \exp\left(-j < \mathbf{k}(\phi_{\mathrm{RX},\ell}, \theta_{\mathrm{RX},\ell}), \left(\mathbf{r}_{\mathrm{RX}}^{(i)} - \mathbf{r}_{\mathrm{RX}}^{(1)}\right) >\right)$$

and $\mathbf{n}(f_k) \in \mathbb{C}^{N_r \times 1}$ is the additive zero-mean Gaussian noise. The elements of the vector correspond to the different RX antenna elements. The aim of the estimator is now to find the parameters that explain the observables. Note that in the MIMO case (compare Section 9.5.7), the data model easily generalizes, to

$$\mathbf{H}(f_k) = \sum_{\ell=1}^{N} \mathbf{H}_\ell(f_k) = \sum_{\ell=1}^{N} a_\ell \widetilde{\mathbf{g}}_{\mathrm{RX}}(\phi_{\mathrm{RX},\ell}, \theta_{\mathrm{RX},\ell}, f_k) \widetilde{\mathbf{g}}_{\mathrm{TX}}^\dagger(\phi_{\mathrm{RX},\ell}, \theta_{\mathrm{RX},\ell}, f_k) e^{-j2\pi f_k \tau_\ell} + \mathbf{N}(f_k) \in \mathbb{C}^{N_r \times N_t}, \tag{9.40}$$

where superscript $\dagger$ indicates "Hermitean transpose."[8] For the sake of notational simplicity, the descriptions for the algorithms CLEAN (Section 9.5.6) and SAGE (Section 9.5.7) below are done for SIMO.

A noteworthy feature of all HRPEs is that their resolution is not limited by the aperture size or bandwidth. Rather, it is the SNR that mainly determines the performance; a more detailed discussion of this effect and performance bounds (in particular the Cramer–Rao Lower Bound) is given in Section 29.2.3.

*9.5.6 Serial Interference Cancellation – The CLEAN Algorithm

The CLEAN algorithm is an iterative algorithm with a very clear intuitive explanation. We first aim to find the strongest MPC. Once we have determined its parameters (we will describe later how), it is possible to compute the contribution that this MPC makes to the observables, i.e., the signals measured at the different antenna elements. We then subtract these contributions. In the thus "cleaned up" observables, we find the next strongest MPC, determine its parameters, subtract its contribution, and so on, until either a maximum number of MPCs is reached, or the power in the remaining signal falls below a specified threshold. From this description, one can identify CLEAN as a serial interference cancellation approach, which we will encounter later on in a variety of other wireless signal processing applications (Sections 16.2.9 and 28.2.4). To be more mathematically explicit, the objective function of CLEAN, i.e., the likelihood of the measured channel, is maximized by estimating the parameters that minimize the Mean Squared Error (MSE):

$$\begin{bmatrix} \hat{a} \\ \hat{\Upsilon} \end{bmatrix} = \begin{bmatrix} \hat{a} \\ \hat{\tau} \\ \hat{\phi}_{\mathrm{RX}} \\ \hat{\theta}_{\mathrm{RX}} \end{bmatrix} = \arg\min_{a,\Upsilon} \sum_{k=1}^{N_f} \left\| \mathbf{h}(f_k) - a\widetilde{\mathbf{g}}_{\mathrm{RX}}(\phi_{\mathrm{RX}}, \theta_{\mathrm{RX}}) e^{-j2\pi f_k \tau} \right\|^2 \tag{9.41}$$

$$= \arg\min_{a,\Upsilon} \left\| \mathbf{h} - \mathbf{u}(\Upsilon; \mathbf{f}) a \right\|^2, \tag{9.42}$$

where $\mathbf{h}$ is the aggregation of the channel measurements, which can be defined as

$$\mathbf{h} \triangleq \left[h_1(f_1), \cdots, h_{N_r}(f_1), \cdots, h_1(f_{N_f}), \cdots, h_{N_r}(f_{N_f}) \right]^T \in \mathbb{C}^{N_r N_f \times 1},$$

---

[8] Depending on the definition of the coordinate system, the Hermitean transposition might need to be replaced by a regular transposition, and/or $\widetilde{G}^{\mathrm{TX}}$ must be replaced by $\left(\widetilde{G}^{\mathrm{TX}}\right)^*$, see also Sec. 6.7.

and the vector $\mathbf{u}(\Upsilon;\mathbf{f})$ containing antenna pattern information and propagation delay can be expressed as

$$
\mathbf{u}(\Upsilon;\mathbf{f}) \triangleq
\begin{bmatrix}
\widetilde{G}_1(\phi_{\mathrm{RX}},\theta_{\mathrm{RX}})e^{-j<\mathbf{k}(\phi_{\mathrm{RX}},\theta_{\mathrm{RX}}),\left(\mathbf{r}_{\mathrm{RX}}^{(1)}-\mathbf{r}_{\mathrm{RX}}^{(1)}\right)>}\,e^{-j2\pi f_1\tau} \\
\vdots \\
\widetilde{G}_{N_{\mathrm{r}}}(\phi_{\mathrm{RX}},\theta_{\mathrm{RX}})e^{-j<\mathbf{k}(\phi_{\mathrm{RX}},\theta_{\mathrm{RX}}),\left(\mathbf{r}_{\mathrm{RX}}^{(N_{\mathrm{r}})}-\mathbf{r}_{\mathrm{RX}}^{(1)}\right)>}\,e^{-j2\pi f_1\tau} \\
\vdots \\
\widetilde{G}_1(\phi_{\mathrm{RX}},\theta_{\mathrm{RX}})e^{-j<\mathbf{k}(\phi_{\mathrm{RX}},\theta_{\mathrm{RX}}),\left(\mathbf{r}_{\mathrm{RX}}^{(1)}-\mathbf{r}_{\mathrm{RX}}^{(1)}\right)>}\,e^{-j2\pi f_{N_{\mathrm{f}}}\tau} \\
\vdots \\
\widetilde{G}_{N_{\mathrm{r}}}(\phi_{\mathrm{RX}},\theta_{\mathrm{RX}})e^{-j<\mathbf{k}(\phi_{\mathrm{RX}},\theta_{\mathrm{RX}}),\left(\mathbf{r}_{\mathrm{RX}}^{(N_{\mathrm{r}})}-\mathbf{r}_{\mathrm{RX}}^{(1)}\right)>}\,e^{-j2\pi f_{N_{\mathrm{f}}}\tau}
\end{bmatrix}
\in \mathbb{C}^{N_{\mathrm{r}}N_{\mathrm{f}}\times 1},
\tag{9.43}
$$

where $\mathbf{f} \triangleq \left[f_1,\cdots,f_{N_{\mathrm{f}}}\right]^{T}$. Based on the assumption of Gaussian noise in the residue at each CLEAN cycle, the Maximum Likelihood Estimator (MLE) is equivalent to a least-square estimator as shown in (9.41). The minimizer can also be interpreted as the parameters corresponding to the maximum correlation to the measured channel given the calibrated antenna pattern. Note that we assume here that the antenna element pattern is independent of frequency – this simplifies the algorithm, but the validity of the assumption has to be checked from case to case.

The transfer function is impacted by the path amplitude in a linear way, but the *structural parameters* τ, ϕ_{RX}, θ_{RX}, which are summarized in the vector Υ, in a nonlinear way. Consequently, it is difficult to derive a closed-form solution to (9.41). Considering the objective function in the vector form (9.42), the problem is quadratic in the complex weight a for fixed structural parameters Υ, and the estimator of the *path weight* can be found as a closed-form expression of the structural parameters,

$$
\hat{a} = \frac{\mathbf{u}^{\dagger}(\Upsilon;\mathbf{f})\mathbf{h}}{\|\mathbf{u}(\Upsilon;\mathbf{f})\|^2}.
\tag{9.44}
$$

Inserting (9.44) into (9.41), the objective function to be optimized is not a function of path weights anymore. The structural parameters are computed as

$$
\begin{aligned}
\hat{\Upsilon} &= \arg\max_{\Upsilon} \frac{1}{\|\mathbf{u}(\Upsilon;\mathbf{f})\|^2}\left|\mathbf{u}^{\dagger}(\Upsilon;\mathbf{f})\mathbf{h}\right|^2 \\
&= \arg\max_{\Upsilon} \frac{1}{N_{\mathrm{f}}\|\widetilde{\mathbf{g}}_{\mathrm{RX}}(\phi_{\mathrm{RX}},\theta_{\mathrm{RX}})\|^2}\left|\sum_{k=1}^{N_{\mathrm{f}}}\widetilde{\mathbf{g}}_{\mathrm{RX}}^{\dagger}(\phi_{\mathrm{RX}},\theta_{\mathrm{RX}})\mathbf{h}(f_k)e^{j2\pi f_k\tau}\right|^2.
\end{aligned}
\tag{9.45}
$$

The summation over frequencies and antennas can be interpreted as the correlation function with the measured channel, where the maximum is achieved as the estimated parameters are approaching the ground truth. The optimum argument Υ can be found, e.g., through a multi-dimensional grid search.

Figure 9.7 depicts the flow chart of CLEAN. As a first step, the residue is initialized as the calibrated measured channel, $\mathbf{h}_{\mathrm{res},0} = \mathbf{h}_{\mathrm{meas}} \triangleq \left[\mathbf{h}(f_1),\cdots,\mathbf{h}(f_{N_{\mathrm{f}}})\right] \in \mathbb{C}^{N_{\mathrm{r}}N_{\mathrm{f}}\times 1}$. The initial solution set is empty, $\ell_s = 0$; here ℓ_s denotes the number of paths that have already been evaluated. At each iteration, the "strongest" path will be extracted by finding the most highly correlated contribution to the residual channel response, where "strongest" corresponds to the largest correlation value. Then the corresponding path weight can be evaluated. The estimated path parameters at ℓ_sth iteration, $\left[\hat{a}_{\ell_s},\hat{\tau}_{\ell_s},\hat{\phi}_{\mathrm{RX},\ell_s},\hat{\theta}_{\mathrm{RX},\ell_s}\right]$, are added to the solution set after updating the iteration numbering $\ell_s \leftarrow \ell_s + 1$. From Eq. (9.39), the contribution of the extracted path, $\hat{\mathbf{h}}_{\ell_s}$, can be reconstructed. The residual channel response is updated by subtracting this corresponding contribution, $\mathbf{h}_{\mathrm{res},\ell_s} = \mathbf{h}_{\mathrm{res},\ell_s} - \hat{\mathbf{h}}_{\ell_s}$. This process will continue until the halting condition is satisfied, which can be either the maximum number of paths to be estimated is reached, or the residual power is lower than a certain threshold level.

The CLEAN algorithm provides useful estimates of the MPC parameters, but it is inherently subject to error propagation. Assume that the estimation of the parameters of the strongest component is imperfect, due to interference from other MPCs, and/or noise. Then the subtraction process will not completely eliminate the true contribution of this first component, the residue in turn increases the errors in the next component, and so on. Due to wrong estimates of the parameters, so-called ghost paths can occur, i.e., MPCs that do not exist physically, but which the algorithm conjectures in order to explain the residuals from the imperfect subtraction.

*9.5.7 Iterative Maximum-Likelihood Estimation – SAGE Algorithm

To resolve the problems of CLEAN, an iterative refinement of the MPC parameters is required. As an intuitive explanation, once we have estimates of the later (weaker) paths, we can subtract their impact on the original observables, and then obtain a more accurate estimate of the parameters for the first path. Now this contribution can be subtracted, more accurately, from the overall observables,

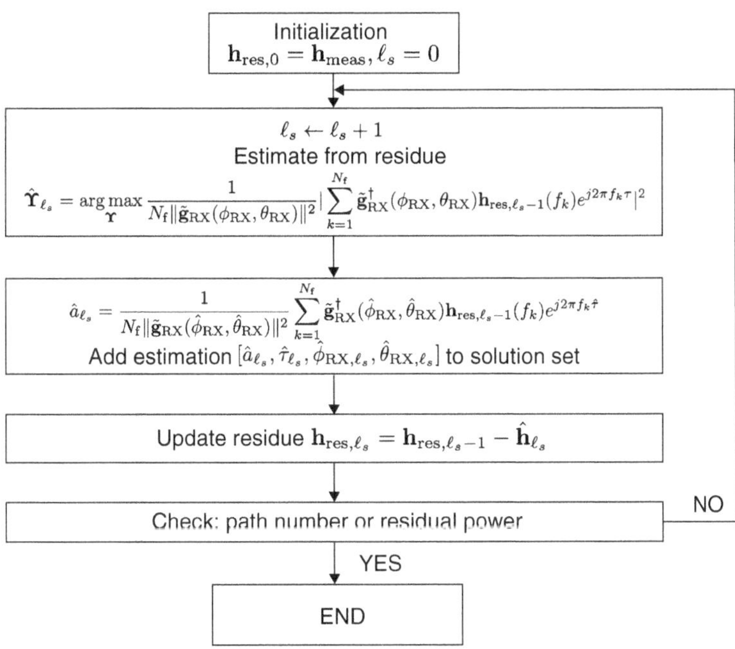

Figure 9.7 CLEAN algorithm flowchart for the SIMO case.

and a better estimate of the parameters for the second path can be obtained. This process is repeated until all parameters have converged, i.e., do not change "significantly" from iteration to iteration, see Figure 9.8.

In a more formal notation, SAGE is an iterative Expectation-Maximization (EM) algorithm, where the expectation performs essentially the interference cancellation, and the maximization computes a correlation function for each parameter. The correlation we are maximizing is:

$$z\left(\tau, \phi_{\mathrm{RX}}, \theta_{\mathrm{RX}}; \hat{\mathbf{h}}_{\ell}\right) = \sum_{k=1}^{N_{\mathrm{f}}} \widetilde{\mathbf{g}}_{\mathrm{RX}}^{\dagger}(\phi_{\mathrm{RX}}, \theta_{\mathrm{RX}})\hat{\mathbf{h}}_{\ell}(f_k)e^{j2\pi f_k \tau}. \tag{9.46}$$

The algorithm then proceeds as follows:

- Iterate until convergence, iteration index n
  - For $\ell = 1, \ldots N$
    - *Expectation Step*: the estimate of the contribution of the ℓth path of the received signal is the total received signal, minus the contributions from all the other paths as computed in the current iteration round (if already computed) or previous iteration

$$\hat{\mathbf{h}}_{\ell}\left(f_k; \hat{\mathbf{Y}}_{\ell}^{\{n+1\}}\right) = \mathbf{h}(f_k) - \sum_{\ell' < \ell}\hat{\mathbf{h}}_{\ell'}\left(f_k; \hat{\mathbf{Y}}_{\ell'}^{\{n+1\}}\right) - \sum_{\ell' > \ell}\hat{\mathbf{h}}_{\ell'}\left(f_k; \hat{\mathbf{Y}}_{\ell'}^{\{n\}}\right). \tag{9.47}$$

- *Maximization Step*: using a one-dimensional grid search, we one-by-one compute the structural parameters that correlate best with the contribution of the ℓth path; finally we compute the amplitude of the path in closed form

$$\hat{\tau}_{\ell}^{\{n+1\}} = \arg\max_{\tau} \frac{1}{N_{\mathrm{f}}\left\|\widetilde{\mathbf{g}}_{\mathrm{RX}}\left(\hat{\phi}_{\mathrm{RX},\ell}^{\{n\}}, \hat{\theta}_{\mathrm{RX},\ell}^{\{n\}}\right)\right\|^2} \left|z\left(\tau, \hat{\phi}_{\mathrm{RX},\ell}^{\{n\}}, \hat{\theta}_{\mathrm{RX},\ell}^{\{n\}}; \hat{\mathbf{h}}_{\ell}\left(f_k; \hat{\mathbf{Y}}_{\ell}^{\{n+1\}}\right)\right)\right|^2 \tag{9.48}$$

$$\hat{\phi}_{\mathrm{RX},\ell}^{\{n+1\}} = \arg\max_{\phi_{\mathrm{RX}}} \frac{1}{N_{\mathrm{f}}\left\|\widetilde{\mathbf{g}}_{\mathrm{RX}}\left(\phi_{\mathrm{RX}}, \hat{\theta}_{\mathrm{RX},\ell}^{\{n\}}\right)\right\|^2} \left|z\left(\hat{\tau}_{\ell}^{\{n+1\}}, \phi_{\mathrm{RX}}, \hat{\theta}_{\mathrm{RX},\ell}^{\{n\}}; \hat{\mathbf{h}}_{\ell}\left(f_k; \hat{\mathbf{Y}}_{\ell}^{\{n+1\}}\right)\right)\right|^2 \tag{9.49}$$

$$\hat{\theta}_{\mathrm{RX},\ell}^{\{n+1\}} = \arg\max_{\theta_{\mathrm{RX}}} \frac{1}{N_{\mathrm{f}}\left\|\widetilde{\mathbf{g}}_{\mathrm{RX}}\left(\hat{\phi}_{\mathrm{RX},\ell}^{\{n+1\}}, \theta_{\mathrm{RX}}\right)\right\|^2} \left|z\left(\hat{\tau}_{\ell}^{\{n+1\}}, \hat{\phi}_{\mathrm{RX},\ell}^{\{n+1\}}, \theta_{\mathrm{RX}}; \hat{\mathbf{h}}_{\ell}\left(f_k; \hat{\mathbf{Y}}_{\ell}^{\{n+1\}}\right)\right)\right|^2 \tag{9.50}$$

$$\hat{a}_{\ell}^{\{n+1\}} = \frac{1}{N_{\mathrm{f}}\left\|\widetilde{\mathbf{g}}_{\mathrm{RX}}\left(\hat{\phi}_{\mathrm{RX},\ell}^{\{n+1\}}, \hat{\theta}_{\mathrm{RX},\ell}^{\{n+1\}}\right)\right\|^2} z\left(\hat{\tau}_{\ell}^{\{n+1\}}, \hat{\phi}_{\mathrm{RX},\ell}^{\{n+1\}}, \hat{\theta}_{\mathrm{RX},\ell}^{\{n+1\}}; \hat{\mathbf{h}}_{\ell}\left(f_k; \hat{\mathbf{Y}}_{\ell}^{\{n+1\}}\right)\right). \tag{9.51}$$

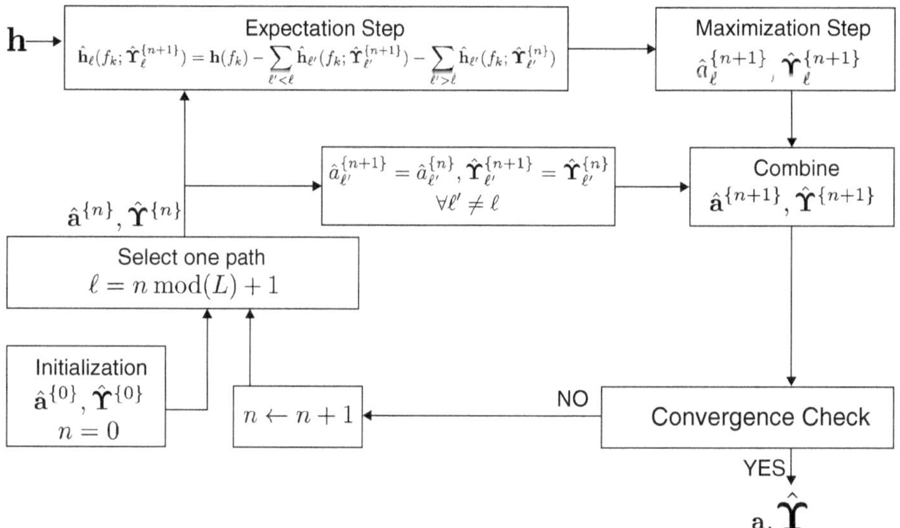

Figure 9.8 SAGE algorithm flowchart for the SIMO case starts at Initialization block.

Convergence is checked through Root Mean Square Error (RMSE) between the reconstructed channel (based on the estimates) and the input measured channel.

*9.5.8 Multiple Input Multiple Output Measurements

The methods described above are intended for getting the directions of arrival at one link end. They can be easily generalized to double-directional or MIMO measurements, where antenna arrays (or equivalent structures) are used at *both* link ends. In this case, it is necessary to use transmit signals in such a way that the RX can determine which antenna they were transmitted from. This can be done, e.g., by sending signals at different times, on different frequencies, or modulated with different codes (see Figure 9.9). Of these methods, using different times requires the least hardware effort, and is thus in widespread use.

Another major point of concern in MIMO measurements is the measurement duration, in particular, when virtual or switched arrays are used. Assume $N_{\rm TX}$ antenna elements (or an equivalent number of horn directions) are used at the TX. Then, the

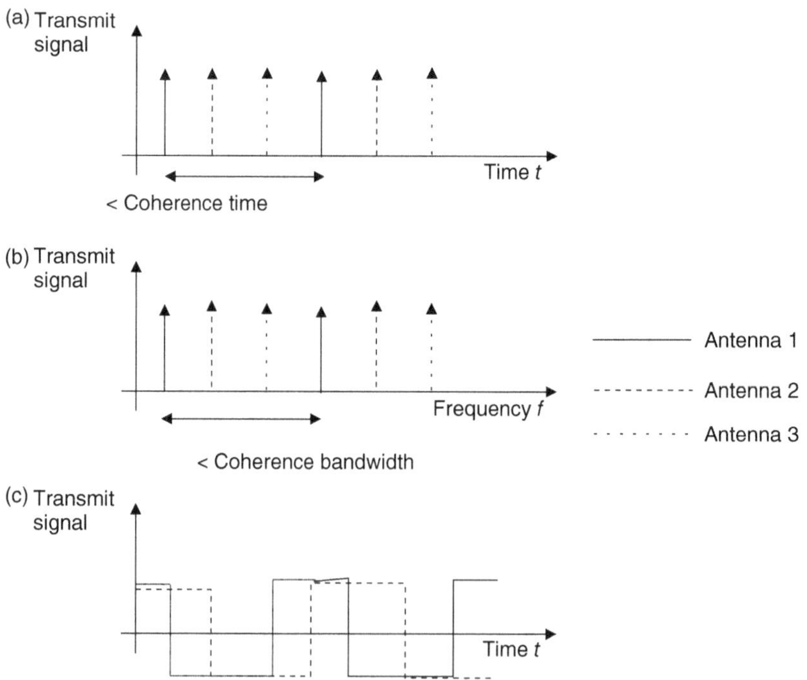

Figure 9.9 Transmission of sounding signals from different antennas: signals orthogonal in time (a), frequency (b), or code (c).
Reproduced with permission from [Molisch and Tufvesson 2005] © Hindawi.

measurement duration is simply N_{TX} times the measurement duration of the SIMO case. In the case of the virtual array, this means that the measurement duration becomes

$$N_{TX}[N_{RX}(T_{meas} + T_{mov}) + (T_{mov})], \tag{9.52}$$

where T_{meas} is the measurement duration at one particular location (antenna orientation), and T_{mov} is the time it takes to move (rotate) the antenna from one location to another. Since the latter is on the order of several seconds, one can easily see that measurements with a large number of position at both link ends can take hours to complete.

For the case of switched arrays, an equation similar to (9.52) holds. Now T_{mov} is replaced by T_{switch}, i.e., the duration it takes to electronically switch from one antenna element to another. This time is typically between 100 ns and 10 μs. Since T_{meas} is typically on the same order (depending on the maximum excess delay of the channel), a MIMO measurement can often be performed within several ms. This duration can be smaller than a typical coherence time of the channel, and thus allow full identification of the directional channel characteristics. It is also possible to identify a channel even if the measurement time is somewhat larger than the inverse Doppler frequency, see [Wang et al. 2019] for a discussion of this approach.

The signal processing techniques for determination of the directions at the two link ends are also fairly similar to the processing techniques for the one-dimensional case. We stress, however, that evaluations for SIMO (array only at RX) and a MISO (array only at TX) are *not* equivalent to a MIMO evaluation. Rather, a *joint* processing of the MIMO data has to be performed, to obtain the double-directional characteristics of the MPCs.

*9.6 Appendices

App. 9.A: The ESPRIT Algorithm

See App9.pdf at wiley.com/go/molisch/wireless3e

App. 9.B: Guidelines for Evaluation of Channel Measurements

See App9.pdf at wiley.com/go/molisch/wireless3e

Further Reading

Overviews of different channel-sounding techniques are given in [Parsons 1992] and [Parsons et al. 1991], the more recent monograph of [Salous 2013], and (specifically for higher frequencies) [Rappaport et al. 2022]. [Cullen et al. 1993] concentrate on correlative channel sounders; the STDCC is also described in detail in [Cox 1972], where it was first introduced. [Matz et al. 2002] discuss the impact of time variations of the channel on measurement results. Measurement procedures are also discussed by most papers presenting measurement results (especially back-to-back calibration and deconvolution). For the measurement of directional properties, the method of using a rotating directional antenna is discussed, e.g., in [Pajusco 1998]. Detailed discussions of sounder calibration for mm-wave channels (many of which carry over to lower frequencies) are in [Rappaport et al. 2022].

The ESPRIT algorithm is described in [Haardt and Nossek 1995] and [Roy et al. 1986]; MUSIC in [Schmidt 1986]; for the MVM (Capon's beamformer), see [Krim and Viberg 1996]; SAGE is described in [Fleury et al. 1999]. The RiMax algorithm, an alternative to SAGE with fast convergence, is described in [Thomae et al. 2005]. The incorporation of diffuse radiation into the extraction is discussed in [Richter 2006]. Other high-resolution algorithms include the CLEAN algorithm that is especially suitable for the analysis of ultrawideband signals [Cramer et al. 2002], the JADE algorithm [Vanderveen et al. 1997], and many others. Tracking of multi-path components obtained from a sequence of measurements by means of Extended Kalman Filters (compare also Sec. 29.11) is discussed in [Salmi et al. 2009].

For updates and errata for this chapter, see wides.usc.edu/teaching/textbook

Exercises

See Sec. 36.9 in Exercises.pdf at wiley.com/go/molisch/wireless3e

Part III

Wireless Communication Over a Single Link

The ultimate performance limits of wireless systems are determined by the wireless propagation channels that we have investigated in the previous parts. The task of practical transceiver design now involves finding suitable modulation schemes, codes, and signal processing algorithms so that these performance limits can be approximated "as closely as possible." In this part, we deal with techniques to get data from *one* transmitter to *one* receiver – in other words, a single-link transmission. Such single-link techniques are well explored, though research is still ongoing, and significant innovations are developed on a continuous basis.

The part starts in Chapter 10 with a description of the various *modulation formats*, by which we mean mappings of data onto particular waveforms that can be sent over the air. Each modulation format has specific advantages and disadvantages in a wireless context; we will describe those, as well as various methods to represent them. Building on such formal mathematical descriptions, Chapter 11 then describes how to evaluate their performance in terms of *bit error probability* in different types of channels. We find that the performance of such systems is mostly limited by two effects: fading and delay dispersion. The effect of fading can be greatly mitigated by *diversity*, i.e., by transmitting the same signal via different paths. Chapter 12 describes the different methods of obtaining such different paths, e.g., by implementing multiple antennas, by repeating the signal at different frequencies, or at different times. The chapter also discusses the effect that the diversity has on the performance of the different modulation schemes.

However, even with diversity, the bit error rate is often too high for practical purposes. In many cases, *coding* can greatly enhance the performance and provide the transmission quality required by a specific application. Chapter 13 thus gives an overview of the different coding schemes that are most popular for wireless communications, including the near-optimum turbo codes and Low Density Parity Check (LDPC) codes that have drawn great attention since the early 1990s, and the more recent polar codes. The chapter also describes the fundamentals of information theory, which establishes the ultimate performance limits that can be achieved with "ideal" codes.

While negative effects of the delay dispersion can also be combated by diversity, it is more effective to use equalization. *Equalizers* do not only combat intersymbol interference created by delayed echoes of the original signal, but they make active use of them, exploiting the energy contained in such echoes. They can thus lead to a considerable improvement of performance, especially in systems with high data rates and/or systems operating in channels with large delay spreads. Chapter 14 describes different equalizer structures, from the simple linear equalizers to the optimum (but highly complex) maximum-likelihood sequence detectors.

While equalizers are efficient ways of combatting delay dispersion, their complexity can become overwhelming in systems that operate at high data rate, or in channels with high delay dispersion, or both. An alternative way is the use of multicarrier modulation, where a highrate stream is split into a number of parallel streams whose spectra are overlapping yet orthogonal to each other. Since each of those channels operates now with a much lower data rate, delay dispersion can be combatted more easily. Chapter 15 describes such multi-carrier techniques, in particular Orthogonal Frequency Division Multiplexing (OFDM), which is used in LTE, 5G, and Wi-Fi, as well as related emerging schemes for Beyond 5G.

One of the most important trends of the past 30 years has been the development of robust and spectrally efficient multi-antenna transmission techniques. These can use multiple antenna elements either only at one link end, or at both link ends. They apply them to achieve diversity and beamforming, or transmit multiple data streams in the same time/frequency resources. Chapter 16 provides an in-depth look at these techniques and discusses their performance and implementation.

Finally, the viability and performance of these techniques are all depending on the hardware on which they are implemented. Chapter 17 thus discusses the hardware structure of wireless transceivers (e.g., heterodyne receivers), as well as the individual components such as Analog to Digital Converters (ADCs), local oscillators, and amplifiers, and the limitations and nonidealities that they introduce in the overall system.

Wireless Communications: From Fundamentals to Beyond 5G, Third Edition. Andreas F. Molisch.
© 2023 John Wiley & Sons Ltd. Published 2023 by John Wiley & Sons Ltd.
Companion website: www.wiley.com/go/molisch/wireless3e

10

Modulation Formats

10.1 Introduction

10.1.1 Basic Principles

The data that we want to transmit over the wireless propagation channel are digital – either because we really want to transmit data files, or because the source coder has rendered the source information (audio, video) into a digital form. On the other hand, the wireless propagation channel is an analog medium, over which analog waveforms have to be sent. *Digital modulation is the mapping of the data bits onto analog signal waveforms that can be transmitted over the (analog) channel.* At the receiver (RX), the demodulator tries to recover the bits from the received waveform. This chapter gives a brief review of digital *modulation formats*, i.e., the specific mappings of bits to waveforms. The chapter concentrates mostly on the *results*; for references with derivation of those results and further details see the "Further Reading" section. Chapter 11 will then describe optimum and suboptimum demodulators and their performance in wireless channels.

An analog waveform can represent either one bit or a group of bits, depending on the type of modulation. The simplest modulation is a binary modulation, where a $+1$ bit value is mapped onto one specific waveform, while a -1 bit value is mapped to a different waveform. More generally, a group of K bits can be subsumed into a symbol, which in turn is mapped to one out of a set of $M = 2^K$ waveforms; in that case, we speak of *M-ary modulation*, *higher-order modulation*, *multi-level modulation*, or modulation with *alphabet size M*. In any case, different modulation formats differ by the waveforms that are transmitted, and by the way the mapping from bit groups to waveforms is achieved. Typically, the waveform corresponding to one symbol is time-limited to a time T_S, and waveforms corresponding to different symbols are transmitted one after the other. Obviously, the symbol duration T_S is K times the bit duration T_B, and the data (bit) rate is K times the transmitted symbol rate (signaling rate).

10.1.2 Goals and Evaluation Criteria

When choosing a modulation format in a wireless system, the ultimate goal is to transmit with a certain energy as much information as possible over a channel with a certain bandwidth, while assuring a certain transmission quality (BER). From this basic requirement, some additional criteria follow logically:

- The *spectral efficiency* of the modulated signal should be as high as possible. This is impacted by two factors:
  - How many bits can be conveyed by each transmitted symbol. This can be best achieved with a higher-order modulation format, which allows to transmit more data bits with each symbol.
  - What bandwidth is occupied by the modulation format. To a first, rough, approximation, this bandwidth is equal to the symbol rate. However, at closer inspection, it turns out that the power spectrum of the signal always shows a gradual roll-off outside the desired band. The faster this roll-off, the closer we can place in the frequency domain the signals from different users while keeping the *adjacent-channel interference* small. Fast roll-off is achieved by strong filtering of the transmit signal.

- *Sensitivity with respect to noise* should be very small. This can be best achieved with a low-order modulation format, where (assuming equal average power) the difference between the waveforms of the alphabet is largest.
- The *robustness with respect to delay- and Doppler dispersion* should be as large as possible. Thus, the transmit signal should be filtered as little as possible, as filtering creates delay dispersion that makes the system more sensitive with respect to channel-induced delay dispersion.
- The waveforms should be *easy to generate* with low-cost hardware that is easy to produce and has high energy efficiency. This requirement stems from the practical requirements of wireless transmitters (see Chapter 2). Generally, modulation formats with low variations of the amplitude (known as peak-to-average-power-ratio, PAPR), are preferable; see Chapter 17 for a detailed discussion.

Wireless Communications: From Fundamentals to Beyond 5G, Third Edition. Andreas F. Molisch.
© 2023 John Wiley & Sons Ltd. Published 2023 by John Wiley & Sons Ltd.
Companion website: www.wiley.com/go/molisch/wireless3e

The above outline shows that some of these requirements are contradictory. There is thus no "ideal" modulation format for wireless communications. Rather, the modulation format has to be selected according to the requirement of a specific system and application.

10.1.3 Baseband–Passband

Modulated wireless signals are generally sent in passband, i.e., a frequency range $[f_c - B/2, f_c + B/2]$, where it is assumed that $B \ll f_c$. For example, the typical bandwidth B for LTE wireless systems is 20 MHz, while the carrier frequency f_c is between 800 and 2,300 MHz. A modulated signal can generally be written as showing changes of the amplitude and the phase of a sinusoidal oscillation

$$s_{\mathrm{BP}}(t) = A(t) \cos\left[2\pi f_c t + \phi(t)\right] \tag{10.1}$$

such that $A(t)$ describes the amplitude modulation and $\phi(t)$ the phase modulation. According to standard formulas for sinusoids, this can be written as

$$s_{\mathrm{BP}}(t) = A(t) \cos\left[\phi(t)\right] \cos\left[2\pi f_c t\right] - A(t) \sin\left[\phi(t)\right] \sin\left[2\pi f_c t\right]. \tag{10.2}$$

In other words, instead of interpreting modulation as changing amplitude and phase of a cosine, we can interpret it as changing the amplitude of a cosine and a sine oscillation (of equal frequency f_c). The first term is known as the in-phase component, and the second term is the quadrature component, or simply I- and Q-component for short.

Using Euler's relationship, the bandpass signal can also be written as the real part of a *complex* modulation signal[1]

$$s_{\mathrm{BP}}(t) = \mathrm{Re}\left\{s_{\mathrm{LP}}(t) \exp\left[j2\pi f_c t\right]\right\}. \tag{10.3}$$

Writing $s_{\mathrm{LP}}(t) = \mathrm{Re}\left\{s_{\mathrm{LP}}(t)\right\} + j\,\mathrm{Im}\left\{s_{\mathrm{LP}}(t)\right\}$, and $\exp[j2\pi f_c t] = \cos[2\pi f_c t] + j\sin[2\pi f_c t]$, we easily see that

$$\mathrm{Re}\left\{s_{\mathrm{LP}}(t)\right\} = A(t) \cos\left[\phi(t)\right] \tag{10.4}$$

$$\mathrm{Im}\left\{s_{\mathrm{LP}}(t)\right\} = A(t) \sin\left[\phi(t)\right] \tag{10.5}$$

where (10.4) and (10.5) are the modulation of the I and Q component, respectively. It is thus mathematically convenient to operate with the equivalent baseband representation $s_{\mathrm{LP}}(t)$ for the description of modulation formats.

This equivalent baseband signal is not physically existing, as no complex signals can exist in reality. However, its I- and Q-component can be treated by digital signal processing as two separate real signals, and then subsequently used to synthesize a real passband signal according to (10.3), and vice versa. The shifting of a signal from one frequency band (e.g., baseband) to another (e.g., passband), is essentially achieved by multiplying the signal with a sinusoidal oscillation of appropriate frequency. Mathematically, a multiplication with a complex exponential leads to a convolution with a delta pulse, i.e., a shift, in the frequency domain. In practice, positive and negative frequency components of the oscillators need to be taken into account; these aspects will be discussed in detail in Section 17.8. For the current chapter, the simple relationship (10.3) is sufficient.

Before going into more mathematical details, we will discuss as a simple example a modulation format called QPSK, or quaternary phase-shift keying (it will be discussed again, in greater detail, in Section 10.3.2). It transmits groups of two bits, such that each group is encoded into the phase of the transmit signal during one symbol duration T_s. Interpreting in terms of the representation (10.1), we can say that $A(t)$ is constant, A, and $\phi(t)$ takes on one of the values $\pi/4 + i\pi/2$, $i = 0, 1, 2, 3$.[2] Thus, the I-component of the low-pass signal is $\pm A/\sqrt{2}$, and similarly, the Q-component is $\pm A/\sqrt{2}$.[3] We can write this signal also in the form

$$s_{\mathrm{LP}}(t) = \sum_{i=-\infty}^{\infty} c_i g(t - iT_{\mathrm{S}}) \tag{10.6}$$

where a sequence of time-shifted *basis pulses* $g(t)$ is multiplied with complex modulation coefficients c_i, so that $c_i = (\pm 1 \pm j)A/\sqrt{2}$. Plots of low-pass and band-pass QPSK signals can be seen in Figures 10.10 and 10.12, respectively.

---

[1] Note that definition (10.3) results in an energy of the bandpass signal that is half the energy of the baseband signal. In order to achieve equal energy, the energy must be defined as $E = \| s_{BP} \|^2 = 0.5 \| s_{LP} \|^2$. Furthermore, it is required that the impulse response of a filter $h_{\mathrm{filter,BP}}(t)$ in passband has the following equivalent baseband representation $h_{\mathrm{filter,BP}}(t) = 2\,\mathrm{Re}\left\{h_{\mathrm{filter,LP}}(t) \exp\left[j2\pi f_c t\right]\right\}$, in order to assure that the output of the signal has the same normalization as the input signal. The used normalization is thus not very logical but is used here because it is the one that is most commonly used in the literature. Alternatively, some authors (e.g., [Barry et al. 2003]) define $s_{\mathrm{BP}}(t) = \sqrt{2}\,\mathrm{Re}\left\{s_{\mathrm{LP}}(t) \exp\left[j2\pi f_c t\right]\right\}$.

[2] Some authors use the phases $i\pi/2$. This is purely for convenience and does not impact the overall interpretation.

[3] Note that the energy contained in one (passband) symbol, according to (10.1), is $T_s A^2/2$. For the baseband, the energy contained in the I-component is $T_s A^2/2$, and of the Q-component is similarly $T_s A^2/2$, so that – by definition – the energy in the equivalent complex baseband representation is twice that in passband. This again relates back to the normalization mentioned in footnote 1.

10.1.4 General Aspects of Bandwidth

The bandwidth of the transmitted signal is a very important characteristic of a modulation format, especially in the context of a wireless signal, where spectrum is at a premium. Before going into a deeper discussion, however, we have to clarify what we mean by "bandwidth," since various definitions are possible.

- The *noise bandwidth* is defined as the bandwidth of a system with rectangular transfer function $|H_{\text{rect}}(f)|$ that receives as much noise as the system under consideration.
- The *3dB bandwidth* is the bandwidth at which $|H(f)|^2$ has decreased to a value that is 3 dB below its maximum value.
- The 90% *energy bandwidth* is the bandwidth that contains 90% of the total emitted energy; analogous definitions are possible for the 99% energy bandwidth or other percentages of the contained energy.

The *bandwidth efficiency* (spectral efficiency) is defined as the ratio of the data (bit) rate to the occupied bandwidth and is usually expressed in bit/(s Hz).

10.1.5 Signal-Space Diagram

One of the most important tools in the analysis of modulation formats is the signal-space diagram. It leads to a graphical representation of the signals that allows an intuitive and unified treatment of different modulation methods. In Chapter 11, we will explain in more detail how the signal-space diagram can be used to compute bit error probabilities. For now, we mainly use it as a convenient shorthand, and a method for representing different modulation formats.

To simplify explanations, we assume in the following that the modulation signal $s(t)$ that is present during the ith interval $iT_S < t < (i+1)T_S$ is only impacted by the ith group of bits, so that there is no memory in the system (in the following, we assume $i = 0$ without loss of generality).[4] We furthermore assume that the possible modulation symbols during that time interval form a finite set of functions; the size of this set is called the size of the modulation alphabet and denoted as M. This representation covers both Pulse Amplitude Modulation (PAM) and multi-pulse modulation, which will be discussed later in this chapter.

We now choose an orthonormal set (of size N) of *expansion functions* $\varphi_n(t)$, which are defined in the range $[0, T_S]$.[5] The set of expansion functions has to be complete, in the sense that all transmit signals $s_{\text{BP},i}(t)$ or $s_{\text{LP},i}(t)$ can be represented as linear combinations of the expansion functions. Such a complete set of expansion functions can be obtained by a *Gram-Schmidt orthogonalization* procedure (see, e.g., [Wozencraft and Jacobs 1965]).

Given the set of expansion functions $\{\varphi_n(t)\}$, any transmit signal $s_m(t)$ can be represented as a vector $\mathbf{s}_m = (s_{m,1}, s_{m,2}, ...s_{m,N})$ where

$$s_{m,n} = \int_0^{T_S} s_m(t)\varphi_n^*(t)dt \tag{10.7}$$

where superscript * denotes complex conjugation. When starting with the *bandpass signal*, both the modulated signal and the expansion functions are real, so that the vector components can be simply computed as

$$s_{\text{BP},m,n} = \int_0^{T_S} s_{\text{BP},m}(t)\varphi_{\text{BP},n}(t)dt. \tag{10.8}$$

Conversely, the actual transmit signal can be obtained from the vector components (both for passband and baseband) as

$$s_m(t) = \sum_{n=1}^{N} s_{m,n}\varphi_n(t). \tag{10.9}$$

As each signal is represented by a vector $\mathbf{s}_m$, we can plot those vectors in the so-called *signal-space diagram*.[6] A graphical representation of those points is especially simple for $N = 2$, which fortunately covers most important modulation formats.

An especially important (and simple) formulation of the signal space diagram occurs for signals that can be written in the form (10.6). In that case, the expansion functions in the *complex baseband representation* are simply normalized rectangles in time, multiplied with 1 (representing the real axis) and j (representing the imaginary axis)

---

[4] A somewhat more general representation can be achieved if we assume that time-shifted versions of the basis pulse are orthogonal to each other, i.e., $\int g(t + iT_s)g^*(t + kT_s) = \delta_{ik}$; note that the rectangular basis pulse is just one special case of that assumption. Then the range of the integrations below has to extend over the whole support (time range at which the pulse is non-zero) of the basis pulse, though implicitly the signal of interest is the one created by the modulation symbol at time $i = 0$.

[5] The expansion functions are often called "basis functions." However, we avoid that name in order to avoid confusion with the "basis pulses." It is noteworthy that the expansion functions are *by definition* orthogonal, while the basis pulses can be orthogonal but are not necessarily so.

[6] More precisely, the endpoints of the vectors starting in the origin of the coordinate system.

$$\varphi_1(t) = \begin{cases} \sqrt{\dfrac{1}{T_S}} \cdot 1 & \text{for } 0 \le t < T_s \\ 0 & \text{otherwise} \end{cases} \tag{10.10}$$

$$\varphi_2(t) = \begin{cases} \sqrt{\dfrac{1}{T_S}} \cdot j & \text{for } 0 \le t < T_s \\ 0 & \text{otherwise} \end{cases}.$$

This implies that $s_{m,1}$ and $s_{m,2}$ are simply the real and imaginary part of c.[7] Consequently, plotting the modulation symbol c "in the complex plane," and "plotting the signal space representation" become the same thing. Note, however, that this does *not* hold for other modulation formats like FSK that we will discuss later on; suitable sets of expansion functions can be established either by educated guess (as we will do in this chapter) or by systematic orthogonalization procedures like Gram-Schmidt.

For the passband representation, the expansion functions for such Quadrature Amplitude Modulation (QAM) are

$$\varphi_{BP,1}(t) = \begin{cases} \sqrt{\dfrac{2}{T_S}} \cos\left(2\pi f_c t\right) & \text{for } 0 \le t < T_s \\ 0 & \text{otherwise} \end{cases} \tag{10.11}$$

$$\varphi_{BP,2}(t) = \begin{cases} \sqrt{\dfrac{2}{T_S}} \sin\left(2\pi f_c t\right) & \text{for } 0 \le t < T_s \\ 0 & \text{otherwise} \end{cases}.$$

Here it is assumed that $f_c \gg 1/T_S$; this implies that all products containing $\cos(2 \cdot 2\pi f_c t)$ are negligible and/or eliminated by a filter.

It is important that the signal-space diagrams that result from bandpass and lowpass representation differ by a factor $\sqrt{2}$ – again this is a consequence of the normalization in (10.3). In the following, we will employ the signal constellations resulting from the bandpass signals, so that for a binary antipodal signal, the signal is, e.g.,

$$s_{BP,\begin{smallmatrix}1\\2\end{smallmatrix}} = \pm \sqrt{\dfrac{2E_S}{T_S}} \cos\left(2\pi f_c t\right) \tag{10.12}$$

and the points in the signal-space diagram are located at $\pm \sqrt{E_S}$, with E_S being the symbol energy.

The *energy* contained in a symbol[8] can be computed from the bandpass signal as

$$E_{S,m} = \int_0^{T_S} s_{BP,m}^2(t)dt = \|\mathbf{s}_{BP,m}\|^2 \tag{10.13}$$

where $\|\mathbf{s}\|$ denotes the L_2- norm (Euclidian norm) of $\mathbf{s}$. When considering equivalent baseband, the signal energy is given as

$$E_{S,m} = \frac{1}{2}\int_0^{T_S} |s_{LP,m}(t)|^2 dt \simeq \frac{1}{2}\|\mathbf{s}_{LP,m}\|^2. \tag{10.14}$$

and the $\mathbf{s}_{LP,m}$ are the signal vectors obtained from the equivalent baseband representation. Note that for many modulation formats (e.g., BPSK), E_S is independent of m.

The *correlation coefficient* between $s_m(t)$ and $s_k(t)$ can be computed as a normalized inner product of signal vectors

$$\operatorname{Re}\left\{\rho_{k,m}\right\} = \frac{\langle \mathbf{s}_{BP,m}, \mathbf{s}_{BP,k}\rangle}{\left\|\mathbf{s}_{BP,m}\right\|\left\|\mathbf{s}_{BP,k}\right\|} \tag{10.15}$$

or (note that $\langle \mathbf{x},\mathbf{y}\rangle = x_1 y_1^* + x_2 y_2^* \dots$)

$$\rho_{k,m} = \frac{\langle \mathbf{s}_{LP,m}, \mathbf{s}_{LP,k}\rangle}{\| \mathbf{s}_{LP,m} \|\| \mathbf{s}_{LP,k} \|}. \tag{10.16}$$

---

[7] Note that the use of *1* and *j* as distinct expansion functions is meaningful only when the expansion coefficients are constrained to be real.
[8] again applying the caveats from footnote 4.

The *Euclidean distance* between the two signals is

$$d_{k,m}^2 = \left[E_{S,k} + E_{S,m} - 2\sqrt{E_{S,k}E_{S,m}} \operatorname{Re}\left\{ \rho_{k,m} \right\} \right]. \tag{10.17}$$

We will see in Chapter 11 that this distance is important for the computation of the bit error probability.

10.2 Pulse Amplitude Modulation

10.2.1 General Formulation

Many modulation formats can be interpreted as PAM formats, where a basis pulse $g(t)$ is multiplied with a modulation coefficient c_i

$$s_{LP}(t) = \sum_{i=-\infty}^{\infty} c_i g(t - iT_S). \tag{10.18}$$

Different PAM formats differ in how the data bits b_i are mapped onto the modulation coefficients c_i. It is noteworthy that those modulation coefficients can be complex, describing the joint modulation of In-phase and quadrature signal (in complex baseband), or – equivalently – of cosine carrier and sine carrier (in passband).

Important subcategories are

- *Quadrature Amplitude Modulation* (QAM): this name is used sometimes for two slightly different things: (i) it may be an alternative name for PAM with general complex c_i, or (ii) it refers to a modulation where the c lie on a rectangular grid in the complex plan, i.e., the complex modulation symbols are $(p\Delta_x + \Omega_x) + j(q\Delta_y + \Omega_y)$, $p = 1, ...P$, $q = 1, ...Q$, $PQ = M$. Examples are QPSK (Section 10.3.2) and 16-QAM (Section 10.3.3)
- *Phase Shift Keying* (PSK): here $|c_i| = const$, and the bits are mapped onto the phase of c_i. Examples are QPSK (Section 10.3.2) and 8-PSK (Section 10.3.4). Note that QPSK can be considered both as PSK and QAM.
- *Amplitude Shift Keying* (ASK): here only the I (or only the Q) signal are modulated, and all information is in $|c_i|$.

10.2.2 Spectrum

General Equation for PAM Spectrum

A cyclostationary process $x(t)$ is defined as a stochastic process whose mean and autocorrelation functions are periodic with period T_{per}

$$E\{x(t + T_{per})\} = E\{x(t)\} \tag{10.19}$$

$$R_{xx}(t + T_{per} + \tau, t + T_{per}) = R_{xx}(t + \tau, t).$$

These properties are fulfilled by most modulation formats; the periodicity T_{per} is the symbol duration T_S. We furthermore assume that the data symbols are zero-mean and uncorrelated, so that the spectral density of the data symbols is constant over frequency. Those properties are fulfilled by most PAM signals; either if they are uncoded, or coded and scrambled.[9] The power spectral density of a modulated signal can then be computed as the product of the energy spectrum of a basis pulse $|S_G(f)|^2$ and the power spectral density of the data σ_S^2 per time interval

$$S_{LP}(f) = \frac{1}{T_{per}} \cdot |S_G(f)|^2 \cdot \sigma_S^2. \tag{10.20}$$

The energy spectrum[10] of the basis pulse is simply the squared magnitude of the Fourier transform $G(f)$ of the basis pulse $g(t)$

$$|S_G(f)|^2 = |G(f)|^2. \tag{10.21}$$

---

[9] *Scrambling* refers to modulo-2 adding a bit sequence with a pseudo-random, known, sequence of 1 and 0, which does not show any correlations between subsequent bits. Consequently, the output of the scrambling operation is again a sequence of uncorrelated 1 and 0.

[10] Note that a single pulse has a finite energy, but zero average power. An (infinitely long) signal consisting of a convolution of a data sequence with the basis pulse has finite average power. However, sometimes – and in abuse of notation – the squared magnitude of the Fourier transform of a single pulse is also called a power spectral density.

The power spectral density in passband is

$$S_{\mathrm{BP}}(f) = \frac{1}{2}[S_{\mathrm{LP}}(f - f_{\mathrm{c}}) + S_{\mathrm{LP}}(-f - f_{\mathrm{c}})].\tag{10.22}$$

Rectangular Basis Pulses

Let us turn to the possible shapes of the basis pulse $g(t)$ and the impact on the spectrum. We will assume that the basis pulses are normalized to unit average power, so that

$$\int_{\infty}^{\infty} |g(t)|^2\, dt = \int_{-\infty}^{\infty} |G(f)|^2\, df = T.\tag{10.23}$$

The simplest basis pulse is a rectangular pulse with duration T. Figure 10.1 shows the pulse as function of time; Figure 10.2 the magnitude of the corresponding spectrum (this is a spectrum in the sense of the Fourier transform of the signal. The *energy spectrum* is its squared magnitude)

$$g_{\mathrm{R}}(t, T) = \begin{cases} 1... & \text{for} \quad 0 \le t \le T \\ 0...\text{otherwise} \end{cases}\tag{10.24}$$

$$G_{\mathrm{R}}(f, T) = \mathcal{F}\{g_{\mathrm{R}}(t, T)\} = T\ \operatorname{sinc}(\pi f T) \exp\left(-j\pi f T\right)$$

where $\operatorname{sinc}(x) = \sin(x)/x$.[11]

Nyquist Pulses and Spectrum Shaping

The rectangular pulse has a spectrum that extends over a large bandwidth; the first sidelobes are only 13 dB weaker than the maximum. This leads to large adjacent channel interference, which in turn decreases the spectral efficiency of a cellular system. It is thus often required to obtain a spectrum that shows a stronger roll-off in the frequency domain. The most common class of pulses with strong spectral roll-offs are Nyquist pulses, i.e., pulses that fulfill the Nyquist criterion, and thus do not lead to Intersymbol Interference (ISI).[12]

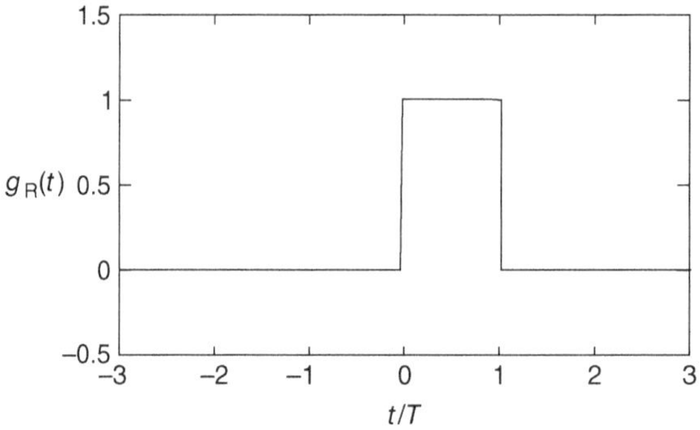

Figure 10.1   Rectangular basis pulse.

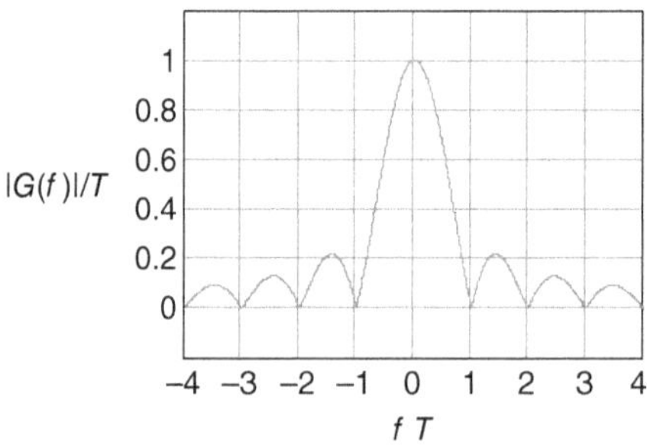

Figure 10.2   Spectrum of a rectangular basis pulse.

---

[11] Note that some books define $\operatorname{sinc}(x) = \sin(\pi x)/(\pi x)$.

[12] More precisely, Nyquist pulses do not create ISI assuming that the channel does not create ISI, and sampling is done at the ideal sampling times.

As an example, we present equations for a raised-cosine pulse (see Figure 10.3). This filter has a roll-off that follows a sinusoidal shape; the parameter determining the steepness of the spectral decay is the roll-off factor α. Defining the functions

$$G_{N0}(f,\alpha,T) = \begin{cases} 1 & 0 \le \mid 2\pi f \mid \le (1-\alpha)\dfrac{\pi}{T} \\[2mm] \dfrac{1}{2}\cdot\left(1-\sin\left(\dfrac{T}{2\alpha}\left(\mid 2\pi f\mid -\dfrac{\pi}{T}\right)\right)\right) & (1-\alpha)\dfrac{\pi}{T} \le \mid 2\pi f\mid \le (1+\alpha)\dfrac{\pi}{T} \\[2mm] 0 & (1+\alpha)\dfrac{\pi}{T} \le \mid 2\pi f\mid \end{cases} \tag{10.25}$$

the spectrum of the raised-cosine pulse is

$$G_{N}(f,\alpha,T) = \frac{1}{\sqrt{1-\frac{\alpha}{4}}}\cdot T\cdot G_{N0}(f,\alpha,T)\exp\left(-j\pi f T_{S}\right) \tag{10.26}$$

where the normalization factors $\dfrac{1}{\sqrt{1-\alpha/4}}$ follow from Eq. (10.23).

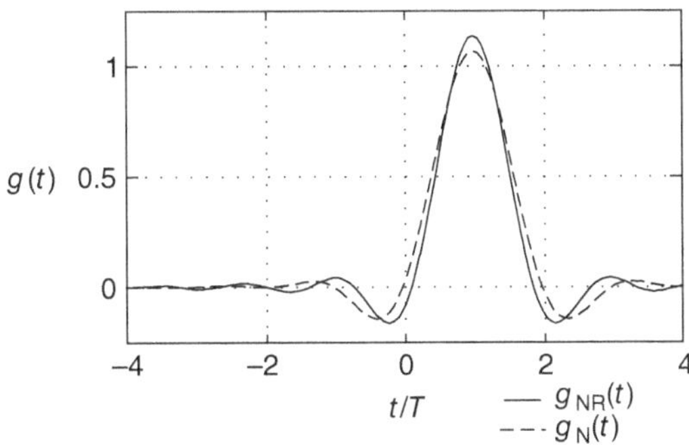

Figure 10.3 Raised-cosine pulse and root raised cosine pulse.

As we will see in Section 11.1.2, an optimum RX performs matched filtering, i.e., sends the received signal through a filter whose impulse response is $g^{*}(T-t)$, i.e., the complex conjugate of the time-reversed impulse response of the basis pulse. Thus, the magnitude of the Fourier transform of the output of the filter is $|G(f)|^{2}$. If we thus want the *concatenation* of basis pulse and RX filter to result in a raised-cosine shape, the spectrum of both the basis pulse and the RX filter should be the square root of a raised cosine filter. Such a filter is known as root-raised-cosine filter and henceforth denoted by subscript *NR*.

The impulse responses for raised cosine and root-raised cosine pulses follow from the inverse Fourier transformation (see Figure 10.4); closed-form equations can be found in [Chennakeshu and Saulnier 1993].

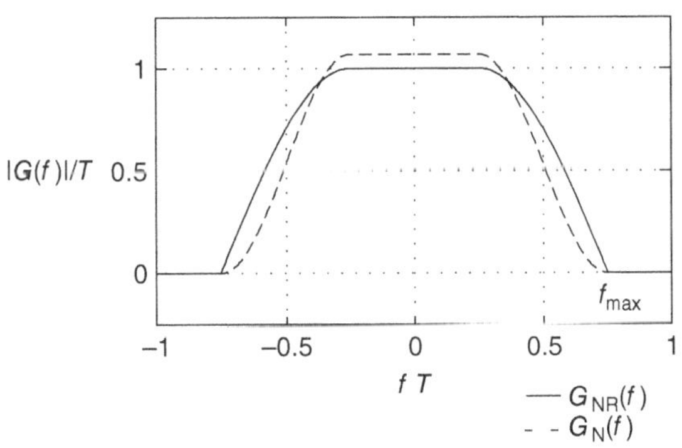

Figure 10.4 Spectrum of raised-cosine pulse G_{N} and root raised-cosine pulse G_{NR}.

Adjacent Channel Interference

An important reason for performing pulse shaping is the improvement of the spectral efficiency of Frequency-Division Multiple Access (FDMA, see also Section 18.3.1). In this scheme, signals of different users are sent on different "frequency channels," i.e., nonoverlapping frequency bands. In order to avoid inter-user interference, the band assigned to a user must be wide enough to contain the largest part of the signal spectrum of this user, such that the spectrum does not spill over into the neighboring band "by a significant amount." The steeper the decay of $|G(f)|$ is for a given T_{s}, the narrower the required bandwidth for one user (for a given symbol rate), and thus the better the spectral efficiency.

Example 10.1 *Compute the ratio of signal power to adjacent-channel interference when using (i) raised-cosine pulses and (ii) root-raised cosine pulses with $\alpha = 0.5$ when the two considered signals have center frequencies 0 and $1.25/T$.*

The two signals overlap slightly in the roll-off region and with $\alpha = 0.5$ the desired signal has spectral components in the band $-0.75/T \le f \le 0.75/T$. There is interfering energy between $0.5/T \le f \le 0.75/T$. Without loss of generality, we assume $T = 1$. In the case of raised cosine pulses and assuming a matched filter at the RX, signal energy is proportional to (dropping normalization factors occurring in both signal and interference)

$$
\begin{aligned}
S &= 2 \int_{f=0}^{0.75/T} |G_{\mathrm{N}}(f, \alpha, T)|^4 \, df \\
&= 2 \int_{f=0}^{0.25} |1|^4 \, df + 2 \int_{f=0.25}^{0.75} \left| \frac{1}{2} \cdot \left(1 - \sin \left(\frac{2\pi}{2\alpha} \left(|f| - \frac{1}{2} \right) \right) \right) \right|^4 \, df \\
&= 0.77.
\end{aligned}
\tag{10.27}
$$

while the interfering signal energy (assuming that there is an interferer at both the lower and the upper adjacent channel) is given by

$$
\begin{aligned}
I &= 2 \int_{f=0.5}^{f=0.75} |G_N(f, \alpha, T) G_{\mathrm{N}}(f - 1.25, \alpha, T)|^2 \, df \\
&= 2 \int_{f=0.5}^{f=0.75} \left| \frac{1}{2} \cdot \left(1 - \sin \left[\frac{2\pi}{2\alpha} \left(|f| - \frac{1}{2} \right) \right] \right) \frac{1}{2} \cdot \left(1 - \sin \left[\frac{2\pi}{2\alpha} \left(|f - 1.25| - \frac{1}{2} \right) \right] \right) \right|^2 \, df \\
&= 9.2 \cdot 10^{-5}.
\end{aligned}
\tag{10.28}
$$

so the signal to interference ratio is $\log_{10}\left(\frac{S}{I}\right) \approx 39\mathrm{dB}$. Using root-raised cosine pulses, the signal energy is given by

$$
S = 2 \int_{f=0}^{f=0.75/T} |G_{\mathrm{N}}(f, \alpha, T)|^2 \, df = 0.875
\tag{10.29}
$$

and the interfering energy is given by

$$
\begin{aligned}
I &= 2 \int_{f=0.5}^{f=0.75} |\, G_{\mathrm{N}}(f, \alpha, T) G_{\mathrm{N}}(f - 1.25, \alpha, T) \,| \, df \\
&= 2 \int_{f=0.5}^{f=0.75} \left\| \frac{1}{2} \cdot \left(1 - \sin \left[2\pi \left(|f| - \frac{1}{2} \right) \right] \right) \right\| \frac{1}{2} \cdot \left(1 - \sin \left[2\pi \left(|f - 1.25| - \frac{1}{2} \right) \right] \right) \right\| df \\
&= 5.6 \cdot 10^{-3}.
\end{aligned}
\tag{10.30}
$$

so the signal-to-interference ratio is $10 \log_{10}(S/I) \approx 22$ dB. Obviously, the root-raised cosine filter, which has a flatter decay in the frequency domain, leads to a worse signal-to-interference ratio.

10.3 Widely Used PAM Modulation Formats

In this and the following sections, we summarize in a very concise form the most important properties of different digital modulation formats. We will give the following information for each modulation format:

- Bandpass and baseband signal as a function of time.
- Signal-space diagram, as well as (in some cases) IQ diagrams, which show the temporal trajectory of the baseband signal in the complex plane.
- Spectral efficiency.

The current section concentrates on the most popular PAM formats; the subsequent section will deal with multi-pulse modulation.

10.3.1 Binary Phase-Shift Keying (BPSK)

Binary Phase-Shift Keying (BPSK) modulation is the simplest modulation method: the carrier phase is shifted by $\pm\pi/2$, depending on whether a $+1$ or -1 is sent.[13] Despite this simplicity, two different interpretations of BPSK are possible. The first one is to see BPSK as a *phase modulation*, in which the data stream influences the phase of the transmit signal. Depending on the data bit b_i, the phase of the transmitted signal is $\pi/2$ or $-\pi/2$.

The second, and more popular, interpretation is to view BPSK as a *PAM* where the basis pulses are rectangular pulses with amplitude 1, so that

$$s_{BP}(t) = \sqrt{2E_B/T_B}\, p_D(t)\, \cos\!\left(2\pi f_c t + \frac{\pi}{2}\right) \tag{10.31}$$

where

$$p_D(t) = \sum_{i=-\infty}^{\infty} b_i g(t - iT) = \sum_{i=-\infty}^{\infty} [b_i \delta(t - iT_s)] * g(t)) \tag{10.32}$$

where $*$ denotes convolution and we write the r.h.s. henceforth simply as $[b_i] * g(t)$, and

$$g(t) = g_R(t, T_B). \tag{10.33}$$

Figure 10.5 shows the signal waveform and Figure 10.6 the signal space diagram. In equivalent baseband, the complex modulation symbols are $\pm j$

$$c_i = j \cdot b_i \tag{10.34}$$

so that the real part of the signal is

$$\text{Re}\{s_{LP}(t)\} = 0 \tag{10.35}$$

and the imaginary part is

$$\text{Im}\{s_{LP}(t)\} = \sqrt{\frac{2E_B}{T_B}} p_D(t). \tag{10.36}$$

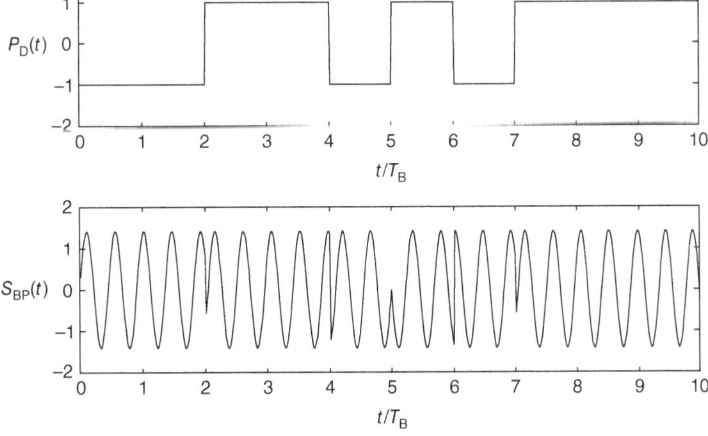

Figure 10.5 BPSK signal as function of time.

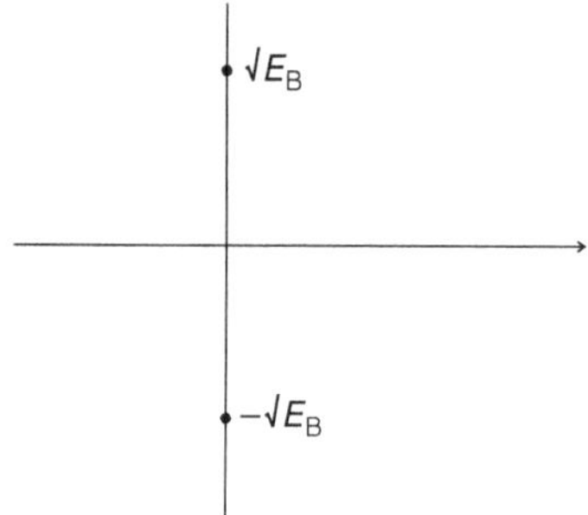

Figure 10.6 Signal-space diagram for BPSK.

The envelope has constant amplitude, except at times $t = iT_B$. The spectrum shows a very slow roll-off, due to the use of unfiltered rectangular pulses as basis pulses. The bandwidth efficiency is 0.59 bit/s/Hz when we consider the bandwidth that contains 90% of the energy but only 0.05 bit/s/Hz when considering the 99% energy bandwidth, see also Figure 10.7.

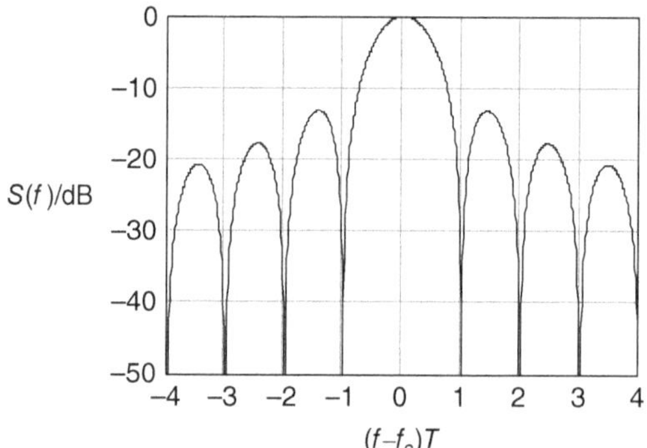

Figure 10.7 Power spectral density for BPSK.

Because of the low bandwidth efficiency of rectangular pulses, practical transmitters often use raised-cosine pulses as basis pulses.[14] Even the relatively mild filtering of $\alpha = 0.5$ leads to a dramatic increase in spectral efficiency: for 90% and 99% energy bandwidth, the spectral efficiency becomes 1.02 and 0.79 bit/s/Hz, respectively. On the other hand, we see that the signal does not have a constant envelope anymore, see Figures 10.8 and 10.9.

10.3.2 Quadrature Phase-Shift Keying (QPSK)

A Quadrature Phase-Shift Keying (QPSK)-modulated signal carries one bit each on the in-phase and quadrature-phase component. The original data stream is split into two streams, $b1_i$ and $b2_i$

$$b1_i = b_{2i} \tag{10.37}$$

$$b2_i = b_{2i+1}$$

each of which has a data rate that is half that of the original data stream

$$R_S = 1/T_S = R_B/2 = 1/(2T_B). \tag{10.38}$$

Let us first consider the situation where the basis pulses are rectangular pulses, $g(t) = g_R(t, T_S)$. Then we can give an interpretation of QPSK as either a phase modulation or as a PAM. We first define two sequences of pulses (see Figure 10.10)

---

[14] Note that there is a difference between a PAM with Nyquist-shaped basis pulses, and a phase modulation with Nyquist-shaped phase pulses. In the following, we will consider only PAM with Nyquist basis pulses, and call it BAM (*binary amplitude modulation*) for clarity.

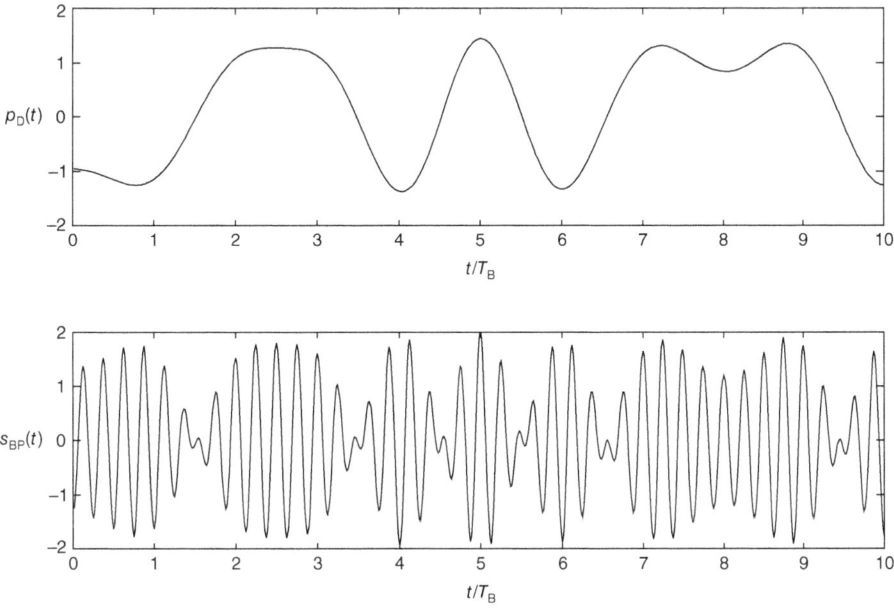

Figure 10.8 BAM signal (with roll-off factor $\alpha = 0.5$) as a function of time.

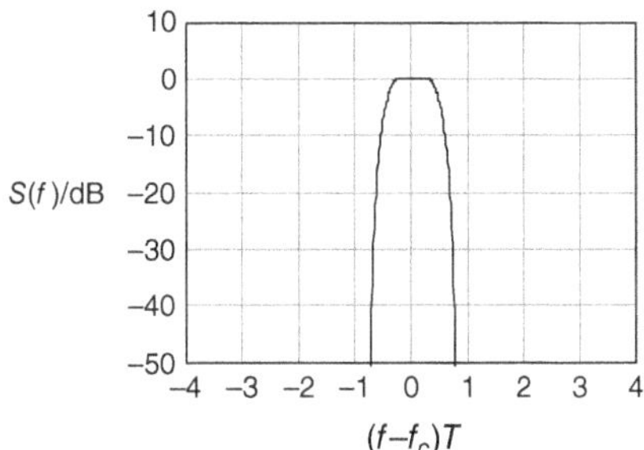

Figure 10.9 Power spectral density of a BAM signal (with roll-off factor $\alpha = 0.5$).

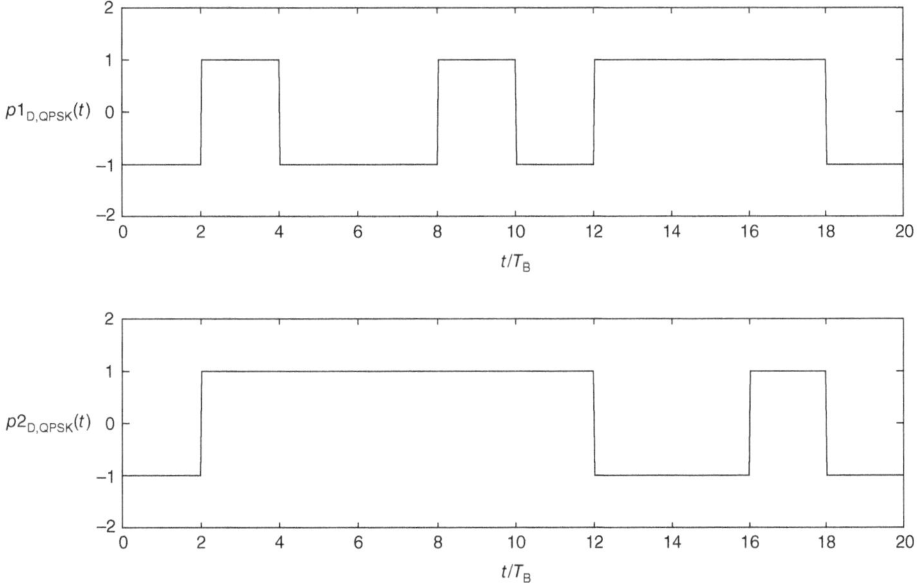

Figure 10.10 Data streams of in-phase and quadrature-phase components in QPSK.

$$p1_D(t) = \sum_{i=-\infty}^{\infty} b1_i g(t - iT_S) = b1_i * g(t)$$

$$p2_D(t) = \sum_{i=-\infty}^{\infty} b2_i g(t - iT_S) = b2_i * g(t).$$

(10.39)

When interpreting QPSK as a *pulse-amplitude modulation*, the bandpass signal reads

$$s_{BP}(t) = \sqrt{E_B/T_B}[p1_D(t)\cos(2\pi f_c t) - p2_D(t)\sin(2\pi f_c t)].$$

(10.40)

Normalization is done in such a way that the energy within one symbol interval is $\int_0^{T_S} s_{BP}(t)^2 dt = 2E_B$, where E_B is the energy expended on the transmission of a bit. Figure 10.11 shows the signal space diagram. The baseband signal is

$$s_{LP}(t) = [p1_D(t) + jp2_D(t)]\sqrt{E_B/T_B}.$$

(10.41)

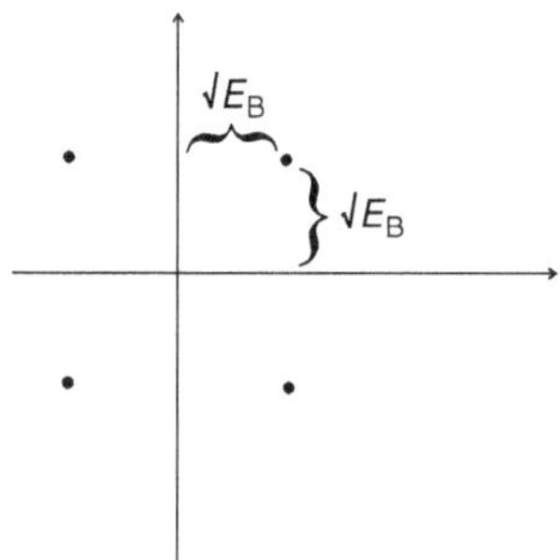

Figure 10.11 Signal space diagram of QAM.

When interpreting QPSK as a *phase modulation*, the lowpass signal can be written as

$$\sqrt{2E_B/T_B}\exp(j\Phi_S(t))$$

(10.42)

with

$$\Phi_S(t) = \pi\cdot\left[\frac{1}{2}\cdot p2_D(t) - \frac{1}{4}\cdot p1_D(t)\cdot p2_D(t)\right].$$

(10.43)

It is obvious from this representation that the signal is constant-envelope, except for the transitions at $t = iT_S$, see Figures 10.12.

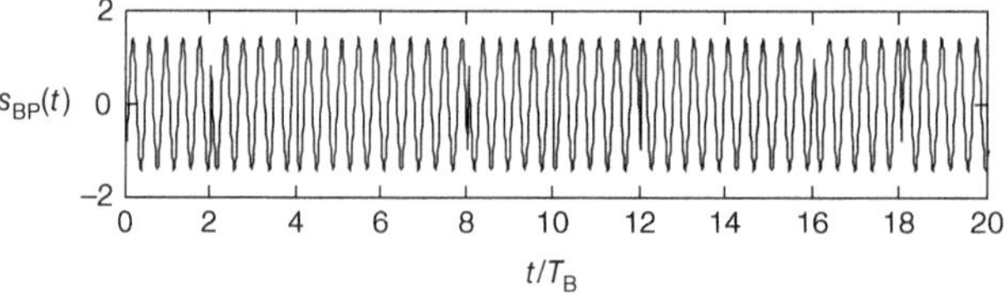

Figure 10.12 QPSK signal as a function of time.

The spectral efficiency of QPSK is twice the efficiency of BPSK, since both the in-phase and the quadrature-phase components are exploited for the transmission of information. That means that when considering the 90% energy bandwidth, the efficiency is 1.1 bits/s/Hz, while for the 99%-energy bandwidth, it is 0.1 bit/s/Hz, see Figure 10.13. The slow spectral roll-off motivates (similarly to BPSK) the use of raised-cosine basis pulses, see Figure 10.14; we will refer to the resulting modulation format as 4-QAM in the following. The spectral efficiency increases to 2.04 and 1.58 bit/s/Hz, respectively, see Figure 10.15. On the other hand, the signal shows strong envelope fluctuations, Figure 10.16.

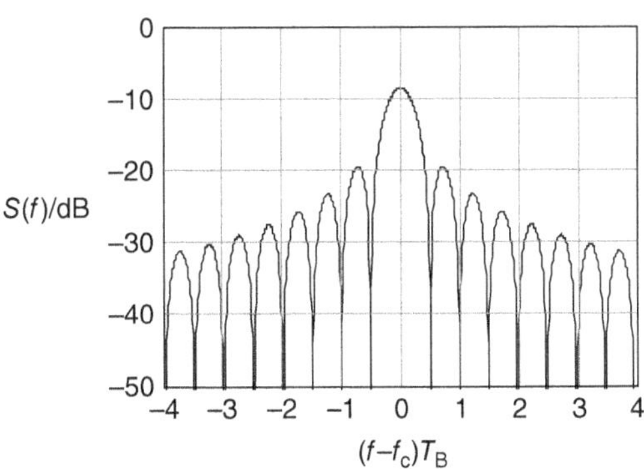

Figure 10.13 Power spectral density of QPSK.

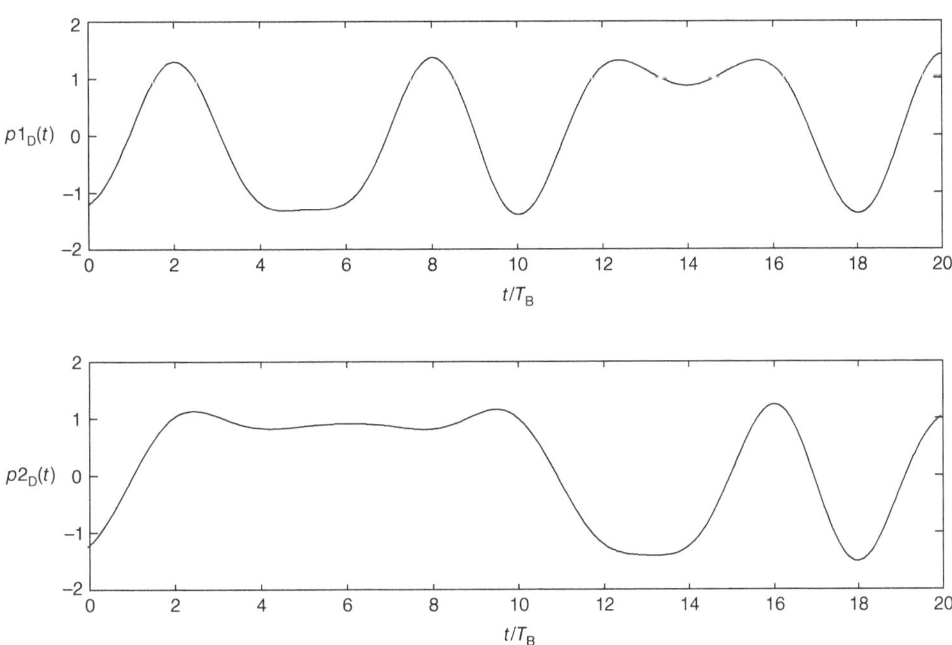

Figure 10.14 4-QAM pulse sequence with raised cosine filter.

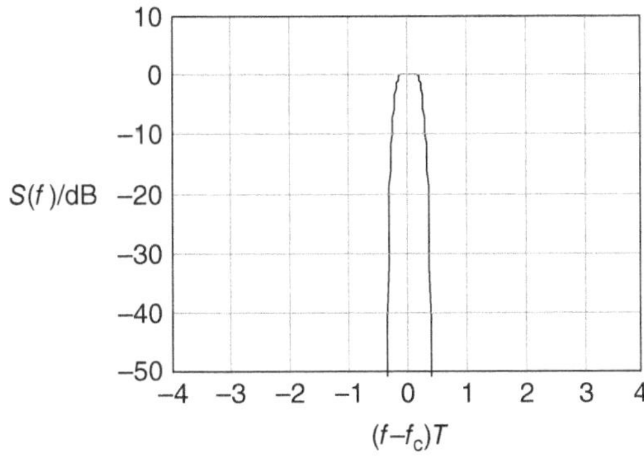

Figure 10.15 Power spectral density of 4-QAM with raised-cosine filters with $\alpha = 0.5$.

10.3.3 Higher-Order QAM

Up to now, we have treated modulation formats that transmit at most two bits per symbol. We next explore how those schemes can be generalized to transmit more information in each symbol interval. Such schemes result in a higher spectral efficiency, but consequently also in a higher sensitivity to noise and interference. They are therefore used for situations with high Signal-to-Interference-and-Noise Ratio (SINR), typically when a User Equipment (UE) is close to a Base Station (BS).

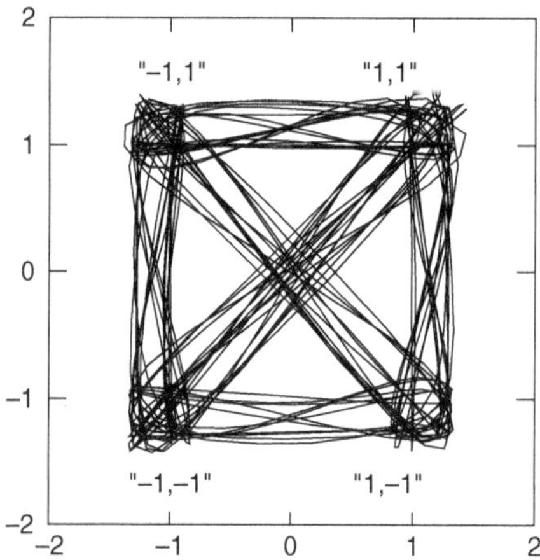

Figure 10.16 IQ diagram of 4-QAM with raised-cosine basis pulses. Also shown are the four points of the signal space diagram, $(1,1), (1,-1), (-1,-1), (-1,1)$.

4-QAM (also known simply as QAM), or QPSK, transmits one bit each on the in-phase and the quadrature-phase component. It does so by sending a signal with a positive or negative polarity, but fixed amplitude, on each component. This scheme can be generalized by allowing multiple amplitude levels. The resulting scheme is called *higher-order QAM*. The mathematical representation is the same as for 4-QAM, just that in the pulse sequences of Eq. (10.40) we not only allow the levels ± 1, but $2m - 1 - \sqrt{M}$, with $m = 1, \ldots \sqrt{M}$.

The signal-space diagram for 16-QAM is shown in Figure 10.17. Larger constellations, including 64 QAM and 256 QAM, can be constructed according to similar principles. Naturally, the peak-to-average ratio of the output signal becomes the larger, the larger the constellation is.

Example 10.2 *Relate the average energy of a 16-QAM signal to the distance between two adjacent points in the signal-space diagram.*

Let us for simplicity consider the first quadrant of the signal-space diagram. Assume that the signal points are located at $d + jd$, $d + j3d$, $3d + jd$, and $3d + j3d$. The average energy is given by $E_s = \frac{d^2}{4}(2 + 10 + 10 + 18) = 10d^2$. The distance between two neighboring points is $2d$, so the ratio is $\frac{10d^2}{2d} = 5d$. Similarly, the squared distance between two points is $4d^2$, so the ratio then becomes 2.5.

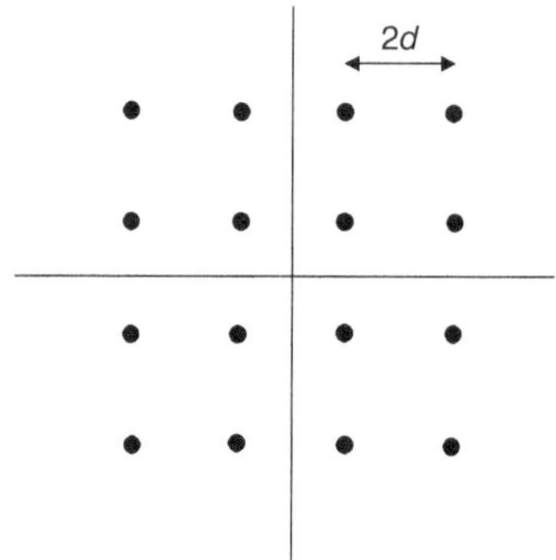

Figure 10.17 Signal-space diagram for 16-QAM.

10.3.4 Higher-Order Phase Modulation

The drawback of the higher-order QAM is the fact that the resulting signals show strong variations of the output amplitude. An alternative is the use of higher-order PSK, where the transmit signal can be written as

$$s_{\text{BP}}(t) = \sum_i \sqrt{2E_S/T_S} \cos\left(2\pi f_c t + \phi_i\right) g(t - iT_s) \qquad \phi_i \in 0, \frac{2\pi}{M}, \ldots \frac{2\pi}{M}(M - 1) \tag{10.44}$$

i.e., the TX picks one of the M transmit phases (see Figure 10.18); note that we normalize the energy here in terms of the *symbol* energy and the *symbol* duration. The equivalent low-pass signal is

$$s_{\mathrm{LP}}(t) = \sum_i \sqrt{2E_\mathrm{S}/T_\mathrm{S}} \, \exp\left(j\phi_i\right) g(t - iT_\mathrm{s}). \tag{10.45}$$

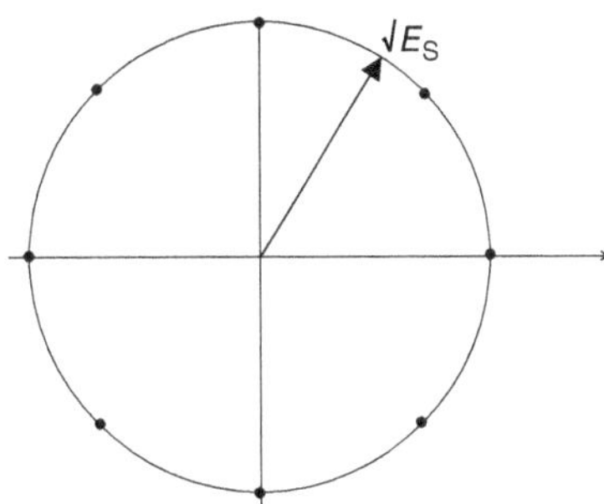

Figure 10.18 IQ diagram of 8-PSK. Also shown are the eight points of the signal space diagram.

The correlation coefficient between two signals is

$$\rho_{km} = \exp\left(j\frac{2\pi}{M}(m-k)\right). \tag{10.46}$$

10.3.5 Differential BPSK

An important variant of BPSK is *differentially encoded PSK (DPSK)*. The basic idea is that the current bit is encoded as the phase difference between the previous and the current symbol. In other words, the transmitted phase is not solely determined by the current symbol; rather, we transmit the phase of the previous symbol added to a phase change that corresponds to the current symbol. For BPSK, this reduces to a particularly simple form: we first encode the data bits (in 0/1 format) according to

$$\tilde{b}_i = b_i \oplus \tilde{b}_{i-1} \tag{10.47}$$

where $\oplus$ denotes modulo 2 addition, and then (after mapping to $-1/1$ format) use $\tilde{b}_i$ instead of b_i in Eq. (10.32).

The advantage of differential encoding is that it enables a differential decoder, which only needs to compare the phases of two subsequent symbols in order to demodulate the received signals. This obviates the need to recover the absolute phase of the received signal, and thus allows to build simpler and cheaper RXs, see Sec. 11.1.

*10.3.6 $\pi/4$-Differential QPSK

Even though QPSK is nominally a constant-envelope format, it has amplitude dips at the bit transitions; this can also be seen by the fact that the trajectories in the signal-space diagram pass through the origin for some of the bit transitions. The duration of the dips is longer when nonrectangular basis pulses are used. Such variations of the signal envelope are undesirable, because they make the design of suitable amplifiers more difficult. One possibility for reducing these problems lies in the use of $\pi/4$-shifted QPSK and its variant ($\pi/4$ Differential Quadrature Phase-Shift Keying ($\pi/4$-DQPSK)). This modulation format was especially popular in the 1990s – it was used in several American standards (IS-54, IS-136, and PWT), as well as the Japanese cellphone (JDC) and cordless (PHS) standards, and the European trunk radio standard (TETRA), and still has some applications today, e.g., in Bluetooth (see Section 34.1.3).

The principle of $\pi/4$-QPSK can be understood from the signal-space diagram, see Figure 10.19. There exist *two* sets of signal constellations: (0°, 90°, 180°, 270°) and (45°, 135°, 225°, 315°). All symbols with an even temporal index i are chosen from the first set, while all symbols with odd index are chosen from the second one. In other words: whenever t is an integer multiple of the symbol duration, the transmit phase is increased by $\pi/4$, in addition to the change of phase due to the transmit symbol. Therefore, transitions between subsequent signal constellation points can never pass through the origin (see Figure 10.22); in physical terms, this means much smaller fluctuations of the envelope. For $\pi/4$ DQPSK, the symbol phase is additionally differentially encoded similar to DBPSK.

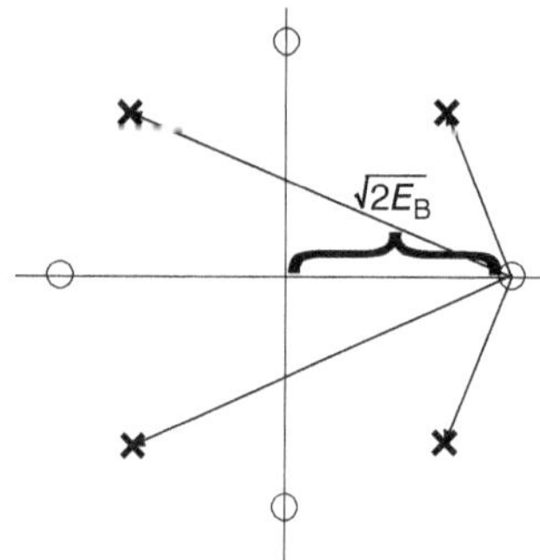

Figure 10.19 Allowed transitions in the signal-space diagram of $\pi/4$-DQPSK.

The signal phase is given by

$$\Phi_{\mathrm{s}}(t) = \pi \left[\frac{1}{2} p2_{\mathrm{D}}(t) - \frac{1}{4} p1_{\mathrm{D}}(t)p2_{\mathrm{D}}(t) + \frac{1}{4} \left\lfloor \frac{t}{T_{\mathrm{S}}} \right\rfloor \right] \tag{10.48}$$

where $\lfloor x \rfloor$ denotes the largest integer smaller or equal to x. Comparing this to Eq. (10.43), we can clearly see the change in phase at each integer multiple of T_{S}. Figure 10.20 shows the underlying data sequences, and Figure 10.21 depicts the resulting bandpass signals when using rectangular or raised cosine basis pulses.

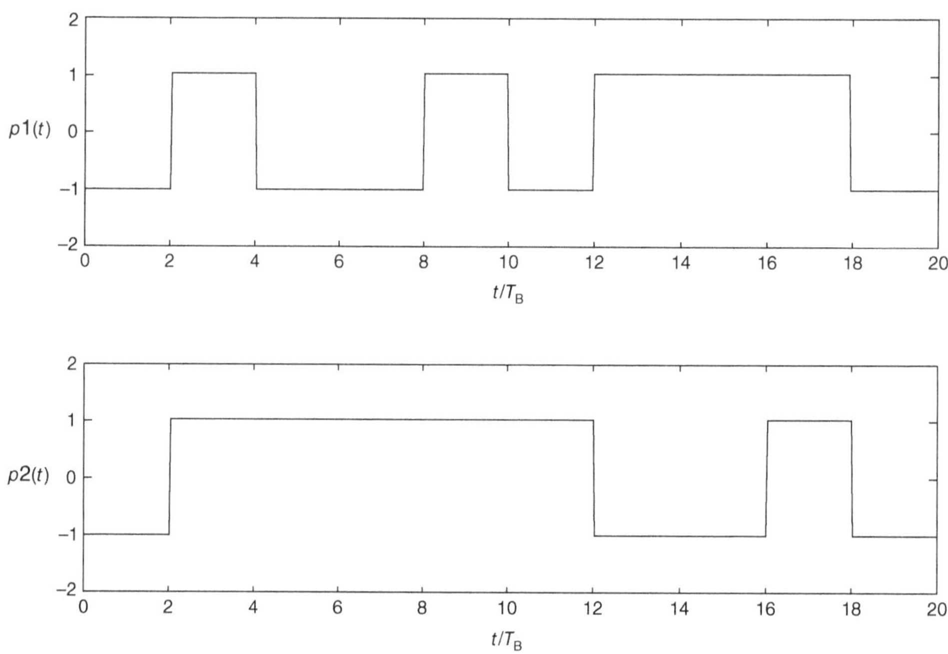

Figure 10.20 Sequence of basis pulses for $\pi/4$-DQPSK.

*10.3.7 Offset-QPSK

Another way of improving the peak-to-average ratio in QPSK is to make sure that the bit transitions for the in-phase and the quadrature-phase components occur at different time instants. This method is called OQPSK (offset QPSK). The bitstreams modulating the in-phase and quadrature-phase components are offset half a symbol duration with respect to each other (see Figure 10.23), so that the transitions for the in-phase component occur at integer multiples of the symbol duration (even integer multiple of the bit duration), while the quadrature-component transitions occur half a symbol duration (one-bit duration) later. Thus, the transmit pulse streams are

$$p1_{\mathrm{D}}(t) = \sum_{i=-\infty}^{\infty} b1_i g(t - iT_{\mathrm{S}}) = b1_i * g(t)$$

$$p2_{\mathrm{D}}(t) = \sum_{i=-\infty}^{\infty} b2_i g\left(t - \left(i + \frac{1}{2}\right)T_{\mathrm{S}}\right) = b2_i * g\left(t - \frac{T_{\mathrm{S}}}{2}\right). \tag{10.49}$$

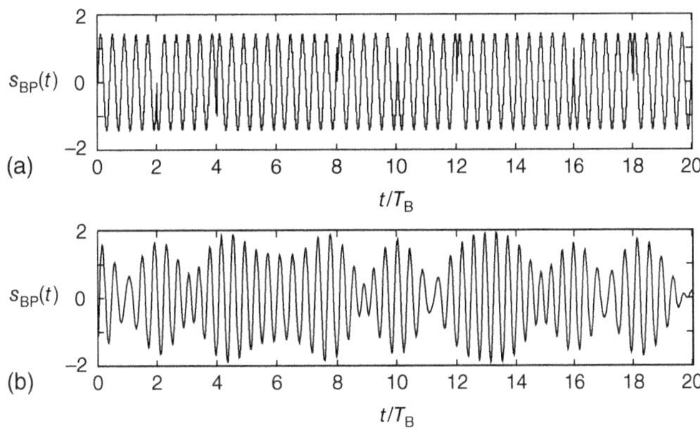

Figure 10.21 $\pi/4$-DQPSK signals as function of time for rectangular basis pulses (a) and raised cosine basis pulses (b).

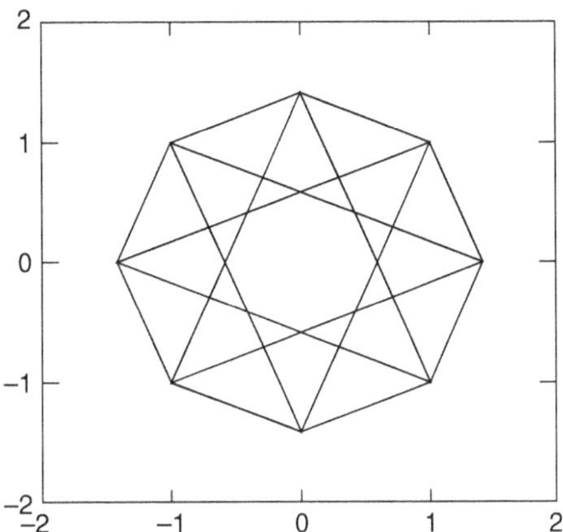

Figure 10.22 IQ diagram of a $\pi/4$-DQPSK signal with rectangular basis pulses.

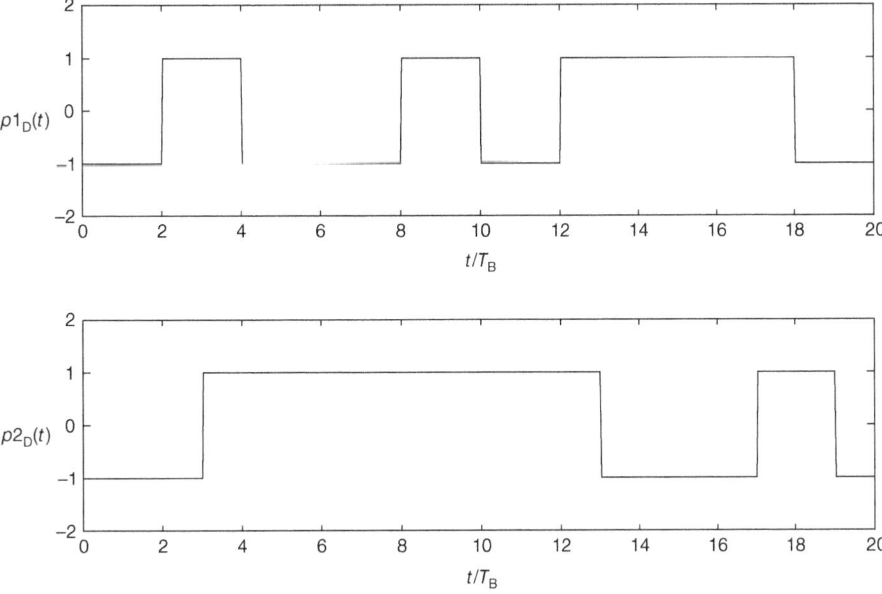

Figure 10.23 Sequence of basis pulses for OQPSK

These data streams can again be used for an interpretation as PAM, Eq. (10.42), or as phase modulation, according to Eq. (10.43). The resulting bandpass signal is shown in Figure 10.24.

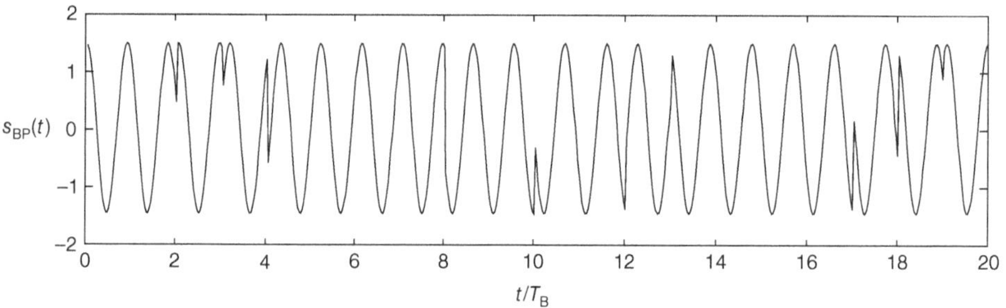

Figure 10.24 OQPSK signal as function of time.

The representation in the signal-space diagram Figure 10.25 makes clear that there are no transitions passing through the origin of the coordinate system; thus also this modulation format takes care of the envelope fluctuations. As for regular QPSK, we can use smoother basis pulses, like raised-cosine pulses, to improve the spectral efficiency. Figure 10.26 shows the resulting IQ diagram.

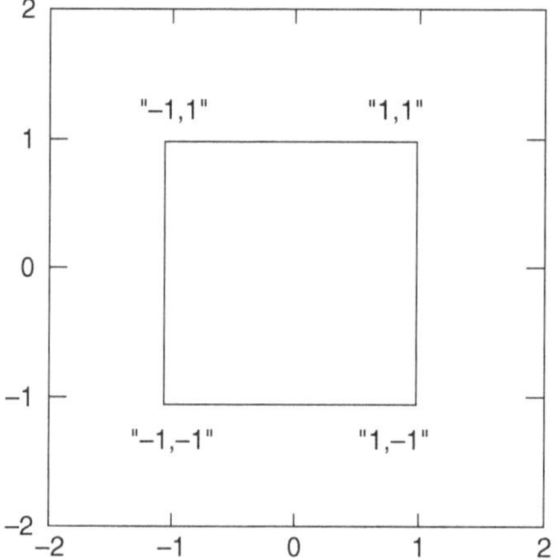

Figure 10.25 IQ diagram for OQPSK with rectangular basis pulses. Also shown are the four points of the signal space diagram, $(1,1)$, $(1,-1)$, $(-1,-1)$, $(-1,1)$.

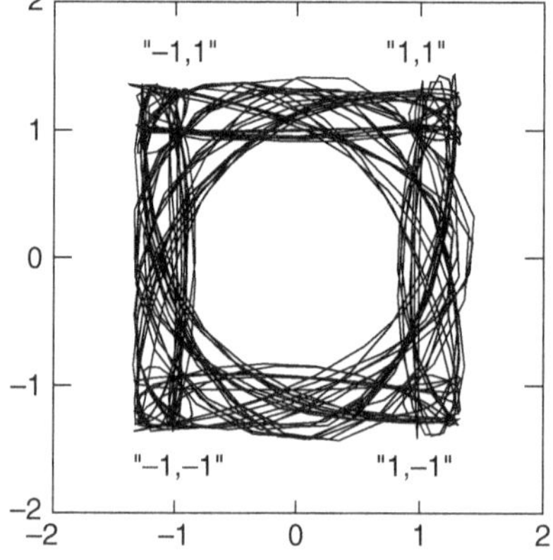

Figure 10.26 IQ diagram for OQAM with raised-cosine basis pulses.

10.3.8   On-Off Keying (OOK)

The simplest modulation scheme is On-Off Keying (OOK), where a cosine carrier signal is transmitted when a "1" is sent, and suppressed when a "−1" is sent. Thus, the bandpass signal during one symbol interval can be written as

$$s_{\mathrm{BP}}(t) = \frac{b_m + 1}{2} \sqrt{4E_{\mathrm{B}}/T_{\mathrm{S}}} \cos\left(2\pi f_{\mathrm{c}} t\right). \tag{10.50}$$

The signal space diagram is shown in Figure 10.27

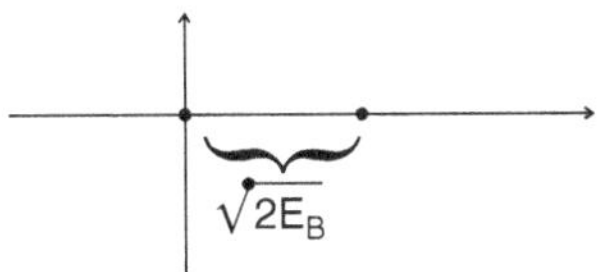

Figure 10.27   Signal space diagram of OOK

OOK falls into the category of orthogonal modulations. It is not usually employed when coherent detection is possible, because it has an unfavorable peak-to-average power ratio, and a smaller distance between signal points than, e.g., BPSK, for the same bit energy. However, it lends itself to easy noncoherent detection, i.e., no knowledge of the phase of the received signal is required, and the detector only needs to determine whether the received signal energy is above a particular threshold; more details will be provided in Section 11.1.3.

10.4   Multi-Pulse Modulation

10.4.1   Multi-Pulse Modulation

PAM, as described in Eq. (10.18), can be generalized to multi-pulse modulation, where the signal is composed of a set of basis pulses; the pulse to be transmitted depends on the modulation coefficient c_i

$$s_{\mathrm{LP}}(t) = \sum_{i=-\infty}^{\infty} g_{c_i}(t - iT). \tag{10.51}$$

While this formulation at first glance looks very similar to regular PAM, there is a key difference: in PAM, every symbol in the modulation alphabet is represented by the same pulse shape, and only the complex factor multiplying it provides a differentiation. In contrast, multi-pulse modulation allows to represent different symbols by different pulse shapes, which might thus be more easily distinguishable.

Multi-pulse modulation is often used to enable noncoherent RXs (compare also Section 10.1.3), such that the RX does not need any information about the phase that is imparted on the received signal by the channel.

10.4.2   Frequency Shift Keying

In *Frequency Shift Keying (FSK)*, each symbol is represented by transmitting (for a time T_{S}) a sinusoidal signal whose frequency depends on the symbol to be transmitted. In M-ary FSK, the basis pulses have an offset from the carrier frequency if_{mod}, where $i = \pm 1, \pm 3, \ldots \pm(M-1)$. It is common to choose a set of orthogonal or bi-orthogonal pulses, as this simplifies the detector; this is assumed henceforth in this section. For ease of exposition, let us specialize in the following to the case of binary FSK. Then the basis pulses are simply (see also Figure 10.28):

$$g_m(t) = \cos\left[(2\pi f_{\mathrm{c}} + b_m 2\pi f_{\mathrm{mod}})t + \psi\right] \quad \text{for } 0 \le t \le T. \tag{10.52}$$

Note that the phase of the transmit signal can jump at the bit transitions. This leads to undesirable spectral properties, as discussed below.

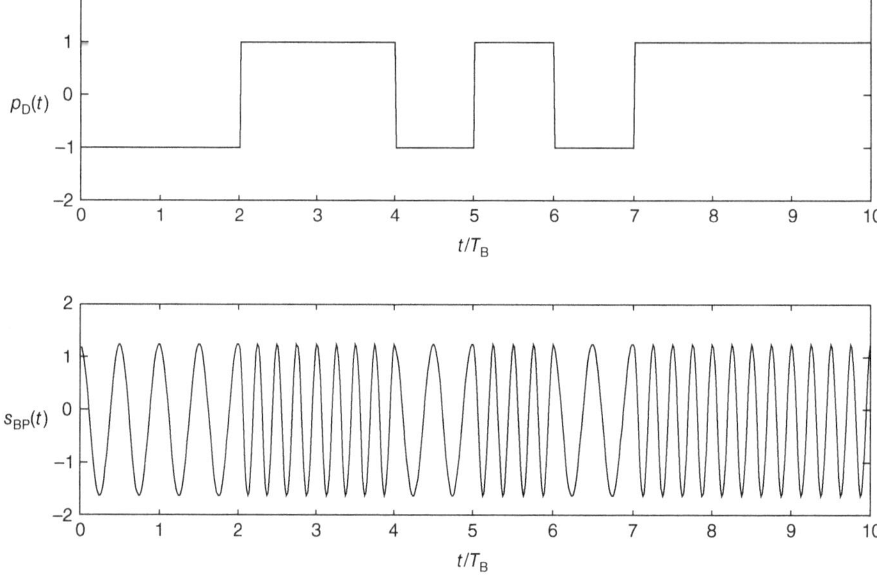

Figure 10.28 FSK signal as a function of time.

Alternatively, we can also interpret FSK as a phase modulation, but with the phase varying all the time, instead of being piecewise-constant as in PSK. Consequently, we can find two different representations:

1. Use sinusoidal oscillations at the possible signal frequencies as expansion functions. For the case of binary FSK, BFSK, this is $f_c \pm f_{mod}$. The expansion functions (in passband) thus read:

$$\varphi_{BP,1}(t) = \sqrt{2/T_B} \cos \left(2\pi f_c t + 2\pi f_{mod} t\right)$$
$$\varphi_{BP,2}(t) = \sqrt{2/T_B} \cos \left(2\pi f_c t - 2\pi f_{mod} t\right). \tag{10.53}$$

In this case, the signal-space diagram consists of two points on the two orthogonal axes, see Figure 10.28. Note that we have made the implicit assumption that the two signals $\varphi_{BP,1}(t)$ and $\varphi_{BP,2}(t)$ are orthogonal to each other.

2. Use the in-phase and quadrature-phase component of the center frequency f_c, i.e., consider the IQ diagram. In that case, the signal shows up not as a discrete point, but as a continuous trajectory, namely a circle (see Figure 10.29). At any time instant, the transmit signal is represented by a different point in the IQ diagram (Figure 10.30).

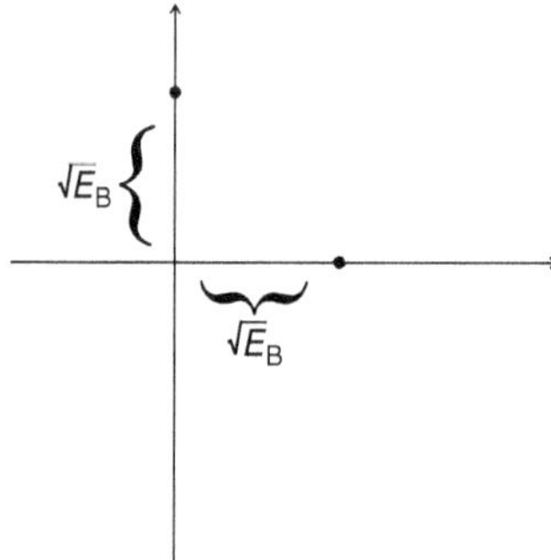

Figure 10.29 Signal-space diagram of FSK when using $\cos[2\pi(f_c \pm f_{mod})t]$ as expansion functions.

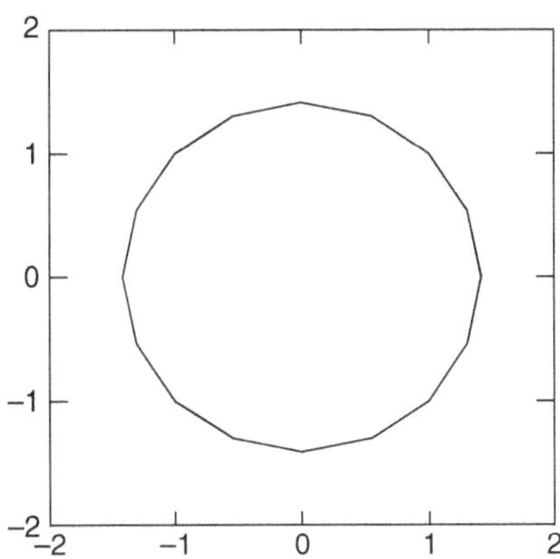

Figure 10.30 IQ diagram of FSK, i.e., using $\cos(2\pi f_c t)$ and $\sin(2\pi f_c t)$ as expansion functions.

Spectrum of FSK

The power spectrum of FSK can be shown to consist of a continuous and a discrete (spectral lines) part

$$S(f) = S_{\text{cont}}(f) + S_{\text{disc}}(f). \tag{10.54}$$

where [Benedetto and Biglieri 1999]

$$S_{\text{cont}}(f) = \frac{1}{2T}\left\{ \sum_{m=1}^{2} |G_m(f)|^2 - \frac{1}{2}\left| \sum_{m=1}^{2} G_m(f) \right|^2 \right\}. \tag{10.55}$$

and

$$S_{\text{disc}}(f) = \frac{1}{(2T)^2}\left| \sum_{m=1}^{2} G_m(f) \right|^2 \sum_{n} \delta\left(f - \frac{n}{T}\right). \tag{10.56}$$

where $G_m(f)$ is the Fourier transform of $g_m(t)$. An example is shown in Figure 10.31.

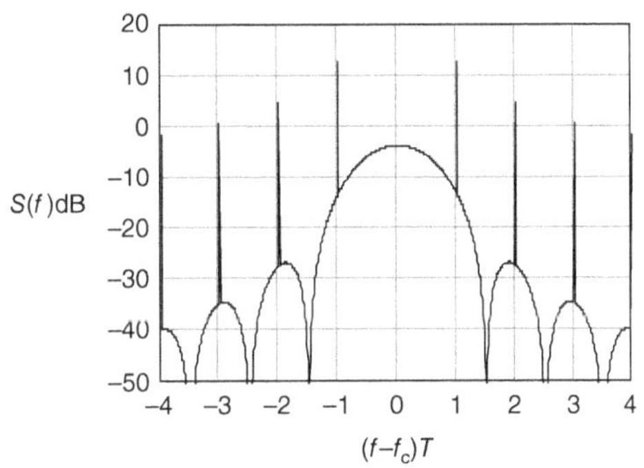

Figure 10.31 Power spectral density of (noncontinuous phase) FSK with $h_{\text{mod}} = 1$.

10.4.3 *Continuous-Phase Modulation*

An important variant of multi-pulse modulation is *Continuous-Phase Modulation, CPM*. In that case, the symbols follow each other in such a way that the phase of the total signal is continuous. Consequently, the transmit waveform at a given time depends not just on one specific symbol that we want to transmit but also on the *history* of the transmit signal. In particular, for *Continuous-Phase Frequency Shift Keying*, CP-FSK, the amplitude of the total signal is chosen as constant; the phase $\Phi(t)$ for binary CP-FSK is

$$\Phi_{\text{CPFSK}}(t) = 2\pi h_{\text{mod}} \sum_{i=-\infty}^{\infty} b_i \int_{-\infty}^{t} \tilde{g}(u - iT)\,du \tag{10.57}$$

where u is the integration variable, h_{mod} is the modulation index, $\tilde{g}(t)$ is the *phase basis pulse*, which is normalized to

$$\int_{-\infty}^{\infty} \tilde{g}(t)dt = 1/2. \tag{10.58}$$

Note that the normalization of the phase basis pulse is fundamentally different from the basis pulse in PAM, and is not related to the energy of the signal.

Gaussian Basis Pulses

A Gaussian basis pulse is the convolution of a rectangular and a Gaussian function – in other words, the output of a filter with Gaussian impulse response that is excited by a rectangular waveform (Figure 10.32). Speaking mathematically, the rectangular waveform is given by Eq. (10.24), and the impulse response of the Gaussian filter is

$$\frac{1}{\sqrt{2\pi}\sigma_G T} \exp\left(-\frac{t^2}{2\sigma_G^2 T^2}\right) \tag{10.59}$$

where

$$\sigma_G = \frac{\sqrt{\ln(2)}}{2\pi B_G T} \tag{10.60}$$

and B_G is the 3 dB bandwidth of the Gaussian filter. The transfer function of the Gaussian filter is

$$\exp\left(-\frac{(2\pi f)^2 \sigma_G^2 T^2}{2}\right) \tag{10.61}$$

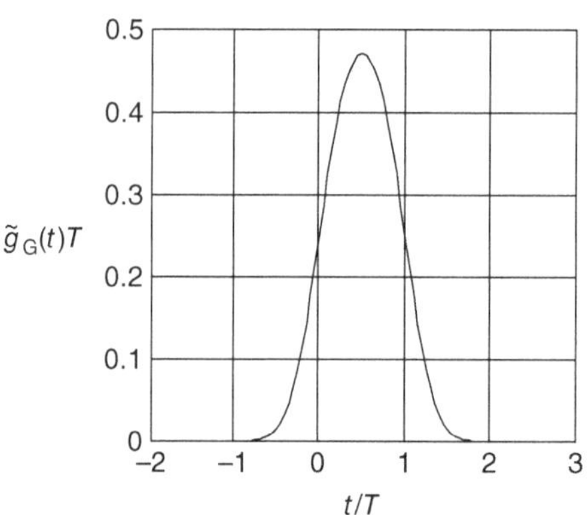

Figure 10.32 Shape of a Gaussian phase basis pulse with $B_G T = 0.5$.

Using the normalization for phase basis pulse (10.58),

$$\tilde{g}_G(t) = \frac{1}{4T}\left[\text{erfc}\left(\frac{2\pi}{\sqrt{2\ln(2)}}B_G T\left(-\frac{t}{T}\right)\right) - \text{erfc}\left(\frac{2\pi}{\sqrt{2\ln(2)}}B_G T\left(1-\frac{t}{T}\right)\right)\right] \tag{10.62}$$

where $\text{erfc}(x)$ is the complementary error function

$$\text{erfc}(x) = \frac{2}{\sqrt{\pi}}\int_x^{\infty} \exp\left(-t^2\right)dt. \tag{10.63}$$

Continuous-Phase FSK

Let us now return to the discussion of FSK. Conventional FSK can have jumps of the phase at the symbol transitions. Such sharp changes lead to a broadening of the spectrum, as we saw. Instead, with CP-FSK we enforce continuity of the phase. Using the normalization for phase pulses Eq. (10.58) and assuming rectangular phase basis pulses

$$\tilde{g}_{FSK}(t) = \frac{1}{2T_B} g_R(t, T_B) \tag{10.64}$$

the phase pulse sequence is

$$\tilde{p}_{D,FSK}(t) = \sum_{i=-\infty}^{\infty} b_i \tilde{g}_{FSK}(t - iT_B) = b_i * \tilde{g}_{FSK}(t). \tag{10.65}$$

The instantaneous frequency is given as

$$f(t) = f_c + b_i f_{mod}(t) = f_c + f_D(t) = f_c + \frac{1}{2\pi} \frac{d\Phi_S(t)}{dt}. \tag{10.66}$$

For a continuous-phase FSK (CPFSK) signal, the phase is given as

$$\Phi_S(t) = 2\pi h_{mod} \int_{-\infty}^{t} \tilde{p}_{D,FSK}(\tau)d\tau \tag{10.67}$$

and the signal has a constant envelope.

Real- and imaginary part of the equivalent baseband signal are

$$\mathrm{Re}\,(s_{LP}(t)) = \sqrt{2E_B/T_B}\,\cos\left[2\pi h_{mod} \int_{-\infty}^{t} \tilde{p}_{D,FSK}(\tau)d\tau\right] \tag{10.68}$$

$$\mathrm{Im}(s_{LP}(t)) = \sqrt{2E_B/T_B}\,\sin\left[2\pi h_{mod} \int_{-\infty}^{t} \tilde{p}_{D,FSK}(\tau)d\tau\right]. \tag{10.69}$$

The resulting signal has memory, as the signal at time t depends on all previously transmitted bits.

*10.4.4 Minimum Shift Keying

Minimum Shift Keying (MSK) is a *CP-FSK* with modulation index

$$h_{mod} = 0.5, \quad f_{mod} = 1/4T. \tag{10.70}$$

This implies that the phase changes by $\pm\pi/2$ during one-bit duration (see Figure 10.33). The bandpass signal is shown in Figure 10.34.

Alternatively, it is also possible to interpret MSK as *offset-QAM*. The basis pulses are sinusoidal half-waves extending over a duration of $2T_B$ (see also Figures 10.34 and 10.35):

$$g(t) = \sin\left(2\pi f_{mod}(t + T_B)\right)g_R(t, 2T_B). \tag{10.71}$$

For proof, see Appendix 10.A.

Due to the use of smoother basis pulses, the spectrum decreases faster than that of "regular" OQPSK:

$$S(f) = \frac{16T_B}{\pi^2}\left(\frac{\cos\left(2\pi f T_B\right)}{1 - 16 f^2 T_B^2}\right)^2 \tag{10.72}$$

see also Figure 10.36. On the other hand, MSK is only a binary modulation format, while OQPSK transmits 2 bits in a symbol duration. As a consequence, MSK has a lower spectral efficiency when considering the 90% energy bandwidth (1.29 bit/s/Hz) but still performs reasonably well when considering the 99% energy bandwidth (0.85 bit/s/Hz).

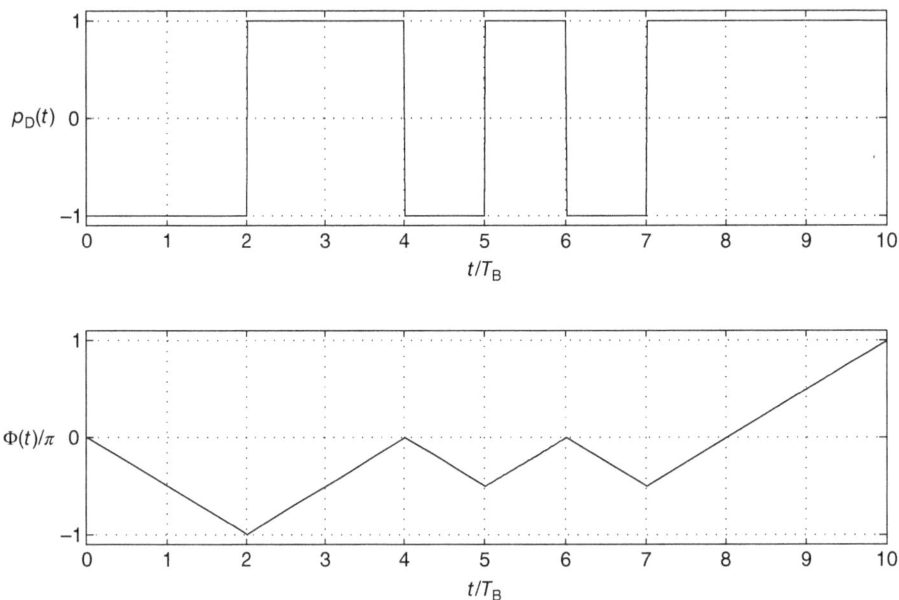

Figure 10.33 Phase pulse and phase as function of time for MSK signal.

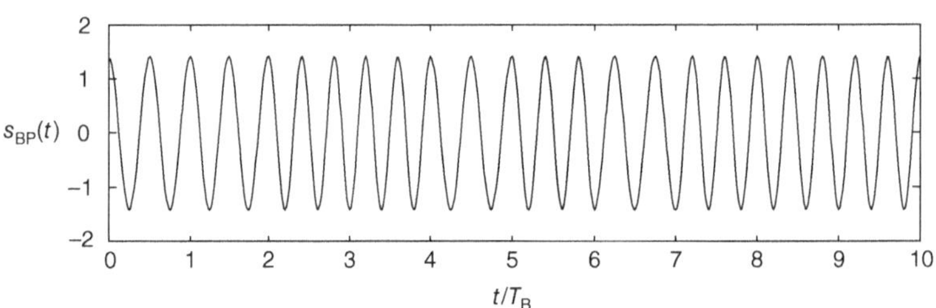

Figure 10.34 MSK modulated signal.

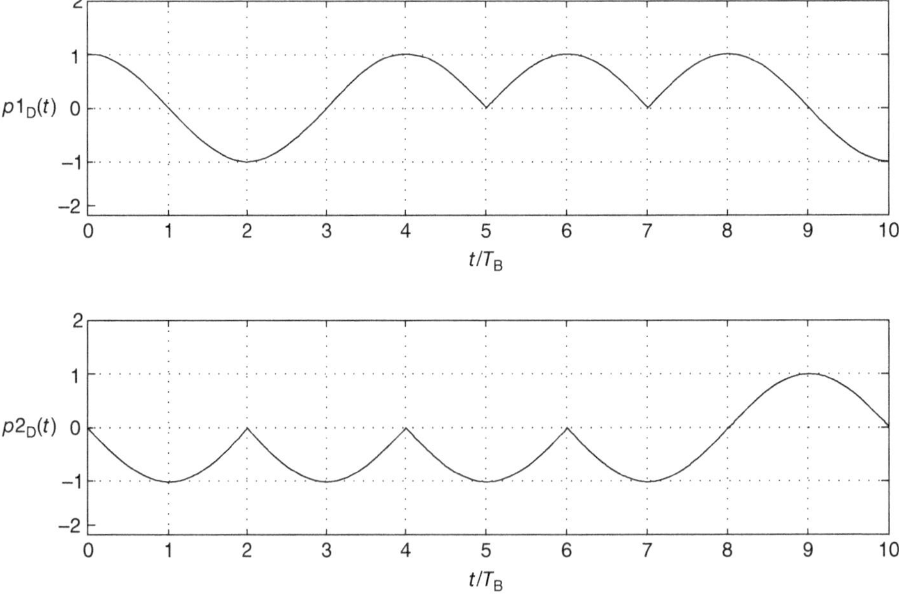

Figure 10.35 Composition of MSK from sinusoidal half-waves.

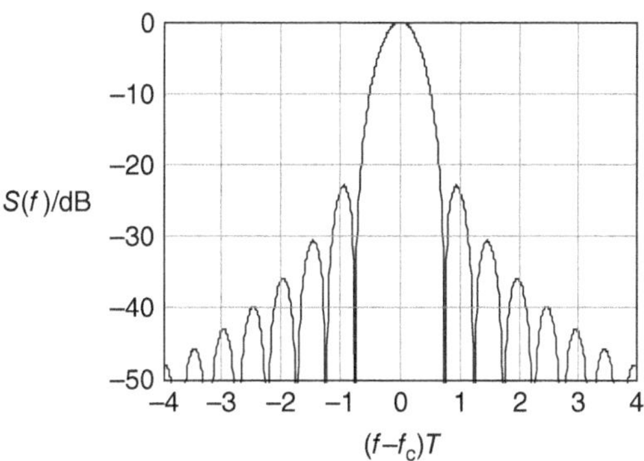

Figure 10.36 Power spectral density of MSK.

Example 10.3 *Compare the spectral efficiency of MSK and OQPSK with rectangular constituent pulses. Consider systems with equal bit duration. Compute the out-of-band energy when the boundary of the band is at $1/T_B$, $2/T_B$, and $3/T_B$.*

The power spectral density of MSK is given by

$$S_{\mathrm{MSK}}(f) = \frac{16 T_B}{\pi^2} \left(\frac{\cos\left(2\pi f T_B\right)}{1 - 16 f^2 T_B^2} \right)^2 \tag{10.73}$$

whereas the power spectral density for OQPSK with rectangular pulses is the same as for ordinary QAM given by (note that we normalize such that the integral over the power-spectral density becomes unity)

$$S_{\mathrm{OQPSK}}(f) = (1/T_S)(T_S \operatorname{sinc}(\pi f T_S))^2 \tag{10.74}$$

where it must be noted that $T_S = 2T_B$ for OQPSK. The out-of-band power is, for MSK and $T_B = 1$, given by

$$P_{\mathrm{out}}(f_0) = 2 \int\limits_{f=f_{\mathrm{boundary}}}^{\infty} S(f) df = 2 \int\limits_{f_{\mathrm{boundary}}}^{\infty} \frac{16}{\pi^2} \left(\frac{\cos\left(2\pi f\right)}{1 - 16 f^2} \right)^2 df$$

$$= \frac{32}{\pi^2} \int\limits_{f_{\mathrm{boundary}}}^{\infty} \frac{\cos^2 2\pi f}{256 f^4 - 32 f^2 + 1} df = \frac{32}{\pi^2} \int\limits_{f_{\mathrm{boundary}}}^{\infty} \frac{\frac{1}{2}\cos 4\pi f + \frac{1}{2}}{256 f^4 - 32 f^2 + 1} df \tag{10.75}$$

and for OQPSK with $T_B = 1$, given by

$$P_{\mathrm{out}}(f_0) = 2 \int\limits_{f_{\mathrm{boundary}}}^{\infty} \frac{1}{2} \left(2 \frac{\sin\left(2\pi f\right)}{(2\pi f)} \right)^2 df. \tag{10.76}$$

Solving these integrals numerically gives the following table

	$1/T_B$	$2/T_B$	$3/T_B$
OQPSK	0.050	0.025	0.017
MSK	$2.4*10^{-3}$	$2.8*10^{-4}$	$7.7*10^{-5}$
ratio (dB)	13.1	19.7	23.4

*10.4.5 Demodulation of MSK

The different interpretations of MSK are not just useful for gaining insights in the modulation scheme but also for building demodulators. Different demodulator structures correspond to different interpretations:

- *Frequency discriminator detection:* since MSK is a type of FSK, it is straightforward to check whether the instantaneous frequency is larger or smaller than the carrier frequency (larger or smaller than 0 when considering equivalent baseband). The instantaneous frequency can be sampled in the middle of the bit, or it can be integrated over (part of the) bit duration in order to reduce the effect of noise. This RX structure is simple, but suboptimum, since it does not exploit the continuity of the phase at the bit transitions.

- *Differential detection:* the phase of the signal changes by $+\pi/2$ or $-\pi/2$, depending on the bit that was transmitted. An RX thus just needs to determine the phases at times iT and $(i+1)T$, in order to make a decision. It is remarkable that no differential encoding of the transmit signal is required; an erroneous estimate of the phase at one sampling time leads to two (but not more) bit errors.
- *Matched-filter-reception:* it is well-known that matched-filter reception is optimum (see also Chapter 11). This is true both when considering MSK as OQPSK, and when considering it as multi-pulse modulation. However, it has to be noted that MSK is a modulation format with memory. Thus, a bit-by-bit detection is suboptimum: consider the signal-space diagram: four constellation points (at $0°$, $90°$, $180°$, and $270°$) are possible. For a bit-by-bit decision, the decision boundaries are thus the first and second main diagonals. The distance between the signal constellations and the decision boundary is $\sqrt{E}/\sqrt{2}$, and thus worse by 3 dB compared to BPSK. However, such a decision method has thrown away the information arising from the memory of the system: if the previous constellation point had been at $0°$, the subsequent signal constellations can only be at either $90°$ or $270°$. The decision boundary should thus be the x-axis, and the distance of the signal constellations to the decision boundary is $\sqrt{E}$, i.e., equal to BPSK. The memory can be exploited, e.g., by a maximum-likelihood sequence estimation, compare Section 13.3.[15]

*10.4.6 Gaussian MSK

GMSK (Gaussian minimum shift keying) is CP-FSK with modulation index $h_{\mathrm{mod}} = 0.5$ and *Gaussian* phase basis pulses

$$\tilde{g}(t) = g_{\mathrm{G}}(t, T_{\mathrm{B}}, B_{\mathrm{G}} T_{\mathrm{B}}) \tag{10.77}$$

where g_{G} is given in (10.62). The sequence of transmit phase pulses is

$$p_{\mathrm{D}}(t) = \sum_{i=-\infty}^{\infty} b_i \tilde{g}(t - iT_{\mathrm{B}}) = b_i * \tilde{g}(t) \tag{10.78}$$

(see Figure 10.37). The spectrum is shown in Figure 10.38. We see that GMSK achieves better spectral efficiency than MSK because it uses the smoother Gaussian phase basis pulses as opposed to the rectangular ones of MSK.

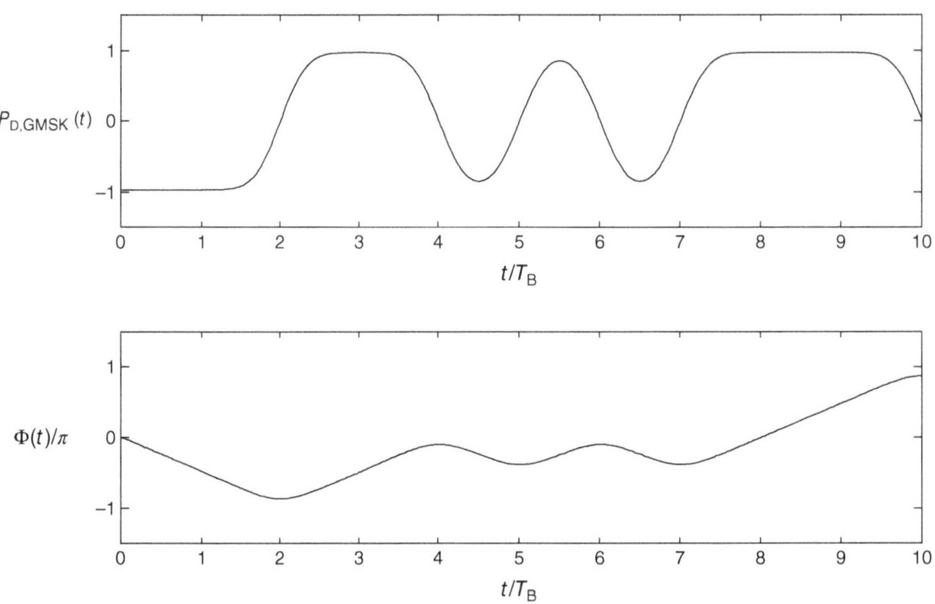

Figure 10.37 Pulse sequence and phase of GMSK signal.

GMSK used to be the most widely used modulation format in wireless communications, because of its use in the cellular GSM standard (with $B_{\mathrm{G}} T = 0.3$), see Appendix 30.A. Its main application nowadays is in legacy deployments of GSM, as well as in the Bluetooth (IEEE 802.15.1) standard for wireless personal area networks, see Chapter 34.

It is noteworthy that – in contrast to regular MSK – GMSK *cannot* be interpreted as PAM. However, [Laurent 1986] derived equations that allow the interpretation of GMSK as PAM with memory.

---

[15] It can be shown that also for a maximum-likelihood sequence estimation, only three sampling values can influence the decision for a specific bit, see [Benedetto and Biglieri 1999].

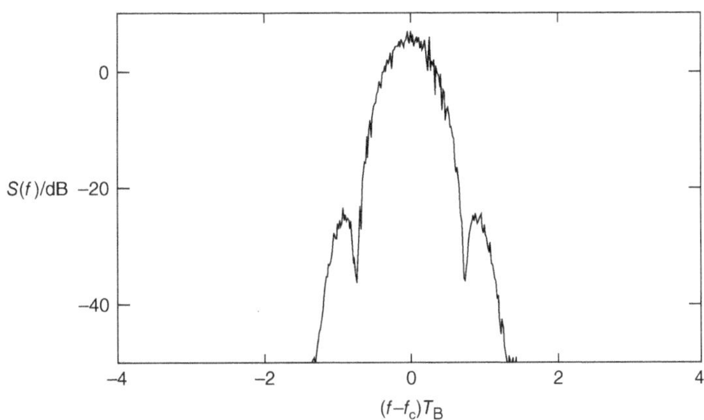

Figure 10.38 GMSK power spectral density (from simulations).

*10.4.7 Pulse Position Modulation

Pulse Position Modulation (PPM) is another form of multi-pulse modulation. Remember that for FSK, we used pulses that had different center frequencies as basis pulses. For PPM, we use pulses that have different delays. In the following we will consider M-ary PPM:

$$s_{LP}(t) = \sqrt{\frac{2E_S}{T_S}} \sum_{i=-\infty}^{\infty} g_{c_i}(t - iT) \tag{10.79}$$

$$= \sqrt{\frac{2E_S}{T_S}} \sum_i g_{PPM}(t - iT - \tau_i) \quad \tau_i \in 0, T_d, \dots (M-1)T_d \tag{10.80}$$

where T_d is the modulation delay (see Figure 10.39), and T_P is the duration of the pulse. Note that the modulation symbols are directly mapped onto the delay of the pulses, and are thus real.

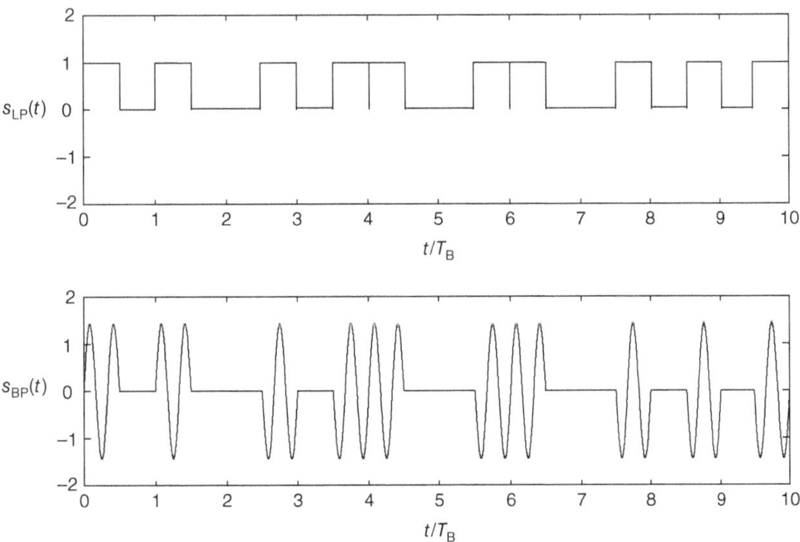

Figure 10.39 Temporal signal in the low-pass and bandpass domain for pulse position modulation.

The envelope shows strong fluctuations; however, it is still possible to use nonlinear amplifiers, since only two output amplitude levels are allowed.

Because PPM is a nonlinear modulation, the spectrum is not given by Eq. (10.20), but rather by the more complicated expression for multi-pulse modulation (see also Eq. (10.54))

$$S(f) = \frac{1}{M^2 T_S^2} \sum_{l=-\infty}^{+\infty} \left(\left| \sum_{m=0}^{M-1} G_m(f) \right|^2 \delta\left(f - \frac{i}{T_S}\right) \right) + \frac{1}{T_S} \left(\sum_{m=0}^{M-1} \frac{1}{M} |G_m(f)|^2 - \left| \sum_{m=0}^{M-1} \frac{1}{M} G_m(f) \right|^2 \right). \tag{10.81}$$

We note that the spectrum has a number of lines.

Up to now, we have assumed that the transmitted pulses are rectangular pulses, and offset with respect to each other by one pulsewidth. However, this is not spectrally efficient, as the pulses have a very broad spectrum. Other pulseshapes can be used as well. The performance of those different pulses depends on the detection method. When coherent detection is employed, then the key quantity determining the performance is the correlation between the pulses representing $+1$ and -1 (assuming binary PPM):

$$\rho = \frac{\int g(t) g^*(t - T_d) dt}{\int |g(t)|^2 dt}. \tag{10.82}$$

If $\rho = 0$, the modulation is orthogonal as is the case for rectangular pulses if $T_d > T_P$. It is actually possible to choose pulseshapes so that $\rho < 0$. In that case, we have not only better spectral efficiency but also better performance. However, when the RX uses *incoherent* (energy) detection, then it is best if the pulses do not have any overlap. The relevant correlation coefficient in that case is

$$\rho_{\text{env}} = \frac{\int |g(t)||g(t - T_d)| \, dt}{\int |g(t)|^2 dt}. \tag{10.83}$$

PPM can be combined with other modulation formats. For example, binary PPM can be combined with BPSK, so that each symbol represents 2 bits – one bit is determined by the phase of the transmitted pulse, and one bit by its position.

Example 10.4 *Consider a PPM system with $g(t) = \sin(t/T)/t$. What is the correlation coefficient ρ and the correlation coefficient of the envelopes when T_d is (i) T, (ii) 5T?*

The correlation coefficient is given by

$$\rho = \frac{\int g(t) g(t - T_d) \, dt}{\int |g(t)|^2 \, dt}. \tag{10.84}$$

Assuming $T = 1$, the denominator is

$$\int |g(t)|^2 \, dt = \int_{-\infty}^{\infty} \frac{\sin^2(t)}{t^2} \, dt = \pi. \tag{10.85}$$

For $T_d = 1$, using numerical integration, the numerator becomes

$$\int g(t) g(t - T_d) \, dt = \int_{-\infty}^{\infty} \frac{\sin(t)}{t} \frac{\sin(t-1)}{t-1} \, dt \approx 2.63 \tag{10.86}$$

so that the correlation coefficient becomes $\rho \approx 0.84$. For $T_d = 5$, using numerical integration, the numerator becomes

$$\int g(t) g(t - T_d) \, dt = \int_{-\infty}^{\infty} \frac{\sin(t)}{t} \frac{\sin(t-5)}{t-5} \, dt \approx -0.61 \tag{10.87}$$

so that the correlation coefficient becomes $\rho \approx -0.2$. For $T_d = 1$, the numerator for the envelope correlation is given by

$$\int_{-\infty}^{\infty} \left| \frac{\sin(t)}{t} \right| \left| \frac{\sin(t-1)}{t-1} \right| \, dt \approx 2.73 \tag{10.88}$$

so that the envelope correlation becomes $\rho \approx 0.87$. For $T_d = 5$, the numerator for the envelope correlation is given by

$$\int_{-\infty}^{\infty} \left| \frac{\sin(t)}{t} \right| \left| \frac{\sin(t-5)}{t-5} \right| \, dt \approx 1.04. \tag{10.89}$$

so that the envelope correlation becomes $\rho \approx 0.33$.

PPM is not used very often for wireless systems. This is due to its relatively low spectral efficiency, as well as to the effect of delay dispersion on a PPM system. Still, there are some emerging applications where PPM is used, especially impulse radio (see Section 19.4).

10.5 Summary of Spectral Efficiencies

See Table 10.1.

Table 10.1 Spectral efficiency for different modulation schemes.

Modulation method	Spectral efficiency for 90% of total energy (bit/s/Hz)	Spectral efficiency for 99% of total energy (bit/s/Hz)
BPSK	0.59	0.05
BAM ($\alpha = 0.5$)	1.02	0.79
QPSK, OQPSK	1.18	0.10
MSK	1.29	0.85
GMSK ($B_G = 0.5$)	1.45	0.97
QAM ($\alpha = 0.5$)	2.04	1.58

*10.6 Appendix

App. 10.A: Interpretation of MSK as OQPSK

See App10.pdf at wiley.com/go/molisch/wireless3e

Further Reading

The description of different modulation formats, and their representation in a signal-space diagram, can be found in any of the numerous textbooks on digital communications, e.g., [Anderson 2005], [Proakis and Salehi 2005], [Sklar 2001], [Barry et al. 2003], [Wilson 1996], [Xiong 2006]. A book specifically dedicated to modulation for wireless communications is [Burr 2001]. QAM is described in [Hanzo et al. 2000]. A description of the multiple interpretations of MSK, and the resulting demodulation structures, can be found in [Benedetto and Biglieri 1999]. GMSK was invented in [Murota and Hirade 1981]. An authoritative description of different forms of CPM is [Anderson et al. 1986].

For updates and errata for this chapter, see https://wides.usc.edu/students.html#textbooks

Exercises

See Sec. 36.10 of Exercises.pdf at wiley.com/go/molisch/wireless3e

11

Demodulation

In this chapter, we describe how to demodulate the received signal, and discuss the performance of different demodulation schemes in *Additive White Gaussian Noise* (AWGN) channels as well as flat-fading channels and dispersive channels.

11.1 Demodulator Structure and Error Probability in Additive White Gaussian Noise Channels

This section deals with basic demodulator structures for the modulation formats described in Chapter 10, and the computation of the Bit Error Rate (BER) and Symbol Error Rate (SER) in an AWGN channel. As in the previous chapter, we concentrate on the most important results; for more detailed derivations, we refer the reader to the monographs [Anderson 2005], [Benedetto and Biglieri 1999], [Proakis and Salehi 2005], and [Sklar 2001].

11.1.1 Model for Channel and Noise

The AWGN channel attenuates the transmit signal, causes phase rotation, and adds Gaussian-distributed noise. Attenuation and phase rotation are temporally constant and are thus easily taken into account. Thus, the received signal (in complex baseband notation) is given by

$$r_{\mathrm{LP}}(t) = \alpha s_{\mathrm{LP}}(t) + n_{\mathrm{LP}}(t) \tag{11.1}$$

where α is the (complex) channel gain $|\alpha| \exp(j\varphi)$, and $n(t)$ is a (complex) Gaussian noise process.

In order to derive the properties of noise, let us first consider noise in the bandpass system. We assume that over the bandwidth of interest the noise power spectral density is constant. The value of the *two-sided* noise power-spectral density is $N_0/2$ (see Figure 11.1a). The (complex) equivalent low-pass noise has a power-spectral density (see Figure 11.1b):

$$S_{\mathrm{n,LP}}(f) = \begin{cases} N_0 & |f| \leq B/2 \\ 0 & \text{otherwise.} \end{cases} \tag{11.2}$$

Note that $S_{\mathrm{n,LP}}(f)$ is symmetric with respect to f – i.e., $S_{\mathrm{n,LP}}(f) = S_{\mathrm{n,LP}}(-f)$.

When considering noise in the time domain, we find that it is described by its autocorrelation function $R_{\mathrm{LP},nn}(\tau) = (1/2)E\{n_{\mathrm{LP}}^*(t) n_{\mathrm{LP}}(t + \tau)\}$.[1] For a system limited to the band $[-B/2, B/2]$, the ACF is

$$R_{\mathrm{LP},nn}(\tau) = N_0 \frac{\sin(\pi B \tau)}{\pi \tau} \tag{11.3}$$

which becomes

$$R_{\mathrm{LP},nn}(\tau) = N_0 \delta(\tau) \tag{11.4}$$

as the bandwidth B tends to infinity. We also note that the correlations of the in-phase and the quadrature-phase components of noise, respectively, are both $R_{\mathrm{LP},nn}(\tau)$, while the cross-correlation between in-phase and quadrature-phase noise is zero, since $S_{\mathrm{n,LP}}(f)$ is an even function.

---

[1] Note that this definition of the AutoCorrelation Function (ACF) differs by a factor of 2 from the one used in Chapter 6. The reason for using this modified definition here is that it is commonly used in communications theory for BER computations.

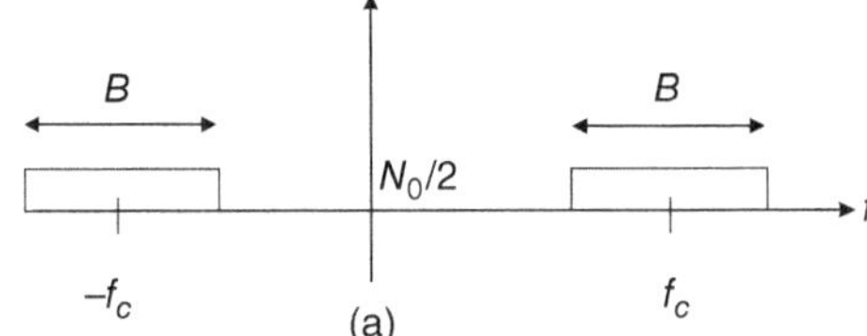

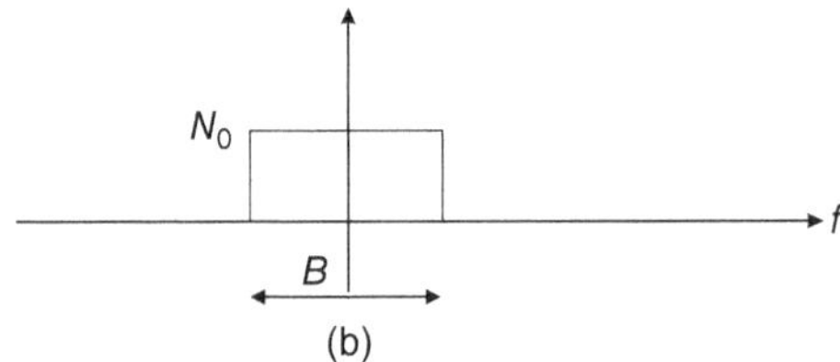

Figure 11.1 Bandpass noise and equivalent low-pass noise power spectrum.

The ACF of the bandpass signal is

$$R_{\mathrm{BP},nn}(\tau) = \mathrm{Re}\left\{R_{\mathrm{LP},nn}(\tau)\,\exp\left(j2\pi f_c\tau\right)\right\}. \tag{11.5}$$

11.1.2 Signal Space Diagram and Optimum Receivers

We now derive the structure of the optimum Receivers (RXs) for digital modulation. In the process, we will also find an additional motivation for using signal space diagrams. For these derivations we make the following assumptions:

1. All transmit symbols are equally likely.
2. The modulation format does not have memory.
3. The channel is an AWGN channel, and both absolute channel gain and phase rotation are completely known. We assume henceforth that phase rotation has been compensated completely, so that the channel attenuation is real, so that $\alpha = |\alpha|$.

The ideal detector, called the *Maximum A Posteriori* (MAP) detector, aims to answer the following question: "If a signal $r(t)$ was received, then which symbol $s_m(t)$ was most likely transmitted?" In other words, which symbol m maximizes

$$\Pr[s_m(t) \mid r(t)]. \tag{11.6}$$

Bayes' rules can be used to write this as (in other words, find the symbol m that achieves)

$$\max_m \Pr[n(t) = r(t) - \alpha s_m(t)]\Pr[s_m(t)]. \tag{11.7}$$

Since we assume that all symbols are equiprobable, the MAP detector becomes identical to the *Maximum Likelihood* (ML) detector:

$$\max_m \Pr[n(t) = r(t) - \alpha s_m(t)]. \tag{11.8}$$

In Section 10.1.5, we introduced the signal space diagram for the representation of modulated transmit signals. There, we treated the signal space diagram just as a convenient shorthand. Now we will show how a transmit and receive signal can be related, and how the signal space diagram can be used to derive optimum RX structures. In particular, we will derive that the ML detector finds in the signal space diagram the transmit symbol that has the smallest Euclidean distance to the receive signal.

Remember that the transmit signal can be represented in the form:[2]

$$s_m(t) = \sum_{n=1}^{N} s_{m.n}\varphi_n(t) \tag{11.9}$$

---

[2] Note that these equations are valid for both the baseband and the bandpass representation – it is just a matter of inserting the correct expansion functions. However, we will henceforth concentrate on the baseband case.

where

$$s_{m,n} = \int_0^{T_S} s_m(t)\varphi_n^*(t)dt. \tag{11.10}$$

Now we find that the received signal can be represented by a similar expansion. Using the same expansion functions $\varphi_n(t)$ we obtain

$$r(t) = \sum_{n=1}^{\infty} r_n \varphi_n(t) \tag{11.11}$$

where

$$r_n = \int_0^{T_S} r(t)\varphi_n^*(t)dt. \tag{11.12}$$

Since the received signal contains noise, it seems at first glance that we need more terms in the expansion – infinitely many, to be exact. However, we find it useful to split the series into two parts:

$$r(t) = \sum_{n=1}^{N} r_n \varphi_n(t) + \sum_{n=N+1}^{\infty} r_n \varphi_n(t) \tag{11.13}$$

and similarly:

$$n(t) = \sum_{n=1}^{N} n_n \varphi_n(t) + \sum_{n=N+1}^{\infty} n_n \varphi_n(t). \tag{11.14}$$

Using these expansions, the expression that the ML RX aims to maximize can be written as

$$\max_m \Pr[\mathbf{n} = \mathbf{r} - \alpha \mathbf{s}_m] \tag{11.15}$$

where the received signal vector $\mathbf{r}$ is simply $\mathbf{r} = (r_1, r_2,...)^T$, and similarly for $\mathbf{n}$. Since the noise components are independent, the Probability Density Function (pdf) of the received vector $\mathbf{r}$, assuming that $\mathbf{s}_m$ was transmitted, is given as

$$p(\mathbf{r}_{LP} \mid \alpha \mathbf{s}_{LP,m}) \propto \exp\left\{ -\frac{1}{2N_0}\|\mathbf{r} - \alpha \mathbf{s}_m\|^2 \right\} \tag{11.16}$$

$$= \prod_{n=1}^{\infty} \exp\left\{ -\frac{1}{2N_0}(r_n - \alpha s_{m,n})^2 \right\}. \tag{11.17}$$

Since $s_{m,n}$ is nonzero only for $n \leq N$, the ML detector aims to find:

$$\max_m \prod_{n=1}^{N} \exp\left\{ -\frac{1}{2N_0}(r_n - \alpha s_{m,n})^2 \right\} \prod_{n=N+1}^{\infty} \exp\left\{ -\frac{1}{2N_0}(r_n)^2 \right\}. \tag{11.18}$$

A key realization is now that the second product in Eq. (11.18) is independent of s_m, and thus does not influence the decision. This is another way of saying that components of the noise (and thus of the received signal) that do not lie in the signal space of the transmit signal are irrelevant for the decision of the detector (*Wozencraft's irrelevance theorem*). Finally, as $\exp(.)$ is a monotonic function, we find that it is sufficient to minimize the metric (writing now explicitly for the baseband case):

$$\mu(\mathbf{s}_{LP,m}) = \|\mathbf{r}_{LP} - \alpha \mathbf{s}_{LP,m}\|^2. \tag{11.19}$$

Geometrically, this means that the ML RX decides for symbol m whose transmit vector $\mathbf{s}_{LP,m}$ has the smallest Euclidean distance to the received vector $\mathbf{r}_{LP}$. Based on this interpretation, we can define the *decision region* of symbol m: it contains all points in the plane that have a shorter distance to $\mathbf{s}_m$ than to any other $\mathbf{s}_k$ with $k \neq m$ (such regions are called the *Voronoi tessellation* of the signal space diagram). The *decision boundaries* are the boundaries of the decision regions (Voronoi cells).

We need to keep in mind that this is an optimum detection method only in memoryless, uncoded systems. Vector $\mathbf{r}$ contains "soft" information – i.e., how sure the RX is about its decision. This soft information, which is lost in the decision process of finding the nearest neighbor, is irrelevant when one bit does not tell us anything about any other bit. However, for coded systems and systems with memory (Chapters 13 and 14), this information can be very helpful.

The decision metric can be rewritten as

$$\mu(\mathbf{s}_{\mathrm{LP},m}) = \|\mathbf{r}_{\mathrm{LP}}\|^2 + \|\alpha\mathbf{s}_{\mathrm{LP},m}\|^2 - 2\alpha\,\mathrm{Re}\left\{\mathbf{r}_{\mathrm{LP}}\mathbf{s}_{\mathrm{LP},m}^*\right\}. \tag{11.20}$$

Since the term $\|\mathbf{r}_{\mathrm{LP}}\|^2$ is independent of the considered $\mathbf{s}_{\mathrm{LP},m}$, minimizing the metric is equivalent to maximizing:

$$\mathrm{Re}\left\{\mathbf{r}_{\mathrm{LP}}\mathbf{s}_{\mathrm{LP},m}^*\right\} - \alpha E_m \tag{11.21}$$

(remember that $E_m = \|\mathbf{s}_{\mathrm{LP},m}\|^2/2$, see Section 10.1).

One important consequence of this decision rule is that the RX has to know the value of channel gain α. This can be difficult in wireless systems, especially if channel properties change quickly (see Part II). Thus, there are situations where modulation and detection methods are preferred that do not require this information. Specifically, the magnitude of channel gain does not need to be known if all transmit signals have equal energy, $E_m = E$. On the other hand, the phase rotation of the channel (argument of alpha) does not need to be known if either incoherent detection or differential detection is used, as will be explained below in more detail.

The beauty of the above derivation is that it is independent of the actual modulation scheme. The transmit signal is represented in the signal space diagram, which gives all the *relevant* information. The RX structure of Figure 11.2 is then optimum for any modulation alphabet represented in a signal space diagram. The only prerequisite is that the conditions mentioned at the beginning of this chapter are fulfilled. This RX can be interpreted as a correlator, or as a matched filter (matched to the different possible transmit waveforms).

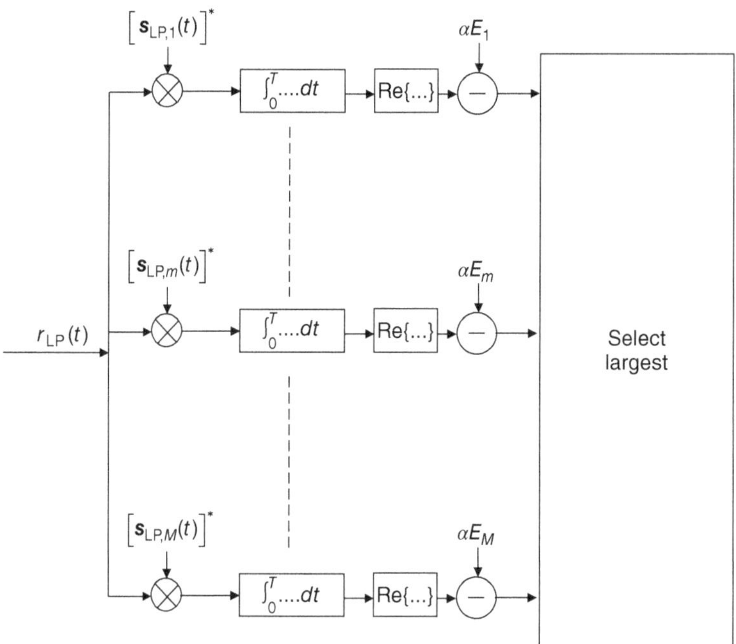

Figure 11.2 Structure of a generic optimum RX.

The term $\mathbf{r}_{\mathrm{LP}}\mathbf{s}_{\mathrm{LP},m}^*$ is the correlation between the representation of the received signal and the mth transmit signal in the signal space representation, i.e., between $\mathbf{r}_{\mathrm{LP}}$ and $\mathbf{s}_{\mathrm{LP},m}$. Equivalently, this can be written as the correlation of the continuous signals $r(t)$ and $s_m(t)$, i.e.,

$$\int_0^{T_s} r_{\mathrm{LP}}(t)s_{\mathrm{LP},m}^*(\tau + t)dt \tag{11.22}$$

evaluated at lag $\tau = 0$.

Alternatively, we can interpret the operation in Eq. (11.22) as a convolution, i.e., sending the received signal through a filter with impulse response $h_{\mathrm{F}}(\tau)$. Since the general equation for such a filtering reads

$$\int_0^{T_s} r_{\mathrm{LP}}(t)h_{\mathrm{F}}(\tau - t)dt \tag{11.23}$$

it is clear that $h_F(t) = s^*_{LP,m}(-t)$, and sampling should occur at $\tau = 0$. Of course such an impulse response is noncausal; we make it causal by modifying the filter impulse response $h_F(t) = s^*_{LP,m}(T_s - t)$ with sampling at $\tau = T_s$.

A third interpretation is for the optimum RX to maximize the Signal-to-Noise Ratio (SNR) at the RX output. This can be most easily seen from Eq. (11.23), where for simplicity we assume that all signals have the same energy. Write $r(t) = \alpha s(t) + n(t)$, so that the output corresponding to the pure signal is $\int \alpha s(t) h_F(\tau - t) dt$, while the output corresponding to the noise is $\int n(t) h_F(\tau - t) dt$. From the Cauchy–Schwarz inequality, we see that the first term is maximized by setting $s(t) = h^*_F(T_s - t)$, or equivalently $h_F(t) = s^*_{LP,m}(T_s - t)$. The expected power of the second term is not impacted by the choice of $h_F(t)$ as long as it is properly normalized.

In the following, we will find that the Euclidean distance of two points in the signal space diagram is a vital parameter for modulation formats. We find that in the bandpass representation for equal-energy signals, which we assumed in Chapter 10, the Euclidean distance is related to the energy of the signal component as

$$d^2_{12} = 2E(1 - \mathrm{Re}\{\rho_{12}\}) \tag{11.24}$$

where ρ_{km} is the correlation coefficient as defined in Section 10.1:

$$\mathrm{Re}\{\rho_{k,m}\} = \frac{\langle \mathbf{s}_k, \mathbf{s}_m \rangle}{|\mathbf{s}_k||\mathbf{s}_m|}. \tag{11.25}$$

11.1.3 Methods for the Computation of Error Probability

In this subsection, we discuss how to compute the performance that can be achieved with optimum RXs. Before going into details, let us define some important notation: the *BER* is, as its name says, a rate. Thus, it describes the number of bit errors per unit time and has dimension s^{-1}. The *bit error ratio* is the number of errors divided by the number of transmitted bits; it thus is dimensionless. For the case when infinitely many bits are transmitted, this becomes the *bit error probability*. In the literature, there is a tendency to mix up these three expressions; specifically, the expression BER is often used when the authors mean bit error probability. This misnomer has become so common that in the following we will use it as well.

Though we have treated a multitude of modulation formats in the previous chapter, they can all be classified easily with the framework of signal space diagrams. For example:

1. Binary Phase Shift Keying (BPSK) signals are antipodal signals.
2. Binary Frequency Shift Keying (BFSK), and Binary Pulse Position Modulation (BPPM), are orthogonal signals. While On–Off Keying (OOK) is an orthogonal modulation format, it is not equal energy.
3. Quadrature-Phase Shift Keying (QPSK), $\pi/4$-DQPSK (Differential Quadrature-Phase Shift Keying), and Offset Quadrature-Phase Shift Keying (OQPSK) are bi-orthogonal signals.

Error Probability for Coherent Receivers – General Case

As mentioned above, coherent RXs compensate for phase rotation of the channel by means of *carrier recovery*. Furthermore, the channel gain α is assumed to be known, and absorbed into the received signal, so that in the absence of noise, $\mathbf{r} = \mathbf{s}$ holds and E denotes henceforth the energy of the *received* signal. The probability that symbol $\mathbf{s}_j$ is mistaken for symbol $\mathbf{s}_k$ that has Euclidean distance d_{jk} from $\mathbf{s}_j$ (*pairwise error probability*) can then be easily computed.

We assume a two-dimensional signal-space diagram and, without loss of generality, that the symbols s_j and s_k are located at $[-d_{jk}/2, 0]$ and $[d_{jk}/2, 0]$, respectively (since the error probability is invariant to translations and rotations in the signal space, any two points can be shifted to these coordinates). With this configuration, the decision region for s_j is the left half plane, i.e., $x < 0$. An error occurs if the Gaussian noise shifts $\mathbf{r}$ into the right half plane. Even though the noise can have multiple dimensions, the only *relevant* noise is the one-dimensional noise along the x-axis. Since the total noise (along x- and y-axes) has variance N_0, the noise along the x-axis has variance $\sigma^2_{n,x} = N_0/2$. The pdf of $r(t)$ component along the x-axis is

$$\frac{1}{\sqrt{2\pi}\sigma_{n,x}} \exp\left[-\frac{[r_x - (-d_{jk}/2)]^2}{2\sigma^2_{n,x}}\right] \tag{11.26}$$

and consequently, the error probability is

$$\frac{1}{\sqrt{2\pi}\sigma_{n,x}} \int_0^\infty \exp\left[-\frac{[r_x + (d_{jk}/2)]^2}{2\sigma^2_{n,x}}\right] dr_x. \tag{11.27}$$

Substituting $t = [r_x + (d_{jk}/2)]/\sigma_{n,x}$, this becomes

$$\frac{1}{\sqrt{2\pi}} \int_{d_{jk}/(2\sigma_{n,x})}^\infty \exp\left[-\frac{t^2}{2}\right] dt. \tag{11.28}$$

Using then the well-known Q-function

$$Q(x) = \frac{1}{\sqrt{2\pi}} \int_x^\infty \exp\left[-\frac{t^2}{2}\right] dt \tag{11.29}$$

the pairwise error probability can be written as

$$\Pr_{\text{pair}}(\mathbf{s}_j, \mathbf{s}_k) = Q\left(\sqrt{\frac{d_{jk}^2}{2N_0}}\right) = Q\left(\sqrt{\frac{E}{N_0}\left(1 - \text{Re}\left\{\rho_{jk}\right\}\right)}\right). \tag{11.30}$$

A useful upper and lower bound for the Q function is

$$\left(\frac{x}{1+x^2}\right)\frac{\exp\left(-x^2/2\right)}{\sqrt{2\pi}} < Q(x) < \left(\frac{1}{x}\right)\frac{\exp\left(-x^2/2\right)}{\sqrt{2\pi}}. \tag{11.31}$$

Alternatively, these expressions can be written in terms of the complementary error function, which is related to the Q-function as:

$$\text{erfc}(x) = 2Q\left(\sqrt{2}x\right) \quad \text{and} \quad Q(x) = \frac{1}{2}\text{erfc}\left(\frac{x}{\sqrt{2}}\right). \tag{11.32}$$

Error Probability for Coherent Receivers – Binary Orthogonal Signals

As we saw in Chapter 10, a number of important modulation formats can be viewed as binary orthogonal signals – most prominently, BFSK and BPPM. Figure 11.3 shows the signal space diagram for this case. The figure also shows the decision boundary: if a received signal point falls into the shaded region, then it is decided that a $+1$ was transmitted, otherwise a -1 was transmitted.

Defining the SNR for one symbol as $\gamma_S = E_S/N_0$, we get

$$\Pr_{\text{pair}}(\mathbf{s}_j, \mathbf{s}_k) = Q\left(\sqrt{\gamma_S\left(1 - \text{Re}\left\{\rho_{jk}\right\}\right)}\right) \tag{11.33}$$

$$= Q(\sqrt{\gamma_S}). \tag{11.34}$$

Note that for binary signaling $\gamma_S = \gamma_B$.

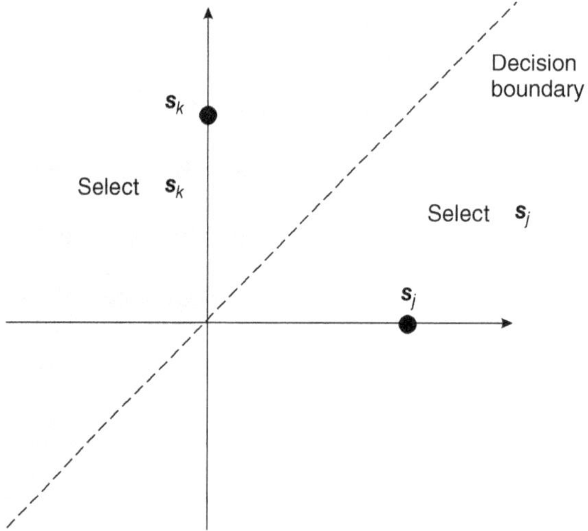

Figure 11.3 Decision boundary for the selection between $\mathbf{s}_k$ and $\mathbf{s}_j$.

Error Probability for Coherent Receivers – Antipodal Signaling

For antipodal signals, the pairwise error probability is

$$\Pr_{\text{pair}}(\mathbf{s}_j, \mathbf{s}_k) = Q\left(\sqrt{\gamma_S\left(1 - \text{Re}\left\{\rho_{jk}\right\}\right)}\right) \tag{11.35}$$

$$= Q(\sqrt{2\gamma_S}). \tag{11.36}$$

For binary signals (including antipodal and orthogonal binary signals) with equal-probability transmit symbols, pairwise error probability is equal to symbol error probability, which in turn is equal to bit error probability. For example, the bit error probability for BPSK, as well as for MSK with ideal coherent detection is given by Eq. (11.35). Note that MSK can be detected like FSK, but then does not exploit the continuity of the phase. In this case, it becomes an orthogonal modulation format, and the BER is given by Eq. (11.34). This means deterioration of the effective SNR by 3 dB.

Union Bound and Bi-Orthogonal Signaling

For M-ary modulation methods, exact computation of the BER and SER is much more difficult; these computations typically require the integration of a non-zero-mean Gaussian distribution over a possibly complicated-shaped Voronoi cell. For this reason, the SER is often upper-bounded by the *union bound* method. The principle of this bounding technique is outlined in Figure 11.4. The region of the signal space diagram that results in an erroneous decision consists of partial regions, each of which represents a pairwise error – i.e., confusing the correct symbol with another symbol. The symbol error probability is then written as the sum of these pairwise probabilities. Since the pairwise-error regions overlap, this represents an upper bound for true symbol error probabilities. The approximation improves as the SNR increases, because the SER is then mainly determined by regions close to the decision boundaries whereas overlap regions have little impact.

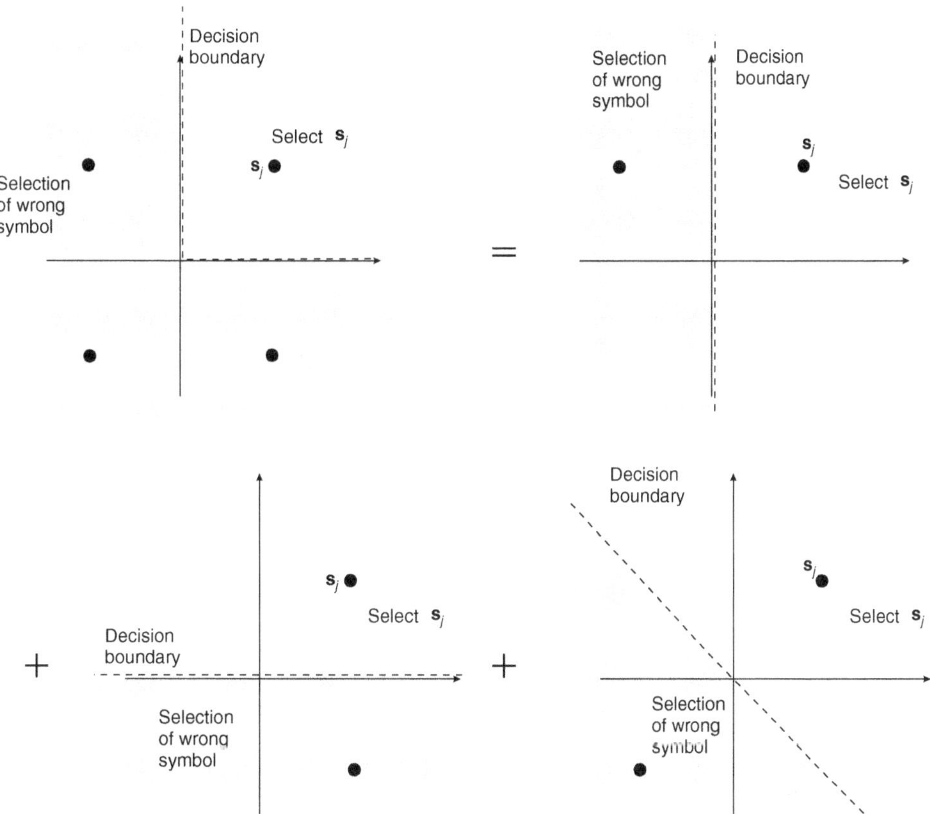

Figure 11.4 Union bound for symbol error probabilities.

We can now consider two types of union bounds:

1. For a "full" union bound, we compute the pairwise error probability with *all* possible signal constellation points and add them up. This is shown in Figure 11.4.

2. For a "tightened" union bound, we compute pairwise error probability using a subset of neighboring points, where the subset is chosen such that the union of their decision regions covers the whole signal space except the decision region of the correct signal. In our example, this means that we omit the last decision region in Figure 11.4 from our computations. As we can see, the union of the first two regions already covers the whole "erroneous decision region." In the following, we will use this type of union bound.

Besides the union *bound*, there is also an *approximation* of the SER in M-ary modulation, namely the *nearest-neighbor* approximation. In this case, we add up the pairwise error probability with the symbols s_k that have the smallest distance to s_j; this approximation is justified by the fact that the nearest neighbors dominate the overall error probability. For some modulations, this approximation coincides with the above-described tightened union bound; however, in other situations, we might not get an upper bound but just an approximation.

For higher-order modulation, it is furthermore important to distinguish BER and SER. Take, e.g., the 4-Quadrature Amplitude Modulation (QAM) of Figure 11.4, and assume furthermore that the signal constellations are Gray-coded, so that the four points represent the bit combinations 00, 01, 11, and 10 (when read clockwise). There is a high probability of errors occurring between two neighboring signal constellations. One symbol error (in one transmitted symbol) then corresponds to one bit error (one error in *two* transmitted bits). Thus, bit error probability is only about half symbol error probability.[3]

Example 11.1 *Compute the BER and SER of QPSK.*

As QPSK is a four-state modulation format (2 bit/symbol), we can invest twice the bit energy for each symbol. The signal points thus have squared Euclidean distance $d^2 = E_\mathrm{S} = 2E_\mathrm{B}$ from the origin; the points in the signal constellation diagram are thus at $\pm\sqrt{E_\mathrm{B}}(\pm 1 \pm j)$, and the distance between two neighboring points is $d_{jk}^2 = 4E_\mathrm{B}$.

From Eq. (11.26) it follows that the pairwise error probability is

$$Q\left(\sqrt{2\frac{E_\mathrm{B}}{N_0}}\right) = Q(\sqrt{2\gamma_\mathrm{B}}). \tag{11.37}$$

According to Figure 11.4, the symbol error probability as computed from the union bound is twice as large:

$$SER \approx 2Q(\sqrt{2\gamma_\mathrm{B}}). \tag{11.38}$$

Now, as discussed above, the BER is half the SER:

$$BER = Q(\sqrt{2\gamma_\mathrm{B}}). \tag{11.39}$$

This is the same BER as for BPSK. Notably, Eq. (11.39) is exact, even though Eq. (11.38), as well as $BER \approx SER/2$ are both approximations. The reason is that the two approximation errors cancel each other out: Eq. (11.38) "double counts" the lower left quadrant; but it is exactly in this region where two bit errors per symbol (instead of the one bit error we assume with $BER \approx SER/2$) occur.

For QPSK, it is also possible to compute the symbol error probability exactly: the probability of a correct decision (in Figure 11.4) is to stay in the right half plane (probability for that event is $1 - Q(\sqrt{2\gamma_\mathrm{B}})$) times the probability of staying in the upper half-plane (the probability for this is independent of the probability of being in the right half plane, and also is $1 - Q(\sqrt{2\gamma_\mathrm{B}})$). The overall symbol error probability in thus $1 - [1 - Q(\sqrt{2\gamma_\mathrm{B}})]^2$, i.e.,

$$SER = 2Q(\sqrt{2\gamma_\mathrm{B}})\left[1 - \frac{1}{2}Q(\sqrt{2\gamma_\mathrm{B}})\right] \tag{11.40}$$

which shows the magnitude of the error made by the union bound. The relative error is $0.5Q(\sqrt{2\gamma_\mathrm{B}})$, which tends to 0 as γ_B tends to infinity.

We stress again that great care must be taken when using equations for the BER of higher-order modulation formats from the literature. There are several possible pitfalls:

- Does the equation give the symbol error probability or the bit error probability?
- Does computation of the SNR use bit energy or symbol energy? A fair comparison between different modulation formats should be based on E_B/N_0.
- Some authors define the distance between the origin and (equal-energy) signal constellation points not as $\sqrt{E}$, but rather as $\sqrt{2E}$. This is related to a different normalization of the expansion functions, which is compensated by different values of the noise variance The final results do not change, but this makes the combination of intermediate results from different sources much more difficult.

---

[3] This demonstrates the drawbacks of mixing up the expressions "bit error probability" and "bit error rate." In our example, the bit error rate, i.e., number of bit errors per second, is the same as the symbol error rate, while the bit error probability is only half the symbol error probability.

Error Probability for Differential Detection

Carrier recovery can be a challenging problem (see [Meyr et al. 1997] and [Proakis and Salehi 2005]), which makes coherent detection difficult for some situations, in particular for low-cost, low-energy devices. Differential detection is an attractive alternative to coherent detection, because it renders irrelevant the absolute phase of the detected signal. The RX just compares the phases (and possibly amplitudes) of two subsequent symbols to recover the information in the symbol; this phase difference is independent of the absolute phase. If the phase rotation introduced by the channel is slowly time varying (and thus effectively the same for two subsequent symbols), it enters just the absolute phase, and thus need not be taken into account in the detection process.

For differential detection of Phase Shift Keying (PSK), the transmitter (TX) needs to provide differential encoding. For binary symmetric PSK, the transmit phase Φ_i of the ith bit is

$$\Phi_i = \Phi_{i-1} + \begin{cases} +\dfrac{\pi}{2} \text{ if } b_i = +1 \\[2mm] -\dfrac{\pi}{2} \text{ if } b_i = -1 \end{cases}. \tag{11.41}$$

Comparison of the difference between phases on two subsequent sampling instances determines whether the transmitted bit b_i was $+1$ or -1.[4]

For Continuous Phase Frequency Shift Keying (CPFSK), such differential encoding can be avoided. Remember that in the case of MSK (without differential encoding), the phase rotation over a 1-bit duration is $\pm\pi/2$. It is thus possible to determine the phases at two subsequent sampling points, take the difference, and conclude which bit has been transmitted. This can also be interpreted by the fact that the phase of the transmit signal is an integral over the uncoded bit sequence. Computing the phase difference is a first approximation to taking the derivative (it is exact if the phase change is linear), and thus reverses the integration.

For Differential Binary Phase Shift Keying (DBPSK), the BER for differential detection is [Proakis and Salehi 2005]

$$BER = \frac{1}{2} \exp\left(-\gamma_b\right). \tag{11.42}$$

For 4-DPSK with Gray-coding, it is

$$BER = Q_M(a, b) - \frac{1}{2} I_0(ab) \exp\left(-\frac{1}{2}\left(a^2 + b^2\right)\right) \tag{11.43}$$

where

$$a = \sqrt{2\gamma_B\left(1 - \frac{1}{\sqrt{2}}\right)}, \quad b = \sqrt{2\gamma_B\left(1 + \frac{1}{\sqrt{2}}\right)} \tag{11.44}$$

and $Q_M(a, b)$ is Marcum's Q-function:

$$Q_M(a, b) = \int_b^\infty x \exp\left[-\frac{a^2 + x^2}{2}\right] I_0(ax)dx \tag{11.45}$$

whose series representation is given in Eq. (5.31).

Error Probability for Noncoherent Detection

When the carrier phase is completely unknown, and differential detection is not an option, then noncoherent detection can be used. For equal-energy signals, the detector tries to maximize the metric:

$$\left|\mathbf{r}_{LP}\mathbf{s}_{LP,m}^*\right| \tag{11.46}$$

so that the optimum RX has a structure according to Figure 11.5.

An actual realization, where $r(t)$ is a bandpass signal, will in each branch of Figure 11.5 split the signal into two subbranches, in which it obtains and processes the I- and Q-branches of the signal separately; the squared outputs of the "absolute value" operation of the I- and Q-branches are then added up (and possibly the square root of the sum is taken) before the "select largest" operation.

---

[4] Theoretically, differential detection could also be performed on a non-encoded signal. However, one bit error could then lead to a whole chain of errors.

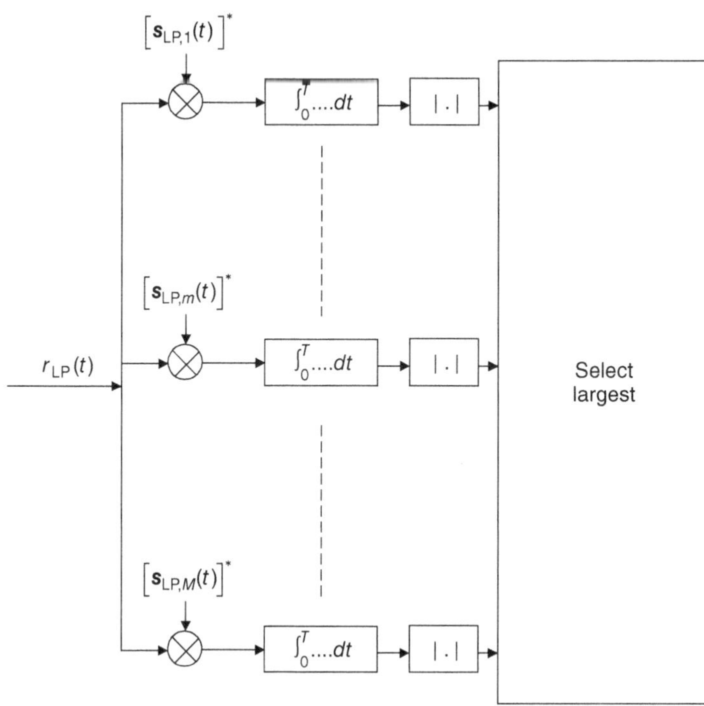

Figure 11.5 Optimum RX structure for noncoherent detection.

In this case, the BER for binary signals can be computed from Eq. (11.43), but with a different definition of the parameters a and b:

$$a = \sqrt{\frac{\gamma_{\text{B}}}{2}\left(1 - \sqrt{1 - |\rho|^2}\right)}, \quad b = \sqrt{\frac{\gamma_{\text{B}}}{2}\left(1 + \sqrt{1 - |\rho|^2}\right)}. \tag{11.47}$$

The optimum performance is achieved in this case if $\rho = 0$ – i.e., the signals are orthogonal; in this case, the BER is $(1/2)\exp(-\gamma_{\text{B}}/2)$. For the case when $|\rho| = 1$, which occurs for PSK signals, including BPSK and 4-QAM, the BER becomes 0.5.

Example 11.2 *Compute the BER of binary FSK in an AWGN channel with $\gamma_{\text{B}} = 5\,dB$, and compare it with DBPSK and BPSK.*

For binary FSK in AWGN, the BER is given by Eq. (11.34) with $\gamma_{\text{S}} = \gamma_{\text{B}}$. Thus, with $\gamma_{\text{B}} = 5\,$dB, we have:

$$\begin{aligned} \text{BER}_{\text{BFSK}} &= Q\left(\sqrt{\gamma_{\text{B}}}\right) \\ &= 0.038. \end{aligned} \tag{11.48}$$

For BPSK, the BER is instead given by Eq. (11.36), and thus:

$$\begin{aligned} \text{BER}_{\text{BPSK}} &= Q\left(\sqrt{2\gamma_{\text{B}}}\right) \\ &= 0.006. \end{aligned} \tag{11.49}$$

Finally, for differential BPSK, the BER is given by Eq. (11.42), which results in:

$$\begin{aligned} \text{BER}_{\text{DBPSK}} &= (1/2)e^{-\gamma_{\text{B}}} \\ &= 0.021. \end{aligned} \tag{11.50}$$

While OOK is an orthogonal modulation format, it is not equal energy. Its noncoherent detection thus requires comparison of the correlator output to a threshold. In the case of high SNR, this threshold is $\sqrt{E/2}$, but at lower SNR, the optimum threshold level actually depends on the SNR; for further details see [Benedetto and Biglieri 1999, Section 4.4.2].

11.2 Error Probability in Flat-Fading Channels

11.2.1 Average Error Probability – Classical Computation Method

In fading channels, the received signal power (and thus the SNR) is not constant but changes as the fading of the channel changes. In many cases, we are interested in the BER in a fading channel averaged over the different fading states. For a mathematical computation of the BER in such a channel, we have to proceed in three steps:

1. Determine the BER for any arbitrary SNR.
2. Determine the probability that a certain SNR occurs in the channel – in other words, determine the pdf of the power gain of the channel.
3. Average the BER over the distribution of SNRs.

In an AWGN channel, the BER decreases approximately exponentially as the SNR increases: for binary modulation formats, a 10-dB SNR is sufficient to give a BER on the order of 10^{-4}, for 15 dB the BER is below 10^{-8}. In contrast, we will see below that in a fading channel the BER decreases only linearly with the (average) SNR. At first glance, this is astonishing: sometimes fading leads to high SNRs, sometimes it leads to low SNRs, and it could be assumed that high and low values would compensate for each other. The important point here is that the relationship between (instantaneous) BER and (instantaneous) SNR is highly nonlinear, so that the cases of low SNR essentially determine the overall BER.

Example 11.3 *BER in a two-state fading channel.*

Consider the following simple example: a fading channel has an average SNR of 10 dB, where fading causes the SNR to be $-\infty$ dB half of the time, while it is 13 dB the rest of the time. The BERs corresponding to the two-channel states are 0.5 and 10^{-9}, respectively (assuming antipodal modulation with differential detection). The mean BER is then $\overline{BER} = 0.5 \cdot 0.5 + 0.5 \cdot 10^{-9} = 0.25$. For an AWGN channel with a 10-dB SNR, the BER is $2 \cdot 10^{-5}$.

Following this intuitive explanation, we now turn to the mathematical details of the abovementioned three-step procedure. Step 1, the determination of the BER of an arbitrary given SNR, was treated in Section 11.1. Point 2 requires computation of the SNR distribution. In Section 5.4, we showed that, e.g., for Rayleigh fading the pdf of the received power becomes

$$pdf_P(P) = \frac{1}{\overline{P}} \exp\left(-\frac{P}{\overline{P}}\right) \text{ for } P \geq 0. \tag{11.51}$$

Since the SNR is the received power scaled with the noise power, the pdf of the SNR is similarly

$$pdf_{\gamma_B}(\gamma_B) = \frac{1}{\overline{\gamma}_B} \exp\left(-\frac{\gamma_B}{\overline{\gamma}_B}\right) \tag{11.52}$$

where $\overline{\gamma}_B$ is the mean SNR. For Rician-fading channels, we saw that:

$$pdf_{\gamma_B}(\gamma_B) = \frac{1 + K_r}{\overline{\gamma}_B} \exp\left(-\frac{\gamma_B(1 + K_r) + K_r\overline{\gamma}_B}{\overline{\gamma}_B}\right) I_0\left(\sqrt{\frac{4(1 + K_r)K_r\gamma_B}{\overline{\gamma}_B}}\right) \tag{11.53}$$

where K_r is the Rice factor. Analogous computations are possible for other amplitude distributions.

In the last step, the BER has to be averaged over the distribution of the SNR:

$$\overline{BER} = \int pdf_{\gamma_B}(\gamma_B) BER(\gamma_B) d\gamma_B. \tag{11.54}$$

For Rayleigh fading, a closed-form evaluation of Eq. (11.54) is possible for many modulation formats. Using the relationship

$$2\int_0^\infty Q\left(\sqrt{2x}\right) a \exp\left(-ax\right) dx = 1 - \sqrt{\frac{1}{1 + a}} \tag{11.55}$$

the mean BER for coherent detection of binary antipodal signals is

$$\overline{BER} = \frac{1}{2}\left[1 - \sqrt{\frac{\overline{\gamma}_B}{1 + \overline{\gamma}_B}}\right] \approx \frac{1}{4\overline{\gamma}_B} \tag{11.56}$$

and the mean BER for coherent detection of binary orthogonal signals is

$$\overline{BER} = \frac{1}{2}\left[1 - \sqrt{\frac{\overline{\gamma}_B}{2 + \overline{\gamma}_B}}\right] \approx \frac{1}{2\overline{\gamma}_B}. \tag{11.57}$$

These BERs are plotted in Figure 11.6.

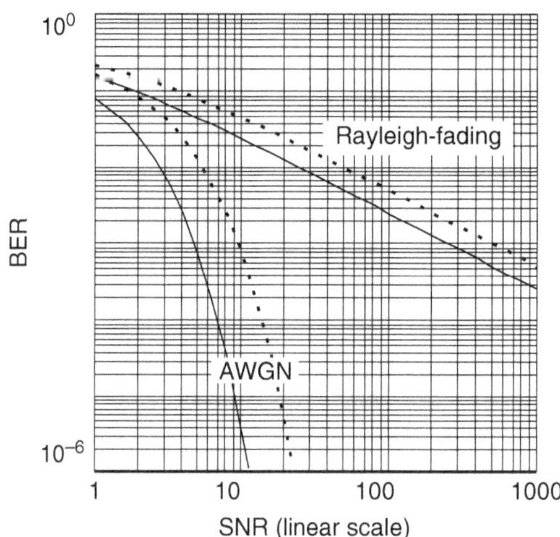

Figure 11.6 BER for binary signals with coherent detection: antipodal (solid) and orthogonal (dashed).

For differential detection, the averaging process is even simpler, as the BER curve is an exponential function of the instantaneous SNR. For binary antipodal signals:

$$\overline{BER} = \frac{1}{2(1 + \overline{\gamma}_B)} \approx \frac{1}{2\overline{\gamma}_B} \tag{11.58}$$

and for noncoherent detection of binary orthogonal signals:

$$\overline{BER} = \frac{1}{2 + \overline{\gamma}_B} \approx \frac{1}{\overline{\gamma}_B}. \tag{11.59}$$

For differential detection, the BER in Rician channels can also be computed in closed form. For antipodal signals:

$$\overline{BER} = \frac{1 + K_r}{2(1 + K_r + \overline{\gamma}_B)} \exp\left(-\frac{K_r \overline{\gamma}_B}{1 + K_r + \overline{\gamma}_B}\right) \tag{11.60}$$

and for orthogonal signals:

$$\overline{BER} = \frac{1 + K_r}{(2 + 2K_r + \overline{\gamma}_B)} \exp\left(-\frac{K_r \overline{\gamma}_B}{2 + 2K_r + \overline{\gamma}_B}\right). \tag{11.61}$$

We note that as $K_r \to \infty$, Eq. (11.60) becomes $(1/2) \exp(-\overline{\gamma}_B)$, i.e., the BER of differential detection of antipodal signals in AWGN, while for $K_r = 0$, it becomes $1/(2 + 2\overline{\gamma}_B)$, i.e., the solution for Rayleigh channels in Eq. (11.58).

Example 11.4 *Compute the BER of DBPSK with $\overline{\gamma}_B = 12\,dB$ and $K_r = -3\,dB$, $0\,dB$, $10\,dB$.*

The BER is given by Eq. (11.56), and hence, with $K_r = -3$ dB:

$$\begin{aligned}\overline{BER} &= \frac{1 + K_r}{2(1 + K_r + \overline{\gamma}_B)} \exp\left(-\frac{K_r \overline{\gamma}_B}{1 + K_r + \overline{\gamma}_B}\right) \\ &= 0.027.\end{aligned} \tag{11.62}$$

Then, with $K_r = 0$ dB we get $\overline{BER} = 0.023$, and with $K_r = 10$ dB we get $\overline{BER} = 0.00056$. Note the strong decrease in BER for high Rice factors, even though we kept the average SNR constant.

*11.2.2 Average Error Probability – Moment-Generating Function Method

In the late 1990s, a new method for computation of the average BER was proposed, and shown to be very efficient. It is based on an alternative representation of the Q-function, and allows easier averaging over different fading distributions [Annamalai et al. 2000, Simon and Alouini 2004].

Alternative Representation of the Q-Function

When evaluating the classical definition of the Q-function

$$Q(x) = \frac{1}{\sqrt{2\pi}} \int_x^\infty \exp\left(-t^2/2\right) dt \tag{11.63}$$

there is the problem that the argument of the Q-function is in the integration limit, not in its integrand. This makes evaluation of the integrals of the Q-function much more difficult, especially because we cannot use the beloved trick of exchanging the sequence of integration in multiple integrals. This problem is particularly relevant in BER computations. The problem can be solved by using an alternative formulation of the Q-function:

$$Q(x) = \frac{1}{\pi} \int_0^{\pi/2} \exp\left(-\frac{x^2}{2\sin^2\theta}\right) d\theta \ \text{ for } x > 0. \tag{11.64}$$

This representation now has the argument in the integrand (in a Gaussian form, $\exp(-x^2)$), and also has finite integration limits. We will see below that this greatly simplifies evaluation of the error probabilities.

It turns out that Marcum's Q-function, as defined in Eq. (11.45), also has an alternative representation:

$$Q_M(a,b) = \begin{cases} \dfrac{1}{2\pi} \displaystyle\int_{-\pi}^{\pi} \dfrac{b^2 + ab\sin\theta}{b^2 + 2ab\sin\theta + a^2} \exp\left(-\dfrac{1}{2}\left(b^2 + 2ab\sin\theta + a^2\right)\right) d\theta \ \text{ for } b > a \geq 0 \\[3ex] 1 + \dfrac{1}{2\pi} \displaystyle\int_{-\pi}^{\pi} \dfrac{b^2 + ab\sin\theta}{b^2 + 2ab\sin\theta + a^2} \exp\left(-\dfrac{1}{2}\left(b^2 + 2ab\sin\theta + a^2\right)\right) d\theta \ \text{ for } a > b \geq 0 \end{cases}. \tag{11.65}$$

Error Probability in Additive White Gaussian Noise Channels

For computation of the BER in AWGN channels, we find that the BER for BPSK can be written as (compare also Eq. (11.32)):

$$BER = Q\left(\sqrt{2\gamma_S}\right) \tag{11.66}$$

$$= \frac{1}{\pi} \int_0^{\pi/2} \exp\left(-\frac{\gamma_S}{\sin^2\theta}\right) d\theta. \tag{11.67}$$

For QPSK, the SER can be computed as (compare Eq. (11.37)):

$$SER = 2Q\left(\sqrt{\gamma_S}\right) - Q^2\left(\sqrt{\gamma_S}\right) \tag{11.68}$$

$$= \frac{1}{\pi} \int_0^{3\pi/4} \exp\left(-\frac{\gamma_S}{\sin^2\theta} \sin^2(\pi/4)\right) d\theta$$

and quite generally for M-ary PSK:

$$SER = \frac{1}{\pi} \int_0^{(M-1)\pi/M} \exp\left(-\frac{\gamma_S}{\sin^2\theta} \sin^2(\pi/M)\right) d\theta. \tag{11.69}$$

For binary orthogonal FSK, we finally find that

$$BER = Q\left(\sqrt{\gamma_S}\right) \tag{11.70}$$

$$= \frac{1}{\pi} \int_0^{\pi/2} \exp\left(-\frac{\gamma_S}{2\sin^2\theta}\right) d\theta. \tag{11.71}$$

Example 11.5   *Compare the SER for 8-PSK as computed from Eq. (11.69) with the value obtained from the union bound for $\gamma_S = 3$ and 10 dB.*

First using Eq. (11.69) with $M = 8$ and $\gamma_S = 3$ dB we have

$$SER = \frac{1}{\pi} \int_0^{(M-1)\pi/M} \exp\left(-\frac{\gamma_S}{\sin^2\theta} \sin^2(\pi/M)\right) d\theta \tag{11.72}$$

$$= 0.442.$$

and for $\gamma_S = 10\,\mathrm{dB}$ we get $SER = 0.087$, where the integral has been evaluated numerically.

The 8-PSK constellation has a minimum distance of $d_{\min} = 2\sqrt{E_S}\sin(\pi/8)$. The nearest neighbor union bound on the SER is then given by

$$
\begin{aligned}
SER_{\text{union-bound}} &= 2{\cdot}Q\left(\frac{d_{\min}}{\sqrt{2N_0}}\right) \\
&= 2{\cdot}Q\left(\frac{2\sqrt{E_S}\sin(\pi/8)}{\sqrt{2N_0}}\right) \\
&= 2{\cdot}Q\left(\sqrt{2\gamma_S}\sin(\pi/8)\right).
\end{aligned}
\tag{11.73}
$$

Thus, with $\gamma_S = 3\,\mathrm{dB}$ we get $SER_{\text{union-bound}} = 0.445$, and with $\gamma_S = 10\,\mathrm{dB}$ we get $SER_{\text{union-bound}} = 0.087$. In this example, the union bound gives a very good approximation, even at the rather low SNR of 3 dB.

Error Probability in Fading Channels

For AWGN channels, the advantages of the alternative representation of the Q-function are rather limited. They allow a simpler formulation for certain higher-order modulation formats but do not exhibit significant advantages for the modulation formats that are mostly used in practice. The real advantage emerges when we apply this description method as the basis for computations of the BER in fading channels. We find that we have to average over the pdf of the SNR $pdf_\gamma(\gamma)$, as described in Eq. (11.55). We have now seen that the alternative representation of the Q-function allows us to write the SER (for a given SNR) in the generic form:

$$
SER(\gamma) = \int_{\theta_1}^{\theta_2} f_1(\theta)\exp\left(-\gamma f_2(\theta)\right)d\theta.
\tag{11.74}
$$

Thus, the average SER becomes

$$
\overline{SER} = \int_0^\infty pdf_\gamma(\gamma)SER(\gamma)d\gamma
\tag{11.75}
$$

$$
= \int_0^\infty pdf_\gamma(\gamma)\int_{\theta_1}^{\theta_2} f_1(\theta)\exp\left(-\gamma f_2(\theta)\right)d\theta\,d\gamma
\tag{11.76}
$$

$$
= \int_{\theta_1}^{\theta_2} f_1(\theta)\int_0^\infty pdf_\gamma(\gamma)\exp\left(-\gamma f_2(\theta)\right)d\gamma\,d\theta.
\tag{11.77}
$$

Let us now have a closer look at the inner integral:

$$
\int_0^\infty pdf_\gamma(\gamma)\exp\left(-\gamma f_2(\theta)\right)d\gamma.
\tag{11.78}
$$

We find that it is the moment-generating function of $pdf_\gamma(\gamma)$, evaluated at the point $-f_2(\theta)$. Remember (see also, e.g., [Papoulis 1991]) that the moment-generating function is defined as the Laplace transform of the pdf of γ:

$$
M_\gamma(s) = \int_0^\infty pdf_\gamma(\gamma)\exp(\gamma s)\,d\gamma
\tag{11.79}
$$

and the mean SNR is the first derivative, evaluated at $s = 0$:

$$
\overline{\gamma} = \left.\frac{dM_\gamma(s)}{ds}\right|_{s=0}.
\tag{11.80}
$$

Summarizing, the average SER can be computed as

$$
\overline{SER} = \int_{\theta_1}^{\theta_2} f_1(\theta)M_\gamma(-f_2(\theta))\,d\theta.
\tag{11.81}
$$

The next step is then finding the moment-generating function of the distribution of the SNR. Without going into the details of the derivations, we find that for a Rayleigh distribution of the signal amplitude, the moment-generating function of the SNR distribution is

$$M_\gamma(s) = \frac{1}{1 - s\bar{\gamma}}$$ (11.82)

for a Rice distribution, it is

$$M_\gamma(s) = \frac{1 + K_r}{1 + K_r - s\bar{\gamma}} \exp\left[\frac{K_r s\bar{\gamma}}{1 + K_r - s\bar{\gamma}}\right]$$ (11.83)

and for a Nakagami distribution with parameter m:

$$M_\gamma(s) = \left(1 - \frac{s\bar{\gamma}}{m}\right)^{-m}.$$ (11.84)

Having now the general form of the SER (Eq. (11.81)) and the form of the moment-generating function, the computation of the error probabilities becomes straightforward (if sometimes a bit tedious).

Example 11.6 *BER of BPSK in Rayleigh fading.*

As one example, let us go through the computation of the average BER of BPSK in a Rayleigh fading channel – a problem for which we already know the result from Eq. (11.56). Looking at Eq. (11.67), we find that $\theta_1 = 0$, $\theta_2 = \pi/2$:

$$f_1(\theta) = \frac{1}{\pi}$$ (11.85)

$$f_2(\theta) = \frac{1}{\sin^2(\theta)}.$$ (11.86)

Since we consider Rayleigh fading:

$$M_\gamma(-f_2(\theta)) = \frac{1}{1 + \frac{\bar{\gamma}}{\sin^2(\theta)}}$$ (11.87)

so that the total BER is, according to Eq. (11.81):

$$\overline{BER} = \frac{1}{\pi}\int_0^{\pi/2} \frac{\sin^2(\theta)}{\sin^2(\theta) + \bar{\gamma}} d\theta$$ (11.88)

which can be shown to be identical to the first (exact) expression in Eq. (11.56) – namely:

$$\overline{BER} = \frac{1}{2}\left[1 - \sqrt{\frac{\bar{\gamma}}{1 + \bar{\gamma}}}\right].$$ (11.89)

Many modulation formats and fading distributions can be treated in a similar way. An extensive list of solutions, dealing with coherent detection, partially coherent detection, and noncoherent detection in different types of channels, is given in the book by [Simon and Alouini 2004].

11.2.3 *Outage Probability versus Average Error Probability*

In the previous section, we computed the average bit error probability, where averaging was done over small-scale fading. Similarly, we could also average this distribution over large-scale fading – the mathematics are quite similar. But what is the physical meaning of these averaged values? In order to understand *what* we are computing here, and what kind of averaging makes sense, we first have to have a look at the different operating scenarios that can occur in a wireless system.

As a first step, we have to compare the length of the "memory" of the system with the coherence time of the channel. The "memory" of the system in this context could, e.g., be caused by

1. perception of the human ear during transmission of speech data (error bursts on the order of a few milliseconds or less tend to be "smoothed out" by the human ear); or
2. buffers at the RX, such as video buffers for streaming video, which store a few seconds or minutes of upcoming video, and thus allow smooth replay even when transmission of new data is temporarily interrupted;
3. the size of typical data structures (file size) that are to be transmitted over the wireless connection;

4. further memory can also come from coding (in which case the block length of the block code or the constraint length of a convolutional code would determine memory duration), and interleaving, in which case the length of the interleaver would determine the memory (see also Chapter 13).

Let us first consider the case where the RX or TX moves, so that the channel changes within a finite time and system memory can thus extend over many channel realizations. Such system memories might typically see many small-scale realizations of the channel, as well as large-scale variations of the channel. When we now want to see the average BER for a file transfer, we have to average the BER over the duration of that file transfer, and thus average it over the distribution of the SNRs seen during that time.[5] The BER is a single, deterministic value.

Next, we consider the case where system memory is much shorter than coherence time. To name but one example: if we wish to transmit a file from a (stationary) laptop, and the objects in the environment are not moving, the channel seen by the wireless connection is static for the duration of the transmission. The SNR seen by the demodulator is thus constant, but random, depending on the position. It is thus meaningful to investigate the pdf of the SNR, in order to see what SNR is available in what percentage of locations. For each of the locations – i.e., for each realization of the SNR – we compute the BER simply as the BER of an AWGN channel with that specific SNR. In such a case, we will also get a *distribution* of the BER; note that this distribution is for a fixed transmit power.

From these considerations, we arrive at the concept of *outage probability*. For many applications, it is not important what the exact value of the BER is, as long as it stays below a certain threshold. For example, a file transfer is successful as long as the raw BER is small enough so that errors can be corrected by the error correction coding – in other words, as long as a certain threshold raw BER (typically on the order of a few percent) is not exceeded. It is then meaningful to determine the percentage of locations where successful file transfer will not be available. This percentage is known as outage probability.

In some situations, the memory is long enough to see multiple realizations of the small-scale fading, but only one realization of the large-scale fading. In this case, an outage occurs if the small-scale averaged BER lies below a certain threshold, and the probability for such an outage to occur is determined by the large-scale fading statistics only.

Computation of outage probability becomes much simpler if we define not a maximum BER but rather a minimum SNR γ_0, for the system to work properly. We can then find the outage probability as

$$\mathrm{Pr_{out}} = \mathrm{Pr}(\gamma < \gamma_0) = \int_0^{\gamma_0} pdf_\gamma(\gamma) d\gamma. \tag{11.90}$$

Outage probability can also be seen as another way of establishing a fading margin: we need to find the mean SNR that guarantees a certain outage.

11.3 Error Probability in Delay- and Frequency-Dispersive Fading Channels

11.3.1 Physical Cause of Error Floors

In wireless propagation channels, transmission errors are caused not only by noise but also by signal distortions. These distortions are created on one hand by delay dispersion (i.e., echoes of the transmit signal arriving with different delays), and on the other hand by frequency dispersion (Doppler effect – i.e., signal components arriving with different Doppler shifts). For high data rates, delay dispersion is dominant; at low data rates, frequency dispersion is the main reason for signal distortion errors. In either of these cases, an increase in transmit power does not lead to a reduction of the BER; for that reason, these errors are often called *error floor* or *irreducible errors*. Of course, these errors can also be reduced or eliminated, but this has to be done by methods other than increasing power (e.g., equalization, diversity, etc.). In this section, we only treat the case when the RX does not use any of these countermeasures, so that dispersion leads to increased error rates. Later chapters discuss in detail the fact that dispersion can actually be a benign effect if specific RX structures are used.

Frequency Dispersion

We first consider errors due to frequency dispersion. For FSK, it is immediately obvious how frequency dispersion leads to errors: random Frequency Modulation (FM) (see Section 5.7.3) leads to a frequency shift $\dot{\psi}$ of the received signal, and can push a symbol over the decision boundary. Assume that $+1$ was sent (i.e., the frequency $f_c + f_{\mathrm{mod}}$). Due to the random FM effect, the frequency $f_c + f_{\mathrm{mod}} + \dot{\psi}/2\pi = f_{\mathrm{inst}}$ is received. If this is smaller than f_c, the RX opts for -1. Note that *instantaneous* frequency shifts can be significantly larger than the maximum Doppler frequency even though the statistics of the random FM are determined by the Doppler spectrum of the channel. Consider the following equation for the instantaneous frequency:

$$f_{\mathrm{inst}}(t) = \frac{\mathrm{Im}\left(r^*(t)\frac{dr(t)}{dt}\right)}{|r(t)|^2}. \tag{11.91}$$

---

[5] For a coded system, where the code length is larger than channel coherence time, computations are a bit trickier (as discussed in Chapter 13).

Obviously, this can become very large when the amplitude becomes very small. In other words, deep fading dips lead to large shifts in the instantaneous frequency, and thus higher error probability.

A somewhat different interpretation can be given for differential detection. As mentioned above, differential detection assumes that the channel does not change between two adjacent symbols. However, if there is a finite Doppler, then the channel *does* change – remember that the Doppler spectrum gives a statistical description of channel changes. Thus, a nonzero Doppler effect implies a wrong reference phase for differential detection. If this effect is strong, it can lead to erroneous decisions. Also in this case, it is true that channel changes are strongest near fading dips.[6]

For MSK with differential detection, [Hirade et al. 1979] determined the BER due to the Doppler effect:

$$\overline{BER}_{\text{Doppler}} = \frac{1}{2}(1 - \xi_s(T_B)) \tag{11.92}$$

where $\xi_s(t)$ is the normalized autocorrelation function of the channel (so that $\xi_s(0) = 1$) – i.e., the Fourier transform of the normalized Doppler spectrum, e.g., $J_0(2\pi v_{\max}T_B)$ in the case of a Jakes spectrum. For a Jakes spectrum with small $v_{\max}T_B$ we then get a BER that is proportional to the squared magnitude of the product of Doppler shift and bit duration:

$$\overline{BER}_{\text{Doppler}} = \frac{1}{2}\pi^2(v_{\max}T_B)^2. \tag{11.93}$$

This basic functional relationship $\overline{BER}_{\text{Doppler}} = K(S_v T_B)^2$ also holds for other Doppler spectra and modulation formats; it is only the proportionality constant K that changes.

From this relationship, we find that errors due to frequency dispersion are mainly important for systems with a low data rate. For example, paging systems and sensor networks exhibit data rates on the order of 1 kbit/s, while Doppler frequencies can be up to a few hundred Hz. Error floors of 10^{-2} are thus easily possible. This has to be taken into account when designing the coding for such systems. For high-data-rate systems with modulation of a single carrier signal, as described in this chapter, errors due to frequency dispersion do not play a noticeable role.[7]

Delay Dispersion

In contrast to frequency dispersion, delay dispersion has great importance for high-data-rate systems. This becomes obvious when we remember that the errors in unequalized systems are determined by the ratio of symbol duration that is disturbed by InterSymbol Interference (ISI) to that of the undisturbed part of the symbol. The maximum excess delay of a channel impulse response is determined by the environment, and independent of the system; let us assume in the following a maximum excess delay of 1 μs. In a system with a symbol duration of 20 μs, the ISI can disturb 5% of each symbol, while it can disturb 20% if the symbol duration is 5 μs.

Many theoretical and experimental investigations have shown that the error floor due to delay dispersion is given by the following equation:

$$\overline{BER} = K\left(\frac{S_\tau}{T_B}\right)^2 \tag{11.94}$$

where S_τ is the rms delay spread of the channel (see Section 6.5.3). Just as for frequency dispersion, errors mainly occur near fading dips. Section 11.3.2 gives an interpretation of this fact in terms of group delay, which reaches its largest values near fading dips (see also Chapter 5).

Equation (11.94) is only valid if the maximum excess delay of the channel is much smaller than the symbol duration, and the channel is Rayleigh fading. The proportionality constant K depends on the modulation method, filtering at TX and RX, the form of the average impulse response, and choice of the sampling instant, as we will discuss in the sections below.

Choice of the Sampling Instant
In a flat-fading channel, the choice of sampling instant is obvious – sampling should always occur at those times where the SNR at the decision device is largest; this usually occurs either at the bit transitions or exactly in the middle between bit transitions; in either case, we call this time henceforth $t_s = 0$.

For channels with delay dispersion, the choice of sampling time is no longer obvious. Most theoretical derivations assume that either $t_s = 0$ (i.e., sampling occurs at the minimum excess delay),[8] or at the *average mean delay*. The latter actually is the optimum sampling time for some Power Delay Profiles (PDPs, see Chapter 6), as demonstrated in Figure 11.7.

---

[6] For general QAM, not only the reference phase but also the reference amplitude is relevant. However, in the following we will restrict our considerations to PSK and FSK, and thus ignore amplitude distortions.

[7] However, this does *not* mean that time variations in the channel are unimportant in such systems. Channel variations can also have an impact on coding, on the validity of channel estimation, etc. Furthermore, frequency dispersion can be important in multicarrier systems, see Chapter 15.

[8] Without restriction of generality, we assume that $\tau_0 = 0$.

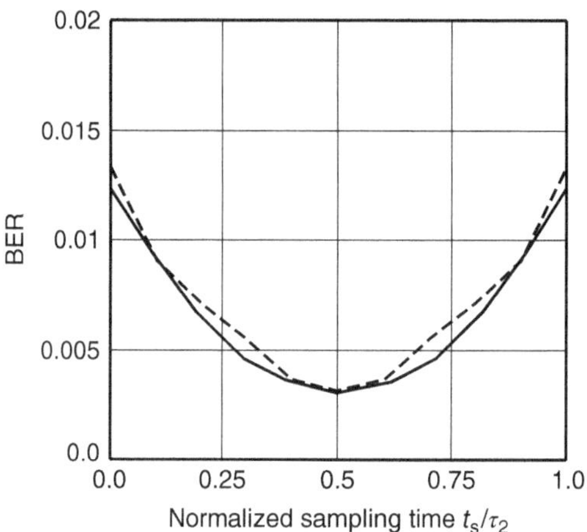

Figure 11.7 Dependence of delay-dispersion-induced error probability BER on choice of sampling instant in a two-spike channel.
Reproduced with permission from [Molisch 2000] © Prentice Hall.

When the sampling instant is chosen adaptively, according to the instantaneous state of the channel, the error floor can be decreased considerably, and – in unfiltered systems – even completely eliminated.

Impact of the Shape of the Power Delay Profile

To a first approximation, only the *rms delay spread* determines the BER due to delay dispersion. A closer look reveals, however, that the actual shape of the PDP also has an impact. The variation of the BER for different PDPs (for equal S_τ) is usually less than a factor of 2 and is thus often neglected. Figure 11.8 shows that a rectangular PDP leads to a slightly larger BER than a two-spike PDP; the difference is 75%.

An exponential PDP leads to an even larger error floor; however, for this shape, sampling at the average mean delay is not optimum.

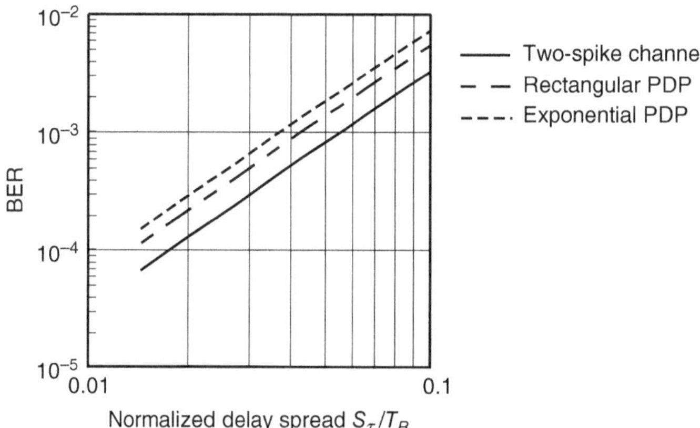

Figure 11.8 Impact of the shape of the PDP on the error floor due to delay dispersion. The modulation method is MSK with differential detection.
Reproduced with permission from [Molisch 2000] © Prentice Hall.

Filtering

Filtering at the TX and/or RX also leads to signal distortion, and thus makes the signal more susceptible to errors by the additional distortions caused by the channel. The narrower the filtering, the larger the error floor. Figure 11.9 shows the effect of filtering on QAM with Raised Cosine (RC) filters.

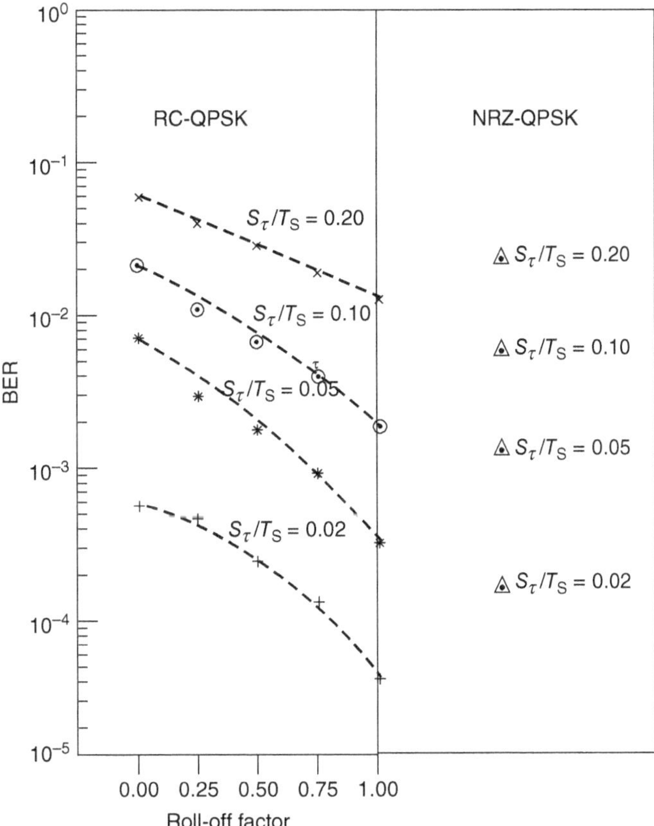

Figure 11.9 Error floor of quadrature-phase shift keying with coherent detection and RC filters as a function of the roll-off factor. For comparison purposes, we also show the BER of conventional QPSK. Note that there is no roll-off factor where RC QPSK reduces to conventional Nonreturn to Zero (NRZ) QPSK.
Reproduced with permission from [Chuang 1987] © IEEE.

The question naturally arises as to the optimum filter bandwidth. Very narrow filters (bandwidth on the order of the inverse symbol duration or less) lead to strong ISIs by themselves. Even if all decisions can be made correctly in the absence of further disturbances, such a filter makes the system more susceptible to channel-induced ISI and noise. For wide filters, a lot of noise can pass through the filter, which also leads to high error rates. The optimum filter bandwidth depends on the ratio of delay dispersion to noise. Figure 11.10 shows an example of such a tradeoff.

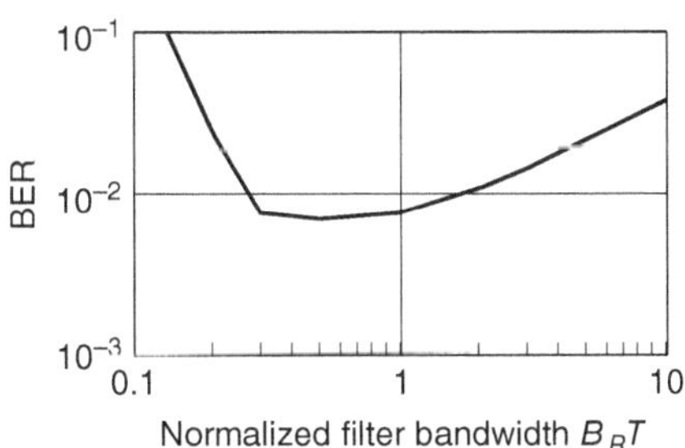

Figure 11.10 BER of filtered minimum shift keying with differential detection. The normalized rms delay spread is 0.1, and $SNR = 12\,\mathrm{dB}$.
Reproduced with permission from [Molisch 2000] © Prentice Hall.

Modulation Method

The modulation method also has an impact on the error floor: obviously, a modulation format is more sensitive to distortions by the channel the closer the signals are in the signal constellation diagram. QPSK shows a higher error floor than BPSK when the rms delay spread is normalized to the same *symbol* duration (see Figure 11.11). Higher-order modulation formats fare better when equal *bit* durations are assumed.

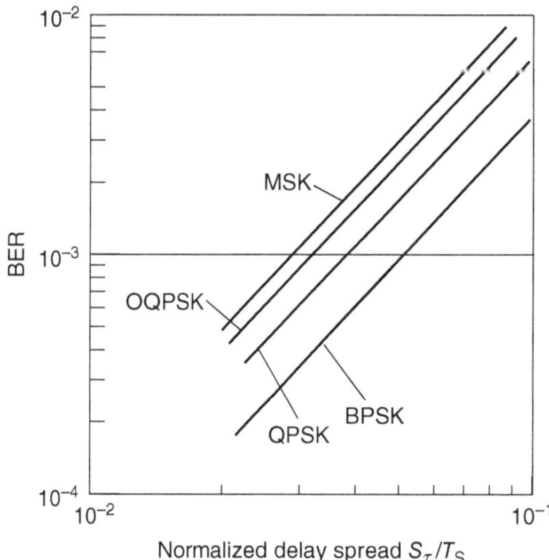

Figure 11.11 Error floor of different modulation formats as a function of normalized rms delay spread (normalized to symbol duration). Reproduced with permission from [Chuang 1987] © IEEE.

*11.3.2 Computation of the Error Floor Using the Group Delay Method

In this section, we present a very simple, approximate method for computing the BER due to delay dispersion. The method was introduced in [Andersen 1991] for PSK and [Crohn et al. 1993] for MSK. In the following, we present the method for differentially detected MSK.

Group delay T_g is defined as:

$$T_g = -\left.\frac{\partial \Phi_c}{\partial \omega}\right|_{\omega = 0} \tag{11.95}$$

where ω is angular frequency. Here, $\Phi_c(\omega)$ is the phase of the channel transfer function, which can be expanded into a Taylor series:

$$\Phi_c(\omega) = \Phi_c(0) + \omega\left.\frac{\partial \Phi_c}{\partial \omega}\right|_{\omega = 0} + \frac{1}{2}\omega^2\left.\frac{\partial^2 \Phi_c}{\partial \omega^2}\right|_{\omega = 0} + \cdots. \tag{11.96}$$

The first term of this series corresponds to the *average mean delay* and can be omitted if the sampling is done at the average mean delay. Terminating the Taylor series after the linear term, we obtain

$$\Phi_c(\omega) = -\omega T_g. \tag{11.97}$$

Obviously, phase distortion at the sampling instant depends on the instantaneous frequency at these instants. If a +1 is transmitted, then the instantaneous frequency is $\pi/(2T_B)$; otherwise, it is $-\pi/(2T_B)$. When the same bits are transmitted (i.e., a +1 followed by a +1, or a −1 followed by a −1), then the difference between the phase distortions is $\Delta\Phi = 0$; when different bits are transmitted, then

$$\Delta\Phi = \pm\frac{\pi}{T_B}T_g. \tag{11.98}$$

A decision error occurs when the size of channel-induced phase distortions (or rather; their difference at the sampling instants) is larger than $\pi/2$ – i.e., when:

$$|T_g| > T_B/2. \tag{11.99}$$

The statistics of group delay in a Rayleigh-fading channel is well known [Andersen et al. 1990] – namely, a *Student's t* distribution:

$$pdf_{T_g}(T_g) = \frac{1}{2S_\tau}\frac{1}{\left[1 + \left(T_g/S_\tau\right)^2\right]^{3/2}}. \tag{11.100}$$

The probability for bit errors can thus be easily computed from Eq. (11.99). Furthermore, when averaging over the different possible bit combinations, we get:

$$BER = \frac{4}{9}\left(\frac{S_\tau}{T_B}\right)^2 \approx \frac{1}{2}\left(\frac{S_\tau}{T_B}\right)^2. \tag{11.101}$$

Example 11.7 *Consider a system using differentially detected MSK with $T_B = 35\,\mu s$, operating at 900 MHz, moving at 360 km/h (high-speed train), and an exponential PDP with $S_\tau = 10\,\mu s$. Compute the BER due to frequency dispersion and delay dispersion.*

The BER due to frequency dispersion can be computed by first calculating the maximum Doppler shift as

$$\begin{aligned}
v_{\max} &= f_c \frac{v}{c} \\
&= 9 \cdot 10^8 \frac{100}{3 \cdot 10^8} \\
&= 300 \text{Hz}.
\end{aligned} \tag{11.102}$$

For MSK with differential detection, and assuming a classical Jake's Doppler spectrum, the BER due to frequency dispersion is given by Eq. (11.93), and hence:

$$\begin{aligned}
\overline{BER}_{\text{Doppler}} &= \frac{1}{2}\pi^2 (v_{\max} T_B)^2 \\
&= 5.4 \cdot 10^{-4}.
\end{aligned} \tag{11.103}$$

The BER due to delay dispersion is given by Eq. (11.101), which means that

$$\begin{aligned}
\overline{BER}_{\text{Delay}} &= \frac{4}{9}\left(\frac{S_\tau}{T_B}\right)^2 \\
&= 3.6 \cdot 10^{-2}.
\end{aligned} \tag{11.104}$$

*11.3.3   General Fading Channels: The Quadratic Form Gaussian Variable Method

A general method for the computation of BERs in dispersive fading channels is the so-called Quadratic Form Gaussian Variable (QFGV) method.[9] The channel is Rayleigh or Rice fading, suffering from delay dispersion and frequency dispersion, and adds AWGN. Mathematically speaking, the QFGV method determines the probability that a variable D:

$$D = A|X|^2 + B|Y|^2 + CXY^* + C^*X^*Y \tag{11.105}$$

is smaller than 0. Here, X and Y are complex Gaussian variables, A and B are real constants, and C is a complex constant. Defining the auxiliary variables:

$$\left.\begin{aligned}
w &= \frac{AR_{xx} + BR_{yy} + CR_{xy}^* + C^*R_{xy}}{4\left(R_{xx}R_{yy} - |R_{xy}|^2\right)\left(|C|^2 - AB\right)} \\
v_{1,2} &= \sqrt{w^2 + \frac{1}{4\left(R_{xx}R_{yy} - |R_{xy}|^2\right)\left(|C|^2 - AB\right)}} \mp w \\
\alpha_1 &= 2\left(|C|^2 - AB\right)\left(|\overline{X}|^2 R_{yy} + |\overline{Y}|^2 R_{xx} - \overline{X}^*\overline{Y}R_{xy} - \overline{X}\,\overline{Y}^* R_{xy}^*\right) \\
\alpha_2 &= A|\overline{X}|^2 + B|\overline{Y}|^2 + C\overline{X}^*\overline{Y} + C^*\overline{X}\,\overline{Y}^* \\
p_1 &= \frac{\sqrt{2v_1^2 v_2(\alpha_1 v_2 - \alpha_2)}}{|v_1 + v_2|} \\
p_2 &= \frac{\sqrt{2v_1 v_2^2(\alpha_1 v_1 + \alpha_2)}}{|v_1 + v_2|}
\end{aligned}\right\} \tag{11.106}$$

where R_{xy} is the second central moment $R_{xy} = \frac{1}{2}E\{(X - \overline{X})(Y - \overline{Y})^*\}$, the error probability becomes

---

[9] This method is based on evaluation of certain quadratic forms of Gaussian variables. It is also often named after the groundbreaking papers of [Bello and Nelin 1963] and [Proakis 1968].

$$P\{D < 0\} = Q_M(p_1, p_2) - \frac{v_2/v_1}{1 + v_2/v_1} I_0(p_1 p_2) \exp\left(-\frac{p_1^2 + p_2^2}{2}\right) \tag{11.107}$$

where Q_M is Marcum's Q-function (Eq. (11.45)). If there is only Rayleigh fading, and no dispersion, then α_1 and α_2 are zero, so that $P(D < 0) = v_1/(v_1 + v_2)$. The biggest problem is often to formulate the error probability as $P\{D < 0\}$. This will be discussed in the following.

Canonical Receiver

It is often helpful to reduce different RX structures to a "canonical" RX [Suzuki 1982], whose structure is shown in Figure 11.12. The received signal (after bandpass filtering) is multiplied by reference signals: once by a "normal" reference signal, and once by the signal shifted by $\pi/2$ to get the quadrature component. The resulting signal is lowpass-filtered and sent to the decision device. For coherent detection, the reference signal is obtained from a carrier recovery circuit; for differential detection, the reference signal is a delayed version of the incoming signal.

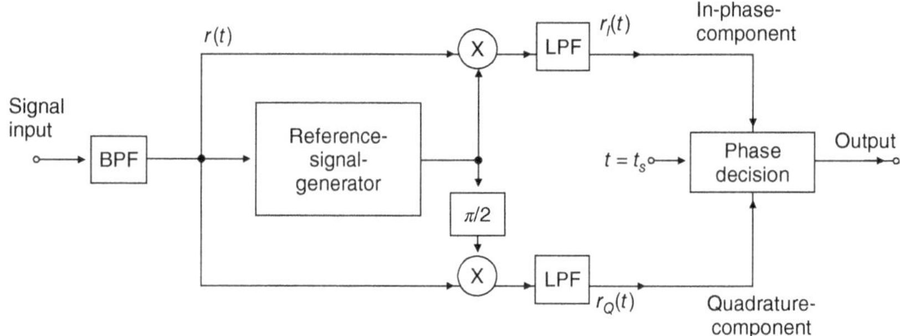

Figure 11.12   Canonical RX structure. *In this figure*: BPF, Bandpass Filter; LPF, Low-Pass Filter. Suzuki [1982] © IEEE.

As an example, we consider the differential detection of MSK. It is fairly easy to describe a bit error in the form $D < 0$. An error is made if $+1$ was transmitted, but the phase difference of the signals at the sampling instants, $X = r(t_s)$ and $Y = r(t_s - T)$, lies between π and 2π.

The condition for an error is thus identical to

$$\text{Re}\left\{b_0 XY^* \exp\left(-j\pi/2\right)\right\} < 0 \tag{11.108}$$

where b_0 is the transmitted bit. Since $\text{Re}\{Z\} = (Z + Z^*)/2$, Eq. (11.108) is a quadratic form D with $A = B = 0$.

Bit Error Probability

A first step in the application of the QFGV method to differential detection is computation of the correlation between the quantities $X = r(t_s)$ and $Y = r(t_s - T)$. For detection using a frequency discriminator, we define X as the sample value $r(t_s)$ and define Y as the derivative $dr(t)/dt$ at time $t = t_s$. Explicit equations for the resulting correlation coefficient are given in [Adachi and Parsons 1989] and [Molisch 2000].

After the correlation coefficient has been found, the mean BER can be computed from Eq. (11.107). For the case of differential detection of binary FSK in Rayleigh fading, these equations can be simplified to

$$\overline{BER} = \frac{1}{2} - \frac{1}{2} \frac{b_0 \text{Im}\{\rho_{XY}\}}{\sqrt{\text{Im}\{\rho_{XY}\}^2 + \left(1 - |\rho_{XY}|^2\right)}}. \tag{11.109}$$

For $\pi/4$-DQPSK we obtain:

$$\overline{BER} = \frac{1}{2} - \frac{1}{4}\left\{\frac{b_0 \text{Re}\{\rho_{XY}\}}{\sqrt{(\text{Re}\{\rho_{XY}\})^2 + \left(1 - |\rho_{XY}|^2\right)}} + \frac{b_0' \text{Im}\{\rho_{XY}\}}{\sqrt{(\text{Im}\{\rho_{XY}\}^2) + \left(1 - |\rho_{XY}|^2\right)}}\right\} \tag{11.110}$$

where b_0 and b_0' are the bits making up a symbol.

Summarizing, the BER can be computed in the following steps:

(i)   reduce the actual RX structure to canonical form;
(ii) formulate the condition for the occurrence of errors as $D < 0$;
(iii) compute the mean and the correlation coefficients of X and Y;
(iv) compute the BER according to the general Eqs. (11.107) and (11.106), or use the simplified Eq. (11.109) or (11.110) for MSK and $\pi/4$-DQPSK in Rayleigh-fading channels.

Further Reading

The literature describing the bit error probability of digital modulation formats encompasses hundreds of papers. For the BER in AWGN, we again just refer to the textbooks on digital communications [Anderson 2005, Barry et al. 2003, Proakis and Salehi 2005, Sklar 2001]. An extremely rigorous mathematical description can be found in [Gallagher 2008]. A discrete-time approach is used in [Rice 2008]. Fundamental computation methods that are applicable both to nonfading and fading channels were proposed in [Pawula et al. 1982], [Proakis 1968], and [Stein 1964]. The computation of the error probability in flat-fading channels is described in many papers: essentially, each modulation format, combined with the amplitude statistics of the channel, results in at least one paper. Some important examples in Rayleigh-fading channels include [Chennakeshu and Saulnier 1993] for $\pi/4$-DQPSK, [Varshney and Kumar 1991], and [Yongacoglu et al. 1988] for Gaussian Minimum Shift Keying (GMSK), and [Divsalar and Simon 1990] for differential detection of M-ary Phase Shift Keying (MPSK).

An important alternative to the classical computation of the BER is computation via the moment generating function described in [Simon and Alouini 2004]. or via the characteristic function (see [Annamalai et al. 2000]). [Simon and Alouini's 2004] monograph gives a number of BER equations for different modulation formats, channel statistics, and RX structures.

For delay-dispersive channels, a number of different computation methods have been proposed: we have already mentioned in the main part of the text the QFGV method [Adachi and Parsons 1989, Proakis 1968], and the group delay method [Andersen 1991, Crohn et al. 1993]. Furthermore, a number of papers use the pdf of the angles between Gaussian vectors as derived by [Pawula et al. 1982]. [Chuang 1987] is a very readable paper comparing different modulation formats. A wealth of papers are also dedicated to modulation formats or detection methods that reduce the impact of delay dispersion in unequalized systems. A summary of these methods, and further literature, can be found in [Molisch 2000].

For updates and errata for this chapter, see wides.usc.edu/teaching/textbook

Exercises

See Sec. 36.11 of Exercises.pdf at wiley.com/go/molisch/wireless3e

12

Diversity

12.1 Introduction

12.1.1 Principle of Diversity

In the previous chapter, we treated conventional transceivers that transmit an uncoded bitstream over fading channels. For non-fading Additive White Gaussian Noise (AWGN) channels, such an approach can be quite reasonable: the Bit Error Rate (BER) decreases exponentially as the Signal-to-Noise Ratio (SNR) increases, and a 10-dB SNR leads to BERs on the order of 10^{-4}. However, in Rayleigh fading the BER decreases only linearly with increasing SNR. We thus would need an SNR on the order of 40 dB in order to achieve a 10^{-4} BER, which is unpractical. The reason for this different performance is the fading of the channel: the BER is mostly determined by the probability of channel attenuation being large, and thus of the instantaneous SNR being low. A way to improve the BER is thus to change the effective channel statistics – i.e., to make sure that the SNR has a smaller probability of being low. Diversity is a way to achieve this.

The principle of diversity is to ensure that the same information reaches the Receiver (RX) on statistically independent channels.[1] Consider the simple case of an RX with two antennas. The antennas are assumed to be far enough from each other that small-scale fading is independent at the two antennas. The RX always chooses the antenna that has instantaneously larger receive power.[2] As the signals are statistically independent, the probability that both antennas are in a fading dip *simultaneously* is low – certainly lower than the probability that one antenna is in a fading dip. The diversity thus changes the SNR statistics at the detector input.

Example 12.1 *Diversity reception in a two-state fading channel.*

To quantify this effect on the (average) BER, let us consider a simple numerical example: the noise power within the RX filter bandwidth is 50 pW, the average received signal power is 1 nW, the SNR is thus 13 dB. In an AWGN channel, the resulting BER is 10^{-9}, assuming that the modulation is differentially detected Frequency Shift Keying (FSK). Now consider a fading channel where during 90% of the time the received power is 1.11 nW, and the SNR is thus 13.5 dB, while for the remainder, it is zero. This means that during 90% of the time, the BER is 10^{-10}; the remainder of the time, it is 0.5; the average BER is thus

$$0.9 \cdot 10^{-10} + 0.1 \cdot 0.5 = 0.05. \tag{12.1}$$

For the case of two-antenna diversity, the probability that the received signal power is 0 at both antennas simultaneously is $0.1 \cdot 0.1 = 0.01$. The probability that the received power is 1.11 nW at both antennas simultaneously is $0.9 \cdot 0.9 = 0.81$; the probability that it is 1.11 nW at one antenna and 0 at the other is 0.18. Assuming selection diversity, in both the latter cases, the SNR at the detector is 13.5 dB. The total BER is thus

$$0.01 \cdot 0.5 + 0.99 \cdot 10^{-10} = 0.005. \tag{12.2}$$

This is approximately the square of the BER for a single-antenna system. If we have three antennas, then the probability that the signal power is 0 at all three antennas simultaneously is 0.1^3; the total BER is then $0.5 \cdot 0.001 + 0.999 \cdot 10^{-10} = 0.0005$; this is approximately the third power of the BER for a single-antenna system.

Later sections will give exact equations for the BER with diversity in Rayleigh-fading channels. The general concepts, however, are the same as in the simple example described above. With N_r diversity antennas, we obtain a bit error probability $\propto BER_{oc}^{N_r}$, where BER_{oc} is the BER with just one receive channel.[3]

---

[1] More precisely, we require that the channels are not completely correlated. Having them independent provides the best diversity, but is not strictly necessary. This will be discussed more in Section 12.1.2.

[2] We will see later on that this scheme is only one of many different possible diversity schemes.

[3] Since in a Rayleigh-fading channel, $BER_{oc} \propto SNR^{-1}$, we find that for a diversity system with N_r independently fading channels, $BER \propto SNR^{-N_r}$. Quite generally in fading channels with diversity, $BER \propto SNR^{-d_{div}}$, where d_{div} is known as *diversity order*.

Wireless Communications: From Fundamentals to Beyond 5G, Third Edition. Andreas F. Molisch.
© 2023 John Wiley & Sons Ltd. Published 2023 by John Wiley & Sons Ltd.
Companion website: www.wiley.com/go/molisch/wireless3e

The above example described the impact of diversity on the average BER. An even more important motivation for diversity stems from considerations of system reliability. Consider, for example, a nomadic device, such as a laptop. Then for a particular placement of the device, its channel to the Base Station (BS) might be in a fading dip, leading to consistently low BER for the whole time that the device is at this location. Thus, the probability of being in a bad channel state must be reduced, which can be achieved through diversity. Assume that with a single antenna, the probability for being located in a catastrophic fading dip is $P_{\text{out,oc}}$, then with N_r diversity antennas, it may be reduced to $P_{\text{out,oc}}^{N_r}$.

In the following, we first describe the characterization of correlation coefficients between different transmission channels. We then give an overview about how transmission over independent channels can be realized – spatial antenna diversity described above is one, but certainly not the only approach. Next, we describe how signals from different channels can best be combined, and what performance can be achieved with the different combining schemes.

12.1.2 Definition of the Correlation Coefficient

Diversity is most efficient when the different transmission channels (also called diversity branches) carry independently fading copies of the same signal. This means that the joint probability density function (pdf) of field strength (or power) $pdf_{r_1, r_2, \dots}(r_1, r_2, \dots)$ is equal to the product of the marginal pdfs for the channels, $pdf_{r_1}(r_1), pdf_{r_2}(r_2), \dots$ Any correlation between the fading of the channels decreases the effectiveness of diversity.

The *correlation coefficient* characterizes the correlation between signals on different diversity branches. A number of different definitions are being used for this important quantity: complex correlation coefficients, correlation coefficient of the phase, etc. An important one is the correlation coefficient of signal envelopes x and y:

$$\rho_{xy} = \frac{E\{x\cdot y\} - E\{x\}\cdot E\{y\}}{\sqrt{\left(E\{x^2\} - E\{x\}^2\right)\cdot\left(E\{y^2\} - E\{y\}^2\right)}}. \tag{12.3}$$

For two statistically independent signals, the relationship $E\{xy\} = E\{x\}E\{y\}$ holds; therefore, the correlation coefficient becomes zero. Signals are often said to be "effectively" decorrelated if ρ is below a certain threshold (typically 0.5 or 0.7).

12.2 Microdiversity

As mentioned in Section 12.1, the basic principle of diversity is that the RX has multiple copies of the transmit signal, where each of the copies goes through a statistically independent channel. This section describes different ways of obtaining these statistically independent copies.

We concentrate first on methods that can be used to combat small-scale fading, which are therefore called "microdiversity." The five most common methods are as follows:

1. *Spatial diversity*: several antenna elements separated in space.
2. *Temporal diversity*: transmission of the transmit signal at different times.
3. *Frequency diversity*: transmission of the signal on different frequencies.
4. *Angular diversity*: multiple antennas (with or without spatial separation) with different antenna patterns.
5. *Polarization diversity*: multiple antennas with different polarizations (e.g., vertical and horizontal).

When we speak of antenna diversity, we imply that there are multiple antennas at the *RX*. Only in Section 16.1.5 will we discuss how multiple *transmit* antennas can be exploited to improve performance, and Section 16.1.6 will explore multiple antennas at both Transmitter (TX) and RX.

The following important equation will come in handy: Consider the correlation coefficient of two signals that have a temporal separation Δt and a frequency separation $f_1 - f_2$. As shown in Appendix 12.A, the correlation coefficient is

$$\rho_{xy} = \frac{J_0^2(k_0 v\tau)}{1 + (2\pi)^2 S_\tau^2 (f_2 - f_1)^2}. \tag{12.4}$$

Note that for moving User Equipments (UEs), temporal separation can be easily converted into spatial separation, so that temporal and spatial diversity become mathematically equivalent. Equation (12.4) is thus quite general in the sense that it can be applied to spatial, temporal, and frequency diversity. However, a number of assumptions were made in the derivation of this equation: (i) validity of the Wide Sense Stationary Uncorrelated Scatterer (WSSUS) model, (ii) no existence of Line Of Sight (LOS), (iii) exponential shape of the Power Delay Profile (PDP), (iv) isotropic distribution of incident power, and (v) use of omnidirectional antennas.

Example 12.2 *Compute the correlation coefficient of two frequencies with separation (i) 30 kHz, (ii) 200 kHz, (iii) 5 MHz, in the "typical urban" environment, as defined in COST 207[4] channel models.*

For zero temporal separation, the Bessel function in Eq. (12.4) is unity, and the correlation coefficient is thus only dependent on rms delay spread and frequency separation. Rms delay spread is calculated according to Eq. (6.40), where the PDP of the COST 207 typical urban channel model is given in App. 7C. We obtain

$$S_\tau = \sqrt{\frac{\int_0^{7\cdot10^{-6}} e^{-\tau/10^{-6}} \tau^2 d\tau}{\int_0^{7\cdot10^{-6}} e^{-\tau/10^{-6}} d\tau} - \left(\frac{\int_0^{7\cdot10^{-6}} e^{-\tau/10^{-6}} \tau\, d\tau}{\int_0^{7\cdot10^{-6}} e^{-\tau/10^{-6}} d\tau}\right)^2} \tag{12.5}$$

$$= 0.977\,\mu s.$$

The correlation coefficient thus becomes

$$\rho_{xy} = \frac{1}{1 + (2\pi)^2 (0.977\cdot10^{-6})^2 (f_1 - f_2)^2}$$

$$= \begin{cases} 0.97, & f_1 - f_2 = 30\,\text{kHz} \\ 0.4, & f_1 - f_2 = 200\,\text{kHz} \,. \\ 1\cdot10^{-3}, & f_1 - f_2 = 5\,\text{MHz} \end{cases} \tag{12.6}$$

From this, we can see that the correlation between two neighboring 30-kHz bands – as used, e.g., in the old IS-136[5] Time Division Multiple Access (TDMA) cellular system – is very high; correlation over 200-kHz bands (two neighboring channels in Global System for Mobile communication (GSM) are 200 kHz apart) is also appreciable; note that both of these systems were second-generation cellular systems designed for speech communication, and thus with low data rate. When considering modern systems, bandwidths on the order of 5 MHz are more common; such channels are uncorrelated in this environment. Furthermore, carriers that are, e.g., tens of MHz apart – used for uplink and downlink communication respectively – are completely uncorrelated.

12.2.1 Spatial Diversity

Spatial diversity is the oldest and simplest form of diversity. Despite (or because) of this, it is also the most widely used. The transmit signal is received at several antenna elements, and the signals from these antennas are then further processed according to the principles that will be described in Section 12.4. But, irrespective of the processing method, performance is influenced by correlation of the signals between the antenna elements. A large correlation between signals at antenna elements is undesirable, as it decreases the effectiveness of diversity. A first important step in designing diversity antennas is thus to establish a relationship between antenna spacing and the correlation coefficient. This relationship is different for BS antennas and UE antennas, and thus will be treated separately.

1. *UE in cellular and WLAN systems*: it is a standard assumption that waves are incident from all directions at the UE (see also Section 5.6.1). Thus, points of constructive and destructive interference of Multi-Path Components (MPCs) – i.e., points where we have high and low received power, respectively – are spaced approximately $\lambda/4$ apart, compare also Example 5.3. This is therefore the distance that is required for decorrelation of received signals. This intuitive insight agrees very well with the results from the exact mathematical derivation (Eq. (12.4), with $f_2 - f_1 = 0$), given in Figure 12.1: decorrelation, defined as $\rho = 0.5$, occurs at an antenna separation of approximately $\lambda/4$. Note, however, that for some environments, e.g., street canyons, the angular power spectrum can be more concentrated (see Section 7.3.2), leading to higher correlation.

The above considerations imply that the minimum distance for antenna elements at 900 MHz is about 8 cm, and systems in the 1,800-MHz band it is about 4 cm. For systems at 2.4 and 5 GHz – mainly Wireless Local Area Networks (WLANs) – the distances are even smaller. It is thus clearly possible to place two sufficiently separated antennas on a UE of a cellular system.

2. *BS for WLANs*: in a first approximation, the angular distribution of incident radiation at indoor BSs is also uniform – i.e., radiation is incident with equal strength from all directions. Therefore, the same rules apply as for UEs.
3. *BSs in cellular systems*: for a cellular BS, the assumption of uniform directions of incidence is no longer valid. Interacting Objects (IOs) are typically concentrated around the UE (Figure 12.2, see also Section 7.3.1). Since all waves are incident essentially from one direction, the correlation coefficient (for a given distance between antenna elements d_a) is much higher. Expressed differently, the antenna spacing required to obtain sufficient decorrelation increases.

---

[4] European COoperation in the field of Scientific and Technical research.
[5] IS-136 is a (now defunct) second-generation cellular system that was used mainly in the USA.

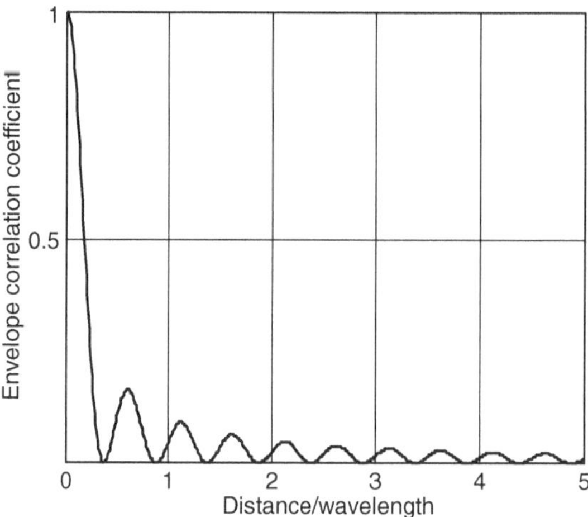

Figure 12.1   Envelope correlation coefficient as a function of antenna separation when MPCs are incident from all directions.

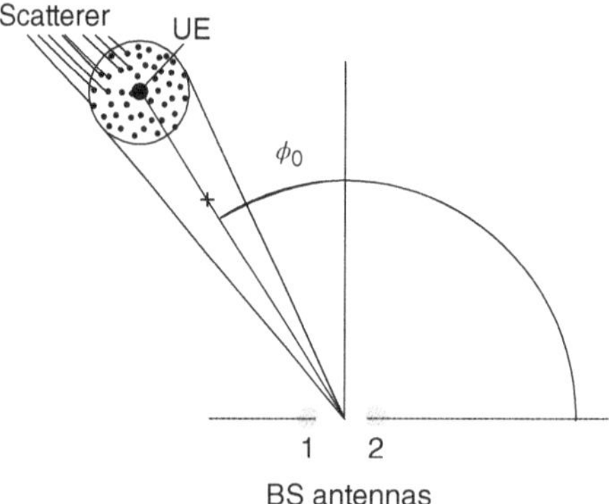

Figure 12.2   Scatterers concentrated around the UE.

To get an intuitive insight, we start with the simple case when there are only two MPCs whose wave vectors are at an angle α with respect to each other (Figure 12.3). It is obvious that the distance between the maxima and minima of the interference pattern is the larger the smaller α is. For vanishing α, the connection line between antenna elements lies on a "ridge" of the interference pattern, and antenna elements are completely correlated.

Numerical evaluations of the correlation coefficient as a function of antenna spacing are shown in Figure 12.4. The first column shows the results for rectangular angular power spectra; the results for Gaussian distributions are shown in the second column. We can see that antenna spacing has to be on the order of 2–20 wavelengths for angular spreads between 1° and 5° in order to achieve

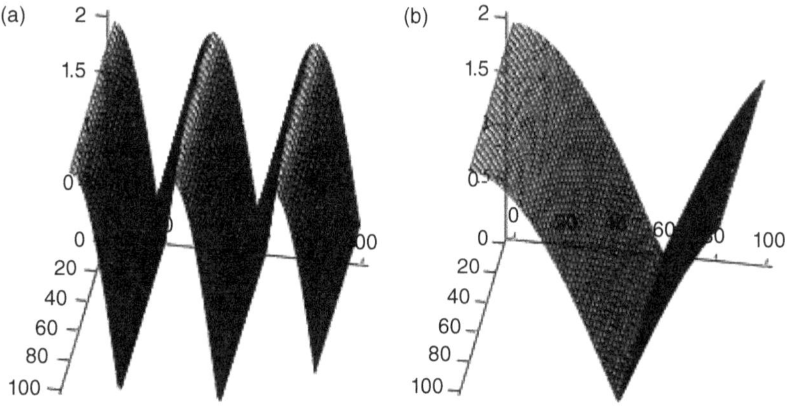

Figure 12.3   Interference pattern of two waves with 45° (a) and 15° (b) angular separation.

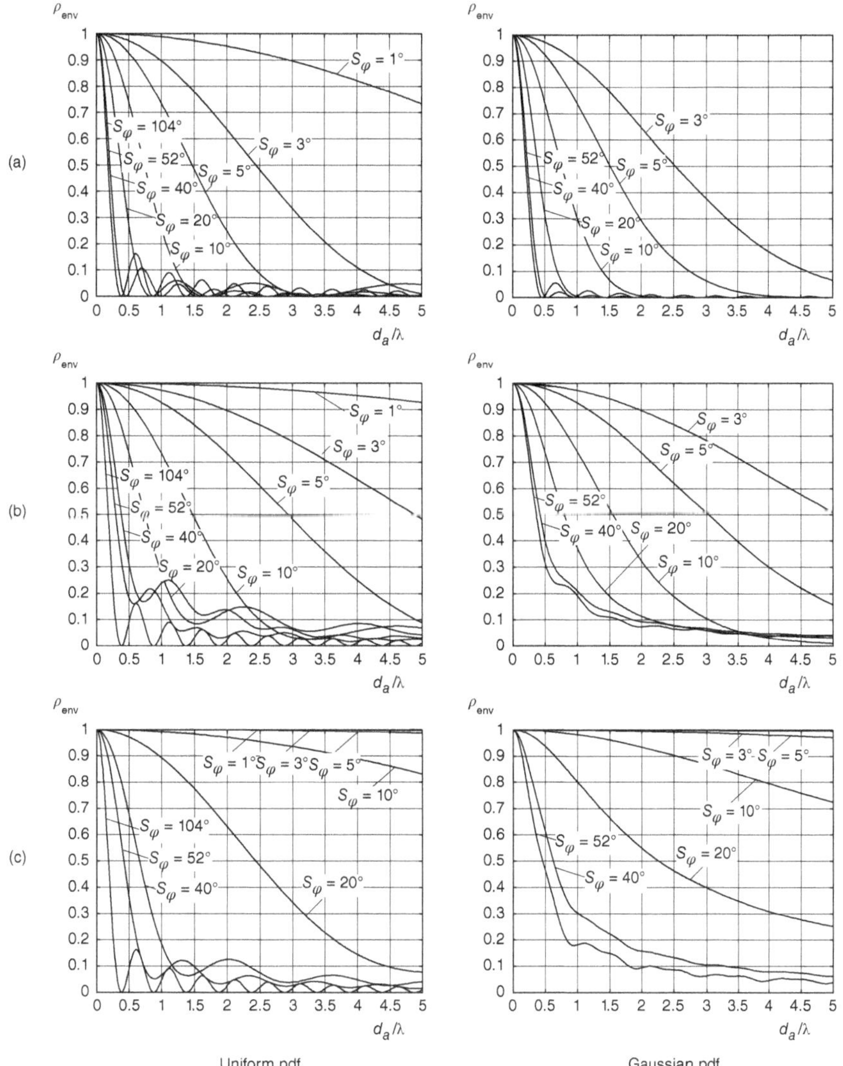

Figure 12.4 Envelope correlation coefficient at the BS for uniform and Gaussian probability density function (pdf) of the directions of arrival (a) $\phi_0 = 90°$, (b) $\phi_0 = 45°$, (c) $\phi_0 = 10°$, and different values of angular spread S_φ.
Reproduced with permission from [Fuhl et al. 1998] © IET.

decorrelation. We also find that it is mostly rms angular spread that determines the required antenna spacing, while the shape of the angular power spectrum has only a minor influence.

12.2.2 Temporal Diversity

As the wireless propagation channel is time variant, signals that are received at different times may be uncorrelated. For "sufficient" decorrelation, the temporal distance must be at least $1/(2\nu_{\max})$, where $\nu_{\max}$ is the maximum Doppler frequency. In a static channel, where neither TX, RX, nor the IOs are moving, the channel state is the same at all times. Such a situation can occur, e.g., for nomadic devices, such as laptops, hotspot devices, and many wireless sensors/actuators. In such a case, the correlation coefficient is $\rho = 1$ for all time intervals, and temporal diversity is useless.

When the channel shows temporal variations, then multiple signal copies using the different diversity branches can be obtained in different ways:

1. *Repetition coding*: this is the simplest form. The signal is repeated several times, where the repetition intervals are long enough to achieve decorrelation. This obviously achieves diversity but is also highly bandwidth inefficient. Spectral efficiency decreases approximately by a factor that is equal to the number of repetitions.

2. *Automatic Repeat reQuest (ARQ)*: here, the RX sends a message to the TX to indicate whether it received the data with sufficient quality (see Section 13.12). If this is not the case, then the transmission is repeated (after a wait period that achieves decorrelation). The spectral efficiency of ARQ is better than that of repetition coding, since it requires multiple transmissions only when the first

transmission occurs in a bad fading state, while for repetition coding, retransmissions occur always. On the downside, ARQ requires a feedback channel.

3. *Combination of interleaving and coding*: an alternative is forward error correction coding with interleaving. The different symbols of a codeword are transmitted at different times (the interleaver ensures that the times are sufficiently different), which increases the probability that at least some of them arrive with a good SNR. The transmitted codeword can then be reconstructed. For more details, see Chapter 13.

Regardless of the specific technique for realizing temporal diversity, an important drawback is the increase in latency: as mentioned above, a two-branch diversity takes up a time of at least $T_{\rm coh} \sim 1/(2\nu_{\rm max})$. In many applications, this might not be acceptable.

12.2.3 Frequency Diversity

In frequency diversity, the same signal is transmitted on two (or more) different frequencies. If these frequencies are spaced apart by more than the coherence bandwidth of the channel, then their fading is approximately independent, and the probability is low that the signal is in a deep fade at both frequencies simultaneously. For an exponential PDP, the correlation between two frequencies can be obtained from Eq. (12.4) by setting the numerator to unity as the signals at the two frequencies occur at the same time. Thus

$$\rho = \frac{1}{1 + (2\pi)^2 S_\tau^2 (f_2 - f_1)^2}. \tag{12.7}$$

This again confirms that the two signals have to be at least one coherence bandwidth apart from each other. Figure 12.5 shows ρ as a function of the spacing between the two frequencies. For a more general discussion of frequency correlation and typical coherence bandwidths, see Secs. 6.5.4 and 7.2, respectively.

It is not common to actually repeat the same information at two different frequencies, as this would greatly decrease spectral efficiency – similar to repetition coding in time diversity. Rather, information is spread over a large bandwidth, so that small parts of the information are conveyed by different frequency components. The RX can then sum over the different frequencies to recover the original information.

This spreading can be done by different methods:

- *Compressing the information in time*: – i.e., sending short bursts that each occupy a large bandwidth – TDMA (see Section 18.3.2).
- *Code Division Multiple Access (CDMA)*: (Section 19.2).
- Coded orthogonal frequency division multiplexing (Section 15.4.3), Multicarrier CDMA (Section 15.11.1), and Orthogonal Time-Frequency-Space Modulation (Section 15.12).
- *Frequency hopping in conjunction with coding*: different parts of a codeword are transmitted on different carrier frequencies (Section 19.1).

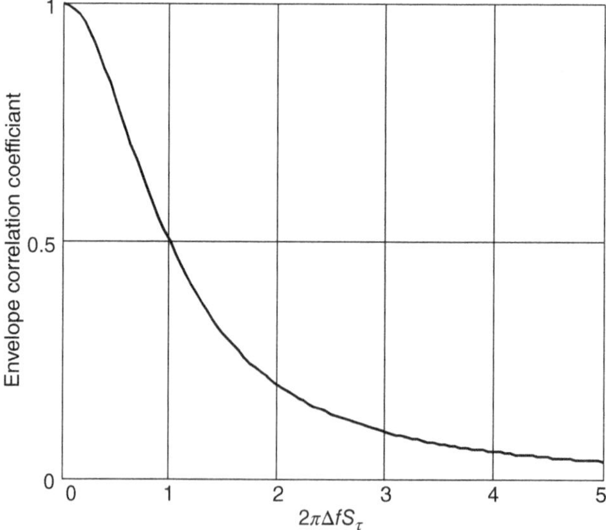

Figure 12.5 Correlation coefficient of the envelope as a function of normalized frequency spacing.

These methods allow the transmission of information without wasting bandwidth and will be described in greater detail in later chapters. For the moment, we just stress that the use of frequency diversity requires the channel to be frequency selective. In other words, frequency diversity (delay dispersion) can be exploited by the system to make it more robust, and decrease the effects of fading. This seems to be a contradiction to the results of Section 11.3, where we had shown that frequency selectivity leads to an *increase* of the BER and even an error floor. The reason for this discrepancy is that Chapter 11 considered a very simple RX that takes no measures to combat (or exploit) the effects of frequency selectivity.

12.2.4 Angular Diversity

A fading dip is created when MPCs, which usually come from different directions, interfere destructively. If some of these waves are attenuated or eliminated, then the location of fading dips changes. In other words, two colocated antennas with different patterns "see" differently weighted MPCs, so that the MPCs interfere differently at the two antennas. This is the principle of *angle diversity* (also known as *pattern diversity*).

Angular diversity is usually used in conjunction with spatial diversity; it enhances the decorrelation of signals at closely spaced antennas. Different antenna patterns can be achieved very easily. Of course, different types of antennas have different patterns. But two directional antennas of the same type also provide pattern diversity when mounted in different directions, e.g., on different parts of the casing. Internal antennas, such as patch antennas and inverted-F antennas (see Chapter 8) can be placed on all parts of the casing (see Figure 12.6). Further decorrelation can occur due to the presence of the head, hand, or body of the user holding the device, since these objects lead to pattern distortion, whose specific form depends on the exact relative placing of objects and antennas.

Angular diversity can be used not only for combatting small-scale fading but also shadowing. Consider, e.g., two antennas with non-overlapping 90° fields of view, covering [0,90] and [90,180] degree in an environment in which MPCs come from a 150 degree angular range [0,150]. Now let an object shadow off MPCs coming from one 90 degree sector [0,90]; the second antenna in this case can still provide significant power. This is in contrast to, e.g., two antennas pointing in the same direction [0,90] but being offset by a wavelength (standard spatial diversity), in which case neither antenna can receive sufficient power.

Even identical antennas can have different patterns when mounted close to each other (see Figure 12.7). This effect is due to *mutual coupling*: antenna B acts as a reflector for antenna A, whose pattern is therefore skewed to the left.[6] Analogously, the pattern of antenna B is skewed to the right due to reflections from antenna A. Thus, the two patterns are different.

12.2.5 Polarization Diversity

Horizontally and vertically polarized MPCs propagate differently in a wireless channel,[7] as the reflection and diffraction processes depend on polarization (see Chapter 4). Even if the transmit antenna only sends signals with a single polarization, the propagation effects in the channel lead to depolarization so that both polarizations arrive at the RX. The fading of signals with different polarizations is statistically independent. Thus, receiving both polarizations using a dual-polarized antenna, and processing the signals separately, offers diversity. This diversity can be obtained without any requirement for a minimum distance between antenna elements.

Let us now consider more closely the situation where the transmit signal is vertically polarized, while the signal is received in both vertical and horizontal polarization. In that case, fading of the two received signals is independent, but the average received signal strength in the two diversity branches is *not* identical. Depending on the environment, the horizontal (i.e., cross-polarized) component is some 3–20 dB weaker than the vertical (co-polarized) component. As we will see later on, this has an important impact on the effectiveness of the diversity scheme. Various antenna arrangements have been proposed in order to mitigate this problem; the most

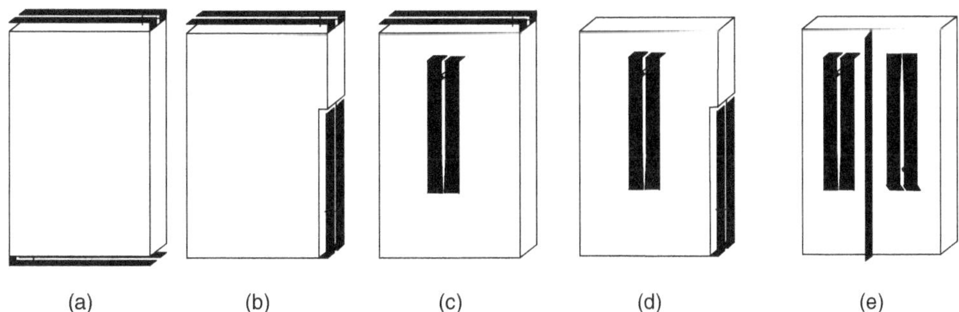

Figure 12.6 Five example configurations (subfigures (a–e)) of diversity antennas at a UE.
Reproduced with permission from [Eratuuli and Bonek 1997] © IEEE.

---

[6] This arrangement can also be considered as a Yagi antenna. It depends on the spacing of the two elements whether antenna B acts as director or reflector, and thus which direction the pattern is skewed into.

[7] For simplicity, we henceforth speak of horizontal and vertical polarization. However, the considerations are valid for any two orthogonal polarizations.

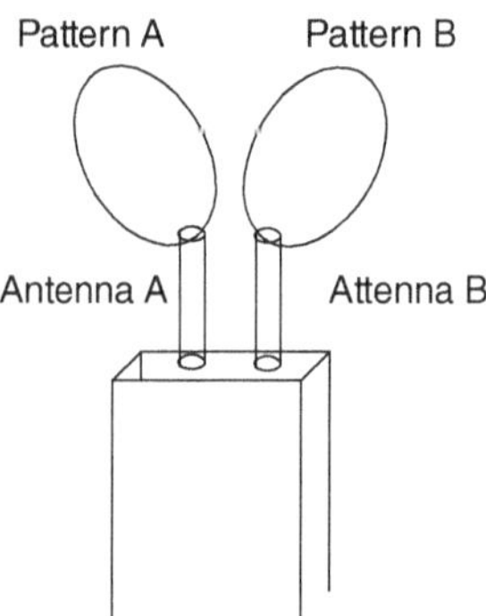

Figure 12.7 Angle/pattern diversity for closely spaced antennas.

common one consists of two antennas arranged at $\pm45°$ with respect to the vertical axis; the two antennas are thus orthogonal to each other (and thus have independent fading) but both receive a mixture of horizontal and vertical polarization and thus have equal average power.

It has also been claimed that the diversity order that can be achieved with polarization diversity is up to 6: three possible components of the E-field and three components of the H-field can all be exploited [Andrews et al. 2001].[8] However, propagation characteristics as well as practical considerations prevent a full exploitation of that diversity order especially for outdoor situations. This is usually not a serious restriction for diversity systems, as we will see later on that going from diversity order 1 (i.e., no diversity) to diversity order 2 gives larger benefits than increasing the diversity order from 2 to higher values. However, it is an important issue for Multiple Input Multiple Output (MIMO) systems (see Section 16.2).

12.3 Macrodiversity and Simulcast

The previous section described diversity methods that combat small-scale fading – i.e., the fading created by interference of MPCs. However, not all of these diversity methods are suitable for combating large-scale fading, which is created by shadowing effects. Shadowing is almost independent of transmit frequency and polarization, so that frequency diversity or polarization diversity is not effective. Spatial diversity (or equivalently, temporal diversity with moving TX/RX) can be used, but we have to keep in mind that the correlation distances for large-scale fading are on the order of tens or hundreds of meters. In other words, if there is a hill between the TX and RX, adding antennas on either the BS or the UE does not help to eliminate the shadowing caused by this hill (note, however, the discussion in 12.2.4 about the use of angular diversity to combat shadowing). Rather, we should use a separate base station (BS2) that is placed in such a way that the hill is not in the connection line between the UE and BS2. This in turn implies a large distance between BS1 and BS2, which gives rise to the word *macrodiversity*.

The simplest method for macrodiversity is the use of *on-frequency repeaters* that receive the signal and retransmit an amplified version of it. *Simulcast* is very similar to this approach; the same signal is transmitted simultaneously from different BSs. In cellular applications, the two BSs should be synchronized, and transmit the signals intended for a specific user in such a way that the two waves arrive at the RX almost simultaneously (timing advance).[9] Note that synchronization can only be obtained if the runtimes from the two BSs to the UE are known. Generally speaking, it is desirable that the synchronization error is no larger than the delay dispersion that the RX can handle. This is especially critical for RXs in regions where the strengths of the signals from the two BSs are approximately equal.

Simulcast is also widely used for broadcast applications, especially digital TV. In this case, the exact synchronization of all possible RXs is not possible – each RX would require a different timing advance from the TXs.

A disadvantage of simulcast is the large amount of signaling information that has to be carried. Synchronization information as well as transmit data have to be transported on cable connections (or microwave links) to the BSs. This used to be a serious problem in the early days of digital mobile telephony, but the current wide availability of fiber-optic links has made this less of an issue.

The use of on-frequency repeaters is simpler than that of simulcast, as no synchronization is required. On the other hand, delay dispersion is larger, because (i) the runtime from BS to repeater, and repeater to UE is larger (compared with the runtime from a second BS), and (ii) the repeater itself introduces additional delays due to the group delays of electronic components, filters, etc.

In modern systems, the most important form of macrodiversity arises from BS cooperation such that different BSs act as parts of a large distributed MIMO system. This approach is motivated by a variety of advantages, but among other benefits also combats shadowing. A detailed discussion is given in Section 22.11.

---

[8] Note that this result is surrounded by some controversy, and there is no general agreement in the scientific community whether diversity order 6 can really be achieved from polarization only.

[9] If the RX cannot easily deal with delay dispersion, it is desirable that the two signals arrive exactly at the same time. For advanced RXs it can be desirable to have a small amount of delay dispersion; if the timing of the BSs and the runtimes of the signals to the RX are exactly known, this can also be achieved.

12.4 Combination of Signals

Now we turn our attention to the question of how to use diversity signals in a way that improves the total quality of the signal that is to be detected. To simplify the notation, we speak here only about the combination of signals from different antenna signals at the RX. However, the mathematical methods remain valid for other types of diversity signals as well. In general, we can distinguish two ways of exploiting signals from the multiple diversity branches:

1. *Selection diversity*, where the "best" signal copy is selected and processed (demodulated and decoded), while all other copies are discarded. There are different criteria for what constitutes the "best" signal.
2. *Combining diversity*, where all copies of the signal are combined (before or after the demodulator), and the combined signal is decoded. Again, there are different algorithms for combination of the signals.

Combining diversity leads to better performance, as all available information is exploited. On the downside, it requires a more complex RX than selection diversity. In most RXs, all processing is done in the baseband. Thus, an RX with combining diversity needs to downconvert all available signals and combine them appropriately in the baseband. Thus, it requires N_r antenna elements as well as N_r complete Radio Frequency (RF) (downconversion) chains. An RX with selection diversity requires only *one* RF chain, as it processes only a single received signal at a time.

In the following, we give a more detailed description of selection (combination) criteria and algorithms. We assume that different signal copies undergo statistically independent fading – this greatly simplifies the discussion of both the intuitive explanations and the mathematics of the signal combination. Discussions on the impact of a finite correlation coefficient are relegated to Section 12.5.

In these considerations, we also have to keep in mind that the gain of multiple antennas is due to two effects: *diversity gain* and *beamforming gain*. Diversity gain reflects the fact that it is improbable that several antenna elements are in a fading dip simultaneously; the probability for very low signal levels is thus decreased by the use of multiple antenna elements. Beamforming gain reflects the fact that (for combining diversity) the combiner performs an averaging over the noise at different antennas. Thus, even if the signal levels at all antenna elements are identical, the combiner output SNR is larger than the SNR at a single-antenna element. More discussions about beamforming versus diversity can be found in Section 16.1.2.

12.4.1 Selection Diversity

Received-Signal-Strength-Indication-Driven Diversity

In this method, the RX selects the signal with the largest instantaneous power (or *Received Signal Strength Indication (RSSI)*) and processes it further. This method requires N_r antenna elements, N_r RSSI sensors, and a N_r-to-1 multiplexer (switch), but only one RF chain (see Figure 12.8). The method allows simple tracking of the selection criterion even in fast-fading channels. Thus, we can switch to a better antenna as soon as the RSSI becomes higher there.

1. If the BER is determined by noise, then RSSI-driven diversity is the best of all the selection diversity methods, as maximization of the RSSI also maximizes the SNR.
2. If the BER is determined by co-channel interference, then RSSI may no longer be a good selection criterion. High receive power can be caused by a high level of interference, such that the RSSI criterion makes the system select branches with a low signal-to-interference ratio. This is especially critical when interference is caused mainly by one dominant interferer – a situation that often occurs in cellular systems.
3. Similarly, RSSI-driven diversity is suboptimum if the errors are caused by the frequency selectivity of the channel. RSSI-driven diversity can still be a reasonable approximation, because we have shown in Section 11.3 that errors caused by signal distortion occur mainly in the fading dips of the channel. However, this is only an approximation, and it can be shown that (uncoded, unequalized) systems with RSSI-driven selection diversity have a BER that is higher by a constant factor compared with optimum (BER-driven) diversity.

For an exact performance assessment (Section 12.5), it is important to obtain the SNR distribution of the output of the selector. Assume that the instantaneous signal amplitude is Rayleigh distributed, such that the pdf of the SNR of the nth diversity branch, γ_n, is (see Eq. (11.52))

$$\mathrm{pdf}_{\gamma_n}(\gamma_n) = \frac{1}{\bar{\gamma}} \exp\left(-\frac{\gamma_n}{\bar{\gamma}}\right) \tag{12.8}$$

where $\bar{\gamma}$ is the mean branch SNR (assumed to be identical for all diversity branches). The cumulative distribution function (cdf) is then

$$\mathrm{cdf}_{\gamma_n}(\gamma_n) = 1 - \exp\left(-\frac{\gamma_n}{\bar{\gamma}}\right). \tag{12.9}$$

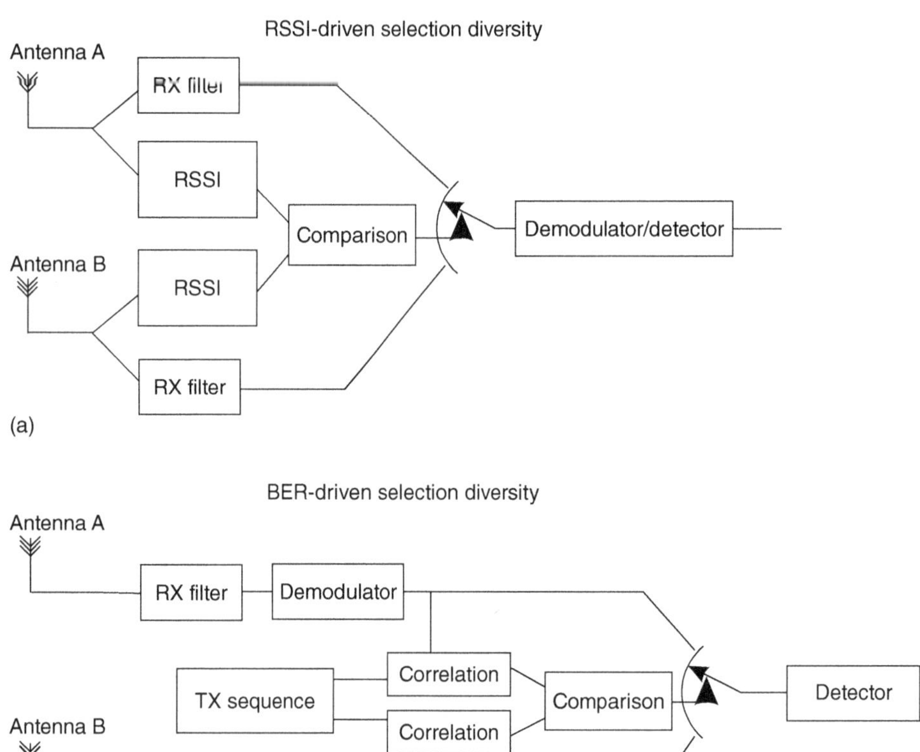

Figure 12.8 Selection diversity principle: (a) Received-signal-strength-indication-controlled diversity. (b) Bit-error-rate-controlled diversity.

The cdf is, by definition, the probability that the instantaneous SNR lies below a given level. As the RX selects the branch with the largest SNR, the probability that the chosen signal lies below the threshold is the product of the probabilities that the SNR at each branch is below the threshold. In other words, the cdf of the selected signal is the product of the cdfs of each branch:

$$\mathrm{cdf}_\gamma(\gamma) = \left[1 - \exp\left(-\frac{\gamma}{\overline{\gamma}}\right)\right]^{N_\mathrm{r}}.$$

(12.10)

Example 12.3 *Compute the probability that the output power of a selection diversity system is 5 dB lower than the mean power of each branch, when using $N_\mathrm{r} = 1, 2, 4$ antennas.*

The threshold is $\gamma_{|\mathrm{dB}} = \overline{\gamma}_{|\mathrm{dB}} - 5\,\mathrm{dB}$, which in linear scale is $\gamma = \overline{\gamma} \cdot 10^{-0.5}$. Using Eq. (12.10), the probability that the output power is less than $\overline{\gamma} \cdot 10^{-0.5}$ becomes:

$$\mathrm{cdf}_\gamma\left(\overline{\gamma} \cdot 10^{-0.5}\right) = \left[1 - \exp\left(-10^{-0.5}\right)\right]^{N_\mathrm{r}}$$
$$= \begin{cases} 0.27, & N_\mathrm{r} = 1 \\ 7.4 \cdot 10^{-2}, & N_\mathrm{r} = 2 \\ 5.4 \cdot 10^{-3}, & N_\mathrm{r} = 4 \end{cases}.$$

(12.11)

Example 12.4 *Consider now the case that $N_\mathrm{r} = 2$, and that the mean powers in the branches are $1.5\overline{\gamma}$ and $0.5\overline{\gamma}$, respectively. How does the result change?*

In this case, the probability is

$$cdf_\gamma\left(\overline{\gamma} \cdot 10^{-0.5}\right) = \left[1 - \exp\left(-\frac{1}{1.5} \cdot 10^{-0.5}\right)\right]\left[1 - \exp\left(-\frac{1}{0.5} \cdot 10^{-0.5}\right)\right]$$

(12.12)

$$= 8.9 \cdot 10^{-2}.$$

(12.13)

This demonstrates that diversity is less efficient when the average branch powers are different.

Bit-Error-Rate-Driven Diversity

For BER-driven diversity, we first transmit a *training sequence* – i.e., a bit sequence that is known at the RX. The RX then demodulates the signal from each receive antenna element and compares it with the transmit signal. The antenna whose associated signal results in the smallest BER is judged to be the "best," and used for the subsequent reception of data signals. A similar approach is the use of the mean square error of the "soft-decision" demodulated signal, or the correlation between transmit and receive signal.

If the channel is time variant, the training sequence has to be repeated at regular intervals, and selection of the best antenna has to be done anew. The necessary repetition rate depends on the coherence time of the channel.

BER-driven diversity has several drawbacks:

1. The RX needs either N_r RF chains and demodulators (which makes the RX more complex), or the training sequence has to be repeated N_r times (which decreases spectral efficiency), so that the signal at all antenna elements can be evaluated.
2. If the RX has only one demodulator, then it is not possible to continuously monitor the selection criterion (i.e., the BER) of all diversity branches. This is especially critical if the channel changes quickly.
3. Since the duration of the training sequence is finite, the selection criterion – i.e., bit error probability – cannot be determined exactly. The variance of the BER around its true mean decreases as the duration of the training sequence increases. There is thus a tradeoff between performance loss due to erroneous determination of the selection criterion, and spectral efficiency loss due to longer training sequences. However, this problem can be overcome by not estimating directly the raw BER, but rather the signal to noise-and-interference ratio and delay dispersion of the channel (which can be done based on a short training sequence) and from that conclude about the performance measure such as BER; this also has the advantage that other performance measures, such as channel capacity (see Chapter 13) can be analyzed.

*12.4.2 Switched Diversity

The main challenge of selection diversity is that the selection criteria (power, BER, etc.) of *all* diversity branches have to be monitored in order to know when to select a different antenna. As we have shown above, this leads to either increased hardware effort or reduced spectral efficiency. An alternative solution, which avoids these drawbacks, is *switched diversity*. In this method, the selection criterion of just the active diversity branch is monitored. If it falls below a certain threshold, then the RX switches to a different antenna.[10] Switching only depends on the quality of the active diversity branch; it does not matter whether the other branch actually provides a better signal quality or not.

Switched diversity runs into problems when both branches have signal quality below the threshold: in that case, the RX just switches back and forth between the branches. This problem can be avoided by introducing a hysteresis or hold time, so that the new diversity branch is used for a certain amount of time, independent of the actual signal quality. We thus have two free parameters: switching threshold and hysteresis time. These parameters have to be selected very carefully: if the threshold is chosen too low, then a diversity branch is used even when the other antenna might offer better quality; if it is chosen too high, then it becomes probable that the branch the RX switches to actually offers lower signal quality than the currently active one. If hysteresis time is chosen too long, then a "bad" diversity branch can be used for a long time; if it is chosen too short, then the RX spends all the time switching between two antennas.

Summarizing, the performance of switched diversity is worse than that of selection diversity; we will therefore not consider it further.

12.4.3 Combining Diversity

Basic Principle

Selection diversity wastes signal energy by discarding $(N_r - 1)$ copies of the received signal. This drawback is avoided by combining diversity, which exploits *all* available signal copies. Each signal copy is multiplied by a (complex) weight and then added up. Each complex weight w_n^* can be thought of as consisting of a phase correction,[11] plus a (real) weight for the amplitude:

- Phase correction causes the *signal amplitudes* to add up, while, on the other hand, noise is added incoherently, so that *noise powers* add up.
- For amplitude weighting, two methods are widely used: *Maximum Ratio Combining* (MRC) weighs all signal copies by their amplitude. It can be shown that (using some assumptions) this is the best combination strategy. An alternative is *Equal Gain Combining* (EGC), where all amplitude weights are the same (in other words, there is no weighting, but just a phase correction). The two methods are outlined in Figure 12.9.

---

[10] The method is mostly applied when two diversity branches are available.

[11] Note that in our notation the signals are weighted with the *complex conjugate* of w; this simplifies notation later on.

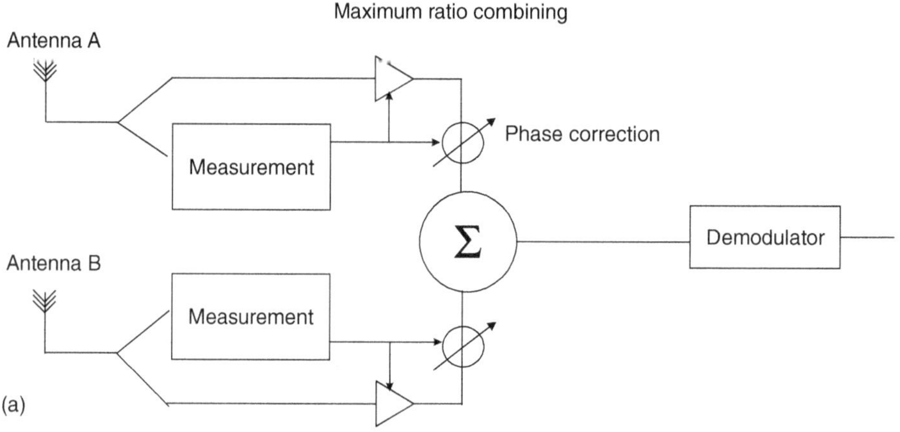

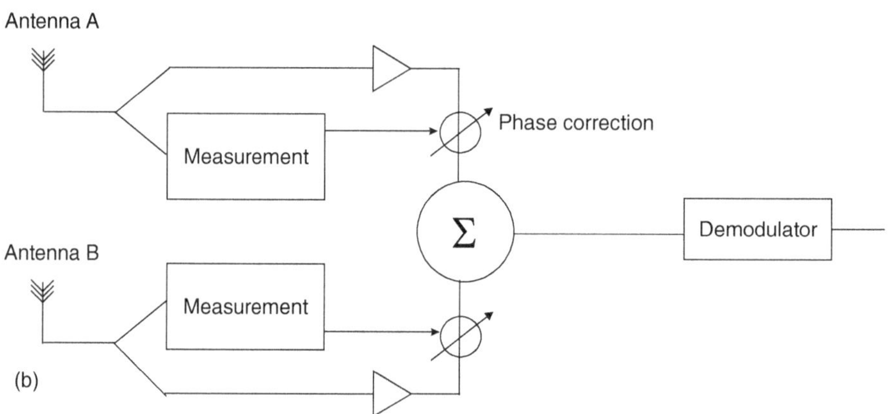

Figure 12.9 Combining diversity principle: (a) maximum ratio combining, (b) equal gain combining.

Maximum Ratio Combining

MRC compensates for the phases and weights the signals from the different antenna branches according to their SNR. This is the optimum way of combining different diversity branches – if several assumptions are fulfilled. Let us assume a propagation channel that is slow fading and flat fading. The only disturbance is AWGN. Under these assumptions, each channel realization can be written as a time-invariant filter with impulse response:

$$h_n(\tau) = \alpha_n \delta(\tau) \tag{12.14}$$

where α_n is the (instantaneous) complex gain of diversity branch n. These signals at the different branches are multiplied with weights w_n^* and added up, so that the SNR becomes (assuming unit transmit power)

$$\frac{\left| \sum_{n=1}^{N_{\mathrm{r}}} w_n^* \alpha_n \right|^2}{\sigma_{\mathrm{n}}^2 \sum_{n=1}^{N_{\mathrm{r}}} |w_n|^2} \tag{12.15}$$

where σ_{n}^2 is the noise power per branch (assumed to be the same in each branch). According to the Cauchy–Schwarz inequality, $\left| \sum_{n=1}^{N_{\mathrm{r}}} w_n^* \alpha_n \right|^2 \leq \sum_{n=1}^{N_{\mathrm{r}}} |w_n^*|^2 \sum_{n=1}^{N_{\mathrm{r}}} |\alpha_n|^2$, where equality holds if and only if $w_n = \alpha_n$. Thus, the SNR is maximized by choosing the weights as

$$w_{\mathrm{MRC},n} = \alpha_n \tag{12.16}$$

i.e., the signals are phase-corrected (remember that the received signals are multiplied with w^*) and weighted by the amplitude. We can then easily see that in that case, the output SNR of the diversity combiner is the *sum* of the branch SNRs:

$$\gamma_{\text{MRC}} = \sum_{n=1}^{N_r} \gamma_n. \tag{12.17}$$

If the branches are statistically independent, then the moment-generating function of the total SNR can be computed as the product of the moment-generating functions of the branch SNRs. If, furthermore, the SNR distribution in each branch is exponential (corresponding to Rayleigh fading), and all branches have the same mean SNR $\bar{\gamma}_n = \bar{\gamma}$, we find after some manipulations that

$$pdf_\gamma(\gamma) = \frac{1}{(N_r - 1)!} \frac{\gamma^{N_r - 1}}{\bar{\gamma}^{N_r}} \exp\left(-\frac{\gamma}{\bar{\gamma}}\right) \tag{12.18}$$

and the mean SNR of the combiner output is just the mean branch SNR, multiplied by the number of diversity branches:

$$\bar{\gamma}_{\text{MRC}} = N_r \bar{\gamma}. \tag{12.19}$$

Figure 12.10 compares the statistics of the SNR for RSSI-driven selection diversity and MRC. Naturally, there is no difference between the diversity types for $N_r = 1$, since there is no diversity.

We furthermore see that the slope of the distribution is the same for MRC and selection diversity, but that the difference in the mean values increases with increasing N_r. This is intuitively clear, as selection diversity discards $N_r - 1$ signal copies – something that increases with N_r. For $N_r = 3$, the difference between the two types of diversity is only about 2 dB.

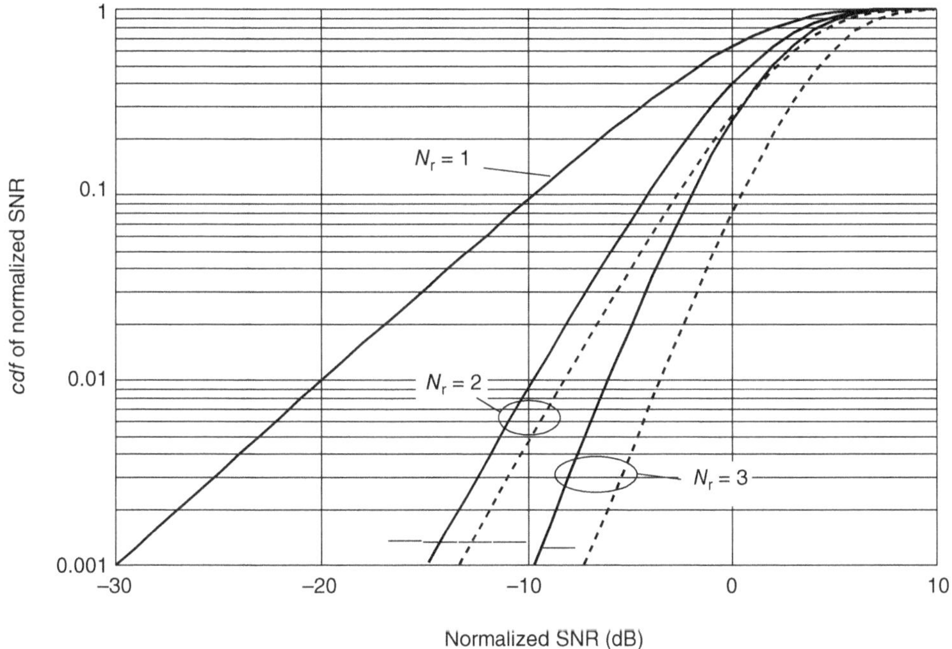

Figure 12.10 Cumulative distribution function of the normalized instantaneous signal-to-noise ratio $\gamma/\bar{\gamma}$ for received-signal-strength-indication-driven selection diversity (solid), and maximum ratio combining (dashed) for $N_r = 1, 2, 3$. Note that for $N_r = 1$, there is no difference between diversity types.

*Equal Gain Combining

For EGC, we find that the SNR of the combiner output is

$$\gamma_{\text{ECG}} = \frac{\left(\sum_{n=1}^{N_r} \sqrt{\gamma_n}\right)^2}{N_r} \tag{12.20}$$

where we have assumed that noise levels are the same on all diversity branches. The mean SNR of the combiner output can be found to be

$$\bar{\gamma}_{\text{EGC}} = \bar{\gamma}\left(1 + (N_r - 1)\frac{\pi}{4}\right) \tag{12.21}$$

if all branches suffer from Rayleigh fading with the same mean SNR $\bar{\gamma}$. Remember that we only assume here that the *mean* SNR is the same in all branches, while instantaneous branch SNRs (representing different channel realizations) can be different. It is quite remarkable that EGC performs worse than MRC by only a factor $\pi/4$ (in terms of mean SNR). The performance difference between EGC and MRC becomes bigger when mean branch SNRs are also different.

Equations for the pdf of the SNR can be computed from Eq. (12.20), but become unwieldy very quickly (results can be found, e.g., in [Lee 1982]).

Optimum Combining

One of the assumptions in the derivation of MRC was that only AWGN disturbs the signal. If it is interference that determines signal quality, then MRC is no longer the best solution. In order to maximize the Signal-to-Interference-and-Noise Ratio (SINR), the weights should then be determined according to a strategy called *optimum combining*, first derived in the groundbreaking paper of [Winters 1984]. The first step is the determination of the correlation matrix of noise and interference at the different antenna elements:

$$\mathbf{R} = \sigma_n^2 \mathbf{I} + \sum_{k=1}^{K} E\{\mathbf{r}_k \mathbf{r}_k^\dagger\} \tag{12.22}$$

where expectation is over a time period when the channel remains constant, and $\mathbf{r}_k$ is the receive signal vector of the kth interferer. We furthermore need the complex transfer function (complex gain, since we assume flat-fading channels) of the desired signal for the N_r diversity branches; these are written into the vector $\mathbf{h}_d$. The vector containing the optimum receive weights is then

$$\mathbf{w}_{\text{opt}} = \mathbf{R}^{-1}\mathbf{h}_d. \tag{12.23}$$

These weights have to be adjusted as the channel changes. It is easy to see that for a noise-limited system the correlation matrix becomes a (scaled) identity matrix, and optimum combining reduces to MRC.

A further interesting observation is that minimizing the Mean Square Error (MSE) is equivalent to maximizing the SINR, which can be written as

$$\text{SINR} = \left((\mathbf{w}^\dagger \mathbf{h}_d \mathbf{h}_d^\dagger \mathbf{w}) / (\mathbf{w}^\dagger \mathbf{R}\mathbf{w}) \right). \tag{12.24}$$

This SINR is a generalized Rayleigh quotient and is maximized by the generalized eigenvector corresponding to the maximum generalized eigenvalue of

$$\mathbf{h}_d \mathbf{h}_d^\dagger \mathbf{w} = \lambda \mathbf{R}\mathbf{w}. \tag{12.25}$$

The weight vector obtained from this generalized eigenvalue problem is the same as $\mathbf{w}_{\text{opt}}$ in Eq. (12.22).

Optimum combining of signals from N_r diversity branches gives N_r degrees of freedom. This allows interference from $N_r - 1$ interferers to be eliminated. Alternatively, $N_s \le N_r - 1$ interferers can be eliminated, while the remaining $N_r - N_s$ antennas behave like "normal" diversity antennas that can be used for noise reduction. This seemingly simple statement has great impact on wireless system design! As we discussed in Chapter 3, many wireless systems are interference limited. The possibility of eliminating at least some of the interferers by appropriate diversity combining opens up the possibility of drastically improving the link quality of such systems, or of increasing their capacity. This fact forms the basis of both single-user MIMO (Chapter 16) and multi-user MIMO (Chapter 22) systems.

Example 12.5 *Consider a desired signal with Binary Phase Shift Keying (BPSK) propagating through a frequency-flat channel with $\mathbf{h}_d = [1, 0.5 + 0.7j]^T$ and a synchronous, interfering BPSK signal with $\mathbf{h}_{\text{int}} = [0.3, -0.2 + 1.7j]^T$. The noise variance is $\sigma_n^2 = 0.01$. Show that weighting the received signal with the weights as computed from Eq. (12.23) leads to mitigation of interference.*

Let the desired transmitted signal be $s_d(t)$, the interfering transmitted signal be $s_{\text{int}}(t)$, and $n_1(t)$ and $n_2(t)$ be independent zero-mean noise processes. It is assumed that desired and interfering signals are uncorrelated, and that $E\{s_d s_d^*\} = 1$, and also $E\{s_{\text{int}} s_{\text{int}}^*\} = 1$.

The noise-plus-interference correlation matrix is computed according to Eq. (12.22):

$$\mathbf{R} = \sigma_n^2 \mathbf{I} + \sum_{k=1}^{K} E\{\mathbf{r}_k \mathbf{r}_k^\dagger\} \tag{12.26}$$

$$= 0.01 \begin{pmatrix} 1 & 0 \\ 0 & 1 \end{pmatrix} + \begin{pmatrix} 0.3 \\ -0.2 + 1.7j \end{pmatrix} (0.3 \quad -0.2 - 1.7j) \tag{12.27}$$

$$= \begin{pmatrix} 0.1 & -0.06 - 0.51j \\ -0.06 + 0.51j & 2.94 \end{pmatrix}. \tag{12.28}$$

Using Eq. (12.23) and normalizing the weights, we obtain:

$$\mathbf{w} = \frac{\mathbf{R}^{-1}\mathbf{h}_\mathrm{d}}{|\,\mathbf{R}^{-1}\mathbf{h}_\mathrm{d}\,|} \tag{12.29}$$

$$= \begin{pmatrix} 0.979 + 0.111j \\ 0.041 - 0.165j \end{pmatrix}. \tag{12.30}$$

Inserting this value for $\mathbf{w}$ into Eq. (12.24), we obtain an SINR of 78. This can be compared to the SNR = 100, which could be obtained if the RX would not have to suppress an interferer, and the SINR of 0.6 that is available at a single antenna.

Hybrid Selection – Maximum Ratio Combining

A compromise between selection diversity and full signal combining is the so-called hybrid selection scheme, where the best L out of N_r antenna signals are chosen, downconverted, and processed. This reduces the number of required RF chains from N_r to L and thus leads to significant savings. The savings come at the price of a (usually small) performance loss compared with the full-complexity system. The approach is called *Hybrid Selection/Maximum Ratio Combining* (H-S/MRC), or sometimes also *Generalized Selection Combining* (GSC).

It is well known that the output SNR of MRC is just the sum of the SNRs at the different receive antenna elements. For H-S/MRC, the instantaneous output SNR of H-S/MRC looks deceptively similar to MRC – namely:

$$\gamma_{\mathrm{H-S/MRC}} = \sum_{n=1}^{L} \gamma_{(n)}. \tag{12.31}$$

The major difference from MRC is that the $\gamma_{(n)}$ are *ordered* SNRs – i.e., $\gamma_{(1)} > \gamma_{(2)} > \dots > \gamma_{(Nr)}$. This leads to different performance and poses new mathematical challenges for performance analysis. Specifically, we have to introduce the concept of *order statistics*. Note that selection diversity (where just one of N_r antennas is selected) and MRC are limiting cases of H-S/MRC with $L=1$ and $L=N_\mathrm{r}$, respectively.

H-S/MRC schemes provide good diversity gain, as they select the best antenna branches for combining. Actually, it can be shown that the diversity *order* obtained with such schemes is proportional to N_r, not to L. However, they do not provide full beamforming gain. If the signals at all antenna elements are completely correlated, then the SNR gain of H-S/MRC is only L, compared with N_r for an MRC scheme.

The analysis of H-S/MRC based on a chosen ordering of the branches at first appears to be complicated, since the SNR statistics of the ordered branches are *not* independent. However, we can alleviate this problem by transforming ordered branch variables into a new set of random variables. It is possible to find a transformation that leads to *independently distributed* random variables (termed *virtual branch variables*). The fact that the combiner output SNR can be expressed in terms of independent identically distributed (iid) virtual branch variables enormously simplifies performance analysis of the system. For example, the derivation of Symbol Error Probability (SEP) for uncoded H-S/MRC systems, which normally would require evaluation of nested N-fold integrals, essentially reduces to evaluation of a *single* integral with finite limits.

The mean and variance of the output SNR for H-S/MRC is thus

$$\overline{\gamma}_{\mathrm{H-S/MRC}} = L\left(1 + \sum_{n=L+1}^{N_\mathrm{r}} \frac{1}{n}\right)\overline{\gamma} \tag{12.32}$$

and

$$\sigma^2_{\mathrm{H-S/MRC}} = L\left(1 + L\sum_{n=L+1}^{N_\mathrm{r}} \frac{1}{n^2}\right)\overline{\gamma}^2. \tag{12.33}$$

H-S/MRC is also related to the concept of hybrid beamforming, see Sections 16.2.10 and 22.10.2.

12.5 Error Probability in Fading Channels with Diversity Reception

In this section, we determine the Symbol Error Rate (SER) in fading channels when diversity is used at the RX. We start with the case of flat-fading channels, computing the statistics of the received power and the SER. We then proceed to dispersive channels, where we analyze how diversity can mitigate the detrimental effects of dispersive channels on simple RXs.

12.5.1 Error Probability in Flat-Fading Channels

Classical Computation Method

Analogous to Section 11.2, we can compute the error probability of diversity systems by averaging the conditional error probability (conditioned on a certain SNR) over the distribution of the SNR:

$$\overline{SER} = \int_0^\infty pdf_\gamma(\gamma)SER(\gamma)d\gamma. \tag{12.34}$$

As an example, let us compute the performance of BPSK with N_r diversity branches with MRC. The SER of BPSK (which equals its BER) in AWGN is (see Eq. (11.36))

$$SER(\gamma) = Q\left(\sqrt{2\gamma}\right). \tag{12.35}$$

Let us apply this principle to the case of MRC. When inserting Eqs. (12.18) and (12.35) into Eq. (12.34), we obtain an equation that can be evaluated analytically:

$$\overline{SER} = \left(\frac{1-b}{2}\right)^{N_r}\sum_{n=0}^{N_r-1}\binom{N_r-1+n}{n}\left(\frac{1+b}{2}\right)^n \tag{12.36}$$

where b is defined as

$$b = \sqrt{\frac{\overline{\gamma}}{1+\overline{\gamma}}}\;. \tag{12.37}$$

For large values of $\overline{\gamma}$, this can be approximated as

$$\overline{SER} = \left(\frac{1}{4\overline{\gamma}}\right)^{N_r}\binom{2N_r-1}{N_r}. \tag{12.38}$$

From this, we can see that (with N_r diversity antennas) the SER decreases with the N_r-th power of the SNR.

***Computation via the Moment-Generating Function Method**

In the previous section, we averaged the SER over the distribution of SNRs, using the "classical" representation of the Q-function. As we have already seen in Section 11.2.2, there is an alternative definition of the Q-function, which can easily be combined with the moment-generating function $M_\gamma(s)$ of the SNR. Let us start by writing the SER conditioned on a given SNR in the form (Eq. (11.74)):

$$SER(\gamma) = \int_{\theta_1}^{\theta_2} f_1(\theta)\exp\left(-\gamma_{\mathrm{MRC}}f_2(\theta)\right)d\theta. \tag{12.39}$$

Since

$$\gamma_{\mathrm{MRC}} = \sum_{n=1}^{N_r}\gamma_n \tag{12.40}$$

this can be rewritten as

$$SER(\gamma) = \int_{\theta_1}^{\theta_2} f_1(\theta)\prod_{n=1}^{N_r}\exp\left(-\gamma_n f_2(\theta)\right)d\theta. \tag{12.41}$$

Averaging over the SNRs in different branches then becomes (assuming independent fading in the branches)

$$\overline{SER} = \int d\gamma_1 pdf_{\gamma_1}(\gamma_1)\int d\gamma_2 pdf_{\gamma_2}(\gamma_2)\cdots\int d\gamma_{N_r} pdf_{\gamma_{N_r}}(\gamma_{N_r})\int_{\theta_1}^{\theta_2} d\theta f_1(\theta)\prod_{n=1}^{N_r}\exp\left(-\gamma_n f_2(\theta)\right) \tag{12.42}$$

$$= \int_{\theta_1}^{\theta_2} d\theta f_1(\theta)\prod_{n=1}^{N_r}\int d\gamma_n pdf_{\gamma_n}(\gamma_n)\exp\left(-\gamma_n f_2(\theta)\right) \tag{12.43}$$

$$= \int_{\theta_1}^{\theta_2} d\theta f_1(\theta) \prod_{n=1}^{N_r} M_\gamma(-f_2(\theta)) \tag{12.44}$$

$$= \int_{\theta_1}^{\theta_2} d\theta f_1(\theta) \left[M_\gamma(-f_2(\theta)) \right]^{N_r}. \tag{12.45}$$

With that, we can write the error probability for BPSK in Rayleigh fading as

$$\overline{SER} = \frac{1}{\pi} \int_0^{\pi/2} \left[\frac{\sin^2(\theta)}{\sin^2(\theta) + \overline{\gamma}} \right]^{N_r} d\theta. \tag{12.46}$$

Example 12.6 *Compare the symbol error probability in Rayleigh fading of 8-PSK (Phase Shift Keying) and four available antennas with 10-dB SNRs when using L = 1, 2, 4 receive chains.*

The SER for M-ary Phase Shift Keying (MPSK) with H-S/MRC can be shown to be

$$\overline{SER}_{e,\text{H-S/MRC}}^{\text{MPSK}} = \frac{1}{\pi} \int_0^{\pi(M-1)/M} \left[\frac{\sin^2\theta}{\sin^2(\pi/M)\overline{\gamma} + \sin^2\theta} \right]^L \prod_{n=L+1}^{N_r} \left[\frac{\sin^2\theta}{\sin^2(\pi/M)\overline{\gamma}\frac{L}{n} + \sin^2\theta} \right] d\theta. \tag{12.47}$$

Evaluating Eq. (12.47) with $M = 8$, $\overline{\gamma} = 10\,\text{dB}$, $N_r = 4$, and $L = 1, 2, 4$ yields:

L	$\overline{SER}_{e,\text{H}-\text{S/MRC}}^{\text{MPSK}}$
1	0.0442
2	0.0168
4	0.0090

*12.5.2 Error Probability in Frequency-Selective Fading Channels

We now determine the SER in channels that suffer from time dispersion and frequency dispersion. We assume here binary FSK (e.g., MSK) with differential phase detection. The analysis uses the correlation coefficient ρ_{XY} between signals at two sampling times that was discussed in Section 11.3.3.

For binary FSK with selection diversity:

$$\overline{SER} = \frac{1}{2} - \frac{1}{2} \sum_{n=1}^{N_r} \binom{N_r}{n} (-1)^{n+1} \frac{b_0 \text{Im}\{\rho_{XY}\}}{\sqrt{(\text{Im}\{\rho_{XY}\})^2 + n(1 - |\rho_{XY}|^2)}}. \tag{12.48}$$

where b_0 is the transmitted bit. This can be approximated as

$$\overline{SER} = \frac{(2N_r - 1)!!}{2} \left(\frac{1 - |\rho_{XY}|^2}{2(\text{Im}\{\rho_{XY}\})^2} \right)^{N_r}. \tag{12.49}$$

where $(2N_r - 1)!! = 1 \cdot 3 \cdot 5 \dots (2N_r - 1)$.

For binary FSK with MRC:

$$\overline{SER} = \frac{1}{2} - \frac{1}{2} \frac{b_0 \text{Im}\{\rho_{XY}\}}{\sqrt{1 - (\text{Re}\{\rho_{XY}\})^2}} \sum_{n=0}^{N_r-1} \frac{(2n-1)!!}{(2n)!!} \left(1 - \frac{(\text{Im}\{\rho_{XY}\})^2}{1 - (\text{Re}\{\rho_{XY}\})^2} \right)^n. \tag{12.50}$$

which can be approximated as

$$\overline{SER} = \frac{(2N_r - 1)!!}{2(N_r!)} \left(\frac{1 - |\rho_{XY}|^2}{2(\text{Im}\{\rho_{XY}\})^2} \right)^{N_r}. \tag{12.51}$$

This formulation shows that MRC improves the SER by a factor $N_r!$ compared with selection diversity. A further important consequence is that the errors due to delay dispersion and random Frequency Modulation (FM) are decreased in the same way as errors due to noise. This is shown by the expressions in parentheses that are taken to the N_r-th power. These terms subsume the errors due to all different effects. The SER with diversity is approximately the N_r-th power of the SER without diversity (see Figures 12.11–12.13).

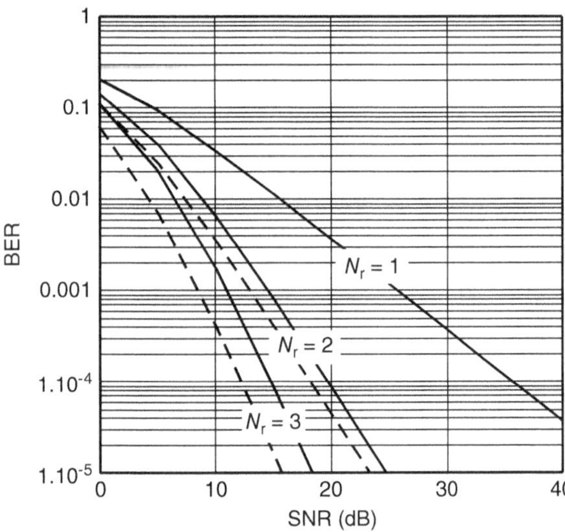

Figure 12.11 Bit error rate of minimum shift keying (MSK) with received-signal-strength-indication-driven selection diversity (solid) and maximum ratio combining (dashed) as a function of the signal-to-noise ratio with N_r diversity antennas. Reproduced with permission from [Molisch 2000] © Prentice Hall.

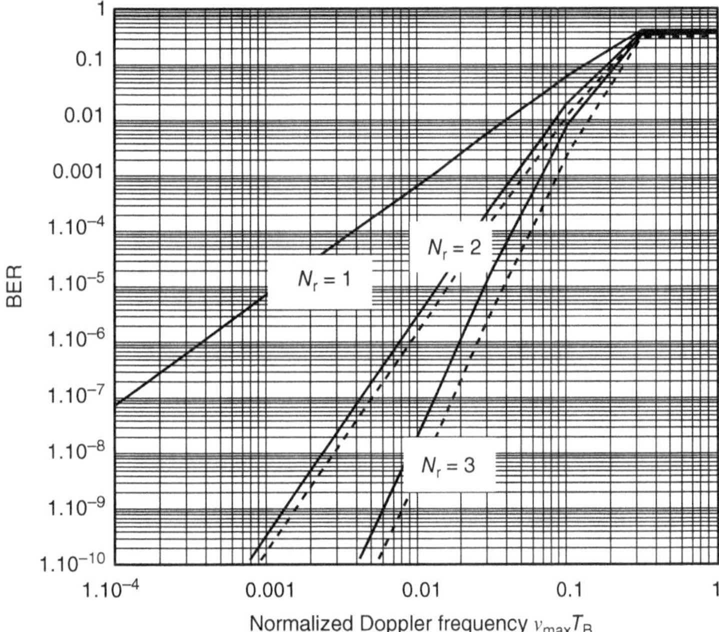

Figure 12.12 Bit error rate of MSK with received-signal-strength-indication-driven selection diversity (solid) and maximum ratio combining (dashed) as a function of the normalized Doppler frequency with N_r diversity antennas. Reproduced with permission from [Molisch 2000] © Prentice Hall.

More general cases can be treated with the Quadratic Form Gaussian Variable (QFGV) method (see Section 11.3.3), where Eq. (11.108) is replaced by

$$P(\mathrm{D} < 0) = Q_M(p_1, p_2) - I_0(p_1 p_2) \exp\left[-\frac{1}{2}\left(p_1^2 + p_2^2\right)\right]$$

$$+ \frac{I_0(p_1 p_2) \exp\left[-\frac{1}{2}\left(p_1^2 + p_2^2\right)\right]}{\left(1 + \nu_2/\nu_1\right)^{2N_r - 1}} \sum_{n=0}^{N_r - 1} \binom{2N_r - 1}{n} \left(\frac{\nu_2}{\nu_1}\right)^n$$

$$+ \frac{\exp\left[-\frac{1}{2}\left(p_1^2 + p_2^2\right)\right]}{\left(1 + \nu_2/\nu_1\right)^{2N_r - 1}} \cdot \sum_{n=1}^{N_r - 1} I_n(p_1 p_2) \sum_{k=0}^{N_r - 1 - n} \binom{2N_r - 1}{k}$$

$$\times \left[\left(\frac{p_2}{p_1}\right)^n \left(\frac{\nu_2}{\nu_1}\right)^k - \left(\frac{p_1}{p_2}\right)^n \left(\frac{\nu_2}{\nu_1}\right)^{2N_r - 1 - k}\right]. \tag{12.52}$$

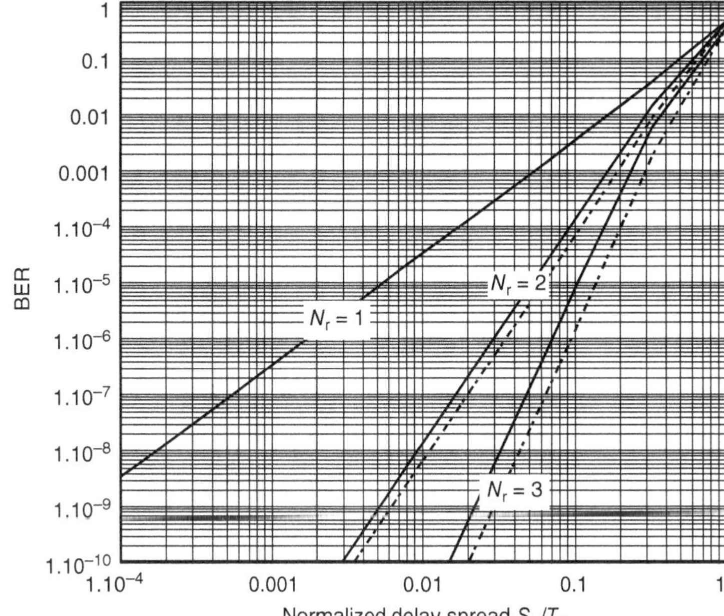

Figure 12.13 Bit error rate of MSK with received-signal-strength-indication-driven selection diversity (solid) and maximum ratio combining (dashed) as a function of the normalized rms delay spread with N_r diversity antennas.
Reproduced with permission from [Molisch 2000] © Prentice Hall.

RSSI-driven diversity is not the best selection strategy when errors are mostly caused by frequency selectivity and time selectivity. It puts emphasis on signals that have a large amplitude, and not on those with the smallest distortion.[12] In these cases, BER-driven selection diversity is preferable. For $N_r = 2$, the BER of Minimum Shift Keying (MSK) with differential detection becomes

$$\overline{BER} \approx \left(\frac{\pi}{4}\right)^4 \left(\frac{S_\tau}{T_B}\right)^4.$$
(12.53)

compared with the RSSI-driven result:

$$\overline{BER} \approx 3\left(\frac{\pi}{4}\right)^4 \left(\frac{S_\tau}{T_B}\right)^4.$$
(12.54)

*12.6   Appendix

App. 12.A: Correlation Coefficient of Two Signals with Frequency Separation

See App12.pdf at wiley.com/go/molisch/wireless3e

Further Reading

Spatial (antenna) diversity is the oldest form of diversity and is discussed in textbooks on wireless propagation channels (see, e.g., [Vaughan and Andersen 2003]). Evaluations of the antenna correlation coefficient for different angular spectra can be found in [Fuhl et al. 1998] and [Roy and Fortier 2004]. For antennas on a handset, results are given in [Ogawa et al. 2001]. [Taga 1993] also shows the effect of mutual coupling on pattern diversity and thus the correlation coefficient, as do a number of papers written in the context of MIMO systems [Waldschmidt et al. 2004; Wallace and Jensen 2004]. Polarization diversity is discussed in more detail, e.g., in [Narayanan et al. 2004] and [Shafi et al. 2006]; joint spatial, polarization, and pattern (angle) diversity is discussed in [Dietrich et al. 2001]. Different combination strategies for antenna signals, and the resulting channel statistics, are discussed in [Proakis and Salehi 2005], and in many papers in the primary literature. The situation is similar to that mentioned in Chapter 11: each type of fading statistics and each combination strategy merits at least one paper. Application of these fading statistics to computation of BERs gives rise to an even greater variety of papers. A nice, unified treatment can be found in [Simon and Alouini 2004], who include different fading statistics and antenna combination strategies for a variety of modulation formats. For Rayleigh or Rice fading, the

---

[12] For a similar reason, MRC is not the best combining strategy.

classical QFGV method of [Proakis 1968] is applicable; [Adachi and Ohno 1991] and [Adachi and Parsons 1989] compute the performance for DQPSK and FSK, respectively. The virtual path method for H-S/MRC was introduced in [Win and Winters 1999]; [Simon and Alouini 2004] also investigate this case. General mathematical aspects of order statistics can be found in [David and Nagaraja 2003]. Discussions on the impact of diversity on systems in frequency-selective channels can be found in [Molisch 2000, ch. 12].

For updates and errata for this chapter, see https://wides.usc.edu/students.html#textbooks.

Exercises

See Sec. 36.12 of Exercises.pdf at wiley.com/go/molisch/wireless3e

13

Channel Coding and Information Theory

13.1 Fundamentals of Coding and Information Theory

13.1.1 History and Motivation of Coding

Chapter 11 demonstrated that Bit Error Rates (BERs) on the order of 10^{-2} can occur for Signal-to-Noise Ratios (SNRs) typically encountered in wireless systems. Those high BERs are mostly due to the effect of multi-path propagation. Advanced Receiver (RX) structures can help to reduce those values: diversity combats fading dips, while equalizers and Rake receivers (see Chapters 14 and 19) improve the performance in frequency-selective channels. However, even those advanced RXs might not sufficiently reduce the BER. Data communications often require BERs on the order of 10^{-6}–10^{-9}. Such low values can only be achieved by employing coding of the data, i.e., introducing redundancy into the transmission. The use of error-correcting codes[1] leads to a reduction of the BER, or equivalently to a coding gain G_{code}, i.e., we have to use G_{code} decibel (dB) less transmit power than in an uncoded system to achieve the target BER.[2]

Coding and information theory builds on the seminal work of Claude Shannon [Shannon 1948] entitled "A mathematical theory of communication." He showed that it is possible to transmit data without errors as long as the bit rate is smaller than the *channel capacity*. The absence of errors is achieved by the use of "appropriate" codes. Shannon showed that (infinitely long) random codes achieve capacity. Unfortunately, such codes cannot be used in practice due to the enormous effort required for their decoding. For more than 50 years, the work of the coding theorists mainly consisted of finding practical codes that come close to the Shannon limit, i.e., allow virtually error-free communications with rates close to the channel capacity.

In the subsequent sections of this chapter, we will give a brief overview of error-correction coding. Basic coding theory is developed in Sections 13.2–13.9 for the Additive White Gaussian Noise (AWGN) channel, laying out theoretical backgrounds for the most important classes of codes and their decoding: block codes, convolutional codes, Trellis Coded Modulation (TCM), turbo codes, Low-Density Parity Check (LDPC) codes, and polar codes. Sections 13.10–13.12 then deal specifically with the idiosyncrasies of fading channels and describe how the coding structures need to be adapted for this case.

13.1.2 Fundamental Concepts of Information Theory

Information theory explores the theoretical performance limits of optimum communication systems – in other words, the limits of what *any* communication system can achieve. Knowledge of information-theoretic limits is useful for system designers because it indicates how far a real system could be improved. Key concepts we will encounter in this regard are (i) a mathematically solid definition of *information*, and (ii) the *channel capacity*, which describes at what rate information can be transmitted, at best, over a given channel. We will find that suitable coding is required for communication to approach the channel capacity.

As a preliminary step, let us define the *Discrete Memoryless Channel* (DMC). Assume there is an alphabet $\mathcal{X}$ of transmit symbols (the size of the alphabet is $|\mathcal{X}|$), where the probability for the transmission of each element of the alphabet is known. Similarly, assume there is a receive symbol alphabet $\mathcal{Y}$, with size $|\mathcal{Y}|$. The DMC defines transition probabilities from each transmit symbol to each receive symbol. The most common example of a DMC is the binary symmetric channel, which is characterized by $|\mathcal{X}| = |\mathcal{Y}| = 2$, and a transition probability p: if we transmit $X = +1$, then with $1 - p$ probability, the receive symbol will be $Y = +1$, and with p probability, it will be $Y = -1$; similarly if we transmit $X = -1$, probability for receiving -1 is $1 - p$, and for receiving $+1$ it is p. Furthermore, a memoryless channel has the property that if we transmit a sequence of transmit symbols $\mathbf{x} = \{x_1, x_2, \ldots x_N\}$ and observe a sequence of output symbols $\mathbf{y} = \{y_1, y_2, \ldots y_N\}$, then

---

[1] In the remainder of this chapter, we will say *coding* when we mean *error-correcting coding*. Note that this is different from Chapters 24 and 25, where coding will refer to source coding.

[2] Note that the coding gain can depend on the target error rate. Coding leads to a change of shape of the BER-over-SNR curve.

Wireless Communications: From Fundamentals to Beyond 5G, Third Edition. Andreas F. Molisch.
© 2023 John Wiley & Sons Ltd. Published 2023 by John Wiley & Sons Ltd.
Companion website: www.wiley.com/go/molisch/wireless3e

$$\Pr(\mathbf{y}\,|\,\mathbf{x}) = \prod_{n=1}^{N} \Pr(y_n\,|\,x_n) \tag{13.1}$$

which implies that there is no Inter Symbol Interference (ISI).

DMCs can describe, e.g., the concatenation of a Binary Phase Shift Keying (BPSK) modulator, an AWGN channel, and a demodulator with hard-decision output. There, the variable X corresponds to the transmit symbols $+1/-1$. The variable Y corresponds to the output of the demodulator/decision device, which is also $+1/-1$. The amount of noise in the channel determines the symbol error probability (as computed in Chapter 11), which in this case is identical to the transition probability p. If we want to avoid the restriction of a hard-decision demodulator, a useful channel model is the discrete-time AWGN channel (as used in Chapters 11), with

$$y_n = x_n + n_n \tag{13.2}$$

where the n_n are Gaussian-distributed random variables with variance σ^2, and the input variables x_n are subject to an average power constraint, $E\{x_n^2\} \leq P$.[3]

We now define the mutual information between two discrete random variables X and Y. If we observe a certain realization of Y, i.e., $Y = y$, then the mutual information is a measure of how much this observation tells us about the occurrence of an event $X = x$. The mutual information between the realizations x and y is defined as

$$I(x;y) = \log \frac{\Pr(x\,|\,y)}{\Pr(x)} \tag{13.3}$$

where $\Pr(x|y)$ is the probability of x, conditioned on y. The logarithm in the above equation can either have base 2, in which case the mutual information is in the unit of *bits*, or it can have base e, in which case the unit of I is *nats*. The mutual information between the random variables X and Y is then the average of $I(x; y)$, namely

$$I(X;Y) = \sum_{x}\sum_{y}\Pr(X = x,\, Y = y) \log \frac{\Pr(x\,|\,y)}{\Pr(x)} \tag{13.4}$$

$$= \sum_{x}\sum_{y}\Pr(X = x,\, Y = y) \log \frac{\Pr(x,y)}{\Pr(x)\Pr(y)}. \tag{13.5}$$

In the case that y and/or x are continuous variables, the summations are replaced by integrations.

An intuitive interpretation of the mutual information follows from the DMC: associate again X with the transmit symbols and Y with the received symbols. If X and Y are statistically independent, i.e., receiving a particular symbol gives us no information about which symbol was transmitted, then the mutual information is 0; this situation occurs in channels with very low SNR; for example, in a binary symmetric channel, this corresponds to $p = 0.5$. Mutual information is high if receiving a certain symbol, say $+1$, tells us with high probability which symbol was transmitted. This situation occurs in channels with small transition probability p, or in other words, in channels with little noise. Note that the mutual information can never become larger than $\min[\log |\mathcal{X}|, \log |\mathcal{Y}|]$. If $X = Y$, then for discrete alphabets

$$I(X;Y) = -\sum_{x}\sum_{y}\Pr(X = x, Y = y) \log\left[\Pr(x)\right] \tag{13.6}$$

$$= -\sum_{x}\Pr(X = x) \log\left[\Pr(x)\right] \tag{13.7}$$

which is called the *entropy* of X, $H(X)$. The entropy characterizes the information content of the random variable X. For discrete memoryless sources, the entropy is maximized if all symbols of the alphabet are equiprobable.

In general, mutual information can also be written in terms of the entropies of the received symbols $H(Y)$ and the conditional entropy $H(Y\,|\,X) = -\sum_{x,y}\Pr(x,y) \log\left[\Pr(y\,|\,x)\right]$ (or, equivalently, the entropies of transmitted symbols and $H(X)$, and the conditional entropy $H(X\,|\,Y)$):

$$I(X;Y) = H(Y) - H(Y\,|\,X) = H(X) - H(X\,|\,Y). \tag{13.8}$$

---

[3] Note that P is power, while p is the transition probability. Furthermore, to avoid confusion about the meaning of subscript (n as index of the symbols versus n as subscript denoting noise), we call the noise variance here simply σ^2, instead of σ_n^2 as in the rest of the book.

13.1.3 Channel Capacity

For any communication over a DMC, the following theorem holds: reliable communications (i.e., communications with error probability tending to zero) is possible only if the communication rate $R < C$, where C is the channel capacity

$$C = \max_{\Pr(x)} I(X;Y) \tag{13.9}$$

i.e., the mutual information of X and Y maximized over all input distributions. If the rate is higher than the capacity, reliable communications are not possible.

The key to achieving, or at least approaching, the channel capacity is the use of coding with long codewords. As mentioned in the introduction, coding introduces redundancy, so that even if some bits in the codeword are received erroneously, it is still possible to recover the original codeword. Consider transmission of a codeword of length N over a binary symmetric channel: as $N \to \infty$, the probability of receiving $\approx Np$ errors tends to unity. There are

$$\binom{N}{Np} = \frac{N!}{(Np)!(N(1-p)!)} \tag{13.10}$$

$$\approx \frac{\sqrt{2\pi N}N^N e^{-N}}{\left[\sqrt{2\pi Np}(Np)^{Np}e^{-Np}\right]\left[\sqrt{2\pi N(1-p)}(N(1-p))^{N(1-p)}e^{-N(1-p)}\right]} \tag{13.11}$$

$$\approx \frac{1}{2^{N[p\log(p)]}\,2^{N[(1-p)\log(1-p)]}} \tag{13.12}$$

$$= 2^{N\,H_b(p)} \tag{13.13}$$

different codewords with Np errors; the first approximation follows from Sterling's formula, and the second is a rearrangement of terms using $p = 2^{\log(p)}$. The last equality is simply the definition of the binary entropy function, $H_b(p) = -p\log(p) - [(1-p)\log(1-p)]$. The overall number of available sequences of length N is 2^N. Thus, the number of clearly distinguishable sequences is $M = 2^{N(1-H_b(p))}$. Therefore, using N symbols, we can transmit M different messages, and thus $\log(M)$ information bits. The possible data rate is thus

$$R = \frac{\log(M)}{N} = 1 - H_b(p) \tag{13.14}$$

For an AWGN channel, a similar argument can be made. For a long codeword, we know that with high probability

$$\frac{1}{N}\sum_n |n_n|^2 \to \sigma^2 \tag{13.15}$$

so that the received signal vector $\mathbf{y}$ lies near the surface of a sphere (called noise sphere) of radius $\sqrt{N\sigma^2}$ around the transmit signal vector $\mathbf{x}$. Reliable communication is possible as long as the spheres associated with the different codewords do not overlap, i.e., each received signal point can be uniquely associated with a particular transmit signal point. On the other hand, due to the average power constraint, we know that all received signal points must lie within a sphere of radius $\sqrt{N(P + \sigma^2)}$. We can thus conclude that the number of different received sequences that can be decoded reliably is equal to the number of noise spheres that fit into a sphere of radius $\sqrt{N(P + \sigma^2)}$. Since the volume of an N-dimensional sphere with radius ρ is proportional to ρ^N, the number of different messages that can be communicated with a codeword of length N is

$$M = \frac{[N(P + \sigma^2)]^{N/2}}{[N\sigma^2]^{N/2}} \tag{13.16}$$

which makes the possible rate of communications

$$R = \frac{\log(M)}{N} = \frac{1}{2}\log\left[1 + \frac{P}{\sigma^2}\right] \tag{13.17}$$

This equals the capacity of the AWGN channel. Without derivation, we state here also that this capacity is achieved when the transmit alphabet is Gaussian (and thus continuous).

It is also noteworthy that Eq. (13.17) is the capacity *per channel use* (i.e., per transmitted underlying symbol) for a *real* modulation alphabet and channel. When using complex modulation, the capacity per unit bandwidth becomes

$$C_{\text{AWGN}} = 2 \cdot \frac{1}{2}\log\left[1 + \frac{P}{N_0 B}\right], \tag{13.18}$$

which can be expressed in terms of the SNR as

$$C_{\mathrm{AWGN}} = \log\left[1 + \gamma\right].$$ (13.19)

It is noteworthy that the capacity increases only logarithmically with the SNR. This implies that increasing transmit power has only limited value for improving the capacity of an AWGN channel. Remember also that this statement is different from our insights in Chapters 11 and 12: there we said that *for a given transmission rate*, increasing the transmit power is very effective for decreasing the symbol error rate of *uncoded* transmission over an AWGN channel. Here, we are stating that for an *ideally coded* system (which by definition has no symbol errors), increasing the transmit power is *not* an efficient way for *increasing the transmission rate*. However, the admissible transmission rate clearly increases linearly with the bandwidth that is available.

13.1.4 Power–Bandwidth Relationship

From the capacity equation, we also see that we can trade off spectral efficiency with energy efficiency of a transmission. We know that the transmission rate has to be smaller than the normalized capacity, multiplied by the invested bandwidth

$$R < B \log\left[1 + \frac{P}{N_0 B}\right].$$ (13.20)

Furthermore, the energy expended per transmitted information bit is

$$E_{\mathrm{B}} = \frac{P T_{\mathrm{seq}}}{\log\left(M\right)} = \frac{P}{\dfrac{\log\left(M\right)}{T_{\mathrm{seq}}}} = \frac{P}{R}$$ (13.21)

where T_{seq} is the duration of the symbol sequence of length N. Denoting R/B as the spectral efficiency r, we can rewrite Eq. (13.20) as

$$r < \log\left(1 + r\frac{E_{\mathrm{B}}}{N_0}\right)$$ (13.22)

so that (assuming the rate is expressed in bit/s)

$$\frac{E_{\mathrm{B}}}{N_0} > \frac{2^r - 1}{r}.$$ (13.23)

This equation demonstrates the trade-off between energy per bit and the spectral efficiency. As the spectral efficiency decreases, so does the requirement for the E_{B}/N_0. In the limit of zero spectral efficiency, $E_{\mathrm{B}}/N_0 \to \ln(2) = -1.59$ dB, as follows from L'Hopital's rule. *This -1.59 dB is the smallest value for the E_{B}/N_0 for which reliable communication is possible.* Note that orthogonal modulation (e.g., Frequency Shift Keying (FSK) or Pulse Position Modulation (PPM)) with a large number of dimensions approximates this case.

Note that the above relationship is between the spectral efficiency when using a complex Gaussian modulation alphabet and the E_{B}/N_0. When considering the minimum SNR in practical cases, we have to take the following aspects into consideration:

- is the modulation format real or complex? When considering a one-dimensional modulation format such as BPSK, the achievable rate is given by Eq. (13.17) instead of (13.19) which is applicable for two-dimensional QAM;
- for a finite code rate R_{c}, i.e., where $1/R_{\mathrm{c}}$ symbols are being transmitted to represent a single source bit (see below for a more detailed description), do we represent the achievable spectral efficiency as a function of the E_{B}/N_0 or the E_{c}/N_0, where E_{c} is the energy of a transmitted symbol; note the relationship $E_{\mathrm{c}} = R_{\mathrm{c}} E_{\mathrm{B}}$;
- the use of a finite modulation alphabet results in the SNR required for error-free transmission increasing beyond what would be anticipated from (13.19), or equivalently, a decrease in achievable rate. For binary signaling using only the real dimension, i.e., BPSK, this becomes

$$C_{\mathrm{BPSK}} = 1 - \frac{1}{\sqrt{2\pi}}\int \exp\left(-\frac{u^2}{2}\right)\log_2\left[1 + \exp\left(-2u\sqrt{\gamma} - 2\gamma\right)\right]du$$ (13.24)

which naturally saturates at 1 bit/s/Hz.

We see from the above derivation that there is a trade-off between spectral efficiency and energy efficiency. For most wireless applications, such as cellular systems, Wi-Fi, etc., spectral efficiency is paramount. However, for sensor networks, IoT, and specialized applications such as deep-space communications, energy efficiency is more important. Generally, research for energy-efficient coding is not as advanced as that for spectrally efficient coding, though recent years have seen a strong uptick in this area.

13.1.5 Impact of Code Length

The above derivations of channel capacity were using infinitely long codes. Furthermore, the results do not suggest a way *how* to construct good codes, which have all the properties discussed above. Shannon's derivations show that random codes achieve channel capacity; however, the only way of decoding truly random codes is through list decoding, and with large N, the lists quickly become unmanageable. In Sections 13.2–13.9, we discuss practical coding methods that have worse performance than random codes but have the advantage of being actually decodable with finite effort. This is particularly true for block codes and convolutional codes, which have been used for many years but which – due to their relatively short codeword length – do not come close to the theoretical performance limits. The 1990s finally saw codes that achieve practical decodability while having large effective codeword length, and thus close-to-optimum performance: *turbo codes* and *LDPC codes* perform within less than 1 dB of the Shannon limit; they are discussed in Sections 13.6 and 13.7. More recently, it has been shown that *polar codes* also reach channel capacity, see Section 13.8.

The channel capacity is defined for the limiting case of very long codewords. Some more recent work has established valuable bounds for finite-length codes. In that case, it is not anymore possible to guarantee error-free transmission, and therefore an admissible bit error probability ε needs to be defined. For this ε, and for codewords of size N, the achievable rate in an AWGN channel is approximated as (see [Polyanski et al. 2010])

$$R(N, \varepsilon) = C_{\mathrm{AWGN}} - \sqrt{\frac{V_{\mathrm{cd}}}{N}} Q^{-1}(\varepsilon) + \mathcal{O}\left(\frac{\log(N)}{N}\right) \tag{13.25}$$

where $Q(x)$ is the Q-function Eq. (11.29), and V_{cd} is called the "channel dispersion" (note that this is different from the delay dispersion or frequency dispersion we encountered in multipath channels), which is defined as

$$V_{\mathrm{cd}} = \frac{\gamma(2 + \gamma)}{(1 + \gamma)^2}. \tag{13.26}$$

This shows that the penalty for having a finite-sized codeword decreases as $1/\sqrt{N}$. These equations are valuable because they give insights into the minimum length of LDPC and turbocodes necessary to achieve a certain performance level, and also provide target rates for the design of short codes.

13.1.6 Classification of Practical Codes

One way of classifying codes is to distinguish between *block codes*, where the redundancy is added to blocks of data and *convolutional codes*, where redundancy is added continuously to sets of bits/symbols. Block codes are well suited for correcting burst errors – something that frequently occurs in wireless communications; however, error bursts can also be converted into random errors by interleaving techniques. Convolutional codes have the advantage that their optimum decoder, the Viterbi algorithm, is of practically feasible complexity. They also offer the possibility of joint decoding and equalization by means of the same algorithm. Turbo codes easily fit into this categorization. As we will find in Section 13.6, turbo codes use two parallel, interleaved, convolutional codes to encode the information; however, the employed decoding structures are different because of the memory length of the encoder. A similar thing can be said about LDPC codes: they are block codes, but their large size necessitates different decoding structures; for that reason, they are treated in a separate section.

Later research has actually led to a "smearing" of the boundary between block codes and convolutional codes. It has been shown that Viterbi decoders (the classical solution for convolutional codes) can be used to decode block codes [Wolf 1978], while belief propagation algorithms [Loeliger 2004], commonly used for decoding block codes can be generalized to decode convolutional codes. However, those are very advanced research topics and will not be treated further here. The interested reader is referred to the cited papers.

Another classification of codes can be based on whether the coder input and/or output use *hard* or *soft* information. Hard information just tells us the (binary) value that we detect for a bit. Soft information tells us also the confidence we have in our decision for this bit.

Example 13.1 *Soft and hard information for repetition coding.*

Let us consider a simple example: a repetition code that repeats a bit three times. Let the transmitted bit sequence be 1 1 1. Let the demodulated signal at the RX be $-0.05, -0.1, 1.0$. One strategy is now to make a hard decision on each bit before decoding. After the slicer, the bits are $-1\ -1\ 1$. The decoder performs a majority decision and decides that -1 was transmitted. A soft decoder, on the other hand, might add up the demodulated signals,[4] giving 0.85, and a hard decision on that results in deciding that $+1$ was transmitted.

---

[4] This is purely for demonstration purposes. We will discuss more intelligent soft combining strategies later on.

For some applications (e.g., iterative and concatenated decoders), it is necessary that the decoder does not only use soft input information but also that it *puts out* soft information, and not just hard bits. In any case, the use of soft input information always leads to a performance improvement compared to the use of hard input information.

13.2 Block Codes

13.2.1 Introduction

Block codes are codes that group the source data into blocks, and – from the values of the bits in that block – compute a longer codeword which is the one actually transmitted. The smaller the code rate – i.e., the ratio of the number of bits in the original datablock to that of the transmitted block – the higher the redundancy, and the higher the probability that errors can be corrected. The simplest codes are repetition codes with blocksize 1: for an input "block" x, the output block is xxx (for a repetition code with rate 1/3).

After this intuitive introduction, let us now give a more precise description. First, we define some important terms and notations:

- *Block coding*: for block coding, source data are parsed into blocks of K symbols. Each of these uncoded datablocks is then associated with a codeword of length N symbols.
- *Code rate:* The ratio K/N is called the *code rate* R_c (assuming the symbol alphabet of coded and uncoded data is the same).
- *Binary codes:* these occur when the symbol alphabet is binary, using only "0" and "1". Almost all practical block codes are binary, with the exception of *Reed–Solomon* (RS) *codes* (see below). If not stated otherwise, the remainder of the chapter always talks about binary codes. Therefore, in the following, "sum" means "modulo-2 sum", and "+" denotes a modulo-2 addition.
- *Hamming distance:* The Hamming distance $d_{\mathrm{H}}(\mathbf{x}, \mathbf{y})$ between two codewords is the number of different bits:

$$d_{\mathrm{H}}(\mathbf{x},\mathbf{y}) = \sum_n |x_n - y_n| \qquad (13.27)$$

where $\mathbf{x}, \mathbf{y}$ are codevectors, $\mathbf{x} = [x_1 x_2 \ldots x_N]$, $x_n \in \{0, 1\}$. For example, the Hamming distance between the codewords 01001011 and 11101011 is 2. Note that it is common in coding theory to use row vectors instead of column vectors commonly used in communication theory. In order to simplify cross-referencing to coding books, we follow this established notation in this chapter.

- *Euclidean distance:* The squared Euclidean distance between two codewords is the squared geometric distance between the codevectors $\mathbf{x}$ and $\mathbf{y}$:

$$d_{\mathrm{E}}^2(\mathbf{x},\mathbf{y}) = \sum_n |x_n - y_n|^2. \qquad (13.28)$$

- *Minimum distance:* the minimum distance $d_{\min}$ of a code is the minimum Hamming distance $\min(d_{\mathrm{H}})$, where the minimum is taken over all possible combinations of two codewords of the code. Note that this minimum distance is equal to the number of linearly independent columns in the parity check matrix (see below).
- *Weight:* the weight of a codeword is the distance from the origin – i.e., the number of 1's in the codeword. For example, the weight of codeword 01001011 is 4.
- *Systematic codes:* in a systematic code, the original, information-bearing bits occur explicitly in the output of the coder, at a fixed location. The parity check (redundant) bits, which are computed from the information-bearing bits, are at different (also fixed) locations. For transmission over an ideal (noise-free, nondistorting) channel, the codeword could be determined without any information from the parity check bits. As an example, a systematic (7, 4) block code can be created in the following way:

k	k	k	k	m	m	m

where k represents information symbols and m represents parity check symbols.

- *Linear codes (group codes):* for these codes, the sum of any two codewords gives another valid codeword. Important properties can be derived from this basic fact:
  - The all-zero word is a valid codeword.
  - All codewords (except the all-zero word) have a weight equal to or larger than $d_{\min}$.
  - The distribution of distances – i.e., the Hamming distances between valid codewords – is equal to the weight distribution of the code.
  - All codewords can be represented by a linear combination of basic codewords (*generator words*).
- *Cyclic codes*: cyclic codes are a special case of linear codes, with the property that any cyclic shift of a codeword results in another valid codeword. Cyclic codes can be interpreted either by codevectors or by *polynomials* of degree $\leq N - 1$ where N is the length of the codeword. The nonzero coefficients correspond to the nonzero entries of the codevector; the variable x is a dummy variable.

As an example, we show both representations of the codeword 011010:

$$\mathbf{x} = [0\,1\,1\,0\,1\,0] \tag{13.29}$$

$$X(x) = 0 \cdot x^5 + 1 \cdot x^4 + 1 \cdot x^3 + 0 \cdot x^2 + 1 \cdot x^1 + 0 \cdot x^0. \tag{13.30}$$

- *Galois Fields*: A Galois field GF(p) is a finite field with p elements, where p is a prime integer. A field defines addition and multiplication for operating on its elements, and it is closed under these operations (i.e., the sum of two elements is again a valid element, and similar for the product); it contains identity and inverse elements for the two operations; and the associative, commutative, and distributive laws apply. The most important example is GF(2). It consists of the elements 0 and 1 and is the smallest finite field. Its addition and multiplication tables are as follows:

+	0	1
0	0	1
1	1	0

×	0	1
0	0	0
1	0	1

Codes often use GF(2) because it is easily represented on a computer by a single bit. It is also possible to define extension fields GF(p^m), where again p is a prime integer, and m is an arbitrary integer.

- *Primitive polynomials:* we define as *irreducible* a polynomial of degree N that is not divisible by any polynomial of degree less than N and greater than 0. A primitive polynomial $g(x)$ of degree m is defined as primitive if it is an irreducible polynomial such that the smallest integer N for which $g(x)$ divides $(x^N + 1)$ is $N = 2^m - 1$.

13.2.2 Encoding

The most straightforward encoding is a mapping table: any K-valued information word is associated with an N-valued codeword; the table just checks the input and reads out the associated codeword. However, this method is highly inefficient: it requires the storing of 2^K codewords.

For linear codes, any codeword can be created by a linear combination of other codewords, and thus it is sufficient to only store a subset of codewords. For example, for a K-valued information word, only K out of the 2^K codewords are linearly independent, and only those have to be stored. It is advantageous to select those codewords that have only a single 1 in the first K positions, as this choice automatically leads to a systematic code. The encoding process can then best be described by a matrix multiplication:

$$\mathbf{x} = \mathbf{u}\mathbf{G}. \tag{13.31}$$

Here, $\mathbf{x}$ denotes the N-dimensional codevector, $\mathbf{u}$ the K-dimensional information vector, and $\mathbf{G}$ the $K \times N$-dimensional generator matrix. For a systematic code, the leftmost K columns of the generator matrix are a $K \times K$ identity matrix, while the right $N - K$ columns denote the parity check bits. The first K bits of $\mathbf{x}$ are identical to $\mathbf{u}$. Note that – as discussed above – we use *row* vectors to represent codewords, and that a vector–matrix product is obtained by *pre*multiplying the matrix with this row vector.

Example 13.2 *Encoding with a (7,4) code.*

To make things more concrete, let us now consider an example of a $(7, 4)$ code. We encode the source word $[1011]$ with a generator matrix:

$$\mathbf{G} = \begin{bmatrix} 1 & 0 & 0 & 0 & 1 & 1 & 0 \\ 0 & 1 & 0 & 0 & 1 & 0 & 1 \\ 0 & 0 & 1 & 0 & 0 & 1 & 1 \\ 0 & 0 & 0 & 1 & 1 & 1 & 1 \end{bmatrix}. \tag{13.32}$$

Computing $\mathbf{x} = \mathbf{u}\mathbf{G}$:

$$\mathbf{x} = [1\ 0\ 1\ 1] \begin{bmatrix} 1 & 0 & 0 & 0 & 1 & 1 & 0 \\ 0 & 1 & 0 & 0 & 1 & 0 & 1 \\ 0 & 0 & 1 & 0 & 0 & 1 & 1 \\ 0 & 0 & 0 & 1 & 1 & 1 & 1 \end{bmatrix} \tag{13.33}$$

the codeword becomes

$$\mathbf{x} = [1\ 0\ 1\ 1\ 0\ 1\ 0]. \tag{13.34}$$

13.2.3   Decoding

In order to decide whether the received word is a valid codeword, we multiply it by a *parity check matrix* $\mathbf{H}$. This results in a $N - K$ dimensional *syndrome vector* $\mathbf{s}_{\text{synd}}$. If this vector is the all-zero vector, then the received word is a valid codeword.

Example 13.3   *Syndrome for (7,4) code.*

Let us demonstrate the computation of the syndrome with our previous example. Assume that the received bit sequence (after hard decision) be

$$\hat{\mathbf{x}} = \begin{bmatrix} 1 & 0 & 0 & 0 & 1 & 0 & 1 \end{bmatrix}. \tag{13.35}$$

A parity check matrix for the code of Example 13.2 is (we will describe later a constructive method for obtaining $\mathbf{H}$ from $\mathbf{G}$):

$$\mathbf{H} = \begin{bmatrix} 1 & 1 & 0 & 1 & 1 & 0 & 0 \\ 1 & 0 & 1 & 1 & 0 & 1 & 0 \\ 0 & 1 & 1 & 1 & 0 & 0 & 1 \end{bmatrix}. \tag{13.36}$$

Let us now compute the expression $\mathbf{s}_{\text{synd}} = \hat{\mathbf{x}}\mathbf{H}^T$. Writing this expression component by component, we obtain three parity check equations, corresponding to the three parity check bits:

$$\mathbf{s}_{\text{synd}}^T = \hat{\mathbf{x}}\mathbf{H}^T = \begin{bmatrix} 0 & 1 & 1 \end{bmatrix}. \tag{13.37}$$

Thus, the received word does not satisfy the last two parity checks and is therefore not a valid codeword.

The computation of the syndrome can also be interpreted the following way:

- The information received in the positions of the systematic bits is 1000
- The parity bits computed from the received systematic bits is 110 (this follows from taking the product [1 0 0 0] with $\mathbf{G}$, and taking the last 3 entries in the resulting 7-element vector
- The actually received parity check bits: 101
- $\Rightarrow$ syndrome: 011.

Next, let us determine how to find an $\mathbf{H}$-matrix. The relationship $\mathbf{H} \cdot \mathbf{G}^T = \mathbf{0}$ has to hold, as each row of the generator matrix is a valid codeword, whose product with the parity check matrix has to be 0.[5] Representing $\mathbf{G}$ as

$$\mathbf{G} = (\mathbf{I} \;\; \mathbf{P}). \tag{13.38}$$

the relationship $\mathbf{H} \cdot \mathbf{G}^T = \mathbf{0}$ reduces to

$$\mathbf{H} \cdot \mathbf{G}^T = (\mathbf{H}_1 \;\; \mathbf{H}_2) \begin{pmatrix} \mathbf{I} \\ \mathbf{P}^T \end{pmatrix} = (\mathbf{H}_1 + \mathbf{H}_2\mathbf{P}^T) = 0. \tag{13.39}$$

If we now select $\mathbf{H}_2 = \mathbf{I}$ and $\mathbf{H}_1 = -\mathbf{P}^T$ then the equation is certainly satisfied: the subtraction of two identical matrices is the all-zero matrix; note that in modulo-2 arithmetic, there is no difference between addition and subtraction. Also note that different parity check matrices can exist for each generator matrix. The above constructive method gives just one possible solution.

Example 13.4   *Computation of parity check matrix for the (7,4) code.*

The procedure for computing the parity check matrix for the (7, 4) code (and actually for any systematic code) can also be described the following way: start with the $N - K$ rightmost columns of the generator matrix:

$$\begin{bmatrix} 1 & 1 & 0 \\ 1 & 0 & 1 \\ 0 & 1 & 1 \\ 1 & 1 & 1 \end{bmatrix} \tag{13.40}$$

and transpose them. Appending now an $(N - K) \times (N - K)$ identity matrix results in $\mathbf{H}$:

$$\mathbf{H} = \begin{bmatrix} 1 & 1 & 0 & 1 & 1 & 0 & 0 \\ 1 & 0 & 1 & 1 & 0 & 1 & 0 \\ 0 & 1 & 1 & 1 & 0 & 0 & 1 \end{bmatrix}. \tag{13.41}$$

---

[5] In other words, $\mathbf{G}$ is the codespace, and $\mathbf{H}$ is the associated nullspace.

Hamming codes are a well-known class of linear codes. A Hamming code can be defined easily through its parity check matrix. The columns of $\mathbf{H}$ contain all possible 2^{N-K} bit combinations of length K, with the exception of the all-zero word. Consequently, all columns of $\mathbf{H}$ are distinct. Note that the parity check matrix of Eq. (13.41) fulfills this condition, as can be easily verified by the reader. The size of a Hamming code is $(2^m - 1, 2^m - 1 - m)$, where m is a positive integer.

13.2.4 Recognition and Correction of Errors

Due to the linear properties of the code, each received word can be interpreted as the sum of a codeword $\mathbf{x}$ and an error word $\boldsymbol{\varepsilon}$. This implies that the syndrome depends only on the error word, not on the transmitted codeword. If a word $\hat{\mathbf{x}}$ is received, then the RX computes the syndrome vector:

$$\mathbf{s}_{\text{synd}} = \hat{\mathbf{x}}\mathbf{H}^T = (\mathbf{x} + \boldsymbol{\varepsilon})\mathbf{H}^T = \mathbf{x}\mathbf{H}^T + \boldsymbol{\varepsilon}\mathbf{H}^T = \mathbf{0} + \boldsymbol{\varepsilon}\mathbf{H}^T = \boldsymbol{\varepsilon}\,\mathbf{H}^T. \tag{13.42}$$

An error ($\boldsymbol{\varepsilon} \neq 0$) is indicated by a nonzero syndrome. Decoding requires that we find the "correct" $\boldsymbol{\varepsilon}$ – by subtracting $\boldsymbol{\varepsilon}$ from $\hat{\mathbf{x}}$, we obtain the correct codeword. On the other hand, it is clear that a number of different error words $\boldsymbol{\varepsilon}$ can lead to the same syndrome; these error words are called a *coset*. The goal is thus to find the most probable $\boldsymbol{\varepsilon}$ corresponding to the observed syndrome. In a "reasonable" channel, the probability that a bit arrives correctly is higher than the probability that a bit arrives with an error. Therefore, the $\boldsymbol{\varepsilon}$ that has the minimum weight is the most probable $\boldsymbol{\varepsilon}$ in a given coset.

A different interpretation, which leads to an estimate of the number of detectable and correctable errors, starts with the representation of the code in the *codespace* (see Figure 13.1): Each of the 2^N possible bit combinations of a binary (N, K) block code corresponds to a point in the codespace. 2^K of these bit combinations correspond to a valid codeword, each of which is separated at least by a distance $d_{\min}$. The other bit combinations can be interpreted as points located in "correction spheres" around the valid codewords. The minimum radius of correction spheres t determines the number of corrigible errors.

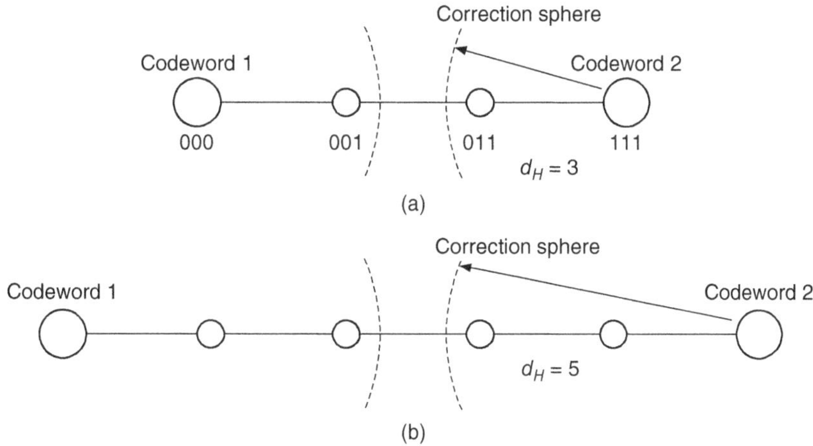

Figure 13.1 Cuts through a codespace (a) for two codewords with $d_H = 3$ and (b) a codespace for two codewords with $d_H = 5$ and a sketch of the correction spheres.

For codes of small size, decoding via lookup tables is possible – in other words, all possible syndromes and their corresponding minimum-weight error words are stored in a lookup table. For each received codeword, this table is then used to determine the transmitted codeword. Unfortunately, codes with small size usually show bad performance. For this reason, alternative decoding schemes that can be used for larger codes as well have been investigated. The most important one, the *belief propagation* algorithm, is described in more detail in Section 13.7. Furthermore, in certain scenarios, special cases of linear codes that allow simplified decoding, like cyclic codes, might be preferable.

From these considerations, the following conclusions can be drawn: a code with minimum distance $d_{\min}$ always allows the *detection* of $d_{\min} - 1$ error in a codeword. Only errors that influence at least $d_{\min}$ bits can lead to another valid codeword. Alternatively, such a code can be capable of *correcting* $\lfloor (d_{\min} - 1)/2 \rfloor$ errors – in other words, the received bit sequence has to lie in the correction sphere of the true codeword ($\lfloor x \rfloor$ denotes the largest integer smaller than x). If a bit combination has the same Hamming distance from its two (or more) nearest neighbor, a random decision needs to be made which codeword was transmitted, which is undesirable; if all bit combinations can be uniquely assigned to a correction sphere, the code is called *perfect*.

For a Hamming code, the location of a single error can be uniquely determined. Since all the columns in the parity check matrix are different and linearly independent, and a Hamming code can correct exactly one error, the syndrome tells us the location of the error.

13.2.5 Concatenated Codes

Error protection can be made stronger by using two codes: a so-called *inner code* protects the data in the usual way, as described above. Some errors can still remain (when the received word lies in the wrong correction sphere); those errors are then corrected with an *outer code* (see Figure 13.2).

Intuitive insights can be obtained by interpreting the combination of channel and inner code as a *superchannel* that exhibits a lower BER than the original channel. The errors of the superchannel usually occur in bursts; if the inner code is an (N, K) block code, then the length of the burst is K bits. Thus, the outer code is usually a code that is especially well suited for combating burst errors. RS codes are efficient for handling burst errors and are thus popular as outer codes in concatenated codes.[6]

Appropriate concatenation of codes is quite a difficult task. In particular, the relative capabilities of the two codes need to be carefully balanced. If the inner code is not strong enough, then it is useless, and might even increase the BER that the outer code has to deal with.

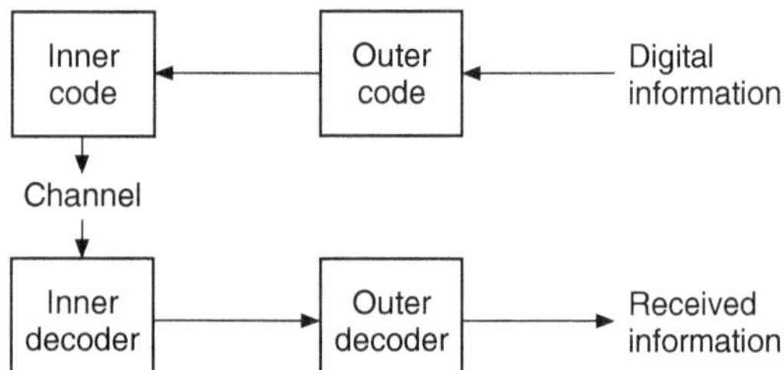

Figure 13.2 Concatenation of codes.

13.3 Convolutional Codes

13.3.1 Principle of Convolutional Codes

Convolutional codes do not divide (source) data streams into blocks, but rather add redundancy in a quasi-continuous manner. A convolutional encoder consists of a shift register with L memory cells and N (modulo-2) adders (see Figure 13.3). Let us assume that at the outset we have a clearly defined state in memory cells – i.e., they all contain 0. When the first data bit enters the encoder, it is put into the first memory cell of the shift register (the other zeros are shifted to the right, and the rightmost zero "falls out" of the register). Then a multiplexer reads out the output of all the adders $n = 1, 2, 3$. We thus get three output bits for one input bit. Then, the next source data bit is put into the register (and the contents of all memory cells are shifted to the right by one cell). The adders then have new outputs, which are again read by the multiplexer. The process is continued until the last source data bit is put into register. Subsequently, zeros are used as register input, until the last source data bit has been pushed out of the register, and the memory cells are again in a clearly defined (all-zero) state.

A convolutional encoder is thus characterized by the number of shift registers and adders. Adders are characterized by their connections to memory cells. In the example of Figure 13.3, only element $\ell = 1$ is connected to the output $n = 1$, so that source data are directly mapped to the coder output (the encoder thus creates a systematic code). For the second output, the contents of memory cells $\ell = 1, 2$ are added. For the third output, the contents of elements $\ell = 1, 2, 3$ are combined.

This coder structure can be represented in different ways. One possibility is via generator sequences: we generate N vectors of length L each. The ℓth element of the nth vector has value 1 if the ℓth shift register element has a connection to the nth adder; otherwise, it is 0. Generator sequences can be interpreted for building an encoder; generator polynomials can be used similarly.

For the decoder, the trellis diagram is a more useful description method. In this representation, the state of the encoder is characterized by the content of the memory cells. The trellis shows which input bits get the shift register into which state, and which output bits are consequently created. As an example, Figure 13.4 shows the trellis of the convolutional encoder of Figure 13.3. Only the states of the cells 2,... L need to be described, since the content of cell 1 is identical to the input (information) bit. For that reason, the number of states that need to be distinguished is $2^{L-1} = 4$. Two lines originate from each state: the upper represents source data bit 0, and the lower source data bit 1 (in general, it can be necessary to label each transition with the information bit it corresponds to). It is not possible to get from each state directly into each other state. For example, from state A we can only get to state A or B (but not C or D). This is the redundancy that can be used for reducing error probability during decoding. We also find that the trellis is repeated periodically. It is thus unnecessary to plot an infinitely long trellis diagram, even if the input data sequence is infinitely long.

Decoding at the RX would be very simple if the received data sequence was identical to the transmit data sequence. As we would then know which coded sequence was received, we would just have to trace in the trellis the source data from which it was created. When there is additive noise, there are errors in the received bit sequence, and we have to discover from the perturbed receive data which sequence is most likely to have been originally transmitted (*Maximum Likelihood Sequence Estimation* (MLSE)). In practice, this estimation is usually done by the Viterbi algorithm that is discussed in the next section. Alternative approaches are the Fano algorithm, the stack algorithm, and the feedback algorithm [Lin and Costello 2004; McEliece 2004].

---

[6] Code concatenation can also be applied when the inner code is a convolutional code, see Section 13.3.

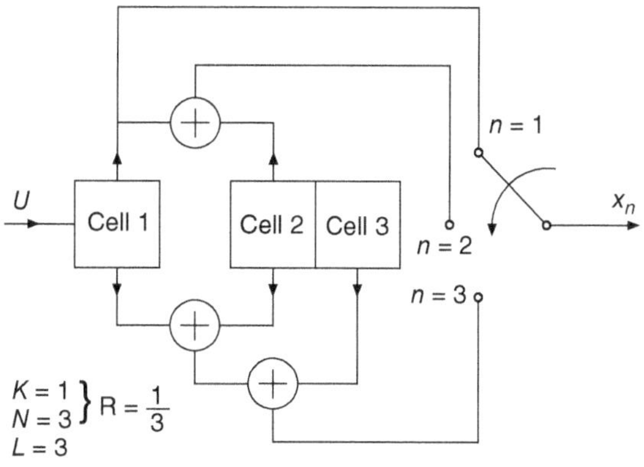

$\left.\begin{array}{l} K = 1 \\ N = 3 \\ L = 3 \end{array}\right\}\ R = \dfrac{1}{3}$

If shift-register initially is in all-zero state

$U_1 = 1 \rightarrow X_1 = 1, 1, 1$

$U_1 = 1 \quad U_2 = 1 \rightarrow X_2 = 1, 0, 0$

Number of states: $2^{L-1} - 1$

State	Cell 2	Cell 3
A	0	0
B	1	0
C	0	1
D	1	1

Figure 13.3 Example of a convolutional encoder.
Reproduced with permission from [Oehrvik 1994] © Ericsson AB.

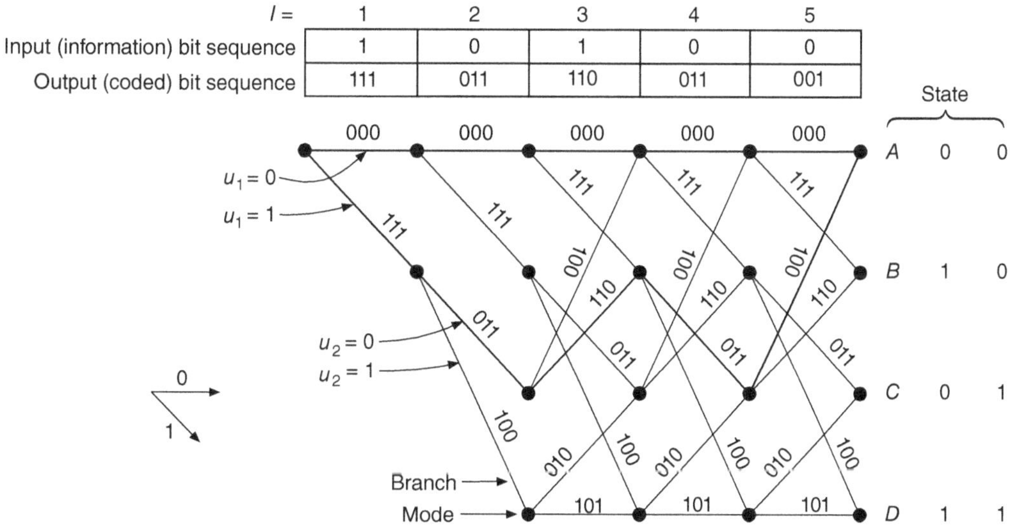

Figure 13.4 Trellis for the convolutional encoder of the previous figure.
Reproduced with permission from [Oehrvik 1994] © Ericsson AB.

13.3.2 Viterbi Decoder – Classical Representation

The Viterbi algorithm [Viterbi 1967] is the most popular algorithm for MLSE. The goal of this algorithm is to find the sequence $\hat{\mathbf{s}}$ that is the most likely sequence to have been transmitted given that the sequence $\mathbf{r}$ was received.[7] This may be represented as

$$\hat{\mathbf{s}} = \max_{\mathbf{s}} \Pr(\mathbf{r} \,|\, \mathbf{s}) \tag{13.43}$$

---

[7] This is the definition for Maximum A Posteriori (MAP) detection. However, it is equivalent to MLSE for equiprobable sequences (see Chapter 11).

where maximization is done over all possible transmit sequences **s**. If perturbations of the received symbols by noise are statistically independent,[8] then the probability for the sequence $\Pr(\mathbf{r}|\hat{\mathbf{s}})$ can be decomposed into the product of the probabilities of each symbol:

$$\hat{\mathbf{s}} = \max_{\mathbf{s}} \prod_i \Pr(r_i \mid s_i). \tag{13.44}$$

Now, instead of maximizing the above product, maximizing its logarithm also finds the optimum sequence – this is true since the logarithm is a strictly monotonous function:

$$\hat{\mathbf{s}} = \max_{\mathbf{s}} \sum_i \log\left[\Pr(r_i \mid s_i)\right]. \tag{13.45}$$

The logarithmic transition functions $\log[\Pr(r_i \mid s_i)]$ are also known as *branch metrics*; in the case that the decoder puts out only "hard decisions" (estimates of transmitted coded bits), $r_i = \pm 1$, and the branch metric is $d_{\mathrm{H}}(r_i, s_i)$, the Hamming distance between r_i and s_i.

The MLSE now determines the total metrics $\sum_i d_{\mathrm{H}}(r_i, s_i)$ for all possible paths through the trellis – i.e., for all possible input sequences. In the end, the path with the smallest metric is selected. Such an optimum procedure requires large computational effort: the number of possible paths increases exponentially with the number of input bits. The key idea of the Viterbi algorithm is the following: instead of computing metrics for all possible paths (working from "top to bottom") in the trellis, we work our way "from the left to the right" through the trellis. More precisely, we start with a set of possible states of the shift register (A_i, B_i, C_i, D_i, where i denotes the time instant, or considered input bit). Let us now consider, for a particular time instant i, all paths that (from the left) lead into state A, denoting them as $\mathbf{s}^{(1)}$, $\mathbf{s}^{(2)}$, $\mathbf{s}^{(3)}$,... We discard any possible path $\mathbf{s}^{(p)}$ if it merges at state A_i with a path $\mathbf{s}^{(q)}$ that has a smaller metric. As the paths run through the same state of the trellis, there is nothing that would distinguish the paths from the point of view of later states. We thus choose the one with the better properties. Similarly, we choose the best paths that run through the states B_i, C_i, and D_i. After having determined the *survivors* for state i, we next proceed to state $i + 1$ (or rather, to the tuple of states A_{i+1}, B_{i+1}, C_{i+1}, D_{i+1}), and repeat the process. All paths in a trellis ultimately merge in a single, well-defined point, the all-zero state.[9] At this point, there is only a single survivor – the most-likely sequence to have been transmitted.

Example 13.5 *Example for Viterbi decoding.*

Figure 13.5 shows an example of the algorithm. The basic structure of the trellis is depicted in Figure 13.5a; the received bit sequence is shown in Figure 13.5b. The metrics shown are the Hamming distances of the received sequence, i.e., *after* the hard decision compared with the theoretically possible bit sequences in the trellis.

We assume that at the outset the shift register is in the all-zero state. Figure 13.5c shows the trellis for the first 3 bits. There are two possibilities for getting from state A_0 to state A_4: by transmission of the source data sequence 0, 0, 0 (which corresponds to the coded bit sequence 000, 000, 000, and thus via states A_2, A_3) or by transmission of the source data sequence 100 (coded bit sequence 111, 011, 001 – i.e., via states B_2, C_3). In the former case, the branch metric is 2; in the latter, it is 6. This allows us to immediately discard the second possibility. Similarly, we find that the transition from state A_0 to state B_4 could be created (with greatest likelihood) by the source data sequence 0, 0, 1, and not by 1, 1, 0. The following subfigures of Figure 13.5 show how the process is repeated for ensuing incoming bits.

The Viterbi algorithm greatly decreases storage requirement by elimination of nonsurviving paths, but it is still considerable. Thus it is undesirable to wait for the decision as to which sequence was transmitted until the last bits of the source sequence. Rather, the algorithm makes decisions about bits that are "sufficiently" in the past. More precisely, during consideration of states A_i, B_i, C_i, D_i we decide about the symbols of the state tuple $A_{i-L_{\mathrm{Tr}}}$, $B_{i-L_{\mathrm{Tr}}}$, $C_{i-L_{\mathrm{Tr}}}$, $D_{i-L_{\mathrm{Tr}}}$, where L_{Tr} is the *truncation depth*. This principle is shown in Figure 13.6. Data within the window of length L_{Tr} are stored. When moving to the next tuple in the trellis, the leftmost-state tuple moves out of the considered window, and we have to make a final decision about which bits were transmitted there. The decision is made in favor of the state that contains the path that has the smallest metric in the *currently* observed state – i.e., at the right side of the window. While this procedure is suboptimum, performance loss can be kept small by judiciously choosing the length of the window. In practice, a duration

$$L_{\mathrm{Tr}} = 6L \tag{13.46}$$

has turned out to be a good compromise.

---

[8] If the noise is correlated between the different samples then the receiver has to use a so-called "noise-whitening filter" (see Chapter 14).

[9] Actually, this only happens if the convolutional encoder appends enough zeros (tail bits) to the source data sequence to force the encoder into the defined state. If this is not done, then the decoder has to consider all possible finishing states, and compare the metrics of the path ending in each and all of them.

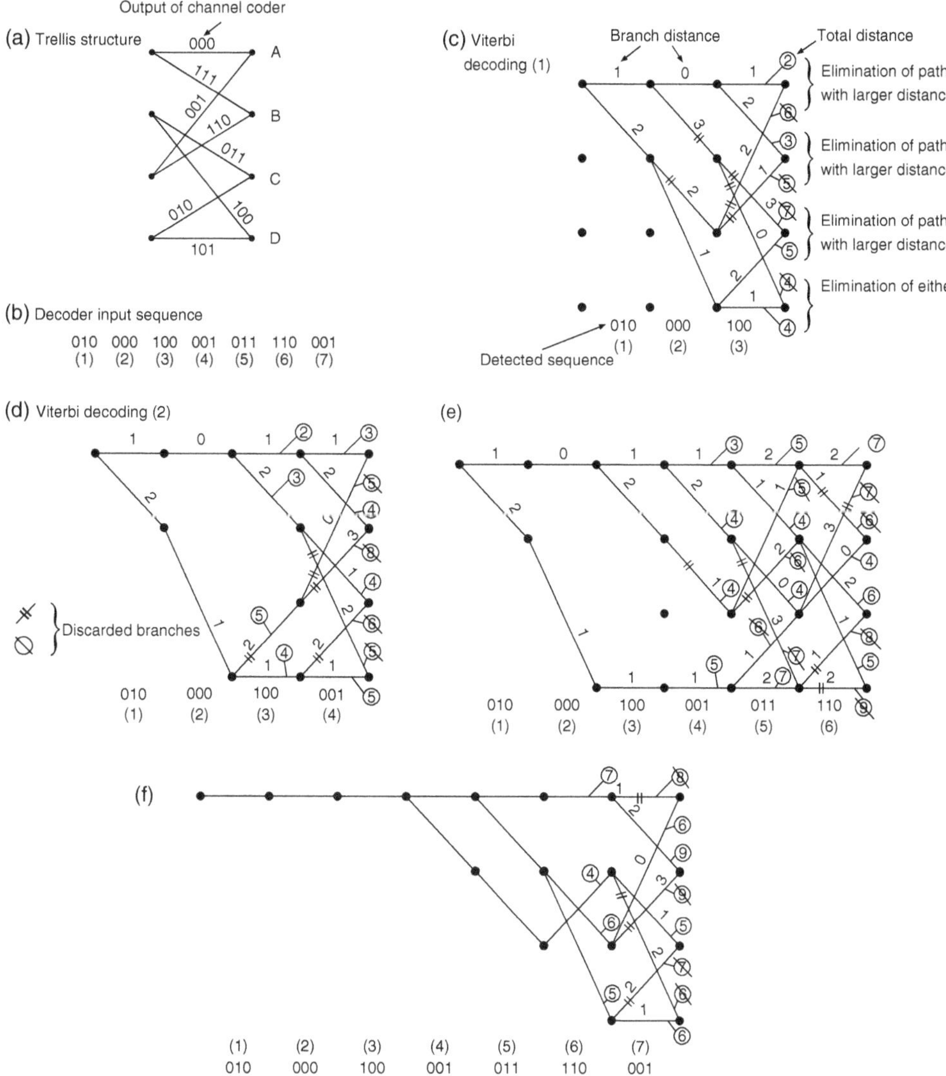

Figure 13.5 Example of Viterbi detection. (a) Structure of the trellis, (b) decoder input sequence, (c–f) progressive steps in the Viterbi decoding. Reproduced with permission from [Oehrvik 1994] © Ericsson AB.

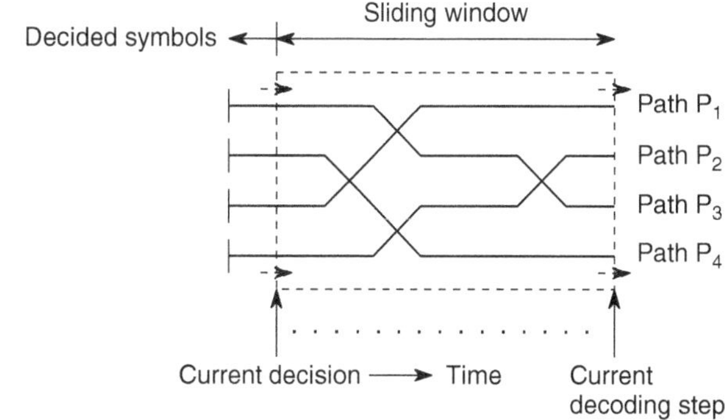

Figure 13.6 Principle of decision using a finite duration sliding window. Reproduced with permission from [Mayr 1996] © B. Mayer.

*13.3.3 Improvements of the Viterbi Algorithm

Soft-Input Decoding

The above example used the Hamming distance between received symbols and possible symbols as a metric. This means that received symbols first undergo a hard decision before being used in the decoding process. This algorithm thus uses the redundancy of the code (not all combinations of bits are valid codewords) but does not use this information about the reliability of the received bits. In some cases, the bits might be close to the decision boundary, in which case they should have less impact on the finally chosen sequence than if they are far away from that boundary. "Soft information" can be taken into account by using a different metric in the Viterbi algorithm. It can be shown that In an AWGN channel as well as in a flat-fading channel the best decision metric is proportional to the squared Euclidean distance $d_i^2 = |r_i - s_i|^2$ in the signal space diagram. Proportionality constants are of no further importance, as they have no impact on finding the optimum metric:

$$\hat{s} = \min_s \sum_i |r_i - s_i|^2. \tag{13.47}$$

This equation assumes that the channel gain is the same for all received symbols and that the channel gain has been compensated by the RX. If this is not the case, then we have to minimize the expression:

$$\min_s \sum_i |r_i - \alpha_i s_i|^2 \tag{13.48}$$

where α_i is the channel gain of the ith symbol. A detailed example of a Viterbi algorithm with soft information is given in Section 14.4 (we will see there that MLSE equalization can also be done by means of this algorithm).

Tail Bits

As mentioned above, it is best if – at the end of transmitting the data sequence – the encoder ends up in a defined state, usually the all-zero state. In that case, it is clear which surviving path has to be chosen. Figure 13.7 shows that this approach significantly decreases the error probability for the data bits at the end of the sequence. The drawback is a loss in spectral efficiency, as noninformation-bearing symbols (the zeros appended at the end) have to be transmitted. This can become a problem in systems with very short sequence durations.

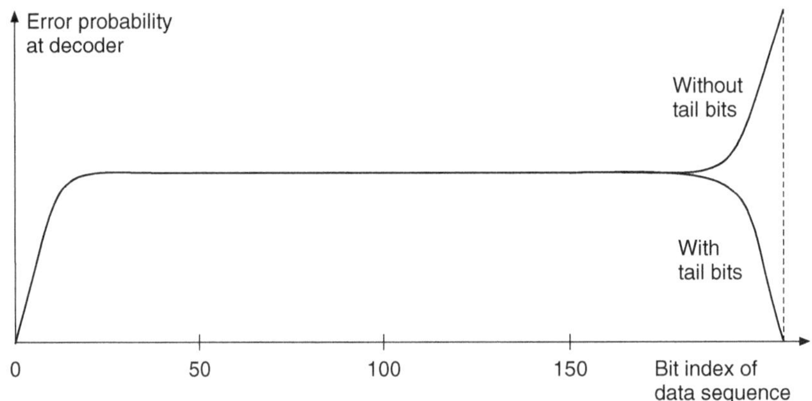

Figure 13.7   Error probability when tail bits are used.
Reproduced with permission from [Oehrvik 1994] © Ericsson AB.

If the sequences that are to be transmitted are very long, the principle of tail bits can be used also within a sequence: at certain, predefined locations, a series of zeros is transmitted, which forces the shift register (and the trellis) into a defined state. Such *waists* in the trellis allow a reduction in error probability.

Puncturing

Normal convolutional codes can realize only a certain subset of coding rates, like $R_c = 1/2, 1/3, 1/4$. However, for many applications, we need different code rates that are determined by the source rate and the available bandwidth, e.g., 5/6 for situations with good SNR.

An easy way of adapting a code rate is puncturing of a code. It starts out with a code that has a rate that is lower than the desired. Then certain bits of the coded sequence are omitted from the transmission. Since the code contains considerable redundancy, this is not a problem – it is similar to the physical situation when codebits are erased due to fading. The only requirement is that the punctured bits are appropriately distributed throughout the codeword, so that not all information about a certain bit is eliminated.

*13.3.4   The BCJR Algorithm

We have seen that the Viterbi algorithm can have either "hard" or "soft" information as the *input*, but in either case, the *output* is hard information, namely the sequence of bits that was transmitted. This makes sense if the convolutional decoding is the last processing step occurring at the RX, and the bit sequence is the ultimate output. However, there are situations where further processing occurs that requires information about the reliability of each bit in the output sequence. This is particularly relevant if the convolutional decoder is only a component in a larger detector scheme, as occurs, e.g., in turbo decoding (see Section 13.6). Obtaining this soft information requires a generalization of the Viterbi algorithm. One possible form is the BCJR algorithm proposed by [Bahl et al. 1974]. We describe it here even though it might not be the most computationally efficient one because it is conceptually easier than most of the alternatives.

We first give a more exact definition of soft-output information: it is the confidence in a specific decision, i.e., the *a posteriori* probability. This confidence is described by the *LLR (Log-Likelihood Ratio)* value. For LLR, we generally distinguish between the following two cases: the *a priori* LLR (for symmetry reasons henceforth write the bits as +1 and −1),

$$L(u_i) = \ln \left[\frac{\Pr(u_i = +1)}{\Pr(u_i = -1)} \right] \tag{13.49}$$

i.e., the logarithm of the ratio of the probabilities that source bit s_i is +1 or −1. For equiprobable transmit symbols and a pure convolutional decoder, $L(u_i) = 0$. The *a posteriori LLR* is the confidence in a decision taking into account the observation at the RX,

$$L(u_i \,|\, \mathbf{r}) = \ln \left[\frac{\Pr(u_i = +1 \,|\, \mathbf{r})}{\Pr(u_i = -1 \,|\, \mathbf{r})} \right]. \tag{13.50}$$

If the LLR is positive, we assume that a +1 was sent, and the larger the LLR, the more confident we are in that decision. Conversely, a negative LLR indicates a −1, and a large magnitude of the LLR again indicates greater confidence.

Consider now the part of the trellis diagram in Figure 13.4 that describes the step from some arbitrary $i-1$ to i. There is a set of possible transitions (four in this case) that are caused by a source bit with value $u_i = -1$, and another four possible transitions associated with $u_i = 1$. We call these sets $\mathcal{T}_0$ and $\mathcal{T}_1$. Furthermore, call the state at time $i-1$ the state Φ' that has the possible values ϕ', and the state at time i as state Φ with possible values ϕ. We can then write the a posteriori LLR as

$$L(u_i \,|\, \mathbf{r}) = \ln \left[\frac{\sum_{\mathcal{T}_1} \Pr(\phi', \phi \,|\, \mathbf{r})}{\sum_{\mathcal{T}_0} \Pr(\phi', \phi \,|\, \mathbf{r})} \right] \tag{13.51}$$

$$= \ln \left[\frac{\sum_{\mathcal{T}_1} \Pr(\phi', \phi, \mathbf{r})}{\sum_{\mathcal{T}_0} \Pr(\phi', \phi, \mathbf{r})} \right]. \tag{13.52}$$

Now we split up the received sequence into past (denoted $\mathbf{r}_{<i}$), present ($\mathbf{r}_i$), and future ($\mathbf{r}_{>i}$)

$$\Pr(\phi', \phi, \mathbf{r}) = \Pr(\phi', \phi, \mathbf{r}_{<i}, \mathbf{r}_i, \mathbf{r}_{>i}). \tag{13.53}$$

Using Bayes' theorem, this can be written as

$$\Pr(\phi', \phi, \mathbf{r}) = \Pr(\mathbf{r}_{>i} \,|\, \phi', \phi, \mathbf{r}_{<i}, \mathbf{r}_i) \; \Pr(\phi', \phi, \mathbf{r}_{<i}, \mathbf{r}_i). \tag{13.54}$$

Since – for a given state ϕ – the future does not depend on the past (ϕ', $\mathbf{r}_{<i}$, $\mathbf{r}_i$), this becomes simply

$$\Pr(\phi', \phi, \mathbf{r}) = \Pr(\mathbf{r}_{>i} \,|\, \phi) \; \Pr(\phi', \phi, \mathbf{r}_{<i}, \mathbf{r}_i). \tag{13.55}$$

The second term can be similarly simplified to $\Pr(\mathbf{r}_i, \phi \mid \phi')\Pr(\phi', \mathbf{r}_{<i})$, so that we finally get

$$
\begin{aligned}
\Pr(\phi', \phi, \mathbf{r}) &= \lambda_{i-1}(\phi')\mu(\phi', \phi)\nu_i(\phi) \\
\lambda_{i-1}(\phi') &= \Pr(\phi', \mathbf{r}_{<i}) \\
\mu_i(\phi', \phi) &= \Pr(\mathbf{r}_i, \phi \mid \phi') \\
\nu_i(\phi) &= \Pr(\mathbf{r}_{>i} \mid \phi).
\end{aligned}
\tag{13.56}
$$

Thus, λ is the probability that we are in state ϕ' at $i-1$, and the received sequence up to that time is $\mathbf{r}_{<i}$, $\mu(\phi', \phi)$ is the probability that we start out in state ϕ', and – with an observation of $\mathbf{r}_i$ – transition to state ϕ, and $\nu_i(\phi)$ is the probability that we start in state ϕ and the subsequently received sequence is $\mathbf{r}_{>i}$.

We then have to consider how to determine the three factors. We start out with $\mu(\phi', \phi)$, which we conveniently rewrite as

$$
\mu(\phi', \phi) = \Pr(\mathbf{r}_i \mid \mathbf{s}_i)\Pr(u_i)
\tag{13.57}
$$

where $\Pr(u_i)$ is the *a priori* probability of source bit i, which can be written as

$$
\Pr(u_i) = B_i \exp\left[u_i L(u_i)/2\right].
\tag{13.58}
$$

Here, B_i does not depend on whether u_i is $+1$ or -1. The term $\Pr(\mathbf{r}_i \mid \mathbf{s}_i)$ represents the probability that the received symbol is $\mathbf{r}_i$ given that the nominal symbol – that should be received over a noise- and distortion-free channel with u_i as the coder input – is $\mathbf{s}_i$. We assume that K coded bits are transmitted for each input symbol (so that $K = 1/R_c$), and we denote the coded symbols associated with u_i as s_{ik} and similarly the corresponding received symbols as r_{ik}, with $k = 1, \dots K$.

The BCJR algorithm now makes some simplifications, which we will show in the following for the example case that the modulation scheme is BPSK. Let $E_c = R_c E_B$ denote the energy per coded bit, and $\tilde{s}$ and $\tilde{r}$ the transmitted and received signals normalized by E_c. Then

$$
\Pr(\tilde{\mathbf{r}}_i \mid \tilde{\mathbf{s}}_i) = \left(\frac{E_c}{\pi N_0}\right)^{K/2} \exp\left(-\frac{E_c}{N_0}\sum_{k=1}^{K}(\tilde{r}_{ik} - \alpha\tilde{s}_{ik})^2\right)
\tag{13.59}
$$

$$
= C_i \exp\left(2\alpha\frac{E_c}{N_0}\sum_{k=1}^{K}(\tilde{r}_{ik}\tilde{s}_{ik})\right)
\tag{13.60}
$$

where

$$
C_i = \left(\frac{E_c}{\pi N_0}\right)^{K/2} \exp\left(-\frac{E_c}{N_0}\sum_{k=1}^{K}\tilde{r}_{ik}^2\right)\exp\left(-\frac{E_c}{N_0}\alpha^2\sum_{k=1}^{K}\tilde{s}_{ik}^2\right)
\tag{13.61}
$$

does not depend on u_i or s_i (note that this is true only for constant-modulus modulation schemes like BPSK; otherwise the energy $\sum_{k=1}^{K}\tilde{s}_{ik}^2$ has to be considered). We will see later that C_k appears both in the numerator and denominator of the LLR and cancels out; we will thus neglect it henceforth; the same applies for B_i in (13.58). Finally, we denote the *channel reliability* as L_c

$$
L_c = 4\alpha\frac{E_c}{N_0}.
\tag{13.62}
$$

Thus $\mu_i(\phi', \phi)$ can be written as

$$
\mu_i(\phi', \phi) = \exp\left[u_i\frac{L(u_i)}{2}\right]\exp\left(\frac{L_c}{2}\sum_{k=1}^{K}(\tilde{r}_{ik}\tilde{s}_{ik})\right).
\tag{13.63}
$$

We next turn to the computation of the λ, which can be done iteratively. It is easy to see that the probability of being in a state ϕ at time i is the probability of being in state ϕ' at time $i-1$ multiplied with the transition probability, summed over all ϕ' that have transitions into ϕ. Thus

$$\lambda_i(\phi) = \sum_{\phi'} \lambda_{i-1}(\phi') \mu_i(\phi', \phi). \tag{13.64}$$

If the encoder starts in the all-zero state, then the initial condition is

$$\lambda_0(\phi') = \begin{cases} 1 & \text{for } \phi' = 0 \\ 0 & \text{otherwise} \end{cases}. \tag{13.65}$$

Similarly, the ν can be computed as

$$\nu_{i-1}(\phi') = \sum_{\phi} \nu_i(\phi) \mu_i(\phi', \phi) \tag{13.66}$$

and the final condition, assuming the coder terminates in the all-zero-state, is

$$\nu_N(\phi) = \begin{cases} 1 & \text{for } \phi = 0 \\ 0 & \text{otherwise} \end{cases}. \tag{13.67}$$

Note that the λ can be computed by a forward iteration in the trellis, while the ν are computed by a backward iteration (starting from the final state) – for this reason, the BCJR algorithm is also called the forward-backward algorithm. Thus, in this simple formulation, we have to wait until all symbols are received before we can start the decoding process, and the "sliding window" approach described for the Viterbi algorithm cannot be used; there are however a number of methods to circumvent this problem.

Once we have computed these terms, we can finally obtain the LLR easily as

$$L = \ln \left[\frac{\sum_{T_1} \lambda_{i-1}(\phi') \mu(\phi', \phi) \nu_i(\phi)}{\sum_{T_0} \lambda_{i-1}(\phi') \mu(\phi', \phi) \nu_i(\phi)} \right]. \tag{13.68}$$

Example 13.6 *Consider the systematic convolutional encoder in Figure 13.8a; note that it is a recursive encoder, i.e., the bit that is pushed out from the storage cell is added to the input bit to create the new cell state. The transmission scheme is BPSK, with modulation alphabet $\{1, -1\}$; the $E_b/N_0 = 4$ dB. Let the input data sequence be $u = [0\ 1\ 0\ 1]$. Let the received signal be $[0.7, -0.1, 1.2, -0.2, -2, 0.8, 1.5, -0.9]$. The encoder begins and terminates in the all-zero state. Decode the transmitted sequence with (i) hard-input Viterbi algorithm, (ii) soft-input Viterbi algorithm, and (iii) BCJR algorithm.*

For the hard-input Viterbi algorithm, we first perform a bit-by-bit decision, i.e., the decided received sequence is $[1\ 0\ 1\ 0\ 0\ 1\ 1\ 0]$. The state transition diagram can be found in Figure 13.8b, and the transmitted and received signals in Figure 13.8c.[10] The trellis diagram, including input bits/branch metrics and total distances, can be found in Figure 13.8d, using exactly the same procedure as in Example 13.5. We note that the decoded sequence is $[0\ 1\ 0\ 1]$ (which coincides with the originally transmitted sequence). A second sequence, $[1\ 1\ 1\ 1]$ gives the same Hamming distance; the decoder must pick one of the two at random.

For the soft-input Viterbi algorithm, we show the trellis diagram in Figure 13.8e. The branch metrics and total distances have changed, because we now use the squared Euclidean distance as branch metric. The result of the decoded sequence is correct.

---

[10] Note that here we explicitly give the input source bit as 1/0, followed by the transmitted BPSK symbol bit combination (e.g., 1 −1); this is done because in contrast to Example 13.5 the transition associated with source bit value 0 is not always the "upper" transition in the diagram.

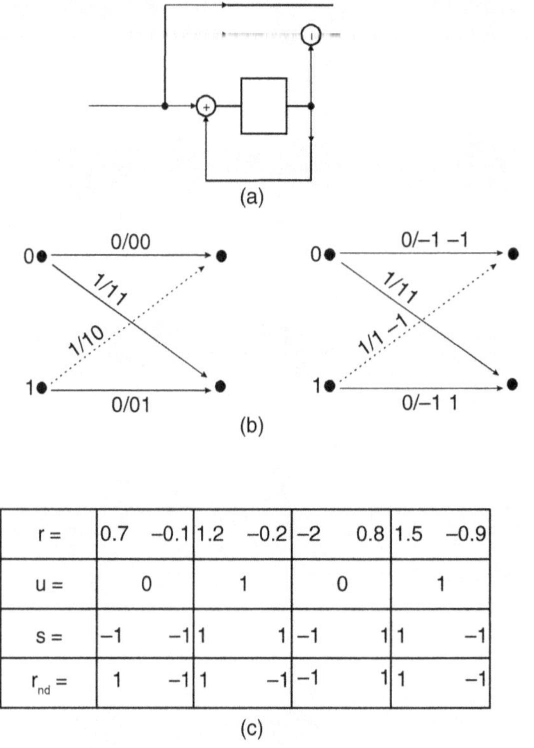

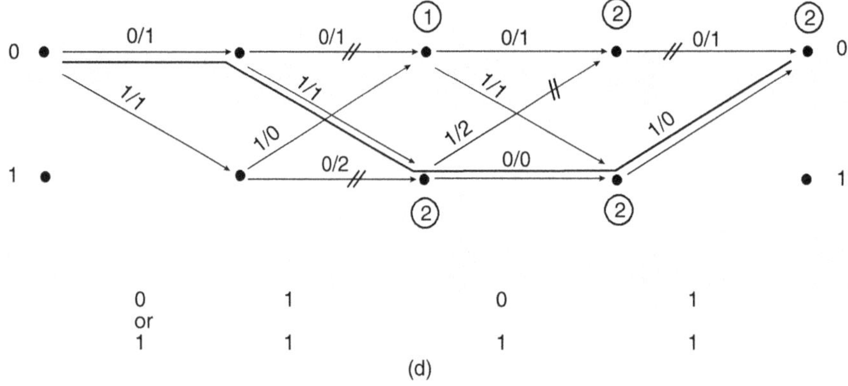

Figure 13.8 Example of the BCJR algorithm. (a) encoder; (b) state transition diagram (left: showing bits; right: showing $+1/-1$ symbols); (c) signals; (d) Viterbi decoding of hard input. (e) Viterbi decoding of soft input. (f–h) Evaluation of μ, λ, ν variables for LLR computation. The computed LLRs and their hard decision are at the bottom.

We notice that while hard- and soft-input Viterbi give the same result in many cases, this does not necessarily have to be fulfilled – soft-input Viterbi is generally superior and thus can correctly decode even some cases where the hard-decision Viterbi fails.

Finally, for the BCJR algorithm, Figures *13.8f–h* show the quantities λ, μ, and ν, as well as the LLRs. Note that the term $\exp\left[u_i \frac{L(u_i)}{2}\right]$ is unity because the transmit symbols are equiprobable. By making a hard decision on the LLRs, we get a decoded sequence [0 1 0 1], which is again correct. Note that for this example, the BCJR algorithm might seem like an unnecessarily complicated alternative to the classical Viterbi algorithm. However, if we need not only the decisions on the bits but also the reliability values of the decisions, the LLRs, available from the BCJR algorithm before the hard decision, provide this information.

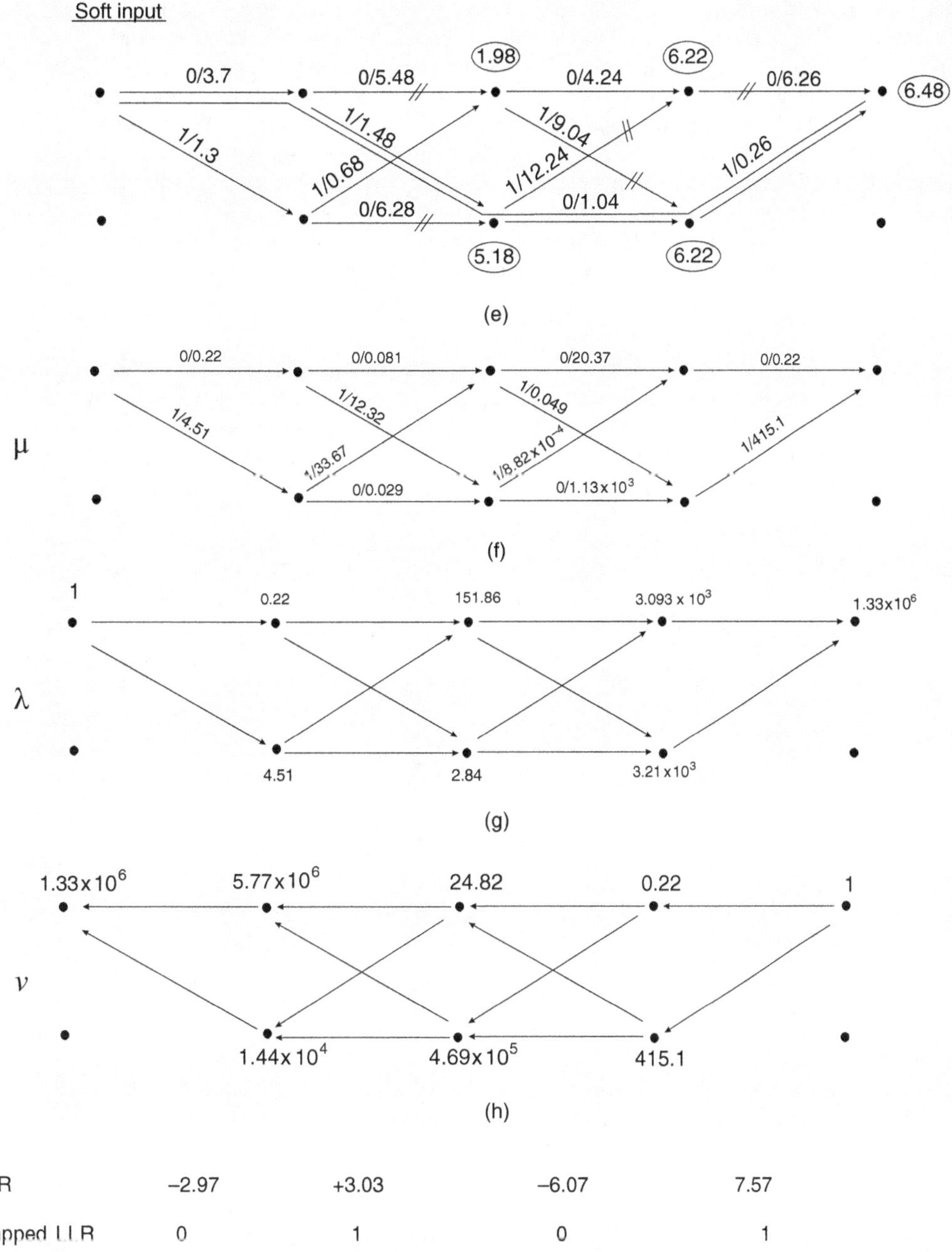

Figure 13.8 (Continued).

Note that the values of λ and ν can become very large. The reason for this is that the expressions we gave above are not true probabilities, since all constant terms have been dropped; rather the computed values are only proportional to the actual probabilities. Proper normalization in each step can improve the numerical behavior.

*13.4 Trellis Coded Modulation

13.4.1 Basic Principle

A main problem of coding is the reduction in spectral efficiency. Since, with the addition of check bits, we have to transmit more bits, the bandwidth requirement increases. This problem can be avoided by the use of higher-order modulation alphabets, which allow the transmission of more bits within the same bandwidth. Compare the transmission of uncoded BPSK with coded (rate 1/3 code) 8-Phase Shift Keying (PSK): using 8-PSK instead of BPSK decreases the symbol rate by a factor of 3, while rate 1/3 encoding increases the symbol rate by a factor 3 compared to the uncoded case; these two effects cancel out. In other words, using a rate 1/3 code, while at the same time changing the modulation alphabet to 8-PSK, keeps the number of *symbols* transmitted per unit time, and thus the bandwidth requirement, at the same level as when transmitted uncoded data with BPSK modulation.

As we discussed in Chapter 11, increasing the symbol alphabet increases the probability of error; on the other hand, introducing coding decreases the error probability. A simplistic approach to solving the spectral efficiency problem would thus be to add parity check bits to the data bits and map the resulting coded data to higher-order modulation symbols. However, this may not give good results. In contrast, *TCM* adds to the redundancy of the code by increasing the dimension of the signal space, while disallowing some symbol sequences in this enlarged space. The important aspect here is that modulation and encoding are designed as a joint process. This allows the design of a modem-plus-codec that shows higher resilience to noise than uncoded systems with the same spectral efficiency.

Example 13.7 *Simple trellis-coded modulation.*

To understand the basic principle of TCM, consider first a simple example in an AWGN channel. We compare an uncoded system with a TCM system, both of which transmit two source data bits per symbol duration. In the uncoded case, Quadrature-Phase Shift Keying (QPSK) is used as the modulation format (see Figure 13.9). Every combination of two bits corresponds to one valid point in the signal constellation diagram. With the energy per bit being E_B, the distance between points in the signal constellation diagram is $2\sqrt{E_B}$ so the squared distance is $4E_B$. The error probability in an AWGN case can thus be represented (see Section 11.1.3) as

$$BER \approx Q\left(\sqrt{2\gamma_B}\right). \tag{13.69}$$

For TCM, we use a larger symbol alphabet, and thus a higher-order modulation format (in our case, 8-PSK). As each symbol can transmit 3 bits, the code rate is $R_c = 2/3$. The allowed symbol sequences are determined by a shift register structure, like in a convolutional code. As with any other convolutional codes, the (coded) transmit symbol depends on the current source bit, as well as the current state of the memory. In our example, the memory is 2 bits long, so that there are four possible states: 00, 01, 10, 11. Depending on the state, as well as the source bit, different 8-PSK symbols are transmitted.

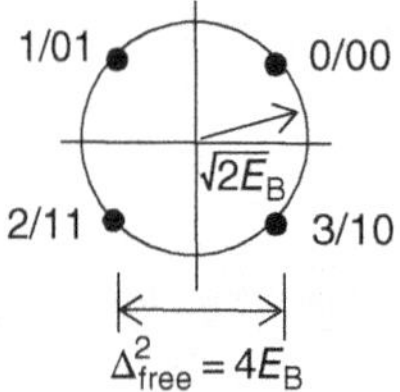

Figure 13.9 Quadrature-phase shift keying signal constellation diagram.

Figure 13.10 shows the possible transitions between states. If, e.g., memory is in state 11, and the information symbol 01 is to be transmitted, then the PSK symbol 3 is transmitted, and memory ends up in state 10. We also find from the diagram that so-called *parallel transitions* are possible. These are transitions where we can get from memory state A to memory state B by transmission of different 8-PSK symbols. However, not all combinations of states and 8-PSK symbols are possible: when in state 11, only symbols 1, 3, 5, 7 are allowed to be transmitted. Figure 13.11a,b shows the encoder structure and the trellis diagram for this system.

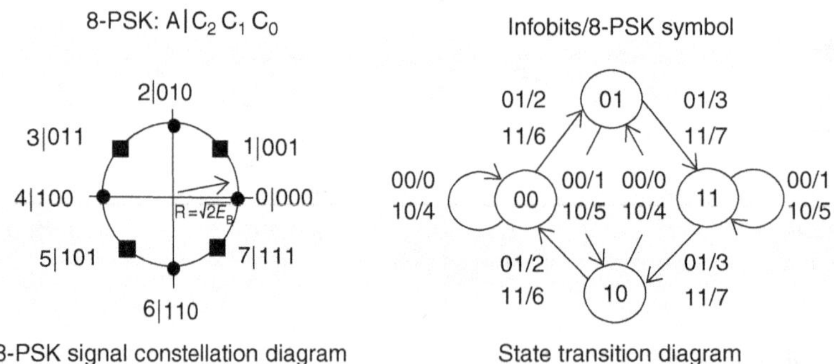

Figure 13.10 8-PSK signal constellation, and state transition diagram for a rate-2/3 8-PSK trellis-coded modulation. Reproduced with permission from [Mayr 1996] © B. Mayer.

The trellis diagram also allows determination of the BER of TCM. As the first step, we determine the smallest squared Euclidean distance between symbol sequences that can lead to errors. Looking at a part of the trellis diagram, we find all allowed paths that lead from state 00 at time 0 to state 00 at time 3, see Figure 13.11c. The first possible pair is the parallel transitions using PSK

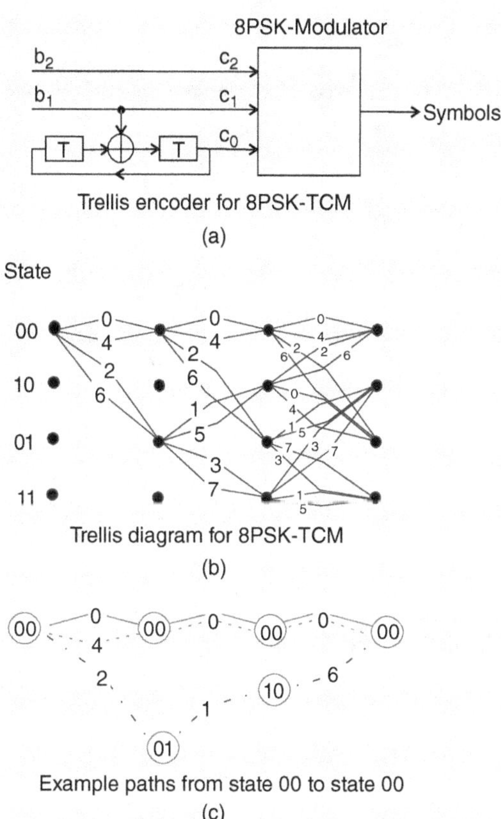

Figure 13.11 Encoder and trellis diagram for a rate-2/3 8-PSK trellis-coded modulation with four trellis states. (a) Trellis-coded modulator structure, (b) trellis diagram, (c) example paths. Color version available at wiley.com/go/molisch/wireless3e.
Reproduced with permission from [Mayr 1996] © B. Mayer.

symbols 0 and 4 (to get from state 00 at time 0 to state 00 at time 1; later symbols need not differ), denoted as the solid and dotted line in Figure 13.11c. The squared Euclidean distance between these two symbols is $d^2 = 8E_B$. Other transitions are not possible by moving just one step to the right in the trellis; rather, a whole sequence is required. For example, a transition from state A_0 (00) to state A_3 (00) is also possible by transmitting the PSK symbol sequence $2 - 1 - 6$, i.e., the dash-dotted line in Figure 13.11c. The squared Euclidean distance between symbols 0 and 2 is $d^2 = 4E_B$, between 0 and 1 it is $2E_B [2\sin(\pi/8)]^2$, and between 0 and 6, it is again $4E_B$. The sum of the squared magnitudes of the distances between the two paths is thus $(4 + 1.17 + 4) E_B = 9.17 E_B$. Other symbol sequences that can lead to errors (not shown in Figure 13.11c for clarity) can be searched for in a similar fashion. It can be shown that the smallest squared Euclidean distance between sequences that begin and end in the same state is $d^2_{\text{coded}} = 8E_B$. This is twice the distance we had for the uncoded case.

Generally, we can define the asymptotic coding gain – i.e., the coding gain at high SNRs – as

$$G_{\text{coded}} = 10 \log_{10} \left(\frac{d^2_{\text{coded}}}{d^2_{\text{uncoded}}} \right). \tag{13.70}$$

The above example achieves an asymptotic coding gain of 3 dB. In other words, for equal bandwidth requirement, equal source data rate, and equal target BER, the system needs 3 dB less transmit power. For finite target error rates, the coding gain is somewhat smaller; for a BER of 10^{-5}, the coding gain is only 2.5 dB (see Figure 13.12). It is, however, also noteworthy that at low SNRs, the performance is *worse* than for an uncoded system.

13.4.2 Set Partitioning

The previous subsection demonstrated the advantages of TCM, but the underlying code was ad hoc. For 8-PSK, such ad hoc construction is still feasible and can lead to good results, but it becomes impossible for higher-order modulation formats. [Ungerboeck 1982] suggested a heuristic method for the construction of good codes, the so-called *set partitioning*. The basic principles of the method are as follows:

1. Double the size of the modulation alphabet.
2. Select the allowable transitions so that the minimal squared Euclidean distance between sequences is maximized.

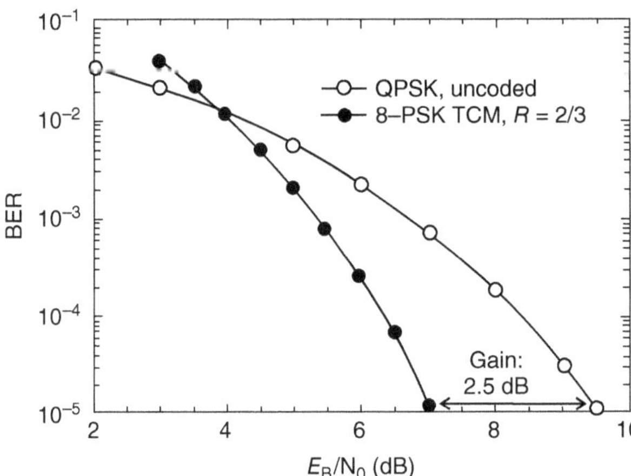

Figure 13.12 Simulated error probability of uncoded quadrature-phase shift keying and rate-2/3 8-PSK trellis-coded modulation in an AWGN channel. Reproduced with permission from [Mayr 1996] © B. Mayer.

For maximization of the distance, the symbol alphabet is partitioned in several steps, and at each step, the minimum distance between symbols of the partitioned sets should increase.

Example 13.8 *Set partitioning.*

This somewhat abstract principle can best be explained with an example. In other words, we show how the symbol alphabet needs to be partitioned for maximizing the minimum distances within each of the sets. Consider the 8-PSK constellation of the previous section: the distance between two neighboring points in the signal constellation diagram is $d = \sqrt{E_B}\left[2\sqrt{2}\sin\left(\pi/8\right)\right] = 1.08\sqrt{E_B}$. As a first step, the existing symbols are partitioned into two sets (see Figure 13.13). In order to maximize the distance between elements within each step, each set is a QPSK constellation that is rotated 45° with respect to the other. This increases the

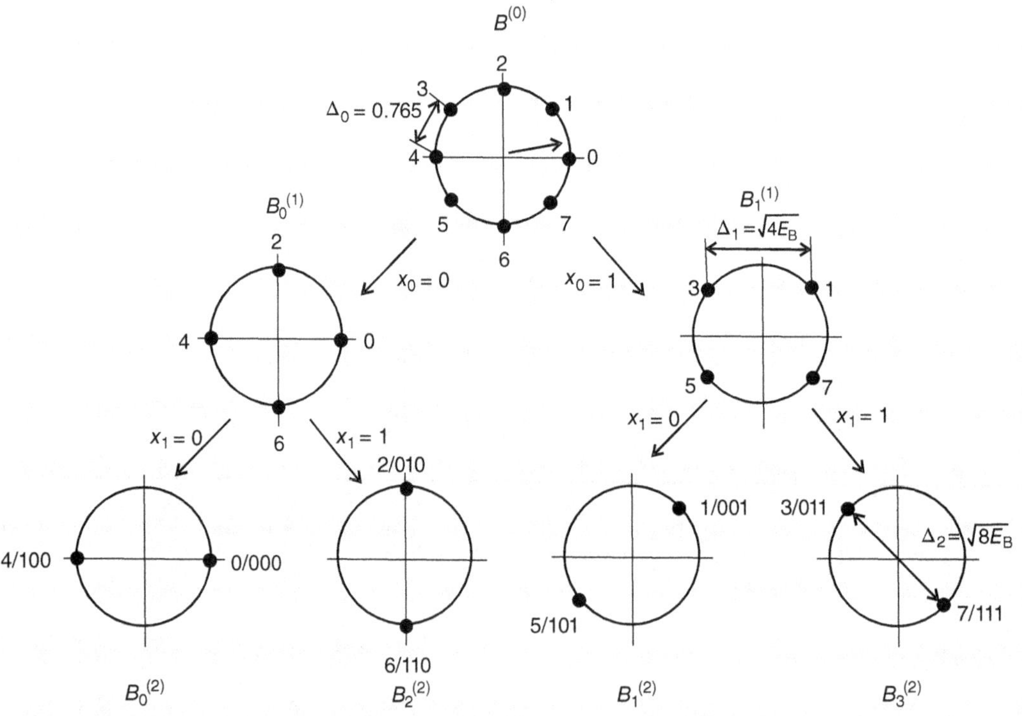

Figure 13.13 Partitioning of 8-PSK.

Euclidean distance within a set to $2\sqrt{E_B}$; there is no constellation that would lead to a larger minimum distance. In the next step, the QPSK constellation is partitioned. This results in two BPSK constellations that are rotated 90° with respect to each other, and the Euclidean distance within the set is $2\sqrt{2E_B}$. This completes the set partitioning.

In the next step, the partitioned sets are used as the basis for a code. Figure 13.14 shows the structure of a TCM encoder with such a set partitioning. We distinguish between N_{symb}, the number of information bits that can be transmitted per symbol, and $\tilde{N}$, the number of bits that are mapped into coded bits by a convolutional encoder of rate $R_c = \tilde{N}/(\tilde{N} + 1)$. During the encoding process, the source bits $u_1 \cdots$ $u_{\tilde{N}}$ are mapped to coded bits $x_0 \cdots x_{\tilde{N}}$. The remaining source bits $u_{\tilde{N}+1} \cdots u_{N_{\mathrm{symb}}}$ are directly mapped to "uncoded" bits $x_{\tilde{N}+1} \cdots x_{N_{\mathrm{symb}}}$. The uncoded bits are used to select a signal constellation point within a subset; such points have a large Euclidean distance, so that errors are less probable. The other bits select the subsets. The fewer bits are uncoded, the fewer parallel transitions are possible.

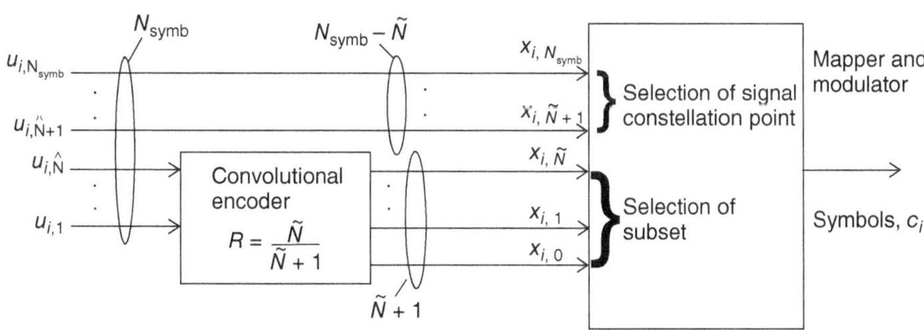

Figure 13.14 Structure of a trellis-coded modulation coder according to Ungerboeck. Reproduced with permission from [Mayr 1996] © B. Mayer.

Continuing Example 13.8, we find that two information bits are mapped onto one of 8 PSK symbols, i.e., $N_{\mathrm{symb}} = 2$, $\tilde{N} = 1$. The number of information bits that remain uncoded and are used, according to Figure 13.14, to select the constellation point within a subset is $N_{\mathrm{symb}} - \tilde{N} = 1$. Specifically, they select within each of the 4 sets at the bottom (0 or 4, 2 or 6, 5 or 1, 3 or 7); these are the parallel transitions we mentioned in Example 13.7. The number of other information bits, which are encoded, is $\tilde{N} = 1$; specifically, those bits are encoded with a rate $R = 1/2$ convolutional encoder, so that one information bit is mapped onto two encoded bits. Using a recursive convolutional encoder then results in the structure of Figure 13.8a. Decoding is done by means of a Viterbi decoder (see Section 13.3.2).

13.5 Bit Interleaved Coded Modulation (BICM)

Another way of combining coding and higher-order modulation is *Bit Interleaved Coded Modulation* (BICM). In this approach, the channel coder and the modulator are separated by an interleaver. As a consequence, encoded bits that are correlated through the coding process are mapped onto temporally separated symbols (which might increase diversity, see Section 13.10), which provides greater robustness with respect to noise spikes.

Figure 13.15 shows a block diagram of a BICM transmission. The bits are encoded (by any binary encoder) and interleaved by the interleaver Π. The coded and interleaved bits are finally mapped onto the modulation symbols, which belong to an alphabet $\mathcal{X}$. The binary labeling scheme (discussed in more detail below) determines which bit combinations (as seen after the interleaver) are associated with which modulation symbol. The symbols are then transmitted over the channel, and the received signal is sent through a de-mapper, which computes reliability measures L (LLRs, compare Section 13.3.4) for the estimated signals. The de-interleaver provides the "soft information" that forms the input of the decoder. Optionally, the output of the decoder can be re-interleaved and fed

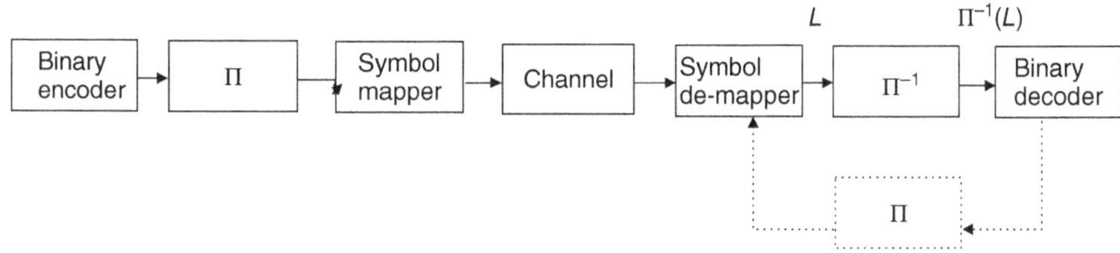

Figure 13.15 Block diagram of a bit-interleaved coded modulator. Feedback in the demodulator is optional.

back to help the symbol de-mapper to refine its estimate of the LLRs. Such iterative decoding improves performance but increases complexity. It is similar in spirit to the turbo codes that will be discussed in Section 13.6 but will not be considered further in the current subsection.

The binary labeling is of critical importance for the performance of BICM. We assume in the following that higher-order modulation (at least QPSK) is used. Numerical investigations, as well as theoretical arguments, show that for moderate-to-high SNR values, binary reflected Gray mapping is optimum. For low SNRs, set partitioning mapping is optimum.

The de-mapper computes the LLR for each particular (coded) bit

$$L(u_i|\mathbf{r}) = \ln \left[\frac{\Pr(b_i = +1 \mid \mathbf{r})}{\Pr(b_i = -1 \mid \mathbf{r})} \right] \tag{13.71}$$

$$= \ln \left[\frac{\Pr(\mathbf{r} \mid b_i = +1)}{\Pr(\mathbf{r} \mid b_i = -1)} \right] + \ln \left[\frac{\Pr(b_i = +1)}{\Pr(b_i = -1)} \right]. \tag{13.72}$$

Just as in Section 13.3.4, if we further assume equal a priori probabilities, and an AWGN model for the channel, the LLRs can be computed as

$$L(u_i|\mathbf{r}) = \ln \left[\frac{\sum_{s \in X_{i,1}} \exp\left(-\gamma|\mathbf{r} - \alpha s|^2 \right)}{\sum_{s \in X_{i,0}} \exp\left(-\gamma|\mathbf{r} - \alpha s|^2 \right)} \right] \tag{13.73}$$

where α is the (complex) channel gain, and $s \in X_{i,1}$ denotes modulation symbols whose associated binary representation (in the binary labeling) has the ith bit equal to $+1$.

It has been shown that BICM, in particular when used in combination with iterative decoding, has excellent performance. It is thus widely used in practical wireless systems. The separation of the coder and the modulator allows a very flexible design (more flexible and easier to implement than TCM) and allows an easy implementation of adaptive modulation and coding where, e.g., the modulation constellation can be changed depending on the quality of the channel over which signaling is happening while keeping the coding scheme unchanged.

*13.6　Turbo Codes

13.6.1　Introduction

Turbo codes are among the most important developments in coding theory since the field was founded. As we have already mentioned in Section 13.1, *very long codes* can approach the Shannon limit (channel capacity). However, the brute force decoding of such long codes is prohibitively complex. Turbo codes were the first practically used codes that came close to the Shannon limit with reasonable decoding effort. [Berrou et al. 1993] created *very long codes* by a combination of several parallel, simple codes. The codes are interleaved by a pseudorandom interleaver. The vital trick now lies in the decoder: because the code is a combination of several short codes, the decoder can also be broken up into several simple decoders that exchange soft information about the decoded bits and thus iteratively arrive at a solution.

The random interleaver in the code approximately realizes the idea of a random code, so that the total codeword has very little structure. This has the following advantages:

- Interleaving increases the effective codelength of the combined code. In other words, it is the interleaver (whose operation can easily be reversed), and not the constituent codes, that determines the codelength. This allows the construction of very long codes with simple encoder structures.
- The special structure of the total code – i.e., the composition of separate constituent codes – makes decoding possible with an effort that is essentially determined by the length of the constituent codes.

13.6.2　Encoder

Turbo codes use a combination of several codes. One method of combining codes, serial concatenation, was already discussed in Section 13.2. In this section, we concentrate on parallel codes. The principle is shown in Figure 13.16. The source data stream is put out directly, and also sent to several encoder branches. Each of the branches initially contains an interleaver that is different from branch to branch. One of the branches (the direct feedthrough), uses the original source data sequence as input, while in the other cases, the sequence is first interleaved.

After the interleaver, the data stream (in each parallel branch) is sent through a convolutional encoder, which maps the information vector $\mathbf{u}$ to the output. The number of these parallel encoders is – in principle – arbitrary. Due to concerns about the data rate, two encoders with rate $R = 1/2$ each are most common. Each of those encoders is systematic; the systematic bits are discarded since the turbo code transmits the original data sequence anyway (see topmost branch in Figure 13.16); only the parity bits of the encoders are

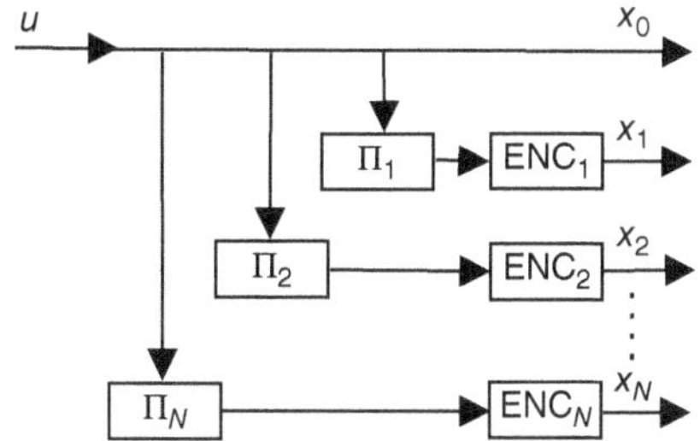

Figure 13.16 Structure of a turbo encoder. Π denotes interleavers.
Reproduced with permission from [Valenti 1999] © M. Valenti.

actually used as output **x**. Therefore, each a constituent encoder usually has code rate 1 – i.e., puts out one coded bit per source bit (see Figure 13.18).[11]

The sequences (original sequences, plus the parity bits from the encoders) are then multiplexed. The resulting code is systematic, as it still contains the original data sequence. Since the systematic part is transmitted only once, and each of the constituent encoders has code rate 1, the code rate of the total system in our example is $R_c = 1/(N + 1) = 1/3$.

It is advantageous to use *Recursive Systematic Convolutional* (RSC) codes for the encoding of each branch. The structure of such a code is outlined in Figure 13.17. A shift register with feedback is used to compute the encoded bits. These codes are systematic, so that there is one output that directly maps the source bits, just as described above. Furthermore, in the shown example, the coderate of the RSC is ½, so that after discarding of the systematic bits, a constituent RSC has coderate R = 1. Fig. 13.18 shows a turbo-encoder based on such RSC.

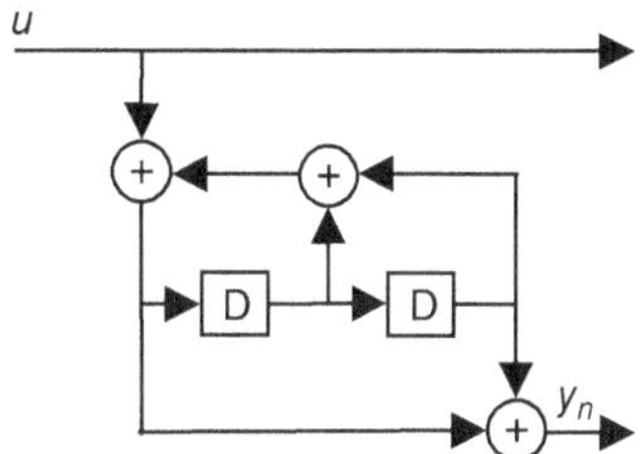

Figure 13.17 Structure of a recursive systematic convolutional encoder. D denotes delay elements.
Reproduced with permission from [Valenti 1999] © M. Valenti.

Besides this basic structure, many other turbo encoders are possible. As research in this area is still active, we refrain from giving a taxonomy of encoders here.

The code rate of the above encoder is $R_c = 1/3$. However, it can be increased by puncturing. For example, outputs x_1 and x_2 can be used as input to a multiplexer that alternatingly uses a bit from the first encoder x_1 and discards a bit x_2 from the second encoder; or uses x_2 and discards x_1. This realizes an encoder with code rate $R_c = 1/2$.

13.6.3 Turbo Decoder

While encoding is always a comparatively simple process, decoding is the area where problems normally occur. Brute-force implementation of a decoder, where the combined code is viewed as a single, very complicated, code, would require prohibitive computational effort. It is here that turbo codes show their great advantage: it is possible to decode two constituent codes separately, and then combine the information from these two decoders.

Turbo codes are decoded iteratively, exchanging soft information between constituent decoders. Figure 13.19 shows a block diagram of a turbo decoder. It consists of two Soft Input Soft Output (SISO) decoders for the constituent codes, an interleaver, and a de-interleaver. The SISO decoder puts out not only the various bits but also the confidence it has in a specific decision. This confidence is described by the *LLR* value which can be computed by the BCJR algorithm as discussed in Section 13.3.4.

The LLR consists of 3 components

- The LLR of the a priori probability for the bits (that LLR is usually 0, as all bits are equally likely).

- The information from the received raw data – i.e., the direct observation. This LLR also depends on the SNR.

---

[11] If we just used this encoder as a normal convolutional encoder, then the code rate would be 1/2, as the encoder would put out the source bit and the redundant bit for each incoming source bit.

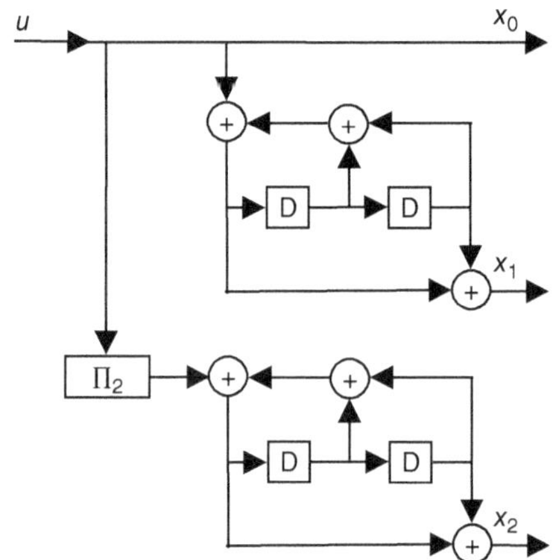

Figure 13.18 Structure of a turbo encoder based on recursive systematic convolutional encoders.
Reproduced with permission from [Valenti 1999] © M. Valenti.

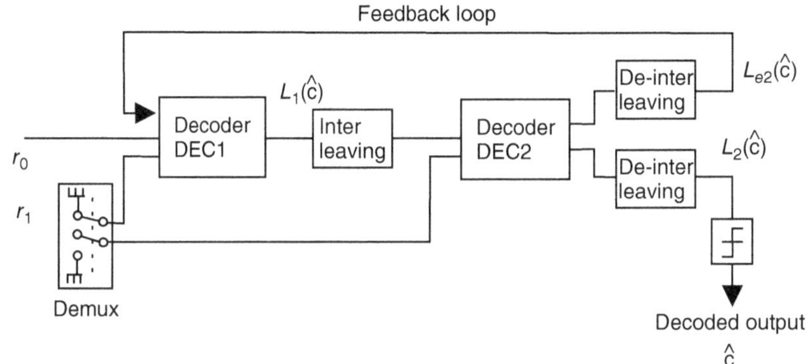

Figure 13.19 Structure of a turbo decoder.
Reproduced with permission from [Sklar 1997] © IEEE.

- The extrinsic log likelihood, which contains the information from decoding. The extrinsic information from the second decoder helps the first decoder, and the first decoder helps the second decoder.

To demonstrate this, we analyze in more detail (13.68). If the considered code is systematic, $s_{i1} = u_i$. Then we can write

$$\mu_i(\phi', \phi) = \exp\left[u_i \frac{L(u_i)}{2}\right] \exp\left(u_i \frac{L_c}{2}\widetilde{r}_{i1}\right)\widetilde{\mu}_i(\phi', \phi) \quad \text{where } \widetilde{\mu}_i(\phi', \phi) = \exp\left(\frac{L_c}{2}\sum_{k=2}^{K}(\widetilde{r}_{ik}\widetilde{s}_{ik})\right). \tag{13.74}$$

Consequently

$$L(u_i \,|\, \mathbf{r}) = \ln\left[\frac{\sum_{\mathcal{T}_1}\exp\left[u_i \frac{L(u_i)}{2}\right]\exp\left(u_i \frac{L_c}{2}\widetilde{r}_{i1}\right)\lambda_{i-1}(\phi')\widetilde{\mu}(\phi', \phi)v_i(\phi)}{\sum_{\mathcal{T}_0}\exp\left[u_i \frac{L(u_i)}{2}\right]\exp\left(u_i \frac{L_c}{2}\widetilde{r}_{i1}\right)\lambda_{i-1}(\phi')\widetilde{\mu}(\phi', \phi)v_i(\phi)}\right]. \tag{13.75}$$

Since the $u_i = 1$ in the numerator, and $u_i = -1$ in the denominator, this becomes

$$L(u_i \,|\, \mathbf{r}) = L(u_i) + L_c\widetilde{r}_{i1} + L_e(u_i) \tag{13.76}$$

where the *extrinsic information* is defined as

$$L_e(u_i) = \ln\left[\frac{\sum_{\mathcal{T}_1}\lambda_{i-1}(\phi')\widetilde{\mu}(\phi',\phi)\nu_i(\phi)}{\sum_{\mathcal{T}_0}\lambda_{i-1}(\phi')\widetilde{\mu}(\phi',\phi)\nu_i(\phi)}\right]. \tag{13.77}$$

The first term in (13.76) thus describes the a-priori likelihood for the symbol, the second term the information we get from the systematic bit, and the last term the extrinsic information; this last term will be exchanged during turbo iteration.

The first two components also exist for uncoded systems, the last one is the contribution from parity bits of the constituent code; it is fed as input to the other constituent decoder. It is important that the extrinsic information of a constituent decoder does not depend on the input of that same decoder; otherwise, the information in the extrinsic information and in the direct observation would be strongly correlated, and the decoder would just confirm its own opinion.

The iteration now starts with the assumption that the extrinsic LLR is zero, and decoder 1 decodes the received signal like a "normal" convolutional decoder (though with soft output). From this soft output, the RX then computes the extrinsic LLR for decoder 2. The extrinsic information computed from the output of decoder 2 is then used for the next iteration step at decoder 1. The extrinsic LLR from decoder 1 obtained in that next iteration is used for the subsequent iteration of decoder 2. This procedure is continued until convergence is achieved.

The BER improves as the number of iterations increases. In an AWGN channel, $N_{it} = 5$ iterations are usually sufficient to achieve convergence, while a fading channel usually requires $N_{it} = 10$ iterations. The number of operations per information bit has been estimated, for typical setups, as

$$N_{op} \leq N_{it}(62 + 8/R_c). \tag{13.78}$$

Typically 400–800 operations per information bit are required. This explains why the widespread adoption of turbo codes happened only after the turn of the century. Transmission rates of 400 kbit/s require up to 300 MIPS (Million Instructions Per Second) for decoding – signal-processing equipment of such computational power has only been available since then.

Turbo codes allow close approximation of channel capacity. As an example, Figure 13.20 shows the BER of a rate-1/2 turbo encoder in an AWGN channel. Six iterations are obviously sufficient to closely approach the converged value. After 18 iterations, the BER is 10^{-5} at a 0.7-dB SNR.

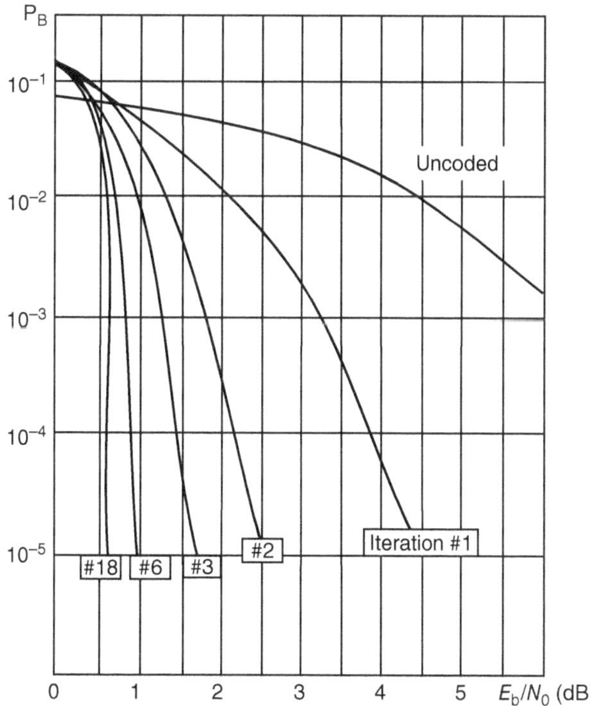

Figure 13.20 Bit error rate of a $R_c = 1/2$ turbo code with interleaver length 64,000 for different numbers of iterations in an AWGN channel. Reproduced with permission from [Sklar 1997] © IEEE.

*13.7   Low-Density Parity-Check Codes

When turbo codes were announced in 1993, they immediately drew a large amount of richly deserved attention. It seemed that for the first time the Shannon bound could be approached by practical codes. Yet it turns out that the problem had already been solved in the early 1960s! Gallagher in his doctoral thesis [Gallagher 1961] had designed linear block codes, called *LDPC codes*, which allow the Shannon limit to be closely approached, and also proposed efficient iterative decoding mechanisms. However, this work was largely overlooked, because iterative decoding exceeded the computational power available at that time. However, several papers in the mid-1990s led to the rediscovery of these codes. And by that time the decoding complexity that once had seemed prohibitive looked quite reasonable. Since then, a large number of papers have been published on LDPC codes, and a considerable number of improvements have been proposed for them. Most importantly, LDPC codes are now used in both the most recent cellular (5G NR) and Wi-Fi (802.11ax) standards, see Chapters 32 and 33, respectively.

13.7.1   Definition of Low-Density Parity-Check Codes

LDPC codes are linear block codes (as discussed in Section 13.2). One interesting aspect of them is that they are not defined via the generator matrix $\mathbf{G}$, but rather via the parity check matrix $\mathbf{H}$. This is a key trick, since it is normally decoding that causes the biggest problems, not encoding. It thus makes eminent sense to define a structure that allows for easy decoding! Blocksize of LDPC codes, and thus the dimensions of the check matrix, are very large. However, the number of nonzero entries in that matrix is kept low. More precisely, the ratio of the number of nonzero elements to the total number of entries is small; this is the reason why the codes are called "low-density." Following Gallagher, let us define an (N, p, q) binary LDPC code as a code of length N, whose parity check matrix has p 1's in each column and q 1's in each row. In order for the code to have good properties, it is necessary that $p \geq 3$. If all rows are linearly independent, then the resulting rate of the code is $(q - p)/q$.

Good parity check matrices can be constructed from a few simple rules. First, subdivide the matrix horizontally into p submatrices of equal size. Then put one "1" into each column of such a submatrix. Let the first submatrix be defined as, e.g., q concatenated identity matrices, or a structure like:

$$\begin{bmatrix} 1 & 1 & 1 & 1 & 0 & 0 & 0 & 0 & 0 & 0 & 0 & 0 & 0 & 0 & 0 & 0 & 0 & 0 & 0 & 0 \\ 0 & 0 & 0 & 0 & 1 & 1 & 1 & 1 & 0 & 0 & 0 & 0 & 0 & 0 & 0 & 0 & 0 & 0 & 0 & 0 \\ 0 & 0 & 0 & 0 & 0 & 0 & 0 & 0 & 1 & 1 & 1 & 1 & 0 & 0 & 0 & 0 & 0 & 0 & 0 & 0 \\ 0 & 0 & 0 & 0 & 0 & 0 & 0 & 0 & 0 & 0 & 0 & 0 & 1 & 1 & 1 & 1 & 0 & 0 & 0 & 0 \\ 0 & 0 & 0 & 0 & 0 & 0 & 0 & 0 & 0 & 0 & 0 & 0 & 0 & 0 & 0 & 0 & 1 & 1 & 1 & 1 \end{bmatrix}. \tag{13.79}$$

Then, let the other submatrices be random column permutations of this first submatrix. For example, using (13.79) as the first submatrix, we arrive at the following example of a (20, 3, 4) code [Davey 1999]:

$$\mathbf{H} = \begin{bmatrix} 1 & 1 & 1 & 1 & 0 & 0 & 0 & 0 & 0 & 0 & 0 & 0 & 0 & 0 & 0 & 0 & 0 & 0 & 0 & 0 \\ 0 & 0 & 0 & 0 & 1 & 1 & 1 & 1 & 0 & 0 & 0 & 0 & 0 & 0 & 0 & 0 & 0 & 0 & 0 & 0 \\ 0 & 0 & 0 & 0 & 0 & 0 & 0 & 0 & 1 & 1 & 1 & 1 & 0 & 0 & 0 & 0 & 0 & 0 & 0 & 0 \\ 0 & 0 & 0 & 0 & 0 & 0 & 0 & 0 & 0 & 0 & 0 & 0 & 1 & 1 & 1 & 1 & 0 & 0 & 0 & 0 \\ 0 & 0 & 0 & 0 & 0 & 0 & 0 & 0 & 0 & 0 & 0 & 0 & 0 & 0 & 0 & 0 & 1 & 1 & 1 & 1 \\ 1 & 0 & 0 & 0 & 1 & 0 & 0 & 0 & 1 & 0 & 0 & 0 & 1 & 0 & 0 & 0 & 0 & 0 & 0 & 0 \\ 0 & 1 & 0 & 0 & 0 & 1 & 0 & 0 & 0 & 1 & 0 & 0 & 0 & 0 & 0 & 0 & 1 & 0 & 0 & 0 \\ 0 & 0 & 1 & 0 & 0 & 0 & 1 & 0 & 0 & 0 & 0 & 0 & 0 & 1 & 0 & 0 & 0 & 1 & 0 & 0 \\ 0 & 0 & 0 & 1 & 0 & 0 & 0 & 0 & 0 & 0 & 1 & 0 & 0 & 0 & 1 & 0 & 0 & 0 & 1 & 0 \\ 0 & 0 & 0 & 0 & 0 & 0 & 0 & 1 & 0 & 0 & 0 & 1 & 0 & 0 & 0 & 1 & 0 & 0 & 0 & 1 \\ 1 & 0 & 0 & 0 & 0 & 1 & 0 & 0 & 0 & 0 & 0 & 1 & 0 & 0 & 0 & 0 & 0 & 1 & 0 & 0 \\ 0 & 1 & 0 & 0 & 0 & 0 & 1 & 0 & 0 & 0 & 1 & 0 & 0 & 0 & 0 & 1 & 0 & 0 & 0 & 0 \\ 0 & 0 & 1 & 0 & 0 & 0 & 0 & 1 & 0 & 0 & 0 & 0 & 1 & 0 & 0 & 0 & 0 & 0 & 1 & 0 \\ 0 & 0 & 0 & 1 & 0 & 0 & 0 & 0 & 1 & 0 & 0 & 0 & 0 & 1 & 0 & 0 & 1 & 0 & 0 & 0 \\ 0 & 0 & 0 & 0 & 1 & 0 & 0 & 0 & 0 & 1 & 0 & 0 & 0 & 0 & 1 & 0 & 0 & 0 & 0 & 1 \end{bmatrix}. \tag{13.80}$$

This structure obviously follows the rules of having p 1's in each column, and q 1's in each row; it also appears that the structure is reasonably random.

13.7.2 Encoding of Low-Density Parity-Check Codes

Since LDPC codes are defined via their parity check matrix, the encoding process is more complicated than for "traditional" block codes. Remember that with those codes, we only have to multiply the message vector by the generator matrix to obtain the codeword. However, for LDPC codes, the generator matrix is not known first-hand. Fortunately, the computation is not very difficult: Using Gaussian elimination and reordering of columns, we can cast the parity check matrix in the form:

$$\widetilde{\mathbf{H}} = \left(-\mathbf{P}^T \; \mathbf{I} \right). \tag{13.81}$$

The corresponding generator matrix is then

$$\mathbf{G} = \left(\mathbf{I} \quad \mathbf{P} \right). \tag{13.82}$$

Note that, due to the process of Gaussian elimination, the generator matrix is typically *not* sparse. This also means that the encoding process requires more operations. Fortunately, all the required operations are simple because they are performed on discrete, known bits. Thus, encoding complexity is usually not an issue.

13.7.3 Decoding of Low-Density Parity-Check Codes

As we have mentioned above, the sparse structure of the parity check matrix is the key to decoding schemes of reasonable complexity. But it is still far from trivial! Performing an exact maximum likelihood decoding is an N-p hard problem (in other words, we have to check all possible codewords, and compare them with the received signal). It is therefore common to use an iterative algorithm called *belief propagation*. It is this algorithm that we describe in more detail in the following.[12]

Let the received signal vector be **r**. Let us next put the parity check equations into graphical form. In the so-called "Tanner graph" (see Figure 13.21),[13] we distinguish two kinds of nodes:

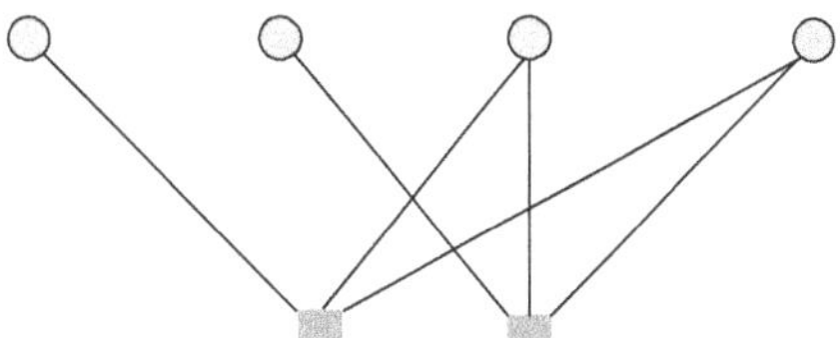

Figure 13.21 Tanner graph for the parity check matrix $H = \begin{bmatrix} 1 & 0 & 1 & 1 \\ 0 & 1 & 1 & 1 \end{bmatrix}$

1. *Variable (bit) nodes*: each variable node corresponds to one bit, and we know that it can be either in state 0, or state 1. Variable nodes correspond to the *columns* of the parity check matrix. We denote these nodes by circles.
2. *Constraint nodes*: constraint nodes (checknodes) describe the parity check equations; we know that if there are no errors present, the inputs to constraint nodes have to add up to 0. This follows from the definition of the syndrome, which is all-zero if no errors are present. Constraint nodes correspond to the *rows* of the parity check matrix. We denote these nodes by squares.

Because there are two different types of nodes, and no connections between nodes of the same type, such a graph is also called a "bipartite" graph. In addition to constraint nodes and variable nodes, there is also external information, obtained by observation of the received signal, which has to influence our decisions.

Constraint nodes are connected to variable nodes if the appropriate entries in the parity check matrix are 1 – i.e., constraint node i is connected to variable node j if $H_{ij} = 1$. "Soft" information from the observed signals – i.e., the external evidence – is associated with the variable nodes. We also need to know the probability density function (pdf) of the amplitude of the variables – i.e., the probability that a variable node has a certain state, given the amplitude value of the received signal.

Decoding on such a graph is done by a procedure called *message passing* or *belief propagation*. Each node collects incoming information, makes computations according to a so-called *local rule*, and passes the result of the computation to other nodes. Essentially, the jth variable node tells each of the constraint nodes it is connected to what it thinks its – i.e., the variable node's – value is, given the external information r_j and the information from the other constraint nodes it is connected to. This message is denoted λ_{ij}.[14] In turn, the ith constraint node tells the jth variable node (assuming they are connected) what *it* thinks the variable node has to be, given the

---

[12] Note that the decoding algorithm can be described based on the syndrome vector (the method we will choose here) as well as the data vector.
[13] There are two types of graphical representations: the "Tanner graph" used here, and the "Forney factor graph" [Loeliger 2004].
[14] Note that the variables μ, λ are different from the variables with the same name in Section 13.3.4. The definitions in this section, Section 13.7, are specific for Section 13.7.

information it has received from all the *other* connected variable nodes; this message is called $\mu_{i,j}$. This is shown in Figure 13.22. The belief that a bit should have a particular value is expressed as an LLR (rather than a simple probability).

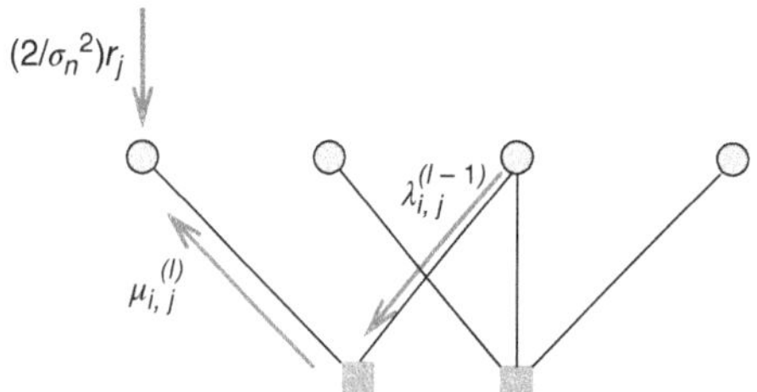

Figure 13.22 Message-passing in a factor graph. Circles: variable nodes (index j); squares: constraint nodes (index i).

Let us formulate the decoding strategy mathematically, for an AWGN channel:

1. First, the data bits decide what value they *think* they are, given the external evidence $\mathbf{r}$ only. Knowing the statistics of the noise σ_n^2, the variable nodes can easily compute their probability to be a 1 or a 0, i.e., the LLR, and pass that information to the constraint nodes. Conversely, the constraint nodes cannot pass a meaningful message to the variable nodes yet. Therefore,

$$\mu_{i,j}^{(0)} = 0, \qquad \text{for all } i \tag{13.83}$$

$$\lambda_{i,j}^{(0)} = \left(2/\sigma_n^2\right)r_j, \quad \text{for all } j. \tag{13.84}$$

2. Then, the constraint nodes pass a different message to each variable node. Elaborating on the principle mentioned above, let us look specifically at constraint node i: assume that a set of edges ends in node i, which originate from an ensemble $A(i)$ of variable nodes.

   Now, each of the checknodes has two important pieces of information: (i) it knows the values (or probabilities) of all data bits connected to it; (ii) furthermore, it knows that all the bits coming into a checknode have to sum up to 0 mod 2 (that is the definition of a parity check equation). From these pieces of information, it can compute the probability for the value that it thinks data bit j has to have. Since we are using LLRs, the message becomes

$$\mu_{i,j}^{(\ell)} = -2\tanh^{-1}\left(\prod_{k \in A\,(i)-j} \tanh\left(\frac{-\lambda_{i,k}^{(\ell-1)}}{2}\right)\right) \tag{13.85}$$

where $A(i)-j$ denotes "all the members of ensemble $A(i)$ with the exception of j" – i.e., all variable nodes that connect to the ith constraint node, with the exception of the jth node. Superscript$^{(\ell-1)}$ denotes the $\ell-1$th iteration – i.e., we use the results from the previous iteration steps.

3. Next, we update our opinion of what the variable nodes are, based on the information passed by the constraint nodes, as well as the external evidence. This rule is very simple:

$$\lambda_{i,j}^{(\ell)} = \left(2/\sigma_n^2\right)r_j + \sum_{k \in B(\,j)-i} \mu_{k,j}^{(\ell)} \tag{13.86}$$

where $B(j)-i$ denotes all constraint nodes that connect to the jth variable node, with the exception of i.

4. From the above, we can compute the pseudoposterior probabilities that a bit is 1 or 0:

$$L_j = \left(2/\sigma_n^2\right)r_j + \sum_i \mu_{i,j}^{(\ell)} \tag{13.87}$$

based on which tentative decision we make about the codeword. If that codeword is consistent – i.e., its syndrome is 0 – then decoding stops.

The above sequence of steps is the simplest conceptually, but in practice step 4 is done before step 3, for two reasons: (i) it saves computation of potentially unnecessary λ, since the iteration stops when the codeword is consistent and (ii) if this is not the case, Eq. (13.86) can be simplified to

$$\lambda_{i,j}^{(\ell)} = L_j - \mu_{i,j}^{(\ell)}. \tag{13.88}$$

Example 13.9 *Decoding of a low-density parity-check code.*

Let us now consider a very simple example for this algorithm. Let the parity check matrix be

$$\mathbf{H} = \begin{bmatrix} 0 & 0 & 1 & 1 & 1 & 1 \\ 1 & 1 & 1 & 1 & 0 & 0 \\ 1 & 1 & 0 & 0 & 1 & 1 \end{bmatrix}. \tag{13.89}$$

Let the codeword:

$$\bar{y} = \begin{bmatrix} 0 & 1 & 1 & 0 & 1 & 0 \end{bmatrix} \tag{13.90}$$

be sent through an AWGN channel with $\sigma_n^2 = 0.237$ corresponding to $\gamma = 6.25$ dB; let the received word be

$$\bar{r} = \begin{bmatrix} -0.71 & 0.71 & 0.99 & -1.03 & -0.61 & -0.93 \end{bmatrix}. \tag{13.91}$$

Then according to step 1 (mentioned above), likelihood values are computed from external evidence as

$$\overline{\lambda^{(0)}} = \begin{bmatrix} -6.0 & 6.0 & 8.3 & -8.7 & -5.2 & -7.9 \end{bmatrix}. \tag{13.92}$$

This is the top row of Figure 13.23. Hard thresholding of the received likelihood values would result in codeword error with an error at bit position 5:

$$\begin{bmatrix} 0 & 1 & 1 & 0 & 0 & 0 \end{bmatrix} \tag{13.93}$$

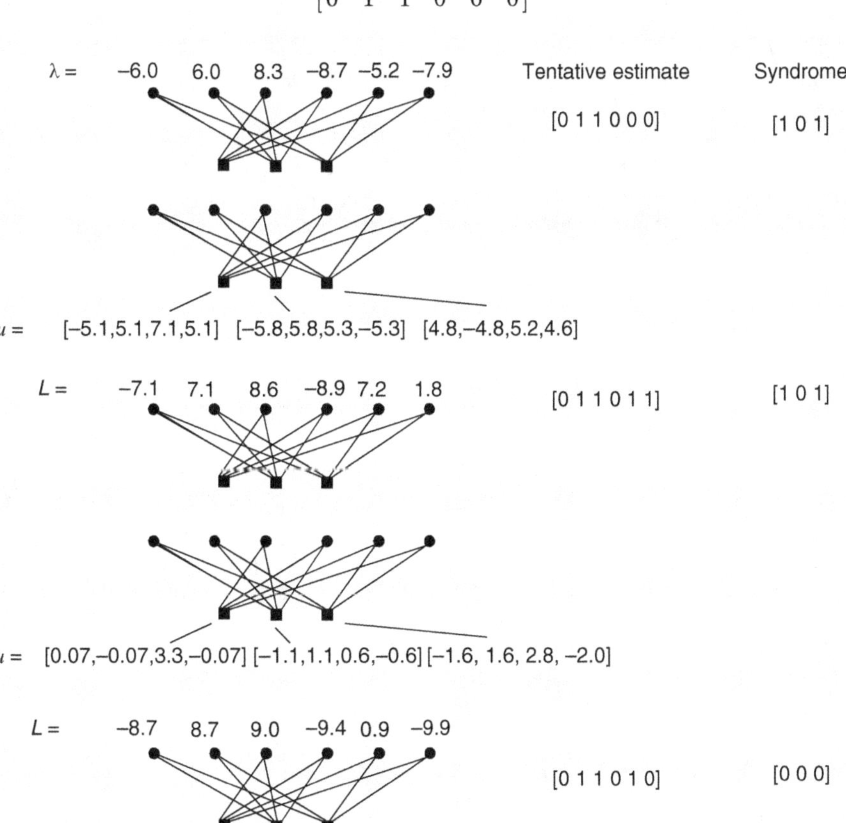

Figure 13.23 Example for the iterations of low-density parity-check message passing. Iterations use accurate numerical values of λ, μ, but displayed results are rounded.

that can be detected by computing the syndrome [1 0 1], which is nonzero. The μ values computed from (13.85) are computed in the next step, shown in the second subfigure of Figure 13.23. From this, the new LLRs are computed from (13.87), but the resulting tentative estimates for the message bits [0 1 1 0 1 1] are still inconsistent (nonzero syndrome), so that another round of iterations (computation of the λ from (13.86), μ from (13.85), and LLR from (13.87)) is required, leading finally to the estimate [0 1 1 0 1 0], which is a consistent message word (syndrome is all-zero), so that the iteration stops.

13.7.4   Performance Improvements

It can be shown that the belief propagation algorithm always converges to the maximum likelihood solution if the Tanner graph can be rolled up into a tree structure – i.e., each node is a "parent" or a "child" of another node, but not both at the same time. In other words, there should be no cycles in the Tanner graph. Short cycles, like the one shown in Figure 13.24, lead to problems with convergence: if the set of nodes creating the cycle start out with a wrong belief, they tend to reinforce it among themselves, instead of being convinced by evidence from other variable nodes that they need to change their self-assessment. Related to the concept of cycles is the *girth* of a code, i.e., the minimum length of any cycle in the graph. The construction of codes with large girth, i.e., without short cycles, is one of the most important, and most challenging, tasks in the design of LDPC codes. Note, however, that codes that lead to a *pure* tree structure are usually not good codes, even though they can be decoded exactly by the belief propagation algorithm.

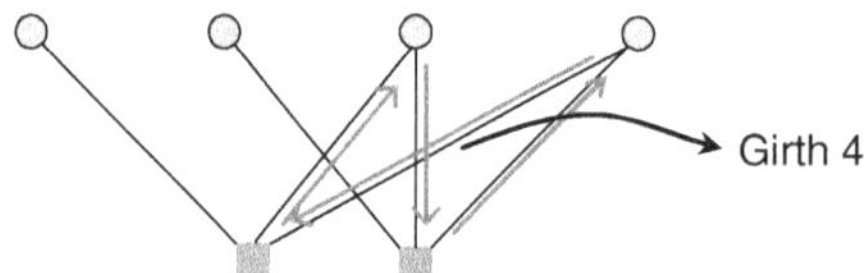

Figure 13.24   Tanner graph from Figure 13.21, showing a short cycle (with "girth" 4).

Another approach to improve convergence and performance is through the use of irregular codes. This means that column and row weights are not fixed, as we have assumed up to now, but that we rather prescribe only the *mean* weights and allow some of the columns to have more entries and some have less. The nodes associated with these "heavier" columns often converge faster and can spread their "secure" knowledge to other nodes, which leads to improved convergence for those nodes as well.

The performance that can be achieved with LDPC codes is very impressive, and (for large blocksizes) can approach the Shannon capacity within a fraction of a dB. There is also an interesting rule for computational complexity in decoding LDPC codes: when the code rate approaches $(1 - \delta)$ of Shannon capacity, then the complexity per bit goes like $(1/\delta)\log_2(1/\delta)$ [Richardson and Urbanke 2008]. For most other codes, the complexity grows with $\exp(1/\delta)$.

*13.8   Polar Codes

13.8.1   Introduction

Polar codes were introduced in the seminal paper of [Arikan 2009]. The basic idea of polar codes is to take a regular AWGN channel, where each "channel use" (transmission of a symbol) is affected by the same amount of noise, and convert it into a set of channels of which some are effectively noise-free, and others that are extremely noisy – such a separation into very good or very bad channels is called "channel polarization", giving the codes their name. From a theoretical point of view, polar codes are the first codes that are actually proven to achieve the capacity of binary symmetric channels. From a practical point of view, polar codes also have a relatively simple and explicit code construction, which has complexity $N\log(N)$, and there also exists a rather simple decoding method (*Successive Cancellation Decoding* (SCD)) that achieves good (though not optimal) performance. In the following, we will describe their implementation in particular for AWGN channels.

13.8.2   Code Construction and Encoding Process

First of all, we have to note that while the *code construction method* for polar codes is independent of the operating SNR, the *actual codes* are strongly dependent. Choosing a good design SNR is thus very important. We find that at a particular SNR, channel polarization results in a percentage p of good channels. We then choose our coderate such that $K/N < p$, i.e., that K out of N bits can be reliably transmitted. Remarkably, instead of having to construct redundant bits from the information bits, we rather just have to set the remaining $N - K$ bits to some fixed (frozen) value, and for convenience, we set them all to 0.

The basis of polar codes is the polar transform. To map two input bits onto two output symbols, consider the matrix

$$\mathbf{F}^{(1)} = \begin{pmatrix} 1 & 1 \\ 0 & 1 \end{pmatrix} \tag{13.94}$$

which is strongly related to the Walsh-Hadamard matrix that we will encounter in Section 19.3.2. This means that if we have the input bits u_1 and u_2, the output of the transform is $u_1 + u_2$ and u_2, where all additions are modulo 2. Note that – keeping with the most widely used notation – for polar codes we write codewords into column vectors, not row vectors as in Section 13.2.

Larger polar transforms are created out of Kronecker products of this fundamental matrix

$$\mathbf{F}^{(2)} = \mathbf{F}^{(1)} \otimes \mathbf{F}^{(1)} = \begin{pmatrix} \mathbf{F}^{(1)} & \mathbf{F}^{(1)} \\ 0 & \mathbf{F}^{(1)} \end{pmatrix} \tag{13.95}$$

where $\otimes$ denotes the Kronecker product, and generally $\mathbf{F}^{(n+1)} = \mathbf{F}^{(n)} \otimes \mathbf{F}^{(1)}$. Consider now a set of K input bits u_i, $i \in \mathcal{I}$, where $\mathcal{I}$ is the set of indices in the codeword that carry information bits (the rest of the codeword entries carries frozen bits). Thus, define

$$d_i = \begin{cases} u_i & i \in \mathcal{I} \\ 0 & \text{otherwise} \end{cases} \tag{13.96}$$

and

$$\mathbf{x} = \mathbf{F}^{(n)}\mathbf{d}. \tag{13.97}$$

Note that the transform from $\mathbf{d}$ to $\mathbf{x}$ can be implemented efficiently: instead of a matrix multiplication (which would require N^2) operations, the Kronecker structure of $\mathbf{F}$ allows to implement it as a butterfly structure (similar to an FFT), as shown in Figure 13.25.

It is now important to explore the choice of the set $\mathcal{I}$. Let the design SNR be $R_c E_B/N_0 = E_c/N_0$. Then define the Bhattacharyya parameter, which assuming BPSK is

$$z = \exp\left(-E_c/N_0\right). \tag{13.98}$$

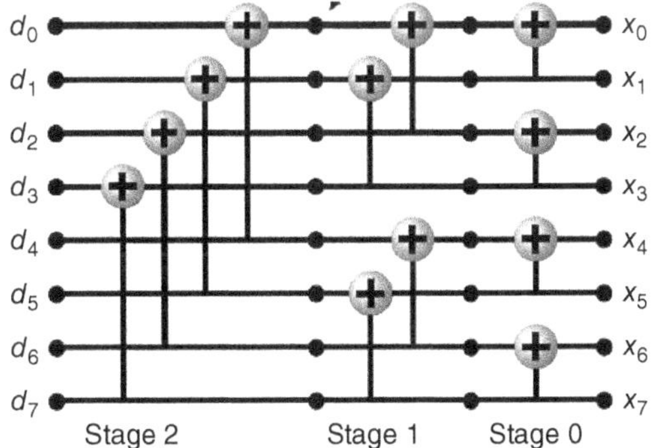

Figure 13.25 Butterfly implementation of an 8×8 polar transform. Reproduced with permission from [Vangala 2017] © H Vangala.

From this, we can construct, via the recursion structure in Figure 13.26, a set of N output parameters – every node splits into two, where to the upper leaf node the mapping $z \rightarrow (2z - z^2)$ is applied, and to the lower leaf node $z \rightarrow z^2$. Among the N leaf nodes found in the end, find the K nodes that hold the least values, and write their indices into a set $\mathcal{J}$. The indices of nodes that carry information

Figure 13.26 Recursive computation of the node reliability from the Bhattacharyya parameter. Reproduced with permission from [Vangala 2017] © H Vangala.

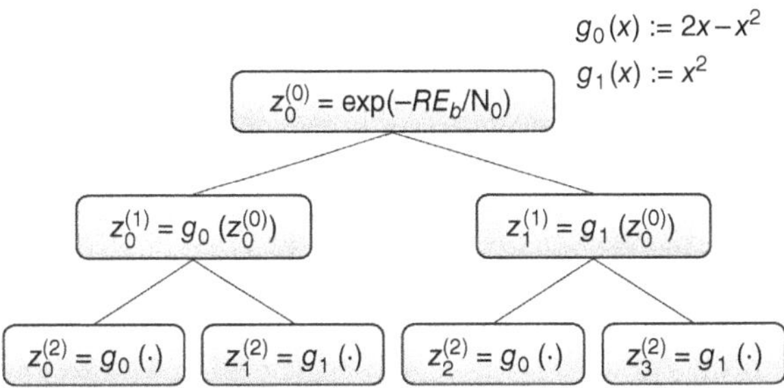

bits (i.e., members of the set $\mathcal{I}$) are then the bit-reversed versions of the indices in set $\mathcal{J}$. Let us consider an example: the bit-reversed numbers counting from 7 down to 0 are

$$111,\ 011,\ 101,\ 001,\ 110,\ 010, 100,\ 000 = 7, 3, 5, 1, 6, 2, 4, 0.$$

Assuming that $\mathcal{J} = \{3, 4, 5, 6, 7\}$, the set $\mathcal{I} = \{1, 3, 5, 6, 7\}$. Note that different operating SNRs, and different code rates, result in different sets of bits that have a predetermined (frozen) state, and thus different codes.

13.8.3 Decoding

An appealing feature of polar codes is that a simple algorithm exists for decoding, the SCD, which furthermore has a fixed complexity. While it is not optimum, it can be useful either by itself for low-complexity decoding, or to form the basis for better-performing but more complex algorithms. The key to the decoding algorithm is the elementary mapping of the likelihood functions (see Figure 13.27)

$$\begin{pmatrix} L_1 \\ L_2 \end{pmatrix} \rightarrow \begin{pmatrix} f(L_1, L_2) \\ g(L_1, L_2) \end{pmatrix} = \begin{pmatrix} \dfrac{L_1 L_2 + 1}{L_1 + L_2} \\ \begin{cases} L_2 L_1 & d_{\text{upper branch}} = 0 \\ L_2/L_1 & d_{\text{upper branch}} = 1 \end{cases} \end{pmatrix}. \tag{13.99}$$

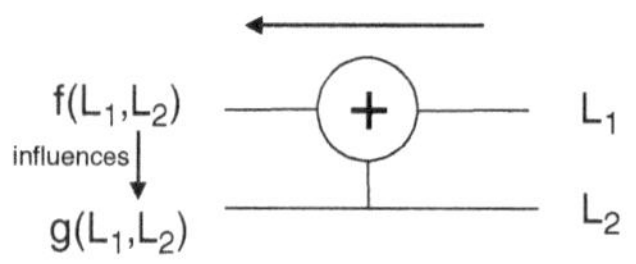

Figure 13.27 Building block for SCD decoder for polar codes: likelihood computations.
Reproduced with permission from [Vangala 2017] © H Vangala.

This implies that based on the likelihoods, we can compute the function f, i.e., the likelihood at the upper branch. However, the function g, i.e., the likelihood at the lower branch requires knowledge of the actual bit decision occurring at the upper branch. Fortunately, an efficient recursive computation is possible, where likelihoods are propagated from the right (observation nodes) to the left (bit decision nodes), and bits are propagated from the left to the right. Note that bit decisions on the left are only effective if the bit in question is one of the source bits; if it is a frozen bit the arriving information is ignored and the bit is always set to the frozen state 0.

Example 13.10 Consider as an example the 8×8 decoding case in Figure 13.28a). We call the leftmost layer the layer 0, and the rightmost layer $\log_2 N = 3$; and we write a function $f(L_i, L_j)$ that takes the nodes i (upper node) and j (lower node) at layer $n + 1$ as input and puts the output into layer n as $f_i^{(n)}\left(L_i^{(n+1)}, L_j^{(n+1)}\right)$, and similarly for $g_j^{(n)}\left(L_i^{(n+1)}, L_j^{(n+1)}\right)$. We now proceed in the following steps

- Figure 13.28b: We start out on the right (layer 3), with the likelihood ratios obtained from the observations and the knowledge of the SNRs. Since we do not know any of the bit decisions yet, we can only compute f-functions. Thus, propagating one layer to the left (layer 2), we get likelihoods at nodes 0, 2, 4, 6, namely $f_0^{(2)}\left(L_0^{(3)}, L_1^{(3)}\right)$, $f_2^{(2)}\left(L_2^{(3)}, L_3^{(3)}\right)$, $f_4^{(2)}\left(L_4^{(3)}, L_5^{(3)}\right)$, $f_6^{(2)}\left(L_6^{(3)}, L_7^{(3)}\right)$. Propagating further to the left gives $f_0^{(1)}\left(L_0^{(2)}, L_2^{(2)}\right)$, $f_4^{(1)}\left(L_4^{(2)}, L_6^{(2)}\right)$ and finally on the leftmost layer (layer 0) only a single likelihood value $f_0^{(0)}\left(L_0^{(1)}, L_4^{(1)}\right)$, is computed for the bit with index 0. Now for this bit, u_0, we can make a hard decision based on the single available likelihood value (if this bit is in the set of frozen bits, then the value is known a priori and no decision is necessary).

- Figure 13.28c: With the knowledge of u_0, $g_4^{(0)}\left(L_0^{(1)}, L_4^{(1)}\right)$ can be computed, because the $d_{\text{upper branch}} = d_0^{(0)} = u_0$ is known. Therefore, a hard decision on $u_4 = d_4^{(0)}$ can be computed.

- Figure 13.28d: This bit decision can be propagated to the right. Then $d_0^{(1)} = d_0^{(0)} + d_4^{(0)}$ and $d_4^{(1)} = d_4^{(0)}$ can be computed. Using these, we can now compute $g_2^{(1)}\left(L_0^{(2)}, L_2^{(2)}\right)$ and $g_6^{(1)}\left(L_4^{(2)}, L_6^{(2)}\right)$ (note that $L_{0,2,4,6}^{(2)}$ were already computed in the previous round). From these in turn $f_2^{(0)}\left(L_2^{(1)}, L_6^{(1)}\right)$ can be computed, based on which $u_2 = d_2^{(0)}$ can be decided.

- Figure 13.28e: With this, $g_6^{(0)}\left(L_2^{(1)}, L_6^{(1)}\right)$ can be computed, and $u_6 = d_6^{(0)}$ decided.

- Figure 13.28f: With the knowledge of $d_2^{(0)}, d_6^{(0)}$, we can now compute $d_2^{(1)} = d_2^{(0)} + d_6^{(0)}$ and $d_6^{(1)} = d_6^{(0)}$, and consequently $d_0^{(2)} = d_0^{(1)} + d_2^{(1)}$, $d_2^{(2)} = d_2^{(1)}$, $d_4^{(2)} = d_4^{(1)} + d_6^{(1)}$, and $d_6^{(2)} = d_6^{(1)}$. We then find $g_1^{(2)}\left(L_0^{(3)}, L_1^{(3)}\right)$, $g_3^{(2)}\left(L_2^{(3)}, L_3^{(3)}\right)$, $g_5^{(2)}\left(L_4^{(3)}, L_5^{(3)}\right)$, $g_7^{(2)}\left(L_6^{(3)}, L_7^{(3)}\right)$, from which $f_1^{(1)}\left(L_1^{(2)}, L_3^{(2)}\right)$, $f_5^{(1)}\left(L_5^{(2)}, L_7^{(2)}\right)$, and finally $f_1^{(0)}\left(L_1^{(1)}, L_5^{(1)}\right)$ follows, based on which u_1 is decided.

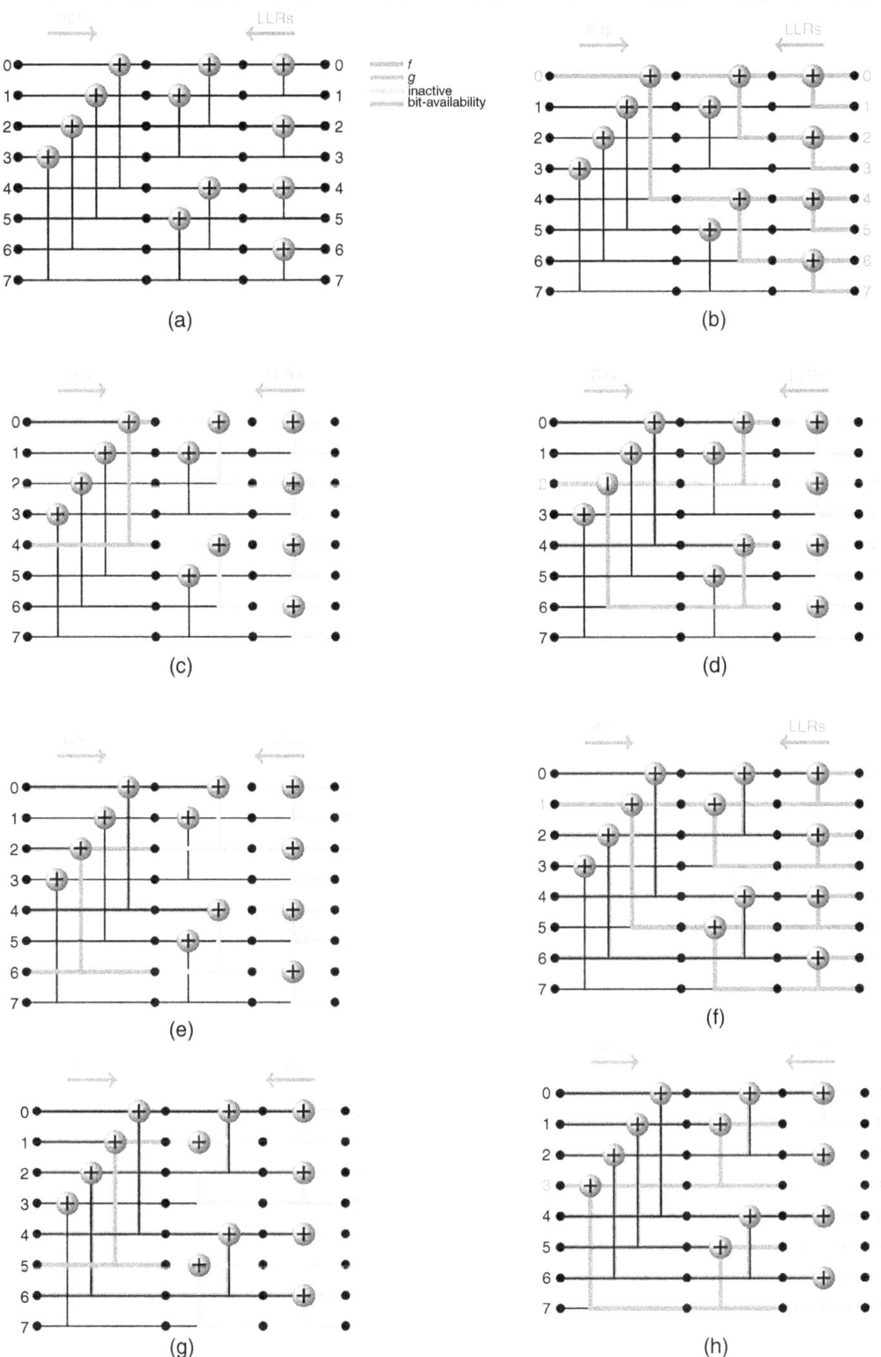

Figure 13.28 Decoding of a polar code by successive cancellation decoding; see text for the description of the steps (a)–(h). Color version available at wiley.com/go/molisch/wireless3e.

Reproduced with permission from [Vangala et al 2016b] © H Vangala.

- Figure 13.28g: Based on u_1, $g_5^{(0)}\left(L_1^{(1)}, L_5^{(1)}\right)$ is computed and therefore u_5 is decided.

- Figure 13.28h: Propagating this bit to the right (i.e., computing $d_1^{(1)}$ and $d_5^{(1)}$), $g_3^{(1)}\left(L_1^{(2)}, L_3^{(2)}\right)$ and $g_7^{(1)}\left(L_5^{(2)}, L_7^{(2)}\right)$, $f_3^{(0)}\left(L_3^{(1)}, L_7^{(1)}\right)$, and u_3 are determined,

- Finally with u_3, $g_7^{(0)}\left(L_3^{(1)}, L_7^{(1)}\right)$ and thus u_7 are determined.

*13.9 Comparison of Capacity-Approaching Codes

We have in the previous sections discussed turbo-codes, LDPC codes, and polar codes. All of them are approaching capacity of AWGN channels, and so it is natural to wonder whether one is better than the other. Unfortunately, there is no simple "the best", as different codes might be optimum for different criteria (such as latency, decoding complexity, and block error rate), and for different block lengths.

In terms of decoding complexity, polar codes have an advantage in that a simple (complexity order $N\log(N)$) decoding algorithm with fixed complexity is available. Depending on the code rate, this algorithm is within 1–3 dB of the results achievable with optimum decoding. When higher complexity decoding is used for polar codes, the performance becomes more similar to LDPC and turbo codes, but at the price that the decoding complexity also becomes comparable. LDPC codes tend to require somewhat smaller computational effort than turbo codes, though this depends very much on the specific implementation of the codes (e.g., LDPC codes with relatively short cycles might need a lot of iterations to converge, or might not converge at all) as well as of the decoding algorithms. Finally, Figures 13.29 and 13.30 show the performance of these codes at small ($K = 256$) and large ($K = 8192$) block length Note that the

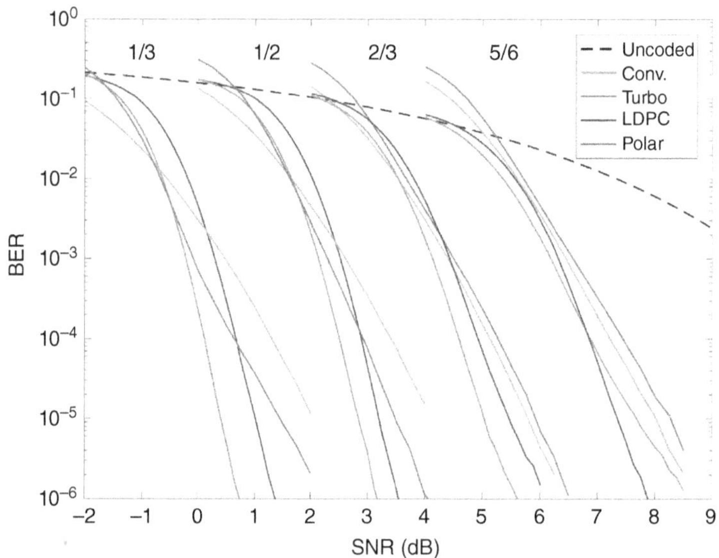

Figure 13.29 Comparison of convolutional, turbo, LDPC, and polar codes at different code rates with source block rate $K = 256$ in AWGN with BPSK modulation. Color version available at wiley.com/go/molisch/wireless3e.

Reproduced with permission from [Tahir et al. 2017] © IEEE.

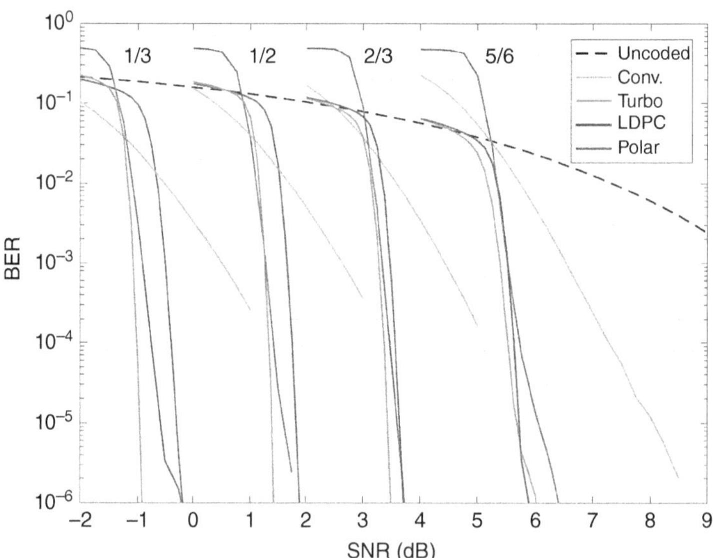

Figure 13.30 Comparison of convolutional, turbo, LDPC, and polar codes at different code rates with source block size $K = 8192$ in AWGN with BPSK modulation. Color version available at wiley.com/go/molisch/wireless3e.

Reproduced with permission from [Tahir et al. 2017] © IEEE.

comparison between LDPC and turbo codes depends on the particular code construction, though both approximate capacity. Polar codes perform similar to turbo and LDPC codes for large block sizes (note that in these figures the polar codes use an improved decoding algorithm with higher complexity, compared to what was discussed in Section 13.8) but generally perform comparatively better at low block lengths (e.g., $K = 64$).

It is noteworthy that different codes are used in different wireless standards. For example, LTE uses turbo codes, while NR and Wi-Fi use (mostly) LDPC codes, and NR uses polar codes for short blocks of control signals. Some of the reasons are performance, technological (certain coding/decoding algorithms are easier to implement in silicon, and/or chip manufacturers have experience in optimizing particular algorithms), while some are based on ownership of patents.

13.10 Coding for the Fading Channel

The structure of errors in a fading channel is different from that in an AWGN channel. The presence of error bursts is noticeable: when the channel shows (instantaneous) high attenuation, e.g., due to the destructive interference of multi-path components, then the error probability is much larger than in constructive interference. Correction of these error bursts can be achieved either by using codes that are especially suited for bursty errors; or, alternatively, interleaving "breaks up" the bursty structure of errors.

13.10.1 Interleaving

Compared with a symbol duration, wireless channels change slowly. Typically, a link between a Base Station (BS) and a User Equipment (UE) has a coherence time of 10–100 ms. As a consequence, neighboring bits are strongly correlated, and in particular, if the UE is in a fading dip, a large number of adjacent bits are strongly attenuated (and thus more susceptible to errors by noise). A normal code usually cannot correct such a large number of errors. To understand the basic principle, consider a simple rate-1/3 repetition code. The probability of a wrong decision in a coded bit is $P_{single} \sim Q(\sqrt{\gamma})$, where γ is the *instantaneous* SNR. Due to the majority rule, the probability of a wrong final decision is approximately $P_{single}^2 \sim Q^2(\sqrt{\gamma})$, where we have made use of the fact that the SNR of two subsequent bits is the same. Averaged over different channel states, the BER is then approximately:

$$BER \sim \int_0^\infty pdf_\gamma(\gamma) Q^2(\sqrt{\gamma}) d\gamma. \tag{13.100}$$

When an interleaver is used, then the three bits associated with one source bit are transmitted at large intervals from each other, so that the SNR for each of these transmissions is different (see Figure 13.31). Therefore, an error occurs only if two of these *independent* transmissions are all in (independent) fading dips, which is very unlikely (see Figure 13.32). Mathematically, this can be written as

$$BER \sim \int_0^\infty \int_0^\infty pdf_{\gamma_1}(\gamma_1) Q(\sqrt{\gamma_1}) pdf_{\gamma_2}(\gamma_2) Q(\sqrt{\gamma_2}) d\gamma_1 \, d\gamma_2$$

$$= \left(\int_0^\infty pdf_\gamma(\gamma) Q(\sqrt{\gamma}) d\gamma \right)^2. \tag{13.101}$$

We stress that an interleaver can reduce the mean BER only in combination with a codec. For an uncoded system, the interleaver still breaks up error bursts (which sometimes can be desirable), but it does not lead to a decrease in the mean BER. Another problem

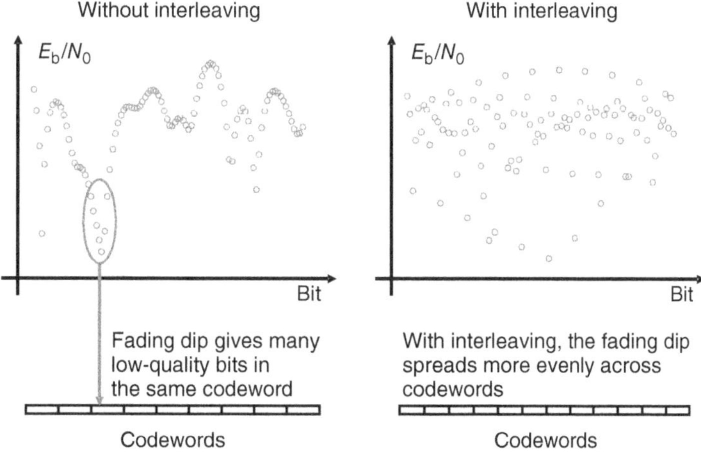

Figure 13.31 Effect of an interleaver.

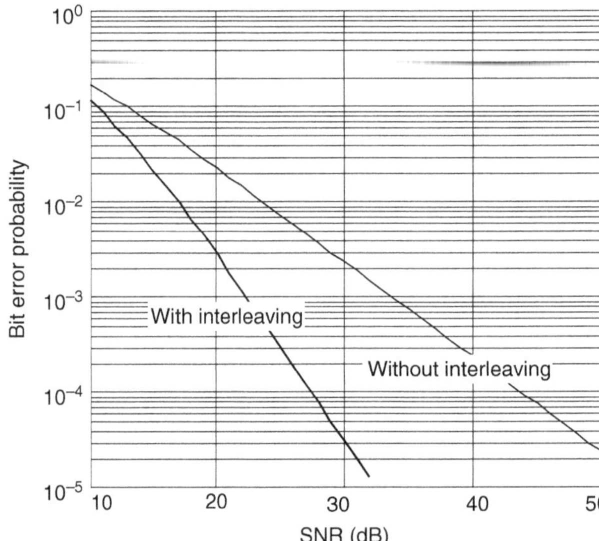

Figure 13.32 Bit error rate of a rate-1/3 repetition code with hard majority decision in a flat Rayleigh-fading channel, with and without interleaving.

with interleavers lies in the fact that they increase latency of transmission – decoding is only possible after the interleaved bits of a codeblock have all been received. Conversely, the maximum admissible latency places an upper limit on the interleaver depth. Thus, if the maximum latency is smaller than the duration of a fading dip, the effectiveness of the interleaver is greatly reduced (interleaving in the frequency domain is discussed in Chapter 15).

After these basic considerations, we now turn to the structure of the interleaver. Basically, two structures are possible: block interleaving and convolutional interleaving. For a block interleaver (Figure 13.33), a block of data of size $N_{\mathrm{itnerleav}}\,D_{\mathrm{interleav}}$ is interleaved at once. The structure of the interleaver is a matrix: bits are read in line-by-line and read out column-by-column. The bits that are originally adjacent are now separated by $D_{\mathrm{interleave}}$. Denoting the symbol duration as T_S, latency due to this interleaver is $N_{\mathrm{interleave}}\,D_{\mathrm{interleave}}\,T_S$ – it has to wait until the matrix has been filled up before it can read out the bits. Latency is thus considerably larger than the separation of bits that can be achieved.

For a convolutional interleaver, the data are interleaved in a continuous stream, similar to a convolutional encoder. This reduces latency compared with the block interleaver (for the same separation of bits). It is common to use block interleavers together with block codes, and convolutional interleavers together with convolutional codes.

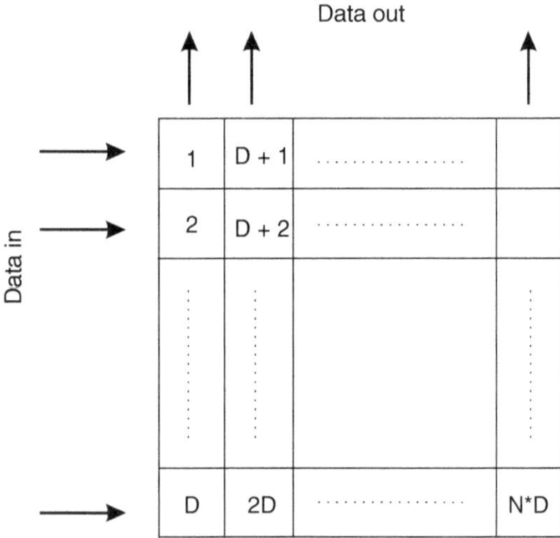

Figure 13.33 Structure of a block interleaver.

13.10.2 Block Codes and Convolutional Codes

We now apply the above-mentioned principles to block codes. In the absence of interleaving, the effectiveness of the code would be much reduced, since all bits of a codeword might fall into a deep fading dip. Due to interleaving, each of the bits (symbols) in a codeword fades independently. The redundancy implicit in the code should thus make it possible to recover the information even when some of the bits are in fading dips and therefore have a very poor SNR.

It is intuitive that in a block code with hard decoding, the achievable diversity order (see Chapter 12) is $\lfloor \frac{d_{min}-1}{2} \rfloor + 1$, where the term 1 reflects the diversity order of an uncoded transmission, and $\lfloor \frac{d_{min}-1}{2} \rfloor$ is the radius of the correction sphere (see Section 13.2.4).

Example 13.11 *Compare the diversity order with and without fading for a block code with hard decoding and $K = 12$, $N = 23$, $d_{min} = 7$, and coding gain 2 dB.*

Due to the slowness of fading, all 23 codebits see the same channel if there is no interleaving. In other words, each and every codeword has a gain of 2 dB. An uncoded system needs a 26-dB SNR to achieve a BER of 10^{-3}, with coding, this reduces to 24 dB – which is still bad. This demonstrates again that even with an encoder, the BER decreases only proportionally with the SNR per bit γ_B, just the proportionality constant changes.

The use of interleaving dramatically changes the situation. Errors occur only when there are errors at least at four locations in the codeword (note that $\lceil \frac{d_{min}-1}{2} \rceil = 3$ can be corrected). The BER is approximately proportional to [Wilson 1996]:

$$\sum_{i=4}^{23} K_i \left(\frac{1}{2 + 2\overline{\gamma}_B} \right)^i \left(1 - \frac{1}{2 + 2\overline{\gamma}_B} \right)^{23-i} \approx \frac{1}{\overline{\gamma}_B^4}. \tag{13.102}$$

where the K_i are constants. Thus, the required BER can be achieved with less than 10 dB SNR.

When interleaving together with soft decoding is used, the resulting diversity order is almost twice as large – namely, d_{min}. To put this into more mathematical terms, we write the metric that we need to minimize as

$$\min_{s} \sum_{i} |r_i - \alpha_i s_i|^2 \tag{13.103}$$

and the pairwise error probability becomes [Benedetto and Biglieri 1999]:

$$P(\mathbf{s} \rightarrow \mathbf{s}_E) \leq \prod_{i \in A} \frac{1}{1 + |s_i - s_{E,i}|^2 / (4N_0)} \tag{13.104}$$

where A is the set of indices so that $s_i \neq s_{E,i}$. For a linear code, the error probability becomes

$$P_e \leq \sum_{w \in W} N_w \left(\frac{1}{1 + R_c \gamma} \right)^w \tag{13.105}$$

where R_c is the code rate, N_w is the number of codewords with weight w, and W is the set of nonzero Hamming weights of the code. This shows clearly that the minimum distance determines the diversity order (slope of the BER versus SNR curve).

The performance measures and design rules for convolutional codes are quite similar to those of block codes. As a matter of fact, the mathematical description given above for block codes with soft decoding can directly be applied to convolutional codes.

These estimates of the error probability also show what properties of the code most impact performance of interleaved codes for fading channels. According to (13.102) and (13.105), the Euclidean distance of the code is no longer an important metric. Rather, we strive to find codes with good minimum (Hamming) distance properties. If the code sequences differ in many positions, then the probability is small that all of the distinguishing bits are in a fading dip simultaneously. Expressed in other words: we wish that two sequences that can be confused with each other (i.e., start and end in the same state of the trellis) are as long as possible. The minimum length of such sequences is often called the *effective length* of the code.

From these simple considerations, we can derive a few rules about code design for flat-fading channels:

- The most important parameter in the fading channel is effective length, which enters exponentially into the BER. Thus, maximization of the effective length is the most important design goal.
- The minimal Euclidean distance does not occur in the equations for the BER in Rayleigh fading channels, but it does dominate performance in AWGN channels, as shown in Section 13.3. Thus, the design criteria for AWGN channels and fading channels are quite different. Since Rayleigh fading and AWGN channels are limiting cases of the Rician channels maximization of Euclidean distance can be used as a secondary goal.

13.10.3 Concatenated Codes

An important class of codes for fading channels is the concatenation of convolutional codes with RS codes. Before 1995, this was actually the most popular code for cases where extremely high coding gains are required. Even today, this method is sometimes used for situations where high data rates or complexity restrictions prevent the use of turbo codes and LDPC codes.

RS codes are good at correcting bursts of bit errors – actually much better than at correcting distributed errors. It would seem that this makes them well suited for fading channels. However, for typical coherence times and bit rates, the length of the fades (in units of bits) in a wireless channel is typically much larger than the correction capability of an RS code. Thus, the way RS codes are applied is the following: first, the data are convolutionally encoded, and the resulting data stream is interleaved, to break up any possible error bursts (long interleavers are much easier to build than codes for long error bursts). While the convolutional code does not see long strings of severely impaired data (as it would without the interleaver), the decoding of a convolutional code results in error bursts, as this is a fundamental property of convolutional codes. These error bursts are then corrected by means of the outer code – namely, the RS code – which is well equipped to deal with these bursts.

*13.10.4 Trellis Coded Modulation in Fading Channels

For TCM, the same design criteria are valid as for convolutional codes. Figure 13.34 shows that the effectiveness of the codes can be very different in different channels. We show a code by [Ungerboeck 1982], designed for AWGN channels, and a code from [Jamali and Le-Ngoc 1991] designed for fading channels. The superior performance of each code in its "natural" environment is obvious.

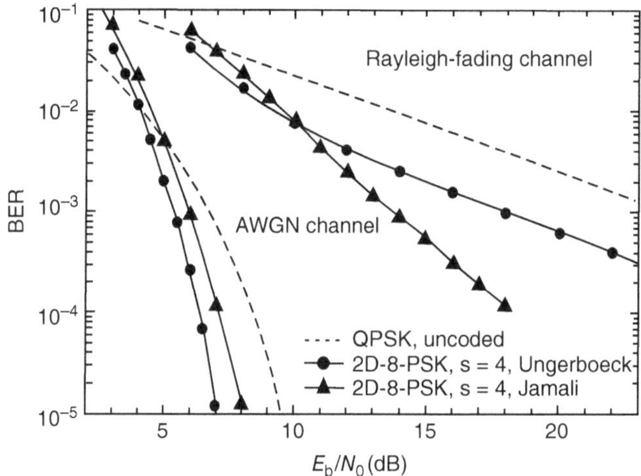

Figure 13.34 Bit error rate in a fading channel: uncoded quadrature-phase shift keying and 8-PSK trellis-coded modulation. Reproduced with permission from [Mayr 1996] © B. Mayer.

TCM is also very helpful in combating the errors due to delay dispersion that was discussed in Section 11.3. [Chen and Chuang 1998] showed that these errors are determined essentially by the Euclidean distance between the codewords. Minimization of this metric gives codes that can combat these types of errors (see Figure 13.35). While the search for codes for flat-fading channels is relatively mature, code designs for frequency-selective channels are more recent.

13.11 Information-Theoretic Performance Limits of Fading Channels

After discussing the performance of practical codes in fading channels, we now analyze the information-theoretic limits in fading channels. We will consider for the most part frequency-flat fading channels, which can be described by a time-variant SNR γ. Only in the last subsection will we turn to frequency-selective channels.

13.11.1 Ergodic Capacity vs. Outage Capacity

We first reiterate two critical facts about AWGN channels: the normalized capacity of a (complex) channel is given by

$$C(\gamma) = \log_2[1 + \gamma] \ \text{bits/s/Hz} \tag{13.106}$$

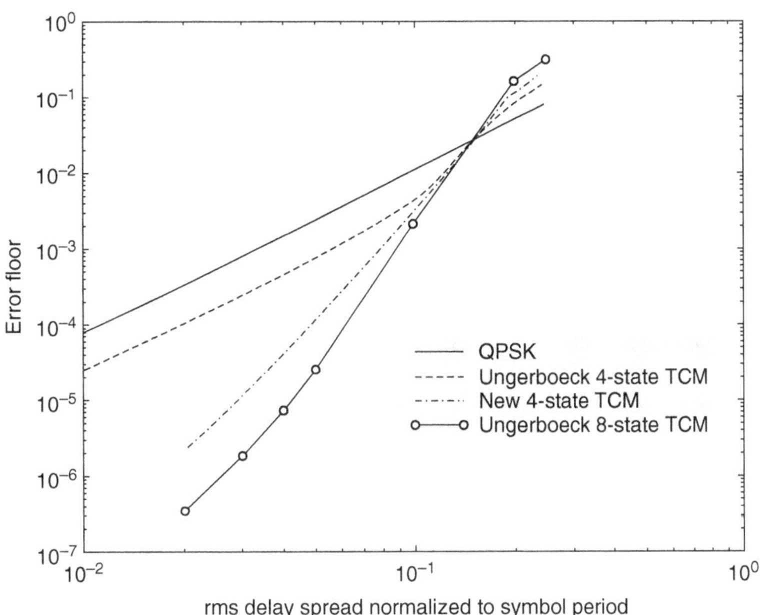

Figure 13.35　Error floor of trellis-coded modulation in frequency-selective channels.
Reproduced with permission from [Chen and Chuang 1998] © IEEE.

and this capacity is derived under the assumption of codes with infinite codeword length.

For time-varying (fading) channels, different definitions of capacity exist, depending on the assumptions about channel state information CSI, as well as the SNR distribution that a codeword "sees".

First and foremost, the *ergodic (Shannon) capacity* in a fading channel is computed under the assumption that a codeword extends over all possible channel realizations; under some assumptions about CSI, it becomes simply the expected value of the capacity, taken over all realizations of the channel. We note that in the definitions of information theory, only the ergodic capacity is a true capacity, while other quantities we discuss next are also called "achievable rates". However, among wireless researchers, a more relaxed use of the word capacity is common, and we will use it as well in the remainder of the book.

In practical situations, we are often faced with the situation where the data are encoded with a near-Shannon-limit-achieving code whose codewords "see" only a single channel realization. More precisely, a codeword extends over a period that is much shorter than the channel coherence time. For example, LDPC codes with reasonable block lengths (e.g., 10,000 bits) get close to the Shannon limit (see Section 13.7), and for a data rate of 10 Mbit/s, such a block can be transmitted within 1 ms. This is much shorter than 10 ms, which is a typical coherence time of wireless channels (see Chapter 5). Thus, each channel realization can be associated with a capacity value. This capacity (sometimes called *instantaneous capacity*) is a random variable (r.v.) with an associated Cumulative Distribution Function (cdf). Thus henceforth we will look at this distribution function, or equivalently the capacity that can be guaranteed for $x\%$ of all channel realizations. The latter quantity is also known *outage capacity*.

Since the channel is time varying, it can also be meaningful to let the transmit power vary with time. Whether, and to what extent, this can be done depends on the amount of Channel State Information at the Transmitter (CSIT), i.e., what the Transmitter (TX) knows about the SNR at the RX, as discussed in the sequel.

13.11.2　Capacity for Channel State Information at the Receiver (CSIR) Only

Without CSIT the TX cannot adapt its transmit power to the channel state, so the best thing it can do is transmit with a constant power. The Shannon capacity (per unit bandwidth) is then the AWGN capacity averaged over the SNR distribution, i.e.,

$$E\{C(\gamma)\} = \int_0^\infty \log_2[1 + \gamma]pdf_\gamma(\gamma)d\gamma \text{ bits/s/Hz.} \tag{13.107}$$

From Jensen's inequality, it follows immediately that $E\{C(\gamma)\} < C(E\{\gamma\})$; in other words, the Shannon capacity in a fading channel is smaller than the capacity in an AWGN channel with the same mean SNR.

We now turn to the computation of the outage capacity, or capacity cdf. The mapping $\gamma \rightarrow \log_2(1 + \gamma)$ constitutes a transformation between two random variables, so that we can derive the pdf of C using transformation of variables with the Jacobian, resulting in

$$pdf_C(C) = pdf_\gamma(2^C - 1)\ln(2)2^C \tag{13.108}$$

For example, in a Rayleigh-fading channel,

$$pdf_C(C) = \frac{1}{\overline{\gamma}} \exp\left[-\frac{2^C - 1}{\overline{\gamma}}\right] \ln(2) 2^C \tag{13.109}$$

and the resulting cdf is

$$cdf_C(C) = 1 - \exp\left[-\frac{2^C - 1}{\overline{\gamma}}\right]. \tag{13.110}$$

In the case that the admissible outage is given, we can compute the corresponding γ from $cdf_\gamma(\gamma)$, and insert this value into the capacity equation to obtain the outage capacity.

From (13.109), we see that there are channel constellations at which the capacity vanishes. Thus, it is not possible to guarantee a zero-outage transmission.

13.11.3 Capacity for CSIT and CSIR – Waterfilling

When the TX knows the channel state, it can adapt its power (subject to a long-term power constraint) to maximize the channel capacity. Thus, the ergodic capacity is given by

$$C = \max_{P(\gamma)} \int_0^\infty \log_2\left[1 + \gamma \frac{P(\gamma)}{\overline{P}}\right] pdf_\gamma(\gamma) d\gamma \tag{13.111}$$

with the constraint

$$\int_0^\infty P(\gamma) pdf_\gamma(\gamma) d\gamma \le \overline{P}. \tag{13.112}$$

This optimization problem can be solved by

$$\frac{P(\gamma)}{\overline{P}} = \begin{cases} \dfrac{1}{\gamma_0} - \dfrac{1}{\gamma} & \gamma > \gamma_0 \\ 0 & \text{otherwise} \end{cases} \tag{13.113}$$

where γ_0 is determined by inserting Eq. (13.113) into the constraint Eq. (13.112). This power allocation strategy is known as "waterfilling"; we will discuss it in more detail in Section 15.8. Essentially, it means that the TX does not transmit if the channel takes on a very bad realization (SNR is below the threshold γ_0). If the SNR is above the threshold, then the TX will increase the transmission power with increasing SNR.

One might try to improve the outage capacity by letting the transmit power compensate for the losses of the RX, i.e., $P(\gamma)/\overline{P} = K_{\text{inv}}/\gamma$ where, K_{inv} is a proportionality constant that is determined from Eq. (13.112), so that $K_{\text{inv}} = 1/E\{1/\gamma\}$. This approach results in an instantaneous capacity $C_{\text{inv}} = \log_2(1 + K_{\text{inv}})$ that is independent of γ; therefore, there is no "outage," i.e., the capacity never falls below C_{inv}; The value C_{inv} is therefore also known as the "zero-outage capacity." Note that in Rayleigh fading, $E\{1/\gamma\} \to \infty$, so that $C_{\text{inv}} \to 0$. When we allow outage in $x\%$ of all cases, then the best strategy for maximizing outage capacity is not to transmit at all for the $x\%$ worst channels, and for the remainder of channel, realizations control the transmit power in such a way that the SNR is a constant.

Last, but not least, we find that in the case of diversity reception, the above equations remain true, with γ representing the SNR at the output of the diversity combiner (see Section 12.4).

13.12 Automatic Repeat Request

While forward error correction is *in principle* capable of eliminating all errors, this cannot be achieved *in practice* – most importantly, the length of the codewords is limited, and therefore a finite error probability has to be anticipated. In typical situations, the remaining BER is on the order of $10^{-3} - 10^{-6}$, which is obviously much better than what can be achieved in the uncoded case, but still not sufficient for certain applications, e.g., financial transactions. Alternatively, one can consider the packet error rate: since typical packet lengths are between $10^2 - 10^4$ bits, packet error rates are around $10^{-3} - 10^{-1}$.

The practical solution for detecting such packet errors is the addition of error *detection* (parity checks) in addition to the error correction with Forward Error Correction (FEC). And when the RX determines that a codeword was received with errors, it provides feedback to the TX, and the transmission is repeated. This procedure, which is known as *backward error correction* (also known as *Automatic Repeat Request (ARQ)*) is simple, yet surprisingly efficient. Still, there are some restrictions that have to be kept in mind.

- ARQ leads to an increase in the latency of data transmission. The feedback on whether the transmission was successful takes time, and if a retransmission is required, even more time is consumed. The resulting delay is acceptable for most data applications (e.g., web browsing, and even video or audio streaming), but not for speech communications and communications for control. For this reason, ARQ is widely used in data communications systems like LTE and Wi-Fi and much of 5G, but not for the low-latency communications in 5G, and some speech communications systems.
- ARQ is efficient only when the average packet error rate is reasonably small, say, 5% or less. Otherwise, most of the time is spent on retransmission of packets. Thus, ARQ is almost always combined with error-correcting codes.
- ARQ requires the presence of an error-detection code. This can be achieved either with a simple block code (cyclic redundancy check) or can be implicit in an error-correcting code (e.g., decoding of an LDPC code establishes whether the output is a valid codeword or not).
- The feedback loop has to be well protected against errors.
- The retransmission is advantageously done over a different channel, or with increased power. For example, a retransmission can occur after a time that is larger than the coherence time of the channel, or it can be done on a different frequency. If retransmission does not change any of the parameters, the chance is high that it will also fail. ARQ is similar in spirit to selection diversity, and the retransmission process should make sure that the different "diversity branches" have low correlation.

ARQ is a suboptimum scheme, because it discards the information in a packet in which an error is detected, and just tries a retransmission. A better way is for a RX to store and exploit *all* the information it gets. Even if the information received during one transmission attempt is not sufficient for successful decoding, it can still help if combined with information from a second transmission attempt. In other words, two or more received signals that each have insufficient SNR for correct decoding can sometimes be combined in such a way that the total signal can be decoded. This approach is called *hybrid Automatic Repeat Request (HARQ)*. Different types of HARQ are distinguished in the way that the signals are combined.

In *Chase combining*, repetitions of the same signal are added up according to the principle of maximum-ratio combining. This implies that the SNRs of the repetitions add up, see Section 12.4. An alternative is *incremental redundancy,* which transmits additional parity-check bits during retransmission. Take, for example, a rate 1/3 convolutional code that is punctured to a rate 2/3 code. During the first transmission, the 2/3-rate coded bits are transmitted. If the SNR at the RX is bad, it requests a retransmission, and this retransmission now sends the bits that were originally punctured out. Thus, after the retransmission, the RX can attempt to decode the rate 1/3 code. This will be more often successful than the decoding of a rate 2/3 code with increased SNR (which is what we would get if the RX did Chase combining).

Another way of implementing incremental redundance is the transmission of so-called *rateless codes*, of which *Fountain codes* are a well-known representative. The basic principle is that a fixed-length message (length K) is encoded into a very long data stream, i.e., the encoder generates constantly new parity check bits. For an erasure channel, the codes have the property that any set of nonerased received bits of size (slightly larger than) K is sufficient for the RX to decode, and recover the message. For an AWGN channel, decoding becomes possible when the received information at the RX is larger than the entropy of the message. When the RX has decoded the message, it sends, on a feedback channel, a brief notification to the TX, so that the TX can stop encoding and transmitting this message word. Note that HARQ that transmits different parity bits, as described above, can be seen as a quantized version of the rateless transmission.

Generally speaking, Chase combining allows a capacity $C = \log(1 + \sum \gamma_i)$, while incremental redundancy provides $C = \sum \log(1 + \gamma_i)$, with i being the index of transmission time. From Jensen's inequality, it is clear that the latter quantity is larger. A more detailed discussion of HARQ will be given in the context of LTE and NR, see Sections 31.5.2 and 32.5.2.

Further Reading

More details on channel coding can be found in the many excellent textbooks on this topic, e.g., [Lin and Costello 2004, MacKay 2002, Moon 2020, McEliece 2004, Sweeney 2002, Wilson 1996]. [Steele and Hanzo 1999, ch. 4] give a detailed description of block codes, including the decoding algorithms for RS codes, while [Sklar and Harris 2004] give an intuitive introduction; further details can also be found in [Lin and Costello 2004]. Convolutional codes are detailed in [Johannesson and Zigangirov 1999]. The original Viterbi algorithm was published in [Viterbi 1967]. Soft decoding of convolutional codes was explored by Bahl et al. in the 1970s [Bahl et al. 1974] and [Hagenauer and Hoeher 1989]. An excellent tutorial description of the BCJR algorithm, which inspired some of the explanations in this chapter, is [Abrantes 2004]. These works also form the basis for turbo decoding, which was introduced in the seminal paper [Berrou and Glavieux 1996]. More details about turbo codes can be found in [Schlegel and Perez 2003], and also in the excellent tutorial of [Sklar 1997].

TCM was first invented in the early 1980s by [Ungerboeck 1982]; excellent tutorial expositions can be found in [Biglieri 1991]. BICM is described in the tutorial booklet [Fabregas et al. 2008]. LDPC codes originally proposed by [Gallagher 1961], were rediscovered by [MacKay and Neal 1997], and described in a very understandable way in [MacKay 2002]. The important improvement of irregular LDPC codes was proposed by [Richardson et al. 2001]. [Richardson and Urbanke 2008] give a detailed description of LDPC codes.

Polar codes were introduced by [Arikan 2009]; analysis in AWGN channels are, e.g., in [Vangala et al. 2015]. An excellent tutorial is [Vangala et al. 2016b], which also provides MATLAB code. Fundamentals of information theory are described, e.g., in [Cover and Thomas 2006, Yeung 2006].

For updates and errata for this chapter, see https://wides.usc.edu/students.html#textbooks.

Exercises

See Sec. 36.13 of Exercises.pdf at wiley.com/go/molisch/wireless3e

14

Equalizers

14.1 Introduction

14.1.1 Equalization in the Time Domain and Frequency Domain

Wireless channels can exhibit delay dispersion – i.e., Multi-Path Components (MPCs) can have different runtimes from the Transmitter (TX) to the Receiver (RX) (see Chapters 6 and 7). Delay dispersion leads to InterSymbol Interference (ISI), which can greatly disturb the transmission of digital signals. We already saw in Section 11.3 that even a delay spread that is smaller than the symbol duration can lead to a considerable Bit Error Rate (BER) degradation. If the delay spread becomes comparable with or larger than the symbol duration, as happens often in wireless systems with high data rate, then the BER becomes unacceptably large if no countermeasures are taken. Coding and diversity can reduce, but not completely eliminate, errors due to ISI (Chapters 12 and 13). On the other hand, delay dispersion can also be a positive effect. Since fading of the different resolvable MPCs is statistically independent, resolvable MPCs can be interpreted as diversity paths. Delay dispersion thus offers the possibility of *delay diversity*, if the RX can separate, and exploit, the resolvable MPCs. Using the fact that the transfer function is the Fourier transform of the impulse response with the Fourier transform pair $\tau \to f$, delay diversity can also be interpreted as frequency diversity (see Section 12.2).

Equalizers are RX structures that work both ways: they reduce or eliminate ISI, and at the same time exploit the delay diversity inherent in the channel. The operational principle of an equalizer can be visualized either in the time domain or the frequency domain.

For an interpretation in the frequency domain, remember that delay dispersion corresponds to frequency selectivity. In other words, ISI arises from the fact that the transfer function is not constant over the considered system bandwidth. The goal of an equalizer is thus to reverse distortions by the channel. In other words, the product of the transfer functions of channel and equalizer should be constant.[1] This can be expressed mathematically the following way: let the original signal be $s(t)$; it is sent through a (quasi-static) wireless channel with the impulse response $h(t)$, received, and sent through a receiver (containing an equalizer) with impulse response $\widetilde{e}(t)$. We furthermore assume that the transmit signal uses Pulse Amplitude Modulation, PAM (see Section 10.2), so that

$$s(t) = \sum_i c_i g(t - iT) \tag{14.1}$$

where the c_i are the complex transmit symbols. We now require that

$$H(\omega)G(\omega)\widetilde{E}(\omega) = \text{const} \tag{14.2}$$

over the bandwidth of interest, where $\widetilde{E}(\omega)$, $H(\omega)$, and $G(\omega)$ are the Fourier transforms of $\widetilde{e}(t)$, $h(t)$, and $g(t)$, respectively.

An equivalent formulation in the *time domain* requires that the received signal be free of ISI at the sampling instants. Define $\eta(t)$ as the convolution of the channel impulse response $h(t)$ with the basis pulse $g(t)$:

$$\eta(t) = h(t) * g(t). \tag{14.3}$$

Then we require that

$$[\eta(t) * \widetilde{e}(t)]_{t = iT_s} = \begin{cases} 1 & i = 0 \\ 0 & \text{otherwise} \end{cases}. \tag{14.4}$$

---

[1] Actually, this is a special form of equalizer, the so-called "Zero-Forcing" (ZF) linear equalizer, that will be discussed below in more detail.

Wireless Communications: From Fundamentals to Beyond 5G, Third Edition. Andreas F. Molisch.
© 2023 John Wiley & Sons Ltd. Published 2023 by John Wiley & Sons Ltd.
Companion website: www.wiley.com/go/molisch/wireless3e

If the channel were known and static, we could build (in hardware) a filter that performs the required equalizations of the transfer function. In wireless communications, however, the channel is (i) unknown and (ii) time variant. The former problem can be solved by transmission of a *training sequence* (also known as pilot signal) – i.c., a sequence of known bits. From the received signal $r(t)$ and knowledge of the shape of the basis pulses $g(t)$, the RX can estimate the channel impulse response $h(t)$. The problem of time variance is solved by repeating the transmission of the training sequence at "sufficiently short" time intervals, so that the equalizer can be adapted to the channel state at regular intervals. The concept is thus known as *adaptive equalization*.

Over the years, many different types of equalizers have been developed. The simplest is the linear equalizer, which is usually a tapped-delay-line filter with coefficients that are adapted to the channel state (Section 14.2). Decision feedback filters make use of the fact that the ISI created by past symbols can be computed (and subtracted) from the received signal (Section 14.3). Finally, the optimum way of detection in a delay-dispersive channel is the Maximum Likelihood Sequence Estimation (MLSE) (Section 14.4). Blind equalizers, which do not need a training sequence, have been intensively investigated by researchers, but are not very popular for practical systems.

We note that the description of this chapter assumes single-carrier modulation, i.e., the type of modulation previously described in Chapter 10. Other modulation methods, such as multi-carrier modulation (Chapter 15) and CDMA (Chapter 19) use similar concepts but have different implementations that will be described in more detail in those chapters.

14.1.2   Modeling of Channel and Equalizer

The following sections describing different equalizer structures will require a discrete-time model of the channel and equalizer. We now give such a model, together with the concept of the *noise-whitening filter* that has importance for optimum RXs.

The first stage of the RX consists of a filter that limits the amount of received noise power. This filter also should make sure that all information is contained in sample values at instances $t_s + iT_S$. This is achieved by a filter that is matched to $\eta(t)$ – i.e., the convolution of channel impulse response and basis pulse. The sequence of sample values at the output of the matched filter is then given by

$$\psi_i = c_i \zeta_0 + \sum_{n \neq i} c_n \zeta_{i-n} + \hat{n}_i \tag{14.5}$$

where ζ_i are the sample values of the AutoCorrelation Function (ACF) of $\eta(t)$, and $\hat{n}_i$ is a sequence of complex Gaussian random variables with ACF $N_0 \zeta_i$ – i.e., the noise filtered by the matched filter.

The z-transform of the sampled ACF ζ_i can be factored as

$$\Xi(z) = F(z)F^*\left(z^{-1}\right). \tag{14.6}$$

This factorization is not unique, but it is advantageous to choose $F^*\left(z^{-1}\right)$ in such a way that all roots are within the unit circle. In such a case, $1/F^*(z^{-1})$ is a stable, realizable filter.

Now, if the matched filter is concatenated with a filter whose z-transform of the impulse response is $1/F^*\left(z^{-1}\right)$, then the noise at the output of this concatenation is spectrally white again, with an ACF given by $N_0 \delta_i$. It is therefore also known as the *noise-whitening filter*. The sample values of the impulse response of concatenation of the basis pulse, wireless channel, matched filter, and noise-whitening filter are henceforth denoted as the *discrete-time channel*. The impulse response of the discrete-time channel is written as f_i and its z-transform as $F(z)$. Note that the impulse response of this channel is causal, so that the output signal can be written as

$$u_i = \sum_{n=0}^{L_c} f_n c_{i-n} + n_i \tag{14.7}$$

where L_c is the length of the impulse response of the discrete-time channel. Therefore, the noise-whitening filter is also often known as the *precursor equalizer*.

We will use the following notation in this chapter:

c_i	*i*th complex transmit symbol;
$\eta(t)$	convolution of basis pulse and channel impulse response;
e_i	*i*th equalizer coefficients;
ζ_i	*i*th sample value of the ACF of $\eta(t)$;
$\hat{n}_i$	*i*th sample of a sequence of complex Gaussian random variables with ACF $N_0 \zeta_i$;
n_i	*i*th sample of a sequence of uncorrelated complex Gaussian random variables;
f_i	*i*th sample of the impulse response of the time-discrete channel;
u_i	*i*th sample of the output signal of the time-discrete channel;
$\hat{c}_i$	estimate of transmit symbol c_i;
ε_i	deviation between estimated and true transmit symbol $c_i - \hat{c}_i$.

14.1.3 Channel Estimation

A common strategy for data detection with an equalizer is to separate the estimation of $\mathbf{f}$ and $\mathbf{c}$. In a first step, a training sequence (i.e., known $\mathbf{c}$) is used to estimate $\mathbf{f}$. During the subsequent transmission of the unknown payload data, we assume that the estimated impulse response is the true one, and solve the above equation for $\mathbf{c}$.

In this subsection, we discuss estimation of the channel impulse response by means of a training sequence. Channel estimation shows strong similarities to the "channel-sounding" techniques described in Chapter 9. A simple estimate can be obtained by means of *Pseudo Noise* (PN) sequences with period N_{per}. The ACF of the PN sequence approximates a Dirac delta function. More precisely, periodic continuation of the sequence $\{b_i\}$, convolved with a time-reversed version of itself,[2] gives a sum of Dirac pulses spaced N_{per} symbols apart:

$$\{b_{-i}\}*_{per}\{b_i\} \approx \sum_{n=-\infty}^{\infty} \delta_{i-n\,N_{per}} \tag{14.8}$$

where $*_{per}$ denotes periodic convolution. If this sequence is sent over a channel with impulse response $\mathbf{f}$, the output of the correlator becomes

$$\left\{\hat{f}_i\right\} = \left(\{b_{-i}\}*_{per}\{b_i\}\right) * \{f_i\}. \tag{14.9}$$

Now, if the duration of the channel impulse response is shorter than N_{per}, then $\hat{\mathbf{f}}$ is simply a periodic repetition of $\mathbf{f}$ (see Figure 14.1).

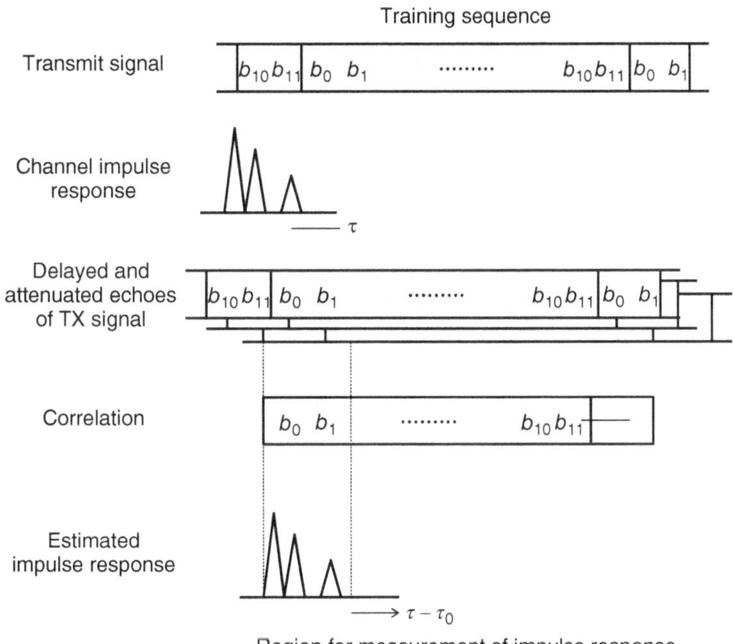

Figure 14.1 The principle of channel estimation by correlation.

In practice, we do not transmit a periodic continuation of the PN sequence, but just a single realization. In order to avoid the possible influence of (unknown) payload bits on correlator output, known "buffer bits" have to be transmitted before and after the sequence.

Channel estimation by means of a training sequence technique has several drawbacks:

1. *A reduction in spectral efficiency*: the training sequence does not convey any payload information. For example, the Global System for Mobile communications (GSM) uses 26 bits in every 148-bit frame for the training sequence (see Appendix 30.A).

2. *Sensitivity to noise*: in order to keep spectral efficiency reasonable, the training sequence has to be short. However, this implies that the training sequence is sensitive to noise (longer training sequences can average out noise), and also to nonidealities in sounding sequences (remember that the peak-to-offpeak ratio of a PN-sequence increases with the length of such a sequence).

---

[2] Strictly speaking, we need to take the complex conjugate of the time-reversed sequence. However, as the transmit sequence is usually real $(+1/-1)$, bits and complex symbols can be considered to be equivalent.

3. *Outdated estimates*: if the channel changes after transmission of the training sequence, the RX cannot detect this variation. Use of an outdated channel estimate leads to decision errors.

Despite these problems, training-sequence-based channel estimation is used in practically every system. The reason for this is that the alternative (blind techniques, see Section 14.7) requires high computational effort and suffers from significant numerical problems.

14.2 Linear Equalizers

Linear equalizers are simple linear filter structures that try to compensate the impact of the channel in the sense that the product of the transfer functions of channel and equalizer fulfills a certain criterion. This criterion can either be achieving a completely flat transfer function of the channel – filter concatenation, or minimizing the mean-squared error at the filter output.

The basic structure of a linear equalizer is sketched in Figure 14.2. Following the system model of Section 14.1.2, a transmit sequence $\{c_i\}$ is sent over a dispersive, noisy channel, so that the sequence $\{u_i\}$ is available at the equalizer input. We now need to find the coefficients of a Finite Impulse Response (FIR) filter (transversal filter, Figure 14.3) with $2K + 1$ taps. This filter should convert sequence $\{u_i\}$ into sequence $\{\hat{c}_i\}$:

$$\hat{c}_i = \sum_{n=-K}^{K} e_n u_{i-n} \tag{14.10}$$

that should be "as close as possible" to the sequence $\{c_i\}$. Defining the deviation ε_i as

$$\varepsilon_i = c_i - \hat{c}_i. \tag{14.11}$$

we aim to find a filter so that

$$\varepsilon_i = 0 \text{ for } N_0 = 0 \tag{14.12}$$

which gives the *ZF equalizer*, or that

$$E\{|\varepsilon_i|^2\} \to \min \quad \text{for } N_0 \text{ having a finite value} \tag{14.13}$$

which gives the *Minimum Mean Square Error (MMSE) equalizer*.

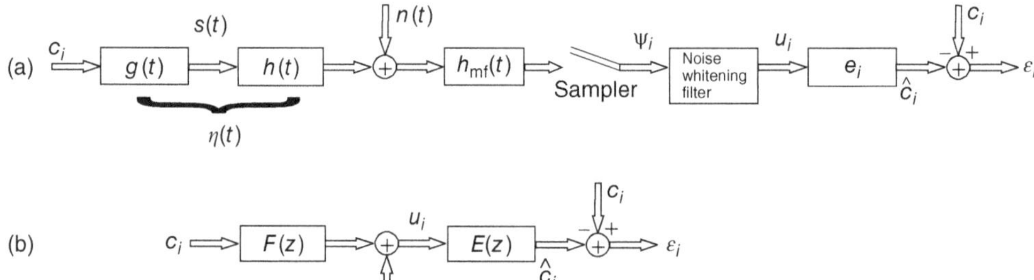

Figure 14.2 Linear equalizer in the time domain (a) and time-discrete equivalent system in the z-transform domain (b).

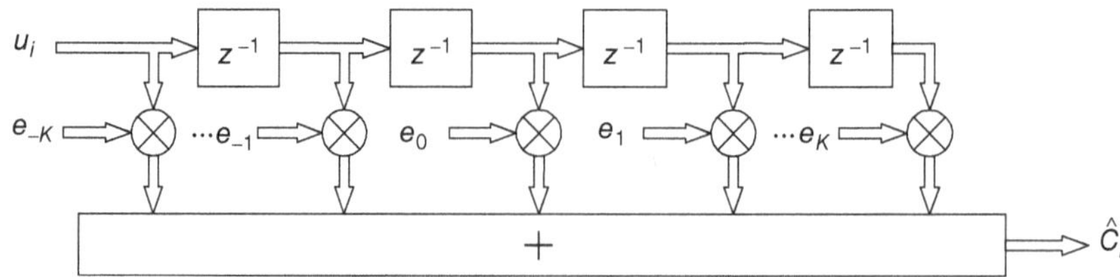

Figure 14.3 Structure of a linear transversal filter. Remember that z^{-1} represents a delay by one sample.

14.2.1   Zero-Forcing Equalizer

The ZF equalizer can be interpreted in the frequency domain as enforcing a completely flat (constant) transfer function of the combination of channel and equalizer by choosing the equalizer transfer function as $E(z) = 1/F(z)$. Alternatively, in the time domain, this can be interpreted as minimizing the maximum ISI (*peak distortion criterion*). Appendix 14.A shows that these two criteria are identical.

The ZF equalizer is optimum for elimination of ISI. However, channels also add noise, which is amplified by the equalizer. At frequencies where the transfer function of the channel attains small values, the equalizer has a strong amplification, and thus also amplifies the noise. As a consequence, the noise power at the detector input is larger than for the case without an equalizer (see Figure 14.4).

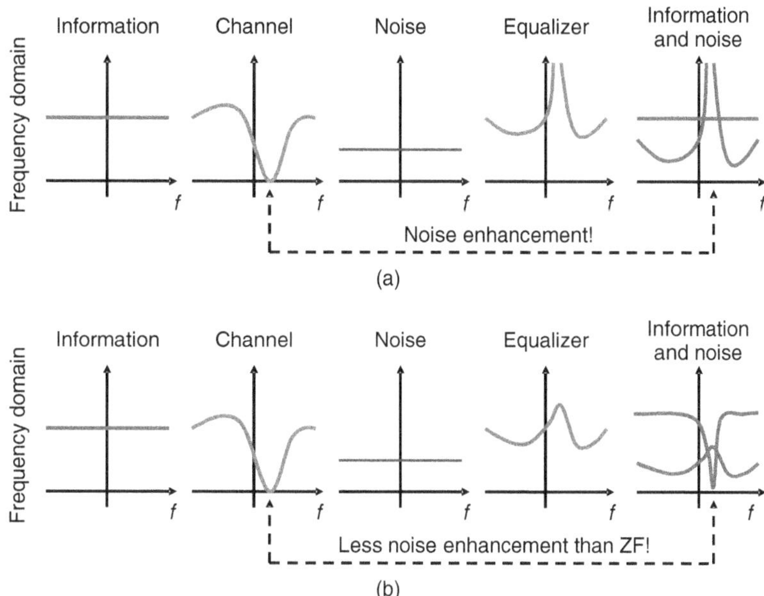

Figure 14.4   Illustration of noise enhancement in zero-forcing equalizer (a), which is mitigated in an MMSE linear equalizer (b).

The Fourier transform $\Xi(e^{j\omega T_s})$ of the sample ACF ζ_i is related to $\hat{\Xi}(e^{j\omega T})$, the Fourier transform of $\eta(t)$, as

$$\Xi\left(e^{j\omega T_s}\right) = \frac{1}{T_S} \sum_{n=-\infty}^{\infty} \hat{\Xi}\left(\omega + \frac{2\pi n}{T_S}\right), \quad |\omega| \leq \frac{\pi}{T_S}. \tag{14.14}$$

The inverse Signal-to-Noise Ratio (SNR) at the detector is (assuming $\sigma_s^2 = 1$)

$$\frac{1}{\gamma_{LE-ZF}} = N_0 \frac{T_S}{2\pi} \int_{-\pi/T_S}^{\pi/T_S} \frac{1}{\Xi(e^{j\omega T_s})} \, d\omega. \tag{14.15}$$

It is finite only if the spectral density Ξ has no (or only integrable) singularities.

14.2.2   The Mean Square Error Criterion

The ultimate goal of an equalizer is minimization, not of the ISI, but of the bit error probability. Noise enhancement makes the ZF equalizer ill-suited for this purpose. A better criterion is minimization of the *Mean Square Error* (MSE) between the transmit signal and the output of the equalizer treating ISI like effective noise.

We are thus searching for a filter that minimizes:

$$MSE = E\{|\varepsilon_i|^2\} = E\{\varepsilon_i \varepsilon_i^*\}. \tag{14.16}$$

As shown in Appendix 14.B, this can be achieved with a filter whose coefficients $\mathbf{e}_{opt}$ are given by

$$\mathbf{e}_{opt} = \mathbf{R}^{-1}\mathbf{p} \tag{14.17}$$

where $\mathbf{R} = E\{\mathbf{u}_i^*\mathbf{u}_i^T\}$ is the correlation matrix of the received signal, and $\mathbf{p} = E\{\mathbf{u}_i^* c_i\}$ the cross-correlation between the received signal and the transmit signal. Considering the frequency domain, concatenation of the noise-whitening filter with the equalizer $E(z)$ has the transfer function (again assuming $\sigma_s^2 = 1$):

$$\breve{E}(z) = \frac{1}{\Xi(z) + N_0} \tag{14.18}$$

which is the transfer function of the Wiener filter. The MSE and the resulting SNR are then

$$MSE_{\mathrm{LE}} = N_0 \frac{T_S}{2\pi} \int\limits_{-\pi/T_S}^{\pi/T_S} \frac{1}{\Xi(e^{j\omega T_s}) + N_0}\, d\omega \quad \text{and} \quad \gamma_{\mathrm{MSE}} = \frac{1 - MSE}{MSE}. \tag{14.19}$$

The second equation holds for both linear equalizers and DFE. Comparison with Eq. (14.15) shows that the noise power of an MMSE equalizer is smaller than that of a ZF equalizer (as illustrated in Figure 14.4).

Example 14.1 *Equalizer coefficients and noise enhancement for linear equalizers: consider a channel with impulse response* $h(\tau) = 0.4\delta(\tau) - 0.7\delta(\tau - T_S) + 0.6\delta(\tau - 2T_S)$, $N_0 = 0.3$, *and* $g(t) = g_R(t, T_S)$. *Compute the noise variance at the output of a ZF equalizer and an MMSE equalizer.*

First we note that $\eta(t) = h(t) * g(t) = 0.4g(t) - 0.7g(t - T_S) + 0.6g(t - 2T_S)$. For the given rectangular pulse, $\sigma_s^2 = 1$. The z-transform of the ACF of $\eta(t)$ is given by

$$\Xi(z) = 0.24z^2 - 0.7z + 1.01 - 0.7z^{-1} + 0.24z^{-2} = F(z)F^*\left(z^{-1}\right).$$

Let us then choose $F(z) = 0.4 - 0.7z^{-1} + 0.6z^{-2}$. Thus, $F^*\left(z^{-1}\right) = 0.4 - 0.7z^1 + 0.6z^2$ has roots at $0.58 \pm 0.57j$, which are inside the unit circle.

The transfer function of the ZF equalizer is then

$$E_1(z) = \frac{1}{F(z)} = \frac{1}{0.4 - 0.7z^{-1} + 0.6z^{-2}}. \tag{14.20}$$

The inverse effective SNR is given by Eq. (14.15):

$$\frac{1}{\gamma_{\mathrm{LE-ZF}}} = N_0 \frac{T_S}{2\pi} \int\limits_{-\pi/T_S}^{\pi/T_S} \frac{1}{0.24e^{\,j2\omega T_s} - 0.7e^{j\omega T_s} + 1.01 - 0.7e^{-j\omega T_s} + 0.24e^{-j2\omega T_s}}\, d\omega$$

$$= N_0 \frac{1}{2\pi} \int\limits_{-\pi}^{\pi} \frac{1}{0.24e^{\,j2\omega} + 0.24e^{-j2\omega} - 0.7e^{j\omega} - 0.7e^{-j\omega} + 1.01}\, d\omega \tag{14.21}$$

$$= N_0 \frac{1}{2\pi} \int\limits_{-\pi}^{\pi} \frac{1}{0.48\cos 2\omega - 1.4\cos\omega + 1.01}\, d\omega$$

$$\approx 2.94.$$

The effective SNR is thus $\gamma_{LE-ZF} = 0.34$.

The MMSE equalizer is given by

$$E_2(z) = \frac{F^*(z^{-1})}{F(z)F^*(z^{-1}) + N_0} = \frac{0.4 - 0.7z + 0.6z^2}{0.24z^2 - 0.7z + 1.31 - 0.7z^{-1} + 0.24z^{-2}}. \tag{14.22}$$

The MSE is given by Eq. (14.19). We note that the only difference from the ZF case is the addition of N_0 in the denominator of the integrand:

$$MSE_{\mathrm{LE}} = N_0 \frac{1}{2\pi} \int\limits_{-\pi}^{\pi} \frac{1}{0.48\cos 2\omega - 1.4\cos\omega + 1.31}\, d\omega \tag{14.23}$$

$$\approx 0.46.$$

As expected, the noise variance is lower for the MMSE equalizer than for the ZF equalizer. The effective SNR is computed according to Eq. (14.19) as

$$\gamma_{\text{LE-MSE}} = \frac{1 - MSE}{MSE} = 1.17 \tag{14.24}$$

compared with $(1/N_0) = 3.33$ if there were no ISI (and thus, no equalizer). Thus, the necessity to equalize decreases the effective SNR (SNR at the output of the equalizer) by 4.5 and 10 dB for the MMSE and ZF equalizer, respectively.

*14.2.3 Adaptation Algorithms for Mean Square Error Equalizers

In order to find the optimum equalizer weights, we can directly solve Eq. (14.17). However, this requires on the order of $(2K + 1)^3$ complex operations. To ease the computational burden, iterative algorithms have been developed. The quality of an iterative algorithm is described by the following criteria:

- *Convergence rate*: how many iterations are required to "closely approximate" the final result? It is usually assumed that the channel does not change during the iteration period. However, if an algorithm converges too slowly, it will never reach a stable state – the channel has changed before the algorithm has converged.
- *Misadjustment*: the size of deviation of the converged state of the iterative algorithm from the exact MSE solution.
- Computational effort per iteration.

In the following, we discuss two algorithms that are widely used – the Least Mean Square (LMS) and the Recursive Least Square (RLS).

Least Mean Square Algorithm

The LMS algorithm; also known as the *stochastic gradient method*, consists of the following steps:

1. Initialize the weights with values $\mathbf{e}_0$.
2. With this value, compute an approximation for the gradient of the MSE. The true gradient cannot be computed, because it is an expected value. Rather, we are using an estimate for $\mathbf{R}$ and $\mathbf{p}$ – namely, their instantaneous realizations. Define now:

$$\mathbf{u}_i = [u_{i+K}, u_{i+K-1}, \dots\dots, u_{i-K}]^T \tag{14.25}$$

$$\varepsilon_i = c_i - \hat{c}_i, \qquad \hat{c}_i = \mathbf{u}_i^T \mathbf{e}_i. \tag{14.26}$$

The gradient is estimated as

$$\hat{\nabla}_i = -\varepsilon_i \mathbf{u}_i^*. \tag{14.27}$$

3. We next compute an updated estimate of the weight vector $\mathbf{e}$ by adjusting weights in the direction of the negative gradient:

$$\mathbf{e}_{i+1} = \mathbf{e}_i - \mu \hat{\nabla}_i \tag{14.28}$$

where μ is a user-defined parameter that determines convergence and residual error.

4. If the stop criterion is fulfilled – e.g., the relative change in weight vector falls below a predefined threshold – the algorithm has converged. Otherwise, increment i and return to step 2.

It can be shown that the LMS algorithm converges if

$$0 < \mu < \frac{2}{\lambda_{\max}}. \tag{14.29}$$

Here $\lambda_{\max}$ is the largest eigenvalue of the correlation matrix $\mathbf{R}$. The problem is that we do not know this eigenvalue (computing it requires larger computational effort than inverting the correlation matrix). We thus have to guess values for μ. If μ is too large, we obtain faster convergence, but the algorithm might sometimes diverge. If we choose μ too small, then convergence is very probable but slow. Generally, convergence speed depends on the condition number of the correlation matrix (i.e., the ratio of largest to smallest eigenvalue): the larger the condition number, the slower the convergence of the LMS algorithm.

The Recursive Least Squares Algorithm

In most cases, the LMS algorithm converges very slowly. Furthermore, the use of this algorithm is justified only when the statistical properties of the received signal fulfill certain conditions. The general Least Squares (LS) criterion, on the other hand, does not require such assumptions. It just analyzes the N subsequent errors ε_i and chooses weights such that the sum of the squared errors is minimized. This general LS problem can be solved by a recursive algorithm as well – known as *RLS* – the details of which are given in Appendix 14.C.

Comparison of Algorithms

There are two classes of algorithms for the determination of equalizer coefficients: direct implementation (Wiener filter, LS criterion) and iterative methods (LMS, RLS).

For the Wiener filter, we first have to determine the correlation matrix; this can be the major part of the numerical effort especially if the number of weights is small. The actual inversion of the matrix requires $(2K + 1)^3$ operations. Alternatively, we can use the data matrix directly, and invert it (LS algorithm). Construction of the data matrix requires less numerical effort than the correlation matrix; on the other hand, the effort for inversion is much larger and depends on the number of used bits.

When comparing iterative algorithms, we find that the LMS algorithm usually converges too slowly. The RLS algorithm converges faster but has a larger residual error. Figure 14.5 shows the typical example of the MSE for a Digital Enhanced Cordless Telecommunications (DECT) cordless telephone (see App. 30.C). We see that the RLS algorithm has converged after 10 bits, while the LMS algorithm requires almost 300 bits. Due to the temporal variance of the channel, as well as spectral efficiency considerations, fast convergence is more important than an extremely low residual error rate.

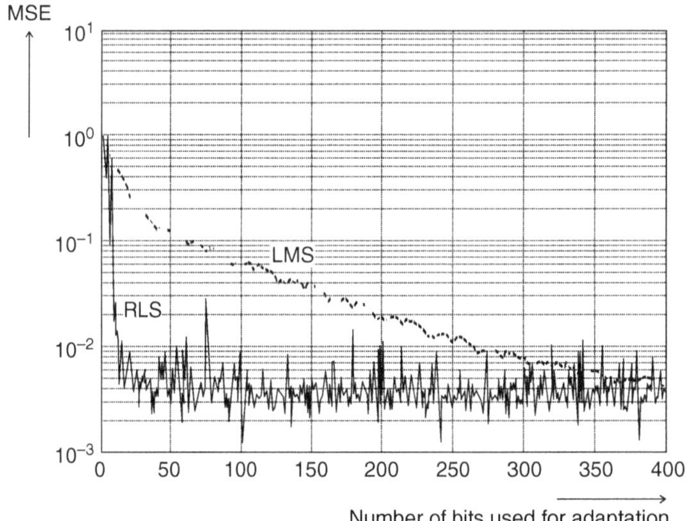

Figure 14.5 Mean-square error as a function of the number of iterations for a decision feedback equalizer (see below). For least mean square: $\mu = 0.03$; recursive least squares: $\lambda = 0.99$, $\delta = 10^{-9}$.
Reproduced with permission from [Fuhl 1994] © J. Fuhl.

On the other hand, the LMS algorithm requires far fewer (complex) operations (see Figure 14.6; the terms "Decision Feedback Equalizer (DFE)" and "gradient lattice" will be explained below). One important conclusion from this figure is that for a small number of weights the complexity of the algorithms does not differ significantly. For up to 5–8 weights, the differences are less than 50% (with the exception of the LMS which is however disqualified because of its slow convergence). In this regime, convergence, stability, and ease of implementation are the dominant criteria.

*14.2.4   Further Linear Structures

Up to now, we have considered transversal FIR filters. However, linear filters can also be realized by means of other structures. One possibility is the use of recursive filters (*Infinite Impulse Response filters* (IIR)). They have the advantage that fewer taps are required to achieve equalization. The major drawback is that these filters can show not only zeros but also poles in the transfer function, such that they can become unstable. For this reason, IIR filters are rarely used in practice.

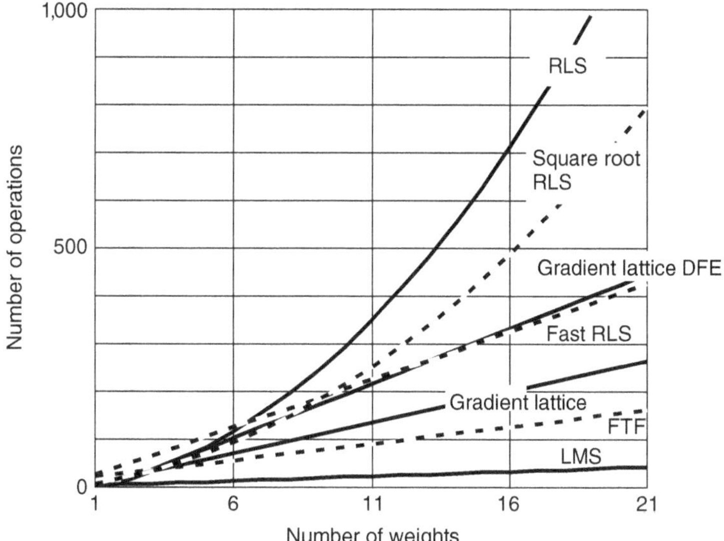

Figure 14.6 Number of operations per iteration step for different algorithms for a decision feedback equalizer (see below). *In this figure*: FTF, Fast Transversal Filter.

A further possible structure is the *lattice filter*. The equations for the equalization algorithms are different from those of transversal filters (details can be found in [Proakis and Salehi 2005]).

14.3 Decision Feedback Equalizers

A *DFE* has a simple underlying premise: once we have detected a bit correctly, we can use this knowledge in conjunction with knowledge of the channel impulse response to compute the ISI caused by this bit. In other words, we determine the effect this bit will have on subsequent samples of the receive signal. The ISI caused by each bit can then be subtracted from these later samples.

The block diagram of a DFE is shown in Figure 14.7. The DFE consists of a *forward filter* with transfer function $E(z)$, which is a conventional linear equalizer, as well as a *feedback filter* with transfer function $D(z)$. As soon as the RX has decided on a received symbol, its impact on all *future* samples (*postcursor ISI*) can be computed, and (via the feedback) subtracted from the received signal. A key point is the fact that the ISI is computed based on the signal *after* the hard decision; this eliminates additive noise from the feedback signal. Therefore, a DFE results in a smaller error probability than a linear equalizer.

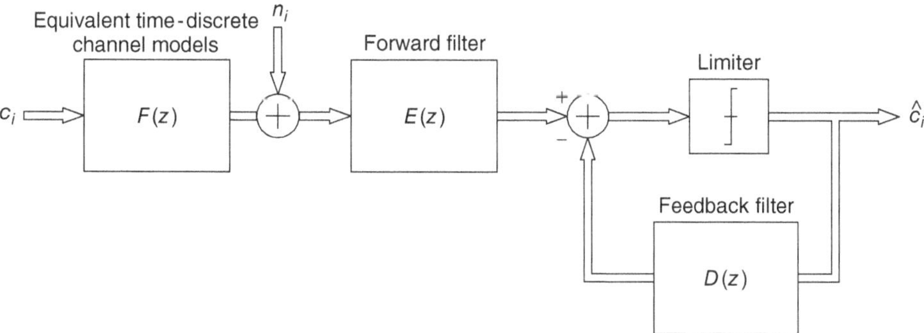

Figure 14.7 Structure of a decision feedback equalizer.

One possible source of problems is *error propagation*. If the RX decides incorrectly for one bit, then the computed postcursor ISI is also erroneous, so that later signal samples arriving at the decision device are even more afflicted by ISI than the unequalized samples. This leads to a vicious cycle of wrong decisions and wrong subtraction of postcursors.

Error propagation does not usually play a role when the BER is small. Note, however, that small error rates are often achieved via coding. It may therefore be necessary to decode the bits, re-encode them (such that the signal becomes a noise-free version of the received signal), and use this new signal in the feedback from the DFE.[3] In the following, we will only consider uncoded systems without error propagation.

14.3.1 MMSE Decision Feedback Equalizer

The goal of the MMSE DFE is again minimization of the MSE, by striking a balance between noise enhancement and residual ISI. As noise enhancement is different in the DFE case from that of linear equalizers, the coefficients for the forward filter are different: as postcursor ISI does not contribute to noise enhancement, we now aim to minimize the sum of noise and (average) *precursor* ISI. Obviously, performance also differs.

The coefficients of the feedforward filter can be computed from the following equation:

$$\sum_{n=-K_{\mathrm{ff}}}^{0} e_n \left(\sum_{m=0}^{-\ell} f_m^* f_{m+\ell-n} + N_0 \delta_{nl} \right) = -f_{-\ell}^* \quad \text{for } \ell, n = -K_{\mathrm{ff}}, ..., 0 \tag{14.30}$$

where K_{ff} is the number of taps in the feedforward filter. The coefficients of the feedback filter are then

$$d_n = - \sum_{m=-K_{\mathrm{ff}}}^{0} e_m f_{n-m} \quad \text{for } n = 1, ..., K_{\mathrm{fb}} \tag{14.31}$$

where K_{fb} is the number of taps in the feedback filter.

Assuming some idealizations (the feedback filter must be at least as long as the postcursor ISI; it must have as many taps as required to fulfill Eq. (14.30); there is no error propagation), the MSE at the equalizer output is

$$MSE_{\mathrm{DFE}} = N_0 \exp\left(\frac{T_S}{2\pi} \int_{-\pi/T_s}^{\pi/T_s} \ln\left[\frac{1}{\Xi(e^{j\omega T}) + N_0} \right] d\omega \right). \tag{14.32}$$

14.3.2 Zero-Forcing Decision Feedback Equalizer

The ZF DFE is conceptually even simpler. As mentioned in Section 14.1.2, the noise-whitening filter eliminates all precursor ISI, such that the resulting effective channel is purely causal. Postcursor ISI is subtracted by the feedback branch. The effective noise power at the decision device is

$$\frac{1}{\gamma_{\mathrm{DFE\text{-}ZF}}} = N_0 \exp\left(\frac{T_S}{2\pi} \int_{-\pi/T_s}^{\pi/T_s} \ln\left[\frac{1}{\Xi(e^{j\omega T})} \right] d\omega \right). \tag{14.33}$$

This equation demonstrates that noise power is larger than it is in the unequalized case but smaller than that for the linear ZF equalizer.

Example 14.2 *Using the channel from Example 14.1, compute the noise enhancement for the MMSE DFE and ZF DFE*

From Example 14.1, remember that $\Xi(e^{j\omega T}) = 0.48 \cos 2\omega T_S - 1.4 \cos \omega T_S + 1.01$. Inserting this in Eq. (14.32), we obtain:

$$MSE_{\mathrm{DFE}} = N_0 \exp\left(\frac{T_S}{2\pi} \int_{-\pi/T_s}^{\pi/T_s} \ln\left[\frac{1}{\Xi(e^{j\omega T}) + N_0} \right] d\omega \right) \tag{14.34}$$

$$= N_0 \exp\left(\frac{1}{2\pi} \int_{-\pi}^{\pi} \ln\left[\frac{1}{0.48 \cos 2\omega - 1.4 \cos \omega + 1.31} \right] d\omega \right) \tag{14.35}$$

$$\approx 0.33 \tag{14.36}$$

---

[3] As the decoder itself shows a delay, this can become a challenging task. Possible solutions to this include the design of joint equalization and decoding, with the exchange of soft information (see also Chapter 13).

so that the output SNR is 2. Thus, the SNR deteriorates by 2 dB compared with the Additive White Gaussian Noise (AWGN) case. For the ZF DFE, the noise variance at the output is 0.83, so that the SNR deteriorates by 4.4 dB.

14.4   Maximum Likelihood Sequence Estimation – Viterbi Detector

The equalizer structures considered up to now influence the decision about which *symbol* has been transmitted. For MLSE, on the other hand, we try to determine the *sequence of symbols* that has most likely been transmitted. This situation shows strong similarities to the decoding of convolutional codes. As a matter of fact, transmission through a delay-dispersive channel can be viewed as convolutional encoding with a code rate $R_c = 1/1$. MLSE estimators give the best performance of all equalizers.

Remember that the output signal of the time-discrete channel can be written as

$$u_i = \sum_{n=0}^{L_c} f_n c_{i-n} + n_i \tag{14.37}$$

where n is AWGN with variance σ_n^2. For a sequence of N received values, the joint Probability Density Function (pdf) of the vector of received signals $\mathbf{u}$ (conditioned on the data vector $\mathbf{c}$ and impulse response vector $\mathbf{f}$) is[4]

$$pdf(\mathbf{u}|\mathbf{c};\mathbf{f}) = \frac{1}{\left(2\pi\sigma_n^2\right)^{N/2}} \exp\left(-\frac{1}{2\sigma_n^2}\sum_{i=1}^{N}\left|u_i - \sum_{n=0}^{L_c} f_n c_{i-n}\right|^2\right). \tag{14.38}$$

The MLSE of $\mathbf{c}$ (for a given $\mathbf{f}$) is the values of the vectors that maximize the joint pdf $pdf(\mathbf{u}|\mathbf{c}, \mathbf{f})$. As the variables only occur in the exponent, it is sufficient to minimize:

$$\sum_{i=1}^{N}\left|u_i - \sum_{n=0}^{L_c} f_n c_{i-n}\right|^2. \tag{14.39}$$

As for convolutional decoding, various algorithms exist for determination of the optimum sequence. The RX first generates all possible sequences that can result from convolution of valid transmit sequences with the channel impulse response. We then try to find the sequence that has the smallest distance (best metric) from the received signal. The most straightforward (but also most computationally intensive) method is the *exhaustive search*. In practice, the Viterbi algorithm is used instead.

MLSE as described above is only optimum if the additive noise at MLSE input is uncorrelated. Therefore, the sample values used at the detector have to be the output of a noise-whitening filter. This filter has to be adapted to the current channel state; and each channel realization requires spectral factorization. Due to these difficulties, the total input filter (matched filter and noise-whitening filter) is often replaced by a simple brickwall filter whose bandwidth is approximately the inverse symbol duration. Note, however, that in this case one sample per symbol no longer provides sufficient statistics.

Example 14.3   *Viterbi equalization.*

This example shows the working of the Viterbi detection of a symbol stream that went through a channel with a discrete-time impulse response:

$$\mathbf{f} = \begin{pmatrix} 1 \\ -0.5 \\ 0.3 \end{pmatrix}. \tag{14.40}$$

The channel can be viewed as a tapped delay line (shift register) with weights 1, −0.5, and 0.3; see the top part of Figure 14.8 (the left top part shows the tapped delay line model, the right part shows the "cell" model analogous to the convolutional codes discussed in Section 13.3). For ease of exposition, we chose a channel with a real impulse response, and Binary Phase Shift Keying (BPSK) as the modulation format. The lower part of Figure 14.8 shows possible transitions in the trellis diagram. We have to consider four states in the trellis, as $L_c = 2$ samples, and the number of possible states in a cell of the equivalent shift register is equal to the size of the modulation alphabet $M = 2$. We assume furthermore that we know the starting state of the trellis –i.e., −1 −1 (e.g., because known bits have been transmitted before the start of our decoding). The bottom part of Figure 14.9 shows the "unfolding" of the trellis diagram. The numbers next to the transitions are metrics of the considered sequence.

---

[4] Here and in the following, we assume that all transmit symbols are equally likely, such that MLSE and maximum-a-posteriori estimation are identical.

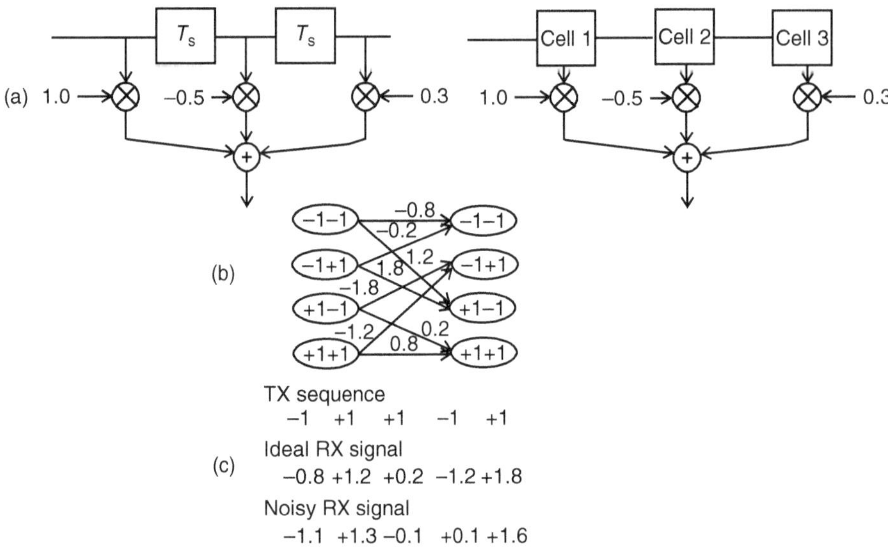

Figure 14.8 Representation of tapped delay line channel (a), transition probabilities (b), and transmitted and received signals (c).

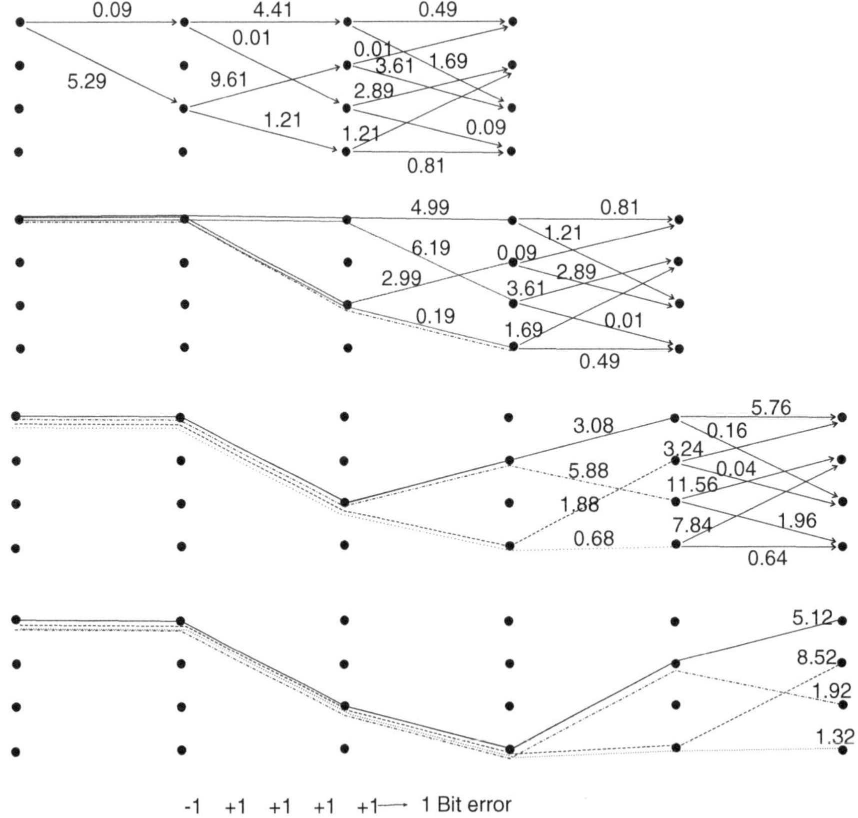

Figure 14.9 Viterbi algorithm for detection of the transmit sequence in a delay-dispersive channel.

When convolutionally coded data are sent over the channel, two possibilities for MLSE exist: (i) we can consider the convolutional encoder and the channel as a "meta-encoder" with a joint impulse response, and the Viterbi decoding is applied to this channel; (ii) we first perform equalization using a soft-Viterbi algorithm such as the BCJR algorithm (see Section 13.3.4), and use the resulting output as input for the FEC decoder. The advantage of the latter technique is that the time-invariant encoding (FEC encoding) and the time-variant "encoding" (convolution with the channel impulse response) are separated, which can ease processing; furthermore, a similar approach can also be used for nonconvolutional FECs.

14.5   Comparison of Equalizer Structures

Figure 14.10 shows a taxonomy of equalizer structures. When selecting an equalizer for a practical system, we have to consider the following criteria:

- *Minimization of the BER*: here MLSE is superior to all other structures. DFEs, though worse than MLSE estimators, are better than linear equalizers. The quantitative difference between the structures depends on the channel impulse response.
- *Can the channel deal with zeros in the channel transfer function?* ZF equalizers have problems, as they invert the transfer function and thus create poles in the equalizer transfer function. Neither MMSE nor MLSE equalizers have this problem.
- *Computational effort*: the effort for linear equalizers and DFEs is not significantly different. Depending on the adaptation algorithm, the number of operations increases linearly, quadratically, or cubically with equalizer length (number of weights). For MLSE, the computation effort increases exponentially with length of the impulse response of the channel. For short impulse responses (e.g., impulse response is at most four symbol durations long, as in GSM), the computational complexity of MLSE is comparable with that of other equalizer structures.
- *Sensitivity to channel misestimation*: due to the error propagation effect, DFE equalizers are more sensitive to channel estimation errors than linear equalizers. Also, ZF equalizers are more sensitive than MMSE equalizers.
- *Power consumption and cost*: these can be deduced from the computational effort.

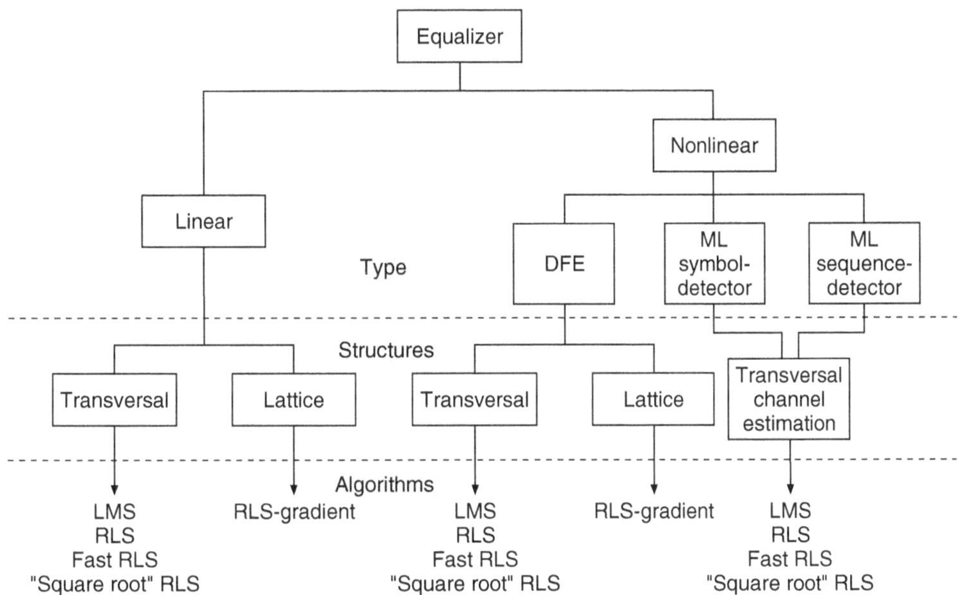

Figure 14.10   Taxonomy of equalizer structures.
Reproduced with permission from [Proakis 1991] © IEEE.

*14.6   Fractionally Spaced Equalizers

In most cases, the RX samples and processes signals at the symbol frequency $(1/T_S)$. This is suboptimum if this sampling rate is lower than the Nyquist rate and the matched filter is matched only to the TX pulse (and not the received, distorted pulse). The spectrum of a signal that was filtered by a raised cosine filter (with roll-off factor α) extends up to a frequency $(1 + \alpha)/2T_S$, such that the Nyquist rate is $(1 + \alpha)/T_S$. A *fractionally spaced equalizer* is based on sampling at no less than the Nyquist rate. The taps of the equalizers are spaced at $T_S a/b$, where $a < b$ and a and b are integers.

Fractionally spaced equalizers can also be interpreted as performing equalization and matched filtering in one step. They are also less sensitive to errors in the sampling time. The drawback is that the number of required taps is larger than for a symbol-spaced equalizer, and thus the computational effort is higher.

*14.7   Blind Equalizers

14.7.1   Introduction

Equalization as described above works in two stages: a training phase and a detection phase. During the training phase, a known bit sequence is transmitted over the channel, and the distorted version at the RX is compared with a (locally generated) undistorted version; this, in turn, gives us information about the channel impulse response that is used for equalization (see Section 14.1.3

for a discussion of the pros and cons). In contrast, blind equalization exploits known statistical properties of the transmit signal to estimate both channel and data. Equalizer coefficients are adjusted in such a way that certain statistical properties of the equalizer output match known statistical properties of the transmit signal.

The advantages of blind equalization include the following:

- The whole timeslot is used for determination of the impulse response, not just a short training sequence.
- Spectral efficiency is improved, because no time is "wasted" on transmission of the training sequence.

The following signal properties can be used for blind equalization:

- *Constant envelope*: for many signals (Frequency Shift Keying (FSK), Minimum Shift Keying (MSK), Gaussian Minimum Shift Keying (GMSK)), the envelope (amplitude) is constant.
- *Statistical properties*: e.g., cyclostationarity.
- *Finite symbol alphabet*: only certain discrete values are valid points in the signal constellation diagram.
- *Spectral correlation*: signal spectra are correlated with shifted versions of themselves.
- A combination of these properties.

Well-studied blind algorithms include (i) *Constant Modulus Algorithm* (CMA), (ii) blind *MLSE*, and (iii) algorithms based on higher-order statistics. In the following, we will discuss these classes of algorithms.

14.7.2   Constant Modulus Algorithm

CMAs are the oldest algorithms for blind equalization. In their simplest form, they use the LMS adaptation. The data-aided LMS algorithm was described in Section 14.2; it is based on minimization of the difference between a desired signal (known as the RX) and the output of an equalizer. In a blind LMS algorithm, the desired signal first has to be generated *from the output of the equalizer*. This is achieved by sending the equalizer output through a nonlinear function. The error signal is then the difference between the output of this nonlinear function and the equalizer output. The difference between a conventional equalizer and the CMA equalizer is shown in Figure 14.11.

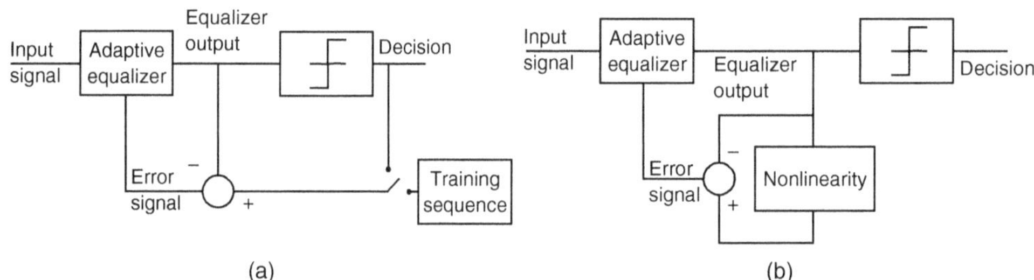

Figure 14.11   Structure of a conventional equalizer (a) and that of a constant-modulus-algorithm-based equalizer (b).

The nonlinear function can either be memoryless, or contain mth order memory. Various types of algorithms (usually named for their inventors) are distinguished by different nonlinearities. The best known are the Sato algorithm [Sato 1975] and the Godard algorithm [Godard 1980].

The LMS algorithm used for blind adaptation has the same drawbacks as the conventional LMS: the convergence rate is slow (by increasing the stepwidth μ, convergence is improved and the error after convergence is worsened). It is also possible that the algorithm converges to a local minimum. These problems can be solved by the "analytic CMA" [Vanderveen and Paulraj 1996]. This is a noniterative algorithm that provides exact solutions in the noise-free case but is also robust with respect to noise.

14.7.3   Blind Maximum Likelihood Estimation

For conventional MLSE, a training sequence is used for determination of the channel impulse response $\mathbf{f}$. This estimate is then used during the actual transmission of data, so that the Maximum Likelihood (ML) estimate only has to solve for $\mathbf{c}$. For a blind estimate, $\mathbf{c}$ and $\mathbf{f}$ have to be estimated simultaneously. This can be achieved by either of the following methods:

1. The channel impulse response is estimated from $pdf\,(\mathbf{u}|\mathbf{f},\,\mathbf{c})$ by averaging over all possible data sequences. This method has two drawbacks: it requires considerable computation time for the averaging, and it is suboptimum.

2. Alternatively, we can determine the ML estimate for $\mathbf{f}$ for all possible data sequences. Then we select the pair $\mathbf{f}$, $\mathbf{c}$ that has the best overall metric. This method is even more computationally intensive than the averaging method; however, it is also more accurate. Furthermore, methods for reducing the computational effort have been proposed [Proakis and Salehi 2005].

14.7.4   Algorithms Using Second- or Higher-Order Statistics

In general, second-order statistics (ACFs) cannot provide information about the phase of the channel impulse response. An exception occurs when the ACF of the received signal is periodic. Cyclostationary properties are thus the basis for blind estimation methods using second-order statistics. Similar to our discussion in Section 6.4, we distinguish between strict-sense and wide-sense cyclostationarity. A process is strict-sense cyclostationary if *all* its statistical properties are invariant to shifts by integer multiples of the sampling period T_{per}. For wide-sense stationarity, only the mean and the ACF have to fulfill this condition:

$$E\{x(t + iT_{\mathrm{per}})\} = E\{x(t)\} \tag{14.41}$$

$$E\{x^*(t_1 + iT_{\mathrm{per}})x(t_2 + iT_{\mathrm{per}})\} = E\{x^*(t_1)x(t_2)\}. \tag{14.42}$$

If the signal is oversampled, then the resulting sequence of sample values is guaranteed to be cyclostationary. The actual channel estimate is then based on different correlation matrices of the received signal (for details, see [Tong et al. 1994, 1995]). The use of higher-order statistics does not require the applicability of cyclostationarity; however, its accuracy is usually much lower.

14.7.5   Assessment

Historically speaking, blind equalization was first developed in the 1970s for multi-terminal computer networks (one central station linked to multiple terminals). During the 1980s and 1990s, a lot of theoretical work was devoted to this topic, as it offers some fascinating mathematical challenges. However, up to now, truly blind equalization has not been able to replace training-sequence-based equalization in practical systems. The main reason seems to be that the difference in computational effort and reliability is significant. In particular, blind equalizers require a long time to converge, and thus do not work well in quickly time-variant wireless channels.

14.7.6   Joint Equalization and Detection

Another approach that has been developed in recent years uses a training sequence to obtain an initial estimate of the channel and then refines these estimates by means of blind (decision-aided) equalization [Loncar et al. 2002]. In that case, the "blind" equalization need not even rely on particular statistical properties of the TX signal, but rather can employ demodulated data symbols.

This can be applied, e.g., in systems in which the channel changes more quickly than can be tracked by means of training sequences. A concrete example is the IEEE 802.11p standard for car-to-car communications: there, data packets have only a training sequence at the beginning of the packet, but the channel might change within one packet duration. The RX can then use the original training sequence as the basis for the detection of the first data symbols. These data symbols can then be used as another training sequence (since the data are now known at the RX), from which an updated channel estimate can be obtained. This updated estimate is then used to detect the next data symbols, and so on. Such an iterative approach can be quite efficient.

More generally, the estimation of channel and data can be viewed as a joint detection problem, where known (training) symbols reduce the uncertainty of the channel estimation, and the presence of constraints on the data symbols (finite alphabet) allows to reduce noise effects. A joint estimation, while theoretically optimum, is computationally rather expensive, and thus currently rarely used in practice.

14.8   Predistortion at the Transmitter

Reception of signals that went through a delay-dispersive channel can be difficult because of the presence of noise. In some situations, the TX can estimate the channel to the RX, and use this information to predistort the signal in such a way that an "ideal" signal arrives at the RX. For example, a zero-forcing equalizer can be replaced by a filter with transfer function $1/(H(f)$ in the TX chain. The signal arriving at the RX then sees an undistorted waveform shape $s(t)$.

It must be emphasized that the TX predistortion does *not* eliminate many of the fundamental problems of equalization. Consider the zero-forcing predistortion mentioned above: while there is obviously no noise enhancement at the RX, the TX suffers a power penalty because it has to invest a lot of power into overcoming the channel attenuation at the dips of the transfer function; a similar statement holds true for MMSE equalizers.

The power penalty can be partly overcome by so-called Tomlinson–Harashima (TH) precoders, which are the TX predistortion equivalents of DFEs. A significant advantage that TH precoding has over DFE reception is that there is no danger of error propagation – since there is no noise at the TX, and all symbols are known.

However, the biggest challenge to TX precoding is the requirement of accurate channel state information at the TX. This topic is discussed in more detail in Section 16.1.7, but it is shown that considerable effort is required, and furthermore, the channel estimates are noisy, thus leading to the transmission of imperfectly predistorted signals.

*14.9 Appendices

App. 14.A: Equivalence of Peak Distortion and Zero-Forcing Criterion
See App14.pdf at wiley.com/go/molisch/wireless3e

App. 14.B: Derivation of the Mean-Square Error Criterion
See App14.pdf at wiley.com/go/molisch/wireless3e

App. 14.C: The Recursive Least Squares Algorithm
See App14.pdf at wiley.com/go/molisch/wireless3e

Further Reading

The first comprehensive description of adaptive equalizers, which is still worth reading today, was given in [Lucky et al. 1968]. The description in this chapter – particularly, that of linear and DFE equalizers – was inspired by the excellent exposition in [Proakis and Salehi 2005], which also gives many more interesting details. [Haykin 1991] describes adaptive filters, which is another way of interpreting equalizers. DFEs are also described in [Belfiore and Park 1977]. Fractionally spaced equalizers were analyzed, e.g., in [Gitlin and Weinstein 1981] and [Ungerboeck 1976]. The Viterbi equalizer is an application of the Viterbi algorithm [Viterbi 1967]; the impact of channel estimation errors is discussed in [Gorokhov 1998]. [Vitetta et al. 2000] give a detailed description of a wide variety of equalizer structures, including the impact of channel-state information (perfect, estimated, or averaged). Spectral factorization techniques are surveyed in [Sayed and Kailath 2001]. Blind algorithms for equalization are too numerous to list here; as examples, we just name [Giannakis and Halford 1997], [Liu et al. 1996], [Sato 1975], and [Vanderveen and Paulraj 1996]. Iterative equalizers are described in [Wymeersch 2007].

For updates and errata for this chapter, see https://wides.usc.edu/students.html#textbooks

Exercises

See Sec. 36.14 of Exercises.pdf at wiley.com/go/molisch/wireless3e

15

Orthogonal Frequency Division Multiplexing (OFDM)

15.1 Introduction

Orthogonal Frequency Division Multiplexing (OFDM) is a modulation scheme that is especially suited for high-data-rate transmission in delay-dispersive environments. It converts a high-rate data stream into a number of low-rate streams that are transmitted over parallel, narrowband channels that can be easily equalized.

Let us first analyze why traditional single-carrier modulation methods become problematic at very high data rates. As the required data rate increases, the system bandwidth increases, and the symbol duration T_s, which is approximately the inverse of the bandwidth, has to become very small.[1] Now, delay dispersion of a wireless channel is imposed by nature; its values depend on the environment, but not on the transmission system. Thus, if the symbol duration becomes very small, then the impulse response (and thus the required length of the equalizer) becomes very long *in units of symbol durations*. The computational effort for such a long equalizer is very large (see Chapter 14), and the probability of instabilities increases. For example, the Global System for Mobile (GSM) communications system (see Appendix 30.A), which was designed for peak data rates up to 200 kbit/s, uses 200 kHz bandwidth, while the LTE system (see Chapter 31), with data rates of $\sim$100 Mbit/s uses 20 MHz bandwidth. In a channel with 10 µs maximum excess delay, the former needs a two-tap equalizer, while the latter would need 200 taps. OFDM, on the other hand, increases the symbol duration on each of its carriers compared to a single-carrier system, and can thus have a very simple equalizer for each subcarrier.

OFDM dates back more than 50 years; a patent for one of its variants was issued in the mid-1960s [Chang 1966]. A few years later, an important improvement – the *Cyclic Prefix* (CP) – was introduced; it helps to eliminate residual delay dispersion. [Cimini 1985] was the first to suggest OFDM for wireless communications. But it was only in the early 1990s that advances in hardware for digital signal processing made OFDM a realistic option for wireless systems. Furthermore, the high-data-rate applications for which OFDM is especially suitable emerged only since the early 2000s. Since that time, however, it has become the dominant modulation waveform for both cellular (LTE, 5G), *wireless Local Area Networks* (LANs) (IEEE 802.11), as well as broadcasting services such as *Digital Audio Broadcasting* (DAB), *Digital Video Broadcasting* (DVB).

In recent years, a variety of generalizations of OFDM have been developed, which show advantages in particular through improved robustness to impairments, more flexibility to accommodate a large number of potentially nonsynchronized users, and reduced interference to adjacent channels. Generally, these types of signals are described as *multi-carrier modulation*, since they are all based on demultiplexing a data stream into parallel streams, and – after some preprocessing – transmitting them on several closely spaced carriers. We will review the most relevant of those variants at the end of this chapter.

15.2 Principle of Orthogonal Frequency Division Multiplexing

OFDM splits a high-rate data stream into N parallel streams, which are then transmitted by modulating N distinct carriers (henceforth called *subcarriers* or *tones*). The symbol duration on each subcarrier thus becomes larger by a factor of N. In order for the Receiver (RX) to be able to easily separate signals carried by different subcarriers, they have to be orthogonal. Conventional Frequency Division Multiple Access (FDMA), as depicted in Figure 15.1 (for more details see Section 18.3.1), can achieve this by having large (frequency) spacing between carriers. This, however, wastes precious spectrum. A much narrower spacing of subcarriers can be achieved. Specifically, let subcarriers be at the frequencies $f_n = nW/N$, where n is an integer, and W is the total available bandwidth; in the simplest case, $W = N/T_s$. We furthermore assume for the moment that modulation on each of the subcarriers is Pulse Amplitude Modulation (PAM) with rectangular basis pulses. We can then easily see that subcarriers are mutually orthogonal, since the relationship

---

[1] This can be compounded by multiple access formals like Time Division Multiple Access (TDMA), which has a high peak data rate because it compresses data into bursts (see Section 18.3.2).

Wireless Communications: From Fundamentals to Beyond 5G, Third Edition. Andreas F. Molisch.
© 2023 John Wiley & Sons Ltd. Published 2023 by John Wiley & Sons Ltd.
Companion website: www.wiley.com/go/molisch/wireless3e

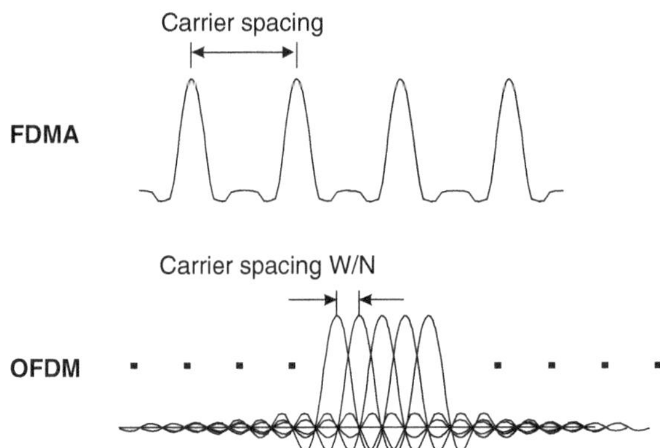

Figure 15.1 Principle behind orthogonal frequency division multiplexing: N carriers within a bandwidth of W.

$$\int_{iT_s}^{(i+1)T_s} \exp\left(j2\pi f_k t\right) \exp\left(-j2\pi f_n t\right) dt = \delta_{nk} \tag{15.1}$$

holds. Therefore, as long as the RX does the appropriate demodulation (multiplying by $\exp(-j2\pi f_n t)$ and integrating over symbol duration), the data streams of any two subcarriers will not interfere with each other.

Figure 15.1 shows this principle in the frequency domain. Due to the rectangular shape of pulses in the time domain, the spectrum of each modulated carrier has a $\sin(x)/x$ shape. The spectra of different modulated carriers overlap, but each carrier is in the spectral nulls of all other carriers.

15.3 Implementation of Transceivers

OFDM can be interpreted in two ways: one is an "analog" interpretation following from the picture of Figure 15.2a. As discussed in Section 15.2, we first split our original data stream into N parallel data streams, each of which has a lower data rate. We furthermore have a number of Local Oscillators (LOs) available, each of which oscillates at a frequency $f_n = nW/N$, where $n = 0, 1, \ldots, N-1$. Each of the parallel data streams then modulates one of the carriers. This picture allows an easy understanding of the principle, but is ill suited for actual implementation – the hardware effort of multiple LOs is too high.

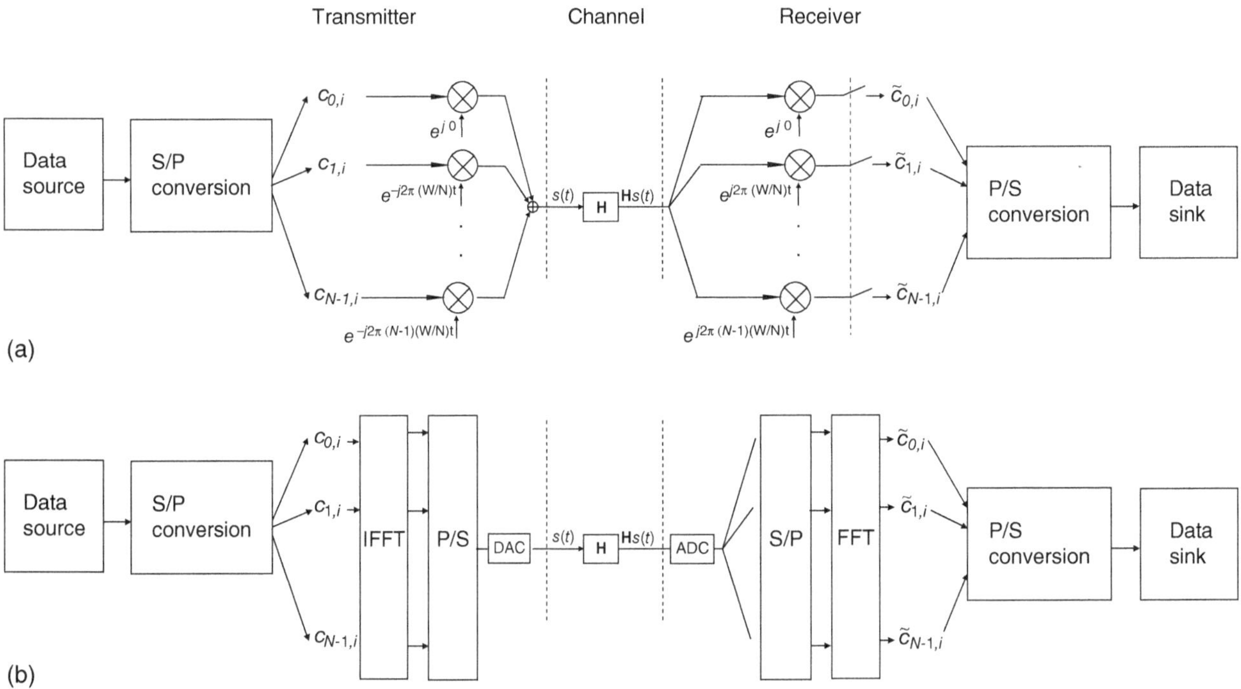

Figure 15.2 Transceiver structures for orthogonal frequency division multiplexing in purely analog technology (a), and using inverse fast Fourier transformation (b).

An alternative implementation is *digital*. It first divides the transmit data into blocks of N symbols. Each block of data is subjected to an *Inverse Fast Fourier Transformation* (IFFT) and then transmitted (see Figure 15.2b). This approach is much easier to implement with integrated circuits. In the following, we will show that the two approaches are equivalent.

We first consider the analog interpretation. Let the complex transmit symbol at time instant i on the nth carrier be $c_{n,i}$. The transmit signal is then:

$$s(t) = \sum_{i=-\infty}^{\infty} s_i(t) = \sum_{i=-\infty}^{\infty} \sum_{n=0}^{N-1} c_{n,i}\, g_n(t - iT_S) \tag{15.2}$$

where the basis pulse $g_n(t)$ is a normalized, frequency-shifted rectangular pulse:

$$g_n(t) = \begin{cases} \dfrac{1}{\sqrt{T_S}}\exp\left(j2\pi n\,\dfrac{t}{T_S}\right) & \text{for } 0 < t < T_S \\ 0 & \text{otherwise.} \end{cases} \tag{15.3}$$

Let us now – without restriction of generality – consider the signal only for $i = 0$, and sample it at instances $t_k = kT_s/N$:

$$s_k = s(t_k) = \dfrac{1}{\sqrt{T_S}} \sum_{n=0}^{N-1} c_{n,0}\exp\left(j2\pi n\,\dfrac{k}{N}\right). \tag{15.4}$$

Now, this is nothing but the *Inverse Discrete Fourier Transform* (IDFT) of the transmit symbols. Therefore, the Transmitter (TX) can be realized by performing an IDFT on the block of transmit symbols (the blocksize must equal the number of subcarriers). In almost all practical cases, the number of samples N is chosen to be a power of 2, and the IDFT is realized as an IFFT. In the following, we will use IDFT and IFFT interchangeably, and similar for Discrete Fourier Transformly (DFT) and Fast Fourier Transform (FFT).

Note that the input to this IFFT is made up of N samples (the symbols for the different subcarriers), and therefore the output from the IFFT also consists of N values. These N values now have to be transmitted, one after the other, as temporal samples – this is the reason why we have a P/S (Parallel to Serial) conversion directly after the IFFT. At the RX, we can reverse the process: sample the received signal, write a block of N samples into a vector – i.e., an S/P (Serial to Parallel) conversion – and perform an FFT on this vector. The result is an estimate $\widetilde{c}_n$ of the original data c_n.

Analog implementation of OFDM would require multiple LOs, each of which has to operate with little phase noise and drift, in order to retain orthogonality between the different subcarriers. This is usually not a practical solution. The success of OFDM is based on the above-described digital implementation that allows an implementation of the transceivers that is much simpler and cheaper. In particular, highly efficient structures exist for the implementation of an FFT (so-called *butterfly structures*), and the computational effort (per bit) of performing an FFT increases only with log (N).

OFDM can also be interpreted in the time–frequency plane. Each index i corresponds to a (temporal) rectangular pulse modulated around a specific subcarrier frequency; each index n to a carrier frequency. This ensemble of functions spans a grid in the time–frequency plane (Figure 15.3).

15.4　Frequency-Selective Channels

In the previous section, we explained how the OFDM TX and RX work in an *Additive White Gaussian Noise* (AWGN) channel. We could take this scheme without any changes, and just let it operate in a frequency-selective channel. Intuitively, we would anticipate that delay dispersion will have only a small impact on the performance of OFDM: since we convert the system into a parallel system of narrowband channels, the symbol duration on each carrier is made much larger than the delay spread. But, as we saw in Section 11.3,

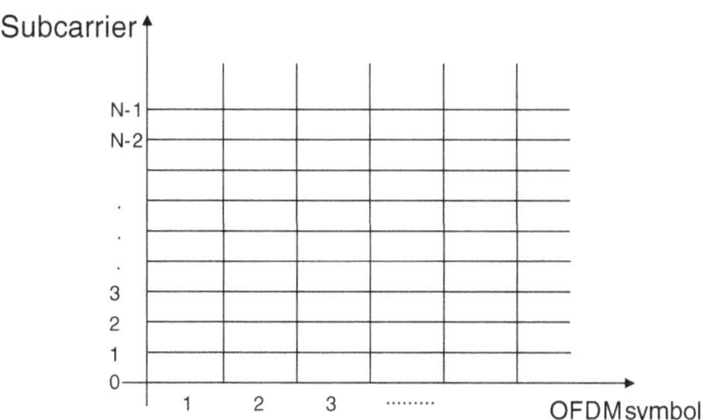

Figure 15.3　Time-frequency plane.

delay dispersion can lead to appreciable errors even when $S_\tau/T_s < 1$. Furthermore, as we will elaborate below, delay dispersion also leads to a loss of orthogonality between the subcarriers, and thus to *Inter Carrier Interference* (ICI). Fortunately, both these negative effects can be eliminated by a special type of guard interval, called the CP. In this section, we show how to construct this CP, how it works, and what performance can be achieved in frequency-selective channels.

15.4.1 Cyclic Prefix

Let us first define a new (unnormalized) basis pulse for transmission:

$$g_n(t) = \exp\left[j2\pi n \frac{W}{N} t\right] \quad \text{for } -T_{cp} < t < \hat{T}_S \tag{15.5}$$

where again W/N is the carrier spacing, and $\hat{T}_S = N/W$. The symbol duration T_S is now $T_S = \hat{T}_S + T_{cp}$. This definition of the basis pulse means that for duration $0 < t < \hat{T}_S$ the "normal" OFDM symbol is transmitted (Figure 15.4). It can be easily seen by substituting in Eq. (15.5) that, $g_n(t) = g_n(t + N/W)$. Therefore, during time $-T_{cp} < t < 0$, a copy of the last part of the symbol is transmitted. From linearity, it also follows that the *total* signal $s(t)$ transmitted during time $-T_{cp} < t < 0$ is a copy of s(t) during the last part, $\hat{T}_S - T_{cp} < t < \hat{T}_S$. This prepended part of the signal is called the CP.

Now that we know what a CP is, let us investigate why it is beneficial in delay-dispersive channels. When transmitting any data stream over a delay-dispersive channel, the arriving signal is the linear convolution of the transmitted signal with the channel impulse response. The CP converts this *linear* convolution into a *cyclical* convolution. During the time $-T_{cp} < t < -T_{cp} + \tau_{max}$, where τ_{max} is the maximum excess delay of the channel, the received signal suffers from "regular" InterSymbol Interference (ISI), as echoes of the last part of the preceding symbol interfere with the desired symbol; thus if $\tau_{max} \leq T_{cp}$, the ISI is restricted to the time interval $-T_{cp} < t < 0$. This "regular" ISI is eliminated by discarding the received signal during this time interval. During the remainder of the symbol, we have *cyclical* ISI; especially, it is the last part of the current (not the preceding) symbol that interferes with the first part of the (nondiscarded part of the) current symbol. In the following, we show how an extremely simple mathematical operation can eliminate the effect of such a cyclical convolution.

For the following mathematical derivation, we assume that the duration of the impulse response is exactly equal to the duration of the prefix; furthermore, in order to simplify the notation, we assume (without restriction of generality) $i = 0$. In the RX, there is a bank of filters that are matched to the basis pulse *without* the CP:

$$\bar{g}_n(t) = \begin{cases} g_n^*(\hat{T}_S - t) & \text{for } 0 < t < \hat{T}_S \\ 0 & \text{otherwise.} \end{cases} \tag{15.6}$$

This operation removes the first part of the received signal (of duration T_{cp}) from the detection process; as discussed above, the matched filtering of the remainder can be realized as an FFT operation. The signal at the output of the matched filter is thus the convolution of the transmit signal with the channel impulse response and the receive filter:

$$r_{n,0} = \int_0^{\hat{T}_S} \left[\int_0^{T_{cp}} h(t,\tau) \left(\sum_{k=0}^{N-1} c_{k,0} g_k(t-\tau) \right) d\tau \right] g_n^*(t)dt + n_n \tag{15.7}$$

where n_n is the noise at the output of the matched filter. Note that the argument of g_k can attain values between $-T_{cp}$ and $\hat{T}_S$, which is the region of definition of Eq. (15.5). If the channel can be considered as constant during the time $\hat{T}_S$, then $h(t, \tau) = h(\tau)$, and we obtain:

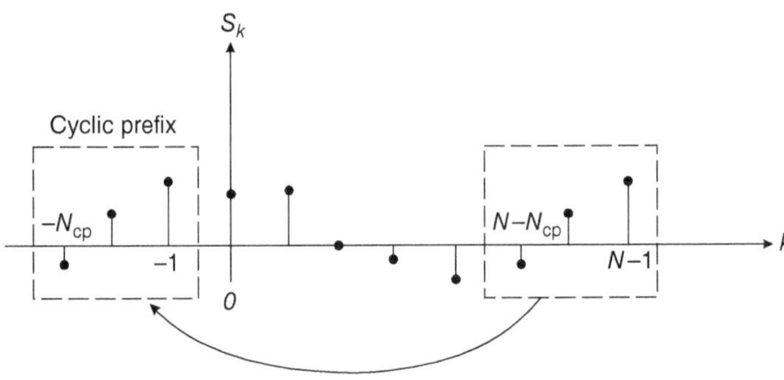

Figure 15.4 Principle of the cyclic prefix. $N_{cp} = NT_{cp}/(N/W)$ is the number of samples in the cyclic prefix.

$$r_{n,0} = \sum_{k=0}^{N-1} c_{k,0} \int_0^{\hat{T}_S} \left[\int_0^{T_{cp}} h(\tau)(g_k(t-\tau))\, d\tau \right] g_n^*(t)\, dt + n_n.$$

(15.8)

The inner integral can be written as

$$\exp\left[j2\pi tk\frac{W}{N} \right] \int_0^{T_{cp}} h(\tau)\exp\left(-j2\pi\tau k\frac{W}{N} \right) d\tau = g_k(t)H\left(k\frac{W}{N} \right)$$

(15.9)

where $H\left(k\dfrac{W}{N} \right)$ is the channel transfer function at the frequency kW/N. Since, furthermore, the basis pulses $g_n(t)$ are orthogonal during the time $0 < t < \hat{T}_S$:

$$\int_0^{\hat{T}_S} g_k(t)g_n^*(t)\, dt = \delta_{kn}(t)$$

(15.10)

the received signal samples r can be written as

$$r_{n,0} = H\left(n\frac{W}{N} \right)c_{n,0} + n_n.$$

(15.11)

The OFDM system is thus represented by a number of parallel *nondispersive* fading channels, each with its own complex attenuation $H\left(n\dfrac{W}{N} \right)$. Equalization of the system thus becomes exceedingly simple: it just requires division by the transfer function at the subcarrier frequency, independently for each subcarrier. In other words, the CP has recovered the orthogonality of the subcarriers.

Two caveats have to be noted: (i) we assumed in the derivation that the channel is static for the duration of the OFDM symbol. If this assumption is not fulfilled, interference between the subcarriers can still occur (see Section 15.7); (ii) discarding part of the received signal decreases the Signal-to-Noise Ratio (SNR), as well as spectral efficiency. For usual operating parameters (CP about 10% of symbol duration), this loss is tolerable.

The block diagram of an OFDM system, including the CP, is given in Figure 15.5. The original data stream is S/P converted. Each block of N data symbols is subjected to an IFFT, and then the last NT_{cp}/T_S samples are prepended. The resulting signal is modulated onto a (single) carrier and transmitted over a channel, which distorts the signal and adds noise. At the RX, the signal is partitioned into blocks. For each block, the CP is stripped off, and the remainder is subjected to an FFT. The resulting samples (which can be interpreted as the samples in the frequency domain) are "equalized" by means of one-tap equalization – i.e., division by the complex channel attenuation – on each carrier.

15.4.2　Performance in Frequency-Selective Channels

The CP converts a frequency-selective channel into a number of parallel flat-fading channels. This is positive in the sense that it gets rid of the ISI that plagues Time Division Multiple Access (TDMA) and Code Division Multiple Access (CDMA) systems. On the downside, an uncoded OFDM system does not show any frequency diversity at all. If a subcarrier is in a fading dip, then the error probability on that subcarrier is very high, and dominates the *Bit Error Rate* (BER) of the total system for high SNRs.

Example 15.1　*Bit error rate of uncoded* OFDM.

Figure 15.6 shows the transfer function and the BER of a *Binary-Phase Shift Keying* (BPSK) OFDM system for specific realization of a frequency-selective channel. Obviously, the BER is highest in fading dips. Note that the results are plotted on a logarithmic scale – while the BER on "good" subcarriers can be as low as 10^{-4}, the BER on subcarriers that are in fading dips are up to 0.5. This also has a significant impact on average error probability; the error probability on bad subcarriers dominates the behavior. Figure 15.7 shows a simulation of the average BER (over many channel realization) for a frequency-selective channel. We find that the BER decreases only linearly as the SNR increases, closer inspection reveals that the result is the same as in Figure 11.6 (i.e., time-varying frequency-flat channel).

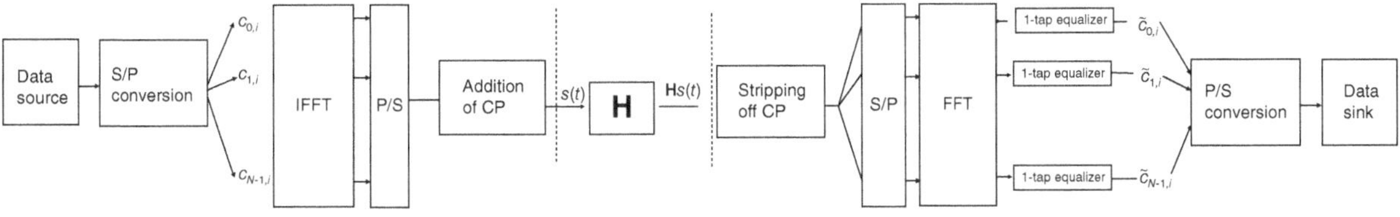

Figure 15.5　Structure of an orthogonal-frequency-division-multiplexing transmission chain with cyclic prefix and one-tap equalization. DAC and ADC not shown for conciseness.

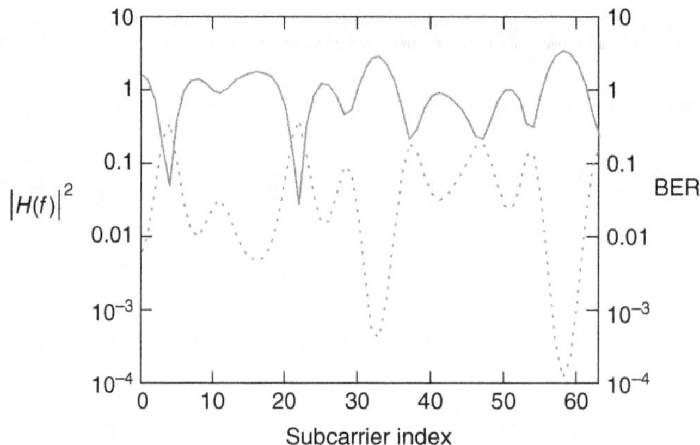

Figure 15.6 Normalized squared magnitude of the transfer function (solid), and bit error rate (dashed), for a channel with taps at [0, 0.89, 1.35, 2.41, 3.1] with amplitudes [1, −0.4, 0.3, 0.43, 0.2]. The average signal-to-noise ratio at the RX is 3 dB; the modulation format is binary-phase shift keying. Subcarriers are at $f_k = 0.05k$, $k = 0, ..., 63$.

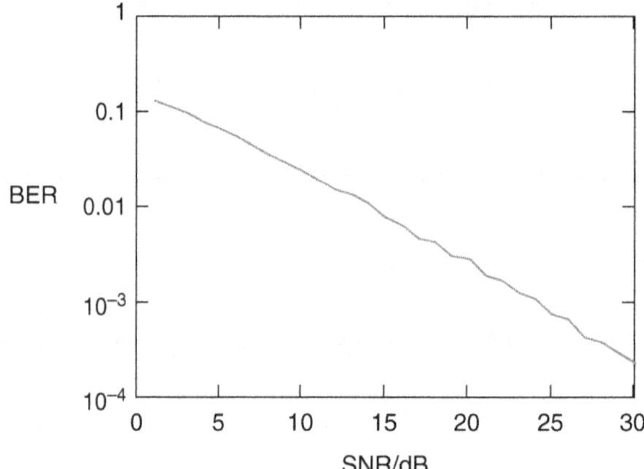

Figure 15.7 Bit error rate for a channel with taps at [0, 0.89, 1.35, 2.41, 3.1] with mean powers [1, 0.16, 0.09, 0.185, 0.04], each tap independently Rayleigh fading. The modulation format is binary-phase shift keying. Subcarriers are at $f_k = 0.05k$, $k = 0, ..., 63$.

More generally, we find that uncoded OFDM has the same average BER irrespective of the frequency selectivity of the channel. This can also be interpreted the following way: frequency selectivity gives us different channel realizations on different subcarriers; time variations give us different channel realizations at different times. Doubly selective channels have different realizations on different subcarriers as well as different times. But, for computation of the average BER, it does not matter how the different realizations are created, as long as the fading has the same statistics (e.g., Rayleigh), and the ensemble is large enough.[2]

From these examples, we see that the main problem lies in the fact that carriers with poor SNR dominate the performance of the system. Any of the following approaches circumvents this problem:

- *Coding across the different tones*: such coding helps to compensate for fading dips on one subcarrier by a good SNR in another subcarrier. This is described in more detail in Section 15.4.3.
- *Spreading the signal over all tones*: in this approach, each symbol is spread across all carriers, so that it sees an SNR that is the average of all tones over which it is spread. This method is discussed in more detail in Sections 15.11 and 15.12.
- *Adaptive modulation*: if the TX knows the SNR on each of the subcarriers, it can choose its modulation alphabet and coding rate adaptively. Thus, on carriers with low SNR, the TX will send symbols using stronger encoding and a smaller modulation alphabet. Also, the power allocated to each subcarrier can be varied. This approach is described in more detail in Section 15.9.

---

[2] Note that in a time-invariant, frequency-selective channel, the number of independent channel realizations depends on the ratio of system bandwidth to coherence bandwidth of the channel. If this value is small, there might not be a sufficiently large ensemble to obtain good averaging.

15.4.3 Coded Orthogonal Frequency Division Multiplexing

Just as coding can be used to great effect in single-carrier systems to improve performance in fading channels, so can it be gainfully employed in OFDM systems. But now we have data that are transmitted at different frequencies as well as at different times. This gives rise to the question of how coding of the data should be applied.

To get an intuitive feeling for coding across different subcarriers, imagine again the simple case of repetition coding: each of the symbols that are to be transmitted is repeated on K different subcarriers. As long as fading is independent of the different subcarriers, K-fold diversity can be achieved. In the simplest case, the RX first makes a hard decision about symbols on each subcarrier, and then makes a majority decision among the K received symbols about which bit was sent. Of course, practical systems do not use repetition coding, but the principle remains the same.

We could now try and develop a whole theory for coding on OFDM systems. However, it is much easier to just consider the analogy between the time domain and the frequency domain. Remember the main lessons from Section 13.10: enough interleaving should be applied such that fading of coded bits is independent. In other words, we just need independent channel states over which to transmit our coded bits; this will automatically result in a high diversity order. It does not matter whether channel states are created by temporal variations of the channel, or as different transfer functions of subcarriers in frequency-selective channels. Thus, it is not really necessary to define new codes for OFDM, but it is more a question on how to design appropriate mappers and interleavers that assign the different coded bits in the time–frequency plane. This mapping, in turn, depends on the frequency selectivity as well as the time selectivity of the channel. If the channel is highly frequency selective, then it might be sufficient to code *only* across available frequencies, without any coding or interleaving along the time axis. This has two advantages: on one hand, this scheme also works in static channels, which occur quite often for wireless LANs and other nomadic transmission scenarios; on the other, the absence of interleaving in the time domain results in lower latency of the transmission and decoding process.

Figure 15.8 shows a performance example. We see that for AWGN, both rate-1/3- and rate-3/4-coded systems exhibit good performance, with approximately a 1-dB difference. In fading channels, performance is dramatically different. While the rate-1/3 code has good diversity, and therefore the BER decreases fast as a function of the SNR, the rate-3/4 code has very little frequency diversity, and thus bad performance.

15.5 Channel Estimation

As for any other coherent wireless system, operation of OFDM systems requires an estimate of the *Channel State Information* (CSI); specifically the channel transfer function, or, equivalently, the channel impulse response. Since OFDM is operating with a number of parallel narrowband subcarriers, it is intuitive to estimate the channel in the frequency domain. More precisely, we wish to obtain estimates of the N complex-valued channel gains on the subcarriers. Let us denote these channel gains as $h_{n,i}$, where n is the subcarrier index and i is the time index.

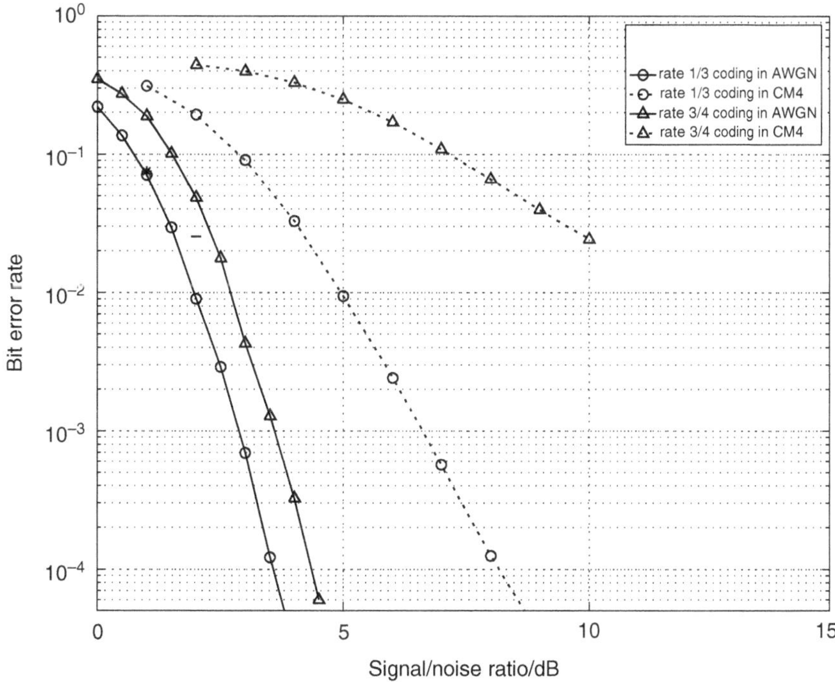

Figure 15.8 Bit error rate as a function of the signal-to-noise ratio for rate-1/3- and rate-3/4-coded OFDM system. Channel is either AWGN or fading channel (model 4 of the IEEE 802.15.3a channels). The OFDM system follows the specifications of the WiMedia standard. Reprinted with permission from [Ramachandran et al. 2004] © JEEE.

In the following, we first discuss "single-shot" channel estimation techniques, based on the transmission of a single pilot symbol. We will discuss estimators that exploit the structure of the channel in a variety of ways, and in particular, show performance can be improved if we know the statistical properties of the transfer functions. We then proceed to the case of time varying channels, where pilot symbols are transmitted at periodic intervals, and discuss how to exploit known temporal properties of the channel. We conclude with an exposition on joint channel estimation and data detection.

15.5.1 Single-Shot Estimation with Pilot Symbol

We first consider the case where we have a dedicated pilot *symbol* containing only known data – in other words, the data on each of the subcarriers is known; we also assume that estimation based on a single pilot symbol is sufficient, and the channel does not change afterward. This approach is appropriate for initial estimation of the channel, e.g., at the beginning of a transmission burst; it is used, e.g., in Wi-Fi. Denoting the known data on subcarrier n at time i as $c_{n,i}$, we can find a least-squares channel estimate as

$$\hat{h}_{n,i}^{LS} = r_{n,i}/c_{n,i} \tag{15.12}$$

where $r_{n,i}$ is the received signal on subcarrier n. Note that this estimates the channel on each subcarrier independently, and thus neglects the correlation between the subcarriers.

The least-square estimate can be improved by exploiting the correlation of the channel at the different subcarrier frequencies. This exploitation can be implemented in different ways:

- Incorporating the second-order statistics of the transfer function, i.e., the frequency correlation function.
- Exploiting the nature of the channel impulse response in the delay domain, i.e., considering it as sum of contributions from Multi-Path Components (MPCs), and thus possibly exhibiting sparsity and limited support in the delay domain. Application of parametric channel estimation can be very useful in this context.

Estimation Based on Second-Order Statistics

We first consider the exploitation of second-order statistics, which are assumed to be known. Intuitively speaking, since the channels at adjacent subcarriers are strongly correlated, fast fluctuations from carrier to carrier are likely caused by noise; the frequency correlation function should be applied like a low-pass filter to the estimates to smooth out the errors the noise induces in the LS estimates.

Mathematically speaking, the knowledge of the frequency correlation function allows to consider the estimation problem as a Bayesian estimation problem in which we know the prior. Applying standard estimation theory, we can obtain a Linear Minimum Mean Square Error (LMMSE) estimator: arranging the LS estimates in a vector

$$\hat{\mathbf{h}}_i^{LS} = \left(\hat{h}_{1,i}^{LS} \ \ \hat{h}_{2,i}^{LS} \ \ \cdots \ \ \hat{h}_{n,i}^{LS} \right)^T \tag{15.13}$$

the corresponding vector of linear MMSE estimates becomes

$$\hat{\mathbf{h}}_i^{LMMSE} = \mathbf{R}_{h\hat{h}^{LS}} \mathbf{R}_{\hat{h}^{LS}\hat{h}^{LS}}^{-1} \hat{\mathbf{h}}_i^{LS} \tag{15.14}$$

where $\mathbf{R}_{h\hat{h}^{LS}}$ is the covariance matrix between the channel gains and the least-squares estimate of the channel gains, $\mathbf{R}_{\hat{h}^{LS}\hat{h}^{LS}}$ is the autocovariance matrix of the least-squares estimates. Given that we have AWGN with variance σ_n^2 on each subcarrier, $\mathbf{R}_{h\hat{h}^{LS}} = \mathbf{R}_{hh}$, i.e., the autocovariance matrix of the channel gains

$$\mathbf{R}_{hh} = E\left\{ \mathbf{h}_i \mathbf{h}_i^\dagger \right\} \tag{15.15}$$

where $\mathbf{h}_i = \left(h_{1,i} \ \ h_{2,i} \ \ \cdots \ \ h_{n,i} \right)^T$. Note that $\mathbf{R}_{hh}$ is independent of time i if the channel is wide-sense stationary. Finally, then

$$\mathbf{R}_{\hat{h}^{LS}\hat{h}^{LS}} = \left(\mathbf{R}_{hh} + \frac{\beta}{\gamma} \mathbf{I} \right) \tag{15.16}$$

where $\beta = E\{|c_{n,i}|^2\}E\{1/|c_{n,i}|^2\}$ is a factor that depends on the modulation format of the pilot tones, and specifically becomes unity if the pilot symbol is constant envelope $|c_{n,i}| = const$; $\gamma = E\{|c_{i,n}|^2\}/\sigma_n^2$ is the SNR (at the TX); we will assume this henceforth. With all these simplifying assumptions, the channel estimator takes on a form that is widely used

$$\hat{\mathbf{h}}_i^{\text{LMMSE}} = \mathbf{R}_{hh}\left(\mathbf{R}_{hh} + \frac{1}{\gamma}\mathbf{I}\right)^{-1}\hat{\mathbf{h}}_i^{\text{LS}}. \tag{15.17}$$

This estimation approach has its strengths and weaknesses. It produces very good estimates, but the computational complexity is high if the number of subcarriers is large: it requires N^2 multiplications, i.e., N multiplications per estimated channel gain (assuming that all correlation matrices and inversions are precalculated). This is quite a large complexity, even if this pilot-symbol-based estimation usually is done only at the beginning of a transmission burst. For that reason, there are several other sub-optimal approaches available, where, e.g., smoothing FIR-filters of limited length (much less than N) are applied across the least-squares estimated attenuations to exploit the correlation between neighboring subcarriers.

We finally note that the estimation of the second-order statistics might sometimes not be possible, because no sufficient number of pilot symbols could be observed in the past. However, the support of the impulse response, i.e., the duration of the channel impulse response (in units of T_s), can often be determined. Using a rectangular power delay profile of this length provides a crude but still sometimes useful approximation that can be used for the computation of the frequency correlation function. It can be shown that when the PDP has N_τ contiguous nonzero samples, and there are N_p pilot tones distributed uniformly across frequency (in other words, the pilot tones determine an impulse response of length N_p), then an SNR improvement of N_p/N_τ can be expected. Intuitively, this is achieved by simply not collecting samples in the "noise-only" part of the impulse response.

*Improved Efficiency Estimation Based on Second-Order Statistics

As discussed above, a naive implementation of the estimator Eq. (15.14) or its variants, has rather high complexity as it implies the inversion and multiplication of matrices of large sizes. However, the structure of OFDM allows for more efficient channel estimator structures. We know that the channel impulse response is short compared to the OFDM symbol length in any well-designed system. This fact can be used to reduce the dimensionality of the estimation problem. In essence, when using the LMMSE estimator in (15.14), we would like to use the statistical properties of the channel to perform the matrix multiplication more efficiently. This can be done using the theory of optimal rank-reduction from estimation theory, where a singular-value decomposition[3] of the channel autocovariance matrix $\mathbf{R}_{hh} = \mathbf{U}\mathbf{\Lambda}\mathbf{U}^\dagger$ results in a new more computationally efficient version of (15.14). The dimension of this space is (assuming the duration of the impulse response equals the CP) approximately $N_{\text{cp}} + 1$, i.e., one more than the number of samples in the CP. We can therefore expect that after the first $N_{\text{cp}} + 1$ diagonal elements in Λ the magnitude should decrease rapidly. Using the Singular Value Decomposition (SVD) to rewrite (15.14) as (assuming no dependence on i)

$$\hat{\mathbf{h}}^{\text{LMMSE}} = \mathbf{U}\mathbf{\Delta}\mathbf{U}^\dagger\hat{\mathbf{h}}^{\text{LS}} \tag{15.18}$$

where Δ is a diagonal matrix containing the values $\delta_n = \lambda_n/(\lambda_n + 1/\gamma)$ on its diagonal. The diagonal elements δ_n will decrease rapidly after the first $N_{\text{cp}} + 1$ since the λ_n's do. By setting all but the p first λ_n's to zero, i.e. assigning $\delta_n = 0$ for $n > p$, we get an optimal rank-p estimator for the channel gains. The computational complexity of this estimator is $2Np$ multiplications, which is $2p$ per estimated transfer function sample. This should be compared with the N multiplications per estimated coefficient in the original estimator (15.14). The estimator principle as a block diagram is illustrated in Figure 15.9.

In the case that the autocorrelation matrix $\mathbf{R}_{hh}$ is a circulant matrix, the resulting optimal transforms $\mathbf{U}^\dagger$ and $\mathbf{U}$ are the IDFT and DFT, respectively, and there are only N_{cp} nonzero singular values. The basic estimator structure stays the same, as shown in Figure 15.10, while the FFT processor already available in the OFDM RX can be used to perform the channel estimation as well.

In many cases, when the channel correlation matrix is not circulant, the computational efficiency of the DFT-based estimators may outweigh the suboptimality of their rank reduction.

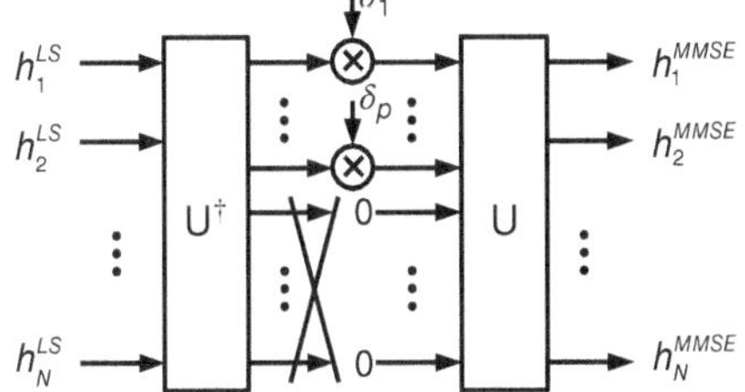

Figure 15.9 The optimal rank-p channel estimator viewed as a transform ($U^\dagger$) followed by p scalar multiplications and a second transform (U).

---

[3] In this particular case the SVD is equivalent to an eigenvalue decomposition, since the matrix is square and Hermitian symmetric.

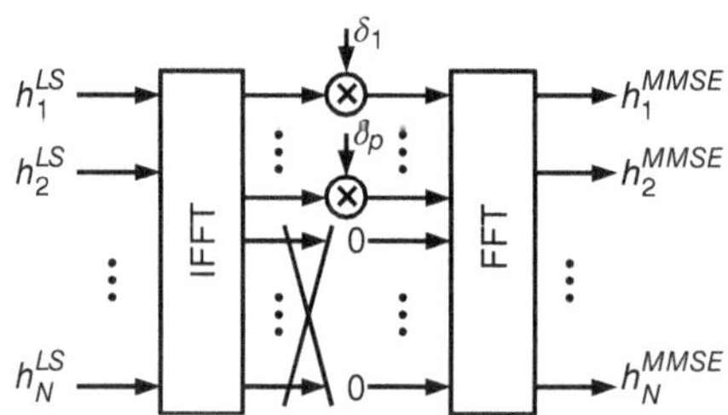

Figure 15.10   Low-rank estimator for channels with circulant autocorrelation, implemented using fast Fourier transforms.

*Parameter-Estimation-Based Methods

We have established in Section 7.2.1 that the impulse response can be approximated as a finite sum of contributions from MPCs

$$h(\tau) = \sum_{\ell=1}^{\tilde{N}} c_{\text{channel},\ell}\delta(\tau - \tau_\ell).$$

Thus, the $2\tilde{N}$ parameters $c_{\text{channel},\ell}$ and τ_ℓ determine the impulse response or transfer function. In other words, if we can estimate these parameters, then the channel transfer function follows immediately. If the number of parameters to be estimated is smaller than the number of subcarriers, a gain in SNR is obtained, with the value $N/\tilde{N}$. Note that we need to estimate only MPCs that are (approximately) an inverse system bandwidth apart, since more closely spaced MPCs can be merged into an "effective" MPC. The key advantage of this method arises when the impulse response is "sparse" in the sense that resolvable delay bins containing MPCs are interlaced with delay bins that do not contain any MPCs.

Estimates of the parameters can be obtained in a variety of ways: the methods described in Section 9.6, such as ESPRIT and SAGE can all be applied. In particular for ultrawideband and mm-wave channels, also sparse estimation methods such as Least Absolute Shrinkage and Selection Operator (LASSO) or matching pursuit, can be used.

*15.5.2   Estimation in Time-Variant Channels

When the channel is not only frequency selective but also time-variant, the two-dimensional transfer function $\mathbf{H}$ needs to be estimated, where the (n,i)th element of $\mathbf{H}$ is the transfer function during the ith OFDM symbol interval on the nth subcarrier. Let us first consider the (unrealistic) case that the TX continuously sends out pilot symbols, each of which occupies the whole system bandwidth – in other words, all time–frequency resources are spent on pilot transmission. We can then write the LS estimates of the channel into a matrix $\hat{\mathbf{H}}^{\text{LS}}$. In principle, the treatment of the temporal correlation can be handled in the same way as the frequency correlation; more generally, we need to consider the joint time–frequency correlation function $\mathbf{R}_{HH}$, which might be *approximated* in some cases by the product of the time correlation and the frequency correlation. Under these circumstances, the LMMSE solution can be written as

$$\hat{\mathbf{H}}^{\text{LMMSE}} = \mathbf{R}_{HH}\left(\mathbf{R}_{HH} + \frac{1}{\gamma}\mathbf{I}\right)^{-1}\hat{\mathbf{H}}^{\text{LS}}. \tag{15.19}$$

Note that the correlation matrix $\mathbf{R}_{HH}$ is a four-dimensional matrix; equivalently we can stack the different columns of $\mathbf{H}$ into a vector of dimension $N \cdot T$ where T is the number of considered symbols, to then obtain a two-dimensional correlation matrix of dimension $N \cdot T \times N \cdot T$. In some cases, the time–frequency correlation matrix might be *approximated* by the product of the time correlation and the frequency correlation.[4]

Obviously, spending all spectral resources on pilot tones is not acceptable in practice, and it is not necessary. Rather, we can obtain time–frequency channel estimates from pilot symbols scattered in the OFDM time–frequency grid as illustrated in Figure 15.11, where the pilot spacing in time (in units of OFDM symbols) is denoted as N_t, and the spacing in frequency (in units of subcarriers) as N_f.[5]

When estimating the channel based on scattered pilots, we can start by performing least-squares estimation of the channel at pilot positions, i.e., $\hat{h}_{n,i}^{\text{LS}} = r_{n,i}/c_{n,i}$, where $r_{n,i}$ is the received value and $c_{n,i}$ is the known pilot data in pilot position (n, i). From these initial estimates at pilot positions, we would like to perform interpolation to obtain estimates of the channel at all other positions. Interpreting the pilots as samples in a two-dimensional space, we can use standard sampling theory to put limits on the required density of our pilot pattern as [Nilsson et al. 1997]

---

[4] In the case of the stacked vector, the total correlation matrix is then a Kronecker product of the temporal correlation matrix and the frequency correlation matrix.

[5] We have used a rectangular pilot pattern in the illustration, but other pilot patterns can be used as well.

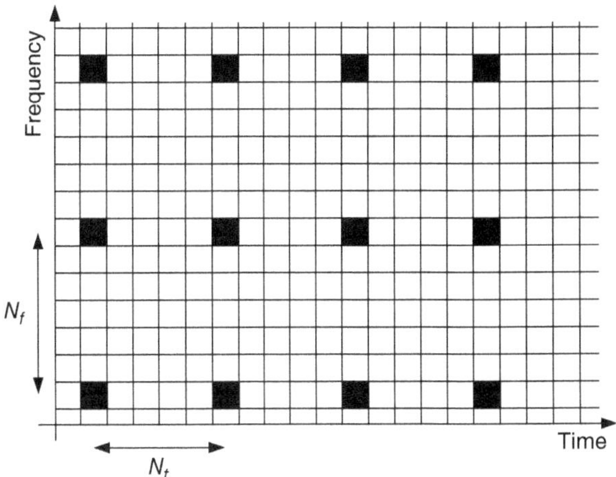

Figure 15.11 Scattered pilots in the ODFM time–frequency grid. In this case, the pattern is rectangular with pilot distances N_f subcarriers in frequency and N_t OFDM symbols in time.

$$N_f < \frac{N}{N_{cp}} \tag{15.20}$$

$$N_t < \frac{1}{2\left(1 + N_{cp}/N\right)\nu_{max}}.$$

Since we need to reduce the effect of noise of the pilots and also help to reduce the complexity of estimation algorithms, it has been argued that a good trade-off is to place twice as many pilots in each direction as required by the sampling theorem [Nilsson et al. 1997].

In principle, the channel estimation with scattered pilots can be done using the same estimation theory as for the all-pilot symbol case. When estimating a certain channel attenuation $h_{n,\,i}$ using a set of K pilot positions $(n_j, i_j), j = 1, ..., K$, we place the least-squares estimates in a pilot-vector $\hat{\mathbf{p}}^{LS} = \left(\hat{h}^{LS}_{n_1,i_1} \quad \hat{h}^{LS}_{n_2,i_2} \quad \cdots \quad \hat{h}^{LS}_{n_K,i_K} \right)^T$ and calculate the LMMSE estimate as

$$\hat{h}_{n,i}^{LMMSE} = \mathbf{r}_{h\hat{p}^{LS}} \mathbf{R}^{-1}_{\hat{p}^{LS}\hat{p}^{LS}} \hat{\mathbf{p}}^{LS} \tag{15.21}$$

where $\mathbf{r}_{h\hat{p}^{LS}}$ is the correlation (row) vector $E\left\{ h_{n,i}\hat{\mathbf{p}}^{LS\dagger} \right\}$ and $\mathbf{R}_{\hat{p}^{LS}\hat{p}^{LS}}$ is the autocovariance matrix of the least-squares estimates at the pilot positions. Channel states for the symbols/subcarriers that do not have a pilot can be determined by interpolation, assuming that Eq. (15.20) is fulfilled.

The complexity of this estimator grows with the number of pilot tones included in the estimation and requires K multiplications per estimated channel coefficient, again assuming that all correlation matrices and inversions are precalculated. To obtain good channel estimates, a quite large number of pilots may have to be used, in particular in connection with multi-antenna techniques. Chapters 31, 32 discusses how this problem is solved in current cellular standards.

An alternative approach to the 2-D filtering above, where we use pilots in both the frequency and time direction at the same time, is to apply separable filters. This implies that we use two 1-D filters, one in the time and the other in the frequency direction. Many more pilots are thus influencing each estimated channel coefficient, for a given estimator complexity. The resulting increase in performance has been shown to dominate over the loss in optimality when going from general 2-D filters to separable ones based on two 1-D filters.

The rank-reduction structure discussed above in Section 15.5.1 has also been used as one of the two 1-D estimators when performing 2-D estimation (see above). The gain here is that the time-direction smoothing can be done between the two transforms, leading to a smaller number of filters that have to be applied in parallel. Instead of N filters (one per subcarrier), only p filters are needed in a rank-p estimator, as shown in Figure 15.12.

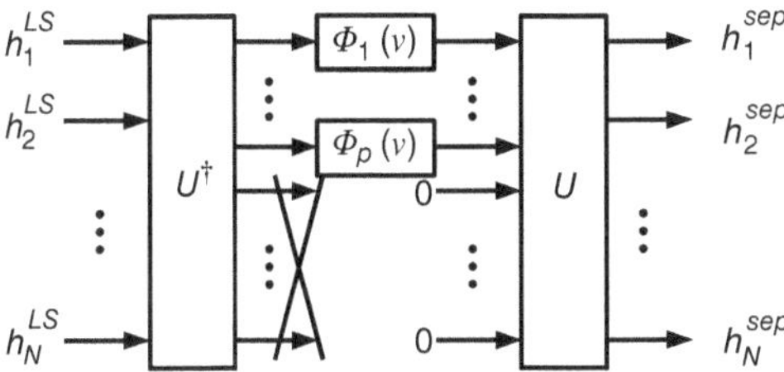

Figure 15.12 2-D (separable) channel estimation where time-domain smoothing is done in the transform domain. This reduces the number of parallel filters needed, from N to p.

*15.5.3　Joint Channel Estimation and Data Detection

We have up to now treated channel estimation as a problem that is completely separate from data detection; in particular, we assumed that we can determine the CSI from the pilots and subsequently use this information for the coherent demodulation/detection of the data. The CSI estimation thus services the data detection. However, it is also possible for the data detection process to help the channel estimation, either to reduce the overhead for pilots or to improve the channel estimates, similar to the approach for single-carrier systems described in Section 14.7.6.

One approach is to exploit the detected data for channel tracking, obviating the need for pilot symbols beyond a first symbol. In essence, the RX operates in a decision-directed mode: it uses the channel estimates from the first pilot symbol (at time $i = 0$) to detect the first OFDM symbol, $i = 1$. After demodulation and decoding, the data of this OFDM symbol are known and can thus be used as an effective pilot symbol. Using this new pilot symbol, it is now possible to obtain an updated channel estimate (valid now at time $i = 1$), which is used to demodulate the OFDM symbol at time $i = 2$, and so on. This approach can be further improved by exploiting the temporal correlation function, and thus refine the channel estimates – in other words, all OFDM symbols transmitted in the past, over the whole bandwidth, act as pilot symbols.

The main challenge of this approach is error propagation: incorrect estimation of a data symbol, $\hat{c}_{n,i} \neq c_{n,i}$ leads to increased error in the channel estimation (since $\hat{h}_{n,i}^{\rm LS} = r_{n,i}/\hat{c}_{n,i}$). This in turn increases the probability of further data detection errors, which further worsen the channel estimation quality, and so on (very similar to the error propagation in decision feedback equalizers, compare Section 14.3). Under typical operating conditions, data detection errors are likely to occur if the raw detected symbols (with hard decision) are used for $\hat{c}_{n,i}$; the error probability is drastically reduced if the symbols as obtained at the output of the error-correction decoder are used. Note, however, that – depending on the error correction code – a chicken-and-egg problem might arise: consider a block code that extends over 20 OFDM symbols' say, $i = 41$, ..., 60. The decoding can only be done based on the (coherently demodulated) soft symbols of all those 20 OFDM symbols. However, the only decoded symbols, and associated channel estimates, that are available are based on the *previous* code block ($i = 21$, ..., 40); if the channel varies rapidly, those estimates will be outdated by the time the codeword ends, $i = 60$. To demodulate the symbol with $i = 60$, we would like to have the channel state at $i = 59$, but we can get that state only *after* the decoding of the current symbol has been done. This chicken and egg problem can be solved by suitable iteration, as explained next.

Instead of performing channel estimation and decoding as separate problems (even if they are done in small steps), we can see them as a *joint* problem: the demodulated/decoded data can provide improved information about the channel states, and improved channel estimation results in better demodulation and decoding. For such a joint problem, application of the turbo principle (compare Section 13.6) has obvious appeal: soft information (which may consist of *a priori* and *a posteriori* information) is fed iteratively between a channel estimator – plus – demapper on one hand, and the channel decoder on the other hand. A variety of implementations for the corresponding turbo RXs have been proposed. Similarly, joint channel estimation and decoding can be tackled via factor graphs and belief propagation (compare Section 13.7).

15.6　Peak-to-Average Power Ratio

15.6.1　Origin of the Peak-to-Average Power Ratio Problem

One of the major problems of OFDM is that the peak amplitude of the emitted signal can be considerably higher than the average amplitude. This *Peak-to-Average Power Ratio* (PAPR) issue originates from the fact that an OFDM signal is the superposition of N sinusoidal signals on different subcarriers. On average the emitted power is linearly proportional to N. However, sometimes, the signals on the subcarriers add up constructively, so that the *amplitude* of the signal is proportional to N, and the power thus goes with N^2. We can thus anticipate the (worst case) PAPR to increase linearly with the number of subcarriers.

We can also look at this issue from a slightly different point of view: the contributions to the total signal from the different subcarriers can be viewed as random variables (they have quasi-random phases, depending on the sampling time as well as the value of the symbol with which they are modulated). If the number of subcarriers is large, we can invoke the central limit theorem to show that the distribution of the amplitudes of in-phase components is Gaussian, with a standard deviation $\sigma = 1/\sqrt{2}$ (and similarly for the quadrature components) assuming that mean power is unity. Since both in-phase and quadrature components are Gaussian, the absolute amplitude is Rayleigh distributed, and the power is exponentially distributed (see Chapter 5 for details of this derivation). Knowing the amplitude distribution, it is easy to compute the probability that the instantaneous amplitude will lie above a given threshold, and similarly for power. For example, there is a $\exp(-10^{6/10}) = 0.019$ probability that the peak power is 6 dB above the average power. Note that the Rayleigh distribution can only be an approximation for the amplitude distribution of OFDM signals: an actual OFDM signal has a bounded amplitude ($N\times$ amplitude of signal on one subcarrier), while realizations of a Rayleigh distribution can take on arbitrarily large values.

There are three main methods to deal with the PAPR:

1. Put a power amplifier into the TX that can amplify linearly up to the possible *peak* value of the transmit signal. This typically requires expensive and power-consuming class-A amplifiers. The larger the number of subcarriers N, the more difficult this solution becomes.

2. Use a nonlinear amplifier, and accept the fact that amplifier characteristics will lead to distortions in the output signal. Those nonlinear distortions destroy orthogonality between subcarriers and also lead to increased out-of-band emissions (*spectral regrowth* – similar to

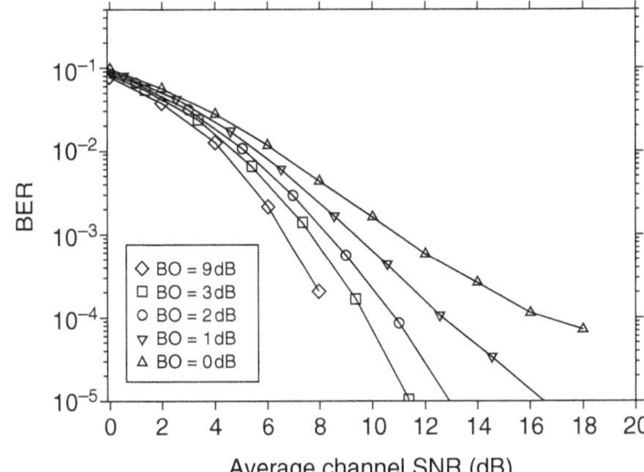

Figure 15.13 Bit error rate as a function of the signal-to-noise ratio, for different backoff levels of the transmit amplifier.
Reproduced with permission from [Hanzo et al. 2003] © J. Wiley & Sons, Ltd.

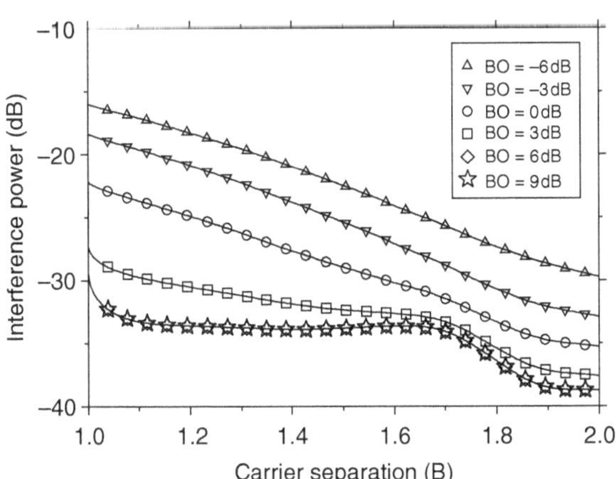

Figure 15.14 Interference power to adjacent bands (OFDM users), as a function of carrier separation, for different values of backoff of the transmit amplifier.
Reproduced with permission from [Hanzo et al. 2003] © J. Wiley & Sons, Ltd.

third-order intermodulation products – such that the power emitted outside the nominal band is increased). The first effect increases the BER of the desired signal (see Figure 15.13), while the latter effect causes interference to other users and thus decreases the cellular capacity of an OFDM system (see Figure 15.14). This means that in order to have constant adjacent channel interference we can trade-off power amplifier performance against spectral efficiency (note that increased carrier separation decreases spectral efficiency).

3. Use PAPR reduction techniques. These will be described in the next subsection.

*15.6.2 Peak-to-Average Ratio Reduction Techniques

A wealth of methods for mitigating the PAPR problem has been suggested in the literature. Some of the promising approaches are as follows:

1. *Coding for PAPR reduction*: under normal circumstances, each OFDM symbol can represent one of 2^N codewords (assuming BPSK modulation). Now, of these codewords, only a subset of size 2^K is acceptable in the sense that its PAPR is lower than a given threshold. Both the TX and the RX know the mapping between a bit combination of length K, and the codeword of length N that is chosen to represent it, and which has an admissible PAPR. The transmission scheme is thus the following: (i) parse the incoming bitstream into blocks of length K; (ii) select the associated codeword of length N; (iii) transmit this codeword via the OFDM modulator. The coding scheme can guarantee a certain value for the PAPR. It also has some coding gain, though this gain is smaller than for codes that are solely dedicated to error correction.

2. *Phase adjustments*: this scheme first defines an ensemble of phase adjustment vectors ϕ_ℓ, $\ell = 1, ..., L$, that are known to both the TX and RX; each vector has N entries $\{\phi_n\}_\ell$. The TX then multiplies the OFDM symbol to be transmitted c_n by each of these phase vectors to get the modified sequences

$$\{\hat{c}_n\}_\ell = c_n \exp\left[j(\phi_n)_\ell \right] \tag{15.22}$$

and then selects

$$\ddot{\ell} = \arg\min_{\ell} \left(PAPR(\{\hat{c}_n\}_\ell) \right) \tag{15.23}$$

which gives the lowest PAPR. The vector $\{\hat{c}_n\}_{\hat{\ell}}$ is then transmitted, together with the index $\hat{\ell}$. The RX can then undo phase adjustment and demodulate the OFDM symbol. This method has the advantage that the overhead is rather small (at least as long as L stays within reasonable bounds); on the downside, it cannot guarantee to keep the PAPR below a certain level.

3. *Correction by multiplicative function*: another approach is to multiply the OFDM signal by a time-dependent function whenever the peak value is very high. The simplest example for such an approach is the clipping we mentioned in the previous subsection: if the signal attains a level $s_k > A_0$, it is multiplied by a factor A_0 / s_k. In other words, the transmit signal is modified to

$$\hat{s}(t) = s(t) \left[1 - \sum_k \max \left(0, \frac{|s_k| - A_0}{|s_k|} \right) \right]. \tag{15.24}$$

A less radical method is to multiply the signal by a Gaussian function centered at times when the level exceeds the threshold:

$$\hat{s}(t) = s(t) \left[1 - \sum_n \max \left(0, \frac{|s_k| - A_0}{|s_k|} \right) \exp \left(-\frac{t^2}{2\sigma_t^2} \right) \right]. \tag{15.25}$$

Multiplication by a Gaussian function of variance σ_f^2 in the time domain implies convolution with a Gaussian function in the frequency domain with variance $\sigma_f^2 = 1/(2\pi\sigma_t^2)$. Thus, the amount of out-of-band interference can be influenced by the judicious choice of σ_t^2. On the downside, we find that the ICI (and thus BER) caused by this scheme is significant.

4. *Correction by additive function*: in a similar spirit, we can choose an additive, instead of a multiplicative, correction function. The correction function should be smooth enough not to introduce significant out-of-band interference. Furthermore, the correction function acts as additional pseudo noise, and thus increases the BER of the system.

When comparing the different approaches to PAPR reduction, we find that there is no single "best" technique. The coding method can guarantee a maximum PAPR value, but requires considerable overhead, and thus reduced throughput. The phase adjustment method has a smaller overhead (depending on the number of phase adjustment vectors), but cannot give a guaranteed performance. Neither of these two methods leads to an increase in either ICI or out-of-band emissions. The correction by multiplicative functions can guarantee performance – up to a point (subtracting the Gaussian functions centered at one point might lead to larger amplitudes at another point). Also, it can lead to considerable ICI, while out-of-band emissions are fairly well controlled.

15.7   Inter Carrier Interference

The CP provides an excellent way of ensuring orthogonality of the carriers in a delay-dispersive (frequency-selective) environment – in other words, there is no ICI due to frequency selectivity of the channel. However, wireless propagation channels are also time varying, and thus time selective (=frequency-dispersive). Time selectivity has two important consequences for an OFDM system: (i) it leads to random Frequency Modulation (FM, see Section 5.7.3), which can cause errors especially on subcarriers that are in a fading dip; and (ii) it creates ICI. A Doppler shift of one subcarrier can cause ICI in many adjacent subcarriers (see Figure 15.15). Another important source of frequency shifts is errors in the LO signals. Such errors can be produced by

- *Synchronization errors*: as we discussed in Section 15.4, synchronization is critical for retaining orthogonality between carriers. Any errors in the frequency synchronization procedure will result in deviations of the RX's LO from the optimum frequency, and thus ICI.
- *Phase noise of the TX and RX*: phase noise, which stems from inaccuracies in the oscillator, leads to deviation of the LO signal from its nominal, strictly sinusoidal shape. The phase noise is characterized by its power-spectral density, see Section 17.6.4. Essentially, a narrow spectrum means that phase only changes very slowly, which can be more easily compensated by various RX algorithms. The effect of phase noise is a spilling of the spectrum of subcarrier signals into adjacent subcarriers, and thus ICI.

As we will see in Section 17.6, errors in the LO increase with the carrier frequency and are thus especially relevant at mm-wave frequencies. Irrespective of the particular source of the frequency dispersion, the impact of time selectivity is mostly determined by the

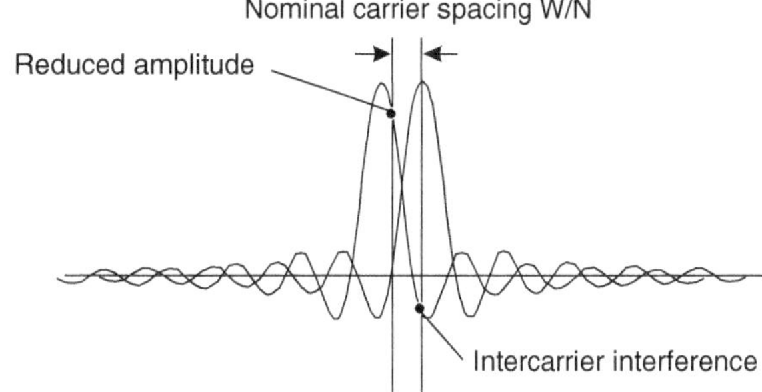

Figure 15.15   Intercarrier interference due to frequency offset.

product of mean-square frequency error and symbol duration of the OFDM symbol. The spacing between the subcarriers is inversely proportional to symbol duration. Thus, if symbol duration is large, even a small frequency dispersion can result in appreciable ICI.

Delay dispersion can be another source of ICI, namely if the CP is shorter than the maximum excess delay of the channel. This situation can arise for various reasons. A system might consciously shorten or omit the CP in order to improve spectral efficiency. In other cases, a system may originally be designed to operate in a certain class of environments (and thus a certain range of excess delays) and is later also deployed in other environments that have a larger excess delay. Finally, for many systems, the length of the CP is a compromise between the desire to eliminate ICI, and the need to retain spectral efficiency – in other words, a CP is not chosen to cope with the worst-case channel situation.

In the following, we mathematically describe the received signal if ICI occurs either as a result of frequency dispersion or insufficient CP. Instead of Eq. (15.11), the relationship between data symbols c_n and receive samples after FFT is now given by

$$r_k = \sum_{n=0}^{N-1} c_n H_{k,n} + n_k \tag{15.26}$$

where

$$H_{k,n} = \frac{1}{N} \sum_{q=0}^{N-1} \sum_{\ell=0}^{L-1} h[q,\ell] \exp\left[j\frac{2\pi}{N}(qn - n\ell - qk) \right] \mathcal{H}\left[q - \ell + N_{\text{cp}} \right] \tag{15.27}$$

where $h(q, \ell)$ is a sampled version of the time-variant channel impulse response $h(t, \tau)$, $\mathcal{H}[\cdot]$ denotes the Heaviside function, and L is the maximum excess delay in units of samples $L = \tau_{\max} N/\hat{T}_S$. Note also that Eq. (15.27) reduces to Eq. (15.11) for the case of a time-invariant channel and a sufficiently long guard interval.

Because ICI can be a limiting factor for OFDM systems, a large range of techniques for fighting it has been developed and can be classified as follows:

- *Optimum choice of carrier spacing and OFDM symbol length*: in this approach, we influence the OFDM symbol length in order to minimize its ICI. It follows from our statements above that short symbol duration is good for reduction of frequency-dispersion-induced ICI. On the other hand, spectral efficiency considerations enforce a minimum duration of T_S: the CP (which is determined by the maximum excess delay of the channel) should not be shorter than approximately 10% of the symbol duration. The following equation gives a useful guideline on how to choose T_S. Let $R(k, \ell) = P_h (\ell T_c, k T_c)$ be the sampled delay cross power spectral density of the channel (see Chapter 6). Define furthermore a function:

$$w(q,r) = \frac{1}{N}
\begin{cases}
N - |r| & 0 \leq q \leq N_{\text{cp}} & 0 \leq |r| \leq N \\
N - q + N_{\text{cp}} - |r| & N_{\text{cp}} \leq q \leq N + N_{\text{cp}} & 0 \leq |r| \leq N - q + N_{\text{cp}} \\
N + q - |r| & -N \leq q \leq 0 & 0 \leq |r| \leq N + q \\
0 & \text{elsewhere}
\end{cases} \tag{15.28}$$

Then the desired signal power can be approximated as [Steendam and Moeneclaey 1999]:

$$P_{\text{sig}} = \frac{1}{N} \sum_{\ell} \sum_{k} w(k,\ell) R(k,\ell) \tag{15.29}$$

and the ICI and ISI powers as

$$P_{\text{ICI}} = \sum_k w(k,0)R(k,0) - P_{\text{sig}} \tag{15.30}$$

$$P_{\text{ISI}} = \sum_k [1 - w(k,0)]R(k,0) \tag{15.31}$$

and the Signal-to-Interference-and-Noise Ratio (SINR) is

$$\text{SINR} = \frac{\dfrac{E_S}{N_0}P_{\text{sig}}\dfrac{N}{N_{\text{cp}}+N}}{\dfrac{E_S}{N_0}P_{\text{sig}}\dfrac{N}{N_{\text{cp}}+N}\dfrac{P_{ISI}+P_{ICI}}{P_{\text{sig}}}+1}. \tag{15.32}$$

The above equations allow an easy trade-off between the ICI due to frequency dispersion, the ICI due to residual delay dispersion, and SNR loss due to the CP.

- *Optimum choice of OFDM basis pulses*: a related approach influences the OFDM basis pulse *shape* in order to minimize ICI. We know that a rectangular temporal signal has a very sharp cutoff in the temporal domain, but has a $\sin(x)/x$ shape in the frequency domain and thus decays slowly. And while in a perfect system each subcarrier is in the spectral nulls of all other subcarriers, the slope of the $\sin(x)/x$ is large near its zeros. Thus, even a small frequency dispersion leads to large ICI. By choosing basis pulses whose spectrum decays faster and gentler, we decrease ICI due to frequency dispersion. On the downside, faster decay in the frequency domain is bought by slower decay in the time domain, which increases delay-spread-induced errors. Gaussian-shaped basis functions have been shown to be a useful compromise, compare Section 15.10.

- *Self-interference cancellation techniques*: in this approach, information is modulated not just onto a single subcarrier but onto a group of them. This technique is very effective for mitigation of ICI but leads to a reduction in spectral efficiency of the system.

- *Frequency domain equalizers*: if the channel and its variations are known, then its impact on the received signal, as described by Eq. (15.26), can be reversed. While this reversal can no longer be done by a single-tap equalizer, there is a variety of suitable techniques. For example, we can simply invert $\mathbf{H}$, or use a minimum mean square error criterion. These inversions can be computationally expensive: as the channel is continuously changing, the inverse matrix has to be recomputed for every OFDM block. However, methods with reduced computational complexity exist. Another approach is to interpret different tones as different users, and then apply multi-user detection techniques (as described in Section 28.2) for detection of the tones. Figure 15.16 shows an example of the effect of different equalization techniques (*Operator Perturbation Technique* (OPT) denotes a linear inversion technique, while *Parallel Interface Cancellation* (PIC) and *Successive Interface Cancellation* (SIC) denote multi-user detection – inspired approaches).

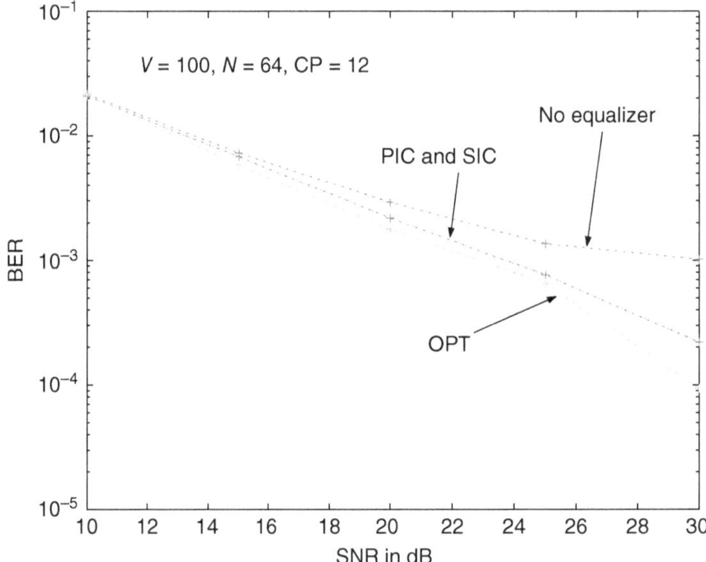

Figure 15.16 Bit error rate as a function of signal-to-noise ratio for an 802.11a-like orthogonal-frequency-division-multiplexing system with 64 carriers and 12 samples CP. Performance is analyzed in channel model F of the 802.11n channel models, with 100-m/s velocity.

Example 15.2   *Consider a system with a 5-MHz bandwidth, 128 tones, and a CP that is 40 samples long. It operates in a channel with an exponential Power Delay Profile (PDP), $\tau_{\mathrm{rms}} = 1\,\mu s$, $\nu_{\mathrm{rms}} = 500\,Hz$, and an E_S/N_0 of 10 dB. What is the SINR at the RX? How do results change when the CP is shortened to 12 samples?*

In a first step, we need to find the sampled delay cross power spectral density. For a bandwidth of 5 MHz, the sampling interval is 200 ns. Therefore, the rms delay spread is five samples, and the sampled PDP is described as $\exp(-k/5)$. The Doppler spectrum is assumed to have a Gaussian shape. Assuming furthermore that the Doppler spectrum is independent of delay, we obtain:

$$R(k,\ell) = \exp\left(-k/5\right)\cdot \exp\left(-\frac{\ell^2}{2\cdot 10,000^2}\right).$$
(15.33)

An accurate solution for interference power can then be found by inserting this sampled delay cross power spectral density into Eqs. (15.28)–(15.32). We obtain:

$$\frac{P_{\mathrm{ICI}}}{P_{\mathrm{sig}}} = 2.46\cdot 10^{-5}$$
(15.34)

$$\frac{P_{\mathrm{ISI}}}{P_{\mathrm{sig}}} = 1.18\cdot 10^{-5}.$$
(15.35)

This shows that ISI and ICI are reasonably balanced, which overall leads to low interference power.

Furthermore, the CP reduces the effective SNR by a factor $128/(128 + 40) = 0.762$. Thus, the total SINR becomes

$$\frac{7.62}{7.62(2.46 + 1.18)\cdot 10^{-5} + 1} = 7.6.$$
(15.36)

This indicates that the major loss of SINR occurs due to the CP. When we shorten it from 40 to 12 samples, the sum of ISI and ICI increases to

$$\frac{P_{\mathrm{ICI}} + P_{\mathrm{ISI}}}{P_{\mathrm{sig}}} = 6.2\cdot 10^{-3}.$$
(15.37)

On the other hand, the SNR becomes $10 \cdot 128/(128 + 12) = 9.14$. Thus, the effective SNR becomes

$$\frac{9.14}{9.14\cdot 6.2\cdot 10^{-3} + 1} = 8.65.$$
(15.38)

This shows that a long CP is not always the best way to improve the SINR. Rather, it is important to correctly balance ISI, ICI, and duration of the CP.

*15.8   Synchronization

The previous derivations have assumed that the RX knows, and can tune its LO to, exactly the frequency of the TX; furthermore, we assume that the start time of each OFDM symbol (or more generally, each block of symbols) is exactly known. But how can the RX obtain this knowledge? In this section, we will review methods that allow to obtain (approximate) synchronization, and also discuss the impact of synchronization errors.

15.8.1   Timing Synchronization

Since modern systems are mostly dealing with packet transmissions, we consider the question: "how can we determine the start time of a packet after a period of silence"? One could, of course, revert to energy detection, and determine the rising edge of the received energy. However, this is not a good solution for a variety of reasons: (i) in the downlink, the Base Station (BS) is likely to transmit continuously – though not continuously for one particular user; (ii) other sources of interference, as well as delay dispersion and fading, make such energy detection unreliable.

A widely used synchronization method is the exploitation of particular periodic signal structures. In particular, the *Schmidl–Cox* synchronization transmits two subsequent copies of the same signal. The RX then forms the correlation of the received signal with a delayed version of itself; if there is a pronounced correlation peak, it is concluded that the synchronization sequence has been found. Note that for this approach to work, the two copies of the *arriving* signal must be identical; this, in turn, implies that the synchronization blocks require a sufficient CP. To be more precise, let us write the arriving signal as

$$r_k = s_k \exp\left(j2\pi\varepsilon_{\mathrm{f}}k/N\right) + n_k \qquad\qquad \theta \leq k \leq \theta + N/2 - 1$$
$$r_{k+N/2} = s_k \exp\left(j2\pi\varepsilon_{\mathrm{f}}k/N\right)\exp\left(j\pi\varepsilon_{\mathrm{f}}\right) + n_{k+N/2} \quad \theta \leq k \leq \theta + N/2 - 1$$
(15.39)

where θ is the (unknown) timing offset in units of samples; it is assumed to be integer, the residual offset is considered part of the channel impulse response. Further, n_k is the noise at the kth sample, and ε_f is the frequency offset normalized to the subcarrier spacing. We now have a sliding-window correlator that identifies the maximum of the correlation

$$\hat{\theta} = \arg\max_{\theta} (|\, ACF(\theta)\,|) \tag{15.40}$$

where

$$ACF(\theta) = \frac{\displaystyle\sum_{m=\theta}^{\theta + N/2 - 1} r_{m + N/2} r_m^*}{\displaystyle\sum_{m=\theta}^{\theta + N/2 - 1} \left| r_{m + N/2} \right|^2} \tag{15.41}$$

is the normalized correlation. Furthermore, $ACF(\hat{\theta})$ must exceed a certain minimum threshold; otherwise, the detected peak might not stem from an actual synchronization block. Figure 15.17 shows an example for the timing metric in a delay-dispersive channel; the peak of the correlation is rather broad. The reason for this is that each MPC leads to its own correlation peak, so that for a channel with significant delay dispersion there are a lot of correlation peaks.

The situation can be somewhat improved by introducing additional synchronization blocks; for example, a fourfold repetition with the pattern $[[A \quad A \quad -A \quad A]$ (where A denotes one block) has been proposed, with a corresponding detector creating a metric

$$ACF(\theta) = \frac{S1 + S2 + S3}{\dfrac{3}{2} \displaystyle\sum_{m=\theta}^{\theta + N - 1} |r_m|^2} \tag{15.42}$$

$$S1 = \left| \sum_{m=\theta}^{\theta + N/4 - 1} \left(r_{m + N/4} r_m^* - r_{m + 2N/4} r_{m + N/4}^* - r_{m + 3N/4} r_{m + 2N/4}^* \right) \right| \tag{15.43}$$

$$S2 = \left| \sum_{m=\theta}^{\theta + N/4 - 1} \left(r_{m + 3N/4} r_{m + N/4}^* - r_{m + 2N/4} r_m^* \right) \right| \tag{15.44}$$

$$S3 = \left| \sum_{m=\theta}^{\theta + N/4 - 1} r_{m + 3N/4} r_m^* \right|. \tag{15.45}$$

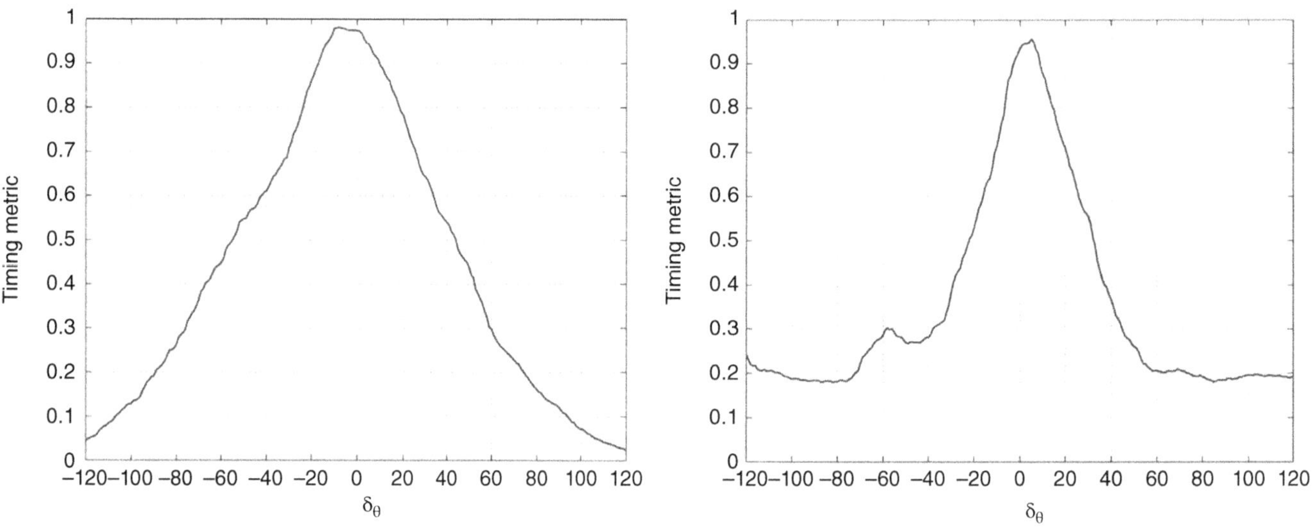

Figure 15.17 Timing metric obtained by a Schmidl–Cox synchronization correlator (left) and four-block correlator (right) in a Rayleigh-fading channel with eight delay taps; $N = 256$.
Reproduced with permission from [Morelli et al. 2007] © IEEE.

Of course, the insertion of a synchronization block increases the overhead. Several works have thus suggested to instead use the inherent autocorrelation of the CP – after all, we know that the CP and the last part of an OFDM symbol are, by design, a repetition of each other. Thus, forming the autocorrelation with a lag of N can provide timing information, though care must be taken concerning the effects of inter-symbol interference, which shows up in the CP.

We finally note that some residual timing errors need not be catastrophic. After all, the RX cannot know whether a misestimation of the timing is due to a synchronization problem, or the inherent propagation delay of the channel. In other words, the channel equalizer will take care of any residual timing error as long as the composite delay (timing error plus maximum excess delay) is smaller than the CP duration.

15.8.2　Frequency Synchronization

Besides the absolute timing, we also have to estimate the frequency of the incoming signal. A typical crystal oscillator that might be used in a low-cost User Equipment (UE) has about 100 ppm accuracy, which translates, at a 1 GHz carrier frequency, to a 100 kHz *Carrier Frequency Offset* (CFO) – considerably larger than the subcarrier spacing in many systems. We can thus divide the CFO into two parts: (i) integer offset of the frequency q, i.e., in multiples of two subcarrier spacings, and (ii) fractional offset $\kappa = \mathrm{mod}(\varepsilon_\mathrm{f}, 2)$, such that

$$\varepsilon_\mathrm{f} = \kappa + 2q \tag{15.46}$$

where the modulo operation is taken such that $\kappa = (-1, 1]$ and $q \in (0, N-1]$ is an integer.[6] Note that the integer offset does not lead to ICI (any shift of frequency that is an integer multiple of subcarrier spacing just shifts a subcarrier from one null of the surrounding subcarriers to another).

Estimation of the fractional part is relatively easy and can be done together with the timing acquisition – it follows from Eq. (15.39) that it can be estimated as

$$\hat{\kappa} = \frac{1}{\pi} \arg\left(\sum_{m=\hat{\theta}}^{\hat{\theta}+N/2-1} r_{m+N/2} r_m^* \right). \tag{15.47}$$

Estimation of the integer part cannot be done from the block with the time-shifted repetitions Eq. (15.39), but rather requires an additional block. For this, we take two pn sequences d_1 and d_2 that are interlaced on even and odd subcarriers (note that they are interlaced on a subcarrier basis, not in time). The RX first compensates the fractional frequency offset. Then the time samples are transformed, via DFT, to the frequency domain, resulting in the sequences $R_{1,n}$ and $R_{2,n}$

$$\begin{aligned} R_{1,n} &= e^{j\varphi_1} H_{n-2q} d_{1,n-2q} + n_{1,n} \\ R_{2,n} &= e^{j\varphi_2} H_{n-2q} d_{2,n-2q} + n_{2,n} \end{aligned} \tag{15.48}$$

where a modulo operation is applied such that $n-2q$ is always in the interval $[0, N-1]$, and $\varphi_i = 2\pi i \varepsilon_\mathrm{f}(1 + N_\mathrm{cp}/N)$. Denoting as $p_n = d_{2,n}/d_{1,n}$ for $n = 0, 2, ...,$ and neglecting the noise, we can write

$$R_{2,n} = e^{j(\varphi_2 - \varphi_1)} p_{n-2q} R_{1,n}. \tag{15.49}$$

We can thus find an estimate for q as

$$\hat{q} = \arg\max_q \frac{\left| \sum_{n=\mathrm{even}} R_{2,n} R_{1,n}^* p_{n-2q}^* \right|}{\sum_{n=\mathrm{even}} |R_{2,n}|^2}. \tag{15.50}$$

The CFO is not fixed, but varies with time, due to phase noise and other impairments in the system. Thus, even after the correct frequency offset has been acquired at the beginning of a packet or frame, further tracking is required. Such tracking can be done through a variety of techniques:

- *Continuous pilot tones:* particular subcarriers can be dedicated to tracking of the frequency. It allows for a simple implementation, but leads to a relatively high overhead, since a certain percentage of resources is always dedicated to providing the information about frequency offset. The Wi-Fi standard uses this approach, where, e.g., in IEEE 802.11a, four pilot tones (out of 56 subcarriers) are dedicated to frequency tracking, see Section 33.2.

---

[6] Taking into account the circular shift nature of OFDM with CP (i.e., any integer shift that lands outside the interval $(0, N-1]$ is subjected to a modulo operation to shift it into this range).

- *Periodic synchronization blocks:* synchronization blocks that can be used for signal acquisition must anyway be inserted at regular times so that RXs that lost sync can reset/reacquire synchronization and/or new users can acquire synchronization. It is then also possible for existing users that have an estimation of the frequency, to employ these blocks for tracking of the CFO.
- *Blind synchronization:* we can follow the evolution of the CFO from each symbol to the next, but using two possible characteristics of the modulation signal:
  - *Finite-alphabet properties:* all transmitted symbols are part of a Quadrature Amplitude Modulation (QAM) alphabet. If the constellation (usually obtained from observing multiple complex symbols) shows a rotation that increases from OFDM-symbol to OFDM-symbol, a positive CFO exists, whose magnitude can be estimated from the amount of rotation per symbol (and similarly for negative CFOs);
  - *Decision feedback adjustment:* the transmitted data symbols can serve as training symbols. After demodulation and decoding, the actual data symbol is known. Comparison of this decoded symbol to the projection of the received symbol into signal space (i.e., the "soft" information, compare Chapter 11) allows determination of the CFO. Compared to the use of the finite-alphabet properties, larger rotations of the constellation can be estimated.
  - *Rotation obtained from the CP:* here we are again exploiting the fact that the CP is a copy of the last part of the OFDM symbol in a time domain representation (after discarding the part impacted by ISI). Thus, a phase rotation between CP and that last part must be due to frequency offset.

Any of these methods provides an "error signal," i.e., the deviation of the instantaneously measured frequency from the nominal value. This error signal needs to be low-pass filtered, in order to avoid wild fluctuations due to noise. Figure 15.18 shows the typical structure of an RX based on those principles. The obtained signal is phase corrected by the output of a controlled oscillator. The resulting signal then undergoes the usual OFDM reception, namely discarding of the CP and FFT. The resulting signal, together with the originally observed signal, is used to obtain an error signal by any of the abovementioned methods. The error signal is then passed through a loop (typically low-pass) filter to reduce rapid fluctuations, and then drives the control of the LO.

15.8.3   Uplink Synchronization

The discussion up to now is related to downlink synchronization. The case of uplink synchronization when multiple UEs talk to a single BS, is somewhat more difficult, for a number of reasons:

- Each signal (UE to BS) goes through a different propagation channel.
- The time interval between transmissions from a particular TX (i.e., UE) may be large, while BSs are usually transmitting continuously. Thus, tracking may be more difficult in the uplink.

For these reasons, it is preferable that the UE derives its synchronization from the BS transmission, and also applies it in the uplink transmission. However, such compensation usually is not perfect, and residual errors remain.

Residual *timing* errors may be compensated by a timing advance (see Section 18.5), and/or absorbed into the CP. Treatment of residual CFO is more difficult, due to the fact that different offsets, stemming from the LOs of the different TXs, need to be estimated. If the different UEs are transmitting one after the other, like in Wi-Fi (see Chapter 33), then the estimation for each separate user is

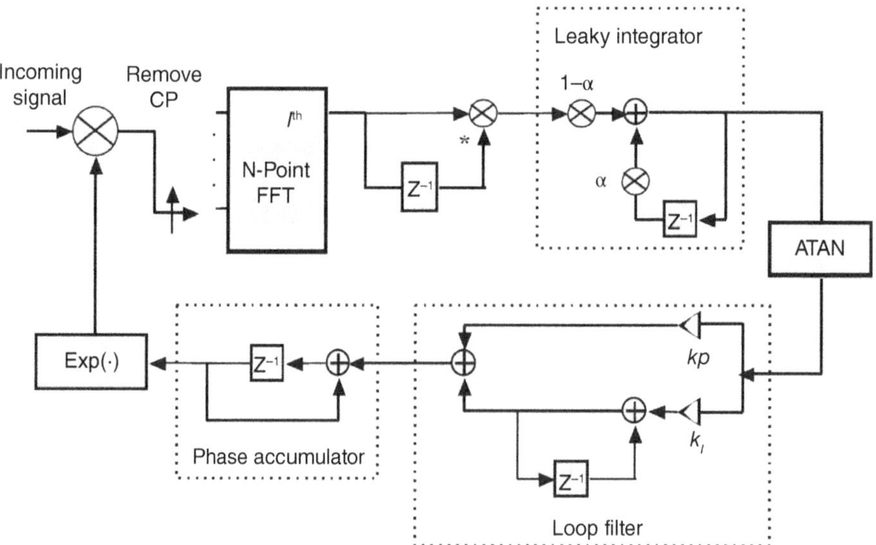

Figure 15.18   Receiver for tracking CFO with closed-loop structure. Color version available at wiley.com/go/molisch/wireless3e.

independent and follows the same principle as for the downlink. However, the situation is more difficult if each UE transmits only on a subset of the subcarriers (for more details, see Section 18.3.3), since the signals in the frequency domain overlap. In case that all subcarriers for a particular UE are next to each other, we can perform a bandpass filtering, and obtain the CFO of the signals within the filter bandwidth; note, however, that there will be residual interference from other UEs because the filtering cannot achieve a perfect separation of the users. For the case that the subcarriers of different users are interleaved, joint maximum-likelihood estimators for the offsets of the different users should be used. Since the number of parameters to be estimated is rather large, iterative procedures can be applied: start with an initialization to estimate CFO for each user (e.g., assuming zero CFO, or assuming the CFO from the previous timestep). Next, assume that the CFOs of all UEs except UE1 are known (i.e., identical to the initialization values), and perform a maximum-likelihood estimation for the CFO of UE 1, i.e., a single unknown parameter. Then repeat this process for all other UEs; these updated estimates form now the basis for the next iteration round, where again we assume those estimates to be true, except for UE 1, for which we estimate the CFO, then UE 2, and so on. Note the similarity of this approach to the SAGE algorithm for channel estimation described in Section 9.6, as well as multi-user detection, which we will describe later in Section 28.2.

15.9 Adaptive Power Allocation, Modulation, and Coding

Adaptive modulation changes the coding scheme and/or modulation method depending on CSI – choosing it in such a way that it always "pushes the limit" of what the channel can transmit. In OFDM, modulation and/or coding can be chosen differently for each subcarrier, and can also change with time. We thus not only accept the fact that the channel (and thus the SNR) shows strong variations but even exploit this fact. On subcarriers with good SNR, transmission is done at a higher rate than on subcarriers with low SNR. In other words, such an adaptive modulation selects a modulation scheme and code rate according to the channel quality of a specific subcarrier.

Let us compare adaptive modulation for OFDM with coded OFDM and multi-carrier CDMA. Both of the latter systems try to "smear" the data symbols over many subcarriers, so that each symbol sees approximately the same average SNR. It can be shown that – at least theoretically – systems using adaptive modulation perform better than systems whose modulation and coding are fixed once and for all. As a matter of fact, OFDM using waterfilling power allocation and capacity-achieving codes on each subcarrier (with a data rate that is adjusted to the SNR on each subcarrier) has been shown to be information-theoretically optimum.

15.9.1 Channel Quality Estimation

Adaptive modulation requires that the TX knows the CSI. This requirement sounds trivial but is quite difficult to realize in practice. Two approaches can be distinguished (for more details see Section 16.1.7): the first is applicable if the channel is reciprocal – e.g., the BS learns the CSI while it is in receiving mode; when it then transmits, it relies on the fact that the channel is still in the same state. This can be fulfilled in systems with Time Domain Duplex (TDD) in slowly time-varying channels. Alternatively, feedback from the RX to the TX can be used to inform the TX about the channel state. It is also noteworthy that the TX has to know the CSI *for the time instant when it will transmit* – in other words, it has to look into the future. Channel prediction is thus an important component for many adaptive modulation systems. The problem is obviously easier if we are in a frequency-selective but time-invariant channel.

15.9.2 Power Adaptation

Once the channel state is known, the TX has to decide how much power should be assigned to each channel, since this determines (together with the channel state of each channel) the SNR. In the following, we will assume that the channel is frequency selective but time invariant. This will make the discussion easier, and the principles can be easily extended to doubly selective channels. Furthermore, the optimum power will be derived under the assumption that capacity-achieving modulation and coding can be used. The actual choice of modulation and coding format will be discussed later.

In order to determine how much power should be assigned to each channel, let us first reformulate the question in a more abstract way: "Given a number of parallel subchannels with different attenuations, what is the distribution of transmission power that maximizes capacity?" The answer to this latter question was given by Shannon in the 1940s, and is known as "waterfilling." The power allocation P_n to the nth subchannel is

$$P_n = \max\left(0, \varepsilon_{\mathrm{WF}} - \frac{\sigma_\mathrm{n}^2}{|\alpha_n|^2}\right) \tag{15.51}$$

where $|\alpha_n|^2$ is the power gain (inverse attenuation) of the nth subchannel, σ_n^2 is noise variance, and the threshold $\varepsilon_{\mathrm{WF}}$ (note that this threshold has nothing to do with the frequency offset ε_f discussed in the previous section) is determined by the constraint of the total transmitted power P as

$$P = \sum_{n=1}^{N} P_n. \tag{15.52}$$

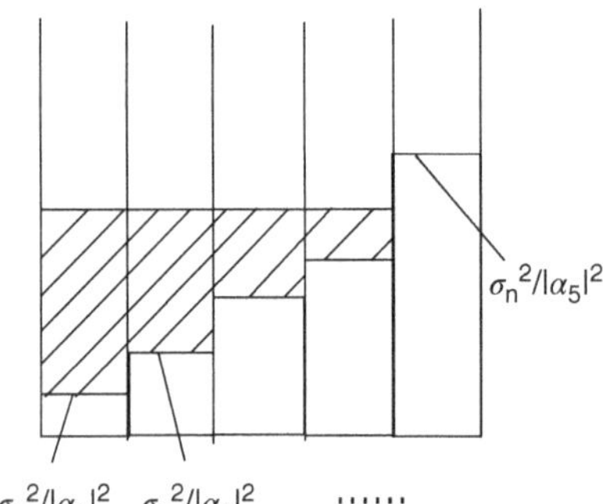

Figure 15.19　Principle of waterfilling.

The optimality of this scheme, in the sense that it maximizes Shannon capacity, can be demonstrated through standard convex optimization techniques, see Exercise 15.11 in the end-of-chapter exercises.

Waterfilling can be interpreted visually, according to Figure 15.19. Imagine a number of connected vessels. At the bottom of each vessel is a block of concrete with a height that is proportional to the inverse SNR, $\sigma_n^2/|\alpha_n|^2$ of the subchannel that we are considering. Then take P units of water, and pour it into the vessels; thus the amount of poured water represents the total transmit power that is available. Because the vessels are connected, the surface level ε of the water is guaranteed to be the same in all vessels. The amount of power assigned to each subchannel is then the amount of water in the vessel corresponding to this subchannel. Obviously, subchannel 1, which has the highest SNR, has the most water in it. It can also happen that some subchannels that have a poor SNR (like channel 5), do not get any power assigned to them at all (the concrete block of that vessel is sticking out of the water surface). Essentially, waterfilling makes sure that energy is not wasted on subchannels that have poor SNR: in the OFDM context, this means not wasting power on subcarriers that are in a deep fade.

With waterfilling, power is allocated preferably to subchannels that have a good SNR. This is optimum from the point of view of theoretical capacity; however, it requires that the TX can actually make use of the large capacity on good subchannels. If this is not possible, e.g., because the modulation alphabet is too small to support high data rates, it is preferable to give the excess power that cannot be exploited by good subchannels to subchannels that do not use the maximum modulation format.

Example 15.3　*Waterfilling: consider an OFDM system with three subcarriers, with $\sigma_n^2 = 1, \alpha_n^2 = 1, 0.4,$ and $0.1,$ and total power $\sum P_n = 15$. Compute the power assigned to different tones according to (i) waterfilling, (ii) equal power allocation, (iii) predistortion (inverting the channel attention), and compute the resulting (Shannon) capacity.*

From Eq. (15.51) we find that $\varepsilon_{\mathrm{WF}} = 9.25$ gives the correct solution: in that case, the power in the different subcarriers are

$$P_1 = 8.25 \tag{15.53}$$

$$P_2 = 6.75 \tag{15.54}$$

$$P_3 = 0 \tag{15.55}$$

that is, no power is assigned to the channel that suffers from the strongest attenuation. The total capacity can be computed as

$$C_{\mathrm{waterfill}} = \sum_{n=1}^{N} \log_2\left(1 + \alpha_n^2 P_n/\sigma_n^2\right) = 5.1 \ \mathrm{bit/s/Hz}. \tag{15.56}$$

For equal power allocation:

$$P_1 = P_2 = P_3 = 5 \tag{15.57}$$

so that capacity becomes:

$$C_{\mathrm{equal\text{-}power}} = 4.8 \ \mathrm{bit/s/Hz}. \tag{15.58}$$

For the predistortion case, the powers become

$$P_1 = 1.1 \tag{15.59}$$

$$P_2 = 2.8 \tag{15.60}$$

$$P_3 = 11.1 \tag{15.61}$$

from which we obtain a capacity of

$$C_{\text{predistort}} = 3.2 \, \text{bit/s/Hz}. \tag{15.62}$$

In this example, equal power allocation gives almost as high a capacity as (optimum) waterfilling, while predistortion leads to significant capacity loss.

This section described the power allocation over a set of subcarriers, assuming that the propagation channel is time-invariant. Somewhat similar principles could be applied, in principle, for time-variant channels as well, i.e., more power is given to the channel at times when it is in a good state, and less when it shows a small channel gain. However, a challenge arises that in order to apply temporal waterfilling we have to know the statistics of the channel, which is only feasible if the channel is stationary (and learning the channel statistics can thus be done from past transmissions). If the channel statistics are not stationary, learning them would require observation of the future.

15.9.3　Adaptive Modulation and Coding

The above derivation was based on the assumption that on each subchannel (subcarrier), transmission as close to capacity as possible is performed. This means that the TX has to adapt the data rate according to the SNR that is available (note that waterfilling increases SNR differences between subcarriers) and pick a modulation and coding scheme that approximates this capacity. Consequently, the coding rate and the constellation size of the modulation alphabet have to be adjusted.

As was shown by Shannon, the optimum modulation alphabet, i.e., the one that can achieve capacity, in an AWGN channel is itself Gaussian. This is, however, not practical – actually used modulation alphabets are QAM modulation, and thus have a finite size. The data rate associated with uncoded transmission on a subchannel is $\log_2 (M)$, where M is the size of the symbol alphabet.[7] Furthermore, in order to approximate capacity, the datastream on each subcarrier needs to be coded, with such a code rate that the resulting data rate is at/below the channel capacity associated with the SNR available on this subcarrier.[8] This then allows to *approximate* the rates

$$R = \min \left[\log_2(1 + \gamma), \, \log_2(M) \right].$$

In other words, the combination of modulation alphabet and code rate determines the actual data rate achieved on each subcarrier. Note also that different combinations of alphabet size and code rate can give the same data rate – for example, BPSK with rate 2/3 coding, and QPSK with rate 1/3 coding provide the same data rate. As we have seen in Chapter 13, adjustment of the code rate, e.g., through puncturing of a common mother code, can be achieved relatively easily. However, for implementation reasons, and to reduce the signaling overhead, only a relatively small number of code rates is implemented, leading to a discrete set of modulation format/code rate combinations that can be selected by the TX. This means that in practice, we can select a number of discrete rates, each of which can be used for a *range* of SNRs. Figure 15.20 shows an example.

Signaling of Chosen Parameters

As outlined above, the available data rates form a discrete set. After the TX has decided which transmission mode – i.e., combination of signal constellation and encoder – to use on each tone, it has to communicate that decision to the RX. There are three possibilities to achieve that task:

- *Explicit transmission*: the TX can send, in a predefined and robust format, the index of the transmission mode it intends to use. Transmission of this information itself should always be done in a format that is independent of the channel state, and care should be taken that the message is well protected against errors during transmission.
- *Implicit transmission*: implicit transmission is possible when the TX gets its CSI from the RX via feedback. In such a case, the RX knows exactly what channel-state information is available to the TX, and thus the basis on which the decision for a transmission mode is being made. Thus, the RX just needs to know the decision rule on which the TX bases its choice of transmission mode. If the RX feeds back the transmission mode that the TX should use, the situation is even simpler.

---

[7] Quadrature Amplitude Modulation (QAM) of alphabet size 4096 currently seems to be the largest constellation size that can be used in practical systems.

[8] We assume in the following that near-capacity-achieving codes are used. If this is not the case, we usually try to choose the data rate in such a way that a certain BER can be guaranteed.

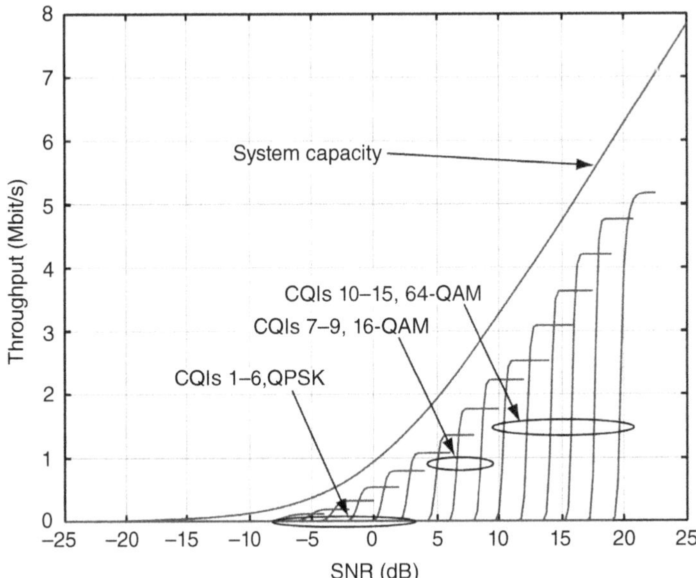

Figure 15.20 System capacity, and throughput achievable with different modulation formats and code rates (choice of which depends on the "channel quality indicator" CQI) in LTE. Color version available at wiley.com/go/molisch/wireless3e. Reproduced with permission from [Mehlführer et al. 2009] © IEEE.

The drawback to this method is that errors in channel-state feedback (from the RX to the TX) not only lead to a wrong choice of transmission mode (which is bad, but usually not fatal) but also to detection and decoding using the wrong mode, which leads to very high error rates.

- *Blind detection*: from the received signal, the RX can try to determine the signal constellation. This can be achieved by considering different statistical properties of the received signal, including the PAPR, autocorrelation functions, and higher-order statistics of the signal.

Limitations of AMC

While Adaptive Modulation and Coding (AMC) provides significant benefits in slowly varying channels, it might either require high feedback overhead or be completely impossible (when the channel coherence time is less than the feedback time) in systems operating in fast-varying channels. When accurate CSI at the TX is not available, then channel variations are detrimental to performance, as the AMC must be chosen to accommodate the worst-case SNR. Thus, for high-mobility situations (vehicle-to-vehicle, high-speed train, etc.), channel averaging through spreading, and thus operating with a constant SNR over extended time periods, is not only the simpler solution but also provides the better performance.

The importance of a robust and fixed-rate channel is increased for applications that – due to latency constraints – do not allow retransmissions. This is especially critical for applications running over the TCP/IP protocol, which dramatically backs off the rate when packet failures occur, and subsequently takes a long time to converge again to a higher rate.

AMC in Other Transmission Systems

We finally note that waterfilling and AMC is not limited to OFDM. The same principles can be applied for single-carrier modulation as well; different modulations can be used when the channel has different SNR due to temporal fading; and – if there is only an average power constraint and not a peak power constraint – the transmit power can be distributed in time according to the waterfilling rules.

*15.10   Generalizations of OFDM

15.10.1   General Framework – Gabor Systems

The OFDM we have considered up to now can be viewed as a special case of the more general class of *multi-carrier modulation*. In recent years, a number of multi-carrier modulations have been proposed, motivated mainly by the changing applications and requirements in 5G. While the NR standard still uses "classical" OFDM as defined in the previous sections, alternatives have been proposed, and might be selected in future versions of the 5G standard and/or other wireless systems. This section provides an overview and dichotomy of such techniques.

General multi-carrier modulation can be related to the theory of *Gabor systems*. Specializing to the case of interest here, a signal is represented as

$$s(t) = \sum_{i=-\infty}^{\infty} \sum_{n=0}^{N-1} c_{n,i} g_{n,i}(t) \tag{15.63}$$

and the basis function $g_{n,i}(t)$ fulfills the condition

$$g_{n,i}(t) = p_{\mathrm{TX}}(t - i\tau_0) \exp(j2\pi n\nu_0 t) \tag{15.64}$$

where $p_{\mathrm{TX}}(t)$ is the transmit pulse shape (also known as transmit filter, synthesis function, or Gabor atom), τ_0 is the symbol spacing, and ν_0 is the subcarrier spacing. Points with coordinates $(i\tau_0, n\nu_0)$ form a *lattice* (see below for more details). For the case of $p_{\mathrm{TX}}(t)$ being a rectangular function of length T_{S}, and $\tau_0 = 1/\nu_0 = T_{\mathrm{S}}$, we obtain conventional, no-CP-, OFDM. The received signal is

$$r(t) = \int h(t,\tau) s(t-\tau) d\tau + n(t) \tag{15.65}$$

and the RX performs a projection of the received signal onto a receive pulse (also known as receive filter or analysis function)

$$d_{\ell,m}(t) = p_{\mathrm{RX}}(t - m\tau_0) \exp(j2\pi\ell\nu_0 t) \tag{15.66}$$

as

$$\hat{c}_{\ell,m} = \langle r(t), d_{\ell,m}(t) \rangle = \int r(t) d_{\ell,m}^{*}(t) dt. \tag{15.67}$$

The Gabor system is thus characterized by symbols, filters, and lattices. Symbols are relatively straightforward: they correspond to the standard modulation symbols used for data transmission and are usually a complex, finite-size set, e.g., complex QAM symbols (other choices are also possible, e.g., the symbols can be constrained to be real). The choice of filters and lattices is more involved; they will be discussed in the subsequent two subsections.

15.10.2   Filters (Pulses)

The fundamental transmit pulse shape $p_{\mathrm{TX}}(t)$ determines the spectrum (or more generally, the time–frequency characteristics) of the signals. The receive filter depends on the transmit filter, through the requirement of reconstructability, and achieving good SNR.

The most common $p_{\mathrm{TX}}(t)$ has a rectangular shape; as a matter of fact, this has been the assumption up to now in this chapter. However, as we have seen in Section 10.2.2, the spectrum of a rectangular pulse is a sinc shape, which decays slowly in frequency and has high side lobes (power of sidelobes decays as $1/f^2$) – actually, the second central moment of its power spectrum is infinite. In order to improve spectral efficiency, and in particular for systems in which different users use different sets of subcarriers (see Section 18.3.3), a better confinement in the frequency domain is desirable. Gaussian pulses have the attractive property of symmetry, in that they are Gaussian in both time and frequency domain. This implies that delay and frequency dispersion have a similar effect, and design of good pulses can be easily related to the channel properties; however, Gaussian pulses do not generally fulfill orthogonality conditions across frequency/time translations. Another family of pulses is *Prolate Spheroidal Wave Functions* (PSWF), which are the finite-duration waveforms that have the best energy concentration in frequency (or can be made to be bandlimited functions with the best energy concentration in time). Other commonly used pulseshapes include such well-known functions as Hamming, Blackman, root-raised cosine, and Kaiser filters, and isotropic orthogonal transform algorithm (IOTA) pulses. Note that in all of the above cases, the filter has to be implemented, in principle, on every subcarrier, which can result in prohibitive computational burden. However, much more efficient implementations, relying on *polyphase filters*, are known.

Pulse designs should take "typical" channel properties into account. In particular, define the σ_{t} and σ_{f} as the square root of the second central moments of $|p(t)|^2$ and $|P(f)|^2$, respectively. Then a good design is

$$\frac{\sigma_{\mathrm{t}}}{\sigma_{\mathrm{f}}} = \frac{S_{\tau}}{S_{\nu}}. \tag{15.68}$$

Of course, such a choice has an impact on the desired lattice as well; this will be discussed below.

Turn now to the receive pulse. As mentioned above, it depends on the transmit pulse. A standard requirement for the pulse shape is that the symbols can be reconstructed from a simple projection operation Eq. (15.67). This is fulfilled if $\langle g_{n,i}(t), d_{\ell,m}(t) \rangle = \delta_{n,\ell}\delta_{i,m}$; this means that the basis signals are either orthogonal to each other if $p_{\mathrm{TX}}(t) = p_{\mathrm{RX}}(t)$, or are bi-orthogonal to each other if $p_{\mathrm{TX}}(t) \neq p_{\mathrm{RX}}(t)$. Note, however, that the (bi-)orthogonality is not simply a matter of the pulse shape, but also connected to the lattice, according to the definition of $g_{n,i}(t)$ and $d_{\ell,m}(t)$; see also the next subsection.

An important quantity to judge the (quasi-)orthogonality of pulses (and lattices) and their time–frequency localization is the cross-ambiguity function

$$A_{p_{\mathrm{TX}},p_{\mathrm{RX}}}(\tau,\nu) \triangleq \int e^{-j2\pi\nu t} p_{\mathrm{TX}}(t + \tau/2) p^*_{\mathrm{RX}}(t - \tau/2)\,\mathrm{d}t. \tag{15.69}$$

When sampled at $\tau = m\tau_0$ and at $\nu = \ell\nu_0$, it should be 1 for $\ell = m = 0$, and 0 elsewhere.

The receive filter output should also have good SNR. The choice $p_{\mathrm{TX}}(t) = p_{\mathrm{RX}}(t)$ implies matched filtering, which maximizes SNR. However, deviation from that condition may not have critical impact on the performance. The most common example is CP-OFDM which relies on bi-orthogonal pulses. Indeed, the CP-OFDM RX pulse discards the CP samples to enable simpler processing at the cost of a reduction in SNR as discussed in Sec. 15.4.

15.10.3 Lattices

The way the symbols are arranged in time and frequency defines the lattice structure and will directly impact the system performance. Mathematically speaking, a lattice Λ is an arrangement of points that is invariant to integer-valued shift operations (where the shifts may be modulo some (larger) integer number). It can be described by basis vectors, and any combinations of integer multiples of basis vectors is again another lattice point. For example, the basis vectors $\begin{pmatrix} T \\ 0 \end{pmatrix}$ and $\begin{pmatrix} 0 \\ F \end{pmatrix}$ provides a rectangular lattice, while $\begin{pmatrix} T \\ 0 \end{pmatrix}$ and $\begin{pmatrix} T/2 \\ F \end{pmatrix}$ provide a hexagonal lattice. The density of a lattice Λ can be described as $D(\Lambda) = 1/|\det(\mathbf{L})|$, where $\mathbf{L}$ is the matrix created by horizontal stacking of the basis vectors.

The density of the grid is related to the spectral efficiency (spectral efficiency is grid density multiplied with number of bits carried by each grid point, i.e., symbol), and also impacts the admissible pulse shapes. A critically sampled grid is defined as $D(\Lambda) = 1$; on a rectangular grid, this means $F = 1/T$, as we are accustomed from (no-CP) OFDM. For a critically sampled grid, the *Balian–Low* theorem tells us that the pulse shapes cannot be well-localized in the time–frequency plane (e.g., a rectangular pulse is not well localized in frequency, because its second moment is infinite) if complex symbols are transmitted. Use of well-localized pulses (which can reduce ISI/ICI) requires an undersampled grid, $D(\Lambda) < 1$, though this leads to a reduction in spectral efficiency.

Depending on the shape of the pulse, the grid needs to be placed in a specific way in the time–frequency plane to ensure orthogonality. In particular, if the pulse shape is adapted to the ratio of delay dispersion and frequency dispersion (e.g., we pick a long pulse in a channel with a lot of delay dispersion and little frequency dispersion, see Eq. (15.68)), then the spacing of the grid points should be similarly chosen

$$\frac{\tau_0}{\nu_0} = \frac{S_\tau}{S_\nu}. \tag{15.70}$$

15.10.4 Dichotomy of Multi-Carrier Schemes

Based on the above characteristics of Gabor systems, we can now construct a variety of multi-carrier schemes, including

- *Orthogonal schemes* $\langle g_{n,i}(t), d_{\ell,m}(t) \rangle = \delta_{n,\ell}\delta_{i,m}$ and $p_{\mathrm{TX}}(t) = p_{\mathrm{RX}}(t)$
  - *Standard OFDM*: as mentioned above, standard OFDM fits into the above framework by setting $p_{\mathrm{TX}}(t) = rect(T)$, and $D(\Lambda) = 1$.
  - *Null suffix OFDM (NS-OFDM)*: instead of a CP, we can also use zero padding at the end of each symbol, to create a guard interval. This implies again $p_{\mathrm{TX}}(t) = rect(T)$, but now $D(\Lambda) < 1$, since the spacing in the time domain between points of the lattice is larger than T.
  - *Filtered Multitone (FMT)*: here the pulses do not overlap in the *frequency* domain – as a matter of fact there is a guard interval between the spectra of the different carriers, making the scheme more robust to frequency dispersion. Typically, $D(\Lambda) < 1$. An example for suitable pulseshapes are raised-cosine or root-raised cosine pulses, but other possibilities exist.
  - *Lattice OFDM*: it uses orthogonalized Gaussian pulses, with various lattice geometries, designed to minimize the total amount of interference. Again, $D(\Lambda) < 1$.
  - *Staggered Multitone (FBMC/OQAM)*: this scheme employs a staggering of the lattices for the real and imaginary parts of QAM symbols used for modulation; this serves to escape the constraint on complex orthogonality and circumvent the Balian–Low theorem. More details will be described in Section 15.10.7. A variant called cosine-modulated multitone (which uses vestigial sideband modulation) has been shown to be actually equivalent.

- *Bi-orthogonal schemes* $\langle g_{n,i}(t), d_{\ell,m}(t) \rangle = \delta_{n,\ell}\delta_{i,m}$ and $p_{\mathrm{TX}}(t) \neq p_{\mathrm{RX}}(t)$
  - *CP OFDM*: in this scheme, $p_{\mathrm{TX}}(t) = rect(T + T_{\mathrm{cp}})$ is a rectangular filter, and $p_{\mathrm{RX}}(t) = rect(T)$ is a shorter rectangular filter, so that, $D(\Lambda) < 1$.

- *Windowed OFDM*: in order to reduce the spectral sidelobes, a (temporally) smoother $p_{\mathrm{TX}}(t)$ is used. Due to the existence of a CP also for this setup, the scheme is bi-orthogonal.
- *Bi-orthogonal FDM*: this scheme uses Gauss–Hermitian pulseshapes, and $D(\Lambda) = 1/2$, with the lattice designed to minimize the errors due to frequency as well as delay dispersion, i.e., $\tau_0/\nu_0 = S_\tau/S_\nu$. The spectral efficiency can be improved through the use of lattice staggering.

- *Nonorthogonal schemes* $\langle g_{n,i}(t), d_{\ell,m}(t)\rangle \neq \delta_{n,\ell}\delta_{i,m}$
  While in this class of schemes, any RX output (result of a projection operation) is impacted by multiple transmit symbols, perfect reconstruction is still possible through suitable RXs. The most important schemes in this category are:
  - *Generalized FDM (GFDM)*: this scheme uses tail biting in the pulse shape (to enable channel shortening) as well as a CP. It will be described in more detail in Section 15.10.6.
  - *Faster-than-Nyquist signaling*: while it seems oxymoronic to transmit signals with a rate higher than the Nyquist rate (which allows perfect reconstruction), allowing some interference while using optimum sequence estimators at the RX can provide as good BER as Nyquist signaling while improving spectral efficiency.

It is important to note that we generally would like to fulfill three conditions on a scheme:

1. A critically sampled lattice, $D(\Lambda) = 1$, to maximize spectral efficiency.
2. Orthogonality in the *complex* domain, $\langle g_{n,i}(t), d_{\ell,m}(t)\rangle = \delta_{n,\ell}\delta_{i,m}$, to allow simple demodulation.
3. Well-localized filters, such that, e.g., the second central moments of $|p(t)|^2$ and $|P(f)|^2$ are finite.

Unfortunately, the Balian–Low theorem shows that the three conditions cannot be fulfilled at the same time. The modulation formats discussed above sacrifice one or the other, achieving various trade-offs.

	Complex orthogonality	Critically sampled lattice	Well-localized filters
OFDM	Y	Y	N
NS-OFDM	Y	N	N
FMT	Depends	N	Y
Lattice OFDM	N	N	Y
Staggered multitone	N	Y	Y
CP-OFDM	Y	N	N
Windowed OFDM	Depends	N	Depends
Bi-orthogonal OFDM	Y	N	Y
GFDM	N	N	Y

A comparison of the power spectral density of several of these schemes is given in Figure 15.21. We also note that FMBC/OQAM shows the smallest sensitivity to synchronization errors, which is to be expected, since inter-carrier interference is suppressed by design in this scheme. Finally, the computational complexity of the different schemes can also vary considerably; Figure 15.22 provides a comparison. Note that complexity at the UE is typically a bigger concern.

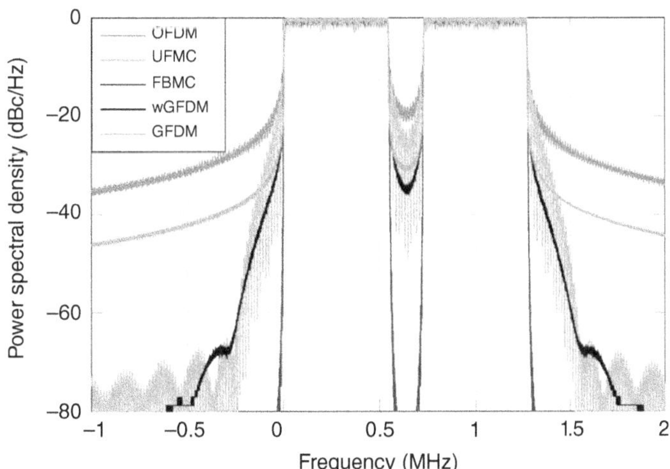

Figure 15.21　Power Spectral Density of different multi-carrier modulation methods. Color version available at wiley.com/go/molisch/wireless3e. Reproduced with permission from [Gerzaguet et al. 2017] © Eurasip.

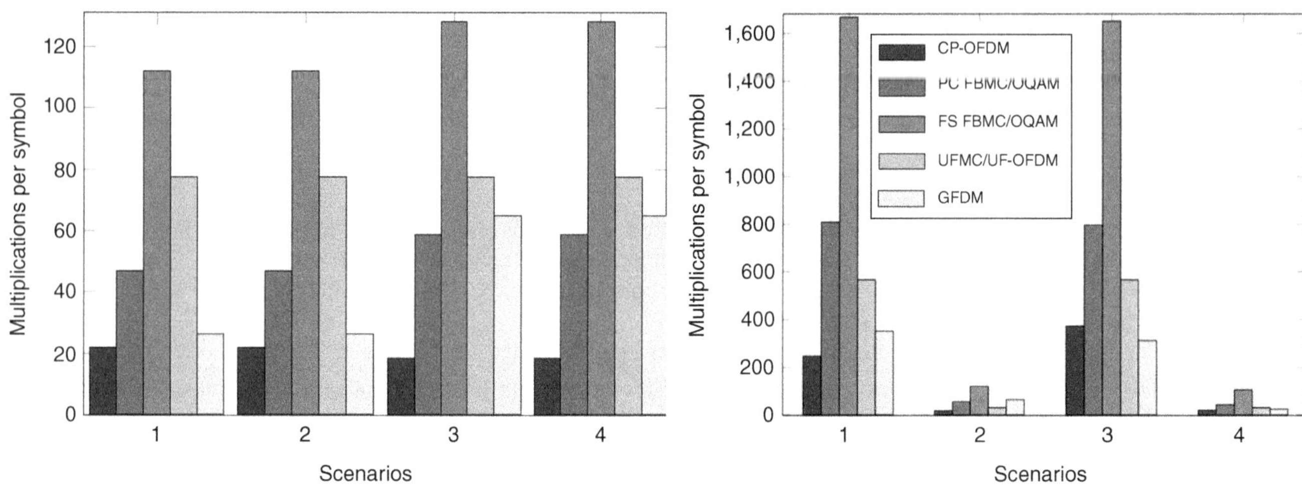

Figure 15.22 Complexity of different multi-carrier schemes in units of real multiplications at BS (left) and UE (right). Scenarios: (1) Downlink with multiple UEs frequency multiplexed, (2) Downlink single UE, (3) Uplink multiple UEs frequency multiplexed, (4) Uplink single UE. Color version available at wiley.com/go/molisch/wireless3e.
Reproduced with permission from [Gerzaguet et al. 2017] © Eurasip.

15.10.5 Filtered Multitone (FMT) and UFMC

We now consider FMT in greater detail. FMT has two essential components:

- Use of a lattice with $\tau_0\nu_0 > 1$. For this reason, the intercarrier interference is naturally reduced.
- Use of pulseshapes that reduce spectral interference. Since $\det(L) > 1$, it holds that $D(\Lambda) < 1$, i.e., we have an undersampled grid. This provides more degrees of freedom, which in turn allows the use of pulses with good spectral roll-off, at the price of a reduced spectral efficiency. Note that the system needs to use an oversampling factor $\rho = \tau_0\nu_0$ for all processing.

The resulting signal at the discrete sampling instances can then be written as

$$s_k = \sum_{i=-\infty}^{\infty} \sum_{n=0}^{N-1} c_{n,i} p_{\text{TX}}(k - i\rho N)\exp\left[j2\pi\frac{kn}{N}\right]. \tag{15.71}$$

The FMT scheme is robust to Doppler and phase noise related to the CFO of a single oscillator. However, in other situations, the interference between subcarriers arises because different subcarriers are assigned to different users (compare Section 15.8.3), which might not be well synchronized. In that case, it makes more sense to filter blocks of subcarriers; this scheme is called *Universal Filtered Multi-Carrier* (UFMC). These blocks could either be all (adjacent) subcarriers that are assigned to a particular user; or it could be a prescribed block size that is the minimum unit that is assigned to a particular user. The former solution has the advantage of better efficiency but needs to be reconfigured whenever the number of subcarriers assigned to a user changes. The latter case is the situation encountered in LTE and 5G-NR (see Chapters 31 and 32), where these minimum-size blocks are called *resource blocks*.[9]

15.10.6 Generalized FDM

Generalized FDM, GFDM, groups multiple points of a lattice into one block. Within each block, subcarrier wise processing, including upsampling, pulse shaping, and tail-biting, is performed, as can be seen from Figure 15.23. Let us assume a block size of N subcarriers and I OFDM symbols. Then the transmit signal is

$$s_k = \sum_{i=0}^{I-1} \sum_{n=0}^{N-1} c_{n,i} \widetilde{p}_{\text{TX}}(k - iN)\exp\left[j2\pi\frac{kn}{N}\right] \tag{15.72}$$

where $\widetilde{p}_{\text{TX}}$ is

$$\widetilde{p}_{\text{TX}}(k) = p_{\text{TX}}[\text{mod}(k, IN)]. \tag{15.73}$$

The periodic filtering creates a circular convolution, which – as we have seen above – simplifies some of the signal processing. Importantly, there are no orthogonality constraints on the filter functions, so that significant flexibility is available; this implies,

---

[9] Note, however, that neither LTE nor the current (Release 16) version of 5G-NR foresee the use of UFMC.

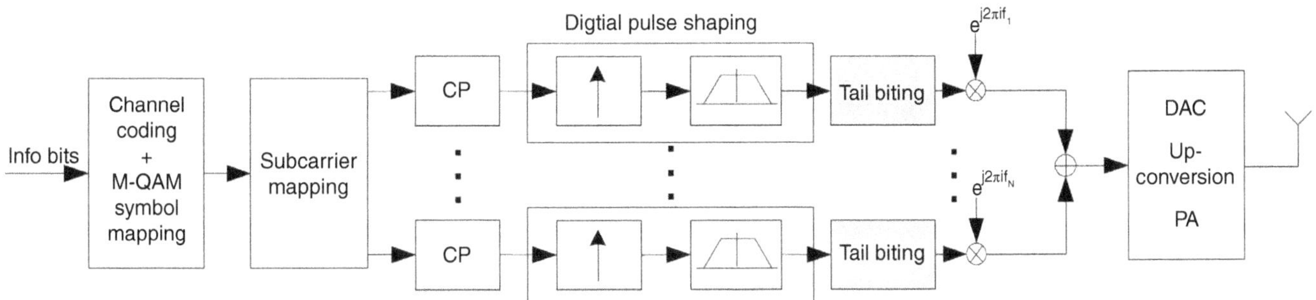

Figure 15.23 Block diagram of GFDM TX.
Reproduced with permission from [Fettweis et al. 2009] © IEEE.

however, also that more complicated RX processing for the equalization is required. Furthermore, the size of the FFT is increased, namely to IN. On the other hand, there is no need for a CP between the symbols in a block (only for the total block), which improves the spectral efficiency.

15.10.7 Staggered Multitone – FBMC/OQAM

An important trick for achieving better pulse localization without sacrificing spectral efficiency is lattice staggering. To be specific, the lattice points are spaced $T/2$ in the time domain, and $1/T$ in the frequency domain (subcarrier spacing).[10] The lattice density is thus $D(\Lambda) = 2$. However, for regular complex signaling, we would not be able to easily reconstruct the signals from such a dense grid. Thus, there is the additional constraint that all modulation symbols are real; alternatively, we can say that the lattices of the real and imaginary parts are shifted with respect to each other, see Figure 15.24. This has some analogies to offset-QPSK, where similarly real and imaginary pulseshapes are staggered in time (compare Section 10.3.7); however, even and odd subcarriers are treated differently. The modulation symbols for each lattice point are

$$\text{for } n \text{ even} \begin{cases} \widetilde{c}_{n,i'} = \text{Re}\{c_{n,i}\} \\ \widetilde{c}_{n,i'+1} = \text{Im}\{c_{n,i}\}j \end{cases} \tag{15.74}$$

$$\text{for } n \text{ odd} \begin{cases} \widetilde{c}_{n,i'} = \text{Im}\{c_{n,i}\}j \\ \widetilde{c}_{n,i'+1} = \text{Re}\{c_{n,i}\} \end{cases} \tag{15.75}$$

so that the $\widetilde{c}$ are *real* symbols, i is the index of the original complex modulation symbols (which occur at rate $1/T$), i' indexes the real symbols which occur on the grid at twice the rate, $2/T$, so that $i' = 2i$. The data symbols are then sent through the transmit filter shape $p_{\text{TX}}(t)$. The transmit signal is thus

$$s_k = \sum_{i'=-\infty}^{\infty} \sum_{n=0}^{N-1} \widetilde{c}_{n,i'} p_{\text{TX}}(k - i'N/2) e^{j2\pi kn/N} e^{-j\pi i'n}. \tag{15.76}$$

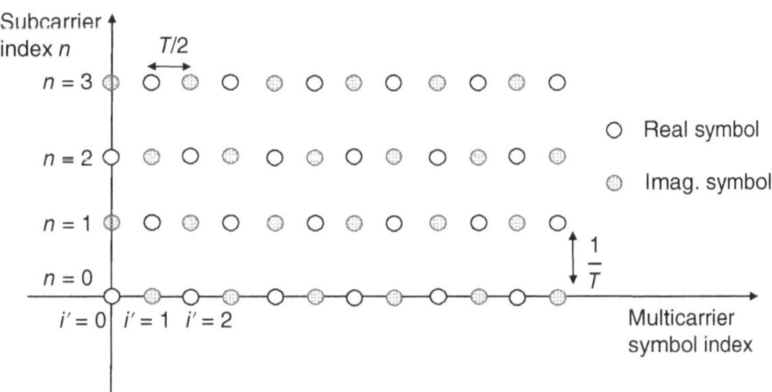

Figure 15.24 Staggered lattice for FBMC/OQAM. Color version available at wiley.com/go/molisch/wireless3e.
Reproduced with permission from [Rottenberg 2017] © F. Rottenberg.

---

[10] A similar lattice can be obtained by spacing the symbols T apart in the time domain and $1/(2T)$ apart in the frequency domain.

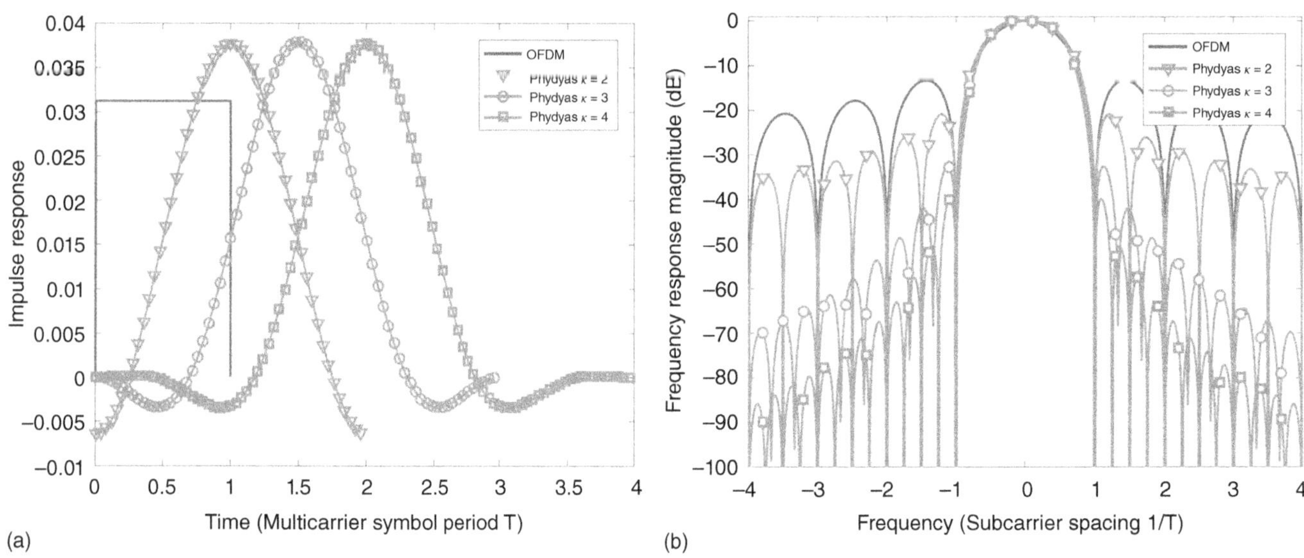

Figure 15.25 Pulseshapes for FBMC/OQAM, compared to standard rectangular pulse. (a) Time domain representation and (b) frequency domain. Color version available at wiley.com/go/molisch/wireless3e.
Reproduced with permission from [Rottenberg 2017] © F. Rottenberg.

When a matched-filter RX is used, we find that orthogonality of the real part, i.e., $\mathrm{Re}\{\langle g_{n,i}(t), d_{\ell,m}(t)\rangle\} = \delta_{n,\ell}\delta_{i,m}$ is achieved, subject to imposing of symmetry conditions on the pulse shape and Nyquist design.

Since the density of the lattice is $D(\Lambda) = 2$, i.e., an oversampled grid, the use of well-localized pulses in the time–frequency plane is possible; typical shapes in time-and-frequency domain are shown in Figure 15.25. At the same time, since real symbols occurring at integer multiples of $T/2$ carry the same amount of information as complex symbols at T, the system reaches the same spectral efficiency as standard OFDM (and a larger one than CP-OFDM since no CP is required). The drawback of this scheme lies in its more complicated signal processing, in particular in dispersive environments.

*15.11   Multi-Carrier Spread Spectrum

Up to now, we have dealt with modulation that concentrates each modulation symbol in the time–frequency domain in order to minimize the effect of dispersion, and simplify RX design. However, this has the drawback of sacrificing diversity (which we handled by either coding, or AMC). We now turn to multi-carrier techniques that consciously use waveforms that are widely spread out in the time–frequency domain; equivalently, we can say that we still use "basis pulses" concentrated in the time–frequency domain, but that the information symbols are spread out over many such basis pulses. This makes each symbol less sensitive to fading on this particular time/frequency resource. In this category, *Multi-Carrier CDMA* (MC-CDMA), DFT-spread OFDM, and Orthogonal Time Frequency Space (OTFS) modulation (see Section 15.12) are particularly prominent.

15.11.1   MC-CDMA

Let us explain the basic idea using the example of MC-CDMA (we will see below that the other formats have a very similar approach). We transmit a data symbol on all available subcarriers simultaneously. In other words, a code symbol c is mapped to a vector $c\mathbf{p}$, where $\mathbf{p}$ is a predetermined vector. This can be interpreted as a repetition code (a symbol is repeated on each tone, but multiplied by a different known constant p_n), or as a spreading action, where each symbol is represented by a code sequence – the sequence is in the frequency domain, instead of the time domain as would occur in CDMA (compare Section 19.2).

Irrespective of the interpretation, we obtain a bandwidth expansion by a factor of N, i.e., the transmission requires a larger bandwidth than one would expect from standard Fourier theory, leading to a loss of spectral efficiency. However, we can eliminate this problem by transmitting N different symbols, and thus N different codevectors, *simultaneously*. The first symbol c_1 is multiplied by codevector $\mathbf{p}_1$, the second symbol c_2 by codevector $\mathbf{p}_2$, and so on, and those spread symbols are added up and transmitted. If all the vectors $\mathbf{p}_n$ are chosen to be orthogonal, then the RX can easily recover the different transmitted symbols.

Let us now put this into a more compact mathematical form by writing the codevectors into a "spreading matrix" $\mathbf{P}$:

$$\mathbf{P} = [\mathbf{p}_1 \, \mathbf{p}_2 \cdots \mathbf{p}_N] \tag{15.77}$$

where we will see later that it is advantageous if $\mathbf{P}$ is unitary. The *symbol spreader* performs a matrix multiplication:

$$\widetilde{\mathbf{c}} = \mathbf{Pc}. \tag{15.78}$$

Now it is this modified signal that is OFDM-modulated – i.e., undergoes an IFFT and has the CP prepended – and sent over the wireless channel.

In the RX, we first perform the same operation as in "normal" OFDM: stripping off the CP and performing an FFT. Thus, we have again transformed the symbols into the frequency domain. At this point, the received symbols are

$$\widetilde{\mathbf{r}} = \mathbf{H}\widetilde{\mathbf{c}} + \mathbf{n} \tag{15.79}$$

where $\mathbf{H}$ is a diagonal matrix with entries $H\left(n\frac{W}{N}\right)$ along the diagonal. The next step is "one-tap equalization." Let us assume for the moment that we use zero-forcing equalization (we will see below why this is more relevant in MC-CDMA than in conventional OFDM). Then employing the unitary properties of the spreading matrix, we can perform despreading by just multiplying the received signal by the Hermitian transpose of the spreading matrix to obtain:

$$\mathbf{P}^{\dagger}\mathbf{H}^{-1}\widetilde{\mathbf{r}} = \mathbf{P}^{\dagger}\mathbf{H}^{-1}\mathbf{HPc} + \mathbf{P}^{\dagger}\mathbf{H}^{-1}\mathbf{n} \tag{15.80}$$

$$= \mathbf{c} + \widetilde{\mathbf{n}}. \tag{15.81}$$

We have thus recovered the transmit symbols. A summary of the transceiver structure can be seen in Figure 15.26.

Note that noise is no longer spectrally white, since the $\widetilde{\mathbf{n}}$ includes noise enhancement from zero-forcing equalization. When the RX uses MMSE equalization instead of zero-forcing, then noise enhancement is not as bad. However, MMSE equalization does not recover the orthogonality of different codewords the way that zero-forcing does. After equalization and despreading, we get

$$\mathbf{P}^{\dagger}\frac{\mathbf{H}^{\dagger}}{|\mathbf{H}|^{2} + \sigma_{\mathrm{n}}^{2}}\widetilde{\mathbf{r}} = \mathbf{P}^{\dagger}\frac{\mathbf{H}^{\dagger}\mathbf{H}}{|\mathbf{H}|^{2} + \sigma_{\mathrm{n}}^{2}}\mathbf{Pc} + \mathbf{P}^{\dagger}\frac{\mathbf{H}^{\dagger}}{|\mathbf{H}|^{2} + \sigma_{\mathrm{n}}^{2}}\mathbf{n}. \tag{15.82}$$

The matrix $\mathbf{P}^{\dagger}\mathbf{H}^{\dagger}\mathbf{H}/(|\mathbf{H}|^{2} + \sigma_{\mathrm{n}}^{2})\mathbf{P}$ is not diagonal. This means that there is residual crosstalk from one codeword to the other.

What spreading matrices should be used for MC-CDMA systems? A popular choice are the so-called Walsh–Hadamard matrices. Define the $n + 1$-order Hadamard matrix $\mathbf{H}_{\mathrm{had}}^{(n+1)}$ in terms of the nth order matrix:

$$\mathbf{H}_{\mathrm{had}}^{(n+1)} = \begin{pmatrix} \mathbf{H}_{\mathrm{had}}^{(n)} & \mathbf{H}_{\mathrm{had}}^{(n)} \\ \mathbf{H}_{\mathrm{had}}^{(n)} & \overline{\mathbf{H}}_{\mathrm{had}}^{(n)} \end{pmatrix} \tag{15.83}$$

where $\overline{\mathbf{H}}$ is the modulo-2 complement of $\mathbf{H}$. The recursive equation is initialized as

$$\mathbf{H}_{\mathrm{had}}^{(1)} = \begin{pmatrix} 1 & 1 \\ 1 & -1 \end{pmatrix}. \tag{15.84}$$

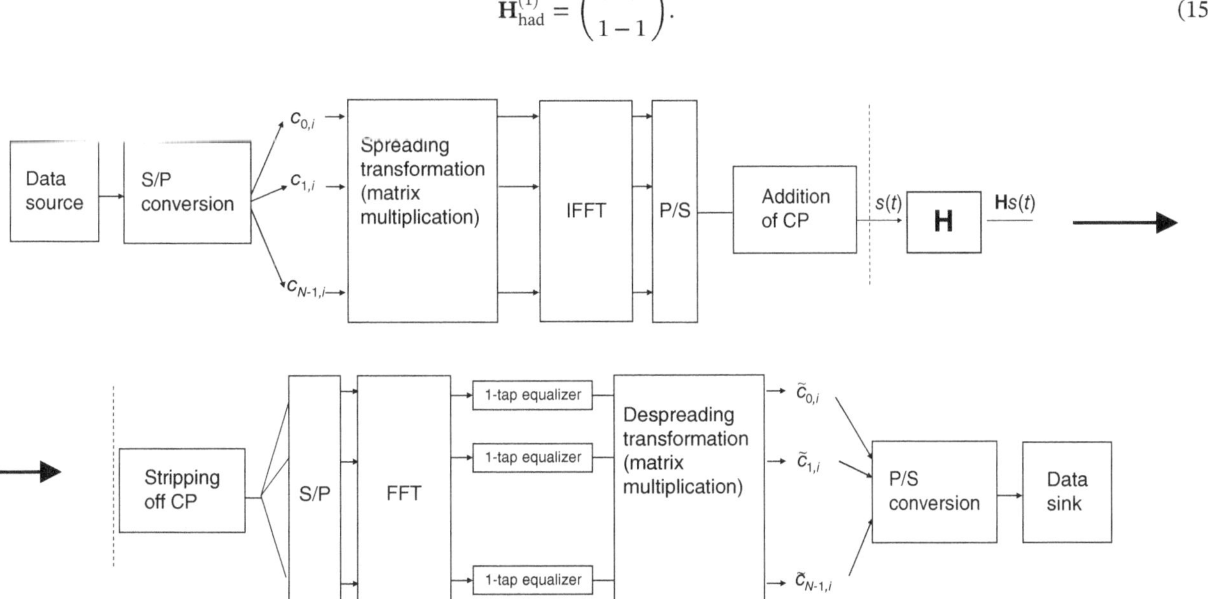

Figure 15.26 Block diagram of a multi-carrier code-division-multiple-access transceiver.

The columns of this matrix represent all possible Walsh–Hadamard codewords of length 2; it is immediately obvious that the columns are orthogonal to each other. From the recursion equation, we find that $\mathbf{H}_{\text{had}}^{(2)}$ is

$$
\mathbf{H}_{\text{had}}^{(2)} = \begin{pmatrix} 1 & 1 & 1 & 1 \\ 1 & -1 & 1 & -1 \\ 1 & 1 & -1 & -1 \\ 1 & -1 & -1 & 1 \end{pmatrix}.
\tag{15.85}
$$

The columns of this matrix are all possible codewords of duration four; again it is easy to show that they are all orthogonal to each other. Further iterations give additional codewords, each of which is twice as long as that of the preceding matrix.

Note that these matrices are unitary and have all the coefficients as ± 1; furthermore, the recursive definition allows implementation of a Walsh–Hadamard transform with a "butterfly" structure, similar to implementation of an FFT.

How does spreading influence the PAPR problem? In most cases, there is no significant impact. The explanation can again be found from the central limit theorem: the output (amplitudes of the I- and Q-components on different subcarriers) of the spreading matrix is approximately Gaussian distributed, as it is the sum of a large number of variables. The IFFT then just weights and sums those Gaussian variables. But the weighted sum of Gaussian variables is again a Gaussian variable. The amplitude distribution of the transmit signal of MC-CDMA is thus the same as the amplitude distribution of normal OFDM.

Besides the WH codes, other types of unitary code vectors are possible as well. We also note that the subcarriers used for the spreading need not cover all subcarriers assigned to a particular user. Rather, we can use sparse coding vectors that use, e.g., only every Kth subcarrier. This does not necessarily reduce the available diversity – as long as there is at least one used subcarrier per coherence bandwidth of the channel, the available diversity is (approximately) exploited.

15.11.2 DFT-Spread OFDM

A variant of MC-CDMA occurs when the unitary transformation matrix $\mathbf{P}$ is chosen to be the FFT matrix; this case is called *DFT-spread OFDM*. In such a case, multiplication by the spreading matrix and the IFFT inherent in the OFDM implementation cancel out. In other words, the transmit sequence that is transmitted over the channel is the original data sequence – plus a CP that is just a prepending of a few data symbols at the beginning of each datablock.

This just seems like a rather contrived way of describing the single-carrier system that has already been discussed in Chapter 10. However, the big difference here lies in the existence of the CP, as well as in how the signal is processed at the RX (see Figure 15.27). After stripping off the CP, the signal is transformed by an FFT into the frequency domain. Due to the CP, there are no residual effects of ISI or ICI. Then, the RX performs equalization on each subcarrier (this can be zero-forcing or MMSE equalization), and finally transforms the signal back into the time domain via an IFFT (this is the de-spreading step of MC-CDMA). The RX thus performs equalization in the frequency domain. Since FFTs or IFFTs can be implemented efficiently, the computational effort (per bit) for equalization goes only like $\log_2(N)$. This is a considerable advantage compared with other equalization techniques discussed in Chapter 14. On the downside, frequency domain equalization is a linear equalization scheme and therefore does not give optimum performance. Also, the extra overhead of a CP has to be taken into account.

Another important advantage of DFT-spread OFDM is the fact that it has a much lower PAPR. Since it is essentially single-carrier modulation, its PAPR is determined by the QAM modulation alphabet used for the transmission, but no extra penalty arises from the superposition of different modulation symbols. Furthermore, complexity of the TX is shifted to the RX, since the TX needs no FFT (but a RX now needs both FFT and IFFT). The combination of lower PAPR and simpler TX makes DFT-spread OFDM especially attractive for uplink transmission, and indeed it is being used for this purpose in LTE (see Chapter 31) and also (optionally) in 5G (see Chapter 32).

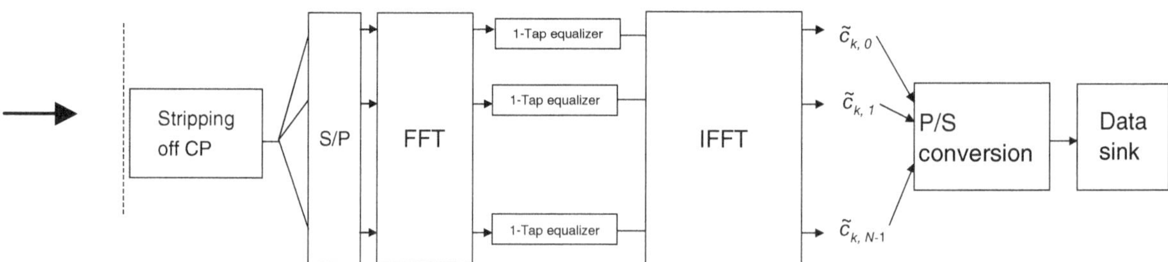

Figure 15.27 Block diagram of a single-carrier frequency domain equalization RX.

*15.12 Orthogonal Time Frequency Spreading (OTFS)

15.12.1 Introduction

After 2015, another scheme related to multi-carrier communication has emerged, OTFS (the acronym is sometimes interpreted as *Orthogonal Time Frequency Spreading*, and – when used in connection with multiple antennas, as *Orthogonal Time Frequency Space* modulation). This format can be interpreted in a variety of ways:

- *Modulation in the delay-Doppler domain*: while OFDM is a PAM where the basis pulses are reasonably concentrated in the time–frequency domain, OTFS modulates pulses that are concentrated in the delay-Doppler domain (essentially, the Fourier dual of the time–frequency domain).
- *Generalization of DFT-spread OFDM*: while DFT-spread OFDM spreads a symbol over all subcarriers, using DFT coefficients as spreading sequence, OTFS *in addition* performs a spreading in the time domain, thus spreading each symbol over a long time interval.
- *Representation via the Zak transform*: the basis pulses of OTFS can also be interpreted as "living" in the Zak domain, where the Zak transform can be viewed as the "square root" of a Fourier transform (the concatenation of two Zak transforms results in one Fourier transform). For details of this interpretation, see [Hadani et al. 2018].

15.12.2 Mathematical Description

We now consider the first interpretation. We have seen in Chapter 6 that the propagation channel can be represented in the time-frequency domain (as the time-variant transfer function $H(t, f)$, or equivalently by the spreading function (Doppler-variant impulse response) $h_s(\nu, \tau)$.[11] Similarly, we can think of signals "living" in the time-frequency domain, as OFDM signals do, or in the delay-Doppler domain. Obviously, the two forms can be transformed into each other by means of Fourier transforms, but for the purpose of algorithm design and analysis, it is often advantageous to think in one or the other form. We also recall the interpretation of what a time-invariant channel does to a signal: a *multiplication* of the signal spectrum with the transfer function, or a *convolution* of the time-domain signal with the impulse response. When now considering time-variant systems, we have a simple generalization of this behavior: again, a multiplication in the time-frequency domain of the signal with the time-variant transfer function, and a two-dimensional convolution of the delay-Doppler signal with the Doppler-variant impulse response. A particular form of this two-dimensional convolution is the *twisted convolution*, denoted by $*_\sigma$

$$h_2(\nu, \tau) *_\sigma h_1(\nu, \tau) = \int\int h_2(\nu', \tau') h_1(\nu - \nu', \tau - \tau') e^{j2\pi\nu'(\tau - \tau')} d\tau' d\nu'. \tag{15.86}$$

Thus, if we send a delay-Doppler pulse into a doubly-dispersive channel, we get an output that is broadened in both delay and Doppler. An example is shown in Figure 15.28. We see the convolution of the pulses in the delay/Doppler domain with the Doppler-variant impulse response. For clarity of the depiction, the signal pulses are widely separated; in a spectrally efficient transmission, they would be placed closely to each other.

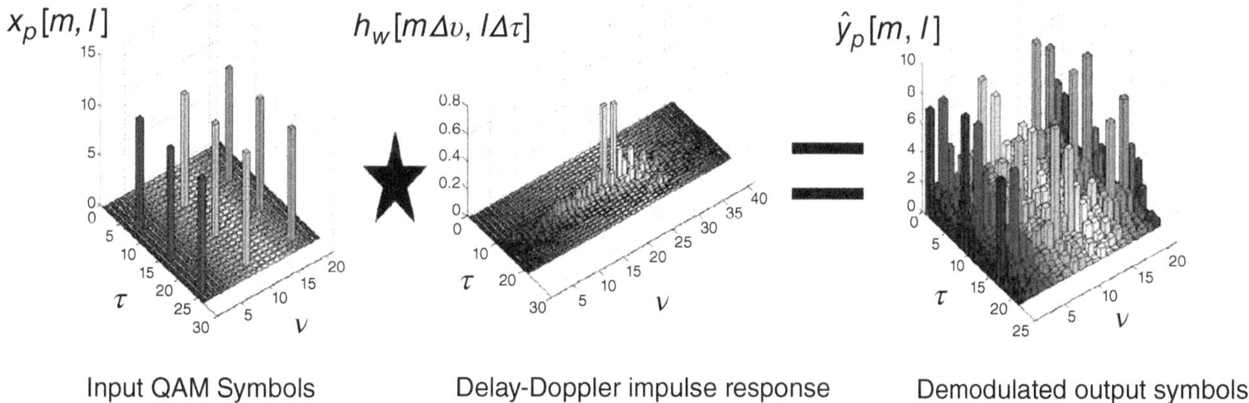

Figure 15.28 Response with modulation symbols in the delay/Doppler domain. Color version available at wiley.com/go/molisch/wireless3e. Reproduced with permission from [Hadani et al. 2017] © IEEE.

---

[11] Note that in Chapter 6 we named this quantity $s(\nu, \tau)$. To avoid confusion with the transmit signal s, we here choose a different symbol h_s.

The signal dispersion leads to intersymbol interference, and has to be reversed by a two-dimensional equalizer. Possible structures include the 2D versions of standard equalizers, namely (i) linear equalizers, (ii) non-linear equalizers such as decision feedback and maximum-likelihood sequence estimators, possibly approximated as turbo equalizers. As we will discuss below, this somewhat higher signal processing effort leads to significant improvement in performance.

The OTFS modulation itself can now be interpreted as a cascade of two two-dimensional transforms. At the TX, in a first step, the information symbols $c_{\ell,m}$, which reside in the delay-Doppler domain, are mapped into the time-frequency domain through the 2D inverse Fourier transform:[12] This mapping also involves windowing (since any practical implementation has finite block size) and periodization with period (I, N): $\widetilde{c}_{n,i} = \mathrm{SFFT}^{-1}(c_{\ell,m})$ for

$$\widetilde{c}_{n,i} = W_{\mathrm{TX},n,i}\frac{1}{NI}\sum_{\ell,m}c_{\ell,m}\exp\left[-j2\pi\left(\frac{n\ell}{N}-\frac{im}{I}\right)\right]\tag{15.87}$$

where $m = 0, ..., I-1$, and $\ell = 0, ..., N-1$; and W_{TX} is the transmit window. Note that a windowing in the time-frequency domain is necessary since the FFT leads to outputs that are periodic. In the following, the combination of windowing and inverse SFFT is called the OTFS transform.

The two-dimensional symbols are then converted into a time-domain signal by means of classical multi-carrier modulation and serial/parallel conversion (this step is also sometimes called a *Heisenberg transform*, with its dual at the RX called the *Wigner transform*:

$$s(t) = \sum_{i=0}^{I-1}\sum_{n=0}^{N-1}\widetilde{c}_{n,i}p_{\mathrm{TX}}(t-iT)\exp\left[j2\pi n\Delta f(t-iT)\right].\tag{15.88}$$

At the RX, the signal is first filtered with the receive basis pulse. The delay-Doppler representation of this filtered signal can be interpreted as the cross-ambiguity function of the received signal with the basis pulse, where the cross-ambiguity function has a slightly modified definition compared to Eq. (15.69): $A_{p_{\mathrm{TX}},p_{\mathrm{RX}}}(\nu,\tau) \triangleq \int e^{-j2\pi\nu(t-\tau)}p_{\mathrm{TX}}(t)p_{\mathrm{RX}}^*(t-\tau)\mathrm{d}t$. It can be shown that the end-to-end channel (without OTFS transform) can be described as

$$r(\nu,\tau) = h_{\mathrm{s}}(\nu,\tau)*_\sigma\widetilde{c}_{n,i}*_\sigma A_{p_{\mathrm{TX}},p_{\mathrm{RX}}}(\nu,\tau).\tag{15.89}$$

This signal is then sampled at $\nu = n\Delta f$, $\tau = iT$, to provide $r_{n,i}$. This signal is then multiplied with a receive window $W_{\mathrm{RX},n,i}$, and periodized (with periods N and I) to provide $\widetilde{r}_{n,i}$. The estimates of the symbols in the delay-Doppler domain are then obtained as $\hat{r}_{\ell,m} = \mathrm{SFFT}(\widetilde{r}_{n,i})$, i.e.,

$$\hat{r}_{\ell,m} = \sum_{n,i}\widetilde{r}_{n,i}\exp\left[j2\pi\left(\frac{n\ell}{N}-\frac{im}{I}\right)\right].\tag{15.90}$$

The estimated sequence $\hat{c}_{\ell,m}$ of information symbols obtained after demodulation is given by the two dimensional periodic convolution

$$\hat{c}_{\ell,m} = \frac{1}{IN}\sum_{p,q}c_{p,q}h_{\mathrm{w}}\left(\frac{m-q}{IT},\frac{\ell-p}{N\Delta f}\right)\tag{15.91}$$

of the input QAM sequence $c_{\ell,m}$ and a sampled version of the windowed impulse response $h_{\mathrm{w}}(\cdot)$,

$$h_{\mathrm{w}}\left(\frac{m-q}{IT},\frac{\ell-p}{N\Delta f}\right) = h_{\mathrm{w}}(\nu',\tau')\big|_{\nu'=\frac{m-q}{IT},\tau'=\frac{\ell-p}{N\Delta f}}\tag{15.92}$$

where $h_{\mathrm{w}}(\nu',\tau')$ denotes the circular convolution of the channel response with a windowing function, which in turn is the symplectic Fourier transform of the product of the transmit and receive window. Importantly, $h_{\mathrm{w}}\left(\frac{m-q}{IT},\frac{\ell-p}{N\Delta f}\right) \approx 0$ for $m \neq q, \ell \neq p$, which means that in a nondispersive channel, we can reconstruct the transmit symbols through projection and Fourier transforms. In dispersive channels, equalizers are required, as discussed above.

---

[12] To stay consistent with the notation of the literature on OTFS, we use here the symplectic FT, implemented as symplectic FFT (SFFT) which has some signs in the exponents different from a "standard" FT.

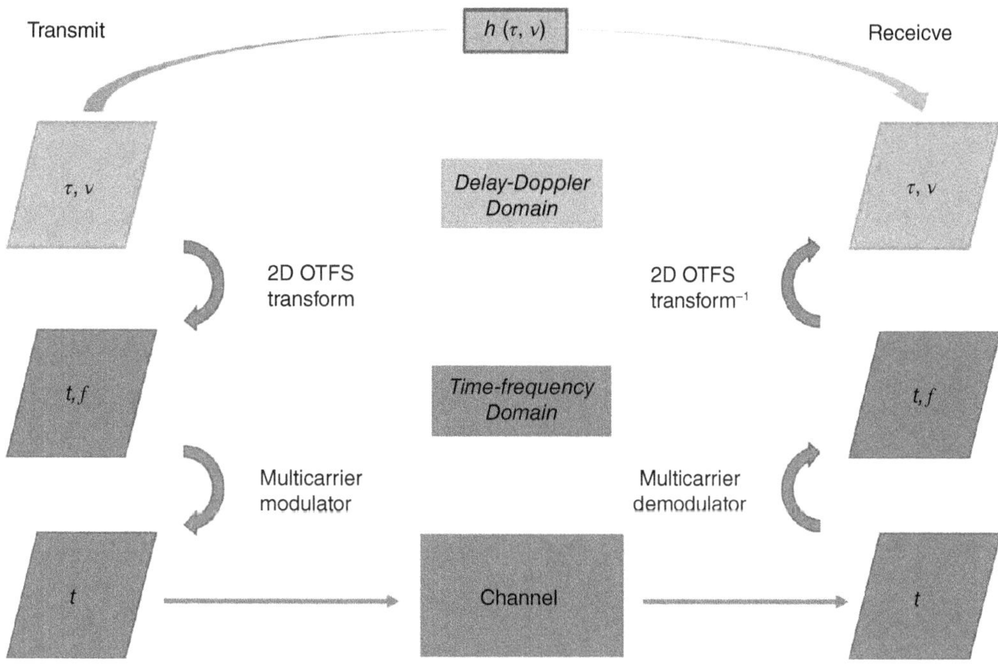

Figure 15.29 Signal flow in an implementation of OTFS as time-frequency overlay. Color version available at wiley.com/go/molisch/wireless3e. Reproduced with permission from [Hadani et al. 2017] © IEEE.

15.12.3 Implementation as Overlay

OTFS can be implemented in a variety of ways. One expedient method is as an overlay of an existing OFDM system, since highly optimized hardware already exists for such systems, especially in the context of cellular (3GPP) and wireless LAN (Wi-Fi) systems. Figure 15.29 shows the flow diagram that makes use of this structure. Current OFDM transceivers already implement a form of the Heisenberg/Wigner transform. At the TX, it is thus sufficient to perform a 2D OTFS transform (which can be implemented as a 2D SFFT), and let the resulting symbols be the input for the existing OFDM modulator. At the RX, the output of the (soft) OFDM demodulator also undergoes a 2D OTFS transform, the results of which are used as input to the OTFS equalizer and demodulator. This can be thought of as a generalization of the approach in single-carrier (SC)-FDMA (also referred to as DFT-spread-OFDM) where a one-dimensional FFT is applied. Note, however, that this implementation is an approximation that substitutes the cyclical convolution for the twisted convolution described in Eq. (15.86); this approximation works well for small Doppler.

15.12.4 Diversity and Channel Gain

From (15.42) we see that over a given frame, each demodulated symbol $\hat{c}_{m,k}$ for a given m and k experiences the same channel gain on the transmitted symbol $c_{m,k}$. This, combined with a suitable equalizer at the RX, allows to extract the full channel diversity. The almost-constant channel offers several important performance benefits. Firstly, it obviates the need for fast AMC while providing high diversity order. This is beneficial in fast-changing environments, see Section 15.9.

A high diversity could also be achieved in OFDM, using suitable interleaving and coding. However, that solution does not uniformly distribute the information over the time–frequency plane (the higher the coderate the more pronounced is the effect); furthermore it is not an effective solution for short codewords, since the diversity is upper bounded by the number of transmitted bits; in contrast, OTFS always provides full diversity.

A second important advantage of the almost-constant channel lies in enabling simplified signal processing, because the channel coefficients and the signal processing derived from them (e.g., precoder coefficients) are the same for all symbols.

Figure 15.30 shows the BER of an uncoded system operating in a time-and-frequency dispersive channel, with a medium Doppler (velocity corresponding to highway vehicle speed) and with different modulation formats. When comparing the results of OFDM modulation to that of OTFS, we see that the slope of the BER-vs-SNR curve is steeper for OTFS than for OFDM, which can be explained by the higher diversity order. While for higher-order modulation OFDM outperforms OTFS with linear (MMSE) equalizer, OTFS outperforms when using nonlinear (e.g., DFE) equalizers.

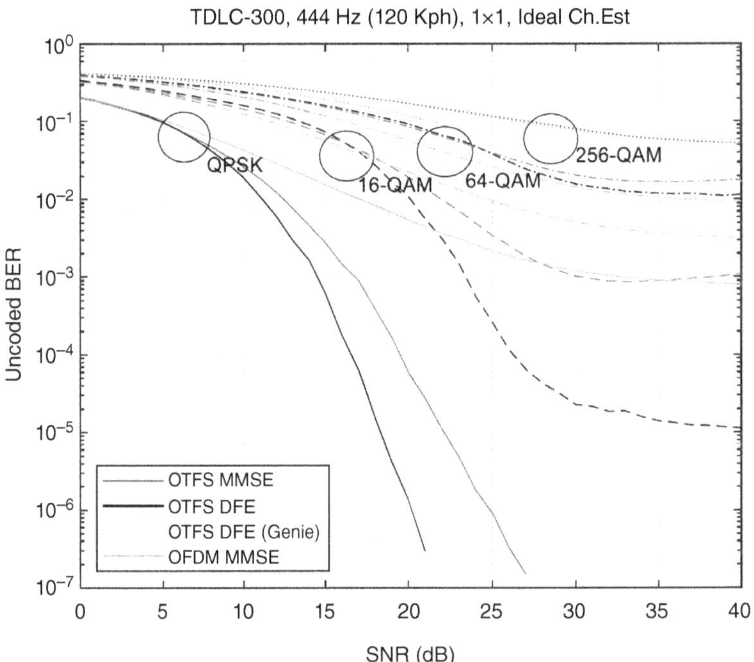

Figure 15.30 Uncoded BER for 4QAM/16QAM/64QAM/256QAM, TDL-C channel model, 120 kmph. Curves show OFDM with MMSE equalizer, and OTFS with (i) MMSE equalizer, (ii) DFE equalizer with error propagation, and (iii) DFE equalizer with genie feedback. Color version available at wiley.com/go/molisch/wireless3e.
Reproduced with permission from Hadani et al. [2017] © IEEE.

Further Reading

Several books describe multi-carrier schemes, including OFDM and multi-carrier CDMA, e.g., [Hanzo et al. 2003] and [Li and Stuber 2006]; another interesting description that also covers applications to various standardized systems is [Bahai et al. 2004]. Review papers on that topic include [Wang and Giannakis 2000], [Jajszczyk and Wagrowski 2005], [Hwang et al. 2009].

Multi-carrier modulation was invented by [Chang 1966], and further explored by Saltzberg in 1967 and the CP was proposed in [Weinstein and Ebert 1971]. A more detailed history of the milestones of multi-carrier modulation is given in Section III.E of [Sahin et al. 2013]. Discussion of the CP and its comparison to zero-padding can be found in [Muquet et al. 2002]. The performance of a coded OFDM system in a multi-path channel is analyzed, e.g., in [Kim et al. 1999]. ICI is discussed in [Cai and Giannakis 2003] and [Choi et al. 2001]; the latter also discussed ICI mitigation techniques, as does a number of other papers (see, e.g., [Schniter 2004] and [Molisch et al. 2007b] and references therein).

Synchronization and channel estimation are very important topics. Channel estimation techniques are discussed in [Edfors et al. 1998] and [Li et al. 1998]. Approximate DFT estimators were introduced in [van de Beek et al. 1995] and later analyzed in detail in [Edfors et al. 2000]. Filtering after eigen transformation was introduced in [Li et al. 1998] and extended to the transmit diversity case in [Li et al. 1999]. A more recent summary, that also includes extensive discussion of joint channel estimation and decoding, is provided in [Liu et al. 2014a]. Some examples of time and frequency synchronization include [Schmidl and Cox 1997], [Speth et al. 1999], and [van de Beek et al. 1999]. An excellent tutorial is provided in [Morelli et al. 2007]. Methods for PAPR reduction are reviewed in [May and Rohling 2000] and [Jiang and Wu 2008]. Waterfilling dates back to the classical work of [Shannon 1948]. Adaptive modulation was introduced by [Goldsmith and Chua 1998], [Chung and Goldsmith 2001], [Keller and Hanzo 2000]. Furthermore, the paper by [Wong et al. 1999], and especially the overview paper by [Keller and Hanzo 2000], gives a good account.

A number of different multi-carrier techniques have been introduced over the years. FBMC/OQAM was already proposed in the 1960s by [Chang 1966]; bi-orthogonal systems with Gaussian pulses for greater robustness were proposed by [Kozek and Molisch 1998], GFDM was introduced by [Fettweis et al. 2009]; an excellent overview of the different multi-carrier techniques, which partly inspired the presentation in Section 15.10, is provided in [Sahin et al. 2013]. Further reviews of multi-carrier techniques can be found in [Gerzaguet et al. 2017] and [van Eeckhaute et al. 2017].

[Yang and Hanzo 2003] review MC-CDMA. [Falconer et al. 2002] and [Benvenuto et al. 2010] discuss single-carrier frequency equalization schemes; a combination of this scheme with diversity is detailed in [Clark 1998]. OTFS was introduced by [Hadani et al. 2017].

For updates and errata for this chapter, see wides.usc.edu/teaching/textbook

Exercises

See Sec. 36.15 of Exercises.pdf at wiley.com/go/molisch/wireless3e

16

Multiple Antenna Systems – SIMO, MISO, and MIMO

Since the 1990s, there has been enormous interest in multiple antenna systems as a way to improve wireless link performance. Work originally focused on *smart antenna systems,* i.e., systems with multiple antenna elements at one link end only. Depending on whether the multiple antennas are at the Receiver (RX) or Transmitter (TX), such systems are called *Single-Input–Multiple-Output (SIMO)* or *Multiple-Input–Single-Output (MISO),* respectively. The multiple antenna elements are used for increasing the Signal-to-Noise Ratio (SNR) and reliability of the system, through diversity and beamforming. Similar benefits can also be obtained from *Multiple-Input–Multiple-Output (MIMO) systems.* However, the core motivation for the study of MIMO is the possibility of *Spatial Multiplexing (SM),* i.e., transmitting multiple data streams from one TX to one RX in the same time-frequency resources. All of these aspects will be discussed in this chapter, concentrating on the single-link (single user) case. The use of multiple antennas in multi-user systems will be discussed in Chapter 22.

16.1 Diversity and Beamforming

16.1.1 *Motivation*

SIMO/MISO systems use multiple antenna elements to achieve the benefits of diversity and beamforming gain. Multiple-antenna arrangements with appropriate signal processing for obtaining diversity or beamforming gain are also often called "smart antennas." In most practical situations, the smart antennas are located at the BS. Depending on whether we consider uplink or downlink, the smart antenna is thus at the RX or TX, respectively, resulting in SIMO or MISO, respectively (many examples in the remainder of this subsection assume the smart antenna at the RX).

The intelligence of a smart antenna is not in the actual antenna, but rather in the signal processing. In the simplest case, the combination of the antenna signals is a linear combination using a weight vector $\mathbf{w}^*$. The methods for determining $\mathbf{w}^*$ are what essentially differentiates various smart antenna systems. We see immediately that there is a strong relationship between multiple-antenna systems and diversity systems – as a matter of fact, an RX with antenna diversity *is* a smart antenna. This chapter will thus reuse many results from Chapter 12 (Figure 16.1).

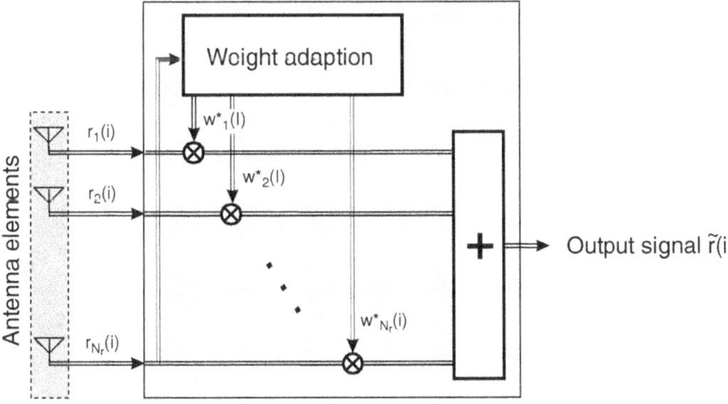

Figure 16.1 Linear combination of antenna signals. Color version available at wiley.com/go/molisch/wireless3e.

Smart antennas can be interpreted in a variety of ways that are equivalent, but provide different insights:

- A smart antenna exploits signals sampled at different *spatial locations*, thus providing different linear combinations of the signal echoes – remember that the total received signal is made up of contributions from different MPCs of a user, and possibly signals from different users. The smart antenna then processes/combines these MPC sums to obtain a favorable total signal.
- A smart antenna exploits the *directional properties of the channel*. Remember from Chapters 8 and 9 that a RX with multiple antenna elements can distinguish MPCs with different directions of arrival. Thus, one way to interpret smart antennas is as an RX that distinguishes between MPCs with different directions of arrival and processes them separately. This allows the RX to, e.g., coherently add up the different MPCs, and thus avoid fading; MPCs from interferers can also be suppressed.
- Yet another interpretation is that the smart antenna adaptively forms a beam pattern that enhances (coherently combines) the desired MPCs, while suppressing the contributions from the interferers.

Application of Smart Antennas

Smart antennas can be used for various purposes:

1. *Increase of coverage*: By coherently combining the signals at the different antenna elements (or, equivalently, forming an antenna pattern that maximizes the receive power), the SNR of the signal at the combiner output is increased. In a noise-limited cellular system, an improvement in the SNR increases the area that can be covered by one BS. Conversely, the coverage range can be kept constant while the transmit power is decreased. In a simple example, with N_r antenna elements and thus an array gain of N_r, the transmit power can be decreased by a factor of N_r.
2. *Robustness to fading:* The different elements of a smart antenna provide spatial diversity, and thus improve robustness to small-scale fading, as discussed in Chapter 12.
3. *Interference suppression:* A smart antenna can suppress interference, e.g., by optimum combining, which creates better sensitivity, and allows transmission over longer ranges, similar to the previous point. Note that if the interferer is another user in the system of interest, further benefits can be reaped, see below.
4. *Increase of number of served users.* Multiple antennas can increase the number of served users by reducing interference and/or serving multiple users located in different directions. These aspects will be discussed in detail in Chapter 22.
5. *Decrease of the delay dispersion:* By suppressing MPCs with large delays, the delay dispersion can be reduced. This feature can be especially useful in systems with a very large bandwidth, such as millimeter-wave systems.
6. *Improvement of user position estimation:* Knowledge of the directions-of-arrival, especially for the (quasi-)line-of-sight component, improves the geolocation. This is of value both for *location-based services* and for emergency (*E911*) location. For a more detailed discussion, see Chapter 29.

It is *not* possible to have all those advantages simultaneously to their fullest extent. For example, we can use the capability of smart antennas to reduce interference in order to *either* improve the quality of a single link *or* have more users in the system *or* to have a trade-off between the two. For the system design, the engineer has to decide which aspect is the most important one.

16.1.2   System Model

We first establish the system model and the notation; to allow a unified treatment of diversity and *SM*, we set up this model to be general enough to cover all cases, even though some components might not be needed in all applications. Figure 16.2 exhibits a block diagram. At the TX, the original source data stream enters an encoder, which puts out N_s different, possibly lower-rate, data streams $\widetilde{s}_m(t)$, $m = 1, ..., N_s$, which can be written as a $N_s \times 1$ vector $\widetilde{\mathbf{s}}(t)$. Each data stream is normalized to have on average unit power, such that $E\left\{\int_{T_s} |\widetilde{s}_m(t)|^2 dt\right\} = T_s$, where the expectation is over the data symbols. The signals then undergo a (linear) preprocessing

$$\mathbf{s} = \mathbf{T}\widetilde{\mathbf{s}} \tag{16.1}$$

i.e., the preprocessing is a multiplication of $\widetilde{\mathbf{s}}(t)$ with the $N_t \times N_s$ matrix $\mathbf{T}$, resulting in an $N_t \times 1$ output signal $\mathbf{s}(t)$, whose components, $s_1(t), s_2(t), ..., s_{N_t}(t)$ are the transmit signals at the different antenna elements.

The preprocessing can be decomposed into a power weighting and a beamforming operation

$$\mathbf{T} = \hat{\mathbf{T}}\begin{bmatrix} [\mathbf{P}]^{1/2} \\ \mathbf{0} \end{bmatrix} \tag{16.2}$$

where the $N_s \times N_s$ diagonal matrix $\mathbf{P}$ has diagonal entries P_m, $\mathbf{0}$ is an $(N_t - N_s) \times N_s$ all-zero matrix, and $\hat{\mathbf{T}}$ is an $N_t \times N_t$ unitary matrix, $\hat{\mathbf{T}}\hat{\mathbf{T}}^\dagger = \mathbf{I}$. To be completely general, this expression can be postmultiplied with a unitary matrix $\hat{\hat{\mathbf{T}}}^\dagger$, so that $\hat{\mathbf{T}}$, power allocation, and $\hat{\hat{\mathbf{T}}}^\dagger$ constitute the singular value decomposition of an arbitrary precoder. However, the presence of $\hat{\hat{\mathbf{T}}}^\dagger$ does not change the performance of the precoder for Gaussian signals, and is thus usually disregarded for theoretical investigations [Heath and Lozano 2018]. In this

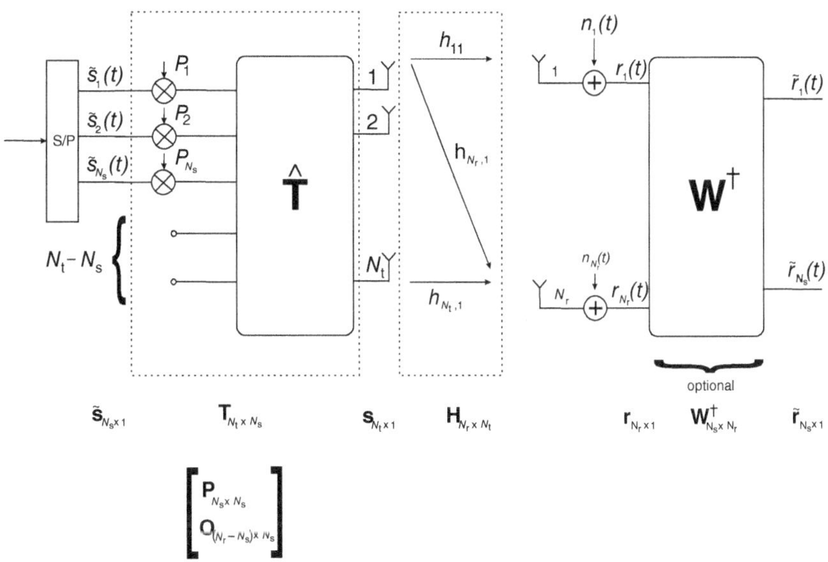

Figure 16.2 Block diagram of a MIMO system.

picture, the matrix $\mathbf{T}$ can be interpreted as the preprocessing consisting of a (power-preserving) beamforming operation $\hat{\mathbf{T}}$, and the assignment of powers P_m to the different data streams, (i.e., the (amplitude) signal is multiplied with $\sqrt{P_m}$).

We generally assume a per-symbol power constraint on the TX

$$tr\left[\mathbf{TT}^{\dagger}\right] = tr[\mathbf{P}] = P_{\text{tot}} \leq P_{\max} \tag{16.3}$$

that can be interpreted as a limit on the transmitted peak power of the antenna array, which could be due to either hardware constraints, or regulatory constraints on the conducted power.

Other possible power constraints (which will not be used in this book) are

- Long-term power constraints, where the temporal average over the possibly time-varying $tr[\mathbf{TT}^{\dagger}]$ is constrained. This constraint is meaningful when the limiting factor for the power expenditure is the battery drainage, since it is determined by the cumulative (or average) power consumption.
- Per antenna power constraint, such that $[\mathbf{TT}^{\dagger}]_{i,\,i} \leq P_{\text{pa}}$: this constraint should be used if each transmit antenna has its own power amplifier whose peak output power is limited.

From the antennas, the signal is sent through the wireless channel, which is characterized by the $N_{\text{r}} \times N_{\text{t}}$ channel matrix of the channel as

$$\mathbf{H} = \begin{pmatrix} h_{11} & h_{12} & . & . & . & h_{1N_{\text{t}}} \\ h_{21} & h_{22} & . & . & . & h_{2N_{\text{t}}} \\ . & & . & . & . & . \\ h_{N_{\text{r}}1} & h_{N_{\text{r}}2} & & & & h_{N_{\text{r}}N_{\text{t}}} \end{pmatrix} \tag{16.4}$$

where entry h_{ij} characterizes the channel from the jth transmit to the ith receive antenna.

If the channel is flat fading, h_{ij} is a complex scalar, namely the complex channel gain for this TX/RX antenna pair. It will often be convenient to factor the channel matrix into a scalar $\beta^{1/2}$, where β describes the combination of path gain and shadowing (which usually is assumed to be the same for all antenna elements), and a matrix $\widetilde{\mathbf{H}}$ that is normalized such that $E\left\{\|\widetilde{\mathbf{H}}\|_F^2\right\} = N_{\text{t}}N_{\text{r}}$, where the expectation is over the small-scale fading. For most of the current chapter, we ignore pathloss and shadowing and treat $\mathbf{H}$ and $\widetilde{\mathbf{H}}$ interchangeably. If the channel is Rayleigh fading, and the fading is independent at the different antenna elements, the (normalized) variables h_{ij} are i.i.d. zero-mean, circularly symmetric complex Gaussian random variables with unit variance, i.e., the real and imaginary part each has variance $1/2$.

Consequently, the power carried by each transmission channel h_{ij} is chi-square distributed with two degrees of freedom. This is one of the simplest possible, and most widely used, channel models; it requires the existence of "heavy multi-path," i.e., many MPCs, all of which have approximately equal strength, as well as a sufficient distance between the antenna elements.

If the channel is delay dispersive, $h(\tau)$ is a function of the delay with finite support, or equivalently $h(f)$ is a function of frequency. The latter description is more common for handling the frequency-selective case, because it is a natural fit to describing Orthogonal Frequency Division Multiplexing (OFDM) systems. Note that in such a wideband system, the power allocation $\mathbf{P}$, the precoding matrix $\mathbf{T}$, and the receive combiner (see below) $\mathbf{W}^\dagger$ may all be frequency dependent, i.e., have different values for the different subcarriers.

The received $N_r \times 1$ signal vector $\mathbf{r}$, i.e., the signal obtained at the N_r receive antenna elements, is

$$\mathbf{r} = \mathbf{Hs} + \mathbf{n}. \tag{16.5}$$

It is received by N_r antenna elements, where the $N_r \times 1$ vector $\mathbf{n}$ describes the noise, where the entries are commonly assumed to be realizations of circularly complex Gaussian i.i.d. random variables. In many RXs, the first stage is a linear combiner, described by the $N_s \times N_r$ matrix $\mathbf{W}^\dagger$ (which is often chosen to be unitary), so that $\tilde{\mathbf{r}} = \mathbf{W}^\dagger \mathbf{r}$. The output of the combiner is then forwarded to the decoder for further processing.

For the case of diversity and beamforming, i.e., the case we consider in this section, the model can be simplified considerably. Since only a single data stream exists, $N_s = 1$, and the power allocation matrix is simply a scalar multiplying the incoming data stream, the TX is simply multiplication of the data stream $s(t)$ with a scalar power P, and a beamforming vector $\hat{\mathbf{t}}$ of dimension $N_t \times 1$ that is normalized as $\left\| \hat{\mathbf{t}} \right\|_F^2 = 1$. Similarly, at the RX, the signal is combined through multiplication with a normalized $N_r \times 1$ vector $\mathbf{w}^\dagger$. For SIMO and MISO, $N_t = 1$ and $N_r = 1$, respectively.

Channel State Information

Algorithms for beamforming, diversity, and SM can be categorized by the amount of *Channel State Information (CSI)* that they require, noting that CSI at the RX (CSIR) is easier to obtain than CSI at the TX (CSIT). To obtain CSIR, we usually have to insert pilot signals into the transmission, and the RX can extract the CSIR from those symbols (compare, e.g., Section 15.5). In order to obtain CSIT, typically this information needs to be fed back from the RX to the TX, which results in additional overhead, or it can be inferred from CSIR if reciprocity holds (see Section 16.1.7).

Thus the following cases can be distinguished:

1. *Full CSIT and full CSIR*: In this ideal case, both the TX and the RX have full and perfect knowledge of the instantaneous CSI. This case obviously results in the highest-possible performance. However, in many instances, it is difficult to obtain the full CSIT, or the acquisition of CSIT requires such a high overhead that it negates the performance advantages that can be achieved. A more detailed discussion will be given in Section 16.1.7.

2. *Second-order CSIT and full CSIR:* In this case, the RX has full information of the instantaneous channel state, but the TX knows only the second-order CSI, e.g., the correlation matrix of the channel impulse responses or the Angular Power Spectrum (APS). As we will discuss in Section 16.1.7, this is easier to achieve.

3. *No CSIT and full CSIR*: This is the case that can be achieved most easily. The TX simply does not use any CSI, while the RX learns the instantaneous channel state from a training (pilot) sequence or using blind estimation (see Section 16.1.4).

4. *Noisy CSI*: When we assume "full CSI" at the RX, this implies that the RX has learned the channel state perfectly. However, any received training sequence will be affected by additive noise as well as quantization noise and possible interference. When taking into account the effect of noisy CSI, the RX becomes a "mismatched RX," where the RX processes the signal based on the *estimated* channel $\hat{\mathbf{H}}$, while in reality, the signals pass through the channel $\mathbf{H}$

$$\hat{\mathbf{H}} = \mathbf{H} + \boldsymbol{\Delta}. \tag{16.6}$$

Some papers have taken this into account by a modification of the noise variance that affects the received data signal (replacing σ_n^2 by $\sigma_n^2 + \sigma_e^2$, where σ_e^2 is the variance of the entries of $\boldsymbol{\Delta}$).

5. *No CSIT and no CSIR*: We have seen in previous chapters that signals can be detected even without any CSI, e.g., when noncoherent or differentially coherent modulation (see Section 10.3.5) is used, or when blind channel estimation is applied. These approaches can be generalized to the multi-antenna case.

Beamforming vs. Diversity

Before proceeding further, it is useful to define more clearly the expressions "beamforming" and "diversity", since those often lead to confusion. We will again use the SIMO case (smart antennas at the RX) for the examples.

Intuitively, beamforming is the adaptive forming of an antenna pattern that enhances signals from a particular direction. This is easily interpreted if there is only a single MPC (e.g., a LOS connection) between the UE and the BS. The gain of the beamforming antenna, similar to the gain of a horn antenna pointed in the right direction, enhances the signal coming from the UE. The adaptiveness of the antenna is only used to point in different directions when, e.g., the UE moves. As we saw in Section 8.4, an antenna array with N_r antenna elements can have a gain of $G_e N_r$ where G_e is the element gain (assumed to be unity for the rest of this subsection). We also found that the antenna elements have to be spaced *no more than $\lambda/2$ apart* from each other, to avoid grating lobes.

Conversely, diversity combining performs a phase-corrected, weighted combining of signals at the different antenna elements. The SNR gain of Maximum Ratio Combining (MRC) with N_r antenna elements is N_r. We also have argued that antenna elements have to be spaced *at least* $\lambda/2$ from each other to obtain independent fading, and larger if the angular spread is limited. We saw in Section 12.4.3 that MRC combining not only provides an improvement in the average SNR, but also the slope of the SNR distribution.

These two descriptions give the impression that diversity and beamforming are two disparate techniques for smart antennas. However, we also see that the general operation performed on the signals received at the different antenna elements is a linear combination – this is true for both diversity and beamforming. We thus can conclude that there really is no distinction between diversity and beamforming if they are based on the same information, namely the instantaneous CSI, or in other words, the vector of transfer functions between the TX and the RX. Furthermore, any linear combining vector can be associated with a beampattern, according to Eq. (8.62). In other words, even a diversity combining is really a beamforming – it is just that the computation of the antenna weights is based on a mathematical criterion, while the shape of the associated beampattern is incidental, and its explicit computation often not even necessary. In contrast, the intuition for beamforming starts out with a beampattern, and from that derives the weight vector that achieves this array pattern. For a small spacing of the antenna elements, $\leq\lambda/2$, the associated beampattern will often look "physically reasonable," i.e., having the main lobe in the direction of the strongest component; as a matter of fact, for the case of pure LOS, the antenna pattern as obtained from MRC just has the sinc-type shape derived in Eq. (8.63). For a rich multi-path environment and larger antenna spacing, the magnitude of the antenna pattern may look very haphazard, but it will still lead to the maximum SNR if obtained via the MRC weights.

A true distinction between diversity and beamforming arises if beamforming is based on second-order channel statistics, instead of the instantaneous CSI (by definition, diversity combining is always based on instantaneous CSI). By second-order channel statistics we typically mean the APS, i.e., power distribution over angle, averaged over small scale fading. A beamformer pointing in the direction of the strongest MPCs will enhance the average SNR, but cannot provide any diversity, e.g., to protect against small-scale fading: as long as multiple MPCs fall within the beampattern of the RX antenna, they might interfere destructively and thus lead to a fading dip. Furthermore, the gain obtained from such a beamformer may be smaller than N_r – this can occur if, e.g., the APS is wider than the main lobe of the beampattern, so that an antenna array with maximum gain cannot collect all energy in the APS, and vice versa. One might then wonder why such a lower-performing processing method is considered at all. The answer to this is the reduced requirement for channel knowledge – instead of having to know the instantaneous channel state (and possibly having to know it at each subcarrier in an OFDM system), it only requires the second-order CSI, which changes much more slowly. This aspect is particularly important for beamforming at the TX, since CSIT is more difficult to obtain than CSIR.

16.1.3 Hardware Structures for SIMO and MISO

In this section, we will describe transceiver structures that can be used for the separation and processing of MPCs. For ease of exposition, we mostly discuss the uplink (SIMO) case; however, the structures are applicable to the downlink as well (see Section 16.1.5 for more details of downlink).

Adaptive Beamforming

The standard approach for beamforming/diversity, as discussed above, is a linear combination of signals (see Figure 16.1). Typically, the antenna weighting and summing are done in baseband (fully digital solution). In that case, on the positive side, there are essentially no constraints on the antenna weights that can be realized; on the downside it requires N_r complete downconversion chains for the N_r antenna elements, resulting in considerable cost and energy consumption in particular when N_r is large.

An alternative realization performs weighting and combining in RF (i.e., at the carrier frequency). The complex weighting can be seen as multiplying with a real weight (implemented through variable-gain amplifiers), and a phase shift (implemented through adjustable phase shifters). Clearly, if adaptive spatial processing is thus performed in RF, it obviates the need for multiple downconversion chains. Drawbacks of this approach are more difficult channel estimation, and larger errors in the weights that occur at higher frequencies (because the amplifier gains often cannot be adjusted as precisely as digital weighting in baseband could be done, and similarly for the phase shifters).

A compromise between the two approaches (baseband or RF) is provided by *hybrid beamforming*. There, the signals from N_r receive antenna elements are combined in RF to provide L output signals, which are then downconverted for further processing. This approach provides a compromise between the fully digital and the fully analog solution. A more detailed description of hybrid beamforming will be given in Section 16.2.10.

Note also that adaptive spatial processing can be done based on different CSI – it could be instantaneous CSI or second-order CSI. The former performs better, while the latter has a smaller overhead for pilot signals. In practice, spatial processing based on second-order statistics is often done for analog implementations, while digital implementations use instantaneous CSI.

Finally, operation of wideband systems needs to take delay dispersion/frequency selectivity into account. The discussion above was restricted to space-only processing; its output is a single signal that can be further processed to deal with delay dispersion. In other words, the OFDM processing, Rake reception, or equalization is done only on the *single* signal that is put out from the spatial processor. This approach works well only if the temporal properties of the signal (delay dispersion) are the same at all the antenna elements, or conversely, if the spatial properties are independent of the frequency or delay. While this situation can occur in some cases (see also Chapter 7), it is certainly not universally valid. In that case, in order to fully exploit the possibilities of the different MPCs, adaptive *space–time processing* has to be used. For OFDM system, each subcarrier, or each group of subcarriers within a coherence bandwidth, can use a different set of combiner weights (or, equivalently, a different beam pattern), based on the CSI for this subcarrier. For Code Division Multiple Access (CDMA) systems (see Sec. 19.2), the optimum linear RX is a two-dimensional Rake

RX, i.e., a linear combiner that weights all *resolvable* (in the space-time domain) MPCs and adds them up, see Figure 16.3. Equivalently we can say that each Rake finger can have a different beampattern associated with it.

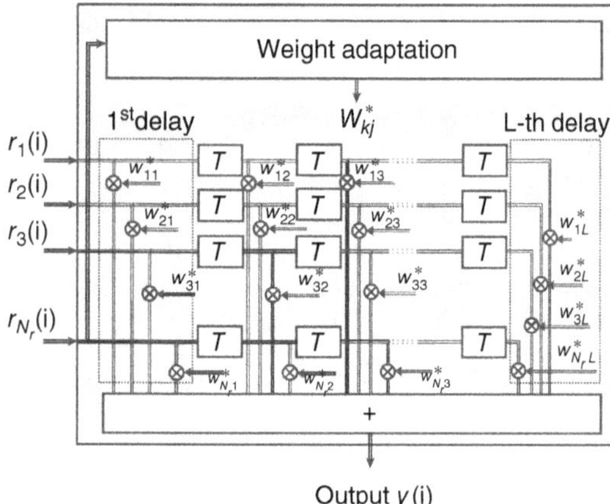

Figure 16.3 Space-time filter. L is the number of delay taps of the channel.

Note that forming different beampatterns on different subcarriers is only possible for a fully digital implementation. Analog amplifiers/phase shifters do not have the flexibility of changing their values within the system bandwidth, and only can adjust to the frequency-averaged (essentially, second-order statistics) CSI (compare also Sections 16.2.10 and 8.4.2).

Space–Time Detection

Space–time processing and decoding/detection have to be done jointly. Special properties of the received signal, including finite-alphabet properties, can be taken into account for the detection. The optimum detector is a generalized *Maximum Likelihood Sequence Estimator, MLSE* RX. Such structures are mainly used in the context of SM, as discussed in Section 16.2.9.

Switched-Beam Receivers

A switched beam antenna is an antenna array that can form only a small set of patterns, which are typically beams pointing into certain discrete directions. A switch then selects one of the possible beams[1] for downconversion and further processing. In other words, the RX selects the beam that is *best* in the sense that it gives the highest SNR or the highest Signal-to-Interference-and-Noise Ratio (SINR). This approach greatly simplifies the RX design. Relating to Chapter 12.4.1, we find switched beam antennas as an implementation of selection diversity, while the above-discussed adaptive beamforming is combining diversity.

There are many different ways of realizing switchable beams, for example

- An array containing multiple directional elements, oriented into different directions. Such setups are sometimes used, e.g., at BSs operating at mm-wave or THz frequencies, since at those high frequencies even directional antennas are reasonably small.
- A linear or rectangular array is followed by a spatial Fourier transformation. The output of such a spatial Fourier transformation is a number of beams that are all orthogonal to each other, and point into different directions. The spatial Fourier transformation can be realized, e.g., as a so-called *Butler matrix*. It has a structure that is essentially the butterfly structure well known from the software implementations of the FFT. The elements of each stage are simple phase-shifters that can be realized in the RF domain.
- A lens antenna, followed by pickup-antennas placed at different locations, so that they "see" different beam directions (compare Section 8.3.8).

The main advantage of the switched-beam approach is its simplicity: all processing (spatial FFT and selection) is done in the RF domain, so that only a single signal has to be downconverted into the baseband and processed there. Since the downconversion circuits are among the most expensive components in today's wireless systems, this is a significant advantage. The drawback of this scheme is its limited flexibility. The main beam can only be pointed into certain fixed directions, so that the gain into the actual direction of the UE might not be the maximum achievable value. Even more importantly, the nulls cannot be pointed into arbitrary directions, so that nulling of interferers is difficult.

Switched beam antennas have mainly the following applications: (i) systems that require feedback for CSIT. Since usually the information about dominant directions is quantized coarsely, and obviously, the TX needs to be able to form beams only into the quantized directions, (ii) systems with beamforming in the RF domain, where the phase shifters can only provide a rough quantization. This situation occurs mostly at very high frequencies.

---

[1] More precisely, it selects the signal associated with one of the possible beams.

Example 16.1 *For a switched-beam antenna, compute the maximum gain, and the maximum relative loss due to mismatch of Direction-of-Arrival (DoA) and beamdirection for a uniform linear array with Butler matrix and number of antenna elements (spaced $\lambda/2$ apart) equal to $N_r = 2, 4, 8$.*

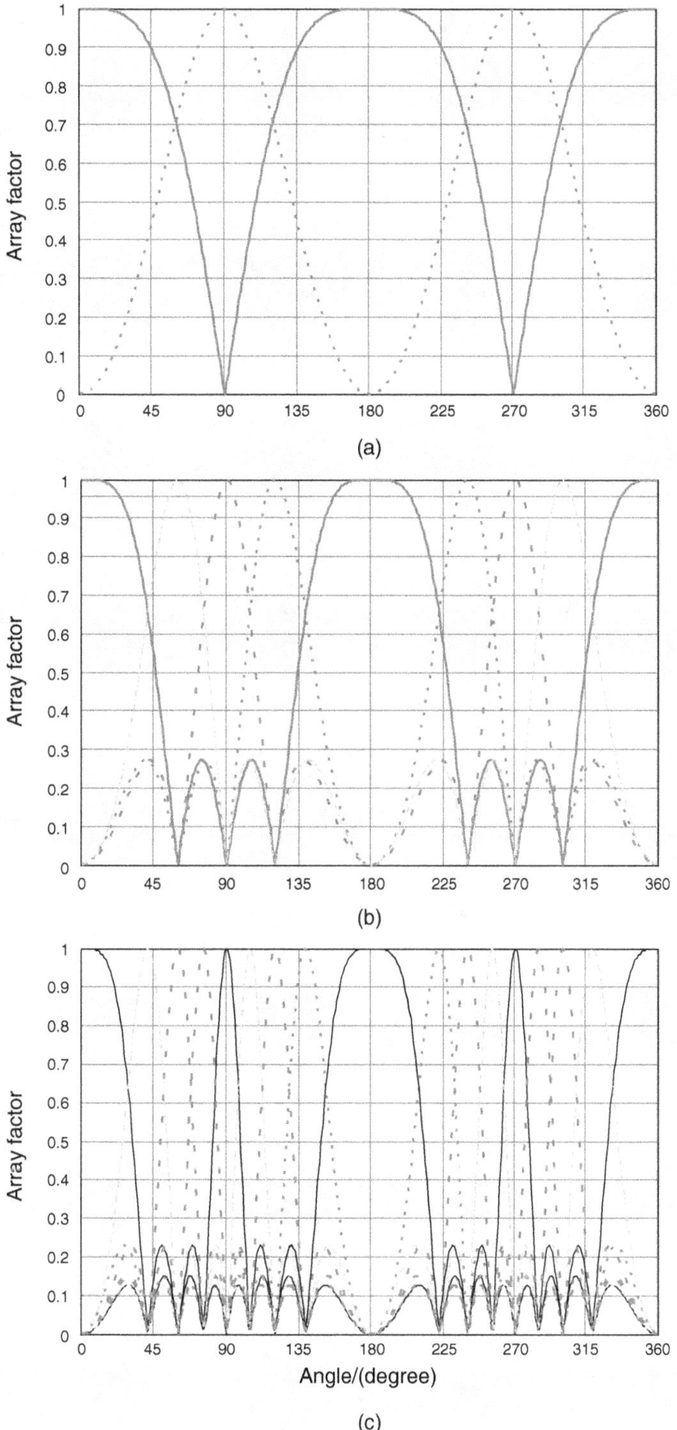

Figure 16.4 Normalized array factor of switched-beam antenna with 2 (a), 4 (b), and 8 (c) elements. Different linetypes denote different beams; note that beamwidth at endfire (0°, 180°) is larger than at broadside.

Consider the RX case. Let us first write down the IFFT matrix for N_r elements, with $k = 0, ..., N_r - 1$ and $n = 0, ..., N_r - 1$

$$w_{\text{IFFT},k,n} = \exp\left(j\frac{2\pi}{N_r} \right)^{kn}. \tag{16.7}$$

The array factor for a uniform linear array was given in (8.62)

$$M(\phi) = \sum_{n=0}^{N_r-1} w_n^* \exp\left[-j\cos(\phi)2\pi d_a n/\lambda\right].$$ (16.8)

With the weights, and thus the array factor, depending on k, and specifically taking on the values $w_{\text{IFFT},k,n}$, and furthermore $d_a = \lambda/2$, this becomes

$$M_k(\phi) = \sum_{n=0}^{N_r-1} \exp\left[-j2\pi n\left(\frac{\cos(\phi)}{2} - \frac{k}{N_r}\right)\right]$$ (16.9)

and the magnitude of the array factor is (compare Eq. (8.63)

$$|M_k(\phi)| = \left|\frac{\sin\left[\frac{N_r}{2}\left(\pi\cos\phi - 2\pi\frac{k}{N_r}\right)\right]}{\sin\left[\frac{1}{2}\left(\pi\cos\phi - 2\pi\frac{k}{N_r}\right)\right]}\right|.$$ (16.10)

The crossover points between patterns at FFT output k and $k+1$ are at

$$\left|\frac{\sin\left[\frac{N_r}{2}\left(\pi\cos\phi - 2\pi\frac{k}{N_r}\right)\right]}{\sin\left[\frac{1}{2}\left(\pi\cos\phi - 2\pi\frac{k}{N_r}\right)\right]}\right| = \left|\frac{\sin\left[\frac{N_r}{2}\left(\pi\cos\phi - 2\pi\frac{k+1}{N_r}\right)\right]}{\sin\left[\frac{1}{2}\left(\pi\cos\phi - 2\pi\frac{k+1}{N_r}\right)\right]}\right|.$$ (16.11)

Figure 16.4 shows the array factor for different values of N_r. We see that in all cases, the value of the array factor of the crossover point is approximately 0.7. The exact values can be found in Table 16.1.

Table 16.1 Gain and relative loss (due to DoA mismatch) of switched beam antennas.

N_r	Max SNR gain (dB)	Max relative loss due to DoA mismatch (dB)
2	3.01	1.51
4	6.02	1.84
8	9.03	1.93

16.1.4 Algorithms for SIMO

Algorithms for the adaptation of antenna weights can be broadly classified into those depending on instantaneous CSI, and those based on second-order channel statistics. The former usually act on the information from the pilot signals, as sampled at the different antenna elements, and are thus called Temporal Reference (TR) algorithms. For the latter, information can be related to the APS, if a suitable knowledge (calibration) of the array structure is known. Since the resulting algorithms are then based on angular (spatial) information, they are often called Spatial Reference (SR) algorithms. Processing based on instantaneous and second-order CSI shows important differences.

Algorithms Based on Instantaneous CSI

TR algorithms are based on the use of a training signal, which serves as a "temporal reference" to which the output of the combiner is matched – thus the name. The antenna weights $\mathbf{w}^*$ are chosen in such a way that the deviation of the combiner output from the (known) training sequence is minimized. As a criterion for the deviation we can use the signal-to-interference ratio, the minimum mean square error, the BER during the training sequence, or any other suitable criterion. These criteria and their advantages and disadvantages have already been discussed in Chapter 12 (we stress again that smart antennas with diversity combining at the RX is the "standard" diversity). As discussed in Sec. 16.1.2, the linear weights lead to the creation of an effective antenna pattern, but this pattern need not have an intuitive interpretation. There are also no assumptions about the existence of discrete DoAs.

In summary, the TR algorithms proceed according to the following steps (see Figure 16.5):

1. During a training phase, the smart antenna receives the signal at all antenna elements and adjusts the weights in such a way that the deviation from the known signal is minimized. This adjustment can be done iteratively, as indicated in Figure 16.5a, or – as is

much more common in modern systems – done as "single shot" based on the channel estimation obtained from the training signal. The determination of the combiner weights may follow the rules of MRC (if only thermal noise is present) or optimum combining (if interference is present).

2. During the transmission of the user payload data, the antenna weights stay fixed and are used to weigh the incoming signals before they are combined and decoded/demodulated.

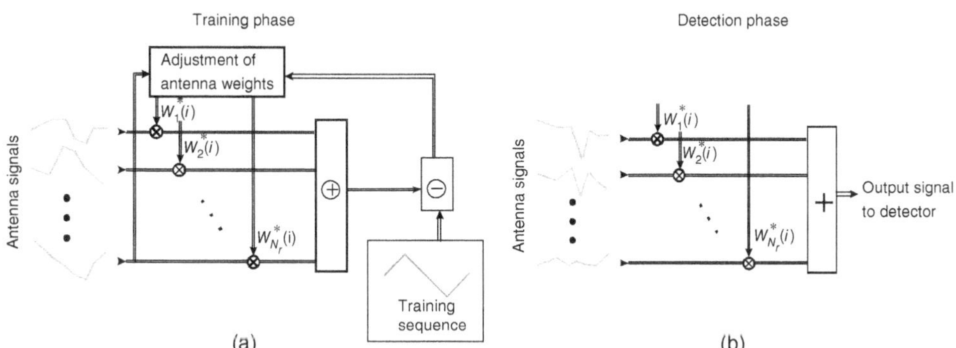

Figure 16.5 Principle of TR algorithms. Weight adjustment based on training sequence (a), and operation during data reception (b).

We finally note that a practical difficulty in TR algorithms is the fact that the RX must be synchronized to the incoming signal *before* the determination of the antenna weights can be done. This is required because the sampling instants for the determination of the training sequence have to be known before the weights can be adapted. However, the SINR for the (spatially unfiltered) receive signal can be poor and may make synchronization to the desired training sequence difficult. If the interferers are caused by other nodes in the same system, suitable coordination of transmission time to allow, e.g., interference-free synchronization, can help the performance. Examples of such approaches are discussed in Section 32.4.2.

Algorithms Based on Second-Order Statistics

In the *SR* case, the antenna tries to form a beam pattern that puts maxima in the direction of the main directions of arrival, and nulls in the direction of the MPCs coming from the interferers. A SR algorithm thus proceeds in the following way (see Figure 16.6):

1. In a first step, it determines the main DoAs of the MPCs, and the power carried by each of them. Note that we assume here that the nonfading MPCs can be isolated, or – if we just analyze the power coming from a particular angular "bin" – averaging over the small-scale fading is performed, compare also Section 6.2; such an approach does not attempt to use the phase information, and thus needs only second-order CSI. The directly observable quantities are the signals at the antenna elements; it is thus necessary to extract the DoAs from those signals. This can be done by the same methods as the ones for channel sounding (see Section 9.6): spatial Fourier transformations (which only provides information in angular bins) or high-resolution algorithms (e.g., MVM, ESPRIT, and SAGE) can be applied. There are some important practical differences from the evaluation of channel sounder data: the SNR for the reception of data can be considerably worse than the one used for channel sounding – this makes the DoA determination more difficult. On the other hand, a lot of data are transmitted within a relatively short time. This allows the use of either averaging or tracking algorithms that observe the change of DoAs. Note that some of the DoA estimation algorithms (e.g., ESPRIT) do not require a training sequence, since they are based exclusively on the correlation matrix of the received signal, and can thus estimate the DoA from user data that are unknown. Other algorithms, like SAGE, can eschew the training sequence only at the price of greatly complicating the algorithm.[2]

2. Association of DoAs with specific users. If multiple UEs are allowed to communicate with the BS at the same time (as in MU-MIMO, Chapter 22), MPCs can stem from different users. It is thus necessary to separate the desired and the interfering users. For this identification problem, it is necessary to have a training sequence, or some other property that is unique for each user (e.g., spreading sequence in a CDMA system).

3. After the DoAs for user and interferer are established, it is possible to form the beam pattern that maximizes the SINR. This beam pattern is used for the actual reception of the data.

We note that many of the DoA estimation algorithms require the validity of a certain channel model (e.g., that the received signal can be written as a finite sum of plane waves). If this model is not valid, the performance of the algorithms, and of the ensuing beamforming, deteriorates.

---

[2] However, when SAGE is used in conjunction with a training sequence, it usually outperforms ESPRIT.

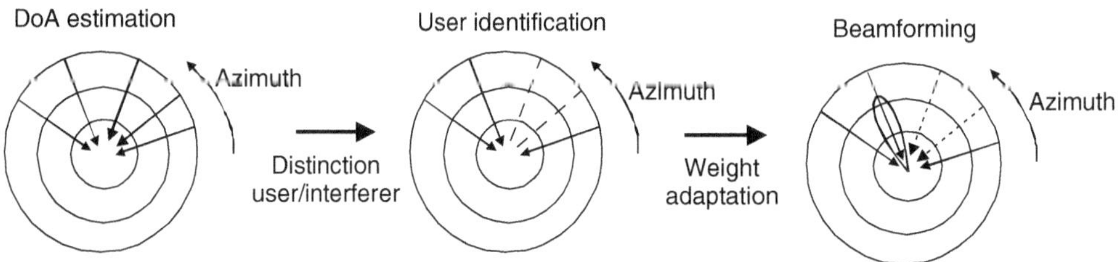

Figure 16.6 Using smart antennas with spatial reference algorithms. Solid lines: MPCS from desired user; dashed lines: MPCs from interfererers.

An alternative approach to exploiting the second-order statistics of the channel is based on the forming of the spatial correlation matrix $E_{\mathrm{ssf}}\{\mathbf{HH}^\dagger\}$, where the expectation is over the small-scale fading. This correlation matrix is related to the second-order statistics, and in the case of perfect knowledge of the array patterns, can be mapped to an angular spectrum. However, a key advantage is that such a calibration and mapping is not necessary, but rather it is obtained directly from observations, and some algorithms can make use of it directly. On the downside, it does not allow the physical interpretation of an angular spectrum.

*Blind Algorithms

The TR algorithms rely on the existence of a training sequence, while the SR algorithms rely on the existence of a certain spatial structure of the arriving signal. Yet another group of algorithms, dubbed *blind* algorithms, makes neither of those assumptions. Rather, those algorithms exploit statistical properties of the transmit signal.

Let us start out by describing the received signal by the equation

$$\mathbf{r}(i) = \mathbf{h}_{\mathrm{d}}s(i) \tag{16.12}$$

where $\mathbf{r}(i)$ is the receive signal vector created by the transmit signal $s(i)$ related to the ith bit, and $\mathbf{h}_{\mathrm{d}}$ is the channel vector, and we omit the noise contribution for convenience. A more general representation, which also includes possible intersymbol interference, is obtained by stacking the signals corresponding to a large number of bits into matrices

$$\mathbf{R} = \mathbf{H}_{\mathrm{stack}}\mathbf{S} \tag{16.13}$$

where the matrix $\mathbf{H}_{\mathrm{stack}}$ is related to the channel impulse responses at the different antenna elements; the exact form of the matrices and their dimensions will be explained below by means of an example. Note also that $\mathbf{R}$ here is a matrix of stacked receive signals, not a correlation matrix. The channel description is thus similar to the TR approach in the sense that there is no assumption about a spatial structure, and the channel is characterized by the impulse response (or transfer function) alone. However, a blind algorithm does *not* learn the values of $\mathbf{H}_{\mathrm{stack}}$ from a training sequence. Rather, it tries to compute a factorization of the matrix $\mathbf{R}$ according to Eq. (16.13), so that the signal matrix $\mathbf{S}$ fulfills certain properties, namely *known* properties of the transmit signal.

Example 16.2 *Consider a system with one transmit antenna and three receive antennas. Let the impulse response for each of the antenna elements last for two symbol durations, denoted as $h_n(\ell)$, $\ell = 0, 1$; $n = 1, 2, 3$. For a 5-symbol long transmit signal, $s(i), i = 1, ...5$, write the explicit form of $\mathbf{R} = \mathbf{H}_{\mathrm{stack}}\mathbf{S}$, assuming that the impulse responses stay constant over the duration of the transmit signal.*

Writing the ith received symbol at the nth antenna element as $r_n(i)$, we can easily see that

$$r_n(i) = \sum_{\ell=0}^{L-1} h_n(\ell)s(i\text{-}\ell) \tag{16.14}$$

where $h_n(\ell)s(i\text{-}\ell)$ is the channel response to the signal to the nth antenna element in the ℓth channel delay tap, L is the duration of the channel impulse response, and I is the number of transmitted symbols. Let us then define a matrix $\mathbf{S}$ as [Laurila 2000]

$$\mathbf{S} = \begin{bmatrix} s(1) & s(2) & \cdot\ , & s(I) \\ s(0) & s(1) & & s(I-1) \\ \cdot & \cdot & & \cdot \\ \cdot & \cdot & & \cdot \\ s(-L+2) & s(-L+3) & & s(I-L+1) \end{bmatrix} \tag{16.15}$$

which in our case becomes

$$\mathbf{S} = \begin{bmatrix} s(1) & s(2) & s(3) & s(4) & s(5) \\ s(0) & s(1) & s(2) & s(3) & s(4) \end{bmatrix}. \tag{16.16}$$

The channel matrix gets the block form

$$\mathbf{H}_{\text{stack}} = \begin{bmatrix} h_{1,0} & h_{1,1} & . & . & h_{1,L-1} \\ h_{2,0} & h_{2,1} & . & . & h_{2,L-1} \\ . & . & & & . \\ . & . & & & . \\ h_{N_r,0} & h_{Nl_r,1} & . & . & h_{N_r,L-1} \end{bmatrix} \tag{16.17}$$

with $h_{n,\ell} = h_n(\ell)$ which in our case becomes

$$\mathbf{H}_{\text{stack}} = \begin{bmatrix} h_{1,0} & h_{1,1} \\ h_{2,0} & h_{2,1} \\ h_{3,0} & h_{3,1} \end{bmatrix}. \tag{16.18}$$

Inserting Eqs. (16.16) and (16.18) into Eq. (16.13), we get

$$\mathbf{R} = \begin{bmatrix} r_1(1) & r_1(2) & r_1(3) & r_1(4) & r_1(5) \\ r_2(1) & r_2(2) & r_2(3) & r_2(4) & r_2(5) \\ r_3(1) & r_3(2) & r_3(3) & r_3(4) & r_3(5) \end{bmatrix} \tag{16.19}$$

$$= \begin{bmatrix} h_{1,0}s(1) + h_{1,1}s(0) & .. & . & h_{1,0}s(5) + h_{1,1}s(4) \\ h_{2,0}s(1) + h_{2,1}s(0) & & & h_{2,0}s(5) + h_{2,1}s(4) \\ h_{3,0}s(1) + h_{3,1}s(0) & & & h_{3,0}s(5) + h_{3,1}s(4) \end{bmatrix} \tag{16.20}$$

which can easily be seen to be identical to Eq. (16.14).

Depending on which signal properties are exploited, we get different algorithms for the determination of $\mathbf{S}$. If there is oversampling in the temporal and/or spatial domain, then the cyclostationarity of the sampled signal can be used. Another group of algorithms exploits higher-order statistics. Also, the finite-alphabet property can be exploited: for example, we know that for a BPSK signal, the elements of $\mathbf{S}$ have to be ± 1. If the signal has constant envelope, the Constant-Modulus Algorithm (CMA) can be used.

Blind algorithms have a number of advantages:

- No assumptions about the channel are required: the method works even for channels that do not exhibit discrete directions of arrival or separable MPCs.
- Blind algorithms do not require any calibration of the antenna array and make no assumptions about the specific structure of the array.
- The detection process can be done in one step, yielding directly the desired user data.
- Blind algorithms eliminate the need for a training sequence, thus increasing the spectral efficiency of the system.

However, those advantages are bought at a price:

- Most of the blind algorithms assume that the channel impulse response does not change over the time it takes to establish statistics of $\mathbf{R}$. For a good estimate of those statistics, we have to collect samples over a long time, and the assumption of a time-invariant impulse response might be violated. The number of samples that forms the basis of the statistics is thus a compromise between the need to have a large-enough basis for obtaining good statistics and the need to use only samples that correspond to the same channel state.
- The initialization of the estimate for the channel is critical. For this reason, so-called *semi-blind* algorithms have been proposed, which use a very short training sequence at the beginning of the transmission. This sequence then allows to obtain an initial estimate, which is then refined by the blind algorithm. The sequence length is chosen to be short enough not to significantly influence the spectral efficiency of the system but helps to avoid possible convergence problems of the blind algorithms.

These drawbacks are essentially the same as for blind equalization (see Section 14.7), and have prevented widespread use of purely blind algorithms.

However, a variant of the semi-blind approach described above is now in widespread use: it uses a training sequence as the (possibly noisy) initialization of the channel estimate, followed by a decision-directed refinement of the channel estimates. In other words, the (possibly very short) training sequence provides an initial estimate of the channel, based on which the first data symbols are estimated in a conventional way. Knowledge of the data symbols allows to then use them like training symbols, i.e., to refine (in the case of noisy channel) and/or update (in the case of time-variant channel) the channel estimates. This process is continued until the end of the data packet or the transmission of the next training sequence. Such an approach can also be viewed as a low-complexity approximation to the ideal approach, namely joint estimation of channel and data.

16.1.5 Algorithms for MISO

For the downlink, the multiple antennas are located at the *TX*, resulting in a MISO system. In this subsection, we will thus discuss methods of how to transmit signals from several TX antennas and achieve a diversity effect and/or beamforming gain with it.[3]

TX Diversity with CSI

The first situation we analyze is the case where the TX has full CSIT. In that case, we find that (at least for the noise-limited case) there is a complete equivalence between transmit diversity and receive diversity. In other words, the optimum transmission scheme linearly weights the signals transmitted from the different antenna elements with the complex conjugates of the channel transfer functions from the transmit antenna elements to the single receive antenna. This approach is known as *Maximum Ratio Transmission (MRT)* [Lo 1999], so that

$$\mathbf{t} = \mathbf{h}^*. \tag{16.21}$$

TX Diversity Without CSI – Delay Diversity

In many cases, CSI is not available at the TX. In that case, the TX cannot simply transmit weighted copies of the same signal from the different antennas, because it cannot know how they would add up at the RX. There is equal likelihood that the addition of the different MPCs would be constructive or destructive; in other words, the RX would just be adding up signal copies with random phases, which results in Rayleigh fading. We thus cannot gain any diversity. In order to give benefits, the transmission of the signals from the different antenna elements has to be done in such a way that it allows the RX to distinguish the different transmitted signal components.

One way to achieve this is *delay diversity*. In this scheme, the signals transmitted from the different antenna elements are differently delayed copies of the same signal. This makes sure that the effective impulse response is delay-dispersive, even if the channel itself is flat fading. For example, in a flat-fading channel, the data stream from the nth antenna is transmitted with a delay *of n* symbol durations, nT_S (relative to the signal at the reference antenna $n = 0$). The effective impulse response of the channel then becomes

$$h(\tau) = \frac{1}{\sqrt{N_t}} \sum_{n=0}^{N_t - 1} h_n \delta(\tau - nT_S) \tag{16.22}$$

where the h_n is the complex channel gains from the nth transmit antenna to the receive antenna, and the expression has been normalized so that the total transmit power is independent of the number of antenna elements. The signals from the different transmit antennas to the RX act effectively as delayed MPCs. If the antenna elements are spaced sufficiently far apart, those coefficients are fading independently. An appropriate RX for delay-dispersive channels (e.g., an equalizer as described in Chapter 14), achieves a diversity order that is equal to the number of antenna elements. If the channel from a single transmit antenna to the RX already is delay dispersive, then the scheme still works, but care has to be taken in the choice of the delays of the different antenna elements. The delay between signals transmitted from different antenna elements should be at least as large as the maximum excess delay of the channel. On the other hand, the length of the effective impulse response must not exceed the length of the equalizer available at the RX, to avoid residual intersymbol interference.

In OFDM systems, a variant of delay diversity called *Cyclic Delay Diversity* (CDD), is commonly used. As discussed in Section 15.4, the special structure of OFDM with cyclic prefix converts linear delay shifts into cyclic shifts. Thus, an OFDM data block is *cyclically* shifted by different amounts at different antenna elements, after which the CP is applied. This retains the property that the effective channel impulse response is no longer than $T_{\rm cp}$ and consequently allows single-tap equalization. Alternatively, we can say that the delay shift imposed in CDD effectively results in a phase shift of $2\pi\Delta f k \tau_{\rm shift}$ for the kth subcarrier. It is in this form that delay diversity has gained widespread use in Wi-Fi (Section 33.3.3) and LTE (Section 31.3.7).

An alternative method is *phase sweeping diversity*. In this method, which is especially useful if there are only two antenna elements, the same signal is transmitted from both antenna elements. However, one of the antenna signals is multiplied with a time-varying

---

[3] Note that antenna diversity encompasses spatial diversity, pattern diversity, and polarization diversity.

phase shift. That means that at the RX, the received signals add up in a time-varying way; in other words, we are artificially introducing temporal variations into the channel. The reason for this is that even if TX, RX, and the Interacting Objects (IOs) are stationary, the signal does not remain stuck in a fading dip. If this scheme is combined with appropriate coding and/or interleaving, it improves the performance.

Yet another alternative is *precoder cycling*. In this approach, the TX has a codebook, consisting of different code vectors (complex coefficients that multiply the signals at the various antenna elements, see also section titled "CSIT based on feedback" in 16.1.7). Each codevector is applied to a different set of subcarriers and/or transmission times; this enforces pattern diversity so that a diversity order on the order of the number of codewords is achieved. If the TX has information about second-order statistics, then it can restrict the codewords through which it cycles, to a smaller set of precoder settings that are good-on-average, sacrificing diversity order for improved average SNR.

TX Diversity Without CSI – Alamouti Codes

Another possibility for achieving diversity is *space-time coding*. This method is generally applicable to an arbitrary number of TX and RX antennas. In this subsection, we only discuss the most popular code, the Alamouti code [Alamouti 1998], in a setup with two TX and one RX antennas. General space–time codes can use not only more antennas but also multiple data streams; their discussion is deferred to Section 16.2.12.

The idea of the Alamouti code is to transmit the data in a way that guarantees high diversity, while allowing a simple decoding process. Consider a frequency-flat channel, where the complex channel gain from TX antenna 1 is given by h_1 and from TX antenna 2 to the RX by h_2. Transmit now the two symbols, c_1 and c_2, from the two transmit antennas at time instant 1:

$$\mathbf{s}_1 = \frac{1}{\sqrt{2}} \begin{pmatrix} c_1 \\ c_2 \end{pmatrix}. \tag{16.23}$$

At the second time instant, transmit the vector

$$\mathbf{s}_2 = \frac{1}{\sqrt{2}} \begin{pmatrix} -c_2^* \\ c_1^* \end{pmatrix}. \tag{16.24}$$

The factor $1/\sqrt{2}$ stems from the necessity to keep the transmit energy constant, as we use now two transmit antennas. Then the received signal can be written as (note that the receive signal at the second antenna is complex-conjugated before further processing):

$$\mathbf{r} = \begin{pmatrix} r_1 \\ r_2^* \end{pmatrix} = \frac{1}{\sqrt{2}} \begin{pmatrix} h_1 & h_2 \\ h_2^* & -h_1^* \end{pmatrix} \begin{pmatrix} c_1 \\ c_2 \end{pmatrix} + \mathbf{n} = \mathbf{H}\mathbf{c} + \mathbf{n}. \tag{16.25}$$

We have thus created a "virtual" MIMO system. It is important to note that the columns of the "virtual" channel matrix $\mathbf{H}$ are orthogonal. Therefore, $\mathbf{H}^\dagger\mathbf{H}$ becomes a scaled identity matrix $\alpha\mathbf{I}$. Thus, for the decoding, we thus first multiply the received signal vector by $\mathbf{H}^\dagger$. Then we get

$$\widetilde{\mathbf{r}} = \mathbf{H}^\dagger\mathbf{r} = \mathbf{H}^\dagger\mathbf{H}\mathbf{c} + \mathbf{H}^\dagger\mathbf{n} = \alpha\mathbf{c} + \widetilde{\mathbf{n}} \tag{16.26}$$

where

$$\alpha = \frac{|h_1|^2 + |h_2|^2}{2}. \tag{16.27}$$

Since the columns of $\mathbf{H}$ are orthogonal, the components of $\widetilde{\mathbf{n}}$ are still uncorrelated zero-mean Gaussian, and have variance $\alpha\sigma_n^2$. Therefore the decoding of the data c_1 and c_2 becomes decoupled, which greatly reduces the computational effort in the RX. The scaling factor α describes the "effective" gain by the channel, and the SNR is given by $\alpha^2/\alpha = \alpha$. This is a factor 2 lower than with MRT. In other words, while the diversity order is two – both h_1 and h_2 would have to be in a fading dip for the effective channel gain to be low – the scheme cannot provide beamforming gain, since it does not have CSI at the TX.

We finally note that instead of transmitting the two symbols at two subsequent times, we can also transmit them on two adjacent subcarriers in an OFDM system – this is a special case of *space-frequency* coding. This implementation, which has the advantage of reduced latency, is used in Wi-Fi and LTE.

TX with Second-Order Statistics

If only second-order statistics of the CSI is available, the TX can realize beamforming gain without diversity. If the APS is available, then the TX forms a beam into the direction of the strongest MPCs, aiming to match the beampattern to the APS to maximize the averaged receive power, within the constraints of how well the TX array can adjust its pattern. This is achieved through the SR algorithms discussed previously.

However, in some cases, the TX does not know the APS, since this requires also knowledge of the (calibrated) TX array response. In this case, the TX can perform beamforming based on the (spatial) correlation matrix of the uplink, $\mathbf{R} = E\{\mathbf{rr}^\dagger\}$, where the expectation is over the small-scale fading. Since the small-scale average channel statistics are the same in uplink and downlink (see Section 16.1.7), the downlink correlation matrix can be deduced from the uplink matrix, and then subsequent matching of the TX signal to this matrix provides good beamforming gain [Farsakh and Nossek 1998].

As mentioned above, TX beamforming based on second-order statistics provides only beamforming gain. It is possible, however, to form two or more beams (sacrificing some beamforming gain in the process), and then apply diversity techniques that do not require CSI to the inputs of those beamformers.

16.1.6 Diversity and Beamforming in MIMO Systems

Since we can obtain diversity and beamforming gain with smart antennas at either RX or TX, the obvious question arises what we can achieve with smart antennas at both link ends, i.e., in MIMO systems. It turns out that in the best case, it is possible to increase the diversity order to $N_r N_t$, and the beamforming gain also to $N_r N_t$ (though these two things cannot be realized simultaneously). Again, we have to distinguish between systems that have CSI at the TX and systems that do not.

Diversity and Beamforming with CSI at the TX

Having the CSI at the TX leads to a conceptually simple transceiver structure. In a diversity system, the vector of transmit symbols is a weighted replica of a single data symbol, $\widetilde{\mathbf{s}} = \mathbf{t}c$. Consider a *singular value decomposition*[4] of the channel

$$\mathbf{H} = \mathbf{U}\boldsymbol{\Sigma}\mathbf{V}^\dagger. \tag{16.28}$$

Then the received signal is

$$\mathbf{r} = \mathbf{U}\boldsymbol{\Sigma}\mathbf{V}^\dagger \mathbf{t}c + \mathbf{n} \tag{16.29}$$

where $\mathbf{n}$ is the noise vector with complex Gaussian distribution with variance σ_n^2.

The RX performs a linear combination (summation) of the signals at the different antenna elements with combining vector $\mathbf{w}^\dagger$, so that the output of the combiner is $\hat{c} = \mathbf{w}^\dagger \mathbf{r}$. Choosing $\mathbf{t} = \mathbf{V}_1$, and $\mathbf{w} = \mathbf{U}_1$ maximizes the SNR at the RX, where $\mathbf{U}_1$ is the left singular vector corresponding to the largest singular value (and similarly for $\mathbf{V}$). In other words, we choose the transmit and receive weights according to the singular value decomposition of the channel.

The output of the receive combiner is thus

$$\hat{c} = \mathbf{w}^\dagger \mathbf{r} = (\mathbf{U}_1)^\dagger \mathbf{U}\boldsymbol{\Sigma}\mathbf{V}^\dagger (\mathbf{V}_1)c + (\mathbf{U}_1)^\dagger \mathbf{n}. \tag{16.30}$$

Again, the multiplication of the noise vector $\mathbf{n}$ by $\mathbf{U}_1$ does not change its statistics, i.e., the entries of the resulting vector are independent Gaussian-distributed scalars with variance σ_n^2. The SNR that can be achieved this way is

$$\gamma = \frac{P}{\sigma_n^2}\widetilde{\sigma}_{\max}^2 = \frac{P}{\sigma_n^2}\lambda_{\max} \tag{16.31}$$

where $\widetilde{\sigma}_{\max}$ is the largest singular value of the matrix $\mathbf{H}$, and $\lambda_{\max}$ is the largest eigenvalue of $\mathbf{HH}^\dagger$. For the case that of only one transmit antenna, the scheme reduces to MRC; for the case of a single receive antenna, it becomes MRT. Figure 16.7 shows the cdf of the SNR that can be achieved that way. The slope of the distribution of the largest eigenvalue is essentially the same as the slope of the distribution of the sum of the eigenvalues. In a 4×4 MIMO system, the diversity order is 16, and the SNR distribution is almost a step function – in other words, the required fading margin is very small. There is also an enhancement of the average SNR, but it is smaller than $N_r N_t = 16$ (12 dB), see also section titled "Diversity-Multiplexing Trade-Off" in 16.2.6.

---

[4] A singular value decomposition is similar to an eigenvalue decomposition, but exists also for rectangular matrices (more rows than columns or vice versa). It decomposes any matrix into a product if three matrices: a unitary matrix corresponding to the row space, a diagonal matrix describing the strength of the different eigenmodes, and a unitary matrix describing the column space.

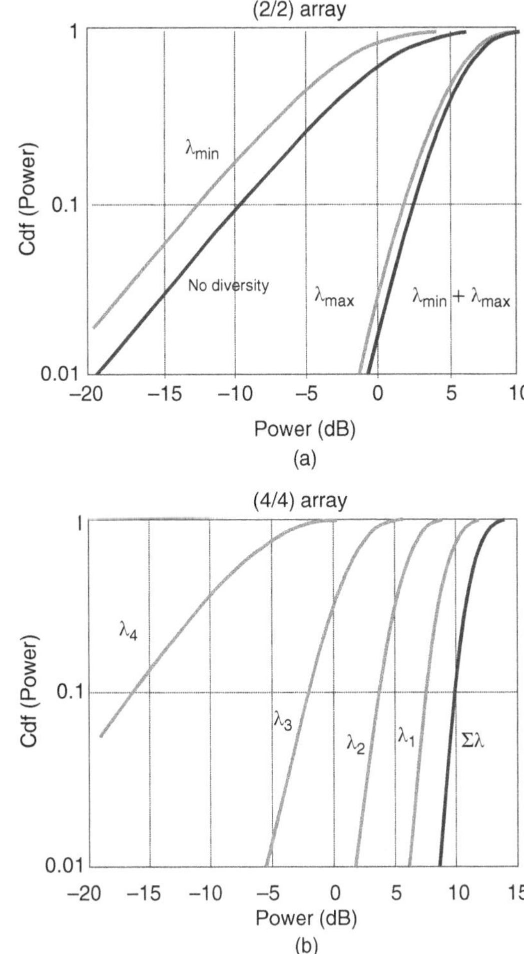

Figure 16.7 Cumulative distribution function of the SNR in a 2×2 (a) and 4×4 (b) MIMO diversity with i.i.d. Rayleigh fading at all antenna elements.

Example 16.3 *Consider a 2×2 MIMO system with 10 dB average SNR at each of the antenna elements. What is the probability that the SNR is smaller than 7 dB?*

As shown in [Andersen 2000], the probability density function for the largest eigenvalue of the matrix $\mathbf{HH}^\dagger$ in a 2×2 system is

$$pdf_{\lambda_{\max}}(\lambda) = \exp(-\lambda)\left[\lambda^2 - 2\lambda + 2\right] - 2\exp(-2\lambda) \tag{16.32}$$

and the cumulative distribution function is

$$cdf_{\lambda_{\max}}(\lambda) = 1 - \exp(-\lambda)\left(\lambda^2 + 2\right) + \exp(-2\lambda). \tag{16.33}$$

Here λ describes the normalized SNR. We now want to identify the probability that the largest eigenvalue is 3 dB below the mean value of $\mathbf{HH}^\dagger$. This is, by definition, *cdf(0.5)*, which evaluates to

$$cdf_{\lambda_{\max}}(0.5) = 3{\cdot}10^{-3}. \tag{16.34}$$

Diversity and Beamforming Without CSI at the TX

For diversity transmission, computation of the achievable performance is rather straightforward: the diversity order is the diversity order achievable with the TX diversity scheme (at a maximum $N_{\rm t}$) multiplied with $N_{\rm r}$. Since the TX does not provide any beamforming gain (due to the absence of CSIT), the total beamforming gain is upper limited by $N_{\rm r}$.

16.1.7   CSIT in TDD and FDD Systems

Up to now, we have assumed that CSIT, when available, is given by a genie, without limitations on accuracy and cost for acquiring it. This is, of course, not realistic. This subsection will discuss how CSIT can be acquired, which overhead this entails, and what the resulting limitations on the accuracy of the CSIT are.

CSIT Based on Reciprocity

Consider the situation that uplink and downlink occur within one coherence bandwidth of the channel of each other, and within one coherence time (we will discuss below when this occurs). Then, for practical purposes, the uplink and downlink channel are reciprocal, i.e., the transfer function from the UE antenna to the m-th BS antenna, $(\mathbf{h}_d)_m$, is the same as the transfer function from the m-th BS antenna to the UE antenna. Assume furthermore that the only impairment is thermal noise, i.e., there is no significant interference. Then a BS can learn the downlink channel state, i.e., the CSIT, by observing the uplink channel. If the BS receives signals from the UE anyway, including pilot signals that allow it to obtain the CSIR, it can then obtain the CSIT without further overhead. Furthermore, transmission might not even require explicit channel estimation: if the BS has chosen, by any means, the antenna weights in such a way that they constructively add the signals from the antenna elements during the uplink, the *same* antenna weights, when used in the downlink, will ensure that the signals from the different BS antenna elements add up constructively at the UE antenna.

This principle sounds very simple; however, there are a number of practical obstacles and constraints that need to be considered.

TDD vs. FDD

Let us first consider when the above conditions for validity of reciprocity are fulfilled: uplink and downlink must operate within one coherence bandwidth, and within one coherence time. The former condition is fulfilled (at least for modern cellular communications) only in *Time Domain Duplexing* (TDD) mode. In a *Frequency Domain Duplexing* (FDD) system, the uplink and downlink use different frequencies with a separation (duplexing distance) that is typically on the order of 100 MHz. This is much larger than the coherence bandwidth of the channel (typically on the order of 1 MHz). As a consequence, the small-scale fading of the channel impulse response (or transfer function) in the uplink and in the downlink are completely decorrelated. In other words: the small-scale fading is created by the superposition of MPCs with different phases. The relative phase shifts depend, among other things, on the carrier frequency. For a large-enough duplexing distance, the phase shifts are sufficiently different in uplink and downlink that the superposition of the MPCs occurs in different ways, and thus the channel impulse response in uplink and downlink bands differ significantly.[5]

In TDD systems, uplink and downlink occur at the same frequency, but at different times. The condition that the transmission is within one coherence time of the channel can often be fulfilled: typical coherence times for standard cellular frequencies lie between 10 ms (when the UE is moving at slow vehicular speeds) and several seconds or even minutes or hours (when the UE is nomadic). However, coherence times can be 1 ms or even less in high-mobility scenarios (highway, high-speed train) and/or for mm-wave systems (remember that Doppler shifts scale linearly with the carrier frequency). The duplex time, which is specific to each system, then needs to be sufficiently short if reciprocity is to hold.

Reciprocity of Interference

In a noise-limited system, reusing uplink antenna weights for the downlink provides optimum receive SNR and thus performance. Furthermore, if the BS antenna weights are chosen to suppress interference from other UEs during the uplink, the BS will also cause little or no interference *to* those other UEs in the downlink transmission phase.

However, the interference is not completely reciprocal for the uplink and downlink: the interference that the UE sees in the downlink stems from other *BSs*, not from other UEs (Figure 16.8). The antenna weights $\mathbf{w}^\dagger$ that are determined during the uplink cannot take this into account – especially if the set of active UEs is different in uplink and downlink. Thus, interference suppression for the

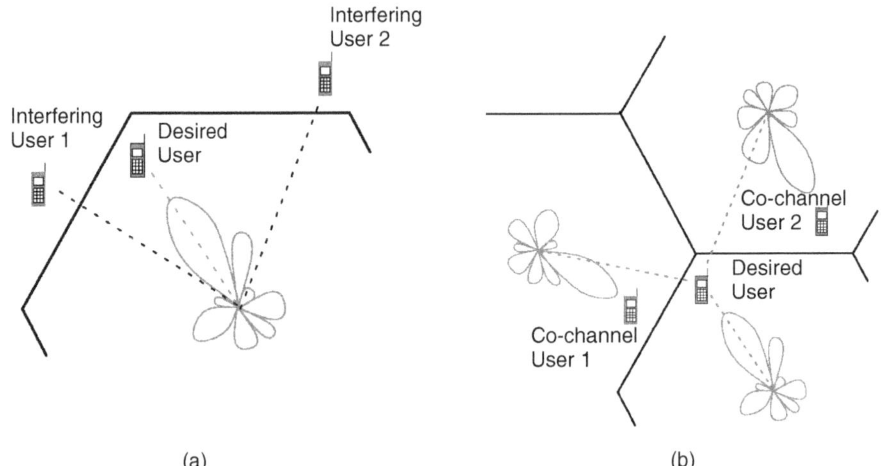

Figure 16.8 Interference situation for the uplink (a) and downlink (b) for a BS with smart antennas.

[5] Research has been done about the extrapolation of the channel in frequency or time beyond the coherence bandwidth/coherence time. These approaches might use high-resolution parameter estimation or machine learning. Yet at the time of this writing, they are not accurate enough for practical deployment.

downlink cannot be as effective as the suppression in the uplink. An exception would be if the UE notifies all the BSs in its surroundings about the details of the interference it sees from them, and the BSs would cooperate in order to provide as little interference as possible to all the users, see also Section 22.11.

*Reciprocity Calibration

We have stated that the *radio channel* (propagation channel plus antennas) is reciprocal. However, since the antenna weights are determined in baseband, it is really required that the baseband-to-baseband "effective radio channel" is reciprocal, which includes not only the radio channel but also the RF elements and up/downconversion chains at the transceivers of BS and UE, i.e., everything from the antenna connectors to the baseband. Since up and downconversion use different electronic components, such reciprocity will not be satisfied for many transceivers. This problem can be eliminated or, at least, mitigated by a calibration procedure: each device determines the difference between the transmit and receive chains and corrects for it during operation. This determination, also known as reciprocity calibration, is of critical importance for operating smart antennas based on reciprocity.

The basic principle of reciprocity calibration is to measure the channel transfer function twice: once based on the assumption of reciprocity, and once with a different method; the difference between the two approaches can then be used for calibration. Define first the effective radio channel $\widetilde{g}$ as the product of transmit hardware μ, radio channel h, and receive hardware ρ transfer functions:

$$\widetilde{g}_{i \to j} = \mu_i h_{i \to j} \rho_j. \tag{16.35}$$

Then a calibration coefficient between radio i (more precisely, the TX hardware associated with the ith antenna element) and radio j can be computed from knowledge of the radio channel in both directions:

$$b_{i \to j} = \frac{\widetilde{g}_{i \to j}}{\widetilde{g}_{j \to i}} = \frac{\mu_i h_{i \to j} \rho_j}{\mu_j h_{j \to i} \rho_i} = \frac{\mu_i \rho_j}{\mu_j \rho_i} = \frac{1}{b_{j \to i}}. \tag{16.36}$$

Then, with the knowledge of the calibration coefficients, we can determine the desired downlink channels after having measured the uplink channels. The remaining question is now how to measure the calibration coefficients.

A first, "brute-force" approach measures the transfer function once in the uplink $\widetilde{g}_{j \to i}$, and once in the downlink $\widetilde{g}_{i \to j}$; in the latter case, the measurement result is fed back to the BS, which can then compute the $b_{j \to i}$. This approach requires feedback of the channels between all BS antenna elements and UEs, resulting in a large overhead especially for massive MIMO systems, where N_{BS} can be 100 or more (see Section 22.9). An approach to reducing this overhead is to first perform an *internal* calibration, so that all BS antenna elements are calibrated to the same reference antenna element. Such internal calibration can be done, e.g., by naming one antenna element in the transmit array the reference element; this element can transmit, with all other elements monitoring their receive signal. It is important that the calibration coefficient between two antennas can be computed if their respective calibration coefficients to a reference antenna are known:

$$b_{i \to j} = \frac{b_{1 \to j}}{b_{1 \to i}}. \tag{16.37}$$

The internal calibration can be further improved by declaring, in subsequent time intervals, each element of the array as the reference antenna, and finally performing appropriate averaging to reduce the effects of noise from the calibration.

To get *absolute* calibration, feedback of the channel from the UE to the reference antenna on the BS is still required. However, it turns out that for many applications such absolute calibration is unnecessary, since a common phase shift (or scaling factor) of the channel coefficients does not impact the overall results, such as SINR at the RX. Thus, we can always assume that $b_{1 \to k} = 1$. Instead of reconstructing the true channel from the mth BS antenna to the kth UE, we can thus get modified channel estimates $\widetilde{g}'_{m \to k}$

$$\widetilde{g}_{m \to k} = \widetilde{g}_{k \to m} \frac{b_{1 \to k}}{b_{1 \to m}} \quad \Rightarrow \quad \widetilde{g}'_{m \to k} = \frac{\widetilde{g}_{k \to m}}{b_{1 \to m}} \tag{16.38}$$

and perform all the beamforming with these modified channel estimates.

Calibration needs to be repeated at time intervals during which the calibration coefficients change considerably. Since the calibration coefficients are independent of the channel, this value is not a function of the coherence time, but only of the temporal changes of the hardware components. The fastest-changing quantity is typically the phase of the Local Oscillator, LO, due to the unavoidable phase noise. However, a common phase factor affecting all antennas equally does not impact the calibration factors, so this becomes only an issue if different antennas use different LOs. Apart from this, amplifier characteristics and other hardware components exhibit thermal drift. During the ramping up of a transceiver, time between calibrations might then be necessary to be on the order of a minute, while in thermal equilibrium, 10 minutes or more might be sufficient (of course, these values depend on the hardware design, thermal tolerances of the components, and cooling).

There are numerous situations where instantaneous CSIT based on calibration is not an option – because the system is FDD, or TDD with a duplexing time much larger than the coherence time of the channel. In that case, the small-scale fading and, thus, the instantaneous impulse responses of the channel, are different for uplink and downlink. What remains constant in uplink and downlink is the *second-order statistics* of the channel state: in other words, the DoAs, delays, and mean powers of the MPCs. The correlation matrix is not the same for uplink and downlink in FDD systems, because the steering vectors are different: antenna elements have a given physical distance between each other, which means that the electrical distance (i.e., in units of wavelength) depends on the carrier frequency, and is thus different at uplink and downlink frequency; however, a suitable transformation of the calibration matrix from uplink to downlink can solve this problem, and thus allow to obtain second-order CSIT.

CSIT Based on Feedback

Feedback for MIMO

If instantaneous CSIT is desired in an FDD system, feedback from the UE to the BS is required, which may lead to considerable overhead. In particular, "ideal" CSI requires quantizing the entries of the transfer function matrix very finely, resulting in a large number of bits in the feedback message. It is, therefore, worthwhile to investigate how a reduction of the feedback overhead impacts the performance of a MIMO system.

Example 16.4   *We consider a typical example for a cellular MIMO system: let the BS have eight antenna elements, and the UE have four antenna elements. The system has 5 MHz bandwidth centered at 2 GHz carrier frequency and operates in a channel with 250 kHz coherence bandwidth. The coherence time is 5 ms, corresponding to typical vehicular speeds. What is the total overhead data rate for the feedback?*

Assume that real and imaginary part are quantized with 6 bits each, and a rate 1/3 code is used to protect the feedback information. The total number of feedback bits per second is

$$
2 \cdot 6 \cdot \frac{1}{1/3} \cdot N_\mathrm{t} \cdot N_\mathrm{r} \cdot \frac{B}{B_\mathrm{coh}} \cdot \frac{1}{T_\mathrm{coh}} \tag{16.39}
$$

$$
= 2 \cdot 6 \cdot \frac{1}{1/3} \cdot 8 \cdot 4 \cdot \frac{5000}{250} \cdot \frac{1}{0.005} \tag{16.40}
$$

$$
= 4.6 \cdot 10^6 . \tag{16.41}
$$

Thus, the total feedback load for a single user in a 5 MHz band is almost 5 Mbit/s. This example shows clearly that feedback reduction techniques are extremely important.

If the TX uses linear precoding (beamforming), an effective way for reducing feedback is the use of *feedback codebooks*. In this approach, the RX does not feed back the channel transfer function matrix itself, but rather the index of a "codeword" (i.e., a specific setting of the precoder, or, equivalently, the index of a beamforming pattern) that it wants the TX to use. Codebook sizes are usually very small (8 or 16 entries, requiring 3 or 4 bits to communicate the index of the entry), and thus the savings can be significant: instead of feeding back $36 \cdot N_\mathrm{t} \cdot N_\mathrm{r}$ bits in each coherence time and each coherence bandwidth (1152 bits in the example above), only 3 or 4 bits need to be transmitted – a saving of more than two orders of magnitude. It is thus important to find good codebook designs.

There are two key reasons why it is more efficient to use codebooks instead of quantized channel coefficients:

- The eigenstructure of a channel matrix is fairly sensitive to changes in the coefficients. In other words, if some of the entries of the channel matrix change even by a small amount (due to the quantization), then the eigenvectors computed from this perturbed matrix can be quite different from the eigenvectors of the original matrix. Since the optimal beamforming vectors are usually along the eigenvectors of the matrix, we thus find a strong sensitivity of beamforming vectors (and resulting SNR) to the quantization.
- The channel coefficients are the entries into a matrix, whose size is proportional to $N_\mathrm{t}N_\mathrm{r}$. On the other hand, the precoding matrix is of size $N_\mathrm{s}N_\mathrm{t}$, i.e., each data stream needs only as many coefficients as there are transmit antennas. Especially for the case of a single transmit data stream, this can result in significant savings.

In the following, we first discuss feedback in flat-fading channels. A generalization to frequency-selective channels is given at the end of this section.

Feedback for Linear Precoding

We first discuss the situation where the BS transmits a single data stream with linear precoding to a UE. Assume a codebook $\mathcal{C}$ (set of quantized precoding vectors $\mathbf{t}$) of size Q, so that feedback of one codebook entry takes $\log_2 Q$ bits; this codebook is known a priori at both TX and RX. For a given $\mathbf{t}^{(q)}$, the SNR available at the UE is

$$
SNR_q = \gamma_\mathrm{TX} \left\| \mathbf{H} \mathbf{t}^{(q)} \right\|_2^2 \tag{16.42}
$$

where γ_{TX} is the transmit SNR (i.e., transmit power divided by noise power at one of the receive antenna elements), and $\mathbf{H}$ is the channel matrix. The RX measures the channel matrix and defines the best possible precoding vector as

$$\widetilde{\mathbf{t}} = \arg\max_{\mathbf{t}^{(q)} \in \mathcal{C}} SNR_q \tag{16.43}$$

where $\widetilde{\mathbf{t}}$ depends on $\mathbf{H}$, even though we do not explicitly write this dependence. The UE can determine the desired $\widetilde{\mathbf{t}}$ for a given (measured) $\mathbf{H}$, e.g., by a brute-force search over the whole codebook.

But before the codebook can be used, we first have to define what its entries should be. We know that the optimum beamforming vector is $\mathbf{v}_{\max} = \mathbf{V}_1$, the eigenvector corresponding to the largest eigenvalue, $\lambda_{\max}$ of $\mathbf{H}\mathbf{H}^{\dagger}$. We thus wish to find a codebook such that – irrespective of the realization of $\mathbf{H}$ – there exists an entry into the codebook that is close to $\mathbf{v}_{\max}$. Note that any multiplication by a real scalar of the matrix $\mathbf{H}$ (i.e., change of pathloss) is irrelevant, since we only deal with normalized precoding vectors $\left\| \mathbf{t}^{(q)} \right\|_2^2 = 1$, since the transmit power is given and just has to be distributed to the TX antennas. Furthermore, the codebook is invariant to a common phase shift.

It is desirable to have a codebook that minimizes – on average – the loss of SNR at the RX, ζ:

$$\zeta = \gamma_{TX} E_{\mathbf{H}} \left\{ \left\| \mathbf{H}\mathbf{v}_{\max} \right\|_2^2 - \left\| \mathbf{H}\widetilde{\mathbf{t}} \right\|_2^2 \right\}. \tag{16.44}$$

Now it can be shown that this can be upper-bounded as

$$\zeta \leq \hat{\zeta} = E_{\mathbf{H}} \left\{ \gamma_{TX} \lambda_{\max} \left[1 - \max_{\mathbf{t}^{(q)} \in \mathcal{C}} \left| \mathbf{v}_{\max}^{\dagger} \mathbf{t}^{(q)} \right|^2 \right] \right\}. \tag{16.45}$$

This is a general bound on the performance loss.

For the case of isotropic channels, i.e., i.i.d. zero-mean complex Gaussian fading,

$$\hat{\zeta} = E_{\mathbf{H}} \{ \gamma_{TX} \lambda_{\max} \} E_{\mathbf{H}} \left\{ 1 - \max_{\mathbf{t}^{(q)} \in \mathcal{C}} \left| \mathbf{v}_{\max}^{\dagger} \mathbf{t}^{(q)} \right|^2 \right\} \tag{16.46}$$

because eigenvalues and eigenvectors of $\mathbf{H}\mathbf{H}^{\dagger}$ are independent if $\mathbf{H}$ is i.i.d. complex Gaussian.

$\hat{\zeta}$ can furthermore be upper-limited as

$$\hat{\zeta} \leq E_{\mathbf{H}} \{ \gamma_{TX} \lambda_{\max} \} \left(\left(\frac{\delta}{2} \right)^2 \left(\frac{\delta}{2} \right)^{2(N_t - 1)} Q + \left[1 - \left(\frac{\delta}{2} \right)^{2(N_t - 1)} Q \right] \right) \tag{16.47}$$

where

$$\delta = \min_{p \neq q} \sqrt{1 - \left| \left[\mathbf{t}^{(q)} \right]^{\dagger} \mathbf{t}^{(p)} \right|^2} \tag{16.48}$$

i.e., the minimum over the subspace distance between $\mathbf{t}^{(p)}$ and $\mathbf{t}^{(q)}$. Finding the optimum codebook thus means that we have to maximize the minimum of the subspace distances – a problem that is related to so-called *Grassmannian line packing*. An example codebook for $N_t = 2$ and $Q = 4$ is given in Table 16.2.

Table 16.2 Codebook entries for limited feedback with $N_t = 2$ and $Q = 4$.

$q =$	$t_1^{(q)}$	$t_2^{(q)}$
1	$-0.1612 - 0.7348j$	$-0.5135 - 0.4128j$
2	$-0.0787 - 0.3192j$	$-0.2506 + 0.9106j$
3	$-0.2399 + 0.5985j$	$-0.7641 - 0.0212j$
4	-0.9541	0.2996

Based on [Love et al. 2003].

For Grassmannian codebooks, the δ is lower limited by

$$\delta \geq \sqrt{1 - \frac{Q - N_t}{N_t (Q - 1)}}. \tag{16.49}$$

Note that the codebook design is not unique – a multiplication with a constant phase (for all entries) does not change the SNR (and that multiplication is undone at the RX anyway). The minimum codebook size to guarantee a relative capacity loss no larger than $\varepsilon = (\mathcal{C}_{\text{ideal}} - \mathcal{C}_{\text{quant}})/\mathcal{C}_{\text{ideal}}$ is

$$Q \geq (1 - \varepsilon)\left(\frac{4N_{\text{t}}}{N_{\text{t}} - 1}\right)^{N_{\text{t}} - 1}. \tag{16.50}$$

Remarkably, this number does not depend on the SNR. An alternative derivation of codebooks for i.i.d. channels can be done via random vector quantization.

It must be stressed that the above derivation makes one big simplification – it separates the channel statistics from the codebook through the upper-bounding technique Eq. (16.47), which is specific to i.i.d. channels. However, most practical MIMO channels are not i.i.d. Another "extreme" case of angular distribution is the pure LOS case, i.e., all radiation is coming from a particular direction. A good codebook for that case is a Fourier codebook, i.e., the beamforming vectors are simply the column vectors of the FFTs, so that each beamforming vector is a Fourier beam pointing into a particular direction.

An "in-between" situation can often be described by a correlated channel. Let us consider the Kronecker model of Section 7.5.3:

$$\mathbf{H} = \mathbf{R}_{\text{RX}}^{1/2}\mathbf{G}_{\text{G}}\mathbf{R}_{\text{TX}}^{1/2} \tag{16.51}$$

where $\mathbf{G}_{\text{G}}$ is a matrix with independent identically distributed (i.i.d.) complex Gaussian entries. While the receive correlation does not have an impact on the precoder design, it is intuitive that the transmit correlation is very important. Keep in mind that (for a uniform linear array at the TX) there is a relationship between the correlation matrix and the angles in which the radiation is transmitted. Clearly, there is no sense to quantize finely in directions which are never used to transmit energy. This statement can be put into a more quantitative form by the following precoder design recipe: the codebook vectors $\hat{\mathbf{t}}^{(q)}$ are the codebook vectors from the uncorrelated case, scaled by the correlation matrix

$$\hat{\mathbf{t}}^{(q)} = \frac{\mathbf{R}_{\text{TX}}^{1/2}\mathbf{t}^{(q)}}{\left\|\mathbf{R}_{\text{TX}}^{1/2}\mathbf{t}^{(q)}\right\|}. \tag{16.52}$$

A general strategy for deriving optimal codebooks follows from the *Generalized Lloyd Algorithm* for vector quantization. We explain it with the example of a MISO channel, so that $\mathbf{H}$ becomes a row vector $\mathbf{h}$. The goal is to partition the input space into Q regions (called *quantization cells*) $\mathcal{R}_q$, $q = 1, 2, ...Q$ and each region is associated with a particular precoder $\mathbf{t}^{(q)}$. We then define the correlation matrix of the qth region as $E_{\mathcal{R}_q}\{\mathbf{h}^\dagger\mathbf{h}\}$, i.e., the expectation of $\mathbf{h}^\dagger\mathbf{h}$ for all the $\mathbf{h}$ that are in region $\mathcal{R}_q$. The algorithm then first initializes the codebook with any valid codebook, and then iterates the following two steps to convergence:

- Given the codebook, find the optimal quantization cells through the nearest-neighbor rule, i.e., any possible $\mathbf{h}$ is mapped to the $\mathbf{t}^{(q)}$ for which $|\mathbf{t}^{(q)}\mathbf{h}| \geq |\mathbf{t}^{(p)}\mathbf{h}|$ for $p \neq q$.
- Given the quantization cells, the optimal quantized precoder $\mathbf{t}^{(q)}$ for quantization cell q is the dominant eigenvector of $E_{\mathcal{R}_q}\{\mathbf{h}^\dagger\mathbf{h}\}$.

A comparison of the performance with different codebooks is given in (Figure 16.9)

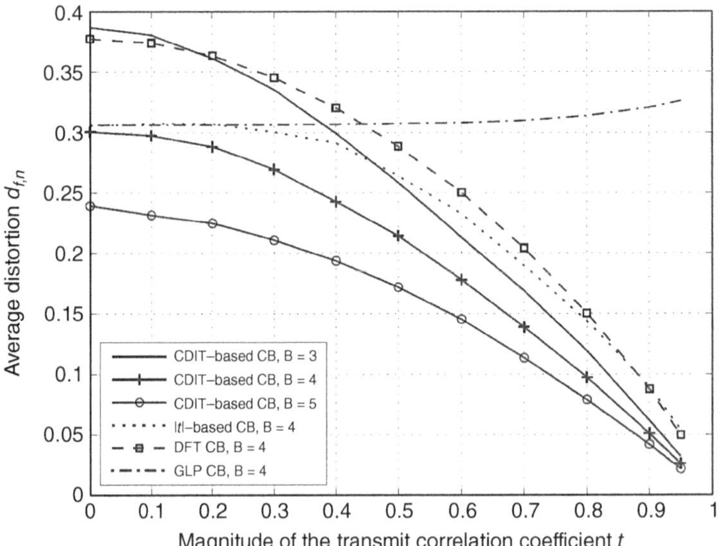

Figure 16.9 Average distortion due to quantized feedback for various codebooks in correlated channels (correlation coefficient t): GLP (Grassmannian Line Packing), DFT (Discrete Fourier transform), CDIT (adaptive codebook Eq. (16.52)).

All of the above considerations are for a "one-shot" feedback, where during each time interval the index of the best beam is sent back, without memory of the beams at previous time instances. However, this might be a waste of resources, since the optimal beam directions have a temporal correlation (depending on the ratio of coherence time of the channel to feedback cycle time). A better way is to start out with the beamforming vector from a standard codebook, but then for the next time step refine this estimate, according to a *differential codebook*. For example, in a rotation-based codebook, the codewords are unitary rotation matrices $\boldsymbol{\Theta}_i$, each of which has dimension $N_t \times N_t$. The code vector at time instant $n = 0$ is taken from a standard (e.g., Grassmannian) codebook, and the code vector at time index $n + 1$ is the code vector at time index n, multiplied with $\boldsymbol{\Theta}_i$; it is the index i, i.e., which of the defined rotation matrices should be chosen, that is fed back to the TX. Other forms of differential feedback also exist.

*Feedback for Space-Time Block Codes

The design of the precoding codebook becomes more complicated when the TX employs space-time coding. In that case, the precoding vector $\mathbf{t}$ has to be replaced by a precoding matrix $\mathbf{T}$ of size $N_t \times \widetilde{L}$, where $\widetilde{L}$ is a dimension of the space-time code (e.g., 2 for an Alamouti code). The optimization goal is still the maximization of the overall SNR at the RX (note that there is only a single data stream, even though it is space-time coded)

$$SNR_q = \gamma_{\text{TX}} \left\| \mathbf{H}\mathbf{T}^{(q)} \right\|_F^2. \tag{16.53}$$

A bound to the SNR loss due to the quantization can be derived for isotropic channels as

$$\zeta = E_{\mathbf{H}} \left\{ \gamma_{\text{TX}} \left\| \mathbf{H}\mathbf{T}^{(\text{opt})} \right\|_F^2 - \gamma_{\text{TX}} \left\| \mathbf{H}\widetilde{\mathbf{T}} \right\|_F^2 \right\}$$

$$\leq E_{\mathbf{H}} \{ \gamma_{\text{TX}} \lambda_{\max} \} \left\{ \widetilde{L} + \left(\frac{\delta}{2\sqrt{\widetilde{L}}} \right)^{2N_t \widetilde{L} + \mathcal{O}(N_t)} Q \left[\left(\frac{\delta}{2} \right)^2 - \widetilde{L} \right] \right\} \tag{16.54}$$

where $\mathbf{T}^{(\text{opt})}$ is the optimum precoder (assuming perfect CSIT and no quantization), while $\widetilde{\mathbf{T}}$ is the best-quantized precoder. The minimum subspace distance is now defined as the minimum *projection two-norm*

$$\delta = \min_{p \neq q} \frac{1}{\sqrt{2}} \left\| \mathbf{T}^{(p)} \left(\mathbf{T}^{(p)} \right)^\dagger - \mathbf{T}^{(q)} \left(\mathbf{T}^{(q)} \right)^\dagger \right\|_F. \tag{16.55}$$

The codebook entries are not unique – clearly, a multiplication with a unitary matrix does not change the characteristics of $\mathbf{T}$. This is similar to the case of a single data stream, where multiplication with a constant phase does not change the optimality.

One might now ask why to consider Space-Time Block Codes (STBC), in particular Alamouti codes, in a setting where feedback is available. After all, Alamouti codes perform worse than maximum-ratio transmission (which is approximated by the beamforming based on the feedback). However, there are situations where CSI becomes uncertain (due to noise) or quickly outdated. In this case, it is advantageous to supplement the partial-CSI-based beamforming with extra diversity from the space-time coding. If second-order CSI is available, then the precoder should be used to appropriately align the transmission signal with the correlation matrix information before transmission.

Feedback in Frequency-Selective Channels

Consider now an OFDM system operating in a frequency-selective channel. The brute-force approach would be to feed back the codebook entry index for each subcarrier separately. However, this would be highly wasteful, since channels on adjacent subcarriers are correlated, and therefore also the optimum beamformers are correlated. An intuitive improvement is thus to group the subcarriers, and just feed back one index for each group (e.g., the index for the subcarrier at the center of the group). This approach is used, e.g., in 5G (Sections 32.3.6 and 32.3.7).

16.2 Spatial Multiplexing

16.2.1 Introduction

Besides diversity and beamforming, multiple antenna elements – if present at *both TX and RX* – can also be used to increase the spectral efficiency by transmitting multiple data streams in the same time-frequency resources. This so-called spatial multiplexing (SM) is the main reason for the great attention MIMO systems have received since the mid-1990s. We stress that SM is only possible in MIMO, not in SIMO or MISO;[6] as a matter of fact, MIMO is often used synonymously with SM, though that usage is not accurate. Originally suggested

---

[6] Theoretically, spatial multiplexing can be done in MISO, i.e., different data streams are sent from the TX antennas, and the single-antenna RX demodulates/decodes all of them, e.g., using the principles of multi-user detection (see Chapter 28). However, the multiplexing gain also in this case is 1. Therefore, in practice spatial multiplexing is used in MIMO systems.

and investigated in [Winters 1987], and in patents by [Paulraj and Kailath 1994], MIMO attracted great attention through theoretical investigations [Foschini and Gans 1998, Telatar 1999]. Since that time, research in those systems has exploded and many current systems, including LTE, 5G NR, and Wi-Fi, incorporate SM in their standards (compare Chapters 31–33).

16.2.2 How does Spatial Multiplexing Work?

SM uses the multiple antenna elements at the TX for the transmission of parallel data streams, see Figure 16.10. An original high-rate data stream is multiplexed into several parallel streams, each of which is sent from one transmit antenna element. The channel "mixes up" those data streams, so that each of the receive antenna elements sees a combination of them. If the channel is well-behaved, the received signals represent *linearly independent* combinations. In that case, appropriate signal processing at the RX can separate the data streams. The ability to separate the data streams can be explained in a number of different ways:

- The RX performs optimum combining (see Section 12.4.3), where N_r receive antenna elements can suppress $N_r - 1$ interferers. Thus, by interpreting one data stream as "desired," and the others as "interferers," a data stream can be properly received when up to $N_r - 1$ other streams are present. Since this statement holds for any of the transmitted data streams, it shows that the RX can separate and demodulate a total of up to N_r data streams.
- The RX performs linear reception: if the different channel realizations are independent, the number of independent linear combinations of N_r TX data streams is equal to N_r. Thus, a suitable linear processing such as matrix inversion can separate out the data streams. While this is not an optimum reception method, it is sufficient to explain the ability to receive the multiple data streams.
- Counting dimensions: a transmit signal with N_s data streams spans N_s (complex) dimensions. A RX sampling at N_r locations obtains an N_r dimensional signal (assuming no degeneracies in the channel that transforms the TX signal to the RX signal), so that (again with the caveats of footnote 6) reconstruction of the transmit signal is possible if $N_r \geq N_s$.

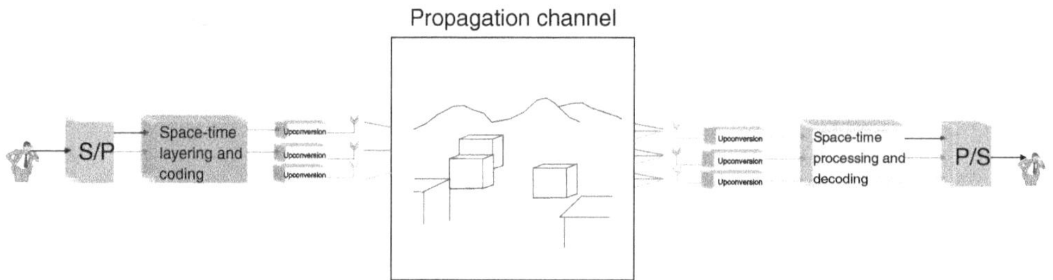

Figure 16.10 Principle of spatial multiplexing.

From all of these (equivalent) representations, the basic condition follows that the number of receive antenna elements must be at least as large as the number of transmit data streams. At the same time, the number of data streams obviously cannot be larger than the number of transmit antennas. Thus, the use of SM approximately increases the data rate by a factor $\min(N_t, N_r)$.

There is, however, also another important condition: since the RX needs to have at least N_s *linearly independent* combinations of the TX data streams, the rank where R_H, the rank of the channel matrix **H**, must at least be N_s. This, in turn, requires a sufficiently "rich scattering," i.e., a sufficient number of MPCs along which the signal can propagate from the TX to the RX. As we will discuss later on, small angular spread, large Rice factor, and small number of IOs can all decrease the channel rank. This leads to the somewhat counterintuitive result that a channel that is usually considered a "bad" channel, namely one with rich multi-path and therefore deep fading, is actually beneficial for SM. However, great care must be taken in the interpretation, in particular, if the rich multi-path is often associated with large pathloss, see section titled "LOS vs. NLOS" in 16.2.6 for a more detailed discussion.

For the case that the TX knows the channel, we can also develop another transmission scheme, see Figure 16.11. With N_t transmit antennas, we can form N_t different beams. We point all those beams at different IOs and transmit different data streams over them. At the RX, the N_r antenna elements are used to form N_r beams that also point at the different IOs. If all the beams from the TX to the IOs, and all the beams from the IOs to the RX, can be kept orthogonal to each other, there is no interference between the data streams; in other words, we have established parallel channels. The IOs (in combination with the beams pointing into their direction) play the same role as wires in the transmission of multiple data streams on multiple wires. In this case, it is not necessary to do a complicated separation of the data streams – each receive beam contains one stream.

The picture of forming orthogonal beams to different IOs is intuitive, but actually overly restrictive. If instantaneous CSI is available, then the formed beams can be singular vectors of **H** – still beams that are orthogonal to each other in signal space, but not necessarily separated in the angular domain anymore. Note that the channel rank R_H is limited not only by the number of TX and RX antennas but also by the number of (significant) IOs, N_{IO}.

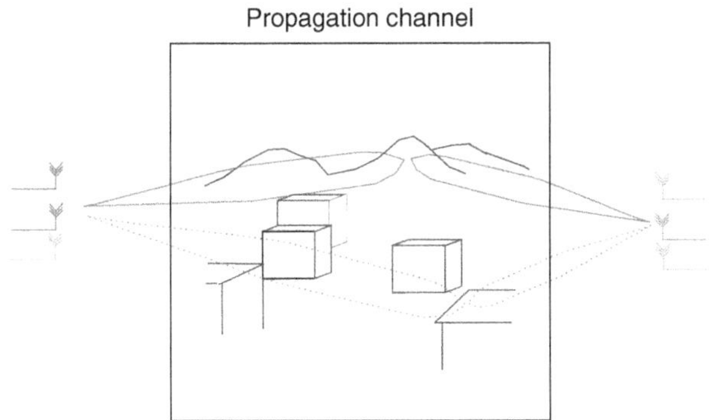

Figure 16.11 Transmission of different data streams via different IOs.

In summary, both with and without CSIT, the number of possible data streams is limited by $\min(N_t, N_r, N_{IO})$. A more exact mathematical treatment follows in the sections below. For later user, we define $N = \min(N_t, N_r)$ and $M = \max(N_t, N_r)$.

In the following subsections, we will compute quantitatively the capacity of MIMO systems in deterministic and fading channels. Note that not all of the resulting expressions might be "capacities" in the information-theoretic definition of the word (in many cases "maximum throughput" might be more appropriate). Yet a more "forgiving" use of the word capacity has become common in the MIMO literature, and we will follow it here as well.

16.2.3 Capacity in Nonfading Channels

General

Before going into a discussion on how to realize SM, let us first discuss the information-theoretic limits of multi-antenna systems. Recall from Section 13.1 that the capacity equation for "normal" (single-antenna) nonfading Additive White Gaussian Noise (AWGN) channels is (expressing normalized capacity in bit/s/Hz)

$$C_{\text{shannon}} = \log_2\left(1 + \gamma_{\text{TX}} \cdot |H|^2\right) \tag{16.56}$$

where γ_{TX} is the SNR at the TX (more precisely, the ratio of signal power at the TX to noise power at the RX), and H is the complex channel gain from the TX to the RX (as we are for now dealing with the frequency-flat case). The key consequence from this equation is that the capacity increases (at high SNR) only logarithmically with the SNR, so that boosting the transmit power is a highly ineffective way of increasing capacity.

We now turn to capacity of MIMO systems, where in this subsection we consider transmission over a fixed (time-invariant) channel described by matrix $\mathbf{H}$. The capacity can then be computed under the assumption of full, partial, or no CSIT.

A basic lemma needed for the subsequent considerations is that the mutual information between TX and RX for a deterministic channel $\mathbf{H}$ is given as

$$\mathcal{I}(\mathbf{H}, \mathbf{R}_{ss}) = \log_2 \det\left[\mathbf{I}_{N_r} + \gamma_{\text{TX}} \mathbf{H} \mathbf{R}_{ss} \mathbf{H}^{\dagger}\right] \tag{16.57}$$

where $\mathbf{R}_{ss}$ is the input covariance matrix with $tr[\mathbf{R}_{ss}] = 1$, and $\mathbf{I}_{N_r}$ is the $N_r \times N_r$ identity matrix; for the proof, see Exercise 16.13.

Full CSIT

From Eq. (16.57), it follows immediately that the capacity for full CSIT is

$$C_{\text{CSIT}} = \max_{\mathbf{R}_{ss} \succeq 0,\, tr[\mathbf{R}_{ss}] = 1} \log_2 \det\left[\mathbf{I}_{N_r} + \gamma_{\text{TX}} \mathbf{H} \mathbf{R}_{ss} \mathbf{H}^{\dagger}\right]. \tag{16.58}$$

To find the $\mathbf{R}_{ss}$ that optimizes $\log_2 \det[.]$, first consider the singular value decomposition, Eq. (16.28), i.e., $\mathbf{H} = \mathbf{U}\boldsymbol{\Sigma}\mathbf{V}^{\dagger}$. The relationship between transmitted and received signal can then be written as

$$\mathbf{r} = \mathbf{U}\boldsymbol{\Sigma}\mathbf{V}^{\dagger}\mathbf{s} + \mathbf{n}. \tag{16.59}$$

Then, choose the TX signal $\mathbf{s}$ such that it can be written as the product of $\mathbf{V}$ with a signal $\tilde{\mathbf{s}}$, and the received signal is multiplied with $\mathbf{U}^{\dagger}$; this diagonalizes the channel,

$$U^\dagger r = U^\dagger U \Sigma V^\dagger V \tilde{s} + U^\dagger n \tag{16.60}$$

$$\tilde{r} = \Sigma \tilde{s} + \tilde{n}. \tag{16.61}$$

Note that – because U is a unitary matrix – $\tilde{n}$ has the same statistical properties as n, i.e., it is i.i.d. white Gaussian noise. Now the computation of the capacity of Eq. (16.61) is rather straightforward. The matrix Σ is a diagonal matrix with R_H nonzero entries σ_k (since R_H is the number of nonzero singular values) and σ_k is the kth singular value of H. Remember that in a rich scattering channel, $R_H = N = \min(N_t, N_r)$. We have therefore N *parallel* channels, and the capacity of those parallel channels just adds up. Equivalently, we can say that the input correlation matrix takes on the form

$$R_{ss} = V \begin{bmatrix} P_1 & & & \\ & P_2 & & \\ & & \cdots & \\ & & & P_{R_H} \end{bmatrix} V^\dagger \tag{16.62}$$

which is to say that the directional precoder is $\hat{T} = V$ and the overall precoder is

$$T = \hat{T} \begin{bmatrix} P^{1/2} \\ 0 \end{bmatrix} = V \begin{bmatrix} P^{1/2} \\ 0 \end{bmatrix}. \tag{16.63}$$

Either formulation leads to the conclusion that the capacity is

$$C_{\text{CSIT}} = \max_{P_k \geq 0,\, \sum P_k = P_{\text{tot}}} \sum_{k=1}^{N} \log_2 \left[1 + \frac{P_k}{\sigma_n^2} \sigma_k^2 \right]. \tag{16.64}$$

This is the problem of optimally allocating power to several parallel channels, each of which has different SNR. This is the same problem that was considered in Section 15.9.2, and therefore the answer is the same: *waterfilling*. This is another nice example of how the same mathematics can be applied to different communications problems: replacing the word "subchannel" or "subcarrier in an OFDM system" by "eigenmode in a MIMO system" allows to apply the whole discussion in Section 15.9.2 to MIMO systems with CSI at the TX. The optimum power allocation $P_{\text{opt},k}$ is thus

$$P_{\text{opt},k} = \left[\varepsilon_{\text{WF}} - \frac{1}{\gamma_{\text{TX}} \sigma_k^2} \right]^+ \tag{16.65}$$

where $[x]^+ = \max[x, 0]$, and ε_{WF} is chosen such that $\Sigma P_{\text{opt},\,k} = P_{\text{tot}}$.

Since the nonzero singular values of H are the same as those of $H^\dagger$, the capacity remains unchanged if we flip the role of TX and RX. This symmetry is predicated on the existence of CSIT and CSIR; we will see below that the symmetry breaks when CSI is available only at the RX.

For the limiting case of high SNR, waterfilling degenerates to assigning the same amount of power to each eigenmode; when there is a lot of water available, the height of the "concrete blocks" in the vessel has little influence on the total amount that ends up in the vessels. Note, however, that for $N_t > N_r$ this is not identical to transmitting one data stream, omnidirectionally, from each TX antenna (the solution that is best for the case of no CSIT, see below). When CSIT is present, the TX will perform beamforming along the singular vectors of H; it is only the amount of power assigned to each of those beams that is uniform. The capacity then becomes

$$C_{\text{CSIT}} = \sum_{k=1}^{N} \log_2 \left[1 + \frac{\gamma_{\text{TX}}}{N} \sigma_k^2 \right]. \tag{16.66}$$

One can also conclude that (16.66) is a lower bound of the capacity.

For the other limiting case of very low SNR, all power is assigned to the dominant eigenmode only. In other words, enhancing the SNR via beamforming, and sticking with a single transmit stream, is the most efficient way of using the multiple antenna elements when the SNR is low. Generally, it is important to note that the optimum solution might transmit a smaller number of streams than N. Thus, so-called *rank adaptation*, i.e., adjusting the number of data streams to the channel conditions, is important in practical systems.

A general upper bound for the capacity can be found from Jensen's inequality

$$C_{\text{CSIT}} = \sum_{k=1}^{N} \log_2 \left[1 + \frac{P_{\text{opt},k}}{\sigma_n^2} \sigma_k^2 \right]$$

$$\leq N \log_2 \left[1 + \sum_{k=1}^{N} \frac{P_{\text{opt},k}}{N \sigma_n^2} \sigma_k^2 \right]$$

$$\leq N \log_2 \left[1 + \frac{\gamma_{\text{TX}}}{N} \sigma_{\max}^2 \right].$$

We can also write, more generally, the capacity in the limit of high SNR as

$$C_{\text{CSIT}} = N\left[\log_2(\beta\gamma_{\text{TX}}) - L_\infty\left(\widetilde{\mathbf{H}}\right)\right] + \mathcal{O}\left(\frac{1}{\beta\gamma_{\text{TX}}}\right) \tag{16.67}$$

where $\mathcal{O}(.)$ is the big-O notation, i.e., $\mathcal{O}(1/x)$ decreases "at least as fast as a constant multiplied by $1/x$," and $\beta\gamma_{\text{TX}} = \gamma_{\text{RX}}$ is the average *receive* SNR at an RX antenna element. The function $L_\infty\left(\widetilde{\mathbf{H}}\right)$ describes the difference at high SNR between N parallel AWGN channels and the results in the actual channel; it can be shown to be

$$L_\infty\left(\widetilde{\mathbf{H}}\right) = \begin{cases} \log_2(N_t) - \dfrac{1}{N_t}\log_2\left|\widetilde{\mathbf{H}}^\dagger\widetilde{\mathbf{H}}\right| & N_t \leq N_r \\[3mm] \log_2(N_r) - \dfrac{1}{N_r}\log_2\left|\widetilde{\mathbf{H}}\widetilde{\mathbf{H}}^\dagger\right| & N_r \leq N_t. \end{cases} \tag{16.68}$$

Equation (16.67) shows clearly the improvement that can be achieved by increasing the number of antennas: as long as adding antennas increases N (i.e., adding antennas on the link end that has fewer antennas), the capacity increases linearly. Increasing the SNR by 3 dB increases the capacity by N bit/s/Hz, i.e., 1 bit/s/Hz for every data stream that can be sent over the air.

No CSIT

If the TX does not have any CSI, it cannot adapt the beamforming vector and power allocation to the channel state. Thus, it is optimum to assign equal transmit power to all the TX antennas, $P_k = P_{\text{tot}}/N_t$, and to use uncorrelated data streams. The capacity thus takes on the form

$$C_{\text{noCSIT}} = \log_2\left[\det\left(\mathbf{I}_{N_r} + \frac{\gamma_{\text{TX}}}{N_t}\mathbf{H}\mathbf{H}^\dagger\right)\right]. \tag{16.69}$$

Comparing Eqs. (16.69) to (16.64) shows that the capacity of a MIMO system increases linearly with $\min(N_t, N_r)$, irrespective of whether the channel is known at the TX or not.

Consider now a few special cases. The capacity of a SISO channel with average receive SNR γ_{RX} (note that in this case there is really no averaging, since we do not have multiple antennas) is $C_{\text{SISO}} = \log_2(1 + \gamma_{\text{RX}})$. If we have multiple antennas, to make the discussion easier, we assume that $N_t = N_r = N$.

1. All transfer functions are identical, i.e., $h_{1,1} = h_{1,2} = \cdots = h_{N,N}$. This case occurs when all antenna elements are spaced very closely together, and the waves are all coming from similar directions. In that case, the rank of the channel matrix is unity. Then, the capacity is

$$C_{\text{MIMO}} = \log_2(1 + N\gamma_{\text{RX}}). \tag{16.70}$$

   We see that in this case, the SNR is increased by a factor of N compared to the single antenna case, due to the beamforming gain at the RX. However, this only leads to a logarithmic increase of the capacity with the number of antennas.

2. Parallel transmission channels (however, *without* CSIT), i.e., the reception at the jth RX antenna is only impacted by the signal transmitted from the jth TX antenna. In that case, the capacity increases linearly with the number of antenna elements. However, the SNR per channel decreases with N, so that the total capacity is

$$C_{\text{MIMO}} = N\log_2\left(1 + \frac{\gamma_{\text{RX}}}{N}\right). \tag{16.71}$$

For the high-SNR case, we can again establish the capacity as

$$C_{\text{noCSIT}} = N\left[\log_2(\beta\gamma_{\text{TX}}) - L_\infty\left(\widetilde{\mathbf{H}}\right)\right] + \mathcal{O}\left(\frac{1}{\beta\gamma_{\text{TX}}}\right) \tag{16.72}$$

where now

$$L_\infty\left(\widetilde{\mathbf{H}}\right) = \begin{cases} \log_2(N_t) - \dfrac{1}{N_t}\log_2\left|\widetilde{\mathbf{H}}^\dagger\widetilde{\mathbf{H}}\right| & N_t \leq N_r \\[3mm] \log_2(N_t) - \dfrac{1}{N_r}\log_2\left|\widetilde{\mathbf{H}}\widetilde{\mathbf{H}}^\dagger\right| & N_r \leq N_t. \end{cases} \tag{16.73}$$

For the case $N_r \geq N_t$, the asymptotic capacity (for the SNR growing to infinity) is the same as for the CSIT case; this is due to the fact that the capacity improvement of waterfilling (compared to equal-power allocation) vanishes for very large SNR. For the case of $N_t > N_r$, the capacity without CSIT is relatively smaller. This is because the lack of CSIT prevents beamforming (which is effective for $N_t > N_r$), and thus worsens the SNR.

Figure 16.12 shows the capacity without CSIT as a function of N for different values of the SNR.

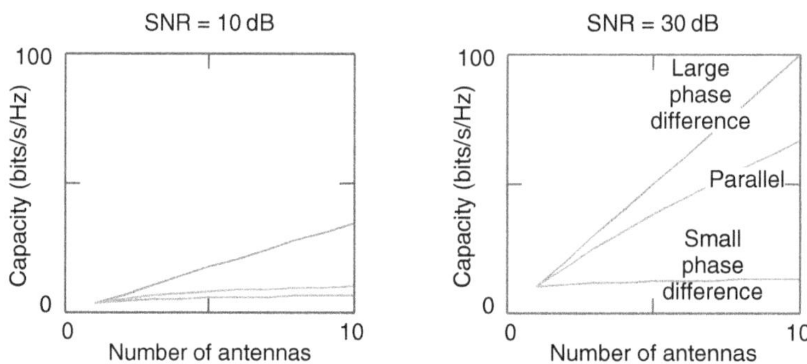

Figure 16.12 Capacity of MIMO systems in AWGN channels.

No CSIR and no CSIT

It is remarkable that the channel capacity is also high when neither the TX nor the RX has CSI. We can, for example, use a generalization of differential modulation. For high SNR, the capacity no longer increases linearly with $N = \min(N_t, N_r)$, but rather increases as $\widetilde{N}\left(1 - \widetilde{N}/T_{\mathrm{coh,norm}}\right)$, where $\widetilde{N} = \min\left(N_t, N_r, \lfloor T_{\mathrm{coh,norm}}/2 \rfloor\right)$, and $T_{\mathrm{coh,norm}}$ is the coherence time of the channel in units of symbol duration.

*16.2.4 Capacity in Flat-Fading Channels

General Concepts

In the previous section, we considered the capacity for one given channel realization, i.e., for one channel matrix $\mathbf{H}$. In wireless systems, we are, however, usually faced with fading of the channel. In that case, the entries in the channel matrix Eq. (16.4) are random variables. In the current subsection, we consider the case of *fast fading* (i.e., time-selective) *channels*: the channel is fading, and the codewords being transmitted over the channel last much longer than the coherence time of the channel (similar considerations may apply in frequency- and space-selective channels). Thus, each data stream will experience many channel realizations. The resulting capacity is the Shannon capacity or *ergodic capacity* of this fading MIMO channel.

The ergodic capacity can be computed simply as the expectation over the ensemble of realizations that the channel can take on

$$C_{\mathrm{ergodic}} = E_{\mathbf{H}}\{C(\mathbf{H})\} \tag{16.74}$$

which explicitly denotes the dependence of the deterministic capacity on the channel realization.

In the presence of full CSIT, we can further distinguish two important cases:

- in the case of an *instantaneous power constraint,* the ergodic capacity can generally be written as

$$
\begin{aligned}
C_{\mathrm{ergodic,CSIT}} &= E_{\mathbf{H}}\left\{ \max_{\mathbf{R}_{ss}\succeq 0,\, tr[\mathbf{R}_{ss}]=1} \log_2 \det\left[\mathbf{I}_{N_r} + \gamma_{\mathrm{TX}}\mathbf{H}\mathbf{R}_{ss}\mathbf{H}^{\dagger}\right] \right\} \\
&= E_{\mathbf{H}}\left\{ \max_{P_k\succeq 0,\, \sum P_k = P_{\mathrm{tot}}} \sum_{k=1}^{N} \log_2\left[1 + \frac{P_k}{\sigma_{\mathrm{n}}^2}\sigma_k^2\right] \right\}.
\end{aligned}
\tag{16.75}
$$

This implies that for each time instant, the constraint $\sum P_k = P_{\mathrm{tot}}$ is active, i.e., the power (summed over all spatial eigenmodes) at every moment is limited. The expectation over the channel is, essentially, the expectation over the eigenvalues, σ_k^2. The unordered eigenvalues all have the same distribution, so that the capacity becomes (using also Eq. (16.65))

$$C_{\mathrm{ergodic,CSIT}} = NE_{\mathbf{H}}\left\{ \log_2\left[\frac{\gamma_{\mathrm{TX}}\sigma_k^2 \varepsilon_{\mathrm{WF}}}{N_t}\right]^{+} \right\}. \tag{16.76}$$

- in the case of a *time-average* (also called long-term) *power constraint*, the σ_k^2 are random variables, and the waterfilling is now done with respect to the singular values in time as well as in space. In other words, assume that different "realizations of the ensemble" are different realizations in time, with time index i. Then the singular values can be written as $\sigma_{i,k}$, and the associated powers as $P_{i,k}$. The capacity is then

$$C_{\text{ergodic,CSIT}} = \sum_{k=1}^{N} E_{\mathbf{H}}\left\{ \max_{P_k \succeq 0} \log_2\left[1 + \frac{P_k}{\sigma_n^2}\sigma_k^2\right]\right\} \qquad \text{s.t.} \qquad E_{\mathbf{H}}\left\{\sum_k P_k = P_{\text{tot}}\right\} \tag{16.77}$$

where the σ_k and P_k are now functions of the channel state $\mathbf{H}$. In other words, the power assignment might reduce the total transmit power in moments where the MIMO channel is bad, and spend more during good times.

It turns out that the difference between the capacity with short-term and long-term power constraint is generally small; in particular it vanishes for high SNR (where power assignment is uniform), and for very large arrays (due to the high diversity order of such arrays), implying that the capacity does not change much even when the small-scale fading realization changes.

For the case of no CSIT, the TX cannot adapt to the channel states, and the optimum input covariance matrix is thus always a scaled identity matrix, $\mathbf{R}_{ss} = \mathbf{I}_{N_t}/N_t$. The ergodic capacity is thus

$$C_{\text{ergodic,noCSIT}} = E_{\mathbf{H}}\left\{ \log_2 \det\left[\mathbf{I}_{N_r} + \frac{\gamma_{\text{TX}}}{N_t}\mathbf{HH}^\dagger\right]\right\}. \tag{16.78}$$

For the case of partial CSIT, the input covariance matrix is adapted to the available CSIT (e.g., the second-order statistics); with this modification, Eqs. (16.75) and (16.77) remain valid.

Full CSIT in i.i.d. Rayleigh Fading Channels

In the case of high SNR, such that the power allocated to each data stream is the same, the ergodic capacity is (16.72) with

$$L_\infty\left(\widetilde{\mathbf{H}}\right) = \log_2(N) + \log_2(e)\left[\gamma_{\text{Euler}} - \sum_{k=1}^{M-N}\frac{1}{k} - \max\left(\frac{N_r}{N_t}, \frac{N_t}{N_r}\right)\sum_{k=M-N+1}^{M}\frac{1}{k} + 1\right] \tag{16.79}$$

where the derivation (see [Heath and Lozano 2018]) is based on the following relation for Wishart matrices

$$E\left\{\ln\left[\det\left(\mathbf{HH}^\dagger\right)\right]\right\} = \sum_{n=0}^{N-1}\psi(M-n) \qquad \psi(k) = -\gamma_{\text{Euler}} + \sum_{m=1}^{k-1}\frac{1}{m} \quad \text{for} \quad k \text{ integer} \tag{16.80}$$

where $\psi(k)$ is the digamma function, and $\gamma_{\text{Euler}} \approx 0.5772$ is Euler's constant.

No CSIT in i.i.d. Rayleigh Fading Channels

We next turn to the case that no CSIT is available. Note that due to the assumption of i.i.d. channels, this is identical to having second-order statistics available – since those second-order statistics just indicate isotropic distribution.

The exact expression for the *ergodic capacity* can be written as

$$C_{\text{ergodic,noCSIT}} = N\int_0^\infty \log_2\left[1 + \frac{\gamma_{\text{TX}}}{N_t}\lambda\right]p_\lambda(\lambda)d\lambda \tag{16.81}$$

where $p_\lambda(\lambda)$ is the pdf of the nonordered eigenvalues of the Wishart matrix $\mathbf{HH}^\dagger$ (for $N_t > N_r$) or $\mathbf{H}^\dagger\mathbf{H}$ (for $N_t < N_r$), which can be evaluated to (see [Telatar 1999])

$$p_\lambda(\lambda) = \frac{1}{N}\sum_{k=0}^{N-1}\frac{k!}{(k+M-N)!}\left[L_k^{M-N}(\lambda)\right]^2 \lambda^{M-N}\exp(-\lambda) \tag{16.82}$$

where $L_k^{M-N}(\lambda)$ are Associated Laguerre polynomials. This integral can be evaluated in closed form in terms of special functions (exponential integrals). For the high SNR regime, the following holds

$$L_\infty\left(\widetilde{\mathbf{H}}\right) = \log_2(N_t) + \log_2(e)\left[\gamma_{\text{Euler}} - \sum_{k=1}^{M-N}\frac{1}{k} - \max\left(\frac{N_r}{N_t}, \frac{N_t}{N_r}\right)\sum_{k=M-N+1}^{M}\frac{1}{k} + 1\right]. \tag{16.83}$$

Comparing to the result of CSIT, we find a difference in the first term: it is $[\log_2(N)]$ if we have CSIT, but $[\log_2(N_t)]$ if we have no CSIT. This makes a difference only if $N_t > N_r$, as discussed previously.

To summarize, the capacity gain by CSIT is rather small when the number of transmit and receive antennas is identical, and in particular at large SNRs. For i.i.d. channels and $N_t \leq N_r$, beamforming is not effective. Furthermore, if the TX has no channel knowledge, then $N_t > N_r$ does not provide any advantages: the number of data streams is limited by the number of receive antennas. Of course, we can transmit the same data stream from multiple transmit antennas, but that does not increase the SNR for that stream at the RX (without channel knowledge at the TX, the streams add up incoherently at the RX). Only in the case that the TX has full channel knowledge and $N_t > N_r$ can the TX perform beamforming, and direct the energy better toward the receive array. In that case, increasing the number of TX antennas improves the SNR, and (logarithmically) the capacity, see Figure 16.13.

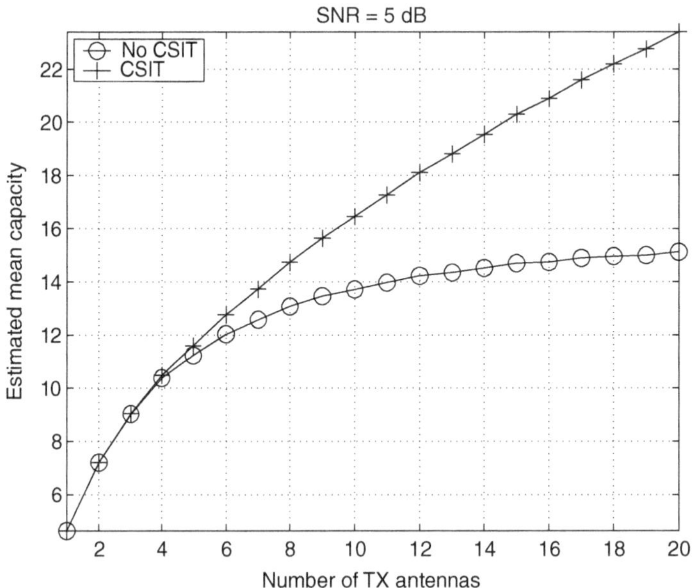

Figure 16.13 Capacity with and without CSI at the TX with $N_r = 8$ antennas and SNR $= 5$ dB.
Reproduced with permission from [Molisch and Tufvesson 2005] © Hindawi.

For the case that the number of antenna elements becomes very large, such that N_t, N_r both go to infinity, but $\zeta = M/N = \max(N_t, N_r)/ \min(N_t, N_r)$ is constant, then random matrix theory tells us that the distribution function of the eigenvalues converges to a deterministic function, so that the variations of the capacity with the different channel realizations vanishes. In this particular case of $\zeta = 1$, an exact expression for the ergodic capacity can be obtained as:

$$\frac{C_{\text{ergodic,noCSIT}}}{N} = 2\log_2\left(1 + \sqrt{4\bar{\gamma}_{\text{RX}} + 1}\right) - \frac{\log_2(e)}{4\bar{\gamma}_{\text{RX}}}\left(\sqrt{4\bar{\gamma}_{\text{RX}} + 1} - 1\right)^2 - 2. \tag{16.84}$$

16.2.5 Capacity Distributions in Flat-Fading Channels

Up to now, we have computed the ergodic capacity, where the transmission extends over many channel states. Another case of interest is the situation where the channel is fading, and we are sending codewords of a capacity-approaching code, but the codewords are shorter than the coherence time of the channel. Thus, each channel realization has its own associated capacity, which means that the capacity becomes a random variable whose distribution we aim to compute (see Section 16.2.2 for the caveats on using the word "capacity" for this). The capacity that is exceeded in $x\%$ of realizations is called the $x\%$ outage capacity.[7]

In general, the capacity distribution can be obtained by using the equations for ergodic capacity given in the previous subsection but removing the expectation over $\mathbf{H}$. The cdf of the capacity distribution is thus given as

$$cdf_C(x) = \Pr\{C_\mathbf{H} \leq x\}$$

where C is a random variable depending on the channel state.

---

[7] There are two ways to look at a capacity distribution. One is to consider a particular power allocation and precoding strategy, and analyze the cdf. Another way is to prescribe a particular outage probability, and then optimize the power allocation and precoding strategy for this particular setting (for example, we might allocate no power at all for the situations in which an outage occurs). The latter method might be useful for truly optimizing a system with minimum requirements, but is rarely used; we will henceforth not consider it further.

Exact computation of this cdf is very difficult; thus numerical simulations are often used instead: generate a number of channel realizations obeying the prescribed statistics (e.g., i.i.d.), and compute the "instantaneous capacity" for each of the realizations; then create a cdf of those results. Figure 16.14 shows the result for some interesting systems at 21 dB SNR, all assuming no CSIT and i.i.d. Rayleigh fading. The (1,1) curve describes a single-input, single-output system. The mean capacity is on the order of 6 bit/s/Hz, but the 5% outage capacity is considerably lower (on the order of 3 bit/s/Hz). When using a (1,8) system, i.e., one transmit antenna and eight receive antennas, the mean SNR increases by only about 50% – from 6 to 10 bit/s/Hz. However, the 5% outage capacity triples, from 3 to 9 bit/s/Hz. The reason for this is the much higher resistance to fading that such a diversity system has. However, when going to a (8, 8) system, i.e., a system with eight transmit and eight receive antennas, both capacities increase dramatically: the mean capacity is on the order of 46 bit/s/Hz, and the outage probability is more than 40 bit/s/Hz.

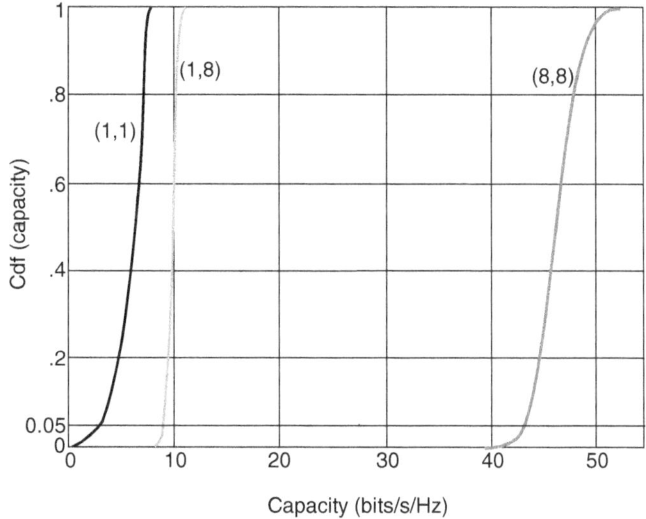

Figure 16.14 Cumulative distribution function of the capacity for a 1×1, 1×8, and the 8×8 optimum scheme. Reproduced with permission from [Foschini and Gans 1998] © Kluwer.

Also, two approximations are in widespread use

- The capacity can be well approximated by a Gaussian distribution, so that only the mean (i.e., the ergodic capacity given above) and the variance need to be computed.
- From physical considerations, the following upper and lower bounds for the capacity distribution have been derived in [Foschini and Gans 1998] for the case $N_t \geq N_r$

$$\sum_{k=N_t-N_r+1}^{N_t} \log_2\left[1 + \frac{\overline{\gamma}_{RX}}{N_t}\chi_{2k}^2\right] < C < \sum_{k=1}^{N_t} \log_2\left[1 + \frac{\overline{\gamma}_{RX}}{N_t}\chi_{2N_r}^2\right] \tag{16.85}$$

where χ_{2k}^2 is a chi-square distributed random variable with $2k$ degrees of freedom, normalized so that $E\{\chi_2^2\} = 1$.[8] Those two bounds have very clear physical interpretations. The lower bound corresponds to the capacity that can be achieved with a BLAST system; this system and its operating principle will be described below in Section 16.2.9. The upper bound corresponds to an idealized situation where there is a separate array of receive antennas for each transmit antenna, and it receives the signal in such a way that there is no interference from the other transmit streams. Figure 16.15 shows some examples. We see that the lower bound is quite tight, while the upper bound can be rather loose, in particular for large N.

16.2.6 Impact of the Channel

Up to now, we have discussed the capacity of a channel with flat fading and independent, identically distributed (i.i.d.) zero-mean complex Gaussian coefficients. Real channels are more complicated, and the deviations of the channel from the idealized assumptions can have a significant impact on the capacity. In the following, we will describe some of the more important effects.

---

[8] The above equation is a slight abuse of notation, indicating that the cdf of the capacity is bounded by the cdfs of the random variables given by the equations on the left and right side.

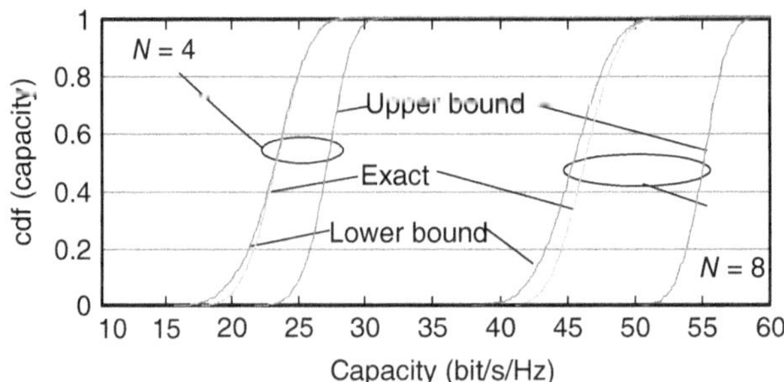

Figure 16.15   Exact capacity, upper bound, and lower bound of MIMO system in an i.i.d. channel at 20 dB SNR, with $N_r = N_t = N$ equal to 4 and 8. Reproduced with permission from [Molisch and Tufvesson 2005] © Hindawi.

Channel Correlation

Correlation of the signals at the different antenna elements can significantly reduce the capacity of a MIMO system. This can be understood in the following way: the capacity is determined by the distribution of the singular values of the channel matrix. Consider now the case of high SNR and $N_r = N_t$: for a given SNR, the maximum capacity is achieved when the channel transfer matrix has full rank and the singular values of $\mathbf{H}$ are equally strong. If the coefficients of the channel matrix are i.i.d. Rayleigh fading, then this situation is *approximately* fulfilled (though the ordered eigenvalues still have different values). But if the fading of the channel coefficients is correlated, then the eigenvalue spread, i.e., the difference between the largest and the smallest eigenvalues (squared singular values), becomes much bigger. Remember that, at high SNR, since the capacity is proportional to the logarithm of the SNR, two eigenvalues (SNRs) that are similar to each other result in a much higher sum capacity than a very high and a very low eigenvalue. Thus, for the sum of the eigenvalues of $\mathbf{H}\mathbf{H}^\dagger$ constant, i.e., constant average SNR, a higher eigenvalue spread results in a lower sum capacity.

The correlation is influenced by the angular spectrum of the channel as well as the arrangement and spacing of the antenna elements, see Section 12.2. For antennas that are spaced half a wavelength apart, a uniform APS leads approximately to a decorrelation of the received signals. A smaller angular spread of the channel leads to an increase of the correlation. Since we are now looking at a MIMO system, we have to consider the correlation both at the TX and at the RX. A popular model (the so-called Kronecker model, see Section 7.5.3) assumes that the correlation at the TX is independent of the correlation of the RX. A realization of the channel matrix can then be obtained as

$$\mathbf{H}_{\mathrm{kron}} = \mathbf{R}_{\mathrm{RX}}^{1/2}\mathbf{G}_{\mathrm{G}}\mathbf{R}_{\mathrm{TX}}^{1/2} \tag{16.86}$$

where $\mathbf{G}_{\mathrm{G}}$ is a matrix with i.i.d. complex Gaussian entries and $\mathbf{R}_{\mathrm{TX}}$ and $\mathbf{R}_{\mathrm{RX}}$ are the correlation matrices at the TX and RX, respectively. The model implies that the correlation matrix at the RX is independent of the direction of transmission and vice versa.

In order to gain analytical insights, we first assume no CSIT and thus, equal power allocation to all TX antennas. When further considering the high-SNR case, the ergodic capacity[9] becomes

$$C_{\mathrm{corr}} = E_{\mathbf{H}}\left\{ \log_2\left[\det\left(\frac{\gamma_{\mathrm{RX}}}{N_t}\mathbf{G}_{\mathrm{G}}\mathbf{G}_{\mathrm{G}}^\dagger \right) \right] \right\} + \log_2[\det(\mathbf{R}_{\mathrm{RX}})] + \log_2[\det(\mathbf{R}_{\mathrm{TX}})] \; . \tag{16.87}$$

Note that $\det(\mathbf{R}_{\mathrm{TX}}) \leq 1$, with equality being reached when $\mathbf{R}_{\mathrm{TX}}$ is an identity matrix. The stronger the correlation (and thus, the more skewed the eigenvalues of $\mathbf{R}_{\mathrm{TX}}$ are), the bigger is the loss in capacity. The same argument holds for $\mathbf{R}_{\mathrm{RX}}$.

Analytical computation of the capacity distribution is even more complicated in the case of correlated channels than for i.i.d. channels. Generally, it is easier to obtain simulation results, by generating realizations of the channel matrix from Eq. (16.86), and then inserting them into Eqs. (16.69) or (16.64). Figure 16.16 shows the ergodic capacity of a 4×4 MIMO system with uniform linear arrays at TX and RX as a function of the angular spread at one link end.[10] We see that a small angular spread leads to a drastic reduction of the capacity. Another topic of importance in particular in correlated channels is the proper rank adaptation, i.e., choice of N_s, since performance suffers if N_s is larger than the "effective" rank.

For the case of correlated channels, there is a significant difference in the capacity between no CSIT and second-order CSIT, as in the latter case, the TX can beamform into the on-average strongest directions. Specifically, let us write an eigendecomposition of the correlation matrix,

$$\mathbf{R}_{\mathrm{TX}} = \mathbf{U}_{\mathbf{R}_{\mathrm{TX}}}\mathbf{\Lambda}_{\mathbf{R}_{\mathrm{TX}}}\mathbf{U}_{\mathbf{R}_{\mathrm{TX}}}^\dagger. \tag{16.88}$$

---

[9] This is under the assumption that the TX transmits uniformly in all directions. When the second-order statistics, but no instantaneous CSIT is available, better performance can be achieved, see below.

[10] Note that this figure is based on the idealized assumption that there is no mutual coupling between the antenna elements; investigations have shown that mutual coupling influences the capacity by introducing pattern diversity as well as changing the average power received in the different antenna elements.

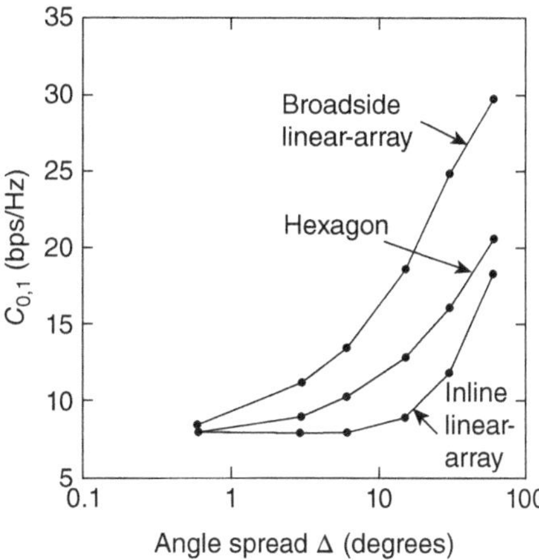

Figure 16.16 10% outage capacity as a function of BS angular spread for one-ring model (maximum angle between direction from UE and direction from IO is Δ. SNR is 18 dB, 7×7 MIMO system, antenna spacing is half a wavelength.
Reproduced with permission from [Shiu et al. 2000] © IEEE.

Then the optimum input covariance matrix can be written as

$$\mathbf{R}_{ss} = \mathbf{U}_{\mathbf{R}_{\mathrm{TX}}} \mathbf{\Lambda}_{\mathrm{bf}} \mathbf{U}_{\mathbf{R}_{\mathrm{TX}}}^{\dagger} \tag{16.89}$$

where $\mathbf{\Lambda}_{\mathrm{bf}}$ is a diagonal matrix whose entries are the power allocations to the different eigenmodes. For the case where $\mathbf{R}_{\mathrm{RX}} = \mathbf{I}$, i.e., correlation exists only at the TX link end, the optimum $\mathbf{\Lambda}_{\mathrm{bf}}$ can be approximated by waterfilling along the entries of $\mathbf{\Lambda}_{\mathbf{R}_{\mathrm{TX}}}$ (while this method is not exact, and numerical optimization can improve results further, it is often good enough). For the case of general $\mathbf{R}_{\mathrm{RX}}$, numerical optimization of the elements of $\mathbf{\Lambda}_{\mathrm{bf}}$ is unavoidable.

At low SNRs, only one eigenmode should be used for transmission. While this is generally true, the existence of transmit correlation raises the corresponding SNR threshold below which such strategies should be applied. This is again intuitive, since for high correlation, the eigenvalue distribution of the channel is more skewed, and therefore the second-strongest eigenmode does not have enough SNR even when the average SNR is higher.

Frequency-Selective Channels

The previous sections assumed frequency-flat channels. Fortunately, the generalization to the frequency-selective case is straightforward. As Shannon has shown, the use of an OFDM-like scheme is optimum for dealing with the frequency-selective channel, converting it into a number of parallel flat-fading channels. Thus, the capacity per unit bandwidth is an integration of flat-fading capacity over frequency. For no CSIT and with frequency-independent TX power, it becomes:

$$C = \frac{1}{B} \int_B \log_2 \left[\det \left(\mathbf{I}_{N_i} + \frac{\gamma_{\mathrm{TX}}}{N_{\mathrm{t}}} \mathbf{H}(f) \mathbf{H}(f)^{\dagger} \right) \right] df \tag{16.90}$$

where B is the bandwidth of the considered system.

This equation also implies that frequency selectivity offers additional diversity that increases the slope of the capacity cdf. If one of the frequency subchannels shows poor capacity, there is a good chance that another subchannel has high quality. Figure 16.17 shows an example of a measured capacity cdf in a microcellular environment. We see that the capacity cdf becomes steeper as the bandwidth increases.

LOS vs. NLOS

In some situations, there is a line-of-sight connection between TX and RX, resulting in different fading statistics. As we have seen in Section 5.5, the fading statistics of a SISO link becomes Ricean instead of Rayleigh. For a MIMO system, the channel matrix can be written as

$$\widetilde{\mathbf{H}} = \sqrt{\frac{K_{\mathrm{LOS}}}{K_{\mathrm{LOS}} + 1}} \hat{\mathbf{H}}_{\mathrm{LOS}} + \sqrt{\frac{1}{K_{\mathrm{LOS}} + 1}} \hat{\mathbf{H}}_{\mathrm{res}} \tag{16.91}$$

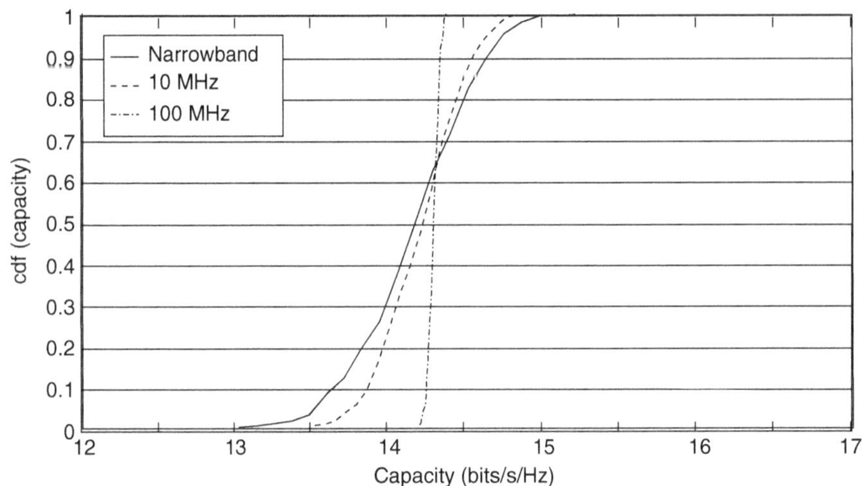

Figure 16.17 Capacity in a measured microcellular channel, for different system bandwidths: 4×4 MIMO system. Reproduced with permission from [Molisch et al. 2002b] © IEEE.

where K_{LOS} is the Rice factor, i.e., ratio of the powers in LOS and residual components, $\hat{\mathbf{H}}_{\mathrm{LOS}}$ is a purely deterministic matrix, and $\hat{\mathbf{H}}_{\mathrm{res}}$ has (uncorrelated or correlated) zero-mean Gaussian entries.[11] If the distance between TX and RX is large (much larger than the Rayleigh distance, see Sections 4.1 and 8.1.4), the LOS gives rise to a matrix $\hat{\mathbf{H}}_{\mathrm{LOS}}$ that has rank one (a single wave can only be associated with one singular value of the channel matrix!). This in turn implies that the singular value spread of the matrix Eq. (16.91) is much larger than for a NLOS matrix. Consequently, the capacity of a LOS channel is lower than for a NLOS channel *when assuming equal SNR*. One should however note that the SNR is often better in the LOS case compared to the NLOS case. For a power limited scenario with realistic channels, the LOS case often gives the highest capacity, despite the imbalance between the singular values.

Example 16.5 *Capacity in a channel with LOS. The transmit and receive arrays are uniform linear arrays with element spacing λ between the elements and $N_r = N_t = 8$. The arrays are perpendicular to the LOS connection. The directions of arrival of the NLOS components is uniformly distributed between 0 and π. Estimate the mean capacity if the TX does not have CSI for an SNR of 20 dB, and $K_{LOS} = 0$ and 20 dB.*

In a first step, we have to determine the channel matrix. Since the transmit and receive arrays are linear arrays that are oriented perpendicular to the LOS, we find that $\hat{\mathbf{H}}_{\mathrm{LOS}}$ is the all-ones matrix if transmit and receive array are sufficiently far apart from each other. Furthermore, $\hat{\mathbf{H}}_{\mathrm{res}}$ has unit-energy, i.i.d. complex Gaussian entries because the angular spectrum is uniformly distributed and the antenna elements are more than $\lambda/2$ apart from each other. From Eq. (16.69), the capacity is then

$$C = \log_2\left[\det\left(\mathbf{I}_{N_r} + \frac{\gamma_{\mathrm{TX}}}{N_t}\mathbf{H}\mathbf{H}^{\dagger}\right)\right] \tag{16.92}$$

$$= \log_2\left[\det\left(\mathbf{I}_{N_r} + \frac{\gamma_{\mathrm{RX}}}{N_t}\left[\frac{K_{\mathrm{LOS}}}{K_{\mathrm{LOS}}+1}\hat{\mathbf{H}}_{\mathrm{LOS}}\hat{\mathbf{H}}_{\mathrm{LOS}}^{\dagger}\right.\right.\right.$$
$$\left.\left.\left. + \frac{\sqrt{K_{\mathrm{LOS}}}}{K_{\mathrm{LOS}}+1}\left(\hat{\mathbf{H}}_{\mathrm{res}}\hat{\mathbf{H}}_{\mathrm{LOS}}^{\dagger} + \hat{\mathbf{H}}_{\mathrm{LOS}}\hat{\mathbf{H}}_{\mathrm{res}}^{\dagger}\right) + \frac{1}{K_{\mathrm{LOS}}+1}\hat{\mathbf{H}}_{\mathrm{res}}\hat{\mathbf{H}}_{\mathrm{res}}^{\dagger}\right]\right)\right]. \tag{16.93}$$

Using Jensen's inequality, the expected value of the capacity can be approximated as (compare [Ayadi et al. 2002])

$$E\{C\} \leq \log_2\left[\det\left(\mathbf{I}_{N_r} + \frac{\gamma_{\mathrm{RX}}}{N_t}\left[\frac{K_{\mathrm{LOS}}}{K_{\mathrm{LOS}}+1}E\{\hat{\mathbf{H}}_{\mathrm{LOS}}\hat{\mathbf{H}}_{\mathrm{LOS}}^{\dagger}\}\right.\right.\right.$$
$$\left.\left.\left. + \frac{\sqrt{K_{\mathrm{LOS}}}}{K_{\mathrm{LOS}}+1}E\{\hat{\mathbf{H}}_{\mathrm{res}}\hat{\mathbf{H}}_{\mathrm{LOS}}^{\dagger} + \hat{\mathbf{H}}_{\mathrm{LOS}}\hat{\mathbf{H}}_{\mathrm{res}}^{\dagger}\} + \frac{1}{K_{\mathrm{LOS}}+1}E\{\hat{\mathbf{H}}_{\mathrm{res}}\hat{\mathbf{H}}_{\mathrm{res}}^{\dagger}\}\right]\right)\right] \tag{16.94}$$

$$= \log_2\left[\det\left(\mathbf{I}_{N_r} + \frac{\gamma_{\mathrm{RX}}}{N_t}\left[\frac{K_{\mathrm{LOS}}N_t}{K_{\mathrm{LOS}}+1}\mathbf{1} + \frac{1}{K_{\mathrm{LOS}}+1}E\{\hat{\mathbf{H}}_{\mathrm{res}}\hat{\mathbf{H}}_{\mathrm{res}}^{\dagger}\}\right]\right)\right] \tag{16.95}$$

$$= \log_2\left[\det\left(\left[1 + \frac{\gamma_{\mathrm{RX}}}{K_{\mathrm{LOS}}+1}\right]\mathbf{I}_{N_r} + \left[\gamma_{\mathrm{RX}}\frac{K_{\mathrm{LOS}}}{K_{\mathrm{LOS}}+1}\mathbf{1}\right]\right)\right] \tag{16.96}$$

---

[11] Both $\hat{\mathbf{H}}_{\mathrm{LOS}}$ and $\hat{\mathbf{H}}_{\mathrm{res}}$ are normalized to $E\{\|\hat{\mathbf{H}}\|_F^2\} = N_t N_r$, where $\|\hat{\mathbf{H}}\|_F$ is the Frobenius norm of $\hat{\mathbf{H}}$.

where $\mathbf{1}$ is a $N_r \times N_r$ matrix where each entry is 1; such a matrix has only a single nonzero eigenvalue, whose magnitude is N_r. Using some further approximations, we finally get at high SNR

$$E\{C\} \simeq \log_2\left[1 + \frac{\gamma_{\mathrm{RX}} K_{\mathrm{LOS}} N_r}{K_{\mathrm{LOS}} + 1}\right] + (N_r - 1)\log_2\left[1 + \frac{\gamma_{\mathrm{RX}}}{K_{\mathrm{LOS}} + 1}\right]. \tag{16.97}$$

Using Eq. (16.97), we find that the capacity for $K_{\mathrm{LOS}} = 0$ and 20 dB is 48 and 17 bit/s/Hz, respectively. Note that the above is a rather crude approximation; Monte Carlo simulations give 40 and 15 bit/s/Hz, respectively.

In case that $\hat{\mathbf{H}}_{\mathrm{res}}$ is i.i.d., and second-order CSIT is available, then the TX should simply beamform along the direction of the LOS. This can be shown mathematically, but it is also intuitive – since no direction is preferable for the $\hat{\mathbf{H}}_{\mathrm{res}}$ part, we should beamform to make a difference in the $\hat{\mathbf{H}}_{\mathrm{LOS}}$, i.e., along the direction of the LOS.

It is also noteworthy that $\hat{\mathbf{H}}_{\mathrm{LOS}}$ has unit rank only if the LOS component is a *plane* wave. A *spherical* wave leads to a transfer function matrix that can have full rank if the antenna elements are spaced appropriately. The curvature of the waves is noticeable up to one Rayleigh distance, i.e., typically a few meters.

For a more detailed investigation, consider the geometry of Figure 16.18. Let the elements at TX and RX be spaced d_{TX} and d_{RX}, respectively, and assume that $N_t = N_r = N$. The distance r_{ji} from the jth TX to the ith RX antenna element can be obtained, for large R, through a Taylor expansion of the actual distances (which follow from elementary trigonometry)

$$\begin{aligned} r_{ji} = R &+ jd_{\mathrm{RX}}\sin(\theta_{\mathrm{RX}})\cos(\phi_{\mathrm{RX}}) - id_{\mathrm{TX}}\sin(\theta_{\mathrm{TX}}) + \\ &\frac{[jd_{\mathrm{RX}}\sin(\theta_{\mathrm{RX}})\sin(\phi_{\mathrm{RX}})]^2 + [jd_{\mathrm{RX}}\cos(\theta_{\mathrm{RX}}) - id_{\mathrm{TX}}\cos(\theta_{\mathrm{TX}})]^2}{2R}. \end{aligned} \tag{16.98}$$

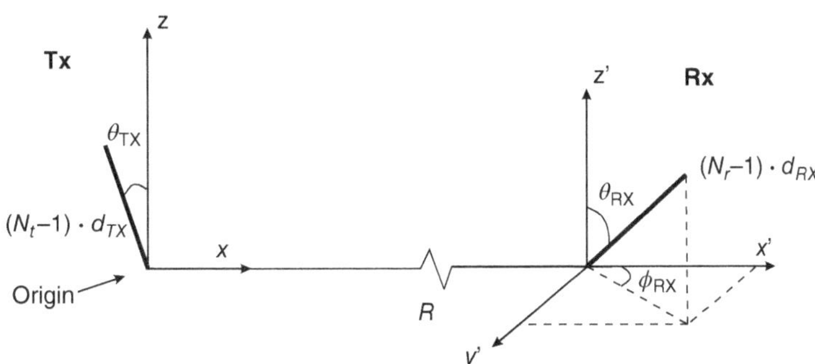

Figure 16.18 Geometry for LOS MIMO. The TX array is a ULA in the x–z plane; the RX array is a ULA with arbitrary direction. Reproduced with permission from [Bohagen et al. 2007] © IEEE.

It then follows that the eigenmodes of the transmission are orthogonal, and the eigenvalues are equal, if

$$d_{\mathrm{TX}} d_{\mathrm{RX}} = \frac{\lambda_c R}{M\cos(\theta_{\mathrm{TX}})\cos(\theta_{\mathrm{RX}})}. \tag{16.99}$$

For the special case $d_{\mathrm{TX}} = d_{\mathrm{RX}}$ and $\cos(\theta_{\mathrm{TX}}) = \cos(\theta_{\mathrm{RX}}) = 1$, this requires $d_{\mathrm{TX}} = \sqrt{\lambda_c R/N}$, or conversely, it can be achieved only for $R \le N d_{\mathrm{TX}}^2/\lambda_c$ (note that this expression differs from the Rayleigh distance as the numerator has only N, not N^2 as would show up in the Rayleigh distance of an array). For 2 GHz carrier frequency and a 0.6 m long array, and $M = 4$ (and thus $d_{\mathrm{TX}} = 0.15$ m), this implies that $R \le 1.2$ m is required. At 60 GHz carrier frequency, 36 m can be achieved, which is somewhat more realistic for, e.g., communications in data centers, or backhaul from picocell BSs. Of course, larger distances can be used, but at the price of a worse ratio of eigenvalues of the different modes, thus reducing the capacity.

Limited Number of IOs

In the intuitive picture of Section 16.2.1, we have already seen that a sufficient number of IOs are required to act as relays for the data streams. Relating this to the mathematics of the previous section, we note that a sufficient number of IOs are needed in order to guarantee that the channel coefficients in the matrix $\mathbf{H}$ are independent. While the number of existing IOs is always large, the number of *significant* IOs might be limited in practice. After all, IOs that are too weak to provide appreciable SNR (and thus capacity) are not useful in carrying data streams.

Figure 16.19 shows some measurement results for the capacity as a function of the number of antenna elements. Measurements were taken in a microcellular scenario where the number of IOs was rather small. The capacity does not increase linearly with the number of antenna elements N when N exceeds 4. This is a clear sign that the number of IOs starts to put a limitation on the achievable capacity.

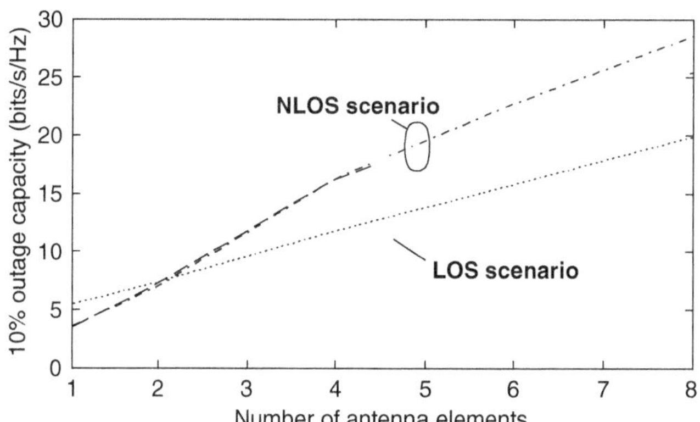

Figure 16.19   10% outage capacity as a function of the number of antenna elements at TX and RX in measured LOS and NLOS scenario. Reproduced with permission from [Molisch et al. 2002b] © IEEE.

Keyhole Channels

There are some special cases where the capacity is low even though the signals at the antenna elements are uncorrelated. These cases are often referred to as *keyholes* or *pinholes*. An example of a keyhole scenario is a rich scattering environment at both the TX and RX side; between them, there is only one propagation path that has only one degree of freedom. Such a scenario can occur when TX and RX are surrounded by IOs, but those IOs groups are separated by a long stretch of empty space (green field). Another scenario is the case where the IO areas are connected by a single-moded waveguide or by a diffraction edge. In all those cases, the total transfer function can be written as

$$\mathbf{H} = \mathbf{R}_{\mathrm{RX}}^{1/2}\mathbf{G}_{\mathrm{G},1}\mathbf{R}_{\mathrm{RX\text{-}TX}}^{1/2}\mathbf{G}_{\mathrm{G},2}\mathbf{R}_{\mathrm{TX}}^{1/2} \qquad (16.100)$$

where $\mathbf{G}_{\mathrm{G},1}$ and $\mathbf{G}_{\mathrm{G},2}$ are both i.i.d. complex Gaussian matrices, and where $\mathbf{R}_{\mathrm{RX\text{-}TX}}$ describes the correlation of the channel matrix between the TX and RX environments; in a keyhole channel, this is a low-rank matrix. It can also be seen from this description that the statistics of the entries of $\mathbf{H}$ are not Gaussian anymore – this explains why it can be possible to have low correlation at the TX, low correlation at the RX, and low capacity at the same time. It should, however, be noted that keyhole channels occur very seldom in practice. While [Almers et al. 2006] measured one in a controlled environment, it seems to be a rare occurrence in "normal" environments.

Diversity-Multiplexing Trade-Off

MIMO systems can be used to achieve SM, diversity, and/or beamforming. However, it is not possible to reach all of those goals simultaneously to their full extent. Firstly, there is a trade-off between beamforming and diversity gain; this trade-off also depends on the operating environment. Consider first a LOS scenario. In that case, it is obvious that the achievable beamforming gain is $N_t N_r$: beams at the TX (with gain N_t) and at the RX (with gain N_r), should point at each other; the gains thus multiply. On the other hand, there is obviously no diversity gain, since there is no fading in a LOS scenario: in other words, the slope of the SNR distribution curve does not change.

In a heavily scattering environment, the slope of the SNR distribution changes drastically due to the use of the multiple antenna elements. As established in Section 16.1.6, it is easy to make sure that *at least one* receive signal has good quality. Define the diversity order as the slope of the uncoded bit error probability for very high SNRs

$$d_{\mathrm{div}} = -\lim_{\bar{\gamma}\to\infty}\frac{\log\left[BER(\bar{\gamma})\right]}{\log\left(\bar{\gamma}\right)}. \qquad (16.101)$$

Alternatively, the diversity can also be defined as the slope of the curve of the outage vs. SNR, which makes it independent of any particular modulation format

$$d_{\mathrm{div}} = -\lim_{\bar{\gamma}\to\infty}\frac{\log\left[P_{\mathrm{out}}(\bar{\gamma})\right]}{\log\left(\bar{\gamma}\right)}. \qquad (16.102)$$

The diversity order in a heavily scattering environment can be shown to be $N_t N_r$. On the other hand, it also turns out that the maximum beamforming gain in such a heavily scattering (i.i.d.) environment is upper-limited by $\left(\sqrt{N_t} + \sqrt{N_r}\right)^2$, for large N_t and N_r. The reason is that it is not possible to form the transmit beam pattern in such a way that the MPCs overlap constructively *at all receive antenna elements simultaneously*. Note the key difference here between achieving full diversity order (which requires to have a good signal at least one receive antenna element) and beamforming gain (which requires good signal quality at all receive antenna elements).

There is also a fundamental trade-off between SM and diversity. [Zheng and Tse 2003] showed that the optimum trade-off curve between the diversity order and the normalized rate r_{MIMO} is piecewise linear, connecting the points

$$d_{\mathrm{div}}(r) = (N_t - r_{\mathrm{MIMO}})(N_r - r_{\mathrm{MIMO}}) \qquad r_{\mathrm{MIMO}} = 0, ..., \min(N_t, N_r) \ . \tag{16.103}$$

This implies that maximum diversity order, $N_t N_r$, and maximum rate, $\min(N_t, N_r)$, cannot be achieved simultaneously.

*16.2.7 Channel Estimation

We have assumed throughout the previous discussion that the RX knows the CSI. Habitually, this information is obtained from pilot signals. We now turn to the question of how many pilot signals are needed, and how that number impacts the spectral efficiency.

First consider the case of flat-fading channels. Since the RX needs to estimate the channel from each TX antenna element, and these pilots need to be orthogonal to each other, N_t/T_{coh} pilot symbols per second are required (more precisely, following the sampling theorem in the time domain, $N_t 2\nu_{\max}$ symbols per second are required). Let these symbols represent a fraction α_{pilot} of all the transmitted symbols, then the capacity is reduced by a factor $(1 - \alpha_{\mathrm{pilot}})$.

Generalizing to the case of frequency-selective channels, and assuming that OFDM with subcarrier spacing Δf is used, a fraction $\Delta f/B_{\mathrm{coh}}$ of subcarriers are needed as pilot signals. Taking into account the relationship between OFDM symbol duration and subcarrier spacing, the overall fraction of spectral resources that is needed for pilot signals is $N_t/(T_{\mathrm{coh}} B_{\mathrm{coh}})$ (or, more precisely, $2N_t \nu_{\max}\tau_{\max}$). Coherence times are typically on the order of 1 ms or more, and coherence bandwidths 100 kHz or more, so that it would seem $1/T_{\mathrm{coh}}B_{\mathrm{coh}}$ is less than 1%, making the overhead small for $N_t \lesssim 10$. However, due to recent trends for increasing the number of antenna elements at base stations to dozens or even hundreds, methods need to be devised to reduce the pilot overhead.

The key for improving the overhead is to exploit the correlation of the signals at the antenna elements, or equivalently, the limited angular spread. In this respect, the channel estimation for MIMO is very similar to that of OFDM. Due to the Fourier duality between angle (or rather directional cosine) and space, a limited support in the angle domain corresponds to correlation of the signals at the antenna elements, just like in OFDM, the limited support of the impulse response corresponds to frequency correlation. Knowing the support, or even better, the complete second-order statistics thus allows to either improve the performance of the channel estimator or reduce the amount of pilot signals.

Consider the situation where the TX has a Uniform Linear Array (ULA) with N_t antenna elements, such that an FFT forms beams into different directions (compare section titled "Switched-Beam Receivers" in 16.1.3). If the signal incident at the TX is "seen" only by two beams, we know that two coefficients can describe the CSI – it is obviously two coefficients in the Fourier (transform) domain, but due to the nature of the Fourier transform, it implies the same thing for the representation in the antenna domain. Thus, instead of sending out N_t pilot signals, it is sufficient to send two. Thus, with a limited angular support (and suitable antenna spacing), we can save on the number of pilot signals, and thus improve spectral efficiency.

Let us now turn to the best estimator when the correlation matrix is known. Writing the estimated channel matrix $\hat{\mathbf{H}}$ as a vector (by stacking all the columns), we get

$$vec(\hat{\mathbf{H}}) = \left(E_p \left[\mathbf{C}_p \mathbf{C}_p^{\dagger} \right]^T \otimes \mathbf{I}_{N_r} + \sigma_n^2 \left(\mathbf{R}_H^T \right)^{-1} \right)^{-1} \sqrt{E_p} \left(\mathbf{C}_p^T \otimes \mathbf{I}_{N_r} \right)^{\dagger} vec(\mathbf{r}_p) \tag{16.104}$$

where E_p is the energy of the pilot symbol, $\mathbf{C}_p$ the pilot sequence, $\mathbf{R}_H$ the correlation matrix, and $\mathbf{r}_p$ the $N_t N_r$ receive signals obtained when the pilot sequence is sent out (remember that a pilot symbol is sent from each TX antenna and received at every RX antenna). This structure is very similar to that of OFDM (see Eq. (15.19)) if the training sequence fulfills $\mathbf{C}_p \mathbf{C}_p^{\dagger} = \mathbf{I}_{N_t}$. The performance can be improved if the training sequences are specifically matched to the correlation matrix [Zhang et al. 2005]; however, this is not done in practice because it would require significant signaling effort between TX and RX, which both need to know the training sequences.

Another important aspect of pilot signals is that they are affected by noise and interference, so that CSI is not perfect. A certain MSE (often approximated as the inverse SNR $1/\gamma_{\mathrm{ce}}$ of the channel estimate) causes the nominal SNR γ to degrade to an effective value γ_{eff}

$$\gamma_{\mathrm{eff}} = \frac{1 - 1/\gamma_{\mathrm{ce}}}{(1/\gamma) + (1/\gamma_{\mathrm{ce}})}. \tag{16.105}$$

If the number of pilot symbols corresponds to critical sampling, i.e., there is one pilot symbol per transmit antenna element, coherence time, and coherence bandwidth, and the energy of a pilot symbol equals the energy of a data symbol, then $\gamma_{\mathrm{ce}} = \gamma$. This leads to an approximate 3dB loss in effective SNR. The loss can be mitigated by increasing the number of pilot symbols (thus leading, effectively, to noise averaging); however, this increases the overhead, i.e., more spectral resources need to be assigned to pilot signals, which is not desirable. Rather, it is preferable to increase the energy of the pilot symbols compared to the data symbols. If the overall power is limited, this implies a reduction of the data symbol SNR, but since (in the high SNR regime) capacity decreases only logarithmically with shrinking SNR, it is preferable compared to the use of more pilot signals. Care must be taken, however, when the pilot impairments are dominated not by noise but by adjacent-cell interference (see Section 21.1). If all cells are synchronized such that pilot symbols are transmitted at the same time and the same subcarriers, then pilot boosting does not improve γ_{eff}. Thus, a suitable "staggering" of the pilot symbols in different cells need to be implemented, so that pilot symbols in one cell are interfered by data symbols in another cell, rather than pilot symbols from another cell.

16.2.8   Feedback for Spatial Multiplexing

Just like in the case for beamforming for smart antennas, i.e., MISO (see Section 16.1.5), also for precoding in SM, we face the problem of obtaining the necessary CSI at the TX. In TDD systems, reciprocity can be used (and the same principles hold as in Section 16.1.7). In FDD, feedback from the RX to the TX is required, and just like in the MISO case, we need to find quantized codebooks to keep the overhead for the feedback within reasonable bounds. Finding good codebooks is obviously a generalization of the MISO problem, since we now need to find a codebook of quantized precoding matrices $N_t \times N_s$ from which the RX selects the desired one, and feeds back the codebook index. The problem thus becomes that of Grassmannian subspace packing (instead of line packing, as in the MISO case). Again, the goal is to maximize the minimum distance between two subspaces; an added difficulty is to define a good measure for this distance; and, furthermore, what constitutes a "good" measure depends on the particular RX type. For Maximum-Likelihood (ML) RXs, it turns out that the optimum codebook design relates to finding precoding matrices with the largest projection two-norm. However, in practice, Grassmannian and especially Fourier codebooks of Section 16.1.5 are still widely used: an example for the performance is shown in Figure 16.20.

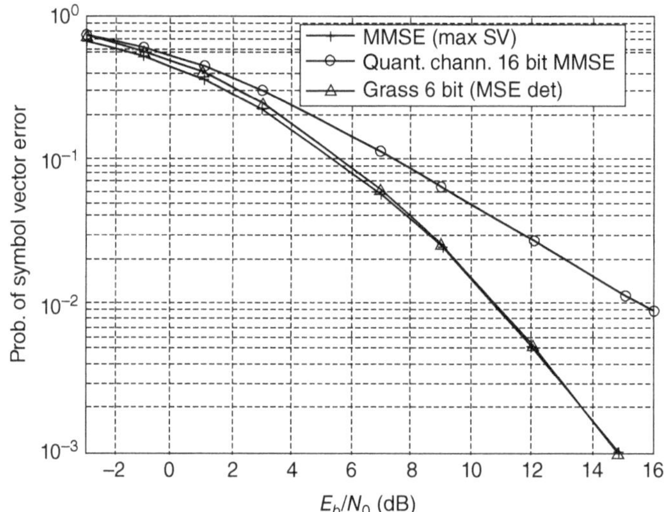

Figure 16.20   Symbol vector error as function of SNR when channel feedback is used, and the detector is MMSE. Ideal feedback (max SV); 16-bit quantization of the channel, and Grassmannian codebook with 6 bits.
Reproduced with permission from [Tsoulos 2006] © CRC Press.

16.2.9   Receiver Structures

The discussion up to now was restricted to the information-theoretic limits of MIMO systems. A big question is how to realize those capacities in practice. This section thus will discuss various RX structures that trade-off complexity vs. accuracy. Interestingly, the basic principles are the same as the ones discussed for temporal equalizers in Chapter 14: linear RXs, decision-feedback equalizers (here called BLAST), and ML. This similarity is not surprising, since there is a strong similarity in the formal mathematical description of frequency-selective and spatially selective channels. We will in this section consider frequency-flat channels; for frequency-selective channels and OFDM signaling, the operations can be performed on every subcarrier.

Linear Receivers

Linear RXs multiply the received signal with a linear filter (represented by a matrix), to separate the different data streams and allow their easy detection.

In the case that full CSI is available at the TX, the transmitted signal vector can be precoded in a way that makes a linear RX optimum. As discussed at length in Section 16.2.3, the transmitted and received signal vectors are multiplied with the right and left singular vectors of the channel matrix, respectively, thus providing a diagonalization of the channel. Therefore, the different data streams at the output of the linear RX do not interfere with each other and can be detected separately.

In the case of no CSIT, linear RXs still can be used, though they are, in general, far from optimal.[12] Assuming that each transmit stream is sent from a particular TX antenna, separation of the streams can be achieved by a linear receive filter $\mathbf{W}^{\dagger}$ that inverts the channel matrix

$$\mathbf{W}^{\dagger} = \mathbf{H}^{+} = \begin{cases} \left(\mathbf{H}^{\dagger}\mathbf{H}\right)^{-1}\mathbf{H}^{\dagger} & \text{if} \quad N_{\mathrm{r}} \geq N_{\mathrm{t}} \\ \mathbf{H}^{\dagger}\left(\mathbf{H}\mathbf{H}^{\dagger}\right)^{-1} & \text{if} \quad N_{\mathrm{r}} \leq N_{\mathrm{t}} \end{cases} \tag{16.106}$$

possibly normalized to ensure correct TX power, where $^{+}$ denotes the Moore–Penrose pseudo-inverse;[13] note that stream separation by linear means is not possible when $N_{\mathrm{r}} < N_{\mathrm{s}}$. If the transmit signal is precoded with a precoding matrix $\widetilde{\mathbf{T}}$, then the receive filter needs to be computed from the "effective" channel $\mathbf{H}\widetilde{\mathbf{T}}$ instead of from $\mathbf{H}$.

This approach is obviously a Zero-Forcing (ZF) RX in the spatial domain. Like all ZF-RXs, it suffers from noise enhancement, and the amount of noise enhancement depends on the condition number of the channel matrix. The SNR at each of the RX antenna elements can be lower bounded by

$$\gamma_{\mathrm{TX}}\sigma_{\mathrm{min}}^{2}\left(\mathbf{H}\widetilde{\mathbf{T}}\right). \tag{16.107}$$

In an i.i.d. Rayleigh fading channel, the expected value of the RX output SNR becomes

$$\frac{N_{\mathrm{r}} - N_{\mathrm{t}} + 1}{N_{\mathrm{t}}}\gamma_{\mathrm{RX}} \tag{16.108}$$

which indicates bad effective SNR in particular when the number of used TX and RX antennas is the same. For $N_{\mathrm{r}} = N_{\mathrm{t}} = N$, the performance penalty compared to an optimum RX can be expressed in the high SNR regime as an increase in L_{∞} in (16.67)

$$\log_{2}(e) \sum_{k=2}^{N} \frac{1}{k}. \tag{16.109}$$

When the SNR is low, the performance penalty is even more significant.

Furthermore, the diversity-multiplexing trade-off is worse, with the achievable diversity order limited to, for $N_{\mathrm{r}} \geq N_{\mathrm{t}}$

$$d = (N_{\mathrm{r}} - N_{\mathrm{t}} + 1)\max\left[1 - \frac{r}{N_{\mathrm{t}}}, 0\right]. \tag{16.110}$$

This is also reflected in the approximate expression for the distribution of the achievable capacity at high SNR in an i.i.d. channel

$$C_{\mathrm{ZF}} \approx N\log_{2}\left(\frac{\overline{\gamma}_{\mathrm{RX}}}{N_{\mathrm{t}}}\right) + N\mathrm{E}_{\mathbf{H}}\left\{\log_{2}\left(\chi_{2(N_{\mathrm{r}} - N_{\mathrm{t}} + 1)}^{2}\right)\right\} \tag{16.111}$$

where the second term on the r.h.s. again indicates the achievable diversity.

A somewhat better performance can be achieved by an LMMSE RX, which minimizes the mean-square error, in the process allowing some inter-stream interference to obtain the advantage of reduced noise enhancement. This principle is completely analogous to the discussion of ZF vs. MMSE equalizers in Sections 14.2 and 14.3. The RX filter is

$$\mathbf{W} = \left(\mathbf{H}\widetilde{\mathbf{T}}\widetilde{\mathbf{T}}^{\dagger}\mathbf{H}^{\dagger} + \frac{1}{\gamma_{\mathrm{TX}}}\mathbf{I}\right)^{-1}\mathbf{H}\widetilde{\mathbf{T}}. \tag{16.112}$$

For a given channel realization, the achievable MSE for the jth stream is

$$MSE_{j} = \left[\left(\left(\gamma_{\mathrm{TX}}\widetilde{\mathbf{T}}^{\dagger}\mathbf{H}^{\dagger}\mathbf{H}\widetilde{\mathbf{T}} + \mathbf{I}\right)^{-1}\right)\right]_{j,j} \tag{16.113}$$

---

[12] There are some specific cases, e.g., certain massive MIMO configurations, where they are optimal, compare Section 22.8.

[13] We are sometimes using $[x]^{+}$ to indicate $[max(x), 0]$; it is clear from the context what is meant.

and the achievable SNR follows as $(1 - MSE_j)/MSE_j$. From this, closed-form equations can be derived for the capacity in i.i.d. Rayleigh fading channels.

While linear RXs were popular in the early days of MIMO, when simple implementation was paramount, the performance disadvantages have made linear RXs to have fewer applications nowadays.

H-BLAST

We next turn to successive interference cancellation techniques, also known under the name of *BLAST (Bell labs LAyered Space Time architectures)*. *Horizontal BLAST (H-BLAST)* is the simplest possible layered space-time structure.[14] The TX first demultiplexes the data stream into N_t parallel streams (assuming $N_s = N_t$), each of which is encoded *separately*. Each encoded data stream is then transmitted from a different transmit antenna. The channel mixes up the different data streams; the RX separates them out by nulling and interference subtraction. More specifically, the RX proceeds in the following steps (Figure 16.21):

- It considers the first data stream as the useful one and looks at the other data streams as interference. It can then use *optimum combining* for the suppression of the interfering streams, see Section 12.4.3. The RX has $N_r \geq N_t$ antenna elements available. If $N_r = N_t$, it can suppress all $N_t - 1$ interfering data streams, and receive the desired data stream with diversity order 1.[15]
- The desired stream can now be demodulated and decoded. The outputs of that process are firm decisions on the bits of stream 1. Note that since we have separate encoding of the different data streams, we need only the knowledge of the first stream to complete the decoding process.
- The bits that have thus been decoded are now re-encoded, and remodulated. Multiplying that symbol stream with the transfer function of the channel determines the contribution that stream 1 has made to the total received signal at the different antenna elements, with the effect of noise on the received data of that stream eliminated.
- The RX subtracts these contributions from the signals at the different antenna elements.
- Now consider this "cleaned-up" signal to detect the second data stream. There are N_r received signals, but only $N_t - 2$ interferers. Using optimum combining again, the RX can now receive the desired data stream with diversity order 2.
- The next step is again the decoding, recoding, and remodulating of the considered data stream (stream 2 now), and subtraction of the associated signal from the total signal at the receive antenna elements. This cleans up the received signal even more.
- The process is repeated until the last data stream is decoded.

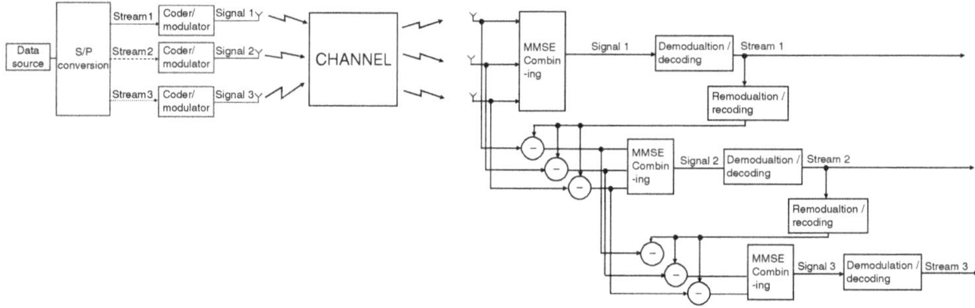

Figure 16.21 Block diagram of an H-BLAST transceiver.

This scheme is actually very similar to a multi-user detection scheme called Serial Interference Cancellation, SIC, (Section 28.2) with the only difference being that in standard SIC the different data streams come from different users. Note also that the encoding scheme does not require "cooperation" between the different antenna elements. This also implies that H-BLAST can be used for the detection of data streams from multiple UEs at the BS, i.e., multi-user MIMO (compare Chapter 22).

If the RX uses ZF or MMSE combining for each of the decoding stages, the ergodic capacity that can be achieved with this scheme is approximately (at high SNR) in a fast-fading channel (i.e., each codeword sees many channel realizations).

$$C_{\text{HBLAST}} \approx N \log_2 \left(\frac{\bar{\gamma}_{\text{RX}}}{N_t}\right) + \sum_{k=1}^{N} \mathrm{E}_{\mathbf{H}} \left\{ \log_2 \left(\chi^2_{2(N_r - N_t + k)}\right) \right\}. \tag{16.114}$$

---

[14] Confusingly, this scheme was originally called *V-BLAST* (for vertical BLAST) but changed that name later [Foschini et al. 2003].

[15] If the RX has more antennas, it can receive the first data stream with higher diversity order, i.e., better quality. But in any case, the interference from the other streams can be eliminated.

Note the important difference to the linear RX: in a linear RX, the second term on the right-hand side has the same degree of freedom of the χ^2 variable for all data streams; in H-BLAST, the degrees of freedom (i.e., diversity order), increase for every stream.

We next consider slow-fading channels. If the RX uses MMSE filtering in every stage (or ZF at high SNR), and Per-Antenna Rate Control (PARC), it can achieve the (open-loop) capacity (Eq. 16.69) in the slow-fading channels. If such separate rate control is not available (e.g., because of insufficient feedback), capacity cannot be achieved in such channels. Since in the absence of PARC all streams have the same rate but different degrees of freedom, the outage probability is higher (or, equivalently, the outage capacity is lower). However, even in this case the simplicity still makes H-BLAST a very attractive scheme.

Similar to decision-feedback equalizers and multi-user SIC detectors, H-BLAST also faces the problem of error propagation, especially since the first decoded data stream has the worst quality. In other words: if the data stream 1 is decoded incorrectly, then we subtract the "wrong" signal from the remaining signals at the antenna elements. Thus, instead of "cleaning up" the receive signal, the process introduces even more interference. This in turn increases the likelihood that the second data stream is decoded incorrectly, and so on. In order to mitigate this problem, stream ordering should be used: the RX should first decode the stream that has the best SINR, then the one with the next best, and so on. Further challenges arise in the presence of forward error correction coding, see the discussion in section titled "Transceivers for Coded Transmission." below.

D-BLAST

The main problem with H-BLAST is that it does not provide diversity, so that – in a quasi-static setting – different data streams experience different fading states, and can handle different rates. If no PARC is possible, this leads to suboptimum results. The first stream, which has diversity order one, dominates the performance at high SNRs. A better performance can be achieved with the so-called D-BLAST scheme. In this approach, the streams are cycled through the different transmit antennas, so that each stream sees all possible antenna elements. In other words, each single transmit stream is subdivided into a number of subblocks. The first subblock of stream 1 is transmitted from antenna 1, the next subblock from antenna 2, and so on (Figure 16.22).

Figure 16.22 Assignment of bitstreams to the different antennas for H-BLAST and D-BLAST.

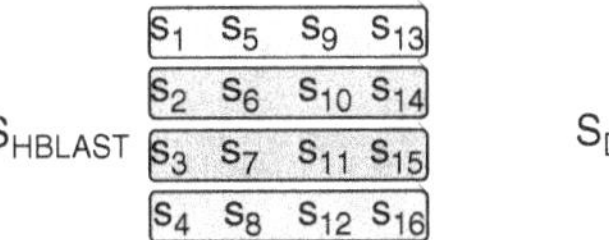

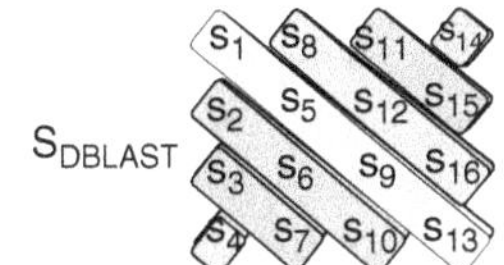

The decoding can be done stream-by-stream; again, each decoded block can be subtracted from the signals at the other antenna elements and, thus, enhances the quality of the residual signal. The difference to H-BLAST is that each stream sometimes is in a "good" position in the sense that the other streams have already been subtracted, and thus the SINR is very high, while sometimes it is in a bad position, in the sense that it suffers full interference. Thus, each stream experiences full diversity. As shown by [Ariyavisitakul 2000], D-BLAST is also related to the lower capacity bound of Eq. (16.85): let us specialize it to the case of $N_t = N_r$:

$$C_{\text{D-BLAST}} = \sum_{k=1}^{N_t} \log_2 \left[1 + \frac{\overline{\gamma}_{\text{RX}}}{N_t} \chi_{2k}^2 \right]. \tag{16.115}$$

The diversity for each stream is higher, and thus the outage capacity, compared to H-BLAST without PARC, is as well: the data streams alternatingly see a channel with diversity order 1 (whose SNR has a pdf that is a chi-square distribution with 2 degrees of freedom χ_2^2), diversity order 2 (chi-square with 4 degrees of freedom, χ_4^2), and so on. In other words, D-BLAST achieves the outage capacity.

Maximum-Likelihood Receiver

For a symbol-by-symbol detection, the optimum RX is (assuming all codewords are equiprobable) the ML RX, which determines an estimate $\hat{\mathbf{s}}$ as

$$\begin{aligned}
\hat{\mathbf{s}} &= \arg\min_{\mathbf{s}} \left\| \mathbf{r} - \sqrt{\frac{E_s}{N_t}} \mathbf{H}\mathbf{s} \right\|^2 \\
&= \arg\min_{\mathbf{s}} \left(\frac{E_s}{N_t} \mathbf{s}^\dagger \mathbf{H}^\dagger \mathbf{H}\mathbf{s} - 2\sqrt{\frac{E_s}{N_t}} \mathrm{Re}\left(\mathbf{s}^\dagger \mathbf{H}^\dagger \mathbf{r} \right) \right)
\end{aligned} \tag{16.116}$$

where the minimization is over all possible symbol vectors (symbols at all the TX antenna elements), i.e., a discrete set.

For several classes of algorithms, it is useful to view the constellation points as forming a lattice. More precisely, let QAM symbols separately be described by their real and imaginary parts, and normalize their distances so that all entries are integer

$$\mathcal{S} = [\, \mathrm{Re}\,\{c_1\}, \ldots \mathrm{Re}\,\{c_{N_t}\}, \mathrm{Im}\{c_1\}, \ldots \mathrm{Im}\{c_{N_t}\}\,]^T \tag{16.117}$$

where we assume henceforth that $N_s = N_t = N_r$. Then all the possible QAM symbols form a lattice, i.e., a set of integer points in a $2N_t$ dimensional space. We similarly define $\mathcal{R}$ as the receive vector with stacked real and imaginary parts. The channel matrix is written as

$$\mathcal{H} = \begin{bmatrix} \mathrm{Re}\,(\mathbf{H}) & \mathrm{Im}(\mathbf{H}) \\ -\,\mathrm{Im}(\mathbf{H}) & \mathrm{Re}\,(\mathbf{H}) \end{bmatrix} \tag{16.118}$$

and the equation for the received signal becomes

$$\begin{aligned} \mathcal{R} &= \mathcal{H}_{\mathrm{eff}}\mathcal{S} + \mathcal{N} \\ &= \mathcal{W} + \mathcal{N} \end{aligned} \tag{16.119}$$

where $\mathcal{H}_{\mathrm{eff}}$ is the normalized version of $\mathcal{H}$, and $\mathcal{W}$ denotes the received lattice. Note that the regular transmit lattice $\mathcal{S}$ gets distorted by the channel. Thus, the $2N_r$-dimensional receive lattice is not necessarily rectangular anymore, but rather skewed. The worse the condition number of the channel, the more distorted the receive lattice is. In any case, the goal of decoding is to find the lattice point $\mathcal{W}$ that is closest to $\mathcal{R}$.

The minimization Eq. (16.116) is np-hard, and can thus only be done for very small constellation sizes and number of data streams. A simple approximation is to do the minimization over the N_t dimensional complex space (which is easy – it is a minimum distance problem) and then round to the nearest symbol. This estimate, $\hat{\mathcal{S}}_{\mathrm{Bab}} = \mathcal{H}_{\mathrm{eff}}^{-1}\mathcal{R}$, also known as the Babai estimate, is not optimal, and in fact is the output of the linear RX discussed previously. It is, however, a starting point for approximate ML RXs we will describe below.

One way to solve the exact problem are lattice reduction techniques, where the minimization problem Eq. (16.116) is to transform the effective lattice $\mathcal{H}_{\mathrm{eff}}$ by postmultiplication with a matrix $\mathcal{P}$, which has only integer entries, and fulfills $|\det(\mathcal{P})| = 1$, so that consequently $\mathcal{P}^{-1}$ also has integer entries. If this operation results in an (almost) orthogonal matrix, the simple unconstrained distance minimization becomes close to optimum. Finding a suitable matrix $\mathcal{P}$ can be computationally expensive, but – in particular when the channel is slowly varying, or when, e.g., OTFS (see Section 15.12) is used as modulation format, the relative effort of finding $\mathcal{P}$ may easily pay off by making the subsequent detection much easier, enabling the use of linear detectors.

Finally, we note that modern RXs usually do not make a symbol-by-symbol decision, but rather compute the log-likelihood ratios of the bits, which are then passed on to the decoders of the (turbo or Low-Density Parity Check (LDPC)) codes, see Sections 13.6 and 13.7 [Studer and Bölcskei 2010].

*Sphere Decoder

The sphere decoder is an approximate ML decoder with significantly reduced complexity. Essentially, the goal is to identify lattice points that lie within a sphere of radius D around the received signal vector. This then allows to search for the ML estimate in a much smaller set. However, the question is how to do this efficiently: if it involves computing the distance of all lattice points from the received vector, then this is as much effort as the (np-hard) search for the ML estimate, which also requires a distance computation for every possible lattice point.

The basic trick of the sphere decoder is as follows: squared distances of an N-dimensional space are the sum of the squared distances in the different dimensions. Thus, if the distance between two points in a lower-dimensional space is already larger than a threshold, then we know for sure that the overall distance is larger than the threshold as well. Applied to our problem, this means that in order to lie within an N-dimensional sphere of radius D around $\mathcal{R}$, it is a *necessary but not sufficient* that it lies within a sphere of radius D in an $n < N$-dimensional space.

The algorithm thus proceeds iteratively: consider a one-dimensional space, i.e., projection of all lattice points onto the lattice points of a line. Each of those projections can be considered the "parent" of a lot of lattice points, since many points in the higher-dimensional lattice have the same projection. Now, points whose projections are at a distance $>D$ in the one-dimensional space *and all their child points* can be discarded from further consideration, because they are known to have a too-large distance also in the higher-dimensional space. For those lattice points that passed the first round, the sphere decoder looks at the projection onto a two-dimensional space and test which of those lie within a two-dimensional sphere (i.e., a circle) around the two-dimensional projection of $\mathcal{R}$. Again, those points with too large of a distance and all their child points are discarded, and for the rest, we proceed to the next round. This process is continued until the test is applied to the full-dimensional distance. An example for such a tree is shown in Figure 16.23. Because each iteration round eliminates a large number of points, the computational complexity is dramatically reduced. It is also noteworthy that the computation of higher-dimensional distances can make use of the (already-computed) lower-dimensional distances, further improving computational efficiency.

Let us now mathematically formulate the above principles. We start out with the Q-R factorization of the channel matrix.[16] Assuming $N_r \geq N_t = N_s$ (note that due to the notation (16.119), N_t, N_r denote twice the number of physical TX, RX antennas, respectively and N_s twice the number of (complex) data streams).

---

[16] For this subsection only, we will use $\mathbf{Q}$ and $\mathbf{L}$ as the matrices of this factorization. While it is common to call the upper triangular matrix $\mathbf{R}$ we do not follow the convention to avoid confusion with the received vector $\mathbf{r}$ and the correlation matrix $\mathbf{R}$.

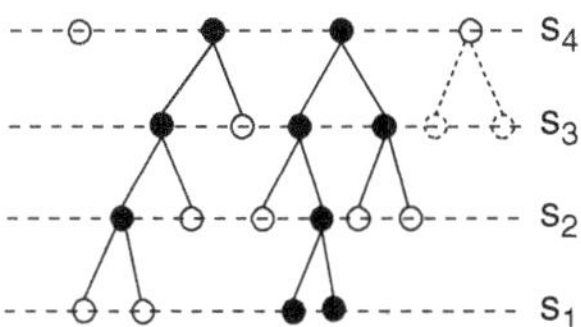

Figure 16.23 Tree structure showing the downselection of possible lattice points: sufficiently low distance in lower-dimensional space (solid), distance exceeds D even in lower-dimensional space (hollow). The dotted lines indicate an example of lower-level lattice points that do not need to be checked.

$$\mathbf{H} = \mathbf{Q} \begin{bmatrix} \mathbf{L} \\ \mathbf{0}_{(N_r - N_t) \times N_t} \end{bmatrix} \tag{16.120}$$

where $\mathbf{Q} = [\mathbf{Q}_1 \ \mathbf{Q}_2]$ is an $N_r \times N_r$ orthogonal matrix (i.e., $\mathbf{Q}^\dagger = \mathbf{Q}^{-1}$), and $\mathbf{L}$ is an $N_t \times N_t$ upper triangular matrix. Points that are within a sphere of radius D fulfill

$$\left| \mathbf{r} - [\mathbf{Q}_1 \ \mathbf{Q}_2] \begin{bmatrix} \mathbf{L} \\ \mathbf{0}_{(N_r - N_t) \times N_t} \end{bmatrix} \mathbf{s} \right|^2 \leq D^2 \tag{16.121}$$

which can be reformulated as

$$\left| \mathbf{Q}_1^\dagger \mathbf{r} - \mathbf{L}\mathbf{s} \right|^2 \leq D^2 - \left| \mathbf{Q}_2^\dagger \mathbf{r} \right|^2 . \tag{16.122}$$

The only unknown here is $\mathbf{s}$. We thus denote the (known) $\mathbf{Q}_1^\dagger \mathbf{r}$ as $\widetilde{\mathbf{r}}$ and $D^2 - \left| \mathbf{Q}_2^\dagger \mathbf{r} \right|^2$ as $\widetilde{D}_{N_t}^2$. Furthermore, $L_{i,j}$ is the i,jth entry of $\mathbf{L}$. Thus, Eq. (16.122) becomes

$$\sum_{i=1}^{N_t} \left(\widetilde{r}_i - \sum_{j=i}^{N_t} L_{i,j} s_j \right)^2 \leq \widetilde{D}_{N_t}^2. \tag{16.123}$$

Now write the left-hand side as

$$\left(\widetilde{r}_{N_t} - L_{N_t,N_t} s_{N_t} \right)^2 + \left(\widetilde{r}_{N_t - 1} - L_{N_t - 1, N_t} s_{N_t} - L_{N_t - 1, N_t - 1} s_{N_t - 1} \right)^2 + \cdots. \tag{16.124}$$

Each of the terms $()^2$ is the distance in one dimension. To check whether a lattice point is within D in a one-dimensional space (indexed by N_t), the following condition needs to be fulfilled

$$\left\lceil \frac{-\widetilde{D}_{N_t} + \widetilde{r}_{N_t}}{L_{N_t,N_t}} \right\rceil \leq s_{N_t} \leq \left\lfloor \frac{\widetilde{D}_{N_t} + \widetilde{r}_{N_t}}{L_{N_t,N_t}} \right\rfloor \tag{16.125}$$

where $\lceil . \rceil$ and $\lfloor . \rfloor$ denote rounding up to the nearest larger, and rounding down to the nearest smaller, integer, respectively. To proceed to the next step, define

$$\widetilde{D}_{N_t - 1}^2 = \widetilde{D}_{N_t}^2 - \left(\widetilde{r}_{N_t} - L_{N_t,N_t} s_{N_t} \right)^2. \tag{16.126}$$

Note that we have several possible values of s_{N_t}, so the computation has to be done for each of them. We furthermore compute a received signal conditioned on the already-discovered dimensions (again, this has to be computed for each allowed s_{N_t}):

$$\widetilde{r}_{N_t - 1 | \widetilde{r}_{N_t}} = \widetilde{r}_{N_t - 1} - L_{N_t - 1, N_t} s_{N_t}. \tag{16.127}$$

The condition for the lattice to be within the two-dimensional sphere can then be written as

$$\left\lceil \frac{-\widetilde{D}_{N_t - 1} + \widetilde{r}_{N_t - 1 | \widetilde{r}_{N_t}}}{L_{N_t - 1, N_t - 1}} \right\rceil \leq s_{N_t - 1} \leq \left\lfloor \frac{\widetilde{D}_{N_t - 1} + \widetilde{r}_{N_t - 1 | \widetilde{r}_{N_t}}}{L_{N_t - 1, N_t - 1}} \right\rfloor. \tag{16.128}$$

This process, i.e., computing $\widetilde{D}_{N_t - k}, \widetilde{r}_{N_t - k | \widetilde{r}_{N_t - k + 1}}$, checking the distance condition, accepting or discarding particular lattice points is continued with $k = 2, 3, \ldots$ until $k = N_t - 1$, i.e., we have checked the full-dimensional sphere.

A critical question is the choice of the radius D. If it is too small, there might not be a point within the sphere. If it is too large, the computational effort becomes very large. Computing the optimal radius is itself an np-hard problem. Various approximations have been proposed; one simple approach is to take the distance between R and the Babai estimate.

The sphere decoder is a "depth-first" search on a tree, and as such can only guarantee a given complexity *on average*. For many practical implementations, a maximum complexity for each separate symbol is required, since otherwise certain symbols, which require more complex computations, could suffer from higher latency. This problem is tackled by a "breadth-first" search, see Figure 16.24. In each cycle of the search, all nodes in a layer are investigated. A fixed number of nodes, i.e., the K best nodes, are retained in each layer. This guarantees a limited computational complexity, but of course can lead to missing the best lattice point within the sphere. The factor K is obviously important for trading off complexity with accuracy.

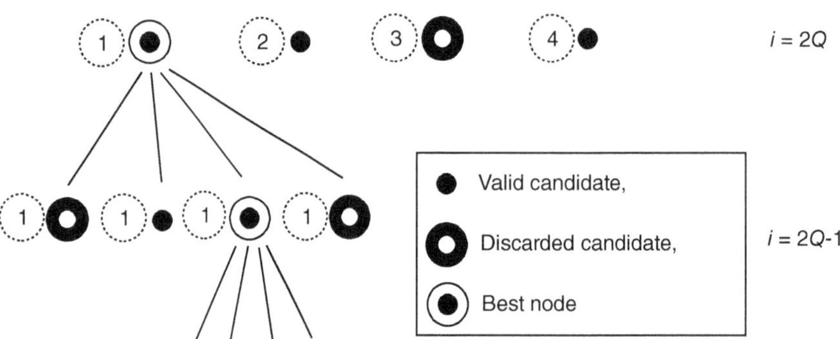

Figure 16.24   Breadth-first search for a sphere decoder.
Reproduced with permission from [Clerckx and Oestges 2013] © Academic Press/Elsevier.

Transceivers for Coded Transmission

The above discussion has concentrated on the symbol-by-symbol transmission of the data streams. Yet, for achieving of capacity, the data streams need to be encoded with error-correcting codes, such as turbo-codes, BICM, or LDPCs (see Chapter 13). For this encoding, there generally are two possibilities:

1. *Single-codeword transmission (joint encoding)*: in this approach, a high-rate datastream is encoded with the Forward Error Correction (FEC), and different coded symbols are transmitted from the different TX antenna elements. Note that while there is only a single coded data stream, it is *not* the same as single-stream transmission that we previously discussed (because in that case, multiple copies of the same coded symbols of a lower-rate data stream are transmitted from the different antenna elements). This approach provides inherent antenna diversity, because the coded data symbols are distributed over different TX antenna elements. However, it may result in higher complexity of the decoding.

2. *multiple-codeword transmission*: an approach that is more intuitive for SM is to separately encode the different data streams with FEC codes, thus allowing separate decoding at the RX. This approach is especially appealing for eigenmode transmission with CSIT, since all the data streams are orthogonal to each other, and decoding operations are thus naturally isolated from each other. Multiple codeword transmission is also an obvious fit to the BLAST architectures, though there is one significant challenge in the interference cancellation step: the "raw" detected symbols in a data stream are error prone (typically on the order of 10%), and thus their subtraction in the SIC step can lead to error propagation. Yet the decoded bits are only available after all symbols of a codeword have arrived, which would entail considerable latency. Further challenges arise if the detection order changes during the codeword.

For the reception of coded data streams, RXs with soft information and turbo iterations are commonly used. This means that the RXs we described above do not provide a particular "hard" symbol as the detector output, but rather log-likelihood ratios. These can then be used as priors in a turbo stage of the RX, similar to the turbo RX outlined in Section 13.6.

16.2.10   Antenna Selection and Hybrid Beamforming

One of the major challenges of MIMO lies in the hardware effort and energy consumption for the up-downconversion and digitization of the signals for the multiple antenna elements. This problem can be mitigated by transceiver structures that have a smaller number of up/downconversion chains and digitizers than they have antenna elements, since antenna elements themselves are cheap and (as they are passive) do not contribute to the energy consumption. The interface between the antenna elements and the RF chains can either be an L out of $N_{t,r}$ switch, resulting in *hybrid antenna selection*, or a $L \times N_{t,r}$ beamforming, resulting in *hybrid beamforming*. Both of these will be discussed here, as well as their combination.

Antenna Selection

The basic principle of hybrid antenna selection is to connect, through a suitable switch, the instantaneously "best" L antennas to the RF chains. In hybrid selection/MRC, which was already discussed in Section 12.4.3, the signals from the RF chains are processed with MRC (at the RX) or MRT (at the TX) to provide high diversity; the achieved diversity order is given by the number of antenna elements, while the beamforming gain is given by the number of RF chains. For hybrid selection SM, henceforth abbreviated H-S/MIMO, we consider first the case without CSIT. Then, the TX uses all antenna elements (selecting TX antennas would require CSIT), while the RX performs hybrid antenna selection. The goal of the antenna selection is to maximize the information-theoretic capacity:

$$C_{\text{H-S/MIMO}} = \max_{S(\breve{\mathbf{H}})} \left(\log_2 \left[\det\left(\mathbf{I}_{L_r} + \frac{\gamma_{\text{TX}}}{N_t} \breve{\mathbf{H}} \breve{\mathbf{H}}^\dagger \right) \right] \right) \tag{16.129}$$

where $\breve{\mathbf{H}}$ is created by striking $N_r - L_r$ rows from $\mathbf{H}$ and $S(\breve{\mathbf{H}})$ is the set of all possible matrices $\breve{\mathbf{H}}$. Practical algorithms for finding the best switch position will be discussed below. It is obvious that the dimension of the channel matrix, and therefore the number of spatial streams, is upper limited by $\min(N_t, L)$. An upper bound for the capacity for i.i.d. fading channels can be derived similar to the principles of the case of i.i.d. channels (assuming a separate RX for each TX stream): For $L_r \leq N_t$, this bound is

$$C_{\text{H-S/MIMO}} \leq \sum_{i=1}^{L_r} \log_2 \left(1 + \frac{\gamma_{\text{RX}}}{N_t} \gamma_{(i)} \right) \tag{16.130}$$

where the $\gamma_{(i)}$ is obtained by ordering a set of i.i.d. N_r chi-square-distributed random variables with $2N_t$ degrees of freedom such that $\gamma_{(1)} > \gamma_{(2)} > \ldots$. For $L_r > N_t$, the following bound is tighter

$$C_{\text{H-S/MIMO}} \leq \sum_{j=1}^{N_t} \log_2 \left[1 + \frac{\gamma_{\text{RX}}}{N_t} \sum_{i=1}^{L_r} \tilde{\gamma}_{(i)} \right] \tag{16.131}$$

where (for each j) the $\tilde{\gamma}_{(i)}$ are ordered (from a set of size N_r chi-square distributed variables with 2 degrees of freedom). Figure 16.25 shows the cdf of the capacity obtained by Monte Carlo simulations for $N_r = 8$, $N_t = 3$, and various L_r. With full exploitation of *all* available elements, a mean capacity of 23 bit/s/Hz can be transmitted over the channel. This number decreases gradually as the number of selected elements L_r decreases, reaching 19 bit/s/Hz at $L_r = 3$. For $L_r < N_t$, the capacity decreases drastically, since a sufficient number of antennas to spatially multiplex N_t independent transmission channels is no longer available.

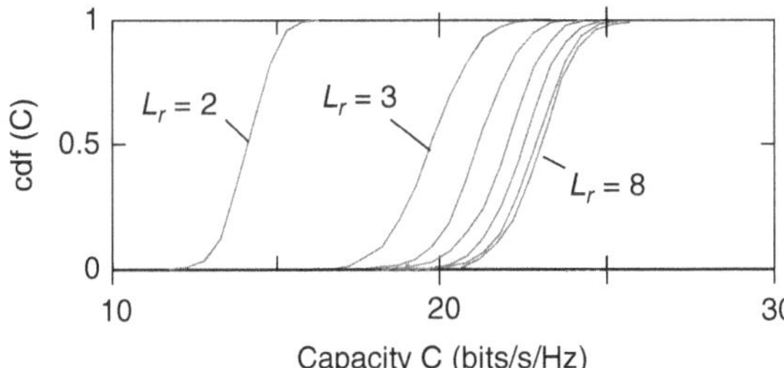

Figure 16.25 Capacity for a spatial multiplexing system with $N_r = 8$, $N_t = 3$, SNR $= 20$ dB, and $L_r = 2, 3, \ldots 8$. Color version available at wiley.com/go/molisch/wireless3e.

We now turn to algorithms for selecting the best antennas. Truly optimum selection requires an exhaustive search. However, computational effort for this is usually prohibitive. When considering suboptimum methods, selecting the antennas with the highest power is *not* a good approach in general. This behavior can be interpreted physically: the goal of the RX is to separate the different data streams. Thus it is suboptimum to use the signals from two antennas that are highly correlated, even if both have high SNR.

Better algorithms select antennas one by one, according to some criterion related to capacity. For example, we can search for two rows with highest correlation and then delete the one of them with the lower power. This process is continued until the remaining number of rows is L.

A similar algorithm adds one by one the antenna that leads to the largest increase in capacity, given the set of antennas that have already been chosen. The capacity that can be achieved in the $k + 1$th step of this loop is

$$C\left(\breve{\mathbf{H}}_{k+1} \right) = C\left(\breve{\mathbf{H}}_k \right) + \log_2 \left(1 + \frac{\gamma_{\text{TX}}}{N_t} \mathbf{h}_j \mathbf{B}_k \mathbf{h}_j^\dagger \right) \tag{16.132}$$

where $\mathbf{B}_k$ is related to the channel matrix of the antenna elements that have already been selected

$$\mathbf{B}_k = \left(\mathbf{I} + \frac{\gamma_{\text{TX}}}{N_t} \breve{\mathbf{H}}_k \breve{\mathbf{H}}_k^\dagger \right)^{-1} \tag{16.133}$$

and $\mathbf{h}_j$ is the (row) channel vector of the antenna we consider to add. We thus select the antenna that maximizes $\mathbf{h}_j \mathbf{B}_k \mathbf{h}_j^\dagger$, since this maximizes the capacity increase. The computation of $\mathbf{B}_k$ does not need to be done from scratch for each iteration step k, but rather can be done using $\mathbf{B}_{k-1}$ and the matrix inversion lemma.

Antenna selection does not reduce, and sometimes even increases, the overhead for training. For transmit antenna selection, we have to send one training sequence per antenna. Different training sequences should be orthogonal, e.g., transmitted at different time instances. The number of such transmissions is thus N_t, irrespective of whether transmit antenna selection is used or not. Receive antenna selection *increases* channel estimation overhead: each training sequence for a TX antenna has to be repeated $\lceil N_r/L_r \rceil$ times, so that the RX can downconvert and sample the signals at all antenna elements.

Another aspect of antenna selection is the noise and quantization error of the channel estimates. For those antennas that are part of the selected set, channel estimation errors impact results in the same manner as for full-complexity systems. However, the selection of the antennas themselves is also afflicted by noise. Fortunately, these errors have very small impact on the actual capacity. The reason is intuitive: if the channel estimation errors are small, then even selected "wrong" antennas provide capacity that is close to the ideal case. Conversely, channel estimation errors that would lead to selection of antennas that would give much smaller capacity would have to be very large, which is unlikely to occur.

Frequency selectivity of the channel greatly decreases the effectiveness of antenna selection when the average characteristics of the channel are the same at all antenna elements. Consider the case of OFDM: the set of optimum antennas is different for each coherence bandwidth, and for a sufficiently large ratio of system bandwidth to coherence bandwidth, all antennas are equally good (or equally bad). However situations can arise where there is different shadowing of the different antenna elements (e.g., different body shadowing depending on how a handset is held); since shadowing is approximately independent of frequency, antenna selection in this case is effective even in frequency-selective channels.

Hybrid Beamforming

Hybrid beamforming uses a combination of analog beamformers in the RF domain, together with digital beamforming in baseband that is connected to the RF with a smaller number of up/downconversion chains. In contrast to antenna selection, it provides beamforming gain. Depending on the beamforming mechanism, it can also provide diversity gain and reduce the overhead in the determination of the CSI at each of the antenna elements. Hybrid beamforming was invented by the author and his collaborators in the early 2000s; it is nowadays widely used both for single-user MIMO as discussed in this section, and for multi-user MIMO in particular in the context of massive MIMO (see Section 22.9) and millimeter-wave systems.

Figure 16.26 shows block diagrams of two hybrid beamforming structures at the BS, where we assume a downlink transmission from the BS (acting as TX) to the UE (RX). In each case, at the TX, a baseband digital precoder $\mathbf{T}_{BB}$ processes N_S data streams to produce N_{RF}^{BS} outputs, which are upconverted to RF and mapped via an analog precoder $\mathbf{T}_{RF}$ to N_{BS} antenna elements for transmission. The structure at the RX is similar: an analog beamformer $\mathbf{W}_{RF}^\dagger$ combines RF signals from N_{UE} antennas to create N_{RF}^{UE} outputs, which are downconverted to baseband and further combined using a matrix $\mathbf{W}_{BB}^\dagger$, producing signal $\mathbf{y}$ for detection/decoding. For a full-complexity structure, Figure 16.26a, each analog RF precoder output can be a linear combination of *all* RF signals. Further complexity reduction at the price of a somewhat reduced performance can be achieved when each RF chain can be connected only to a subset of antenna elements, as in Figure 16.26b. We will henceforth consider the full-complexity structure unless stated otherwise.

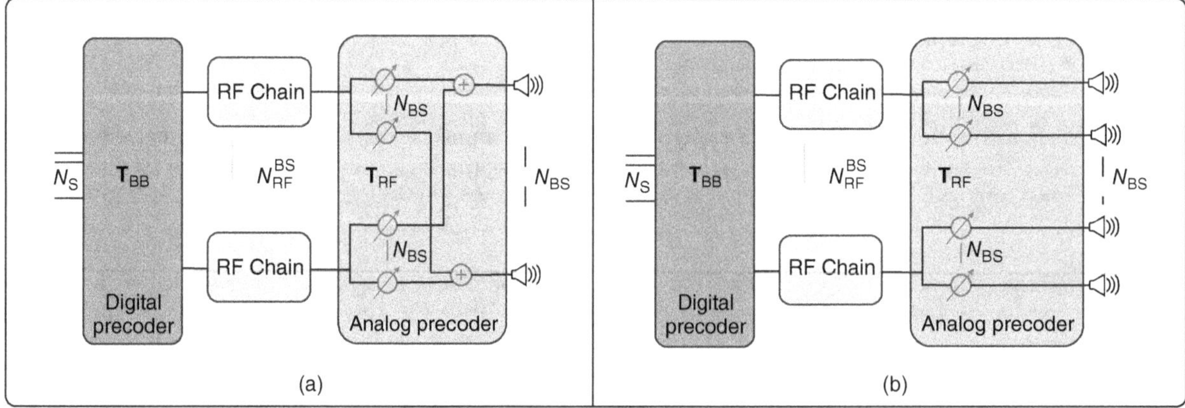

Figure 16.26   Block diagrams of hybrid beamforming structures at BS for a downlink transmission, where structures A and B denote the full-complexity and reduced-complexity structures, respectively. Color version available at wiley.com/go/molisch/wireless3e.

A further classification of the hybrid beamforming is based on whether the analog beamforming is based on instantaneous or average (second-order) statistics. In the former case, the analog beamformer can adjust to the instantaneous channel realizations, and thus provide the maximum diversity (i.e., diversity determined by the number of antenna elements). On the downside, this approach requires that the channel states at all antenna elements are sounded at least once per coherence time, thus leading to a high overhead of the CSI acquisition, similar to the case of antenna selection. Alternatively, analog beamforming can adjust to the second-order statistics of the channel, such as the APS. It provides the beamforming gain that can be obtained from such knowledge, but the diversity order is limited by the number of RF chains; thus overall capacity and other performance measures tend to be worse. On the positive side, the CSI for the analog beamforming only needs to be updated once per *stationarity time*, which is considerably longer than the coherence time, often on the order of seconds or minutes. We also note that analog beamformers do not allow different shaping of the beampatterns at different subcarrier frequencies (assuming here an OFDM system). Optimum beamforming even in the presence of instantaneous CSI then essentially averages over different small-scale fading states. Thus, for highly frequency-selective channels analog beamforming based on instantaneous CSI converges to that of second-order CSI.

Consider now the case that only the RX performs hybrid beamforming based on instantaneous CSI, while the TX is operating with full complexity, and the system is operating in a frequency-flat channel. Then the hybrid beamforming can perform as well as full-complexity beamforming, as long as $L \geq N_s$. This can be explained intuitively as follows: the N_s data streams have N_s dimensions in the N_r dimensional signal space of the received signal, The analog beamforming is a linear operation that allows to transform the received signal such that the signal stream components are contained within an $L \geq N_s$ dimensional subspace of the received signal, i.e., no signal energy is contained along other signal space dimensions. Then, downconversion and digitization of only the L signals carrying the subspace does not entail any loss of information. Note that in the preceding we assumed that the analog beamformer can adjust both amplitude and phase; if only phase shifters are available, then the number of required RF chains is doubled.

For more general cases, it is very difficult to find the analog and digital beamforming matrices that optimize, e.g., the net data rate. The main difficulties include:

- Analog and digital beamformers at each link end, as well as combiners at the different link ends, are coupled, which makes the objective function of the resulting optimization nonconvex.
- Typically the analog precoder/combiner is realized as a phase-shifter network, which imposes additional constraints on the elements of $\mathbf{W}_{\mathrm{RF}}^{\dagger}$ and $\mathbf{T}_{\mathrm{RF}}$.
- Moreover, with finite-resolution phase shifters, the optimal analog beamformer lies in a discrete finite set, which typically leads to NP-hard integer programming problems.

For sparse channels, i.e., channels that have a small number of MPCs that are resolvable with the (analog) beamformer, the following approach shows good results: first consider the optimum beamforming for the fully digital case with $N_{\mathrm{RF}}^{\mathrm{BS}} = N_{\mathrm{BS}}$ and $N_{\mathrm{RF}}^{\mathrm{UE}} = N_{\mathrm{UE}}$, where the solution is known, as discussed in Section 16.2.3. Then, find an (approximate) optimum hybrid beamformer by minimizing the Euclidean distance to the fully digital one; the finding of the approximation can incorporate additional constraints such as phase-shifter only for the analog beamformer.

In nonsparse channels, an alternating optimization of analog and digital beamformer can be used, i.e., we "freeze" the digital precoder and then develop the optimum analog precoder, then freeze that analog beamformer and obtain the best digital beamformer for it, and so on. Closed-form solutions or efficient approximations for each of the alternating optimization steps exist for various types of architectures and/or constraints. Even simpler decoupled solutions can be found, e.g., by assuming that the digital beamformer is a linear MMSE beamformer, which then allows a closed-form determination of the analog beamformer. Such simplified solutions can also be used as an initialization for an iterative approach. Joint optimization of the beamformers is even more difficult for the case that the analog beamformer is based on second-order CSI. In that case, the optimization is occurring on two different timescales: the digital beamformer adjusts to the instantaneous CSI while the analog is based on second-order CSI. For hybrid beamforming at the RX, the closed-form solution is simple if the channel obeys a Kronecker model; in this case, the optimum beamformer is the L largest eigenvectors of the correlation matrix $E_{\mathbf{H}}\{\lambda_1^2 \mathbf{u}_1 \mathbf{u}_1^{\dagger}\}$.

Finally, since the main benefit of antenna selection is enhanced diversity, while for hybrid beamforming it is beamforming gain, it is sometimes advantageous to combine the two. This is particularly true for the case where the analog beamformer is based on second-order CSI and thus does not provide enhanced diversity for small-scale fading.

*16.2.11 MIMO Systems with Low-Resolution ADCs

As MIMO systems become larger and operate with larger bandwidth, the power consumption of the ADCs is becoming a major bottleneck, compare Section 17.3. A possible way to reduce this problem is the use of low-resolution ADCs. This leads, however, to a reduction of the overall capacity, since the quantization noise increases. The performance analysis of such systems is quite difficult due to the nonlinear nature of quantization, and almost all closed-form obtained results are for 1-bit ADCs.

In the following, we consider the case that the TX has infinite-resolution DACs (remember that DACs are easier to produce and do not need as fine a quantization since they send out signals belonging to a discrete constellation), and has full CSIT. The RX uses 1-bit ADCs for both the in-phase and the quadrature-phase components.

For the SISO case, the capacity can be shown to be

$$C_{\mathrm{SISO}} = 2\left[1 - H_{\mathrm{b}}\left(Q\left(\sqrt{\gamma_{\mathrm{RX}}}\right)\right)\right] \tag{16.134}$$

where $H_{\mathrm{b}}(p)$ is the binary entropy function (compare Section 13.1)

$$H_b(p) = -p\log_2(p) - (1-p)\log_2(1-p).\tag{16.135}$$

Equation (16.134) can be interpreted such that (i) the optimum modulation format is QPSK, i.e., binary orthogonal signaling on each of the in-phase and quadrature branches, and the 1-bit quantization at the RX makes the channel on each branch into a real-valued binary symmetric channel. The transition probability (i.e., probability of getting into an erroneous state) is $Q(\sqrt{\gamma_{RX}})$. Note that since we assume perfect CSIT, the TX can compensate any phase rotation that the channel imposes.

For the MISO case, the TX performs maximum-ratio transmission to create an "effective" channel whose resulting receive SNR can then be used in the above equation.

While for the infinite-resolution ADC/DAC SIMO and MISO with perfect CSIT are completely equivalent, the situation changes dramatically in the setup considered here. Since the 1-bit resolution is only at the RX, having multiple RX antennas changes not only the capacity equation but also the optimum modulation format. With multiple antenna elements, each with its own 1-bit ADC, and with phase shifts between the different antenna elements, we can now detect higher-order modulation formats. Figure 16.27 shows the fundamental principle: each RX antenna element can distinguish between the four quadrants, but each with a different "reference" phase determined by the channel phase shift. Since in this example there are three antenna elements, the complex plane is divided into 12 sectors; in an ideal case, each sector contains exactly one signal constellation point.

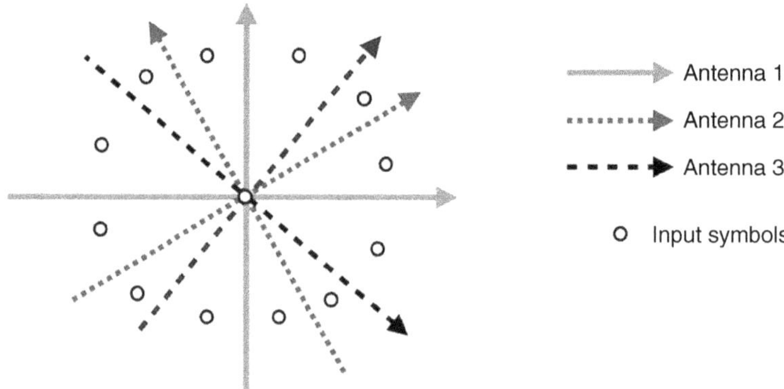

Figure 16.27 Transmitted symbol constellation, and decision regions for three antenna elements. The optimal constellation contains 12 nonzero symbols. Reproduced with permission from [Mo and Heath 2015] © IEEE. Color version available at wiley.com/go/molisch/wireless3e.

Generally, the number of constellation points that can be distinguished is $4N_r$ (an additional point can be used if the zero-point is included in the constellation). the capacity for the SIMO case can thus be bounded as

$$\log_2(4N_r) \le C_{\mathrm{SIMO}} \le \log_2(4N_r + 1).$$

For the MIMO case, consider infinite SNR, where the capacity is limited by the resolution of the ADCs and no other effects. Define a function

$$K(N_r, N_t) = 2 \sum_{k=0}^{2N_t - 1} \binom{2N_r - 1}{k}.\tag{16.136}$$

This function is related to the number of orthants that a space is divided into by N_r hyperplanes passing through the origin of the space. Then for $N_t < N_r$ the capacity of a MIMO system at infinite SNR is upper and lower bounded by

$$\log_2(K((N_r, N_t)) \le C_{\mathrm{MIMO}} \le \log_2(K((N_r, N_t) + 1).\tag{16.137}$$

If the channel has a limited rank (such as when there is a finite number of MPCs), then (at infinite SNR) the upper bound still holds when replacing N_t with the channel rank in the above equation.

An important sub-problem for the use of low-resolution ADCs is channel estimation, which we discuss in the following for the frequency-flat case. A widely used approach is the so-called Bussgang estimator, based on the Bussgang decomposition that finds equivalent linear operators for nonlinear functions of a Gaussian signal; in the case of a one-bit ADC, received and transmitted signal are related as $\mathbf{z} = \mathbf{A}\mathbf{r} + \mathbf{q}$ where $\mathbf{r}$ is the received signal, $\mathbf{z}$ the quantized signal, $\mathbf{A}$ is the linear operator, and $\mathbf{q}$ the quantization noise; to make the latter uncorrelated with $\mathbf{s}$, we choose $\mathbf{A} = \mathbf{C}_{rz}^{\dagger}\mathbf{C}_{rr}^{-1}$, where $\mathbf{C}_{rr}$ is the autocorrelation matrix of $\mathbf{r}$ and the cross-correlation matrix between transmitted and quantized signal is, for single-bit quantization and Gaussian inputs,

$$\mathbf{C}_{rz} = \sqrt{\frac{2}{\pi}}\mathbf{C}_{rr}\mathbf{\Sigma}_{rr}^{-1/2} \qquad \mathbf{\Sigma}_{rr} = diag(\mathbf{C}_{rr}).\tag{16.138}$$

Assume that we have a set of pilot sequences of length M_p (in units of symbols), which we write into a $M_p \times K$ pilot matrix $\mathbf{\Phi}^\dagger$ where K is the number of "sources" (streams, antennas, or - in Chapter 22 - users); we assume $M_p \geq K$ in order to ensure that the pilot signals from all sources can be distinguished. The matrix $\mathbf{A}$ can then be computed as

$$\mathbf{A} = \alpha_p \mathbf{I}_{N_r M_p} \quad \text{with } \alpha_p = \sqrt{\frac{2}{\pi} \frac{1}{K\gamma_p + 1}} \tag{16.139}$$

if the pilot signals form a semi-unitary matrix $\mathbf{\Phi}^\dagger \mathbf{\Phi} = M_p \mathbf{I}_K$, e.g., as K columns of the FFT matrix.

The channel estimate and covariance matrix of the estimate of the so-called Bussgang LMMSE estimator becomes

$$\widehat{\mathbf{h}}^{\text{BLSM}} = \mathbf{B}^\dagger \mathbf{C}_{\mathbf{zz}}^{-1} \mathbf{z} \qquad \mathbf{C}_{\widehat{\mathbf{h}}^{\text{BLSM}} \widehat{\mathbf{h}}^{\text{BLSM}}} = \mathbf{B}^\dagger \mathbf{C}_{\mathbf{zz}}^{-1} \mathbf{B} \tag{16.140}$$

with $\mathbf{B} = \left[\mathbf{A} \left(\mathbf{\Phi}^* \otimes \sqrt{\gamma_p} \mathbf{I}_N \right) \right]$. For a single-bit ADC, the correlation matrix of the quantized signal is

$$\mathbf{C}_{\mathbf{zz}} = \frac{2}{\pi} \left[\arcsin \left(\mathbf{\Sigma}_{\mathbf{rr}}^{-1/2} \text{Re}(\mathbf{C}_{\mathbf{rr}}) \mathbf{\Sigma}_{\mathbf{rr}}^{-1/2} \right) + j \arcsin \left(\mathbf{\Sigma}_{\mathbf{rr}}^{-1/2} \text{Im}(\mathbf{C}_{\mathbf{rr}}) \mathbf{\Sigma}_{\mathbf{rr}}^{-1/2} \right) \right]. \tag{16.141}$$

For the case of high SNR, the channel estimation error for $M_p = K$ converges to $1 - 2/\pi = -4.4$ dB, which is usually not acceptable. However, longer pilot sequences allow to get better channel estimates even with low-resolution ADCs, at the price of increased overhead. Improvement is also possible when the ADC has more than one bit resolution [Jacobsson et al. 2017].

*16.2.12 Space Time Coding for MIMO

If the channel is unknown at the TX and $N_t > N_s$, it is advantageous to transmit different versions of the data stream(s) from the different transmit antennas. Section 16.1.5 already described some methods that achieve this, such as delay diversity and the *Alamouti* code. The latter is the most popular example of a space-time code. Generally, space-time codes introduce redundancy by sending from each transmit antenna differently encoded versions of a basic signal vector. There are multiple ways how that encoding can be done. In the following, we will first describe STBC, essentially, generalizations of the Alamouti code, followed by the basic principles of *Space-Time Trellis Codes (STTC)*. Note that while space-time codes might provide some coding gain, we treat them separately from FEC, i.e., the input to the space-time coder might be an already-FEC-encoded symbol stream.

Space-Time Block Codes

The idea of STBC is to transmit the data in a way that guarantees high diversity, while allowing a simple decoding process. For STBCs, we always assume that the channel is constant over the duration of the codeword. Even when using space-frequency block codes (i.e., the code symbols are transmitted on different subcarriers, instead of different times), we assume that the subcarriers are spaced less than a coherence bandwidth apart, so that symbols transmitted from the same antenna but on different subcarriers "see" the same channel.

The most popular STBC is the Alamouti code discussed in Section 16.1.5. We saw that with 2 TX antennas it can achieve a SM rate of 1 (which is optimum for a single RX antenna), and achieve diversity order 2, but no coding gain. For convenience, we repeat here the codewords

$$\mathbf{C} = \frac{1}{\sqrt{2}} \begin{pmatrix} c_1 & -c_2^* \\ c_2 & c_1^* \end{pmatrix}. \tag{16.142}$$

There have been many attempts to generalize the Alamouti code to more than two transmit antennas. Unfortunately, it can be shown that *orthogonal complex* STBCs for more than two antennas have a SM rate smaller than one – in other words, they cannot even achieve the rate that we could have with a single-antenna system [Tarokh et al. 1999]. For four antennas, a rate 3/4 code is given as

$$\mathbf{C} = \frac{1}{\sqrt{3}} \begin{pmatrix} c_1 & -c_2^* & c_3^* & 0 \\ c_2 & c_1^* & 0 & c_3^* \\ c_3 & 0 & -c_1^* & -c_2^* \\ 0 & c_3 & c_2 & -c_1 \end{pmatrix}. \tag{16.143}$$

This sacrificing of rate is generally not desirable. Thus, STBCs for larger N_t typically sacrifice orthogonality, with the benefit of allowing rate-1 codes. Without orthogonality, the decoding becomes somewhat more complicated. A typical example is the code

$$\mathbf{C} = \frac{1}{2}\begin{pmatrix} c_1 & -c_2^* & c_3 & -c_4^* \\ c_2 & c_1^* & c_4 & c_3^* \\ c_3 & -c_4^* & c_1 & -c_2^* \\ c_4 & c_3^* & c_2 & c_1^* \end{pmatrix} \tag{16.144}$$

which has the property

$$\mathbf{CC}^\dagger = \frac{1}{4}\begin{pmatrix} a & 0 & b & 0 \\ 0 & a & 0 & b \\ b & 0 & a & 0 \\ 0 & b & 0 & a \end{pmatrix} \tag{16.145}$$

where a, b can be computed from the c_i. The structure shows that the pairs $\{c_1, c_3\}$ and $\{c_2, c_4\}$ can be decoded independently of each other, but within the symbols, each of those pairs is coupled. Other codes with these same properties are also possible.

Another alternative is to design STBCs with full rate but less than full diversity. For example, a 4-antenna TX can transmit an Alamouti codeword from antennas 1 and 2 in the first two timeslots, and another Alamouti codeword on antennas 3 and 4 in the next two timeslots. Each of the codewords achieves only diversity 2 but can be decoded with a standard Alamouti decoder. Coding across the codewords with an FEC then further increases the diversity. This and similar schemes are popular in standardized systems and described in Sections 31.3.6 and 33.3.3.

We finally note that STBCs can also be combined with other transmission schemes, such as SM. For example, when three streams are to be transmitted from four antenna elements, then two streams can be transmitted from one antenna element each, while the third stream can be sent, via Alamouti coding, from the remaining two antenna elements. Such combinations have been used in commercial MIMO systems such as Wi-Fi.

General Design Principles

Any space-time code can be assessed according to some general guidelines that are related to the pairwise error probability between codewords. We will see that some guidelines are more helpful for enhancing diversity, while others can provide coding gain. For STBC, diversity and simple decoding are paramount, so those principles are somewhat less directly applicable (though they still hold); for STTC they are especially relevant. Note, however, that these general space-time codes are rarely used in commercial systems, as they are competing with other methods of obtaining coding gain.

Due to an assumed absence of CSIT, no linear precoding is used (one could combine knowledge of second-order channel statistics with space-time codes, as discussed previously, but we will not treat this case here). Assuming that the space-time coder converts N_b bits into N_Q symbols, which are transmitted over a time duration T (in units of symbol durations) over N_t transmit antennas, the signaling rate is N_b/T and the SM rate is N_Q/T. It is noteworthy that this rate can be larger, equal, or even smaller than unity.

Consider now the case of a slow-fading channel, such that each space-time codeword experiences the same channel. Assume furthermore an i.i.d. Rayleigh fading channel and high SNR. Then the average pairwise error probability, i.e., the probability of confusing one codeword $\mathbf{C} = (\mathbf{c}_1, \mathbf{c}_2, ...\mathbf{c}_L)$ of length L_c with another codeword $\widetilde{\mathbf{C}}$ is upper-bounded at high SNR by

$$P\left(\mathbf{C} \to \widetilde{\mathbf{C}}\right) \leq \left(\prod_{i=1}^{R_e} \lambda_i\right)^{-N_r}\left(\frac{E_S}{4N_0}\right)^{-R_e N_r} \tag{16.146}$$

where R_e and λ_i are the rank and eigenvalues, respectively, of the error matrix

$$\left[\mathbf{E} = \left(\mathbf{C} - \widetilde{\mathbf{C}}\right)\left(\mathbf{C} - \widetilde{\mathbf{C}}\right)^\dagger\right]. \tag{16.147}$$

The term

$$\left(\prod_{i=1}^{R_e} \lambda_i\right)^{-N_r} \tag{16.148}$$

represents the coding gain, while the term

$$\left(\frac{E_S}{4N_0}\right)^{-R_e N_r} \tag{16.149}$$

describes the diversity gain. In order to achieve full diversity, the rank of the error matrix should thus be as high as possible (*rank criterion*); in order to achieve high coding gain, the determinant of the error matrix should be maximized (*determinant criterion*).

In the low-SNR regime, or when N_t is large, the determinant criterion needs to be replaced by the *trace criterion*, i.e., maximize

$$\min_{C,\tilde{C}\neq C} Tr\{\mathbf{E}\}. \tag{16.150}$$

Since the trace is equal to the Frobenius norm, this criterion essentially states that the minimum Euclidean distance should be maximized, similar to code design in AWGN. The rank criterion and the determinant (or trace) criterion provide important guidelines on how to design space-time codes in slow-fading channels.

Consider next the case of fast-fading channels, such that each codeword experiences multiple fading states. Then the average error probability is

$$P\left(\mathbf{C}\to\widetilde{\mathbf{C}}\right) \leq \left(\prod_{i\ \text{s.t.}\mathbf{c}_i\neq\tilde{\mathbf{c}}_i} |(\mathbf{c}_i-\widetilde{\mathbf{c}}_i)|^{-2N_r}\right)\left(\frac{E_S}{4N_0}\right)^{-L_{C,\tilde{C}}N_r} \tag{16.151}$$

where the product is taken over all time instances in which $\mathbf{c}$ and $\tilde{\mathbf{c}}$ are different, and the number of such time instances is called the "effective length" of the pair of codewords $L_{C,\tilde{C}}$. This leads then to the following design criteria for the codes. The distance

$$L_{\min} = \min_{C,\tilde{C}\neq C} L_{C,\tilde{C}} \tag{16.152}$$

describes the worst-case diversity gain. The coding gain is described by

$$\left(\prod_{i\ \text{s.t.}\mathbf{c}_i\neq\tilde{\mathbf{c}}_i} |(\mathbf{c}_i-\widetilde{\mathbf{c}}_i)|^{-2N_r}\right) \tag{16.153}$$

and the goal of the optimization is to minimize the product distance

$$\min_{C,\tilde{C}\neq C;L_{\min}} \left(\prod_{i\ \text{s.t.}\mathbf{c}_i\neq\tilde{\mathbf{c}}_i} |(\mathbf{c}_i-\widetilde{\mathbf{c}}_i)|^2\right)$$

for the code word with a minimum effective length. Maximization of the distance and the product criteria (in that order of importance) then provides guidelines for the code design. It is noteworthy that the "fast fading" situation occurs not only in the case of actual fast fading of the channel but also in frequency-selective channels, when the symbols are distributed over subcarriers that experience different fading.

Space-Time Trellis Codes

STBCs provide full diversity order, but they do not give any beamforming gain, nor do they provide coding gain. Such coding gain can be obtained from STTC. Given N_t transmit antennas, the STTC maps each symbol from the information source onto a *vector* of N_t transmit symbols that are sent from the different antenna elements. This mapping is done similar to a conventional convolutional encoder, in that the output is a function of the input bits and the state of the encoder. The difference now is that the input is a group of N_b bits, and the output is a vector of symbols, which are sent to the different TX antennas. The encoder has a memory size, corresponding to the shift register length of the traditional convolutional encoder. Conversely, the decoding requires a vector Viterbi decoder. Two examples for such STTCs are given in Figure 16.28.

There is a large variety of codes that can be implemented, but they follow general guidelines that can be obtained from theoretical considerations above.

The main drawback of the STTCs is their complexity. The requirement for a vector Viterbi RX has proven to be a major obstacle in their application in practical systems. Rather, STBCs combined with conventional (SISO) FECs are commonly used.

*16.2.13 Spatial Modulation

Antenna selection as presented in Section 16.2.10 aims to select the best antenna elements, and then communicate over them in a "conventional" way, such as diversity or SM. A different way for exploiting transmit selection is *spatial modulation*, where the choice of the antenna element itself conveys information. Write the number of TX antenna elements as $N_t = 2^{Q_a}$, and assume for the moment that only a single TX antenna is used at a given point in time, implying that only a single RF chain is used as well. Then split an

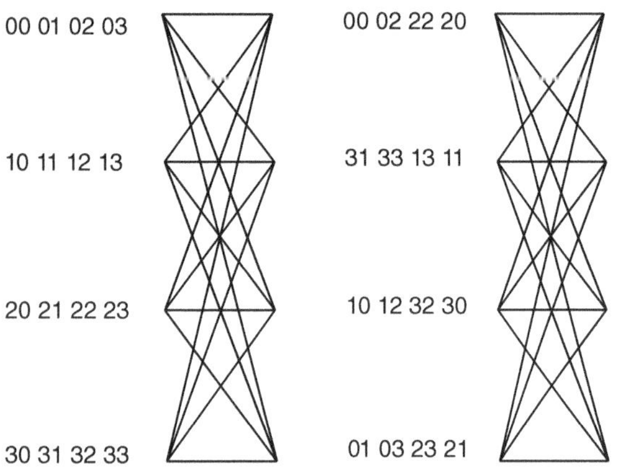

Figure 16.28 Examples for space-time trellis codes, using two antennas and QPSK. The QPSK symbols are represented by 0, 1, 2, 3. The duplet 01 denotes that a 0 is transmitted from the first antenna element and a 1 from the second. The trellis on the left-hand side implements delay diversity; the one on the right-hand side is designed according to the rank-determinant criterion.
Reproduced with permission from [Clerckx and Oestges 2013] © Academic Press/Elsevier.

information symbol containing Q bits as $Q = Q_a + Q_m$, where Q_a bits are encoded in the index of the antenna that is used for transmission, while the Q_m bits are transmitted via "normal" QAM modulation. This principle is outlined in Figure 16.29. Since the channels from the different TX antennas to the RX (assumed single-antenna for the moment) are all different, these channel characteristics allow a discrimination of which antenna served for the transmission, and thus the first Q_a bits. Generally, the RX uses standard demodulation techniques; for example in the case of per-symbol uncoded transmission, performing a search for the *nearest neighbor* of the received signal among all possible combinations of TX antennas and TX QAM symbols.

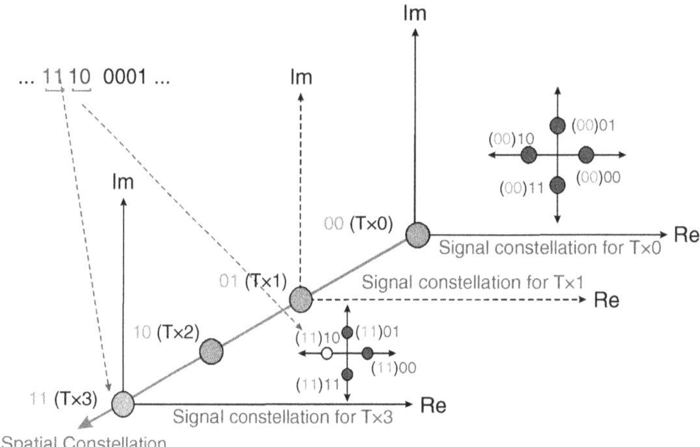

Figure 16.29 Principle of spatial modulation. Color version available at wiley.com/go/molisch/wireless3e.
Reproduced with permission from [di Renzo et al. 2013] © IEEE.

Spatial modulation depends on favorable propagation conditions: imagine the use of 4QAM and 2 TX antenna elements, whose complex channel gains are 1 and −1. It is immediately obvious that in this case, demodulation is not uniquely possible, since transmit symbol 00 from antenna A would result in the same receive signal as transmit symbol 11 from antenna B. While this exact situation is unlikely, generally any situation that results in "somewhat similar" RX signals for different antenna/QAM symbols is a disadvantage, resulting in an increased sensitivity to noise and interference. Note also that the RX needs to learn the channels from all TX antenna elements to the RX, which implies significant training overhead.

Compared with single-RF chain antenna selection, spatial modulation has a higher throughput, since it can provide additional Q_a bits/symbol; though the actual gain might be less since the average SNR is lower, and thus the size of the modulation alphabet, and thus Q_m, might have to be reduced. Compared to SM, spatial modulation offers lower hardware complexity (due to reduced number of RF chains) as well as reduced decoding complexity (since there is no inter-stream interference at the RX that needs to be resolved). On the other hand, the throughput is lower: while in SM the throughput may grow linearly with N_t, is grows at most with $\log_2(N_t)$ in spatial modulation. Furthermore, the scheme requires very fast switches (switching times need to be much shorter than a symbol duration); fortunately switches with nanosecond switching times are available.

Spatial modulation can be generalized by selecting not just one antenna at a time, but rather a group of antennas, i.e., hybrid selection. The set of selected antennas then still can carry payload, while the multiple active antennas can be used for SM or space/time codes. The number of bits carried by the "spatial information," i.e., the choice of antenna sets, can become quite large, since there are $\binom{N_t}{L}$ combinations of antennas.

*16.2.14 Intelligent Reflective Surfaces

Another implementation of MIMO that has recently emerged is *Intelligent Reflective Surfaces, IRS,* also known as reconfigurable reflective surfaces.[17] In brief, an IRS is a two-dimensional area made of a large number of *passive,* reflective (scattering) elements, where the properties of the scattering, such as the incurred phase shift, can be modified through control signals. This then creates a *metasurface* with reflection characteristics that can be controlled by software. At lower frequencies, IRSs occupy a large area – this is necessary in order to provide a large radar cross section, and is also in line with the concept of having a massive array. They would thus typically be coating the walls of buildings, or be integrated into the surface of aerial vehicles. At high frequencies, even surfaces with relatively small (in absolute dimensions) area can provide strong reflections.

Before going into further details, it is helpful to draw a distinction between IRS and massive MIMO relays (see Chapter 27). A relay receives the signal, downconverts it, possibly cleans it up (if the relay is of the decode-and-forward type), and then re-transmits it using a beamformer into the desired direction. The relay is thus an active network component, requiring significant power, and having to solve the duplexing problem. The IRS, on the other hand, solely redirects the signal via *passive* reflectors, with the only energy consumption arising from the reconfiguration of the scattering elements. The greatly reduced power consumption can be a major advantage for terrestrial deployments and is even more important for aerial deployments. Compared to a purely passive reflector (similar to mirrors sometimes placed at street intersections with bad visibility), the IRS has the advantage of reconfigurability, so that it can, e.g., redirect a beam in a way that tracks the position of a UE it wants to reach. It can furthermore even redirect a signal from a BS into multiple directions or synthesize any almost arbitrary beam pattern. A key limitation is, however, that the beam pattern is the same over the signal bandwidth; in other words, it is not possible to impose different beampatterns on different OFDM subcarriers (compare also the discussion about hybrid beamformers above). This may inherently reduce the flexibility and performance.

The main goal of an IRS is to increase the coverage in areas that cannot be reached directly by a BS. This may be particularly important at high frequencies, since diffraction is an inefficient process there. A region that is shadowed off by a large building could be covered, instead, by reflections. In contrast to a reliance on conventional reflectors such as building surfaces, which have a fixed reflection angle following Snell's law (compare Section 4.2.1), the IRS can vary the reflection angle and thus cover a larger area. Figure 16.30 shows the principle of coverage extension. Besides improvement of the coverage, an IRS can also help to avoid interference, by creating destructive interference between the signal coming directly from the BS, and the signal reflected by the IRS. Besides these tasks, IRSs have also been suggested for goals such as "invisibility cloaks" (mainly for optical radiation), analog computing, sensing for diagnostic devices, and more; we will not discuss those cases in this section.

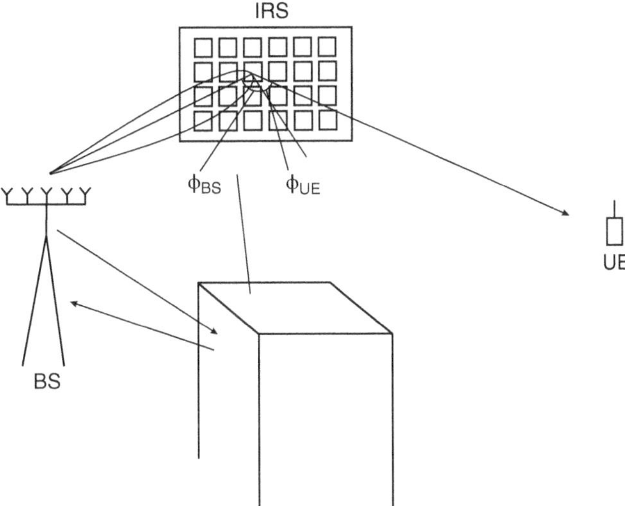

Figure 16.30 Principle of IRS: the IRS reflects a signal from the BS to a UE that has insufficient coverage from the BS. Note that the angle of incidence ϕ_{BS} and departure ϕ_{UE} at the IRS need not be identical.

The basic structure of an IRS is shown in Figure 16.31. It consists of a large number of passive scattering elements, whose reflection phase is determined by diode switches, varactor diodes, or other tunable elements that adjust the (resistive and/or capacitive) load of the reflector elements, and thus the phase shift. The states of the tunable elements are governed by a controller, which is informed by communication from the BS (and possibly the UE) about the required reflection angles. This communication may be wired or wireless. Consider now the downlink of a cellular system (uplink works completely analogously): The link from the BS to the IRS is a fixed wireless link, and will typically be LOS. If the IRS is in the far-field of the BS, only a single mode can be transmitted on this link, and is

---

[17] We note that the term "reconfigurable intelligent surfaces" is also being employed in the literature for very large active arrays, e.g., at the BS. In our notation, this would simply be a case of massive MIMO, see Sections 22.9 and 22.10.

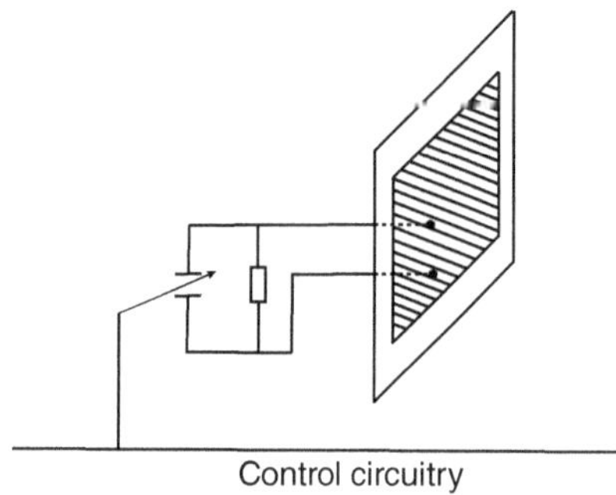

Figure 16.31 Typical building block of an IRS: a metallic patch is terminated with an adjustable RC-circuit.

reflected to the location of the UE. If we thus assume that the reflected component propagates only on a single path from the IRS to the UE, and via $\ell = 1, ...N_{\mathrm{MPC}}$ MPCs from the IRS to the UE, then the overall signal is

$$\sum_q \widetilde{G}_{\mathrm{BS}}\left(\breve{\phi}_{\mathrm{BS}}\right) \frac{1}{d_{\mathrm{BS\text{-}IRS},q}} e^{-jk_0 d_{\mathrm{BS\text{-}IRS},q}} \sum_\ell \rho_{\mathrm{IRS}}\left(\phi_{\mathrm{BS}}, \phi_{\mathrm{UE},\ell}\right) e^{j\Delta_q} e^{-jk_0 d_{\mathrm{IRS\text{-}UE},q,\ell}} a_\ell e^{j\phi_\ell} \widetilde{G}_{\mathrm{UE}}\left(\breve{\phi}_{\mathrm{UE},\ell}\right) \tag{16.154}$$

where $\widetilde{G}_{\mathrm{BS}}$ and $\widetilde{G}_{\mathrm{UE}}$ are the complex (amplitude) antenna patterns at the BS and UE, respectively, $\breve{\phi}_{\mathrm{BS}}$ the angle under which the LOS departs the BS in the direction of the IRS, $d_{\mathrm{BS\text{-}IRS},q}$ the distance from the BS to the qth IRS element (which, for the amplitude term $1/d_{\mathrm{BS\text{-}IRS},q}$ can be approximated to be independent of q), ρ_{IRS} is the reflection coefficient of a single reflector element when the direction of incidence is ϕ_{BS} and we are analyzing the signal departing in the direction $\phi_{\mathrm{UE},\ell}$, namely the angle at which the ℓth MPC departs from the IRS, a_ℓ and φ_ℓ are the amplitude gains and excess phase shifts experienced by the ℓth MPC. The angle $\breve{\phi}_{\mathrm{UE},\ell}$ is the angle under which the MPC arrives at the UE. The Δ_q is the phase shift for the qth reflector, i.e., the adaptive component. This is the core equation from which both the design of beampatterns (via the Δ_q) and the performance analysis can be obtained.

At the time of this writing, IRS is still very much in its developing stage. Topics of current investigation range from channel estimation, design of beamforming algorithms, use in multi-user scenarios, coordination between different cells, physical-layer security, impact of finite-resolution phase shifters, and many more. Details can be found in some of the review papers mentioned in "Further Reading."

*16.2.15 Orbital Angular Momenta

SM is, basically, the transmission of multiple data streams over different spatial modes. One can argue that generally, any electromagnetic field can be expanded into different sets of orthogonal modes, each of which can carry a different data stream. Thus, SM can be seen as (the most widely used) special case of Mode Division Multiplexing (MDM).

Another form of MDM that has found considerable attention over the past decade is Orbital Angular Momentum (OAM) multiplexing, which is useful in particular for pure line-of-sight connections at high frequencies (it was originally pioneered by Willner et al. for free-space optical links). The following will first present an intuitive description of the most common form of OAM multiplexing, followed by a mathematical description of the most general case.

Due to the wave/particle duality, an electromagnetic wave corresponds to a photon – while commonly photons are associated with visible light, they of course also exist for lower-frequency electromagnetic waves. A photon is characterized by various quantum numbers, including the spin, which takes on only values, namely $\pm 1/2$, and corresponds to the polarization of the wave. Another quantum number is the OAM, which can take on any integer value. In "normal" propagation, we only encounter the photon with OAM number $\ell = 0$,[18] but photons/waves with other ℓ are physically reasonable as well. Furthermore, photons/waves with different ℓ are orthogonal to each other, and thus can carry independent data streams.

Further in the interpretation of waves, the OAM gives a "twist" to the phase of the propagating wave, resulting in a helical transverse phase. In other words, when considering propagation along the z-axis, then the field on any $x - y$ cut (i.e., fixed z) shows a phase rotation of $2\pi\ell$ over any circle centered at $(0//0/z)$. Note that this is different from circular polarization, where the orientation of the E vector rotates as the wave propagates along the z-axis (compare Section 8.2.4), see Figure 16.32. Related to the helical phase structure, the intensity profile of a wave with $|\ell| > 0$ is ring-shaped, with a central null, and a maximum that is the farther from the center, the larger the $|\ell|$ is, see Figure 16.33.

---

[18] Following the convention in the OAM literature, we denote the OAM number as ℓ. Note that this has nothing to do with the ℓ that we otherwise use in this book to index MPCs.

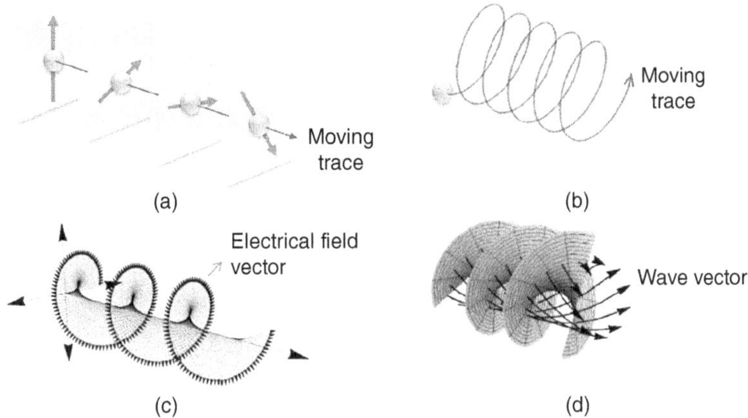

Figure 16.32 Two cases of angular momenta: (a) a spinning object carrying spin and (b) an orbiting object carrying OAM. (c) A circularly polarized wave carrying spin, and (d) the phase structure of an OAM-carrying wave. Color version available at wiley.com/go/molisch/wireless3e. Reproduced with permission from [Willner et al. 2015] © Optical Society of America.

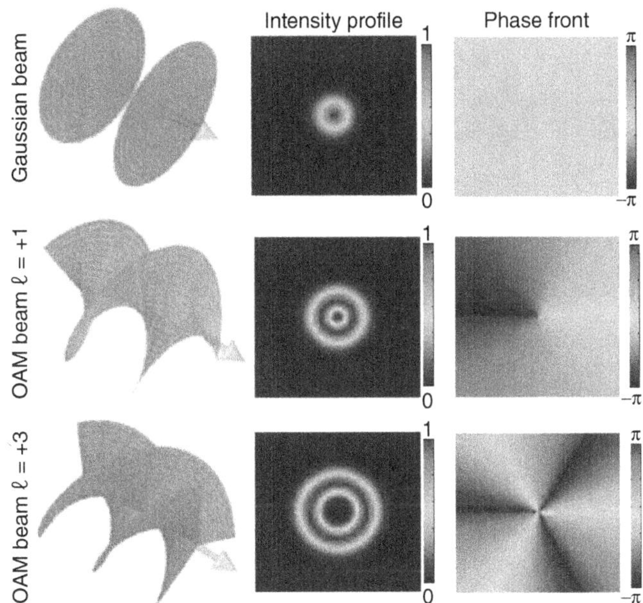

Figure 16.33 Intensity (left) and phase (right) figure for different OAM modes. Color version available at wiley.com/go/molisch/wireless3e. Reproduced with permission from [Ren et al. 2014] © IEEE.

There are a number of different ways in which OAM waves can be generated. The conceptually simplest one is a Spiral Phase Plate (SPP) as shown in Figure 16.34. A dielectric plate is shaped in such a way that it induces a phase shift that depends on the azimuthal angle. Different OAM modes are then created by using different SPPs, and the modes can be combined with beam combiners at the TX. At the RX, the incident beam is split up, and each of the components is, via a SPP, converted back to a fundamental Gaussian mode $\ell = 0$ that is then detected with a regular horn antenna. Note that the received beam that has passed a particular SPP still contains multiple modes, but horn antennas are "blind" to nonfundamental modes, since such modes, when integrated over the aperture of the antenna, result in zero energy. Thus, imagine a signal incident at the RX that contains modes $\ell = -3, -1, 1, 3$. When it is sent through

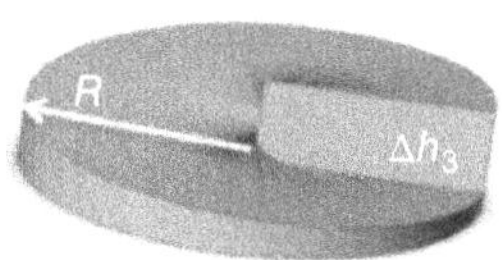

Figure 16.34 Spiral phase plate. Δh_3 is chosen to induce an extra phase shift of $\ell \cdot 2\pi$.

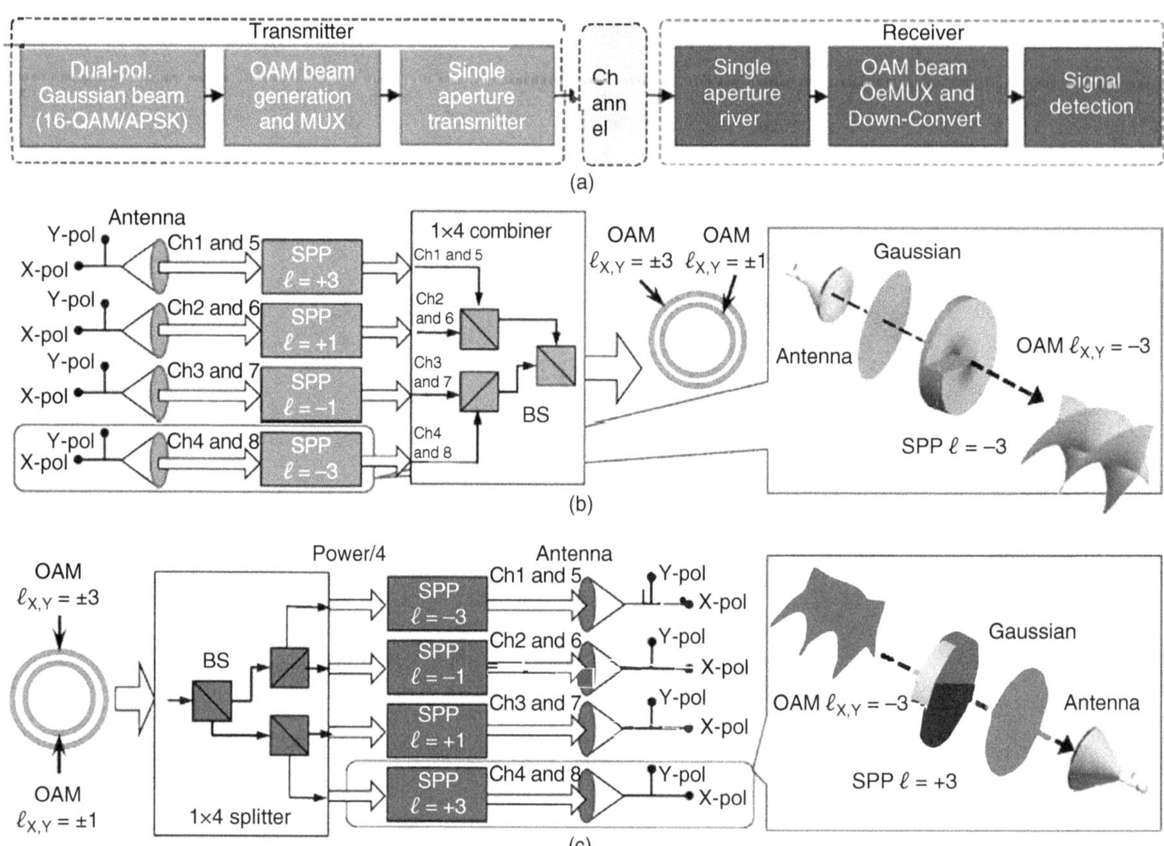

Figure 16.35 System for OAM multiplexing combined with polarization multiplexing. (a) Block diagram, (b) TX multiplexing. (c) RX demultiplexing. Color version available at wiley.com/go/molisch/wireless3e.

an SPP with $\ell = 3$, it is converted to modes $\ell = 0, 2, 4, 6$, of which only the first mode (the former $\ell = -3$ mode) creates a nonzero signal at the horn antenna output. A block diagram of a transceiver is shown in Figure 16.35.

Of course, SPPs are only one of many possibilities for generating OAM modes. An important alternative is the use of circular antenna arrays, where the phase difference between different antenna elements can be used to generate different OAM modes. It is also noteworthy that OAM multiplexing can be combined with other forms of multiplexing, such as polarization multiplexing or frequency multiplexing.

As is well known, a "beam" (to use optical notation) excited by a finite aperture diverges as it propagates. It is important to note that the beam divergence is different for different OAM modes. The divergence angle is asymptotically

$$\alpha_{\text{diverge}} = \sqrt{\frac{|\ell| + 1}{2}} \frac{2}{k_0 w_0} \tag{16.155}$$

where k_0 is the wavenumber and w_0 the beam waist. This indicates that higher-order modes spread out in space more quickly than lower-order modes, and that therefore a larger-aperture receive antenna is required to collect a required amount of energy. The size of the "beamspot," i.e., the radius of the disc that contains most of the energy, depends on the following variables: (i) the distance between TX and RX, (ii) the wavelength, via k_0, (iii) the TX aperture, via the beamwaist w_0, and (iv) the OAM index ℓ. For reasonably sized antennas (30 cm), a mm-wave link can transmit up to $|\ell| = 3$ over distances of tens of meters; higher frequencies can cover longer distances. For traditional cellular frequencies, very large antenna apertures would be required. It is also noteworthy that the ratio of the powers between a higher-order mode and the fundamental Gaussian mode $\ell = 0$ decreases with distance, so that at large distances (far beyond the Rayleigh distance), only a single mode effectively exists. This situation is analogous to LOS-MIMO, where in the farfield of an antenna array only a single mode exists. In other words, the scaling laws for these two types of multiplexing are the same: as the RX moves far beyond the Rayleigh distance, the number of supported modes converges to 1. The differences thus lie more in the practical implementation aspects, in particular, the fact that OAM demultiplexing can be easily achieved by analog means, which has advantages especially at very high-speed communications (compare the computational effort for spatial-multiplexing RXs discussed in Section 16.2.9).

We have above assumed that OAM mode separation operates ideally. However, there are a number of effects that can create inter-modal interference:

- *Nonideal SPPs and other components*: Due to manufacturing tolerances, the generated beams are not pure, but contain components from other modes (and similarly, at the RX, an SPP might create intermodal interference).
- *Wide-band signals*: SPPs and similar devices are designed to create the ideal phase shift at the nominal center frequency of the signal. For very wideband signals, the upper and lower band edges might have significant deviations and thus create modal impurities.
- *Misalignment*: both a tilt of the TX or RX apertures with respect to the nominal propagation axis and a lateral displacement of the antennas from the propagation axis, lead to mode impurities.
- *Multi-path*: multi-path affects modal purity in two ways: firstly, any reflection changes the sign of the OAM mode, e.g., a $\ell = 1$ mode becomes a $\ell = -1$. Secondly, since an MPC is incident at an angle from the nominal propagation axis, the interference pattern with the direct component creates modal impurities. It is noteworthy that for OAM, multi-path is usually considered a drawback, not an advantage as in SM.
- *Atmospheric disturbances and shadowing*: turbulence, as well as shadowing by objects, can distort the field and thus create modal impurities.

There are two main types of countermeasures: firstly, it is possible to use not each OAM mode in a particular range, but "space them out"; e.g., using only modes $-3, -1, 1, 3$ (and not $-2, 0, 2$). This, obviously, reduces the achievable capacity. Secondly, one can use digital signal processing to counteract the modal impurities at the RX; however, this has the drawback of increased computational effort. The particular choice of the trade-off depends on the application.

Let us now turn to a mathematical description, which covers a more general case than the description above. Consider an electromagnetic wave propagating along the z-axis. Then, assuming a paraxial approximation, the electrical field in the cylindrical coordinates (r, ϕ), is

$$E_{\ell,p}(r, \phi, z, w_0) = K_{\ell,p} \frac{1}{w(z)} \left(\frac{\sqrt{2}r}{w(z)} \right)^{|\ell|} \exp\left(-\frac{r^2}{w^2(z)} \right) L_p^{|\ell|} \left(2\frac{r^2}{w^2(z)} \right) \exp\left(j\left[-k_0 z + \ell\phi + k_0 \frac{r^2}{2R(z)} - \psi(z) \right] \right) \qquad (16.156)$$

where $L_p^{|\ell|}$ is an associated Laguerre polynomial, and $p = 0, 1, 2, ...$, and $\ell = ... -2, -1, 0, 1, 2, ...$ Furthermore, $K_{\ell, p}$ is normalization constants such that $\int d\phi \int r dr |E_{\ell, p}(r, \phi, z, w_0)|^2 = 1$;

$$K_{\ell,p} = \sqrt{\frac{2p!}{\pi(p + |\ell|)!}} \qquad (16.157)$$

the so-called Gouy phase is

$$\psi(z) = (|\ell| + 2p + 1) \arctan(z/z_R) \qquad (16.158)$$

the Rayleigh range is

$$z_R = \pi w_0^2 / \lambda \qquad (16.159)$$

and beam waist and curvature radius are

$$w(z) = w_0 \sqrt{1 + (z/z_R)^2} \qquad (16.160)$$

$$R(z) = z\left[1 + \left(\frac{z_R}{z} \right)^2 \right]. \qquad (16.161)$$

Note that all these modes are propagating along the same axis.

We can thus see that the field can be expanded into a doubly infinite set of modes (p, ℓ), which are all orthogonal to each other, and which each can carry a separate data stream. Most important is the interpretation of the mode numbers p and ℓ. The fundamental (Gaussian) beam has $p = 0, \ell = 0$. For a fixed z, the field strength is a Gaussian function of the radius (and, as can be shown, so is the intensity given by the Poynting vector). Next let us consider $p > 0, \ell = 0$. It can be immediately seen from Eq. (16.156) that p impacts the dependence of the amplitude, but note the phase, on r, via the term $L_p(r)$, which shows oscillations. Phase and intensity plots of the 01, 11, and 21 modes are shown in Figure 16.36. Obviously, by setting $p = 0$ and only considering the variations of ℓ, we obtain the OAM modes discussed above.

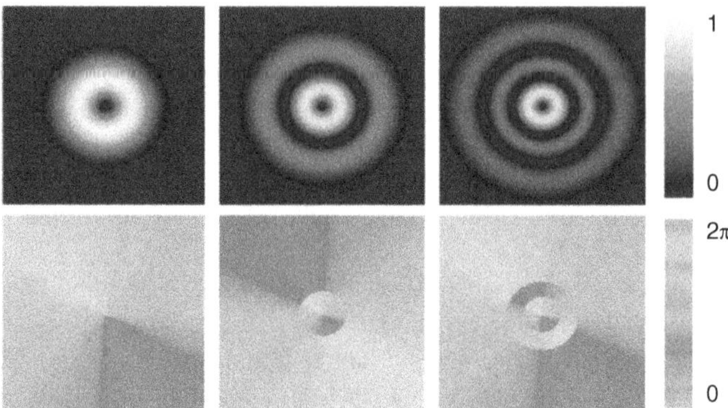

Figure 16.36   Normalized intensity (top) and phase (bottom) plots of Laguerre-Gaussian modes: LG01, LG11, and LG21 (left to right) showing the $p + 1$ concentric rings and the effect on the phase pattern. Color version available at wiley.com/go/molisch/wireless3e.
Reproduced with permission from [Yao and Padgett 2011] © Optical Society of America.

Further Reading

An overview of early smart antennas is given in the paper by [Godara 1997], and the paper collections [Rappaport 1998, Tsoulos 2001]. The topic of space-time processing for SIMO systems, which forms a basis for smart antenna systems, involves many topics we have already discussed in previous chapters, and we refer to those references. Maximum-ratio transmission was first suggested in [Lo 1999]; delay diversity in [Seshradi and Winters 1994]. Alamouti codes were developed in [Alamouti 1998]. The exposition of diversity in MIMO systems in [Vaughan and Andersen 2003] is very clear and didactic. Reciprocity calibration is described, e.g., in [Shepard et al. 2012], while further developments and detailed experimental results can be found in [Vieira et al. 2017]. Limited-feedback considerations for MIMO were initiated by the seminal paper of [Love et al. 2003]; a comprehensive review of limited feedback can be found in [Love et al. 2008]. Generally, the use of SIMO, MISO, and MIMO for diversity and beamforming is often subsumed in the description of MIMO in general, see the corresponding references for this.

MIMO including SM has been one of the dominant research topics in wireless in the past 30 years, and a plethora of books and survey papers have been written. We mention here only some personal favorites. Among the earliest- and still very worth reading – books are [Paulraj et al. 2003] and [Tse and Visvanath 2005], as well as the edited books of [Tsoulos 2006] and [Biglieri et al. 2007]. More recently, the books of [Clerckx and Oestges 2013] and [Heath and Lozano 2018] are very worth reading. Also, the review papers by [Gesbert et al. 2003], [Diggavi et al. 2004], and [Paulraj et al. 2004] are excellent introductions to the topic.

For understanding the basic concept of capacity in MIMO sytems and the derivations of the capacity distributions and bounds, the original papers by [Foschini and Gans 1998] and [Telatar 1999] are still worth reading. For the case of channels that are known at the TX, [Raleigh and Cioffi 1998] is the first paper, and still very interesting to read. [Goldsmith et al. 2003] review the information-theoretic capacity for different assumptions about the CSI. The book chapter [Lozano et al. 2008] debunks several common myths about MIMO capacity. The possibility for a signaling scheme that allows MIMO communications without CSI at the RX was first pointed out by [Marzetta and Hochwald 1999].

The impact of channel correlation on the capacity is described in [Shiu et al. 2000] and [Chuah et al. 2002]. In frequency-selective channels, MIMO is most often combined with OFDM, see [Stuber et al. 2004] and [Jiang and Hanzo 2007]. Keyhole channels are elucidated in [Gesbert et al. 2002], [Chizhik et al. 2002], and [Almers et al. 2006]. For LOS MIMO, [Driessen and Foschini 1999] and [Bohagen et al. 2007] provide important insights. Channel estimation techniques for MIMO are surveyed in [Ganesh and Kumari 2013].

[Foschini et al. 2003] is devoted to reviewing different layered space-time structures. [Kailath et al. 2001] give an excellent overview of receive algorithms, and in particular of sphere decoding, the description of which has inspired the exposition in this chapter. The impact of low-resolution ADCs is discussed in [Mo and Heath 2015], [Li et al. 2017a], and surveyed in [Liu et al. 2019]. MIMO systems with antenna selection are reviewed in [Molisch and Win 2004], and their capacity investigated in detail in [Molisch et al. 2005]. Hybrid beamforming was introduced in a series of papers [Molisch and Zhang 2004, Zhang et al. 2005, and Sudarshan et al. 2006]; hybrid beamforming for sparse channels is discussed in [El Ayach et al. 2014]. A survey of hybrid beamforming is provided in [Molisch et al. 2017].

For space-time coding [Diggavi et al. 2004] gives an extensive introduction to space-time coding while [Tarokh et al. 1998] and [Tarokh et al. 1999] give more mathematical details for STTC and STBC respectively. The books of [Jafarkhani 2005] and [Larsson and Stoica 2008] are mostly dedicated to space-time codes.

An excellent survey of spatial modulation is given in [Renzo et al. 2013]. IRS is reviewed in [Gong et al. 2020] and [Liu et al. 2021]; the concept of IRSs is also closely related to adaptive reflectarrays, for which there is a rich literature in the antenna community. MDM via OAM was pioneered by Willner and collaborators, e.g. [Willner et al. 2012]. For an extensive survey for OAM for optical communications see [Willner et al. 2015] while other applications are described in [Yao and Padgett 2011]. First experiments of OAM multiplexing with two single-polarized streams in the RF (2 GHz) domain were shown in [Tamburini et al. 2012]. Transmission of four dual-polarized OAM modes at mm-wave frequencies was demonstrated in [Yan et al. 2014].

For updates and errata for this chapter, see https://wides.usc.edu/students.html#textbooks

Exercises

See Sec. 36.16 of Exercises.pdf at wiley.com/go/molisch/wireless3e

17

Hardware Aspects

17.1 Introduction

17.1.1 Motivation

Up to now, we discussed the transmission of information under the assumption that the transceivers, and thus data generation and reception, are ideal – in other words, the only impairments are the noise and distortion introduced by the channel. In fact, the Transmitter (TX) was idealized to the point where we assumed that the baseband signal is shifted in frequency to bandpass and amplified perfectly linearly, and similarly, at the receiver (RX), the signal was amplified linearly, shifted to baseband, and converted to digital representation with infinite precision.[1] However, such ideal transceivers do not exist in reality, and the nonidealities of the hardware can be a major factor in wireless system design.

A further motivation for reviewing wireless hardware arises from the use of Software-Defined Radio (SDR) in both system development and teaching; they allow to flexibly create and process, in real time, essentially arbitrary complex baseband signals. However, interfacing those SDR outputs with actual antennas and thence wireless channels still requires the implementation of hardware. Suitable so-called "Radio Frequency (RF) daughterboards" may or may not be commercially available, but in either case, any experiment with such SDRs must investigate and understand the hardware constraints.

In this chapter, we thus give a brief overview of the hardware of wireless transceivers, and the constraints that it puts on the modulation, demodulation, and other processing. The material in this chapter is intended as a brief primer or refresher; the emphasis is on the description of the macro-characteristics of the blocks, and not how to design the individual components. Those interested in more details will find suitable references in the "Further Reading" section.

17.1.2 Structure of RF Transceivers

Impairments of wireless signals can generally be categorized as follows:

- *Noise*: as discussed in Chapter 3, thermal noise, as well as man-made noise, that falls into the band of interest, can impair the desired signal. Additional noise may also be created by the RX, where we distinguish between the following categories:
  - *Thermal noise*:
    - Signal-to-Noise Ratio (SNR) deterioration because of noise created by active components, e.g., amplifiers and active mixers.
    - SNR deterioration due to attenuation by passive components.
    - SNR deterioration due to imperfect matching between components (in the sense that the impedance matching is not optimized for SNR maximization).
    - SNR deterioration due to mixing of noise at the image frequency and Local Oscillator (LO) harmonics.

    The impact of a particular noise source depends on its placement within the transceiver – the same absolute amount of noise has much less impact when added to an-already strong signal.
  - *1/f noise*: this is noise created in many electronic components and – as indicated by the name – is strongest at low frequencies and decreases with increasing frequency.
  - *Phase noise*, in particular of the LO, creates a broadening of the received signal spectrum and temporal drift of the phase of this signal over time. It impacts, e.g., the intercarrier interference in Orthogonal Frequency Division Multiplexing (OFDM) (see Section 15.7). In the downconversion process, the widened spectrum of the LO may also lead to mixing of out-of-band interference into the band of interest, this is known as reciprocal mixing.

---

[1] We have briefly alluded to such problems as Peak-to-Average Power Ratio (PAPR), and difficulty of creating narrowband filters.

- o *Quantization noise*: the analog received signal has to be quantized before the digital baseband processing that we discussed in Chapters 10–16 can happen. The quantization itself introduces noise (in the sense that the quantized signal is only an approximation of the unquantized signal), which should be minimized.

- *Interference*: a second key source of signal impairment is interference. Here we distinguish:

  - o *Intra-channel interference*: interference signals that are in the same band at the desired signal. They are an unavoidable part of cellular systems (and all other systems where multiple TXs operate on the same frequency channel, i.e., any systems applying frequency reuse), see Chapter 21. While this type of interference is important, it has relatively little impact on transceiver hardware, because it needs to be handled by the digital signal processing.

  - o *Adjacent-channel interference*: this is interference stemming from signals typically generated by the same system, but operating in different frequency channels. Such signals can have a significantly larger signal strength than the desired signal and are operating at a frequency very close to that of the desired channel. They strongly impact the design of filters in the system, as well as linearity of RXs, phase noise requirements, etc.

  - o *Out-of-band interference*: this is interference stemming from TXs belonging to a different system. Such signals are generally farther away from the band of interest but may have much higher power than the desired signal. Their impact is governed both by the filter design and the linearity of the RX.

- *Signal distortions*: while we have previously assumed that the signal received in the RF domain is simply frequency-translated into baseband, in reality, the signal may be distorted during the reception process, in particular by nonlinearities of the RX.[2]

TXs can create the following impairments:

- *Noise*:

  - o thermal noise generated by TX components has little impact on the receive SNR, since the power levels of all signals are high compared to thermal noise. However, it can lead to wideband emissions that create adjacent-channel and out-of-band interference.

  - o Phase noise: creates adjacent-channel and out-of-band emissions.

  - o Quantization noise: this is determined by the capabilities of the Digital-to-Analog Converter (DAC), and impacts the Error Vector Magnitude (EVM) of the transmitted signal. Depending on the design of the digital processing of the TX, quantization effects may be avoided or mitigated.

- *Interference*: the TX might create signal components both in other channels of the same service and outside the band of service. The transceiver must be designed to fulfill the following limits:

  - o Adjacent-channel interference: typically limited by design specifications of the standard.

  - o Out-of-band interference: typically limited by the frequency regulator; needs to be mitigated by filters.

- *Signal distortion*: the TX can create strong signal distortions because nonlinearities particularly in the TX amplifiers are stronger than in the RX. These signal distortions lead to additional detection errors, and can furthermore increase the adjacent-channel and out-of-band interference.

To summarize, transceivers can be characterized by the amount of noise they add, the ability to suppress interference, and the signal distortions they introduce. Key characteristics are:

- *Noise figure*: degradation of the SNR due to noise created by the hardware. Related to this is the RX sensitivity, i.e., the required power arriving at the antennas so that signals can be demodulated and decoded.

- *Maximum input/output power level*: for the RX/TX, it describes what is the maximum input/output power level that can be received/created without undue distortions or damage to the hardware. It is mostly determined by nonlinear distortions. Related to this is the maximum dynamic range, which describes the range between the lowest meaningful RX sensitivity, and the maximum input level.

- *Channel selectivity*: for the RX, this is the amount by which adjacent-channel interference can be suppressed. Furthermore, out-of-band rejection can be characterized similarly. For the TX, power spectral density of adjacent-channel transmissions and out-of-band transmissions is an important characteristic.

- *Phase noise*: this can be characterized by the phase noise spectrum, see Section 17.6.

We now give a very brief description of the working of a typical wireless transceiver, based on the block diagram in Figure 17.1, and how it relates to the above "macro" characteristics. A more detailed characterization of the individual components follows in Sections 17.3–17.7.

An antenna, or – as shown – an antenna array[3] is connected via a duplex filter (in the case of FDD) or a switch (in the case of TDD) to the TX and RX branches of the transceiver. This ensures that the transmit signal is connected to the antennas (and not the

---

[2] According to this definition, quantization could also be seen as a signal distortion. However, it is more common to treat it as noise.

[3] As can be seen, the array considered here is a phased array, consisting of multiple antenna elements, phase shifters, and a splitter/combiner; we show this structure because it contains several non-ideal RF components, and is widely used for mm-wave systems. Alternatively, we can consider arrays where each of the elements has a separate up/downconversion chain, i.e., multiple copies of the structure outside the dashed boxes.

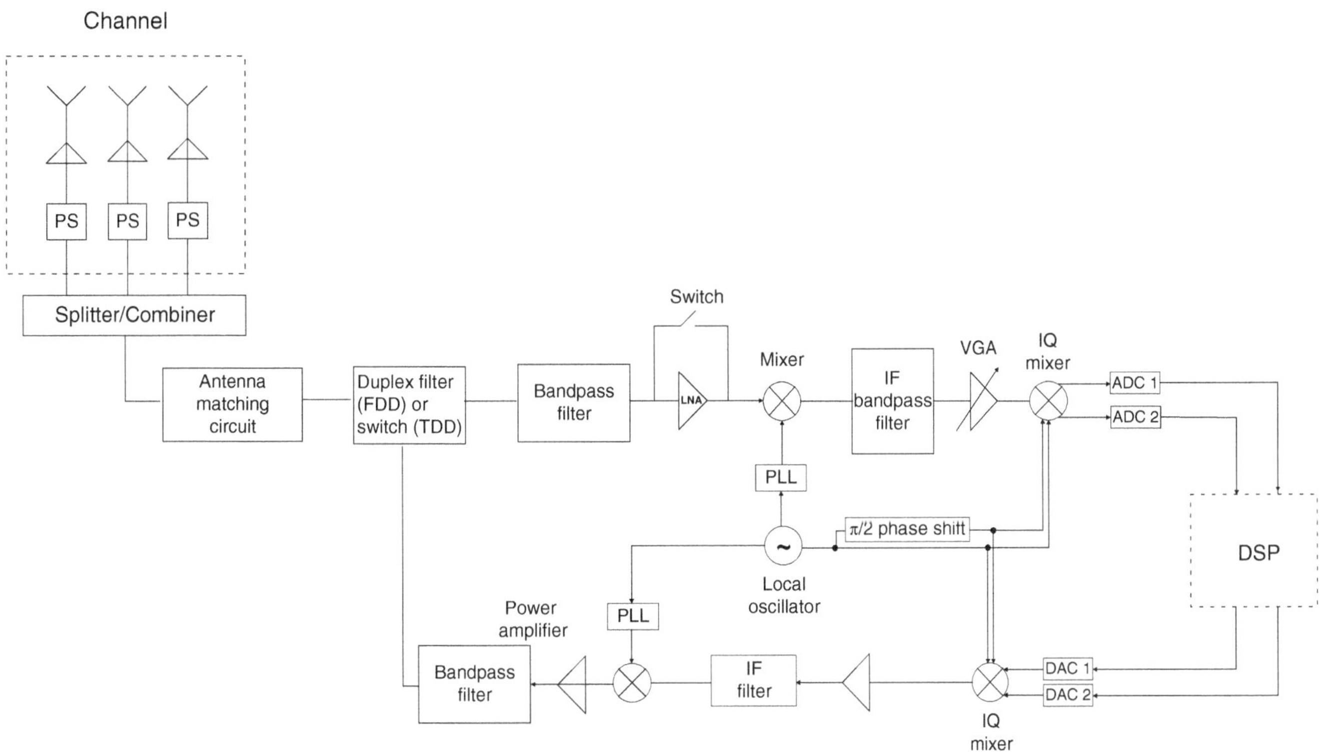

Figure 17.1 Block diagram of a typical wireless transceiver. For TDD, TX, and RX can share a PLL.

RX branch), while the signals coming *from* the array are connected to the RX branch only. Between the array and the filter/switch, there is a matching circuit that avoids incoming radiation being reflected back to the antenna, as this would reduce the signal energy available to the RX, and thus deteriorate the SNR (and similarly reduce the power available from the TX to be transmitted).

Consider now first the RX branch. The first component is a bandpass filter, which is typically designed to let through all frequency channels of the desired system (e.g., the multiple 5 MHz channels of an Long-Term Evolution (LTE) system in the 2.1 GHz band). This filter is followed by a Low-Noise Amplifier (LNA); this amplifier might be bypassed if the received signal is very strong. The reason why the filter is *before* the LNA is to reduce the out-of-band interference, which otherwise might be strong enough to saturate the LNA, thus making any signal reception impossible (a good example for the impact of signal distortions on the RX design). Since the filter is before the first amplifier, it is important that it has only small attenuation of the desired signal (to avoid SNR degradation). The design of the bandpass filter must trade off the interference suppression with the desired-signal attenuation.

The signal is then downconverted in the frequency domain to an Intermediate Frequency (IF). This process, in which it is mixed with the signal from an LO, which might in turn drive a Phase-Locked Loop (PLL), not only affects the desired frequency shift, but might also lead to a broadening of the spectrum (due to phase noise), nonlinear distortions (due to mixer nonlinearities), and additional interference (due to so-called image frequencies, and reciprocal mixing). This downconverted signal is then sent through another filter, which now selects the desired frequency channel within the system bandwidth; it again trades off suppression of interference (especially adjacent-channel interference) with attenuation of the passband signal; however, since the signal has already been amplified, desired-signal attenuation is less of a concern compared to the RF filter.

The IF signal is then further amplified by a variable gain amplifier (VGA) that ensures that its output always has a similar voltage level independent of the power received at the antenna elements. This is important because the resulting signal will be either directly sampled by an Analog-do-Digital Converter (ADC), or downconverted to baseband and sampled by an ADC there. Since ADCs are designed for a fixed range of received voltage, the output of the VGA has to be matched to it. An In-Phase–Quadrature Phase (IQ) mixer creates the real and imaginary part of a complex baseband signal from a real passband signal, by mixing with two LO signals that are 90 degree out of phase. Note that if the signal is converted via an IQ mixer to baseband, further signal distortions can arise because the two signal branches (see Section 17.8) might experience different amplifications in the mixer. The digitized signal is sent to the Digital Signal Processor (DSP) for all the processing (equalization, decoding, demodulation, etc.) that we have discussed in Chapters 11–16.

At the TX (lower part of the figure), the baseband signal created by the DSP is first converted from the digital to the analog domain. If the baseband signal is complex, DACs are required both for in-phase and quadrature-phase components. Those signals are then upconverted to an IF via an IQ mixer; similarly to the RX case, imbalance in the IQ mixer can lead to signal distortions. Alternatively, the baseband processing can create a real (passband at IF) signal. In either case, the resulting signal is filtered and then upconverted to an even high frequency with an LO signal that might be affected by phase noise. The upconverted signal is then amplified by a Power Amplifier (PA); this is the step that usually generates the largest signal distortions due to the nonlinear characteristics of PAs. The distortions in turn give rise to adjacent-channel and out-of-band emissions, which have to be reduced by further bandpass filters. Also in this case, there is a trade-off between the signal attenuation and the suppression of the undesired signal components.

In the remainder of this chapter, we will first review some fundamental concepts related to the characteristics of transceivers, namely noise, nonlinearities, and matching between components. We then describe in detail, one by one, the individual components that make up a transceiver, as discussed above (Sections 17.3–17.7). This will be followed by a discussion of the finer points of the up/downconversion, and how the properties in particular of filters and oscillators impact the choice of the RF architecture (Section 17.8). Further system aspects revolve around the definition of the admissible emission spectrum of a TX (Section 17.9), and the question of whether simultaneous transmission and reception at the same time, and at the same frequency, is possible (Section 17.10).

17.2 General Concepts

17.2.1 Noise

As mentioned in Chapter 3, an unavoidable impairment of the received signal is noise. In particular, thermal noise stems from the environment that is usually at a temperature around 300 K. The power stemming from additional noise sources, and even interference, is often described by an equivalent increased noise temperature. In any case, the noise power spectral density is related to the noise temperature as

$$N_0 = k_{\mathrm{B}} T_0 \tag{17.1}$$

which at 300 K is -174 dBm/Hz. The noise power is

$$P_{\mathrm{n}} = N_0 B \tag{17.2}$$

or, on a logarithmic scale (all logarithms in this chapter are for base 10),

$$10 \log(k_{\mathrm{B}} T_0) + 10 \log(B) \text{ dBm.} \tag{17.3}$$

Note that this is the power actually delivered to a load (the voltage of the equivalent noise source with internal (real) impedance R that delivers power to a matched load R is $V = \sqrt{4Rk_{\mathrm{B}}T_0 B}$). This thermal noise, which stems from the environment, is further increased by the components of the RX chain; in the following, we calculate this increase.

First note that an ideal amplifier will, to an equal degree, amplify signal and noise, so that it does not change the SNR. However, a real amplifier adds internal noise to the amplified input noise, so that the overall SNR is decreased. This effect is described by the noise factor F, which is defined as the SNR at the input divided by the SNR at the output

$$F = \frac{P_{\mathrm{s,in}}/P_{\mathrm{n,in}}}{P_{\mathrm{s,out}}/P_{\mathrm{n,out}}} \tag{17.4}$$

which is always ≥ 1. When the noise factor is expressed in dB, it is commonly called "noise figure" (though sometimes noise figure is used for F on a linear scale as well).

Not only amplifiers but also attenuators have a noise factor: an attenuator decreases the strength of the desired signal and the thermal noise to an equal degree; however, the attenuator itself has a temperature of around 300 K, and thus generates output noise at the same level at the input noise; it thus also decreases the SNR. The noise factor of an attenuator with loss L is thus itself L; or if we prefer to describe the attenuator by its gain $G = 1/L < 1$, the noise factor $F = 1/G$. The additional noise from electronic components can be represented formally by an equivalent noise temperature T_{e}, which is the temperature at which a matched resistor at the input of a circuit would have to be in order to deliver P_{n} to the circuit, i.e., $T_{\mathrm{e}} = P_{\mathrm{n,excess}}/(k_{\mathrm{B}}B)$, so that $F = 1 + T_{\mathrm{e}}/T_0$. The total noise output power is then $Gk_{\mathrm{B}}T_0 B + Gk_{\mathrm{B}}T_{\mathrm{e}}B$.

In most RXs, there are a number of (cascaded) components, such as amplifiers, attenuators (e.g., lossy transmission lines, filters), etc. We thus derive in the following the effective noise factor of a cascade of two components. Let each of the components be characterized by its gain G_i, and noise factor F_i, $i = 1, 2$. By definition, the noise power at the output of the first component is

$$P_{\mathrm{n,1}} = P_{\mathrm{n,in}} G_1 F_1 = kT_0 B G_1 F_1 \tag{17.5}$$

and from this, we can conclude that the excess noise created by the first component is

$$kT_0 B G_1 (F_1 - 1). \tag{17.6}$$

Analogously, the excess noise power created by the second component is

$$kT_0 B G_2 (F_2 - 1). \tag{17.7}$$

The total noise power at the output of the second component is thus

$$P_{\mathrm{n,out}} = kT_0 B G_1 G_2 + kT_0 B G_1 (F_1 - 1) G_2 + kT_0 B G_2 (F_2 - 1) \tag{17.8}$$
$$= kT_0 B [F_1 G_1 G_2 + (F_2 - 1) G_2] \tag{17.9}$$

where the first term of the rhs in the first line is the external noise multiplied by both amplifier gains, the second term is the excess noise created in the first amplifier multiplied by the gain of the second amplifier, and the third term is the excess noise created in the second amplifier. The signal power at the output is the signal power at the input of the first component, multiplied by $G_1 G_2$, the total gain of the cascade. The total noise factor of the cascade is thus, following the definition of (17.4),

$$F_{\text{cascade}} = F_1 + \frac{F_2 - 1}{G_1}. \tag{17.10}$$

It is obvious that for a general cascade, the total noise factor is

$$F_{\text{eq}} = F_1 + \frac{F_2 - 1}{G_1} + \frac{F_3 - 1}{G_1 G_2} + \cdots \tag{17.11}$$

where the F_i and G_i are noise factors and gains of the individual stages in linear units (not in dB). We can see immediately that it is important to put an amplifier with a high gain as close to the RX antenna as possible, since in this case, the excess noise created by later components has little impact. Furthermore, this first amplifier should have a low noise factor. In a link budget, we then have to add $\log_{10}(F_{\text{eq}})$ to the RX power required to achieve a certain system performance (compared to the ideal-RX case), see Chapter 3.[4]

Example 17.1 *Consider an RX that has in cascade an input amplifier ($G_1 = 5$ dB, $F_1 = 3$ dB), followed by a lossy transmission line ($G_2 = -2$ dB, $F_2 = 2$ dB), and another amplifier ($G_3 = 20$ dB, $F_3 = 3$ dB). What is the noise factor of the cascade?*

We first need to convert the gains and noise figures into linear units, resulting in $G = 3.2, 0.63, 100$ and $F = 2, 1.6, 2$. Inserting into (17.11), we get $F_{\text{eq}} = 2.68$. If we could swap the first and second amplifier, the noise factor would reduce to $F_{\text{eq}} = 2.02$, confirming that it is – all other things being equal – preferable to put the component with the highest gain closest to the input. If the lossy transmission line is the first component (because it is, e.g., impractical to bring the amplifier closer to the RX antenna), the noise factor becomes 3.2.

17.2.2 Linearity and Dynamic Range

Communication-theoretic system investigations often assume that RF components have linear relationships between input and output voltage. However, this is an (over)simplification in particular for active components like amplifiers but also diodes and some other passive components. Those generally have a nonlinear relationship, typically (roughly) exponential or quadratic, between input voltage and output current, which can only be *approximated* by a linear description, and that only if the "voltage swing," i.e., range of the input signal, is sufficiently small. Furthermore, active components experience saturation as the output power increases – the output voltage amplitude is limited by the supply voltage. Saturation is, of course, also a nonlinearity. The following discussion will use, for concreteness, amplifiers as the quintessential nonlinear components.

Nonlinearities are generally undesirable, because they lead to intermodulation between different frequency components in the signal of a single user, and – in a multi-user system (see Part IV) – to the mixing of signals from multiple users. Either of these effects is difficult to undo at the RX and thus leads to an increase of the bit error probability of the system.

We first consider the issue of amplifier saturation: as the input power increases, the output power does not increase quite as strongly. A widely used parameter to describe this effect is the 1-dB compression point, i.e., the power level at which the actual output power is 1 dB lower than the output power one would obtain if truly linear amplification would hold, see Figure 17.2. Care must be taken to distinguish between the input 1-dB compression point, and the output 1-dB compression point (these are, of course simply related as $P_{\text{1dB,out}} = P_{\text{1dB,in}} + G - 1$; in data sheets for power amplifiers, one finds mostly the latter.

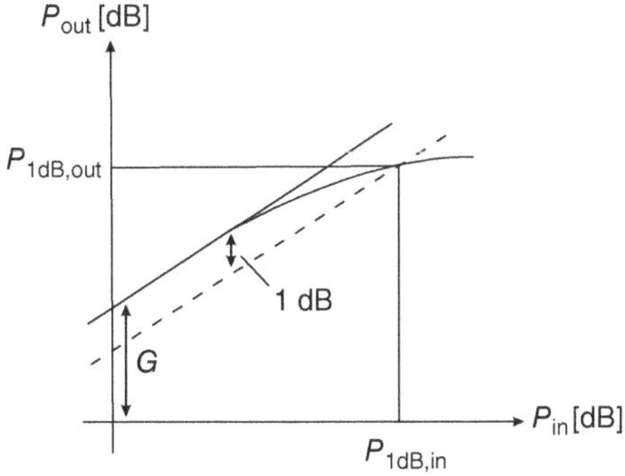

Figure 17.2 Principle of the 1 dB compression point.

[4] All of the above considerations are based on the RX antenna "seeing" an environment with T_0 temperature; for different temperatures, computations based on noise temperature (and not noise figures) are preferable [Pozar 2005].

Let us now turn to a more mathematically detailed explanation of amplifier nonlinearities. The input–output relationship for any amplifier can be written as a Taylor series expansion,

$$y(t) = b_1 x(t) + b_2 x(t)^2 + b_3 x(t)^3 \tag{17.12}$$

and we neglect the higher-order terms $\mathcal{O}(x(t)^4)$, as well as any Direct Current (DC) offset. When the input is a pure sinusoid with amplitude A, the output becomes

$$y(t) = \left[b_1 + \frac{3}{4}b_3 A^2\right] A \cos(\omega t) + \left[\frac{1}{2}b_2 A\right] A \cos(2\omega t) + \left[\frac{1}{4}b_3 A^2\right] A \cos(3\omega t). \tag{17.13}$$

Usually, the terms involving $\cos(2\omega t)$ and $\cos(3\omega t)$ can be ignored, because they are far outside the band of interest and can be easily filtered out. This also motivates to henceforth ignore the quadratic nonlinearities, i.e., set $b_2 = 0$. The third-order nonlinearity however impacts the band of interest, by changing the amplification factor from b_1 (the ideal linear amplification factor, which is independent of the amplitude of the input signal), to

$$\left[b_1 + (3/4)b_3 A^2\right]. \tag{17.14}$$

According to this model, the input 1 dB compression point and the nonlinear model parameters of Eq. (17.12) are related as

$$A_{1\mathrm{dB}}^2 = 0.145 \left|\frac{b_1}{b_3}\right|. \tag{17.15}$$

A second, and more destructive, effect of nonlinearities is intermodulation, i.e., the mixing of different frequency components to create new components. Let us start out with the simple case of two sinusoids whose frequencies ω_1 and ω_2 are different, though both are in the passband of the system. Inserting into Eq. (17.12) provides

$$y(t) = b_1[A_1 \cos(\omega_1 t) + A_2 \cos(\omega_2 t)] + b_2[A_1 \cos(\omega_1 t) + A_2 \cos(\omega_2 t)]^2 + b_3[A_1 \cos(\omega_1 t) + A_2 \cos(\omega_2 t)]^3 \tag{17.16}$$

where we again have neglected DC component and all higher-order terms. This equation can be evaluated to provide

$$y(t) = \left[b_1 + \frac{3}{4}b_3 A_1^2 + \frac{3}{2}b_3 A_2^2\right] A_1 \cos(\omega_1 t) + \left[b_1 + \frac{3}{4}b_3 A_2^2 + \frac{3}{2}b_3 A_1^2\right] A_2 \cos(\omega_2 t) \tag{17.17}$$

$$+ \frac{3}{4}b_3 A_1^2 A_2 \cos((2\omega_2 - \omega_1)t) + \frac{3}{4}b_3 A_1 A_2^2 \cos((2\omega_1 - \omega_2)t)$$

as components in the passband, as well as a "beat" signal (signal in the baseband) at $\omega_2 - \omega_1$ with amplitude $b_2 A_1 A_2$, and a DC component $(1/2)b_2(A_1^2 + A_2^2)$, which depends on the second-order nonlinearity. The latter components might be problematic if they "talk through" to the baseband circuitry, e.g., due to insufficient insulation between passband and baseband circuitry. Furthermore, there are a number of components at approximately 2 or 3 times the carrier frequency, which are easily filtered out by bandpass filters.

Let us now turn to the components in the passband, Eq. (17.17). We see, first of all, components at the desired frequencies, ω_1 and ω_2, though their amplitudes are distorted, similar to what we observed in (17.14) – note that the amplitude distortion depends on the amplitudes of both (or, more generally, all) frequency components. More importantly, the nonlinear interaction of the two frequency components has given rise to two new components, at $2\omega_2 - \omega_1$ and $2\omega_1 - \omega_2$. These components greatly complicate communications applications: for example, for OFDM modulation, the signal at a particular subcarrier ω_3 might be interfered by any combination of two subcarriers that fulfill $2\omega_2 - \omega_1 = \omega_3$. As a consequence, we either have worse Symbol Error Rate (SER) (if we ignore this interference) or have to create new detection algorithms that take such interference into account (highly undesirable, due to the involved complexity). The situation is even worse in multi-user systems when the intermodulation products depend on the signals of two or more users. Another undesirable effect is "spectral regrowth": consider a signal that has been shaped to fit well into a spectral mask, and has negligible contributions outside the frequency range $[\omega_c - \Delta/2, \omega_c + \Delta/2]$. Now consider two subcarriers at $\omega_c + 0.4\Delta$, $\omega_c + 0.2\Delta$. Their mixing products include a component at $\omega_c + 0.6\Delta$, which is outside the admissible band. Such a signal thus creates adjacent-band interference, which cannot be easily filtered out by a passive filter, since PAs operate in passband, and creating narrowband filters in passband is difficult, see Section 17.5.[5] Avoiding such intermodulation is the motivation for a number of design choices for modulation formats that were discussed in Chapter 10.

The intermodulation behavior of amplifiers is typically characterized by the *intercept point for third-order modulation, IP-3*. This point is determined as follows: let the amplitudes of two sinusoidal signals be equal, $A_1 = A_2 = A$. Then the amplitude of the

---

[5] Note also that in reality, the considered input signals will consist of many spectral components (usually a continuum), so that the intermodulation products also create a continuous spectrum of interference.

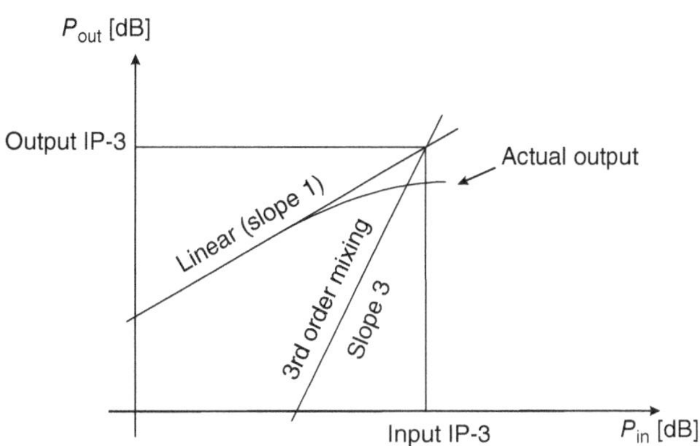

Figure 17.3 Principle of the IP-3 point.

intermodulation products scale as A^3. We now determine the A such that the intermodulation product has the same power as the ideal response of the amplifier, i.e., $b_1 A$. It then follows that the IIP-3 point is at

$$A_{\text{IP3}}^2 = \frac{4}{3} \left| \frac{b_1}{b_3} \right|. \tag{17.18}$$

The power at this operating point is called the IP-3, see Figure 17.3; by comparing to (17.15), it can be easily seen that it is related to the 1 dB compression point as $A_{\text{IP3}}^2 / A_{\text{1dB}}^2 = 4/0.435 \approx 9.2$. Actual operation of an amplifier has to occur far below the IP-3. As a matter of fact, we would like to operate in such a fashion that the intermodulation products are no stronger than noise and other external error sources (this is often called *spurious-free operation*). It is also worth noting that the IP-3 of a cascaded system can be computed from the properties of the constituent components, similar to the computation of the noise figure of a cascaded system. In particular, the cascade of two devices has an (input) IP-3 point (assuming noncoherent relationship of the stages):

$$\frac{1}{A_{\text{IP3,casc}}^2} = \left| \frac{1}{\left[A_{\text{IP3}}^{(1)} \right]^2} + \frac{G^{(1)}}{\left[A_{\text{IP3}}^{(2)} \right]^2} \right| \tag{17.19}$$

where superscript $^{(1)}$ denotes the first device in the cascade (and similarly for the second).

Another important nonlinear effect is Amplitude Modulation–Phase Modulation (AM–PM) conversion. This occurs if the amplifier contains, e.g., capacitors whose capacitances depend on the input signal magnitude. Then, as the input signal changes in magnitude, it changes the phase response to the system, and even for a constant-phase input signal, the output signal shows phase variations. This effect is most significant for modulations that use both amplitude and phase variations (constant-envelope signals like Phase Shift Keying (PSK) do not exhibit this effect by definition, and pure Amplitude Shift Keying (ASK) might show phase variations due to AM/PM conversion, but they do not impact the error probability).

The nonlinear distortions discussed above create an upper limit of the admissible input signal power. A lower limit is provided by what signal is detectable with the amplifier, e.g., due to noise. The *Minimum Detectable Signal (MDS)*, thus describes the sensitivity of the amplifier (or, more generally, of a RX). The dynamic range of an amplifier is then the difference between the MDS and the maximum input signal level that can be tolerated without "excessive" nonlinear distortions. If the amplifier is mandated to operate in spurious-free range, then we call this difference the *spurious-free dynamic range.*

17.2.3 Matching

A wireless transceiver can be represented as a source (e.g., the DSP outputs) with a particular impedance, a load (the antenna with its radiation impedance, see Chapter 8), and a cascade of other components in between. Those other components can be described as *two-ports*, i.e., having an input port and an output port. Different components might be connected via cables or microstrip lines, which themselves can be interpreted as two-ports (e.g., the inner and outer connector of a coaxial line on one end of the cable are the input port, and on the other end of the line the output port).[6]

In microwave circuits, there is often a need to match a particular load impedance to the impedance of the preceding component of the circuit. Matching is generally achieved by putting a two-port network between the load and the other component, see Figure 17.4a, with two example implementations in Figure 17.4b,c. This matching can be used for the elimination of reflections, maximization of

---

[6] Note that a two-port that has a load impedance across its output port can be interpreted as a one-port, whose load impedance is the input impedance seen at the remaining port.

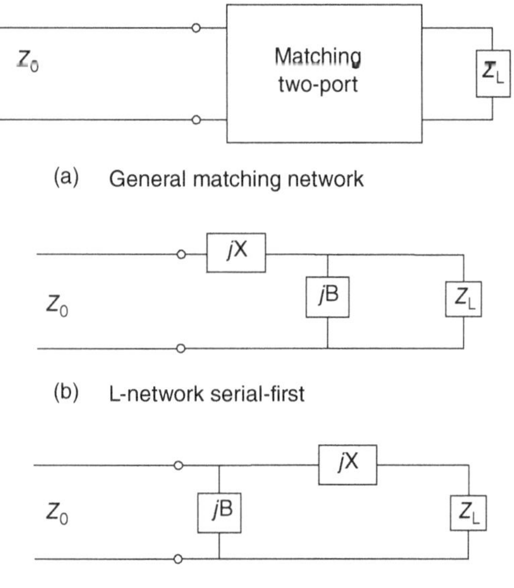

(a)   General matching network

(b)   L-network serial-first

(c)   L-network shunt-first

Figure 17.4   Two-element passive lossless matching network.

transferred power, minimization of noise, or decorrelation of antennas. A more extensive discussion of two-port networks, reflection coefficients, and matching is given in Appendix 17.

In most microwave applications, commercially available modules of components contain matching networks such that the input and output impedance is 50 Ω (in rarer cases 75 Ω). This is also the impedance of most cables. This enables easier construction of wireless transceivers from such modules, because they can be connected without having to worry about reflections and designs of matching networks. However, power matching on chips is rare, due to the limited physical dimensions available there; instead used are voltage signals (load is high impedance compared to source), and current signals (vice versa).

In some circumstances, matching to completely eliminate reflections is not possible, or not desirable, see above; yet existing reflections might constitute a significant problem. For example, a sensitive source might be destroyed by too-large amplitude of reflections. In those cases, the use of *isolators* might be beneficial. Such isolators are two-ports characterized by the S-matrix

$$\mathbf{S} = \begin{bmatrix} 0 & 0 \\ 1 & 0 \end{bmatrix} \tag{17.20}$$

so that only transmission from port 1 to port 2 is possible, while the reverse direction is blocked. Such isolators require the use of ferrites, and thus cannot be easily integrated into an Integrated Circuit (IC). Still, they are occasionally used, e.g., near a discrete (transistor or tube) PA.

*17.3   ADCs and DACs

17.3.1   ADC

Antennas receive analog signals (which in turn consist of distorted attenuated transmit signals, noise, and interference). Even after filtering and downconversion, usually to baseband (for exceptions see Section 17.8), these signals are still analog. However, the processing of the baseband signal (demodulation, equalization, decoding, etc.) is done digitally. Thus there must be a conversion of the analog waveform to a digital signal by an *ADC*.

An ADC performs two functions: (i) time discretization, and (ii) amplitude quantization. These two tasks are reflected by the fundamental structure of an ADC: its first stage is a Sample and Hold (S/H) device, while the second is a quantizer and output-word generator. The necessary sampling rate is generally determined by the Nyquist criterion.

Consider first the S/H stage in more detail, see Figure 17.5a. A continuous-time input signal is amplified by a buffer amplifier, and charges (or discharges) the capacitor to the value of the input voltage at the current moment; this is achieved by putting the switch into the "connected" position. Then the switch is opened; the Operational Amplifier (OpAmp) at the output ensures that the voltage at the output remains constant. During that time, the amplitude value is discretized and read out. After one clock interval, the process is repeated, see Figure 17.5b.

We next turn to the amplitude quantizer. Q bits of quantization can represent 2^Q amplitude levels. The (continuous) sampled value at the S/H needs to be mapped to one of these amplitude values. The simplest way of achieving this is the *flash architecture*, which compares the S/H output, in parallel, to each of 2^Q reference voltages. To achieve this, 2^Q comparators are implemented, see Figure 17.5c, and each comparator compares the input signal to one of the reference voltages (which in this example are obtained

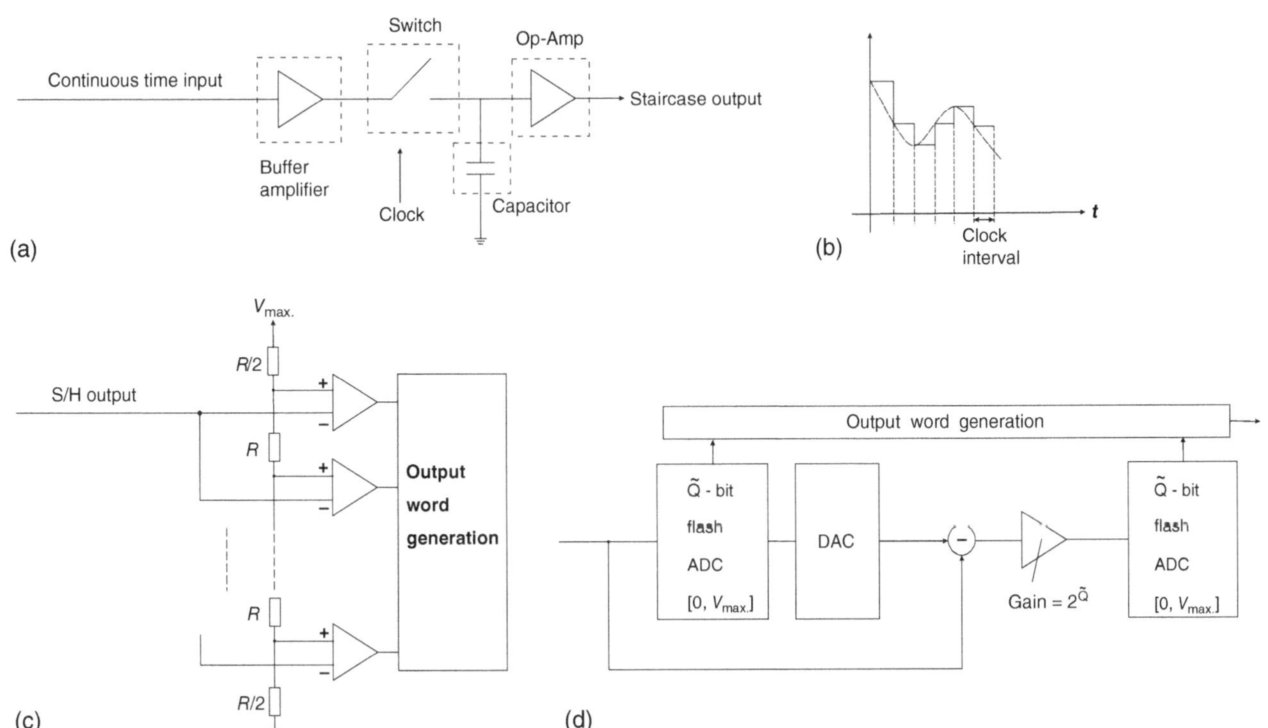

Figure 17.5 Principle of an ADC: (a) sample and hold, (b) output waveform of a sample-and-hold, (c) quantizer for flash ADC, (d) pipeline ADC.

by dividing a single "golden" voltage with a series of resistors). The output is then in the form 00..0111.....1, where the transition from 0 to 1 occurs at the level of the input voltage; this 2^Q long sequence is then translated into standard binary code by the output word generator. The advantage of the flash architecture is its fast operation (all operations of level comparison are done simultaneously); the drawback is the large number of comparators that needs to be implemented. Typically, converters with up to eight bit resolution can be implemented in flash architecture. For especially fast ADCs (beyond several Gsample/s), multiple flash ADCs can be interleaved, so that one ADC processes time samples 1,5,9,..the next 2,6,10,... and so on.

One possibility for achieving finer amplitude quantization is the *pipeline architecture*. It concatenates several stages, each of which consists of a $\tilde{Q}$ flash ADC (with $\tilde{Q} < Q$), followed by a DAC. The first stage determines a low-resolution estimate of the input voltage, and (after converting it back to analog with the DAC) subtracts this estimated voltage from the actual input voltage, resulting in a signal with possible input range that is equal to the quantization stepwidth of the first ADC. This input signal is now sent into a second ADC that provides a $\tilde{Q}$ bit quantized estimate of the difference signal (if the same type of ADC should be used in the first and second stage, the difference signal needs to be amplified, see Figure 17.5d). The combination of the first and second quantization thus provides a $2\tilde{Q}$ bit resolution estimate, using only $2 \cdot 2^{\tilde{Q}}$ comparators, instead of $2^{2\tilde{Q}}$ comparators that would be necessary with a flash architecture. This hardware saving is bought at the price of slower operation, since quantization takes two clock cycles. Besides the pipeline architecture, *successive approximation registers* are used, where in a binary search process the input voltage is compared against a series of successively smaller voltages ($V_{max}/2, V_{max}/4,. . . ...V_{max}/2^Q$) that correspond to the bits in the output codeword, and for each of those comparisons subtract the result from the input voltage. The architecture requires only a single comparator but needs Q comparison actions. Another approach, which uses also just a one-bit ADC, is the *sigma–delta modulator*, which compares, with a high oversampling factor, the integrated voltage based on past decisions with the current voltage. These ADCs are mainly used for high-precision, low-bandwidth applications.

A key measure of ADCs is their resolution, and thus the resulting quantization noise. Generally, the *Quantization SNR (QSNR)*, is

$$QSNR_{dB} = 4.77 + 6.02Q + 10\log\left[\frac{E\{V(t)^2\}}{V^2_{max}}\right] \tag{17.21}$$

which in the case of sinusoidal signals filling out the available range of the ADC becomes

$$QSNR_{dB} = 1.76 + 6.02Q. \tag{17.22}$$

For a low number of bits, the quantization noise can become the dominant source of noise. Since QSNR is related to the maximum possible input signal, a signal with lower input amplitude automatically has a worse SNR. It is thus essential that an input signal is amplified such that it (approximately) exploits the full dynamic range of the ADC, while not exceeding $V_{\max}$ (which would lead to clipping of the signal). This requirement motivates an *Automatic Gain Control (AGC)*, as discussed in Section 17.4.3, especially since wireless signals can have amplitude fluctuations on the order of many tens of dB.

Besides the quantization error, an ADC can also have other error sources (e.g., internal noise, nonlinear distortions) that lead to a worsening of the SNR. These are often subsumed in the *Effective Number of Bits (ENOB)*, i.e., the resolution that an ideal ADC would have that results in the same SNR. Typically, the ENOB is 1–2 bits lower than the nominal number of bits (for large bandwidth, the degradation can be up to three bits).

A further impairment comes from *timing jitter* (sometimes also called aperture uncertainty). If the ADC does not sample at the prescribed time, the output is obviously not a true representation of the signal at the nominal sampling time. The resulting SNR can be approximated (for sinusoidal signals) as

$$SNR = \frac{1}{\left[2\pi f_\mathrm{s} \sigma_\tau\right]^2}. \tag{17.23}$$

Here f_s is the sampling frequency and σ_τ the rms timing deviation.

ADCs trade off resolution, bandwidth, and power consumption. Roughly speaking, power consumption scales linearly with the sampling rate, since it is proportional to the number of switching operations that have to be performed. It also scales with 2^Q, which is intuitive for the flash architecture. The actual proportionality constants depend, of course, on the particular architecture, as well as the technology used for implementation (e.g., 28 nm Complementary Metal Oxide Semiconductor (CMOS)).

This trade-off has important consequences for wireless system design. To give but one example: Chapter 15 showed that the dynamic range (PAPR) of OFDM signals is much larger than those of single-carrier systems. Thus, ADCs with larger ENOBs are required. On the other hand, ADCs with larger number of bits and very high sampling rate (>1 GSample/s) are difficult to produce, and even if they exist, are power hungry. This has led to a renaissance of single-carrier transmission for extremely high bandwidth applications, in particular for mm-wave systems at 60 GHz and beyond, where >10 GHz bandwidth is available.

17.3.2 DAC

The baseband signal processing generally creates a digital signal[7] that can be interpreted as a sampled version of the analog waveform that we actually want to transmit over the air. The *DAC* takes this digital signal and converts it to the analog waveform that can be upconverted to passband and transmitted over the air.

The operation of a DAC is, in essence, a reversal of that of a DAC. The first step is the mapping of the DAC input into a decoder that translates the Q input bits onto a code for one of the 2^Q amplitude levels that should be put out. These levels then serve as the input of a waveform assembly, which translates the digital input into a time-continuous, but amplitude-discrete output. The simplest form is a current source array, which has 2^Q identical current sources (each of which just requires a single transistor), and each bit of the 2^Q-element code determines whether the current source provides a contribution to the total current or not, see Figure 17.6a. Alternatively, one can also build a current array with unequal weighting, see Figure 17.6b, which can then directly use the binary data (with length Q) as input: the Least Significant Bit (LSB) determines whether the current source with weight 1 is active, the next one for a current source with weight 2, and the Most Significant Bit (MSB) for a current source with weight 2^{Q-1}. The advantage of this structure is the smaller number of components; the drawback is that it is more sensitive to errors in the component values: while the uniform-weighted current source array always puts out more current when the index of the input code takes on a higher value, this need not be the case for the weighted array. When, e.g., the current source for the third-LSB has an actual value that is smaller than the nominal value by at least one unit current source strength, the output current will actually decrease when input code increases from 011 to 100. Finally, the output of a DAC has a "staircase" shape. Proper low-pass filtering of this shape is thus necessary.

Many other aspects of DACs are similar to those of ADCs, such as considerations of quantization noise, dynamic range, etc.

*17.4 Amplifiers

17.4.1 General Considerations

An amplifier is, abstractly speaking, a device that can increase the power of a signal, using power from a power supply. It is generally a nonlinear device, but often can be approximated as linear for a limited range of the input signal. Amplifiers in wireless circuitry are almost exclusively transistor amplifiers.[8] Transistors use structures made of doped semiconductor materials, especially silicon (Si), to change the current (provided by the power supply) between two terminals according to the much smaller voltage (or current) at a third terminal. Amplifiers thus consist of both an active element, the transistor, and auxiliary circuitry such as power supplies. In the following we discuss amplifiers from a phenomenological point of view, without going into the details of semiconductor physics or the design of the auxiliary circuitry; we refer to the books in the "Further Reading" section for those aspects.

---

[7] Note that this signal is *not* the complex modulation symbols; rather it is the discrete version of the actual modulated signal.
[8] For high-power radar applications, tube amplifiers might still sometimes be used.

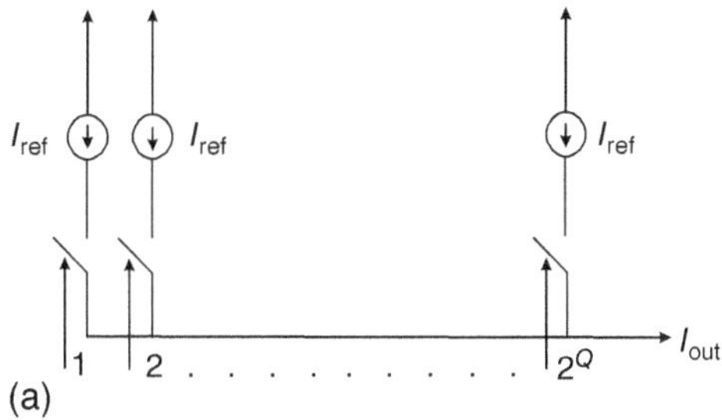

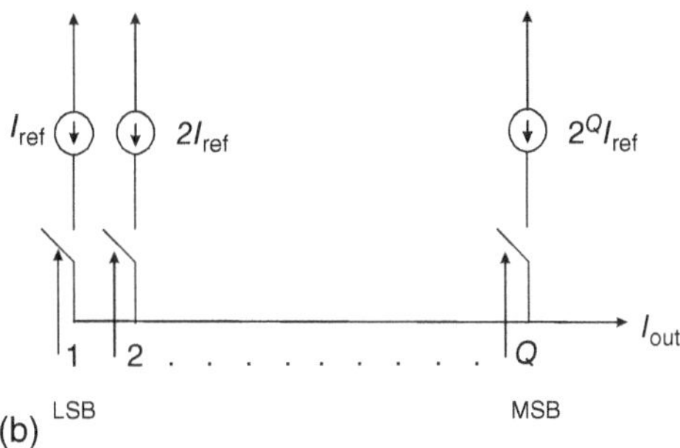

Figure 17.6 Principle of a current array for a DAC: (a) uniform weighting, (b) binary weighting. Color version available at wiley.com/go/molisch/wireless3e.

The most important parameter of an amplifier is, obviously, the amplification factor. Since amplifiers are two-ports, they can be described by their S-parameters (compare the summary of S-parameters in Appendix 17.A). S_{21} describes the amplification, i.e., the (complex) wave amplitude emanating from the output port, divided by the wave amplitude entering the input port. Obviously, in proper operation, $|S_{21}| > 1$. However, S_{21} is not the only parameter determining the power delivered to the load; rather reflections at the input and the output of the two-port are important, which are related to the S_{11} and S_{22} and how the amplifier is matched to the preceding and following components of the chain; this will be discussed in Section 17.4.4.

We must further note that all the parameters mentioned above are frequency-dependent. Amplifier gain is often characterized by the 3-dB bandwidth (i.e., the frequency at which the power gain decreases by 3 dB). The amplifier bandwidth is itself an important quantity, which of course needs to fulfill the requirements of the particular application at hand.

The gain of a transistor generally shows a low-pass characteristic and is often described by the *transit frequency*, i.e., the frequency where the gain falls to 1. Achievable transit frequencies have changed significantly over the years – for example, it is now possible to create amplifiers operating at 60 GHz with silicon-based CMOS technology – something that was impossible 20 years ago. This trend has allowed the use of (low-cost) CMOS in frequency bands that previously required much more expensive III–V semiconductors.

However, the gain of an amplifier is not just determined by the gain of the transistor, but also of the auxiliary circuitry, including the matching circuits. Since designing broadband matching is difficult, one thus has to determine the actual band of operation, and try to provide suitable circuitry only for this band. In particular, not only the magnitude of the gain but also the phase, as well as the complex input and output reflection coefficient are frequency-dependent.

Another key factor of amplifiers is the nonlinearity; this has already been discussed in detail in Section 17.2.2. Finally, the noise characteristics of an amplifier (in particular the noise figure) are of importance, as discussed in Section 17.2.1.

17.4.2 Power Amplifiers

All of the above discussions were valid for any amplifier. We now turn to a specific application, namely the *PA*, in the TX. Its purpose is to amplify the transmit signal to high power levels – the main criterion here is the output power, not the gain, since a high enough input power level can be achieved by preamplifiers. PAs need to pay particular attention to the *power added efficiency*, i.e., the ratio between added (output minus input) RF power and the power put into the amplifier by the power supply

$$\eta_{\mathrm{PAE}} = \frac{P_{\mathrm{RF,out}} - P_{\mathrm{RF,in}}}{P_{\mathrm{DC}}}. \tag{17.24}$$

Power efficiency is so important for PAs simply because of the large absolute amount of power dissipated in a PA – remember that the output power of a Base Station (BS) can easily be 40 W and more, so that $\eta = 0.4$ would mean approx. 60 W of power dissipated as

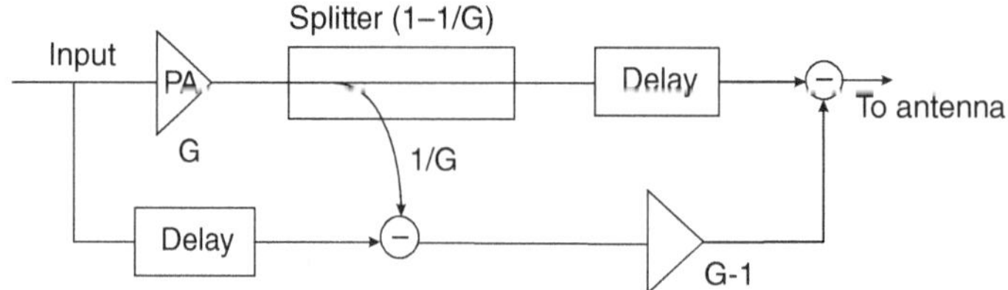

Figure 17.7 Block diagram for feedforward linearization of a power amplifier.

heat. This is not only a significant waste of energy but also creates thermal problems – the heat might have to be carried away by air conditioning in order to avoid thermal meltdown of the BS. For LNAs and other small-signal amplifiers, which often require output powers on the order of 1 mW or less, this problem is obviously much less relevant.

Unfortunately, increasing the power efficiency of an amplifier usually implies sacrificing linearity. Particular ways of biasing and choosing or combining different operating points allow highly linear operation but have low power efficiency. The literature defines different classes of amplifiers, that range from class A (linear), to class AB and B (quasi-linear), to class C (nonlinear), and class D, E, F (essentially switched between on and off state, and thus highly nonlinear). While class A has $\eta << 0.5$, class AB and B can reach up to 0.6, while the nonlinear amplifiers can reach up to 0.8 in practice (and close to 1 in theory). Details about the different amplifier classes and their designs can be found, e.g., in [Ellingson 2016, Section 17.4]. For the purposes of this chapter, it is enough to say that the trade-off between power efficiency and linearity exists.

In order to benefit from high η, one can also combine a high-efficiency amplifier with a linearization circuit. In this approach, the nonlinearities of the output signal are compensated through appropriate circuitry. Imagine first a situation in which the nonlinear characteristics of the amplifier are known $y(t) = Gf(x(t))$. It is then straightforward to predistort the input signal as $x'(t) = f^{-1}(x(t))$, so that the output of the nonlinearity becomes $Gx(t)$. Note that the predistortion can be done either through analog components, or as *digital predistortion*; the latter is more popular in modern solutions. The distortion can be done either in a static manner (based on amplifier characteristics that have been measured once), or, preferably, in a dynamic manner where the attenuated output signal of the amplifier is compared to the ideal signal $x(t)$, and deviations adaptively impact the predistortion; this approach relies on the assumption that changes in the nonlinear characteristics occur on a much longer timescale than the time it takes to adapt the digital predistortion. Yet another alternative is feedforward linearization, whose principle can be seen from Figure 17.7. Note, however, that delay elements can be difficult to implement in integrated circuits.

17.4.3 Gain Control

The power of the received signal in a wireless transceiver can undergo extreme variations: the channel attenuation can vary by up to 70 dB. The dynamic range of a single receive amplifier is not capable of accommodating such large variations. Furthermore, we have seen that the ADCs available for higher bandwidths could not digitize such a signal with good SNR, even if the amplifier were available. Thus, to compensate the difference in RX power level the standard RX structure is to have multiple amplifiers in the RX chain, several of which have variable gain. Such variable gain can be controlled either digitally or by mains of an analog signal. The actual implementation might be through a bank of attenuators that can be selectively bypassed according to a control signal, or by directly varying the gain of the transistor amplifier through adjusting the voltages in the power supply.

The VGA is part of an *AGC*, which ensures that the average output power is always at the same level, irrespective of the input power. The principle is shown in Figure 17.8: the output signal of the amplifier is split into two parts, the actual output, and a signal whose power is detected and used to control the amplifier gain. This signal is first sent through a lowpass filter, since we only wish to compensate for slow signal variations (see below for more details). This filtered control signal is then compared to a preset desired reference level, and any difference serves to adjust the amplifier gain.

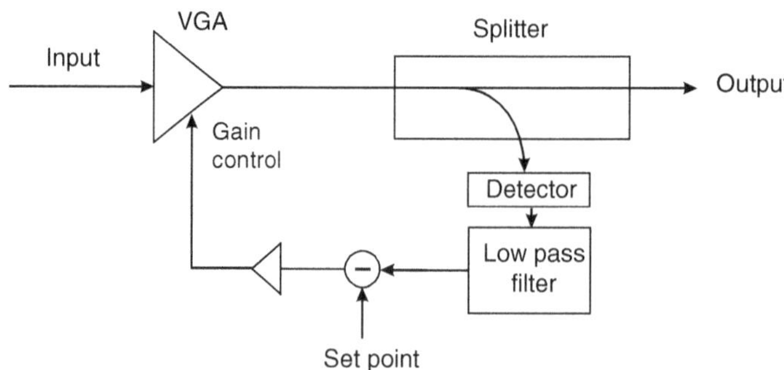

Figure 17.8 Block diagram of an automatic gain control.

Thus, generally, we encounter three types of amplification stages in a corresponding RX RF chain:

1. Fixed-gain LNA: this is the first amplifier, as close to the RX antenna as possible, to boost the signal level and ensure sufficient SNR. As indicated in Section 17.2.2, the gain and noise figure of the first amplifier in a cascade mostly determine the overall behavior, so that the first LNA concentrates on these characteristics. For cases of very high RX power, it might be possible to bypass this LNA; such bypass can be seen as a crude form of gain control.
2. AGC to adjust to slow variations: this serves to compensate for the large variations of the channel gain due to pathloss and shadowing.
3. AGC to adjust to fast variations: here the filter bandwidth is on the order of the inverse coherence time of the channel, and can thus track the small-scale fading.

17.4.4 Amplifier Matching

Since amplifiers are two-ports, they are described by their S-parameters:

- S_{21} describes the amplification, i.e., the (complex) wave amplitude emanating from the output port, divided by the wave amplitude entering the input port. Obviously, in proper operation, $|S_{21}| > 1$. From this one might think that the power gain is equal to $|S_{21}|^2$. However, this is true only if input and output are matched for zero reflection. For the case that this is not fulfilled, the relevant quantity to consider is the transducer power gain G_T, which is the ratio of the delivered power over the power that would be delivered (without the amplifier) to a perfectly matched load, see also Appendix 17.A. For the case that S_{12} is very small (i.e., the output does not impact the input – which is often true for amplifiers), it can be approximated as

$$G_T = \frac{\left(1 - |\Gamma_S|^2\right)|S_{21}|^2\left(1 - |\Gamma_L|^2\right)}{|1 - S_{11}\Gamma_S|^2|1 - S_{22}\Gamma_L|^2} \tag{17.25}$$

where Γ_S and Γ_L are reflection coefficients of source and load. The G_T for arbitrary S_{12} is given in Appendix 17.A.

- S_{11} describes the reflection at the input port, and thus the requirements for the matching.
- Similarly, S_{22} describes the reflection at the output port.

As for every two-port, matching of the amplifier is of great importance (a brief primer on matching is given in Appendix 17.B). It must be noted that matching can be done for different purposes:

- *Matching for maximum gain*: in this approach, Γ_{in}, i.e., the reflection coefficient "looking into" port 1, must be the conjugate of the source reflection coefficient, $\Gamma_{in} = \Gamma_S^*$, and similarly, the reflection coefficient looking into port 2 must be conjugate matched to the load, $\Gamma_{out} = \Gamma_L^*$. This conjugate matching is known to optimize power delivery to the load. This requires in turn that

$$\Gamma_S^* = S_{11} + \frac{S_{12}S_{21}\Gamma_L}{1 - S_{22}\Gamma_L}$$

$$\Gamma_L^* = S_{22} + \frac{S_{12}S_{21}\Gamma_S}{1 - S_{11}\Gamma_S} \tag{17.26}$$

which is a system of equations that can be easily solved simultaneously.
- *Matching for stability*: amplifiers can easily become unstable and start to oscillate if care is not taken in the design of the matching. Imagine that a signal at port 1 is amplified by S_{21} and sent from port 2 toward the load. If the load is not matched for zero reflections, part of that signal gets reflected and becomes an incoming wave at port 2. This wave, scaled by S_{12}, is an outgoing wave at port 1, part of which gets reflected by the source impedance, and is thus again incident into port 1. The above process is thus a possibly positive feedback loop, and even a small signal can quickly lead to an undesired instability of the system. Proper design of the matching has to ensure that such instabilities cannot occur. Note that there are some amplifiers that are unconditionally stable, i.e., no matter what the source and load impedances are, the amplifier can never get unstable. However, in most cases, the matching design is critical.
- *Matching for noise minimization*: the additional noise created by an amplifier also depends significantly on the matching. Every amplifier has a source impedance for which the noise figure is optimized. In particular, for LNA design, it is desirable to create a matching that lets the amplifier "see" this source impedance.

- *Matching for no reflections:* reflections are deleterious in many contexts – they can lead to delayed components that create intersymbol interference; can possibly destroy sensitive electronic components, etc. Matching to minimize or eliminate reflections is thus an obvious choice.
- *Matching for PA efficiency:* this is done to optimize the energy efficiency of a transmitter.

These different goals cannot be all satisfied at the same time. Stability is a requirement that has to be fulfilled in all cases; as for the other criteria (gain, reflections, noise), the system designer can either make a choice, or try to find a suitable compromise. For example, it is not possible to have maximum gain and minimum noise figure at the same time. Rather, one can, e.g., pick the desired gain, and then find the matching that results in the minimum noise figure for this gain.

*17.5 Filters, Power Dividers, and Phase Shifters

17.5.1 Principles and Characteristics of Filters

Filters, by definition, are devices that change the (complex) spectrum of a signal. Thus, in principle, any device that has a nonconstant (in amplitude and/or phase) transfer function is a filter. In practice, the name filter is reserved for devices whose main (or only) purpose is to enhance the magnitude of one part of the spectrum compared to another.[9] We usually distinguish between the following categories:

- *Low-pass filters* allow transmission of low frequencies (the passband extends from 0 to ω_L) and suppress higher frequencies. These are the most common filters in baseband processing.
- *High-pass filters* allow transmission of high frequencies and suppress low frequencies, so that the passband from ω_H to ∞.
- *Band-pass filters* allow frequencies in a specified band extending from $\omega_{B,L}$ to $\omega_{B,H}$ to pass. These filters are applied to RF and IF signals to suppress noise and interference outside the band of interest (at the RX) and suppress spurious emissions (at the TX).
- *Band-stop filters* block frequencies in a specified band extending from $\omega_{S,L}$ to $\omega_{S,H}$. They are useful for suppressing strong interference at a particular frequency/band.

Ideally, each of those filters would have a "brickwall" shape, so that it allows the frequencies in the passband to pass without attenuation, and completely block all other frequencies. However, this is not possible; deviations from this ideal picture are characterized by (see also Figure 17.9):[10]

- *Passband attenuation:* any real filter will lead to attenuation even in the desired frequency band. The (average) passband attenuation is thus an important filter parameter.
- *Passband ripple:* the magnitude of the transfer function in the passband is not completely flat, but rather shows some changes in amplitude. Those changes can be either monotonous (as in a simple RC-filter, where the magnitude of the transfer function is decreasing according to $\sqrt{1 + (\omega/\omega_L)^2}$), or showing actual ripples, as indicated in Figure 17.9. The maximum amplitude of those ripples is a filter parameter.
- *Stopband attenuation:* an ideal suppression of the spectral components falling into the stopband is not possible. The minimum attenuation taken over the stopband is another important filter parameter.
- *Width of the transition band:* the transition of the transfer function from passband to stopband cannot occur abruptly, but rather requires a band of width $\Delta\omega$.
- *Phase linearity:* all the above parameters discussed the magnitude of the transfer function. However, the phase as a function of frequency $\phi(\omega)$ (or, equivalently, the group delay, which is defined as $-d\phi(\omega)/d\omega$) plays an important role as well. Nonlinearities of the phase (or, equivalently, variations of the group delay) lead to signal distortions and intersymbol interference that require equalization, and thus increase complexity or reduce performance of a system. Thus, filters generally aim to provide linear phase.
- *Number of components:* when the filters are implemented in hardware, the number of components (such as inductors, capacitors, or transmission lines) plays an important role for space requirements and cost.

Unfortunately, not all of these filter characteristics can be optimized simultaneously; rather a design choice, or a trade-off, is required. Obviously, an increase in the number of components allows to synthesize more complex (and thus closer to ideal) filter characteristics. For a given number of components, different filter designs (often named after their inventors) are optimal for achieving particular characteristics.

---

[9] One exception from this definition is the all-pass filter, which aims to only manipulate the phases of the different frequency components.

[10] For the sake of simplicity, we discuss in the following low-pass filters; other filter types can be characterized similarly.

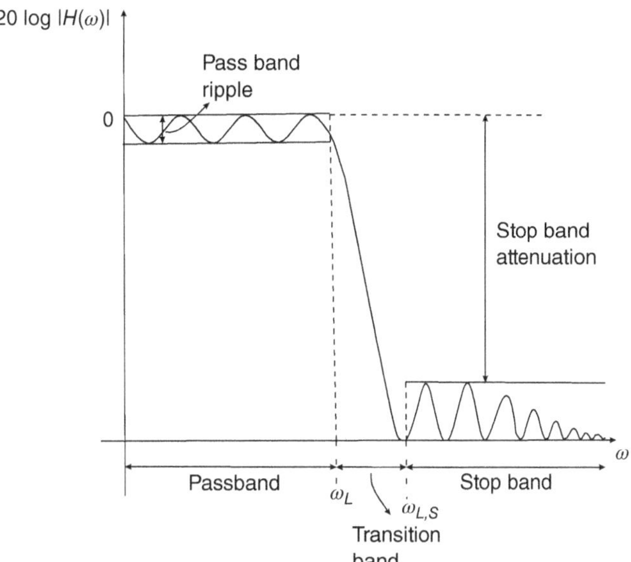

Figure 17.9 Definition of passband, transition band, and stopband for a low-pass filter. Also shown are passband ripple and stopband attenuation.

- *Butterworth filters* are maximally flat, i.e., show no ripples, and the smallest slope, in the passband.
- *Chebyshev filters* have a faster transition from passband to stopband, but buy this advantage with a ripple in the passband or stopband. The magnitude of those ripples is a design parameter – the larger the ripples that are allowed, the smaller is the transition band.
- *Cauer filters* have ripples in both passband and stopband and the smallest transition band.
- *Bessel filters* have maximally linear phase response, so that in the passband the phase has no inflection points.

We do not discuss here the actual techniques for synthesizing these filters; details can be found in the references of the "Further Reading" section.

These filters can be implemented in a variety of technologies. For lower frequencies implementations, the use of lumped elements (inductors and capacitors) has long been the standard; on-chip implementations use active filters, where inductors are replaced by active circuitry to create a compact size. Generally, analog filter implementations are required for all filters placed before the ADC (or after the DAC). Baseband filters mostly use digital implementations of filters, which can realize very steep filter characteristics. Note that for digital filters the importance of "number of components," i.e., complexity of operation, is generally less important, since the required operations are much less than for other baseband tasks.

At higher frequencies (typically a few hundred MHz and above), lumped elements are not available, or show strong deviations from ideal behavior. In those frequency ranges, implementation of filters is preferably done by means of transmission lines. Two techniques are generally considered: use of (open-circuit or short-circuit) stubs, or stepped-impedance filters, where transmission lines with small and high impedance are interleaved.

Most of the filters at higher frequencies are bandpass filters. These can also be interpreted as exploiting a resonance of the filter structure – remember that a lossless structure has one or more resonance frequencies that allow passing of signals, while the remainder is completely blocked. A resonator is characterized by its Q-factor, which describes the ratio between the stored energy versus the dissipated energy. In other words, a resonator with very small losses is best suited to provide a sharp transition between passband and stopband. Filters made of microstrip transmission lines are usually not well suited in this context, as they incur losses due to (lossy) dielectric substrate on which the transmission line is located, and possible radiation losses. Helical filters, cavity filters, crystal filters, Surface Acoustic Wave (SAW) filters, Bulk Acoustic Wave (BAW) filters, and dielectric resonators are all technologies that are used for various parts of the >100 MHz frequency regime. In particular in the range <7 GHz, SAW and BAW filters dominate.

In any case, it is worth remembering that the Q-factor describes a *relative* bandwidth. Thus, a filter with small (in absolute terms) bandwidth at high carrier frequency is very difficult to realize; it is even more difficult to make such a filter tunable, though there is ongoing research in particular using active structures, to realize such filters. This is a major motivation for the superheterodyne RX structure that will be discussed in Section 17.8.

17.5.2 Switches and Phase Shifters

Another important class of components used in RF circuitry is switches. A number of techniques discussed in Part III, such as switched antenna diversity, depend on the availability of suitable switches. Some switches are used as ON/OFF switches, while other applications require switching between multiple ports. A variety of technologies are available for creating switches, such as electromechanical, pin diode, transistor, and MEMS (Micro-Electro-Mechanical Switch). Without going into the details of the actual technology, the main characteristics of switches are as follows:

- *Switching speed*: the time required to switch between the ON and OFF states (or between two output ports) of a switch can range from nanoseconds (for most semiconductor switches) to a few microseconds (typical MEMS), to tens of milliseconds (electromechanical).
- *Number of switching operations*: mechanical switches, including MEMS, might wear out after a certain number of switching operations. The frequency at which switching operations are necessary (together with the desired lifetime, which usually is several years) thus becomes a determining factor in the selection of a suitable switch.
- *Power handling capability*: semiconductor switches usually have a limited power handling capability, often on the order of mW (higher-power switches are available, but very expensive). Switches for TXs (when used after PAs) can thus be difficult to realize.
- *Insertion loss*: a switch in the ON state should ideally not have any attenuation, but in reality, it does. The amount of attenuation depends both on the switch technology, and the number of switching ports, see below.
- *Isolation*: this describes the amount of leakage power when a switch is in the OFF position (or connected to a different port).
- *Number of switching ports*: the "canonical" switch is a two-throw switch, i.e., connecting an input to one of two possible outputs (or vice versa). Larger switches are often created as a cascade of such canonical switches. This implies also that larger switches have a larger attenuation, as they add up the insertion losses of the different stages of the cascade.

All these characteristics are functions of frequency. The higher the operating frequency, the larger the deviations from ideal state, and in particular, the lower the power handling capability.

Switches can also be used to create adaptive phase shifters. By bypassing or adding (depending on the switch position) delay lines or other circuit elements, phase shifts can be imposed on signals. This approach is particularly useful at higher frequencies. It is noteworthy that these phase shifts are discretized, with typical step widths of 22.5°. At lower frequencies, phase shifters can be realized by lumped elements, e.g., using RLC circuits in which the capacitor is realized by a varactor diode, whose capacitance can be tuned by the bias voltage.

The most important application of phase shifters is in phased arrays, where the adjustment of the phase shifters changes the antenna pattern. The fact that phase shifters might only work in discrete steps needs to be taken into account in the algorithms synthesizing the antenna patterns; in particular, the location of nulls might be sensitive to the quantization effects.

17.5.3 Power Dividers, Circulators, and Directional Couplers

We now turn to multi-port devices. A first important component is the *power divider*, which takes an input signal and splits it into two (not necessarily equal) signals. Such splitting is important, e.g., for all monitoring of RF signals and feedback loops based on such monitoring; we obviously have to split off a (usually small) part of the signal for the monitoring circuit. Power dividers can be lossless (in which case their S-matrix is unitary) or lossy. Lossless power dividers have the drawback that the output ports are not isolated from each other, and the impedances looking into the output ports are not matched. Lossy power dividers are thus more commonly used. A particularly popular version is the Wilkinson power divider. In its simplest form it provides equal division of power to the two output ports, while providing isolation between the output ports, and matched output impedance, see Figure 17.10. Generalizations to arbitrary dividing ratio are possible.

Another important device is the *circulator*, which is characterized by an S-matrix

$$\mathbf{S} = \begin{bmatrix} 0 & 0 & 1 \\ 1 & 0 & 0 \\ 0 & 1 & 0 \end{bmatrix} \tag{17.27}$$

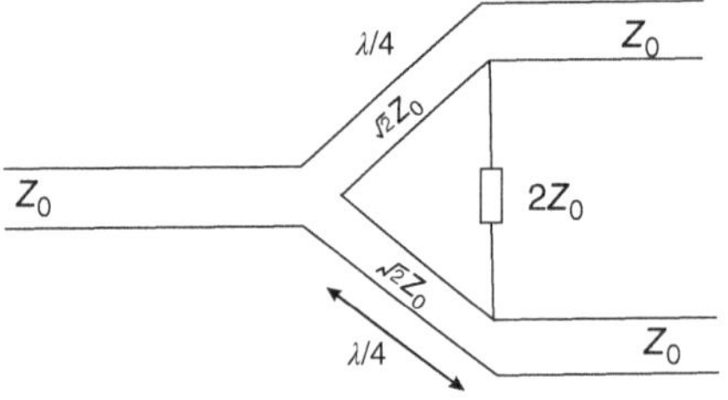

Figure 17.10 Wilkinson power divider.

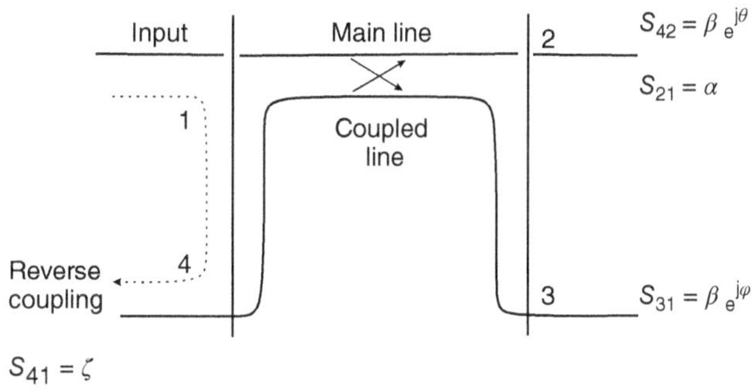

Figure 17.11 Directional coupler.

which means power from port 1 goes to port 2, power from port 2 to port 3, and power from port 3 to port 1 (note that terminating one of the ports with a matched load converts a circulator to an isolator). The most common implementations of circulators are passive, using ferrites to allow the implementation of the nonreciprocity; active designs are based on transistors. The advantage of the latter lies in the more compact design; the drawback is the power consumption and the reduction in SNR.

Finally, it is worthwhile to mention *directional couplers*, see Figure 17.11. They have an S-matrix[11]

$$
\mathbf{S} = \begin{bmatrix}
0 & \alpha & \beta e^{j\phi} & \zeta \\
\alpha & 0 & \zeta & \beta e^{j\theta} \\
\beta e^{j\phi} & \zeta & 0 & \alpha \\
\zeta & \beta e^{j\theta} & \alpha & 0
\end{bmatrix}
\tag{17.28}
$$

where ζ is assumed to be very small and

$$
\alpha^2 + \beta^2 + \zeta^2 = 1
\tag{17.29}
$$

$$
\phi + \theta = \pi \pm 2n\pi.
$$

From this, we can define the key quantities, namely the coupling

$$
C = 10\log\left(\frac{P_1}{P_3}\right) = -20\log(\beta)
\tag{17.30}
$$

the directivity

$$
D = 10\log\left(\frac{P_3}{P_4}\right) = 20\log\left(\frac{\beta}{\zeta}\right)
\tag{17.31}
$$

and the isolation

$$
I = 10\log\left(\frac{P_1}{P_4}\right) = -20\log(\zeta).
\tag{17.32}
$$

Ideally, we wish for $\zeta = 0$, which leads to infinite directivity and isolation. The phases ϕ and θ are often chosen as $\phi = \theta = \pi/2$, so that there is a phase shift of $\pi/2$ between the two ports. When furthermore the coupling $C = 3$ dB, we have a quadrature hybrid, which are useful, e.g., in IQ up/downconverters.

*17.6 Oscillators

Oscillators provide the sinusoidal signals used in up/downconversion of signals. Various methods have been proposed for generating sinusoids – all are based on the principle of feedback circuits containing resonant elements, but differ in what resonant elements they use (e.g., LC circuits, crystals, ceramic resonators, etc.), and how they are placed in the feedback circuitry. Also, different methods have been developed for tuning the oscillator frequency.

---

[11] We ignore the phases of S_{23}, S_{14}, S_{32}, and S_{41}, which are irrelevant for the subsequent considerations.

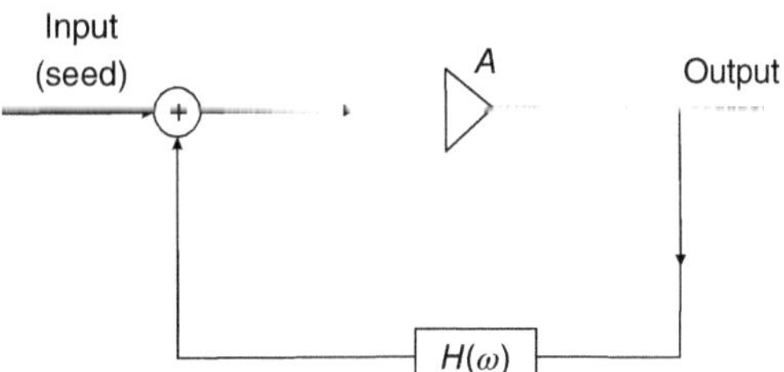

Figure 17.12 Block diagram of a feedback circuit that forms the basis of an oscillator.

From a phenomenological point of view, oscillators are mainly characterized by the phase noise, which describes the deviations from an ideal sinusoidal output signal. More specifically, we can distinguish between short-time variations and long-term drift; two aspects that have different implications for different aspects of system design.

17.6.1 Feedback Oscillators

Generation of oscillations is usually done with a feedback circuit that contains a frequency-selective element in the transfer loop, see Figure 17.12. In this circuit, stable oscillations occur for the ω for which

$$|AH(\omega)| = 1 \quad \text{and} \quad \angle AH(\omega) = 2\pi n \quad n = 0, 1, 2, \ldots \tag{17.33}$$

In a self-sustained oscillator, nonlinearity is needed to ensure stable oscillations. This nonlinearity can be, for instance, due to the gain saturation in amplifiers. The other condition is for $H(\omega_0)$ to have a phase of $2\pi n$ at the desired frequency ω_0; this is fulfilled by placing a resonant circuit in the feedback loop.

A widely used oscillator circuit is the *Colpitts oscillator*, which employs a transistor in common-base configuration as amplifier. In the simplest case, it furthermore uses a parallel LC circuit (tank circuit) as resonator, see Figure 17.13. Defining the series capacitance

$$C_{\mathrm{T}} = \frac{1}{\dfrac{1}{C_1} + \dfrac{1}{C_2}} \tag{17.34}$$

the resonance frequency is

$$\omega_0 = \frac{1}{\sqrt{LC_{\mathrm{T}}}}. \tag{17.35}$$

In many cases, it is desirable to adapt the frequency of the oscillator. This can be realized by having adjustable capacitance in the resonant circuit (theoretically, adjustable inductors are equivalent, but they are much more difficult to realize in practice). The very first radio RXs used mechanically tunable capacitances, but nowadays varactor diodes, whose capacitance can be adjusted by changing a bias voltage, are the most common form. When the capacitance needs to be changed only with a certain stepsize (and within a limited range), the use of multiple capacitors that can be switched in and out of the resonator by means of switches can be helpful. In any case, the use of an adjustable capacitance in the resonator circuit results in a *Voltage-Controlled Oscillator (VCO)*, i.e., an oscillator

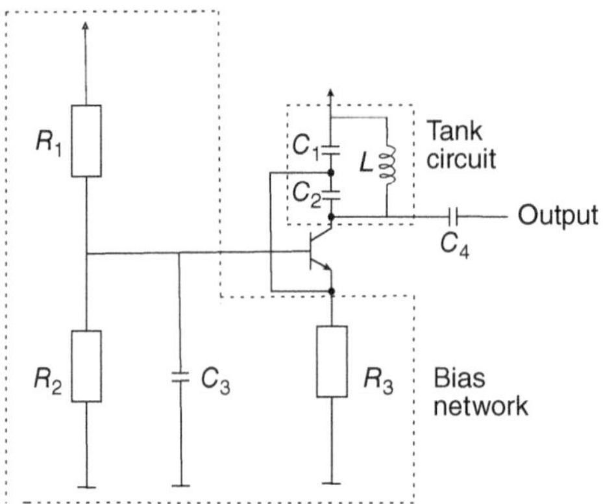

Figure 17.13 Simplified circuit diagram of a Colpitts oscillator, showing the tank circuit, the bias network, and the amplifier. Note that the base of the transistor is grounded with respect to AC through C_3, and that only AC is coupled to the output, due to C_4.

whose resonance frequency can be adjusted according to a control voltage. For the case that the oscillation frequency is controlled by a digital signal, we speak of a Digitally Controlled Oscillator (DCO).

17.6.2   Resonators for Oscillators

In order to ensure that oscillations occur only at one frequency, Eq. (17.33) should be fulfilled only at one specific frequency ω_0. This in turn requires a resonator with high Q as part of the circuitry.

The simplest example, discussed already in the previous section, uses lumped inductors and capacitors in the circuit. At low frequencies, such resonators can have good Q. However, the oscillation frequency depends on the actual value of the reactances, whose deviations from the nominal values (and thus deviation of the oscillation frequency from its desired value) can be on the order of percent – something that is completely unacceptable for most applications. Even when components are carefully selected, aging and temperature drift will lead to detuning. VCOs can compensate for these effects, but the need to find a suitable reference to determine the "correct" frequency, and the effort in tuning over a large range, make this a suboptimum solution.

Thus, most wireless transceivers use crystal oscillators; these employ the resonances of piezoelectric crystals, usually quartz, mounted between two metallic plates. These crystals are inserted into the resonant circuitry in the feedback loop, e.g., as part of a Colpitts oscillator. Typical quartz oscillator frequencies are in the range of $3 - 30$ MHz. Quartzes can achieve $Q-$ values on the order of $10^4 - 10^6$, and provide a frequency accuracy on the order of $10^{-8} - 10^{-4}$, depending on the implementation. However, they still suffer from some error sources: (i) *aging*: due to outgassing of the crystal and similar effects, the resonance frequency might change, (ii) *sensitivity to "hard" radiation,* such as gamma rays: this is important mainly for systems mounted on satellites or airplanes, (iii) *sensitivity to mechanical shock*, and (iv) *temperature drift*: the resonance frequency changes with the ambient temperature; this is the most important source for frequency detuning. It can be combatted by the use of oven-heated quartz, i.e., placing the quartz in a miniature oven (not much larger than the quartz itself) and heating it to a constant temperature that is higher than the highest ambient temperatures (typically 75°C). This setup achieves the highest frequency accuracy and stability ($\sim 10^{-8}$), but has as a drawback the increased size and power consumption for the oven.

Even higher accuracy can be achieved with *atomic clocks*, most frequently *Rubidium (Rb)-clocks*, which exploit the hyperfine transition in certain atoms as reference. Such clocks are usually bulky and expensive (though some progress has been recently made to create miniaturized atomic clocks). They are used for controlling the frequency of cellular BSs, Global Positioning System (GPS) TXs, and precision measurement equipment such as channel sounders and vector network analyzers. In many situations, atomic clocks are combined with oven-heated quartz oscillators to combine the high frequency stability of the former with the low phase noise of the latter. Atomic clocks are almost always used to create a 10 MHz standard frequency.

If frequencies >100 MHz are desired, different types of resonators can be used, such as surface acoustic wave resonators, ceramic resonators, or – if higher Q is necessary – dielectric resonators or Yttrium Iron Garnet (YIG) resonators. However, in many cases, higher frequencies are generated by creating a lower-frequency sinusoid using quartz oscillators, followed by frequency multiplication and/or a PLL, discussed below.

In some circumstances, it is possible to improve the accuracy of resonant oscillators by means of GPS signals (see Chapter 29). GPS provides not only location, but also (i) a precise frequency reference, and (ii) accurate timing information (in particular, a 1 pulse per second (1 pps) signal that can be used to adjust any timing drift that might accumulate due to inaccurate oscillator frequency). Such GPS-disciplined oscillators thus have the precision of GPS (which is based on atomic clocks), while only requiring relatively simple, low-cost oscillators. However, this approach is only feasible for those situations where GPS is available, which might not be the case indoors, in urban street canyons, etc.

17.6.3   PLL

Up to now, we have analyzed how to generate a single sinusoid with high precision. Yet, in wireless transceivers, there often is the need for generating a sinusoid that is locked, in frequency and phase, to another *reference oscillation*. A key application is to generate a sinusoid with selectable output frequency: while using VCOs enables tuning of this frequency, this happens at the expense of accuracy and stability. On the other hand, most modern wireless systems, including cellular and Wi-Fi, can operate on a variety of different frequency channels, and thus need a whole set of possible LO frequencies for up/downconversion (see below for details). Generating such a set of frequencies can be realized by a PLL. This device locks the phase of one oscillation to the phase of a reference oscillator that might operate at a different frequency.

The fundamental block diagram of a PLL is shown in Figure 17.14. Assume for the moment that the reference oscillation stems from a quartz oscillator, operating at, say, $f_{\rm ref} = 10$ MHz, and the division factor $N_{\rm ref} = 1$. The reference signal is thus one input of a phase comparator (or a phase frequency detector); the other input is the output signal of a VCO that is operating at approximately $f_{\rm out} = N_{\rm mult}f_{\rm ref}$, and is then divided in an appropriate circuit by a factor $N_{\rm mult}$ – thus, the two inputs of the phase comparator oscillate with the same frequency, and comparing the phase between them is meaningful. If there is a phase difference between the two inputs, the phase comparator generates a signal that (filtered through a loop filter) drives a charge pump that pushes a proportionate charge to the input of the VCO. This detunes the VCO output until the phases at the phase comparator are in agreement. Various types of phase comparators can be used for this purpose; digital PLLs (which replace analog phase comparator, charge pump, and loop filter by digital circuitry followed by a DAC) are rapidly gaining in popularity. Similarly, the frequency dividers with division factors $N_{\rm ref}$ and $N_{\rm mult}$ can be implemented in various ways, with digital (flip-flop) dividers the most common method for lower frequencies.

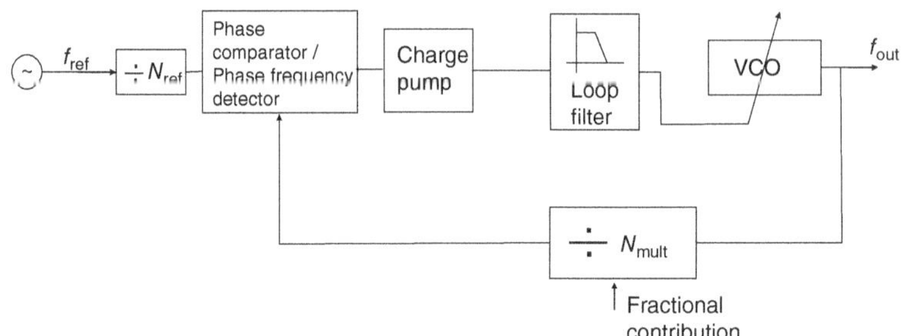

Figure 17.14   Block diagram of a PLL.

Changing the output frequency in steps of f_{ref} simply requires the changing of the division factors. In the case that $N_{\text{ref}} = 1$, the output frequency can be an integer multiple of f_{ref}, according to the settings of the divisor N_{mult}. If a finer quantization of the available frequencies is desired, then $N_{\text{ref}} > 1$ must be chosen (of course N_{ref} is an integer); the possible frequencies are then in steps of $f_{\text{ref}}/N_{\text{ref}}$.

In practice, the frequency resolution achievable by integer PLLs is either too coarse, or would require division by a very large factor such that $f_{\text{ref}}/N_{\text{ref}}$ is sufficiently small. Instead, it is common to use *fractional* PLLs, which adds a signal (typically the output of a Sigma-Delta modulator is added) in the division block, thus creating a division by a fractional factor; for details see, e.g., [Su and Pamarti 2009].

Up to now, we have assumed that the oscillations are essentially noise free and stable, so that the PLL serves just as a means of generating selectable frequencies. However, PLLs have more general applications that arise from the presence of a loop filter in the system. If this loop filter is a low-pass, then it essentially serves to limit the speed with which the output frequency can follow variations of the input frequency. Such variations might arise either from conscious changes, or from noise, giving rise to the following two applications for PLLs, which might use $N_{\text{ref}} = N_{\text{mult}} = 1$.

- *Carrier recovery*: we want to be able to lock a LO to the frequency and phase of the carrier of a modulated RF signal. In this case, the input is not a clean sinusoid, but rather a modulated input signal, and the loop filter should suppress the frequency components caused by the modulation. In this case, the PLL output follows the average frequency, i.e., the carrier (this is a reason why, e.g., long runs of 0 or 1 are undesirable as input to an FSK modulator, motivating the use of scrambling sequences to break them up). On the other hand, the loop bandwidth must be large enough to pass frequency components that the PLL wishes to track, such as time variations of the Doppler shift of a signal.
- *Reduction of phase noise*: a denoising of an external oscillation and/or reduction of noise generated in the PLL is desired. On one hand, a low loop bandwidth suppresses external (phase) noise, as well as transients in the signal; on the other hand, it enhances the impact of noise created within the feedback loop.

PLL loop bandwidth usually has to be determined empirically. Bandwidths on the order of $10 - 1000$ kHz are typical.

An important nonideality of PLLs is the *cycle slip*, which may occur in the presence of transients. Imagine a situation where the (divided) VCO frequency, which was previously locked, is driven by an external influence to be 2% smaller than f_{ref}. Assume furthermore that the loop bandwidth is $10^{-2} f_{\text{ref}}$. Then the PLL will not immediately react to the difference of the phases due to the small cycle bandwidth. After about 50 periods of the reference frequency, the two signals have the same phase again, but the frequency (number of cycles) experienced by the two oscillators differs by one. The PLL will however only aim to correct the phase imbalance it "sees" after about 100 cycles, and ignore the fact that the number of cycles experienced by the VCO has "slipped" behind the reference signal.

The theory of PLLs is quite involved, and we again refer to the references of the "further reading" section for a mathematical treatment and design guidelines for PLLs. From a system point of view, a PLL can be characterized by the following parameters:

- *Pull-in range* (capture range, acquisition range): this is the range of oscillations that the PLL can synchronize to when starting up. More precisely, the pull-in range is the largest frequency deviation (for arbitrary phase) such that the PLL acquires lock.
- *Lock-in range*: if the PLL is a locked state, then any sudden frequency jump that is smaller than the lock-in range can be followed *without cycle slip*.
- *Hold-in range*: the hold-in range describes the frequency range over which the PLL can follow a sequence of small perturbations (as opposed to the abrupt steps considered in the lock-in range) of the frequency.
- *Transient response*: this describes the overshoot and settling time to a certain defined accuracy.
- *Steady-state errors*: phase errors created in the PLL (as opposed to the phase errors implicit in the reference oscillator).

Signals generated from a single reference generator by multiple PLLs are not completely phase synchronous – both due to phase noise generated in the PLL, temperature drift, cycle slip, and different states of the delta-sigma modulators in fractional PLLs. This has important consequences for systems that require multiple oscillator signals such as antenna arrays or distributed antenna systems: it is usually easiest to distribute a single reference clock to the different locations where the oscillator signal is needed, since such signals are low-frequency and suffer less from cable attenuations. However, the most precise phase alignment between the signals is created if a single high-frequency local-oscillator signal is generated and then distributed to the different locations. Careful trade-offs are required from the system designers.

If only a single high-frequency oscillator signal is needed, it is often preferable to use a simple frequency multiplier instead of a PLL. In essence, any nonlinear component, such as a diode, creates oscillations at integer multiples of the input oscillations, as can be easily seen by expanding $[\cos(\omega_0 t)]^n$ into a Fourier series. The advantage of this approach is the absence of cycle slips; the drawback is the reduced flexibility in the generated output frequency.

17.6.4 Phase Noise

The key characteristic of all oscillators is the purity of the created sinusoidal wave. Description of the deviations is thus the essential characterization of real-world oscillators. These deviations can be described in the time domain, as timing jitter (usually, deviations of the zero-crossings of the sinusoid from their nominal position), or in the frequency domain, as phase noise.

Denote, generally, the output of an oscillator as

$$v(t) = V_0[1 + \alpha(t)] \cos[\omega_0 t + \varphi(t)] \tag{17.36}$$

where $\alpha(t)$ describes the time variations of the amplitude (which we henceforth ignore), and the phase $\varphi(t)$, whose time variations describe the deviations from the ideal oscillator frequency, as $d\varphi(t)/dt$ describes an instantaneous frequency. These phase variations are a random process, which is characterized by its power spectral density $S_\varphi(f)$; it describes the power in a single-sided spectral sideband (typically, of 1 Hz bandwidth, at a particular offset from the carrier frequency), divided by the power concentrated in the nominal carrier frequency. The unit of this power spectrum is thus dBc (which stands for dB carrier).

It is important to note that phase noise is not an additive noise like Gaussian noise. It can be described as a Wiener process, which describes essentially a random walk: at every time instance, a random change is imposed on the total phase. Though the process is zero-mean (assuming there is no frequency offset of the oscillator from the nominal frequency), the variance of the process increases linearly with time.

When high-frequency oscillations are created by frequency-multiplying a lower-frequency oscillator like a quartz oscillator, then the phase noise increases (see also (17.37) below).

Figure 17.15 shows a typical phase noise spectrum of a crystal oscillator. We first observe that the phase noise is largest near zero offset (i.e., near the carrier), and then decreases, with changing slope, as the offset increases. Theoretical models have described the phase noise as $1/f^3$ closest to the carrier (where f by convention denotes the offset from the carrier frequency, not the absolute frequency), then for larger frequencies changing to $1/f$, while far from the carrier it shows essentially frequency-flat characteristics. Particularly, Leeson's equation for the single-sideband phase noise (in dBc/Hz) reads

$$L(f) = 10 \log \left[\frac{1}{2} \left(\left(\frac{f_0}{2Qf} \right)^2 + 1 \right) \left(\frac{f_{\text{corner}}}{f} + 1 \right) \left(\frac{Fk_B T}{P_s} \right) \right]. \tag{17.37}$$

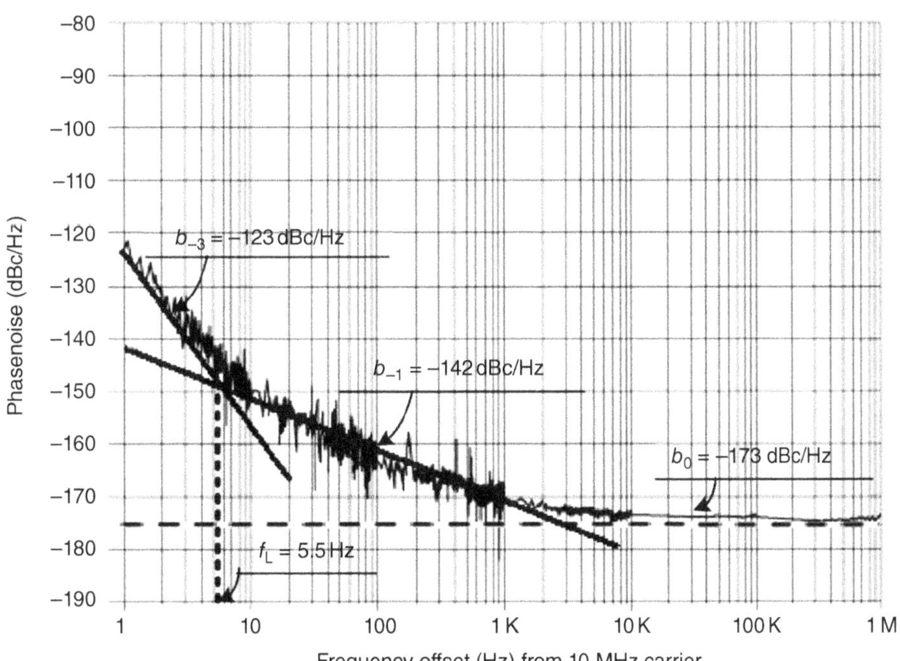

Figure 17.15 Phase noise spectrum of quartz oscillators at 10 MHz.

Here f_0 is the carrier frequency, f_{corner} the so-called corner frequency, Q the loaded quality factor, F is the noise figure, $k_B T$ the noise power spectral density, and P_s the power at the oscillator input. Figure 17.15 shows a transition between the different slopes, with a slope of 3 from 0 to 5 Hz, changing continuously to a slope of 0 above 10 kHz. For assessing the impact of phase noise spectrum on the system performance, it is useful to remember that the inverse of the frequency offset indicates the timescale over which the phase noise leads to significant changes in the phase; for example, the 1 kHz phase noise is relevant for processes lasting for 1 ms.

Another way of describing phase noise is by means of the *Allan variance*, also called two-sample variance. It is defined as

$$\sigma^2_{\text{Allen}}(\tau) = \frac{1}{2\tau^2} E\{[\varphi(2\tau) - 2\varphi(\tau) + \varphi(0)]^2\} \tag{17.38}$$

where the expectation is over the random process (with ergodicity often assumed) and τ is the observation period. Allan variances (or Allan deviations, i.e., the square root of the Allan variance) for different observation periods describe therefore the phase errors that are relevant for different aspects of the system design. Short-term Allan deviations describe the phase noise occurring, e.g., between the beginning and the end of a modulation symbol, which impacts the Bit Error Rate (BER) of coherent RXs; Allan deviations over 1 ms impact the timing jitter of frames in LTE systems (see Chapter 31), and so on.

Phase noise has negative influence on a variety of system aspects, for example:

- *Reduction of the SNR of a received symbol*: since a coherent received integrates the received signal over the symbol duration, phase noise during that integration period reduces the output of the integrator, and thus the effective SNR of the signal. Furthermore, considering a time scale of multiple symbol periods, the phase noise can push the received signal constellation points away from their ideal position toward the decision boundary; Figure 17.16 shows a 16-Quadrature Amplitude Modulation (16-QAM) constellation in the presence of phase noise.
- *Intercarrier interference in OFDM systems*: we have seen in Section 15.7 that broadening of the spectrum of a subcarrier destroys the orthogonality between the signals of the subcarriers. We mostly discussed such broadening due to the Doppler effect, but phase noise can create the same problems. This becomes especially relevant at mm-wave frequencies (remember that the phase noise gets increased more when the frequency of the fundamental oscillation is multiplied by a larger factor).
- *Violation of spectral masks*: the actually emitted spectrum in passband is the convolution of the baseband spectrum with the LO spectrum. If the LO spectrum is not a delta pulse, the emitted spectrum is broadened and thus may violate the spectral mask imposed by the frequency regulator, even if the baseband spectrum is designed to properly adhere to it.
- *Insufficient suppression of adjacent-channel interference*: due to reciprocal mixing. Consider a desired signal with support in the frequency range $[f_0 - \Delta/2, f_0 + \Delta/2]$. This gets downconverted by a noisy LO and filtered by a lowpass filter to the range $[-\Delta/2, \Delta/2]$. Consider now an arbitrary frequency component of the phase noise spectrum, at $f_0 + f_1$. Any adjacent-channel interference (i.e., signal with $f_0 + f_{adj}$ outside the range $[f_0 - \Delta/2, \, f_0 + \Delta/2]$) that fulfills $|f_{adj} - f_1| \leq \Delta/2$ will be downconverted into the passband of the lowpass filter. All frequency components that can lead to a mixing product within the lowpass filter bandwidth have to be considered and added up.

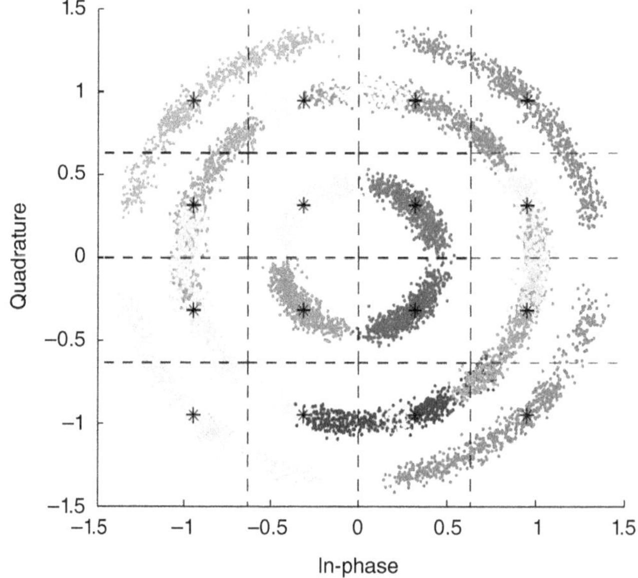

Figure 17.16   16-QAM constellation in the presence of phase noise. Color version available at wiley.com/go/molisch/wireless3e.

*17.7 Mixers and Frequency Conversion

Downconversion is the shifting of the spectrum of a signal from a higher to a lower frequency. It needs three essential ingredients: (i) the RF signal, (ii) a *LO*, and (iii) a mixer, i.e., a device outputting the product of the RF signal and the oscillator signal, which is usually at the *IF*. Upconversion works, of course, similarly.

Mixers usually exploit nonlinearities of semiconductor devices. A simple example is a single-ended diode mixer, where the two signals are added together by means of a coupler, and then used to drive a (biased) diode.

From a phenomenological point of view, mixers are designed to produce *only* the mixing products, but those with the highest possible efficiency. They furthermore should provide good matching at the inputs, and isolate the input ports from each other. A variety of mixer configurations, such as balanced mixers, double-balanced mixers (ring modulator), etc., exist. Without going into the details of the implementations, they are generally characterized by

- *Conversion loss or gain*: this measures the ratio of the output (IF) power over the RF input power; note that there is also a requirement that the LO power falls within a specified range. Passive mixers usually have a conversion loss on the order of $5 - 10$ dB, while active mixers have a conversion gain.
- *Noise figure*: a mixer generates internal noise that is added to the input noise; see Section 17.2.2 for a more detailed discussion. Further characteristics, namely dynamic range and IP 3 are defined similar to those of amplifiers discussed previously.
- *Isolation*: defines the amount of leakage from one port to another; thus each mixer is characterized by multiple isolation values.
- *Spurious*: this describes signals from undesired frequency bands that are converted into the IF-band.
- *Quadrature accuracy*: for IQ mixers, this describes the accuracy of the separation (combining) of the signal into I and Q components.

17.8 Transceiver Structures

17.8.1 Up/Downconversion

Up/downconversion is a key part of any wireless transceiver, and in the preceding subsections, we have investigated the components necessary to make it work. We now proceed to the investigation of the whole transceiver structure that implements it; for ease of explanation, all descriptions will refer to downconversion unless stated otherwise.

The simplest downconversion architecture is the single-conversion superheterodyne RX. In it, the RF signal is converted from RF to a (fixed) IF. If the RF channel we are interested in might change, such as in cellular or Wi-Fi systems, then the LO must be tunable. The IF signal is then directly digitized (using a suitably high sampling rate), and further processed by digital processing algorithms. Note that these steps only describe the actual frequency conversion; a number of amplifiers and filters are also included to adjust signal power levels and suppress undesired signals. Figure 17.17 shows a more detailed block diagram:

- Directly after the antenna is a bandpass filter whose passband encompasses the system bandwidth of interest (e.g., all Wi-Fi channels in the 2.45 GHz frequency band), while it suppresses out-of-band signals. This suppression both serves to eliminate signals at the image frequencies (see below), and to avoid strong interference signals that could drive the LNA into saturation.
- The LNA amplifies the RF signals falling within the bandpass filters. It might be followed by other amplifiers, ensuring that the output power level of this amplifier cascade is within the dynamic range of the mixer. After the amplifier, further filters might be implemented.
- The mixer has an LO that can be tuned such that the mixing product with the desired frequency channel (e.g., a specific 20 MHz Wi-Fi channel within the 2.45 GHz band) is in the IF band.

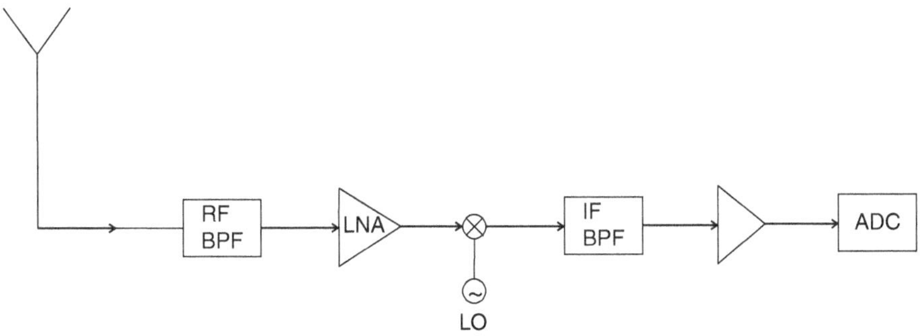

Figure 17.17 Block diagram of a single-conversion superheterodyne receiver. BPF, bandpass filter.

- Next follows the IF filter, which has a bandwidth of the desired frequency channel (e.g., 20 MHz in the example above). This filter needs to suppress well all signals outside the range of interest, since – after the subsequent sampling – any aliased out of band signal will be folded into the processed signal.
- The IF signal is then amplified further and sent to an ADC for digitization.

17.8.2 Image Rejection and Multi-Stage Conversion

Ideally, downconversion would simply mean a shifting of the center frequency from RF to IF. However, mixers create *products* of signals. Consider thus the simplest case where both the RF and the LO signal are sinusoids, $A_{\mathrm{RF}} \cos(\omega_{\mathrm{RF}}t)$ and $A_{\mathrm{LO}} \cos(\omega_{\mathrm{LO}}t)$. Then the mixer output is their product, which – using standard trigonometric identities – can be written as

$$ s_{\mathrm{mixer\text{-}out}}(t) = \frac{A_{\mathrm{RF}}A_{\mathrm{LO}}}{2} \cos\left[(\omega_{\mathrm{RF}} + \omega_{\mathrm{LO}})t\right] + \frac{A_{\mathrm{RF}}A_{\mathrm{LO}}}{2} \cos\left[(\omega_{\mathrm{RF}} - \omega_{\mathrm{LO}})t\right]. \tag{17.39} $$

We generally want the IF to be lower than the RF, so that the first term on the r.h.s. is eliminated through lowpass filtering, so that $s_{\mathrm{mixer\text{-}out}} = s_{\mathrm{IF}}(t) = A_{\mathrm{IF}} \cos(\omega_{\mathrm{IF}}t)$, with $A_{\mathrm{IF}} = A_{\mathrm{RF}}A_{\mathrm{LO}}/2$ (assuming the mixer gain to be unity).

Now depending on the relationship between ω_{LO} and ω_{IF}, we obtain different RX architectures. Most common is *low-side injection,* i.e., $|\omega_{\mathrm{LO}}| < |\omega_{\mathrm{RF}}|$, so that

$$ \omega_{\mathrm{IF}} = \omega_{\mathrm{RF}} - \omega_{\mathrm{LO}}. \tag{17.40} $$

An alternative approach is *high-side injection*, where $|\omega_{\mathrm{LO}}| > |\omega_{\mathrm{RF}}|$, so that the IF becomes

$$ \omega_{\mathrm{IF}} = -[\omega_{\mathrm{LO}} - \omega_{\mathrm{RF}}], \tag{17.41} $$

This implies that the IF is shifted to negative frequencies; however since we are dealing with real signals, any negative frequency component has a corresponding positive component. For a continuous spectrum, higher frequencies are shifted to "less negative" frequencies, which, in the positive-frequency half of the spectrum, means lower frequencies. In other words, the spectrum gets not only shifted but also mirrored around its new center frequency when using high-side injection. Such mirroring can, if required, easily be undone by the digital signal processing of the received signals. The other mixing product, at $\omega_{\mathrm{RF}} + \omega_{\mathrm{LO}}$ is now located at more than twice the RF frequency and thus can be easily suppressed.

Image Frequency and Image Rejection Filters

It follows from (17.39) that not only can two different LO frequencies mix an RF signal into the desired IF band but also do two different RF bands get mixed, by the same LO, into the desired IF band. Specifically, any RF frequency such that

$$ |\,\omega_{\mathrm{LO}} \pm \omega_{\mathrm{IF}}\,| \approx \omega_{\mathrm{RF}} \tag{17.42} $$

gets downconverted into the IF band. Figure 17.18 shows an example for low-side injection, i.e., the band at $\omega_{\mathrm{LO}} + \omega_{\mathrm{IF}}$ is the band of interest. In addition to this band, also the frequency band centered on $\omega_{\mathrm{LO}} - \omega_{\mathrm{IF}}$ gets converted to the IF; we call this the *image frequency* band. Thus, the (interfering) signal from a TX operating in the image band might show up at IF unless bandpass filtering before the mixer suppresses the image frequency.[12] This image suppression filter is centered on ω_{RF}, and has a transition from passband to stopband within less than $2\omega_{\mathrm{IF}}$, i.e., the separation between the desired band and the image band.

Thus, the choice of the IF involves, among other things, a trade-off between image rejection and channel selection. A low IF makes construction of the image rejection filter difficult, because it requires a very small *relative* bandwidth of the transition region, which is difficult to achieve with on-chip components. On the other hand, a low IF makes it easier to build channel selection filters, which pick a subband, such as a Frequency Division Multiple Access (FDMA) channel, out of the total signal available at IF; furthermore, if the IF signal is sampled directly (see Section 17.8.3), a high IF poses high requirements to the ADC. A way out of this dilemma is to use two (or even more) downconversion stages: we first do downconversion to a high IF, such that the image rejection filter can be easily realized. From there, a second downconversion is done, where the second IF can be quite low, e.g., such that the lower band edge of the IF band is just barely above $\omega = 0$. This then facilitates the use of highly selective channel selection filters, as well as a lower sampling frequency. It must be noted, however, that this second downconversion can give rise to a secondary image frequency, which needs to be filtered out too.

---

[12] There are also filter-less image rejection methods based on Hilbert transforms, see [Razawi 2011, Chapter 4].

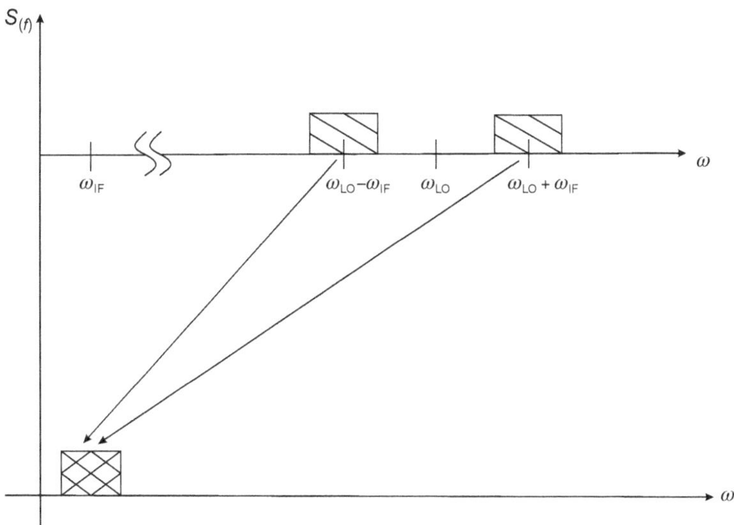

Figure 17.18 How image frequencies map into the IF band.

Zero IF and Direct Conversion Receivers

From this, the question naturally arises whether the (second) IF center frequency can be made equal to zero, i.e., the signal can be converted into baseband. In this case, the signal becomes complex, i.e., the RX needs to convert the signal in a way that allows extraction of the I and Q part of the signal (compare Chapter 11). Thus, the downconverter has to be a *quadrature downconverter* that provides the mixing product of the IF with both $\sin(\omega_{\mathrm{IF}}t)$ and $\cos(\omega_{\mathrm{IF}}t)$, see Figure 17.19. Obviously, this approach needs two ADCs for the digitization of the mixing products, but those can operate at a low sampling rate. Practical challenges arise from IQ imbalance, where the amplification of the I and Q branch are not identical, and the phase shift between the two branches is not exactly $\pi/2$.

IQ downconversion can also be used directly on the RF signal; such RXs are called *direct conversion RXs* (homodyne RXs). They have the advantage of simplicity, reducing the number of analog RF components, and eliminating the problem of image frequencies. On the downside, direct conversion RXs face the following challenges:

- LO leakage: the signal from the LO might couple directly into the antenna and get transmitted, thus interfering with other RXs in the vicinity; note that this is not an issue in heterodyne RXs.
- DC offset: the LO leakage also creates a DC offset, which can result in saturation of RF components, and difficulty for the ADC, which usually has a dynamic range that is the same in positive and negative amplitude direction.
- Flicker noise, which is proportional to $1/f$, can be problematic in particular for narrowband RXs (bandwidth <1 MHz)
- Higher sensitivity to IQ mismatch in the IQ downconversion, since, e.g., a timing offset between the I and Q branch translates into a larger phase offset if the downconverted frequency is larger; the mismatch problem can be partly solved through calibration procedures.
- For a direct-conversion transmitter, oscillator pulling can be a problem. The output of the power amplifier has a frequency very similar to that of the LO; thus when part of it couples into the LO, It can modulate the LO frequency.

Despite these challenges, direct-conversion RXs have become very popular especially for frequency range <7 GHz.

Transmitter Design

TX designs are mostly analogous to the RX architectures, though noise and interference are less of an issue, while linearity is a larger concern.

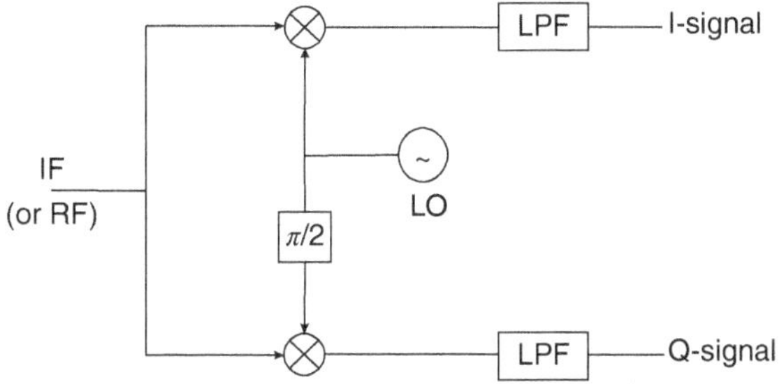

Figure 17.19 IQ demodulator.

For heterodyne TXs, the input frequency of a mixer is at IF, and the output frequency is at RF; also in this case, we can consider low-side injection and high-side injection. Similar to the downconversion case, there exists an image frequency, in other words, frequency bands around $\omega_{LO} + \omega_{IF}$ and $\omega_{LO} - \omega_{IF}$ are created, with one of them suppressed by an RF filter.

Just like for downconversion, the output of the mixer contains – in addition to the two mixing products – the LO leakage, i.e., the LO signal that reaches (in attenuated form) the output port. This is undesirable and often is further suppressed by filtering. Note that typically the IF is relatively small, so that ω_{RF} and ω_{LO} are similar. This means that suppression of the LO leakage is much more difficult for upconversion, since the leakage signal lies close to the desired output band, and furthermore bandpass filters with small relative bandwidth (such that ω_{LO} lies in the stopband and ω_{RF} lies in the passband) are difficult to realize. LO leakage is especially problematic for direct-conversion TXs. Multi-step upconversion can help to mitigate these problems.

17.8.3 Sampling and Digitization

Once the signal has been converted to IF, digitization can be done as follows:

(i) Nyquist sampling: in this case, the sampling rate must be at least twice the frequency of the upper band edge of the IF band. For IF frequencies on the order of tens of MHz, this is quite feasible, although it leads to unnecessarily high energy consumption, as energy consumption of ADCs increases with the sampling frequency. Oversampling can be used to relax the requirements on the anti-aliasing filters.

(ii) Undersampling: in this case, the sampling frequency is consciously chosen lower than the Nyquist frequency, such that the IF band lies in one of the periodic repetitions (aliasing) of the Nyquist band, see Figure 17.20. However, this approach is sensitive to jitter in the sampling clock, and suffers from noise folding, where out-of-band noise is folded into the sampled signal.

(iii) Downconversion to zero-IF, and IQ-sampling, as discussed in Section 17.8.2.

17.9 Spectrum Masks

The transmission power of a wireless user is usually limited by the frequency regulator of a country. These limits can be related to protection of health (excessive absorption of electromagnetic waves can lead to heating of tissues), and/or can be rooted in limiting interference to other users.

The first and simplest form of such emission regulations is a limit of the total emitted power (often called the conducted power). However, the impact of interference is governed not only by the total amount of power but also by how that power is distributed over frequency, i.e., the power spectrum of the emission. It is thus common to prescribe – instead of, or in addition to, the limits on conducted power – a limit on the emitted *power spectral density*. For this purpose, the power is measured over a sliding window with defined bandwidth, typically between 10 kHz and 1 MHz.

Spectral lines and spectral ripples thus lead to a reduction of the total allowed emission power when power spectral density is limited. Consider the situation of Figure 17.21, which shows a signal with a 5 MHz channel bandwidth and 100 kHz measurement bandwidth: if its power spectrum is flat, then a limit of 1 mW per 100 kHz bandwidth results in a total conducted power of 50 mW. However, imagine that the signal does not properly suppress the carrier, so that the power in the central 100 kHz is 10 times stronger than in any other 100 kHz band: then the power in the central band is only allowed to be 1 mW as per the regulation, which implies that the power in any other 100 kHz intervals is only allowed to be 0.1 mW, leading to a total conducted power of 5.9 mW.

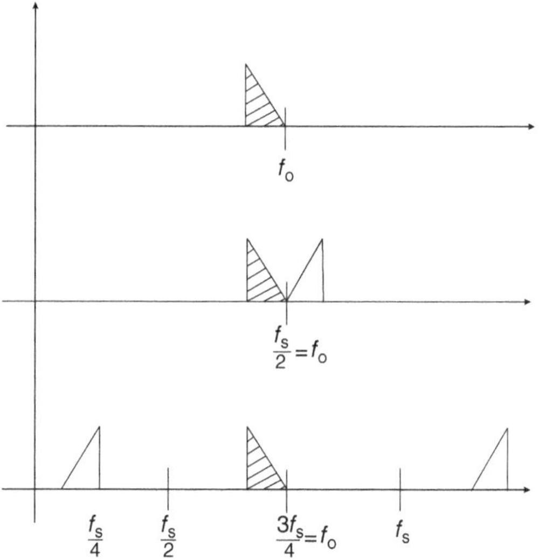

Figure 17.20 (Top) Bandpass signal of interest. (Middle) Critical sampling, obtaining signal of interest (shaded) and aliasing (outline). (Bottom) Using low sampling frequencies to obtain desired band (shaded) and aliasing products (outline).

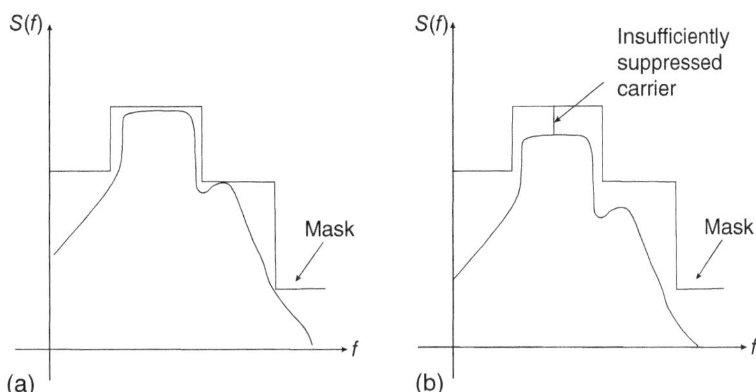

Figure 17.21   Emission spectrum closely fitting spectral mask (a). Emission spectrum with backoff to accommodate the insufficiently suppressed carrier signal (b).

Regulations might also limit the *area power density*, such that the power is *limited in every direction*. The interpretation of this limitation is quite similar to that of the power spectral density: uniform emission into all directions allows to maximize the transmitted total power while staying within the limits of the spectral mask. In such a situation, directional transmit antennas (fixed or adaptive) cannot help to improve the receive SNR, since any gain due to antenna gain will be compensated by the need to back off the total transmitted power in order to stay within the emission regulations (note, however, that directionality might be still helpful to reduce interference to other users, and reduce power consumption). Finally, the *area power spectral density* may be limited (i.e., limitations are imposed for every frequency measurement band in every direction).

Frequency regulators usually also prescribe a measurement (averaging) duration for the measurement of the transmitted power (or power spectral density); such durations are usually on the order of milliseconds. The motivation for this is that "instantaneous spectra" might have quite a different shape than average spectra, and staying within the spectral mask for any possible realization of the instantaneous spectrum would be difficult; more importantly, most systems that might be affected by interference (or thermal heating) are mainly impacted by the average power, and not short-term spikes. This limitation has important consequences in multi-user Multiple Input Multiple Output (MIMO) systems (see Chapter 22): if a BS forms a beam in the direction of a particular user and continuously transmits to this user over a duration that is longer than the averaging duration, it needs to back off the conducted power. If, however, it transmits for a time that is much shorter and then changes to a different direction to transmit to a different user, it might not need to back off. This in turn impacts user scheduling.

Besides limitations on the intentionally transmitted power, regulators also impose limitations on unintentional emissions. When discussing about frequency bands for different users, we have up to now used the assumption that the spectra of the different users are completely orthogonal; in other words, that the spectrum of one user does not overlap with the spectrum of another user. This assumption does not hold in reality, as it is essentially impossible to create signals whose support in the frequency domain is truly finite. A user who is assigned a particular frequency band in which to transmit it thus also assigned a spectral mask, within which all emissions need to be contained. This spectrum mask limits the power spectral density outside the intended band, i.e., the maximum adjacent-channel interference or spurious emissions that a TX may cause. Typically, the spectrum mask imposes lower emissions at frequencies that are farther away from the nominal band edge, since those can be filtered out more easily. A typical example is shown in Figure 17.22. In addition, there is usually a limit (in the US, it is called the Part-15 regulation) on spurious emissions at *any* frequency range, even far away from the assigned band, to ensure that such emissions do not disturb other wireless operations.

The nominal emission spectrum is determined by the choice of the modulation format and the burst structure and possible carrier frequency changes of the signal (compare Chapters 18 and 19). However, as we could see throughout this chapter, hardware non-idealities lead to increase in out-of-band emissions, so that in most cases a physical testing of the emissions is required. Commercial transceivers generally have to go through type approval to show that they follow the emission regulations.

*17.10   Full Duplex

In traditional wireless systems, transmission and reception occur either at different frequencies so that the signals can be separated by suitable filters (Frequency Domain Duplex FDD)) or at different times (Time Domain Duplex (TDD)), compare Section 18.5. However, both of these schemes have the disadvantage that each link direction uses only a fraction (e.g., half) of the available time-frequency resources. It would be more advantageous if each device could transmit and receive at the same time, a concept known as full duplex.[13] In the most straightforward case, the spectral efficiency is doubled: for example, instead of using a separate band for uplink and downlink, both bands can be used for uplink and downlink simultaneously. Simultaneous TX and RX are also essential for in-band relays (compare Chapter 27).

---

[13] "Full duplex" in this sense is different from the "full duplex" of Chapter 1, which refers to quasi-simultaneous communication but encompasses, e.g., standard FDD. Both meanings are habitually used in the literature.

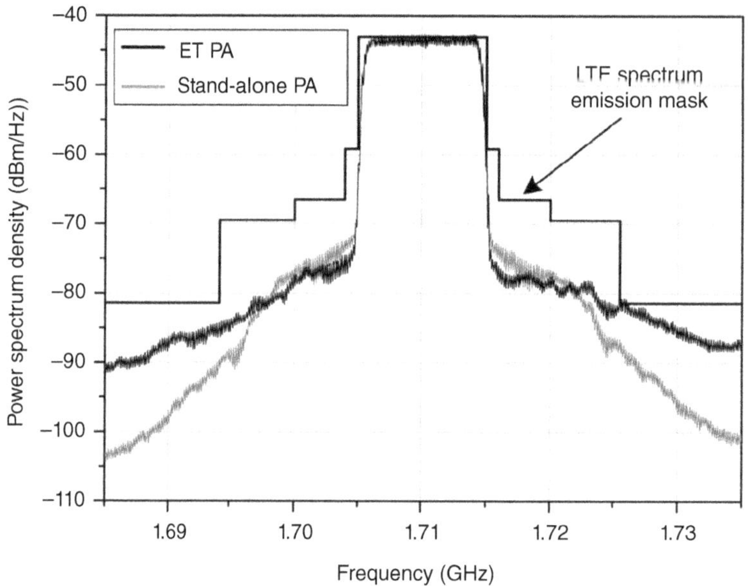

Figure 17.22 LTE spectral mask and emission spectrum with two power amplifiers, indicating spectral regrowth. Reproduced with permission from [Kim et al. 2014] © IEEE.

The main challenge of full-duplex is the interference between two data streams – obviously, the received data (and the received signal) are different from the transmitted data/signal. In principle, any interference from the TX to the RX (on the same device) can be eliminated through serial interference cancellation (compare Section 28.2), which can be implemented in digital or analog. The difficulty lies in the enormous power differences between the TX and RX signals. Consider a Wi-Fi system: the TX power is 20 dB, while the typical RX power levels are on the order of −60 dB. Since furthermore a Signal-to-Interference Ratio (SIR) of 20 dB is required for optimum operation, interference from the TX to the RX of the device should be suppressed by 100 dB. In the cellular case, the requirements are even more stringent, since the BS has higher TX power levels, and RX power levels can be −80 to −100 dB. Thus interference suppression by 120–140 dB might be required.

Consequently, it is usually necessary to suppress the interference in multiple stages, so that the suppression effects add up: (i) antenna, (ii) analog, and (iii) digital, see Figure 17.23. We will now discuss these stages one after the other.

Firstly, discrimination between TX and RX signals can be achieved if the device has separate antennas (one for TX, one for RX). Assuming that the two antennas are in the far-field of each other, the power gain between them is

$$\frac{\lambda^2}{(4\pi d)^2}\left(G_{TX,\phi}G_{RX,\phi} + G_{TX,\theta}G_{RX,\theta}\right). \tag{17.43}$$

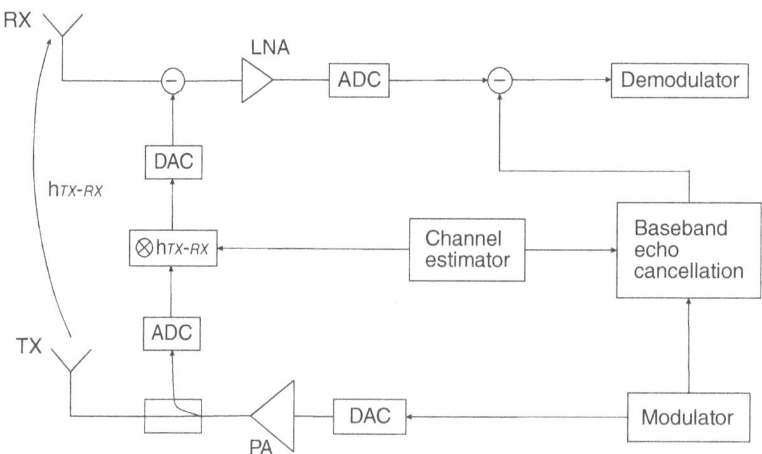

Figure 17.23 Block diagram of a full-duplex transceiver.

This tells us that there are three ways of increasing the attenuation:

- increase the distance between the TX and the RX antenna (this is limited by the device size),
- let the RX antenna pattern have a null in the direction of the TX antenna, and vice versa. This can be achieved by (i) the antenna design itself, or (ii) by having antenna arrays (minimum two elements) that can create suitable nulls, or (iii) have absorbing material between TX and RX antennas,
- have the TX and RX antenna elements with different polarization. If, e.g., the TX antenna emits with the ϕ polarization and the RX antenna receives on the orthogonal (θ) polarization, no signal power is coupled to the RX.

Method 2 and 3 can be implemented either in a channel-adaptive, or a nonadaptive way. Note that the signal coupled from the TX to the RX antenna can occur not just due to direct transmission but also through reflections from surrounding objects. A nonadaptive approach cancels the direct transmission, by designing the antennas (and possibly calibrating them in an anechoic chamber) such that direct transmission is suppressed. However, reflected signal components may still act as interference. An adaptive approach can be implemented, e.g., through antenna arrays with adaptive phase shifters, so that the array pattern can be changed.

These measures help to suppress the interference between TX and RX but are not able to completely suppress it. Typical attenuations are on the order of 40 dB, much lower than the 80–140 dB required in practical systems. Thus additional suppression is required.

In most circumstances, this additional suppression occurs in two stages, in an analog and a digital stage. First, the analog cancellation can be obtained by taking the transmit signal, weigh it both in amplitude and phase such that it compensates the residual transmit signal (i.e., what remains after the suppression by the antenna). This compensation usually assumes the residual interference is frequency-independent; any frequency-dependent residual is handled in a further digital stage.

The digital interference suppression is the most flexible, and possibly most accurate. The suppression requires an accurate model of the whole transmission chain from TX baseband to RX baseband, including antenna and analog suppression. Such a model can be obtained, e.g., through suitable training. Then, knowing the transmit data, the digital compensation can suppress any residual TX contributions in the received signal. However, the effectiveness of the digital suppression is limited by the dynamic range of the ADC – this is the main reason why the suppression cannot be done through digital means only. As discussed in Section 17.3.1, the ENOB strongly depends on the bandwidth – while it is easy to build an ADC with $ENOB = 16$ for a bandwidth <100 kHz (this is done in every Compact Disk (CD) player), the ENOB for bandwidths of 1 GHz are typically $5 - 6$. Thus, digital suppression is least effective for large-bandwidth signals, but it is exactly for this case that it is most needed, due to the frequency selectivity of the interference.

*17.11   Appendices

App. 17.A: Two-port Network and S-parameters

See App17.pdf at wiley.com/go/molisch/wireless3e

App. 17.B: Matching

See App17.pdf at wiley.com/go/molisch/wireless3e

Further Reading

A detailed description of S-Parameters, matching, and noise, as well as amplifiers, RF filters, couplers, etc. can be found in [Pozar 2005]; an overview of various aspects of RF circuit design is provided in [Ellingson 2016] and [Sayre 2008]. RF transceiver design including component characteristics is also described very clearly in [Razawi 2011]. Taken together, those four books cover most of the topics of this chapter.

There are also a number of excellent books on particular aspects. The list below has no claim to completeness, but just contains some personal favorites.

ADCs: Extensive description of ADC characteristics and types can be found in [van der Plaasche 2003] and [Pelgrom 2017].

Amplifiers: A classical, and easy to read, book on the topic is [Gonzales 1996].

Filter design is discussed in detail in [Schaumann et al. 2008]. For a handbook on RF filters, couplers, etc. see [Matthaei et al. 1980].

Oscillators and mixers: Design aspects for amplifiers, mixers, and oscillators, as well as models for semiconductor components, are described in [Rohde and Rudolph 2013]. For PLLs, the classical (and still worth reading) text is [Gardner 2005]. Another good description is [Best 2007].

Hardware for millimeter-wave systems is discussed in [Rappaport et al. 2015b].

Synchronization can be considered to lie on the border of hardware and signal processing. Synchronization in OFDM was discussed in Chapter 15. More general discussions can be found in [Mengali and D'Andrea 1997, Meyr and Ascheid 1990, and Meyr et al. 1997].

For updates and errata for this chapter, see https://wides.usc.edu/students.html#textbooks.

Exercises

See Sec. 36.22 of Exercises.pdf at wiley.com/go/molisch/wireless3e

Part IV
Wireless Communication with Multiple Users

In Part III, we described how a single transmitter can communicate with a single receiver. However, for most wireless systems, multiple devices should communicate simultaneously in the same area. It is therefore necessary to provide means for *multi-user communications*, in particular *multiple access*, where multiple User Equipments (UEs) can talk to the same Base Station (BS), and *multicell communications*, where multiple BSs are deployed in a geographical area, with at least some of them using the same time/frequency resources. The area of multi-user communications shows significant differences between the speech communications systems that were dominant until the mid-2000s and modern systems that are dominated by data transmission. The greater flexibility of data packet transmission, as well as the wider variety of services with different rate requirements and priorities, has a major impact on the multi-user system designs.

The most straightforward way to enable multiple access is to have different devices communicate on different frequencies (Frequency Division Multiple Access, FDMA), different times (Time Division Multiple Access, TDMA), or both. When combined with Orthogonal Frequency Division Multiplexing (OFDM) modulation, a very flexible assignment of time-frequency resources called Orthogonal Frequency Division Multiple Access (OFDMA) can be realized as well. Finally, resource assignment can either be determined by a central controller or done at random. All these topics are discussed in Chapter 18.

A different type of multiple access is based on spreading the signal over a large bandwidth, and to make that spreading unique for each user. This allows multiple users to be on the air simultaneously, and the receiver can determine from the spreading which part of the "on-air" signal comes from which user. This *spread-spectrum* approach, described in Chapter 19, is used, e.g., for third-generation cellular systems in the form of *Code Division Multiple Access* (*CDMA*). The chapter also describes the Rake receiver, a device that enables CDMA systems to deal with the delay dispersion of the channel, and how multi-user detection can be exploited to significantly increase the performance in a multiple-access environment.

Not only the amount of resources allocated to each user but also *which* specific resources (at what time, and on which frequency) need to be determined. This problem called *scheduling* is particularly important for data transmission where services with different latency constraints might all use the same infrastructure. Latency, total throughput, and fairness need to be traded off against each other. All of this is discussed in Chapter 20.

While Chapters 18–20 deal with multiple UEs communicating with a single BS, Chapter 21 then extends the discussion to multiple BSs. The cellular principle, which allows a reuse of spectrum for different BSs, is described. While the classical cellular networks assign a fixed percentage of the system bandwidth to each BS, we also discuss schemes that use adaptive assignments of frequencies, possibly depending on where in the cell the UE is located, as well as heterogeneous networks.

Chapter 22 generalizes the analysis of Chapters 18–21 to the case when the BS has multiple antenna elements. In such multi-user Multiple Input Multiple Output (MIMO) situations, the BS can communicate with multiple UEs at the same time on the same time/frequency resources. It also incorporates a discussion of massive MIMO, where the BS has a very large number of antenna elements. Finally, cell-free massive MIMO (also known as cooperative multipoint or C-RAN) allows the elimination of inter-cell interference.

All of the above topics assume the existence of infrastructure nodes (base stations) that have fixed locations, are powered by power mains and connected to a backbone network. While this is an efficient setup, there are situations where wireless devices just need to talk amongst each other, so that direct communications between UEs is the only operation mode, and no infrastructure is necessary. Such ad-hoc networks provide much lower cost; yet their communications require new approaches since no central control mechanism is available anymore. Chapter 23 outlines the challenges and solutions for this type of communication.

18

Multiple Access

18.1 Introduction

In this chapter, we will discuss the situation where multiple User Equipments (UEs) communicate with a single Base Station (BS). We will consider both the downlink (i.e., BS to multiple UEs) and the uplink of the system (i.e., the UEs transmitting to the BS). The considered system setup thus represents infrastructure-based wireless systems, including cellular and Wireless Local Area Network (WLAN) systems. This chapter discusses a variety of methods for keeping the transmissions from/to different UEs separable. We will start out by considering the performance limits of various Multiple-Access (MA) methods, as obtained from information theory, in Section 18.2.

To provide a rough categorization of MA protocols, we can distinguish between contention-free and contention-based access protocols. In a contention-free protocol, Section 18.3, a certain percentage of "spectral resources" (to be defined more accurately later on) is set aside for the exclusive use of one particular UE. This UE thus never has to compete with other UEs for these particular resources; they are always available to it alone – hence the name contention-free access. This technique has the advantage of simple design and usually provides easy performance guarantees. The most important categories of these protocols are

- *Frequency Division Multiple Access (FDMA)*, where different frequency subbands are assigned to different users.
- *Time Division Multiple Access (TDMA)*, where different time periods are assigned to different users.
- *Orthogonal Frequency Division Multiple Access (OFDMA)*, where different symbols on various subcarriers are assigned to the different users.

It is now clear that all of these methods assign spectral resources, i.e., use of a particular part of the spectrum for a particular amount of time. FDMA and TDMA are discussed in Sections 18.3.1 and 18.3.2, respectively. The fact that those sections are rather brief should not detract from the importance of TDMA and FDMA. Firstly, they serve to explain important concepts. Secondly, while not often used in their pure form in modern systems, they are often combined with other approaches. Section 18.3.3 then describes OFDMA (not to be confused with the modulation format OFDM), which – due to its use in the fourth and fifth-generation cellular standards Long-Term Evolution (LTE) and New Radio (NR) (see Chapters 31 and 32) has gained increased relevance in recent years. Another important contention-free multiple access format, *Code Division Multiple Access (CDMA)*, where each user is assigned a different code, has been studied mainly in the context of second- and third-generation cellular systems. We will discuss this scheme, as well as other spread-spectrum methods, in Chapter 19.

A second category of MA formats are contention access protocols. In these methods, no resources are a priori assigned to (reserved for) a particular user. Rather, the users try to "grab" them when they have data to transmit in the uplink (note that for the downlink, the BS has the data for all users available, and can thus schedule transmissions to them without contention). Contention access is particularly suitable for systems in which a large number of UEs are present, and the requested data rates are highly time-variant, e.g., bursty data traffic. In that situation, a contention-free protocol would be inefficient: it would have to reserve a lot of resources for each UE, to accommodate the data when they are ready, but the UE would make use of it only at rare occurrences. Rather, in contention access, each UE competes for resources when it actually has data to transmit. However, contentions (conflicts) arise when two or more UEs want to "grab" the same spectral resources. A variety of methods have been developed to handle such conflicts – from a simple retransmission to sophisticated exchange of control information. In Section 18.4 we will review these different methods and the performance that can be achieved with them.

It is common to associate contention-free access with cellular communications systems, and contention access with wireless LANs (and ad hoc networks). This association is historic, and correct for the dominant standards. Early cellular communications systems concentrated on voice transmission, where a user that was to be served would be active consistently for many minutes; thus reserving certain resources for such a user was eminently meaningful. In contrast, the Wi-Fi standard for wireless LAN was intended for transmission of bursty packet traffic (e.g., web pages), so that contention access was most meaningful. However, these associations are not "laws of nature." Every cellular system has at least a contention-access component, namely when new UEs associate themselves with the BS (since the UE does not yet known to the BS, it has to contend for channel resources to make itself heard), and random

Wireless Communications: From Fundamentals to Beyond 5G, Third Edition. Andreas F. Molisch.
© 2023 John Wiley & Sons Ltd. Published 2023 by John Wiley & Sons Ltd.
Companion website: www.wiley.com/go/molisch/wireless3e

(grant-free) access is an emerging topic for 5G. On the other hand, the Wi-Fi standard also introduced a contention-free access method that is intended for data that require certain quality of service guarantees, like voice.

*18.2   Performance Limits for Multiple Access

The ultimate performance limits of transmission between a BS and multiple UEs can be investigated through information-theoretic considerations (we will in Chapter 23 separately consider the performance limits of ad hoc networks, where multiple Transmitters (TXs) and Receivers (RXs) are present). We separately consider

1. *broadcast channel*, which describes a channel where a single TX sends (different) messages to multiple RXs. This situation corresponds to the downlink of a cellular system.
2. *multiple access channel*, where multiple TXs send their messages to a single RX. This situation corresponds to a cellular uplink.

While we consider broadcast and multiple-access channels separately, there are strong dualities between the two.

We have seen in Section 13.1 that the performance limits of a single link are given by the channel capacity, which forms an upper limit for the data rate at which (for a particular channel) reliable communication can be achieved by any arbitrary scheme. When dealing with multiple users, we aim to find the *capacity region*, i.e., the set of tuples of rates for the different users that can be achieved simultaneously. The capacity region is multidimensional (one dimension per considered user); therefore it is sometimes preferable to have a more compact representation: the *sum capacity*, which adds up the capacities of the different users. (we again use a relaxed definition of capacity, sometimes meaning "achievable rate", compare Section 13.11.2).

In line with the general theme of this chapter, the equations below consider only communications between a single BS and multiple UEs (i.e., one cell) without interference. We note that the information-theoretic capacity of interference channels (as would be relevant in multi-cell systems (see Chapter 21) is not known exactly (only upper and lower bounds are known).

18.2.1   Downlink

Consider a system with a single TX and K RXs. As a first constraint, we find that the rate per unit bandwidth for a specific user k, $k = 1, ..., K$ cannot exceed the single-user capacity

$$R_k \leq \log_2\left(1 + \frac{\overline{P}}{\sigma_n^2}|h_k|^2\right) \tag{18.1}$$

where $\overline{P}$ is the overall (mean) power available to the TX, h_k is the channel coefficient (complex amplitude channel gain) to the kth user, and $\sigma_n^2 = N_0 B$ is the noise power.

We next consider the capacity for the different users in the case of TDMA, where signals for different users are distinguished by transmitting them at different times. With such a scheme, the users are completely orthogonal, i.e., are ensured not to interfere with each other. We can distribute time and power to the different users at will. If we give the kth user a fraction τ_k of the available time, and let it use power P_k, the capacity (for unit bandwidth) for its transmission is

$$C_k = \tau_k \log_2\left(1 + \frac{P_k}{\tau_k \sigma_n^2}|h_k|^2\right) \tag{18.2}$$

and the rate region is characterized by

$$\{R_1 \leq C_1, R_2 \leq C_2, ...\} \quad \text{with} \quad \sum_k \tau_k = 1 \quad \text{and} \quad \sum_k P_k = \overline{P}. \tag{18.3}$$

The largest capacity region is indeed achieved by varying both allocated power and time for each user; exceptions are found only when the channels for the different users are identical, i.e., $|h_k|^2 = |h_{k'}|^2, \forall k, k'$. Furthermore, in asymmetric channels, the sum rate is maximized by assigning all resources to the user with the best channel (though this might violate fairness constraints). This important insight is exploited further in time- or frequency-selective channels through multi-user diversity, see Chapter 20.

Higher rates (i.e., rate combinations that lie outside the region Eq. (18.3)) can be achieved by superposition coding, combined with serial interference cancellation. To realize this, order the users according to the strength of their channels, such that $|h_1|^2 \leq |h_2|^2 \leq \cdots |h_K|^2$. Therefore, a user k can decode all signals that can be decoded by user $i < k$. The rate for the first (weakest) user is

$$R_1 \leq \log_2\left(1 + \frac{P_1|h_1|^2}{\sigma_n^2 + \sum_{i>1} P_i|h_1|^2}\right) \tag{18.4}$$

since the signals assigned to the users with the better channels appear as noise to user 1. Note that those signals are designed to be transmitted to the ith user and transmitted with power P_i, but the interference power arriving at UE 1 is impacted by the channel h_1, which describes the channel gain from the BS to UE 1. Next consider UE 2: because it has a better SNR than UE 1, it can certainly decode message 1. After decoding this message, it can subtract its contribution from the overall received signal (serial interference cancellation, see also Section 28.1). Consequently, the only interference present in the decoding at UE 2 are the signals intended for users 3, 4, ... Generally, the rates for superposition coding must fulfill

$$R_k \leq \log_2\left(1 + \frac{P_k|h_k|^2}{\sigma_\mathrm{n}^2 + \sum_{i > k} P_i|h_k|^2}\right). \tag{18.5}$$

It turns out that this approach achieves the broadcast capacity, so that the capacity region is the union of all the tuples of rates that satisfy Eq. (18.5) with power constraint $\sum_k P_k = \overline{P}$ (Figure 18.1).

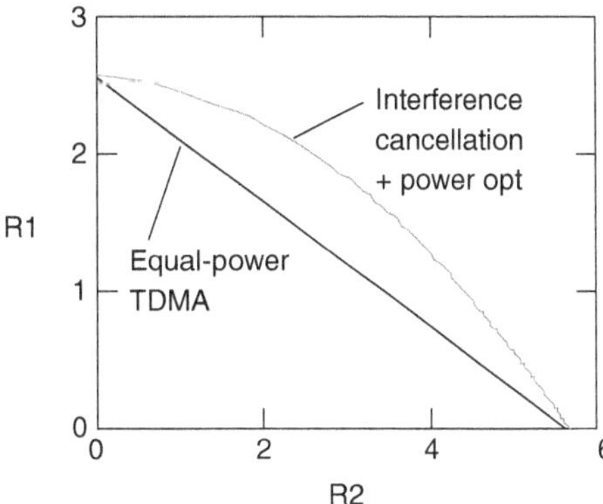

Figure 18.1 Rate/capacity region for two-user broadcast channel: $N_0 = 0.1$, $|h_1|^2 = 0.5$, $|h_2|^2 = 5$, $P = 1$. Color version available at wiley.com/go/molisch/wireless3e.

It is remarkable that the same capacity can be achieved without requirement of interference cancellation at the RX. Instead, the TX can encode the transmit signal in such a way that the RX does not "see" the interference. More precisely, we design the signal for the first RX. Since this signal is now known, the TX can design the signal for the next RX such that it is not interfered by the first signal. The signal for the third RX is designed to be not interfered by the first and second transmitted signal, and so on. This precancellation at the TX can be seen as the dual of the interference cancellation at the RX and leads to the same capacity region.[1] It is also known as "writing on dirty paper," or, in short "dirty paper coding." In practice, it is often approximated by Tomlinson–Harashima precoding.

Let us now investigate how the capacity changes in the presence of fading. We have to distinguish here between the cases with and without Channel State Information at the TX (CSIT), while we assume that the RX always knows the Channel State Information (CSI). Let us first consider the case without CSIT. Then, for a symmetric channel (i.e., where the fading statistics are the same for each UE), the sum rate is the same as for the single-user case, namely

$$\sum_k R_k < E\left\{\log_2\left(1 + \frac{|h|^2\overline{P}}{\sigma_\mathrm{n}^2}\right)\right\} \tag{18.6}$$

which shows that orthogonalization of the different users (i.e., each user transmits at a different time, of over different parts of the frequency band) achieves the capacity. Unfortunately, no such simple solutions are available for the asymmetric channel (i.e., different average channel strength). This is because the superposition coding scheme described above for the nonfading case is based on a sorting of the channels according to their SNR. In the fading case, however, different channels are the best at different times, so that (without CSIT) proper ordering cannot be done.

For the case that CSIT is present, different powers can be assigned for the signals to the different users. In this case, the capacity is the time average of the capacities of the different fading states of the channel. The power allocation is generalized (to multiple users) waterfilling (see Section 15.9) in time.

---

[1] In practice, there are important differences: dirty paper coding requires channel state information at the TX, which might be difficult to obtain (see Section 16.1.7). On the other hand, interference cancellation can suffer from error propagation (similar to decision feedback equalizers, see Section 14.3), while this does not occur in dirty paper coding, because the TX knows all the signals perfectly (there is no noise at the TX).

18.2.2 Uplink

We now turn to the capacity of the uplink, where K UEs send messages to a single BS. The critical difference to the downlink case is that power cannot be distributed arbitrarily between different users, but rather each user (each UE) has its own power constraint. Maximum rates are achieved by each UE transmitting at its maximum power. Transmission at maximum power is optimum in terms of information-theoretic limits of a multiple-access channel. Note, however, all the considerations in this section are for a multiple-access system that does *not* suffer inter-cell interference. In a multi-cell system, increasing transmit power for one user increases not only desired signal power at the intended BS but also increases inter-cell interference and thus might actually reduce the cellular capacity (see Chapter 21).

Just like in the downlink, the capacity for each UE is limited by the single-link communication capacity – i.e., a UE cannot communicate with a higher rate than in the situation where it is alone in the cell

$$R_k \leq \log_2\left(1 + \frac{P_k}{\sigma_n^2}|h_k|^2\right). \tag{18.7}$$

Furthermore, the sum rate of the users is limited by the capacity that would be achieved by a "superuser" that has all resources available

$$\sum_k R_k \leq \log_2\left(1 + \frac{\sum_k P_k|h_k|^2}{\sigma_n^2}\right). \tag{18.8}$$

The rates fulfilling conditions Eqs. (18.7) and (18.8) can again be achieved by successive interference cancellation (see also Chapter 28), as shown in the following.

It is remarkable that if user 1 transmits at its maximum (single-user) rate, it is possible for other users to still transmit with nonzero rate. Consider the case of two UEs, and let UE 1 transmit at its maximum rate $\widetilde{R}_1 = \log_2\left(1 + \frac{P_1}{N_0}|h_1|^2\right)$, while UE 2 transmits with a lower rate $\widetilde{R}_2$. When the BS decodes the signal from UE 2, it will "see" the signal from UE 1 as interference. Therefore, $\widetilde{R}_2$ is limited by

$$\widetilde{R}_2 \leq \log_2\left(1 + \frac{P_2|h_2|^2}{\sigma_n^2 + P_1|h_1|^2}\right). \tag{18.9}$$

After decoding this signal, the BS can subtract its contribution from the overall received signal, so that it can then actually decode the "cleaned up" signal (containing only contributions from UE 1) received at a rate $\widetilde{R}_1$. We also find that the rate pair $\widetilde{R}_1, \widetilde{R}_2$ fulfills Eq. (18.8) with equality, see Figure 18.2

$$\begin{aligned}
\widetilde{R}_1 + \widetilde{R}_2 &= \log_2\left(1 + \frac{P_1}{\sigma_n^2}|h_1|^2\right) + \log_2\left(1 + \frac{P_2|h_2|^2}{\sigma_n^2 + P_1|h_1|^2}\right) \\
&= \log_2\left(1 + \frac{P_1|h_1|^2 + P_2|h_2|^2}{\sigma_n^2}\right)
\end{aligned} \tag{18.10}$$

indicating that the successive interference cancellation actually achieves capacity (for rigorous proofs see the references cited in the section titled "Further Reading").

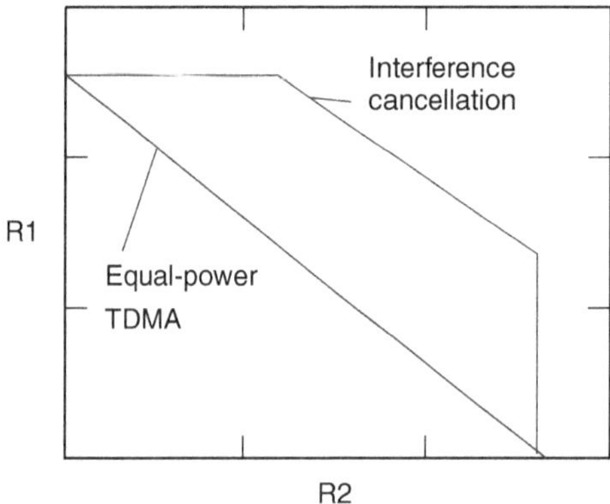

Figure 18.2   Rate/capacity region of the multiple access channel.

Let us now turn to the fading channel for the uplink. The sum capacity is given by taking the expectation over Eq. (18.8), i.e.,

$$\sum_k R_k \leq E\left\{ \log_2\left(1 + \frac{\sum_k P_k |h_k|^2}{\sigma_{\mathrm{n}}^2} \right) \right\} \tag{18.11}$$

where the expectation is over the fading. From Jensen's inequality, it is clear that this is always smaller than the sum capacity in a nonfading channel with equal average channel gain.

For orthogonal access of the users, and assuming each user is given $1/K$ of the time/frequency resources, the achievable sum rate is simply

$$\sum_k \frac{1}{K} E\left\{ \log_2\left(1 + \frac{K P_k |h_k|^2}{\sigma_{\mathrm{n}}^2} \right) \right\}$$

where the factor K in the SNR arises from the fact that the signal power is concentrated in a part of the time/frequency domain (compare also 18.2). This is less than (18.11).

18.3 Contention-Free Multiple Access

We now turn to practical ways of realizing *orthogonal* multiple access. In the previous section, we pointed out the performance limits that such a system has (and that it is not necessarily optimum in terms of the maximum rates that can be achieved). In this section, we deal in particular with contention-free multiple access, i.e., the system reserves particular resources for the different UEs (for both uplink and downlink).

18.3.1 *Frequency Division Multiple Access (FDMA)*

FDMA is the oldest, and conceptually simplest, multiple access method. Each user is assigned a frequency (sub)band, i.e., a (usually contiguous) part of the available spectrum, see Figure 18.3. The assignment of frequency bands is done during the call setup (for a phone call) or the establishment of a data connection (for data transmission) and retained during the whole session. FDMA is usually combined with frequency-domain duplex (see Section 18.5), so that two frequency bands (with a fixed duplex distance, i.e., frequency offset) are assigned to each user: one for the downlink (BS-to-UE) and one for the uplink (UE-to-BS) communication.

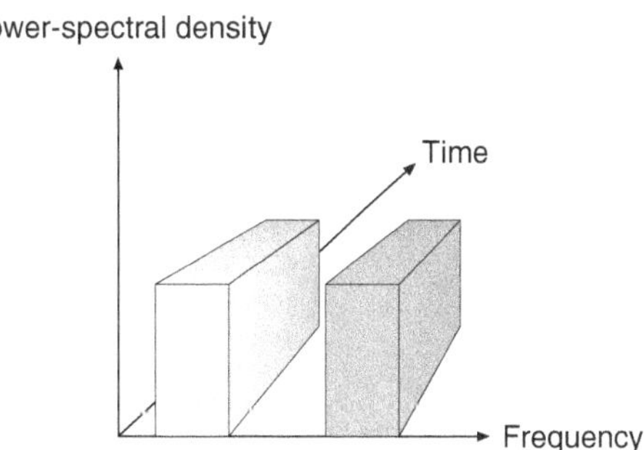

Figure 18.3 Principle of FDMA.

Pure FDMA is conceptually very simple, and has some advantages for implementation:

- TX and RX require much *less digital signal processing*. However, this is not so important in practice anymore, as the costs for digital processing are continuously decreasing.
- The *(temporal) synchronization is simple*. Once synchronization has been established in the call setup, it is easy to maintain by a simple tracking algorithm, as transmission occurs continuously.

However, pure FDMA also has significant disadvantages. These problems arise from spectral efficiency considerations, as well as from the sensitivity to multi-path effects:

- *Frequency synchronization and stability are difficult*. For speech communications, each frequency subband is quite narrow (typically between 5 and 30 kHz). The local oscillators thus must be very accurate and stable; jitters in the carrier frequency result in adjacent-channel interference. High spectral efficiency also requires the use of steep filters to extract the desired signal. Both accurate oscillators and steep filters are expensive and thus undesirable; to avoid them, guard bands can be used to mitigate the filter requirements. This, however, reduces the spectral efficiency of the system.

- *Sensitivity to fading.* Since each user is assigned a distinct frequency band, those bands are narrower than for other multiple access methods (compare TDMA, CDMA). For such narrow subbands, the fading is flat in practically all environments. This has the advantage that no equalization is required; the drawback is that there is no frequency diversity. Note that frequency diversity is mainly provided by signal components that are more than one channel coherence bandwidth apart: in that case, they are fading independently, so that the fading of the total signal is reduced, see Chapters 12 and 14.
- *Sensitivity to random Frequency Modulation (FM).* Due to the narrow bandwidth, the system is sensitive to random FM: the Bit Error Rate (BER) due to random FM is proportional to $(\nu_{\max}T_S)^2$, see Section 11.3. Thus, it is inversely proportional to the square of the bandwidth. On the positive side, appropriate signal processing schemes can not only mitigate these effects but even exploit them for obtaining time diversity. Note that the situation here is dual to wideband systems, where delay dispersion can be a drawback, but equalizers can turn them into an asset by exploiting the frequency diversity.
- *Intermodulation.* The BS needs to transmit multiple FDMA channels, each of which is active the whole time. Typically, a BS uses 20–100 frequency channels. If those signals are amplified by the same power amplifier, third-order modulation products can be created, which lie at undesirable frequencies, see Section 17.2.2. We thus need either a separate amplifier for each FDMA channel, or highly linear amplifiers – each of those solutions makes a BS more expensive.
- *Increased latency:* due to the narrower bandwidth, the transmission of a data packet takes longer. This effect is mostly compensated, however, by the fact that the channel is always available to the TX (in contrast, e.g., to TDMA, where a TX has to wait for its assigned timeslot until it can transmit.

Due to these reasons, FDMA is mostly used for the following applications:

- *Analog communications systems.* Here, FDMA is the only practicable MA method.
- *Combination of FDMA with other MA methods.* The spectrum allocated for a service (or a network operator) is divided into larger subbands, each of which is used for serving a *group* of users. Within this group, multiple access is done by means of another MA method, e.g., TDMA or CDMA. Most current wireless systems use FDMA in that way.
- *High data rate systems:* the disadvantages of FDMA are mostly relevant if each user requires only a small bandwidth, e.g., 20 kHz. The situation can be different for wireless LANs. where a single user requires on the order of 20–160 MHz bandwidth (and even higher for Personal Area Networks) and only a few frequency channels are available at each access point (BS). Note, however, that data networks usually do not assign a frequency band to a user on a permanent basis, but rather combine it with packet radio (see Chapter 33).

18.3.2 Time Division Multiple Access (TDMA)

For TDMA, different users transmit not at different frequencies, but rather at different times, see Figure 18.4. In the simplest form of TDMA, a time unit is subdivided into K timeslots of fixed duration, and each user is assigned one such time slot (in more advanced implementations, the time percentage and power can differ between different users), see Section 18.1. During the assigned timeslot, the user can transmit with a high data rate (as it can use the whole system bandwidth); subsequently, it remains silent for the next $K-1$ timeslots, when the other users take their turn. This process is then repeated periodically.[2] At first glance, this approach has the same performance as FDMA: a user transmits only during $1/K$ of the available time, but then occupies K times the bandwidth. However, there are some important practical differences:

- The users occupy a larger bandwidth. This allows to exploit the frequency diversity available within the bandwidth allocated to the system; furthermore, the sensitivity to random FM is reduced. On the flip side, equalizers are required to combat ISI for most operating environments; this increases the effort needed for the digital signal processing.
- There is no need for frequency guard bands, as each user completely fills up the assigned band. However, temporal guard intervals are now required. A TX needs a finite amount of time to ramp up from 0 W output power to "full power." Furthermore, there has to be sufficient guard time to compensate for the runtime of the signal between UE and BS, as discussed below.
- Each timeslot might require a new synchronization and channel estimation, as the transmission is not continuous. The optimization of the time slot duration is a challenging task. If it is too short, then a large percentage of the time is used for synchronization and channel estimates (in the second-generation cellular standard GSM (Global System for Mobile communications, see Appendix 30.A), 17% of a time slot are used for this purpose). If the timeslot is too long, latency becomes too long (which users find annoying especially for speech communications), and the channel starts to change during one timeslot. In that case, the equalizer has to track the channel during transmission, which increases the implementation effort. If the time between two timeslots assigned to one user is larger than the coherence time, the channel has changed between those two time slots, and a new channel estimate is required.

---

[2] A more general system can also change the assignment from timeslot to timeslot, see also Section 18.3.3.

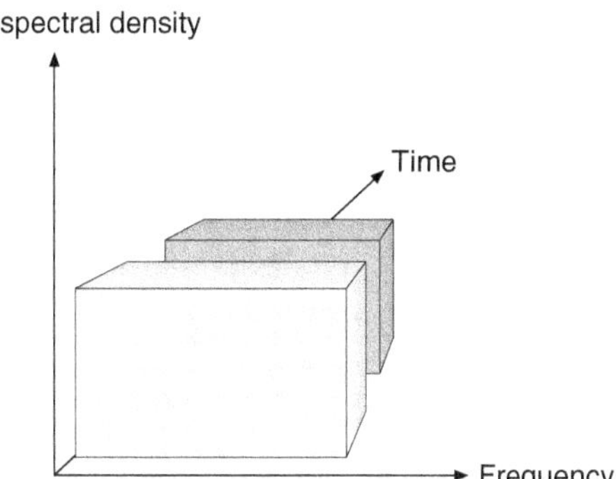

Figure 18.4   Principle of TDMA.

- For interference-limited systems, TDMA has a major advantage: during its period of inactivity, the UE can "listen" to the transmissions on other timeslots.[3] This is especially useful for the preparation of handovers from one BS to another, when the UE has to find out whether a neighboring BS would offer better quality and has available communications channels, as well as for other forms of BS cooperation.

In practice, TDMA is often combined with FDMA, see Figure 18.5, which increases the flexibility and allows to combine some of the advantages of the two schemes.

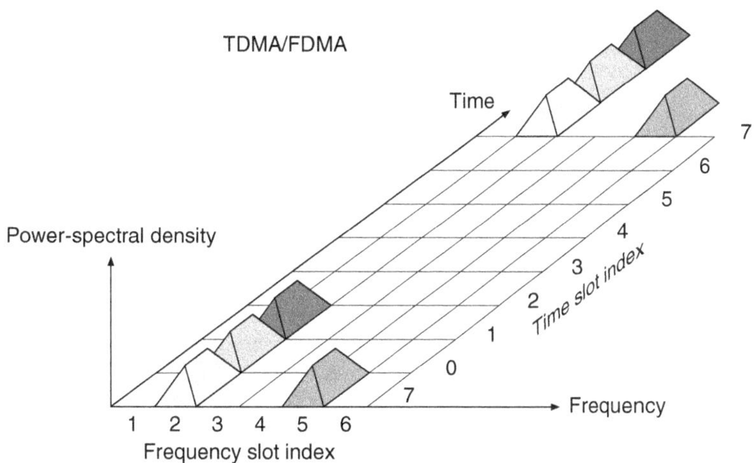

Figure 18.5   Combination of TDMA with FDMA, using the example of the GSM system.

A challenge in TDMA systems arises from the fact that signals to/from different UEs have different runtimes. Consider the situation that transmission in the first timeslot is by a UE that is far away from the BS, while the UE that transmits in the subsequent timeslot is very close to the BS and thus has negligible runtime. As the signals from the two users must not overlap at the BS, the second UE must not transmit during the time it takes the first signal to propagate to the BS. Take the following example: user A is at 30 km distance from the BS, and transmitting bursts in the timeslot TS 3 of every frame. User B is located close at the BS and accesses the timeslots TS 4. The propagation delay of user A is around 100 μs, whereas the propagation delay of user B is negligible. Therefore, without a guard interval of suitable length (i.e., 100 μs), the end of a burst from user A partly overlaps with the beginning of a burst of user B at the BS, this situation is illustrated in Figure 18.6.

This problem can be largely overcome by the so-called *timing advance*. In this approach, the propagation delay from UE to BS is estimated by the BS during the initial phase of establishing a connection. The result is transmitted to the UE, which then sends its bursts *advanced* (with respect to the regular timing structure) to ensure that the bursts *arrive* within the dedicated timeslots at the BS. In the example above, UE A would start and end its transmission 100 μs earlier than what is shown in Figure 18.6, thus avoiding overlap with the signal from UE B, and eliminating the need for a long guard interval.

TDMA had been used in a large number of wireless standards: the worldwide second-generation cellular standard GSM as well as the US standards IS-54 and IS-136 and the Japanese PHS standard, and the worldwide cordless standard Digital Enhanced Cordless

---

[3] In a mixed TDMA/FDMA system, the RX can also listen to the activity on other frequencies.

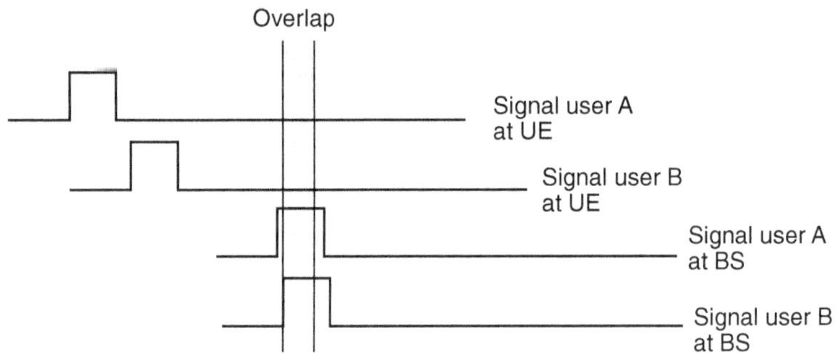

Figure 18.6   Overlapping bursts assuming uncompensated propagation delay.

Telecommunications (DECT) are based on it. However, it is noteworthy that all of these are older standards that were intended for speech communications, and exist today mostly as legacy systems, compare Sec. 30.3.

18.3.3   OFDMA

OFDM, as discussed in Chapter 15, is a modulation format that allows the transmission of high data rates for a single user; it is not (as sometimes erroneously stated) a multiple access format. In order to realize multiple access, it can be combined with various multiple access methods, e.g.,

- TDMA: each user occupies the whole system bandwidth, and different users are served at different times. At a minimum, one user transmits for one symbol duration, but it is also possible that one user transmits/receives multiple symbols before the system switches to the next user.
- Packet radio: in this mode, every user transmits complete data packets, where the modulation format for each packet is OFDM. The access of the users to the channel is regulated by the packet access techniques discussed in Section 18.4, like Aloha, and CSMA. This approach is used in IEEE 802.11 (see Chapter 33).

A more flexible approach is known as OFDMA. Remember that in OFDM, the frequency axis is divided into subcarriers, and the symbol duration on each subcarrier is an OFDM symbol (see Chapter 15); the basic time-frequency unit (i.e., one subcarrier for the duration of one OFDM symbol), is also often called a *resource element*. OFDMA now assigns different resource elements to different users. General OFDMA has no specific restrictions on how the assignment of resource elements can be done. Note that it is possible to assign multiple *noncontiguous*, resource elements to one user; the motivation for this will be explained below.

The assignment of resource elements is afflicted with administrative overhead – the finer grained, the more overhead. This is because each user has to be informed which resource is assigned to it. Thus, in most practical cases, whole "chunks" of such time-frequency resources are assigned. The choice of granularity of the chunks presents a trade-off between flexibility and overhead.

We can also think of OFDMA as "slicing" the time-frequency plane into pieces, each of which can be assigned to different users. Now clearly this principle has similarities to TDMA/FDMA. The main difference is that in TDMA/FDMA, the slicing method, and assignment of the slices, has to follow a very restrictive rule, namely that periodic slices are assigned. In contrast, the slicing in OFDMA gives much more flexibility, at the price of higher overhead.

There are essentially three ways of assigning the subcarriers to different users (see Figure 18.7):

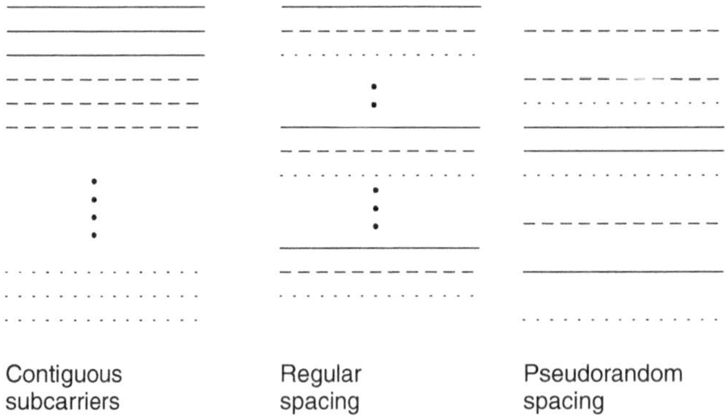

Figure 18.7   Assignment of subcarriers to different users (indicated as solid, dashed, and dotted lines).

1. *Assigning adjacent subcarriers* to one user, i.e., a set of $N^{(k)}$ carriers f_n, with $n = n_{\text{low},k}, n_{\text{low},k} + 1, \ldots n_{\text{low},k} + N^{(k)} - 1$. The advantage of the method is that channel estimation is simplified and/or pilot overhead is reduced. This occurs because adjacent subcarriers are correlated – they are close in frequency, and they are assigned to the same link. Furthermore, it is possible to perform "intelligent scheduling": each UE is assigned the group(s) of frequencies that provide the best channel quality for it. The fading of the channels for the different users is independent. Thus if $h^{(k)}(f_{\tilde{n}})$ is very small (i.e., the channel for the kth user is in a fading dip at frequency $\tilde{n}$), then it is unlikely that $h^{(j)}(f_{\tilde{n}})$ is also small for $j \neq k$. In other words, we obtain a high degree of selection diversity without suffering a loss of spectral efficiency – every subcarrier is assigned to somebody. This form of diversity is also known as "multi-user diversity," compare also Section 20.3.2, because it requires the presence of multiple users to take effect. Note, however, that the BS has to first know the propagation channel (over the full system bandwidth) from/to each user; and as channels change, the channel assignments for the users have to change.

2. *Assigning regularly spaced subcarriers* to one user, i.e., the set f_n with $n = n_{\text{low},k}, n_{\text{low},k} + Q, \ldots n_{\text{low},k} + Q(N^{(k)} - 1)$, where $n_{\text{low},k}$ is the index of the lowest subcarrier used for UE k, and Q is the subcarrier spacing. This assignment has the advantage of a high degree of frequency diversity – actually, essentially the same as "standard" OFDM (with the same occupied bandwidth) that has every single carrier assigned to one user, as long as Q is at most one coherence bandwidth. Furthermore, the use of a smaller number of subcarriers for each user allows a more flexible assignment of the data rate to the different users. Finally, the scheme does not require knowledge of all the channels at the BS for scheduling, because (due to the high frequency diversity) all assignments provide the same effective channel quality. Thus, it is sufficient to prescribe the *fraction* of resources that a user is assigned, instead of having to search for the *specific* subcarriers that should be used. The drawback (compared to the adjacent subcarrier assignment) is that there is no multi-user diversity.

3. *Assigning randomly spaced subcarriers* to one user, so that the indices n of the subcarriers f_n are part of a random sequence $b(i)$, i.e., $n = b(i)$, with $i = i_{\text{low},k}, i_{\text{low},k} + 1, \ldots i_{\text{low},k} + N^{(k)} - 1$. Largely, this assignment strategy has the same pros and cons as the regularly spaced subcarriers: (i) assignment of subcarriers to users can stay constant over time, (ii) no full channel knowledge is required for assignment strategy, (iii) full frequency diversity, (iv) no multi-user diversity. The main advantage of this scheme is with respect to the adjacent-cell interference. If adjacent cells use different random sequences, then the adjacent-cell interference is randomized: interference on different subcarriers comes from different users in the neighboring cells (see Chapter 21). Such interference randomization can reduce the worst-case interference.

As indicated above, OFDMA can have a large overhead. For the downlink, this is due to the fact that the BS has to communicate to the UE which time-frequency resources are assigned to it. Typically, the BS will broadcast from time to time an "information block" that tells each UE exactly this information (obviously, each UE has to listen to this block and decode this information). The smaller the "chunks" are into which the time-frequency plane can be broken, the larger the overhead for this information block. For the uplink, the UE has to tell the BS how many data it wants to transmit – and then the BS has to send a reply, telling the UE which time-frequency resources are assigned to it for the uplink. This is because a UE cannot by itself determine on which time-frequency resource it wants to send in the uplink – this has to be coordinated between all the UEs, and only the BS can provide this coordination. Note again that all this overhead exists only in OFDMA, and not in TDMA/FDMA where resources are assigned on a (semi-) permanent basis. This is the price OFDMA pays for the enhanced flexibility allowing for its data transmission. In many situations, this is a worthwhile price to pay, and both LTE and 5G-NR consequently use this approach.

In addition to the overhead for the assignment, there may also be an increased channel estimation overhead, in particular when a distributed subcarrier assignment is used. Note that a pilot tone is required for every coherence bandwidth, at least once per coherence time, for each user transmitting in the coherence frequency/time block. Thus, multiple users transmitting in interlaced fashion means more users transmitting in the same coherence time/frequency block and thus larger number of required pilots.

As long as the channel is time-invariant and frequency-flat, it does not matter which subcarriers, and which OFDM blocks, are assigned to a particular user *as long as the energy per symbol is independent of the scheduling*. However, in particular for the uplink of cellular systems, it often occurs that the transmit *power* is limited. In that case, it is disadvantageous for a UE to transmit for a short time, over a large bandwidth; rather transmission with few subcarriers over multiple timeslots is preferable (and, using interleaved subcarriers, even relatively small number of subcarriers can provide sufficient frequency diversity). Note that such a situation generally does not occur in cellular downlinks, since the BS transmits (almost) continuously anyway, as it sends signals to multiple users, see Chapter 21. On the other hand, for transmissions that require low latency, use of large bandwidth over a short time is preferable.

We finally note that OFDMA can be used not only in conjunction with OFDM but also with other multi-carrier-based schemes. An important practical example is a combination with DFT-spread OFDM, where a particular user gets assigned a group of contiguous subcarriers, and then transmits a DFT-spread OFDM signal on this group, see Figure 18.8. This is used, e.g., in the uplink of LTE, see Section 31.3.3.

18.4   Contention Multiple Access

Contention-free access allows centralized planning and guaranteed quality of service. However, it also has afflicted by high overhead and/or low spectral efficiency, and cannot be used at all in certain situations, e.g., initial access of a UE to a BS. There are thus situations where a different approach, called contention access, is used.

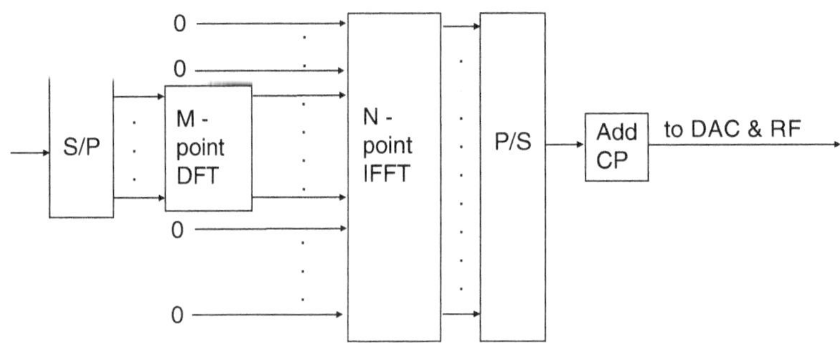

Figure 18.8   Combination of OFDMA as multiple access with DFT-spread OFDM modulation.

In contention access, each data packet, irrespective of its source, has to fight for its "own" resources. When two users try to access the channel at the same time, and transmit packets, *collisions* can occur, i.e., the signals from the two users arrive at the intended RX simultaneously, and create such high level of interference for each other that neither of them can be decoded. The resource contention thus consists of two parts, the *Channel Access Algorithm* (CAA), which tries to not let collisions happen in the first place, and the *Collision Resolution Algorithm* (CRA), which deals with what happens when such a collision actually occurs. While this chapter considers the case that multiple UEs try to transmit packets to a given BS (remember that for the downlink the BS has all the information and thus can perform contention-free access), the principle also applies to ad hoc networks (see Chapter 23).

18.4.1   ALOHA

The first wireless packet radio system was the ALOHA system of the University of Hawaii; it was used to connect the computer terminals in the different parts of this archipelago to the central computer in Honolulu. The CAA of ALOHA is very simple: the starting time of a packet transmission is chosen completely at random by the TX (strictly speaking, in this case, the system is called a *pure* or *unslotted* ALOHA system). A TX does not take any consideration whether other users are already transmitting and thus an additional transmission might result in a collision; a collision is discovered only after the fact, by the RX, who sees a bad SINR. Furthermore, the CRA is also very simple: a packet that suffers a collision, and thus cannot be decoded at the RX, is retransmitted. From a protocol point of view, during retransmission, such a packet is treated like a new packet, and transmitted with a random start time. The obvious advantage of such a system is simplicity; the drawback is that it becomes inefficient when the traffic load is large, because the probability for collisions becomes large, and most of the spectral resources are spent on unsuccessful transmissions/retransmissions.

Let us now determine the possible throughput of an unslotted ALOHA system. For that purpose, we first determine the possible collision time, i.e., the time during which collisions with packets of other users are possible; we assume that all packets have the same length T_p. Figure 18.9 shows that the packet A (from TX 2) can suffer collisions with packets that are transmitted by TX 1 either before or after packet A starts. We assume here that even a short collision (small overlap time) leads to such strong interference that the packet has to be retransmitted. In order to completely avoid collisions, a packet from TX 1 must start its transmission at least T_p seconds before packet A, or has to start its transmission after packet A has finished (i.e., must start no sooner than T_p seconds after the start of packet A). The total possible collision time is thus $2T_p$.[4]

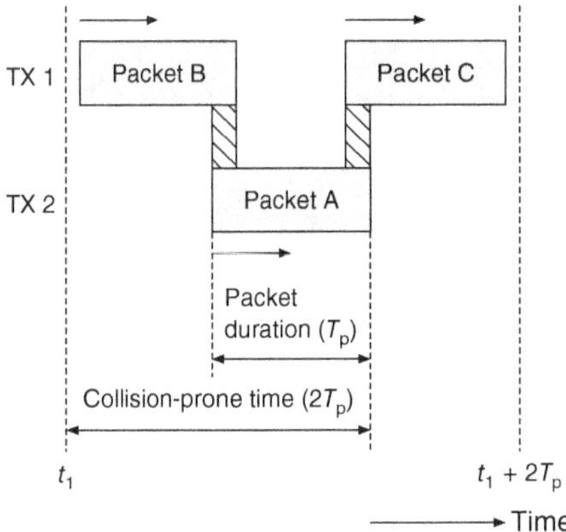

Figure 18.9   Possible collision time in unslotted ALOHA.
Reproduced with permission from [Rappaport 1996] © Prentice Hall.

---

[4] In the following, we also assume that T_p contains a guard period that accounts for the different physical runtimes of a packet in a cell.

For mathematical convenience, we assume now that all transmission times are random, and the different TXs are independent of each other. The average transmission rate of all TXs is denoted as λ_p packets per second; the offered rate $R = \lambda_p T_p$ is the normalized channel usage, which has to lie between 0 and 1. The probability that n packets are transmitted within a time duration t is given by a Poisson distribution [Papoulis 1991]

$$\Pr(n, t) = \frac{(\lambda_p t)^n \exp\left(-\lambda_p t\right)}{n!}. \tag{18.12}$$

The probability that during the time t zero packets are generated is thus

$$\Pr(0, t) = \exp\left(-\lambda_p t\right). \tag{18.13}$$

The effective throughput is thus the percentage of time during which the channel is used in a meaningful way, i.e., packets are offered, and transmitted successfully. As we have seen above, the possible collision time is twice the packet length, so that the probability of not having a collision is $\exp(-2\lambda_p T_p)$. The effective channel throughput is thus

$$\lambda_p T_p \exp\left(-2\lambda_p T_p\right). \tag{18.14}$$

It can be easily shown that the maximum effective throughput is $1/(2e)$, where e is Euler's number.

In a *slotted* ALOHA system, the BS prescribes a certain slot structure. Each TX has a synchronized clock that makes sure that the start of the transmission time coincides with the beginning of a slot. Thus, partial collisions cannot occur anymore: either two packets collide completely, or they do not collide at all. It is immediately obvious that the possible collision time in such a system is T_p, so that the effective throughput is

$$\lambda_p T_p \exp\left(-\lambda_p T_p\right) \tag{18.15}$$

and the maximum achievable throughput is $1/e$, i.e., twice as large as in an unslotted ALOHA system.

18.4.2 Carrier Sense Multiple Access (CSMA)

A more advanced CAA determines whether the channel is already occupied by another user, and does not transmit if that is the case. In other words, the TX determines (*senses*) whether the channel is currently occupied by another user (*carrier*). Such a method is called *Carrier Sense Multiple Access* (*CSMA*). It is more efficient than ALOHA, because it does not disturb other users that are already on the air.

A key parameter of CSMA is the *sensitivity threshold*, i.e., the receive power level above which a device considers the channel to be "occupied." A high threshold implies that a UE might transmit even though that transmission leads to collisions; a low threshold could mean that it foregoes transmission opportunities even when it would not create lethal interference. Since the device does not know the power at which the currently on-air transmission will arrive at its intended RX, the threshold can at best be set according to some average criterion.

Two other important parameters of a CSMA system are the detection delay and the propagation delay. The *detection delay* is a relative measure for how long it takes a UE to determine whether the channel is currently occupied. It depends essentially on the hardware of the system, but also on the desired false-alarm probability and the SNR. The *propagation delay* is the measure of how long a data packet takes to get from one UE to another. It can happen that at time t_1, UE 1 determines that the channel is free, and thus sends off a packet. At time t_2 another UE senses the channel. If $t_2 - t_1$ is shorter than the time it takes the data packet A to get from UE 1 to UE 2, then UE 2 determines that the channel is free, and sends off a data packet B. In that case, a collision occurs.

A major challenge for CSMA is the so-called *hidden node* problem. It essentially means that the node (typically a UE) that wants to transmit data cannot necessarily hear all other nodes transmitting data to the RX it is targeting. Figure 18.10a shows that principle: UE 1 is in the process of transmitting to the BS, and has a line-of-sight (LOS) connection to it. At the same time, UE 2 wants to transmit a packet to the BS. It also has a LOS connection to it, but a thick concrete wall is between UE 1 and 2, so that UE 2 cannot hear the transmission from UE 1 (more precisely, the power received at UE 2 from UE 1 lies below the detection threshold of the sensing mechanism). Thus, UE 2 thinks that the channel is not occupied, and transmits, leading to a collision.

The reverse problem is known as *exposed node*, shown in Figure 18.10b. Here, UE 1 is transmitting to BS A, while UE 2 wants to transmit to BS B. Even though the link UE 1 – BS B (and from UE 2 to BS A) is weak enough that no collision would occur, UE 2 will not transmit, because it hears UE 1, and thus thinks the channel is occupied. Note that the exposed node problem normally does not occur when a single BS is present and is also less of a concern as it only leads to a missed transmission opportunity, which is usually easier to handle than a collision.

An elegant solution to the hidden and exposed node problems is the RTS/CTS mechanism. In this approach, a UE that wants to send a packet first sends a short *Request to Send* (RTS) message. The intended BS determines whether it senses the channel to be free. If yes, it replies with a *Clear to Send* (CTS) message; if the BS gets multiple RTSs, it makes an autonomous decision on which of these to reply to with a CTS. In such an RTS/CTS approach, collisions occur only during the transmission of the RTS/CTS messages, which are very short, and thus less prone to collisions. Furthermore, if a collision occurs, the CTS will not arrive at the prospective transmitting UE, so that no data packet will be sent off. The RTS/CTS mechanism is used in the Wi-Fi standard (Chapter 33), and thus of great

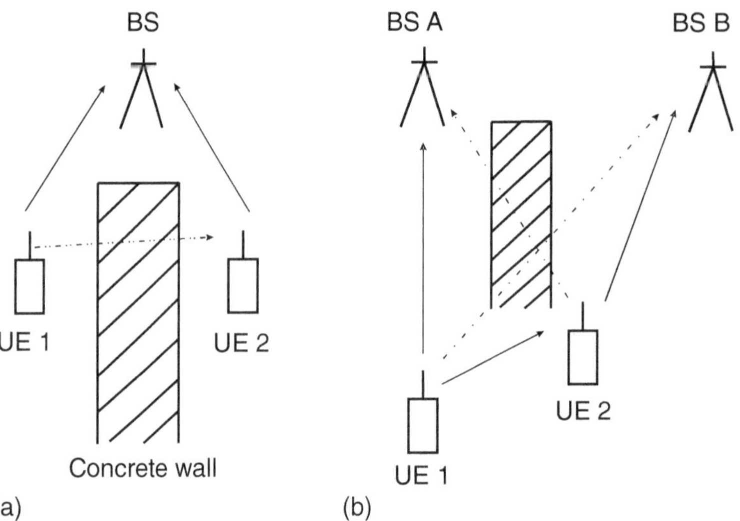

Figure 18.10   Principle of the hidden node problem (a) and exposed node problem (b).

practical importance. In addition, Wi-Fi foresees that the duration of a packet is announced at the beginning of a packet, so that the surrounding nodes know how long the channel will remain occupied.

*18.4.3   Conflict Resolution Algorithms – Random Backoff

As we have seen above, collisions cannot be avoided completely. An important part of contention algorithms is thus how to resolve collisions when they do occur. There are two fundamental types of *CRA*: (i) *random backoff*, and (ii) *splitting algorithms*. In this section, we will deal with the former category.

An important approach to reducing collisions are P-persistent algorithms, which sense the channel, and if it is free, they transmit; if it is occupied they continue sensing the channel until it is free and then transmit with a certain probability. In particular, we can distinguish the following types:

- *p-persistent CSMA*: once a TX senses that a channel is not occupied anymore, it transmits with probability p; otherwise, it waits for one timeslot and again makes a random decision (with probability p), and so on; it continues this process until it either has transmitted the packet or the channel becomes busy; in the latter case, it restarts the whole process (sensing until the channel becomes free, making a random decision on whether to transmit in the first free timeslot, etc.).
- *1-persistent CSMA*: the TX constantly senses the channel, until it realizes that the channel is free; then it immediately sends off the packet. This is obviously a special case of p-persistent transmission, with $p = 1$. If a collision occurs, the TX waits a random time until it attempts the whole procedure again. In most cases, 1-persistent CSMA has lower throughput than p-persistent with $p < 1$; generally, throughput increases with decreasing p, but also the latency increases.
- *Nonpersistent CSMA*: the TX senses the channel. If the channel is busy, the TX waits a random number of unit times and restarts the algorithm (i.e., sensing the channel, transmitting the packet if the channel is free, ...). The name non-persistent comes from the fact that the TX does not continuously check until the channel is free, but rather just waits a random time. This is more energy-efficient, but can significantly increase the latency of the transmission.

Thus, whenever a collision occurs, the UEs have to wait for a (random) time period until they can try to transmit again.

Alternatively, a UE can insert preemptively a random wait time before transmission. Specifically, each UE that wants to send a packet senses the channel; if it is free it transmits; otherwise it senses until the channel is free, then initializes a *backoff timer* with a random start value, and starts a countdown. As long as the channel is free, the countdown is continued; when the channel is occupied, the countdown is put on hold (but the time is not reset). When the timer has expired, another transmission attempt is made. If the timers of the different UEs expire at sufficiently different times, collisions are avoided. The way that the retransmission times are determined is the main distinction between the different backoff algorithms. Typically, the value of the backoff time is chosen within the so-called *contention window*, i.e., there is a minimum and a maximum value of the backoff time. The size of the contention window might depend on the priority of the packet (for a discussion of packet/user priorities, see Section 20.1), and how many previously failed transmission attempts there were. This approach is used in Wi-Fi, see Section 33.4.

Figure 18.11 shows a flow diagram of a backoff algorithm: we start with the attempt of a packet transmission. If it is successful, then the backoff stage is decreased (e.g., decrease in the contention window). If the transmission is unsuccessful, then the backoff stage is increased, and the backoff counter is set, at random, according to the current backoff stage (contention window size). The algorithm then starts to sense the channel at unit time intervals: whenever the channel is free, the counter is decreased by one; if

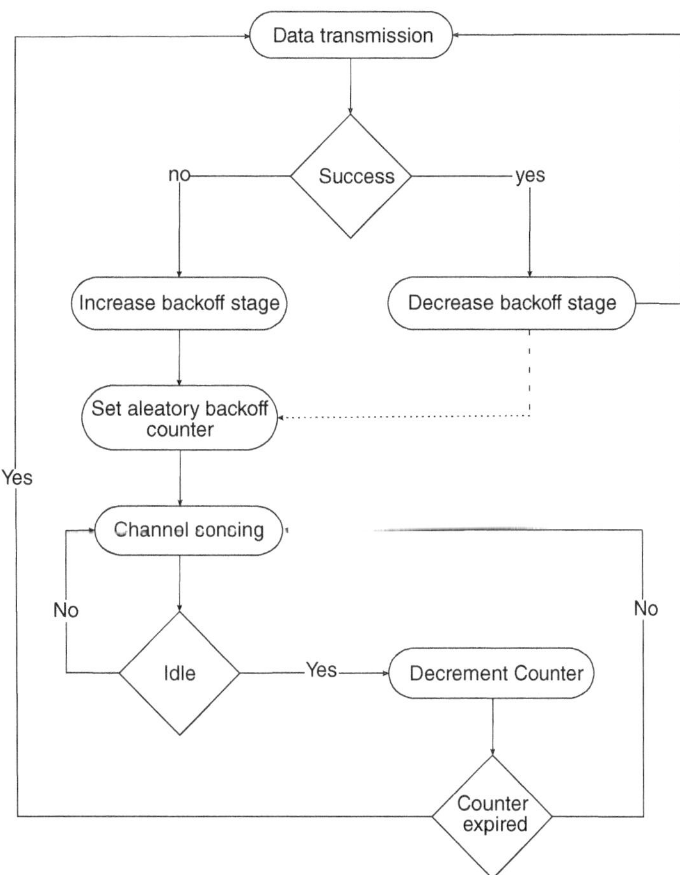

Figure 18.11 Flow diagram of backoff algorithm.
Reproduced with permission from [Wu and Pan 2008] © Nova Publisher.

the channel is busy, the counter value is retained. It is important that the counter is decreased only while the channel is free; otherwise, the counters of multiple UEs might all reach zero while a packet is on the air, and they would all try to transmit at the first time that the channel becomes available again. When the counter reaches zero, another attempt is made to transmit the data.

A method for determining the contention window size is the *binary exponential backoff*, which is used in Wi-Fi (see Section 33.4). In this method, the contention window CW starts out with a minimum value $CW_{\min}$, and is doubled after each failed transmission attempt (and reset to $CW_{\min}$ in case of a successful transmission): obviously, if there had been a collision in the previous attempt, then the range of backoff times was not sufficient, and a larger contention window should be used. On the other hand, one does not want to choose the backoff time needlessly large, as this increases transmission delays and reduces the spectral efficiency of the system (no payload is transmitted during the backoff times). We therefore define a maximum value of the contention window size that is never exceeded. Formally, the contention window size for the mth transmission attempt thus becomes

$$CW_m = \min\left[CW_{\max}, 2^{m-1}CW_{\min}\right]. \tag{18.16}$$

The backoff timer is then chosen uniformly within this contention window. The parameters $CW_{\min}$ and $CW_{\max}$ are the minimum and maximum size of the contention window.

In addition, there are a number of variants that change the contention window in various ways depending on whether the previous packet transmission was successful or not. For example, the contention window can be decreased by a constant amount when a transmission is successful, and increased by a (possibly different) constant amount when a collision occurs, all within the bounds of a minimum and maximum window size; this algorithm is called *Linear Increase Linear Decrease* (LILD). Another algorithm, called MILD does a multiplicative increase of the contention window (when a collision occurs), but a linear decrease (without collisions). These are but two examples, a wide variety of algorithms exist that all have their respective advantages and disadvantages in various scenarios, but none are generally optimum.

The contention window size should, in principle, be adapted to the expected collision probability. If all users employ the same constant value for CW, the optimum value can be shown to be

$$CW = K\sqrt{2T} \tag{18.17}$$

where K is the number of active users, and T the packet transmission time (including overhead) in units of slot times. The *Dynamic Tuning Backoff* (DTB) determines the optimum window size for the binary exponential backoff, also based on an estimate of the number of active users. In either case, the main challenge for implementing such a protocol is the determination of K. This is usually achieved through feedback of ternary information, namely whether the channel state is idle, transmission was successful, or a collision occurred. However, the requirement for such feedback, and the need to compute the optimum parameters in real-time, leads to it being not in very widespread use.

*18.4.4 Conflict Resolution Algorithms – Splitting Algorithms

Splitting algorithms for conflict resolution first determine whether there is a collision in a large group of users; if yes, the group is divided successively into finer and finer subgroups before retransmission until the collision is resolved. To give more details, the algorithm works as follows: all UEs that want to transmit in a given timeslot send out their data packets. If a collision occurs, the set of transmitting UEs is split into subsets (we call them level-1 subsets), and the first subset is allowed to transmit in the next timeslot. If there is no collision, the next level-1 subset can transmit. If there is a collision, the subset is further split (into level-2 subsets), and the first of the level-2 subsets transmits. If that is successful, the next level-2 subset transmits, and so on. If not, further splitting occurs, until no collision occurs (after sufficient splitting, every TX is in its own subgroup, so that no collision can happen). The splitting can also be visualized as a tree, with the complete set of TXs at the root, and each splitting giving rise to two or more branches. A tree node (corresponding to a subset) might be empty (in which case the channel is idle during its transmission), have one entry (in which case successful transmission occurs), or multiple entries (in which case there is a collision). A collision is considered fully resolved when all members of the original transmit set have successfully sent their packets; the time required for this is called the *Collision Resolution Interval* (CRI).

The following parameters may differ according to the particular form of the algorithm:

- The number of subsets created at each splitting process can be 2, 3, or generally Q (with Q an integer ≥ 2), and the CRA is correspondingly called binary, ternary, or Q-ary.
- The splitting is done according to the addresses of the nodes, so that the addressing scheme determines the partitioning. Addresses can be assigned deterministically, randomly, or according to the arrival time. Note, however, that the assignment of the nodes to the subsets always occurs randomly, since the RX cannot determine (due to the collisions) which nodes are trying to talk to it.
- The feedback from the RX can be (i) binary, i.e., "collision" or "no collision," (ii) ternary, i.e., collision, success, or idle, or (iii) of higher multiplicity, namely indicating (in addition to idle and success) how many packets have collided during an interval. This latter type of information allows a more efficient splitting, but it is often difficult to determine from the measurement at the RX.
- Incorporation of capture effects: when a collision occurs where one signal is much stronger than the other, the RX might be capable of decoding both of them through multi-user detection (compare Section 28.2). Most of the existing algorithms do not incorporate this effect, but it can dramatically increase the efficiency of splitting algorithms.
- The channel access: in *Free Access* algorithms, newly arrived packets are transmitted right away, without taking into account whether a collision resolution is going on at this time. In case they give rise to a collision, they become part of the collision resolution. In *Gated Access*, users with new packets wait until the end of the current collision resolution until they attempt to transmit; similarly, in *Window Access*, only packets arriving within a given window are transmitted after the end of the CRI.

The *Binary Tree Algorithm* is the first, and most straightforward of these various implementations. It implements (i) a binary split, (ii) binary feedback, (iii) no capture effects, and (iv) gated access. An example is shown in Figure 18.12 and discussed in the following.

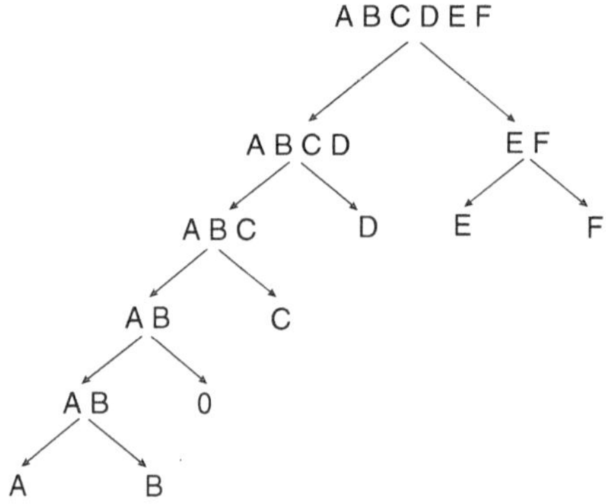

Figure 18.12 Example of a splitting algorithm.

At each collision, members of a set are assigned randomly to one of the two subsets. In the first step, every user is assigned a random bit (i.e., zero or one), and thus assigned to the first subset (A,B,C,D) or the second subset (E,F). The feedback conveys that in both sets collisions occur, so both have to undergo the splitting. The set ABCD is divided, through random assignment of bits 0 or 1 to each member, into subsets ABC and D. ABC gives rise to a collision and is further split into AB and C. AB undergoes another split, but in that step, it happens that both members are assigned to the same subset, so that again a collision occurs, while the other subset is empty. Finally, a further splitting attempt results in splitting A and B into separate subsets in which no collisions occur. Next, the subset containing C only is tested, and the transmission is successful because C is alone in its subset. The same happens to the subsets containing D. Then, transmission of the level-1 subset EF is done, and a collision occurs. A splitting divides the nodes into a subset containing only E, and one containing only F, each of which can successfully transmit in their timeslots. At that point in time, the transmission of all packets is successfully completed, and the CRI has ended. We see that 13 timeslots were required to transmit six packets, so that the efficiency was slightly less than 0.5. This is just one random example; a more detailed mathematical analysis shows that the average throughput per timeslot is 0.346 packets. It is remarkable that this number of actually *smaller* than the throughput of slotted ALOHA, which is $1/e = 0.367$.

Example 18.1   *Derive the expected throughput of the binary splitting algorithm.*

Denote the number of users whose packets are colliding as k. Let furthermore T_k denote the (random) number of timeslots required to resolve a collision, and $\overline{T}_k$ the expected value of this random variable, i.e., the average number of timeslots for collision resolution with k users. Obviously, $\overline{T}_0 = \overline{T}_1 = 1$, i.e., with zero or one TXs, there is no collision, and the timeslot is either idle, or a successful transmission occurs. For any larger number of users, a collision resolution occurs. Define as $T_{k,m}$ the time for collision resolution when exactly m users in the splitting process are assigned to the first group, and $k - m$ to the second group. Then, obviously

$$T_{k,m} = 1 + T_m + T_{k-m} \tag{18.18}$$

since one timeslot is used up to determine that there is a collision, and then each of the subsets has to resolve their collisions, which by definition lasts T_m and T_{k-m}, respectively. Due to the linearity of the expectation, Eq. (18.18) also holds for the $\overline{T}$. Let now p denote the probability of assigning a UE to the first subset (note that this is a design parameter that we wish to optimize). Then the probability that m out of k users are assigned to the first group is given by the binomial distribution

$$P_{m,k} = \binom{k}{m} p^m (1-p)^{k-m} \qquad 0 \le m \le k. \tag{18.19}$$

Thus, we can now write $\overline{T}_k$ for $k \ge 2$ as

$$\overline{T}_k = \sum_{m=0}^{k} P_{m,k} T_{k,m}. \tag{18.20}$$

Inserting Eqs. (18.18) and (18.19) into (18.20), we get

$$\overline{T}_k = \frac{1 + \sum_{m=0}^{k-1} P_{m,k} T_m + \sum_{m=1}^{k} P_{m,k} T_{k-m}}{1 - P_{0,k} - P_{k,k}} \tag{18.21}$$

which can be further shown to be

$$\overline{T}_k = 1 + \sum_{m=2}^{k} \binom{k}{m} \frac{2(m-1)(-1)^m}{1 - p^m - (1-p)^m} \qquad k \ge 2 \tag{18.22}$$

and, as mentioned above, $\overline{T}_k = 1$ for $k = 0, 1$. The limit of the efficiency of the algorithm for large number of users is $\lim_{k \to \infty} \overline{T}_k/k$, which in the strict mathematical sense does not exist, but where numerical evaluations show that it oscillates with small amplitude around the value 2.886 ($=1/0.346$) when $p = 0.5$ is chosen.

The Binary Tree Algorithm can be improved through various measures, such as skipping transmissions that are known to lead to collisions, adapting the splitting to the traffic conditions, etc.; for details, we refer to the references mentioned in the section titled "Further Reading".

*18.4.5   Multi-Channel Transmission

The previous discussions all assumed that a user occupies the whole bandwidth for the transmission of a packet, and when that packet transmission has finished, another user gets the whole bandwidth (possibly after some time during which the channel is idle).

However, it is also possible to divide the available bandwidth into multiple channels. The latter approach can either serve to have one channel that can serve as control channel and thus ease the allocation of spectral resources to the actual users, and/or to allow several payload data transmissions in parallel.

Multi-Channel Transmission with Single Payload Channel

The first suggestion for such a system was *Multi-channel with Busy Tone*. In this method, UEs send out a narrowband busy-tone signal on the control channel to indicate a busy channel. When a UE is ready to transmit, it first senses whether the control channel is idle for a certain amount of time. If yes, it transmits; if no, it waits for a random time and then schedules the transmission again. Note that the busy tone also suffers from the hidden/exposed node problem discussed in Section 18.4.2.

Several variants of the Multi-channel with Busy Tone protocol exist. For example, it is possible that the RX sends out the busy tone, in response to a request from a possible transmitting UE. The busy tone thus has a role similar to the CTS in the RTS/CTS exchange described in Section 18.4.2. At the same time, it tells the surrounding UEs how long the packet is, since it is sent continuously (this has the same effect as the announcement of the packet duration in Wi-Fi protocols, though of course, the actual implementation is different). Yet another implementation has both TX and RX transmit a busy tone.

Multi-Channel Transmission with Multiple Payload Channels

The use of multiple payload channels can decrease the number of collisions (on each of the subchannels), and thus ease the requirements for the RTS/CTS exchanges. Essentially, each UE listens to all available subchannels. If the subchannel is free on which a previous successful transmission was made, the same channel is used. Otherwise, another free channel is looked for. Blockage/collision properties of multi-channel systems are discussed in Section 20.6.2.

*18.4.6   Reservation Mechanisms

The challenges in contention access arise from the fact that a data packet does not have a timeslot available in which it can be sure to be free from interference. On the other hand, reserving the same timeslot for a long time (as done in TDMA) is inefficient. An obvious compromise is thus to find a way to reserve a timeslot for a shorter period of time.

Packet Reservation Multiple Access

In a *Packet Reservation Multiple Access* (PRMA), we consider a frame/slot structure, i.e., the time-axis is divided into frames, each of which contains a constant number of slots. A slot can be in one of three states: busy, reserved, or idle. At startup, each UE that has data to transmit can compete for an idle timeslot (through any of the above-mentioned contention schemes, such as ALOHA). After a successful transmission in timeslot k, the UE can "reserve" the corresponding timeslot k in the next frame. This is the slot that would be assigned to it if it used TDMA, and the UE will actually keep using the slot until it has no more data to transmit. At this time it releases the resource, and all other UEs can compete for it.

PRMA also allows an effective prioritization of delay-critical traffic (such as voice) over best-effort traffic (like email). Firstly, one can stipulate that only voice traffic (or voice + video, in the example of video conferencing) can reserve traffic. Secondly, if a UE contends for an idle timeslot and a collision occurs, a UE with delay-critical traffic is allowed to retransmit with probability that is higher than the UE with best-effort traffic. With these measures, voice traffic can transmit with almost the same success probability as TDMA, while the main drawback of TDMA, namely timeslots that are reserved even when the UE has nothing to transmit, are eliminated. We notice that this technique is related to semi-persistent scheduling, which is extensively used in LTE and NR (Chapters 31 and 32).

Split-Channel Reservation Multiple Access

Split-channel Reservation Multiple Access (SRMA) can send a request to transmit a data packet. A control mechanism (which can be centralized or noncentralized) answers by telling the UE when it is allowed to send off the packet. This eliminates the risk of collisions of data packets; however, the signals that carry the requests for time can collide. Furthermore, the system sacrifices a part of the transmission capacity for the transmission of the reservation requests. In order to maintain reasonable efficiency, the requests for time must be much shorter than the actual data packets. RTS/CTS and Multi-channel with Busy Tone can be seen as special cases of that approach. Equally importantly, the assignment of time-frequency resources to users in the OFDMA of LTE can be seen as such a method: reservation requests are sent on specific time/frequency resources in the uplink, and the BS provides, in the downlink, on specific time/frequency resources the assignment that it decided to provide to each of the requesters, before sending the actual payload on the announced time/frequency resources.

Polling

In *Polling*, a BS asks (polls) one UE after the other whether it wants to transmit a data packet. The shortest polling cycle occurs when no UE wants to transmit information; this is also the most inefficient case, as capacity is sacrificed for the polling, and no payload is transmitted. The approach is thus suitable mostly in the case of a small number of users, each of which has a relatively high utilization rate. It is less suitable for "internet of things" type of applications, where thousands of potential users, each of which sends only a few data packets at large intervals, communicate with a BS.

Both SRMA and Polling allow to take quality of service into account, since a centralized algorithm assigns the timeslots, and thus can take QoS (as well as other considerations, such as fairness and backlog, see Chapter 20), into account.

18.4.7 Comparison of the Methods

Figure 18.13 shows the efficiency of various packet-switched methods. The abscissa shows the channel usage, the ordinate the average packet delay. We see that ALOHA methods are only suitable if efficient channel use is of no importance. The maximum achievable capacity is only 36% in slotted ALOHA. CSMA, SRMA, and polling achieve better results.

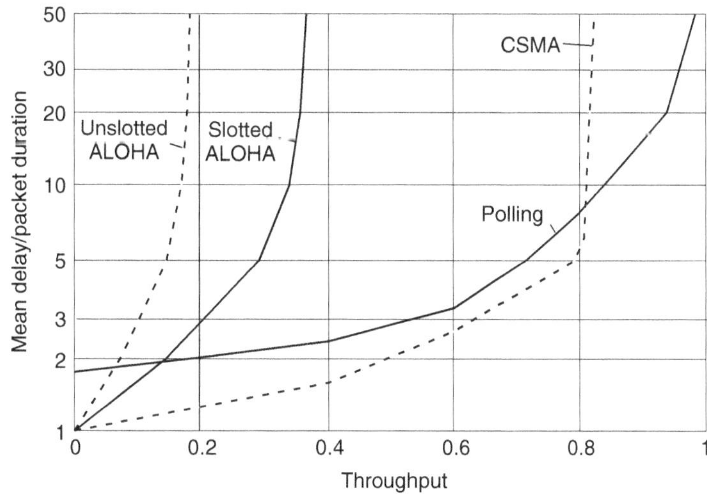

Figure 18.13 Channel usage and mean packet delay for different contention access methods. Reproduced with permission from [Oehrvik 1994] © Ericsson AB.

When comparing packet radio to contention-free systems, we find that the latter schemes are very useful for applications that require low latency (1–100 ms) as long as the parameters (timeslot duration, etc.) are chosen correctly. Each TX can be certain to be able to transmit its data to the RX without significant blocking or delay on the line, since it has a channel (frequency or timeslot) exclusively reserved; this is critical for guaranteed QoS. On the other hand, for bursty traffic with large number of users, contention-free multiple access tends to waste resources. This is particularly true if a channel (time slot) is *always* reserved for a single user becomes a major drawback -even if that user does not have any data to transmit, the multiple access scheme does not allow anybody else to use that slot in the meantime.

18.5 Duplexing

18.5.1 FDD and TDD

Duplexing serves to separate the uplink and downlink. Most systems cannot transmit and receive simultaneously in the same frequency band, as the strong transmit signal would also be seen by the co-located RX, and overwhelm any (weak) signal that comes from a far-away TX. The transmit and receive signals of a device (UE or BS) should thus be separated by time or frequency, so that we correspondingly speak of *Time Domain Duplex (TDD)* and *Frequency Domain Duplex (FDD)*; the first "D" in the acronyms may also be interpreted as *division* instead of *domain*. In TDD, the uplink data are sent at different times than the downlink data, see Figure 18.14a. In FDD, uplink and downlink data are sent in different frequency bands, see Figure 18.14b.

FDD can be used in combination with any multiple-access method. In most cases, there is a fixed duplexing distance, i.e., a fixed frequency difference between the transmit and the receive band. The separation between these bands is mainly given by hardware constraints, namely the necessity to build filters that sufficiently suppress the cross talk from the transmit signal into the receive chain. Since filters with steep flanks are more difficult and expensive to build, larger separation of the bands allows the use of cheaper hardware. On the other hand, the separation should not be so large as to necessitate different antennas for transmit and receive case; since the relative bandwidth of antennas is typically <10%, this poses a natural limit on the duplexing distance.

The isolation of the users can be further improved in so-called semi-duplex systems, where transmit and receive slots are offset in both time and frequency. Consider, as an example, the GSM standard, which uses FDD in conjunction with TDMA. There, a particular user occupies 1 out of 8 timeslots in each frame, but the corresponding slots of one user for uplink and downlink are not only in

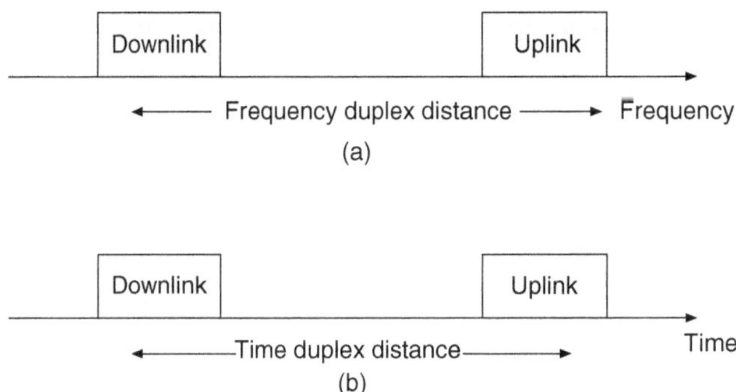

Figure 18.14 (a) Frequency and (b) time duplexing.

different frequency bands, but also offset in time by two slot durations. Thus, the joint effect of bandpass filter and power up/power down status of the transmit amplifier suppresses the transmit signal. A similar approach is taken for some device types in LTE and NR systems (see Chapters 31 and 32).

TDD is often used in conjunction with TDMA. In the simplest case, the available time is divided not into K, but rather $2K$ timeslots, where each of the K users is assigned one timeslot for the uplink, and one for the downlink. We also find that TDD can be used well in conjunction with packet transmission schemes for which the transmission occupies the whole available system bandwidth. On the other hand, the use of TDD in an FDMA system would counteract many of the advantages inherent in FDMA (continuous transmission and reception, which simplifies synchronization), while retaining the disadvantages.

The previous paragraph described a "symmetric" TDD system, where the number of timeslots assigned to each user is the same and is furthermore the same for uplink and downlink. This need not be the case: any of the timeslots can, in principle, be assigned for any of the users, and for either uplink or downlink. This is important in particular for data transmission, since in most cases the download traffic is much heavier than any uplink traffic. Care must be taken, however, when considering such an approach in a cellular system, since problems can arise with the interference levels when one cell is operating in downlink while a neighboring cell is in uplink, compare Chapter 21. We also note that no such flexible assignment is possible for FDD: the bands for uplink and downlink are pre-assigned (and have almost always the same width), so that the ratio of uplink to downlink spectral resources is fixed as 1 : 1 (deviations from this principle can occur in systems with so-called carrier aggregation, i.e., transmission on multiple bands, see Sections 31.6 and 32.6).

Care has to be taken when designing TDD systems with large cell sizes: there needs to be a *dead time* between transmission and reception. Imagine that the BS sends data to a user at distance d (and thus runtime d/c_0) stopping the transmission at time T. Then data will arrive at the UE until time $T + d/c_0$. Only now can the UE switch to a transmission mode, and send out its own block of data. Since the data have to propagate back to the BS, they start arriving there at time $T + 2d/c_0$. Consequently, the BS has experienced a dead time of $2d/c_0$. This can result in a considerable loss of efficiency for the system. Note that this effect cannot be combatted through timing advance (Section 18.3.2), since the UE cannot simply start early transmission – it would then have to transmit while still listening to downlink data.

For both TDD and FDD, it is interesting to consider the question whether the channel for uplink and downlink are identical. We find that for TDD, the requirement is that the duplexing time is much smaller than the coherence time, while for FDD systems, the requirement is that the frequency duplexing distance is much smaller than the coherence bandwidth; more details are given in Section 16.1.7. When considering practical system parameters, we find that the former condition can be fulfilled quite well, especially for wireless LANs that are operating in a quasi-static environment where neither BS nor UE are moving. The latter condition (for FDD) is practically never fulfilled. However, there are considerable advantages when the channel for uplink and downlink are identical; remember that we found in Chapter 15 that it simplifies the implementation of adaptive modulation, and in Chapter 16 that it is beneficial for the implementation of various smart antenna systems. For this reason, there is a general trend to assign newly available spectrum to TDD usage, even though FDD retains importance due to the large amount of corresponding legacy frequency assignments and systems designed for it.

18.5.2 Full-Duplex Systems

In recent years, interest has increased in full-duplex transceivers, i.e., devices that transmit and receive at the same time at the same frequency (see Section 17.9 for a more detailed definition). As we mentioned above, the transmit signal is much stronger than the received signal (typically by 100–140 dB), so that it overwhelms the received signal if no special measures are taken. On the other hand, the transmit signal is known, so that it can be subtracted from the total received signal, and the self-interference of the device can thus be eliminated. More details about how such a self-interference cancellation can be achieved are given in Section 17.9. From a system design point of view, one can obviously expect the spectral efficiency to improve, since the available spectral resources for transmission are effectively doubled (note, however, that the spectral efficiency will not be doubled, because there is more interference on the air when all transceivers are transmitting all the time).

18.6   Broadcast and Multi-Cast

Up to now, we have discussed the case of *unicast*, where – even though we have multiple TXs and/or RXs – the communication of a particular data stream occurs between one TX and one RX. This is the situation we are most familiar with from cellular telephony and Wi-Fi. However, there are also situations where a BS wants to send a message to multiple RXs simultaneously, i.e., *broadcast* this message (do not confuse the use of broadcast with the "broadcast channel" discussed in Sec. 18.2).

Fundamentally, a message sent to one user can be overheard by any other user in the cell; thus broadcasting a message does not require more resources than transmission of a unicast message. However, the following needs to be kept in mind:

- The signals transmitted by broadcast can either be data packets that are of interest to all users in the cell (e.g., earthquake warnings), or they can be control information that all users in the cell need, such as synchronization information, identification of the BS, etc. More examples for such control signals will be discussed in Chapters 31 and 32.
- The required transmit energy for a broadcast is determined by the *weakest* user that shall be reached by the message (in case that there is interference, then the user with the worst ratio of channel power gain to interference-plus-noise power is the determining one). Thus, even just one user that is in a deep fading dip can lead to requirements of high broadcast energy.
- No Automatic Repeat Request (ARQ) can be used to achieve a desired reliability. ARQ would require transmission of confirmation messages from all the users the broadcast tries to reach; such overhead is not acceptable.
- For transmission of user data, improvement of reliability (or reduction of required power) can be achieved if users with good reception act as relays, i.e., forward the information to users that have a bad channel to the BS.
- Broadcast is difficult in systems with multiple antenna elements: the beamforming inherent in multiple-antenna transmission is antithetical to the principle of broadcasting. The situation can thus arise that a UE can communicate well with a BS in unicast (exploiting SINR enhancement through beamforming) but cannot hear the broadcast message.

Multi-cast is similar to broadcast, just with a smaller set of target UEs. While in the case of broadcast, *all* the UEs in the cell should be reached by the transmitted signal, in the case of multi-cast it is *some* of them – more specifically, a predetermined set (e.g., subscribers of a particular information service). Most of the other considerations written above for broadcast also hold in this case.

We finally note the concept of *anycast*. Here, it is sufficient that the signal reaches one of the possible RXs – it does not matter which one. This situation rarely occurs in the cellular downlink; rather, certain situations in sensor networks, e.g., a temperature sensor sends out its reading, and it is sufficient that any of the UEs in the vicinity receives it, because it can forward to a suitable central point.

Further Reading

The analysis of multiple access (more precisely, the multiple-access channel and the broadcast channel) is one of the fundamental problems of information theory and thus explored in many of the related textbooks, e.g., [Cover and Thomas 2006]. A very readable summary is provided in [Goldsmith et al. 2003]. Also the descriptions in [Goldsmith 2005] and [Tse and Viswanath 2005] provide very clear expositions.

FDMA and TDMA systems are "classical" multiple access schemes, and more recently mostly analyzed in the context of specific systems, and generically in classical textbooks; there is also a nice summary in [Falconer et al. 1995]. An interesting performance analysis in interference is given in [Xu et al. 2000]; message delays in TDMA and FDMA systems are analyzed in [Rubin 1979]. The transition from TDMA/FDMA to CDMA is discussed in [Sari et al. 2000]. Various aspects of OFDMA, going beyond the single-cell multiple access discussed in this chapter, are described in [Jiang et al. 2010].

Contention-based access has been widely discussed in the context of computer networks. The ALOHA system was first suggested in [Abramson 1970] and analyzed for mobile radio applications in [Namislo 1984]. A classical description of CSMA is [Kleinrock and Tobagi 1975] and [Tobagi 1980]. PRMA is described in [Goodman et al. 1989]. Random backoff and splitting algorithms are described in the book chapter [Kartsakli et al. 2008 in Wu and Pan 2008]. Many of the concepts are also discussed in [Kumar et al. 2006], though the emphasis is on ad hoc networks. Full duplex is discussed, e.g., in [Sabharwal et al. 2014].

For updates and errata for this chapter, see https://wides.usc.edu/students.html#textbooks.

Exercises

See Sec. 36.18 of Exercises.pdf at wiley.com/go/molisch/wireless3e

19

Spread Spectrum Systems

Spread spectrum techniques spread information over a very large bandwidth – specifically, a bandwidth that is much larger than the data symbol rate. In this chapter, we discuss various ways of providing multiple access by spreading the spectrum. We start out with the conceptually simplest approach, Frequency Hopping (FH). We then proceed to the most popular form of spread spectrum, Direct Sequence-Code Division Multiple Access (DS-CDMA). Finally, we elaborate on time-hopping impulse radio, a scheme that has gathered interest because of its application to ultrawideband systems.

The concept of "spread spectrum" originated in the national security/military area, where the main interest lies in keeping communications stealthy, safe from intercept, and safe from jamming efforts by hostile Transmitters (TXs) – we will show below how spread spectrum can help to reach these goals. However, spread spectrum has also manifold applications in the commercial space. Yet in previous chapters, we have stressed how important spectral efficiency is: we want to transmit as much information per available unit bandwidth as possible. Thus, it might seem like a strange idea to spread information over a large bandwidth in a commercial wireless system.[1]

This seeming paradox can be resolved when we recognize that different users can be spread across the spectrum in different ways. This allows multiple users to transmit in the same frequency band simultaneously; the Receiver (RX) can determine which part of the total contribution comes from a specific user by looking only at signals with a specific spreading pattern. Thus, capacity (per unit bandwidth) is not necessarily decreased by using spread spectrum techniques, and can even be increased by exploiting its special features.

19.1 Frequency Hopping Multiple Access (FHMA)

19.1.1 Principle of Frequency Hopping – Single User Case

The basic thought underlying FH is to change the carrier frequency of a narrowband transmission system so that transmission is done in one frequency band only for a short while. The ratio between the bandwidth over which the carrier frequency is hopped and the narrowband transmission bandwidth is the spreading factor.

FH originated from military communications; it was invented by Hedy Lamarr during the Second World War. It was inspired by the problem that emissions from radio TXs could be used by the enemy to triangulate the position of TXs, or that transmission could be jammed by the enemy with powerful (narrowband) TXs. By changing the carrier frequency frequently, the signal is in the vulnerable (observed or jammed) band only for a short while. The FH pattern has to be known to the desired RX, but unpredictable for the enemy, making them unable to "follow" the FH.

FH is used nowadays in commercial systems for a variety of reasons. Firstly, it can suppress narrowband interference: for example, Bluetooth (Section 34.1) uses it to mitigate the effect of interference from Wi-Fi as well as from microwave ovens. More accurately, FH provides *interference diversity* such that the RX sees different amounts of interference at different times. Secondly, the FH also helps to mitigate the effect of deep fading dips through *frequency diversity*. Sometimes the system transmits on a "good" frequency – i.e., one with low attenuation between TX and RX – and sometimes on a "bad" frequency – i.e., in a fading dip; this is used, e.g., in the uplink of LTE (Section 31.3.4).

There are two basic types of FH: "slow" and "fast." *Fast FH* changes the carrier frequency several times during transmission of one symbol; in other words, transmission of each separate symbol is spread over a large bandwidth. Consequently, the effects of fading or interference can be combated for each symbol separately. It follows from elementary Fourier considerations that transmission of each *part* of a symbol requires more bandwidth than that of a narrowband system. Fast FH can be used to average out the amount of interference encountered in the different parts of the band; this is advantageous to avoid worst-case situations. However, more advanced signal processing that takes particular advantage of the different Signal-to-Interference-and-Noise Ratio (SINRs) at different frequencies can obtain better performance. Note that combining of the different contributions belonging to one symbol has to use processing that works faster than at the symbol rate. Furthermore, in channels with maximum excess delay comparable to, or larger than, the symbol duration, a large number of MPCs (signal echoes) at a particular carrier frequency keep arriving even while the signal

---

[1] While security (insensitivity to intercept operations) is important, it can be achieved by cryptographic means, and does not require spectral spreading.

Wireless Communications: From Fundamentals to Beyond 5G, Third Edition. Andreas F. Molisch.
© 2023 John Wiley & Sons Ltd. Published 2023 by John Wiley & Sons Ltd.
Companion website: www.wiley.com/go/molisch/wireless3e

has already "moved on" to the next frequency in the hopping sequence. Fast FH is not in widespread use in commercial wireless systems; it has been mostly edged out by Code Division Multiple Access (CDMA).

Slow FH transmits one or several symbols on each frequency (this is used in a number of systems, including Bluetooth, LTE, and GSM). This method is often used in conjunction with Time Division Multiple Access (TDMA): each timeslot is transmitted on a given carrier frequency; the next slot then changes to a different frequency. In order for the FH to be effective, interleaving and coding should distribute information belonging to one source bit over several timeslots. Imagine simple repetition coding, where each bit is sent twice, in different timeslots (and thus on different carrier frequencies). If the first timeslot is transmitted in a deep fading dip, chances are that the second is at a frequency where channel attenuation is small; thus the information can be recovered. Alternatively, slow FH can be used in conjunction with ARQ, such that groups of bits that have not been received with sufficient quality are simply repeated in a subsequent timeslot according to the feedback from the RX.

19.1.2 Frequency Hopping for Multiple Access (FHMA)

In the previous section, we looked at FH for the suppression of interference, and increasing frequency diversity. The price that has to be paid is that a larger bandwidth has to be used for transmission, which seems wasteful. In the following, we will show that FH can be used as a multi-access method that is as spectrally efficient as TDMA and Frequency Division Multiple Access (FDMA). For these considerations, we distinguish between synchronized and unsynchronized systems.

Let us consider first the *synchronized* case – e.g., in the downlink of a cellular system – where the Base Station (BS) can always make sure that it emits to all User Equipments (UEs) at the same time. Figure 19.1 shows an example with three available bands (carrier frequencies). Clearly, during one time interval, the BS can transmit to three users simultaneously, but for ease of exposition, we assume just two active users (users A and B). During the first time interval (hopping period), the BS transmits to UE A in band 2. At the same time, band 3 is free, so the BS can transmit to UE B in that band. In the next timeframe, the BS now transmits to UE A in band 1, and UE B in band 2. In the third timeslot, UE A is serviced in band 3, and UE B in band 1. Then the whole sequence repeats. The signals for all the UEs use the same hopping sequence, and it can be made sure that there is never a collision between them, i.e., perfect orthogonalization between the users. Thus, clearly, we have the same capacity as FDMA, with the added benefit of frequency diversity. In order to apply the same concept for the uplink, all UEs have to send their signals in such a way that they arrive at the BS synchronously, and thus recover the situation of Figure 19.1. This requires information about the runtime from each UE to the BS, which tells each UE exactly when to start transmission (timing advance, see also Section 18.3.2).

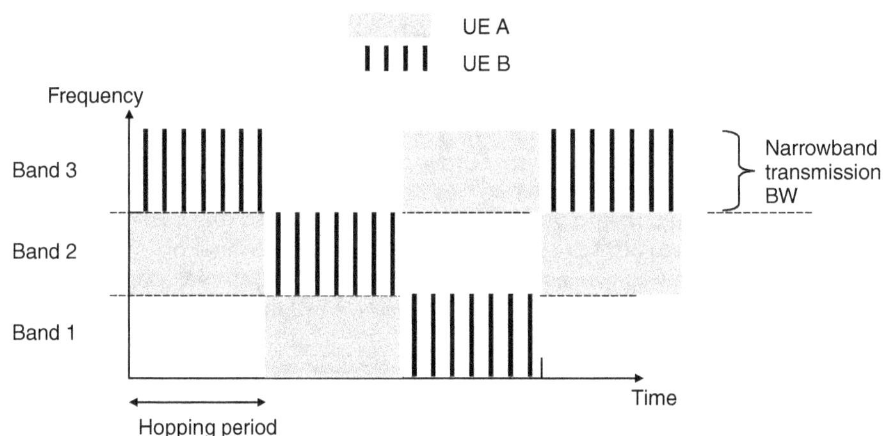

Figure 19.1 Principle behind frequency hopping for multiple access for synchronized users. *In this figure*: BW, bandwidth.

The situation is different when users are not synchronized – such a situation can either occur in simple networks where timing advance is not foreseen, for consideration of intercell interference,[2] or in ad hoc networks. For such a case, it is not a good idea to use the same hopping sequence for all users. Remember that, due to the lack of synchronization, any delay between the signals of different users is possible, including a zero-delay. If all users use the same hopping sequence, then such a zero-delay leads to *catastrophic collisions*, where different users interfere with each other all the time. In order to circumvent this problem, different hopping sequences are used for each user (see Figure 19.2). These sequences are designed in such a way that during each hopping cycle (i.e., one repetition of the hopping sequence; in our example, three times the hopping period), the duration of exactly one timeslot is disturbed, while the remainder of the time is guaranteed to be collision free.[3] Obviously, the performance of such a system is worse than that of a synchronized system (or an FDMA system). The design of hopping sequences that guarantee the low probability of collisions is also nontrivial.

---

[2] Users that design their uplink signals to arrive synchronously at one BS cannot be strictly synchronous for another BS.

[3] By collision free we mean "free of collisions from user B" – other users might still interfere.

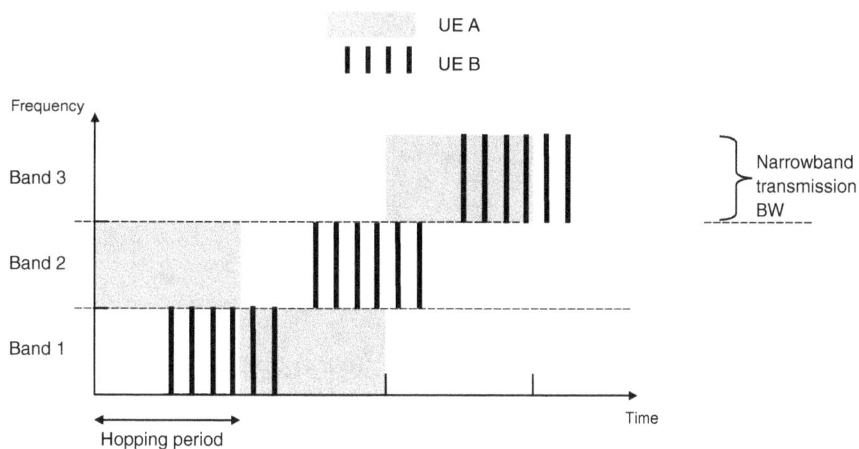

Figure 19.2   Principle behind frequency hopping for multiple access for unsynchronized users. *In this figure*: BW, bandwidth.

19.2   Direct Sequence Spread Spectrum – Single-User Case

An alternative for spreading a signal over a larger bandwidth is *Direct Sequence-Spread Spectrum* (DS-SS), which forms the basis of the CDMA method and 2G and 3G commercial cellular systems (IS-95, cdma2000, WCDMA) that built on it. While the commercial significance of these systems has decreased, as 3G has been supplanted by 4G and 5G (which use OFDMA instead of CDMA), the fundamental operating principles of CDMA are still important, and often used as part of other systems. The current subsection describes the DS-SS principle, i.e., how to obtain spectral spreading for a single user, while Section 19.3 will discuss the multiple access and related questions on how to incorporate it into cellular systems.[4]

19.2.1   Basic Principle

DS-SS spreads the signal by multiplying the transmit signal by a second signal that has a very large bandwidth. The bandwidth of this total signal is approximately the same as the bandwidth of the wideband spreading signal. The ratio of the bandwidth of the new signal to that of the original signal is again known as the *spreading factor*. As this spreading factor increases, and the transmit power stays constant, the *power spectral density* of the transmitted signal decreases – depending on the spreading factor and the distance between TX and RX, it can lie below the noise power spectral density. This is important in military applications, because an unauthorized RX then cannot determine whether a signal is being transmitted. Authorized listeners, on the other hand, can invert the spreading operation and thus recover the narrowband signal (whose power-spectral density lies considerably *above* the noise power).

Figure 19.3 shows the block diagram of a DS-SS TX. The information sequence (possibly encoded by a forward error correction encoder) is multiplied by a broadband signal that was created by modulating a sinusoidal carrier signal with a spreading sequence $p(t)$. This can be interpreted alternatively as multiplying each information symbol of duration T_S by a spreading sequence $p(t)$ before modulation.[5] We call the duration of a "symbol" of the sequence $p(t)$ a *chip* (the name *symbol* will remain reserved for the information symbols); we assume in the following that the spreading sequence is M_C *chips* long, where each chip has the duration $T_C = T_S/M_C$.[6] Since the bandwidth is the inverse of the chip duration, the bandwidth of the total signal is now also $W = 1/T_C = M_C/T_S = M_C B$; i.e., larger than the bandwidth of a narrowband-modulated signal by a factor M_C. Assuming that the spreading operation does not change the total transmit power, it also implies that the power-spectral density decreases by a factor M_C.

The RX now has to invert the spreading operation. This can, under some conditions, be achieved by correlating the received signal with the spreading sequence. This process reverses bandwidth spreading, so that after correlation, the desired signal again has a bandwidth of $1/T_S$. In addition to the desired signal, the received signal also contains noise, other wideband interferers, and possibly narrowband interferers. Note that the effective bandwidth of noise and wideband interferers is not significantly affected by the despreading operation, while narrowband interferers are actually spread over a bandwidth W. After the despreading, the signal passes through a low-pass filter of bandwidth $B = 1/T_S$. This leaves the desired signal essentially unchanged but reduces the power of noise, wideband interferers, and narrowband interferers by a factor M_C. At the symbol demodulator, DS-SS thus has the same Signal-to-Noise Ratio (SNR) as a narrowband system: for a narrowband system, the noise power at the demodulator is N_0/T_S. For a DS-SS system, the noise power at the RX input is $N_0/T_C = N_0 M_C/T_S$, which is reduced by narrowband filtering by a factor of M_C; thus, at the detector input, it is N_0/T_S. A similar effect occurs for wideband interference.

---

[4] It has become common to call the spectral spreading method of DS-SS as "CDMA" even when multiple access does not come into play.

[5] This interpretation is valid if the narrowband signal and the wideband signal use the same modulation method, e.g., BPSK.

[6] This interpretation assumes a "short spreading sequence," where each symbol is spread by the same sequence; more details and alternatives will be described below.

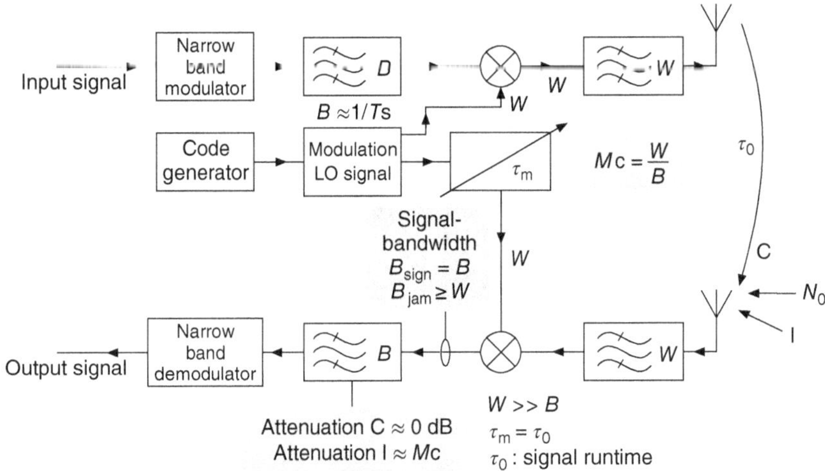

Figure 19.3   Block diagram of a direct sequence-spread-spectrum TX and RX. *In this figure*: LO, Local Oscillator.
Reproduced with permission from [Oehrvik 1994] © Ericsson AB.

As a matter of fact, the SNR at the RX can be computed in two ways:

(i) *After the despreader:* the spreading is undone by the despreading (correlation) operation, and a narrowband filtering limits the noise and interference power to what can pass through the narrowband receive filter.

(ii) *Before the despreader:* in this case, we count the full noise power and interference power (i.e., collected over the system bandwidth W). However, we argue that the desired signal is enhanced by the correlator by a factor M_C.

The two interpretations give the same SNR; one just has to take care not to mix the two approaches – enhancing the desired signal by M_C *and* performing narrowband filtering on the noise power would give incorrect results.

Let us next discuss the spreading signals for DS-SS systems. Remember that (assuming an ideal channel), the transmitted signal is $s(t)p(t)$, which is correlated at the RX with $p(t)$. Therefore, in order to perfectly reverse the spreading operation in the RX by means of a correlation operation, we want $ACF(i)$, the AutoCorrelation Function (ACF) of $p(t)$ at times iT_C to be a Dirac delta function

$$ACF(i) = \begin{cases} M_C & \text{for } i = 0 \\ 0 & \text{otherwise} \end{cases}. \tag{19.1}$$

These ideal properties can only be approximated in practice. One group of suitable code sequences is a type of Pseudo Noise (PN) sequences called *maximum length sequence* (*m-sequence*). These sequences have the following ACF:

$$ACF(i) = \begin{cases} M_C & \text{for } i = 0 \\ -1 & \text{otherwise} \end{cases} \tag{19.2}$$

see Figure 19.4. We assume here $p(t)$ to be real.

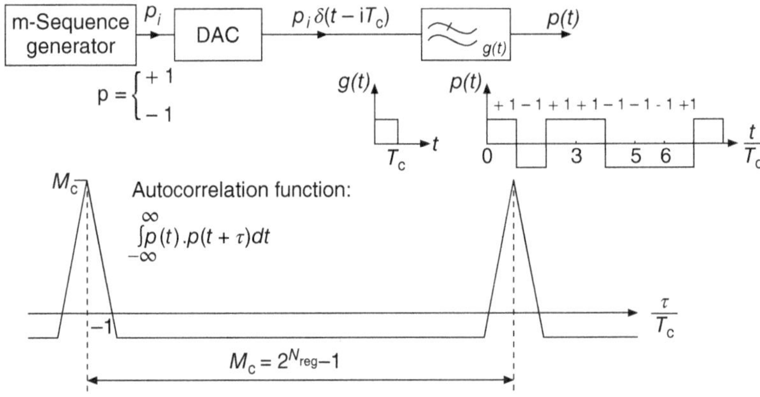

Figure 19.4   Autocorrelation function of a m-sequence. *In this figure*: DAC, Digital to Analog Converter.
Reproduced with permission from [Oehrvik 1994] © Ericsson AB.

We will see later that – for the purpose of user separation in multiple access – different users are assigned different sequences and there are requirements on the cross-correlation between these sequences, for details see Section 19.3. For the purpose of this section, we will treat signals of other users as wideband interferers.

*19.2.2 Error Probability and SINR

In the following, we put the above qualitative description into a mathematical framework [Viterbi 1995]. We assume for this that Binary Phase Shift Keying (BPSK) modulation is used, and that perfect synchronization between the TX and RX is available (discussed below in more detail). We then obtain four signal components at the RX:

- *Desired (user) signal*: let the kth user be the desired user, let $c_{i,k}$ represent the ith information symbol of this user, and $r_{i,k}$ the corresponding receive signal. Assuming that *InterSymbol Interference* (ISI), *interchip interference* (to be defined later), and noise are zero-mean processes, then the expected value of the received signal is proportional to the transmit symbol:

$$E\{r_{i,k} \mid c_{i,k}\} = \sqrt{(E_C)_k} c_{i,k} \int_{-\infty}^{\infty} |H_R(f)|^2 df \tag{19.3}$$

where $(E_C)_k$ is the chip energy of the kth user, and $H_R(f)$ is the transfer function of the receive filter normalized to $\int_{-\infty}^{\infty} |H_R(f)|^2 df = 1$.

- *Interchip interference*: the receive filter has a finite-duration impulse response, so that the convolution of a chip with this impulse response lasts longer than the chip itself. Thus, the signal after the receive filter exhibits interchip interference (we will see later on that delay dispersion of the channel also leads to interchip interference). If the spreading sequences are zero-mean, then the interchip interference increases the variance of the received signal by

$$(E_C)_k \sum_{i \neq 0} \left[\int_{-\infty}^{\infty} \cos(2\pi i f T_C) |H_R(f)|^2 df \right]^2 . \tag{19.4}$$

- *Noise* increases the variance by $N_0/2$.
- *Co-Channel Interference (CCI)*: this is the interference by other users that use different spreading sequences (compare Section 19.3). We assume that the mean of the received signal is not changed by the CCI, as the transmitted chips of interfering users are independent of the data symbols and chips of desired users. The CCI increases the variance by

$$\sum_{j \neq k} \frac{(E_C)_j}{2T_C} \int_{-\infty}^{\infty} |H_R(f)|^4 df. \tag{19.5}$$

If we now assume that interchip interference and CCI are approximately Gaussian, then the problem of computing the error probability reduces to the standard problem of detecting a signal in Gaussian noise:

$$BER = Q\left(\sqrt{\frac{(E_C)_k M_C}{\text{Total variance}}} \right). \tag{19.6}$$

In a fading channel, the energy of the desired signal varies when the transmit power is kept constant. However, for the uplink, power control is generally used to make sure that these variations are compensated, see Section 19.3.3.

19.2.3 Effects of Multi-Path Propagation

Principle of Rake RX

The above, strongly simplified, description of a DS-SS system assumed a frequency-flat channel. This assumption is violated under most practical circumstances. The basic nature of a DS-SS system is to spread the signal over a large bandwidth; thus, it can be anticipated that the transfer function of the channel exhibits variations over this bandwidth.

The effect of frequency selectivity (delay dispersion) on a DS-SS system can be understood by looking at the impulse response of the concatenation spreader–channel–despreader. If the channel is slowly time variant, the effective impulse response can be written as[7]

$$h_{\text{eff}}(t_i, \tau) = \widetilde{p}(\tau) * h(t_i, \tau) \tag{19.7}$$

where the effective system impulse response $\widetilde{p}(\tau)$ is the convolution of the transmit and receive spreading sequence:

$$\widetilde{p}(\tau) = p_{\text{TX}}(\tau) * p_{\text{RX}}(\tau) = ACF(\tau). \tag{19.8}$$

In the following, we assume an ideal spreading sequence (Eq. 19.1). The despreader output then exhibits multiple peaks: more precisely, one for each Multi-Path Component (MPC) that can be resolved by the RX – i.e., spaced at least T_C apart. Each of the peaks contains information about the transmit signal. Thus, all peaks should be used in the detection process: just using the largest correlation peak would mean that we discard a lot of the arriving signal. A RX that can use multiple correlation peaks is the so-called *Rake RX*, which collects ("rakes up") the energy from different MPCs. A Rake RX consists of a *bank of correlators*. Each correlator is sampled at a different time (with delay τ), and thus collects energy from the MPC with delay τ. The sample values from the correlators are then weighted and combined.

Alternatively, we can interpret the Rake RX as a tapped delay line, whose outputs are weighted and added up, as shown in Figure 19.5. The tap delays, as well as the tap weights, are adjustable, and matched to the channel. Note that the taps are usually spaced at least one chip duration apart, but there is no requirement for the taps to be spaced at regular intervals. The combination of the RX filter and the Rake RX constitutes a filter that is matched to the receive signal: the RX filter is matched to the transmit signal, while the Rake RX is matched to the channel.

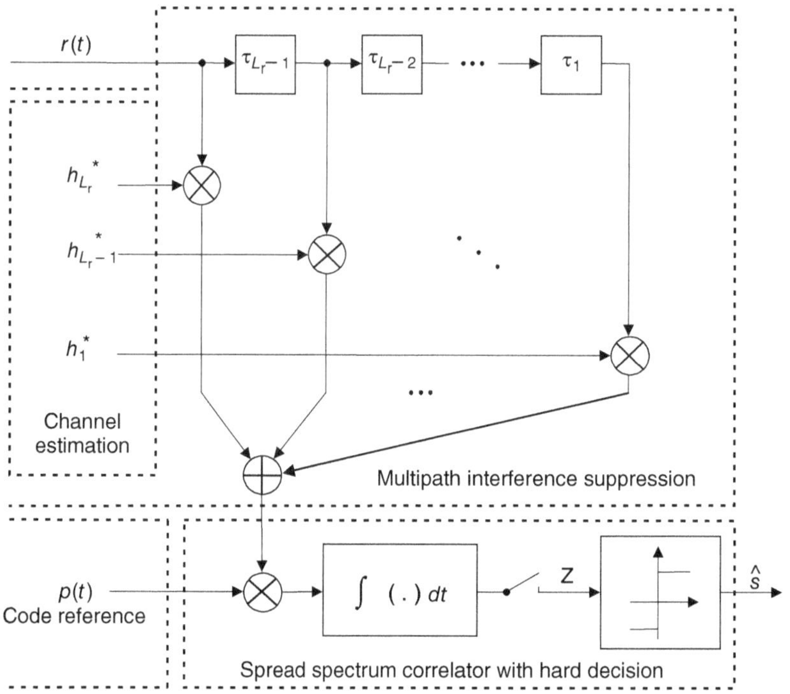

Figure 19.5 Rake receiver.
Reproduced with permission from [Molisch 2000] © Prentice Hall.

*Performance Analysis of Rake RX

Independent of the above-mentioned interpretation, the RX adds up the (weighted) signal from the different Rake fingers in a coherent way. As these signals correspond to different MPCs, their fading is (approximately) statistically independent – in other

---

[7] Note that this expression is identical to that in correlative channel sounders (see Section 9.2.2). Correlative channel sounders and DS-SS systems have the same structure, it is just the goals that are different: the channel sounder tries to find the channel impulse response from the received signal and knowledge of the transmit data, while the DS-SS RX tries to find the transmit data from knowledge of the received signal and the channel impulse response.

words, they provide delay diversity (frequency diversity). A Rake RX is thus a diversity RX, and all mathematical methods for the treatment of diversity remain valid. As for the performance of Rake RX systems, we can now simply refer to the equations of Sections 12.4–12.5.

Example 19.1 *Performance of a Rake RX: Compute the Bit Error Rate (BER) of BPSK in (i) a narrowband system and (ii) with a DS-SS system that can resolve all multi-paths, using a six-finger Rake RX at a 15-dB SNR, in an International Telecommunications Union (ITU) Pedestrian-A channel.*

The tapped delay line model of an ITU Pedestrian-A channel has the following (resolvable) tap weights:

$$|h(n)|_{\mathrm{dB}} = [0 - 9.7 - 19.2 - 22.8] \tag{19.09}$$

which on a linear scale is

$$|\,h(n)\,| = [1\ \ 0.3273\ \ 0.1096\ \ 0.0724]. \tag{19.10}$$

The average channel gain of the flat-fading channel is

$$\Sigma|h(n)|^2 = 1 + 0.33^2 + 0.11^2 + 0.07^2 = 1.1 \tag{19.11}$$

and the transmit SNR has to be

$$\overline{\gamma}_{\mathrm{TX}} = \frac{10^{1.5}}{1.1} = 28.8 \tag{19.12}$$

so that a receive SNR of 15 dB is achieved.

As can be found from Section 11.2.2, the BER for the flat-fading channel is

$$\overline{BER} = E\,[\,P_{\mathrm{BER}}(\gamma_{\mathrm{Flat}})] = \int_0^{\pi/2} \frac{1}{\pi} M_{\gamma_{\mathrm{Flat}}}\left(-\frac{1}{\sin^2\theta}\right) d\theta \tag{19.13}$$

and

$$\overline{BER} = \int_0^{\pi/2} \frac{1}{\pi} \frac{\sin^2\theta}{\sin^2\theta + \overline{\gamma}_{\mathrm{TX}}\Sigma|h(n)|^2}\, d\theta = 7.72 \times 10^{-3}. \tag{19.14}$$

When combining the signals from the different Rake fingers with maximum-ratio combining,

$$\gamma_{\mathrm{Rake}} = \gamma_1 + \cdots + \gamma_6. \tag{19.15}$$

Since only four MPCs carry energy, only four Rake fingers are effectively used. If the $\gamma_1, ..., \gamma_4$ are independent, the joint pdf of $f_{\gamma 1,...,\gamma 4}(\gamma_1, ..., \gamma_4) = f_{\gamma 1}(\gamma_1) \cdot ... \cdot f_{\gamma 4}(\gamma_4)$ and (see also Eq. 12.42):

$$\overline{BER} = \int d\gamma_1 pdf_{\gamma_1}(\gamma_1) \int d\gamma_2 pdf_{\gamma_2}(\gamma_2) \cdots \int d\gamma_4 pdf_{\gamma_4}(\gamma_4) \int_0^{\pi/2} d\theta f_1(\theta) \prod_{k=1}^{N_{\mathrm{r}}} \exp\left(-\gamma_k f_2(\theta)\right)$$

$$= \int_0^{\pi/2} \frac{1}{\pi} \prod_{k=1}^{4} \int_{\gamma_k} f_{\gamma_k}(\gamma_k) \exp(-\gamma_k/\sin^2\theta)\, d\gamma_k d\theta \tag{19.16}$$

$$= \int_0^{\pi/2} \frac{1}{\pi} \prod_{k=1}^{4} M_{\gamma_k}\left(-\frac{1}{\sin^2\theta}\right) d\theta.$$

Thus,

$$\overline{BER} = \int_0^{\pi/2} \frac{1}{\pi} \prod_{k=1}^{4} \left[\frac{\sin^2(\theta)}{\sin^2(\theta) + \overline{\gamma}_k}\right] d\theta. \tag{19.17}$$

For the same transmit SNR as above, $\widetilde{\gamma}_{\mathrm{TX}} = 28.8$, we then get:

$$\overline{BER} = \int_0^{\pi/2} \frac{1}{\pi} \frac{\sin^2\theta}{\sin^2\theta + \overline{\gamma}_{\mathrm{TX}}} \frac{\sin^2\theta}{\sin^2\theta + 0.33^2\overline{\gamma}_{\mathrm{TX}}} \frac{\sin^2\theta}{\sin^2\theta + 0.1^2\overline{\gamma}_{\mathrm{TX}}} \frac{\sin^2\theta}{\sin^2\theta + 0.07^2\overline{\gamma}_{\mathrm{TX}}}\, d\theta \tag{19.18}$$

$$= 9.9 \times 10^{-4}.$$

Another consequence of the delay diversity interpretation is the determination of the weights for the combination of Rake finger outputs. The optimum weights are the weights for maximum-ratio combining – i.e., the complex conjugates of the amplitudes of the

MPC corresponding to each Rake finger. However, this is only possible if we can assign one Rake finger to each resolvable MPC (the term *all Rake* has been largely used in the literature for such a RX). Up to $L_r = \tau_{max}/T_C$ taps, where τ_{max} is the maximum excess delay of the channel (see Chapter 6), are required in this case. For a typical system bandwidth, say 5 MHz, and delay dispersion encountered in outdoor environments, this number can easily exceed 20 taps. However, the number of taps that can be implemented in a practical Rake combiner is limited by power consumption, design complexity, and channel estimation. A Rake RX that processes only a *subset* of the available L_r resolved MPCs achieves lower complexity, while still providing a performance that is better than that of a single-path RX. The *Selective Rake* (*SRake*) RX selects the L_b best paths (a *subset* of the L_r available resolved MPCs) and then combines the selected subset using maximum-ratio combining. This combining method is "hybrid selection/maximum ratio combining" (as discussed in Section 12.4.3); however, note that the average power in the different diversity branches is different. It is also noteworthy that the SRake still requires knowledge of the instantaneous values of *all* MPCs so that it can perform appropriate selection. Another possibility is the *Partial Rake* (*PRake*), which uses the first L_f MPCs. Although the performance it provides is not as good, it only needs to estimate L_f MPCs.

Another generally important problem for Rake RXs is interpath interference. Paths that have delay τ_i compared with the delay the Rake finger is tuned to are suppressed by a factor $ACF(\tau_i)/ACF(0)$, which is infinite only when the spreading sequence has ideal ACF properties. Rake RXs with nonideal spreading sequences thus suffer from interpath interference.

Finally, we note that in order for the Rake RX to be optimal there must be no ISI – i.e., the maximum excess delay of the channel must be much smaller than T_S, though it can be larger than T_C. If there is ISI, then the RX must have an equalizer (working on the Rake output – i.e., a signal sampled at intervals T_S) in addition to the Rake RX. An alternative to this combination of Rake RX and symbol-spaced equalizer is the chip-based equalizer, where an equalizer works directly on the output of the despreader sampled at the chip rate. This method is optimum but very complex. As we showed in Chapter 14, the computational effort for equalizers increases quickly as the product of sampling frequency and channel maximum excess delay increases.

*19.2.4   Synchronization

Synchronization is one of the most important practical problems of a DS-SS system. Mathematically speaking, synchronization is an estimation problem in which we determine the optimum sampling time out of an infinitely large ensemble of possible values – i.e., the continuous time. Implementation is facilitated by splitting the problem into two partial problems:

- *Acquisition*: a first step is the coarse synchronization, which determines in which time interval (of duration T_C or $T_C/2$) the optimum sampling time lies. This is a hypothesis-testing problem: we test a finite number of hypotheses, each of which assumes that the sampling time is in a certain interval. The hypotheses can be tested in parallel or serially.
- *Tracking*: as soon as this interval has been determined, a control loop can be used for fine synchronization, i.e., to fine-tune the sampling time to its exact value and follow any variations over time.

For the acquisition phase, we use a special synchronization sequence that is shorter than the spreading sequence used during data transmission. This decreases the number of hypotheses that have to be tested, and thus decreases the time that has to be spent on synchronization. Furthermore, the synchronization sequence is designed to have especially good autocorrelation properties. For the tracking part, the normal spreading sequence used for data communications can be employed.

It is also often important to know whether signals from different users are synchronized with respect to each other:

- *Synchronization within a cell*: the signals transmitted by a BS are always synchronous, as the BS has control over when to transmit them. For the uplink, synchronous arrival of the signals at the BS is usually not achieved. It would require that all UEs arrange their timing advance – i.e., when they start transmitting a code sequence – in such a way that all signals arrive simultaneously. The timing advance would have to be accurate within one chip duration. This is too complicated for most applications, especially since movement of the UE leads to a change in the required timing advance.
- *Synchronization between BSs*: BSs can be synchronized with respect to each other. This is usually achieved by means of timing signals provided by GPS (*Global Positioning System*, see Section 29.4), so that each BS requires a GPS RX and free line of sight to several GPS satellites. While the former is not a significant obstacle, the latter is difficult for microcells and picocells. Synchronization via the backbone (e.g., internet timing) is a possible alternative.

19.3   Code-Division-Multiple-Access Systems

19.3.1   Principle of Code Division Multiple Access

The Direct Sequence (DS) spreading operation itself – i.e., multiplication by the wideband signal – can be viewed as a modulation method for stealthy communications, and is as such mainly of military interest. CDMA is used on top of it, exploiting the spreading to achieve multi-access capability. Each user is assigned a different spreading code, which determines the wideband signal that multiplies the information symbols. Thus, many users can transmit simultaneously in a wide band (see Figure 19.6).

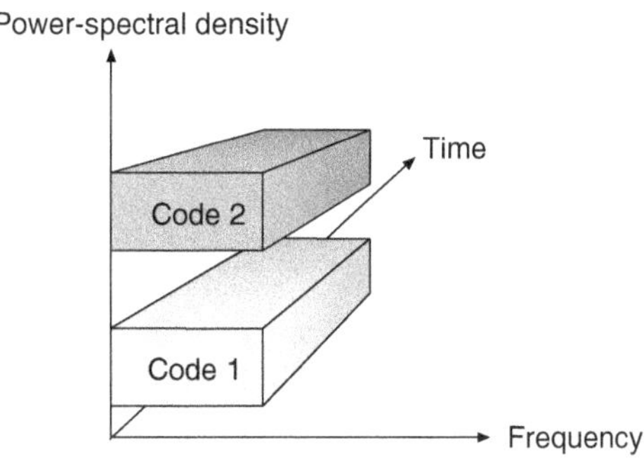

Figure 19.6 Principle behind code division multiple access.

At the RX, the desired signal is obtained by correlating the received signal with the spreading signal of the desired user. Other users thus become wideband interferers; after passing through the despreader, the amount of interference power seen by the detector is equal to the *Cross-Correlation Function* (CCF) between the spreading sequence of the interfering user and the spreading sequence of the desired user. In particular, we can distinguish between *orthogonal spreading sequences* and *nonorthogonal* sequences. Orthogonal sequences fulfill

$$CCF_{j,k}(t) = 0 \text{ for } j \neq k \tag{19.19}$$

for all users j and k. In other words, we require code sequences to be orthogonal. Perfect orthogonality can be achieved for at most M_C spreading sequences; this can be immediately seen by the fact that M_C orthogonal sequences span an M_C-dimensional space, and any other sequence of that duration can be represented as a linear combination.

If the spreading sequences are not orthogonal, the RX achieves finite interference suppression – namely, suppression by a factor ACF/CCF. If the different spreading sequences are shifted m-sequences, then this suppression factor is M_C.

With these prerequisites, we can now proceed to explain multiple access via DS-spreading. We need to distinguish here the uplink and downlink.

The downlink typically uses *orthogonal* spreading codes. Thus, the signals intended for different users do not interfere with each other,[8] but the number of users is limited by M_C – once that number is reached, no further users can be accepted into the system. This case is very comparable to TDMA/FDMA, where the number of users is limited by the number of available timeslots/frequencies.

In the uplink, the mechanism is subtly different. The uplink uses codes that are *approximately* orthogonal, but not perfectly so, i.e., the CCF is small but finite. The reason for choosing such sequences is that they retain low CCF even when different signals are not synchronized (remember that exact synchronization in the uplink is difficult). Furthermore, it is possible to trade off the number of available codes and the maximum CCF, allowing up to a number of codes that is larger than M_C.

As user separation is not perfect, each user in the cell contributes interference to all other users. Thus, as the number of users increases, the interference for each user increases as well. Consequently, transmission quality decreases gradually (*graceful degradation*), until users find the quality too bad to place (or continue) calls.[9] Consequently, CDMA puts a soft limit on the number of uplink users, not a hard limit like TDMA. Therefore, the number of users in a system depends critically on the SINR required by the RX. It also implies that any increase in the obtained SINR at the RX, or reduction in the required SINR, can be immediately translated into higher capacity.

We can draw some important conclusions from the above description:

- The choice of spreading sequences is an essential factor for the quality of a CDMA system. Sequences have to have good ACFs (similar to a Dirac delta function) and a small cross-correlation. One possible choice is the above-mentioned m-sequences. Alternative sequences include Gold and Kasami sequences, which will be discussed below.
- CDMA requires accurate power control in the uplink. If ACF/CCF has a finite value, the RX cannot suppress interfering users perfectly. When the received interference power becomes much larger than the power from the desired TX, it exceeds the interference suppression capability of the despreading RX. Thus, each UE has to adjust its power in such a way that the powers of all the signals arriving at the BS are approximately the same. Experience has shown that power control has to be accurate within about ± 1 dB in order to avoid significant performance loss compared to the theoretical capacity of CDMA systems.

Most interference stems from within the same cell as the desired user and is thus termed *intracell interference*. Total intracell interference is the sum of many independent contributions, and thus behaves approximately like Gaussian noise. Therefore, it causes effects that are similar to thermal noise. It is often described by *noise rise* – i.e., the increase in "effective" noise power (sum of noise

---

[8] In an AWGN channel. Inter-user interference in the presence of delay dispersion will be discussed below.

[9] CDMA for cellular systems was used initially in second- and third-generation systems where voice calls played an important role. For data transmission, an equivalent mechanism is that the packet error rate, and associated repeat requests, become so high that the scheduler reduces the number of transmissions, or the data rate is reduced.

and interference power) compared with the noise alone $(N_0 + I_0)/N_0$. Figure 19.7 shows an example of noise rise as a function of system load; here *system load* is defined as the number of active users, compared with M_C. We see that noise rise becomes very strong as system load approaches 100%. A cell is thus often judged to be "full" if noise rise is 6 dB. However, as mentioned above, there is no hard limit to the number of active users.

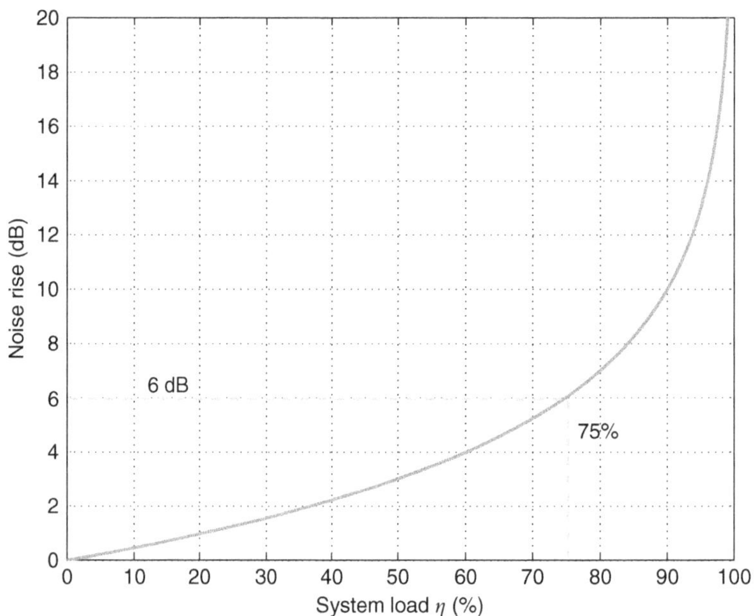

Figure 19.7   Noise rise as a function of system load in a code-division-multiple-access system.
Reproduced with permission from [Neubauer 2001] © T. Neubauer.

Many of the advantages of CDMA are related to the fact that interference behaves almost like noise, especially in the uplink. This noise-like behavior is due to several reasons:

- The number of users (and therefore, of interferers) in each cell is large.
- Power control makes sure that all intracell signals arriving at the BS have approximately the same strength.
- Interference from neighboring cells is also sometimes be approximated as noise-like, though this approximation might be crude, see Section 21.5.1. As we will outline further in Section 21.5.1, codes in different cells need to be different.

Due to the above effects, total interference power in the uplink shows very little fluctuations. At the same time, the power control makes sure that the signal strength from the desired user is always constant. Since SINR is thus constant, no fading margin has to be used in the link budget. However, note that making interference as Gaussian as possible is not always the best strategy for maximizing data throughput; multi-user detection (Section 28.2) actively exploits structure in interference and works best when there are only a few strong interferers.

In the downlink, there is no intra-cell interference (for non-dispersive channels) when orthogonal codes are used, see Sec. 19.3.2. Inter-cell interference does exist, see Section 21.5.2.

19.3.2   Spreading Codes for Multiple Access

Selection Criteria

The selection of spreading codes has a vital influence on performance on a CDMA system. In general, the quality of spreading codes is determined by the following properties:

- *Autocorrelation*: ideally, $ACF(0)$ should be equal to the number of chips per symbol M_C, and zero at all other instances. For m-sequences $ACF(0) = M_C$, and $ACF(n) = -1$ for $n \neq 0$. Good properties of the ACF are also useful for synchronization. Furthermore, as discussed above, autocorrelation properties influence interchip interference in a Rake RX: the output of the correlator is the sum of the ACF_S of the delayed echoes; a spurious peak in the ACF looks like an additional MPC.[10]

---

[10] As the ACF is known to the RX, spurious peaks belonging to the first-detected MPCs can be subtracted from the remaining signal, and thus their effect can be eliminated, compare Sec. 28.1.

- *Cross-correlation*: if all codes are orthogonal to each other, interference from other users can be completely suppressed. Nonzero cross-correlation leads to residual inter-user interference, which of course should be as small as possible. For unsynchronized systems, the cross-correlation magnitude should either be averaged over arbitrary delays between the different users, or the worst-case cross-correlation should be used for the assessment of the sequence.

- *Number of codes*: a CDMA system should allow simultaneous communications of as many users as possible. This implies that a large number of codes have to be available. The number of orthogonal codes is limited by M_C. If more codes are required, worse cross-correlation properties have to be accepted. The situation is complicated further by the fact that codes in adjacent cells have to be different: after all, it is only the codes that distinguish different users. Now, if cell A has M_C users with all orthogonal codes, then the codes in cell B cannot be orthogonal to the codes in cell A; therefore, intercell interference cannot be completely suppressed. The codes in cell B can, however, be chosen in such a way that the interference to users in the original cell becomes noise-like; this approach then requires code planning instead of frequency planning. An alternative approach is the creation of a large number of codes with suboptimum CCFs (see the next subsection), and assigning them to wherever they are needed. In such a case, no code planning is required; however, system capacity is lower.

Note that bandwidth spreading and separation of users can be done by different codes; in this case, we distinguish between spreading codes and scrambling codes.

Pseudo Noise Sequences

The spreading sequences most frequently used for the uplink of CDMA systems are m-sequences, Gold sequences, and Kasami sequences. The autocorrelation properties of m-sequences are excellent; shifted versions of an m-sequence are again valid codewords with almost ideal cross-correlation properties. However, the same m-sequence with different shifts cannot be used in unsynchronized settings, or settings with significant delay dispersion. On the other hand, when considering *different* m-sequences (i.e., with different shift-register settings), only a small number exist (e.g., 18 sequences for a 127-bit sequence), and the cross-correlation for suitably selected pairs of sequences can be as low as $\approx -3N_{\mathrm{reg}}/2 + 1.5$ dB.

Gold sequences are created by the appropriate combination of m-sequences; the ACFs of Gold sequences can take on three possible values; both the offpeak values of the ACF and the CCF can be upper-bounded.

An even more general family of sequences is the Kasami sequences. We distinguish between S (*small*), L (*large*), and VL (*very large*) Kasami sequences. The letters describe the number of codes within the family. S-Kasami sequences have the best CCFs, but only a rather small number of such sequences exists. VL-Kasami sequences have the worst CCF, but there is an almost unlimited number of such sequences. Table 19.1 shows the properties of the different code families.

Table 19.1 Properties of widely used code division multiple access codes. Number of codes and CCF are approximate; exact values may depend on specific length of register and construction method.

Sequence	Number of codes	Maximum CCF/dB	Comment
Gold	$2^{N_{\mathrm{reg}}} + 1$	$\approx -3N_{\mathrm{reg}}/2 + 1.5$	
S-Kasami	$2^{N_{\mathrm{reg}}/2}$	$\approx -3N_{\mathrm{reg}}/2$	Best CCF of all Kasami sequences
L-Kasami	$2^{N_{\mathrm{reg}}/2}(2^{N_{\mathrm{reg}}} + 1)$	$\approx -3N_{\mathrm{reg}}/2 + 3$	
VL-Kasami	$2^{N_{\mathrm{reg}}/2}(2^{N_{\mathrm{reg}}} + 1)^2$	$\approx -3N_{\mathrm{reg}}/2 + 6$	Almost unlimited number

N_{reg} is the size of the shift register used to create the sequences.

Walsh–Hadamard Codes

In the downlink, signals belonging to different users can be made completely synchronous as they are all emitted by the same TX (the BS). Under these circumstances, a family of codes that are all completely orthogonal to each other is given by the columns of Walsh–Hadamard matrices defined in Section 15.11.1, and repeated here for convenience: the matrices are computed from a recursive equation,

$$\mathbf{H}_{\mathrm{had}}^{(1)} = \begin{pmatrix} 1 & 1 \\ 1 & -1 \end{pmatrix} \text{ and } \mathbf{H}_{\mathrm{had}}^{(n+1)} = \begin{pmatrix} \mathbf{H}_{\mathrm{had}}^{(n)} & \mathbf{H}_{\mathrm{had}}^{(n)} \\ \mathbf{H}_{\mathrm{had}}^{(n)} & \overline{\mathbf{H}}_{\mathrm{had}}^{(n)} \end{pmatrix}. \tag{19.20}$$

Orthogonal codes lead to perfect multi-user suppression at the RX if the signal is transmitted over an Additive White Gaussian Noise (AWGN) channel. Delay dispersion destroys the orthogonality of the codes. The RX can then either accept the additional interference (described by an *orthogonality factor*), or send the received signal through a chip-spaced equalizer that eliminates delay dispersion before correlation (and thus user separation) is performed.

Example 19.2   *Orthogonality of Walsh–Hadamard codes in frequency-selective channels: the codewords $[1\ 1\ 1\ 1]$ and $[1\ -1\ 1\ -1]$ are sent through a channel whose impulse response is $(0.8, -0.6)$. Assuming that $[1\ 1\ 1\ 1]$ was sent, compute the correlation coefficient of the signal arriving at the RX with the possible transmit signals.*

Assume that the BS communicates with two UEs and transmits spreading codes, $t_1(n) = [1\ 1\ 1\ 1]$ and $t_2(n) = [1\ -1\ 1\ -1]$. $t_1(n)$ and $t_2(n)$ are orthogonal to each other; e.g.,

$$t_1(n){\cdot}t_2(n)^T = 0. \tag{19.21}$$

However, the time dispersive channel (multi-path channel) $h(n) = [0.8 - 0.6]$ will affect the orthogonality between the two codes. The received signal is the linear convolution of the impulse response and the transmitted signal (assuming that the channel is linear and time invariant); e.g.,

$$r(n) = t(n) * h(n) = \sum_{k=-\infty}^{\infty} t(k)h(n-k). \tag{19.22}$$

The received signal when transmitting $t_1(n)$ is then

$$r_1(0) = 1{\cdot}0.8 = 0.8 \tag{19.23}$$
$$r_1(1) = 1{\cdot}(-0.6) + 1{\cdot}0.8 = 0.2 \tag{19.24}$$
$$r_1(2) = 1{\cdot}(-0.6) + 1{\cdot}0.8 = 0.2 \tag{19.25}$$
$$r_1(3) = 1{\cdot}(-0.6) + 1{\cdot}0.8 = 0.2 \tag{19.26}$$
$$r_1(4) = 1{\cdot}(-0.6) = -0.6. \tag{19.27}$$

Thus, by correlating the received signals with $t_1(n)$ and $t_2(n)$ we get for $r_1(n)$:

$$\rho_{r_1 t_1} = \sum_{n=0}^{3} r_1(n)t_1(n) \tag{19.28}$$

$$= [0.8\ 0.2\ 0.2\ 0.2][1\ 1\ 1\ 1]^T = 1.4 \tag{19.29}$$

$$\rho_{r_1 t_2} = \sum_{n=0}^{3} r_1(n)t_2(n) \tag{19.30}$$

$$= [0.8\ 0.2\ 0.2\ 0.2][1\ -1\ 1\ -1]^T = 0.6 \neq 0 \tag{19.31}$$

In a similar manner, the signal that is received when $t_2(n)$ is transmitted is $r_2(n) = [0.8\ -1.4\ 1.4\ -1.4\ 0.6]$. The correlation coefficients with the two transmit signals are:

$$\rho_{r_2 t_1} = \sum_{n=0}^{3} r_2(n)t_1(n) = -0.6 \neq 0 \tag{19.32}$$

$$\rho_{r_2 t_2} = \sum_{n=0}^{3} r_2(n)t_2(n) = 5. \tag{19.33}$$

Correlation between the two received signals is

$$\rho_{r_1 r_2} = \sum_{n=0}^{3} r_1(n)r_2(n) = 0.36 \neq 0. \tag{19.34}$$

An additional challenge arises if different users require different data rates, so that codes of different length need to be used for the spreading. *Orthogonal Variable Spreading Factor (OVSF)* codes are a class of codes that fulfills these conditions; they are derived from Walsh–Hadamard codes.

Let us first define what we mean by orthogonality for codes of different duration. The chip duration is the same for all codes: it is given by the available system bandwidth, and independent of the data rate to be transmitted. Consider a code A that is two chips long $(1, 1)$, and a code B that is four chips long $(1, -1, -1, 1)$. The output of correlator A has to be zero if code B is at the input of the correlator. Thus, the correlation between code A and the first part of code B has to be zero, which is true: $1 \cdot 1 + 1 \cdot (-1) = 0$. Similarly, correlation of code A with the second part of code B has to be zero $1 \cdot (-1) + 1 \cdot 1 = 0$.

Let us now write all codewords of different Walsh–Hadamard matrices into a "code tree" (see Figure 19.8). All codes within one level of the tree (same duration of codes) are orthogonal to each other. Codes of different duration A, B are only orthogonal if they are

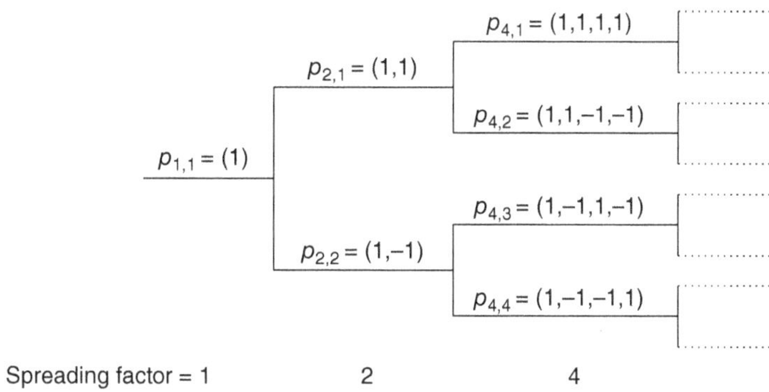

Figure 19.8 Code tree of orthogonal-variable-spreading-factor codes.

in different branches of the tree. They are *not* orthogonal to each other if one code is a "mother code" of the second code – i.e., code A lies on the path from the "root" of the code tree to code B. Examples of such codes are $p_{2,2}$ and $p_{4,4}$ in Figure 19.8, whereas codes $p_{2,2}$ and $p_{4,1}$ are orthogonal to each other.

19.3.3 Power Control

As we mentioned above, power control is important to make sure that the desired user has a time-invariant signal strength, and that the interference from other users becomes noise-like. For further considerations, we have to distinguish between power control for the uplink and that for the downlink:

- *Power control in the uplink*: for the uplink, power control is vital for the proper operation of CDMA. Power control is done by a closed control loop: the UE first sends with a certain power, the BS then tells the UE whether the power was too high or too low, and the UE adjusts its power accordingly. The bandwidth of the control loop has to be chosen so that it can compensate for small-scale fading – i.e., has to be on the order of, or larger than, the Doppler frequency. Due to time variations of the channel and noise in the channel estimate, there is a remaining variance in the powers arriving at the BS; this variance is typically on the order of 1.5–2.5 dB, while the dynamic range that has to be compensated is 60 dB or more. This variance leads to a reduction in the capacity of a CDMA cellular system of up to 20% compared with the case when there is ideal power control.

  Note that an open control loop (where the UE adjusts its transmit power based on its own channel estimate) cannot be used to compensate for small-scale fading in a Frequency Domain Duplexing (FDD) system: the channel seen by the UE (when it receives signals from the BS) is different from the channel it transmits to. However, an open loop can be used in conjunction with a closed loop. The open loop compensates for large-scale variations in the channel (path loss and shadowing), which are approximately the same at uplink and downlink frequencies. The closed loop is then used to compensate for small-scale variations.

  We finally note that uplink power control is not necessary if multi-user detection is used, in particular to handle the situation where interferers have much larger power than the desired signal. This approach is further discussed in Section 28.2; however, commercial CDMA systems like 3GPP WCDMA do not use it.

- *Power control in the downlink*: for the downlink, power control is not necessary for CDMA to control intra-cell interference: all signals from the BS arrive at one UE with the same power (the channel is the same for all signals). However, it can be advantageous to still use power control in order to keep the total transmit power low. Decreasing the transmit power for all users within a cell by the same amount leaves unchanged the ratio of desired signal power to intracell interference – i.e., interference from signals destined for other users in the cell. However, it does decrease the power of total interference to other cells. On the other hand, we cannot decrease signal power arbitrarily, as the SNR must not fall below a threshold. The goal of downlink power control is thus to minimize the total transmit power while keeping the BER or SINR level above a given threshold. The accuracy of downlink power control need not be as high as for the uplink; for many cases, open-loop control is sufficient.

  Interference power from all users is the same only if all users employ the same data rate. Users with higher data rates contribute more interference power; high-data-rate users can thus be a dominant source of interference. This fact can be understood most easily when we increase the data rate of a UE by assigning multiple spreading codes to it. In this case, it is obvious that the interference this user contributes increases linearly with the data rate. While this situation did not occur for second-generation cellular systems, which had only speech users, it became relevant for third-generation cellular systems, which provide high-data-rate services.

  It should also be noted that power control is not an exclusive property of CDMA systems; it can also be used for FDMA, TDMA, or OFDMA systems, where it decreases intercell interference and thus improves capacity.

*19.3.4 Methods for Capacity Increases

- *Quiet periods during speech transmission:* for speech transmission, CDMA makes implicit use of the fact that a person does not talk continuously, but rather only about 50% of the time, the remainder of the time they listen to the other participant. In addition, there are pauses between words and even syllables, so that the ratio of "talk time" to "total time of a call" is about 0.4. During quiet periods, no signal, or a signal with a very low data rate, has to be transmitted.[11] In a CDMA system, not transmitting information leads to a decrease in total transmitted power, and thus interference in the system. But we have already seen above that decreasing the interference power allows additional users to place calls. Of course, there can be a worst-case scenario where all users in a cell are talking simultaneously, but, statistically speaking, this is highly improbable, especially when the number of users is large. Thus, pauses in the conversation can be used very efficiently by CDMA in order to improve capacity.

- *Flexible data* rate: in an FDMA (TDMA) system, a user can occupy either one frequency (timeslot), or integer multiples thereof. In a CDMA system, arbitrary data rates can be transmitted by an appropriate choice of spreading sequences. This is not important for speech communications, which operate at a fixed data rate. For data transmission, however, the flexible data rate allows for better exploitation of the available spectrum. Note, however, that OFDMA provides even better flexibility in the data rate.

- *Soft capacity*: the capacity of a CDMA system can vary from cell to cell. If a given cell adds more users, it increases interference to other cells. It is thus possible to have some cells with high capacity, and some with lower; furthermore, this can change dynamically, as traffic changes. This concept is known as *breathing cells*.

- *Error correction coding*: a drawback of error correction coding is that the data rate that is to be transmitted is increased, which decreases spectral efficiency in systems with fixed modulation format. On the other hand, CDMA consciously increases the amount of data to be transmitted. It is thus possible to include error correction coding without decreasing spectral efficiency; in other words, different users are distinguished by different error correction codes. Note, however, that commercial systems (such as WCDMA) did not use this approach; they have separate error correction and spreading.

*19.3.5 Combination with Other Multi-Access Methods

CDMA has advantages compared with TDMA and FDMA, especially with respect to flexibility, while it also has some drawbacks, like complexity. It is thus obvious to combine CDMA with other multi-access methods in order to obtain the "best of both worlds." The most popular solution is a combination of CDMA with FDMA: the total available bandwidth is divided into multiple subbands, in each of which CDMA is used as the multi-access method. Clearly, frequency diversity in such a system is lower than for the case where spreading is done over the whole bandwidth. On the positive side, the processing speed of the TXs and RXs can be lower, as the chip rate is lower. The approach was used, e.g., in IS-95 (which used 1.25-MHz-wide bands) and WCDMA, which used 5-MHz-wide subbands.

Another combination is CDMA with TDMA. Each user can be assigned one timeslot (as in a TDMA system), while the users in different cells are distinguished by different spreading codes (instead of different frequencies).

*19.4 Time Hopping Impulse Radio

When the desired spreading bandwidth W is on the order of 500 MHz or higher, it becomes interesting to spread the spectrum by transmitting short pulses whose position or amplitude contains the desired information. This method, often called *Impulse Radio* (IR), allows the use of simple TXs and RXs. It is mostly used for so-called "ultrawideband" communications systems [diBenedetto et al. 2005], see also Section 26.8.

19.4.1 Simple Impulse Radio

Let us start out with the simplest possible IR: a TX sending out a single pulse to represent one symbol. For the moment, assume that the modulation method is orthogonal pulse position modulation (see Section 10.4.7). A pulse is either sent at time t, or at time $t + T_d$, where T_d is larger than the duration of a pulse T_C. The detection process in an AWGN channel is then exceedingly simple: we just need an energy detector that determines during which of the two possible time intervals

$$[t + \tau_{run}, t + \tau_{run} + T_C] \tag{19.35}$$

or

$$[t + \tau_{run} + T_d, t + \tau_{run} + T_C + T_d] \tag{19.36}$$

we get more energy. Here, τ_{run} is the runtime between TX and RX, which is determined via the synchronization process.

Obviously, a pulsed transmission achieves spreading, because the bandwidth of the transmit signal is given by the inverse of pulse duration $1/T_C$. And "despreading" is done in a very simple way: a decision about which bit was sent is made by only recording and

---

[11] Actually, most systems transmit *comfort noise* during this time – i.e., some background noise. People speaking into a telephone feel uncomfortable (think the connection has been interrupted) if they cannot hear any sound while they talk.

using the arriving signal in the intervals given by Eqs. (19.35) and (19.36). The SNR occurring at the RX after the despreading is $E_{\rm B}/N_0$: if the RX is an energy detector, the observed peak power is $\overline{P}T_{\rm S}/T_{\rm C}$, and noise power is $N_0/T_{\rm C}$, resulting in an SNR of $\overline{P}T_{\rm S}/N_0 = E_{\rm B}/N_0$. This can be interpreted as suppressing the (wideband) noise by a factor $T_{\rm S}/T_{\rm C}$.

However, there are two main drawbacks to such a simple scheme:

1. The peak-to-average ratio (crest factor) of the transmitted signal is very high, giving rise to problems in the design of transmit and receive circuitry.
2. The scheme is not robust to different types of interference. In particular, there are problems with multi-access interference. In other words, what happens if the desired TX sends a 0 such that the signal would arrive at $t + \tau_{\rm run,desired}$, while another user sends a 0 that arrives at $t + \tau_{\rm run,interfere}$? It can happen that this interfering pulse arrives precisely at the time the RX expects a pulse representing a 1 from the desired user. More exactly, this occurs when $t + \tau_{\rm run,interfere} = t + \tau_{\rm run,desired} + T_{\rm d}$. Since it is very difficult to influence the runtimes from different users in a highly accurate way, such a situation occurs with finite probability. We term it "catastrophic collision," since the RX has no way of making a good decision about the received symbol, and error probability will be very high; compare also the catastrophic collisions in frequency-hopping systems (see Section 19.1).

In order to solve these problems, we need to transmit multiple pulses for each symbol. The first idea that springs to mind is to send a regular pulse train, where the duration between pulses is $T_{\rm f}$. This solves the peak-to-average problem. However, we can easily see that it does not decrease the probability of catastrophic collisions: if the first pulse of the train collides with a pulse from an interfering pulse train, then subsequent pulses collide as well. A solution to this problem is provided by the concept of time-hopping impulse radio, described in the following section.

19.4.2 Time Hopping

The basic idea of Time-Hopping Impulse Ratio (TH-IR) is to use a sequence of *irregularly spaced* pulses to represent a single symbol. More precisely, we divide the available symbol time into a number of so-called *frames* of duration $T_{\rm f}$ and transmit one pulse within each frame (see Figure 19.9).[12] The key idea is now to vary the position of the pulses *within* the frames. For example, in the first frame, we send the pulse in the third chip of the frame; in the second frame, we send the pulse in the eighth chip of the frame, and so on. The positions of the pulses within a frame are determined by a pseudorandom sequence called a "time-hopping sequence." Mathematically speaking, the transmit signal is

$$s(t) = \frac{\sqrt{E_{\rm S}}}{\sqrt{N_{\rm f}}} \sum_i \sum_{j=1}^{N_{\rm f}} g\big(t - jT_{\rm f} - c_j T_{\rm C} - b_i T_{\rm d} - iT_{\rm S}\big) = \frac{\sqrt{E_{\rm S}}}{\sqrt{N_{\rm f}}} \sum_i p(t - b_i T_{\rm d} - iT_{\rm S}) \tag{19.37}$$

where $g(t)$ is the transmitted unit energy pulse of duration $T_{\rm C}$, $N_{\rm f}$ is the number of frames (and therefore also the number of pulses) representing one information symbol of length $T_{\rm S}$, and b is the information symbol transmitted. The time-hopping sequence provides an additional timeshift of $c_j T_{\rm C}$ seconds to the jth pulse of the signal, where c_j are the elements of a pseudorandom sequence, taking on integer values between 0 and $N_{\rm c} - 1$ (see also Figure 19.9). The RX performs matched filtering (matched to the total transmitted waveform $p(t)$), and samples this output at times $t = T_{\rm s} + \tau_{\rm run}$, and $t = T_{\rm s} + \tau_{\rm run} + T_{\rm d}$. A comparison of the sample values at these two times determines which sequence had been sent. As in the case of CDMA, the matched filter operation can also be interpreted as a correlation with the transmit sequence.

This TH-IR has the same suppression of noise as the "simple" IR system – namely, $T_{\rm S}/T_{\rm C}$; however, now the gain comes from two different sources. One part of the gain stems from the fact that the RX observes the noise only over a short period of time – i.e., the same type of gain as in the simple IR. Its value is now $T_{\rm f}/T_{\rm C}$ and is thus smaller than in simple IR. The other type of gain stems from

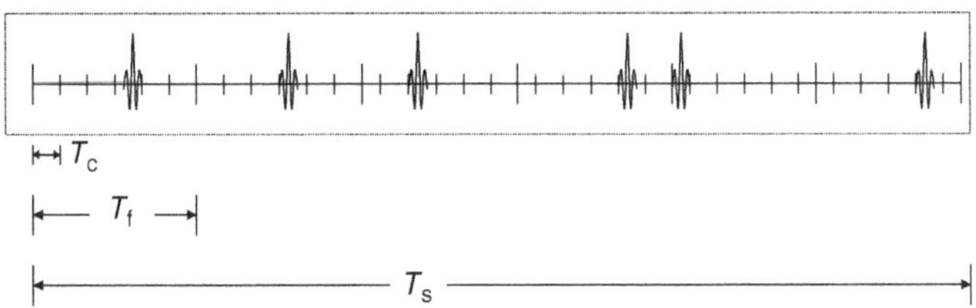

Figure 19.9 Transmit waveform of time-hopping impulse radio for one symbol, $p(t)$, indicating chip, frame, and symbol duration.

---

[12] Note that the notation "frame," while established in the IR literature, can give rise to confusion. In IR, one data symbol contains several frames. For TDMA, Packet Reservation Multiple Access (PRMA), and in LTE and 5G-NR, a block of data is also often called a "frame"; when used in this context, a frame contains several symbols.

combining the pulses in the different frames: the desired signal components from the different frames add up coherently, while the noise components add up incoherently. Taken together, those two gains provide a total gain of T_S/T_C.

Now, what is the advantage of this approach, compared with a regular pulse train? Different users can be assigned *different* time-hopping sequences. As we will show below, these sequences can be constructed in such a way that pulses collide only in a few (N_{collide}) frames, but not the others. And as the RX correlates the incoming signal with the time-hopping sequence of the desired user, it sees interference only in those frames where there is actually a collision. This leads to a suppression of interference by a factor N_{collide}/N_f.

We have already mentioned that signals between desired user and interferer are not synchronized, and thus can have an arbitrary timeshift against each other. Thus, time-hopping sequences are constructed according to the following criterion: irrespective of the relative shift between the different sequences, the number of collisions between pulses must not exceed a threshold λ (usually, we choose $\lambda = 1$).[13] This way the system designer does not have to worry about runtime effects or synchronization between users; a good suppression of Multiple Access Interference (MAI) is always guaranteed. Designing such sequences is difficult, especially if we want a large number of sequences with small collisions; exhaustive computer searches are often the best method.

Example 19.3 *The time-hopping sequence of a TH-IR system with $N_f = 6$ is [1, 2, 4, 6, 3, 5]. Find another hopping sequence that has at most one collision for arbitrary shifts.*

A systematic search reveals that the sequence [1, 3, 5, 2, 6, 4] fulfills the requirements; this is shown in Figure 19.10.

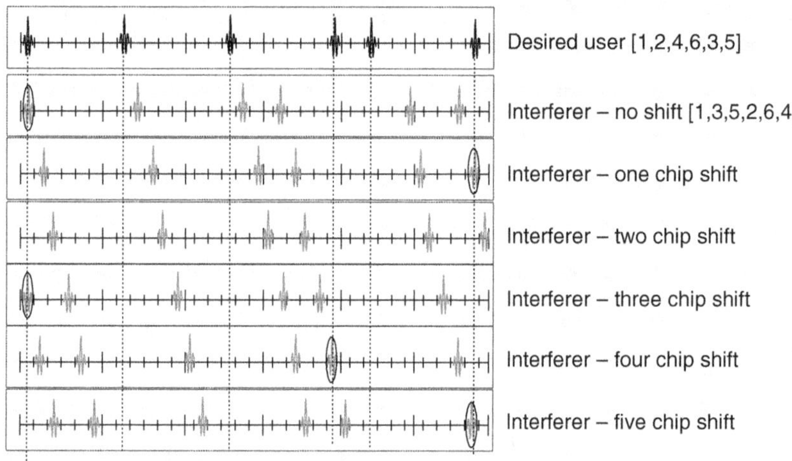

Figure 19.10 Interference between two users, for all possible (integer) shifts between the two users. Circles around a pulse indicate collisions.

Impulse radio can also be given a different, very useful, interpretation: it is a direct-sequence CDMA system (like the one in Section 19.2), where the spreading sequence has a large number of 0s, and a small number of 1s. Compare this with the conventional DS-CDMA systems, where there is an almost equal number of +1s and −1s. The interpretation of TH-IR as a DS-CDMA system allows the adaptation of many research results of conventional DS-CDMA to IR.

Finally, we note that TH-IR can be used not only in conjunction with pulse position modulation but also with pulse amplitude modulation. Furthermore, the polarity of transmitted pulses within a symbol can be randomized, so that the transmit signal then reads:

$$s(t) = \frac{\sqrt{E_S}}{\sqrt{N_f}} \sum_i b_i \sum_{j=0}^{N_f-1} d_j g\left(t - jT_f - c_j T_C - iT_S\right) \tag{19.38}$$

where each pulse is multiplied by a pseudorandom variable d_j that can take on the values +1 or −1 with equal probability. Such a polarity randomization has advantages with respect to spectral shaping of the transmit signal. For such a system, the resemblance to DS-CDMA is even more striking: the spreading sequence is now a *ternary* sequence, with an (almost) equal number of +1's and −1's, and a large number of 0's in between. The suppression of interference now occurs not just because of a small number of pulse collisions but also because the interference contributions from different frames (but the same symbol) might cancel out.

19.4.3 Impulse Radio in Delay-Dispersive Channels

Up to now, we have discussed IR in AWGN channels. We have found that the TX as well as the RX can be made very simple in these cases. However, TH-IR almost never works in AWGN channels. The purpose of such a system is the use of a very large bandwidth

---

[13] It is interesting that this problem of constructing good time-hopping sequences has strong similarities to constructing good frequency-hopping sequences for FH spread systems (see Section 19.1).

(typically 500 MHz or more), which in turn implies that the channel will certainly show variations over that bandwidth. It is then necessary to build a RX that can work well in a dispersive channel.

Let us first consider coherent reception of the incoming signal. In this case, we need a Rake RX, just like the one discussed in Section 19.2.3. Essentially, the Rake RX consists of multiple correlators or fingers. In each of the fingers, correlation of the incoming signal is done with a delayed version of the time-hopping sequence, where the delay is equal to the delay of the MPC that we want to "rake up." The output of the Rake fingers is then weighted (according to the principles of maximum-ratio combining or optimum combining) and summed up. Because Impulse Radio (IR) systems usually use very large bandwidth, they always use structures whose number of fingers is smaller than the number of available MPCs (SRake, PRake, see Section 19.2.3). In order to get reasonable performance, it might be required to have 20 or more Rake fingers even in indoor environments. Differentially coherent (transmitted reference) or noncoherent schemes thus become an attractive alternative.

In order to understand the principle of *Transmitted Reference* (TR) schemes, remember that the ideal matched filter in a delay-dispersive channel is matched to the convolution of the time-hopping sequence with the impulse response of the channel. For coherent reception, we first have a filter matched to the time-hopping sequence, followed by the Rake RX, which is matched to the impulse response of the channel. A so-called TR scheme creates this composite matched filter in a different way. A TR TX sends out two pulses in each frame: one unmodulated (reference) pulse, and, a fixed time T_{pd} later, the modulated (data) pulse. The sequence of reference pulses is convolved with the channel impulse response when it is transmitted through the wireless channel; this signal thus constitutes a noisy version of the system impulse response (convolution of transmit basis waveform and channel impulse response). In order to perform a matched filtering on the data-carrying part of the received signal, the RX just multiplies the received signal by the (suitably delayed) received reference signal and integrates the result.

Let us now have a look at the mathematical expression of the transmit signal for one symbol:

$$s(t) = \frac{\sqrt{E_S}}{\sqrt{2N_f}} \sum_i b_i \sum_{j=0}^{N_f} d_j \big[g(t - jT_f - c_j T_C) + b_i \cdot g(t - jT_f - c_j T_C - T_{pd}) \big]. \tag{19.39}$$

Inspection shows that correlation with the received reference signal can be done by just multiplying the total received signal by a delayed (by time T_{pd}) version of itself and integration. To be more precise: the first step at the RX involves filtering by a receive filter $h_R(t)$; this filter should have a wide enough transfer function not to introduce any signal distortions and just limit the available noise. The filtered received signal $\hat{r}(t)$ is multiplied by a delayed version of itself, and integrated over a finite interval T_{int}, which should be long enough to collect most multi-path energy, but short enough not to collect too much noise energy.

TR schemes have the advantage of being exceedingly simple (though implementation of the delay in the RX may be nontrivial). On the downside, they show poorer performance than coherent schemes for two reasons: (i) reference pulses waste energy, in the sense that they do not carry any information (this results in a 3-dB penalty); (ii) the reference part of the signal is noisy, as is the data-carrying part of the signal. Multiplication of the two signals in the RX gives rise to noise–noise cross-terms that worsen the SNR. Now, remember that IR is a spread spectrum system, so that the SNR of the received signal is usually negative. Noise–noise cross-terms can thus become large.

Finally, we note that noncoherent detection can be an attractive alternative as well. This is especially true if only a single user is to be served. However, noncoherent RXs have problems with multi-access interference. Due to multi-path propagation, energy is dispersed over several adjacent chips. Thus, energy detection will see much more interference (as exemplified in Figure 19.11).

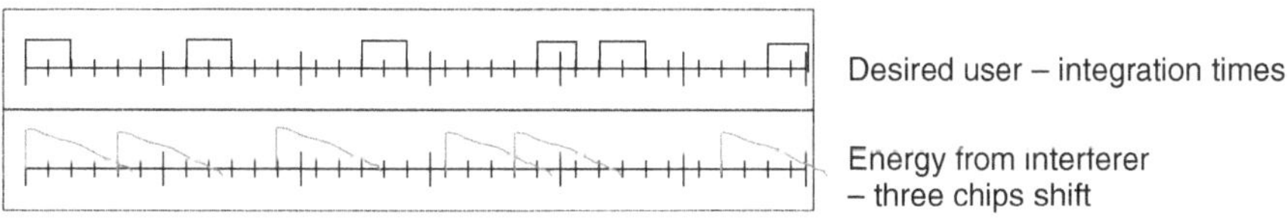

Figure 19.11 Interference from a delay-dispersed interferer signal (lower row) to the desired signal, whose integration time (two-chip durations per frame) is sketched in the upper row.

Further Reading

An overview of spread spectrum communications in general can be found in the classic monographs of [Dixon 1994] and [Simon et al. 1994], which provide good coverage of FH and DS systems. Books that are more tuned to cellular systems, especially CDMA, are [Glisic and Vucetic 1997, Goiser 1998, Lee and Miller 1998, Viterbi 1995, and Ziemer et al. 1995]; the overview paper [Kohno et al. 1995] gives a shorter description. [Scholtz 1982] gives an overview of the history of the spread spectrum (though more from a military perspective, since cellular applications were not yet considered at the time of that article). [Milstein 1988] discusses the interference rejection capabilities of various spread spectrum techniques.

Frequency-hopping codes are designed in [Maric and Titlebaum 1992]. Estimates for the capacity of CDMA systems were first published in the widely cited paper of [Gilhousen et al. 1991]. The effect of multi-path on CDMA – namely, the use of Rake RXs – is reviewed in [Goiser et al. 2000], [Swarts et al. 1998], and [Win and Chrisikos 2000]. Other papers on Rake reception include [Holtzman and Jalloul 1994] and [Bottomley et al. 2000], among many others. For synchronization aspects, the papers by [Polydoros and Weber 1984] are still interesting to read. Spreading codes are reviewed in [Dinan and Jabbari 1998]. An authoritative description of sequence design (for radar as well as communications) is [Golomb and Gong 2005]; sequence design and particular cross-correlation properties are also discussed in the classical paper by [Sarwate and Pursely 1980]. While the Gaussian approximation is widely used for intercell interference, it can give significant deviations from reality when shadowing is present; a more refined model is presented in [Singh et al. 2010].

Time-hopping impulse radio was pioneered by Win and Scholtz in the 1990s. For a description of the basic principles, see [Win and Scholtz 1998], and more detailed results in [Win and Scholtz 2000]. An overview of the different aspects of ultrawideband system design can be found in the monographs [Shen et al. 2006, Ghavami et al. 2006, Reed 2005, diBenedetto et al. 2005, Roy et al. 2004]; TR RXs and other simplified RX structures are discussed in the overview paper of [Witrisal et al. 2009].

For updates and errata for this chapter, see https://wides.usc.edu/students.html#textbooks

Exercises

See Sec. 36.19 of Exercises.pdf at wiley.com/go/molisch

20

Resource Allocation: Scheduling, Power Control, and Admission Control

We have seen in Chapters 18 and 19 that the spectral resources can be shared between different users. In the current chapter, we will elaborate on that concept, and explain *what amounts of resources* should be assigned to the different users at any given time, and *which specific resources* that should be – as we will see below, not all subbands (or timeslots) are of equal value to a particular user. This allocation of time/frequency resources is also known as scheduling, and good algorithms for these tasks are required for performance optimization of wireless transmission systems. Scheduling, together with power allocation, i.e., how much transmit power can be used by each user, is also known more generally as *resource allocation*. While it did play some role in earlier (voice-centric) systems, it has become particularly important in modern data systems, and determines the *Quality of Experience* (QoE) for a variety of services.

A topic intimately related to resource allocation is *admission control*, i.e., determining how many users are admitted to be served in a system. Obviously, as the number of users grows, fewer resources can be assigned to each user, and thus the QoE will decrease. If too many users would have unacceptable QoE, the system will simply refuse to admit further users into the system.

20.1 Rate and Latency Requirements for Different Kinds of Traffic

The "optimum" allocation of resources is very much dependent on the objective function that should be maximized – should it be the highest total data throughput, or should it maximize the minimum data rate that every user obtains, or some other function? To find such meaningful objective functions, we have to first analyze the nature of the data traffic: the criteria will be different, e.g., for speech communication and data file transmission.

20.1.1 Speech, Video, and Data

We first consider a few practical examples to show the differences between various kinds of traffic (see also Chapter 2). Consider, first, a speech communication system. In this case, the traffic is generated at a constant rate and needs to arrive at the Receiver (RX) with little delay (latency). Thus, a scheduler has to make available a constant data rate to the user, with the requirement that this rate matches the rate required by the speech codec. Lower instantaneous data rates lead to bad speech quality or outages in the communication, while data rates beyond the requirements of the speech codec are wasted (compare Chapter 24).

The opposite situation occurs during file downloading: the traffic is generated only once (when the request for a particular file arrives at the device that stores it, which starts transmission). The transmission is successful when all bits have been transmitted. In this situation, it is the *cumulative* data rate over the transmission time that counts; it does not matter whether at a particular time the rate is very high or very low. The scheduler might thus assign resources in such a way that the data rate shows strong variations, if such an assignment is beneficial from an overall throughput point of view.

The scheduling requirements for *video streaming* lie between those extremes: firstly, the exogenous data rate created by video is time-varying, since at different times different amounts of compression are feasible – for example, shots of a static landscape can be compressed much more than action scenes, see Chapter 25. Secondly, due to the particular implementation of video transmission, a certain amount of mismatch between the current exogenous rate and the scheduled data rate is acceptable: a RX has a playback buffer, and when the video transmission starts, this buffer is filled over the time of a few seconds with the first moments of the video. Subsequently, while the user views the video, and thus empties these bits from the buffer, new bits, representing the next moments of the video, are transmitted and fill the buffer. As long as the playback buffer is not empty, there are no *stalls*, i.e., interruptions of the playback. Consequently, the system can handle some variations of the data rate (which result in the playback buffer becoming more or less full), but there are limits to how long a low data rate can be tolerated. From all these examples we see that the requirements on the scheduler differ significantly depending on the application.

Wireless Communications: From Fundamentals to Beyond 5G, Third Edition. Andreas F. Molisch.
© 2023 John Wiley & Sons Ltd. Published 2023 by John Wiley & Sons Ltd.
Companion website: www.wiley.com/go/molisch/wireless3e

Let us next turn to a general system model for data transmission, see Figure 20.1. Assume that the time is divided into slots, and during each slot, a data packet (with time-varying amount of data) is received/transmitted. At the Transmitter (TX), data intended for a particular user are arriving from an exogenous source at a particular (possibly time varying) rate $a(t)$.[1] These data are written into the *transmit queue* $Q(t)$. Data in the queue are then sent, depending on the scheduling/resource allocation at a (possibly time-varying) rate $R(t)$.[2] Note that $R(t)$ need not be identical to $a(t)$; if the exogeneous source rate $a(t)$ is smaller than the allocated transmission data rate $R(t)$, then the size of the queue $Q(t)$ decreases during a transmission time interval; if the queue is empty at time t_i and only the new data from the exogeneous source need to be transmitted, assigning resources to achieve $R(t_i) > a(t_i)$ is a waste, since the maximum achievable data rate is $a(t)$. If $a(t_i) > R(t_i)$, then the size of the transmit queue $Q(t)$ increases at $t = t_i$. Also note that the transmit queue need not stick to transmitting the packet sizes that it received from the queue. Rather, the process of *fragmentation/ defragmentation* can adapt the transmission data rate to the channel conditions – for example, a large packet from the source is divided into five packets of smaller size, each of which is sent on the channel in one time slot.

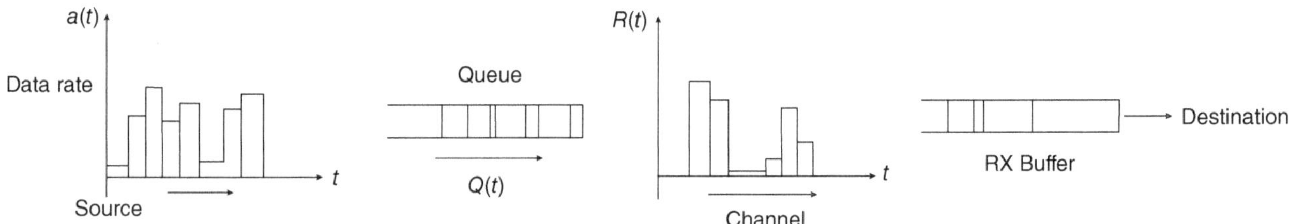

Figure 20.1 Queue setup for scheduling in wireless systems.

For theoretical considerations, we often aim for *queue stability* such that the size of the transmit queue should not become infinitely large; in other words, the transmit data rate should be at least as large as the exogeneous arrival data rate *on average*.[3] Also note that the size of the queue is a measure for the delay experienced by a data packet; for example in a FIFO (First-In-First-Out) queue, a long queue means that many packets have to be transmitted before the one created now by the exogeneous source can be sent. At the RX, the arriving data are either written into another buffer (often called *playback buffer*, a name that derives from music and video streaming) or directly written to the output (e.g., the screen of the device).

As we will discuss below, the assignment of resources to different users can occur on an *opportunistic* basis, i.e., when resources can be used most efficiently, or on a *persistent* basis. The former case is most efficient from the point of view of utilization of resources. The latter is most meaningful if the particular application is known to require a constant data rate, and will transmit over a period of time. Such persistent resource allocation was actually the first widely used scheme, since it applies for voice telephony, and particular frequencies were assigned to each user, resulting in a perfectly persistent resource allocation.[4] A variant of this is a semi-persistent scheduling, where a particular resource is assigned for a time that is shorter than a call or a data session.

20.1.2 Traffic Characteristics

Generally speaking, different types of traffic distinguish themselves by the following (compare also Section 1.3):

- *Absolute data rate*: as we have seen in Chapter 2, different types of traffic might require data rates that are orders of magnitude different from each other. In most cases, the mean required data rate does not have a big impact on the *type* of scheduling, but just influences the amount of the required resources, and the final (numerical) result.
- *Variability of data rate*: different types of services create different temporal variability of their exogenous data rate, with speech being fairly constant, video showing higher variability, and websurfing creating strong but short bursts. There are also a number of other services that can create constant data rates, such as the transmission of sensing and control data in automated production machinery (essentially, the data sent on a control loop), or short, rare bursts (e.g., environmental sensing data).
- *Admissible latency*: this describes the maximum amount of delay that data can have in order to be still useful. These constraints can vary widely. For signals that are part of control loops (such as tactile feedback in video games, or feedback in industrial power plants), maximum latency might be on the order of 1 ms. For speech communication and real-time video (such as video conferencing), 100 ms is a typical maximally admissible value. For data files, there is usually no specific upper limit of the latency, which greatly increases the flexibility of scheduling.

---

[1] In the case of relaying and multi-hop transmission, the data might be arriving from another node instead of, or in addition to, exogeneous sources. This will be discussed more in Chapter 23.

[2] To ease discussion, we henceforth quantize time into transmission time intervals of length T_{TTI}, so that the number of bits transmitted at time t_i is $R(t_i) T_{\mathrm{TTI}}$, and $R(t_i)$ is the average data rate between times $[t_i, ... t_i + T_{\mathrm{TTI}}]$.

[3] If the buffer that stores the transmit queue has a finite size (this is always fulfilled in practice), then buffer overflows might occur even if the queue is stable, leading to loss of data.

[4] In OFDMA systems, it is particular timeslot/subcarrier combinations that are assigned to a particular user.

- *Duration of transmission*: in many situations, we assume that the transmission of a "commodity" (file, video stream, etc.) lasts for a very long time (much larger than the scheduling intervals or times during which the channel changes), which is thus approximated as infinite. However, there are also a number of situations where the transmission is short (e.g., a text message, a small data file, a short voice call, transmission of sensor data in the internet of things). As we will see later on, in many situations the scheduling is heavily influenced by the queue lengths at the TX, where commodities with larger queue lengths get preferential treatment. Short data files, which by definition cannot have a large queue length, might be disadvantaged in such a scenario, so that appropriate countermeasures have to be taken.
- *Playback buffer*: a special case arises for wireless on-demand video streaming (nonrealtime video). The traffic is generated continuously, but need not be played back immediately. Rather, as outlined above, a playback buffer can smooth out mismatches between the exogenous data rate and the scheduled transmission rate over the channel, and only a prolonged insufficient data rate leads to outages (rebuffering). Thus, the existence of, and the size of, a playback buffer is an important descriptor.

While real-life traffic statistics can be very complicated, theoretical treatment of system analysis often makes idealized assumptions to enable tractable formulations. The two most common are:

- *Infinite-backlog traffic*: under this assumption, the transmit queue is never empty, no matter what transmit data rates have been assigned to the particular TX in the past. Thus, it can never happen that a particular TX might not have data to transmit. As a consequence, the scheduling is essentially determined by the channel state and the notion of *fairness* between the users.
- *Poisson arrival process*: here it is assumed that the generation (start) times of data packets (or possible whole voice calls) form a *Homogeneous Poisson Point Process* (HPPP). This means that the generation of packets is independent at different TXs and that the times between generations (inter-generation times) are exponentially distributed random variables. The average generation rate is the parameter of this process. Depending on the packet generation rate and transmission data rates, queues for the individual users might be full (with different lengths) or empty.

20.1.3 Optimization Goals

Resource allocation is ultimately an optimization problem, whose solution first requires a criterion for *what* we actually want to optimize. Typically, we want to maximize the sum of the "satisfactions," or *utilities*, of the different users; the utility in turn is a function of data rate, delay, and errors. More concretely, the following criteria are often used, and algorithms are specifically designed to satisfy them:

- *Spectral efficiency*: the simplest way to describe spectral efficiency in a multi-user setting is the sum rate. For theoretical analysis, it is common to assume that each user can transmit with the Shannon rate $\log_2(1 + \text{SINR})$, though for algorithms used, e.g., in real-world 5G Base Stations (BSs), the limitations of the physical layer, the overhead for communicating channel state information, effort for retransmissions, etc., all have to be taken into account. An alternative criterion for maximizing spectral efficiency is the number of users that can be supplied, where typically each user has to fulfill a certain minimum quality criterion.
- *Fairness:* maximizing the sum rate typically leads to a "winner takes all" solution, where the users closest to the BS, which have the highest Signal-to-Interference-and-Noise Ratio (SINR), get assigned all resources. This is an unfair resource allocation, where cell edge users (who are already disadvantaged, see Chapters 3 and 21) get fewer or no resources assigned. To ensure a better balancing of the different users, alternative optimization criteria that provide various forms of fairness can be used. The most important fairness criteria are *resource fair*, where all users get the same amount of resources, and *rate fair*, where resources are assigned in such a way that the rate for all users is the same. We will see below that the implications of fairness are different for static and dynamic systems. In a static system, "instantaneous" and "on average" fairness are the same thing, so that, for example, resource fairness means that within a short time window, all users get the same amount of resources assigned (e.g., through a round-robin scheduler). On the other hand, in a dynamic system, we could either enforce instantaneous fairness (which might again lead to round-robin), or "on-average" fairness, such that a user might get preferential treatment at one time, but will then receive fewer resources in the future. We will see that this latter case allows opportunistic resource allocation, leading to higher spectral efficiency in time-varying channels.
- *Quality of Service (QoS) provisioning*: the above schemes stressing spectral efficiency and fairness fall into the category of *best effort* services: depending on the number of users and the channel conditions, the assigned resources and resulting data rates might give better or worse quality; there is no specific *guarantee* for the utility. This is not acceptable for many real-time services such as voice calls and video conferences. There, a certain minimum QoS should be guaranteed. Different definitions for QoS exist (minimum data rate; maximum latency of each data packet, etc.). The QoS constraints can be combined, e.g., with spectral efficiency goals. A typical example would be to maximize the sum throughput in the system, subject to a guarantee that each user has a certain minimum data rate. We also note that a system might carry traffic with QoS requirements and best-effort traffic at the same time, so that only a part of the users (those with particular needs, or those who particularly pay for the privilege), get guaranteed QoS. One practically important case for QoS occurs in voice networks, where users require a fixed data rate, and we need to ensure that the system has enough resources for providing for them (see Section 20.6).
- *Priority*: Related to the topic of QoS is the handling of priority of different services. Besides the services that need to get a guaranteed rate, there might be other services that receive resources on a best-effort basis, but which have different priorities. In this case, the

easiest way is to assign resources to the services is in sequence of priority; however, this can lead to starvation of lower-priority services. Alternatively, each service can be assigned a "priority rate" that might provide some QoS, though not necessarily the best possible one (e.g., video streaming with low resolution, e.g., 480p). Then the system first assigns resources up to the priority rate (in sequence of priority); that way every service gets at least some rate. If there is capacity left over, it is distributed to the services in sequence of priority for transmission of additional data that might enhance the QoS.

- *Energy efficiency*: a topic that has gained increased importance in the past years is the energy efficiency of the transmission system. For Internet of Things (IoT) and sensor network applications, energy efficiency might determine lifetime of the nodes, while for the downlink of cellular systems, energy efficiency might impact the cost of operation for the network provider. In either case, a trade-off between the spectrum efficiency/QoS and energy efficiency might be desirable for a scheduler to strive for.

20.1.4 Network Utility Function

In many cases, we wish to find a scheduling of the users that maximizes the *network utility*, i.e., the sum of the utilities over all the users. For this purpose, we make the following assumptions:

- There are K users that all should be served.
- The system has a set of time-frequency resources available that can be assigned to the different users. The quantization of those time/frequency (and possibly power) resources is independent of the data packets to be transmitted; in other words, a large packet may be fragmented into smaller units that are transmitted on different time/frequency resources.
- There is no notion of blockage – data packets might be delayed longer if the transmission queue serving them is very long, but they *are* delivered if the system is stable (for treatment of blockage, see Section 20.6).
- The optimization is for the sum utility of all users; there is no guarantee on the individual utilities. Such guarantees could be introduced if desired, but we do not consider this case, to keep the treatment reasonably simple.

Under the assumption that the utility U is a function of the data rate R_i only, the optimization goal thus becomes

$$\max_{\mathbf{R}} \sum_{k=1}^{K} U(R_k) \quad \text{with} \quad \mathbf{R} \in \mathcal{R} \tag{20.1}$$

where $\mathbf{R} = [R_1, R_2, \ldots]^T$ and the maximization is over the *rate region* $\mathcal{R}$: depending on the channel condition and the transmission scheme, certain combinations of data rates for individual users are possible (for details see Section 18.2).

The function $U(R)$ describes how much a user benefits from the assigned data rate R. We commonly assume that $U(R)$ is a concave function, i.e., that increasing the rate by a factor x increases the utility by *less than* a factor x. This is in line with the user experience for most common services: e.g., in video streaming, doubling the rate might allow a higher-resolution picture, but the user does not feel that the benefit has doubled. In other words, two users that can both receive videos in standard resolution are, taken together, happier than one user that receives an HD video and another user that gets nothing.

Any rate assignment should be Pareto-efficient, i.e., it should not be possible to further increase the utility of one user without harming any other user. The question of how to trade-off utility improvements between different users is more difficult and depends on the utility function itself. A common family of utility functions, sometimes called *alpha-fair*, is

$$U(R) = \begin{cases} \ln(R) & \alpha = 0 \\ \dfrac{1}{\alpha} R^\alpha & \alpha \leq 1, \alpha \neq 0. \end{cases} \tag{20.2}$$

We consider that as a consistent "family" because $\ln(R)$ is the limit of $(R^\alpha - 1)/\alpha$ for $\alpha \to 0$; since that differs from R^α/α only by a constant, a scheduling that optimizes one also optimizes the other. Generally, α controls the "fairness" of the distribution. A value of $\alpha = 1$ means that increasing the rate by a given *amount* (not factor) for one user and decreasing it for another user by the same amount does not change the sum utility. Using this utility in Eq. (20.1) thus leads to a *sum-rate maximization*. On the other hand, for $\alpha = 0$, increasing the rate for user A by *a factor* (not an amount) x and decreasing the rate for user B by at least a factor x leaves the sum utility unchanged. Consequently, the optimal rate vector $\mathbf{R}^*$ (in this chapter, superscript * indicates optimality) has the property that deviating from it decreases the sum of the *normalized* rates, i.e., any small perturbation δR_k leads to:

$$\sum_{k=1}^{K} \frac{\delta R_k}{R_k^*} \leq 0. \tag{20.3}$$

Using this utility function results thus in the so-called *proportional-fair scheduling*. Finally, $\alpha \to -\infty$ leads to *max-min fairness*. Note that choice of the fairness parameter a may not only be determined by the type of file being transmitted but also possibly the history of the users, the amount of money they pay for the service, etc. Figure 20.2 shows some examples for the shape of utility functions.

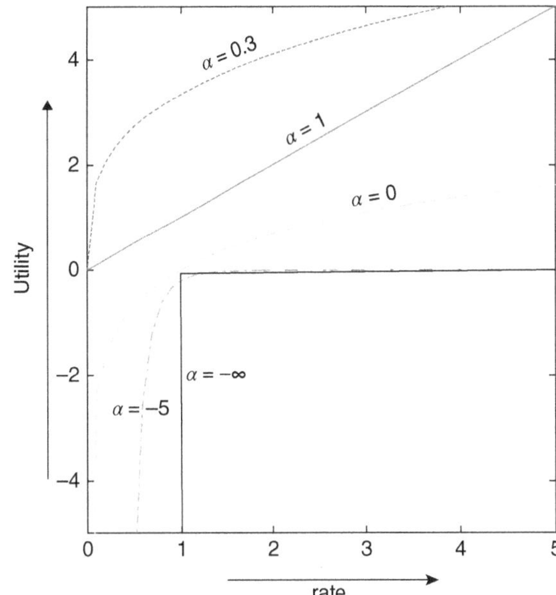

Figure 20.2 α-Fair utility functions with $\alpha = 1, 0.3, 0, -5, -\infty$. Color version available at wiley.com/go/molisch/wireless3e.

20.1.5 Power Control

While resource allocation mostly deals with assigning *time-frequency* resources to particular data streams, a wireless system has other resources as well, in particular spatial resources (beams, antennas) that will be discussed in Chapter 22, and *power*. Judicious power allocation is thus an important part of resource allocation.

Just like for time-frequency resources, power control can serve to achieve different purposes, such as optimization of the network utility function versus achieving target QoS. Furthermore, the constraints are often different in the uplink and the downlink: in the downlink, the BS has a sum power constraint for transmitting signals. In other words, the power amplifier (and sometimes the frequency regulators) limit the total power that the BS can transmit. In the uplink, the User Equipments (UEs) introduce a per-user power constraint – in other words, the power must be below the admissible value for each separate TX.

20.2 Dichotomy of Resource Allocation

There is a large number of resource allocation problems, resulting in the following core distinctions:

- *Infinite backlog versus random arrivals*: these two assumptions (Section 20.1.2) lead to very different mathematical frameworks.
- *Orthogonal versus nonorthogonal users*: in the case of single-cell Orthogonal Frequency Division Multiple Access (OFDMA),[5] users scheduled on different resources are orthogonal, i.e., do not interfere with each other. The rate that each user achieves thus depends only on the particular resources assigned to this user. In Code Division Multiple Access (CDMA) (and also multicell systems, see Chapter 21), the data rate for each user is determined by the ratio of signal power to noise-plus-interference power, where the interference is determined by the resources assigned to the *other* users. This obviously complicates the mathematical formulations.
- *With versus without power control*: in the simplest case, the assigned resources are only time/frequency (in the case of OFDMA) or code (in the case of CDMA) resources, while the transmit powers to/from each user are fixed. Additional flexibility is obtained by adding power control.
- *Frequency-flat and time-invariant channels versus selective channels*: in frequency-flat, time-invariant (over the period of the scheduling) channels without interference, it does not matter which particular time/frequency resources are assigned to each user; all that matters is the percentage of resources that are assigned. If the channels are selective, it becomes important which particular resources are assigned, because some are better for a particular user than others. Again the latter case leads to a more complicated mathematical formulation.
- *Achieving target rate versus optimizing network utility function*: in some cases, all we wish to achieve is to provide a certain QoS (data rate) to each user; this usually maps onto providing a minimum SINR for each user. However, in most cases, we wish to maximize a function that reflects the overall network utility function.
- *Uplink versus downlink*: these two cases differ in the constraints on the transmit power (sum power constraint for the downlink versus per-user power constraint in the uplink), as well as the ease with which centralized resource allocation can be done. It must be noted, however, that for the dominant cellular standards, LTE (Chapter 31) and New Radio (NR) (Chapter 32), resource allocation for both the uplink and the downlink is done centrally, at the BS.

---

[5] And also TDMA/FDMA; for ease of notation we henceforth just talk about OFDMA.

Just using the above dichotomy (further finer specialization is possible), there are 32 different cases, many of which require different mathematical tools to tackle. Obviously, this book cannot treat all of them, but rather selects some important examples. In particular, we will discuss

- Maximization of network utility function with infinite backlog in OFDMA
  - No power control
    - Nonselective channels: Section 20.3.1
    - Selective channels: Section 20.3.2
  - With power control
    - Nonselective channels: Section titled "Power Control in Nonselective Channels"
    - Selective channels: Section titled "Power Control in Selective Channels"
- Maximization of network utility function with infinite backlog in CDMA
  - With power control, in selective or nonselective channels: Section 20.4
- Throughput maximization for random packet arrivals: Section 20.5

These three main types of problems all require different mathematical tools. Solutions are often difficult to obtain, especially when further constraints (e.g., maximum latency constraints) are introduced. For this reason, Section 20.7 will discuss Machine Learning (ML) methods for handling resource allocation.

20.3 Resource Allocation in OFDMA with Infinite Backlog

We start out by considering resource allocation for the case where each user has an *infinite backlog*, i.e., always wants to transmit data, no matter what rates were assigned in the past. This section treats resource allocation in systems without interference (orthogonal users), for example, OFDMA.

20.3.1 Network Utility Maximization in Nonselective Channels

We wish to maximize the network utility function defined in Section 20.1.3 in the case where the channel is nonselective in time and frequency. Assume that during one "scheduling interval" N time/frequency resources are available (and let $N \gg K$). As we are dealing with nonselective channels, it does not matter whether resources are divided in time or frequency; the discussion applies to Time Division Multiple Access (TDMA), Frequency Division Multiple Access (FDMA), and OFDMA. For concreteness, we consider the downlink of a cellular system with TDMA, where different timeslots are assigned to different users. Finally, let the transmit power be fixed, i.e., no power control, so that each user has a Signal-to-Noise Ratio (SNR) γ_k that is determined by the propagation channel.

In such a setup, all the scheduler can impact is what *fraction* β_k of the N radio resources are assigned to each user.[6] Consider the downlink, such that the available power of the BS is distributed uniformly over time and frequency. Then the goal is to maximize the sum utility. For α-fairness, assuming that the SNRs are independent of the β_k and assuming that the achievable spectral efficiency for each user equals the Shannon capacity, the optimization problem becomes

$$\max_{\beta} \sum_{k=1}^{K} \frac{1}{\alpha} [\beta_k \log_2(1 + \gamma_k)]^{\alpha}$$

$$\Sigma \beta_k = 1$$

$$\beta_k \geq 0 \quad \forall k. \tag{20.4}$$

This is a convex optimization problem with equality constraints.[7] Such optimization problems can be solved by means of the Lagrangian:

$$L(\boldsymbol{\beta}, \lambda) = \sum_{k=1}^{K} \frac{\beta_k^{\alpha}}{\alpha} [\log_2(1 + \gamma_k)]^{\alpha} + \lambda \left[1 - \sum_{k=1}^{K} \beta_k \right] \tag{20.5}$$

---

[6] In practical systems, only discrete values of $\beta = n/N$ with n integer can be realized. However, such a formulation would be much harder to solve mathematically, so that β is often assumed to be continuous for the computations, and then rounded to the nearest admissible discrete value in the end.

[7] Strictly speaking, inequality constraints apply, since the requirement is that each timeslot is assigned to *at most* one user. However, it is obvious that no timeslot should be left unassigned – since the utility function is monotonous, the best approach is to assign the timeslot to *somebody*. Note that the situation would be different in the presence of a power constraint.

where λ is the Lagrange multiplier and $\boldsymbol{\beta} = [\beta_1, \beta_2, ...]^T$. We now need to find the $\boldsymbol{\beta}^*$ and λ^* such that

$$\frac{\partial L}{\partial \beta_1}(\boldsymbol{\beta}^*, \lambda^*) = \frac{\partial L}{\partial \beta_2}(\boldsymbol{\beta}^*, \lambda^*) = ... \frac{\partial L}{\partial \beta_K}(\boldsymbol{\beta}^*, \lambda^*) = 0$$

$$\frac{\partial L}{\partial \lambda}(\boldsymbol{\beta}^*, \lambda^*) = 0.$$

(20.6)

It can be easily seen that the following set of allocations fulfills (20.6) and thus provides the optimum resource allocation:

$$\beta_k^* = \frac{[\log_2(1 + \gamma_k)]^{\frac{\alpha}{1-\alpha}}}{\sum\limits_{k=1}^{K} [\log_2(1 + \gamma_k)]^{\frac{\alpha}{1-\alpha}}}.$$

(20.7)

For different values of α, this corresponds to the following scheduling policies in nonselective channels:

- For $\alpha = 1$, i.e., *sum-rate maximization*, the optimum allocation is to give all resources to the user with the highest rate, since giving more resources to the user with higher SNR (and thus higher throughput) increases the utility more. This is intuitive but also follows mathematically from Eq. (20.7) when $\alpha \to 1$ and thus $\frac{\alpha}{1-\alpha} \to \infty$.
- For $\alpha = 0$, i.e., *proportional fairness*, each user gets the same amount of resources, i.e., $\beta_k^* = 1/K$. In other words, the scheme becomes *resource-fair*.
- For $\alpha \to -\infty$, the percentage of assigned timeslots is inversely proportional to the rate, as can be seen from inserting $\frac{\alpha}{1-\alpha} \to -1$ in Eq. (20.7). With this assignment, each user is guaranteed the same rate, i.e., we have *equal-rate assignment*.

The sum rate for those three cases are

$$\max_k \log_2(1 + \gamma_k) \geq \frac{1}{K}\sum_k \log_2(1 + \gamma_k) \geq \frac{K}{\left(\sum\limits_k \frac{1}{\log_2(1 + \gamma_k)}\right)}.$$

(20.8)

In other words, the sum rate that can be achieved with sum rate maximization (the first term) is always larger or equal to the sum rate for proportional fair (middle term), which is greater than the sum rate for max–min fairness.

Example 20.1 *Consider three users in the downlink of a TDMA cellular system with bandwidth 3 MHz, where the first has a (time-invariant) SNR of 30 dB; the second has SNR of 15 dB, and the third 0 dB. Compute the individual rates and the sum rates for the cases of sum rate maximization, proportional fairness, and equal-rate assignment.*

Solution:
The maximum rates (per unit bandwidth) $\log_2(1 + \gamma)$ that can be sustained on each of the links are 10, 5, and 1 bit/s/Hz. For sum rate maximization, all resources are assigned to the best user, so that it obtains a rate of 30 Mbit/s – this is equal to the sum rate in this case, since the other two users get no resources assigned and obtain a zero rate. For the proportional-fair case, each user is assigned 1/3 of the time (equivalently, in an FDMA system we would assign each user 1 MHz bandwidth), resulting in user rates of 10, 5, and 1 Mbit/s, and a sum rate of 16 Mbit/s. Finally, for the equal rate assignment, the time fraction assignments are $0.1/1.3 = 0.077$ for user 1, $0.2/1.3 = 0.154$ for user 2, and $1.0/1.3 = 0.769$. This leads to the same rate for each user, namely 2.31 Mbit/s, and a sum rate of 6.93 Mbit/s. As expected, the sum throughput decreases, and the minimum data rate increases, as we move from sum rate maximization to proportional fair to equal-rate assignment.

The above resource allocation can also be reformulated as weighted fair queuing,

$$\max_{\boldsymbol{\beta}} \Sigma_k w_k R_k$$

$$\sum_k \beta_k = 1$$

(20.9)

$$\beta_k \geq 0 \quad \forall k$$

where the weights w_k are determined by the fairness criterion. We will make use of this formulation in later sections. The min–max problem can also be rewritten as

$$\max_{\boldsymbol{\beta}} t$$

subject to

$$R_k(\boldsymbol{\beta}) \geq t$$

$$\Sigma_k \beta_k = 1,$$

$$\boldsymbol{\beta} \geq 0.$$

A similar approach can also be used for solving the transmit power control problem discussed in Section 20.4.

20.3.2 Network Utility Maximization in Selective Channels

In most practical scenarios, the rate for each user is time-varying (time-selective) and/or frequency selective (for ease of exposition we first treat a flat-fading, time-selective channel). This has important consequences for how to schedule in an optimum way. We will start out by describing round-robin scheduling; in this approach, the time/frequency resources are given to each user in a predetermined way, independent of the channel state. We then proceed to *opportunistic scheduling*, which schedules the users according to their channel states. For this latter case, we first give an intuitive picture and then derive a general mathematical framework that can be used to give a more mathematically exact justification of this type of scheduling.

Round-Robin

Consider the downlink (the uplink behaves similarly) where all links undergo flat Rayleigh fading; the mean SNR for the kth link is $\bar{\gamma}_k$. In a *round-robin* scheme, each user is assigned the same amount of resources and is furthermore assigned a *fixed* (periodically repeating) time/frequency resource. This system is resource-fair. Due to the time variation of the channel, the SNRs are different at different times, and the probability for user k to "see" a certain SNR γ_k is then given by the exponential distribution (see Section 5.4.3):

$$pdf_{\gamma_k}(\gamma) = \frac{1}{\bar{\gamma}_k} \exp\left(-\frac{\gamma}{\bar{\gamma}_k}\right). \tag{20.10}$$

This probability is independent from user to user and is also independent of the *number* of users K. Each user can thus realize the rate

$$\frac{1}{K} \int \frac{1}{\bar{\gamma}_k} \exp\left(-\frac{\gamma}{\bar{\gamma}_k}\right) \log_2(1 + \gamma) d\gamma. \tag{20.11}$$

The scheme is the best we can do if the scheduler does not know anything about the channel.

Opportunistic Scheduling – Intuitive Picture

However, we can do better if the scheduler knows the *Channel State Information* (CSI) for the downlink of all users. At any given point in time, chances are that some users have a good SNR, while others have a bad SNR. The sum-rate-maximizing strategy for the BS is now to communicate always with the user who has the best link, i.e., the highest instantaneous SNR. Note that in a time-invariant channel, always talking to the best user is inherently unfair as discussed in Section 20.3.1. However, in a time-variant channel, different users become the *instantaneously best* at different times, since the channel changes. As a matter of fact, in the simple case that all users have the same average SNR $\bar{\gamma}_k = \bar{\gamma}$, each user has the same probability of becoming the instantaneously best. We can then create a scheduling that *on average* is resource-fair, but exploits the time variations of the channel and at any given time assigns all resources to the instantaneously best user, see Figure 20.3.

Since the scheduler selects the best out of K channels, the multiple users act like diversity branches – without the need for additional antennas. Opportunistic scheduling thus gives rise to *multi-user diversity*; it is optimum in the sense that the sum rate is maximized. The cumulative distribution function for the SNR of the active user (i.e., the instantaneously best user) is the same as for selection diversity, Eq. (12.10).

Consider now the case that different users have different average SNRs (and thus feasible average rates), e.g., due to different distances from the BS. Thus, selecting the users with the instantaneously best channel leads to the UE closest to the BS getting selected most often, while UEs that are far away are almost never the instantaneously best, and thus are chosen much less frequently. Thus, not only the data rate but also the fraction of assigned resources for users at the cell boundary becomes lower than for UEs near the BS, making the scheduler unfair. This problem can be solved by having the BS communicate not with the *absolutely* best user, but rather with the user whose ratio of instantaneous to average rate is the highest. In that case, each user is assigned the same amount of resources if the *normalized* fading statistics of the different users are identical (e.g., all channels are Rayleigh fading), leading to resource-fair scheduling.

Multi-user diversity is mainly useful for data communications. For voice and other systems with stringent delay requirements (total delay less than about 100 ms), the long time it may take until a user becomes the best, precludes the use of "conventional" multi-user

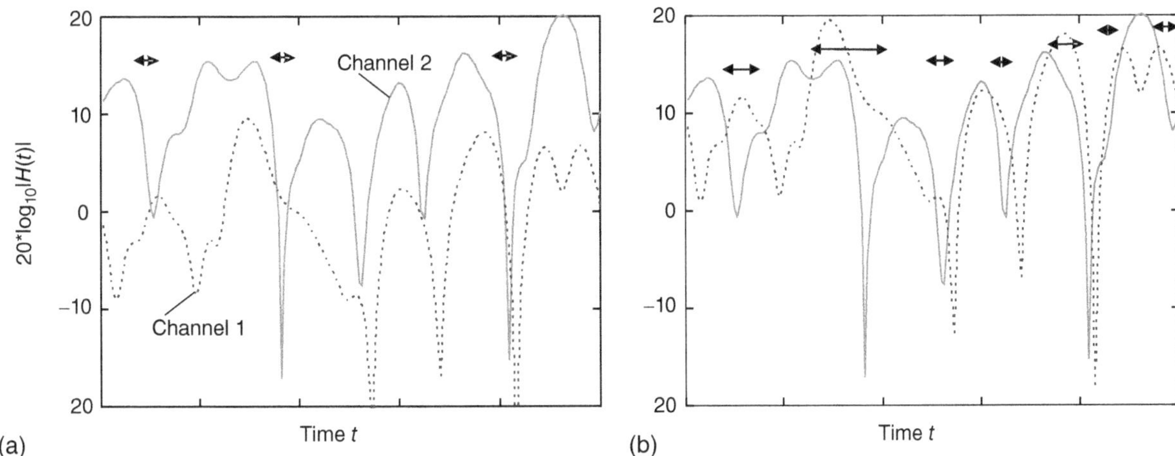

Figure 20.3 SNR as function of time for two users, and the selection of users to get resources for max-rate scheduling (a) and proportional fair (b). Channels are the same, but in figure (b) are normalized to same average power to show principle of proportional fair scheduling. Arrows indicate times at which user with Channel 1 is chosen. Color version available at wiley.com/go/molisch/wireless3e.

diversity, especially when users are stationary or slowly moving (pedestrian speeds) – though note that when mutliple antenna elements are available, random beamforming (see Section 22.8.1) can alleviate that problem somewhat. On the other hand, opportunistic scheduling is well suited for data communications with larger acceptable delays.

It is noteworthy that opportunistic scheduling leads to a paradigm change in the physical-layer design. Many of the transceiver structures and signal processing methods aim at combatting fading by *reducing* the variations of the SNR for a specific link (e.g., using diversity antennas), since with round-robin scheduling, transmission quality is determined mainly by the timeslots in which a UE experiences bad SNR (compare Section 11.2). However, with opportunistic scheduling, large variations of the SNR become desirable, since the scheduler simply does not assign the resources with bad SNR to the affected user.

*Opportunistic Scheduling – Mathematical Treatment

Turning now to a more detailed mathematical description, let the vector containing the channel coefficients of all users be written as $\mathbf{h}(t)$, with the associated rate region $\mathcal{R}_\mathbf{h}$. Now interpret this vector as a realization of an ergodic stochastic process (compare Chapter 5), having a probability $\pi_\mathbf{h}$ to occur, such that $\Sigma_\mathbf{h}\pi_\mathbf{h} = 1$. The feasible steady-state rate region $\overline{\mathcal{R}}$, i.e., the set of all achievable steady-state long-term rate vectors, can then be defined as (see [Li et al. 2013])

$$\overline{\mathcal{R}} = \left\{ \overline{\mathbf{R}} = \left[\overline{R}_1, \overline{R}_2, ...\overline{R}_K\right]^T \quad \text{s.t.} \quad \overline{R}_k = \sum_\mathbf{h}\pi_\mathbf{h}R_{\mathbf{h},k}, \mathbf{R}_\mathbf{h} \in \mathcal{R}_\mathbf{h} \right\} \tag{20.12}$$

where $R_{\mathbf{h},k}$ is the rate achievable by user k when in state $\mathbf{h}(t)$. Eq. (20.1) then becomes

$$\overline{\mathbf{R}}^* = \arg\max_{\overline{\mathbf{R}}} \sum_{k=1}^{K} U(\overline{R}_k) \quad \text{with} \quad \overline{\mathbf{R}} \in \overline{\mathcal{R}}. \tag{20.13}$$

Instantaneous rates $\mathbf{R_h}^*$ need to be chosen such that $\overline{\mathbf{R}}^* = \sum_\mathbf{h}\pi_\mathbf{h}\mathbf{R_h}^*$. It is noteworthy that the seemingly obvious choice for the instantaneous rates

$$\mathbf{R}^* = \arg\max_{\mathbf{R}} \sum_{k=1}^{K} U(R_k) \quad \text{with} \quad \mathbf{R} \in \mathcal{R} \tag{20.14}$$

does in general *not* optimize the average utility.[8]

The above maximization requires the channel statistics $\pi_\mathbf{h}$ as input, i.e., the scheduler needs to know them. This implies that the scheduler would have to first observe the channel for a long time before it can make any scheduling decision (and, in practice, the channel statistics might start to change by the time it is ready to make any decisions); clearly, this is inefficient. This problem can be

---

[8] Despite the suboptimality, some practical schemes do use instant-by-instant optimization of the utility.

avoided by using an iterative algorithm with exponential forgetting factor: Let $\mathbf{T}(t)$ be the measured throughput at time t, obtained by the following exponential averaging

$$\mathbf{T}(t+1) = \mathbf{T}(t) + \delta_{\text{sch}}(\mathbf{R_h}(t) - \mathbf{T}(t)) \qquad (20.15)$$

where δ_{sch} is the scheduling stepsize, and $\mathbf{T}(0) = 0$. Denote now the sum utility (over the users) as $\widetilde{U}$. Then, as time progresses, let the scheduler maximize at each time instant

$$\widetilde{U}(\mathbf{T}(t+1)) - \widetilde{U}(\mathbf{T}(t)) \simeq \delta_{\text{sch}} \nabla \widetilde{U}(\mathbf{T}(t))^T [\mathbf{R_h}(t) - \mathbf{T}(t)] \qquad (20.16)$$

where $\nabla \widetilde{U}(\mathbf{T}(t))$ is the gradient of the utility function. The maximization is achieved by choosing the rate assignment as

$$\mathbf{R_h}^* = \arg \max_{\mathbf{R}} \nabla \widetilde{U}(\mathbf{T}(t))^T \mathbf{R} \qquad \text{with} \qquad \mathbf{R} \in \mathcal{R} \qquad (20.17)$$

i.e., maximizing the projection of the rate (within the rate region) onto the gradient of the utility. The gradient of the utility can also be interpreted as the weights in *weighted fair queuing*, i.e., optimizing a (linearly weighted) sum rate $\Sigma w_k R_k$, where $w_k = dU(T_k(t))/dT_k(t)$. Note that this is different from maximizing the utility (within the rate region), which is the suboptimum approach described by Eq. (20.14).

This framework directly applies to the proportional-fair scheduling, since the utility function is $U(r) = \ln(r)$, and consequently $U'(r) = 1/r$. Assume that in a scheduling interval each user is assigned β_k time-frequency resources, so that $\Sigma \beta_k = 1$. The rate should thus maximize $\Sigma \beta_k R_k(t)/T_k(t)$; obviously, the best strategy is to give, in this scheduling interval, all resources to the user with the largest ratio of instantaneous rate to average rate, $R_k(t)/T_k(t)$. This is exactly the opportunistic scheduling strategy described in section titled "Opportunistic Scheduling – Intuitive Picture". Thus, while there are two seemingly different definitions of "proportionally fair", namely (i) maximizing the sum of the logarithms of the rates, and (ii) scheduling the user with the "instantaneously relatively best" channel, the above derivation shows that these two definitions are (under many circumstances) identical.

Challenges of Scheduling in Time-Varying Channels

A key requirement for scheduling is that the BS has knowledge of the instantaneous downlink channel. As discussed in Section 16.1.7, in a Frequency Domain Duplexing (FDD) system this requires that the UEs feed back their instantaneous SNRs to the BS. The rate of this feedback depends on the coherence time of the channel. Furthermore, it also involves a system design trade-off. If the feedback is done too rarely, then the BS picks a suboptimum user. If the feedback is done too often, then the overhead for this feedback becomes prohibitive. It is also noteworthy that *all* UEs have to feed back their SNRs, even though only one will ultimately be picked for communication.[9]

Another challenge for opportunistic scheduling is latency introduced by the scheme. The BS retains communication with the chosen UE for approximately one coherence time of the channel. If the user mobility is low, the coherence time can be quite large. Remember that each user is not assigned resources until its feasible instantaneous rate becomes the largest; this waiting time (latency) is proportional to T_{coh} and grows with the number of users. In a system with many users and large coherence times, this latency can become prohibitively large.

All those above considerations were for the downlink; however, the scheme works for the uplink with minor variations of the protocol: the BS determines the quality of the uplink channels, and then broadcasts which UE is allowed to transmit. Note that the overhead may be different in the uplink and the downlink: in the uplink, the system requires K pilot signals and one feedback message, while for the downlink case, one broadcast pilot signal and K feedback messages are required.

Frequency-Selective Channel

OFDMA can provide opportunistic scheduling in frequency-selective channels even if the channel is time-invariant, by assigning subcarriers to the UEs that have good SNRs on them. In other words, instead of giving the whole band at different times to different users, the scheduler gives different subcarriers to different users; to visualize, just consider Figure 20.3 with the x-axis label being frequency.

However, there is an important difference to the frequency-flat, time-selective channel: there is no inherent delay until a user becomes the best for a particular frequency. Furthermore, the scheduler just needs to determine once which user is best for which subcarrier (assuming that the set of users that want to access the channel stays the same). The available diversity order in the frequency domain is the ratio of system bandwidth to coherence bandwidth.

More generally, let the channel be both time- and frequency-selective. Assume that the transmit power spectral density is constant, so that the SNR of a user on a particular subcarrier does not depend on the amount of assigned resources. Using a weighted fair queueing formulation, the optimization problem becomes

---

[9] Some tricks can be used to alleviate this problem: for example, a UE that sees an SNR below a certain threshold does not need to feed back its information since the chance that it will be picked is very low. Again, the choice of this threshold involves a tradeoff: if the threshold is too low, it does not lead to significant savings; if it is too high, then there are times when no UE provides feedback, and the BS thus does not know which user to pick.

$$\max_{\boldsymbol{\beta}} \Sigma_k w_k \left(\Sigma_n \beta_{k,n} \log_2 \left(1 + \gamma_{k,n} \right) \right) \tag{20.18a}$$

$$\Sigma_k \beta_{k,n} = 1, \qquad \text{for all } n \tag{20.18b}$$

$$\beta_{k,n} \in [0, 1] \qquad \text{for all } k, n \tag{20.19}$$

where n indexes the subcarriers/subbands. Note that $\beta_{k,n}$ does not stand anymore for the percentage of the resources, but actually indexes the specific resources. For this reason, $\beta_{k,n}$ must be either 0 or 1, and the fact that any subcarrier can only be assigned to a single user is reflected in (20.18b). However, (20.18a, 20.18b, 20.19) is an (np-hard) mixed-integer optimization problem. For mathematical convenience, (20.19) can be relaxed such that $\beta_{k,n}$ can take on any value in the range [0, ... 1], instead of being forced to the values 0 or 1, which allows easier solution.

However, while with that simplification the solution can be found with polynomial complexity (as for every convex problem), it might be too time-consuming to be executed in real-time. Furthermore, we need to take into account the fact that the subcarriers are actually discrete. For this reason, a number of approximate algorithms have been suggested, most of which are based on a greedy approach (i.e., find the subcarrier/user combination that gives the highest reward, take this subcarrier out of consideration, look for the next-best combination, and so on, until all subcarriers have been assigned).

One possible greedy approach is:

- Iterate $i = 1, \dots I$ times, such that in each iteration (indexed by i) one subcarrier is assigned to a specific user. In each iteration
  - update the "tone index" $\tau_k^{(i)}$ that indicates the subcarrier that user k would like to be assigned if it gets a subcarrier assigned in this, the ith, iteration round. The update is done for all users k. One possible way of doing this assignment is by sorting the tones according to the channel conditions for each individual user; in other words, for all k choose $\tau_k^{(i)} = \arg\max_n \left(\gamma_{k,n} \right)$, where the argmax is over the subcarriers that have not yet been assigned;
  - update a metric g_k that indicates for each user the utility increase if it gets assigned the subcarrier in this iteration round. This could for example be simply the utility of this particular tone, $U\left(R\left(\gamma_{k,\tau_k^{(i)}} \right) \right)$. Possible limits in the utility (e.g., a maximum rate) can be taken into account as well. If subcarriers can interfere with each other, it is possible to take into account the degradation that this assigned tone creates for other users. The subcarrier is then assigned to the user that can derive the highest utility increase; ties are broken by making arbitrary decisions. If the maximum utility gain is not positive, then the subcarrier is not assigned to anybody.

Various investigations in the literature have shown that this approach gives almost optimum results.

We finally note that further generalizations of the weighted fair queuing formulation are possible. For example, the weights can be made time-variant, and thus account for different delay requirements for data from different users. Furthermore, additional constraints on the individual utilities, e.g., minimum required data rates for some of the users, can be added. Then it is possible, e.g., to treat a system with different classes of data, where some users have guaranteed QoS, while for the rest we aim to optimize the sum of the utility functions.

20.3.3 *Network Utility Maximization with Power Control*

Power Control in Nonselective Channels

Consider the downlink of a system in which we can perform independent power control for the users, and a fraction of the orthogonal time/frequency resources is assigned to each user. This corresponds to the case of OFDMA in a time invariant, frequency-flat channel. When specializing to a utility that is the weighted sum rate of the different users, this problem can be formulated as weighted fair queuing:

$$\begin{aligned} \max_{\mathbf{p},\,\boldsymbol{\beta}} &\ \Sigma_k U_k(p_k, \beta_k) \\ &\ \Sigma_k p_k = 1, \\ &\ \Sigma_k \beta_k = 1, \\ &\ \mathbf{p}, \boldsymbol{\beta} \geq 0 \end{aligned} \tag{20.20}$$

where β_k and p_k are the fraction of bandwidth resources and power that are assigned to the kth user, and γ_k the SNR that is obtained when all bandwidth resources and power are assigned to the kth user. If there is only a sum power constraint for all the users, the situation corresponds to a downlink where the total power of a UE is distributed to the signals intended for the different users. Assume further that the utility function is a weighted sum of the rates R_k

$$U_k = w_k R_k = w_k \beta_k \log_2 \left(1 + \frac{p_k \gamma_k}{\beta_k} \right) \tag{20.21}$$

and the weights w_k are determined by a fairness criterion. To simplify the math, do not assign discrete subcarriers but rather allow p_k and β_k to be continuous; in that case, the optimization in (20.20, 20.21) is convex and can thus be solved by the well known and efficient tools for solving convex optimization problems [Boyd and Vandenberghe 2004].

Power Control in Selective Channels

In this case, the problem formulation becomes

$$\max_{\mathbf{p},\boldsymbol{\beta}} \Sigma_k \Sigma_n w_k R_{k,n} \tag{20.22}$$

$$R_{k,n} = \beta_{k,n} \log_2\left(1 + \frac{p_{k,n}\gamma_{k,n}}{\beta_{k,n}}\right)$$

$$\Sigma_k \Sigma_n p_{k,n} = 1 \tag{20.23}$$

$$\Sigma_k \Sigma_n \beta_{k,n} = 1$$

$$\mathbf{p},\boldsymbol{\beta} \geq 0$$

where n indexes the subcarriers/subbands (a stricter constraint on subband allocation is $\Sigma_k \beta_{k,n} = 1\ \forall n$). The utility function is jointly convex in p and β. The theoretically optimum solution is the joint optimization of carrier allocation and power control. While this is in principle possible [Huang et al. 2010], the computational effort is very high. A popular sub-optimum technique is then to perform subcarrier assignment (i.e., solving the scheduling problem) under the assumption of uniform power distribution, and then perform the power optimization for the given allocation; that latter part of the power control can be implemented by waterfilling.

*20.4 Resource Allocation in CDMA with Infinite Backlog

Resource allocation in CDMA is somewhat different from OFDMA. Due to the inherent frequency diversity that each user experiences, there is little multi-user diversity in the frequency domain, and as well as the time domain (since the small-scale fading is essentially eliminated through frequency diversity). Also, it does not matter which specific codes are assigned to each user, since the SNR or utility is independent of it. Thus, scheduling in CDMA is *much* simpler than in OFDMA systems, since only the amount of resources, but not the specific resources, have to be identified.

On the other hand, users impact each other through interference and changing the transmit power impacts not only the received signal power of the data stream whose power is adjusted but also the SINR of all the other users. Thus, power control plays a more central role in CDMA – it is noteworthy that OFDMA works (though somewhat worse) without power control, but CDMA does not work at all.

The case we consider here adapts the transmit power such that the overall transmit power is minimized while each user reaches a target SINR $\widehat{\gamma}_k$. We consider this in the context of the uplink in a CDMA system. For the case that all users require the same SINR, the solution was already discussed in Chapter 19: each UE adjusts its transmit power P_k such that the power received at the BS is the same; each user gets (assuming negligible noise) at the RX input an SINR of $1/(K\text{-}1)$ (which is afterward enhanced by the Spreading Factor (SF)). For the general case, the optimization problem is

$$\text{minimize} \sum_k P_k \tag{20.24}$$

$$\text{s.t.} \quad \gamma_k \geq \widehat{\gamma}_k \quad \forall k \tag{20.25}$$

$$\mathbf{p} \geq 0$$

where $\mathbf{p}$ is the vector of transmit powers P_k.

This can be rewritten as a linear program, and can be solved efficiently in a centralized way: define $\mathbf{D}(\widehat{\gamma})$ as a diagonal matrix with entries $\widehat{\gamma}_k$ on its diagonals, $\mathbf{F}$ as a matrix with entries $F_{ki} = |h_i|^2/|h_k|^2$ for $i \neq k$, which describes the relative power gain of an interfering link i to that of the considered link k and $\mathbf{v}$ as a vector with entries $\widehat{\gamma}_k \sigma_n^2/|h_k|^2$. Then the optimization becomes

$$\text{minimize } \mathbf{1}^T\mathbf{p} \tag{20.26}$$

$$\text{s.t. } (\mathbf{I} - \mathbf{D}(\widehat{\gamma})\mathbf{F})\mathbf{p} \succeq \mathbf{v} \tag{20.27}$$

which has a solution as

$$\mathbf{p}^{\text{opt}} = (\mathbf{I} - \mathbf{D}(\widehat{\gamma})\mathbf{F})^{-1}\mathbf{v} \tag{20.28}$$

if the spectral radius of $\mathbf{D}(\widehat{\gamma})\mathbf{F}$ is smaller than unity; if that condition is not fulfilled, the problem is not feasible, i.e., no combination of powers can be found that fulfills all the SINR requirements.

Interestingly, Eq. (20.24, 25) can also be solved through a simple iterative algorithm in a distributed fashion, by having each UE apply an update in the $t + 1$ timestep, such that

$$P_k[t + 1] = \frac{\widehat{\gamma}_k}{\gamma_k[t]} P_k[t]$$ (20.29)

If the UEs also have a maximum power constraint, then the update rule is simply

$$P_k[t + 1] = \min \left[\frac{\widehat{\gamma}_k}{\gamma_k[t]} P_k[t], \; P_{\max} \right].$$ (20.30)

Note that the users that hit the power limit might not fulfill their respective target SINRs – the power control problem becomes infeasible. In other words, if the requirements are too stringent and/or too many interferers are present, there simply is no solution that satisfies all the constraints. If that happens, the system must decide on whether to limit access to the system or allow some users to operate below their target SINRs. It is the admission control (see Section 20.6) that might handle these situations, i.e., not admit users whose addition to the system would lead to the targets not be fulfilled.

Generally, if the execution of (20.29) is actually done in a distributed way, with time-slotted updates, then considerable time may pass until the algorithm has converged. For it to be useful, several conditions have to be fulfilled: (i) the transmission from each user must be longer than the convergence time. This condition might be violated in typical data transmission, where only a short packet is transmitted from one user, while in the next timeslot a different user (with different channel and SINR goal) might be active. (ii) the convergence time has to be shorter than the channel coherence time (otherwise, the channel, and thus the SINR, has changed before convergence is reached).

Another challenge for the algorithm is a lack of robustness. Assume that the P_k have converged to a steady state. Then, when a new user enters the cell and starts transmitting according to Eq. (20.29), the SINR for all users falls under the target thresholds until the algorithm has converged again. This problem can be resolved by a robust version of the algorithm, which consists of two – related – parts: (i) giving the users that have reached their targets an extra safety margin, so that their SINRs are $(1 + \varepsilon)\widehat{\gamma}$, and letting new users ramp up slowly, so that the percentage increase of their power in each step is limited to ε. Thus, the update equations become

$$P_k[t + 1] = \begin{cases} \dfrac{\widehat{\gamma}_k(1 + \varepsilon)}{\gamma_k[t]} P_k[t] & \gamma_k[t] \geq \widehat{\gamma}_k \\ (1 + \varepsilon)P_k[t] & \text{else.} \end{cases}$$ (20.31)

The steady-state energy consumption is higher than the one of Eq. (20.29) by a factor $(1 + \varepsilon)$. The factor ε allows to trade-off energy consumption and ramp-up time for new users. Further improvements can be made by having ε not as a fixed constant, but rather adaptive to the state of the network: when the network is uncongested, ε should be larger, so that ramp-up happens faster, while in a congested state ε should be smaller, to reduce power fluctuations.

Summarizing, the problem of minimizing power for fixed SINR targets can be considered essentially solved; both centralized and distributed algorithms that reach the global optimum exist, and even the transient behavior in the distributed implementation is under control in the robust implementation. This is noteworthy insofar that this is one of the few resource allocation problems for which such optimal solutions are known.

*20.5 Scheduling with Random Data Arrivals

20.5.1 Queue Definitions

The previous sections assumed that each user always has data to transmit, no matter what rates are assigned. We next turn to the case where the data are generated as a random process, i.e., there is a random *exogeneous* process that generates data packets that are to be transmitted. The data packets are written into a buffer (queue), and from there (typically in a FIFO) manner transmitted over the channel.

In this context, the overriding concern for the scheduling policy is *queue stability*, which is defined as the length of the queue staying finite. As an idealization, it is often assumed that the buffer size can be very large (so that computations do not depend on a specific buffer size). Then, queue stability means intuitively that the transmission opportunities assigned for a certain device should result in an average data rate that is larger than, or equal to, the average data rate of the exogeneous data; a mathematical definition will be given below.

We treat here the simple case of an uplink or downlink in a single cell.[10] Consider a slotted system with unit slot duration. Then the length of the queue $Q_k(t + 1)$ for the kth user in the $t + 1$ slot is given by

$$Q_k(t + 1) = \max \left(Q_k(t) - R_k(t), 0 \right) + a_k(t) \qquad \text{for } t = 0, 1, \dots.$$ (20.32)

---

[10] The general case of network optimization will be treated in Chapter 23.

where t indicates the time index of the slot, $a_k(t)$ is the exogenous input rate, and $R_k(t)$ is the rate at which packets depart the queue (and are thus transmitted to their destination); this rate is usually called transmission rate or *service rate*. For future use, we assume that the time averages $\bar{a}_k$ and $\bar{R}_k$ are well defined. The max operation is present because a queue cannot have negative length. Rate stability of $Q(t)$ means now that

$$\lim_{t \to \infty} \frac{Q(t)}{t} = 0 \quad \text{with probability 1} \tag{20.33}$$

while the less stringent condition for mean rate stability is

$$\lim_{t \to \infty} \frac{E\{|\, Q(t)\, |\}}{t} = 0. \tag{20.34}$$

Note that this section treats the comparatively simple case of interference-free channels that are frequency-flat (though they can be time-variant).

20.5.2 Rate Stability and Stationary Randomized Policies

The rate stability theorem [Neely 2010] states that $Q(t)$ is rate stable if $\bar{a}_k \le \bar{R}_k$. This is intuitive: on average, the queue must transmit with a rate that is at least as large as the rate at which data are incoming. To be feasible, the transmission rate (and thus the exogeneous rate) should lie in the steady-state capacity region $\bar{\mathcal{R}}$ (i.e., $\bar{a}_k \le \bar{R}_k\ \forall k$). This theorem suggests that any strategy that ensures well-defined time-averaged exogeneous rate and service rate and ensures that $\bar{a}_k \le \bar{R}_k$ is rate stable. This can be achieved through a stationary randomized policy, which probabilistically selects which queue to serve, based on the channel state information only. Such a strategy requires knowledge of the achievable average service rate, which in turn requires knowledge of the channel statistics, giving rise to the same issues discussed in section titled "Challenges of Scheduling in Time-Varying Channels".

Example 20.2 *Consider packet transmission in a wireless downlink, where the channels from the BS to the UEs are either in a "good" or "bad" state, i.e., it can either support the transmission of a packet or not. Channel states for the transmissions to the two UEs are independent of each other, and probability for a "good state" on channel 1 is 0.7, and for a good state on channel two is 0.6. (i) Compute the rate stability region, and (ii) Develop a transmission policy that stabilizes the transmission queues.*

Since the probabilities for good and bad channels are independent for the two UEs, the joint probabilities are

$$\begin{aligned} p_{11} &= 0.42 \\ p_{10} &= 0.28 \\ p_{01} &= 0.18 \\ p_{00} &= 0.12. \end{aligned} \tag{20.35}$$

We first need to determine the rate region; a sketch is shown in Figure 20.4. Obviously, the point (0.7/0) belongs to the rate region: in this case, we always transmit to UE1 when the opportunity arises (when channel 1 is in a good state), and never to UE2.

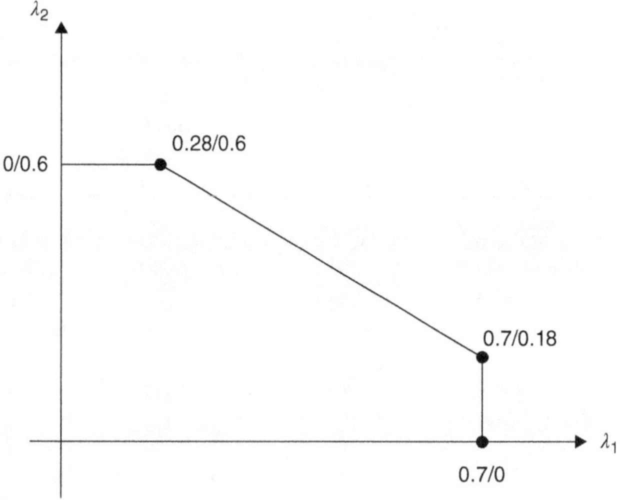

Figure 20.4 Rate region example.

Similarly, the point (0/0.6) is in the rate region. Furthermore, we can increase the rate for UE2 without decreasing the rate for UE1 up to a point: in $p_{01} = 0.18$ cases, the channel for UE1 is bad while the channel for UE2 is good; transmitting a packet to UE2 in those cases does not decrease the rate for UE1.[11] If we want to increase the rate for UE2 further, we have to sacrifice some of the rate for UE1: this occurs when both channels are in a good state, and we have to decide which UE to send the packet to. Thus, the rate region is bounded by a line connecting (0.7/0.18) and (0.28/0.6). Stability can be achieved if the average exogeneous input rates lie within the rate region.

A transmission policy that stabilizes the rates is then very simple: (i) the BS transmits to UE1 if the channel to it is good, and the channel to UE2 is bad; (ii) the BS transmits to UE2 if BS-UE2 is good and BS-UE1 is bad, and (iii) when both channels are good, the BS transmits i.i.d. to UE1 or UE2, with the probability of transmission to UE1 such that this probability multiplied by p_{11}, added to p_{10}, supports the input data rate.

In most cases, there are a number of different policies that stabilize the queues. We usually aim to pick the one that is the farthest away from the boundary of the feasible rate region, as this is the most robust solution and (as we will see later on) is preferable from a delay point of view.

20.5.3 Dynamic Scheduling

Queue Stabilization/Throughput Optimization

To avoid having to know the channel statistics, we would like to develop a scheduling method that can determine which packets to transmit based on instantaneous information only. The following derives such a policy, based on the mathematical theory of *Lyapunov drift*, using the example of a cellular downlink with K UEs. The resulting optimum policy is also known as the backpressure algorithm: in essence, the scheduling favors users whose product of queue length and achievable rate on their link is the largest at a given point in time. This achieves both fairness (users that have not been served in a long time, and thus built up a long queue, are preferentially served) and efficiency (users that have a very good channel and thus a high rate available are preferentially served).

The Lyapunov function $L(\mathbf{Q}(t))$ is defined as a scalar measure of the congestion in the network, namely the average of the squared queue lengths

$$L(\mathbf{Q}(t)) = \frac{1}{2}\sum_{k=1}^{K} Q_k^2(t). \tag{20.36}$$

The Lyapunov drift, which is the change of the Lyapunov function over one timeslot, can be upper-bounded as

$$L[\mathbf{Q}(t+1)] - L[\mathbf{Q}(t)] = \frac{1}{2}\sum_{k=1}^{K}\left[Q_k^2(t+1) - Q_k^2(t)\right]$$

$$\leq \frac{1}{2}\sum_{k=1}^{K}\left[a_k^2(t) + R_k^2(t)\right] + \sum_{k=1}^{K} Q_k(t)[a_k(t) - R_k(t)] \tag{20.37}$$

where the inequality holds because $(\max[Q - R, 0] + a)^2 \leq Q^2 + a^2 + R^2 + 2Q(a - R)$. Define further the conditional Lyapunov drift (i.e., $L(t)$ conditioned on a queue length vector $\mathbf{Q}(t)$)

$$\Delta(\mathbf{Q}(t)) = E\{L[\mathbf{Q}(t+1)] - L[\mathbf{Q}(t)]|\mathbf{Q}(t)\} \tag{20.38}$$

where the expectation is over the random channel states $\mathbf{h}$ and the transmission decisions made depending on those channel states. We next upper-bound the following term by some constant B

$$E\left\{\frac{1}{2}\sum_{k=1}^{K}\left[a_k^2(t) + R_k^2(t)\right]|\mathbf{Q}(t)\right\} \leq B. \tag{20.39}$$

The constant can be obtained as follows: firstly, we know that the exogeneous arrival rates are independent of the queue length, and thus

$$E\left\{\frac{1}{2}\sum_{k=1}^{K}\left[a_k^2(t)\right]|\mathbf{Q}(t)\right\} = \frac{1}{2}\sum_{k=1}^{K} E\{a_k^2(t)\} \tag{20.40}$$

---

[11] Again, analogously, we can transmit with maximum possible rate to UE2 whenever possible and still transmit to UE1 in p_{10} of the cases.

where often a homogeneous Poisson process is assumed for the arrival process. Secondly, we can be upper-bound $R_k^2(t)|\mathbf{Q}(t)$ by the square of the highest transmission rates that can occur for any queue lengths.

The conditional Lyapunov drift can thus be upper-bounded

$$\Delta(\mathbf{Q}(t)) \le B + \sum_{k=1}^{K} \nu_k Q_k(t) - E\left\{\sum_{k=1}^{K} Q_k(t)R_k(t)|\mathbf{Q}(t)\right\} \tag{20.41}$$

where $\nu_k = \overline{a}_k$. Note that the transmission rate $R_k(t)$ depends on the scheduling decision (also known as control action) $I(t)$ as well as the channel state $\mathbf{h}(t)$. In a dynamic transmission policy, we wish to minimize the Lyapunov drift at each time instant, since the smaller the drift, the less the network congestion. The only term on the right-hand side that a scheduling decision impacts is the term in curly brackets. If we want to maximize the expectation of $\sum_{k=1}^{K} Q_k(t)R_k(t)|\mathbf{Q}(t)$, then at each time we should maximize the term itself opportunistically. In other words, during each timestep, assign the following transmission rates to the users:

$$\mathbf{R}^*(t) = \arg\max_{\mathbf{R}\in\mathcal{R}(t)} \mathbf{Q}(t)^T\mathbf{R}. \tag{20.42}$$

As mentioned above, this approach ensures a certain fairness: users that have a large queue length (possibly because they have not been served in a long time) have a higher probability of getting scheduled, even if the channel conditions (and the associated rates) are not so advantageous. This scheduling policy is a special case of the *backpressure* algorithm. Remarkably, it does not need to know the fading statistics or arrival statistics, but simply schedules in each timeslot according to Eq. (20.42). Despite this simplicity, it can be shown to perform as well (in terms of achievable throughput while maintaining queue stability) as the stationary randomized policy that requires knowledge of the channel statistics and exogeneous rate statistics.

Figure 20.5 shows simulations of a two-user system in which the arrival rates at each time slot are described by a Poisson distribution with rate parameter ν for both users. Both channels are Rayleigh fading, though their average SNRs are different (20 and 10 dB, respectively). Simulations are performed over 20,000 timeslots; data rates are according to Shannon capacity. We can see that a high arrival rate (starting at slot 9993) of user 2 leads to increase in the (already large) queue length, and thus scheduling of this user for transmission even though user 1 can transmit with higher data rate. Only when the instantaneous rate of user 2 becomes particularly small (at slot 9996) does the system switch to user 1, before switching back to user 2 in the subsequent timeslot.

It can furthermore be shown that when the arrival rates are a minimum distance ε away from the boundary of the capacity region, then the average queue length can be bounded as

$$\limsup_{T\to\infty}\frac{1}{T}\sum_{t=0}^{T-1}\sum_{k=1}^{K} E\{Q_k(t)\} \le \frac{B}{\varepsilon}. \tag{20.43}$$

However, the above scheduler is not the only throughput-optimum one. It has actually been shown [Eryilmaz et al. 2005] that *any* scheduler that fulfills

$$\mathbf{R}^*(t) = \arg\max_{\mathbf{R}\in\mathcal{R}(t)} \mathbf{f}(\mathbf{Q}(t))^T\mathbf{R} \tag{20.44}$$

is throughput-optimum, as long as the function f fulfills some mild conditions: (i) f is a nondecreasing, continuous function with $\lim_{x\to\infty} f(x) = \infty$, and (ii) given any $M_1 > 0, M_2 > 0$, and $0 < \varepsilon < 1$, there exists a finite constant such that if x is larger than this constant, and for all k,

$$(1-\varepsilon)f_k(x) \le f_k(x-M_1) \le f_k(x+M_2) \le (1+\varepsilon)f_k(x). \tag{20.45}$$

The set of functions fulfilling these conditions contains, e.g., $f_k(x) = (K_k x)^\alpha$ for all positive a, but not the exponential function. Different functions f lead to different delay characteristics of the scheduler.

Queue Stability Plus Power Minimization

We now turn to the case where we want to not only find a stable solution but also want to minimize another quantity, say, power consumption. This can be solved by the *drift plus penalty* approach: instead of minimizing the Lyapunov drift, it minimizes a weighted sum of the drift and a penalty for power consumption:

$$\Delta(\mathbf{Q}(t)) + V\cdot E\{P(t)|\mathbf{Q}(t)\} \tag{20.46}$$

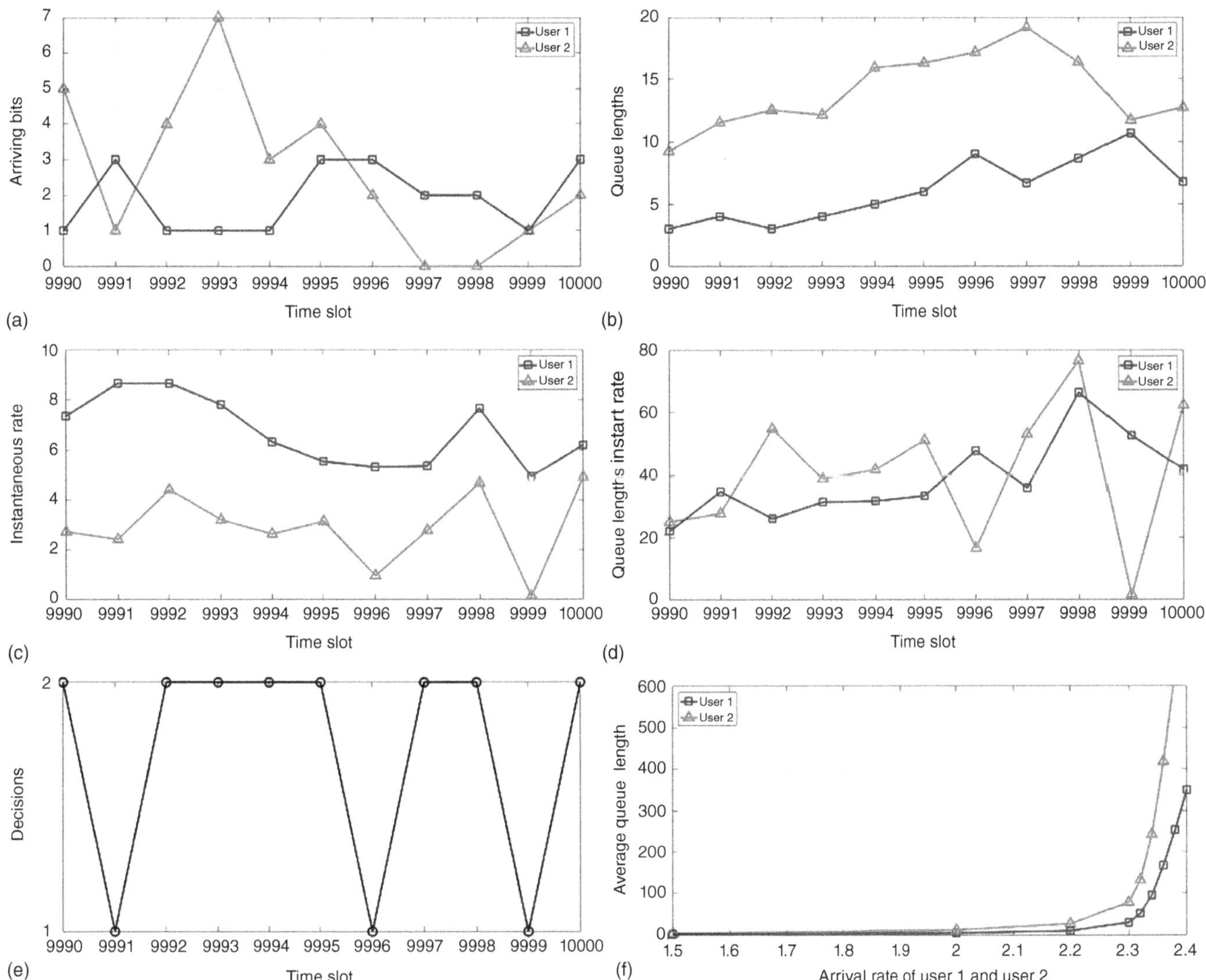

Figure 20.5 Scheduling of a two-user system with random arrivals with arrival rate $\nu = 2, \bar{\gamma}_1 = 20$ dB, $\bar{\gamma}_2 = 10$ dB. Achievable rate is Shannon capacity; only one user scheduled per timeslot. 20,000 timeslots simulated, results shown for part of that range in steady-state. (a) Number of arrivals. (b) Queue evolution. (c) Instantaneous rate evolution. (d) Evolution of the backpressure (e) Decisions of which user signals to send. (f) Queue length as function of arrival rates of the users. From M.C. Lee, private communication. Color version available at wiley.com/go/molisch/wireless3e.

where the parameter V shows how much importance is assigned to the power consumption, compared to keeping the queues short. In the simplest case, the power $P(t)$ is a step function, meaning that a power P is used when a transmission occurs, or zero power is used when the TX remains idle at time t. Following a similar procedure as above, the drift-plus-penalty can be upper-bounded by

$$\Delta(\mathbf{Q}(t)) + V \cdot E\{P(t)|\mathbf{Q}(t)\} \le B + V \cdot E\{P(t)|\mathbf{Q}(t)\} + \sum_{k=1}^{K} \nu_k Q_k(t) - E\left\{\sum_{k=1}^{K} Q_k(t)R_k(t)|\mathbf{Q}(t)\right\}. \tag{20.47}$$

The optimum scheduling decision is then at each time to pick the user/power combination that minimizes

$$VP(t) - \sum_{k=1}^{K} Q_k(t)R_k(t) \tag{20.48}$$

where it is noteworthy that both $P(t)$ and $R_k(t)$ depend on the decision of the scheduler. One can see that when the queues are almost empty, the scheduler might decide for all users to remain idle (since this requires zero power). As the queues are getting longer, transmitting on a channel with good SNR (and thus requiring little power) becomes more attractive; the larger V, the longer Q has to be in order to make transmission attractive.

The choice of the parameter V involves a trade-off between the queue length and how close we get to the minimum power consumption: the power consumption is within an additive constant B/V of the optimum, while the average queue length is bounded as

$$\lim_{T \to \infty} \frac{1}{T} \sum_{t=0}^{T-1} \sum_{k=1}^{K} E\{Q_k(t)\} \le \frac{B}{\varepsilon} + 2V \tag{20.49}$$

which implies that as V gets large, the power consumption approaches optimum (for achieving stability) as $1/V$, but average queue length grows linearly with V, see Figure 20.6. Improvements can be achieved by the so-called placeholder technique, see [Neely 2010].

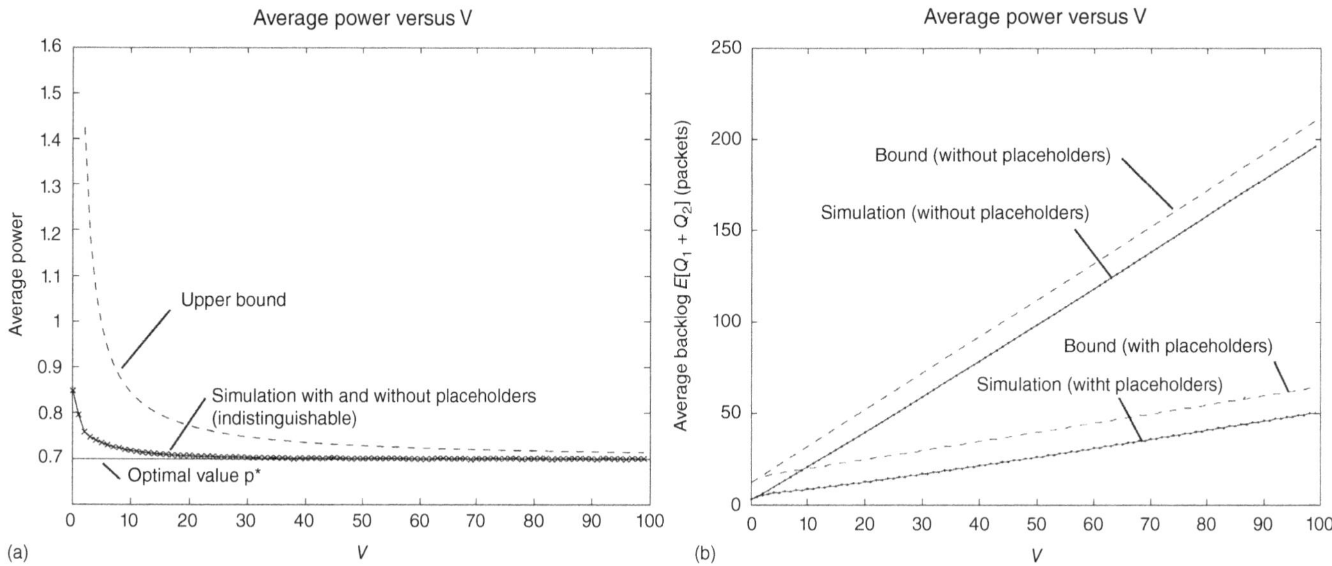

Figure 20.6 Trade-off of power consumption (a) and queue length (b) via the parameter V. Exogenous arrival rates 0.3 and 0.7. Reproduced with permission from [Neely 2010] © Morgan Claypool.

20.5.4 Scheduling of Short Files

The previous derivations assumed that the exogeneous rates are time-invariant, i.e., that a particular user just keeps on generating data as time goes on. In a number of cases, however, users just want a relatively short file. With a backpressure scheduling policy, these users are disadvantaged because they cannot build up the necessary "pressure" (i.e., backlog) to obtain high priority even when the channel conditions are good. As a matter of fact, such users might never be able to transmit.

An intuitive way to solve this problem is to schedule a particular user when it can transmit with the maximum possible data rate, without considering queue length. The question is now what the "maximum data rate" is. For an ideal transmission system, data rate is unbounded, since it just is $\log_2(1 + \gamma)$. In a practical system, however, the highest-order modulation/coding scheme (e.g., 64 QAM with rate 5/6 coding) places an upper limit on the achievable data rate. Clearly, a user with a short file should transmit under these circumstances. However, if the user always has a suboptimum SNR, such a data rate might be never achieved, and the user would have to wait forever to transmit. We could thus also estimate the maximum data rate from the history (i.e., maximum in a window of length T), and transmit when the user (again) reaches this maximum. Choice of the window length results in a trade-off between data rate and delay.

Yet another possibility is to modify the backpressure algorithm by weighing the "backpressure factor" with a monotonically increasing function of the file delay.

20.6 Multi-Channel Systems and Admission Control

As long as we are only maximizing a utility, we can in principle consider as many users as we want. However, the utility per user decreases with increasing number of users. Thus, if all users require a minimum data rate, then the system requires *admission control*: there is a maximum number of users that is allowed to enter the system – but once they have entered, a certain QoS (subject to an outage) is guaranteed for them. In other words, if a system has a certain number of resources, we want to guarantee that communication is successful in all but $1 - \varepsilon$ situations.

20.6.1 System Model for Speech Communications

We start out by analyzing a situation that is not admission control by itself but forms the basis for its analysis: blocking probability in a circuit-switched communications system, e.g., for speech communication.[12] Specifically, there is a fixed number of *speech channels* (where in this section by channel we mean a set of time/frequency resources),[13] and the task is to compute how many users can be supplied by one BS. The statistics of the amount of traffic each user generates are known, and the system wants to guarantee that attempts of a user to make a call are successful in all but $1 - \varepsilon$ of situations. The following considerations also carry over to some data packet transmissions, which will be discussed at the end of this section. Furthermore, they will be important for the discussion of handover failure probability, as also discussed later on.

To simplify the discussion, we make the following assumptions:

- *All users generate data with the same, constant, rate.* This holds, e.g., in simple voice telephony systems; variable-rate speech codecs and exploitation of silence periods are not taken into account;
- *The data of each user are transmitted over a channel that provides a fixed data rate with a certain maximum BER.* No adaptive modulation/coding is taken into account. We assume that each user has sufficient SINR to communicate at, or below, the required BER – in other words, no outages occur due to the wireless channel conditions. On the other hand, having a lower BER than the required one does not lead to any advantages, and in particular does not allow to reduce the required bandwidth per speech channel.
- *The rate of call generation ν is fixed, and the call generation (start) times are a Poisson Point Process (PPP).* Typically, the considered rate is the one that occurs during *the busy hour*, which is defined as the hour when most calls are made (the specific time of day when this busy hour occurs might be different, e.g., in business districts and in entertainment districts). While the number and duration of calls depend on the time of the day, designing for the busy hour is a worst-case design.
- *The duration of calls is an exponentially distributed random variable;* the decay time constant (which equals the inverse of the average call duration) of this distribution is denoted as μ. The validity of this assumption might depend on the considered country: in places where prepaid accounts and charging by the minute are common, the model is quite reasonable. In countries where "unlimited calls" plans are the norm, long call times are the norm (this is compounded by the fact that the customer service numbers of many companies and government agencies involve hold times in excess of 30 minutes), while short phone calls have to a large degree been replaced by text messaging.

We define from the above the average *offered traffic* T_{tr}, which is the number of call attempts (summed over all users), multiplied by the average duration of calls, per unit time. This quantity is dimensionless but is often given as having the unit *Erlang* (i.e., Erlang is a dimensionless unit). It can obviously be computed as

$$T_{tr} = \frac{\nu}{\mu}. \tag{20.50}$$

We can further define a "system loading" factor $\rho = T_{tr}/N_C$, where N_C is the number of channels per cell.

With these assumptions, a wireless system simply provides a fixed number N_C of communications channels, and each channel can be used by exactly one user at a time. The channels constitute a "pool," and a user requiring a channel can be assigned to any of the free channels in that pool. Demands for a channel by users are independent of each other, and each user occupies a channel for an exponentially distributed time. The system can then admit calls as long as there are free channels; when no channel is available a new incoming call can either be rejected/blocked (leading to the Erlang-B system in Section 20.6.2) or kept waiting until a channel becomes available (leading to Erlang C, Section 20.6.3). In the former case, a blocked request is cleared from the system and is assumed to have no impact on the future requests of that user; rather that user will require a channel again according to the regular call time statistics. While this may not be a realistic description of typical user behavior, it simplifies the mathematical treatment. In the latter case, the user is assumed to stay on hold until a channel has freed up.

20.6.2 Call Blocking – The Erlang B Model

System Model

To provide some intuition, we can imagine two extreme cases in the planning of a cellular network:

1. The *worst-case design*. It is assumed that all users want to call simultaneously. If the network operator wants to serve 720 users per cell, it has to provide 720 speech channels. Of course, such a network should never be built in practice – this would be like designing a hospital that can treat all inhabitants of a city at the same time.

---

[12] Note that most modern wireless systems, including LTE (Chapter 31), use "Voice over IP," where the speech is transmitted through data packets. Circuit-switched systems include GSM (which is still widely used for speech in parts of the world, see Appendix 30.A). Note also that the formulation applies equally to any system that requires a fixed, continuous data rate.

[13] E.g., for GSM, this number follows immediately from the number of GSM frequencies and the fact that each frequency can support eight speech channels.

2. The *best-case design*: if a typical user uses a phone only 20 minutes per day and there are 720 potential users per cell, then 14,400 minutes of call time are actually used during one day. Normalizing by the considered timeframe (in our example, the number of minutes per day, i.e., 1440) the *offered traffic* is 10 Erlang. A system with 10 speech channels per cell could thus supply the required number of users. However, this computation assumes that all users call sequentially, i.e., a new user dials in as soon as old ones have finished their calls, and they do that evenly distributed over a 24 hour period. In the notation defined above, this corresponds to a system loading factor $\rho = 1$.

Obviously, neither of the two extreme cases is realistic. The art of network design is, to a considerable degree, to predict the call behavior of the users, and derive the physical infrastructure (available number of speech channels) that guarantees an acceptable QoS. More precisely, we determine the *probability* that a user gets blocked if, through a statistical fluke, more users want to telephone simultaneously than channels are available. In order to achieve a certain (small) blockage probability, we have to include a safety margin, i.e., the system loading factor ρ must be smaller than 1. We will see in the following that not only does the average throughput increase as N_C increases but also the required safety margin shrinks.

*Derivation of the Erlang-B Equation

By assumption, call start times that are independent of each other, so that the time between the call starts is exponentially distributed, and the number of calls created within a time interval D is Poisson-distributed

$$\Pr[n \text{ calls generated in } D] = \frac{(\nu D)^n}{n!} \exp(-\nu D) \qquad n \geq 0. \tag{20.51}$$

The probability of starting (finishing) a call in an infinitesimal time interval dt is νdt (μdt) respectively. The probability that two or more calls are starting within such an infinitesimal interval is vanishing.

Furthermore, assume that the rates at which calls are placed do not depend on the current number of calls; this neglects the fact that people that are already in the middle of a call cannot at the same time make another call, and thus reduce the number of potential callers (this is a good approximation if the number of possible callers is much larger than the number of channels – a condition that is fulfilled in most cellular systems). The probability that at time t there are k calls in the system is denoted as $\Pr[k, t]$. Clearly, the probability of zero calls at time $t + dt$ is the probability that there were no calls at time t, and no call was added (probability for *not* adding a call is $(1 - \nu dt)$, plus the probability that there was one call at t, and one call terminated during the time interval of duration dt (the probability for terminating a call during this interval is μdt)

$$\Pr[0, t + dt] = (1 - \nu dt)\Pr[0, t] + \mu dt \Pr[1, t]. \tag{20.52}$$

No additional terms occur since there is vanishing probability for starting or terminating more than one call during the infinitesimal interval dt.

Having one call in the system at time $t + dt$ can arise if (i) there is no call at time t and one call starts during dt, (ii) one call exists at time t, and no calls are started or terminated, or (iii) two calls exist at time t and one of the calls is terminated (note that the probability for that is $2\mu dt$, since either of the two calls might be terminated):

$$\Pr[1, t + dt] = \nu dt \Pr[0, t] + (1 - \nu dt - \mu dt)\Pr[1, t] + 2\mu dt \Pr[2, t]. \tag{20.53}$$

Similarly, for any $1 \leq k < N_C$, the following holds

$$\Pr[k, t + dt] = \nu dt \Pr[k - 1, t] + (1 - \nu dt - k\mu dt)\Pr[k, t] + (k + 1)\mu dt \Pr[k + 1, t]. \tag{20.54}$$

However, when the system is fully loaded ($k = N_C$), then

$$\Pr[N_C, t + dt] = \nu dt \Pr[N_C - 1, t] + (1 - N_C \mu dt)\Pr[N_C, t]. \tag{20.55}$$

In steady state, the *probabilities* (not the actual number of calls) for having k calls should be independent of time, so that $\Pr[k, t + dt] = \Pr[k, t] = \Pr[k]$. Dividing by μ, and remembering that $T_{tr} = \nu/\mu$, the above set of equations reduces to the following linear system of equations

$$
\begin{aligned}
0 &= -T_{tr}\Pr[0] + \Pr[1] \\
0 &= T_{tr}\Pr[0] - (T_{tr} + 1)\Pr[1] + 2\Pr[2] \\
&\quad\quad\quad \cdots\cdots \\
0 &= T_{tr}\Pr[k - 1] - (T_{tr} + k)\Pr[k] + (k + 1)\Pr[k + 1] \\
&\quad\quad\quad \cdots\cdots \\
0 &= T_{tr}\Pr[N_C - 1] - N_C\Pr[N_C].
\end{aligned}
\tag{20.56}
$$

It can be easily seen that this system of equations is solved by $\Pr[k] = T_{tr}^k/k!$ for $0 \leq k \leq N_C$. To obtain proper probabilities, the normalization requirement $\Sigma \Pr[k] = 1$ needs to be fulfilled. Thus,

$$\Pr[k] = \begin{cases} \dfrac{T_{tr}^k/k!}{\displaystyle\sum_{k=0}^{N_C} T_{tr}^k/k!} & \text{for } 0 \leq k \leq N_C \\[4pt] 0 & \text{otherwise.} \end{cases} \tag{20.57}$$

Based on the statistical knowledge of the user habits, we can now design a system that *with a certain probability* allows a given number of users per cell to make calls.

Erlang-B Blocking Probability and Interpretation

The *probability of call blocking* is the probability that the system is full and cannot accommodate more users, $\Pr[N_C]$, i.e.,

$$\Pr_{\text{block}} = \frac{T_{tr}^{N_C}/N_C!}{\displaystyle\sum_{k=0}^{N_C} T_{tr}^k/k!}. \tag{20.58}$$

Figure 20.7 shows the relationship graphically. Define the admissible loading factor as the ratio of traffic that can be offered for a given outage probability and number of channels, divided by the number of channels. This quantity is low if N_C is small, especially for low required blocking probabilities. For example, for a required blocking probability of 1%, the admissible loading factor is less than 0.1 if $N_C = 2$. If N_C is very large, then the loading factor is only slightly less than unity and becomes almost independent of the required blocking probability. For example, assuming again a required blocking probability of 1%, the admissible loading factor is about 0.9 for $N_C = 50$. This effect can be understood essentially as *diversity effect* (or, mathematically, as concentration of the pdf around its mean): when there is a large number of channels, then the probability is low that the instantaneous number of users deviates significantly from the mean. In other words, when N_C increases, the admissible traffic (for a given blockage probability) increases due to two factors: the increase in average traffic, which is linearly proportional to N_C, and the increase in admissible loading factor, which is also known as the *trunking gain*. An example is shown in Figure 20.7.

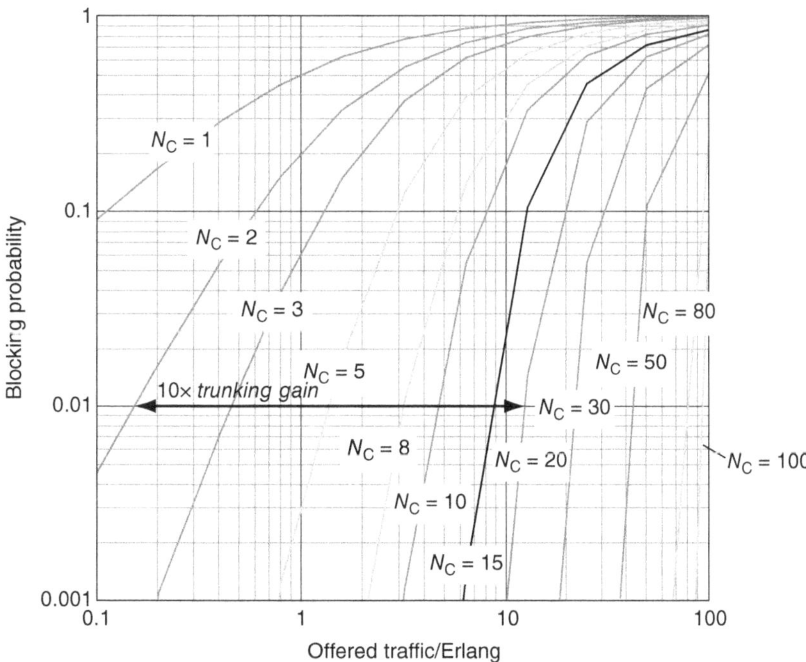

Figure 20.7 Blocking probability in an Erlang B system. N_C is the number of available speech channels.

The above derivation assumed that a blocked caller is simply cleared from the system, and the next attempted call from this user is independent of the blocked call; in other words, the user simply "forgets" that it wanted to make the previous, blocked, call.

An alternative model is that such a user tries to make the same call again after a random time. Thus, the statistical user behavior, namely placing calls according to a HPPP, remains the same, but the effective rate at which the traffic is offered is increased to

$$\nu' = \nu + \nu \mathrm{Pr}_{\mathrm{block}} + \nu \mathrm{Pr}_{\mathrm{block}}^2 + \nu \mathrm{Pr}_{\mathrm{block}}^3 + \dots = \nu \frac{1}{1 - \mathrm{Pr}_{\mathrm{block}}}.$$

This means that the offered traffic $T_{\mathrm{tr}} = \dfrac{\nu'}{\mu}$ is itself a function of the blocking probability, so that the computation has to be done iteratively. If the blocking probability is reasonably small ($<10\%$), one or two iterations are sufficient.

Summarizing, we find that the number of users that can be accommodated with a given QoS increases faster than linear with the number of available channels. The difference between the actual increase and the linear increase is called the *trunking gain*. From a purely technical point of view, it is thus preferable to have a large pool of channels that serves all users, rather than a priori dividing the resources (and the users) into several groups, and performing scheduling only within each group. This is also known as the benefit of a *fat pipe*. This situation would be fulfilled, e.g., if there is only a single operator for cellular systems, owning the complete spectrum assigned to cellular services. The reasons for *not* choosing this approach are political/economic (pricing, monopoly), not technical.

Example 20.3 *In an Erlang-B system, 30 channels are available. A blocking probability of less than 2% is required. What is the traffic that can be served if there is one operator or three operators?*

1. By inserting the required blocking probability $P_{\mathrm{block}} = 0.02$ and the number of channels $N_C = 30$ into Eq. (20.58), we get

$$0.02 = \frac{T_{\mathrm{tr}}^{30}/30!}{\displaystyle\sum_{k=0}^{30} T_{\mathrm{tr}}^{k}/k!}. \tag{20.59}$$

Solving this equation for T_{tr} gives

$$T_{\mathrm{tr}} = 21.9. \tag{20.60}$$

2. Similarly, sharing the 30 speech channels among the three operators, each having $N_C = 10$ speech channels, results in an admissible average traffic T_{tr} of each operator of

$$T_{\mathrm{tr}} = 5.1. \tag{20.61}$$

Hence, the total average traffic that can be handled by all three operators together is

$$T_{\mathrm{tr,tot}} = 3 \cdot 5.1 = 15.3. \tag{20.62}$$

Call Dropping During Handover

The above derivations considered a system where there is no mobility of the calls, i.e., the call blocking was depending only on whether the cell that serves the user has available channels or not. In mobile systems, the situation can occur that a call is initiated in one cell where a channel is available, but then wants to move to a different cell; if the new cell does not have a channel available, then the call is dropped. In this case, the above derivations hold with minor modifications:

- The inverse average call duration μ is replaced by $\mu + 1/T_{\mathrm{cell}}$, where T_{cell} is the average time a UE stays in a cell (it is assumed that this time is also exponentially distributed). This means that the connection of a UE to a particular BS is either terminated because the call is terminated, or because the UE moves to a different cell.
- Call arrival rate ν is replaced by the sum of the "new call generation rate" in the cell (i.e., the traditional call generation rate), plus the handoff arrival rate. The handoff arrivals are typically modeled as a Poisson process, so that the overall new incoming calls are also a Poisson process (the sum of two Poisson processes is again a Poisson process). However, computation of the actual handoff probabilities is nontrivial [Ghaderi and Boutaba 2006, p. 75].

With these modifications, the probability of call blocking plus call dropping (due to handover failure) can be computed. Note that this approach does not distinguish between these two quantities, although users sometimes consider a dropped call as more annoying than a blocked call.

20.6.3 Call Waiting – The Erlang C Model

An alternative model assumes that any user that is not immediately assigned a channel is transferred to a waiting loop, and assigned a channel as soon as it becomes available. The *probability that a user is put on hold* is

$$\mathrm{Pr}_{\mathrm{wait}} = \frac{T_{\mathrm{tr}}^{N_{\mathrm{C}}}}{T_{\mathrm{tr}}^{N_{\mathrm{C}}} + N_{\mathrm{C}}!\left(1 - \dfrac{T_{\mathrm{tr}}}{N_{\mathrm{C}}}\right)\sum\limits_{k=0}^{N_{\mathrm{C}}-1}\dfrac{T_{\mathrm{tr}}^{k}}{k!}} \tag{20.63}$$

and the average wait-time (for all customers, i.e., users that are not placed in a wait loop are included in the average) is

$$t_{\mathrm{wait}} = \mathrm{Pr}_{\mathrm{wait}}\frac{T_{\mathrm{call}}}{N_{\mathrm{C}} - T_{\mathrm{tr}}} \tag{20.64}$$

where $T_{\mathrm{call}} = 1/\mu$ is the average duration of the call.

Example 20.4 *Consider an Erlang C system where users are active 50% of the time, and the average call duration is 5 minutes. It is required that no more than 5% of all calls are put into a waiting loop. How many channels are required for K = 1, 8, 30 users? What is the average wait time in each of those cases?*

Since $T_{\mathrm{tr}} = 0.5 \cdot K$ is the average offered traffic, we need to find N_{C} that fulfills

$$0.05 \geq \frac{T_{\mathrm{tr}}^{N_{\mathrm{C}}}}{T_{\mathrm{tr}}^{N_{\mathrm{C}}} + N_{\mathrm{C}}!\left(1 - \dfrac{T_{\mathrm{tr}}}{N_{\mathrm{C}}}\right)\sum\limits_{k=0}^{N_{\mathrm{C}}-1}\dfrac{T_{\mathrm{tr}}^{k}}{k!}}. \tag{20.65}$$

This equation needs to be solved numerically; the results are given in the table below. With $T_{\mathrm{call}} = 5$ minutes, the average wait time is

$$t_{\mathrm{wait}} = \mathrm{Pr}_{\mathrm{wait}}\frac{5}{N_{\mathrm{C}} - T_{\mathrm{tr}}}. \tag{20.66}$$

The required number of channel to fulfills the inequality and the resulting average wait time is

T_{tr}	0.5	4	15
N_{C}	3	9	23
$\mathrm{Pr}_{\mathrm{wait}}$	0.0152	0.0238	0.0380
t_{wait}	0.0304	0.0238	0.0238

20.6.4 Applications to Data Transmission

The above discussion considered a speech communication system. However, almost all the considerations can be applied to data communication systems as well. Let us replace the words "phone calls" by "data packets": most of the above assumptions can be adapted to this new interpretation: data packets are generated according to a homogeneous Poisson process (this is a often assumed for data transmission problems for mathematical convenience, compare Section 18.4.1); the length of the data packets is distributed according to an exponential distribution (this is obviously an approximation, since most data transmission protocols allow only discrete packet sizes); the data generation rate is independent of the current traffic and is independent between users. The numerical values of duration and generation rates are obviously different between speech calls and data packets, but the mathematical model remains valid. Thus, many of the conclusions that can be drawn from the above derivations carry over to packet data systems, in particular, the advantages of multi-channel systems, as outlined in the seminal work of Kleinrock. It is important to keep in mind, however, that the model is based on the fundamental assumption that every channel has the same, time-invariant, data rate – an assumption that is often fulfilled in wired communications systems, but not always in wireless systems.

A major simplification that does not hold in practice is the "forgetting" of a blocked call/packet. While this might hold sometimes for human (speech) communication, it rarely is applicable in data systems. Rather, the various retransmission mechanisms will ensure that a blocked packet is transmitted at the next possible opportunity and that a packet that has been blocked for a considerable time will get preferential treatment (see Sections 31.5.2 and 32.5.2 for the implementation in LTE and NR, respectively). However, incorporation of this effect into the system model leads to a considerable complication of the resultant mathematical treatment, and will thus not be discussed here further.

20.6.5   Admission Control

Armed with these performance evaluations, we can now analyze admission control. The QoS for the admitted users (which may include minimum average data rate as well as maximum delay and other criteria) can only be fulfilled if the number of users is limited. The following situations are of interest:

- Admission control has to limit the probability that calls are dropped during a handover. This is achieved by making sure that each cell has a sufficient number of free channels to handle (with a certain probability) the calls that will be handed into this particular cell. The amount of such resources might be determined by the loading of the cell of interest, as well as the surrounding cells. Users joining a cell via handover should have higher priority for admission than newly joining users; however, already-existing calls in the cell have equally high priority. Admission control usually has to make a pessimistic estimate of how much resources will be needed in the future to accommodate existing and handover users; clearly, this leads to a trade-off between reliability and efficiency of the system.
- Admission control has to handle time variations of the systems: due to channel variations, or variations in the required data rates of the users (e.g., video), the effective number of available channels might change over time. Thus, the admission control can only determine the admissible number of users according to the expected traffic, plus some safety margin. In this context, the evolution of traffic can be modeled as a Markov decision process; however, note that optimization based on this model is computationally expensive.

The system flexibility can be somewhat improved when real-time traffic as well as delay-insensitive traffic has to be accommodated. The real-time traffic will then be scheduled with a higher priority, while the delay-insensitive traffic can be sent whenever there are resources not used up by the real-time traffic. For the case of multiple classes of traffic with different (but finite) delay constraints, different utilities, and constraints for the different classes, can be defined and the scheduling performed according to the sum utility.

Finally, we note the special nature of video (and music) streaming, in which the TX can choose between several possible transmission qualities (higher quality requiring higher data rate), but where actual interruptions in the transmission (leading to a stalling of the video at the RX) need to be avoided. This leads to special requirements for the design of the admission policies, since the video quality can be traded-off with the number of users.

20.7   Machine Learning for Resource Allocation

As all of the above examples have shown, resource allocation is a very complicated problem that in most cases leads to formulations for which no optimal solutions can be found in polynomial times. Furthermore, the above examples only aimed to optimize a relatively simple utility function – when hard delay constraints (e.g., for signals arising from control circuitry or from edge computing) have to be taken into account, the situation becomes even more complicated. For this reason, the design of resource allocation algorithms has been heuristic, and somewhere between an art and a science.

Over the past decade, ML algorithms have found usage in a wide array of engineering problems. Generally, ML excels at problems for which no theoretically-optimum deterministic algorithms are known, or are too complicated to derive. Consequently, they are well suited for finding good resource allocation. Generally, there are three ways in which ML can be applied to resource allocation problems: (i) accelerating a computation step in a traditional resource allocation algorithm, (ii) supervised learning, and (iii) unsupervised learning.

The first method makes use of the property of deep neural networks as a *universal function approximator*. In other words, there are mathematical subtasks in traditional resource allocation algorithms that are very computationally intensive and thus often make real-time operation of an algorithm impractical; a typical example is the branching step in a "branch and bound" algorithm. This task can be handled by a neural network that is trained offline, and then provides a fast solution during the online operation. The advantage of such an approach, compared to other ML implementations, is that it uses the domain expertise and ensures that results are explainable; the drawback is that the problem solution still requires an analytical formulation that might be based on restrictive assumptions.

Another approach is to use supervised learning as a "black box" to obtain a scheduling solution. A neural network is trained by means of a training data set where labeled data, i.e., both the input parameters and the optimum output (e.g., the optimum schedule), are known. A suitable training algorithm, e.g., backpropagation, determines the neural network parameters in such a way that the difference between input and output is minimized. The main obstacle for the application of this method is that optimum (or at least very good) solutions must be known for the training process. This is not an option if the optimum solution is, per se, unknown. However, it can be done for the case that the optimum solution is known, but is too slow for online execution of the scheduling. In this case, ways to generate a sufficient amount of training data, or methods to reduce the need for training data, are required.

The most natural type of ML for scheduling is for the network to adjust its parameters to maximize a reward or minimize a loss function. This loss function can be as simple as the sum rate of the users, but can also be defined in a much more complicated way that includes minimum rates for each user, penalties for packets that are delivered after a deadline, reward for delivering video packets with higher quality, and so on. Consider a formulation of the resource allocation problem as a decision process, where the algorithm observes the system state (e.g., channel states, queue lengths, etc.), and selects a resulting action, which provides a certain reward. Furthermore, for the next time step, the system moves, with certain probabilities, into a new state, which is impacted by the decision. The goal is to find a policy that – based on the state – determines the probability for choosing a particular action so that the reward is maximized.

Reinforcement Learning (RL) is a natural way for optimizing this policy. There are a number of different RL algorithms, which can be categorized into value-based, policy-based, and actor-critic. Among the value-based algorithms, Q-learning is popular. It defines action-value functions $Q_\pi(s, a)$ that describe the reward for starting in state s, taking action a, and following policy π. Instead of actually solving for the optimum policy in closed form (which would require knowledge of the dynamics of the state transitions), Q-learning uses an iterative update at time t

$$Q(S_t, A_t) \leftarrow (1 - \Delta)Q(S_t, A_t) + \Delta\left[B_{t+1}(S_t, A_t) + \lambda \max_{a'} Q(S_{t+1}, a')\right] \tag{20.67}$$

where B_t is the reward, Δ is the step size, and λ the discount factor for future rewards. Note that the Q-functions defined here are not related to the queue lengths Q defined in Section 20.5. The action A usually follows a soft greedy approach, e.g., choosing the action that provides the highest estimated value with a certain probability p_0, and a random action with probability $1 - p_0$. The output is a policy

$$\pi^*(s) = \arg\max_a Q^*(s, a). \tag{20.68}$$

The Q-learning approach requires to store the action-value functions. For a large state and action space, this can be prohibitive. Deep-Q learning thus uses a deep neural network to approximate the value functions.

In the policy-based RL methods, the neural network aims to directly learn the optimum policy. The policy is represented by a function computed via a deep neural network, parameterized by θ. The parameter is then updated by gradient descent, such that in each step it improves the discounted reward factor $E\{\sum \lambda^t B_{t+1}\}$.

In the actor-critic method, there are two neural networks, the actor network that uses as an input the state, and provides as output an action, and a critic network, which takes as the input state, and puts out the reward that follows when using the policy suggested by the actor. It can thus be seen as a combination of value-based and policy-based RL algorithms.

RL-based algorithms have been applied to a wide variety of resource allocation problems, from scheduling, to power allocation, to dynamic spectral access. Yet there remain many open questions, in particular about suitable training data, stationarity (are the policies learned at a particular time of day, with certain traffic pattern, applicable during a different time of day), transfer learning, robustness, and so on. Still, given the difficulties of obtaining closed-form scheduling solutions, RL promises to play a major role in future wireless networks.

Further Reading

Resource allocation in wireless networks has a long history. An excellent exposition of resource allocation in OFDMA, from which the current chapter partially draws, is provided in [Li et al. 2013]. [Huang et al. 2010] describe both optimum and suboptimum methods for resource allocation including power control. Other interesting reviews include [Sadr et al. 2009] which concentrates on the downlink, and [Abu-Ali et al. 2014], which covers the uplink in LTE. Opportunistic scheduling was first analyzed in [Knopp and Humblet 1995] and further extended and popularized in [Viswanath et al. 2002]. For CDMA, the distributed power control algorithm of Eq. (20.27) was proposed in [Foschini and Miljanic 1993]. An extensive description of power control for CDMA, including several of the cases described in this chapter, is given in [Chiang et al. 2008]. Power control in systems with different types of services is discussed in [Lee et al. 2005]. The trade-off between rate and reliability is discussed in [Lee et al. 2006].

The backpressure algorithm was first proposed in the seminal paper of [Tassiulas and Ephremides 1990]. An excellent summary of Lyapunov drift techniques is found in [Neely 2010].

The Erlang-B equation is derived, in a very understandable form, in [Miller 2002]. An extensive description of general queuing theory can be found in [Shortle et al. 2018]. Further useful surveys on call admission include [Ahmed 2005, Ghaderi and Boutaba 2006].

A discussion of Reinforcement Learning can be found in [Luong et al. 2019]; the application to scheduling is described in [Liang et al. 2019].

For updates and errata for this chapter, see https://wides.usc.edu/students.html#textbooks.

Exercises

See Sec. 36.20 of Exercises.pdf at wiley.com/go/molisch/wireless3e

21

Principles of Cellular Networks

21.1 Frequency Reuse

Let us now turn to the question of how an infrastructure-based wireless system can cover a large area, and provide service to as many user equipments (UEs) as possible within this area (for wireless networks without dedicated infrastructure, so-called ad-hoc networks, see Chapter 23). The first such radio systems were noise-limited systems with few users; the spacing between Base Stations (BSs) was so large that interference was not an issue. Therefore, it was advantageous to put each BS on top of a mountain or high tower, so that it could provide coverage for a large area.[1] However, this approach severely limited the number of UEs per unit area that could communicate simultaneously. The cellular principle, which we will describe in this section, provides the solution to this problem.

In a cellular system, the coverage area is divided into many small areas, so-called cells. In each of those cells, there is one BS that provides coverage for this (and only this) cell area; assume for the moment that each BS uses Frequency Division Multiple Access (FDMA) to provide multiple access within a cell, employing multiple frequency sub-bands. Importantly, each frequency sub-band can be used in *multiple* cells. Imagine for the moment that such a reuse is possible in every single cell – this is called "reuse-1." The number of users that can be served *per cell* is then approximately independent of the cell size, or, in other words, the number of users *per unit area* increases linearly with the number of cells. A densification of the cells thus leads to increased throughput (call handling capability, sum of the data rates of all the users in the network). This throughput is often called the *cellular capacity*.[2] A related measure is the *area spectral efficiency*, which states how many bits per second can be transmitted in a unit area per unit bandwidth (i.e., dimension bit/s/Hz/m^2). Historically, cell radii in urban areas have decreased from 10 to 30 km in the 1960s to 100–300 m today, leading to an increase in area spectral efficiency by four orders of magnitude.

The question that naturally arises is: can we really use each frequency sub-band in each cell?[3] The short answer is "generally no, though it might be possible with special techniques." Consider, for example, the downlink of an FDMA system where UE 1 is at the boundary of its assigned cell, so that the distances to its *serving BS* (i.e., the BS with which it communicates) and to a neighboring BS are the same. If now the neighboring BS transmits in the same frequency sub-band (in order to communicate with a UE 2 in its own cell), then the Signal-to-Interference Ratio (SIR) seen by UE 1 is $C/I = 0$ dB. This is certainly not enough to sustain reliable communications, especially if 0 dB is the *median* SIR, and due to fading, the actual SIR is worse 50% of the time.[4]

A way to improve the SIR is to re-use a frequency sub-band not in every cell, but only in cells that have a certain minimum distance from each other. The normalized distance between two cells that can use the same frequency sub-bands is called *reuse distance*, D/R, where D is the distance between two BSs that use the same frequency subband, and R is the cell radius. This reuse distance can be computed from the link budgets as described in Section "Link Budget for Interference-Limited Systems" in Chapter 3. As a consequence of ensuring this minimum reuse distance, there is a *cluster* of cells that all use different frequencies; therefore, there can be no Co-Channel Interference (CCI) within such a cluster. The number of cells in a cluster is called the *cluster size*. However, frequencies can be reused *across* clusters.

Obviously, a cluster size N_{cluster} reduces the area spectral efficiency by a factor of N_{cluster} compared to the reuse-1 case *if the data rate for each user is unchanged*; a large cluster size would thus seem disadvantageous. For example, an operator that has licenses for 35 frequency sub-bands, and uses a cluster size 7, can support 5 simultaneous users in each cell. Maximization of the capacity thus requires minimization of the cluster size. On the other hand, increasing the cluster size improves the Signal-to-Interference-and-Noise Ratio (SINR), in particular for cell-edge users, which in systems with adaptive modulation and coding can increase the throughput per user. Thus, a cellular system design has to carefully trade off these two effects.

Typical cluster sizes have decreased significantly over the years. First-generation (analogue) FDMA systems required an *SIR* of 18 dB, which results in a cluster size of 21 (see below). Second-generation digital systems like Global System for Mobile

---

[1] Some of those systems, like police radio, often were designed only for a single city anyway.

[2] Note that we use "capacity" in this chapter for the number of users or communication devices, or the total throughput of the system, that can be supported simultaneously. We do *not* refer to the information-theoretic capacity.

[3] We assume here and in the rest of this chapter a full-buffer model, so that all BSs always have data to transmit (compare Chapter 20).

[4] For some fading distributions, the probability density function of the SIR can be computed analytically, in particular if both desired and interfering signal have the same distribution; for Rayleigh fading, see (5.25).

---

communications (GSM) required less than 10 dB, which decreases the reuse distance to 7 or less. This allowed a dramatic increase of the capacity and was one of the most important reasons for the shift from analogue to digital cellular systems in the early 1990s. Fourth- and fifth-generation systems like Long-Term Evolution (LTE) and New Radio (NR) have very small reuse distances (three or less), and use additional adaptive techniques to improve the reuse distance, as will be explained later on.

This chapter explains cell planning by the example of (persistently scheduled) users in an FDMA system. This is done for ease of explanation, and furthermore because very similar principles hold for Orthogonal Frequency Division Multiple Access (OFDMA) in data transmission; firstly, the frequency sub-bands we will talk about here are similar to groups of subcarriers in OFDMA systems; secondly, while data users might require different amounts of subcarriers (while users in FDMA systems get assigned sub-bands of equal size), users with larger bandwidth requirements can be seen as a group of co-located users each of which is requiring a unit block of subcarriers. In summary, while the mathematics and nomenclature become more involved in OFDMA-based data communications systems, the principles remain the same. Code Division Multiple Access (CDMA) systems, which are based on a somewhat different principle, will be discussed in Section 21.5.

21.2 Cell Planning with Symmetric BS Deployment

What shapes do cells normally take on? First consider the idealized situation where the pathloss depends only on the distance from the BS, but not on the direction. The most natural choice would be a disk (circle), as it provides a constant power on the cell boundary. However, disks cannot fill a plane without either gaps or overlaps. Hexagons, on the other hand, have a shape similar to a circle, and they *can* fill up a plane, like in a beehive pattern. Thus, hexagons are usually considered the "basic" cellshape, especially for theoretical considerations.

21.2.1 Cell Planning with Hexagonal Cells

In the following, we will derive the frequency reuse that can be obtained in hexagonal cells. The current section makes some simplifying assumptions, namely that only interference between two cells is relevant, and that cell edge users are considered. In this case, results are identical for downlink and uplink. The subsequent sections will then relax those assumptions, and find that uplink and downlink require separate treatment.

For the case of hexagonal cells, some interesting conclusions can be drawn about the relationship between link margin and reuse distance. Consider the hexagon whose center is at the origin of the coordinate system. Proceed now l_1 hexagons in the y direction, turn $60°$ counterclockwise, and proceed l_2 hexagons in that new direction (see Figure 21.1). This gets us to the cell whose center has the following distance from the origin:

$$D = \sqrt{3}\sqrt{(l_1 R + \cos(60°)l_2 R)^2 + (\sin(60°)l_2 R)^2}$$
$$= \sqrt{3}R\sqrt{l_1^2 + l_2^2 + l_1 l_2}. \tag{21.1}$$

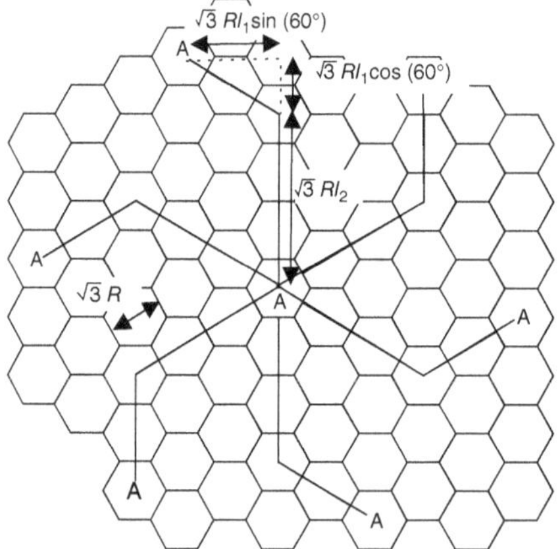

Figure 21.1 Minimum reuse distance.

Note that the distance between the centers of two adjacent hexagons is $\sqrt{3}R$, where R is the distance from the center of a hexagon to its farthest corner. Also note that only integer values of l_1 and l_2 are possible.

The task of frequency planning is to find those values of l_1 and l_2 that make sure that the distance from Eq. (21.1) is larger than the required reuse distance. Of course, there is an infinite manifold of such pairs – large values of l_1 and l_2 certainly satisfy the condition. What we want to find, however, is the pair of values that *minimizes* the cluster size, and thus maximizes the spectral efficiency, while still satisfying the minimum reuse distance.

The relationship between the cluster size N_{cluster} and the parameters l_1 and l_2 is

$$N_{\mathrm{cluster}} = l_1^2 + l_1 l_2 + l_2^2, \qquad l_1, l_2 = 0, 1, 2, \ldots \tag{21.2}$$

which is the area of a cluster, divided by the area of a cell, see also Figure 21.2.

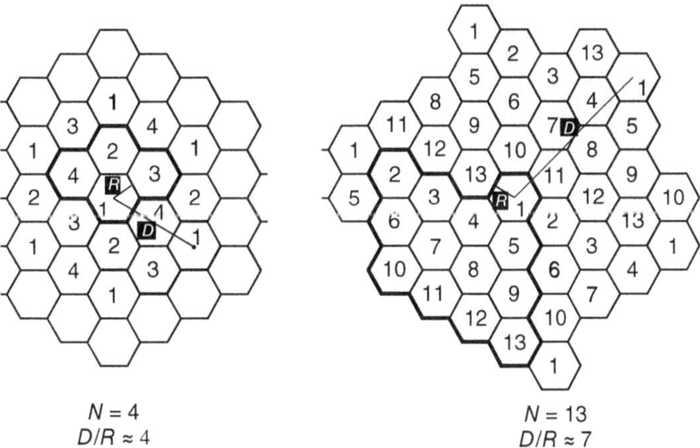

Figure 21.2   Interdependence between reuse distance D/R and cluster size N.
Reproduced with permission from [Oehrvik 1994].

Equation (21.2) also establishes that not all integers are possible cluster sizes. The cluster size can only take on the numbers $N_{\mathrm{cluster}} = 1, 3, 4, 7, 9, 12, 13, 16, 21, \ldots$. The relationship between reuse distance D/R and cluster size results from Eqs. (21.2) and (21.1) as

$$D/R = \sqrt{3 N_{\mathrm{cluster}}}. \tag{21.3}$$

Table 21.1 shows typical cluster sizes and reuse distances.

Table 21.1   Typical cluster sizes and reuse distances, assuming omni-antennas

N_{cluster}	D/R	
1		CDMA
3	3	
4	3.46	
7	4.58	TDMA-system (GSM)
9	5.2	
12	6.0	
13	6.24	
16	6.93	
21	7.94	Analogue system (NMT, AMPS)

Cell planning thus proceeds in the following steps: starting from the specifications for the minimum transmission quality, the link budget establishes the minimum distances between desired BS and interferer. From this relationship, Eq. (21.3) provides the cluster size; note that it has to be the smallest integer number out of the set defined by Eq. (21.2). Using the procedure for obtaining nearest neighbors (move l_1 cells into one direction, turn 60°, move another l_2 cells), the frequencies for each cell can be determined.

Example 21.1 Cell Planning in an Interference-Limited System

Consider a cellular system where each frequency sub-band is 30 kHz wide, and the link margin is 18 dB for satisfactory transmission (speech) quality.[5] The fading margin (for shadowing plus Rayleigh fading) is set to 15 dB. How many sub-bands (speech channels) per cell can an operator with a license for 5 MHz of spectrum provide?

Given the link margin and the fading margin, the median values of the signal power must be 33 dB ($2 \cdot 10^3$) stronger than that of the interference power at the cell edge. The distance between the desired BS and the farthest corner of the hexagon is R, the distance between the interfering BS and that corner is (approximately) $D - R$. Assuming further that the power decreases with d^{-4}, we require that

$$\frac{D - R}{R} = \left(2 \cdot 10^3\right)^{1/4} = 6.7 \tag{21.4}$$

so that the reuse distance must be $D/R = 7.7$. From Eq. (21.3), the reuse distance is $(3 N_{\mathrm{cluster}})^{0.5}$, so that the cluster size is 19.8. The smallest $N_{\mathrm{cluster}} \geq 19.8$ is $N_{\mathrm{cluster}} = 21$. Having a total of $5000/30 = 167$ possible frequency sub-bands, only $167/21 \simeq 8$ can be used in each cell.

The above description provides a *normalized* cluster size, namely in units of cell radii. As indicated in Section 21.1, the area spectral efficiency is determined by the number of clusters that can be fitted into a unit-size area on an absolute scale.

Another implicit simplification made in the above derivation is that the interference from a single interferer dominates the overall behavior. This provides a lower bound on the overall interference. A more conservative estimate can be achieved by assuming that a fixed number (usually chosen as 6) of equally strong interferers are present. More detailed derivations will be provided in the following subsections.

The simplest form of improving area spectral efficiency is a sectorization of the cell. Specifically, a hexagonal (or similarly shaped) cell can be divided into several (typically three) sectors. Each sector is served by one sector antenna. Thus, the number of "distinct areas" (areas that can be covered individually) has tripled, as has the number of BS *antennas*. However, the number of BS *locations* has remained the same, because the three antennas are at the same location, see Figure 21.3. Note that the number of BS locations (and not the number of antennas or transceivers) dominates the cost of deployment for the network operator.

Sectorization can thus significantly enhance the area spectral efficiency for a given number of BSs. It can also increase the range of the cells, since with sectorization, antennas with a larger gain are naturally used – remember that the integral of the gain over all directions is constant (see Section 8.2.1), so that if the antenna has zero gain in some directions (angles outside its assigned sector), it shows higher gain in the remaining directions.

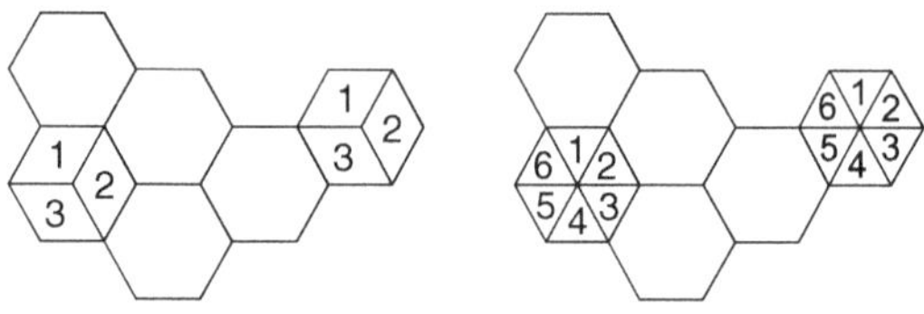

(a) 120°-sectorization   (b) 60°-sectorization   **Figure 21.3** Principle of sector cells.

*21.2.2 Spectral Efficiency for the Downlink

Let us now turn to the computation of the spectral efficiency in the downlink when the interference from multiple BSs is taken into account, and shadowing is considered as well. Furthermore, we take into account that the UEs are not all at the cell edge, but distributed throughout the cell. To remain closer to the reality of a modern cellular system mainly providing data services, we assume use of adaptive modulation and coding.

Consider a cellular system with K users per cell or sector. Each UE suffers from CCI from L neighboring BSs. The number of interferers depends on the geometric layout of the cellular system and sectorization. For example, for the hexagonal layout with only first-tier interferers, $L = 6$ without sectorization, $L = 2$ for three sectors per cell, and $L = 1$ for six sectors per cell; when the second-tier interferers are also considered, the corresponding values are $L = 18$, $L = 7$, and $L = 4$.

Each fading state and interference realization of a UE results in a specific instantaneous SINR, and thus instantaneous capacity (see Section 13.11.1); the planning of cellular layout and the determination of service coverage area necessitate the evaluation of fading-averaged capacity, which for the kth user can be written as

---

[5] In order to work with a real system that uses pure FDMA, we consider the numbers of the historic Advanced Mobile Phone System (AMPS) system.

$$
\begin{aligned}
C_k &= \int_0^\infty \min\{\log_2(1+\gamma), C_{\max}\}pdf(\gamma)d\gamma \\
&= \log_2(e)\int_0^{\gamma_{\mathrm{T}}} \frac{1}{1+\gamma}[1-F(\gamma)]d\gamma
\end{aligned}
\tag{21.5}
$$

where $C_{\max}$ is the maximum capacity that a link can provide (due to the finite modulation order) and $\gamma_{\mathrm{T}} = 2^{C_{\max}} - 1$; the step from the first to the second line is obtained through integration by parts. The function $F(\gamma)$ is the Cumulative Distribution Function (cdf) of γ. It depends on the type of averaging over fading: if the scheduling is based on the small-scale fading state, we get different scheduling realizations over a time frame over which the pathloss and shadowing can be considered constant. Then, assuming that the small-scale fading is Rayleigh, and denoting $F(\gamma) = F_{\gamma_k|\beta}(\gamma)$ is the conditional cdf of γ_k, conditioned on pathloss and shadowing between the kth UE and the lth BS that is subsumed in the coefficients, β_{kl}, it can be shown to be

$$
F_{\gamma_k|\beta}(\gamma) = 1 - \exp\left(-\frac{\gamma}{\overline{\gamma}}\frac{1}{\beta_{k0}}\right)\prod_{l=1}^{L}\left(1+\frac{\beta_{kl}}{\beta_{k0}}\gamma\right)^{-1}
\tag{21.6}
$$

where $\overline{\gamma}$ is the transmit Signal-to-Noise-Ratio (SNR), and β_{k0} is the pathloss-plus-shadowing coefficient between the kth UE and its desired BS.

On the other hand, we might consider the spectral efficiency when the ensemble of scheduling decisions extends over different shadowing realizations. Then, the long-term spectral efficiency, $F(\gamma) = F_{\gamma_k}(\gamma)$ is the corresponding unconditional cdf, i.e., the cdf of the SINR that includes both Rayleigh fading and lognormal shadowing. An exact closed-form expression is not available; however, for the case that noise is negligible, the SIR, γ_k, can be accurately approximated by a lognormal RV, $\widetilde{\gamma}_k$, whose parameters are determined, e.g., by the Moment Generating Function (MGF) matching method, see Section 5.8. The cdf of the lognormally distributed SIR, $\widetilde{\gamma}_n$, can then be written as

$$
F_{\widetilde{\gamma}_k}(\gamma) = 1 - Q\left(\frac{10\log_{10}\gamma - \mu_{\widetilde{\gamma}_k}}{\sigma_{\widetilde{\gamma}_k}}\right).
\tag{21.7}
$$

Short-Term – Fading-Averaged Spectral Efficiency

Consider first a Round-Robin (RR) scheduler. It gives the same percentage of time-frequency resources to each user and does not take into account the channel states of the users. Thus, the average spectral efficiency is

$$
C_{\mathrm{RR}} = \frac{1}{K}\sum_{k=1}^{K} C_k.
\tag{21.8}
$$

The actual values for each UE can be computed by inserting (21.6) into (21.5). In the presence of Rayleigh fading, and a noise-limited system, this becomes

$$
C_k = \log_2(e)e^{1/(\overline{\gamma}\beta_{k0})}\left[\Gamma\left(0, \frac{1}{\overline{\gamma}\beta_{k0}}\right) - \Gamma\left(0, \frac{2^{C_{\max}}-1}{\overline{\gamma}\beta_{k0}}\right)\right].
\tag{21.9}
$$

Similar expressions can also be derived for the max-SINR and the proportional-fair scheduler; see [Wu et al. 2011] for further details.

As an example, Figure 21.4 plots the spectral efficiencies of the Max-SINR, Proportional Fair (PF), and RR schedulers for different numbers of UEs, K, in the cell. As expected, the spectral efficiency for the Max-SINR Scheduler increases the most as K increases, since it benefits the most from multi-user diversity, followed by PF scheduler and RR scheduler.

Long-Term Spectral Efficiency Analysis Under Composite Channels

We now investigate the spectral efficiency when the averaging is done over both small-scale Rayleigh fading and long-term lognormal shadowing, restricting to the case of interference-limited systems. As outlined above, the SIRs $\widetilde{\gamma}_k$ can be approximated by a Q-function whose parameters can be obtained from moment-matching. The average spectral efficiency, $\overline{C}_{\mathrm{RR}}$, with a RR scheduler is now obtained as

$$
\overline{C}_{\mathrm{RR}} = \frac{\log_2 e}{K}\sum_{k=1}^{K}\int_0^{\gamma_{\mathrm{T}}}\frac{1}{1+\gamma}Q\left(\frac{\xi\ln\gamma - \mu_{\widetilde{\gamma}_k}}{\sigma_{\widetilde{\gamma}_k}}\right)d\gamma
\tag{21.10}
$$

where $\xi = 10/\ln(10)$.

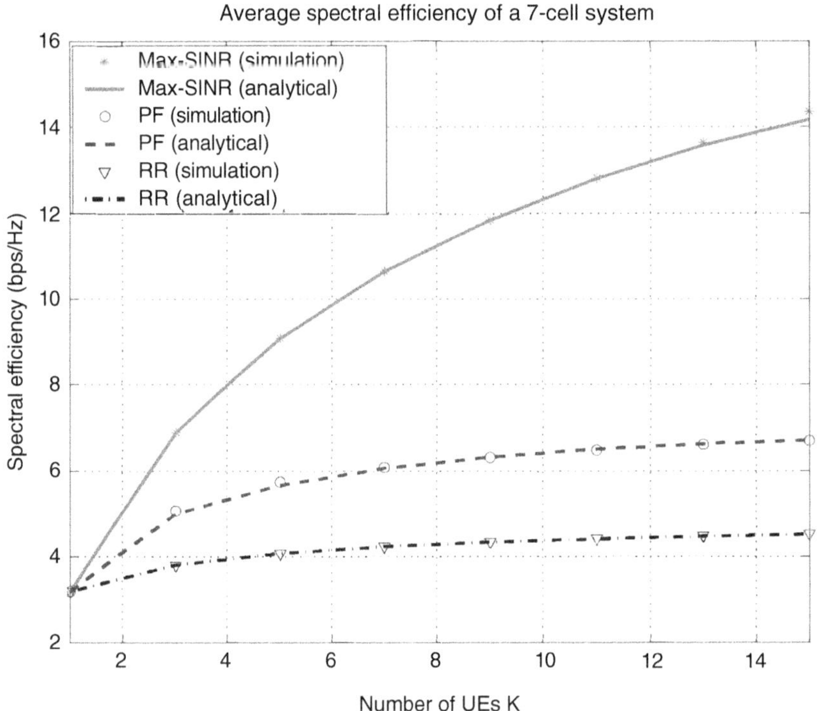

Figure 21.4 Short-term Rayleigh fading-averaged spectral efficiencies of systems with various schedulers. Pathloss exponent is 3.7, the dB standard deviation of all lognormal RVs is $\sigma = 8$ dB. The SNR at the cell corner is $\bar{\gamma}_{\mathrm{edge}} = 10$ dB. The kth UE is placed at a distance of $\frac{k}{K}R$ from its serving BS, where R is the cell radius, and at an azimuth of $\frac{2k\pi}{K}$. Color version available at wiley.com/go/molisch/wireless3e.
Reproduced with permission from [Wu et al. 2011] © IEEE.

For a Max-SIR scheduler, which serves the UE with the highest SIR, the cdf of $\tilde{\gamma}_{\max}$ is the product of the cdfs of the individual SIRs of the UEs, and thus can be written as

$$F_{\tilde{\gamma}_{\max}}(\gamma) = \prod_{k=1}^{K} \left[1 - Q\left(\frac{\xi \ln \gamma - \mu_{\tilde{\gamma}_k}}{\sigma_{\tilde{\gamma}_k}} \right) \right] \tag{21.11}$$

assuming that the SIRs $\tilde{\gamma}_k$ are independent. The spectral efficiency of the Max-SIR scheduler upon averaging over composite fading is

$$\overline{C}_{\mathrm{MSIR}} = \log_2(e) \int_0^{\gamma_{\mathrm{T}}} \frac{1}{1+\gamma} \left\{ 1 - \prod_{k=1}^{K} \left[1 - Q\left(\frac{\xi \ln \gamma - \mu_{\tilde{\gamma}_k}}{\sigma_{\tilde{\gamma}_k}} \right) \right] \right\} d\gamma. \tag{21.12}$$

A similar equation can be derived for the proportional-fair scheduler.

Figure 21.5 shows the effects of constellation limits on the system spectral efficiency. While the Max-SIR scheduler always outperforms the PF and RR schedulers, limitations on the constellation size undercut its throughput advantage as the Max-SIR scheduler quickly reaches the spectral efficiency cap. While all schedulers benefit from having larger constellation size for adaptive modulation and coding, the Max-SIR scheduler benefits the most and the RR scheduler benefits the least.

*21.2.3 Spectral Efficiency for the Uplink

The computation of the spectral efficiency is, in general, more complicated for the uplink. This is due to the fact that the interference now stems from the UEs in the neighboring cells, not the BSs.

For a given scheduling and a given location of the interfering UEs, the computation of the previous subsection for the downlink can be reused, with just a reinterpretation of the channel coefficients. However, since the interference is stemming from UEs in other cells, the key question is which UEs are actively contributing interference for the UE of interest. The interference thus depends on the actual scheduling of the UEs in their respective cells, and the distribution of the distance of those interferers from the victim BS. Furthermore, if the UEs have power control, then their power is regulated by the channel *to their serving BS*, while the attenuation of their transmit power as affecting the victim BS depends on the channel between interfering UE and victim BS. These aspects are discussed in more detail in Section 21.5, where they are discussed in the context of CDMA.

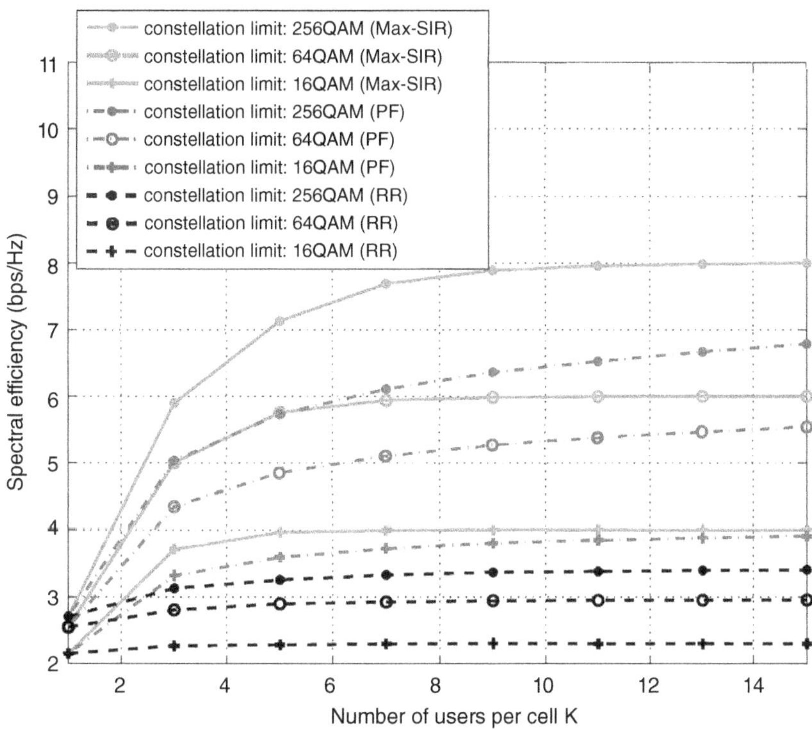

Figure 21.5 Composite channels: effects of constellation size limits on spectral efficiency for RR, PF, and Max-SIR schedulers. Color version available at wiley.com/go/molisch/wireless3e.
Reproduced with permission from [Wu et al. 2011] © IEEE.

*21.2.4 Considerations for TDD Systems

In Time Domain Duplexing (TDD) systems, the timing of the switching from uplink to downlink and vice versa needs to be considered. Typically, neighboring cells need to synchronize their switching, such that when the BS in cell A is in reception mode, the BS in cell B is also in reception mode, not in transmission mode. If this were not fulfilled, BS-A would be interfered by transmissions from another BS, which is problematic because (i) transmit power of BSs can be orders of magnitude larger than that of a UE, (ii) path gain between BSs can be much higher than path gain from BS to UE at the same distance (since BSs are generally elevated), see Chapter 7, and (iii) the power control in a BS is not designed to reduce interference to other BSs. For all these reasons, switching between uplink and downlink is generally synchronized. Exceptions may occur in small cells (Section 21.7), where the difference between the TX powers of BS and UE is small.

The switching between uplink and downlink also needs to consider that the runtime from an adjacent-cell UE to a BS is larger than from an in-cell UE to a BS. The design of guard intervals and dead times for the switching needs to take this into account.

21.3 Inter-Cell Interference Reduction

21.3.1 Taxonomy of Inter-Cell Interference Reduction Techniques

The frequency reuse approach described above is very simple and has the advantage of allowing closed-form analysis of the performance. However, it does not necessarily lead to the best possible spectral efficiency. Rather, a range of measures both with respect to frequency reuse methodology, and with respect to transceiver structures, can be taken to improve the area spectral efficiency further. These techniques can generally be categorized as follows:

- *Interference Avoidance:* in this class of schemes, interference to victim receivers (RXs) – which can be BS or UE, depending on whether we are considering uplink or downlink – is *avoided*, in the sense that transmissions from other devices are arranged in a way that little/ no interference is arriving from them at the victim RX. Interference avoidance can be further subdivided into (i) schemes based on frequency reuse, such as sectorization and Fractional Frequency Reuse (FFR), which statically assign a certain frequency band to a particular part of the cell, and (ii) cell coordination based techniques, where the transmission scheduling is adapted dynamically to reduce the interference at victim RXs in the neighboring cells. The effectiveness of the cell coordination can be enhanced by the use of multiple antennas, an approach that will be described in Section 22.8.4 in more detail. A further division into centralized (where a central controller determines the resource allocation in all cells) vs. distributed schemes is possible.

- *Interference Mitigation:* in this approach, no attempt is made to reduce the interference power at the victim RX; rather, measures are taken to reduce the *effect* of the interference on the reception quality for the desired signal. This can be achieved, e.g., through multi-user detection at the RX (see Section 28.2), or through the use of BS cooperation, such that the antenna elements at different BSs cooperate, and thus form a distributed array that sees all incoming signals as "useful" (for more details, see Section 22.11). Since the principles of interference mitigation schemes are discussed elsewhere in this book, we will not further consider them in this chapter.

21.3.2 Fractional Frequency Reuse (FFR)

The fundamental idea of FFR is to employ a reuse factor that is not necessarily limited to those listed in Table 21.1. There are a variety of methods that can achieve this; all of which are based on using different frequency and/or power assignments in different parts of the cell, specifically in areas with different distances from the BS (annular rings). Note that these techniques can be combined with sectorization, which gives different frequencies to areas with different *angles* as seen from the BS. In the following, we will use the terms "cell" and "sector" interchangeably.

Partial Frequency Reuse

Partial Frequency Reuse (PFR), also known as *Fractional Frequency Reuse with Full Isolation* (FFR-FI), uses different parts of the spectrum in the inner part of the cell, and near the cell edge, respectively, and employs different reuse factors in them. This allows a better utilization of the overall spectrum. This can also be interpreted as follows: the reuse factor we normally compute is for a cell-edge user. However, for users inside the cell, it is an overkill, leading to much better SIR than the required minimum value.[6] Thus, it is desirable to have a more aggressive frequency reuse in the cell center than at the cell edge.

For a more concrete discussion, take the example of Figure 21.6, in which the starting point is a sectorized cell with standard frequency reuse, cluster size 3. The total available spectrum is divided now into *four* parts of possibly unequal size: band 1 that is used in the cell centers, and bands 2, 3, and 4, which are used at the cell edges. Since the cell edges are susceptible to adjacent-cell interference, frequency reuse must be applied there (remember that the situation at the cell edges is the main motivation for cluster sizes larger than 1). However, for the inner part of the cells, the same frequency band can be used in every cell – by definition, a UE within this region has a shorter distance to the serving BS than to any interfering BS, and by suitable selection of the size of the "inner part," it can be assured that for all UEs in them the (small-scale averaged) SIR is better than required for a given modulation and coding scheme. Say that the inner part of the cell gets assigned a fraction κ_0 of the available spectrum, and the outer parts the remainder, which is split into three parts. The effective cluster size is then given as

$$\frac{1}{\kappa_0 + \frac{1 - \kappa_0}{3}} . \tag{21.13}$$

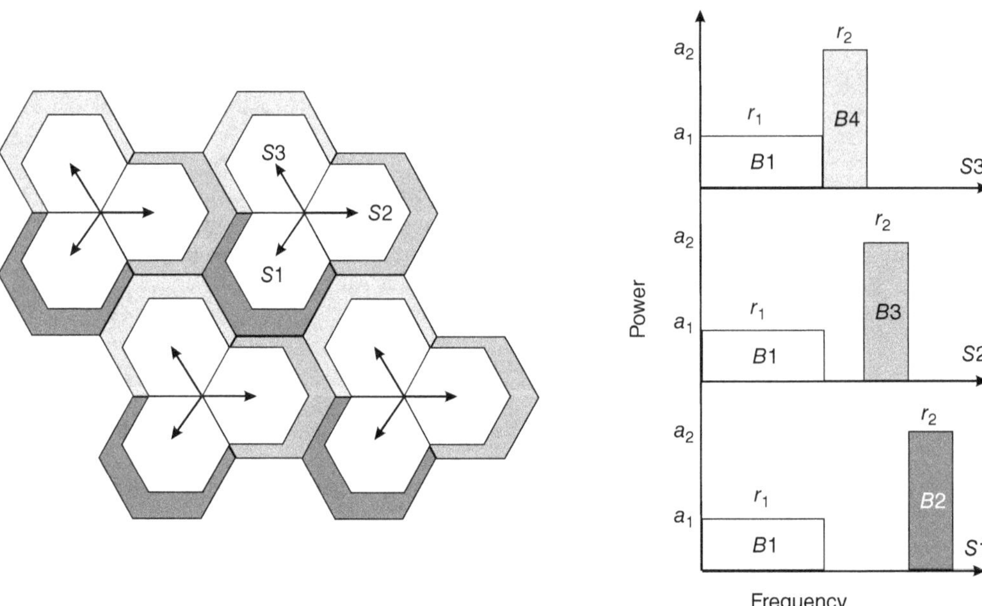

Figure 21.6 Principle of partial frequency reuse and sectorization. Color version available at wiley.com/go/molisch/wireless3e. Reproduced with permission from [Hamza et al. 2013] © IEEE.

In the limiting case that the "inner part" is zero, we return to standard cluster size 3. In the case that the inner part is very large, we approach frequency reuse 1.

---

[6] This assumes that SIR better than a minimum value goes to waste. For data transmission with adaptive modulation and coding, this need not be the case, as higher SIR actually leads to higher capacity; yet the benefits of higher SIR are limited by the maximum modulation order.

Soft Frequency Reuse

Soft Frequency Reuse (SFR) modifies not only the frequency allocation scheme but also the power assignments in different parts of the cell.[7] Every cell can use all parts of the spectrum, but with different power levels, as shown in Figure 21.7. A particular sub-band is used by a certain cell for its cell edge users and is assigned the largest power, e.g., sub-band 1 in cell 1. The other bands are used with lower power. Since sub-band 1 is used with low power in cells 2 and 3, the BSs of those cells do not create significant interference at the cell edge of cell 1; thus *all* users in cell 1 that are close to the cell boundary should use sub-band 1. Users in the interior can use any band, since the shorter distance to the BS allows reasonable SINR even with lower power levels. Completely analogous considerations are valid in cells 2 and 3, with the only variation that different sub-bands are used there near the cell edges. Thus, there is no fixed band that is *always* used in the inner part.

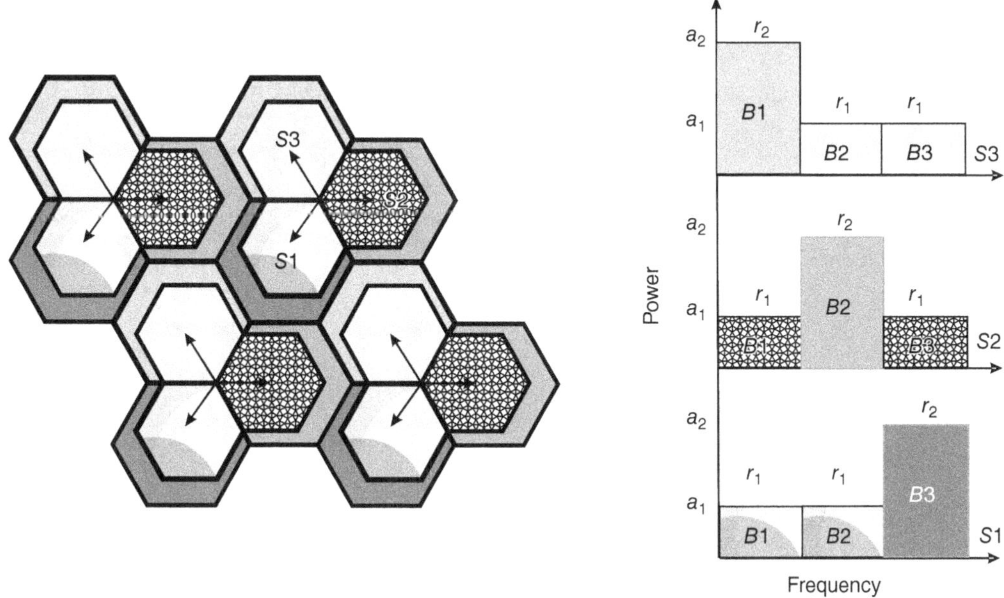

Figure 21.7   Principle of soft frequency reuse. Color version available at wiley.com/go/molisch/wireless3e.
Reproduced with permission from [Hamza et al. 2013] © IEEE.

The reuse factor is adaptively changed with the power ratio between the band used in the inner and the one used in the outer part of the cell. Obviously, there is a tradeoff between frequency reuse (a smaller power ratio means that the sub-bands are used over a larger part of the cell) and interference level (smaller power ratio means worse SIR, in particular near the cell edge). Furthermore, increasing the power ratio provides higher SIR in the cell edge, while at the same time increasing interference in the center of the neighboring cells, and thus reducing the throughput there.

Incremental Frequency Reuse

Incremental Frequency Reuse (IFR) is an attempt to improve the SINR in situations with low traffic load. It does this by starting out like a system with larger frequency reuse factor when the load is light. Assume that the system is capable of reuse-1, but early in the day, there is only demand for 1/3 of the frequency sub-bands. At this time, it may then be advantageous to treat the system as a reuse-3 system, i.e., a user in cell 1 is assigned a frequency sub-band from the first third of the available spectrum, a user in cell 2 from the next third, and so on. This works until the system load is at 1/3. At this point, sub-bands start to be occupied that are also used in other cell types – in other words, the system starts to become similar to a reuse-1 system.

The principle of IFR can also be combined with PFR. In that case, each cell has an edge band that is used only in a particular cell, while the other sub-bands are assigned according to traffic load, in an interference-aware manner. This scheme is also known as *Enhanced Fractional Frequency Reuse* (EFFR). Note that IFR and EFFR require measurements of interference levels and coordinations between cells, and can thus also be considered a form of dynamic Inter-Cell Interference Coordination (ICIC), see Section 21.3.4.

Invert Selective Frequency Reuse

The fundamental idea of *Invert Frequency Reuse* (also known as selective interference avoidance) is to improve the channel quality in the border region by restricting the interference power in certain parts of the cell. Consider Figure 21.8, which shows an example with BSs that transmit into three sectors each. Cells are divided into different groups, and each group transmits with reduced power in one particular sub-band. For example, cell group 1 transmits with reduced power in sub-band 1. That means now that in a neighboring cell, e.g., cell 5, users near the border to cell 1 can operate in sub-band 1 without experiencing significant interference. Similarly, cell 1

---

[7] Different papers have different definitions of SFR.

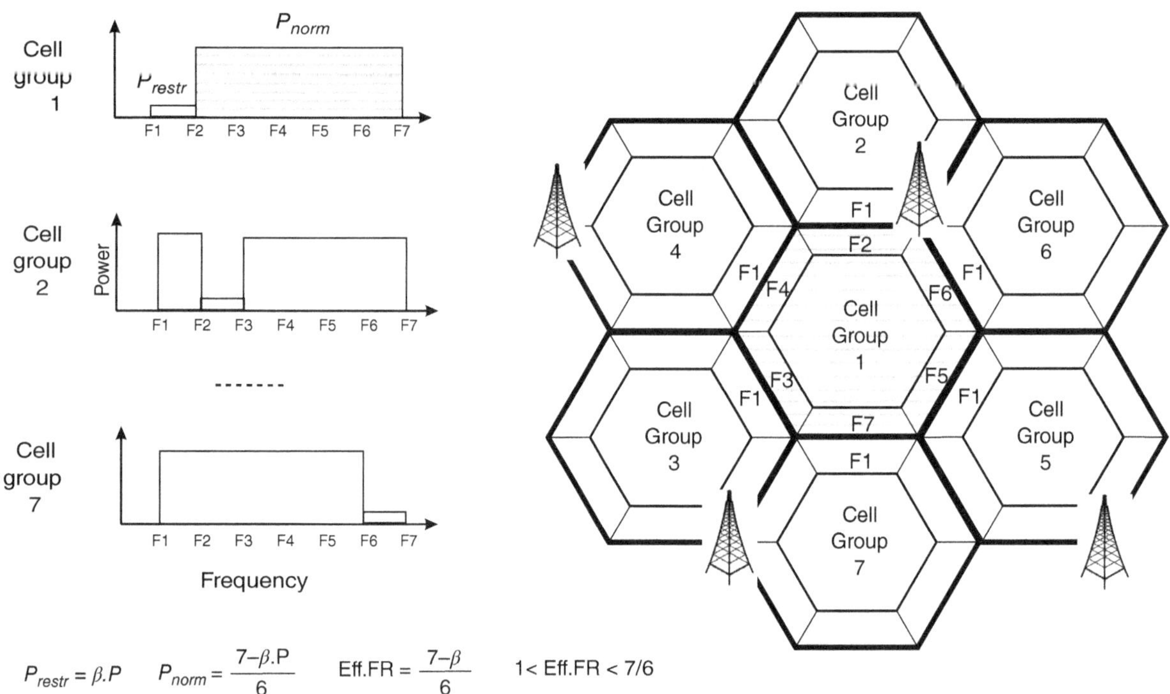

$$P_{restr} = \beta.P \qquad P_{norm} = \frac{7-\beta.P}{6} \qquad \text{Eff.FR} = \frac{7-\beta}{6} \qquad 1 < \text{Eff.FR} < 7/6$$

Figure 21.8 Invert frequency reuse.
Reproduced with permission from [Kosta et al. 2012] © IEEE.

can schedule its associated users on sub-bands 2–7 without significant interference. Note, though, that users in different parts of the cell edge have to be scheduled on different parts of the spectrum.

21.3.3 Cell Breathing and BS Sleeping

Up to now, we have assumed that there is the same (or at least similar) amount of user traffic in every cell. This need not be the case, however. Not only are there random fluctuations arising from the fact that users are active at random times, but there can also be changes on a larger timescale, e.g., a cell covering a conference center might experience heavy traffic during the day, while traffic might shift to the cell covering a nearby train station once the conference has finished. These changing traffic conditions lead to a decreased per-user throughput (in the case of data traffic) or higher probability of blocked calls (in the case of voice traffic) in the more heavily loaded cell.

A method to adapt the cell to changing traffic conditions is *cell breathing*. Depending on the occurring traffic situation, one cell might have fewer active users in it, while a neighboring has more. It is then advantageous to ensure that not the *cell area*, but rather the *traffic load* is approximately the same for all cells. The change of cell size can be affected by variations of the transmit power.[8] If we define as the "cell edge" the point where an UE has an SINR that exceeds (possibly averaged over fading) a required operational threshold, then increasing the transmit power obviously increases the size of the cell. At the same time, such power increase creates additional interference for users in the adjacent cells.

When pushing this approach to the limit, we can imagine situations where a BS is turned off completely (BS sleeping), and the neighboring BSs increase their power to the point where they can cover all UEs in that cell area. Obviously, this cannot be used for enhancing the capacity (since one BS has gone off the air, and thus cannot provide any UE with suitable capacity). However, such BS sleeping can be advantageous for saving energy. The additional transmit power required at the neighboring cells is often minor compared with the savings at the sleeping BS, especially when taking into account that the overall power consumption of a BS is governed not just by the actual Radio Frequency (RF) transmit power, but mostly by the circuit power for baseband and RF components, as well as the interface to the backhaul network, and possibly the cooling system for the BS.

21.3.4 Dynamic Inter-Cell Interference Coordination (ICIC)

The frequency assignment schemes described above are suboptimum because they are based on *average* traffic behavior and *homogeneous* user distributions, without taking into account the *instantaneous* variations of these quantities.[9] In dynamic ICIC, spectral

---

[8] Alternatively, we can simply assign UEs to cells with lighter load even if the link to the associated BS does not provide the highest receive power.
[9] With the exception of IFR and EFFR, which are actually schemes that are at the border between conventional frequency assignment and ICIC.

efficiency is improved by adapting the frequency band assignments to the users dynamically – either according to small-scale averaged channel states, or possibly even according to instantaneous channel states, and traffic demand.

In principle, dynamic ICIC is a large scheduling and power allocation problem. Given the users that need to communicate in the different cells, we wish to determine on which frequency sub-band, and with what power, each user is to be served. Such a problem formulation is especially relevant in OFDMA, where the operating principle allows very a fine-grained assignment of frequency resources. Before going into a mathematical formulation, let us first get an intuitive understanding why scheduling and power control can help to reduce interference between cells.

Consider the situation in Figure 21.9, which depicts the downlink (similar considerations hold for the uplink): UEs A and B belong to cell α; UEs C and D to cell β. UEs A and C are near their respective BSs, while B and D are near the cell edges. The BS uses downlink power control to compensate for the different pathloss to the assigned UEs (though very similar considerations hold without power control). Assume further that only two frequency sub-bands available, and that all channels are frequency-flat. Then let without loss of generality UE A be on sub-band 1, and UE B on sub-band 2. For the scheduling in cell β, we now have two possibilities, namely putting UE C on sub-band 1, or on sub-band 2 (and conversely for UE D). So which is the better choice? Since UE C is a user near the cell center and thus has low pathloss, its associated BS (BS β) can transmit to it with low power. So let us first assume that BS β transmits to UE C on sub-band 1. Then UE A will experience low interference: (i) little power is emitted by the interference source, and (ii) UE A is located far away from the interference source BS β, and thus the interference is strongly attenuated. At the same time, since UE D is a cell-edge user, BS β has to transmit to it with full power. The transmission is in sub-band 2 (the only one left available), and thus in the same sub-band as the BS α – UE B link. Since UE B is close to the cell edge, it will experience strong interference, and have low SIR. Thus one pair of UEs (UE A and C) has excellent SIR and rate, and another pair of UEs (UE B and D) have low SIR and rate. From a sum throughput point of view, this arrangement might be advantageous or not; however, it is certainly intolerable from a fairness point of view, as user satisfaction for cell-edge users will be very low.

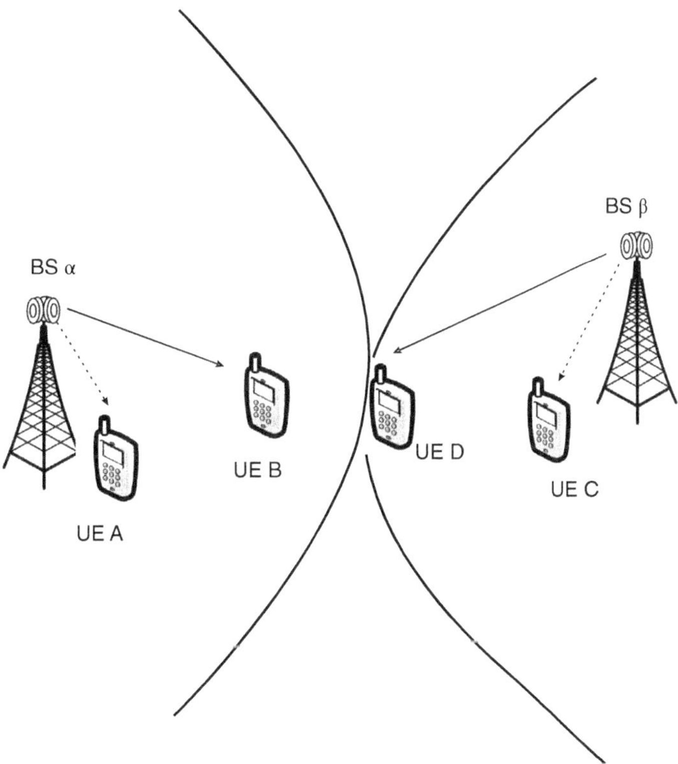

Figure 21.9 Principle of dynamic ICIC.

On the other hand, scheduling UE C on sub-band 2 results in it interfering with the cell edge user UE B. Due to power control, interference will be low (remember that BS β needs to transmit only with small power to UE C). The interference from BS β to user A will also be moderate: it is the link BS β – UE D that creates interference, which has high transmit power, but this power is attenuated more strongly because the distance BS β – UE A is large. This scheduling thus results in two UE pairs that experience reasonably low interference (worse than UE A and UE C, but better than UE B and D, in the previously-mentioned scheduling approach). This scheduling is thus better from a fairness point of view.

But the advantages of ICIC can go further. Consider a situation with multiple cell-edge users that each has the same channel state to their serving BS, but exhibit frequency-selective fading on the channel to the interfering BS. Consider UE E and UE F, both in cell α, with channel of equal strength to their serving BS α. Assume further that UE E has a strong channel to BS β in sub-band 1 (but a weak channel in sub-band 2), while it is vice versa for UE F. Obviously, in this case, it is advantageous to schedule UE E on sub-band 2 (so that it sees little interference) and UE F on sub-band 1.

For a more general and quantitative treatment, we formulate the problem mathematically. Again, consider the downlink: let there be $K(l)$ users in the lth cell. Each user is scheduled on a sub-band (more generally, an orthogonal time-frequency resource), so that there is no intra-cell interference. Let there be N sub-bands, indexed by $n = 1, \dots, N$. The indicator function $\mathbb{1}_{k,l}(n)$ is set to 1 if the nth sub-band is assigned to the kth user in the lth cell, and 0 otherwise; the associated power is $P_{n,l}$ (note that this power is uniquely defined by the sub-band and the cell in which it is transmitted; the dependence on the user index is indirect and follows from the scheduling assignment). The SIR at the kth user in the lth cell on the nth sub-band, denoted as $\gamma_{k,l,n}$ is then

$$\gamma_{k,l,n} = \frac{\mathbb{1}_{k,l}(n)P_{n,l}|h_{k,l,n}|^2}{\sum\limits_{l' \neq l, k'} \mathbb{1}_{k',l'}(n)P_{n,l'}|h_{k,l',n}|^2 + \sigma_n^2}. \tag{21.14}$$

The scheduling constraints are that for each cell the total transmitted power must be upper limited

$$\sum_{n, k'} \mathbb{1}_{k',l'}(n)P_{n,l'} \leq P_{\max} \text{ for all } l' \tag{21.15}$$

and that no more than one user within each cell can be assigned a sub-band

$$\sum_{k} \mathbb{1}_{k,l'}(n) \leq 1 \text{ for all } n, l'. \tag{21.16}$$

The rate that a particular user can then achieve is just the summation of the rates on the different subcarriers, and in the case of capacity-achieving transmission simply becomes

$$R_{k,l} = \sum_{n} \log_2\left(1 + \gamma_{k,l,n}\right). \tag{21.17}$$

Just as in the single-cell case (compare Section 20.1.4), we usually aim to maximize the sum throughput, i.e., $\sum_{l}\sum_{k} R_{k,l}$, or some other (usually convex) measure of the throughput, such as the sum of the logarithms of the rates. We can also have additional constraints such that each user must be guaranteed a minimum rate.

It is clear from the above formulation that optimum ICIC is a mixed-integer optimization problem; such problems are generally very hard to solve exactly (they are usually Non-deterministic Polynomial Time (NP)-hard, except in special circumstances). Furthermore, optimum ICIC would require knowledge of all Channel State Information (CSI) both within a cell, and between cells, at a central controller. The overhead for acquiring and distributing such information is very high, and thus often not practicable. For these reasons, suboptimal algorithms for ICIC have been developed that use approximate methods to solve the optimization problem, use reduced amount of CSI, or both; alternatively, use of machine learning is an option (see Sec. 20.7).

Simplified optimization methods mainly fall into three categories:

- *Greedy Algorithms.* This approach is similar to the one described in Section "Frequency-Selective Channel" in Sec. 20.3.2. Assume first that all signals are transmitted with the same power. Then start out by assigning the sub-band to the user that has the maximum $\gamma_{k,l,n}$ among all sub-band/user/cell combinations. We then select the next best, where "best" now has to incorporate not only the utility the assignment gives to the selected user, but also the impact it has on the previously selected user (if it is in the same sub-band, it might create interference, and thus reduce the utility). This process is repeated until all bands are assigned (sub-bands can be left empty if their use would reduce the overall utility function).

  If power control is also foreseen, then we can first assign the Time Frequency (TF)-blocks assuming equal power for all users, and then optimize the power assignments based on the chosen TF-block scheduling. Obviously, this approach is not optimum, but it does provide a fairly simple (and often surprisingly effective) method of interference coordination.

- *Relaxation Techniques.* The main difficulty in solving the general ICIC problem is the mixture of continuous and discrete constraints. In other words, the problems arise because we specify that assignment of a sub-band to a particular user can only be "full," or not all, in other words, the assignment variables are restricted to the discrete values 0 and 1. A way around this problem is the so-called "relaxation" technique, where the assignment variables are allowed to take on *any continuous value* between 0 and 1. This converts the mixed integer-continuous optimization problem into a purely continuous problem that often is convex (and thus can be solved optimally). For a final result, the assignment variables might be "rounded" to 0 or 1, to recover again a solution that can be implemented with FDMA/TDMA/OFDMA scheduling.

  Scheduling also includes the problem of BS association, i.e., which BS should serve a particular UE. While a standard solution is to associate the UE with the BS to which it has the (instantaneously, or on average) strongest link, this might not be optimum, since it might connect too many users with one particular BS, which then becomes overloaded. The scheduling formulation given above allows to include the BS association as a variable. In most cases, the association variable is binary, i.e., one UE can be connected to one particular BS at a time. However, in the case of BS cooperation (see Section 22.11), fractional association is feasible. This not only makes the mathematical solution easier but also improves the overall system performance.

- *Breaking into sub-problems.* One of the challenges in solving the general ICIC problem is that it requires taking into account CSI and scheduling over many different cells. However, interference from far-away cells usually is not a major factor in the performance. Complexity can thus be reduced by dividing the interferers into "dominant" and "nondominant," and performing ICIC only for dominant interferers.

21.4 Cell Planning with Irregular Deployment

21.4.1 Motivations for Irregular Deployment

Hexagonal cell structures are only ideal under certain assumptions that are often not fulfilled in practice:

- *The requested traffic density is independent of the location.* This condition is obviously violated whenever the population density changes. As a matter of fact, traffic densities are extremely inhomogeneous both with respect to time and to space. For example, during office hours, traffic in downtown areas is much higher than in suburban areas, and even within a downtown, traffic density varies greatly. Figure 21.10 shows an example of a traffic density map in Shanghai, China.
- *The pathloss is influenced only by the distance between BS and UE*; this in turn usually requires that the terrain is completely flat, and there are no high edifices. This is rarely fulfilled. Figure 21.11 shows the terrain, and the power obtained from one BS in a typical hilly terrain in Vienna, Austria. According to the simplified picture above, lines of equal power should be concentric circles around the BS. We see that the actual result is anything but – the power is not even necessarily a monotonic function of the distance in some directions.
- *It is permissible to locate the BSs at those places that are ideal from the coverage/capacity point of view.* In many situations this is not fulfilled, since building owners might either refuse to have a BS placed on their building, local building codes might prohibit it (e.g., on historical buildings), or placement on certain buildings might not be economically viable (high rent demands from the owners, difficulty in connecting to the backbone networks, etc.).

For all these reasons, realistic deployments are not on a hexagonal grid, but rather in a manner that looks almost random. Consequently, it is mathematically convenient to model the BS locations as realizations of a random spatial process, since this allows the application of powerful tools of stochastic geometry to the problem of interest.

A *Poisson Point Process* (PPP) can be viewed as the most extreme case of a point process, in the sense that all points are placed independently of each other. In real life, there is of course a certain amount of cell planning. Yet, the PPP provides often one extreme case of the system performance, just like the hexagonal grid provides another. Actually, the difference in performance of these two deployment scenarios is often surprisingly small – Figure 21.12 compares the outage probability in a grid and a PPP deployment.

Other point processes have been identified that provide a compromise – "repulsive" processes describe situations where it is very unlikely that two points are located very closely to each other; this situation obviously occurs in the placement of BSs. On the other hand, "attractive" point processes describe situations where points occur in clusters – this describes often the location of UEs, since users tend to be located close to each other.

In the following, we will first introduce some mathematical foundations of stochastic geometry, and then use those to analyze the system performance in a PPP environment. The derivations are restricted to the simplest cases (homogeneous PPP, pathloss plus Rayleigh fading), to work out the key properties of such networks.

*21.4.2 Random Point Processes

Since the theory of random point processes is the foundation of the mathematical treatment of random BS deployment, and this theory is not covered in typical engineering mathematics courses, this section provides a *very* brief introduction. Much more extensive descriptions can be found in the references cited in "Further Reading."

Owing to its remarkable tractability, the most popular point process is the PPP. It is created by dropping, at random, points on a (typically two-dimensional) map; it is characterized by the density λ of the process, i.e., how many points are dropped on average in a unit area. Let this process be denoted as $\mathbf{\Phi}(\mathbf{X}_1, \mathbf{X}_2, \ldots)$ where $\mathbf{X}_1, \mathbf{X}_2, \ldots$ are the locations. It can be shown that for any region $\mathbf{A}$, the probability that there are exactly n points in this region is given as

$$P(\Phi(\mathbf{A}) = n) = \exp(-\lambda A)\frac{(\lambda A)^n}{n!} \tag{21.18}$$

where Φ is the counting measure (i.e., number of points) of the process $\mathbf{\Phi}$, and A is the size of region $\mathbf{A}$.[10] Furthermore, the number of points in regions $\mathbf{A_1}$ and $\mathbf{A_2}$ are independent RVs if $\mathbf{A_1}$ and $\mathbf{A_2}$ do not overlap. Generally, the point density λ can vary with the location

---

[10] The expected number of points $\Lambda(\mathbf{A})$ in an area $\mathbf{A}$ is generally known as the intensity measure. In the case of the homogeneous PPP, it becomes $\lambda(\mathbf{A})$.

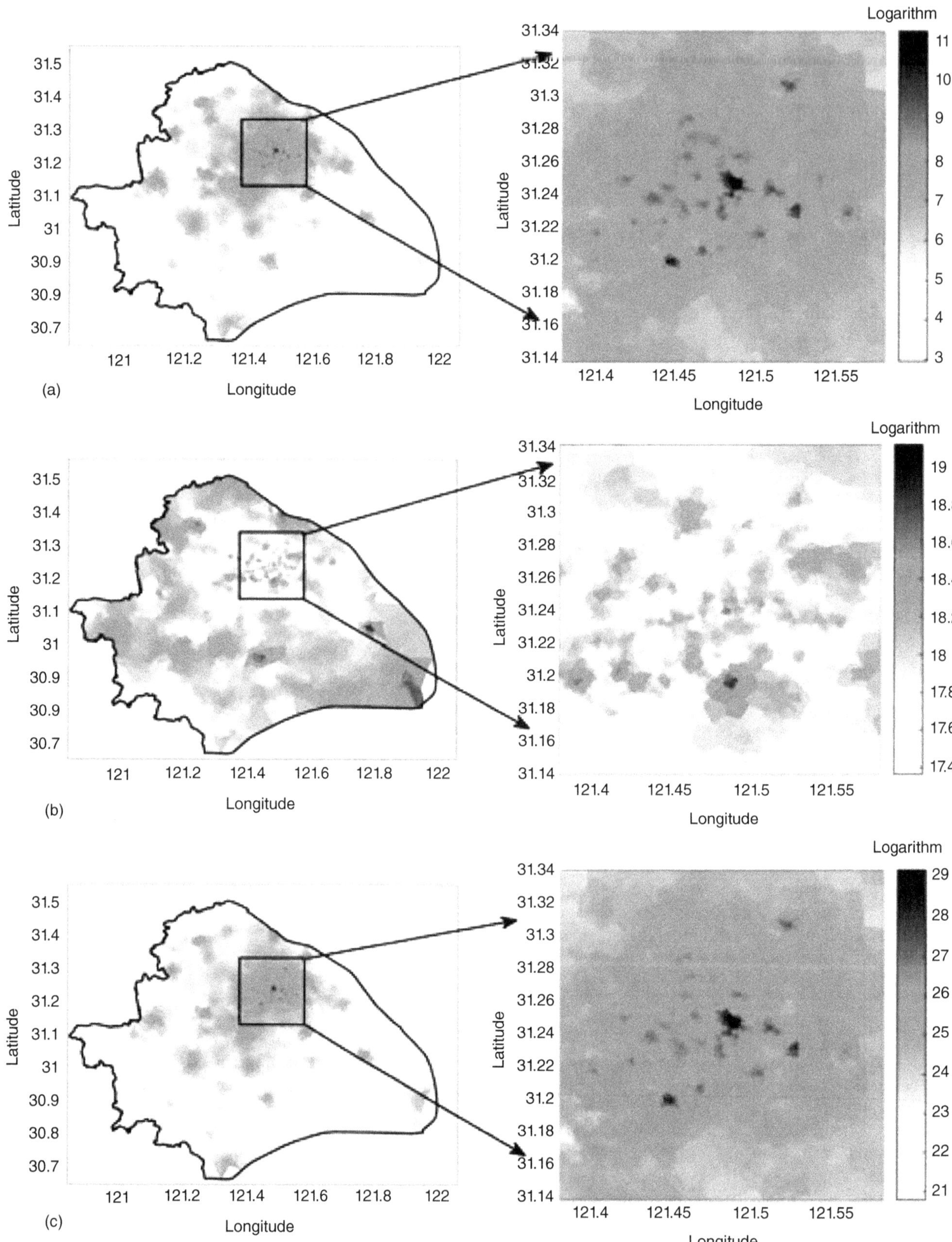

Figure 21.10 Geographical distributions of subscriber density (a), per-subscriber demand (b), and traffic density (c) in Shanghai, China. Color version available at wiley.com/go/molisch/wireless3e.

Reproduced with permission from [Ding et al. 2017] © IEEE.

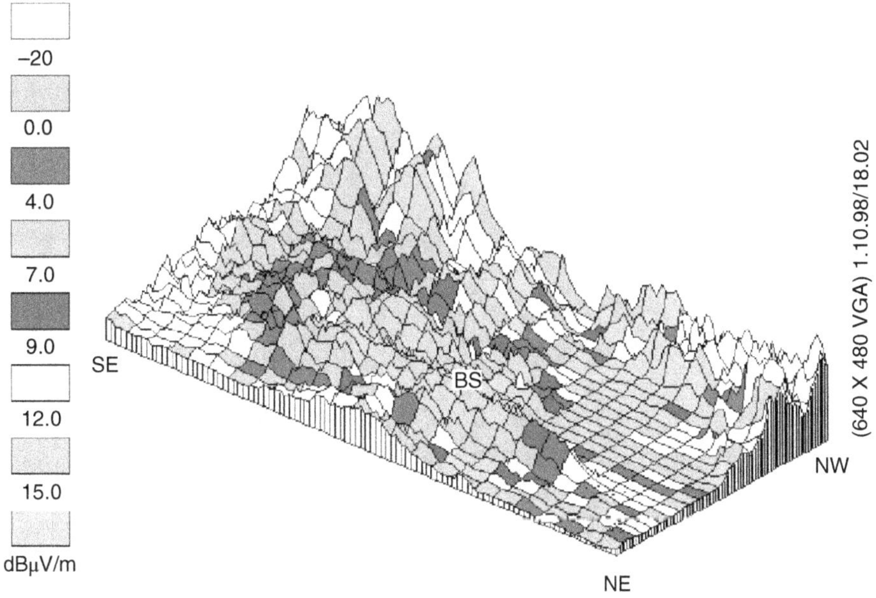

Figure 21.11 Example of received field strength in hilly terrain (western part of Vienna, Austria).
Reproduced with permission from [Buehler 1994] © H. Buehler.

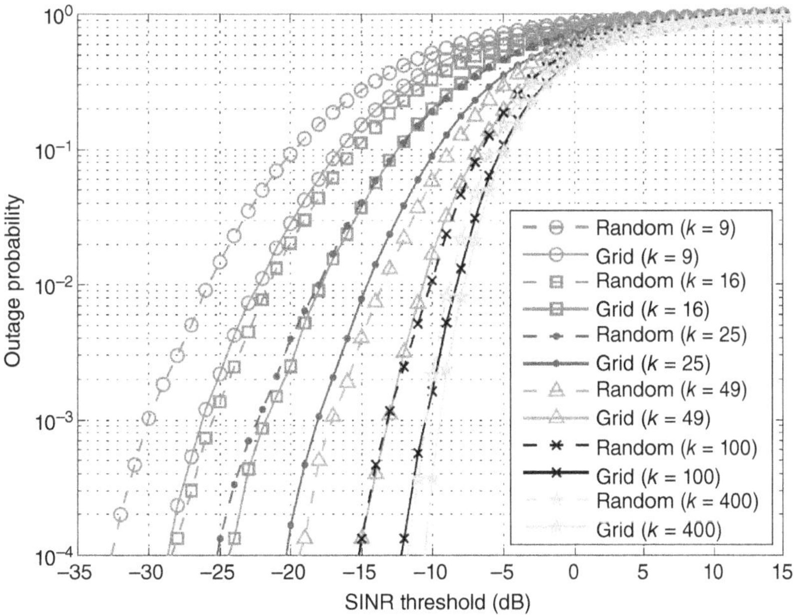

Figure 21.12 Outage probability versus SINR in grid and random topologies. k is the number of BSs per 1 km^2. Color version available at wiley.com/go/molisch/wireless3e.
Reproduced with permission from [Chen et al. 2012a] © IEEE.

(and thus strictly speaking should be written as $\lambda(\mathbf{A})$ – which we don't do to keep the notation simple). In the case that λ is independent of the location, we speak of a *homogeneous* PPP. Most of our considerations below are done for this important special case.

Obviously, the distance between two points of the process becomes a RV. The Probability Density Function (pdf) of the distance to the nth nearest neighbor can be obtained from first considering the complementary cdf: the probability that the nth nearest neighbor points is at a distance larger than r is the probability that there are fewer than n points in a circle of radius r around the considered location. But from (21.18), this is

$$\exp\left(-\lambda\pi r^2\right)\sum_{k=0}^{n-1}\frac{\left(\lambda\pi r^2\right)^k}{k!}. \tag{21.19}$$

The pdf then follows as the negative derivative with respect to r, which after some simplifications becomes

$$pdf_{R_n}(r) = r^{2n-1}(\lambda\pi)^n \frac{2}{\Gamma(n)} \exp\left(-\lambda\pi r^2\right) \quad \text{for} \quad r \geq 0. \tag{21.20}$$

For the special case of the nearest neighbor, $n = 1$, this becomes a Rayleigh distribution.

The *mapping theorem* allows to derive a number of other important properties. As a specific example, let $f(\mathbf{x})$ be a mapping function that maps the two-dimensional space (of locations) onto a one-dimensional space. Then $\hat{\Phi}(f(\mathbf{X_1}), f(\mathbf{X_2}), ...)$ is a one-dimensional Poisson process with intensity function $\Lambda^*(A) = \Lambda(f^{-1}(A))$. For example, consider the density of the distances of points (taken from a homogeneous PPP) from the origin: the mapping function $f(\mathbf{x}) = |\mathbf{x}|$ and the inverse function maps all points within a given distance r onto $b(r)$, a circle of radius r centered at the origin, $f^{-1}(A) = b(r)$. Consequently, $\Lambda^*(A) = \int_{b(r)}\lambda dx = \lambda\pi r^2$, and the density is the derivative

$$\lambda^*(r) = 2\pi\lambda r. \tag{21.21}$$

In other words, the distances form a nonhomogeneous PPP with linearly increasing density.

Another important formula for a PPP is *Campbell's theorem*, which allows to compute the expectation of a function, summed over a point process. It states that

$$E\left\{\sum_{\mathbf{X}_i \in \Phi} f(\mathbf{X}_i)\right\} = \int f(\mathbf{x})\Lambda(d\mathbf{x}). \tag{21.22}$$

Finally, we introduce the *Probability Generating Functional* (PGF), $G(v)$. For a function $v(x)$ that is defined in the interval $[0, 1]$, the PGF is defined as

$$G(v) = E\left\{\prod_{\mathbf{x} \in \Phi} v(\mathbf{x})\right\}. \tag{21.23}$$

For a PPP, this can be computed as

$$G(v) = \exp\left(-\int[1 - v(\mathbf{x})]\Lambda(d\mathbf{x})\right). \tag{21.24}$$

The PGF can be used to compute the moment-generation function, and the Laplace transform of the pdf of the variable of interest.

A few other important properties of PPPs include:

- If we create a "thinned-out" process by starting out with a PPP, and independently making a binary decision (keep/discard) for every point, the resulting process is again a PPP, with a density that is the product of the original density and the probability of keeping a point. Similarly, adding two independent PPP results in another PPP, whose density is the sum of the densities of the original processes.
- Another important statement is that due to the independence of all the points, conditioning on a point at $\mathbf{x}$ does not change the distribution of the remainder of the process; i.e., there is no change in the distribution whether or not we condition on having a point at $\mathbf{x}$ (Slivnyak's theorem).
- If we start from a PPP and displace each point by a random vector, the resulting point process will also be a PPP. A key application of the displacement theorem mentioned above is the incorporation of large-scale fading into computations in wireless networks. In particular, assuming a pathloss law $r^{-\alpha_{\text{PL}}}$, and independent shadowing on the different links, all with the same distribution of the shadowing variable S, the "effective" PPP seen for computations is a PPP with density $\lambda E\{S^\delta\}$ with $\delta = 2/\alpha_{\text{PL}}$.

*21.4.3 Received Signal Power from a Homogeneous PPP

The above mathematical preliminaries can be used to compute the total signal power received at the origin (where we assume the UE to be located) from sources that are distributed according to a homogeneous PPP. Assume that pathloss (with path loss coefficient α_{PL}) is present, as well as fading with a *power* distribution $\tilde{h}_{\mathbf{r}}$, so that the ratio of RX power to TX power is $h = \tilde{h}_{\mathbf{r}} r^{-\alpha_{\text{PL}}}$ where $\mathbf{r}$ is the location vector of the point and r the distance from the origin. Thus, assuming unit transmit power,

$$P_{\text{PPP}} = \sum_{\mathbf{r} \in \Phi} \tilde{h}_{\mathbf{r}} r^{-\alpha_{\text{PL}}}. \tag{21.25}$$

In order to obtain the expected value of the interference, note first that the expectation consists actually of two (independent) expectations: expectation over the fading, and expectation over the locations, i.e., the PPP. The expectation over the fading is easily handled if we assume that the fading has Rayleigh amplitude distribution, i.e., the power (variance) is $E\{\widetilde{h}_r\} = 1$, see Section 5.4.3. For the expectation over the locations, apply Eqs. (21.22) and (21.21), so that

$$E\{P_{PPP}\} = \int r^{-\alpha_{PL}} 2\pi\lambda r dr = \begin{cases} \frac{2\pi\lambda}{2-\alpha_{PL}} r^{2-\alpha_{PL}} \Big|_{r_a}^{r_b} & \alpha_{PL} \neq 2 \\ 2\pi\lambda\ln(r)\big|_{r_a}^{r_b} & \alpha_{PL} = 2 \end{cases}. \tag{21.26}$$

This expression becomes infinite when we set $r_a = 0$ and $r_b = \infty$, as would be intuitive for interference in an unbounded plane. This is of course not physical, but its source lies in certain mathematical simplifications: for $\alpha_{PL} < 2$, we get too much interference from far-away nodes; however in practice a pathloss coefficient smaller than 2 occurs only in Line Of Sight (LOS), and even then only until the breakpoint (compare Section 7.1), so that such a situation would never occur in practice. For $\alpha_{PL} > 2$, the problem arises from inter-ferers at very small distance from the origin; which (due to the oversimplified pathloss model) can receive more power than the inter-fering node transmits. A practical solution is to bound the smallest distance so that the received interference power is never higher than a node's transmit power, $\min(1, r^{-\alpha_{PL}})$. For $\alpha_{PL} < 2$, a breakpoint model should be introduced to avoid problems with far-away interferers, or we can simply limit the size of the network to D. Under the latter assumption, the expectation of the interference becomes

$$\lambda \frac{2\pi}{2-\alpha_{PL}} D^{2-\alpha_{PL}}. \tag{21.27}$$

We next proceed to the determination of the pdf of the power; more precisely the Laplace transform of the pdf:

$$E\{\exp(-sP_{PPP})\} = E_{\hat{\Phi},\widetilde{h}}\left\{\exp\left(-s\sum_{\mathbf{r}\in\hat{\Phi}} \widetilde{h}_{\mathbf{r}} r^{-\alpha_{PL}}\right)\right\}. \tag{21.28}$$

Again, assume that the fading statistics are independent of the location; furthermore, according to the definition of the PPP, the loca-tions of the points are independent of each other. Therefore, Eq. (21.28) can be written as

$$E_{\hat{\Phi}}\left\{\prod_{\mathbf{r}\in\hat{\Phi}} E_{\widetilde{h}}\left\{\exp\left(-s\widetilde{h}_{\mathbf{r}} r^{-\alpha_{PL}}\right)\right\}\right\}. \tag{21.29}$$

We can now identify $E_{\widetilde{h}}\left\{\exp\left(-s\widetilde{h}_{\mathbf{r}} r^{-\alpha_{PL}}\right)\right\}$ as $v(r)$ in Eq. (21.23). Using (21.24), Eq. (21.29) becomes

$$E\{\exp(-sP_{PPP})\} = \exp\left(-\int_0^\infty \left[1 - E_{\widetilde{h}}\left\{\exp\left(-s\widetilde{h}_{\mathbf{r}} r^{-\alpha_{PL}}\right)\right\}\right]\lambda^*(dr)\right) \tag{21.30}$$

where $\lambda^*(dr) = 2\pi\lambda r dr$. Swapping expectation and integration, and conditioning on a fading realization $\widetilde{h}$, and noting that the statistics of the fading are assumed to be independent of location, the integral can be evaluated to

$$\lambda\pi\left(s\widetilde{h}\right)^\delta \Gamma(1-\delta) \qquad \text{for } 0 < \delta < 1 \tag{21.31}$$

where again $\delta = 2/\alpha_{PL}$. Taking the expectation over $\widetilde{h}$, we finally get

$$E\{\exp(-sP_{PPP})\} = \exp\left\{-\lambda\pi E\{\widetilde{h}^\delta\}\Gamma(1-\delta)s^\delta\right\} \quad 0 < \delta < 1. \tag{21.32}$$

This is known as a *stable distribution*. It is characterized by the *characteristic exponent* which in our case is δ, and the *dispersion*, which in our case is $\lambda\pi E\{\widetilde{h}^\delta\}\Gamma(1-\delta)$. For Rayleigh fading, $E\{\widetilde{h}^\delta\} = \Gamma(1+\delta)$. The Laplace transform then becomes

$$E\{\exp(-sP_{PPP})\} = \exp\left\{-\lambda\pi\Gamma(1+\delta)\Gamma(1-\delta)s^\delta\right\} = \exp\left\{-\lambda\pi \frac{\pi\delta}{\sin(\pi\delta)} s^\delta\right\}. \tag{21.33}$$

The situation is more complicated when considering the interference power in a cellular network. Assume that the closest BS to the UE of interest is at distance r_0; this BS functions as the serving BS. Interference can only come from BSs that are farther away than that, since by definition the signal from the closest BS is the desired one. We thus wish to compute

$$P_I = \sum_{\mathbf{r} \in \Phi \backslash b(r_0)} \widetilde{h}_{\mathbf{r}} |\mathbf{r}|^{-\alpha_{PL}}.$$ (21.34)

We can again start out from Eq. (21.29). However, we note that for the next step, the integration range can only extend from r_0 to ∞. Thus,

$$E\{\exp(-sP_I)\} = \exp\left(-\int_{r_0}^{\infty}\left[1 - E_{\widetilde{h}}\left\{\exp\left(-s\widetilde{h}_{\mathbf{r}} r^{-\alpha_{PL}}\right)\right\}\right]2\pi\lambda r dr\right).$$ (21.35)

Using the fact that $\widetilde{h}_{\mathbf{r}}$ is an exponential RV,

$$\begin{aligned}
E\{\exp(-sP_I)\} &= \exp\left(-\int_{r_0}^{\infty}\left(1 - \frac{1}{1 + sr^{-\alpha_{PL}}}\right)2\pi\lambda r dr\right) \\
&= \exp\left(-\int_{r_0}^{\infty}\left(\frac{1}{1 + s^{-1}r^{\alpha_{PL}}}\right)2\pi\lambda r dr\right)
\end{aligned}$$ (21.36)

using $1 - \dfrac{1}{1+a} = \dfrac{1}{1+1/a}$. This integral needs to be evaluated numerically.

*21.4.4   Performance in Random Deployment – Downlink

With the above derivations, we can now tackle an analysis of the interference levels; we start out with the downlink. Let the UE of interest be located at the origin, let the (geographically) nearest BS be the serving BS, and all other BSs be interfering BSs (note that this implies a scenario in which we have frequency reuse 1, i.e., there is no clustering that reduces the interference as discussed in Section 21.2).

We now assume that a UE has coverage if the receive SINR is larger than a particular threshold θ. The probability of coverage is

$$p_c(\theta) = \int_0^{\infty} \Pr(\gamma > \theta \mid r_0) pdf_{r_0}(r_0) dr_0$$ (21.37)

$$\begin{aligned}
&= \int_0^{\infty} \Pr\left(\frac{\widetilde{h}_0 r_0^{-\alpha_{PL}}}{\sigma_n^2 + P_I} > \theta\right) 2\pi\lambda r_0 \exp\left(-\lambda\pi r_0^2\right) dr_0 \\
&= 2\pi\lambda \int_0^{\infty} \Pr\left[\widetilde{h}_0 > \theta r_0^{\alpha_{PL}}\left(\sigma_n^2 + P_I\right)\right] r_0 \exp\left(-\lambda\pi r_0^2\right) dr_0
\end{aligned}$$ (21.38)

where $\widetilde{h}_0$ is the fading on the desired link. Since it is an exponentially distributed variable, $\Pr()$ is

$$E_{P_I}\left[\exp\left(-\theta r_0^{\alpha_{PL}}\left(\sigma_n^2 + P_I\right)\right)\right] = \exp\left(-\theta r_0^{\alpha_{PL}}\sigma_n^2\right) E_{P_I}\left[\exp\left(-\theta r_0^{\alpha_{PL}} P_I\right)\right]$$ (21.39)

where the second term equals $E\{\exp(-sP_I)\}$ evaluated at $s = \theta r_0^{\alpha_{PL}}$. Using Eq. (21.36), $p_c(\theta)$ becomes

$$p_c(\theta) = 2\pi\lambda \int_0^{\infty} \exp\left(-\theta r_0^{\alpha_{PL}}\sigma_n^2\right) r_0 \exp\left(-\lambda\pi r_0^2\right) \exp\left(-\int_{r_0}^{\infty}\left(\frac{1}{1 + \theta^{-1}r_0^{-\alpha_{PL}} r^{\alpha_{PL}}}\right)2\pi\lambda r dr\right) dr_0.$$ (21.40)

Defining $u = (r_0/r)^2 \theta^{-2/\alpha_{PL}}$, $v = r_0^2$ and

$$\Xi(\theta) = \theta^{2/\alpha_{PL}} \int_{\theta^{-2/\alpha_{PL}}}^{\infty} \frac{1}{1 + u^{\alpha_{PL}/2}} du$$ (21.41)

the coverage probability can be written as

$$p_c(\theta) = \pi\lambda \int_0^{\infty} \exp\left[-\lambda\pi v(1 + \Xi(\theta)) - \theta v^{\alpha_{PL}/2}\sigma_n^2\right] dv.$$ (21.42)

For the special case of $\alpha_{PL} = 4$ and noise negligible compared to the interference, this becomes

$$p_c(\theta) = \frac{1}{1 + \sqrt{\theta}\arctan\left(\sqrt{\theta}\right)}.$$ (21.43)

The coverage probability of such a system is much too low for practical purposes – even for a threshold of $\theta = 1$, i.e., SIR of 0 dB, transmission is successful only at about half of the locations. Obviously, also in the random deployment case, we need a reuse distance larger than 1. If we silence the $n - 1$ interferers that are strongest on average, then the success probability becomes

$$p_s = (n-1)\delta \int_0^1 \frac{(1-r^\delta)^{n-2} r^{\delta-1}}{({}_2F_1(1, -\delta; 1 - \delta; -\theta r))^n}\, dr \tag{21.44}$$

where F is the hypergeometric function [Abramowitz and Stegun 1965].

*21.4.5 Performance in Random Deployment – Uplink

Computation of the uplink case is considerably more complicated. This seemingly surprising problem arises from the fact that the interference, and resulting capacity, depends on the user associations with a particular BS. A number of approximations have been developed in the literature to deal with this complication.

Let us consider the following simple situation: The UEs are distributed according to a homogeneous PPP with density λ_U. They are associated with the BSs that are closest to them (the area of all points that are closest to a particular BS is called a *Voronoi cell*) and thus have the best SNR. We assume OFDMA/TDMA/FDMA, so that on one time-frequency resource only one user is active in each Voronoi cell. We then make the *approximation* that the resulting distribution of active users is also a PPP.[11] We consider a typical UE located at the origin of the coordinate system, associated with its serving BS at $\mathbf{B}_0$, denoting $\|\mathbf{B}_0\| = r_0$, the distance from the kth interfering UE to its serving BS as r_k, and the distance from the interfering UE to the BS at $\mathbf{B}_0$ as $\breve{r}_k$. We will also consider fractional power control, i.e., the transmit power is adjusted as $r^{\alpha_{PL}\epsilon}$ with $\epsilon = [0, 1]$, so that $\epsilon = 0$ means fixed TX power, and $\epsilon = 1$ means full power control; note, however, that the power control does not adapt to the instantaneous channel state. Thus, the fading state (but not the distance!) of an interfering UE with respect to its own serving BS is irrelevant for the subsequent interference computation; however, the channel state $\widetilde{h}_k$ with respect to the BS at $\mathbf{B}_0$ is important, as we will see below.

It is noteworthy that an interfering UE can be closer to $\mathbf{B}_0$ than the desired UE; the only condition for an interfering UE is that its own serving BS is closer to it than $\mathbf{B}_0$. In other words, it is possible that $\breve{r}_k < r_0$, but we always must have $\breve{r}_k > r_k$. Generally, we want to find the probability that a UE located at distance $\breve{r}_k$ from $\mathbf{B}_0$ actually is an interfering UE. This happens when $\mathbf{B}_0$ is *not* the closest BS – the probability of which is, as seen in (21.20) with $n = 1$, is $1 - \exp\left(-\pi\lambda\breve{r}_k^{\,2}\right)$. We can thus approximate the density of interfering UEs as a PPP Φ_t with a location-dependent thinning,

$$\lambda_I = \lambda\left(1 - \exp\left(-\pi\lambda\breve{r}^{\,2}\right)\right) \tag{21.45}$$

the total interference can thus be computed as

$$P_I = \sum_{\mathbf{X}_k \in \Phi_t} r_k^{\alpha_{PL}\epsilon} \widetilde{h}_k \breve{r}_k^{\,-\alpha_{PL}}. \tag{21.46}$$

The SINR can then be written as

$$\gamma = \frac{\widetilde{h}_0 r_0^{\alpha_{PL}(\epsilon-1)}}{\sigma_n^2 + P_I} \tag{21.47}$$

where we have again set the nominal transmit power (without power control) to unity. The distance r_0 is Rayleigh distributed, since it is simply the probability that there are no other BSs closer to the considered UE (which is at the origin) than the one at $\mathbf{B}_0$. The distribution of the distances between the interfering UEs and the BSs they are associated with is somewhat more complicated. As mentioned above $r_k < \breve{r}_k$ (since otherwise, the kth UE would connect to $\mathbf{B}_0$). We thus use as the marginal distribution for r_k a *truncated Rayleigh distribution*

$$pdf\left(r \mid \breve{r}_k\right) = \frac{2\pi\lambda r \exp\left(-\lambda\pi r^2\right)}{1 - \exp\left(-\pi\lambda\breve{r}_k^{\,2}\right)} \quad \text{for} \quad r < \breve{r}_k. \tag{21.48}$$

---

[11] In reality, the "thinning out" of the users for each BS is not independent, since only one user per BS can be selected.

Furthermore, though the r_i are all correlated with each other, it is mathematically convenient (and a reasonable approximation, as simulations have shown) to neglect the correlation.

We next proceed to the determination of the Laplace transform of the pdf. Similar to Eq. (21.29)

$$E\{\exp(-sP_{\mathrm{I}})\} = E_{\Phi_{\mathrm{I}}, r_k, \tilde{h}_k}\left\{\prod_{\mathbf{X}_k \in \Phi_{\mathrm{I}}} \exp\left(-s\tilde{h}_k r_k^{\alpha_{\mathrm{PL}}\epsilon}\breve{r}_k^{-\alpha_{\mathrm{PL}}}\right)\right\} \tag{21.49}$$

where by $E_{\Phi_k}\{\}$ we mean the expectation over the interfering UEs (and thus the $\mathbf{X}_k$, and thus the $\breve{r}_k$), while E_{r_k} is the expectation over their serving BSs, and $E_{\tilde{h}_k}$ is the expectation over the fading states. Since the $\tilde{h}_k$ are exponentially distributed variables and independent of the r_k, this becomes

$$E\{\exp(-sP_{\mathrm{I}})\} = E_{\Phi_{\mathrm{I}}}\left\{\prod_{\mathbf{X}_k \in \Phi_{\mathrm{I}}} E_{r_k}\left\{\frac{1}{1 + sr_k^{\alpha_{\mathrm{PL}}\epsilon}\breve{r}_k^{-\alpha_{\mathrm{PL}}}}\right\}\right\} \quad . \tag{21.50}$$

Now applying Eqs. (21.24) and (21.45), we get

$$E\{\exp(-sP_{\mathrm{I}})\} = \exp\left[-2\pi\lambda\int_0^\infty \left(1 - \exp\left(-\pi\lambda x^2\right)\right)\left(1 - E_{r_i}\left\{\frac{1}{1 + sr_i^{\alpha_{\mathrm{PL}}\epsilon}x^{-\alpha_{\mathrm{PL}}}}\right\}\right)xdx\right]. \tag{21.51}$$

Using $1 - \frac{1}{1+a} = \frac{1}{1+1/a}$ and the pdf of r Eq. (21.48), we finally get

$$E\{\exp(-sP_{\mathrm{I}})\} = \exp\left[-2\pi\lambda\int_0^\infty\int_0^{x^2} \frac{1}{1 + s^{-1}u^{-\alpha_{\mathrm{PL}}\epsilon/2}x^{\alpha_{\mathrm{PL}}}}\pi\lambda\exp(-\lambda\pi u)dudx\right] \quad . \tag{21.52}$$

The probability of uplink coverage, i.e., the probability that the uplink SINR exceeds a threshold θ, needs to be averaged over the distance distribution of the UEs

$$\begin{aligned}p_{\mathrm{c}}(\theta) &= \int_0^\infty \Pr(\gamma > \theta \mid r_0)pdf_{r_0}(r_0)dr_0 \\ &= \int_0^\infty \Pr\left(\frac{\tilde{h}_0 r_0^{\alpha_{\mathrm{PL}}(\epsilon-1)}}{\sigma_{\mathrm{n}}^2 + P_{\mathrm{I}}} > \theta\right)2\pi\lambda r_0\exp\left(-\lambda\pi r_0^2\right)dr_0.\end{aligned} \tag{21.53}$$

Since $\tilde{h}_0$ is exponentially distributed,

$$p_{\mathrm{c}}(\theta) = \int_0^\infty E_{P_{\mathrm{I}}}\left\{\exp\left(-\theta r_0^{\alpha_{\mathrm{PL}}(1-\epsilon)}P_{\mathrm{I}}\right)\right\}\exp\left(-\theta r_0^{\alpha_{\mathrm{PL}}(1-\epsilon)}\sigma_{\mathrm{n}}^2\right)2\pi\lambda r_0\exp\left(-\lambda\pi r_0^2\right)dr_0 \quad . \tag{21.54}$$

We see that the first term in the integral is Eq. (21.52) evaluated at $s = \theta r_0^{\alpha_{\mathrm{PL}}(1-\epsilon)}$. Substitution thus gives the final result

$$\begin{aligned}p_{\mathrm{c}}(\theta) &= 2\pi\lambda\int_0^\infty dr_0\exp\left(-\theta r_0^{\alpha_{\mathrm{PL}}(1-\epsilon)}\sigma_{\mathrm{n}}^2 - \lambda\pi r_0^2\right)r_0\cdot \\ &\quad \cdot\exp\left[-2\pi\lambda\int_0^\infty\int_0^{x^2} \frac{1}{1 + \theta^{-1}r_0^{-\alpha_{\mathrm{PL}}(1-\epsilon)}u^{-\alpha_{\mathrm{PL}}\epsilon/2}x^{\alpha_{\mathrm{PL}}}}x\pi\lambda\exp(-\lambda\pi u)dudx\right].\end{aligned} \tag{21.55}$$

Figure 21.13 shows probability of coverage for various power control strategies. Again we find rather low coverage probability for typical SINR thresholds – on the order of 50% for an SINR threshold of 0 dB. Again, this means that blanking of BSs (i.e., frequency reuse) is required in order to obtain reasonable coverage probabilities.

For the special case of full power control ($\epsilon = 1$) and no noise, coverage probability can be shown to reduce to

$$p_{\mathrm{c}}(\theta) = \exp[-\Xi(\theta)] \tag{21.56}$$

with the definition of $\Xi(\theta)$ from Eq. (21.41). When comparing this with the downlink case, we see that the coverage probability decreases much faster (exponentially, instead of as $1/(1 + \Xi)$) with the SINR threshold. In other words, in an uplink system with power control, obtaining very high SINR is difficult.

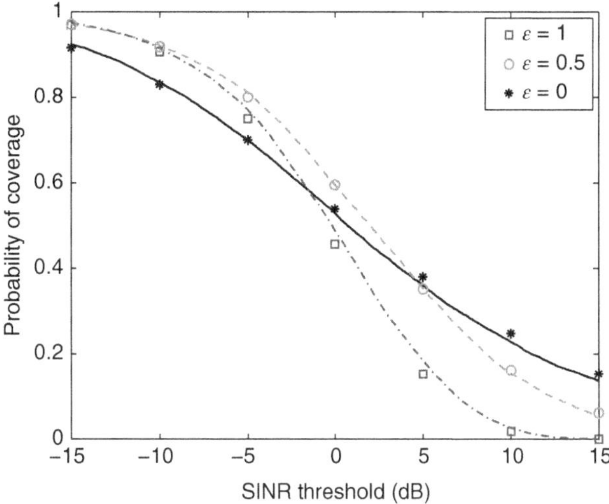

Figure 21.13 Coverage probability as function of required SINR for different power control policy when using stochastic geometry analysis. Color version available at wiley.com/go/molisch/wireless3e.
Reproduced with permission from [Andrews et al. 2016] © H. Dhillon.

21.5 CDMA-Based Cellular Systems

21.5.1 Uplink

Intuition

CDMA-based systems have fundamentally different interference characteristics, in particular in the uplink. While an FDMA system has no intra-cell interference, and the inter-cell interference is dominated by one (or a few) UEs in neighboring cells, in CDMA there is a large number of both intra- and intercell interferers. This effect is compounded by the fact that CDMA has frequency reuse-1, so that the whole system bandwidth is used in every single cell. Inter-cell interference is thus mainly determined by the characteristics of the spreading codes.

These special characteristics of inter-cell interference depend, however, on the way that the codes in the different cells are constructed. Imagine, for a moment, that the system reused the same codeset in every cell. In this case, if the BS aims to detect user k_1, which uses spreading code 1, the interference from a user in the neighboring cell that uses the same spreading code 1 would dominate the total interference, similar to the situation in FDMA. It is often advantageous if the characteristics of the interference are similar to those of noise, which is achieved if the inter-cell interference is effectively an average over the different adjacent-cell users. Thus, codesets for adjacent cells should be constructed in such a way that the interference suppression by the despreading is approximately the same for all users from an adjacent cell. In order to achieve this, the codesets must have a much larger cardinality than the value of the spreading factor *SF*. A discussion of various codesets that achieve this condition, either by themselves, or by multiplication with cell-specific scrambling codes, is given in Section 19.3.2.

However, due to the effects of power control, the *power* contributed by each inter-cell interferer is not the same even if the interference suppression is the same. Remember that power control is essential even in single-cell CDMA systems, to ensure appropriate mitigation of intra-cell interference; due to the mechanism of power control, each UE contributes exactly the same amount of intra-cell interference. Furthermore, the power of the arriving desired signal is constant as well. Now in the multi-cell case, the desired signal still has constant power, since this is how the power control algorithm is designed. However, the inter-cell interference is *not* constant anymore. A user in an adjacent cell is power-controlled by its own serving BS – in other words, the power is adjusted in such a way that the signal arriving at its desired BS is constant. However, it "sees" a completely different channel to the undesired BS, with temporal fluctuations that the desired BS neither knows nor cares about. Consequently, the inter-cell interference from different users is different, and furthermore temporally variant. It is also not necessarily Gaussian. One intuitive explanation is that only users close to the cell edge create a strong inter-cell interference (since those transmit with high power). Thus for the summation over the different interferers the central limit theorem is not applicable.

*Derivation of the SINR Distribution

Let us now analyze this aspect mathematically. We use the following notation. Let D_l denote the distance between BS l and the reference BS 0, which is located at the origin of the coordinate system. Let $r_{k,\,l}$ denote the distance of the kth UE from the lth BS. A UE in cell l is served by BS l, which is located at the cell's center. When a UE k located at $\mathbf{x}_k$ transmits a signal with power P_k, the received power at BS l is $const \cdot P_k(d_0/r_{k,l})^{\alpha_{\mathrm{PL}}} \exp(\kappa y_k(l))\widetilde{h}_k(l)$, where y is a zero-mean Gaussian variable, and κ is $0.1\ln(10)$. The number of UEs in a cell and their locations is modeled as a PPP with density λ.

Consider first the case where short-term fading has been averaged out by spatial or multi-path diversity. We set P_k in such a way that the receive SNR is equal to γ. Hence, the interference power $P_{\text{I},\,k}$ that BS 0 sees from UE k is equal to

$$P_{\text{I},k} = const \cdot P_k \left(\frac{d_0}{r_{k,0}}\right)^{\alpha_{\text{PL}}} e^{\kappa y_k(0)} = \sigma_{\text{n}}^2 \gamma e^{\kappa(y_k(0) - y_k(l))} \left(\frac{r_{k,l}}{r_{k,0}}\right)^{\alpha_{\text{PL}}}. \tag{21.57}$$

The total interference $P_{\text{I},l}$ at the reference cell from all the K_l UEs served by cell l equals

$$P_{\text{I},l} = \sum_{k=1}^{K_l} P_{\text{I},k} = \sigma_{\text{n}}^2 \gamma \sum_{k=1}^{K_l} e^{\kappa(y_k(0) - y_k(l))} \left(\frac{r_{k,l}}{r_{k,0}}\right)^{\alpha_{\text{PL}}}. \tag{21.58}$$

When the power control reacts to short-term fading as well, the short-term fading gain $\tilde{h}_k(l)$ of the channel between UE k and BS l needs to be considered as well. The expression in (21.58) for the aggregate interference from a cell changes accordingly.

Now we derive the MGF of the aggregate interference. Let $\varphi_l(s)$ denote the MGF of the interference from an arbitrary UE in cell l. Then, the MGF $\Psi_l(s)$ of the aggregate uplink interference from all UEs served by cell l equals

$$\Psi_l(s) = e^{\lambda A \varphi_l(s) - 1} \tag{21.59}$$

where A is the cell area.

Since the interferers of different cells are Independent Identically Distributed (i.i.d.), the MGF of $P_{\text{I},\Sigma}$ can be computed as the product of the interference MGF in the different cells. Furthermore, $\Psi_{P_{\text{I},\Sigma}}(s)$ directly follows from the equations for MGFs in shadowing, which can be obtained, e.g., from the MGF-matching method described in Section 5.8, or other suitable techniques. While the approximation of the interference as Gaussian is very popular, it is not accurate. Figure 21.14 compares the cdfs of the aggregate interference power of various models when the average number of UEs per cell is 20.

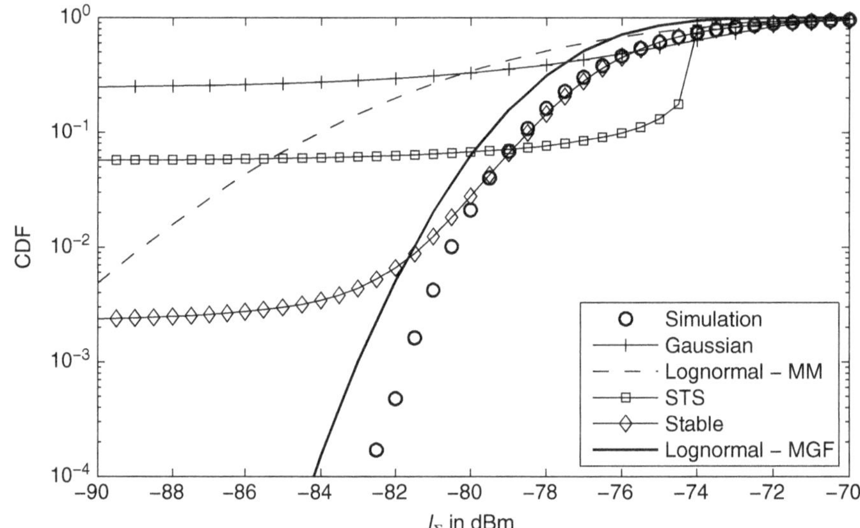

Figure 21.14 Lower UE density (20 UEs/cell): Comparison of cdf of aggregate interference with path-loss and shadow fading. Sum of shadowing computed via lognormal-MGF method (compare Section 5.8) with $(s_1, s_2) = (0.01, 0.1)/\sigma_{\text{n}}^2$. [N. B. Mehta, private communication].

21.5.2 Downlink

In the downlink, the intercell interference does *not* come from a large number of independent sources. All the interference comes from the BSs in the vicinity of the considered UE, i.e., a few (at most 6) BSs constitute the dominant source of interference. The fact that each of those BSs transmits signals to a large number of users within their cell does not alter this fact – the interfering signal still comes from a single geographical source that has a single propagation channel to the victim UE. Consequently, the downlink might require a fading margin in its link budget.

The downlink of CDMA systems also usually employs orthogonal codes, so that no intra-cell interference occurs (at least with suitable RXs). Since the number of Walsh–Hadamard (WH) codes is limited by the spreading factor, the same set of codes needs to be used in every cell, leading to a situation where one adjacent-cell interferer can have a dominant effect on a user.

A signal transmitted from the lth BS, intended for the kth user in the associated cell, contributes the following amount of interference power to the k'th user:

$$P_{l,k}|h_{l,k'}|^2/SF. \tag{21.60}$$

This is thus (apart from the spreading factor) the same as in Section 21.2.2. If there is no power control, then the interference observed at a specific UE is determined by the total transmit power of the neighboring BSs, and the channels between those BSs and the UE under consideration. Also, the stochastic-geometry analysis of Sec. 21.4 can be extended to CDMA.

It is advantageous to use power control in the downlink as well, in order to reduce inter-cell interference. The power control is usually chosen based on large-scale fading (to reduce feedback overhead), and the transmit power is chosen such that the SINR of the signal arriving at the UE is larger than some threshold. If such power control is used, then the level of inter-cell interference at the UE depends on the UE in the neighbor cell that shares the same WH code (again, similar to the situation in FDMA).

To mitigate worst-case situations, it can be useful to have interference averaging in the downlink as well. This can be realized by multiplying the WH codes with a *scrambling code* that is specific for each cell. WH codes that are multiplied by the same scrambling code remain orthogonal; codes that are multiplied by different scrambling codes do not interfere catastrophically.

21.6 Handover

One of the key features of cellular systems is that the UE can move over a large area, while retaining a seamless connection to the network. At different times during a call or data session, the UE is connected to different BSs. To ensure that the user experiences no call drops or interruptions, the procedure for changing the serving BS, called *handover*, has to work fast, and without interrupting the actual service. The detailed steps of the procedure depend on the specifics of the system; Chapters 31 and 32 discuss it for 4G and 5G systems (and Appendix 30.A and B for 2G and 3G systems). In this section we discuss some general principles, distinguishing between two main approaches: *hard handover*. where the connection to the current serving BS is terminated before a connection with a new BS can be taken up, and *soft handover*, where a UE can maintain over a period of time connection to two BSs simultaneously.

21.6.1 Hard Handover

In a hard handover, the data connection of a UE to one BS is halted before the connection to a new BS is started. It must be stressed that this is very different from terminating a connection to one BS and building up a completely new one to the other BS. Setting up a completely new connection to the network takes up a lot of time and resources, and would lead to noticeable breaks in the user experience. Rather, in a handover, the old and the new BS cooperate in preparing the handover. And while the data connection to the old BS is halted before the one to the new one is started, control information to/from and between the two BSs may flow even at times where the specific BS is not the serving BS.

A handover starts with the UE measuring the receive power (or some other quality measure) from both the currently serving, and the neighboring BS. The ability to listen to the strength of neighboring BSs can be achieved in different ways: (i) in an FDMA system with frequency reuse, the UE can simply listen to the power of the BSs that are transmitting on a different frequency (note that this either requires some time in which the UE is idle, or extra circuitry). Also note that when the BS uses downlink power control, then listening to the power-controlled data transmissions might give a wrong impression of the channel quality; rather specific signals such as beacons, sent out without power control, should be observed. (ii) In an FDMA system with reuse-1, specific beacon signals need to be sent out by the different BSs at different times to allow the UE to assess the associated channel qualities. (iii) In a TDMA and OFDMA system, the UE can listen to the beacons from neighboring BSs during the times that it is not receiving its desired signal; (iv) in CDMA systems, the signals from neighboring BSs typically will have different spreading codes. The UE can simultaneously detect the total arriving signal with different despreading codes, and thus determine payload or beacon signals from different BSs. Note that in principle, the BSs could measure an uplink beacon from the UE, and communicate the results to the serving BS (or a central decision unit) about the results, so that full CSI is now available at the serving BS; however, this can result in considerable overhead and delay on the backbone.

After measuring the signals from the different BSs, the UE can decide whether it wants to change its serving BS; alternatively, it can inform the BS about the current CSI to all the surrounding BSs and let the BS make a decision of whether a handover is warranted. In any case, it is the serving (source) BS that then initiates a handover, by starting communication with the target BS, i.e., the BS to which the handover is done. This communication can be direct, or through a "switching center" or similar entity. In this process, the source BS informs the target BS about the details of the UE it wants to hand over, such as identifiers, service requirements, what type of service it has paid for, etc., as well as the information about the channel quality to the surrounding BSs the source BS had received from the UE. The target BS might communicate how many resources it has available – in the most extreme case, its admission control might decline to accept the UE because it has insufficient resources available.

Once the preparatory work is done, the actual data link is rerouted from the source to the target BS. For the downlink, this means that the source packets that come from the core network are now routed to the target BS instead of the source BS. For the uplink, the transmissions are now received and decoded at the target BS. Note that the switching of the serving BS does require some physical-layer procedures: for example, the UE needs to synch itself up to the new BS. Yet most of the time-consuming procedures of a call setup need not be performed. For example, the UE does not need to send access requests to the BS to indicate that it wants to join, because the target BS already knows that this UE has joined. Once the new link, between UE and target BS is established, the connection to the old (source) BS can be broken off, and the resources are freed up for other purposes in that cell.

21.6.2 Soft Handover

If all cells use the same frequency band, a UE can have contact with two (or more) BSs at the same time; this is particularly relevant for CDMA systems. Figure 21.15 shows a UE close to a cell boundary that receives signals from three BSs and also transmits to all of those BSs. The signals coming from the different UEs are time-shifted with respect to each other, but that can be compensated by the Rake RX, and the signals from the different BSs can be added coherently.[12]

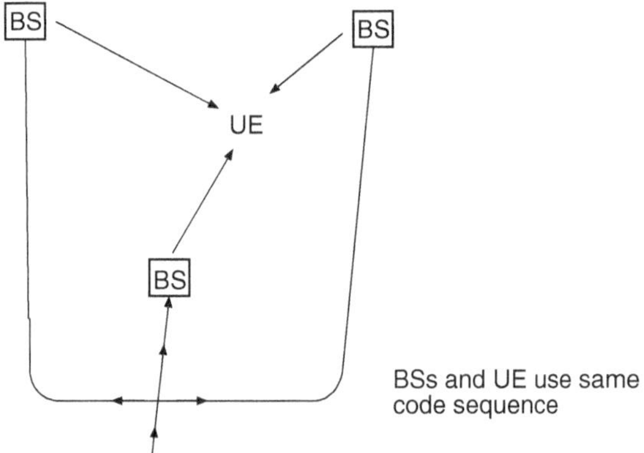

Figure 21.15 Principle of soft handover in CDMA.
Reproduced with permission from [Oehrvik et al. 1994] © Ericsson AB.

Consider now a UE that starts in cell A but has already established a link to BS B as well. In the beginning, the UE gets the strongest signal from BS A. As it starts to move toward cell B, the signal from BS A becomes weaker, and the signal from BS B becomes stronger, until the system decides to drop the link to BS A. The soft handover dramatically improves performance while the UE is near the border of the two cells, as it provides diversity (macrodiversity) that can combat the large-scale fading as well as small-scale fading. On the downside, soft handover decreases the available capacity in the downlink: one UE requires resources (WH codes) in two cells at the same time, while the user talks – and pays – only once. Furthermore, soft handover increases the amount of signaling that is required between BSs.

While soft handover has been traditionally associated with CDMA systems, it can also be applied to OFDMA systems. In particular, 5G NR foresees the possibility of a UE being connected to multiple BSs simultaneously, either to increase the total amount of available bandwidth (if the two BSs transmit different information), or in order to provide macrodiversity, see Sec. 32.7.

21.7 Heterogeneous Networks

21.7.1 Motivation for Heterogeneous Networks

The most important method for increasing capacity in a cellular network is a densification of the BSs. Assuming operation in the interference-limited regime, the area spectral efficiency simply scales linearly with the number of deployed BSs. This effect can also be understood intuitively that the number of UEs associated with a BS scales with the size of the cell; smaller cells thus means fewer UEs that have to share the resources of the BS.

Due to the nonhomogeneous nature of traffic density (see Section 21.4.1), it is not useful to deploy BSs with very high density in all parts of a city or country. Rather, the deployment density should be proportional to the expected density of users/traffic. This means that certain "hotspots" should be covered with especially high area spectral efficiency. These hotspots often have a very small area, and only a few, or possibly even one, BS is sufficient to cover it. For example, a popular coffee shop might contain a large number of users, and a dedicated BS can cover all the users in it. To enforce that a BS covers a very small geographical area only, this small coverage area can be achieved by a combination of (i) lower transmit power and (ii) small height of the BS antenna, so that over-the-rooftop propagation becomes strongly attenuated, and only points close to the BS (often with LOS) receive significant power. Such small-area cells are often called *femtocells*.

On the other hand, peak capacity is not the only thing that counts for a good cellular deployment; coverage (i.e., probability of being able to be connected to the network) is important as well. Thus, femtocells must be combined with macrocells that provide large-area

---

[12] Note that the different cells might use different codes. This is not a major problem; it just means that (for the downlink) the RX has to use different correlators that use different spreading sequences.

coverage.[13] In many urban environments, we see a combination of macrocells not only with femtocells, but also microcells, where the BS is below rooftop, but typically higher than for femtocells, and having a larger area; Table 21.2 shows typical operating parameters for macro-, micro-, and femtocells. Note that the cell radii in macrocells can vary significantly, depending on whether they are deployed in urban or suburban/rural areas.

Table 21.2 Operating parameters for macro-, micro-, and femtocells

Parameter	Macro	Micro	Femtocell
Cell radius	500–5000 m	200 m	50 m
TX power	40 W	10 W	1 W
BS height	Above rooftop	Below rooftop	2–3 m, or indoor
Number of potential users	10,000	1000	1–100
Open to all users	Always	Always	depends
Frequency planning	Yes	Yes	No
Capability for ICIC and CoMP	Yes	Sometimes	No

Another important motivation for a heterogeneous network structure is the different amount of mobility of users. Moving users require handovers between neighboring cells. A user moving with 10 m/s (a typical driving speed in a city) would move through the coverage area of a femtocell in 10 s. This implies that the high overhead for handover has to be spent for having a connection for a very short time. It is therefore preferable to have such a user connected to a macro-BS,[14] with its much larger coverage range, in particular, if those driving UEs run applications that require smaller data rates anyway. On the other hand, stationary (nomadic) users, which do not need handovers, tend to require high data rates, since they might be doing intensive web browsing, or stream videos; such users are thus advantageously connected to a femto-BS. This solution can be implemented by defining a "minimum time of stay" for a UE in a femtocell; if that time cannot be provided, a handover into this cell does not happen.

Another approach for heterogeneous networks separately considers the control signaling and the actual data transmission. In other words, the control signaling is communicated always between UE and macro-BS, while the data can be transferred either via the macro- or the femto-BS. This approach does not impact the spectral efficiency significantly (since control signaling should take up only a small part of the time-frequency resources anyway) but ensures better reliability for the control signals, and in particular helps when frequent handovers are required.

Additional challenges occur when (as is often the case in practice) the femto-BS is not directly connected to the core network and thus does not have a low-latency connection – a fact that can lead to considerable delays in the handover.

Heterogeneous networks combining macro- and microcells are widely used. The emergence of femtocells has been predicted since the early 2010s, but up to now, they have been used relatively little. This is expected to change with the advent of millimeter-wave systems in 5G (compare Chapter 32), which have a relatively small coverage range, but large available bandwidth, and are thus naturally suited for implementation of femtocells.

21.7.2 Types of Heterogeneous Networks

Femtocells can be categorized into "open" and "closed" access cells. An open femtocell BS allows any cellular user to associate with it; a closed BS restricts connections to either one particular UE, or a small group of UEs. The former case occurs when the femtocell is owned by the network operator and is truly a part of the network. For example, the operator might decide to deploy femtocells in an airport, a busy shopping mall, or other areas where a large number of subscribers are usually present. A closed femtocell usually belongs to a subscriber and is set up to improve coverage and capacity in the home of that particular subscriber. The subscriber usually pays for the BS, and also pays for the backhaul (through a wired broadband internet connection) as well as operating costs such as electricity. Obviously, the subscriber wants to reap the benefits of such a BS for themselves and thus will restrict the access to either one UE (the cellphone of this particular user), or a small number of phones (phones, tablets, etc. of family, friends, etc.). The price of femto-BSs for personal use is often subsidized by the network operator, since it increases the coverage and user satisfaction of a group of customers and may also increase the overall spectral efficiency of the system (since the user does not have to connect to the macro-cell). However, it also makes network planning more difficult, since the femto-BS is essentially outside the control of the network operator.

Another way to categorize heterogeneous networks is (i) networks where there is a separate frequency band for the femtocells and (ii) networks where the femtocells and the macrocells share the same band. The former approach makes frequency planning easier but is not spectrally efficient. There are usually large areas that are not within the coverage region of any of the femtocells, and in those

---

[13] One might now ask: why not simply let the macrocell handle the traffic from the hotspot? The answer is that the macro-BS would be overwhelmed if it had to deal with the UEs in this hotspot area in addition to the UEs in the remainder of the surrounding area. Note that as per the table below, the coverage area of a macrocell might contain, e.g., 10,000 potential users.

[14] To simplify notation, we will henceforth only consider a network with two layers, namely macrocells and femtocells; generalization to more layers is straightforward.

areas spectrum reserved for femtocells is "wasted." Still, the situation occurs, e.g., in 5G when the femtocells are using the millimeter-wave bands while the macrocells are operating in the classical cellular bands below 6 GHz. Network planning is relatively simple in this case, as the signals of macro- and femto-cells do not influence each other; each user simply associates with the BS that provides the best overall throughput. Of course, this user association influences the per-user throughput: the more users associated with a femtocell, the smaller the per-user throughput in the femtocell, and the larger the per-user throughput in the macrocell. It can be shown that optimal results can be achieved with a very simple association scheme in which the *UE* determines with which BS it associates.

The latter approach (joint spectrum for macro- and femtocell) leads to a much more complicated problem, namely which resources should be provided to the femto-BS for its operation, and which users should be associated with it. As a matter of fact, it is not even obvious whether (in an FDD system) the downlink from the femtocell should happen in the downlink band or the uplink band of the BS. Another question is whether the UE should simply connect to the BS that provides the strongest signal, or other criteria should be applied. We will analyze these questions in the subsequent sections.

21.7.3   Interference and Cell Association

Also in heterogeneous networks, inter-cell interference is of critical importance. The performance analysis is challenging and, at the time of writing, still very much an area of ongoing research. As the deployment of femto-BSs is not centrally planned, a random deployment (in the simplest form modeled as a PPP) is a suitable model. Consequently, two fundamental approaches are popular for the modeling of hetnets: (i) modeling the macrocellular network as a hexagonal grid (as in Section 21.2), and the femtocells as a random point process (as in Section 21.4); (ii) modeling the macro-cell and femto-cell locations as point processes: a single random point process is generated, and a random subset of the points are designated as macro-BS, while the remainder represents the locations of the femto-BSs. Most of the analysis has concentrated on the downlink, because (i) it is usually considered the bottleneck in providing cellular data, and (ii) it is easier to analyze.

In a heterogeneous network, inter-cell interference can be categorized into co-layer and cross-layer interference; these can furthermore be different in the uplink and the downlink. For the downlink, the transmit power of the BS has a major impact on the range of the cell, and the SINR that can be achieved at the associated UEs. Since the transmit power of a macro-BS can be two orders of magnitude or more larger than that of a femto-BS, the large majority of UEs will be associated with the macro-BS. There may be many UEs that have better downlink SINR to the macro-BS even though geographically they are closer to the femto-BS. For the uplink, on the other hand, there is much less difference between macro- and femto-BS.[15] It is thus possible that a UE has its best uplink SINR to a femto-BS, but its best downlink SINR from the macro-BS.

We note here that the stochastic-geometry analysis of Section 21.4 can also be applied to heterogeneous networks, though the details are considerably more complicated. In this case, we first have to compute the probability that a UE associates with a BS from a certain "layer" (macro-, micro-, or femto) – a computation that depends on the particular association rules. This is followed by the computation of the pdf of the distance to the serving BS as well as the interference statistics. For details refer to [Andrews et al. 2016] and references therein.

Co-Layer Interference

The interference from one macrocell to a neighboring one follows the discussions in Section 21.2.

The interference between femto-cells in the uplink depends on the access policy. For open access, each UE connects to the strongest associated BS (either on average, or instantaneously). Since a frequency reuse one would result in too-large interference, a guard spacing (dead zone) has to be provided, again following the principles of Section 21.2. For closed access, the interference might be stronger, since a UE very close to a particular femto-BS might be connected to a *different* BS, and thus create strong interference. For OFDMA as multiple access, the effect of such interference might be mitigated by sensing the strong interference and placing the desired users on different time-frequency resources; however, this can only be effective if the time-frequency-resource assignment is constant in time (persistent or semi-persistent scheduling).

For the downlink, similar considerations hold. The interference can, in principle, be computed from the equations in Section 21.4.4. Interference can be reduced, among other methods, by suitable power control.

Cross-Layer Interference

Uplink

In the uplink, femto-UEs (i.e., UEs connected to the femto-BS) can create interference to the macro-BS, and vice versa (Figure 21.16). The former case usually creates less of a problem, since a femto-UE is close to its associated femto-BS, and thus (assuming power control) transmits with relatively small power. In the latter case, strong inter-layer interference is possible: the macro-UE can be far away from its associated BS, and thus transmit with high power. This implies that the BS the UE is associated with (i.e., the macro

---

[15] Macro-BSs might have somewhat better noise figures (since there are fewer of them, they can use better, and thus costlier, LNAs). Also, the higher mounting of the macro-BS might reduce the pathloss somewhat. But none of these effects is as significant as the power difference between macro- and femto-BS in the downlink.

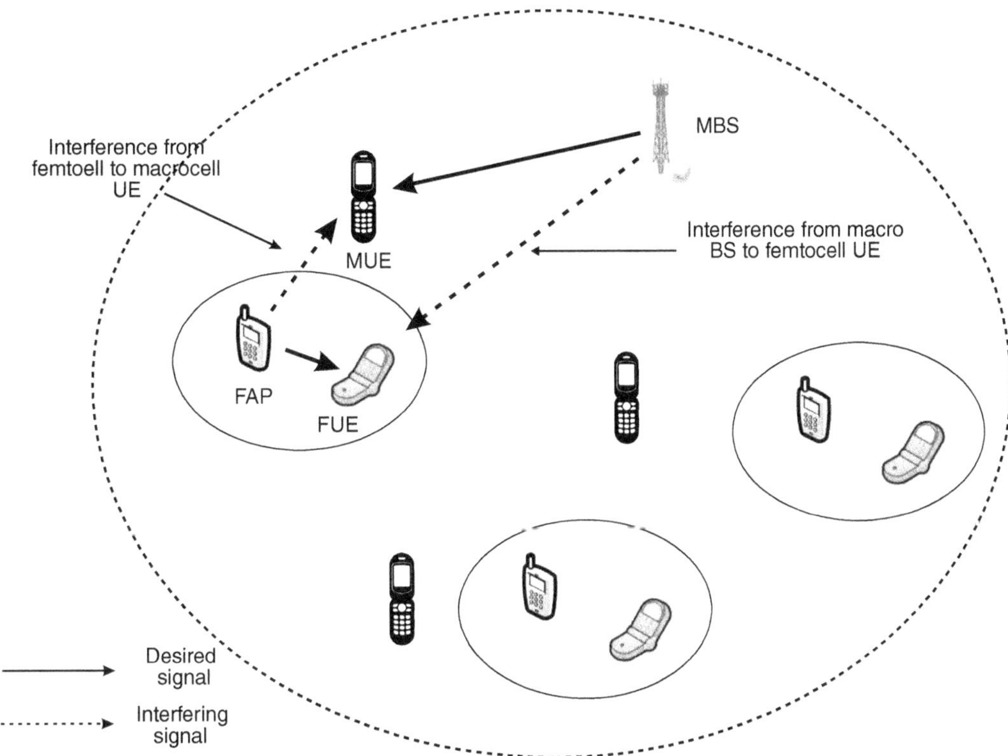

Figure 21.16 Inter-layer interference between macrocells and femtocells. In this figure: FAP means femto BS; MBS means macro BS. Reproduced with permission from [Zahir et al. 2013] © IEEE.

BS) is not the BS with the best connection. Such a situation can occur in particular when the femto-BS has closed user groups – in which case a UE might be right next to the femto-BS, and yet not be connected to it because it does not have the permissions.

Furthermore, the interference created between femtocell layers and macrocell layers strongly depends on the multiple access format. In the case of CDMA, strong interference from a macrocell user can effectively destroy femtocell performance, since interference suppression relies on effective power control.

In OFDMA, TDMA, and FDMA, a reduction of the interference can be achieved through suitable scheduling: in essence, signals from macro- and femto-BSs that might strongly interfere with each other can be simply put on different time/frequency resources. In other words, we can use the ICIC techniques discussed in Section 21.3.4. The drawback of this approach is that it requires a considerable degree of coordination. Since femtocells are often user-operated and connected to the main network through slow and unreliable connections (best-effort internet connections of the users owning the femtocell), ICIC might not be possible to its fullest extent. However, semi-static assignments of the time-frequency resources are possible.

Downlink

For the downlink, first consider the case of open-access femtocells. Then, under certain circumstances, the use of a two-layer network does not change the interference statistics of the downlink (compared to the single-layer case). To be more precise, consider a network where (i) both macro- and femto-BS are independent PPPs, (ii) there is the same pathloss coefficient for all users, and channel gain consists only of pathloss and Rayleigh fading, and (iii) all users connect to the BS that is strongest on average. Then it can be shown that the SIR distribution is *independent of the number of layers, their densities, and their power levels*.[16] It must be stressed that the fact that the SIR distribution does not change when adding BSs does not mean that additional BSs are useless: on the contrary, the more BSs, the more capacity per user is available for each BS. The success probability (which is equivalent to the cdf of the SIR), is given by

$$p_\mathrm{s}(\theta) = \frac{1}{{}_2F_1(1, -\delta; 1-\delta; -\theta)} \ . \tag{21.61}$$

As discussed in Sec. 21.4, this leads to a success probability of about 50% for practical threshold levels, so that interference reduction methods such as frequency reuse have to be applied.

The interference situation can again be considerably worse in systems with closed access. Consider the interference created by a femto-BS to a macro-UE in its vicinity. The UE might receive much more power from the femto-BS (due to the close proximity) than

---

[16] Note that this result applies to the *SIR* distribution, not the SINR – the relative impact of noise changes with the density of BS and power levels.

from the macro-BS; however, because it is forbidden to connect to the femto-BS, it has to operate essentially in negative SINR conditions. Thus, some interference reduction by switching to a different sub-bands, or some form of ICIC, is required.

Besides the standard methods for interference management, a popular approach in fourth generation cellular networks (LTE, see Chapter 31) are *Almost Blank Subframes* (ABS). In this method, the macro-BS transmits with reduced power, or not at all, during certain time durations (subframes), so as to not generate significant interference. The femto-BS can then schedule during these times those UEs near the cell edge that would suffer most from interference (Figure 21.17). UEs that are very close to the femto-BSs can be scheduled when the macro-BS is on as well, since they have very high received power from the (desired) femto-BS anyway. The ABS approach thus creates a TDMA-like reuse, but mainly for those UEs that need it; this approach is thus spectrally more efficient than providing separate time-frequency resources for *all* femto-UEs. Also, multiple femto-BSs within the reach of the same macro-BS can exploit the same ABS, as long as they are sufficiently separated so that they do not create intra-layer interference.

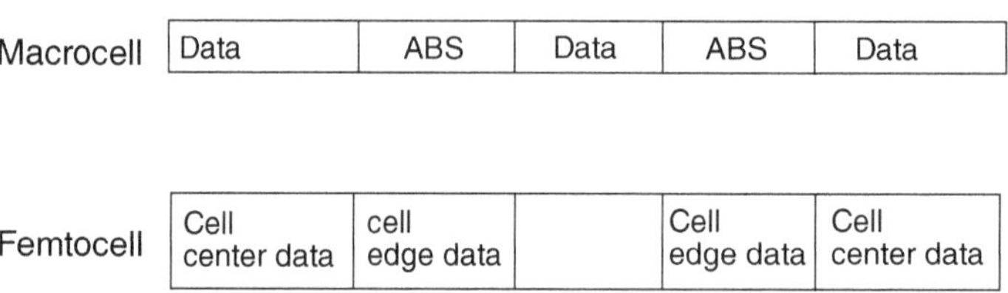

Figure 21.17   Principle of Almost Blank Subframes (ABS). Femto-cells schedule cell-edge users in those subframes (time intervals) that the macro-BS is silent.

As discussed above, the most straightforward way for associating a UE with a BS is to find the one that provides the best SINR. However, this approach is often not optimum from an overall system point of view: because the femtocell has only a small range, the number of users associated with a femto-BS is extremely small. While this is good for the femto-users, it means that the traffic load of the macro-BS is reduced only very little. It is preferable to divide users in such a way that the *network utility* – which can be essentially interpreted as the sum rate under some fairness constraints, see Section 20.1.4 – is optimized. A mathematical framework for such optimization can become complicated; for this reason, *biasing* has become popular as an ad-hoc but effective method. In biasing, a UE is associated with a femtocell if the resulting SINR is no worse than b_{bias} dB compared to the SINR achieved by connecting to the macrocell, where the bias b_{bias} depends on the system load.

21.7.4   Self-Organizing Networks

Traditional macrocells undergo a very detailed planning process: both the location, and the assigned frequencies, are designed by experts, supported by suitable computer simulations. However, as networks grow larger and larger, such manual planning becomes difficult. Furthermore, for femtocells, manual planning is not even an option: firstly, the number of deployed femtocells is too large; secondly, the femtocells are often under the control of a user, so that the network operator cannot determine a priori where the femto-BS will be deployed, and when it is moved to a different location. Thus, all the network planning functionality must be automated and adjusted automatically whenever the environment changes. Networks that perform such functionality are called *Self-Organizing Networks* (SON).

SON functionality can be separated into the following categories/steps:

- *Self-configuration*: when a BS is added to a network, various network parameters need to be set. Most important is the cell-ID, and parameters that are derived from it (e.g., certain scrambling codes; the details depend on the particular system). Also, the relations to the neighbor cells need to be established, both to minimize interference, and to support handovers. For femto-cells, the self-configuration has to occur every time the femto-cell BS is moved to a new location.
- *Self-optimization*: after the initial configuration, the BS has to continuously monitor the environment and adapt its operating parameters (such as admission control, access modes, transmit power, etc.) to the environment. Typical examples include mobile load balancing, adaptation of the handover parameters, as well as BS sleeping. Self-optimization bears considerable connections to "cognitive radios" (compare Chapter 26), as it observes the environment and adjusts the radio parameters to it.
- *Self-healing*: the BS should be able to recover from failures. Furthermore, the network should also be able to adapt to the failure of a BS, and "take up the slack," i.e., re-configure the network parameters so that the remaining BSs can cover the area sufficiently.

21.7.5   Economic Aspects of Femtocells

Femtocells have been promoted both by network operators and equipment manufacturers. For network operators, the clear benefit lies in offloading a considerable amount of traffic from the (expensive) macro-BSs to local points that typically are connected to an existing internet connection. The situation is especially stark in the case that the femtocell is on customer premises and connected to the wired internet of the customer; in this case, the network operator has a small capital investment (subsidy or full cost of the femto-BS), but no operating expense, and in particular no cost for establishing a backhaul.

For customers, the main benefit lies in a better performance, though it must be noted that this is not always guaranteed (see the above discussion about strong interference to closed femtocells). When the customer provides the backhaul on a conventional (low-speed cable or Asymmetric Digital Subscriber Line (ADSL)) internet connection, that connection might become the bottleneck; furthermore, if the cable provider imposes a data cap for the user, the backhaul traffic (in conjunction to the regular internet traffic of the customer) might cause the customer to exceed the cap.

Finally, it is worthwhile to compare femto-cells and WiFi offloading. Femto-cells in LTE and other current systems are in-band femto-cells, and thus share the available spectrum with the macro-BSs. Control signaling is following the cellular standards, and the UE can be handed over from macro-BS to femto-BS within a very short time. In WiFi offloading, the signal transmission occurs in a different band (the assigned Wifi bands, see Chapter 33). The handover between cellular system and WiFi can take a long time (several seconds) and use considerable resources, since it requires tearing down the connection in one system and creating a new connection in the other system; it thus should happen as rarely as possible. For the cellular operators, WiFi offloading has the advantage that no licensed spectrum is used up. However, the traffic on the WiFi connection cannot be easily managed, as WiFi systems are typically designed as "best effort" systems. While there have been recent activities to establish "carrier grade WiFi," i.e., WiFi with as good a quality-of-experience as cellular, the principle of unlicensed bands means that there can never be a guarantee against the presence of strong interferers. Operating cellular systems jointly in licensed and unlicensed bands constitutes a compromise solution; its implementation in LTE and NR is discussed in Sections 31.6.2 and 32.6.2, respectively.

21.8 Backhaul

Besides the connection between BS and UE, another important question is how are data transferred between the core network and the BS? In the past, BSs were typically connected to the core network by optical fiber connections. However, as the BS density increases, it becomes more and more expensive to lay fiber to each of the BSs. In other words, the main deployment costs are not the costs of the BSs, but rather the costs of the backhaul.

This makes wireless backhaul an attractive alternative, which can take on two forms:

1. *Directional LOS links*: this approach is useful when a LOS exists between a remote BS (e.g., a femto- or micro-BS) and the connection point to the core network (e.g., a macro-BS with fiber connection to the internet). Directional antennas at the two link ends are pointed at each other, thus providing high gain, such that a higher-order modulation and coding scheme can be applied. In order to further enhance the data rate, polarization multiplexing, and possibly LOS-MIMO (Multiple Input Multiple Output system) (Section 16.2.6) or Orbital Angular Momenta (Section 16.2.15) can be used. As operating frequency, either microwave or millimeter-wave frequencies are suitable. These links are somewhat similar in their operating principle to the directional links that have been used for long-distance wireless signaling for many years.

2. *In-band, non-LOS links*: a remote BS can be considered by the macro-BS as just another user (similar to a macro-UE). Thus, physical-layer transmission and scheduling can be done in the same manner as for other users; most importantly, there is no specific requirement that the femto-BS have a LOS to the macro-BS. Obviously, such a link requires much higher capacity than a regular UE, and might thus constitute an overall system bottleneck. Furthermore, if the remote BS transmits at the same time as the macro-UE, this means that remote downlink has to happen at the same time as the macro-uplink, and vice versa. This in turn has consequences for the interference statistics. The in-band scheme is especially efficient if the BS has multiple antenna elements, so that it can spatially multiplex the transmission to the remote BS with other transmissions. The implementation in 5G NR is discussed in Section 32.8.

When the density of the BSs increases, also the requirement for wireless backhaul increases. It can be shown that it is possible to achieve scalability in the wireless-only backhaul network if we increase the number of antennas per node as the network size increases.

Finally, femto-BSs are often envisioned to use existing wired internet connections as backhaul. The problem with this approach is that often these wired connections have slower speed than the wireless air interface. In this case, it is the backhaul that becomes the bottleneck; a fact that has to be included in the mathematical formulations of the network optimization.

21.9 Other Methods for Increasing Capacity

The system capacity is the most important measure for a cellular network. Methods for increasing the capacity are thus an essential area of research. In the following, we give a brief overview, often referring to other chapters in this book.

1. *Increasing the amount of spectrum used*: this is the "brute-force" method. While the amount of spectrum that is available for both cellular communications and WiFi has increased significantly over the years, adding new spectrum is expensive and time consuming, as discussed in Chapters 1 and 2. Despite these obstacles, over the past years, a number of new frequency bands have been opened for cellular and WiFi systems: (i) the digital dividend, i.e., frequencies in the sub-1 GHz band that became available when terrestrial television transmission moved from analogue to the more bandwidth-efficient digital transmission; (ii) additional bands

in the range 2–4 GHz, as well as (for WiFi) 6–7 GHz (iii) millimeter-wave bands. For wireless Local Area Network (LAN) applications, the frequency range from 58 to 65 GHz has been opened up – this in itself is more spectrum than all previous bands for cellular and WiFi taken together. Further bands in the 27, 38, and 72 GHz range are currently being discussed for 5G cellular applications on a worldwide basis and have been made available in some countries; for more details see Chapter 32.

2. *More efficient modulation formats and coding*: modulation formats that require less bandwidth (higher-order modulation) and/or are more resistant to interference. The former allows to increase the data rate for each user (or increase the number of users in a cell while keeping the data rate per user constant). However, higher-order modulation is also more sensitive to noise and interference (see Chapter 10 and 11), so that the reuse distance might have to be increased. In any case, the choice of the modulation format (and the coding rate) should be adapted to the channel state, to make better use of the available power, and, among other effects, reduce the interference, see Section 15.9.

   The introduction of near-capacity-achieving codes (turbo-codes and low-density parity check codes, see Chapter 13) is another way to achieve better immunity to interference, and thus increases the system capacity. Note that modern systems such as LTE use adaptive modulation and near-capacity-achieving codes, so that further improvements in this field will be limited.

3. *Better source coding*: Compression of data files and music/video compression also allows to serve more users. For example, uncompressed High Definition TeleVision (HDTV) transmission requires more than 1 GBit/s, while suitable compression can reduce this to less than 10 MBit/s, see Chapter 25.

4. *Multi-user detection:* it greatly reduces the effect of interference, and thus allows more users per cell (for CDMA systems) or smaller reuse distances (for FDMA systems); for details see Chapter 28.

5. *Multiple antennas:* they can be used to enhance capacity via different scenarios:
   (a) Diversity (Chapter 13 and Section 16.1) increases the quality of the received signal, which can be exploited to increase capacity, e.g., by use of higher-order modulation formats, or reduction of the reuse distance.
   (b) Spatial Multiplexing (Section 16.2) increases the capacity of each link.
   (c) Multi-user MIMO (Chapter 22) allows to serve several users in the same frequency sub-band in the same cell. Furthermore, cooperation between the BSs can decrease the amount of interference, see Section 22.11.

6. *Caching:* In case that the content is reused by many users (even though asynchronously), like in video on demand, caching at the wireless edge, or directly on the devices, can also greatly increase the capacity.

Further Reading

The cellular principle, and the basic concepts of cellular radio, are described in detail in [Lee 1995]. Though the description is somewhat outdated and tries to cover both analog and digital systems, the principles have remained valid; [Alouini and Goldsmith 1999] discuss more advanced aspects, including the effect of fading. [Chan 1992] discusses the impact of sectorization.

Various methods for frequency reuse and intercell interference coordination are discussed in [Necker 2008, Boudreau et al. 2009, Hamza et al. 2013, and Kosta et al. 2012].

[Frullone et al. 1996] give an overview of advanced cell planning techniques; [Woerner et al. 1994] describe issues concerning the simulation of cellular systems and network planning.

Excellent introductions to stochastic geometry can be found in [Haenggi 2012] and [Andrews et al. 2016]. Multi-tier networks are analyzed with stochastic geometry in [Dhillon et al. 2011].

Heterogeneous networks and femtocells are discussed in [Andrews et al. 2012, Andrews 2013]; their implementation in LTE in [Damnjanovic et al. 2011]; the specific aspects of interference coordination in [Lopez-Perez et al. 2011] and [Zahir et al. 2013], and SON in [Peng et al. 2013]. The scalability of wireless backhauls was analyzed in [Dhillon and Caire 2015].

For updates and errata for this chapter, see https://wides.usc.edu/students.html#textbooks

Exercises

See Sec. 36.21 of Exercises.pdf at wiley.com/go/molisch/wireless3e

22

Multiple Antennas for Multi-User Systems – MU-MIMO, Massive MIMO, and CoMP

22.1 Introduction and Intuition

We now turn our attention to the question of how Multiple Input Multiple Output (MIMO) systems work in a cellular scenario where the Base Station (BS) communicates with multiple users. As we will see in the following, this situation requires some new paradigms for the usage of the degrees of freedom provided by the multiple antenna elements.

The idea of using multiple antennas at the BS to enhance the throughput of cellular systems goes back more than 30 years, and was generally known as "smart antennas" in the 1990s. Since then, these approaches have been improved and generalized, and in the form of *multi-user MIMO (MU-MIMO)* and *Massive MIMO (mMIMO)* form a cornerstone of 5G (compare Chapter 32).

As we will expound below, there are different ways in which multiple BS antennas can be used in cellular systems, ranging from simultaneous communication of the BS with multiple users within the same cell, to reduction of interference in adjacent cells. The current section will provide a brief intuitive overview, while details will be discussed in the subsequent sections.

Generally, two different operating principles for throughput enhancement in cellular systems can be distinguished:

1. *Space Division Multiple Access (SDMA)*: In this approach, different User Equipments (UEs) can communicate with the BS on the *same* time/frequency slot, because the BS distinguishes them by means of their different spatial signatures. SDMA is also known under the name of *multi-user MIMO, MU-MIMO*. This name is used even if each of the UEs has only a single antenna element, since the signals from the different UEs can still be interpreted as "multiple inputs" or outputs. Such single-antenna-user situations are practically relevant, and often simplifies the theoretical treatment. Most of the subsequent discussion will concentrate on this case. There are different ways in which the operating principle of SDMA can be intuited:

   (a) *Pointing of antennas:* Imagine a situation where there are multiple directional horn antennas at the BS, each of which can be pointed mechanically at a different UE, independently of each other. An actual system realizes those directional antennas as the different beams of an antenna array (remember that an antenna array can form multiple beams simultaneously). Importantly, antenna arrays can form not only maxima in the antenna patterns, but also notches. The latter is important for suppressing co-channel interference, i.e., signals to/from another UE in the cell.

   (b) *Spatial multiplexing:* Consider the uplink of a cellular system. The different UEs all send out data streams that are independently encoded. This is similar to spatial multiplexing of Single-User MIMO (SU-MIMO) as discussed in Section 16.2. In other words, the Receiver (RX), i.e., the BS, can perform a joint detection of the signals and separate them as long as the number of BS antenna elements is larger than the total number of data streams. However, while the capacity equations *for a given channel* are similar to the SU-MIMO case, the channel that the BS "sees" is often very different. In the MU-MIMO case, the different UE antennas are distributed over a much larger angular range, thus usually leading to a larger angular spread and a smaller correlation between the signals at the BS antennas. Furthermore, while in the SU-MIMO case the average signal characteristics (more precisely, the second-order statistics) of different streams are often the same, this does not hold true in the MU-MIMO case, since different UEs experience different pathloss and shadowing. Finally, note that in the downlink there is a key difference between SU-MIMO and SDMA: in the former case, Channel State Information at the Transmitter (CSIT) provides only minor advantages, while in the latter, it is absolutely vital. The reason for this is that in the SDMA downlink the RX antennas are distributed among the different UEs, and no joint processing (like in the SU-MIMO case) can be performed. Thus, the elimination of inter-stream interference has to be performed at the Transmitter (TX), and that requires CSIT. Furthermore, in MU-MIMO, the question of user scheduling takes on a bigger role, as we will discuss in Section 22.8.

Wireless Communications: From Fundamentals to Beyond 5G, Third Edition. Andreas F. Molisch.
© 2023 John Wiley & Sons Ltd. Published 2023 by John Wiley & Sons Ltd.
Companion website: www.wiley.com/go/molisch/wireless3e

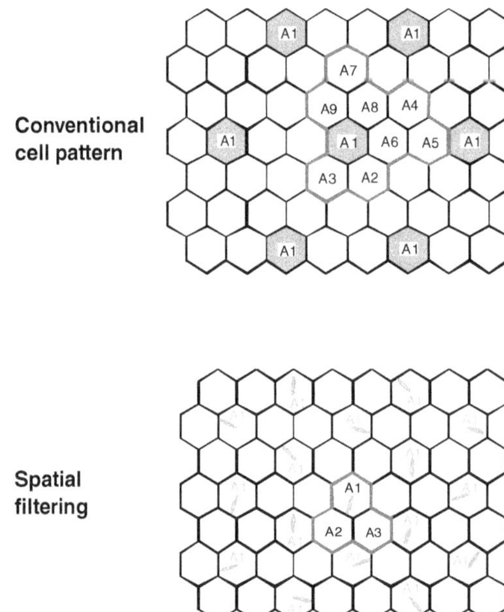

Figure 22.1 Principle of spatial filtering for interference reduction. Color version available at wiley.com/go/molisch/wireless3e.

2. *Spatial Filtering for Interference Reduction (SFIR)*: is used in Time Division Multiple Access/Frequency Division Multiple Access/ Orthogonal Frequency Division Multiple Access (TDMA/FDMA/OFDMA) systems to reduce the impact of inter-cell interference, and thus improve the area spectral efficiency, as shown in Figure 22.1.[1] The improvement in spectral efficiency can work via two mechanisms:

(i) in a system with a fixed *Modulation and Coding Scheme (MCS),* reduction of the inter-cell interference allows to reduce the cellular cluster size (as discussed in Section 21.2, cells belonging to the same cluster cannot use the same spectral resources; the smaller the cluster size the better the area spectral efficiency). Consequently, the same spectral resources can be used in more cells, and the number of admissible users per area increases proportionately.

(ii) In a system with *Adaptive Modulation and Coding (AMC)*, the reduction of the interference improves the Signal-to-Interference-and-Noise Ratio (SINR), and thus enables to operate with an MCS that has a higher spectral efficiency (usually close to the Shannon capacity). SFIR is strongly related to Inter-Cell Interference Coordination (ICIC) and network MIMO, see Section 22.11. Use of AMC and cluster size reduction can also be combined.

When comparing SDMA with SFIR, the required modifications within a system are considerably smaller for SFIR; for this reason it was preferred in the early days of smart antennas. However, the capacity improvement is usually greater for SDMA: in SDMA, doubling the number of antennas at the BS allows to double the number of users that can communicate at the same time. In SFIR, doubling the number of antennas helps to improve the SINR; the amount of improvement and the corresponding increase in spectral efficiency depends on a variety of factors including amount of noise, and available MCSs. Furthermore, while SDMA is beneficial in both noise- and interference-limited systems, SFIR displays its advantages only in interference-limited systems. In the remainder of this chapter, the discussion will thus focus on SDMA.

Forming the beams at the BS to separate the signals from/to the different UEs can be based on: (i) instantaneous Channel State Information (CSI), and (ii) second-order CSI. The former provides better performance, but has a larger overhead for the acquisition and possible feedback of CSI. Information theory provides the performance limits, which usually can be approached through non-linear signal processing. In practice linear processing is the dominant implementation method. A plethora of methods in particular for precoding at the BS is available, trading off complexity and performance. While we will mainly discuss narrowband situations, the methods and equations directly carry over to the frequency-selective case when Orthogonal Frequency Division Multiplexing (OFDM) is used as modulation format; in other words, the precoders and combiners can be applied on a subcarrier-by-subcarrier basis.

In the case where user separation is achieved based on second-order CSI, the number of UEs that can be handled simultaneously might be smaller – in the most extreme case, when Multi-Path Components (MPCs) are arriving at the BS from all directions for every UE, no user separation based on second-order CSI is possible, even though user separation might be feasible based on the instantaneous CSI.

Figure 22.2 explains the difference between these two principles in greater detail. On top we see the (more intuitive) system based on second-order statistics. We assume here that the MPCs from the two users, UE 1 and UE 2, are separated in the angle domain. The multi-antenna BS can thus form a beam into the direction of the MPCs from one user (we assume here that UE 1 is the desired user), and suppresses the MPCs from UE 2. The suppression might not be perfect if the beams have sidelobes, or are so broad that some

---

[1] Note that SDMA also uses spatial filtering in order to reduce interference. However, it has become common to use the term SFIR specifically for reducing *adjacent-cell* interference, while SDMA deals with interference from other users in the same cell.

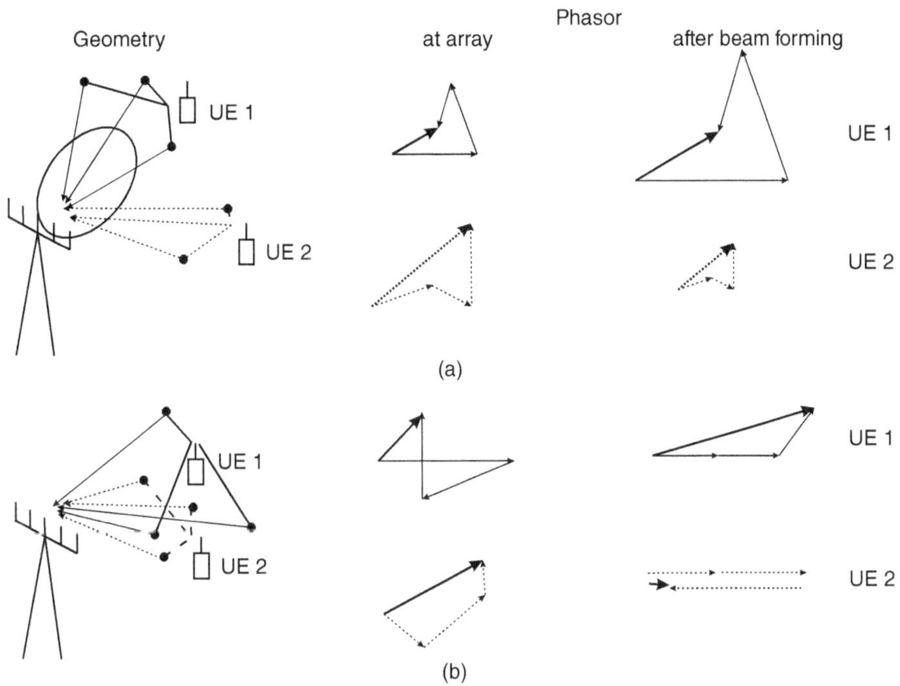

Figure 22.2 Principle of SDMA. (a) Based on second-order statistics. (b) Based on instantaneous CSI.

MPCs of UE 2 are only attenuated but not eliminated. Also note that the MPCs from UE 1 add up incoherently, so that there can be small-scale fading dips in the received power, but on average, signals coming from the direction of UE 1 are enhanced, and those from UE 2 are suppressed. Finally, remember that an array with N elements can simultaneously form N beams, so that the BS can simultaneously form a beam that enhances UE 1 and suppresses UE 2, and vice versa. Thus, in situations where the Angular Power Spectra (APSs) of the different users are well separated, the above approach gives good performance.

Now consider the situation in the lower figure. The angular ranges for the MPCs of the different users overlap, so that simply suppressing/nulling certain MPCs according to their directions might become impossible. Rather, what we do in this case is to manipulate the *complex* antenna characteristics. We do not care about the actual directional characteristics of the MPCs anymore, but rather concentrate on the complex signal from UE 1 and UE 2 at the different antenna elements (of course, those signals are the complex superposition of the different MPCs). The signals of the antenna elements are then adjusted, by means of the precoder (downlink) or combiner (uplink) in such a way that the desired signal gets a constructive interference of all the contributions for UE 1, and a destructive one for UE 2 (of course we cannot have both of these conditions fulfilled ideally at the same time; we will aim to either find a compromise between the two, as is done in Minimum Mean Square Error (MMSE) algorithms, or simply prioritize interference suppression, leading to Zero-Forcing (ZF) algorithms). Now with this approach it is possible to separate users even if their APS are completely overlapping. This approach furthermore gives the better overall performance since small-scale fading is greatly reduced as well.

22.2 System Model

The system model we consider in the following is outlined in Figures 22.3 and 22.4. A single BS with N_{BS} antenna elements communicates with K UEs with $N_{\mathrm{UE},k}$ antenna elements each, the sum of the number of all antennas at the UEs is denoted $N_{\mathrm{UE}} = \sum_k N_{\mathrm{UE},k}$.

Each of the UEs is transmitting/receiving $N_{\mathrm{S},k}$ data streams (with $N_{\mathrm{S},k} \leq N_{\mathrm{UE},k}$), with the sum of the number of data streams denoted as N_{S}. We assume henceforth that $N_{\mathrm{BS}} > N_{\mathrm{S},k}$ for all k, and even more that $N_{\mathrm{BS}} \geq N_{\mathrm{S}}$. This is the case normally occurring in a cellular network, and also the one with the most interesting effects for multi-user MIMO. A typical fourth-generation cell would have $N_{\mathrm{BS}} = 8$, $N_{\mathrm{S},k} = 2$, and $K = 4$. Note that K here means the UEs that are communicating with the BS on a particular time/frequency slot; the total number of UEs in the cell, $\hat{K}$, can be much larger (and the sum of all antennas of these potential UEs can be larger than N_{BS}); particular users can be scheduled by the BS according to their channel state and the traffic conditions.

Depending on whether we consider uplink or downlink, the UE is the TX or the RX. We consequently also have to consider two different block diagrams of the transmission chain. For self-consistency, we repeat here the various definitions and normalizations of the parameters previously defined in Chapter 16.

Figure 22.3 shows the uplink block diagram, which is similar to Section 16.1.2, though the single TX is now replaced by multiple TXs, indexed with k. At each UE, the original source data stream enters an encoder, which puts out $N_{\mathrm{S},k}$ different, possibly lower-rate,

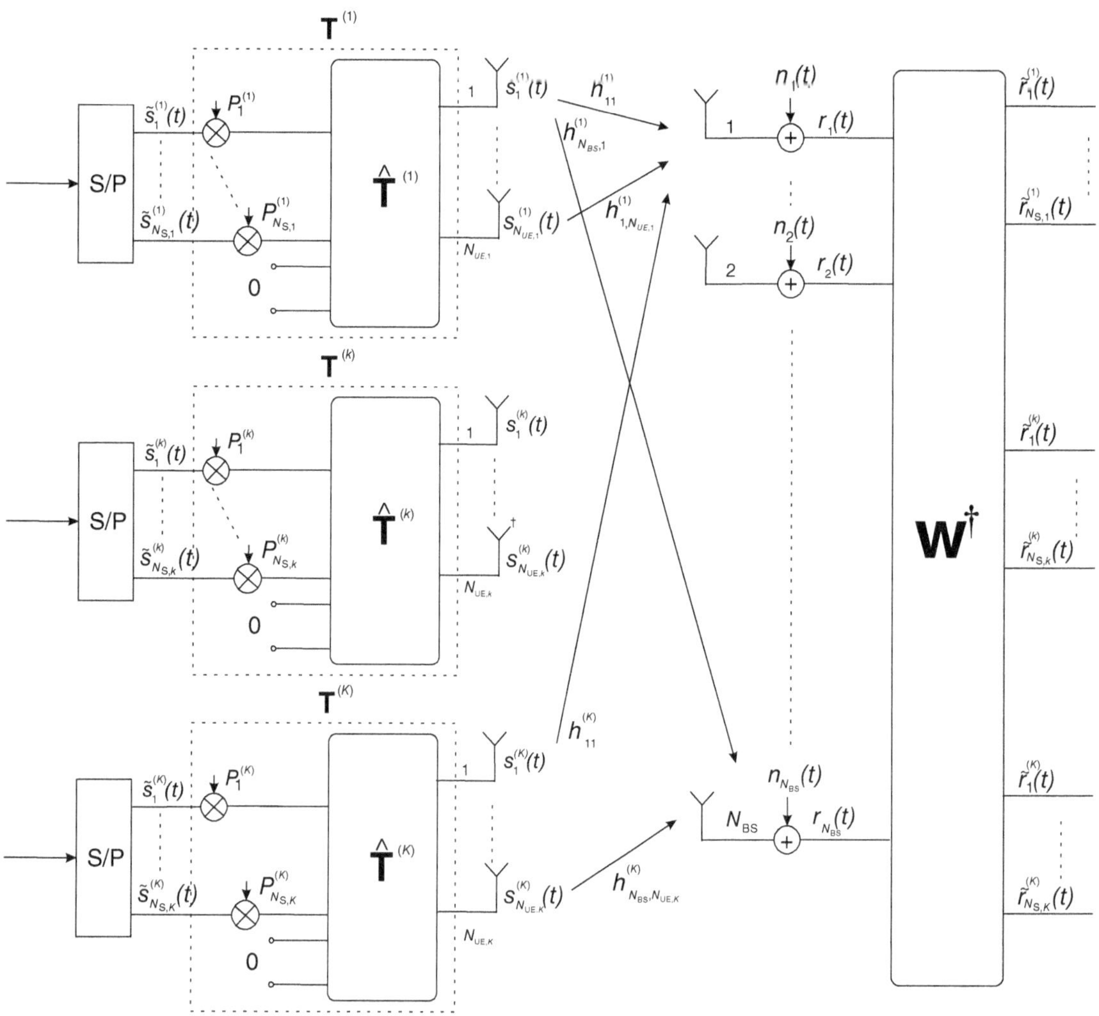

Figure 22.3 Multi-user MIMO system setup uplink.

data streams $\widetilde{s}_m^{(k)}(t)$, $m = 1, \ldots N_{S,k}$, which can be written as a vector $\widetilde{\mathbf{s}}_k(t)$.[2] Each data stream is normalized to $\int_{T_s} \left|\widetilde{s}_m^{(k)}(t)\right|^2 dt = T_s$. The different data streams then each get different powers assigned to them, $P_m^{(k)}$, which make up the entries in a diagonal matrix $\mathbf{P}_k$. The vector is augmented by $N_{UE,k} - N_{S,k}$ zeros and then sent through a linear precoder $\hat{\mathbf{T}}_k$ of dimension $N_{UE,k} \times N_{UE,k}$, resulting in an output signal $\mathbf{s}^{(k)}(t)$, where the components of the vector, $\left[s_1^{(k)}(t), s_2^{(k)}(t), \ldots, s_{N_{UE,k}}^{(k)}(t)\right]$ are the transmit signals at the different antenna elements. The precoder (steering matrix) can generally be written as

$$\mathbf{T}_k = \hat{\mathbf{T}}_k \begin{bmatrix} [\mathbf{P}_k]^{1/2} \\ \mathbf{0} \end{bmatrix}$$

where $\mathbf{P}_k$ is thus a diagonal matrix whose entries are the powers for the different streams, and $\hat{\mathbf{T}}_k$ is unitary $\hat{\mathbf{T}}_k \hat{\mathbf{T}}_k^{\dagger} = \mathbf{I}$.[3] More compactly, we can consider $\hat{\mathbf{T}}_k$ as the normalized precoder matrix (fulfilling a unit power constraint); we can (under some assumptions) interpret

---

[2] The fact that k is written as superscript $^{(k)}$ of (most) scalars but as subscript $_k$ for vectors and matrices has no deep mathematical meaning, but is just for easier readability. Note also that $\widetilde{\mathbf{s}}_k$ does not denote the kth element of the vector $\widetilde{\mathbf{s}}$, but rather indicates that this is the vector $\widetilde{\mathbf{s}}$ for the kth user.

[3] Just as in the system model in Section 16.2, this expression can be postmultiplied with a unitary matrix $\hat{\mathbf{T}}^{\dagger}$ whose presence does not change, however, the performance for Gaussian signals, and is thus usually disregarded [Heath and Lozano 2018].

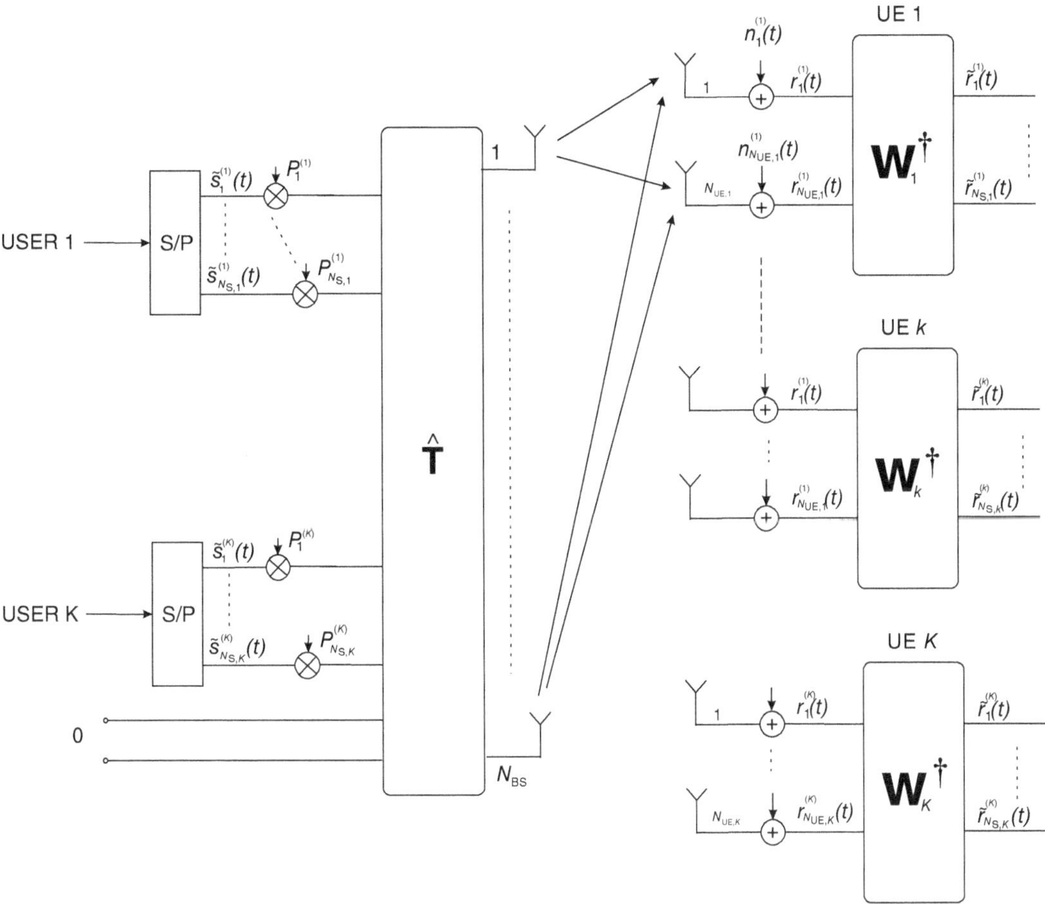

Figure 22.4 Multi-user MIMO system setup downlink.

a column as determining the direction in which a data stream will be sent. The **P** provides the power allocation for the different data streams. The precoder outputs are forwarded to the $N_{\mathrm{UE},k}$ transmit antennas of the UE, so that the transmit signal is

$$\mathbf{s}_k = \mathbf{T}_k \widetilde{\mathbf{s}}_k \tag{22.1}$$

The power constraints are generally different from the SU-MIMO case. In SU-MIMO we typically constrain the total transmitted power, or have a per-antenna power constraint. For the uplink of MU-MIMO, it is often meaningful to constrain the total power transmitted from each UE, such that

$$tr\left[\mathbf{T}_k \mathbf{T}_k^\dagger\right] = tr\left[\mathbf{T}_k^\dagger \mathbf{T}_k\right] = tr[\mathbf{P}_k] = P_{\mathrm{tot},k} \le P_{\max,k}. \tag{22.2}$$

Note that there can be very large variations between the actual transmitted powers of the different UEs – typically much more than between the powers of the data streams in SU-MIMO. This might be due to uplink power control that compensates the pathloss.

The channel from the kth UE to the BS is then written as

$$\mathbf{H}_k = \beta_k^{1/2}\widetilde{\mathbf{H}}_k$$

where β_k^2 describes the combination of path gain and shadowing, while $\widetilde{\mathbf{H}}_k$ is normalized such that $\left\{\left\|\widetilde{\mathbf{H}}_k\right\|_F^2\right\} = N_{\mathrm{UE},k}N_{\mathrm{BS}}$.

Now the different transmit signals can be stacked into an aggregate vector $\widetilde{\mathbf{s}}$, with an aggregate precoding matrix $\mathbf{T}$, and going through an aggregate channel $\mathbf{H}$. Denoting the total number of streams on the air as $N_{\mathrm{s}} = \Sigma_k N_{\mathrm{s},k}$, the total number of UE antenna elements as $N_{\mathrm{UE}} = \Sigma_k N_{\mathrm{UE},k}$

$$\tilde{\mathbf{s}} = \begin{pmatrix} \tilde{\mathbf{s}}_1 \\ \tilde{\mathbf{s}}_2 \\ \cdots \\ \tilde{\mathbf{s}}_K \end{pmatrix} \qquad \mathbf{s} = \begin{pmatrix} \mathbf{s}_1 \\ \mathbf{s}_2 \\ \cdots \\ \mathbf{s}_K \end{pmatrix} \tag{22.3}$$

$$\mathbf{T} = \begin{pmatrix} \mathbf{T}_1 & \mathbf{0} & .. & \mathbf{0} \\ \mathbf{0} & \mathbf{T}_2 & .. & \mathbf{0} \\ .. & .. & .. & .. \\ \mathbf{0} & \mathbf{0} & .. & \mathbf{T}_K \end{pmatrix} \tag{22.4}$$

and the total channel as

$$\mathbf{H} = (\mathbf{H}_1 \ \ \mathbf{H}_2 \ \ ... \ \ \mathbf{H}_K). \tag{22.5}$$

For the case that each UE has only a single antenna, $\hat{\mathbf{T}} = \mathbf{I}$ and thus $\mathbf{T} = \mathbf{P}$.

With this notation, the overall input–output relationship can be written as

$$\mathbf{r} = \mathbf{Hs} + \mathbf{n} \tag{22.6}$$

where the vector $\mathbf{r}$ contains the signals received at the BS antenna elements and $\mathbf{n}$ contains the noise. Just like in the SU-MIMO case, a first processing stage is often a linear matrix filter $\mathbf{W}^\dagger$ providing linearly filtered outputs $\tilde{\mathbf{r}}$.

For the downlink, the TX has the same characterization as in the SU-MIMO case: it first creates a number of data streams; though now groups of data streams are intended for particular UEs, and we might index such a group accordingly with a subscript k, see Figure 22.4. Importantly, the (single) precoder $\mathbf{T}$, which is located at the BS, can now have a general form

$$\mathbf{T} = \hat{\mathbf{T}} \begin{bmatrix} \mathbf{P}^{1/2} \\ \mathbf{0} \end{bmatrix} \tag{22.7}$$

in other words, for the downlink case we do not have a restriction to the block-diagonal structure of Eq. (22.4). Furthermore, we now again have a total power limit $tr[\mathbf{TT}^\dagger] = tr[\mathbf{P}] = P_{\text{tot}} \le P_{\text{max}}$. Note that we can write the downlink channel as the transpose of the uplink channel $\mathbf{H}^{(\text{DL})} = [\mathbf{H}^{(\text{UL})}]^T$.[4] Each of the RXs has a separate matrix filter $\mathbf{W}_k^\dagger$, and the outputs from those filters $\mathbf{r}_k$ at the different UEs are processed separately.

Just like for the SU-MIMO case, MU-MIMO distinguishes between situations with full CSIT, second-order CSIT, and no CSIT. Note that the no-CSIT case is only meaningful in the uplink, while for the downlink the achievable capacity is extremely low, as we face the following choice: (i) the BS transmits only $\min_k(N_{\text{UE},k})$ data streams, which drastically limits the capacity of the downlink; (ii) the BS transmits N_{BS} data streams, but since the UEs cannot decode that many data streams, the SINR at each UE is extremely bad.

22.3 Performance Limits

In a first step, we want to explore the fundamental limits of MU-MIMO, i.e., how many users can be supported, and at what rate. We will again start with an intuitive description that draws from the analogy to SU-MIMO, and then summarize key results from information theory.

22.3.1 Uplink

Consider first the case that each UE has only a single antenna: then the system is identical to an Horizontal BLAST (H-BLAST) system (Section 16.2.9); it is known that as long as the transmit rates from the different antenna elements are chosen suitably, an H-BLAST system (with per-antenna rate control) can achieve capacity. To assess the degrees of freedom, i.e., the number of orthogonal streams that can be transmitted over the air (see Chapter 28 for non-orthogonal transmissions), we have to distinguish a number of different cases:

- Single-antenna UEs, $K \le N_{\text{BS}}$: in this case, the number of data streams is clearly limited by K. For a fixed K, increasing N_{BS} beyond K does not impact the degrees of freedom, but only helps to obtain additional diversity and beamforming gain, and thus increases the data throughput logarithmically. The situation is different if we consider K to be variable. Assume that $K \le N_{\text{BS}} \le \hat{K}$, i.e., there is a

---

[4] Some references write the channel as conjugate transpose $\left[\mathbf{H}^{(\text{UL})}\right]^\dagger$.

large "pool" of $\hat{K}$ UEs that can be scheduled in the same time/frequency slot. In that case, we can increase K as we increase N_{BS}. In other words, in this situation, increasing N_{BS} leads to a linear scaling of the system throughput.

- Multi-antenna UEs, $N_{\mathrm{UE}} \leq N_{\mathrm{BS}}$: in this case, each UE can send out multiple data streams; apart from this, the situation is the same as in the single-antenna case.
- Multi-antenna UEs, $K \leq N_{\mathrm{BS}} \leq \sum N_{\mathrm{UE},k}$: The overall number of transmitted streams N_s can be up to N_{BS}. Even if we consider the number of active UEs, K, to be fixed, the system throughput scales linearly as we increase N_{BS}, because we can increase the number of streams *per UE* when N_{BS} increases (note, however, the condition $N_{\mathrm{BS}} \leq \sum N_{\mathrm{UE},k}$). The actual throughput (as opposed to the number of admissible streams) depends on how many data streams are assigned to which UE, and what the channels from these UEs are. Furthermore, even if scheduling from a larger pool of UEs is taken into account, the total number of streams remains upper-limited by N_{BS}. Allowing streams from more different users (as opposed to many streams from a few users each) may increase the actual throughput because streams with better SINR might be selected. Yet the scaling law, namely linear increase with N_{BS}, remains the same.

After these intuitive considerations, we now turn to information-theoretic treatment of the performance limits. As shown in Section 18.2, the performance limits are given by a capacity region, i.e., the set of all achievable rate tuples. In the case of two users (where the capacity region can be most easily represented graphically), and for the case of single-antenna UEs, the rate region is almost completely analogous to that of Section 18.2.1:

$$R_k \leq \log_2\left(1 + \frac{P_k}{\sigma_\mathrm{n}^2}\|\mathbf{h}_k\|^2\right) \tag{22.8}$$

where due to the single-antenna assumption, $\mathbf{h}_k$ is an $N_{\mathrm{BS}} \times 1$ vector; P_k is the power assigned to the kth user, which is upper limited as $P_k \leq P_{\max,k}$. The sum rate of all users is limited by

$$\sum_k R_k \leq \log_2 \det\left(\mathbf{I} + \frac{\sum\limits_k P_k \mathbf{h}_k \mathbf{h}_k^\dagger}{\sigma_\mathrm{n}^2}\right). \tag{22.9}$$

The optimum RX structure is an H-BLAST RX, i.e., Successive Interference Cancellation (SIC), with Per Antenna Rate Control (PARC). The achieved rates for the different users depend on the decoding order.

For $K = 2$ and single-antenna UEs, the rate for the first user is

$$R_1 \leq \log_2\left(1 + P_1 \mathbf{h}_1^\dagger \left(\sigma_\mathrm{n}^2 \mathbf{I} + P_2 \mathbf{h}_2 \mathbf{h}_2^\dagger\right)^{-1} \mathbf{h}_1\right) \tag{22.10}$$

since it suffers from the interference from the second user. When decoding the second user, the contribution from the first user has already been subtracted, so that the rate is

$$\begin{aligned}
R_2 &\leq \log_2\left(1 + P_2 \mathbf{h}_2^\dagger \left(\sigma_\mathrm{n}^2 \mathbf{I}\right)^{-1} \mathbf{h}_2\right) \\
&= \log_2\left(1 + P_2 \frac{\|\mathbf{h}_2\|^2}{\sigma_\mathrm{n}^2}\right).
\end{aligned} \tag{22.11}$$

When flipping the decoding order, the achievable rates have to be changed accordingly. Note that H-BLAST with this particular decoding order corresponds to one corner of the capacity region.

In the case of multiple antenna elements at the UE, the precoder at the UE and the propagation channel form an "effective channel," so that $\mathbf{H}_1$ needs to be replaced by $\mathbf{H}_1 \mathbf{T}_1$ (note that the power allocation is part of $\mathbf{T}$, so it does not show up explicitly in this formulation). Specifically,

$$R_k \leq \log_2 \det\left(\mathbf{I} + \frac{1}{\sigma_\mathrm{n}^2} \mathbf{H}_k \mathbf{T}_k \mathbf{T}_k^\dagger \mathbf{H}_k^\dagger \left(\mathbf{I} + \sum_{k'=k+1}^{K} \frac{1}{\sigma_\mathrm{n}^2} \mathbf{H}_{k'} \mathbf{T}_{k'} \mathbf{T}_{k'}^\dagger \mathbf{H}_{k'}^\dagger\right)^{-1}\right). \tag{22.12}$$

Again we clearly see from the summation term that the decoding order plays an important role in what rates can be achieved. For optimization of weighted rates, $\kappa_k R_k$ (where the κ_k are the weights), the highest-weight user should be decoded last.

Computation of the true capacity region requires a maximization of the rate regions over all possible precoders. The resulting region is, of course, larger than the rate region corresponding to the restrictive assumption $\hat{\mathbf{T}}_k \propto \mathbf{I}$.

Optimization of the precoders to provide maximum sum rate can be achieved by *iterative waterfilling* [Yu et al. 2004]. Considering the kth user, fix power and precoders of all other users, and treat their signals as noise/interference. The optimum precoder can be then computed from the SU-MIMO principles. Specifically, iterate the following algorithm to convergence

- for $k = 1$ to K

  ○ for a given user k, compute the covariance matrix $\widetilde{\mathbf{R}}_{\mathrm{ni},k}$ of the noise and the interference from all other users. Compute an "effective" channel matrix $\widetilde{\mathbf{R}}_{\mathrm{ni},k}^{-1/2}\mathbf{H}_k$.

  ○ obtain the optimum signal correlation matrix from singular value decomposition of this effective matrix, with the optimum precoder derived from the right singular vectors and the powers according to waterfilling, as described in Section 16.2.3.

Up to now we have assumed perfect CSIT and a static channel. In the case of fading channels where the codewords extend over multiple fading realizations, i.e., when considering ergodic capacity, Eqs. (22.8) and (22.9) are modified by taking the ensemble average of the right hand side.

22.3.2 Downlink

Let us now turn to the downlink. As mentioned in Section 22.1, the downlink (Broadcast Channel (BC)) differs from the SU-MIMO case in that CSIT is essential. The received signal $\mathbf{r}_k$ at the kth UE is

$$\mathbf{r}_k = \mathbf{H}_k\mathbf{s} + \mathbf{n}. \tag{22.13}$$

Here the transmit signal is the superposition $\mathbf{s} = \sum_k \mathbf{s}_k$, and the $\mathbf{s}_k$ are linearly or nonlinearly encoded versions of the signal streams intended for the different users. As for the uplink case, the UEs cannot cooperate – but now they cannot cooperate for the *decoding* of the data streams. Consequently, the only way to avoid inter-user interference is via a *TX precoding* that eliminates interference at the RX (and which requires CSIT).

However, *if* CSIT is present, then there exists a duality between the uplink (*Multiple-Access, MAC*) and the downlink (*BC*) case (i.e., the boundary of the capacity region is the same) when the following additional conditions are fulfilled:

- the power constraint at the BS is the same as the sum power constraint of the UEs, and the noise is the same at UE and BS antennas. Note that while the UEs can pool their powers, they cannot jointly precode.
- the propagation channels are reciprocal.

This result has two important interpretations: if MAC and BC channel fulfill those conditions, then uplink and downlink have the same data rates. More importantly, it is relatively easy to compute the capacity region for the MAC case, as done in the previous subsection. Thus, for the computation of the BC capacity region, first construct a "dual MAC" setup in which these conditions are fulfilled by definition (e.g., only a sum-power constraint, but no per-UE power constraint is imposed on the uplink transmission). Then compute the capacity region for this equivalent MAC channel, and finally conclude from the duality theorem that the BC capacity region is the same.

The duality holds not only for the capacity boundary, but also for the transceiver structures. While for the uplink the BS, as the RX, uses serial interference cancellation, in the downlink it should use *dirty paper coding*. This is based on the fact that if interference is known at the TX, then the desired signal can be precoded in such a way that the RX does not "see" any interference. The name derives from the principle that if one knows the position of dirt on a piece of paper, one can write letters in such a way that the dirt does not hinder the reading of the letters [Peel 2003]. Codes that achieve such interference suppression are called "*dirty paper codes*" or "*Costas codes*." There are numerous ways on how such dirty-paper codes can be approximated, e.g., through Tomlinson–Harashima precoding.

When performing the encoding of multiple data streams, the order in which the streams are encoded is very important: The encoder for the kth UE treats the interference from data streams for UE $1, \dots k - 1$ as known (which can therefore be completely suppressed), while the interference for users $k + 1, \dots K$ has the same effect as noise. Consequently, the first encoded stream has a poor SINR, while later encoded streams (which are derived from the "cleaned-up" signals) have good SINR:

$$R_k \leq \log_2 \det\left(\mathbf{I} + \frac{1}{\sigma_{\mathrm{n}}^2}\mathbf{H}_k\mathbf{T}_k\mathbf{T}_k^\dagger\mathbf{H}_k^\dagger\left(\mathbf{I} + \sum_{k'=k+1}^{K}\frac{1}{\sigma_{\mathrm{n}}^2}\mathbf{H}_k\mathbf{T}_{k'}\mathbf{T}_{k'}^\dagger\mathbf{H}_k^\dagger\right)^{-1}\right). \tag{22.14}$$

In contrast to the uplink case, only the channel $\mathbf{H}_k$ and not $\mathbf{H}_{k'}$ appears in (22.14), since all the signals decoded at the kth UE go through the same channel.

The computation of the capacity region according to (22.14) requires optimization of precoders and is complicated by the fact that it is defined as the maximum over all possible decoding orders. Computation of the capacity region is thus preferable via the dual MAC; similarly the computation of the precoders can be done via the dual MAC [Vishwanath et al. 2003].

We close this discussion by noting that the dual MAC is a computational tool, not a physical MAC setup. This can be seen from the fact that the power constraints in the dual MAC are generally different from the power constraints that would occur in an actual physical uplink channel, where there are per-UE power constraints, and the admissible sum power is typically smaller than the power used by the BS in the downlink. Furthermore, in a Frequency Domain Duplex (FDD) system the physical uplink operates on a different frequency than the downlink, and thus has different channel coefficients, while the dual MAC assumes the same channel as the BC channel.

In the remainder of this section, we will often use the term "capacity" even though strictly speaking we refer to a "maximum achievable throughput under certain constraints." This nomenclature (which we also used in earlier chapters) is not accurate from an information-theoretic point of view, but widely used in the MIMO literature, so that we adopt it here as well.

22.4 Linear Processing for Uplink

We now turn to a special case of processing in MU-MIMO, namely linear processing. For the uplink case, this is very similar to the SU-MIMO case, as we previously discussed. The main challenges arise in the case that the UEs have multiple antenna elements and we wish to perform a precoding for each of them.

22.4.1 Processing for Single-Antenna UEs

We first review the solution for the case of UEs with a single antenna. Without loss of generality, denote the kth user as the user of interest. Section 12.4.3 derived the optimum antenna weights, i.e., the ones that maximize the SINR. Define the channel correlation matrix (note that due to the SIMO configuration the channel matrix $\mathbf{h}$ is a $N_{\mathrm{BS}} \times 1$ column vector)

$$\mathbf{R}_k = \mathbf{h}_k \mathbf{h}_k^\dagger \tag{22.15}$$

and the interference matrix

$$\mathbf{R}_{\mathrm{intf},k} = \sum_{\ell \neq k} \mathbf{h}_\ell \mathbf{h}_\ell^\dagger.$$

We can then define the ZF RX as any receive filter $\mathbf{w}_k^\dagger \perp \mathbf{R}_{\mathrm{intf},k}$, i.e., the desired combiner weight lies in the nullspace of the interference signal, where $\mathbf{w}_k^\dagger$ is a $1 \times N_{\mathrm{BS}}$ row vector. There are multiple solutions to this problem that may all be equivalent, since suppression of interference is the only optimization criterion; a more practically useful solution also requires at the same time that the Signal-to-Noise Ratio (SNR) at the filter output is maximized.

For this reason, it is more common to optimize the mean-squared error of the signal, treating noise and interference as equally detrimental to decoding. We thus first define the noise-plus-interference matrix for equal-power transmission as

$$\mathbf{R}_{\mathrm{ni},k} = \frac{1}{\gamma_{\mathrm{TX}}}\mathbf{I} + \sum_{\ell \neq k} \mathbf{h}_\ell \mathbf{h}_\ell^\dagger. \tag{22.16}$$

More generally, let the transmit power be P_k. Let furthermore the power-weighted versions of these channel and interference matrices be

$$\widetilde{\mathbf{R}}_k = P_k \mathbf{h}_k \mathbf{h}_k^\dagger, \qquad \widetilde{\mathbf{R}}_{\mathrm{ni},k} = \sigma_{\mathrm{n}}^2 \mathbf{I} + \sum_{\ell \neq k} P_\ell \mathbf{h}_\ell \mathbf{h}_\ell^\dagger, \qquad \widetilde{\mathbf{R}}_{\mathrm{tot}} = \sigma_{\mathrm{n}}^2 \mathbf{I} + \sum_k P_k \mathbf{h}_k \mathbf{h}_k^\dagger. \tag{22.17}$$

Then the optimum antenna weights $\mathbf{w}_k^\dagger$ for the kth user are obtained by maximization of the SINR, i.e., (compare (12.24))

$$\mathbf{w}_{\mathrm{opt},k} = \arg\max_{|\mathbf{w}|^2 = 1} \mathrm{SINR}_k = \arg\max_{|\mathbf{w}|^2 = 1} \frac{\mathbf{w}^\dagger \widetilde{\mathbf{R}}_k \mathbf{w}}{\mathbf{w}^\dagger \widetilde{\mathbf{R}}_{\mathrm{ni},k} \mathbf{w}} \tag{22.18}$$

which is the solution of a generalized eigenvalue problem. The ZF RX or *decorrelating RX*, is a special case obtained by setting $\sigma_{\mathrm{n}} = 0$.

Now let us turn to the case of multiple users. In principle, the above derivation still holds, assuming that the interference for a specific user is fixed and known. For the ZF RX, we now have to ensure that the RX weight vector for each user signal is in the null space of all the other users. Since we can design the RX filter for each user independently of the filters for all other users, the RX weight matrix has to be the pseudoinverse of the effective (stacked) channel matrix

$$\mathbf{W}^\dagger = \mathbf{H}^+ = \begin{cases} \left(\mathbf{H}^\dagger \mathbf{H}\right)^{-1} \mathbf{H}^\dagger & \text{if } N_{\mathrm{BS}} \geq K \\ \mathbf{H}^\dagger \left(\mathbf{H} \mathbf{H}^\dagger\right)^{-1} & \text{if } N_{\mathrm{BS}} \leq K. \end{cases} \tag{22.19}$$

Note that this RX filter leads to noise enhancement, and that (in an Independent Identically Distributed (i.i.d.) Rayleigh-fading channel) the statistics of the SNR are χ^2 distributed with the degrees of freedom equal to $2(N_{\mathrm{BS}} - K + 1)$, assuming $N_{\mathrm{BS}} \geq K$. This solution is independent of the transmit powers of the UEs, since the optimization criterion is to eliminate all interference to the data streams from other users. Additional criteria, such as minimization of the total TX power while each UE reaches a target error rate, or maximization of the sum throughput of the BS, can be used to obtain a unique solution.

For the MMSE RX, we can similarly state that for a given TX power of each UE, the RX filters for these users can be computed independently of each other. Thus, the optimum receive filter is simply a stacked version of Eq. (22.18) or equivalently

$$\mathbf{W} \propto \widetilde{\mathbf{R}}_{\text{tot}}^{-1}\,\mathbf{H} \tag{22.20}$$

which needs to be scaled to ensure proper normalization.

However, we can get better overall results if the transmit powers of the different users are not fixed, but rather can be optimized. This requires feedback commands from the BS to the UE (to instruct the UEs about which power level to use), as well as power control capabilities of the UEs. Finding the optimum UE transmission powers together with the optimum receive antenna weights is complicated because adjusting the power for one UE affects the SINR for all users. There is therefore no closed-form solution; rather, we iterate an outer and two inner loops to convergence: in the outer loop, we alternatingly keep the beamforming weights fixed and optimize the power allocation, or vice versa. In each step of the inner loops we optimize one UE while keeping the parameters of the other ones constant. Concretely:

1. Fix the transmit powers of the UEs, and obtain the BS antenna weights for the reception of the different users by Eq. (22.18).

2. Then, leave the set of antenna weights $\mathbf{w}_k^\dagger$ unchanged and compute the optimum power allocations P_k. This computation itself does not have a closed-form solution but can be obtained through various numerical optimization methods that depend on the target of the optimization. For example, for optimizing the sum throughput, iterate the following equation to convergence (where superscript (n) is the iteration counter) [Chrisanthopoulou and Tsoukatos 2007]:

$$P_k^{(n+1)} = \min\left[P_{\max}, P_k^{(n)} + \delta_{\text{step}}\left(\frac{1}{P_k^{(n)}} - \sum_{\ell\neq k}\frac{\mathbf{w}_\ell^\dagger \mathbf{R}_k \mathbf{w}_\ell}{\sigma_{\text{n}}^2 + \sum_{m\neq\ell} P_m \mathbf{w}_\ell^\dagger \mathbf{R}_m \mathbf{w}_\ell}\right)\right] \tag{22.21}$$

where $P_{\max}$ is the maximum power a UE can transmit, and δ_{step} is a stepsize parameter. Do this for all k. In essence, this step is a gradient projection method, i.e., updating the power along the gradient of the capacity.

3. Update the $\widetilde{\mathbf{R}}_k$ and $\widetilde{\mathbf{R}}_{\text{ni},k}$, and recompute the $\mathbf{w}_k$ from Eq. (22.18). Return to step 2.

Besides the above capacity optimization, other relevant optimization criteria exist, such as minimization of the total transmit power of the UEs while guaranteeing the quality of service (minimum SINR) for every user. In that case, in each iteration round, each user reduces its transmit power so that it just meets the SINR threshold. This reduces interference to other users in the cell, so that in the next iteration round, every user can reduce its transmit power even more.

Of course, linear reception is not optimum. Different power allocations are desirable when nonlinear reception is used; in this case the equations for the MAC capacity serve as guidelines for the power allocation design.

*22.4.2 Processing for Multi-Antenna UEs

The simplest practical implementation of multi-user MIMO processing with multi-antenna UEs is based on linear processing at TX and RX. The RX (the BS) has all signals available, and can thus perform optimum (linear) processing for interference suppression. The UE (the TX) on the other hand has only its own signal available, and thus can perform precoding only with respect to this signal, and over the limited set of TX antennas mounted on this UE. Obviously, whenever one UE changes its precoder, it impacts the interference seen at the BS, which leads to a change at the RX filter, which in turn might require a change in the precoder of another UE. In other words, the optimum transmit and receive weights depend on each other. Therefore an iterative determination of the optimum weights is generally used.

We first re-define the desired signal correlation matrix and the noise-and-interference correlation matrix: it now becomes

$$\widetilde{\mathbf{R}}_k = \mathbf{H}_k \mathbf{T}_k \mathbf{T}_k^\dagger \mathbf{H}_k^\dagger, \qquad \widetilde{\mathbf{R}}_{\text{ni},k} = \sigma_{\text{n}}^2 \mathbf{I} + \sum_{\ell\neq k}\mathbf{H}_\ell \mathbf{T}_\ell \mathbf{T}_\ell^\dagger \mathbf{H}_\ell^\dagger \qquad \widetilde{\mathbf{R}}_{\text{tot}} = \sigma_{\text{n}}^2 \mathbf{I} + \sum_k \mathbf{H}_k \mathbf{T}_k \mathbf{T}_k^\dagger \mathbf{H}_k^\dagger. \tag{22.22}$$

Those expressions are implicitly power-weighted – remember that $\mathbf{T}$ encompasses the power allocation. Furthermore, we note that each RX filter $\mathbf{W}_k^\dagger$ (for the kth user) is now a matrix, because it might process multiple data streams.

The details of the weight optimization depend on the optimization criterion; in the following, we consider the minimization of the overall Mean-Square Error (MSE). The MSE matrix for the kth user (whose diagonal elements are the MSE for the various streams of the user) can be written as

$$\mathbf{MSE}_k = \mathbf{I} - \mathbf{W}_k^\dagger \mathbf{H}_k \mathbf{T}_k - \mathbf{T}_k^\dagger \mathbf{H}_k^\dagger \mathbf{W}_k + \mathbf{W}_k^\dagger \widetilde{\mathbf{R}}_{\mathrm{tot}} \mathbf{W}_k. \tag{22.23}$$

The optimum RX filter (note that this expression is not normalized) for the kth user then fulfills

$$\mathbf{W}_k = \widetilde{\mathbf{R}}_{\mathrm{tot}}^{-1} \mathbf{H}_k \mathbf{T}_k \tag{22.24}$$

and the MSE becomes

$$\mathbf{MSE}_k = \mathbf{I} - \mathbf{T}_k^\dagger \mathbf{H}_k^\dagger \widetilde{\mathbf{R}}_{\mathrm{tot}}^{-1} \mathbf{H}_k \mathbf{T}_k. \tag{22.25}$$

The goal is then to minimize this MSE under the power constraints

$$tr\{\mathbf{T}_k \mathbf{T}_k^\dagger\} \le P_k. \tag{22.26}$$

This formulation confirms that the TX weights are functions of the RX weights of *all* users, while the optimum RX weights depend on the TX weights of *all* users. Thus, the RX weights and TX weights can be computed with the following iteration [Serbetli and Yener 2004]

1. Update for all users ($k = 1, \dots K$) the RX weight, which is simply the MMSE RX filter of the "effective channel" as given in Eq. (22.24).
2. Update for all users ($k = 1, \dots K$) the TX weights, which again are MMSE filters, with updated power assignments.

$$\mathbf{X}_k(\mu_k') = \left[\mu_k' \mathbf{I} + \sum_{j=1}^{K} \mathbf{H}_k^\dagger \mathbf{W}_j \mathbf{W}_j^\dagger \mathbf{H}_k \right]^{-1} \mathbf{H}_k^\dagger \mathbf{W}_k \tag{22.27}$$

$$\mu_k = \arg_{\mu_k'} \left(tr\{\mathbf{X}_k(\mu_k') \mathbf{X}_k(\mu_k')^\dagger\} = P_k \right) \tag{22.28}$$

$$\mathbf{T}_k = \left[\mu_k \mathbf{I} + \sum_{j=1}^{K} \mathbf{H}_k^\dagger \mathbf{W}_j \mathbf{W}_j^\dagger \mathbf{H}_k \right]^{-1} \mathbf{H}_k^\dagger \mathbf{W}_k. \tag{22.29}$$

Typically, 10–20 iterations are sufficient for convergence.

22.5 Linear Processing for the Downlink

Determination of the optimum transmission strategy for the downlink is complicated because changing the transmit strategy (e.g., the antenna weights) for one user influences the SINR for every other user. This section only considers algorithms for linear (beamforming) transmission schemes, which are most widely used in practice. The BS has a constraint on the *total* transmit power, which is reasonable, because total transmit power is limited by the admissible interference to other cells, limitations by the frequency regulator, and the power amplifiers.[5] Finally, the BS is assumed to have complete CSI, i.e., it knows $\mathbf{H}_k$ for all the k users.

22.5.1 Processing for Single-Antenna UEs

The situation is again somewhat simplified if each UE has only a single antenna. Thus, there is no joint processing or spatial filtering at the RX side, and all processing (including power allocation) has to be done at the TX.

Zero-Forcing

Let us first establish how many users can communicate simultaneously with the BS when linear processing is used – as we will show momentarily, it is the same as the "theoretically optimum" degrees of freedom obtained from capacity considerations. We know that in the uplink with linear processing, the number of users is limited by the number of antenna elements in the BS, N_{BS}. This follows from the principle of "optimum combining" that each additional antenna element can suppress one interfering signal; thus, one set of BS antenna weights can deal with 1 desired user and $N_{\mathrm{BS}} - 1$ interferers simultaneously. The key insight is now that this admissible number of users is the same for the downlink, i.e., a BS with N_{BS} antenna elements can transmit to N_{BS} different UEs at the same time.

---

[5] If each antenna has its own power amplifier, then a per-antenna power constraint is more realistic. However, this leads to considerable mathematical complications and is thus not treated here.

Intuitively, this result follows from reciprocity: if – by adapting its weights – the BS in the uplink forms an antenna pattern that nulls out an interferer, then the same BS will not create any interference at this user when the BS transmits.[6]

We now formulate these insights mathematically. Denote again the directional precoder as $\hat{\mathbf{T}}$, and the precoder including power allocation as $\mathbf{T}$. The precoders are at the BS, while the UEs don't have receive spatial filters since by assumption they have a single-antenna. If the kth UE should see no interference intended for other users, the transmit signal for the kth UE must lie in the nullspace of the channel vectors for all other users. The precoding vector for the kth UE is then the kth column of the pseudoinverse of $\mathbf{H}$, so that the transmit signal must be (assuming for notational convenience $N_t = N_s$)

$$\mathbf{s} = \mathbf{H}^+ \mathbf{P}^{1/2}\widetilde{\mathbf{s}} = \mathbf{H}^\dagger \left(\mathbf{H}\mathbf{H}^\dagger\right)^{-1}\mathbf{P}^{1/2}\widetilde{\mathbf{s}}. \tag{22.30}$$

In other words, choosing $\hat{\mathbf{T}} = \mathbf{H}^+$ ensures that there is no interference signal arriving at any of the UEs, while the $\mathbf{P}$ independently allocates the power. Thus, the beamforming can be combined with a variety of power allocation strategies. For example, maximization of throughput is obtained by water filling. Another example is to optimize the sum SNR, while guaranteeing a certain minimum SNR for each user. Clearly, zero interference can only be achieved if no more than N_{BS} users are served at one given time, so we assume henceforth $K \le N_{\mathrm{BS}}$. The key disadvantage of ZF TXs is that in the case of ill-conditioned channel matrices, the transmit power along a particular dimension can become very large; conversely, since the total transmit power is limited, the transmit power is used very inefficiently. In other words: if two users are "effectively" close together (in the sense that their channel vectors are almost linearly dependent – they do not have to be close together geographically), then it is difficult to transmit to one user while simultaneously suppressing transmissions to the other. This situation is dual to the effect of noise enhancement in the RX. Note that in systems with a large number of users, of which only a subset transmits on the same time-frequency resources via MU-MIMO, this problem can be mitigated through smart scheduling, see Section 22.8.

Regularization

The performance can be improved by so-called *regularization* that trades off noise enhancement and interference. The transmit signal becomes

$$\mathbf{s} = \mathbf{H}^\dagger \left(\mathbf{H}\mathbf{H}^\dagger + \zeta\mathbf{I}\right)^{-1}\mathbf{P}^{1/2}\widetilde{\mathbf{s}} \tag{22.31}$$

where ζ is the so-called "regularization parameter." This is obviously a dual of an MMSE RX. We can now aim to optimize the sum rate of the transmission, or guarantee a minimum rate for each user; for either of those goals, iterative algorithms exist [Stojnic et al. 2006].

Furthermore, the random beamforming discussed below can be used to provide greater fairness. There are many similar, but subtly different, optimization problems. For example, the "power control problem" aims to minimize the total transmitted power while guaranteeing a certain minimum SINR (and thus a certain rate) for every user. A number of such resource allocation problems can be tackled by convex optimization techniques [Boyd and Vandenberghe 2004].

Optimum Beamforming

An optimum beamforming can be shown to be computationally hard, yet the solution has a simple structure, as shown by [Bengtsson and Ottersten 2001]. The starting point is the problem of minimizing the sum of the transmit weights (and thus the transmit powers)

$$\underset{\mathbf{t}_1, \mathbf{t}_2, \dots \mathbf{t}_K}{\mathrm{minimize}} \sum_k \|\mathbf{t}_k\|^2$$

$$\text{s.t.} \quad SINR_k \ge \breve{\gamma}_k$$

where $\breve{\gamma}_k$ is the target SINR for the kth user. This problem can be reformulated as a convex optimization problem; the optimum beamformers are

$$\mathbf{t}_k = \sqrt{P_k}\,\hat{\mathbf{t}}_k \qquad \hat{\mathbf{t}}_k^* = \frac{\left[\mathbf{I}_{N_{\mathrm{BS}}} + \sum_{i=1}^{K} \dfrac{\lambda_{\mathrm{Lag},i}}{\sigma_\mathrm{n}^2}\mathbf{h}_i\mathbf{h}_i^\dagger\right]^{-1}\mathbf{h}_k}{\left\|\left[\mathbf{I}_{N_{\mathrm{BS}}} + \sum_{i=1}^{K} \dfrac{\lambda_{\mathrm{Lag},i}}{\sigma_\mathrm{n}^2}\mathbf{h}_i\mathbf{h}_i^\dagger\right]^{-1}\mathbf{h}_k\right\|} \qquad \text{for } k = 1, \dots K$$

---

[6] Note, however, that the optimum power assignments are different for uplink and downlink, because noise and interference levels are not necessarily reciprocal.

where P_k is the power assigned for the kth user stream, which is related to the beamforming vectors as

$$
\begin{bmatrix} P_1 \\ \vdots \\ P_K \end{bmatrix} = \mathbf{M}^{-1} \begin{bmatrix} \sigma_{\mathrm{n}}^2 \\ \vdots \\ \sigma_{\mathrm{n}}^2 \end{bmatrix} \quad \text{with} \quad [\mathbf{M}]_{ij} = \begin{cases} \dfrac{1}{\breve{\gamma}_i} \left| \mathbf{h}_i^T \hat{\mathbf{t}}_i \right|^2 & i = j \\ -\left| \mathbf{h}_i^T \hat{\mathbf{t}}_j \right|^2 & i \neq j. \end{cases}
$$

The Lagrangian multipliers $\lambda_{\mathrm{Lag},i}$ can be computed from the fixed point equations

$$
\lambda_{\mathrm{Lag},k} = \frac{\sigma_{\mathrm{n}}^2}{\left(1 + \frac{1}{\breve{\gamma}_k}\right) \mathbf{h}_k^\dagger \left[\mathbf{I}_{N_{\mathrm{BS}}} + \sum_{i=1}^{K} \frac{\lambda_{\mathrm{Lag},i}}{\sigma_{\mathrm{n}}^2} \mathbf{h}_i \mathbf{h}_i^\dagger \right]^{-1} \mathbf{h}_k} \quad \forall k.
$$

The above solution applies for known $\breve{\gamma}_k$.

In the case that we wish to optimize a utility function, the problem is more complicated, as the set of $\breve{\gamma}_k$ that optimizes this utility function is unknown (though the structure of this solution still applies, and provides useful insights). For this problem, the weighted MMSE algorithm described in Section 22.5.2 (which shows some relations to the optimum beamforming) provides a concrete prescription on how to optimize utility functions.

*22.5.2 Processing for Multi-Antenna UEs

Block Diagonalization

How can we now generalize from the single-antenna-UE case to the more general case? A simple way would be to treat each UE antenna as a separate (single-antenna) user, and the BS preprocessing makes sure that data streams arriving at the antennas are well separated. This eases the design of the UE, which does not have to do any additional processing. However, it results in a constraint on the number of users, $\sum_k N_{\mathrm{UE},k} \leq N_{\mathrm{BS}}$, that is actually too strict. According to this constraint, adding receive antennas at the UEs decreases the number of users that can be transmitted to at a given point in time; this is obviously not a reasonable constraint, as additional receive antennas serve to *improve* the reception quality. Rather, it is the *number of data streams* intended for the UEs that is limited, and data streams intended for different UEs have to be kept apart by appropriate beamforming at the BS side.

More precisely, it is not necessary that the signals arriving at each separate antenna element are interference-free. Rather, the signals arriving at the kth UE should only contain data streams intended for this UE – the signals from multiple data streams that are all intended for this UE may be intermixed (the UE can separate them with the usual SU-MIMO processing), but signals intended for *other* UEs should be suppressed by the transmit beamforming. Keeping the data streams of different UEs apart is achieved by a *block diagonalization* technique. Define for each user an interference channel matrix $\overline{\mathbf{H}}_k$

$$
\overline{\mathbf{H}}_k = \left[\mathbf{H}_1^T, \ldots \mathbf{H}_{k-1}^T, \mathbf{H}_{k+1}^T, \ldots \mathbf{H}_K^T \right]^T. \tag{22.32}
$$

Any precoding matrix $\mathbf{T}_k$ that lies in the nullspace of $\overline{\mathbf{H}}_k$ leads to a situation where the data streams intended for the kth UE do not influence the other UEs. Consequently, there is now a new dimensionality constraint: Let J_k be the rank of $\overline{\mathbf{H}}_k$. Then the block diagonalization can be achieved if

$$
N_{\mathrm{t}} > \max_k \left(J_1, J_2, \ldots J_K \right). \tag{22.33}
$$

If the propagation channels are all full rank, this means that the number of transmit antennas must be as large as the number of data streams. Define then the Singular Value Decomposition (SVD) of $\overline{\mathbf{H}}_k$

$$
\overline{\mathbf{H}}_k = \overline{\mathbf{U}}_k \overline{\boldsymbol{\Sigma}}_k \left[\overline{\mathbf{V}}_{\mathrm{fc},k} \ \ \overline{\mathbf{V}}_{\ell\mathrm{c},k} \right]^\dagger \tag{22.34}
$$

where $\overline{\mathbf{V}}_{\mathrm{fc},k}$ contains the first J_k and $\overline{\mathbf{V}}_{\ell\mathrm{c},k}$ the last $N_{\mathrm{t}} - J_k$ right singular vectors; $\overline{\mathbf{V}}_{\ell\mathrm{c},k}$ thus forms an orthonormal basis of the nullspace of $\overline{\mathbf{H}}_k$. We define now an "overall effective channel matrix" $\hat{\mathbf{H}}$, which can be seen as the projection of the channel to each UE onto the nullspace of the interference to this UE:

$$
\hat{\mathbf{H}} = \begin{bmatrix} \mathbf{H}_1 \overline{\mathbf{V}}_{\ell\mathrm{c},1} & \mathbf{0} & \mathbf{0} \\ \mathbf{0} & .. & \mathbf{0} \\ \mathbf{0} & \mathbf{0} & \mathbf{H}_K \overline{\mathbf{V}}_{\ell\mathrm{c},K} \end{bmatrix}. \tag{22.35}
$$

Now we need to find the best beamformer and power allocation for this matrix, which is based on familiar elements: SVD and waterfilling. Thus, we next find the SVD of $\hat{\mathbf{H}}$; due to its block-diagonal structure, this can be done in an efficient (block-by-block) manner. For the kth block $\mathbf{H}_k$,

$$\hat{\mathbf{H}}_k = \hat{\mathbf{U}}_k \begin{bmatrix} \hat{\boldsymbol{\Sigma}}_k & \mathbf{0} \\ \mathbf{0} & \mathbf{0} \end{bmatrix} \begin{bmatrix} \hat{\mathbf{V}}_{\text{fc},k} & \hat{\mathbf{V}}_{\ell\text{c},k} \end{bmatrix}^{\dagger}. \tag{22.36}$$

The total beamforming/power allocation matrix $\mathbf{T}$ is then

$$\mathbf{T} = \begin{bmatrix} \overline{\mathbf{V}}_{\ell\text{c},1} \hat{\mathbf{V}}_{\text{fc},1} & \overline{\mathbf{V}}_{\ell\text{c},2} \hat{\mathbf{V}}_{\text{fc},2} & \cdots\cdots & \overline{\mathbf{V}}_{\ell\text{c},K} \hat{\mathbf{V}}_{\text{fc},K} \end{bmatrix} \hat{\boldsymbol{\Lambda}}^{1/2} \tag{22.37}$$

where $\boldsymbol{\Lambda}$ is a diagonal matrix that performs waterfilling on the elements of

$$\begin{bmatrix} \hat{\boldsymbol{\Sigma}}_1 & & & \mathbf{0} \\ & \hat{\boldsymbol{\Sigma}}_2 & & \\ & & \ddots & \\ \mathbf{0} & & & \hat{\boldsymbol{\Sigma}}_K \end{bmatrix}. \tag{22.38}$$

Coordinated Beamforming

If an RX has multiple antennas, then it can have a linear filter $\mathbf{W}_k^{\dagger}$ for its receive signals. For a given transmit precoder, the optimum RX filter matrix is given by the Wiener filter

$$\mathbf{W}_k = [\mathbf{J}_k]^{-1} \mathbf{H}_k \mathbf{T}_k \tag{22.39}$$

where

$$\mathbf{J}_k = \mathbf{H}_k \mathbf{T}_k \mathbf{T}_k^{\dagger} \mathbf{H}_k^{\dagger} + \left[\sigma_{\text{n}}^2 \mathbf{I} + \sum_{j \neq k} \mathbf{H}_k \mathbf{T}_j \mathbf{T}_j^{\dagger} \mathbf{H}_k^{\dagger} \right]. \tag{22.40}$$

If such a filter is present, $\mathbf{H}_k$ should be replaced by an "effective" channel that the TX sees, namely $\mathbf{W}_k^{\dagger} \mathbf{H}_k$. Changing this effective channel indirectly requires changes in the transmit beamforming matrix as well. The computation of the weights $\mathbf{T}$ and $\mathbf{W}$ thus has to be done in an iterative fashion: start out with a set $\mathbf{W}_k^{\dagger}$ (e.g., the weights for maximum-ratio combining at the kth UE), and compute the optimum transmit weights according to one of the strategies described above. With the new $\mathbf{T}$ compute the signal statistics at the UEs, and re-compute the optimum linear weights according to a certain criterion (e.g., MMSE). These new RX weights create a new set of effective channels, which then forms the basis for a recomputation of $\mathbf{T}$, and so on. The iteration is stopped when the weights (or the performance measures) do not change appreciably anymore.

In FDD systems, coordinated beamforming can be implemented in one of two ways: (i) the beamforming is performed, iteratively, at TX and RX. In other words, the TX (BS) sets beamforming weights, the RXs (UEs) observe the resulting effective channel and feeds this back to the TX, which adjusts its beamforming weights appropriately, and so on, until convergence; (ii) the RXs observes the full (matrix) channels and feed this information to the TX; the TX then performs the iterative computation *locally*, i.e., without need for further iterations with the RXs, and then transmits (the RX beamformer weights can either be communicated to the UEs, or the UEs can compute them locally). The former approach may require less feedback overhead, since there is no need to feed back matrix channels; however, each iteration step requires a time that is at least as long as the packet duration or duplexing time of the system, so that the computation overall might take longer than the coherence time of the channel, and thus never converge. The situation is easier in Time Domain Duplex (TDD) systems that can exploit reciprocity, since with suitable pilot signal transmission, the BS can obtain the full CSI between every TX and every RX antenna element (compare Section 16.1.7).

Figure 22.5 shows the capacity for various types of precoding. With a pure channel inversion (ZF), the capacity first increases as the number of users increases, but then starts to decrease again as K starts to approach the number of antenna elements at the BS. This is due to the increased probability of an ill-conditioned channel matrix. When the RX has more antennas, the difference between channel inversion and regularized channel inversion becomes smaller (see Figure 22.6).

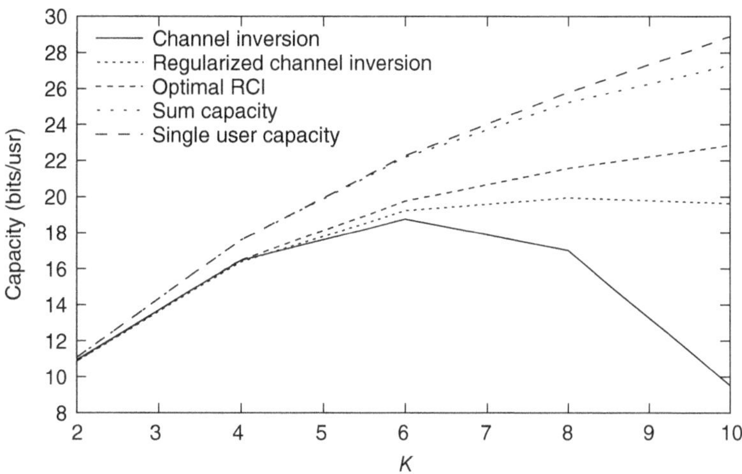

Figure 22.5 Multi-user-MIMO spectral efficiency as function of the number of users K when each UE has only one antenna; $N_t = 10$, $\gamma = 10$ dB. Different precoding methods are shown.

Reproduced with permission from [Tsoulos 2006] © CRC Press.

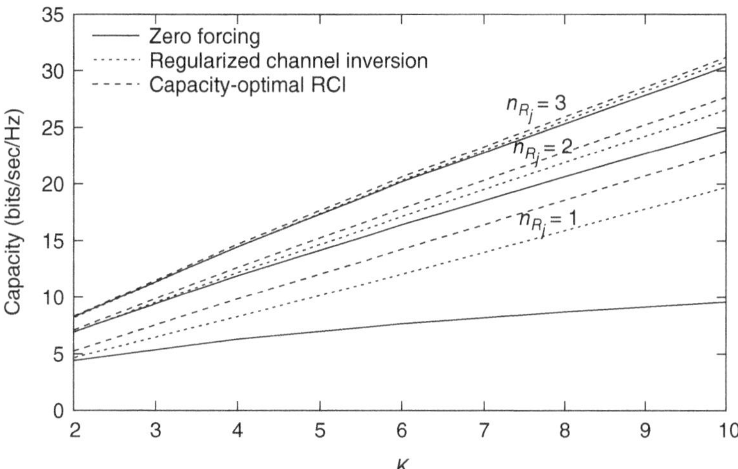

Figure 22.6 Multi-user-MIMO capacity as function of the number of users K when each UE has multiple antennas; number of data streams per users is fixed to 1. n_R indicates number of antennas per UE. $\gamma = 10$ dB, $N_t = K$. Different precoding methods are shown.

Reproduced with permission from [Tsoulos 2006] © CRC Press.

Weighted MMSE Algorithm

We now turn to the problem of weighted sum rate maximization

$$\max_{\mathbf{T}_k} \sum_{k=1}^{K} a_k R_k \quad \text{s.t.} \quad \sum_{k=1}^{K} tr\left(\mathbf{T}_k \mathbf{T}_k^\dagger\right) < P_k$$

where the a_k are arbitrary weights and

$$R_k = \log_2 \det\left[\mathbf{I} + \mathbf{H}_k \mathbf{T}_k \mathbf{T}_k^\dagger \mathbf{H}_k^\dagger \left(\sigma_n^2 \mathbf{I} + \sum_{j \neq k} \mathbf{H}_k \mathbf{T}_j \mathbf{T}_j^\dagger \mathbf{H}_k^\dagger\right)^{-1}\right]$$

is the rate of user k. Then it can be shown [Shi et al. 2011] that this problem is equivalent to a weighted MMSE problem that can be solved through an efficient distributed algorithm. Specifically, define a weighted MSE matrix as

$$\mathbf{E}_k = \mathbf{I} - \mathbf{T}_k^\dagger \mathbf{H}_k^\dagger \mathbf{J}_k^{-1} \mathbf{H}_k \mathbf{T}_k. \tag{22.41}$$

where $\mathbf{J}_k$ is defined in (22.40). Then the weighted sum rate maximization is equivalent to the problem

$$\min_{\mathbf{T},\mathbf{W},\mathbf{B}} \sum_{k-1}^{K} a_k [tr(\mathbf{B}_k \mathbf{E}_k) - \log_2 \det(\mathbf{B}_k)] \quad \text{s.t.} \quad \sum_{k=1}^{K} tr(\mathbf{T}_k \mathbf{T}_k^\dagger) \leq P_k. \tag{22.42}$$

This can be solved by the following algorithm:

Initialize the $\mathbf{T}_k$ such that $tr(\mathbf{T}_k \mathbf{T}_k^\dagger) = P_k/K$

Repeat until convergence

- Update $\mathbf{W}_k$ for all k according to (22.39)
- Update $\mathbf{B}_k$ for all k

$$\mathbf{B}_k = \left[\mathbf{I} - \mathbf{W}_k^\dagger \mathbf{H}_k \mathbf{T}_k\right]^{-1}.$$

- Update $\mathbf{T}_k$ for all k according to

$$\mathbf{T}_k = \left[\sum_{k'=1}^{K} a_{k'} \mathbf{H}_{k'}^\dagger \mathbf{W}_{k'} \mathbf{B}_{k'} \mathbf{W}_{k'}^\dagger \mathbf{H}_{k'} + \mu \mathbf{I}\right]^{-1} a_k \mathbf{H}_k^\dagger \mathbf{W}_k \mathbf{B}_k \quad \forall k = 1, \dots K \tag{22.43}$$

where $\mu_k \geq 0$ is a constant chosen to satisfy the power constraint and can be found by a simple numerical search.

The resulting solutions for $\mathbf{T}_k$ and $\mathbf{W}_k$ also maximize the weighted sum rate.

Joint Leakage Suppression

In joint leakage suppression, the precoding matrices are designed to maximize *Signal-to-Leakage and Noise Ratio* (SLNR), i.e., the ratio of the power of the desired signal received by the kth UE, and the sum of the noise and the total interference power (leakage) due to the signal intended for the kth user at all the other UEs (note that this is different from the SINR at the kth UE). The SLNR of the kth user is [Sadek et al. 2007]

$$SLNR_k = \frac{E\left\{\mathbf{s}_k^\dagger \mathbf{T}_k^\dagger \mathbf{H}_k^\dagger \mathbf{H}_k \mathbf{T}_k \mathbf{s}_k\right\}}{N_{r,k}\sigma_n^2 + E\left\{\sum_{i \neq k}\sum_{j \neq k} \mathbf{s}_k^\dagger \mathbf{T}_k^\dagger \mathbf{H}_i^\dagger \mathbf{H}_j \mathbf{T}_k \mathbf{s}_k\right\}} \tag{22.44}$$

where the expectations are over the symbols. The key motivation for this criterion is that it allows an easier optimization (often in closed form) than the SINR at the RX.

Equation (22.44) can be simplified to

$$SLNR_k = \frac{tr\left\{\mathbf{T}_k^\dagger \mathbf{H}_k^\dagger \mathbf{H}_k \mathbf{T}_k\right\}}{tr\left\{\mathbf{T}_k^\dagger \left[N_{r,k}\sigma_n^2 \mathbf{I} + \overline{\mathbf{H}}_k^\dagger \overline{\mathbf{H}}_k\right] \mathbf{T}_k\right\}}. \tag{22.45}$$

The BS thus needs to find precoding matrices $\mathbf{T}_k$ that maximize the $SLNR_k$ subject to a transmit power constraint $tr\left\{\mathbf{T}_k \mathbf{T}_k^\dagger\right\} = P_{t,\max}$. Note that we assume here an absence of power control, so that the powers for each user are fixed to a certain value – this decouples the optimization of the different users. Then define an auxiliary matrix $\mathbf{A}$ as a matrix that fulfills

$$\begin{aligned}
\mathbf{A}_k^\dagger \mathbf{H}_k^\dagger \mathbf{H}_k \mathbf{A}_k &= \mathbf{\Lambda}_k \\
\mathbf{A}_k^\dagger \left[N_{r,k}\sigma_n^2 \mathbf{I} + \overline{\mathbf{H}}_k^\dagger \overline{\mathbf{H}}_k\right] \mathbf{A}_k &= \mathbf{I}
\end{aligned} \tag{22.46}$$

where $\mathbf{\Lambda}_k$ is a diagonal matrix with (arbitrary) nonnegative entries, sorted in descending order. The matrix $\mathbf{A}_k$ describes thus the generalized eigenspace of the pair $\left\{\mathbf{H}_k^\dagger \mathbf{H}_k, N_{r,k}\sigma_n^2 \mathbf{I} + \overline{\mathbf{H}}_k^\dagger \overline{\mathbf{H}}_k\right\}$, and it can be shown that the precoding matrix $\mathbf{T}$ is then

$$\mathbf{T}_k = \mathbf{A}_k \begin{bmatrix} \mathbf{I}_{N_{s,k} \times N_{s,k}} \\ \mathbf{0} \end{bmatrix}.$$

For the special case that each user has only one data stream, $L_k = 1$, the beamforming matrix becomes a vector, and can be obtained in a simple way: perform an eigendecomposition of the matrix

$$\left[N_{r,k}\sigma_n^2 \mathbf{I} + \overline{\mathbf{H}}_k^\dagger \overline{\mathbf{H}}_k\right]^{-1} \mathbf{H}_k^\dagger \mathbf{H}_k. \tag{22.47}$$

The optimum precoding vector is the eigenvector associated with the largest eigenvalue of this matrix. Since the performance is ultimately governed by the SINR at the RX, performance of systems that maximize SNLR is worse than systems that maximize SINR, paying a price for the greater simplicity of the algorithm.

22.6 Beamforming Based on Second-Order Statistics

Up to now, we assumed that the TX has, by some means (feedback or reciprocity) the *instantaneous* CSI. However, this might be impossible or undesirable in some situations – e.g., because of the overhead, or the required speed of the feedback being too high. Alternatively, MU-MIMO can be based on the *second-order* CSI, which requires much less training, though it also has inferior performance. This approach is strongly related to the *spatial reference* algorithms described in Section 16.1.4.

Generally, second-order statistics refer to knowledge of the channel covariance matrices, which can be used for beamforming without any knowledge of the array geometry, and no relation to the angular characteristics of the incident radiation. However, in many cases the BS arrays are linear, rectangular, or cylindrical arrays, in which case the second-order statistics can be immediately related to the angular power spectrum of the channel. This is the case we discuss below, but it is important to remember that it is not the only feasible, or interesting, case.

Figure 22.7 a shows the fundamental principle of the simplest implementation: the BS forms a number of beams in fixed directions. Assume now that each UE is surrounded by a ring of Interacting Objects (IOs), so that the BS "sees" the MPCs associated with the UE in a small angular sector. Then the BS determines the beams in which each UE is located. This determination occurs through a training phase called beam sweeping: For example, the BS transmits towards the different beams consecutively; each UE then determines its best beam, and feeds back that information to the BS. For the actual data transmission, the BS picks a set of K nonoverlapping beams to communicate with K (out of the totally available $\hat{K}$) UEs. The beam sweeping can occur at intervals that are much larger than the typical coherence time of the channels, since the location of the UE, and thus which beam it belongs to, changes much more slowly than the instantaneous CSI.

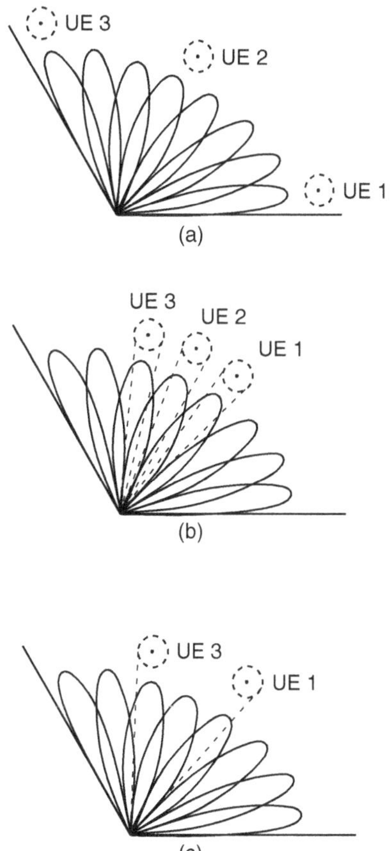

Figure 22.7 Principle of beamforming based on second-order statistics. (a) UEs are in well-separated beams. (b) Interference between the downlink signals intended for different UEs. (c) Uplink situation for the same locations of UEs, but UE 2 is not scheduled to transmit.

The above description provides only a crude picture of the procedure and its challenges. Points that need to be taken into consideration for practical implementation are:

- *How can the BS form different beams*: the two most popular methods involve (i) the use of uniform linear or planar arrays combined with a spatial Fourier transform, where this Fourier transform can be implemented in digital, or in analog through a so-called

Butler matrix (see section titled on "Switched-Beam Receivers" in Chapter 16). (ii) the use of multi-beam antennas, such as lens antennas (compare Section 8.3.8). In either case, the beams tend to overlap, so that even a UE that is in the center of one beam creates, in the uplink, contributions in multiple beams. Thus, if UEs in adjacent beams are active simultaneously, there is inherent inter-user interference (the same holds for the downlink).

- *Are the beams fixed or adaptive*: in most cases, the centers of the beam directions, as well as the beam shapes, are fixed, and the adaptivity of the system is only related to which beams are activated at a particular time. However, beamforming in an array enables in principle the adaptation of the beam centers and/or having some beams with wider and others with narrower beamwidths (though, of course, a lower limit of the beamwidth is given by the physical aperture size of the array).

- *What is the separation of the APSs*: while the location of the UE usually tells us where the direction of the highest energy is, the question remains how widely this radiation is dispersed in the angular domain. Imagine a situation where the IO circles around two UEs are overlapping. It is then immediately obvious that – even with fully adaptive beams – in this case we have to trade-off inter-user interference with the collection of all available energy. To avoid inter-user interference, a beam should only point towards those IOs that "carry" MPCs from one particular UE. However, this reduces the overall energy collected from that user. The situation is made worse by the fact that beams do not have a rectangular shape, but rather taper off slowly: in that case, even an IO circle that is not overlapping, but "close by," can create interference in the beam receiving signals from another UE. For example, in Figure 22.7b, the beam pointing towards UE 2 can see IOs in the circle around UE 3. Consequently, a tradeoff is required to maximize the SINR.

  The requirements on the APS also imply that beamforming based on second-order CSI is better suited for some environments than others. Outdoor BSs, in particular for macrocells, tend to see fairly concentrated APSs, with the mean direction in the Quasi-Line-of-Sight (LOS) to the UE (see Chapter 7); a similar trend holds for millimeter-wave and THz channels, both in macrocells and even in microcells. On the other hand, indoor environments lead to high angular spread also at the BS, so that separation of users based on the beam direction is much more difficult. At the same time, in indoor environments there is less motivation to use second-order statistics beamforming, because the instantaneous channels change more slowly (due to lower movement speed), so that acquiring and using instantaneous CSI for MU-MIMO does not require prohibitive overhead.

- *Impact of common IOs*: the tradeoff between energy collection and interference becomes even more pronounced when far IO clusters, such as high-rise buildings (compare Chapter 7) are conduits of significant energy. Since those clusters carry energy from multiple users, the decision on whether to collect energy from them or not needs to be considered.

- *How to perform the channel estimation*: The determination of the best beam combination can be done either in the uplink or the downlink. For the downlink, the BS sends training signals to the UEs, and each UE determines the best beam index and feeds this information back to the BS. However, the best index does not give all the important information, for example, it does not allow determination of the inter-user interference that would occur if a particular user combination would be selected. A more complete feedback would provide information about the power levels observed by each UE for every single beam the BS sweeps through. With this information, the BS can then determine the SINRs that result from choosing different beam combinations. Alternatively, the channel estimation can also be done in the uplink, with the different UEs transmitting orthogonally, while the BS receives on all beams simultaneously. It is important that the APS are reciprocal for uplink and downlink. Thus, even for FDD, information from the uplink measurement can be used for the determination of the downlink beam.

- *Uplink and downlink beams*: despite the reciprocity of the APSs, it is not a given that the same beam is optimum for uplink and downlink. This is due to the asymmetry of the data traffic; for example, a UE might only require downlink traffic so that its associated transmissions create a lot of interference in the neighboring beams during the downlink phase (making transmissions in neighboring beams to other UEs inefficient during that time), while those same neighboring beams might work well during the uplink phase; see also Figure 22.7b,c.

- *Scheduling*: overall, the use of second-order statistics needs to be combined with smart user selection and scheduling, see also Section 22.8. In particular, the performance of second-order-statistics based MU-MIMO is more sensitive to scheduling than the beamforming based on instantaneous CSI.

Despite its lower performance, second-order-CSI based MU-MIMO is a method of choice for many cellular standards. This is due to its simpler implementation and reduced overhead for the channel estimation.

Second-order CSI for beamforming is also beneficial when combined with instantaneous CSI. Knowledge of the second-order CSI can reduce the overhead for instantaneous CSI estimation (Section 22.7), and can be used in the analog part of a hybrid beamformer (Section 22.10.2).

22.7   Channel Estimation and Feedback

In all of the above discussion we have assumed the existence of CSIT, since (downlink) MU-MIMO is based on this assumption. We mentioned that for the case of TDD systems, CSIT can be obtained from the uplink pilots in conjunction with proper reciprocity calibration. However, in FDD systems, this is not available, and we have to use downlink pilots and feedback, where quantized codebooks provide a good way of reducing the feedback required for transmit beamforming (note that the beamsearch described previously can be interpreted as feeding back the index of a Fourier codebook, compare Section 16.1.7). While this approach works well for SU-MIMO, as discussed in Sections 16.1.7 and 16.2.8, it is more complicated in MU-MIMO.

MU-MIMO requires that the quantization of the transmit precoding vector has to be much finer than in the single-user case. This can be explained the following way: in a single-user case the BS forms a beampattern that shows a maximum in the direction of the targeted UE. Slight deviations from the optimum beampattern, due to the quantization effects, do not lead to a significant loss of SNR. In the multi-user case, in order to achieve good SINR at the UEs, we need to guarantee low interference by signals from other users; in other words, each transmit beampattern needs to place nulls toward the users it should *not* cover. But nulls are much more sensitive to perturbations of the antenna weights than "main beams." Thus, quantization in multi-user settings must be finer.

Another complication for codebook-based systems arises from the fact that the optimum settings of the precoder depend on the channels of *all* users, and the UE therefore cannot easily compute the precoding matrix settings that it needs to optimize its performance. In other words, the UE cannot simply observe transmissions by the BS with different precoder settings and identify which one gives the best SINR; it can only identify which one provides the best SNR; furthermore, different users might identify the same precoder setting as optimal for them, creating a dilemma for the BS which UE to give this setting to.

It is possible to mitigate these issues through *projection techniques*. Here the BS first sends out a number of training signals on $Q = K$ different beams $\mathbf{t}^{(q)}$. The kth UE then forms the quantities

$$\gamma_k = \max_q \frac{\left|(\mathbf{h}_k)^T (\mathbf{t}^{(q)})\right|^2}{\sigma_\mathrm{n}^2 + \sum_{p \neq q} \left|(\mathbf{h}_k)^T (\mathbf{t}^{(p)})\right|^2} \tag{22.48}$$

which is simply the maximum SINR it can achieve if the BS transmits to this UE on the beam that is optimum for it, and to all the other UEs on different beams, without any power control. The kth UE then just feeds back the index q with which it achieves this SINR. To avoid the problem of two beams claiming the same precoder, each UE transmits not just the index for the best beam, but also for the second-best, so that the BS is aware of a "fallback position" that gives a certain UE an acceptable (though not optimum) SINR.

When taking into account the signaling overhead, information-theoretic results show that for both TDD and FDD systems, the Spatial Multiplexing Gain (SMG) of a MIMO downlink with full digital structure equals to

$$\widetilde{N}\left(1 - \frac{\widetilde{N}}{T_{\mathrm{coh,norm}}}\right) \quad \text{with } \widetilde{N} = \min\left(N_{\mathrm{BS}}, K, \frac{T_{\mathrm{coh,norm}}}{2}\right). \tag{22.49}$$

$K = N_\mathrm{s}$ is the number of single-antenna UEs, and $T_{\mathrm{coh, norm}}$ is the coherence time of the channel in units of symbol durations, or, more generally, the number of channel uses in a coherence time-frequency block. It is evident that for massive MIMO systems (Sec. 22.9) relying on full CSI from all antenna elements of the BS to users, the maximal achievable SMG is limited by the size of coherence block of channels because N_{BS} and K are generally large in massive MIMO systems. This therefore necessitates the design of reduced-dimensional CSI acquisition strategies for massive MIMO systems in order to reduce signaling overhead.

The use of second-order CSI allows such a reduction of the channel estimation overhead. Fundamentally, we can partition the UEs into groups with approximately similar covariances; in the case that all the IOs are in a disk around the UE, that means we cluster almost-co-located UEs together. Then we can split the beamforming into two stages: a first stage consisting of a pre-beamformer that depends only on the second order statistics, i.e., the covariances of the user channels, and a second stage comprising a standard MU-MIMO precoder for spatial multiplexing on the effective channel obtained after pre-beamforming. The instantaneous CSIT of such an effective channel is easier to acquire thanks to the considerable dimensionality reduction produced by the pre-beamforming stage. In other words, the degrees of freedom needed to describe the channel are greatly reduced, because we know that the instantaneous CSI must lie in a subspace of the second-order CSI. This is the channel estimation scheme of Joint Spatial Division and Multiplexing (JSDM) [Adhikary et al. 2013, 2014], which will be discussed in more detail in Section 22.10.2.

JSDM is not the only way in which the second-order statistics can be exploited for channel estimation. Firstly, there are alternative methods in which the required overhead for pilot tones can be reduced. Secondly, the correlation matrix can be used for noise smoothing, thus improving the SNR of the CSI estimates, in a manner that is analogous to SU-MIMO, compare Section 16.2.7.

22.8 Scheduling for MU-MIMO

The previous part of this chapter assumed that there is a fixed number, K, of users that need all be scheduled on the same time/frequency resources. However, this is not the typical situation in a cellular setup. Rather, there is a larger set of users, $\hat{K}$, from which the system selects subsets that get scheduled on particular time/frequency resources.

Good scheduling can dramatically increase the performance of a MU-MIMO system. For example, derivations in Section 22.4 and 22.5 showed that linear precoding leads to bad performance when the channel vectors of two users A and B are almost collinear (a situation that can occur quite often in random-fading channels). Under the assumption that the number of users K is fixed, we either have to live with this performance degradation, or drop one of those users and thus violate the quality-of-service requirement. However a good scheduler will put UE B into a different time-frequency resource, and instead put another UE C (that is not collinear with the channel of user A) from the available pool onto the same resource as UE A. This results then in an improved spectral efficiency.

Just as in the single-antenna case discussed in Chapter 20, the scheduling is done with the goal of maximizing a utility function. This utility function can be the sum rate, an α-fairness rate, and/or fulfill the constraint of a minimum rate (quality of service) for each of the users. Most of the subsequent examples will use the sum rate as a criterion, but the principles apply to other target functions as well.

Scheduling in MU-MIMO is considerably more difficult than in single-antenna systems, because an additional dimension, namely the spatial (or beam, or precoding) vector is available. The situation becomes even more complicated when incorporating inter-cell interference. We will in the following proceed from the simpler to the more complicated setups:

- single-cell scenario with only one UE scheduled on a particular time/frequency resource;
- multi-cell scenario with only one UE scheduled on a particular time/frequency resource;
- single-cell scenario with multiple users scheduled on a time-frequency resource;
- multi-cell scenario with ICIC.

The case of a multi-cell scenario with joint signal transmission from the different BSs (compare Section 22.11) is formally similar to item 2 above, and will thus not be treated separately. To keep the discussion reasonably simple, we only consider the case where the UEs have a single antenna; for the more general case see the references in the "Further Reading" section. Besides the (deterministic) algorithms discussed here, scheduling in MU-MIMO systems might also use stochastic algorithms such as simulated annealing, or machine-learning algorithms, in particular based on reinforcement learning.

22.8.1 Single-Cell with One UE at a Time – Random Beamforming

We first deal with SFIR, i.e., the situation where many potential users are present in a cell, but only one is used on a particular time/frequency resource, and we wish to exploit the multiple antenna elements at the BS for improving the SINR and thus capacity. In this case we can exploit multi-user diversity as discussed in Section 20.3.2. As we have seen there, a major challenge is the inherent delay. If the BS always uses proportionally fair scheduling, it retains communication with the chosen user for approximately one coherence time of the channel. If the user mobility is low, the coherence time can be quite large. Remember that each user has to wait until its $\gamma/\bar{\gamma}$ (where $\bar{\gamma}$ is the average SNR) becomes the instantaneously largest and it can communicate; this waiting time (latency) is roughly KT_{coh}. In a system with many users and large coherence times, this latency can become prohibitively large. An ingenious solution for this problem, called *random beamforming*, uses multiple antennas at the BS [Viswanath et al. 2002]. The signals transmitted from the different antenna elements of the BS are multiplied with time-varying complex coefficients. This can be interpreted in two (equivalent) ways:

1. Each vector of transmit weight coefficients can be associated with a beampattern. No matter what beam is formed – it will enhance the channel to *some* UE. The scheduling then makes sure that the enhanced UE is chosen for communication. Every time the coefficients are changed, the beampattern changes, and a different UE gets enhanced.
2. The combination of multiple antennas (with weighting coefficients) and physical channel can be viewed as an "equivalent" channel. Varying the coefficients for the antenna elements enforces time variations of the equivalent channel, and thus reduces the effective coherence time T_{coh}. In other words, the effective channel exhibits small-scale fading with a coherence time that is approximately the smaller of the coherence time of the physical channel and the time over which the antenna weights change.

The main difference from the transmit diversity/beamforming for the downlink is the following:

- For *conventional transmit diversity with CSIT*,[7] the BS needs to know the amplitude and phase of the transfer function to all UEs. It then chooses the weights for the signals at the different antenna elements in such a way that the different transmit signals add up in an optimum way at the destined UE. Coefficients are changed only if the transfer function between the BS and the chosen UE changes, or if the BS decides to communicate with another UE.
- For *random beamforming*, the BS chooses the coefficients completely at random and changes them according to system parameters (related to the latency), but independently of the channel.

The time over which the weighting coefficients are kept constant constitutes a compromise between the resulting latency and the system overhead for feeding back the channel quality. Random beamforming can also be combined with SDMA. For this purpose, the BS forms not just one beampattern at a time, but rather a set of orthogonal patterns. The UEs then feed back the signal quality that they receive on the different beams; in order to reduce feedback, they sometimes just notify the BS about the index of the beam on which they receive the optimum SINR. The BS then schedules one user on each of the orthogonal beams. This technique enhances the spectral efficiency approximately by a factor that is equal to the number of beams; note that the method is related to the projection technique mentioned above.

---

[7] Transmit diversity without CSIT does not have beamforming gain.

22.8.2 Multi-cell with One UE at a Time

The combination of beamforming capability at the BS and coordinated user scheduling allows the reduction of *Inter-Cell Interference (ICI)*. Generally, the existence of beamforming reduces the probability of strong ICI. Consider the downlink example in Figure 22.8. The interference from BS 1 to UE B is strong only if UE B happens to lie in the direction of the beam that BS 1 forms to communicate with its intended target, UE A. However, while the probability for such an event is reduced compared to the case of omni- or sector-antennas at the BS, it can still occur. In this situation, coordinated scheduling needs to come into play: the two BSs can agree on a joint schedule, such that when BS 1 transmits to UE A, then BS 2 will transmit to UE C, and schedule UE B later, or on a different frequency. Note that this reduction of interference can be done in addition to interference reduction techniques such as fractional frequency reuse that were discussed in Section 21.3.

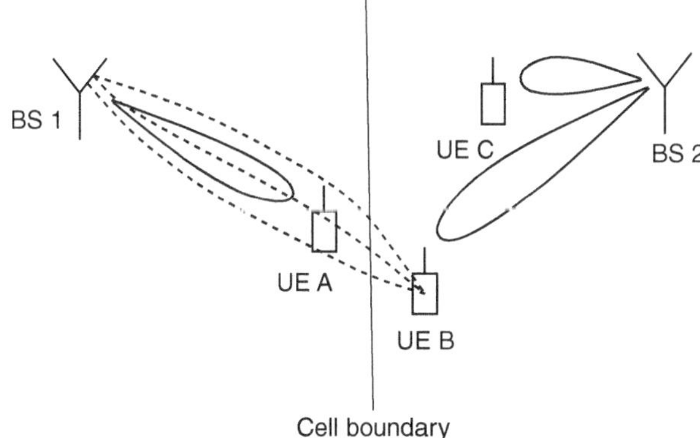

Figure 22.8 Example for scheduling in a multi-cell system with beamforming capability.

First consider the case that the beamforming in each cell is done based on the signal to/from the desired user only, and then scheduling is used to minimize interference (for simplicity we assume that each user has a fixed data rate and enough data to occupy exactly one time/frequency resource). There is an enormous number of ways in which users can be placed onto those resources, and the optimum scheduler needs to test all of them and select the best combination. This is not practically feasible, so that suboptimum methods are employed.

One such approximate method is to iterate the scheduler between cells: BS 1 determines the best scheduling under the assumption that no ICI occurs; BS 2 then computes the resulting ICI its UEs receive at the different times/frequencies, and computes an optimum scheduler based on that. This now creates the interference seen by BS 1, which thus re-computes the scheduler for it, which forms the basis for BS 2 to recompute its own scheduler, and so on. In some cases, this algorithm converges to a stable schedule. However, it can also occur that the scheduler simply oscillates between different solutions. The computational effort, feedback overhead, and problems with oscillations can be avoided by having a hierarchy between cells, in that BS 1 does the scheduling under the assumption of no interference, and then BS 2 schedules according to the interference from BS 1. Then BS 1 does not adjust its scheduler, but only takes the resulting ICI into account when it determines the rate at which communications with the scheduled UEs occurs. The order in which the cells are scheduled can either be changed from time to time in the interest of fairness, or can be determined by some global interference criterion (i.e., which cell generally creates more interference to its neighbors).

The scheduling can be further improved by adding adaptive beamforming and power control. Firstly, for the uplink, the BS can perform optimum beamforming to suppress interference from the scheduled UE in the neighboring cell. This may completely eliminate the ICI; scheduling then mainly aims to minimize the noise enhancement that occurs due to the need for interference suppression; the downlink works analogously. Of course, the beamforming and the scheduling interact, and the exact optimization of such a system becomes very difficult – it is a mixed integer optimization problem. Commonly, this issue is circumvented by an iterative approach: freeze the scheduler and optimize the beamformer, followed by freezing the beamformer and optimizing the scheduler, and so on.

22.8.3 Single-Cell with Multiple UEs at a Time

Now we turn to the case where multiple users are scheduled simultaneously on the same time-frequency resources. We first determine a scheduler that maximizes the downlink capacity. To repeat and rewrite the equation for the achievable rate:

$$R_{\pi(k)} = \log_2 \frac{\left| \mathbf{I} + \mathbf{H}_{\pi(k)} \left[\sum_{k' \geq k} \mathbf{R}_{\mathbf{T},\pi(k')} \right] \mathbf{H}_{\pi(k)}^{\dagger} \right|}{\left| \mathbf{I} + \mathbf{H}_{\pi(k)} \left[\sum_{k' > k} \mathbf{R}_{\mathbf{T},\pi(k')} \right] \mathbf{H}_{\pi(k)}^{\dagger} \right|} \qquad k = 1, ...K \qquad (22.50)$$

where $\pi(k)$ is a particular permutation in which the Dirty Paper Coding (DPC, see Sec. 18.2) is performed, and $\mathbf{R}_{\mathrm{T},\pi(k')}$ is the covariance of the transmit signal, $\mathbf{TT}^\dagger$, for that permutation, which has to satisfy the total power constraint $tr(\mathbf{R}_{\mathrm{T},1} + ...\mathbf{R}_{\mathrm{T},K}) \leq P_{\mathrm{tot}}$, where tr is the trace of a matrix. The capacity region is the convex hull of the rate vectors over all permutations π and all covariance matrices satisfying the constraints. Alternatively, use the dual MAC channel to compute the capacity region as union over all capacity regions of MAC channels that satisfy the sum power constraint.

We see from Eq. (22.50) that the capacity region depends on the order of the encoding (or decoding, for the MAC), as well as on the set of users $\mathcal{K}$ that is selected from the overall available set $\mathcal{K}'$. This means again that the cardinality of the search space of the optimization is very large. As often in such situations, a low-complexity solution can be found based on the greedy principle, i.e., we pick in subsequent iteration rounds the user that locally provides the largest improvement of the utility function, given the already-chosen users, without consideration of what further selections might happen in the future. Such a greedy approach is suboptimum – it might pick, e.g., a user that "blocks" several good users in the future – yet the simplification of the scheduling is often worth this tradeoff. So, generally speaking a greedy algorithm will select as first user the one with the best SNR, and thus highest rate. It will then next pick the user among all remaining users that provides the highest increase in utility (or some – possibly approximate – proxy criterion). This process is continued until adding further users does not increase the utility function anymore – for example, once the number of chosen users has reached N_{BS}, any further increase in scheduled users increases interference and thus reduces the utility. We next give two concrete examples of this principle.

For DPC-based scheduling, one example algorithm is based on a Q-R decomposition of the channel matrix [Tu and Blum 2003]. Let $\mathbf{H} = \mathbf{FQ}$, where $\mathbf{F}$ is a $K \times N_{\mathrm{BS}}$ lower triangular matrix, and $\mathbf{Q}$ is a $N_{\mathrm{BS}} \times N_{\mathrm{BS}}$ matrix of orthonormal rows. Let the transmit signal be precoded, $\mathbf{s} = \mathbf{Q}^\dagger \widetilde{\mathbf{s}}$ According to this decomposition, the received signal is

$$r_k = f_{k,k}\widetilde{s}_k + \sum_{j < k} f_{k,j}\widetilde{s}_j$$

i.e., each user sees interference from the subsequently decoded channels, which, however, can be eliminated by DPC. In other words, there are N_{BS} interference-free channels with gains $f_{k,\,k}$. However, the capacity still depends on the coding order, whose exact optimization would result in high complexity. The greedy principle can provide the ordering of rows in the QR decomposition process:

1. Set $\mathcal{L}$ as the null space and label all rows of the channel matrix as unprocessed.
2. Project all unprocessed rows onto the complement of $\mathcal{L}$.
3. Select the row with the highest 2-norm projection and label it as processed.
4. Set $\mathcal{L}$ to be the span of all processed rows.
5. Repeat 2–4 as long as there are non-zero projections.

When considering ZF linear precoding for the downlink, a somewhat different greedy approach can be used [Yoo and Goldsmith 2006]. It starts out with the user that gives the best SNR. Now comes an iteration loop where each step takes into account the interference of a newly selected user onto the already selected users (this does not need to be considered in above-described algorithm). The first step within the loop computes the set of users that is quasi-orthogonal to the already-selected users, i.e., the inner product of the channel vector with the already-scheduled users is below a threshold; this keeps the interference to already-selected users in acceptable limits. For each of the users in this set, we then select the one whose channel vector has the largest projection onto the nullspace of the previously selected users. This is continued until there is no more user available whose channel vector fulfills the interference-threshold criterion. Note that this algorithm uses a proxy criterion for the impact of new users on the already-established users, namely the projection onto the signal space (instead of the actual rate reduction due to interference). This is, of course, not the only such possible criterion, but rather an example for a wide class of such possible algorithms.

22.8.4 MU-MIMO Inter-Cell Interference Coordination

We now turn to the general case of MU-MIMO in each cell, and different cells can coordinate with each other. If there is full exchange of CSI and scheduling information between the cells, then the multiple BSs form a "super-BS" that performs a joint scheduling and beamforming for all users. Importantly, the beamforming is constrained by assuming that the transmission for a particular UE may only employ the antennas of the BS with which the user is associated, as we assume that the payload signal is only available there (this is in contrast to the Coordinated Multiple Point Transmission Joint Processing (CoMP-JP) approach discussed in Section 22.11, which assumes that not only CSI but also payload data are shared between the BSs). Because of this formal analogy of scheduling in the "super-BS" with scheduling in single-cell MU-MIMO, the considerations of Section 22.8.3 mostly carry over. Of course, the complexity of the decisions increases because the pool of UEs that can be scheduled, and therefore the dimensionality of the problem, is larger.

Another possible approach is a sequential scheduling (and beamforming) in the different BSs. In this approach, the first BS makes its scheduling decisions that are optimum for it. The next BS is informed about those decisions and can thus compute the interference its own users will experience. It then makes scheduling decisions either based on its own interest, or trying to maximize the utility for each user while keeping interference to BS 1 at a minimum. BS 3 then does a similar scheduling based on the decisions of BS 1 and BS 2, and so on. Once the last BS has done its scheduling, BS 1 can either adjust its schedule, starting a new round of iterations, or stick to its original scheduling and just recompute its beamformers based on the interference from all the other BS; the latter version is suboptimum but faster.

22.8.5   General Scheduling Considerations

Most scheduling algorithms are designed to provide a good sum utility *in the long run*. The MCS used for the communication between BS and UE depends on the SINR, and thus can be determined only *after* the scheduling and beamforming is done, and thus the number of time/frequency resources assigned to each user is established. This can create problems if a certain minimum number of packets needs to be transmitted in a particular timeslot in order to sustain a particular quality of service. In a system intended, e.g., for video streaming, the ability of transmitting a packet in a particular timeslot is not critical; rather the data rate averaged over a window of hundreds of milliseconds or more is what counts. Thus, if the achievable rate in a given timeslot is low due to low SINR, the system can make up for it in subsequent timeslots by assigning more resources. However, for extremely time-critical information, such as for self-driving cars, such delay is not acceptable. In that case, a scheduler might have to sacrifice efficiency for guaranteed performance.

Another challenge for the implementation of scheduling lies in the delay between the time that the CSI has been determined and number and priority of incoming data packets is known, and the time the actual transmission of the packets occurs. Since the scheduling algorithm itself requires time to compute, and exchange of information between the BSs is needed, such delay can be several milliseconds. This might be problematic in highly mobile environments because the CSI can change within that time frame. Furthermore, there are services (again, take the example of a self-driving car) for which a delay of several milliseconds is not acceptable. Such services might need dedicated time-frequency resources. Alternatively, the BS can try to predict the channel several milliseconds into the future, and perform the scheduling plus beamforming for those future channels.

An alternative to the deterministic scheduling and beamforming algorithms described above is machine-learning based beamforming. The fundamental principles are similar to those expounded in Section 20.7 for the single-antenna case; in particular due to the difficulty of finding optimum solutions that can be used as labeled data, unsupervised reinforcement learning is the most promising approach. The methods can either compute the scheduler and then have the associated beamformers computed via classical techniques, or can provide joint scheduling and precoding solutions.

22.9   Massive MIMO Theory

For a multi-user MIMO system, the capacity is essentially determined by $\min(N_{BS}, K)$ users (assuming that each UE has only one antenna element – an assumption that we will make for the rest of this section). This obviously suggests that going to a large number of BS antenna elements would provide very large capacity. Such an approach is now generally known as *massive MIMO*, and has received great attention both in the academic literature and the wireless industry since 2010.

However, massive MIMO is not just a "souped-up" MU-MIMO system, in which all the algorithms of the previous sections should be reused without modifications – such an approach would be possible, but inefficient and – since the algorithm complexity increases quickly with the number of antenna elements – often lead to prohibitive system complexity. Rather, the large number of antenna elements leads to a number of important physical effects that need to be considered in the system design. The key insights for massive MIMO are as follows:

- With a large number of BS antenna elements, the performance of maximum ratio transmission (for the downlink) or Maximum Ratio Combining (MRC) (for the uplink) approaches that of ZF transmission (downlink) or optimum combining (uplink). This dramatically simplifies signal processing, since now the precoding and combining weights can be determined for each user separately, instead of requiring a joint design that involves the inversion of a large matrix. Note, however, that this convergence is a theoretical property that holds under some special circumstances, such as single-cell with good enough SNR; in many practical situations there is significant performance difference.
- The beamforming gain for each user becomes very large, since it is proportional to the number of BS antenna elements. This implies that the required Radio Frequency (RF) transmission energy becomes small, and/or cell-edge coverage (or range) can be improved significantly.
- The effective channel gain for each user (including beamforming effects) becomes essentially time- and frequency-invariant, i.e., small-scale fading effects disappear. This *channel hardening* not only simplifies theoretical analysis, but also makes scheduling (that normally depends on the small-scale fading state) much easier.
- An effect that – at least according to early investigations – gets great importance (compared to the traditional MU-MIMO) is pilot contamination, i.e., the fact that during the training phase the received signal suffers from co-channel interference. Since during the training there is no beamforming yet, this effect is not decreased when the number of antenna elements increases. Erroneous channel estimates affect the beamforming for the actual data transmission, and thus have an important impact on the capacity. It must be noted, though, that recent theoretical advances have shown how to mitigate this effect.

There are also practical concerns with increasing the number of antenna elements:

- Each BS antenna element needs a down/upconversion chain if digital (baseband) beamforming is to be performed. This may be not only cost-prohibitive, but might lead to excessive overall power consumption despite the fact that the actual RF transmit energy is very small as outlined above; remedies will be discussed in Section 22.10.
- Arrays containing a large number of antenna elements might have a very large size, depending on the wavelength and the array configuration. For Uniform Linear Arrays (ULAs) with element spacing $\lambda/2$, a 128-element array at 2 GHz carrier frequency would

be 8 m long, which might cause problems both from an aesthetics point of view, and result in an excessive wind load for the mounting. However, a 16×8 dual-polarized rectangular array might have dimensions of less than 1×0.5 m, certainly comparable to current BS antennas. Pros and cons of different array configurations will be discussed below.

We now turn to a more detailed mathematical description to refine and quantify these insights.

22.9.1 Basic Performance

Consider a cellular system with L cells, each of which has K single-antenna users in the cell. Each BS has $N_{\rm BS}$ antenna elements; we will assume (unless stated otherwise) that BSs in different cells do not coordinate, either by exchange of CSI, scheduling, or data (see Section 22.11 for a general discussion of that case). Let now $h_{n\ell jk}$ denote the complex instantaneous channel gain between the nth BS antenna in the ℓth cell and the kth UE in the jth cell. Keeping with our notation of Section 22.2, we define the $N_{\rm BS} \times K$ matrix $\mathbf{H}_{\ell j}$ that contains the fast-fading coefficients from the K UEs to the $N_{\rm BS}$ BS antenna elements, where the UEs are in the jth cell and the BS in the ℓth cell.[8] Assuming now for the moment that the array is configured in such a way that shadowing is the same for all antenna elements (i.e. the array is small enough, and all elements have the same pattern), we can now rewrite the channel gain as

$$h_{n\ell jk} = \widetilde{h}_{n\ell jk}\beta_{\ell jk}^{1/2} \tag{22.51}$$

where the β describe path gain and shadowing, while $\widetilde{h}$ is the normalized small-scale fading. We define a diagonal matrix $\mathbf{D}_{\ell j}$ with entries $\beta_{\ell j1}, \beta_{\ell j2}, \ldots$ Finally, we define $\mathbf{H}_{\ell j} = \widetilde{\mathbf{H}}_{\ell j}\mathbf{D}_{\ell j}^{1/2}$ as the total channel matrix. Note that this matrix is defined based on an uplink transmission, so that in the downlink the relevant channel matrix is $\mathbf{H}^T$; reciprocity is assumed implicitly.

When considering the limit of very large arrays, we need to distinguish the following two regimes:

- fixed number of users K, number of antenna elements $N_{\rm BS}$ going to infinity
- fixed ratio $\eta = K/N_{\rm BS}$, $N_{\rm BS} \to \infty$. We always assume $N_{\rm BS} \geq K$.

Note that the limiting regimes mainly serve to gain simple analytical insights. Much of the performance analysis that exists (and is discussed below) is valid for arbitrary array sizes.

Now let us assume that the following relationship holds for a pair of channel vectors $\widetilde{\mathbf{h}}_i$ and $\widetilde{\mathbf{h}}_j$ as $N_{\rm BS} \to \infty$

$$\frac{1}{N_{\rm BS}}\widetilde{\mathbf{h}}_i^{\dagger}\widetilde{\mathbf{h}}_i \to \sigma_i^2 \tag{22.52a}$$

$$\frac{1}{N_{\rm BS}}\widetilde{\mathbf{h}}_i^{\dagger}\widetilde{\mathbf{h}}_j \to 0 \tag{22.52b}$$

$$\frac{1}{\sqrt{N_{\rm BS}}}\widetilde{\mathbf{h}}_i^{\dagger}\widetilde{\mathbf{h}}_j \to \mathcal{CN}\left(0, \sigma_i^2\sigma_j^2\right) \tag{22.52c}$$

where $\mathcal{CN}$ denotes the circularly symmetric complex Gaussian variable. These relations are fulfilled when the entries of $\widetilde{\mathbf{h}}_i$ and $\widetilde{\mathbf{h}}_j$ are zero-mean i.i.d. complex Gaussian random variables with variance for each entry σ_i^2 and σ_j^2, respectively. As we will see below, a set of channels that fulfills these conditions constitutes favorable conditions for massive MIMO. The conditions are not just fulfilled for i.i.d. Rayleigh fading, but also if the UEs have LOS to the BS but are in random directions.

These mathematical relationships provide one of the key insights of massive MIMO: since the beamforming vectors for MRC are just the complex conjugates of the channel vectors (Section 12.4), MRC inherently provides orthogonality between the users, in other words, by performing MRC for one user, the interference from the other user is automatically suppressed when $N_{\rm BS}$ becomes large. Similarly, the effect of the noise vanishes, and the only factor limiting the SINR is the pilot contamination.

Another important consequence is that the effect of fading vanishes. The power of the output signal from the antenna combiner (in the uplink) or the power arriving at the UE (in the downlink) becomes independent of the small-scale fading state of the channel. We stress here that the fading of the channel between a UE and a BS antenna does *not* disappear – it is, after all, an innate property of multi-path channels. Rather, it is the *effect* of fading at the RX that disappears due to a very high degree of diversity.

The following sections present some of the key results for massive MIMO under various assumptions. By necessity, we eschew detailed derivations, and refer the readers to [Marzetta et al. 2016] and [Björnson et al. 2017] for those.

---

[8] Note that this is a matrix, with $j\ell$ serving to further define for which cell combination the matrix is valid – it is *not* the $j\ell$–th element of a matrix; entries of a matrix are written as $[\mathbf{H}]_{j\ell}$.

22.9.2 Single-Cell Massive MIMO with Perfect CSI

Uplink

Let us first assume that there is only a single cell (i.e., no pilot contamination); for the analysis of the single-cell case, we henceforth drop subscripts j and ℓ. Now consider the uplink of a massive MIMO system, with linear detection, where the received signal $\mathbf{r}$ is multiplied with a matrix $\mathbf{W}^\dagger$. Let the BS have perfect CSI. In that case, the BS uses the following (un-normalized) receive filters:

$$\mathbf{W} = \begin{cases} \mathbf{H} & \text{for MRC} \\ \mathbf{H}\left(\mathbf{H}^\dagger\mathbf{H}\right)^{-1} & \text{for zero-forcing} \\ \mathbf{H}\left(\mathbf{H}^\dagger\mathbf{H} + \dfrac{1}{\gamma_{\mathrm{TX},k}}\mathbf{I}_K\right)^{-1} & \text{for MMSE} \end{cases} \tag{22.53}$$

where $\gamma_{\mathrm{TX},k} = P_k/\sigma_\mathrm{n}^2$. If written without index, it is assumed to be identical for all users.

Letting $\mathbf{w}_k$ and $\mathbf{h}_k$ be the kth column of $\mathbf{W}$ and $\mathbf{H}$, respectively, the achievable rate of the kth user is:

$$R_k = E\left\{ \log_2\left(1 + \frac{P_k\left|\mathbf{w}_k^\dagger\mathbf{h}_k\right|^2}{\sum_{i\neq k}P_i\left|\mathbf{w}_k^\dagger\mathbf{h}_i\right|^2 + \|\mathbf{w}_k\|^2\sigma_\mathrm{n}^2}\right)\right\}. \tag{22.54}$$

It is challenging to exactly evaluate the average rate for a particular RX based on this equation, since we have to take an expectation over the small-scale fading. It is considerably easier to derive the "effective SINR," and compute a lower bound on the rate as $\log_2(1 + \gamma_\mathrm{eff})$. It turns out that for the limiting case of $N_\mathrm{BS} \to \infty$, this bound becomes tight. We will demonstrate this equivalence in the following using the example of MRC with perfect CSI; similar results can also be obtained for other situations.

Let us first determine the SINR in the MRC case. Without normalization, the RX filter would be $\mathbf{w}_k = \mathbf{h}_k$, but we introduce a normalization $\mathbf{w}_k = \mathbf{h}_k\beta_k^{-1/2}N_\mathrm{BS}^{-1/2}$ so that (assuming equal power for all users) the power of the desired signal is $P_k\left(|\mathbf{h}_k|^2|\mathbf{w}_k|^2\right) = P_k\left(|\mathbf{h}_k|^2\right)^2\beta_k^{-1}N_\mathrm{BS}^{-1}$, and its expectation is, according to Eq. (22.52a), $P_k\beta_k N_\mathrm{BS}$. The power of the noise is σ_n^2, while the intra-cell interference is $\sum_{i\neq k}P_i\left|\mathbf{w}_k^\dagger\mathbf{h}_i\right|^2$, the expectation of this converges to $\sum_{i\neq k}P_i\beta_i$. Thus, the SINR becomes

$$\begin{aligned} SINR_k &= \frac{P_k\beta_k N_\mathrm{BS}}{\sum_{i\neq k}P_i\beta_i + \sigma_\mathrm{n}^2} \\ &= \frac{\gamma_\mathrm{TX}\beta_k N_\mathrm{BS}}{\sum_{i\neq k}\gamma_\mathrm{TX}\beta_i + 1}. \end{aligned} \tag{22.55}$$

We thus see that the desired signal grows linearly with N_BS, while the interference grows with the number of users in the cell.

Example 22.1 *Derivation of lower bound on rate.*

Now let us establish a lower bound on the actual rate. Again, we start by substituting $\mathbf{w}_k = \mathbf{h}_k\beta_k^{-1/2}N_\mathrm{BS}^{-1/2}$ in (22.54) to get

$$R_k = \mathbb{E}\left\{ \log_2\left[1 + \frac{P_k|\mathbf{h}_k|^2|\mathbf{w}_k|^2}{\sum_{i\neq k}P_i\left|\mathbf{w}_k^\dagger\mathbf{h}_i\right|^2 + \|\mathbf{w}_k\|^2\sigma_\mathrm{n}^2}\right]\right\}. \tag{22.56}$$

We then again assume that all transmit powers are identical, and exploit the fact that $\log(1 + 1/x)$ is a convex function. Multiplying numerator and denominator with $\beta_k N_\mathrm{BS}$ and using Jensen's inequality $E\{f(x)\} \geq f(E\{x\})$ for convex $f(x)$, and assuming for convenience equal transmit power for all users, we obtain the following lower bound:

$$R_k \geq \log_2\left(1 + \left(E\left\{\frac{\gamma_\mathrm{TX}\sum_{i\neq k}\left|\mathbf{h}_k^\dagger\mathbf{h}_i\right|^2 + \|\mathbf{h}_k\|^2}{\gamma_\mathrm{TX}\|\mathbf{h}_k\|^4}\right\}\right)^{-1}\right) \tag{22.57}$$

$$= \log_2\left[1 + \left(\mathbb{E}\left\{\frac{\sum_{i\neq k}^K|\widetilde{g}_i|^2 + \frac{1}{\gamma_\mathrm{TX}}}{\|\mathbf{h}_k\|^2}\right\}\right)^{-1}\right] \tag{22.58}$$

where $\widetilde{g}_i = \dfrac{\mathbf{h}_k^\dagger\mathbf{h}_i}{\left\|\mathbf{h}_k\right\|}$ and we can observe that $\widetilde{g}_i$ is a zero-mean Gaussian random variable with variance β_i that is independent of $\mathbf{h}_k$ when conditioned on $\mathbf{h}_k$. Consequently, the expectation in Eq. (22.58) can be rewritten as

$$\mathbb{E}\left\{\frac{\gamma_{TX}\sum_{i\neq k}|\widetilde{g}_i|^2 + 1}{\gamma_{TX}\|\mathbf{h}_k\|^2}\right\} = \left(\gamma_{TX}\sum_{i\neq k}\mathbb{E}\{|\widetilde{g}_i|^2\} + 1\right)\mathbb{E}\left\{\frac{1}{\gamma_{TX}\|\mathbf{h}_k\|^2}\right\} = \left(\gamma_{TX}\sum_{i\neq k}\beta_i + 1\right)\mathbb{E}\left\{\frac{1}{\gamma_{TX}\|\mathbf{h}_k\|^2}\right\} . \tag{22.59}$$

Now further exploiting the following identity for $m \times m$ central complex Wishart matrices $\mathbf{A}$ with unit variance and n degrees of freedom

$$E\{tr(\mathbf{A}^{-1})\} = \frac{m}{n-m} \tag{22.60}$$

we get

$$\mathbb{E}\left\{\frac{1}{\gamma_{TX}\|\mathbf{h}_k\|^2}\right\} = \frac{1}{\gamma_{TX}(N_{BS}-1)\beta_k} \qquad N_{BS} \geq 2. \tag{22.61}$$

Thus the rate can be bounded as

$$\widetilde{R}_k^{mrc} = \log_2\left(1 + \frac{\gamma_{TX}(N_{BS}-1)\beta_k}{\gamma_{TX}\sum_{i\neq k}\beta_i + 1}\right). \tag{22.62}$$

Note that this bound is valid for arbitrary array sizes. In the large-array limit

$$R_k = \log_2\left(1 + \frac{\gamma_{TX}\beta_k N_{BS}}{\gamma_{TX}\sum_{i\neq k}\beta_i + 1}\right) \qquad \text{for } N_{BS} \to \infty. \tag{22.63}$$

Note that in this case the rate expression is exact (i.e., the exact value coincides with the lower bound). This can be explained by the channel hardening effect: since for every channel realization the small-scale fading does not have an effect, the expectation is taken over an ensemble that takes on only a single value, and Jensen's inequality (on which the lower bounding was based) becomes an equality.

For a ZF RX, the derivation of the rate bounds follows very similar principles:

$$R_k = \mathbb{E}\left\{\log_2\left(1 + \frac{\gamma_{TX}}{\left[(\mathbf{H}^\dagger\mathbf{H})^{-1}\right]_{kk}}\right)\right\} \geq \log_2\left(1 + \frac{\gamma_{TX}}{\mathbb{E}\left\{\left[(\mathbf{H}^\dagger\mathbf{H})^{-1}\right]_{kk}\right\}}\right). \tag{22.64}$$

This bound is only good for $N_{BS} \gg K$. Now we can evaluate the expectation as

$$\begin{aligned}
\mathbb{E}\left\{\left[(\mathbf{H}^\dagger\mathbf{H})^{-1}\right]_{kk}\right\} &= \frac{1}{\beta_k}\mathbb{E}\left\{\left[(\widetilde{\mathbf{H}}^\dagger\widetilde{\mathbf{H}})^{-1}\right]_{kk}\right\} = \\
&= \frac{1}{K\beta_k}\mathbb{E}\left\{tr\left[(\widetilde{\mathbf{H}}^\dagger\widetilde{\mathbf{H}})^{-1}\right]\right\} = \\
&= \frac{1}{(N_{BS}-K)\beta_k} \qquad \text{for} \quad N_{BS} \geq K+1
\end{aligned} \tag{22.65}$$

where we again use Eq. (22.60) for the transition from the second to the third line. Finally the rate bound becomes

$$R_k \geq \log_2(1 + \gamma_{TX}(N_{BS}-K)\beta_k). \tag{22.66}$$

This result can be easily interpreted in light of the discussion of optimum combining in Section 12.4. We established there that each antenna element provides a degree of freedom that can be used either for diversity/beamforming, or the suppression of an interferer. With K users, the bound (22.66) says that we have to expend K degrees of freedom on the suppression of the intra-cell interference, leaving the remaining $N_{BS} - K$ antenna elements for beamforming, and thus allowing to improve the SNR or reduce the energy consumption. The corresponding SINR is

$$SINR_{\text{ZF,UL},k} = \gamma_{\text{TX}}(N_{\text{BS}} - K)\beta_k. \tag{22.67}$$

We see from the above that the throughput goes to infinity as N_{BS} goes to infinity. In the case that K is constant, the SINR for each user goes to infinity, so we have infinite capacity for a finite number of users. In the case that K/N_{BS} is constant, the SINR per user still goes to infinity, and we have an infinite number of users with infinite rate each. In the case that the SINR were to converge to a constant (we will encounter such cases in later subsections), we have an infinite number of users with finite (non-zero) throughput, again providing infinite sum throughput.

However, whenever dealing with "going-to-infinity" situations, some caution is necessary. Infinitely many antennas are not feasible, and once the antenna array becomes very large, various approximations, such as equal shadowing and pathloss for all antenna elements, start to break down. Also, various other error sources, like finite-precision phase shifters for the beamformers, crosstalk between user signals due to nonlinearities, etc., all might come into play. The importance of the above results lies in the fact that they constitute a good approximation at large-but-finite antenna sizes, and that they give an intuitive insight into the effects and parameters that determine the system performance.

Downlink

Starting out with the ZF case, the precoder is

$$\hat{\mathbf{T}} = \sqrt{\frac{N_{\text{BS}} - K}{K}}\widetilde{\mathbf{H}}^{+} \tag{22.68}$$

which provides the correct normalization in Rayleigh fading (note that the channel matrix $\mathbf{H}$ is defined for the uplink channel). The SINR that can be achieved with this is

$$SINR_{\text{ZF,DL},k} = \gamma_{\text{TX},k}(N_{\text{BS}} - K)\beta_k \tag{22.69}$$

which is identical to the uplink case. However, note that in the uplink case γ_{TX} is proportional to the power per UE, while in the downlink case, it is the power that the BS assigns to the data streams intended for the kth UE; there is therefore a sum power constraint. For the case of equal transmit power for each user, $\gamma_{\text{TX},k} = P_{\text{BS}}/\left(K\sigma_{\text{n}}^2\right)$.

For the MRT case, there is a difference between the uplink and the downlink case, because there is inter-user interference (remember that interference is not reciprocal). The SINR in this case is

$$SINR_{\text{MRT,DL},k} = \frac{\gamma_{\text{TX},k}\beta_k N_{\text{BS}}}{\beta_k \sum_{i \neq k}\gamma_{\text{TX},i} + 1} \tag{22.70}$$

because the various interference powers now go through the same channel, namely the channel for user k.

*22.9.3 Single-Cell Massive MIMO with Noisy CSI

Uplink

Perfect CSI is impossible to obtain in a realistic setup. In conventional MIMO systems, noisy CSI plays a minor role because other effects such as inter-user interference dominate the performance. However, for massive MIMO, many of the traditional limitations can be eliminated, so that noisy CSI becomes a key factor that fundamentally changes the scaling behavior of the capacity.

Assume that we have a set of pilot sequences of length M_{p} (in units of symbols), which form the columns of a $M_{\text{p}} \times K$ pilot matrix $\mathbf{\Phi}$ (where $M_{\text{p}} \geq K$ in order to ensure that the pilot tones from all users can be distinguished). There is no pilot boosting, which means that the peak power during pilot transmission is the same as during data transmission. For notational convenience, we assume that the pilot sequences form a semi-unitary matrix $\mathbf{\Phi}^{\dagger}\mathbf{\Phi} = \mathbf{I}_K$. Then we transmit $\mathbf{\Phi}^{\dagger}$, and an MMSE estimate of the channel matrix can be obtained by postmultipling the received signal $\mathbf{r}_{\text{p}}$ with $\mathbf{\Phi}$, to get a vector $\mathbf{r}'$, from which the channel estimate is obtained

$$\hat{h}_{nk} = \frac{\sqrt{M_{\text{p}}\gamma_{\text{TX},k}}\beta_k}{1 + M_{\text{p}}\gamma_{\text{TX},k}\beta_k}\left[r'_{\text{p},n,k}\right] \tag{22.71}$$

so that the mean-square of the channel estimate is

$$\hat{\beta}_k = \frac{M_{\mathrm{p}}\gamma_{\mathrm{TX},k}\beta_k^2}{M_{\mathrm{p}}\gamma_{\mathrm{TX},k}\beta_k + 1} \tag{22.72}$$

which obviously converges to the true mean-square value β_k as the pilot SNR $M_{\mathrm{p}}\gamma_{\mathrm{TX}}$ increases. The channel estimation error for the kth user, ε_k, has zero mean and variance

$$\sigma_{\varepsilon,k}^2 = \beta_k - \hat{\beta}_k = \frac{\beta_k}{M_{\mathrm{p}}\gamma_{\mathrm{TX},k}\beta_k + 1}. \tag{22.73}$$

We start by writing the estimated channel as the true channel plus the estimation error, $\hat{\mathbf{H}} = \mathbf{H} + \varepsilon$ in the usual input/output relationship

$$\mathbf{r} = \mathbf{HP}^{1/2}\mathbf{s} + \mathbf{n} \tag{22.74}$$

$$= \left(\hat{\mathbf{H}} - \varepsilon\right)\mathbf{P}^{1/2}\mathbf{s} + \mathbf{n} \tag{22.75}$$

$$= \hat{\mathbf{H}}\mathbf{P}^{1/2}\mathbf{s} + \left[\mathbf{n} - \hat{\mathbf{H}}\varepsilon\mathbf{P}^{1/2}\mathbf{s}\right]. \tag{22.76}$$

The term in square brackets is thus the "effective noise," with a covariance of

$$\mathbf{I}_{N_{\mathrm{BS}}}\left(\sigma_{\mathrm{n}}^2 + \sum_k P_k\sigma_{\varepsilon,k}^2\right). \tag{22.77}$$

For MRC, we use the receive filter for the kth user $\mathbf{w}_k = \hat{\mathbf{h}}_k\beta_k^{-1/2}N_{\mathrm{BS}}^{-1/2}$, i.e., already normalized, where $\hat{\mathbf{h}}_k$ is again the kth column of $\hat{\mathbf{H}}$. The signal power for the kth user is thus

$$P_k\left|\hat{\mathbf{h}}_k\right|^2|\mathbf{w}_k|^2 = P_k\hat{\beta}_kN_{\mathrm{BS}} \tag{22.78}$$

while the standard deviation of the Gaussian term is

$$\left(\sigma_{\mathrm{n}}^2 + \sum_i P_i\sigma_{\varepsilon,i}^2 + \sum_{i\neq k}P_i\hat{\beta}_i + P_k\hat{\beta}_k\right) \tag{22.79}$$

where the first term in brackets denotes the noise, the second the channel estimation errors, the third the inter-user interference, and the last the variations of the beamforming gain due to noisy channel estimates. Again assuming equal power for all users, we thus get an effective SINR of

$$SINR_{\mathrm{MRC,UL},k} = \frac{\gamma_{\mathrm{TX}}\hat{\beta}_kN_{\mathrm{BS}}}{1 + \gamma_{\mathrm{TX}}\sum_i\beta_i}. \tag{22.80}$$

We note that the impact of the channel estimation error is a reduction in the desired signal, i.e., the numerator is $\hat{\beta}$ instead of β that occurs in the ideal-CSI case.

For a ZF RX, the SINR can be shown to be

$$SINR_{\mathrm{ZF,UL},k} = \frac{\gamma_{\mathrm{TX},k}\hat{\beta}_k(N_{\mathrm{BS}} - K)}{1 + \sum_i\gamma_{\mathrm{TX},i}\sigma_{\varepsilon,i}^2} \tag{22.81}$$

which has a similar interpretation as for the MRC case; however, the array gain in the numerator is reduced from N_{BS} to $(N_{\mathrm{BS}} - K)$ because of the reduced degrees of freedom for eliminating inter-user interference; on the other hand the denominator only retains the terms for noise and channel estimation error.

Figure 22.9 shows the spectral efficiency versus the number of BS antennas N_{BS} for MRC, ZF, and MMSE processing at the RX, with perfect CSI and with imperfect CSI (obtained from uplink pilots). The figure confirms that the scaling for the different detection methods is the same, though there is a constant offset in the spectral efficiency. Fig. 22.10 shows experimental results.

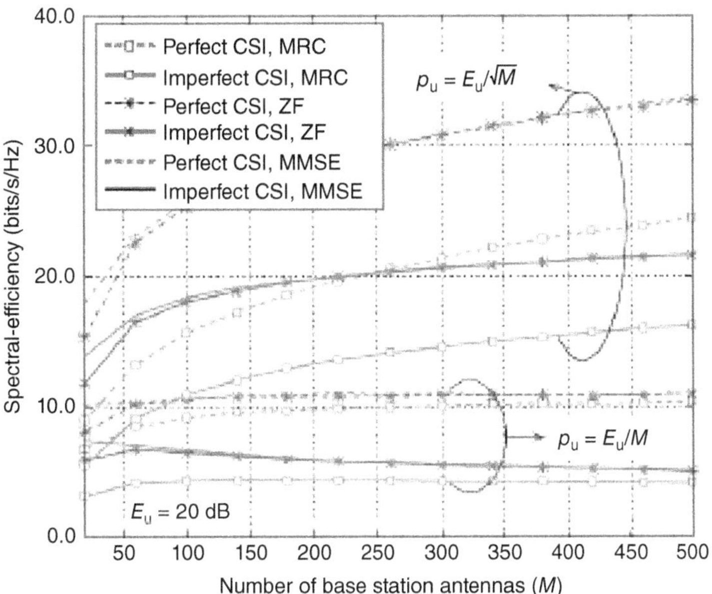

Figure 22.9 Spectral efficiency of massive MIMO versus number of BS antennas. $K = 10$ UEs are served simultaneously. Both the case that transmit power scales as $p_{\mathrm{u}} = E_{\mathrm{u}}/\sqrt{N_{\mathrm{BS}}}$ and as $p_{\mathrm{u}} = E_{\mathrm{u}}/N_{\mathrm{BS}}$ is shown. Note that M in the figure means N_{BS} in the notation of this book. Color version available at wiley.com/go/molisch/wireless3e.
Reproduced with permission from [Ngo et al. 2013a] © IEEE.

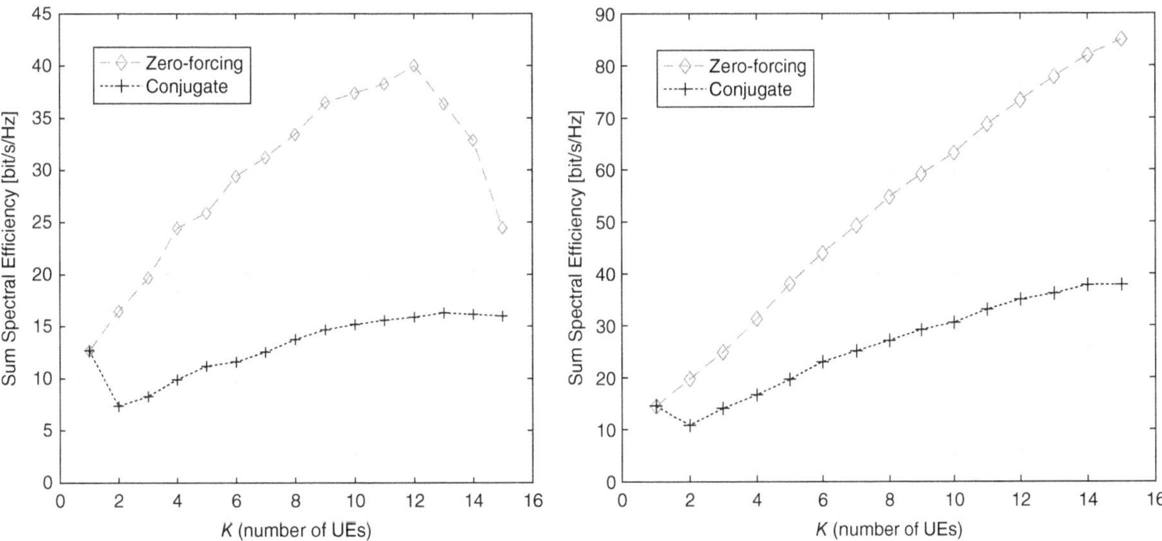

Figure 22.10 Capacity measured in an indoor channel with a massive MIMO testbed (Argos). (a) $N_{\mathrm{BS}} = 15$. (b) $N_{\mathrm{BS}} = 64$. From C. Shepard, private communication (2021). Color version available at wiley.com/go/molisch/wireless3e.

Downlink

The derivation of the effective SINR is very similar to the one for uplink. For the MRC case, the signal power for the kth user is

$$P_k \hat{\beta}_k N_{\mathrm{BS}} \tag{22.82}$$

while the standard deviation of the Gaussian term is

$$\left(\sigma_{\mathrm{n}}^2 + \sigma_{\varepsilon,k}^2 \sum_i P_i + \hat{\beta}_k \sum_{i \neq k} P_i + P_k \hat{\beta}_k \right) \tag{22.83}$$

where the first term in brackets denotes the noise, the second the channel estimation errors, the third the inter-user interference, and the last the variations of the beamforming gain due to noisy channel estimates. We thus get an effective SINR of

$$SINR_{\mathrm{MRC,DL},k} = \frac{\gamma_{\mathrm{TX},k}\hat{\beta}_k N_{\mathrm{BS}}}{1 + \beta_k \sum_i \gamma_{\mathrm{TX},i}}. \tag{22.84}$$

For ZF, we get

$$SINR_{\mathrm{ZF,DL},k} = \frac{\gamma_{\mathrm{TX},k}\hat{\beta}_k (N_{\mathrm{BS}} - K)}{1 + \sigma^2_{\varepsilon,k} \sum_i \gamma_{\mathrm{TX},i}}. \tag{22.85}$$

It is now interesting to interpret the above results in the limiting case of large N_{BS}. Consider first a fixed K, and $N_{\mathrm{BS}} \to \infty$. Then we can see that in all of the above situations, the effective SINR tends to infinity. In other words, the increasing number of BS antenna elements allows to form sharper and sharper beams, which helps to eliminate inter-user interference (which may occur in MRC for finite N_{BS}) as well as increase the beamforming gain such that noise starts to become irrelevant. If, on the other hand, we keep N_{BS}/K constant while letting both N_{BS} and K go to infinity, then the "estimation-error-free" ZF approach is the only for which the SINR goes to infinity. The reason for this is that only noise during the data transmission is a relevant error source, and the *degrees of freedom* available to suppress noise, namely $(N_{\mathrm{BS}} - K)$ go to infinity. In the case of errors in the channel estimate, there is residual inter-user interference, so that the SINR converges to a constant value. Now it must be stressed that while in this case the SINR stays finite, the sum rate of the BS still goes to infinity, because we have an infinite number of users with finite rate each.

*22.9.4   Inter-Cell Interference and Pilot Contamination

We now finally analyze the case that includes ICI, which occurs in any cellular system. This gives rise to two effects: (i) ICI for the desired data signal. The effect is similar to what we experienced in the single-cell system: its effect can be combatted by increased N_{BS} – the beams become sharper and sharper, and the ICI can be suppressed. (ii) Pilot contamination: this is the effect that channel estimation is impacted by ICI, and the associated estimation error leads to a limitation of the capacity. This is an effect that cannot be overcome by simply increasing N_{BS} (we will below discuss some other mitigation methods), and thus has drawn great attention as a limit for massive MIMO performance.

To determine the MMSE estimate and MSE of the channel, use

$$\hat{h}_{n\ell jk} = \frac{\sqrt{M_{\mathrm{p}}\gamma_{\mathrm{TX}}}\beta_{\ell jk}}{1 + M_{\mathrm{p}}\gamma_{\mathrm{TX}} \sum\limits_{j' \in \mathcal{P}_\ell} \beta_{\ell j'k}} \left[r'_{\mathrm{p},n,\ell,k} \right] \tag{22.86}$$

where $\mathcal{P}_\ell$ is the set of cells that use the same pilot sequence, and we assumed equal transmit power for all users. In other words, the signals from the kth UE in the jth cell, and from the kth UE in the $j' \in \mathcal{P}_\ell$ cell cannot be distinguished by the RX, resulting in an estimate of $\hat{h}$ that is "contaminated." The mean-square estimation value $\hat{\beta} = E\left\{\left|\hat{h}\right|^2\right\}$ then becomes

$$\hat{\beta}_{\ell jk} = \frac{M_{\mathrm{p}}\gamma_{\mathrm{TX}}\beta^2_{\ell jk}}{1 + M_{\mathrm{p}}\gamma_{\mathrm{TX}} \sum\limits_{j' \in \mathcal{P}_\ell} \beta_{\ell j'k}} \tag{22.87}$$

and the MSE is $\sigma^2_{\varepsilon,\ell jk} = \beta_{\ell jk} - \hat{\beta}_{\ell jk}$. Comparing to the estimate (22.72) we find that even as the pilot SNR $M_{\mathrm{p}}\gamma_{\mathrm{TX}}$ goes to infinity, the estimate $\hat{\beta}_{\ell jk}$ does not converge to $\beta_{\ell jk}$. This is the essence of pilot contamination.

It is worth noting that the pilot contamination leads to two distinct effects: (i) it reduces the beamforming gain of the desired user, since the formed beam is based on somewhat erroneous CSI, and (ii) it prevents a perfect suppression of the interference. Both of these effects are relevant, and occur for all array sizes. The common association of pilot contamination with very large arrays stems from the fact that it plays a relatively bigger role in those cases, but it is important to keep in mind that it is not *only* playing a role there.

The SINRs for our considered cases can be derived analogously to the previous discussion, i.e., establish the received signal, set the receive filter to follow from the estimated channel $\hat{\mathbf{H}}$, while the signal between UE goes through a channel $\hat{\mathbf{H}} - \varepsilon$, and finally taking the ratio of the power in the desired signal component and the variance of the deviations. The power in the desired signal component is formally the same as for the single-cell case, namely

$$P_{\ell k}\hat{\beta}_{\ell\ell k}\Xi \tag{22.88}$$

where $\Xi = N_{\mathrm{BS}}$ in the case of MRC, and $\Xi = N_{\mathrm{BS}} - K$ in the case of ZF. The actual value of $\hat{\beta}_{\ell\ell k}$ in the multi-cell case is different from $\hat{\beta}_k$ in the single-cell case, due to the pilot contamination. In terms of noise and interference, we now have:

1. Noise with variance σ_n^2,
2. Channel estimation errors with variance (again we see the formal similarity to the single-cell case; now we just have to consider the estimation errors from signals in multiple cells)

$$\sum_{j' \in \mathcal{P}_\ell} \sum_{k'} \sigma_{\varepsilon,\ell j'k'}^2 P_{j'k'}. \tag{22.89}$$

3. ICI from cells that do not have the same pilot tones as the considered cell, i.e., do not give rise to pilot contamination (this is a type of error not arising in the single-cell case), with a variance

$$\sum_{j' \notin \mathcal{P}_\ell} \sum_{k'} \beta_{\ell j'k'} P_{j'k'}. \tag{22.90}$$

4. ICI from contaminating cells, with variance

$$\Xi \sum_{j' \in \mathcal{P}_\ell \setminus \{\ell\}} \hat{\beta}_{\ell j'k} P_{j'k}, \tag{22.91}$$

i.e., summation over the cells that have the same pilot sequences, but not the home cell ℓ. We stress again that these terms increase in power as a function of N_{BS} in the same fashion as the numerator; this is due to the following effect: due to pilot contamination, the BS does not form a beam into the direction of the desired user, but into an aggregate channel that contains directions towards all users with contaminating pilots (after all, these pilots cannot be distinguished). Thus, the "sharpening" of the beams for data reception due to increasing N_{BS} does not help – it enhances signals from desired UE and contaminator in the same way.

5. ICI, with variance

$$\sum_{j' \in \mathcal{P}_\ell} \sum_{k'} \hat{\beta}_{\ell j'k'} P_{j'k'}. \tag{22.92}$$

which is a term that occurs only in the MRC case, since it is suppressed by design in the ZF RX.

We finally get the SINRs as

$$SINR_{\mathrm{MRC,UL},\ell k} = \frac{N_{\mathrm{BS}} \hat{\beta}_{\ell\ell k} P_{\ell k}}{\sigma_n^2 + \sum_{j' \in \mathcal{P}_\ell} \sum_{k'} \beta_{\ell j'k'} P_{j'k'} + \sum_{j' \notin \mathcal{P}_\ell} \sum_{k'} \beta_{\ell j'k'} P_{j'k'} + N_{\mathrm{BS}} \sum_{j' \in \mathcal{P}_\ell \setminus \{\ell\}} \hat{\beta}_{\ell j'k} P_{j'k}} \tag{22.93}$$

$$SINR_{\mathrm{ZF,UL},\ell k} = \frac{\hat{\beta}_{\ell\ell k} (N_{\mathrm{BS}} - K) P_{\ell k}}{\sigma_n^2 + \sum_{j' \in \mathcal{P}_\ell} \sum_{k'} \sigma_{\varepsilon,\ell j'k'}^2 P_{j'k'} + \sum_{j' \notin \mathcal{P}_\ell} \sum_{k'} \beta_{\ell j'k'} P_{j'k'} + (N_{\mathrm{BS}} - K) \sum_{j' \in \mathcal{P}_\ell \setminus \{\ell\}} \hat{\beta}_{\ell j'k} P_{j'k}}. \tag{22.94}$$

For the downlink, we obtain in a similar fashion

$$SINR_{\mathrm{MRC,DL},\ell k} = \frac{N_{\mathrm{BS}} \hat{\beta}_{\ell\ell k} P_{\ell k}}{\sigma_n^2 + \sum_{j' \in \mathcal{P}_\ell} \sum_{k'} \beta_{j'\ell k} P_{j'k'} + \sum_{j' \notin \mathcal{P}_\ell} \sum_{k'} \beta_{j'\ell k} P_{j'k'} + N_{\mathrm{BS}} \sum_{j' \in \mathcal{P}_\ell \setminus \{\ell\}} \hat{\beta}_{j'\ell k} P_{j'k}} \tag{22.95}$$

$$SINR_{\mathrm{ZF,DL},\ell k} = \frac{(N_{\mathrm{BS}} - K) \hat{\beta}_{\ell\ell k} P_{\ell k}}{\sigma_n^2 + \sum_{j' \in \mathcal{P}_\ell} \sum_{k'} \sigma_{\varepsilon,j'\ell k}^2 P_{j'k'} + \sum_{j' \notin \mathcal{P}_\ell} \sum_{k'} \beta_{j'\ell k} P_{j'k'} + (N_{\mathrm{BS}} - K) \sum_{j' \in \mathcal{P}_\ell \setminus \{\ell\}} \hat{\beta}_{j'\ell k} P_{j'k}}. \tag{22.96}$$

With pilot SNR and N_{BS} going to infinity, the uplink (for equal UE TX powers) SINR converges to:

$$SINR_{\ell,k} \to \frac{\beta_{\ell\ell k}^2}{\displaystyle\sum_{\substack{i=1 \\ i \neq \ell}}^{L} \beta_{\ell i k}^2}. \tag{22.97}$$

This shows that even with infinite beamforming gain, we actually cannot achieve infinite SINR. Rather, the pilot contamination puts a limit on it. Interestingly, the SINR is proportional to the *square* of the channel (power) gain ratios – this can be explained that the channel gains enter for the pilot SINR, and again (and thus multiplying with itself) for the data transmission.

*22.9.5　Pilot Contamination Countermeasures

Since pilot contamination is a key factor limiting the performance of massive MIMO systems, it is obviously important to find ways to decrease its impact. A number of different methods have been proposed.

Increased Reuse Distance

An obvious method for decreasing pilot contamination is to introduce frequency reuse for the pilot tones. Thus, the other-cell pilots that could interfere with the desired pilots would originate from farther-away cells, and thus be more strongly attenuated. According to Eq. (22.97), this increases the SINR, since now the summation in the denominator is only taken over cells that use the same pilot tones as the desired cell, and might be sufficient for practical purposes, even though it does not change the fundamental scaling law – in other words, pilot contamination will always be a limiting factor (for very large N_{BS}) unless every cell in the network uses orthogonal pilots.

The drawback of an increased reuse distance for pilots is the reduction of the spectral efficiency, since more orthogonal time/frequency resources need to be assigned to the pilot tones. We can also use scrambled pilot sequences in the different cells: in other words, instead of having one set of completely orthogonal pilots, where the set is reused in different cells, every cell has a different set of pilot sequences that are orthogonal within the cell, but partially correlated between cells.

Time-Delayed Pilot Sequences in Different Cells

One might be tempted to reduce pilot contamination by time-staggering the pilot tones (that use the same time-frequency resources). Then the BS in the desired cell receives a pilot tone only from the desired UE, while the neighboring cell is actually transmitting user data, see Figure 22.11. This does not, however, fundamentally solve the problem: while it clearly eliminates pilot contamination, now the pilots are contaminated by data (instead of pilots) from UEs in adjacent cells.

This approach mainly has advantages in quasi-static channels, as they allow an inherent reduction of interference in this case. Let the pilot tones for one cell be interfered by data from another cell. When pilot tones are repeated at a later time, they will be interfered by different data (though they might be from the same interfering user). Since the transmitted data symbols are usually zero-mean (e.g., in Quadrature Amplitude Modulation (QAM)), the data interference will average out, while the (desired) pilot tones add up. Of course, one might question why pilot tones should be frequently repeated in a quasi-static channels (one transmission should be enough since the channel does not change). The reason might lie in the fact that the wireless standard prescribes a certain pilot repetition frequency (as is the case in most modern standards, such as Long-Term Evolution (LTE) or NR, see Chapters 31 and 32). The same principle holds, to a lesser degree, for slowly time-varying channels where pilot signals are repeated more often than necessary from a sampling-theory point of view.

Use of Second-Order Statistics

If MPCs from a desired UE are restricted to a certain angular range $[\varphi_1, \varphi_2]$, and the MPCs from a pilot-contaminating UE in a neighboring cell are in a non-overlapping range $[\varphi_3, \varphi_4]$, then the pilot contamination can be eliminated. This is done essentially by a "beamforming" of the BS antennas in the pilot reception mode, where the beam pattern is based on the support of the power angular spectrum.

More generally, we can eliminate pilot contamination if the second-order statistics of all the channels are known. We can then take the received signal, and project it onto the nullspace of the pilot sequences from all possible interferers (as determined by the second-order statistics). This projection operation reduces the desired signal power (because the part of the desired signal that lies in the interference gets eliminated), but also gets rid of pilot contamination.

Blind Channel Estimation

Another interesting approach to channel estimation is based on blind estimation. The correlation matrix $\mathbf{R}_r$ of the received signal can then be approximated by the sample correlation matrix. Assuming that the pathloss and shadowing is known a priori, the fading channel matrix $\mathbf{H}_{\ell\ell}$ can be estimated from an eigendecomposition of the correlation matrix – up to a multiplicative ambiguity.

Summary

While pilot contamination for a long time was considered the most important aspect of massive MIMO performance, it is now accepted that (i) for practical operating points (SINRs, N_{BS}), other effects are at least as relevant, and (ii) the impact is most pronounced when straightforward MMSE channel estimation is used, while methods like subspace separation are actually effective in eliminating pilot contamination.

22.10 Massive MIMO Implementation Aspects

22.10.1 Antenna Configurations and Propagation Channels

The mathematical interpretation of massive MIMO relies on the fact that the channel vectors from the different users to the BS become orthogonal as the number of antenna elements increases to large values. This is a convenient mathematical approximation, but we now have to raise the question of how that can be achieved in practice. Furthermore, there are also situations where a physical separability of the MPC directions from the different users is required, i.e., the second-order statistics are orthogonal of each other. The answer to these questions depends both on the particular antenna structure, and the propagation channel characteristics.

A mathematically convenient formulation for a massive MIMO array is as a ULA with a large number of antenna elements. From a mathematical point of view, this has the advantage that a spatial Fourier transformation of the signals at the antenna elements provides beams pointing into essentially equidistantly spaced azimuth angles (strictly speaking, the directional cosines are equidistant, see Section 8.4). It can be shown that with such a structure, independent fading of the propagation channel occurs in two idealized scenarios: (i) in rich scattering, where there is an unlimited number of uniformly distributed scatterers, and (ii) when the angular spread of each UE approaches zero, and the directions under which the BS sees the UEs are uniformly distributed, see Figure 22.12. This latter case can be interpreted as follows: as N_{BS} increases, the resolution of the beams becomes finer and finer, so that the contributions of all the UEs except the desired one are completely suppressed. However, in this case only a single data stream per UE can be transmitted, even if the UE has multiple antenna elements. Generally, it is remarkable that for multi-user MIMO (and in particular massive MU-MIMO) high capacity can be achieved even if the propagation channel to each user is low-rank, as long as a spatial separability between the users is given. This is in striking contrast to the SU-MIMO case, where the number of spatial streams that can be meaningfully transmitted is upper limited by the channel rank, and in particular the angular spread and number of significant scatterers in the channel.

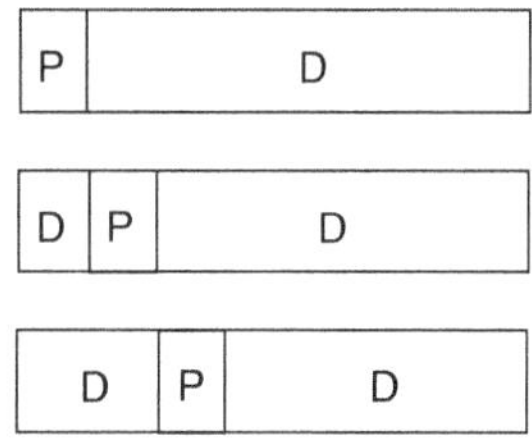

Figure 22.11 Principle of staggered pilot tones for pilot contamination reduction. P...pilots, D...data.

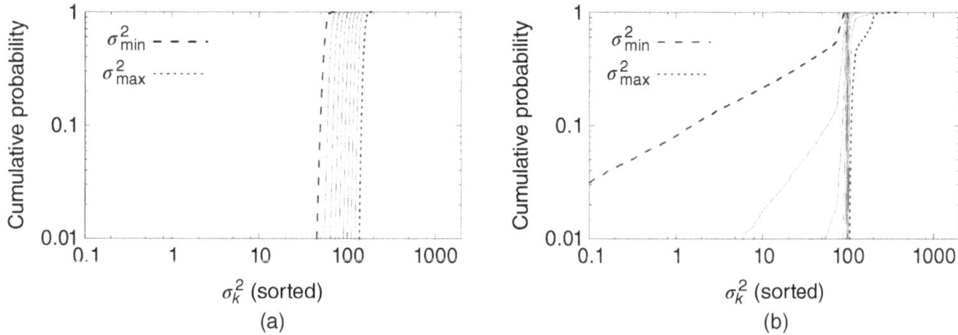

Figure 22.12 Sorted eigenvalue distribution for massive MIMO in i.i.d. Rayleigh fading (a) and pure LOS in random directions (b). $N_{\mathrm{BS}} = 100$, $K = 10$. Reproduced with permission from [Marzetta et al. 2016] © Cambridge University Press.

In most practical situations in outdoor environments, the propagation channel will be somewhere between those extremes. According to the "ring of scatterers" model (compare Section 7.6), scatterers are located on a ring around the UE. As long as the scatterers for the different users are different, the instantaneous channels will be orthogonal as $N_{\mathrm{BS}} \to \infty$.

There might even be orthogonality of the second-order statistics (i.e., the APSs, which are averaged over the small-scale fading) of at least some UEs. Increasing N_{BS} helps to increase the orthogonality when the number of UEs stays fixed: while the beams formed by the antenna array become sharper, the angular spread per user stays the same. If now the APSs of two users are partially overlapped, then by selecting some (sharp) beams that point towards the non-overlapped region of the APSs, the SIR of the users can be improved. However, if the number of UEs goes to infinity as well, this approach does not work anymore: the BS will "see" contributions from multiple users at each possible angle, so that user separation based on the APS is not possible (note that this is different from the ability to separate users based on instantaneous CSI).

At microwave frequencies, ULAs for massive MIMO become very large – for example, at 2.5 GHz carrier frequency, a 256-element array is 15 m long.[9] Such large physical dimensions have two major effects: (i) they might not be sustainable in a practical deployment, because of aesthetic considerations, and the wind load that such large antenna arrays would create. (ii) The assumption of identically distributed signals at the antenna elements might not be true anymore, in other words the channel is not (spatially) wide-sense stationary. This is because different parts of the signal suffer from different shadowing, see Figure 22.13. Both the directions of the MPCs, and the power coming from the MPCs, varies over the array. It also becomes clear that not all parts of the antenna array contribute equally to the capacity – it is not very helpful to process signals from those parts of the array that get little power anyway; rather, antenna selection should be combined with massive MIMO.

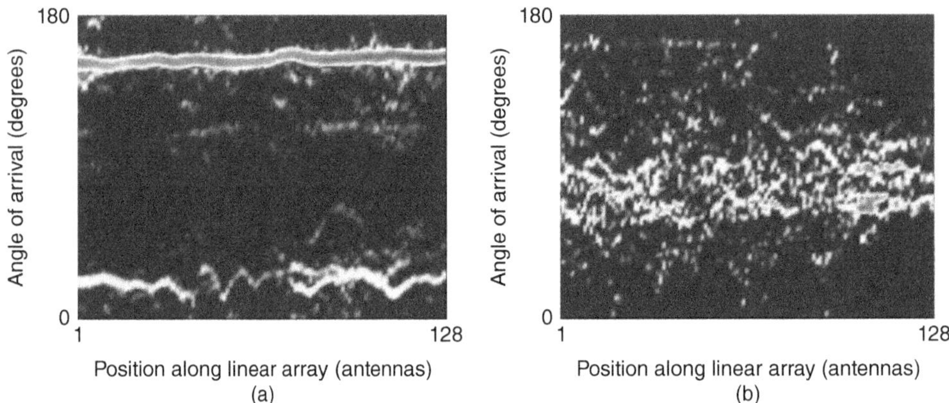

Figure 22.13 Angular power spectrum as a function of the position along a 128-element linear array. (a) A LOS scenario (b) a NLOS scenario. Color version available at wiley.com/go/molisch/wireless3e.
Reproduced with permission from [Gao et al. 2012] © Asilomar.

In order to reduce the size of the antenna arrays, two-dimensional array structures, such as rectangular or cylindrical arrays, or the one shown in Figure 22.14, are used. The resulting MIMO structure is often called *Full-Dimensional MIMO* (FD-MIMO).

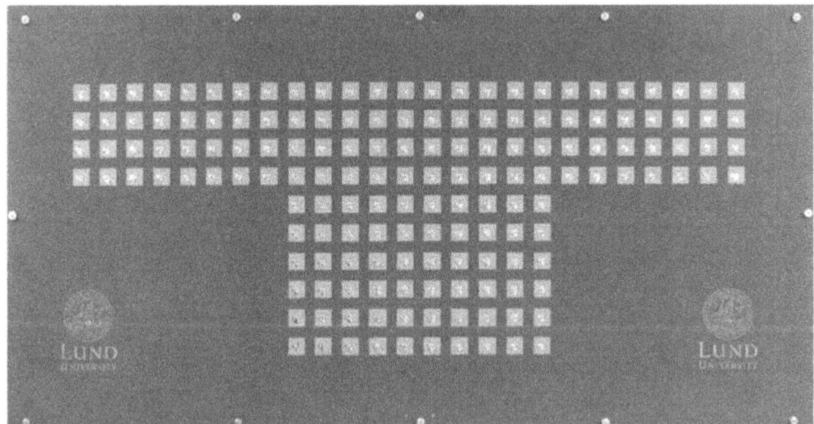

Figure 22.14 Example of two-dimensional array at microwave frequencies (3.7 GHz). Color version available at wiley.com/go/molisch/wireless3e.
Reproduced with permission from [Björnson et al. 2016] © IEEE.

In order to make use of the adaptive beamforming in the elevation domain, UEs that are located in the same azimuthal direction, but have different elevation spectra, should be separable. This occurs mainly under two circumstances: (i) in rural environments, when the scatterers are located close to the UE, the different UEs must be at different distances, see Figure 22.15a; (ii) in microcells near high-rise buildings, different floors can be seen under different elevation angles by the BS, again allowing a separation of the different users, see Figure 22.15b. The situation is more complicated in urban macrocells, where "over-the-rooftop" propagation is dominant, see Figure 22.15c. Consider two UEs in the same building, but on different floors. The MPCs from each of the UEs to the BS are diffracted by the roof edge near the UEs, and from there propagate to the BS. Thus, the BS "sees" the UEs under the same elevation angle, and second-order statistics cannot be easily separated. A similar problem occurs for azimuthal separation for wave-guided MPCs (mostly of importance for microcells, but also for many MPCs in macrocells): MPCs from different UEs that are guided down the same street canyon in which the BS is located lead to overlapping APS, so that a separation of users in that domain is not easily possible.

---

[9] At millimeter-wave frequencies, realization of such large arrays can be achieved more easily.

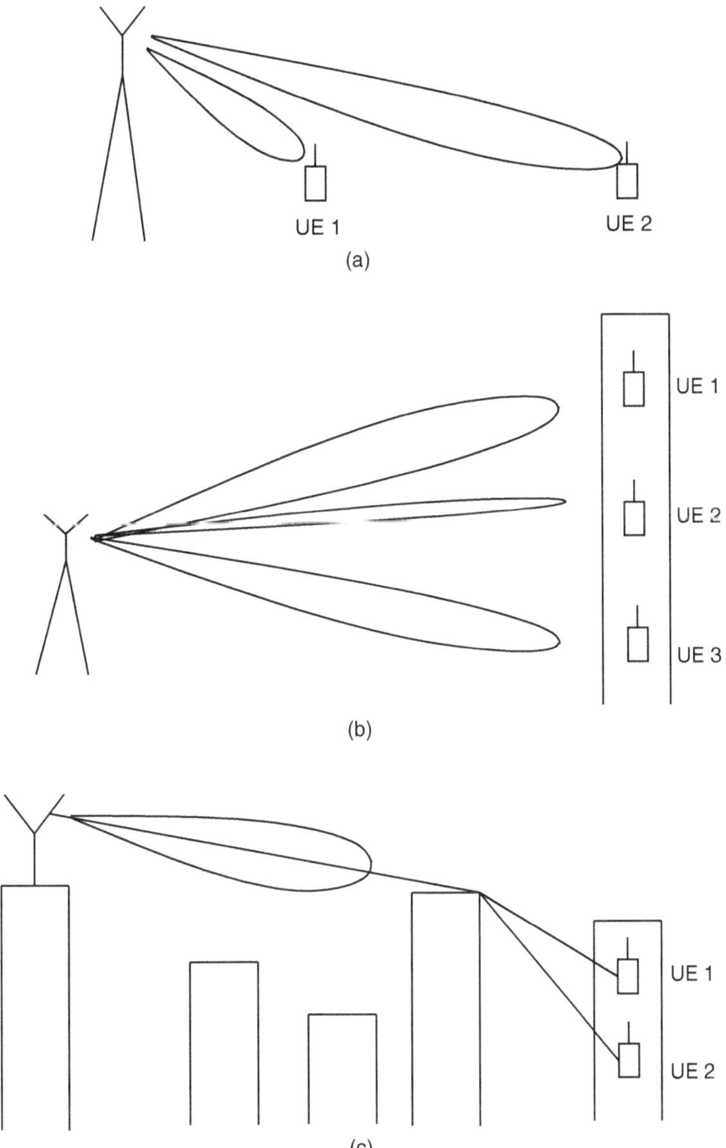

Figure 22.15 MU beamforming in the elevation domain in (a) rural environment, (b) highrise building, (c) urban macrocell.

22.10.2 Hybrid Beamforming Transceivers

The large number of antenna elements in massive MIMO also poses major hardware and efficiency challenges: (i) *Cost.* While antenna elements themselves are not expensive (usually consisting of a metal patch or rod), each antenna element requires an RF up/down-conversion chain. These analog elements do not fully benefit from Moore's law, and thus constitute a considerable part of chip area and cost. (ii) *Energy consumption.* The bias current of the RF chains can be significant, and energy wasted on those and similar effects often outstrips the energy savings for the RF transmission, making massive MIMO *less* energy efficient than traditional systems. (iii) *Channel estimation overhead.* The determination of CSI between each TX and RX antenna requires the usage of a considerable part of the spectral resources.

A widely used solution to these problems are *hybrid beamforming* transceivers, already introduced in Section 16.2.10. They are generally suitable for MU-MIMO, but particularly relevant for massive MIMO. Fundamentally, hybrid transceivers need at least K RF chains, where K is the number of *simultaneously* operating users. Beamformers that can adapt to the fast changes of the wireless propagation channel provide higher gains, but also require a higher effort for training and faster RF components. Thus, we will use the following a rough categorization based on the CSI that is used as the basis for the analog beamformers: *instantaneous CSI*, and *second-order CSI*. Within each of those categories, we can distinguish between "full complexity" architectures, in which the analog beamformers allow to fully exploit all degrees of freedom provided by the RF chains and antenna elements, and "reduced complexity" architectures, that achieve lower hardware effort at the cost of reduced flexibility – in complete analogy to the SU-MIMO case.

Block diagrams of the transceiver structures for these architetures are shown in Fig. 22.16; also shown is a structure that is suitable for the implementation of JSDM mentioned in Sec. 22.7 and further elaborated below.

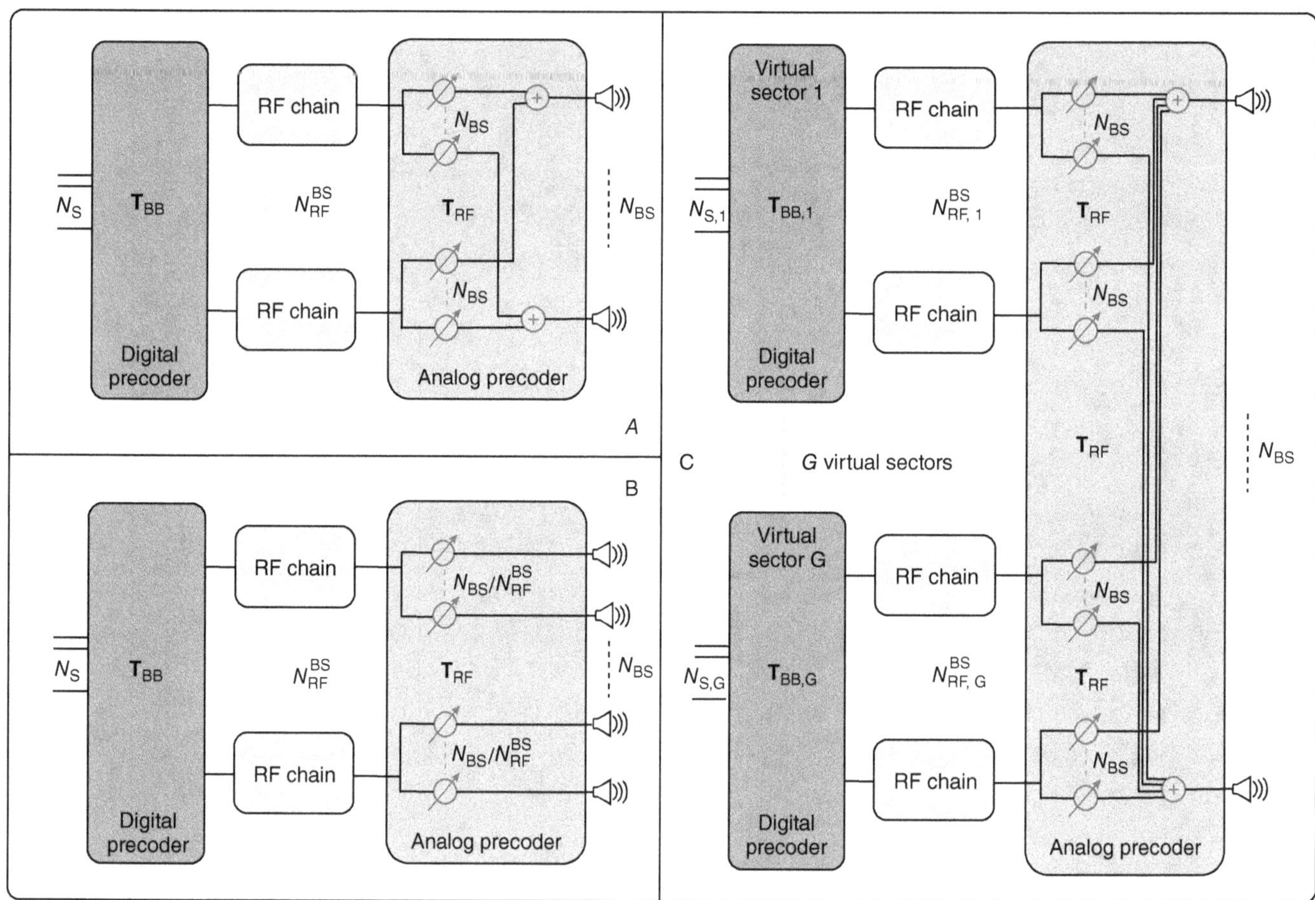

Figure 22.16 Block diagrams of hybrid beamforming structures at BS for a downlink transmission, where structures *A*, *B*, and *C* denote the full-complexity, reduced-complexity, and virtual-sectorization structures, respectively. Color version available at wiley.com/go/molisch/wireless3e. Reproduced with permission from [Molisch et al. 2017] © IEEE.

Based on Instantaneous CSI

There are two main methodologies to find good beamformers

- *Approximation of full-digital beamformer:* this approach first directly optimizes the one-stage beamformer, i.e., $\mathbf{W}_{os}$, $\mathbf{T}_{os}$ by assuming that both BSs and UEs have full-digital structure, i.e., the number of RF chains equals to the number of antenna elements. Then, the approximation of optimal beamforming by the hybrid one to minimize the matrix distance between them, e.g., $\|\mathbf{T}_{os} - \mathbf{T}_{RF}\mathbf{T}_{BB}\|_F$ and $\|\mathbf{W}_{os} - \mathbf{W}_{RF}\mathbf{W}_{BB}\|_F$, under different constraints for hardware structures and operating bands.
- *Decoupling of analog and digital beamformers:* Decoupling the interacting impact of analog and digital precoder/combiner is another approach. As for the SU-MIMO case, we can freeze the analog beamformer and optimize the digital one, followed by freezing the digital beamformer and re-optimizing the analog one. If the UEs also have a hybrid transceiver, iteration of BS and UE optimization can be done as a further loop; and since a change of the beamformer for one user might impact the interference to all other users, looping over the beamformers of the different users might provide yet another layer of iteration. By introducing appropriate additional assumptions, e.g., fixing the analog precoder to be eigen-beamforming and digital precoder to be ZF, a simple practical hybrid precoder design can be developed without iterative optimization.

Figure 22.17 compares the performance of the three structures for downlink transmission. The full-complexity structure of Figure 22.16a performs the same as the fully digital structure when the number of RF chains is no smaller than the number of users (or streams). Performance loss of structure B is rather large for the considered MU case, though it is much smaller for SU-MIMO (not shown here). For structure C, the employed JSDM algorithm divides the users into four or eight groups, which might lead to a performance floor due to inter-group interference.[10]

---

[10] Note that the reduced training overhead of JSDM is not shown here.

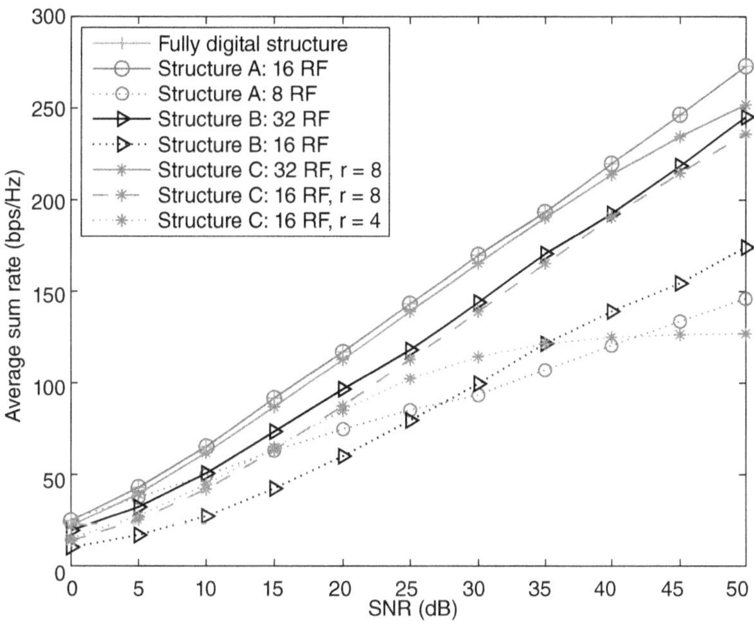

Figure 22.17 Performance comparison of the three hybrid structures with MU-MIMO; $N_{\mathrm{BS}} = 64$, $N_{\mathrm{UE}} = 1$, four groups of users located in a sector with mean directions $[-45°, -15°, 15°, 45°]$, and each group has four users. Color version available at wiley.com/go/molisch/wireless3e. Reproduced with permission from [Molisch et al. 2017] © IEEE.

The previously introduced hybrid beamforming designs focus on narrowband (i.e., single-subcarrier) massive MIMO systems for both SU-MIMO and multi-user MIMO cases. When applied to wideband OFDM systems, hybrid beamformers, in particular those with analog beamformers designed based on spatial correlation information, can work well if every user can occupy all subcarriers, i.e., frequency-domain scheduling is not employed. This is due to the fact that an analogue beamformer can only provide a single beam shape for all of the subcarriers.

Frequency-domain scheduling was believed unnecessary for full digital massive MIMO systems because the sufficiently large number of antennas can harden the channels and provide plenty of spatial degrees of freedom for multiplexing users. However, when considering the hybrid structure and practical array size limitation, channel hardening might not occur to a sufficient degree. Thus, there may be still the need for frequency-domain scheduling, so that different users are served over different subcarriers, complicating the precoder design.

Another important issue that is often ignored in the design of hybrid beamforming is control signaling coverage. Different from the transmission of user-specific data, for which analog precoders need to form multiple narrow beams towards the users, wide beams are preferred for control signaling broadcasting to ensure the cell coverage. This problem may be solved by using a signaling and data splitting architecture, where control signals and user-specific data are transmitted on different carrier frequencies, or by repeating the control information on multiple narrow beams. Practical solutions to these issues are implemented in 3GPP NR, see Chapter 32.

Based on Average CSI

One of the approaches for hybrid beamforming based on average CSI is JSDM, which groups users into "virtual sectors," i.e., groups that have a similar correlation matrix. This helps to reduce the overhead of channel estimation. But beyond that, it forms the basis for a reduced-complexity processing scheme that is well aligned with the structure of hybrid transceivers. Essentially, the analog beamformer creates beams towards the virtual sectors, and the digital domain is used for separating the signals within such a sector.

Specifically, using the Karhunen–Loeve representation, the N_{BS}-by-1 channel vector can be modeled as $\mathbf{h} = \mathbf{U}\boldsymbol{\Lambda}^{\frac{1}{2}}\mathbf{z}$, where $\mathbf{z} \in \mathbb{C}^{r \times 1} \mathcal{CN}(\mathbf{0}, \mathbf{I}_r)$, $\boldsymbol{\Lambda}$ is an r-by-r diagonal matrix, which has the eigenvalues of the channel covariance $\mathbf{R}$ on its diagonal, $\mathbf{U} \in \mathbb{C}^{M \times r}$ indicates the eigenmatrix of $\mathbf{R}$, and r denotes the rank of the channel covariance. Dividing UEs into G groups and assuming that UEs in the same group g exhibit the same channel covariance $\mathbf{R}_g$ with rank r_g, the JSDM analog precoder is

$$\mathbf{T}_{\mathrm{RF}} = [\mathbf{T}_{\mathrm{RF},1}, ..., \mathbf{T}_{\mathrm{RF},G}] \text{ with } \mathbf{T}_{\mathrm{RF},g} = \mathbf{E}_g \mathbf{G}_g. \tag{22.98}$$

By selecting $r_g^\star \leq r_g$ dominant eigenmodes, the matrices $\mathbf{E}_g$ and $\mathbf{G}_g$ are related to the nullspace of the interference covariance matrix of the gth group, and the dominant eigenmodes of the projection of $\mathbf{R}_g$ onto that eigenspace, respectively (compare Sec. 22.5.2).

A variety of other methods exist that use second-order CSI as the basis of the analog beamformer, and that – in contrast to JSDM – do not perform a subdivision of the users into groups. Rather, an iterative optimization of the analog beamformer and the digital beamformer for the maximization of the SINR and/or the sum rate can be performed. For more details see the "Further Reading" section.

The presence of RF transceiver imperfections degrades the spectral efficiency in various ways: e.g., it is harder to accurately generate desired transmit signals when higher beamformer gain is aimed for; non-linear distortion at the RX depends on the instantaneous channel gain and hence the SNR. In particular, transceiver imperfections are more pronounced at mm-waves, so that the performance of hybrid precoder/combiners no longer scales well with the number of RF chains. Figure 22.18 compares the spectral efficiency of spatially sparse hybrid precoding, including RF imperfections, with that from fully-digital beamforming. The coarsely quantized phase shifters and the transceivers' imperfections significantly degrade the spectral efficiency.

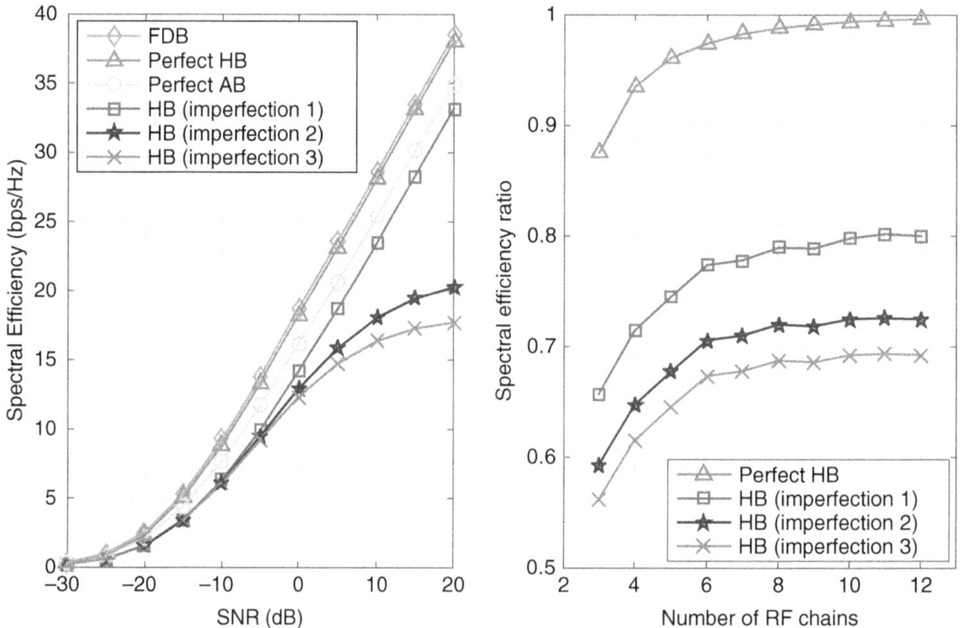

Figure 22.18 SE comparison of Fully-Digital Beamforming (FDB), Hybrid Beamforming with perfect RF hardware (Perfect HB), Analog-only beam steering with perfect RF hardware (Perfect AB), and HB with three different cases of RF hardware imperfection: (a) quantization error caused by limited-resolution phase shifters; (b and c) additional residual transceiver impairments at BS and at both BS and UE, respectively. Right subfigure, spectral efficiency normalized to FDB. Color version available at wiley.com/go/molisch/wireless3e.
Reproduced with permission from [Molisch et al. 2017] © IEEE.

*22.10.3 Implementation Aspects – Load Modulators

Another interesting way to reduce the number of RF chains is to modulate not the baseband signal, but rather the load in the RF domain. To understand this approach, let us first consider traditional modulation. The equivalent circuit of almost any practical wireless modulation system is a time-variant voltage source attached to a time-invariant impedance, namely the RF circuitry which is in series with the radiation impedance of the antenna, compare Section 8.2.2. This setup provides proper matching between Z and R; specifically matching circuits are present to ensure that no reflections are caused by impedance steps. The information is then contained in a modulated baseband signal that is mixed with a sinusoidal local oscillator signal and sent via the time-invariant matching circuit to the antenna.

An alternative way of modulating the transmitted waveform would be to have a constant voltage source (i.e., a pure sinusoidal generator at the carrier frequency), amplified by a power amplifier, followed by circuits with time-variant impedance, see Figure 22.19. This approach is normally not used in standard transceiver design because of the (time-variant) impedance mismatch and the resulting reflections in the system. However, for a massive MIMO system, the many parallel loads average out by the law of large numbers. Specifically, the total admittance at the output of the circulator is

$$Y(t) = \sum_n \frac{1}{Z_n(t) + R} \tag{22.99}$$

which tends to a constant value as $N_{\mathrm{BS}} \rightarrow \infty$. Consequently, it is possible to directly modulate the load (matching circuit) of each antenna. At the time of this writing, prototypes based on this principle are under development.

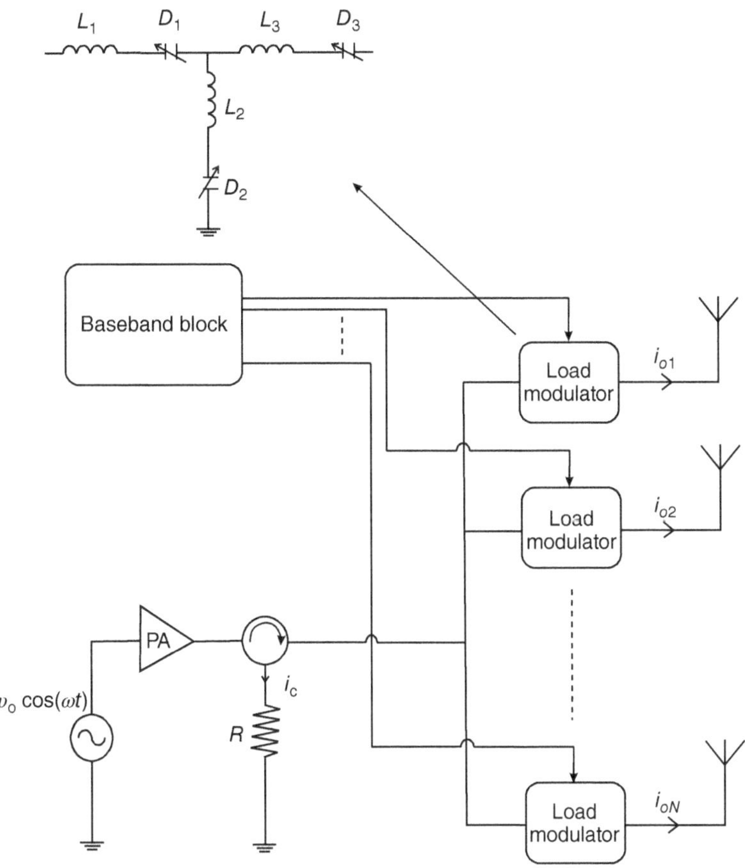

Figure 22.19 Equivalent circuit for load modulation.

*22.10.4 Low-Resolution ADCs

Another approach to reduce cost and energy consumption is the employment of low-resolution, possibly 1-bit Analog-do-Digital Converters (ADCs), as already discussed in Section 16.2.11. The quantization noise impacts both the channel estimation quality (which in turn leads to a reduction of the desired signal and an increase of the interference signal), and creates a quantization noise that needs to be taken into account in the total capacity expressions.

Similar to the channel estimation procedure of Section 16.2.11, we can define a Bussgang decomposition that describes the quantization as a linear operation plus an uncorrelated quantization noise, $\mathbf{z} = \mathbf{AHs} + \mathbf{An} + \mathbf{q}$, where

$$\mathbf{A} = \sqrt{\frac{2}{\pi}}diag\left[\gamma\mathbf{HH}^{\dagger} + \mathbf{I}_N\right]^{-1/2} = \alpha_\mathrm{d}\mathbf{I}_N \quad \text{where } \alpha_\mathrm{d} \approx \sqrt{\frac{2}{\pi}\frac{1}{K\gamma + 1}} \tag{22.100}$$

where the second equality holds under the assumption of i.i.d. channel coefficients and in the limit of large number of users K. The general equation for the SINR of the kth user in the uplink is

$$SINR_k = \frac{\gamma\left|\mathbf{w}_k^{\dagger}\mathbf{A}\hat{\mathbf{h}}_k\right|^2}{\gamma\sum_{i\neq k}\left|\mathbf{w}_k^{\dagger}\mathbf{A}\hat{\mathbf{h}}_i\right|^2 + \gamma\sum_{i=1}^{K}\left|\mathbf{w}_k^{\dagger}\mathbf{A}\left(\mathbf{h}_i - \hat{\mathbf{h}}_i\right)\right|^2 + \left\|\mathbf{w}_k^{\dagger}\mathbf{A}\right\|^2 + \mathbf{w}_k^{\dagger}\left(\mathbf{C}_{\mathbf{zz}} - \mathbf{AC}_{\mathbf{rr}}\mathbf{A}^{\dagger}\right)\mathbf{w}_k} \tag{22.101}$$

where $\mathbf{C}_{\mathbf{zz}}$ is the correlation matrix of the quantized signal and can be obtained, e.g., from (16.141) for the output of one-bit ADCs. In the limit of either low SNR, or large number of users, the following approximations hold

$$SINR_{\mathrm{MRC},k} = \frac{\gamma_\mathrm{d}\alpha_\mathrm{d}^2\sigma_{\mathrm{eq}}^2 N_{\mathrm{BS}}}{\gamma_\mathrm{d}\alpha_\mathrm{d}^2 K + \alpha_\mathrm{d}^2 + 1 - \frac{2}{\pi}} \tag{22.102}$$

$$SINR_{\mathrm{ZF},k} = \frac{\gamma_{\mathrm{d}}\alpha_{\mathrm{d}}^2\sigma_{\mathrm{eq}}^2(N_{\mathrm{BS}}-K)}{\gamma_{\mathrm{d}}\alpha_{\mathrm{d}}^2\left(1-\sigma_{\mathrm{eq}}^2\right)K + \alpha_{\mathrm{d}}^2 + 1 - \frac{2}{\pi}} \tag{22.103}$$

where we defined an equivalent channel estimation variance related to (16.139)

$$\sigma_{\mathrm{eq}}^2 = \frac{\alpha_{\mathrm{p}}^2 M_{\mathrm{p}}\gamma_{\mathrm{p}}}{\alpha_{\mathrm{p}}^2 M_{\mathrm{p}}\gamma_{\mathrm{p}} + \alpha_{\mathrm{p}}^2 + 1 - \frac{2}{\pi}}. \tag{22.104}$$

For the special case that $M_{\mathrm{p}} = K$ and no pilot boosting (i.e., pilot SNR γ_{p} equals data SNR γ_{d}), it can be shown that 2.5 as many antennas are needed for 1-bit ADC compared to the "infinite-resolution" case. However, this ratio can be somewhat reduced when optimization of resource allocation, e.g., more pilot tones, or larger pilot power, is incorporated. Figure 22.20 shows an example.

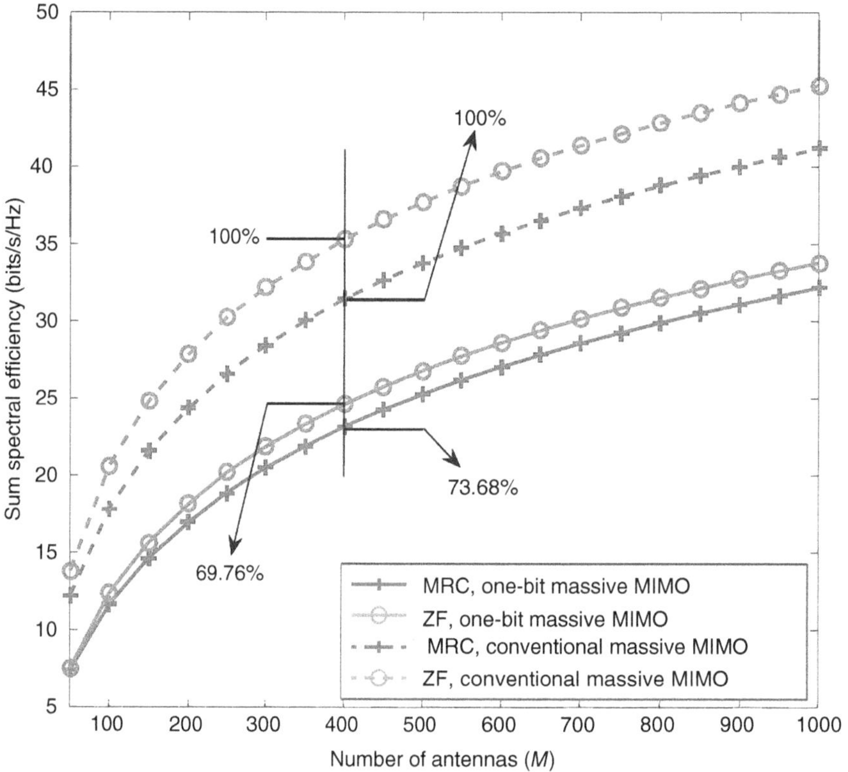

Figure 22.20 Comparison of uplink sum spectral efficiency versus number of BS antennas, with average SNR $\gamma = -10\mathrm{dB}$. Color version available at wiley.com/go/molisch/wireless3e.
Reproduced with permission from [Li et al. 2017b] © IEEE.

22.11 Base Station Cooperation and Distributed Antenna Systems

22.11.1 Principle of Capacity Increase

In any cellular system, the existence of multiple cells – and thus ICI – influences the capacity, and decreases the data rate that is possible for a single user. A first investigation by [Catreux et al. 2001] of a MIMO-based a cellular TDMA system with MMSE detection showed that the *cellular* capacity of a MIMO system (i.e., each BS and each UE has multiple antenna elements) is hardly larger than that of a system with multiple antennas at the BS only. The reason for this somewhat astonishing result is that in a cellular system with multiple antennas at the BS only, those antennas can be used to suppress ICI, and thus decrease the reuse distance, or increase the cell edge user capacity (see Section 22.1). For a cellular MIMO system with $N_{\mathrm{BS}} = N_{\mathrm{UE}}$, the degrees of freedom created by the multiple BS antennas are all used for the separation of the multiple data streams from a single user, or the intracell interference, and none for the suppression of interfering users from other cells. The ICI limits the overall performance.

To avoid the ICI, multiple BSs can be connected via backhaul links, so that they can collaborate and thus act as a single huge MIMO BS [Molisch 2001]. This effectively eliminates ICI and furthermore enhances the desired signal (similar to macro diversity in soft handover), thus ensuring high SINR even at the cell-edge. This concept can improve both cell-average and cell-edge throughput. It nowadays is mostly known as *CoMP*, though *BS cooperation, network MIMO*, and *cloud RAN* (Radio Access Network) are often used synonymously. CoMP is generally divided into two categories: CoMP-JP, and CoMP Coordinated Beamforming (CoMP-CB), depending on what kinds of information are shared among BSs. An illustration of CoMP-JP and CoMP-CB is shown in Figure 22.21.

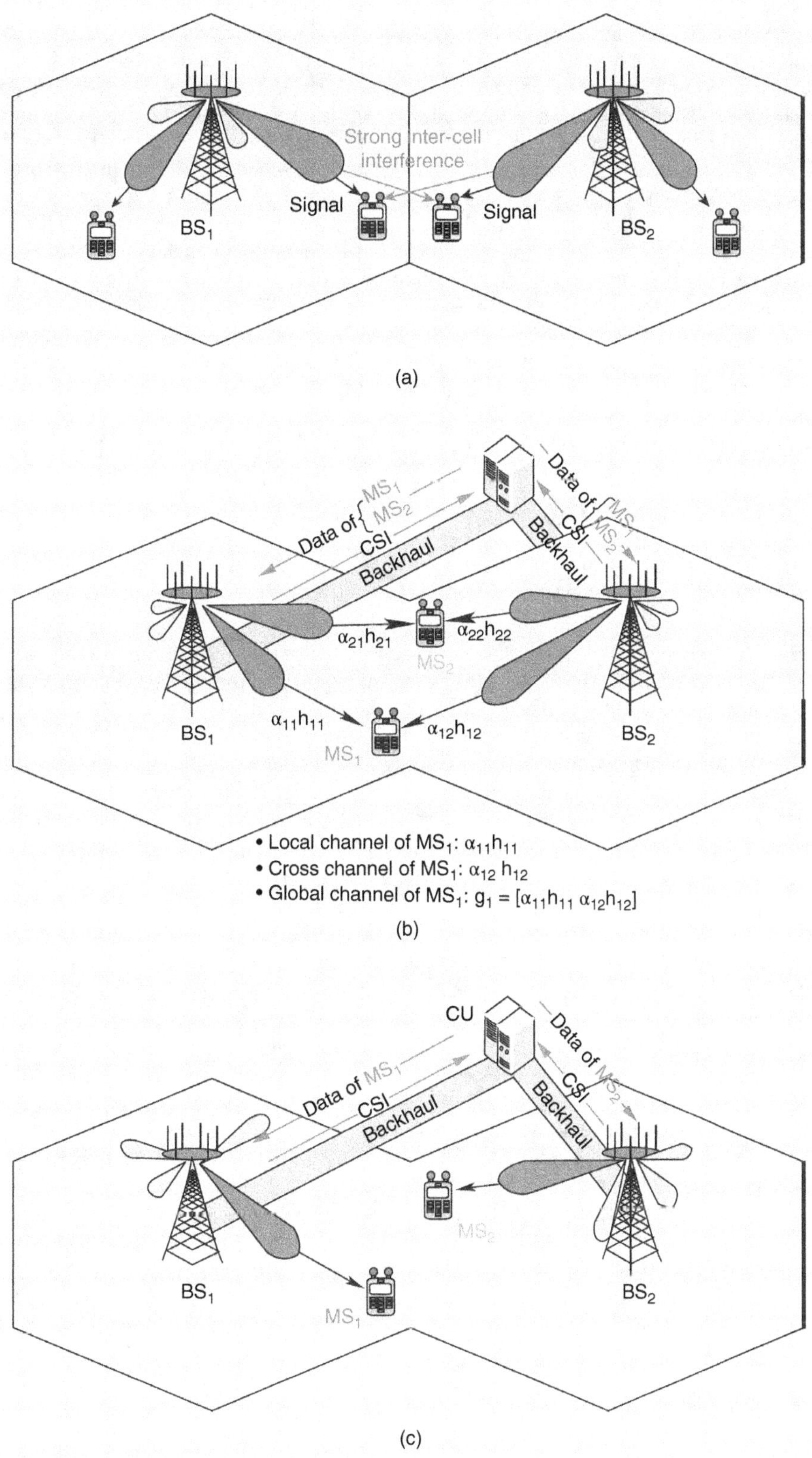

Figure 22.21 Illustration of Non-CoMP, CoMP-JP, and CoMP-CB. For the CoMP-CB systems, BS_1 is the serving BS of MS_1, from which the average channel gain is stronger. (a) Non-CoMP: no coordination between BSs, and strong interference might be caused to adjacent cell edge users. (b) CoMP-JP: BSs cooperate by jointly serving multiple users in their covered area, (c) CoMP-CB: BSs cooperate by avoiding interference to adjacent cell-edge users. The acronym MS in the figure is UE in the notation of this book. Color version available at wiley.com/go/molisch/wireless3e. Reproduced with permission from [Yang et al. 2013] © IEEE.

For CoMP-JP, both data and CSI need to be shared via backhaul links between each BS and a *Central Unit (CU)*. The CU could be a separate entity, or possibly co-located with one of the BSs. Alternatively, each BS can serve as a CU – a concept called decentralized cooperation. This approach allows the cooperating BSs to behave like a single large multi-antenna BS with distributed antenna elements. The distributed array forms beams towards all the users in its coverage area (i.e., the cells covered by all the cooperating BSs) simultaneously, employing all available BS antennas for each beam. Consequently, ICI is turned into desired signals. On the downside, this approach places high demands on the backhaul links and requires signal level synchronization as well as data level synchronization among BSs.

In contrast, CoMP-CB only needs the coordinated BSs to share CSI and the scheduling information. CoMP-CB, also called spatial ICIC, retains the concept of cells: in each cell, the BS forms beams towards the users in such a way that it not only increases the desired signal strength towards the desired user in its own cell, but also reduces interference towards the users in the adjacent cells. Since each BS only needs the CSI and scheduling information of the adjacent cells, the demands for backhaul links are greatly reduced. However, CoMP-CB only passively avoids ICI rather than proactively exploiting it. Further discussion is in Sec. 22.8.4.

Since the above description pointed out great similarities of downlink CoMP-JP to single-cell MU-MIMO, one might wonder why the manifold and well-explored techniques for implementing MU-MIMO cannot be applied in a straightforward manner. As a matter of fact, many of the same principles can be applied, but there are three important obstacles for such a simplistic approach: (i) single-cell MIMO differs in some subtle but important aspects from a true CoMP-JP setup, thus requiring changes in the transmission strategies; (ii) the acquisition of CSI is more difficult in CoMP systems, and (iii) there are restrictions on the sharing of information between the cooperating BSs due to the limitations of the backhaul network. These points will be discussed next.

22.11.2   Single-Cell MIMO versus CoMP-JP

CoMP-JP differs from single-cell MIMO system even if the BSs are synchronized and connected with perfect backhaul, which come from the distributed BSs and practical limitations.

- *Per-BS Power Constraint:* CoMP is subject to a *per-BS power constraint*, since the power limits are mostly given by regulatory constraint for a specific site, as well as hardware limitations (e.g., maximum power of the power amplifiers). On the other hand, most of the optimization of beamforming, scheduling and power allocation for single-cell MIMO is subject to a sum power constraint. Since power cannot be shared among BSs, directly applying the transmit strategies optimized under sum power constraint will lead to optimistic results, especially for heterogeneous networks including Macro, Micro, and Pico BSs with very different transmit powers. In homogeneous networks where multiple BSs have the same transmit power, the transmission schemes designed for maximizing the sum rate under per-BS power constraint generally perform close to those under the sum power constraint.
- *Asynchronous Interference:* Since BSs are not co-located, interference from different BSs received at each UE are asynchronous. This cannot be compensated by timing-advance technique conventionally applied in cellular systems, because the "degree of freedom" of the timing advance is used up by ensuring that the signals from the BSs arrive synchronously at a *desired* UE. For OFDM systems, this problem can simply be solved by prolonging the cyclic prefix. Considering that cell sizes nowadays are fairly small, in particular since CoMP is more desirable for high density networks, even such a prolonged cyclic prefix does not need to be very long.
- *Dynamic Clustering:* Due to the prohibitive complexity and overhead, it is not possible to allow all BSs (which might span a whole city) to cooperate. Moreover, a CoMP system with many BSs in a network exhibits negligible performance gain compared to one where only a few BSs are cooperating. As a pragmatic tradeoff, cooperative clusters can be formed, within which several adjacent BSs jointly transmit. The drawback of this technique is that there are now users at the "cluster edge" that show especially poor performance, and inter-cluster interference takes the role that ICI has in conventional systems. Dynamic clustering outperforms fixed clustering, but it leads to a dynamic overall number of transmit antennas. Considering the flexibility and scalability, channel training, and feedback mechanisms need to be redesigned for such a cooperative network. A related concept, *cell-free MIMO*, will be discussed in more detail in Section 22.11.5.
- *Non-i.i.d. Global Channels:* The global channel for each UE in a CoMP-JP system is a stacking of multiple single-cell channel vectors. The antennas at the different BSs yield a special non-i.i.d. global channel, where the average channel energies of the links between multiple BSs and each UE differ. As a result, the statistics of the global channel depend on the UE location, i.e., path loss and shadowing. This implies that algorithms for single-cell MU-MIMO that rely on the assumption of identical small-scale-averaged channel statistics between UE and the different BS antenna elements, might not be directly applicable.

22.11.3   Challenges Related to Channel Information Acquisition

As discussed above, the performance gain of MU-MIMO is largely dependent on the CSIT quality, and the same holds true for CoMP. To facilitate downlink spatial precoding and scheduling, the CU needs to gather CSIT from all coordinated BSs to all UEs in their serving cells. In TDD systems, the CSI is estimated at each BS by uplink training via exploiting channel reciprocity; in FDD, the CSI is estimated at each UE by downlink training, and then is fed back to the UE's serving BS via uplink channels, see Figure 22.22.

TDD Systems

TDD relies on reciprocity for CSIT acquisition, just like in the SU-MIMO and single-cell MU-MIMO case. In order to perform hardware calibration, self calibration (see Section 16.1.7) is a popular antenna calibration method in single cell systems. It ensures a

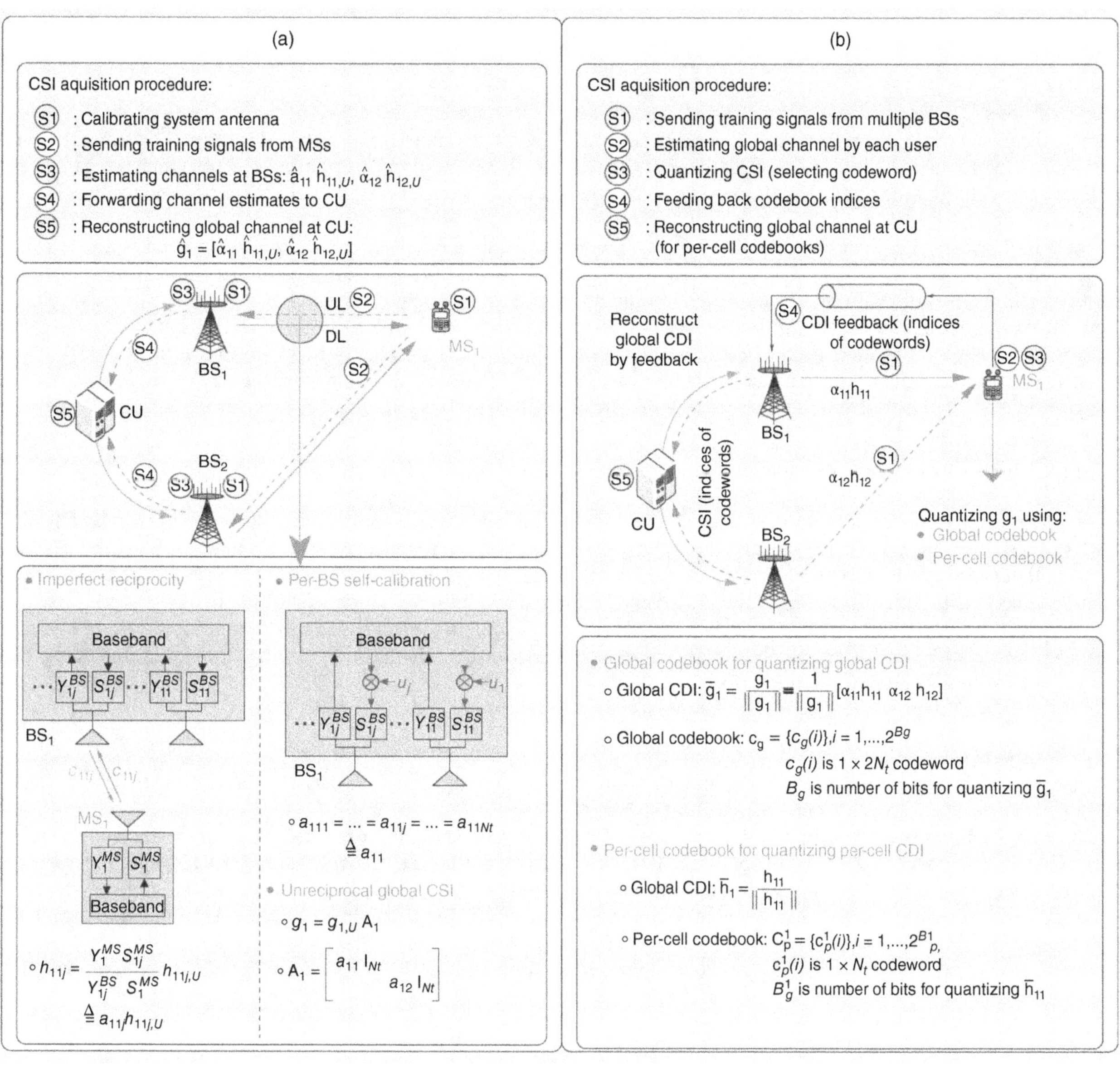

Figure 22.22 CSI acquisition procedures in TDD (a) and FDD (b) CoMP systems, where the specific steps from *S1* to *S5* are respectively shown in each of the procedure. The acronym MS in the figure is UE in the notation of this book. Color version available at wiley.com/go/molisch/wireless3e. Reproduced with permission from [Yang et al. 2013] © IEEE.

complex scalar ambiguity between the uplink and downlink channels for all antennas of this BS. In other words, as shown in Figure 22.22, the equivalent downlink and uplink channels between BS1 and UE1 are related by $\mathbf{h}_{11} = a_{11}\mathbf{h}_{11,U}$, where a_{11} is a complex ambiguity factor. As discussed in Section 16.1.7, such a scalar ambiguity does not affect the performance of single-cell single-user systems. However, when self calibration is employed at each BS in a CoMP system, it will lead to N_b ambiguity factors between the uplink and downlink channels at different coordinated BSs, e.g., a_{11} and a_{12} for a two BSs CoMP illustrated in Figure 22.22. Then the global equivalent uplink channel $\mathbf{g}_{1,U}$ and downlink channel $\mathbf{g}_1$ of UE1 is related by $\mathbf{g}_1 = \mathbf{g}_{1,U}\mathbf{A}_1$, where $\mathbf{A}_1$ is the diagonal ambiguity matrix.

The multiple ambiguity factors in the global equivalent channels lead to imperfect downlink global CSIT even if the uplink channel estimation is perfect. Such an ambiguity is more detrimental than channel estimation errors. Because it is a kind of multiplicative noise rather than additive noise, it will hinder co-phasing of coherent CoMP transmission.

One possible way to avoid this dilemma is over-the-air calibration, where the channel is estimated simultaneously through two different approaches: through reciprocity of propagation channels and through measurement on the downlink and feedback; knowledge of those two channel estimates allows to estimate multiple ambiguity factors. The performance of this method depends on the accuracy of the channel estimation. To improve the performance, the calibration should be obtained from the measurements of multiple uplink and downlink frames or of multiple users.

Furthermore, the interference experienced at the BSs and UEs are also not reciprocal. As a result, the actual SINR based on which the supported downlink *MCS* is selected, should be estimated at the UE then fed back to the BSs.

Training Overhead

To estimate the global channel for downlink cooperative transmission, the uplink training overhead is roughly N_b times the overhead of single-cell systems, where N_b is the number of the coordinated BSs. To ensure that the downlink spectral efficiency gain over Non-CoMP is not "eaten up" by the uplink training overhead, it is paramount to reduce the training overhead especially for relatively fast fading channels.

The trade-off between the performance gain of CoMP-JP and the required overhead for channel estimation is non-trivial. For a given coherence time, the training length as well as the number of cooperative BSs can be optimized such that they maximize the net throughput. Of course, the overhead reduction schemes discussed in Section 22.7 can be used for this case as well.

A simplified form of this parameter optimization is a switching between CoMP and Non-CoMP transmission modes. For users that require asymptotically zero training overhead (e.g., static users), CoMP-JP always outperforms Non-CoMP transmission. For users requiring high training overhead, the net data rate of some cell-center UEs under CoMP transmission may be lower than Non-CoMP. This suggests to develop transmission mode selection either from a system perspective, or independently from each UE's perspective.

FDD Systems

In order to save uplink resources for channel feedback, the feedback is done using a quantization codebook known at both BS and UE. In CoMP systems, the dimension of the global CSI may vary dynamically and the statistics of the CSI depends on each UE's location. This implies that every UE needs a unique codebook, and some attempts at adaptive codebook design using machine learning have shown promising results. Still, considering network scalability and compatibility, it is highly desirable to design a per-cell codebook based feedback strategy, where the existing codebooks can be reused to quantize each single-cell channel as shown in Figure 22.22. Though such a structured codebook is suboptimal, its performance can be enhanced by various techniques, including iterative approaches similar to those discussed in Sec. 16.1.7.

Feedback Overhead

To increase the network spectral efficiency, a fundamental question is: how much uplink overhead is required to achieve the downlink performance gain of CoMP-JP over single-cell MU-MIMO?

To reduce the feedback overhead, the non-i.i.d. feature of CoMP channels should be exploited. This can be realized by optimizing the sizes of the per-cell codebooks. To a first approximation, the feedback overhead of CoMP-JP is about N_b times of that of single-cell MIMO, which depends on user location. When a user has equal average per-cell channel gains, its overhead is largest. When a user is in the cell-center, its overhead is less because only the channels to one of the BSs is strong enough to need a fine quantization (large size codebook).

Alternatively, we can switch the transmission modes between CoMP-JP and Non-CoMP, or design spatial scheduling exploiting the channel statistics, as in the TDD case.

22.11.4   Imperfect Backhaul

In reality, the backhaul links among BSs are not perfect but limited by the capacity of the backhaul connection (which can be large if the backhaul is done via optical fiber, but much smaller for wireless backhaul or Asymmetric Digital Subscriber Line (ADSL) connections). The limited-capacity backhaul might not allow BSs to share a large amount of data. The CSI shared among BSs may have quantization error and severe latency if the BS-to-BS interface shows high latency, as is the case, e.g., in LTE (compare Chapter 31). The backhaul imperfection is even more severe in parts of heterogeneous networks, such as femto-cells, which hinders the application of CoMP. While most backhaul links could be upgraded by high speed optical fiber without technical challenges, this creates higher costs to the operators.

The impact of limited-rate backhaul is largest for uplink CoMP, since in that case (quantized) analogue signals need to be conveyed on the backhaul. Though not as stringent as in the uplink case, also downlink CoMP-JP might have to reduce its throughput to accommodate the capacity limitation of the backhaul links.

Switching between Different Transmission Modes

CoMP-CB needs much less backhaul capacity than CoMP-JP, since it only shares the CSIT, but not the user data. Considering that both of these schemes are able to mitigate interference, CoMP-CB may outperform CoMP-JP under stringent backhaul capacity constraints. Consequently, mode switching between these two CoMP transmission modes – adaptive to location and number of UEs – will provide better overall throughput.

Another natural way is to switch between CoMP-JP and Non-CoMP. Since cell-center users experience lower ICI, they will not benefit as much from CoMP as cell-edge users. Intuitively, we can simply divide the users in each cell with a threshold based on their average channel gains. Since only the users to be served by CoMP need to share their data among the BSs, the backhaul load can be controlled by judiciously selecting the threshold. Figure 22.23 shows simulation results of the mode switching, where the results for pure CoMP-JP and pure Non-CoMP schemes are shown as a baseline, either with or without considering the training overhead. It can be seen that for a severely limited backhaul, CoMP-JP performs worst, while becoming best for very high backhaul capacities.

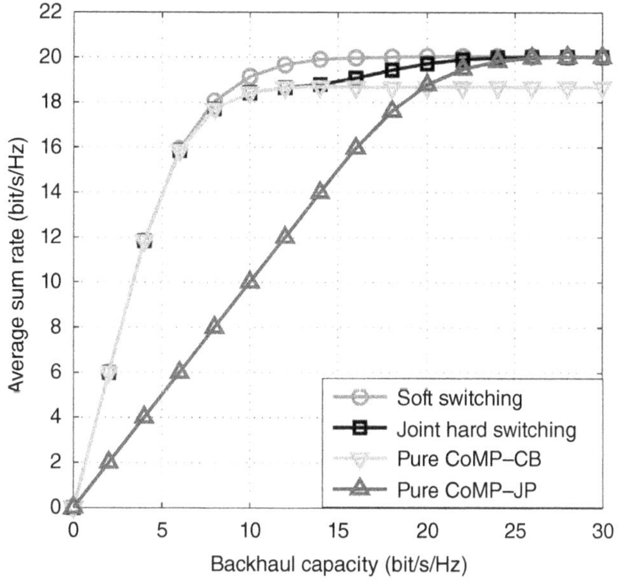

Figure 22.23 Average sum rate achievable in a ComP system with finite backhaul capacity, comparing pure CoMP-CB, pure CoMP-JP, and switching between the two. Color version available at wiley.com/go/molisch/wireless3e.
Reproduced with permission from [Zhang et al. 2013] © IEEE.

22.11.5   Cell-Free MIMO

In an ideal CoMP system, the concept of "cell" loses its meaning: the whole area covered by the (connected) BSs provides useful signals, and no signal to/from any BS constitutes interference. Furthermore, the number of antennas distributed throughput such a system is very large. For these reasons, it has recently (in the late 2010s) become common to call such systems *Cell-Free Massive MIMO* (CF-mMIMO). Figure 22.24 shows a more detailed architecture diagram, with a more hierarchical structure of the BSs. *Access Points* (APs), i.e., groups of distributed antenna elements are connected via a so-called "front-haul" to a CU, which in turn are connected via their backhaul to the core network. Some processing can be done at each AP, and further processing is implemented at the CUs. It must be noted that in case of infinite fronthaul/backhaul capacity and processing capability, the distinction between APs and CUs is irrelevant, and one can think of all antennas as being connected to an infinitely powerful processor. However, in the presence of constraints on those quantities, the detailed structuring starts to play a major role.

When comparing CF-mMIMO with traditional single-cell massive MIMO with the same number of antenna elements, we notice significant performance differences: (i) the path gain between a UE and the infrastructure antenna elements has a different structure:

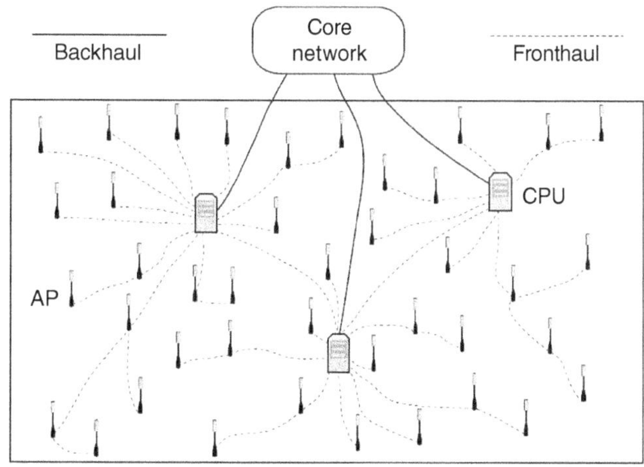

Figure 22.24 Architecture of a CF-mMIMO system. Color version available at wiley.com/go/molisch/wireless3e.
Reproduced with permission from [Demir et al. 2021] © NOW Publisher.

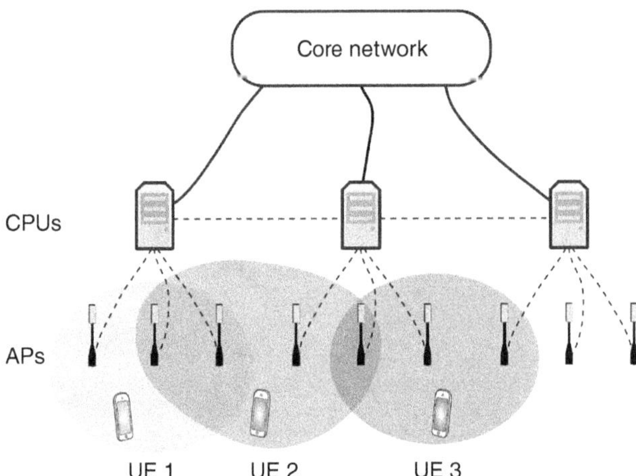

Figure 22.25 Principle of dynamic AP association in a cell-free network. Each UE is associated with the APs in its shaded region. Color version available at wiley.com/go/molisch/wireless3e.
Reproduced with permission from [Demir et al. 2021] © NOW Publisher.

in single-cell massive MIMO, the path gain and shadowing is the same for all antenna elements; it is large for UEs close to the BS and small for far-away UEs, while in CF-mMIMO there is always at least one infrastructure antenna element that is reasonably close to the UE; (ii) the probability for a low-rank channel between the UE and the CF-mMIMO system is much lower than for a concentrated massive MIMO system, even in LOS situations, due to the spatially distributed nature of the infrastructure array.

It follows from basic MIMO theory that the only fundamental limit to the number of UEs that can communicate with the infrastructure simultaneously is the total number of antenna elements. However, in practice there are three main limitations:

- Channel estimation effort, which grows linearly with the number of UEs in the system, as already discussed in Section 22.11.3. We stress that accurate CSI is critical for the proper working of CF-mMIMO, and that essentially all envisioned systems will be based on TDD and acquisition of CSI through reciprocity.
- Backhaul/fronthaul effort, which increases as signals from more and more antenna elements need to be hauled to a central processing location.
- Processing effort, as the dimension of the observed signal matrix (and thus the associated effort of linear or nonlinear processing) increases with the number of antenna elements.

In order to limit overhead and processing effort, there is still a requirement of limiting the number of APs a UE communicates with; in other words, each UE is associated with a *group of APs*; a network is often called "scalable" if the effort (per UE or AP) for channel estimation, the signal processing for data reception and transmission, and the backhaul/fronthaul effort remain finite as the number of UEs goes to infinity.

The traditional way of achieving this is to group the APs into fixed clusters, and associate each UE with a cluster; yet as we have seen above, the inter-cluster interference may limit the performance. Instead, CF-mMIMO selects dynamically the APs it communicates with, see Fig. 22.25; this selection can, e.g., be based on a maximum number of APs (e.g., the seven best ones); this method has the advantage that the processing effort and fronthaul/backhaul load is predictable. Alternatively, a UE might select all APs whose signals are above a certain power threshold, or ensure that the signals of the selected APs have a particular power ratio compared to the signals to the non-selected APs (resulting in a desired SIR).

Much of the mathematical treatment of massive MIMO, Section 22.9, can be applied directly to CF-mMIMO, to wit, all equations that do not rely on equal average channel gain between BS antennas and UEs; note, however that this makes strong assumptions about the channel knowledge, such as all UE-to-AP path losses. The impact of using only the signals from a subset of APs can be taken into account formally simply by forcing the combining weights for the signals from the non-participating APs to zero. Define a matrix

$$\mathbf{D}_{k\ell} = \mathbf{I}_{N^{(\mathrm{AP})}} \quad \text{for } \ell\text{-th AP processing signal to/from } k\text{-th UE}$$
$$= \mathbf{0}_{N^{(\mathrm{AP})}} \quad \text{otherwise}$$

that indicates when the signal from an AP is actually used for the processing or not; this is mainly introduced for notational convenience. The MMSE channel estimates are then a straightforward generalization of (22.71): we now aim to compute the channel estimate between the kth UE and the ℓth AP

$$\hat{\mathbf{h}}_{k\ell} = \sqrt{M_{\mathrm{p}} P_{\mathrm{TX},k}} \mathbf{R}_{k\ell} \boldsymbol{\Psi}_{k\ell}^{-1} \left[\mathbf{r}'_{\mathrm{p},k,\ell} \right] \tag{22.105}$$

where $\mathbf{R}_{k\ell} = E\{\mathbf{h}_{k\ell}\mathbf{h}_{k\ell}^{\dagger}\}$ is the correlation matrix between the kth UE and the ℓth AP (and can be seen as a generalization of the β_k), while the $\boldsymbol{\Psi}_{k\ell}$ is $\boldsymbol{\Psi}_{k\ell} = \sum_{i\in\mathcal{P}_k} M_{\mathrm{p}} P_i \mathbf{R}_{i\ell} + \sigma_{\mathrm{n}}^2 \mathbf{I}_{N^{(\mathrm{AP})}}$ is the received signal correlation matrix (where the summation is over all UEs that use the same pilot tone as UE k (including $i = k$); it can be interpreted as the generalization of the $1 + M_{\mathrm{p}}\gamma_{\mathrm{TX},k}\beta_k$ term in (22.71). From this, the error correlation matrix of the channel estimate

$$\mathbf{C}_{k\ell} = \mathbf{R}_{k\ell} - M_{\mathrm{p}}P_k\mathbf{R}_{k\ell}\boldsymbol{\Psi}_{k\ell}^{-1}\mathbf{R}_{k\ell}$$

can be computed. The normalized MSE of the estimate can be computed as $tr(\mathbf{C}_{k\ell})/tr(\mathbf{R}_{k\ell})$.

The channel estimation suffers from similar issues as the classical massive MIMO described in Section 22.9, including pilot contamination. UEs employing the same pilots create interference that distorts the channel estimates, and the resulting worsening of the MSE and capacity can be computed from the above equations. However, there is a new and challenging problem in the CF-mMIMO case, namely the assignment of pilot resources to the different UEs. Finding the optimum sets of UEs that use the same pilot resources, and can thus cause pilot contamination, is in itself a hard optimization problem. This problem is typically solved in an iterative manner, with a variety of possible algorithms. In any case, the quality of the pilot estimates differs with the path gain between the UE and the considered APs, with the highest-gain link having the best estimation quality. The signal processing for the reception or transmission of the payload signal thus needs to take into account the different estimation qualities of those channel estimates.

For the uplink with centralized processing, the desired signal power for the kth user is (in contrast to (22.78), the weights $\mathbf{w}$ are not normalized)

$$P_k\left|\mathbf{w}_k^{\dagger}\mathbf{D}_k\hat{\mathbf{h}}_k\right|^2 \tag{22.106}$$

while the noise and interference terms are Gaussian with standard deviation (compare (22.79))

$$\left(\sigma_{\mathrm{n}}^2\|\mathbf{D}_k\mathbf{w}_k\|^2 + \mathbf{w}_k^{\dagger}\left[\sum_i P_i\mathbf{D}_k\mathbf{C}_i\mathbf{D}_k\right]\mathbf{w}_k + \sum_{i\neq k}P_i\left|\mathbf{w}_k^{\dagger}\mathbf{D}_k\hat{\mathbf{h}}_i\right|^2\right) \tag{22.107}$$

where $\mathbf{C}_i$ is the correlation matrix of the collective channel $\mathbf{h}_i$ and the weights can be determined according to MRC, ZF, or MMSE criteria. Similar equations also hold for the downlink; for details we refer to [Demir et al. 2021].

There are numerous techniques for complexity reduction, i.e., keeping the computation scalable. For the computation of the combining weights, only the participating APs need to be taken into account, since the total channel matrix can be written as a block diagonal matrix where the serving APs receive the desired signals while the non-serving APs receive no useful signal from the considered UE. Furthermore, the number of UEs whose interference contribution is considered can be limited. This latter simplification is not optimal, but sometimes a reasonable approximation. Another form of complexity reduction can be obtained in the uplink by computing estimates of the data from each user at each AP; these estimates are then combined, at the Central Processing Units (CPUs), to give the final estimate. The simplest way of obtaining the total estimate is to simply add up the estimates from each AP; however, that might give too much weight to error-prone (due to noise and interference) APs. The combining weights thus should take the large-scale statistics into account.

Another topic of interest is how many antenna elements should be placed on each AP. Assuming that the total number of antenna elements is fixed, the two extreme cases are N distributed single-antenna APs, and a single N-element concentrated BS. A fully distributed setup enables the most spatially uniform SNR, since the distance to the nearest AP is smallest; however, the benefits of massive MIMO such as channel hardening are reached faster in a concentrated setup. Figure 22.26 shows some example results with randomly distributed AP and UE locations.

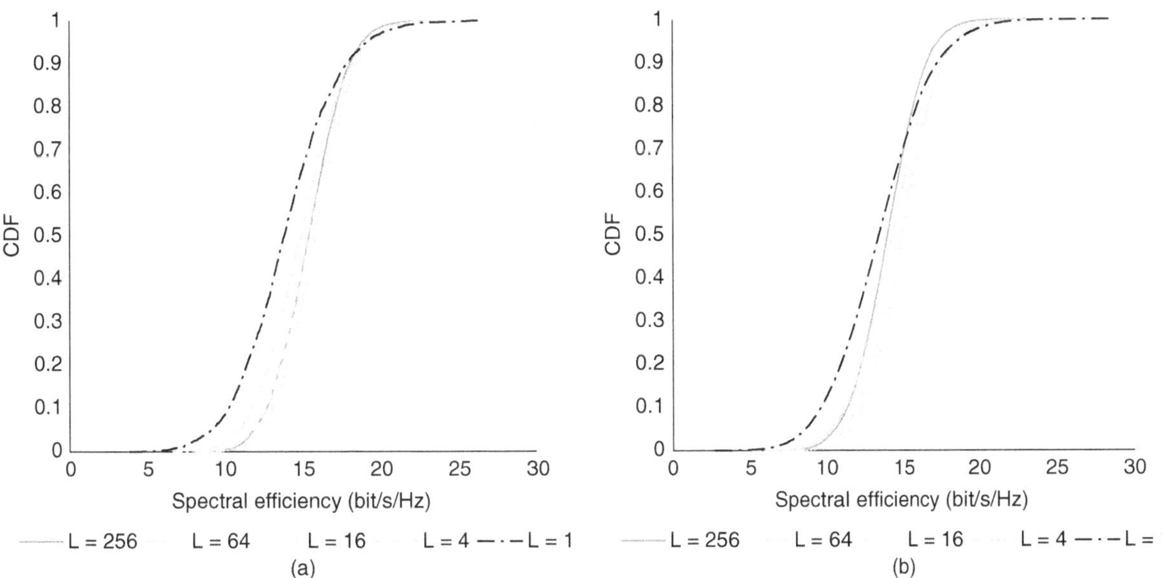

Figure 22.26 Spectral efficiency for 256 antenna elements distributed in a 1×1 km^2 area, for various numbers L of APs. (a) K = 8; (b) K = 64. Both APs and UEs are distributed as binomial point process (compare Section 21.4).
Reproduced with permission from [Ito et al. 2021] © IEEE.

Care must also be taken when selecting the channel model. While concentrated massive MIMO generally uses a "correlated Rayleigh" fading assumption, CF-mMIMO assumes that the fading at antenna elements of the same AP is correlated, but independent at the different APs, resulting in a block-diagonal structure of the correlation matrix. Furthermore, the commonly assumed combination of correlated Rayleigh fading, together with a power pathloss law, need not hold in practice – firstly, the dense deployment of APs can result in high LOS probability; secondly, the deployment of multiple APs along a street canyon in an urban environment can lead to large-scale pathloss that depends on the street orientation as well as the distance (compare Section 7.1). However, not many alternative models are available at the time of this writing, since the topic of channel modeling for CF-mMIMO is still in its infancy.

*22.12 Appendix

App. 22A: Smart Antennas for CDMA

See App22.pdf at wiley.com/go/molisch/wireless3e

Further Reading

An overview of early smart antenna techniques is given in the paper by [Godara 1997], and the paper collections [Rappaport 1998] and [Tsoulos 2001]. The topic of space-time processing, which forms a basis for smart antenna systems, involves many topics we have already discussed in previous chapters (diversity, Rake RXs, SU-MIMO); an overview can be found in [Paulraj and Papadias 1997]. Smart antennas for CDMA systems are discussed in [Liberti and Rappaport 1999].

The information-theoretic basis of MU-MIMO was laid in [Caire and Shamai 2003] and the work of Goldsmith and coworkers, which is summarized in the very readable overview article [Goldsmith et al. 2003]. An overview of both the information theory and the linear algorithms for MU-MIMO can be found in the excellent monographs of [Heath and Lozano 2018] and [Clerckx and Oestges 2013], Overviews of the multi-user downlink can be found in [Spencer et al. 2004a, 2004b, 2006] and [Gesbert et al. 2007]. A tutorial discussion of optimum beamforming is given in [Björnson et al. 2014], Leakage-based precoding was introduced by Sayed and coworkers, see, e.g., Sadek et al. [2007]. Iterative waterfilling was first suggested in the context of ADSL in [Yu et al. 2002]; see also the discussion in [Kobayashi and Caire 2006]; for MMSE-based multi-user downlinks, see [Shi et al. 2007] and references therein. WMMSE was introduced in [Shi et al. 2011].

The topic of scheduling for MU-MIMO is reviewed in [Castaneda et al. 2016] and – with particular emphasis on LTE – in [Li et al. 2014].

Massive MIMO was first suggested in the seminal paper by [Marzetta 2010]. Reviews and surveys of the topic are provided in [Rusek et al. 2012], [Larsson et al. 2014], and [Lu et al. 2014], and in particular the textbook [Marzetta et al. 2016] as well as [Björnson et al. 2017]. Considerable work was done in a series of papers by [Ngo et al. 2013a, 2013b, 2014]. Pilot contamination is reviewed in [Elijah et al. 2015]. A pioneering massive MIMO testbed is described in [Shepard et al. 2012]. Propagation channels for massive MIMO with long linear arrays were first measured in [Gao et al. 2012]. FD-MIMO is surveyed in [Kim et al. 2014]. Hybrid beam forming was first proposed in several papers by Molisch and co-workers ([Molisch and Zhang 2004], [Zhang et al. 2005], and [Sudarshan et al. 2006]); algorithms particularly for mm-Wave systems are discussed in [El Ayach et al. 2014]. JSDM was proposed in [Adhikary et al. 2013], and average-CSI based MU-MIMO systems in [Li et al. 2017a]. A survey is [Molisch et al. 2017]. Load modulation was proposed in [Sedaghat et al. 2014]. MIMO systems with low resolution ADCs are discussed in [Li et al. 2017b] and a comparison with hybrid beamforming is in [Roth et al. 2018].

An early (though non-public) suggestion of BS cooperation/CoMP is [Molisch 2001]. A number of algorithms for BS cooperation are reviewed in [Zhang and Dai 2004]. A pioneering paper on network MIMO is [Foschini et al. 2006]. A comprehensive survey is given in the book edited by [Marsch and Fettweis 2011], and the paper [Gesbert et al. 2010], as well as in [Yang et al. 2013]. Synchronization and calibration of distributed systems is discussed in [Rogalin et al. 2014]. Readers interested in the higher-layer aspects of CoMP (or Cloud RAN) find a comprehensive survey in [Checko et al. 2014]. CF-mMIMO is reviewed in [Demir et al. 2021].

For updates and errata for this chapter, see https://wides.usc.edu/students.html#textbooks

Exercises

See Sec. 36.22 of Exercises.pdf at wiley.com/go/molisch/wireless3e

23

Ad hoc Networks, Device-to-Device Communications, and Mesh Networks

23.1 Introduction and Motivation

A wireless ad hoc network, also often called a *Mobile Ad hoc NETwork* (*MANET*) is a decentralized network that does not use fixed infrastructure, such as Base Stations (BSs). Instead, each node can function as source, relay, or destination, depending on the requirements of particular data packets. In other words, "all nodes are created equal," and associated communications are *Peer-to-Peer* (P2P) communications. Ad hoc networks are complementary to the infra-structure-based cellular networks that are at the center of most of this book (and in particular Chapter 21). From this fundamental distinction, a number of important consequences arise

- The deployment of an ad hoc network can be considerably faster than an infrastructure-based network. Building up a cellular network can take years, especially since permits for BSs have to be obtained, but also the planning of the BS locations, providing the backhaul, power supply, cabling, etc., requires a significant amount of time and cost. In ad hoc networks the nodes are small and usually battery powered; locations of the nodes are not planned in advance.
- While the network topology of an infrastructure-based network is usually static and planned in advance (with the exception of self-organizing networks, see Section 21.7.4), ad hoc networks usually have a time-variant topology, though exceptions exist, e.g., mesh networks (Section 23.11).
- Infrastructure-based networks usually have a fairly strict requirement for the quality of service for the different users. High probability of coverage is a key goal for network operators, often mandated by frequency regulators. In ad hoc networks, on the other hand, reliability can rarely be guaranteed, since the location of the nodes is not planned. Again, exceptions to this rule exist, either in static planned networks like mesh networks or in networks where the location of the nodes can be controlled and adapted, e.g., in some drone networks.
- The distance over which direct transmission can happen between two nodes is usually much smaller in ad hoc networks, since both nodes are at street level (if in an outdoor environment), or at the same (low) height above a building floor (when indoors). As discussed in Chapter 7, this usually results in a higher pathloss than between a User Equipment (UE) and an elevated BS. Due to this shorter transmission distance, connections between a data source and a sink usually have to be routed, via multi-hop transmission, through the network.
- Resource allocation, scheduling, etc., usually has to be done in a decentralized way in ad hoc networks. This might reduce the administrative overhead but also reduces the spectral efficiency.
- Due to the lack of infrastructure, ad hoc networks are much cheaper than infrastructure based networks. Furthermore, since there is no single point of failure, they are more robust. This explains why ad hoc networks are popular for military applications (and as a matter of fact much of the theory grew out of the work of DARPA (the Defense Advanced Research Project Agency) in the 1970s.

An important variant of ad hoc networks, which has emerged since 2010, are Device-to-Device communications (D2D), where devices, i.e., noninfrastructure wireless nodes, are talking directly with each other, but they do so in coordination with, and possibly under instruction of, the infrastructure nodes. The infrastructure can interact with the D2D communication by providing an alternative transmission method – i.e., a device can communicate with another device either directly, or through a cellular BS. Alternatively, the BS can collect central *control* information, and determine resource allocation and scheduling, while only the payload transmission is done directly between the devices.

23.2 Applications

Ad hoc networks have a wide range of applications, and we mention here only a few key examples. In principle, all applications for wireless communications can be fulfilled with ad hoc networks, with all the pros and cons described in Section 23.1. However, the following applications are especially relevant:

23.2.1 Sensor Networks

Sensor networks in general are networks of nodes that have the ability to sense physical data (temperature, pressure, but also images), and communicate them to a data sink; this data sink often is a computer-human interface, or an automated control/monitoring system. Sensor networks can in general be implemented either using infrastructure-based, or ad hoc network architectures. Specifically, when using ad hoc implementation, the following key differences from "standard" ad hoc networks are commonly present:

- There is just a single data sink, even though there are multiple data sources.
- Multi-hop relaying is required, due to the shorter range of P2P communication. At the same time, the information obtained from the different sensing nodes is correlated; for example, temperature-sensing nodes in one room will record similar temperatures. For this reason, data aggregation and compression can be performed during the data-forwarding action.
- Sensor networks tend to be especially energy constrained (depending on the application), since the sensor nodes often have to survive for years on a single battery charge.

23.2.2 Health Care Applications and Body Area Networks

The use of sensor nodes on the human body is especially interesting in the context of health care applications. Simple applications include tracking of heart rate, exercise, etc. More advanced monitoring would include blood pressure, blood sugar levels, gait monitoring, and other information that could help with the monitoring and treatment of specific diseases. A specific aspect of such sensor networks may be that information is transmitted only between nodes mounted on the human body.[1] Furthermore, main concerns for health care applications are reliability and latency. It is nontrivial to trade-off these characteristics against energy efficiency.

The properties of such *Body Area Networks* (*BANs*) are mainly determined by the propagation characteristics to/from different parts of the human body. On one hand, the covered distances are small. On the other hand, there is a large lossy dielectric object, i.e., the human torso, between two nodes mounted on front and back of a person (similar, though smaller, losses occur, e.g., between shoulder and hip, etc.). The pathloss can thus be large, in particular at high frequencies. In indoor environments, the most significant propagation path is often not directly through or around the human body, but rather via reflections on the ceiling or walls of the room that the person is located in.

23.2.3 Gaming and Communication in Social Networks

In many situations, users that want to interact with each other are located in close physical proximity. Such a situation can easily occur in social networks, such as at parties (and while it might seem counterintuitive that people who can talk to each other in person would instead do so via electronic means, this is actually a frequent situation). Also, people playing video games against each other are often in the same room or at least close by. Cellular connections might not offer the data rate and latency required for some of these applications, and in any case, the spectral efficiency for sending information over a long-distance link to the BS, and from there back to another user (that is close to the first user) is much lower than a direct communication. Thus, ad hoc or D2D seems a suitable solution.

23.2.4 Emergency Communications

Natural disasters, such as earthquakes, tsunamis, etc., have shown the need for information and communication without infrastructure for emergency situations. In such large-scale disasters, communication infrastructures are generally useless due to the damage to BSs and backbone networks. Therefore, ad hoc networking needs to be used. It is noteworthy that Long-Term Evolution (LTE) Direct or Sidelink (a mode of the dominant 4G standard, which mostly deals with D2D communications) was motivated by a mandate of government regulators for infra-structure-less communications (though it ultimately included also more efficient infrastructure-facilitated communications for nondisaster communications), see Sec. 31.10.

23.2.5 Distributed Storage Systems

As many mobile devices, in particular cellphones and tablets, have considerable storage space on them, data files could be stored and retrieved from the mobile users themselves in order to offload traffic from the infrastructure network. Consider a cellular network

---

[1] It is popular nowadays to uses a smartphone as a "hub" (data sink). The processing and storage of the data may occur on the smartphone, or the phone might relay it to the cellular infrastructure.

where mobile users roam freely in and out of a geographically limited area. Assume that the mobile users themselves can be used to store (cache) data and they can, upon request, transmit data to one another. A set of mobile users that are within a specified distance from each other forms a storage community, or a local network. The local mobile users can communicate with each other through ad hoc networking or D2D communications.

23.2.6 Video Distribution

As video is becoming the most popular and data-hungry application of wireless communications, network structures have to be modified for optimizing the delivery. A variety of ad hoc network schemes can help with that goal. Obviously, groups of people that are in close physical proximity and want to exchange video files (e.g., a movie clip from a recent vacation) can do so via ad hoc networking; in this sense, video distribution is no different from other communications in social networks as discussed above. A second application for D2D is the use of mobile devices as caches for popular video files, where groups of users in a geographical region form a storage community, and exchange files through ad hoc communications (files that might be required by users but are not cached by the community could still be downloaded from an infrastructure node). Finally, D2D communications can also be used to extend the range of video broadcasts, or enhance the efficiency of such broadcasts, in that the transmission from the infrastructure nodes is optimized to reach a smaller group of nodes, which then relay the video to other nodes to which they have a good connection.

23.2.7 Vehicular Communications

Self-driving cars have garnered great attention over the past years. While much of the interaction with other cars is based on sensors such as radar and cameras, the direct communication via ad hoc networking is also very important. This is particularly true for interactions in non-line-of-sight situations, e.g., when a car can communicate to a number of following cars (some of which might not see the first one) that it needs to make an emergency brake, or when a car approaching an intersection can communicate to cars in the cross street.

Another aspect of vehicular communications is an efficient and scalable delivery system to distribute location-aware data for users in cars. For example, as a vehicle passes by a roadside infostation, it can download movies, maps, and brochures of local entertainment options. This part of the download could be considered either as infra-structure based or D2D link. Subsequently, this vehicle can distribute the information to other vehicles through ad hoc links.

23.3 Node Types and Hierarchical Structure

The conceptually simplest networks occur when all nodes have the same hardware, and the same functionality. This allows an easy design of transmission protocols, as well as a simplified computation of the network performance. However, such a setup is not always suitable for all applications. Neighbor discovery, scheduling, and other network functionalities can be greatly simplified if the network has a central controller. In many situations, some nodes can be equipped with more expensive hardware and/or have better network connectivity. For example, the widely used Zigbee standard for Personal Area Communications (Section 34.2) foresees *full function devices* and *reduced function devices*. Full function devices are obviously well suited to play a leading role in network formation. They often start up a network by sending out beacons (which give other nodes the possibility to link up to them), and act like a BS, but without themselves being connected to a wired backhaul network (and being small and lightweight enough so that they can be transported anywhere).

But even if all the devices have the same hardware, it can be advantageous to establish a "hierarchical" system, in which some nodes are assigned coordination roles. Typically the first node in an ad hoc network that actually wants to establish a connection (e.g., because it has data to transmit) will acquire the role of such controller (called *piconet controller* in some standards), as is done, e.g., in Bluetooth, see Section 34.1. The role of controller might either remain with this node for the duration of the network being active, or it can rotate among the different nodes in the network. The reason for the latter arrangement can be that the controller has a larger energy consumption, since it cannot go into a sleep mode as often as the other nodes in the network. To prevent the controller node from running out of battery too quickly, the controller role then needs to be rotated between different users.

Larger networks can have multiple controllers, each of which is controlling a subset of nodes, so-called *clusters*. In this case, the controller is called a *cluster head*; the grouping of nodes into clusters is usually based on the network layout and propagation conditions – nodes within a cluster should have strong and reliable connections to their cluster heads; furthermore, the cluster heads should also have strong connections with each other. The problem of clustering is a well-explored one in the computer science literature, and a wide variety of algorithms are available. When nodes are clustered, communications are performed in a hierarchical manner: a node sends its information first to the cluster head, which will forward it (directly, or via multi-hop) to the cluster head closest to the destination node, which in turn sends it on to the destination. Routing protocols for clustered networks are discussed in more detail in Section 23.6.8.

*23.4　Neighbor Discovery and Channel Estimation

23.4.1　General Considerations

Neighbor discovery is the first task in an ad hoc network, since it is a prerequisite for all other operations such as channel estimation, scheduling, adaptation of modulation and coding, etc. This is also required in BS-assisted D2D communications: even though devices have to make themselves known to the BS (as always in cellular networks), this does not obviate the need for the devices to discover their neighbors.

In ad hoc neighbor discovery, each device has to find all of its neighbors – this means that if there are K devices in the cell, each device has to discover which of the other $K-1$ devices are its neighbors. In total, there could be up to $K \times (K-1)$ potential links, so that fast discovery of neighbors is essential. The situation is further complicated by the mobility of devices, which changes the neighbor structure. Due to the short range of ad hoc links, the neighborhood of a device changes much faster than, e.g., the association to different BSs in a cellular link. Thus, neighborhood discovery must be repeated sufficiently often such that knowledge of the network topology is always up to date.

To provide a more precise mathematical description, let us define neighbor relation as follows: Two nodes are neighbors if each can hear the other with power greater than a certain threshold $\Gamma = \theta\sigma_n^2$, i.e.,

$$N(v_i) = \left\{ v_j : P_{\text{TX}}G_{\text{TX}}G_{\text{RX}}|h_{ij}|^2 \geq \Gamma \right\} \tag{23.1}$$

where P_{TX} is the transmission power (equal for all nodes), G_{TX}, G_{RX} are Transmitter (TX) and Receiver (RX) antenna (power) gains, and h_{ij} is the complex amplitude channel gain between nodes v_i and v_j; θ is the Signal-to-Noise Ratio (SNR) threshold (compare Section 21.4), and σ_n^2 is the noise power. The choice of the threshold has important implications for the efficiency of the discovery process, as well as the accuracy of subsequent operations. In principle, even very far away nodes have a finite link strength $|h_{ij}|^2$, and can thus later be either communication partners or effective interferers. However, detecting all "potential" neighbors, even if they are very far away, is very time consuming.[2] Setting the detection threshold thus involves a trade-off between missing potential neighbors and accepting inefficiencies in the discovery process.

Neighbor discovery algorithms can be categorized in different ways. We first distinguish between randomized vs. deterministic protocols. In the randomized algorithm, the nodes randomly choose to either transmit or listen in each time slot so that each node gets a chance to hear its neighbors as well as be heard by its neighbors, or – in order to save energy – to do nothing.[3] In deterministic protocols, each node has a fixed transmission sequence assigned to it and repeats this sequence multiple times.

We also distinguish between protocols that transmit Identifiers (IDs) as part of their *discovery packet*, and those whose identity is determined from the sequence of discoverable signals being sent out. In the former case, each node transmits a sequence of bits in each *time unit*, each of those sequences contains information like a Medium Access Control (MAC) address, IP address, mobile device ID, etc. Thus, if the RX can receive one such discovery packet, it can completely identify the neighbor that had transmitted such a packet. The alternative is that the pattern of transmissions (e.g., the times at which pulses are transmitted) itself identifies the transmitting node, not the bits sent out during that transmission. The latter case is less used in practice, since establishment of an information source ID is usually part of protocols of commercial network standards, but is often convenient for theoretical analysis.

23.4.2　Randomized Protocols

One of the most influential early approaches was the *birthday protocol*, which is based on randomly transmitting/receiving discovery packets or beacon signals. The algorithm draws its name from the fact that even for a rather small group of people, the probability is high that at least two of them have the same birthday (for a group of 23 randomly selected people, this probability is 50%). Let us now consider how that translates to a neighbor discovery problem. Assume a slotted transmission structure (similar to slotted ALOHA, see Section 18.4.1).[4] Now let one node transmit randomly in k out of n available timeslots; while the other listens in randomly selected k slots; both nodes remain idle for the remaining slots (e.g., to save energy). The probability that one node hears another is

$$1 - \frac{\dbinom{n-k}{k}}{\dbinom{n}{k}}. \tag{23.2}$$

Even with a fairly small ratio k/n, the probability of hearing each other is quite high.

---

[2] As outlined below, a threshold can be determined below which it is information-theoretically optimum to consider interference as noise; however, determination of the threshold requires knowledge of the channels – this approach is thus suitable for setting thresholds in scheduling, but not for neighbor discovery.

[3] Almost all currently existing devices cannot listen and transmit at the same time on the same frequency.

[4] In a pure ad hoc network, establishing a slotted structure before neighbor discovery is not easy, since no common time basis exists. Explaining discovery algorithms based on slotted structures thus mainly serves to simplify the mathematics and provide easier intuitive explanations.

More generally, consider now a network where each node transmits with probability p_t, listens with probability p_r, and stays idle (sleep) with probability p_s. Assume now that there are $K \times (K-1)$ links, and a total of n timeslots are available for discovery. We furthermore assume that a discovery can only take place if exactly one discovery packet is transmitted by a neighbor, i.e., there are no collisions between packets coming from different neighbors.

The number of TXs in a K-node neighborhood that transmits is a random variable with binomial distribution. Therefore, the probability that exactly one node is transmitting is

$$\binom{K}{1} p_t (1-p_t)^{K-1} \tag{23.3}$$

and the number of listeners (conditioned on having one transmitting node) follows from the fact that there are $K-1$ possible listeners, each of which has a probability of $\frac{p_r}{p_r + p_s}$ to be in receiving mode. This results in a binomial distribution of the number of listeners, which has an expectation

$$(K-1) \frac{p_r}{p_r + p_s}. \tag{23.4}$$

Since $p_r + p_s = 1 - p_t$, the probability of having a neighbor discovery within a certain timeslot, D_{ts}, is

$$D_{ts} = K(K-1) p_t p_r (1-p_t)^{K-2}. \tag{23.5}$$

One might now be tempted to think that the number of discoveries in n timeslots is simply nD_{ts}. However, discoveries in different timeslots might be duplicates of each other. Rather, each discovery is the discovery of one of $K(K-1)$ bidirectional pairings. For a large number of timeslots, the overhearing event on the available links becomes a Poisson process with mean

$$\lambda = \frac{nD_{ts}}{K(K-1)} = np_t p_r (1-p_t)^{K-2} \tag{23.6}$$

and the fraction of discovered links is

$$1 - \exp(-\lambda) = 1 - \exp\left[-np_t p_r (1-p_t)^{K-2} \right]. \tag{23.7}$$

The variables p_t and p_r can now be optimized, using various constraints. Firstly, as the number of nodes K increases, the probability of transmission should decrease as $p_t = 1/K$; otherwise, the probability of collision becomes too large. Secondly, the best choice of p_r is $1 - p_t$, i.e., when a node is not transmitting, it should be listening to the transmissions of other nodes. This is obviously the best choice from a discovery point of view, but it is not necessarily energy efficient. Constraints can be put on the sleep probability to guarantee certain energy savings.

23.4.3 Deterministic Protocols

The main drawback of randomized neighborhood discovery protocols is that they cannot provide performance guarantees: while *on average* a high percentage of links can be discovered quickly, the *worst-case* time it can take for detection can be large. For this reason, deterministic protocols have become popular, where nodes transmit their IDs according to a predetermined transmission pattern. There are two aspects to the deterministic patterns: (i) how to ensure detection between two nodes while minimizing energy, and (ii) how to handle or avoid collisions between different nodes.

For the first problem, call slots in which either transmission or reception is happening as *active*, and other slots as *idle*, and assume that detection happens when the active slots of two nodes overlap. The main difficulty in optimization of the sleep pattern is that the nodes are not synchronized, so that their active and idle times are not naturally aligned (however, they *may* be synchronized by chance – a situation that also has to be incorporated by the discovery protocols).

A simple solution to guarantee detection is to have each node be active in more than half of all slots (the "51%" solution, see Figure 23.1); however, this is not an energy-efficient solution. The question is thus how a lower duty cycle can be achieved while still guaranteeing detectability; in other words, how can we ensure that for any two given nodes, there is at least one slot in which both nodes are active. In the *Searchlight* protocol, a group of slots is divided into cycles. During each cycle, a node is active in both the anchor slot (which is always the same, denoted as "A" in Figure 23.2), and in the probe slot (denoted as "P" in Figure 23.2), whose position changes from cycle to cycle. Assuming there are N slots, the offset between two anchor slots cannot be larger than $N/2$ (assuming N even), so that it is sufficient to send out $N/2$ probe slots to ensure discovery, i.e., overlap between active slots of two nodes, and the total detection time is $N^2/2$. The active slots are stretched out over a longer period of time, which reduces the duty cycle; at the same time detectability is still guaranteed, see Figure 23.2.

Another possible approach is the *Quorum* protocol: consider a group of m^2 slots. Arrange those now in a matrix, and let each node pick one row and one column of entries as the slots in which they are active; in the remaining slots, the node is idle. Obviously, a duty cycle of

	0	1	2	3	4	5	6	0	1	2
A	x	x	x	x				x	x	x
B		x	x	x	x				x	x
C				x	x	x	x			
D	x				x	x	x	x		

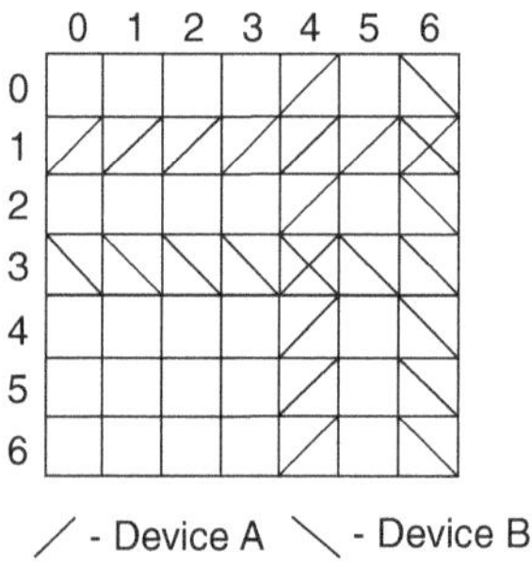

Figure 23.1 Guaranteed detectability when nodes are on more than half of the time.

	0	1	2	3	4	5	6	7
Cycle 1	A	P						
2	A		P					
3	A			P				
4	A				P			

Figure 23.2 Active slots in the searchlight algorithm.

$(2m-1)/m^2$ is obtained. As can be seen from Figure 23.3, the rows/columns of the two nodes intersect in two slots (unless the two nodes happened to select the same rows/columns, in which case there is more overlap). In other words, no matter what the relative offset between the starting times of the different nodes; discovery is guaranteed (assuming periodic repetition of the group of slots). Drawbacks of this algorithm are that (i) the mean discovery latency is large, and (ii) it assumes that all nodes in the network use the same duty cycle.

Figure 23.3 Operating principle of quorum discovery.

Another set of algorithms is based on the principle that the length of a transmission cycle is a prime number, and only transmits in the first slot of a cycle. As long as two nodes picked different primes (i.e., different cycles), detection is guaranteed. However, problems can occur if two nodes pick the same prime p, but the starting time of their cycle is different – in that case, the nodes are never active at the same time. A refinement of that idea is the "Disco" protocol, where each node picks two primes p, q and is active at integer multiples of both primes. Now even if both nodes happen to pick the same primes (and have a timing offset), detection is guaranteed since $p \neq q$, so sooner or later the two nodes will be active at the same time.

Figure 23.4 shows the cumulative distribution function (cdf) of the discovery latency for a 5% duty cycle for the different protocols for active/idle slots. We see that the (randomized) birthday protocol provides a very low mean discovery time, but has no guaranteed performance, meaning discovery times for perfect discovery can become very large.

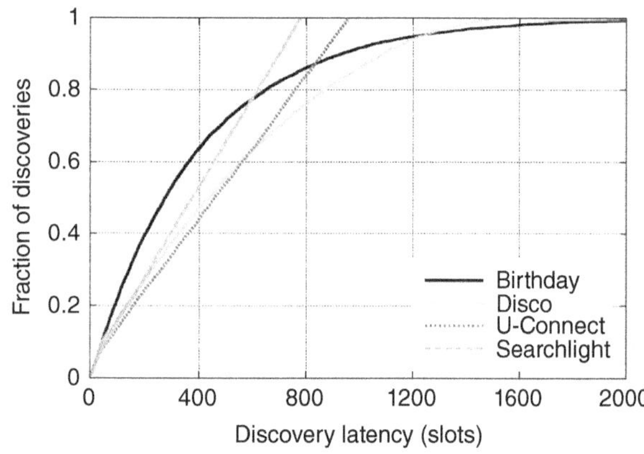

Figure 23.4 Latency of discovery with various methods with 5% duty cycle. Color version available at wiley.com/go/molisch/wireless3e.
Reproduced with permission from [Sun et al. 2014] © IEEE.

Turning now to the problem of collision handling, an example is the *"ZigZag"* algorithm. Nodes transmit and receive according to a predetermined *on–off* pattern. Thus, the nodes may cancel the known IDs (encoded as signal waveforms) from received signals to retrieve IDs of other neighbors. The principle is shown in Figure 23.5, where Nodes v_1, v_2, and v_3 are neighbors of v_0 with IDs s_1, s_2, and s_3. In Figure 23.1, node n_0 can discover the three neighbors: from the transmission in the third time slot, it can determine s_3. This also allows to use the signal received in timeslot 2: by subtracting the now-known s_3 from the total received signal, the collision can be resolved, which allows to determine s_2. This in turn enables to consider the signal in the first slot, subtract s_2, and determine s_1.

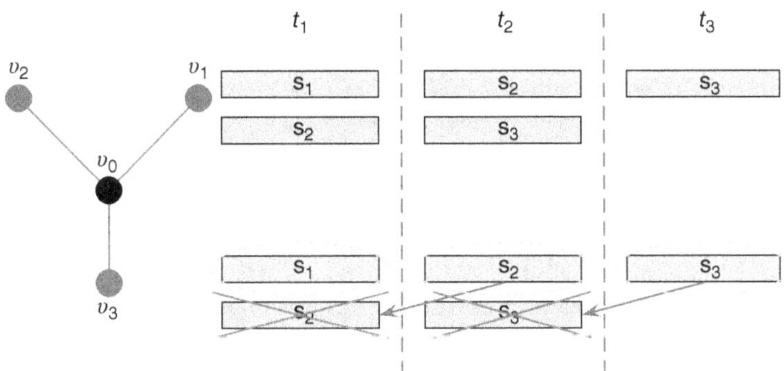

Figure 23.5   Operating principle of Zig-Zag discovery. Color version available at wiley.com/go/molisch/wireless3e.

The challenge consists of designing the transmission patterns so that the discovery time remains short while all nodes manage to detect all their neighbors with high probability. Construction methods that rely on the connection between Low Density Parity Check (LDPC) codes (compare Section 13.7) and on/off patterns have been suggested in [Tehrani and Caire 2014].

23.4.4   Channel State Information Acquisition

In contrast to cellular networks, where only the Channel State Information (CSI) between devices and BS is required, in ad hoc and D2D networks, the CSI *between* devices is needed. This poses new challenging problems; even if a BS exists, it cannot infer the state of the links between devices from direct observations. For example, the link between two devices could be bad due to shadowing, while their links to the BS are fine. Thus, it is necessary to perform estimation of the channels between devices, and subsequently report the results to the BS. Yet the question remains how often, and in what form, pilot transmission and the associated CSI training should occur; temporal variability of the channel variation and the broadcast effect makes "straightforward" solutions inefficient. For instance, Time Division Multiple Access (TDMA), where each device in the cell would get a separate timeslot for transmission of its pilot sequence, reduces the interference between devices but increases channel outdatedness due to the long times between pilot sequences of each particular device. On the other hand, contention-based techniques (e.g., Carrier Sense Multiple Access (CSMA)) might reduce the delay of CSI acquisition but the quality of estimated channels might poor due to interference. For these reasons, the acquisition of CSI between *all* devices is rather difficult.

To provide a general formulation, the acquisition of CSI can also be viewed as a scheduling problem, where every device has to transmit a pilot signal to its neighbors. Scheduling the pilot tones would mean to determine the optimum sets of devices that can transmit simultaneously without exceeding an interference threshold. This problem is similar to the traditional scheduling problem (i.e., when which node should transmit their data) in ad hoc networks, which is known to be NP hard. Approximate solutions have been studied thoroughly for the purpose of maximizing the throughput of data communication, see Section 23.5. However, those heuristics cannot be easily reused for pilot scheduling, because the quality of the CSI has an impact on the *subsequent* communication, leading to a change in the objective function. Also, many scheduling schemes discussed in the literature assume the knowledge of interference levels between devices, information that is not available during the CSI acquisition process, because we know it only *after* CSI is acquired/known.

One solution proposed in the literature is the Location Aware Training Scheme (LATS) scheme, which exploits the geographical information in order to optimize the scheduling of training sequences for CSI acquisition. LATS uses the spatial reuse concept (Section 21.1) to trade off interference and latency. Increasing distance leads to reduction of interference and thus enables reusing the same time/frequency resources for the pilots for far-away node pairs. To simplify the problem, LATS groups devices into segments of equal size, and performs scheduling of those segments. Note that it requires knowledge of the location of the devices, which could be obtained, e.g., from Global Positioning System (GPS) (compare Chapter 29).

The detailed setup of the scheme can be seen in Figure 23.6. The coverage area is divided into equal size square segments with L_s side length. The communication disk, i.e., the area centered around a device within which neighbors lie, is assumed to consist of an integer number of segments. In order to reduce interference, LATS defines the minimum separation distance between devices that are scheduled simultaneously through use of a guard distance.

By writing the Normalized Mean Square Error (NMSE) as a function of the geometry parameters (segment size, communication disk length, guard distance) as well as the mobility parameters, those geometry parameters can be optimized.

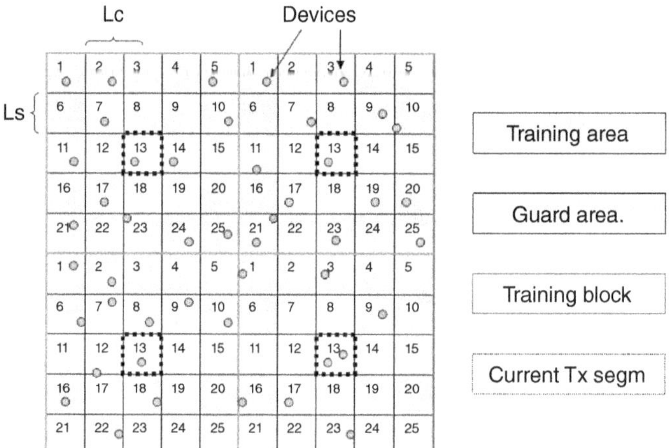

Figure 23.6 Simplified segmented cell and LATS cell components. Devices in training set 13 are allowed to transmit. Color version available at wiley.com/go/molisch/wireless3e.

Reproduced with permission from [Burghal and Molisch 2016] © IEEE.

23.5 Scheduling of Single-Hop Transmissions

23.5.1 Problem Formulation

Once the network topology is known (neighbor discovery) and the CSI for all links has been acquired, the network needs to determine the scheduling, i.e., which nodes are allowed to transmit at what time. Scheduling in ad hoc networks is complicated by the fact that the transmission from any link (i.e., transmission between two nodes) might interfere with the transmission of other links. One typical objective of these scheduling algorithms is to find a maximal feasible subset of links, i.e., a set of transmissions that can simultaneously co-exist with each other while maintaining a sufficiently large Signal-to-Interference Ratio (SIR), among all the given one-hop links that have data to transmit. One popular way to make the problem more tractable (and draw on a wealth of algorithms from the computer science literature) is a binary interference (protocol) model. In this model, if a given link A interferes "too strongly" with another link B, then link B cannot be scheduled at all; by "too strongly" we define that if the D2D link A transmits, then the received signal power at the RX of link B from the TX of link A is higher than the interference threshold Γ.

The interference situation is pictorially described by an interference graph, in which links are denoted as vertices $\{A, B, \cdots\}$; while edges $\{(A, B), \cdots\}$ represent "too strong" interference between two links, see Figure 23.7. Note that generally a conflict graph can be directed; for example, link B can interfere too strongly with link C (because the TX of link B is too close to the RX of link C), while activation of link C does not interfere with link B. However, if we require that a link is activated only if it *both* does not interfere with a prior link, *and* does not suffer from interference from another link, C cannot be scheduled when B is active.

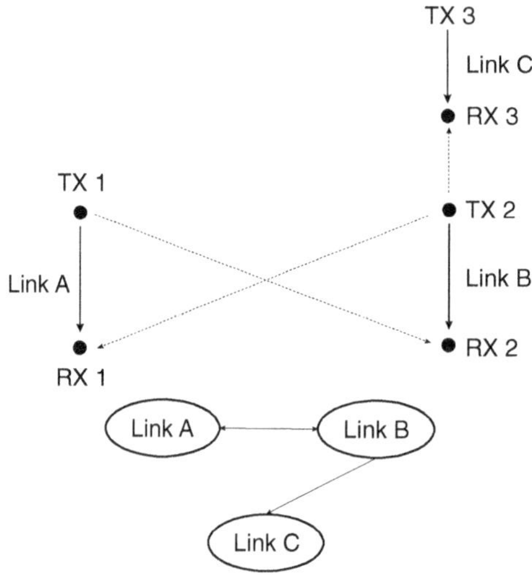

Figure 23.7 Conflict graph construction.

A main challenge in this formulation is the determination of the interference threshold; the choice of different thresholds results in different algorithms and performance. Furthermore, setting a threshold for each single link ignores the fact that interference from different concurrently active TXs accumulates, and the interference from each TX depends on the transmit power as well as the channel state. Thus, it might seem preferable to compute the actual interference level under the scheduling schemes and then choose the scheduling that optimizes the capacity resulting with this interference. While in principle more exact, this leads to computationally extremely hard problems, and suitable approximations need to be used.

23.5.2 FlashLinQ

Currently, the best-known D2D scheduling protocol is *FlashLinQ*. It is a distributed algorithm that schedules D2D links according to their priorities, such that the higher-priority links do not suffer from significant interference of possibly scheduled lower-priority links. Theoretically, it can guarantee the maximum number of activated D2D links. The FlashLinQ scheduler is designed to find a feasible subset of nodes that can transmit at the same time in a distributed manner, under the assumption that all channel states are known, and no power control is used.

The algorithm works in the following steps:

1. Each link is assigned a priority. These priorities can be derived from the priority of the application (e.g., an emergency call would automatically get highest priority), or they can be randomized and made time-variant (such that fairness, averaged over a longer time period, is provided), or some mixture thereof.
2. The highest-priority link is scheduled for sure.
3. The second-highest priority link is scheduled only if (i) it does not create excessive interference to the RX of the higher-priority link, and (ii) it does not suffer excessive interference from the TX of the higher-priority link.
4. Then the next link (in order of priority) is considered according to a similar criterion, and so on, until the last link is scheduled or rejected.

In other words, for a given TX–RX pair (as denoted by Tx-A and Rx-A in Figure 23.8), we need to determine which other links (Tx-B and Rx-B in Figure 23.8) are allowed to simultaneously transmit without generating too much interference at Rx-A. The FlashLinQ mechanism can be interpreted as each RX having a guard zone, which is a circular region where no other TX can simultaneously transmit. Link A conflicts with (i.e., is a neighbor to) link B if and only if link A's TX falls into the guard zone of link B's receiver or link B's TX falls into the guard zone of link A. Clearly, it is critical to define the right radius of the guard zones to maximize the spatial reuse of the system; heuristics have been discussed in the literature.

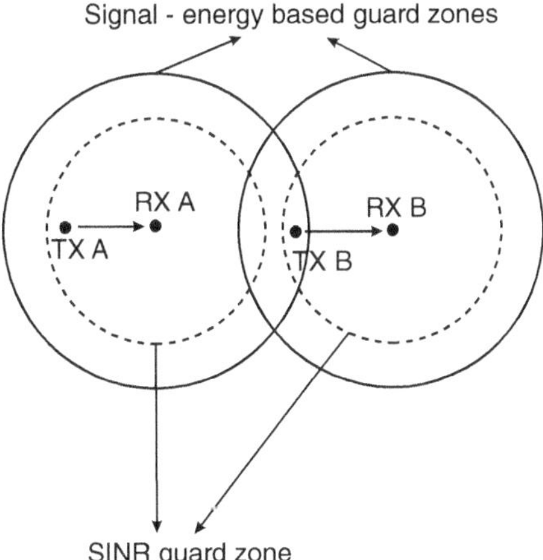

Figure 23.8 SINR-based Guard Zone vs. Energy-based Guard Zone.

23.5.3 ITLinQ

The ITLinQ algorithm exploits insights for when it is information-theoretically optimum to treat interference as noise and uses it to schedule users simultaneously only if it results in interference that is optimally treated as noise. For this to occur, it is required that

$$\mathrm{SNR}_i \geq \max_{j \neq i} \mathrm{INR}_{ij} \max_{k \neq i} \mathrm{INR}_{ki} \quad \text{for} \quad \forall i = 1, \dots. K \tag{23.8}$$

where SNR_i is the SNR on the ith link, and INR_{ij} is the interference-to-noise ratio created by source j at destination i. An optimum subset S for transmission in ITLinQ thus requires that for any link i in that set

$$\mathrm{SNR}_i \geq \max_{j \in S \setminus \{i\}} \mathrm{INR}_{ij} \max_{k \in S \setminus \{i\}} \mathrm{INR}_{ki}. \tag{23.9}$$

However, this requires central knowledge of all the SNRs and INRs. A distributed approximation can be done with the same algorithm foreseen with FlashlinQ, though the test of whether a link is admissible uses different criteria. To verify that a destination j does not receive excessive interference from already-scheduled links

$$\mathrm{INR}_{ji} \leq \mathrm{SNR}_j^\kappa \ \text{ for } \ \forall i < j$$

where κ is a design parameter; $\kappa = 0.5$ is a possible, but often too pessimistic, solution. To verify that the link j must not create excessive interference to existing links i

$$\mathrm{INR}_{ij} \leq \mathrm{SNR}_i^\kappa \ \text{ for } \ \forall i < j.$$

For comparison, remember that the conditions of distributed scheduling of FlashLinQ are of the form:

$$\frac{\mathrm{SNR}}{\mathrm{INR}} \geq \Gamma. \tag{23.10}$$

While FlashLinQ performs well in many scenarios, it requires heuristic adaptations of its threshold parameters. ITLinQ provides a theoretically justified choice of the scheduling that does not need parameter adaptation [Mungara et al. 2015], and provides significant gains over FlashLinQ in a number of situations.

23.5.4 CSMA, TDMA, and ALOHA

Besides the above-mentioned optimized protocols, the contention-based protocols discussed in Section 18.4 can be used in ad hoc networks. CSMA, as discussed in Section 18.4.2, is also very popular for ad hoc networks, especially due to its use in IEEE 802.11 (Wi-Fi) standards. However, the efficiency in ad hoc networks is questionable. Request to Send (RTS)/Clear to Send (CTS) (at least as implemented in 802.11) is often too conservative and reduces the efficiency of the multiple access, but switching off this functionality (as often done in mesh networks) leads to problems with hidden nodes.

A very simple method of scheduling is ALOHA, as discussed in Section 18.4.1. Due to the lack of required sensing or coordination, the efficiency is not good (similar to the case of multiple access to a BS) especially when the overall throughput is large. However, for small offered traffic, the simplicity of ALOHA can still be attractive.

TDMA-based scheduling is generally not suitable for ad hoc networks because of the bursty nature of the traffic (compare Chapter 20). However, for mesh networks (see Section 23.11), traffic is more stable over time, so that TDMA can make sense. Scheduling in TDMA is somewhat similar to the scheduling problems described above, i.e., we need to find independent sets in the conflict graph so that interference is limited, and different sets are assigned different timeslots.

23.6 Routing and Resource Allocation for Multi-Hop Networks

We now turn our attention to larger networks, where a single transmission is not sufficient to get the information from the source to the destination. Rather, we have to use multiple relays sequentially. In particular, this section considers multi-hop systems, where a packet of data is transferred via multiple relays using decode and forward (compare Chapter 27 for other forms of relaying), in the manner of a bucket brigade: the packet is transmitted from the source to the first relay node, where it is decoded, re-encoded, and sent to the next relay, which also does decode/re-encode, then transmits it to the next relay that lies on the route from source to destination, and so on, until finally, the last relay transmits to the destination. Each of the hops is operated as a point-to-point link, i.e., the broadcast effect is not exploited.

The first task is the determination which nodes should act as relays (i.e., forward the message) and in what sequence; in other words, what is the *route* of the packet to its destination? Related to this issue is the question what resources (power, bandwidth) should be allocated to those nodes. The combined *routing and resource allocation* is a cross-layer design problem, i.e., involves physical layer, MAC layer, and networking. However, many of the studies in the literature treat only subaspects: routing algorithms tend to assume a fixed physical layer, for which an optimal route is constructed. Conversely, other papers assume a given route and try to optimize the PHY layer for this route.

The subsequent discussions assume unicast situations, i.e., each packet has exactly one source and one destination. In many cases, the algorithms can be generalized to cover multi-cast or broadcast (i.e., multiple destinations), but for the sake of simplicity, we will not deal with this, except briefly in Sec. 23.7.5.

23.6.1 Overview

We first give a high-level overview of routing in packet switched ad hoc networks. A packet switched system builds up a *logical* connection between TX and RX, but – in contrast to circuit-switched systems – not necessarily a constant *physical* connection. It is only important to get the packets from the TX to the RX in some way; the actual choice of the route (the physical connection) can change with time. For this purpose, the message is broken into small pieces (packets), each of which can take a different route to the RX, depending on what transmission paths are currently available, see Figure 23.9. The packets can thus take routes via different network nodes; each of the nodes can pass on the packet right away, or it may buffer it; the latter case arises if transmission is undesirable (bad channel conditions), or other packets with higher priority need to be transmitted. This buffering can lead to considerable delays in the transmission, and it can also happen that the sequence how the data arrive is different from the sequence with which they were transmitted. For this reason, packet radio with routing over many nodes cannot easily be used for applications with guaranteed maximum latency, though there is active research toward this goal. Voice-over-IP is a typical example for this development: when first introduced, voice quality of these systems was far inferior to that of circuit-switched voice transmission; nowadays the quality is comparable at most times.

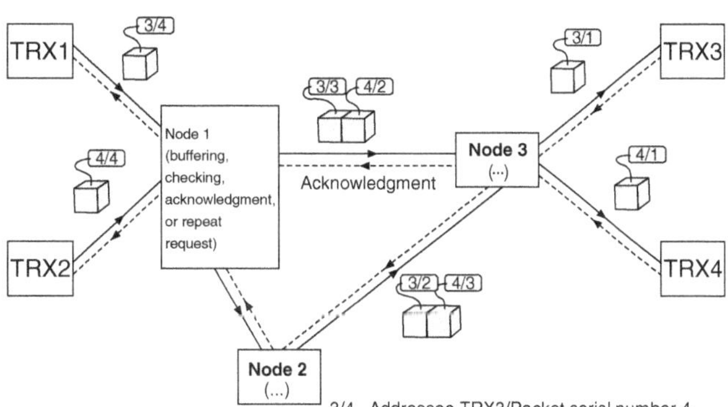

Figure 23.9   Principle of a packet radio network.
Reproduced with permission from [Oehvrik 1994] © Ericsson AB.

Turning now to the content of each packet: a data packet must contain not only payload data but also *routing information*. The payload data have to contain error detection coding (and usually error correction coding), so that transmission errors can be recognized by the receiving nodes, and a retransmission can be requested if necessary. The routing information spells out the origin and the destination of the message; furthermore, additional information about the path of the packet can be included. If a very low packet error rate is required, each node that acts as relay stores the packets in a buffer and deletes them only after receiving an acknowledgment of successful transmission from the node it forwarded the packet to (see Figure 23.10).

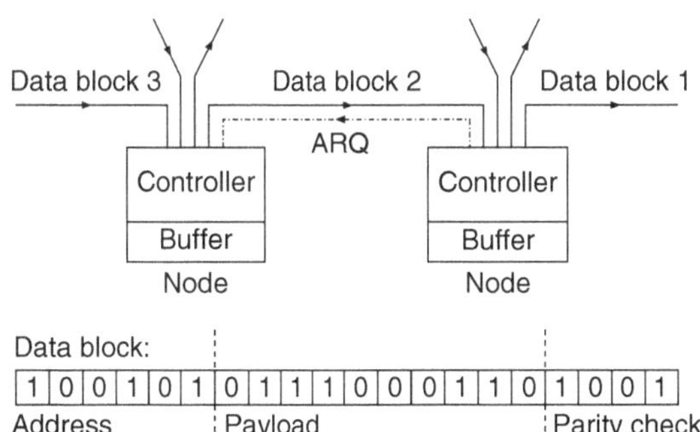

Figure 23.10   Structure of a data block, and principle of the buffering mechanism.
Reproduced with permission from [Oehvrik 1994] © Ericsson AB.

Generally, routing methods can be classified into the following two categories:

- *source-driven routing*: in that case, the header of the packet includes the complete route, and the nodes just follow the instructions for forwarding. The drawback is that the header can become quite long; especially for packets with little payload, this leads to a significant decrease of spectral efficiency.
- *table-driven routing*: in this approach, each node stores in a table the nodes to which it should forward packets (depending on the destination address, and the node the packet came from). This method has a better spectral efficiency; the drawback is that the tables can become quite large, especially at nodes in the middle of a network.

A related topic is also the *route discovery*, i.e., determination which route a packet should take from the TX to the receiver. Route discovery is typically done by means of special packets that are broadcast in the network, and record the quality of the links between different nodes. In order to achieve optimum performance, the routing has to be changed whenever the channel between nodes changes significantly.

Besides the above-mentioned approaches, there are also *backpressure-type* algorithms that perform scheduling and forwarding without establishing an explicit route, thus obviating both the need for route discovery and routing tables. Such algorithms will be discussed in more detail in Section 23.6.12.

23.6.2 Mathematical Preliminaries

We start out with networks where each link between two nodes can be treated as essentially "binary": it can either support error-free transmission of a packet (in which case there is a valid connection between the nodes), or it cannot. The network is then represented as a *graph*: each node corresponds to a vertex, and each (working) link corresponds to an edge. Assume furthermore that the links are reciprocal, i.e., that the transfer function from node A to node B is the same as from node B to node A; in that case, the graph is undirected. Each edge has a *weight*: for the simple case of fixed transmit power, and assuming a link either works or does not work, the edge weights of all working links are unity, so that the cost of transmission is the *hop count*.[5] In the case that different nodes can use different power (to ensure that certain links work), the edge weight could represent the power expenditure for sending a packet over a certain link. In either case, the routing problem becomes a shortest-path problem – more exactly, the problem of finding the path with the smallest "distance" (sum of the edge weights) between two vertices in a graph. These shortest-path problems have been considered in the computer-science literature, and several algorithms have been developed to solve them. The *Dijkstra* algorithm provides a fast solution if all edge weights are positive (as they usually are); in those cases where negative edge weights need to be considered, the *Bellman–Ford* algorithm can be used.

Dijkstra Algorithm

The Dijkstra algorithm finds the shortest-path weights from a source node to every node in the network, based on the principle of "greedy relaxation." It proceeds in the following steps:

1. Assign the source node the distance $d_0 = 0$, and all other nodes the distance $d_i = \infty$ (note that d is the distance from the source node, and as such is a property of the nodes, not of the edges). Furthermore, mark all nodes as "unvisited," and declare the source node the "current" node.
2. Loop until all nodes are visited
   (a) Consider all unvisited neighbor nodes i of the current node c, i.e., nodes with (direct) links to the current node. For each neighbor node i, compute the $\hat{d}_i = d_c + w_{c,i}$ where $w_{c,i}$ is the edge cost between nodes c and i. If $\hat{d}_i < d_i$, replace d_i by $\hat{d}_i$. Let the node store a pointer to the previous node on the route that led to the lowest cost.
   (b) Mark the current node as visited.
   (c) Pick the unvisited node with the smallest distance as current node.

The algorithm is very efficient; depending on the particular implementation, the runtime is proportional to $\mathcal{O}(|V|^2 + |E|)$ or $\mathcal{O}(|V| \log(|V|) + |E|)$, where $|V|$ is the number of vertices, and $|E|$ is the number of edges.

Bellman–Ford Algorithm

The Bellman–Ford is an alternative algorithm for finding the shortest path. It proceeds in the following steps

1. We again start by assigning the source node the distance $d_0 = 0$, and all other nodes the distance $d_i = \infty$.
2. Repeat $|V| - 1$ times
   (a) do for all edges in the graph
      (i) Call the starting point of the edge j, and the end point i; then compute the $\hat{d}_i = d_j + w_{j,i}$; if $\hat{d}_i < d_i$, replace d_i by $\hat{d}_i$. Let the node store a pointer to the previous node on the route that led to the lowest cost.
3. Check whether further reductions in the distances are still possible. If yes, this indicates that the graph contains "negative cycles," and the weights will not converge. Otherwise, the algorithm is finished. This step can be skipped if we know that all edge weights are positive.

Example 23.1   Figure 23.11 shows a network with positive edge weights; a message should be transmitted from node A to node C. The left side of the figure shows for the various stages of the Dijkstra algorithm. The right side shows the working of Bellman–Ford algorithm, where each graph corresponds to one iteration of the outer loop; the tables indicate the evolutions of the weights as the inner loop (stepping through all the edges in the graph). Note that for the Bellman–Ford algorithm, the sequence of the edges in the inner loops does not play a role; in other words, the numbering of the edges as (1,2,3, ...) is arbitrary.

---

[5] A somewhat better metric is the Expected Number of Transmissions (ETX), which takes into account that not all packet transmissions over a link are successful, and therefore require a retransmission.

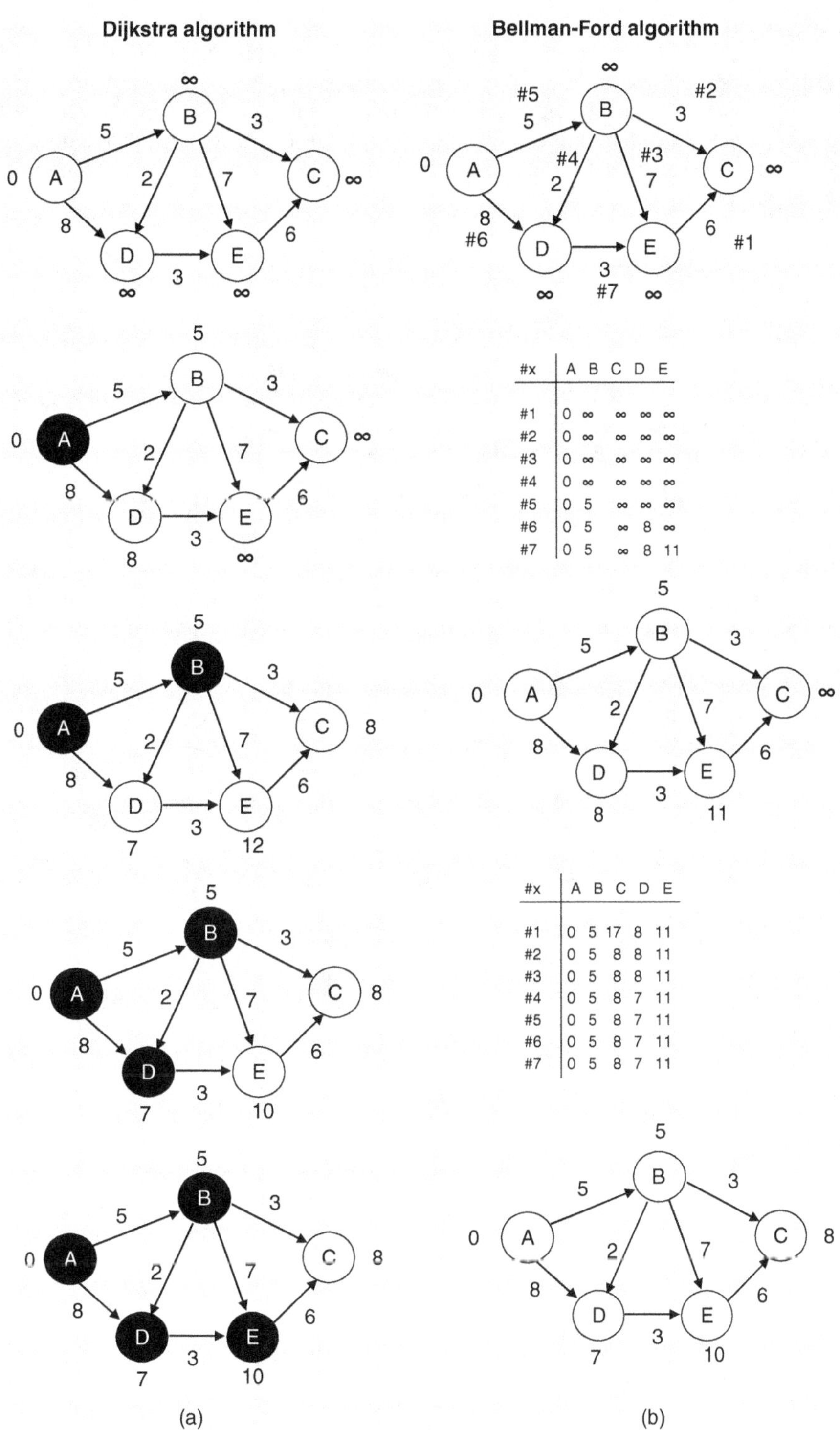

Figure 23.11 Dijkstra (a) and Bellman–Ford (b) algorithm.

23.6.3 Goals and Classifications of Routing Protocols

Routing protocols have been investigated extensively in the context of wired packet networks, in particular the internet. There, the main goal is to transmit the information to the destination in the shortest possible time. In wireless ad hoc networks, additional constraints have to be considered. We can thus have a combination of the following goals (i) the energy consumption for the forwarding of the information should be kept as small as possible, (ii) the lifetime of the network, i.e., the time until the first node runs out of battery power, should be maximized, (iii) the protocol should be distributed, i.e., not require centralized control, (iv) the protocol should be able to react quickly to changes in the topology or link states, (v) the protocol should be bandwidth-efficient, (vi) the end-to-end transmission time should be minimized.

Taking the changing topology and link states into account can be done in one of the following two ways: (i) *proactive*: in this case, the network keeps track of the optimum routes from all possible sources to all possible destinations at all times. Thus, the actual transmission of packets can be done very quickly, as the optimum route is immediately available. On the downside, the overhead required to keep track of all the routes can be significant. (ii) *reactive*: in this approach, a route to a destination is only determined when there is actually a packet to be sent to that particular destination, i.e., *on demand*. This approach is more efficient but clearly leads to a slower delivery of packets.

23.6.4   Source Routing

In *source routing*, each information-generating node can determine, e.g., from a lookup table, the sequence of nodes that this packet should take through the network. The sequence of nodes is added to the data packet, so that each node on the route can learn to which node the packet should be transmitted next. A key advantage of source routing is that it is loop-free, i.e., there is no danger that a packet returns to an intermediate node that it had already visited. This absence of loops can be guaranteed since the source determines the route through the network, and therefore can make sure during the route setup that there are no loops. This point may sound trivial, but we will see below that other routing methods can suffer from loops.

In *proactive source routing*, link state advertisements are sent out periodically by each node to neighboring nodes. Those nodes compare the received link state to the one they have stored in their local tables; if it is fresher (i.e., has a higher sequence number), then they update their table, and forward the information to their neighbors, and so on. Thus, refreshed link state information is propagated throughout the network. It is obvious that the amount of information that has to be distributed grows enormously as the number of nodes in the network increases. For this reason, proactive source routing algorithms are not well-suited for large networks.

Dynamic Source Routing (DSR) can greatly reduce the overhead by performing on-demand routing. The routing procedure consists of two steps: an initial *route discovery*, followed by *route maintainance* that reacts to changes in the link states in the network. During the route discovery, the network is flooded with so-called *route request packets*. The route request contains the ID of the intended information destination, a unique packet ID, as well as a list of nodes that have already been visited by the message. When a node receives a route request packet, it checks whether it is either the intended destination or has a path to the destination stored in its own routing table. If that is *not* the case, the node re-broadcasts the route request, adding its own address to the list of visited nodes in the message. If the node *is* the destination (or has a path to the destination), then the node answers with a *route reply packet*, which tracks back along the identified path to the source, and finally informs the source about the sequence of nodes that have to be taken from source to destination.[6] The route request packets can record the quality of the links, which then can be used for the route determination through the edge weights as described above for the Dijkstra algorithm.

During route maintenance, the protocol observes whether links in the established route are "broken" (which might include that the throughput of a particular link has gone below a certain threshold). It then either uses an alternative stored route for the destination or initiates another full route discovery process.

A problem in DSR is the so-called reply storm, which occurs when a lot of neighbors know the route to the target and try to simultaneously send the information. This wastes network resources. Route reply storms are prevented by making each node that wants to transmit wait by an amount of time that is inversely proportional to the quality of the route it can offer. In other words, a node that wants to suggest a good route is allowed to transmit sooner than a node that suggests an alternative, worse, route. Furthermore, nodes listen to route replies transmitted by other nodes and do not transmit if another node has already transmitted a better route.

23.6.5   Link-State-Based Routing

In link-state routing, each node gathers information about the states of the links in the whole network. Based on this information, a node can then construct the most efficient routes through the network to all other nodes, e.g., by means of the Dijkstra algorithm.

Link-state-based routing requires acquisition and distribution of the link states throughout the network. A node first has to learn its link state, e.g., from training sequences transmitted by other nodes. The distribution of the link-state information to other nodes is then achieved by means of short messages called *link-state advertisements*. Those messages contain the following pieces of information: (i) the ID of the node that is creating the advertisement, (ii) the nodes to which the advertising node is connected, as well as the link quality (edge weights) of that link, (iii) the sequence number, which indicates how "fresh" the information is (the sequence number is incremented every time the node sends out a new advertisement).

A popular algorithm for implementing link-state-based routing for wireless networks is the *Optimized Link State Routing Protocol* (OLSR). It is a proactive protocol, where messages between nodes are exchanged on a regular basis, to create routing tables at every node. While classical link-state protocols flood the link-state information throughout the network, OLSR limits this information. This fact makes it particularly suitable for wireless networks, since the amount of link-state information in wireless

---

[6] Intermediate nodes on that route might also store the route from them to the destination nodes; this can accelerate future route searches.

ad hoc networks can be particularly large. OLSR therefore uses the concept of *Multi-Point Relays* (MPRs). The set of MPRs of a particular node is defined as a subset of the one-hop neighbors, where the subset must be chosen in such a way that the node can reach all two-hop neighbors via an MPR. In other words, we do not need to reach each two-hop neighbor by all routes that would be possible, but reaching it via one route is deemed sufficient. In order to sense the environment of a node, OLSR periodically lets the node transmit "Hello" messages. Those messages are sensed by the surrounding nodes, but not re-broadcast. Since a "Hello" message contains information about all known links and neighbors of a particular node, every node can find *two-hop neighbors* by passively listening to the Hello messages.

As mentioned above, each node selects a subset of its neighbors as MPRs. The set of MPRs is regularly broadcast by means of the *Topology Control* (TC) messages, to other MPRs, which can then forward that information. Thus, any node in the network can be accessed through a series of MPRs. By reducing the ways that a node can be reached, the overall amount of link-state information that has to be sent through the network is reduced.

*23.6.6 Distance-Vector Routing

In distance vector routing, each node maintains a list of all destinations that *only* contains the cost of getting to that destination, and the *next* node to send the message to, but not the overall network topology. Thus, the source node only knows to which node to hand the packet, which in turn knows the next node, and so on. This approach has the advantage of vastly reduced storage costs compared to link-state algorithms (remember that in link-state routing, every node has to store, for each destination, the complete sequence of nodes via which to send the message). It follows that distance vector algorithms are easier to implement and require less storage space. The actual determination of the route is based on the Bellman–Ford algorithm, which was discussed in Section 23.6.2.

However, distance-vector routing also has a number of drawbacks:

1. Slow convergence. Compared to the Dijkstra algorithm, Bellman–Ford requires multiple passes of the cost information. In a network with quickly-changing topology, this can lead to situations where the link states have changed before an optimum route has been set up.
2. "Counting to infinity": in the extreme case that part of the network becomes separated, the network can form loops, so that information comes back to a node that it had already passed (remember that such a situation cannot occur in source routing). The essence of the problem is that even if node B tells node A that it has a route to the destination, node A does not know if that route contains node A (which would make it a loop). Under "normal," converged, circumstances this is not a problem, because a route containing a loop has a higher total cost than a route that "cuts out" that loop. However, in case that a node "goes down" (either because of node failure or because of fading of links), a loop can be created. Consider a linear network, connected as A–B–C–D–E–F, and let the edge cost be unity for each hop. Now if node A goes down, node B learns during the update process of Bellman–Ford that its link to node A is down, because it does not receive an update. However, at the same time, node B gets an update from node C, which tells node B that node A is only two hops from node C (since node C does not know about the outage yet). This misinformation thus propagates through the network, until the nodes have increased their costs to infinity.

A solution of the counting-to-infinity problem can be obtained by means of *destination sequences*, resulting in the *DSDV (Destination-Sequenced Distance Vector)* algorithm. For this case, the nodes store not only the cost of getting to the destination and the next node on the route but also a sequence number. Each node then periodically advertises routes to other destinations; the destination increases the sequence number and propagates the route to other nodes in the network. The route that is selected is the one with the largest sequence number in the network. If a link is broken, the sequence number is increased, and the node cost is increased to infinity. This change is immediately propagated to other nodes.

The *Ad hoc On-demand Distance Vector (AODV)* routing algorithm is the reactive version of DSDV. Routes are built up using route requests and route replies (similar to DSR): when a source needs to send a packet to a destination for which it does not already have a route, it broadcasts a route request. Nodes that hear this request update their information for the source node and set up backward pointers to the source node in the route tables. If the node has a route to the destination, it replies to the source; otherwise, it broadcasts the information to further nodes.

An important practical form of a proactive distance vector routing is *RPL*, the IPv6 routing protocol for multi-hop low-power wireless sensor networks. The key element of RPL is a Destination-Oriented Directed Acyclic Graph (DODAG), which is rooted in one (or several) gateway connections, and establishes a loop-free connection to each node; each route supports bi-directional communications. Each node sends out messages that advertise routing metrics, based on which the surrounding nodes select parent nodes (different criteria for that choice are possible). Mechanisms are foreseen for reverse route construction, adding of nodes into the network, etc. Connections between nodes and multiple gateways are possible, to enhance route diversity (compare also Section 23.7).

*23.6.7 Geography-Based Routing

In a number of cases, the nodes in the network know their geographic position. This can be achieved either by GPS or other localization mechanisms, see Chapter 29. Routes can be designed based on this geographic information (in particular using the protocol model mentioned in Section 23.5.1) – though it is noteworthy that geographic distance between two nodes forming a link and SNR of that link need not be monotonic – a fact that is sometimes ignored in the networking literature.

Geographical routing schemes are based on the concept of "progress" toward the destination. "Greedy" schemes, which are the most popular geographical schemes, pick the node within the coverage disk that has the smallest distance to the destination. Alternatively, it can maximize the projection of the link connecting the transmitting node to a particular receiving node, onto the line connecting transmitting node to the destination; this approach is called "most forward within radius." An example how those two algorithms behave is given in Figure 23.12; however, it must be emphasized that in most cases the two algorithms find the same path.

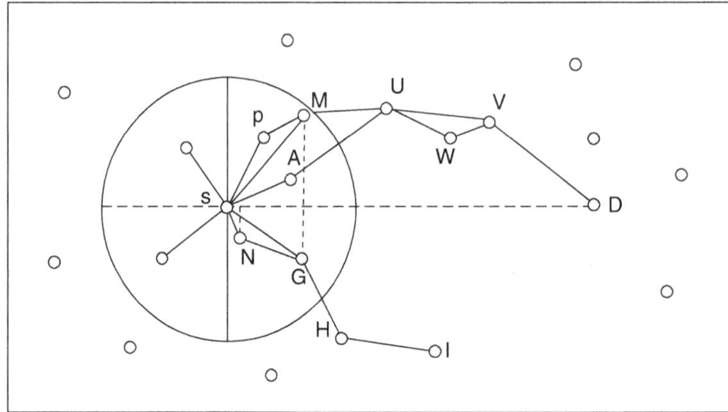

Figure 23.12 Criteria for choosing nodes in geographic routing: a greedy algorithm picks the path SGHI (which fails to deliver); the "most forward within radius" algorithm picks SMUVD.
Reproduced with permission from [Stojmenovic 2002] © IEEE.

When greedy approaches fail to advance the message to the destination, networks can go into "recovery mode." This occurs when the message arrives at so-called "concave" nodes, which have no neighbor that is closer to the destination than themselves. An easy way out is to momentarily use flooding, i.e., concave nodes flood their neighbors with the message, and subsequently reject any further copies of the message (to avoid that the neighbor sends the message back to the concave node, which after all is close to the destination). After this flooding step, the routing continues using a greedy mode.

23.6.8 Hierarchical Routing

Up to now, we have assumed that all nodes are equal with respect to forwarding messages. However, this need not be the case. In hierarchical routing, the nodes of the network are subdivided into clusters, i.e., group of nodes, see Section 23.3. When a message is to be passed between clusters, this can only be done by communications via the clusterheads. One popular hierarchical algorithm (mainly used for data-driven routing, see below) is the LEACH protocol, see Figure 23.13.

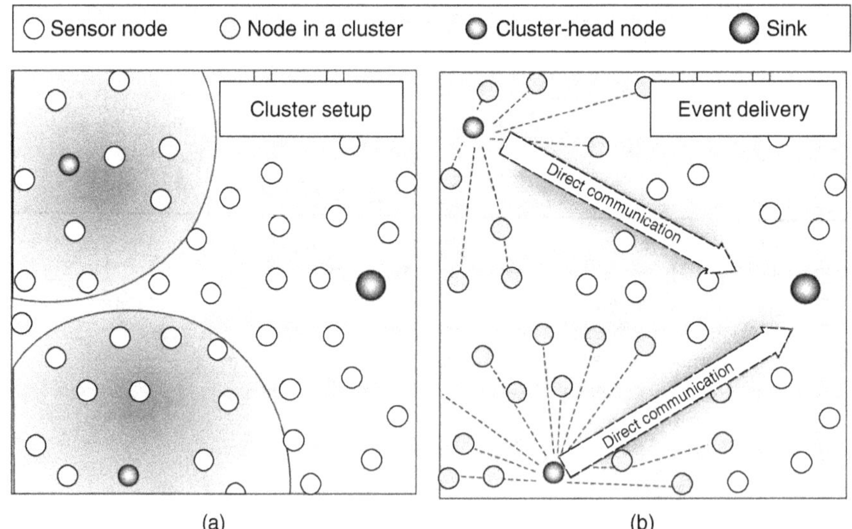

Figure 23.13 Clustering of nodes in hierarchical routing.
Reproduced with permission from [Martirosyan et al. 2008] © IEEE.

23.6.9   Impact of Node Mobility

In most ad hoc networks, the nodes are fairly static during operation. However, there are also some networks where nodes show a very high degree of mobility – e.g., in vehicular ad hoc networks. This mobility has both advantages and drawbacks:

- The key advantage is that data packets can "hitchhike" on nodes that are moving. Imagine that a data packet has to be sent from node A to node B, both of which are static, and widely separated. Node A then transmits the packet to a highly mobile node C when it is passing by node A. Node C stores the message, and when it comes into the vicinity of node B, transmits it. Thus, the large distance between nodes A and B can be bridged without either high transmit power for direct transmission, or multiple transmissions. Note, however, that the latency of the packet transmission is significant.
- The key drawback of high node mobility is that the network can become temporarily disconnected, especially in sparse networks, where there are only few possible routes between a source and a destination. If a few nodes are moving, it can easily happen that there is no valid path for a multi-hop connection between source and destination anymore. On the other hand, there are also sparse networks that are never connected in a static sense (i.e., at no point in time is there a route between TX and receiver); still, exploiting node mobility by the "hitchhiking" process described above can allow packet delivery to the destination.

A routing algorithm that takes account of the node mobility is *epidemic routing*. Whenever two nodes come into range of each other, they exchange all the packets that they do not have in common. In this way, every packet is ultimately distributed to every node in the network. This approach is therefore quite wasteful of resources, in particular, if it is mostly unicast messages that are to be distributed. A more efficient algorithm is *spray and wait*, where L copies of the message are transmitted, and then we just wait until one of the nodes that received a copy comes into range for direct transmission to the destination.

23.6.10   Data-Driven Routing

The ultimate task of sensor networks is different from those in ad hoc networks. While ad hoc networks have to transmit data files from one node to another, the purpose of sensor networks is to get sensing data (typically data about the environment in which the sensor is located), to a data sink. In this process, it is not important from which particular node the data originated; all that matters is that the underlying environmental information arrives correctly. As a concrete example, take temperature measurements in a factory hall. Say that the hall has 100 temperature sensors in it. Yet all that matters is that we get the correct picture of the temperature distribution to a central monitoring station. The data from all the temperature sensors are correlated, so it might be overkill to ensure that each single sensor can communicate with the monitoring station. Thus, routing in sensor networks is *data-driven* (or application-driven), not connection-driven.

The most popular data-driven routing approach is *directed diffusion*, see Figure 23.14. The monitoring station requests data by sending out an *interest* message, which contains specifics of the type of information that it wants, intervals at which the data are to be collected, and geographical area. The message is propagated throughout the network; during this propagation, the nodes also set up gradients, i.e., reply links toward the nodes from which the interest statement was received. Note that multiple paths can be set up back to monitoring station. Particular paths can be reinforced during the actual transmission of the data. During the data transmission, each node can cache and process data and can aggregate data from different sources.

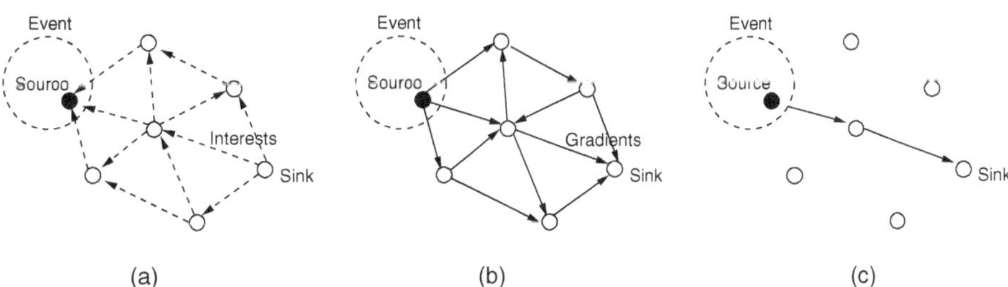

Figure 23.14   Directed diffusion. (a) Interest propagation. (b) Initial gradients set up. (c) Data delivery along reinforced path.
Reproduced with permission from [Intanagonwiwat et al. 2003] © IEEE.

Further refinements of the protocol can include "meta-data negotiations," which essentially make sure that nodes only transmit new data to the clusterhead. The transmission of data occurs in three steps: advertising, request, and data transmission. In the advertising step, a node uses meta-data to advertise the existence of new data; in the request stage, the recipient requests the data if they are useful (e.g., not already collected from another node that has similar/correlated data), and in the data transmission stage, the actual data are sent.

23.6.11 Power Allocation Strategies

In the description of the above algorithms, we have assumed that the transmit power of the nodes is fixed – an assumption that is often fulfilled in ad hoc networks, where the nodes should be as simple as possible. However, there are cases where the nodes can adapt their power and/or their transmission rate. In that case, we should try to optimize the transmission power as much as possible. We can distinguish the following cases:

1. *Routes fixed, transmission rate fixed*: in this case all we can achieve with power control is a reduction of the expended power. The transmit power at each node should be lowered as much as possible with the limitation that the SNR at the receiving node has to be high enough to guarantee decoding. As a side benefit of the power control, the interference to the rest of the network is reduced (this becomes important when multiple messages are to be sent through the network at the same time).
2. *Routes fixed, but transmission rate variable*: in this case, reducing the transmit power increases the necessary transmission time. If minimization of the energy expenditure is the ultimate goal of the optimization process, then transmit power should be lowered as much as possible. If the message has to be delivered to the destination within a certain deadline, then an optimization of the power allocation for a given delivery time can be done.
3. *Routes and transmission rates variable*: by changing the power and/or the rate, we can adapt the edge weights of the graph representing the network. Thus, a route that is optimal for a certain set of transmit powers might not be optimum for a different set. Thus, routing and power allocation have to be done in one joint step. Actually, there is an even more general and harder problem, which occurs when there are *multiple* messages for which routing, transmission rate control, and power control needs to be done. This general problem will be discussed in Section 23.6.12.

*23.6.12 Joint Scheduling and Routing for Multi-Hop Transmission

Up to this point, we have assumed that either (i) there is only a single packet that needs to be transmitted, so that routing (and possibly power allocation) is the only problem that needs to be solved, or (ii) routing and scheduling can be treated as separable, so that the multi-hop routing is done according to the algorithms described above, while each of the transmissions in the route is scheduled according to the methods of Chapter 20. However, when multiple messages are transmitted, separate routing and scheduling is not necessarily desirable, since each transmission creates interference on other active links. Thus, a link that provides high capacity when only a single message is routed over it might become a bottleneck when multiple messages are trying to be transmitted on nearby links, thus creating interference. In other words, the routes that are found to be optimum for the different messages separately (e.g., by the Dijkstra algorithm) are not necessarily the best (or even good) routes when multiple messages are being transmitted in the network simultaneously. Conversely, scheduling is critical in avoiding bottlenecks: different messages can be sent over the same or nearby nodes; we just have to ensure that this happens at different times – similar to road traffic at an intersection: the scheduling algorithm corresponds to a traffic light that tells the traffic from one road to stop while traffic on an intersecting road is using the intersection, so that no collisions occur. The general problem of finding routes, scheduling, and power/rate control for multiple messages is thus a very complicated, but also practically very important problem. A number of different algorithms have been designed to solve this. We will consider in particular the stochastic optimization approach leading to the *backpressure algorithm*.

The backpressure algorithm for ad hoc networks is a generalization of the algorithm described in Section 20.5 for the scheduling of packets in single-hop broadcast/multiple access nodes. We first reformulate the problem in a suitably generalized notation. The network is described (i) by its states, and (ii) by control actions that determine what and how much data are transmitted with what power. Each node has a buffer (queue) in which it stores arriving data, and it tries to transmit data to empty the buffer. The algorithm then computes link weights that take into account (i) the difference of the queue size between the transmitting and receiving node of a link and (ii) the data rate that can be achieved over this link.

To quantify: a network contains N nodes, connected by L links. The transmission of the messages occurs in a slotted manner, where t is the slot index. The link between two nodes a and b is characterized by a transmission rate $\mu_{ab}(t)$; the rates are summarized in the transmission matrix $\boldsymbol{\mu}(t) = \mathbf{C}(\mathbf{I}(t), S(t))$, where C is the transmission rate function that depends on the network topology state $S(t)$, which describes all the effects that the network cannot influence (e.g., fading) and the link control action $\mathbf{I}(t)$, which includes all the actions of the network that can be influenced, like power control, etc. The most important example of a transmission rate function is capacity-achieving transmission, so that on the lth link

$$C_l(\mathbf{P}(t), S(t)) = \log_2\left[1 + \frac{P_l(t)|h_{ll}(S(t))|^2}{\sigma_{\mathrm{n}}^2 + \sum_{k\neq l}P_k(t)|h_{kl}(S(t))|^2}\right] \tag{23.11}$$

where $|h_{kl}(S(t))|^2$ is the power gain of the signal transmitted by the intended TX of link k to the intended RX of link l, when the network topology is in the state $S(t)$, and $P_l(t)$ is the power used for transmission along the lth link.

During each timeslot, an amount of data R_q^{out} is being transmitted by a node. At the same time, data are arriving from an external source (e.g., sensors, or computers that want to transmit data); and the node is receiving data via the wireless links from other sources;

the total amount of data arriving during one timeslot is R_q^{in}.[7] Each node has an infinitely large buffer, which can store the arriving messages before they are sent out over the wireless links. The amount of data in the buffer (also called queue backlog) is written as $Q_q(t)$, where q is the index of the considered queue; all queues in the network are stacked into a vector $\mathbf{Q}(t)$. During one timeslot, the backlog of a queue at a particular node changes as

$$Q_q(t+1) \leq \max\left[Q_q(t) - R_q^{\text{out}}(I(t), S(t)), 0\right] + R_q^{\text{in}}(I(t), S(t)) \tag{23.12}$$

where the max[., 0] operator is used to ensure that the length of the queue cannot become negative. We can define a general "penalty function" $\bar{\xi}$, which is defined as the time-average of an instantaneous penalty function – for example, the overall power consumption of the network. An additional set of constraints $\bar{x}_i$ can be defined. We then aim to solve the following optimization problem:

$$\text{Minimize } \bar{\xi} \tag{23.13}$$

$$\text{subject to } \bar{x}_i \leq x_{\text{av}} \text{ for all } i$$
$$\text{and network stability.} \tag{23.14}$$

For example, we can aim to e.g., to constrain the (time-averaged) power consumption per node, so that $\bar{x}_i = \bar{P}_i$ and $x_{\text{av}} = P_{\text{max}}$ is the maximum admissible mean power.

Here, network stability means that the size of the queues remains bounded – in other words, that on average not more data flow into the queue than can be "shoveled out." Note that it is important to ultimately deliver the data to their final destination, since such data delivery is the only way of making the data "vanish" from the network. As long as data are sent on their way through a multi-hop network, they are part of some queue, and therefore detrimental to achieving network stability. We also note that a possible optimization goal is simply the achievement of network stability without any further constraints (this is formally achieved by setting $\bar{\xi} = 1$).

This problem can be solved with the "Lyapunov drift" technique described in Section 20.5. The first step is to convert the additional constraints $\bar{x}_i \leq x_{\text{av}}$ into "virtual queues" (these are not true data queues, but simply update equations):

$$Z_i(t+1) = \max\left[Z_i(t) - x_{\text{av}}, 0\right] + x_i(I(t), S(t)) \tag{23.15}$$

which are stacked into a vector $\mathbf{Z}(t)$. The optimization problem is then converted into the problem of minimizing $\bar{\xi}$ while stabilizing both the actual queues $\mathbf{Q}(t)$ and the virtual queues $\mathbf{Z}(t)$, combined into vector $\boldsymbol{\Theta}(t)$. Defining now the Lyapunov drift

$$\Delta_Z(\boldsymbol{\Theta}(t)) = E\{L_Z(\boldsymbol{\Theta}(t+1)) - L_Z(\boldsymbol{\Theta}(t)) \mid \boldsymbol{\Theta}(t)\}$$
$$\Delta_Q(\boldsymbol{\Theta}(t)) = E\{L_Q(\boldsymbol{\Theta}(t+1)) - L_Q(\boldsymbol{\Theta}(t)) \mid \boldsymbol{\Theta}(t)\} \tag{23.16}$$
$$\Delta(\boldsymbol{\Theta}(t)) = \Delta_Z(\boldsymbol{\Theta}(t)) + \Delta_Q(\boldsymbol{\Theta}(t))$$

where

$$L_Z(\boldsymbol{\Theta}(t)) = \frac{1}{2}\sum_i [Z_i(t)]^2, \qquad L_Q(\boldsymbol{\Theta}(t)) = \frac{1}{2}\sum_q [Q_q(t)]^2. \tag{23.17}$$

Upper bounds for the Lyapunov drift are given by

$$\overline{\Delta}_Z(\boldsymbol{\Theta}(t)) = B_Z - \sum_i Z_i(t)E\{x_{\text{av}} - x_i(I(t), S(t)) \mid \boldsymbol{\Theta}(t)\} \tag{23.18}$$

$$\overline{\Delta}_Q(\boldsymbol{\Theta}(t)) = B_Q - \sum_q Q_q(t)E\left\{R_q^{\text{out}}(I(t), S(t)) - R_q^{\text{in}}(I(t), S(t)) \mid \boldsymbol{\Theta}(t)\right\} \tag{23.19}$$

where B_Z and B_Q are finite constants.

The "generalized max-weight policy," i.e., the determination for the optimum weights for the backpressure algorithm, in each step, greedily minimizes

$$\overline{\Delta}_Z(\boldsymbol{\Theta}(t)) + \overline{\Delta}_Q(\boldsymbol{\Theta}(t)) + VE\{\xi(I(t), S(t)) \mid \boldsymbol{\Theta}(t)\} \tag{23.20}$$

where V is a control parameter. This can be written as minimizing

$$V\hat{\xi}(I(t), S(t)) + \sum_i Z_i(t)\hat{x}_i(I(t), S(t)) - \sum_q Q_q(t)\left[\hat{R}_q^{\text{out}}(I(t), S(t)) - \hat{R}_q^{\text{in}}(I(t), S(t))\right] \tag{23.21}$$

---

[7] The half-duplex constraint can be included, e.g., by assuming that two orthogonal channels exist for each node, one on which data can be received, and one on which only transmission can happen.

where $\hat{\xi}(I(t), S(t)) = E\{\xi(I(t), S(t)) \mid I(t), S(t))\}$ and similarly for the other "hat" functions in the equation above. These functions of $I(t)$ and $S(t)$ are assumed to be known. For a given slot t, the network state $S(t)$ and queue backlogs $Q(t)$ and $Z(t)$ are known, though the probability density function (pdf) of S does not need to be known.

The "network stabilization" only guarantees that – as a time average – the length of the queues does not grow to infinity. There is, however, no guarantee about the delay of a particular data packet. Even the average delay of the data packets can be quite large. The control parameter V allows a trade-off between this average delay and how close the solution of Eq. (23.21) comes to the theoretical minimum of the penalty function, similar to the behavior obtained in Section 20.5.

After this rather general description, let us turn to simpler special cases: we wish to minimize the average network power (without additional constraints on the per-node power). In that case, there are no "virtual" queues. We thus define now for each node Ω_n as the set of links for which node n acts as TX; we furthermore realize that at each node there can be multiple queues, each for one particular commodity (packet stream for a particular destination). We thus aim to choose the power vector $\mathbf{P}(t)$ that maximizes the expression

$$\sum_n \left[\sum_{l \in \Omega_n} 2C_l(\mathbf{P}(t), S(t)) W_l^*(t) - V \sum_{l \in \Omega_n} P_l \right] \tag{23.22}$$

where P_l is the temporal average over $P_l(t)$, and the weights W_l^* are

$$W_l^*(t) = \max \left[Q_l^t - Q_l^r, 0 \right]$$

and Q_l^t and Q_l^r are the backlogs of the queue at the transmitting node and receiving node, respectively, of link l for the commodity that has the biggest backlog differential at this particular link.

An even simpler case occurs when we wish to perform network stabilization only. In this case, the optimization reduces to the following simple rule: for each queue, and each possible receiving node, compute the product of the "queue length difference" at the two link ends, and the transmission rate over that link. Then use the timeslot to transmit (only) packets from the particular queue to its corresponding receiving node.[8] This is the celebrated *backpressure* algorithm. Like its more general cousins described above, it is a completely distributed algorithm (the decision at every node requires only knowledge of the characteristics of the node and its immediate neighbors), and yet can be proven to be throughput-optimal.

It is also noteworthy that the backpressure algorithm provides an inherent routing; data packets will ultimately end up in their intended sinks, since this is the only way to get "out of the network." However, there is no guarantee that the routes taken by the packets follow a shortest path; especially in lightly loaded networks, data packets can take very circuitous routes. This is especially pronounced if we try to simply stabilize the network (without energy minimization). This problem can be alleviated by introducing a "bias" for shorter routes in the penalty function. Alternatively, the expected number of hops (how many transmissions will be necessary on the remainder of a route to the destination) can be used as utility function. Furthermore, use of a first-in-last-out principle for packets in the queues (instead of the traditional first-in-first-out) can drastically reduce the average delay.

Figure 23.15 shows the average backlog as a function of the generation rate of packets at the nodes, when no power minimization is attempted. We see that the shortest-path algorithm becomes unstable (backlog becomes infinite) at lower packet generation rates. The "enhanced" backpressure algorithm (one that includes a bias term) performs better than the regular algorithm at low packet generation rates.

*23.7   Routing and Resource Allocation in Collaborative Networks

Transmission of signals in an ad hoc network can also involve cooperation between different nodes. Instead of the "decode and forward" approach described in Section 23.6.1, the transmission during each hop might involve multiple nodes that have the message available. In particular, a receiving node might perform *energy accumulation* when it gets multiple copies of the same signal, and adds up their energies, e.g., through maximum-ratio combining. Or a node might get multiple, differently encoded versions of the same signal, and then perform *accumulation of the mutual information* from those signal copies. More details on *how* such energy or information accumulation can be obtained are provided in Section 27.3; for this section, it is sufficient to state that it is possible to perform it.

Obviously, the change in how the information transmission occurs during each hop of a multi-hop forwarding also changes the optimum routing. In contrast to the decode-and-forward case, where efficient algorithms for routing have been proven to be optimal, the routing problem with cooperative communications relies on a number of heuristic algorithms, in particular for the case when multiple messages are to be transmitted through the network simultaneously.

23.7.1   Edge-Disjoint Routing and Anypath Routing

A simple form of cooperative routing is obtained if the goal of the cooperation is increased robustness. In that case, it is advantageous to route the message on several parallel routes through the network; these routes should have as few links and nodes in common as

---

[8] For one-hop problems we have $Q_l^r = 0$, and the algorithm reduces to maximizing a weighted sum of transmission rates over each link, minus a power cost, compare Sec. 20.5.

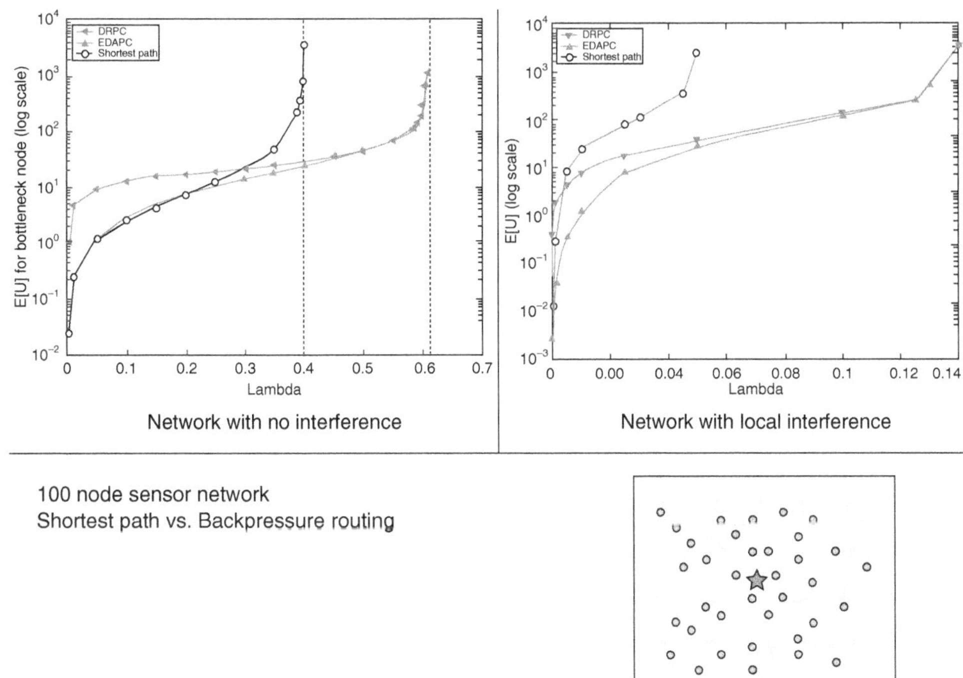

Figure 23.15 Performance of different routing algorithms in an ad hoc network: DRPC (backpressure algorithm), EDRPC (enhanced backpressure), and shortest-path.
Reproduced with permission from [Georgiadis et al. 2006] © NOW publishing.

possible, so that the failure of a particular link cannot lead to a blockage of all routes to a destination. *Edge-disjoint shortest-path routing* is a way of identifying routes that do not share any links; a suitable algorithm is a minor modification of the Bellman–Ford algorithm. Note, however, that this approach does not make significant use of the broadcast effect.

Anypath routing exploits the broadcast effect to achieve diversity: each node broadcasts the data packet to a group of neighbors, called the *forwarding set*. As long as at least one of the nodes in the forwarding set receives the message, a retransmission is not required. The successful node then acts as the next relay on the route, see Figure 23.16. Note, however, that only a single node from a forwarding set retransmits. In other words, the broadcast effect is used to obtain selection diversity. Thus, even if a particular link on the "nominal" shortest path goes down, the data packet can reach its destination, without necessity of finding a new route. Anypath routing is thus especially useful for links that frequently change in quality.

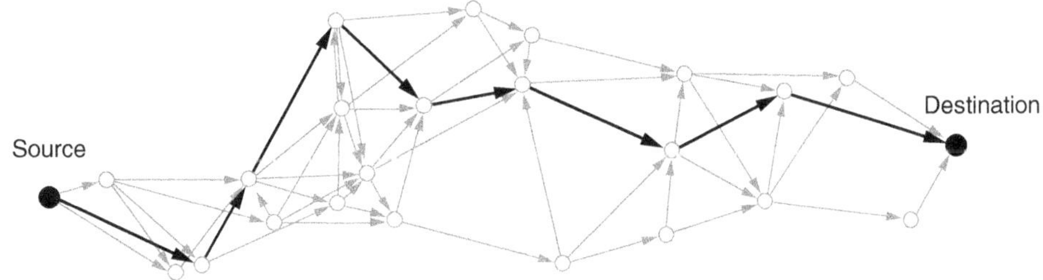

Figure 23.16 An anypath route (grey), and one possible trajectory taken by a packet (bold).
Reproduced with permission from [Dubois-Ferriere 2006] © EPFL Switzerland.

Instead of a specific route, anypath routing results in an ensemble of routes; depending on the outcomes of the transmissions from the various nodes, a packet can take different routes through the network. Formally speaking, an anypath route is the union of all possible trajectories along which a packet can travel from the source to the destination. Finding the best anypath route involves a trade-off: increasing the candidate set gives a better robustness (and might therefore help to decrease the required link margin, and thus transmit power, while retaining the same outage probability for message delivery). On the other hand, a larger forwarding set increases the danger that the packet is routed farther away from the true shortest path (which we would take if we had perfect, completely current, knowledge of all the network states). Additional complications arise when the data rate is also allowed to vary – changing the data rate changes the set of nodes that are, in principle, able to correctly receive the packet from a certain preceding node. In any case, however, variations of the Bellman–Ford algorithm can find the optimum anypath route (and associated rates) in polynomial time.

23.7.2 Routing with Energy Accumulation

Another way of exploiting diversity is energy accumulation at the relay nodes. This occurs when a node stores a received signal of a packet that is too weak for decoding and combines it with another signal of the same packet that arrives later. When using energy accumulation at the nodes instead of simple multi-hopping, the optimum route changes. A simple example is given in the following.

Example 23.2 *Consider a linear network with three nodes, where direct transmission from node A to node C has a path gain of 0.1, while transmission from node* A *to node B has a pathgain of 0.19, and similarly from node 2 to node 3.*

Let the threshold of the received signal energy for decodability of the signal be 1 J. Then in a classical multi-hop scenario, the optimum route is direct transmission, with transmit energy of node A equal to 10; multi-hop with route $A \to B \to C$ would require $2(1/0.19) = 10.53$ J. However, when the destination node can perform energy accumulation, the route $A \to B \to C$ becomes preferable: the source node 1 uses $1/0.19 = 5.26$ J for its transmission (so that node B can decode). During that transmission, node C receives $0.1 \cdot 5.26 = 0.53$ J energy from "overhearing" the packet. Thus, it requires only 0.47 J during the second transmission, which it obtains if node B transmits with $0.47/0.19 = 2.47$ J. Thus, total transmission energy is 7.63 J; less than the 10 J required for direct transmission.

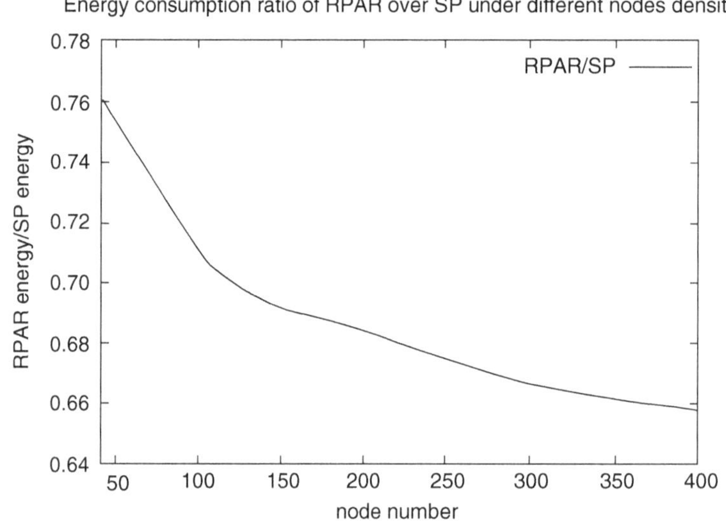

Figure 23.17 Energy savings of routing taking into account energy accumulating compared to shortest-path algorithm when network density increases. Reproduced with permission from [Chen et al. 2005] © IEEE.

The problem of finding the optimum route involves the issues of finding (i) what nodes should participate, and (ii) in what sequence, and with what power. Unfortunately, the former problem is NP-hard, i.e., it can be solved exactly only by trying out all possible node combinations, and picking the best one.[9] A number of heuristic algorithms exist for finding the best route. Some of those algorithms start out with the optimum multi-hop route (which is determined by the Dijkstra or Bellman–Ford algorithm) and then add on nodes that reduce the overall energy consumption. Another type of algorithm builds up the route from scratch, starting from the source node. When it adds the next relay on the path, it reduces the signal energy still required at all the other nodes; this energy reduction depends on how much energy those other nodes can "overhear" when the new node transmits. In either case, the energy savings from the energy accumulation (and associated routing) increase as the density of nodes increases: the closer the nodes, the more energy a node can "overhear," see Figure 23.17.

A somewhat different case of energy accumulation occurs when multiple nodes transmit, synchronously, in parallel, to effect higher receive power:

1. parallel nodes use orthogonal transmission (Section 27.3.3) or distributed space-time coding (Section 27.3.4) for transmission;
2. parallel nodes use distributed beamforming (Section 27.3.2).

Also, in this case, finding the optimum route is NP-hard. A heuristic method for finding a route is to subsume the nodes acting in parallel into a "super-node," and then try to find the best route of supernodes.

23.7.3 Information Accumulation

When appropriate coding is used, relay nodes can exploit "overhearing" the signals intended for other relay nodes in an even more efficient manner: they accumulate mutual information, instead of energy, as discussed in Section 27.3.6. Still, routing with mutual-

---

[9] For the case of cooperative broadcasting, the first problem is easy (since all nodes participate in the transmission), but the determination of the correct order of node participation is NP-hard.

information accumulation shares two important properties with energy accumulation: (i) finding an optimum route is NP-hard, and (ii) for heuristic algorithms, it is useful to break down the problem into two sub-problems: determination of the physical route or order of nodes through which packets propagate, and the allocation of resources (time, power) among the nodes. Under the assumption that each node has a fixed transmission power, the determination of the optimum resource allocation (time) can be done by a Linear Program (LP) for a specific routing order. A simple algorithm then revises the routing order based on the results of the LP. Iterating between the two sub-problems (resource allocation and routing order) yields an efficient approach to good route-finding even in large networks. Furthermore, the Lyapunov-drift method described in Section 23.6.12 can be modified to include mutual information accumulation.

23.7.4 Other Collaborative Routing Problems

Different Optimization Criteria

In the previous parts of this section, we often used "overall energy consumption" as criterion for optimization of a route. However, other criteria can be used in practice. To give but a few examples:

- *Network lifetime maximization:* network lifetime is usually defined as the time during which all nodes have sufficient energy to operate properly. Nodes in the center of a network are in particular danger of running out of energy, since they have the highest likelihood of acting as relays.
- *Message delay*: while the use of many hops through the network can decrease the energy consumption, it also increases the latency; in particular when the communication rate in the network is fixed.[10]
- *Network throughput*: when multiple messages are being transmitted, then the resulting interference decreases the overall throughput of the network. The amount of interference depends on the route as well as on the particular collaboration scheme.

Routing with Selfish Nodes

Up to now, we have considered routing with the goal of decreasing the *overall* energy consumption of the network. For either centralized or distributed routing algorithms, it is assumed that the nodes will obey an algorithm that maximizes the social benefits, not the individual benefits of the nodes. This assumption works well in ad hoc networks in industrial or military/security settings, where all nodes are under control of a single operator. However, in ad hoc networks made up of, e.g., laptops of individual users, the situation is different: every user asks "what is the benefit for me," or, in other words "why should I exhaust my battery in order to forward messages for somebody else?" There have to be proper incentives for users (typically, that their messages will also be forwarded by somebody else). Designing proper rules that maximize benefits for each user, while at the same time discouraging "rule breaking," is thus an interesting problem of networking.

It is easy to see from the above description that *game theory* can be applied to this problem. One class of game-theoretic approaches uses "virtual payments and credits": whenever a node acts as relay for somebody else's message, it receives a "virtual credit"; it can then spend it as payment to other nodes for forwarding its own message when the need arises. A second class of game-theoretic algorithms uses "enforcement" of good behavior either by a watchdog (centralized) or by "reputation-based" algorithms: a node that does not forward messages gets a bad reputation, which negatively affects its own ability to ask other nodes to forward messages for it. In either case, selfish behavior is discouraged by appropriate punishment through the other nodes.

Clustering and Partitioning

A cluster of nodes can use cooperative communications to bridge larger distances (via collaborative beamforming) than a single node can achieve. Thus, there is a better connectivity of a network, i.e., the probability that a node is isolated (cannot be reached by any route) is smaller for a network allowing collaborative beamforming than for a noncollaborative (multi-hop) network. An example for this is shown in Figure 23.18

Another case where clustering of nodes comes in handy is in networks where the number of hops is restricted (e.g., to two hops), but there are still a large number of nodes. In that case, it is required to find suitable cooperation nodes, in other words, which nodes should be "paired up" for the forwarding of a message. Choosing such node pairs can be considered a special case of so-called "matching problems on graphs," for which there is a rich literature in computer science and operations research. Particular examples include (i) minimal weighted matching, (ii) greedy matching, and (iii) random matching [Scaglione et al. 2006].

---

[10] With adaptive modulation and coding, a short link allows the use of a higher communication rate, so that the overall time for a message to reach the destination might actually decrease when many short (instead of one long) link is used.

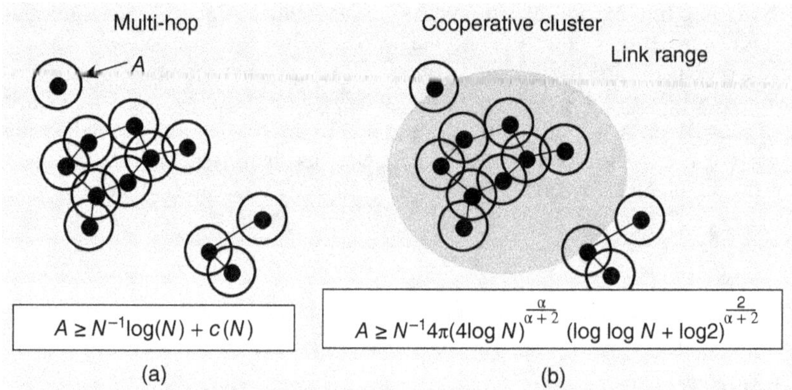

$$A \geq N^{-1}\log(N) + c(N)$$

$$A \geq N^{-1}4\pi(4\log N)^{\frac{\alpha}{\alpha+2}}(\log\log N + \log 2)^{\frac{2}{\alpha+2}}$$

(a)   (b)

Figure 23.18 Improved connectivity through use of cooperative clusters (b) compared to traditional multi-hop (a): collaborative beamforming of the cluster in the center increases the possible range. A denotes the required range of radio coverage to achieve connectivity with high probability; α is the pathloss exponent.

Reproduced with permission from [Scaglione et al. 2006] © IEEE.

23.7.5   Broadcasting in Ad hoc Networks

In a number of situations, e.g., emergency communications, data need to be communicated to all other nodes in the network. In contrast to infrastructure networks, where this would simply require an omni-directional transmission from the BS, the problem is more difficult in ad hoc networks since a single transmission cannot reach all nodes in the network. On the other hand, treating the problem as a sum of unicast transmissions (one to each node in the network) would be wasteful of resources. Rather, relaying via multiple nodes is required, though each relay transmission needs to be carefully coordinated.

A widely used protocol for this purpose is the *Barrage Relay Network* (BRN). It is a time-slotted protocol: in the first timeslot, a source node sends out a packet that is overheard by any node that is within range (first tier). Each first-tier node then broadcasts, in the next timeslot, the packet, such that the nodes in the vicinity of those nodes (second tier) can hear the packet, and then re-broadcast in the next timeslot, and so on. Critically, the retransmissions of all the nodes in a timeslot have to occur in such a way that an RX that hears multiple nodes sees those transmissions as a benefit, not as interference/collision. This requires several important ingredients: (i) alignment of the transmissions at least on a timeslot level (residual misalignment must be handled, e.g., by the cyclic prefix if OFDM transmission (Chapter 15) is chosen, (ii) ensuring that all transmitted packets are identical (implying, e.g., that no specific information about the transmit node is contained in the header), (iii) no node transmits a packet twice (i.e., a tier-1 node does not re-transmit when it hears a packet from a tier-2 node) and (iv) using a cooperative transmission method that does not require CSI at the TX (see Section 27.3). In the BRN protocol, the method chosen for the latter purpose is burst-wise phase dithering combined with LDPC codes. Since no node transmits twice, messages "ripple outward" from the source, which also means that a source can send off another packet while the first one is still propagating through far-away tiers, a process called "spatial pipelining" in BRN.

23.8   Scaling Laws

23.8.1   Definitions and Assumptions

As the number of nodes that want to communicate with each other in a network increases, the interference between the messages becomes worse and worse. A very active area of information-theoretic research is to investigate scaling laws, i.e., the functional dependence of the overall throughput of the network on the number of nodes, in the limiting case of a large number of nodes. Such a scaling law does not quantify the absolute throughput that can be achieved but rather only how much the throughput can be increased by, e.g., doubling of the number of nodes. Yet, this is a very important method for understanding whether increasing the number of nodes is worthwhile, or even beneficial. It must be noted, however, that there exist a large number of scaling law results that seem contradictory at first glance, because they were derived under different assumptions.

A random network is defined as a network where nodes are placed at random locations within a bounded area (typically a disk). Each node has a randomly chosen destination. The transmission can be done through multi-hop, over a distance d. For the transmission model, we need to distinguish between the protocol model and the physical model for transmission. The protocol model uses the idea of "guard zones," where the communication distance (with assumed fixed rate R) is called d and the guard zone extends from d to $d(1 + \Delta)$. In the physical model, the SINR at the receiver is computed based on the actually existing interference, summed up over all other transmitting nodes.

Let us next define the throughput capacity, which is the main quantity that scaling laws analyze: a per-node throughput $C(N)$ for a network with N nodes is feasible if there exists a (spatio-temporal) scheduling scheme such that each node sends C bits/s to its intended destination without errors. Note that the capacity requires an achieving scheduling scheme to exist, but in most cases, the scaling laws do not provide an explicit construction of the scheduling scheme. The capacity is said to be order $\Theta(f(N))$[11] if $C(N)$ is asymptotically upper and lower bounded by functions of shape $f(N)$; more precisely, there exist constants $c_1 > 0$ and $c_2 < \infty$ such that [Lu and Shen 2013]

$$\lim_{N \to \infty} \Pr[C(N) = c_1 f(N) \text{ is feasible}] = 1$$
$$\lim_{N \to \infty} \Pr[C(N) = c_2 f(N) \text{ is feasible}] < 1.$$

(23.23)

Another model distinction that has to be made is how the number of nodes is increased: in one approach, the density of nodes is kept constant, while the area in which the nodes are deployed (typically a disk) is scaled up. In an alternative approach, the size of the disk is kept constant, while the density of nodes increases. In that latter approach, great care must be taken to ensure a physically reasonable propagation model: if a d^{-n} law is used for the pathloss, then at very short distances (that become likely as the node density becomes large) the receive power might be higher than the transmit power (physically speaking, this may occur when the distance between nodes becomes smaller than the Rayleigh distances, compare Section 4.1). In this case, a limit on the received power must be introduced, otherwise, the results are not physically reasonable. Fading is not considered in either of these models.

23.8.2 Throughput Capacity of Random Networks

For a random network where the nodes are on a unit disk, and the transmission from each node has a fixed rate, only short-distance transmissions are possible. It can then be shown that for the protocol model, the feasible throughput is of the order

$$C(N) = \Theta\left[\frac{1}{\sqrt{N \log(N)}}\right].$$

(23.24)

This result holds both if the nodes are distributed on a disk, and on the surface of a unit sphere. The latter fact indicates that the limitation does not come from a "bottleneck" node at the center of a sphere, but from the interference that exists all over the network area. For the physical model, the lower bound of the feasible data rate per source scales as $1/\sqrt{N \log(N)}$, while the upper bound is slightly looser, $1/\sqrt{N}$. This result can be interpreted as follows: for multi-hop transmission, where each hop covers only a small distance, the guard zone can be smaller (which allows more nodes to be on simultaneously). On the other hand, a message needs to be sent over several hops, thus creating interference in the network for a longer time. If we now furthermore assume that all nodes are equal, generating traffic that needs to be forwarded to destination nodes, the overall transport capacity of a network covering an area A scales like $\sqrt{AN}$, which also implies that the capacity per node *decreases* like $1/\sqrt{N}$, and thus vanishes as N becomes large.

Performance improvements can be achieved by (i) power control, (ii) multi-packet reception, and (iii) directional antennas. The capacity gains compared to (23.24) that can be achieved this way in the protocol model are $\Theta\left[\sqrt{\log(N)}\right]$, $\Theta[\log(N)]$, and $\Theta[\log(N)^2]$, respectively. This is somewhat helpful, but since the logarithm increases less quickly than the square root function, the per-node throughput still goes to zero as the number of nodes increases to infinity.

23.8.3 Throughput Laws with Cooperation

Cooperation between nodes leads to a change in the scaling of the network throughput as the node density increases. For cooperative communications, the feasible throughput per node is at least constant, in other words, the aggregate network throughput increases linearly. More precisely, it has been shown that the network throughput cannot increase faster than $N \log(N)$; and furthermore there is a known, constructive scheme that achieves a network throughput that scales as N, i.e., the throughput per node is constant.

This well-scaling scheme is hierarchical cooperation, which as its basic building block contains a three-phase cooperation scheme based on clustering:

1. in the first phase, the source node transmits the information to surrounding nodes. To be more precise, we divide the area containing nodes into clusters, and the source node sends the information to the nodes located in the cluster. Note that by application of the cellular principle, and an appropriate reuse distance, transmission from source node to surrounding nodes can happen in many clusters in parallel. The source then divides the information into M blocks and sends one such block to a particular node in the cluster (this does not exploit the broadcast effect).[12]

2. In the second phase, the cluster of nodes performs MIMO transmission to the cell (cluster) in which the destination node is located. Each node independently encodes the information block it received in the first phase, thus providing (distributed) spatial multiplexing. The nodes in the receiving cluster quantize the received signal.

3. In the third phase, the nodes in the receiving cluster send this quantized information within the cluster; by appropriate decoding of the spatial-multiplexing signal, a node can recover the original information.

---

[11] The order notation $\Theta(N)$ employed in Section 23.8 has nothing to do with the queue vector $\Theta(t)$ used in Section 23.6.12.

[12] Some special cases arise when the source node and the destination node are in adjacent clusters; for details see [Ozgur et al. 2007].

This three-phase cooperation scheme can now be applied in a recursive manner, if distances need to be covered that are larger than what can be achieved (under given power constraints) with a single application of the three-phase scheme. The recursion starts with a small area and is applied over consecutively larger areas until it can encompass the whole network area, see Figure 23.19.

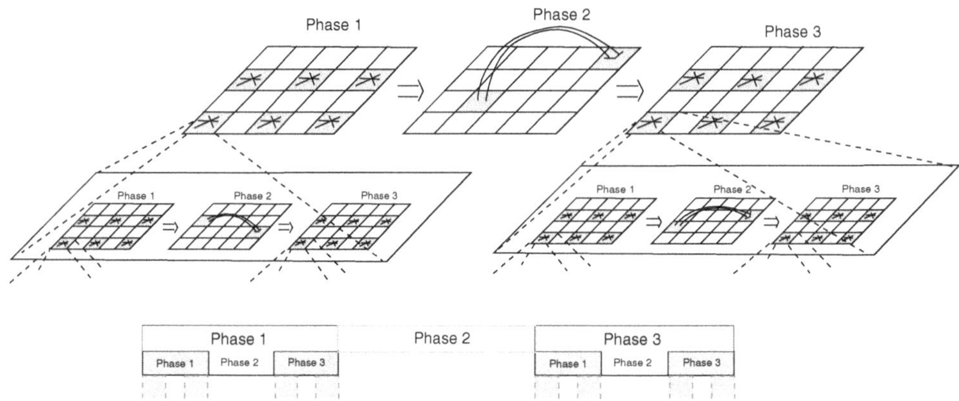

Figure 23.19 Principle of hierarchical cooperation.
Reproduced with permission from [Ozgur et al. 2007] © IEEE.

Another case in which the capacity can increase is in the presence of mobility. If we allow arbitrary (though finite) delay, then a two-hop strategy can greatly increase the throughput capacity, in that nodes can act as mobile data storage, i.e., receive data when they are close to the TX, and then retransmit them when they are close to the receiver. Also, in this case, the throughput per node can be constant.

Another case in which higher throughput scaling can occur is for wireless video distribution, or more generally when the content that a user requests follows a peaked popularity distribution, so that a limited number of files (e.g., videos) accounts for a majority of the traffic. For certain types of popularity distributions, the per-node throughput is independent of the network density even with simple single-hop transmission, the reason being that popular files are cached on many devices, and can thus be transmitted to the requesting nodes via short-distance, and thus spectrally very efficient, Device-to-Device (D2D) links. A more detailed consideration needs to trade-off the total throughput with the outage, i.e., the probability that a file is not cached at all in the network. Still, even at very small admissible outage probabilities dramatic increases of the throughput can be achieved, compare Figure 23.20.

23.9 Energy Management

23.9.1 Optimization Goals

Since nodes in ad hoc networks are typically battery operated, energy management is a very important aspect of ad hoc networks. While energy efficiency is *important* for infrastructure (to keep operating costs low), it is *vital* for mobile devices, since they become nonfunctional when the battery runs out. Energy management often pursues one of two goals in an ad hoc network:

(i) *minimization of the overall transmitted energy* – this is the optimization criterion that can be most easily handled mathematically, and it is also related to the total amount of interference generated by the network – or

(ii) *maximization of the network lifetime*, which can be defined either as the time until the first device becomes nonfunctional, or as the time until the network becomes disconnected.

To appreciate the distinction between the latter two quantities, consider a network that has two nodes very close to each other, and which act only as relay nodes (not as source or destination). The first relay node is the *sole* relay node that can also forward packets from C to C′, and the second relay node is the sole node that can forward packets from D to D′. Furthermore, both nodes can forward packets from A to A′; however, the first node has slightly better channels to A and A′. Then in order to optimize total energy consumption, the first node will always act as relay node between A and A′, while the other remains idle for these packets. Clearly, the first node will sooner run out of energy; network lifetime is much worse than if the two nodes are relaying those packets alternately. Now for the relaying from A to A′, the event that the first relay node runs out of battery is not catastrophic, since the second relay node can forward all packets afterward. However, since the first node is the only one that can forward from C to C′, the network is now partly disconnected.

23.9.2 Processes Influencing Energy Consumption

It is also important to understand the main sources of energy consumption in ad hoc networks. In many theoretical formulations, only the transmit power is considered. However, this does not agree with reality, for various reasons:

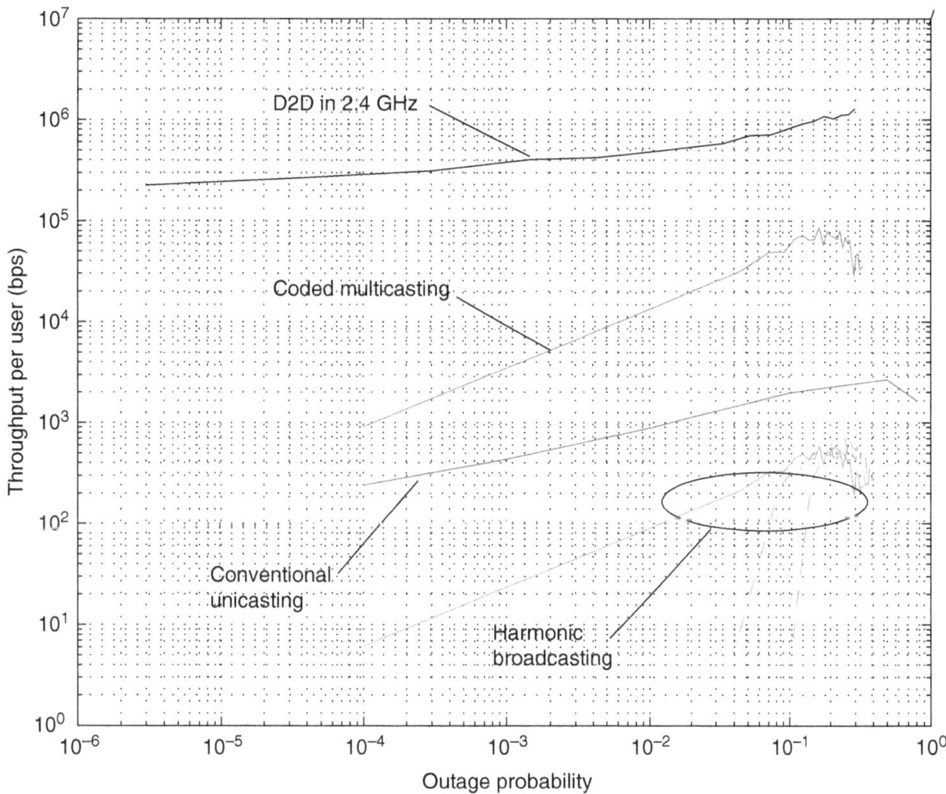

Figure 23.20 Simulation results for the throughput-outage trade-off for conventional unicasting, coded multi-casting, harmonic broadcasting (with different numbers of transmitted files), and the caching-plus-D2D-communication scheme under 2.4 GHz office channel models. Color version available at wiley.com/go/molisch/wireless3e.

(i) When the TX is active, there is a "power overhead" in addition to the transmit power, associated with the inefficiency of the power amplifier, as well as for all the other digital and analog functionality required to generate the transmit signal before amplification, compare Chapter 17.

(ii) The reception process may require considerable energy, due to the inefficiency of the amplification of received signals, as well as the signal processing required for detection and decoding. Obviously, the energy consumption for this processing increases with the transmission data rate.

(iii) The standby power: keeping the circuitry for reception and transmission up and running between the times that actual transmission/reception happens also requires some amount of power. While the power per unit time is much lower than the actual transmit or receive power, the time-integrated energy consumption can be significant especially in systems with low duty cycle. Consider, for example, a network that transmits one byte of temperature data every hour. The transmission energy of this system will be minuscule, while the standby power will be dominant.

23.9.3 Methods for Reducing Energy Consumption

To deal with the transmit energy consumption, power control can be used. Obviously, the transmit power should only be large enough so that the receiver can properly decode the data packet; any larger power is wasted. For a given modulation/coding format, this implicitly provides a rule for the power adaptation. Furthermore, even when adaptive modulation and coding are used, a trade-off between delay and energy consumption is possible, since transmission with a high bit rate requires more energy, as can be seen from Eq. (13.23). Power control can also be worked into the stochastic network optimization discussed in Section 23.6.12.

Another important aspect of energy consumption in multi-hop ad hoc networks is energy-efficient routing. Clearly, the energy consumption can be made as the optimization criterion (edge weight) when determining "shortest path" routes. In clustered structures, the cluster head has a significantly higher energy consumption, both because it has to forward more packets, and cannot easily go into sleep mode, because it has to listen to possible transmissions from the nodes in its cluster, as well as from other cluster heads. Thus, the cluster head should either be a node with outlet power, or the role of cluster head should rotate among the nodes in a cluster.

An important method for reducing the standby power is the introduction of a *sleep mode*, in which the circuitry is powered down during the times where no transmission/reception is expected, and only woken up upon actions. The question arises how the wakeup should occur: (i) at regularly scheduled intervals (in which case the device has to keep at least its clock circuitry up and running during sleep time, or (ii) upon receiving a wakeup signal; in this case, there must be a receive circuitry suitable for detecting the wakeup signal with very small power consumption (much smaller than for receiving payload signals), since obviously such circuitry must remain active all the time.

A further interesting method for increasing the lifetime of ad hoc networks is energy harvesting, where nodes collect energy from surroundings, be it sunlight, or electromagnetic radiation in other bands, e.g., from TV transmission, cellular TXs, etc. This energy can be used to replenish the battery of the node, and thus lead to an increased lifetime. A challenge in network optimization with energy-harvesting nodes is that the energy sources are usually time-varying and nondeterministic (e.g., whether the sun will shine), so that stochastic optimization is necessary.

23.10 Cellular vs. D2D Mode in Hybrid Networks

In D2D communications, a device might communicate either with another device, or the BS. We can thus distinguish between D2D mode and cellular mode (details about D2D implementation in LTE and 5G NR can be found in Sections 31.10 and 32.10, respectively). A key question in D2D systems is which links should be handled by D2D connections, and which should be handled by cellular connections – a problem often called *mode selection*. Generally, the goal of mode selection (and other optimizations in the D2D networks) is to maximize the sum rate or minimize the individual power for D2D communications subject to the constraint of a SINR threshold for the cellular users and/or D2D users, or a constraint of the maximum outage probability of the cellular users.

Furthermore, the D2D links can be categorized based on the spectrum in which D2D communication occurs, i.e., *In-band D2D* and *Out-of-band D2D*.

In-band D2D: In-band D2D technologies propose to use the cellular spectrum for both D2D and cellular links. The motivation for choosing in-band communication is usually the high level of control the operator has over licensed cellular spectrum. In-band D2D can be further divided into non-orthogonal and orthogonal categories. In non-orthogonal D2D, cellular networks and D2D networks are sharing the same spectrum whereas orthogonal D2D networks allow to communicate on cellular links on one hand, and on D2D links on the other hand. using orthogonal spectrum. In-band D2D can improve the spectrum efficiency of cellular networks by reusing spectrum resources (in the non-orthogonal case) or allocating dedicated cellular resources to D2D links that show better performance (in the orthogonal case). The key disadvantage of non-orthogonal D2D is the interference caused by D2D links to cellular networks (and vice versa). This interference can be mitigated by introducing suitable resource allocation methods, though these might be computationally intensive and have high complexity.

Out-of-band D2D: Out-of-band D2D technologies exploit unlicensed spectrum. The motivation behind this is to eliminate the interference issue between D2D and cellular link and retain all licensed spectrum for cellular links. Using unlicensed spectrum requires an extra interface and sometimes adopts other wireless technologies such as Wi-Fi Direct. It can be advantageous to give the control of the second interface/technology to the cellular network, since this provides best coordination between the different radios. On the other hand, less control overhead is achieved by keeping only cellular communications *controlled* while leaving the D2D communications to the users (named *autonomous*). While out-of-band D2D does not create interference between D2D and cellular users, it may suffer from the uncontrolled nature of unlicensed spectrum.

23.11 Mesh Networks

Mesh networks are very similar to ad hoc networks, but consist of *infrastructure nodes* between which multi-hop routing is performed. Figure 23.21 shows the structure of a mesh network and the different types of nodes: (i) BSs[13] provide the connectivity to the UEs (also called *Mobile Clients*, MC). The BSs forward their traffic through the Mesh Routers (MRs) to one of the Mesh Gateways (MG), which has a (typically wired) connection to the internet. Note that the transmission from UE to BS follows the traditional cellular principles, as described in Chapter 21. The Mesh Network has its unique properties due to the way how the information is transmitted between the BSs and the wired backbone network (traditional cellular networks assume the BSs to be directly connected to the backbone).

The similarity of the transmission *between the mesh nodes* (BS, MR, and MG) to ad hoc networks are obvious: the BSs are the data sources, the MGs the data sinks, and the transmission between sources and sinks occurs through multi-hop transmission. For this reason, much of the theory of ad hoc networks, including routing algorithms, can be applied with minor modifications to Mesh Networks. However, there are also some aspects that are special to Mesh Networks,

- *No mobility*: the mesh nodes do not move; once they have been established, their location remains unchanged. Also, since they are typically located above the average height of persons and cars, the propagation channels between the mesh nodes show only minor temporal variations (compare Section 5.6.2 for "fixed wireless" channels).
- *Multiple destinations*: the goal of the mesh network is to haul the data from the BSs to one of the MGs (or vice versa); it does not matter which MG it is, since all of them are wired to the same backbone network. This is somewhat different from both the unicast

---

[13] If mesh networks are based on WiFi infrastructure, it is common to talk about access points (APs) as constituent infrastructure nodes, not BSs. However, to stay consistent with the notation in the rest of the book, we use the expression BS.

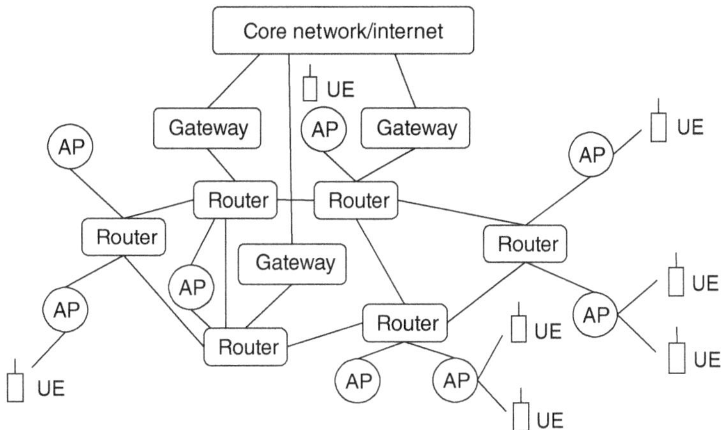

Figure 23.21 Mesh network structure.

situation (where there is only a single admissible destination node) or the multi-cast situation (where a data packet should reach all desired destination nodes, not just a single one of them). However, it must also be considered that no single MG should have too high a load, as it would lead to a congestion of the wired network at this location.

- *No energy constraints*: since mesh nodes are typically connected to power mains, energy management is not as important as for ad hoc networks.
- *Planability of the network*: while some mesh networks simply try to connect existing locations to extend coverage, in many cases the locations of the mesh nodes can be influenced by the network planner. It must be noted that the location of the BSs is mostly restricted by conventional cell planning restrictions (e.g., how to space nodes to get sufficient coverage, sufficient capacity for covering all users, etc.). The location of the MGs is limited by the requirement to access high-speed backhaul. The possibility of consciously selecting node locations (even with limitations) distinguishes mesh networks from ad hoc networks, where we assume by default that the nodes are randomly placed (the deployer of the network has no control) as well as possibly mobile.
- *Channelization*: while most of the ad hoc literature assumes that a packet transmission occupies the whole bandwidth, this is rarely assumed in mesh networks. Typically, the available spectrum is divided into a number of subchannels, which can then be assigned for different transmissions within the network. For example, a particular BS might transmit a data packet on subchannel 1 to mesh relay A, while transmitting a different data packet on subchannel 2 to mesh relay B, and so on. It is noteworthy that there is no *fundamental* reason why a channelization is commonly used in mesh networks but not ad hoc nodes. Rather, the additional hardware requirements (filters, digital signal processing, better synchronization) required for a frequency channelization make it an undesirable option in ad hoc networks but are not major obstacles in the more expensive mesh nodes. The use of efficient multi-channel routing algorithms is however critical for making efficient use of this capability.

Routing protocols are, in general, similar to the case of ad hoc networks. However, the better stability of the mesh networks should be taken into account when considering the trade-offs. In particular, proactive protocols are more competitive in a mesh setting, since the routes discovered in the initial discovery phase will remain valid for a long time. On-demand protocols should avoid repetitive route discovery, but reuse routes that have been discovered before. Mesh routing also has to consider the average load associated with each traffic flow, and ensure that there are no areas through which a disproportionate percentage of data flows, as this would create congestions. Distributing the traffic load among different areas/nodes can be done either in an ad hoc manner or can be incorporated into the channel assignment and routing.

If traffic and/or channels are strongly time-varying, dynamic channel assignment should be used. Dynamic assignments can be further classified according to whether assignments are changed on a per link, per packet, or per time-slot basis. Most of these assignments are based on CSMA-like schemes, possibly with RTS/CTS. The presence of multiple channels leads to a number of additional challenges; one example is the multi-channel hidden terminal problem, where a receiver does not hear the RTS/CTS exchange of another node pair in its range, because it is listening in a different channel. A variety of MAC protocols can be used to circumvent these issues.

The topology of the mesh network can be controlled by adjusting the transmit power of the nodes – the larger the transmit power of a node, the more connections to other nodes can be established. For this reason, the expressions "power control" and "topology control" are often used interchangeably.[14] However, all nodes using the same (static) power have been shown to give capacity close to the optimum. Another power control method that has been suggested is to adapt power just enough that each node can reach its intended destination in a single hop.

When multiple channels are available, the multi-channel assignment problem needs to be solved. We can firstly distinguish between static and dynamic assignments. Static assignments are easier to realize because optimized assignments require considerable overhead and computation, so that it is preferable to have to do them only once (or at least at very long intervals). A number of

---

[14] This is different from controlling the actual location of the mesh nodes, which of course also influences the network topology.

methods are based on graph-theoretical models, since each link (both desired link and created interference) can be considered to be an edge in a graph. Assigning channels to the various links can then be seen as a *graph coloring problem*, a well-known problem in computer science, where we aim to assign colors to edges (or vertices) in a graph such that no two adjacent edges (vertices) use the same color. Computation in polynomial time of the optimum graph coloring is possible only in some special cases. A number of approximate algorithms exist, and we refer here only to the literature (see section titled "Further Reading") for the treatment of these issues.

The descriptions above assume that each link in the multi-hop network has the same capacity, so that the "color" (channel assignment) is the only relevant quantity. In wireless channels, we can however expect that some channels have better quality than others. In this case, a more general description is required. Essentially, we wish to maximize the overall throughput, while fulfilling a set of constraints:

- the scheduling constraints: only one data stream on each subchannel entering or leaving a node,
- power allocation constraints: transmit power, summed over all subchannels, of each node must remain below the maximum transmit power,
- data rate of one data stream coming into a node (possibly summed over subchannels) has to equal the outgoing rate of that stream, and
- link capacity constraints: data rate cannot exceed the capacity of the wireless link, where notably this capacity is impacted by the amount of interference experienced from other transmitting nodes, and thus their scheduling assignment and power assignment.

The resulting optimization problem can be solved by various approaches, such as iterative convexification or greedy algorithms.

Further Reading

A general overview of wireless ad hoc networks, which discusses at length several of the topics mentioned in this chapter (multi-hop routing, energy management), is given in the monograph [Sarkar et al. 2013]; a short overview article is [Conti and Giordano 2014]. Aspects of the multiple access that are particular to ad hoc networks can be found in [Kumar et al. 2006]. Vehicular ad hoc networks are surveyed in [Al-Sultan et al. 2014]. Neighbor discovery in ad hoc networks is surveyed in [Kumar et al. 2006] and [Sun et al. 2014]. The FlashlinQ scheduling protocol was proposed in [Wu et al. 2013]; ITLinQ in [Naderializadeh and Avestimehr 2014].

There is a rich literature on routing in multi-hop networks. Many of the basics, which relate to computer networks in general, are explained in textbooks of computer science and operations research literature, e.g., [Peterson and Davie 2003], though the specifics of wireless routing are mostly described in research papers. A taxonomy of various protocols is given in [Boukerche et al. 2009, 2011]. Source routing is discussed in [Johnson et al. 2001]. AODV routing was introduced in [Perkins and Royer 1999]. Geography-based routing is discussed in [Stojmenovic 2002] and [Cadger et al. 2012], and cluster-based routing in [Martirosyan et al. 2008]. The spray-and-wait algorithm was proposed in [Spyropoulos et al. 2005] and directed diffusion in [Intanagonwiwat et al. 2003]. Joint routing and resource allocation with the backpressure algorithm are described in tutorial form in [Georgiadis et al. 2006] and backpressure with average-power optimization is treated in [Neely 2006]. A practical implementation of the backpressure algorithm is described in [Moeller et al. 2010], which contains a number of improvements such as LIFO queues and Expected Number of Transmissions (ETX) as utility function. A large number of convex-optimization-based papers exist for this problem as well (see, e.g., [Cruz and Santhanam 2003; Chiang 2005]).

A Bellman–Ford approach to anypath routing is developed in [Lott and Teneketzis 2006]. Anypath routing is described in [Laufer et al. 2009], and backpressure with anypath is in [Neely and Urgaonkar 2009]. Routing on cooperative networks for unicast is discussed in [Khandani et al. 2003, Chen et al. 2005]. Cooperative routing for multi-cast with energy accumulation is analyzed in [Maric and Yates 2004]. Routing in networks with selfish nodes is described in [Han and Poor 2009]. Routing protocols for multi-cast situations are surveyed in [Junhai et al. 2009]. The barrage routing protocol for broadcasting is described in [Blair et al. 2008] and [Halford and Chugg 2010].

The scaling laws for (noncooperative) multi-hop networks were derived in the landmark paper of [Gupta and Kumar 2000]. A number of further investigations have considered the impact of the pathloss model, and the possible impact of fading on the throughput capacity. The survey paper [Lu and Shen 2013] provides a good overview of the various results. The constructive method for achieving higher throughput was proposed in [Ozgur et al. 2007]. Ad hoc networks that allow caching on the devices, and where the desired content has a peaked popularity distribution, were first considered in [Golrezaei et al. 2013], and their scaling laws are derived in [Ji et al. 2015] and surveyed in [Lee et al. 2022].

Energy management in wireless sensor networks is surveyed in [Anastasi et al. 2008] For energy harvesting, the survey of [Ulukus et al. 2015] is recommended. D2D communications, in particular the questions of in-band and out-of-band, and scheduling, are surveyed in [Fodor et al. 2012, Asadi et al. 2014]. Mesh networks are discussed in [Pathak and Dutta 2010, Benyamina et al. 2011]. Mesh networks are also important in combination with mobile edge computing, i.e., combining communications and computations, see [Mao et al. 2017]. A more general form of edge computing, called "Augmented Information Services", is surveyed in [Cai et al. 2022].

For updates and errata for this chapter, see https://wides.usc.edu/students.html#textbooks.

Exercises

See Sec. 36.23 of Exercises.pdf at wiley.com/go/molisch/wireless3e

Part V

Advanced Transmission Techniques and Special Features

The previous sections have provided the fundamentals of sending information between transmitters and receivers. The topics in the current part fall into two main categories: (i) creation and handling of special types of information occurring in wireless systems, such as speech, video, and location, and (ii) advanced forms of information transmission, such as techniques that go beyond the techniques described in Parts III and IV.

We start out with a discussion on how to encode two types of information that have dominated earlier and current wireless systems, respectively, namely speech and video, which are covered in Chapters 24 and 25, respectively. While brute force quantization of the "source information" would lead to very large quantities of data that need to be transmitted, smart exploitation of the special structure of speech and video allows a much more efficient source coding (compression), and thus more efficient information transmission. Both the basic compression methods and their interaction with wireless transmission are discussed.

Spectrum is a precious, and thus expensive, resource. In order to ensure a better utilization, adaptive spectrum use, in which different systems share the spectrum, or so-called "secondary" systems might use spectrum not needed by the primary system, have been considered. Chapter 26 provides an overview of those techniques and their impact on current and possible future frequency regulations.

We have seen already in Chapter 23 that relaying information from one node to another can be an effective way of extending coverage. While the relaying scheme itself that we discussed there was very simple, namely decoding and forwarding, Chapter 27 discusses more advanced forms of relaying that are more spectrally efficient. The cooperation of different nodes in the transmission of the information can also enhance performance. Finally, this chapter will give a brief introduction to *network coding*, where the information intended for different destinations can be combined during the forwarding process, increasing the sum data rate that can be realized in the network.

Chapter 28 discusses advanced forms of handling interference. While traditionally interference has been treated like noise, and either is accepted, or reduced by exploiting its spatial structure (compare Chapters 12, 16, and 22), we can also exploit the fact that interference actually aims to provide information to somebody. By detecting this information and taking it into account in the detection of the desired information, i.e., multi-user detection, performance can be improved. Related to multi-user detection, NOMA (Nonorthogonal Multiple Access) also provides improved throughput. Finally, interference alignment allows to constrain interference into only a part of the signal space, such that the desired signal can be communicated in the remaining part without seeing interference.

Location information is of great importance in modern wireless systems, either by itself (e.g., for navigation and maps), or to support other features. Chapter 29 discusses the various methods to obtain location information, from time-of-arrival to fingerprinting of the received power. Also included is a discussion of the Global Positioning System (GPS) and the E911 localization features of cellphones.

24

Speech Coding

Gernot Kubin

Signal Processing and Speech Communication Laboratory, Graz University of Technology, Graz, Austria

24.1 Introduction

24.1.1 Speech Telephony as Conversational Multi-Media Service

When O. Nußbaumer succeeded in the first wireless transmission of speech and music in the experimental physics lab at Graz University of Technology in 1904, nobody would have predicted the tremendous growth in wireless multi-media communications 100 years after this historical achievement. Many new media types have emerged such as text, image, and video, and modern services range from essentially one-way media download, browsing, messaging, application sharing, broadcasting, and real-time streaming to two-way, interactive, real-time conversations by text (chat), speech, and videotelephony. Still, speech telephony is the backbone of all conversational services and continues as an indispensable functionality of any mobile communications terminal. It is for this reason that we will focus our discussion of source-coding methods on speech signals – i.e., on their efficient digital representation for transmission (or storage) applications.

The success story of digital speech coding started with the introduction of digital switching in the Public Switched Telephone Network (PSTN) using Pulse Code Modulated (PCM) speech at 64 kbit/s and continued with a cascade of advanced compression standards from 32 kbit/s in the early 1980s over 16 to 8 kbit/s in the late 1990s, all with a focus on long-distance circuit multiplication while maintaining the traditional high-quality level of wireline telephony (*toll quality*). For wireless telephony, the requirements on digital coding were rather stringent with regard to bit rate and complexity from the very beginning in the mid-1980s whereas some compromises in speech quality seemed acceptable because users either had never made the experience of mobile telephony before or, if so, their expectations were biased from the relatively poor quality of analog mobile radio systems. This situation changed quickly, and the first standards introduced at the beginning of the 1990s were completely overturned within 5 years with significant quality enhancements through new coding algorithms, maintaining a bit rate of about 12 kbit/s where the accompanying complexity increase was mitigated by related advances in microelectronics and algorithm implementation.

24.1.2 Source-Coding Basics

The foundations for source coding were laid by [Shannon 1959], who developed not only channel-coding theory for imperfect transmission channels but also *rate–distortion theory* for signal compression. The latter theory is based on two components:

- a stochastic source model which allows us to characterize the *redundancy* in source information; and
- a distortion measure which characterizes the *relevance* of source information for a user.

For asymptotically infinite delay and complexity, and certain simple source models and distortion measures, it can be shown that there exists an achievable lower bound on the bit rate necessary to achieve a given distortion level and, vice versa, that there exists an achievable lower bound on the distortion to be tolerated for a given bit rate. While complexity is an ever-dwindling obstacle, delay is a substantial issue in telephony, as it degrades the interactive quality of conversations severely when it exceeds a few 100 ms. Therefore, the main insight from rate–distortion theory is the existence of a three-way tradeoff among the fundamental parameters *rate, distortion*, and *delay*. Traditional telephony networks operate in circuit-switched mode where transmission delay is essentially given by the electromagnetic propagation time and becomes only noticeable when dealing with satellite links. However, packet-switched networks are increasingly being used for telephony as well – as in Voice over Internet Protocol (VoIP) systems – where substantial delays can be accumulated in router queues, etc. In such systems, delay becomes the most essential parameter and will determine the achievable rate–distortion tradeoff.

Wireless Communications: From Fundamentals to Beyond 5G, Third Edition. Andreas F. Molisch.
© 2023 John Wiley & Sons Ltd. Published 2023 by John Wiley & Sons Ltd.
Companion website: www.wiley.com/go/molisch/wireless3e

Source coding with a small but tolerable level of distortion is also known as *lossy coding* whereas the limiting case of zero distortion is known as *lossless coding*. In most cases, a finite rate allows lossless coding only for discrete amplitude signals which we might consider for *transcoding* of PCM speech – i.e., the digital compression of speech signals which have already been digitized with a conventional PCM codec. However, for circuit-switched wireless speech telephony, such lossless coders have two drawbacks: first, they waste the most precious resource – i.e., the allocated radio spectrum – as they invest more bits than necessary to meet the quality expectations of a typical user; second, they often result in a bitstream with a *variable rate* – e.g., when using a Huffman coder – which cannot be matched efficiently to the *fixed rate* offered by circuit-switched transmission.

Variable-rate coding is, however, a highly relevant topic in packet-switched networks and in certain applications of joint source–channel coding for circuit-switched networks (see Section 24.4.5). While Shannon's theory shows that, under idealized conditions, source coding and channel coding can be fully separated such that the two coding steps can be designed and optimized independently, this is not true under practical constraints such as finite delay or time-varying conditions where only a joint design of the source and channel coders is optimal. In this case, a fixed rate offered by the network can be advantageously split into a variable source rate and a variable channel code rate.

24.1.3 Speech Coder Designs

Source-coding theory teaches us how to use models of source redundancy and of user-defined relevance in the design of speech-coding systems. Perceptual relevance aspects will be discussed in later sections; however, the use of the source model gives rise to a generic classification of speech coder designs:

1. *Waveform coders* use source models only *implicitly* to design an adaptive dynamical system which maps the original speech waveform on a processed waveform that can be transmitted with fewer bits over the given digital channel. The decoder essentially inverts encoder processing to restore a faithful approximation of the original waveform. All waveform coders share the property that an increase of the bit rate will asymptotically result in lossless transcoding of the original PCM waveform. For such systems, the definition of a coding error signal as the difference between the original and the decoded waveform makes sense (although it is no immediate measure of the perceptual relevance of the distortion introduced).
2. *Model-based coders* or *vocoders* rely on an *explicit* source model to represent the speech signal using a small set of parameters which the encoder estimates, quantizes, and transmits over the digital channel. The decoder uses the received parameters to control a real-time implementation of the source model that generates the decoded speech signal. An increase of the bit rate will result in saturation of the speech quality at a nonzero distortion level which is limited by systematic errors in the source model. Only recently, model-based coders have advanced to a level where these errors have little perceptual impact, allowing their use for very-low-rate applications (2.4 kbit/s and below) with slightly reduced quality constraints. Furthermore, due to the signal generation process in the decoder, the decoded waveform is not synchronized with the original waveform and, therefore, the definition of a waveform error is useless to characterize the distortion of model-based coders.
3. *Hybrid coders* aim at the optimal mix of the two previous designs. They start out with a model-based approach to extract speech signal parameters but still compute the modeling error explicitly on the waveform level. This model error or *residual waveform* is transmitted using a waveform coder whereas the model parameters are quantized and transmitted as *side information*. The two information streams are combined in the decoder to reconstruct a faithful approximation of the waveform such that hybrid coders share the asymptotically lossless coding property with waveform coders. Their advantage lies in the explicit parameterization of the speech model which allows us to exploit more advanced models than is the case with pure waveform coders which rely on a single invertible dynamical system for their design.

The model-based view of speech-coding design suggests that description of a speech-coding system should always start with the decoder that typically contains an implementation of the underlying speech model. The encoder is then obtained as the signal analysis system that extracts the relevant model parameters and residual waveform. Therefore, the encoder has a higher complexity than the decoder and is more difficult to understand and implement. In this sense, a speech-coding standard might specify only the decoder and the format for transmitted data streams while leaving the design of the best matching encoder to industrial competition.

Further Design Issues

The reliance on source models for speech coder design naturally results in a dependence of coder performance on the match between this model and the signal to be encoded. Any signal that is not clean speech produced from a single talker near the microphone may suffer from additional distortion which requires additional performance testing and, possibly, design modifications. Examples of these extra issues are the suitability of a coder for *music* (e.g., if put on hold while waiting for a specific party), for severe *acoustic background noise* (e.g., talking from the car, maybe with open windows), for *babble noise* from other speakers (e.g., talking from a cafeteria), for *reverberation* (e.g., talking in hands-free mode), etc.

Further system design aspects include the choice between *narrowband* speech as used in traditional wireline telephony (e.g., an analog bandwidth from 300 Hz to 3.4 kHz with 8 kHz sampling frequency) and *wideband* speech with quality similar to Frequency Modulation (FM) radio (e.g., an analog bandwidth from 50 Hz to 7 kHz with 16 kHz sampling frequency), which substantially enhances user experience with clearly noticeable speech quality improvements over and above the Plain Old Telephone Service (POTS).

Second, even with the use of sophisticated channel codes, the *robustness of channel errors* such as individual bit errors, bursts, or entire lost transmission frames or packets is also a source-coding design issue (as discussed below in Section 24.4.5). Finally, within the network path from the talker to the listener, there may be several *codec-tandeming* steps where transcoding from one coding standard to another occurs, each time with a potential further loss of speech quality.

24.2 The Sound of Speech

While the "sound of music" includes a wide range of signal generation mechanisms as provided by an orchestra of musical instruments, the instrument for generating speech is fairly unique and constitutes the physical basis for speech modeling, even at the acoustic or perception levels.

24.2.1 Speech Production

In a nutshell, *speech* communication consists of information exchange using a natural *language* as its code and the human *voice* as its carrier. Voice is generated by an intricate oscillator – the vocal folds – which is excited by sound pressure from the lungs. In the view of wireless engineering, this oscillator generates a nearly periodic, Discrete Multi-Tone (DMT) signal with a fundamental frequency f_0 in the range of 100–150 Hz for males, 190–250 Hz for females, and 350–500 Hz for children. Its spectrum slowly falls off toward higher frequencies and spans a frequency range of several 1,000 Hz. From its relative bandwidth, it should be considered an *ultrawideband* signal,[1] which is one of the reasons why the signal is so robust and power-efficient (opera singers use no amplifiers) under many difficult natural environments.

The voice signal travels further down the *vocal tract* – the throat and the oral and nasal cavities – which can be described as an acoustic waveguide shaping the spectral envelope of the signal by its resonance frequencies known as *formant frequencies*. Finally, the signal is radiated from a relatively small opening (mouth and/or nostrils) in a large sphere (our head) which gives rise to a high-pass radiation characteristic such that the far-field speech signal has a typical spectral roll-off of 20 dB per decade.

Sound Generation

Besides the oscillatory voice signal ("phonation"), additional sound sources may contribute to the signal carrier, such as turbulent noise generation at narrow flow constrictions (in "fricative" sounds like [s] in "lesson") or due to impulse-like pressure release after complete flow closures (in "plosive" sounds like [t] in "attempt"). If the vocal folds do not contribute at all to the sound generation mechanism, the speech signals are called *unvoiced*, otherwise, they are *voiced*. Note that in the latter case, nearly periodic and noise-like excitation can still coexist (in "voiced fricative" sounds like [z] in "puzzle"), such sounds are often referred to as *mixed excitation* sounds.[2] Besides the three above-mentioned sound generation principles (phonation, frication, explosion) there is a fourth one which is often overlooked: *silence*. As an example, compare the words "Mets" and "mess." Their major difference lies in the closure period of the [t] which results in a silence interval between the vowel [ε] and the [s] in "Mets" which is absent in "mess." Otherwise, the two pronunciations are essentially the same, so the information about the [t] is carried entirely by the silence interval.

Articulation

While the sound generation mechanisms provide the speech carrier, the *articulation* mechanism provides its modulation using a language-specific code. This code is both sequential and hierarchical – i.e., spoken language utterances consist of a sequence of phrases built of a sequence of words built of a sequence of syllables built of a sequence of speech sounds (called *phonemes* if discussed in terms of linguistic symbols). However, the articulation process is not organized in a purely sequential way – i.e., the shape of the vocal tract waveguide is not switched from one steady-state pose to another for each speech sound. Rather, the articulatory gestures (lip, tongue, velum, throat movements) are strongly overlapping and interwoven (typically spanning an entire syllable), mostly asynchronous, and continuously evolving patterns, resulting in a continuous modulation process rather than in a discrete shift-keying modulation process. This *co-articulation* phenomenon is the core problem in automatic speech recognition or synthesis where we attempt to map continuously evolving sound patterns on discrete symbol strings (and vice versa).

24.2.2 Speech Acoustics

Source Filter Model

The scientific study of speech acoustics dates back to 1791 when W. von Kempelen, a high-ranked official at Empress Maria Theresa's[3] court, published his description of the first mechanical speaking machine as a physical model of human speech production. The

---

[1] This analogy becomes even more striking if we consider that the phase velocities of light and sound are related by a scale factor of approx. 10^6, showing that 1-kHz soundwaves have about the same wavelength as 1-GHz electromagnetic waves.
[2] Note that from our definitions, mixed excitation speech sounds are clearly a subclass of voiced speech.
[3] Empress Maria Theresa was Archduchess of Austria and Queen of Hungary in the 1700s.

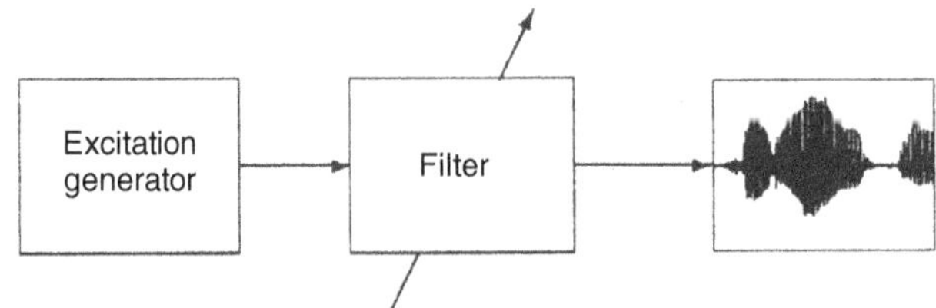

Figure 24.1 Source filter model of speech: The excitation signal generator provides the source which drives a slowly time-varying filter that shapes the spectral envelope according to formant frequencies to produce a speech waveform.

foundations of modern speech acoustics were laid by G. Fant in the 1950s [Fant 1970] and resulted in the *source filter* model for speech signals (see Figure 24.1).

While this model is still inspired from our understanding of natural speech production, it deviates from its physical basis significantly. In particular, all the natural sound sources (which can be located at many positions along the vocal tract and which are often controlled by the local aerodynamic flow) are collapsed into a single source which drives the filter in an independent way. Furthermore, there is only a single output whereas the natural production system may switch between or even combine the oral and nasal branches of the vocal tract. Therefore, the true value of the model does not lie in its accuracy in describing human physiology but in its flexibility in modeling speech acoustics. In particular, the typical properties of speech signals as evidenced in its temporal and spectral analyses are well represented by this structure.

Sound Spectrograms

An example of the typical time–frequency analysis for speech is shown in Figure 24.2.

The lower graph shows the time domain waveform for the phrase "*is the clear spring,*" spoken by a male speaker and limited to an analog bandwidth of 7 kHz for 16 kHz sampling. This graph shows the marked alternation between the four excitation source

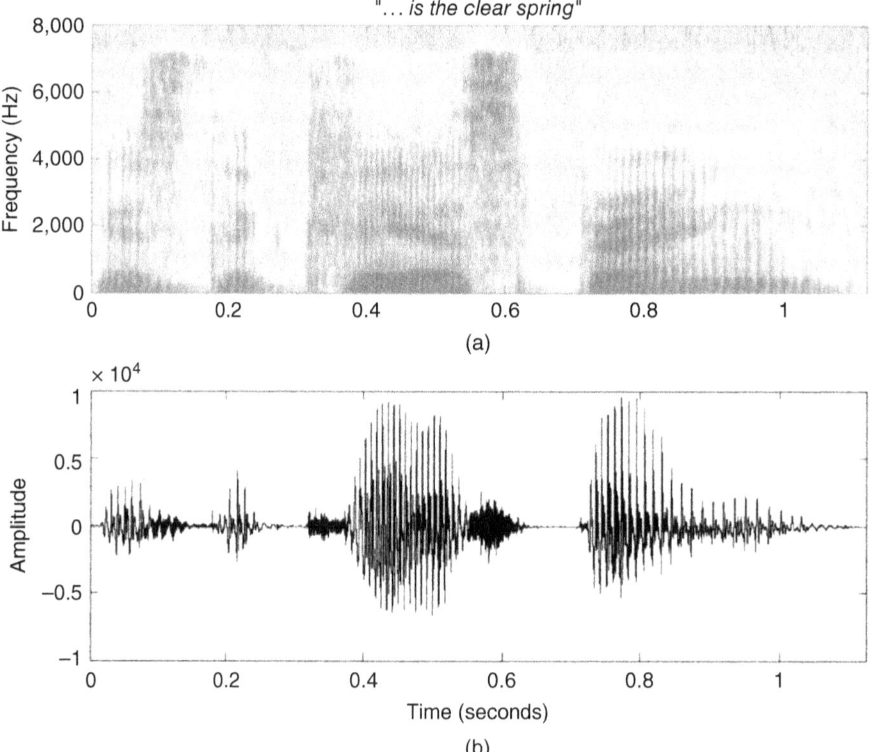

Figure 24.2 Spectrogram of the phrase "*is the clear spring*" spoken by a male speaker, analog bandwidth limited to 7 kHz, sampled at $f_s = 16$ kHz, horizontal axis = time, vertical axis = frequency, dark areas indicate high-energy density (a). Time-domain waveform of the same signal, vertical axis = amplitude (b).

mechanisms where we can note that the "nearly periodic" source of voiced speech can mean an interval of only three fundamental "periods" which show a highly irregular pattern in the case of the second voiced segment corresponding to "*the.*" Furthermore, strong fluctuations of the envelope are visible, which often correspond to 20–30 dB.

The upper graph shows a *spectrogram* of the same signal which provides the time–frequency distribution of the signal energy where darker areas correspond to higher energy densities. This representation can be obtained either by means of a filterbank or short-time Fourier analysis. It illustrates global signal properties – like anti-aliasing low-pass filtering at 7 kHz – and the observation that a significant amount of speech energy is found above 3.4 kHz (in particular for fricative and plosive sounds) suggesting that conventional telephony is too narrow in its bandwidth and destroys natural speech quality in a significant way.[4] Local properties are found in:

- noise-like signal components visible as broadband irregular energy distributions;
- impulse-like components visible as broadband spikes;
- nearly periodic components visible as narrowly spaced (around 10 ms for a male voice) vertical striations, corresponding to the pulse trains seen in the time domain;
- formant frequencies visible as slowly drifting energy concentration due to vocal tract resonances which apparently are very context-dependent as they differ even for the three occurrences of an [i] sound in this utterance.

The first three properties are related to the source or excitation signal of the source–filter model. The fourth property is related to the filter and its resonance frequencies or poles. The slow temporal variation of these poles models the temporal evolution of the formant frequencies.

24.2.3 Speech Perception

The ultimate recipient of human speech is the human hearing system, a remarkable receiver with two broadband, directional antennas shaped for spatiotemporal filtering (the outer ear) in terms of the individual, monaural *Head-Related Transfer Functions*, HRTFs (functions of both azimuth angle and frequency) which along with interaural delay evaluation give rise to our spatial hearing ability. Second, a highly adaptive, mechanical impedance-matching network (the middle ear) covers a dynamic range of more than 100 dB and a 3,000-channel, phase-locking, threshold-based receiver (hair cells in the inner ear's cochlea) with very low self-noise (just above the level where we could hear our own blood-flow-induced noise) converts the signal to an extremely parallel, low-rate, synchronized representation useful for distributed processing with low-power and imprecise circuits (our nervous system).

Auditory Speech Modeling

Auditory models in the form of *psychoacoustic* – i.e., behavioral – models of perception are very popular in audiocoding (like those for the ubiquitous MP3 standard) because they allow us to separate relevant from irrelevant parts of the information. For instance, certain signal components may be masked by others to such an extent that they become totally inaudible. Naturally, for high-quality audiocoding, where little prior assumptions can be made about the nature of the signal source and its inherent redundancy, it is desirable to shape the distortion of lossy coders such that it gets masked by the relevant signal components.

In speech coding, the situation is reversed: we have a lot of prior knowledge about the signal source and can base the coder design on a source model and its redundancy, whereas the perceptual quality requirements are somewhat relaxed compared with audiocoding – i.e., in most speech coders some unmasked audible distortion is tolerated (note that we have long learnt to live with the distortions introduced by 3.4 kHz band limitation which would never work when listening to music). Therefore, perceptual models play a lesser role in speech coder design, although a simple *perceptual weighting filter*, originally proposed by [Schroeder et al. 1979], has wended its way into most speech-coding standards and allows a perceptually favorable amount of noise shaping. More recent work on invertible auditory models for transparent coding of speech and audio is reviewed in [Feldbauer et al. 2005].

Perceptual Quality Measures

The proof of a speech coder lies in listening. Till today, the best way of evaluating the quality of a speech coder is by controlled listening tests performed with sizable groups of listeners (a couple of dozens or more). The related experimental procedures have been standardized in International Telecommunications Union (ITU-T) Recommendation P.800 and include both *absolute category rating* and *comparative category rating* tests. An important example of the former is the so-called *Mean Opinion Score* (MOS) test which asks listeners to rate the perceived quality on a scale from 1 = poor to 5 = excellent where traditional narrowband speech with logarithmic PCM coding at 64 kbit/s is typically rated with an MOS score in the vicinity of 4.0. With a properly designed experimental setup, high reproducibility and discrimination ability can be achieved. The test can be calibrated by using so-called anchor conditions generated with artificially controlled distortions using the ITU's Modulated Noise Reference Unit (MNRU).

Besides these one-way listening-only tests, two-way *conversational tests* are important whenever speech transmission suffers from delay (including source-coding delay!). Such tests have led to an overall planning tool for speech quality assessment, the so-called

---

[4] The impact of limited POTS bandwidth on intelligibility is evidenced from the need to use "alpha/bravo/charlie ..." spelling alphabets when communicating an unknown proper name over the phone.

E-model standardized by ITU-T Recommendation G.107 which covers various effects from one-way coding and transmission losses to the impact of delay on conversations or the subjective user advantage obtained from mobile service access.

As a means of bypassing tedious listening and/or conversational tests, objective speech quality measures have gained importance. They are often based on perceptual models to evaluate the impact of coding distortions, as observed by comparing the original and the decoded speech signal. Such measures are also called *intrusive* because they require transmission of a specific speech signal over the equipment (or network connection) under test in order to make this original signal in its uncoded form available for evaluation at the receiver site. One standardized example is the *Perceptual Evaluation of Speech Quality* (PESQ) (compare ITU-T Recommendation P.862).

Nonintrusive or single-ended speech quality measures are still under development. They promise to assess quality without having access to the original signal, just as we do as human listeners when we judge a telephone connection to have bad quality without direct access to the other end of the line. A first attempt (from May 2004) at standardization is the *Single-Sided Speech Quality Measure* (3SQM) (ITU-T Recommendation P.563).

24.3 Stochastic Models for Speech

24.3.1 Short-Time Stationary Modeling

With all the physical insights obtained from speech production and perception, speech coding has only the actual acoustic waveform to work with and, as this is an information-carrying signal, a stochastic modeling framework is called for. At first, we need to decide how to incorporate the time-varying aspects of the physical signal generation mechanism which suggests utilization of a *nonstationary*[5] *stochastic process* model. There are two "external" sources for this nonstationarity, best described in terms of the time variations of the acoustic wave propagation channel:

- The movements of the articulators which shape the boundary conditions of the vocal tract at a rate of 10–20 speech sounds/second. As the vocal tract impulse response typically has a delay spread of less than 50 ms, a *short-time stationary representation* is adequate for this aspect of nonstationarity.
- The movements of the vocal folds at fundamental frequencies from 100 to 500 Hz give rise to a nearly periodic change in the boundary conditions of the vocal tract at its lower end – i.e., the vocal folds. In this case, comparison with delay spread suggests that the time variation is too fast for a short-time stationary model. Doppler-shift-induced modulation effects become important and only a *cyclostationary representation* can cope with this effect (compare Section 24.3.5).

Most classical speech models neglect cyclostationary aspects; so, let us begin by discussing the class of short-time stationary models. For these models, stochastic properties vary slowly enough to allow the use of a subsampled estimator – i.e., we need to reestimate the model parameters only once for each speech signal *frame* where in most systems a new frame is obtained every 20 ms (corresponding to $N = 160$ samples at 8 kHz). For some applications – like signal generation which uses the parameterized model in the decoder – the parameters need to be updated more frequently (e.g., for every 5 ms subframe) which can be achieved by interpolation from available frame-based estimates.

Wold's Decomposition

Within a frame, the signal is regarded as a sample function of a stationary stochastic process. This view allows us to apply *Wold's decomposition* [Papoulis 1985] which guarantees that *any stationary stochastic process* can be decomposed into the sum of two components: a *regular component* $x_n^{(r)}$ which can best be understood as *filtered noise* and which is not perfectly predictable using linear systems and a *singular component* $x_n^{(s)}$ which is essentially a *sum of sinusoids* that can be perfectly predicted with linear systems:

$$x_n = x_n^{(r)} + x_n^{(s)}. \tag{24.1}$$

Note that the sinusoids need not be harmonically related and may have random phases and amplitudes. In the following three subsections, we will show that this result – already established in the theory of stochastic processes by 1938 – serves as the basis for a number of current speech models which only differ in the implementation details of this generic approach. These models are the Linear Predictive voCoder (LPC), sinusoidal modeling, and Harmonic + Noise Modeling (HNM).

---

[5] An alternative view would introduce a hypermodel that controls the evolution of time-varying speech production model parameters. A stationary hypermodel could reflect the stationary process of speaking randomly selected utterances such that the overall two-tiered speech signal model would turn out to be stationary.

24.3.2 Linear Predictive voCoder (LPC)

The first implementation of Wold's decomposition emphasizes its relationship to the Linear Prediction (LP) theory as first proposed by [Itakura and Saito 1968]. The model derives its structure from the representation of the regular component as white noise run through a linear filter, which is causal, stable, and has a causal and stable inverse – i.e., a *minimum-phase filter*. The same filter is used for the generation of the singular component in voiced speech where it mostly consists of a number of harmonically related sinusoids which can be modeled by running a periodic pulse train through the linear filter. This allows flexible shaping of the spectral envelope of all harmonics but does not account for their original phase information because the phases are now determined by the phase response of the minimum-phase filter which is strictly coupled to its logarithmic amplitude response (via a Hilbert transform). Furthermore, as there is only one filter for two excitation types (white noise and pulse trains), the model simplifies their additive superposition to a *hard switch in the time domain*, requiring strict temporal segregation of noise-like and nearly periodic speech (compare the decoder block diagram shown in Figure 24.3). This simplification results in a systematic model mismatch for mixed excitation signals.

Linear Prediction Analysis

The LPC encoder has to estimate the model parameters for every given frame and to quantize and code them for transmission. We will only discuss estimation of the LP filter parameters here; estimation of the fundamental period $T_0 = 1/f_0$ is treated later in this text (Section 24.4.3). As a first step, specification of the minimum-phase filter is narrowed down to an *all-pole filter* – i.e., a filter where all transfer function zeros are concentrated in the origin of the z-plane and where only the poles are used to shape its frequency response. This is justified by two considerations:

- The vocal tract is mainly characterized by its resonance frequencies and our perception is more sensitive to spectral peaks than to valleys.
- For a minimum-phase filter, all the poles and zeros lie inside the unit circle. In this case, a zero at position z_0 with $|z_0| < 1$ can be replaced by a geometric series of poles, which converges for $|z| = 1$:

$$\left(1 - z_0 z^{-1}\right) = \frac{1}{1 + z_0 z^{-1} + z_0{}^2 z^{-2} + \cdots}. \tag{24.2}$$

This allows us to write the signal model in the z-transform domain as

$$X(z) = \frac{U(z)}{A(z)} \tag{24.3}$$

where the speech signal is represented by $X(z)$, the excitation signal by $U(z)$, and the filter transfer function by $1/A(z)$. In the time domain, this reads as

$$x_n = -\sum_{i=1}^{m} a_i x_{n-1} + u_n \tag{24.4}$$

where the filter or predictor order is chosen as $m = 10$ for speech sampled at 8 kHz and the parameters a_i are known as *predictor coefficients*, with normalization[6] $a_0 = 1$.

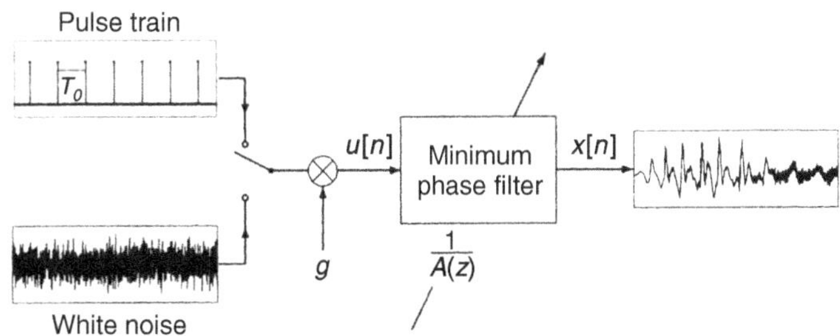

Figure 24.3 Linear predictive vocoder signal generator as used in decoder.

---

[6] The gain g can be included in the excitation signal amplitude.

For a given speech frame, we estimate model parameters $\hat{a}_i$ such that model mismatch is minimized. This mismatch can be observed through the prediction error signal e_n:

$$e_n = x_n - \hat{x}_n = x_n - \left(-\sum_{i=1}^{m} \hat{a}_i x_{n-i} \right) = \sum_{i=1}^{m} (\hat{a}_i - a_i) x_{n-i} + u_n. \tag{24.5}$$

For uncorrelated excitation u_n, the prediction error power achieves its minimum iff $\hat{a}_i = a_i$, for $i = 1, ..., m$. In this case, the prediction error signal e_n becomes identical to the excitation signal u_n. Note that we use the prediction framework only for model fitting, not for forecasting or extrapolation of future signal samples. To apply this estimator to short-time stationary speech data, for every frame with an update rate of N samples, a window of $L \geq N$ samples is chosen, where the greater sign means that the windows have some overlap or lookahead. Typically, a special window function w_n is applied to mitigate artifacts due to the data discontinuity introduced by the windowing mechanism. Asymmetric windows with their peak close to the most recent samples allow a better compromise between estimation accuracy and delay. The window can be applied to the data in two different ways, giving rise to two LP analysis methods:

1. The *autocorrelation method* defines the prediction error power estimate based on a windowed speech signal:

$$\tilde{x}_n = \begin{cases} w_n \cdot x_n, & \text{for } n = 0, 1, ..., L-1 \\ 0, & \text{for } n \text{ outside the window} \end{cases} \tag{24.6}$$

as

$$\sum_{n=-\infty}^{\infty} e_n^2 = \sum_{n=-\infty}^{+\infty} \left(\tilde{x}_n + \sum_{i=l}^{m} \hat{a}_i \tilde{x}_{n-i} \right)^2 \tag{24.7}$$

where data windowing results in an implicit limitation of the infinite sum.

2. The *covariance method* defines the prediction error power estimate via a windowing of the error signal itself without explicit data windowing:

$$\sum_{n=m}^{L-1} (w_n \cdot e_n)^2 = \sum_{n=m}^{L-1} w_n \left(x_n + \sum_{i=1}^{m} \hat{a}_i x_{n-i} \right)^2 \tag{24.8}$$

where the summation bounds are carefully chosen to avoid the use of speech samples outside the window.

In both methods, minimization of the quadratic cost function leads to a system of linear equations for the unknown parameters $\hat{a}_i$. In the autocorrelation method, the system matrix turns out to be a proper correlation matrix with *Toeplitz* structure which allows a computationally efficient solution using the order-recursive *Levinson–Durbin algorithm*. This algorithm reduces the operation count of the estimator from $\mathcal{O}(m^3)$ to $\mathcal{O}(m^2)$ and guarantees that the roots of the polynomial $\hat{A}(z)$ lie inside the unit circle. In the covariance method, no such structure can be exploited such that higher complexity and additional mechanisms for stabilizing $\hat{A}(z)$ are required. Its advantage is the significantly increased accuracy of estimated coefficients because the absence of explicit data windowing avoids some systematic errors of the autocorrelation method. Finally, to enhance the numerical properties of LP analysis, additional (pre-)processing steps are routinely included such as high-pass prefiltering of input speech to remove unwanted low-frequency components, a bandwidth expansion applied to correlation function estimates and a correction of the autocorrelation at lag 0 which corresponds to the addition of a weak white noise floor (at $-40\,$dB of the data).

24.3.3 Sinusoidal Modeling

The second implementation of Wold's decomposition emphasizes the idea of spectral modeling using sinusoids as proposed by [McAulay and Quatieri 1986]. In this model, both signal components are produced by the same *sum of sinusoids*. The noise-like regular component can well be approximated by such a sum (as suggested by spectral representation theory) if the relative phases of the sinusoids are randomized frequently, at least once for each frame. This scheme offers the advantage of retaining the original phase information in the singular component (if the available bit rate allows us to code it, of course) and it is more flexible in combining regular and singular components. Typically, even voiced speech contains a noise-like component in the higher frequency bands which can be modeled by using a fixed-phase model for the lower harmonics and a random-phase model for the higher harmonics. This *hard switch in the frequency domain* assumes the segregation of noise-like and nearly periodic signal components along the frequency axis. While this allows the modeling of mixed excitation speech signals, the transitions between excitation signal types are now a priori restricted to the frame boundaries (whereas the LPC allows in principle higher temporal resolution for voicing switch update).

A further development of sinusoidal modeling is the Multi-Band Excitation (MBE) coder which allows separate voicing decisions in multiple frequency bands, thereby achieving a more accurate description of mixed excitation phenomena. This also establishes a relationship to the principles of subband coding or transform coding which do not rely on specific source signal models and which are most popular in coding of generic audio.

24.3.4 Harmonic + Noise Modeling

The third implementation of Wold's decomposition strives at a full realization of the additive superposition of regular and singular components *without enforcing a hard switch in either the time or frequency domains*. Thus, the HNM is maybe the simplest in its decoder structure which just follows Eq. (24.1), whereas the encoder is the most difficult as it has to solve the problem of simultaneous estimation of superimposed continuous spectra (the regular component) and discrete spectra (the singular component, assumed to be harmonically related). So far, use of this model has been confined to applications in speech and audio synthesis, whereas for speech coding it has been recognized that the level of sophistication achieved in the HNM-spectral representation needs to be complemented by a more detailed analysis of time-domain variations, too, leading us beyond the conventional short-time stationary model.

24.3.5 Cyclostationary Modeling

The short-time stationary model of speech neglects the rapid time variations induced by vocal fold oscillations. For voiced speech, this nearly periodic oscillation not only serves as the main excitation of the vocal tract but also as a cyclic modulation of all signal statistics. Stochastic processes with periodic statistics are known as *cyclostationary processes*. For our purposes, where the fundamental period T_0 and the associated oscillation pattern may slowly evolve over time, we need to adapt this concept to short-time cyclostationary processes. For a given speech frame, the waveform will be decomposed in a cyclic mean signal (corresponding to the singular component of Wold's decomposition) and a zero-mean noise-like process with periodically time-varying correlation function. Most importantly, the variance of this noise-like component is a periodic function of time representing the periodic envelope modulation of the noise-like component of voiced speech. This effect is clearly audible for lower-pitched male voices and constitutes one of the main improvements of cyclostationary speech models over HNMs. These models were originally introduced as (Prototype) Waveform Interpolation (PWI) coders by [Kleijn and Granzow 1991] where the cyclic mean is interpreted as the "slowly evolving waveform" and the periodically time-varying noise-like component is termed the "rapidly evolving waveform." More recent developments work with filterbanks whose channels are adapted to multiples of f_0 by prewarping the signal in the time domain. A key aspect of cyclostationary signal modeling is the extraction of reliable "pitch marks" that allow explicit time-domain *synchronization* of subsequent fundamental periods. As all speech properties evolve over time, this synchronization problem is a major challenge and leads to the characterization of speech signals in terms of an underlying self-oscillating nonlinear dynamical system with the added benefit of explaining certain nonadditive irregularities of speech oscillation such as jitter or period-doubling phenomena.

24.4 Quantization and Coding

Once the signal model and the parameter estimation algorithms have been selected, the encoder has to quantize the parameters and, in hybrid coders, samples of the modeling residual waveform as well.[7] To achieve source-coding efficiency, several techniques beyond the simple uniform quantization scheme found in most analog-to-digital converters have been developed.

24.4.1 Scalar Quantization

Scalar quantization refers to the quantization of a single random variable which can be either a waveform sample or a single model parameter. If there is more than one variable to be quantized then they are treated independently of each other. As a general example, the continuous (or high-resolution discrete) value x of the random variable X is quantized to the ith quantization interval $[t_i, t_{i+1}]$ if $t_i \leq x < t_{i+1}$ and the index $i = Q(x)$ will be transmitted using a (binary) codeword. The receiver decodes the codeword back to the quantizer index and uses it to address a lookup table that stores a high-resolution reconstruction value $x^{(q)} = Q^{-1}(i)$. The latter

---

[7] Note that this is the case for *forward-adaptive* speech coders. For extremely low delay, *backward-adaptive* coders avoid the transmission of parameters as side information and rather estimate model parameters making exclusive use of speech samples which have already been decoded. This requires implementation of both encoder and decoder algorithms in the transmitter and works under the assumption of low channel error rates. The parameters estimated from decoded data only implicitly contain some delay relative to those parameters found from the current frame in forward-adaptive schemes. However, due to the short-time stationarity assumption, this delay results in an acceptable loss in modeling performance which is approximately compensated by the bit rate reduction obtained from suppression of the side information channel.

operation is sometimes referred to as "inverse quantization" although the static nonlinearity of a quantizer is a many-to-one map and, as such, has no unique inverse function. A typical performance measure for scalar quantizers is the Signal-to-Quantization Noise Ratio (SQNR) based on the mean-square value of the quantization error $q = x^{(q)} - x$:

$$SQNR_{[dB]} = 10\log_{10}\frac{E(x^2)}{E(q^2)} = 10\log_{10}\frac{E(x^2)}{E\left((x^{(q)}-x)^2\right)} \qquad (24.9)$$

where the operator E is the expectation operator performing an average weighted by the probability measure of the variable X. This definition includes both overflow distortions when the input value reaches the boundaries of the quantizer range and granular distortions related to the individual quantizer steps. In most designs, overflow will be avoided by using proper scaling such that granular distortions will be the quality-determining effect. Normalization of error power by input power makes the measure somewhat more relevant to human perception when dealing with waveform quantization.

In uniform quantization, all intervals have the same width $t_{i+1} - t_i = q = $ constant. The most prominent nonuniform scalar quantization law is 8-bit *logarithmic* quantization for PCM speech signals with two variants of the logarithmic characteristic, the μ-law in the U.S.A and Japan and the A-law in Europe. Logarithmic quantization has the remarkable property that the SQNR becomes nearly independent[8] of the single power $E(x^2)$ as long as overflows are avoided and the signal does not fall in the close vicinity of 0 (where the log function is approximated with a linear function, corresponding to the resolution of a 12-bit uniform quantizer). Thus, logarithmic quantizers are designed to cope with the strong temporal variations of the speech signal envelope which are both intrinsic (short-time stationary sequence of highly different speech sounds, soft versus loud voice) and extrinsic (variable distance to the microphone).

Adaptive quantization extends the dynamic range of a quantizer even further than logarithmic quantization such that input power fluctuations of 40 dB and more show no pronounced effect on SQNR performance. It is implemented with an adaptive gain control mechanism, operating on a sample-by-sample backward-adaptive basis in one of the most popular variants.

If the random variable X itself has a nonuniform amplitude probability distribution, the uniform quantizer does not achieve the minimal SQNR performance for a given number of quantization intervals I. The set of *optimal quantizer* thresholds t_i, $i = 1, ..., I$ and reconstruction values $x^{(q)}(i)$, $i = 1, ..., I$ can be found using the iterative Lloyd–Max algorithm which alternates between the two following implicit conditions found from minimizing the mean-square quantization error for fixed input power:

1. The best quantization thresholds lie at equal distance between adjacent reconstruction values:

$$t_i = \frac{x^{(q)}(i) - x^{(q)}(i-1)}{2}. \qquad (24.10)$$

2. The best reconstruction values lie in the centers of mass of the quantization intervals, computed as a conditional expected value using the input amplitude probability density $f_X(x)$:

$$x^{(q)}(i) = E(x \mid t_i \le x < t_{i+1}) = \int_{t_i}^{t_{i+1}} f_X(x)x\, dx. \qquad (24.11)$$

In practice, probability distributions are often not known such that the center-of-mass computations are done by evaluating the class-conditional means of a large set of training data samples collected in real-world experiments.

24.4.2   *Vector Quantization*

Vector Quantization (VQ) combines d scalar random variables X_k into a d-dimensional random vector **X**. In the d-dimensional space, the scalar quantization intervals turn into multi-dimensional, convex *Voronoi cells* V_i, $i = 1,..., I$. If an input vector $\mathbf{x} \in V_i$, then $Q(\mathbf{x}) = i$ and the value i is mapped to a binary codeword for transmission. [Shannon 1949] showed that using VQ with asymptotically increasing dimension d can achieve the rate distortion bound for source coding – i.e., just like in channel coding the best coding performance is obtained when considering large blocks of signal samples simultaneously. VQ is known to have three different advantages over scalar quantization. For this comparison, we will study the number of bits per dimension $\log_2(I)/d$ needed to achieve a certain quantization distortion.

---

[8] Actually, for logarithmic quantization, the SQNR is essentially independent of the amplitude probability distribution of the input, thereby making it maximally *robust* to unknown or strongly varying signal properties.

1. The *memory advantage* – i.e., the handling of statistical dependences: in general, a random vector exhibits statistical dependences among its components, which may go beyond linear correlations. Therefore, even if we use linear transforms (such as the Discrete Cosine Transform, DCT, or principal component analysis) to decorrelate the components of the random vector, the joint quantization of all components is still more efficient due to the redundancy contained in the remaining (nonlinear) dependences.

2. The *space-filling advantage*: in scalar quantization, there is not much choice in selecting the shape of quantization intervals, whereas, in multiple dimensions, Voronoi cells can be shaped to achieve the best space-filling degree – e.g., according to a dense sphere-packing principle. This allows coverage of a certain volume in space with the same maximal distortion (given by a linear dimension of the Voronoi cell) but fewer cells – i.e., a lower number $\log_2(I)$ of codeword bits. This advantage approaches 0.25 bit/dimension for large vector dimension d. For low-rate speech coding at 4 kbit/s or less (which means less than 0.5 bit/sample), this becomes a very significant contribution.

3. The *shape advantage*: this is very similar to the gains achieved for optimal nonuniform quantizers of scalar variables where the shape (and size) of the quantizer cells are matched to the probability distribution of the random vector. This advantage is only relevant for resolution-constrained quantizers with a predetermined number of cells I. For entropy-constrained quantizers (where we only care about the average entropy of all codewords but we neither constrain the number of cells or codewords nor the length of the codewords), there is no shape advantage as this can equally be obtained by lossless entropy coding of the codewords themselves.

The design of optimal VQ proceeds along the same lines as the design of nonuniform quantizers for scalar variables. The generalization of the Lloyd–Max algorithm to multiple dimensions is known as the Linde–Buzo–Gray or LBG algorithm.

Suboptimum Vector Quantization

The larger the vector dimension d, the better the VQ performance. However, if we design a VQ for a certain *given number of bits per dimension*, this means that the complexity of the VQ grows exponentially with dimension d because I, the number of Voronoi cells, grows exponentially with d. As multi-dimensional Voronoi cells may have complicated shapes, we rather define them by their reconstruction vectors $\mathbf{x}_q(i)$ and perform quantization by selecting the index $i = Q(\mathbf{x})$ which minimizes the quantization distortion $D(\mathbf{x}_q(i), \mathbf{x})$ over all $i = 1,..., I$. A typical distortion measure would be the quadratic distortion or Euclidean distance $\|\mathbf{x}_q(i) - \mathbf{x}\|^2$. Thus we need to memorize a table or *codebook* of I vectors of dimension d and we need to make a full search through this codebook to minimize distortion, which requires I distortion measure computations where each has a cost proportional to d in case of Euclidean distance. So even per dimension, the memory cost and the number of operations is directly proportional to I and, therefore, suffers from the curse of (= exponential growth with) dimensionality.

To address this issue, simplified schemes for VQ design and implementation have been developed. Surprisingly, it turns out that the suboptimal search for a good vector in a codebook filled with suboptimal entries can still outperform an optimal VQ system of lower dimension. While early attempts at VQ-based waveform quantization worked at 1 bit/dimension with a full search over $I = 2^{10} = 1,024$ entries of dimension $d = 10$, the new suboptimal systems work at 0.875 bit/dimension with a greedy search over $I = 2^{35} = 34.4 \cdot 10^9$ entries of dimension $d = 40$ and still achieve better performance.[9] Examples of such simplified, suboptimal designs are as follows:

- *Gain shape quantization* where the codevector is the product of a scalar gain and a unit vector (norm 1).
- *Multi-stage VQ* where a first codebook provides a coarse approximation and the subsequent $K - 1$ codebooks successively refine the quantization.
- *Split VQ* where the high-dimensional vector is split into K pieces of dimension d/K each, which are then quantized independently.

In the two latter cases, complexity is approximately reduced to that of a d/K-dimensional VQ, as nonexponential factors do not matter much for such comparison.

As one of the most successful simplifications – having made it into several international standards – we discuss *algebraic codebooks* where codevector amplitudes are restricted to a ternary alphabet $\{-1, 0, +1\}$ and where, furthermore, only a small percentage of nonzero amplitudes is allowed for (e.g., 10 in a vector of dimension 40). This design is inspired from the earlier techniques of "multi-pulse excitation" and "Regular Pulse Excitation" (RPE) where the source excitation of a source–filter model was pruned to a few nonzero amplitudes. As Euclidean distance computations consist in the evaluation of inner products, the algebraic structure reduces each evaluation from 40 multiply–add operations to 10 additions only. Furthermore, the selection of nonzero pulses follows a greedy scheme that does not try all possible combinations. Finally, they are encoded efficiently using several interleaved combs of subsampled pulse trains or polyphases (Interleaved Single Pulse Permutation, ISPP). Note that this codebook is built with a systematic structure in mind, which achieves a fairly uniform coverage of the entire vector space, so no training or optimization is done for codevector entries.

---

[9] Note that a full search through these 40-dimensional codebooks would result in an increase in complexity *per dimension* by a factor of $2^{23} = 8.4 \cdot 10^6$ when compared with the earlier 10-dimensional codebooks!

Line Spectrum Pairs (LSP)

When applying VQ to the parameters of a speech model (and not to waveform samples), special considerations regarding the distortions induced by parameter quantization errors are needed. A common application is the simultaneous quantization of the $m = 10$ LP coefficients a_i, which raises the following issues:

1. The quantized parameter vector should still correspond to a minimum-phase filter allowing stable operation of the filter and its inverse.
2. Quantization error is best measured in terms of induced *spectral distortion* which is the mean-square difference between the original spectral envelope and the one described by the quantized parameters, both expressed in decibels (dBs). From experience, this distortion should be less than 1 dB for most of the parameter vectors to achieve transparent coding quality.
3. Within the short-time stationary model, the transition from one frame to the next should allow simple interpolation mechanisms between two parameter vectors which still maintain stability at all times and avoid unexpected spectral excursions of the interpolated spectral envelopes.

All these issues are best resolved before VQ by applying a nonlinear transform from the predictor coefficients to a vector of line-spectral frequencies which occur in distinct pairs and, therefore, are also referred to as Line Spectrum Pairs (LSPs). They are derived by introducing two polynomials $F_1(z)$ and $F_2(z)$ for order $m + 1$ with even and odd symmetries in their coefficients, respectively:

$$F_1(z) = A(z) + z^{-1} \cdot z^{-m} A(z^{-1}) \tag{24.12}$$

$$F_2(z) = A(z) - z^{-1} \cdot z^{-m} A(z^{-1}) \tag{24.13}$$

where $z^{-m} A(z^{-1}) = \sum_{i=0}^{m} a_{m-i} z^{-i}$ is the *time-reversed predictor polynomial*. We recognize that this first step is invertible as we have $A(z) = (F_1(z) + F_2(z))/2$. However, for a minimum-phase filter $A(z)$, the mirror poynomials $F_1(z)$ and $F_2(z)$ have their roots not inside but exactly on the unit circle such that each of their locations can be fully specified by a single real-valued frequency. Furthermore, the root location sets of the two polynomials follow a joint sorting relationship such that, with increasing frequency, the two sets are strictly interleaved on the unit circle. These properties simplify the search for the polynomial roots significantly and also suggest the combination of pairs of adjacent line spectrum frequencies (one from each mirror polynomial). Furthermore, this sorting relationship helps in interpolation of estimated LSP vectors over time. Finally, the VQ for this vector of ten real numbers is often implemented as a split VQ with three parts sorted by increasing frequencies which contributes to the relative independence of vector parts and minimizes the effects of suboptimal quantization.

24.4.3 Noise Shaping in Predictive Coding

Hybrid speech coders do not rely entirely on the source signal model but observe and transmit the modeling error or residual waveform as well. If we use a predictive signal model, the *residual waveform* is obtained as the prediction error $e_n = x_n - \hat{x}_n$. If we quantize the prediction residual e_n for transmission, it results in $e_n^{(q)} = e_n + q_n$ from which the decoder reconstructs $x_n^{(d)} = \hat{x}_n + e_n^{(q)}$. The *decoder* computes the prediction $\hat{x}_n$ from the delayed reconstructed signal samples $x_{n-i}^{(d)}$:

$$\hat{x}_n = -\sum_{i=1}^{m} a_i x_{n-i}^{(d)} \tag{24.14}$$

which results in a recursive or *closed-loop* reconstruction filter:

$$x_n^{(d)} = -\sum_{i=1}^{m} a_i x_{n-i}^{(d)} + e_n^{(q)} \tag{24.15}$$

$$A(z)X^{(d)}(z) = E^{(q)}(z) = E(z) + Q(z). \tag{24.16}$$

However, when computing the predicted signal sample $\hat{x}_n$ in the *encoder*, we can do this in two different ways:

1. *Open-loop prediction* makes use of the delayed original signal samples as available in the transmitter only:

$$\hat{x}_n = -\sum_{i=1}^{m} a_i x_n - i \tag{24.17}$$

$$E(z) = A(z)X(z) \tag{24.18}$$

from which we get the decoder output with Eq. (24.16):

$$X^{(d)}(z) = X(z) + \frac{Q(z)}{A(z)}. \tag{24.19}$$

2. *Closed-loop prediction*, however, uses exactly the same reconstructed samples for prediction in the encoder as in the decoder (Eq. (24.14)) to compute:

$$e_n = x_n - \hat{x}_n = x_n + \sum_{i=1}^{m} a_i x_{n-i}^{(d)}.$$
(24.20)

This results in the decoder output as

$$x_n^{(d)} = -\sum_{i=1}^{m} a_i x_{n-i}^{(d)} + e_n^{(q)}$$
(24.21)

$$= -\sum_{i=1}^{m} a_i x_{n-i}^{(d)} + x_n + \sum_{i=1}^{m} a_i x_{n-i}^{(d)} + q_n$$
(24.22)

$$X^{(d)}(z) = X(z) + Q(z).$$
(24.23)

In this setup, the encoder predictor duplicates the *closed-loop* structure of the decoder, hence its name.

Quantization noise $Q(z)$ is modeled with a white spectrum. According to Eq. (24.19), open-loop prediction results in an overall coding distortion $X^{(d)}(z) - X(z)$ with a spectrum shaped like the speech model spectrum $1/A(z)$ which reduces the audibility of this noise due to the masking properties of human hearing. However, filtering of the noise by $1/A(z)$ also results in an overall gain that makes the total noise power identical to that observed in direct quantization of the speech waveform without using prediction. Closed-loop prediction keeps a white noise spectrum without noise amplification according to Eq. (24.23). However, as quantization noise is not shaped like the speech spectrum, it may become audible in the spectral valleys of the latter. A compromise between these two extremal positions is found in the use of a *perceptual noise-shaping filter* [Schroeder et al. 1979]. To this end, quantization noise is observed as the difference between quantizer output and input, $q_n = e_n^{(q)} - e_n$, and a filtered version of this noise is fed back to the input of the quantizer $E(z) \rightarrow E(z) + (W(z) - 1)Q(z)$ (where the zero-delay gain is $w_0 = 1$). Thereby, the open-loop prediction system results in a decoder output equal to $X^{(d)}(z) = X(z) + \frac{W(z)}{A(z)} Q(z)$. If we choose the weighting filter $W(z) = A(z/\gamma)$, with $0 < \gamma \leq 1$, we can control the spectral shape of the decoder distortion to be anywhere between a white spectrum ($\gamma = 1$) and the speech model spectrum ($\gamma = 0$). Note that this *bandwidth expansion* for the perceptual weighting filter can be simply obtained by setting $w_i = a_i \gamma^i$.

Long-Term Prediction

The *Short-Term Predictors* (STPs) discussed so far only address the statistical dependences among neighboring signal samples (short-term correlations) which can be related to a nonflat spectral envelope – i.e., the formant structure of speech. The harmonic structure of voiced speech results in an additional spectral fine structure which corresponds to long-term correlations in the speech signal. These correlations are visible in the repetitive nature of the waveform cycles for a nearly periodic signal. The simplest *Long-Term Predictor* (LTP) $B(z)$ predicts the current sample x_n from a sample that lies one period T_0 back into the past:

$$\hat{x}_n = b \cdot x_{n-T_0}$$
(24.24)

As both the STP and LTP are stable linear systems, they can be cascaded in any sequence, but it often is more advantageous to first perform STP and then LTP. For LTP, closed-loop prediction is the preferred option. The best parameters b and T_0 can be obtained from analyzing the signal autocorrelation over a range of lags that spans from the shortest fundamental periods (high-pitched female voices) to the longest (low-pitched male voices). Even if the true pitch is not found in some cases (e.g., period doubling), the LTP will still contribute to coder performance by extracting any redundancy related to the autocorrelation maximum.

A specific problem in periodicity modeling with LTP occurs whenever the true fundamental period T_0 is not an integer multiple of the sampling interval T_s ("fractional pitch"). As a workaround, signal interpolation by a factor of 3–6 can be used to increase the effective sampling resolution. Another approach is to increase the order of the LTP filter $B(z)$ such that it simultaneously solves the problem of signal interpolation *and* prediction.

24.4.4   Analysis by Synthesis

The analysis-by-synthesis procedure has been introduced in the context of multipulse-excited linear predictive coding [Atal and Remde 1982] and, later on, it was combined by [Atal and Schroeder 1984] and [Schroeder and Atal 1985] with VQ to result in *Code Excited Linear Prediction* (CELP). Essentially it consists in reformulation of the open-loop/closed-loop prediction principle with perceptual weighting, which replaces the LP analysis step of the encoder with a linear predictive synthesis step. Thereby, the basic structure of the underlying signal model becomes more directly visible and certain computational advantages can be realized when combined with VQ. Note, however, that this analysis-by-synthesis formulation is still exactly equivalent to the noise-shaping predictive coding ideas presented in the previous subsection.

The starting point is to generate or synthesize one (or more) decoded signal sample by selecting a quantized residual signal sample $e_n^{(q)}$ from an inventory of allowed values (codebook). To this end, all entries from the inventory are run through the LP synthesis filter $1/A(z)$ to produce a set of all decoded signal candidates $e_n^{(d)}$. The coding distortion can be computed from the difference $x_n^{(d)} - x_n$. From the above, it is evident that this distortion is proportional to quantization error q_n via the filters $W(z)/A(z)$. To select the best quantized residual signal, the coding distortion is filtered by the inverse weighting filters $A(z)/W(z)$ to obtain the quantization noise $q_n = e_n^{(q)} - e_n$. If we minimize the squared magnitude of this variable we effectively perform a nearest neighbor decision among all possible residual candidates $e_n^{(q)}$ from the codebook to find the entry closest to e_n.

The power of this analysis-by-synthesis interpretation is that it lends itself to any description of the residual waveform inventory as long as it can be enumerated. This is, in particular, the case for VQ of the residual waveform where, typically, instead of a single sample a number of $d = 40$ samples are collected into a subframe. The squared magnitude estimate of the perceptually weighted synthesis error is now computed by summing the squares of the d samples. The VQ application suggests the following two computational reorganizations that reduce numerical complexity without compromising the solution:

1. As we process blocks or subframes of d samples at a time, we have to make sure that all filters carry over their filter states from one subframe to the next. This means that the filter output for one subframe will be the sum of a *zero-input response* (the decaying filter states) and a *zero-state response* (the component driven by the current input). Only the latter component provides the information for selecting the best codevector from the VQ codebook. Therefore, it is convenient to first subtract the zero-input response from the speech signal x_n such that codevector selection is done on zero-state responses only.
2. For the zero-state response of each codevector, we have to run the vector through both the LP synthesis filter $1/A(z)$ and the (inverse) weighting filters $A(z)/W(z)$. This filter cascade can be simplified to the overall transfer function $1/W(z)$, resulting in the block diagram shown in Figure 24.4.

Adaptive Codebook

The analysis-by-synthesis principle can also be carried over to the LTP which tries to explain the current subframe with predictions from (scaled) samples delayed by approximately one fundamental period T_0. As the fundamental period can vary in a certain range, it is advantageous to store subframe-length vectors taken from this range of the residual signal itself in a buffer memory that is called the *adaptive codebook*. One residual subframe is then represented by the weighted sum of two vectors, one selected from this adaptive codebook and another one selected from the *fixed codebook*. In most systems, the two selections are made sequentially such that first the adaptive codebook entry is selected to model the periodicity of the signal and, based on this choice, the fixed codebook entry is selected to represent the remaining "stochastic" part of the residual (Figure 24.4). This approach is essentially identical to the use of an LTP, with minor differences for high-pitched voices where the fundamental period is shorter than a subframe length. Just as for LTP, good quality requires interpolation of the signal to increase the temporal resolution of the adaptive codebook, at least in the vicinity of the long-term correlation maximum.

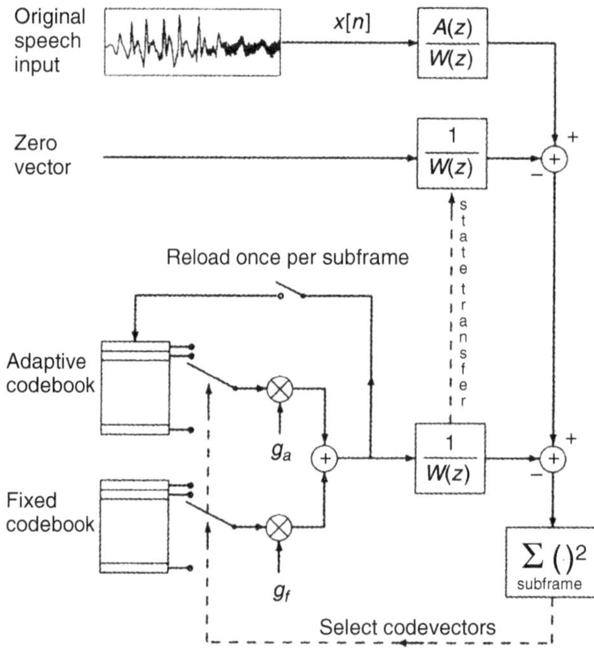

Figure 24.4 Code-excited linear predictive coding of speech; simplified block diagram showing separate treatment of zero-input and zero-state response with adaptive and fixed codebook excitation; weighting filter $W(z) = A(z/\gamma)$.

24.4.5 Joint Source Channel Coding

Unequal Error Protection

Once all the source parameters have been computed for an entire speech frame, their codewords need to be assembled for transmission and extra bits will be added for channel-coding purposes. Channel coding usually works with the assumption that the codeword sequence is independent identically distributed (iid) and that all codewords bear the same relevance for the receiver. The last condition is certainly not the case for hybrid speech coders where some codewords represent model parameters such as fundamental period, gains, or spectral envelopes (LSPs) whereas other codewords represent the residual waveform by entries in the fixed VQ codebook. A channel error in one of the more relevant parameters (where T_0 is maybe the most important one) will result in significant waveform errors in the decoded signal as, e.g., the reconstructed voice may sound much lower or higher than the original. A channel error in one of the less relevant codewords may hardly be noticed as a properly parameterized speech model will generate rather good speech even if the details of the residual waveform are wrong. In conclusion, "*all source bits are equal, but some are more equal than others.*" This results in classification of the codewords in different importance or channel sensitivity classes which are handled by different levels of channel error protection, ranging from sophisticated convolutional codes to no error protection at all.

Adaptive Multi-Rate Coding

Both channel and source statistics vary over time. Therefore, the optimum coding strategy has to adapt itself to the time-varying channel state and source state. For circuit-switched communication, fixed overall rates are desired on the channel, so the task of the rate adaptation mechanism is to determine the best tradeoff between the rate budget spent on source coding and on channel error protection such that the total rate remains constant (Adaptive Multi-Rate AMR coding). Because source statistics change much faster (essentially they are different for each frame – i.e., every 20 ms) than channel-state information is updated, current systems will essentially select the best source versus channel rate split based on channel-state information only. Depending on this information, the source coder produces a different number of source bits per frame and the channel coder will use the remaining bits to perform adequate unequal error protection as described in the previous paragraph.

Source-Coding Optimization

There are two approaches to optimization of source coding in the light of joint source channel coding:

- *Encoder optimization*, where, for instance, the *index assignment* of the VQ is optimized such that bit errors in the codewords representing the VQ index will result in the minimum mean perceptual impact, very much like the Gray coding strategy for scalar index assignment. A further encoder technique that strives for equal relevance of all codewords is found in *multiple-description coding*.
- *Decoder optimization*, where residual redundancy among the received codewords is used to assist the channel decoder to increase the reliability of the overall decoding process.

24.5 From Speech Transmission to Acoustic Telepresence

The ultimate goal of speech transmission or telephony has always been to recreate the acoustic presence of a human talker at a geographically distant location. In that sense, telephony is just a special case of *telepresence* where we attempt to augment our local reality with the virtual presence of one or more remote persons, preferably recreating the remote environment to some extent. Our short discussion starts out from simple add-on functionalities for speech transmission and progresses to the most advanced services of three-dimensional virtual audio.

24.5.1 Voice Activity Detection

In a symmetric conversation between two persons, each of the participants is silent for about 50% of the time. This fact has long been observed and exploited for multiplexing telephone conversations over the same transmission channel in a technique known as Digital Speech Interpolation (DSI). For wireless communications, multiple access to the same shared radio spectrum requires the reduction of interference created among the users. This can be achieved by transmitting speech frames over the air interface only when the talker is actively speaking, a state which is detected by a Voice Activity Detector (VAD). For an inactive speaker, Discontinuous Transmission (DTX) results in:

1. less power consumption on the mobile terminal (saving of baseband-processing and radio transmitter power), thereby extending battery life;
2. less multi-user interference on the air interface, thereby enhancing mobile access network performance;
3. less network load if packet switching is used, thereby increasing backbone network capacity.

In a typical realization, speech transmission is cut after seven nonactive frames, and a "SIlence Descriptor" (SID, only 35 bits for 20 ms) frame is sent as a model for acoustic background noise which is used to perform Comfort Noise Generation (CNG) at the receiver end. This noise model is updated at least every 24 frames and can be seen as a first step to virtual reality rendering of the auditory scene, thereby maintaining coherence between the communication partners by letting the other know implicitly that the call was made while driving a car, etc.

While this DTX method is sometimes referred to as source-controlled rate adaptation, it operates at a very high source description level (i.e., only source on versus source off) and, e.g., is not able to rapidly vary the source rate according to the phonetic contents (e.g., voiced versus unvoiced speech). This is also not the same as AMR coding described above (with VAD, the source rate is either zero or the currently allowed maximum source rate, the latter still being adapted according to the channel state).

24.5.2   Receiver End Enhancements

If an entire speech frame is lost due to severe fading on the air interface, the resulting error cannot be corrected with channel codes targeted for individual or bursty bit errors. Rather, *error concealment* methods are required to interpolate such lost frames from received neighboring frames. This usually works very well for substitution of the first lost frame but may lead to unnatural sounding speech if continued over several frames lost in a row. In the latter case, later substituted frames will receive a gradual damping to achieve slow fadeout of the signal amplitude (typically over six frames = 120 ms).

A further receiver end signal enhancement is *adaptive postfiltering* which constitutes filters shaped according to the spectral envelope and LTP filters which will make noise introduced by lossy coding less audible to the listener. *Adaptive playout buffers* or *jitter buffers* help to conceal lost frames on packet-switched networks. If a terminal is equipped with wideband speech Input/Output (I/O) functionality (analog bandwidth from 50 Hz up to 7 kHz), speech received over a narrowband network may be augmented by artificially created frequency bands above 3.4 kHz and below 300 Hz using a synthesis mechanism known as *bandwidth extension.*

24.5.3   Acoustic Echo and Noise

The acoustic environment of a speaker may not only contain useful background information that adds to the presence of the auditory scene but also noise sources that are considered annoying, or reverberation and (acoustic) echo due to the fact that both speaking parties share a common acoustic space when using hands-free phones. In such situations, echo and noise need not only be controlled for the comfort of the human users but also for the speech-coding systems which, if strongly based on a speech signal model, will fail to handle background noise or echo appropriately. Solutions that perform *joint echo and noise control* will often achieve the best compromise in reducing the two impairments. An uncontrolled echo can have drastic consequences on the stability of the entire communication system; strict performance requirements have been set as standards by ITU-T in recommendations G.167 and G.168.

24.5.4   Service Augmentation for Telepresence

Speech-Enabled Services

In a telepresence framework, users will access a much wider range of services than in conventional telephony. Given the shrinking size of mobile terminals, service access benefits a lot from spoken language dialog with the machine offering the service. This can range from simple name dialing by voice (possibly activated by a spoken magic word) to Text To Speech (TTS) synthesis for email reading or *Distributed Speech Recognition* (DSR) where feature vectors for speech recognition are extracted on the mobile terminal, encoded with sufficient accuracy for pattern recognition (but not necessarily for signal regeneration at the central server end), and *transmitted as data* over the wireless link. This allows us to transmit recognition features based on higher quality signals than those used in telephony and it also allows us to use specific source- and channel-coding mechanisms that maximize benefits for the remote speech recognition server (rather than a human listener).

While DSR carries speech information over a data channel, the dual application of *Cellular Text Telephony* (CTT) allows the carriage of textual data over the speech channel so as to provide augmentative communication means for people with hearing or speaking impairments who still can exchange interactive text (not just short messages) in a chat-like style using a modem operating over the digital speech channel.

Personalization

For personalized services, *talker authentication* will be a must which should be distinguished from authentication of the mobile station or the infrastructure itself. The identity of a talker can be established via voice identification or verification techniques and, if needed, an additional personalized watermark might be inserted in the speech signal prior to encoding it.

For *talker privacy,* the traditional phone booth might once be replaced by a virtual talker sphere outside of which active speech cancellation (using wearable loudspeaker arrays) would make the phone conversation hardly audible to bystanders.

Three-Dimensional Audio

The virtual talker sphere has already introduced the concept of advanced audioprocessing for speech telephony which has seen a tremendous boost from the use of microphone and loudspeaker arrays and virtual/augmented audio techniques that allow the spatial rendering (e.g., ambisonic) of three-dimensional sound fields, potentially converted to binaural headphone listening. The best effects are expected if personalized HRTFs can be used together with real-time tracking of head movements to place virtual sound sources in the context of the real environment, creating the immersive telepresence which blends the presence of virtual participants and local participants for successful teleconferencing.

Ultimately, this will require a significant shift beyond today's speech-coding standards to allow for high-quality, multi-channel audio including metadata information. A first move in this direction has been made by standardization of the AMR wideband speech codec which has become the first speech-coding standard ever to be accepted almost simultaneously for wireline communication (ITU), wireless communication (European Telecommunications Standards Institute/Third Generation Partnership Project (ETSI/3GPP)), and Internet telephony (Internet Engineering Task Force (IETF)).

Further Reading

The classical textbooks on speech coding are [Jayant and Noll 1984] and [Kleijn and Paliwal 1995], which, at the time of this writing, are unfortunately out of print; however [Vary and Martin 2006] and [Kondoz 2004] are available. Another excellent overview of various aspects of speech processing is [Rabiner 1994]. Source-coding theory is treated in [Berger 1971], [Gray 1989], and [Kleijn 2005] which is maybe best suited for the needs of speech coding. For speech signal processing in general, we recommend the classic [Rabiner and Schafer 1978] and the more recent textbooks of [Deller et al. 2000], [O'Shaughnessy 2000], and [Quatieri 2002]. For extensions to nonlinear speech modeling, see [Chollet et al. 2005] and [Kubin 1995]. In [Gersho and Gray 1992] an extensive discussion of VQ is provided. [Hanzo et al. 2001] address joint solutions for source and channel coding for wireless speech transmission. The wider context of acoustic signal processing for telecommunications is treated in [Gay and Benesty 2000], [Haensler and Schmidt 2004], and [Vaseghi 2000]. Bandwidth extension is the main topic of [Larsen and Aarts 2004]. An excellent reference for spoken language processing is [Huang et al. 2001] and [Jurafsky and Martin 2000]. [Gibson et al. 1998] treat the compression of general multi-media data.

For updates and errata for this chapter, see https://wides.usc.edu/students.html#textbooks.

Exercises

See Sec. 36.24 of Exercises.pdf at wiley.com/go/molisch/wireless3e

25

Video Coding

Anthony Vetro

Mitsubishi Electric Research Labs, Cambridge, MA, USA

25.1 Introduction

Digital video is being generated and consumed at an increasing rate. It has been reported that 300 h of video are uploaded to YouTube every minute and more than 80% of all Internet traffic is video [Cisco Systems 2019]. While video streaming over the Internet and wireless networks account for a vast majority of digital video content, digital video is also available from over-the-air broadcast services; many countries around the world have fully transition from analog to digital broadcasting, allowing for better quality video, and a wide range of services to be offered to the public.

The means for communicating video will generally be supported by a specific infrastructure and associated communication techniques, each providing their own quality of service. An important task for service and network providers is to match their offerings with the capabilities of their network. When video is involved, the overall system design including the characteristics of the digital video signal, the compression scheme, and its capabilities, as well as the robustness to transmission errors, must all be considered.

In contrast to other digital signals such as speech, digital video requires a relatively large bandwidth. For instance, a typical high-definition video signal has a raw data rate in the range of 600–800 Mbit/s depending on the specific spatial resolution and frame rate. To transmit video in a practical manner over existing networks, compression of the digital video signal is necessary. This section introduces the digital video representation and describes the fundamental compression architecture. The major components for video compression are elaborated on in subsequent sections. A summary of major video coding standards is then provided. The chapter concludes with sections devoted to robust video transmission and video streaming.

25.1.1 Digital Video Representation and Formats

Video is a multi-dimensional signal that represents emitted and reflected light intensity from objects in a scene over time. The samples of a digital video signal are typically recorded by an imaging system, such as a camera. The capture system will typically record two-dimensional (2D) sample values in a particular color space, which correspond to a specific set of wavelengths, at discrete instants of time. The number of samples at each time instant is referred to as the spatial resolution, while the rate at which samples are taken is referred to as the frame rate.

It is well established that most colors can be produced with three properly chosen color primaries. The RGB primary, which includes red, green, and blue colors, is perhaps the most popular set for both capture and display. However, video coding and transmission systems utilize a color coordinate system that is based on luminance (Y) and chrominance components (Cb and Cr). One reason is historical: in older analog systems, the luminance signal was backward compatible with the monochrome black-and-white television signals. The other advantage of this color space for transmission is that some of the information in the chrominance samples could be discarded since the human visual system is less sensitive to the color components relative to the luminance. There is a well-defined linear relation between the RGB and YCbCr color spaces.

The color sampling formats specified by the ITU-R recommendation BT.601 [International Telecommunications Union 1998] are illustrated in Figure 25.1. These formats are utilized by all international video coding standards. In the 4:4:4 color sampling format, the luminance and chrominance samples have the same resolution. In the 4:2:2 and 4:2:0 color sampling formats, the chrominance samples have a reduced resolution relative to the luminance samples. The 4:4:4 and 4:2:2 color sampling formats are primarily used in studio and production environments, where it is important to maintain the fidelity of the color components. For distribution to the consumer, current systems utilize the 4:2:0 sampling format. As display technology and distribution capabilities increase, it is expected that video with higher color fidelity will be transmitted to the consumer, e.g., in a 4:4:4 color sampling format.

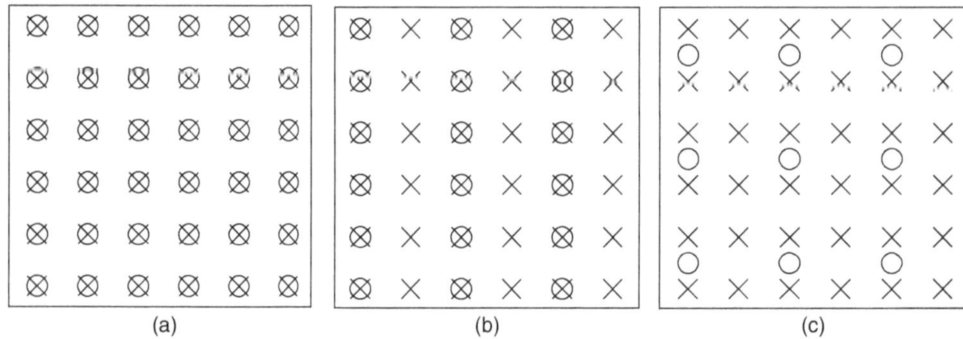

(a)   (b)   (c)

Figure 25.1 Color sampling structures with the location of luminance samples represented by a × and the location of chrominance represented by a ○. 4:4:4 color sampling in which chrominance samples have the same resolution as luminance samples (a). 4:2:2 color sampling in which chrominance samples have half the horizontal resolution as luminance samples (b). 4:2:0 color sampling in which chrominance samples have half the horizontal and vertical resolution as luminance samples (c).

25.1.2 Video Coding Architecture

There is a great deal of redundancy in a video signal, in both spatial and temporal dimensions. Video compression schemes exploit that redundancy and determine a compact binary representation. The most popular and effective scheme for video compression is known as a block-based hybrid video coder, which uses a combination of both transform coding and temporal prediction to code the video signal. Since this scheme has been adopted by all international video coding standards to date, this chapter will focus exclusively on this scheme. A common architecture for the block-based hybrid video coder is shown in Figure 25.2.

There are several types of pictures that are commonly referred to in the video coding literature. Intracoded pictures, or I-pictures, refer to pictures that are coded without reference to any other picture in the video. Pixels within the I-picture may still be predicted from other neighboring pixels to exploit spatial redundancy. Due to their independence, they are often inserted periodically in the bitstream to facilitate random access since the decoding can start from these pictures. Another type of picture is referred to as a P-picture, which utilizes a unidirectional temporal prediction. Since pixels in neighboring pictures are correlated with the current picture, temporal prediction is a very effective means to reduce the energy of the signal to be coded. Biprediction, or the prediction of the current picture from two references, is another effective means of temporal prediction. Pictures that use biprediction are referred to as B-pictures. An example prediction structure is shown in Figure 25.3, where P-pictures are predicted from the previous I/P-picture and the B-pictures are predicted from neighboring I/P-pictures.

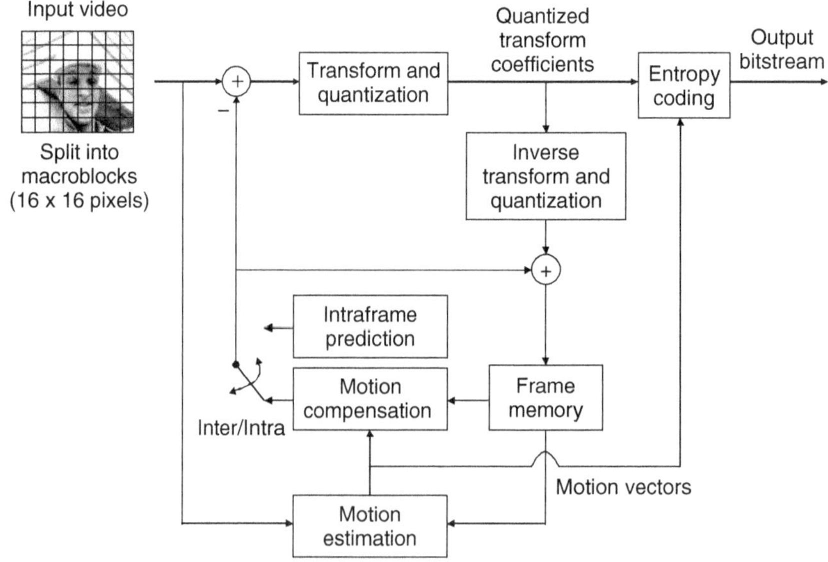

Figure 25.2 Block diagram of a typical video encoder.

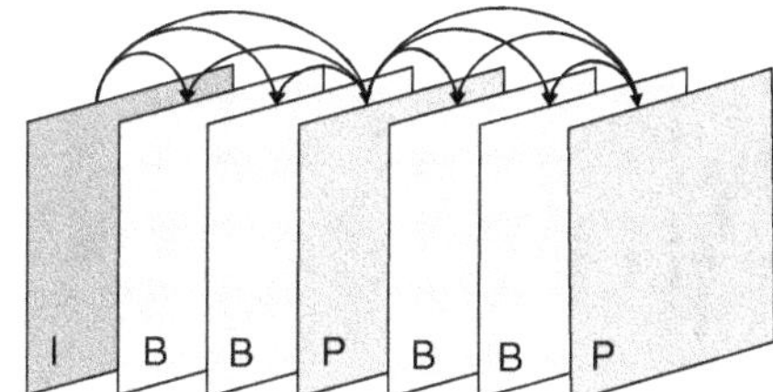

Figure 25.3 Sample video prediction structure.

The first step in the video coding process is to divide each picture into fixed-size blocks of pixels or macroblocks (MBs). The most common size of an MB in most earlier video coding standards was 16×16, although larger block sizes have been adopted in more recent standards. Then, on a block basis, a prediction is performed. For I-pictures, an intraframe prediction is utilized, where neighboring pixels are used to predict pixels in the current block. For P- and B-pictures, a motion-compensated temporal prediction is performed. The result of the prediction is a residual signal, which is then subject to a transform and quantization. The quantized transform coefficients are then entropy coded to produce the resulting video bitstream. Since the motion-compensated prediction is based on previously coded pictures that must also be available at the decoder, the inverse operations to reconstruct those reference pictures are also performed. These reference pictures are stored in a frame memory.

The remainder of this chapter describes the major components of a video coder in more detail and discusses the particular set of coding tools that are supported by different video coding standards. Extensions to accommodate scalable and multiview video, as well as efficient coding of screen content and 360° video, are also briefly described. Finally, a brief review of error-resilient video transmission and video streaming is given.

25.2 Transform and Quantization

In the previous section, a basic video coding architecture was introduced. In the following, further discussion on the transform and quantization modules are discussed. We specifically focus on block-based transforms and the Discrete Cosine Transform (DCT) in particular, as well as scalar quantization, since these methods have proven to be especially effective in modern image and video codecs.

25.2.1 Discrete Cosine Transform

The use of a transform in an image or video codec attempts to modify the input signal in two ways: (i) compact the energy of the signal and (ii) decorrelate the signal. The energy compaction property generally leads to a set of transform coefficients with few coefficients having a large magnitude and many coefficients becoming small or zero, which has obvious benefits for compression. The decorrelation property is advantageous since each coefficient could be quantized independently in an optimal manner, e.g., to either minimize Mean Square Error (MSE) or according to a perceptual sensitivity.

The transform of an image or video block represents the signal as a linear combination of basis functions. The coefficient that corresponds to a particular basis function denotes the weighting or contribution of that function in the overall signal representation.

Let x_n, $n = 0, 1, ..., N - 1$, be an input block of length N. Further, assume that x is a Markov-1 process with correlation coefficient, ρ. The covariance matrix of this random process is then given by

$$R_x(i,j) = \rho^{|i-j|}. \tag{25.1}$$

The input signal can be transformed by matrix A to yield an output signal in the transform-domain, $y = Ax$, which has a covariance matrix given by

$$R_y = AR_xA^T. \tag{25.2}$$

The gain of the transform, which is a typical measure of transform performance relative to Pulse Code Modulation (PCM), is defined as

$$G_T = \frac{\frac{1}{N}\sum_{i=0}^{N-1}\sigma_{y_i}^2}{\left(\prod_{i=0}^{N-1}\sigma_{y_i}^2\right)^{1/N}}. \tag{25.3}$$

This metric is the ratio of the arithmetic mean of the transform coefficient variances to the geometric mean of these variances, and essentially measures the reduction in MSE that the transform provides relative to PCM.

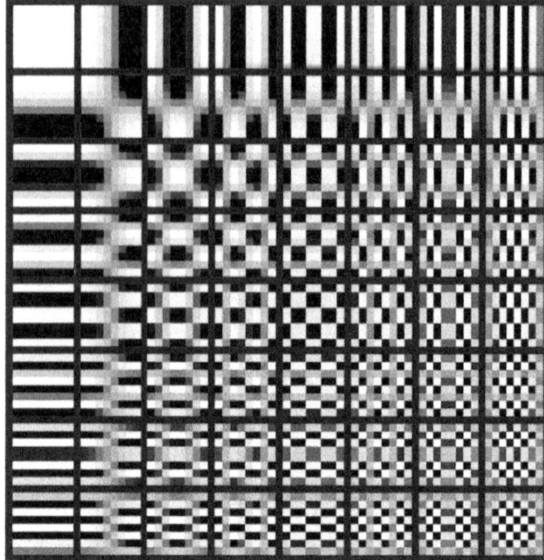

Figure 25.4   Basis functions of the 2D 8×8 DCT.

A transform that maximizes the transform coding gain is the Karhunen Loève Transform (KLT). Its basis functions are the eigenvectors of the covariance matrix of the input signal, R_x. The transformed signal not only achieves optimum energy concentration, but the transform coefficients are also fully decorrelated, i.e., the covariance matrix R_y is diagonal. Despite these excellent properties of the KLT, there are notable disadvantages that have made it unusable for image and video coding in practice. Most importantly, it is dependent on the signal statistics and can only be computed for stationary sources with known covariance matrix. Also, it is not a separable transform, so additional complexity would be incurred when applying it to 2D image blocks.

Fortunately, the DCT has been shown to perform very close to the KLT for natural image and video data and has a basis function defined as follows [Rao and Yip 1990]:

$$a_{i,j} = \alpha_i \cos \frac{(2j + 1)\pi i}{2N} \tag{25.4}$$

for $i, j = 0, 1, ..., N - 1$, with $\alpha_0 = \sqrt{1/N}$ and $\alpha_i = \sqrt{2/N}$ for $i \neq 0$. An illustration of the 2D basis function of the DCT is given in Figure 25.4. The DCT achieves similar transform gain to that of the KLT, while not being dependent on signal statistics.

The 8-point DCT has been specified as the transform in most of the earlier image and video coding standards, but more recent standards have employed a wider range of sizes from 4-point up to 64-point. To ease implementation on a range of platforms, there exist various factorizations and fixed-point implementations of the DCT [Arai et al. 1988], [Reznik et al. 2007] as well as scaled integer approximations of the DCT [Malvar et al. 2003]. Further information about the transform sizes supported by different standards is described in Section 25.5.

25.2.2   Scalar Quantization

In contrast to the transformation process, which is generally invertible and does not incur loss, quantization is inherently a lossy process and introduces distortion into the reconstructed signal. The basics of scalar quantization have already been introduced in the chapter on speech coding. A very brief review is provided here for completeness and to introduce some specifics on quantization related to video coding.

In most video coding schemes, quantization is applied to the transform coefficients. The dynamic range of the transform coefficients will vary with the bit depth of the input video and the length of the transform[1]. Assuming transform coefficients in the range $[t_{min}, t_{max}]$ with dynamic range B, the quantization process will assign a quantization index $i = Q(x)$ to an input value x subject to a quantization function $Q(\cdot)$. A sample quantization function is shown in Figure 25.5, which is defined by the number of quantization indices, i.e., reconstruction values, and the boundary of each interval.

Consider a uniform quantizer with L quantization indices over the range of possible values and equal distance between adjacent boundary values, and let the distance of each interval be denoted by $\Delta = B/L$. With a fixed-length binary representation, each

---

[1] For 8-bit image samples and 8-point DCT, the transform coefficients have a dynamic range of 11 bits with coefficient values between $[-1024, 1023]$. When the input to the transform is a prediction signal that may be in the range $[-255, 255]$, the dynamic range of the transform coefficients increases to 12 bits.

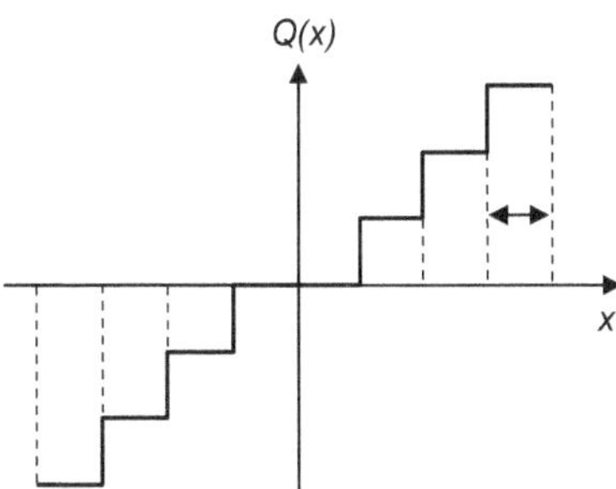

Figure 25.5 Illustration of scalar quantizer with uniform quantization step size Δ.

quantization index requires $R = \lceil \log_2 L \rceil$ bits. For a uniformly distributed source, it can be shown that the variance of quantization errors is equal to $\sigma_q^2 = \Delta^2/12 = \sigma_x^2 2^{-2R}$. The Signal-to-Noise Ratio (SNR) of the quantizer can then be computed as

$$SNR = 10 \log_{10} \frac{\sigma_x^2}{\sigma_q^2} = 20 \log_{10} 2R = 6.02R. \tag{25.5}$$

which reveals a classic result in coding theory that every additional bit in a uniform quantizer yields a 6.02-dB gain in SNR for a uniform source.

The quantization scheme utilized in most state-of-the-art video codecs is essentially uniform quantizers with some minor adjustments. One such adjustment adds a *dead-zone* which expands the interval of the quantization index that corresponds to a zero-level reconstruction. This has the effect of forcing more transform coefficients to zero. Second, a weighting matrix is often used to account for the fact that the human visual system is less sensitive to higher frequencies than lower frequencies. In this way, a courser quantization step size is applied to higher frequencies. Most standards allow the quantization matrix used for coding a picture to be signaled as part of the bitstream. Finally, a quantization offset could be utilized to shift the location of the reconstruction value within the quantization interval, i.e., different than the center of that interval. The reconstruction offset has been shown to provide benefits when quantizing sources with a Laplacian distribution, which are typical of transform coefficients in video coding.

25.3 Prediction

Prediction is perhaps one of the most powerful and effective means of reducing redundancy in a video signal. This section briefly introduces two commonly used approaches. Intraprediction refers to the prediction of data from data within the same picture, while interprediction or motion-compensated prediction refers to temporal predictions that utilize data from neighboring pictures.

25.3.1 Intraframe Prediction

Intracoded pictures are important for random access, which requires that they be decoded independently of any other pictures, i.e., there are no temporal dependencies. To code such pictures more effectively, intraprediction techniques that make use of neighboring data within the current picture are typically utilized to reduce the signal energy, thereby reducing the number of bits required to represent that portion of the signal. There exist both transform-domain prediction schemes as well as spatial domain prediction schemes. A few commonly used block-based intraprediction schemes are shown in Figure 25.6.

For natural images and video, the average values of pixels within a block are highly correlated within a given picture. Therefore, an intuitive and effective way to exploit this correlation is in the transform domain. When a block-based transform, such as the DCT discussed in Section 25.2.1, is applied to a block in the picture, a DC (zero frequency) or average component of the block is computed. As shown in Figure 25.6, the DC component can be predicted from the preceding block in the same row. This prediction results in a set of residual DC components, which follow a Laplacian-like distribution around zero. Without prediction, the distribution of DC components for natural images is closer to a uniform distribution. The DC prediction has the effect of reducing the entropy, thereby reducing the required bits to represent that signal component.

Rather than always predicting from the preceding block, one can also adaptively select either the DC value of the left block or above block as a reference. A method to adaptively select the direction would then be needed. For example, the DC prediction direction could be selected based on a comparison of the horizontal and vertical DC gradients around the block to be coded.

For many natural images, the low-frequency components of neighboring transformed blocks are also correlated. As shown in Figure 25.6, the first rows and first columns of a transformed block, which represent low-frequency AC (nonzero frequency) coefficients, could also be predicted from neighboring blocks. When combined with adaptive DC prediction, i.e., prediction from either the left or above block, the prediction of AC coefficients would typically use the same direction. In this way, AC prediction is applied only to the rows or the columns of any block.

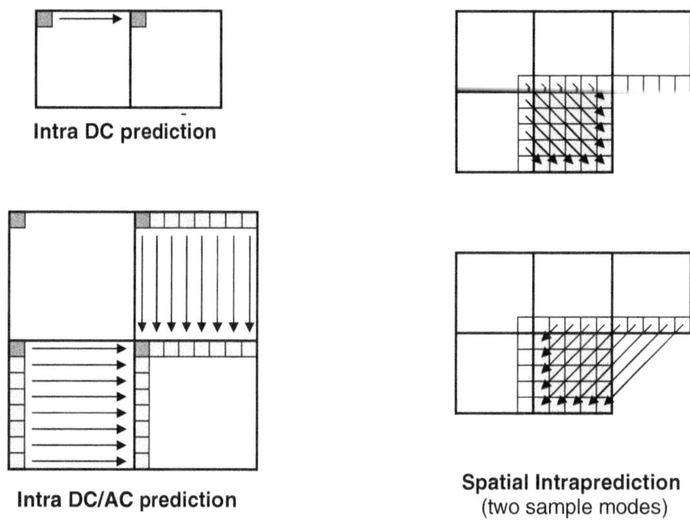

Figure 25.6 Various types of block-based intraprediction schemes.

Another common method of intraprediction is to apply prediction in the spatial domain. Optimal linear predictors could be derived based on the correlation of the source, but in practice, a fixed set of predictors is typically used. The latest video coding standards employ directional predictors, where multiple prediction directions could be selected adaptively on a block basis. Two sample prediction modes for spatial prediction are shown in Figure 25.6. In these examples, the neighboring pixels from nearby blocks are used to predict the pixels in the current block according to the chosen direction. The are many variations and different ways to form such predictors. When presented with such options, a method to determine the optimal mode is needed and the selected mode must then be signaled as part of the bitstream. This overhead is usually what limits one from devising too many different prediction modes to select from.

25.3.2   Interframe Prediction

The previous subsection discussed methods to predict data within a frame of the video. However, video also has a very high correlation in the temporal direction. In contrast to intraframe prediction in which both transform-domain and spatial domain predictors have shown to be effective, most interframe prediction techniques tend to focus on spatial domain methods.

The simplest type of temporal or interframe predictor is to use the data from a colocated block in the previous frame to predict data in the current frame. For stationary scenes or parts of the scene, this is actually quite effective. It is so effective and commonly used in practice that special signaling is used to indicate that a block to be coded in the current frame can simply be copied from a colocated position in a reference frame. The corresponding coding mode is referred to as a *skip* mode.

Most natural video scenes do have motion, including global motion that would result from camera pans and zooms as well as local motion that results from objects moving within the scene. Under such conditions, motion-compensated temporal prediction becomes an effective means to predict frames in the video from neighboring frames. The basic concept of motion-compensated prediction is to determine the displacement of pixels in one frame, which we will refer to as the current frame, relative to the set of pixels in a neighboring frame, which we refer to as reference frame. Note that there may be multiple reference frames used for the prediction of the current frame.

Determining the way that pixel intensities move from one time instant to another has been a rich area of study among the video coding community. One can imagine trying to estimate the movement of each pixel, but this is an ill-defined problem since there could be many pixels that match the current pixel intensity in neighboring frames, especially considering regions of a scene with constant intensity. Two common approaches to get around this problem are (i) use regularization techniques to enforce smoothness constraints on the motion field, and (ii) assume that pixels within a small neighborhood have the same motion. One drawback of the first option is that there would be significant overhead in signaling the motion for each pixel. Due to the block-based nature of most video coding standards, the second option has thus become a more popular and practical approach for compression. Specifically, for any block of a frame, a translational motion model is assumed and a motion vector that indicates the 2D displacement is determined.

Motion vectors are determined through a block matching process, which is illustrated in Figure 25.7. There is a search region associated with each block in the current frame, from which the best match between the current block and candidate blocks in the search region of the reference frame are found. There is typically not a perfect match between the current block and the reference block, so the residual signal that represents the difference after prediction must also be coded to reconstruct the video. Additionally, the motion vector must be coded and signaled as part of the bitstream so that the prediction can be formed at the decoder based on previously decoded frames, which serve as the reference frame.

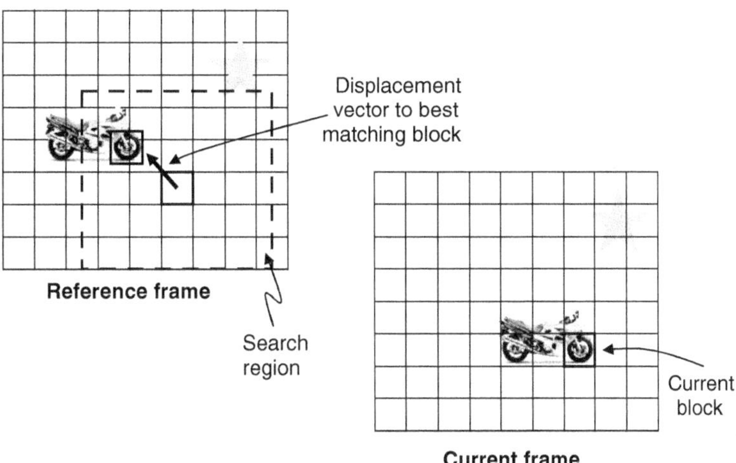

Figure 25.7 Illustration of block-based matching for interframe prediction.

25.4 Entropy Coding

The entropy of a given source is a limit on the data compression rate that could be achieved. If all the symbols of the source data have equal probability, then a fixed-length binary representation for each symbol would achieve the lower limit on the rate. However, it is typical in video coding and other applications that the symbols have a nonuniform probability distribution. Therefore, a lower bit rate could be achieved by assigning shorter codewords to higher probability symbols and longer codewords to lower probability symbols. This is the basic principle of entropy coding, which is also referred to as Variable Length Coding (VLC).

VLC has two important properties: (i) the code should be uniquely decodable and (ii) the code should be instantaneously decodable. The first property requires that there is only one possible set of source symbols that the codeword represents, while the second implies that no codeword is the prefix of any other codeword. Codes that satisfy this second property are referred to as prefix codes.

In the case of video coding, the symbols to be encoded include the quantized transform coefficients and any side information that is required to reconstruct the video signal, e.g., motion vectors, block coding modes, etc. It is noted that in contrast to quantization, which introduces distortion to the signal, entropy coding is a lossless process. Huffman coding and arithmetic coding are two popular entropy coding schemes that are widely used for video coding. The basic principles of each are described in the following subsections.

25.4.1 Huffman Coding

[Huffman 1952] published an algorithm to construct an optimal prefix code for a given distribution of symbols. Consider a finite alphabet source $\chi = x_1, x_2, ..., x_N$ with probabilities $p(x_i)$. The optimal binary code should assign longer codewords to higher probability symbols and shorter codewords to those that appear more frequently. The procedure is outlined as follows:

1. Let each symbol be a leaf node of a tree with an assigned probability, $p(x_i)$. Arrange the symbols so that their occurrence probabilities are in a decreasing order.
2. While there is more than one node,
    (a) identify the two nodes with smallest probabilities and arbitrarily assign a binary 1 and a binary 0 to these nodes;
    (b) merge the two nodes to create a new node with probability equal to the sum of those nodes.
3. The last remaining node is the root node. The codeword for each symbol can be traced from the root node to each leaf node.

An example of a Huffman code construction is shown in Figure 25.8. This example illustrates the successive assignment of 0 and 1 to each of the leaf nodes with lowest probability and merging of those nodes until the root node is reached. This example demonstrates that none of the resulting codewords are the prefix of any other codeword. The code shown in this example has an average length of 2.4 bits.

A proof of the optimality of the Huffman code as well as other properties and examples of Huffman code constructions can be found in [Cover and Thomas 2006]. It can be shown that the average length of the Huffman code, L_{avg}, satisfies the following condition, where $H(X)$ is the entropy of the source.

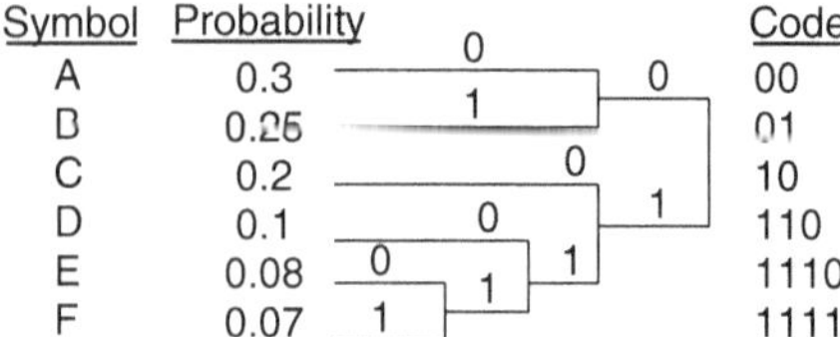

Figure 25.8 Example of Huffman coding based on probabilities of symbols to be coded.

$$H(X) \leq L_{avg} < H(X) + 1. \tag{25.6}$$

One well-known drawback of Huffman codes is that when coding individual symbols, each symbol always requires a minimum of 1 bit. One way to overcome this is to consider the coding for blocks of symbols, i.e., a vector. In this way, each vector would be assigned a probability, which is a joint probability of each symbol in the block. The procedure to construct a code follows the same as that outlined above, only that the symbols represent vectors. Another way to code sets of symbols more efficiently is to condition them on a context. A simple example would be to condition on the previous sample, but more complex contexts are considered in practical video codecs.

25.4.2 Arithmetic Coding

It was noted in the previous section that Huffman coding could suffer from some coding loss relative to the optimal representation since each symbol must be coded with at least 1 bit. In an extreme case, consider an alphabet with two symbols, one with probability that is close to 1 and the other with probability close to 0. Since there is little uncertainty, the entropy of this source would be very small and close to 0. However, Huffman coding would still require 1 bit for each symbol, which shows that there is still further room for improvement. While blocks of symbols could be used to improve the coding efficiency, the complexity grows exponentially with the block length.

Rather than representing a symbol or block of symbols by a sequence of bits, arithmetic coding represents a sequence of symbols by a subinterval in the interval between 0 and 1, where the size of each subinterval is proportional to the probability of each symbol. Starting with an initial division, the first subinterval is selected based on the first symbol to be coded. This subinterval is then recursively divided based on each new symbol to be coded. A subinterval is represented by its lower and upper boundaries and a binary representation of those boundaries is used to code the sequence of symbols at each stage. In particular, when the most significant bit of the lower and upper boundaries are the same, that bit is written to the output. In this way, higher probability symbols will be associated with larger intervals, which require fewer bits to represent.

An example of the arithmetic coding process is shown in Figure 25.9. The source data includes three symbols, "A," "B," and "C," each with their respective probabilities given as $p(A) = 1/2$, $p(B) = 1/3$, and $p(C) = 1/6$. To begin the encoding process, the initial interval $[0, 1]$ is first divided according to these probabilities. Assume we would like to code the sequence of symbols "ABAC." Since the first symbol is "A," the subinterval associated with this symbol $[0, 1/2]$ is carried to the next stage and again subdivided according to the symbol probabilities. The next symbol to be coded is "B" so we take the second subinterval in the range $[1/4, 5/12]$ and further subdivide it according to the probabilities. Following this process for each symbol in the sequence, we can arrive at the final subinterval that represents the sequence of symbols and its binary representation.

At first glance, it seems that arithmetic encoding would require extremely high precision to encode a long sequence of symbols. However, in practice, a fixed-point precision based on integer arithmetic can be maintained by renormalization of the upper and lower boundaries when bits are written out.

One advantage of arithmetic coding relative to Huffman coding is that it is incremental, i.e., additional symbols can be coded based on the code for the previous sequence of symbols. This property allows it to closely approach the entropy rate in practice since longer sequences can be effectively coded without maintaining large codebooks. Another merit is that it can easily adapt to changes in the statistics of the source data. It is only required that the encoder and decoder update their probabilities tables in a synchronized manner. In practical video coding schemes, context-based arithmetic coding is also used to improve the coding efficiency. In this case, a set of probability tables are maintained for each context. Due to the higher coding efficiency and capability to adapt better to the signal statistics, arithmetic coding is generally favored over Huffman coding for platforms that could tolerate the computational requirements.

25.5 Video Coding Standards

Since the early 1990s, a number of video coding standards have been developed to satisfy industry needs. These standards have been developed by two major international standardization organizations: the Moving Pictures Expert Group (MPEG) of International Standards Organization/International Electrotechnical Commission (ISO/IEC), and the Video Coding Expert Group (VCEG) of ITU-T, the Telecommunications Standardization Sector of the International Telecommunications Union.

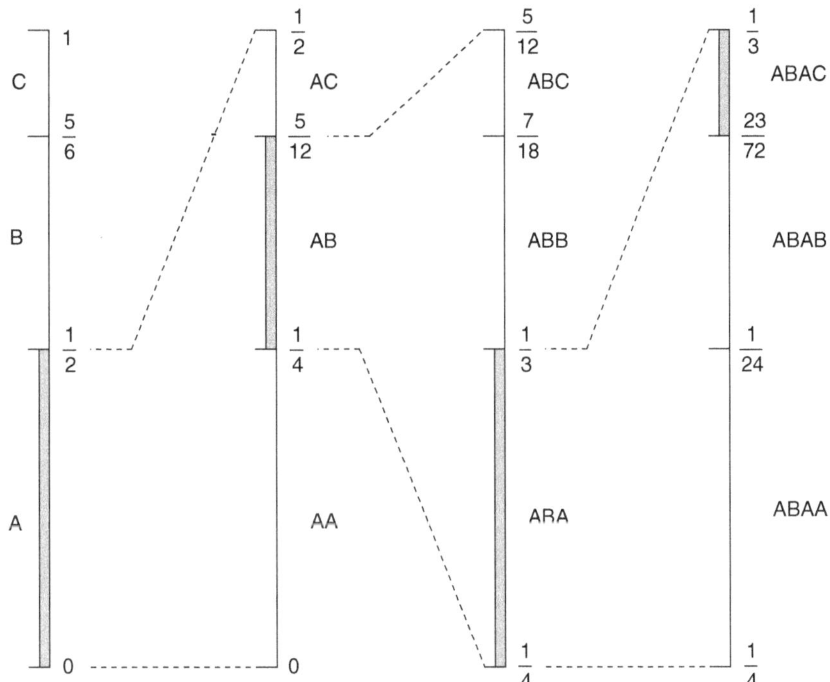

Figure 25.9 Example of arithmetic coding of the source sequence "ABAC" according to the probabilities of each symbol.

The video coding standards developed by ISO/IEC include the MPEG-x series of standards, while the standards developed by ITU-T fall under the H.26x series of recommendations. It should be noted that Recommendation H.262 is the same as MPEG-2; this standard was jointly developed by both organizations. In fact, most major video coding standards that followed were developed jointly by experts from both standardization bodies, including the AVC standard, known as H.264 and MPEG-4 Part 10, which was developed by the Joint Video Team (JVT), the High-Efficiency Video Coding (HEVC) standard, known as H.265 and MPEG-H Part 2, which was developed by the Joint Collaborative Team on Video Coding (JCT-VC) and the latest Versatile Video Coding (VVC) standard, known as H.266 and ISO/IEC 23090-3, which was developed by the Joint Video Experts Team (JVET). With ever-increasing requirements on coding higher resolution video, each generation of a video coding standard has generally improved compression efficiency by a factor of 2, i.e., the new standard achieves similar perceptual quality with half the bit rate relative to the prior standard. These standards have found many successful applications such as digital television broadcast, optical disc storage including CD, DVD, and Blu-ray Disc, digital telephony, video streaming, and mobile video. In the following, we provide a short review of the main coding tools used by each standard, focusing on the notable changes from prior standards.

The first version of H.261 was completed in 1990 with later amendments published in 1993 [International Telecommunications Union 1993] and was mainly designed for low-delay video conferencing over Integrated Services Digital Network (ISDN) lines with bit rates between 64 and 320 Kbps. It was the first standard to define the basic video coding architecture as shown in Figure 25.2. This standard utilized 16×16 MBs, 8×8 DCT with uniform quantization, and supported unidirectional forward motion-compensated prediction with integer-pixel precision. A VLC scheme known as run-level coding was used to code quantized transform coefficients. With run-level coding, the 2D transform coefficients are first scanned into a one-dimensional (1D) vector. Huffman coding is then applied to symbols that comprise a pair of numbers indicating the *run* of zeros followed by the *level* of the next nonzero coefficient. A special symbol is used to indicate the last nonzero coefficient in a block. H.261 also defined an optional loop filter that applied a low-pass filtering on the motion-compensated prediction to decrease prediction error and blocking artifacts at high compression ratios.

The MPEG-1 standard was completed in 1991 [MPEG-1]. The target application of MPEG-1 was digital storage media on CD-ROM at bit rates between 1 and 2 Mbps. MPEG-1 specified motion compensation with half-pixel accuracy as well as the use of B-frames and bidirectional prediction. The DC coefficients in MPEG-1 are predicted from the left neighbor.

MPEG-2 was completed in 1994, with later amendments in 2000 [MPEG-2]. The standard was developed jointly with ITU-T and is also known as H.262. MPEG-2 is an extension of MPEG-1 and allows for greater input format flexibility and higher data rates that include support for standard-definition and high-definition resolutions. Target bit rates are in the range of 4–30 Mbps. This standard has been widely used for television broadcast and DVD applications. MPEG-2 adds coding tools that specifically support the efficient encoding of interlaced material. It also defined various modes of scalability, which are outlined briefly in the next section.

The first version of H.263 was completed in 1996 [International Telecommunications Union 1996], and it is based on the H.261 framework. It defined more computationally intensive and efficient algorithms to increase the coding performance in telecommunication applications. New technical features included advanced prediction, which supported overlapped block motion compensation and optional use of four motion vectors per MB, motion vector prediction to code motion vector data with less rate, and improved entropy coding that integrates the end of block symbol into the run-level coding. H.263 also allowed for arithmetic coding in place of Huffman coding.

A second version of H.263 referred to as H.263+ was approved in 1998. Several new optional features were added to improve coding efficiency. Most notably, advanced intraprediction, in which spatial prediction from neighboring blocks was employed, and an in-loop deblocking filter that is applied to block boundaries of 8 × 8 blocks of the reconstructed images that is used for reference. Additionally, significant improvements in error resilience were realized through a variety of tools including flexible resynchronization marker insertion, reference picture selection, data partitioning, reversible VLC, and header repetition.

A first version of MPEG-4 Part 2 was completed in 2000, with later editions in 2004 [MPEG 4]. It was the first object-based video coding standard and is designed to address the highly interactive multimedia applications. Specific profiles of MPEG-4 Part 2, targeting lower bit rate video, have been used for some mobile and Internet streaming applications. Among the new technical features are adaptive DC/AC prediction for intrablocks and quarter-pixel motion compensation. It also adds support for improved error resilience and generally has coding performance similar to that of H.263.

H.264 was the next generation of video coding, with the first edition approved in 2003 and second edition with more advanced capabilities in 2005. The standard was developed jointly between MPEG and ITU-T and is also known as MPEG-4 Part 10 [International Telecommunications Union 2009]. Another name for this standard is Advanced Video Coding, or simply AVC. H.264 has significantly improved the coding performance over both MPEG-2 and H.263. The standard improves intraprediction with directionally adaptive spatial prediction, it defines a 4 × 4 and 8 × 8 integer transform that could be selected adaptively, there are more powerful motion-compensated prediction capabilities including support for various block partitioning schemes and multiple reference pictures, as well as a context-adaptive binary arithmetic coding scheme.

In addition to coding efficiency, H.264 also targeted a design that would ease transport system integration and data loss resilience. To achieve this, a high-level syntax was designed with the following features: a parameter set structure that provides a robust mechanism for conveying data that are essential to the decoding process; a Network Abstraction Layer (NAL) unit to readily identify the purpose of the associated payload data; and a slice data structure that can be decoded independently from other slices of the same picture to enable resynchronization in the event of data losses.

H.264 is capable of achieving the same quality as prior standards such as H.263 and MPEG-4 Part 2 with approximately half the bit rate. It has been deployed widely for television broadcast, Blu-ray Disc, and Internet/mobile streaming applications. For further details on the H.264 standard, interested readers are referred to the overview paper by [Wiegand et al. 2003].

In 2013, a decade after the H.264 standard was first approved, the joint team of MPEG and ITU-T experts produced the first version of the HEVC standard, also referred to as H.265. This standard was mainly developed to address the needs for higher coding efficiency required by an increasing diversity of services and the emergence of formats beyond traditional high-definition (HD) resolutions, including 4K × 2K or 8K × 4K resolutions. While the HEVC coding architecture largely follows that of prior standards, a few notable changes were adopted. The most significant change was the introduction of larger blocks sizes and much more flexible partitioning of these blocks. The traditional 16 × 16 MB specified in all prior video coding standards was replaced by a Coding Tree Unit (CTU), which could be up to 64 × 64 in size and partitioned in very flexible ways using a tree structure and quadtree-like signaling as shown in Figure 25.10. The smaller partitions within a CTU could then be coded using a variety of prediction and transform options. For the prediction, each coding unit could use either intraprediction or interprediction and could be further split into prediction blocks with varying height and width. For the transform, HEVC defines integer basis functions similar to those of a DCT for square blocks of larger size, including 4 × 4, 8 × 8, 16 × 16, and 32 × 32. In order to reduce signaling overhead, methods of advanced motion vector prediction were also adopted, which derive a motion vector prediction from adjacent blocks or simply inherit the motion vector of a temporally or spatially neighboring block. Intraprediction efficiency was also improved with support for 33 directional modes, compared to 8 directional modes in H.264.

The high-level syntax features of H.264 have generally been retained in HEVC with some additions to improve flexibility and robustness to data losses. To enhance the parallel processing capability, HEVC introduced several new features.

1. Tiles: these are independently decodable regions of a picture that are encoded with some shared header information. Segmenting a picture into multiple tiles can provide parallelism in the decoding process.
2. Wavefront parallel processing: When enabled, a slice is divided into rows of CTUs, and each row can be processed with a two-CTU processing lag relative to the preceding row. This mode provides a form of processing parallelism at finer level of granularity within a slice and may offer better compression efficiency than tiles.

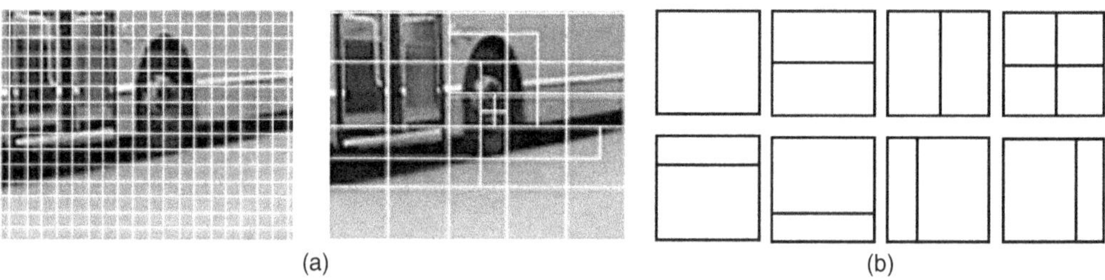

(a)   (b)

Figure 25.10 Illustration of (a) larger block size and flexible block partitioning, and (b) motion partitions, including asymmetric motion partitions, supported by HEVC.

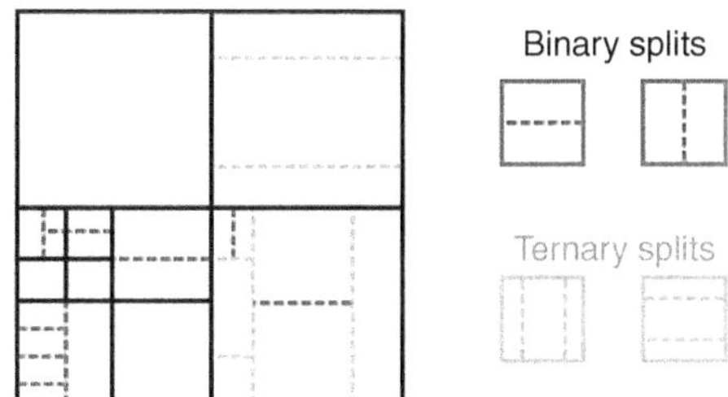

Figure 25.11 Block partitioning in VVC using recursive quadtree partitioning and nested recursive multi-type tree partitioning with binary and ternary splits.

3. Dependent slice segments: This type of structure allows data associated with a particular wavefront entry point or tile to be carried in a separate NAL unit. The main benefit is being able to make that data available to a system for fragmented packetization with lower latency than if it were all coded together in one slice.

The HEVC design has been shown to achieve approximately a 50% bit-rate savings for equivalent perceptual quality relative to the performance of the prior H.264 standard. Further details about the HEVC standard can be found in [Sullivan et al. 2012].

The latest generation video coding standard is referred to as VVC and was formally approved as ITU-T H.266 and ISO/IEC 23090-3 in 2020. This standard provides yet another 50% reduction in bit rate compared to HEVC and offers a wide range of additional functionalities. In addition to improved compression efficiency for high-resolution video, new features designed into the first version of this new coding standard include support for coding of High Dynamic Range (HDR) content and 360° video content, multi-layered coding, streaming with adaptive picture resolution, and support for compressed-domain bitstream extraction and merging.

A number of new coding tools and refinements to existing tools have been adopted into VVC. Some select changes include:

- The maximum CTU size has been increased to 128×128 and the minimum size has been set to 32×32.
- There is greater flexibility in block partitioning using recursive quadtree partitioning and nested recursive multi-type tree partitioning with binary or ternary splits, as illustrated in Figure 25.11.
- The maximum transform size has been extended to 64×64 to have better energy compaction for the residual signals of large-sized smooth areas. Also, nonsquare transforms are supported by applying different length transform kernels in horizontal and vertical directions.
- The number of directional intra modes that can be used for prediction has been increased to 85 with wider prediction angles beyond $45°/135°$. Additionally, two additional reference lines could be used and a matrix-weighted intra prediction could be formed based on multiple pixel values in the reference line.
- Inter prediction has been improved through a newly introduced geometric partitioning mode that enables motion compensation on nonrectangular partitions of blocks. Also, an affine motion mode has been introduced to represent nontranslational motion more efficiently, and decoder-side motion vector refinement is used to improve the accuracy of the motion vectors.

Further details about the VVC standard can be found in [Bross et al. 2021].

25.6 Video Coding Extensions

25.6.1 Scalable Video Coding

The traditional dimensions of scalability include quality scalability, temporal scalability, and spatial scalability. A key objective of a scalable representation is to encode the video source signal once, then decode many times according to specific delivery, and receiver capabilities. Such functionality is highly desirable in any dynamic and heterogeneous communication environment, and especially for mobile video delivery. Scalable video coding typically incurs some loss relative to single nonlayered video coding. One challenge is to minimize this loss, another is to keep the complexity to a minimum.

Scalable video coding was first introduced in the MPEG-2 standard and revisited in MPEG-4 Part 2. These scalable extensions were not very successful since the coding efficiency loss relative to nonlayered video was relatively high, and there was a notable increase in complexity to support such modes. The scalable video coding extension of the H.264/AVC standard has overcome these drawbacks, so we will focus on fundamental aspects of scalable video coding with an emphasis toward the architecture and features introduced in AVC and later standards. For further details about the support for scalable video coding in various standards, readers are referred to [Schwarz et al. 2007], [Boyce et al. 2015], and [Bross et al. 2021].

Temporal scalability is very easily supported in the context of current standards with a hierarchical prediction structure. In older standards such as MPEG-2, B-frames were not used as reference and were at the bottom of a simple hierarchy between I- and P-frames, so they could easily be dropped without any impact on the decoding of other frames. In H.264/AVC, the prediction dependency is more flexible so deeper hierarchies and hence more temporal layers could be supported. An example of a hierarchical prediction

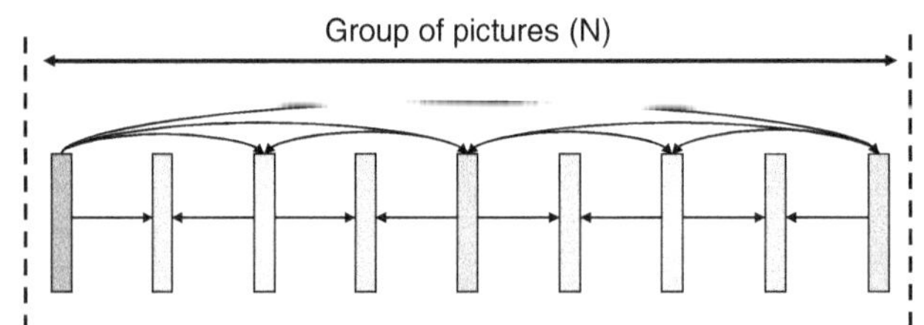

Figure 25.12 Hierarchical coding structure supporting temporal scalability.

structure is shown in Figure 25.12. Interestingly, it has been found that such hierarchical prediction structures actually improve coding efficiency provided that the quantizers for each level are selected appropriately, i.e., finer quantizers should be used at lowest temporal layers and coarser quantizers at the highest layers.

To support spatial scalability, one performs a multi-layered coding of each spatial scale, where each spatial scale supports conventional motion-compensated prediction as well as an interlayer prediction. There are a number of ways to perform the interlayer prediction in a multi-scale framework. One method is to simply up-sample the reference data in the lower layer and use the up-sampled data for prediction. Another method is to infer block-level data such as the motion vectors from lower reference layers. Finally, the residual from the lower reference layer could also be used to predict the residual that is derived at the higher spatial layer. All three forms of interlayer prediction are supported by the H.264 standard. Another major innovation in this standard to overcome the complexity issues that have plagued past standards was to constrain the interlayer prediction so that single-loop decoding could be enabled rather than having decoding loops for each scale. Finally, combined spatial/temporal scalability is also possible since lower layer pictures need not be present at every time instant. A figure illustrating this is provided in Figure 25.13.

The main purpose of quality scalability is to refine the SNR of the video with increasing quality layers. This form of scalability is often referred to as SNR scalability as well. To achieve a coarse-grained scalability, a similar multi-scale architecture as used for spatial scalability could be used without the up-sampling operation. In this way, each layer would use a different quantizer to achieve the desired level of quality at each layer. The base layer would be coded with a coarse quantization and finer levels of quantization would then be used for higher layers. A finer level of quality control could also be imposed within each layer by coding fragments of transform coefficients. This enables successive refinement of quality within a layer.

In both the HEVC and VVC designs, scalability is based on a multi-layer architecture that relies on multiple single-layer decoders. In this way, there are no modifications to any of the block-level decoding tools. The reconstruction of a higher enhancement layer from a lower layer, e.g., reconstructing Ultra HD video from an HD base layer for spatial scalability, is enabled through picture referencing with added interlayer reference picture-processing modules, including texture and motion resampling and color mapping. While this allows reusing the base decoder cores, such an architecture increases processing requirements by needing multiple decoder cores plus the additional modules.

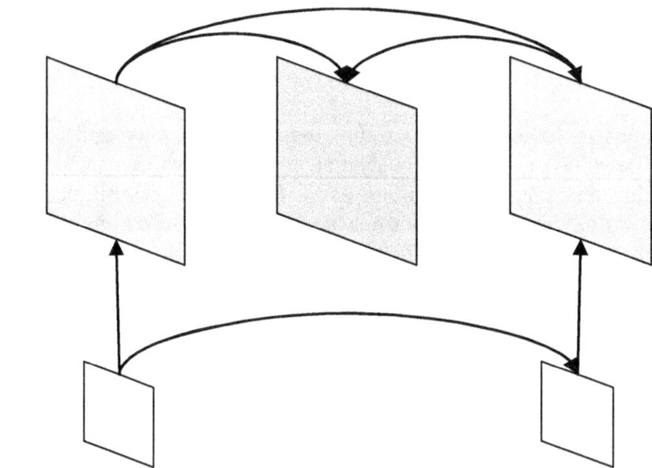

Figure 25.13 Combined spatial and temporal scalability including interlayer prediction and motion-compensated prediction.

25.6.2 Multiview Video Coding

Multiview video is used to support three-dimensional (3D) video applications. A special case of multiview video is stereo video in which only two views are present, a left view and right view corresponding to each eye. For most stereoscopic displays, glasses are needed to view the 3D scene. Auto-stereoscopic displays are capable of rendering multiple views of a 3D scene simultaneously and so do not require glasses. Three-dimensional services are becoming more popular for home entertainment systems and in mobile environments.

Performing efficient compression relies on having good predictors. While the correlation between temporally neighboring pictures is often very strong, including spatially neighboring pictures offers some advantages. For example, spatially neighboring pictures are useful predictors in uncovered regions of the scene, during fast object motion, or when objects appear in one view that are already present in neighboring views at the same time instant. Interview prediction is employed in all related works on efficient multiview video coding and aims to exploit both spatial and temporal redundancy for compression. The prediction is adaptive, so the best predictor among temporal and interview references is selected on a block basis. It is also noteworthy that there exists a base layer that can be independently decoded and possibly used as a 2D representation of the 3D scene.

It has been shown that coding multiview video with inter-view prediction does give significantly better results compared to independent coding of each view. For instance, improvements of more than 2 dB have been reported relative to independent encoding of views using the multiview extensions of H.264/AVC. Furthermore, subjective testing has indicated that the same quality could be achieved with approximately half the bit rate.

It is noted that the layered coding approach in HEVC and VVC supports both scalable and multiview video coding capabilities in a unified manner by utilizing the mechanisms for referencing pictures to include pictures in other views without any changes to the block-level coding tools. However, there is a version of HEVC, referred to as 3D-HEVC, that extends the 3D coding capabilities of HEVC to achieve a more efficient representation of the Multiview video and its associated depth map. The depth map enables a receiver to render additional intermediate views and can also be used to further exploit correlation between views as well as statistical dependencies between video texture and depth maps. The main drawback of the 3D-HEVC design is that it introduces new block-level coding tools, which requires new and modified decoder cores relative to the base HEVC design.

For further details about the support for multiview and 3D video coding in various standards, readers are referred to [Vetro et al. 2011], [Tech et al. 2016], and [Bross et al. 2021].

25.6.3 360° Video Coding

Immersive video experiences have become more popular in recent years, and one of the key representation formats is 360° video, which enables a look-around viewing experience for virtual reality applications through head-mounted displays. To encode the 360° video with existing coding tools, the captured content is first mapped from a 3D sphere to a 2D picture representation. This can be done in numerous ways depending on the type of geometry that is considered. For instance, using an EquiRectangular Projection (ERP) format, the sphere is projected onto a rectangular picture with some geometric distortion, especially at the poles. Another mapping is the Cube Map Projection (CMP), where the sphere is mapped onto the six faces of a cube, which are then packed together into one picture. These projections are illustrated in Figure 25.14.

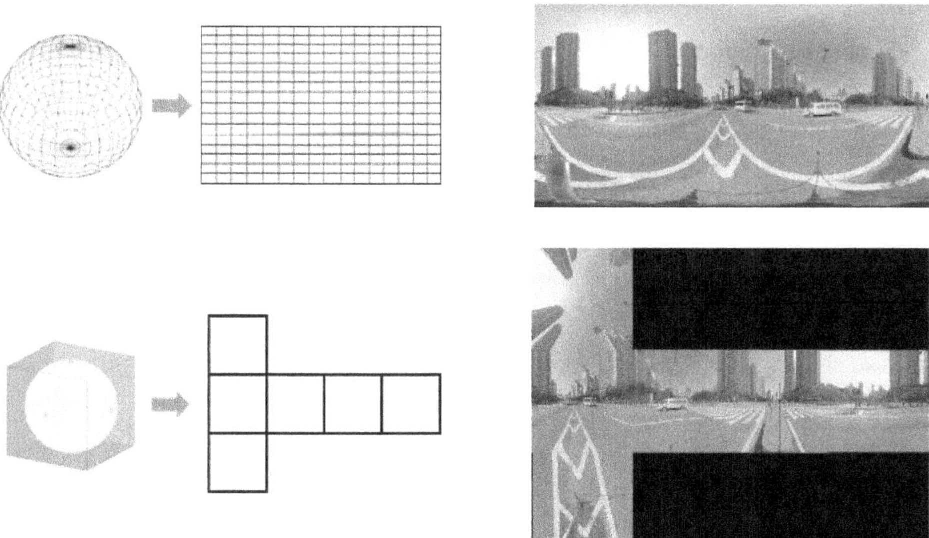

Figure 25.14 Illustration of equirectangular projection (top) and cube map projection (bottom) to map 360° video into a 2D picture representation format.

HEVC defined signaling to enable remapping of coded video samples onto a 3D coordinate space in both angular and Cartesian coordinates. In addition to signaling the type of projection, signaling of a region-wise packing was also specified, which enables manipulation of the image, e.g., scaling, translation, rotation, and mirroring, of any rectangular region of the projected picture in order to overcome weaknesses of a particular projections, e.g., oversampling toward the poles in ERP.

VVC extended the number of projections that could be signaled and improved on the individual packing formats of each projection in order to minimize visual discontinuities between adjacent surfaces. Furthermore, to increase the coding efficiency for video pictures using these projection formats, a wrap-around technique was defined to allow prediction samples to wrap around from the opposite left or right boundary in cases where motion vector points outside of the coded area. Also, virtual boundaries for in-loop filtering are used to prevent applying in-loop filtering across certain virtual boundaries, e.g., corresponding to the CMP face boundaries.

25.6.4 Screen Content Coding

The need to compress and transmit computer-generated content has substantially increased in recent years through the rise in screen sharing and gaming applications. Since the signal characteristics present in screen content are very different than those found in camera-captured content, new coding tools have been designed to better exploit those characteristics. A number of tools, including intra-picture block copy, palette mode, adaptive color transform and transform skip, first appeared in an HEVC extension for screen content coding, which was finalized in 2016. These tools were retained in the VVC standard with some simplifications and refinements.

Intra-picture block copy makes use of repeated patterns inside a picture. It can be seen as a very basic form of motion-compensated prediction with block vectors referencing previously coded regions of the same picture instead of previously coded reference pictures. This tool was first defined in HEVC and later simplified in the VVC design.

Palette mode is used to represent the sample values by a set of representative color values, which is referred to as the palette. To use this coding mode, the palette must first be signaled, and then, a palette index is signaled for each sample. Escape symbols are used to signal values that are not represented in the palette. For blocks with very few colors present, this mode can provide a very compact representation. The key principles of this coding mode are illustrated in Figure 25.15.

Recognizing that much of the screen content is captured in a 4:4:4 chroma format and in the RGB color space, an adaptive color transform can be applied to reduce the correlation between the three color components. The transform essentially performs an in-loop color-space conversion of the prediction residual by adaptively converting the residuals from the input color space (presumed to be RGB) to the YCgCo-R luma–chroma color representation, thereby compacting the overall energy of the signal.

In some cases, the spatial transform may not be beneficial when coding some types of screen content. Therefore, an option to skip the transform is provided. Since the quantization and entropy coding stages would be unchanged, some scaling of the spatial domain prediction residuals would be needed to approximate the dynamic range of the transform coefficients.

The screen content coding tools provide substantial gains for a wide variety of computer-generated and mixed content, with bit rate reductions up to 75% on certain types of content. Further details about screen content coding tools can be found in [Peng et al. 2016] and [Bross et al. 2021].

25.7 Error Control

The previous sections focused mainly on the techniques to efficiently compress video and discussed related standards. These compression standards rely heavily on predictive coding and VLC techniques, both of which are not necessarily favorable characteristics when transmitting the compression video bitstream over error-prone channels. Prediction creates a dependency in the bitstream, whereby errors in one segment propagate to other segments. Errors in the VLC could also lead to synchronization issues and ultimately decoding failure.

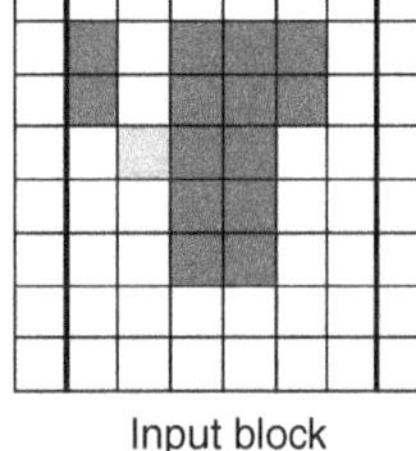

Figure 25.15 Illustration of palette coding mode for screen content coding.

This section covers three basic levels of error control that could be utilized to overcome errors during transmission: mechanisms that are available at the transport layer to protect the video, error-resilient features within the video layer, and techniques to conceal errors in a reconstructed video. We consider random bit errors that may result from characteristics of the physical channel, as well as the loss of packets that typically impact a greater portion of the bitstream. A more thorough treatment of this subject could also be found in [Wang and Zhu 1998].

25.7.1 Transport Layer Mechanisms

A well-known method for error detection and correction is Forward Error Correction (FEC). FEC could be applied directly to the compressed bits to protect against bit errors or across data packets to recover from erasures. When applied to compressed bits, the FEC code is typically capable of correcting a single bit error within a frame comprising several hundred bits. When applied across data packets, a typical approach is to combine Reed–Solomon (RS) coding with block interleaving, where the RS code is first applied to blocks of data and then those blocks of data are interleaved into packets. In this way, a loss of one packet is essentially dispersed over multiple blocks of data and could be recovered. Since FEC increases the data transmission rate, which in turn reduces the available rate for the coded video, an appropriate balance between rate for source and channel coding must be considered in the overall design.

Unequal error protection is another effective means for increasing the robustness of the video transmission at the transport layer. It must first be noted that not all bits of the video bitstream is equally important. For instance, certain header information and other side information are much more critical to the final picture quality than some of the other block data. Also, in layered coding schemes described in Section 25.6, the base layer is much more critical than the enhancement layers since the enhancement layers are of no use without the base layer. Therefore, when error correction is used, important parts of the video bitstream should be coded with greater levels of protection. For networks that allow prioritization of data, higher prioritization can also be assigned accordingly so that such aspects including congestion control, retransmission, and power control can be optimized based on the priority of the data.

25.7.2 Error-Resilient Encoding of Video

While coding efficiency is the most important aspect in the design of any video codec, the transmission of compressed video through noisy channels must be considered. There are many error-resilience tools available in today's video coding standards. A brief review of some of the most relevant tools is provided in the following.

Localization

The basic principle of localization is to remove the spatial and temporal dependency between segments of the video to reduce error propagation. Such techniques essentially break the predictive coding loop so that if an error does occur, then it is not likely to affect other parts of the video. Obviously, a high degree of localization will lead to lower compression efficiency. There are two methods for localization of errors in a coded video: spatial localization and temporal localization; these methods are illustrated in Figure 25.16.

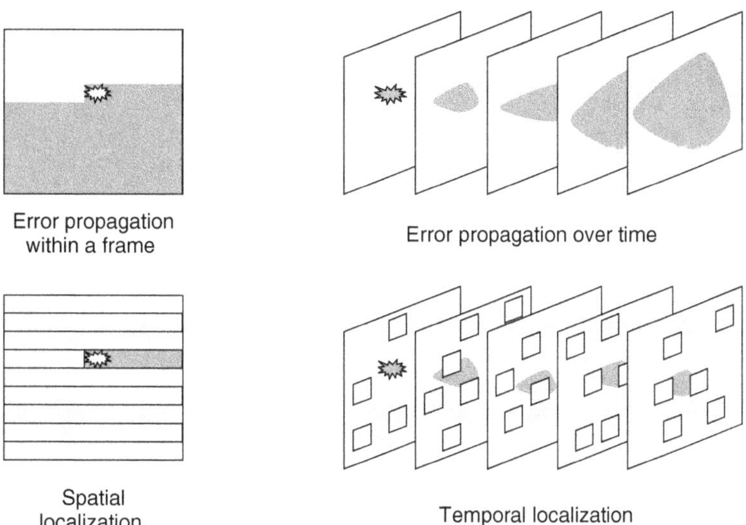

Figure 25.16 Illustration of spatial and temporal localization to minimize error propagation within a frame and over time, respectively. Spatial localization is achieved by means of resynchronization markers, while temporal localization is achieved by means of intrablock coding.

Spatial localization considers the fact that most video coding schemes make heavy use of VLC to reach high coding performance. In this case, even if 1 bit is lost or damaged, the entire bitstream may become undecodable due to the loss of synchronization between the decoder and the bitstream. To regain synchronization after a transmission error has been detected, resynchronization markers are added periodically into the bitstream at the boundary of particular MBs in a frame. This marker would then be followed by essential header information that is necessary to restart the decoding process. When an error occurs, the data between the synchronization point prior to the error and the first point where synchronization is reestablished are typically discarded. For portions of the image that have been discarded, concealment techniques could be used to recover the pixel data, e.g., based on neighboring blocks that have been successfully decoded. For resynchronization markers to be effective in reducing error propagation, all predictions must be contained within the bounds of the marker bits. The restriction on predictions results in lowered compression efficiency. In addition, the inserted resynchronization markers and header information are redundant, and they lower the coding efficiency. Spatial localization is supported in a number of standards through a *slice* or *Tile* structure, which is essentially a group of independently decodable MBs or CTUs.

While resynchronization marker insertion is suitable to provide a spatial localization of errors, the insertion of intracoded MBs is used to provide a temporal localization of errors by decreasing the temporal dependency in the coded video sequence. While this is not a specific tool for error resilience, the technique is widely adopted and recognized as being useful for this purpose. The higher percentage of intrablocks used for coding the video will reduce the coding efficiency, but reduce the impact of error propagation on successively coded frames. In the most extreme case, all blocks in every frame are coded as intrablocks. In this case, there will be no temporal propagation of errors, but a significant increase in bit rate would be expected. The selection of intracoded blocks may be cyclic, in which the intracoded blocks are selected according to a predetermined pattern; the intracoded blocks may also be randomly chosen or adaptively chosen according to content characteristics.

Another form of temporal localization is reference picture selection, which was introduced in the H.263 and MPEG-4 standards for improved error resilience. Assuming a feedback-based system, the encoder receives information about corrupt areas of the picture from the decoder, e.g., at the slice level, and then alters its operation by choosing a noncorrupted reference for prediction or applying intracoding to the current data. In a similar spirit, the support for multiple reference pictures in H.264/AVC and later standards could be used to achieve temporal localization as well.

Data Partitioning

The objective of data partitioning is to group coded data according to relative importance to allow for unequal error protection or transport prioritization as discussed in the previous subsection. Data partition techniques have been developed to group together coded bits according to their importance to the decoding such that different groups may be more effectively protected or handled. For example, during the bitstream transmission over a single channel system, the more important partitions can be better protected with stronger channel codes than the less important partitions. Alternatively, with a multi-channel system, the more important partitions could be transmitted over the more reliable channel.

In MPEG-2, data partitioning divides the coded bitstream into two parts: a high-priority partition and low-priority partition. The high-priority partition includes picture type, quantization scale, and motion vector ranges, without which the rest of the bitstream is not decodable. It may also include some MB header fields and DCT coefficients. The low-priority partition contains everything else. In MPEG-4, the data partitioning is achieved by separating the motion and MB header information away from the texture information. This approach requires that a second resynchronization between motion and texture information, which may further help localize the error. For instance, if the texture information is lost, the motion information may still be used to conceal these errors. Additional means of partitioning and signaling critical data for decoding have been devised in later standards, including the introduction of the NAL unit, as well as a variety of Supplemental Enhancement Information (SEI) messages, which could carry timing of the video pictures or describe various properties of the coded video or how the video can be used or enhanced.

Redundant Coding

With this approach, segments of the video signal or syntactic elements of the bitstream are coded with added redundancy to enable robust decoding. The redundancy may be added explicitly, such as with the Redundant Slices tool, or implicitly in the coding scheme, as with Reversible Variable Length Codes (RVLC) and Multiple Description Coding (MDC).

RVLC has been developed for the purpose of data recovery at the receiver. Using this tool, the VLC are designed so that they can be read both in the forward and reverse directions. This allows the bitstream to be decoded backward from the next synchronization marker until the point of error. Examples of 3-bit codewords that satisfy this requirement include 111, 101, 010. It is obvious that this approach will reduce the coding efficiency compared with using normal VLC due to the constraints imposed in constructing the RVLC tables, which is the primary reason we classify the RVLC approach as a redundant coding technique. It also shares the benefit with other tools that robust decoding could be performed. However, since this tool is designed to recover from bit errors, it is not helpful for packet-erasure channels. RVLC has been adopted to both H.263 and MPEG-4 Part 2 standards.

MDC encodes a source with multiple bitstreams such that a basic-quality reconstruction is achieved if any one of them is correctly received, while enhanced-quality reconstructions are achieved if more than one of them is correctly received. With MDC, the redundancy may be controlled by the amount of correlation between descriptions. Generally, MDC video streams are suitable for delivery over multiple independent channels in which the probability of failure over one or more channels is likely. Some limited forms of MDC can be achieved with H.264/AVC.

Redundant slice was a new tool adopted into the H.264/AVC standard that allows for different representations of the same source data to be coded using different encoding parameters. For instance, the primary slice may be coded with a fine quantization, while the redundant slice may be coded with a coarse quantization. If the primary slice is received, the redundant slice is discarded, but if the primary slice is lost, the redundant slice would be used to provide a lower level of reconstructed quality. In contrast to MDC, the two slices together do not provide an improved reconstruction.

25.7.3 Error Concealment at the Decoder

In any video transmission system, there will inevitably be errors in the received bitstream. If they are not corrected by transport layer mechanisms or suppressed during the decoding of the video bitstream, there could be severe damage to the reconstructed video. If the errors are appropriately localized, it is possible to conceal the effects of the transmission loss in the video signal. In most cases, it is assumed that the locations of errors have been detected and erroneous data have been discarded.

Perhaps the most widely studied error concealment approach is the recovery of texture information from neighboring data. By exploiting the inherent spatial and temporal redundancy in the video signal, it is often possible to recover missing blocks or even larger slices of the picture from neighboring data. A straightforward approach recovers missing data from one picture by copying colocated data from a neighboring picture, e.g., the previously decoded picture. The method could be applied when motion data is also lost and provides reasonable results for static parts of the video but does not provide satisfactory quality when there is significant motion in the scene. When the motion vector data is available, a much better recovery of the missing texture could be performed from reference pictures that are used for prediction of the current picture.

Another means to recover texture information is by utilizing the valid data in the same picture. Spatial interpolation methods recover missing texture in damaged blocks based on pixel values in neighboring blocks. There are also more sophisticated methods that optimize the recovery by imposing smoothness constraints and accounting for any nonerroneously received DCT coefficients.

Since the motion vector data could also be viewed as smooth field for natural scenes, there also exist methods that attempt to recover motion information for damaged blocks so that the temporal recovery schemes discussed above could be applied. For example, the motion vector for a damaged block could be estimated based on an average or median of motion vectors in neighboring blocks, or the motion vector from the corresponding MB in the above row of MBs could be copied.

25.8 Video Streaming

Video streaming refers to the real-time transport of video to a receiving device. The video itself may be live or stored on a server. In either case, the delivery of video over wireless and wired networks poses a number of unique challenges to ensure that the best Quality of Service (QoS) could be achieved. There are of course unique challenges to transmitting and receiving video over a wireless network including multi-path propagation, interference, energy, and power management as well as user mobility. Further discussion on these characteristics of a wireless communication system is provided in Chapter 2.

Generally speaking, QoS requirements are typically specified in terms of bandwidth, delay, and error rates. These parameters are likely to be time varying depending on the channel characteristics. The bandwidth is essential to ensure that the video at a particular resolution could be represented with high enough quality given the rate constraints. Since video must be played out continuously, there are also strict timing constraints imposed on the delivery and decoding processes. As discussed in the previous section, error loss also has a notable impact on the final reconstructed quality of the video.

Since the bandwidth and loss characteristics over a given channel are often fluctuating, the rate of the compressed video that is being streamed should ideally change based on such dynamics. There are a variety of techniques that could be used to regulate the rate of the video bitstreams depending on the bitstream characteristics, system level constraints, and application requirements. For instance, given a nonlayered MPEG-2 video bitstream, one may apply transcoding operations so that the source rate matches the channel bandwidth. This could be achieved by requantizing the transform coefficients, which requires a partial decoding of the bitstream, or by simply dropping frames. If the video is encoded using a scalable format, adjustments to the rate could be made with simpler operations as described in Section 25.6. In multi-cast networking, receiver based strategies could also be used to regulate the stream data to be processed locally.

Synchronization is another important aspect of video streaming. In most applications, the video is accompanied by an associated audio stream, and in some cases, there are other graphics and text that are delivered as separate streams that correspond to the video as well. It is very critical that the temporal relationship between the different streams should be maintained, and if lost, for them to be resynchronized at a later point in time. For instance, in broadcasting applications, even minor differences between the audio and video stream can lead to what is known as lip synchronization problems, which are annoying to the viewer.

25.8.1 Networking Protocols

There is a very rich set of networking protocols that supports the streaming of video. A brief discussion of relevant transport, session, and application-layer protocols is given in the following.

- *Transmission Control Protocol (TCP)* is the dominant transport-layer protocol for Internet Protocol (IP)-based data transfer and handles such functions as multiplexing, error control, and flow control. While TCP could be used for video streaming, there are several aspects that prevent it from providing reliable and good-quality video streaming. For one, it utilizes retransmission for packet loss so end-to-end delay may be relatively large. Also, TCP does not handle variability in the data rate well. These problems can be addressed by buffering the data. An optimal buffer size could be determined based on a target delay, smoothness of playback, and data loss. In general, a small buffer size implies smaller delay. On the other hand, a larger buffer will provide for smoother playback since larger variations in the bit rate and transmission time could be tolerated.

- *User Datagram Protocol (UDP)* has become a preferred network protocol for video streaming. In contrast to TCP, UDP allows damaged or lost packets to be dropped. While this feature allows for reduced delay, it does not guarantee packet delivery; therefore, packet loss should be expected and error concealment techniques described in Section 25.7.3 would be needed to recover those losses.
- *Real-time Transport Protocol (RTP)* is a protocol for the transport of real-time data, including audio and video. RTP consists of two parts, a data part and a control part which is called RTCP. The data part of RTP supports real-time transmission for continuous media such as video and audio. It provides timing reconstruction, loss detection, security, and content identification. The control part provides source identification and support for gateways like audio and video bridges as well as multicast-to-unicast translators. While it offers QoS feedback from receivers to the multicast group as well as support for the synchronization of different media streams, it does not provide QoS guarantees. RTP/RTCP is commonly built on the top of UDP.
- *Real Time Streaming Protocol (RTSP)* is a session control protocol for media streaming. This protocol is designed to initiate a session and direct the delivery of the video stream. One of its main functions is to select the delivery channel and mechanism. It also controls the playback of the video stream with support of so-called trick-play operations such as pause, fast-forward, and reverse play.
- *Hyper Text Transfer Protocol (HTTP)* is an application-layer protocol that facilitates the fetching of resources and is the basic foundation for data transfer on the Internet. It functions as a request-response protocol, where the client submits a request message to the server, and then the server provides a response message back to the client with the requested resources. The response also contains completion status information about the request. It presumes an underlying transport-layer protocol, such as TCP or UDP. HTTP has primarily been used for the transfer of relatively static content, such as web pages that include text and image data. Progressive download can be used for video streaming, but it has several disadvantages including being bandwidth inefficient since fetched content will be wasted if the user stops watching, it is not rate adaptive to network conditions, and cannot be used for live media services. The Dynamic Adaptive Streaming over HTTP (DASH) standard has been developed by MPEG to overcome these shortcomings and will be described in the next subsection.

The above protocols are used in a variety of mobile delivery standards including Third Generation Partnership Project (3GPP), 1-seg in Japan, Digital Video Broadcasting – Handheld (DVB–H) in Europe and Advanced Television Systems Committee – Mobile/Handheld (ATSC–M/H) in North America. An overview of some important mobile video application standards, including Multimedia Messaging Service (MMS), streaming, video telephony, and multicast and broadcast is provided by [Wang et al. 2007].

25.8.2   Dynamic Adaptive Streaming over HTTP

Relative to RTP, streaming over HTTP has several benefits. For one, HTTP is very widely supported, and the Internet infrastructure has evolved to efficiently support HTTP, e.g., Content Delivery Networks (CDNs) provide localized edge caches that reduce long-haul traffic. Additionally, the client manages the streaming without having to maintain a session state on the server, which makes it easier to provision a large number of streaming clients. As a result, HTTP streaming has become a popular approach in commercial deployments, including Apple's HTTP Live Streaming, Microsoft's Smooth Streaming, and Adobe's HTTP Dynamic Streaming. However, these platforms are not interoperable with each other since they use different manifest and segment formats, which was a key motivation to develop the DASH standard, which was first published in 2012 with later editions adding more features and functionalities.

Figure 25.17 illustrates the basic DASH streaming framework between an HTTP server and a DASH client. The media presentation is stored on HTTP server and delivered using HTTP. In addition to the segments, which contain the actual multimedia bitstreams in the form of chunks, in single or multiple files, the other important part of the content is the Media Presentation Description (MPD), which describes a manifest of the available content, its various alternatives, their URL addresses, and other characteristics.

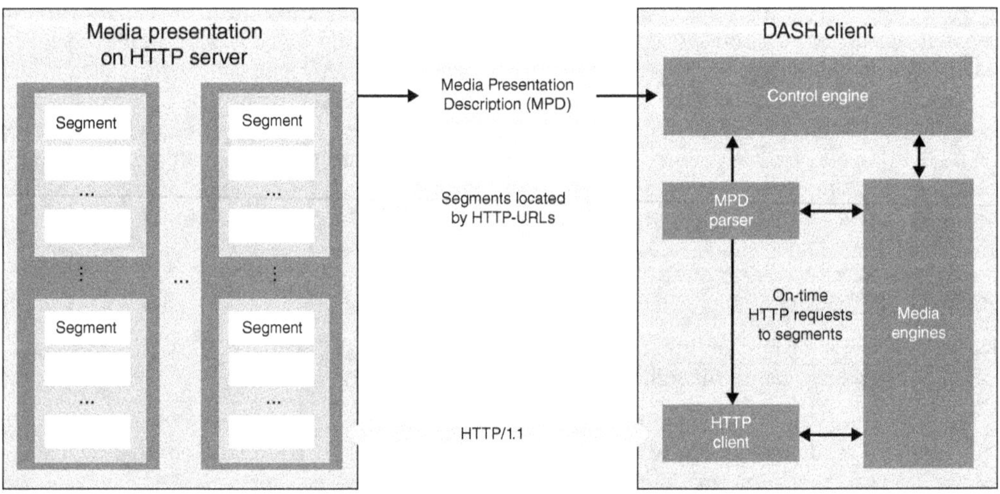

Figure 25.17   Illustration of DASH-based media delivery framework, including media presentation on HTTP composed of segments, media presentation description, and key functions of the DASH client. Color version available at wiley.com/go/molisch/wireless3e.

To play the content, the DASH client first obtains the MPD. The MPD can be delivered using HTTP or other transports. By parsing the MPD, the DASH client obtains information about the program timing, media-content availability, media types, resolutions, minimum and maximum bandwidths, and the existence of various encoded alternatives of multimedia components, media-component locations on the network, and other content characteristics. Using this information, the DASH client selects the appropriate encoded alternative and starts streaming the content by fetching the segments using HTTP GET requests.

After appropriate buffering to allow for variations in the network bandwidth, the client continues fetching the subsequent segments and also monitors the network bandwidth fluctuations. Depending on its measurements, the client decides how to adapt to the available bandwidth by fetching segments of different alternatives (with lower or higher bitrates) to maintain an adequate buffer. In contrast to other streaming frameworks, the client fully controls the streaming session by managing the on-time requests to the server to ensure a smooth playout of the sequence of segments.

The MPEG-DASH specification only defines the MPD and the segment formats; further details on the specification of these formats can be found in [Sodagar 2011]. It is noted that the delivery of the MPD and the media-encoding formats containing the segments, as well as the client behavior for fetching, adaptation heuristics, and playing content, are outside the scope of the standard.

Further Reading

Further information about image and video processing can be found in the classic textbooks by [Jain 1989], [Gonzalez and Woods 2008], and [Tckalp 1995]. For a more comprehensive overview of the principles of video compression, as well as corresponding algorithms and video compression standards, we recommend the textbooks by [Wang et al. 2002] and [Shi and Sun 2000]. The edited book by [Sun and Reibman 2001] provides an excellent collection of works on networking and transport of compressed video.

For updates and errata for this chapter, see https://wides.usc.edu/students.html#textbooks.

Exercises

See Sec. 36.25 of Exercises.pdf at wiley.com/go/molisch/wireless3e

26

Cognitive Radio

26.1 Types of Cognitive Radio

The efficient use of available spectrum is a key requirement for wireless system design, since spectrum is a finite resource. Due to the propagation characteristics of electromagnetic waves (see Part II), it is mostly the frequencies between 10 MHz and 7 GHz that are used for wireless services with wider-area coverage. While this might sound like a lot of spectrum, we have to keep in mind that it is used for a large variety of wireless services, the demand for which is constantly growing. A listing of the main frequency bands for different services is given in Section 2.3.1.

The dominant regulatory approach for spectrum usage has been to assign a spectral band to a particular service, and even a particular operator, with all users in the band thus employing the same transmission technique (PHYsical layer (PHY)). Such a scheme allows detailed network planning and good quality of service. The drawback of this approach is that if – at any given time – there are fewer users of this particular system than could be sustained by the available bandwidth, spectral resources are wasted. An alternative approach is *cognitive radio*, in which users adapt to the environment, including existing spectrum usage. For the further discussion, it is useful to distinguish between two different definitions for cognitive radio:

- A "spectrum-sensing cognitive radio" only adapts the transmission frequency, bandwidth, time, and power, according to the current spectrum usage in the environment. Such cognitive radio is also often called *Dynamic Spectrum Access* (DSA).
- A fully cognitive radio (also known as "Mitola radio," in recognition of the engineer who first proposed it), adapts *all* transmission parameters to the environment, i.e., modulation format, multiple access method, coding, as well as center frequency, bandwidth, transmission times, and so on. While a fully cognitive radio is interesting from a scientific point of view, it currently seems too complicated for practical purposes.

26.1.1 Dynamic Spectrum Access

A key motivation for DSA comes from the fact that with the current, fixed, assignments, spectrum is not exploited to its full extent at all times. If, e.g., no active users are present in a cell of a cellular system, the spectrum is unused in this particular area. In general, this problem of fallow spectrum (called *white-space* spectrum particularly for spectrum in the TV transmission bands) is quite pervasive, though the exact percentages depend on the specific location and what counts as "fallow". Figure 26.1a shows the maximum, minimum, and average spectrum usage in an outdoor environment, demonstrating enormous variations of interference power. Figure 26.1b shows an example of an indoor environment where the spectrum usage is even smaller, and even on average, mostly thermal noise is present.

Models for DSA can be classified as follows:

- *Geography-based sharing*: geography-based assignments of spectrum can be either time-invariant or time-variant. As an example of the former category, consider areas in which sensitive radio telescopes for astronomy are located. Any electromagnetic emissions in a large geographical area around them should be avoided, even if they are nominally outside the band of interest, because the spurious emissions can interfere with the radio astronomical observations. For this reason, there are large "radio quiet zones" located around radio telescopes throughout the world. Similarly, emissions that can interfere with naval equipment should be avoided specifically near the coast. For the time-variant case, consider the frequency band normally reserved for wireless stage microphones. In large parts of the country, this band will be used for this purpose only a few days per year when a touring singer or musical is in town. By providing a suitable reservation system, this band can be opened to other usage for the rest of the time.
- *Dynamic exclusive model*: here, a frequency band is still reserved for the exclusive use of a particular *service*, but different providers can share the spectrum. As we have seen in Section 20.6, due to trunking gain, joint usage of a single large frequency band (either by

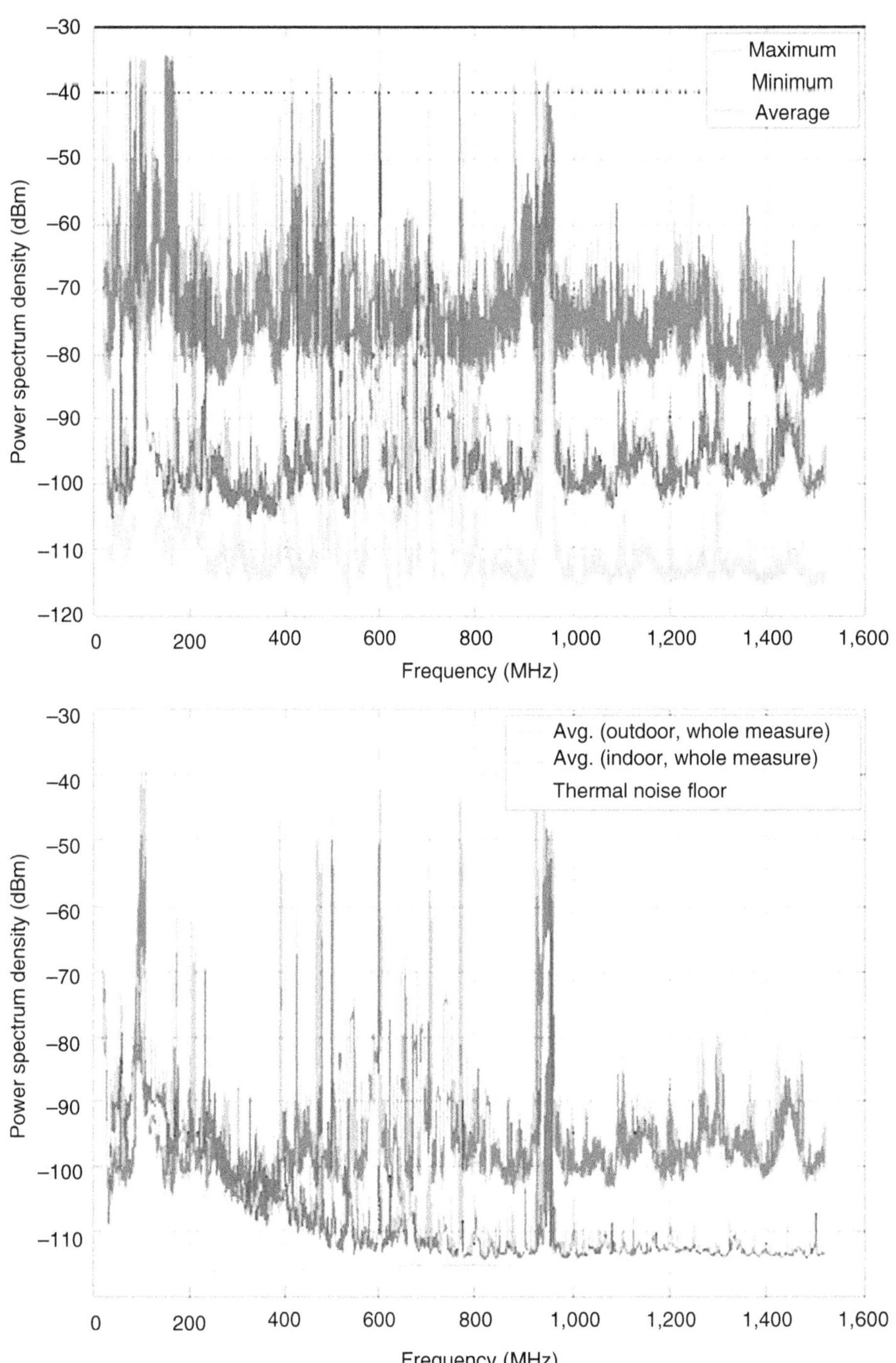

Figure 26.1   Maximum, minimum, and average received power spectral density in the frequency band 20–1, 520 MHz with a 200-kHz resolution bandwidth of the receiver. Outdoor location: on top of 10-storey building in Aachen, Germany. Indoor location: inside the office building in Aachen. Reproduced with permission from [Wellens et al. 2007] © IEEE.

a single cellular operator or by a consortium of operators) results in higher spectral efficiency than the use of N smaller bands by N separate operators. The sharing of the spectrum can be done by a trading (buying/selling or auctioning approach). Another possibility is that a regulator – depending on the usage statistics in a given location – assigns spectrum to a particular user or service on an exclusive but time-varying basis. For example, a cellphone provider might have the right to use 50 MHz of spectrum in the morning and only 20 MHz at noon.[1]

---

[1] Note that this approach is somewhat similar to the partial frequency reuse described in Section 21.3. In partial frequency reuse, the spectrum assigned to a service and a provider is constant, but the assignment to specific cells changes depending on the traffic conditions.

- *Open sharing model*: here, all users can access the spectrum equally, subject to certain constraints on the characteristics of the transmit signal. Such an approach is used today, e.g., in the Industrial, Scientific, and Medical (ISM) bands. The proliferation of Wireless Fidelity (Wi-Fi) devices in this band shows both the advantages and disadvantages of this approach. The free access and ease of type approval contributed to the popularity of Wi-Fi when it was first introduced. However, due to the large number of Wi-Fi devices, it has become almost impossible for other services (especially medical and industrial services) to operate in this frequency band with a reasonable Signal-to-Interference Ratio (SIR).
A way to bring some order and discipline to the "open sharing" is a restriction on the duty cycle and maximum transmission time of each user, combined with the requirement of Listen Before Talk (LBT). The limitation of the duty cycle forces each user to make the spectrum available to other users for a certain percentage of time; the limitation of the maximum transmission time prevents blocking of a spectrum for such a long time that latency of the packets of other users becomes excessive. Furthermore, in LBT, a device has to listen to the possible transmissions of other users and stay silent if someone else is using the spectrum. This prevents collisions between the packets of users. Critical here is the detection threshold above which silence is imposed: it is a compromise between too many collisions, and overcautious, and thus inefficient, usage of the spectrum. If all users follow a particular standard, then LBT can be augmented with such techniques as Carrier Sense Multiple Access (CSMA)/Collision Avoidance (CA) (compare Section 18.4). If different systems (e.g., LTE and Wi-Fi) operate in the same band, then pure energy detection at the device is used to assess whether the channel is free. It must be kept in mind, however, that in an open sharing model, no guarantees for the quality of service can be provided. Further discussions of this issue, as well as the interaction between Wi-Fi and LTE and New Radio (NR), are found in Sections 31.6.2 and 32.6.2.
- *Hierarchical access model*: this model assigns different priorities to different users. *Primary users* should be served in such a way that they experience the same service quality as if the spectrum were reserved exclusively for their usage. *Secondary users* are allowed to transmit, but only in such a way that they do not (or only "insignificantly") affect the performance or service quality of the primary users. The secondary user *adaptively* decides whether they might use parts of a spectrum that is assigned by default to primary users. In other words, a cognitive radio exploits spectrum that is assigned to a primary user but is temporarily available. Therefore it has to sense the current channel usage first and then determine a transmission strategy that does not disturb the current primary users (spectrum management). This approach has drawn much of the research in cognitive radio.

One might now ask why primary users would agree to allow secondary users to employ "their" spectrum. Possible reasons are as follows:

- *Profit*: spectrum owners might be able to charge secondary users. Auction systems have been proposed in the literature where secondary users could buy – in real time – the right to specific parts of spectrum for a short time. This is promising for some applications, but might not always be practical as the costs for monitoring and billing could become higher than the revenue from the auctioning of the spectrum.
- *Regulatory requirements*: the frequency regulator can mandate that a certain spectrum range can be used by cognitive devices as long as they do not interfere with primary users. Such an approach is likely for parts of the spectrum that primary users never paid for – especially TV. In many countries, TV stations did not buy the spectrum they use but rather got it for free because they are deemed to perform a public service. This makes it easy for the frequency regulator to demand that TV stations coexist with other services "in public interest."
- *Emergency services*: another form of cognitive radio occurs in times of emergency, when services that normally count as "primary users" have to give up spectrum for emergency services.

26.1.2 Overlay and Underlay

The key principle of hierarchical cognitive radio is that the secondary users do not disturb the primary users. Such nondisturbance can be achieved by three fundamental approaches: *interweaving*, *overlay*, and *underlay*. In the interweaving approach, the radio first identifies those parts of the spectrum that are not being used at a certain time, and transmits in those; i.e., the dynamic spectrum access we discussed above. In an overlay approach, the cognitive radio detects the actually transmitted signal of the primary user and adjusts its own signal in such a way that it does not disturb the primary Receiver (RX), even though it transmits in the same band.[2] In the underlay approach, the secondary radio actually does not adapt to the current environment, but always keeps its transmit Power Spectral Density (PSD) so low that its interference to primary users is insignificant. Those concepts are discussed in more detail in the remainder of this chapter.

26.1.3 System Co-design

A more efficient use of spectrum can be achieved when different systems are designed to work together. This co-design can take on many different forms:

---

[2] A number of papers actually use the expression "overlay" for what is "interweaving" in the notation of this book.

- Design of waveforms that minimize the interference with each other.
- Co-design of systems with different functionalities. An area of particular interest is the co-design of systems for communications and localization/radar; e.g., pilot tones and synchronization signals used in communications signals can be used for purpose of localization and/or radar. Conversely, radar signals can be designed such that they can carry additional data.
- Coordination of systems with the same functionalities. In many cases, different systems exist to perform the same functionality, namely data communication. Most User Equipments (UEs) might have the capability to communicate with 5G, Wi-Fi, and Bluetooth. If there are infrastructure nodes [Base Stations (BSs)] for these systems available, a backbone network connected to all those infrastructure nodes can selectively use BSs from those different modalities to communicate with the different nodes.
- Use of opportunistic signals. In some cases, signals that are transmitted by other TXs, for other purposes, can be used as "signals of opportunity" for sensing purposes. For example, signals sent from FM radio broadcast towers can be used for radar or localization purposes.

26.2 Cognitive Transceiver Architecture

We next consider the basic structure of a cognitive transceiver. Figure 26.2 divides the usual transceiver structure (compare Chapter 17) into three parts: Radio Frequency (RF), Analog to Digital Converter (ADC), and baseband processing. Depending on the particular type of cognitive radio, one or more of those components is made adaptive. In a spectrum-sensing cognitive radio, only the RF front end is different from a conventional RX. Figure 26.3 shows a typical RF RX front end – a standard heterodyne RX, as discussed in Chapter 17. Its components, including antennas, Low Noise Amplifier (LNA), local oscillator, and automatic gain control, all have to be wideband enough so that they can operate at all possible frequencies at which the cognitive radio might want to operate. The front end also must be able to select the channel which the cognitive radio is to use, by having an adjustable local oscillator (e.g., a Voltage Controlled Oscillator, VCO).

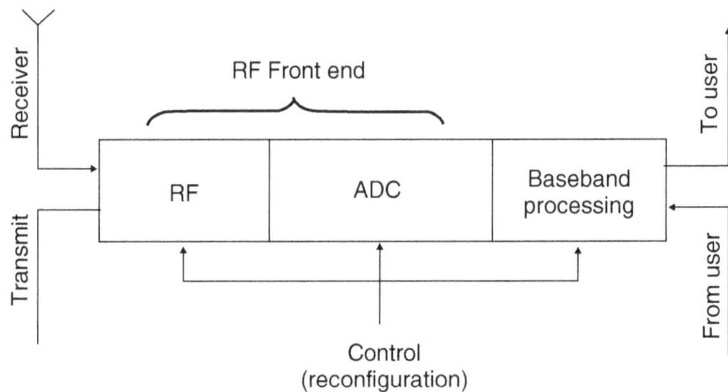

Figure 26.2 Basic structure of a cognitive transceiver. *In this figure:* ADC, Analog to Digital Converter; RF, Radio Frequency. Reproduced with permission from [Akyildiz et al. 2006] © Elsevier.

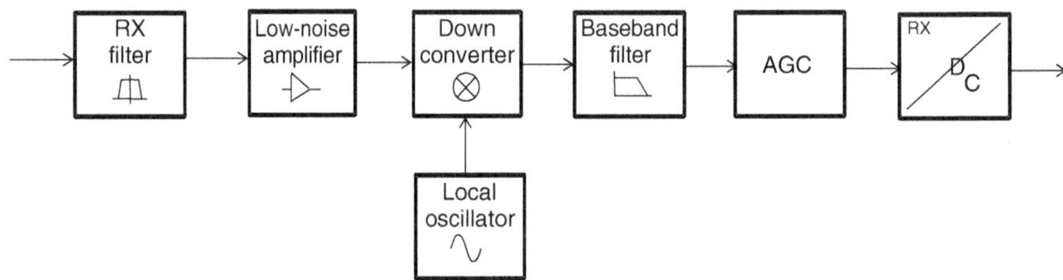

Figure 26.3 RF receiver front end. *In this figure:* AGC, Automatic Gain Control.

Since the channel selection occurs only after the downconverter (mixer), a major challenge in cognitive radio lies in the possibility of RX saturation through strong out-of-band signals. RF components like the LNA can be saturated by a signal that is not in the band that the RX wants to demodulate at a specific time, but still within the overall receive band of the cognitive radio. One possibility of reducing such strong interference is tunable RF notch filters that can be placed before the LNA. However, the hardware cost of such filters is high, and the tunability is rather limited, though there is research to improve the situation.

For a fully cognitive radio, the baseband processing also has to be adaptive. This can be most easily achieved by implementing the baseband processing as software on a Digital Signal Processor (DSP). Consequently, fully cognitive radio has also been called *software radio* by many of its proponents. While general-purpose processing units are indeed capable of providing all the functionalities for radio processing in some cases, 5G cellular systems with sampling rates on the order of hundreds of Msamples/s, and complicated processing, cannot be handled in this way. Furthermore, even for simpler systems, the energy efficiency and cost of such general-purpose architectures cannot compete with chips specifically designed for a system.

26.3   Principles of Interweaving

In an interweaving system, a secondary user tries to identify, and transmit in, the times/locations/frequency resources where primary users are not active; see Figure 26.4. This approach can be seen as "filling the holes" in the time–frequency plane. Clearly, the strategy involves three steps (which, with small modifications, also apply for open-sharing DSA):

1. *Spectrum sensing*: the cognitive radio has to identify which parts of the time–frequency plane are not used by primary users. The sensing has to be done in the presence of noise, so that the sensing cannot be completely reliable. Furthermore, it is important to determine the signal level of the detected radiation. Spectrum sensing is discussed in more detail in Section 26.4.

2. *Spectrum management*: here, the system decides when, and in which part of the spectrum, to transmit. This decision is difficult for several reasons: (i) the secondary system has only causal knowledge of the spectrum occupation, i.e., it knows only which parts of the spectrum were free in the past and the present, but it has to make *assumptions* (that are not necessarily correct) about how the primary system will work in the future, i.e., at the time when the secondary system will actually transmit; (ii) the sensing information on which the decisions of the secondary system is based are not perfect. Spectrum management is discussed in Section 26.5.

3. *Spectrum sharing*: a decision on how to divide up the free spectrum that the secondary system uses. It must be noted that spectrum sharing and spectrum management are strongly related – spectrum might be usable for a particular secondary Transmitter (TX) but not for another.

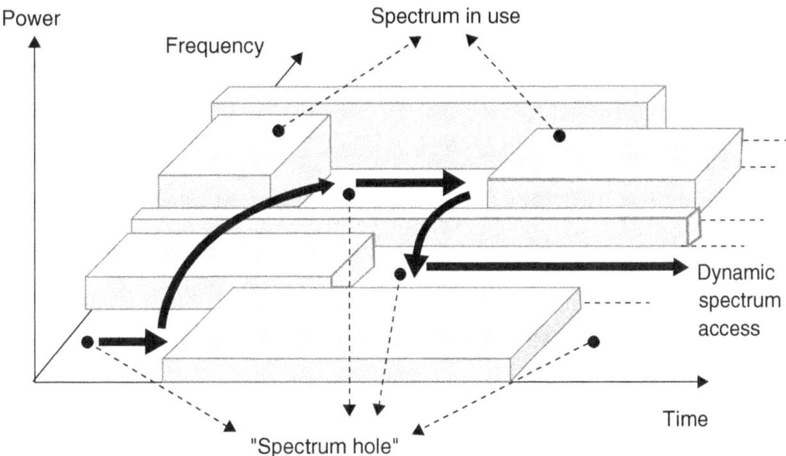

Figure 26.4   Concept of spectrum holes.
Reproduced with permission from [Akyildiz et al. 2006] © Elsevier.

In open sharing systems, there is no concept of primary/secondary users. In these systems, users of the spectrum may agree (or the frequency regulator may impose) on certain behavioral rules, e.g., LBT, and maximum transmission times; the latter allows some prediction of when currently-occupied spectrum becomes available. Note that such rules could also be prescribed for spectrum management in hierarchical systems, but demanding LBT for primary users would be incompatible with a basic tenet of such systems, namely that incumbent primary systems must not suffer a performance degradation, and must not require modification of their operation.

26.4   Spectrum Sensing

A key requirement for cognitive radio is spectrum sensing, i.e., detecting the presence of other users. Interference at a specific RX is often characterized by the interference temperature, which describes the interfering PSD divided by Boltzmann's constant k_B.

26.4.1   Spectrum Sensing in a Hierarchical System

One of the key problems of spectrum sensing in a hierarchical cognitive radio system is that any device can only sense *transmitted* radiation – a completely passive RX is invisible to any sensor. This important dilemma has not been solved completely, though a number of possible solutions have been proposed:

1. Assume that any primary device acts as both TX and RX. Then the presence (and possibly the location) of the primary device can be deduced from the radiation transmitted by this device. Note that a secondary system needs knowledge about the duplexing policies of the primary device. For example, if a frequency domain duplexing primary transmits in a particular band, the secondary system would need to know in which band such a primary system device would receive.

2. Observe spurious emissions of the primary RX. For example, the local oscillators of TV RXs are not perfectly isolated from the antenna, and thus their emissions can be observed by the secondary system. This approach has the drawback that there is no straightforward relationship between the level of the observed signal and the location of the RX and that it actually punishes good RX design (i.e., designs that do not have significant spurious emissions).

3. Assume that primary RXs are in all possible locations where primary transmissions can be heard at a level sufficient for demodulation. Such a policy is extremely conservative: RXs that need to be protected are assumed to be everywhere in the coverage region of the primary system.

4. Let the secondary device start to transmit in any band, and monitor whether the radiation of the primary system changes. For example, an increase in the transmit power of a primary Code Division Multiple Access (CDMA) TX would indicate that a secondary device is interfering with the primary RX (the primary system increases transmit power in order to keep the Signal-to-Interference-and-Noise Ratio (SINR) constant). This approach has, however, three major drawbacks: (i) it interferes, at least initially, with a primary user, (ii) it is difficult to separate changes in the primary TX characteristics due to secondary interference from changes that are caused by other environmental effects, (iii) it does not work in systems that do not change TX characteristics according to interference (and just provide bad transmission quality).

5. A common control channel that allows dissemination of information about spectrum usage. This scheme allows highly efficient utilization of the spectrum but has the drawback that existing devices need to be modified to allow signaling on this channel.

6. A database that makes geographic locations of primary RXs available to secondary systems, compare also Section 26.1.

Like most of the literature, the remainder of this section will ignore the problem of sensing RXs, and just assume that detecting the primary TXs is sufficient.

26.4.2 Types of Detectors

Three types of detectors are commonly used for sensing, depending on the amount of *a priori* knowledge that a sensor has about the transmitted waveform. Energy detection is used when the sensor has no information about the signal structure. Matched filters provide better performance but can only be used when the transmit signal waveform is known. Detection of cyclostationary properties is a compromise solution that does not require knowledge of the waveform and gives a performance that is somewhat better than that of energy detection. It must be noted that a fully cognitive radio requires a matched-filter detection.

Energy Detection

If the sensor knows nothing about the waveform transmitted by other users in a frequency band of interest, it can only measure the energy present in that band. If no other user is present, the sensor measures only thermal noise energy; otherwise, the sensor measures signal-plus-noise energy. The decision of whether a signal is present or not is thus a classical detection problem that can be formulated as hypothesis testing. For simplicity, we only consider a single narrowband channel, where the received signal is

$$r_n = n_n \quad \mathcal{H}_0 : \text{noise-only received} \tag{26.1}$$

$$r_n = hs_n + n_n : \quad \mathcal{H}_1 : \text{noise-plus-signal received} \tag{26.2}$$

where r_n and s_n are the (normalized) received and transmitted signal at time n. Furthermore, h is the channel amplitude gain (assumed time independent), and n_n is the observed noise. $\mathcal{H}_0$ and $\mathcal{H}_1$ are the two hypotheses between which the sensor needs to decide. The received signal samples can be averaged over a number of time instances N, so that the decision variables

$$y = \sum_{n=1}^{N} |r_n|^2 \tag{26.3}$$

are used. If the number of samples is large, y are Gaussian random variables, with expected value and variance[3]

$$E\{y\} = \begin{cases} N\sigma_n^2 & \mathcal{H}_0 \\ N\left[|h|^2 + \sigma_n^2\right] & \mathcal{H}_1 \end{cases} \tag{26.4}$$

---

[3] Keep in mind, though, that the y can never take on negative values, as they are sums of energy.

$$\sigma_y^2 = \begin{cases} 2N\sigma_n^4 & \mathcal{H}_0 \\ 2N\sigma_n^2 \left[2|h|^2 + \sigma_n^2\right] & \mathcal{H}_1 \end{cases} . \tag{26.5}$$

The decision rule is then

$$y \underset{\mathcal{H}_1}{\overset{\mathcal{H}_0}{\lessgtr}} \theta \tag{26.6}$$

where θ is a threshold. Because the noise is random, it is not possible to set the threshold in such a way that the presence of a signal is detected perfectly. There can be *false alarms*, i.e., even though there is only noise, the decision variable exceeds the threshold, and consequently, the sensor thinks that a signal is present. The probability for such an event is

$$P_{\mathrm{f}}(\theta) = \Pr(Y > \theta \mid \mathcal{H}_0) = Q\left(\frac{\theta - N\sigma_n^2}{\sigma_n^2 \sqrt{2N}}\right) \tag{26.7}$$

where $Q(x)$ is the Q-function defined in Eq. (11.29). Similarly, there can be missed detections, which happen when the decision variable remains below the threshold even if a signal is present. The probability for this is

$$P_{\mathrm{md}}(\theta) = \Pr(Y < \theta \mid \mathcal{H}_1) = 1 - Q\left(\frac{\theta - N\left[|h|^2 + \sigma_n^2\right]}{\sigma_n \sqrt{2N\left[2|h|^2 + \sigma_n^2\right]}}\right) . \tag{26.8}$$

By adjusting the threshold, we can then trade off false-alarm probability and missed-detection probability. In a hierarchical cognitive system, the missed-detection probability will usually be prescribed by the frequency regulator, because a missed detection of the spectrum sensor implies that the secondary system will transmit even though a primary user is active in the considered band.

Another interesting question is how many sensing values N should be obtained before a decision is made. If the number is too small, then the decision variable has a large variance, and a high P_{f} (or P_{md}) must be accepted. However, if the number is too large, the decision process takes too much time, reducing the spectrum usage and increasing the chance that the primary system starts to transmit even though the channel was free in the beginning. This problem can be treated by means of "optimum stopping theory."

Matched Filter

The performance of the spectrum sensing can be improved considerably if the sensor knows the transmitted waveform. In that case, matched filtering is the best way of improving the Signal-to-Noise Ratio (SNR) of the detection process, similar to the arguments made in Chapter 11. For a spectrum-sensing cognitive radio, the output of the matched filter is used as a test statistics to detect whether signal energy is present in the considered band or not. A fully cognitive radio might have to demodulate/detect the symbols transmitted by the primary user.

In some circumstances, a sensor might want to detect/decode only pilots or beacons, since those have usually better SNR than the actual payload data, and might also contain information about spectrum usage (e.g., the duration of a packet) that can be exploited by a secondary system.

Cyclostationarities

Modulated signals usually have cyclostationary statistics, i.e., certain statistical properties are periodic; see also Section 14.7.4. This fact can be employed to enhance the performance of an energy detector. In other words, the incoming signal is correlated with itself before being sent to an energy detector.

Wavelet Detection

If the PSD of the signal that is to be detected is smooth within a frequency band but shows steep decay at an edge, a wavelet transform can expose the edges of the PSD. The drawbacks of this technique are the high computational costs and the fact that the method does not work well for spread-spectrum signals [Letaief and Zhang 2007].

26.4.3 Multi-Node Detection

In many cases, the secondary system consists of multiple nodes, which then have the possibility of helping each other reach a better sensing decision. In the simplest case, each of the secondary nodes listens and makes an independent decision concerning whether it can sense a certain channel to be occupied. The nodes then exchange this binary information and come to a joint ("fused") decision as to whether the spectrum is available or not. In order to best protect the primary system, the secondary system should decide that a frequency band is occupied if at least one node senses occupation. Clearly, the overall false-alarm probability increases, becoming

$$P_{\text{f, network}} = 1 - \prod_{k=1}^{K} [1 - P_{\text{f},k}] \tag{26.9}$$

while the probability of missed detection will decrease

$$P_{\text{md, network}} = \prod_{k=1}^{K} P_{\text{md},k}. \tag{26.10}$$

A more sophisticated joint decision can be made if the nodes communicate not just a binary decision about spectrum occupancy (yes/no) but perform a linear combination of the observed averaged samples y (Eq. 26.3). Introducing linear weights w_k for the signal from the kth node, we form a total decision variable

$$z = \sum w_k y_k = \mathbf{w}^T \mathbf{y}. \tag{26.11}$$

Because the decision variable is a weighted sum of Gaussians (see above for a discussion why y is Gaussian), it is itself Gaussian, and its mean and variance is

$$E\{z\} = \begin{cases} N\sigma_{\text{n}}^2 \mathbf{w}^T \mathbf{1} & \mathcal{H}_0 \\ N\mathbf{w}^T \left[\mathbf{g} + \sigma_{\text{n}}^2 \mathbf{1}\right] & \mathcal{H}_1 \end{cases} \tag{26.12}$$

$$\sigma_z^2 = \begin{cases} 2N\sigma_{\text{n}}^4 \mathbf{w}^T \mathbf{w} & \mathcal{H}_0 \\ 2N\sigma_{\text{n}}^2 \mathbf{w}^T \left[2\text{Diag}(\mathbf{g}) + \sigma_{\text{n}}^2 \mathbf{I}\right] \mathbf{w} & \mathcal{H}_1 \end{cases} \tag{26.13}$$

where $\mathbf{g} = [|h_1|^2, |h_2|^2, ..., |h_K|^2]^T$, and $\text{Diag}(\mathbf{g})$ is a diagonal matrix whose entries are the $|h_k|^2$. The probabilities for false alarm and missed detection are then

$$P_{\text{f}}(\theta, \mathbf{w}) = Q\left(\frac{\theta - N\sigma_{\text{n}}^2 \mathbf{w}^T \mathbf{1}}{\sigma_{\text{n}}^2 \sqrt{2N\mathbf{w}^T \mathbf{w}}}\right) \tag{26.14}$$

$$P_{\text{md}}(\theta, \mathbf{w}) = 1 - Q\left(\frac{\theta - N\mathbf{w}^T \left[\mathbf{g} + \sigma_{\text{n}}^2 \mathbf{1}\right]}{\sigma_{\text{n}} \sqrt{2N\mathbf{w}^T \left[2\text{Diag}(\mathbf{g}) + \sigma_{\text{n}}^2 \mathbf{I}\right] \mathbf{w}}}\right). \tag{26.15}$$

Now we have to specify the optimum weights and thresholds for the detection. What is "optimum" under the circumstances depends on the system goals. We might wish, e.g., to maximize the throughput of the secondary system $(1 - P_{\text{f}}(\theta, \mathbf{w}))$, under the constraint that the probability of interference to the primary system $P_{\text{md}}(\theta, \mathbf{w})$ stays below a certain threshold. Standard optimization techniques can then be used to find out the optimum parameters $\theta, \mathbf{w}$.

Clearly, communicating the full decision variables from each sensing node requires more overhead than just communicating a binary decision. On the other hand, it must be noted that it is rarely possible to really just send a single "bit" over a wireless channel. Every transmitted packet needs headers, synchronization sequences, etc. Compared to this necessary overhead, it does not make much difference whether we want to transmit, say, 1 bit or 8 bits of information. Communicating the full decision variable from each node thus might not be a significantly higher effort than a binary decision.

26.4.4 Cognitive Pilots

The sensing of primary (and other secondary) users would be greatly facilitated by the introduction of Cognitive Pilot Channels (CPCs). The procedure for using the CPC can include the following three phases:

- The User Equipment (UE) first listens to the CPC (usually transmitted from a BS) at the initialization.
- The UE gets the information about available channels, selects the most suitable one to setup its communications, and conveys the choice to the network.
- The updated CPC is broadcasted to a wide area.

However, a CPC has to be agreed upon by a wide range of standardized devices and thus constitutes a formidable logistics/standardization problem.

26.5 Spectrum Management

26.5.1 Spectrum Opportunity Tracking

A cognitive system has to make decisions about which frequency band and bandwidth to use, and how long to transmit in this band. However, spectrum sensing only gives information about the spectrum occupancy at a given point in time. It is thus important for the cognitive system to keep a record of the history of spectrum usage, and to have detailed models for the traffic statistics. Ideally, a secondary system observes the environment at all possible frequencies and at all times in order to get the best possible statistics; however, energy and hardware constraints might prevent such an approach.

A simple, analytically tractable model is a Markov model, which just assumes that each subchannel is in one of two states (occupied or unoccupied), and has a certain transition probability (which is obtained from observations) from one state to another. A more detailed model might include variations of the spectrum occupancy with time-of-day, or take typical durations of packets or voice calls into account.

If the cognitive radio recognizes that the current band in which it operates has to be released (either because of worsening propagation conditions or because the band is needed by other users), it either moves to a different band or (if no other band is available) ceases transmission. In the former case, the *spectrum handover* should be done in such a way that the performance of the link is affected as little as possible.

In a hierarchical system, the secondary system has to cease transmission when a primary user reappears and wants to use the spectrum. The duration for which a secondary system transmits in a certain band before doing another round of sensing is thus an important parameter of the system, which depends on the previously measured statistics of the primary channel.

26.5.2 Machine Learning for Spectrum Sensing and Tracking

Machine Learning (ML) is an approach that is well suited to cognitive radio. Essentially, the observables in the environment are the sensing observations in the different bands, and the output of the algorithms are the decisions on when, in which band, and with what power to transmit. As always with ML, a key advantage is that no complicated analytical models for the spectral occupancy need to be established, but rather the raw observations can be used as input.

In order to train the network, two main approaches are available: supervised, and unsupervised. In the former case, labeled training data (i.e., data where the best decision is known) are used to train, e.g., a neural network or a support vector machine. Obviously, those training data should show similar characteristics to the spectral occupancy data that will be experienced by the node (and the ML network) later on. Obtaining a large set of labeled training data can be difficult, as the objective "ground truth" (i.e., which decisions would be the best) has to be known for each data point.

An alternative is thus unsupervised algorithms such as reinforcement learning, so that the network learns from the reaction of the overall system what effect a particular decision had, without requiring to know what the optimum decision in such a situation would have been.

At the time of this writing, ML for spectrum management and spectrum sharing is very much in flux; early work is surveyed in [Bkassiny et al. 2012] for a recent (at the time of this writing) survey see [Kaur and Kumar 2020].

26.5.3 Privacy and Security Considerations

Spectrum tracking for cognitive radio, and in particular in the context of ML, requires the acquisition of a large amount of data about UEs and their use of the spectrum. While no user data are demodulated and decoded, the metadata contained in the spectrum usage are themselves sensitive and may lead to unwitting compromising of privacy. Thus, care needs to be taken in determining which data should be aggregated, with what degree of centralization. Federated learning is generally seen as a way of reducing privacy implications.

26.6 Spectrum Sharing

26.6.1 Introduction

A key question is now how to distribute the detected spectral resources to the different users of a secondary system. A large variety of methods exist, depending on the degree of coordination between the users. At the one extreme, a centrally controlled scheme, where a leader node assigns the spectral resources to the different users, provides the best performance of the system at the price of significant overhead. On the other hand, uncoordinated competition between the different users does not require any message exchange but has lower efficiency.

A key mathematical tool in the analysis of spectrum sharing is *game theory*. A game consists of a number of players, a strategy for each player, and a payoff (utility function); each user adjusts its strategy in such a way that the payoff is maximized. It is noteworthy that the optimum strategies depend on the amount of coordination, i.e., how much each user knows about the other users, and whether there is a central authority that can enforce certain rules of the game.

In many cases, the ideal outcome of a game is to achieve a *Pareto optimum* (also known as *social optimum*); in that case, there is by definition no other outcome that has a better payoff for at least one player and at the same time does not reduce the payoff for any other player.

26.6.2 Noncooperative Games

A noncooperative game is defined as "one in which any player is unable to make enforceable contracts outside of those specifically modeled in the game" [Han et al. 2007]. Consequently, any cooperation must be self-enforcing; there is no outside party that can enforce a particular policy. An important quantity in the context of such games is the *Nash equilibrium*, which is defined as an operating point where no user has a motivation to unilaterally change its current strategy.

Noncooperative games can lead to Nash equilibria that are very inefficient. A typical example is a heavily loaded ALOHA system (compare Section 18.4.1). If every user sends out packets at a high rate, there is a high number of collisions, and the system will become congested. However, a user that tries to behave in a "socially responsible" manner and sends out fewer packets will achieve an even lower throughput as long as the other users keep up their high channel access rate. Since each of the users faces the same dilemma (similar to the famous "prisoners' dilemma"), none will reduce their channel access rate, and the congestion will persist. The users are then in a Nash equilibrium but have low efficiency.

26.6.3 Games with Partial Coordination

In cooperative games, good behavior of the users can be enforced through various means. This enforcement can be done, e.g., through a central authority, or through "punishment" by other players. In the following, we will outline some of the most widespread methods [Han et al. 2007].

Referee-Based Solution

The inefficiency of noncooperative games can be mitigated if a *referee* enforces certain types of behavior and alleviates the collisions and high-interference situations that plague standard competitive behavior. The amount of intervention depends on the amount of information that the referee has, and the amount of control information that is to be sent to the users. In the extreme case that the referee "micromanages" the resource allocation to all users, the referee-based solution becomes equivalent to the centralized solution described below. An example for a low-overhead referee algorithm is given in Figure 26.5.

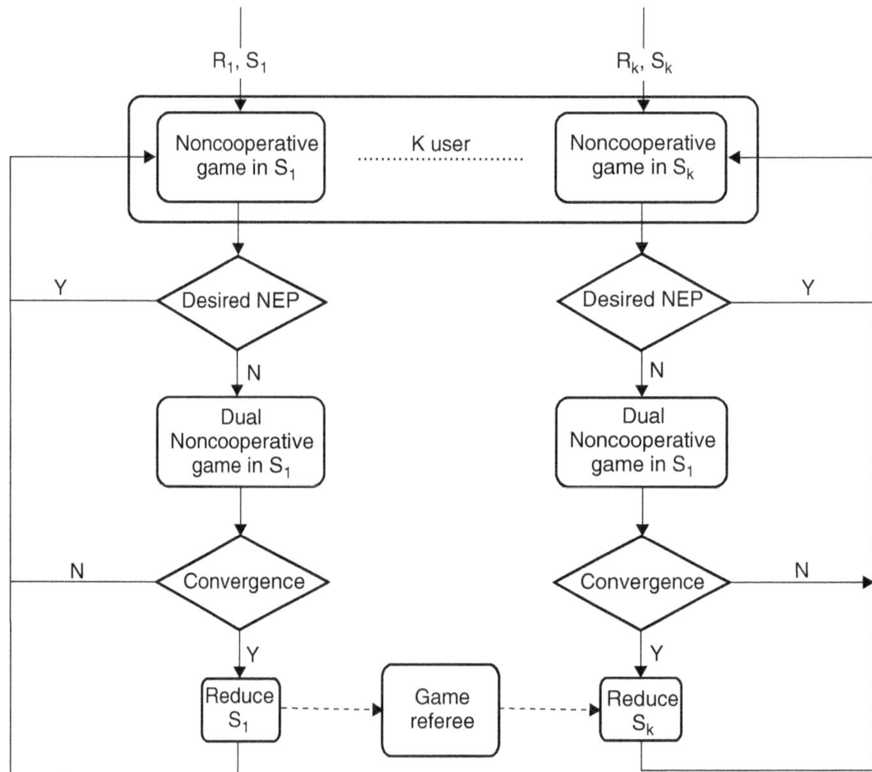

Figure 26.5 Flow diagram of a noncooperative game with referee. *In this figure:* NEP, Nash Equilibrium Point. Reproduced with permission from [Han et al. 2007] © IEEE.

Threat and Punishment in Repeated Games

The Nash equilibrium normally assumes a "static game," i.e., players play only once. Better results can be achieved in repeated games, where the players have the possibility to learn from the past, i.e., make their decisions conditioned on the results of the outcomes of previous games. Consequently, a user will not sacrifice, for its short-term advantage, its long-term benefits if it is going to be severely punished by the other users. The total payoff is the weighted average over time

$$V = \sum_{t=1}^{T} \beta^{t-1} u_t \tag{26.16}$$

where β is the "discount factor" and u_t is the payoff at time t; T is the total time that the game is played. It can be shown that as the discount factor approaches 1, any individually rational (and achievable) payoff can be enforced by an equilibrium.

An implementation of a punishment scheme is the "trigger punishment strategy" that proceeds in the following way:

- All players start in a cooperative mode.
- At the end of the first phase $t = 1$ of the game, the result of the game (e.g., the total payoff for all users) is made public. More cooperation will result in a higher total payoff; on the other hand, if one user behaves selfishly, the overall system behavior will become bad.
- Thus, if the total payoff is below a threshold, users switch to a noncooperative mode and use as their strategy the Nash equilibrium strategy for a time T. If the total payoff is above the threshold, users stay with the cooperative strategy.

Problems can arise if the users are mobile. In that case, a nonsocial user can get high payoff in a certain area by behaving badly, and then escape the punishment by simply moving to a different area.

Spectrum Auction

A spectrum auction in a cognitive radio system works, in principle, like any auction in real life. The players (users) are bidding a certain price for the available spectrum; clearly, a user that pays more will get a larger assignment of resources. It must be noted that the "price" need not be actual money and that it can depend on the quality of the user channel. For example, it has been proposed that each user is charged according to SINR or received power.

Bargaining Solutions

Another assignment algorithm that draws its inspiration from economic theory is bargaining. In this approach, different users can negotiate with each other (not with a central authority) about who gets which spectral resources assigned. In one-on-one bargaining, two neighboring users can exchange time/frequency resources. Since the transmission qualities on those resources are not necessarily the same for different users, an exchange of channels might be beneficial for both users. When more users are involved, a bargaining with one buyer and multiple sellers can be applied.

26.6.4   Centralized Solutions

The conceptually simplest case is one where a central authority has all the information about the bandwidth requirements of the different users as well as the available bandwidth and assigns the resources in such a way as to optimize the spectral efficiency of the (secondary) system; it is assumed that the secondary users obey commands from the central authority. Such a solution involves a high overhead for sending all the required information to the central authority and for the channel control information.

Mathematically, the spectrum allocation can be written as a (constrained) optimization problem similar to Sec. 20.3

$$\min_{\mathbf{x} \in \Omega} f(\mathbf{x}) \tag{26.17}$$

$$g_i(\mathbf{x}) \leq 0 \text{ for } i = 1, \dots, I$$
$$h_j(\mathbf{x}) = 0 \text{ for } j = 1, \dots, J$$

where $\mathbf{x}$ is the parameter vector for the optimization of $f()$ and has to lie in the space Ω. The $g_i(\mathbf{x})$ and $h_j(\mathbf{x})$ are inequality constraints and equality constraints for the parameter vector. Depending on the application and the optimization goal, these quantities can mean different things. To give one example, $\mathbf{x}$ can be a vector containing the power and bandwidth assignment for the different users. The inequality constraints limit the power of the different users, and possibly the bandwidth assigned to the different users. Different mathematical techniques exist for solving constrained optimization problems. If, e.g., the functions f, g, and h are linear functions of $\mathbf{x}$, then the optimization problem is a *linear program* that can be readily solved by standard mathematical packages. Similarly, if the function $f(\mathbf{x})$ is convex, the problem can be solved by numerical optimizations that rapidly converge to global optima.

The situation is somewhat more difficult when the optimization function is nonconvex – in this case, numerical techniques usually can converge to local optima. If the admissible parameter space Ω is discrete, the optimization problem usually cannot be solved in polynomial time. More details are discussed in Sec. 20.3.

We now turn to one concrete example [Acharya and Yates 2007]. We have a situation with multiple secondary systems (BSs) i, where the ith system provides bandwidth b_{ij} to the jth user. A user then transmits its data to the ith BS using that bandwidth. The channel between BS and user has a channel gain $|h_{ij}|^2$. Communication is assumed to happen at Shannon capacity, so that the data rate of a link is

$$R_{ij} = b_{ij} \log_2 \left[1 + \frac{|h_{ij}|^2 P_{ij}}{b_{ij} \, \sigma_{\mathrm{n}}^2} \right] \tag{26.18}$$

where P_{ij} is the power on each link. The goal is the maximization of the overall throughput. The optimization problem can then be written as

$$\max_{b_{ij}, P_{ij}, X_i} \sum_j \sum_i R_{ij} \tag{26.19}$$

$$\sum_j b_{ij} \leq B_i \tag{26.20}$$

$$\sum_i P_{ij} \leq P_j \tag{26.21}$$

$$\sum_i B_i \leq B_{\mathrm{tot}} \tag{26.22}$$

$$b_{ij} \geq 0, P_{ij} \geq 0, B_i \geq 0 \tag{26.23}$$

where Eq. (26.19) says that we want to optimize the total throughput in the system; Eq. (26.20) states that the total bandwidth allocated to the different users within a secondary system is limited to the total bandwidth B_i assigned to that secondary system by a central authority, and Eq. (26.22) limits the sum of the bandwidths assigned to the secondary systems to the total available bandwidth. Eq. (26.21) limits the sum power that each user j can employ to a certain value P_j. Finally, all bandwidth assignments and powers must be positive. The solution of the above-described problem can be obtained via standard Lagrangian optimization techniques.

26.6.5 Spectrum Sharing in Open Access Systems

The most important open-access bands are the ISM bands, e.g., at 2.45 GHz. For these bands, the frequency regulators have not provided any rules beyond (i) maximum total transmit power, (ii) maximum power spectral density of the emissions, and (iii) requirement for spectrum spreading (though that latter rule does not apply, e.g., for Wi-Fi in the 2.45 GHz band). In these bands, spectrum sharing between different systems is simple but inefficient: every user transmits at the maximum power that the regulator allows (it can be seen as a form of underlay, as discussed in Section 26.8). Spectrum sharing *within* one system can, of course, be done in any way the system wants (in the case of Wi-Fi, using CSMA). For some more recently available bands, such as around 3.5 GHz, frequency regulators are imposing Listen Before Talk (LBT) rules, maximum duty cycles, and maximum transmission times, as discussed in Sections 26.1.2, 31.6.2 and 32.6.2.

26.7 Overlay

A very different approach to cognitive radio is the *overlay* principle. Here, we do not try to avoid transmission at the same time and frequency by primary and secondary user, but rather we exploit it. By using appropriate types of cooperation, the secondary user can help the primary user, and still at the same time transmit information of its own.

Consider a very simple cognitive system: a primary TX/RX pair, as well as a secondary TX/RX pair, tries to communicate over an Additive White Gaussian Noise (AWGN) channel. We furthermore assume that the secondary TX has full knowledge of the message of the primary user. This latter assumption is clearly unrealistic in almost any wireless setting, but it can be seen as a reasonable approximation in the following two cases:

1. The primary user uses a special type of code ("rateless code"), which allows an RX to decode the message as soon as it has gathered enough mutual information. In other words, the RX does not have to wait with the decoding until the TX has finished sending out a packet of data. Rather, the better the channel between primary TX and secondary TX, the sooner the secondary TX can decode. Thus, if the secondary TX is very close to the primary TX, it can have almost instantaneous knowledge of the primary message (this is similar to coded cooperation, compare also Section 27.3.5).
2. The primary user employs Automatic Repeat reQuest (ARQ), i.e., it retransmits the same message multiple times, until the primary RX finally gets it (it takes repeated attempts because primary RX has poor SINR). The secondary TX can get the primary message at

the first attempt (again, assuming that it has a very good channel to the primary TX) and then has noncausal knowledge of the primary message for the retransmission.

A secondary TX can now pursue two possible strategies:

1. *Selfish approach*: the secondary TX uses all its power to transmit the secondary message. It uses its knowledge of the primary message to make sure that there is no effective interference to the secondary RX. Such elimination of interference at the TX side can be achieved by "dirty paper coding" [Peel 2003]. It is noteworthy that in such a system there is still interference from the secondary TX to the primary RX, which violates the basic tenet of hierarchical cognitive radio.

2. *Selfless approach*: the secondary TX uses part of its power to help convey the primary message to the primary RX; the remainder of the power is used to send the secondary message to the secondary RX. Depending on the fraction of power used for sending the primary message, the secondary system can actually boost the rate of the primary message. An especially interesting case occurs when the secondary system makes sure that the primary rate remains unchanged (compared to the case when there is no secondary system) [Devroye et al. 2007]. In that case, the primary system is completely oblivious to the fact that a secondary system is present at all, and its capacity in an AWGN channel is [Devroye et al. 2007], [Jovocic and Viswanath 2009]

$$R_1 = \log_2\left[1 + \frac{P_1}{\sigma_n^2}\right] \tag{26.24}$$

and still, the secondary system can transmit some of its data, which can be done at a rate

$$R_2 = \log_2\left[1 + (1 - \alpha')\frac{P_2}{\sigma_n^2}\right] \tag{26.25}$$

where

$$\alpha' = \left[\frac{\sqrt{P_1}\left(\sqrt{1 + |h_{21}|^2\frac{P_2}{\sigma_n^2}\left(1 + \frac{P_1}{\sigma_n^2}\right)} - 1\right)}{|h_{21}|\sqrt{P_2}\left(1 + \frac{P_1}{\sigma_n^2}\right)}\right]^2 \tag{26.26}$$

where we have assumed $|h_{11}| = |h_{22}| = 1$, and $|h_{21}| < 1$. In other words, as long as the crosstalk channel from the secondary TX to the primary RX is weaker than the primary channel, it is possible for the secondary system to transmit its own information "for free," i.e., without reducing the rate of the primary system.

Overlay systems are fascinating from a theoretical point of view; however, at least at the time of this writing, they seem quite far away from practical implementation: (i) the question of how the secondary system obtains the noncausal knowledge of the primary system is at least partially unanswered (though progress has been made on cognitive systems with causal message knowledge); (ii) even more importantly, complete channel state information for *all* relevant channels (including the one from primary TX to primary RX) has to be available at the secondary TX; such knowledge seems to be difficult to obtain without explicit cooperation from the primary system.

26.8 Underlay Hierarchical Access – Ultra Wide Bandwidth System Communications

In an underlay system, the secondary users have severe restrictions on the transmit PSD, so that the effect on a primary RX is a "minor" increase of the noise floor that the RX sees. Such low PSD can be achieved by keeping the transmit power low (which is feasible only if the secondary users communicate only over short distances) and/or by spreading the signal over a very large bandwidth. Only the part of the secondary signal that falls within the RX bandwidth of a primary user acts as interference. Underlay radios are actually not "cognitive" in the sense that they adapt their transmission parameters to the environment, but because of their use as secondary radios, they are still often mentioned in the cognitive category.

26.8.1 *Frequency Regulations and Transmit Power Constraints of UWB Signals*

The underlay principle is realized in Ultra Wide Bandwidth system (UWB) communications, where signals have extremely large bandwidth. Such a large bandwidth offers the possibility of very large spreading factors: in other words, the ratio of the signal bandwidth to the symbol rate is very large. For a typical sensor network application with 5 ksymbol/s throughput, a spreading factor of 10^5 to 10^6 is achieved for transmission bandwidths of 500 MHz and 5 GHz, respectively. Spreading over such a large bandwidth means that the PSD of the radiation, i.e., the power per unit bandwidth, can be made very low while still maintaining good SNR at the secondary RX. A primary (narrowband) RX will only see the secondary-signal power within its own system bandwidth, i.e., a small part of the total secondary transmit power; see Figure 26.6. This implies that the interference to primary (narrowband) systems is small.

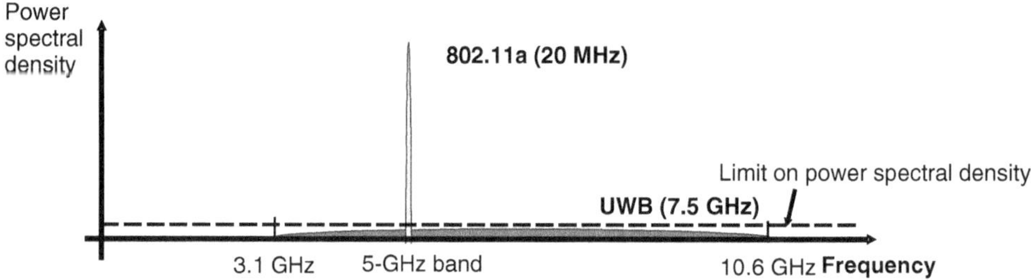

Figure 26.6   Interference between a UWB system and a narrowband (IEEE 802.11a) local area network.

Frequency regulators have defined UWB signals as signals with a minimum of 500 MHz absolute bandwidth.[4] They stipulate that UWB can be used as *unlicensed* underlay systems as long as the PSD is limited to -41.3 dBm/MHz EIRP (Equivalent Isotropically Radiated Power; see Chapter 3). This PSD is so low that it does not significantly disturb primary RXs that are more than approximately 10 m away from a UWB TX. Take, e.g., a signal with 6 GHz carrier frequency. It suffers a free-space attenuation of 68 dB at a 10 m distance, resulting in a received PSD of approximately -109 dBm/MHz, which is comparable to the PSD of thermal noise. Consequently, even a primary RX operating at the thermal-noise sensitivity limit would hardly see an impact on its performance. RXs with nonideal noise figures, and/or impacted by co-channel interference, e.g., from neighboring cells, would not be affected even if the UWB TXs were somewhat closer to the primary RX. It is also noteworthy that transmission at -41 dBm/MHz is allowed only in certain frequency ranges, while others (e.g., the Global Positioning System, GPS band, and also most cellular bands) are more strongly protected.

In the USA, the Federal Communications Commission (FCC) allows emission between 3.1 and 10.6 GHz as shown in Fig. 26.7a. Limits for indoor and outdoor communication systems differ; for outdoor systems, UWB devices are required to operate without a fixed infrastructure. In Europe, the Radio Spectrum Committee (RSC) of the European Commission (EC) imposes a spectral mask shown in Figure 26.7b. Emission between 6 and 8.5 GHz with EIRP of -41.3 dBm/MHz is allowed for devices without some type of additional interference mitigation techniques (techniques similar to those of interweaving systems). Systems *with* interference mitigation techniques or low duty cycle operation are allowed to transmit at -41.3 dBm/MHz in the 3.1–4.8 GHz band. In Japan, operation between 3.4 and 4.8 GHz is admissible as shown in Figure 26.7c, if the UWB TX uses interference mitigation. Operation between 7.25 and 10.25 GHz is admissible also without special techniques.

The high spreading factor of UWB systems helps not only in mitigating interference to primary users but also enables a UWB RX to suppress narrowband (primary-user) interference by a factor that is approximately equal to the spreading factor. These principles are well understood from the general theory of spread-spectrum systems, see Chapter 19. The distinctive feature of UWB is that it goes to extremes in terms of the spreading factor. It must be kept in mind that the spreading factor is a function of both the transmission bandwidth and the data rate. Consequently, UWB systems with high data rates (>100 Mbit/s) exhibit a rather small spreading factor at and can thus be used to communicate over very short distances.

26.8.2   Methods of UWB Signal Generation

There are a number of different ways to spread signals to large bandwidths.

1. *Frequency Hopping (FH)*: FH uses different carrier frequencies at different times. In slow FH, one or more symbols are transmitted on a given frequency; in fast FH, the frequency changes several times per symbol; see also Section 19.1. The bandwidth of the resulting signal is determined by the range of the oscillator, not the bandwidth of the original signal that is to be transmitted. Implementation of an FH TX is fairly simple: it is just a conventional narrowband modulator followed by a mixer with the output of a frequency-agile oscillator. An FH RX can be constructed in a similar way; such a simple RX is efficient as long as the delay spread of the channel is shorter than the hopping time (otherwise, multi-path energy is still arriving on one subcarrier while the RX has already hopped to a different frequency). Consequently, a hopping rate of approximately 1 MHz or less would be desirable. However, such slow FH can lead to significant interference to primary RXs, since – at a given time – a victim RX "sees" the full power of the UWB signal. For this reason, FH for UWB has been explicitly prohibited by several frequency regulators.

2. *Orthogonal Frequency Division Multiplexing (OFDM)*: in OFDM, the information is modulated onto a number of parallel subcarriers (in contrast to FH, where the carriers are used one after the other); see Chapter 15. For this reason, OFDM has no innate spectral spreading. Rather, spreading can be achieved by low-rate coding, e.g., by a spreading code similar to CDMA, or by a low-rate convolutional code. The bandwidth of the resulting signal is determined by the employed code rate and the data rate of the original (source) signal. In modern implementations, the subcarriers are produced by a fast Fourier transformation; see

---

[4]Signals are also defined as UWB if they have more than 20% relative bandwidth. However, for the subsequent discussion, we assume large absolute bandwidth.

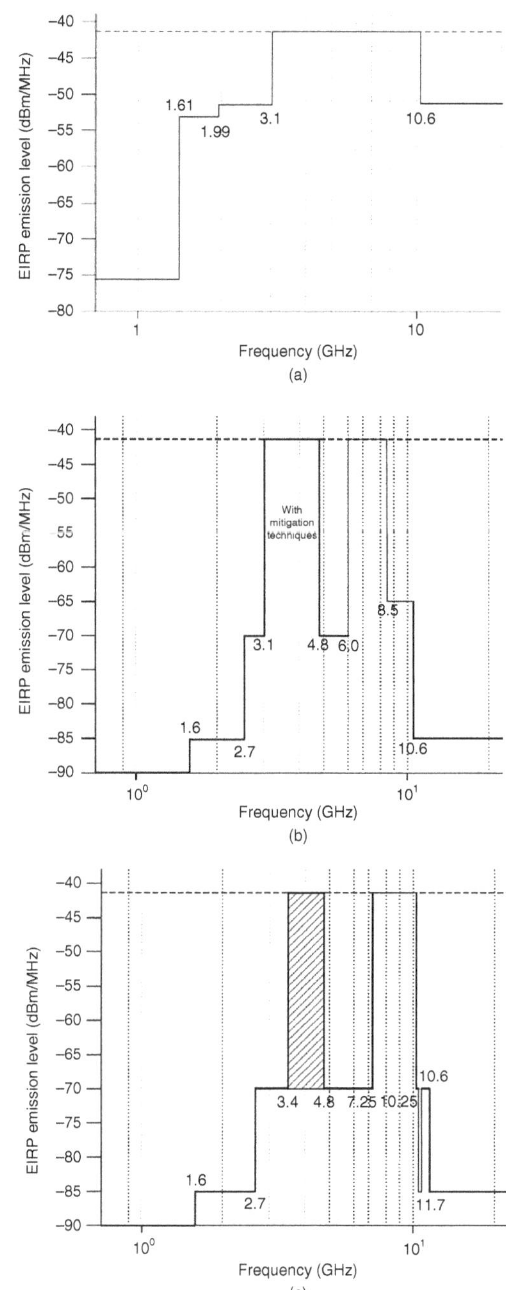

Figure 26.7 Indoor spectral masks for UWB transmission in the USA (a), Europe (b), and Japan (c).
Reproduced with permission from [Decawave 2015] © Decawave.

Section 15.3. However, this implies that signal generation at the TX, as well as sampling and signal processing at the RX, has to be done at a rate that is equal to the employed bandwidth, i.e., at least 500 MHz.

3. *Direct Sequence–Spread Spectrum (DS–SS)*: this multiplies each bit of the transmit signal with a spreading sequence. The bandwidth of the overall signal is determined by the product of the bandwidth of the original signal and the spreading factor. At the RX, despreading is achieved by correlating the received signal with the spreading sequence; see Section 19.2. The key implementation challenge lies in the speed at which the RX has to sample and process (despread) the signal.

4. *Time Hopping Impulse Radio (TH-IR)*: TH-IR represents each data symbol by a sequence of pulses with pseudorandom delays. The duration of the pulses essentially determines the width of the transmit spectrum; see also Section 19.4. The key implementation challenge lies in building coherent RXs that keep complexity low while still maintaining adequate performance. It is used, e.g., in the IEEE 802.15.4a/z standard (Section 34.2).

Summarizing, we find that there is a strong duality between FH and TH-IR. FH sequentially hops in the frequency domain, while TH-IR hops in the time domain. Similarly, OFDM and DS–SS are dual, in that they perform low-rate coding operations in the frequency and time domains, respectively.

26.8.3 Further Advantages of UWB Transmission

In addition to the low interference to primary users, UWB signals also offer a number of other advantages.

1. A UWB RX can suppress narrowband interference by a factor that is approximately equal to the spreading factor.
2. A large absolute bandwidth can result in a high resilience to fading. First of all, a large absolute bandwidth allows to resolve a large number of (independently fading) Multi-Path Components (MPCs), and thus a high degree of frequency diversity, i.e., the fading at sufficiently separated frequencies is independent. We can also give an alternative interpretation that is especially useful for TH-IR and CDMA systems. An RX with a large absolute bandwidth has a fine delay resolution, and can thus resolve many MPCs. The number of resolvable, and independently fading, MPCs can be up to τ_{max}/B, where τ_{max} is the maximum excess delay of the channel and B is the system bandwidth. By separately processing the different MPCs, the RX can make sure that all those components add up in an optimum way, giving rise to a smaller probability of deep fades. As an additional effect, we have seen in Section 6.6 that in UWB, the number of *actual* MPCs that constitute a *resolvable* MPC is rather small; for this reason, the fading statistics of each resolvable MPC does not have a complex Gaussian distribution anymore but show a lower probability of deep fades.
3. A large absolute bandwidth also leads to high accuracy of ranging and geolocation, see Chapter 29. Most ranging systems try to determine the flight time of the radiation between TX and RX. It follows from elementary Fourier considerations that the accuracy of the ranging improves the bandwidth of the ranging signal. Thus, even without sophisticated high-resolution algorithms for the determination of the time-of-arrival of the first path, a UWB system can achieve centimeter accuracy for ranging.
4. The large spreading factor and low PSD also provide increased protection against eavesdropping.

26.8.4 UWB Dynamic Spectrum Access

Despite the low interference created by a UWB underlay system, the residual interference to nearby victim RXs can still be too large. It is thus often necessary to combine UWB with an LBT scheme. Such a strategy is potentially capable of enabling coexistence, compatibility, interference avoidance, and compliance with regulation. The good ranging (and thus geolocation) capabilities of UWB nodes can help to determine whether nodes are close to potential victim nodes, in particular, if the location of such victim nodes is kept in a database that can be accessed by the UWB nodes.

Further Reading

The field of cognitive radio is still very much in flux, in particular as new bands are freed up for cellular usage and incumbents aim to find now coexistence approaches. The following papers can be especially recommended: The edited monographs [Hossain and Barghava 2008], [Xiao and Hu 2008], [Zhang et al. 2016] give a broad overview of all the aspects of cognitive radio, including spectrum sensing, spectrum allocation and management, OFDM-based implementations of cognitive systems, and UWB-based cognitive systems as well as protocol- and Medium Access Control (MAC) design. Concise overviews of these topics are also in [Akyildiz et al. 2006], [Haykin 2005], [Zhao and Sadler 2007]. Spectral sensing is discussed in [Quan et al. 2008] and [Yucek and Arslan 2009]; cooperative sensing in [Akyildiz et al. 2011], and spectrum opportunity tracking in [Zhao et al. 2007]. Application of game theory to cognitive radio system is reviewed in [Han et al. 2007] as well as in [Ji and Liu 2007], while Machine Learning techniques are reviewed in [Bkassiny et al. 2012] and [Kaur and Kumar 2020]. Overlay systems are described in [Jovocic and Viswanath 2009] as well as in [Devroye and Tarokh 2007]. UWB communications and its many applications are described, e.g., in [di Benedetto et al. 2006]; "detect-and-avoid" methods for UWB are described in [Zhang et al. 2008]. Coexistence and cognitive assignment of spectrum for specific applications are often driven by proposed rulemakings of frequency regulators and the comments on those; see the websites of the national frequency regulators (see Chapter 30) for those.

For updates and errata for this chapter, see https://wides.usc.edu/students.html#textbooks.

Exercises

See Sec. 36.26 of Exercises.pdf at wiley.com/go/molisch/wireless3e

27

Relaying, Cooperative Communications, and Network Coding

27.1 Introduction and Motivation

27.1.1 Principle of Relaying

Traditional wireless communications are based on point-to-point communication, i.e., only two transceivers (one Transmitter (TX), and one Receiver (RX)) are involved in the communication of data. These two transceivers are, e.g., the Base Station (BS) and User Equipment (UE) in a cellular setting, Access Point (AP) and laptop in wireless Local Area Networks (LANs), or two UEs in peer-to-peer communications. Other transceivers that are in the surroundings *compete* for the same (spectral) resources, giving rise to interference.

In Chapter 23, we encountered an alternative approach, where a transceiver helps in the transmission of a packet from the message source to the intended destination, by receiving a data packet and retransmitting it in the manner of a bucket brigade. We discussed the various advantages that such a *relaying approach* can have. In this chapter, we will discuss more advanced forms of relaying, including cooperative communication and network coding.

To set the scene, we will first paraphrase, and expand on, the results from Chapter 23. Relaying can be provided either by

- *Peer devices acting as relays*. Such peer devices, e.g., mobile handsets or sensor nodes, can change their roles depending on the situation at hand – sometimes they help to forward information and sometimes they act as a source or destination – this is the situation discussed in Chapter 23.
- *Dedicated relay devices,* i.e., relays that never act as source or destination of the information, but whose sole purpose is to facilitate the information exchange of other nodes; this situation is typically applicable in cellular systems.

Following established notation in the literature, we will henceforth call any transceiver in the network a "node," irrespective of whether it is source, destination, peer device acting as relay, or dedicated relay.

The introduction of relays creates more degrees of freedom in the system design, which can help to improve the performance, but also complicates the design process. As a starting point, consider the three-node network in Figure 27.1a, where, for simplicity, we assume that only free-space attenuation (and no fading) occurs on each link. Imagine a situation where node A does not have enough transmit power to send a data packet directly to node C. However, it can, in a first step, transmit the data packet to the intermediate node B. Node B retransmits the packet, and this retransmission can then be received by node C. This simple two-hop approach doubles the range of the network. Extending this approach to larger networks, the range that the network covers can be increased even more by multi-hopping, i.e., sending the message to a first relay, from there to a second relay, and so on, until it finally arrives at the destination.

A key feature of the wireless propagation channel is the *broadcast effect*: when one node transmits a signal, it can be received by any node in the vicinity – a fact that *can* be positively exploited in multi-node networks.[1] While the multi-hopping strategy of Section 23.1–23.6 does *not* make use of the broadcast effect, the more advanced cooperative communications approach *does* use it. Consider Figure 27.2a, which just slightly redraws Figure 27.1a. When node A transmits, the signal reaches not only node B but also (in weaker form) node C. This weak signal might not be enough by itself for node C to decode, but it can be used to *augment* the signal received in a subsequent transmission from node B to node C, similar to the situation in Hybrid Automatic Repeat Requests (HARQ), see Section 13.12. The broadcast effect has even more significant impact in larger networks, e.g., the situation depicted in Figure 27.2b: if the first node transmits, the signal reaches nodes B and D at about equal strength. Reaching those two nodes in the network thus does not "cost" anything more (i.e., does not require more transmit power) than reaching a single node. The two nodes B and D can now cooperate

---

[1] The flip side of this coin is that the transmission of a message from a node creates interference to other nodes that want to receive a different message. This negative effect is always present, regardless of whether the positive aspect of the broadcast effect is exploited or not.

Wireless Communications: From Fundamentals to Beyond 5G, Third Edition. Andreas F. Molisch.
© 2023 John Wiley & Sons Ltd. Published 2023 by John Wiley & Sons Ltd.
Companion website: www.wiley.com/go/molisch/wireless3e

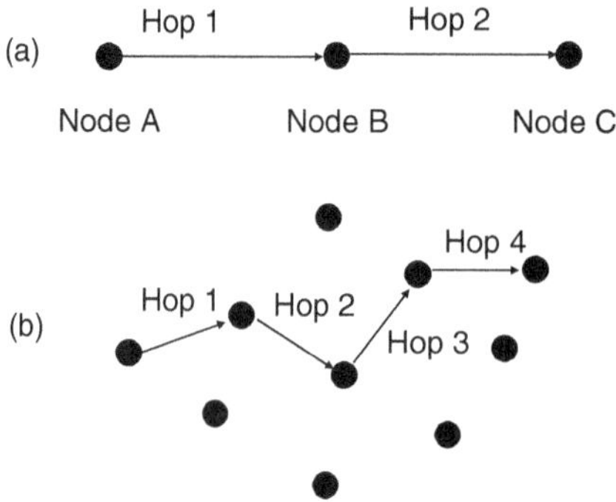

Figure 27.1 Two-hop network (a) and multi-hop ad hoc network (b).

to forward the information to node *C*, and – as we will show later on – such a cooperative transmission can be more efficient than if only a single node transmits. The same principle holds in even larger networks, like the one depicted in Figure 27.2c.

In the subsequent sections, more detailed descriptions of the scenarios of Figure 27.2 are given. We start out with the three-node relaying network, which is the most fundamental cooperative system, while Section 27.3 considers multiple parallel relays. For discussion of larger networks, the reader is referred to Chapter 23.

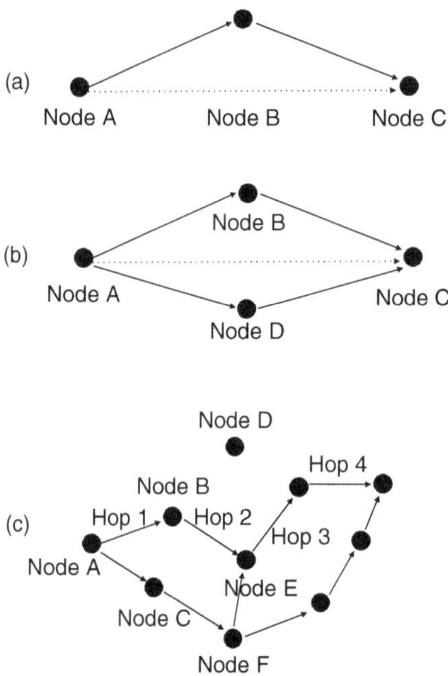

Figure 27.2 Demonstrating the broadcast effect in relay networks.

27.2 Fundamentals of Relaying

27.2.1 Fundamental Protocols

Consider the three-node network shown in Figure 27.3. A source is connected to a relay and a destination, with the complex amplitude of the (frequency-flat) channels between the nodes given by h_{sr}, h_{sd}, and h_{rd}, respectively. The relay can now help in various ways in forwarding the information.

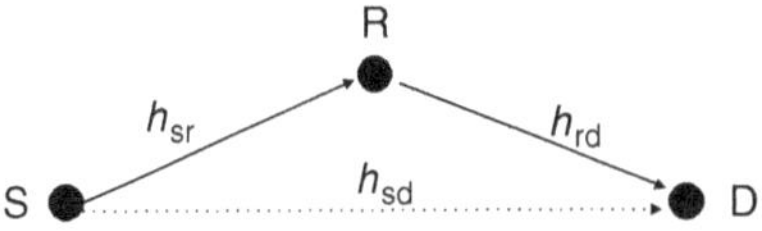

Figure 27.3 The fundamental relay channel.

In *Amplify-and-Forward* (AF), the relay amplifies the received signal by a certain factor and retransmits it. In *Decode-and-Forward* (DF), the relay decodes the packet and then subsequently re-encodes and retransmits it. In *Compress-and-Forward* (CF), the relay creates a quantized (compressed) version of the signal it obtains from the source and forwards that to the destination; the destination combines this compressed signal with the directly transmitted signal from the source. We assume in the following that all relaying nodes operate in a *half-duplex mode*, i.e., they cannot transmit and receive in the same frequency band at the same time. This is reasonable because the transmit and receive levels of wireless signals are so different that the transmitted signal would "swamp" the RX and make it impossible to detect the receive signal.[2]

These relay processing methods can now be combined with various transmission protocols that prescribe when what information blocks are transmitted from which nodes. We list them here in order of increasing performance (and at the same time of increasing complexity).

- *Multi-hop xF[3] (MxF)*: in the first timeslot, the source transmits, and *only* the relay is listening. In the second timeslot, only the relay is transmitting and the destination is listening. MDF (multi-hop with decode and forward) is a typical example.
- *Split-Combine xF (SCxF)*: in the first timeslot, the source transmits and only the relay is listening (just as in MxF). In the second timeslot, both source and relay transmit, and the destination is listening.
- *Diversity xF (DxF)*: in the first timeslot, the source transmits and both relay and destination are listening. In the second timeslot, only the relay is transmitting and the destination is listening. Thus, the destination gets two copies of the original signal.
- *Nonorthogonal Diversity xF (NDxF)*: in this scheme, the destination listens in both timeslots. The relay sends *differently encoded* information in the second timeslot, similar to incremental redundancy in HARQ.
- *Intersymbol Interference xF (IxF)*: this is a scheme that works only if the relay has full-duplex capability (contrary to our assumption above). In timeslot i, the source sends an information block to the relay, and in timeslot $i + 1$, the relay forwards this block to the destination, while at the same time, the source sends the next information block to the relay. The destination is continuously listening, and in each timeslot hears the superposition of the "current" information block directly from the source and a "previous" information block from the relay.

The above-mentioned protocols are by no means the only ones possible, and a number of variations have been proposed that are either aimed at reducing the penalty of half-duplex operation and/or are intended to allow simplified encoding and decoding.

	Phase 1			Phase 2		
	S	R	D	S	R	D
MxF	TX	RX	–	–	TX	RX
SCxF	TX	RX	–	TX	TX	RX
DxF	TX	RX	RX	–	TX	RX
NDxF	TX	RX	RX	TX	TX	RX
IxF	TX	RX + TX	RX	TX	TX + RX	RX

A further subclassification of the fundamental protocols can be done according to the following criteria:

- *Fixed vs adaptive power*: is the power expended by the source and by the relay fixed, or is it considered to be adaptive to the channel state? If it is adaptive, the power can be adapted:

  - over the nodes, i.e., some node is assigned more power, while another is assigned less;
  - over time, i.e., according to the temporal changes of the channel state, a node can transmit with more power at some time, and less power at another;
  - over frequency: in a wideband system using, e.g., Orthogonal Frequency Division Multiplexing (OFDM) (Chapter 15), a node can transmit more power at some frequencies, and less on others;
  - any combination of the above.

Constraints are then usually imposed on the sum (or average) transmit power, where the summation can be over nodes, time, and/ or frequency. For example, constraining the power summed over all the nodes (and imposing that constraint for any arbitrary time) imposes a restriction on the instantaneous power of the network; this is important because that instantaneous power is a measure for the interference *to other networks*. A summation over nodes and time (and frequency, if applicable) is a constraint on the energy expenditure of the communication. A summation over time (and frequency) only is a measure for the energy consumption of a particular node, which determines its battery lifetime.

---

[2] For a discussion of full-duplex systems, see Section 18.5.2.
[3] xF stands for either AF, DF, or CF.

- *Fixed vs. adaptive allocation of time and spectral resources*: the most common form of half-duplex protocols assumes that the network spends half the time sending a packet from the source to the relay, and the other half of the time forwarding the relay to the destination. This might not, however, be the most effective use of the available time. If the relay power is fixed, then transmission should occur at a rate close to the capacity of the relay-destination channel, and the transmission time chosen accordingly. If the relay power can be varied (as discussed above), then power and time (or bandwidth) should be optimized simultaneously.

27.2.2 Decode-and-Forward

The most important of all relaying schemes is DF. The relay receives a packet and decodes it, thus eliminating the effects of noise, before re-encoding and retransmitting the packet. In the following, we will analyze the capacity of several of the implementations.

The simplest scheme to analyze is Multi-hop Decode-and-Forward (MDF). For fixed transmit power, and equal splitting of the available time between the two phases, the overall data rate per unit bandwidth in bit/s/Hz is

$$R = \frac{1}{2} \min \left[\log_2 \left(1 + \frac{P_s |h_{sr}|^2}{P_n} \right), \ \log_2 \left(1 + \frac{P_r |h_{rd}|^2}{P_n} \right) \right] \tag{27.1}$$

where P_s and P_r are the powers used by source and relay, and $P_n = \sigma_n^2$ is the noise power. In other words, the link with the smallest Signal-to-Noise Ratio (SNR) becomes the "bottleneck" that determines the overall capacity; the factor $1/2$ arises from the half-duplex constraint. To wit: the terms inside the min operation are the capacities of the source-relay and the relay-destination link. In order for a transmission to be successful, a data packet has to pass through *both* links; the link with the smaller capacity is therefore the bottleneck that determines the achievable transmission rate.

The values of P_s and P_r can be fixed or can be optimized given the power constraints and the values of the channel coefficients h_{sr} and h_{rd}. In the latter case, the powers should be adjusted in such a way that the capacity of the source-relay link is the same as the one for the relay-destination link, i.e.,

$$P_s = P_0 \frac{|h_{rd}|^2}{|h_{sr}|^2 + |h_{rd}|^2}, \ P_r = P_0 \frac{|h_{sr}|^2}{|h_{sr}|^2 + |h_{rd}|^2} \tag{27.2}$$

where P_0 is the total transmit power of the system. Further optimizations for this scheme (as well as the ones we discuss below) can be achieved by dividing an available timeslot for data transmission into unequal parts and optimizing the duration of those two parts [Stankovic et al. 2006].

In Diversity-Decode-and-Forward (DDF),[4] the destination listens during both the phases and thus can add up the signals it received from the source (in phase 1) and the relay (in phase 2). We then have to distinguish two important cases:

1. *Transmission from the relay using repetition coding*: in this case, the relay uses the same encoder as the source. Consequently, the destination can add up the received signals before the decoding, resulting in an improved SNR. Assume now the further restriction that transmission is successful only if the relay can correctly decode (irrespective of the total received power at the destination). In that case, the maximum achievable rate is

$$R_{DDF} = \frac{1}{2} \min \left[\log_2 \left(1 + \frac{P_s |h_{sr}|^2}{P_n} \right), \ \log_2 \left(1 + \frac{P_r |h_{rd}|^2}{P_n} + \frac{P_s |h_{sd}|^2}{P_n} \right) \right] \tag{27.3}$$

and its optimum power allocation is

$$P_s = P_0 \frac{|h_{rd}|^2}{|h_{sr}|^2 + |h_{rd}|^2 - |h_{sd}|^2}$$
$$P_r = P_0 \frac{|h_{sr}|^2 - |h_{sd}|^2}{|h_{sr}|^2 + |h_{rd}|^2 - |h_{sd}|^2} \ . \tag{27.4}$$

A more intelligent protocol would employ the relay only if it can actually help, or otherwise keep it idle. Such a protocol is called adaptive DDF. An even better performance can be achieved by "incremental relaying," where the relay does not transmit if the destination can already decode the packet after the first transmission phase.

---

[4] According to our notation, DDF is "Diversity-Decode-and-Forward." However, sometimes the acronym DDF is used in the literature for "dynamic decode and forward," which is a different protocol.

2. *Transmission from the relay using incremental-redundancy encoding*: in this case, the relay decodes the packet and re-encodes it with a different coder. Then, the RX can add up the mutual information from the two transmission phases. The capacity of such a protocol is

$$R_{\mathrm{DDE,IR}} = \frac{1}{2} \min \left[\log_2 \left(1 + \frac{P_s |h_{sr}|^2}{P_n} \right), \log_2 \left(1 + \frac{P_r |h_{rd}|^2}{P_n} \right) + \log_2 \left(1 + \frac{P_s |h_{sd}|^2}{P_n} \right) \right] \quad . \tag{27.5}$$

In the following, when we speak about DDF, we mean the "regular" DDF as described in Eq. (27.3).

Using the concept of outage capacity (see Section 13.11.1), we define the whole system to be in outage if the achievable rate falls below the desired threshold R_{th}. For nonadaptive DDF, the probability for such an outage is

$$\begin{aligned}
\Pr(R_{\mathrm{DDF}} < R_{\mathrm{th}}) = {} & \Pr \left[|h_{sr}|^2 < \left(2^{2R_{\mathrm{th}}} - 1 \right) \frac{P_n}{P_s} \right] \\
& + \Pr \left[|h_{sr}|^2 > \left(2^{2R_{\mathrm{th}}} - 1 \right) \frac{P_n}{P_s} \right] \Pr \left[|h_{sd}|^2 P_s + |h_{rd}|^2 P_r < \left(2^{2R_{\mathrm{th}}} - 1 \right) P_n \right]
\end{aligned} \tag{27.6}$$

where the first term on the right-hand side corresponds to the case where the source-relay connection is too weak (according to the protocol the relay transmits in the second phase anyway, even if it creates an outage), and the second term corresponds to the case when the source-relay connection is strong enough, but both the relay-destination and source-destination connections are too weak to sustain sufficient information flow to the destination. Since the various links in the system are independent, those probabilities add up. If all the links are Rayleigh fading, then in the limit of high SNRs, the first term dominates because it provides only diversity order 1; the second term vanishes if *either* the source-destination *or* the relay-destination link provides sufficient quality. The overall diversity order 1 is linked to the fact that the protocol absolutely requires the source-relay link to have sufficient strength. For adaptive DDF, on the other hand, diversity order 2 can be achieved: two independent paths (source-destination, or source-relay-destination) are available, and transmission is successful if *either* of those paths provides sufficient quality.

27.2.3 Amplify-and-Forward

The basic principle of AF is that the relay takes the (noisy) signal y_r that it receives and amplifies it with a gain β; there are no other manipulations of y_r (like decoding, demodulating, etc.). It is assumed that the AF processing at the relay leads to a delay of half a timeslot.[5] Thus, in the first phase, the signal received at the destination is simply the (attenuated) signal from the source, plus noise (additional terms occur for Intersymbol interference Amplify-and-Forward (IAF), which we do not consider further). In the second phase, the signal is the sum of the direct signal from the source, and the signal from the relay, which is the amplified source word of the *previous* phase of that slot:

$$\begin{aligned}
y_d^{(2)} &= h_{sd} x_s^{(2)} + h_{rd} x_r^{(2)} + n_d^{(2)} \\
&= h_{sd} x_s^{(2)} + \beta h_{sr} h_{rd} x_s^{(1)} + \beta h_{rd} n_r^{(1)} + n_d^{(2)}
\end{aligned} \tag{27.7}$$

where x_s and x_r are the transmit signals from source and relay, respectively, superscript $^{(1)}$ and $^{(2)}$ denote the first and second phase of the transmission, respectively, and n_r and n_d are noise at relay and destination, respectively ($x_s^{(2)}$ can be zero, depending on the protocols outlined above).

The amplification factor is limited by power constraints. In the case of an instantaneous power constraint, we require that

$$|\beta|^2 \leq \frac{P_r}{P_n + P_s |h_{sr}|^2} \tag{27.8}$$

while in the case of an average power constraint, the expression has to be averaged over the fading realizations.

Despite the seeming simplicity of the scheme, there are a number of possible protocol realizations, as indicated in the classification above. Let us first analyze the performance of Multi-hop Amplify-and-Forward (MAF) with $P_r = P_s = P$. The signal arriving at the RX during the first phase is simply the source signal multiplied with the channel coefficient h_{sd}, and the noise has power P_n. In the second phase, the signal arriving at the destination is the source signal multiplied with $\beta h_{sr} h_{rd}$, and the noise has power

$$P'_n = \left(|\beta|^2 |h_{rd}|^2 + 1 \right) P_n. \tag{27.9}$$

---

[5] In the practical case of a purely analog repeater, the delay is actually much smaller.

The SNR is thus simply given by $\gamma = P|\beta h_{sr}h_{rd}|^2/P'_n$. Note, however, that the optimum power allocation in MAF is given by

$$\frac{P_s}{P_r} = \sqrt{\frac{|h_{rd}|^2 P_0 + P_n}{|h_{sr}|^2 P_0 + P_n}}. \tag{27.10}$$

In Diversity-Amplify-and-Forward (DAF), we wish to combine the signals from the two phases such as to maximize the SNR, i.e., we want to perform maximum-ratio combining. Therefore, the signal from the first phase has to be multiplied with

$$\sqrt{\frac{P_s}{P_n}}h_{sd}^* \tag{27.11}$$

i.e., phase adjusted, and multiplied with the square root of the receive SNR during that phase. Using a similar motivation, the signal from the second phase has to be multiplied by

$$\sqrt{\frac{P_s\beta^2}{P'_n}}(h_{sr}h_{rd})^*. \tag{27.12}$$

Assuming $P_s = P_r = P$, the corresponding SNRs for the two phases are

$$\gamma_1 = \frac{P|h_{sd}|^2}{P_n} \tag{27.13}$$

and

$$\gamma_2 = \frac{\left(\dfrac{P}{P_n}\right)^2 |h_{sr}h_{rd}|^2}{\dfrac{P}{P_n}|h_{sr}|^2 + \dfrac{P}{P_n}|h_{rd}|^2 + 1} \tag{27.14}$$

and the overall SNR is, as always with Maximum Ratio Combining (MRC), $\gamma = \gamma_1 + \gamma_2$. The resulting capacity is $\frac{1}{2}\log_2(1 + \gamma)$, where the factor $1/2$ stems from the half-duplex constraint of the relay. We also note that the scheme has diversity order 2, because the signal can reach the destination on two independent paths: directly, or via the relay. However, note that for coverage-extension relays with practical SNRs (as opposed to the infinite SNRs used for diversity order computations), the direct path is not really helpful – if we could achieve good reception with the direct path, there would be no need for the relay in the first place.

27.2.4 Compress-and-Forward

CF is similar to AF in the sense that the relay does not decode a message, but forwards whatever it receives (including noise). The main *difference* compared to AF is that the forwarded signal is a *quantized, compressed* version of the signal received at the relay. The process of quantization and compression can be seen as a source encoding problem, i.e., the received signal is an (analog) source, and its information should be encoded into a digital signal with as few distortions as possible – but on the other hand, the rate at which that signal can be transmitted is limited (and depends on the relay-destination channel). At the destination, this signal is used together with the signal received directly from the source node to reconstruct the original signal.

CF has been shown to provide higher capacity than DF or AF for some specific channel configurations. However, it is considerably more complicated than either of those formats and thus will not be considered further in this book. Further details can be found in the references cited in "further reading".

27.3 Relaying with Multiple, Parallel Relays

In many situations, more than one relay is available for forwarding the information. In that case, a cooperation between the different relays can be used to greatly enhance the performance of the relaying scheme, in particular in fading channels. Essentially, the multiple relays provide diversity paths that bring better robustness with respect to fading and interference. At the same time, the cooperation between the relays necessitates the exchange of Channel State Information (CSI) and control information. There are thus a number of different schemes, which trade off the overhead with the system performance in different ways.

Also for the arrangement with multiple relays, various transmission schemes do exist – AF, DF, CF, with the various protocols discussed in the previous section. However, to keep the discussion focused, we restrict the discussion to DF (in most cases, restricted to MDF) with semi-duplex relays. In the current section, we will only deal with two-hop networks, since in this case the relaying problem can be viewed as a physical-layer problem. Networks with more hops require, in addition, routing, which is discussed in Chapter 23.

Figure 27.4 shows the fundamental setup considered in this section. Transmission occurs in two phases: in phase 1, the source broadcasts the information. This phase takes advantage of the broadcast effect, i.e., the signal arrives at several relays (possibly with different strengths), even though only a single node (i.e., the source) transmits. In the second phase, one or more of the relays forward the information to the destination. We see that this second phase strongly resembles a smart-antenna system, specifically the transmission from a multi-antenna TX to a single-antenna destination – the main difference being that the antennas are distributed over a larger area in space. This analogy to multi-antenna systems will be helpful in the discussion below.

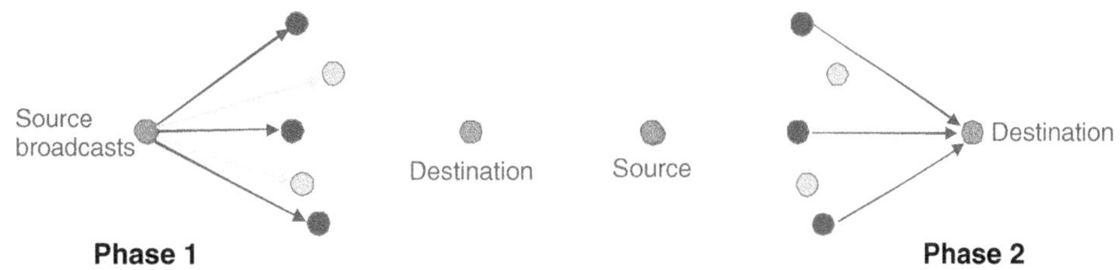

Figure 27.4 Two-phase transmission with parallel relays.

For phase 1, we always assume that the kth relay knows the channel from the source to it (i.e., CSI at the RX [Channel State Information at the Receiver (CSIR)], is available). CSI at the TX [Channel State Information at the Transmitter (CSIT)] is only useful if the source can either adapt its power or its transmission time, in accordance to the channel states. For phase 2, we again assume CSIR (i.e., the destination knows the channels between the kth relay and the destination). However, we distinguish different cases with respect to knowledge of CSIT, i.e., CSI that is available at the relays. Depending on the type of CSIT, different transmission schemes can be used:

- *Full CSIT available*: relays know both amplitude and phase of the channel to the destination. In that case, "virtual beamforming," similar to maximum-ratio transmission in a multiple-antenna system, can be used. This method ensures the maximum SNR at the RX for a given sum power expenditure at the relays. This case will be discussed in Section 27.3.2.
- *Amplitude CSIT available*: in that case, the relays know the amplitude (strength) of the channel to the destination, but not the phase. In this case, the best strategy (for a sum-power constraint) is to select a single relay that provides the best transmission quality (see Section 27.3.1).
- *No CSIT available*: in this case, the relays can transmit Space–Time (ST) encoded versions of the data packet – they act like antennas in a transmit-diversity system without CSIT (see Section 27.3.3). Alternatively, the relays can send out incremental-redundancy encoded bits of the same codeword (see Section 27.3.4). Note that for a sum-power constraint, the SNR for ST codes is worse than for relay selection: transmit diversity provides an effective channel whose SNR is the *average* of the individual relay-destination channels, while relay selection provides a channel with an SNR that is the *maximum* of that of the individual channels.
- *Average CSIT available*: in this case, only the mean channel gain but not the instantaneous realization is available. This case is interesting because average CSIT can be acquired much more easily than instantaneous CSIT, particularly in fast-varying channels. Modifications of the no-CSIT schemes can be used.

27.3.1 Relay Selection

In relay selection, we simply pick the "best" of all the available relays and then perform relaying the same way as described in Section 27.2. This approach sounds deceptively simple; the challenges are (i) defining what we mean by "best relay," (ii) actually finding the best relay for a given set of channel states.

First, consider criteria for the "best" relay, distinguishing two cases: if the source has fixed transmit power and data rate, (i.e., modulation format and coding of the packet are fixed), then we cannot influence which relay will receive the packet correctly during transmission phase 1; rather, we simply consider the set of relays that *do* get the packet, and select the one for forwarding that has the strongest channel to the destination.

If the source can adapt to the channels, then we can make sure that a specific relay (which is selected a priori as the one that will do the forwarding) gets the message in phase 1. Choosing this relay requires a balancing between the source-relay and relay-destination channel strengths: in MDF, where the rate is given by Eq. (27.1), we should aim to avoid bottlenecks, and thus pick the relay that provides the best

$$\eta_k = \min\left[|h_{\mathrm{s},k}|^2, |h_{k,\mathrm{d}}|^2\right]. \tag{27.15}$$

An alternative criterion considers a "smoothed-out" version of (27.15)

$$\eta_k = \frac{2}{\dfrac{1}{|h_{\mathrm{s},k}|^2} + \dfrac{1}{|h_{k,\mathrm{d}}|^2}}. \tag{27.16}$$

It is often assumed that a central control node knows all the $|h_{s,k}|^2$ and $|h_{k,d}|^2$. In practice, this would require considerable signaling overhead, and is therefore undesirable. It is therefore preferable to use algorithms similar to the ones that are used to control multiple access for packet radio systems (see Section 18.4), in a distributed manner:

1. In a first step, the destination sends out a brief broadcast signal that allows the relays to determine the $|h_{k,d}|^2$ (assuming channel reciprocity, see Section 16.1.7).
2. Next, the source sends the data packet, as well as a "Clear To Send" (CTS) message after it is finished. Each relay tries to receive the packet, and also determines its $|h_{s,k}|^2$, and from that, the η_i, according to one of the criteria Eqs. (27.15) or (27.16).
3. Each relay starts a timer with an initial value K_{timer}/η_k (where K_{timer} is a suitably chosen constant) and counts down, while listening for possible on-air signals from the other relays. When the timer reaches 0, the relay starts to transmit – unless another relay has already started to transmit (and thus occupies the channel).

Clearly, the relay with the "best" channel (highest η_k) is the first – and therefore only – relay node to transmit. In practical networks, the performance is not ideal, since a second node might start to transmit between the time that the first node transmits and the time that signal actually arrives at the second node (again, compare Chapter 18), but such collisions can be resolved by repeating step 3 with different K_{timer}.

Relay selection performs remarkably well and provides the same diversity order as other, more complicated relaying schemes discussed below. This is analogous to antenna selection (see Section 12.4.1): antenna selection provides the same diversity order (i.e., slope of the Bit Error Rate (BER) vs. SNR curve) as MRC. Define the outage probability as the probability that the capacity (or more precisely, the mutual information using Gaussian codebooks) I, is below a threshold R_{th}. For K relays, the outage probability can be computed for the case of Rayleigh fading on all links as

$$\Pr[I < R_{\text{th}}] = \prod_{k=1}^{K} \left[1 - \exp\left[-\frac{2^{2R_{\text{th}}} - 1}{1} \left(\frac{1}{\overline{\gamma}_{s,k}^2} + \frac{1}{\overline{\gamma}_{k,d}^2} \right) \right] \right] \tag{27.17}$$

where $\overline{\gamma}_{s,k}^2$ and $\overline{\gamma}_{k,d}^2$ are the mean SNRs of the source-relay and relay-destination channels.

27.3.2   Distributed Beamforming

This protocol consists of two phases: in phase 1 the source broadcasts the information and a set $\mathcal{D}$ (with size $|\mathcal{D}|$) receives the packet in good order. For an MDF protocol, the optimal transmission weight at each selected relay k in phase 2 can then be shown to be proportional to

$$\frac{h_{k,d}^*}{\left(\sum_{k \in \mathcal{D}} |h_{k,d}|^2 \right)^{1/2}} \tag{27.18}$$

once the CSIT at the relays is available. The $|\mathcal{D}|$ nodes cooperate, i.e., transmit coherently, to send data to the destination. This is similar to beamforming or maximum ratio transmission in transmit diversity systems.

In the case that the relays are using AF, the optimum gain w_k applied at relay k is

$$w_k = K^{\text{AF}} \frac{|h_{sk}||h_{kd}|}{P_{\text{n}} + P_{\text{s}}|h_{sk}|^2 + P_k|h_{kd}|^2} \frac{h_{sk}^*}{|h_{sk}|} \frac{h_{kd}^*}{|h_{kd}|} \tag{27.19}$$

where the constant K^{AF} is chosen such that the total power constraint $\sum_k |w_k|^2 \left(1 + P_{\text{s}}|h_{sk}|^2 \right) = P_{\text{r}}$ is met. The power on the kth relay is

$$P_k \propto \frac{|h_{sk}|^2 |h_{kd}|^2 \left[P_{\text{s}}|h_{sk}|^2 + P_{\text{n}} \right]}{\left[P_{\text{n}} + P_{\text{s}}|h_{sk}|^2 + P_k|h_{kd}|^2 \right]^2}. \tag{27.20}$$

Obtaining the CSIT at the relays is nontrivial: not only does each relay need to know its channel to the destination but it also needs to know the sum of the channel gains from all the relays that will be active in the forwarding of the data (i.e., the denominator in Eq. 27.18). This can be implemented through consecutive transmission of training sequences (pilots) from the relays, followed by a feedback from the destination.

The situation is, again, more complicated if the source can adjust its power, because then it can influence the set of possible active relays, $\mathcal{D}$. This leads to a trade-off between the two phases in the relaying: if the source expends little energy on the broadcast, then $|\mathcal{D}|$

is too small, and the diversity order available in phase 2 is low – in other words, there is a risk that all the relays that received the packet have a bad channel to the destination, and thus have to expend a large amount of power to get the packet to the destination. On the other hand, spending too much energy on the broadcast is wasteful. An exact optimization of the best power allocation is somewhat complicated, but as a rule of thumb, $|\mathcal{D}|$ should be 3.

The above discussion assumes that the various relay nodes can co-phase their transmit signals in such a way that they superpose constructively at the intended destination. This is a very difficult endeavor in practice, since the relay nodes are not co-located, yet still have to be phase synchronous (in addition to being frequency and time synchronous). Typically, one node in the network would work as a leader that periodically sends out synchronization signals and forces all other nodes to adapt their frequency and phase to this synchronization signal. Adjusting for the phase shift created by the runtime between nodes has to be done on a link-by-link case. Often it is advantageous if the destination node is this leader node.

An alternative way of dealing with the problem of phase adjustment is the use of random beamforming (compare Section 22.8). If no special measures are taken (i.e., no specific phase adjustment), the beams created by the relays point into random directions, and by changing the relative phases of the nodes, the main direction of the beams changes. In the spirit of opportunistic beamforming, a destination node that finds itself in the main lobe of the beam sends a feedback signal and asks for the relays to send payload data intended for this node.

27.3.3 Transmission on Orthogonal Channels

When CSI is not available at the TX, one possible solution is to have each relay transmit on an orthogonal channel, e.g., a different timeslot; see, e.g., the top row of Fig. 27.5. This clearly eliminates the interference between the different relay channels; however, it also leads to a drastic reduction of the spectral efficiency. In particular, consider a DDF scheme where every relay has a reserved channel – whether it can decode the message or not. Its capacity per unit bandwidth is (or more precisely, the mutual information using Gaussian codebooks) is

$$I = \frac{1}{(K+1)} \log_2 \left[1 + \gamma_{s,d} + \sum_{k \in \mathcal{D}} \gamma_{k,d} \right] \tag{27.21}$$

where $\mathcal{D}$ is the set of the relays that can decode the message from a particular source. When all links are Rayleigh fading, the outage probability conditioned on a particular decoding set for this scheme is for high SNR

$$\Pr[I < R_{th}|\mathcal{D}] \sim \left[2^{(K+1)R_{th}} - 1 \right]^{|\mathcal{D}(s)| + 1} \frac{1}{\overline{\gamma}_{s,d}} \prod_{k \in \mathcal{D}} \frac{1}{\overline{\gamma}_{k,d}} \frac{1}{[|\mathcal{D}| + 1]!} \ . \tag{27.22}$$

The probability of obtaining a particular decoding set is given by

$$\Pr[\mathcal{D}] \sim \left[2^{(K+1)R_{th}} - 1 \right]^{[K - |\mathcal{D}|]} \prod_{k \notin \mathcal{D}} \frac{1}{\overline{\gamma}_{s,k}} \ . \tag{27.23}$$

The overall outage probability is then Eq. (27.22) unconditioned by Eq. (27.23); this expression can be bounded by

$$\left[\frac{2^{(K+1)R_{th}} - 1}{\overline{\gamma}^{lb}} \right]^{(K+1)} \sum_{k \in \mathcal{D}} \frac{1}{[|\mathcal{D}| + 1]!} \ \lesssim \Pr[I < R_{th}] \lesssim \left[\frac{2^{(K+1)R_{th}} - 1}{\overline{\gamma}^{ub}} \right]^{(K+1)} \sum_{k \in \mathcal{D}} \frac{1}{[|\mathcal{D}| + 1]!} \tag{27.24}$$

where

$$1/\overline{\gamma}_k^{lb} = \min \left\{ 1/\overline{\gamma}_{s,k}, 1/\overline{\gamma}_{k,d} \right\} \quad 1/\overline{\gamma}_k^{ub} = \max \left\{ 1/\overline{\gamma}_{s,k}, 1/\overline{\gamma}_{k,d} \right\} \quad \overline{\gamma}_s^{lb} = \overline{\gamma}_s^{ub} = \overline{\gamma}_{s,d} \tag{27.25}$$

and $\overline{\gamma}^{lb}$ is the geometric mean of the $\overline{\gamma}_k^{lb}$, and similarly for $\overline{\gamma}^{ub}$. To again draw an analogy with multiple antenna systems, transmission on orthogonal channels can be compared to *antenna cycling*, where only one antenna element is used (for one particular message) at each time.

We now turn to the situation where the nodes all act as sources as well as relays and which can transmit at different frequencies as well as times. Consider the situation depicted in Figure 27.5. Also, in that case, each node is transmitting information from a particular source only for $1/(K+1)$ of the available time. In other words, the spectral efficiency is not improved compared to the situation discussed above.

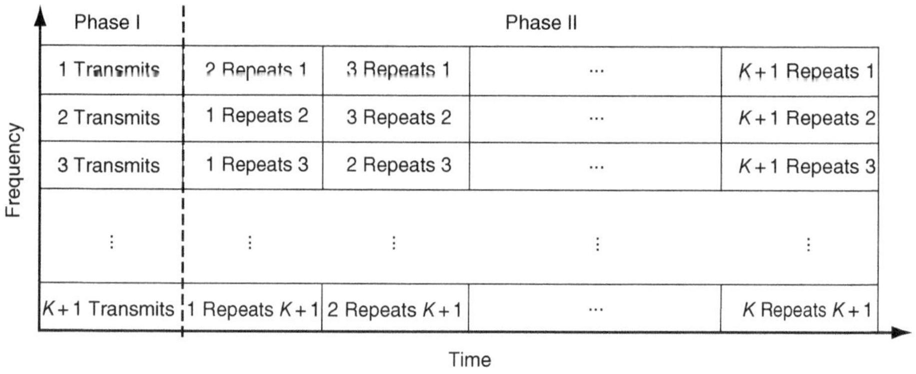

Phase I	Phase II				
1 Transmits	2 Repeats 1	3 Repeats 1		...	$K+1$ Repeats 1
2 Transmits	1 Repeats 2	3 Repeats 2		...	$K+1$ Repeats 2
3 Transmits	1 Repeats 3	2 Repeats 3		...	$K+1$ Repeats 3
⋮	⋮	⋮	⋮		⋮
$K+1$ Transmits	1 Repeats $K+1$	2 Repeats $K+1$		...	K Repeats $K+1$

Figure 27.5 Multiplexing of multiple signals on multiple relays. $K+1$ nodes transmit messages. Reproduced with permission from [Laneman and Wornell 2003] © IEEE.

27.3.4 Distributed Space–Time Coding

An alternative approach for the no-CSIT case is to have the relays use ST codes during the transmission. Consider the following situation: in phase 1, the source sends the information to the relays. In phase 2, the relays now perform a ST-coded transmission to the destination. In other words, each relay node acts as a "virtual antenna," and sends out the signal that – in a Multiple Input Single Output MISO setting – would be sent out by one of antenna elements of the transmit antenna array. For example, if two relays are used, then the used ST code could be an Alamouti code: two symbols, c_1 and c_2, are transmitted from the two relays at time instant 1:

$$\mathbf{s}_1 = \frac{1}{\sqrt{2}} \begin{pmatrix} c_1 \\ c_2 \end{pmatrix} \tag{27.26}$$

where $\mathbf{s}$ is the vector containing the symbols sent from the relays. At the second time instant, the signal vector

$$\mathbf{s}_2 = \frac{1}{\sqrt{2}} \begin{pmatrix} -c_2^* \\ c_1^* \end{pmatrix} \tag{27.27}$$

is transmitted (compare Section 16.1.6). Of course, the communication protocol must have a means to assign to each relay which "antenna" it is, and therefore, which sequence of data $(c_1 - c_2^*....)$ or $(c_2 \, c_1^*....)$ it should send out.

Since the Alamouti code is a rate-1 code, the spectral efficiency of the transmission is better than for the relaying on orthogonal channels, where the rate (during the second phase of the relaying) is only 1/2 (for the case of two relays). When using more relays, the spectral efficiency of relaying with orthogonal ST codes decreases somewhat: for $K > 2$, no rate-1 orthogonal ST codes exist. For $K = 3$ or 4, the achievable rate decreases to 3/4. Still, this is much better than orthogonal relaying, where the rate decreases as $1/K$.

A further practical problem arises from the fact that the number of participating relays changes, depending on how many relays are able to decode the message from the source. Fortunately, this does not impact the operation of distributed ST codes significantly: if a relay does not receive a message from the source, it simply does not transmit (which for the RX looks like that particular "antenna" is in a deep fade). The decoding operation of the RX is therefore not impacted.

An alternative to distributed space-time coding is phase dithering combined with Low-Density Parity Check (LDPC) codes. All the relay nodes transmit a codeword approximately simultaneously, but – since they have no CSIT – without phase synchronization. This might lead to destructive interference of the various transmission at the destination node. To combat this, an LDPC codeword is divided into different bursts, each of which has a different phase dither. Consequently, the superposition of the same symbol transmitted by different nodes at the RX is random but changes from burst to burst, leading to some parts of a codeword with good SNR and some with bad SNR. The LDPC code can then recover the originally transmitted information.

*27.3.5 Coded Cooperation

In coded cooperation, relaying and error correction coding are integrated, leading to enhanced diversity. A data packet from a source is encoded with a Forward Error Correction (FEC) code (see Chapter 13), and different parts of the codewords are sent via two (or more) different paths in the network.

To give more details, consider the example of Figure 27.6. At node 1, a source data block is encoded with an FEC, and the resulting codeword is split up into two parts, with N_1 and N_2 bits, respectively. It is important that it is possible to reconstruct the source data from the first N_1 bits alone. For example, the FEC can be a rate 1/3 convolutional code, which is then punctured to result in a rate 2/3 code that is transmitted in the first N_1 bits; the bits that are punctured out are transmitted in the second N_2 bits. A similar encoding and splitting is done for a different block of source data at node 2.

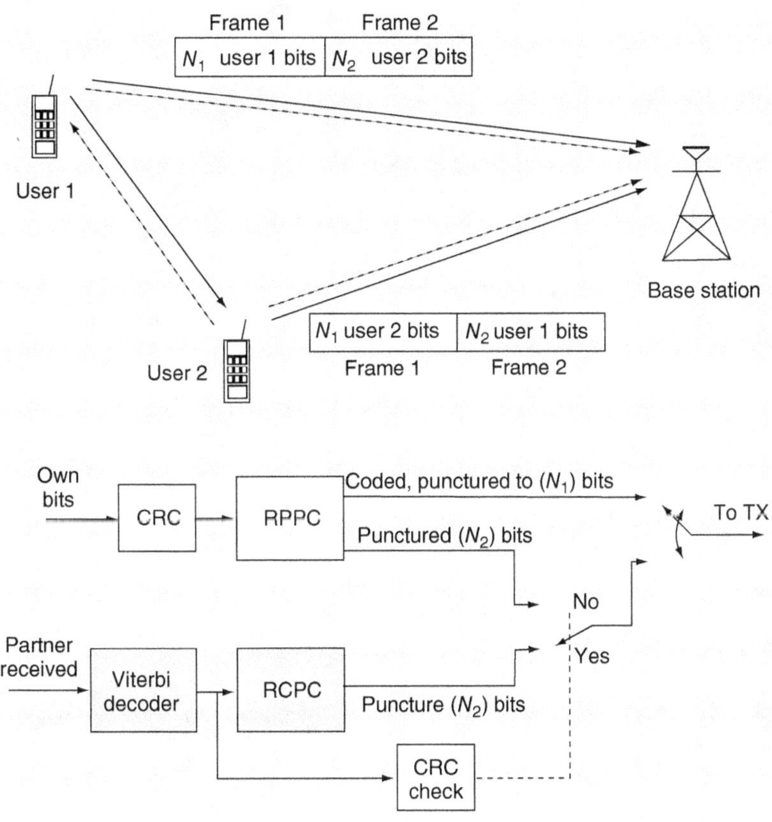

Figure 27.6 Principle of coded cooperation. Solid (dashed) lines: bits associated with the payload user 1 (2) has generated. Reproduced with permission from [Nosratinia et al. 2004] © IEEE.

Now the transmission interval available for one block of source data is divided into two parts. During the first interval, node 1 broadcasts the first N_1 bits. They are received by the destination, as well as by node 2. At the same time, node 2 transmits (on an orthogonal channel, e.g., a different frequency channel) its first N_1 bits. If node 1 can successfully decode the source word of node 2 (this is checked with a Cyclic Redundancy Check (CRC)), then it computes the second N_2 bits associated with that source data of node 2 and transmits it in the second time interval. If it cannot decode successfully, then it sends the N_2 bits associated with its own codeword. Node 2 behaves in a completely analogous way. Since there is no feedback between nodes 1 and 2, the four situations depicted in Figure 27.7 can arise. Summarizing, each node always sends $N_1 + N_2$ bits; if the channel between the two nodes is good, then part of the transmitted bits are helping a partner node; this is the case that we will consider in the following (the other case, where each node just transmits $N_1 + N_2$ of its own data, is a regular coded Frequency Division Multiple Access (FDMA) transmission of two users to an RX, see Section 18.3.1).

Since the different parts of the codeword are transmitted from different locations, the transmission has a diversity order of 2 (if the two nodes 1 and 2 are sufficiently widely separated, they might provide macrodiversity as well as microdiversity (see Chapter 12)). The diversity order is reflected in the asymptotic expressions for the outage probability. Assuming that all channels are Rayleigh fading, $N_1 = N_2$, and the channels from node 1 to node 2 and node 2 to node 1 are independent (as is usually the case in a frequency duplexing), the outage probability is approximated in the high-SNR regime by

$$\Pr[I < R_{\text{th}}] = \frac{(2^{2R_{\text{th}}} - 1)^2}{\gamma_{A,d}\gamma_{A,B}} + \frac{R_{\text{th}} \ln(2)2^{2R_{\text{th}}+1} - 2^{2R_{th}} + 1}{\gamma_{A,d}\gamma_{B,d}}. \tag{27.28}$$

For the case of reciprocal inter-user channels (e.g., if transmission from node A to node B and node B to node A is done on the same frequency channel, e.g., using Time Division Multiple Access (TDMA)), the outage probability is

$$\Pr[I < R_{\text{th}}] = \frac{(2^{R_{\text{th}}} - 1)(2^{2R_{\text{th}}} - 1)}{\gamma_{A,d}\gamma_{A,B}} + \frac{R_{\text{th}} \ln(2)2^{2R_{\text{th}}+1} - 2^{2R_{\text{th}}} + 1}{\gamma_{A,d}\gamma_{B,d}}. \tag{27.29}$$

The transmission of the data packet for one particular user can be considered as DF relaying with incremental redundancy; this is more efficient than conventional DF, where the relay repeats the originally transmitted bits, as discussed in Section 27.2 above. This is

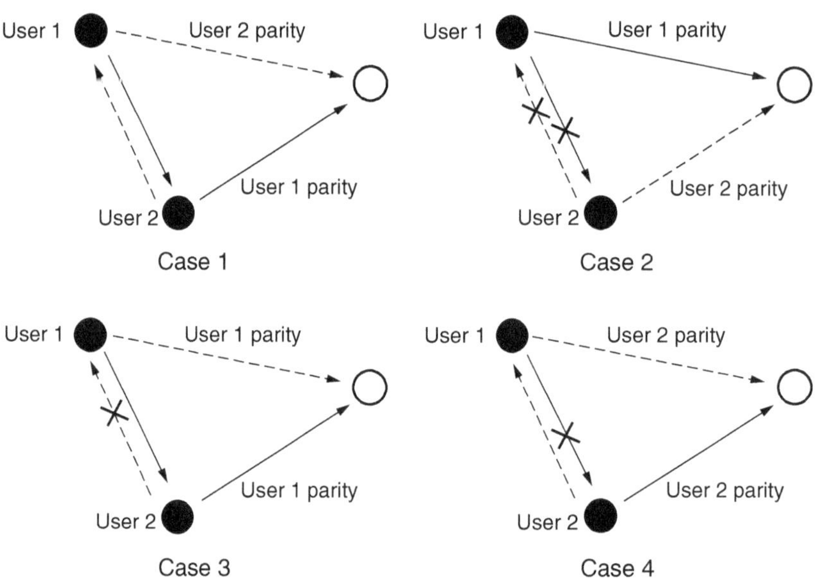

Figure 27.7 Four cases of messaging that can arise in two-user cooperative coding.
Reproduced with permission from [Nosratinia et al. 2004] © IEEE.

also visible in Figure 27.8, which compares the block error rate of coded cooperation with rate 1/4 encoding to AF and DF with rate 1/2 encoding (i.e., all schemes have the same spectral efficiency).

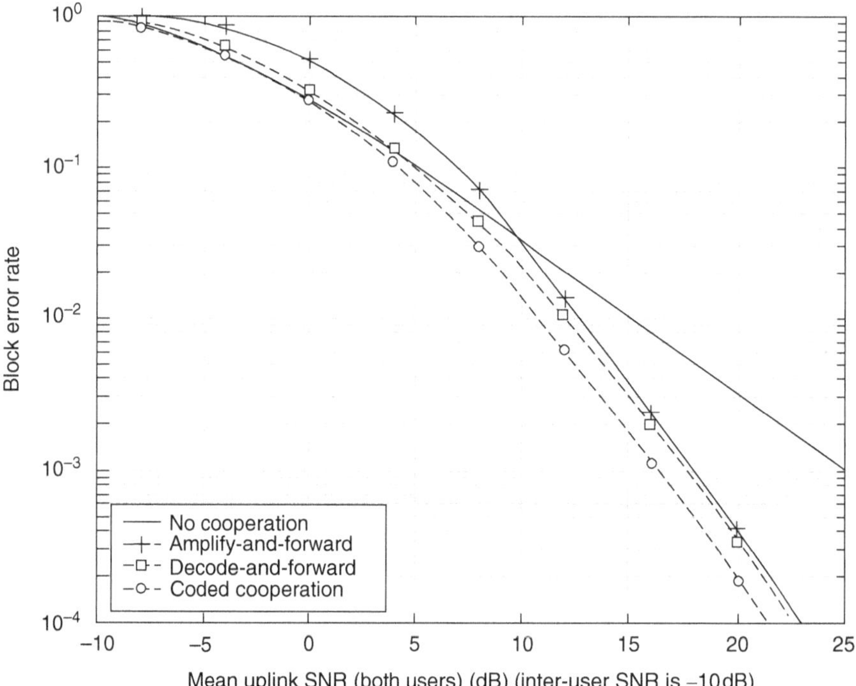

Figure 27.8 Performance comparison of coded cooperation with AF, DF, and direct transmission.
Reproduced with permission from [Nosratinia et al. 2004] © IEEE.

*27.3.6 Fountain Codes

The virtual-MIMO techniques described in the previous section suffer from a number of drawbacks, including the necessity to coordinate simultaneous transmissions to achieve cooperative gain and comparatively low efficiency of the collaboration (RXs accumulate energy from the cooperating nodes). An alternative is the use of rateless codes. Unlike conventional codes, rateless codes do not have a fixed coding rate (hence, the name) and are not optimized for a specific SNR. Rather, they work well for all possible SNRs. The most popular form of rateless codes, called Fountain codes, was originally designed to be applied to combat the erasure of data packets on the Internet (and have been used for that purpose in cellular systems). However, Fountain codes can also be made to work on a bit-by-bit basis, and then work as follows in an erasure channel: the TX creates an (infinitely long) bitstream from a finite-length block of source data. Receiving nodes observe the bitstream, and accumulate the bits that were not erased in the channel. RXs can recover the original information from the observed, unordered subset of the codestream, just as long as the total received number of bits is larger than the number of bits in the original source word. It turns out that bit-by-bit Fountain codes also work well in Additive White Gaussian Noise (AWGN) channels and fading channels. In that case, successful reception of a source word is achieved if the received mutual information equals the source entropy.

Since Fountain codes are "universal" codes that work at *any* SNR rate, the same code design can be used for broadcasting from one source to multiple RXs whose links to the source have different attenuations. At the same time, Fountain codes are useful in relay networks because they allow each node in the network to accumulate mutual information (and not just energy) from multiple transmitting nodes (source and other relays that received the same information at earlier times).

Summarizing, cooperative communications with Fountain codes is different from "regular" cooperative communications in the following way:

1. The transmission duration of a packet is a random variable, depending on the channel state.
2. Successful "one-shot" transmission can be guaranteed, without the need to know the SINR (Signal-to-Interference-and-Noise Ratio) at the TX. The TX just keeps transmitting until it gets a 1-bit feedback from the RX that the message was successfully decoded.
3. A destination node can accumulate mutual information (not just energy) from multiple relays.
4. Different parallel relays can be active for different amounts of time. This is in contrast to, e.g., distributed ST coding, where the parallel nodes transmit for the same amount of time.

27.4 Applications

Relaying and multi-hopping can be applied either in an infrastructure-based (cellular) setting, facilitating the communication between a BS and a UE, or they are an integral part of ad hoc networks. We describe in this section the former application; the latter is discussed in Chapter 23.

Dedicated Relay Stations (RSs) in a cellular network might be beneficial in one or more of the following respects:

- *Increase of coverage area*: since the relay improves the SNR, UEs that are far away from the BS can still receive a decodable signal. This helps to extend the effective radius of a cell, or eliminate "coverage holes," i.e., provide coverage in areas within a cell that due to the peculiarities of the topography are not covered by the BS. Dedicated RSs are usually placed at locations where they have good connection to the BS, e.g., on rooftops. This allows the RS to receive the signal with good quality, and forward it (after amplifying and/or "cleaning up" by decoding and re-encoding) to the destination; in either case, the signal arriving at the UE has improved SNR.
- *Improvement of indoor coverage*: similarly, relays can help to improve the indoor coverage. While coverage on the streets of urban/ metropolitan areas is usually very good, indoor coverage on the same streets is often spotty or nonexistent due to the additional path loss suffered by signals when they penetrate into the building – this is especially pronounced at high carrier frequencies, e.g., millimeter-wave bands. Thus, relays (which are usually placed close to the building perimeter) often are necessary to fully cover the inside of office or residential buildings. In a related application, relays can be used to improve connections to the inside of trains, buses, and other vehicles: a single relay can connect to the BS, using complex signal processing to compensate for the high Doppler shifts encountered, e.g., in high-speed trains. The connection from the relay to the UEs inside the train, on the other hand, is fairly simple.
- *Increase of reliability*: by improving the SNR, relays help to improve the reliability. Furthermore, relays can add diversity, and thus increase reliability.
- *Increase of throughput*: if both the BS-RS and RS-UE links have good SNR, then higher throughput can be achieved, e.g., by using a higher-order modulation alphabet and higher coding rate. On the flip side, most relays lose rate because of the half-duplexing constraint, as discussed in Section 27.2. The trade-off between the duplexing loss and the gains in per-link data rate determines whether a relay can help to increase the capacity and throughput of a cellular system or not. A different way of enhancing throughput is the use of relays to redirect traffic from overloaded BSs to less-congested BSs. Imagine a situation where a cell is temporarily overloaded (e.g., due to a large number of people congregating in that cell because of a special event). Then calls from some of the UEs can be connected via relays to BSs in neighboring cells that are momentarily underutilized.

Many of the practical implementation issues of dedicated RSs are related to the MAC-layer protocol and the control information that is being transmitted. In the downlink of *transparent relays*, the RS receives the signal from the BS and repeats it (with exactly the same control information); the uplink operates analogously. Care must be taken that the retransmission is consistent with the timing and frequency information contained in the associated control information. In *nontransparent relays*, the RS adds control information of its own, and thus appears to the UE like a distinct BS. Implementations in the LTE and NR standards are discussed in Sections 31.8 and 32.8, respectively.

While, in principle, multiple hops through a network are possible in infrastructure-based systems, this is rarely used in practice. Two-hop relaying (i.e., use of a single relay) is by far the most common situation. Of course, one physical relay can serve multiple UEs.

27.5 Network Coding

The basic principle of network coding is that nodes in the network form combinations (functions) of data packets or symbols that they receive, and forward those combinations; the destination then recovers the original messages from different combinations it receives. This is in contrast to relaying, where a node forwards a packet it has received in unadulterated form (at most performing noise clean-up). Thus, network coding can be seen as the epitome of collaboration: not only the resources (power, airtime) are shared between the nodes but also the packets.

Network coding is most easily described for multi-cast situations, i.e., where multiple sources transmit packets, and *several* nodes want to learn those messages (broadcast and unicast are special cases of multi-cast). We will in the following discuss two different cases:

- Networks without broadcast effect: these are networks in which a node can send different packets toward different destinations, and can receive packets from multiple sources. This situation occurs, of course, in wired networks, but also in wireless networks with highly directional antennas and small angular spread.
- Networks with broadcast effect: this is the situation we are more accustomed to in wireless networks, where the transmission from a single node can be heard by multiple RX nodes, which can be an advantage, but also a possible drawback since it gives rise to interference.

27.5.1 Network Coding Without Broadcast Effect

We will in the following first discuss the networks without broadcast, since this has been historically the first situation in which network coding was explored, and several concepts are easier to explain. We will also assume at first that noise plays no significant role. Later we will consider erasure channels, which arise when noise and/or interference overwhelm the FEC of the transmission and packets are not received in good order.

The Butterfly Network

We start out with a famous example for network coding, the "butterfly network," see Figure 27.9. Two messages, a and b, are to be transmitted from the source node S to two receiving nodes D and E.

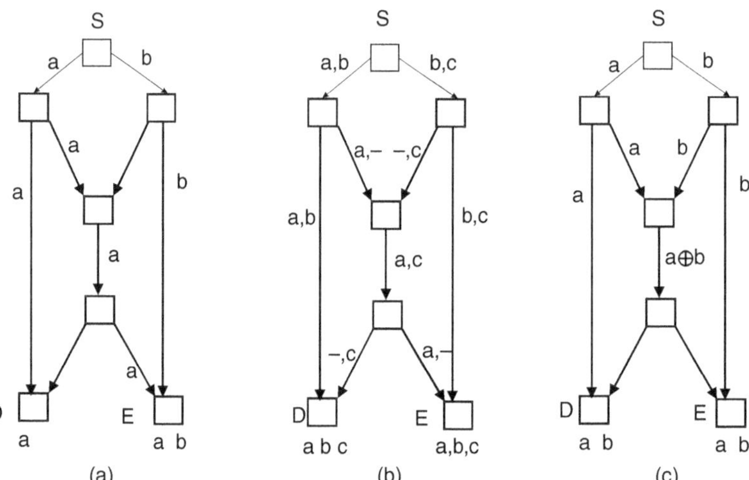

Figure 27.9 Butterfly network: (a) simple routing through the bottleneck link; (b) routing of three packets in two timeslots, (c) network coding transmission of two packets in one timeslot. S, source node; D,E, destination nodes.

The standard way of transmitting the packets would be via routing (Figure 27.9a): imagine that the source transmits message a via the left branch to the left RX, and message b via the right branch. The question is now what to do with the center branch, which can carry only one message. With routing, we could send, e.g., packet a over this link, so that a and b can be transmitted to destination E; however, only packet a arrives at destination D. We thus need more than one "channel use" to transmit the packets. Similarly, we could send packet b over the bottleneck link, so that now node D gets both packets, but node E gets only b. Thus, in either case, we would have to spend an extra timeslot to unicast the missing packet to the RX that did not get it, leading to a rate 1 (2 packets per 2 channel uses). A somewhat better routing method, that transmits 3 packets in 2 timeslots is shown in Figure 27.9b (i.e., in the first timeslot S sends packet a on the left link and packet b on the right link, in the next timeslot it sends b on the left link and c on the right link, etc.). But also in this case we have to decide whether in a given timeslots we want to send one packet *or* another over a given link.

The key difference between routing and network coding is that in the latter we send a *function of two packets* over a link; in our example, we specifically send the modulo-2 addition $a \oplus b$ over the central link (note that we are again just considering a single timeslot). Then, RX D obtains packet a from the left branch, and $a \oplus b$ from the center branch; from this, it can then easily compute b as well. Similarly, RX E obtains packet b from the right branch and $a \oplus b$ from the center branch, from which it computes a. Thus, with one "channel use," two packets have been transmitted to both the RXs; we thus have increased the throughput compared to the routing case by 33% (in general, the improvement can be – depending on the network structure – arbitrarily large). It is easy to see that in this example, throughput of two packets per channel use is actually the optimum: a destination node has at most two connections coming in, and thus cannot receive more than two packets per channel use.

Linear Network Coding

The butterfly network is just one example of a general network coding problem, where each node can create linear combinations of the incoming packets. The elements of each data packet are taken from an alphabet that is a finite field F_q (e.g., a binary field, such that $q = 2$). In a finite field F_q, additions and multiplications are well defined. To simplify notation, we henceforth assume that each packet has only one element (in the case of $q = 2$, that means a single bit). Thus each packet $p_i, i = 1, \ldots I$ is described by a scalar (generalization to multi-element packets is straightforward).

Consider now a particular node n. It has a number of incoming packets, which are written into a column vector $\mathbf{x}^{(n)}$ whose elements are the packets coming in on the different edges leading into node n, and a vector of outgoing packets $\mathbf{y}^{(n)}$ that is similarly defined. The formation of a linear combination at the node is then described by a coefficient matrix $\mathbf{L}^{(n)}$; a row associated with a particular outgoing edge is called the local encoding vector corresponding to that edge. The encoding process is now simply

$$\mathbf{y}^{(n)} = \mathbf{L}^{(n)}\mathbf{x}^{(n)} \tag{27.30}$$

where all operations are in the arithmetic of the finite field F_q (e.g., for $q = 2$, additions are XOR, and multiplications are AND). Since the concatenation of linear operations is again a linear operation, we can describe the packet carried along any edge in the network as a linear combination of the source packets. Writing the source packets into a vector $\mathbf{p}$,

$$\mathbf{x}^{(n)} = \mathbf{G}^{(n)}\mathbf{p}^{(n)} \tag{27.31}$$

where the matrix $\mathbf{G}^{(n)}$ now describes the global encoding process. This formulation is particularly important for the RX nodes, because it provides the direct relationship between what those nodes receive in terms of the source packets. It is obvious that the global matrix has to have rank I at each receiving node: only in this case can we find a left-inverse $[\mathbf{G}^{(n)}]^{-1}$ (such that $[\mathbf{G}^{(n)}]^{-1}\mathbf{G}^{(n)} = \mathbf{I}$, so that packets can be decoded as

$$\mathbf{p}^{(n)} = \left[\mathbf{G}^{(n)}\right]^{-1}\mathbf{x}^{(n)}. \tag{27.32}$$

There are constructive ways in which good linear network codes can be found, based, e.g., on the "linear information flow" algorithm [Jaggi et al. 2005]. Such a construction might be useful for fixed topologies. However, in many cases, a random choice of the encoding vector is preferable, as it is easy to realize. Let η denote the number of edges at which the local encoding vectors are chosen at random. Then it can be shown that successful decoding at all T destinations is possible with probability $1 - \delta$, where

$$1 - \delta \geq 1 - \frac{\eta T}{q}. \tag{27.33}$$

This indicates that larger size of the Galois field is important for making random network coding work more efficiently. Note that the above equation is a bound; some practical simulations are shown in Figure 27.10.

Key advantages of random network coding are that it can adjust to different network configurations, that it does not require the network to operate synchronously, and does not even require the network to be acyclic. The drawbacks are the increased overhead (which stems from the need to convey the coefficients with the RXs), and possibly the requirement of a feedback from RX to TX whether packets have been received.

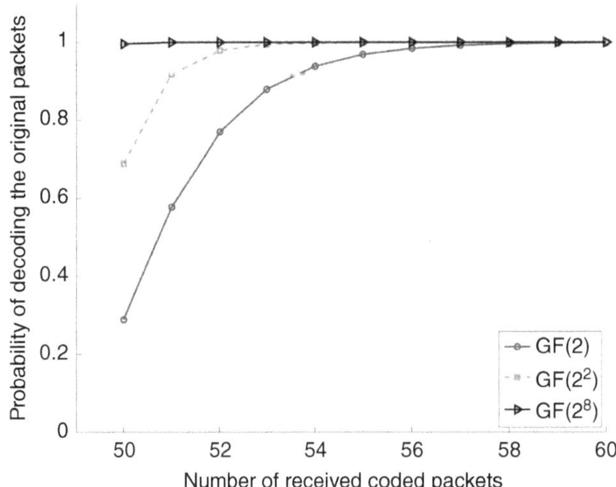

Figure 27.10 For 50 source packets, number of packets that need to be transmitted with random network coding for Galois fields of various sizes. Color version available at wiley.com/go/molisch/wireless3e.

Reproduced with permission from [Fitzek and Medard 2020] © F. Fitzek.

27.5.2 Theoretical Performance

In order to assess the performance of network coding, we first need to know the network capacity, i.e., the maximum throughput that could be achieved by *any* possible scheme. The answer to this question is given by the *max-flow min-cut theorem*. First define the meaning of a *cut*: it is a collection C of edges, such that every path from the source node to the destination node contains at least one edge that is part of the collection C. Thus, if we would eliminate the edges contained in C from the graph, then there would be no possible path from source to destination anymore. Assuming now that one packet per channel use can be sent over one edge, a *minimum cut* is then a cut with the smallest possible number of edges. It is easy to see that the transmission rate between source and destinations (in units of packets per channel use) is upper bounded by the minimum cut between source and destination; this value is the same as the maximum number of pairwise edge-disjoint paths from source to destination. In the case of multiple destinations, i.e., multi-cast, the rate is upper bounded by the smallest min-cut between source and any of the destinations (this is also termed the *multi-cast capacity*).

A key theorem of network coding states that this upper bound can be achieved by linear network coding for sufficiently large q. In other words, this theorem holds only when operations are happening in a large-size Galois field, not in a field with $q = 2$, which indicates again the usefulness of operating with a large q. However, practical investigations of network coding have shown that the penalty for using $q = 2$ is relatively small (typically on the order of 10% capacity loss).

While the above considerations were written for the case of a single source and multiple destinations, generalization to multiple sources is straightforward: introduce a virtual source from which all transmitted packets emanate, and simply connect this virtual source via parallel edges to the different actual sources. Then, both the assessment of the theoretical performance and the methods for linear network coding discussed above can be retained. This also implies that network coding is well suited for mesh networks.

Network Coding for Multi-Hop Transmission

Network coding is also useful for simple multi-hop unicast transmission: imagine that a relay node forwards packets from a source to a destination. Network coding now means that we allow that relay to create various linear combinations of the transmitted packets (this approach is known as *recoding*). This approach is useful when dealing with channels that have packet erasures due to noise, interference, etc. (remember that bit errors lead to packet erasures if the FEC of the packet is not strong enough to correct the errors). Consider now a standard relay channel in which the erasure rate of the first hop is ε_1, and of the second hop is ε_2. In this case, the total erasure rate is $1 - (1 - \varepsilon_1)(1 - \varepsilon_2)$. Now remarkably, with recoding, $1 - \min[(1 - \varepsilon_1), (1 - \varepsilon_2)]$ can be achieved. The intuitive reason behind this is that in a conventional system, if the relay does not get a useful packet from the source, it cannot forward any useful information. However, with network coding, the relay will still send out potentially useful information, namely a linear combination of the packets that have already arrived at the relay. Some of those might not have yet been decoded by the destination, so that receiving such a new linear combination helps in the overall reception. Obviously, the effectiveness of the scheme depends on I, i.e., over how many packets such linear combinations might be taken. Figure 27.11 shows simulation results from a more advanced example, where packets can be transmitted directly to the destination, as well as over two relays. Results are shown both for the case where the relays simply forward the packets, and for relays performing recoding; obviously, a significant performance improvement can be achieved with recoding. Note that given the erasure rates, the min-cut law tells us that transmission with rate 1 is possible, i.e., 64 time-slots is the minimum transmission time. The remaining inefficiencies are tied to the "startup" and ending of the transmission process and can be further reduced by suitable protocol design.

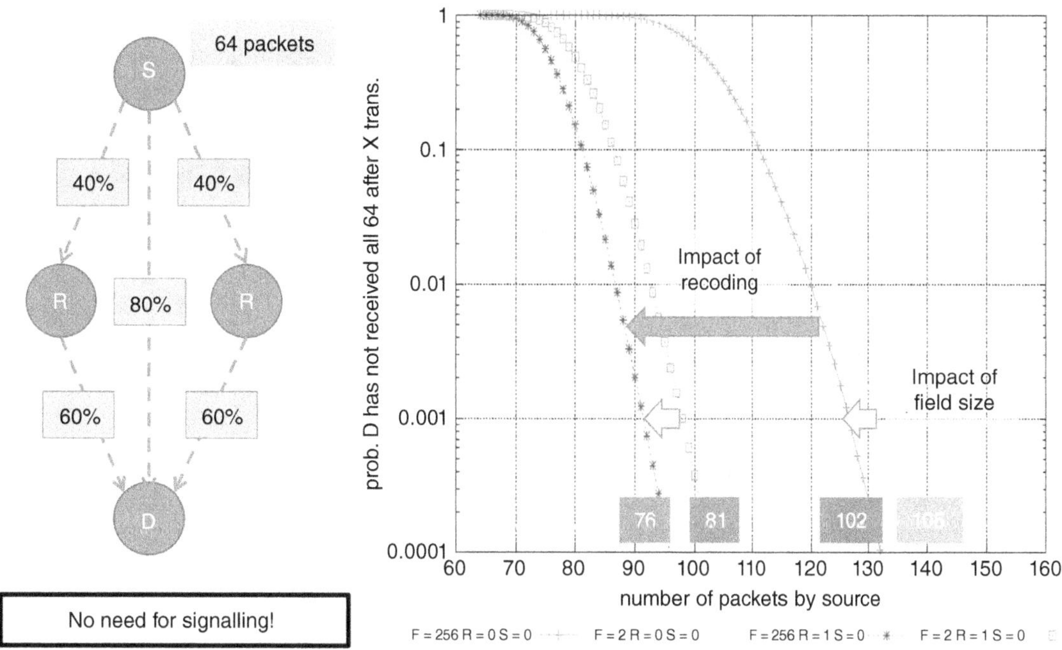

Figure 27.11 Number of required transmissions for sending 64 packets over the shown network with and without recoding. Color version available at wiley.com/go/molisch/wireless3e.
Reproduced with permission from [Fitzek and Medard 2020] © F. Fitzek.

No matter whether unicast, multi-cast, or mesh, network coding provides considerable robustness. As long as the destination nodes correctly receive a sufficient number of linearly independent combinations of packets, they can do a decoding – irrespective of which packets get lost or irreparably damaged in transit.

Practical Issues

A first practical challenge in network coding is how the RX knows how to interpret the received packets. For this, it has to know which packets have been combined in what form, i.e., it has to know the matrix **G** described above. For a network in which the structure is fixed and packets are always combined in the same way, this is just information shared at the beginning of a session between the TXs and the RXs. However, in practice, it is preferable to retain more flexibility. This is achieved by prepending the sequence of local coding vectors to the actual packets. In such an approach, the size of the header increases as the packets propagate through the network, yet for small networks the overhead is manageable.

An alternative approach is to prepend, at the source, a unit vector of size I to each packet. These headers then undergo the same network coding combinations as the payload bits; consequently, the first I columns of the received packets **Y** are the global encoding matrix **G**, which can then be used for decoding. While also this method induces some overhead, the overhead does not increase for networks with many hops. Rather, the header size is $\log_2(q)I$, which, e.g., is on the order of 100 Byte. Compared to the typical packet sizes (1 kB in many applications), this is quite acceptable.

Besides the overhead, another important challenge of network coding is the computational complexity. We have seen above that the robustness increases with I, but at the same time, the computational effort for inverting **G** goes with I^3. Various complexity reduction techniques have been suggested. For example, if the matrix **G** is sparse, well-known techniques from sparse signal processing can be brought to bear. The "1"s in the matrix can be either placed randomly or can be concentrated within a sliding window, making the decoding matrix D-diagonal, i.e., only elements offset by at most D from the diagonal can take on nonzero values. All these approaches trade-off computational efficiency with performance.

27.5.3   Network Coding With Broadcast Effect

Now we turn to the case with broadcast effect, i.e., a packet sent out by a node can be heard by multiple other nodes simultaneously. As mentioned before, this is the more typical case for wireless systems, in particular at lower frequencies and/or when the nodes are in unpredictable locations.

Two-Way Relaying

The simplest example of a wireless network code occurs in the bi-directional relay. Consider the situation sketched in Figure 27.12. There are two messages a and b, which are to be exchanged between nodes A and B via relay node R. The conventional way of relaying uses, e.g., TDMA to separate the messages, and thus requires four packet durations (time slots) for the exchange: (i) in slot 1, node A sends message a to the relay, (ii) in slot 2, node B sends message b to the relay, (iii) in slot 3, the relay sends message a to node B (though, strictly speaking, this transmission is a broadcast – in a wireless setting, a relay cannot help but send messages to multiple nodes), (iv) in slot 4, the relay sends message b to node A.[6]

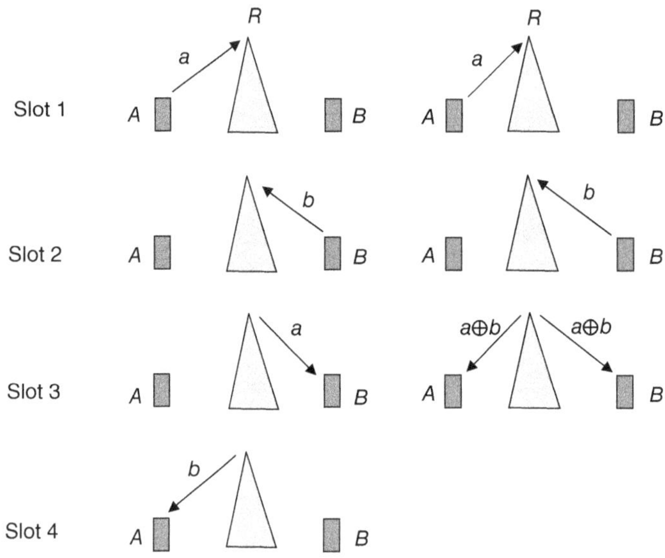

Figure 27.12   Bidirectional relaying: conventional method (left side) and network coding (right side).

A more efficient approach is the following: slots (i) and (ii) are as above, i.e., nodes A and B separately send their messages to the relay. However, in the third slot, the relay broadcasts the *sum* of the two messages, $s = a \oplus b$. Since node A already knows message a, it can easily determine message b from the sum signal. Similarly, node B can determine message a from the sum signal s. This approach thus improves the spectral efficiency of the transmission: it needs only 3 timeslots instead of 4 to relay two messages, resulting in a 33% throughput improvement. Note that the sum signal s really is again summation of the messages, not a concatenation of the two messages, and thus has the same length as, e.g., the individual message a.

This principle is not limited to the simple relay channel. Consider the cross topology in Figure 27.13. Here, nodes A and B want to exchange packets, as well as node C and D. We could apply the above method of the bidirectional relay twice, thus using 6 timeslots to

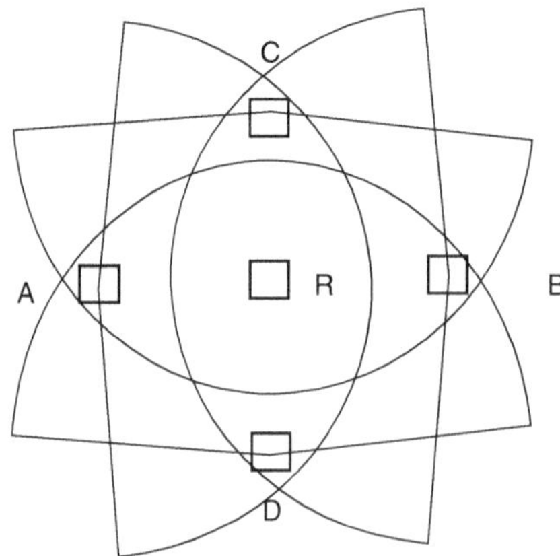

Figure 27.13   Cross topology for network coding.

---

[6] Of course, slot 2 and 3, or slot 3 and 4, can be interchanged.

transmit 4 packets. However, we can do even better, due to the fact that nodes can "overhear" packets that are not intended for them. For example, when node A transmits its packet a, not only the intended relay node R but also nodes C and D can hear this packet. Similarly, when packet b is transmitted from node B, nodes R, C, and D hear the packet. Then packet c is heard at R and also overheard at A and B, and ditto for packet D. Thus, at the end of 4 timeslots, the relay knows all packets, and each node knows all the packets *except the one it wants*. Consequently, it is sufficient that in timeslot 5 the relay broadcasts $a \oplus b \oplus c \oplus d$, from which each node can obtain its missing packet. Thus, we have exchanged four packets in five timeslots, instead of the 6 that would be required if we applied the double-directional relay method twice.

Broadcast in the Presence of Erasures

The discussion up to now ignored the presence of noise, and thus errors (and consequently, packet erasures). Now consider how we can multi-cast packets over channels that suffer from erasures. Imagine a situation where a BS uses three timeslots to broadcast three packets to five different UEs: two of them get all three packets, one of them gets packets 1 and 2, one gets 1 and 3, and one gets only 2. Clearly, we want to perform re-transmissions so that all RXs get all packets. Simple repetition coding is inefficient; rather we need to transmit properly coded combinations of packets. This problem is a special case of the general problem of *index coding*, i.e., how to code broadcasts when the RXs have different amounts of side information. Index coding has been an intensely studied problem for many years, though an exact solution of even the capacity region in the general case is unknown. Again, transmissions of linear combinations of the packets until the last node in the network has received sufficient information can be used as a practical method.

Analog Network Coding

In systems with broadcast effect, and thus interference, different sources have to transmit to the intended RX (e.g., the relay in our above example) in different timeslots (or on some other orthogonal resources); the RX then performs the creation of the linear combinations by computation in the digital domain. To improve the efficiency, we can exploit the fact that the receive antenna *inherently adds up the signals coming from different TXs*. Such an "over-the-air combination" can thus form the basis for what is called *physical-layer network coding* or *analog network coding*.

Consider again a simple example, where two nodes are transmitting Quadrature Phase Shift Keying (QPSK) symbols to the RX, called $a^{(I)} + ja^{(Q)}$ and $b^{(I)} + jb^{(Q)}$, respectively (where a, b are now taken from ± 1, not $\{0, 1\}$. Assume furthermore that the two signals arrive at the RX without any phase shifts due to their channels. Then the in-phase and quadrature-phase components of the total signal are

$$y^{(I)} = a^{(I)} + b^{(I)}; \quad y^{(Q)} = a^{(Q)} + b^{(Q)} \quad . \tag{27.34}$$

This could, in principle, be used as an appropriate linear combination, if one were to do network coding in the complex plane. This approach leads, however, to higher power values of the sum signal.

An alternative is to use the same approach as employed previously, namely adding bits (or, more generally, symbols in an F_q field) using an appropriate modulo-operation. The RX thus has to perform a suitable mapping. In our example,

$$a_{\text{out}}^{(I)} = \begin{cases} -1 & \text{if} \quad y^{(I)} = 0 \\ 1 & \text{if} \quad y^{(I)} = \pm 2 \end{cases} \tag{27.35}$$

and similarly for $a_{\text{out}}^{(Q)}$.

Note that the addition can be done directly in passband, or using downconversion and digital processing. In the former case, the TXs have to perform linear preprocessing (similar to linear precoding for distributed arrays), so that the different gains and phase shifts of the TX1-RX and TX2-RX links are compensated before transmission. As discussed in Section 16.1.7, obtaining such channel state information at the TX might require significant overhead particularly in the presence of mobility. On the other hand, combining in baseband requires to first separately estimate the two channels, and then performing a mapping similar to Eq. (27.35), but taking into account the different channel coefficients. It is also required (for RF combining) or advantageous (for baseband combining) if the two signals are synchronized to within a fraction of the duration of a modulation symbol (for an OFDM system, a small synchronization error can be absorbed into the cyclic prefix).

The mapping described above can be seen as a special example of an operation on a two-dimensional lattice. More generally, linear combination of signals combined with appropriate lattice codes, also known as *compute and forward* operation, has been shown to provide excellent performance in wireless systems.

Further Reading

General overviews of relaying and cooperative communications are found in the monograph [Liu et al. 2009] and the review papers [Hong et al. 2007, Stankovic et al. 2006].

Repeaters, which are a primitive form of relays, have been used for almost 100 years. Studies grounded in information theory started with the work of [van der Meulen 1971] that introduced the relay channel and [Cover and El Gamal 1979] which explored its information-theoretic limits in greater detail. A number of relaying protocols were introduced, and their performance analyzed in [Kramer et al. 2005], [Laneman and Wornell 2003], and [Laneman et al. 2004]. A summary of all these facts can be found in the monograph [Kramer et al. 2007]. SCxF and NDxF have more complicated forms for the capacity equations; they are discussed in [Nabar et al. 2004] (called protocols III and I, respectively). Optimum power allocations for the two-hop case with multiple relays and average CSIT only are derived for various relaying methods in [Annavajjala et al. 2007].

For relaying with multiple parallel relays, relay selection was analyzed in [Bletsas et al. 2006, 2007]. Practical aspects of distributed beamforming, including the critical synchronization issues, are discussed in [Mudumbai et al. 2009]. The impact of training overhead on beamforming is analyzed in [Madan et al. 2009]. Transmission on orthogonal channels and distributed ST coding are analyzed in detail in [Laneman and Wornell 2003]. Various forms of user cooperation were introduced in [Sendonaris et al. 2003], [Hunter et al. 2006], and [Nosratinia et al. 2004]. Fountain codes for relaying are discussed in [Molisch et al. 2007a].

Network coding was introduced in the seminal paper of [Ahlswede et al. 2000]. The primer of [Fragouli et al. 2005] gives an excellent introduction. Compute and forward was proposed in [Nazer and Gastpar 2008]. Algebraic network coding was developed in [Koetter and Médard 2003], with random network codes being proposed in [Ho et al. 2006]. Physical layer network coding is surveyed in [Liew et al. 2013]. Tutorial expositions of network coding can be found in [Médard and Sprintson 2011], [Fragouli and Soljanin 2007, 2008], and [Fitzek and Médard 2020]. The latter also provides many examples and references to software libraries for executing network coding.

For updates and errata for this chapter, see https://wides.usc.edu/students.html#textbooks.

Exercises

See Sec. 36.27 of Exercises.pdf at wiley.com/go/molisch/wireless3e

28

Advanced Interference Processing: Multi-User Detection, Nonorthogonal Multiple Access, and Interference Alignment

28.1 Introduction and Motivation

Up to now, we have considered situations where the signals to/from different users are orthogonal, or quasi-orthogonal, to each other. This orthogonalization can be imposed in time/frequency (Orthogonal Frequency Division Multiple Access (OFDMA)/Time Division Multiple Access (TDMA)/Frequency Division Multiple Access (FDMA)), space/angle (Space Division Multiple Access (SDMA)/Multi-User Multiple-Input Multiple-Output (MU-MIMO)), or code (Code Division Multiple Access (CDMA)/Frequency Hopping Multiple Access (FHMA)). However, the question arises whether one can transmit multiple signals in such a way that they overlap in all (time/frequency/code/space) domains, and still demodulate/decode them. The answer to this question is a (qualified) yes. By employing particular types of Receiver (RX) algorithms, *multi-user detection*, i.e., joint detection of the signals from several users operating in the same domains, can be achieved. The different signals may have to fulfill certain relationships to each other, e.g., different receive signal powers and/or specific data rates. In Section 28.2, we will discuss the principles of multi-user detection. We also note that capacity-achieving schemes for the multi-access and broadcast channels, which were discussed in Section 18.2, make use of some form of multi-user detection, thus indicating the importance of such schemes for maximizing data rates.

Based on the concept of multi-user detection is *Nonorthogonal Multiple Access* (NOMA). As we will discuss in Sections 28.3 and 28.4, NOMA allows to increase both throughput and fairness, though at the cost of complexity of the transceivers. We will distinguish two types of NOMA approaches, namely (i) power-based NOMA, which is usually combined with OFDMA, such that multiple users with different power levels are on the same time/frequency resources, and (ii) code-based NOMA, such that the codes assigned to the different users are partially overlapping by design, and more users than chips are incorporated in the system.

A further important idea to enhance multi-user capacity is *interference alignment*, which means manipulating interference in such a way that it can be more easily separated from the desired signal. This approach will be discussed in Section 28.5.

28.2 Multi-User Detectors

28.2.1 Basic Idea of Multi-User Detection

Multi-user detection is based on the idea of detecting the interference, and exploiting the resulting knowledge to mitigate its effect onto the desired signal. Up to 1986, it had been an established belief that interference from other users cannot be mitigated, and that in the best case, interference behaves like additive Gaussian noise. Correct detection and demodulation in the presence of strong interference were thus considered impossible, just as it is impossible to correctly detect signals in strong noise. The work of Poor and Verdu demonstrated that it is actually possible to exploit the *structure* of multi-user interference to combat its effect. Using such a strategy, interference is *less* detrimental than Gaussian noise. Multi-user detection was intensely researched during the 1990s, in particular in the context of CDMA. Principles related to multi-user detection have also been used extensively in spatial-multiplexing systems, since various MIMO receivers treat the different data streams as different "users" (compare Section 16.2), even when they come from the same UE. More recently, the use of multi-user detection in conjunction with Orthogonal Frequency Division Multiplexing (OFDM)/OFDMA systems has gathered interest in the context of implementation in 5G systems.

The conceptually simplest version of multi-user detection is *serial interference cancellation*. Consider a two-user system where the interfering signal is much stronger than the desired signal. The RX detects and demodulates first this strongest signal. This signal has a good Signal-to-Interference-and-Noise Ratio (SINR), and can thus hopefully be detected, i.e., demodulated and decoded, without errors. Its effect on the total received signal can be computed by re-encoding and remodulating the signal and filtering it by the channel impulse response. This is then subtracted (canceled) from the total received signal. The RX then detects the desired signal by demodulating/decoding the "cleaned-up" signal. As this cleaned-up signal consists only of the desired signal and the noise, the SINR is good, and detection can be done correctly. This example makes clear that detection at a Signal-to-Interference Ratio (SIR) of less than 0 dB is possible with multi-user detection.

Other multi-user structures include Maximum Likelihood Sequence Estimation (MLSE) detectors, which try to perform an optimum detection for the signals of *all* users simultaneously. These RXs show very good performance, but their complexity is usually prohibitive, as the effort increases exponentially with the number of users to be detected; approximative ML RXs such as sphere decoders (see Section 16.2.9) are possible. The performance of MLSE can also be approximated by RXs that use the turbo-principle (see Section 13.6). Serial interference cancellation is another attractive alternative that has lower complexity than MLSE and performance that (under certain circumstances) is optimum from an information-theoretic point of view.

A general classification of multi-user detection distinguishes between linear and nonlinear detectors. The former class includes the decorrelation RX and the Minimum Mean Square Error (MMSE) RX; the latter includes MLSE, interference cancellation, decision-feedback RXs, and turbo RXs.

28.2.2 Assumptions

In our basic description here, we make a number of simplifying assumptions:

- The RX has perfect knowledge of the channel from the interferer to the RX. This is obviously a best case. Consider the situation in a serial interference cancellation RX: only if the RX has perfect channel knowledge for the interfering signal can its effect be perfectly computed and subtracted from the total signal. The stronger the interference, the larger is the impact of any channel estimation error on the subtraction process.
- For ease of exposition, the examples in this section assume that all users employ CDMA as a multiple access scheme. We stress, however, that multi-user detection is also possible for other multiple access methods, as discussed in the introduction.
- All users are synchronized. This is helpful for the analysis, but might give worse performance, since absence of synchronization might lead to situations where a particular user is the only one on the air; this can serve as a "hook" to decode this user, whose later transmissions can be subtracted more easily from the total arriving signal, helping the detection of other users.
- This section only treats multi-user detection at the RX. A related topic is the design of transmission schemes so that the interference at the different RXs is mitigated. Precoding at the Transmitter (TX), such as *dirty paper coding* [Peel 2003], is discussed in Section 18.2.1.

28.2.3 Linear Multi-User Detectors

As mentioned above, we consider the example of a CDMA system with nonorthogonal spreading sequences. The block diagram of a linear multi-user RX for such a system is sketched in Figure 28.1. It first estimates the signals from the different users by de-spreading with the spreading sequences of the different users.[1] Note that this requires a number of parallel despreaders, each operating with a different spreading sequence. The outputs from those de-spreaders are then linearly combined. This combination step can be considered as a filtering with a matrix filter and is used for the suppression of interference. This approach shows a strong similarity to linear equalization for the suppression of intersymbol interference. Therefore, concepts like zero-forcing, Wiener filtering, etc., which we discussed in Chapters 14, 16, and 22, are encountered also in this context.

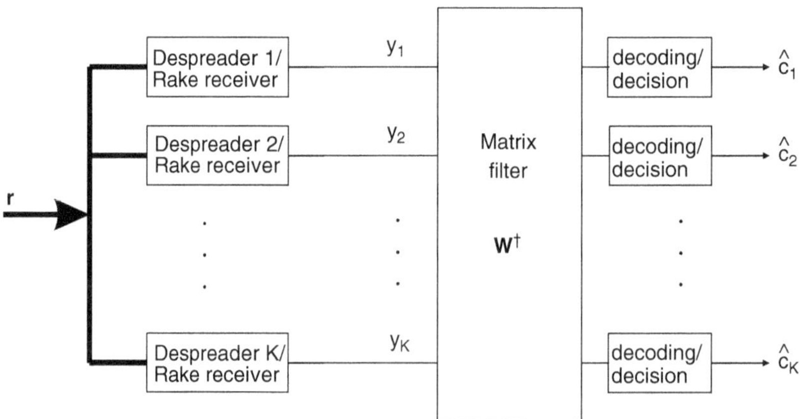

Figure 28.1 Linear multi-user detector.

---

[1] This step might also include the combination of signals from multiple Rake fingers (see Section 19.2.3) and the elements of an antenna array.

Decorrelation Receiver

The decorrelation RX is the simplest multi-user detection; it is the equivalent of a zero-forcing equalizer. We write the output of the RX filter (after despreading) as

$$\mathbf{y} = \mathbf{R}\mathbf{c} + \mathbf{n} \tag{28.1}$$

where the $K \times 1$ vector $\mathbf{c}$ contains the symbols of the K users; the correlation matrix $\mathbf{R}$ includes the effect of channel, matched filtering, and possible antenna and/or delay diversity; $\mathbf{n}$ is the noise vector with standard deviation σ_{n}. The estimation of the symbols is now obtained simply by filtering with a matrix filter $\mathbf{W}^{\dagger} = \mathbf{R}^{-1}$:

$$\hat{\mathbf{c}} = \mathbf{R}^{-1}\mathbf{y} = \mathbf{c} + \mathbf{R}^{-1}\mathbf{n}. \tag{28.2}$$

The advantage of this approach is its simplicity and the fact that it is not necessary to know the received amplitudes. Only the correlation matrix $\mathbf{R}$ needs to be determined. The drawback lies in the noise enhancement (compare, again, the zero-forcing equalizer). The worse the conditioning of the correlation matrix, the more the noise is increased.

MMSE Receiver

Just like for the MMSE equalizer, the MMSE multi-user detector strikes a balance between interference suppression and noise enhancement. A measure for the total disturbance is the mean quadratic error $E\{|\mathbf{c} - \hat{\mathbf{c}}|^2\}$. The matrix filter is thus

$$\mathbf{W}^{\dagger} = \left[\mathbf{R} + \sigma_{\mathrm{n}}^2\mathbf{I}\right]^{-1}. \tag{28.3}$$

The MMSE detector does not lead to a complete suppression of the interference, but due to its smaller noise enhancement, the signal distortions are still smaller than for the decorrelation RX.

Notes on Applicability

The linear multi-user RXs are helpful and simple concepts for improving performance in conventional CDMA systems, like those discussed in Chapter 19. However, they also have significant limitations. Most importantly, they are not suitable for OFDMA-type situations (different users on the same time/frequency resources), since the signals there cannot be separated by decorrelation – they are completely linearly dependent in the assigned resources. Furthermore, even for CDMA, the RXs cannot handle more users than chips/symbol, since that situation would lead to a system of equations that cannot be solved by matrix inversion.

28.2.4   Nonlinear Multi-user Detectors

Linear multi-user detectors ignore part of the structure of the transmit signals: they allow any continuous value for the estimate $\hat{c}$ of the transmit signal. Thus, they are ignoring the fact that the transmit signal can only contain elements of a finite transmit alphabet. Nonlinear detectors exploit also this information and are useful for the detection in OFDMA-like situations as well as for CDMA.

Successive Interference Cancellation

Successive Interference Cancellation (SIC) detects the users one after the other. The contributions of the signal of each user are subtracted from the total signal before the next user is detected, see Figure 28.2. The SIC is thus a special case of a decision-feedback RX. The sequence of detection is governed by both the received signal strengths for the different users, and the data rates; in the following examples we assume equal data rate for all users.

The detailed operation of a SIC-RX for CDMA is as follows: the sum of all signals is received, and it is despread with the different spreading codes of each user. Then, the strongest signal is detected, demodulated and decoded, so that we get the original bitstream, unaffected by noise or interference. This bitstream is then re-encoded, re-spread, and filtered by the estimated propagation channel from the TX to the RX; the thus-estimated contribution of this user signal is then subtracted from the total signal. The resulting "cleaned-up" signal is then sent through the de-spreaders again; now the strongest user within this new signal undergoes the same detection and re-spreading operation as described above. The process is repeated until the last user has been detected. Note that error propagation can seriously affect the performance of SIC: if the RX decides wrongly about one bit, it subtracts the wrong contribution from the total signal, and the residual signal, which is further processed, suffers from more, not less, interference.

There are two possibilities for subtracting the interference: "hard" and "soft." For a hard subtraction, the interference is subtracted completely; for a soft subtraction, only a scaled-down version of the signal is subtracted. This does not lead to a complete cancellation of the signal but makes error propagation less of an issue. Another aspect that impacts error propagation is whether the RX actually performs decoding/re-encoding to compute the contribution to the total signal, or only works off the "raw" (i.e., demodulated but not decoded) symbols. As the error probability after the decoder is much lower than before, decoding/re-encoding obviously reduces error propagation.

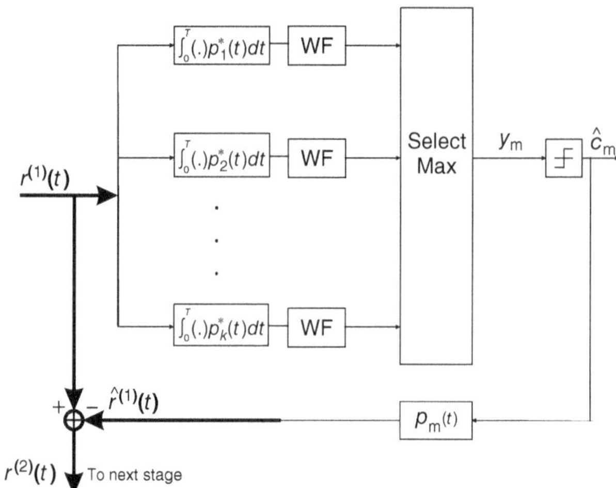

Figure 28.2   Successive interference cancellation. Received signal $\mathbf{r}$ from the K users is correlated with spreading sequence of the kth user; the largest signal is selected and its impact (i.e., the re-spread signal) is subtracted from the received signal. WF: noise-whitening filter (compare Chapter 14).

On the downside, the decoding process increases the latency of the detection process. These issues are analogous to the SIC RXs for MIMO systems, also known as BLAST, see Section 16.2.9.

Parallel Interference Cancellation

Instead of subtracting the interference in a serial (user-by-user) fashion, we can also cancel all users simultaneously. To achieve this, a first step makes a (hard or soft) decision for all users based on the total received signal. The signals are then respread, and the contributions from all interferers to the total signal are subtracted. Note that a different group of interferers is active for each user: For user 1, users 2, ..., K act as interferers; for user 2, users 1 and 3, ..., K act as interferers, and so on. The next stage of the canceller then uses the "cleaned-up" signals as a basis for a decision and again performs remodulation and subtraction. The process is repeated until the decisions do not change anymore from iteration to iteration, or until a certain number of iterations is reached, see Figure 28.3.

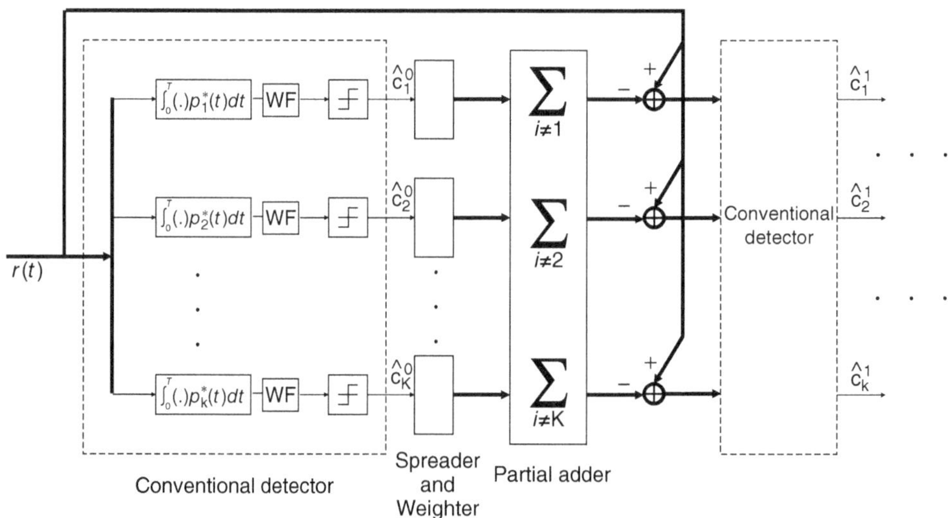

Figure 28.3   Parallel interference cancellation.

    Error propagation can be a significant problem also for parallel interference subtraction. Note that only the first stage would be necessary if all decisions there would be correct. However, each signal in the first stage has a bad signal-to-interference ratio, so that with high probability, only the strongest signal is decided correctly. Moreover, the wrong decisions distort the signal even more, so that later stages perform even worse than the first.

    One approach to mitigate those problems is the subdivision of the signals into power classes; only signals that are detected reliably are used for interference cancellation. The arriving signals are divided into several groups, depending on their SINR. For a decision in class m, we use the feedback from the classes $1, ..., m-1$, i.e., the more reliable decisions, but not those of classes $m+1, m+2, ...$

It is also possible to use partial cancellation. At each stage, the signals are sent through a mapping $x \rightarrow \tanh(\lambda x)$ (instead of a hard decision device), where the steepness of the mapping curve, i.e., λ, increases from stage to stage. Thus, only the last stages use de-facto hard decisions. The *turbo-multi-user detector* further improves on this principle by feeding back the log-likelihood ratios for the different bits, see below.

Multi-user ML Detector

The structure of the multi-user ML detector is the same as for a conventional system; if we have K users, then M^K combinations of transmit symbols can be sent out by the users, with M the size of the modulation alphabet. The number of possible states thus also increases exponentially with the number of users. For this reason, a brute-force multi-user ML detector is not used in practice.

However, there are approximate ML detectors that can achieve good performance with considerably reduced complexity. The underlying principles of those RXs fall mainly into three categories:

1. *Sphere decoders*: following the discussion of Section 16.2.9, the number of potential constellations that is to be investigated is decreased to a set of "most likely" ones – thus trading off potential decoding errors with system complexity.
2. *Belief propagation*: these structures are particularly well suited for sparse CDMA spreading codes: in other words, a symbol of a particular user is only mapped onto a sparse subset of chips. Such a mapping is similar in spirit to a Low-Density Parity Check (LDPC) code discussed in Section 13.7, and thus the decoding can use similar principles.
3. *Turbo detection*: in SIC or PIC, error propagation occurs if a wrong decision is made on a "raw" symbol. Turbo detection mitigates this problem by making not hard decisions for each symbol, but rather assigning likelihoods to each symbol, which can be iterated between the detection processes of the different users. This greatly decreases the probability of error propagation: bits that are decided wrongly usually have a considerable uncertainty to them (it is unlikely that the noise is so large that we make a wrong decision with great confidence). The turbo detector does exploit this fact, by assigning a small log-likelihood ratio. Furthermore, the process can be extended to improve the channel estimate during the iteration; this is important because multi-user detection is very sensitive to errors in the channel state information.

28.3 NOMA in the Power Domain

28.3.1 Basic Principle

We now turn to the application of multi-user detection in the power domain, which is particularly important for NOMA. The basic principle can be explained by the simple example of a cellular downlink with two users, where user 1 has a bad channel (small channel gain $|h_1|^2$) and user 2 has a good channel (large channel gain $|h_2|^2$). The BS transmits a combination

$$s = \sqrt{P_1}s_1 + \sqrt{P_2}s_2 \tag{28.4}$$

of the signals intended for the two users, s_1 and s_2, where the signal intended for user 1, i.e., the user with the bad channel, is assigned a *larger* power, $P_1 > P_2$. Now consider the decoding at the RXs: from the standpoint of RX 1, both signals go through the same channel, so that it receives $h_1 s$. RX 1 can decode s_1 as long as

$$\frac{|h_1|^2 P_1}{|h_1|^2 P_2 + \sigma_n^2} \tag{28.5}$$

is sufficiently large, where σ_n^2 is the noise power. There is no need to decode s_2 at RX 1, and it might not even be possible, since $|h_1|^2 P_2 / \sigma_n^2$ may be small.

The reception is more complicated at RX 2: it can apply SIC (described above) as follows: first considering s_2 as noise-like interference, it can decode s_1 assuming the effective SINR,

$$\frac{|h_2|^2 P_1}{|h_2|^2 P_2 + \sigma_n^2} \tag{28.6}$$

is sufficiently high for the used modulation and coding scheme; note that this condition for P_1 is less stringent than the one for decoding s_1 at RX 1. The RX then subtracts the contribution of s_1 from the total received signal, and then decodes s_2, which is the signal it is really interested in; this will be successful assuming that $|h_2|^2 P_2 / \sigma_n^2$ is sufficiently large.

From this simple example, we draw a number of important conclusions: (i) decoding of the desired signals at all intended RXs is possible, and (ii) the distribution of power at the BS to the signals of the different users is critical for the SINR, and thus rates, $\log_2(1 + SINR_k)$, that can be achieved for all users. Different trade-offs are possible. As a comparison with Section 18.2 shows (and will be elaborated below), the superposition of the signals realizes different points on the boundary of the capacity region of the channel;

in other words, NOMA is a realization of the superposition coding that forms the basis of the derivation of the capacity region. Compared to orthogonal MA, NOMA provides a larger rate region and greater flexibility.

In addition to the improved capacity, (we again use the word "capacity" loosely in this chapter, such as for achievable rate, compare the discussion in Chapter 16) NOMA also provides improved fairness, since larger power is assigned to the users with the worse channels.

28.3.2 Downlink Rate Analysis

Let us now generalize those results. First consider the downlink in a cell with K users, which are characterized by the channel gains $|h_k|^2$, and which are transmitted with power P_k from the TX, see Figure 28.4. Index the users, without loss of generality, such that $|h_1|^2 \leq |h_2|^2 \leq \cdots |h_K|^2$ and assign $P_1 \geq P_2 \geq \cdots P_K$.

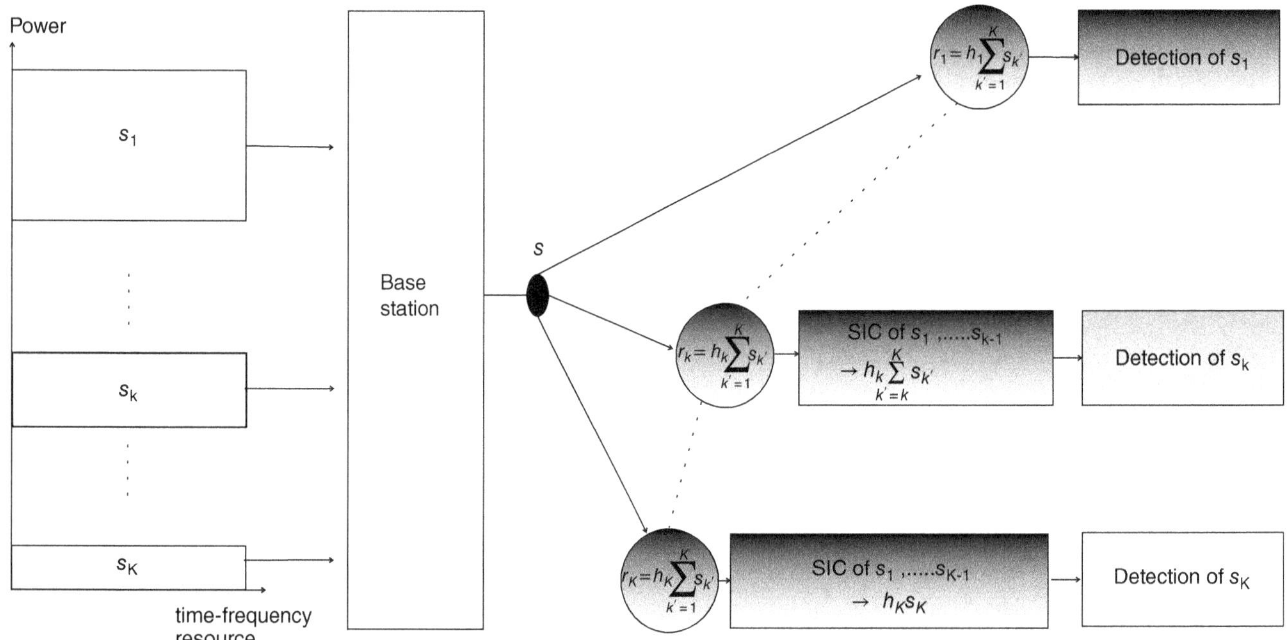

Figure 28.4 NOMA setup for the downlink.

Write $P_k = \alpha_k P_{\text{TX}}$, so that α_k is the percentage of the total transmit power assigned to the signal intended for user k, and define $\gamma_{\text{TX}} = P_{\text{TX}}/\sigma_n^2$ be the transmit SNR. The total received signal at the kth User Equipment (UE) is then

$$r_k = h_k \sqrt{P_{\text{TX}}} \sum_{k'} \sqrt{\alpha_{k'}} s_{k'} + n_k \tag{28.7}$$

and the SINR for the decoding of the kth signal at the kth UE is

$$\gamma_k = \frac{\alpha_k \gamma_{\text{TX}} |h_k|^2}{\gamma_{\text{TX}} |h_k|^2 \sum_{k'=k+1}^{K} \alpha_{k'} + 1} \tag{28.8}$$

where the sum in the denominator is 0 for $k = K$. Eq. (28.8) reflects that the signals that have lower transmit power than the considered one act as interference, while the signals with higher power ($k' < k$) are decoded and subtracted. The rates associated with γ_k are, as usual, assumed to approximate the Shannon capacity $R_k = \log_2(1 + \gamma_k)$. The sum capacity is then

$$R_{\text{sum}} = \sum_{k=1}^{K} \log_2 \left(1 + \frac{\alpha_k \gamma_{\text{TX}} |h_k|^2}{\gamma_{\text{TX}} |h_k|^2 \sum_{k'=k+1}^{K} \alpha_{k'} + 1} \right). \tag{28.9}$$

In the limit of high SNR, this becomes

$$R_{\text{sum,DL}}^{\text{NOMA}} = \sum_{k=1}^{K-1} \log_2\left(1 + \frac{\alpha_k}{\sum_{k'=k+1}^{K} \alpha_{k'}}\right) + \log_2\left(\alpha_K \gamma_{\text{TX}} |h_K|^2\right). \tag{28.10}$$

Optimization of the sum rate with respect to the α_k is then a standard optimization problem. To perform it, the BS needs the channel gains, which it can obtain either from reciprocity, or via feedback from the UEs, compare Section 16.1.7. Note that instead of the sum rate, we could also consider other utility functions, compare Section 20.1.4. It is thus possible to impose, e.g., min–max fairness or proportional fairness.

28.3.3 Uplink Rate Analysis

Consider now the uplink, as sketched in Figure 28.5.

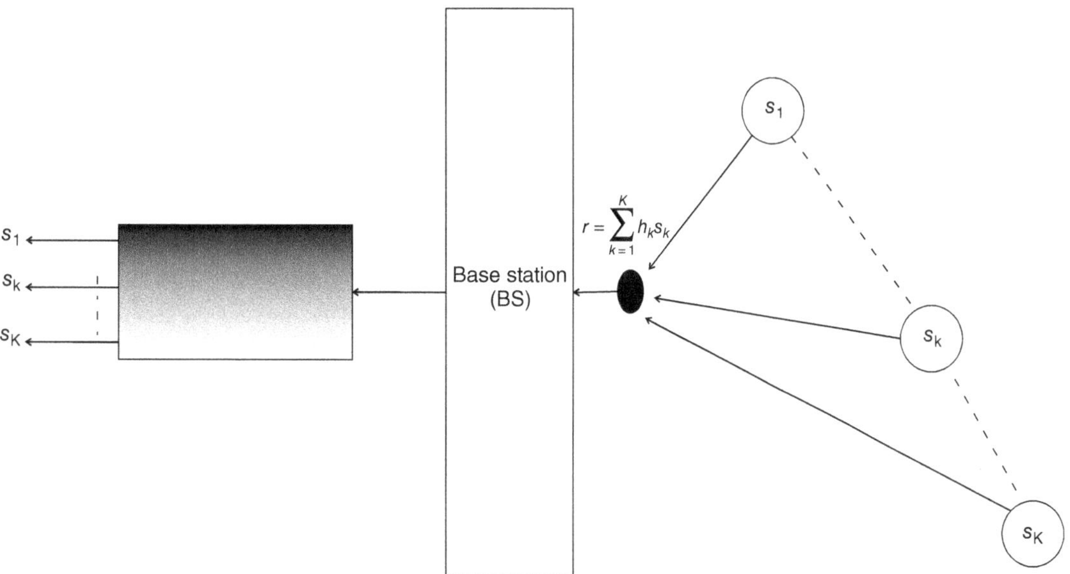

Figure 28.5 NOMA setup for the uplink.

The uplink performance is somewhat different, since the signal from every TX goes through a *different* channel. Defining now γ_{TX} as the maximum transmit SNR at each separate user, we adjust the transmit powers by factors $\alpha_k \leq 1$ in a way that allows joint detection with the desired SINR at the BS. Of course, the required α_k depend on each other, so they have to be computed centrally at the BS and sent to the UEs via control signaling. The users are now assigned a possibly different index, namely according to their decoding order (often the user with the best channel is decoded first). Then the SINRs for the different users become

$$\gamma_k = \frac{\alpha_k \gamma_{\text{TX}} |h_k|^2}{\gamma_{\text{TX}} \sum_{k'=k+1}^{K} \alpha_{k'} |h_{k'}|^2 + 1}. \tag{28.11}$$

Note the subtle but important difference that the $|h_{k'}|^2$ in the denominator is now inside the summation. The sum rate then follows as

$$R_{\text{sum}} = \sum_{k=1}^{K} \log_2\left(\frac{\alpha_k \gamma_{\text{TX}} |h_k|^2}{\gamma_{\text{TX}} \sum_{k'=k+1}^{K} \alpha_{k'} |h_{k'}|^2 + 1}\right) \tag{28.12}$$

which for the high SNR case becomes

$$R_{\text{sum,UL}}^{\text{NOMA}} = \sum_{k=1}^{K-1} \log_2\left(\frac{\alpha_k|h_k|^2}{\sum\limits_{k'=k+1}^{K} \alpha_{k'}|h_{k'}|^2}\right) + \log_2\left(\alpha_K\gamma_{\text{TX}}|h_K|^2\right). \tag{28.13}$$

For comparison, in both the uplink and the downlink, the rate achievable by orthogonal multiple access is

$$R_{\text{sum,UL}}^{\text{OMA}} = \sum_{k=1}^{K} \beta_k \log_2\left(1 + \frac{\alpha_k\gamma_{\text{TX}}|h_k|^2}{\beta_k}\right) \tag{28.14}$$

where β_k is the percentage of spectral resources assigned to the kth user. It can be shown that with suitable optimization of a, NOMA is as good as, or better than, orthogonal multiple access. For the special case of two users in the downlink, the difference in the sum rate at high SNR can be expressed nicely as $(1/2)\log_2(|h_1|^2/|h_2|^2)$, which shows that the difference in performance becomes the larger, the more different the channels are.

We finally note that NOMA can also be used in conjunction with cooperative communications. In particular, a node that needs to decode the signal of the second user anyway, can act as a Decode-and-Forward relay, and thus extend the range of coverage in a weak cell, without having to dedicate separate spectral resources for the first hop of the relay.

28.3.4 MIMO–NOMA

NOMA can be combined with multiple antenna techniques. Firstly, we can combine NOMA with single-user (SU)-MIMO. The operating principle, in this case, is relatively straightforward: for each UE, the BS sends "standard" SU-MIMO signals, which are decoded at the corresponding UEs. Signals for multiple UEs are separated in the power domain. If the number of transmitted streams is two, polarization multiplexing is a common way of ensuring that both streams intended for a single UE have comparable SNR.

An alternative for increasing sum throughput in cellular networks is MU-MIMO, as discussed in Chapter 22. The most common approach of combining it with NOMA is to divide users into clusters that are covered by different beams formed by the BS, and then apply NOMA separately in each of those beams; consequently, the approach is sometimes called *beam division NOMA*. The clustering needs to trade-off two requirements: (i) there should be as little inter-cluster interference as possible; and (ii) the users within one cluster should have particular differences in the path gain, since NOMA is exploiting those differences. The principle is outlined in Figure 28.6. Note, however, that in the case of MIMO, NOMA is not capacity achieving. How well it performs depends on the configuration and the channel. It is most useful when traditional MIMO encounters difficulties, such as when the channels to the different users are similar (as in the case of beam division), and/or in overloaded systems. A generalized method that encompasses both traditional MU-MIMO and NOMA is *rate-splitting multiple access,* see [Mao et al. 2018].

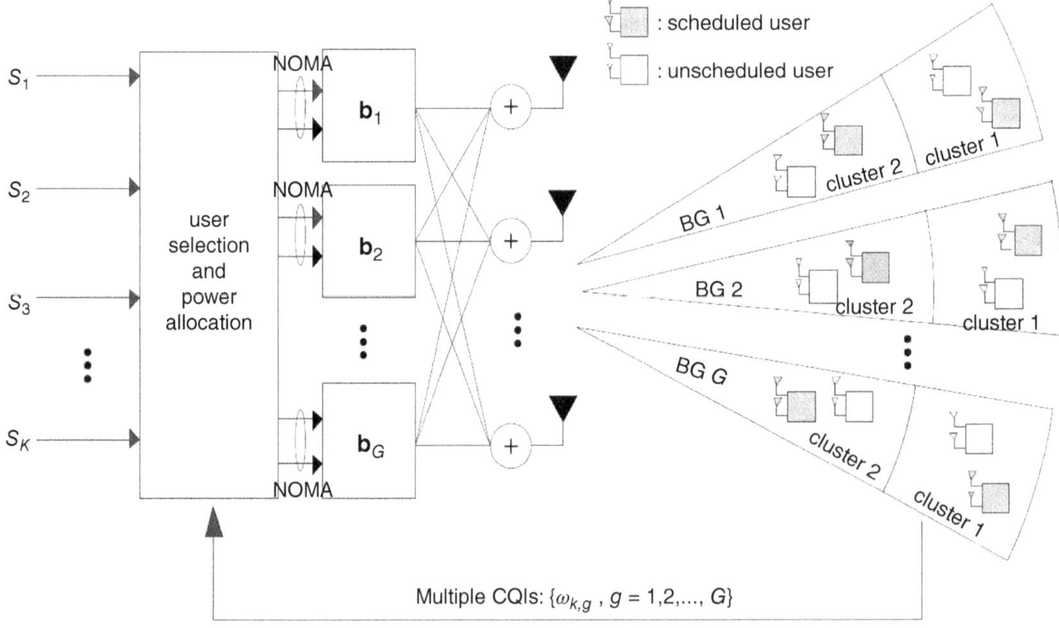

Figure 28.6 Principle of beam-division NOMA. CQI is channel quality indication.
Reproduced with permission from [Choi et al. 2017] © IEEE.

Even for a given clustering, finding the optimum (downlink) beamformers and power allocation is complicated, and usually done in an iterative manner: for a given set of beamforming vectors and total power assigned to each beam, the inter-beam interference can be computed (this interference can be treated as equivalent noise), and the computation of the optimum power allocation reduces to the standard problem of NOMA. On the other hand, once the ratios of the powers in one beam are established (and assuming noise can be neglected), we can easily determine the power $P_{\mathrm{TX},n}$ that each beam should use, considering the overall constraint $\sum_k P_{\mathrm{TX},k} \leq P_{\mathrm{BS}}$.

Figure 28.7 shows some sample results of a simulation with a 128-element uniform linear array. We can see that for small angular spreads, NOMA provides significant capacity gains compared to standard massive MIMO; however, as the angular spread increases, those gains decrease as well. This reflects the conflict mentioned above – interbeam interference starts to dominate the NOMA performance, so that separating multiple users in the power domain becomes very difficult.

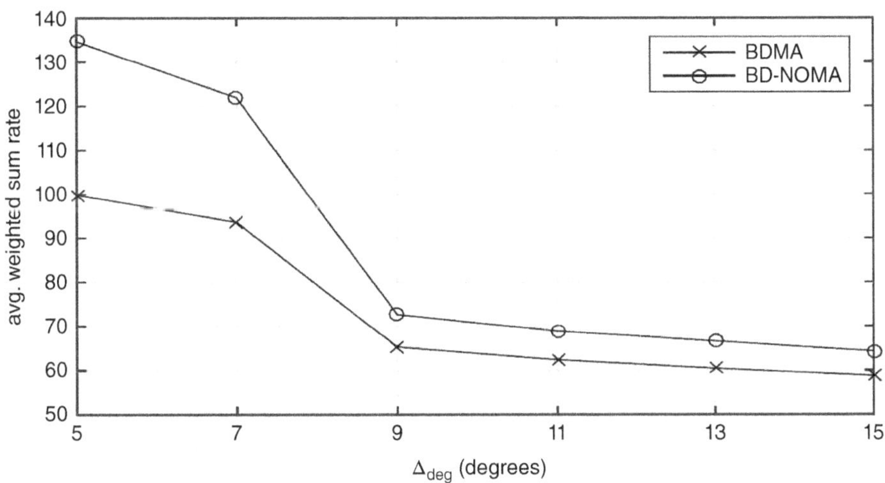

Figure 28.7 Weighted sum rate for NOMA system with 128-element uniform linear array, as a function of angular spread. Reproduced with permission from [Choi et al. 2017] © IEEE.

A further challenge is the extension of NOMA to the multi-cell case. Since more power is assigned for users near the cell edge, significant interference can be expected. This can be mitigated by Coordinated Multiple Point Transmission (CoMP) (see Section 22.11), though at the price of requiring the signals to be present at multiple BSs and making the design of the precoding vectors difficult. A possible approximate solution is to have only the cell-edge users (which suffer most from the interference) use CoMP. Consider the setup in Figure 28.8. Since U1 and U4 are near the cell edge, the signals intended for them are assigned higher power. This implies, however, strong inter-cell interference between those users, which can be turned into an advantage if the two BSs use joint precoding for those users.

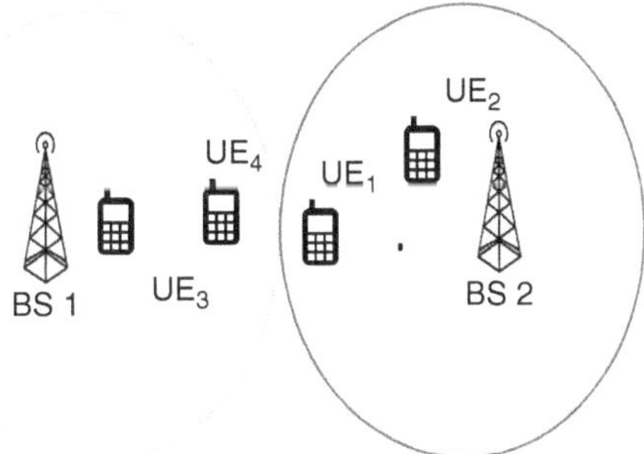

Figure 28.8 Inter-cell interference for NOMA. Color version available at wiley.com/go/molisch/wireless3e.

28.3.5 Implementation Aspects

While the above discussion has shown the advantages that NOMA can, in principle, obtain, there are also some significant challenges. Among them are:

- *Complexity*: The most obvious challenge is the implementation complexity, since with NOMA, RXs have to be capable of multi-user detection. This capability has to be realized both at the BS and at the UEs; this can be a major challenge since it has an impact on cost and energy consumption, which are sensitive issues in particular for UEs (compare Chapter 2). The complexity issue becomes particularly significant if more than two users are covered by a NOMA transmission, though this might not occur frequently in practice. For this reason, a combination of NOMA and orthogonal access, e.g., using OFDMA, with only two users performing NOMA in each orthogonal resource, is the most practical approach. Generally, the complexity can be more easily accommodated for uplink-NOMA, since the BS needs to detect all signals anyway. In the downlink, the detection of the signals not intended for a particular UE is pure overhead; together with the restrictions on computational effort at a UE, it motivates more stringent limits on NOMA signals on a particular time-frequency resource.

- *Intercell interference*: The use of NOMA might increase the overall power usage in each cell, and thus increase the inter-cell interference. Thus, the power allocation to the users should take into account the impact on the neighboring cells.
- *Channel State Information (CSI) acquisition*: As mentioned before, CSI is critical for the proper working of the interference cancellation. However, for the uplink, CSI for the different UEs has to be obtained by *orthogonal* pilots, since interference during the channel estimation phase cannot be tolerated. This leads to a significant increase in the pilot overhead; this becomes even more pronounced when NOMA is combined with MIMO. The situation is somewhat better in the downlink, since a single pilot broadcast by the BS can be used by all the UEs for estimating their channels. An additional challenge arises from the fact that the power assignment has to be cognizant of the CSI. Thus, for the downlink, the BS has to be aware of the instantaneous downlink CSI, which is nontrivial particularly in fast-varying channels (see the discussion in Section 16.1.7). Similarly, the BS has to assign the transmit powers to the UEs for their uplink transmission, based on (possibly outdated) uplink CSI the BS has from a previous pilot tone transmission of the UE. Use of the average CSI can decrease these problems but at the price of overall performance loss. For MIMO–NOMA that requires feedback from the UEs to the BS, traditional MIMO feedback techniques, such as predefined codebooks, and combinations of second-order and instantaneous CSI (compare Section 16.2), can be used to limit the feedback overhead.
- *Synchronization between the users*: In particular for the uplink, SIC is made more difficult if the received signals have offsets in time and frequency. At a minimum, those offsets have to be estimated and compensated during the SIC procedure, which increases complexity.
- *Clustering/scheduling*: We have implicitly assumed, up to now, that the users that are to be supplied with NOMA have a given location/channel state, and the system can only influence the power it can assign to it. However, an OFDMA cellular system in which there are many active users can decide which users are assigned to the same time/frequency resources, and thus are to be separated by NOMA. In the Single Input Single Output (SISO) case, the clustering of the users serves to minimize the used power while ensuring Quality of service (QoS) constraints for the far-away users and separability (in the power domain) of users on the same time/frequency resource. For beam division NOMA, it is desirable that scheduling is done in such a way that there is as little inter-beam interference as possible; yet the scheduling best fulfilling those constraints might not be the best for separating users in the power domain [Choi et al. 2017]. In either case, optimum clustering is an NP-hard problem, so that polynomial-complexity approximations have to be found. Besides the ubiquitous relaxation techniques, a mathematical method called *matching theory* also seems well suited.
- *Types of interference cancellers and encoders*: Error propagation is one of the most significant error sources for SIC. It is rooted in the fact that errors can occur in a symbol-by-symbol decoding even if the signal is transmitted at a rate that falls within the capacity region, and with a near-capacity-achieving code – i.e., errors are only eliminated by the error-correction code. The proper way of decoding is thus a codeword-level SIC. This not only greatly increases complexity, but also leads to reduced flexibility in the user assignments (e.g., the two users may have to overlap for the whole duration of the codeword). An alternative is superposition with Gray mapping, Similar to standard Gray coding (compare Chapter 10), it makes sure that two "super-symbols" (symbols encompassing the superposition of the symbols for both users in the NOMA scheme) differ by only one bit for either of the two users. Figure 28.9 shows a typical assignment, in which the bit combinations for the lower-powered user are "flipped" along the x- and y-axis to achieve a Gray constellation. Notably, the resulting NOMA scheme with symbol-level SIC can achieve a performance similar to that of codeword-level SIC without Gray coding, see Figure 28.10. Note that for proper choice of power ratios, the resulting superconstellations are still regular Quadrature Amplitude Modulation (QAM) constellations, which can thus be implemented by standard modulators.

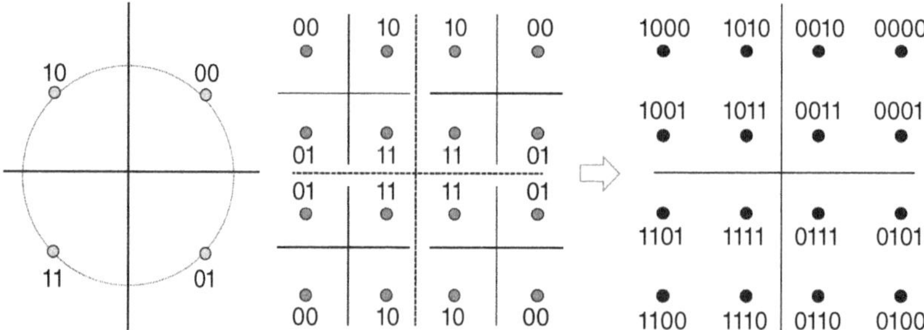

Figure 28.9 Gray encoding for two-user NOMA. Each user uses QPSK; power ratio is 4:1. Color version available at wiley.com/go/molisch/wireless3e. Reproduced with permission from [Yuan et al. 2016] © J. Wiley and Sons, Ltd.

Figure 28.11 shows a recent example of a system-level simulation. We can see that the improvement in throughput is on the order of 20–30% compared to the orthogonal case. Other investigations have shown advantages of up to 70%. At the time of this writing, only a few experimental results have shown somewhat similar improvements, but have been done only for sample locations; this area is anticipated to develop fast in the near future.

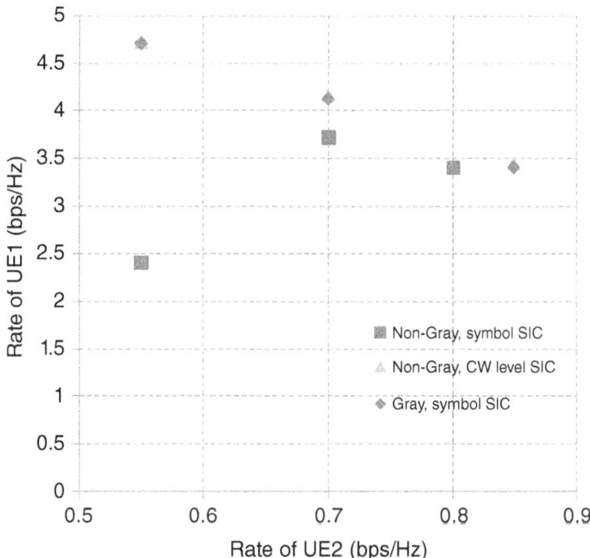

Figure 28.10   Achievable rates for two-user NOMA SNR = 20 dB for UE1, SNR = 0 dB for UE 2. Color version available at wiley.com/go/molisch/wireless3e. Reproduced with permission from [Yuan et al. 2016] © J. Wiley and Sons, Ltd.

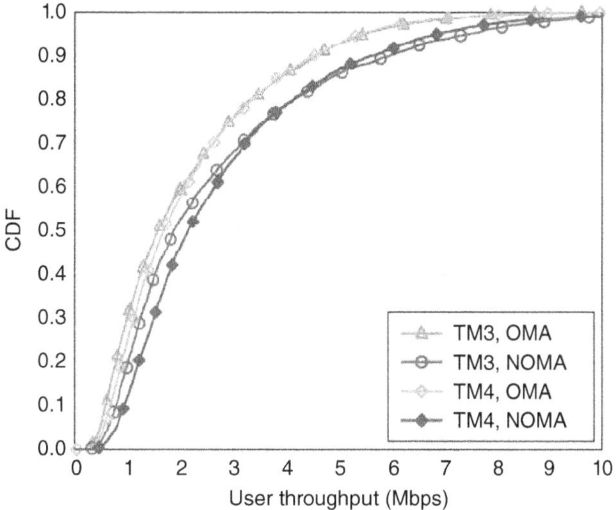

Figure 28.11   Simulations comparing NOMA with OMA with 2 × 2 MIMO; TM3: open loop; TM4: closed loop. Color version available at wiley.com/go/molisch/wireless3e. Reproduced with permission from [Benjebbour et al. 2015] © IEEE.

*28.4   NOMA in the Code Domain

Multi-user detection is naturally suited for CDMA, as outlined in Section 28.2. It is envisioned that through improvements in the spreading sequence, CDMA can become suitable for massive connectivity, i.e., handle a large number of connections at once, without excessive overhead for resource allocation; the main application for this is machine-type communications, where many devices are sending short messages to a BS. Importantly, the transmissions are intended to be *uplink* transmissions.

In order to achieve a simultaneous number of users that is higher than the spreading factor, several different designs have been developed, including: Low-density spreading (LDS) CDMA, low-density spreading OFDM, sparse code multiple access, multi-user shared access, and pattern division multiple access.

28.4.1   Low-Density Spreading (LDS) CDMA

While conventional CDMA uses spreading sequences in which the number of ones is approximately half of all the chips, LDS-CDMA uses sparse sequences (large percentage of zeros). Those sparse sequences allow the use of belief propagation for the decoding, thus improving the efficiency of a ML-type of decoding. LDS-CDMA allows overloading, i.e., having a larger number of users than there are chips in a symbol period. The design of the spreading sequences follows the same guidelines as the design of LDPC codes, such as the avoidance of cycles in the factor graph. Note that this belief propagation serves to distinguish between the users; a separate belief propagation (or turbo decoding) is required for the error correction decoding. Decoding and user separation can also be done jointly and/or iteratively, further improving performance but also increasing complexity.

28.4.2 Low-Density Spreading OFDM

LDS-OFDM relates to LDS-CDMA in the same way as multi-carrier CDMA relates to direct-sequence CDMA; in other words, the spreading with the LDS sequences is done over subcarriers instead of chips. Symbols are spread with the LDS sequences and then mapped onto the subcarriers.

28.4.3 Sparse Code Multiple Access (SCMA)

Sparse Code Multiple Access (SCMA) does not use traditional spreading codes but rather directly maps bitstreams to different sparse codewords. Codewords in the same codebook have zeros in the same place, while codewords from different codebooks are different, so that collision avoidance is made easier. An example is shown in Figure 28.12, using a loading of 150%. The key goal in the design of the codebooks is to create constellations that are close to a sphere, since this allows for the best exploitation of the signal energy. Design of these codebooks is rather involved,; the achieved gain is typically on the order of 1.5 dB, though.

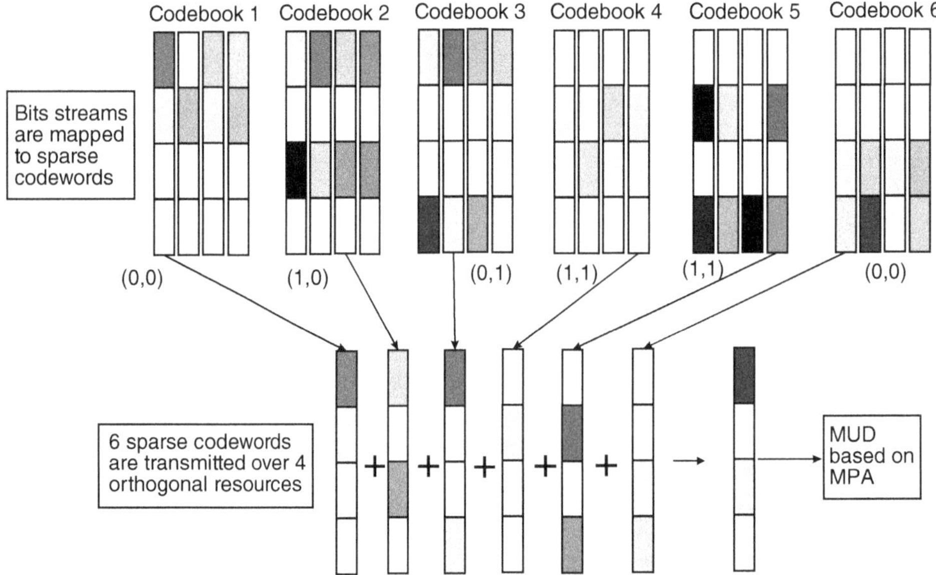

Figure 28.12 Principle of SCMA. Color version available at wiley.com/go/molisch/wireless3e. Reproduced with permission from [Dai et al. 2015] © IEEE.

28.4.4 Multi-User Shared Access (MUSA)

Multi-User Shared Access (MUSA) is, in a sense, traditional CDMA with overloading, i.e., use of multi-user detectors at the RX. Short spreading codes, which are not sparse, are used. Since the scheme needs a large set of spreading codes with low cross-correlation, *complex* spreading codes are preferable. Each user randomly picks a spreading sequence and transmits it in an unsynchronized manner. The RX needs to perform code-level SIC; the associated complexity is acceptable because the RX is the BS, where high complexity is admissible, and where all signals need to be decoded anyway. Figure 28.13 shows the performance for different user load factors and using real pn sequences and complex sequences (real and imaginary parts taken with equal probability from the set $\{-1, 0, 1\}$). It can be seen that for complex spreading sequences of length 8 and load factor of 300%, a block error rate of less than 0.01 can be achieved.

28.4.5 Pattern Division Multiple Access (PDMA)

Pattern Division Multiple Access (PDMA) aims to design spreading over different resources (time, frequency, and/or space) in a way that maximizes diversity while at the same time simplifying decoding with belief propagation methods through minimization of the overlap between the different users. It can thus be viewed as a generalization of SCMA.

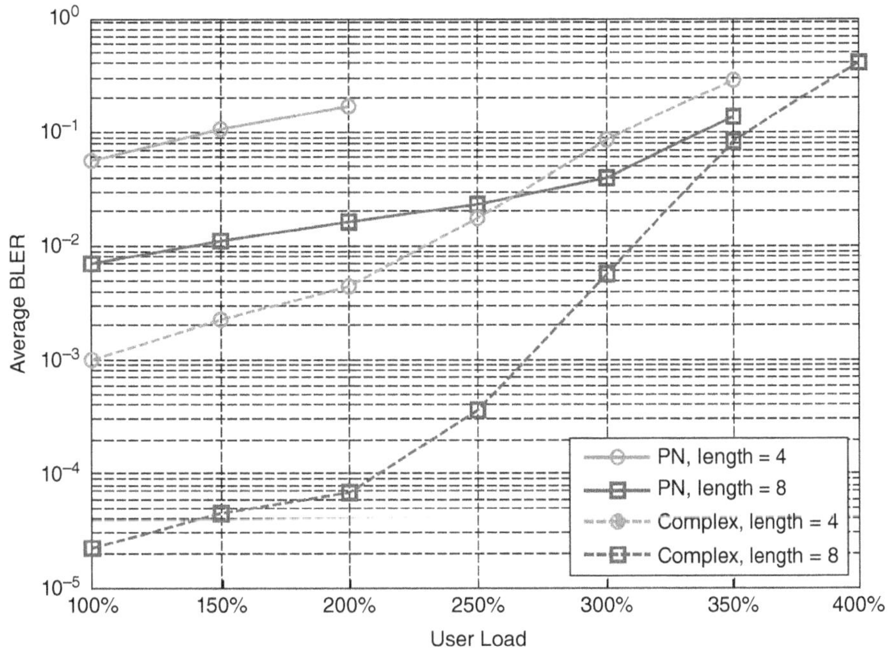

Figure 28.13 Example performance (Block Error Rate, BLER) of MUSA. Color version available at wiley.com/go/molisch/wireless3e. Reproduced with permission from [Yuan et al. 2016] © J. Wiley and Sons, Ltd.

28.5 Interference Alignment

28.5.1 Principle

Another revolutionary idea that emerged in 2008 is *interference alignment*, i.e., manipulating interference in such a way that it can be more easily separated from the desired signal. With this approach, it becomes possible to provide each (unicast) user with *half* the capacity it would obtain in an interference-free environment, irrespective of the number of users. This is a remarkable performance improvement compared to "standard" systems, where the capacity normally decreases with the number of users. Much of the work since 2010 has been dedicated to finding methods to make this great theoretical promise practically realizable. The description here draws on several examples from [Jafar 2013].

To understand the principle better, let us first review the way interference is handled in "standard" wireless systems, where K TXs wish to communicate with K RXs; assume all signals are in an M-dimensional signal space. In order to avoid interference, we usually have to orthogonalize the users, i.e., each TX sends a signal in M/K dimensions in which no other TX is sending (compare Chapter 18). The propagation channel does not change the occupied signal dimensions (e.g., timeslot, frequency band). Thus, a RX can eliminate interference by simply looking at the received signal in the M/K dimensions assigned to its desired user, and ignoring everything else. The method is effective for eliminating interference, but the capacity decreases as $1/K$, since each user gets fewer dimensions when more users are present.

Interference alignment now aims to have the signal sent from each TX occupy $M/2$ dimensions, and the received signal should occupy $M/2$ dimensions, while the interference also occupies $M/2$ dimensions, regardless of the number of users K. This means that the interference needs to be aligned in those limited dimensions. It is instructive to compare interference alignment with interference cancellation (e.g., SIC, see Section 28.2): in interference cancellation, the RX correctly decodes each of the interfering data symbols, though it has no use for them. In interference alignment, the RX can only detect some (typically linear) combination of the interfering symbols; this is not a relevant loss of information because the RX is not interested in them anyway.

A simple example is shown in Figure 28.14. In this particular channel realization, all channels for the desired user have the transfer function j, and all interfering links have transfer function 1. Thus, if each user transmits complex symbols ($M = 2$), but uses only $M/2 = 1$ dimensions by transmitting only real symbols, then each RX can determine interference-free desired symbols by discarding the real part of the desired symbol and making the decision only based on the imaginary part. Of course, this is a very simple example in which the channel is naturally suited to enable interference alignment. In the following, we will see that aligning interference can be done by either (i) manipulating the channels or (ii) manipulating the transmit symbols.[2]

---

[2] Besides the methods discussed here, there are specific interference alignment methods for wired networks. Wired networks can implement network coding in such a way that interference alignment is achieved, i.e., combine packets with suitable weights to enforce this interference alignment.

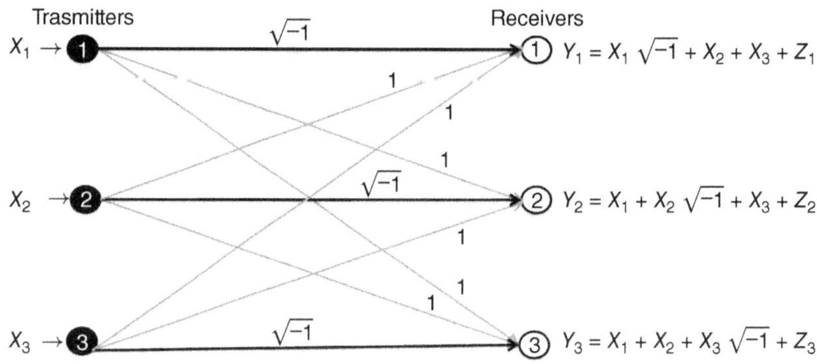

Figure 28.14 Channel in which interference alignment occurs. Color version available at wiley.com/go/molisch/wireless3e. Reproduced with permission from [Jafar 2013] © S. Jafar.

28.5.2 Signal-Space Interference Alignment with Full CSIT

Let us first consider the case of shaping the transmit symbols. For this case, we need to assume that the TX has full channel state information. To explain the principle, we again use a very simple example: two TXs and two RXs, and each of the TXs has one symbol to transmit to each of the RXs. Thus, a total of four transmit bits is to be transmitted, and we wish to transmit them with three channel uses (time instances or carrier frequencies in an OFDM system).[3] We assume that the channel is different for each channel use. Thus, let $h^t_{j,i}$ denote the channel from the ith TX to the jth RX during the tth channel use. Then let the transmitted symbol vector (i.e., stacking the three channel uses into a vector) from the first TX be

$$\mathbf{s}_1 = \begin{bmatrix} 1/h^1_{2,1} \\ 1/h^2_{2,1} \\ 1/h^3_{2,1} \end{bmatrix} a_1 + \begin{bmatrix} 1/h^1_{1,1} \\ 1/h^2_{1,1} \\ 1/h^3_{1,1} \end{bmatrix} b_1 \tag{28.15}$$

and from the second TX

$$\mathbf{s}_2 = \begin{bmatrix} 1/h^1_{2,2} \\ 1/h^2_{2,2} \\ 1/h^3_{2,2} \end{bmatrix} a_2 + \begin{bmatrix} 1/h^1_{1,2} \\ 1/h^2_{1,2} \\ 1/h^3_{1,2} \end{bmatrix} b_2 \tag{28.16}$$

where the symbols a are intended for the first RX and the symbols b for the second RX. Then the signal arriving at the first RX is

$$\mathbf{s}_1 \odot \mathbf{h}_{1,1} + \mathbf{s}_2 \odot \mathbf{h}_{1,2} = \begin{bmatrix} h^1_{1,1}/h^1_{2,1} \\ h^2_{1,1}/h^2_{2,1} \\ h^3_{1,1}/h^3_{2,1} \end{bmatrix} a_1 + \begin{bmatrix} h^1_{1,2}/h^1_{2,2} \\ h^2_{1,2}/h^2_{2,2} \\ h^3_{1,2}/h^3_{2,2} \end{bmatrix} a_2 + \begin{bmatrix} 1 \\ 1 \\ 1 \end{bmatrix} (b_1 + b_2) \tag{28.17}$$

where $\mathbf{h}_{1,\,1}$ is $\begin{bmatrix} h^1_{1,1} & h^2_{1,1} & h^3_{1,1} \end{bmatrix}^T$ (and similarly for $\mathbf{h}_{1,\,2}$), $\odot$ denotes the element-wise product. The signal at the second RX is

$$\mathbf{s}_1 \odot \mathbf{h}_{2,1} + \mathbf{s}_2 \odot \mathbf{h}_{2,2} = \begin{bmatrix} h^1_{2,1}/h^1_{1,1} \\ h^2_{2,1}/h^2_{1,1} \\ h^3_{2,1}/h^3_{1,1} \end{bmatrix} b_1 + \begin{bmatrix} h^1_{2,2}/h^1_{1,2} \\ h^2_{2,2}/h^2_{1,2} \\ h^3_{2,2}/h^3_{1,2} \end{bmatrix} b_2 + \begin{bmatrix} 1 \\ 1 \\ 1 \end{bmatrix} (a_1 + a_2) \tag{28.18}$$

Each RX now has three equations for a_1, a_2, b_1, b_2, which normally would not be sufficient to get a solution. But we see that RX 1 really has only three unknowns: a_1, a_2, and $(b_1 + b_2)$. It can thus solve for these three variables – and is actually only interested in two of them, namely a_1 and a_2, which are intended for it. The fact that it can see the b values only in a certain linear combination is irrelevant. Similarly, RX 2 can decode b_1 and b_2 from the equations it has available.

---

[3] This is slightly different from the $M/2$ dimensions mentioned above; a more extensive analysis shows that in larger-dimensional systems the $M/2$ rule is recovered.

For the general case of K single-antenna users, the problem becomes more difficult because every TX must satisfy multiple alignment conditions. For example, in the three-node case, the signal from TX 1 needs to align with the signal from TX 2 at RX 3, and with the signal from TX 3 at RX 2. The solution for this problem lies in finding a signal space $\mathbf{V}$ that is invariant to all undesired channels, but not to the desired channels; such subspaces can be found, e.g., through iterative algorithms [Cadambe and Jafar 2008].

An especially interesting application is the use of interference alignment to suppress intercell interference. In other words, assume that users within a cell are separated in the usual way (orthogonalization), while inter-cell interference is combatted by interference alignment. Assuming K users per cell, and single-antenna BSs, it can be shown that the "degrees of freedom"[4] (DoF) are

$$DoF = \frac{K}{K+1} \tag{28.19}$$

per cell, while it would be 1 if there were no interference (the latter fact follows simply from the fact that a single-antenna BS has only one DoF, i.e., can serve only one user per time/frequency unit). Thus, for a large number of users, the effect of intercell interference can be de-facto eliminated.

28.5.3 Blind Interference Alignment

The above methods assumed that every TX has CSI not only for its desired link but for interfering links as well. This is not necessarily easy to achieve in practice. It is thus interesting to consider "blind" interference alignment, by which the system shapes the propagation channel to enable alignment of the interference. The basic principle of this approach is to repeat symbols while at the same time manipulating the channel in such a way that interfering channels stay constant while desired channels change.

We again explain by means of an example, in which we send four messages over three channel uses. Consider RX1 that "sees" a channel that changes between channel use 1 and 2, and RX2 seeing a channel that is constant between channel use 1 and channel use 2, but changes from channel use 2–3 (we will discuss below how to create such channels). Then let TX1 and 2 send the following symbol vectors

$$\mathbf{s}_1 = \begin{bmatrix} x_1^1 \\ x_1^1 + x_1^2 \\ x_1^2 \end{bmatrix}, \qquad \mathbf{s}_2 = \begin{bmatrix} x_2^1 \\ x_2^1 + x_2^2 \\ x_2^2 \end{bmatrix} \tag{28.20}$$

where x_i^j is a symbol sent from the ith TX intended for the jth RX. Let the channel from the ith TX to the jth RX at the tth channel use again be written as $h_{j,i}(t)$, where we note that $h_{1,i}(2) = h_{1,i}(3)$ and $h_{2,i}(1) = h_{2,i}(2)$. The total received signal at RX 1 can then be written as

$$\mathbf{r}_1 = \begin{pmatrix} h_{1,1}(1)x_1^1 + h_{1,2}(1)x_2^1 \\ h_{1,1}(2)x_1^1 + h_{1,1}(2)x_1^2 + h_{1,2}(2)x_2^1 + h_{1,2}(2)x_2^2 \\ h_{1,1}(2)x_1^2 + h_{1,2}(2)x_2^2 \end{pmatrix}. \tag{28.21}$$

It can be easily seen that interfering signals (i.e., signals intended for RX 2, with the notation x_i^2) occur in the same linear combination, namely $h_{1,1}(2)x_1^2 + h_{1,2}(2)x_2^2$, in the signal during the second and third channel use. Thus, the interference can be canceled simply by subtracting the signal during the third channel use from the one in the second channel use. This leaves us with two equations for the two unknown variables x_1^1 and x_2^1. Note that those two equations are independent because $h_{1,i}(1) \neq h_{1,i}(2)$. Very similar equations can be derived for the received signal at RX2.

Now let us turn to the question how we can create channels that change in the desired manner. The most obvious choice is antenna selection: let each RX have two antenna elements, but only one RF chain. Then it can only receive one symbol at a time; however, by switching the antennas at specific times (after the first channel use for RX1 and the second channel use for RX2), the desired channel characteristics can be realized. Note that this is a method in which antenna selection can improve the degrees of freedom (i.e., number of data streams on the air), and not just improve the SNR and thus data rate for a given stream, as we assumed in Chapter 12 and Section 16.1). Other possibilities for channels staying constant over different intervals may occur if the coherence time and bandwidth of channels are different at different RXs, but it may be very difficult to "design" a channel in this manner, as it depends on the physical environment characteristics.

The degrees of freedom that can be achieved with this scheme depending on the number of TXs and RXs. With M TXs and N RXs,

$$DoF = \frac{MN}{M+N-1} \tag{28.22}$$

---

[4] DoF characterizes the capacity in the high SNR regime. Specifically, the capacity is proportional to $DoF \log_2 (SNR)$.

can be achieved. Remember that this is achieved without any TX cooperation or CSIT. It is remarkable that it is not possible to improve on those degrees of freedom by either (i) allowing TX cooperation but no CSIT, or (ii) CSIT, but no TX cooperation. However, it is possible to achieve higher DoFs when both TX cooperation and CSIT are allowed.

28.5.4 Ergodic Interference Alignment

The simplest, and intuitively clearest, interference alignment exploits time variations of the wireless propagation channel: assume $M=N$, and characterize the channels between the M TXs and the M RXs with a $M \times M$ matrix (transfer function matrix, analogous to a MIMO system):

$$
\begin{pmatrix}
h_{11} & h_{12} & h_{13}. & . & . & h_{1M} \\
h_{21} & h_{22} & h_{23} & . & . & h_{2,M} \\
. & . & . & . & . & . \\
h_{M1} & h_{M2} & & & & h_{MM}
\end{pmatrix}.
\tag{28.23}
$$

The signals transmitted when the channel has one particular realization are repeated when the channel is in its *complementary* state

$$
\begin{pmatrix}
h_{11} & -h_{12} & -h_{13} & . & . & -h_{1M} \\
-h_{21} & h_{22} & -h_{23} & . & . & -h_{2M} \\
. & . & . & . & . & . \\
-h_{M1} & -h_{M2} & & & & h_{MM}
\end{pmatrix}.
\tag{28.24}
$$

The RXs then just have to add up the signals from those two transmissions, to get signals that are effectively transmitted over a diagonal (i.e., interference-free) channel

$$
2\begin{pmatrix}
h_{11} & 0 & 0 & . & . & 0 \\
0 & h_{22} & 0 & . & . & 0 \\
. & . & . & . & . & . \\
0 & 0 & & & & h_{MM}
\end{pmatrix}.
\tag{28.25}
$$

The signal for each user is thus interference-free (allowing transmission with capacity of the noise-only channel). Due to the required signal repetition, DoF is decreased by a factor 2.

As mentioned above, the concept can be implemented with very little hardware effort. However, it can lead to large latency in the signal transmission, because the TX has to wait until the channel takes on the complementary state. The slower the temporal channel variations, the longer that delay can be.

*28.5.5 Interference Alignment in Static Channels

Interference alignment in static channels is practically very important (many communications systems use nomadic TXs and RXs); however, the above-described methods require changes in the channel. A method to circumvent this problem is interference alignment in the amplitude (power) domain. The interference alignment then tries to ensure that interference aligns at the same signal levels.

To explain the concept in more detail, consider a deterministic discrete channel model in Figure 28.15. Assume, for simplicity, real-valued inputs, and the signal is interpreted as occupying a succession of levels. The example of Figure 28.15 represents the transmit signal as bits b_1 to b_5, where b_1 corresponds to the most significant bit, and is transmitted with the highest energy. The number of bits that can be resolved in the presence of noise is $\log_2(SNR)$. When the signal is transmitted over the channel, the SNR is reduced, so that lower-level bits "vanish in the noise" (see the right side of Figure 28.15). The stronger the attenuation, the fewer bits can be transmitted in one channel use, i.e., the lower the channel capacity.

Consider now the many-to-one interference channel in Figure 28.16. This is a special type of channel in which all TXs interfere with link from TX1 to RX1, but the transmission from TX1 does not interfere with other RXs (such a channel is physically feasible, though not necessarily of great practical importance. But it serves well to explain the principle of signal-level alignment). The right-hand figure shows how the contributions from the different interferers add up at the different signal levels of RX0.[5] The RX can only

---

[5] Note that in order to simplify the explanations, all additions of signal contributions are modulo 2. Thus, two contributions coinciding on level 0 add up to a contribution of level 0, and so on.

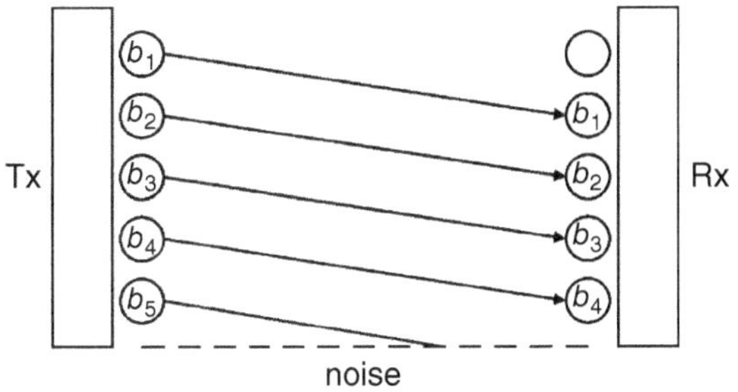

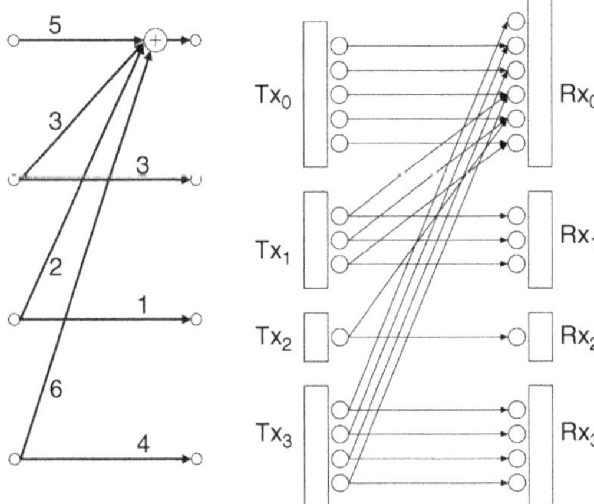

Figure 28.15 Deterministic channel model for an AWGN channel. The dashed line corresponds to the noise level.
Reproduced with permission from [Bresler et al. 2010] © IEEE.

Figure 28.16 Many-to-one interference channel, and its deterministic channel representation.
Reproduced with permission from [Bresler et al. 2010] © IEEE.

decode bits on whose levels there is no incident interference; however, it does not matter whether one or more interferers contribute interference on this level. For this reason, it is best if each TX sends bits in such a manner that all their signals are arriving on a particular level (or groups of levels). Consider signal level 0 of TX0: when it transmits, level 0 of TX1 has to be quiet; the other TXs have no impact. Thus, level 0 of TX0 and level 0 of TX1 have to transmit at orthogonal times (i.e., time-share). Level 1 of TX0 has to time-share with level 1 of TX1 and level 0 of TX2, and so on. All these constraints together provide the limits on the overall capacity.

An actual realization of the above-described scheme requires the use of nested lattice codes, which are a useful but rather complicated class of codes; the interested reader is referred to the references in the "further reading" section.

*28.5.6 Interference Alignment in Partially Connected Networks

Up to now, we have implicitly assumed that every TX is connected to every RX with approximately the same channel gain. This is clearly not fulfilled in practice, as the different distances between nodes (as well as the fading) will lead to vast variations of the channel gains, and thus the received powers. As a matter of fact, the interference from most nodes in a wireless network will be rather weak. Since the complexity of interference alignment increases dramatically with the number of interferers, it is important to determine which interferers should be eliminated by interference alignment, and which should be ignored. The answer to this question is discussed in Section 23.5.3.

Once we have eliminated the "sufficiently weak" links, the network is characterized to a large degree by its connectivity graph, i.e., which node can talk to which other nodes; this is fundamentally an interference alignment problem. Consider the network in Figure 28.17. With traditional orthogonalization (scheduling users that interfere with each other in different timeslots), four messages can be transmitted in two timeslots, resulting in DoF = 2. However, with interference alignment, better performance can be achieved. As shown in Figure 28.18, five information symbols can be transmitted over a period of two timeslots. The "trick" is to schedule transmission from TX2 and TX5 simultaneously in timeslot 1, and then cancel the interference through transmission from TX2 in timeslot 2.

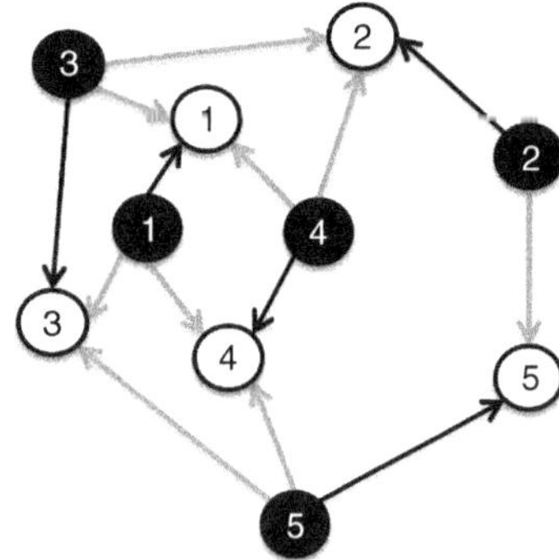

Figure 28.17 Partially connected network. Black circles: TXs. White circles: RXs. Black arrows: desired connections. Red arrows: interference. Color version available at wiley.com/go/molisch/wireless3e.
Reproduced with permission from [Jafar 2013] © S. Jafar.

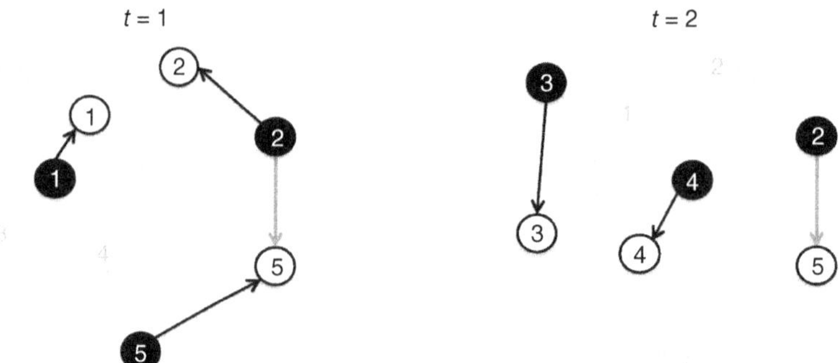

Figure 28.18 Scheduling in partially connected network. Two channel uses (timeslots) are required for five data transmissions. Color version available at wiley.com/go/molisch/wireless3e.
Reproduced with permission from [Jafar 2013] © S. Jafar.

For the general case, the optimum scheduling for achieving such interference alignment is still an unsolved problem at the time of this writing. It has been shown that this problem is strongly related to the index-coding problem, which has only admitted partial solutions and solutions for some special cases up to now.

Further Reading

For multi-user detection, the book [Verdu 1998] is the standard reference. Also the overview papers [Duel-Hallen et al. 1995], [Moshavi 1996], and [Poor 2001] give useful insights. Advanced concepts like blind multi-user detection are described extensively in [Wang and Poor 2003] and [Honig 2009]; the interaction between multi-user detection and decoding is explored in [Poor 2004].

A widely cited paper for NOMA is [Saito et al. 2013]. A combination of beam division and NOMA was explored in [Choi et al. 2017]. Experimental results can be found in [Benjebbour et al. 2015]. Overall, a wide variety of NOMA schemes have been established. A good starting point is the survey papers/chapters of [Makki et al. 2020]. [Vaezi et al. 2019], [Dai et al. 2018], [Islam et al. 2016], [Benjebbour et al. 2016] and [Yuan 2016].

Interference alignment was introduced in the seminal paper of [Cadambe and Jafar 2008]. Ergodic interference alignment is described in [Nazer et al. 2012]. Blind interference alignment with antenna switching is described in [Gou et al. 2011]. Signal-level interference alignment is discussed in [Bresler et al. 2010]. A comprehensive overview can be found in the book [Jafar 2011], which also has constant updates on the website of S. Jafar at UCI. An excellent tutorial is also [Jafar 2013].

For updates and errata for this chapter, see https://wides.usc.edu/students.html#textbooks.

Exercises

See Sec. 36.28 of Exercises.pdf at wiley.com/go/molisch/wireless3e

29

Localization

29.1   Introduction and Motivation

The ability to accurately determine the location of a wireless device has changed from a niche application that was mostly relevant for military devices, to one of the dominant applications of wireless technology. Accurate location knowledge opens the market for a host of new applications. In this chapter, we review the main technologies that are employed, emphasizing *active localization*, where the device that is to be localized is participating in the process of localization. The related issue of *passive localization*, where the object of interest is illuminated by an external device, but not doing anything itself, is strongly related to the classical radar problem, and we refer the interested reader to the vast literature on this topic (for a few samples, see the "Further Reading" section). Here we will discuss fundamental localization techniques, categorized as time-of-arrival based, direction-based, and field-strength-based (Sections 29.2–29.5), localization systems such as satellite, cellular, and Radio Frequency Identification (RFID) (Sections 29.6–29.8), and advanced techniques such as cooperative localization, tracking, and Machine-Learning (ML)-based methods (Sections 29.9–29.11).

Localization was originally driven by military applications. It is obviously of great advantage for armed forces if every soldier knows their position to a high degree of accuracy, as this allows better coordination of forces, reduces the risk of friendly fire, etc. The US military thus decided to establish a satellite-based system called *Global Positioning System* (GPS), which became operational in the 1990s. It was decided at this point to also make GPS available for civilian users, a step that has had enormous impact on localization systems and created completely new market categories. GPS can be used for essentially any application that requires precise positioning *outdoors*; however, it has limited effectiveness in indoor environments. It is also useful for providing high-precision timing information. Other countries built similar systems later on, see Sec. 29.6.5.

In the early years, the main commercial applications of (satellite-based) localization were *vehicle navigation systems*; they quickly became ubiquitous in cars, either built-in by the car manufacturer (usually in high-end cars), or as after-market products, and were widely used by the mid-2000s. As the price for GPS Receivers (RXs) decreased due to advances in technology, GPS units were built into smartphones, which quickly became the dominant devices for satellite navigation. Nowadays, most people rely on GPS route guidance for driving to new destinations, saving significant amounts of time, frustration, and gas as they are not getting lost on their way to the destination anymore. Furthermore, improved route guidance systems obtain traffic information (either historic, real-time, or a combination of both) for various possible routes to a destination and make recommendations, thus often alleviating traffic congestion. GPS is also used for train control systems (e.g., ensuring that trains do not go faster than a certain track allows, that they are actually on the correct track, etc.). Similarly, fleet management systems for trucking companies either use GPS or other proprietary systems. This improves efficiency, reduces risk of highjacking, etc.

A second major application is *emergency localization of cellphone users*, originally motivated by requirements from the US frequency regulator, the Federal Communications Commission (FCC). As more and more people started to use cellphones, an increasing percentage of emergency calls to police, ambulance, and firefighters (collectively known as "911" calls), However, no possibility existed in the 1990s to automatically localize a phone from which such a call was made. The FCC thus mandated that cellular operators should be able to localize a certain percentage of emergency calls made from outdoors locations with a given accuracy: 67% within 150 m, and 90% within 300 m. Heavy fines were foreseen for operators that did not meet the targets by 2004 (the date was actually shifted several times); other countries including the European Union later introduced similar requirements. The mandates were made more stringent over time, and requirements for indoor localization, including vertical localization (i.e., which floor a caller is on) were added. These mandates motivated a lot of research in cellular-based localization, which will be discussed in Section 29.7. Note that in particular in the early days, GPS was not seen as a suitable method for localization for two reasons: (i) few phones contained GPS RXs in those days, as they were too costly and battery-hungry, and (ii) GPS was not able to localize indoor callers. While the former reason is mostly gone, and the latter has been mitigated, GPS still cannot be used as sole method of localization.

A third category of applications is so-called *location-based services*. Many of the functionalities of modern smartphones can be improved if they are operated in a suitable geographical context, i.e., the location at which the smartphone is located. Take as an example a search for a restaurant. In all likelihood, the person performing the search is not interested in restaurants across town from the current location, but rather in a place close by; thus the search results and recommendations can be prioritized based on the current location. Obviously, a prerequisite is for the phone to actually know its current location and communicate that to the search engine. From a commercial point of view, these location-based services are the most important application. A number of companies have thus invested large amounts of money into localization and the associated services. Creation of digital maps has itself been transformed into a multi-billion dollar business, and leading smartphone companies are constantly working on improving their digital maps so as to provide better services based on them.

Location-based services also provide a strong motivation for indoor localization. As mentioned above, satellite systems are not capable of reliable indoor localization, but alternative methods such as Received Signal Strength Indication (RSSI)-based localization or ranging to/from cellular Base Stations (BSs) or Wi-Fi access points (discussed in Section 29.7) can be used instead. The applications for which such indoor localization is used are partly the same as for outdoor (e.g., determine the location to ensure more relevant search results), while other applications are unique to indoor applications (e.g., guiding a user to a certain product in a store or warehouse). Of particular importance for indoor localization is to find not only the horizontal coordinates but also the floor on which a device is located. This can be relatively easy when the location is obtained by measurements with respect to other indoor nodes (such as signal strength from Wi-Fi access points) but is very challenging when done by measurements with respect to outdoor nodes such as cellular BSs. Also, the requirements for localization accuracy are usually much more stringent for indoor localization: e.g., one wishes to localize the specific room or office in which a device is located, while for outdoor localization it is often sufficient to localize the correct street block.

A particular form of (usually) indoor localization is RFID. In this service, the location of a tag is to be determined by measurements from dedicated readers. These tags are attached to merchandise and other products, and RFID serves for inventory management, localizing merchandise when requested by a customer, and theft prevention. As the tags often cannot be reused or are damaged during usage, they have to be very cheap. For this reason, tags are usually passive (without their own power source), though active tags have been considered for specialty applications. Large warehouse management systems are strongly dependent on RFID.

Last but not least, localization greatly helps in the setup and administration of wireless networks, i.e., the location and trajectory of wireless nodes is a side information that a network can use to improve network efficiency. For example, the relative location between wireless nodes is a good indicator of their pathloss; thus location information indicates which node might best act as relay for another node, and which node might create strong interference. The trajectory of a node indicates that it might soon become ready for a handover; "breathing cells" (see Section 21.3) can adjust their cell radius based on geographical information, and so on.

The above description covers only some of the most important applications of localization. As a matter of fact, the field of localization has obtained its own momentum, and a huge variety of applications is constantly emerging. When it comes to localization *methods*, we find that they are usually based on the following quantities: (i) runtime of the signal (also known as *Time Of Arrival* (TOA)) – this is the most important and widely used method, (ii) received signal strength, (iii) direction of arrival. Note also that the information from these electromagnetic measurements can be combined with measurements from other sensors such as *Inertial Measurement Units* (IMUs), for tracking the distance during movement, compass, and gyroscope to track orientation of the devices, barometer for height, and so on. Such hybrid localization methods are widely used in practice as they provide greater reliability, and one type of sensor might be able to cover "gaps" or inaccuracies arising in the measurements of another sensor in a particular location.

29.2   Principles of TOA/TDOA

One of the most important forms of localization is based on measuring the time it takes a signal to traverse the distance from the Transmitter (TX) to the RX. This can be achieved by transmitting a signal at a predetermined time and measuring the TOA at the RX.

Before proceeding further, let us establish some notation. We will in the following call a wireless node whose position is precisely known as *anchor*, and a node whose position is to be determined as *agent*. The problem of determining the distance between anchor and agent is known as *ranging*. When the ranges of an agent to three or more anchors are known, the (two-dimensional) location of the agent can be determined, in what constitutes the actual localization. Corresponding to these two steps, we also find that the errors arise in two forms: (i) errors in the determination of the range between anchor and agent, and (ii) *Geometric Dilution Of Precision* (GDOP), i.e., depending on the location of the agent with respect to the set of anchors, even small ranging errors can translate into large location errors in certain directions.

29.2.1   Determination of the Runtime

In this subsection, we will describe how the runtime, i.e., the time it takes the signal to propagate between TX and RX, can be determined, and what physical effects induce ranging errors. Consider first the case that TX and RX are both time- and frequency synchronized, and that signals are sent out at specific times that are known to all nodes. Furthermore, assume for now that there is no multi-path propagation, and only the Line-Of-Sight (LOS) component can reach the RX. In the absence of noise or interference, the RX can determine the arrival time of the signal, and thus the runtime and range, precisely and no errors are incurred.

However, in practice there are a number of effects that can lead to erroneous measurement of the arrival time:

- noise in the received signal;
- the LOS component propagates through dielectric material and therefore suffers longer delay;
- multi-path propagation and/or suppression of the LOS component by blocking objects;
- interference; and
- errors in time and/or frequency synchronization.

Figure 29.1 outlines these error sources. In the following, we will discuss them one by one.

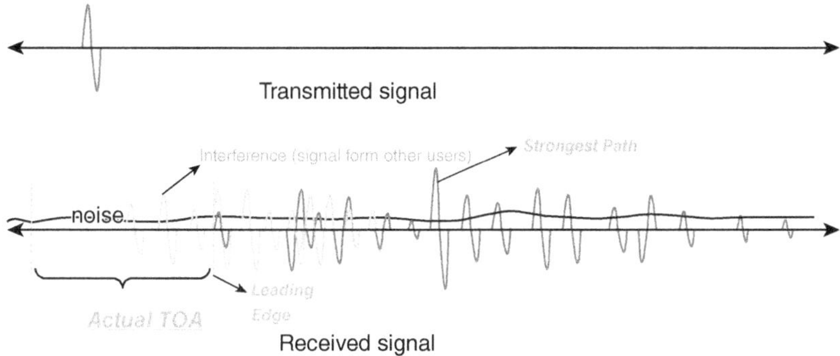

Figure 29.1 Various types of error sources in ranging measurements. Color version available at wiley.com/go/molisch/wireless3e.

TOA Estimation in the Presence of Noise

First, consider the case that there is noise but no other error sources for the TOA estimation, i.e., estimation of the arrival time in an Additive White Gaussian Noise (AWGN) channel. Two types of RX are commonly used (similar to communication RXs): matched filters (correlation receivers), and energy detectors. The former provide the better performance but require knowledge of the pulse shape and frequency synchronization. Essentially, this RX determines the peak of the cross-correlation of a local template with the arriving signal. In the absence of noise, this peak occurs at the (nominal) arrival time of the signal; noise can disturb the received signal and shift the peak from its ideal location. Figure 29.2 shows the shape of the autocorrelation function for narrowband and wideband signals.

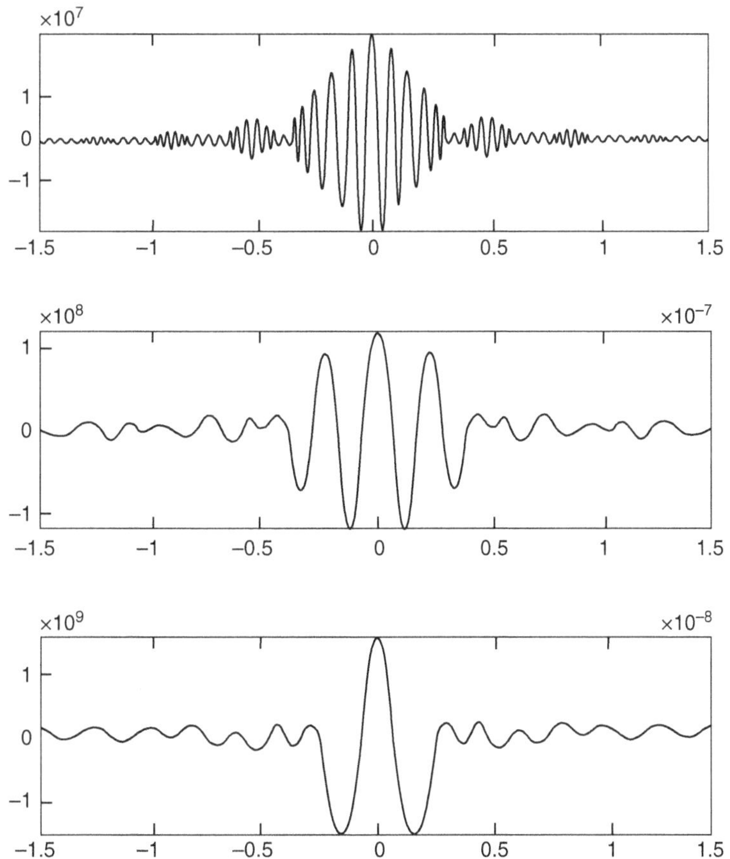

Figure 29.2 Autocorrelation for signals with (i) $f_c = 0.4$ GHz, $B = 60$ MHz, (ii) $f_c = 0.9$ GHz, $B = 400$ MHz, (iii) $f_c = 7$ GHz, $B = 6$ GHz. The x-axis is time in seconds; note the different scales of the x-axis.

For a narrowband signal, the central correlation peak is very sharp (determined by the carrier frequency), but a number of almost equally large sidelobes exist; the decay of the sidelobes is determined by the bandwidth (and also the used waveform). For very high Signal-to-Noise Ratio (SNR), the variance of the signal amplitude due to noise is much smaller than the difference in the height between the central correlation peak and the neighboring peaks. In that case, it is possible to determine the TOA with a precision that is essentially determined by the carrier frequency. However, at intermediate SNRs, which are most relevant in practice, the accuracy is determined by the probability of confusing one of the neighboring peaks with the central peak; this probability is determined by the bandwidth, which governs the amplitude ratio between the different peaks, and the SNR.

We now determine the limit of what can be achieved with the best-unbiased estimator, also known as *Cramer–Rao Lower Bound* (CRLB). As we will discuss in more detail in Section 29.2.3, this bound can be derived from the Fisher Information Matrix (FIM). For the moment, suffice it to say that the variance of the estimation error can be lower bounded by

$$var\left(\hat{d}\right) \geq \frac{c_0^2}{8\pi^2\gamma\beta^2} \tag{29.1}$$

where γ is the SNR and β is the effective bandwidth

$$\beta = \left[\frac{\int f^2|S(f)|^2 df}{\int |S(f)|^2 df}\right]^{1/2} \tag{29.2}$$

with $S(f)$ the Fourier transform of the TX signal; we can see that β of a narrowband signal converges to the carrier frequency. It must be emphasized that the CRLB provides a good bound only at high SNR. At intermediate frequencies, the Ziv–Zakai lower bound and the Baranakin bound provide more realistic bounds for the mean-square error; for example, the Baranakin bound is larger by a factor $12(f_c/B)^2$ than the CRLB. This is in line with the above discussion on the impact of correlation sidelobes at different SNRs (Figure 29.3).

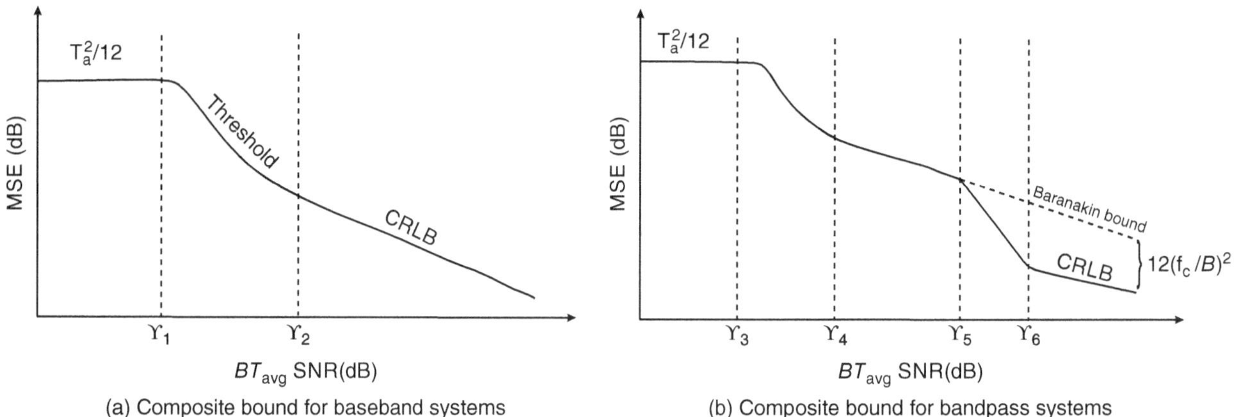

Figure 29.3 Composite bounds for TOA estimation. T_{avg} denotes the averaging time. Reproduced with permission from [Sahinoglu et al. 2008] © Cambridge University Press.

Many ranging algorithms do not consider the location of the correlation maximum, but rather sample the output of a matched filter once per symbol duration, and compare the output of the sampler to a threshold – this method is suboptimum but simple. The probability that the output energy exceeds a threshold χ is (compare Chapter 11)

$$P\left(z^{MF} > \chi\right) = 2Q\left(\frac{\chi}{\sigma}\right) \tag{29.3}$$

where $\sigma^2 = N_0/2N_r$ and N_r is the number of transmitted ranging symbols (note that averaging over multiple ranging symbols provides improvement in the effective SNR, and thus decreases the variance). The choice of χ allows to trade-off missed detection and false alarm probability; this becomes particularly relevant in multi-path environments, where detection of the first (in time) sample exceeding the threshold provides the range estimate, see also [Bartoletti et al. 2015].

Besides the coherent (matched filter) RX, also energy detectors can be used, with the advantage that the transmit pulseshape does not need to be known, and no precise frequency synchronization is required. In the case of a noise-only signal, the probability of exceeding a threshold is given by

$$P\left(z^{ED} > \chi\right) = \exp\left(-\frac{\chi N_r}{N_0}\right) \sum_{k=0}^{M/2-1} \frac{1}{k!} \left(\frac{\chi N_r}{N_0}\right)^k \tag{29.4}$$

where the degrees of freedom of the system $M = N_r(2BT_s + 1)$ are assumed to be an even number. If there is both the ranging pulse and noise present, the probability of exceeding the threshold can be expressed in terms of $Q_K(a,b)$, the generalized Marcum's Q function of parameter K

$$P\left(z^{\mathrm{ED}} > \chi\right) = Q_{M/2}\left(\frac{E}{\sigma}, \frac{\sqrt{\chi}}{\sigma}\right) \quad \text{where } Q_K(a,b) = \frac{2}{a^{K-1}} \int_b^\infty x^K \exp\left[-\frac{a^2 + x^2}{2}\right] I_{K-1}(ax) dx \tag{29.5}$$

where $I_K(x)$ is the modified Bessel function Kth order, first kind and $\sigma^2 = N_0/(2N_r)$.

TOA in Multi-Path Situations

The simplest form of "propagation-induced" errors occurs when the signal propagates along the direction of the line-of-sight but traverses dielectric materials on the way from the TX to the RX. In dielectric material, the speed of light is $c_0/\sqrt{\varepsilon_r}$, so that the runtime of the signal is increased by

$$\Delta\tau = (\sqrt{\varepsilon_r} - 1)d_{\mathrm{diel}}/c_0. \tag{29.6}$$

where d is the distance the wave traverses in the dielectric. If, for example, the wave traverses a wall of thickness d_w at an angle α, $d_{\mathrm{diel}} = d_w/\cos(\alpha)$. In any case, the error introduced by this effect is a bias, i.e., it can only increase the TOA, it cannot lead to a smaller estimate than what would be observed in a free-space environment.

Let us next turn to a situation where the LOS component might be attenuated and additional, delayed, Multi-Path Components (MPCs) exist. This is a case that often occurs in practice. The correct range can then be determined by measuring the TOA of the *first* MPC, i.e., the LOS connection. However, determination of this LOS is not trivial. Finding the MPC with the highest amplitude in the impulse response may not provide the true TOA, since the strongest MPC need not be the one to arrive first. In other words, determining the strongest arriving component will usually result in a positive bias of the ranging measurement. Similarly, using the center of gravity of the squared impulse response or power delay profile will provide a ranging estimate with positive bias.

The simplest detector of the first MPC is a *threshold-crossing* detector: it monitors the amplitude or energy of the received impulse response, and the delay at which this amplitude first crosses a given threshold is considered to be the arrival time of the first MPC. If there were no noise, and the threshold is set sufficiently low, then this method would provide the correct result. However, noise peaks might be mistaken for MPCs, so that a noise peak occurring at a small delay might be mistaken for the first MPC. The probability for such an event to occur depends on (i) the threshold setting: the higher the threshold χ, the lower the probability that such an event will occur (but also the higher the probability that the actual first MPC is below the threshold and thus fails to be detected), (ii) the size of the delay window within which the RX searches for this first MPC: the range of possible delays for the first MPC is constrained by the fact that the runtime cannot be negative, and additional side information about the possible distance range between TX and RX.

An alternative method is the "searchback" approach, see Figure 29.4: it starts from the strongest MPC (which possibly has a delay larger than the first MPC). We then search backward, i.e., in the direction of smaller and smaller delays, until the impulse response magnitude falls below a certain threshold, and remains below a threshold within a delay window of length W_D. The latter condition is intended to ensure that the backward search does not stop just because there is a fading dip at a particular delay. The parameter W_D trades off (i) stopping the backward search too early (i.e., at a delay larger than τ_0), and (ii) mistaking a noise peak as an MPC, thus identifying a $\tau < \tau_0$ as the delay of the first MPC. Alternatively, the searchback can happen in a window of size W_{sb} from the strongest component.

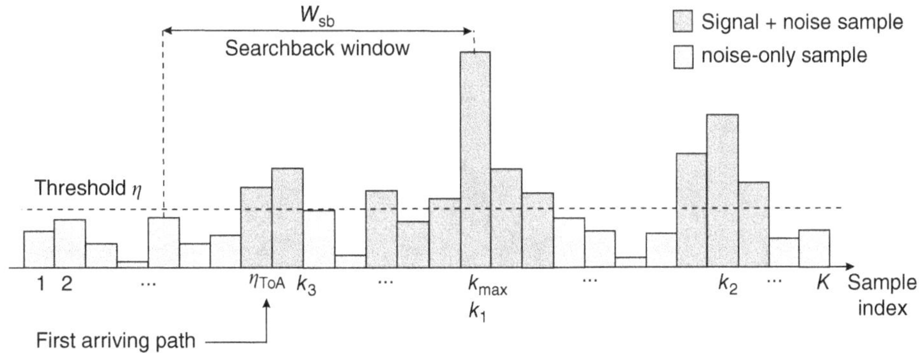

Figure 29.4 Principle of searchback algorithm and thresholding algorithm. Color version available at wiley.com/go/molisch/wireless3e. Reproduced with permission from [Dardari et al. 2009] © IEEE.

Besides these two simple search methods, there are a variety of other methods for determining the first MPC (see the "Further Reading" for references). The performance of these algorithms often depends on the amount of available side information, e.g., whether the average Power Delay Profile (PDP) is known. Generally, it is found that in the presence of MPCs, the range estimate has a positive bias on average, but that negative errors are possible as well. It is also remarkable that the CRLB for range estimation in the presence of MPCs is (in many circumstances) the same as the one for AWGN channels, just with the total signal energy replaced by the energy carried in the first MPC.

Much larger errors can occur if the first MPC (corresponding to the LOS) is cut off completely (i.e., attenuated so strongly that it cannot be detected in the noise), and only much-delayed MPCs can be identified. In this case, the positive bias of the range estimate can be very large, and no amount of signal processing can improve the actual ranging estimate, see Figures 29.5 and 29.6. Instead, it is necessary to identify the fact that an non-LOS (NLOS) situation is occurring, and take this into account during the localization procedure as discussed further in Section 29.2.2.

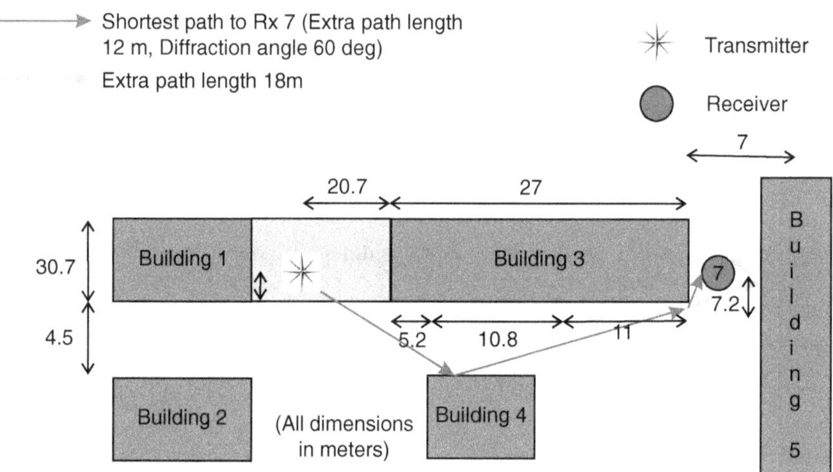

Figure 29.5 Outdoor geometry for measurements. Color version available at wiley.com/go/molisch/wireless3e. Reproduced with permission from [Kristem et al. 2014] © IEEE.

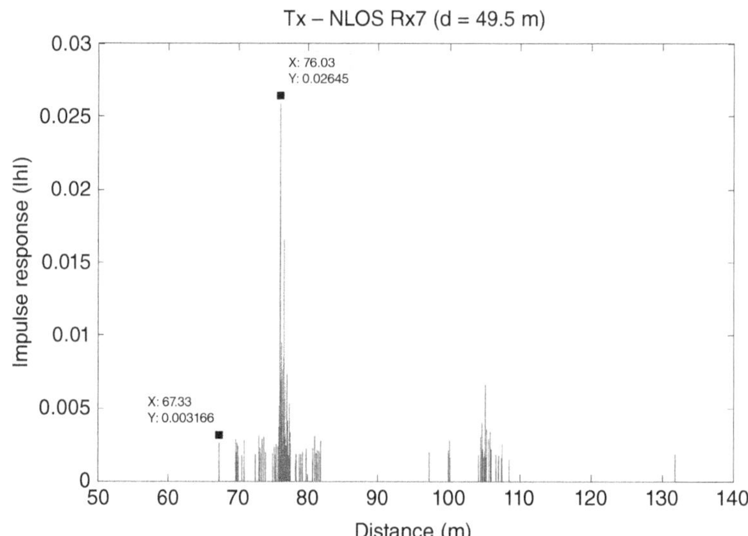

Figure 29.6 Impulse response measured at position 7 of Figure 29.5. Note the large error (nominal distance 49.5 m, pseudorange of first arriving MPC 67 m due to blockage by building; strongest MPC at 76 m). Color version available at wiley.com/go/molisch/wireless3e. Reproduced with permission from [Kristem et al. 2014] © IEEE.

TOA in the Presence of Interference

Co-channel interference can lead to wrong delay estimates because MPCs corresponding to interfering signals might be mistaken for desired signals. In particular interfering signals arriving earlier than the MPCs of the desired signal can lead to large errors, as the system will identify those as the LOS signals, and thus lead to a strong negative bias of the range estimate. One way of reducing the impact of interference is to treat it like noise. Assume that the interfering signal has some characteristic that changes over time, e.g., in a time-hopping impulse radio system (compare Section 19.4), the signal from the desired TX is sent out at $t_k = kT_f + c_k T_c$, while the interfering signal is sent out at $kT_f + c'_k T_c$. Then if we de-hop with the desired sequence c_k and average the signals over multiple repetitions of the transmit signal, the desired signals will not be impacted (it is always the same, and thus not impacted by averaging), while the interfering signal will be suppressed by a factor that is equal to the number of signal repetitions; furthermore, it occurs at more and more delays and thus becomes similar to a constant floor. A similar statement holds true when the ranging signal is, e.g., a multi-tone signal, and the interfering signal is a multi-tone signal whose subcarriers change from transmission to transmission (relative to the desired signal).

In some situations, even averaging might not suppress interference sufficiently. In this case, outlier detection can be used to identify the ranging signals that lead to large deviations, e.g., through median filtering. Alternatively, multi-user detection can be used advantageously, compare Section 28.2. In serial interference cancellation, the strongest ranging signal is detected first (irrespective of whether it is the desired signal or not), and its contributions are subtracted from the overall received signal. Then the next-strongest-ranging signal is detected, and so on, until the desired ranging signal(s) can be identified.

Impact of Synchronization Errors

The previous discussion assumed that TX and RX are perfectly synchronized, i.e., have both the same time base and the same frequency. In reality, this is not possible, as the oscillators driving the clocks at TX and RX show timing offsets and clock drift (frequency offset). Let δ denote the clock drift (relative to the correct, nominal rate) and μ the timing offset at the start of the procedure. Then the estimate $G(t)$ of the true time t can be modeled (over short time intervals, where a linear drift model holds), as

$$G(t) = (1 + \delta)t + \mu. \tag{29.7}$$

Now consider a one-way ranging protocol, in which the last frequency synchronization (either between each other, or each separately to an external clock) between TX and RX occurred at a nominal zero time; timing offset correction at that instant is generally imperfect, with a residual at TX and RX of μ_{TX} and μ_{RX}, respectively. The actually estimated runtime is

$$\hat{\tau} = \tau_0 + \delta_{\mathrm{RX}}(t_0 + \tau_0) - \delta_{\mathrm{TX}}t_0 + \mu_{\mathrm{RX}} - \mu_{\mathrm{TX}} \tag{29.8}$$

where t_0 is the time at which the pulse was sent out.

To obtain meaningful ranging estimates, the estimation error $|\hat{\tau} - \tau_0|$ has to be much smaller than the runtime that we wish to measure. This may be difficult in many networks. For this reason, two-way ranging protocols are popular. Such a protocol measures the round-trip time between two nodes, so that offsets between the timings at the nodes become irrelevant. The protocol is shown in Figure 29.7. A ranging signal is sent from node i to node j. Node j determines the arrival time of the signal, waits a predetermined time τ_{d}, and then sends a signal back. Note that since Node j waits for τ_{d} according to its own clock, the correct (absolute) time that expires during that wait time is $\tau_{\mathrm{d}}/(1 + \delta_j)$. Node i thus receives the signal back after a time (according to the timescale on its clock)

$$\hat{\tau}_{\mathrm{RT}} = \left(2\tau_0 + \frac{\tau_{\mathrm{d}}}{1 + \delta_j}\right)(1 + \delta_i). \tag{29.9}$$

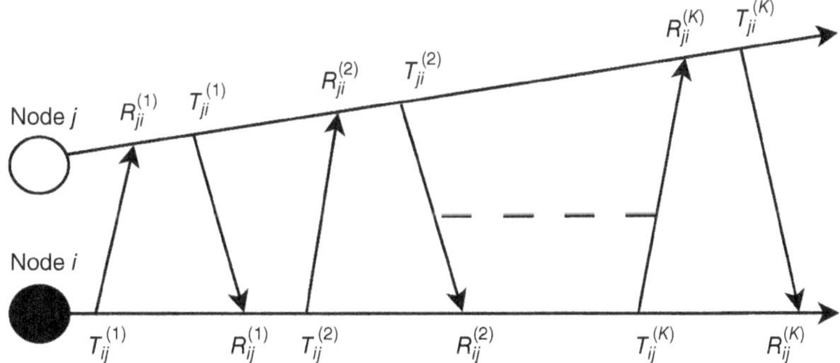

Figure 29.7   Two-way ranging protocol. Node i is the reference with and also initiates the communication with node j whose clock skew, clock offset, and distance from node i are unknown. There are K two way communications between the node pair during which 2K time markers are recorded at the respective nodes.
Reproduced with permission from [Rajan and VanderVeen 2011] © IEEE.

Using this estimate as the true roundtrip time $(2\tau_0 + \tau_{\mathrm{d}})$ results in a relative error

$$\varepsilon = \delta_i + \frac{\delta_i - \delta_j}{(1 + \delta_j)} \frac{\tau_{\mathrm{d}}}{2\tau_0}. \tag{29.10}$$

This can be significantly smaller than the error in one-way ranging, in particular when the t_0 and/or $\mu_{\mathrm{RX}} - \mu_{\mathrm{TX}}$ is large.

29.2.2 Determination of the Location

TOA Formulation

Once the runtimes between TX and RX have been estimated, the location of the agent node can be found. In this context, we need to distinguish between two cases: (i) anchors and agents are all synchronized to the same time base. In this case, the absolute runtime between agent and all anchors can be determined, resulting in the TOA as a basis for localization. (ii) anchors are synchronized among each other, but the agent is not synchronized to them. In this case, the *Time Difference Of Arrival* (TDOA) of the agent signal received at the various anchor nodes forms the basis of the localization (or conversely, the TDOA of anchor transmissions at the agent node).

In the first case, a particular runtime τ_i between the ith anchor and the agent means that the agent lies on a circle of radius $d_i = \tau_i/c_0$ centered at the anchor. When measurements to multiple anchors are available, then the agent must lie on all corresponding circles, i.e., on the intersection of them, see Figure 29.8a. To frame this mathematically, let $\mathbf{x} = [x, y]^T$ denote the position of the agent, and $\mathbf{u}_i = [u_i, v_i]^T$ the position of the ith anchor. Using n_i to denote the ranging error with respect to the ith anchor, a range measurement can be written as

$$r_i = d_i + n_i = \sqrt{(x - u_i)^2 + (y - v_i)^2} + n_i, \quad i = 1, ...I. \tag{29.11}$$

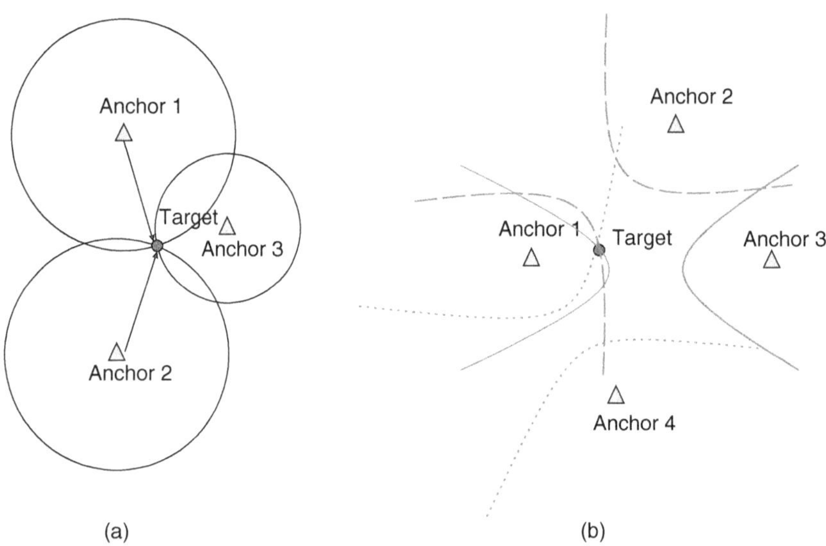

Figure 29.8 Principle of TOA localization (a) and TDOA localization (b). Color version available at wiley.com/go/molisch/wireless3e.

Note that r is the estimated range, not the received signal envelope. The vector of estimated distances can be written in compact form as

$$\mathbf{r} = \mathbf{f}(\mathbf{x}) + \mathbf{n}. \tag{29.12}$$

where

$$\mathbf{f}(\mathbf{x}) = \begin{pmatrix} \sqrt{(x - u_1)^2 + (y - v_1)^2} \\ \sqrt{(x - u_2)^2 + (y - v_2)^2} \\ \vdots \\ \sqrt{(x - u_I)^2 + (y - v_I)^2} \end{pmatrix} \tag{29.13}$$

Also note that n is the error in the range, not AWGN added to the received signal. This error is often modeled as zero-mean Gaussian (though, as discussed in Section 29.2.1, it need not be so in particular in the presence of bias from possibly existing MPCs as well as errors induced by noise). For the Gaussian-error case the pdf of $\mathbf{r}$ is

$$pdf(\mathbf{r}) = \frac{1}{\sqrt{(2\pi)^I |\mathbf{R}|}} \exp\left(-\frac{1}{2}(\mathbf{r} - \mathbf{d})^T \mathbf{R}^{-1}(\mathbf{r} - \mathbf{d})\right) \tag{29.14}$$

where the error correlation matrix[1] is

$$\mathbf{R} = E\{\mathbf{n}\mathbf{n}^T\} = \begin{pmatrix} \sigma_1^2 & 0 & .. & 0 \\ 0 & \sigma_2^2 & 0 & 0 \\ 0 & 0 & .. & .. \\ 0 & 0 & .. & \sigma_I^2 \end{pmatrix} \tag{29.15}$$

if the range measurements to the different anchors have uncorrelated errors (which is reasonable for noise-induced errors, but which might not hold for multi-path-induced errors).

TDOA Formulation

Now turn to the case of TDOA. To simplify the discussion, we assume that the agent is sending out a signal whose arrival times at the anchors are measured. Since the agent is not synchronized to the anchors, the anchors do not know the actual time at which the agent sent out its signal; all information lies in the difference of the runtimes. Consider the measurement of the runtime from one particular anchor as the reference time measurement, and relate all other arrival time measurements to it; without loss of generality, we index this anchor as $i = 1$. By definition, the set of all locations that have a fixed distance difference from two points constitutes a hyperbola. Thus, the runtime difference of the signal to the ith ($i > 1$) anchor and the reference anchor ($i = 1$) defines a hyperbola with the two anchor locations as foci, and the runtime difference as the major axis. Thus, the agent must lie on the intersection of the hyperbolas corresponding to the multiple TDOAs of the anchor nodes with respect to the reference anchor, see Figure 29.8b. Thus, the hyperbolae take on the role in TDOA that the circles have in TOA

$$\mathbf{f}(\mathbf{x}) = \begin{pmatrix} \sqrt{(x-u_2)^2 + (y-v_2)^2} - \sqrt{(x-u_1)^2 + (y-v_1)^2} \\ \sqrt{(x-u_3)^2 + (y-v_3)^2} - \sqrt{(x-u_1)^2 + (y-v_1)^2} \\ \vdots \\ \sqrt{(x-u_I)^2 + (y-v_I)^2} - \sqrt{(x-u_1)^2 + (y-v_1)^2} \end{pmatrix} \qquad i = 2, 3, ...I \tag{29.16}$$

Thus, TDOA needs at least four anchors for unique localization in two dimensions, while TOA needs only 3. Also, note that while the signal model (29.11) remains valid, the errors n_i are not uncorrelated anymore, because each error is the difference between the ith anchor measurement and the reference anchor measurement; in other words, the reference anchor measurement induces correlation.

*Nonlinear Solution Methods

Based on these formulations, we now can find different ways of determining the best location vector $\mathbf{x}$ by minimizing the error associated with this location. We first consider a direct solution, which results in a nonlinear optimization problem. Let the cost function (which we wish to minimize) be the mean square error between the measured range information and the "true" one:

$$C(\mathbf{x}) = (\mathbf{r} - \mathbf{f}(\mathbf{x}))^T (\mathbf{r} - \mathbf{f}(\mathbf{x})) \tag{29.17}$$

then the location estimate is $\check{\mathbf{x}} = \arg\min_{\check{\mathbf{x}}} C(\check{\mathbf{x}})$. Now the search variables, namely, the coordinates (x, y) are in the cost function in a nonlinear manner. For example, for TOA,

$$C(\mathbf{x}) = \sum_i \left(r_i - \sqrt{(x-u_i)^2 + (y-v_i)^2} \right)^2 . \tag{29.18}$$

Thus, the solution also requires nonlinear processing; specifically, an iterative solution can be obtained by means of the Newton–Raphson algorithm. Each step of this algorithm advances in the direction of the gradient. More precisely, in the $n + 1$th step,

$$\hat{\mathbf{x}}^{(n+1)} = \hat{\mathbf{x}}^{(n)} - \mathbf{H}^{-1}\left[C\left(\hat{\mathbf{x}}^{(n)}\right)\right] \nabla\left[C\left(\hat{\mathbf{x}}^{(n)}\right)\right] \tag{29.19}$$

where $\mathbf{H}(C(\hat{\mathbf{x}}))$ is the Hessian matrix

---

[1] Since the errors are assumed here to be zero-mean, we do not distinguish between correlation and covariance.

$$\mathbf{H}[C(\mathbf{x})] = \begin{bmatrix} \dfrac{\partial^2 C(\mathbf{x})}{\partial x^2} & \dfrac{\partial^2 C(\mathbf{x})}{\partial x \partial y} \\[2ex] \dfrac{\partial^2 C(\mathbf{x})}{\partial y \partial x} & \dfrac{\partial^2 C(\mathbf{x})}{\partial y^2} \end{bmatrix} \tag{29.20}$$

and the gradient

$$\nabla[C(\mathbf{x})] = \begin{bmatrix} \dfrac{\partial C(\mathbf{x})}{\partial x} \\[2ex] \dfrac{\partial C(\mathbf{x})}{\partial y} \end{bmatrix}. \tag{29.21}$$

By inserting (29.18) in (29.21), (29.20)

$$\frac{\partial C(\mathbf{x})}{\partial x} = -2\sum_i \frac{\left(r_i - \sqrt{(x-u_i)^2 + (y-v_i)^2}\right)(x-u_i)}{\sqrt{(x-u_i)^2 + (y-v_i)^2}} \tag{29.22}$$

$$\frac{\partial C(\mathbf{x})}{\partial y} = -2\sum_i \frac{\left(r_i - \sqrt{(x-u_i)^2 + (y-v_i)^2}\right)(y-v_i)}{\sqrt{(x-u_i)^2 + (y-v_i)^2}} \tag{29.23}$$

$$\frac{\partial^2 C(\mathbf{x})}{\partial x^2} = 2\sum_i \frac{(x-u_i)^2}{(x-u_i)^2 + (y-v_i)^2} - \frac{\left(r_i - \sqrt{(x-u_i)^2 + (y-v_i)^2}\right)(y-v_i)^2}{\left[\sqrt{(x-u_i)^2 + (y-v_i)^2}\right]^3} \tag{29.24}$$

$$\frac{\partial^2 C(\mathbf{x})}{\partial x \partial y} = \frac{\partial^2 C(\mathbf{x})}{\partial y \partial x} = 2\sum_i \frac{(r_i)(x-u_i)(y-v_i)}{\left[\sqrt{(x-u_i)^2 + (y-v_i)^2}\right]^3} \tag{29.25}$$

$$\frac{\partial^2 C(\mathbf{x})}{\partial y^2} = 2\sum_i \frac{(y-v_i)^2}{(x-u_i)^2 + (y-v_i)^2} - \frac{\left(r_i - \sqrt{(x-u_i)^2 + (y-v_i)^2}\right)(x-x_i)^2}{\left[\sqrt{(x-u_i)^2 + (y-v_i)^2}\right]^3}. \tag{29.26}$$

A related solution is obtained from a maximum-likelihood formulation. Taking the log-likelihood function, i.e., the logarithm of Eq. (29.14), we see that we need to minimize

$$C(\mathbf{x}) = (\mathbf{r} - \mathbf{f}(\mathbf{x}))^T \mathbf{R}^{-1} (\mathbf{r} - \mathbf{f}(\mathbf{x})) \tag{29.27}$$

(note that $\mathbf{R}$ denotes the noise correlation matrix, while C is the scalar cost function). Assuming uncorrelated measurements, this can be re-written as

$$C(\mathbf{x}) = \sum_i \frac{1}{\sigma_i^2}\left(r_i - \sqrt{(x-u_i)^2 + (y-v_i)^2}\right)^2 \tag{29.28}$$

which is simply a weighted version of Eq. (29.18). The weights are related to the noise variance, and measurements that have a large noise variance get a small weight for the location computation, as is intuitive. Since the mathematical problem formulation is very similar to Eq. (29.17) the Newton–Raphson iterative approach described above can be used for solving this ML problem as well.

*Linear Solution Methods

The nonlinear relationship between target location and cost function can be handled by the iterative solution methods described above. An alternative approach is to linearize the equations. The following shows the example for TOA; TDOA can be treated in a similar fashion. Start out by taking the square of Eq. (29.11)

$$r_i^2 = (x - u_i)^2 + (y - v_i)^2 + n_i^2 + 2n_i\sqrt{(x - u_i)^2 + (y - v_i)^2}. \tag{29.29}$$

Introducing two new variables, $R = x^2 + y^2$, and $\varsigma_i = n_i^2 + 2n_i\sqrt{(x - u_i)^2 + (y - v_i)^2}$ gives

$$r_i^2 = (x - u_i)^2 + (y - v_i)^2 + \varsigma_i \tag{29.30}$$

from which

$$-2u_i x - 2v_i y + R + \varsigma_i = r_i^2 - u_i^2 - v_i^2 \tag{29.31}$$

follows. Define now further

$$\mathbf{A} = \begin{pmatrix} -2u_1 & -2v_1 & 1 \\ -2u_2 & -2v_2 & 1 \\ \vdots & \vdots & \vdots \\ -2u_I & -2v_I & 1 \end{pmatrix} \tag{29.32}$$

$$\boldsymbol{\theta} = \begin{pmatrix} x \\ y \\ R \end{pmatrix}$$

$$\mathbf{b} = \begin{pmatrix} r_1^2 - u_1^2 - v_1^2 \\ r_2^2 - u_2^2 - v_2^2 \\ \vdots \\ r_I^2 - u_I^2 - v_I^2 \end{pmatrix}.$$

If the noise is sufficiently small that $E\{\varsigma_i\} \simeq 0$, Eq. (29.31) can be written as

$$\mathbf{A}\boldsymbol{\theta} = \mathbf{b}. \tag{29.33}$$

The cost function is then

$$\begin{aligned} C(\boldsymbol{\theta}) &= (\mathbf{A}\boldsymbol{\theta} - \mathbf{b})^T (\mathbf{A}\boldsymbol{\theta} - \mathbf{b}) \\ &= \boldsymbol{\theta}^T \mathbf{A}^T \mathbf{A}\boldsymbol{\theta} - 2\boldsymbol{\theta}^T \mathbf{A}^T \mathbf{b} + \mathbf{b}^T \mathbf{b}. \end{aligned} \tag{29.34}$$

Minimizing the cost function is done by taking the derivative with respect to $\boldsymbol{\theta}$ and setting it to zero, so that we obtain

$$\hat{\boldsymbol{\theta}} = \left(\mathbf{A}^T \mathbf{A}\right)^{-1} \mathbf{A}^T \mathbf{b} \tag{29.35}$$

and the target location is often taken as the first two components of $\boldsymbol{\theta}$.

The above approach works only well if the errors for the different measurements are identically distributed. Otherwise, we can define an error correlation matrix

$$\mathbf{R} = \begin{pmatrix} 4r_1^2\sigma_1^2 & 0 & .. & 0 \\ 0 & 4r_2^2\sigma_2^2 & 0 & 0 \\ 0 & 0 & .. & .. \\ 0 & 0 & .. & 4r_I^2\sigma_I^2 \end{pmatrix} \tag{29.36}$$

and the location estimate becomes

$$\hat{\boldsymbol{\theta}} = \left(\mathbf{A}^T \mathbf{R}^{-1} \mathbf{A}\right)^{-1} \mathbf{A}^T \mathbf{R}^{-1} \mathbf{b}. \tag{29.37}$$

Note that the above derivation treats (x, y) and R as independent variables (and R essentially as a dummy variable that is discarded in the end). This is of course not correct as R contains important information about (x, y), but this approach is often chosen because of its relative simplicity. Alternatively, the estimate can be refined in an iterative procedure that is known as two-step weighted least squares.

To summarize, errors on the localization are mostly determined by the error in the underlying ranging signal but also by the geometrical relation between anchor and agents. The best precision is obtained if the circles that correspond to the range measurements with respect to the different anchors intersect at right angles. In that case, the localization error is on the same order as the ranging error. However, if circles intersect at "grazing angle," the error in the localization can be much larger; this effect is known as GDOP, see Figure 29.9. To be precise, GDOP is defined as follows: let $\mathbf{e}_i = [e_{i,1}\ e_{i,2}\ e_{i,3}]^T$ contain the x, y, and z coordinates of a unit-norm vector pointing from the agent to the ith anchor. Then define a geometry matrix

$$\mathbf{H}_{\mathrm{GDOP}} = \begin{bmatrix} e_{1,1} & e_{1,2} & e_{1,3} & 1 \\ e_{2,1} & e_{2,2} & e_{2,3} & 1 \\ \vdots & \vdots & \vdots & \vdots \\ e_{I,1} & e_{I,2} & e_{I,3} & 1 \end{bmatrix}$$

and the GDOP is (assuming equal error variances on all ranging measurements; a weighted version can be used otherwise)

$$GDOP = \sqrt{tr\left(\mathbf{H}_{\mathrm{GDOP}}^T \mathbf{H}_{\mathrm{GDOP}}\right)^{-1}}. \tag{29.38}$$

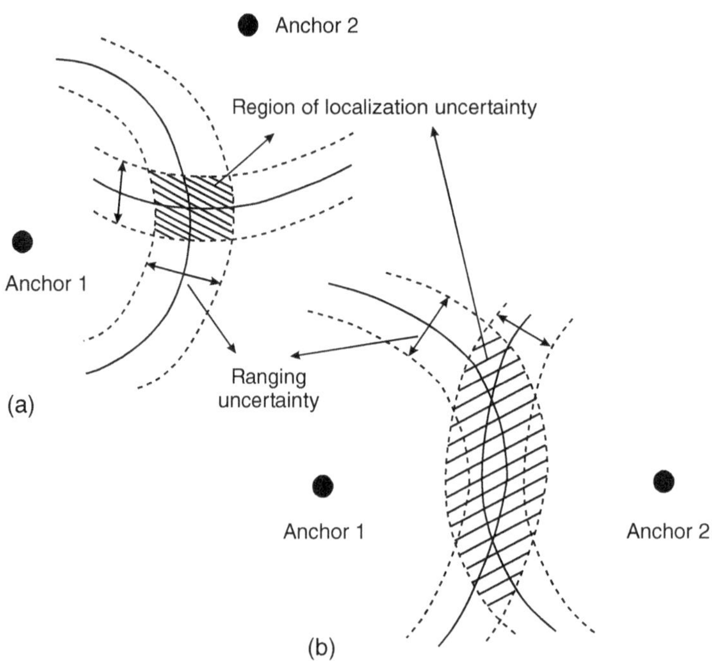

Figure 29.9 Principle of GDOP: small localization error when ranging "circles" intersect at almost normal angles (a); large localization for grazing intersection (b).

29.2.3 Cramer–Rao Lower Bound (CRLB)

In any estimation problem, it is important to know what the best achievable accuracy is. A very important tool in this context is the CRLB, which specifies the smallest variance that can be achieved by an unbiased estimator (note that the variance of a biased estimator might be lower). The CRLB essentially states that the variance cannot be smaller than the inverse of the Fisher information.

Let us thus start out by defining the Fisher information. We wish to determine a vector of deterministic parameters – in our case the location – $\mathbf{x}$ from measurements $\mathbf{z}$. Write the probability density function as $pdf(\mathbf{z}\,|\,\mathbf{x})$. The FIM is then defined as having the entries

$$\mathbf{I}(\mathbf{x})_{i,j} = E\left\{ \left(\frac{\partial}{\partial x_i} \ln\left(pdf(\mathbf{z}\,|\,\mathbf{x})\right) \right) \left(\frac{\partial}{\partial x_j} \ln\left(pdf(\mathbf{z}\,|\,\mathbf{x})\right) \right) \right\}. \tag{29.39}$$

The covariance matrix of the estimate is lower bounded by the inverse of the FIM

$$cov_{\mathbf{x}}(\hat{\mathbf{x}}(\mathbf{x})) \geq \mathbf{I}(\mathbf{x})^{-1} \tag{29.40}$$

where $\hat{\mathbf{x}}(\mathbf{x})$ denotes an unbiased estimator of $\mathbf{x}$. A bound of the mean square error MSE of the position estimate can be obtained from the FIM as

$$\text{MSE} \geq \text{trace}\left[\mathbf{I}(\mathbf{x})^{-1}\right]. \tag{29.41}$$

From (29.39), the CRLB for localization with zero-mean Gaussian errors can be easily obtained. The FIM becomes

$$\mathbf{I}(\mathbf{x}) = \left[\frac{\partial \mathbf{f}(\mathbf{x})}{\partial \mathbf{x}}\right]^{T} \mathbf{R}^{-1} \left[\frac{\partial \mathbf{f}(\mathbf{x})}{\partial \mathbf{x}}\right]. \tag{29.42}$$

Inserting, e.g., the definition of $\mathbf{f}(\mathbf{x})$ for the TOA case, Eq. (29.11), results in

$$\mathbf{I}_{\text{TOA}}(\mathbf{x}) = \begin{bmatrix} \displaystyle\sum_i \frac{(x-u_i)^2}{\sigma_i^2 r_i^2} & \displaystyle\sum_i \frac{(x-u_i)(y-v_i)}{\sigma_i^2 r_i^2} \\ \displaystyle\sum_i \frac{(x-u_i)(y-v_i)}{\sigma_i^2 r_i^2} & \displaystyle\sum_i \frac{(y-v_i)^2}{\sigma_i^2 r_i^2} \end{bmatrix}. \tag{29.43}$$

The bound on the MSE can then be obtained by inserting (29.43) into (29.41).

29.2.4 Soft Information

The above derivations determined a range estimate based on the individual measurement, and from this established the location, assuming that disturbances are Gaussian. However, combining those "hard" estimates of the ranges is not necessarily the best approach. Rather, it is preferable to characterize the *likelihood* that the range takes on various values (i.e., a "soft estimate" of the range), and combine the likelihoods of the ranges to the different anchors. Considering that errors can be mainly from multi-path, and can have distinctly non-Gaussian characteristics, the result of such a location estimate based on soft information can be significantly different from the methods described above. Figure 29.10 outlines the differences: in the left figure, each ranging measurement is evaluated with a hard thresholding technique, providing a discrete range estimate; which is combined with other such estimates to give the location. in the rightmost figure, the likelihood functions (see middle figure) of the range estimates are combined.

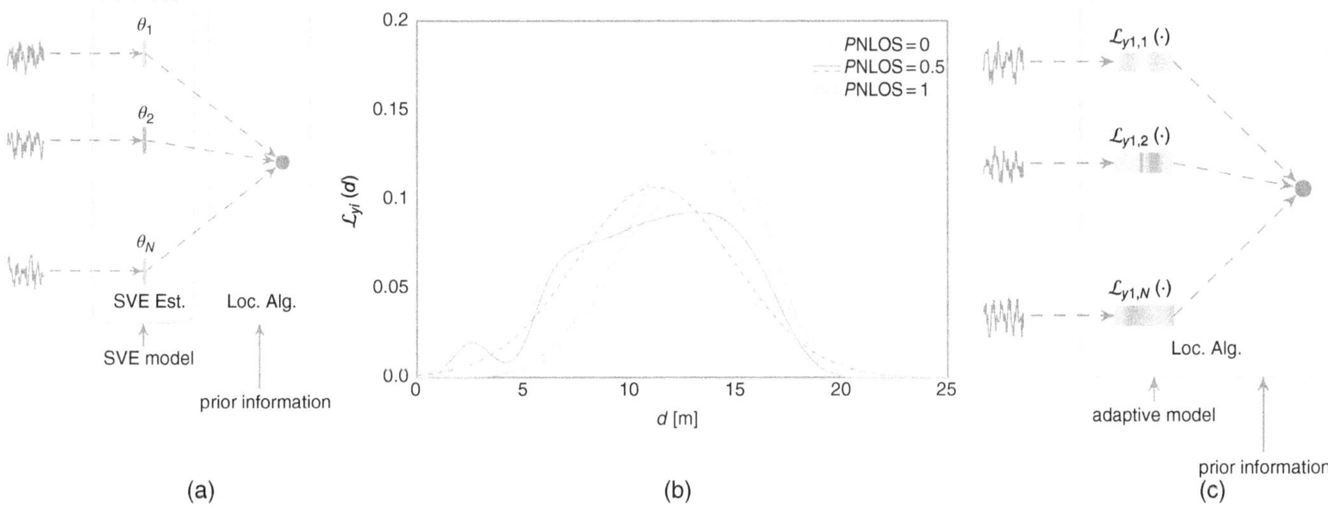

Figure 29.10 Extraction of location by hard thresholding or range estimates (a); likelihood functions of range estimate (b), and location estimate from combining likelihood functions (c). Color version available at wiley.com/go/molisch/wireless3e. Reproduced with permission from [Conti et al. 2019] © IEEE.

29.3 NLOS Detection, Mitigation, and Exploitation

TOA-based localization is commonly based on the assumption that there is a LOS path between TX and RX, so that the signal runtime corresponding to this path serves as a measure for the distance. However, as discussed in Sec. 29.2.1, this assumption need not be true. When TX and RX are separated by one or more walls, ceilings, or whole buildings, then the assumption breaks down. There might be still a quasi-LOS component that propagates along the geometrical LOS and is attenuated and delayed by the wall; or the LOS component might be for all practical purposes be completely suppressed (i.e., attenuated so strongly that it cannot be extracted from the noise). In particular in the latter case, the measured runtime gives a *biased* estimate of the distance. The question is how a localization algorithm should deal with this situation. In the following, we describe three steps and approaches (i) NLOS detection, (ii) NLOS mitigation, and (iii) multi-path exploitation.

29.3.1 NLOS Detection

The first step in dealing with a NLOS situation is to detect that a certain range measurement actually suffers from being performed in NLOS. For this task, we distinguish two approaches: (i) single-link techniques, where we consider the characteristics of the link between a particular TX and RX, and determine just from those whether an NLOS situation exists or not. (ii) multi-link (collaborative) approaches, where an analysis of the relationship between the different links, e.g., consistency checks, are used.

Single-Link Techniques

NLOS testing for a single link is, in principle, a hypothesis testing problem. We have certain observables, like receive power, Rice factor, etc. (for details see below), and know the probability density function of these observables. Then given a particular observation, the task is to make a decision, namely whether the hypothesis "LOS" is more likely to be fulfilled than the hypothesis "NLOS" or not. A different decision criterion is to allow only a certain "false alarm probability," where by false alarm we mean that a situation is judged NLOS even though it is LOS. Note that some knowledge of the environment is necessary in order to obtain the pdf of the observables – for example, knowledge of the average Rice factor either from the type of operating environment, or from measurements at nearby locations.

Observables that can be used as a basis for the hypothesis test include

- *Received power:* in a LOS situation the pathloss coefficient is smaller or equal to 2, and shadowing variance is low.
- *Rice factor:* in an NLOS situation, the Rice factor will be low (typically <3 dB), while it can be much higher in LOS situations.
- *Delay spread:* LOS scenarios typically lead to much smaller delay spread than NLOS. Note, however, that the delay spread also depends on the environment (e.g., urban outdoor, hilly terrain, indoor office, ...) as well as the distance between TX and RX. Thus, delay spread is usually a good measure for LOS/NLOS only when the statistics of the operating environment are well known.
- *Shape of the power delay profile:* in a LOS situation, the first MPC is the strongest one. This holds true even in situations where there is rich multi-path, e.g., because there are many metallic reflectors that might lead to low Rice factors and long delay spreads. However, note also that there might be many situations in which a monotonous decay of the PDP occurs even though the scenario is NLOS. Single-parameter measures for the PDP, such as kurtosis and skewness, can be considered as well.

Closer analysis of measurement results shows that identification of NLOS based on a single parameter is difficult and error-prone. Consider, e.g., the identification by means of the Rice factor. The first question that arises is "where to put the threshold"? The answer to this might depend on the specific environment and thus needs to be learned from training data where the ground truth is known. Secondly, after the threshold has been established, there are still many situations where, e.g., the Rice factor is above the threshold even though the link is in a NLOS situation, and vice versa, see Figure 29.11. It is thus advantageous to combine multiple parameters, such as Rice factor, kurtosis, and delay spread, to increase the probability of correct identification. Such a joint identification requires, however, more complex decisions than simple thresholding. Various types of Machine Learning, such as Support Vector Machines (SVMs), Neural Networks (NN), etc., can be used for this purpose.

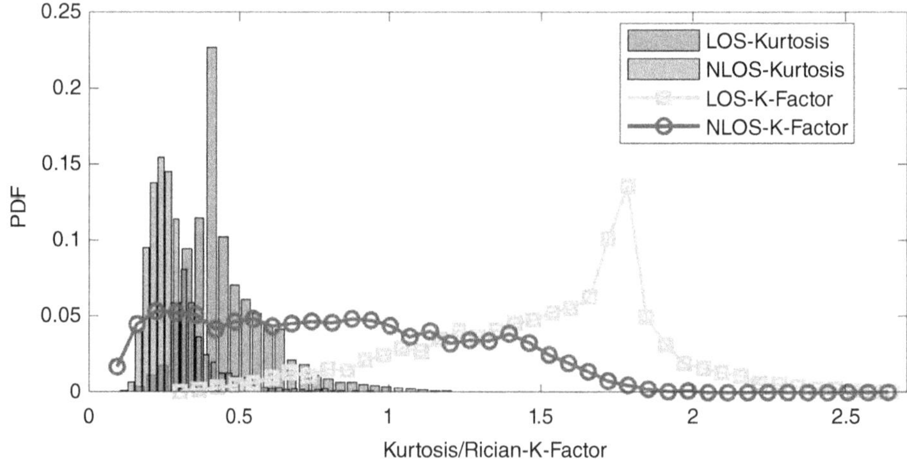

Figure 29.11 Distribution of Rice factor and Kurtosis in LOS and NLOS situations in a vehicle-to-vehicle scenario. It can be seen that in neither case there exists a threshold that allows clear distinction of LOS/NLOS based on one parameter. Color version available at wiley.com/go/molisch/wireless3e. Reproduced with permission from [Huang et al. 2019b] © IEEE.

Another single-link technique checks the consistency of Direction-of-Arrival (DoA) measurements, and/or compares DoA and Direction-of-Departure (DoD). We know that a LOS component has to leave the TX in the direction of the RX; if it does not do so, then clearly the measured component is not LOS. Equivalently, we can say that the DoD and the DoA have to point at each other. Obviously, this technique can only be applied when the DoA/DoD can be measured, e.g., by means of antenna arrays.

Multi-Link Techniques

Multi-link techniques test the consistency of the estimates on multiple links to determine which ones (if any) are in a LOS situation. The principle is the following: imagine that the true agent location is known. Now the circles corresponding to the different ranging measurements might not pass exactly through that point – due to noise, some deviation from the ideal location can be expected. The circles thus do not intersect in a single point, but rather there are multiple intersections (typically within the standard deviation of the noise-induced ranging error). If one or more of the ranging measurements is associated with a NLOS connection, the deviation of some of these intersections from the true agent location can be much larger than in the case of noise-only.

Of course, in reality, we do not know the location of the agent node (this is, after all, what we want to find out), and thus cannot say for sure which measurements are the NLOS ones. We can, however, hypothesize that a certain set of range measurements are all LOS (and the rest NLOS), and test their consistency; of course, the assumed NLOS measured ranges have to be *larger* than the distance between the hypothesized agent location and the considered anchor. Such an approach can be generalized to a maximum-likelihood estimation (assuming that the statistics for LOS and NLOS reception are known). Irrespective of the specific identification procedure, the localization algorithm then discards (disregards) all measurements that are deemed to be NLOS, and determines the location from the remainder.

*29.3.2 NLOS Mitigation

Identifying and discarding NLOS situations is not easy (resulting in a possible residual bias), and throws away possible useful information. It is thus preferable to find approaches that mitigate the impact of the positive NLOS bias without discarding the corresponding measurements completely. One of the best-known such algorithms is *Projection Onto Convex Sets* (POCS), which interprets the localization problem as a convex feasibility problem. The algorithm can be briefly summarized as follows:

1. Initialization: choose an arbitrary initial agent position $\mathbf{x}^{(0)}$
2. For $n = 0$ to convergence, update

$$\mathbf{x}^{(n+1)} = \mathbf{x}^{(n)} + \lambda^{(n)}\left(\mathcal{P}_{\mathcal{D}_j}^{(n)}\left(\mathbf{x}^{(n)}\right) - \mathbf{x}^{(n)}\right) \tag{29.44}$$

where the projection of a point $\mathbf{x}$ onto a closed convex set Ω is the solution to

$$\mathcal{P}_\Omega(\mathbf{x}) = \arg\min_{\mathbf{x}'\in\Omega} \left\|\mathbf{x} - \mathbf{x}'\right\|. \tag{29.45}$$

When Ω is a disc whose center is located at $\mathbf{u}_j$, the projection becomes

$$\mathcal{P}_{\mathcal{D}_j}(\mathbf{x}) = \begin{cases} \mathbf{x}_j + \dfrac{\mathbf{x} - \mathbf{u}_j}{\left\|\mathbf{x} - \mathbf{u}_j\right\|}\hat{r}_j & \left\|\mathbf{x} - \mathbf{u}_j\right\| \geq \hat{r}_j \\ \mathbf{x} & \text{otherwise} \end{cases}. \tag{29.46}$$

Here, $\hat{r}_j$ is the radius of the jth disk. The sequence $\lambda^{(k)}$ is a "steering sequence" that determines the stepsize in the iteration.

What is remarkable about POCS is that it is insensitive to even large positive biases, and thus well suited for obtaining location estimates in NLOS situations, see Figure 29.12. However, the algorithm does not work well if the agent location is outside the convex hull of the anchor locations – then the error can be large even in the noiseless case.

29.3.3 Multi-Path Exploitation

Last but not least, MPCs can be actively exploited for improving localization accuracy. The basic idea can be easily understood from the image principle that we discussed in Section 4.7: an MPC reflected from a flat wall looks to the RX as if it came from an "image source," i.e., a source location that is the original source location mirrored at the reflecting wall.

Consider an environment consisting of a single anchor having LOS to a target, as shown in Figure 29.13. In addition to the LOS path, suppose that the agent also receives four paths from single-bounce reflections off each of the four walls enclosing the anchor and the target. In this case, the target cannot be localized using the LOS path alone. However, according to the image principle, the path reflected on wall 1 has the same TOA as that of a path from a Virtual Anchor (VA) on the other side of wall 1. On repeating the same process for the remaining walls, it is easily seen that there are a sufficient number of (real + virtual) anchors to localize the agent unambiguously.

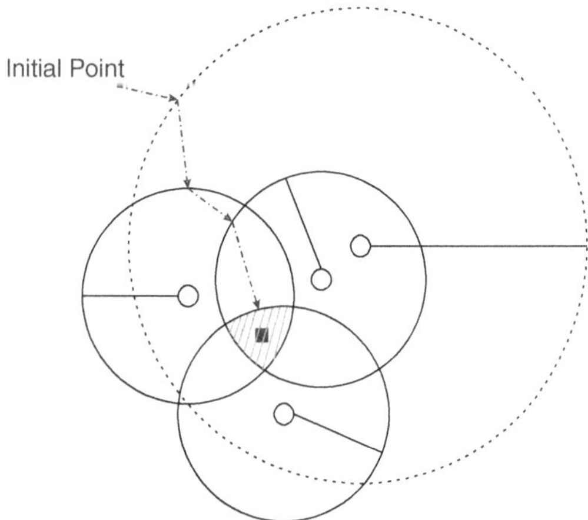

Figure 29.12 POCS is able to suppress a very large positive bias. Color version available at wiley.com/go/molisch/wireless3e.
Reproduced with permission from [Gholami et al. 2011] © EuraSIP.

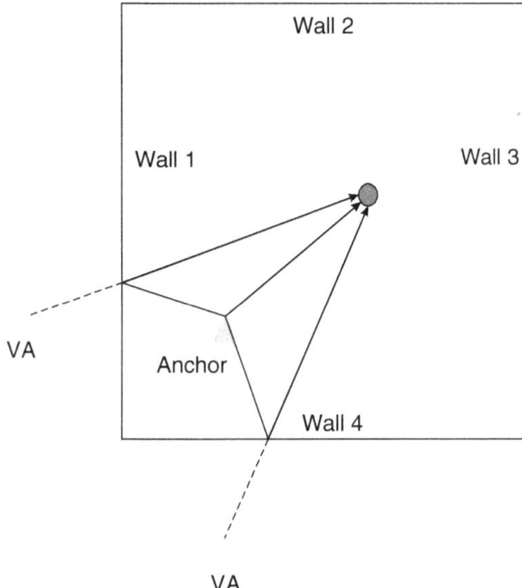

Figure 29.13 Virtual anchors for improved localization. Color version available at wiley.com/go/molisch/wireless3e.

The principle can be extended from single-reflection to other processes as well. In general, if the map of the environment is known, then the propagation path of any indirect path can, in principle, be determined, which in turn renders it equivalent to a LOS from a virtual anchor. In practice, the location of the walls might not be known accurately, or not known at all. In that case, a joint localization and detection of the environment are required; however, this usually requires multiple observations as, e.g., the agent moves around on a trajectory (see Section 29.10). This approach is related to the well-known Simultaneous Localization and Mapping (SLAM) problem.

The principle of using VAs is useful even when there are enough real anchors available for localization. Each VA provides additional information, thereby increasing the overall ranging information. Even if the positions of the VAs are not precisely known (e.g., due to inaccuracies of the underlying map), the large number of VAs makes their contributions useful, and often can overcome an unfavorable GDOP.

29.4 Direction-Of-Arrival (DoA)

The direction of arrival is also a quantity that can greatly help in localizing wireless nodes. The combination of TOA and DoA determines a node location almost completely (it provides one or two intersections between the circle describing the TOA and the line describing the DoA). A DoA encompassing both azimuth and elevation furthermore helps in 3D localization.

Methods for the determination of the DoA have been discussed in Section 9.5. Consider DoA determination at the anchor. The anchor then needs a calibrated antenna array with known orientation; the phase shifts of the signal at the different antenna elements allow the determination of the DoA. Just like for TOA, the main information that is useful for the localization is the DoA of the LOS component (though if the User Equipment (UE) is far away from the BS, the angle error created by reflections near the UE is minor). Thus, joint determination of DoA and TOA of the LOS component is often desirable, which can be achieved through maximum-likelihood estimation methods like SAGE (compare Section 9.5.7).

A major drawback of DoA is that even small errors in the angle estimation translate into large errors in the location of the agent when the distance between anchor and agent is large. Another challenge is that high-accuracy determination of the DoA requires large arrays with good calibration. Finally, arrays on a portable device (e.g., a cellphone) rarely allow DoA determination: (i) the orientation of the cellphone with respect to the absolute coordinate system is usually not known (though it could be estimated from gyroscopes or compass), (ii) the presence of the human user distorts the radiation patterns, and thus destroys a possibly existing calibration of the antennas.

Despite all these drawbacks, DoA can still be used in conjunction with other techniques, to refine existing estimates, or to provide rough estimates in the situations where other methods break down (e.g., an agent not having connections to enough anchor nodes to locate itself from TOA measurements only). DoA can also be very useful in indoor scenarios, where distances between agent and anchor are short, and even a relatively rough estimation of the angle can significantly improve the overall accuracy of a system.

The mathematical framework of Section 29.2.2 still applies. Define the measured angles as[2]

$$r_i = \phi_i + n_i = \arctan\left(\frac{y - v_i}{x - u_i}\right) + n_i, \quad i = 1, ..., I \tag{29.47}$$

and a vector

$$\mathbf{f}(\mathbf{x}) = \phi = \begin{pmatrix} \arctan\left(\dfrac{y - v_1}{x - u_1}\right) \\ \arctan\left(\dfrac{y - v_2}{x - u_2}\right) \\ \vdots \\ \arctan\left(\dfrac{y - v_I}{x - u_I}\right) \end{pmatrix} \tag{29.48}$$

and the distribution of the error-prone angles follows (29.14) with $\mathbf{d}$ replaced by ϕ. The resulting nonlinear and linear solution methods then again follow the principles of Section 29.2.2; for example, the cost function is defined abstractly as (29.17), now with the new definition of $\mathbf{f}(\mathbf{x})$, whose gradients can be easily computed. For details see [Zekavat and Buehrer 2019, Chapter 2].

An alternative to the use of explicit DoA is the direct use of the Channel State Information (CSI) at different antenna elements of an array. In other words, instead of using a single transfer function for the determination of a location, the set of transfer functions at the various array elements is used. Such an approach does not provide geometrically interpretable data; however, it can be applied for a fingerprinting approach (possibly in conjunction with Machine Learning), and in this case, improve the accuracy and remove ambiguities, see Sections 29.5 and 29.11.

29.5 RSSI and Fingerprinting

Besides the signal runtime, the received signal power is also an important quantity that varies with the distance between TX and RX. Thus, RSSI can form the basis for localization. It is particularly suitable for relatively narrow band systems (the definition of which depends on the desired precision), which often face challenges in obtaining sufficiently accurate TOA measurements. However, as we will see, there are other challenges in measuring distance/location on the basis of RSSI. In the following, we will discuss the following two types of systems: (i) range estimation based on RSSI, and (ii) Radio Frequency (RF) fingerprinting.

29.5.1 RSSI-Based Range Estimation

The principle of RSSI-based range estimation can be easily explained from the basic equation for received signal power, discussed in Section 7.1, and repeated here for convenience. On a logarithmic scale, the received power is often modeled as

$$P(d) = P(d_0) - \alpha_{\mathrm{PL}} 10 \log_{10}(d/d_0) + S + \xi \tag{29.49}$$

where S is a random variable describing the shadowing, and ξ describes the small-scale fading. RSSI-based range estimation is now based on averaging out (as much as possible) the small-scale and large-scale fading, and estimating the offset $P(d_0)$ and pathloss coefficient α_{PL}. Once that has been achieved, computation of the distance is obviously a simple task. With the given distance values, we can then use the trilateration techniques of Section 29.2 to determine the location of the desired node.

While this approach is appealing in its simplicity, it faces great challenges in practice:

- It requires multiple measurements, in order to be able to average out the small-scale and large-scale fading. Small-scale fading might be eliminated by averaging over frequency (but that is only possible in a system that is wideband in the sense of Chapter 6). Averaging out the large-scale fading might require a large number of spatially widely separated measurement points.

---

[2] Note that r does not describe a distance anymore, but the variable name is used to retain compatibility to Section 29.2.2.

- It requires calibration points (i.e., points with known distance from the BS) in order to obtain the intercept $P(d_0)$.
- It assumes that the model Eq. (29.49) is valid. However, it has been shown that for many situations a two-slope model is a better fit to the measurements. Furthermore, pathloss coefficients in different directions (as seen from the BS) can be different.
- The pathloss is influenced by the antenna pattern, including distortions of the pattern by objects within the near field (e.g., the human holding the device).
- The offset and pathloss coefficient can change quickly when moving from a LOS to an NLOS situation.

Due to all of these constraints, RSSI-based ranging is not in widespread practical use.

A variation on the theme of RSSI-based localization is cell-ID (Section 29.7.2), which only determines whether a node is within the radio range of a BS or not. This (rough) estimate is used sometimes in cellular, WiFi, and PAN networks.

In ad hoc networks, where many nodes can be interconnected, the *connectivity graph* can provide quite detailed information about the relative locations. In a graph with N nodes, on the order of $N!$ links can exist in principle. Assume a simple "protocol model," where each node is surrounded by a disc within which it can establish communications but cannot communicate to a node outside the disc. The goal is then to find a set of node locations that fulfills all the constraints (i.e., nodes with a connection to each other are within the communication radius, and vice versa).

29.5.2 Fingerprinting

Due to the strong fading (shadowing and small-scale), the received signal might vary considerably even at points that have very similar distance. While this fact is a problem for RSSI-based ranging, it can be actively exploited for high-precision localization by the so-called *fingerprinting* approach. Consider a situation where the UE measures the received power from several different BS. Let the measured value from the first BS be s_1. Due to noise, antenna effects, etc., we actually have to consider not a particular discrete value, but rather a range $[s_1 - \Delta, s_1 + \Delta]$. There will be many regions in the coverage area that have signal power in this particular range; and since the received signal varies over relatively small distances in realistic cases, those regions will be disjoint, and small in size. Now take the measurement to the next BS, s_2. Clearly, the intersection of the areas whose powers are in the range around s_1 from the first BS and s_2 from the second BS is a much smaller set. Now continue this process by measurements with respect to other BSs. With a sufficient number of measurements, we can narrow down the location of the UE very accurately. Since the set of measured RSSIs from the different BSs constitutes the "signature" of the location, localization from this signature is often called "fingerprinting."

It is interesting to compare this to the RSSI-based ranging. In that case, we aim to eliminate shadowing and other fluctuations, so that the area associated with one ranging measurement is an annular ring. The ring is relatively wide, because even small variations of the receive power, like 3 dB, translate into a fairly large distance range. We can then localize the UE at the intersection of the annular rings, but the width of the identified area in which the UE is will be much larger than the area identified by fingerprinting.

From the above description, it is clear that the UE (or the system) needs to know the set of locations a particular RSSI value from a BS corresponds to. For this reason, fingerprinting consists of two major steps:

1. *Training phase*, in which the RSSI created by a particular anchor node is measured or computed; these values are stored in a database.
2. *Matching phase*, during which the observed RSSI value is compared to the database, and the best match is determined.

The set of RSSIs or full-CSI fingerprints can also be used as input to supervised machine learning, as discussed in Section 29.11. Such an approach can be seen as a generalization of fingerprinting, and the below considerations about training phase, matching phase, and use of full CSI largely carry over.

Training Phase

The number of training points constitutes a tradeoff between effort and accuracy: measurement at more points will later on allow a more precise localization, but it requires more time during the training phase, and also requires more time during the (online) process of matching the RSSI measured by the UE to the stored values. The requirement for number of training points is greatly reduced if small-scale fading can be eliminated through temporal or frequency averaging. The coherence length of small-scale fading is on the order of a wavelength; having a database with such fine resolution (which is necessary to match the particular small-scale fading state) is very costly. Stationarity distance, i.e., coherence length of shadowing, on the other hand, is meters (indoor) or tens of meters (outdoor), see Section 7.1.

The recent popularity of smartphones with GPS has opened the possibility for "crowd-sourced" establishment of fingerprinting databases. As users with GPS-enabled phones drive (or walk) through an area, they can record the RSSI as well as the GPS coordinates at their various locations, and feed it to a database. These values can then later on be used by other UEs that do not have GPS capability. Of course, this method does not work in GPS-denied environments, where it is most urgently needed.

Due to the large effort inherent in measurement-based databases, computer-based models (such as ray tracing, see Section 4.7) are widely considered as an alternative for creating large-scale databases. The accuracy is lower, but the system can react more flexibly to changes in morphology (e.g., a new building being erected), which might change the RSSI from all surrounding BSs.

A factor that can lead to discrepancies between RSSI values in the database and during the measurements is the actual measurement device, and the human operator holding it. If the UE that wants to locate itself has a different antenna pattern or antenna gain compared to the device that was used during the establishment of the database, a mismatch can occur. Also, the human operator leads to distortions of the received signal, and thus leads to mismatch between recorded signature and observed one at the same location. Machine Learning is the most promising approach for overcoming these challenges.

Matching Phase

A large number of training points also leads to an increased computational effort during the matching phase. However, this problem can be greatly mitigated by smart searching strategies. A simple linear search through the database of training points is not an efficient method. Rather, a hierarchical approach can be more efficient. Assume that the fingerprints are sorted by the signal strength of the strongest connection. Then – even assuming some measurement errors – the UE can be assumed to lie on points whose signal strength of the strongest BS are within a range of values (note that the corresponding geographical locations need *not* be in a contiguous area). Then, the field strength of the second-strongest component of points within that set of locations is investigated, and the range of possible locations can be further narrowed down, and so on.

This way, the location algorithm can either find the location in the database whose fingerprint is closest to the measured one, or it can determine a set of K nearest neighbors, and then estimate the location as the center of gravity between them.

Fingerprinting with More Detailed CSI

RSSI is the most frequently used basis for fingerprinting, both because of its compact description, and because it is available on various externally accessible interfaces of the BSs and/or UEs, thus enabling third-party programs to perform localization. However, to the network operator or handset manufacturer, much more detailed CSI is available from the pilot signals, which provide the complex transfer function, possibly at multiple antenna elements. This additional information allows improved fingerprinting. In particular, it can resolve ambiguities that could arise in RSSI-only fingerprinting. Consider a situation where only training signals from a single BS can be received with reasonable strength (this can occur, e.g., in rural areas, or for Wi-Fi). Then there may be a multitude of locations that see identical or at least very similar RSSI, leading to a large standard deviation of the location estimate. If, on the other hand, impulse responses and directions (i.e., the Fourier transforms of the transfer functions at different antenna elements) are available, inability to distinguish two locations would occur only if the two locations have the similar CSI on all the subcarriers and antenna elements. Thus, it is much more likely that a unique location (or small area) can be identified.

On the downside, more detailed CSI also leads to a higher effort in training, as well as a more complicated matching phase, since more than just a single number characterizes each particular location. Hybrid localization techniques, where RSSI-based fingerprinting is used when it gives low standard deviation, and full-CSI fingerprinting covers the situations in which RSSI-fingerprinting is insufficient, are a possible way out of this dilemma.

29.6 Global Positioning System (GPS)

29.6.1   History and Types of System

The Navstar-GPS (NAVigation System with Timing And Ranging – GPS), commonly known simply as GPS, is based on a network of satellites that act as "flying anchor stations," emitting extremely precisely timed signals.[3] A GPS RX measures the runtime of those signals from the satellite to itself and also acquires the correct clock timing from the satellites. It can then perform a TOA localization. Note that GPS is a one-way system (i.e., signals only are transmitted by the satellites, they do not receive return signals), so that two-way ranging is not a possibility. The operating principle is thus comparatively simple, though the implementation details can be quite complicated.

Before going into the technical details, we first discuss the evolution of satellite navigation systems in general. The GPS system was the first widely used GNSS (Global Navigation Satellite System), though it built on some older, very application-specific, systems. GPS was originally intended for military applications, in particular, to provide accurate location information to mobile launch platforms for intercontinental ballistic missiles (such as submarines). When civilian applications became apparent, it was decided to open the system to civilian use.

GPS reached initial operational capability (i.e., 24 satellites aloft) in 1993; it has since increased to 32 satellites. At first, GPS provided lower accuracy (100 m vs. 20 m) for civilian users – in other words, the version available to the public contained consciously introduced errors, while the military users had additional information to obtain full accuracy. In 2000, this conscious reduction of precision was eliminated. The capabilities of GPS are being continuously enhanced particularly through the use of new signal structures.

---

[3] While this book generally does not concern itself with satellite communications systems, GPS has become such a vital component of cellular and other terrestrial-based systems that this chapter on localization would not be complete without at least a brief discussion.

The success of GPS has motivated other countries to deploy similar systems. Russia has deployed the GLONASS, the European Union is in the process of deploying the Galileo system, China the Compass system, and India and Japan have deployed smaller regional systems. Some of these systems provide interoperability (in particular GPS and Galileo), so that RXs are able to improve localization by exploiting the signals from both of these satellite systems. For simplicity, we will in the following only talk about GPS, though most of the material applies (sometimes with minor modifications) to other systems as well.

29.6.2   System Structure

The GPS consists of three segments: the space segment (i.e., the satellites), the control segment (ground stations controlling all aspects of the system), and the user segment (i.e., the GPS RXs).

Space Segment

There were originally 24 active satellites, arranged in six orbital planes with four satellites each, each plane having a 55° inclination to the equator. The orbital radius is about 26,500 km (20,200 km above ground), and a satellite makes two complete orbits in a day. Since 2012, there are 32 satellites, now providing a nonuniform constellation. The additional satellites provide redundant measurements and thus improve the accuracy. As we will see below, a GPS RX on the ground needs LOS to at least four satellites in order to get a navigational fix. With the 24-satellite constellation, typically, 6–10 satellites are visible if there are no obstacles; in street canyons, a much smaller number of satellites may be visible.

Control Segment

The control segment consists of a Master Control Station (MCS) (with an alternate), four dedicated ground antennas, and six monitor stations. The monitor stations track the trajectories of the satellites (which can drift, and which particularly can be impacted by imprecise repositioning maneuvers). It sends the information about the new trajectories (ephemerides), as well as clock timing that forces the precision clocks (hydrogen masers) aboard the satellites into tight synchronization. The MCS performs all the administration, software, control, etc.

User Segment

The user segment consists of the actual GPS RXs. This part of the system has undergone the largest changes, from bulky, expensive devices in the 1990s, to cheap RX modules of size less than $1\,\text{cm}^2$, which can be found in almost any cellphone. As we will see below, the GPS satellites transmit spread-spectrum signals with precise timing, so that the RX has to perform acquisition of CDMA signals (which provides the timing) and reception of the data (such as satellite position) modulated onto those signals. These operations are similar to those described in Chapter 19, so that we will not further discuss the associated RX structures.

As we know from Section 29.2, a RX needs TOA measurements with respect to three anchors to obtain a unique position fix. However, a GPS RX does not have its own precision clock, but rather has to obtain the timing information from the GPS satellites as well. This means that there is another unknown variable in the system, namely the bias between the local clock and the constellation clock. This in turn means that we need measurements with respect to one more satellite, for a total of 4; this can also be thought of as performing TDOA. When more satellites are available, the RX can either select the four "best ones" (this is used in simple GPS RXs that have very limited computational power), or combine the signals from all the satellites for improving the overall accuracy. Note that in addition to the range errors, the location error is also impacted by GDOP. Depending on the satellite constellation, the GDOP can increase the location error by a factor of 10 or more compared to the ranging error. This is especially critical in street canyons, where the GPS RX not only sees a smaller number of satellites, but also these satellites all tend to be in one slice of the sky, so that GDOP errors are considerable worse. Figures 29.14 and 29.15 compare the error for an open-sky scenario and an urban canyon scenario.

The determination of the user location can employ the fundamental TOA/TDOA locations discussed in Section 29.2. However, either iterative or linearization methods strongly depend on the initialization of the search. If the GPS RX has to search through a large variety of possible locations, then the computation of the location can take a long time.

*29.6.3   Signal Structure

Satellite Frequencies

GPS transmits its signal on three frequencies: L1 (1575.42 MHz carrier frequency, band 1563–1587 MHz); L2 (1227.60 MHz, band 1217–1237 MHz), and L5 (1176.45 MHz, band 1164–1188 MHz). These frequencies can be derived by multiplication from a 10.23 MHz oscillator (a standard crystal frequency), by multiplication with 154, 120, and 115, respectively. Galileo and GLONASS have frequency bands that partly overlap with GPS.

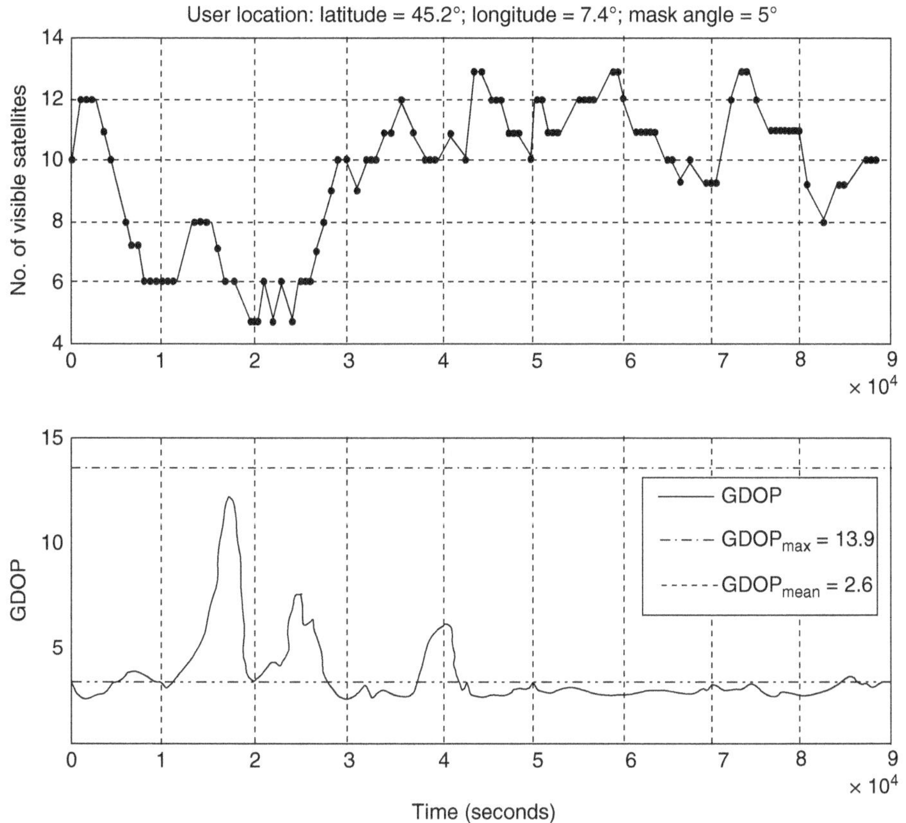

Figure 29.14 GDOP for a GPS constellation over Italy, open sky environment.
Reproduced with permission from [Zekavat and Buehrer 2019] © J. Wiley and Sons, Ltd.

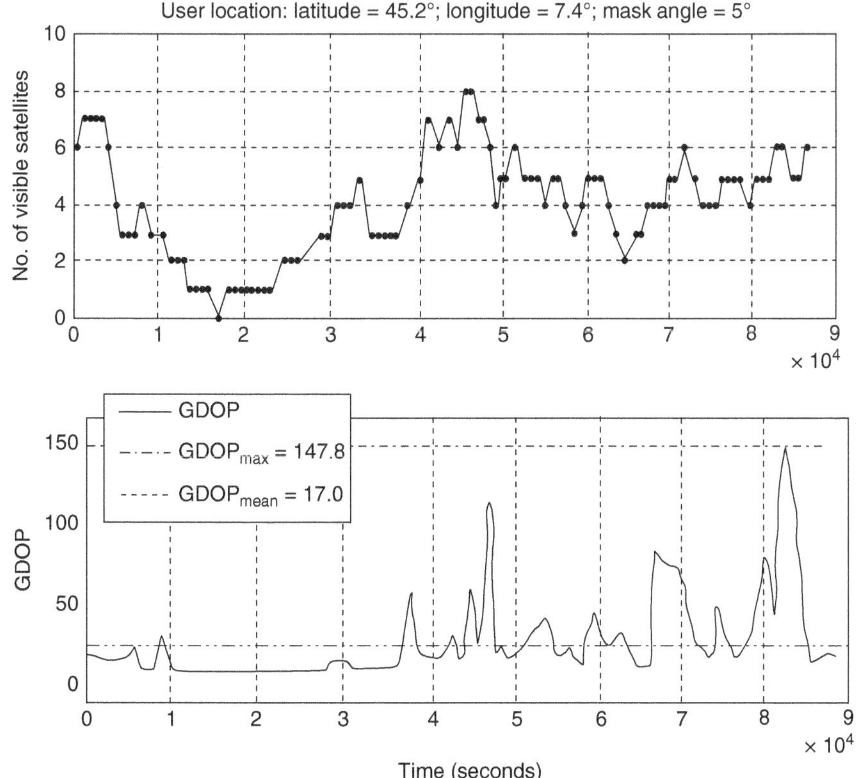

Figure 29.15 GDOP for GPS signal in a street canyon in Italy.
Reproduced with permission from [Zekavat and Buehrer 2019] © J. Wiley and Sons, Ltd.

Modulation Format

The modulation format of the GPS signal is either *Binary Phase Shift Keying* (BPSK), or *Binary Offset Carrier* (BOC), modulation. A BOC signal is characterized by a subcarrier frequency m and a chip rate n. Both are expressed in multiples of $f_R = 1.023$ MHz, i.e., $1/10$ of the standard oscillator frequency. The BOC signal is

$$s(t) = \sum_i c_i d_i \text{rect}\left[t - \frac{i}{f_R n}\right] \text{sign}[\sin(2\pi m f_R t)] \tag{29.50}$$

where c_i are the code chips, and d_i the data.

As we will see below, most of the spectral occupancy is due to the acquisition signals which provide the timing to the RXs; actual payload data are transmitted with a rate of 50 bit/s. Figure 29.16 shows the transmission parameters of all of the different types of codes; Figure 29.17 shows a simplified block diagram concentrating on the most important codes and carriers.

Band	Service	Signal	Modulation Scheme	Spectral Occupation (MHz)	Code Rate (cps)	Navigational Data Rate (bps)	Minimum Rx Power (dBW)	Status
L1	C/A	C	BPSK (1)	2.046	1.023×10^6	50	−157.0	T
L1/L2	P	M	BPSK (10)	20.46	10.23×10^6	50	−161.5	T
L1	L1C	C	TMBOC (6, 1, 4/33)	4.092	CP 1.023×10^6 CD 1.023×10^6 CO 100	No data 50 or 75 No data	−157.0	F
L2	L2C	C	BPSK (1)	2.046	CM 511.5×10^3 CL 511.5×10^3	25 No data	−160.0	T
L1/L2	M	M	BOC (10,5)	30.69	c.g.	N/A	−158.0	T
L5	L5	C	QPSK (10)	20.46	I5 10.23×10^6 Q5 10.23×10^6	50 N/A	−154.5	F

Figure 29.16 Types of GPS signals and their modulation format. c.g., cryptically generated; M, military; C, civilian; T, transmitted; F, foreseen. BOC(m,n) is s(t) in (29.50) with m,n. TMBOC: time-multiplexed BOC.
Reproduced with permission from [Zekavat and Buehrer 2019] © J. Wiley and Sons, Ltd.

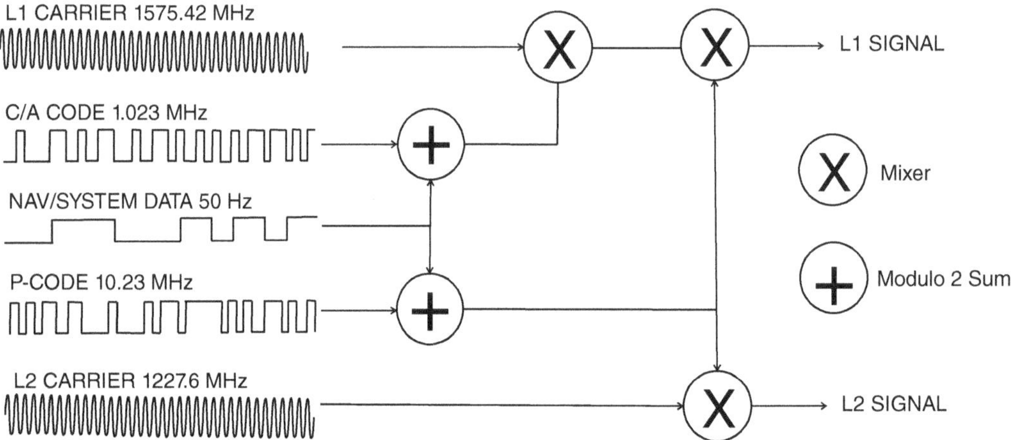

Figure 29.17 Modulation of GPS coding signals onto carriers L1 and L2.
Reproduced with permission from [Dana 1998] © P.H. Dana.

Codes

There are two spreading sequences that are transmitted in GPS: the first is the *coarse acquisition sequence*, C/A, which is a 1023 bit pseudorandom sequence, transmitted at 1.023 Mbit/s, and thus repeating every millisecond. The C/A sequences transmitted from the different satellites are different, and thus allow ranging to each of the satellites. The second is the precision (P) code, which is another pseudorandom sequence, but with a length of $6.2 \ 10^{12}$ bits, so it repeats only once per week. Signals transmitted from different satellites are different parts of the same fundamental code. As it is transmitted with a chiprate of 10.23 Mchip/s, it provides higher accuracy. The code is encrypted with a further code to prevent spoofing. The encryption code is secret. However, since it is applied with a frequency of only about 500 kHz (i.e., much slower than the chiprate of the precision code), RXs can be built that partly exploit this code without needing to know the encryption code.

GPS Message Format

Besides the timing, a GPS RX needs to know the precise position of the satellites, as well as some messaging information. This information is transmitted at a rate of 50 bit/s, as described above. The navigation message consists of three parts:

(i) GPS date and time, which is required for proper synchronization of all components.
(ii) Ephemerides, i.e., orbital data that allow the GPS RX to compute the position of the satellite. A RX cannot obtain a position fix without the ephemeris data of a satellite.
(iii) Almanac information, which also contains information about location of satellites and pseudo-random sequences. Almanac information is very detailed, and valid for a long time (which means a GPS RX usually has to download it only during a cold start; it allows the RX to determine which satellites to look for).

A navigation message consists of 1,500 bit frames (30 s), which is divided into five subframes. The first subframe provides the GPS time, and information on timing correction; the second and third subframes contain the ephemeris data; and subframes four and five contain 1/25th of the almanac data (i.e., a RX needs to download 25 frames before it has the complete almanac).

29.6.4 Localization Methods and Accuracy Enhancements

Differential GPS

GPS errors fall into two categories: noise-induced errors, which are independent between different locations of the RX, and systematic errors. In the second category, the most important is signal delays in the ionosphere but also inaccuracies in the ephemeris data; these errors are usually identical over a larger region. One way of reducing the impact of those errors is to have a GPS RX at a known location but also determine the location through GPS localization from it. The difference between the location obtained from GPS, and the known correct location, allows to get a correction for the systematic errors, which can then be applied to other devices in the vicinity; note that the correction has to be transmitted through a terrestrial system such as cell phones.

Assisted GPS (A-GPS)

A slight variation on the theme of error correction is *Assisted GPS* (A-GPS). As we have seen above, GPS RXs starting from a "cold start" can take a long time to acquire satellites, in particular when the received signal is weak. The acquisition time, and even the acquisition probability, can be greatly reduced if the RX knows its approximate location. In many cellular systems, the BS can provide this information (e.g., through cell ID), at least to the accuracy of a cell diameter. Due to the wide availability of A-GPS, even indoor localization is now possible with GPS in more than 50% of all cases.

29.6.5 Alternative Systems

As mentioned in the introduction, several other countries or regional organizations have launched Navigation Satellite Systems. An important example is the GALILEO system, which consists of 30 satellites (intended), and is designed to provide accuracy that is improved compared to GPS (1 vs. 3 m). The basic operating principles are the same as for GPS, but there are some important technical differences, often arising from the fact that it was specified considerably later, when more complicated designs had become technically feasible.

Galileo operates on frequencies that are similar to, but somewhat offset to, the GPS frequencies, namely 1575.42, 1191.795 MHz (which is split into two subbands with 1176.45 and 1207.14 MHz), and 1278.75 MHz. Similar to the L1 codes in GPS, it uses long spreading codes, but while GPS uses Gold codes, Galileo uses hand-selected memory codes (i.e., stored codewords), which are claimed to provide better tracking performance and improved interference mitigation. Modulation formats are BPSK or various forms of BOC, which is again somewhat similar to GPS, though the details of the BOC differ. Details of these formats can be found in [Lohan et al. 2015]. The Galileo satellites have, for the purpose of redundancy, multiple Rb clocks on board, as well as two passive microwave masers, which provide enhanced accuracy. Use of Galileo RXs is becoming mandatory in the European Union for emergency localization.[4] Most smartphones intended for worldwide distribution thus have Galileo (as well as GPS) RXs built-in.

29.7 Localization in Cellular Systems

29.7.1 Introduction

While GPS RXs are nowadays present in most cellphones, and GPS is the standard "first line of defense" for localization, it is not sufficiently reliable to serve as a sole means of localization. As mentioned in Section 29.6, GPS often cannot acquire a sufficient

---

[4] In this book, like in most of the technical literature, we call emergency localization E911 localization. However, European documents usually call it E112 localization, after the emergency phone number used throughout Europe

number of satellites when the UE is indoor, or in a deep street canyon. For this reason, alternative localization methods need to be provided. The most important of those are:

(i) RSSI fingerprinting: in this approach, the UE observes channel characteristics, in particular the RSSI, of the connection to various BSs. The combination of such values constitutes a "fingerprint" characteristic for a particular location. By comparing this fingerprint to a database of stored values, the location of the UE can be identified. This method need not be included in any wireless standards, but it is widely used, see Section 29.5.

(ii) For BSs and/or Wi-Fi access points: if a UE is associated with a particular cell, then this means automatically that it is within the coverage region of a particular BS. Listening to the strength of the beacon signal from this and other BSs can provide additional useful information. Similarly, observation of beacon signals from Wi-Fi access points provides implicit location information.

(iii) Runtime measurements between UE and BSs. Since UEs usually do not have precise (i.e., better than within signal runtime) timing synchronization to the BSs, TOA is not possible, so that TDOA has to be used. The runtimes can be measured either in the uplink (UTDOA) or in the downlink (OTDOA). Also, the timing advance measurements (see Section 18.3.2) provide some rough information about the runtime between UE and serving BS.

(iv) Machine Learning: based in signal strength, wideband channel characteristics, or even multi-antenna channels, machine learning can be used to determine UE positions; this is discussed in detail in Section 29.11.

In the following, we describe these methods in more detail. They can either be used by themselves, in conjunction with each other, or even combined with data from IMUs, and/or GPS data.

29.7.2   Cell ID

The simplest form of localization is just to determine which BS is currently serving the UE. Obviously, the UE is then within the coverage region of this BS. Since in particular in urban environments cell radii are becoming smaller and smaller, it is often possible to obtain UE location accuracy on the order of 300–500 m purely by this information, called *cell ID*.

Further improvements can be made when the UE listens to which Wi-Fi access points it can hear (it does not matter if those Wi-Fi networks are password protected, as the UE does not need to connect, but just needs to hear the "beacon" signal). Since the coverage region of access points is usually below 100 m, high localization accuracy can be achieved by such identification. However, note that access points are usually outside the control of network operators, so that the location of the access points needs to be found by secondary means (e.g., devices that record observed beacons together with GPS-derived coordinates).

Estimates based on cell ID can be improved by additional measurements, leading to the so-called *enhanced cell ID*, E-CID. Such additional measurements include

- Round-trip time between BS and UE, which can be deduced, e.g., from the timing advance. The UE is then in an annular ring whose mean radius corresponds to the measured timing advance, and whose width is determined by the uncertainty of the timing advance measurement. Overall, the location is then known to be in the intersection between annular ring and cell coverage area, which leaves a large uncertainty in the azimuthal coordinate.
- Receive power, typically using simple models for the relation between path gain and distance to obtain an estimate of the distance BS to UE.
- Angle of arrival: if multiple antenna elements are at the BS, and some (approximate) calibration has been done, a determination of the angle of arrival can be performed in a number of situations. This is a good indication of the direction in which the UE is located; this info can then be combined with info from the received power and/or timing advance to obtain a more accurate estimation of the location.

29.7.3   OTDOA

Consider first the case that all BSs are perfectly synchronized to each other. Then location can be obtained if the runtime of the signal between the UE and at least four BSs can be measured: obviously, the runtime from the UE to the serving BS will be measured; in addition, the runtime to several of the neighboring BSs needs to be determined – at least 3, but more is better as it increases the measurement precision and reduces the risk of strong GDOP.

In Observed TDOA (OTDOA), it is the BSs that transmit signals and the UE that determines their arrival times. The UE can then either feedback those arrival times to its serving BS (which then does the localization), or it can find its position by doing the TDOA localization itself, and feed that back to the serving BS. The two main challenges are:

(i) *Finding a suitable signal for ranging*: either an existing signal component needs to be used (i.e., a signal component already foreseen for other purposes), or a dedicated ranging signal.

(ii) *Interference*: since the runtime has to be measured, inter alia, from BSs that are farther away from the UE than the serving BS, the signal power from those far-away BSs is lower, while the interference is usually higher. Thus measures to improve the signal-to-interference ratio need to be found.

Let us first consider the ranging signal. A prearranged signal (such as a beacon or pilot), has to be transmitted from each of the BSs. The signal has to be unique to each BS, so that the UE can differentiate between the signals from the different BSs. As mentioned above, the ranging signal could be a signal that is already used for other purposes. A cell-wide (downlink) pilot signal, which is used for channel estimation, would in principle be a suitable signal: after all, it is designed to allow an estimation of the impulse response, and determination of the range is (as discussed above in Section 29.2), a sub-problem of impulse response estimation. However, there is a major difference: in channel estimation, the UE only aims to obtain the impulse response from the serving BS. The required SIR is thus similar to the Signal-to-Interference Ratio (SIR) of the payload data.[5] For localization, the UE needs good SIR even for signals that stem from far-away BSs. For this reason, modern systems (e.g., the Long-Term Evolution (LTE) standard) foresee a separate signal that is used exclusively for location.

The main advantage of this separate signal is that it can be assigned a much larger spatial reuse factor than the regular pilots (and the payloads). Such a larger frequency reuse factor reduces the area spectral efficiency, and would thus not be advisable for regular pilots. However, the location pilot is sent out only rarely, and thus does not influence the overall area spectral efficiency significantly; it can thus be acceptable to use the large frequency reuse, while reaping the benefit of higher SIR.

Use of the downlink signals for ranging also has another important advantage (compared to the use of uplink signals, see below): the transmit power of the BS is quite large, so that in a noise-limited situation, OTDOA can provide ranging with many BSs.

As the name implies, OTDOA is a TDOA scheme, so that the BSs and the UE do not have to be synchronized to each other. The BSs, however, do have to be accurately synchronized among themselves. This is relatively easy to achieve when each BS has GPS connectivity and can thus be synchronized by the GPS clocks; in those cases where that is not possible (e.g., indoor BS), accurate internet timing protocols, such as White Rabbit, can be used to achieve below-ns accuracy.

The most important practical example for OTDOA is in the LTE/NR standards, which employs localization pilots, or – in LTE parlance – *Position Reference Signals (PRS)*. These signals show some similarity to other types of reference signals (compare Sections 31.3.5 and 32.3.5), but are sent at less frequent intervals. Their bandwidth can be up to the system bandwidth. Their transmission can be triggered either by the UE, or the network (more precisely, the location server of the network).[6] The PRS is then sent over a number of consecutive subframes, and usually arranged in such a way that only a small subset of the BSs is transmitting simultaneously, in order to minimize interference. Furthermore, payload data transmission that might interfere with the PRS may also be reduced or completely avoided. The network also sends some assistance data to the UE, such as transmission bandwidth, neighbor cell lists, and number of consecutive subframes, which helps the UE with the detection process. The UE then measures the incoming signal and sends the results to the location server, which then performs the estimation of the location.

29.7.4 UTDOA

An alternative approach is Uplink TDOA (UTDOA), where the UEs send out ranging signals, and the BSs determine their arrival times. This approach, while also using TDOA, has some important implementation differences from OTDOA.

The first relates to the signal that can be used for ranging. Clearly, a dedicated signal could be foreseen in the standards, just like there is for OTDOA. However, this has not been included in any of the major cellular standards, because the number of distinct ranging signals would have to be proportional to the number of UEs that want to perform ranging simultaneously (as opposed to OTDOA, where the number of distinct ranging signals is given by the number of BSs that want to send out ranging signals without overlap, i.e., the reuse factor for ranging). Rather, in LTE, a particular type of uplink training sequence, the SRS (see Section 31.3.5) is used. The Sounding Reference Signal (SRS) is a training sequence (pilot sequence) that extends over the whole bandwidth and is different for different users, and in different cells. The structure of the SRS in each cell is known a priori to all other cells, so that the BS can easily determine the runtime at each BS (which still includes the timing offset between the UE and the BS), and from that the TDOAs (where the difference is usually taken with respect to the time at the serving BS).

UTDOA also offers the possibility of using the payload signal itself for ranging: the serving BS receives and "cleans up" the signal (i.e., demodulates and decodes, and then re-encodes, which eliminates noise and channel distortion).[7] From the serving BS, the clean signal can then be forwarded, via the backbone connections, to the other BSs in the vicinity. Each of those other BSs then correlates this signal with the signal that they receive over the air. This allows to determine the runtime of the signal from the UE to it (of course, taking into account the processing time and transmission time via the backhaul links), using the signal essentially like a pilot sequence.

While UTDOA based on payload data shows good performance in systems for speech transmission, it faces challenges in data-heavy systems, because data transmission is bursty, and might be assigned different time–frequency resources for different data bursts.

29.7.5 Wi-Fi, UWB, and Bluetooth

For indoor localization, cellular ranging is not precise enough, partly due to the fact that the BSs are far away from the target UEs, and the elevation (i.e., the floor on which the UE is) cannot be easily found. More precise localization can be obtained by ranging

---

[5] Though an improvement by some dB is often advantageous, leading to techniques such as pilot boosting.

[6] In the case of network-triggered localization, the UE can reject the request depending on the privacy settings.

[7] In OTDOA the payload cannot be used for determination of the correlation, because each BS sends out a different payload signal, and there is no possibility of communicating to the UE the "correct" payload data of a nonserving BS.

with Wi-Fi access points, since those are typically much closer, and also provide a larger bandwidth (up to 160 MHz in the latest incarnation of the Wi-Fi standard, with impending further increases, see Chapter 33). The principles of localization remain the same – measurement of the runtime with respect to the anchor nodes, and TDOA localization. The TDOA localization can be combined with (or replaced by) RSSI-based methods as discussed below. A challenge for Wi-Fi localization is that the access points are not necessarily synchronized, or only synchronized with fairly low accuracy through the internet-provided timestamps. Still, Wi-Fi-based systems may provide accuracy within a few meters, which is sufficient to identify the room or office in which a UE is located. The Advanced Mobile Location (AMS) system mandated in Europe uses a combination of satellite localization and WiFi.

Even better accuracy can be achieved with *Ultrawideband* (UWB) signals. Since the accuracy of TOA measurements is partly determined by the bandwidth of the ranging signal, the 500 MHz or more provided by UWB signals allow very high precision localization. In LOS situations, ranging accuracy on the order of millimeters has been reported, though for NLOS and larger distances between TX and RX, experimental results are considerably worse (on the order of tens of cm or more). The IEEE 802.15.4a standard is explicitly designed for UWB-based localization, see Section 34.2.

Yet another alternative is Bluetooth, a standard for personal-area networks. Since it is designed mainly for short-range communications, ability to detect a beacon signal from a Bluetooth TX can be used for proximity detection (further refinements of the distance are possible based on the observed RSSI), see Section 34.1.

29.8   Radio Frequency Identification (RFID)

29.8.1   History

RFID is a system that identifies an object from a signature it returns when polled (typically by illumination with a probe signal). According to this general definition, any communications system in which a node gives a predefined answer to a polling (where the answer is different for different wireless nodes) is an RFID system. In this section, we will particularly concentrate on passive RFID systems, where each identifying node or "tag" does not have its own power source, but rather reflects probing radiation (which comes from an active device called *reader*) in a way that is unique for this particular tag. In other words, RFID is based on the principle of backscattering, where the characteristics of the backscatter identify a certain device. The obvious advantage of such passive backscattering is that the tag does not need its own battery, so it has no expiration date and can be manufactured in a cheaper way than active tags. The drawback of using backscatter is the much smaller range over which identification (and localization) can be achieved, since the backscattered power is very low.

Identification based on backscattering goes back at least to the Second World War, when the IFF (Identification Friend or Foe) was based on modulating incoming radiation from a radar TX in a manner that identified friendly airplanes. During the 1960s and 1970s, relatively simple and inexpensive backscatter modulators were invented, and various applications emerged: automated toll collection systems, tracking of people, etc. During the 1990s, various standards evolved for tracking of animals, contactless proximity cards, and finally product labeling. This latter application, which was enabled by the global Electronic Product Code (EPC) made RFID a vital part of supply chain management. Many of the advances in logistics since 2000, e.g., in e-commerce, are due to the automated supply administration enabled by RFID.

29.8.2   Operating Principle

The fundamental principle of a passive RFID tag is as follows: the reader sends out a polling signal, which is coupled into the tag. The coupled signal is either directly modified by the tag, or rectified and used to power internal circuitry that creates a return signal. Depending on the frequency emitted by the reader, and the distance between TX and RX, the tag is in the near-field or the far-field of the reader, with the coupling mechanism changing accordingly (Figure 29.18).

Near-Field Coupling

For near-field coupling, the transmit antenna is usually a coil, which inductively couples its energy into a coil on the tag. The tag communicates its identity through load modulation, i.e., changing of the load, and thus the current, on the tag. Since the tag coil and the reader coil are inductively coupled, a change in the tag current affects a small change in the reader current, which can be detected by the reader.

Far-Field Coupling

In the far field, the modulation is slightly different: the tag antenna receives the incoming signal, and a part (which depends on the antenna impedance) of the radiation is reflected back. Changing the load impedance thus changes the backscattered radiation; this effect is used to convey information.

Surface Acoustic Wave Filters

Yet another approach is Surface-Acoustic-Wave (SAW) filters, where the tag converts (through the piezo-electric effect) the incoming electromagnetic radiation to acoustic waves, which are then reflected by acoustic reflectors (unique for each device); they are then re-converted to electromagnetic radiation and radiated back toward the reader.

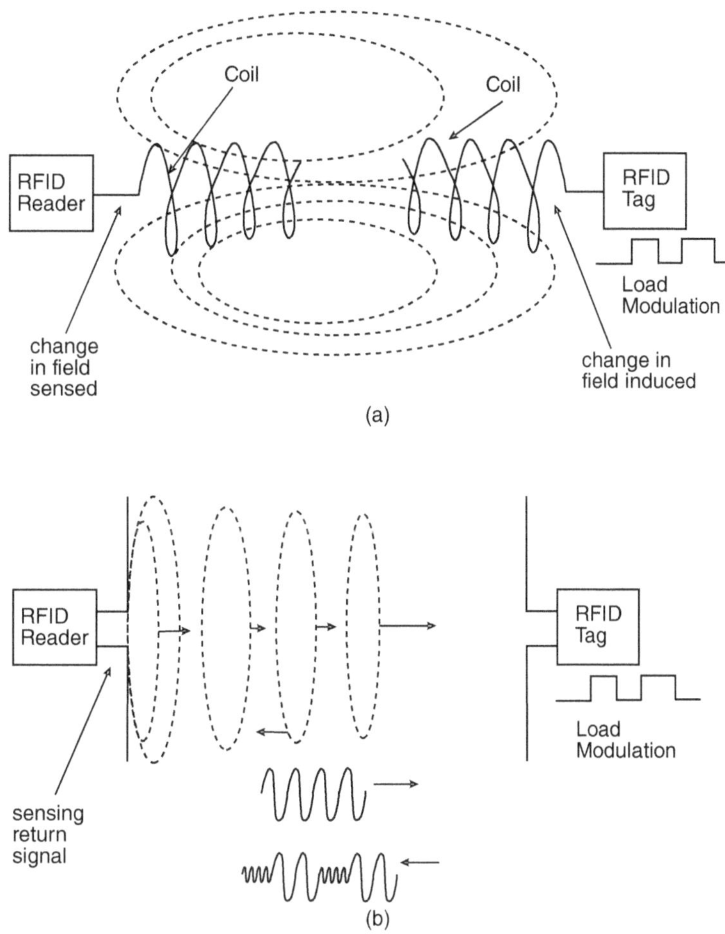

Figure 29.18   Principle of near-field (a) and far-field (b) RFID.

Operating Frequency

Nearfield RFID readers mostly work in the 13.56 MHz band, while far-field devices are in the 900 MHz and 2.45 GHz bands. This means that near-field coupling works at distances up to about 3 m in the 13.56 MHz band, while 2.45 GHz devices almost always use far-field coupling.

29.8.3   Localization Using RFID

The simplest form of localization with RFID is proximity sensing. Take the example of a contactless access system: a gate will open only when the RFID tag will come within close distance to the reader; this automatically establishes the location of the tag. By establishing "gateways" (readers) in a storage hall, tags (and the products they are mounted on) can be localized based on which gateways they passed through.

*29.9   Cooperative Localization

29.9.1   Intuitive Picture

Up to this point, we have assumed that agent nodes can only determine their range with respect to the anchor nodes but not with respect to other agents. This seems reasonable at first glance, since the location of other agent nodes is unknown, so that knowing the range from them does not appear to carry useful information. However, Figure 29.19 shows intuitively why agent cooperation is useful:

Agent node 2 and Agent node 4 each try to determine their locations from the anchor nodes (nodes 1, 3, 5). However, since node 2 has connection only to two anchor nodes (nodes 1 and 3), the location cannot be determined exactly – it could be either of the two intersections of the circles corresponding to the runtime between agent and anchor; a similar situation holds for agent node 4. In a cooperative scheme, the agent nodes then additionally determine the distance between each other. Only one set of node locations is consistent with that measured distance, while the "ghost locations" (marked by filled circles) are not consistent with those estimates.

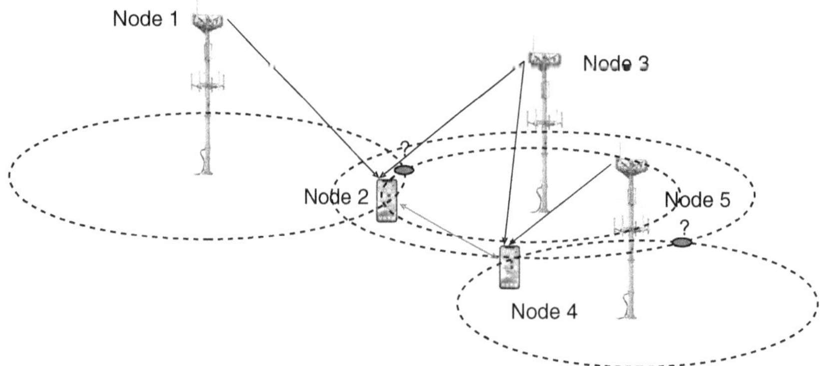

Figure 29.19 Principle of cooperative localization. Color version available at wiley.com/go/molisch/wireless3e. Reproduced with permission from [Wymeersch et al. 2009] © IEEE.

Thus, a cooperative scheme can obtain accurate localization where the noncooperative scheme fails. Similar results hold if the errors in the location in each node are caused by noise.

29.9.2 Fundamental Limits

The fundamental limits of the localization accuracy can be determined from the CRLB, where the FIM is defined in Eq. (29.39). In a conventional localization system, the FIM is a block-diagonal matrix, where each block characterizes the (2D) location of a particular agent; in other words, each of the blocks has the form Eq. (29.43). In a cooperative localization system, on the other hand, there are additional measurements between the agents available, so that the off-diagonal blocks are also filled with nonzero values, see Figure 29.20.

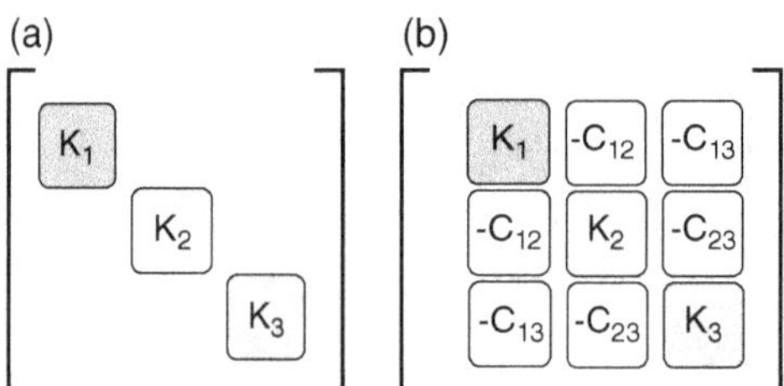

Figure 29.20 Structure of Equivalent Fisher Information Matrix for noncooperative (a) and cooperative (b) case. Color version available at wiley.com/go/molisch/wireless3e.

The *Square Position Error Bound* (SPEB), of a particular agent k, can now be computed as

$$E\left\{|\hat{\mathbf{x}}_k - \mathbf{x}_k|^2\right\} \geq SPEB(\mathbf{x}_k) = tr\left[\mathbf{I}(\boldsymbol{\theta})^{-1}\right]_{2 \times 2,k} \tag{29.51}$$

which is a 2×2 submatrix of the inverse FIM. The FIM itself consists of two terms, describing the localization information from the anchors, and from the agents' cooperation. Computation of the SPEB according to Eq. (29.51) requires the inversion of a high-dimensional matrix; if we are interested only in a subset of parameters (as is usually the case), then we can invert the Equivalent Fisher Information Matrix (EFIM) instead. Say that the parameter vector $\boldsymbol{\theta}$ consists of the interesting part $\boldsymbol{\theta}_1$ and the part $\boldsymbol{\theta}_2$ that we are not interested in, $\boldsymbol{\theta} = \left[\boldsymbol{\theta}_1^T \, \boldsymbol{\theta}_2^T\right]^T$; in that case, the FIM is written as

$$\mathbf{I}(\boldsymbol{\theta}) = \begin{bmatrix} \mathbf{A} & \mathbf{B} \\ \mathbf{B}^T & \mathbf{C} \end{bmatrix} \tag{29.52a}$$

then the EFIM, which is defined as

$$\mathbf{I}_e(\boldsymbol{\theta}_1) = \mathbf{A} - \mathbf{B}\mathbf{C}^{-1}\mathbf{B}^T \tag{29.52b}$$

contains all the necessary information for computing the SPEB. The EFEM itself has an interesting structure: it is built up of 2×2 blocks (each block corresponding to an x and y position). The kth block along the main diagonal contains the information from ranging of the k-th agent with the anchors, while the i, k (off-diagonal) block contains the information from the cooperation, see Figure 29.20.

Each block in the entry can be written as the product of a scalar (which represents the "quality" of the range estimate), with a matrix of the form

$$\begin{bmatrix} \cos^2(\phi) & \cos(\phi)\sin(\phi) \\ \cos(\phi)\sin(\phi) & \sin^2(\phi) \end{bmatrix} \tag{29.53}$$

which indicates the direction along which the ranging information helps to improve the position information. This so-called *information intensity* is related to the GDOP – ideally, the information intensity in two orthogonal directions is equally strong, which corresponds to low GDOP.

29.9.3   Algorithms

The implementation of cooperative localization can be done in a centralized manner, when a central control node receives the ranging estimates from all nodes. However, this requires a considerable communication overhead. Alternatively, a distributed iterative implementation is possible. The most straightforward (though not best-performing) method works as follows: each node broadcasts its current location estimate; then receives the location estimates from all the neighbors, updates its location estimate, and continues this process until convergence is achieved. Faster convergence can be obtained through belief propagation algorithms, similar to what is used in the decoding of Low-Density Parity Check (LDPC) codes (see Section 13.7). Note that in any case, a round of iterations has to be finished before nodes move significantly. For a more detailed survey of algorithms, see [Buehrer et al. 2018].

*29.10   Tracking

29.10.1   Motivation for Tracking

The previous sections showed how to localize an agent based on a single snapshot, i.e., determine a specific location. This is useful, e.g., for localization of sensors, laptops, and other fixed or nomadic devices; it is also the most frequent situation for E911 localization, since it starts the determination of the location only when the call begins (after all, the system does not know beforehand that such a call would be made).

There are, however, also many situations where we want to track the location of an agent over time, i.e., follow the locations of a UE as it moves around. Of course, it would be possible to see such a problem just as a sequence of isolated location estimation problems, in other words, apply the techniques from the previous sections to one snapshot at a time, determine the location for each time instant, and then assume that the trajectory as the curve connecting those "snapshot locations." However, this would be suboptimum whenever noise and uncertainty affect the underlying measurements – in other words, almost always. A better solution is based on Kalman filters and their variants and extensions. The following will present a very brief summary of these filters without a derivation or justification; see the "Further Reading" section at the end of this chapter for more detailed material.

29.10.2   Linear Kalman Filters

Kalman filtering implements, iteratively, a two-step process, where first the future position is *predicted* from the past, and then *corrected* from a current measurement. Both the prediction and the measurement are noisy, and the Kalman filter finds a good way of combining these two pieces of information to get a better estimate. Figure 29.21 shows this fundamental principle.

We now turn to a mathematical description. The key starting points for all computations are the *state equation* and the *measurement equation*. The state vector $\mathbf{x}$ contains, in the case of localization, the position variables (e.g., x and y for a two-dimensional localization), and possibly velocity components and acceleration components (see below). The state at time t_k, also simply written as $\mathbf{x}_k$, is modeled as a linear function of the state at time t_{k-1}, plus some noise

$$\mathbf{x}_k = \mathbf{A}\mathbf{x}_{k-1} + \mathbf{B}\mathbf{u}_{k-1} + \mathbf{D}\mathbf{n}_{\mathrm{s},k-1} \tag{29.54}$$

where $\mathbf{A}$ is the state transition matrix, $\mathbf{u}$ is the external input, and $\mathbf{n}_\mathrm{s}$ describes the uncertainties of the dynamics, i.e., the deviation from the simple linear model, and the matrices $\mathbf{B}$ and $\mathbf{D}$ describe how the external input and the state uncertainty map into the state variables. Furthermore, the measurement results are related to the state as

$$\mathbf{z}_k = \mathbf{C}\mathbf{x}_k + \mathbf{n}_{\mathrm{m},k} \tag{29.55}$$

where $\mathbf{z}$ are the measurement variables (e.g., runtime), and $\mathbf{n}_{\mathrm{m},k}$ is the measurement noise at time k. Note that both the state equation and the measurement equation assume linear relationships; for the cases that this is not fulfilled, generalized versions such as the *Extended Kalman Filter (EKF)* need to be used as discussed below. The two equations (29.54, 29.55) are fairly general. In many location

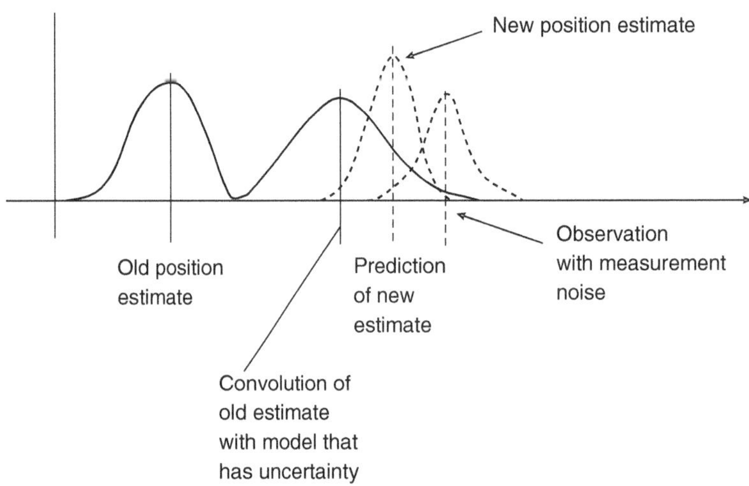

Figure 29.21 Principle of Kalman filter for tracking.

tracking applications, there is no external input (such as a known acceleration imposed by an external control), so $\mathbf{u} = 0$. Furthermore, the state uncertainty is assumed to be Gaussian distributed with covariance matrix $\mathbf{S}_k$ (and $\mathbf{D} = \mathbf{I}$, since the effect of $\mathbf{D}$ can be absorbed into the covariance matrix), and the measurement noise is also assumed to be zero-mean Gaussian distributed, with covariance matrix $\mathbf{M}_k$.

We now describe the workings of the prediction and correction steps. Define the state at time k after the prediction (but before the correction) as $\mathbf{x}_k^-$, and after the correction as $\mathbf{x}_k^+$. The estimated state after the prediction is simply

$$\hat{\mathbf{x}}_k^- = \mathbf{A}\hat{\mathbf{x}}_{k-1}^+ + \mathbf{B}\mathbf{u}_{k-1} \tag{29.56}$$

since we assume that the deviations from the dynamics are zero-mean. We furthermore define the errors $\boldsymbol{\varepsilon}$ and their covariance matrices $\mathbf{P}$ as

$$\begin{aligned} \boldsymbol{\varepsilon}_k^- &= \hat{\mathbf{x}}_k^- - \mathbf{x}_k & \boldsymbol{\varepsilon}_k^+ &= \hat{\mathbf{x}}_k^+ - \mathbf{x}_k \\ \mathbf{P}_k^- &= E\left\{ \boldsymbol{\varepsilon}_k^- \left(\boldsymbol{\varepsilon}_k^- \right)^T \right\} & \mathbf{P}_k^+ &= E\left\{ \boldsymbol{\varepsilon}_k^+ \left(\boldsymbol{\varepsilon}_k^+ \right)^T \right\}. \end{aligned} \tag{29.57}$$

The covariance matrices are related as

$$\mathbf{P}_k^- = \mathbf{A}\mathbf{P}_{k-1}^+ \mathbf{A}^T + \mathbf{D}\mathbf{S}_{k-1}\mathbf{D}^T. \tag{29.58}$$

The correction equation becomes

$$\begin{aligned} \hat{\mathbf{x}}_k^+ &= \hat{\mathbf{x}}_k^- + \mathbf{K}_k \left[\mathbf{z}_k - \mathbf{C}\hat{\mathbf{x}}_k^- \right] \\ \mathbf{P}_k^+ &= (\mathbf{I} - \mathbf{K}_k\mathbf{C})\mathbf{P}_k^- \end{aligned} \tag{29.59}$$

where $\mathbf{K}_k$ is the Kalman gain, which can be shown to minimize the overall error if

$$\mathbf{K}_k = \mathbf{P}_k^- \mathbf{C}^T \left[\mathbf{C}\mathbf{P}_k^- \mathbf{C}^T + \mathbf{M}_k \right]^{-1}. \tag{29.60}$$

To summarize, based on the results from timestep t_{k-1}, we first compute the updated state $\hat{\mathbf{x}}_k^-$ from (29.56) and covariance $\mathbf{P}_k^-$ from (29.58), then update the Kalman gain from (29.60), and finally get the new state estimate $\hat{\mathbf{x}}_k^+$ from (29.59).[8]

---

[8] The Kalman filter is based on a Markov model, i.e., it is assumed that the update in every step only depends on the current state, and not on the history and/or the future. However, knowledge of future states would allow us to improve the estimates of the current states. While this process, which is called "smoothing" is not a realistic option in real-time systems, it can be realized, e.g., in non-real-time reconstruction of trajectories.

Example 29.1 *Consider a tracking system with a known BS location and a UE that should be tracked, where movement occurs only along the x-axis. The velocity vector is assumed to have a constant speed* v *plus random variations. At each step, an independent TOA measurement is done to provide an estimation of the location. Let the error of the state model (deviation from the constant speed) be Gaussian with variance $\sigma_{\rm s}^2$ and the error of the pseudorange (runtime times speed of light) as $\sigma_{\rm m}^2$. Derive the Kalman filter for this case.*

The state equation becomes

$$x_k = x_{k-1} + {\rm v} + n_{{\rm s},k} \tag{29.61}$$

and the measurement equation

$$z_k = x_k + n_{{\rm m},k}. \tag{29.62}$$

Assume we have an estimate of the initial location $\hat{x}_0^+$ that has a variance σ_0^2. The state update equation, (29.56), for the first step is then

$$\hat{x}_1^- = \hat{x}_0^+ + {\rm v}. \tag{29.63}$$

The covariance matrix $\mathbf{P}$ becomes a scalar σ, so that

$$\left(\sigma_1^-\right)^2 = \left(\sigma_0\right)^2 + \left(\sigma_{\rm s}\right)^2. \tag{29.64}$$

This is very intuitive; there are two independent Gaussian sources of errors (the error in the initial estimate, and the error in the translation vector), and the variances of Gaussians add up. We now come to the update step. From (29.59)

$$\hat{x}_1^+ = \hat{x}_1^- [1-K] + Kz_1 \tag{29.65}$$

where

$$K = \left(\sigma_1^-\right)^2 \left[\left(\sigma_1^-\right)^2 + \left(\sigma_{\rm m}\right)^2\right]^{-1} \tag{29.66}$$

so that

$$\hat{x}_1^+ = \frac{\left(\sigma_{\rm m}\right)^2}{\left(\sigma_1^-\right)^2 + \left(\sigma_{\rm m}\right)^2} \hat{x}_1^- + \frac{\left(\sigma_1^-\right)^2}{\left(\sigma_1^-\right)^2 + \left(\sigma_{\rm m}\right)^2} z_1. \tag{29.67}$$

This is essentially a maximum-ratio combining of the location based on prediction, and the location based on the measurement.

Another important component in Kalman tracking is the choice of the state model. The example above assumed a known constant velocity as baseline, with zero-mean random variations around it. However, this is not the only possible choice. We could, for example, assume that the agent follows a Brownian motion, so that the velocity vector changes randomly at every time instant (this could occur, e.g., for a person moving around in a small area while reading on their phone). We could also assume that there is a slowly changing, but unknown, velocity. In that case, the state vector would not only contain the coordinates of the location but also the velocity components, $\mathbf{x} = (x, y, dx/dt, dy/dt)^T$. Clearly, the new position estimate would be a function of the velocity vector estimated in the previous timestep; in other words:

$$\mathbf{A} = \begin{pmatrix} 1 & 0 & T & 0 \\ 0 & 1 & 0 & T \\ 0 & 0 & 1 & 0 \\ 0 & 0 & 0 & 1 \end{pmatrix} \tag{29.68}$$

where T is the duration of the timestep.

29.10.3 Extended Kalman Filters

The relationships in the state equations, as well as the measurement equations, need not be linear. For example, an angle measurement is related nonlinearly (via cos and sin) to the position. We thus need to be able to modify the Kalman filter to handle nonlinear relationships

$$\mathbf{x}_k = f(\mathbf{x}_{k-1}) + \mathbf{Dn}_{s,k-1} \tag{29.69}$$

$$\mathbf{z}_k = h(\mathbf{x}_k) + \mathbf{n}_{m,k}. \tag{29.70}$$

The most popular of these forms is the EKF, which linearizes the nonlinear relationships around the current estimate of the state. In other words, we start out with a linearized relationship that is a Taylor expansion around the initial estimate of the position. The position is then updated, and the nonlinear relationship of the state equation is expanded into a new Taylor series (and truncated after the linear term), this time around the estimate x_1, and so on. It is clear that as long as the intervals between the different updates are sufficiently small, this gives a good estimate of the overall solution. This means that the prediction and correction equations are updated as follows:

The estimated state after the prediction is simply

$$\hat{\mathbf{x}}_k^- = f\big(\hat{\mathbf{x}}_{k-1}^+\big). \tag{29.71}$$

Defining then the Jacobian

$$\mathbf{A}_k = \frac{\partial f(\mathbf{x})}{\partial x}\bigg|_{\mathbf{x}=\hat{\mathbf{x}}_{k-1}^+} \tag{29.72}$$

that contains the partial derivatives, the update of the covariance matrix becomes

$$\mathbf{P}_k^- = \mathbf{A}_k \mathbf{P}_{k-1}^+ \mathbf{A}_k^T + \mathbf{DS}_{k-1}\mathbf{D}^T. \tag{29.73}$$

The correction equation becomes

$$\hat{\mathbf{x}}_k^+ = \hat{\mathbf{x}}_k^- + \mathbf{K}_k\big[\mathbf{z}_k - h\big(\hat{\mathbf{x}}_k^-\big)\big] \tag{29.74}$$

$$\mathbf{P}_k^+ = (\mathbf{I} - \mathbf{K}_k\mathbf{H}_k)\mathbf{P}_k^-$$

where

$$\mathbf{H}_k = \frac{\partial h(\mathbf{x})}{\partial x}\bigg|_{\mathbf{x}=\hat{\mathbf{x}}_k^-} \tag{29.75}$$

and

$$\mathbf{K}_k = \mathbf{P}_k^- \mathbf{H}_k^T\big[\mathbf{H}_k\mathbf{P}_k^- \mathbf{H}_k^T + \mathbf{M}_k\big]^{-1}. \tag{29.76}$$

There are a number of improvements and variations on this general theme, such as iterative EKF, unscented Kalman filters, and Particle filters, which we do not discuss here. However, the above equations provide a sufficient framework for first implementations. It must also be noted that the formulations allow information fusion, i.e., incorporating information from different sensors such as IMUs, TOA measurements, DoA measurements, and so on.

29.10.4 Accuracy Improvements of Kalman Filters

Kalman filters and EKFs can be combined with various accuracy-enhancing measures previously discussed in this chapter. Firstly, an incorporation of the NLOS detection/mitigation can be important. Figure 29.22 shows the result of an EKF tracking in an environment where the LOS is frequently blocked, and the ranging is not very precise. The red curve shows the true track of the agent, while green points show the snapshot-based localization results. A straightforward application of the EKF gives the blue curve, which clearly has large deviations from the ground truth, see Figure 29.22a. The first step for dealing with multipath is, as discussed in Section 29.3.1, the detection of NLOS situations, and throwing them away. This results in a considerable improvement, see Figure 29.22b. However, considerable errors can still occur, since the system does not have *any* information about the trajectory during the NLOS situation. A further improvement can be achieved by mitigating the NLOS effect. In the example shown in Figure 29.22c, a biased Kalman filter is used.

EKF can also be combined with VAs (see Section 29.3.3), both for the case that a map of the environment is available, and that it is not. In the former case, the application is straightforward, since the VA locations are known, and thus the EKF can be applied without further modifications. In the case that the environment map is unknown, the EKF can be used for Simultaneous Localization and

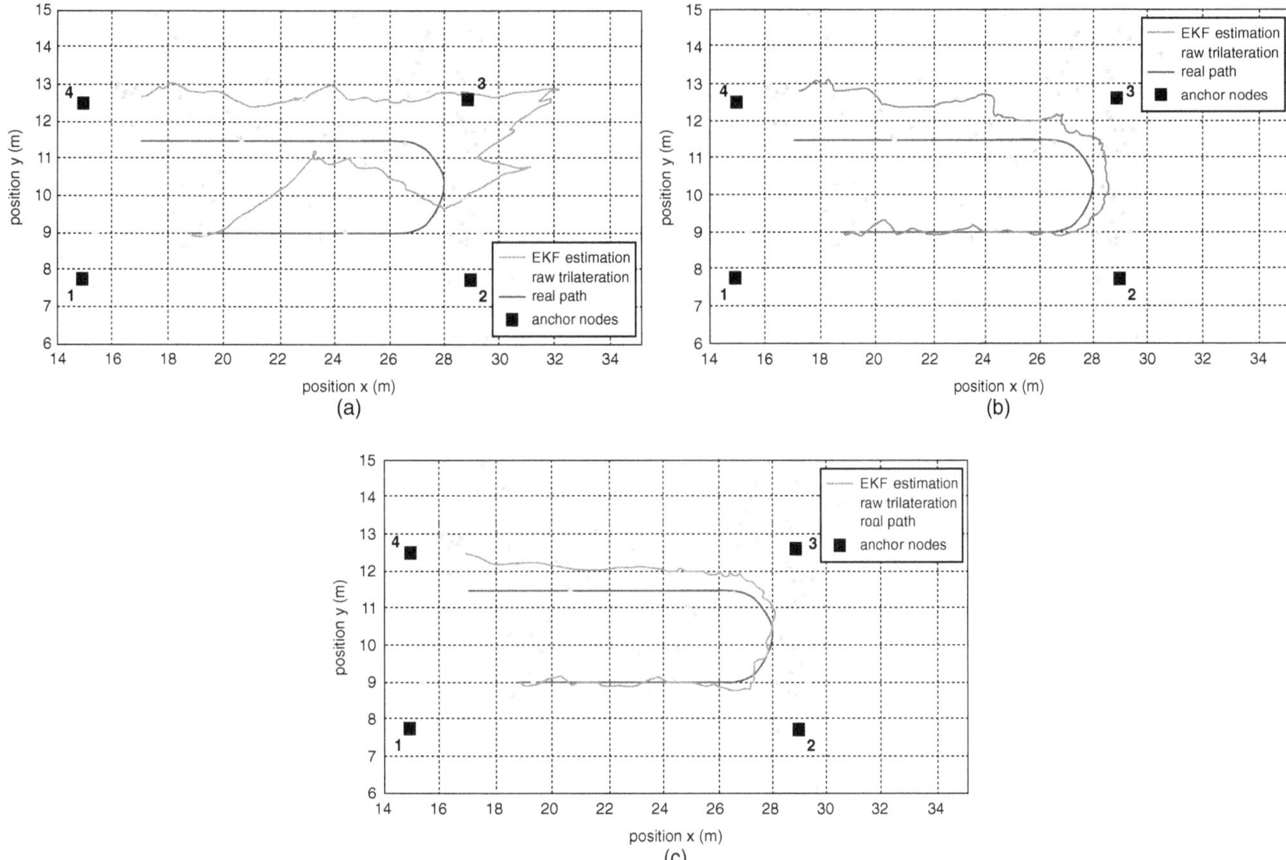

Figure 29.22 EKF tracking in a scenario with some NLOS locations. Black dots signify anchor locations, the black curve is the ground truth trajectory of the agent. Grey points are snapshot-based locations, the grey curve is the EKF output, for (a) standard EKF, (b) omission of all NLOS points, and (c) biased EKF. Color version available at wiley.com/go/molisch/wireless3e.
Reproduced with permission from [Roehrig and Mueller 2009] © IEEE.

Mapping (SLAM). The location of the walls can then be treated as unknowns, but since they are known to be static, the part of the transition matrix describing them is an identity matrix.

Last, but not least, the EKF can be combined with cooperative communications to provide network navigation. Refinement of the anchor-based location estimates then occurs both due to the inter-agent ranging measurements, and the temporal evolution of the location along the track.

*29.11 Machine Learning for Localization

Over the past 10 years, ML-based localization has been investigated intensely. As a matter of fact, localization is one of the applications in wireless communications best suited for the application of ML. For this reason, we in the following give a description of generic ML algorithms (that is also useful for other ML applications discussed in this book) and point out under what circumstances these algorithms can be best applied to localization problems.

29.11.1 Types of ML Problems

Depending on the type of available data, a number of different algorithms have been proposed. They can be generally divided into the following categories.

- In supervised learning, training data (so-called labeled data) are available where a connection between the labels (the location, or the parameters of a location model) on one hand, the observed *features* (such as RSSI) on the other hand exists. Within the category of supervised learning, we distinguish
  - Classification problems, where the input–output relationship maps the features onto a discrete set of labels (classifiers); a typical example would be to establish in which room a UE is.
  - Regression problems, where the features at the input relate to a continuous function at the output, such as the distance from a BS.

- In unsupervised learning, the collected dataset has no labels.
- Semi-supervised learning, where the data contain a mixture of labeled and unlabeled data.
- Reinforcement learning, where the system provides feedback values about the quality of the system's decisions (location estimates).
- Transfer Learning (TL), where training occurs in one environment, but the actual operation (testing) occurs in a different environment.

29.11.2 Supervised Learning

ML is usually based on a NN or similar structure. The structures mostly used for localization are feedforward NN, convolutional NNs, recurrent NNs, autoencoders, and LSTM networks; the latter are well suited for data that show both short-term and long-term variations and are thus inherently well fitted to wireless signals. Figure 29.23 shows some key examples.

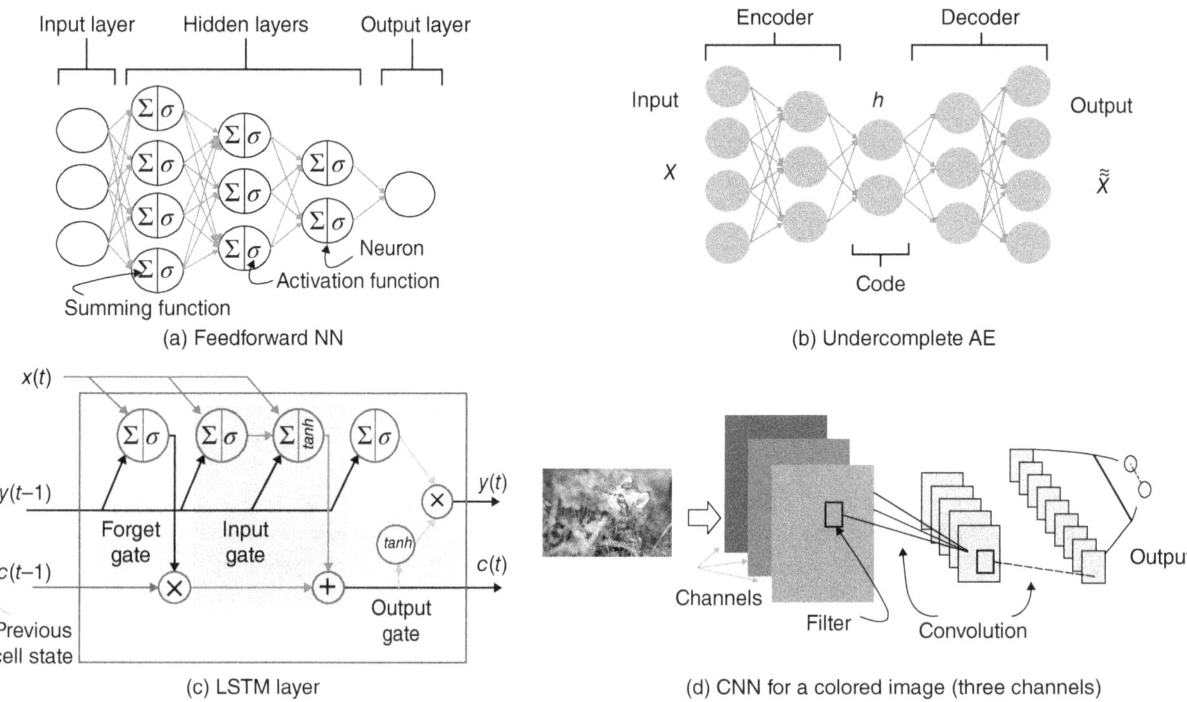

Figure 29.23 Examples of different neural network architectures. Color version available at wiley.com/go/molisch/wireless3e. Reproduced with permission from [Burghal et al. 2021] Cat image from imazite/Pixabay.

Neural Networks are powerful structures designed to imitate the human brain. An NN consists of one or more layers, each of which has a number of parallel neurons (nodes), see Figure 29.23a. The neuron computes a weighted combination of the input features and then passes it through a (usually nonlinear) transformation, a.k.a. "activation function," such as a sigmoid function. When NNs have no feedback loops they are referred to as Feedforward NN. The simplest architectures are sometimes referred to as fully connected Multi-Layer Perceptrons (MLP), where each node (perceptron) in a layer is connected to all other nodes in the following layer. One reason for their popularity can be attributed to the universal approximation theorem, which states that with at least one hidden layer and non-linear activation functions (such as sigmoid), NNs can approximate any function with arbitrary accuracy. However, to achieve such results with single hidden layer, it could be necessary to have a very large number of neurons. Increasing the depth of the network can alleviate this problem, which partly explains the good performance of "deep learning." NN training commonly uses backpropagation, which consists of forward pass of the training input values. Then, utilizing the chain rule, we back-propagate the gradient of the loss function over the layers; subsequently, algorithms such as gradient descent can be used to update the parameters. Since many deep learning problems use large datasets, the parameters update is usually done over mini-batches (compared to using all the data at once to calculate the gradient in each iteration as a single batch); this is a version of Stochastic Gradient Descent (SGD).

NNs and deep NNs have been widely used in localization. The number of layers is typically on the order of 3–5, though this is a function of both the available computational power, and the size of the training set. The features used for the training range from RSSI, to TOA, to full CSI; when more than one type of those features is available, it is possible to either train a large network with them, or obtain separate location estimates from networks using only one feature type, and then combining them by some kind of soft combining.

Convolutional NNs (CNNs) are efficient Feedforward NNs architectures, see Figure 29.23d, that have shown outstanding performance in computer vision, and all problems that can be reformulated as working on a "picture." Compared to a fully connected NN, CNNs use parameter sharing, which allows building deeper networks with a much smaller number of parameters. Key components in CNNs are the "filters" and the convolution operation. The filters have tunable weights that are multiplied with the input data during the convolution process. However, different from the fully connected NN, the size of the filter could be much smaller than the size of the input data.

Many problems in localization can be re-framed as image classification. In the simplest case, RSSI values are placed into a 2D grid and the output is the corresponding fingerprint. Of particular importance is the normalization of the data, including whether RSSI should be represented on a logarithmic (generally preferable) or linear scale. When observations from multiple BSs are used, and/or full CSI is used, the images become three- or higher dimensional. However, this does not impact the fundamental operating principle, and CNNs retain their ability of pattern recognition (and thus localization) even in those higher-dimensional images. When wideband and/or multi-antenna CSI is available, it is usually advantageous to represent the information in the delay (or delay/angle) domain – while this is nominally equivalent to frequency/antenna domain, the resulting pictures can be more easily processed by CNNs.

Recurrent Neural Networks (RNNs) are a type of NN widely used in modeling sequential data. They have shown great promise in time series data processing and natural language processing. RNNs are capable of performing tasks in which the current output is dependent on both the previous outputs and the current input. RNNs can be thought of having the ability to "remember" what has been calculated so far. RNNs have a hidden state, which can store information about the past computations. The hidden state is a function of the previous hidden state and the current input. The weights in RNNs are shared across time. At each point in time, the Back Propagation Through Time (BPTT) algorithm is used for training and adjusting the weights in each iteration. However, RNNs usually suffer from gradient vanishing or explosion, where the gradient values decay (or grow) as we propagate the gradient through time. An even bigger problem is the decaying influence over time i.e., the model cannot remember too far in the past. Hence vanilla RNNs are not widely used in practical applications.

In practice, decaying influence or degradation over time is addressed by using "gates" in the RNN cells. Long Short Term Memory (LSTMs) and Gated Recurrent Units (GRUs) are the two most popular gated RNN models, see Figure 29.23c for an example of an LSTM architecture, where the network uses three gates to control what to remember in the memory cell, when to use it, and when to forget it. LSTMs are a natural choice in many wireless problems, since they are naturally suited to the combined small-scale and large-scale fading.

Sequences of input features can occur in a variety of localization problems. The most obvious arises from the movement of the UE on a trajectory, and this is also where the majority of work has occurred up to now. Sequences can also occur in the frequency domain, in particular when extremely wideband measurements are taken.

Auto-Encoders (AEs) are NN architectures where the output of the network is a "copy" of the input data, see Figure 29.23b. We can distinguish three main components of the AEs, (i) the encoder (ii) the latent variable, and (iii) the decoder. The encoder maps the input data to the hidden state (latent variable), the decoder maps that back to the input. In practice, we are interested in ensuring that the output equals the input. The training procedure follows the standard backpropagation algorithm that we described earlier. We are typically interested in the latent variable, which is sometimes referred to as "code." This can be seen as dimensionality reduction so that the latent variable provides a robust concise representation or a projection on the underlying manifold. Note that when the neurons of the AEs use linear activation functions, AEs can be shown to be equivalent to principal component analysis. With nonlinear activation functions, AEs can provide powerful nonlinear dimensionality reduction.

By their operating principle, AEs are mainly used in unsupervised learning, which is not widely used in localization (though see below for some counter-examples). However, the dimensionality reduction that can be obtained from AEs makes feature representation more robust, so that the output of the AE can be used advantageously as input for supervised learning. Thus, AEs are mainly used for preprocessing in localization problems. In an alternative approach, the AE could be used in fingerprint-based localization to assess the similarity between the observed features and the collected fingerprints.

Kernel machines use a kernel $\mathcal{K}(.,.)$ to capture the similarity between vectors; they are generally used for pattern analysis. Instead of having to operate explicitly in a high-dimensional space, the kernel allows to determine the important characteristics in a low-dimensional representation that describes the high-dimensional space implicitly. One natural way to capture the similarity between two vectors $\mathbf{x}_i$ and $\mathbf{x}_j$ is based on the distance between them. The Radial Basis Function (RBF) is a kernel that is a function of the distance, e.g., the Gaussian Kernel

$$\mathcal{K}(\mathbf{x}_i, \mathbf{x}_j) = \exp\left(-\frac{\|\mathbf{x}_i - \mathbf{x}_j\|^2}{2\sigma^2}\right) \tag{29.77}$$

where σ controls the speed of similarity decay between $\mathbf{x}_i$ and $\mathbf{x}_j$.

One of the powerful algorithms that uses kernels is the SVM. It aims at finding the best hyperplane to separate different classes. In particular, the hyperplane is chosen such that it maximizes the margin between the different classes. In SVM the predictions usually appear in the form of inner products between training instances and observed feature; this allows SVM to use the kernels to find efficiently this product in higher (possibly infinite) dimensional space, which can increase the separability of the features.

SVMs are a natural method for determining location of UEs from training data. For example, the measurements of the RSSI to anchors establish a region in which the UE can be located. The hyperplanes are essentially combinations of RSSI thresholds for the different links, whose values are obtained from the training data. Other features besides RSSI can and have been used for SVM localization as well. Another aspect that needs to be considered is the choice of the kernel – while the Gaussian kernel mentioned above has been widely used in the localization literature, other kernels, such as sum-of-exponentials, can be considered as well.

29.11.3 Training and Preprocessing

A NN trains the network parameters, such as the combiner weights inside the network, based on the labeled data. In this context, it aims to adjust the network parameters to minimize the *loss function*, which measures the deviation of the learned mapping from the true label; a popular loss function is the mean-squared error. A major challenge in training neural networks is overfitting, where the network has so many different free parameters that it can fit all variations of the observed features, even the noise. In this case, the trained model will have large error values when applied to unobserved data. This can be combatted by validation, where the training data set is divided into two subsets: one for training (parameter tuning), and the other for hyperparameter and model selection. For example, for the fitting of a polynomial with degree n, the choice of the degree can be selected with validation. The fundamental idea is that overfitting during the training phase will drive up the error during validation. The final choice of the neural network will then be *tested* on another set of labeled data, called (unsurprisingly) the test data. Note that obtaining the labeled data set is one of the biggest challenges, and has been discussed in Section 29.5.2.

Another important aspect of ML is the preprocessing of the data, which improves training and model robustness. Such preprocessing includes normalization and filtering. Noise reduction and outlier filtration could reduce the overfitting. Dimensionality reduction and reducing co-linearity make the learning easier. Also, proper formatting of the output values impacts the choice of the algorithm (e.g., representation of the different classes). Feature normalization also speeds up training and allows the model to capture the impact of feature variation correctly.

These NN types can now be applied to a variety of features that we already mentioned in Section 29.5: (i) RSSI, (ii) TOA, (iii) full (wideband and/or multi-antenna CSI). These features can be furthermore easily combined with other input data, such as from IMUs, maps of the environment, etc.

29.11.4 Other Learning Solutions

Unsupervised Learning: Unlabeled data can still be useful as they contain the structure and distribution of the features. For this reason, many solutions pretrain the model with unlabeled data. This step can be used in combination with any other ML solution, as discussed above in the context of AEs. In addition to pretraining, unlabeled data could be used for radio-map construction, based on clustering of the observed RSSI or other features.

Transfer Learning (TL): In many situations, it is desirable to exploit the knowledge acquired in the source domain region for the learning task in the target region. This is particularly true for localization, because of the effort in creating labeled data. TL enables to train a network in one region where creation of such labels has been done already (or is easily feasible), and apply the results in a region where it is difficult. In most cases, a "naïve" TL gives bad results, since the building structures and propagation conditions are never exactly comparable. However, TL can also be used to reduce the training effort, e.g., by leaving most of the parameter settings of the source region the same, while retraining, on a small set of data, a single layer of the network with data from the target region.

Federated learning: In Federated Learning (FL), the training is carried out at the target or distributed computing edges. This collaborative framework has many advantages, such as reduced training load and maintaining users' privacy, as the user does not need to share private data with the network. FL usually assumes the presence of a centralized entity that updates the model by combining the locally (at the user side) trained models and broadcasts it back. There have been limited research papers that use FL for localization. However, this is expected to change as location information is one of the basic private pieces of information.

Further Reading

There is a rich literature on wireless localization. A very comprehensive source is [Zekavat and Buehrer 2019], which covers aspects of ranging, localization algorithms, RFID, GPS, LTE-based localization, etc. and provides many further references. Two special issues [Buehrer et al. 2014, Win et al. 2018] also provide a good survey of recent trends. For radar, the textbook [Skolnik 2002], as well as the handbook [Skolnik 2008], provide comprehensive overviews; [Richards 2014] concentrates on the signal processing aspects.

TOA-based localization is reviewed in [Guvenc and Chong 2009]; UWB localization (which is essentially based on TOA) is reviewed in [Gezici et al. 2005] and [Sahinoglu et al. 2008]. The impact of multi-path on localization and various mitigation techniques are reviewed in [Dardari et al. 2009], [Aditya et al. 2018], [Witrisal et al. 2016]. The use of soft information for improving accuracy is presented in [Conti et al. 2019]. NLOS detection and mitigation techniques are also surveyed in [Khodjaev et al. 2010] and [Huang et al. 2020]. The POCS algorithm is described in [Gholami et al. 2011].

GPS is described, e.g., in [Kaplan and Hegarty 2005], [El-Rabbany 2013]. A survey of localization using cellular systems is provided in [del Peral-Rosado et al. 2017], while indoor systems, including Wi-Fi and Bluetooth, are at the center of [Zafari et al. 2019]. Fingerprinting for outdoor systems is reviewed in [Vo and De 2015]. RFID is surveyed in [Chawla and Ha 2007].

The theory of cooperative localization was developed in a series of papers from MIT, in particular [Wymeersch et al. 2009, Shen et al. 2010, Win et al. 2011, 2018]; algorithms are discussed in [Buehrer et al. 2018], and experiments in [Conti et al. 2012]. The principles of Kalman filters can be found in many textbooks on digital signal processing and adaptive filters, e.g., [Sayed 2008]. The application to localization is described, e.g., in [Negenborn 2009], [Zekavat and Buehrer 2019, Part V]. The use of machine learning for localization is surveyed in [Burghal et al. 2020].

For updates and errata for this chapter, see https://wides.usc.edu/students.html#textbooks

Exercises

See Sec. 36.29 of Exercises.pdf at wiley.com/go/molisch/wireless3e

Part VI

System Design and Standardization

One of the main reasons for the success of wireless systems has been the development of widely accepted standards. Those standards make sure that the same type of equipment can be used all over the world, and also for different operators of wireless networks within a country – be those operators nationwide cellular operators or a private person who installed an access point for a Wireless Local Area Network (WLAN) in their house, and lets friends access it with their laptops. In this part of the book, we will describe the most important standards for wireless systems.

We start out with a discussion of overall wireless system design. The previous parts of the book described a variety of transmission methods and algorithms, but did not discuss in detail the signaling that is required to send control information of those algorithms to the other link end; for example to let the receiver know which modulation alphabet an adaptive transceiver uses at a certain point in time. This discussion is provided in Chapter 30. Furthermore, this chapter discusses the basic principles of standardization and the key organizations that govern the international standardization process. Finally, some older standards are described in the Appendices of Chapter 30.

The subsequent chapters then provide overviews of the most important current standards: the cellular standards LTE (Chapter 31) and 5G New Radio (Chapter 32), as well as the wireless LAN standard Wi-Fi (Chapter 33), and the standards for personal area networks and internet-of-things applications, Bluetooth and Zigbee (Chapter 34). An outlook to what systems and technologies might dominate Beyond 5G is provided in Chapter 35.

30

System Design and Standardization

30.1 From Components to Systems

A wireless communications system is a very intricate machinery that needs a lot of moving parts to work together in order to function properly. The previous parts of the book have described many of those key components. This section will briefly describe how they work together (with references to the appropriate chapters) and enumerate additional required components, using a top-down approach.

30.1.1 Backbone, Core Network, and Gateways

The previous parts of this book have talked about the wireless links, i.e., the connections between Base Stations (BSs) and User Equipments (UEs). But, considering the uplink, what happens to the data once they have reached the BS? And similarly, how do the data for the downlink transmission get to the BS? This is achieved through the core network, a – typically – wired network that has the following functionalities:

- It connects the different BSs to each other. Calls between users in different cells require that the call is sent between the BSs; this happens via the core network. A call (data session) between two users in the same cell does not require the core network for transmission of data between the users, but might require some core network functionality like authentication and billing.
- It contains control elements that determine routing, mobility management, call admission, etc., for all the data to/from the UEs.
- It provides the interface to the internet, as well as the traditional (landline) phone system. Specific "gateways" perform this interfacing. The core network may also provide caches for popular content so that, e.g., fetching a popular video does not require going through the gateway every time the video is requested by user. Obviously, the more often the video is replicated in the caches, and thus the closer a video is to the BS that will need it, the smaller the load on the network.

Core networks can be circuit-switched or packet switched. The former was used mainly for earlier generations of cellular systems, which were predominantly dedicated to speech communications. Due to the cost of replacement, some such legacy systems still exist. The core networks of 4G [Long-Term Evolution (LTE)] and 5G [New Radio (NR)] are packet-switched. This not only makes them efficient for handling wireless traffic that nowadays is almost exclusively packet-based but also more seamlessly provide the interfaces and gateways to the internet.

The core network also has to provide key administrative functions. Those include

- Registry for all users: the network operator creates databases that provide the association of a user (person paying the bills) with the device information, the rights that this user has (e.g., whether roaming is allowed), billing information (how are calls or data packets charged), and so on.
- Mobility management: the core network needs to determine which BS a user is associated with. This BS can change as the UE moves, and handover from one BS to the other is necessary.

The situation is somewhat different in (nonenterprise) Wireless Local Area Networks (WLANs). Here, the BS (also known as Access Point, AP) is directly connected to the internet, so that no core network is required. Furthermore, Personal Area Networks (PANs) and similar types of networks do not need core networks.

30.1.2 System Synchronization, Joining of Users, and Mobility Management

Most of the communication-theoretic treatments of wireless networks start with the assumption that Transmitter (TX) and Receiver (RX) have an established connection and are now sending payload data over it. But how does a UE join the network of a BS? This occurs in multiple steps:

Wireless Communications: From Fundamentals to Beyond 5G, Third Edition. Andreas F. Molisch.
© 2023 John Wiley & Sons Ltd. Published 2023 by John Wiley & Sons Ltd.
Companion website: www.wiley.com/go/molisch/wireless3e

- *Beacon signals by the BS*: the BS sends, at regular intervals, and at a known frequency, a beacon with a known signal structure. Thus, any UE that wants to join this type of network (say, an LTE network), searches for that beacon signal by correlating whatever signal it sees with the known beacon signal. When a sufficiently strong correlation is observed, the UE concludes that it has found an LTE BS, and can take further steps. It is noteworthy that the UE needs to search not only over time but also at different frequencies. Firstly, the BS might be operating at one of multiple possible carrier frequencies (due to the need for frequency reuse, not every BS is operating in all sub-bands available to a carrier; similarly for Wi-Fi, a BS is allowed to operate only on one of several available carriers). Thus, the UE needs to search at a number of widely separate carrier frequencies. Furthermore, the local oscillator at the UE might be detuned from that of the BS at that point, so that a search over a frequency offset range also has to be performed. The beacon signals need to be (i) robust in multi-path environments, (ii) convey some fundamental information about the network that the UE will need in the next steps of the network-joining (initial access) procedure, (iii) be frequent enough so that the UE can join in a reasonable amount of time, (iv) be robust to interference (e.g., for Wi-Fi, the beacon must be easily distinguished from a beacon of other networks/BSs operating in the same area), and (v) result in low overhead. Designing beacon signals that fulfill all those criteria is not trivial. Note that the beacon design and network joining is required not only for cellular networks and WLANs but also for PANs and similar networks. Even for device-to-device communications, discovering and connecting to neighboring nodes is essential (see Section 23.4).

- *Joining request by the UE*: next the UE needs to send a message to the BS that it wishes to join the network. At this point, the BS does not know of the existence of the UE and cannot provide contention-free access (assigning dedicated time/frequency resources to the UE); thus, the UE has to use random access even if the network generally uses contention-free access. The BS then takes this message, assigns an Identifier (ID) to the requesting UE, and starts the authentication process – after all, the BS needs to verify whether the UE is allowed to join the network or not. Depending on the type of network, assignment of ID and authentication can be quite complex. In a cellular system, it may be required for the BS to communicate, via the core network, with network servers that store information about all admissible users and their devices. For a WLAN system, the situation is easier as the BS just needs to check whether the UE knows the correct password. During the random-access process, the BS also determines the flight time between UE and BS and thus obtains the timing advance (compare Section 18.3.2). At the end of the process, the UE is connected to the BS and can start to send and receive payload data.

- *Sleep and reconnecting*: while the above procedure allows a UE to connect to a BS, this does not mean that it will stay connected the whole time. Firstly, the connection might "break off" as the UE moves in and out of the coverage region of the system. Secondly, a UE might go into a sleep mode (where it does not listen to possible signals from the BS for some time). Upon waking or coming into the coverage region again, the process of re-connecting should now be much faster than the original joining process, since the BS already "knows" the UE – i.e., has already authenticated the device, know (an estimate of) the timing advance, etc. Still, some additional signaling needs to be exchanged.

- *Mobility management*: In some systems, the mobility is limited – as long as the UE is within the coverage region of the BS, communication can take place; when it moves outside, the connection breaks off. However, in cellular systems, mobility is handled in a more advanced form. Firstly, the UE monitors the signal quality it obtains not only from its currently serving BS but also what it could get from another BS, and informs the system when it thinks a handover is required. Following this, the BSs coordinate at what time which of them should be the serving BS and/or whether they should both serve at the same time, see also Sec. 21.6). Further mobility management involves the system keeping track of where a UE is located even when it is, e.g., in sleep mode or disconnected mode, so that incoming transmissions can be sent to the correct BS for forwarding to the UE.

- *Capabilities exchange*: in most systems, UEs can have different capabilities, e.g., due to a trade-off between cost and performance. For example, a UE might be able to receive four parallel data streams (using spatial multiplexing), while another can support only two. These capabilities need to be communicated to the BS when the UE joins the network. This also impacts the control information discussed below.

30.1.3   Payload Data and Control Information

Any wireless system needs to send two types of data over the air: payload data, and control information. The former type of data is easily understood – it is the data a user wants for their application – a website, video stream, or speech. The control information, on the other hand, is all the information that is needed to enable the payload data transmission; for example, for whom is the data intended?[1] Control information can be bundled together with the user data (e.g., included at the beginning of a data packet), or it can be sent on a separate physical channel. But in either case, there needs to be a mechanism for the RX to recognize which of the received bits are control information and which are payload. Furthermore, control information has to be very carefully protected against transmission errors – if control information is wrong, then all the associated payload data are meaningless, even if they are received correctly.

We start out with the simplest possible way of multiplexing the payload and control information: they are assigned separate, fixed, time–frequency resources, and are thus distinguished easily by the de-multiplexer at the RX. A typical case of this implementation

---

[1] In the spirit of the OSI layer model, we can generally distinguish between two types of control data: control of the physical/MAC/RLC (radio link control) -layer transmission, and control data for the higher layers, which include, e.g., TCP/IP headers or control data for the particular application. It is the former that we generally call "control data" in our context, while the latter are – in the eyes of the wireless system – similar to application data in that they are not needed for performing the wireless part of the transmission properly, but rather are used by the higher layers of the end device.

occurs in GSM, a second-generation wireless system (see Appendix 30.A), where a certain number of bits, at periods spaced $26 \cdot 4.6$ ms apart, are used for transmission of certain control information; these bits are called the control channel (more precisely, the slow associated control channel). In this setup, it is purely the timing of bits that determines whether they are payload or control information; obviously, this solution is most suitable for a system with low adaptivity. Furthermore, an agreed-upon timing needs to be available to both BS and UE, which in turn requires synchronization (see Section 30.1.2). A variant of this multiplexing occurs in Wi-Fi packet data transmission. Here, the first bits of every packet are control information, followed by the payload bits. Within the control channel, the meaning of the bits is also predetermined and agreed upon between all devices in the network. For example, a set of bits can correspond to the ID of the intended RX, another set of bits can describe the modulation format that is used in the transmission, and so on. The association between the different types of data and the various transmission resources is sometimes called the mapping of logical to physical channels (note that here and in most of this Part of the book, a "channel" is a set of time–frequency resources, and not a *propagation* channel; we will explicitly say "propagation channel" when we mean that).

The above-described static control channels are simple and work well in a system that shows a low level of adaptivity. For systems with many different types of control info, much of which is not needed for a specific implementation, a different approach is needed. For example, the time–frequency resources for a control channel can be agreed upon to always start at a particular time, but they do not need to last a predetermined time. Rather, the first bits in a particular control block can be used to indicate how many of the following bits are control information (and which types of control info are transmitted), after which transmission of payload data can resume. This type of approach is used, e.g., in LTE and NR, see Chapters 31 and 32 for details.

Control information in the forward transmission. The actual control information that needs to be transmitted in the forward link depends, of course, on the specifics of the considered system. Not all of the types of data we mentioned in the following are required in all cases. Still, this general categorization gives an indication of the wealth of information that needs to be communicated together with the payload data.

- User address, i.e., for whom is the information intended, and the time–frequency resources used for this UE. For the downlink, the address information can range from using the full IP address of the user, to a (shorter) temporary ID discriminating the UEs connected to a particular BS, to implicit characterization (the user is indicated by the time/frequency resource on which the information is sent). The time–frequency resources for a particular user can be assigned on a semi-permanent basis (e.g., for phone calls), or re-assigned frequently.
- Physical-layer transmission parameters, such as modulation format, coding rate, spatial multiplexing settings.
- Control information for retransmission of data packets.
- Pilot signals, so that the RX can learn the channel.
- Synchronization signals.

Feedback control information. For the reverse direction, i.e., transmission of feedback, the following types of control information are important

- Which UE and/or data packet is the feedback about.
- Desired physical-layer transmission parameters, e.g., which modulation format is robust enough for the Signal-to-Noise Ratio (SNR) that the UE currently experiences, or which beamformer should be used by the BS.
- Did the previous packet arrive ok at the RX, or is a retransmission required?

Detailed examples for the different types of control information will be presented in Chapters 31–34.

30.2 Motivation and Operation of Standards

30.2.1 What Is a Standard?

In wireless systems, different devices need to have agreement on a number of transmission parameters so that they can talk to each other. To use the example of a cellular system, a UE needs to know the structure of the beacon signal that a BS sends out, so that it can even start the process of joining the network; the BS needs to know the structure (signal waveform, frequency) of the random access request it will receive from the UE, and so on. The common knowledge/agreement between the different devices can be achieved in two possible ways:

- *Proprietary solutions*: all devices in the network are created by the same company, which in this case can easily enforce that all devices use the same assumptions and settings. A related approach arises when companies make specifications public (or at least share them with (select) other companies), such as for Open Source software, so that others can build compatible devices. An example for the latter is the Android(TM) operating system for smartphones.
- *Standardized solutions*: here, an industry-wide agreement specifies the various transmission parameters and other settings and assumptions that every device has to follow, thus guaranteeing *interoperability* between devices from different manufacturers.

In other words, a standard (more precisely, an air interface standard, which is the topic of this section) defines the interfaces, e.g., between BS and UE, such that the two devices have a common understanding of what waveforms and bits are to be transmitted, and what they mean. It does *not* imply that all devices that follow the standard have the same performance. To provide a simple example: a standard may specify transmission using Quadrature Phase Shift Keying (QPSK) with a particular rate 1/3 convolutional code, so that the TX knows how to encode the raw data, and the RX knows how to interpret the received signal stream. However, the standard does not prescribe how the RX should perform the demodulation. It is up to the individual manufacturer to decide whether to use a hard-decision demodulation followed by a Fano decoder, a soft decision demodulator followed by Viterbi decoding, or some other demodulation and decoding means. The resulting bit error probability of the demodulation process may thus differ from device to device. Note that while for most wireless air interface components the transmitter is specified while the receiver is up to the implementer, the situation is often reversed in speech, audio, and video codecs.

In addition to the interoperability requirements, a standard may also define minimum performance requirements. For example, for the TX, the standard can define the admissible deviation of a transmitted waveform from the nominal "ideal" waveform. This is an obvious necessity, since without a limit on the deviation, the prescription of a particular waveform is meaningless. On the RX side, a standard may require that a device can demodulate an incoming (standards-compliant) signal with a certain maximum Bit Error Rate (BER) if the SNR is at a particular value. This serves a dual purpose: it prevents low-quality devices from being marketed as "standards compliant" and thus damaging the public perception of a particular standard, and it avoids a situation where a low-quality device requires higher TX power or more retransmissions, thus taking up a disproportionate part of the spectral and/or power resources. Such performance standards obviously impact both the hardware and the signal processing of a device.

Another important aspect of standards development may be *backward compatibility*:[2] a new system with improved technical specifications should allow participation of devices that follow an older standard. A classical example for that is the Wi-Fi standard, where devices that follow the old (year 2000) 802.11a standard can interoperate with, e.g., a BS that follows the most recent (2021) 802.11ax standard. This is done to ensure that users with an existing base of older devices do not need to discard their existing devices, but can upgrade or replace them as desired or necessary. For example, computers with Wi-Fi are sometimes upgraded; appliances and other IoT devices with embedded Wi-Fi are typically upgraded only at end-of-life. Backward compatibility has, however, also considerable drawbacks, as it constrains the freedom of design that is inherent in "greenfield" solutions. As a result there is often lower performance, such as lower capacity. A solution widely used for cellular systems is a "multi-mode" approach, where a UE might include modems for several standards (e.g., GSM, WCDMA, and LTE), and use a particular modem depending on the capabilities of the BS it is connected to (an exception occurs for LTE and NR, which can be used in conjunction, see Chapter 32). While such an approach may increase manufacturing costs, it is often acceptable in particular for smartphones, where the modems constitute only a fraction of the overall device costs.

*30.2.2 Reading Standards Documents

One of the biggest challenges for engineers working in standardization is the reading of standards documents, which are written in a way that is very different from most other technical documents. While academic papers or technical reports are (hopefully) written in a way that aims to elucidate technical concepts and explain the reasoning for taking a particular approach, standards documents are prescriptive documents that aim to specify in an unambiguous way the settings and interfaces, often without giving any reasoning. For example, a textbook or technical paper might discuss the topic of transmit antenna selection by explaining its motivation of increasing robustness in fading, give criteria for which antenna should be selected under which circumstances, derive performance measures, and present simulation results. A standard, in contrast, would define a maximum number of antenna elements that a device can have, how this number (as part of device capability) is communicated, and define a group of bits (transmitted on specified time/frequency resources) whose values indicate which antenna is selected for the transmission of a packet; no discussion of the motivation or performance will be given since this is irrelevant for standard-compliance, and no description of the selection algorithm is provided, because that is up to the manufacturer to choose. A further reason for not providing motivations or rationales in the standard is that they might give the impression of favoring particular implementations, and thus it would be very difficult for multiple companies to find generally agreeable formulations.

The difficulty of reading standards is made worse by the fact that some standards are several thousand, or even tens of thousands, pages long. Trying to understand a wireless system from reading its associated standard is thus exceedingly difficult; it is recommended to first read suitable explanatory books and magazine articles. The various chapters of this book aim to provide some first insights; more detailed expositions are recommended in the "Further Reading" sections of the individual chapters. It may also be helpful to read the "standards submissions" (see below) that led to the adoption of the particular techniques, and which often provide more background and motivation; and of course attendance of the actual standards meeting allows one to follow the discussions and better understand both the decisions and reasons for them.

Another challenge of wireless standards is their extensive use of acronyms, which is necessary for conciseness, but which turns many sentences into alphabet soups. They sometimes also use words in an unusual way, for example, the word "numerology," which is defined in the dictionary as "the study of the occult significance of numbers" [Merriam-Webster], is used in some standards as the set of symbol durations and intercarrier spacings of Orthogonal Frequency Division Multiplexing (OFDM) modulation. Worse, technical terms that have generally accepted meaning are sometimes redefined in a standard. Anybody working with the standards documents simply has to learn this language.

30.2.3 Advantages and Drawback of Standards Versus Proprietary Solutions

Both proprietary and standardized solutions exist in the wireless world, and it is natural to inquire about their relative advantages and drawbacks. Proprietary solutions typically can be more efficient, because all algorithms are optimized to work together, while in a

---

[2] Backward compatibility might even be important for different generations of proprietary solutions.

standardized solution interfaces are defined that should be as general as possible, and enable algorithms from different manufacturers, to interact. Standardized solutions have, in principle, the advantage that the ideas and expertise of a larger number of researchers enters the design of the system, which may lead to technically superior solutions. Further advantages of standards include a reduced risk for developers (because standards help to create a market), reduced risk for network operators (if one manufacturer goes out of business, there are others producing compatible products to fill the resulting gap), and convenience in particular for international customers, as the same product usually works in large parts of the world.

The biggest advantage of standards is, however, economical. Standards allow network operators and consumers to benefit, at the same time, from competition between manufacturers, and from economies of scale. By having, e.g., a worldwide standard for cellular communications, any standards-compliant product has a potential market of billions of users. Yet any company is motivated to find technically sound and low-cost solutions because any other manufacturer beating them in those key numbers could take away all their market share.

An example for the usefulness of standards is the history of cellular systems in Europe in the 1980s. Almost every European operator had its own proprietary system for (first-generation) cellular systems, which led to high development cost and absence of economies of scale (especially given the small size of many of the European countries). The CEPT (European Conference of Postal and Telecommunications Administrations) therefore established a group that developed a common standard for the second-generation cellular systems, which was called GSM, and which became ultimately accepted (almost) worldwide. It allowed the network operators to buy infrastructure from different manufacturers, and UE manufacturers to create devices that could be marketed to hundreds of millions of potential users. The resulting boom in cellular telephony was a major motivator for further international standardization activities.

Standardized systems may use technology that is patented by a particular company. Such patents are called Standards-Essential Patents (SEPs), i.e., it is not possible to build a standards-compliant device without using such a patent. Not having a license for an SEP could exclude a company from the market; for this reason, participants in standardization are often obligated to make their essential patents available to other companies on a *Reasonable and Nondiscriminatory* (RAND) basis.[3] This means that a patent holder cannot claim "unreasonable" license fees from other companies, and that it has to offer the patent to other companies on the same basis, i.e., cannot charge 10 times the license fees to one company compared to another. In other cases, companies might agree to license their patents on a royalty-free basis, or even not agree to a RAND policy. The details of the IP policies can be quite intricate, and strongly depend on the particular standardization organization. Still, ownership of SEPs might guarantee a considerable revenue stream, allowing to recoup research and development cost.

30.2.4 The Role of ITU and National Organization

Standards are being developed on multiple scales: international treaty organizations (such as the ITU, see below), internationally recognized organizations (such as the International Standards Organization and the International Electrotechnical Commission, though neither of them is very active in wireless air interface standards), regional/government organizations (such as ETSI in Europe and ANSI in the US), as well as transnational industry consortia (such as 3GPP) and regional/national industry consortia. In the area of wireless air interfaces, international treaty organizations and government organizations are mainly concerned with spectrum regulations and the setting of parameters (e.g., desired performance) for wireless systems, while industry consortia do the actual technical development.

The most significant international treaty organization is the *International Telecommunications Union*, ITU, a specialized agency of the United Nations, and in particular its radio communications office, the ITU-R. It performs mainly the following functions:

- It organizes the World Radio Conference (WRC), an event taking place every 3–4 years, which provides guidelines for the frequency regulators in different countries. In particular, it may establish worldwide bands for particular types of wireless systems.
- It defines the requirements for various generations of cellular systems, in terms of data rate, latency, and other high-level parameters. In particular, IMT-2000 defined requirements for 3G (third-generation cellular systems), IMT-Advanced for 4G (fourth-generation systems), and IMT-2020 for 5G (fifth-generation cellular systems). It must be noted that while ITU-R creates technical reports in WP5D (Working Party 5D) with approval in SG5 (Study Group 5), the formal votes happen in the WRC by the national delegates.
- It approves system proposals as being compliant with the requirements specifications. For example, Universal Mobile Telephone System (UMTS, also known as WCDMA) was approved as a system compliant with the requirements of IMT-2000. There is no restriction to one particular system as being an ITU standard, in other words, the ITU generally certifies fulfillment of performance specifications, but does not select a unique worldwide, interoperable standard. Even though ITU called, e.g., UMTS and the rival cdma-2000 different "modes" of the IMT-2000 "standard," it is more accurate to call them incompatible standards that both fulfill a certain requirements profile.

Spectrum regulation is also done on a national basis. For example, in the USA, the national frequency regulator is the *Federal Communications Commission* (FCC), in the UK, it is *OFCOM*, in China, it is the *State Radio Regulation of China*, in India, the *Telecom*

---

[3] Some standards organizations use the term Fair Reasonable and Nondiscriminatory (FRAND) instead of RAND; in practice there is little difference in the meaning.

Regulatory Authority of India (TRAI), in Japan the *Ministry of Internal Affairs and Communications*, and in the European Union, the *European Commission* in coordination with the regulating bodies of the individual member states. Such regulators determine permissible use of spectrum, which can be assigned to a particular operator, a particular service, or available without license to any user obeying certain rules, e.g., in terms of transmit power spectral density and other criteria. The frequency assignments are impacted by the decisions of the WRC, but it is noteworthy that there is no requirements for every frequency band to be used for the same service, or in the same manner (licensed vs unlicensed) in different countries. Due to the multitude of bands in different countries, Radio Frequency (RF) design for products that are sold worldwide can be quite challenging.

The national telecom regulators can also impose requirements beyond spectrum rules. Most importantly, they can assign or sell (often via an auction) exclusive usage of the spectrum to a particular operator. They can also require the use of a particular system for a particular band, or various RF technical parameters are often included (for cellular systems, Europe often adopts some of the 3GPP parameters). Regulators can also impose operating rules (e.g., coverage of rural areas, or net neutrality) on wireless operators, etc. Different countries have taken very different approaches in the handling of these questions.

The transnational industrial organizations are discussed in the following subsections.

30.2.5　3GPP and the Cellular Development

3GPP (Third-Generation Partnership Project) has over the years evolved into the main standardization organization for cellular telephony. It started as an organization for the standardization of third-generation systems (hence the name) and drew significantly from the companies driving GSM, as well as Japanese organizations (Japan was mainly using a different 2G system). For much of the 2000s, it competed with a rival organization (3GPP2), most of whose members (after developing their own standard Ultra Mobile Broadband (UMB)), joined 3GPP for the development of a unified 4G systems. After another rival 4G system (WiMax) became essentially defunct, it was left as the only major organization to develop advanced 4G and 5G systems.

3GPP is an industry consortium that mainly consists of infrastructure manufacturers, handset and modem manufacturers, and network operators; it has a total of more than 700 members at the time of this writing (2022). It is divided into three main areas – RAN (Radio Access Network), CT (core network and terminals), and SA (service and system aspects). The RAN Technical Specification Group (TSG), which covers the main wireless aspects, is divided into five working groups (WGs), which handle physical-layer aspects (RAN 1), Medium Access Control (MAC) and radio resource management (RAN 2), UTRAN/E-UTRAN/NG-RAN architecture and related network interfaces specifications (RAN 3), radio performance and procotol aspects (RAN 4), and mobile terminal conformance testing (RAN 5). Companies contribute standards submissions, which specify proposed technologies, simulation results, and possibly text to be used for the standard; these documents are (partly) discussed at bi-monthly meetings,[4] with the aim of achieving consensus between the participants. At regular intervals (typically quarterly), there are also so-called RAN plenaries, where the different RAN WGs meet and approve work and study items, coordiante the work, resolve issues that can't be resolved in the WGs, and approve final documents.

There is limited backward compatibility of new generations of 3GPP standards: the radio interface of each new cellular generation is essentially incompatible to the previous one, with the exception of 5G NR being partially compatible with LTE (compare Chapter 32). 3GPP also specifies core networks, and certain core networks can operate with BSs of different generations. However, there are also different improved versions (called *releases*) within each generation; for example, LTE covers 3GPP releases 8–14. Backward compatibility of those releases is fulfilled.

30.2.6　IEEE and Its Standards

IEEE (Institute of Electrical and Electronics Engineers) is an international professional society, which – in addition to many other activities – develops a number of wireless standards. Historically, IEEE standards development is rooted in computer communications, which included the development of the Ethernet standard as IEEE 802.3. The most important wireless standards are known as IEEE 802.11 (WLAN) and IEEE 802.15 (PAN), on which this section (and Chapters 33-34) focus. Standardization activity of these groups is usually considered to be the realm of individual contributors, though there are some company-based standards groups within IEEE as well. Even though in practice most participants attend on behalf of their companies, the underlying philosophy guides the process: individuals obtain voting rights by attending a certain minimum number of meetings, and selection of technologies occurs through voting. In particular, the process for the development of a standard works as follows:

- The process starts with a "call for interest" for a standards topic within a working group (e.g., 802.11), and interested members form a study group and draft a Project Authorization Request (PAR). If the 802 plenary approves the PAR, it is forwarded to a board that is responsible for authorizing such requests (NesCom).
- Upon approval of the PAR by NesCom, "task group" status is given to the project.
- The working group develops a draft standard. During this stage, technical proposals are submitted, and preliminarily approved by the voting members. Note that a working group (such as 802.11) is a superset of the different tasks groups. Most of the work such as discussion of proposals and preliminary votes is done in the task groups, but the decisions must be brought to the working group, which has the final say of whether to concur with, or overturn, the task group recommendations.

---

[4] Over time, the number of contributions has swelled to the point that discussion of all submissions is not feasible anymore. The number of submissions to a single meeting can exceed 1,000 within one of the groups.

- The various contributions are then merged into a complete draft. The draft standard is approved with a supermajority (75%); if it cannot gather enough votes, modifications are made and the process is continued until the threshold is reached.[5]
- The draft standard is submitted to the 802 executive committee for approval and forwarding to the sponsor ballot, where a broader community is allowed to vote on the draft, possibly requiring further modifications. Usually by the time of the draft standard the major issues have been resolved; however, sometimes a small issue is uncovered that needs to be resolved.
- After passing of the sponsor ballot, two other committees (RevCom and SASB) approve the standard, and it is finally published.

As can be seen, the process can last for a long time; in particular, the various refinements and approvals after the working group has approved a draft standard can take more than a year. For this reason, many companies introduce "prestandard" products that are likely to fulfill the specifications of a future standard, but this cannot be guaranteed as this standard is not official yet.

IEEE standards tend to have the following challenges: (i) a large number of optional features, and (ii) incompatibilities even between standards-compliant devices from different manufacturers. For these reasons, industry consortia have been formed that – on the basis of an IEEE standard – provide a modified standard avoiding these problems. The most famous of these is the Wi-Fi Alliance, which defines which capabilities in an IEEE 802.11 standard have to be supported to become a "certified Wi-Fi" device. The Alliance also performs compatibility tests to ensure complete interoperability between all such certified devices.[6] Furthermore, since IEEE 802 standards are mainly concerned with physical and MAC layer, some industry groups develop standards for the higher layers based on an existing IEEE 802 layer. In the remainder of this book, we will often speak of "Wi-Fi" or "Zigbee" standards even when strictly speaking "802.11" or "802.15.4" would be the correct nomenclature.

30.2.7 The Creation of a Standard

The first step in the creation of a standard is, typically, the determination of the scope and goals. These are governed by the applications that the standard aims to achieve. The scope can range from fairly narrow for certain types of application-specific standards, to extremely wide (in the case of new generations of cellular systems).

The next step for an air interface standard is the determination of suitable rules and models for assessing performance (e.g., through simulations). In the process of standardization, different proponents (companies) will present their proposals for the system design, and demonstrate the performance to support their suggestions. If it were left up to the individual proponents to select the simulation methodology and conditions, then proposals from different companies could not be compared on an apples-to-apples basis. Therefore, before standardization activities start out, the standards body determines guidelines or rules about how these simulations have to be performed. For example, it might be mandated that simulations assume hexagonal cells, that the link of interest is in the center cell, and that two tiers of interfering cells surround the center cell. The placement of the UEs (e.g., as Poisson point process) could be described. Importantly, a standardized model for the propagation channel has to be prescribed as well, and the simulation procedure has to specify which environments (e.g., urban, indoor office, etc.) have to be simulated. It is important that the simulation procedures and channel models are established *before* the actual proposal presentation begins.

During the development of a standard, a lot of compromise solutions and mergers between different proposals occur. This is a natural consequence of both the nature of technical development work and the realities of the standards process. With respect to the former, clearly, the performance can be improved if the best aspects of the different ideas and proposals are selected and combined into the standard. With respect to the latter, the voting (in IEEE) or consensus approach that could result in voting (in 3GPP) suggests that sufficient support for the eventual solution can only be obtained through compromise. Exceptions to this rule occur if a standard is developed essentially as a proprietary solution by one company but then becomes so popular that it is adopted as a standard. In this case, the proprietary solution often serves as the base for future revisions of the standard done in a normal open committee structure. Often the initial version of the proprietary solution is a detailed review to verify that the text is sufficiently unambiguous.

30.3 Some Important Standards

In this section, we will discuss a very brief evolution of standards in cellular communications, WLAN, PAN, and Internet of Things (IoT). This serves as an introduction and "setting the scene" for the subsequent chapters, which discuss details of the currently most important standards: LTE, NR, Wi-Fi, Bluetooth, Zigbee, and an outlook to Beyond 5G standards. This list is clearly not exhaustive, as there are literally hundreds of different wireless standards.

The discussion here and in the rest of this Part focuses on the MAC and PHYsical (PHY) layer of these standards. These aspects are most closely related to what we discuss in this book. We note, however, that many of those standards also have a network layer and application layer component (compare, e.g., the different profiles in Bluetooth and Zigbee), which build on the MAC and PHY layers. We refer to the "Further Reading" section of those chapters for references on those aspects.

---

[5] It is possible that the process does not converge; in that case a task group may be terminated and the PAR retracted.

[6] Usually plugfests are done before the certification process begins. It helps developers to resolve compatibility issues and find interpretation issues in the standards.

30.3.1 Cellular Standards

The most important standards worldwide are the cellular standards, since they cover the largest market segment of the wireless industry. This current chapter will provide, however, only an extremely short survey, since the current standards, LTE and 5G NR are discussed in Chapters 31 and 32, respectively, and description of some older standards can be found in App. 30.A-30.B. A detailed and authoritative history of cellular standardization is given by E. Tiedeman in the foreword of [Chen et al. 2022].

First-generation cellular phones were based on analog technology and served purely for voice communications. A large number of different "standards" existed – from Advanced Mobile Phone System (AMPS) to C450 to Nordic Mobile Telephone (NMT), and many more. Each of those standards was usually the result of development by one specific company – AMPS was developed by AT&T in collaboration with Motorola, NMT by Ericsson, C450 by Siemens, and so on. Different countries tended to use different standards, and different operators within a country would use different standards as well. It is noteworthy, though, that the AMPS standard was very successful not only in the USA, but was also adopted by several other countries; furthermore, several countries adopted minor modifications of it, e.g., the variant in the U.K. was called TACS.

The situation changed with the introduction of second-generation cellular systems. Firstly, operators recognized the need to change to a digital system, both for the improved spectral efficiency, for ease of operation, and because of the possibility of introducing new products such as text messaging. The problem was that economies of scale are essential for digital products. The European operators, which were the main drivers of this development, each had too small a market for achieving such economies. Thus, CEPT decided to develop a unified cellular standard for Europe. A number of different proposals were analyzed and by the mid-1980s the standard that became known as GSM was specified, see Appendix 30.A for a detailed description. The success of this standard exceeded all expectations, dramatically driving down costs for handsets, and making digital cellphones truly mass market products. This was further supported by European frequency regulators, which assigned new frequency bands to cellular operators only on the condition that they would deploy GSM.

It is noteworthy that two other big markets used a different approach. In Japan, the incumbent operator established essentially its own system. The comparatively lower performance was one of the main reasons that Japan became a driver in the development of 3G cellular technology. In the USA, no particular system was prescribed by the regulatory authorities. This led to a three-way competition between GSM, another Time Division Multiple Access (TDMA)-based standard called IS-54/IS-136 (see [Coursey 1999]), and the IS-95 CDMA-based system developed by Qualcomm (see [Liberty and Rappaport 1999]). Different operators within the USA used different standards.

For the development of 3G systems, it was clear that data communications capabilities would be essential, and universally, CDMA was adopted as the key underlying technology. Yet, two different, incompatible, standards were developed. On one hand, there was Wideband Code Division Multiple Access (WCDMA), which was developed by the companies supporting GSM and the Japanese operators, in the organization 3GPP, see Appendix 30.B. The other was cdma-2000, an evolution of the IS-95 standard, developed by 3GPP2 [Garg 2000, Vanghi et al. 2004, Eternad 2004, Tiedemann 2001]. Ultimately, the latter was used by two major operators in the USA, as well as in Korea, by one of the Chinese operators, by one of the Japanese operators, and by some operators in India, Australia, South America, and some other parts of the world; WCDMA was deployed by almost all of the other operators in the world. The continued split of the mobile standards world slowed down the overall development and furthermore complicated the Intellectual Property Rights (IPRs) situation. China also started widespread deployment of cellular technology, with one operator using CDMA, another used TD-SCDMA, and a third WCDMA.

The first versions of 3G standards were still voice-centric. The cdma-2000 standard then created a data-centric improvement 1xEV-DO, while WCDMA was later improved by the HSPA mode for data transmission. Yet, it became obvious that in the long run, CDMA could not fulfill the constantly increasing demands of data transmission. This motivated the development of the 4th generation of cellular systems. Initially, three main competitors existed: WiMax, also known as IEEE 802.16, which proposed Multiple Input Multiple Output-Orthogonal Frequency Division Multiplexing (MIMO-OFDM), see [Eklund et al. 2006], [Andrews et al. 2007]. Not being developed by the traditional cellular organizations, it used the internet as backbone network. MIMO-OFDM was also used as the technical basis by 3GPP for the development of its own 4G standard, LTE (see Chapter 31) which similarly could be considered as "data first": voice transmission was treated almost as an afterthought and often introduced years after the first deployment of LTE (with 2G and 3G networks handling voice calls in the meantime). 3GPP2 started to develop its own 4G standard, Ultra Mobile Broadband (UMB). Even though the standard provided good performance, it was never deployed on a large scale. 3GPP2 mainly focused on maintenance of the cdma-2000 standard, while its member companies joined 3GPP and contributed to the development of LTE. In the early 2010s, Wimax was seen as a strong competitor to LTE but ultimately could not gain significant market share. Thus, LTE became the first worldwide-accepted standard for cellular technology.

In order to satisfy the demand for new applications that require lower latency, higher reliability, and higher datarates, 3GPP started to develop the 5G standard, "NR," in 2015, see Chapter 32. In contrast to previous generation changes, the air interface of 5G is not a complete break from the past, but rather an evolution of LTE, and LTE and NR BSs can operate with the same backbone network. A first version of this standard was published in 2018, while further refinements are included in updated versions called "releases."

While academic research has started on the "next generation," also known as "beyond 5G" or "6G," the standards organizations have not yet started any official activities. Some possible topics for beyond 5G standardization are discussed in Chapter 35.

As outlined above, cellular standards have a rich history, and some of the older standards may still be of interest for historical reasons and/or for engineers maintaining legacy systems; GSM and WCDMA are thus described in App. 30.A and 30.B. However, the main interest for most readers is expected to be LTE, NR, and B5G, which are thus covered in the main part of the book (Chapters 31, 32, 35).

30.3.2 WLAN Standard

The development of the WLAN standards occurred in a much more straightforward way. WLANs started out as wireless replacement of ethernet cables that used to tether every computer to a modem or ethernet socket in the wall. For this reason, the IEEE Computer Society decided to develop a wireless standard, which became the original IEEE 802.11. As required for operation in the unlicensed frequency band around 2.45 GHz, the standard used spectrum spreading techniques to fill a bandwidth of 20 MHz but was not widely used. Deployment increased dramatically with the next generation of the standard, 802.11b, which increased the datarate to 11 Mbit/s using direct sequence spread spectrum. This was followed by 802.11a, which changed the modulation format to OFDM and was designed to operate in the 5 GHz band. A major step was 802.11g which essentially brought 802.11a into the 2.4 GHz band. All later versions of the standard build on the approach of 802.11a/g, with the technologies in the two bands being unified. As a matter of fact, modifications and enhancements followed in rapid succession, some of which enhanced the physical layer, while others improved the MAC layer. Compatibility between devices is ensured by an industry organization called the Wi-Fi Alliance (see Section 30.2.6 and 33.1 for a more detailed description of the relationship between IEEE 802.11 and Wi-Fi). The main physical-layer standards following 802.11a were 802.11n, 802.11ac, and 802.11ax (also known as Wi-Fi 6). All these standards are described in Chapter 33.

802.11 faced serious competition only once in its long history: in the late 1990s, the European HiperLAN standard was designed as competition to 802.11. While arguably technically superior, it lost the battle for market share, and many of its features were later subsumed into revised version of Wi-Fi. Around 2003, another standard, IEEE 802.15.3a, aimed to provide higher data rates, though over shorter distances, using ultrawideband technology. However, the proponents of two different system proposals within the standard blocked each other for years (remember that IEEE standards need 75% of the votes to be approved), until the project was abandoned.

Today, 802.11 is the undisputed standard (or better, family of standards) for WLANs. Competition arises mainly from 5G NR, which aims to fulfill similar functionality as 802.11, and also operates now in unlicensed bands, as well as from IoT standards, which aim to replace Wi-Fi as dominant standard for home automation and machine-to-machine communication.

30.3.3 Personal Area Networks and Internet of Things (IoT) Standards

While 802.11 was designed for data communications between a computer and a modem, covering distances up to 100 m, there was also demand for cordless communications between computers and computer peripherals such as a mouse, or a phone and a headset, with the required distances naturally much shorter (on the order of <10 m). The development of a suitable standard was already started in the mid-1990s, and became known as 802.15.1, or, more popularly, as Bluetooth (see Chapter 34 for a more detailed discussion of the relationship between 802.15.1 and Bluetooth). Since then, the standard was refined, adding additional usage cases (and the associated user profiles), and increasing the data rate. Since 2015, the main emphasis has been on the reduction of the energy consumption and concomitant increase in battery lifetime; this revised standard is known as Bluetooth Low Energy (BLE), and it is deployed in parallel with the classical Bluetooth standard. This standard is, in particular, useful for IoT applications, where battery-powered sensors communicate with a hub. Both classical Bluetooth and BLE are described in Section 34.1.

While 802.15.1 initially concentrated on short distances and computer peripheral applications, another PAN standard called 802.15.4 aimed to provide low data rate communications for PANs. Defining full-function devices and reduced-function devices that can be set up in a mesh architecture, it is also well suited for machine-to-machine communications. Under the name of Zigbee, it has proliferated for applications like home automation. An ultrawideband version of the standard is particularly suitable for localization. All this is discussed in detail in Section 34.2.

Bluetooth and Zigbee both compete in the market for short-distance IoT applications. Another area of application is IoT connections over long distances, e.g., from a single sensor to a far-away BS. For those applications, popular applications include the IoT modes of LTE and NR (see Sections 31.9 and 32.9), and proprietary standards such as LoRa and Sigfox. For ultra-short distance communications (a few centimeters), the Near-Field Communication (NFC) standard is currently dominant, which is based on inductive coupling between two devices.

*30.4 Appendices

App. 30.A: 2G Cellular – GSM

See App30A.pdf at wiley.com/go/molisch/wireless3e

App. 30.B: 3G Cellular – WCDMA/UMTS

See App30B.pdf at wiley.com/go/molisch/wireless3e

App. 30.C: Cordless Telephony – DECT

See App30C.pdf at wiley.com/go/molisch/wireless3e

For updates and errata for this chapter, see https://wides.usc.edu/students.html#textbooks

Exercises

See Sec. 36.30 of Exercises.pdf at wiley.com/go/molisch/wireless3e

31

4G Cellular – 3GPP Long-Term Evolution (LTE)

31.1 Introduction

31.1.1 History

In 2004, just as the first widespread rollout of the 3G system called *Wideband Code Division Multiple Access* (WCDMA) systems was happening, the Third Generation Partnership Project (3GPP) industry consortium started to work on Fourth-Generation (4G) systems. It was predicted at that time (and borne out by later developments) that the data rates and spectral efficiencies of WCDMA would not meet the demand of future applications; therefore, a new system had to be developed. In a somewhat bold move, it was decided to completely change both the air interface and the core network. The air interface was to move to Orthogonal Frequency Division Multiplexing (OFDM)/ Orthogonal Frequency Division Multiple Access (OFDMA), with (limited) support for Multiple-Input Multiple-Output (MIMO) antenna technology. The core network was to evolve into a pure packet-switched network. The new standard became known as *3GPP Long-Term Evolution*, or simply *LTE*.

The development of LTE originally took place in parallel to the further evolution of WCDMA. Around 2007/2008, LTE started to take the center stage of the 3GPP meetings. The basic parameters of the air interface were soon agreed on, but the implementation details required an enormous effort to achieve compromises yet stay reasonably simple and self-consistent. The first version of the standard, Release 8, was finalized at the end of 2008, and formed the basis of the first deployments. A number of minor improvements, such as support for broadcast, were included in Release 9; Release 8 and 9 are thus often lumped together as the "original" LTE. A major jump in the technology, and achieved data rates, occurred with Release 10, which was published in 2011 under the name *LTE-Advanced*. Including technology such as Carrier Aggregation (CA) and enhanced MIMO led to a significant increase in average throughput and peak data rates. Further steps in that direction were included in Release 11 (cooperative multipoint, enhanced hetnets) and Release 12 (small cell enhancements, D2D, dual connectivity). A further major step occurred again in 2015, with Release 13, called *LTE Advanced Pro*, which includes license-assisted access, narrowband Internet-of-Things modes, and full-dimensional MIMO, and Release 14. Starting in 2016, attention mainly shifted to the development of 5G, known in 3GPP parlance as "New Radio (NR)," which will be described in Chapter 32. The current chapter describes LTE without making too many distinctions of which release a particular feature was first published in. It must be kept in mind, though, that not all deployments have all features included.

LTE received strong support from the vast majority of cellphone and infrastructure manufacturers. Most notably, the *3GPP2* alliance (which had promoted the cdma2000 systems, a rival to 3GPP's WCDMA system) decided to terminate the development of its own incipient 4G standard; rather, its members participated in the development of LTE. China, where some operators pursued their own 3G standard (TD-SCDMA), also moved to LTE (with an emphasis on the TDD-mode of LTE, see below). By the time of this writing, LTE has largely supplanted 3G and is the by far dominant cellular technology in the world, though some replacement by 5G has started. Worldwide, the number of LTE subscriptions was more than 4 billion in 2020.

31.1.2 Goals and Applications

The original design of LTE did not have any particular applications in mind. Rather, the goal was to provide a general increase in data rate and spectral efficiency, under the assumption that suitable applications would be found if the standard would provide sufficiently high data rates. 3GPP originally set as the goals of LTE to achieve a *peak* data rate of 100 Mbit/s in the Downlink (DL) and 50 Mbit/s in the Uplink (UL), respectively, with a 20-MHz spectrum allocation for *each* of the DL and UL. Thus, the required spectral efficiency was 5 and 2.5 bit/s/Hz for the DL and UL, respectively. However, due to the wide range of applications and requirements, LTE defined a number of different types of User Equipments (UEs) that present a trade-off between complexity and performance (see Table 31.1).

For operation under realistic circumstances, LTE defined performance requirements relative to the performance of WCDMA systems: generally, user throughput should improve two to four times. LTE is intended to be optimized for low speeds (0–15 km/h), since the main usage, especially for data services, was expected to be for nomadic terminals. Slight performance degradation is allowed for speeds up to 120 km/h, while for truly high-speed applications (up to 500 km/h), only basic connectivity needs to be retained. Later releases of LTE also include a support for *Multimedia Broadcast and Multicast Services* (MBMS) applications, though intended applications such as cellphone-based broadcast video services never became popular.

Wireless Communications: From Fundamentals to Beyond 5G, Third Edition. Andreas F. Molisch.
© 2023 John Wiley & Sons Ltd. Published 2023 by John Wiley & Sons Ltd.
Companion website: www.wiley.com/go/molisch/wireless3e

Table 31.1 Performance requirements for different UE classes. Categories 1-5 were defined in Release 8, 6–8 in release 10, 9–10 in release 11, and 11–15 in release 12. Following [Dahlman et al. 2016].

Category	1	2	3	4	5
Peak data rate DL (Mbit/s)	10	50	100	150	300
Max DL modulation	64 QAM	64 QAM	64 QAM	64 QAM	64 QAM
Peak data rate UL (Mbit/s)	5	25	50	50	75
Max UL modulation	16 QAM	16 QAM	16 QAM	16 QAM	64 QAM
Max number of layers for DL MIMO	1	2	2	2	4

Category	6	7	8	9	10
Peak data rate DL (Mbit/s)	300	300	3000	450	450
Max DL modulation	64 QAM	64 QAM	64 QAM	64 QAM	64 QAM
Peak data rate UL (Mbit/s)	50	100	1500	50	100
Max UL modulation	16 QAM	16 QAM	64 QAM	16 QAM	16 QAM
Max number of layers for DL MIMO	4	4	8	4	4

Category	11	12	13	14	15
Peak data rate DL (Mbit/s)	600	600	400	4000	800
Max DL modulation	256 QAM	256 QAM	256 QAM	256 QAM	256 QAM
Peak data rate UL (Mbit/s)	50	100	150	1000	225
Max UL modulation	16 QAM	16 QAM	64 QAM	64 QAM	64 QAM
Max number of layers for DL MIMO	4	4	4	8	4

The first rollouts of LTE coincided with the beginning of the smartphone revolution, which acted synergistically with LTE deployment: the faster data rates in LTE enabled many new applications, which in turn increased the demand for data and accelerated the LTE rollout. If one can speak of a "killer application" for LTE, it is wireless video streaming: by the time of this writing, more than 70% of all wireless data traffic is video. Both long-form (e.g., Netflix™) and short-form (e.g., YouTube™) video services have greatly contributed to this increase.

In addition to high data rate and high spectral efficiency, as needed for video and smartphone apps, the area of Internet-of-Things (IoT) started to emerge after 2010. Therefore, later releases incorporated components tuned to this application. In particular, energy efficiency and wide coverage were considered important; while low latency was considered, its implementation was left mostly for 5G, see Chapter 32.

31.2 System Overview

In Part 3 of this book, we have discussed how, on an already-established link, a transmitter (TX) can send data to a receiver (RX), and in Part 4, how multiple such links can coexist. A considerable part of the LTE standard deals with the specifications for the implementations of these functionalities. However, we also have to keep in mind that – as discussed in Section 30.1 – a cellular system has many more components that need to be similarly defined.

Some of the key steps that a cellular system has to perform are:

- *Initial access:* before any communication can take place, a UE has to associate with the network. When turned on, or moving into range of the network for the first time, it has to recognize the network, send a message that it wants to join the network, and get admitted. These steps require, in turn, that the network sends out some type of "beacon" signal so that the UE can recognize its existence. Furthermore, we can anticipate an extensive exchange of messages between the network and the UE to establish the identity of the UE, and verify permissions and possible restrictions of what it can do (e.g., limited data rate when roaming abroad). Finally, the network has to convey to the UE configuration information, such as the used bandwidth, that will be needed for all future data exchanges.
- *Handover:* when a UE moves from the coverage area of one cell into the next, this fact needs to be made known to the network, and the network needs to react accordingly. This "reaction" can be different depending on whether there is an active data transmission going on, or the network just needs to know where the UE is located.
- *Paging:* when the network has data or emergency notifications for the UE, it needs to inform the UE of this fact. This is relatively simple if the network knows the Base Station (BS) in whose coverage region the UE is. However, there are situations where – for energy-saving reasons – the network does not track the UE locations; this occurs in particular when the UE moves over large distances without sending or receiving data. Then the network has to send messages to multiple BSs, to identify where exactly the UE is located.
- *Random access:* even when a UE is already connected to a BS, it might not have the "right" to transmit data, since such transmission might interfere with transmission from other UEs. Rather, the network needs to assign time/frequency resources to the UE (this can

be on a long-term basis, e.g., for a voice call, or for just a single data packet). In any case, the UE has to tell the network that it wants to transmit. In modern packet data systems, this signaling usually is done through random access or, for UEs with an active connection[1] to the BS, through a Scheduling Request (SR).

Once the connection between BS and UE has been established, the actual data transmission can occur. For this purpose, a lot of control information needs to be exchanged. This control information partly is related to the actual transmission parameters of the PHYsical layer (PHY) packets. For example, we have mentioned previously that adaptive modulation can be used to match the modulation format to the capacity of the channel. But how does the TX know what modulation format can be supported by the current channel and the RX capability, and conversely, how does the RX know which modulation format the TX chose? All of this requires the exchange of control information between the UE and the BS; in LTE, the control information related to the PHY and Medium Access Control (MAC) of the transmission is called L1/L2 control information.

Another aspect that must be kept in mind is that data do not originate, or end, at the BS (for simplicity we discuss in the following the DL). Rather, the BS is just a step on the way, and data have to get to it from the source, which often is an internet-connected computer or cloud center. Thus, data arrive at the BS in the form of internet packets and have to be "translated" into a format that is suitable for transmission to the UE.

As outlined in Chapter 30, standardized systems usually do not prescribe the use of particular algorithms; rather they prescribe how the data and control information must be formatted so that the network and UE can understand it. What they do with this information is up to the implementer. This also holds for LTE. Consequently, this chapter concentrates on the formal description, while we refer to the earlier Parts of this book for a description of algorithms and their performance.

31.2.1　Network Structure

The network structure in LTE is quite simple in principle (and actually, simplified with respect to the structure of earlier cellular standards, Global System for Mobile Communications (GSM) and WCDMA), see Figure 31.1: there is only a single type of access point, namely, the eNodeB (or BS, in our notation).[2] Each BS can supply one or more cells, providing the following functionalities:

- air interface communications and PHY functions;
- radio resource allocation/scheduling;
- retransmission control.

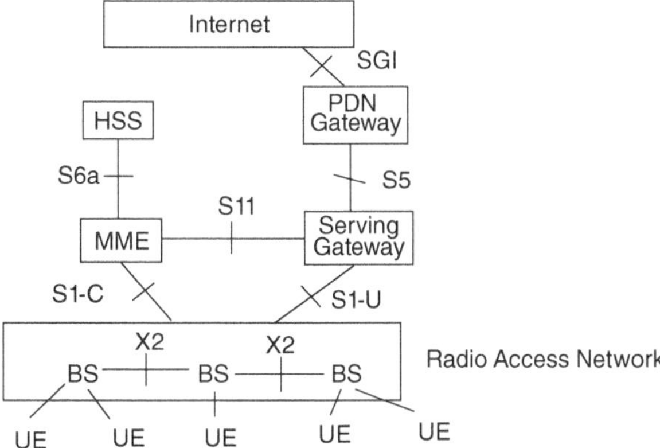

Figure 31.1　Network structure and interface definitions of LTE. *In this figure*: HSS, Home Subscriber Server; PDN, Packet Data Network.

The X2 interface is the interface between different BSs. Information that is important for the coordination of transmissions in adjacent cells (e.g., for intercell interference reduction or for dual connectivity) can be exchanged on this interface. Each BS is connected by the S1 interface to the core network. The BSs and UEs together constitute the Radio Access Network (RAN).

While this book really concerns itself only with the RAN (both in the current and all the previous chapters), we mention a few things about the core network in the following paragraphs. Up to 3G, core networks were mostly circuit-switched, having evolved from the network for the landline telephone systems to which cellphones connected. With the increased importance of connecting cellphones to the internet, it was decided to develop a new packet-switched core network, called Evolved Packet Core (EPC), for LTE. It consists of (i) a Mobility Management Entity (MME), that, as the name states, manages the movement of the UEs between the cells, (ii) the serving gateway (connecting the network to the RAN), and (iii) the Packet Data Network (PDN) gateway, which connects the network to the Internet. In addition, the Home Subscriber Server (HSS) is defined as a separate entity that has a directory of registered

---

[1] More precisely, a Radio Resource Controller connection.

[2] While there can be some subtle differences between the official definition of eNodeB and a BS (depending on the definition of BS), we will use them henceforth interchangeably.

users and listings of the operations the particular UE is allowed to do (e.g., whether the UE has hit its data cap), or may even register to a network when turned on abroad. The core network thus fulfills, *inter alia*, the following functions:

- mobility management;
- subscriber management and charging;
- quality of service provisioning and policy control of user data flows;
- connection to external networks.

The network functionality can be divided into two parts – user plane and control plane. The user plane processes the data packets created by the applications, while the control plane carries the signaling traffic, is responsible for the routing, and carries all the information required for the delivery of the payload packets.

The control-plane functionality is handled by either the MME or the *Radio Resource Controller* (RRC). The MME assigns IP addresses to devices and handles authentication, security, paging, etc. The RRC, which is located in the BS, handles certain control functions relevant for the RAN: broadcasting of system information, connection management, measurement and reporting, and handling of device capabilities.

31.2.2 Protocol Structure

For the user plane, the RAN protocol provides to the higher layers one or more *radio bearers* onto which the internet packets are mapped. The transmission protocol of LTE is divided into several layers that translate the IP packets, via intermediate steps, into *transport blocks* that can be sent over the physical layer link. Thus, for the DL, the data packets are handed from the upper to the lower layers, transmission occurs via the physical layer, and at the UE the packets are handed from the lower to the upper layers; the reverse process happens in the UL. The various layers are shown in Figure 31.2.

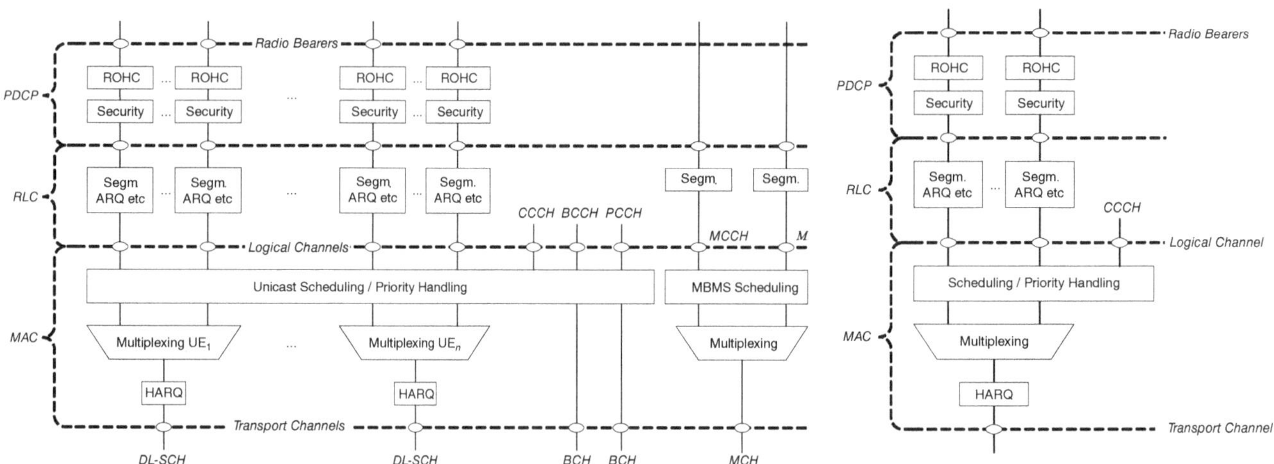

Figure 31.2 Protocol structure of LTE at BS (left) and UE (right). See glossary for explanation of acryonyms.
Reproduced from [3GPP TS 36.300] © 2009. 3GPP™ TSs and TRs are the property of ARIS, ATIS, CCSA, ETSI, TTA and TTC, who jointly own the copyright in them. They are subject to further modifications and are therefore provided. to you "as is" for information purposes only. Further use is strictly prohibited.

The data packets of the radio bearers first enter the Packet Data Convergence Protocol (PDCP) layer, which performs functions related to data integrity, enciphering, and robust header compression (ROHC); it also handles the in-sequence delivery of packets during a handover. The data are then handed to the Radio Link Control (RLC) layer, which performs segmentation/concatenation of packets (see below). The RLC also makes sure that at the RX, all packets arrive (and arrange for retransmission if they do not) and hands them to the PDCP in their correct order. The RLC also has an Automatic Repeat reQuest (ARQ) retransmission protocol (which is separate from, and in addition to, the Hybrid Automatic Repeat reQuest (HARQ) protocol on the MAC (see below)); it is meant to "catch" residual errors that the MAC HARQ does not correct.

The MAC provides a set of logical channels to the RLC, each of which carries a particular type of data, such as specific types of control information, or payload; more detailed discussions can be found in Section 31.4. It handles the multiplexing of the logical channels for a UE (using input from the scheduler that is not part of the standard) and the different logical channels, as well as the HARQ for retransmissions on the PHY.[3]

---

[3] See Section 31.5.2 for the relationship between HARQ and RLC retransmission.

Finally, the *PHY* provides transport channels to the MAC. It handles all the processes of actually transmitting data over the air, including coding and modulation. Note that the PHY not only interfaces with the MAC layer (layer 2), but also the RRC in layer 3. MAC layer and PHY are at the center of this chapter.

We now show how packets are handed over from one layer to the next. As can be seen from Figure 31.3, the input to the transmission system is IP packets, containing the payload and the headers related to internet transmission, namely IP (Internet Protocol), TCP (Transport Control Protocol), or UDP (User Datagram Protocol) headers. These packets go first to the PDCP, which performs the header compression and security as mentioned above, adds a PDCP header, and then hands its packets, called PDCP *Protocol Data Units* (PDUs), to the *Radio Link Control* (RLC).

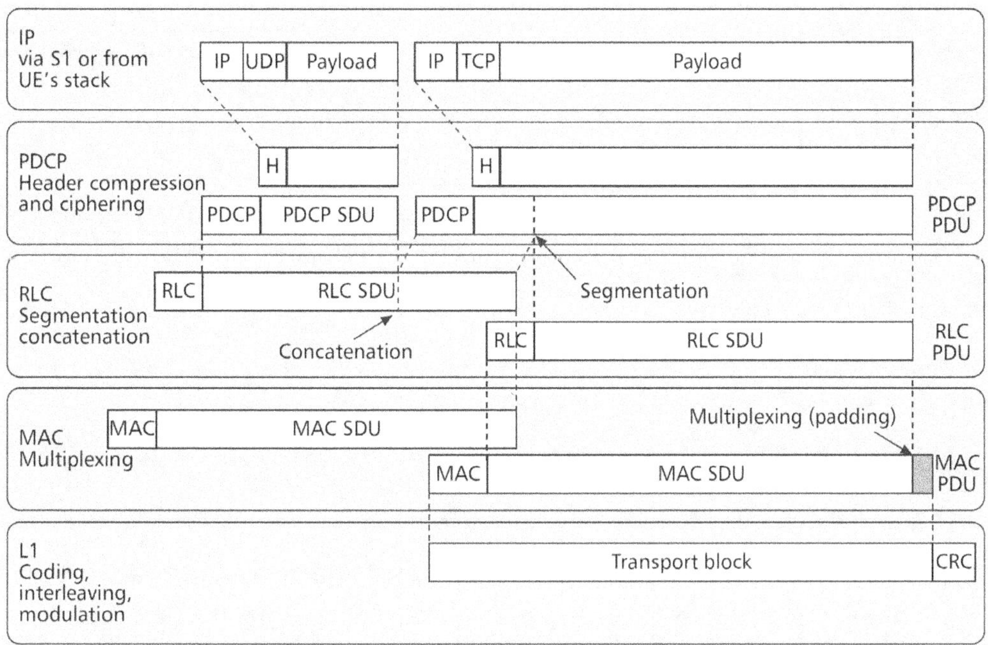

Figure 31.3 Data flow through the protocol stack in LTE. Color version available at wiley.com/go/molisch/wireless3e. Reproduced with permission from [Larmo et al. 2009] © IEEE.

The RLC segments and/or concatenates the PDCP PDU into packets that are more suitable for transmission over the radio channel, the RLC *Service Data Units* (SDUs). To these it prepends the RLC header, creating RLC PDUs. The segmenting/concatenation is, inter alia, necessary because – due to the large dynamic range of the transmission data rates – the amount of data that can be transmitted from the RLC buffer can vary dynamically, which in turn requires the size of the RLC PDUs to be adjusted dynamically. Thus, an RLC PDU can contain bits from one or more PDCP PDUs; conversely, one RLC SDU might be segmented, and its segments transmitted in multiple RLC PDUs. The packets are then handed to the MAC, which might concatenate multiple RLC PDUs into a MAC SDU, which (together with the MAC header) is called a *transport block*. The size of the transport block depends on the instantaneous data rate. Each transport block is protected by Cyclic Redundancy Check (CRC) code for error detection.

LTE defines different types of "channels," i.e., ways to transport bits, see Figure 31.2. In particular, there are *logical channels* (defined by the type of information they carry), *transport channels* (defined by how the information is transmitted), and the *physical channels* (which are defined as a certain set of time/frequency resources; not shown in Figure 31.2). The logical channels can be divided into *traffic channels* that carry payload data, and *control channels*. These are mapped, as mentioned, onto transport channels, most importantly the DownLink Shared Channel (DL-SCH) and UpLink Shared Channel (UL-SCH), which carry the user data, as well as most of the control information. Finally, these transport channels are mapped onto physical channels; there are also physical channels that do not carry any transport channel but are purely used for PHY functionality. Again the most important ones are *Physical Downlink Shared CHannel (PDSCH)*, which carries the DL-SCH, i.e., user data and some control data for the DL, and the *Physical Downlink Control CHannel (PDCCH)*, which carries control information, such as for scheduling, that is required for reception of the PDSCH; this information does not have a related transport or logical channel. Similarly, for the UL the most important channels are *Physical Uplink Shared Channel (PUSCH)* and *Physical Uplink Control Channel (PUCCH)*. The preceding is just a brief overview that helps in the discussion of the PHY/MAC layer transmission techniques in Section 31.3; a more extensive description of all these channels will be given in Section 31.4.

31.2.3 PHY and MAC Layer Overview

Although the details of the PHY are quite intricate, the key features can be summarized in the following straightforward bullet points:

- In the DL, LTE uses OFDM as modulation (Chapter 15); in the UL, it uses DFT-spread OFDM, i.e., OFDM precoded with a Discrete Fourier Transform (DFT), see Section 15.11.2.

- The multiple access format for both UL and DL is OFDMA (Sec. 18.3.3). In other words, the spectral resources, as represented in the time/frequency plane, are assigned in a flexible manner to the different users. Furthermore, different users can have different data rates. The transmissions for a specific user can be scheduled to happen on those subcarriers that offer the best propagation conditions, thus exploiting multi-user diversity, or spread over the whole available bandwidth, thus providing classical frequency diversity.

- *Multicast/Broadcast over Single Frequency Network* (MBSFN), i.e., transmission of the same information from different BSs, can be realized in a straightforward manner using OFDM, as long as the runtime differences of the signals from the different BSs are less than the cyclic prefix.

- LTE provides means for intercell interference coordination, i.e., making sure that signals transmitted in specific time–frequency resources on one cell experience controlled interference, including no interference, from signals transmitted in the time–frequency resources on neighboring cells. While principles of Chapters 21 and 22 are generally applied, most of the implementation details are vendor specific.

- Support for multiple antennas, including receive diversity, various forms of transmit diversity, and spatial multiplexing (see Chapter 16).

- Adaptive modulation and coding (Sec. 15.9.3), together with advanced coding schemes (Chapter 13).

- FDD or TDD can be used for duplexing (Sec. 18.5), depending on the assigned frequency bands. FDD and TDD modes are very similar, though a number of signaling details and parameter choices differ. To keep the description compact, this chapter only considers the FDD mode unless otherwise stated.

- LTE is, by design, completely under control of the infrastructure. UEs should only act upon instruction of the BS. For example, the time–frequency resources that the UE should use for transmission are assigned to it by the BS. Similarly, the modulation and coding scheme, and other transmission parameters, that the UE uses are set by the BS. The UE can send feedback and requests to the BS, but the BS always can decide whether to use or ignore those.

31.2.4 Frequency Bands and Device Classes

LTE can be operated in a variety of frequency bands that are assigned by national frequency regulators, based on the decisions of the World Radio Conference. This spectrum can be, subject to national restrictions, used for any member of the IMT-2000 and IMT-Advanced family. Tables 31.2 and 31.3 show the bands available by the time of LTE Advanced Pro). It originally encompassed the frequency bands of 3G systems (mostly around 2 GHz); later, additional frequencies were assigned, which became available through the so-called "digital dividend," – i.e., spectrum that was freed up when TV was converted to digital transmission techniques that required less spectrum than the old analog techniques. However, not all bands are available in all countries. Overall, the available spectrum is constantly expanding, either globally, or for specific regions. This trend also continues with the deployment of 5G, see Chapter 32.

Table 31.2 Bands for FDD operation of LTE, following 3GPP TS 36.101-14.5.0. Main operating regions at time of writing. Af: Africa, Am: America, As: Asia, Eu: Europe, CSAm: Central and South America, NAm: North America, Oc: Oceania.

E-UTRA operating band	Uplink operating band F_{UL_low}–F_{UL_high} (MHz)	Downlink operating band F_{DL_low}–F_{DL_high} (MHz)	Main operating region
1	1920–1980	2110–2170	Eu, As, Oc, CSAm, Af
2	1850–1910	1930–1990	Am
3	1710–1785	1805–1880	Eu, Af, As, Oc, CSAm
4	1710–1755	2110–2155	Am
5	824–849	869–894	NAm, (As)
6[1]	830–840	875–885	
7	2500–2570	2620–2690	Eu, As, Af, CSAm
8	880–915	925–960	Eu, As,Oc, CSAm, Af
9	1749.9–1784.9	1844.9–1879.9	Japan
10	1710–1770	2110–2170	Am
11	1427.9–1447.9	1475.9–1495.9	Japan
12	699–716	729–746	NAm
13	777–787	746–756	Am
14	788–798	758–768	Am
15	Reserved	Reserved	
16	Reserved	Reserved	
17	704–716	734–746	Am
18	815–830	860–875	Japan
19	830–845	875–890	Japan
20	832–862	791–821	Eu, Af

Table 31.2 (*Continued*)

| E-UTRA operating band | Uplink operating band | Downlink operating band | Main operating region |
	F_{UL_low}–F_{UL_high} (MHz)	F_{DL_low}–F_{DL_high} (MHz)	
21	1447.9–1462.9	1495.9–1510.9	Japan
22	3410–3490	3510–3590	Eu
23[1]	2000–2020	2180–2200	
24	1626.5–1660.5	1525–1559	
25	1850–1915	1930–1995	
26	814–849	859–894	
27	807–824	852–869	NAm
28	703–748	758–803	CSAm, As, Oc
29	N/A	717–728	NAm
30	2305–2315	2350–2360	NAm
31	452.5–457.5	462.5–467.5	Am
32	N/A	1452–1496	Eu
...			
64	Reserved		
65	1920–2010	2110–2200	Eu
66	1710–1780	2110–2200	NAm
67	N/A	738–758	Eu
68	698–728	753–783	
69	N/A	2570–2620	
70	1695–1710	1995–2020	

Table 31.3 Bands for TDD operation of LTE, following 3GPP TS 36.101-14.5.0. Main operating regions at time of writing. LAA: license-assisted access.

E-UTRA operating band	Operating band (MHz)	Main operating region
33	1900–1920	Europe, Asia
34	2010–2025	Europe, Asia
35	1850–1910	
36	1930–1990	
37	1910–1930	
38	2570–2620	Europe, CSAm, China
39	1880–1920	China
40	2300–2400	Europe, Asia, Oceania, Africa
41	2496–2690	USA, China, Japan
42	3400–3600	Europe, Canada, Japan
43	3600–3800	
44	703–803	
45	1447–1467	
46	5150–5925	USA (LAA)
47	5855–5925	
48	3550–3700	USA

LTE can also be operated with various bandwidths. The most common bandwidths are 5 and 10 MHz, but lower bandwidths (1.4 and 3 MHz) as well as higher bandwidths (15 and 20 MHz) are also possible. When peak data rates are mentioned, they usually refer to usage of 20-MHz bandwidth. Due to the use of OFDM as modulation format, bandwidths can be adjusted with great ease by changing the number of subcarriers, without changing any of the other parameters in the system.

Another interesting development is the use of license-assisted access. In this approach, part of the information (including the control information) is sent in the licensed bands, while the actual payload data are transmitted in unlicensed bands, see Section 31.6 for a discussion of the implementation and the advantages and drawbacks.

LTE furthermore defines different classes of BSs and UEs, in particular with respect to output power. For the BS, the admissible output power depends on the application: while BSs for wide-area coverage are mainly limited by the requirements of the national spectrum regulator, there are limits of the standard for medium-range BSs at 38 dBm, local-area (e.g., office) BSs at 24 dBm, and home BSs at 20 dBm. UE transmit power is generally limited to 23 dBm (exceptions for machine-type communications will be discussed in Section 31.9). The standard also defines emission spectral masks, such that the interference into other bands is limited. These limits are in addition to any limits that frequency regulators might impose.

31.3 Physical Layer

31.3.1 Overview of the Transmission Steps

Generating the PHY signal for the DL consists of the following steps (see Figure 31.4):

- Coding and scrambling (Section 31.3.2):
  - *Error correction encoding*: provides forward error correction coding of the PHY-layer bits. The coding can depend on the type of information communicated by the bits, including a distinction between user plane and control plane, and also depends on the channel state.
  - *Scrambling of coded bits*: the bits of all transport channels are scrambled by multiplication with a Pseudo Noise (PN) (Gold) sequence, compare Chapter 19. Note that it is the bits, and not the complex-valued symbols, that are scrambled.

- *Modulation (Section 31.3.3)* of scrambled bits to generate complex-valued modulation symbols: the modulation format used for data transmission is Quadrature Amplitude Modulation (QAM) with various sizes of modulation alphabets.
- *Multi-antenna processing (Section 31.3.6)*:
  - *Mapping of the modulation symbols onto the transmission layers*: LTE foresees multiple layers (roughly equivalent to "spatial streams" of Section 16.2) for the transmission with multiple antennas.
  - *Precoding of the symbols on each layer for transmission on the antenna ports*: this step is essentially single-stream or multi-stream beamforming, or transmit diversity.

- Mapping of symbols to Resource Elements (REs):
  - Assignment of which symbols are to be transmitted in which time/frequency resource (i.e., time and subcarriers), see *Section 31.3.4*. In the case of multiple transmit antennas, this mapping is done at each antenna port separately. Finally, note that the BS multiplexes different users into different REs.
  - This step also involves the insertion of the reference signals (pilot signals); *Section 31.3.5*.

- *Generating the time domain OFDM signal:* (again, for each antenna port separately), through IFFT and insertion of the Cyclic Prefix (CP).

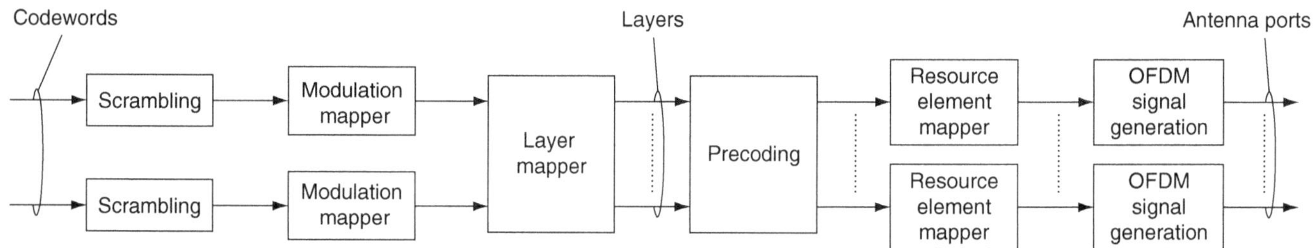

Figure 31.4 Overview of the physical layer procedure.
Reproduced from [3GPP TS 36.211] © 2009. 3GPP™ TSs and TRs are the property of ARIS, ATIS, CCSA, ETSI, TTA and TTC, who jointly own the copyright in them. They are subject to further modifications and are therefore provided. to you "as is" for information purposes only. Further use is strictly prohibited.

For the UL, the steps are almost identical, except that:

- The transmission is based on DFT-spread-OFDM, i.e., signals are DFT encoded before being sent to the OFDM modulator. This is done to maintain phase continuity, minimize amplifier back-off, and improve coverage.
- The assignment of symbols to time/frequency resources is different; only contiguous subcarriers can be used by one UE; this is related to the use of DFT-spread OFDM.
- Data scrambling is done with sequences that depend on the UE.

31.3.2 Coding

The encoding of the information is performed in several steps (see Figure 31.5).

Cyclic Redundancy Check

Data are handed from the MAC layer to the physical layer as transport blocks (see Section 31.2.2); one such transport block lasts for one Transmission Time Interval (TTI).[4] For each transport block, we first compute a CRC of length 24 bits (or, 16 or 8 bits under some

---

[4] A TTI in LTE is typically a 1 ms time interval. Under specific circumstances, such as certain multi-antenna transmissions, two transport blocks per TTI can be transmitted. We do not consider this situation here any further.

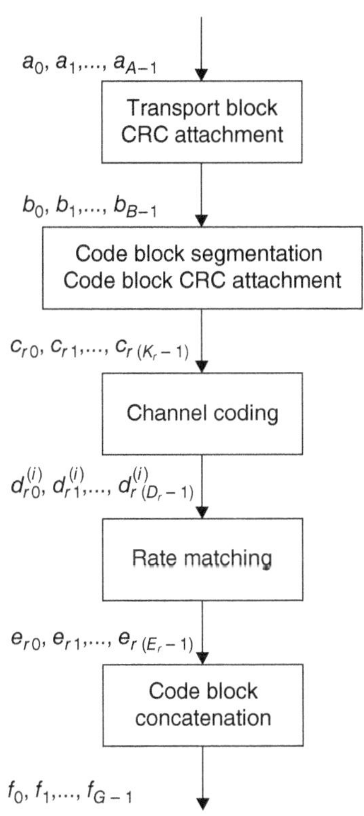

Figure 31.5   Encoding procedure in LTE.
Reproduced from [3GPP TS 36.212] © 2009. 3GPP™ TSs and TRs are the property of ARIS, ATIS, CCSA, ETSI, TTA and TTC, who jointly own the copyright in them. They are subject to further modifications and are therefore provided. to you "as is" for information purposes only. Further use is strictly prohibited.

circumstances). The CRC is computed by dividing the transport block (written as a binary polynomial, see Chapter 13) by the code polynomial and use the remainder of the division as the CRC. The code polynomials are

$$G(D) = D^8 + D^7 + D^4 + D^3 + D + 1 \quad \text{for 8-bit CRC} \tag{31.3}$$

$$G(D) = D^{16} + D^{12} + D^5 + 1 \qquad \text{for 16-bit CRC} \tag{31.4}$$

$$\begin{aligned} G(D) = D^{24} + D^{23} + D^{18} + D^{17} + D^{14} + D^{11} + D^{10} + D^7 + \cdots \\ + D^6 + D^5 + D^4 + D^3 + D + 1 \ \text{ for 24-bit CRC} \end{aligned} \tag{31.5}$$

and are attached at the end of the block. The CRC helps to determine whether a transport block was received correctly, and triggers the HARQ if that is not the case.

If the transport block is too large, data are parsed into code blocks, where each block has a maximum length of 6144 bits, and – if necessary – filler bits are inserted such that each code block has a permissible length (only certain discrete values of code block length are allowed). Then for each code block, another 24-bit CRC is computed according to

$$G(D) = D^{24} + D^{23} + D^6 + D^5 + D + 1 \tag{31.6}$$

and turbo coding (see below) is applied separately to each code block. The division of transport blocks into code blocks is done to reduce memory requirements for the turbo interleaver. However, the codeblock should not be made too small, since near-optimum codes require large codeblocks, see Section 13.6. The codeblock size is thus a compromise between those two requirements.

Convolutional Codes

Convolutional codes are used in LTE only for the encoding of control information, not for the payload data. In particular, the standard defines a length-7 tail-biting convolutional code with the following code polynomials.

$$G1(D) = 1 + D^2 + D^3 + D^5 + D^6 \tag{31.7}$$

$$G2(D) = 1 + D + D^2 + D^3 + D^6 \tag{31.8}$$

$$G3(D) = 1 + D + D^2 + D^4 + D^6 \tag{31.9}$$

This code is used for the following information (see Section 31.4):

- Broadcast CHannel (BCH),
- Downlink Control Information (DCI);
- Uplink Control Information (UCI).

Turbo Codes

Turbo codes are the main type of codes used in LTE. They are applied to the UL-SCH and DL-SCH, as well as the Paging CHannel (PCH), and Multi-Cast CHannel (MCH) (see Section 31.4). There is no option to encode payload data with convolutional codes; only turbo codes are allowed.

This is because turbo-codes provide better performance for payloads above ~100 bits with the performance benefit increasing as the payload increases. At the same time, turbo codes are by now so well established, and RXs are sufficiently optimized, that the higher complexity (compared to the equally well established convolutional codes) is not considered prohibitive.

The principles of turbo coders were already discussed in Section 13.6, namely the interleaving of several convolutional encoders. Specifically, in LTE, two recursive systematic convolutional encoders are employed, as shown in Figure 31.6. The data stream is fed into the first one directly, and into the second one after passing an interleaver. Both encoders have a coding rate of 1/2. Thus, the output is the original bit X or X' and the redundancy bits Y or Y', which are the output of the recursive shift registers. However, as X contains the same bits as X' only X, Y, and Y' are transmitted. Thus, the code rate of the turbo encoder is 1/3.

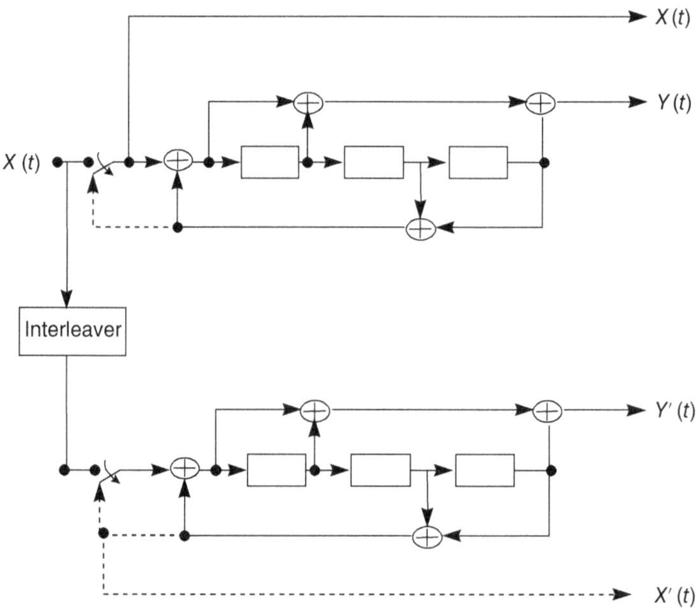

Figure 31.6 Structure of the turbo encoder.

In LTE, the interleaver is a Quadrature Permutation Polynomial (QPP) interleaver that moves bits from location

$$\left(f_1 i + f_2 i^2 \right) \bmod K \tag{31.10}$$

to location i. Here, K is the blocksize, and f_1 and f_2 are tabulated constants that depend on K.

After the encoding, a rate matching is done, such that the number of bits that are put out by the encoder matches the time–frequency resources and modulation format assigned to this UE.

HARQ

LTE employs various forms of HARQ for the DL-SCH and UL-SCH (it is not useful for other channel types). When an RX receives a transport block, it attempts to decode and sends a single feedback bit to indicate whether a retransmission is necessary (more details about the feedback process are given in Section 31.5.2). If a retransmission is indeed required, LTE transmits differently encoded versions (known as "redundancy versions") of the same source bits. For the RX, the accumulation of multiple such transmissions looks like a lowering in the code rate (i.e., increase in the available redundancy); it is thus called "incremental redundancy," see also Section 13.12. The choice of the specific redundancy versions in the different retransmissions is important: for example, the first transmission should include the systematic bits (which are especially important in turbo-codes). More details about HARQ will be described in Sections 31.4.6 and 31.5.2.

Scrambling

The bits of the coded signals are scrambled (undergo an XOR operation with a random bit sequence) before mapping onto modulation symbols. The scrambling sequences are different in neighboring cells, which enhances the capability of the Forward Error Correction (FEC) decoder to suppress interference.

Scrambling is applied to the DL transport channels and most control signaling, but *not* to channels for single-frequency networks, since in that case, signals from different BSs should usually convey the same information.

31.3.3 Modulation

UL and DL use different modulation formats: while the DL employs "classical" OFDM, the UL uses DFT-spread OFDM (Figure 31.7).

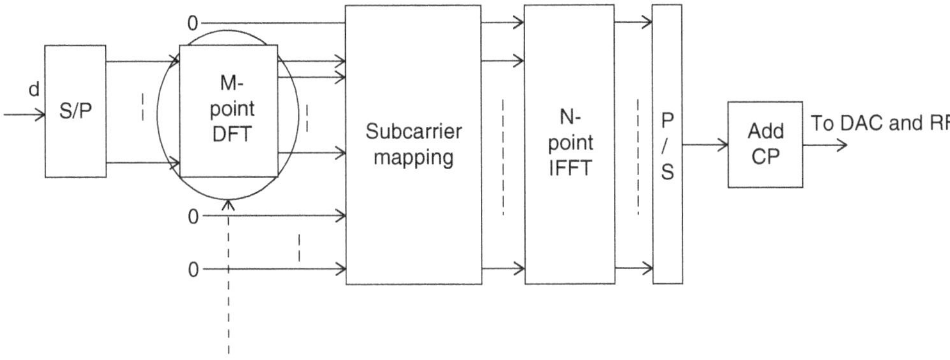

Figure 31.7 Block diagram of DFT-spread OFDM. *In this figure*: DAC, Digital to Analog Converter; P/S, Parallel to Serial; RF, Radio Frequency; S/P, Serial to Parallel.

The implementation of the DL transmission is straightforward: the modulation is OFDM with 15-kHz subcarrier spacing.[5] Data intended for different UEs are multiplexed onto different Resource Blocks (RBs), i.e., groups of subcarriers, see Section 31.4. The data for different users can employ a different modulation format (strictly speaking each transport block can use a different modulation format). For each OFDM symbol, the overall signal is then subjected to an Inverse Fast Fourier Transform (IFFT) transformation; the cyclic prefix is prepended, and the signal is upconverted to passband for transmission (the more complicated case of multi-antenna transmission is described in Section 31.3.6).

For the UL, a UE has a number of contiguous subcarriers available for signaling. The UE maps the symbols onto the input of a DFT whose size equals the number of subcarriers. The sizes of the DFT have to be a product of 2, 3, and 5, in order to allow a simple implementation (radix-2, radix-3, and radix-5 FFTs are well developed in hardware). The output of this DFT is then mapped onto the subcarriers, which are processed like in the DL (IFFT, cyclic prefix, and upconversion to passband). The combination of the DFT with the IFFT inherent in the OFDM implementation results in a single-carrier signal with cyclic prefix, which can be effectively equalized by frequency domain equalization (see Section 15.11.2). The bandwidth of the signal corresponds to the number of subcarriers used in the transmission, multiplied with the subcarrier spacing of 15 kHz.

In either case, the modulation format on each subcarrier is QAM, specifically Quadrature-Phase Shift Keying (QPSK), 16-QAM, and 64-QAM, with Gray mapping (i.e., signal points located next to each other are distinguished only by 1 bit), see Figure 31.8; for later releases, 256-QAM was introduced as additional, optional, format. The choice of the modulation format depends on the quality of the propagation channel: for better signal-to-interference- and Signal-to-Noise Ratio (SNR), higher-order modulation can be employed.

---

[5] There is also an option of 7.5 kHz subcarrier spacing. This is mainly intended for MBSFN and large cells, as it increases the symbol length and thus allows larger cyclic prefix without loss of spectral efficiency. However, it is rarely used in practice.

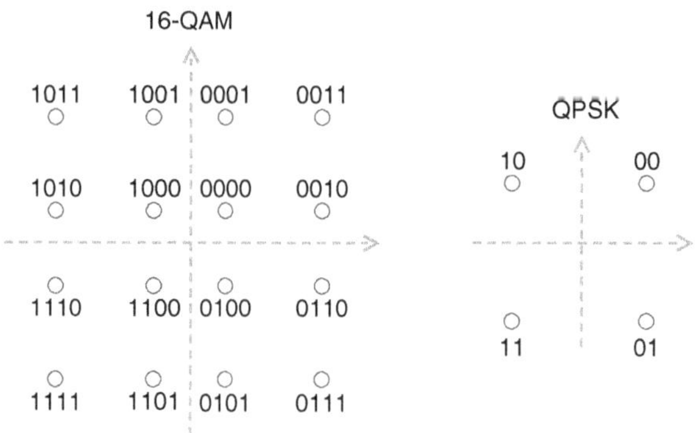

Figure 31.8 Mapping of bit combinations onto symbols.
Reproduced from [3GPP TS 36.211] © 2009. 3GPP™ TSs and TRs are the property of ARIS, ATIS, CCSA, ETSI, TTA and TTC, who jointly own the copyright in them. They are subject to further modifications and are therefore provided. to you "as is" for information purposes only. Further use is strictly prohibited.

31.3.4 Mapping of Modulation Symbols to Time/Frequency Resources

Frames, Slots, and Symbols

Before discussing how symbols are mapped, we first need to establish the time–frequency grid used in LTE, and the various quantities defined therein.

The time axis of LTE is divided into entities that play an important role in the transmission of different channels. These time entities have the following hierarchy (see Figure 31.9):

- The fundamental time unit of LTE transmission is a *radio frame*, which has a duration of 10 ms. Each frame is indexed by its System Frame Number, which distinguishes different parts of processes with periodicity larger than 10 ms.
- Each radio frame is divided into ten *subframes* (each being 1 ms long). Subframes are the fundamental time unit for most LTE processing, like scheduling, and are thus also a Transmission Time Interval (TTI).
- Each subframe consists of two *slots*, which are each 0.5 ms long.
- Each slot consists of seven (for normal CP) or six (for extended CP) OFDM symbols.

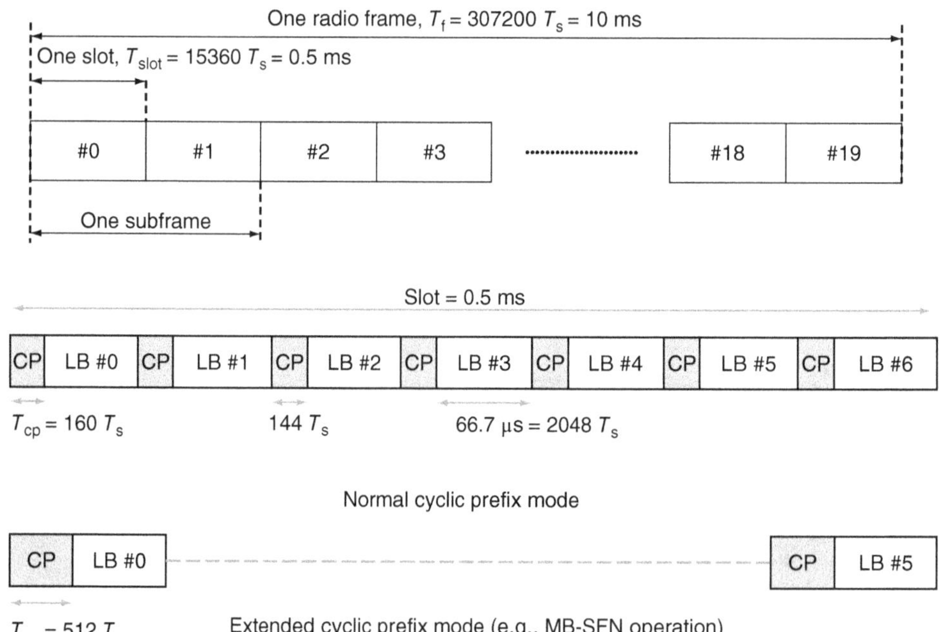

Figure 31.9 Structure of the time units in LTE.
Reproduced from [3GPP TS 36.211] © 2009. 3GPP™ TSs and TRs are the property of ARIS, ATIS, CCSA, ETSI, TTA and TTC, who jointly own the copyright in them. They are subject to further modifications and are therefore provided. to you "as is" for information purposes only. Further use is strictly prohibited.

Duration of the different units is often given in terms of the sampling time $T_s = 1/30{,}720{,}000$ s. Note that this "sampling time" is a bookkeeping unit; RXs are not obligated to actually sample at the corresponding rate. In particular, for bandwidths <15 MHz, a larger sampling time (lower sampling frequency) is feasible.

Duplexing

As mentioned previously, the standard duplexing method is FDD, where one band is assigned for the UL and a paired band for the DL, see Section 31.2. Since the DL often requires higher data rate than the UL, supplemental DL bands have been introduced, which provide larger bandwidth to the DL.

LTE also foresees half-duplexing, where transmission and reception for a specific UE are distinguished both by different times and by different frequencies; this eases implementation of UEs without affecting spectral efficiency (see Section 18.5). The standard foresees two different half-duplex modes, where in mode A one OFDM symbol used as a guard interval between the two transmission directions, while in mode B a whole subframe is used. The latter approach obviously allows lower peak data rates, but allows cheaper implementation with only a single Local Oscillator (LO), since there is sufficient time for it to be tuned from the DL to the UL frequency and vice versa during the guard interval.

Finally, LTE foresees TDD, where subframes can be assigned flexibly to UL and DL, based on a corresponding UL–DL TDD configuration, with the exception of subframes 0 and 5, which are always used for the DL, and subframe 2, which is always used for the UL. Typically, UL/DL assignments are changed only rarely (such a change is signaled by system information). Furthermore, the same assignment is used in neighboring cells, in order to avoid severe cross-link interference situations. The switching from DL to UL occurs in a special subframe, whose structure is described at the end of this subsection. A timing advance is used to combat differences in the runtime from different UEs, see Section 18.3.2.

OFDM Structure

Let us now turn to the details of a symbol. Since the modulation format is OFDM (regular OFDM for the DL, and DFT-spread OFDM in the UL), multiple subcarriers are present. The regular spacing between the subcarriers is $\Delta f = 15$ kHz, leading to an OFDM symbol duration (without cyclic prefix) of 67 μs $= 2048\ T_{\rm s}$. One subcarrier, for the duration of 1 OFDM symbol, is called a *Resource Element* (RE). We can fit 6 or 7 OFDM symbols into one slot, depending on the duration of the CP. In the "regular" case, with 7 OFDM symbols per slot, the duration of the CP is $144\ T_{\rm s}$, i.e., 4.7 μs (except for the first OFDM symbol, where it is $160\ T_{\rm s}$, so that the 7 OFDM symbols and CPs add up to exactly 0.5 ms). If we use only six OFDM symbols per slot, a long CP of $512\ T_{\rm s} = 16.7$ μs is used; this is useful for environments with large delay spread, such as large cells in rural environments, and/or for single-frequency networks. To simplify the notation, the description in the remainder of this chapter assumes "normal-length" CP unless otherwise stated.

Time/frequency resources are assigned to different users as integer multiples of an RB (Figure 31.10). An RB is 12 subcarriers (180 kHz) over the duration of one slot. For the UL, only contiguous RBs can be assigned to one UE in accordance with the DFT-spread-OFDM transmission. Furthermore, the size of the DFT has to be decomposable into factors of 2, 3, and 5; this is again related to efficient implementation with radix-2, radix-3, and radix-5 butterfly structures. Note that while RBs are generally treated as "fundamental units," they extend only over one slot, while scheduling is done on the basis of a subframe (two slots). Thus, a scheduler assigns an RB pair.

Mapping to Physical Resources – Downlink

The complex symbols from the modulator (possibly assigned to particular layers and antenna ports) have to be mapped onto the Physical Resource Blocks (PRBs), which are the basic units for the generation of the OFDM signal. In order to facilitate the implementation, the mapping proceeds in two steps: (i) map the symbols (in the sequence they occur) onto *Virtual Resource Blocks* (VRBs), (ii) map the VRBs onto the *PRBs*.

For the mapping of the symbols onto VRBs, the BS has to assign which VRBs are to be used for the signal intended for a particular UE. This assignment can either be contiguous, or noncontiguous. The following three types of allocation exist: types 0 and 1, which support noncontiguous allocation, and type 2, which only supports contiguous allocation:

- *Type 0*: it groups RBs (the size of the groups depends on the bandwidth), and then provides a bitmap to indicate which elements of that group are assigned to a particular UE. If there are eight groups, and the bitmap reads 10011000, then groups 1, 4, 5 are used for the UE in question. The motivation for assigning RB groups (instead of RBs) is to reduce the size of the bitmap codeword.
- *Type 1*: here the RBs are grouped into interlaced subsets. A particular UE is then allocated to a particular subset. Within that subset, again a bitmap indicates the RBs that are actually used.
- *Type 2*: indicates simply the starting point, and the length of the block allocation. It is thus much shorter than types 0 and 1, which need bitmaps.

In a second step, the VRBs are mapped to PRBs. For this mapping, two different approaches are possible, corresponding to two choices of VRBs: localized or distributed. A *localized* VRB with location *i* is directly mapped onto a PRB with location *i*; in other words, the mapping of symbols to PRBs is purely determined by the mapping of symbols to VRBs as described above. For *distributed* VRBs, we map RBs that are contiguous in the virtual domain onto noncontiguous PRBs. More precisely, the mapping proceeds in two steps (see Figure 31.11):

1. *RB-pair interleaving*: since each subframe (=scheduling interval) consists of two slots, scheduling involves the assignment of RB pairs. In RB-pair interleaving, RB pairs that are adjacent in the virtual domain are separated in the PRB domain.
2. *RB distribution*: the two elements of a VRB pair are mapped onto two different PRBs that are separated in the frequency domain by approximately half the system bandwidth.

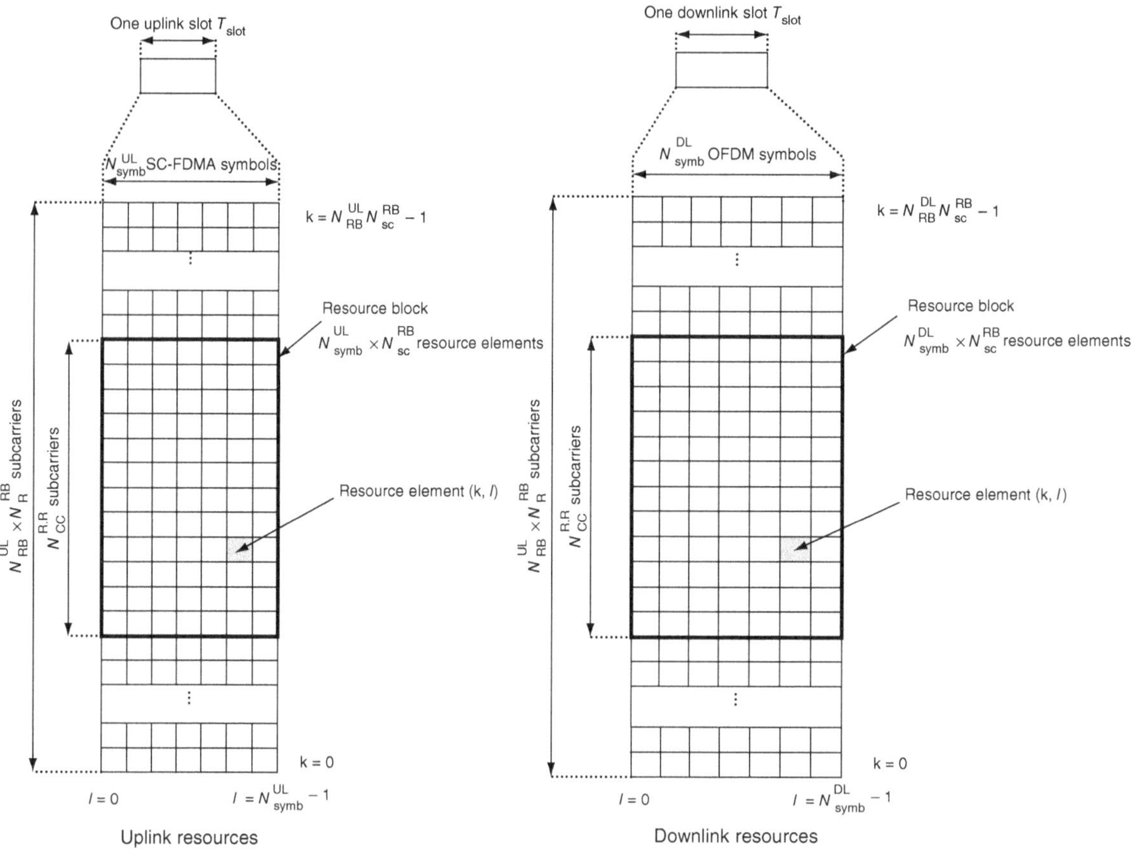

Figure 31.10 Resource blocks for uplink and downlink. *In this figure*: SC-FDMA, Single-Carrier Frequency Division Multiple Access.
Reproduced from [3GPP TS 36.211] © 2009. 3GPP™ TSs and TRs are the property of ARIS, ATIS, CCSA, ETSI, TTA and TTC, who jointly own the copyright in them. They are subject to further modifications and are therefore provided. to you "as is" for information purposes only. Further use is strictly prohibited.

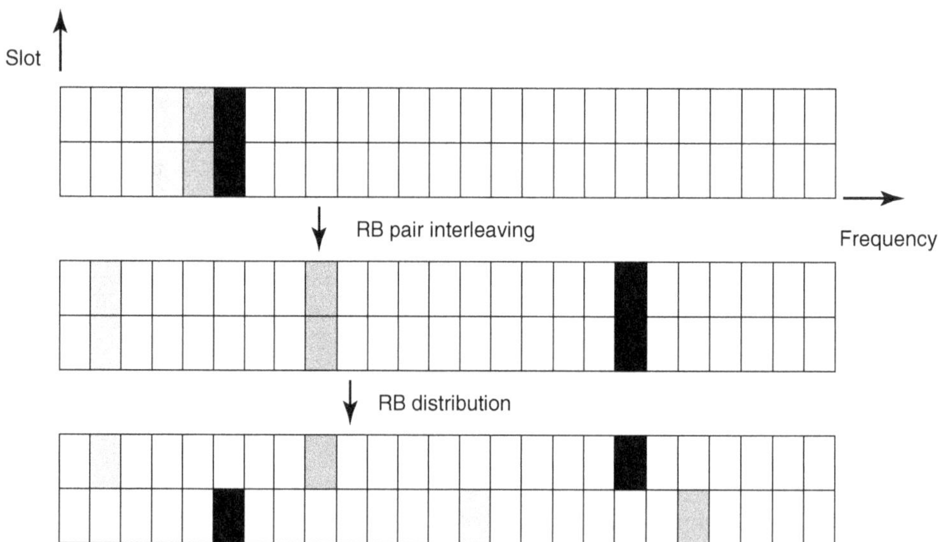

Figure 31.11 Mapping from VRBs to PRBs. Color version available at wiley.com/go/molisch/wireless3e.
Reproduced with permission from [Dahlman et al. 2008] © Academic Press.

The different types of resource allocation provide flexibility to adjust the allocation method to the amount of channel state information and the admissible overhead. If the BS has full channel state information, then a fully flexible approach employing type-0 allocation encoding and localized RBs allow full exploitation of the available multi-user diversity. Using localized RBs with contiguous assignment (type 2) has a smaller overhead, since the assigned RBs can be described in a very simple manner. However, it might not be optimum, particularly when many RBs are assigned to a particular UE, because some of the assigned resources might be on subcarriers that have bad channel quality for this particular UE.

The main reason for using distributed VRBs is to achieve frequency diversity, which is useful when scheduling according to the channel state is not feasible (e.g., in fast-changing environments). Of course, a distributed assignment can also be achieved by using type-0 or type-1 allocation with a suitable bitmap allocation. However, the use of the VRBs allows to provide frequency diversity with type-2 resource allocation notification (small overhead), i.e., assigning a *contiguous* block of *VRBs*.

Mapping to Physical Resources – Uplink

As mentioned above, the modulation for the UL is single-carrier transmission also known as DFT-spread OFDM. The mapping of symbols to the specific frequency resources in the UL is somewhat simpler, since there are no distributed RBs. The scheduler thus assigns a block of adjacent subcarriers, for the duration of one subframe (i.e., two slots, or 14 OFDM symbols; however, since 2 of the OFDM symbols are occupied by pilots, only 12 OFDM symbols are available for data transmission). For one slot, a particular user can only occupy contiguous subcarriers.[6] However, in order to provide frequency diversity, frequency hopping from slot to slot is possible (this can be interpreted in the abovementioned framework of VRB–PRB, namely that the mapping from VRB to PRB can change from slot to slot). The hopping can be done according to one of two methods: (i) use of a cell-specific hopping pattern or (ii) explicit pre-scription of the hopping pattern.

For cell-specific hopping patterns, a part of the available cell bandwidth is subdivided into a number of subbands. During each new hop, the position of the RBs is (cyclically) shifted by the width of the subband multiplied with an integer number provided by the cell-specific random sequence. In the case of explicit hopping patterns, the information about how many RB widths the RB in the second slot should hop compared to the first one is transmitted by the BS in the scheduling grants.

Frame Format

Each DL subframe contains control information, payload data, and pilots. Typically, the control information is transmitted at the beginning of the subframe, since this allows the UE to determine whether part of the payload is intended for it, and – if yes – start the demodulation process as the data are coming in (otherwise, it would have to store the whole frame before it could start processing). In addition to this standard subframe structure, there are also *Multi-cast Broadcast Single Frequency Network* (MBSFN) subframes that contain the control information and pilots for unicast scheduling, concentrated in the first one or two symbols of the subframe, as well as another part of the subframe whose use depends on the particular application. It is noteworthy that MBSFN subframes are not only used for MBSFN but can be applied for a variety of purposes.

A further complication arises for subframes in TDD. For every transition from DL to UL, there is a guard interval, to avoid collisions between the packets "on the air" (see Section 18.5). This necessitates the definition of a different type of subframe that contains three distinct parts: a Downlink Pilot Time Slot (DwPTS), Uplink Pilot Time Slot (UpPTS), and a guard interval between them. The UpPTS is not used for data transmission, but rather for ancillary transmissions such as random access or SRS (see Section 31.3.5); this is because the duration of the UpPTS is too short to be used efficiently. Note that a guard interval is not necessary for a transition from UL to DL.

31.3.5   Pilots or Reference Signals

Downlink

Pilots, or *Reference Signals* (RSs) in the nomenclature of 3GPP, are used to estimate the channel. As outlined in Section 15.5.2, scattered pilots are used to provide sufficient sampling in the time and frequency domains, while limiting the amount of overhead and thus detrimental effects on the spectral efficiency.

Cell-specific Reference Signal, CRS: For DL channel estimation, it is usually sufficient to have a cell-wide pilot CRS, i.e., a pilot that is broadcast by the BS. From this broadcast signal, any UE can estimate the channel from the BS to it. Since the pilot can be heard by all UEs, it has to cover the whole bandwidth used in the cell. The UE can use the channel estimates for (i) coherent demodulation, and/or (ii) channel quality estimation to be fed back to the BS (note that a UE can obtain channel estimates outside the RBs assigned to it).

The pilots are transmitted on the following locations (REs) within each RB (see Figure 31.12): first OFDM symbol, $1 + i$ subcarrier; first OFDM symbol, $7 + i$ subcarrier; fifth OFDM symbol, $4 + i$ subcarrier, fifth OFDM symbol, $10 + i$ subcarrier, where $i = 0, ..., 5$ is a frequency shift parameter specific to a cell, and all additions are to be taken modulo 12. The frequency shifts depend on the cell identity, which is assigned by the system to each cell, and which is detected by the UE as part of the cell-search procedure (see Section 31.4.2). By choosing appropriate cell IDs, and their associated frequency shifts, the system can orthogonalize the pilots of adjacent cells. Of course, there is still interference from data subcarriers in an adjacent cell, but that can be mitigated either by averaging (in channels with large coherence time/bandwidth) or power boosting of the pilot subcarriers.

---

[6] Starting with release 10, it is also possible to assign two *clusters* of subcarriers, which are separated in frequency, in order to increase flexibility.

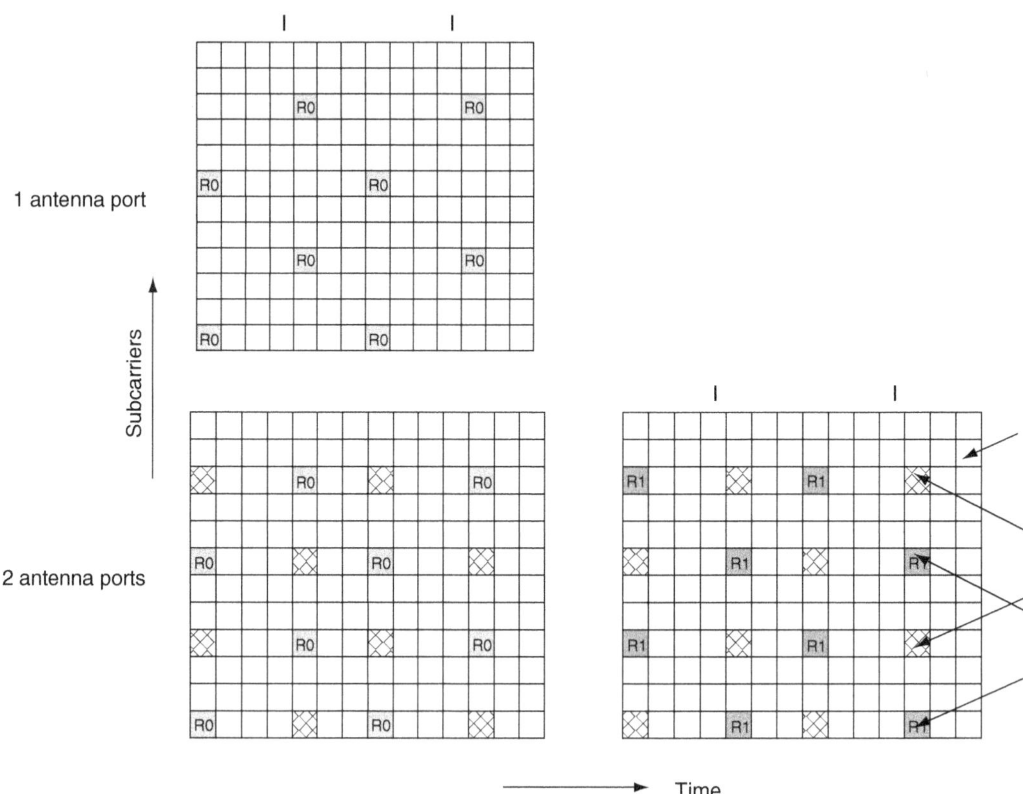

Figure 31.12 Cell-specific Reference Signal.
Reproduced from [3GPP TS 36.211] © 2009. 3GPP™ TSs and TRs are the property of ARIS, ATIS, CCSA, ETSI, TTA and TTC, who jointly own the copyright in them. They are subject to further modifications and are therefore provided. to you "as is" for information purposes only. Further use is strictly prohibited.

The complex symbols that are to be transmitted on the pilot locations are also a function of the cell ID. For each cell ID, a separate sequence with a length of 8800 symbols is defined. This sequence is 10 ms long and independent of the assigned bandwidth: it is computed assuming a 20-MHz bandwidth; if some subcarriers do not exist because the available bandwidth is smaller, those symbols are simply not transmitted.

If multiple antenna elements are used by the BS, then the number of channels that need to be estimated is larger. The pilots for the different antennas are orthogonal, i.e., an RE carrying a pilot for one antenna element is completely empty (no pilot or data assigned) on the other antenna elements. In the case of two antenna elements, the BS uses two different frequency shifts (offset by three subcarriers). In the case of four antenna elements, antenna elements 3 and 4 transmit on the same subcarriers as antenna elements 1 and 2, respectively, but on the second OFDM symbol (instead of the first and fifth), and thus have a lower time-domain density of pilots as their use is mainly targeted to low-mobility UEs experiencing small Dopper shifts. Details are shown in Figure 31.12.

Demodulation Reference Signal, DM-RS: The cell-wide pilots cannot be used when the BS applies non-codebook-based beamforming (see Section 31.3.6), i.e., beamforming in which the UE does not know the beamforming coefficients. In that case, the pilots have to undergo the same beamforming as the user data, so that all the UE sees is an "effective" channel (the concatenation of beamformer and physical channel) that it can estimate from the pilot and use for coherent demodulation. Clearly, this effective channel is specific for one UE. Note that as LTE progressed through the various releases, and in particular evolving into 5G-NR (compare Chapter 32), the trend is to move toward those UE-specific RSs and thus avoid an "always-on" cell-specific RS.

Such a UE-specific DL RS is the demodulation RS, DM-RS.[7] Since it is intended for a specific UE, it is only transmitted in the RBs carrying data for this specific UE. Furthermore, DM-RSs are defined for systems with up to eight layers (spatial data streams), which makes them suitable to be used for antenna ports in the range 5–14 (details depend on the number of DM-RSs). DM-RS are used in multi-antenna transmission modes 7–10 (see Section 31.3.6).

The multi-layer DM-RSs, introduced since Release 9 of the standard, have a structure that is different from the CRS: pilots for different antenna ports are multiplexed not in time or frequency, but rather by orthogonal "cover codes," similar to the Walsh Hadamard codes discussed in Section 19.3.2. We first consider the case that there are one or two antenna ports. On each of the RBs, pilot signals are sent on subcarriers 2, 7, 12, on the sixth and seventh OFDM symbol of each slot. Port 0 transmits the pilot tones multiplied by 1 in the former of those symbols and 1 in the latter, while Port 1 multiplies with 1 in the former and −1 in the latter. When there are

---

[7] The structure of the DM-RS described above relates to the PDSCH. The structure for the EPDCCH (see Section 31.4) is very similar, except that (i) only four antenna ports are allowed, and (ii) the DM-RS symbols are always device-specific.

more than two antenna ports (up to 8), then up to 24 REs are used for pilots in every RB pair: one set of ports uses the REs mentioned above, while the remaining ports use subcarriers 0, 4, 8 (i.e., shifted by one subcarrier).

There are two possibilities for setting the sequence of symbols to be transmitted in the DM-RS (before multiplication by the cover code). The sequence can be cell-specific, which helps with the suppression of inter-cell interference. Up to release 10 of LTE, this was the only possible method. Starting with release 11, there is also the possibility of device-specific sequences, which makes it easier to separate the pilot tones for multiple signals on the same RB in the same cell, which could for example be helpful for distributed MIMO.

MBSFN Reference Signal: In the case of MBSFN, the cell-wide RSs cannot be applied either. Rather, a special MBSFN reference signal is defined. The mapping is shown in Figure 31.13; we see that the spacing in the frequency domain is closer than for the pilots we discussed above. The reason for this is that MBSFN channels usually exhibit stronger frequency selectivity than unicast channels: in addition to multi-path propagation, the MBSFN channels also exhibit the effects of the different runtimes of the signals from the different BSs.

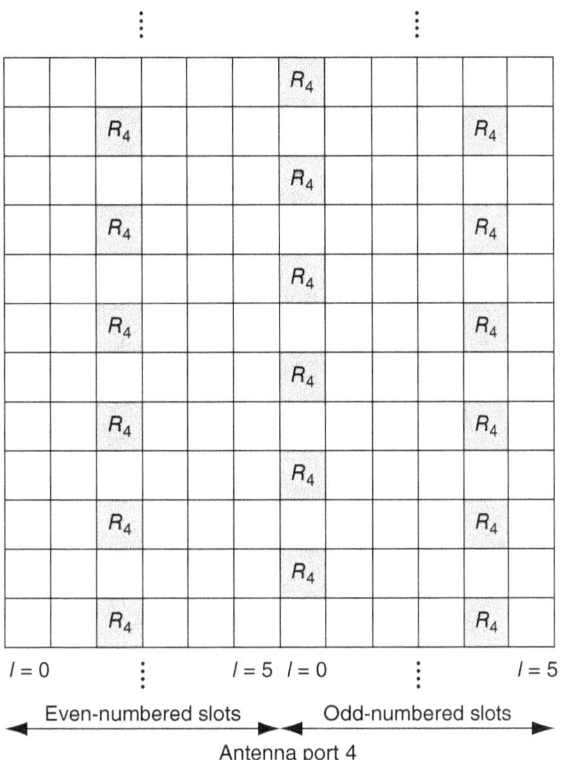

Figure 31.13 Pilots for MB-SFN.
Reproduced from [3GPP TS 36.211] © 2009. 3GPP™ TSs and TRs are the property of ARIS, ATIS, CCSA, ETSI, TTA and TTC, who jointly own the copyright in them. They are subject to further modifications and are therefore provided. to you "as is" for information purposes only. Further use is strictly prohibited.

Channel State Information Reference Signal, CSI-RS: A further type of pilot structure is the Channel State Information (CSI) RS. CSI-RSs are intended to let the RX determine the channel over the whole available bandwidth, e.g., for scheduling considerations. This concept is most important for the UL, and therefore described there in more detail. For the DL, the standard CRS already provides all required information for systems with up to four antenna ports at the BS. However, such a number of signals might not be sufficient, and there could also be other reasons for separate CSI pilots. Thus, the LTE standard (beginning with release 10) also foresees CSI-RS in the DL. Up to 8 (release 10–12) or 16 (release 13 and beyond) antenna ports[8] are supported. The periodicity of the tones can range from 5 to 80 ms, and the pilots for the different ports are separated by a combination of time-frequency orthogonality and cover codes.

Positioning RS: Last but not least, there are positioning RSs, i.e., pilot tones intended for localization of the UEs. The main purpose is for the UE to measure the time of arrival of the signals from different BSs, which then forms the basis for the trilateration that discovers the UE position (compare Chapter 29). In order to minimize the inter-cell interference, the pilots can be configured in a way that surrounding BSs are silent on the REs in which the positioning RSs of one BS are sent.

Uplink

Demodulation RS: The requirements for pilot signals in the UL are significantly different from the pilot signals in the DL. Firstly, *multiple* pilots need to be transmitted, since the BS has to learn the channel from multiple UEs to the BS. Secondly, the data that need to be

---

[8] The antenna ports for CSI-RS are typically actual antennas, while the antenna ports for DM-RSs are usually beamformer ports, see Sec. 31.3.6.

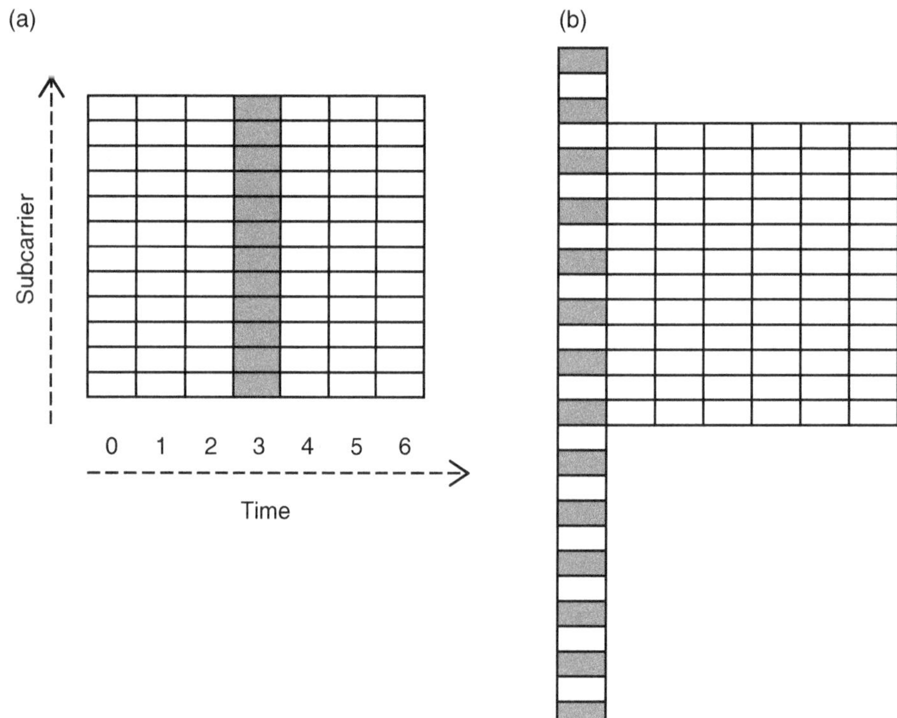

Figure 31.14 Structure of uplink demodulation RS (a) and sounding RS (b).

demodulated are always well localized in the frequency domain. Thirdly, the existence of the pilot should not upset the single-carrier properties of the transmit signal. For these reasons, the pilots in the UL show a very different structure from those of the DL (Figure 31.14).

The main pilot is the *demodulation pilot* (DM-RS), which is used to determine – with high accuracy – the channel transfer function in the specific RB(s) used by a UE for data transmission. In order to retain the single-carrier properties of the transmit signal, the pilot is not frequency multiplexed with the data signal. Rather, a complete OFDM symbol (over the bandwidth used by this particular UE) is assigned to the pilot. In particular, of the seven OFDM symbols in one slot, the fourth symbol is reserved for demodulation pilot.

The exact form of the pilot is defined in the frequency domain (i.e., after having passed through the DFT spreader). It is based on Zadoff–Chu sequences, which are defined as

$$X^{(u)}(k) = e^{-j\pi uk(k+1)/M_{ZC}} \quad 0 \le k < M_{ZC}$$

where M_{ZC} is the length of the sequence, and u is the index of the sequence. Zadoff–Chu sequences have the remarkable property that they have constant amplitude in the time domain (important for the power amplifiers) and the frequency domain (so that the transmit SNR at all measured frequencies is the same); the autocorrelation function of such a sequence is a delta function. For this reason, it belongs to the class of Constant Amplitude Zero AutoCorrelation (CAZAC) sequences. The number of Zadoff–Chu sequences is limited to the number of integers that are relative-prime to the sequence length. For this reason, it is preferred that M_{ZC} is a prime number, since this provides the largest number of distinct sequences, namely $M_{ZC} - 1$.

Choosing M_{ZC} equal to the number of used subcarriers would result in a low number of available sequences, because the number of subcarriers is always an integer multiple of 12 (number of subcarriers in an RB). LTE thus uses for M_{ZC} the largest prime number that is smaller than the desired sequence length. For example, for a signal occupying 4 RBs (48 subcarriers), M_{ZC} is chosen as 47, so that there are 46 available sequences. Since the length of the sequences is not long enough to cover all available subcarriers, the sequence is simply cyclically extended (i.e., after the last element, we start again with the first, then the second, ...). These periodically extended Zadoff–Chu sequences are then the *base sequences* for the UL DM-RS. Note that the minimum number of available sequences is 30: if the number of RBs is ≥ 3, then the above rule provides at least 30 sequences; for the case of 1 and 2 RBs, the standard tabulates a set of 30 different computer-generated CAZAC sequences (that are not ZC sequences), which are to be used as base sequences.

However, the number of base sequences is usually too small for practical purposes. LTE uses the following two modifications of Zadoff–Chu sequences:

1. *Phase rotations*: in order to generate additional sequences, each (cyclically extended) Zadoff–Chu sequence can be multiplied with a linear phase shift in the frequency domain, i.e.,

$$X'(k) = X(k)e^{j\alpha k}$$

where different shifts α lead to different sequences (this corresponds to a cyclic shift in the time domain). With a suitable choice of phase shifts, the resulting sequences can be made completely orthogonal to each other: when the phase shift is an integer multiple of $2\pi/12$, then orthogonality is obtained over 1 RB. The orthogonality can be destroyed by two effects: (i) Timing misalignment of the two considered sequences. Since signals within a cell are typically well time aligned, pilots based on the same Zadoff–Chu sequence, but with different phase shifts, can be used within one cell. (ii) Frequency selectivity of the channel that is so strong that the channel varies significantly over the RB. This is equivalent to saying that the impulse response should not extend beyond the cyclic delay shift; if this condition is violated, a subset of phase shifts corresponding to larger delay shifts (e.g., $2\cdot 2\pi/12$, $4\cdot 2\pi/12$, ...) can be used.

2. *Cover codes:* Further orthogonal sequences can be obtained by multiplying the sequences in the two slots of a subframe by orthogonal cover codes (i.e., $(1/1)$ and $(1/-1)$), though this is valid only if the channel does not change significantly over the duration of a subframe.

Having these orthogonal training sequences in the same time-frequency resources is useful for SU-MIMO (where the sequences are assigned to different data streams/layers) and MU-MIMO, where they are assigned to different devices. They can be used in different cells only if there is very tight synchronization between the cells.

The different ZC sequences are then assigned to cells, such that each cell has 1 periodically extended sequence[9] and can derive orthogonal sequences based on phase shifts. The sequences are assigned to cells based on cell ID (for the *Physical Uplink Control CHannel* (PUCCH), see Section 31.4.10) or explicitly signaled (for *Physical Uplink Shared CHannel* (PUSCH), see Section 31.4.11). The standard also foresees the possibility of group hopping (i.e., the "fundamental" sequence for each cell is changed on a regular basis). Finally, since release 11, there exists the possibility of a device-specific base sequence assignment.

Sounding RS: A second type of pilot, the *sounding pilot* (Sounding Reference Signal (SRS)), is intended to enable efficient scheduling. The BS uses it to estimate the frequency domain channel response over substantially the *entire system bandwidth*; based on this knowledge, it can then assign the users to the most suitable subcarriers. Therefore, the sounding pilot transmitted by each user typically occupies the entire system bandwidth. This large bandwidth can be achieved by (i) sending a DFT-OFDM symbol that spans the whole system bandwidth, or (ii) sending a sequence of frequency-hopped signals that each occupy a smaller bandwidth, but, taken together, occupy the whole bandwidth. The latter approach is more complicated but has the advantage that the power-spectral density of the sounding pilot is higher. Various bandwidths (have to be multiples of 4 RBs) are admissible.

The sounding pilot is based on Zadoff–Chu sequences, having a shift $\alpha = 2\pi n_{SRS}/8$, where n_{SRS} is a parameter that can take on values between 0 and 7. That sequence is then mapped onto a frequency comb, i.e., every second subcarrier (or, since release 13, every fourth subcarrier), in order to enable orthogonal multiplexing of SRS transmissions from different UEs over different bandwidths using different combs. The transmission always occurs in the last symbol of a subframe, but not in every subframe: the transmission periodicity can be chosen in a range from 2 to 160 ms. This accommodates a wide variety of coherence times of the channel. In order to avoid collisions between sounding pilot and payload (PUSCH) transmissions, the BS ensures that there are no PUSCH/PUCCH transmissions in the last subframe symbol in the cell whenever *any* UE transmits a sounding pilot. Orthogonality between sounding pilots from different UEs is handled by a combination of two methods: (i) transmission on different frequency combs, and (ii) transmission of the sequences with different linear phase shifts (similar to the demodulation pilot). Note that the channel estimates used for scheduling need not be as accurate as those used for demodulation (compare the discussion of CSI for antenna selection in Sec. 16.2.10). In addition to the periodic transmission, SRS transmissions can also be "single-shot" aperiodic.

Finally, we note that for multi-antenna situations, the SRS transmitted from each antenna port are *not* precoded, i.e., they provide the true channel from a particular antenna port.

31.3.6 Multiple-Antenna Techniques

LTE makes extensive use of multiple-antenna techniques, including transmit diversity, beamforming, and spatial multiplexing. Let us first establish some notations:

- Transport blocks are mapped onto *layers*, which contain different symbols. Note that a transport block can be mapped onto several layers. In the absence of space-time coding, "layers" correspond to the "spatial streams" discussed in Section 16.2.
- The layers are mapped onto *antenna ports* by means of a precoding. Antenna ports are not necessarily antenna connectors for specific antennas, but rather are beamformer inputs/outputs. To be more precise, signals transmitted from the same antenna port experience the same channel while signals from different antenna ports have (usually) different channels. If we are sending the same signal from two antenna elements, then the sum of these two elements is a "port." Conversely, signals sent from the same set of antenna elements with different precoders originate from different ports.
- *Quasi-colocated antennas*: in brief, quasi-colocated antennas are antennas that experience the same large-scale channels (though the small-scale fading on them might be different). As indicated by the name, this might occur if antennas are at the same location, while it does not apply when antennas are widely separated, e.g., in a distributed MIMO or CoMP system. However, it must be noted

---

[9] And two sequences of length 72 or longer.

that antennas at the same physical location, but, e.g., with different directions (e.g., patch antennas on panels that have 90-degree angle between them) need not experience the same large-scale channel characteristics, and thus would not fall into the category of "quasi-colocated."

Overall, 10 different multi-antenna transmission modes are defined in the standard, see Table 31.4.

Table 31.4 Multi-antenna transmission modes.

Transmission mode index	Transmission type
1	Single-antenna transmission
2	Transmit diversity
3	Open-loop codebook-based precoding
4	Closed-loop codebook-based precoding
5	MU-MIMO version of mode 4
6	Mode 4 specialized to single-layer
7	Noncodebook precoding with 1 layer
8	Noncodebook precoding with ≤ 2 layers
9	Noncodebook precoding with ≤ 8 layers
10	Extension of mode 9 to CoMP

Transmit Diversity

For the case of two transmit antennas, LTE foresees Space Frequency Block Coding (SFBC), i.e., Alamouti encoding on adjacent sub-carriers. This can be interpreted as mapping the symbols of a code block onto the layers according to

$$\mathbf{A} = \begin{bmatrix} s_1 & -s_2^* \\ s_2 & s_1^* \end{bmatrix}. \tag{31.11}$$

For four antennas, the mapping of the code blocks onto the four layers is done according to the following mapping:

$$\mathbf{A} = \begin{bmatrix} s_1 & -s_2^* & & \\ s_2 & s_1^* & & \\ & & s_3 & -s_4^* \\ & & s_4 & s_3^* \end{bmatrix} \tag{31.12}$$

so that, e.g., the transmission vector on frequency 1 is $\left[s_1, 0, -s_2^*, 0\right]$. This scheme gives a rate of 1. The layers are then directly mapped onto the antenna ports.

Spatial Multiplexing

In precoded spatial multiplexing, the symbols in transport blocks are mapped into multiple layers, by a straightforward procedure. To wit, the transport blocks are mapped onto the layers as follows: a single transport block is used in conjunction with a single layer. For a larger number of layers, two transport blocks are used. In particular, if both number of transport blocks and number of layers equals 2, then each layer simply contains the symbols from one transport block. If there are two transport blocks and four layers, then symbols from transport block 1 are alternatingly assigned to layers 1 and 2, while symbols from transport block 2 are alternatingly assigned to layers 3 and 4. In the case of two transport blocks and three layers, one transport block is assigned to the first layer, while the symbols from the second transport block are alternatingly written to layer 2 and 3; in this case, the second transport block has to have twice as many symbols as the first, since each layer contains the same *number* of symbols.

Layers are then multiplied with a matrix $\mathbf{T}$ to provide the signals at the antenna ports; more specifically at a given time instant, $\mathbf{T}$ is multiplied with a vector containing one symbol from each layer; this provides the symbols to be sent from the antenna elements at this time instant.

Closed-Loop Spatial Multiplexing: In closed-loop spatial multiplexing, the UE selects, from a codebook, a precoder matrix that, if used by the BS, will result in good received signal quality at that UE (see Sections 16.1.7 and 16.2.8); it then feeds a few bits indicating this precoder to the BS. Specifically, the UE selects how many layers it wants to the BS to use (Rank Indicator (RI)), and the index of the entry in the codebook that corresponds to the chosen precoder (Precoder Matrix Indicator (PMI)).

The codebooks are Fast Fourier Transform (FFT)-based for the case of two antenna ports, i.e., the entries are given in Table 31.5.

Similar codebooks are defined for transmissions with four antenna ports. An implicit assumption for the design of the codebooks was that the codebook design is governed by small-scale fading characteristics and uniform linear arrays.

Table 31.5 Codebook entries for two antenna port.

Codebook index	Number of layers	
	1	2
0	$\frac{1}{\sqrt{2}}\begin{bmatrix} 1 \\ 1 \end{bmatrix}$	$\frac{1}{\sqrt{2}}\begin{bmatrix} 1 & 0 \\ 0 & 1 \end{bmatrix}$
1	$\frac{1}{\sqrt{2}}\begin{bmatrix} 1 \\ -1 \end{bmatrix}$	$\frac{1}{2}\begin{bmatrix} 1 & 1 \\ 1 & -1 \end{bmatrix}$
2	$\frac{1}{\sqrt{2}}\begin{bmatrix} 1 \\ j \end{bmatrix}$	$\frac{1}{2}\begin{bmatrix} 1 & 1 \\ j & -j \end{bmatrix}$
3	$\frac{1}{\sqrt{2}}\begin{bmatrix} 1 \\ -j \end{bmatrix}$	

For eight-element arrays (defined since release 10), this assumption might not be valid anymore. Thus, the codebook is written as a product $\mathbf{W} = \mathbf{W}_1 \cdot \mathbf{W}_2$, where $\mathbf{W}_1$ models long-term channel effects (average Angular Power Spectrum (APS)), while $\mathbf{W}_2$ models the small-scale fading. For even larger arrays (full-dimensional MIMO, defined since release 13), two methods are defined for codebooks. The first case is a straightforward extension of the eight-element codebook, though now different rectangular, dual-polarized array configurations are assumed. The alternative approach assumes beam-formed CSI signals, and a UE measures in a relatively small number of such formed beams and feeds back the index of the best one.

While the UE feeds back information of what beamformer it wants the BS to use, this is not binding. Rather, it is the BS that ultimately selects the precoder matrix. It can override the recommendation of the UE, in which case it has to signal to the UE which precoder it chose. The UE estimates the channel coefficients from its estimation of the radio channel from the CRS and/or CSI-RS, combined with the knowledge of the precoding matrix.

Non-Codebook-Based Precoding: Non-codebook-based precoding means that the UE is not being told explicitly what precoder the BS is using. Due to this, the BS has to insert DM-RSs that are precoded in exactly the same way as the payload data; the UE can then estimate the "effective" channel (concatenation of precoder and physical radio channel), and use that knowledge for demodulation.

The design of the precoding matrix at the BS can be done in two ways (compare Section 16.1.7): (i) the BS estimates the DL channel based on observations of the UL channel and determines the precoder from that information. This usually requires channel reciprocity and is thus only applicable to TDD systems. (ii) the UE feeds back the DL channel. Typically, such feedback is based on a codebook, and the UE just feeds back the codebook index. It is somewhat confusing that such a scheme is called "non-codebook-based." The term, as used in LTE, thus only relates to the fact that the used codebook entry is unknown to the UE, not to the absence of an actual codebook.

Open-Loop Spatial Multiplexing: In the case that a feedback of the desired beamforming is not available (e.g., because the channel changes too rapidly), an open-loop spatial multiplexing can be used. In that case, the layers are first multiplied by a (subcarrier-independent) matrix $\mathbf{U}$, then a (subcarrier-dependent) diagonal matrix $\mathbf{D}(i)$, and finally by a matrix $\mathbf{W}$. For the case of two antenna ports and two layers,

$$\mathbf{W} = \frac{1}{\sqrt{2}}\begin{bmatrix} 1 & 0 \\ 0 & 1 \end{bmatrix} \quad \mathbf{U} = \frac{1}{\sqrt{2}}\begin{bmatrix} 1 & 1 \\ 1 & \exp(-j2\pi/2) \end{bmatrix} \quad \mathbf{D}(i) = \begin{bmatrix} 1 & 0 \\ 0 & \exp(-j2\pi i/2) \end{bmatrix} \tag{31.13}$$

where the $\mathbf{D}(i)$ effectively leads to a large cyclic shift of the considered block. Open-loop precoding helps to introduce diversity into a spatial multiplexing transmission.

Multi-user MIMO: MU-MIMO implementation in 3GPP essentially follows the principles outlined in Chapter 22. One aspect that is different from the usual academic treatment of the topic is that the time–frequency resource allocation can be *partially* overlapping, i.e., a particular UE (say, UE A) might share a particular RB with UE B, but share another RB with UE C.

The MU-MIMO transmission is essentially transparent to the UE, since non-codebook-based precoding is used, so that the effective channel for each UE can be estimated from a device-specific DM-RS.[10] The RSs for the different users in MU-MIMO are orthogonal in order for channel estimation to be interference-free.

Uplink Multi-Antenna Transmission: the principles for the UL multi-antenna transmission are fairly similar to the DL transmission, though there are a few important differences. Firstly, since the control about UL transmissions should still rest with the BS, the choice of the precoding matrix is done by the BS and communicated to the UE as part of the UL scheduling grant. It also implies that noncodebook beamforming is not possible. Secondly, the beamforming matrices are chosen such that only one layer (data stream) is mapped onto each antenna port, i.e., the precoding matrix has at most one nonzero entry per row. This is done such that the low-PAPR properties of the UL transmit signal is retained.

---

[10] Besides the approach described here, there are some older methods that use either CRS (transmission mode 5) or cell-specific (instead of device-specific) DM-RS.

Finally, MU-MIMO is easier in the UL than in the DL, because the RX (i.e., the BS) can do joint detection of the signals from the different UEs. The only requirement for successful implementation is to have different/orthogonal DM-RSs for the UEs operating on the same time–frequency resources, which is obtained through the use of phase rotations and cover codes, as described in Section 31.3.5.

31.3.7 Feedback for Adaptive Modulation and Beamforming

For the DL transmission, the BS has to know the CSI that the UE sees. This CSI is communicated, in a very specific format, through feedback in the UL control channel. Generally, the information consists of the following pieces:

- *Channel Quality Indicator (CQI)*: this index actually represents the modulation and coding scheme that should be used so that the packet error rate does not exceed 10%; this is the error level deemed acceptable in regular data transmission, such that the overhead for HARQ retransmissions does not become excessive. The reason for reporting a desired MCS and not an SINR is that different RXs might have different quality and processing ability so that the same SINR can represent different packet error rates for different UEs; the UE thus reports back the MCS that is best for their specific RX type.
- *RI*: this is the rank (number of layers) in a MIMO transmission.
- *PMI*: this is the index of the codebook entry that should be used for precoding by the BS.
- *CSI-RS Resource Indicator (CRI)*: in the case that the UE monitors multiple beams and beamformed CSI-RS signals are used, the CRI indicates which beam it would prefer.

The RI, PMI, and CSI-RS are relevant for multi-antenna systems only.

The information is sent in what is called a CSI report. We can distinguish between *periodic* CSI reports and *aperiodic* CSI reports (see Section 31.4.9 about how they are sent). The former are configured by the network to update the BS about the CSI of the different users at periodic intervals, which can be as short as 2 ms. Since this leads to considerable overhead, the reports are kept as small as possible. The aperiodic reports are requested by the network, and can provide much more detailed information – the underlying assumption is that the network can judge when the extra effort for those detailed reports is worth the overhead.

It is important that the parameters transmitted by the UE are only recommendations, and the ultimate decision about which ones to use lies with the BS, and the BS therefore needs to communicate the final decision to the UE. This is in particular true for the CQI, since the BS might take aspects such as scheduling strategy for an actual packet error rate that can be different from the target one that the CQI is based on, buffer size, requirements from other UEs and BSs, etc., into account when deciding on an MCS. For the PMI, the BS can communicate a single bit if it decided to accept the recommendation of the UE; if it decides for a different PMI, it has to send that information to the UE.

Periodic CSI Reports: Those reports must be very concise, because they are transmitted periodically and the corresponding overhead needs to be constrained (see Section 31.4.9). Thus, the UE typically reports *wideband* parameters, i.e., parameters that are valid over the whole bandwidth. For example, the CQI reports a desired MCS that is applied across the whole bandwidth. There is, however, also the possibility of UE-selected reports, in which the total bandwidth is divided into a certain number of *bandwidth parts* (1–4, depending on the cell bandwidth); and each bandwidth part is divided into subbands consisting of 4–8 RBs. The UE then cycles through the bandwidth parts (i.e., reports for one bandwidth part in one CSI report), reporting the wideband CQI and PMI, as well as the (UE-determined) best subbands within the bandwidth part considered in the current report and its CQI (but *not* its PMI, to save overhead).

Aperiodic CSI Reports: Aperiodic reports are on-demand and can provide more detailed information than periodic CSI reports. We distinguish here generally between (i) wideband reports, (ii) UE-selected reports, and (iii) configured reports.

In the wideband reports, only a single CQI value is provided over the whole bandwidth. However, for the PMI, the available cell bandwidth is divided into subbands of size 4–8 RBs, and the PMI is reported per subband.

In the UE-selected reports, the device selects the best M subbands and reports one CQI value reflecting the average channel quality over those bands, in addition to the wideband CQI value. M can range from 1 to 6, and the subband size from 2 to 4 RBs. PMI and RI can also be reported.

In the configured reports, the UE reports CQI per subband, as well as possibly PMI and RI per subband, but the reporting is configured by the higher layers.

CSI for Massive MIMO: since release 13, provisions are foreseen for reporting of CSI with a large number of antenna elements. Two types of reports are foreseen: *per-antenna-element*, and *per-beam* reporting. The per-antenna-element reporting is the traditional approach where an RS is associated with a particular antenna element, and the signal received from this RS is the basis for the feedback. Obviously, the overhead is determined by the number of antenna elements. In contrast, in the per-beam reporting, the BS performs beamforming, and associates an RS with each beamformer input. The number of formed beams can be lower than the number of antenna elements. Thus, the overhead can be lower than for the per-antenna-element reporting when the number of antenna elements is very large.

For per-antenna-element reporting, the codebooks assume that the antenna structure takes on one of a set of admissible rectangular dual-polarized structures, leading to a factorization of the codebook (see discussion on codebooks above). Reporting is limited to a maximum of 16 elements.

For per-beam reporting, there is in principle no limitation on the number of antenna elements. Rather, the number of beams that is sounded and have feedback will be limited. The UE can report the preferred beam, which can possibly be different in different subbands. In the case of mobility, it is common to sound the strongest beam observed in previous time instances, plus beams to either side of it (to account for smooth changes in the direction of the beam), but ultimately the selection of the beams that are sounded lies with the BS.

Interference Estimation: Earlier releases of LTE left it up to the implementation to decide how the interference level is to be measured. This often meant that the interference from neighbor-cell RSs was measured, leading to an estimation of strong interference even if the cell was not heavily loaded. Later releases (after release 11) thus foresee a CSI interference measurement configuration, which prescribes on what time/frequency resources the interference measurements are to be done. Essentially, it defines REs in which the serving BS does not transmit, so that the UE can measure the level of power it receives from other BSs.

31.4   Logical and Physical Channels

31.4.1   Channel Definitions and Mapping

LTE defines different types of "channels" in the standard (note that those have nothing to do with propagation channels). Channels are ways/containers to transport bits. From the point of view of the higher layers of the OSI model, we are mostly interested in *logical channels*, which are defined by the type of information that they carry. Most importantly we can distinguish between traffic channels (i.e., carrying the payload data, and thus constituting the user plane), and the control channels, which constitute the control plane; however, within this dichotomy there are still further distinctions, as described below. These logical channels are then mapped to *transport channels* (which are defined by how, and with which characteristics the information is transmitted). Note that a transport channel can contain information from multiple logical channels, and in particular may carry a mixture of payload and control bits. From the transport channels, the information is mapped onto the *physical channels*. These are defined by their physical properties, i.e., time, subcarrier, etc.), and are thus most similar to what we would imagine a channel from a PHY-layer point of view (one could create, e.g., multiple channels in an FDMA system by just assigning different subbands to different such channels).

The following logical channels are defined in LTE for UL and DL (see Section 31.10 for Device-to-Device channels):

- Traffic channels
  - *Dedicated Traffic CHannel (DTCH)*: it carries the user data for all ULs, as well as for those DL data that are not multi-cast/broadcast.
  - *Multi-cast Traffic CHannel (MTCH)*: it carries the user data for multi-cast/broadcast DL transmission over multiple cells.
  - *Single-Cell Multi-cast Traffic Channel (SC-MTCH)*: transmits the user data for multi-cast/broadcast DL transmission over a single cell.

- Control channels
  - *Broadcast Control CHannel (BCCH)*: it carries system information data that are broadcast to the UEs in a cell. Note the difference from the MTCH, which also broadcasts to UEs, but carries user data.
  - *Paging Control CHannel (PCCH)*: it pages UEs in multiple cells (i.e., when it is not known exactly in which cell the UE currently is located).
  - *Common Control CHannel (CCCH)*: it transmits control data for the Random Access (RA), i.e., when a connection is started.
  - *Dedicated Control CHannel (DCCH)*: it is used for the transmission of control information that relates to a specific UE (as opposed to the system information relevant for all UEs, which is broadcast in the BCCH).
  - *Multi-cast Control CHannel (MCCH)*: it carries the control information related to multi-cell multi-cast/broadcast services.
  - *Single-Cell Multi-cast Control CHannel (SC-MCCH)*: it transmits control information for single-cell multi-cast.

These channels are mapped onto the following transport channels:

- *DownLink Shared CHannel (DL-SCH) and UpLink Shared CHannel (UL-SCH)*: they carry the user data, as well as most of the control information (except the one mentioned below).
- *Broadcast Channel (BCH)*: it carries part of the BCCH (the remainder is on the DL-SCH described above). It has a fixed format, so that any UE can listen to it easily.
- *Paging CHannel (PCH)*: it carries the PCCH.
- *Random Access CHannel (RACH)*: it is also defined as a transport channel even though it does not carry transport blocks. It serves for executing random access of the UEs to the BS.
- *Multicast CHannel (MCH)*: it is used to support broadcast/multi-cast transmission.

The data on transport channels are organized into transport blocks; in each TTI (usually a subframe), one transport block is transmitted (two transport blocks are possible when spatial multiplexing is used). A transport format is associated with each transport block that defines how the data is transmitted.

Finally, these transport channels are mapped onto physical channels; there are also physical channels that do not carry any transport channel but are purely used for PHY functionality.

- Downlink
  - *Physical Downlink Shared CHannel (PDSCH)*: it carries the DL-SCH, i.e., user data, some control data for the DL, as well as the PCH.
  - *Physical Downlink Control CHannel (PDCCH)*: it carries control information, such as scheduling that is required for reception of the PDSCH. This channel does not carry any transport channel.

- *Enhanced PDCCH (EPDCCH)*: it serves the same purpose as the PDCCH, but allows more flexible transmissions.
- *Physical Control Format Indicator CHannel (PCFICH)*: it carries control information about the number of symbols used for PDCCH transmission in a subframe. This channel does not carry any transport channel.
- *Physical HARQ Indicator CHannel (PHICH)*: it carries the feedback bits indicating whether a retransmission of transport blocks from a UE is necessary. This channel does not carry any transport channel.
- *Physical Broadcast CHannel (PBCH)*: it carries the BCH.
- *Synchronization Signal (SS)*: does not carry any transport channel.
- *Physical Multi-cast CHannel* (PMCH): it carries the MCH, which contains the multi-cast payload, as well as some of the control information for multi-cast.
- *Machine-type-communication Physical Downlink Control CHannel (MPDCCH)* is a variant of the EPDCCH, intended for machine-to-machine type communication.
- *Relay Physical Downlink Control CHannel (R-PDCCH)* carries the control information for relaying.

- Uplink
  - *PUSCH*: it is the uplink counterpart to the PDSCH.
  - *PUCCH*: it carries mainly three types of information: (i) channel state information reports, (ii) SRs, and (iii) HARQ feedback bits.
  - *Physical Random Access CHannel (PRACH)*: it is primarily used for initial access by a UE to a BS before an RRC connection for scheduling has been established, but it can also be used by a UE that has established RRC connection to a BS in order to refine UL time synchronization and for other purposes.

Figure 31.15 summarizes the mapping between the channels.

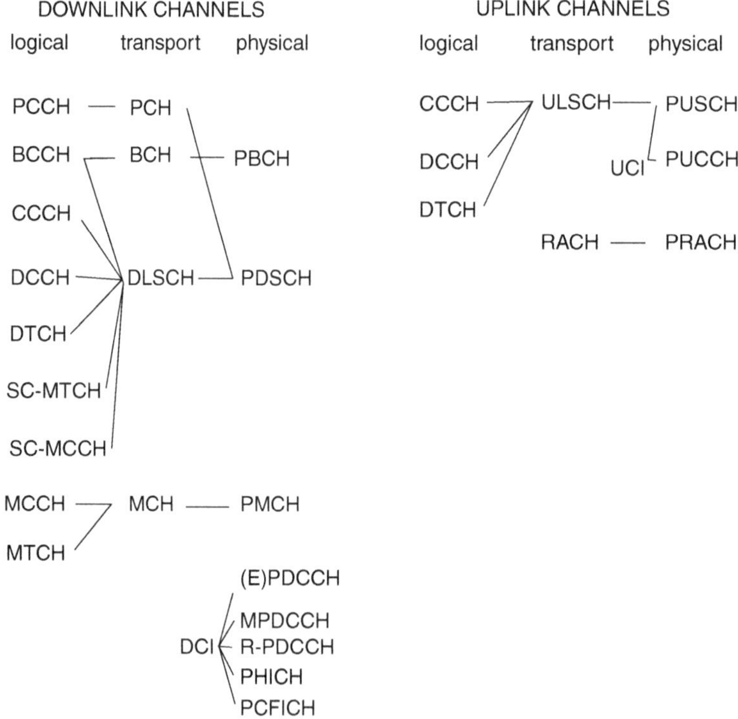

Figure 31.15   Mapping between logical, transport, and physical channels.

In the following, we now describe these channels in more detail. We start with the channels required for establishing a connection between BS and UE, namely the synchronization signal and broadcast channel. We then move to the DL-SCH and the control channels that are associated with it. We finally describe the UL channels, including the random access channel.

31.4.2 Synchronization Signals

When a UE is turned on or enters a cell area, the first thing it has to do is to synchronize itself to the timing of the cell and obtain some basic transmission parameters used in that cell. This is achieved by the SS and the BCH. In particular, the *SS* carries information about the timing of the cell, as well as the *cell ID*. LTE actually provides two SSs, the *Primary Synchronization Signal* (PSS) and the *Secondary*

Synchronization Signal (SSS). In contrast to other systems, these signals are not called "channels," but perform similar functions. To understand the functionality of the SS, keep in mind that there are 504 cell IDs defined for LTE, which are divided into 168 ID groups.

The PSS is transmitted in the last symbol of the first slot of subframes 0 and 5 of every frame,[11] extending over the center 72 sub-carriers (6 RBs) of the band (not including the nulled-out DC subcarrier). The waveform transmitted during that slot is one of three allowed Zadoff–Chu sequences of length 63, extended with five zeroes at the lower and five zeroes at the upper band edge (the ZC sequence element assigned to the DC subcarrier is not transmitted). Which of the three sequences is transmitted depends on the cell ID within a group (note, however, that the actual group ID is transmitted in the SSS). The UE can obtain the fine-resolution timing from the PSS by correlation with the Zadoff–Chu sequence. However, there is still a timing ambiguity by multiples of 5 ms, i.e., the periodicity of the PSS signal; therefore, frame timing is not available from the detection of the PSS alone.

The SSS signal is transmitted in the symbol directly before every PSS signal.[12] The signal also extends over 72 subcarriers. It is an interleaved combination of two *m*-sequences (see Section 19.3.2) of length 31. The resulting length-62 sequence is extended with zeroes at the band edges, just like for the PSS. Only 168 sequences are valid and signify the cell ID group. In contrast to PSS, the signal transmitted in the first half of a frame is different from the signal in the second half: while it contains the same *m*-sequences, the interleaving in the second half is different from that in the first half. This allows the RX to acquire frame timing (i.e., the remaining ambiguity of timing is multiples of 10 ms, which can be resolved from the system frame number transmitted in the BCH, see below), as well as the cell ID, from a single observation of an SSS.

The reason that both PSS and SSS use 72 subcarriers is that this is the smallest bandwidth permissible for an LTE system. At the time of the reception of those signals, the UE does not know yet the actual system bandwidth in the considered cell; the SSs thus use the same bandwidth for all cells, namely the smallest possible.

Once a signal has acquired cell ID and frame timing, it can receive the pilots (remember DL pilots depend on the cell ID), which are needed for the reception and decoding of the BCH.

31.4.3 Broadcast Channel

The BCH contains the *Master Information Block* (MIB), which contains critical information about the cell system information:

- System bandwidth in the cell.
- *PHICH configuration*: some bits describing the specific configuration (number of groups) of the PHICH channel. Knowledge of this configuration is necessary to receive the control information by excluding the PHICH channel.
- System frame number.

The baseband processing and transmission of the MIB bits on the BCH proceed in the following steps:

- Addition of a 16-bit CRC.
- Encoding with tail-biting rate 1/3 convolutional code.
- Scrambling.
- QPSK modulation.
- *Antenna mapping*: this is trivial if only a single antenna is used. If the BS has two antennas, then space-frequency block codes *must* be used. If the BS has four antennas, then the combination of antenna hopping and space-frequency block codes (see Section 31.3.6) *must* be used. This is done so that the UE can learn from the BCH how many antennas the BS has.
- *Demultiplexing*: the signal is mapped onto four consecutive frames; particularly onto the first subframe of each of those four frames. Within each such subframe, the signal is transmitted on the first four symbols of the second slot, over 72 subcarriers (the choice of 72 subcarriers is motivated the same way as for the PSS and SSS). Repetition coding is used to "fill up" this available resource, which is much larger than the number of bits that need to be transmitted.

Note that the BCH extends over 40 ms, while the timing acquired from the PSS and SSS has an ambiguity of multiples of 10 ms. The UE must therefore try to decode the BCH with four different timing shifts, and determine from the CRC which is the correct one. This decoding also provides the two least significant bits of the system frame number; for this reason, those two bits are not included in the MIB.

Other information in the *logical* broadcast channel, the BCCH, is transmitted in the DL-SCH. This information, called the *System Information Block* (SIB) is discussed in more detail in Section 31.5.1.

31.4.4 General Aspects of Control Channels Associated with a DL-SCH

There are four physical channels associated with the DLSCH, namely PCFICH, PHICH, PDCCH, and EDPCCH.[13] The control information contained by these channels is also called L1/L2 signaling, because it is relevant for both the PHY and the MAC layer. It is

---

[11] In TDD systems, it is on the third symbol of subframes 1 and 6.

[12] In TDD, it is transmitted three symbols ahead of the PSS; this difference in spacing allows the UE to determine whether it is in an FDD or TDD cell.

[13] Additional control signaling carried by the MPDCCH and R-PDCCH is only relevant for multi-cast and relaying, respectively.

transmitted, except for the EPDCCH, in the *control region* located at the beginning of each subframe; this control region occupies all the subcarriers (with the exception of the pilots) of the signal, and either (i) the first, (ii) the first two, or (iii) the first three OFDM symbols (or four in the case of very narrow bandwidth); the amount of occupied OFDM symbols can be changed from one subframe to the next and be indicated by the PCFICH. The reason for transmitting control information at the beginning of a subframe is that information contained in it is required for the decoding of the user data; in particular, if a UE sees from the scheduling information that none of the subsequent user data are intended for it, the UE can power down part of the RX circuitry. Also, latency for scheduled receptions or transmissions is reduced.

The mapping of the control channels to particular REs is done in units of RE groups which consist of four REs each, which are in the same OFDM symbol and adjacent subcarriers (REs carrying cell-specific pilots are disregarded).

31.4.5 Physical Control Format Indicator CHannel

The PCFICH carries the information of how many OFDM symbols are dedicated to the control region. Since the control regions can comprise up to three symbols, this information requires 2 bits. The reason for having a dynamical choice of the control region size is that the usage in a cell can vary dramatically. Sometimes, resources are taken up by a few high-data-rate users (in which case the control region, whose size is mainly determined by the resource allocation information, can be shorter). At other times, there are a lot of users (e.g., when most users voice users), which requires a longer control region.

If the PCFICH information is decoded incorrectly, all the remainder of a subframe is interpreted incorrectly. Therefore, the information uses a strong error correction, namely a rate 1/16 block code and may also use power boosting. The encoded bits are then scrambled with a scrambling sequence that depends on the subframe number and the cell ID, and thus reduces the effect of interference. Since the information is needed for the interpretation of the remainder of the subframe, it is always at the same location in the first symbol. In order to enhance frequency diversity, the 16 QPSK symbols are mapped onto 4 RE groups that are each spaced 1/4 of the available bandwidth apart; the starting subcarrier depends on the cell ID.

31.4.6 Physical HARQ Indicator CHannel

The PHICH carries the feedback information for the HARQ of UL transmissions; for more details on the operation of the HARQ process see Section 31.5.2. Groups of eight feedback bits are formed in the following way: first, each feedback bit is repeated three times (repetition coding), and Binary Phase Shift Keying (BPSK) modulated. The resulting waveforms are multiplexed through a combination of In-phase – Quadrature phase (IQ) multiplexing and code multiplexing with length-4 orthogonal spreading sequences. The scrambled output of the multiplexer is mapped to three RE groups. The PHICH group number, IQ branch, and spreading sequence are related to the index of the first RB on which the data block that is to be acknowledged was originally transmitted.

31.4.7 Physical Downlink Control CHannel

The PDCCH occupies the largest part of the control zone. For each of the users, it carries the control information, which consists of DCI, i.e., information that is required for receiving/transmitting and decoding/encoding the payload information in the DL/UL.

*Downlink Control Information

The DCI contains information about the resource allocation, i.e., which RBs in the data region are assigned to which UE (see also Section 31.3.3), as well as, for each user, the transport format, i.e., the modulation and coding scheme, power control information, and control information for spatial multiplexing.

There are many different formats for encoding the DCI, reflecting the trade-off between size of the control information and performance. They are as follows:

- *Format 1*: the fundamental DCI format, using type 0 or 1 resource allocation notification (compare section "Mapping to physical resources" in 31.3.4). It does not assume multiple antennas at the BS.
- *Format 1A*: similar to Format 1, but more compact, since it employs type 2 resource allocation notification.
- *Format 1B*: like Format 1A, but intended when the BS uses codebook-based precoding based on CRS.
- *Format 1C*: extremely compact format with fixed modulation (QPSK); to be used only for special system messages.
- *Format 1D*: like format 1B, but for MU-MIMO.
- *Format 2*: using Type 0 or 1 resource allocation notification, this format is intended for systems with closed-loop spatial multiplexing.
- *Format 2A*: like Format 2, but for systems with open-loop spatial multiplexing with CRS.
- *Format 2B-D*: like format 2, but with DM-RS. Specifically, 2B is associated with transmission mode 8, 2C with 9, and 2D with 10.

The various DCI formats contain some or all of the following:

- *Resource allocation information*: which RBs are allocated to which users. This is the key information required for the UE to determine what it needs to decode.
- *Modulation/coding*: those 5 bits indicate the modulation scheme and code rate used for the transport block. Of the possible 32 combinations, 29 are used to actually indicate combinations of modulation format and code rate; the remaining three can be used to indicate modulation format during retransmission.
- *HARQ-related parameters:*
  - *New Data Indicator (NDI)*: it indicates whether the data are a retransmission and thus need to be combined with previously received soft information for HARQ.
  - *Redundancy version*: it indicates the type of redundancy used in the HARQ transmission.
  - A similar combination of MCS, NDI, and redundancy version might be present for the second transport block, if it exists.
  - *HARQ process number*: this 3-bit message (4 bit for TDD) is the index of the HARQ process associated with the transport block. As explained in Section 31.5.2, several HARQ processes are active simultaneously.
  - *DL assignment index*: it is related to the HARQ operation in the TDD mode.
  - *ACK/NACK offset*: indicates the resource elements on which a PUCCH with ACK/NACK is transmitted.

- VRB-to-PRB mapping
  - *Localized/distributed VRB*: flag indicating whether localized or distributed VRBs are used.
  - *Gap value for VRB*: the spacing between the two parts of the VRB.

- Multi-antenna information
  - *PMI for precoding*: the bits in this field contain the index of the precoding codebook for closed-loop multiple-antenna systems (see Section 31.3.6).
  - *PMI confirmation*, indicating whether the codebook entry requested by the UE is used, or a different value, in which case the information is given in a separate PMI field. Both PMI confirmation and PMI field only exist in modes using CRS.
  - *Transport block swap flag*: indicates whether the two transport blocks need to be swapped before HARQ processing. Such a transport block swap during retransmission might be done to provide additional diversity.
  - *Power offset between PDSCH and CRS*: to enable the UE to perform demodulation for symbols modulated by 16-QAM or 64-QAM.
  - *Reference signal scrambling sequence* for generating multiple DM-RS sequences (see Section 31.3.5).
  - *Number of layers and the set of antenna ports*.
  - *Open-loop precoding info*: this field indicates the type of open-loop precoding (e.g., details of the cyclic delay diversity) used in the transmission with multiple-antenna systems.

- Other parameters
  - *PDSCH resource element mapping* and quasi-colocation indicator.
  - *Transmit power control for PUCCH*.
  - *SRS request*: to trigger a single transmission of an SRS.
  - *Carrier indicator*: this field indicates the carrier for which the information is intended, it exists only if cross-carrier scheduling is enabled for CA (see Section 31.6).
  - *Transport block size index*: only used in the 1C Format, it indicates the size of the data block to be transmitted. Note that in the other formats, the size of transport blocks follows implicitly from the number of allocated RBs and the code rate (the standard contains a table providing this mapping, since there are minor deviations from the nominal rates obtained by closed-form computations).
  - *Flag for 1A/0 differentiation*: this is a single bit indicating whether Format 1A or Format 0 (which is used for UL scheduling grants, see below) is used because these two DCI formats have same size.
  - *Identity* of the device for which the PDSCH is intended: transmitted implicitly, through the use of a specific Radio Network Temporary Identifier (RNTI) that is used to scramble the CRC of the DCI format.

- *Padding*: to resolve some other ambiguities resulting from specific sizes of DCI formats or to align the sizes of specific DCI formats.

The above shows the large amount of control information that has to be conveyed to a UE. This can lead to a significant control overhead and associated reduction of the effective data rate, in particular when multiple PDSCHs/PUSCHs are scheduled in a subframe and the maximum of 3 symbols, out of the 14 symbols, is occupied for PDCCH transmissions (~21.5% control overhead). However, we stress again that not all of the different DCI formats contain all of the above components. Rather, there is obviously a trade-off between the size of the DCI format and the conveyed information, and the different DCI formats are intended to provide flexibility to provide only the information that is needed in a particular case.

Control information related to the UL transmissions is also carried on the DL control channel, i.e., in the PDCCH. Remember that the control of all transmission parameters rests with the BS, while the UE can only make requests. Thus, the PDCCH carries UL scheduling grants, which tell the UE when, and with which format, to transmit. Just like for the DL, that includes information on which RBs to transmit, the transport format, power control, and spatial multiplexing information. This information for the UL is transported in message of DCI format 0, which employs Type 2 resource allocation format in accordance with the use of DFT-spread-OFDM for UL transmission, or DCI format 4, which is used when spatial multiplexing is applied and to provide additional scheduling information that is not included in the compact DCI format 0.

Most of the information for scheduling PUSCH transmissions is similar to the information for scheduling PDSCH receptions. Just like for scheduling PDSCH receptions, not all of these components are transmitted in each frame format.

- *Resource allocation type*: indicates whether one or two groups of contiguous subcarriers are used.
- *Resource block allocation.* This field not only defines the allocated resources for the PUSCH but also indicates whether frequency hopping is used in the UL (which can be beneficial for the frequency diversity).
- *MCS and NDI*: like for the PDSCH. NDI here indicates whether grant is for a retransmission or new transmission.
- *Phase rotation of uplink DM-RS*: indicator for such phase rotation, which can be useful for suppression of interference between DM-RSs from different devices in MU-MIMO.
- *Precoding information*: to signal the precoder to be used in the UL.
- *Channel state request flag*: requests a channel state report to be sent in the PUSCH by the UE to the BS.
- *SRS request*: requesting the transmission of an SRS, to help scheduling decisions by the BS.
- *Uplink index*: for TDD only, it indicates for which subframe the grant is valid.
- *Transmit power control for PUSCH.*

In addition, DCI format 3/3A is used for the transmission of power control commands for PUCCH and PUSCH with 2-bit/1-bit power adjustments. We finally note that DCI format 5 is used for Sidelink, and DCI format 6-0/6-1 is used for the UL/DL of Enhanced Machine-Type Communications (eMTC).

Transmission and Reception of PDCCH

Once all the control information is assembled, the actual transmit signal is created in the following way:

1. A 16-bit CRC is computed. The CRC is scrambled by a Cell RNTI (C-RNTI) that serves to identify the UE for which the DCI is intended. Thus, each UE determines whether information is intended for it by checking the CRC after descrambling with the C-RNTI: if the CRC checks, then the UE concludes that (i) the DCI is intended for it and (ii) was decoded correctly. Consequently, there is no *explicit* signaling of whom the DCI is intended for.
2. The CRC-protected DCI is encoded with a rate 1/3 tail-biting convolutional code (i.e., a convolutional code that ensures that starting state and end state in the trellis are identical).
3. After rate matching, multiple PDCCHs (for different UEs, and even different PDCCHs per UE, see below) are multiplexed.
4. The resulting data are scrambled by a PN sequence (Gold sequence, see Section 19.3.2), and QPSK modulated. The QPSK symbols are grouped into groups of four symbols each. In contrast to the PCFICH, these symbols are not directly mapped onto RE groups, but first interleaved, and then subjected to a cell-specific cyclic shift. The purpose of these measures is to ensure (i) full exploitation of the frequency diversity, and (ii) avoiding consistent collisions between the same signals in neighboring cells.

A BS might transmit multiple PDCCHs per UE, e.g., for scheduling PDSCH reception or PUSCH transmissions or for providing control information such as TPC commands by DCI formats 3/3A. A UE might thus receive multiple PDCCHs. Please see above for the different formats used for DL and UL information.

For the reception, the UE has to *blindly* determine the DCI format (note that this is different from the control format, which *is* explicitly signaled in the PCFICH). This is done through the different sizes of the different DCI formats. The blind search can be quite cumbersome since there are several different sizes of DCI formats, and the starting position of the PDCCHs for a UE is also variable. Through some restrictions about where the PDCCH (or EPDCCH, see below) may be located, the search space is restricted, which allows to accelerate the identification. Overall, the mapping of the PCFICH, PHICH, and the various PDCCHs, is rather complicated, and we refer to Dahlman et al. [2016, Section 6.4.3] and the standards documents for further details.

Enhanced Physical Downlink Control CHannel

The EPDCCH is an alternative control channel that has the purpose of enabling (i) demodulation with DM-RS, (ii) frequency-domain scheduling and ICIC for the control information, and to increase control signaling capacity, when needed. The EPDCCH is transmitted in the *data* region of the subframe, not in the control region.

Only DCI formats scheduling unicast PDSCH receptions or PUSCH transmissions are supported by EPDCCH while information that is relevant to multiple UEs, such as system information, and certain paging or power control commands, is still provided by PDCCH. The processing of the DCI, such as CRC and convolutional encoding, are the same between PDCCH and EPDCCH. The main difference lies in the mapping of the encoded QPSK signals onto the QPSK resource elements. This mapping occurs in a way that is generally very similar to that of data, i.e., can be either localized or distributed. Of course, exploitation of frequency-domain scheduling would use the localized approach, while the distributed approach requires no CSI. Since the EPDCCH is part of the data region, the results are not available until the decoding of the subframe has been completed.

31.4.8   Physical Random Access CHannel

The PRACH is primarily intended for transmissions from UEs that do not yet have resources assigned to them. This can be, e.g., in the situation that the UE has turned on or entered the cell, and needs to establish connection with the BS for the first time.

The PRACH is transmitted on 72 subcarriers (just like the SS). This is because – even though the UE knows the cell bandwidth by the time it uses the PRACH – it still eases the implementation to have the same PRACH in all systems. In the time domain, the BS usually reserves frequency resources in 1-ms long blocks/subframes for the PRACH during which no UL transmissions occur in frequency resources that may be used for PRACH transmission, so that there is no possibility of collision. The periodicity of the PRACH intervals can be configured, e.g., depending on how many possible users are in the cell, and what the latency requirements for random access are.

The most common configuration, a 1-ms PRACH, consists of (i) 0.1-ms cyclic prefix, (ii) 0.8-ms long Zadoff–Chu sequence, and (iii) 0.1-ms guard interval. Note the requirement for long guard intervals because when the PRACH is used, the necessary timing advance (see Section 18.3.2) has not yet been established. There are 64 different sequences that can be used in the "main part" of the PRACH: either distinguished by phase shifts or sequences with different index u (see Section 31.3.5). The advantage of using different phase shifts is that those sequences are truly orthogonal; however, the amount of phase shift is lower bounded by the runtime and delay dispersion of the signal in the cell. If the cell size is below 1.5 km, all 64 sequences can be produced by phase shifting of one base sequence; otherwise, sequences with different indexes have to be used. Different UEs pick, at random, the sequence that they want to use for the PRACH. Due to the large number of sequences, it is unlikely (though not impossible) that two UEs pick the same sequence and try to use it in the same timeslot. If that occurs, later stages of the random access procedure (contention resolution messages) resolve this collision (see Section 31.5.1).

The PRACH channel also foresees power ramping. The power for the first transmission is based on an estimate of the pathloss as obtained from CRS receptions. If the first attempt at reaching the BS is not successful, the transmit power of the random access preamble is increased, and transmission is repeated.

31.4.9   General Aspects of Control Signals Associated with PUSCH

In LTE, there is little information that needs to be signaled in the UL because the BS has central control of almost all transmission parameters. While the amount of information is small, the way that it is communicated is rather complex. The complexity is partly due to the fact that the information should be transmitted both when there is payload being sent for the particular UE (i.e., it has an active PUSCH) and in the case where there is no PUSCH; in the latter case, the information is sent on a separate channel called the PUCCH.

The few pieces of information that need to be sent in UL direction (UCI) are as follows:

- *Scheduling Request (SR)*: indications that the UE wants to send data; the actual resource assignment is then sent by the BS in the DCI. If the SR is sent on the PUCCH, then certain restrictions apply: there is only one bit of information transmitted via on–off signaling, i.e., a request for resources, without stating how much is needed; also the request can only be transmitted every nth subframe according to a configured periodicity, since there is limited capacity for the PUCCHs from different devices in the time/frequency resources assigned for transmission of PUCCHs. If control information is sent together with payload data on the PUSCH, then the buffer status of the UE (how many data it wants to transmit) is reported as part of the MAC header, so that an explicit SR need not be sent.
- *HARQ ACKnowledgements (HARQ-ACKs)*. Depending on the number of transport blocks, one or more bits need to be sent. Of course, the BS also needs the information about which payload data the HARQ-ACK refers to. However, this information does not need to be signaled explicitly, because the DL control channel tells the UE on which resources it should transmit PUCCH with the HARQ-ACK for which data.
- *Channel state information*: this is essential for the BS to determine the appropriate transmission format and scheduling, see Section 31.3.7. Periodic CSI reports are transmitted on the PUCCH (since they occur at fixed times, there is no guarantee that an active PUSCH is present), while aperiodic CSI reports are triggered by a DCI format scheduling a PUSCH transmission and are multiplexed on the PUSCH.

31.4.10   PUCCH

Since few bits typically need to be multiplexed on the PUCCH, it would be difficult to transmit them in the "standard" UL fashion and still retain good spectral efficiency. Rather, for SR or for 1–2 bits HARQ-ACK information, the control information from multiple users is multiplexed onto the same RBs by two measures:

1. The information from different UEs is used to modulate orthogonal sequences called "base sequences"; those orthogonal sequences have the same form as UL pilots, i.e., phase-shifted versions of cyclically extended Zadoff–Chu sequences (see Section 31.3.5). Typically, six different phase shifts, $2\pi n/6$, $n = 0, ..., 5$ are used. Each sequence is multiplied with a BPSK or a QPSK signal, so that each sequence carries 1 or 2 bits of information. In order to randomize intercell interference, the phase shifts

of the sequences are changed from symbol to symbol, where the index of the used phase is determined by a PN sequence whose initialization depends on the cell ID, as well as on the slot in which transmission takes place.

2. Block-wise spreading (i.e., Orthogonal Cover Code (OCC)) over the symbols in a slot is used to further enhance the multiplexing capacity. This block spreading is achieved by assigning different Walsh–Hadamard sequences (see Section 19.3.2) to different users, if the desired spreading factor is 2 or 4, and DFT sequences if the spreading factor is 3. The blockwise spreading is only used for select PUCCH Formats. For randomization of intercell interference, the assigned spreading codes change (in a pseudorandom way, depending on the cell ID) from slot to slot.

The PUCCH signals are transmitted in the RBs at the lower and upper band edges of the system, so that they cannot disturb the contiguous assignment of RBs to the different users for PUSCH transmissions according to DFT-spread-OFDM. From slot to slot, the PUCCH hops between the lower and upper band edge, thus providing maximum frequency diversity (see Figure 31.16).

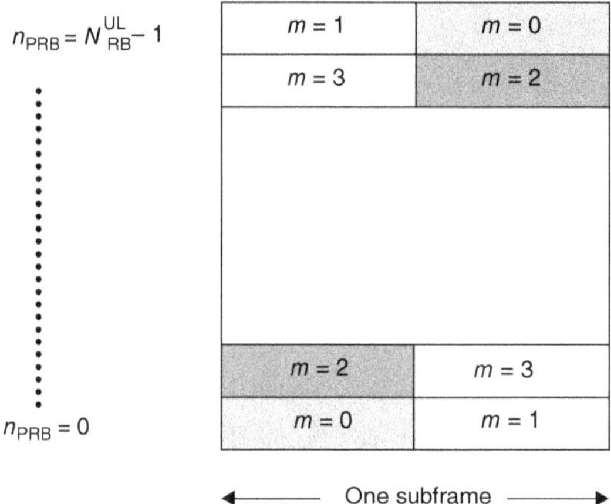

Figure 31.16   Resource assignment for the PUCCH.
Reproduced from [3GPP TS 36.211] © 2009. 3GPP™ TSs and TRs are the property of ARIS, ATIS, CCSA, ETSI, TTA and TTC, who jointly own the copyright in them. They are subject to further modifications and are therefore provided. to you "as is" for information purposes only. Further use is strictly prohibited.

Depending on the amount of information that needs to be transmitted, the PUCCH uses different formats (and possibly additional spreading).

- Format 1 is for the transmission of HARQ acknowledgments and SRs, transmitting 0, 1, or 2 bits of information. It employs the base sequences discussed above, modulated with BPSK (1 bit), or QPSK (2 bits). Out of the 7 OFDM symbols per slot, three are used for channel estimation, while the remaining four are used for transmission of the bits. Both the data and the reference signals use orthogonal cover codes to further distinguish different devices – length 4 OCCs for the data, and length-3 for the RSs. Thus, each base sequence can transmit the PUCCHs of up to three devices. Since there are 12 base sequences, one time/frequency resource can carry the PUCCHs of up to 36 devices.

- Format 2 is used for the transmission of CSI, possibly combined with HARQ and/or SRs. Up to 11 information bits are encoded with a punctured Reed–Muller code and modulated into 10 QPSK symbols, which are transmitted over 10 OFDM symbols in a subframe (the remaining four are used for pilot signals). Since Format 2 requires the transmission of many more bits, it does not use orthogonal cover codes, but rather only one PUCCH can be transmitted per base sequence, i.e., at most 12 UEs can transmit a PUCCH in one time-frequency resource.

- Format 3 is used for the case of CA, when the PUCCH has to be able to carry up to 22 bits of information. A different transmission method, namely DFT-spread OFDM, just like for the PUSCH, is used in this case; multiplication with length-5 orthogonal sequences (DFT sequences) allows transmission of five different UEs to transmit PUCCH simultaneously in same frequency resources (RB). Different cyclic shifts, which depend on the cell ID, are applied for the different OFDM symbols to randomize intercell interference.

- Formats 4 and 5, introduced in release 13, allow the transmission of even more bits.

31.4.11   PUSCH

If there is an active PUSCH for a given UE that overlaps in time with a PUCCH for the UE, then the control information is multiplexed with the payload data in the PUSCH, unless the UE is capable of and is configured for simultaneous PUSCH and PUCCH transmissions. The rules for multiplexing are motivated by the following concerns:

- The HARQ ACK has to be robust; it is therefore transmitted close to a pilot. Thus, there is no degradation of the channel estimate even for fast-varying channels. Similarly, the RI is sent close to the pilot.

- The rate matching should not depend on whether HARQ information is transmitted or not. Therefore, HARQ ACKs are punctured into the data stream.
- The modulation scheme is the same as that for the payload data, in order to ensure the single-carrier properties of the signaling. However, the code rate can be adapted to the channel state. Different codes are available for different pieces of information as corresponding target reception reliabilities can be different: repetition coding and simplex coding for ACKs and RIs, and Reed–Muller block codes or convolutional codes for CQI/PMI.

There is, in principle, also the possibility of simultaneous transmission of PUCCH and PUSCH. This provides additional flexibility, though the benefits are rather limited as single-carrier transmission cannot be maintained and power back-off is required when the transmissions are in a same band, and there are also implementation challenges and various difficulties due to regulatory requirements.

31.5 Physical Layer Procedures

There are a number of procedures that have to be executed in a cell, which make use of the different channels described above. In this section, we discuss the most important of them.

31.5.1 Establishing a Connection

Scanning and Synchronization

The very first thing a UE has to do is to acquire the timing of the signals in the cell it is in; this is done via the SSs (described in Section 31.4.2). Only after that can it learn about vital cell information (transmitted on the BCH), and perform the other functions it needs to do for communication (see Section 31.4.3).

After the acquisition of the timing and reception of the BCH, additional cell system information is communicated in the SIBs transmitted via the DL-SCH, similar to the reception of payload data.

There are a number of different SIBs defined, depending on the type of information that is to be transmitted, see Table 31.6. Note that not all of the SIBs need to be sent, depending on the cell configuration.

Table 31.6 System Information Blocks (SIBs).

SIB	Content
1	(i) Fundamental information about the cell, e.g., network operator, usage restrictions, etc. (ii) TDD configuration information. (iii) Repetition period of other SIBs
2	Parameters for cell access: UL cell bandwidth, random access parameters, etc.
3–8	Information related to handover, including intersystem, interfrequency, and intrafrequency handover
9	Identifier for the home BS
10–12	Information for earthquake and tsunami warnings
13, 15	Information about MBMS
14	Information about access barring
16	Information about GPS and universal time
17	Information about the interaction between LTE and Wi-Fi
18, 19	Information about sidelink
20	Information about point-to-multipoint transmission

A SIB1 is scheduled periodically every 80 ms with a fixed timing. The repetition period for the other SIBs can be configured by the network operator, and multiple SIBs might be mapped together onto a *system information message*, that is transmitted at intervals specified in SIB1.

Random Access (RA)

If a UE wants to join the network, it has to let the BS know its request. Obviously, the UE does not yet have resources assigned to it at this time; it must therefore make a contention-based access, see Section 18.4, also known as random access: it is possible that its request might collide with requests from other UE. LTE specifies a procedure for such contention resolution of a random access in four steps:

1. UE transmits an RA preamble (see Section 31.4.8), also referred to as Msg1, which allows the BS to compute the required timing advance. This is the only step using different PHY signaling. The subsequent steps are transmitted like normal data (except that HARQ is not used).

2. BS transmits an RA Response (RAR), also referred to as Msg2: this contains (i) the index of the RA preamble for which the response is valid (note that at this time the BS does not know/have assigned the ID of the UE yet, only the type of RA preamble that was used), (ii) the timing advance to be used by the UE, (iii) the resources to be used by the UE for a PUSCH transmission, also referred to as Msg3, in the subsequent step, and (iv) a temporary ID. Note that if several UEs used the same RA preamble, then the RAR is valid for all those UEs, and the resulting collision has to be resolved in the subsequent steps.

3. UE transmits Msg3 PUSCH with Radio Resource Control (RRC) signaling information which contains its ID. The details of the information depend on how much information the BS already has about this UE, e.g., whether the UE accesses the network for the first time, or just reestablishes a link after the connection has been interrupted.

4. BS transmits a contention resolution message in a PDSCH, also referred to as Msg4. As described above, ambiguities could have occurred if multiple UEs try to access the system with the same random access preamble. In the contention resolution message, the BS transmits explicitly the ID of the UE to which it assigns resources.

In addition to the messages when joining a network, RA is also used for

- Requesting UL transmission resources when no PUCCH is available to the device.
- Reestablishing a radio link after a link failure.
- Establishing UL synchronization, when a device is connected but unsynchronized, e.g., during handover to a new cell.
- For positioning purposes.

For the last four usages, there is also a contention-free mechanism defined in LTE, which uses only the first two of the above-described four steps of the RA procedure.

Paging

Paging is being sent if the network has a message for a UE that is in an idle or disconnected state.[14] Each UE is assigned (in the DL-SCH) a "paging interval," and a certain subframe in which a paging message might be transmitted. Thus, a UE that is in a sleep state needs to wake up from the sleep state only once per paging interval and listen whether there are data for it. The paging interval can be configured, representing a trade-off between energy saving and latency.

In order to page a UE, the network needs to transmit the paging message in all cells in which the UE *could* be. In case the network has no knowledge of the UE location at all, then this leads to paging in thousands of cells, and thus high overhead. On the other hand, tracking each idle UE as it moves (and thus knowing where it is) also leads to high overhead. The compromise lies in defining cell groups (RAN areas), which comprise multiple cells, so that tracking is only required when the UE moves between such areas; at the same time, the paging message needs to be sent only from the few BSs that are within such an area.

31.5.2 Retransmissions and Reliability

Retransmission is an important part of LTE to ensure transmission quality. There are two components of the retransmission:

1. *HARQ*: it provides an integrated PHY/MAC approach for retransmission of data blocks that were not received successfully the first time, in such a way that the data from the multiple transmissions can be combined. The retransmissions occur quickly. LTE uses "incremental redundancy" (see Section 13.12): different parity check bits are transmitted, so that when retransmission happens, a more strongly coded version of the payload bits is available at the RX.

2. *ARQ on the RLC*: it is a higher layer retransmission protocol that arranges for the retransmissions of all the data blocks that fail even after HARQ. This mechanism is quite a bit slower but will be invoked only rarely if HARQ is configured the right way. Being a "fallback" solution, it provides the extra reliability required for some applications (e.g., file transfer). The retransmission function can be switched off whenever it is not necessary/helpful (e.g., for voice calls, when the delay introduced by the protocol would be larger than what can be tolerated in phone conversations (see Chapter 24)). We note that there is a further layer of retransmission, namely at the PDCP level, which mostly is relevant for handovers of the connection between BSs. All errors not caught by these procedures may trigger the retransmission protocol of TCP, though it is desirable not to invoke this step, since failure of packet delivery is interpreted by TCP as network congestion, resulting in a lowering of the data rate that can persist for several minutes.

The principle of HARQ is the same for UL and DL: (i) the TX sends a transport block,[15] (ii) the RX tries to decode. If successful, it sends an ACK. (iii) if the previous transmission was successful, the TX sends a new packet for the corresponding HARQ process (and indicates this fact in the "new data" field); if the previous transmission was unsuccessful, the transport block is retransmitted (the

---

[14] There are also some special cases where paging can be sent to UEs that are in an RRC-connected state.

[15] Note that in the case of spatial multiplexing, two transport blocks can exist, so that the ACK has to have 2 bits. Also, as discussed in Section 31.3.2, very long transport blocks might be divided into code blocks, which have their own CRCs and ACKs.

"new data" field showing that it is a retransmission, and the "redundancy version" field indicating which type of retransmission is used for the HARQ process).

Since the checking whether a packet arrived o.k., and feedback of this information takes time, the TX would have a lot of "downtime" while waiting for permission to transmit the next transport block. To circumvent this problem, there are multiple HARQ processes, i.e., streams of data that are time-interleaved, such that data from process 2, 3, 4, ..., 8 are sent while the TX is waiting for the feedback from process 1. The principle is outlined in Figure 31.17.

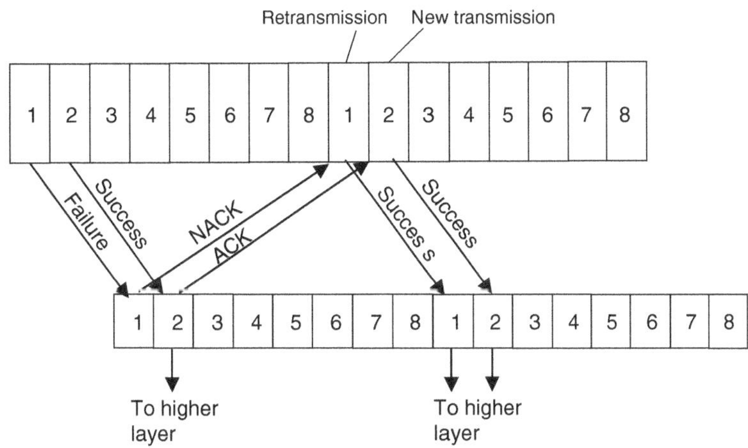

Figure 31.17 Scheduling for HARQ.

There is one key difference between UL and DL: retransmission in the DL is asynchronous and might occur using arbitrary RBs, and special retransmission bits indicate to which original transport block the retransmission belongs. The structure of retransmission in the UL is synchronous/fixed: it is always eight subframes after the first transmission attempt, and the RBs used for retransmission are the same as for the first transmission.[16] For TDD-based LTE, the distance between retransmissions depends on the UL–DL allocation, for details see the standard or [Dahlman et al. 2016]. Also, the interval between reception of a transport block and transmission of the ACK depends on the operating mode: in FDD it is always four subframes, while in TDD it depends on the UL/DL configuration of the setup.

31.5.3 Scheduling

Scheduling in the context of LTE is the question at what time, and on which subcarriers, information for/of which UE is transmitted. Furthermore, it also involves the choice of the transport format, i.e., transport block size, modulation and coding scheme, and the multiple antenna scheme. All scheduling decisions – for both DL and UL – are made by the serving BS.

The standard does not define the scheduling *algorithm*, i.e., on which resources to schedule which users. In general, the scheduler will follow the principles outlined in Chapter 20, but it is ultimately up to the infrastructure manufacturer to decide on the scheduling policy.

Downlink Scheduling: DL scheduling is made easier by the fact that the scheduler, located at the BS, has immediate access to all data, priorities of the data, and all buffer statuses, that are needed for the scheduling. There is thus no need for communicating this information within the standards framework. The BS does need to know the DL CSI, such as CQI and UE-recommended beamformers, which are communicated in the CSI reports that can form part of the UCI. Remember, though, that in particular for feedback-based CSI determination, there is an inherent latency. Once the scheduling decisions have been made, the BS can directly place data from the various physical channels onto the thus associated time/frequency/space resources; these associations have to be communicated to the UEs (in the DCI), so that they know on which resources to listen for which data.

Under normal circumstances, the scheduling interval is 1 ms, i.e., a scheduling decision can change from subframe to subframe, and thus scheduling information is transmitted afresh in every subframe. However, in the case of voice calls (and other applications that have a low data rate but continuous data), Semi-Persistent Scheduling (SPS) is used to reduce control overhead as data arrival (voice traffic) is periodic. Essentially, the BS tells the UE that (until further notice) it has the same resources allocated to it in every *n*th subframe; after that message, only MAC-layer info (telling the UE whether data will actually be sent on the configured resource or not) needs to be transmitted to that UE, thus reducing the overhead.

Uplink Scheduling: While UL scheduling is very similar to the DL, there are some key differences: (i) The scheduler (located at the BS) might base its decisions on information it received from the UEs and from other BSs, in order to provide a better quality. In

---

[16] This is if the retransmission is in response to a PHICH reception and not in response to a PDCCH providing a DCI format scheduling the retransmission. Retransmission on a different RB can be achieved through a clever combination of the "new data indicator" and the scheduling grant.

particular, this information is related to the channel quality of the UE, the buffer status (which the UE tells the BS in a buffer status report), and information from neighboring BSs for inter-cell interference coordination. (ii) The decisions of the scheduler cannot be executed immediately: the UL scheduling grant does not refer to the current subframe, but rather to a later subframe. In FDD, there is a fixed offset of four subframes between the grant and the actual transmission. In TDD, the delay depends on the configuration which subframes are used for UL or DL. Note that also for the UL, a SPS can be established.

Even though scheduling for the UL is performed by the BS, a UE that has multiple radio bearers to transmit decides by itself which of the radio bearers to transmit on the resources assigned to it by the BS via a scheduling grant; these decisions also follow rules that can be configured.

Under some circumstances (transmission on the PUCCH, see Section 31.4.9), the UE only sends a SR to the BS, to let the BS know that it has data it wants to send. This request does not include the *amount* of data it wants to transmit, which leaves the BS with the dilemma of assigning either a small amount of resources (in which case latency might be increased because a large part of the data have to wait for a later subframe) or assign significant resources (which might turn out to be inefficient if the UE has only few data to send). Such PUCCH-based requests only happen if the UE does not have resources assigned to it for data transmission; otherwise, the control information transmitted on the PUSCH includes a report for the buffer status, allowing the BS to make a better-informed decision. Note, finally, that a UE that has not been assigned resources for a PUCCH needs to use the random access mechanism (see Section 31.5.2) as a SR to transmit data.

31.5.4 Power Control

Power control in the UL of LTE consists of an open-loop component and a closed-loop component. The open-loop component establishes a baseline for the desired transmit power: the UE determines the DL path loss from a pilot with known transmit power, and from that, and some other configured parameters, it computes the necessary UL power (including necessary margins in the process). This estimate is improved by the closed-loop component that is intended for tracking short-term fading: a power control command that is provided in a specific DCI format in a PDCCH "fine-tunes" the transmit power. A one-bit power control command indicates a power adjustment of ± 1 dB while a 2-bit power control command indicates an adjustment of $[-1, 0, 1, 3]$ dB. No DL power control is specified, though a BS can adjust the power at will (since such adjustments do not contradict the standard).

The general consideration for power control is that the UL channels should fulfill certain requirements on the SINR (not the received power), subject to constraints on the transmit power. Furthermore, the power of PUCCH and PUSCH can be controlled separately because of separate reliability requirements. For simultaneous PUSCH and PUCCH transmissions, the necessary power of the PUCCH is first determined to ensure that the SINR for this channel is sufficient for decoding within a target error rate. The remaining power (considering a limitation on the maximum power) is then used for the PUSCH transmission (PUCCH is prioritized for power allocation over the PUSCH). Similar, if control information is multiplexed on a PUSCH, and there are multiple PUSCHs transmitted in parallel, the PUSCH with the control information is assigned sufficient transmit power first (prioritized for power allocation over the remaining PUSCHs).

Generally, in order to achieve fairness, the UE tries to compensate the pathloss, so that all UEs have a chance for high data rate transmission. However, the standard foresees also a "partial pathloss compensation" for PUSCH transmissions such that devices with higher pathloss do not increase their power to fully compensate for that loss; this is done to reduce the amount of interference cell-edge users create in adjacent cells. Due to the importance of control information and since it cannot benefit from HARQ retransmissions, full pathloss compensation always apply for PUCCH transmissions.

While the mechanisms of power control are somewhat similar to that of CDMA (compare Chapter 19), their importance and motivation are very different. For CDMA, power control is essential for proper functioning of the data transmission. In LTE, as in any orthogonal-access scheme, power control is used for increasing battery lifetime and for reducing intercell interference.

31.5.5 Handover

Handovers in LTE are "hard handovers," in the sense that communication occurs between UE and one BS at one time, not to two simultaneously, unless the UE is configured for operation with dual connectivity (see Section 31.7.3). On a more detailed level, the handover from a source BS to a target BS proceeds in the following three phases:

1. Handover preparation
   (a) The source BS configures the measurements the UE has to perform and report. Specifically, it sets the thresholds such that reports from the UE to the BS are required if certain measurement results (e.g., signal quality to neighboring BS) exceed those thresholds. Alternatively, the BS requires periodic reports.
   (b) UE sends its (periodic or aperiodic) measurement results.
   (c) BS makes a handover decision based on the measurement results (e.g., to a cell with larger path gain).
   (d) The source BS sends a handover request to the target BS, usually via the X2 interface (the interface between two BSs).
   (e) The target BS performs an admission control. If the target cell has no resources available, the connection might have to be terminated.
   (f) If it admits the handover, the target BS sends a "handover request acknowledgment" to the source BS.

2. Handover execution
   (a) The source BS sends a handover command to the UE, and at the same time starts to forward DL packets (i.e., packets it receives from the network for this UE) to the target BS. Transmission of those packets by the target BS has to wait until the target BS can actually communicate to the UE (see below).
   (b) The source BS tells the target BS which packets were already acknowledged by the UE.
   (c) The UE synchronizes itself to the target BS via the RACH (a preliminary synchronization was already achieved during the cell identification process, when the UE did its measurements).
   (d) The target BS transmits the UL resource allocation and timing advance to the UE.
   (e) The UE sends a "handover confirm" message to the target BS. From that time on, target BS and UE can communicate with each other.
3. Handover completion
   (a) The target BS sends a "path switch" message to the MME, (see Section 31.2.1), requesting that data for the UE are henceforth sent to the target BS.
   (b) The MME forwards this message to the serving gateway.
   (c) The serving gateway switches to the target BS the route the data for the UE have to take.
   (d) The serving gateway confirms the switch to the MME.
   (e) The MME confirms the "path switch" message to the target BS.
   (f) The target BS sends a message to the source BS, telling it to release the resources still reserved for the UE.
   (g) The source BS releases the resources.

*31.6 Carrier Aggregation and License-Assisted Access

31.6.1 Carrier Aggregation

In order to increase the data rates and improve flexibility, later releases of LTE specify CA, i.e., transmission on multiple so-called *component carriers*. Each component carrier has a maximum bandwidth of 20 MHz (i.e., the maximum bandwidth of "classical" LTE). The component carriers can be contiguous or noncontiguous, and in the latter case can be in the same LTE band (see Section 31.2) or in different bands, so that even fragmented spectrum can be used to provide high data rates.

The general structure of CA transmission is as follows: for every device, there is a *primary* component carrier, which performs all idle-mode actions, such as cell search and random access; for the UL it may also carry all the L1/L2 control information. The primary component carrier and even the number of component carriers may be different for the UL and the DL; however, the number of DL carriers must be equal to or larger than the number of UL carriers. Generally, DL CA is much more widely deployed than UL CA. All configurations may be different for different devices (among other reasons, this is important for backward compatibility, as well as enhanced flexibility for the assignments). Different component carriers can be activated and deactivated, depending on traffic requirements, in order to save energy.

The distribution of the data to the different component carriers operates as follows: There is a single RLC that receives the data from the upper layers. The MAC distributes the data for transmission onto the different component carriers, with the PHY transmission aspects on the different component carriers being essentially independent from each other. The scheduling can be done independently for each component carrier and each component carrier has its own MAC entity.

For the DL, the UE learns the scheduling assignment from the DCI; each component carrier has its own scheduling information. However, the scheduling assignments *may* be transmitted on the same component carrier the data are on (self-scheduling), or it may be transmitted on another component carrier (cross-scheduling). For the UL, there is an association (communicated in the system information) about the DL carrier that the UL carrier is paired with in FDD. Another piece of information required for the scheduling is the start time of the data region. For FDD, this time is configured semi-statically in cross-component scheduling. For TDD, there is rather involved signaling about the timing for which the assignments are valid.

There are a number of additional fine points in the allowed configurations, which furthermore might depend on the particular release of LTE. Release 10 allows to aggregate up to five component carriers in FDD, where all carriers have the same configuration. Release 11 enabled TDD aggregation, where different UL/DL configurations may be used in different bands, even though this may imply that simultaneous transmission and reception (in different bands) needs to be implemented. Release 12 enabled a combination of FDD and TDD carriers, while release 13 increased the number of possible component carriers to 32, enabling up to 640 MHz bandwidth. While it is not anticipated that a single network operator has that much licensed spectrum, the use of license-free spectrum (i.e., license-assisted access, see Section 31.6.2) might be in line with these large bandwidths.

All of these aggregations are subject to the capabilities of the UEs. As we have seen in Chapter 17, building transceivers and antennas that span a large bandwidth is rather difficult, so it is a question whether it is worthwhile to enable such a high peak data rate for a particular device; the answer depends on the applications and the price of the device.

31.6.2 License-Assisted Access

While LTE is normally intended to operate in licensed spectrum, 3GPP has also investigated the potential use of unlicensed spectrum. Such usage is attractive for operators, because they do not have to pay the high prices normally paid at spectrum auctions for exclusive use of the band. Since unlicensed bands are, by definition, open to anybody who fulfills certain elementary rules on transmit power, etc., see Chapter 2, the data throughput can be increased with minimal capital expense for the operators. On the downside, unlicensed spectrum does not allow to provide any quality-of-service guarantees, since there is always the possibility that users of a different service, e.g., Wi-Fi, lead to high interference or completely block the channel for a while. For these reasons, LTE provides a *license-assisted* mode, where the control signaling and a minimum data rate is provided in the licensed band, while additional data transmissions in the unlicensed band help to increase the data rate when possible. *License-Assisted Access LAA* is of interest mainly for the DL (since typically more data are sent on the DL than on the UL), and up to release 13 it is the only admissible form; we will in the following concentrate on this as well.[17]

A crude form of LAA has existed for many users for a considerable time – since Wi-Fi is available on their phones, they switch (manually or automatically) to Wi-Fi when an allowed Wi-Fi network is available, while using the LTE connectivity in other cases. However, such a scheme might not be efficient since the user might be connected to Wi-Fi even when LTE provides better service quality; furthermore, the switching can take considerable time (several seconds). With LAA, a faster, more adaptive form of access is provided.

The first mechanism for adaptive spectrum access is *Dynamic Frequency Selection (DFS)*, where the LTE BS determines the frequency bands that have little or no traffic on them, and only uses those. While the mechanism is called dynamic, it operates on a fairly slow timescale. For example, if LTE detects radar usage in a particular band, it has to stop operating for at least 30 min before it can try again. The specific type of DFS algorithm is, however, not prescribed in the standard.

A faster adaptation is provided by *Listen Before Talk* (LBT), which is mandatory in some regions of the world (Europe, Japan), and is therefore implemented in LTE generally, to provide a worldwide compatible solution. The LBT mechanism is somewhat similar to Wi-Fi (see Chapter 33) and needs to ensure fair coexistence with Wi-Fi: before transmission, the BS determines whether the channel is free; after determination of a free channel, the BS waits for 16 μs, and then observes the channel for a certain multiple of 9 μs (to see, e.g., whether any ACK signals are being sent); the exact observation time depends on the priority class of the data. After the observation time, the BS initiates a backoff timer with a random number and waits until the expiration of the timer until transmission. If the channel becomes occupied by somebody else during that time, transmission is deferred. If a collision occurs during the data transmission (which is detected by the HARQ feedback), then for the next transmission the contention window is increased.[18] One problem with LBT is that transmissions in neighboring cells can be interpreted as the channel being busy, thus leading to a reuse factor of less than one. This effect can be combatted by proper synchronization between the cells.

The transmission format for LAA is significantly different from "normal" LTE transmission. Continuous transmission of CRSs, for example, is not allowed in a LBT system, so they can only be sent in frames in which also data are allowed. The duration of the burst is limited due to frequency regulations that want to avoid one user hogging the channel for too long a time; burst durations between 2 and 10 ms are foreseen. Scheduling and HARQ are handled like in CA, i.e., both self-scheduling and cross-carrier scheduling are allowed.

*31.7   CoMP, Dual Connectivity, and Hetnet Support

LTE foresees the support of different forms of BS cooperation, though in very simplified form. For a discussion of the basic principles of ICIC and joint transmission/reception, we refer to Sections 21.3 and 22.11, while here we discuss only the specific implementations. Overall, CoMP is implemented in a rather suboptimum way in LTE, leaving room for significant improvements in 5G NR and later standards.

31.7.1   Intercell Interference Coordination

We discuss here the simplest case of single-antenna BSs, i.e., without beamforming. Consider furthermore the case where there is no *negotiation* between different BSs, just information about what a cell is planning to do; this is the situation covered by release 8. Two pieces of information are communicated for the UL:

- The *High-Interference Indicator (HII)* lists which RBs are especially sensitive to interference. This type of information would allow a neighboring BS to be a "good neighbor" and avoid scheduling cell-edge users (which create the strongest interference) on those RBs.
- The *Overload Indicator (OI)* indicates, *post factum*, the level of intercell interference experienced by various RBs, with a rough quantization (low, medium, high).

For the DL, the *Relative Narrowband Transmit Power (RNTP)* indicates the amount of power that will be sent on a particular RB, and thus lets the neighboring cells anticipate the level of interference; again this is helpful in particular for preventing placing

---

[17] This is compounded by the fact that the unlicensed bands are usually unpaired. Thus, if both UL and DL transmissions are desired, TDD should be used, which complicates implementation for devices that are otherwise intended for FDD use.

[18] There is considerable debate in particular with the Wi-Fi community about the fairness of the access. Wi-Fi is, as discussed in Chapter 33, designed to ensure that a particular user does not "hog" the available spectrum for a long time, and rather easily ceases the spectrum to another user/service that wants to transmit. In particular, the spectrum access mechanisms of LTE LAA do not seem to provide any mechanisms to deal with the "hidden node" problem, see Chapter 18. Some Wi-Fi manufacturers thus claim that there is a danger that LTE LAA "crowds out" Wi-Fi in an unfair manner, while some 3GPP members claim that it enables a fair access to the unlicensed spectrum.

cell-edge users on RBs on which they will see excessive interference. It must be noted that all these information types may be transmitted via the X2 interface, which is relatively slow (typically 10 ms latency).

31.7.2 Multi-Point Coordination/Transmission

Downlink Multi-Point Coordination

Better performance can be achieved if the BSs coordinate between each other the transmissions on the different RBs; this has been realized since release 11 in LTE. An assumption here is that the BSs have a fast way of communicating to each other, e.g., via optical links or direct wireless links (such fast links, if they exist, are outside the scope of the standard, but could be implemented by an infrastructure manufacturer that produces all the involved BSs).

We can distinguish here two types of coordination:

- In *coordinated link adaptation*, a BS uses the information of the transmission decisions in the neighboring cells to perform link adaptation. In order to execute such adaptation, the BS needs to know the impact that various scheduling decisions have on the link quality. It thus becomes necessary to obtain CSI under various hypotheses of neighbor transmission, which is achieved by configuring multiple CSI processes. For example, if just two BSs coordinate, then one BS needs to know the channel quality to its user if there either is, or is not, a transmission on that time/frequency resource in the neighbor cell. Unfortunately, the scheme does not scale easily – the number of hypotheses that need to be measured increases exponentially with the number of BSs that need to be taken into account.
- In *coordinated scheduling*, the BSs coordinate with each other decisions such as whether to transmit at all on a particular time/frequency resource (*dynamic point blanking*), dynamically changing the transmit power (*coordinated power control*), or adjusting the beamforming directions (*coordinated beamforming*). In all of these cases, the decisions are based on the multiple CSI processes as in the coordinated link adaptation, which do not provide the full CSI from/to all BSs. Consequently, the BSs can only make suboptimum decisions.

Downlink MultiPoint Transmission

The simplest form of multi-point transmission is *dynamic point selection*, where the transmitting BS is selected adaptively.[19] This means that the transmitting BS can be different from the serving BS. If DM-RSs are used for channel estimation, then the UE could in principle demodulate the received data without worrying about which is the transmitting BS. However, there are a number of other parts of the PDSCH that are unique to the particular BS, such as the CRS structure. In order to deal with these issues, it is thus foreseen that there are up to four different BS configurations that can communicate with the UE, and a PDSCH *mapping-and-quasi-colocation-indicator* indicates which of those is currently relevant for this UE.

A more advanced form of multi-point transmission is *joint transmission*. LTE does not have mechanisms to enable coherent joint transmission. However, noncoherent joint transmission, which allows beamforming from different BSs, and thus adding up the powers from the BSs, can be used.

Uplink Multi-Point Coordination/Reception

It is possible for multiple BSs to coordinate with each other, or combine the signals received at different BSs. Such approaches are outside the LTE standard, because the UE does not need to know what steps the network takes to coordinate or combine.

31.7.3 Dual Connectivity

While CoMP assumes that multiple transmission points can transmit/receive the *same* data to/from a UE, LTE also considers (since release 12/13) the possibility that *different* data are communicated. In other words, a UE has connection to multiple BSs at the same time. This can be motivated by a number of scenarios:

- *Data aggregation*: a UE can obviously receive more data if it is connected to multiple BSs at a time.
- *Separation of UL/DL*: the best BS for the DL might not be the best BS for the UL. Such a situation can occur, e.g., in a heterogeneous deployment: a macro-BS might be the best BS for the DL because it has a much higher TX power; yet for the UL a pico-BS is better because it is closer to the UE. Even in the case of homogeneous deployment different BSs might be optimum for UL and DL, due to interference and/or load situations.
- *Separation of user plane and control plane*: control data usually should be transmitted via very reliable links, even if they do not offer high data rates, while payload data can be transmitted on a different link with higher peak capacity but lower reliability. In LTE,

---

[19] Strictly speaking, a BS can have multiple transmission points associated with it, leading to an architecture with distributed antenna elements. Transmission points used in multipoint transmission (and similar for multipoint reception) often belong to the same BS. This architecture will be discussed in greater detail in Chapter 32, where a BS consists of a Central Unit and a Distributed Unit.

such situation can occur in heterogeneous deployments; but can also occur in multi-band deployment where, e.g., the 700 MHz band offers high-reliability control links while user data can be transmitted, e.g., in the 3.5 GHz band. For example, a macro-cell operating at the 700 MHz band can be used for mobility support and coverage while a pico cell operating at 3.5 GHz can be used for data offloading.

An LTE system establishes, for each UE with dual connectivity, a *primary BS* and a *secondary BS* (note that the assignment can vary from UE to UE, so that a BS that is the primary BS for one UE can be the secondary BS for another UE). Data are distributed by the network either to the two BSs directly, or first, all data are sent to the primary BS, and the appropriate part of them is then forwarded from there to the secondary BS. The latter case is more demanding on the connection between the two BSs but has the advantage that it makes the dual connectivity "transparent" to the network, i.e., from a network point of view, it is indistinguishable from a normal operation. The primary BS and secondary BS do their own scheduling and PHY/MAC control signaling. Normally, there are separate radio bearers for primary and secondary BS, though there is also a *split bearer*, which has a single PDPC control unit even though the data are routed via different BSs.

Dual connectivity can also be combined with CA (see Section 31.6.1) where CA can be used separately within each BS. In this case, either of the BSs defines a primary component carrier, and one or more secondary component carriers, creating a primary and secondary set of carriers, respectively. The two sets are assumed to use nonoverlapping bands, to simplify the UE design.

From a PHY layer point of view, dual connectivity does not introduce a lot of changes. The main challenge lies in the power control for the UL, since power limits are defined per device, not per link, and scheduling decisions are independent and uncoordinated between the two BSs. For synchronous operation of the two links, modifying the power control is fairly straightforward. However, for asynchronous operation, the situation is more complex, in particular because the subframe boundaries, at which power changes occur, are not aligned between the connections to the two BSs.

31.7.4   Implementation of Heterogeneous Networks

While the LTE standard generally does not say anything about the deployment but leaves this up to the network operator, there are some tools in the standard that help with the implementation of hetnets, following some of the principles described in Section 21.7. Most of the challenges are related to the fact that the DL transmit power of a macrocell is significantly larger than that of a picocell. This is not a problem as long as a UE connects to the BS with the strongest power, but as discussed in Section 21.7, this is not always desirable, in which case the SINR can become smaller than 0 dB. Consequently, a partitioning of macro- and picocell signals, either in the time or the frequency domain, needs to be provided.

Static frequency reuse, as done in classical cellular networks (see Sections 21.1 and 21.2) is not spectrally efficient, and can be wasteful in particular when traffic demand in the picocell is time-varying. The use of CA provides a much more flexible solution here.

Some of the challenges of strong interference from the macro- to a pico-cell can be handled by the standard interference coordination mechanisms discussed in Section 31.7.1. Alternatively, the macro and the pico BS can also be seen as part of a CoMP system, and the corresponding techniques (see above) be applied.

If a time-domain separation of the signal is used, it is desirable that the macro BS transmits with reduced power during certain times, so as to not interfere too strongly with the pico-cell signal. LTE therefore foresees the *Almost Blank Subframes, ABS*, during which the BS transmits with reduced power. Then the picocell can transmit to users at its cell edge during those ABSs, while transmitting to closer users when the macro BS transmits with full power. Information about when ABSs are scheduled can be transmitted in a standards-defined format on the X2 interface. Reporting of the interference level (see above) needs to be done separately for the ABS and the non-ABS situations.

*31.8   Relaying

31.8.1   General Architecture

Relaying, whose principles were explained in Chapter 27, is also foreseen in LTE. Let us first establish nomenclature: LTE calls the coverage range of the (donor) BS the *donor cell*, and the coverage range of the relay the *relay cell*. The connection between BS and relay is the *backhaul link*, and the relay to UE link is the *access link*. Access link and relay link cannot be operated on the same time/frequency resources, since this would create excessive self-interference (see also Section 18.5.2).

If access link and backhaul link operate in different frequency bands, then there is sufficient separation in the frequency domain to allow simultaneous transmission. If access link and backhaul link are in the same spectrum, then more sophisticated mechanisms, operating in the time domain, need to be used for separation. The earliest version (release 8/9) assumed a repeater, so that the relaying operation would be completely transparent to the UE. Due to the requirements for backward compatibility, this also means that relaying must be transparent for later releases as well.[20] Thus, to a UE, the relay must look like a regular BS. Conversely, to a BS, the relay must look like just another UE. Furthermore, the BS acts as a proxy between the core network and the relay, so a relay thinks it is directly connected to the core network; thus the BS appears like an MME to the S1 interface, and a BS via the X2 interface (compare Section 31.2.1)

---

[20] This introduces a number of constraints on the implementation that would not be seen in a "greenfield" implementation of such a system.

31.8.2   Frame Structure and Timing

We now discuss in more detail how this "transparency" can be achieved in systems in which the backhaul link and access link use the same band. The access link must be quiet while the backhaul transmission occurs. Specifically, the access UL has to be quiet while the backhaul UL is active, and the access DL needs to be quiet while the backhaul DL is active. The former can be enforced relatively easily by the relay: it can blank out certain subframes (it is, after all, up to the relay when to give resources to the UE for UL), so that the relay can use these subframes for backhaul UL transmission to the BS.

However, such a simple approach is not possible for the DL, because a normal DL subframe always has some transmissions (at a minimum, cell-specific RS signals), and these parts of the access DL would collide with the backhaul DL transmission. The solution for this dilemma is that the access DL uses MBSFN subframes, which by definition have CRS and other signaling only in the very first part of the subframe, followed by long time periods in which no RS or control signals are to be sent.

Because the first part of the MBSFN still has some signaling, it is occupied by access transmission, and DL control signaling from the donor BS to the relay cannot be transmitted via the regular PDCCH. Instead, a distinct relay control channel, R-PDCCH is defined (note that only the relay must recognize this channel, it is invisible to the UEs, including the release 8/9 UEs). The R-PDCCH carries DL and UL scheduling information: due to relatively narrow frequency region in which the transmission happens, it extends over most of the subframe (except in the first two symbols, which are reserved for the DL control signaling of PCFICH/PHICH/PDCCH). The DL assignments are transmitted in the first part of the subframe (to reduce latency), while the UL grants (which refer to future subframes anyway) are transmitted later.

For determination of the timing, we have to distinguish two cases: (i) the relay derives its timing from the BS, and (ii) the BS and the relay are time-aligned. In the former case, the backhaul UL has the normal timing advance (remember the relay looks like a UE to the BS); in the backhaul DL the first OFDM symbol is left empty of data, to provide a guard interval for the switching of the relay from backhaul DL to access DL. A guard interval may also be necessary for switching between access UL and backhaul UL; switching usually takes longer than a cyclic prefix duration; this guard interval can be enforced by configuring the UEs for transmitting sounding RSs in the last OFDM symbol, but then not actually transmitting them. The latter case (time aligning through a global time reference) is mostly used in TDD systems. Also in this case a guard interval is necessary but handled slightly differently.

31.8.3   HARQ

The challenge for the ARQ operation arises from two requirements: (i) the access link must not show any difference from a "standard" access link between a BS and a UE, and (ii) due to time multiplexing of backhaul link and access link, not all subframes might be available for retransmissions on the access link. In particular, for the UL, the standard procedure would be that a retransmission occurs eight subframes after the original transmission; yet that subframe might be assigned for backhaul. If that occurs, then the retransmission has to be delayed by one of more multiples of 8 ms. For the DL, the availability of the access link for retransmission is not a concern, since those retransmissions are asynchronous anyway, see Section 31.5.2.

For the backhaul link, the challenge is again that backward compatibility requires the 8 ms-based structure for ARQ, while the MBFSN frames have a 10 ms-based structure, and certain subframes cannot be configured for backhaul transmission.

*31.9   LTE for Machine-Type Applications

31.9.1   General Principles

Machine-type Communications (MTC) or *IoT* refer to the communication of sensors, actuators, and other "machines" or "things." In LTE, MTC or IoT refers specifically to transmissions with (i) very low data rate, (ii) from/to devices with low complexity and/or low energy consumption, and (iii) facing possibly very large pathloss (deployment in the wild, in basements, etc.). Other machine-type requirements, such as low latency (which is critical, e.g., in the control of fast-spinning machinery) are *not* targeted by LTE, and the framework for MTC must be seen in this context.

The key components for achieving these goals are:

- *Reduction of the maximum data rate:* by allowing only smaller transport block sizes (1 kbit), and corresponding lower data rate (1 Mbit/s maximum), the complexity of the RX can be reduced.
- *Restriction of transmission modes:* by eliminating the possibility for spatial multiplexing, a number of costly processing steps can be eliminated from the devices. Furthermore, in NB-IoT-based MTC, convolutional codes are used for the DL (such that the UE can do an easy decoding) and turbo codes for the UL (since the UE in this case performs *encoding,* which is simple even for turbo codes). Furthermore, the UL modulation formats are limited to 4-QAM and 16-QAM, reducing PAPR and thus requirements for the power amplifier at the UE.
- *Restriction of duplex operation:* in FDD systems, half-duplex operation is defined as not allowing simultaneous transmission and reception even though TX and RX are operating at different frequencies (see Section 18.5.1). This simplifies the operation and allows to reuse the same RF components for TX and RX. In LTE MTC, this simplification is pushed to the limit by providing long guard intervals (up to one subframe) between TX and RX operation, thus providing a long time for the reconfiguration of the hardware.

- *Power saving mode:* to reduce energy consumption, UEs would often like to power down; however, this eliminates them from the list of UEs registered in the network, so that after wakeup a long and resource-consuming reregistration is necessary. LTE defines a power-saving mode that allows a UE to retain its registration while being powered down. Note that wakeup from that mode can be initiated only by the UE.
- *Operation on narrower bandwidth:* this simplifies the required hardware, and can at the same time improve range.
- *Repetition coding:* this increases the range that can be covered.

The first four items were introduced in release 12, and do not require extensive modifications of the standard. The latter two form the core of Enhanced Machine Type Communications eMTC in release 13, and will be discussed in more detail in the next subsection.

31.9.2   Enhanced Machine-Type Communications

eMTC aims to handle at least 15 dB worse pathloss than standard LTE systems, even though it does not have at its disposal such performance-enhancing methods such as multiple antennas, and the UE has a transmit power of 20 dBm, i.e., 3 dB worse than standard LTE UEs. The main mechanism to achieve this is repetition coding. Two Coverage Enhancement (CE) modes are foreseen, with CE Mode A providing mild improvements mainly aiming to compensate for the lower TX power and single-antenna operation, while CE Mode B provides larger gains. The improvement is achieved by simply repeating the PDSCH/PUSCH transmission over a number of subframes (up to 32 for CE Mode A and 2048 for CE Mode B), where the repetitions are occurring in contiguous subframes that were configured for transmissions from eMTC UEs. If certain subframes are not available (the indices of those subframes are configured semi-statically and are communicated in the system information), repetition continues after the invalid subframes (the presence of invalid frames does not reduce the number of repetitions, but rather increases the transmission time). Frequency diversity can be provided by frequency hopping. Each cell can also define frequency hopping intervals (with number of subframes between hops ranging from 1 to 8 for CE Mode A and 2 to 16 for CE Mode B), and the hopping is synchronized for all UEs in the cell. The modulation format and coding for LTE is 4-QAM or 16-QAM.

The more complicated aspects of eMTC arise from the fact that it is operating with a narrow bandwidth, namely 6 RBs (1.4 MHz), while many of the LTE channels are designed to operate over a wider bandwidth, namely up to 20 MHz. Thus, non-backward-compatible channels are introduced. In particular, because the PDCCH and PCFICH are both inherently wideband, they are replaced by a new control channel called MPDCCH, which is similar to the EPDCCH, but extended to support narrowband operation and repetition coding. A new DCI format with small size to improve coverage and address the reduced scheduling requirements of eMTC UEs is defined as well. Because the MPDCCH can have repetition coding to ensure appropriate coverage, and the UE does not know the number of repetitions in advance and may be able to decode the DCI format with a smaller number of repetitions than the actual one, the number of repetitions is signaled as part of the channel itself in order for the UE to determine the start of a scheduled PDSCH reception or PUSCH transmission. DL scheduling assignments are *inter-subframe:* a scheduling assignment carried by an MPDCCH ending on subframe n is valid for subframe $n + 2$. Note also that due to this scheduling, the round-trip time for HARQ is now $8 + 2 = 10$ ms. Since the UEs are only required to deal with eight HARQ processes, this implies a maximum duty cycle of 80%.

The synchronization signals transmitted by the BS have only a bandwidth of 1.4 MHz, so in principle need no modification except possibly for coverage extension which is instead achieved by tolerating a longer time for initial synchronization. The SIB gets a new format, since the "standard" SIB is extending over a larger bandwidth, and also has a larger size (up to 2.2 kbit) than the maximum size (1 kbit) used in eMTC.

The situation is somewhat easier for the UL, since lower bandwidths are built into the regular LTE standard. The only changes are in the enabling of repetition coding (similar to the DL) and modified HARQ. UL scheduling information is carried by a DCI format on the MPDCCH; the resource allocation of the scheduling grant is split into two parts, a *narrowband indicator* defining in which narrowband channel the resource is located, and a *resource-block indicator* that specifies the exact set of resource blocks to use for the PUSCH in the narrowband channel.

31.9.3   Narrowband IoT

While eMTC is defined for a bandwidth of 1.4 MHz, even narrower bandwidths can be considered. A system that was considered for MTC is GSM (see the Appendix 30.A), because – having been developed in the 1980s – is very simple, has been optimized over 30 years, and excellent economies of scale exist. 3GPP thus had a separate working group to establish MTC based on GSM, but ultimately decided to fold it into the LTE activities. As a legacy of this genesis, the considered bandwidth is close to 200 kHz, giving rise to the name *Narrowband Internet of Things, NB-IoT.* Such systems can be deployed in GSM bands, or in LTE guard bands, or within an LTE band.

In the DL, NB-IoT uses OFDM with 15 kHz subcarrier spacing (i.e., same as in LTE), so that each NB-IoT band has 12 subcarriers (one RB in the frequency domain). Due to the different bandwidth, both the control channel and the traffic channel need to be changed. For the DL, either the narrowband PDCCH carrying the scheduling information or the narrowband PDSCH carrying the payload, are transmitted in each RB. The scheduling is inter-subframe, so that the scheduling information sent in one subframe is valid for a later subframe (the amount of the offset is transmitted as part of the scheduling assignment). The coding is tailbiting convolutional coding, and the modulation is QPSK; both of those choices simplify RX construction. Transmission from either one or two antenna ports is permitted; in the latter case, space-frequency coding (Alamouti codes) is used.

In the UL, two transmission modes exist, one with 15 kHz subcarrier spacing, and one with 3.75 kHz. In the former case, transmission can be done using 1, 3, 6, or 12 subcarriers. In the latter case, transmission can only be done on a single subcarrier, implying that the data rate for this transmission mode is very low. Coding is turbo coding (the encoder for turbo codes is simple), and modulation is QPSK (for transmission with multiple subcarriers), $\pi/4$ shifted QPSK or $\pi/2$ shifted BPSK to minimize peak-to-average power ratio).

The system information is split between an MIB and an SIB. The MIB is transmitted on the Narrowband PBCH, which occupies subframe 0 in every frame, with the information distributed over 64 frames. The Narrowband PBCH omits the first three symbols of the subframe, and all possible locations of the LTE CRS, even when deployed in a dedicated band, so that the UE does not need to know whether it is operating in-band in an LTE system or not. The SIB is transmitted like payload data.

*31.10 Device-to-Device Communications – Sidelink

31.10.1 Motivation, Architecture, and Channel Structure

LTE foresees not only communication from UEs to BSs (and vice versa), but also *between* UEs. There are two main applications for that capability:

- *Improved efficiency communication between UEs that are connected to BSs:* In many situations, communication occurs between devices that are in close physical proximity. It is not effective, from standpoints of spectral efficiency, energy consumption, or latency, to have them communicate via the BS. Rather, direct communication is enabled, though BS signals can be used to facilitate the communications.
- *Communication between UEs that do not have connection to the BSs:* when UEs are out of range of the infrastructure, communication with each other is the only meaningful possibility. This scenario is especially important for emergency communication, e.g., after a natural disaster, when the cellular infrastructure has been destroyed, and first responders have to communicate with each other.

LTE has set up a mechanism for such UE-to-UE communication, which it calls *sidelink* (in analogy to UL and DL), and/or *proximity service, ProSe*. Sidelink is designed to work in both of the above scenarios, which somewhat complicates the overall structure.

The overall architecture is shown in Figure 31.18. For the UE-to-UE-only communication, only the PC5 interface between the UEs is relevant. If a connection via the RAN (PC3 interface) is possible, then the ProSe function is responsible for all sidelink functionality in the core network.

Any sidelink system consists of three main functionalities: (i) synchronization, (ii) discovery, and (iii) communication; these steps will be described in detail in the next subsections. They also motivate the structure of the logical, transport, and physical channels specific to Sidelink, see Figure 31.19.

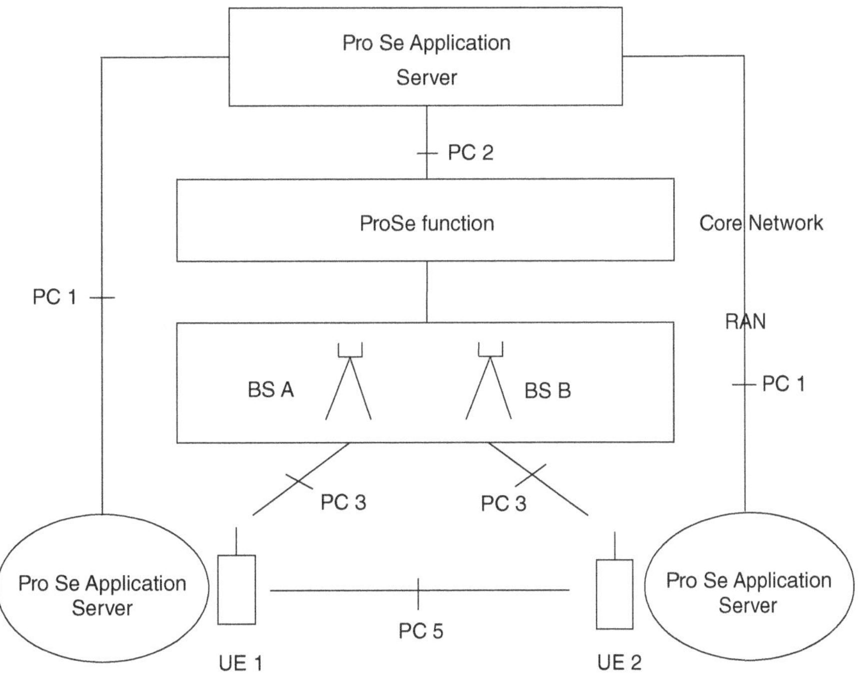

Figure 31.18 Basic sidelink structure.

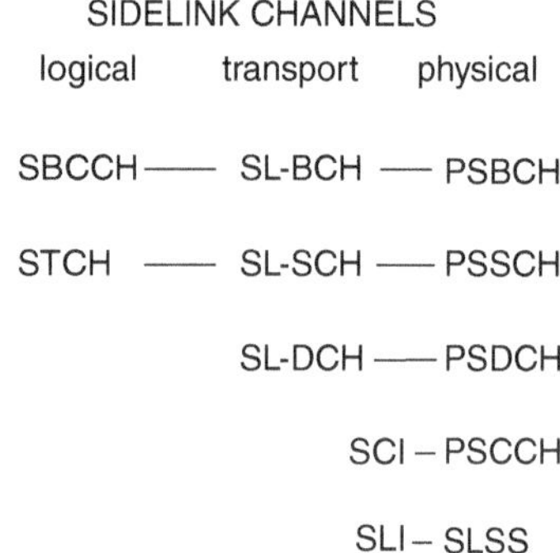

Figure 31.19 Mapping between logical, transport, and physical channels in Sidelink. SCI: Sidelink Control Information

- Logical channels
  - *Sidelink Traffic CHannel (STCH)*, which is used for the payload transmission of Sidelink.
  - *Sidelink Broadcast Control CHannel (SBCCH)*: is used for Sidelink synchronization.

- Transport channels
  - *Sidelink Shared CHannel (SL-SCH)*: the main channel for Sidelink, handling payload data as well as most control information.
  - *Sidelink Broadcast CHannel (SL-BCH)*: is used for Sidelink synchronization.
  - *Sidelink Discovery CHannel (SL-DCH)*: this is used for the Sidelink discovery process.

- Physical channels
  - *Physical Sidelink Shared Channel (PSSCH)*: carries the payload for Sidelink communications.
  - *Physical Sidelink Control CHannel (PSCCH)*: carries the control information for Sidelink.
  - *Physical Sidelink Discovery CHannel (PSDCH)*: is used for the Sidelink discovery process.
  - *Physical Sidelink Broadcast CHannel (PSBCH)* transmits basic information related to Sidelink between devices.
  - *Sidelink Synchronization Signal (SLSS)*: exists only as physical channel, and carries the specific *Sidelink Identity SLI*, and part of the synchronization information (the other part is on the Sidelink broadcast channel).

Sidelink can be viewed as "transmit-centric." A UE sends out information in a broadcast mode to a group of users (of course the group size can be 1). There is no easy bi-directional link; rather communication *between* UE A and UE B would be a transmit operation from UE A (with UE B receiving), and a separate transmit operation from UE B (with UE A receiving). All transmissions occur in the UL band (for the case of FDD) or UL timeslots (for the case of TDD), since this is the "natural" location for transmissions from a UE. Consequently, the transmission format closely hews to the UL format, in particular, DFT-spread OFDM is used for modulation.[21]

31.10.2 Synchronization

The principles of synchronization are different for the cases that (i) all UEs are within coverage of a BS, (ii) some UEs are within coverage while some are out of coverage, and (iii) all UEs are out of coverage.

- In the case that all UEs are within the coverage of a single BS, synchronization is simple: the UEs use the standard synchronization signals of LTE (PSS and SSS). The situation is more complicated when the UEs are associated with different BSs (see Figure 31.20). Since BSs nccd not bc synchronizcd, thc transmitting UE needs to convey its synchronization information to the RX; this is done via the SLSS mentioned above. The SLSS is transmitted by a UE together with the data and allows the RX to obtain the appropriate timing. Due to the UL-centric nature of sidelink mentioned above, both UEs have to send out an SLSS when they transmit.
- For the case that a UE that is within coverage region of a BS transmits to a UE outside coverage, the SLSS can relay the timing of the BS to the out-of-coverage UE. This case is very similar to the case discussed above, where two UEs are in coverage of two different BSs.
- When all devices are out of coverage, a device will start the process of sending out SLSS (it does so when it hears neither a PSS/SSS nor an SLSS from another device), and other devices can synchronize themselves to it.

In summary, the UE will search for synchronization signals in the following order: (i) SS from a BS, (ii) SLSS from an in-coverage UE (as recognized from the in-coverage indicator in the SL-BCH, see below), (iii) SLSS from an out-of-coverage UE that has synched

---

[21] In contrast to the "regular" UL, the last OFDM symbol of each subframe remains unused, in order to create a guard interval that enables switching between sidelink transmission, sidelink reception, and UL transmission.

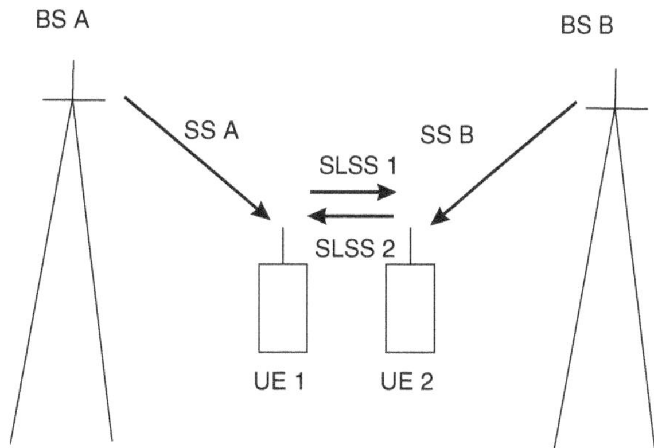

Figure 31.20 Sidelink synchronization signal. Note that SLSS 1 is aligned with SS A, and SLSS 2 is aligned with SS B.

itself to a UE with in-coverage SynchRef signal, (iv) any other SLSS. Note that a receiving UE might synchronize itself to different SLSSs (which might be associated with different groups of UEs) depending on where it receives data from.

We now turn to the detailed structure of the SLSS. Similar to the regular synchronization signals, it consists of a primary (P-) and secondary (S-) SLSS. These may only be transmitted in special SLSS subframes, which exist only with a periodicity of 40 subframes. Both P-SLSS and S-SLSS cover 72 subcarriers, with the P-SLSS consisting of second and third OFDM symbol, and the S-SLSS of the fifth and sixth OFDM symbol in a subframe. The two OFDM symbols in the P-SLSS are identical, and are derived from one of two possible Zadoff–Chu sequences, and are associated with a specific cell identity (*SLIs*). There are two groups of such SLIs (corresponding to the two Zadoff–Chu sequences), one for in-coverage UEs, and one for out-of-coverage. Each group contains 168 possible SLIs, similar to PSS in regular LTE. The two S-SLSS symbols are created the same way as for the SSS.

A UE that sends out an SLSS may also send out the SL-BCH, which includes the *SL-MIB (Master Information Block)*. The SL-MIB in turn contains some basic configuration information, such as bandwidth, TDD configuration, frame number in which the SL-BCH is transmitted, and an in-coverage indicator (i.e., whether the UE sending out the SL-BCH is within coverage of a BS).

31.10.3 Discovery

The "discovery" process in sidelink serves two purposes. The first is, as the name implies, to let a UE discover its neighbors to identify, e.g., whether friends are nearby. The second is to receive announcements, e.g., from local businesses. The common thread in those applications is that a UE broadcasts short, fixed-sized messages that can be detected/discovered by nearby UEs (note that the transmitted message is only a message code, which is linked to the content in the ProSe function in the core network).

The messages consist of 232 bits, which are transmitted with the same approach as UL messages, namely CRC insertion, turbo coding, rate matching, bit-level scrambling, and modulation (with QPSK only). The transmission happens on time/frequency resources that are selected from *resource pools*. Such a resource pool defines a set of subframes, each of which contains a set of RBs that are available for discovery messages. A more extensive discussion of resource pools will be given in the subsection below.

31.10.4 Communications

Since sidelink communications is based on broadcast to a user group, transmission is done without any CSI. A *group identity* is included in the broadcast that tells other UEs whether they are among the intended recipients of the message.

The way time/frequency resources are allocated for a transmission depends on the transmission *mode*. In mode 1, the BS assigns, in a specific scheduling grant, the resources for the transmission. This mode is obviously restricted to the situations where the UE is connected to the network (in-coverage, and in "connected" state). In mode 2, the UE selects by itself the resources from a resource pool. It is noteworthy that this mode is not restricted to the case of out-of-network UEs, but can be used in either in-coverage or out-of-coverage devices.

A resource pool consists of a set of resources for the PSCCH and one for the PSSCH. For each of those, there is a subframe pool (defining in which subframes resources are available), and a resource-block pool (defining the RBs in the available subframes). Resource pools can be configured via dedicated signals from the BS (if the UE are in the coverage region of the BS) or be preconfigured for out-of-coverage UEs.

PSCCH

The PSCCH is transmitted once every *PSCCH period*, which can be configured to be 40, 80, 160, or 320 ms for FDD. The subframes that form the subframe pool are identified to the UEs through a bitmap (of length 40) transmitted as part of the sidelink configuration.

The PSCCH carries the SCI, which is encoded and modulated in the same way as the DCI (see Section 31.4.7). The information includes the TRPI (an indicator for the used subframes, see below), a flag to indicate whether the PSSCH uses frequency hopping, an RB and hopping-resource allocation, an MCS indicator, a group destination ID, and a timing advance indicator. Note that similar information is provided on the PDCCH (using DCI format 5) by the network to a UE when transmission mode 1 is used.

PSSCH

The actual data are transmitted on the PSSCH. The data are prepared in the same way as for UL-SCH, with CRC, rate 1/3 turbo encoding, rate matching, bit-level scrambling, modulation with QPSK or 16-QAM, and DFT precoding (single-carrier transmission). Each encoded transport block is sent on four consecutive subframes of the resource pool. The actual subframes on which the transmission happens are defined by a *Time Repetition Pattern (TRP)*, which is communicated (by means of a *TRP Index (TRPI)*, which is included in the SCI) in the scheduling grant. Note that there is no ARQ in the PSSCH; as mentioned above, all transmissions in sidelink are one-way only.

In the case of transmission mode 1, where the network assigns the resources, collisions between different UEs can be largely avoided or at least limited by proper scheduling. For that reason, it is possible, e.g., to give all RBs, and all subframes, to just one UE. For mode 2, coordination of the scheduling is not possible. Thus, there is a limit on the resources that each UE can use in this mode, which is achieved by allowing only a subset of the TRP indices for each device, and each element of that subset uses less than 50% of the available subframes. Also, the RBs available in this mode are limited, similar to the PSCCH.

Glossary for LTE

ABS	Almost Blank Subframes
BCCH	Broadcast Control CHannel
BCH	Broadcast Channel
CA	Carrier Aggregation
CCCH	Common Control CHannel
CE	Coverage Enhancement
CQI	Channel Quality Indicator
CRI	CSI-RS Resource Indicator
C-RNTI	Cell Radio Network Temporary Identifier
CRS	Cell-Specific Reference Signal
CSI-RS	CSI Reference Signal
DCCH	Dedicated Control CHannel
DCI	Downlink Control Information
DFS	Dynamic Frequency Selection
DL-SCH	Downlink Shared Channel
DM-RS	DeModulation Reference Signal
DTCH	Dedicated Traffic CHannel
DwPTS	Downlink Pilot Time Slot
eMTC	Enhanced Machine-Type Communications
eNodeB	Evolved Node B
EPC	Evolved Packet Core
EPDCCH	Enhanced Physical Downlink Control CHannel
HII	High-Interference Indicator
HSS	Home Subscriber Server
IP	Internet Protocol
LAA	License-Assisted Access
LBT	Listen Before Talk
LTE	Long-Term Evolution
MBSFN	Multicast/Broadcast over Single Frequency Network
MBMS	Multimedia Broadcast and Multicast Services
MCCH	Multicast Control CHannel
MCH	Multicast CHannel
MIB	Master Information Block
MME	Mobility Management Entity
MPDCCH	Machine-Type-Communication Physical Downlink Control CHannel
MTC	Machine-Type Communications
MTCH	Multicast Traffic CHannel
NB-IoT	NarrowBand Internet of Things
NDI	New Data Indicator

NR	New Radio
OI	Overload Indicator
PBCH	Physical Broadcast CHannel
PCCH	Paging Control CHannel
PCDP	Packet Data Convergence Protocol
PCH	Paging CHannel
PCFICH	Physical Control Format Indicator CHannel
PDCCH	Physical Downlink Control CHannel
PDCP	Packet Data Convergence Protocol
PDSCH	Physical Downlink Shared CHannel
PDU	Protocol Data Units
PHICH	Physical HARQ Indicator CHannel
PMCH	Physical Multicast CHannel
PMI	Precoding Matrix Indicator
PRACH	Physical Random Access CHannel
ProSe	Proximity Service
PRB	Physical Resource Block
PSS	Primary Synchronization Signal
PSBCH	Physical Sidelink Broadcast CHannel
PSCCH	Physical Sidelink Control CHannel
PSDCH	Physical Sidelink Discovery CHannel
PSSCH	Physical Sidelink Shared Channel
PUCCH	Physical Uplink Control CHannel
PUSCH	Physical Uplink Shared CHannel
OCC	Orthogonal Cover Code
QPP	Quadrature Permutation Polynomial
RACH	Random Access CHannel
RAN	Radio Access Network
RAR	Random Access Response
RB	Resource Block
RE	Resource Element
RI	Rank Indicator
RLC	Radio Link Control
RNTI	Radio Network Temporary Identifier
RNTP	Relative Narrowband Transmit Power
R-PDCCH	Relay Physical Downlink Control CHannel
RRC	Radio Resource Control
RS	Reference Signal
SAE	System Architecture Evolution
SBCCH	Sidelink Broadcast Control CHannel
SCI	Sidelink Control Information
SC-MCCH	Single-Cell Multicast Control CHannel
SC-MTCH	Single-Cell Multicast Traffic CHannel
SDU	Serivce Data Units
SIB	System Information Block
SL-BCH	Sidelink Broadcast CHannel
SL-DCH	Sidelink Discovery CHannel
SLI	SideLink Identity
SL-MIB	Sidelink Master Information Block
SL-SCH	Sidelink Shared CHannel
SLSS	SideLink Synchronization Signal
SPS	Semi-Persistent Scheduling
SR	Scheduling Request
SRS	Sounding Reference Signal
SS	Synchronization Signal
SSS	Secondary Synchronization Signal
STCH	Sidelink Traffic CHannel
TCP	Transport Control Protocol
TRP	Time Repetition Pattern
TRPI	Time Repetition Pattern Index
TTI	Transmission Time Interval

UCI	Uplink Control Information
UDP	User Datagram Protocol
UL-RS	Uplink Reference Signals
UL-SCH	UpLink Shared Channel
UpPTS	Uplink Pilot Time Slot
VRB	Virtual Resource Block
WCDMA	Wideband Code Division Multiple Access

Further Reading

The official resource for studying the LTE standard is, of course, the standards documentation, which is available from www.3gpp.org. Of particular interest is the 36.2xx series. Two excellent summaries of the standard, which concentrate on the PHY and MAC layer, are [Dahlman et al. 2016] and [Sesia et al. 2011]. Both of these books describe not only *what* has to be implemented, but also *why*, and are therefore recommended as go-to references for any deeper study of LTE (before looking at the actual standards documents). A number of other text/reference books on LTE exist, e.g., [Holma and Tokola 2011, Ahmadi 2013]. A very useful resource for system simulation is the Vienna LTE simulator, which is partly included in MATLAB, and which is described in detail in the book [Rupp et al. 2016].

Furthermore, a number of good surveys exist for specific aspects of LTE. Resource allocation is discussed in [Ku and Walsh 2015], multicell coordinated scheduling in [Li et al. 2014], relays in [Yang et al. 2009], Coexistence of LTE-LAA and Wi-Fi in [Chen et al. 2016], femtocells in [Xenakis et al. 2014], MTC in [Elsaadany et al. 2017], and D2D communication in [Liu et al. 2014].

For updates and errata for this chapter, see https://wides.usc.edu/students.html#textbooks.

Exercises

See Sec. 36.31 of Exercises.pdf at wiley.com/go/molisch/wireless3e

32

5G Cellular – 3GPP New Radio (NR)

32.1 Introduction

In contrast to the other chapters of this book, which are written to be as independent of each other as possible, this chapter should be read in conjunction with Chapter 31. The reason for this is that 5G *New Radio (NR)* is de facto an evolution of LTE,[1] and shares many of the same characteristics. Thus, to avoid duplication that would not only be tedious but also tend to drown out the new aspects, we will here concentrate on the differences between NR and LTE. To ease comparison and cross-reading, the chapter also duplicates the formal structure of Chapter 31 (consequently, some sections consist of little more than a headline).

32.1.1 History

While LTE developed into the dominant cellular standard after 2010, it was already recognized by the industry that further enhancements would be necessary – not just the gradual improvements that come with new Releases of the LTE standard, but rather order-of-magnitude improvements in the system performance. Firstly, as wireless video, gaming, and similar high-data-rate consuming applications became more and more popular, the *average throughput per user* would have to increase by an order of magnitude or more. Secondly, new sets of applications were envisioned, such as massive machine-type communications for the *Internet of Things (IoT)* that would lead to dramatic increases in the required *connection density*, as well as reduction of *latency*. For this reason, research into 5G started around 2010, both as company-internal projects, and large-scale collaborative projects such as the European-Union-funded METIS project. In 2015, a workshop of 3GPP set the scope for the next-generation cellular systems, and technical work started in the spring of 2016. By the end of 2017, a preliminary version of the 5G standard was available (called 3GPP Release 15), which became official in spring 2018. An enhanced version, called Release 16, was published in the fall of 2020. The first major application was Fixed Wireless Access (FWA) using mm-wave frequencies; deployment started in late 2018. More widespread deployment of 5G infrastructure started in 2019 and is proceeding at the time of this writing.

It can be seen that the timetable for the development of the standard was extremely aggressive, with just two years for the development from scope to first standards document. This was achieved by a combination of three measures:

- *A dense schedule of meetings.* The standards groups met every month. Each meeting had up to 2,000 submissions, of which of course only a small part could be discussed in the meetings themselves.
- *Building on LTE.* While previous generational changes (e.g., from GSM to WCDMA, or WCDMA to LTE) involved a complete redesign of the physical layer, the 5G standard took a more evolutionary approach, and thus many of the aspects from LTE could be reused directly.
- *Moving many of the goals to later releases of the standard.* Release 15 mostly concentrated on techniques for mobile broadband. Other topics and requirements, in particular for industrial applications (which had some preliminary support in Release 15) and device-to-device communications, were included in Release 16. Further enhancements are expected for later releases.

32.1.2 Goals and Applications

LTE was a system mainly driven by the need for higher data rates and higher spectral efficiency, as described in Chapter 31. For the development of 5G, a range of new applications was envisioned, see Figure 32.1.

---

[1] Formally, NR is a distinct standard. However, it inherits so many characteristics from LTE that from a practical perspective it can be advantageous to view it as an evolution.

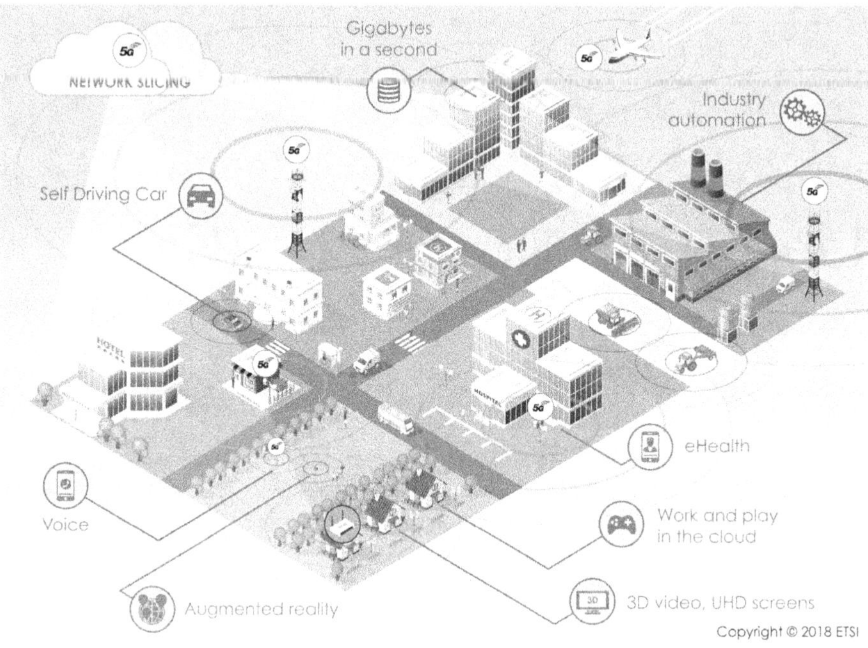

Figure 32.1 Use cases for 5G envisioned by ETSI. From [ETSI 2020, https://www.etsi.org/technologies/5g]. Color version available at wiley.com/go/molisch/wireless3e.

This led to a range of new requirements, in particular in

- *Enhanced Mobile Broadband (eMB)*: the number of broadband users is constantly increasing, as are their requirements for data throughput. This is in particular due to the increase in video traffic, which constitutes almost 80% of wireless data traffic, and is growing about 60% per year (in 2020, it was around 30 ExaByte per month). This necessitates significant improvements in spectrum efficiency, and area traffic capacity. This development continues a trend that started in LTE.
- *Large number of low-power connected devices:* the IoT will lead to many sensors and actuators in the home, in "smart cities," shopping malls, warehouses, etc. These many billions of devices typically will not need high data rates, but create high connection density, and send small data packets with a low duty cycle. This requires a redesign of many system aspects, as discussed in Chapter 30.
- *Ultra-Reliable Low Latency Communications (URLLC):* the tactile internet, autonomous cars, and factory automation in industry 4.0 are often summarized as "mission-critical" applications, because their failure could have catastrophic consequences. In order to function properly, they need to provide not only ultra-high reliability for the packet delivery but also short latency.

We next describe the targets for 5G systems as set by ITU and adopted by 3GPP. The progress from 4G (IMT-Advanced) to 5G (IMT-2020), as also indicated in Figure 32.2, should include:

- *Peak data rate*: increase from 1 to 20 Gbit/s. This is important for new applications such as virtual reality and gaming, where very high peak data rates are needed.
- *User experienced data rate*: increase from 10 to 100 Mbit/s. This follows the general trend of increasing data rates, e.g., for watching of videos, where higher resolution (including 4k) is demanded even from mobile links.
- *Spectrum efficiency*: improve by factor 3. Spectrum efficiency is obviously a major factor driving cost, and thus has a continuous need for improvement.
- *Mobility*: increase from 350 to 500 km/h. This is becoming necessary due to increasing velocity of high-speed trains. These goals have recently increased to 1,000 km/h to account for some airborne systems.
- *Latency*: improve from 10 to 1 ms. This is important for tactile internet and industrial applications. While early systems struggled with this goal, Release 16 foresees features to realize it.
- *Connection density*: increase from 10^5 to 10^6 devices per km^2 This becomes necessary through the deployment of IoT and other massive machine-type applications.
- *Network energy efficiency*: this quantity, defined as number of transmitted bits per Joule of consumed energy, should be increased by a factor 100. Since communications applications use up to 3% of electricity in the world, energy efficiency improvement is important for sustainability reasons. Furthermore, operating costs of networks are dominated by energy costs, and battery lifetime for IoT devices determine the device lifetime.
- *Area traffic capacity*: increase from 0.1 to 10 Mbit/s/m^2 This follows from the requirement of both higher data rate per device, and increased number of devices.

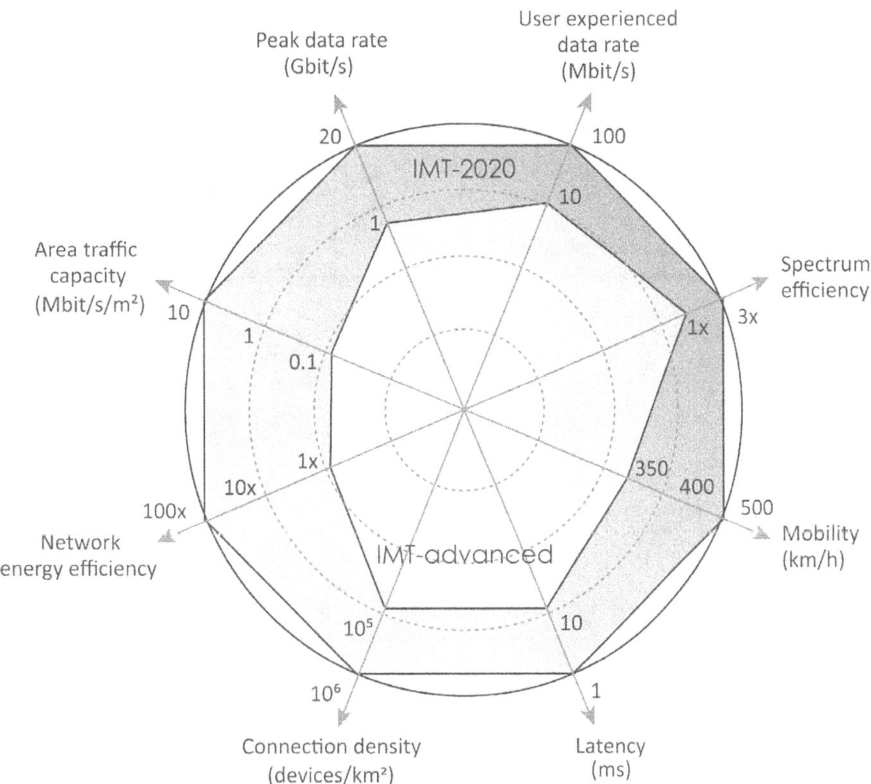

Figure 32.2 Goals of 5G (IMT-2020) compared to 4G (IMT-Advanced). Color version available at wiley.com/go/molisch/wireless3e.
Reproduced with permission from [International Telecommunications Union 2015] © ITU.

32.2 System Overview

32.2.1 Network Structure

Just like LTE, a 5G system consists of a core network and a Radio Access Network (RAN). These two components were developed independently, and it is required that any combination of 4G/5G RAN and 4G/5G core network can work. This includes, but is not limited to, an LTE core network working with both 4G and 5G Base Stations (BSs). More details about the various combinations will be discussed below.

NR Core Network
The functionality of a 5G core network is similar to that of Enhanced Packet Core (EPC) (the core network of LTE), but provides support for a few new functionalities:

- *Service-based architectures*: the network specifications are for services, instead of particular network nodes. This is motivated by the use of NR for a number of different services, and that many network nodes use generic processors on which software-defined networks are implemented.
- *Network slicing*: related to the emphasis on services, NR supports network slicing, so that a logical network can be constructed that encompasses all functionalities from physical layer to service layer. Each such logical network or network slice thus serves to support a particular functionality. Different slices might run on the same hardware.
- *Control plane – user plane split*: this split allows separate implementation and scaling of control plane and user plane, including that the different planes can communicate via different BSs. For example, the control plane can communicate via a LTE BS, while the user data are sent (partly or completely) on a 5G BS. This is especially important for the early deployment phase of NR.

The control plane includes the following components:

- *Access and Mobility Management Functions (AMF),* which handles interaction between the core network and the device.
- *User Plane Function (UPF),* which controls interaction between RAN and the "external" network (usually the internet).
- *Session Management Function (SMF)* that manages a particular data session.

Table 32.1 lists the functionalities of these components. Figure 32.3 shows a more complete figure of the different functionalities of the core network, whose acronyms (which are usually self-explanatory) are defined in the caption. Additional components (not shown in the figure) are Unstructured Data Storage Function (UDSF), Unified Data Repository (UDR), UE radio Capability Management Function (UCMF), 5G-Equipment Identity Register (5G-EIR), NetWork Data Analytics Function (NWDAF), CHarging Function (CHF). None of these functions will be discussed further here, but their names and listings give an impression about the variety and complexity of functions that need to be implemented in a core network.

Table 32.1   Control plane components and functionalities

Function	Role
AMF	Control signaling between UE and core network, data security, and authentication
SMF	Creating and removing PDU sessions, assigning IP addresses to UEs, control policy enforcement
UPF	Packet routing, packet inspection, QoS handling, and traffic measurements

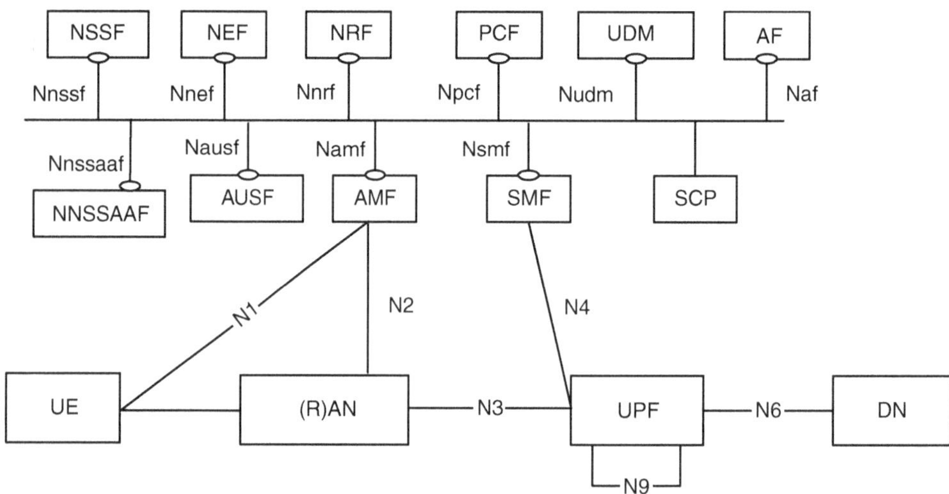

Figure 32.3   5G system architecture. Explanation of acronyms: AUSF, Authentication Server Function; DN, Data Network; NEF, Network Exposure Function; NRF, Network Repository Function; NSSAAF, Network Slice Specific Authentication and Authorization Function; NSSF, Network Slice Selection Function; PCF, Policy Control Function; UDM, Unified Data Management; AF, Application Function; SCP, Service Communication Proxy. Reproduced from [3GPP TS 23.501] © 2021. 3GPP™ TSs and TRs are the property of ARIS, ATIS, CCSA, ETSI, TTA and TTC, who jointly own the copyright in them. They are subject to further modifications and are therefore provided. to you "as is" for information purposes only. Further use is strictly prohibited.

Radio Access Network

The RAN has in principle a similar structure as LTE, consisting of BSs and User Equipments (UEs). The exact name for an infrastructure functional unit (what we always call BS in this book) is now gNodeB.[2] As a new aspect, distributed architectures are defined more explicitly, with a *Central Unit (CU)* that handles the higher layers (Radio Resource Control (RRC), Packet Data Convergence Protocol (PDCP), and Service Data Application Protocol (SDAP), see below), and *Distributed Units (DU)* that handle Radio Link Control (RLC), MAC, and PHY layers. The various interfaces between infrastructure components are shown in Figure 32.4. In addition, the interface between BS and UE is called the *Uu interface.*

As mentioned above, the network can contain a mixture of LTE and NR BSs, each of which can be connected to both LTE and NR UEs. Thus, there need to be, e.g., variants of the Xn interface that are responsible for connecting an NR BS with an LTE BS, as well as one for NR-to-NR BS connections. Another complication arises from the fact that a UE can be connected to one or more BSs. Dual connectivity (i.e., multiple connections, compare Section 31.7.3) can be used to increase data rates, by exchanging data on multiple links simultaneously. It can also be useful for having the control connection to one BS, and the data connection to another; again this is especially important during the first phase of NR where nonstandalone deployments are made so that connection of the UE to the LTE control plane is needed.

---

[2] A *gNB* is defined as a *logical* unit that contains all the functionalities of the RAN. A BS, on the other hand, is a physical implementation. It can be identical to a gNB, but need not be.

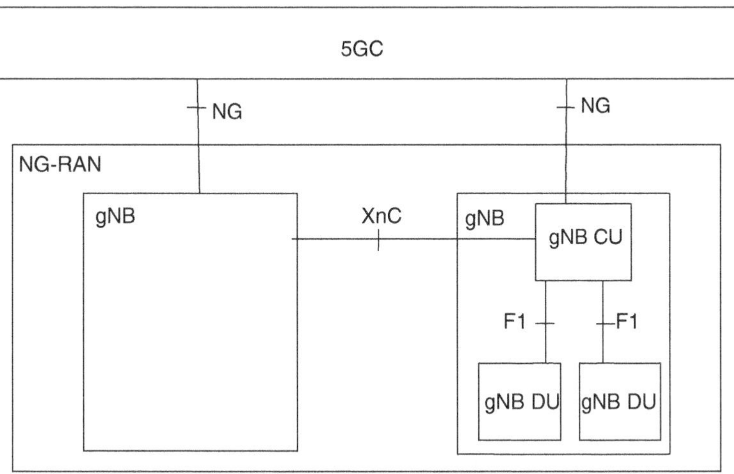

Figure 32.4 RAN architecture. In addition to the shown interfaces, there is also a user-plane interface between gNodeBs (called Xn-u). The Next Generation (NG) interface can be split into NG-c and NG-u.

Reproduced from [3GPP TS 38.401] © 2021. 3GPP™ TSs and TRs are the property of ARIS, ATIS, CCSA, ETSI, TTA and TTC, who jointly own the copyright in them. They are subject to further modifications and are therefore provided. to you "as is" for information purposes only. Further use is strictly prohibited.

32.2.2　Protocol Structure

The protocol structure of NR, shown in Figure 32.5, is very similar to that of LTE. However, a new component is the *SDAP,* which maps *Quality-of-Service (QoS) bearers* (i.e., data streams that need a particular QoS) coming from the core network, to radio bearers. The SDAP is introduced due to the new handling of QoS requirements in the NR core network. If the BS is connected to an LTE core network, then the SDAP is not used.

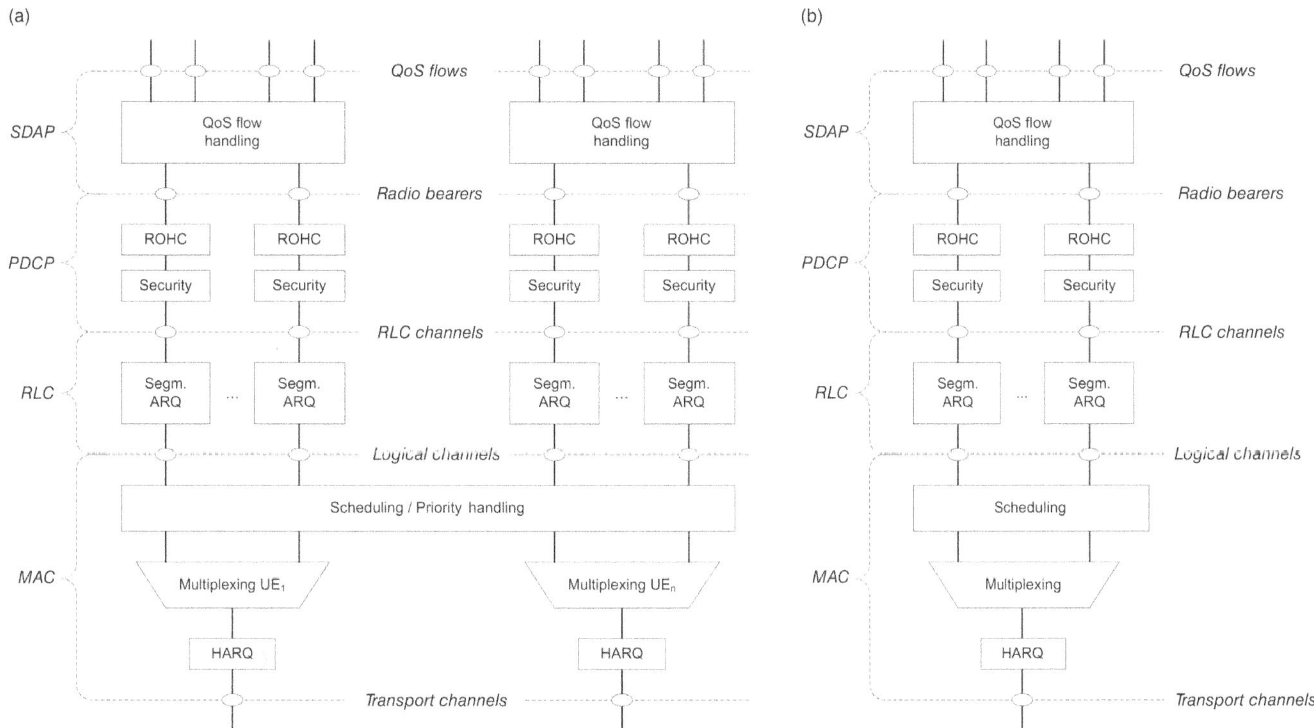

Figure 32.5 Protocol structure for layer 2 of DL (a) and UL (b) in NR.

Reproduced from [3GPP TS 38.300] © 2021. 3GPP™ TSs and TRs are the property of ARIS, ATIS, CCSA, ETSI, TTA and TTC, who jointly own the copyright in them. They are subject to further modifications and are therefore provided. to you "as is" for information purposes only. Further use is strictly prohibited.

The packets from the SDAP are handed to a radio bearer. The subsequent steps are very similar to LTE (see Section 31.2.2): for each radio bearer, the *PDCP* entity performs its functions like header compression, hands the packets to the *RLC* (for segmentation and retransmission), onward to the MAC (handles the multiplexing of logical channels, hybrid ARQ, etc.), and finally the PHY for transmission. The *RRC* handles the control-plane procedures in the RAN, including broadcast of system and paging information, connection management such as setting up of bearers, mobility management, measurement configuration and reporting, and handling of device capabilities. Some differences of the functions compared to LTE arise from specific NR features (e.g., dual connectivity, which requires additional functionality in the PDPC). Also, the RLC does not ensure in-sequence deliveries anymore. This is motivated by a desire to improve overall latency, since if the in-sequence delivery is happening, then when one packet is not sent successfully over the PHY, all packets with higher packet sequence number have to wait at the Receiver (RX) for their handing over to higher layers until this packet has been retransmitted.

32.2.3 PHY and MAC Layer Overview

NR has strong similarities to the LTE physical layer, but also a few important differences. The motivation for changes comes generally from the more stringent system requirements, and in particular from the following design philosophies:

- *Exploiting the mm-wave frequency spectrum*, to achieve larger data rates and higher traffic capacity. The availability of several GHz of bandwidth in the mm-wave spectrum, see Section 32.2.4, enables very data rates. While NR does not specify separate *transmission methods* for the mm-wave band, many of the *parameter choices* (e.g., subcarrier spacing) are either explicitly prescribed for the mm-wave band, or make sense mostly in the context of mm-wave transmission.[3]

- *Greater flexibility in the spectrum assignment*, to accommodate the heterogeneity of many services. While LTE provided bands of 5, 10, or 20 MHz, whose width stays constant over time, NR creates time-varying *bandwidth parts* that allow to change the effective system bandwidth over time (those bandwidth parts are unrelated to the ones in Sec. 31.3.7). The temporal structure is also more flexible to accommodate low-latency transmission and adapt to the requirements of unlicensed bands. Finally, the TDD duplexing structure is more flexible, allowing dynamic choices for Uplink (UL), and Downlink (DL) configuration in each separate cell. This is especially useful in small cells, where (due to the small number of active users), the temporal fluctuations of the different users are not "averaged out," and are also more feasible due to the smaller power differences of UL and DL transmission.

- *Use of massive MIMO*, leading to a *beam-centric design*, to improve area spectral efficiency. Massive MIMO (Sec. 22.9), i.e., using large array sizes, is an essential part of NR. It is mostly exploited to enable multi-user MIMO, and thus enhance the area spectral efficiency. In particular, at higher frequencies, large arrays employ hybrid beamforming (Section 16.2.10) to create beams pointing into various directions. Transmission techniques, including initial access, are modified to take such beamforming into account.

- *Low-latency transmission*, to allow for tactile internet and other control operations. To enable this, the temporal structure is made more flexible, eliminating the requirement that transmissions last for 0.5 ms, and start at the slot boundary. Furthermore, requirements for processing times are tightened (e.g., faster acknowledgments in Hybrid Automated Repeat Request (HARQ) are required). All of this leads to latencies that are much lower than in LTE.

- *Ultra-lean design*, to reduce interference and energy consumption. This is partly achieved by minimizing the always-on signals such as the cell-specific reference and broadcast signals. Furthermore, more flexible configurability of transmission parameters (e.g., pilot density) allows to better match the transmission parameters to the application and channel condition, and thus avoid wasting energy.

- *Forward compatibility*, to allow easier introduction of future improvements. NR provides a mechanism to reserve parts of the time-frequency plane for different transmission methods. The reduction of always-on signals also makes such reservations easier.

These design philosophies lead, *inter alia,* to the following new aspects of NR compared to LTE (the basic transmission structure, as described in Section 31.2.3, has remained the same).

- NR uses OFDM for both DL and UL (DFT-spread OFDM is an option for the UL as well, but only for single-stream transmission); this is in contrast to LTE where OFDM was used for DL and DFT-spread OFDM (as the only option) for UL.

- Support for multi-antenna signaling is greatly expanded. Large antenna arrays, multi-user MIMO with large number of users, and beamforming are all supported, while they were either absent, or severely restricted, in LTE.

- Inter-cell interference coordination is possible, just like in LTE. Furthermore, the system foresees mechanisms (such as indications of when antennas are "quasi-co-located") that allow an easier implementation of remote radio heads, hauling back the signals from distributed antenna locations to a central place for processing. However, the basic interference reduction mechanism through use of inter-cell interference coordination has remained the same.

- Modulation formats have been expanded. While QAM is still being used, the maximum constellation size has been expanded to 256 QAM for both UL and DL (it had been defined for the DL in LTE since Release 12).

- Error-correction coding has been changed from LTE. While previously turbo codes were used for payload and convolutional codes for control, now LDPC codes are used for payloads, and polar codes for control signaling.

---

[3] More specifically, the standard prescribes certain subsets of parameters that are applicable for the different frequency bands. However, those subsets might overlap.

Table 32.2 FDD frequency bands of NR, following [3GPP 38.101]. Main operating regions at time of writing. Af: Africa, Am: America, As: Asia, Eu: Europe, CSAm: Central and South America, NAm: North America, Oc: Oceania.

NR operating band	Uplink (UL) *operating band* (MHz) $F_{UL\_low}-F_{UL\_high}$	Downlink (DL) *operating band* (MHz) $F_{DL\_low}-F_{DL\_high}$	Main operating region
n1	1920–1980	2110–2170	Eu, As, Oc, CSAm, Af
n2	1850–1910	1930–1990	Am
n3	1710–1785	1805–1880	Eu, Af, As, Oc, CSAm
n5	824–849	869–894	NAm, (As)
n7	2500–2570	2620–2690	Eu, As, Af, CSAm
n8	880–915	925–960	Eu, As, Af, CSAm
n12	699–716	729–746	NAm
n14	788–798	758–768	Am
n18	815–830	860–875	Japan
n20	832–862	791–821	Eu, Af
n25	1850–1915	1930–1995	
n26	814–849	859–894	
n28	703–748	758–803	CSAm, As, Oc
n29	N/A	717–728	NAm
n30[3]	2305–2315	2350–2360	NAm
n65	1920–2010	2110–2200	Eu
n66	1710–1780	2110–2200	NAm
n70	1695–1710	1995–2020	
n71	663–698	617–652	
n74	1427–1470	1475–1518	
n75	N/A	1432–1517	
n76	N/A	1427–1432	
n80	1710–1785	N/A	
n81	880–915	N/A	
n82	832–862	N/A	
n83	703–748	N/A	
n84	1920–1980	N/A	
n86	1710–1780	N/A	
n89	824–849	N/A	
n91	832–862	1427–1432	
n92	832–862	1432–1517	
n93	880–915	1427–1432	
n94	880–915	1432–1517	
n95	2010–2025	N/A	

Following [3GPP 38.101].

32.2.4 *Frequency Bands and Spectrum Flexibility*

NR defines two main *Frequency Ranges (FR)*: (i) FR1, which we will call in the following <6 GHz, or cm-wave, and (ii) FR2, which covers 24.25–52.6 GHz, henceforth called the mm-wave band (which is expected to be expanded to 71 GHz in Release 17). In both bands, regulators have made significant new spectrum available. In the cm-wave range, the most important new spectrum is the 3.5 GHz band, which is reserved for NR and is to be used in TDD configuration. In the mm-wave regime, frequency regulators around the world have made several GHz of licensed spectrum available. Most of those regulations have been done on a national basis; e.g., in the US, the frequency regulator FCC has made the frequency bands 27.5–28.35 GHz, 37.0–38.6 GHz, and 38.6–40.0 GHz available for licensed operation (there is also 14 GHz of unlicensed spectrum available between 57 and 71 GHz, but those are not considered in releases 15 or 16 of NR). Tables 32.2–32.4 show the assignments for FDD and TDD in both FR1 and FR2 at the time of this writing. Many of these bands are overlapping with the frequencies for LTE; bands covering the same spectrum have the same band number, just with an "n" as a prefix. It can be anticipated that over the years, more bands that are currently assigned to LTE will be made available to NR, and/or that systems deployed in dual-use bands will migrate from LTE to NR. The maximum bandwidth in FR1 can be 5, 10, 15, 20, 25, 30, 40, 50, 60, 70, 80, 90, or 100 MHz; in FR2 it can be 50, 100, 200, or 400 MHz.

NR distinguishes between different categories of BSs and UEs, similar to LTE, with similar power limitations, see Sec. 31.3.4.

Table 32.3 TDD frequency bands of NR following [3GPP 38.101] and main operating region at the time of writing, in Frequency Range 1

NR operating band	Operating band (MHz)	Main operating region
n34	2,010–2,025	Eu, As
n38	2,570–2,620	Eu
n39	1,880–1,920	China
n40	2,300–2,400	Eu, As, Oceania, Af
n41	2,496–2,690	USA, China, Japan
n46	5,150–5,925	USA (LAA)
n47	5,855–5,925	
n48	3,550–3,700	USA
n50	1,432–1,517	
n51	1,427–1,432	
n53	2,483.5–2,495	
n77	3,300–4,200	
n78	3,300–3,800	
n79	4,400–5,000	
n90	2,496–2,690	
n96	5,925–7,125	

Following [3GPP 38.101].

Table 32.4 TDD frequency bands of NR following [3GPP 38.101] and main operating region at time of writing, in Frequency Range 2

Operating band	Operating band (MHz)	Main operating region
n257	26,500–29,500	As, Am
n258	24,250–27,500	Eu, As, Oc, CSAm
n259	39,500–43,500	
n260	37,000–40,000	
n261	27,500–28,350	

Following [3GPP 38.101].

32.3 Physical Layer

32.3.1 Overview of the Transmission Steps

The basic transmission structure of NR is the same as in LTE; the reader is thus simply referred to Section 31.3.1.

32.3.2 Coding and Scrambling

Coding is an area in which NR is significantly different from LTE. The key changes are that LDPC codes (instead of turbo codes) are used for payload and most control signaling (more precisely, for the PDSCH and PUSCH, see Section 32.4.1), while polar codes (instead of convolutional codes) are used for shorter control signals (the PDCCH and PUCCH).

However, while each of the steps has changed, the general flow of the encoding procedure is similar to LTE, as described in Section 31.3.2. NR sends one transport block[4] per *Transmission Time Interval (TTI)* and component carrier. The first step in the coding procedure is the attachment of a *Cyclic Redundancy Check (CRC)* to the transport block. The CRC is 16 bits long for transport block sizes up to 3,824 bits, and 24 bits for larger transport blocks.

The next step is to segment transport blocks into multiple, equal-sized code blocks (of sizes up to 8,424 or 3,840 bits, see below). The reason for this segmentation is that larger codeblocks would require large storage, be difficult to decode, and possible retransmissions become too costly. If such a segmentation occurs, each segment is appended with its own 24-bit CRC (but using a different CRC polynomial compared to the transport block); this is similar to LTE. As a change to LTE, NR also defines CodeBlock Groups (CBGs), which allow a more efficient retransmission request when a transport block contains a large number of codeblocks; see Section 32.5.2 for details; but these are only relevant for the (H)ARQ part of the error protection.

Each code block is then encoded with an LDPC code, whose principles are described in Section 13.7. The available codes allow code rates from 1/5 to 22/24; even lower rates can be achieved by repetition coding. In particular, NR uses quasi-cyclic LDPC codes with a dual-diagonal structure of the kernel part of the parity check matrix. These codes are versions of so-called *protograph codes*, whose principle can be visualized from the Tanner graph (factor graph), see Figure 32.6. As in Section 13.7, variable nodes are shown on top, and check nodes on the bottom; note that a protograph is called *Base Graph (BG)* in 3GPP parlance. The set of variable and check

---

[4]Two transport blocks per TTI are foreseen for spatial multiplexing with more than 4 streams to one user.

nodes of the protograph is replicated multiple times, while the edges between the variable and check nodes are permuted. The process of protograph copying and edge permutation is called *lifting*, the number of used copies is the lifting size Z. When expressing this as matrix operations, we start out with the base matrix (corresponding to the BG). Then, a "0" in the base matrix is replaced by a $Z \times Z$ all-zeroes matrix, and a "1" in the base matrix is replaced by a $Z \times Z$ permutation matrix. This construction of LDPC codes has the advantage that like nodes from every copy of the protograph can be processed in parallel, and the decoding complexity of this structure scales *linearly* with the number of bits.

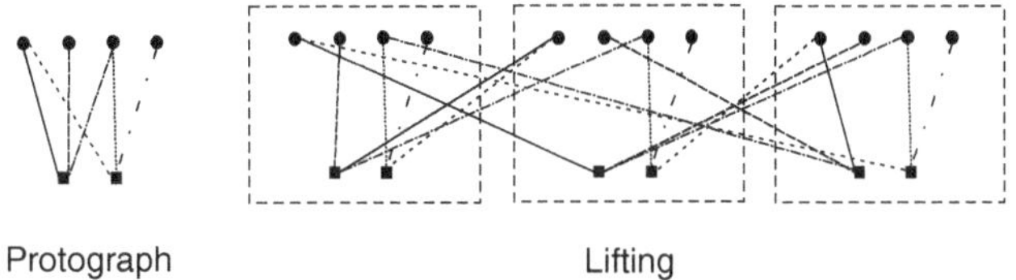

Figure 32.6 Principle of the construction of protograph codes. Permutation is applied for edges with like linetype. Reproduced with permission from [Bae et al. 2019] © APSIPA.

In NR, there are two different BGs, where BG 1 has size 46×68 and mother codes rate 1/3, and BG 2 has size 42×52 and mother code rate 1/5. To further simplify description and signaling, the permutation matrices are cyclically shifted identity matrices (8 different shifts are defined), so that the permutation matrix can be simply identified by the size of the shift. The different BGs are optimized, and should be used, for different block sizes and code rates, see Table 32.5. There is a total of 51 different lifting sizes, corresponding to the different codeblock lengths that can be encoded. If the payload has a size that does not fit into any of these, then it is augmented with dummy bits before encoding. Note that coderates <1/5 can be obtained by repetition coding.

The next step in the encoding process is the rate matching (bit selection, bit interleaving), such that the number of bits that are put out by the encoder matches the time-frequency resources and modulation format assigned to this UE.[5] Furthermore, suitable redundancy bits for the HARQ have to be provided, so that retransmissions can be sent. The generating process is rather complicated: it starts by puncturing some (up to 1/3) of the systematic bits; the remaining systematic bits are then written into the first positions of a circular buffer. Subsequently, the remaining (punctured) systematic bits are written into the buffer, followed by parity bits. The circular buffer also defines different starting points (corresponding to "redundancy versions") that determine which bits are being transmitted. Redundancy versions 0 and 3 are defined to be decodable if *only* this version is received (assuming, of course, sufficiently low transmission errors). A graphic representation of the principle is shown in Figure 32.7. Note that the requirement for soft combining

Table 32.5 Base graph selection for LDPC codes

Coderate\block size (bits)	≤292	293–3824	>3824
≤1/4	BG 2	BG 2	BG 2
1/4–2/3	BG 2	BG2	BG 1
>2/3	BG 2	BG 1	BG 1

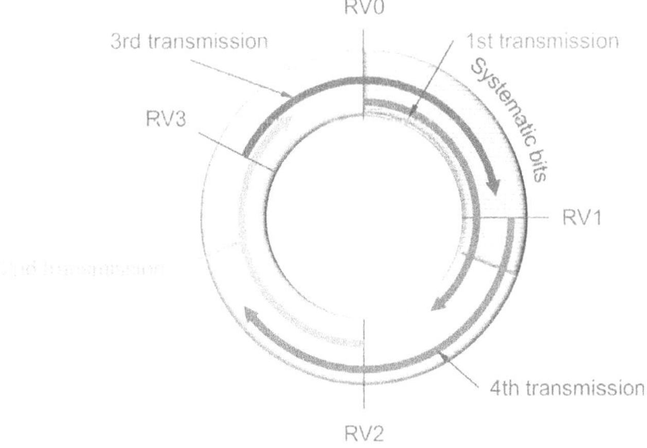

Figure 32.7 Use of the circular buffer to generate different incremental redundancy versions. Color version available at wiley.com/go/molisch/wireless3e. Reproduced with permission from [Dahlman et al. 2020] © Academic Press.

---

[5] Note that not all OFDM symbols in the assigned RBs are actually usable resources, since some of them may be reserved, be assigned to pilot signals, control signals, or synchronization signals.

implies that the Rx has to buffer all received bits until the decoding can be achieved. This might require a large storage, which – in particular at the UE – might not be desirable. The standard thus foresees for the DL a "limited-buffer rate matching," where the UE is only required to store soft bits that correspond to the largest codeblock size at rate 2/3 (which implies that for such large blocks, not all soft bits can be stored, though for smaller blocks, it might still be possible to store all soft bits); obviously, this will lead to a performance loss. Finally, the bits are written (rows first) into a block interleaver, where the number of rows is equal to the modulation alphabet size. Consequently, the systematic bits are spread over the different modulation symbols, which increases reliability.

Finally, the data are scrambled with a scrambling sequence that depends on the *data scrambling identity* that is configured for each device, or the *physical layer cell identity*. The choices of scrambling sequence ensure that neighboring devices use different scrambling sequences, which helps to avoid worst-case interference.

The HARQ aspects of the error correction are discussed in Section 32.5.2.

32.3.3   Modulation

The default modulation method for NR in both DL and UL is OFDM. The modulation formats on each subcarrier are QAM with modulation alphabet size 4, 16, 64, and 256. The modulation symbols are then possibly mapped onto different transmission layers (1st symbol mapped onto the first layer, 2nd symbol onto the second layer,...) for up to 4 layers (with layers being related to spatial multiplexing, see Section 32.3.6); in the DL, also mapping of up to 8 layers is allowed, with a second transport block carrying layers 5–8, and the mapping following the same rules.

In addition, NR also foresees DFT-spread OFDM as an option for the UL. In this case, the modulation formats are the same as for OFDM, except that also $\pi/2$-BPSK is allowed, to reduce the PAPR. The DFT-spread operation is the same as in LTE, see Section 31.3.3. Also, just as in LTE, contiguous Resource Blocks (RBs) must be used in this case. Finally, only a single transmission layer is allowed.

32.3.4   Mapping of Modulation Symbols to Time/Frequency Resources

Frames, Slots, and Symbols

Principles and Comparison to LTE

The time/frequency structure in LTE is relatively straightforward: 10 ms *frames* are divided into 1 ms *subframes*, which in turn consist of two *slots* of 0.5 ms each. Each slot contains 7 OFDM symbols. The OFDM subcarriers are spaced 15 kHz apart. A set of 12 contiguous subcarriers, over a duration of 7 OFDM symbols, constitutes a RB. The admissible bandwidth ranges from 1.4 to 20 MHz.

The structure in NR is quite different, due to the following main reasons:

- *Possible availability of larger bandwidths*: since also mm-wave frequency bands are part of NR, up to 400 MHz of bandwidth should be foreseen as part of the standard.
- *Requirement for shorter latency*: to achieve short latencies, transmissions should be allowed to last shorter durations.
- *Operation in more diverse environments and more applications*: depending on the different environments, different subcarrier spacings should be allowed. This needs to be done to provide higher robustness to intercarrier interference (especially at mm-wave frequencies), as well as shorter OFDM symbols (for lower latency). On the other hand, 15 kHz subcarrier spacing still needs to be part of the standard to ensure backward compatibility; furthermore, OFDM symbol durations should not be made too short in, e.g., macrocells, since the relative overhead for the *Cyclic Prefix* (CP) would become too large.

Time-Frequency Grid

Following from these requirements, NR foresees a number of different subcarrier spacings, namely 15, 30, 60, 120, and 240 kHz.[6] It is obvious that they are all related to each other by a factor of 2. As the subcarrier spacing increases, the OFDM symbol duration scales down (from 66.7 to 4.17 µs), and the CP duration scales down proportionately (i.e., stays a constant fraction of the symbol duration), from 4.7 to 0.29 µs.[7] Different subcarrier spacings are intended for different deployment cases: 15 kHz subcarrier spacing provides backward compatibility with LTE and narrowband IoT LTE; furthermore, the relatively long CP makes it suitable for deployments in macrocells. Larger subcarrier spacings are mainly intended for mm-wave deployment (where the phase noise and higher Doppler would lead to excessive inter-carrier interference with 15 kHz subcarrier spacing), and systems with small required latency, since larger subcarrier spacings imply shorter OFDM symbol duration. Since such large subcarrier spacing is associated with a shorter CP, it is only suitable for cells with small delay dispersion; fortunately, for mm-wave frequencies, the small coverage area and reduced delay spread due to beamforming often are associated with such small delay spread (note, however, that all other things being equal, higher carrier frequency does *not* generally lead to reduced delay spread, see Chapter 7). NR also defines a mode for 60 kHz subcarrier spacing with an extended CP, which allows operation with the higher subcarrier spacing also in environments with larger delay spreads. While the core specifications do not prescribe the use of a particular subcarrier spacing for a particular application or frequency band, the radio requirements specify overlapping sets of subcarrier spacings for the bands (15, 30, 60 kHz for FR1, and 60, 120 kHz for FR2). Figure 32.8 shows some typical use cases.

The different subcarrier spacings lead to different resource grids. As discussed below in more detail, different resource grids (also called *numerologies*) can be combined within a subframe. Finally, we define a basic timing unit $T_c = 1/(480000 \cdot 4096)$, which is 1/64 of the basic timing unit in LTE. Just like in LTE, this is not a required sampling frequency at the RX, but rather a book-keeping unit.

---

[6] 240 kHz subcarrier spacing is only supported for the synchronization signal and only in FR2. 60 kHz is only used for data.

[7] The CP for the first and eight OFDM symbols are slightly longer to retain backward compatibility with LTE.

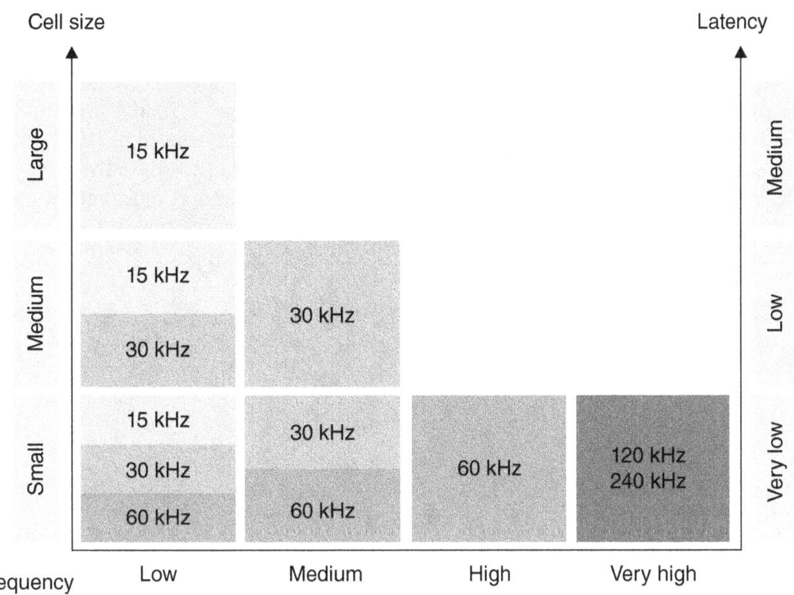

Figure 32.8 Typical use of different subcarrier spacings in a variety of application scenarios. Color version available at wiley.com/go/molisch/wireless3e. Reproduced with permission from [Keysight 2019] © Keysight.

Frames, Subframes, Slots, and Resource Blocks

Frames and subframes in NR have the same definition as in LTE (10 and 1 ms, respectively), so that their absolute durations are independent of the numerology. On the other hand, the *slot* has been redefined to be the time unit containing 14 OFDM symbols, independent of the numerology, which conversely means that the absolute slot duration (e.g., in units of ms) *does* depend on the numerology.[8]

It is also noteworthy that a transmission can occupy only part of a slot, often referred to as *mini-slot*, and need not start at the slot boundary (specifically, in the DL, the mini-slot duration can be 2, 4, or 7 OFDM symbols, while in the UL it can be 1–14. This is useful in multiple respects: (i) it helps to reduce the latency in the transmission, (ii) it allows to time-multiplex more users, which is especially useful in the mm-wave regime, where very large bandwidths are available, but analog beamforming constraints might preclude frequency multiplexing of multiple users, and (iii) when implementing listen-before-talk in unlicensed spectra (which is required by many frequency regulators), it allows to exploit spectral resources when they become free, and not run the risk that a user from a different system might grab them while the NR user is waiting for the next slot to start.

The *RB* has also been redefined in NR. It now consists of 12 subcarriers in the frequency domain, and the duration of 1 OFDM symbol in the time domain. This means that both absolute frequency- and time-extent of the RB depend on the subcarrier spacing. Furthermore, we see that even for the same subcarrier spacing (15 kHz), the RB is much shorter than in LTE. This again is motivated by the desire for enabling lower latencies.

Since the subcarrier spacing may vary, the numbering of the frequency resources is also more complicated. NR defines a *Reference Point A*, which might lie outside the band, as subcarrier 0 from which all RBs, independent of their subcarrier spacing, are counted. This reference point is communicated in the broadcast system information of the synchronization procedure (see Section 32.4.2). The *Common Resource Blocks (CRBs)* are located relative to the reference point, and their offset is communicated in units of their particular subcarrier spacing. Thus, the lowest subcarrier in CRB 0 is always at the frequency of the reference point. However, a frequency offset by 930 kHz from Point A might lie in CRB 0 for subcarrier spacing of 120 kHz, CRB 1 for 60 kHz, CRB 2 for 30 kHz, and CRB 5 for 15 kHz. *Physical Resource Blocks (PRBs)* on the other hand, are counted starting from the lower edge of the bandwidth part used in the transmission, though again the index of a subcarrier at a particular frequency also depends on the used subcarrier spacing.

Bandwidth Constraints and Carrier Aggregation

In NR, the carrier (center) frequency can be on a very fine raster, with resolution of 5 kHz at frequencies up to 3 GHz, 15 kHz for 3–24.5 GHz, and 60 kHz above that. Note, however, that the synchronization sequence is on a much coarser raster, see Section 32.4.2.

The upper limit on the usable bandwidth is 275 RBs (i.e., 3,300 subcarriers), which translates to different bandwidths for different subcarrier spacings. Inherently it implies that the bandwidth can never be larger than 400 MHz, using 120 kHz subcarrier spacing.[9]

NR also defines *bandwidth parts*, which are a set of consecutive RBs using a particular numerology. If, e.g., there were a "standard" bandwidth of 20 MHz, then larger bandwidths would require carrier aggregation; conversely, smaller bandwidths would be difficult to implement. With the bandwidth parts, the network can configure a suitable bandwidth, to (i) adjust the size of the active bandwidth to the capabilities of a UE without having to do carrier aggregation, (ii) enable devices with bandwidth less than the maximum bandwidth

---

[8] Confusingly, this means that an NR slot with 15 kHz subcarrier spacing (same spacing as LTE), lasts 1 ms, which is *twice* the duration of a "slot" in LTE. It is thus best to think of an NR slot to have no connection at all to an LTE slot.

[9] The smallest bandwidth that a device might have according to the RF requirements is 11 RBs; however, the SS block is 20 RBs wide, and it is mandatory for a device to find and receive it, so this truly defines the minimum bandwidth.

(and thus lower cost), (iii) allow RX bandwidth adaptation (e.g., RX adaptively monitoring narrow-band control channels and receiving on a narrow bandwidth, and changing to wider bandwidth when required). A connected device can be configured with up to 4 bandwidth parts in the UL, one of which is active at a time (and similarly in DL); for unpaired spectrum (TDD), the center frequencies of the active UL and DL bandwidth parts are the same. Since only one bandwidth part can be active at a time, the UE can only transmit or receive with one numerology; the currently active bandwidth part is selected via control signaling. Note that there is an important difference between assignment of RBs in a narrow band, and establishment of a bandwidth part: in the former case, the UE has to anticipate that transmission *could* occur anywhere in the channel bandwidth. In contrast, when a narrow bandwidth part is used, the UE can rely on all signaling being only in that narrow frequency band. The use of the narrow bandwidth part might be time-invariant (if, e.g., the UE has no capability for a wider bandwidth), or temporary (to reduce the energy consumption of the UE by requiring it only to monitor a narrower bandwidth).

Still, carrier aggregation is also allowed in NR, just like in LTE. It fulfills a similar functionality as the bandwidth parts, namely dynamic increase/decrease of the bandwidth, but also allows the simultaneous use of noncontiguous spectrum. It will be discussed in more detail in Section 32.6.2.

Duplexing Schemes

NR allows (like LTE) for FDD, TDD, and half-duplex FDD (transmission and reception separated in both time and frequency). However, unlike LTE, the frame structure is the same for all these duplexing schemes.

We next turn to the *slot format*. Depending on the duplexing mode and device capability, a slot might not be used completely for UL or DL. Therefore, NR defines a variety of slot formats, each of which denotes which OFDM symbols in a slot are UL, DL, or flexible. We can thus imagine, e.g., a slot that uses the first 8 symbols for the DL, and the subsequent 6 for the UL. However, we have to keep in mind that due to the timing advance (compare Section 18.3.2), the timing of UL and DL transmissions is different. Thus, e.g., the switch from OFDM symbol 8 to 9 occurs at a different absolute time instant in the DL than in the UL. Appropriate timing of the switching and use of guard intervals is therefore required. In case of strong adjacent-cell interference, the guard interval planning also has to take into account the runtime of the interference.

In NR, an important change from LTE is the introduction of dynamic TDD, where the slots, or parts of slots, can be dynamically assigned for UL or DL. This is not useful for macrocells, where UL/DL split should be coordinated between cells to avoid extreme interference situations.[10] However, it can be useful in small cells (in particular when noise- limited), where the load distribution between UL and DL can change quickly (since there are only few UEs in such a cell, no "averaging out" occurs). Furthermore, in small cells, the imbalance between the Transmitter (TX) powers of BS and UE is smaller; consequently, it is easier to handle situations where one cell might be in DL while the other one is in the UL, resulting in UE-to-UE and BS-to-BS interference. The slot format defines which symbols are used for UL and DL; furthermore, a symbol might also be announced as "flexible."

The slot format can be signaled in three different ways: (i) semi-static signaling, using the RRC, to all devices in a cell; (ii) dynamic signaling for a particular device, and (iii) dynamic signaling for a group of devices. For signaling on the RRC, two slot formats can be used: one is cell-specific, and one is particular for a UE. The UE-specific pattern might constrain the possible use of symbols further – in other words, only symbols that are marked "flexible" in both the cell-specific and UE-specific pattern will be considered as flexible. The RRC can configure a table that constitutes a set of UL/DL/flexible assignments. The BS can then transmit dynamically to a group of devices a *Slot Format Indicator (SFI)*, which is an index for a row of the table and thus shows the configuration of the device. The SFI cannot override semistatically configured UL or DL assignment, but it *can* override a flexible assignment. This is useful to override, e.g., periodic SRS transmissions. Furthermore, the SFI can be used to reserve time-frequency resources for "forward compatibility."

Mapping to Physical Resources

Once the time-frequency resources have been assigned, the modulation symbols need to be mapped onto those resources. For the DL. the transmission to a particular UE is assigned a set of RBs and the modulation symbols are written "frequency first" into those RBs (this reduces latency compared to a "time first" approach). Just like in LTE, this is done in a two-step procedure, first mapping to *Virtual Resource Blocks (VRB)* and from there to *PRBs* though in NR there is a third mapping, from PRBs to *carrier resource blocks* (which counts the RBs from the lower edge of the carrier, not the assigned bandwidth part, though of course the actually used resources still must lie in within the assigned bandwidth part).

For the DL, the mapping from VRB to PRB foresees two possibilities: noninterleaved and interleaved. In the former case, there is simply a direct (1:1) mapping; note that an arbitrary combination of RBs can be assigned to a user, depending on the signaling for the allocation scheme. For the interleaved case, a block interleaver with two rows is used, into which pairs (or quadruplets) of RBs are written column-first, and read out row-first.[11] Interleaving is of obvious importance to gain frequency diversity when the codeblock size is small. However, in contrast to LTE, it is important to gain full frequency diversity even when the codeblock size is large: due to the limit on the size of the codeblock and the (in some cases) large bandwidth, it is possible that the number of VRBs corresponding to one codeblock is only a small fraction of the allocated bandwidth.

For the UL, in contrast to LTE, both direct and interleaved mapping is allowed, and the RBs do not have to be contiguous. Frequency diversity can be obtained through frequency hopping, where the offset between the resources used in different OFDM symbols is configurable, and indicated in the *downlink control information* (see Section 32.4.7); this is similar to LTE.

---

[10] This can still be implemented in NR (after all, if dynamic changes are possible, restricting the change rate is easy).

[11] Note the somewhat simpler structure of the interleaved mapping compared to LTE, where the mapping was different for the first 7, and the second 7, OFDM symbols in a subframe.

Another new aspect of NR is the reserving of resources to enable forward compatibility, in other words, they are "blocked out" from the use by conventional NR, and can thus be used by a future system that might use, e.g., different modulation. To enable such reservation, it has to be ensured that no signals (including control signals and pilots) are transmitted in a specific time/frequency block. This can be achieved by (i) operating in LTE/NR coexistence mode, and referring to an LTE carrier configuration to avoid Cell-specific Reference Signals (CRSs). (ii) referring to a configured CORESET, indicating dynamically whether the configured resources may be used for the DL traffic or not (see section titled "Transmission and Reception of PDCCH" in Section 32.4.7). (iii) use of a bitmap to indicate reserved resources. The bitmap indication operates on three layers: (a) bitmap-3 has a length of 40 slots, and a granularity that equals the length of bitmap-2 (see below), indicating for each granule whether the resource set defined by bitmap-2 and bitmap-1 is valid. (b) bitmap-2 indicates a set of OFDM symbols within a slot, or a pair of slots (this is the granularity of bitmap-3). (c) bitmap-1, which indicates the RBs within one OFDM symbol. Implicitly, the same RBs are reserved in different OFDM symbols and the same OFDM symbols in different (pairs of) slots. The combination of bitmap-1 and bitmap-2 only configures a set of resources that are *potentially* reserved. Whether they are actually reserved is determined by bitmap-3.

32.3.5 Pilots or Reference Signals

NR has dramatically changed the structure of pilot signals (*Reference Signals (RS)*), due to the need to (i) minimize "always on" transmissions, and (ii) handle massive MIMO. Thus, the cell-specific RSs transmitted from each BS antenna element for the DL are eliminated in NR; instead, precoded pilots are now sent. Specifically, NR now contains the following RSs:

- *DeModulation Reference Signal (DM-RS)*: those are precoded with the same precoder settings as the payload channels (PDSCH and PUSCH, respectively, see Section 32.4) and are only transmitted in the RBs in which PDSCH/PUSCH are being transmitted. Note that there are also DM-RSs for the Physical Downlink Control CHannel (PDCCH) and Physical Broadcast CHannel (PBCH).
- *Phase Tracking Reference Signal (PT-RS)*: an extension/modification of DM-RS that is denser in time and sparser in frequency. They are used to help combat phase noise.
- *Sounding Reference Signal (SRS)*: an UL RS, used for channel estimation over the whole bandwidth. Similar to the SRS in LTE.
- *Channel State Information – Reference Signal (CSI-RS)*: the DL equivalent of the SRS. Since there is no cell-specific RS anymore that would allow the UE to estimate the (nonprecoded) DL channel, this RS has been expanded compared to LTE.
- *Tracking Reference Signal (TRS)*: a special CSI-RS that helps frequency tracking.
- *Positioning Reference Signal (PRS)*: serving for localization purposes. While the philosophy is similar to LTE, the signals themselves are different and allow more precise localization.

While in LTE the CRS density was fixed,[12] in NR the density of *all* RSs is adaptive, so that the temporal repetition frequency of the pilots is not a fixed number determined by the worst-case (e.g., high-speed train).

DeModulation Reference Signal

A DM-RS is assigned to a particular *antenna port*; up to 12 orthogonal antenna ports can be configured. A DM-RS can support transmissions over a duration of 2 to 14 OFDM symbols and can be transmitted at up to 4 instances within that duration.

The exact location of the transmission instances within a slot is quite flexible but falls generally into one of two categories: (i) *Mapping Type A*: the first DM-RS is in the second or third OFDM symbol of a slot (i.e., after the CORESET, see section titled

Table 32.6 Location of DM-RS (in units of OFDM symbols away from slot boundary) as function of duration of the transmission (rows) and number of DM-RS elements (columns); for mapping-type A with single symbols in the PUSCH. Index of first symbol is symbol in a slot is 1 here.

Duration\no of symbols	1	2	3	4
4–7	3	–	–	–
8–9	3	3/8	–	–
10–11	3	3/10	3/7/10	–
12	3	3/10	3/7/10	3/6/9/12
13–14	3	3/12	3/8/12	3/6/9/12

Table 32.7 Location of DM-RS (in units of OFDM symbols away from slot boundary) as function of duration of the transmission (rows) and number of DM-RS elements (columns); for mapping-type A with double symbols in the PUSCH

Duration\no of symbols	2	4
4–9	3/4	–
10–12	3/4	3/4/9/10
13–14	3/4	3/4/11/12

---

[12] While, e.g., SRS density was flexible.

Table 32.8 Location of DM-RS (in units of OFDM symbols away from transmission start) as function of duration of the transmission (rows) and number of DM-RS elements (columns); for mapping-type B with single symbols in the PUSCH

Duration\no of symbols	1	2	3	4
2–4	1	–		
5–7	1	1/5	–	–
8–9	1	1/7	1/4/7	–
10–11	1	1/9	1/5/9	1/4/7/10
12–14	1	1/11	1/6/11	1/4/7/10

Table 32.9 Location of DM-RS (in units of OFDM symbols away from transmission start) as function of duration of the transmission (rows) and number of DM-RS elements (columns); for mapping-type B with double symbols in the PUSCH

Duration\no of symbols	2	4
5–7	1/2	–
8–9	1/2	1/2/6/7
10–11	1/2	1/2/8/9
12–14	1/2	1/2/10/11

"Transmission and Reception of PDCCH" in Section 32.4.7); (ii) *Mapping Type B*: the first DM-RS is in the first OFDM symbol of the data allocation for a transmission, independent of where that is compared to the slot boundary. This is obviously meaningful if transmission starts in the middle of a slot. The exact placement for the DL can be configured and signaled in the Downlink Control Information (DCI); for the detailed rules see [3GPP 38.211, Tables 7.4.1.1.2-1 to 7.4.1.1.2-5]. For the UL, the placement is described in [3GPP 38.211, Tables 6.4.1.1.3-1 to 6.4.1.1.3-6]. An example for possible locations with mapping-type A when the first DM-RS is located on the third OFDM symbol is given in Tables 32.6–32.7; examples for mapping type B and first DM-RS on the first OFDM symbol is given in Tables 32.8–32.9.

It can be seen that in either case the DM-RS is located early in the transmission, which helps to reduce latency, as the decoding can only start once a channel estimate is available. Furthermore, contrary to LTE, the density of the RSs in the time domain is configurable, so that anywhere between 1 and 4 DM-RSs are transmitted in one slot. Multiple DM-RSs can either be spaced apart (this is the best solution in quickly time-varying channels) or are transmitted as "double-symbols," which enables to handle more antenna ports than the single symbols. This flexibility helps to greatly improve the spectral efficiency of the pilot signal transmission.

Two different configuration types (types 1 and 2)[13] of DM-RSs are defined, which allow to provide 4 and 6 orthogonal RSs for the single-symbol transmission (and twice that for the double-symbol). Figures 32.9 and 32.10 show the structures. We see that sets of 2 or 4 RSs are transmitted on the same Resource Elements (REs); they are distinguished by *Orthogonal Cover Codes (OCC)* in frequency and/or time domain (i.e., REs that are adjacent in time and/or frequency are multiplied with elements of a Walsh-Hadamard matrix, which is used as OCC); orthogonality is maintained as long as the channels in the REs used for the OCC transmission is identical. This explains how the double symbols can handle twice as many antenna ports as the single-symbol transmissions. Note that both types of DM-RS occupy the same amount of time-frequency resources, but type 1 is denser in frequency (and thus better suited for more frequency-selective channels), while type 2 enables more antenna ports. The actual type that is used is conveyed to the UE in the control signaling.

The symbols for the DM-RS are generated by a $2^{31}-1$ length Gold sequence, where the sequence is generated across all CRBs in the frequency domain, but of course, only transmitted in the RBs that are sent out; this ensures that DM-RSs used from different antenna ports (and distinguished by OCCs) employ the same coding symbols, since only then is orthogonality retained. The Gold sequence itself can, however, be configured, typically based on the cell ID.

The above structure is used for DM-RS for both DL and UL. However, for the UL it only applies when the UL uses OFDM modulation. For DFT-spread OFDM, the DM-RS has a different structure: whole OFDM symbols, filled with Zadoff–Chu sequences, are used as RS (note that DFT-spread OFDM is not intended to use MIMO); this is similar to LTE. The DM-RS then uses the time structure of configuration type A. Since Release 16, it is also possible to instead use pseudorandom sequences with $\pi/2$-shifted BPSK instead of the Zadoff–Chu sequences.

Phase Tracking Reference Signal

For the OFDM case, the PT-RS is similar to the DM-RS (and only sent within the bandwidth and duration of the PDSCH/PUSCH) but denser in the time domain. It is transmitted in the first OFDM symbol, and then repeats every Lth OFDM symbol, where L is configurable. The counter is reset every time a DM-RS occurs, since such a DM-RS can be used for phase tracking as well, and thus no PT-RS is needed in the vicinity.

In the frequency domain, the transmission is sparse: firstly, the PT-RS is transmitted only on one subcarrier in each RB (that subcarrier varies from device to device, to reduce collisions), and secondly, it is only transmitted in every second or fourth RB.

---

[13] Not to be confused with the Mapping Type A and B.

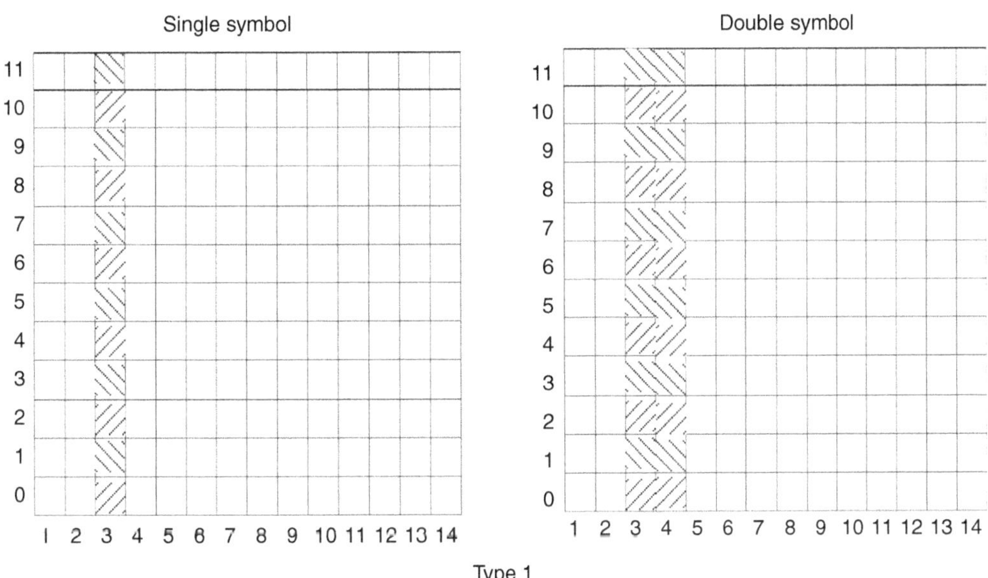

Figure 32.9 DM-RS Type 1 structure assuming mapping-type A, Single-symbol with 1 DM-RS in a 14-RB slot. REs belonging to the same DM-RS show the same shading. For the single-symbol case, groups of 2 REs belonging to the same DM-RS (i.e., same shading) carry one Code Division Multiplexing (CDM) group (2 elements), distinguished by orthogonal cover codes of length 2 (1/1 and 1/−1). For double-symbol case, groups of 4 REs (two in the time, 2 in frequency domain) carry a CDM group with 4 elements.

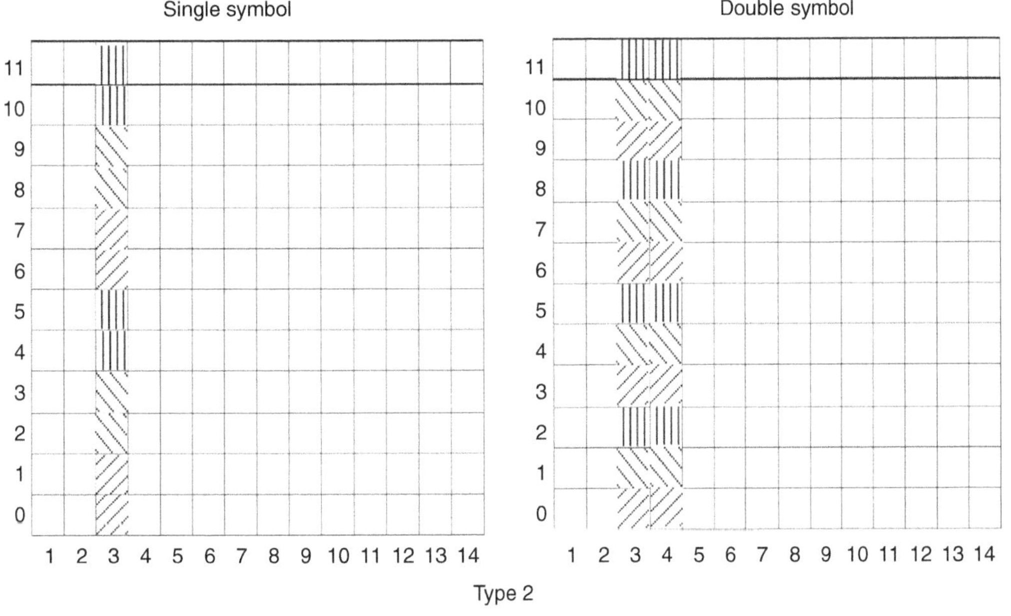

Figure 32.10 DM-RS Type 2 structure assuming mapping-type A, Single-symbol with 1 DM RS in a 14-RB slot. REs belonging to the same DM-RS show the same shading. For the single-symbol case, groups of 2 REs carry one CDM group (2 elements), distinguished by orthogonal cover codes of length 2 (1/1 and 1/−1). For double-symbol case, groups of 4 REs (two in the time, 2 in frequency domain) carry a CDM group with 4 elements.

For the DFT-precoded case in the UL, the PT-RS symbols are inserted before the DFT-precoding, i.e., in the time domain.

The motivation for this PT-RS comes from tracking the phase noise, in particular in mm-wave systems. Since the phase noise is the same for all subcarriers but can have fast time variations (phase noise spectrum extending to tens or hundreds of kHz), a dense pattern in the time domain (and low density in the frequency domain) is essential.

Sounding Reference Signal

The SRS plays a fairly similar role as in LTE, namely to measure the channel over the whole considered bandwidth. These measurements might be useful for two purposes: (i) let the BS determine on which time/frequency/space resources to schedule the UL transmission of the UE, and (ii) use it for reciprocity-based estimation of the DL channel. The RS can span 1, 2, or 4 OFDM symbols, which are located within the last 6 symbols of a slot, and are transmitted on either every second or every fourth subcarrier. Due to the sparsity in the frequency domain, 2 or 4 SRSs can be frequency-multiplexed. The actual sounding signals are – similar to LTE – extended Zadoff–Chu sequences (for the case that the sequence length is ≥ 36; for shorter sequences, sequences found from a computer search and specified in the standard are to be used). An SRS is sent from an "antenna port." Since the BS measures thus an effective channel consisting of the precoder of the UE concatenated with the propagation channel, the subsequent transmissions of user data should be received by the UE with the same antenna weights (compare also Section 32.3.6).

Since a UE can have multiple antenna ports, multi-port SRSs need to be defined. SRSs sent from different ports of the same UE are distinguished by cyclic shifts (compare the discussion of Zadoff–Chu sequences in Section 31.3.5). A UE can be configured with one or more *SRS resource sets*, where each resource set contains one or more SRSs, which can be used for different purposes such as UL and DL precoding and beam management (see Section 32.3.6).

An SRS resource set can be configured to be transmitted (i) periodic, (ii) semi-persistent (where the location of the SRS is configured like in the periodic case, but the actual SRS transmission is activated or deactivated dynamically by MAC signaling, and (iii) aperiodic, where the SRS transmission is triggered by the *SRS request* in the *DCI,* see Section 32.4.7.

NR also introduces a modified SRS that can be used for localization based on UL transmission; the structure of this signal is similar to the PRS (see below), i.e., has a frequency comb structure whose starting position in the frequency domain changes from OFDM symbol to OFDM symbol. Not only time of arrival but also angle of arrival and signal power can be measured at the BS and used for improving the UL localization.

CSI-RS

A major change (compared to LTE) in pilot signals occurs for the DL. Since the cell-specific RSs are abolished, determination of the DL channel for establishing a precoder, measuring interference, etc., requires the use of a different sounding signal called CSI-RS,[14] which is transmitted from a particular antenna port, and thus might include precoding. A CSI-RS is configured in principle on a per-device basis, though it is allowed that multiple devices make use of the same CSI-RS.

A single-port CSI-RS[15] consists of a single RE within a time-frequency area that extends 12 subcarriers (1 RB) in the frequency domain and 14 OFDM symbols (1 slot) in the time domain. It can be anywhere within that area as long as it does not collide with a CORESET or DM-RSs configured for this UE, or an SS block.

A multi-port (up to 32 ports are possible) CSI-RS consists of a set of REs, and the CSI-RSs for the different ports are distinguished by a combination of time, frequency, and code (OCC); as a matter of fact, multiple combinations are possible to create CSI-RSs for the same number of ports. In any case, consecutive port numbers are associated to the CSI-RSs first in the code, then the frequency, and then in the time domain (e.g., port 0 and 1 are on the same REs, distinguished by OCCs, ports 2 and 3 would be in the same OFDM symbol but on different subcarriers, and ports 4–7 in a different OFDM symbol).

In the frequency domain, the CSI-RS extends over the whole, or part of, the bandwidth part in which it is configured. In either case, it uses the numerology of that bandwidth part. It can occur either in every RB, or every second RB. In the time domain, the CSI-RS[16] can be configured to be transmitted in the following configurations:

- *Periodic*: here the CSI-RS is transmitted in every Nth slot, where N can range from 4 to 640. Both the periodicity and the slot offset are configured.
- *Semi-persistent*: also in this case, periodicity and slot offset are configured, but the CSI-RS transmission can be indicated as activated or deactivated by MAC control elements.
- *Aperiodic*: every transmission of the CSI-RS is conveyed by signaling in the DCI.

CSI-RSs for different devices can be on the same, or on different resources. In other words, a particular CSI-RS can be configured for multiple UEs; but it is also possible that the CSI-RSs for different UEs lie on different REs. In the latter case, it is important to let a UE know that it should not interpret the REs containing CSI-RS for other UEs as data. This is done by the *zero-power CSI-RS* (ZP-CSI-RS). Configuration of a ZP-CSI-RS lets the UE know which REs do not contain the PDSCH (though it cannot assume that the RE is unoccupied; other information, e.g., the CSI-RSs for other devices, can be sent on it).

Another application of ZP-CSI-RS lies in the measurement of interference. The BS might decide not to send *any* signal on a particular RE, so that the received power measured on this RE can be interpreted as the interference power; this is then called Channel State Information – Interference Measurement (CSI-IM). Similar to the "normal" CSI-RS (also known as NZP-CSI-RS), the CSI-IM can be configured periodic, semi-persistent, or static.

The UE needs to report the measurement results back to the BS. The method of reporting is also imposed by the BS: in the *CSI-report-config*, it mandates: (i) the set of quantities to be reported (similar to LTE: *Rank Index (RI), Precoder Matrix Indicator (PMI),* and *Channel Quality Indication (CQI)*), (ii) the DL resources on which those quantities should be measured (CSI-RS resource set), and (iii) when/on what UL resources/how the report should be sent. Again, this can be periodic, semi-persistent, or aperiodic, similar in spirit to the discussion above; however, note that reports for periodic CSI-RSs can be sent periodically, semi-persistently, or aperiodically and similarly reports for semi-persistent CRS can be sent not only semi-persistently but also aperiodically.

---

[14] CSI-RS in some basic form existed also in LTE, namely for SU-MIMO with more than 4 layers
[15] Strictly speaking, the non-zero-power CSI-RS, NZP-CSI-RS. The reason for this name will become clear below when we discuss the zero-power CSI-RS.
[16] Strictly speaking, what is configured is a CSI-RS resource set that can contain one or more CSI-RS.

Tracking Reference Signal

The TRS can be seen as a special set of multiple periodic CSI-RS that are intended for oscillator tracking, in particular at mm-wave frequencies. It contains four one-port CSI-RSs located within 2 adjacent slots, and repeated with a periodicity of $T_{\text{rep}} = 10, 20, 40$, or 80 ms, see Figure 32.11. The exact location within the RB is variable, but there is a fixed separation of 4 OFDM symbols between the symbols in a particular slot.

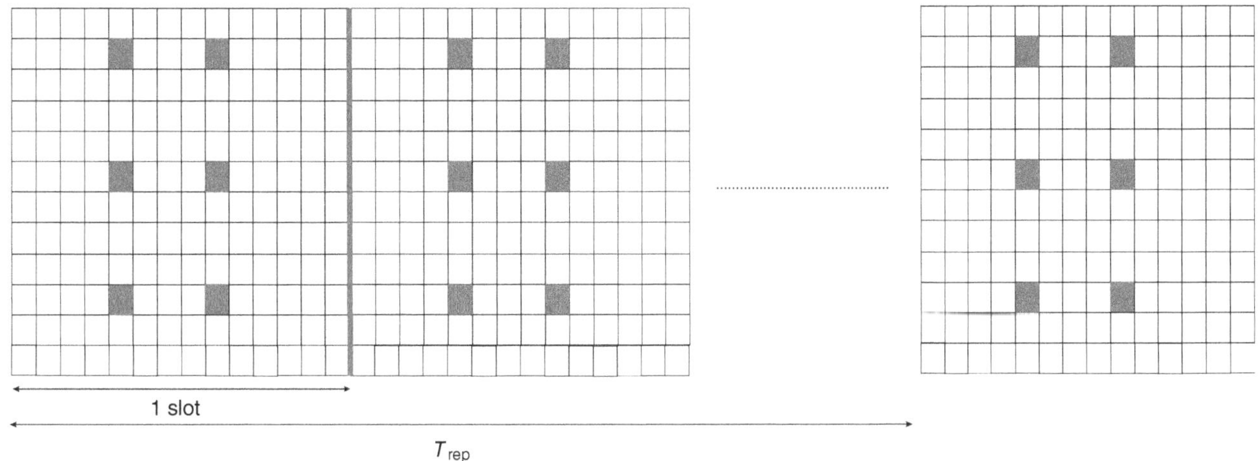

Figure 32.11 Tracking RS structure.

Positioning Reference Signal

Just like in LTE, there are also separate pilot signals for positioning in the DL, called *Positioning Reference Signal (PRS)*. Time of arrival from multiple such signals is measured and reported to the infrastructure. Each BS can be assigned multiple PRSs, each of which corresponds to a beam from that site (which also provides directional information that can be used to refine the position estimate, see Chapter 29). During one OFDM symbol, the PRS is a frequency comb, where every nth subcarrier (where n can be configured to be 2, 4, 6, or 12) is used, while the offset of the comb from the edge of the RBs differs from OFDM symbol to OFDM symbol, using a permutation of a simple $1, 2, 3,....,n-1$ sequence. The total bandwidth of the PRS can be up to 272 RB, and the temporal duration is 2, 4, 6, or 12 OFDM symbols. On these resources, a QPSK-modulated m-sequence is transmitted. Furthermore, multiple PRSs can be set up, increasing the possible accuracy even further.

The PRS is transmitted either on demand or repeated periodically with a configurable repetition period that can range from a few ms to 10 s. The frequency comb structure allows the simultaneous transmission of multiple PRSs. Furthermore, just like in LTE, BSs in the surroundings can be muted (not transmit on the resources used by the PRS) to improve the SINR of the received signal; this is especially beneficial when signals from far-away BSs need to be received by the UE.

As mentioned in the discussion of SRS, there is also the possibility of UL-based localization, where a frequency comb signal sent out by the UE is used as the basis of the localization.

32.3.6 Multiple-Antenna Techniques

Multi-antenna technology is an essential part of NR. Before going into further details, we first define some nomenclature (this is actually similar to LTE, see also Section 31.3.6). An *antenna port* at the TX is not a physical antenna connector, but rather an input to a precoder. In NR, the antenna ports are also numbered depending on which channels (e.g., SRS, PUCCH, etc.) are sent over them. Another extension has occurred in the definition of *Quasi-Co-Located (QCL)*. While the definition from LTE, namely that two QCL antenna ports will "see" the same *large-scale* channel characteristics, has been retained, NR now defines a sub-categorization depending on which large-scale characteristics are meant (e.g., Doppler, delay spread, spatial characteristics, etc.).

Precoding – General Considerations

In NR, multi-antenna precoding can occur both at the BS and at the UE. The mechanisms for DL and UL are, however, somewhat different, due to the following reasons: (i) the network should be able to control all transmissions from the UE, (ii) array configurations at the UE are less predictable, and (iii) UEs might have less capabilities to enforce phase coherence between different antenna elements. A further distinction (both for DL and UL transmission) is between codebook-based and non-codebook-based transmissions. In the former case, feedback from the RX is used as the input for the TX; for the latter case, the standard does not explicitly state the basis on which the precoding is done, but in practice, it is often based on channel reciprocity (see Section 16.1.7).

A further complication is that NR splits what we generally in this book call "precoding" into a multi-antenna precoding and a spatial filtering. DM-RS and payload data undergo both of these steps, thus ensuring that this process is essentially transparent to the UE, i.e., for the purpose of demodulation, the UE neither knows nor cares what precoding is done at the BS; it estimates the "effective" channel, i.e., the concatenation of precoder and propagation channel. However, the CSI-RS signals undergo only the spatial filtering, and the CSI reports the UE sends back to the BS are based on this particular part of the mapping. As a consequence, the precoder specifications in the standard relate to the measurement and feedback of information from the UE to the BS, while the BS can ultimately design any precoder that it wants (it does not have to be contained in the codebook).

Due to the possibility that the precoder, and thus the effective channel, can change between different scheduling occasions, the UE cannot make an assumption about the temporal correlation between the RSs at these occasions. However, in the frequency domain, the UE may assume that the precoder is constant over a particular set of RBs, given by the *Precoding Resource-block Group (PRG)*. The choice of the size of the PRG trades off precoder flexibility (e.g., to create nulls into the directions of different UEs, which might be scheduled on different RBs) with channel estimation performance (which benefits from noise averaging over a larger number of subcarriers, see Section 15.5).

Also for the UL, the precoding is transparent to the RX (which in this case is the BS), though the SRS only goes through the spatial filter and not the precoder matrix. However, there are also some fundamental differences from the DL that arise from the fact that the UL is controlled by the network; this will be discussed below. Finally, it is not assured that a UE can create phase coherence between different antennas even if it has all the required CSI – these limits could arise from the low-quality hardware of a UE; the standard thus foresees the categories of *full coherence*, *partial coherence* (coherence only within pairs of antennas – typically the two polarizations on one physical antenna – but not from pair to pair), and *no coherence*. Depending on these categories, only certain subsets of the codebook entries can be used meaningfully. Finally, the UL precoder has to distinguish between OFDM-based and DFT-spread-OFDM-based transmission. In the former case, only one layer (spatial stream) can be transmitted, while in the latter, up to 4 are used.

CSI-Feedback-Based Precoding in the DL

Similar to LTE, NR provides beamforming that is based on feedback from the UE.[17] The feedback from the UE consists not only of the PMI, but also the RI (number of layers for this UE) and the CQI. As discussed in Sections 16.2.8 and 22.7, SU-MIMO does not require as fine a spatial resolution of the codebook as MU-MIMO. To accommodate both of those applications, NR foresees two different types of codebooks, *type-1 CSI* and *type-2 CSI*.

Within type-1 CSI, there is a further distinction for CSI that is intended for an array structure that is placed on a single panel, and one that is on multiple panels. For a single panel, we can assume that all antennas are of the same type, are placed close to each other at regular intervals (with N_1 and N_2 antenna elements along horizontal and vertical axis, respectively), and are dual-polarized. The precoder matrix $\mathbf{W}$ can be written as a product of a matrix $\mathbf{W}_1$ that depends only on the second-order statistics of the channel (average *Angular Power Spectrum (APS)*), and a matrix $\mathbf{W}_2$ that depends on the instantaneous channel state. $\mathbf{W}_1$ changes relatively slowly; plus, since it depends only on the second-order statistics, it is the same for all subcarriers. Furthermore, since the average ADPS is the same for both polarizations,

$$\mathbf{W}_1 = \begin{bmatrix} \mathbf{B} & \mathbf{0} \\ \mathbf{0} & \mathbf{B} \end{bmatrix}.$$

If the transmission rank is 1 or 2, then $\mathbf{W}_1$ either defines a single beam or four neighboring beams; in the latter case the matrix $\mathbf{W}_2$ (see below) allows to select among those beams. For a larger rank, $\mathbf{W}_1$ describes a set of orthogonal beams whose dimension is half the transmission rank, and $\mathbf{W}_2$ provides the co-phasing between the different polarizations. The matrix $\mathbf{W}_2$ characterizes the instantaneous channel state and thus can change more quickly, and different values can be optimum for different subbands, and different polarizations. Thus, there are no particular restrictions on the structure of the matrix.

Another situation arises when the BS array consists of multiple panels; such a structure could arise, e.g., if the panels are oriented in different directions, or are widely separated from each other. In that case, $\mathbf{W}_1$ defines one beam per panel (and polarization), and $\mathbf{W}_2$ provides the co-phasing between both panels and polarizations – since different panels might be fed with upconverters using different local oscillators, inherent phase coherence between the signals at antenna elements on different panels cannot be assumed.

Type-2 CSI is based on similar principle as type-1 but reports CSI from up to four orthogonal beams. For each beam and each polarization, amplitude and phase are reported (essentially, capturing the parameters of the main MPCs). This reporting uses a combination of wideband and per-subband-based feedback. While release 15 assumed independent small-scale fading for the reported coefficients, release 16 enables to exploit correlation between the subbands. It defines frequency-domain units (which can be either a subband or half a subband), for which the reporting is done. It furthermore introduces a compression of the feedback information. This is mostly achieved by transforming from the frequency domain into the delay domain, exploiting possible (approximate) channel sparsity in the delay domain, and thus reducing the overhead by a configurable parameter p. Furthermore, a matrix that relates the delay domain to the beam domain (i.e., which beams are relevant for which delay bin) can also be viewed as approximately sparse, so that only a subset of its coefficients is reported, where the sparsity factor β can be configured. The parameter p lies between 1/8 and 1/2, the parameter β between 1/4 and 3/4.

---

[17] Though, note that – again as in LTE – the UE provides a recommendation for the beamformer, but the BS can choose to follow that recommendation or not.

Codebook-Based Precoding in the UL

A key aspect to keep in mind for the UL precoding is that it is the BS that selects transmission rank and precoding matrix for the UE (and the UE is mandated to use these settings, as opposed to the DL precoding where the BS can follow the recommendations of the UE or not). To provide the BS with the necessary CSI, the UE has to send out one or more multi-port SRSs (see Section 32.3.5). The UE might have 2 or 4 antenna ports, and the BS selects entries from a codebook whose size corresponds to the number of ports. While LTE allows only such codebooks that one layer is applied to one antenna element, the codebooks in NR are somewhat more general. For 2 ports and transmission rank 1, the precoding vectors can be

$$
\begin{bmatrix} 1 \\ 0 \end{bmatrix} \quad \begin{bmatrix} 0 \\ 1 \end{bmatrix} \quad \begin{bmatrix} 1 \\ 1 \end{bmatrix} \quad \begin{bmatrix} 1 \\ -1 \end{bmatrix} \quad \begin{bmatrix} 1 \\ j \end{bmatrix} \quad \begin{bmatrix} 1 \\ -j \end{bmatrix}.
$$

In a UE, the signals at the different antenna elements might be phase-coherent or not (the latter case occurs if they are using different LOs, or even different PLLs based on the same low-frequency oscillator, see Chapter 17). For the case of noncoherence, only the first two vectors (i.e., antenna selection) may be used. For rank 2, the codebook is

$$
\begin{bmatrix} 1 & 0 \\ 0 & 1 \end{bmatrix} \quad \begin{bmatrix} 1 & 1 \\ 1 & -1 \end{bmatrix} \quad \begin{bmatrix} 1 & 1 \\ j & -j \end{bmatrix}
$$

where for the noncoherent case, only the first setting is allowed (one layer from each antenna port). For the 4-port case, and rank-1 transmission, the noncoherent precoding vectors are simply a row vector with a single "1" (and the rest is "0"). The precoding vectors for partially coherent transmission are

$$
\begin{bmatrix} 1 \\ 0 \\ 1 \\ 0 \end{bmatrix} \quad \begin{bmatrix} 1 \\ 0 \\ -1 \\ 0 \end{bmatrix} \quad \begin{bmatrix} 1 \\ 0 \\ j \\ 0 \end{bmatrix} \quad \begin{bmatrix} 1 \\ 0 \\ -j \\ 0 \end{bmatrix} \quad \begin{bmatrix} 0 \\ 1 \\ 0 \\ 1 \end{bmatrix} \quad \begin{bmatrix} 0 \\ 1 \\ 0 \\ -1 \end{bmatrix} \quad \begin{bmatrix} 0 \\ 1 \\ 0 \\ j \end{bmatrix} \quad \begin{bmatrix} 0 \\ 1 \\ 0 \\ -j \end{bmatrix}
$$

which shows how coherence between two different ports can be achieved, but only selection between port combinations $1 + 3$ and $2 + 4$ is possible. This would typically occur if the same LO is used to receive the signals from the same antenna, but different polarization ports; while different LOs are used for different antennas. For full coherence, the entry for port 1 is always $+1$, while all other ports run through all combinations of ± 1 and $\pm j$.

As a change from LTE, a UE may send out multiple multi-port SRSs, e.g., corresponding to different antenna panels pointing into different directions; the BS will then send an SRS resource indicator, which is an additional control bit which says which of the SRSs (i.e., which panel) should be used, and the PMI should be interpreted in the context of this panel.

Non-Codebook-Based Precoding in the UL

The UE can also design a precoder based on DL channel estimates, if reciprocity applies. There are no particular constraints on which values the precoder matrix might take, as there is no quantized codebook. However, the BS has to retain the right to modify the precoder matrix, since all transmissions have to be controlled by the BS. This is achieved by first having the UE transmit SRSs, one for each of the beams defined by the UE precoder. The BS can then communicate in the scheduling grant (see below) a subset of beams that the UE might use.

32.3.7   Beam Management

The discussion above did not yet say anything about how the second-order statistics are acquired, and the beams used for the transmission are chosen.[18] The beam management procedures of NR deal with exactly this problem. Remember that the second-order statistics are independent of frequency (at least within the bandwidth of a carrier), and change relatively slowly; and that the beam direction is the same for transmission and reception even in FDD. The main steps of beam management are (i) beam establishment, (ii) beam adjustment, and (iii) beam recovery. All these procedures are especially important at mm-wave frequencies, where high-gain beams are essential for establishing and maintaining connectivity.

Beam Establishment

When a connection is established, the best (or at least good-enough) beam directions at BS and UE have to be found. This is usually done in a way that the BS sweeps its transmit beams as it transmits multiple copies of the *Synchronization Signal (SS)* block, whose structure will be explained in more detail in Section 32.4.2. The UE sends a random-access signal when it receives the SS

---

[18] While in general "beam" can refer to an antenna pattern based on instantaneous channel states, in this subsection we refer to beams as related to the second-order statistics. Note, however, that the NR standard does not specify explicitly on what basis the beams are formed.

on a strong-enough beam. By associating each beam with a different random-access opportunity, the initial connection setup also implicitly establishes which is the correct beam from a BS point of view. Since the UE also has beamforming capability, it also needs to try out different beam directions; obviously, the set of UE beam directions has to be tried out for each BS beam direction. The UE beam direction that allows best reception of the SS, and which is then used for the random access signal transmission, is then also used for the subsequent data transmission.

Beam Adjustment

While the connection remains active, the beam directions need to be continuously adjusted, both at the BS and the UE. A change of the beam direction can either be gradual, e.g., when the UE moves and therefore "sees" the BS (or dominant reflectors) under continuously changing angles. Or it can be fairly abrupt, e.g., when a large object such as a truck interrupts the line-of-sight, and a reflected MPC suddenly becomes the strongest MPC (the beam should be directed toward that new strongest direction).

The beam adjustment in the DL occurs similar to the beam establishment, i.e., the BS sweeps through different beam directions, though now pilot signals can be used instead of *SS*s. For the adjustment of the beam direction at the BS, the UE beam direction remains fixed. The UE reports the channel quality, in particular, the *Layer-1 Reference Signal Received Power (L1-RSRP)*. The set of RSs on which the measurements are done is included in the NZP-CSI-RS resource set; during one reporting instance, the results of up to 4 RSs can be reported. For the adjustment of the beam direction at the UE, the beam direction at the BS has to stay fixed (a repetition flag indicates whether the RSs sent within a resource set use the same BS beam). In contrast to the BS, no reporting is needed, since the adjustment is happening in the UE that is doing the measurement.

For the UL, generally, no special beam adjustment is necessary, since the beam directions are reciprocal, as discussed above. If an explicit UL beam adjustment is to be made, it can be performed along similar principles as the DL adjustment, just that the RS to be used as the basis is the SRS.

There is also the possibility for the BS to explicitly inform the UE about which beam it is using; this is useful in particular in the context where the optimal beam direction might "jump." The NR standard establishes *Transmission Configuration Indication (TCI)* states, which include information about the RSs. When a PDSCH (or PDCCH) is associated with a particular TCI, then this implies that it uses the same precoding as the RS associated with the TCI. Thus, let a UE be configured with "candidate" TCI states (there can be up to 64 of them); then the UE knows which RX beam it should use for each of the TCI states. The MAC signaling indicates dynamically which TCI state is used. Consequently, the UE can dynamically adjust its beam direction without having to do a fresh search.

Beam Recovery

When the beam direction changes too quickly, a beam failure might occur from which the system needs to recover. The recovery operation proceeds in the following steps:

- *Beam failure detection*: this is based on measurements of signal strength of the CSI-RS, usually the one that has the same beam (is QCL) with the PDCCH. If the signal strength in too many such measurements falls below a threshold, a beam failure is declared.
- *Candidate beam identification*: the process of identifying a new beam pair with which communication can be restored is fairly similar to the beam establishment procedure.
- *Recovery-request transmission and response*: this is a transmission to the BS that notifies it that a beam failure has occurred, and might also inform it about the newly identified beam. The beam recovery request is transmitted as a contention-free random access request (see Section 32.4.9), where of course the random access request should be transmitted on the BS beam that the UE has identified. The BS replies with a random-access response. Since different beam pairs might have quite different pathlosses, power ramping (see Section 32.4.9) is foreseen.

Multi-TRP Transmission

In release 16, multi-TRP (Transmit-Receive Point) transmission was introduced. A TRP can be interpreted as a BS in the notation of this book,[19] so that multi-TRP implies a cooperation of the BSs in sending multiple data streams to a single UE. The data streams from the different BSs arrive at the UE from different directions, necessitating the forming of multiple simultaneous beams at the UE for reception. The transmission can then be done in one of two ways: (i) single-DCI, or (ii) multi-DCI.

In the single-DCI case, the two transmissions (more exactly, the two PDSCHs – the transmission might contain a single transport block) have to be tightly synchronized in time, as the data streams from the different BSs are interpreted as different data streams in a distributed SU-MIMO transmission. Note, however, that synchronization in phase is not required, as this is not joint transmission where the same data stream is sent from the different BSs. The DCI is allowed to indicate two simultaneous TCIs that are active, such that the UE knows into which directions to form the two simultaneous beams.

In the multi-DCI case, each of the PDSCHs has its own DCI associated, which will usually (but not always) stem from the same BSs from which the PDSCHs are sent. The multi-DCI case always transmits two transport blocks.

---

[19] In the framework of the NR notation, the multiple TRPs belong to the same base station, with the antennas from which they emanate being non-quasi-colocated.

32.4 Physical and Logical Channels

32.4.1 *Mapping of Data onto Logical Channels*

Just like in LTE, NR distinguishes between *logical channels, transport channels,* and *physical channels.* The structure has been simplified, with several channels eliminated, either because the functionality is not (yet) defined, such as multi-cell broadcast, or because certain information that previously was on a separate channel has been subsumed into other channels.

The following logical channels are defined:

- *Broadcast Control CHannel (BCCH)*: for the transmission of the system information throughout the cell.
- *Paging Control CHannel (PCCH)*: for paging information.
- *Common Control CHannel (CCCH)*: for control information related to random access (for UE without RRC connection).
- *Dedicated Control CHannel (DCCH)*: for control information for individual UEs.
- *Dedicated Traffic CHannel (DTCH)*: for transmitting the payload data from/to individual UEs.

These logical channels are then mapped onto the following transport channels:

- *Broadcast CHannel (BCII)*: for transmitting part of the broadcast system information (other broadcast information is sent on other channels).
- *Paging CHannel (PCH)*: for paging information.
- *DownLink Shared CHannel (DL-SCH)*: the main transport channel for the DL carrying *both* data and control information.
- *UpLink Shared CHannel (UL-SCH)*: the equivalent main transport channel for the UL data and control.
- *Random Access CHannel (RACH)*: for random access information.

Multiple logical channels, each coming from its own RLC entity, can be multiplexed by the MAC onto one transport channel. A MAC header is added that enables the de-multiplexing at the RX. Compared to LTE, the MAC header construction has been changed to allow lower latency operation. The MAC layer can also place *MAC control elements* into the transport blocks, such as buffer status reports, timing information, and activation/deactivation of previously configured elements (e.g., semi-persistent scheduling).

The transport channels are then mapped onto the physical channels, which include

- *Physical Broadcast CHannel (PBCH)*.
- *Physical Downlink Control CHannel (PDCCH)*: carrying downlink control information, such as scheduling grants.
- *Physical Uplink Control CHannel (PUCCH)*: for sending HARQ acknowledgments, CSI reports, and for requesting UL grants.
- *Physical Downlink Shared CHannel (PDSCH)*: main physical channel for payload, as well as some control information such as paging, random access responses, and part of the system information.
- *Physical Uplink Shared CHannel (PUSCH)*: the equivalent for the UL.
- *Physical Random Access CHannel (PRACH)*: for random access signals.

The mapping between the different channels is shown in Figure 32.12.

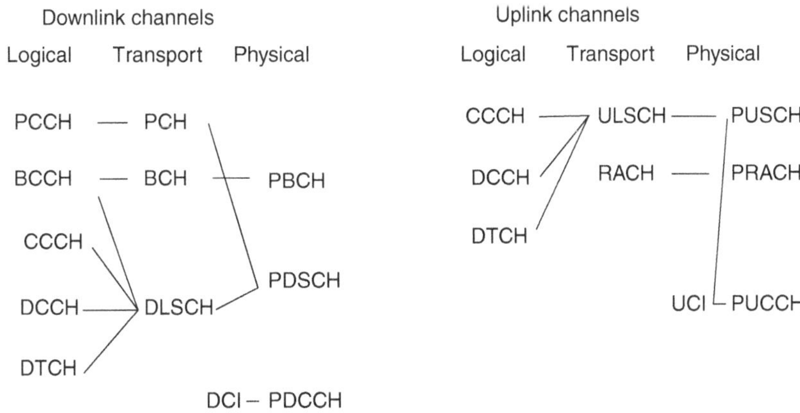

Figure 32.12 Mapping between logical, transport, and physical channels.

32.4.2 Synchronization Signals

There are two SSs in NR: a *Primary Synchronization Signal (PSS)*, and a *Secondary Synchronization Signal (SSS)*. They are transmitted, together with the PBCH, in the *Synchronization Signal Block (SSB)*. This fundamental structure is the same as in LTE. However, there are several important differences that relate to the requirements for NR. Firstly, the periodicity is much lower, to reduce the amount of "always-on signals" that create interference. Secondly, the way the SSB is placed in the time-frequency plane is different, to account for the fact that NR has a more flexible carrier and bandwidth structure.

The basic structure of the SSB is shown in Figure 32.13. The PSS and SSS are sent on the first and third OFDM symbol of the SSB, respectively. They are both 127 subcarriers wide. The PBCH covers 240 subcarriers on the second and fourth OFDM symbol, as well as 96 subcarriers on the third OFDM symbol (48 on each side of the SSS). The subcarrier spacing of the SS block can be 15 kHz for <3 GHz, 15 or 30 kHz for 3–6 GHz carrier frequency, and 120 or 240 kHz for mm-wave bands. The 240 kHz subcarrier spacing allows for very short SS blocks, which are useful for beamsweeping (which is most relevant at mm-wave frequencies), since beamsweeping requires that many SS blocks are sent out, in different directions.

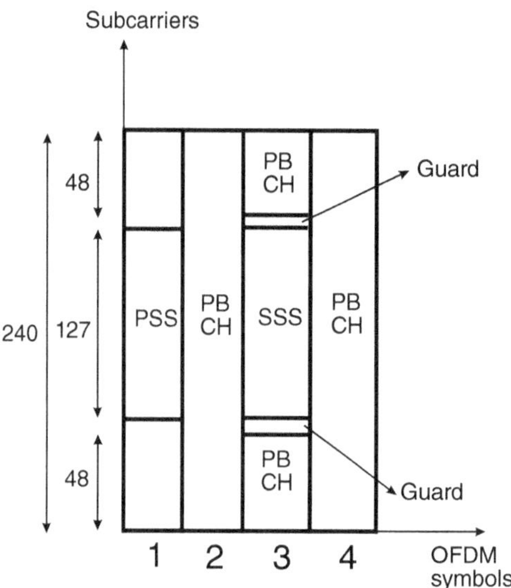

Figure 32.13 Structure of the synchronization signal block. Guard bands are 8 and 9 subcarriers.

The actual transmit signal of the PSS is obtained by modulating the 127 subcarriers with a length-127 pn sequence. There are 3 different pn-sequences, which are obtained by cyclic shifting of a basis sequence. Detecting that sequence allows the UE to obtain timing and frequency synchronization, as well as identify which of the 3 cell groups the BS belongs to. The SSS is obtained from two basic m-sequences, which are then added up with different shifts. There are a total of 336 different resulting sequences, which allow the determination of the cell ID within the cell group. Together with the information from the PSS, this thus allows the determination of the *Physical Cell Identity (PID)* (there are 1,008 different PIDs).

The location of the SS block in time and frequency is a major difference from LTE. In the frequency domain, the blocks can only be located on a sparse *synchronization raster*, so that the SS block may not even lie in the band of the actual channel; the associated band is then communicated separately in the PBCH. This is in contrast to LTE, where the SS block was always at the center of the band. The reason is that in NR there are so many possible carrier frequencies that searching for the SS blocks would take a UE too long. Different bands might use different OFDM numerologies (subcarrier spacings), but within one band, only a single numerology is allowed. In the time domain, the default periodicity has been increased from 5 to 20 ms, to reduce the frequency of always-on signals; however shorter repetition periods (down to 5 ms) are allowed, to reduce the cell search time. Longer periodicities are also allowed, even though it might prevent new UEs from discovering the BS; this makes sense when the BS only wants to maintain connection with already-connected UEs.

In both <6 GHz and mm-wave bands, it might be necessary for the BS to perform beamforming in order to reach the cell edge with sufficient power.[20] Consequently, the BS has to sweep through all beam directions to cover UEs in all possible positions. This is done by transmitting repetitions of the SS block with different beam orientations, see Figure 32.14. Such repetitions are sent out in an SS-burst, which covers a 5 ms interval. It is then such SS bursts that are transmitted with the above-mentioned periodicity (typically 20 ms). The number of SS blocks within an SS burst depends on the frequency band, ranging from 4 (for <3 GHz) to 8 (for 3–6 GHz) to 64 (for mm-wave). Since the PSSs transmitted within one burst are identical (and similar for the SSS), the UE cannot determine the relative timing (with respect to the burst start) from the observation of the PSS; it thus has to be signaled explicitly in the PBCH.

---

[20] Alternatively one could transmit without beamforming for a longer time, but the resulting SNR may suffer if the channel does not remain static during the transmission time.

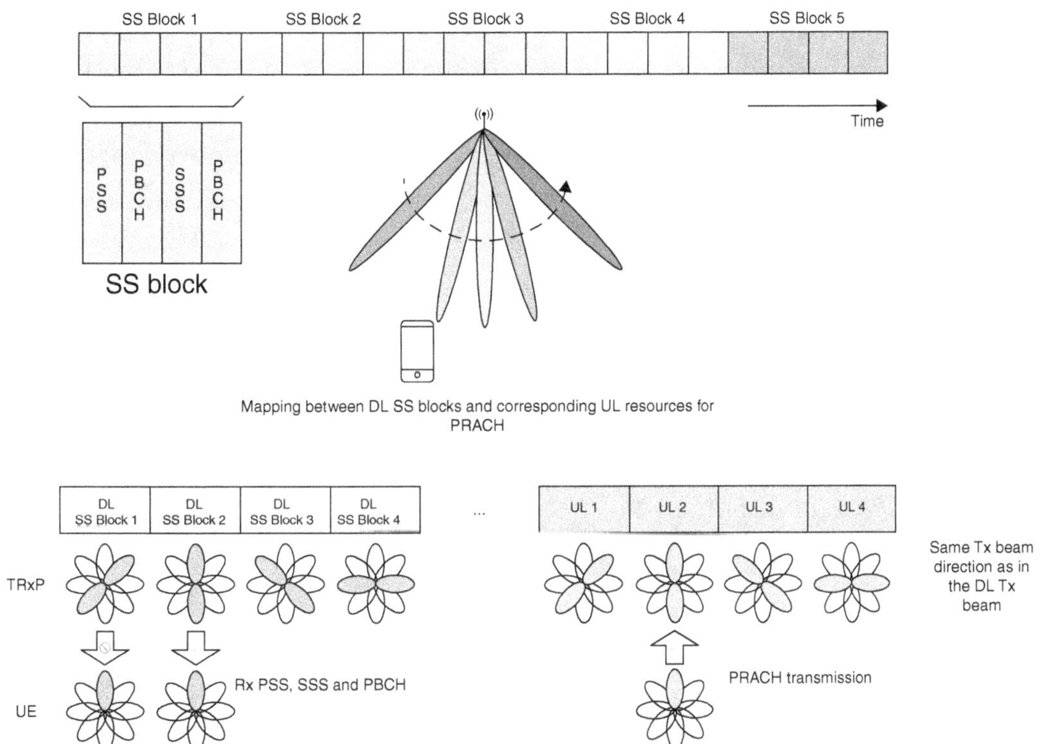

Figure 32.14 Establishment of beams at BS (TRxP) and UE. UE determines the best BS beam by monitoring different repetitions of the SS Block, which are transmitted into different directions. UE then transmits PRACH when the BS points the beam to it again, thus indicating to BS the best beam direction. Color version available at wiley.com/go/molisch/wireless3e.
Reproduced with permission from [Keysight 2019] © Keysight.

32.4.3 Broadcast CHannel

As mentioned above, the BCH is part of the SS block. It contains a total of 576 REs (240+96+240), including the DM-RSs. It contains the *Master Information Block (MIB)*, which – compared to LTE – has changed content, containing among other:

- *SSB time index*: denotes the location of the current SSB within the SS burst. This information is required during beam sweeping: the UE might hear only one of the several SSBs transmitted in a burst, namely the one where the BS beam is pointing in the "right" direction; the UE then has to have a way to identify the timing of that block within the burst, both because it is required for general timing acquisition, and because it indicates the index of the right beam. Since all PSS are the same, the information needs to be transmitted separately and is done in the SSB time index. The MIB contains some explicit bits, namely 0 (for FR1) or 3 (for FR2), which need to be combined with further implicit information that is hidden in the scrambling of the PBCH (8 scrambling patterns, resulting in 3 bit information). This allows to index 8 (for FR1) or 64 (for FR2) locations.
- *Cell barred flag*: indicates whether access is prohibited to UEs. A first bit describes access to the current cell; a second bit indicates whether access to other cells on the same frequency is allowed. This is useful for non-stand-alone NR deployments (where control signaling is only on LTE and data can be sent via both LTE and NR) that need to force devices into accessing the cell via the LTE carrier.
- *First PDSCH DMRS position:* shows the location (in time) of the first DM-RS symbol.
- *SIB1 numerology* (1 bit) *and SIB 1 configuration* (8 bits) describe the numerology, search space, CORESET, and other parameters of the *System Information Block (SIB)* 1.
- *CRB grid offset*: it indicates the frequency offset between the SS block and the CRB grid (see Section 32.3.4).
- *Half-frame bit*: indicates whether the SSB is within the first or second 5 ms part of a 10 ms frame (the SSB can be transmitted in either of those).
- *System frame number* (10 bits).
- *CRC* (24 bits).

Further UE-independent information that the UEs need to know is transmitted in SIBs, where SIB 1 is periodically (160 ms) broadcast over the whole cell area (similar to the PBCH), though this transmission is done as a regularly scheduled PDSCH transmission. Note that the numerology and other information for receiving the SIB 1 are provided in the PBCH. Other SIBs can be either periodically broadcast, or just transmitted on demand, where the requests for such information are sent by the UEs via RACH preambles (the configuration for this request is conveyed in SIB 1).

32.4.4   General Aspects of Control Channels Associated with a DL-SCH

The control channels for the DL have been simplified significantly, such that there is only a single control channel, the PDCCH, anymore.

32.4.5   Physical Control Format Indicator CHannel (PCFICH)

This channel does not exist anymore in NR.

32.4.6   Physical HARQ Indicator CHannel

This channel does not exist anymore in NR: in LTE, it served in the context of synchronous HARQ retransmissions in the UL; however, since all HARQ is asynchronous in NR, there is no need for it anymore.

32.4.7   Physical Downlink Control CHannel

The PDCCH carries the DCI, i.e., the key control information for the DL.

*Downlink Control Information (DCI)

Just like in LTE, the NR standard foresees a number of different formats for communicating the DCI, depending on the particular type of information that needs to be conveyed (though the content and the types of formats are changed considerably compared to LTE). They can be divided into

- *Downlink scheduling assignment* (types 1-0, 1-1, and 1-2)
- *Uplink scheduling grants* (types 0-0, 0-1, and 0-2)
- *Special commands* (types 2-0 to 2-6)
- *Sidelink scheduling* (types 3-0 and 3-1)

A number of these format types were only introduced in release 16. In the following, we do not make a distinction between those and the ones already existing in release 15.

A first key part of the DCI is the scheduling assignment, i.e., which time-frequency resources contain information intended for which UE. There are three formats defined for this:

- *Format 1-1* (also called the non-fallback-format): it supports *all* NR features, but only includes them if they are actually configured for the device. As a consequence, the format is (i) variable in size and (ii) leads to a high overhead.
- *Format 1-0* (the fallback format): it only contains a subset of the data, and is fixed in size (with a small size). This allows easier blind decoding and works also in cases in which the UE feature set has not yet been communicated, e.g., due to transmission errors. In the following listing of the features, we will denote by (F) the features contained in the fallback format. Note that format 1-0 (but with different content) is also used for paging, random access response, and system information delivery.
- *Format 1-2* is mainly intended for IoT applications. It contains a larger subset of the 1-1 parameters, which will be indicated by an I in the listing below but can use more flexible number of bits for each of the information fields. In addition, it foresees a priority indicating field, which is unique to this format, and describes the priority of related UL transmissions such as HARQ acknowledgment; this can be used to reduce latency.

The DCI carries the following DL scheduling information (for some of these parameters, their detailed use is not discussed in this chapter, but they are listed here for completeness):

- *Format identifier* (F,I): indicates whether the DCI is a DL or UL assignment (they might have the same size, and thus cannot be distinguished by blind decoding).
- *Resource information*:
  - *Frequency-domain resource allocation* (F,I): indicates the RBs on which the PDSCH is sent. The following types of allocation are possible (with the first two similar to LTE, though some details differ): (i) *type-0* provides a bitmap that explicitly enumerates the frequency resources to be used. However, to save overhead, it is not on a "per RB" basis, but rather on the basis of groups of RBs, where the group size depends on the size of the bandwidth part, (ii) *type-1* provides the starting RB and the size of the allocation. In either case (i) and (ii), the assignment refers to the VRBs; the mapping to the PRBs occurs either interleaved or noninterleaved, according to another indicator described below. Note that type-0 and type-1 can be signaled in the nonfallback format, while only type 1 can be used for the fallback format. NR also foresees, as a new option, a dynamic switching between the two.

- ○ *Time-domain allocation* (F,I): the time allocation means the slot number (relative to the slot where the DCI is located), the OFDM symbol at which the data transmission starts, the length of transmission in units of OFDM symbols, and (for the DL) the location of the DM-RS. Since this is a lot of information, NR avoids signaling this information explicitly. Rather, it allows the higher layers to configure a table with up to 16 (in release 15) or 64 (release 16) rows, in which various combinations of these informations are defined. The time allocation contained in the DCI is then simply the index of the table row that is to be used. Note that release 16 also allows to add information in the table that is related to operation in unlicensed spectrum, and number of times a URLLC transmission can be repeated.
  - ○ *Other resource information: Bandwidth-part indicator* (I): activates one of the configured bandwidth parts (see section titled "Frames, Slots, and Symbols" in Section 32.3.4). *VRB-to-PRB mapping* (F,I): indicates whether interleaved or noninterleaved mapping (see section titled "Mapping to Physical Resources" in Section 32.3.4) is used. *PRB bundling size indicator* (I): provides PDSCH bundling size (related to PRGs, see section titled "DeModulation Reference Signal" in Section 32.3.5). *Reserved resources* (I): says whether reserved resources can be used for PDSCH. *Zero-power CSI-RS trigger* (I): see Section 32.3.5. *Scheduling offset indicator*: used for cross-carrier scheduling in carrier aggregation (Section 32.6.1). *Channel access type* (F): indicates channel access procedure in unlicensed bands (Section 32.6.2). *Dormancy indicator*: used for power-saving mode. *Carrier indicator* (I): used in carrier aggregation, to indicate which component carrier the DCI describes. *Scheduling offset* (I): to control cross-slot scheduling.
- *Transport-block related info*: these pieces of information are very similar to LTE
  - ○ *Modulation and Coding Scheme (MCS)* (F,I).
  - ○ *New Data Indicator (NDI)* (F,I): for HARQ, see Section 32.5.2.
  - ○ *Redundancy version* (F,I): for HARQ, see Section 32.5.2.
  - ○ *Indicators for second transport block*: provide MCS, NDI, and redundancy version for the second transport block, if applicable.
  - ○ *Priority indication* (I): this is a new quantity (since release 16) that exists *only* in format I and indicates the priority of the UL transmissions related to this transport block.
- *HARQ info* (see also Section 32.5.2):
  - ○ *HARQ process number* (F,I).
  - ○ *Downlink Assignment Index (DAI)* (F,I): for dynamic HARQ codebook.
  - ○ *HARQ feedback timing* (F,I): states when feedback for a packet should be sent.
  - ○ *CBG transmission indicator*: indicates which CBGs are retransmitted.
  - ○ *CBG flush information*: for buffer flushing when CBGs are used.
  - ○ *PDSCH group index, number of requested PDSCH groups, one-shot hybrid ARQ request, new feedback indicator*: related to HARQ in unlicensed spectrum
- *Multi-antenna info*:
  - ○ *Antenna ports* (I): on which ports the data for the desired user, and the data for other users, are transmitted.
  - ○ *Transmission configuration indication* (I): see Section 32.3.7.
  - ○ *SRS request* (I): see Section 32.3.5.
  - ○ *DM-RS sequence initialization* (I): selects between two possible, preconfigured, sequences for DM-RS.
- *PUCCH-related info*
  - ○ *PUCCH power control* (F,I).
  - ○ *PUCCH resource indicator* (I): selects resources (from set of preconfigured options) for transmitting PUCCH.

The DCI also carries control information that governs the uplink transmission. Just like the downlink scheduling, the UL scheduling grant has a nonfallback, a fallback format, and an IoT-related format (called 0-1, 0-0, and 0-2 respectively), with the same motivation and pro/con as the DL schedule information. The priority indicator in format 0-2 now describes the priority of the UL transmission.

- *Identifier* (F,I): same meaning as for DL.
- *DFI flag*: in unlicensed spectrum, indicates whether the DCI is an UL grant or a request for *Downlink Feedback Information (DFI)*.
- *Resource information:*
  - ○ *UL/SUL* (F,I): indicates whether the grant is for the UL or the Supplementary Uplink (SUL) (see Section 32.6.1).
  - ○ *Frequency-hopping flag* (F): instructs whether frequency-hopping is to be used.
  - ○ *Carrier indication* (I), *Bandwidth-part indicator* (I), *Frequency-domain resource allocation* (F,I), *Time-domain resource allocation* (F,I): *Scheduling offset, dormancy indication, channel access type* (F): similar meaning as for DL. However, note that for the frequency allocation, Release 16 also introduced a type-2 allocation that supports interlaced frequency allocation in the UL. Also note that there are two separate tables for time-domain resource allocation, one for the DL, and one for the UL.
- *Transport-block-related information* (F,I): MCS, NDI, and redundancy version, all the same as for DL allocation (no option for second transport block exists for UL).
  - ○ *UL-SCH indicator* (I): indicates whether the PUSCH should contain data from the UL-SCH or just Uplink Control Information (UCI) feedback.
  - ○ *Priority* (I): refers to the priority of the actual UL transmission.

- *HARQ-related info*: Process number (F,I), DAI (I), and CBG transmission indicator; similar as for DL allocation.
- *Multi-antenna info*:
  - *Antenna ports* (I): same as for DL (though fewer bits are used, since the number of possible ports is smaller at the UE).
  - *DM-RS sequence initialization* (I): same meaning as for DL.
  - *SRS resource indication* (I): used for determination of antenna ports for PUSCH transmission, see Section 32.3.6.
  - *Precoding information* (I): selection of precoding matrix and number of layers, for codebook-based precoding.
  - *PTRS-DMRS association* (I): shows the association of the ports used for DM-RS and PT-RS.
  - *SRS request* (I).
  - *CSI request* (I): requests transmission of a CSI report.
- *Power control information*
  - *PUSCH power control* (F,I): to change PUSCH power.
  - *Power control parameter set* (I): for boosting PUSCH power for UL preemption; exists only in signaling format 0-2.
  - *Beta offset*: related to a specific dynamic signaling scheme (not discussed here further).

Several other DCI formats are defined for conveying other control information

- *Slot format indication (format 2-0)*: for transmitting the SFI, as discussed in section titled "Frames, Slots, and Symbols" in Section 32.3.4.
- *Preemption indication (format 2-1)*: see Sections 32.5.3 and 32.9.
- *Uplink power control commands* (format 2-2): same power control info as conveyed in DL assignment or UL grant. It is used in semi-persistent scheduling, which does not have DL assignment or UL grant.
- *SRS control commands* (format 2-3): for power control of SRS signals.
- *Uplink cancelation indicator* (2-4): for preemption (Section 32.9).
- *Soft resource indicator* (2-5) describes whether a particular resource is available for the access UL in an *Integrated Access and Backhaul (IAB)* configuration.
- *DRX activation* (2-6): for control signals related to sleep/wakeup.
- *Sidelink scheduling commands for NR* (3-0): to convey the scheduling when sidelink scheduling is controlled by the network.
- *Sidelink scheduling commands for LTE* (3-1): when an NR network schedules LTE sidelink transmissions.

Transmission and Reception of PDCCH

In NR, there is only a single PDCCH. It is more similar to the EPDCCH in LTE rather than the PDCCH; a few basic design principles are as follows: (i) the channel does not cover the whole carrier bandwidth, since not every UE in NR is capable of receiving the full bandwidth, (ii) it can be transmitted using beamforming, since beamforming in NR is transparent to the UE (see Section 32.3.6); (iii) just like in LTE, there is no explicit signaling of the PDCCH location for a user, but rather a blind decoding in a set of search spaces is used; more details are described in the following.

The general physical-layer processing of the PDCCH is somewhat similar to the PDSCH but also shows some important differences. The length of its payload, the DCI, is at most 140 bits. To this is added a 24-bit CRC (longer than the 16 bit in LTE, to give stronger protection) that is scrambled with a scrambling code that is based on a device-specific scrambler setting (this could be the Cell-Radio Network Temporary Identifier (C-RNTI), i.e., the identity of the device, or some other Radio Network Temporary Identifier (RNTI)). Similar to LTE, this allows the UE to determine whether the information is intended for it. In another difference to LTE, the CRC bits are not simply appended but rather distributed throughout the packet. The resulting data are then encoded with a polar code (see Section 13.8), – as opposed to the convolutional codes used in LTE – resulting in up to 512 coded bits. These coded bits are then rate-matched (using a combination of shortening, puncturing, and repetition), scrambled, and modulated with QPSK symbols. Transmission occurs only from a single antenna port.

Now we turn to the mapping of these symbols onto the time/frequency resources. Here, NR introduces a number of new definitions.[21] A *Resource Element Group* (REG) consists of 12 subcarriers for the length of one OFDM symbol and is thus very similar to a RB, and just like the RB, it carries its own DM-RSs in it. A set of 6 REGs is a *Control Channel Element (CCE)* and carries 54 QPSK symbols (there are 3 DM-RS symbols per REG, i.e., every 4th subcarrier). The mapping of the symbols in a CCE to the REG can either be interleaved or noninterleaved, similar in spirit (and with the same motivation) as the mapping from VRB to PRB in the PDSCH, though the actual details of the mapping are different (and quite complicated). A PDCCH consists of 2^n contiguous CCEs, with $n = 0, ...4$, where 2^n is also called the *aggregation level*.

After these definitions, let us now turn to the locations of those contiguous CCEs, i.e., the location of the control information, in the time-frequency plane; this location is called the *COntrol REsource SET (CORESET)*, and it shows major differences to LTE. Remember

---

[21] The names REG and CCE exist also in LTE, but are defined differently there.

that in LTE the first one, two, or three OFDM symbols of a subframe, extending over the whole carrier bandwidth, are assigned to the control information. In NR, on the other hand, the bandwidth can be much more limited, with as little as 72 ($12 \cdot 6 \cdot 1$) subcarriers, a quarter of which are DM-RS. The CORESET can be located anywhere in the slot and in the active bandwidth part; however, in practice, it will often be located at the beginning of the slot. The duration of the CORESET can be 2 or 3 OFDM symbols, depending on where the DM-RSs for the PDSCH are placed: generally, it is preferred that the CORESET transmission finishes before the PDSCH-DM-RSs are placed. However, to enhance spectral efficiency, PDSCH data can be transmitted before the CORESET finishes.

There may be up to three CORESETS configured for each configured bandwidth part. The location of CORESET 0 is conveyed in the MIB, as this information is required during the connection setup. Once the connection has been established, additional CORE-SETs can be configured through RRC signaling. The different CORESETS can have different aggregation levels and different CCE-to-REG mappings. Their DM-RSs are determined by their absolute position in frequency (again, compare the DM-RSs for the PDSCH).

The DM-RSs for the PDCCH are, in principle, the same as for the PDSCH; they are precoded with the same precoder settings as the DCI, again resulting in a transparent precoding in which the UE simply sees an effective channel. Consequently, the DM-RS are distinct for each PDCCH. Every fourth subcarrier in a REG is used for the DM-RS, thus leading to a relatively high pilot overhead. Similar to the PDSCH, it is possible to define REG group "bundles" within which the precoder settings are assumed to be constant, so that a better noise averaging can be achieved. Note that both narrowband and wideband precoding of the PDCCH is supported. Furthermore, the TCI shows for a particular CORESET whether it is QCL with a particular CSI-RS; this can also be used for improving the noise averaging in the channel estimation.

As seen above, there are different DCI formats. The UE has to *blindly* detect which format is being used. It does so by, essentially, testing all possible locations and DCI sizes. This would be, in principle, an enormous number of combinations, despite the fact that the CORESET structure described above imposes some limitations. NR thus defines some further restrictions; in particular, a set of *search spaces*. A search space is a set of candidate CORESETS using a given aggregation level, which have to be tested by the UE whether they carry the PDCCH. These search spaces are specific for each UE and are tied to the *C-RNTI* of the UE. Since there are multiple aggregation levels, there are multiple search spaces for each UE; the UE has to monitor up to 4 DCI formats, corresponding to the fallback DCI formats, the DL scheduling assignments, the UL scheduling grants, and slotformat indication. The standard defines a limit on the number of blind decoding attempts (and also on the channel estimates), to limit the complexity of the whole process.

Enhanced Physical Downlink Control CHannel

This LTE channel does not exist in NR anymore; as a matter of fact, the above-discussed PDCCH shows great similarity in its design philosophy with the EPDCCH in LTE.

32.4.8 Physical Random Access CHannel

The system reserves certain time/frequency resources for use by the PRACH, with the configuration communicated in SIB 1. It first defines a (configurable) subset of slots, called *RACH slots*. These slots are repeated every *RACH configuration period* (between 10 and 160 ms). Within these slots, multiple configurable *RACH occasions* (each covering time-and-frequency contiguous RBs) are defined, which carry the PRACH.

The signals on the PRACH are called preambles, and – like in LTE – are based on length-L Zadoff–Chu sequences, which are then DFT-precoded. Not only Zadoff–Chu-sequences with different roots but also the same sequence with different shifts can be used to increase the pool of sequences (again, similar to LTE). The admissible shifts are determined by the system, indicated in SIB 1; the underlying physical reasoning is that delayed echoes from one sequence must not be confused with signals from another sequence.

NR defines "long" and "short" preambles; a cell can pick the type of preamble, but then has to use this consistently.

- *Long preambles* are based on the LTE preambles. They have $L = 839$ and can use 1.25 kHz (same as LTE) or 5 kHz subcarrier spacing, which is outside the usual numerology. There are 4 different formats of the long preambles; between 1 and 4 repetitions of the sequences are sent, and the CP ranges from 15 to 680 µs. The resulting preambles might be longer than a slot, so the RACH occasions have to be planned accordingly.
- *Short preambles* have $L = 139$, and use the standard numerology (15 and 30 kHz for FR1 and 60 or 120 for FR2); they always occupy 12 RBs in the frequency domain. There are 9 different formats of the short preamble, using between 1 and 12 repetitions, and a CP ranging from 7.0 to 66.7 µs.

32.4.9 General Aspects of Control Signals Associated with the PUSCH Uplink Control Signals

As outlined above, and similar to LTE, there is relatively little UL control info, i.e., mainly info related to HARQ feedback, CSI reports, and scheduling-request-related information. The information is transmitted in one of two ways: (i) piggy-backed onto the PUSCH, or (ii) as a standalone PUCCH. The former is beneficial from an RF point of view, but cannot always be realized since a PUSCH might not always exist. Thus, provisions for a separate PUCCH have to be made as well; however, simultaneous transmission of PUCCH and PUSCH is not allowed (i.e., the PUSCH *must* be used for the UCI if available).

When *UCI* is sent on the PUSCH, it might include HARQ feedback, CSI reports, scheduling requests, and buffer status reports (which help the BS to make scheduling decisions).

An important type of the UL control info is the reporting of the CSI, based on observation of the CSI-RS. The BS can instruct the UE what to report, and in which format. Generally, there are 3 types of reports: (i) periodic, (ii) semi-persistent (as usual, configured to be periodic, but whether transmission occurs is determined by MAC signaling), and (iii) aperiodic, triggered by a DCI command (there are up to 63 different commands, requesting 63 different types of aperiodic reports).

The UCI can be transmitted either using OFDM or DFT-spread OFDM. The transmit power is kept constant over the PUSCH, for both the data and the UCI part, but the UCI can use different code rates to adapt to different channel conditions. The most critical bits of the UCI (HARQ feedback) are mapped to the OFDM symbols after the DM-RS, so that they suffer least from channel outdatedness.

32.4.10   PUCCH

If the UE has no PUSCH assigned to it, it transmits the UCI on the PUCCH. A number of different formats exist for this channel, depending on the amount of information that needs to be transmitted, and the required robustness/range. The fundamental principles are similar to those for the PUCCH in LTE, see Section 31.4.10. The PUCCH can be beamformed, typically using the same beam configuration on which it receives the corresponding DL information. It is also possible to configure multiple spatial (beam) relations between DL RSs such as CSI-RS on one hand, and the PUCCH signals on the other hand, and using MAC signaling to select which one of them is active.

The PUCCH transmission aims to minimize the PAPR, which is meaningful if the PUSCH also uses DFT-spread OFDM to minimize PAPR. For a UE that normally uses pure OFDM for UL transmission, a PUCCH with low PAPR is not very helpful but was chosen anyway to reduce the number of PUCCH types that need to be defined.

In case of carrier aggregation, the amount of UCI signaling can become very large. In this case, two PUCCHs can be configured, where the first is transmitted on the principal carrier, providing info about one group of component carriers, and the second PUCCH provides info about the other group.

Every PUCCH transmission requires certain time-frequency resources, as well as a set of parameters that govern the transmission. The system thus configures for a UE up to 4 PUCCH *resource sets*, each of which may contain multiple PUCCH configurations, each of which contains the time-frequency resources and parameters. The different resource sets correspond to different ranges of UCI feedback. During the transmission, the size of the UCI then determines which resource set to use, and the *PUCCH resource indicator* in the DCI determines which resource configuration within the set should be used; thus, the BS has control on which time-frequency resources the information is sent.

There are a number of different formats for transmission:

(a) *Format 0* is used for transmitting up to 2 bits of control information (SR and/or ACK/NACK), using up to 2 OFDM symbols. It encodes the information into tabulated CAZAC sequences (the equivalent of ZC sequences for short duration, compare "Uplink" in Section 31.3.5) of length 12. More specifically, a base sequence is determined according to a cell ID, and can further be varied from slot to slot to achieve interference diversity. A linear phase rotation, encoding the 1 or 2 bit information, is then imposed on this sequence to convey the information. The phase shifts are integer multiples of $2\pi 6/12$ and $2\pi 3/12$ for 1 and 2 bit of information, respectively. The phase rotation also depends on a reference phase rotation that is communicated on the higher layers. Since different UEs can have different reference phases, multiple UEs can be multiplexed onto the same time-frequency resource. The PUCCH is normally transmitted on the last OFDM symbol(s) in a slot; however, it is possible to transmit it also sooner if, e.g., required for latency minimization.

(b) *Format 1* is also used for transmission of 1 or 2 bits but uses 4–14 OFDM symbols to achieve that. As a further contrast to Format 0, it requires DM-RS, and actually about half of the transmitted symbols are for DM-RS. The data symbols are BPSK or QPSK modulations onto the same length-12 sequences that are also used for Format 0. These 12 symbols are then block-spread with an orthogonal cover code. The DM-RSs are the same sequences (but unmodulated) as the information-bearing sequences. Frequency hopping is possible to further enhance diversity. An overview of the structure can be found in Figure 32.15. The use of different orthogonal cover codes for different UEs enables multiplexing of more UEs onto the same time-frequency resources.

(c) *Format 2* uses at most 2 OFDM symbols for more than 2 bits of information. Instead of directly transmitting the bits, as in formats 0 and 1, here the symbols are encoded and scrambled. To be specific, the first step is computing and appending a CRC (though only for larger payload sizes). The data are then encoded with either (i) a Reed-Muller code (for ≤ 11 bits of payload), or (ii) a polar code (for larger payloads). The data are then scrambled with a sequence that depends both on device ID and cell ID, and QPSK modulated. The modulated symbols are then distributed over multiple RBs (that also contain DM-RS on every third subcarrier) in 1 or 2 OFDM symbols (typically at the end of a slot). Similar to format 0, transmission normally occurs on the last OFDM symbols but can be chosen to be sooner.

(d) *Format 3* is similar to format 2, but transmission occurs over a larger number of symbols. The first steps of the processing, including the QPSK modulation, are like for format 2. The resulting symbols are then distributed to the different OFDM symbols, and DFT-precoding is applied to each of them. Interspersed with those OFDM symbols are RSs, namely of the form that is used for DFT-spread OFDM transmission. Again, frequency hopping within a slot is possible, in which case there needs to be at least one RS per hopping period.

(e) *Format 4* is similar to format 3, but using block-spreading the payload, and multiplexing different users onto the same time-frequency resources.

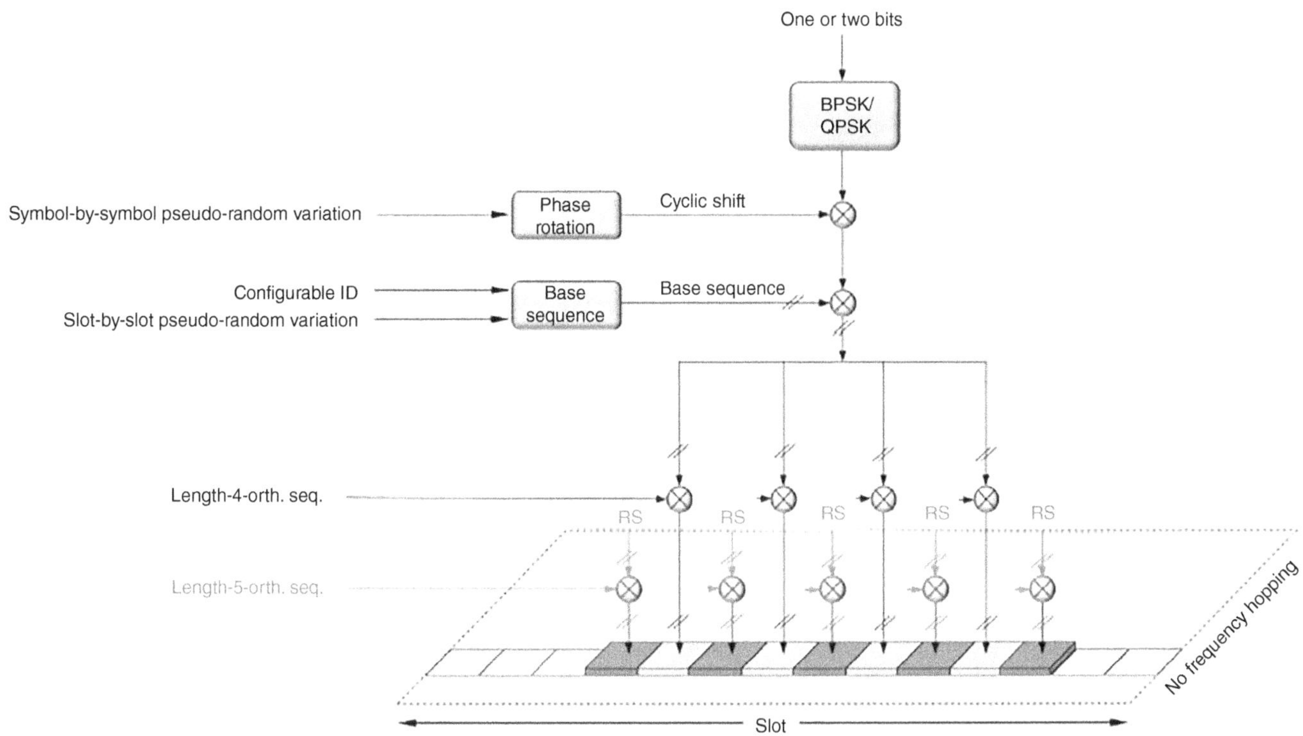

Figure 32.15 PUCCH format 1. Color version available at wiley.com/go/molisch/wireless3e.
Reproduced with permission from [Dahlman et al. 2020] © Academic Press.

32.4.11 PUSCH

If a PUSCH exists for a particular UE, the control info can be sent on it, by time-multiplexing with payload data. This information could be HARQ acknowledgments and CSI reports; there is no need to send a request for an UL scheduling grant if a PUSCH is already scheduled.

Compared to LTE, the HARQ acknowledgments are more complicated: if the number of ACKs is small, then they are punctured onto the data stream (like in LTE, compare Section 31.4.11). However, if the number of ACKs becomes large, rate matching is used, with the DAI field indicating the number of resources intended for this feedback.

Similarly to LTE, the most important bits, namely the HARQ acknowledgments, are transmitted closest to the first DM-RSs.

32.5 Physical Layer Procedures

32.5.1 Establishing a Connection

Scanning and Synchronization

The first step in establishing a connection is the synchronization of the UE to the BS, which occurs as described in Section 32.4.2. The reply from the UE occurs on the PRACH, through transmission of a preamble. The basic principles of this approach are similar to those in LTE, see Section 31.5.1.

An important difference from LTE is the incorporation of beamsweeping. Since the BS can perform beamsweeping during the transmission of the SS, it is important that the reply also indicates during which beam the best DL SS was obtained. This feedback is enabled by associating a specific set of preambles with a particular time index of an SS block (which, as explained in Section 32.4.2, is associated with the beam direction). This is enabled, in turn, by the BS indicating in the random-access configuration information the SS-block time index that is associated with a RACH occasion.

The initial synchronization provides some of the system information that the UE needs for the next steps as part of the PBCH. Other important information, including the SIB1, which contains information about the configuration of the RACH, is broadcast periodically (every 160 ms) through a normally scheduled PDSCH transmission (the parameters required for receiving this transmission are already sent out as part of the MIB). Since PDSCH transmissions are encoded depending on the RNTI (a form of device ID), and the BS does not know the ID of all the devices it broadcasts to (those IDs are only exchanged later, as part of the random access procedure), it uses a special system information RNTI to encode the SIB1. The remaining SIBs are either broadcast regularly or sent out on demand; in either case, they are sent after the UE has achieved connection with the BS.

Random Access

Random access is used when a UE does not (yet) have any scheduled resources assigned to it, which occurs not only during the initial connection process but also at other occasions such as handover and request for UL resources, as discussed in Section 31.5.1.

The random access procedure consists, like in LTE, of 4 steps: (i) the UE transmits a preamble on the PRACH, (ii) the BS sends a random access response, (iii) the UE sends a *message 3*, and (iv) the BS replies with a *message 4*; the latter two steps serve for collision resolution. These steps are now described in more detail.

The transmission of the preamble occurs on the PRACH, which is discussed in Section 32.4.9. If a transmission on the PRACH is not successful, the UE will repeat, ramping up the TX power in the process; this process is very similar to LTE. The main difference from LTE lies in the handling of beam sweeping. The RACH occasions that the BS announces depend on the SSB time index, and thus on the beam direction. Consequently, the RACH occasion in which the RA response by the UE implies for which beam direction it heard the SS transmission. However, note that the number of SSB time indices per RACH occasion can be larger than one; in that case, further specifications about the SSB index are required.

The RA response is then transmitted as a regular PDCCH transmission. It contains information about the preamble sequence that the BS had detected (and to which this response is answering), a timing correction so that the proper timing advance can be set up, a scheduling grant for Message 3, and a temporary ID to be used for the next communication steps. If beamforming is used, then the BS should use the same beamformer setting as when it transmitted the SS block to which the UE is responding; in other words, assuming that beam reciprocity holds, SS block transmission, preamble reception, and random access response should all happen on the same BS beam. In either beamformed or nonbeamformed case, if the BS detects multiple RACH transmissions (that are distinguishable because they use different preambles) it can combine the responses to all of them into a single response. Possible collisions, i.e., multiple preambles that the BS cannot distinguish, on the other hand, have to be resolved in the next steps.

In Message 3, the UE transmits to the BS its device identity. From this, the network determines a C-RNTI. Depending on whether the UE is already known to the network, this is already known to the BS or must be determined via the core network. In the last step, Message 4, the BS transmits, on the PDCCH, a message to resolve contention, essentially identifying now uniquely the UE it was talking to.

Since release 16, there is also a two-step random access procedure, where steps 1 and 3 are combined into a single message (that contains both a RACH preamble and a PUSCH for the message transmission), and similarly steps 2 and 4 are combined. However, for backward compatibility reasons, the four-step procedure must always be supported.

32.5.2 Retransmission and Reliability

Just like in LTE, the retransmission procedures in NR occur on 3 levels: (i) physical-layer, (ii) RLC, and (iii) PDCP. The physical-layer technique can react the fastest, and is thus the "first line of defence." Residual errors are handled by the RLC, which should achieve packet error rates 10^{-5} or better, and PDCP-level retransmissions mainly deal with inter-BS handovers. Any still remaining packet errors are handled by the TCP/IP protocol, and are very costly, as discussed in Section 31.5.2.

Physical-layer HARQ has the same basic principles as in LTE. Firstly, the retransmissions are based on incremental redundancy. Secondly, since the decoding attempts at the RX, and sending of the feedback, take some time, several HARQ processes have to be executed in parallel.

However, there are also some important differences to LTE. In NR, *both* the DL *and* the UL use asynchronous HARQ. For the DL, the asynchronous protocol is simple (and the same as in LTE): the BS transmits in the DCI whether the attached packet is a new packet or a different redundancy version of a previously transmitted packet; it indicates this with the NDI. For the UL, NR also uses an asynchronous protocol (in contrast to LTE, which uses a fixed delay between retransmissions). This is done partly in order to accommodate dynamic TDD, i.e., different UL/DL ratio in different slots, which could preclude a constant offset because the thus-prescribed time might not be available for an UL transmission. Furthermore, it enables to reduce the time between retransmission for latency-critical services. The timing for when the UE should send acknowledgements is determined by the HARQ timing field in the DCI. The system also configures an 8-element table whose entries are different delays between DL and requested ACK; the HARQ timing index indicates dynamically for each slot the request delay by indicating the table entry whose value should be used. Consequently, multiple ACKs might occur at the same time and are multiplexed by a special procedure.

Up to 16 HARQ processes *can* be used in NR (compared to 8 in LTE) if necessary; however, they need not be used, so that the availability of more processes does not create larger latency. Generally, processing times (and thus time between reception and feedback) are much shorter (<1 ms) in NR than in LTE.

A further difficulty for NR lies in the possibly very large transport blocks; to alleviate the resulting problems for retransmissions, each transport block may be divided into CodeBlock Groups (CBGs). Both the transport block and the codeblocks have their own CRC (see Section 32.3.2), and retransmissions may be based on CBGs. Specifically, a transport block may be divided into 2, 4, 6, or 8 CBGs; this serves as a compromise between keeping small the number of bits required, e.g., in retransmissions of CBGs, and in keeping low the amount of control signaling that is necessary for addressing the retransmission units. The DCI contains a *CBG transmission indicator*, which contains a bitmap indicating the presence of a certain CBG, and a *CBG flush indicator*, which tells the RX whether the soft combining with previous bits from the process should be performed, or the CBGs listed in the CBGTI should be flushed (since this is a decision of the scheduler, the information only need to be communicated in the DL). Note that in the case that CBGs are used, also the HARQ feedback must contain a bitmap with one bit per CBG to indicate which of them arrived correctly.

A second level of retransmissions occurs on the RLC. There exists an RLC entity for each logical channel, which provides segmentation (but – in contrast to LTE – no concatenation) of the SDUs, and a retransmission protocol.[22] Retransmissions can occur at different component carriers, because the RLC entity is for the logical channel, which at different times can use different physical carriers (in contrast to the physical-layer HARQ mechanism, where retransmissions occur on the same carrier). The RLC can operate in (i) *Transparent Mode (TM),* where it does not take any actions, (ii) *Unacknowledged Mode (UM),* where it performs segmentation but no retransmissions, and (iii) *Acknowledged Mode (AM),* where it performs both segmentation and retransmission. TM is mainly used for BCHs, where no adaptation to different channel conditions is feasible and thus segmentation is not very relevant, while retransmission is not possible either. UM is mainly used for unicast of services that – due to latency constraints – cannot use retransmissions anyway, such as Voice over IP. AM is the standard mode for DL-SCH and UL-SCH.

Generally, the RLC segmentation process is similar to LTE (see Section 31.2.2): it takes an incoming RLC SDU, and – if there is no segmentation – attaches an RLC header that contains a sequence number. However, if the size of the last RLC Protocol Data Units (PDU) in a transport block does not match the RLC SDU size, then the SDU can be segmented. This segmentation simply involves modification of the header for the first segment and adding of a new header for the second segment. The fact that a PDU contains segments is indicated in the segmentation information field in the RLC header. The RLC retransmission mode is a conventional ARQ (i.e., complete retransmission), again similar to LTE.

32.5.3   Scheduling

Just like for LTE, the scheduling algorithms are mostly out of the scope of the standard. However, there are some new ways in which a scheduling decision can be communicated, as well as in conveying the information based on which the scheduling decision is made. These aspects are discussed in the following. Note that DL scheduling and UL scheduling are essentially decoupled, except for such constraints imposed by normal system operation, such as unavailability of certain timeslots because of half-duplex constraints.

Different schedulers, e.g., located at different BSs, might coordinate their decisions with each other. Again, the exact mechanism for determining the joint schedules is not prescribed; the information exchange communicating the underlying data generally occurs via the Xn-c interface.

Downlink Scheduling

The DL scheduling decisions are taken, and communicated to the UEs, for a time interval, e.g., a slot. However, there might be situations in which an urgent data packet arrives that should be communicated even at the cost of interrupting an ongoing communication. NR foresees a mechanism for it, such that the BS simply substitutes the old data, intended for a particular UE, with the new data and new target UE. The interrupted transmission will then return a failure of reception and request a HARQ retransmission. It is also possible for the BS to send a *preemption indicator* (format 2-1 of the DCI), indicating which bits have been substituted. This information can help the UE in the request for retransmission, and the actual decoding process. While the standard does not specify what the UE does with the information, the obvious choice is to purge all the "substituted" information, since having no information on bits is better than having wrong information.

Uplink Scheduling

Again the principles for UL scheduling are similar as those for LTE. The BS makes scheduling decisions, partly based on the information it receives about channel states and buffer status of the UEs, and communicates the decision with a scheduling grant to the UE; the grant is for a time in the future (in contrast to LTE, where the offset between grant and transmission is usually fixed, it is indicated in a dynamic way in the timing field in NR). The times between the reception of the grant by the UE and the actual transmission in NR are much shorter than in LTE, namely 0.08–0.71 ms (depending on the subcarrier spacing, and the device capability) compared to 3 ms in LTE, implying that a NR UE has to be able to prepare the transmit packets more quickly when it receives the scheduling grant.

Just like in LTE, a UL grant is per UE, so that a UE that has data for several logical channels can determine which data to send, compare Fig. 32.5. Thus, when a UE receives a grant, it has to first determine which of those data it wants to transmit on that grant. The philosophy of assigning resources to the different logical channels, with different priorities, is similar to LTE, but the details of the process are different. NR considers latency constraints implicitly, through constraints on the allowed transmission configurations. Most important among those is the maximum PUSCH duration that is possible to schedule for a particular logical channel, since long PUSCH duration indicates traffic that is not latency critical. This information is combined with information about the allowed subcarrier spacings (which also relate to the transmission duration). There can also be restrictions on the uplink carriers related to reliability considerations (e.g., forcing transmission of critical data on low-frequency channels). Once the set of admissible channels has been established, the scheduling is done based on priority, prioritized bit rates, and the queue backlog (*bucket size duration*).

If a UE does not have a PUSCH associated with it, it sends a scheduling request on the PUCCH. New in NR is that each UE can send multiple scheduling requests, indicating different data types and different priorities. The opportunity for sending such requests occurs

---

[22] In contrast to LTE, there is no mechanism to ensure in-sequence delivery; this was eliminated to reduce latency (note that sometimes even data delivered out of sequence are useful to the higher layers). Furthermore, the elimination of concatenation reduces latency, since concatenation can only be done after a scheduling grant is received and the transport block size for the UL is known.

periodically, with the periodicity ranging from every second OFDM symbol to 80 ms; the actual choice depends on the type of service that the UE is anticipated to send. A UE that has not been configured to have such periodic request opportunities can send a scheduling request using a random-access mechanism; this is mainly useful for IoT applications with a large number of devices and low traffic loads.

If a UE has a PUSCH associated, it sends buffer status reports, i.e., more information than simply "I have data to send." The buffer status reports group logical channels into a maximum of 8 groups, and report the status per group (grouping, instead of sending the status for each logical channel, is done for reduction of the feedback overhead). The reports can be either scheduled periodically or can be triggered by the arrival of data with higher priority than the current ones transmitted, since this might impact the scheduler decisions. The UE also sends power headroom reports, in which it reports the difference between the maximum possible power on a carrier and the power that it would have used to fulfill the rate scheduled by the BS if there were no restrictions on the UE power (thus, the power headroom can be negative, indicating that the rate scheduled by the BS would require more power than the UE can provide).

Semi-Persistent Scheduling

Overhead for scheduling can be reduced when a constant data stream can be anticipated. In this case, semi-persistent scheduling can be used, i.e., assigning particular time-frequency resources to a channel, with the details being configured by the RRC, and where further signaling activates or de-activates the transmission on that scheduled resource. This type of scheduling exists for both the DL and the UL, though the details of the configuration and control differ slightly.

Discontinuous Reception and Other Power Saving Scheduling

Energy consumption of the UE is even more important in NR than it is in LTE, in particular for IoT applications. Furthermore, the energy and computational load for a UE to monitor the PDCCH whether there are data scheduled for it are even higher (inter alia, PDCCH can be transmitted more frequently in NR). Thus, NR foresees several power-saving mechanisms. The first is *Discontinuous Reception, DRX*, which is similar to the same functionality in LTE.

In addition to it, NR also foresees a number of new approaches. Firstly, there are wake-up signals, such that the UE checks a wake-up signal before the beginning of a long DRX cycle, and goes to sleep if being told so by the wake-up signal. While this has the same functionality as checking the PDCCH whether it has data scheduled for the UE, decoding the wakeup signal requires less power, as it uses a special DCI format (format 2-6, see Section 32.4.7) that can be more easily detected.

Another innovation in NR is an indicator that tells the UEs whether all scheduling commands in a slot are intended for a later slot, which reduces the burden on the buffering of possible payload data.

Another way to save energy in NR is to configure a narrow bandwidth part for those times where little or no data are sent, since this implies that the UE only needs to monitor a narrow bandwidth. To enable reception with wider bandwidth, the bandwidth part indicator in the DCI points out that at a certain time in the future, the active bandwidth part will change. The newly chosen bandwidth part will either stay active until a timer expires (and then revert back to the previous value), or until another DCI command changes the active bandwidth part again.

32.5.4   Power Control

Similar to LTE, NR provides mechanisms for UL power control, by combining an open-loop control (for a coarse compensation, mostly of pathloss and shadowing) with a closed-loop control for fine adjustment, see Section 31.5.4.

The main difference to LTE lies in the beam-based power control. The pathloss is obviously different for different beams. As long as there is beam correspondence, i.e., the UE transmits on the same beam pair (TX beam at the UE, RX beam at the BS) in which it measured the DL pathloss, the standard power control algorithm holds. However, a PUSCH can be sent via different beam pairs at different times. Thus, NR foresees that the UE records and stores the pathloss for up to 4 beampairs, as obtained from different CSI-RS or SS transmissions. When the UE then gets a scheduling grant, together with the instruction of which beam to use, it uses the corresponding pathloss estimate for the power adjustment. It is furthermore also possible to have different values of the fractional pathloss compensation parameter and target power parameter P_0, for different PUSCHs.

*32.6   Carrier Aggregation and License-Assisted Access

32.6.1   Carrier Aggregation

Carrier aggregation allows to (dynamically) increase the transmission bandwidth. Just like in LTE, different devices, and even the UL and DL of the same device, can have different component carriers as their primary component, and can have different (and different number of) secondary component carriers. Control signaling is normally carried in the primary component carrier, but for the UL, a second component carrier can be configured for control signaling to avoid overloading one such carrier with too much control signals such as HARQ feedback.

There are different realizations of carrier aggregation, namely in-band aggregation with frequency-contiguous component carriers (which is very similar in spirit to just having a larger active bandwidth part), in-band aggregation with noncontiguous component carriers, and inter-band aggregation. Up to 16 component carriers can be aggregated, so that overall 6.4 GHz bandwidth can be used (although in practice a network operator might not have enough spectrum to fully exploit this). Activating and de-activating secondary component carriers can be done dynamically through MAC signaling.

The underlying scheduling decisions will typically be coordinated between the different carriers. However, a UE that is intended to receive data on multiple carriers will also receive multiple PDCCHs, where each PDCCH can point to its own carrier (self-scheduling), or a different one (cross-scheduling).

Besides carrier aggregation, NR also defines a *SUL*. This is used mainly for increasing the coverage of the UL (e.g., by assigning it a lower carrier frequency) – remember that the UL has a smaller TX power and thus typically a smaller coverage than the DL. The SUL carrier might also overlap with an existing LTE UL carrier; since the latter usually is not fully used anyway, additional information can be transmitted on it. A SUL differs from carrier aggregation in that a SUL is associated with a conventional DL/UL *pair*, while in carrier aggregation, each UL is associated with a particular DL carrier. Note that a supplementary DL already exists since LTE – it emerges implicitly from the fact that the number of DL component carriers can be larger than that for the UL.

32.6.2 License-Assisted Access and Stand-Alone Unlicensed Operation

Just like LTE, NR foresees license-assisted access, in order to exploit the availability of spectrum for which the operators do not need to pay for a license. Going beyond this, it also foresees stand-alone operation in unlicensed bands. One of the major differences between License Assisted Access (LAA) and standalone operation is that in the former case the control signaling can be transmitted in a licensed band, while in the latter case all signals have to be transmitted in the unlicensed band. Given the regulatory constraints, this implies that NR (mainly in release 16) adds a number of new mechanisms, in addition to the basic mechanisms already established in LTE, see Section 31.6.2. It is noteworthy that LTE in unlicensed bands is not widely used, but that NR tries to change this, and in particularly exploit the recently-available 6–7 GHz band, where it will compete with Wi-Fi 6E (see Section 33.6) for the spectrum.

As partly discussed in Section 31.6.2, there are some fundamental rules for operation in unlicensed spectrum that impact the system operation in unlicensed bands:

- *Dynamic Frequency Selection (DFS)*: an NR system must be ready to adjust its operating frequency, e.g., if a spectrum band is needed for another user, such as radar, wireless microphones, etc. That occupation of the band might last for comparatively long times (minutes to months).
- *Listen Before Talk (LBT)*: for many bands, it is mandated that a TX (be it BS or UE) determines whether the channel is occupied by another transmission before it transmits. For LTE and NR, that leads to the dynamic channel access procedure described briefly in Section 31.6.2, which is also very similar to the corresponding Wi-Fi procedure (see Section 33.4).
- Limited *Channel Occupancy Time (COT)*: to avoid one device "hogging" the channel, the COT is limited (depending on regulatory region) to a time between 2 and 10 ms. This, together with the dynamic channel access, determines transmissions of bursts and control signals.
- Limited power spectral density: many frequency regulators impose a limit on the power spectral density, so that greater TX power is allowed when the signal extends over a larger bandwidth (in many cases, a minimum of 20 MHz bandwidth is desirable). For narrower bandwidths, NR foresees a frequency-interleaved transmission, such that only every Nth RB is occupied (with $N = 10$ being typical), which nominally spreads the transmission bandwidth over larger bandwidths. Note that in this case, a victim RX should not treat the interference as spectrally white, requiring it to use more complicated RX algorithms to obtain optimum performance.

For the dynamic channel access, NR foresees two different mechanisms. The first is the above-mentioned LBT, where a device (UE or BS) transmits if the channel is free; if it is occupied, it waits until the other transmission has ended, and then observes the channel for $16 + 9n$ µs to see whether other transmissions occur, where the integer parameter n depends on the priority of the data. If the channel remains free during that time, it waits another random time (whose statistics depend on the priority, and the history of the transmissions), and if the channel still is not occupied, transmits. The threshold (in terms of received signal power) to declare the channel "available" differs depending on whether NR is the only system that may transmit in that band, of whether other systems, such as Wi-Fi, might contend for the channel as well.

Once the channel has been obtained by the network, it may use the remainder of the COT for additional transmissions from different nodes. In particular, if the next transmission occurs with a gap of less than 16 µs after the end of the previous one, no new channel sensing is required, though in this case the duration of the subsequent burst is limited to 584 µs (it is usually control information that is transmitted in this case). For longer bursts, channel sensing is required, though it does not perform a random backoff, and will thus yield to the above-described short bursts, but not "new" users. We stress again that this is allowed only within one COT, to avoid channel "hogging."

Another way to schedule users is semi-static scheduling. While the principle is the same as for licensed bands (i.e., periods of opportunities to transmit data), there is one key difference: transmission may only occur if the channel is free (it has to be free for at least 9 µs before the nominal start time of a scheduled opportunity – in other words, the access to the transmission period is similar to the dynamic channel access described above. The periodicity of the transmissions can be configured to be between 1 and 10 ms, and there has to be a gap of at least 5% of the period, so that other devices can grab the channel for dynamic access.

When the transmission bandwidth is larger than 20 MHz, then in all of the above-described methods it is required to sense the channel in every 20 MHz subband that creates the overall bandwidth. The system can then either transmit only in the subbands that are free, or wait until all subbands are free, and transmit only then.

The basic design of NR is well-suited for unlicensed bands, and most of the aspects of the transmission do not need to be modified compared to the discussions in previous sections. However, there are several modifications that do need to be implemented:

- Hybrid ARQ: in a licensed band, the BS informs the UE when it has to transmit the ACK. However, in an unlicensed band, the channel might not be free at that time. Early code blocks within a COT might be acknowledged at the end of the COT, but later blocks might not be processed fast enough such that the result of the checking of correct reception is available in time for transmission within the COT. For this (and other) reasons, NR allows to postpone the transmission of an ACK to a later, unspecified, time.

- Definition of a number of new parameters in DCI (in the formats 1-0 and 1-1), and UL scheduling grants (DCI formats 0-0 and 0-1) required for unlicensed transmission.

- Change of the PUCCH signaling: to accommodate the interlaced transmission mentioned above.

- Initial access: the biggest change here is that one cannot rely on a periodic transmission of the SS signals, since the channel might be occupied at the "normal" time, so that instead a window is defined within which to expect these signals. Furthermore, the channels required for initial access, namely the SSB (containing PSS, SSS, and PBCH), as well as the PDSCH carrying SIB1, and the PDCCH scheduling this PDSCH, are called a discovery burst, and all sent together.

- Random access: while the general procedure is the same as for licensed bands, two longer preamble sequences (length 1,151 for 15 kHz subcarrier spacing, and 571 for 30 kHz subcarrier spacing) are defined, to ensure that the RACH preambles fill out the whole 20 MHz spectrum, making it more likely that they are detected by other unlicensed users in the band.

*32.7 CoMP, Dual Connectivity, and HetNet Support

32.7.1 Inter-Cell Interference Coordination

For these aspects, there has not been much activity yet, compared to LTE. The section heading mainly exists to retain parallelism to the numbering in Chapter 31.

32.7.2 Multi-Point Coordination/Transmission

NR provides a variety of mechanisms for transmission from non-QCL TXs, which enables a richer set of multi-point transmissions in addition to those foreseen in LTE. For a discussion of these non-QLC methods, see Sections 32.3.6 and 32.3.7.

32.7.3 Dual Connectivity

Dual connectivity retains similar mechanisms (compared to LTE) in NR, but the importance of them has actually *increased*, in particular in the early stages of deployment. Many of those early deployments will run control signaling through an LTE BS, while the data plane will be carried by an NR BS – a classical example of dual connectivity. The backward-compatible structure of the NR waveform (15 kHz subcarrier spacing) enables a dynamic sharing of spectrum between LTE and NR. In principle, collisions between LTE and NR signals can just be avoided by suitable scheduling. There are, of course, the "always-on" signals in LTE such as cell-specific RS; those can be avoided in NR by the mechanisms for "reserved resources" (see also Downlink Scheduling Assignment in Section 32.4.7).

*32.8 Relaying

While the emphasis of relaying in LTE was on coverage extension, NR focuses more on the application of *integrated access and backhaul (IAB)*. The fundamental architecture makes use of the possible split of the BS functionality into a centralized and a distributed unit (DU), which are connected through the standardized F1 interface. The node that is directly connected to the core network through a non-IAB backhaul is called a *donor node*, while the "remote BS node" is called the IAB node. Each of those nodes can be thought of consisting of two functionalities: (i) facing the core network, the donor node looks like a standard CU. Facing the IAB node, it looks like a DU, with all the PHY/MAC functionality of such a node (also, this node can have direct communication with a UE within its range through the DU). (ii) facing the UE, the IAB node looks like a standard DU. Facing the donor node, it looks essentially like another UE (and for this reason this component is called Mobile Terminal (MT) functionality[23]). A typical setup is shown in Figure 32.16. Note that multiple IAB nodes can be concatenated, allowing multi-hop IAB.

---

[23] An unfortunate name, since IAB nodes are usually at a fixed location.

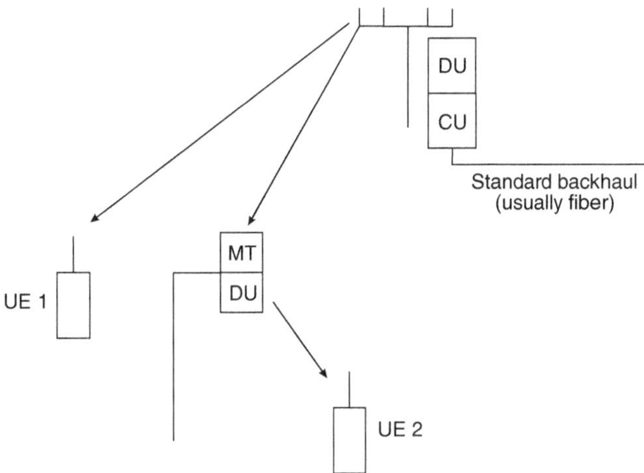

Figure 32.16   Architecture for IAB.

The transmission protocol does not require large changes. In particular, the PHY, MAC, and RLC layer are (almost) unchanged, but there is now a *Backhaul Adaptation Protocol* (BAP), on top of the RCL (and below the PDCP). It adds a header that essentially tells the PDCP unit how to get from the donor BS to the finally transmitting IAB node, and vice versa. Also, initial access and random access do not change significantly – an IAB node has to perform cell search and initial access like any other UE connecting to a donor node (though, with the IAB node typically being fixed and always-on, this happens more rarely than for a UE). The random access preambles are also similar. The main difference between a UE and an IAB node is that the latter can be farther away than the cell edge (this is possible due to more favorable propagation conditions, and possibly higher antenna gain and powers). Thus, the runlength considered in the RACH preamble should be larger than for standard UEs; NR allows to separately configure them.

Returning to the IAB nodes, the reception/transmissions of the MT and DU parts need to be carefully coordinated. Except for those cases where these two parts have antennas that are spatially separated and have exceptionally good isolation (e.g., inside and outside a steel-concrete building operating at mm-wave frequencies) or operate in different frequency bands, the DU should not transmit while the MT is receiving, and vice versa, as this would require true full-duplex operation (see Section 18.5 for a discussion of the challenges). However, it *is* possible that the IAB node simultaneously receives from the donor node and the UE, or simultaneously transmits to them, assuming that the data are on orthogonal resources, e.g., when access link and backhaul link operate in different frequency bands. The separation of these transmissions follows a different logic than in LTE, where the always-on signals constituted the major problem. Rather, in NR, the MT part and the DU part of the IAB node can each configure the radio resources as DL, UL, or flexible, according to the standard NR approach (see section titled "Duplexing Schemes" in Section 32.3.4). As a further refinement, the DU assignment can be hard or soft: in the former case, the allocation is fixed, thus the parent node has to schedule such that it does not conflict with this allocation. If the assignment is soft, then, e.g., the DU part of the IAB will not actually schedule a transmission if it would interfere with a reception of the MT, and the donor node knows this and can rely on it. The configuration of the resources can be done on a slot-by-slot basis.

*32.9   NR for Machine-Type Communications

Machine-type applications are of great interest for NR, in particular in the context of industrial automation. The topic has already been explored in LTE, see Section 31.9. However, the emphasis of the works is quite different. In LTE, the main emphasis was on enabling transmissions in a bandwidth that is much narrower than the standard 20 MHz channel, either through enhanced Machine Type Communications (eMTC) or Narrowband IoT systems.[24] For NR, this is less of a concern, since it already foresees a flexible adaptation of the bandwidth through the definition of bandwidth parts (which can be quite narrow). Rather, the emphasis is on provisioning of low latency, possibly combined with ultra-high reliability. Low energy consumption through so-called Reduced Device Capability will be considered in Release 17.

The low latency is achieved partly through the fundamental design approach, and partly through small enhancements in release 16 that in combination improve latency:

- Flexible slot structure: as mentioned previously, NR has a flexible slot structure. By choosing large subcarrier spacing, the duration of a slot can be made shorter. The fact that transmission need not take up a complete slot, and can start at times different from the slot boundary, further helps. Since release 16, it is even possible to have transmissions extend across slot boundaries.
- The slots themselves are "front-loaded," meaning containing all information needed for decoding in the first OFDM symbols of a slot, so that decoding can start while the transmission is still ongoing. Furthermore, standard-compliant UEs have to be able to perform the processing quite quickly, enabling a faster turnaround.

---

[24] The same eMTC and NB-IoT can be deployed in 5G as well, with some minimal physical changes.

- Various feedback structures have been added to enable fast feedback for low-latency services. For example, HARQ feedback delay can now be configured for both DL and UL, whereas in LTE a fixed delay was mandated.
- Preemption: the standard foresees that transmissions for services with relaxed latency requirements can be interrupted (preempted) by urgent data (transmissions with a low-latency requirement). For the DL, this mechanism was already described in Section 32.5.3. For the UL, there are two possibilities: (i) the BS sends out a cancellation message (DCI format 2-4); this has to be done sufficiently long before the originally scheduled PUSCH) that the UL scheduling assignment has been canceled; the drawback is that a UE whose transmission might be cancelled has to frequently (several times per slot) monitor the channel for possible cancellation messages. (ii) UL power boosting, i.e., transmitting the urgent message with higher power. The amount of power boosting is communicated to the UE with the urgent message through the DCI. This method is simpler because it does not require the other UEs to do anything special, but it has the drawback that it does not work if the UE with the urgent message is at the cell edge and already transmits with the highest power.
- High-priority transmission: while the scheduling of various logical channels from a UE follows complicated rules, see Section 32.5.3, it can happen that a high-priority message is not scheduled. Such a message might be, e.g., the HARQ feedback for a high-priority DL. Since release 16, there is the possibility for the scheduler to mark a message as high priority, using a single bit; in this case, it will be scheduled if there are *any* resources available for the UL.
- For semi-persistent scheduling, DL repetition frequencies as low as 0.125 ms (with 120 kHz subcarrier spacing) can be configured since release 16, while in the UL they can have periodicities as short as every second OFDM symbol. This essentially allows to give a guaranteed low-latency channel to a particular device. It is furthermore possible to restrict those grants to particular services. A typical example would be a machine that sends regular control information containing only a few bits and then at larger intervals large status reports that are less latency critical. The scheduler then can make sure that the less-critical information does not grab the configured semi-static UL grant.

There are also several provisions that help to increase reliability:

- Multi-site connectivity increases the resilience with respect to large-scale fading and pathloss. Furthermore, PDCP duplication is allowed, i.e., multiple (up to 4) versions of the same PDCP PDU are transmitted in multiple logical channels.
- The CQI feedback from the UE requests a particular modulation format, so that a UE can incorporate a particular "safety margin" when requesting a certain MCS.

*32.10   Device-to-Device Communications – Sidelink

32.10.1   Motivation, Architecture, and Channel Structure

Sidelink in NR shows significant differences to sidelink in LTE. These differences stem mainly from three sources:

1. Change in underlying philosophy: Sidelink in LTE was mainly a transmission-oriented system, i.e., communication went only in one direction; bi-directional communication had to be pieced together from two such single-directional systems. Sidelink in NR establishes "normal" bi-directional communication channels between devices, which necessitates new aspects such as HARQ feedback and CSI reports.
2. Change in the "normal" standard: Sidelink aims to be compatible, and reuse as many components as possible, of the standard cellular ULs and DLs. Since many of these components have changed in NR (e.g., use of OFDM as the default mode for the UL), the corresponding parts in the sidelink standard have changed as well.
3. The main envisioned application for Sidelink has changed from emergency communication and brief location-specific announcements to vehicle-to-vehicle communications. This has implications not only on the communications structure mentioned above but also in particular on the discovery process (note the absence of a discovery channel in NR).

Sidelink retains most of the basic mapping between logical, transport, and physical channels (except that now also a logical control channel exists). However, some of the structure carried by these channels changes. The list of NR Sidelink physical channels is thus:

- *Physical Sidelink Broadcast Channel (PSBCH)* and *Sidelink Synchronization Signal (S-SS)*: carries the information required for synchronization, as well as the *Sidelink Master Information Block (S-MIB)*. These structures, which will be described in more detail below, fulfill a similar role as in standard NR systems.
- *Physical Sidelink Control Channel (PSCCH)*: contains part of the *Sidelink Control Information (SCI)*, in particular the parts that are required for the demodulation of the PSSCH (see below), and the resource reservation. This channel needs to be broadcast with a known format, since not only the devices involved in the communication but also the devices within range of the transmission need to hear it, to avoid scheduling colliding transmissions.
- *Physical Sidelink Shared Channel (PSSCH)*: contains other parts of the SCI (the ones that are only needed at the intended RX), as well as the payload data, i.e., the sidelink traffic channel is mapped onto it.
- *Physical Sidelink Feedback Channel (PSFCH)*: this is a new physical channel in NR, and carries the HARQ feedback information.

A UE can be configured with a Sidelink resource pool – the concept is similar to LTE, while the specifics are different. In the-frequency domain, there is a contiguous band, which is divided into contiguous subchannels (with a configurable width of 10, 15, 20, 25, 50, 75, or 100 RBs). In the time domain, a set of – possibly noncontiguous – slots is defined, which are repeated with a configurable resource-pool period. Within each slot, a contiguous block of (7–14) OFDM symbols is configured to be used for sidelink communication, while the remainders of the symbols are available for UL and DL; the starting symbol of this block may be configured as well.

32.10.2 Synchronization

Sidelink synchronization distinguishes between the cases where the UE obtains synchronization information from the network directly, or via another UE that is synched to the network (multi-hop synchronization is possible), or without any network infrastructure in place. These cases and their synchronization approach are similar to LTE, see Section 31.10.2.

The structure of the synchronization block is similar to that of the standard NR DL, consisting of a Sidelink Primary Synchronization Signal (S-PSS), Sidelink Secondary Synchronization Signal (S-SSS), and the PSBCH. In the frequency domain, the S-PSS and S-SSS extend over 127 subcarriers, while the PBSCH extends over 132 subcarriers (note that this is a narrower bandwidth than that used for the DL synchronization). In the time domain, the synchronization block extends over one slot, where the S-PSS is sent on symbol 2 and 3, S-SSS on symbol 4 and 5, and PSBCH on symbols 0 and 6–12; symbol 13 is left empty. The transmit signals on the S-PSS and S-SSS are length 127 m-sequences and Gold sequences, respectively, just like for the DL SSs. There are two groups of sidelink identities (one for in-network-range devices or those that have (possibly indirect) connections to in-network-range devices, and one for out-of-network-range devices, i.e., when there is no infrastructure around), with 336 identities each, which are communicated by the combination of S-PSS and S-SSS signals.

32.10.3 Discovery and Resource Allocation

The allocation of resources for a sidelink transmission can happen in one of two ways:

- Mode 1: the resources are assigned by the BS. In other words, the UE is assigned a scheduling grant, which is communicated via DCI format 3-0 (see Section 32.4.6). Resources are allocated within a window that extends over 32 slots in the time domain; see Sec. 32.10.1 for the frequency domain properties. During this window, up to 3 sidelink resources are scheduled, each of which consists of one slot in the time domain and a configurable extent in the frequency domain. While the bandwidth of the 3 scheduled resources has to be the same, the location in frequency can be different. It is also possible to have (semi-persistently) configured grants, again following the same approach as for the UL. It is noteworthy that also, in this case, the prospective transmitting UE has to make an announcement about the resources it will use in a broadcast announcement to the devices around it, so that devices operating in mode 2 can sense and appropriately react.
- Mode 2: alternatively, a device might decide by itself when, and on what resources, to transmit. However, it has to base this decision on a sensing process, followed by an announcement to surrounding devices about what resources it will use. Based on the announced resource claims of all the other devices, a UE can then decide on what resources it wants to transmit; depending on its priority, it will honor the reservations from other devices (if the resulting SINR at the intended RX would be below a particular threshold) or transmit anyway. If a device gets too few resources with the current power settings, it can increase the power and see whether it now gets more resources (because the SINR at the prospective RX is better).

A potentially receiving UE does not need to know whether the TX uses Mode 1 or 2; it has to monitor the resource reservation announcements and be ready to receive.

32.10.4 Communications

We have discussed above the general resource structure, in particular, that the available resources are (i) in the time domain, a block of N_{SLS} OFDM symbols within a slot, starting at time i_{start} and (ii) in the frequency domain, M contiguous subchannels, indexed $j = 0,$ ...$M - 1$. The PSCCH occupies then the OFDM symbols $i_{\mathrm{start}} + 1$, $i_{\mathrm{start}} + 2$, and possibly $i_{\mathrm{start}} + 3$ in subchannel $j = 0$ (bandwidth can be smaller than a subchannel). The OFDM symbol i_{start} is, for all j, a copy of the symbol at $i_{\mathrm{start}} + 1$, and is used as a preamble to set the AGC. The last OFDM symbol, i.e., $i = i_{\mathrm{start}} + N_{\mathrm{SLS}} - 1$ is a guard interval to allow switching from sidelink to UL/DL operation, or between sidelink TX and RX. The rest of the resources are used for the PSSCH; see Figure 32.17.

The information on the PSCCH is encoded with the same polar codes as DCI in regular NR transmissions, modulated with QPSK, and then transmitted via OFDM. The SCI information on the PSSCH is also encoded with polar codes and transmitted with QPSK, while the payload data are encoded with LDPC codes and modulated with up to 256 QAM, just as in regular NR transmissions.

The feedback channel, PSFCH, which carries the HARQ acknowledgments, has the same basic structure as the PUCCH, i.e., modulating a frequency domain base sequence of length 12 with different phase rotations. The resulting sequence is mapped to a single RB. If this happens, the regular sidelink transmission is only sent up to OFDM symbol $i_{\mathrm{start}} + N_{\mathrm{SLS}} - 5$. The symbol $i_{\mathrm{start}} + N_{\mathrm{SLS}} - 4$ (across all subbands) is used as a guard interval, while $i_{\mathrm{start}} + N_{\mathrm{SLS}} - 3$ and $i_{\mathrm{start}} + N_{\mathrm{SLS}} - 2$ carry two copies of the PSFCH signal in one RB (the first copy is used for setting the AGC – remember that the PSFCH is transmitted in the opposite direction as the original data

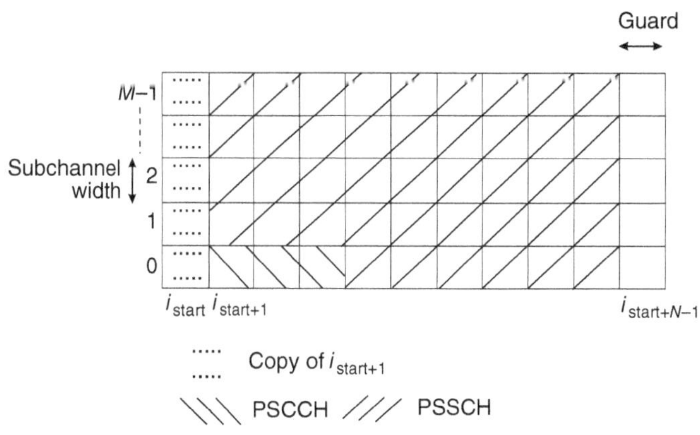

Figure 32.17 Structure of PSCCH and PSSCH within a slot.

transmission). Symbol $i_{start} + N_{SLS} - 1$ is then used as the "regular" guard interval as described above, see Figure 32.18. Since transmission of the PSFCH costs 3 OFDM symbols across many RBs, while providing little information, PSFCH can be configured to be transmitted only in every second or fourth slot. Generally, the HARQ feedback can be either just a NACK, or ACK/NACK. In the case of groupcast, a NACK-only feedback has the advantage that all devices can transmit on the same resources – as long as the original TX can hear that *somebody* sent a NACK, it knows that it has to retransmit. If ACK/NACK feedback is used, each device needs its own feedback resource.

In addition to HARQ feedback, sidelink can also transmit CSI reports. However, this is not done on a separate channel, but rather as part of a PSSCH transmission (with appropriate indications in the MAC control signaling).

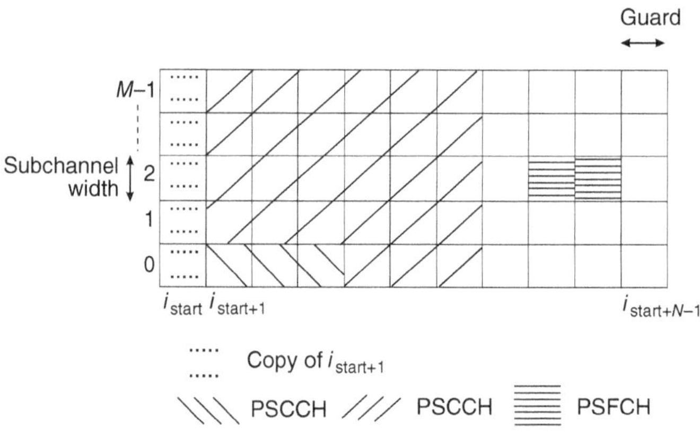

Figure 32.18 Structure of PSCCH, PSSCH, and PSFCH in a slot.

Glossary for 5G-NR

5G-EIR	5G-Equipment Identity Register
AM	Acknowledged Mode
AMF	Access and Mobility management Functions
BAP	Backhaul Adaptation Protocol
BCCH	Broadcast Control CHannel
BCH	Broadcast CHannel
BG	Base Graph
CBG	CodeBlock Group
CBGTI	Code Block Group Transmit Indicator
CCCH	Common Control CHannel
CCE	Control Channel Element
CDM	Code Division Multiplexing
CHF	CHarging Function
CORESET	COntrol REsource SET

COT	Channel Occupancy Time
CQI	Channel Quality Indication
CRB	Common Resource Block
C-RNTI	Cell Radio Network Temporary Identifier
CRS	Cell-specific Reference Signal
CSI-IM	Channel State Information – Interference Measurement
CSI-RS	Channel State Information – Reference Signal
CU	Central Unit
DAI	Downlink Assignment Index
DCI	Downlink Control Information
DL-SCH	DownLink Shared CHannel
DM-RS	DeModulation Reference Signal
DCCH	Dedicated Control CHannel
DFI	Downlink Feedback Information
DFS	Dynamic Frequency Selection
DTCH	Dedicated Traffic CHannel
DU	Distributed Unit
eMTC	enhanced Machine-Type Communications
EPC	Enhanced Packet Core
FR	Frequency Range
IAB	Integrated Access and Backhaul
LAA	License Assisted Access
MIB	Master Information Block
MT	Mobile Terminal
NDI	New Data Indicator
NG	Next Generation
NWDAF	NetWork Data Analytics Function
NZP	Non-Zero Power
OCC	Orthogonal Cover Code
PBCH	Physical Broadcast CHannel
PCCH	Paging Control CHannel
PDCP	Packet Data Convergence Protocol
PCH	Paging CHannel
PCFICH	Physical Control Format Indicator CHannel
PDCCH	Physical Downlink Control CHannel
PDSCH	Physical Downlink Shared CHannel
PDU	Protocol Data Units
PID	Physical cell IDentity
PMI	Precoding Matrix Indicator
PRACH	Physical Random Access CHannel
PRB	Physical Resource Block
PRG	Precoding Resource-Block Group
PRS	Positioning Reference Signal
PSBCH	Physical Sidelink Broadcast CHannel
PSCCH	Physical Sidelink Control CHannel
PSFCH	Physical Sidelink Feedback CHannel
PSS	Primary Synchronization Signal
PSSCH	Physical Sidelink Shared CHannel
PT-RS	Phase Tracking Reference Signal
PUCCH	Physical Uplink Control CHannel
PUSCH	Physical Uplink Shared CHannel
QCL	Quasi-Co-Located
RACH	Random Access CHannel
RAN	Radio Access Network
RB	Resource Block
RE	Resource Element
REG	Resource Element Group
RI	Rank Index
RLC	Radio Link Control
RNTI	Radio Network Temporary Identifier
RRC	Radio Resource Control

RS	Reference Signal
RSRP	Reference Signal Received Power
SCI	Sidelink Control Information
SDAP	Service Data Application Protocol
SFI	Slot Format Indicator
SIB	System Information block
S-MIB	Sidelink Master Information Block
SMF	Session Management Function
S-PSS	Sidelink Primary Synchronization Signal
SRS	Sounding Reference Signal
SS	Synchronization Signal
SSB	Synchronization Signal Block
SSS	Secondary Synchronization Signal
S-SS	Sidelink Synchronization Signal
S-SSS	Sidelink Secondary Synchronization Signal
SUL	Supplementary Uplink
TCI	Transmission Configuration Indication
TM	Transparent Mode
TRP	Transmit-Receive Point
TRS	Tracking Reference Signal
TTI	Transmission Time Interval
UCI	Uplink Control Information
UCMF	UE radio Capability Management Function
UM	Unacknowledged Mode
UPF	User Plane Function
UDR	Unified Data Repository
UL-SCH	UpLink Shared CHannel
URLLC	Ultra-Reliable Low Latency Communications
VRB	Virtual Resource Block
ZP	Zero Power

Further Reading

The most important resource for 5G-NR is, of course, the standard of 3GPP itself; this is especially true as there are constantly new features evolving in the releases (remember that the current chapter describes up to Release 16). That being said, the same holds true for the 5G-NR standards documents as for LTE, namely that they are very difficult to read and so voluminous that they are mainly useful as reference. An excellent description of the standard can be found in the monograph of [Dahlman et al. 2020], which describes the standard in a didactic yet accurate way; similar to the book by the same authors on LTE, it describes not only the "what" but also the "why" of the standard. Another excellent reference that similarly describes both what and why is [Chen et al. 2021], which also gives an outlook on future releases. Several other books, including [Ahmadi 2019], and [Holma et al. 2020], also describe the whole standard. [Asplund et al. 2020] focus on the multiple-antenna aspects, while [Rommer et al. 2019] concentrate on the core network (which we have consciously omitted in this chapter). [Polese et al. 2020] describes IAB, and [Lien et al. 2020] Sidelink. [Shafi et al. 2017] summarize both propagation aspects and early testbed results. When studying academic papers with the words "5G" in the title, great care needs to be taken. In particular, early papers (when the standard was not yet finished) tended to use the moniker 5G as a synonym for advanced technology, and even later papers often develop suggested improvements on the standard that are not stand-ard-compliant. It is thus necessary to carefully check whether a description or simulation actually describes a 3GPP NR compliant setup or not.

For updates and errata for this chapter, see https://wides.usc.edu/students.html#textbooks.

Exercises

See Sec. 36.32 of Exercises.pdf at wiley.com/go/molisch/wireless3e

33

Wireless Local Area Networks – 802.11/Wi-Fi

33.1 Introduction

33.1.1 History

In the late 1990s, wired fast internet connections became widespread both in office buildings and in private residences. For companies, a fast intranet as well as fast connections to the internet became a necessity. Furthermore, private consumers became frustrated with long download times of dial-up connections for elaborate webpages, music, etc., as connection speeds were limited to 56 kbit/s. They therefore opted for cable connections (several Mbit/s) or DSL (up to 1 Mbit/s in the United States, and more than 20 Mbit/s in Asia) for their computer connections. At the same time, laptop computers started to be widely used in the workplace. This combination of factors spurred demand for wireless data connections – from the laptop to the nearest wired Ethernet port – that could match the speed of the new wired connections.

In the following years, two rival standards were developed. The ETSI (European Telecommunications Standards Institute) started to develop the HIPERLAN (*HIghPERformance Local Area Network*) standard, while the IEEE (*Institute of Electrical and Electronic Engineers*) established the 802.11 standard (the number 802 refers to all wireless standards of the IEEE, and the suffix 11 was assigned for *Wireless Local Area Networks* [*WLANs*]). In the subsequent years, the 802.11 standard gained widespread acceptance, while HIPERLAN became essentially extinct.

Actually, it is not correct to speak of *the* 802.11 standard. The standard encompasses some 50 different variants, amendments, and extensions, see Table 33.1, which are not all interoperable. To understand the terminology, we first have to summarize the history of the standard. The "original" 802.11 standard was intended to provide data rates of 2 Mbit/s[1]; since it operated in the 2.45 GHz ISM (*Industrial, Scientific, and Medical*) band, the frequency regulator in the USA (FCC) required that spectrum spreading techniques be used. For this reason, the original 802.11 standard defined two modes: (i) frequency hopping, and (ii) direct-sequence spreading; those two modes were incompatible with each other.

It soon became obvious that higher data rates were demanded by the users. Two subgroups were formed: 802.11a, which investigated Orthogonal Frequency Division Multiplexing (OFDM)-based schemes, and 802.11b, which attempted to retain the direct-sequence approach. 802.11b became popular first; it defined a standard that allowed 11 Mbit/s data rate in a 20 MHz channel. Though the scheme was not really spread-spectrum anymore, the FCC approved its use. The standard was later adopted by an industry group called Wi-Fi (Wireless Fidelity), which was formed to ensure true interoperability between all Wi-Fi-certified products.[2] Wi-Fi products based on 802.11b gained widespread market acceptance after 2000.

However, the data rate of 11 Mbit/s still was not sufficient for many applications – especially in light of the fact that the actual throughput in practical situations was closer to 3 Mbit/s. For this reason, the work of the 802.11a group became of greater interest. 802.11a specified an alternative physical layer that uses OFDM and higher-order modulation alphabets, allowing up to 54 Mbit/s data transfer rate (again, this rate is nominal, and the true throughput is lower by about a factor of three). This mode uses, for regulatory reasons, a different frequency band (above 5 GHz), which furthermore has the advantage of being less "crowded," i.e., has to deal with fewer interferers, though it has worse propagation conditions. When frequency regulators declared OFDM-based WLAN transmissions to be acceptable also in the 2.45 GHz band, a further standard, 802.11g, was introduced, which – apart from the different band – is essentially identical to 802.11a. Further modifications of the 802.11a standard are provided by the 802.11h and 802.11j standards, which adapt it to the European and Japanese regulations, respectively.

Also, the original MAC (Medium Access Control) was amended: the 802.11e standard provides modifications to the MAC that allow to better ensure certain levels of *Quality of Service* (QoS). Additionally, a number of further subgroups of the 802.11 standardization group have been formed, all dealing with "amendments" and "additions" to the original standard. Realistically speaking, though, an 802.11a device, using the 802.11e MAC, bears no resemblance to the original 802.11 standard.

---

[1] Strictly speaking there were three different transmission methods in this standard: a direct-sequence standard with 2 Mbit/s that was later on evolved to provide 11 Mbit/s in 802.11b, as well as frequency hopping and infrared methods that never caught on and were eventually removed from the standard.

[2] Not all 802.11 products, even if following a particular version like 802.11b, are completely interoperable. Wi-Fi imposes additional requirements and constraints, often using only a subset of the 802.11-functionality, and – in contrast to 802.11 – tests interoperability before certifying products.

Table 33.1 The IEEE 802.11 standards and their main focus

Standard	Scope
802.11 (original)	WLAN standard that includes both MAC and PHY functions
802.11a	High-speed (up to 54 Mbps) PHY supplement in the 5 GHz band
802.11b	High-speed (up to 11 Mbps) PHY extension in the 2.4 GHz band
802.11d	Operation in additional regulatory domains
802.11e	Enhancement of the original 802.11 MAC to support QoS (applies to 802.11a/b/g)
802.11f	Recommended practice for interaccess point protocol (applies to 802.11a/b/g)
802.11g	Higher rate (up to 54 Mbps) PHY extension in the 2.4 GHz band
802.11h	Define MAC functions that allow 802.11a products to meet European regulatory requirements
802.11i	Enhancement of 802.11 MAC to provide improvement in security (applies to 802.11a/b/g)
802.11j	Enhancement of 802.11 MAC and 802.11a PHY to operate in Japanese 4.9 GHz and 5 GHz bands
802.11n	Enhancement of 802.11a and 802.11 PHY to operate at data rates up to 600 Mbit/s
802.11p	Modification of 802.11a standard for car-to-car communications
802.11s	Protocol for autoconfiguring mesh networks
802.11w	Provision of data integrity and authentication
802.11ac	Enhancements to PHY and MAC layer to increase throughput to 6.9 Gbit/s multi-link. Also known as Wi-Fi 5
80.11ad	New physical layer operating in the 60 GHz frequency band
802.11ah	Definition of WLAN operating at sub-1 GHz license-exempt bands
802.11ax	Further throughput enhancements compared to 11ac (theoretical data rate up to 9.6 Gbit/s), and provisions for OFDMA multiple access. Also known as Wi-Fi 6
802.11ay	Enhancement of 802.11ad standard for data rates of up to 100 Gbit/s
802.11bd	Enhancement 802.11p for higher throughput
802.11be	Future enhancement of 802.11ax

Since the early 2000s, the requirement for data rates in WLANs has steadily increased. This was partly motivated by the increase in data rates of the cabled connections, which obviously required that the wireless "last 10 m" would not become a bottleneck. In parts of the world, internet connection speed of 1 Gbit/s are by now common. However, even larger data rates are required for communications *between* devices in an office or home, e.g., file exchange between computers, backups, or uncompressed 4k video streaming. Furthermore, Voice-over-IP (VoIP) applications, which require lower data rates, but high reliability and consistent QoS, were also becoming widespread. The 802.11 standard started to accommodate this development with the introduction, in 2009, of 802.11n, which provides several hundred MBit/s through the use of Multiple-Input–Multiple-Output (MIMO) technology as well as larger bandwidths. In order to achieve even higher data rates, two separate approaches were taken in the years after: one was the increase of the number of antenna elements to achieve more data streams both with Single-User (SU)- and Multi-User (MU)-MIMO; the resulting standard, named 802.11ac, was approved in 2013, retroactively called *Wi-Fi 5*. Further enhancements along this avenue are in the 802.11ax standard, which was ratified in 2021; it is also known under the acronym *Wi-Fi 6;* further improvements are currently being deliberated in the 802.11be group. The other approach is the use of mm-wave spectrum, specifically the unlicensed spectrum around 60 GHz. Due to the large available bandwidth there, high data rates can be achieved. The 11ad standard was approved in 2012, and its enhancement 802.11ay in 2021.

Due to the multitude of 802.11 standards, this chapter only presents the most important. In the following, we give an overview of the 802.11a/g, 802.11n, 802.11ac, and 802.11ax physical layers, and the 802.11 and 802.11e MAC layer (and their enhancements in 802.11ac and ax). The description will be done in the sequence of their development, which has not only historical reasons but also is helpful to explain a lot of the technical aspects – many details can only be understood by considering them as modifications of the previous standard version, with a requirement for backward compatibility.

More details can be found, as always, in the official standards publications [www.802world.com] and a multitude of books that has been published on that topic. An excellent summary of the earlier versions of the standard can be found in [O'Hara and Petrick 2005], while the 802.11n and 802.11ac standards are described in [Perahia and Stacey 2015].

IEEE 802.11 vs. Wi-Fi Alliance: The IEEE 802.11 working group is a standards organization created by IEEE, an international professional organization, see Section 30.2.6. While it develops standards, experience has shown that those suffer from two drawbacks: (i) lack of prescribed testing procedures, and (ii) large range of options; both are obstacles to interoperability of devices. For this reason, a number of companies founded in the late 1990s the Wireless Ethernet Compatibility Alliance, which was later rebranded Wi-Fi Alliance. It defines the set of options that a device has to offer to gain the "Wi-Fi-certified" label, and also establishes testing procedures for compatibility that every certified device has to pass. Naturally, companies that are members of the Wi-Fi Alliance tend to be active in IEEE 802.11. Still, the goals, and procedures of the two organizations are not identical. More details can be found in Chapter 30.

Nomenclature: The nomenclature of WLAN deviates somewhat from that of cellular systems. An infrastructure node is called an Access Point (AP), instead of Base Station (BS). In 802.11 notation, a STA denotes a generic device, with a client device (i.e., a UE) called non-AP STA; however, in the Wi-Fi Alliance parlance and in the general public usage, a Station (STA) is a client device. In order to retain a consistent notation, we will retain the terminology of the other chapters, i.e., BS and UE.

33.1.2 Applications

The main markets for WLANs are

- *Wireless networks in office buildings and private homes*, to allow unhindered internet access from anywhere within a building. BSs for wireless LANs need to follow the specifications, but also allow a certain amount of vendor distinction. For example, different numbers of antenna can be used, without leading to incompatibilities. On the UE side, WLAN chipsets have turned into a mass market, often with little vendor distinction, and are built into laptops, cellphones, and many household appliances. As a consequence, research tends to focus on methods for reduction of production cost (implementation with smaller chip area, low-cost semiconductor technology), and energy consumption, while research on BSs includes a broader field of topics.
- *"Hot spots,"* i.e., WLAN BSs that allow the public to connect to the internet. These hot spots are often set up in coffee shops, hotels, airports, etc. Several providers also have a "nationwide" or even "continent-wide" network of hot spots, so that subscribers can log in at many different locations.[3] However, it must be stressed that the coverage of those networks is much lower than for cellular networks. For this reason, research is ongoing on how to seamlessly integrate WLANs with cellular networks or large-area networks.
- *Enterprise WLANs*, i.e., covering a larger campus with a number of coordinated BSs, see also Chapter 1. While 802.11 foresees very few methods for active coordination between different BSs, suitable deployment planning and assignment of frequency channels can enhance SINR. A number of companies have specialized in setting up and managing enterprise WLANs.

Wi-Fi has had an enormous success. By 2020, there were 13 billion installed devices, with accelerating growth.

33.1.3 Relationship Between MAC and PHY Layer

Before going into the details of the MAC and PHY layer, we first have to establish some notation used by the 802.11 community. The data payload received from the upper layers is attached to headers and trailers at both the MAC and PHY layer before it gets transmitted on the air. For example, each *MAC Service Data Unit* (MSDU) received from the *Logic Link Control* layer is appended with a 30-byte-long MAC header and a 4-byte-long *Frame Check Sequence* (FCS) trailer to form the *MAC Protocol Data Unit* (MPDU). The same MPDU, once handed over to the physical layer, is then called *Physical Layer Service Data Unit* (PSDU). Then a *Physical Layer Convergence Procedure* (PLCP) preamble and header, and proper tail bits and pad bits are attached to the PSDU to finally generate the *Physical Layer Protocol Data Unit* (PPDU) for transmission. The relationships among MSDU, MPDU, PSDU, and PPDU are illustrated in Figure 33.1.

From just this brief paragraph, the reader will have seen that – as with most standards – the alphabet soup of the numerous acronyms is a major hurdle for understanding that standard. For this reason, the appendix provides a list of acronyms and their meaning.

33.1.4 Spectrum Considerations

802.11 WLANs operate primarily in three frequency bands, the 2.45 GHz ISM band, the 5 GHz band, and – more recently – the 6 GHz band.[4] In the following, we describe the rules and limits for those bands in more detail. It must be noted that emission limits and other spectrum rules can differ from country to country; most of the discussion here concentrates on the rules in the USA (as set by the frequency regulator there, the Federal Communications Commission [FCC]) as an example, though rules in several other countries are mentioned as well.

The 2.45 GHz band has more favorable propagation conditions when single-antenna transceivers are used, due to the lower isotropic pathloss (compare Section 4.1) and the better penetration through walls and diffraction around obstacles. However, this band is significantly more crowded, and also offers only a relatively small bandwidth. The 5 and 6 GHz bands do provide much wider frequency channels, namely a maximum of 160 MHz (for the next version of the standard, 320 MHz are expected, see Section 33.6.6). In the 2.45 GHz band, a maximum of 40 MHz is foreseen, though in practice this is restricted to 20 MHz because the larger bandwidth negatively impacts coexistence with Bluetooth (see Section 34.1).

The 2.45 GHz ISM band extends from 2.4 to 2.5 GHz. For the purpose of Wi-Fi, a number of channels are defined, as outlined in Table 33.2. The spectrum mask requires that the signal is 20 dB attenuated at ± 11 MHz offset from the center frequency compared to the peak power spectral density that occurs at the center frequency (in short, the 20 dB bandwidth is 22 MHz). In the USA, only channels 1–11 are allowed, while most of the rest of the world allows also channels 12 and 13. Two 20 MHz channels may be combined, resulting in 40 MHz maximum bandwidth, though this is not often used, see above.

---

[3] In most cases, users are charged a per-minute fee or a flat rate for 24 hours of usage. Nationwide networks often have an option for monthly or annual subscriptions.

[4] Further bands, ranging from < 1 GHz to 60 GHz and even visible light, have been defined or are currently explored. However, the three mentioned bands have the greatest practical importance.

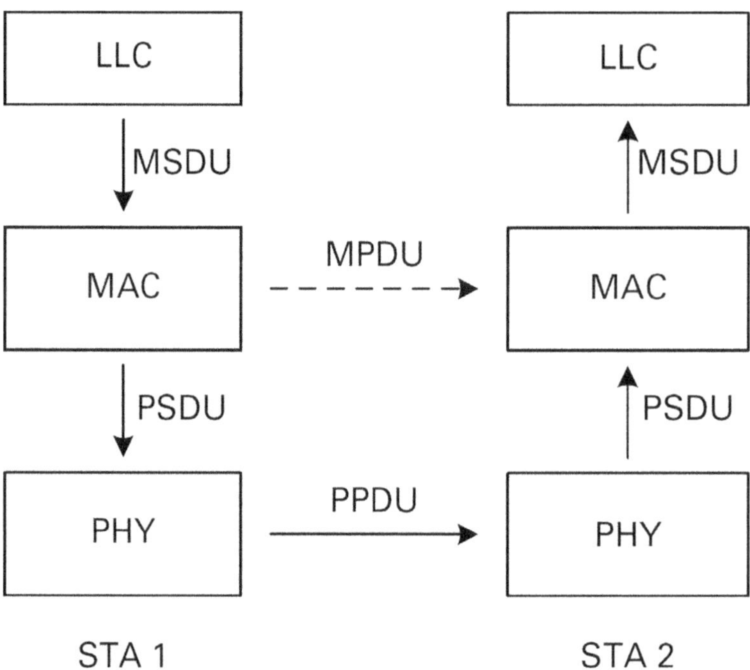

Figure 33.1 Relations among the MAC service data unit, MAC protocol data unit, physical layer service data unit, and physical layer protocol data unit. STA 1 and STA 2 denote generic devices that are transmitter and receiver, respectively.
Reproduced with permission from [Perahia and Stacey 2013] © Cambridge University Press.

Table 33.2 Frequencies for 802.11 in the 2.45 GHz ISM band

Channel ID	Center frequency (MHz)
1	2412
2	2417
3	2422
4	2427
5	2432
6	2437
7	2442
8	2447
9	2452
10	2457
11	2462
12	2467
13	2472
14	2484

With a channel spacing of 5 MHz, 802.11 networks must be separated by five channels to prevent interference; thus channels 1, 6, and 11 are by far the most commonly used. The maximum allowed transmit power is 1000 mW in the United States, 100 mW in Europe. In Japan, the power spectral density (and not the power) is prescribed, and must not exceed 10 mW/MHz.

The 5 GHz band provides a considerably larger bandwidth, though the propagation characteristics are not as favorable. In the United States, these bands are called the *Unlicensed National Information Structure* (U-NII) bands, and are divided into four ranges: UNII-1 (5.15–5.25 GHz), UNII-2A (5.25–5.35 GHz), UNII 2B (5.35–5.47 GHz), UNII-2C (5.47–5.725), UNII-3 (5.725–5.85), and UNII4 (5.85–5.925). Different bands have different power limitations, see Table 33.3. The channels are numbered starting every 5 MHz according to the formula

$$\text{Channel center frequency} = 5000 + 5 \times n_{ch}(\text{MHz}) \tag{33.1}$$

where $n_{ch} = 0, 1, ..., 200$. Obviously, each 20 MHz channel used by 802.11a occupies four channels in the U-NII band.

Table 33.3 UNII-band power limitations for BS and UE. Fixed point-to-point links can have different limits.

Band	Frequency (MHz)	TX power	TX EIRP	DFS required?
U-NII-1	5150–5250	30 dBm (BS) 24 dBm (UE)	36 dBm; 21 dBm for outdoor elevation $>30°$	No
U-NII-2A,C	5250–5350, 5470–5725	24 dBm	30 dBm	Yes
U-NII-3	5725–5850	30 dBm	36 dBm	No
U-NII-5,7	5925–6425, 6525–6875	30 dBm (BS) 18 dBm (UE)	36 dBm (BS) 24 dBm (UE)	
U-NII-6,8	6425–6525, 6875–7125	24 dBm (BS) 18 dBm (UE)	30 dBm (BS) 24 dBm (UE)	

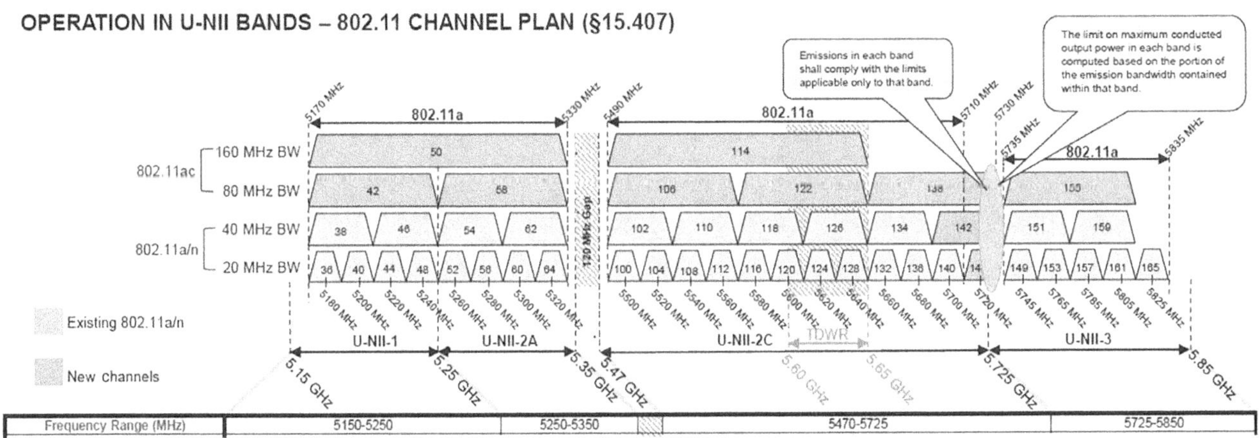

Figure 33.2 802.11 channel plan in the USA. Color version available at wiley.com/go/molisch/wireless3e.
Reproduced with permission from [FCC 802.11] © Federal Communications Commission.

Part of the channel plan for 802.11a in the United States is shown in Figure 33.2. The same channels can be used for 802.11n and 802.11ax. Since those standards allow larger bandwidths, namely 40 and 80, respectively, additional definitions are required. Finally, 160 MHz channels for 802.11ax have been defined as well.

In April 2020, the FCC released a "report and order" making the 6 GHz band (5.925–7.125 GHz) available for Wi-Fi as well,[5] adding additional 14 channels of 80 MHz width and/or 7 channels with 160 MHz. Indoor low-power operation is allowed over the whole bandwidth, considerably enhancing the available peak data rate. Similar rules apply in Canada and much of Central and South America. Europe established rules in 2021, though the assigned frequency band is considerably smaller, namely 5.945–6.425 MHz. Incumbent services in this frequency band are to be protected by constraints on the emitted power, in particular in the vicinity of existing fixed microwave links. Some other details of this allocation are still in flux at the time of this writing.

33.2 802.11a/g – OFDM-Based LANs

The 802.11a/g standard is the oldest and simplest of the 802.11 standards employing OFDM. Its simplicity allows to easily explain the basic principles. In the subsequent sections, we will then discuss the extensions 802.11n, 802.11ac, and 802.11ax.

The main features are (see also Table 33.4):

- Use of the 5 GHz band for 802.11a, and the 2.45 GHz band for 802.11g
- 20 MHz channels.
- Data rates include 6, 9, 12, 18, 24, 36, 48, and 54 Mbps, where support of 6, 12, and 24 Mbps is mandatory.
- OFDM with 64 subcarriers, out of which 52 are used (modulated with binary or quadrature phase-shift keying (BPSK/QPSK), 16-quadrature amplitude modulation (QAM), or 64-QAM.
- Forward Error Correction (FEC), using convolutional coding with coding rates of 1/2, 2/3, or 3/4.

---

[5] Note that the "6" in Wi-Fi 6 refers to the generation of technology, not the frequency band; IEEE 802.11ax operating in the 6 GHz frequency band is called Wi-Fi 6E.

Table 33.4 Important parameters of the 802.11a PHY layer

Information data rate	6, 9, 12, 18, 24, 36, 48, 54 Mbit/s
Modulation	BPSK, QPSK, 16-QAM, 64-QAM
FEC	$K = 7$ convolutional code
Coding rate	1/2, 2/3, 3/4
Number of subcarriers	52
OFDM symbol duration	4 μs
Guard interval	0.8 μs
Occupied bandwidth	16.6 MHz

33.2.1 Modulation and Coding

802.11a uses OFDM as modulation format, enabling high data rates. The principles of OFDM were already described in Chapter 15, so that we here just analyze the details specific to 802.11a. A typical block diagram of a transceiver is shown in Figure 33.3.

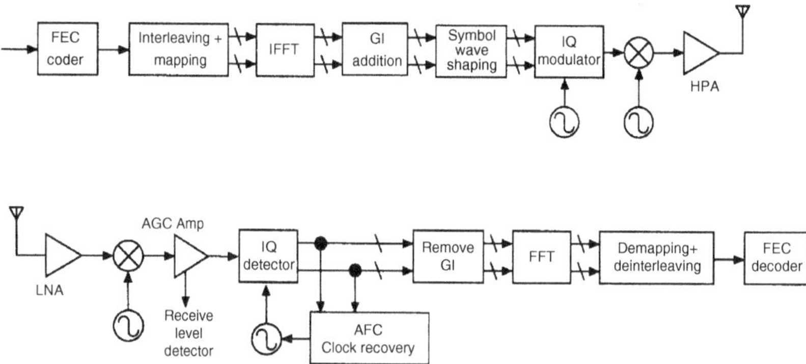

Figure 33.3 Block diagram of a 802.11a transceiver.
Reproduced with permission from [IEEE 802.11] © IEEE

The duration of an OFDM symbol is 4 μs, with a guard interval (cyclic prefix) of 0.8 μs. This is sufficient to accommodate the maximum excess delay of most indoor propagation channels, including factory halls and other challenging environments.

The number of OFDM subcarriers is specified as 64. Powers of 2 are habitually used as number of OFDM carriers, as they allow the most efficient implementation via FFTs. However, only 52 out of the 64 of tones are actually used (modulated and transmitted), while the other 12 tones are null carriers that do not carry any useful information; the useful tones are indexed from -26 to 26, without a DC component. Among those 52 tones, 4 are used as pilot tones, namely tones number $-21, -7, 7, 21$. The pilots are BPSK-modulated by a pseudo binary sequence to prevent the generation of spectral lines. The other 48 tones carry the PSDU data.

The data rate for the PSDU data is chosen depending on the channel state; however, that the standard does not foresee truly adaptive modulation in the sense that the modulation alphabet can differ from tone to tone. Rather, the system uses an average "transmission quality" criterion to adapt the data rate to the current channel state. The rate adaptation is achieved by modifying either the modulation alphabet or the rate of the error correction code, or both. The admissible modulation formats are BPSK, QPSK, 16-QAM, or 64 QAM.

For FEC, 802.11a uses a convolutional encoder with coding rates 1/2, 2/3, or 3/4, depending on the desired data rate. The generator vectors are G1 = 133 and G2 = 171 (in octal notation), for the rate 1/2 coder shown in Figure 33.4. Higher rates are derived from this "mother code" by puncturing.

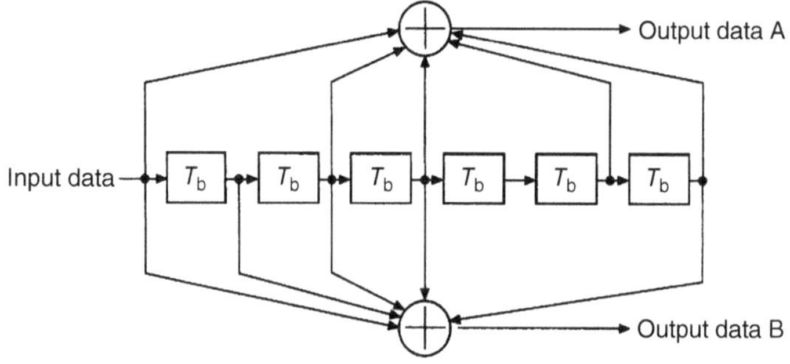

Figure 33.4 Convolutional encoder ($K = 7$).
Reproduced with permission from [IEEE 802.11] © IEEE

Table 33.5 Data rates in 802.11a

Data rate (Mbit/s)	Modulation	Coding rate	Coded bits per subcarrier	Coded bits per OFDM symbol	Data bits per OFDM symbol
6	BPSK	1/2	1	48	24
9	BPSK	3/4	1	48	36
12	QPSK	1/2	2	96	48
18	QPSK	3/4	2	96	72
24	16-QAM	1/2	4	192	96
36	16-QAM	3/4	4	192	144
48	64-QAM	2/3	6	288	192
54	64-QAM	3/4	6	288	216

All encoded data bits are interleaved by a block interleaver with a block size equal to the number of bits in a single OFDM symbol. The interleaver is defined by a two-step permutation. The first permutation ensures that adjacent coded bits are mapped onto non-adjacent subcarriers. The second ensures that adjacent coded bits are mapped alternately onto less and more significant bits of the constellation and, thereby, long runs of low reliability bits are avoided.

Table 33.5 summarizes the rates that can be achieved with different combinations of alphabets and coding rates, as well as the OFDM modulation parameters.

Note that before encoding, the payload data are scrambled; the scrambler initialization is chosen in a pseudorandom way for each packet and is communicated in the service field (see Section 33.2.2). After the payload data, (nonscrambled) zero-value tail bits, and pad bits as required, are added. The scrambled data packet is then encoded.

33.2.2 Preamble and Header

A packet (more precisely, a PPDU) sent out by the TX consists of a PLCP preamble that serves for synchronization and channel estimation, and a PCLP header, which contains signaling information. At the Receiver (RX), the PLCP preamble and header are used to aid in demodulation and data delivery.

Synchronization and channel estimation: Synchronization is achieved by means of the PLCP preamble field. It consists of 10 short symbols and two long symbols, i.e., the first 16 µs shown in Figure 33.5.

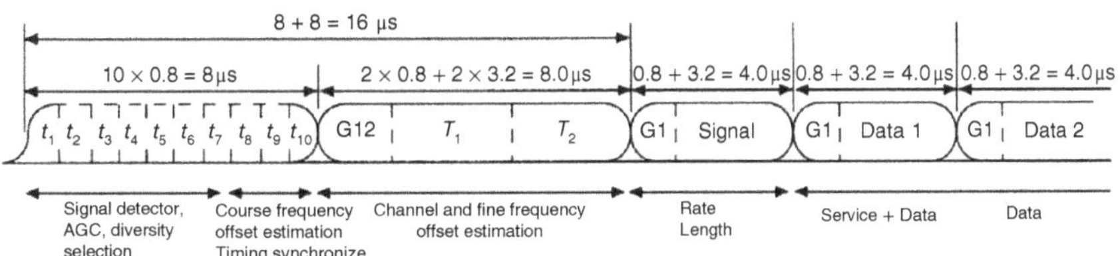

Figure 33.5 PLCP preamble structure.
Reproduced with permission from [IEEE 802.11] © IEEE

The preamble starts out with the *Short Training Field* (*STF*) which consists of 10 short symbols, each of duration 0.8 µs, that allow the RX to detect the signal, adjust the Automatic Gain Control (AGC), and perform a coarse frequency offset estimation. Those short symbols consist only of 12 tones, which are modulated by the elements of the following sequence

$$S_{-26,26} = \sqrt{(13/6)}\{0, 0, 1 + j, 0, 0, 0, -1 - j, 0, 0, 0, 1 + j, 0, 0, 0, -1 - j, 0, 0, 0, -1 - j, 0, 0, 0, 1 + j,$$

$$0, 0, 0, 0, 0, 0, 0, -1 - j, 0, 0, 0, -1 - j, 0, 0, 0, 1 + j, 0, 0, 0, 1 + j, 0, 0, 0, 1 + j, 0, 0, 0, 1 + j, 0, 0\}. \tag{33.2}$$

Multiplication by a factor of $\sqrt{(13/6)}$ is done so as to normalize the average power of the resulting OFDM symbol, which utilizes 12 out of 52 subcarriers. In the time domain, this sequence shows a good autocorrelation function, with more than 10 dB sidelobe suppression, which allows easy synchronization. Furthermore, it allows the estimation of a frequency offset of up to 625 kHz $((1/2)(1/0.8\,\mu s))$.

The STF is followed by the *Long Training Field* (*LTF*), which consists of two OFDM symbols preceded by a guard interval. An OFDM training symbol consists of 53 subcarriers (including a zero value at dc), which are modulated by the elements of the sequence L, given by

$$L_{-26,26} = \{1, 1, -1, -1, 1, 1, -1, 1, -1, 1, 1, 1, 1, 1, 1, -1, -1, 1, 1, -1, 1, -1, 1, 1, 1, 1, 0,$$

$$1, -1, -1, 1, 1, -1, 1, -1, 1, -1, -1, -1, -1, -1, 1, 1, -1, -1, 1, -1, 1, -1, 1, 1, 1, 1\}. \tag{33.3}$$

The LTF allows a more accurate timing, because it shows high correlation peaks, spaced 3.2 µs apart, and a somewhat lower peak 1.6 µs before the main peak (due to the cyclic prefix). It also allows a finer frequency offset estimation, though only up to ±156.25 kHz. Most importantly, the LTF serves as training sequence for channel estimation.

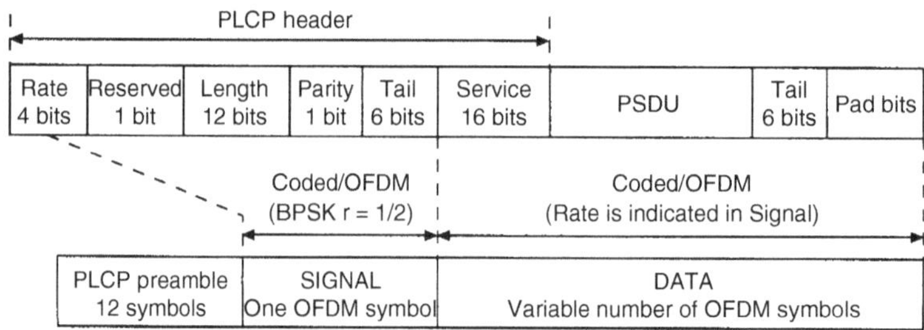

Figure 33.6 PPDU frame format.
Reproduced with permission from [IEEE 802.11] © IEEE

Signaling field: The PLCP preamble is followed by the PLCP header, i.e., the SIGNAL field, see Figure 33.6. It incorporates the RATE field, a LENGTH field, a TAIL field, and so on:

- *Rate (4 bits)*: indicates transmission data rate.
- *Reserved (1 bit)*: future use.
- *Length (12 bits)*: indicates the number of octets in the PSDU.
- *Parity (1 bit)*: parity check.
- *Tail (6 bits)*: convolutional coding tail.
- *Service (16 bits)*: initialization of the scrambler (first 7 bits), and 9 all-zero bits.

The PLCP header is followed by the actual data, transmitted with the modulation and coding discussed above. Table 33.6 summarizes the most important parameters for the 802.11a packet transmission.

Table 33.6 Parameters of 802.11a

Parameter	Value
Number of data subcarriers	48
Number of pilot subcarriers	4
Subcarrier spacing	0.3125 MHz
IFFT/FFT period	3.2 µs
Preamble duration	16 µs
Duration of OFDM symbol	4.0 µs
Guard interval for signal symbol	0.8 µs
Guard interval for training symbol	1.6 µs
Short training sequence duration	8 µs
Long training sequence duration	8 µs

33.3 802.11n – High-throughput Transmission

33.3.1 Overview

The 802.11n standard offers up to (nominal) 600 Mbit/s throughput. These high data rates, as well as improved reliability, were necessitated by a number of new applications, as discussed in Section 33.1.

The 802.11n group was established in 2002. In September 2004, a number of different technical proposals were presented, which were subsequently consolidated into two proposals supported by a major industry alliances each: TGnSync, and WWise. After more than a year of negotiations and fights, a compromise draft proposal was approved by 802.11n in January 2006. After that, the draft was in the process of revisions and corrections, and a final version of the 11n standard was approved in 2009. But even before then, several companies already sold "pre-n" products that followed the 11n draft standard. Note that a similar situation more recently occurred with 802.11ax, as discussed in Section 33.6.

802.11n achieves high data rates mainly by two methods: use of multiple antenna techniques (see Chapter 16), and increase of the available bandwidth from 20 to 40 MHz. The generic structure of an 11n transceiver is shown in Figure 33.7. The source data stream is

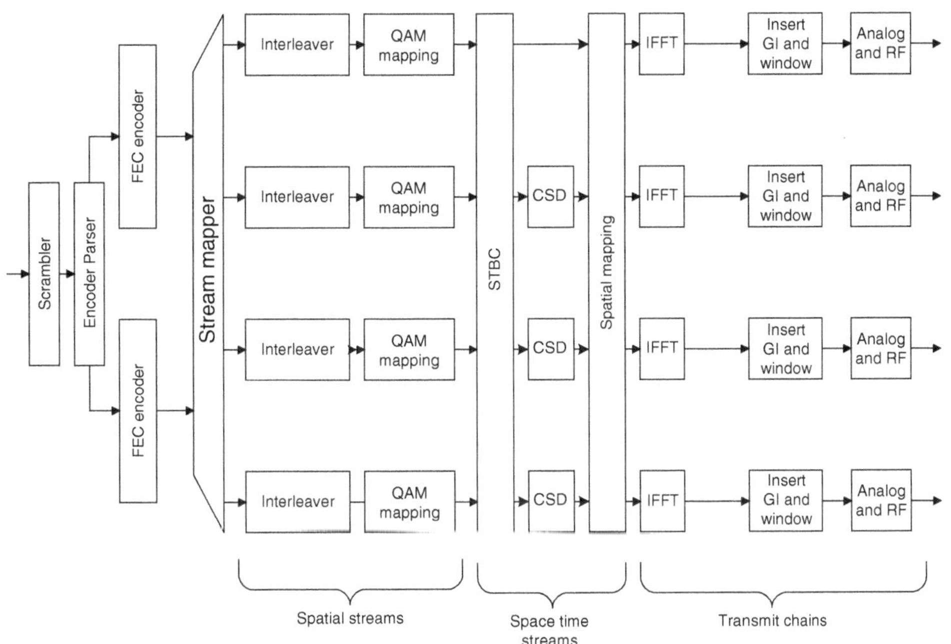

Figure 33.7   Block diagram of an IEEE 802.11n transmitter.
Reproduced with permission from [IEEE 802.11] © IEEE

first scrambled and then (for very high data rates) divided into two parallel data streams, to reduce the processing speed requirements of the encoder/decoder (this situation occurs only for 40 MHz channels, see Section 33.3.4). A number of different codecs are available: binary convolutional codes are the default solution, while LDPC codes are an option for high-performance transmission (compare Section 13.7). The thus-encoded bits are divided into a number of spatial streams (compare Section 16.2), which are to be transmitted in parallel from the antennas. Each of the spatial streams is then interleaved, mapped onto complex modulation symbols, and grouped into OFDM symbols. Next, the spatial streams can be modulated by Alamouti codes and/or cyclic shift delay (compare Section 16.1.5). Next, the "spatial mapping" distributes the spatial streams onto the modulation/upconversion chains. In each of the chains, the symbols are submitted to an IFFT, a guard interval is inserted, and the signal is upconverted to the passband; this part of the processing is identical to 802.11a.

33.3.2   Modulation and Coding

For a single spatial stream, the modulation and coding schemes are very similar to those of 802.11a. The number of subcarriers in a 20 MHz band is slightly increased, from 48 to 52, by replacing 2 null carriers at each of the band edges with data subcarriers. The modulation formats are BPSK, QPSK, 16-QAM, and 64-QAM. When convolutional coding is used, code rates 1/2, 2/3, and 3/4 are the same as for 802.11a; an additional coderate of 5/6 (for higher throughput in very good channel conditions) was introduced as well. This results in a total of 8 MCS schemes with data rates from 6.5 to 65 Mbit/s. When multiple antennas are present, the Transmitter (TX) can either use the same MCS for all spatial streams (this makes sense if the TX has no channel state information), or it can have different MCSs for different streams. A total of 32 MCSs are defined for the case of equal modulation (the eight "fundamental" MCSs equally used on 1, 2, 3, or 4 spatial streams). For unequal modulation, further 44 MCSs (different combinations of the existing MCSs on the various spatial streams) are defined, though not all of them are mandatory.

There is now also the option of MCS feedback, i.e., the RX sends back a suggestion for the MCS that the TX should be using. This feedback can be initiated by the TX (where the feedback can be either immediately after the query, or with some delay), or can be sent by the RX without solicitation.

802.11n also introduces the concept of a short guard interval: the system can adaptively decide whether the length of the cyclic prefix is "normal" 800 µs or shortened to 400 µs; the latter is used to increase the spectral efficiency in environments where the delay spread is so small that a short cyclic prefix is sufficient.

The LDPC codes achieve lower error probability than convolutional codes, at the price of higher decoding complexity. The parity-check matrices can be partitioned into square subblocks (submatrices), which are either cyclic permutations of the identity matrix, or all-zero matrices. Twelve different codes are defined, which are all based on the same codestructure. The codeword sizes and submatrix sizes are 648 (27), 1296 (54), and 1944 (81). A major challenge in the LDPC encoding is to fit the payload bits such that they (i) fit an integer multiple of LDPC blocks and (ii) an integer number of OFDM blocks, where the latter depends on the used MCS. The encoding thus proceeds in a number of steps; for details, we refer to [Perahia and Stacey 2015, Section 6.4.1] and the standard. Finally, note that for LDPC codes, no interleaving is done due to the large diversity of the code.

33.3.3   Multiple Antenna Techniques

The key to the 802.11n standard is the use of multiple antenna techniques. The standard foresees a number of different techniques, in particular (i) spatial multiplexing, (ii) space-time block coding, (iii) eigen beam forming, and (iv) antenna selection. The basics of most of those techniques are outlined in Chapter 16; here we only deal with the specific implementation in 802.11n.

Space-time coding: Space-time block codes can be used in an 802.11n system to increase the robustness of the system. In particular, Alamouti codes are used and can be combined with spatial multiplexing. If there are two transmit Radio Frequency (RF) chains, then one spatial stream is mapped onto the two RF chains by means of the standard Alamouti code. If there are three transmit antennas and two spatial streams, then one stream is mapped to two RF chains by means of Alamouti encoding, and one stream is mapped directly to the remaining RF chain. For four transmit chains, either three streams (where one of them is Alamouti-encoded), or two streams (with each of them Alamouti-encoded) can be used. Different modulation schemes can be used on the different streams; this is motivated by the fact that Alamouti-encoded streams can sustain a higher modulation scheme than nonencoded schemes.

Another way of achieving transmit diversity is the use of *Cyclic Shift Delay* (CSD). As discussed in Section 16.1.5, in CSD the OFDM symbols are *cyclically* shifted, i.e., the signal on the kth subcarrier is multiplied with $\exp[-j2\pi k \Delta_F \tau_i]$, where k indexes the subcarrier frequency, Δ_F is the spacing of the subcarriers, and τ_i is the cyclic shift applied to the ith signal. The cyclic shifts are 0, -400, -200, and -600 ns for the first, second, third, and fourth spatial stream, respectively.

Spatial multiplexing and beamforming: The dividing of the original data stream results in a total number N_S spatial streams, which can be smaller than, or equal to, the number of available RF chains N_{RF} for the upconversion.[6] In any case, linear combinations of the spatial streams are assigned to the RF chains; these combinations are described by means of the so-called "spatial mapping" matrix $\mathbf{Q}$, so that for each time instant the vector of spatial-stream vectors $\mathbf{x}$ is mapped onto the signals for the RF chains $\mathbf{y}$ as $\mathbf{y} = \mathbf{Q}\mathbf{x}$. The following possibilities are defined in the standard:

- *Direct mapping*: this method is used if $N_S = N_{RF}$. In the simplest case, $\mathbf{Q}$ is either an identity matrix, or it is a diagonal matrix in which the elements perform CSD, so that $Q_{i,i} = \exp[-j2\pi k \Delta_F \tau_i]$. The CSD serves to avoid inadvertent beamforming when similar signals occur in the spatial streams.
- *Spatial mapping for the case $N_S = N_{RF}$*: in this case $\mathbf{Q}$ is the product of a CSD matrix with a square matrix with orthogonal columns, such as a Fourier matrix or a Hadamard matrix
- *Spatial mapping for the case $N_S < N_{RF}$*: in this case, some of the spatial streams are duplicated (so that the total number of streams becomes equal to N_{RF}). All the streams are then power-adjusted (so that the total power stays constant) and mapped onto the transmit antennas by means of a CSD matrix.
- *Beamforming steering matrix*: any matrix that improves the overall BER can be used as matrix $\mathbf{Q}$. Realistically speaking, the matrix is based on channel state information at the TX (see Section 33.3.6). In particular, if the TX knows the instantaneous channel transfer matrix $\mathbf{H}$, it can perform an eigenvalue decomposition of $\mathbf{H}$ on each subcarrier, precode with the right singular matrix (see Section 16.2.3), and possibly weight the streams according to the waterfilling rules.

Antenna selection: There are situations where the number of available antenna elements is larger than the number of RF chains – either because of cost reasons, or because the maximum number of RF chains foreseen in the 802.11n standard is 4. In that case, antenna selection allows improvement of the system performance (see Sec. 16.2.10).

The available RF chains are connected to the "instantaneously best" antenna elements via electronic switches. However, in order to determine the "best" antenna elements, the complete channel (from each TX to each RX antenna element) has to be sounded. This is achieved in two or more subsequent packets. For example, for TX antenna selection, the first packet is transmitted from the first subset of antennas, the next packet from a second subset of antennas, and so on. After all subsets have been tried out, the RX sends information to the TX about which subset to use. RX antenna selection works analogously.

33.3.4   20 MHz and 40 MHz Channels

While 802.11n with 20 MHz bandwidth uses 56 subcarriers (52 data and 4 pilot), for 40 MHz bandwidth where two *adjacent* 20 MHz channels are bonded, 114 subcarriers (108 data and 6 pilot) are used. Subcarriers are added in the middle (where the upper guardband of the lower 20 MHz band and the lower guardband of the upper 20 MHz band was), and in the place of the DC subcarriers, Figure 33.8. At the beginning of the packet, the legacy fields, as well as the HT-SIG are duplicated in the two constituent 20 MHz bands. This allows legacy RXs, as well as 20 MHz 11n devices, to decode the preamble, realize that a 40 MHz transmission follows, and defer their own actions. Note that the phases of signals in the upper channel in 40 MHz are rotated by $+90°$ in reference to the lower channel, in order to reduce *Peak to Average Power Ratio, PAPR*. The remaining fields of the preamble are transmitted across the 40 MHz bandwidth, as are the actual data. The training sequence in the HT-STF and HT-LTF have a similar structure as those for 20 MHz, though they are of course adapted to the larger bandwidth. Again, the upper 20 MHz subcarriers are phase-rotated by $90°$. During the data transmission, the pilot tones for a 40 MHz transmission are on subcarriers $-53, -25, -11, 11, 25, 53$.

---

[6] We speak here of the number of RF chains (and not antennas), because the determining factor is the number of signals that can be upconverted. Using antenna selection (see below), the number of antennas can be larger.

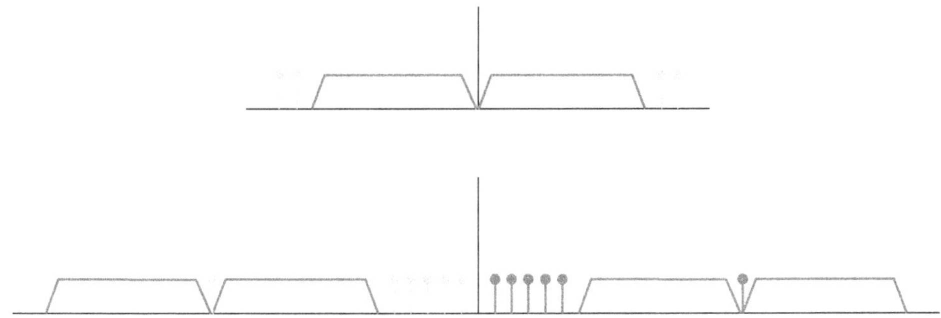

Figure 33.8 20 MHz and 40 MHz subcarriers.

33.3.5 Preamble and Header

802.11n aims to retain backward compatibility to 802.11a/g. Networks with either 11a/g or 11n BSs, and a mixture of 11a/g and 11n UEs have to work. Many of the details of the 11n standard, in particular the design of preamble, can only be understood in the light of this requirement for backward compatibility. The preamble fields that are identical to those of the 802.11a/g standard are called "legacy" fields (signified with a prepended "L"), while the fields added specifically for 802.11n are called "High Throughput," HT.

There are three types of PLCP preambles (i.e., the part of the preamble that is used for synchronization and channel estimation, compare 33.2.2), see Figure 33.9. The first type is the legacy preamble, which is identical to the 802.11a preamble; this is to be used if only legacy (802.11a) devices are used at a given time. If a TX has multiple antenna elements, the transmission of the legacy preamble must be done from multiple antennas to provide an appropriate *total* power (in a multi-antenna system, the power amplifier for each antenna typically has a lower peak power than that of a single-antenna system). In order to allow multiple-antenna transmission without loss of backward compatibility, CSD is used.

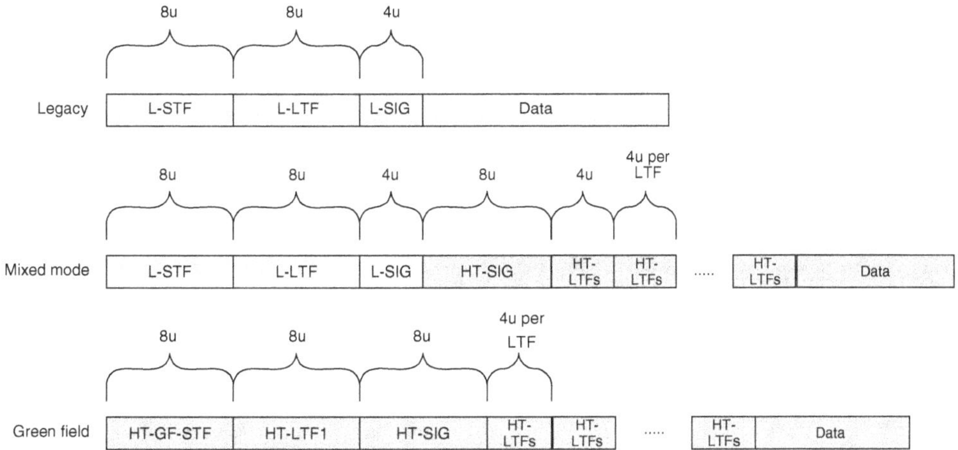

Figure 33.9 Types of preambles in IEEE 802.11n. In this figure: HT-GF-STF, High Throughput – GreenField – Short Training Field; L-SIG, Legacy SIGnalling field; L-STF, Legacy Short Training Field.
Reproduced with permission from [IEEE 802.11] © IEEE

If both 802.11a and 802.11n devices are present in a Local Area Network (LAN), then the mixed-mode preamble has to be used. It starts with the same fields as the legacy preamble, which are then followed by the High-Throughput Signal Field (HT-SIG). This contains information about a number of MIMO parameters, bandwidth allocations, etc., that are unique to 11n devices. In particular, it contains information about (see also Figure 33.10):

- modulation and coding scheme.
- bandwidth indication (20 or 40 MHz).
- length: indicating the length of the packets, allowing up to 64 kBytes.
- smoothing: indicates whether frequency-domain smoothing is recommended as part of channel estimation.
- not-sounding: set to 0 if the current PPDU is a sounding PPDU, see Section 33.3.6.
- aggregation: whether the data portion of the packet is part of a data aggregation transmission.
- STBC: indication of space-time coding.

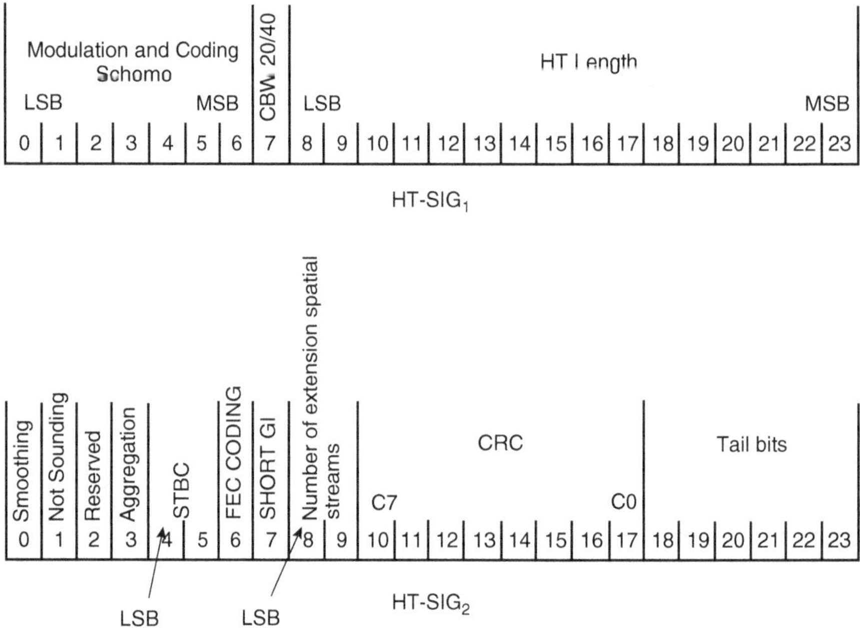

Figure 33.10 High-throughput signaling field.
Reproduced with permission from [IEEE 802.11] © IEEE

- FEC encoding: convolutional or LDPC.
- short GI: whether long or short cyclic prefixes are used.
- number of extension spatial streams.
- CRC: error detection for HT-SIG field.
- tail bits for terminating convolutional code.

The HT-SIG field is encoded with a rate $1/2$ convolutional code, interleaved, and BPSK modulated, to ensure high robustness in the transmission.

The HT-SIG is followed by the HT-STF, which is the same as the L-STF, except that larger values for the CSD (up to 600 ns) are used. This allows a more accurate estimation of the received power. Then follows the HT-LTF, which will be discussed in more detail in the next subsection.

If there are only 11n devices in a LAN, then the greenfield preamble can be used, which omits all the "legacy" fields, and thus provides a shorter and more efficient preamble.

33.3.6 Channel Estimation

The MIMO channel between TX and RX is estimated from the HT-LTFs. If the TX provides training for exactly the available spatial streams, then the preamble uses exactly N_S training symbols (except for the case of three spatial streams, which require four training symbols). In that case, all antennas transmit training symbols simultaneously, where the symbols from each antenna have different cyclic shifts, just as for the HT-SIG. One symbol per stream is multiplied with -1 (it is the $(k+1) \mod (N_S)$ symbol on the kth antenna) thus allowing to extract the channel estimates by adding/subtracting the signals received during different HT-LTFs.

If the TX is providing training for more spatial streams, more spatial dimensions can be estimated, which in turn enables, e.g., beamforming for eigenvalue decomposition. In this case, the PPDU is called a *sounding PPDU*. Thus, the HT-LTF portion has one or two parts. The first part consists of N_S LTFs (known as Data-HT-LTFs) that are necessary for demodulation of the HT-Data portion of the PPDU. The optional second part (the one occurring only in sounding PPDUs) consists of the HT-LTFs that may be used to sound extra spatial dimensions. Figure 33.11 shows an example.

Another way of sounding all spatial dimensions is to use a PPDU that contains no payload data. The preamble of such a *Null Data Packet* (NDP) then contains only data HT-LTFs, but the (nominal) N_S is chosen such that all required spatial dimensions can be sounded. Since the NDP cannot provide MAC information, i.e., whom it is addressed to, it can only be used in sequence with another packet.

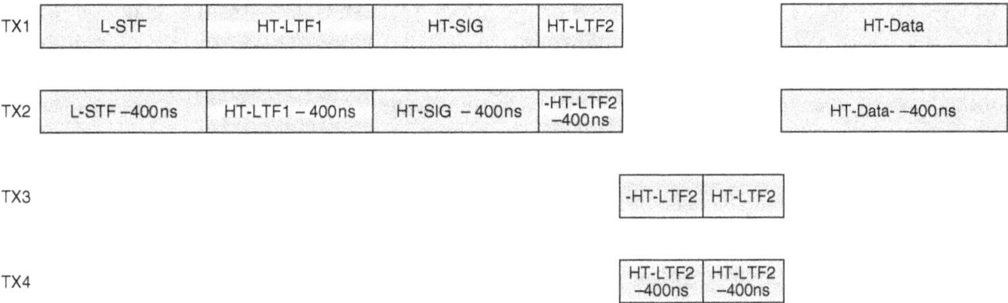

Figure 33.11 Example for sounding PPDU with extension HT-LTFs.
Reproduced with permission from [IEEE 802.11] © IEEE

The HT-LTFs provide the Channel State Information (CSI) that is required for the reception of the signals but can also be used at the TX to enable an appropriate precoding (i.e., creation of a suitable matrix $\mathbf{Q}$ for each subcarrier). The CSI at the TX can either be obtained through explicit feedback or from the principle of reciprocity.

- In the case of explicit feedback, the RX determines either the effective channel matrix $\mathbf{H}_{\text{eff}}$ (which is the product of the matrix $\mathbf{Q}$ with the channel matrix $\mathbf{H}$) or the beamforming matrix. The real and imaginary parts of the channel matrix coefficients are quantized to 4, 5, 6, or 8 bits. Since this can result in rather large channel matrices that need to be fed back a compressed feedback is foreseen as well for the feedback of the beamformer weights.

  The compression is based on the principle of Givens rotations. An $M \times N$ matrix $\mathbf{V}$ describing the right singular vectors of the channel is decomposed as

$$\mathbf{V} = \left(\prod_{i=1}^{min\,(N,M-1)} \left[\mathbf{D_i} \prod_{l=i+1}^{M} \mathbf{G}_{li}^T\left(\psi_{l,i}\right) \right] \times \mathbf{I}_{M \times N} \right) \widetilde{\mathbf{D}} \tag{33.4}$$

  where $\mathbf{I}_{M \times N}$ is an identity matrix with extra rows or columns of zeros if $M \neq N$, and $\widetilde{\mathbf{D}}$ is a diagonal matrix with unit-magnitude entries, $\widetilde{D}_{i,i} = \exp\left(j\angle(V_{M,i})\right)$. The important quantities are diagonal matrices $\mathbf{D_i}$, which ensure that the elements of the ith column of $\mathbf{D_i}\mathbf{V}$ are real nonnegative numbers, and are given as

$$\mathbf{D_i} = \begin{bmatrix} \mathbf{I}_{i-1} & 0 & .. & ... & 0 \\ 0 & \exp\left[j\phi_{i,i}\right] & 0 & ... & 0 \\ ... & 0 & ... & 0 & ... \\ ... & ... & 0 & \exp\left[j\phi_{M-1,i}\right] & 0 \\ 0 & 0 & ... & 0 & 1 \end{bmatrix} \tag{33.5}$$

  where the $\phi_{l,\,i} = \angle\left(\mathbf{V}_{l,i}\right)$ and $\mathbf{I}_i$ is an $i \times i$ identity matrix. The Givens rotation matrices are

$$\mathbf{G}_{li}\left(\psi_{l,i}\right) = \begin{bmatrix} \mathbf{I}_{i-1} & 0 & 0 & 0 & 0 \\ 0 & \cos\left(\psi_{l,i}\right) & 0 & \sin\left(\psi_{l,i}\right) & 0 \\ 0 & 0 & \mathbf{I}_{l-i-1} & 0 & 0 \\ 0 & -\sin\left(\psi_{l,i}\right) & 0 & \cos\left(\psi_{l,i}\right) & 0 \\ 0 & 0 & 0 & 0 & \mathbf{I}_{M-l} \end{bmatrix}. \tag{33.6}$$

  The angles $\phi_{l,j}$ are then quantized with $b + 2$ bits, between 0 and 2π, and $\psi_{l,i}$ is quantized between 0 and b bits, between 0 and $\pi/2$, where b can range from 1, ..., 4. Depending on the size of the beamforming matrix, up to 12 angles need to be transmitted. The angles can be transmitted either for each subcarrier or for groups of 2 or 4 subcarriers. In addition, the average SNR for each stream is transmitted. Note that since 802.11n does not allow a per-subcarrier adaptation of the MCS, a per-subcarrier SNR is irrelevant.

- In implicit feedback, the TX employs the channel reciprocity to obtain knowledge about the channel. As discussed in Section 16.1.7, the upconversion/downconversion RF chains are *not* necessarily reciprocal. The 802.11n standard thus foresees a procedure for calculating a set of calibration matrices that can be applied at the TX side of a device (BS or UE) to correct the amplitude and phase differences between the TX and RX chains of that device. The procedure, which has to be performed only at very large intervals, involves essentially determining channel coefficients implicitly as well as explicitly; the difference between the results can then be used to establish the calibration matrices.

  The actual calibration procedure works as follows: one device (the initiator) sends a "calibration start" signal, after which the other device (the responder) sends a sounding PPDU that allows the initiator to estimate the uplink (responder-to-initiator) channel.

The initiator then sends a sounding PPDU that includes a CSI feedback request: this causes the responder to estimate the downlink channel. Note that the three packets (containing calibration start, sounding response, and CSI feedback request) follow fast upon each other, so that no significant changes of the channel occur during that time. At some later time, the responder provides explicit CSI feedback to the initiator, using one of the methods described above. A slight variation of this approach uses NDP packets instead of sounding PPDUs.

33.4 The MAC Layer in 802.11a/g and 802.11n

The first MAC was developed as part of the original 802.11 standard and was mostly intended as a very straightforward porting of the IEEE 802.3 Ethernet standard to the wireless world. Due to its simplicity, it did not fulfill many of the requirements of more advanced WLANs, such as security, efficiency, etc., and a number of amendments were published, most notably 802.11e. These MAC modifications were mostly *not* part of the discussions in 802.11b, a, and n. We thus describe here the state of the MAC as implemented for 802.11n, but which is nowadays also for 802.11a devices. Starting with 802.11ac, and in particular for 802.11ax the development of MAC and PHY are more aligned, because MU-MIMO is not only a physical-layer technique but also requires significant modifications of the MAC. These developments will thus be reported in the respective sections below.

There are nine MAC services specified by IEEE 802.11. These include distribution, integration, association, reassociation, disassociation, authentication, deauthentication, privacy, and MSDU delivery. Six of the services are used to support MSDU (MAC Service Data Unit) delivery between devices. Three of the services are used to control 802.11 WLAN access and confidentiality. Each of the services is supported by one or more MAC frame types. The 802.11 MAC uses three types of messages – *data, management,* and *control.* Some of the services are supported by MAC management messages and some by MAC data messages. All of the messages gain access to the WM (Wireless Medium) via the IEEE 802.11 MAC medium access method that include both contention-based and contention-free channel access methods – *Distributed Coordination Function* (DCF) and *Point Coordination Function* (PCF). In the following, the 802.11 MAC functions and services will be described.

33.4.1 Network Structure and Purpose of the MAC

The network structure of 802.11 is fairly flexible and can either be an ad-hoc network, where several UEs talk to each other without infrastructure, or the typical, infrastructure-based, WLAN scenario, where a single BS communicates to several UEs. In either case, the network thus formed is called the *Basic Service Set* (*BSS*).[7] In the following, we will consider only the infrastructure-based case, since this is used in the overwhelming majority of cases, but almost all considerations can be transferred directly to the ad-hoc case as well; this is because even in ad-hoc networks, a UE becomes the *group owner,* and takes on most of the functionalities of the BS.

Despite the fact that 802.11 networks in most cases are infrastructure-based, the structure of the MAC is much more "distributed" than it is in cellular networks. This is based on the history, since 802.11 was intended to be a simple wireless extension of the wired Ethernet standard, which is highly decentralized. As we will see in later sections, the more recent versions of 802.11, in particular 802.11ax, are moving toward stronger centralization.

33.4.2 Joining and Leaving a Network

Each BS broadcasts, at regular intervals, a *beacon* signal. This signal contains information necessary for UEs to join the network, such as the name of the network (*Service Set Identifier, SSID*), as well as BS capabilities such as transmit power, RF channels, etc. After hearing that information, a UE will *authenticate* itself to the BS. Such authentication could either be an *open system authentication* if the network is open to any device or might require the exchange of security keys via one of several protocols. These include WEP (which is highly unsecure, even though it is still used in a number of networks), WPA and its successor WPA2 (which is slightly less vulnerable though also for it vulnerabilities have been pointed out and might be exploited), and WPA 3 (approved in 2018, and designed to make brute-force attacks harder). Following the authentication, the UE sends an association request, to which the BS responds. During the association request/response exchange, UE and BS also exchange information about their capabilities.

33.4.3 Multiple Access Methods

Multiple access can be controlled in two different ways: contention-based (i.e., random access), or contention-free (where the BS assigns the resources for the transmission). Each of those has advantages under specific circumstances, which is why 802.11 implements both of them, at least as options. It does so by dividing time into separate intervals for handling them, *Contention Period* (CP)[8] and *Contention-Free Period* (CFP).

The basis of the time separation is the temporal *superframe* structure shown in Figure 33.12. During the CP, access is done according to the *Distributed Coordination Function, DCF* function (or the *Enhanced Distributed Channel Access, EDCA*), while during the CFP, the *Point Coordination Function, PCF* is used. These two control methods are discussed in more detail in the following subsections.

---

[7] It is also possible that several BSs are connected via a *distribution system*, typically a cabled connection, to form an *extended BSS.*

[8] Note that in this section only, we use the abbreviation CP for contention period (and not for cyclic prefix). Since we are talking about the MAC layer only, no confusion can arise.

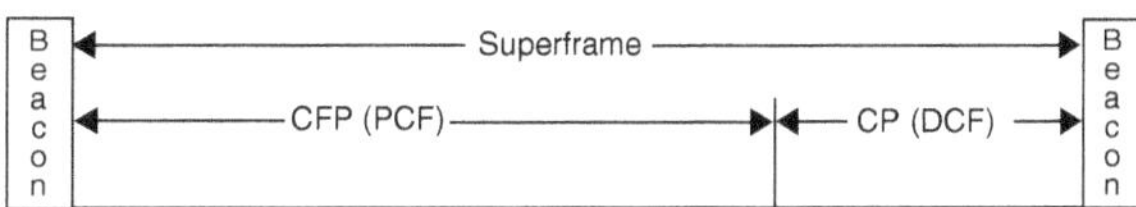

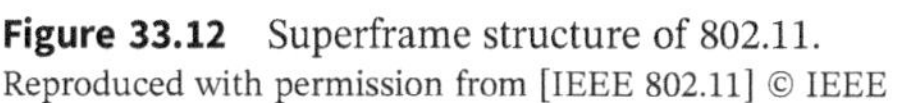

Figure 33.12 Superframe structure of 802.11.
Reproduced with permission from [IEEE 802.11] © IEEE

Distributed Coordination Function The DCF employs *Carrier Sense Multiple Access with Collision Avoidance* (CSMA/CA), as described in Sec. 18.4, plus a random backoff mechanism. In brief, each device performs *Clear Channel Assessment, CCA*, i.e., checks whether the channel is idle before attempting to transmit. This sensing occurs through a physical channel sensing, (which uses energy detection) and virtual carrier sensing (which is based on detection of the preamble of a packet on the air, which contains information about how long the packet will last), which is discussed below in more detail. The channel is considered to be free only if both physical and virtual carrier sensing say that the channel is free.

If the channel is free at the time an MSDU arrives from the higher layers, the device transmits immediately. If the channel is busy, the device waits until the channel is clear. The problem with simple CSMA is that once the channel becomes available, multiple devices might want to transmit. To avoid collisions, each device waits for a random time until it makes a transmission attempt. If another device (that happens to pick a shorter wait time) "grabs" the channel in the meantime, then the device will wait until the channel becomes clear again. This waiting mechanism is the "collision avoidance" in the CSMA/CA.

Let us now consider the waiting ("backoff") mechanism in more detail. 802.11 defines different inter-frame gaps, denoted as *Inter-Frame Spaces* (IFS), in order to control the medium access, i.e., to give devices a higher or lower priority in specific cases. These IFSs are (in the order shortest to longest):

- *Short Inter-Frame Space* (SIFS): In the standards considered here, it is 16 µs in the 5 GHz band and 10 µs in the 2.4 GHz band. This is the minimum delay time between two frames, for example, a data frame and its acknowledgment. This means that an RX has to be capable of finishing all processing of the data frame within one SIFS after the end of the frame. Generally, the SIFS is used between transmission of a frame, and the response triggered by this frame, Also, when a device sends multiple frames in sequence, it spaces them apart by a SIFS (and knows that no other device can grab the channel in the meantime).

- *Point Inter-Frame Space* (PIFS): this equals a SIFS plus one "slot time". The slot time, which is usually 9 µs, is the basic unit for the additional backoffs, and it allows devices to check whether the channel is occupied, and also absorb any imperfections of the time synchronization between the different devices. A PIFS can be used, e.g., to regain medium access following a transmission failure.

- *Distributed Inter-Frame Space* (DIFS): a DIFS equals a SIFS plus two slot times. It is the inter-frame spacing used for "standard" frames, such as data and management frames.

- *Extended Inter-Frame Space* (EIFS): this frame is used if a previously received frame contains an error. The longer wait time is necessary because erroneous decoding prevents the device to properly set its NAV (see below) and any transmissions it makes might collide with the transmissions of other devices, due to the hidden node problem.

802.11n also introduced a *Reduced Interframe Spacing* (RIFS) of length 2 µs for transmission from the same UE if no response is expected between the transmission. However, RIFS was abolished in later versions of the standard.

As mentioned above, the backoff time is not just a deterministic value like a DIFS, but added to that is a *random* backoff count that each device chooses independently. In particular, each device has for itself a *Contention Window* (CW). The device then initializes a backoff counter with a value that is a pseudorandom integer n that is uniformly distributed in $[0, ..., CW]$; the backoff time is then SIFS + $n *$ SlotTime; the transmit times are aligned to slot boundaries so that transmission on one slot can be detected before the start of the next slot. The backoff counter decrements each time the medium is detected to be idle for one slot time. As soon as the backoff counter expires, the device begins to transmit. It is noteworthy that if another device "grabs" the channel, then the considered device does not reset the counter, but just suspends the backoff procedure until the channel becomes free again, and then continues the countdown. This ensures a certain amount of fairness, as a frame that has been on hold for a long time will ultimately have a short backoff time, and then be able to transmit. The basic access method and back-off procedure are shown in Figures 33.13 and 33.14, respectively.

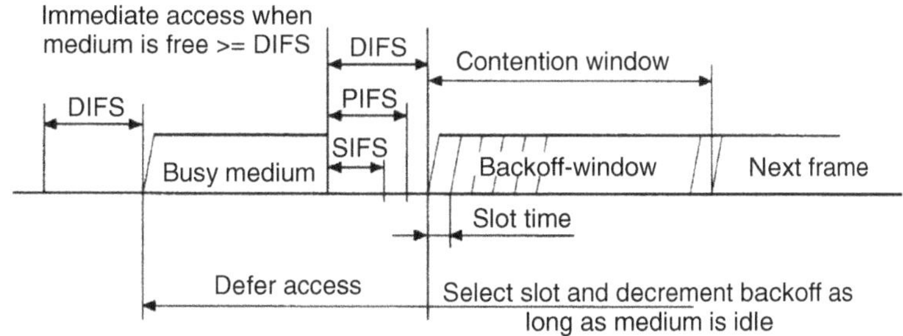

Figure 33.13 Basic access method.
Reproduced with permission from [IEEE 802.11] © IEEE

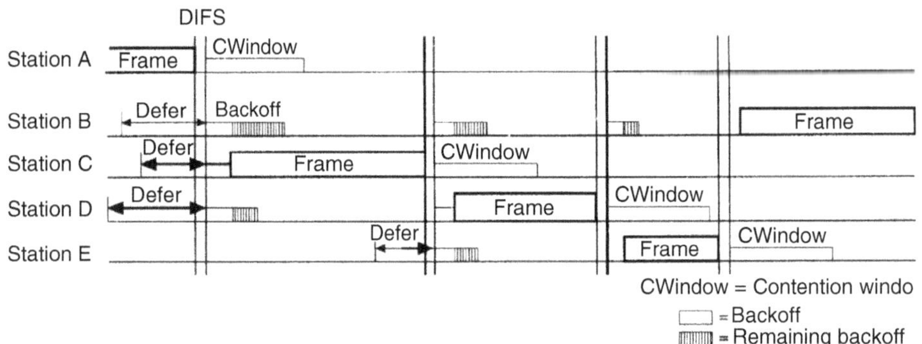

Figure 33.14 Timing backoff procedure.
Reproduced with permission from [IEEE 802.11] © IEEE

The system defines a range of allowed CWs, i.e., $[CW_{\min}, ..., CW_{\max}]$. After every successful transmission, the CW is set to $CW_{\min}$. If the transmission is not successful, the device thinks that a collision has occurred. In this case, the CW is doubled (until it reaches $CW_{\max}$), and a new backoff procedure starts again. The process will continue until the transmission is successful (or discarded). The reason for this increase of the backoff window is that particularly in a dense network, multiple devices could pick the same backoff value. The way to combat this is to reduce the probability for picking a particular value, which is achieved by picking values in a larger range.

The hidden node problem, discussed in Section 18.4.2, is a major problem of CSMA. 802.11 uses thus a virtual carrier sensing mechanism to determine whether the channel is free. It is referred to as the *Network Allocation Vector* (NAV). Essentially, the NAV contains the information how long the channel is expected to be occupied. Information in the NAV is updated by the length field of the MAC header for any frame that is *not* intended for a particular device. More particularly, if a device decodes a MAC header of a packet intended for another device, it can see how long the channel will be occupied. If it receives a frame that indicates a larger NAV value, it increases its own NAV (but it does not decrease it if the second NAV is smaller).

The NAV is particularly effective when combined with the RTS/CTS mechanism discussed in Chapter 18. Essentially, the TX sends an RTS that contains in the MAC header not just the length of the RTS signal, but of the data transmission that will follow (plus the time required for the CTS). In other words, any device that can overhear the RTS knows how long the channel will be occupied by the upcoming frame. Similarly, the CTS contains the information about the length of the frame, allowing any device that can overhear the device that sent out the CTS to determine how long the channel will be occupied. For devices that can overhear both RTS and CTS, the information is simply redundant, but the devices that normally would be "hidden nodes" hear the information from one node only.

Enhanced distributed channel access: EDCA is an enhancement to DCF and provides prioritized access to the channel. The QoS support of EDCA is realized by considering an *Access Category* (AC) mechanism that provides support for the priorities at the devices. A device accesses the medium based on the AC of the frame that is to be transmitted. The mapping from priorities to access categories is defined in Table 33.7.

In essence, EDCA executes parallel versions of the DCF, one for each AC. An AC with higher priority is assigned a shorter CW in order to ensure that in most cases, a higher-priority AC will be able to transmit before the lower-priority ones. For further differentiation, different IFSs are introduced, corresponding to different ACs. Instead of the DIFS, a so-called *Arbitration IFS* (AIFS) is used. The AIFS is at least as long as a DIFS and can be enlarged individually for each AC.

Similarly to DCF, after the end of an existing transmission on the air, the device waits for the duration of the AIFS associated with the current AC to start a backoff procedure. The value of this AIFS increases as the priority of the AC decreases, as discussed above. Each AC within a single device behaves like a virtual device. It contends for access to the wireless medium and independently starts its backoff time after sensing the medium is idle for at least the AIFS. Collisions between ACs within a single device are resolved within the device. The timing relationship for EDCA is shown in Figure 33.15.

Table 33.7 Priority to access category mappings

Priority (same as 802.1D priority)	Access category (AC)	Designation
1	AC_BK (1)	Background
2	AC_BK (1)	Background
0	AC_BE (0)	Best effort
3	AC_BE (0)	Best effort
4	AC_VI (2)	Video
5	AC_VI (2)	Video
6	AC_VO (3)	Voice
7	AC_VO (3)	Voice

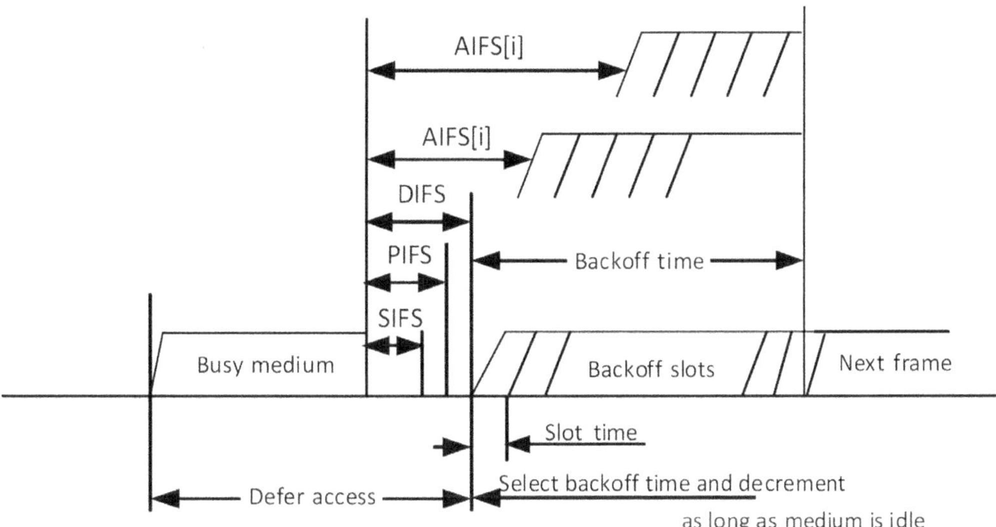

Figure 33.15 Timing relations for EDCA.
Reproduced with permission from [IEEE 802.11] © IEEE

While originally the channel access in DCF was done "frame by frame," 802.11e introduced the concept of *Transmission Opportunity* (*TXOP*) in the context of EDCA. This is a time period during which a device has the right to transmit (and receive responses for) data frames, control frames, and management frames, and not give up the channel. This introduces a degree of resource fairness, i.e., every device gets, on average, to access the channel for a particular duration of time (in contrast per-frame access tends to provide "throughput-fairness," since each device gets to transmit approximately the same number of *frames* and thus bits); consequently, with TXOP better links obtain higher throughput. Since two different devices starting a TXOP at the same time would be especially detrimental (leading to collisions for a long time), a short exchange of frames (RTS/CTS or similar) has to be performed at the beginning of a TXOP. The TXOP information is encapsulated in the QoS Control field of all QoS Data frames, so the other devices can use that information to set up their NAVs to prevent their channel access during the TXOP. We finally note that TXOPs can also be assigned to a device through polling (a contention-free process described in the next subsection).

Point Coordination Function (PCF) is a centralized, contention-free MAC for 802.11, based on polling (see Chapter 18). The *Point Coordinator* (PC) resides in the BS. It is used during the CFP, whose beginning and duration are announced by the beacon signal. All UEs obey the access rules of the PCF, since each UE sets its NAV at the beginning of each CFP, i.e., considers the channel to be occupied for random access purposes, until the end of the CFP.

The PCF relies on the PC to perform polling and enables the polled UEs to transmit without contending for the channel, see Figure 33.16. The PC sends a polling signal to a UE (this signal can either be a stand-alone polling frame, or the polling is piggy-backed on a data frame that the BS sends to the UE). When thus polled by the PC, a UE may transmit only one MPDU, which can be to any destination. Acknowledgments (see below) must be piggy-backed onto data frames as well; this can lead to difficult selection of the MCS if the recipient of the data and the recipient of the acknowledgment are different devices.

The PCF was never widely used, partly because of the considerable latency – packets of a UE operating under PCF that are generated during the CP have to wait until the next CFP, while conversely, packets of a UE operating under DCF during the CFP have to

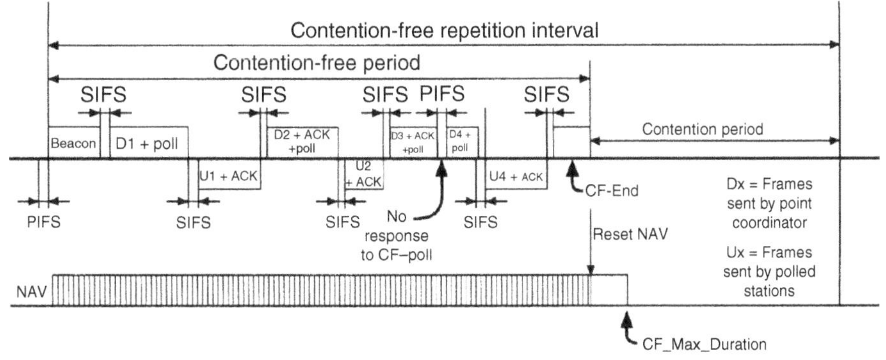

Figure 33.16 PCF frame transfer.
Reproduced with permission from [IEEE 802.11] © IEEE

wait until the CP. It has thus been abolished in later versions of the standard; we mention it here because it is of historical interest, and because it impacted the development of the Hybrid Coordination Function (HCF) that is discussed next.

Hybrid Coordination Function (HCF): An improvement of the PCF was introduced in 802.11e in the form of *HCF Controlled Channel Access, HCCA*, which greatly reduces the latency problem. Like PCF, HCCA uses polling but can employ that in both the CP and the CFP, which reduces the latency. The HC grabs the channel, and starts a *Controlled Access Phase (CAP)*, by waiting a shorter time period than all of the other devices using an EDCA for its first transmission, and not releasing it while active (i.e., the response frames are only a SIFS after the frames the HC sends out. Furthermore, the HC grants TXOPs, and not just MPDUs, thus greatly enhancing the efficiency of the system.

The key difference between HCCA and PCF are:

- HCCA uses a *Hybrid Coordinator* (HC) that resides in the BS to centrally manage the medium access through TXOPs in order to provide parameterized QoS – the capability of providing QoS flows from applications with specific QoS parameters, such as rate, latency, and so forth (the so-called *Traffic Specifications* [TSPEC]).
- HCCA can poll UEs during both CFP and CP. Thus, it is not mandatory for the HC to use CFP for QoS data transfers.
- HC may also operate as a PC, providing (non-QoS) CF-Polls to associated CF-pollable UEs obeying the PCF rules.

The *HCF*, see Figure 33.17, includes both EDCA and HCCA. 802.11e uses a temporal structure with the channel access alternating between CFP and CP as shown in Figure 33.18. EDCA is used for prioritized or differentiated QoS services, while HCCA is used by parameterized QoS services.

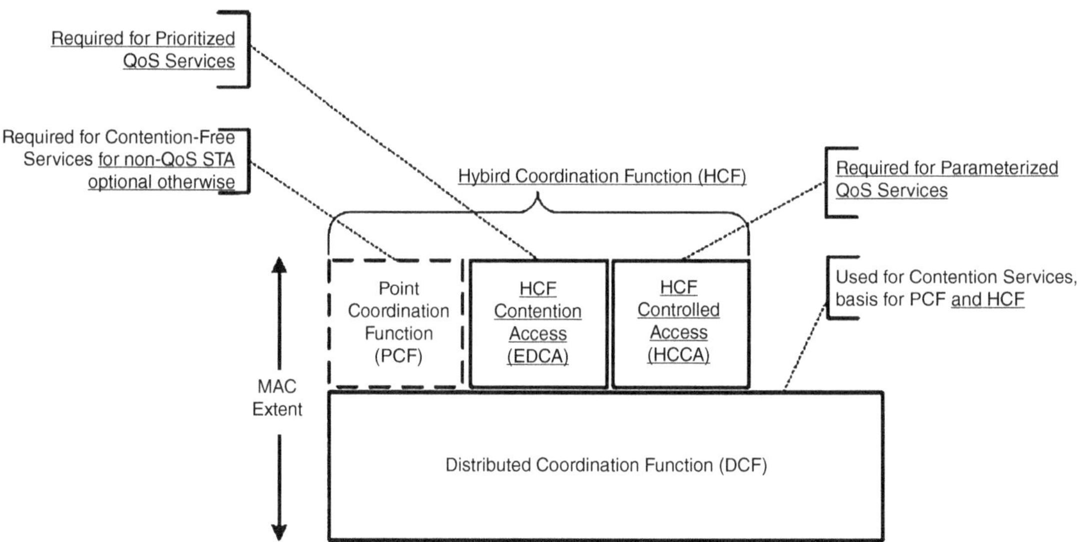

Figure 33.17 Structure of the 802.11e MAC.
Reproduced with permission from [IEEE 802.11] © IEEE

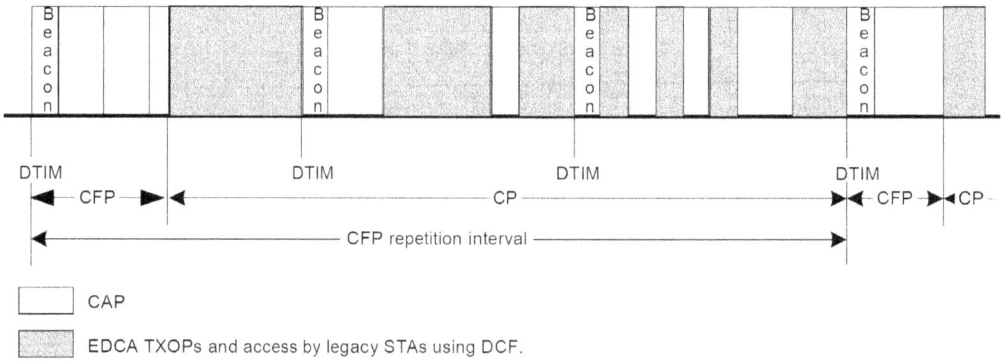

Figure 33.18 Superframe structure of the 802.11e MAC.
Reproduced with permission from [IEEE 802.11] © IEEE

Admission control: The admission control function decides whether a new connection can be granted or not, depending on the status of the network resources and the level of service called for by the new request. As discussed in Section 20.6, the purpose of any admission control is to ensure that admittance of a new data flow into a resource-limited network does not degrade (below the promised level) the QoS of the already-admitted data flows while optimizing the network resource usage. Thus, admission control is a key component of QoS–based resource management scheme. There are two distinct admission control mechanisms in 802.11e – one for contention-based and the other for controlled-access admission control.

- *Contention-based admission control*: Admission control in contention-based systems is realized through the exchange of *Add Traffic Stream* (ADDTS) request and reply management frames between QSTAs and QAP (QoS UE and BS, respectively). QAPs always support admission control, while QSTAs may choose to employ admission control or not. However, if a QSTA decides to use admission control for a specific AC, all ACs with higher priority than that have to use admission control. If a flow is admitted, QSTA maintains two variables – admitted time and used time – to track the communication. If the QAP cannot admit the required traffic, the QSTA may lower the priority of that traffic so that admission control is not mandatory for that traffic flow.
- *Controlled access admission control*: When the HC provides the CAP to QSTAs, it is responsible for granting or denying polling service to a *Traffic Stream* (TS) based on the parameters provided by the QSTAs. If the TS is admitted, HC will keep it active unless the STA requests to tear it down or it has been inactive for a predefined period of time. The polling service provides a "guaranteed channel access" from the scheduler in order to have its QoS requirement met.

33.4.4　Frame Format

The MAC frame format comprises a set of fields that occur in a fixed order in all frames. A frame consists generally of a MAC header, a body, and a frame check sequence that includes a CRC.

Figure 33.19 depicts the general MAC frame format, though not all fields are present in all types of frames. In the following, we will discuss some aspects of the most important content fields. A completely listing would take too much space; please see the standards documents and/or [Peraiha and Stacey 2015] for details.

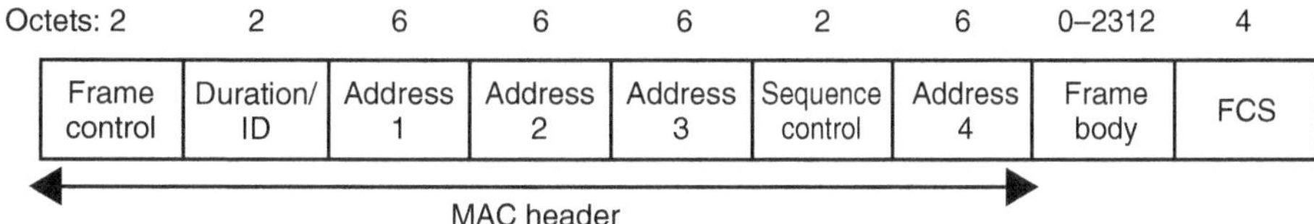

Figure 33.19　MAC frame format (a typical MPDU).
Reproduced with permission from [O'Hara and Petrick 2005] © IEEE.

The MAC header contains, *inter alia*, the following fields:

- *Frame control:* this contains a number of subfields, pertaining to fragmentation, power management, and routing in systems with a distribution system (DS).
- *Duration field:* it indicates the duration in μs, and is used to set the NAV.
- *Address field:* it contains the address of the RX in address 1, the address of the TX in address 2 (absent in ACK and CTS frames), and addresses related to the network (only in specific frame types).
- *Sequence control field:* containing the sequence number of the MSDU, as well as the fragment number if fragmentation is used.
- *QoS control field:* this field contains all parameters related to QoS control, including the traffic identifier, ACK policy, TXOP parameters, and buffer status.
- *HT control field:* this lists parameters related to multi-antenna operation, such as sounding, antenna selection, transmit beamforming, and calibration.

A data frame typically contains most or all of the fields mentioned above. Management frames can be considerably shorter.

33.4.5　Fragmentation, Aggregation, and Acknowledgements

Since the frame transmission is subject to errors, the RX has to inform the TX of whether the data arrived in good order, or a re-transmission is required. This is done by sending an *Acknowledgement* (ACK) frame, which contains a short positive message that uses a more robust MCS than the original data transmission, thus minimizing the probability of an error during the ACK transmission.

The ACK is sent after a SIFS (without a random backoff), so that it is the highest-priority message on the air. If the TX does not receive such an ACK, it assumes that the data was lost in transmission, and retransmits the data frame. This makes clear why the ACK is sent with high priority and robust MCS: the penalty for a missed ACK is high – namely the retransmission of the whole frame.

When the MSDUs handed down to the MAC become too large, it becomes difficult to transmit them in one block: obviously, the probability of a block error (i.e., that one of the bits in the block is in error) increases with the duration of the block.[9] As each block error might lead to the necessity of retransmission, this is highly undesirable. Thus, the MSDUs have to be fragmented in order to increase transmission reliability. This fragmentation is done when the MSDU size exceeds the fragmentation threshold. In that case, the MSDU will be broken into multiple fragments, and each fragment is sent as a separate MPDU. The size of the MPDUs is equal (with the exception of the last fragment), and is equal to the fragmentation threshold. The MPDU also contains a special data field (the "more_data" field) set to 1 in all but the last fragment. The receiving device acknowledges each fragment individually. The channel is not released until the complete MSDU has been transmitted successfully or an acknowledgment is not received for a fragment. In the latter case, the TX will re-contend for channel following the normal rules and retransmit the nonacknowledged fragment, as well as all of the following ones.

However, acknowledging each fragment at a time is quite inefficient. This is why 802.11e introduced *Block Acknowledgement (BA)* frames. For a BA, the devices talking to each other have to first establish a "session"; during the setup of this session, parameters such as the size of the RX buffer, which determine the efficiency of the process, are exchanged.

For the actual BA of data, the TX sends out data frames in which a special field in the MAC header indicates that BA should be used for acknowledgment. The RX writes the received frames into a buffer. It is noteworthy that the frames do not need to arrive in the proper sequence – the sequence number in the MAC header provides the proper information, and the RX is responsible for ordering them and passing them to the higher layers. Since the TX knows (from the session setup) the size of the RX buffer, it can make sure to send only a limited number of frames before sending out a *BA Request* (BAR) frame. This BAR frame first of all contains the sequence number of the oldest MSDU for which a BA is requested; older MSDUs are passed on to the higher layers or discarded. Furthermore, the RX responds to the BAR frame with a BA response that contains a bitmap with the status of all data frames starting with the starting sequence number communicated in the BAR frame. When BA is used, the different frames can be sent with a SIFS spacing between them (according to 11e), or a RIFS (for 11n). It is also possible to send the frames without spacing (aggregation, see below).

The RX can send the BA response using one of two variants: in the first, dubbed *immediate BA*, the BA response is sent within the same TXOP, namely a SIFS after the BAR. In the *delayed BA*, the BA response is sent at a later time, possibly after a new contention for the channel.

The dual to fragmentation is aggregation. This should occur when the payload is too small to be transmitted effectively; multiple frames are then aggregated into a larger one that can be transmitted more effectively. Two mechanisms are foreseen in 802.11: MSDU aggregation, i.e., aggregation at the top of the MAC, and MPDU aggregation, which is at the bottom of the MAC (just before passing data to the physical layer). The structure of how the aggregate MPDU is encapsulated is shown in Figure 33.20.

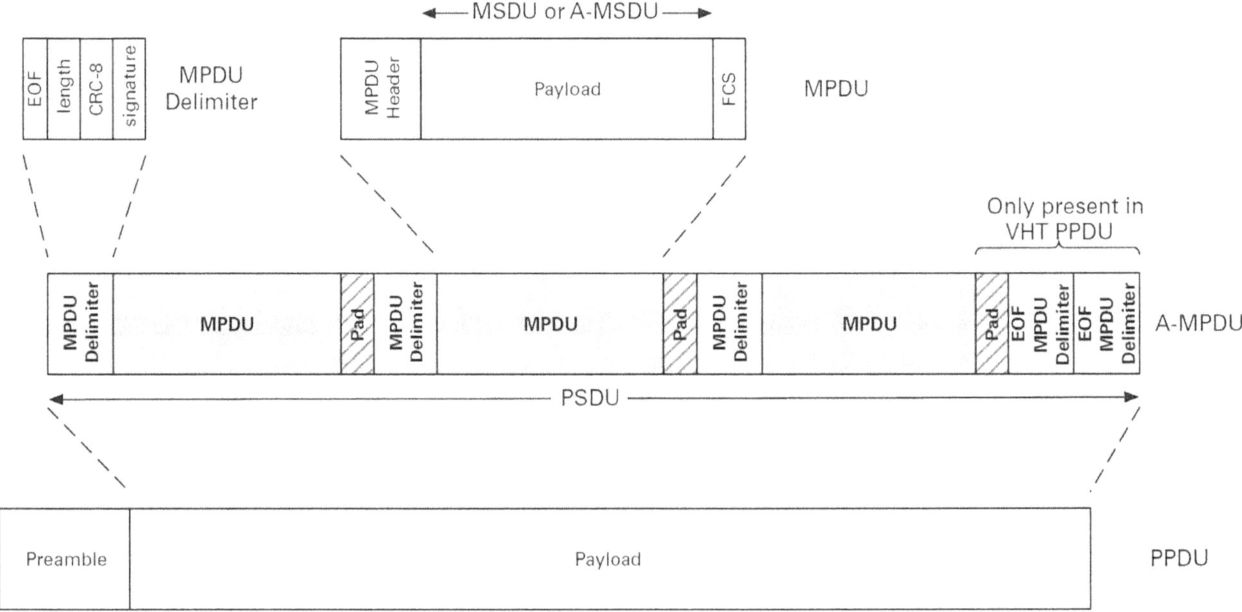

Figure 33.20 A-MPDU encapsulation. VHT refers to 802.11ac, see Sec. 33.5.

Reproduced with permission from [Perahia and Stacey 2013] © Cambridge University Press.

---

[9] Note that in general, this effect may be partly offset by the fact that larger blocks allow the use of better codes, like highly efficient LDPC codes, see Chapter 13).

33.5 IEEE 802.11ac

33.5.1 Overview

Even as 802.11n was being approved, the need for higher data rates became obvious. While 600 Mbit/s is amply sufficient to distribute content received from most cabled internet connections (the average internet speed in the United States in 2010 was 5 Mbit/s, and peak speeds around 20 Mbit/s), much higher data rates are desirable for wireless data exchange within a given enterprise or private network. For example, the exchange of uncompressed high-definition video requires data rates in excess of 1.5 Gbit/s. It was thus decided to form a further study group to study *very high throughput VHT* Wi-Fi. This group developed what became the 802.11ac standard, which was approved and published in 2013. It achieves the increased throughput by a combination of the following main approaches:

- *Increased bandwidth*: channels with a BW of up to 160 MHz are defined.
- *Larger-dimensional MIMO*: up to eight data streams are allowed at the BS, and MU-MIMO is enabled.
- *Higher-dimensional modulation format*: 256 QAM now forms part of the modulation alphabet.

With these improvements, data rates up to 6.9 Gbit/s at the BS (in multi-user mode, i.e., when transmitting to at least two UEs) can be achieved (this is, of course, an idealized rate that is not obtained in reality). At the same time, a number of simplifications were introduced that can reduce performance, but were expected to simplify implementation and thus reduce cost:

- *No unequal MCS*: all spatial streams have to carry the same MCS, even if their SNRs are different.
- *Fewer options for CSI feedback:* implicit feedback, as well as uncompressed explicit feedback, are disallowed.
- *Fewer beamforming options:* various options for block coding and spatial multiplexing were eliminated. For example, either all streams have Alamouti codes, or none do.

33.5.2 Channelization

We first discuss the increase in bandwidth. 802.11ac foresees channel bandwidths of 20, 40, 80, and 160 MHz. Section 33.3.4 discussed how 802.11n increases the bandwidth from 20 to 40 MHz: it does not just bond two channels, but also uses those null-carriers that lie between the two bands for data transmission, since now no guardband is necessary. The same principle holds for the increased bandwidths in 802.11ac: for 20 and 40 MHz, the channelization is the same as for 802.11n. 80 MHz has 256 subcarriers, with null carriers as guard bands at $-128, ..., -123$ and $123, ..., 127$, and null carriers for eliminating DC components at $-1, 0, 1$. Eight pilot tones are on subcarriers $\pm 11, \pm 39, \pm 75, \pm 103$, while all other subcarriers are used for data transmission.

The approach is slightly different for 160 MHz bandwidth. There are two ways of generating it: either by using two disjoint 80 MHz bands, or one contiguous 160 MHz band. In order to allow a simpler design, both approaches have the same subcarrier assignment, i.e., consider a lower and an upper 80 MHz band; even when there is a contiguous 160 MHz bandwidth, the null-carriers in the "middle" guard bands (upper guard band of the lower 80 MHz, and lower guard band of the upper 80 MHz) are not used as data carriers. This inefficiency is the price paid for having a single 160 MHz signal structure. Thus, 160 MHz has 512 subcarriers, with subcarriers $-256, ..., -250, -129, ..., -127, -5, ..., 5, 127, ...129, 251, ..., 255$ being null carriers. The pilot tones are at $\pm 25, \pm 53, \pm 89, \pm 117, \pm 139, \pm 167, \pm 203, \pm 213$.

33.5.3 Preamble

We now turn to the SU (single-user) preamble. It starts out the same way as the 802.11n preamble, namely with the legacy fields: L-STF, L-LTF, and L-SIG. Just like in the 802.11n case, the signals are replicated in frequency when the bandwidth is larger than 20 MHz, phase rotations are put on signals in the upper-frequency ranges ($\pi/2$ for 20–40 MHz when 40 MHz BW is used, and π for 20–80 MHz when 80 MHz BW is used). Furthermore, different cyclic delay shifts are put on signals transmitted from different antennas, again in analogy to 802.11n.

The L-SIG field has the same structure as in 802.11n. Among other things, it carries the length information for the packet. Importantly, while the length information in the L-SIG field can be ignored by 11n RXs because the information is repeated in the HT-SIG field, in 802.11ac the length information *has to* be learned from the L-SIG field, because the VHT-SIG A field does not contain it (though it *is* contained in the VHT-SIG B field).

The signaling field specific to VHT is split into two parts, VHT-SIG A and VHT-SIG B. The former contains most of the information that is critical for detecting/decoding the packet, including (Figure 33.21):

- *Bandwidth*: two bits indicating 20, 40, 80, or 160 MHz.
- *STBC*: one bit stating whether Alamouti coding is used.
- *Group ID and partial AID*: the group ID is mostly used for MU-MIMO (see below); the partial AID (association ID) gives the RX an early indication of whether the packet is intended for it (final confirmation is obtained from the MAC header).

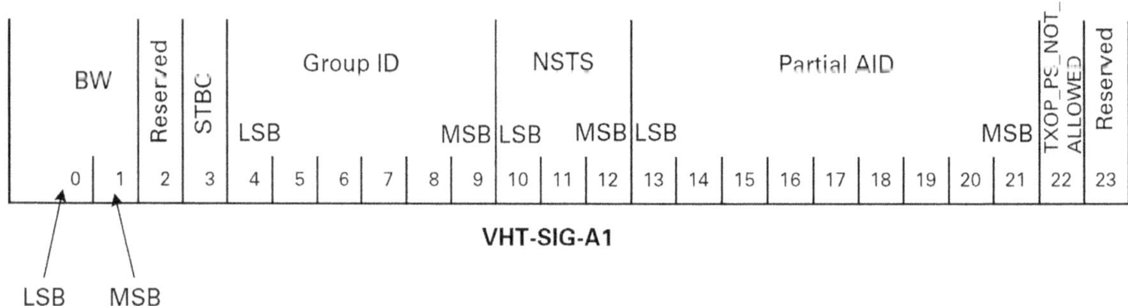

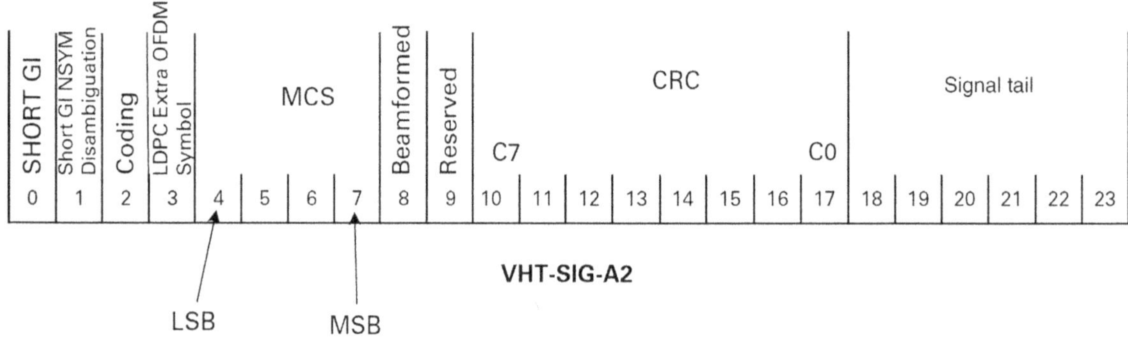

Figure 33.21 VHT-SIG-A structure for SU packet.
Reproduced with permission from [Perahia and Stacey 2013] © Cambridge University Press.

- *Number of spatial streams*: three bits, since there can be up to eight spatial streams. Note that while this index runs from 0 to 7, it indicates number of spatial streams from 1 to 8.
- *TXOP_PS*: related to a rarely used power save mode.
- *Short GI*: this indicates whether a cyclic prefix interval (400 ns instead of 800 ns) is used. A cyclic prefix can also lead to some ambiguities in the duration (since the L-SIG field announces durations in multiples of $4\,\mu s$), which are resolved by the value of a disambiguation field.
- *Coding*: 0 for convolutional coding and 1 for LDPC. Another bit is used for indication of extra OFDM symbols inserted in LDPC encoding.
- *MCS*: indication which of the available 10 MCS schemes is used. It is noteworthy that this applies to all spatial streams.
- *Beamforming*: a bit set to 1 if beamforming is applied and 0 otherwise (due to the reduced number of beamforming options in 802.11ac compared to 802.11n, the number of signaling bits related to beamforming could be reduced).
- *CRC and tail*: same as in 802.11n.

The 48 signaling bits are encoded with a convolutional code, BPSK modulated and mapped onto two OFDM symbols, where for the second symbol, the BPSK symbols have a $\pi/2$ phase shift. This offset in modulation between first and second symbol allows the RX to detect whether the signaling field is an 802.11a, 802.11n, or an 802.11ac field.

The VHT short training field is essentially the same as the legacy training field, though provisions for up to eight spatial streams are made. The VHT long training field is essentially a replication of the LTF in 802.11a (and 802.11n), with some additional "filler" values for the additional subcarriers that are used at larger bandwidths (those carriers that are null-carriers at smaller bandwidths that become data carriers at larger bandwidths, see above). It also allows phase tracking during the training sequence, since it is necessary to transmit up to eight LTFs, which takes up to $32\,\mu s$ – a duration over which phase noise can have a significant effect.

The VHT-SIG B field is intended for user-specific signaling in MU-MIMO but is transmitted also in the SU-MIMO case, just to have a uniform transmission format. The information contained in this packet in the SU case is the length of the packet and some tail bits. For larger bandwidths, the signaling bits are repeated.

Next follows the data field, which contains not only the payload data but starts with the *service field*. The service field contains the CRC for the VHT-SIG B as well as the scrambler seed. The payload data are treated in a similar fashion as in 802.11n, though there are some important differences: (i) for convolutional coding, the raw data are parsed into up to 12 streams before encoding (instead of up to two streams in 802.11n; this is due to the higher data rates that are possible in 802.11ac); (ii) the way the number of padding bits in LDPC codes is computed is slightly different; (iii) there is special parsing for the 160 MHz bandwidth case; (iv) there is an interleaver for LDPC codes (as opposed to 11n[10]), since for large bandwidths, the LDPC code might not spread the bits sufficiently over the whole bandwidth.

As mentioned above, the MCS schemes are significantly simplified. A total of 10 MCSs ranging from BPSK with rate 1/2, to 256-QAM with rate 5/6 are allowed, but unequal MCSs on different streams is not.

33.5.4 Transmit Beamforming

For transmit beamforming, 802.11ac offers *fewer* options than 11n. This happened because adoption of the various beamforming options in 11n was slow, and products from different manufacturers often implemented incompatible options, so it was decided to simplify the approach, even at the cost of performance. Firstly, only explicit feedback is supported (no implicit feedback based on reciprocity), and only compressed feedback of the beamformer (i.e., the **V** matrix) is allowed, though the dimensions of the admissible matrices have been extended to up to 8×8. The quantization for the angles is done with $b = 2$ or $b = 4$. Sounding is now only allowed with NDP packets, while the sounding PPDUs were eliminated.

The sounding sequence has also been simplified. Instead of sending a sounding request as part of a data sequence, there is now a separate VHT NDP announcement, followed by the NDPs, which are immediately followed by the explicit feedback.

33.5.5 Downlink Multi-User MIMO

One of the key new aspects of 802.11ac (compared to 11n) is the enabling of downlink MU-MIMO. For each packet, a part of the preamble (LTF, L-SIG, and VHT-SIG A) is transmitted omnidirectionally, but the subsequent transmissions (VHT-STF, VHT-LTF, VHT-SIG B, and VHT data, are all sent with appropriate precoding (see below on how to obtain that) to the different users simultaneously. Up to eight antennas at the BS are foreseen. For the MU case, up to four spatial streams can be transmitted per user.

Preamble: We next analyze the differences in the preamble between SU-MIMO and MU-MIMO. Consider first the VHT-SIG A, Figure 33.22: it is transmitted omnidirectionally, i.e., is read by all users. We have then different types of information: (i) information intended for all users, such as bandwidth, and STBC configuration (which implies that all users have to have the same bandwidth and the same STBC configuration), (ii) information intended for each particular user (such as the number of spatial streams

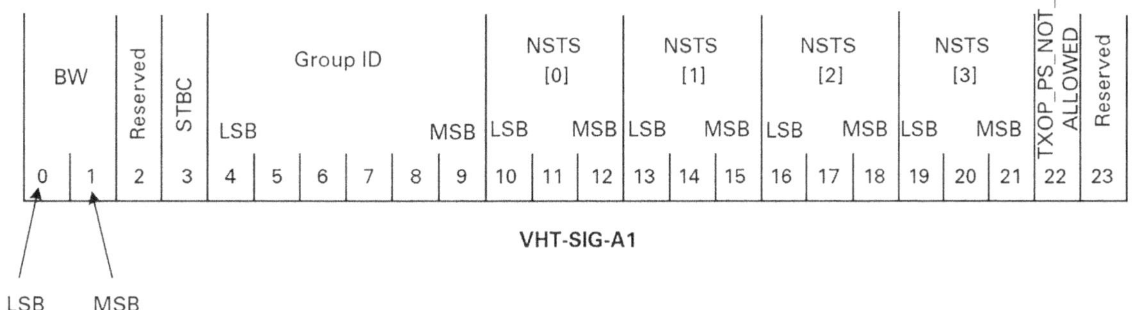

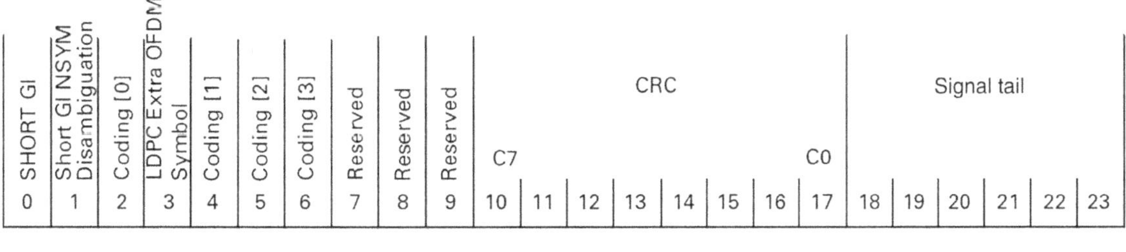

Figure 33.22 VHT-SIG-A field for MU-MIMO.
Reproduced with permission from [Perahia and Stacey 2013] © Cambridge University Press.

---

[10] 11n does have an interleaver for the convolutional code, but not the LDPC code, because a coding block for the latter is so large that it is distributed over all subcarriers and spatial streams anyway.

and the coding-type indicator, where there is a different set of fields for the number of spatial streams intended for each user), (iii) CRC and tailbits. The VHT-SIG A field also contains the *group ID*, which is an index referring to a table entry. This table works as follows: if the group ID is 0 or 63, the following packet is a SU MIMO packet. If the group ID is between 1 and 62, it refers to a table entry that contains the IDs, *in a particular sequence*, of the devices that are part of the group. Thus, every UE looks at the table entry corresponding to the group ID index, and first of all determines whether it is part of the group (if not, it can stop trying to decode all subsequent parts of the packet). Secondly, if it is part of the group, the position in the group gives it information about other parameters in the VHT-SIG field. For example, if it is the zeroth user in the group, then it knows that the Coding[0] bit indicates the coding type for it. The group ID lookup tables are managed by the BS and sent to the devices in a special management frame.

The VHT-STF is essentially the same as for the SU case; one should note that the "number of spatial streams" is the sum over the spatial streams for all users and that the shifts for the cyclic prefixes are not computed "per user," but rather for the total number of spatial streams transmitted by the BS. Similar considerations hold for the VHT-LTF, and of course, the number of LTF fields that need to be transmitted is determined by the total number of spatial streams (summed over all users).

The VHT-SIG-B field carries the information about the length of the packet (length of that field is 16, 17, or 19 bits for 20, 40, and 80 MHz bandwidth), as well as the MCS (remember that for SU-MIMO, the MCS was transmitted in the VHT-SIG A field).

Data field: The data are prepared and encoded similar to the SU-MIMO case, but separately for each user. This means that the convolutional encoding or LDPC encoding is applied. A combination of MAC and PHY padding is used to ensure that the packets for all users transmitted simultaneously have the same length. The different data are then combined in the spatial mapper. From a MAC perspective, every MDPU sent is an aggregated MPDU (A-MPDU). The acknowledgments sent by the different users cannot all happen at the same time. Thus, the BS will typically solicit an immediate feedback only for one user, while asking for later block acknowledgment for the other users.

Beamforming and sounding: In terms of beamforming, MU-MIMO requires that the BS has knowledge of the channels to all the devices. That means that the feedback now has to include not only the matrix of the right singular vectors $\mathbf{V}$ but also the singular values $\mathbf{\Sigma}$; the latter are transmitted by providing the per-tone SNRs. Furthermore, for MU-MIMO, $b = 5$ or 7; this is because MU-MIMO is more sensitive to quantization errors of the CSI, compare Chapter 22.

The sounding is also similar to the SU-MIMO case. The procedure starts with a VHT NDP announcement that identifies to which devices the request for feedback is addressed (the sequence in which these devices are listed is important). Then the BS sends out the NDP, and all addressed UEs measure the channel. Following that, the first UE on the list sends a compressed beamforming feedback (that has the same form as for SU-MIMO), while all other UEs are quiet. This is then followed by a beamforming report poll that solicits the feedback from the other devices one by one.

33.6   802.11ax/Wi-Fi 6

While the peak data rates achievable with 802.11ac seem sufficient for a majority of applications, the actual user-experienced throughput is significantly less. Furthermore, the 802.11ac system lacks flexibility in the assignment of the resources. For these reasons, a new standard has been developed that can be seen as becoming more similar to a "mini-cellular" system. This standard, called 802.11ax (or, more popularly, Wi-Fi 6), was ratified in 2021.

The catchphrase for 802.11ax is "high efficiency," so that the acronym for 802.11ax specific preambles, etc., is "HE" (similar to the use of "very high throughput VHT" for 802.11ac). This also reflects the fact that the maximum theoretical throughput only increases from 6.9 Gbit/s (for 802.11ac) to 9.6 Gbit/s (note that both the step from 11a to 11n and 11n to 11ac was a factor 10 in data rate). Rather it is the spectral efficiency and area spectral efficiency, as well as the energy efficiency, that is at the center of the changes.

The main changes in 802.11ax can be summarized as follows:

- Addition of 1024 QAM to increase the data rate in scenarios with very high SNR.
- Longer symbol period, which reduces the overhead from the cyclic prefix
- Enabling MU-MIMO in the uplink. This is enabled by the introduction of trigger frames that give the BS more control over the transmissions from the UE. Note that DL-MU-MIMO was already introduced in 802.11ac.
- Use of Orthogonal Frequency-Division Multiple Access (OFDMA), including OFDMA random access.
- Spatial reuse, by distinguishing packets to/from different BSs, and setting of thresholds for interference avoidance.
- Special power-saving mode with microsleep and Target Wake Time (TWT).
- Station-to-station (STA2STA, S2S) operation.
- Provisions for outdoor, long-distance transmissions.

It is noteworthy that as 3GPP is adopting approaches from the Wi-Fi space, such as femto-cells and use of unlicensed bands, 802.11 is moving toward solutions that have been traditionally been associated with the cellular approach, such as interference coordination, BS-control of uplink transmissions, and MU MIMO.

33.6.1 Modulation and Channel Assignment

802.11ax changes the subcarrier spacing to 78.125 kHz, making it four times narrower than the 312.5 kHz in the previous 802 standards, and making the OFDM symbol duration four times longer. Furthermore, the cyclic prefix can now take on three different values: 3.2 µs (four times longer than the previous 802.11 CP, which leaves the relative overhead constant), 1.6, and 0.8 µs. The reasons for the change are: (i) the larger number of subcarriers allows a more flexible implementation of OFDMA, (ii) the option for the longer CP enables deployments outdoors, where larger delay spreads occur, while due to slow movement, the intercarrier interference will remain small even with the reduced subcarrier spacing, (iii) perhaps most importantly, an increase in the spectral efficiency, since the ratio of CP duration to symbol duration (for the "standard" CP length) is reduced.

In terms of modulation formats, 802.11ax introduces 1024 QAM, which becomes possible because of technological advances enabling better power amplifiers and other RF components (compare Chapter 17), as well as signal processing methods such as digital predistortion that can compensate for hardware nonidealities. Since each symbol represents 10 bits, the maximum data rate is increased (the combination of this modulation format, and possible lower relative overhead of the CP, leads to a maximum data rate of 9.6 Gbit/s). However, such a high-order modulation can only be used for very high SINR, and thus will not be exploited in most deployments. The associated MCSs (MCS 10 and 11) are optional for both BS and UE.

Enhanced robustness can be achieved through *Dual-Carrier Modulation* (*DCM*) such that the same information is transmitted on two subcarriers that are offset by half the size of the resource unit (see below). To reduce the PAPR, the signal at the upper subcarrier is phase-shifted to that of the lower subcarrier. DCM is only allowed for low-order MCS, namely MCS 0, 1,3, and 4, because it reduces the spectral efficiency, and its sole purpose is to increase robustness.

In terms of coding, both convolutional codes and LDPC codes are now mandatory. Convolutional codes must be applied when the transmission bandwidth is ≤20 MHz, while for larger bandwidths, LDPC codes are mandatory. However, 802.11ax also defines a special class of low-complexity devices that operate only in 20 MHz bandwidth, and thus need not be capable of LDPC decoding, though they have to meet all other mandatory requirements of 11ax. This is mainly intended for IoT devices, which should be low-cost and low-complexity.

802.11ax introduces groups of subcarriers, called *Resource Units (RUs)* as the basis for multiple access. These RUs are similar to the resource blocks in LTE, compare Chapter 31. The narrowest RU consists of 26 subcarriers; larger blocks of 52, 106, 242, 484, and 996 are available as well, depending on the bandwidth, Figure 33.23 shows the possible RU setups for 20, 40, and 80 MHz bandwidth (again, 160 MHz is just seen as two contiguous or noncontiguous 80 MHz bands); Figure 33.24 shows an example of how signals for different users can be transmitted on different RUs in a DL packet.

802.11ax also changes the way in which channels can be bonded. While 802.11n and ac required contiguous bands for transmission (with the exception of the 160 MHz channel), 802.11ax allows a more flexible way of bonding multiple frequency channels by means of *preamble puncturing*. By extending the bandwidth field in the preamble from 2 to 3 bits, four additional bandwidth modes are defined, in which particular 20 MHz channels are eliminated from both the transmission of the preamble and for allocations of the RUs.

33.6.2 Frame Formats

Four new frame formats (more exactly, formats for PPDUs) were established for 802.11ax: one for "regular" single-user transmission, one for multi-user (OFDMA and/or MU-MIMO) transmission, one for transmission for extended range between TX and RX, and one for a trigger-based transmission, which is mostly used for uplink MU-MIMO. Figure 33.25 shows the structure. The modifications of the 11ax preamble follow the pattern established in earlier standards: at the beginning, the legacy fields (L-STF, L-LTF, and L-SIG) is transmitted, followed by fields that are special to the new standard (denoted with the prefix HE), and then finally the data. The preambles are designed for 20 MHz and duplicated into all 20 MHz channels that make up the total used bandwidth of the packet.

A new invention is the RL-SIG, a repetition of the legacy signaling field to enhance the robustness, and at the same time indicates that the PPDU is indeed an 802.11ax PPDU. This is then followed by the HE-SIG A, which is transmitted over all spatial streams and RUs. Its duration is 8 µs, except for the extended-range frame, where it is repeated to increase the robustness. It contains (for the SU-case) information about the MCS, bandwidth, number of spatial streams, and other parameters needed for decoding. New for 11ax, it also contains some information previously contained in the MAC header, such as BSS color and remaining TXOP duration.

In the MU frame format, the HE-SIG A field is followed by a second signaling field, the HE-SIG B, which has variable duration (more precisely, variable number of symbols with 4 µs duration each. It contains two blocks of information: one with common information, such as the OFDMA resource allocation, and one with information that differs between the various users, such as MCS, bandwidth, etc. Note that while a trigger-based frame is also used for the MU case, namely the uplink, there is no need for a HE-SIG B field there, because the BS has already instructed the UEs (in a preceding trigger frame) which resources to use, so that the trigger-based uplink transmission does not need the UEs to state again which resources they use – the BS knows that they will follow the instructions.

Then follow the HE-STF, which lasts for 4 µs, except in the trigger-based frame, where it lasts 8 µs. This is followed by multiple HE-LTFs, which can have variable duration. This is followed by the transmission of the multiple data streams on the different resources. In order to provide some additional time for the RX to process the data, a packet extension may be appended at the end of the PPDU.

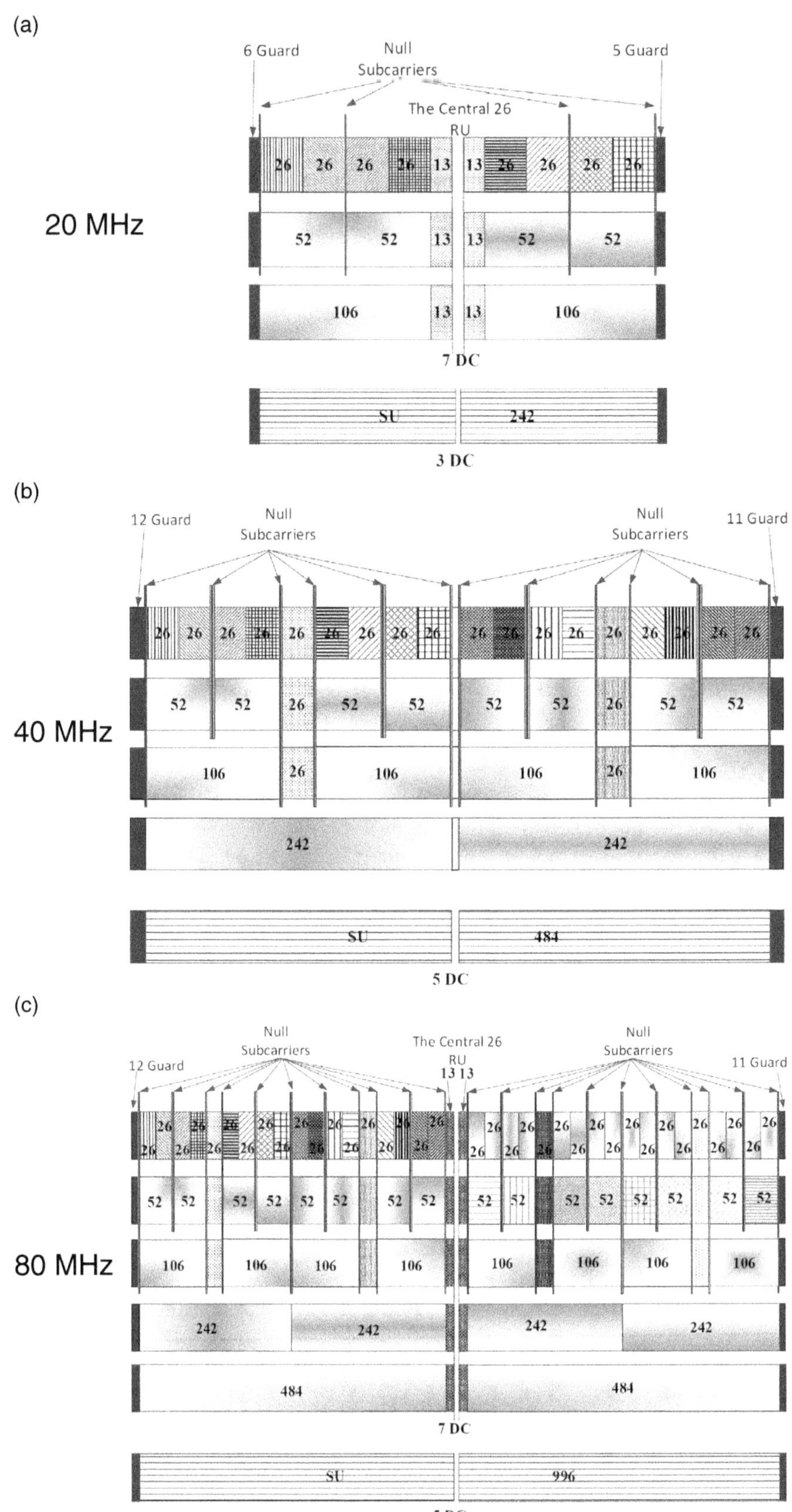

Figure 33.23　RU locations and maximum number of RUs for each channel width: (a) RU locations in a 20 MHz HE PPDU; (b) RU locations in a 40 MHz HE PPDU; (c) RU locations in an 80 MHz HE PPDU.
Reproduced with permission from [IEEE 802.11] © IEEE.

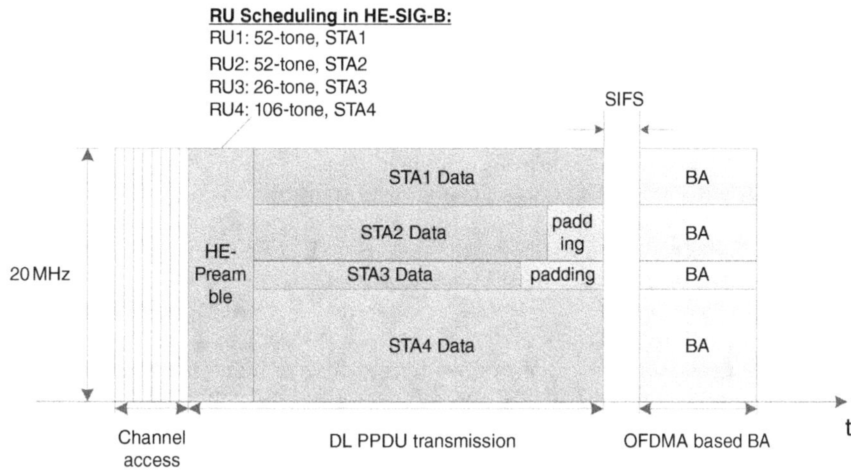

Figure 33.24 Example of a DL MU-MAC. Dark/medium grey: DL frame. Light grey: UL frame. Color version available at wiley.com/go/molisch/wireless3e. Reproduced with permission from [Qu et al. 2017] © Springer.

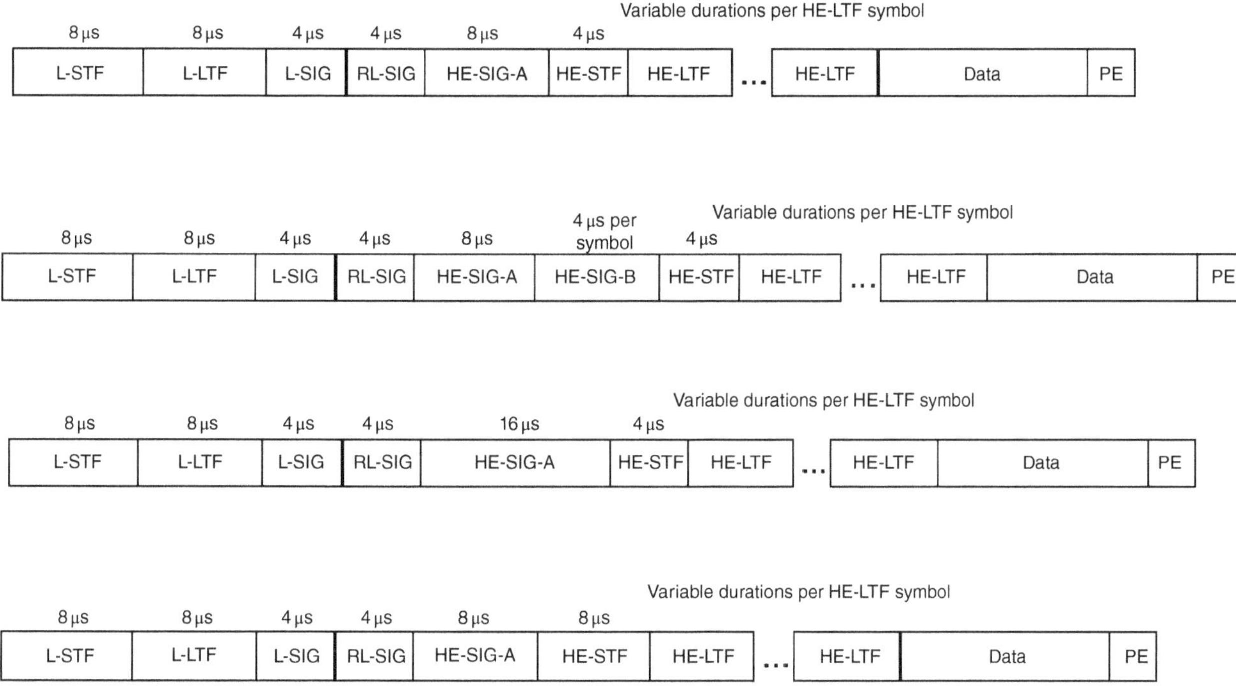

Figure 33.25 Frame formats for 802.11ax transmission. From top to bottom: SU, MU, extended-range SU, and trigger-based. Reproduced with permission from [IEEE 802.11] © IEEE.

33.6.3 Multiple Access

While 802.11ac uses DL-MU-MIMO and requires that each user employ the whole bandwidth, 802.11ax uses a combination of OFDMA and MU-MIMO, for both uplink and downlink, within each packet. This allows high sum throughput while at the same time providing considerable flexibility for handling many users simultaneously.

OFDMA is restricted to an assignment of RUs to a specific user for the whole duration of a frame. Thus, padding or aggregation has to make sure that all transmissions among RUs (and spatial streams, see below) have the same length. The OFDMA is furthermore combined with the traditional CSMA-based multiple access, and every frame can have a different allocation of RUs to users.

MU-MIMO is done separately for different RUs, though it is only allowed if the RU bandwidth is ≥ 106 subcarriers. In each RU, up to eight users can be spatially multiplexed by the BS. Each user can contribute up to four spatial streams, but the total number of streams cannot exceed 8.

For the downlink, the combination of OFDMA and MU-MIMO is relatively straightforward, since for the BS there is no fundamental difference whether to provide the beamformers on the subcarriers for the same or different users. Thus, the fundamental operating principle of the MU-MIMO remains the same as in 802.11ac. The main modification is that the preamble (specifically, the HE-SIG-B field) contains an RU allocation map and the transmission parameters for the individual users. The feedback of the acknowledgments can be done in one of two possible ways: (i) during the downlink frame, informing each UE on which resources to transmit the BA, (ii) transmission of a MU BA request frame (this is a special trigger frame that requires the UE to answer with a BA).

For the uplink, the situation is significantly different. The principle of multiple access in Wi-Fi is generally not suited for OFDMA or MU-MIMO in the uplink: each UE decides autonomously (and asynchronously), based on the clear channel assessment, whether to transmit or not, whereas OFDMA and MU-MIMO require coordinated transmissions. For this reason, 802.11ac has no UL-MU-MIMO. 802.11ax solves this problem by introducing an UL-MU *hybrid MAC*, which allows the BS more control of scheduling of users, without disturbing the legacy random access features. This scheduled access is achieved by so-called *trigger frames*, i.e., packets that tell the UEs when, and on which RUs, to transmit.

We now consider in greater detail the uplink multiple access. As a first step, the BS needs to know which UEs have data to transmit; in other words, the UEs have to send their requests for scheduled transmission resources to the BS (similar to the approach in LTE, see Chapter 31). This happens through the transmission of *Buffer Status Reports* (*BSRs*). These can be delivered either being "piggy-backed" onto other data transmissions of the UE to the BS, or they can be a stand-alone transmission. In the former case, the UE can add subfields in the MAC header of another transmission, or use frame aggregation. The latter case is used when the UE has no opportunity to send other data. If the BS initiates the BSR, it sends out a special frame that requires the addressed UEs to immediately send a BSR. The BS can also schedule UEs even if it does not have BSRs, just based on the assumption that a UE might have accumulated data since the last transmission.

As mentioned above, the UL-MU transmission relies in the concept of trigger frames that contains information how the UEs shall perform the uplink transmission. A trigger frame coordinates start and end of the uplink transmission, assigns resources (time, space, frequency), controls transmit power, and adjusts frequency offset to maintain orthogonality. It also contains information like the length of the frames and guard intervals, the allocation of the different RUs to the UEs, as well as transmission parameters such as MCS. The trigger frame can be padded, in order to give the UEs more time to prepare, since the uplink transmission has to be performed a SIFS after the trigger frame.

Upon receipt of a trigger frame, every UE has to do the following steps: (i) check whether it is one of the target UEs, (ii) synchronize to the trigger frame (both in time and frequency offset), (iii) check the CCA and NAV of the channels mentioned in the trigger frame, to ensure whether channel sensing is required, (iv) prepare the PPDU according to the transmission parameters instructed by the trigger frame, and apply power control according to the instructions of the trigger frame. Concerning the power control, we have to keep in mind that MU-MIMO in practice works best if the power levels received from the different users are similar, since this avoids residual inter-user interference from a strong user to overwhelm the signal from a weak user. The BS thus sends out its power level used for the downlink, and the power level it wants to receive from the UE; based on channel reciprocity, the UE can adjust its TX power level accordingly.

After the trigger frame, the UEs answer with their simultaneously transmitted packets. All the pre-HE parts of the packets are identical. Also, the HE SIG-A field is the same, namely copied from the HE-SIG A field in the trigger frame. The HE-LTFs, on the other hand, are orthogonal (more specific, a base LTF sequence is multiplied by distinct orthogonal codes), thus enabling the BS to separate the data from different UEs. The PPDUs for all users have to have the same length; this is achieved by adding padding bits on the MAC, plus possible filler bits on the PHY to round to the nearest byte (since the MAC works with complete bytes). The ACK to the different UEs can be sent in one of two different ways: (i) individual ACKs, using a DL MU frame, or (ii) a single frame containing the ACKs for all users. The latter method has higher error probability but has less overhead, and – since it is sent in 11a format, can be decoded by all devices in the network; this helps to synchronize channel access across all UEs on the network.

There are, however, still significant components of the CSMA approach of traditional 802.11 even in this approach. Firstly, the transmission of the trigger frame itself only follows after the BS has obtained the channel through the usual CSMA/backoff procedure. Secondly, the trigger frame allows to assign RUs for a mixture of scheduled and random access. For each RU, the trigger frame contains the AID field, which indicates the ID of the UE that should transmit in that RU (similar to 802.11ac). If, however, the AID is set to 0, then it indicates that the RU is available for random access in the following UL PPDU (AID 2045 indicates availability for random access by *unassociated* UEs, i.e., for joining the BS). Note that such a transmission can happen simultaneously with other UEs in different RUs; however, only a single UE may transmit in that band, since the BS lacks the CSI to separate multiple users (due to lack of synchronization, the LTFs of different UEs would not be orthogonal).

The OFDMA random access is somewhat similar to the CSMA procedure. Each UE maintains an *OFDMA backoff (OBO)* counter. When it receives a trigger frame that indicates that n RUs are available for OFDMA random access, then it reduces the OBO by n. If that resulting number is zero or less, then the UE will attempt a transmission, and will do it randomly in one of the available RUs. To ensure that no collisions occur, the UEs have to do carrier sensing after a SIFS interval. If the transmission fails, the size of the CW is doubled, and the process is repeated, similar to the "traditional" CW approach discussed in Section 33.4. It can be seen that in 802.11ax, a UE has two possibilities of access: waiting for trigger frames, or via random access. To ensure fairness, 802.11ax foresees that a UE that has transmitted data via the latter cannot grab the channel for a certain amount of time. Generally, random access is recommended for short packets and BSRs, while longer packets should be transmitted via scheduled transmissions,

A further step toward a "cellular-like" protocol is the introduction of the *cascaded MU MAC*. In it, multiple *scheduled* packets are transmitted in sequence, alternating between uplink and downlink. Through the use of packet aggregation, additional info is included in each of the packets: the downlink packets contain the BA pertaining to the previous uplink transmissions as well as the scheduling information for the next uplink frame – in other words, the functionalities of the BA frame and the trigger frame are aggregated into the downlink data frame. Similarly, the BAs for the previous downlink transmission are transmitted as part of the uplink data frame. It is noteworthy that the assignment of the different RUs does *not* to stay the same from one frame to the next in such a sequence.

The RTS/CTS procedure also needs to be amended in the MU case. 802.11ax introduces a special MU-RTS, which is sent out when the BS wants to send out a trigger frame or a MU-DL transmission. Upon receipt of that MU-RTS, *multiple* UEs, namely the ones addressed by that RTS, send out a CTS. Since they all send out the same information, the collisions are not a problem for the BS to properly receive. Equally importantly, any other UE within range of one of the target UEs (even the ones that might not have heard the RTS from the BS) hear the CTS, and those go into silent mode. This avoids the "hidden node" problem that otherwise might affect the MU setup.

33.6.4 Spatial Reuse

A major problem for WLANs is the interference between different networks, especially in dense deployments. 802.11ax aims to improve this by a combination of various mechanisms, in particular improved channel sensing and power control.

A first step is the determination whether a sensed packet belongs to the same, or a different BS. This is achieved by the *BSS coloring* mechanism. It uses a six-bit sequence in the SIG field to distinguish different BSs. The color ID is picked independently by each BS, which of course means that two neighboring BSs can have the same color. There are two mechanisms to detect such color collisions: (i) the BS detects a packet transmitted from the other BS and sees that it has the same color, or (ii) UEs associated with a BS consistently inform this BS that they experience color collisions (i.e., packets with the same color that belong to a different BS). In that case, the BS may initiate a color change process.

Secondly, the NAV now also distinguished between *inter-BSS* and *intra-BSS* NAV. The problem with having a single NAV is that when a device receives an "end of frame" message, it sets the NAV = 0, allowing it to start transmitting signals. However, there can be situations where a transmission would collide with *both* transmissions within its own network, and that of another BS. Then, if the device recognizes "end of frame" for one of those packets, it might start transmitting, leading to collisions. For this reason, 802.11ax introduces two counters, an *intra-BSS NAV*, and an *inter-BSS NAV*. Transmissions from a device are only allowed when both of them are 0. However, the threshold for the CCA might be different for the two cases, as explained next.

Thirdly, adaptivity is introduced into the channel sensing and packet transmission, through a mechanism called *Dynamic Sensitivity Control* (DSC). Every 802.11 device has a sensitivity threshold that it uses to determine whether the channel is busy or not, by measuring the received power "on the air" and comparing it to this threshold. For intra-BSS packets, this threshold should be low to avoid collisions. However, if a device recognizes that a packet is being transmitted to/from a different BS, it can adopt a higher power threshold for CCA. Furthermore, the admissible level of power sensed from the other device depends also on the power setting for one's own transmission: if a UE will transmit with low power, the threshold for CCA can be higher. This can be motivated by the fact that a higher threshold means that the interfering device is closer but if the UE transmits with low power, it will probably not cause a deadly interference at the RX of the other packet. Essentially, the standard enables a linear adjustment of the threshold level with the transmit power, while imposing lower (−82 dBm, i.e., the standard threshold) and upper (−62 dBm) limits on the threshold, see Figure 33.26.

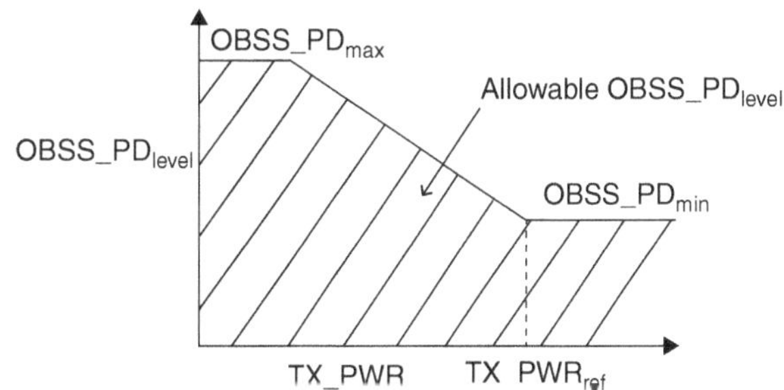

Figure 33.26 Adjustment rules for *OBSS_PD* and *TX_PWR*.
Reproduced with permission from [IEEE 802.11] © IEEE.

Additionally, 802.11ax now also specifies that a device can ignore a nonzero intra-BSS NAV if the device is scheduled by the BS for transmission. This is obviously useful because the BS controls all intra-BS transmissions, and thus would not schedule a UE for uplink transmission if this would lead to collisions. The UE might still decide to not follow the transmission instructions if the *inter-BSS NAV* is nonzero, i.e., if a transmission would lead to collisions with transmission in another network.

An alternative implementation of spatial reuse is the *SRP* mechanism. In it, a BS declares during the trigger frame that it is willing to tolerate interference, and to what level. The devices in the other network that hear the message can then set their NAV = 0 after the UL MU transmission has started, and start their own transmission process to their BS. A different form of interference reduction is achieved by defining a *quiet period* during which only D2D type transmissions are allowed, thus reducing the probability of collision between D2D transmissions and transmissions involving the BS.

33.6.5 Power Saving and Reservation Access

The 802.11ax MAC also makes a further step toward a "cellular" structure by introducing the possibility for services to "reserve" time in advance. The mechanisms for this approach are overlapping, and sometimes identical, with power-saving measures (important for

IoT applications), and will thus be presented here together. Control for these mechanisms lies with the BS, so that a stronger centralization is achieved.

The first important concept is the *Target Wakeup Time* (*TWT*), which tells a UE to wake up at a particular time. In previous versions of the standard, namely 802.11ah, the concept of TWT was already introduced in a rudimentary form, namely that two devices (a BS and a UE) negotiate a wakeup time. This is achieved by the UE sending a TWT request frame, and the BS answers with a response, after which the UE can go to sleep until the wakeup time. This mechanism is retained in 802.11ax.

As an alternative, a broadcast TWT is introduced. The information about TWT is sent out as part of the beacon signal by the BS (a specific TWT element is then configured in the beacon). By assigning different TWT to different UEs, the BS can now reserve parts of the time for a particular UE/service. In particular, the BS divides the time between two beacons into *Service Periods* (*SPs*). More detailed information about transmission during the SP can be set, e.g., whether OFDMA random access will be allowed during the SP. When the UE wakes up at the beginning of its SP, it can then listen to a trigger frame, which would be the method by which the BS ensures orderly transmission of the devices intended for this SP.

The TWT and the setting up of SPs also naturally improve the power efficiency. Since the devices can go to sleep until their scheduled wakeup time, the power consumption is drastically reduced. Furthermore, the BS can also transmit a *Traffic Indication Map* (*TIM*), frame, which indicates the scheduling information for the SP. Any UE that is not mentioned in that TIM can go to sleep. The concept can be taken further: e.g., the trigger frame at the beginning of the SP can announce whether OFDMA random access is allowed during the following SP; if not, the UEs that want to access the channel via this method have to wait until a later SP (and can go to sleep in the meantime). Furthermore, the UEs are allowed to sleep during the transmission of a PPDU that is intended for another UE – while those sleep periods are short, they occur often, and thus greatly improve the energy efficiency.

33.6.6 Beyond 802.11ax

The next version of the IEEE standard is called 802.11be (and the corresponding standard of the Wi-Fi alliance will be called, most likely, Wi-Fi 7). The standardization group started work in September 2020. The actual finalization of the standard is not anticipated before 2024; for this reason, two "releases" (similar to the approach in 3GPP, see Chapter 31) will be established.

While no detailed technical information is available at the time of the writing, some key goals and anticipated technical avenues have been established: the throughput should be increased to 30 Gbit/s, i.e., a factor of 4 higher than 802.11ax. This will be achieved mainly by increasing the maximum bandwidth to 320 MHz (this makes sense due to the availability of the 6–7.125 GHz band) and increasing the number of spatial streams to 16. The latter will be mostly used for MU-MIMO, since the number of antennas at each UE, and thus the maximum number of spatial streams per UE, is expected to remain one or two in most cases. Also, modulation up to 4096 QAM will be enabled. Another important method for increasing throughput is Multi-Link Operation (MLO), which is similar to carrier aggregation in cellular systems. Multiple flavors of this technology are anticipated to be included in the standard; much of the current discussion focuses on this aspect.

Further anticipated features will include HARQ, and BS cooperation (for scheduling and beamforming, as well as joint transmission).

Glossary for WiFi

AC	Access Category
ACK	ACKnowledgement
AID	Association ID
AIFS	Arbitration Inter Frame Spacing
A-MPDU	Aggregate MPDU
AP	Access Point
BA	Block Acknowledgement
BAR	BA Request
BSR	Buffer Status Request
BSS	Basic Service Set
CAP	Controlled Access Period
CCA	Clear Channel Assessment
CCK	Complementary Code Keying
CFB	Contention Free Burst
CFP	Contention Period
CF-Poll	Contention-free Poll
CP	Contention Period (Section 33.4)
CP	Cyclic Prefix (Section 33.6)
CSD	Cyclic Shift Delay
CSMA/CA	Carrier Sense Multiple Access with Collision Avoidance
CTS	Clear to Send

CW	Contention Window
DCF	Distributed Coordination Function
DCM	Dual Carrier Modulation
DIFS	Distributed Inter-frame Space
DLP	Direct Link Protocol
DS	Distribution System
DSC	Dynamic Sensitivity Control
EDCA	Enhanced Distributed Channel Access
EIFS	Distributed Inter-frame Space
ETSI	European Telecommunications Standards Institute
FCS	Frame Check Sequence
GF	Green Field
HCCA	HCF (Hybrid Coordination Function) Controlled Channel Access
HC	Hybrid Coordinator
HCF	Hybrid Coordination Function
HE	High Efficiency (prefix for 802.11ax-related components)
HIPERLAN	HIghPERformance Local Area Network
HT	High Throughput (prefix for 802.11n-related components)
IEEE	Institute of Electrical and Electronic Engineers
IFS	Inter-frame Spaces
ISM	Industrial, Scientific, and Medical
LTF	Long Training Field
MAC	Medium Access Control
MCS	Modulation and Coding Scheme
MLO	Multi-Link Operation
MPDU	MAC Protocol Data Unit
MSDU	MAC Service Data Unit
NAV	Network Allocation Vector
NDP	Null Data Packet
OBO	OFDMA Backoff Counter
PAN	Personal Area Network
PC	Point Coordinator
PCF	Point Coordination Function
PHY	Physical (layer)
PIFS	Point Inter-frame Space
PLCP	Physical Layer Convergence Procedure
PPDU	Physical Layer Protocol Data Unit
PSDU	Physical Layer Service Data Unit
QAP	QoS-Access Point
QoS	Quality of Service
QSTA	QoS-Station
RIFS	Reduced InterFrame Spacing
RL-SIG	Repeated Legacy SIG
RTS	Request to Send
RU	Resource Unit
S2S	Station to Station
SIFS	Short Infer-Frame Space
SIG	Signaling
SP	Service Period
SR	Spatial Reuse
SRP	Spatial Reuse Parameter
SSID	Service Set Identifier
STA	Station
STF	Short Training Field
TC	Traffic Categories
TIM	Traffic Indication Map
TS	Traffic Streams
TXOP	Transmission Opportunity
TSPEC	Traffic Specifications
TWT	Target Wake Time
U-NII	Unlicensed National Information Structure

VHT	Very High Throughput (prefix for 802.11ac-related components)
VoIP	Voice over Internet Protocol
WEP	Wired Equivalent Privacy
Wi-Fi	Wireless Fidelity
WLAN	Wireless Local Area Network
WM	Wireless Medium
WPA	Wi-Fi Protected Access

Further Reading

The official standards documents for the 802.11 standard can be found online at ieee802.org. Excellent summaries of the older 802.11 versions (802.11, 11b, 11a/g) and the more recent 802.11n and 802.11ac standards are given in [O'Hara and Petrick 2005], and [Perahia and Stacey 2015], respectively. A historically interesting comparison of IEEE 802 and HIPERLAN is found in [Doufexi et al. 2002]. Tutorials of the 802.11ax standard are [Khorov et al. 2018], [Deng et al. 2017], and [Qu et al. 2019]. For the mm-wave versions, 802.11ac is surveyed in [Nitsche et al. 2014], and 802.11ay in [Ghasempour et al. 2017] and [Zhou et al. 2018].

For updates and errata for this chapter, see wides.usc.edu/teaching/textbook

Exercises

See Sec. 36.33 of Exercises.pdf at wiley.com/go/molisch/wireless3e

34

PAN and Internet of Things – Bluetooth and Zigbee

While WLANs enable communication over medium distance with high data rate, there are also many applications where only a lower data rate and/or shorter distance is required. At the same time, those applications require lower cost and lower energy consumption than what is typically required from WLANs. This stems from the fact that the involved devices either have a small rechargeable battery that should last for several hours of continuous use, or have to survive for years on a one-way battery. Most important examples for the former are communications between computer peripherals and computer, between wireless headset and smartphone, and so on. For the latter case the Internet of Things (IoT), where various machines communicate status, sensing results, etc., to other machines, is an important example. Therefore, suitable standards have been developed in this space, the most important of which are *Bluetooth* and *Zigbee*.

To be more precise, Bluetooth is an industry standard that includes physical layer, MAC layer, as well as networking and application layer, together with a number of "profiles" that describe how these layers play together for particular applications. We note that these higher layer protocols are as important to the ecosystem as physical and MAC layer.

The development of the Zigbee standard shows some similarities to the approach taken in Wi-Fi. It is based on the physical and MAC layer specified in the *IEEE 802.15.4* standard but adds networking and application layer standards, as well as interoperability requirements, developed by the Zigbee Alliance.

As this book is essentially about physical and MAC layers, this chapter will mostly stick to those layers as well, but the reader should keep in mind that for a full system design the higher layers have to be taken into account at every stage; more details about these aspects can be found in the references listed in the "Further Reading" section.

34.1 Bluetooth

34.1.1 Overview and Applications

Bluetooth was originally devised in the 1990s by the Bluetooth Special Interest Group (SIG) under the leadership of Ericsson, it consists of a physical and MAC layer,[1] as well as higher-layer specifications. Bluetooth is now universally used for short-distance streaming of voice and audio (e.g., between smartphone and wireless headset), as well as for data (e.g., the connection of computer peripherals such as wireless keyboards and trackpads); for example, almost all cars manufactured today have built-in Bluetooth. Versions (so-called core specifications) 1, 2, and 3 of the standard concentrated on increasing the data rate.

A variant, *Bluetooth Low Energy* (BLE), was introduced in 2010 with version 4.0 (and further improved later on; at the time of this writing version 5.2 is the most recent). It provides only a subset of the functionalities, and in particular is intended for intermittent information transmission with extremely low energy consumption, such as IoT applications.

The general Bluetooth architecture is shown in Figure 34.1. The lower layers[2] handle data transmission and related aspects, such as device discovery; those are typically implemented on a Bluetooth chip. The lower layers consist of the Bluetooth radio, the link controller/baseband resource manager, and the Link Manager Protocol (LMP). They interface via the Host Controller Interface (HCI) with the upper layers, which provide functionalities such as fragmentation/defragmentation, music streaming, etc. Together with the actual applications, they are executed on a host device such as a smartphone or computer.

Furthermore, Bluetooth defines a large number of profiles, which are cross-layer definitions of specific parts of the Bluetooth protocol stack that are used for a specific application. The profiles range from Advanced Audio Distribution Profile to Wireless Application Protocol Bearer. It is noteworthy that in addition to the *core protocols*, Bluetooth also adopts protocols from other standards for specific purposes. For example, the USB HID protocol, which was designed for wired keyboard and mouse, is adopted in Bluetooth.

---

[1] Part of the physical layer and MAC layer are standardized in *IEEE 802.15.1*. However, this IEEE group is now dormant, and newer versions of PHY and MAC are published by the Bluetooth SIG.

[2] The lower layers are sometimes named "controller"; in order to not create confusion with the piconet controller described below, we stick with the term "lower layers."

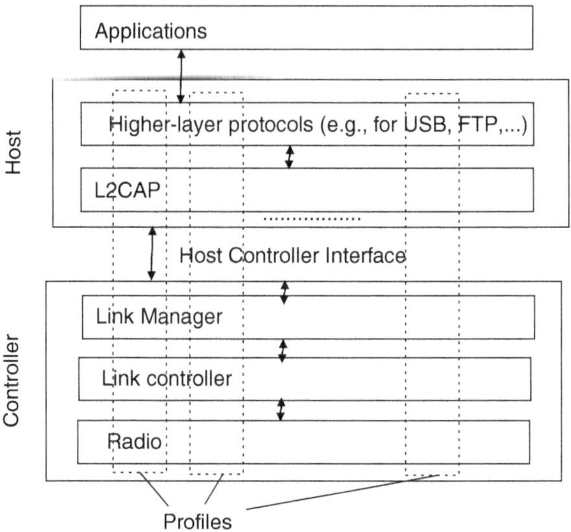

Figure 34.1　Fundamental architecture of Bluetooth.

34.1.2　Bluetooth Network Structure and Link Control

Piconet and Scatternet

The Bluetooth network structure is called a *piconet*, which uses a leader-follower configuration. The leader ("central" node in Bluetooth parlance) is the *piconet controller*, and it has up to seven followers ("peripheral" in Bluetooth parlance).[3] Multiple piconets can share devices; in this case, they make up a *scatternet*. A device may be a follower in both nets, or a leader in one net and a follower in another, but it cannot be leader for two nets. Examples for different cases are shown in Figure 34.2.

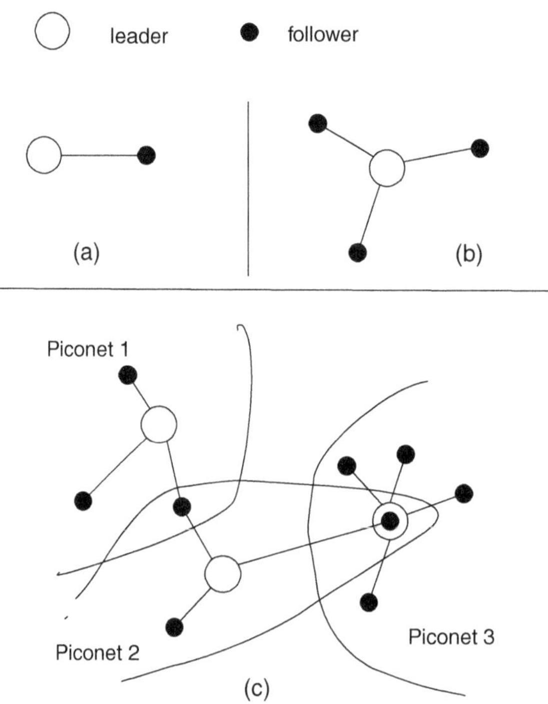

Figure 34.2　Bluetooth network configuration: (a) single follower, (b) piconet with multiple followers, (c) scatternet.

---

[3] Until recently, Bluetooth used different terms (which can still be found in many versions of the standard).

As we will detail below, Bluetooth uses TDD for duplexing, TDMA for multiple access (i.e., different followers transmit at different times), and frequency hopping for robustness against interference, co-existence of multiple piconets, and narrowband fading. Time is divided into slots, which determines both the duration of the packets, and the times at which frequency hopping occurs. A physical channel is characterized by a frequency hopping pattern, on which a leader and possibly multiple followers communicate; a physical link is between a leader and a particular follower. The physical link can be used to transport one or more logical links, where different logical links can be time-multiplexed onto the physical link. Besides other types of data, also the LMP is carried on such logical links.

Associated with the Bluetooth structure shown in Figure 34.1, there is a hierarchy of channels, which goes (from bottom to top) as physical channel, physical link, logical transport, logical link, and L2CAP channel.

Link Controller and Piconet Formation

A link controller has two major states, standby and connection, as well as several substates, see Figure 34.3. The link controller moves from one state to the other by link controller commands or internal signals (e.g., correlator outputs). By default, the device is in standby state, where it consumes little power. It may leave the state to scan for page or inquiry messages, or to send out such messages. Each device has a globally unique 48-bit device identifier, also known as Bluetooth Device Address (BD_ADDR). It is used before a connection is established (afterward, a much shorter identifier, valid within the piconet, can be used).

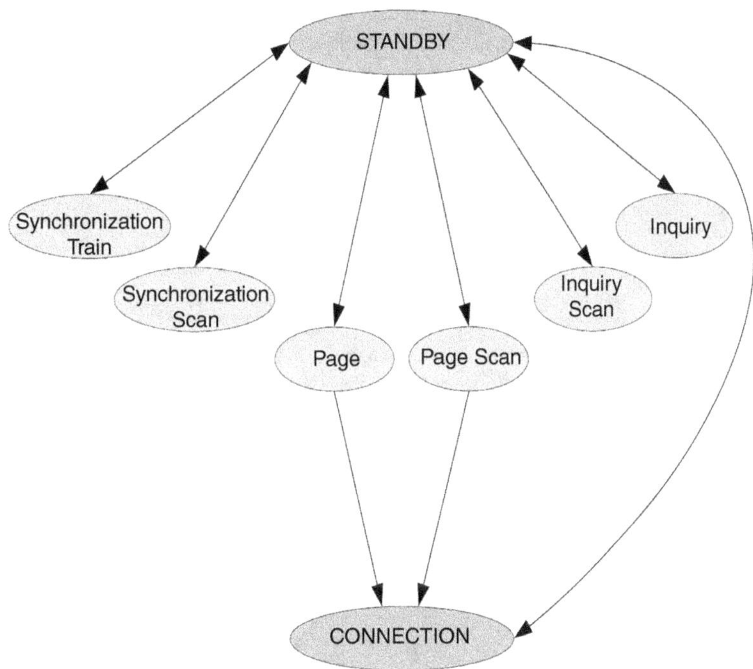

Figure 34.3 State diagram of link controller. Color version available at wiley.com/go/molisch/wireless3e.
Reproduced with permission from [Bluetooth SIG 2019] © Bluetooth SIG.

In the page substate, a prospective leader sends out a paging message containing the desired follower's *Device Access Code* (DAC), which is derived from the global identifier. Since the leader does not know when, and on what frequency, a prospective follower listens, it has to send the paging message multiple times until the follower responds. A device initiating a connection by default is the leader, though a role switch procedure is foreseen such that leader and follower can change roles at a later time.

Conversely, in the page scan substate, a device listens for a message that contains its DAC. The duration of the window during which the listening occurs needs to be sufficient to scan multiple frequencies. Note that a device might enter a page scan substate not only from the standby state but also the connection state; this mechanism allows multiple piconet connections on one device.

When a page message is successfully received, a coarse frequency-hop synchronization is done between leader and follower. Once the connected state is entered, the leader sends a poll packet and receives a reply (any packet) to ensure that the follower has synchronized itself to the clock of the leader. Note also that page, page response, and data transmissions have different frequency hopping patterns. Details about the packet exchange are shown in Figure 34.4.

In order to discover other devices, a device can enter an inquiry substate, where it sends out inquiry messages and collects information (including information from which the DAC can be derived) of the devices that respond. A device that allows itself to be discovered (typically this is determined by the user), regularly enters into the inquiry scan substate, where it periodically scans for inquiry messages; it is allowed to respond to such messages but does not have to. Generally, inquiry, inquiry scan, and inquiry response are fairly similar to page, page scan, and page response. Besides the inquiry response, there is also the extended

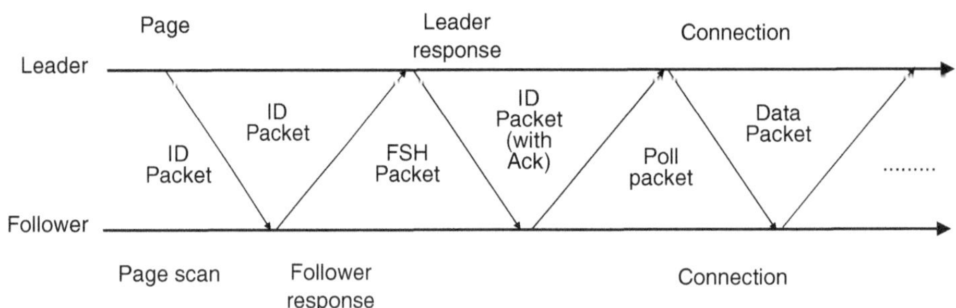

Figure 34.4   Packet exchange for setting up a Bluetooth connection. The different packet types will be explained in Section 34.1.4. Reproduced with permission from [Bluetooth SIG 2019] © Bluetooth SIG.

inquiry response, which contains many more details about the device. This information helps to determine whether a connection should be made at all (e.g., a device might not have the required capabilities for the application the other device wants to set up). Note also that the data format of the extended inquiry response is different from the inquiry response, namely a standard data packet type.

For the connected state, the standard further defines a number of substates, such as "active," "hold," and "sniff" for the connection mode, which serve, e.g., to conserve energy.

Finally, there is also a *park* state, where a follower device needs to remain synchronized with a piconet but does not need to participate; however, this mode turned out to be little used and was removed in version 5 of the standard.

Types of Connections

There are different types of connections, also known as *logical transports*, in Bluetooth, the most important of which are

- *Synchronous Connection-Oriented* (SCO), which is a circuit-switched mode for voice, so that the leader uses a set of reserved timeslots. An SCO connection is set up with certain parameters that stay unchanged until the connection is terminated or re-set. A leader may have up to three SCO links to one or more followers. SCO does not use ARQ (Section 13.12), since in voice channels with standard speech coders, retransmissions are useless due to the induced delay.
- *Extended SCO* (eSCO) is similar to SCO, but considerably more flexible in its configuration. It allows asymmetric data rates in uplink and downlink and allows a limited number of retransmissions directly after the slot used for the original transmission. Since the number of retransmissions is limited, data "expire" if they have not been transmitted within a given deadline, which is in line with the requirements for voice.
- *Asynchronous Connection-Less* (ACL) where a follower responds to requests from a piconet controller, or the piconet controller sends packets to a follower. If such an ACL packet is not addressed to a specific device, then it is a broadcast packet that should be received by all followers in the piconet.

Note that music is transmitted via ACL, not (e)SCO, since the latter does not offer a sufficient data rate for high-quality music transmission, and music streaming (which includes buffering) is less delay sensitive than voice.

Link Manager Protocol

The link managers of two devices involved in a communication talk with each other using the LMP, which employs either DM1 or DV packets (see Section 34.1.4) for the transmission. Functionality enabled by the LMP includes

- Information exchange about device capabilities, since a number of functionalities are optional.
- Connection control, such as connection establishment, termination of a connection, and link supervision, as well as signaling for Adaptive Frequency Hopping (AFH), power control, and QoS requirements.
- Security, including pairing, authentication, and encryption.
- Information requests.
- Role switching (where leader and follower switch roles).
- Mode of operation.

The link manager interfaces with the Logical Link Control and Adaptation Protocol (L2CAP), which is located on the host device. The L2CAP in turn performs such tasks as flow control, streaming, quality of service, segmentation, and reassembly of large packets.

34.1.3   Bluetooth Radio

Transmission Power and Sensitivity

Bluetooth defines different classes of devices, which can have different transmit powers. Class 1 devices can transmit with up to 20 dBm, providing a theoretical range of up to 100 m. Class 2 (the most widely used) has up to 4 dBm, class 3 up to 0 dBm. A minimum transmit power of – 30 dBm is recommended but not mandated. Closed-loop power control with a step width between 2 and 8 dB is foreseen (note the "up to" in the transmission power definitions) in order to save energy; it is mandated for class-1 devices, and optional for class 2 and 3. The required Receiver (RX) sensitivity is -70 dBm for a packet error rate of 0.1%, though many commercial devices provide much better sensitivity (note that the theoretical noise floor is at -114 dBm).

Modulation

First consider the basic mode (*Basic Rate*, BR), which was the mode used in the first version of the standard, and is also the mandatory mode for later versions. The modulation format is GFSK (Gaussian-filtered Frequency Shift Keying); remember that (see also Chapter 10) in this format the transmit signal is a sinusoidal signal with a phase

$$\phi(t) = h_{\text{mod}} 2\pi \int_{-\infty}^{t} \sum_i c_i \widetilde{g}(\iota - iT_{\text{s}}) d\tau \tag{34.1}$$

where $\widetilde{g}(t)$ is a Gaussian pulse with time-bandwidth product BT_{s} and h_{mod} is the modulation index and c_i is ± 1. In Bluetooth, $BT_{\text{s}} = 0.5$ and the modulation index is required to be between 0.28 and 0.35. The symbol rate is 1 MBaud, i.e., $T_{\text{s}} = 1$ μs. As we will see later on, this results in up to 721 kbit/s data rate due to the transmission overhead.

In addition to the BR, later versions (2.0 and higher) define modulations to achieve *Enhanced Data Rate* (EDR). This is achieved by using $\pi/4$-shifted DQPSK or 8-DPSK as modulation, resulting in data rates that are higher by a factor 2 and 3, respectively. For spectral efficiency reasons, pulse shaping with a square-root raised cosine filter with roll-off factor 0.4 is used. The EDR modulation formats follow the descriptions of Chapter 10.

It is noteworthy that none of the Bluetooth modulation formats requires coherent RXs: GFSK, as a frequency shift keying, can be received with a noncoherent RX, while DPSK can be received with a differential RX. Eliminating the need for a carrier recovery circuitry can simplify the hardware significantly. Furthermore, the loosely defined modulation index (0.28 ... 0.35) in BR makes it very hard to create a reliable coherent RX.[4]

Starting with Bluetooth 3.0, even higher data rates are enabled that use Bluetooth for link setup, but then employ 802.11 for the actual data transfer, enabling up to 24 Mbit/s data rate. However, this approach is rarely used in practice.

Packet Duration and Frequency Channels

Slotting: Transmission is done in slots of duration 625 μs; one slot is used for at most one packet, i.e., a new packet can only start at the next slot. Conversely, however, a packet can last longer than one slot, namely up to five slots. The slots are numbered according to the 27 most significant bits of the Bluetooth clock of the piconet leader, which therefore has a periodicity of approximately one day. A basic clock unit is 312.5 μs (i.e., half a slot duration), and all radios in a piconet are synchronized to it.

Duplexing: Transmissions between the leader and the followers are separated by Time Domain Duplex (TDD) with the leader starting a transmission in even-numbered slots and the followers in odd-numbered slots; however, a multi-slot transmission occupies contiguous slots, so that, e.g., a 3-slot transmission starting in slot 4 also occupies slots 5 (an odd slot) and slot 6. Note that it is the application that determines which device is the leader, and which are the followers, implicitly in the process of setting up the links.

Frequency band and hopping: Bluetooth operates in the 2.45 GHz ISM band. Specifically, it defines 79 channels, each 1 MHz wide, with center frequencies $2402 + k$ MHz, $k = 0, ..., 78$. In order to gain robustness to fading and interference, Bluetooth uses frequency hopping over these 79 frequency channels.

The hopping sequence is determined by a cyclic code of length $2^{27} - 1$, which for any typical transmission duration emulates a uniformly distributed random variable. Hops occur at the slot boundaries. If, however, a packet lasts longer than a slot (see Figure 34.5), the band is not changed for the duration of the packet. Consequently, the effective hopping rate is ≤ 1600 hops/s. Furthermore, since version 1.2, there is the possibility of adaptive frequency hopping, where channels with strong interference are marked and henceforth excluded from the frequency hopping, though the number of actually used frequency channels is not allowed to fall below 20. The marking is done by the leader (either based on its own observation, or based on information from the followers), and communicated to the followers to ensure that all participants of the piconet use the same assumptions about which frequencies shall be used. With adaptive hopping, the follower responds on the same frequency as the leader, which reduces the maximum hopping rate, to 800 hops/s.

---

[4] The same statement does not apply, however, to certain high-integrity modulation modes in BT Low Energy (see Section 34.1.5).

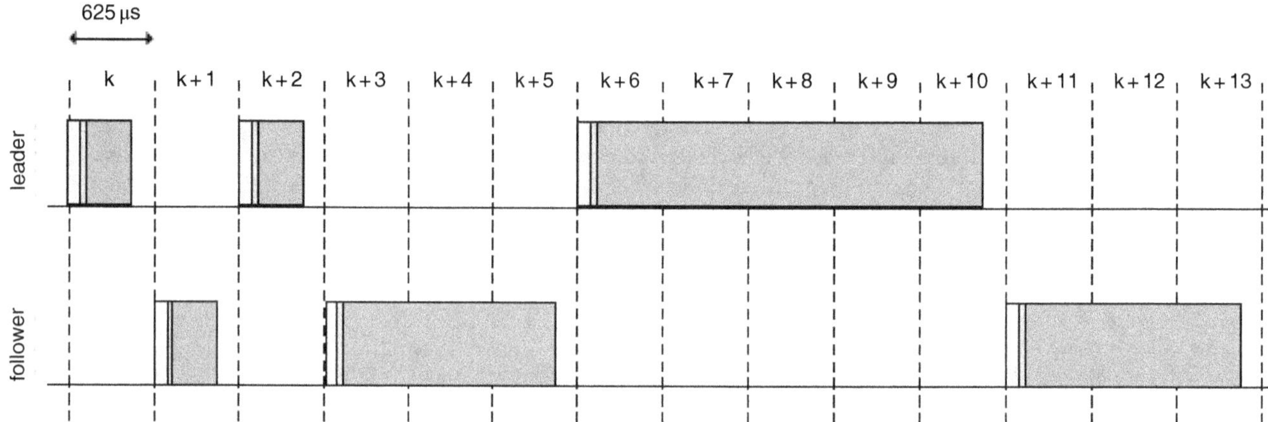

Figure 34.5 Slot structure in Bluetooth. Leaders transmit only packets starting in even timeslots, followers only in odd. Color version available at wiley.com/go/molisch/wireless3e.
Reproduced with permission from [Bluetooth SIG 2019] © Bluetooth SIG.

Physical channels: A "physical channel" is defined in the specifications as a particular frequency hopping pattern, packet timing, and an access code (see also below); it is thus considered part of the baseband (MAC) specifications. There are several different physical channels:

- *Basic piconet channel*: the basic physical channel using the standard frequency hopping described above.
- *Adapted piconet channel*: a physical channel using adaptive frequency hopping described above.
- *Page scan channel*: during paging, a device (which will become the leader) sends out paging messages and listens to answers from prospective followers, see (Figure 34.6). Since the paging message is very short, a faster hopping (3200 hops/s) is used, and the hopping pattern is based on a shorter (length 32) pseudorandom sequence. The hopping sequences used for page and page response have a 1:1 correspondence.
- *Inquiry scan channel*: in this channel, a prospective leader sends inquiry messages with a general or dedicated inquiry access code. The timing and hopping approach for inquiry scan are the same as for page scan, though the actual set of used frequencies is different.
- *Synchronization scan channel*: there is a set of three fixed frequencies that are used for synchronization of the RF channels. The leader may send out synchronization packets that allow followers to obtain a coarse clock synchronization.

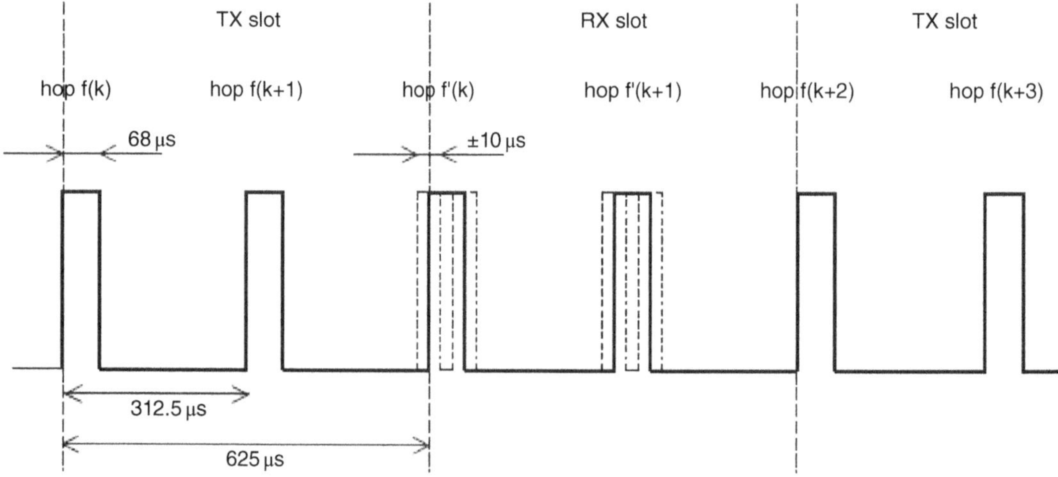

Figure 34.6 RX/TX cycle in paging mode.
Reproduced with permission from [Bluetooth SIG 2019] © Bluetooth SIG.

The hopping sequence for each physical channel is determined (in a rather complicated manner) from the device address, the type of physical channel, and (in the case of adaptive frequency hopping) the AFH map, i.e., the list of available channels.

A Bluetooth device can only use one "physical channel" at a time. If it needs to perform multiple tasks, e.g., data transmission and discovery, it needs to time multiplex several such channels.

Coding: Bluetooth foresees three types of coding: (i) rate 1/3 repetition coding, i.e., every bit gets repeated three times. This method is used for all packet headers, as well as the body of packets of type HV1 (see below). (ii) rate 2/3 Hamming coding, in particular, a

(15/10) shortened Hamming code with generator polynomial $(D + 1)(D^4 + D + 1)$; each block of 10 bits is appended with a block of five redundancy bits; (iii) ARQ, where a single-bit ACK/NACK flag is sent in the header of a subsequent packet. As we will discuss below, ARQ is not used in all packet types.

34.1.4 Bluetooth Packet Structure

Each BR Bluetooth packet consists of three parts:

1. *Access Code*, which is a 72-bit block for packet synchronization, DC offset compensation, and identification. It consists of a 4-bit preamble, a 64-bit synchronization word, and a 4 bit trailer.[5] The preamble is a fixed 0/1 pattern used for DC compensation. The synchronization word is a 64-bit code word that is derived from (part of) the 48-bit address of a Bluetooth device. Special addresses are used for packets during "inquiry" (a device searching for other devices in the vicinity).
2. *Header*, which consists of 54 bits that contain (i) piconet member address (3 bit), (ii) packet type, as described below (4 bit), (iii) flow control bit to stop transmissions when the buffer is full, (iv) ACK/NACK flag, (v) sequence numbering, and (vi) a HEC (header error check), i.e., a CRC for the header bits (8 bit). These 18 bits are encoded with a rate 1/3 repetition code.
3. *Payload*, which may include a payload header, followed by the actual payload data. The payload can be encrypted (either including the payload header, or just the payload data, depending on the encryption type); the encryption keys are exchanged during the initialization of the network.

Both packet header and payload are scrambled to minimize DC bias.

For EDR, the packet structure is somewhat more complicated. It also starts with the access code and header, which are always transmitted using BR modulation, even when the payload uses EDR modulation. This is done so that an RX that sees a packet coming in can always use the same demodulation process. The header also contains information about the modulation format for the payload. After a guard interval follows a synchronization word, payload, and trailer bits, all of which are transmitted with the DPSK (either $\pi/4$-DQPSK or 8-DPSK) modulation of choice for this packet.

There are different types of data packets, some of which are exclusive to SCO, some only for ACL, and some that are used in common, and mainly serve for link control.

Link control packet types include

1. *ID packets* are used before connection establishment to access a particular device/group of devices. They only consist of the DAC, or inquiry access code.
2. *Null packets* carry no payload and are mainly used for returning an ARQ flag, or the status of the receive buffer.
3. *Poll packets* are similar to the Null packets in that they do not have a payload. They serve for the leader to poll the followers for data, and the follower has to send a response.
4. *FHS packets*, which contain real-time clock information and the device address of the sender. The payload contains 144 information bits (see Figure 34.7), 16-bit CRC, and is encoded with a rate 2/3 code to give a total of 240 bits. The packet is used for frequency hop synchronization before a piconet is established.
5. *DM1 packets*, for sending payload data and/or control data.

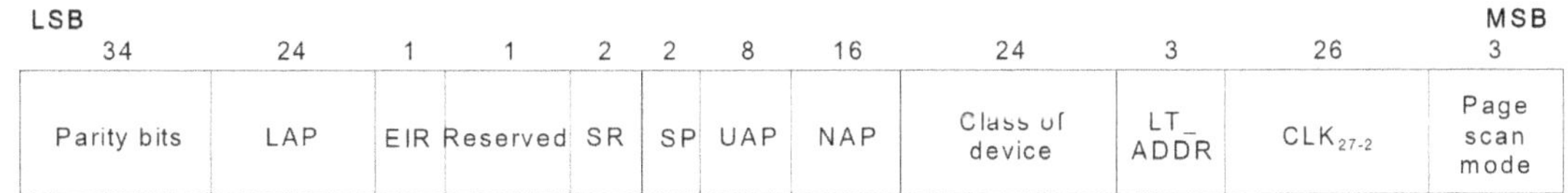

Figure 34.7 Structure of an FHS packet. LAP: lower 24 bits of the 48-bit address of the transmitting device. UAP, NAP: other bits of the address. EIR indicates an extended inquiry response packet to follow. SR: scan repetition mode field. LT_ADDR: 3 bit logical address of the intended receiver. CLK: value of the native clock of the transmitter at beginning of the FHS packet. Page scan model: default scanning mode of the TX.

All link control packets are only a single slot long.

The SCO types of packets carry speech information at 64 kbit/s, but with different levels of quality (HV1, HV2, HV3): different packet types carry different number of source bits, with HV1 having the smallest number of bits but the strongest FEC protection (and thus being most robust for a noisy link), and vice versa for HV3. These two effects cancel out, so that the payload has a fixed size of 240 bytes. Note that there is no CRC, since no retransmissions are foreseen. There is a packet header, but no payload header (compare below for ACL packets) because the speech information constitutes a constant stream whose associated information, that would normally be in the payload header, does not change from packet to packet.

---

[5] If the access code is not followed by a header, i.e., for paging and inquiry, it is shortened to 68 bits by omitting the trailer.

The DV packet type is intended for transmitting a combination of voice and data (DV); it can be used to replace an HV1 packet (like that one, it contains 10 bytes of voice data, though in contrast to HV1 it does not protect them with an FEC).[6] Overall, the structure of the packet is more complicated: after access code and packet header, there are the 10 bytes of (unprotected) voice, followed by the data field. That field consists of 1 byte of payload header, $0-9$ byte of payload data, and 2 byte of CRC; the whole data field is protected by a rate 2/3 FEC. While the voice data are never retransmitted even when delivery fails, the data might be retransmitted (as a DM1 packet) if their proper reception is not acknowledged. A summary of the key parameters of the SCO packet types is given in Table 34.1.

Table 34.1 Key parameters of SCO-type packets. All sizes are in Bytes. D denotes data in mode DV. Data rate is in kbit/s.

Type	PayHeader	Payload	FEC	CRC	Slots	Max sym. data rate
HV1	0	10	1/3	N	1	64
HV2	0	20	2/3	N	1	64
HV3	0	30	N	N	1	64
DV	1	$10+0-9$ D	2/3 (D)	Y (D)	1	$64+57.6$ D

The eSCO type of packets are also for synchronous operation but includes a (16-bit) CRC and the possibility of retransmission. The packet types EV3, EV4, and EV5 are using the BR modulation; while packets 2-EV3, 2-EV5 use $\pi/4$-DQPSK, and 3-EV-3 and 3-EV5 use 8-DPSK. Also for these packets, there is no payload header; the payload length is set during the setup of the eSCO link and then remains fixed. Range of data size, FEC, and number of used slots are given in Table 34.2.

Table 34.2 Key parameters of eSCO-type packets

Type	PayHeader	Payload	FEC	CRC	Slots	Max sym. data rate
EV3	0	1–30	N	Y	1	96
EV4	0	1–120	2/3	Y	3	192
EV5	0	1–180	N	Y	3	288
2-EV3	0	1–60	N	Y	1	192
2-EV5	0	1–360	N	Y	3	576
3-EV3	0	1–90	N	Y	1	288
3-EV5	0	1–540	N	Y	3	864

The ACL type of packets (DM, DH) carry different amounts of data. There is a payload header (1 byte for DM1 and DH1, 2 byte for the other packets), and then the payload with a maximum length given in Table 34.3 (the actual length is indicated in the payload header); if encryption is enabled, a 32-bit *Message Integrity Check, MIC*, might be added. A 16-bit CRC protects the information (retransmission is foreseen).

Table 34.3 Key parameters of ACL-type packets

Type	PayHeader/bytes	PayData/bytes	FEC	CRC	Slots	Symm. max rate	DL max rate	UL max rate
DM1	1	0–17	2/3	Y	1	108.8	108.8	108.8
DM3	2	0–121	2/3	Y	3	258.1	387.2	54.4
DM5	2	0–224	2/3	Y	5	286.7	477.8	36.3
DH1	1	0–27	N	Y	1	172.8	172.8	172.8
DH3	2	0–183	N	Y	3	390.4	585.6	86.4
DH5	2	0–339	N	Y	5	433.9	723.2	57.6
2-DH1	2	0–54	N	Y	1	345.6	345.6	345.6
2-DH3	2	0–367	N	Y	3	782.9	1174.4	172.8
2-DH5	2	0–679	N	Y	5	869.1	1448.5	115.2
3-DH1	2	0–83	N	Y	1	531.2	531.2	531.2
3-DH3	2	0–552	N	Y	3	1177.6	1766.4	265.6
3-DH5	2	0–1021	N	Y	5	1306.9	2178.1	177.1

---

[6] Since HV1 packets are sent every frame, there is a need to define DV so that data can be transmitted in the link as well. HV2 and HV3 are not sent in every frame, so that separate data frames can be interlaced, obviating the need for a DV type of frame.

34.1.5 Bluetooth Low Energy

Basic Philosophy

While the evolution of Bluetooth from version 1 to version 3 was characterized by increase of the data rate and enabling more applications, the steps taken in version 4 and 5 focused on reduction of energy consumption, under the name *Bluetooth Low Energy,* BLE. The motivation for this is IoT applications where a battery charge might have to last years. Typical BLE use cases, such as unlocking of a garage door, require setting up a connection, transferring a small amount of data (a few kByte or less), and disconnecting; this is much more energy efficient than maintaining a continuous connection where a device needs to constantly listen to potentially incoming packets. The advanced connection process of BLE allows making faster on-demand connection while still maintaining low power consumption on the peripheral's side. BLE is distinct from classical Bluetooth (also known as BR/EDR) as defined in standards up to 3.0. A device may support only BR/EDR, only BLE, or both (dual-mode).

In BLE, the physical channel consists of time units called *events,* namely advertising events, extended advertising, periodic advertising, connection, and isochronous events. Data are sent in packets within those events.

Network Structure and Link Control

BLE has a similar network structure as BR/EDR, but the link states are somewhat different, as shown in Figure 34.8. A link layer implementation need not support all of the shown states, though.

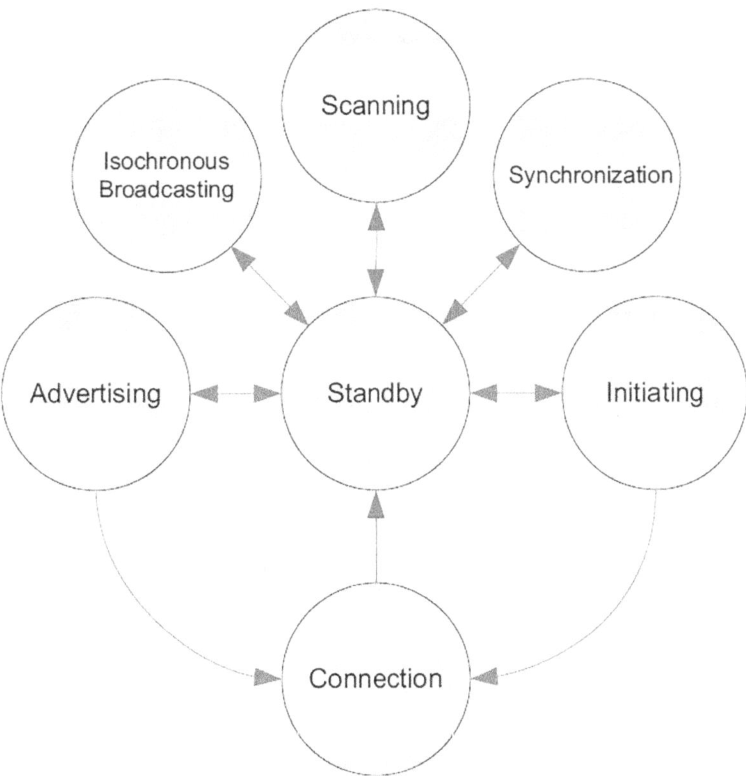

Figure 34.8 State diagram of the BLE link layer state machine. Color version available at wiley.com/go/molisch/wireless3e.
Reproduced with permission from [Bluetooth SIG 2019] © Bluetooth SIG.

Again the standby state is the default state. In the advertising state, a device can send advertising events, which consist of one or more advertising PDUs. Note that "advertising" is not meant in the commercial sense, but rather that the device announces certain information to the surroundings, without having a connection established. Scanning has a similar meaning as in BR/EDR; scanning can be passive (only reception of packets) or active (where a device might send, in response to an advertising packet, a request for more information).

A device can in particular send out "connectable" advertising events. If this is the case, a responder might request to set up a connection, and thus enter a connected mode.

Devices that need to set up an ACL connection to another device have to listen to connectable advertising packets (not all advertising packets fulfill this). Isochronous transmission allows the sending of data streams (at regularly scheduled intervals) to one or more RXs; it is mostly intended for audio transmission. An isochronous connection can be set up via an ACL link. The logical transport is called Connected Isochronous Stream (CIS), which consists of CIS events that occur at regular intervals; a CIS event consists of one or more subevents that contain one leader transmission and one follower response. The structure of the different event types is shown in Figure 34.9.

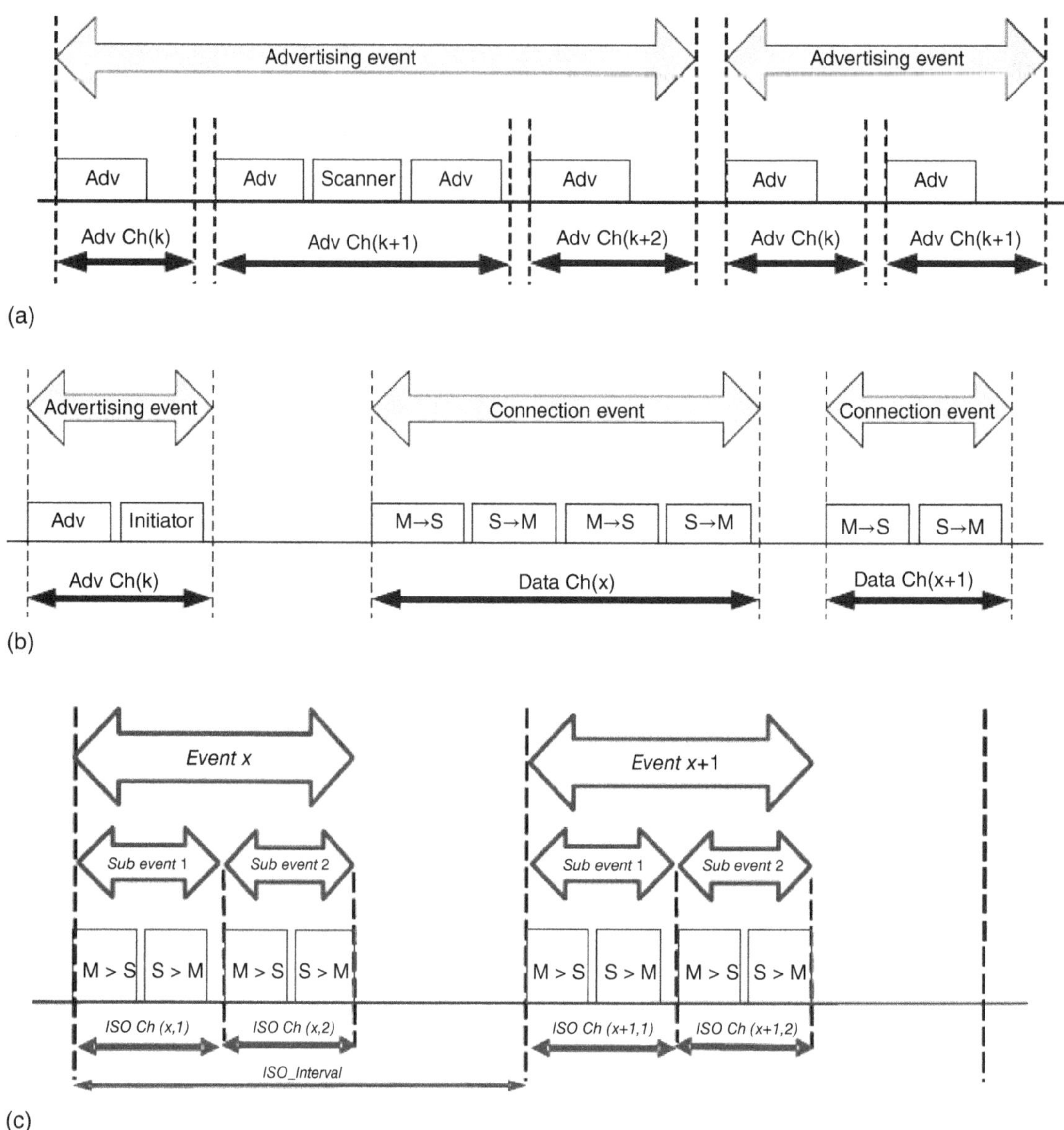

Figure 34.9 (a) Advertising events; (b) connection events; (c) CIS events and subevents; M.leader, S...follower.
Reproduced with permission from [Bluetooth SIG 2019] © Bluetooth SIG.

The topology of the network is also different in BLE compared to BR/EDR. Firstly, in BLE each follower communicates on a separate physical channel (instead of sharing a physical channel with the leader, as in BR/EDR). Secondly, a device might send out advertisements without enabling connections to it. An example configuration is shown in Figure 34.10.

BLE Radio

BLE uses a similar physical layer as BR/EDR, though there are some important distinctions.

Power: The maximum output power is still 20 dBm, but the minimum power is now mandatory and set at −20 dBm. More precisely, there are now four device classes, with class 1 having a minimum power of +10 dBm and a maximum power of +20 dBm. Classes 1.5, 2, and 3 all have a minimum power of −20 dBm, and a maximum power of +10, +4, and 0 dBm, respectively. Power control is not mandatory (though recommended), and if available has a stepsize of no more than 8 dB. The minimum RX sensitivity is −70 dBm for uncoded PHYs, and −75 to −82 dBm for coded PHYs, depending on the amount of coding.

Modulation: the modulation is GFSK with $BT_s = 0.5$, but the modulation index now should be in the range $h_{\mathrm{mod}} = [0.45...0.55]$. The symbol rate can be either 1 Mbaud (mandatory) of 2 Mbaud (optional). By default, the 1 Mbaud mode is uncoded, though an optional coded mode is defined as well, where control information (e.g., access address, Coding Indicator [CI]) are coded at 125 kbit/s, and the payload is coded at either 125 or 500 kbit/s. BLE now also defines a mode of operation referred to as "stable modulation index," where the uncertainty of the modulation index is only 1% (0.495 ... 0.505). The more precise modulation variant makes it substantially easier to create coherent RX implementations.

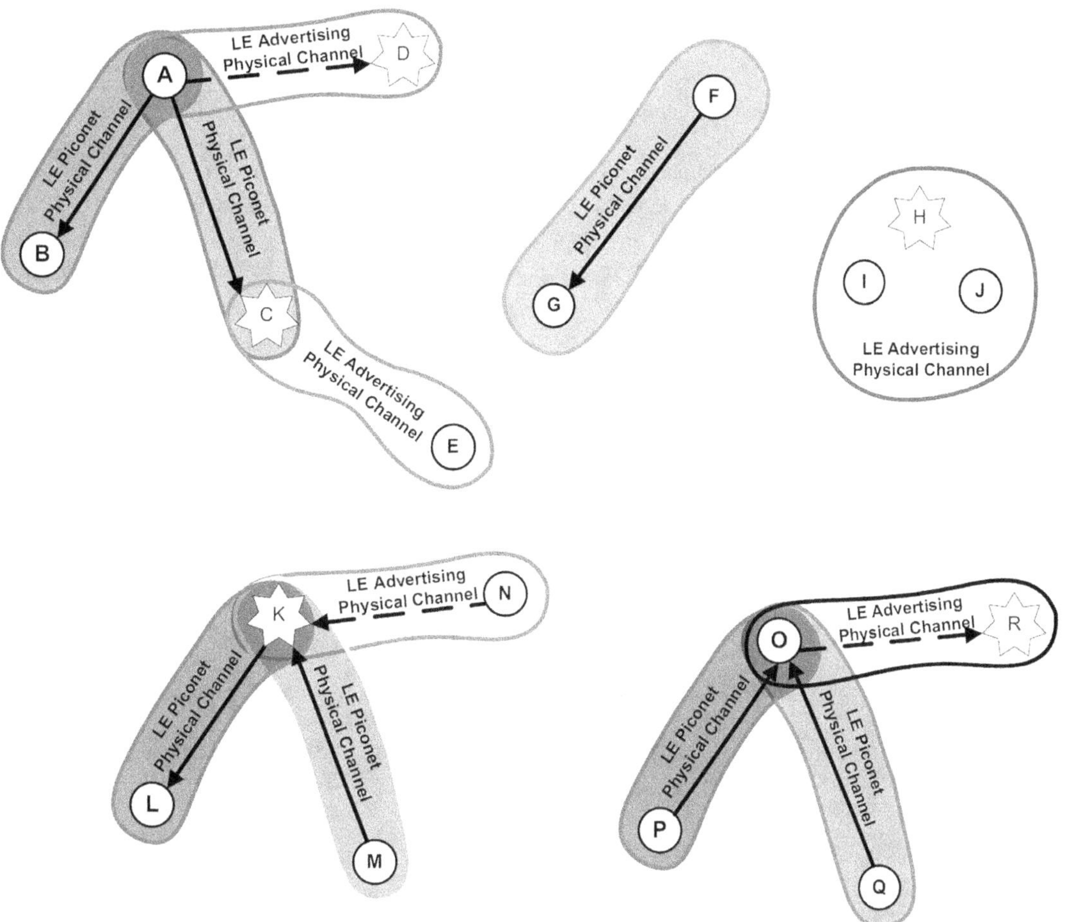

Figure 34.10 Solid arrows indicating connection leader to follower; dashed arrows, indicating a connection initiation, point from initiator to advertiser using a connectable advertising event; devices that are advertising are indicated using stars. Color version available at wiley.com/go/molisch/wireless3e. Reproduced with permission from [Bluetooth SIG 2019] © Bluetooth SIG.

BLE also supports multiple antennas at TX and/or RX with antenna switching; switching patterns of up to a length of 12 (depending on the number of antennas) are supported.

Frequency hopping: The number of frequency channels is reduced to 40, and the channels are spaced 2 MHz apart. There are three dedicated channels that a device can use for advertising its presence (called primary advertising channels), which reduces the connection time from some 20 ms to about 3 ms. The other 37 channels are called general-purpose channels. Just like BR/EDR, BLE allows the use of adaptive frequency hopping that excludes channels that have too much interference.

Error checking and coding: All payloads are protected with a CRC and scrambled to reduce DC offset sensitivity. In the case of the coded PHY, the data are protected by a nonsystematic, nonrecursive convolutional encoder with constraint length 4, as shown in Figure 34.11. Each bit output bit of the convolutional encoder is then mapped onto a symbol or symbol sequence. In the mode called $S = 2$, bits are just mapped 1:1 onto symbols. In the $S = 8$ case, each bit is mapped onto a 4-symbol long sequence.

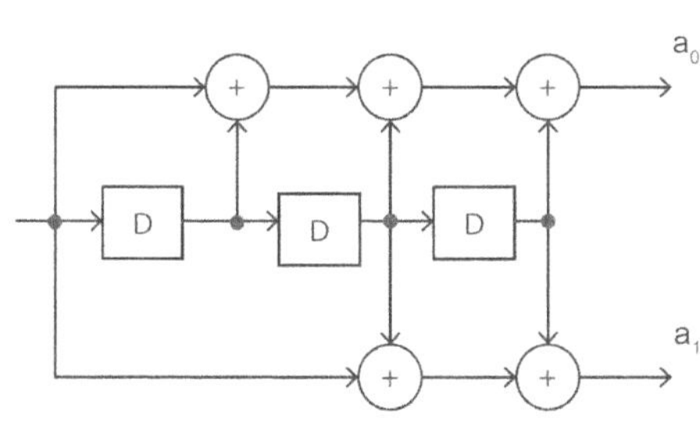

Convolutional encoder output	Transmitted symbol sequ. $S = 2$	Transmitted symbol sequ. $S = 8$
0	0	0011
1	1	1100

Figure 34.11 Convolutional encoder for coded PHY of BLE.
Reproduced with permission from [Bluetooth SIG 2019] © Bluetooth SIG.

Channels: like in BR/EDR, a physical channel is determined by a frequency hopping sequence. There are three main types of channels:

1. *Advertising*, which uses both the primary advertising channels and the general-purpose channels for discovering devices, initiating a connection, and broadcasting. An advertising physical channel PDU consists of a 16-bit PDU header and a variable-size payload (whose length is indicated in the header). Specific forms of advertising PDUs are used for scanning requests and scanning responses.
2. *Data*. A data physical channel PDU consists of a (16 or 24 bit) header, a variable-length payload (up to 251 Bytes), and – in case that encryption is used – a MIC.
3. *Isochronous*: the PDUs of this channel have a structure that is similar to that of data channels, again containing a 16-bit packet header, followed by the payload of up to 251 Bytes and a MIC (in case encryption is used). The payload header contains some modified (compared to the data type) fields.

BLE Packet Structure

The packet structure in BLE is simplified considerably. There is only one packet format for the uncoded PHY and another one for the coded PHY.

For the uncoded PHY, a packet starts with a preamble, which is 1 Byte long for the 1 Mbaud PHY, and 2 Byte for the 2 MBaud PHY; the structure and purpose are the same as in BR/EDR. Next comes the 32-bit access address, which has a new role compared to BR/EDR. It is randomly generated, and each link layer connection between any two devices, each broadcast isochronous stream, as well as each advertising train, should have a different access address. Correlation with the access address tells a device whether a packet is intended for it.

This is followed by the PDU (2–258 octets), which contains either the data of the Advertising Physical Channel, the Data Physical Channel, or the Isochronous Physical Channel. The PDU is followed by a CRC computed from it. All packets contain a PDU header that contains the type of payload, the address of the intended device, and the length of the payload, as well as other control information depending on the packet type.

For the coded PHY, the structure and field sizes of the packet are shown in Figure 34.12. Note that the preamble is not coded. Access address, CI and term1 are encoded with the $S = 8$ coding scheme, while the PDU, CRC, and term2 are either coded with $S = 2$ or $S = 8$ scheme, depending on the value of the CI. Term1 and Term2 are 3-bit termination sequences for the codes.

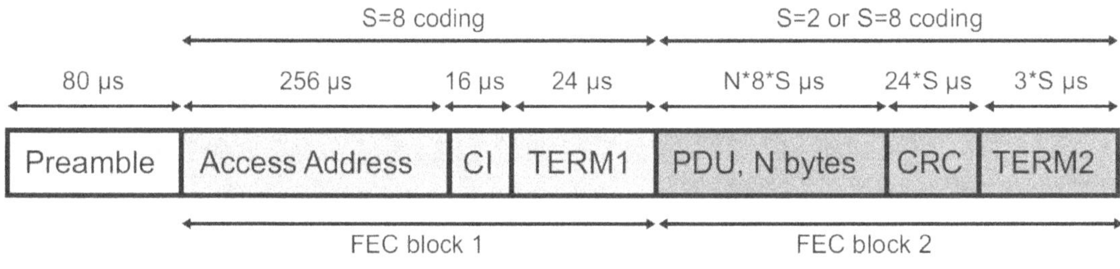

		Fields					
	Preamble	Access Address	CI	TERM1	PDU	CRC	TERM2
Number of Bits	Uncoded	32	2	3	16 – 2056	24	3
Duration when using S=8 coding (µs)	80	256	16	24	128 – 16448	192	24
Duration when using S=2 coding (µs)	80	256	16	24	32 – 4112	48	6

Figure 34.12 Packet format and field sizes and durations (in µs) for BLE coded PHY. Color version available at wiley.com/go/molisch/wireless3e. Reproduced with permission from [Bluetooth SIG 2019] © Bluetooth SIG.

An uncoded-PHY packet may also contain a constant tone extension (between 16 and 160 µs), which is essentially a known signal that can be used for determination of direction of arrival or direction of departure. The length of the extension is long enough that, e.g., an RX can switch between different RX antennas and from the resulting samples determine the DoA, similar to the switched channel sounders discussed in Chapter 9. If the RX switches, the TX must stick to a single antenna, and vice versa; for this reason a special field signals whether a constant tone extension is intended for DoA or DoD determination.

Furthermore, a number of reductions of functionality are implemented to reduce energy consumption, e.g., (i) no support for voice channels, (ii) it is not mandatory to implement both TX and RX, (iii) no support for role switch between leader and follower), (iv) no support for sniff and park modes (since BLE is already energy saving, such energy-saving modes would just serve to complicate the situation).

34.2 Zigbee

34.2.1 Overview and Applications

Already in the early 2000s, the importance for low-energy machine-to-machine communications was realized – at that time mainly in the context of factory automation – and appropriate standards were developed. Similar to the development in WLANs, the physical and MAC layer standard was developed by IEEE, in the 802.15.4 group, while the networking and higher layers, as well as interoperability, were handled by an industry consortium, the Zigbee group.[7] The emphasis is on transmission with low rate (typically a few hundred kbit/s peak rate), over relatively short distances (around 100 m) with low energy consumption. A number of different physical layers are defined as options. Generally, Zigbee uses spread-spectrum, in order to comply with the regulations for ISM bands at the time of its introduction. The specific modulations will be described below in more detail.

There are, of course, numerous applications for Zigbee. Typical examples include:

- Industrial automation: e.g., asset tracking and plant monitoring.
- Home automation: such as light control systems, heating, and air conditioning.
- Smart energy: to monitor and control the delivery and usage of energy and water.
- Healthcare.

As we can see, some of these applications overlap with those of Wi-Fi as well as Bluetooth.

This section will only provide a very brief description of the Zigbee/802.15.4 standard – the description is less detailed than in the case of the other standards in this Part of the book. For details, see the references in the "further reading" section.

34.2.2 Physical Layer

While most systems use the 2.45 GHz ISM band for transmission, there are also several lower-frequency bands that are used in different regions around the world: $868 - 868.6$ MHz in Europe, $902 - 928$ MHz in the United States, $950 - 956$ MHz in Japan, and $779 - 787$ MHz in China. While the 2.45 GHz band has many advantages in terms of available bandwidth, the interference from Wi-Fi is a significant issue. Furthermore, propagation conditions can be better in the lower frequency bands, in particular in terms of diffraction around obstacles and transmission through walls, thus possibly making those bands more robust. Furthermore, the UWB version of the standard (see below) also defines the frequency ranges <1 GHz, 3–5 GHz, and 6–10 GHz; however, due to the severe constraints on transmit power, signaling can be less robust.

Zigbee uses different spreading and modulation modes in different frequency bands. For the 2.45 GHz band, it uses direct-sequence spread-spectrum; specifically, every 4 bits are grouped together, and mapped to a 32-chip sequence, where the mapping is defined in a lookup table. The 16 defined sequences are chosen to have minimum cross-correlation. This means that a signal with a bit rate of 250 kbit/s is spread to a 2 Mchip/s sequence, which is then used as the input of an offset-QPSK (O-QPSK) modulator, as described in Section 10.3.2.

In the 868/915 MHz band, there is a mandatory and two optional spreading and modulation methods. The mandatory mode works as follows: first, the data are differentially encoded. Then the system uses regular (binary) direct-sequence spread spectrum with a spreading gain of 15, i.e., a bit rate of 20 kbit/s (for 868 MHz) or 40 kbit/s (for 915 MHz) is spread with a 15-chip long pseudorandom sequence, resulting in 300 (or 600) kchip/s, which are then BPSK modulated. The optional modes use either amplitude shift keying and parallel sequence spread spectrum (not described here), or direct-sequence spreading and O-QPSK just as in the case of the 2.45 GHz band, though with lower spreading factor. Both of those optional modes provide a higher bit rate than the mandatory mode: 250 kbit/s in the case of ASK, 100 kbit/s (for 868 MHz), or 250 kbit/s (for 915 MHz) for O-QPSK.

In the 780 MHz band, the same spreading method as for the 2.45 GHz band is used, with a bit rate of 250 kbit/s, but mapping to a 16-chip sequence, thus resulting in a 1 Mchip/s chip rate. The chip sequence is then used as input for an O-QPSK modulator, or an MPSK modulator. In the 950 MHz band, binary direct-sequence spread spectrum is used, spreading a 20 kbit/s to 300 kchip/s, which are then modulated with BPSK (a GFSK version exists as well).

As can be seen from the above description, a large number of modulation and spreading methods exist, which are different in different regions of the world (and thus different frequency bands). This makes Zigbee transceivers more complicated than would be necessary with a unified approach, yet the different regulatory environment, in particular with respect to the available bandwidth, prevents such unification.

The mandated RX sensitivity for a 1% packet error rate is -85 dBm in the 2.45 GHz band, as well as for the 868/915 MHz bands when the optional modulations (OQPSK or ASK) are used; for the case of BPSK in the 868/915 MHz band, sensitivity must be better than -92 dBm.

---

[7] There are actually several other standards that also use the 802.15.4 PHY as basis, but we speak here only of Zigbee because it is the most widely used.

34.2.3 Packet Structure and MAC

A Zigbee packet at the PHY layer consists of a synchronization header (which allows the receiving device to obtain time and frequency synchronization), a PHY header (which contains length information), and a PHY payload. The PHY payload itself is a MAC frame, which contains a MAC header (containing address and security information), the MAC payload, as well as a CRC, also known as Frame Check Sequence (FCS), see Figure 34.13. The length of the MAC payload can vary (including length 0), as discussed below.

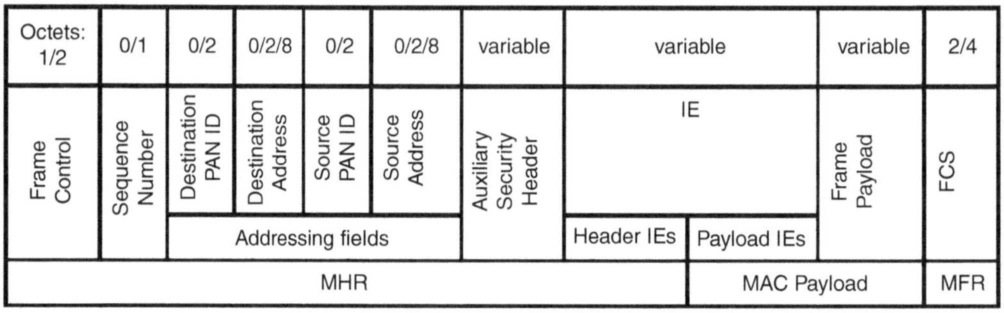

Octets: 1/2	0/1	0/2	0/2/8	0/2	0/2/8	variable	variable		variable	2/4
							IE			
Frame Control	Sequence Number	Destination PAN ID	Destination Address	Source PAN ID	Source Address	Auxiliary Security Header	Header IEs	Payload IEs	Frame Payload	FCS
		Addressing fields					Header IEs	Payload IEs		
MHR								MAC Payload		MFR

Figure 34.13 General MAC frame format for 802.15.4. MHR: MAC Header, MFR: MAC Footer, IE: Information Element, FCS: Frame Check Sequence. Reproduced with permission from [IEEE 802.15.4] © IEEE.

There are four different MAC frame types, which have slightly different structures, for

- *Beacon frame*, which is used for both synchronizing the network, and to let devices know that there are messages waiting for them.
- *Data frame*, which is the main method of data payload transmission.
- *Acknowledgement frame*, which has no MAC payload and simply confirms receipt of a packet.
- *Command frame*, which is used to transport commands such as association requests.

In terms of multiple access, Zigbee defines frequency channels. Different frequency bands have different number of channels and different bandwidths for the channels. For example, the 868 MHz band has only one channel; the 915 MHz band has 10, and the 2.4 GHz band has 16. In 802.15.4e, time-slotted frequency hopping was introduced to provide frequency diversity.

Zigbee uses both contention-based and contention-free multiple access. In the contention-based access, it employs CSMA-CA, similar to 802.11 (see Sections 18.4.2 and 33.4). Whether a channel is clear or not can be determined by energy detection, i.e., measuring the received power in this channel, irrespective of whether this energy comes from another Zigbee signal or from some other transmitter operating in this band. Alternatively, it can perform carrier sense, where the type of signal is determined; in this case, the threshold for transmission might be more stringent than for energy detection, since a Zigbee device should be considerate of other possible Zigbee devices around it.

In the contention-free case, the PAN coordinator sends out a beacon signal that ensures that all devices are synchronized (though this has the drawback of devices having to listen to beacon signals and thus consuming more energy). Then the PAN coordinator dedicates a particular timeslot (called guaranteed timeslot) to a particular device.

Because a coordinator is always awake, while a device may be asleep, data transmission from device to coordinator, and vice versa, have a different structure. In the former case, the device simply sends the data, and the coordinator sends back an acknowledgment if requested. In the latter case, the coordinator first sends with the beacon a message that data are available for a device. If the device is awake, it then sends a data request, indicating that it is awake. The coordinator acknowledges this, indicating to the device that it should stay awake. Only after this does the coordinator send the data. In a nonbeacon network, a coordinator cannot send data unrequited but has to wait until the device sends a request for data.

34.2.4 Network Structure

Zigbee defines Full-Function Devices (FFDs) and Reduced-Function Devices (RFD). The former can perform all the mandatory tasks of the 802.15.4a standard, while a RFD device has reduced capabilities but is cheaper to manufacture and has lower energy consumption. An FFD can talk to both FFDs and RFDs and can act as PAN coordinator (piconet controller), coordinator (router, i.e., performing relaying), or just a regular device.

Zigbee networks are ad hoc networks, i.e., there is no predefined infrastructure. The network formation starts when a first device starts communicating; it then becomes the PAN coordinator, and other devices will send association requests. Zigbee networks can be arranged in a star topology, where every device communicates with the PAN coordinator, or a peer-to-peer (mesh) topology, where every device can talk to any other device within its coverage, Figure 34.14. In a peer-to-peer network, any FFD can become PAN coordinator. Messages can be relayed via the routers, significantly extending the range of the network.

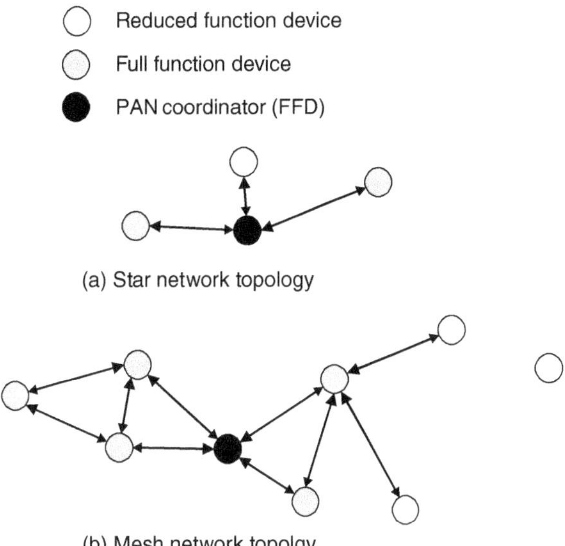

Figure 34.14 Zigbee network structures. Note that RFDs cannot relay messages.

The PAN coordinator selects a PAN ID for the network, assigns unique addresses to all devices in the network, and initiates, terminates, and routes the messages through the network. Since PAN coordinators have these important roles and need to be active most of the time, they are preferably mains-powered; alternatively, the PAN-coordinator role can be rotated among FFDs in the network.

In addition, a Zigbee gateway can enable a connection between a Zigbee network and another type of network, e.g., Wi-Fi.

34.2.5 Ultrawideband Modes

Introduction and Frequency Bands

The IEEE 802.15.4a standard is based on direct-sequence spreading UWB signaling. It has data rates in excess of 1 Mbit/s, and furthermore provides the possibility of ranging. More recently, the IEEE 802.15.4z standard, which enhances this UWB physical layer, has drawn attention for ranging and proximity sensing applications (similar proprietary standards by smartphone manufacturers are either already in use or imminent at the time of this writing). The following description focuses mainly on the original 802.15.4a standard, but at the end describes the enhancements provided in more recent versions.

The UWB mode operates either below 1 GHz, between 3 and 5, or between 6 and 10 GHz, with a bandwidth of either 500 or 1500 MHz, see Figure 34.15.

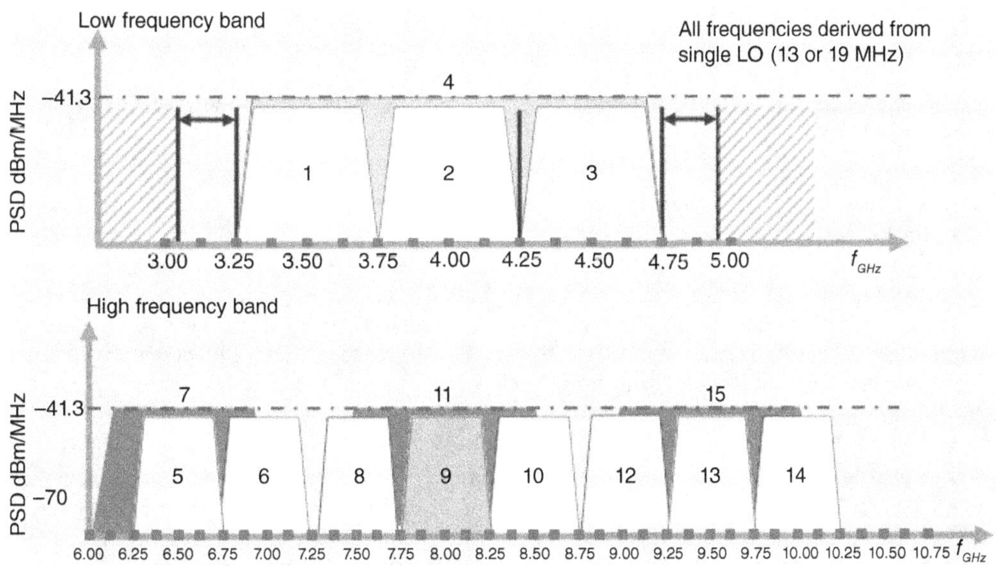

Figure 34.15 UWB bandplan for 802.15.4. Color version available at wiley.com/go/molisch/wireless3e.

Reproduced with permission from [Zhang et al. 2009] © IEEE.

Modulation and Multiple Access

In 802.15.4, the UWB PHY layer, which includes Modulation, Coding, and Multiple access schemes (MCM), has been designed in such a way that it allows both FFDs and RFDs to achieve optimum performance, such as allowing the FFD devices to employ coherent reception (enhanced performance at the cost of energy consumption and complexity), while RFDs use simple energy detectors (noncoherent RXs) for reduced current drain and design simplicity. Furthermore, such a flexible MCM scheme does not deteriorate the possible performance of the FFDs, i.e., the performance of FFDs with flexible MCM is (almost) as good as with an MCM that is designed for homogeneous coherent-RX networks.

In particular, the transmit signal is

$$s^{(k)}(t) = \sum_i \sum_{n=0}^{N-1} \widetilde{b}_i^{(k)} p\left(t - nT_\mathrm{c} - c_i^{(k)} T_\mathrm{b} - iT_\mathrm{s} - b_i^{(k)} T_\mathrm{PPM}\right) d_{i,n}^{(k)} \tag{34.2}$$

where superscript $^{(k)}$ denotes the kth user, b_i is the ith data bit to be transmitted, $\widetilde{b}_i$ is a parity check bit associated with the ith data bit, which is also to be transmitted and modulated onto the phase of the pulses. Furthermore, T_c is the chip (pulse) duration of approximately 2 ns, T_b is the burst-hopping duration, which equals $T_\mathrm{b} = NT_\mathrm{c} = 32$ ns, n indexes the $N = 16$ pulses that are transmitted during each data burst, $c_i^{(k)}$ is the time (burst)-hopping sequence for multi-user access, T_PPM is the modulation interval for the pulse position modulation, $T_\mathrm{PPM} = 16T_\mathrm{b} = 512$ ns, and $T_\mathrm{s} = 2T_\mathrm{PPM}$ is the symbol duration, leading to a 1 Mbit/s data rate. The $d_{i,n}^{(k)}$ denotes a pseudorandom scrambling sequence drawn from $\{-1, 1\}$. The pulse $p(t)$ is the basis pulse that is a raised-cosine pulse.

Figure 34.16 shows a graphical representation of an example. The waveform constitutes a burst of pulses that lasts 32 ns. Depending upon the data bit to be transmitted, the burst of pulses will be in either the first half or the second half of symbol duration. For a noncoherent RX, it provides a PPM modulation signal with a 32 ns excitation signal and a 512 ns modulation interval. The modulation interval is chosen much larger than the typical channel delay spreads, so that a noncoherent RX can detect the PPM even in channels with heavy delay dispersion. On the other hand, the duration of the burst waveform is on the order of, or shorter than, typical delay spreads. Thus, the duration over which a noncoherent RX has to integrate the received signal is essentially determined by the propagation channel. Shortening the duration of the burst waveform would not significantly reduce the optimum integration duration (and thus, the time over which the RX collects noise). Generally, the burst duration can change between 2 and 1025 ns, resulting in symbol rates between 0.11 and 27.24 MBit/s, and average pulse repetition frequencies of 3.9, 15.6, and 62.4 MHz, see Table 15.3 in [IEEE 802.15.4 2020] for a complete listing.

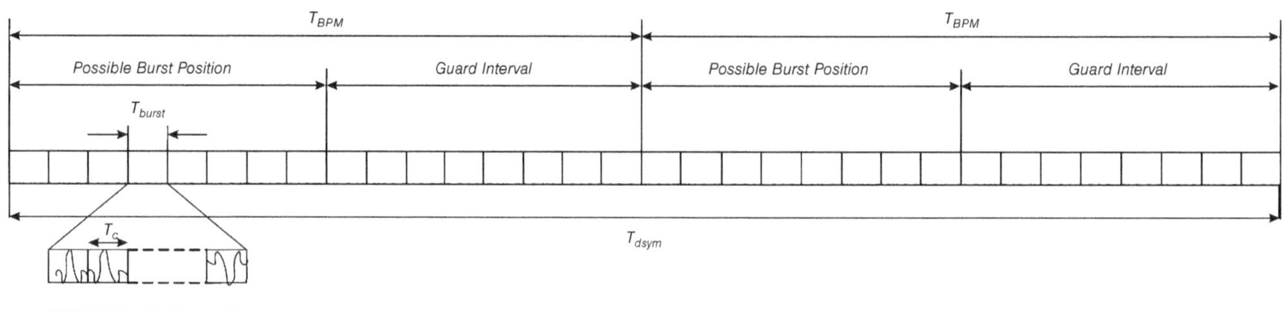

Figure 34.16 Modulation and time-hopping in UWB-802.15.4. T_{dsym} is T_S in the notation of this book. BPM means Binary PPM. Reproduced with permission from [IEEE 802.15.4] © IEEE.

A coherent RX may de-spread the excitation signal by correlating with $d_{i,n}^{(k)}$ resulting in an SNR gain. Furthermore, the coherent RX can extract the parity check bit, $\widetilde{b}_i^{(k)}$, from which it can obtain additional coding gain. Specifically, in order to generate some FEC protection even in the case of noncoherent demodulation, 802.15.4 UWB uses a concatenated encoder. In a first step, the payload is encoded with a systematic Reed-Solomon (RS) code that adds 48 parity bits to the original bits. The output of this encoder is then fed into a systematic rate 1/2 convolutional encoder with generator polynomials $G0 = [010]$ and $G1 = [101]$. The systematic bits of this encoder are used to modulate the position (and can thus be recovered by a noncoherent RX), while the parity bits change the polarity. Thus, a coherent RX can recover those polarity bits and gain the error protection of the concatenation of RS and convolutional code, while a noncoherent RX only gets the benefit of the RS code.

The multiple-access is based on time hopping, as described in Section 19.4: the position of the burst waveform is shifted by multiples of T_b in a pseudorandom way by $c_i^{(k)}$; the shifts are different for different users. Note that the maximum possible shift is $8T_\mathrm{b}$, while the time shift for the PPM is $16T_\mathrm{b}$. Thus, a duration of $8T_\mathrm{b}$ serves as a guard interval for channels with heavy delay dispersion.

The coherent RX obtains additional multi-user separation by the de-spreading of the burst waveform. As each user has a different burst waveform, the matched filtering at the RX input provides multi-access interference suppression. The amount of suppression depends on the cross-correlation between the burst waveforms; it is noteworthy that the spreading sequence, and thus the burst waveform, changes from symbol to symbol.

Preamble, Synchronization, and Ranging

In 802.15.4 UWB a specific preamble, detectable by both coherent and noncoherent RXs, is used for acquisition, synchronization, and channel estimation purposes. It is based on the principle of "Perfectly Balanced Ternary Sequences" (PBTS). For the PBTSs both the periodic autocorrelation function for coherent RXs

$$ACF_k = \sum_n \sum_j \sum_m c_{i+mN} c_{k-i+jN} \tag{34.3}$$

and the periodic autocorrelation function as observed by noncoherent RXs

$$\widetilde{ACF}_k = \sum_n \sum_j \sum_m |c_{i+mN}| \left(2|c_{k-i+jN}| - 1\right) \tag{34.4}$$

are perfect, i.e., proportional to a delta comb $\sum_i \delta_{k+iN}$. The standard foresees the use of either length-31 or length-127 PBTS. Multiple repetitions of those sequences are used to improve the SNR.

The preamble is also useful for ranging. According to the 802.15.4 ranging protocol, an original ranging node A first transmits a signal called range request packet to target ranging node B. After reception of the request, node B prepares and sends an acknowledgment packet, also referred to as a range reply packet, back to node A. In a separate packet, B also reports to A the arrival time of the range request packet and the departure time of the ACK. Node A can then compute the range, since it knows the total round-trip time and the turnaround time of the ACK.

In order to detect the end of the preamble, a Start of the Frame Delimiter (SFD) is placed prior to the PHY header; it consists of 8 or 64 preamble symbols. Upon detection of the SFD of a received range request packet, the ranging timing counter is started. Similarly, at the time instant that the SFD of a range reply packet leaves the transmit antenna, the ranging timing counter is stopped. The difference in these two counter values corresponds to the turnaround time.

Evolution of the Standard

The original 802.15.4a standard was developed in the late 2000s. While the fundamental structure has remained unchanged, a number of enhancements have been provided since then. Firstly, some minor editorial changes were done when its documentation was merged with that of the 802.15.4 (Zigbee) description, to provide the 802.15.4-2011 standard.

The first major enhancement occurred in the 802.15.4f standard in the early 2010s, introducing the so-called Low-Rate Pulse repetition frequency (LRP) mode. It uses on/off keying. In base mode, information is sent by the presence/absence of pulses; specifically, in the case of a logical "1," a short pulse (with at least 500 MHz bandwidth) is transmitted in the middle of a symbol; otherwise, nothing is sent, see Figure 34.17. In the extended mode, the information is encoded with a rate ¼ convolutional code. The 1 μs symbol duration is subdivided into four parts of 250 ns each, and a pulse is sent (or not) according to OOK in each of those parts. The long-range mode employs PPM: it considers groups of 64 subperiods, each of which is 500 ns long. For a logical "1," pulses are sent in the first 32 subperiods, and nothing in the second 32, and vice versa for a logical "0." The effective data rate is thus only 31.25 kbit/s. The possibility for energy accumulation/averaging enables communication over larger distances. The LPR mode was originally mainly intended for RFID and similar applications. The 802.15.4f standard was folded into the overall 802.15 standard in the 802.15.4-2015 document.

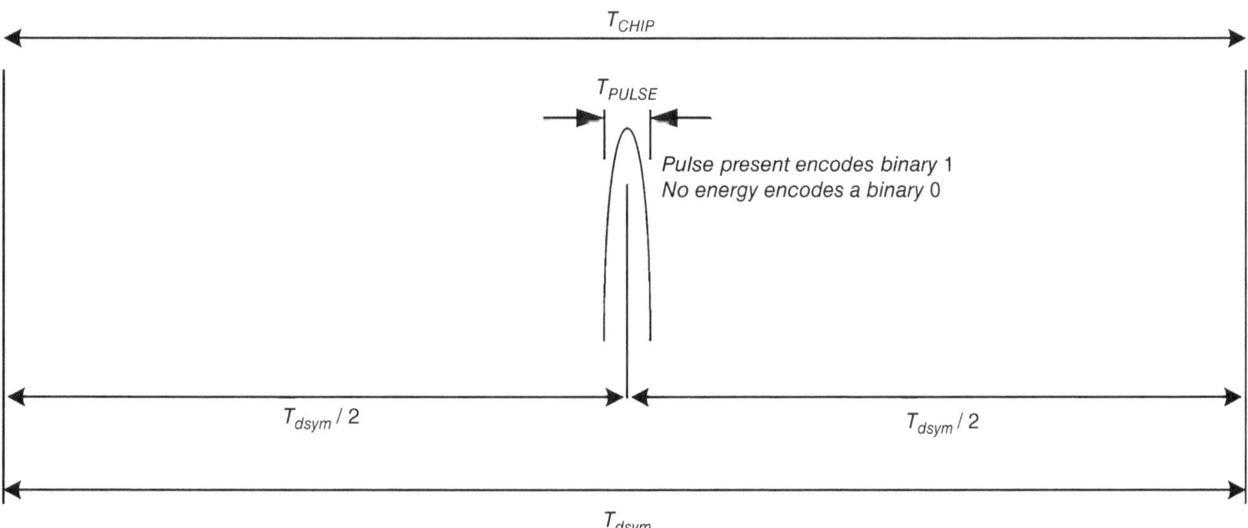

Figure 34.17 Symbol structure of the LRP base mode.
Reproduced with permission from [IEEE 802.15.4] © IEEE.

Further enhancements were made in the 802.15.4z standard. Firstly, ranging capabilities were introduced for the LPR mode, which did not exist in the original 802.15.4f standard. Secondly, provisions were made for one-way ranging, such that Node A sends a ranging signal to Node B, which (possibly at a later time) simply replies with a timestamp when it received the ranging signal. This method is less accurate than the two-way ranging (where both nodes send out ranging signals) assumed in previous versions, but it saves energy because of the reduced number of signals that need to be exchanged.

The 802.15.4z standard also introduces changes in the High-Rate Pulse repetition frequency (HRP) mode, i.e., the mode originally designed in 802.15.4a. It adds even higher pulse repetition frequencies. Furthermore, the security of the ranging process was enhanced, by defining cryptographic sequences that govern the ranging signal that is being sent out.

An industry consortium FiRa has been established, which aims to ensure interoperability of devices following the 802.15.4z standard, and perform overall a similar role for 802.15.4z as the Wi-Fi Alliance does for 802.11.

Glossary

Acronyms Bluetooth

ACL	Asynchronous Connection-Less
AFH	Adaptive Frequency Hopping
BD_ADDR	Bluetooth Device Address
BLE	Bluetooth Low Energy
BR	Basic Rate
CI	Coding Indicator
CIS	Connected Isochronous Stream
DAC	Device Access Code
DM	Data Medium Rate
DV	Data and Voice (packet type)
EDR	Enhanced Data Rate
EIR	Extended Inquiry Response
eSCO	Extended SCO
FHS	Frequency Hop Synchronization
HCI	Host Controller Interface
HEC	Header Error Check
L2CAP	Logical Link Control and Adaptation Protocol
LMP	Link Manager Protocol
MIC	Message Integrity Check
PDU	Protocol Data Unit
SCO	Synchronous Connection-Oriented
USB-HID	Universal Serial Bus Human Interface Device

Acronyms Zigbee

FCS	Frame Check Sequence
FFD	Full Function Device
HRP	High Rate Pulse repetition frequency
LRP	Low Rate Pulse repetition frequency
MCM	Modulation Coding and Multiple access
PBTS	Perfectly Balanced Ternary Sequences
RFD	Reduced Function Device
SFD	Start of Frame Delimiter

Further Reading

For Bluetooth, the official standard [Bluetooth 2019] provides detailed information and is (compared to other standards) surprisingly readable. Also, the book of [Gupta 2016] provides a good overview. The IEEE 802.15.4 standard (Zigbee) is described in [Farahani 2011], while the UWB version IEEE 802.15.4a is described in [Zhang et al. 2009]. MAC innovations in 802.15.4e are surveyed in [Kurunathan et al. 2018]. The current version of the standard is [IEEE 802.15.4 2020].

For updates and errata for this chapter, see https://wides.usc.edu/students.html#textbooks.

Exercises

See Sec. 36.34 of Exercises.pdf at wiley.com/go/molisch/wireless3e

35

Beyond 5G

35.1 Motivation and Process

5G systems – encompassing not only cellular 5G in the form of Third Generation Partnership Project (3GPP) New Radio (NR) but also the latest versions of Wi-Fi and Bluetooth – have brought unprecedented high data rates (up to 10 Gbit/s), low latencies (1 ms), and excellent energy efficiency $(10^{-7}\,\text{J/bit})$. However, a host of current and future applications require to push these numbers even further. Even richer multimedia applications in the form of high-fidelity holograms and immersive reality, tactile/haptic-based communications, as well as the support of mission-critical applications for connecting all things, will require even larger data rates and lower latencies. These developments in turn will require not only improvements in the physical layer and exploitation of new frequency bands but also improvements in the higher layers for the Beyond 5G (B5G) standards, often also called 6G.

At the time of this writing, official standardization has not started yet. Various interested parties have presented their visions to the International Telecommunications Union (ITU), but the actual standardization bodies like 3GPP and IEEE 802 are still busy with the improvement of 5G systems. Any discussion about what will be included in B5G standards is thus necessarily speculative. Thus, ironically, this most forward-looking chapter of this book might be outdated the soonest. Still, I consider it worthwhile to present the perspective of 2022 – for the readers who look at this within a few years it might give an interesting survey, and for later readers, it might provide an object lesson in the old saying "predictions are difficult, especially about the future."

35.2 Applications

A first step in the development of future systems is the determine what applications will run over them, and what are the resulting system requirements. The following lists some major development areas.

35.2.1 Holography, Extended Reality, and Other High-Data-Rate Applications

Video streaming has been the killer application for 4G and 5G, with constantly increasing demand for data rate (due to demand for higher quality, e.g., 4K), increased viewing time (watching complete movies, not just short clips), and increasing number of users. The rise of remote working, accelerated by the COVID-19 pandemic, has furthermore increased the demand for video conferencing, to the point where many people spend more than half their day in such telemeetings. The next step will be extended reality (encompassing virtual and augmented reality), holograms, and multi-sense communications, which will enable remote users as a rendered local presence. For instance, technicians performing remote troubleshooting and repairs, doctors performing remote surgeries, as well as improved remote education in classrooms, could all benefit from hologram renderings.

The data transmission rates for holograms are very substantial (at least for today). Besides the standard video properties such as color, depth, resolution, and frame rate, holographic images will need transmission from *multiple* viewpoints to account for variation in tilts, angles, and observer positions relative to the hologram. This may result in data rate requirements above 1 Tbit/s. Data compression techniques may reduce the data rates needed for the transmission of holograms, but even with compression, holograms will require massive bandwidths, typically far in excess of what 5G can provide. Furthermore, truly immersive scenarios require ultra-low latency – less than 1 ms – to avoid simulator sickness feelings in the user.

Furthermore, synchronization requirements are very tight: as different senses get integrated, the different sensor feeds may be sent over different paths or flows and will require synchronization, as well as coordinated delivery. When streams involve data from multiple sources, such as video, audio, and tactile, precise/stringent inter-stream synchronization is required to ensure timely arrival of the packets. Coordinated delivery of the flows needs synchronization at the physical layer but also the network has to have knowledge of the co-flows. Finally, the system needs to be resilient. At the system level, resilience is about minimizing packet loss, jitter, and latency. At the service level, relevant quality-of-experience metrics are availability and reliability.

Current holograms are mainly restricted to head-mounted displays, where the data rates are lower, or holographic playback (e.g., a virtual concert), where strong compression techniques can be used, because the latency caused by the compressing/uncompressing

Wireless Communications: From Fundamentals to Beyond 5G, Third Edition. Andreas F. Molisch.
© 2023 John Wiley & Sons Ltd. Published 2023 by John Wiley & Sons Ltd.
Companion website: www.wiley.com/go/molisch/wireless3e

process does not impact the quality of experience. However, for future applications, a combination of stronger computational power and higher data rates will be required to make interactive holograms a reality. This market will be targeted both by the cellular industry (3GPP), as well as Wi-Fi, and possibly proprietary systems

Another high-data-rate application will be information kiosks or data showers – a concept that was originally proposed under the name of infostation in the 1990s – where extremely high data speeds are available in a very small area. The kiosks should provide fiber-like speeds, on the order of 1 Tbit/s, and will therefore use very large bandwidth, which is available at high carrier frequencies (see below). Co-existence with contemporaneous cellular services, as well as data security, seems to be the major issue requiring further research in this area. Such systems clearly will need interoperability with cellular or Wi-Fi systems.

Yet another application where high data rates are required are on-chip, inter-chip, and inter-board communications. While those are done nowadays through wired connections, those links are becoming bottlenecks when the data rates exceed 100–1,000 Gbps. There have thus been proposals to employ either optical or THz wireless connections to replace wired links. The development of such *nanonetworks* constitutes another promising area for B5G. Important criteria for such networks – besides the data rate – are energy efficiency (which needs to incorporate possible required receiver (RX) processing), reliability, and latency. While obviously intra-chip communications do not need any standardization, communication between chips or boards of different manufacturers will require new standards, which in all likelihood will be independent of existing wireless standards.

35.2.2   Real-Time Information and Control

Humans and machines are both sensitive to delays in the delivered information (albeit to varying degrees). *Timeliness* of information delivery will be critical for the vastly interconnected society of the future, as new applications that intelligently interact with the network will demand guaranteed capacity and timeliness of arrivals. As we incorporate gadgets in our life, quick responses and real-time experiences are critical and late arrival of information may even be catastrophic. Time sensitivity also has a deep impact on other modes of communications in the future, such as those relying on *tactile and haptic control*. This is rooted in the fact that the human brain requires only 1 ms to receive and process a tactile signal, considerably faster than for audio and visual signals. Furthermore, many control applications for machines need sub-ms latencies. This applies to stationary machinery, where, e.g., fast reaction of control circuitry on a milling machine is required to retain tight tolerances in the output, as well as to highly mobile machines likely autonomous vehicles, as discussed below. The strict latency requirements also make requirements for synchronization of data streams from different sensors more stringent. Networks also need the capability to prioritize data streams based on their criticality.

An important example for such applications is *robotic and industrial automation:* the so-called *Industry 4.0* envisions manufacturing entities that can be reconfigured on short notice, such that per-piece manufacturing costs of small production runs are not significantly larger than those for mass production. This requires the ability to re-arrange location and tasks of robots and other manufacturing devices on short notice, necessitating wireless connectivity for all of them. Advanced robotics scenarios in manufacturing need a maximum latency target in a communication link of 100 μs and round-trip reaction times of 1 ms. For the updating of control information, the concept of *Age of Information* has recently received significant attention.

Another example is *autonomous driving*. Enabled by V2X (encompassing Vehicle-to-Vehicle (V2V) and/or Vehicle-to-Infrastructure (V2I)) communication and coordination, autonomous driving can result in a large reduction of road accidents and traffic jams. However, a latency on the order of a few ms will likely be needed for collision avoidance and remote driving. Thus, advanced driver assistance, platooning of vehicles, and fully automated driving, are the key application areas that B5G aims to support. While V2X communications are part of NR, the performance there is not sufficient for fully autonomous driving. Further research and anticipated battles between the cellular (3GPP) and Wi-Fi (802.11 p/bd) visions will shape the development of the standards in this field.

Tactile applications with the concomitant short latencies also play a key role in *health care,* with telediagnosis, remote surgery, and telerehabilitation are some of the many potential applications. While traditional teleconsultations (as became popular during the COVID-19 pandemic) do not need to meet any special latency limits, remote and robotic surgery have very strict requirements. In those applications, a surgeon gets real-time audio-visual feeds of the patient that is being operated upon in a remote location, as well as haptic information transmitted to/from the robot. End-to-end latencies of better than 1 ms are required in this context.

35.2.3   Network and Computing Convergence

During the 2010s, the main trend in information processing was cloud computing, where local nodes offload their computational tasks to larger server farms that might be far away from the location of the user. This approach has significant advantages in terms of the efficient use of processing resources but increases the load on the communication network and more importantly introduces significant latencies that might be unacceptable in a number of applications such as the control systems mentioned above. Furthermore, privacy and confidentiality are concerns when large sets of data are uploaded to a central server. There has thus been great interest in *Mobile Edge Computing (MEC)*, where computations are executed on edge servers that are located much closer to the user. In other words, when a client requests a low latency service, the network may direct this to the nearest edge computing site. For computation-intensive applications, and due to the need for load balancing, a multiplicity of edge computing sites may be involved, but the computing resources must be utilized in a coordinated manner.[1] The optimization of such MEC systems and the optimization of

---

[1] A more general form is *Augmented Information Services*, where computations are performed on data streams that are transmitted in a multihop fashion from a transmitter to the receiver, and the computations can be performed at intermediate nodes.

the network, which involve trade-offs between the wireless communication aspects and the networking and computing aspects, is an overall very active research area and will be an important part of B5G networks.

Augmented reality/virtual reality (AR/VR) rendering, autonomous driving, and holographic type communications are all candidates for edge – cloud coordination. The key network requirements are computing awareness of the constituent edge facilities, joint network and computing resource scheduling (centralized or distributed), flexible addressing (every network node can become a resource provider), fast routing, and re-routing (traffic should be able to route or re-route in response to load conditions). Figure 35.1 demonstrates this vision via edge-to-edge coordination across local edge clouds of different network and service types, as well as edge coordination with the core cloud architecture. While such systems can be built with existing wireless components, such as Wi-Fi, optimization would benefit from a standardization that has a unified approach to the communication and computation (as well as caching/storage) aspects.

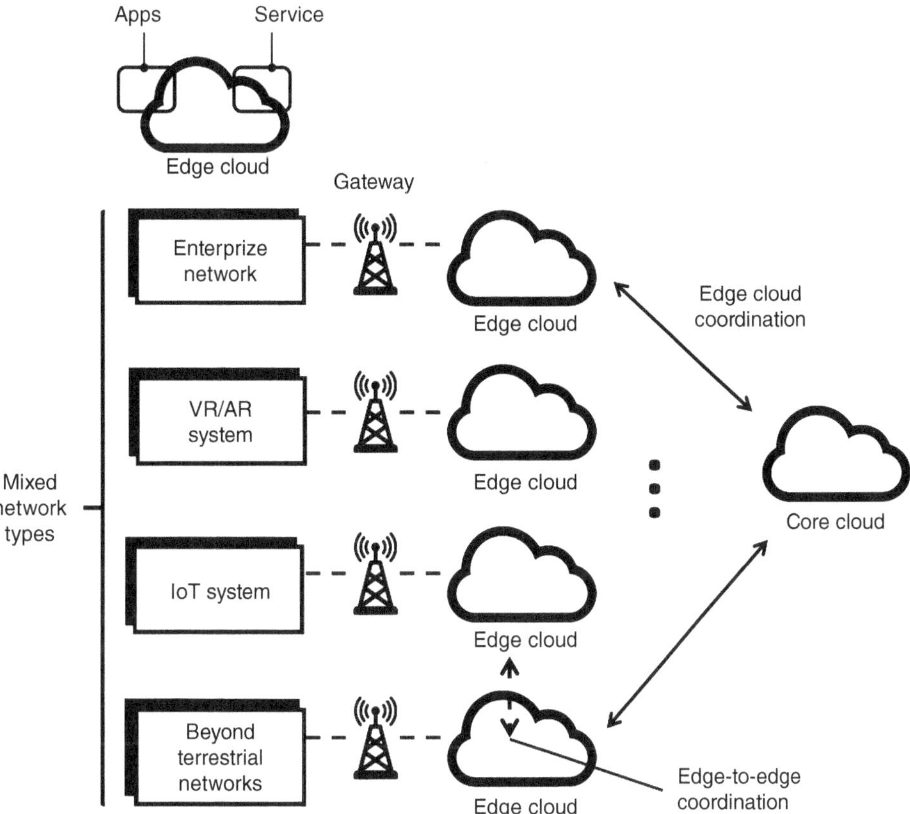

Figure 35.1 Cloud coordination between local edges driven by different network types and services, as well as across the local edge cloud and core cloud. Reproduced with permission from Tataria et al. [2021] © IEEE.

35.2.4 Connectivity for All Things and 3D Networks

While the Internet of Things (IoTs) is already drawing attention and some communication modes for it are part of 5G, where it is called mMTC (see Chapter 32), the trend will be amplified in the future. Applications will include real-time monitoring of buildings, cities, environment, agricultural entities, cars and transportation, roads, critical infrastructure, water, power, etc. Besides these use cases, the internet of bio-things through smart wearable devices, or intra-body communications achieved via implanted sensors, will drive the need of connectivity much *beyond mMTC*. The key network requirements for these use cases are large aggregated data rates due to vast amounts of sensory data, high security, and privacy in particular when medical data are being transmitted, as well as possibly low latency when a fast intervention (e.g., heart attack) is required. This brings the necessity to operate *multiple network types*, going well beyond the standard terrestrial networks of today. There are significant attempts to develop uninterrupted global broadband access via integration between the terrestrial networks and many planned satellite networks – especially for Low Earth Orbit (LEO) satellites – as well as Unmanned Arial Vehicle (UAV)-based systems.

The idea of providing internet from space using large constellations of LEO satellites has regained popularity in the last years (previous attempts such as the Iridium project in the late 1990s had failed), because if offers: ubiquitous internet access on a global scale including on moving platforms (airplanes, ships, etc.), enriched internet paths due to the border gateway protocols across domains relative the terrestrial internet, and ubiquitous edge caching as well as computing. The mobile devices for these integrated systems will be able to have satellite access without relying on ground based infrastructures. The key network requirements for this capability are

(i) Flexible addressing and routing; with thousands of LEO satellites, there are new challenges for the terrestrial internet infrastructure to interact with the satellites. (ii) Satellite bandwidth capability: the inter-satellite links and terrestrial internet infrastructure in some domains could be a bottleneck for satellite capacity. (iii) Admission control by satellites: when a satellite directly acts as an access point, this requires each satellite to have knowledge about the traffic load in the space network to make admission control decisions. (iv) Edge computing and storage: the realization of edge computing and storage will incur challenges on the satellite due to on-board limitations. Latency will also be a challenge as the physical distance between the satellite and end node will set a limit on the minimum delay introduced by the link.

Three-dimensional networks will also include UAVs, commonly known as drones. UAVs can act as mobile Base Stations (BSs), or be mobile User Equipments (UEs) that can, e.g., collect sensor data (including video) and transmit this to the ground, and finally act as relay stations. In the case that UAVs are used as infrastructure nodes, they can operate as temporary BSs, or can be placed quasi-permanently in an area in which coverage provision by such drones (possibly helped by multi-hop relaying to a regular node) is the most cost-effective way of coverage. Research challenges will include the coordination of UAV flight paths, energy expenditure and lifetime of the drone flight, and of the wireless transmitters (TXs), together with coverage optimization. For relaying, the topic of beamforming, as well as choice of suitable bandwidths, needs to be investigated. All of these issues also fall within the purview of cellular standardization.

35.2.5 Summary of Key Requirements

Collectively, in view of the above, the key requirements for B5G systems may be summarized as follows:

- *Peak data rate*: ≥ 1 Tbps catering to holographic communication, multi-sensory extended reality, and extremely high rate information showers. This is at least $50\times$ larger than that of 5G systems.
- *User experience data rate*: At least be $10\times$ that of the corresponding value of 5G.
- *User plane latency*: This is application dependent, yet its minimum should be a factor $40\times$ better than in 5G.
- *Mobility*: It is expected that B5G systems will support mobility of up to 1,000 km/h to include mobility values encountered in commercial airplanes (this has also been recently discussed in 5G).
- *Connection density per-km^2*: Given the desire for B5G systems to support an internet-of-everything, the connection density could be $10\times$ that of 5G.

The above capabilities and more are summarized in Table 35.1, relative to the corresponding values in 5G and 4G systems. Realizations of the technical capabilities as discussed in this are significant challenges which must be overcome, possibly by the approaches described in the subsequent sections.

Table 35.1 Technical performance requirements of B5G systems and a comparison of the B5G KPIs relative to those for 5G and 4G systems.

KPI	4G	5G	B5G
Operating bandwidth	Up to 400 MHz (band dependent)	Up to 400 MHz for sub-6 GHz bands (band dependent) Up to 3.25 GHz for mm-wave bands	Up to 400 MHz for sub-6 GHz bands 1 – 10 GHz for mm-wave bands Indicative value: 10–100 GHz for THz bands
Mac carrier bandwidth	20 MHz	400 MHz	To be defined
Peak data rate	1 Gbit/s (Release 10)	20 Gbps	≥ 1 Tbps (Holographic, VR/AR, and tactile applications)
User experience rate	10 Mbps	100 Mbps	>1 Gbps
Average spectral efficiency	15 bps/Hz (DL) and 6.75 bps/Hz (UL)	30 bps/Hz (DL) and 15 bps/Hz (UL))	$2\times$ that of 5G
Connection density	N/A	10^6 devices/km^2	10^7 devices/km^2
User plane latency	50 ms	4 ms (eMBB) and 1 ms (uRLLC)	25 µs to 1 ms (Holographic, VR/AR and tactile applications)
Control plane latency	50 ms	20 ms	10 ms
Mobility	350 km/h	500 km/h	1,000 km/h Handling multiple moving platforms
Mobility interruption time	N/A	0 ms (uRLLC)	0 ms(Holographic, VR/AR and tactile applications)

35.3 Network Design in B5G

B5G will continue many of the new trends in 5G networking, the most important of which are network slicing and control/user plane separation. The service-driven architecture, together with the trend to Software-Defined Networking (SDN) and Network Function Virtualization (NFV), will lead to a more atomized software structure with faster, more incremental, modifications and improvements, instead of the "generational jumps" that dominated the core network development until 5G. At the same time, there will be new approaches to the underlying transport architecture. All this will have ripple effects on the MAC and PHY.

35.3.1   B5G Network Design Principles

Among the trends in protocol and architecture approaches, the following are anticipated to play a major role in B5G:

1. *Super-Convergence*: non-3GPP systems will be integrated into the B5G cellular ecosystem. From a networking point of view, the details of the physical layer are less important for the interfacing and integration than the security and authentication that ensures that the network is not compromised by incorporation of nonnative components. The convergence of the multiple standards like Wi-Fi and Bluetooth with 3GPP will greatly aid with traffic balancing, and thus resilience, due to the ability to onboard and offload traffic between networks of different loads.
2. *Non-IP Based Networking Protocols*: while Internet Protocol version 6 (IPv6) is the dominant networking protocol, it is now decades old. Various entities have started deliberations on new protocols that take into account the fact that the majority of network traffic now originates from, or terminates at, the wireless edge.
3. *Information-Centric and Intent-Based Networks (ICNs)*: a major shift in the networking architecture is to move away from the source/target-centric approach. Rather, ICNs are a step toward the separation of content and its location identifier. Rather than IP addressing, content is addressed using an abstract naming convention. Different proposals exist today for the protocol realization of ICNs, e.g., by the European Telecommunications Standards Institute (ETSI), and ITU. Furthermore, to bridge latest developments in networking design and operational management, intent-based networking as well as intent-based service design have emerged.
4. *Subnetworking*: while a continuous connectivity between all nodes in a network remains an important goal, there are also many future applications where a semi-closed subnetwork needs to be established. A first application case arises for industrial networks that need to be walled off (with the possible exception of a single gateway) from the general internet, since (almost) all communication is between the machinery and computers within the industrial campus, and connection to the outside world is both unnecessary and a security risk. There are also subnetworks that are created out of the necessity for low latency – for example, for a V2X network over a stretch of road, where any communication outside the subnetwork would lead to too large latency for control application. This network will still be connected to the general network for downloading, e.g., traffic data, but it needs to be functional even if the connection to the outside world breaks down.
5. *Caching at the Wireless Edge*: caching of information close to the end user, such that it does not need to be retrieved from a far-away central repository every time it is requested, has long been used in wired networks. Such *content distribution networks* have been a major enabler of services like video streaming. They not only prevent overload of the central server when a large number of users request the content almost simultaneously but also reduce the latency significantly. However, at this point, caching is mainly done within the core network. In the progression of 5G to B5G, caching at the wireless edge will become more and more dominant. This might mean caching at the central units of a BS (compare Chapter 32), the distributed units, or even the devices themselves. Efficient distribution technologies in particular for video that are based on caching at the UEs have been introduced in the early 2010s (such as caching with D2D communications and coded caching). Optimization of the caching that takes into account content popularity, user density, as well as backhaul capacity and storage space at the various nodes, needs to be developed further. Machine learning is a promising avenue for these very complex optimization tasks.
6. *Cybersecurity and Privacy-by-Engineering-Design*: while security has been taken very seriously in 5G from a protocol and architectural point of view, the underlying embedded code which embodies and executes the various system components has never been part of the standardization efforts. Most security vulnerabilities however have been due to poorly written code. Thus, future efforts will not only focus on a secure end-to-end solution but will also encompass a top (architecture, protocols) – down (embedded software) approach. Furthermore, while security-by-design is now a well-understood design approach, privacy is still being solved at "consent" level. Privacy-by-Engineering-design will ensure that mechanisms are natively built into the protocols and architecture which would, e.g., prevent the forwarding of packets/information if not certified to be privacy-vetted. For instance, a security camera will only be allowed to stream the video footage if certain privacy requirements are fulfilled at networking level and possibly contextual level, i.e., understanding who is in the picture and what privacy settings they have enabled. For sub-networks, much of the security also needs to be handled within the subnetwork.
7. *Emerging Technologies*: a number of novel networking technologies are currently in an early research stage, and might be introduced over a longer time horizon. Examples include (i) quantum communications, which can provide provably secure links; (ii) Distributed Ledger Technologies (DLT), which enables storing and distributing the provenance of data, contracts, etc. in an immutable way; this technology is currently used mainly in the financial world (e. g., digital currency); (iii) Artificial Intelligence (AI), which is already nowadays used in customer service (chatbots), and problems like load balancing, will obtain an even more impactful role. Consumer-facing decisions made by AI will be impacted by the need to be compliant with various consumer-facing policies around the world. As a result, novel paradigms such as *Explainable AI (xAI)* will need to natively sit within B5G; xAI is a set of methods and techniques allowing the results of the solution to be understood by humans, in contrast to traditional AI that operates like a "black box" where even the design team cannot explain why the AI arrived at a specific decision.

The underlying infrastructure, including the transport networks, will need to undergo substantial changes as the amount of traffic to be carried in B5G networks will be orders-of-magnitude larger than what we will see in the next years with 5G networks. We expect the following fundamental changes:

1. *Removal/Reduction of the Transport Network*: The development of the core network for cellular systems as a separate network has its roots in 2G systems [back then the internet was not able to provide the required Quality of Service (QoS)] and has developed

from there due to this history. However, today the transport fiber infrastructure is really well developed and there is less reason for operators to maintain their own private networks. A complete rethink may thus give the opportunity for the cellular community to solely focus on the wireless edge (air interface + radio access network + control plane to support it all); and simply use a sliced Internet fiber infrastructure to carry the cellular traffic. While this requires some policy and operational changes, the technologies to support such a modus operandi do exist.

2. *Flattened Compute-Storage-Transport*: A flattened transport-storage-compute paradigm will be enabled by a powerful B5G air interface and a complete re-think of the core and transport networks as suggested above. A possible scenario is where transport is virtualized over existing fiber but isolated using modern SDN and virtualization methodologies. At the same time, the Core Network functions are packaged into a microservice architecture and enabled on the fly using containers or server-less compute architectures. To underpin novel applications such as gaming, we will also see a clearer split between Central Processing Unit (CPU) and Graphics Processing Unit (GPU) instructions sets, allowing each to be virtualized separately; for instance, the GPU instructions are handled locally on the phone while the CPU instructions are executed on a nearby virtual MEC node.

35.4 Spectrum Usage for B5G

Traditionally, new generations of wireless systems have exploited new spectrum in order to satisfy the increased demands for data rates. 5G systems are characterized to a significant degree by the use of the mm-wave spectrum. A further expansion to higher frequencies for B5G seems almost unavoidable. In this spirit, the spectrum from 100 GHz to 1 THz is being considered as a candidate for B5G systems.[2]

Not all of this spectrum will be made available by frequency regulators, and also not allocated in a contiguous manner, due to already-existing applications. In particular, over the range of 140 to 275 GHz, there are various blocks containing existing services that have co-primary allocation status by the ITU. These services include fixed, mobile, radio astronomy, passive earth exploration satellite services, passive space research systems, inter-satellite communications, radio navigation, and mobile satellite systems. Amongst the above, the passive services are much more sensitive to interference, and their protection will require guard bands, limits on out-of-band emissions as well as in-band transmit power, and restrictions on terrestrial beams and side lobes pointing upward. All of these aspects are critical for the co-existence of terrestrial systems with space-based networks.

The use of the above-mentioned frequency windows is dependent upon the specific use case; naturally, not all the windows will be suitable for all use cases. The various absorption bands in the THz, as well as the mm-wave, band have been discussed in Section 4.6, though absorption is mainly an obstacle for "long-range" applications like hotspots and vehicular radars, and less for short-distance links like inter-chip communications. A window of great interest covers the frequency range from 100 to 350 GHz. The combination of currently unused bandwidth, available semiconductor technology, and ability to develop ultra massive Multiple-Input Multiple-Output (MIMO) antenna arrays with a reasonable form factor makes this band very attractive. The frequency range from 0.35 to 10 THz provides even larger bandwidths, though the hardware challenges become formidable. Even higher frequency ranges bring us to free-space optical (including infrared) links, either through the use of laser diodes, or Light Emitting Diodes (LEDs) commonly assumed for Visible Light Communications (VLCs). Both of these approaches have been explored for a number of years, but it is only recently that an integration into cellular and other wireless systems seems to increasingly become a realistic option, see Section 35.5.4.

Besides the use of THz spectrum, there will also be a significant refarming of spectrum in the lower frequency ranges, including mm-wave bands. Firstly, the frequencies currently used for 4G and 5G systems will slowly but surely migrate to B5G. Secondly, additional spectrum that is currently used for other purposes will be made available to cellular and other wireless services. This re-dedication might occur in one of two methods (i) the existing users will give up their spectrum and either migrate to different spectrum bands or use more spectrally efficient techniques to be able to perform their tasks with less spectrum; usually, such users get significant financial compensation for such freeing-up of spectrum. The subsequent resale or auctioning of such freed-up spectrum in particular to cellular operators can provide significant revue for the government coffers – an auction of 3.5 GHz spectrum in the US in 2021 netted more than $ 80 billion; (ii) spectrum with existing primary users is opened up to secondary users. Such cognitive-radio techniques, which were discussed in greater detail in Chapter 26, were originally investigated in the early 2000s, but have gained fresh interests in the late 2010s. However, great care needs to be taken to avoid interference in particular to existing passive sensing services, such as weather radars and radio telescopes.

35.5 Physical and MAC Layer Aspects

35.5.1 Propagation Channels

As for all wireless systems, the performance of B5G systems will ultimately be limited by the propagation channels they will operate over. It is thus of vital importance to investigate the propagation channels relevant for B5G systems, in particular those that have not already been explored in the context of earlier-generation systems. The following briefly lists topics of special interest, for more details please see Chapter 7.

---

[2] However, we note that not all B5G services will be suitable to be offered in the new bands. The existing bands for 4G and 5G will continue to be used and may be re-farmed for B5G.

- *mm-Wave and THz channels*: with the move to higher frequency bands in B5G, exploration of the associated propagation channels is required. At the time of this writing, a number of measurements exist in the 100–300 GHz frequency band for indoor (in particular office) scenarios as well as information kiosks and chip-to-chip communications; the number of measurements in outdoor hotspots and microcells is small though the pace of new measurements is picking up. In terms of mm-wave channels, which will remain relevant for B5G, there is a much larger number of measurements, yet there remain a lot of unanswered questions in particular for outdoor environments, including UAV and V2X channels.

- *UAV channels*: as UAVs are becoming an important part of B5G networks, the propagation channels between them and ground BSs and UEs are becoming important [channels between UAVs are mostly Line-Of-Sight (LOS)]. Some measurements have been performed in lower frequency ranges, in particular assuming single-antenna transceivers at both UAV and ground station, but much more extensive exploration is required going forward.

- *V2X and railway channels*: propagation channels between vehicles, between vehicles and road-side infrastructure, and between trains and infrastructure, will be required for design of V2X and railway communications. The high dynamics of such channels make measurements more challenging, and models need to take into account not just the time variations but also the changes of the channel statistics. Most existing experimental work for these applications has been done in the <6 GHz range (though more is needed). Since mm-wave and THz systems are envisioned for the future, the associated channels need to be measured.

- *Ultra-massive and distributed MIMO channels*: as the electrical dimensions of massive MIMO arrays are becoming larger and larger, the channel effects that impact such large arrays have to be accounted for. For example, the average power might become different in different parts of an array, due to different shadowing effects of the array. Another important area of research is distributed MIMO channels, since currently the correlations between the different antenna elements are not measured and modeled in a sufficiently accurate way.

- *Industrial channels*: the typical industrial environment is unlike residential or other indoor environments, due to larger dimensions of buildings, reflections from (possibly moving) heavy machinery, as well as interference from electrically powered machinery and ignition systems. While some measurements have been performed over the years in industrial settings, much more needs to be done, in particular in the >6 GHz frequency ranges, and for multi-antenna systems.

- *On-body and in-body channels*: on-body communications are gaining importance, both for entertainment purposes (e.g., VR headset to a processing device carried in a backpack), and for wireless healthcare, such as communication from an on-body sensors to a cellphone. Furthermore, in-body communication, e.g., from a swallowed capsule equipped with a camera or other sensors to recording devices outside the body, can play an important role in medical diagnostics. More extensive measurements of the propagation channels in such scenarios need to be developed. The IEEE 802.15.6 standard for BANs has established several simple models mainly for channels at 900 MHz and 2.4 GHz, though they are based on only a small number of measurements. Furthermore, the scenarios and frequencies covered by the model are insufficient for many applications. Further work is thus needed.

For all of these channels, the first step is to obtain sufficient measurements or detailed ElectroMagnetic (EM)-simulation-based results to obtain sufficient understanding of the propagation to allow system design. Then, improvements and generalization of existing channel modeling approaches will be required: many of the existing channel models are based on "drops," i.e., realizations of the channel that might include temporal evolution within a stationarity distance, but do not cover larger-scale evolutions. Furthermore, many of the existing models are not spatially consistent, i.e., do not correctly account for the fact that a device moving on a closed loop should see the same channel at the end point as on the starting point. Generic models that are suitably improved to take these effects into account need to then be parameterized by the above-mentioned new measurements. These models need then be adapted for standardization purposes. While the 5G channel models were dominated by the need to retain backward compatibility to the 4G simulations, the multitude of new applications and frequency ranges may force a more thorough redesign of standardized models for B5G.

35.5.2   Modulation and Coding

As outlined in Chapters 31–33, Orthogonal Frequency-Division Multiplexing (OFDM) has become the dominant modulation format of modern wireless systems. OFDM's popularity is rooted in four key factors (compare also Chapter 15): (i) its information-theoretic optimality for the maximization of link capacity over frequency-selective channels, (ii) simple equalization, (iii) hardware and algorithms for OFDM processing that has been optimized over several decades, (iv) backward compatibility – OFDM was chosen as a modulation method for 4G, and as a result has also been employed in 5G; yet note the trend in 5G to adaptability of the numerology and frame structure, which required additional standardization effort. On the other hand, the downsides of OFDM discussed in Chapter 15, namely sensitivity to frequency dispersion, spectral efficiency loss due to the cyclic prefix, become more pronounced with increasing carrier frequency, and the Peak-to-Average Power Ratio (PAPR) problem more relevant with increasing bandwidth [since Analog to Digital Converter (ADC)/Digital to Analog Converters (DACs) with the required dynamic range are more difficult to build for large bandwidths]. The inherent signal processing requirements of OFDM also make it less suitable for extremely low energy applications. These considerations are particularly relevant in light of the increased heterogeneity of applications in B5G, in particular, various corner cases such as massive access from IoT devices and Tbps directional links.

For all these reasons, it can be expected that new modulation formats will be explored for B5G. A first area of interest will be modulation for extremely high data rates that balance capacity and ADC/DAC requirements, while keeping the equalization effort low. A promising candidate is temporally oversampled zero-crossing modulation, where information is encoded in the temporal distance between two zero crossings. Secondly, modulation with better robustness to inter-user and inter-carrier interference will be investigated further. For example, the variations of OFDM, such as filtered multi-tone, OTFS, etc., which were discussed in Sections 15.10 and 15.12 will be considered for B5G. Yet another alternative to the above techniques is the use of noncoherent or differentially coherent detection, where recent efforts have been devoted to develop suitable detection methods for multiple antenna systems with large antenna arrays. While the detection method itself will not be part of the standard, the associated encoding at the TX will have to be incorporated. Another area of research is modulation with extremely low energy consumption. Often in such applications, a remote "node" operates on *energy harvesting*, or must survive for years in a single battery charge. While theoretical investigations have shown that *flash* signaling (pulse position modulation with a very high number of bits in every symbol) to be optimal, it is practically infeasible, since it requires high PAPR and precise synchronization. For such applications, new modulation methods that minimize the *total* energy consumption for the TX (for the uplink) and RX (for the downlink) are required. Research in this line will include *clock-free receivers*, since the clock and clock distribution can constitute a significant "floor" in the overall energy consumption, especially when the device is in sleep mode, yet requires the clock to determine when the device needs to wake up. Together with additional approaches discussed in Section 35.5.5, these approaches may drive developments in B5G system development and standardization.

A major challenge of all modulation formats is the acquisition of Channel State Information (CSI) at the RX and at the TX. As discussed in Chapters 31–33, current systems use pilot signals for acquiring CSI at the RX, while for the TX, the system either relies on *reciprocity*, or *feedback* from the RX. However, the pilot overhead can become prohibitive, particularly in systems employing a massive number of transmit and/or receive antennas. Hence, for B5G, better CSI acquisition schemes need to be determined. Promising research efforts in this area include adaptation of the pilot signal spacing in time and frequency domain, exploitation of limited angular spread of the channel, and advanced signal processing methods for reduction of pilot contamination. A particular challenge occurs for Frequency Domain Duplexing (FDD) systems, where feedback-free Channel State Information at the Transmitter (CSIT) requires extrapolation from the uplink to the downlink band; this is an area where in the late 2010s and early 2020s a lot of research has been performed. A related problem is the quantization and feedback of CSI. While again the processing of the signals to obtain CSI estimates is outside the scope of the standards, the pilot format and the feedback types foreseen by the standards will have strong impact on the overall performance.

In terms of coding, the major open problems revolve around (i) finding better codes for short code block lengths (note that in this case there is a residual error rate, as discussed in Section 13.1.5). Another important topic is reliable transmission schemes for IoT and low-latency applications. These aim to improve on, or eliminate the need for, Automatic Repeat Request (ARQ), since ARQ increases the latency, and increases energy consumption by requiring the device to remain in a nonsleep mode for an extended period of time.

35.5.3 Multiple Antenna Techniques

The use of large antenna arrays is one of the defining features of 5G systems. This trend can be expected to continue in B5G systems, where the number of antenna elements will be scaled up by a further order-of-magnitude. Since the relevant theory has been discussed in Chapters 16 and 22, we just list some topics that may be expected to be of particular interest in B5G. First, there is a question of transceiver architecture. At the time of this writing, hybrid beamforming is almost universally used for >6 GHz systems, while for <6 GHz, both hybrid beamforming and fully digital beamforming are employed. It can be anticipated that fully digital beamforming might become more competitive in the higher frequency ranges, due to recent progress in high frequency electronics. Secondly, as the dimensions of MIMO arrays increase beyond the values that are typical at the time of this writing (8×8 wavelengths), effects such as wavefront curvature due to scattering in the near-field of the array, shadowing differences in different parts of the array, and beam squinting due to the nonnegligible run time of the signal across the array, start to become much more pronounced. All of these physical artifacts need to be taken into account in the design and implementation of beamforming architectures and signal processing algorithms at the TX and RX. Spatial modulation (Section 16.2.13) has also drawn interest at very high frequencies, such as those used for Visible Light Communications (VLC).

The concept of *cell-free massive MIMO* has been pointed out as a promising way to realize distributed antenna systems below 6 GHz which can scale to large physical areas. The architecture of 5GNR BSs is well suited for such systems, and the spectral and energy efficiency improvements brought by such systems are now well understood in theory. Future deployments will show to what degree the promised theoretical gains can be retained in practice for realistic scenarios with distances spanning up to hundreds of meters and variations in UE/IO mobility, and what improvements must be implemented in B5G.

Another important development is *Intelligent Reflective Surfaces (IRSs)* which aim to have large physical apertures that are electromagnetically active, see Section 16.2.14. This is at the time of this writing one of the hottest topics in physical-layer research, though challenges remain, such as whether the cost of deployment might reduce the cost advantage compared to active relays, and the performance of channel estimation. Furthermore, novel algorithms for re-calibrating on-the-fly need to be developed, or the IRSs need to be designed a-priori to work without any calibration, i.e., purely based on online pilot tones. In the context of standardization, the control protocols to implement efficient signaling between the BS and the IRS, as well as UEs and the IRS, need to be developed.

Besides conventional spatial multiplexing, which is the fabric of existing multiple antenna systems, *Orbital Angular Momentum (OAM)* (Section 16.2.15) is an alternative spatial multiplexing method that has shown great potential for B5G systems. OAM is especially suitable for LOS propagation, such as in data centers, and for wireless backhaul, as its range is limited due to the fact that the underlying principle for multiplexing best works in the radiating near-field of the antenna. Since OAM performs better with electrically larger antennas, it is better suited for high mm-wave and THz systems, and in particular for free-space optics applications.

35.5.4   Free-Space Optical Communications

Free-space optical communications have great promise for extremely high data rate communications over small-to-medium distances, as long as LOS can be guaranteed. While some operations are possible also in non-LOS (NLOS) situations, the achievable data rates, and required modulation as well as signal processing structures, can be quite different. To this end, more investigations are required into architectures that provide the right complexity-cost-performance trade-off. We can generally distinguish between laser-based and LED-based techniques. The latter (a.k.a. VLC or LiFi) is mostly intended for exploiting LEDs that already exist as lighting source to also transmit information. Furthermore, the optical transmission is intended for the downlink, while the uplink needs to be provided by traditional radio links. This raises interesting challenges in the integration with B5G cellular and Wi-Fi, which need more attention. Furthermore, the adaptation to mobility constitutes an important challenge.

Laser-based systems allow much higher data rates, yet having small beamwidths, they are mainly suitable for fixed wireless scenarios. Furthermore, they are extremely sensitive to blockage of the LOS paths, since no multi-path diversity is available. Modulation and detection methods that are suitable in environments with fast variations of channel conditions also require further investigations.

35.5.5   Backscatter Communication and Wirelessly Powered Communications

Communication from battery-powered sensor nodes to a BS or other collection point faces as its major challenge the lifetime of the battery in the sensor. While deep sleep modes can reduce the overall energy drain, these measures might not be sufficient, in particular, if the admissible latency is not large. Two main avenues that are emerging for B5G communications are backscatter communication and wirelessly powered communications.

Backscatter communications rely on the passive or semi-passive reflection of incident waves by a device. As a fundamental principle, the backscatter tag reflects an incident probing signal with an additional modulation of amplitude, phase, or frequency of the signal; this allows the active transceiver to distinguish the backscattered signal from reflections by other objects in the environment. This modulation is typically achieved by the load impedance of the tag antenna, since the signal reflected by the tag antenna depends on the matching of the antenna impedance to the load (compare Section 17.2.3).[3] Modulation of the impedance based on, e.g., sensor input, does require some energy, but it is much less than required for Radio Frequency (RF) TXs. Despite these fundamental advantages, backscatter systems are not yet ready for practical applications in IoT systems [Xu et al. 2018] due to low achievable range, and a lack of security (the simplicity of the device typically does not allow proper encryption and other security measures). Yet significant development effort and associated standardization activities can be expected for the future, due to the compelling energy efficiency and battery lifetime advantages.

For wireless powering, we can generally distinguish two main forms: targeted powering, and opportunistic harvesting. In the former case, RF radiation is sent intentionally toward the device of interest (typically this would be beamformed, in order to enhance the efficiency of the transfer), and the device uses the received energy to charge its battery. The energy transfer can either occur by itself, or jointly with transmission of downlink information, in which case it is often known under the acronym SWIPT (Simultaneous Wireless Information and Power Transfer). On the other hand, energy harvesting can also exploit EM energy in the air that does not stem from the associated BS. This is particularly advantageous when high-power TXs for, e.g., TV broadcasts are in the vicinity, but also cellular BSs and even Wi-Fi access points can be helpful; even powering by solar cells can be viewed as energy harvesting.

35.5.6   Physical-Layer Security and Encryption

Security from eavesdroppers is a key goal for wireless communications. Traditionally, it has been handled by cryptographic methods, since the broadcast nature of the wireless channels enables eavesdropping of the transmitted signals without any special measures such as tapping into a physical phone line. However, recent years have shown that measures can be taken to also create *Physical-Layer Security* (PLS). It can be demonstrated that communication can be achieved with a secrecy rate that depends on the channel condition between the TX and desired RX, and between TX and the eavesdropper. Both theoretical investigations and construction of codes that achieve the secrecy capacity have been performed. Better secrecy can be achieved in massive MIMO systems, since the ability to focus the transmission to the desired RX improves the relative channel quality of the desired channel versus the eavesdropper channel; however, active eavesdropping exists in which the eavesdropper can contaminate the pilot tones and thus channel estimation, so that more energy is directed toward it. All of these developments are currently in the stage of fundamental research, and no specific provisions exist in the standards for applying PLS. However, further research on this topic will be pursued and may well enter B5G standards.

Besides the PLS, also the traditional cryptography needs to be adjusted for B5G. This requirement is mainly driven by the potential arrival of quantum computers that are mature enough to successfully attack the currently used public-key systems that are based on certain number-theoretic problems that are hard to solve, such as factorization into integers. Since quantum computers can solve such algorithms efficiently, completely new ways of encryption will be required, possibly based on quantum communications.

---

[3] In the simplest case, the impedance is a set of delay lines with different impedances, so that the impulse response is a sequence of pulses with different amplitudes; this binary or M-ary pattern is characteristic for the tag, and thus allows to identify and possibly localize it. This structure is used for RFID.

35.5.7 Multiple Access Techniques

Multiple access techniques require a re-think in B5G, especially due to the integration of massive connectivity and extremely low energy applications. Currently used multiple access schemes, such as Carrier Sense Multiple Access (CSMA) (Section 18.4.2) or OFDMA (Section 18.3.3) do not scale well to scenarios where thousands of devices or more aim to access a single BS, but with a low duty cycle. Current work in this regime concentrates on spread-spectrum type approaches, and tends to have low spectral efficiencies. Hence, new structures that allow for better scaling and possibly further reduce latency need to be studied and standardized.

Another direction of future research is improvement of multiple access in the traditional high spectral efficiency approaches. Here, *Nonorthogonal Multiple Access, NOMA* (Chapter 28) was originally intended to be part of 5G systems, yet was left out of the early releases due to the rush to finish the specifications. Another promising approach is known as *Rate Splitting, RS*, which splits UE messages into common and private parts, and encodes the common parts into one or several common streams, while encoding the private parts into separate streams. The streams are precoded using the available (perfect or imperfect) CSIT, superposed and transmitted. All the RXs then decode the common stream(s), perform Successive Interference Cancellation, and decode their private streams. Each RX reconstructs its original message from the part of its message embedded in the common stream(s) and its intended private stream. The key benefit of RS relative to other techniques is to flexibly manage interference by allowing it to be partially decoded and partially treated as noise. Possibly simplified versions of NOMA or RS will be in contention for B5G systems.

35.5.8 Vehicular Communications

Modern vehicles are equipped with up to 200 sensors, such as video cameras, infrared cameras, automotive radars, light detection, and ranging systems, as well as Global Positioning Systems (GPSs). The sensors and additional devices provide an opportunity to collaborate and share information in order to facilitate accurate and safer automated driving; particularly in congested scenarios. The raw aggregate data rate from the above sensors could be up to 1 Gbps, which is well beyond the capability of the current protocols for connected vehicles. Thus, future research will need to address ways to enhance data rates possibly by going to higher frequency ranges while retaining reliability. On the network side, the current 5G network architecture may not meet the latency needs of reliable autonomous driving until MEC is fully integrated. Related aspects include the processing of the sensor data, including sensor fusion, due to the combination of large amounts of data and tight processing deadlines. The optimal trade-offs between processing at the point of origin, at the BS (if involved), and at the end-point need to be determined, taking into account its relationship with a given level of vehicular traffic density, the amount of available infrastructure, as well as the real-time computational capabilities of involved cars. With high mobility of cars and blockages by intervening vehicles, *beam management* is another aspect that needs more research. In particular, the beam adjustment mechanisms designed for 5G are often too slow in adapting to vehicular scenarios, calling for new methods. For V2I systems, the fast association/disassociation with the various road-side units may require a distributed antenna deployment, and its implications on PHY need to be studied.

35.6 Real-Time Processing and RF Transceiver Design

35.6.1 Implications of Increasing Carrier Bandwidths

The move to extremely large instantaneous bandwidths, started in 5G and accelerated in B5G, has important consequences for transceiver design. For bandwidths that span tens of GHz, building a radio with a single carrier over the entire bandwidth is extremely difficult, especially if one wants to maintain equally high performance and energy efficiency across the band by retaining the linearity of RF front-end circuits. In recognition of this, for 5G systems in case of mm-wave bands, the maximum permissible carrier bandwidth has often been restricted (in the case of 3GPP NR to 400 MHz). Similarly, it might be hardware constraints and not spectrum constraints that limit bandwidth and data rate in THz systems. Higher bandwidths can be obtained by aggregating component carriers. On the downside, the number of carriers that need to be aggregated, and – associated with it – the number of RF transceivers that each handle one carrier – might become impractically large. Furthermore, with such wide bandwidths, the radio performance at the lower end of the band can be expected to be entirely different from the upper end of the band. Thus, the maximum number of carriers, and in turn the maximum operational bandwidth, will be a compromise.

35.6.2 Hardware Challenges for mm-wave and THz Frequency Bands

As already discussed in Section 35.5.3, both hybrid beamforming and fully digital technology will be considered for high frequency ranges. Unlike for cm-wave and mm-wave frequencies, for the THz bands, the phased array processing architecture needs to be redesigned due to the complexities in antenna fabrication, high speed/high power mixed-signal components, RF interconnects, and heat dissipation. The most common type of antenna implementation, namely microstrip patch elements, do not operate efficiently at THz frequencies due to the high dielectric and conductor losses at the RF substrate level. As such, phased arrays fabricated with *nanomaterials*, such as graphene, have been extensively discussed to build miniature plasmonic antennas with dynamic operational modes to reap the benefits of spatial multiplexing and beamforming. On the other hand, metamaterial-based antennas, hypersurfaces, and RF front-end solutions are also emerging as a key technology. To increase the beamforming gain, the concept of *metasurface lenses* was introduced, which acts as a RF power splitting, phase shifting, and power combining network that is applied to the signal radiated from an antenna array.

Using the THz band will impose major challenges on the transceiver hardware design. First and foremost, operating at such high frequencies puts stringent requirements on the semiconductor technology. Even when using state-of-the-art technology, the frequency of operation will approach, or in extreme cases even exceed, the frequency, f_{max}, where the semiconductor is able to successfully provide a power gain. The achievable RX noise figure, as well as TX efficiency, will then be severely degraded compared to operation at lower frequencies. To maximize the high frequency gain, the technology must use scaled-down feature sizes, requiring low supply voltage to achieve reliability, reducing the achievable TX output power. Combined with the degraded RX noise figure, the reduced antenna aperture, and the wide signal bandwidth will naturally result in very short link distances, unless ultra massive number of elements are combined coherently with sharp beamforming. Thousands to tens of thousands of antenna elements may be required for THz BSs. The antennas may be implemented on or off-chip, where on-chip antennas generally have less efficiency, yet they eliminate the loss in chip-to-carrier interfaces. In addition, heat dissipation becomes a major problem. THz transceivers have low efficiency, and the area for heat dissipation will be very small. If heat dissipation becomes too problematic, more sparse arrays may have to be considered, e.g., using compressive sensing-based array thinning principles with more than half wavelength element spacing. However, this would cause side lobes that need to be managed which in turn may pose constraints on spectrum sharing with existing or adjacent services.

To create, e.g., 10,000 transceivers with high level of integration, a silicon-based technology must be used. While silicon Metal Oxide Semiconductor Field Effect Transistor (MOSFET) transistors are predicted to have reached their peak speed, and will actually *degrade* with further scaling, silicon germanium (SiGe) bipolar transistors are predicted to reach an f_{max} of close to 2 THz within a 5 nm device. In such a technology, amplifiers and oscillators up to about 1 THz could be realized with high performance and integration. With today's silicon technology, however, 500 GHz amplifiers and oscillators cannot be realized, and to operate at such frequencies, frequency multiplication in a nonlinear fashion is necessary.

From a real-time processing viewpoint, the major challenge at both mm-wave and THz frequencies is in the dynamic control and management of RF interconnects of the array elements and the associated beamforming networks. While this problem was present in the mm-wave bands, the challenge is elevated even more at THz due to the even shorter channel coherence times (for a fixed Doppler spread), higher phase noise, and higher number of antenna elements. Even with hybrid beamforming, to manage the processing complexity as well as the cost, fully connected architectures which require dedicated phase shifters per-RF signal path may be cost-prohibitive, so that the "array of sub-arrays" principle (see Section 16.2.10) becomes preferable.

Another important challenge is the generation of coherent and low noise Local Oscillator (LO) signals for ten thousand or more transceivers. The generation of a central 500 GHz signal to be distributed to all transceivers on a chip seems impractical, as it would consume very large power in the buffers. As such, a more distributed solution with *local* Phase-Locked Loops (PLLs) is more appealing, since a lower frequency reference can then be distributed over the chip. The phase noise of different PLLs will then be noncorrelated. Regardless of LO architecture, another challenge is frequency tuning of oscillators, since the quality factor of variable reactances (varactors) is inversely proportional to the operating frequency. As such, at THz frequencies, other tuning mechanisms should be investigated, like using resistance for tuning. All of these challenges call for substantial research efforts, since they must be overcome to realize the high data rates that are envisioned for B5G networks.

35.6.3   *Energy Consumption and Efficiency*

As the amount of data to communicate and process is increased by orders of magnitude, energy efficiency becomes critical, especially in battery-powered devices. While large arrays, which reduce the required RF transmit energy, can be more easily implemented at higher frequencies, PA efficiency and UE noise figure will degrade with frequency, counteracting some of the gains of using more directed transmissions. With advances in semiconductor technology, however, like scaled SiGe bipolar technology, the noise figure and power efficiency are predicted to become attractive even at THz frequencies. The power consumption of a large array transceiver may still be high, due to the many transceiver up/down-conversion channels, but the data rate can be extremely high and the energy per bit is expected to drop by orders-of-magnitude compared to existing cellular systems. The bottleneck for energy efficiency may then become processing the data. Thus, technology scaling needs to be supplemented with significant and coordinated advances at all levels of abstraction. Both voltage scaling, which has been extensively used over the past decades, and parallelism, are reaching saturation points in terms of energy gains. Looking ahead toward the next decade, novel design dimensions will be needed to reduce energy consumption.

35.7   Use of Machine Learning

Machine Learning (ML) and Artificial Intelligence (AI) have been topics of intense research since the 1950s. However, it has only been since the mid-2010s that the computational power required for implementing deep neural networks, combined with large sets of labeled (training) data, as well as some new theoretical breakthroughs, has turned ML from an interesting research topic to a widely used tool for solving complicated real-world problems. Initially used in image processing and language processing such as automated translation, ML is now envisioned to be used in a staggering variety of applications. Among those, many aspects of wireless communications can benefit from ML, though great care must be taken to apply ML with a judicious combination with expert knowledge, to select suitable methods (e.g., supervised learning vs re-enforcement learning) depending on the application, properly select training and test data, etc. It also must be kept in mind that ML is best suited for highly complex problems with no known analytical solution. For example, NP-hard problems are well-suited for ML, because exact solutions cannot be obtained within polynomial time; and if heuristic solutions need to be obtained anyway, then using ML is often advantageous. To provide another example: demodulation of signals in a highly nonlinear channel can be done advantageously with ML; demodulation in an Additive White Gaussian Noise (AWGN) channel where exact solutions can be computed easily is usually better handled by those classical methods.

At the physical layer, ML has been used for beamforming and beamtracking, in particular when channel information in different bands is available, or other additional information such as GPS location or video of the surroundings is available. Related to this, channel prediction and channel extrapolation in the frequency domain can be handled by ML. Furthermore, ML can be used advantageously for localization purposes, where – in real environments – it usually outperforms classical localization algorithms. There have also been suggestions to use ML for an end-to-end design where there is no explicit design of waveform, signal constellation, pilot tones, or codes, and the system adapts all these parameters to the environment based on ML. However, such an approach does not seem to be well suited or too complex for the typical dynamically changing multi-user environments at least in the near future.

At the Medium Access Control (MAC) layer, the problems of traffic prediction, scheduling, resource allocation, and load balancing can all be handled by ML. The exact formulations of these problems are usually mixed-integer optimization problems, making exact solutions NP hard. Furthermore, the combination of these problems with physical-layer issues such as antenna selection, BS selection, and precoding is even harder, and again can be solved advantageously with ML.

ML will also be used extensively in deployment planning. Ranging from the placement of the BSs to operating frequencies to the parameter settings of the BSs and the backhaul capacities, the number of parameters in B5G is very large, and the service-centric or data-centric architecture, where multiple services compete for the resources at the BSs, increases the numbers even further. The optimization of all of these parameters cannot be done by hand anymore, so that ML techniques will likely be used for the planning and updating. Furthermore, ML can be used for dynamic spectrum usage, as discussed in Chapter 26.

ML can also be used for further cross-layer optimization by incorporating context awareness. If the system knows the purpose of a communication, e.g., providing video links for surveillance or data for robotic control, the communication parameters can be adjusted accordingly. Such context awareness beyond the usual QoS classification for prioritization (compare Chapter 20) can improve the overall efficiency of the network.

A major challenge for ML in wireless will be the acquisition of suitable training data, in particular in light of privacy concerns. Firstly, different network operators (be it operators of cellular or Wi-Fi networks) might not be willing to share data with each other, complicating the study of interference and spectrum management. Secondly, consumers might not be willing to share their data such as location data with the network operator. A possible way out of these dilemmas is offered by federated learning, where not actual data, but just learned models, are shared. More investigations into the effectiveness and privacy protections in the wireless context will be required.

The ML techniques used for wireless systems span the gamut from deep neural networks to auto-encoders to Q-learning. However, a discussion of those is beyond the scope of this book, and we refer the reader to the "Further Reading" section at the end of this chapter.

35.8 A Final Word on New Technologies

When looking to the future of technological developments, it is helpful to keep in mind the famous *hype cycle* for technology (note, however, that there is no prediction of the duration of the different phases):

- *Technology trigger*: early breakthroughs, often coming from academia, fuel interest in a technology. First proof-of-concepts trigger interest from media and technology funders. At this stage, practical usefulness and implementability are often unknown.
- *Peak of inflated expectations*: based on early success stories, a positive narrative takes hold in the media, often wildly exaggerating the potential of the technology and staying silent about obstacles and implementation issues. Phrases like "this will solve all problems in wireless," or "this will change life as we know it" are the hallmark of inflated expectations.
- *Trough of disillusionment*: new proofs-of-concepts and first products naturally cannot meet the inflated expectations. Interest quickly vanes, and work on the technology is dropped – both in small companies that concentrated on this topic (and often dissolve), and large companies, which simply change direction.
- *Slope of enlightenment*: based on more realistic expectations, and with the emergence of more mature products, the possibilities and limitations of the technology are better understood. However, it is noteworthy that not all technologies reach this stage. Many simply disappear.
- *Plateau of productivity*: mainstream adoption starts to take off.

It is noteworthy that a similar cycle exists even in academic research. "Hot" topics might draw scores of researchers that provide progress within a very short time. Funding from industry and research agencies follows almost inevitably, drawing even more attention. Yet at the end of the development, some of the explored topics prove to be not practically useful or not fulfilling the initial expectations, thus mostly vanishing from the picture. Notably, some of those topics can be revived again decades later, if it was technological (not fundamental) obstacles that had previously constituted the roadblocks, and general progress removes those obstacles. A typical case for this is neural networks, which was a very popular topic in the wireless community in the 1990s, before vanishing in the 2000s and re-emerging in the 2010s when more powerful computers made them practically useful. Other technologies, like

Low-Density Parity Check (LDPC) coding or hybrid beamforming might receive little attention at introduction, and languish for many years with scarce attention before sudden widespread adoption – e.g., LDPC coding was referenced 300 times between 1962 and 2000, and 13,000 times between 2000 and 2020. However, there are also other techniques such as space-time coding where the progression from initial discovery to hot research topic to widely used adoption in products has been continuous and taken only a few years.

Summarizing, it is impossible to say with certainty which of the more futuristic technologies discussed in this chapter will ultimately be successful as mainstream product, which will become niche products, and which will fall by the wayside. It is, however, a fairly safe bet that at least some technologies discussed in this chapter will be employed in B5G and will ensure that B5G will have significantly enhanced capabilities and will enable many new applications.

Further Reading

Much of my thinking for this chapter was impacted by discussions with my colleagues with whom I co-authored a survey paper [Tataria et al. 2021]; this is reflected in the structure of this chapter which draws on this paper, as well as the use of text parts and paraphrases from this paper (with kind permission of all the other authors). A number of other surveys on B5G have been published in the past years, including [Bariah et al. 2020, Chowdhury et al. 2020, Jiang et al. 2021].

For the specific topics mentioned in this chapter, we mostly refer to either the "further reading" section of the chapters in which those areas were already discussed. Additionally, we mention references for some specific topics not covered in the remainder of the book: A brief tutorial on backscatter communication can be found in [Xu et al. 2018]. Energy harvesting is reviewed in [Ulukus et al. 2015]. Information-centric networking is surveyed in [Fang et al. 2018]; Augmented Information services in [Cai et al. 2022], and Age of Information in [Yates et al. 2021]. Security and privacy for IoT are discussed by [Alhirabi et al. 2021]. LED-based VLC is surveyed in [Karunatilaka et al. 2015]. SWIPT is reviewed in [Perera et al. 2017]. V2X communications are surveyed in [Liu et al. 2020], while UAV-based systems are discussed in [Fotouhi et al. 2019], and arial networks including LEO satellites are discussed in [Dao et al. 2021]. Rate splitting is surveyed in [Mao et al. 2022]. A general discussion of THz communications including hardware aspects is given in [Rappaport et al. 2019], while signal processing techniques for THz communications are discussed in [Sarieddeen et al. 2021]. PLS is surveyed in [Wu et al. 2018]. A general introduction of machine learning for engineers can be found, e.g., in [Simeone 2018], and the specific applications in wireless communications in [Chen et al. 2019].

For updates and errata for this chapter, see https://wides.usc.edu/students.html#textbooks.

Exercises

See Sec. 36.35 of Exercises.pdf at wiley.com/go/molisch/wireless3e

References

3GPP 23.501 3GPP, "System architecture for the 5G System", TS 23.501, V16 (2021).

3GPP 36.201 3GPP, "LTE Physical layer - general description", TS 36.201 (2009).

3GPP 36.211 3GPP, "Physical channels and modulation", TS 36.211 (2009).

3GPP 36.212 3GPP, "Multiplexing and channel coding, TS 36.212 (2009).

3GPP 36.300 3GPP, "Overall description", TS 36.300 V11 (2011).

3GPP 38.101 3GPP, "User Equipment (UE) radio transmission and reception", TS 38.101 V16 (2021).

3GPP 38.300 3GPP, "NR and NG-RAN Overall Description", TS 38.300, V16 (2021).

3GPP 38.401 3GPP, "Architecture description", TS 38.401 V16 (2021).

3GPP 38.901 3GPP, "Study on channel model for frequencies from 0.5 to 100 GHz" TR 38.901 v16.0.1, (2019).

Abdi et al. 2000 A. Abdi, K. Wills, H. A. Barger, M. S. Alouini, and M. Kaveh, "Comparison of the level crossing rate and average fade duration of Rayleigh, Rice and Nakagami fading models with mobile channel data", *Proc. VTC Fall*, 2000, 1850–1857 (2000).

Abramowitz and Stegun 1965 M. Abramowitz and I. A. Stegun, *Handbook of Mathematical Functions*, National Bureau of Standards, Washington (1965).

Abramson 1970 N. Abramson, "The ALOHA system – another alternative for computer communications", *Proceedings of Fall 1970 AFIPS Computer Conference* (1970).

Abrantes 2004 S. A. Abrantes, "From BCJR to turbo decoding: MAP algorithms made easier", https://repositorio-aberto.up.pt/bitstream/10216/19735/2/41921.pdf (2004).

Abu-Ali et al. 2013 N. Abu-Ali, A. E. M. Taha, M. Salah, and H. Hassanein, "Uplink scheduling in LTE and LTE-advanced: Tutorial, survey and evaluation framework", *IEEE Commun. Surv. Tutor.*, 16(3), 1239–1265 (2014).

Acharya and Yates 2007 J. Acharya and R. D. Yates, "A framework for dynamic spectrum sharing between cognitive radios", *IEEE International Conference on Communications (ICC)*, pp. 5166–5171 (2007).

Adachi and Ohno 1991 F. Adachi and K. Ohno, "BER performance of QDPSK with postdetection diversity reception in mobile radio channels", *IEEE Trans. Veh. Technol.*, 40, 237–249 (1991).

Adachi and Parsons 1989 F. Adachi and J. D. Parsons, "Error rate performance of digital FM mobile radio with postdetection diversity", *IEEE Trans. Commun.*, 37, 200–210 (1989).

Adhikary et al. 2013 A. Adhikary, J. Nam, J. Y. Ahn, and G. Caire, "Joint spatial division and multiplexing? The large-scale array regime", *IEEE Trans. Inf. Theory*, 59, 6441–6463 (2013).

Adhikary et al. 2014 A. Adhikary, E. Al Safadi, M. K. Samimi, R. Wang, G. Caire, T. S. Rappaport, and A. F. Molisch, "Joint spatial division and multiplexing for mm-wave channels", *IEEE J. Sel. Areas Commun.*, 32(6), 1239–1255 (2014).

Aditya et al. 2018 S. Aditya, A. F. Molisch, and H. M. Behairy, "A survey on the impact of multipath on wideband time-of-arrival based localization", *Proc. IEEE*, 106(7), 1183–1203 (2018).

Ahlswede et al. 2000 R. Ahlswede, N. Cai, S. R. Li, and R. W. Yeung, "Network information flow", *IEEE Trans. Inf. Theory*, 46, 1204–1216 (2000).

Ahmadi 2013 S. Ahmadi, *LTE-Advanced: A Practical Systems Approach to Understanding 3GPP LTE Releases 10 and 11 Radio Access Technologies*, Academic Press, (2013).

Ahmadi 2019 S. Ahmadi, *5G NR: Architecture, Technology, Implementation, and Operation of 3GPP New Radio Standards*, Academic Press, (2019).

Ahmed 2005 M. H. Ahmed, "Call admission control in wireless networks: A comprehensive survey", *IEEE Commun. Surv. Tutor.*, 7(1), 49–68 (2005).

Akyildiz et al. 2006 I. F. Akyildiz, W. Y. Lee, M. C. Vuran, and S. Mohanty, "Next generation/dynamic spectrum access/cognitive radio wireless networks: A survey", *Comput. Netw.*, 50(13), 2127–2159 (2006).

Akyildiz et al. 2011 I. F. Akyildiz, B. F. Lo, and R. Balakrishnan, "Cooperative spectrum sensing in cognitive radio networks: A survey", *Phys. Commun.*, 4(1), 40–62 (2011).

Alhirabi et al. 2021 N. Alhirabi, O. Rana, and C. Perera, "Security and privacy requirements for the internet of things: A survey", *ACM Transactions on Internet of Things*, 2(1), 1–37 (2021).

Alamouti 1998 S. M. Alamouti, "A simple transmit diversity technique for wireless communications", *IEEE J. Sel. Areas Commun.*, 16, 1451–1458 (1998).

Aliakbari and Lau 2021 H. Aliakbari and B. K. Lau, "Low-band MIMO antenna for smartphones with robust performance to user interaction", *IEEE Antennas Wirel. Propag. Lett.*, 20, 1195–1199 (2021).

Aliakbari et al. 2021 H. Aliakbari, L. Y. Nie, and B. K. Lau, "Large screen enabled tri-port MIMO handset antenna for low LTE bands", *IEEE Open J. Antennas Propag.*, 2, 911–920 (2021).

Almers et al. 2006 P. Almers, F. Tufvesson, and A. F. Molisch, "Keyhole effect in MIMO wireless channels: Measurements and theory", *IEEE Trans. Wirel. Commun.*, 5, 3596–3604 (2006).

Almers et al. 2007 P. Almers, E. Bonek, A. Burr, et al., "Survey of channel and radio propagation models for wireless MIMO systems", *EURASIP J. Wirel. Commun. Netw.*, Article ID 19070, 19 (2007).

Alouini and Goldsmith 1999 M. S. Alouini and A. J. Goldsmith, "Area spectral efficiency of cellular mobile radio systems", *IEEE Trans. Veh. Technol.*, 48, 1047–1066 (1999).

Al-Sultan et al. 2014 S. Al-Sultan, M. M. Al-Doori, A. H. Al-Bayatti, and H. Zedan, "A comprehensive survey on vehicular ad hoc network", *J. Netw. Comput. Appl.*, 37, 380–392 (2014).

Anatasi et al. 2008 G. Anastasi, M. Conti, M. Di Francesco, and A. Passarella, "Energy conservation in wireless sensor networks: A survey", *Ad hoc Netw.*, 7 (3), 537–568 (2008).

Andersen 1991 J. B. Andersen, "Propagation parameters and bit errors for a fading channel", *Proceedings of Commsphere'91*, paper 8.1 (1991).

Andersen 1997 J. B. Andersen, "UTD multiple-edge transition zone diffraction", *IEEE Trans. Antennas Propag.*, 45, 1093–1097 (1997).

Andersen 2000 J. B. Andersen, "Antenna arrays in mobile communications: Gain, diversity, and channel capacity", *IEEE Antennas Propag. Mag.*, 42, 12–16 (2000).

Andersen 2002 J. B. Andersen, *Power Distributions Revisited*, COST 273 TD(02)004 (2002).

Andersen and Hansen 1977 J. B. Andersen and F. Hansen, "Antennas for VHF/UHF personal radio: A theoretical and experimental study of characteristics and performance", *IEEE Trans. Veh. Technol.*, VT-26, 349–357 (1977).

Andersen et al. 1990 J. B. Andersen, S. L. Lauritzen, and C. Thommesen, "Distribution of phase derivatives in mobile communications", *Proc. IEEE, Part H*, 137, 197–204 (1990).

Andersen et al. 1995 J. B. Andersen, T. S. Rappaport, and S. Yoshida, "Propagation measurements and models for wireless communications channels", *IEEE Commun. Mag.*, 33(1), 42–49 (1995).

Anderson 2003 H. R. Anderson, *Fixed Broadband Wireless System Design*, Wiley, (2003).

Anderson 2005 J. B. Anderson, *Digital Transmission Engineering*, 2nd edition, Prentice Hall, (2005).

Anderson et al. 1986 J. B. Anderson, T. Aulin, and C. E. Sundberg, *Digital Phase Modulation*, Plenum, (1986).

Andrews 2013 J. G. Andrews, "Seven ways that HetNets are a cellular paradigm shift", *IEEE Commun. Mag.*, 51(3), 136–144 (2013).

Andrews et al. 2001 M. R. Andrews, P. P. Mitra, and R. de Carvalho, "Tripling the capacity of wireless communications using electromagnetic polarization", *Nature*, 409, 316–318 (2001).

Andrews et al. 2007 J. G. Andrews, A. Ghosh, and R. Muhamed, *Fundamentals of WiMAX: Understanding Broadband Wireless Networking*, Prentice Hall, (2007).

Andrews et al. 2012 J. G. Andrews, H. Claussen, M. Dohler, S. Rangan, and M. C. Reed, "Femtocells: Past, present, and future", *IEEE J. Sel. Areas Commun.*, 30(3), 497–508 (2012).

Andrews et al. 2016 J. G. Andrews, A. K. Gupta, and H. S. Dhillon, *A primer on cellular network analysis using stochastic geometry*, arXiv preprint arXiv:1604.03183. (2016).

Annamalai et al. 2000 A. Annamalai, C. Tellambura, and V. K. Bhargava, "A general method for calculating error probabilities over fading channels", *Proceedings of the International Conference in Communication 2000*, pp. 36–40 (2000).

Annavajjala et al. 2007 R. Annavajjala, P. C. Cosman, and L. B. Milstein, "Statistical channel knowledge-based optimum power allocation for relaying protocols in the high SNR regime", *IEEE J. Sel. Areas Commun.*, 25, 292–305 (2007).

Arai et al. 1988 Y. Arai, T. Agui, and M. Nakajima, "A fast DCT-SQ scheme for images", *Trans. IEICE*, E71, 1095 (1988).

Arikan 2009 E. Arikan, "Channel polarization: A method for constructing capacity-achieving codes for symmetric binary-input memoryless channels", *IEEE Trans. Inf. Theory*, 55, 3051–3073 (2009).

Ariyavisitakul 2000 S. L. Ariyavisitakul, "Turbo space-time processing to improve wireless channel capacity", *IEEE Trans. Commun.*, 48, 1347–1359 (2000).

Asadi et al. 2014 A. Asadi, Q. Wang, and V. Mancuso, "A survey on device-to-device communication in cellular networks", *IEEE Commun. Surv. Tutor.*, 16(4), 1801–1819 (2014).

Asplund et al. 2006 H. Asplund, A. A. Glazunov, A. F. Molisch, K. I. Pedersen, and M. Steinbauer, "The COST 259 directional channel model – II. Macrocells", *IEEE Trans. Wirel. Commun.*, 5, 3434–3450 (2006).

Asplund et al. 2020 H. Asplund, D. Astely, P. von Butovitsch, T. Chapman, M. Frenne, F. Ghasemzadeh, M. Hagstrom, B. Hogan, G. Jongren, J. Karlsson, and F. Kronestedt, *Advanced Antenna Systems for 5G Network Deployments: Bridging the Gap Between Theory and Practice*, Academic Press, (2020).

Atal and Remde 1982 B. Atal and J. Remde, "A new model of LPC excitation for producing natural-sounding speech at low bit rates", *Proceedings of the IEEE International Conference on Acoustics, Speech, and Signal Processing, ICASSP'82'*, Paris, pp. 614–617 (1982).

Atal and Schroeder 1984 B. S. Atal and M. R. Schroeder, "Stochastic coding of speech signals at very low bit rates", *Proceedings of the IEEE International Conference in Communication ICC'84'*, Amsterdam (The Netherlands), pp. 1610–1613 (1984).

Ayadi et al. 2002 J. Ayadi, A. A. Hutter, and J. Farserotu, "On the multiple input multiple output capacity of Rician channels", *Proceedings of the International Symposium on Wireless Personal Multimedia Communications 2002*, pp. 402–406 (2002).

Badsberg et al. 1995 M. Badsberg, J. Bach Andersen, and P. Mogensen, *Exploitation of the Terrain Profile in the Hata Model*, COST 231 TD(95)9 (1995).

Bae et al. 2019 J. Bae, A. Abotabl, H. Lin, K. Song, and J. Lee, "An overview of channel coding for 5G NR cellular communications", *APSIPA Trans. Signal Inf. Process.*, 8, e17 (2019).

Bahai et al. 2004 A. R. S. Bahai, B. R. Saltzberg, and M. Ergen, *Multi-Carrier Digital Communications: Theory and Applications of OFDM*, 2nd edition, Springer, (2004).

Bahl et al. 1974 L. Bahl, J. Cocke, F. Jelinek, and J. Raviv, "Optimal decoding of linear codes for minimizing symbol error rate (corresp.)", *IEEE Trans. Inf. Theory*, 20, 284–287 (1974).

Balanis 2005 C. A. Balanis, *Antenna Theory: Analysis and Design*, 3rd edition, Wiley, (2005).

Barclay 2002 L. W. Barclay, *Propagation of Radiowaves*, 2nd edition, IET Press, (2002).

Bariah et al. 2020 L. Bariah, L. Mohjazi, S. Muhaidat, P. C. Sofotasios, G. K. Kurt, H. Yanikomeroglu, and O. A. Dobre, "A prospective look: Key enabling technologies, applications and open research topics in 6G networks", *IEEE Access*, 8, 174792–174820 (2020).

Barry et al. 2003 J. R. Barry, D. G. Messerschmidt, and E. A. Lee, *Digital Communications*, 3rd edition, Kluwer, (2003).

Bartoletti et al. 2015 S. Bartoletti, W. Dai, A. Conti, and M. Z. Win, "A mathematical model for wideband ranging", *IEEE J. Sel. Top. Signal Process.*, 9, 216–228 (2015).

Bas et al. 2019 C. U. Bas, R. Wang, S. Sangodoyin, T. Choi, S. Hur, K. Whang, J. Park, C. J. Zhang, and A. F. Molisch, "Outdoor to indoor propagation channel measurements at 28 GHz", *IEEE Trans. Wirel. Commun.*, 18, 1477–1489 (2019).

Bass and Fuks 1979 F. G. Bass and I. M. Fuks, *Wave Scattering from Statistically Rough Surfaces*, Pergamon, (1979).

Bates 2008 R. J. B. Bates, *GPRS: General Packet Radio Service*, McGraw Hill, (2008).

Beckman and Lindmark 2007 C. Beckman and B. Lindmark, "The evolution of base station antennas for mobile communications", *2007 International Conference on Electromagnetics in Advanced Applications*, pp. 85–92 (2007).

van de Beek et al. 1995 J.-J. van de Beek, O. Edfors, M. Sandell, S. K. Wilson, and P. O. Börjesson, "On channel estimation in OFDM systems", *Proceedings of the IEEE Vehicle Technology Conference*, Vol. 2, pp. 815–819, Chicago, IL, July (1995).

van de Beek et al. 1999 J.-J. van de Beek, P. O. Borjesson, M.-L. Boucheret, et al., "A time and frequency synchronization scheme for multiuser OFDM", *IEEE J. Sel. Areas Commun.*, 17, 1900–1914 (1999).

Belfiore and Park 1977 C. A. Belfiore and J. H. Park, "Decision feedback equalization", *Proc. IEEE*, 67, 1143–1156 (1977).

Bello 1963 P. A. Bello, "Characterization of randomly time-variant linear channels", *IEEE Trans. Commun.*, 11, 360–393 (1963).

Bello and Nelin 1963 P. Bello and B. D. Nelin, "The effect of frequency selective fading on the binary error probabilities of incoherent and differentially coherent matched filter receivers", *IEEE Trans. Commun.*, 11, 170–186 (1963).

Benedetto and Biglieri 1999 S. Benedetto and E. Biglieri, *Principles of Digital Transmission: With Wireless Applications*, Kluwer, (1999).

di Benedetto et al. 2006 M. G. di Benedetto, T. Kaiser, A. F. Molisch, I. Oppermann, C. Politano, and D. Porcino (eds), *UWB Communication Systems – A Comprehensive Overview*, Hindawi Publishing, (2006).

Bengtsson and Ottersten 2001 M. Bengtsson and B. Ottersten, "Optimal and suboptimal transmit beamforming", in L. C. Godara (ed.), *Handbook of Antennas in Wireless Communications*, CRC Press, (2001).

Benjebbour et al. 2015 A. Benjebbour, K. Saito, A. Li, Y. Kishiyama, and T. Nakamura, "Non-orthogonal multiple access (NOMA): Concept, performance evaluation and experimental trials", *International Conference on Wireless Networks and Mobile Communications (WINCOM)*, pp. 1–6 (2015).

Benjebbour et al. 2016 A. Benjebbour, K. Saito, A. Li, Y. Kishiyama, and T. Nakamura, "Non-orthogonal multiple access (NOMA): Concept and design", in F. L. Luo, C. J. Zhang (eds), *Signal Processing for 5G: Algorithms and Implementations*, John Wiley & Sons, (2016).

Benvenuto et al. 2010 N. Benvenuto, R. Dinis, D. Falconer, and S. Tomasin, "Single carrier modulation with nonlinear frequency domain equalization: An idea whose time has come – again", *Proc. IEEE*, 98, 69–96 (2010).

Benyamina et al. 2011 D. Benyamina, A. Hafid, and M. Gendreau, "Wireless mesh networks design: A survey", *IEEE Commun. Surv. Tutor.*, 14(2), 299–310 (2011).

Berg 1995 J. E. Berg, "A recursive method for street microcell path loss calculations", *In Proceedings of 6th International Symposium on Personal, Indoor and Mobile Radio Communications, IEEE*, Vol. 1, pp. 140–143 (1995).

Berger 1971 T. Berger, *Rate Distortion Theory: A Mathematical Basis for Data Compression*, Prentice Hall, Englewood Cliffs (1971).

Bergljung 1994 C. Bergljung, *Diffraction of Electromagnetic Waves by Dielectric Wedges*, Ph.D. thesis, Lund Institute of Technology, Lund, Sweden (1994).

Berrou and Glavieux 1996 C. Berrou and A. Glavieux, "Near optimum error correcting coding and decoding: Turbo-codes", *IEEE Trans. Commun.*, 44(10), 1261–1271 (1996).

Berrou et al. 1993 C. Berrou, A. Glavieux, and P. Thitimajshima, "Near Shannon limit error-correcting coding and decoding: Turbo-codes", *Proceedings of the IEEE International Conference on Communications, ICC'93* (1993).

Bertoni 2000 H. L. Bertoni, *Radio Propagation for Modern Wireless Systems*, Prentice Hall, (2000).

Best 2007 R. Best, *Phase Locked Loops*, 6th edition, McGraw Hill, (2007).

Bettstetter et al. 1999 C. Bettstetter, H. J. Voegel, and J. Eberspecher, "GSM phase 2+ general packet radio service GPRS: Architecture, protocols, and air interface", *IEEE Commun. Surv., Third Quarter 1999*, 2(3), 2–14 (1999).

Bi et al. 2001 Q. Bi, G. L. Zysman, and H. Menkes, "Wireless mobile communications at the start of the 21st century", *IEEE Commun. Mag.*, 39(1), 110–116 (2001).

Biglieri 1991 E. Biglieri, *Introduction to Trellis-Coded Modulation with Applications*, MacMillan, New York (1991).

Biglieri et al. 2007 E. Biglieri, R. Calderbank, A. Constantinides, A. Goldsmith, A. Paulraj, and H. V. Poor, *MIMO Wireless Communications*, Cambridge University Press, (2007).

Biglieri et al. 2013 E. Biglieri, A. J. Goldsmith, L. J. Greenstein, H. V. Poor, and N. B. Mandayam, *Principles of cognitive radio*, Cambridge University Press (2013).

Björnson et al. 2014 E. Björnson, M. Bengtsson, and B. Ottersten, "Optimal multiuser transmit beamforming: A difficult problem with a simple solution structure", *IEEE Signal Process. Mag.*, 31(4), 142–148 (2014).

Björnson et al. 2016 E. Björnson, E. G. Larsson, and T. L. Marzetta, "Massive MIMO: Ten myths and one critical question", *IEEE Commun. Mag.*, 54(2), 114–123 (2016).

Björnson et al. 2017 E. Björnson, J. Hoydis, and L. Sanguinetti, "Massive MIMO networks: Spectral, energy, and hardware efficiency", *Found. Trends Signal Process.*, 11(3–4), 154–655 (2017).

Bkassiny et al. 2012 M. Bkassiny, Y. Li, and S. K. Jayaweera, "A survey on machine-learning techniques in cognitive radios", *IEEE Commun. Surv. Tutor.*, 15(3), 1136–1159 (2012).

Blair et al. 2008 A. Blair, T. Brown, K. M. Chugg, T. R. Halford, and M. Johnson, "Barrage relay networks for cooperative transport in tactical MANETs," *MILCOM 2008-2008 IEEE Military Communications Conference*, pp. 1–7 (2008).

Blanz and Jung 1998 J. J. Blanz and P. Jung, "A flexibly configurable spatial model for mobile radio channels", *IEEE Trans. Commun.*, 46, 367–371 (1998).

Blaunstein 1999 N. Blaunstein, *Radio Propagation in Cellular Networks*, Artech House, (1999).

Bletsas et al. 2006 A. Bletsas, A. Khisti, D. P. Reed, and A. Lippman, "A simple cooperative diversity method based on network path selection", *IEEE J. Sel. Areas Commun.*, 24, 659–672 (2006).

Bletsas et al. 2007 A. Bletsas, H. Shin, and M. Z. Win, "Cooperative communications with outage-optimal opportunistic relaying", *IEEE Trans. Wirel. Commun.*, 6, 3450–3460 (2007).

Bluetooth SIG 2019 Bluetooth SIG, *Bluetooth Core Specifications*, v. 5.2 (2019).

Bohagen et al. 2007 F. Bohagen, P. Orten, and G. E. Oien, "Design of optimal high-rank line-of-sight MIMO channels", *IEEE Trans. Wirel. Commun.*, 6, 1420–1425 (2007).

Bottomley et al. 2000 G. E. Bottomley, T. Ottosson, and Y. P. E. Wang, "A generalized RAKE receiver for interference suppression", *IEEE J. Sel. Areas Commun.*, 18, 1536–1545 (2000).

Boudreau et al. 2009 G. Boudreau, J. Panicker, N. Guo, R. Chang, N. Wang, and S. Vrzic, "Interference coordination and cancellation for 4G networks", *IEEE Commun. Mag.*, 47, 74–81 (2009).

Boukerche et al. 2009 A. Boukerche, M. Z. Ahmad, D. Turgut, and B. Turgut, "A taxonomy of routing protocols for mobile ad-hoc networks", in A. Boukerche (ed.), *Algorithms and Protocols for Wireless and Mobile Ad-hoc Networks*, Wiley, (2009).

Boukerche et al. 2011 A. Boukerche, B. Turgut, N. Aydin, M. Z. Ahmad, L. Bölöni, and D. Turgut, "Routing protocols in ad hoc networks: A survey", *Comput. Netw.*, 55(13), 3032–3080 (2011).

Bowman 1987 J. J. Bowman (ed.), *Electromagnetic and Acoustic Scattering by Simple Shapes*, Hemisphere, New York (1987).

Boyce et al. 2015 J. M. Boyce, Y. Ye, J. Chen, and A. K. Ramasubramonian, "Overview of SHVC: Scalable extensions of the high efficiency video coding standard", *IEEE Trans. Circuits Syst. Video Technol.*, 26, 20–34 (2015).

Boyd and Vandenberghe 2004 S. Boyd and L. Vandenberghe, *Convex Optimization*, Cambridge University Press, (2004).

Braun and Dersch 1991 W. R. Braun and U. Dersch, "A physical mobile radio channel model", *IEEE Trans. Veh. Technol.*, 40, 472–482 (1991).

Bresler et al. 2010 G. Bresler, A. Parekh, and D. N. Tse, "The approximate capacity of the many-to-one and one-to-many Gaussian interference channels", *IEEE Trans. Inf. Theory*, 56, 4566–4592 (2010).

Bross et al. 2021 B. Bross, J. Chen, J. R. Ohm, G. J. Sullivan and Y. K. Wang, "Developments in international video coding standardization after AVC, with an overview of versatile video coding (VVC)", *Proceedings of the IEEE* (2021).

Buehler 1994 H. Buehler, *Estimation of Radio Channel Time Dispersion for Mobile Radio Network Planning, Dissertation*, Technical University Vienna, (1994).

Buehrer et al. 2014 R. M. Buehrer, C. R. Anderson, R. K. Martin, N. Patwari, and M. G. Rabbat, "Introduction to the special issue on non-cooperative localization networks", *IEEE J. Sel. Topics Signal Process.*, 8(1), 2–4 (2014).

Buehrer et al. 2018 R. M. Buehrer, H. Wymeersch, and R. M. Vaghefi, "Collaborative sensor network localization: Algorithms and practical issues", *Proc. IEEE*, 106, 1089–1114 (2018).

Burghal and Molisch 2016 D. Burghal and A. F. Molisch, "Efficient channel state information acquisition for device-to-device networks", *IEEE Trans. Wirel. Commun.*, 15, 965–979 (2016).

Burghal et al. 2020 D. Burghal, A. T. Ravi, V. Rao, A. A. Alghafis, and A. F. Molisch, "A comprehensive survey of machine learning based localization with wireless signals", arXiv preprint arXiv:2012.11171 (2020).

Burr 2001 A. Burr, *Modulation and Coding for Wireless Communications*, Prentice Hall, (2001).

Cadambe and Jafar 2008 V. R. Cadambe and S. A. Jafar, "Interference alignment and degrees of freedom of the K-user interference channel", *IEEE Trans. Inf. Theory*, 54, 3425–3441 (2008).

Cadger et al. 2012 F. Cadger, K. Curran, J. Santos, and S. Moffett, "A survey of geographical routing in wireless ad-hoc networks", *IEEE Commun. Surv. Tutor.*, 15(2), 621–653 (2012).

Cai and Giannakis 2003 X. Cai and G. B. Giannakis, "Bounding performance and suppressing intercarrier interference in wireless mobile OFDM", *IEEE Trans. Commun.*, 51, 2047–2056 (2003).

Cai and Goodman 1997 J. Cai and D. J. Goodman, "General packet radio service in GSM", *IEEE Commun. Mag.*, 35(10), 122–131 (1997).

Cai et al. 2022 Y. Cai, J. Llorca, A. Tulino, and A. F. Molisch, "Compute-and Data-Intensive Networks: The Key to the Metaverse," arXiv preprint arXiv:2204.02001 (2022).

Caire and Shamai 2003 G. Caire and S. Shamai, "On the achievable throughput of a multiantenna Gaussian broadcast channel", *IEEE Trans. Inf. Theory*, 49, 1691–1706 (2003).

Calcev et al. 2007 G. Calcev, D. Chizhik, B. Goransson, et al., "A wideband spatial channel model for system-wide simulations", *IEEE Trans. Veh. Technol.*, 56, 389–403 (2007).

Cassioli et al. 2002 D. Cassioli, M. A. Win, and A. F. Molisch, " "The ultra-wide bandwidth indoor channel: from statistical model to simulations," *IEEE J. Selected Areas Comm.* 20, 1247–1257 (2002).

Cardieri and Rappaport 2001 P. Cardieri and T. Rappaport, "Statistical analysis of co-channel interference in wireless communications systems", *Wirel. Commun. Mobile Comput.*, 1, 111–121 (2001).

Castaneda et al. 2016 E. Castaneda, A. Silva, A. Gameiro, and M. Kountouris, "An overview on resource allocation techniques for multi-user MIMO systems", *IEEE Commun. Surv. Tutor.*, 19(1), 239–284 (2016).

Catreux et al. 2001 S. Catreux, P. F. Driessen, and L. J. Greenstein, "Attainable throughput of an interferencelimited multiple-input multiple-output cellular system", *IEEE Trans. Commun.*, 48, 1307–1311 (2001).

Chan 1992 G. K. Chan, "Effects of sectorization on the spectrum efficiency of cellular radio systems", *IEEE Trans. Veh. Technol.*, 41, 217–225 (1992).

Chang 1966 R. W. Chang, "Synthesis of band-limited orthogonal signals for multichannel data transmission", *Bell Syst. Techn. J.*, 45, 1775–1796 (1966).

Chawla and Ha 2007 V. Chawla and D. S. Ha, "An overview of passive RFID", *IEEE Commun. Mag.*, 45(9), 11–17 (2007).

Checko et al. 2014 A. Checko, H. L. Christiansen, Y. Yan, L. Scolari, G. Kardaras, M. S. Berger, and L. Dittmann, "Cloud RAN for mobile networks: A technology overview", *IEEE Commun. Surv. Tutor.*, 17(1), 405–426 (2014).

Chen and Chuang 1998 Y. Chen and J. C. I. Chuang, "The effects of time-delay spread on unequalized TCM in a portable radio environment", *IEEE Trans. Veh. Technol.*, 46, 375–380 (1998).

Chen and Luk 2009 C. N. Chen and K. M. Luk, *Antennas for Base Stations in Wireless Communications*, McGraw Hill, (2009).

Chen et al. 2005 J. Chen, L. Jia, X. Liu, G. Noubir, and R. Sundaram, "Minimum energy accumulative routing in wireless networks", *IEEE INFOCOM*, 2005, 1875–1886 (2005).

Chen et al. 2012a C. S. Chen, V. M. Nguyen, and L. Thomas, "On small cell network deployment: A comparative study of random and grid topologies", *IEEE Vehicular Technology Conference*, pp. 1–5 (2012).

Chen et al. 2012b X. Chen, Z. Cao, F. Harris, and B. Rao, "OFDM carrier frequency offset correction based on Type-2 control loop," *2012 IEEE International Conference on Acoustics, Speech and Signal Processing (ICASSP)*, pp. 3161–3164 (2012).

Chen et al. 2016 B. Chen, J. Chen, Y. Gao, and J. Zhang, "Coexistence of LTE-LAA and Wi-Fi on 5 GHz with corresponding deployment scenarios: A survey", *IEEE Commun. Surv. Tutor.*, 19(1), 7–32 (2016).

Chen et al. 2019 M. Chen, U. Challita, W. Saad, C. Yin, and M. Debbah, "Artificial neural networks-based machine learning for wireless networks: A tutorial", *IEEE Commun. Surv. Tutor.*, 21(4), 3039–3071 (2019).

Chen et al. 2021 W. Chen, P. Gaal, J. Montojo, and H. Zisimopoulos, *Fundamentals of 5G Communications: Connectivity for Enhanced Mobile Broadband and Beyond*, McGraw Hill Professional, (2021).

Chennakeshu and Saulnier 1993 S. Chennakeshu and G. J. Saulnier, "Differential detection of Pi/4-shifted DQPSK for digital cellular radio", *IEEE Trans. Veh. Technol.*, 42, 46–57 (1993).

Chiang 2005 M. Chiang, "Balancing transport and physical layers in wireless multihop networks: Jointly optimal congestion control and power control", *IEEE J. Sel. Areas Commun.*, 23, 104–116 (2005).

Chiang et al. 2008 M. Chiang, P. Hande, T. Lan, and C. W. Tan, "Power control in wireless cellular networks", *Found. Trends Netw.*, 2(4), 381–533 (2008).

Chiu et al. 2020 C. Y. Chiu, S. Shen, B. K. Lau, and R. Murch, "The design of a trimodal broadside antenna element for compact massive MIMO arrays: Utilizing the theory of characteristic modes", *IEEE Antennas Propag. Mag.*, 62(6), 46–61 (2020).

Chizhik et al. 2002 D. Chizhik, G. J. Foschini, M. J. Gans, and R. A. Valenzuela, "Keyholes, correlations, and capacities of multielement transmit and receive antennas", *IEEE Trans. Wirel. Commun.*, 1, 361–368 (2002).

Chizhik et al. 2021 D. Chizhik, J. Du, and R. A. Valenzuela, "Universal Path Gain Laws for Common Wireless Communication Environments", *IEEE Transactions on Antennas and Propagation* 70, 2928–2941 (2022).

Choi et al. 2001 Y.-S. Choi, P. J. Voltz, and F. Cassara, "On channel estimation and detection for multicarrier signals in fast and frequency selective Rayleigh fading channel", *IEEE Trans. Commun.*, 49, 1375–1387 (2001).

Choi et al. 2017 Y. I. Choi, J. W. Lee, M. Rim, and C. G. Kang, "On the performance of beam division nonorthogonal multiple access for FDD-based large-scale multi-user MIMO systems", *IEEE Trans. Wirel. Commun.*, 16, 5077–5089 (2017).

Choi et al. 2021 T. Choi, J. Gomez-Ponce, C. Bullard, et al., "Using a drone sounder to measure channels for cell-free massive MIMO systems", IEEE WCNC 2022 (2022).

Chollet et al. 2005 G. Chollet, A. Esposito, M. Faundez-Zanuy, and M. Marinaro, *Nonlinear Speech Modeling and Applications*, Vol. 3445, *Lecture Notes in Computer Science*, Springer-Verlag, Berlin, New York (2005).

Chowdhury et al. 2020 M. Z. Chowdhury, M. Shahjalal, S. Ahmed, and Y. M. Jang, "6G wireless communication systems: Applications, requirements, technologies, challenges, and research directions", *IEEE Open J. Commun. Soc.*, 1, 957–975 (2020).

Chrisanthopoulou and Tsoukatos 2007 M. P. Chrisanthopoulou and K. P. Tsoukatos, "Joint beamforming and power control for CDMA uplink throughput maximization", *18th IEEE International Symposium on Personal, Indoor and Mobile Radio Communication*, Athens (2007).

Chuah et al. 2002 C. N. Chuah, D. Tse, J. M. Kahn, and R. Valenzuela, "Capacity scaling in MIMO wireless systems under correlated fading", *IEEE Trans. Inform. Theory*, 48, 637–650 (2002).

Chuang 1987 J. Chuang, "The effects of time delay spread on portable radio communications channels with digital modulation", *IEEE J. Sel. Areas Commun.*, 5, 879–888 (1987).

Chung and Goldsmith 2001 S. T. Chung and A. J. Goldsmith, "Degrees of freedom in adaptive modulation: A unified view", *IEEE Trans. Commun.*, 49, 1561–1571 (2001).

Cimini 1985 L. J. Cimini, "Analysis and simulation of a digital mobile channel using orthogonal frequency division multiplexing", *IEEE Trans. Commun.*, 33, 665–675 (1985).

Cisco Systems 2019 Cisco Systems, "Cisco visual networking index: Forecast and trends, 2017–2022", *Cisco Systems*, White Paper (2019).

Clarke 1968 R. Clarke, "A statistical theory of mobile radio reception", *Bell Syst. Techn. J.*, 47, 957–1000 (1968).

Clark 1998 M. V. Clark, "Adaptive frequency-domain equalization and diversity combining for broadband wireless communications", *IEEE J. Sel. Areas Commun.*, 16, 1385–1395 (1998).

Clerckx and Oestges 2013 B. Clerckx and C. Oestges, *MIMO Wireless Networks: Channels, Techniques and Standards for Multi-Antenna, Multi-User and Multi-Cell Systems*, Academic Press, (2013).

Collin 1985 R. E. Collin, *Antennas and Radiowave Propagation*, McGraw Hill, (1985).

Collin 1991 R. E. Collin, *Field Theory of Guided Waves*, IEEE Press, Piscataway (1991).

Conti et al. 2012 A. Conti, M. Guerra, D. Dardari, N. Decarli and M. Z. Win, "Network Experimentation for Cooperative Localization," *IEEE J. Selected Areas Comm.* 30, 467–475 (2012).

Conti and Giordano 2014 M. Conti and S. Giordano, "Mobile ad hoc networking: Milestones, challenges, and new research directions", *IEEE Commun. Mag.*, 52(1), 85–96 (2014).

Conti et al. 2019 A. Conti, S. Mazuelas, S. Bartoletti, W. C. Lindsey, and M. Z. Win, "Soft information for localization-of-things", *Proc. IEEE*, 107(11), 2240–2264 (2019).

Coursey 1999 C. C. Coursey, *Understanding Digital PCS: The TDMA Standard*, Artech, (1999).

Cover and El Gamal 1979 T. Cover and A. El Gamal, "Capacity theorems for the relay channel", *IEEE Trans. Inform. Theory*, 25, 572–584 (1979).

Cover and Thomas 2006 T. M. Cover and J. A. Thomas, *Elements of Information Theory*, 2nd edition, John Wiley & Sons, New York (2006).

Cox 1972 D. C. Cox, "Delay-doppler characteristics of multipath propagation at 910 MHz in a suburban mobile radio environment", *IEEE Trans. Antennas Propag.*, 20, 625–635 (1972).

Cramer et al. 2002 R. J. Cramer, R. A. Scholtz, and M. Z. Win, "Evaluation of an ultra-wide-band propagation channel", *IEEE Trans. Antennas Propag.*, 50, 541–550 (2002).

Crohn et al. 1993 I. Crohn, G. Schultes, R. Gahleitner, and E. Bonek, "Irreducible error performance of a digital portable communication system in a controlled time-dispersion indoor channel", *IEEE J. Sel. Areas Commun.*, 11, 1024–1033 (1993).

Cruz and Santhanam 2003 R. L. Cruz and A. V. Santhanam, "Optimal routing, link scheduling and power control in multihop wireless networks," *IEEE Infocom.*, pp. 702–711 (2003).

Cullen et al. 1993 P. J. Cullen, P. C. Fannin, and A. Molina, "Wide-band measurement and analysis techniques for the mobile radio channel", *IEEE Trans. Veh. Technol.*, 42, 589–603 (1993).

Dahlman et al. 2008 E. Dahlman, S. Parkvall, J. Skold, and P. Beming, *3G Evolution: HSPA and LTE for Mobile Broadband*, Academic Press, (2008).

Dahlman et al. 2016 E. Dahlman, S. Parkvall, and J. Skold, *4G, LTE-Advanced Pro and The Road to 5G*, 3rd edition, Academic Press, (2016).

Dahlman et al. 2020 E. Dahlman, P. Stefan, and S. Johan, *5G NR: The Next Generation Wireless Access Technology*, Academic Press, (2020).

Dai et al. 2018 L. Dai, B. Wang, Z. Ding, Z. Wang, S. Chen, and L. Hanzo, "A survey of non-orthogonal multiple access for 5G", *IEEE Commun. Surv. Tutor.*, 20(3), 2294–2323 (2018).

Dam et al. 1999 H. Dam, M. Berg, R. Bormann, et al., "Functional test of adaptive antenna base stations for GSM", *3rd EPMCC European Personal Mobile Communications Conference*, Paris (1999).

Damnjanovic et al. 2011 A. Damnjanovic, J. Montojo, Y. Wei, T. Ji, T. Luo, M. Vajapeyam, T. Yoo, O. Song, and D. Malladi, "A survey on 3GPP heterogeneous networks", *IEEE Wirel. Commun.*, 18(3), 10–21 (2011).

Damosso and Correira 1999 E. Damosso and L. Correira, *Digital Mobile Radio – The View of COST 231'*, Commission of the European Union, Luxemburg (1999).

Dana 1998 P.H. Dana, "Global Positioning System Overview", http://gisweb.massey.ac.nz/topic/webreferencesites/gps/danagps/gps.html (1998).

Dao et al. 2021 N. N. Dao, Q. V. Pham, N. H. Tu, T. T. Thanh, V. N. Q Bao, D. S. Lakew, and S. Cho, "Survey on aerial radio access networks: Toward a comprehensive 6G access infrastructure", *IEEE Communications Surveys & Tutorials*, 23(2), 1193–1225 (2021).

Dardari et al. 2009 D. Dardari, A. Conti, U. Ferner, A. Giorgetti, and M. Z. Win, "Ranging with ultrawide bandwidth signals in multipath environments", *Proc. IEEE*, 97(2), 404–426 (2009).

Davey 1999 M. C. Davey, *Error Correction Using Low-Density Parity Check Codes*, Ph.D. thesis, Cambridge University (1999).

David and Nagaraja 2003 H. A. David and H. N. Nagaraja, *Order Statistics*, Wiley, (2003).

Decawave 2015 Decawave, "UWB Regulations A Summary of Worldwide Telecommunications Regulations governing the use of Ultra-Wideband radio" Application note APR001 (2015).

Deller et al. 2000 J. R. Deller Jr., J. H. Hansen, and J. G. Proakis, *Discrete-Time Processing of Speech Signals*, IEEE Press, New York (2000).

Demir et al. 2021 Ö. T. Demir, E. Björnson, and L. Sanguinetti, "Foundations of user-centric cell-free massive MIMO", *Found. Trends Signal Process.*, 14(3–4), 162–472 (2021).

Deng et al. 2017 D. J. Deng, Y. P. Lin, X. Yang, J. Zhu, Y. B. Li, J. Luo, and K. C. Chen, "IEEE 802.11 ax: Highly efficient WLANs for intelligent information infrastructure", *IEEE Commun. Mag.*, 55(12), 52–59 (2017).

Devroye and Tarokh 2007 N. Devroye and V. Tarokh, "Fundamental limits of cognitive radio networks", in F. H. P. Fitzek, M. Katz (eds), *Cognitive Wireless Networks: Concepts, Methodologies and Vision*, Springer, (2007).

Devroye et al. 2007 N. Devroye, P. Mitran, M. Sharif, S. Ghassemzadeh, and V. Tarokh, "Information theoretic analysis of cognitive radio systems", in V. Bhargava, E. Hossain (eds), *Cognitive Wireless Communications*, Springer, (2007).

Deygout 1966 J. Deygout, "Multiple knife edge diffraction of microwaves", *IEEE Trans. Antennas Propag.*, 14, 480–489 (1966).

Dhillon and Caire 2015 H. S. Dhillon and G. Caire, "Wireless backhaul networks: Capacity bound, scalability analysis and design guidelines", *IEEE Trans. Wirel. Commun.*, 14, 6043–6056 (2015).

Dhillon et al. 2011 H. S. Dhillon, R. K. Ganti, F. Baccelli, and J. G. Andrews, "Modeling and analysis of K-tier downlink heterogeneous cellular networks", *IEEE J. Sel. Areas Commun.* (2011).

Di Renzo et al. 2013 M. Di Renzo, H. Haas, A. Ghrayeb, S. Sugiura, and L. Hanzo, "Spatial modulation for generalized MIMO: Challenges, opportunities, and implementation", *Proc. IEEE*, 102(1), 56–103 (2013).

Dietrich et al. 2001 C. B. Dietrich, K. Dietze, J. R. Nealy, and W. L. Stutzman, "Spatial, polarization, and pattern diversity for wireless handheld terminals", *IEEE Trans. Antennas Propag.*, 49, 1271–1281 (2001).

Diggavi et al. 2004 S. N. Diggavi, N. Al-Dhahir, A. Stamoulis, and A. R. Calderbank, "Great expectations: The value of spatial diversity in wireless networks", *Proc. IEEE*, 92, 219–270 (2004).

Dinan and Jabbari 1998 E. H. Dinan and B. Jabbari, "Spreading codes for direct sequence CDMA and wideband CDMA cellular networks", *IEEE Commun. Mag.*, 36(9), 48–54 (1998).

Ding et al. 2017 J. Ding, R. Xu, Y. Li, P. Hui, and D. Jin, "Measurement-driven modeling for connection density and traffic distribution in large-scale urban mobile networks", *IEEE Trans. Mobile Comput.*, 17, 1105–1118 (2017).

Divsalar and Simon 1990 D. Divsalar and M. K. Simon, "Multiple-symbol differential detection of MPSK", *IEEE Trans. Commun.*, 38, 300–308 (1990).

Dixon 1994 R. C. Dixon, *Spread Spectrum Systems with Commercial Applications*, Wiley, (1994).

Dobkin 2005 D. Dobkin, "Short Wire Antennas: A Simplified Approach; Part I: Scaling Arguments and Part 2: Detailed Estimates.", enigmantic-consulting.com (2005).

Doufexi et al. 2002 A. Doufexi, S. Armour, M. Butler, et al., "A comparison of the HIPERLAN/2 and IEEE 802.11a wireless LAN standards", *IEEE Commun. Mag.*, 40(5), 172–180 (2002).

Driessen and Foschini 1999 P. F. Driessen and G. J. Foschini, "On the capacity formula for multiple input-multiple output wireless channels: A geometric interpretation", *IEEE Int. Conf. Commun.*, 3, 1603–1607 (1999).

Dubois-Ferriere 2006 H. Dubois-Ferriere, *"Anypath Routing"*, Ph.D. thesis, Ecole Polytechnique Federale de Lausanne, Switzerland (2006).

Duel-Hallen et al. 1995 A. Duel-Hallen, J. Holtzman, and Z. Zvonar, "Multiuser detection for CDMA systems", *IEEE Pers. Commun. Mag.*, 2(2), 46–58 (1995).

Durgin 2003 G. Durgin, *Space-Time Wireless Channels*, Cambridge University Press, (2003).

Durgin et al. 2002 G. D. Durgin, T. S. Rappaport, and D. A. De Wolf, "New analytical models and probability density functions for fading in wireless communications", *IEEE Trans. Commun.*, 50, 1005–1015 (2002).

Eberspaecher et al. 2009 J. Eberspaecher, H. J. Voegel, C. Bettstetter, and C. Hartmann, *GSM Architecture, Protocols, and Services*, 3rd edition, Wiley, (2009).

Edfors et al. 1998 O. Edfors, M. Sandell, J. J. van de Beek, S. K. Wilson, and P. O. Borjesson, "OFDM channel estimation by singular value decomposition", *IEEE Trans. Commun.*, 46, 931–939 (1998).

Edfors et al. 2000 O. Edfors, M. Sandell, J. J. van de Beek, S. K. Wilson, and P. O. Borjesson, "Analysis of DFT-based channel estimators for OFDM", *Wirel. Pers. Commun.*, 12, 55–70 (2000).

van Eeckhaute et al. 2017 M. Van Eeckhaute, A. Bourdoux, A. P. De Doncker, and F. Horlin, "Performance of emerging multi-carrier waveforms for 5G asynchronous communications", *EURASIP J. Wirel. Commun. Netw.*, 2017(29), 1–15 (2017).

Eklund et al. 2006 C. Eklund, R. B. Marks, S. Ponnuswamy, K. L. Stanwood, and N. J. M. Van Waes, *Wireless-MAN: Inside the IEEE 802.16 Standard for Wireless Metropolitan Area Networks*, IEEE Press, (2006).

El Ayach et al. 2014 O. El Ayach, S. Rajagopal, S. Abu-Surra, Z. Pi, and R. W. Heath, "Spatially sparse precoding in millimeter wave MIMO systems", *IEEE Trans. Wirel. Commun.*, 13, 1499–1513 (2014).

Elijah et al. 2015 O. Elijah, C. Y. Leow, T. A. Rahman, S. Nunoo, and S. Z. Iliya, "A comprehensive survey of pilot contamination in massive MIMO 5G system", *IEEE Commun. Surv. Tutor.*, 18(2), 905–923 (2015).

Ellingson 2016 S. W. Ellingson, *Radio Systems Engineering*, Cambridge University Press, (2016).

El-Rabbany 2013 A. El-Rabbany, *Introduction to GPS: The Global Positioning System*. Artech House, 2nd edition, (2013).

Elsaadany et al. 2017 M. Elsaadany, A. Ali, and W. Hamouda, "Cellular LTE-A technologies for the future Internet-of-Things: Physical layer features and challenges", *IEEE Commun. Surv. Tutor.*, 19(4), 2544–2572 (2017).

Epstein and Peterson 1953 J. Epstein and D. W. Peterson, "An experimental study of wave propagation at 850 MC", *Proc. IEEE*, 41, 595–611 (1953).

Erätuuli and Bonek 1997 P. Erätuuli and E. Bonek, "Diversity arrangements for internal handset antennas", *8th IEEE International Symposium on Personal, Indoor and Mobile Radio Communications (PIMRC'97)*, Helsinki, Finland, pp. 589–593 (1997).

Erceg et al. 1999 V. Erceg, L. J. Greenstein, S. Y. Tjandra, S. R. Parkoff, A. Gupta, B. Kulic, A. A. Julius, and R. Bianchi, "An empirically based path loss model for wireless channels in suburban environments", *IEEE J. Sel. Areas Commun.*, 17, 1205–1211 (1999).

Erceg et al. 2004 V. Erceg, L. Schumacher, P. Kyritsi, A. F. Molisch et al., *TGn Channel Models*, IEEE document 802.11-03/940r4, mentor.ieee.org, May (2004).

Ertel et al. 1998 R. B. Ertel, P. Cardieri, K. W. Sowerby, T. S. Rappaport, and J. H. Reed, "Overview of spatial channel models for antenna array communication systems", *IEEE Pers. Commun.*, 5(1), 10–22 (1998).

Eryilmaz et al. 2005 A. Eryilmaz, A. R. Srikant, and J. R. Perkins, "Stable scheduling policies for fading wireless channels", *IEEE/ACM Trans. Netw.*, 13, 411–424 (2005).

Esmail Hosseini et al. 2018 S. Esmail Hosseini, S. Saied Shojaeddin, and H. Abiri, "Theoretical investigation of an ultra-low phase noise microwave oscillator based on an IF crystal resonator-amplifier and a microwave photonic frequency transposer", *J. Opt. Soc. Am. B*, 35, 1422–1432 (2018).

Eternad 2004 K. Eternad, *CDMA2000 Evolution: System Concepts and Design Principles*, Wiley, (2004).

ETSI 300 175-1 1992 ETSI 300 175-1, *Radio Equipment and Systems (RES); Digital European Cordless Telecommunications (DECT) Common Interface Part 1: Overview, Part 2: Physical Layer ETSI*, October (1992).

Evans 1990 G. E. Evans, *Antenna Measurement Techniques*, Artech House, (1990).

Fabregas et al. 2008 A. G. Fabregas, A. Martinez, and G. Caire, "Bit-interleaved coded modulation", *Found. Trends Commun. Inf. Theory*, 5, 1–2 (2008).

Failii 1989 E. Failli (ed.), *Digital Land Mobile Radio Communications – COST 207*, European Union, Brussels, Belgium (1989).

Falconer et al. 1995 D. D. Falconer, F. Adachi, and B. Gudmundson, "Time division multiple access methods for wireless personal communications", *IEEE Commun. Mag.*, 33(1), 50–57 (1995).

Falconer et al. 2002 D. Falconer, S. L. Ariyavisitakul, A. Benyamin-Seeyar, and B. Eidson, "Frequency domain equalization for single-carrier broadband wireless systems", *IEEE Commun. Mag.*, 40(4), 58–66 (2002).

Fang et al. 2018 C. Fang, H. Yao, Z. Wang, W. Wu, X, Jin, and F. R. Yu, "A survey of mobile information-centric networking: Research issues and challenges", *IEEE Communications Surveys & Tutorials*, 20(3), 2353–2371 (2018).

Fant 1970 G. Fant, *Acoustic Theory of Speech Production*, Mouton, The Hague (1970).

Farahani 2011 S. Farahani, *ZigBee Wireless Networks and Transceivers*, Newnes, (2011).

Farsakh and Nossek 1998 C. Farsakh and J. A. Nossek, "Spatial covariance based downlink beamforming in an SDMA mobile radio system", *IEEE Trans. Commun.*, 46, 1497–1506 (1998).

FCC 802.11 Federal Communications Commission, 802.11 channel plan (§15.407)

Featherstone and Molkdar 2002 W. Featherstone and D. Molkdar, "Capacity benefits of GPRS coding schemes CS-3 and CS-4", *3G Mobile Communications Technologies*, pp. 287–291 (2002).

Feldbauer et al. 2005 C. Feldbauer, G. Kubin, and W. B. Kleijn, "Anthropomorphic coding of speech and audio: A model inversion approach", *EURASIP J. Appl. Signal Process.*, 9, 1334–1349 (2005).

Felhauer et al. 1993 T. Felhauer, P. W. Baier, W. König, and W. Mohr, "Optimized wideband system for unbiased mobile radio channel sounding with periodic spread spectrum signals", *IEICE Trans. Commun.*, E76-B, 1016–1029 (1993).

Felsen and Marcuvitz 1973 L. B. Felsen and V. Marcuvitz, *Radiation and Scattering of Waves*, Prentice Hall, (1973).

Fernandes et al. 2014 C. Fernandes, E. Lima, and J. Costa, "Dielectric lens antennas", in *Handbook of Antenna Technologies*, Z. N. Chen (ed.), Springer (2014).

Fettweis et al. 2009 G. Fettweis, M. Krondorf, and S. Bittner, "GFDM-generalized frequency division multiplexing", *69th IEEE Vehicular Technology Conference,* pp. 1–4 (2009).

Fitzek and Médard 2020 F. Fitzek and M. Médard, "Network coding for 5G systems", *Tutorial at IEEE ICC* (2020).

Fleury 1990 B. Fleury, *Characterisierung von Mobil- und Richtfunkkanälen mit Schwach Stationären Fluktuationen und Unkorrelierter Streuung (WSSUS)*, Dissertation, ETH Zuerich, Switzerland (1990).

Fleury 1996 B. H. Fleury, "An uncertainty relation for WSS processes and its application to WSSUS systems", *IEEE Trans. Commun.*, 44, 1632–1634 (1996).

Fleury 2000 B. H. Fleury, "First- and second-order characterization of direction dispersion and space selectivity in the radio channel", *IEEE Trans. Inf. Theory*, 46, 2027–2044 (2000).

Fleury et al. 1999 B. H. Fleury, M. Tschudin, R. Heddergott, D. Dahlhaus, and K. I. Pedersen, "Channel parameter estimation in mobile radio environments using the SAGE algorithm", *IEEE JSAC*, 17, 434–450 (1999).

Fodor et al. 2012 G. Fodor, E. Dahlman, G. Mildh, S. Parkvall, N. Reider, G. Miklós, and Z. Turányi, "Design aspects of network assisted device-to-device communications", *IEEE Commun. Mag.*, 50(3), 170–177 (2012).

Fontan and Espineira 2008 F. P. Fontan and P. M. Espineira, *Modelling the Wireless Propagation Channel: A Simulation Approach with MATLAB*, Wiley, (2008).

Foschini and Gans 1998 G. J. Foschini and M. J. Gans, "On limits of wireless communications in a fading environment when using multiple antennas", *Wirel. Pers. Commun.*, 6, 311–335 (1998).

Foschini and Miljanic 1993 G. J. Foschini and Z. Miljanic, "A simple distributed autonomous power control algorithm and its convergence", *IEEE Trans. Veh. Technol.*, 42, 641–646 (1993).

Foschini et al. 2003 G. J. Foschini, D. Chizhik, M. J. Gans, C. Papadias, and R. A. Valenzuela, "Analysis and performance of some basic space-time architectures", *IEEE J. Sel. Areas Commun.*, 21, 303–320 (2003).

Foschini et al. 2006 G. J. Foschini, K. Karakayali, and R. A. Valenzuela, "Coordinating multiple antenna cellular networks to achieve enormous spectral efficiency", *IEE Proceedings-Communications*, 153, 548–555 (2006).

Fotouhi et al. 2019 A. Fotouhi, H. Qiang, M. Ding, M. Hassan, L.G. Giordano, A. Garcia-Rodriguez, and J. Yuan, "Survey on UAV cellular communications: Practical aspects, standardization advancements, regulation, and security challenges", *IEEE Communications Surveys & Tutorials*, 21(4), 3417–3442 (2019).

Fragouli and Soljanin 2007 C. Fragouli and E. Soljanin, "Network coding fundamentals", *Found. Trends Netw.*, 2(1), 1–133 (2007).

Fragouli and Soljanin 2008 C. Fragouli and E. Soljanin, "Network coding applications", *Found. Trends Netw.*, 2(2), 135–269 (2008).

Fragouli et al. 2005 C. Fragouli, J. Y. Le Boudec, and J. Widmer, *Network Coding: An Instant Primer*, EPFL *LCA-REPORT-2005-010* http://infoscience.epfl.ch/record/58339 (2005).

Frullone et al. 1996 M. Frullone, G. Riva, P. Grazioso, and G. Falciasecca, "Advanced planning criteria for cellular systems", *IEEE Pers. Commun.*, 3(6), 10–15 (1996).

Fuhl 1994 J. Fuhl, Diploma thesis, Technical University Vienna, Austria (1994).

Fuhl et al. 1998 J. Fuhl, A. F. Molisch, and E. Bonek, "Unified channel model for mobile radio systems with smart antennas", *IEEE Proc. Radar, Sonar Navigation*, 145, 32–41 (1998).

Fujimoto 1987 K. Fujimoto, "A review of research on small antennas", *J. Inst. Electron. Inform. Commun. Eng.*, 70, 830–838 (1987).

Fujimoto 2008 K. Fujimoto, *Mobile Antenna Systems Handbook*, 3rd edition, Artech, (2008).

Gahleitner 1993 R. Gahleitner, *Radio Wave Propagation in and into Urban Buildings*, Ph.D. thesis, Technical University, Vienna (1993).

Gallagher 1961 R. Gallagher, *Low Density Parity Check Codes*, Ph.D. thesis, Massachusetts Institute of Technology (1961).

Gallagher 2008 R. Gallagher, *Principles of Digital Communication*, Cambridge University Press (2008).

Ganesh and Kumari 2013 R. S. Ganesh and J. J. Kumari, "A survey on channel estimation techniques in mimo-ofdm mobile communication systems", *Int. J. Sci. Eng. Res.*, 4(5), 1850–1855 (2013).

Gao et al. 2012 X. Gao, F. Tufvesson, O. Edfors, and F. Rusek, "Measured propagation characteristics for very-large MIMO at 2.6 GHz", *Forty Sixth Asilomar Conference on Signals, Systems and Computers (ASILOMAR)*, pp. 295–299 (2012).

Gardner 2005 F. M. Gardner, *Phaselock Techniques*, 3rd edition, originally published 1966, reprint Wiley, (2005).

Garg 2000 V. K. Garg, *IS-95 CDMA & cdma2000: Cellular/PCS Systems Implementation*, Prentice Hall, (2000).

Gay and Benesty 2000 S. L. Gay and J. Benesty, *Acoustic Signal Processing for Telecommunication*, Kluwer Academic Publishers, (2000).

Gentile 2022 C. Gentile et al., "Pathloss and shadowing", in T. S. Rappaport, K. S. Remley, C. Gentile, A. F. Molisch, A. Zajic (eds), *Radio Propagation Measurements and Channel Modeling: Best Practices for Millimeter-Wave and Sub-Terahertz Frequencies*, Cambridge University Press, (2022).

Georgiadis et al. 2006 L. Georgiadis, M. Neely, and L. Tassiulas, "Resource allocation and cross layer control in wireless networks", *Found. Trends Netw.*, 1, 1–144 (2006).

Gersho and Gray 1992 A. Gersho and R. M. Gray, *Vector Quantization and Signal Compression*, Kluwer Academic Publishers, (1992).

Gerzaguet et al. 2017 R. Gerzaguet, N. Bartzoudis, L. G. Baltar, V. Berg, J. B. Doré, D. Kténas, O. Font-Bach, X. Mestre, M. Payaró, M. Färber, and K. Roth, "The 5G candidate waveform race: A comparison of complexity and performance", *EURASIP J. Wirel. Commun. Netw.*, 2017, 1–14 (2017).

Gesbert et al. 2002 D. Gesbert, H. Boelcskei, D. A. Gore, and A. Paulraj, "Outdoor MIMO wireless channels: Models and performance prediction", *IEEE Trans. Commun.*, 50(12), 1926–1935 (2002).

Gesbert et al. 2003 D. Gesbert, M. Shafi, D. S. Shiu, P. J. Smith, and A. Naguib, "From theory to practice: An overview of MIMO space-time coded wireless systems", *IEEE J. Sel. Areas Commun.*, 21, 281–302 (2003).

Gesbert et al. 2007 D. Gesbert, M. Kountouris, R. W. Heath, C. B. Chae, and T. Saelzer, "Shifting the MIMO paradigm", *IEEE Signal Process. Mag.*, 24(9), 36–46 (2007).

Gesbert et al. 2010 D. Gesbert, S. Hanly, H. Huang, S. S. Shitz, O. Simeone, and W. Yu, "Multi-cell MIMO cooperative networks: A new look at interference", *IEEE J. Sel. Areas Commun.*, 28(9), 1380–1408 (2010).

Gezici et al. 2005 S. Gezici, Z. Tian, G. B. Giannakis, H. Kobayashi, A. F. Molisch, H. V. Poor, and Z. Sahinoglu, "Localization via ultra-wideband radios: A look at positioning aspects for future sensor networks", *IEEE Signal Process. Mag.*, 22(4), 70–84 (2005).

Ghaderi and Boutaba 2006 M. Ghaderi and R. Boutaba, "Call admission control in mobile cellular networks: A comprehensive survey", *Wirel. Commun. Mob. Comput.*, 6(1), 69–93 (2006).

Ghasempour et al. 2017 Y. Ghasempour, C. R. da Silva, C. Cordeiro, and E. W. Knightly, "IEEE 802.11 ay: Next-generation 60 GHz communication for 100 Gb/s Wi-Fi", *IEEE Commun. Mag.*, 55(12), 186–192 (2017).

Ghassemzadeh et al. 2004 S. S. Ghassemzadeh, R. Jana, C. W. Rice, W. Turin, and V. Tarokh, "Measurement and modeling of an ultra-wide bandwidth indoor channel", *IEEE Trans. Commun.*, 52, 1786–1796 (2004).

Ghavami et al. 2006 M. Ghavami, L. Michael, and R. Kohno, *Ultra Wideband Signals and Systems in Communication Engineering*, Wiley, (2006).

Gholami et al. 2011 M. R. Gholami, H. Wymeersch, E. G. Ström, and M. Rydström, "Wireless network positioning as a convex feasibility problem", *EURASIP J. Wirel. Commun. Netw.*, 2011(1), 161 (2011).

Ghosh and Sen 2019 S. Ghosh and D. Sen, "An Inclusive Survey on Array Antenna Design for Millimeter-Wave Communications," in *IEEE Access* 7, 83137–83161 (2019).

Giannakis and Halford 1997 G. B. Giannakis and S. D. Halford, "Blind fractionally spaced equalization of noisy FIR channels: Direct and adaptive solutions", *IEEE Trans. Signal Process.*, 45, 2277–2292 (1997).

Gibson et al. 1998 J. D. Gibson, T. Berger, T. Lookabaugh, R. Baker, and D. Lindbergh, *Digital Compression for Multimedia: Principles and Standards*, Morgan Kaufman, (1998).

Gilhousen et al. 1991 K. S. Gilhousen, I. M. Jacobs, R. Padovani, A. J. Viterbi, L. A. Weaver, and C. E. Wheatley, "On the capacity of a cellular CDMA system", *IEEE Trans. Veh. Technol.*, 40, 303–312 (1991).

Gitlin and Weinstein 1981 R. D. Gitlin and S. B. Weinstein, "Fractionally-spaced equalization: An improved digital transversal equalizer", *Bell System Tech. J.*, 60, 275–296 (1981).

Glassner 1989 A. S. Glassner, *An Introduction to Ray Tracing*, Morgan Kaufmann, (1989).

Glisic and Vucetic 1997 S. Glisic and B. Vucetic, *Spread Spectrum CDMA Systems for Wireless Communications*, Artech House, London (1997).

Godara 1997 L. C. Godara, "Applications of antenna arrays to mobile communications. I. Performance improvement, feasibility, and system considerations", *Proceedings of IEEE* 85, 1031-1060 and Application of Antenna Arrays to Mobile Communications. II. Beam-forming and Direction-of-arrival Considerations, *Proceedings of IEEE* 85, pp. 1195–1245 (1997).

Godara 2001 L. C. Godara, *Handbook of Antennas in Wireless Communications*, CRC Press, Boca Raton (2001).

Godard 1980 D. N. Godard, "Self-recovering equalization and carrier tracking in two-dimensional data communication systems", *IEEE Trans. Commun.*, 28, 1867–1875 (1980).

Golser 1998 A. Golser, *Handbuch der Spread-Spectrum Technik*, Springer, Wien (1998).

Goiser et al. 2000 A. Goiser, M. Z. Win, G. Chrisikos, and S. Glisic, "Code division multiple access", in A. F. Molisch (ed.), *Wireless Wideband Digital Communications*, Prentice Hall, USA (2000).

Goldsmith 2005 A. Goldsmith, *Wireless Communications*, Cambridge University Press (2005).

Goldsmith and Chua 1998 A. J. Goldsmith and S. G. Chua, "Adaptive coded modulation for fading channels", *IEEE Trans. Commun.*, 46, 595–602 (1998).

Goldsmith et al. 2003 A. Goldsmith, S. A. Jafar, N. Jindal, and S. Vishwanath, "Capacity limits of MIMO channels", *IEEE J. Sel. Areas Commun.*, 21, 684–702 (2003).

Golomb and Gong 2005 S. Golomb and G. Gong, *Signal Design for Good Correlation: For Wireless Communication, Cryptography, and Radar*, Cambridge University Press, (2005).

Golrezaei et al. 2013 N. Golrezaei, A. F. Molisch, A. G. Dimakis, and G. Caire, "Femtocaching and device-to-device collaboration: A new architecture for wireless video distribution", *IEEE Commun. Mag.*, 51(4), 142–149 (2013).

Gomez-Ponce et al. 2022 J. Gomez-Ponce, N. Abbasi, A. E. Willner, J. Zhang, and A. F. Molisch, "Directionally Resolved Measurement and Modeling of THz Band Propagation Channels (invited paper)", *IEEE Open Journal on Antennas and Propagation* 3, 663–686 (2022).

Gong et al. 2020 S. Gong, X. Lu, D. T. Hoang, D. Niyato, L. Shu, D. I. Kim, and Y. C. Liang, "Towards smart wireless communications via intelligent reflecting surfaces: A contemporary survey", *IEEE Commun. Surv. Tutor.*, 22, 2283–2314 (2020).

Gonzales 1996 G. Gonzales, *Microwave Transistor Amplifiers: Analysis and Design*, 2nd edition, Pearson, (1996).

Gonzalez and Woods 2008 R. C. Gonzalez and R. E. Woods, *Digital Image Processing*, 3rd edition, Prentice Hall, (2008).

Goodman et al. 1989 D. J. Goodman, R. A. Valenzuela, K. T. Gayliard, and B. Ramamurthi, "Packet reservation multiple access for local wireless communications", *IEEE Trans. Commun.*, 37, 885–890 (1989).

Gorokhov 1998 A. Gorokhov, "On the performance of the Viterbi equalizer in the presence of channel estimation errors", *IEEE Signal Process. Lett.*, 5, 321–324 (1998).

Gou et al. 2011 T. Gou, C. Wang, and S. A. Jafar, "Aiming perfectly in the dark-blind interference alignment through staggered antenna switching", *IEEE Trans. Signal Process.*, 59, 2734–2744 (2011).

Gray 1989 R. M. Gray, *Source Coding Theory*, Kluwer Academic Publishers, (1989).

Greenstein et al. 1997 L. J. Greenstein, V. Erceg, Y. S. Yeh, and M. V. Clark, "A new path-gain/delay-spread propagation model for digital cellular channels", *IEEE Trans. Veh. Technol.*, 46, 477–485 (1997).

Gupta 2016 N. K. Gupta, *Inside Bluetooth Low Energy*, Artech House, (2016).

Gupta and Kumar 2000 P. Gupta and P. R. Kumar, "The capacity of wireless networks", *IEEE Trans. Inf. Theory*, 46, 388–404 (2000).

Guvenc and Chong 2009 I. Guvenc and C. C. Chong, "A survey on TOA based wireless localization and NLOS mitigation techniques", *IEEE Commun. Surv. Tutor.*, 11(3), 107–124 (2009).

Haardt and Nossek 1995 M. Haardt and J. A. Nossek, "Unitary ESPRIT: How to obtain increased estimation accuracy with a reduced computational burden", *IEEE Trans. Signal Process.*, 43, 1232–1242 (1995).

Hadani et al. 2017 R. Hadani, S. Rakib, M. Tsatsanis, A. Monk, A. J. Goldsmith, A.F. Molisch, and R. Calderbank, "Orthogonal time frequency space modulation", *2017 IEEE Wireless Communications and Networking Conference (WCNC)*, pp. 1–6 (2017).

Hadani et al. 2018 R. Hadani, S. Rakib, S. Kons, M. Tsatsanis, et al., "Orthogonal time frequency space modulation", arXiv:1808.00519, (2018).

Haenggi 2012 M. Haenggi, *Stochastic Geometry for Wireless Networks*, Cambridge University Press, (2012).

Haensler and Schmidt 2004 E. Haensler and G. Schmidt (eds), *Acoustic Echo and Noise Control*, John Wiley & Sons, (2004).

Hagenauer and Hoeher 1989 J. Hagenauer and P. Hoeher, "A Viterbi algorithm with soft-decision outputs and its applications", *IEEE Globecom*, 1680–1686 (1989).

Halford and Chugg 2010 T. R. Halford and K. M. Chugg, "Barrage relay networks", *Proc. Information Theory and Applications, UC San Diego*, (2010).

Hamza et al. 2013 A. S. Hamza, S. S. Khalifa, H. S. Hamza, and K. Elsayed, "A survey on inter-cell interference coordination techniques in OFDMA-based cellular networks", *IEEE Commun. Surv. Tutor.*, 15(4), 1642–1670 (2013).

Han and Poor 2009 Z. Han and V. H. Poor, "Impact of cooperative transmission on network routing", in Y. Zhang, H. H. Chen, M. Guizani (eds), *Cooperative Wireless Communications*, CRC Press, (2009).

Han et al. 2007 Z. Han, Z. Ji, and K. J. R. Liu, "Non-cooperative resource competition game by virtual referee in multi-cell OFDMA networks", *IEEE J. Sel. Areas Commun.*, 25, 1079–1090 (2007).

Han et al. 2022 C. Han, Y. Wang, Y. Li, Y. Chen, N. A. Abbasi, T. Kürner, and A. F. Molisch, "Terahertz Wireless Channels: A Holistic Survey on Measurement, Modeling, and Analysis", *IEEE Commun. Surv. Tutor.*, 24, 1670–1707 (2022).

Haneda et al. 2012 K. Haneda, J. Poutanen, L. Liu, and C. Oestges, "COST 2100 multi-link MIMO channel model", in R. Verdone, A. Zanella (eds), *Pervasive Mobile and Ambient Wireless Communications: COST action 2100*, Springer, (2012).

Haneda et al. 2016 K. Haneda, J. Zhang, L. Tan, et al., "5G 3GPP-like channel models for outdoor urban microcellular and macrocellular environments", *Proceedings of 83rd Vehicular Technology Conference*, pp. 1–7, IEEE (2016).

Hansen 1998 R. C. Hansen, *Phased Array Antennas*, Wiley, (1998).

Hanzo et al. 2000 L. Hanzo, W. Webb, and T. Keller, *Single- and Multi-Carrier Quadrature Amplitude Modulation: Principles and Applications for Personal Communications, WLANs and Broadcasting*, Wiley, (2000).

Hanzo et al. 2001 L. Hanzo, F. C. A. Somerville, and J. P. Woodard, *Voice Compression and Communications*, IEEE Press Wiley Interscience, New York (2001).

Hanzo et al. 2003 L. Hanzo, M. Muenster, B. J. Choi, and T. Keller, *OFDM and MC-CDMA for Broadband Multi-User Communications, WLANs and Broadcasting*, Wiley, (2003).

Harrington 1993 R. F. Harrington, *Field Computation by Moment Method*, Wiley/IEEE Press, (1993).

Harryson et al. 2010 F. Harryson, J. Medbo, A. F. Molisch, A. Johansson, and F. Tufvesson, "Efficient experimental evaluation of a MIMO handset with user influence", *IEEE Trans. Wirel. Commun.*, 9, 853–863 (2010).

Hashemi 1979 H. Hashemi, "Simulation of the urban radio propagation channel", *IEEE Trans. Veh. Technol.*, 28, 213–225 (1979).

Haslett 2008 C. Haslett, *Essentials of Radio Wave Propagation*, Cambridge University Press, (2008).

Hata 1980 M. Hata, "Empirical formula for propagation loss in land mobile radio services", *IEEE Trans. Veh. Technol.*, 29, 317–325 (1980).

Haykin 1991 S. Haykin, *Adaptive Filter Theory*, Prentice Hall, Englewood Cliffs (1991).

Haykin 2005 S. Haykin, "Cognitive radio: Brain-empowered wireless communications", *IEEE J. Sel. Areas Commun.*, 23, 201–220 (2005).

Heath and Lozano 2018 R. W. Heath Jr. and A. Lozano, *Foundations of MIMO Communication*, Cambridge University Press, (2018).

Heavens 1965 O. S. Heavens, *Optical Properties of Thin Film Solids*, Dover, New York (1965).

Hewlett-Packard 1994 Hewlett-Packard, *Schulungsunterlagen GSM*, Hewlett-Packard, in German (1994).

Hirade et al. 1979 K. Hirade, M. Ishizuka, F. Adachi, and K. Ohtani, "Error-rate performance of digital FM with differential detection in land mobile radio channels", *IEEE Trans. Veh. Technol.*, 28, 204–212 (1979).

Hirasawa and Haneishi 1991 K. Hirasawa and M. Haneishi (eds), *Analysis, Design, and Measurements of Small and Low-Profile Antennas*, Artech House, (1991).

Hlawatsch and Matz 2011 F. Hlawatsch and G. Matz, *Wireless Communications Over Rapidly Time-Varying Channels*, Academic Press, (2011).

Ho et al. 2006 T. Ho, M. Médard, R. Koetter, D. R. Karger, M. Effros, J. Shi, and B. Leong, "A random linear network coding approach to multicast", *IEEE Trans. Inf. Theory*, 52, 4413–4430 (2006).

Hoeher 1992 P. Hoeher, "A statistical discrete-time model for the WSSUS multipath channel", *IEEE Trans. Veh. Technol.*, 41, 461–468 (1992).

Hofstetter et al. 2006 H. Hofstetter, A. F. Molisch and N. Czink, "A twin-cluster MIMO channel model", *First European Conference on Antennas and Propagation, IEEE*, pp. 1–8 (2006).

Holma and Tokola 2011 H. Holma and A. Tokola, *LTE for UMTS: Evolution to LTE-Advanced*, Wiley, (2011).

Holma and Toskola 2007 H. Holma and A. Toskola (eds), *WCDMA for UMTS: Radio Access for Third Generation Mobile Communications*, 4th edition, Wiley, (2007).

Holma et al. 2020 H. Holma, A. Toskola, and T. Nakamura (eds), *5G Technology: 3GPP New Radio*, John Wiley & Sons, (2020).

Holtzman and Jalloul 1994 J. M. Holtzman and L. M. Jalloul, "Rayleigh fading effect reduction with wideband DS/CDMA signals", *IEEE Trans. Commun.*, 42, 1012–1016 (1994).

Hong et al. 2007 Y. W. Hong, W. J. Huang, F. H. Chiu, and C.-C. Jay Kuo, "Cooperative communications in resource-constrained wireless networks", *IEEE Signal Proc. Mag.*, 5, 47–57 (2007).

Honig 2009 M. L. Honig (ed.), *Advances in Multiuser Detection*, Wiley, (2009).

Hoppe et al. 2003 R. Hoppe, P. Wertz, F. M. Landstorfer, and G. Woelfle, "Advanced ray optical wave propagation modelling for urban and indoor scenarios including wideband properties", *Eur. Trans. Telecommun.*, 14, 61–69 (2003).

Hossain and Barghava 2008 E. Hossain and V. K. Barghava (eds), *Cognitive Wireless Communication Networks*, Springer, (2008).

Huang et al. 2001 X. Huang, A. Accro, and H. W. Hon, *Spoken Language Processing*, Prentice Hall, (2001).

Huang et al. 2010 J. Huang, V. Subramanian, R. Berry, and R. Agrawal, "Scheduling and resource allocation in OFDMA wireless systems", *Orthogonal Frequency Division Multiple Access Fundamentals and Applications*, Auerbach Publications, pp. 147–180 (2010).

Huang et al. 2019a C. Huang, A. F. Molisch, R. He, R. Wang, P. Tang, and Z. Zhong, "Machine-learning-based data processing techniques for vehicle-to-vehicle channel modeling", *IEEE Commun. Mag.*, 57(11), 109–115 (2019a).

Huang et al. 2019b C. Huang, R. He, A. F. Molisch, Z. Zhong, and B. Ai, "Machine learning enabled channel modeling", in R. He, Z. Ding (eds), *Applications of Machine Learning in Wireless Communications*, IET Press, (2019b).

Huang et al. 2020 C. Huang, A. F. Molisch, R. He, R. Wang, P. Tang, and Z. Zhong, "Machine learning-enabled LOS/NLOS identification for MIMO system in dynamic environment", *IEEE Trans. Wirel. Commun.*, 19, 3643–3657 (2020).

Huang et al. 2022 C. Huang, R. He, B. Ai, A. F. Molisch, et al, "Artificial intelligence enabled radio propagation for communications—Part I: Channel characterization and antenna-channel optimization", and "Part II: Scenario identification and channel modeling." *IEEE Trans. Antennas Prop.*, 70, 3939–3969 (2022).

Huffman 1952 D. A. Huffman, "A method for the construction of minimum redundancy codes", *Proc. IRE*, 40, 1098–1101 (1952).

Hunter and Nosratinia 2006 T. Hunter and A. Nosratinia, "Diversity through coded cooperation", *IEEE Trans. Wirel. Commun.*, 5(2), 1–7 (2006).

Hunter et al. 2006 T. E. Hunter, S. Sanayei, and A. Nosratinia, "Outage analysis of coded cooperation", *IEEE Trans. Inf. Theory*, 52, 375–391 (2006).

Hwang et al. 2009 T. Hwang, C. Yang, G. Wu, S. Li, and Y. G. Li, "OFDM and its wireless applications: A survey", *IEEE Trans. Veh. Technol.*, 58, 1673–1694 (2009).

IEEE 802.11 IEEE Standard for Information Technology--Telecommunications and Information Exchange between Systems - Local and Metropolitan Area Networks--Specific Requirements - Part 11: Wireless LAN Medium Access Control (MAC) and Physical Layer (PHY) Specifications, *IEEE 802.11 - 2020* (2020). This mostly subsumes also 802.11e, 802.11n, 802.11-16, and 802.11ax.

IEEE 802.15.4 2020 IEEE Standard for Low-Rate Wireless Networks 802.15.4-2020 (2020).

Ikegami et al. 1984 F. Ikegami, S. Yoshida, T. Takeuchi, and M. Umehira, "Propagation factors controlling mean field strength on urban streets", *IEEE Trans. Antennas Propag.*, 32, 822–829 (1984).

Intanagonwiwat et al. 2003 C. Intanagonwiwat, R. Govindan, D. Estrin, J. Heidemann, and F. Silva, "Directed diffusion for wireless sensor networking", *IEEE/ACM Trans. Netw.*, 11, 2–16 (2003).

International Telecommunications Union 1993 International Telecommunications Union, *Video Codec for Audiovisual Services at px64 Kbit/s*, Recommendation H.261 (1993).

International Telecommunications Union 1996 International Telecommunications Union, *Video Coding for Low Bit Rate Communication*, ITU-T Recommendation H.263 (1996).

International Telecommunications Union 1998 International Telecommunications Union, *Studio Encoding Parameters of Digital Television for Standard 4:3 and Wide-screen 16:9 Aspect Ratios*, ITU-R, Recommendation BT.601-5 (1998).

International Telecommunications Union 2009 International Telecommunications Union, *Information Technology – Coding of Audio-visual Objects – Part 10: Advanced Video Coding (AVC)*, 5th edition, ITU-T Recommendation H.264\ISO/IEC 14496-10:2009 (2009).

International Telecommunications Union 2015 International Telecommunications Union, "IMT Vision – Framework and overall objectives of the future development of IMT for 2020 and beyond", Recommendation ITU-R M.2083-0 (2015).

Islam et al. 2016 S. R. Islam, N. Avazov, O. A. Dobre, and K. S. Kwak, "Power-domain non-orthogonal multiple access (NOMA) in 5G systems: Potentials and challenges", *IEEE Commun. Surv. Tutor.*, 19(2), 721–742 (2016).

ISO/IEC 11172-2:1993 1993 ISO/IEC 11172-2:1993, *Information Technology – Coding of Moving Pictures and Associated Audio for Digital Storage Media at up to about 1.5 Mbit/s – Part 2: Video*. (1993).

ISO/IEC 14496-2:2004 2004 ISO/IEC 14496-2:2004, *Information Technology – Coding of Audio-Visual Objects – Part 2: Visual*, 3rd edition, (2004).

IST-WINNER D1.1.2 et al. 2007 IST-WINNER D1.1.2 P. Kyösti, J. Meinilä, L. Hentilä, et al., *WINNER II Channel Models*, ver 1.1, September 2007. Available: https://www.ist-winner.org/WINNER2-Deliverables/D1.1.2v1.1.pdf (2007).

Itakura and Saito 1968 F. Itakura and S. Saito "Analysis synthesis telephony based on the maximum likelihood principle", *Proceedings of the 6th International Congress on Acoustics*, Tokyo, Japan, pp. C17–C20 (1968).

Ito et al. 2021 M. Ito, I. Kanno, T. Ohseki, K. Yamazaki, Y. Kishi, T. Choi, and A. F. Molisch, "Effect of antenna distribution on spectral and energy efficiency of cell-free massive MIMO", *94th IEEE Vehicular Technology Conference (VTC2021-Fall)*, pp. 1–5 (2021).

Jacobsson et al. 2017 S. Jacobsson, G. Durisi, M. Coldrey, U. Gustavsson, and C. Studer, "Throughput analysis of massive MIMO uplink with low-resolution ADCs", *IEEE Trans. Wirel. Commun.*, 16, 4038–4051 (2017).

Jafar 2011 S. A. Jafar, *Interference Alignment: A New Look at Signal Dimensions in a Communication Network*, Now Publishers Inc, (2011).

Jafar 2013 S. A. Jafar, "Interference alignment - a unified view of signal dimensions across wireless and wired communication networks", *Tutor. IEEE Globecom* (2013).

Jafarkhani 2005 H. Jafarkhani, *Space-Time Coding: Theory and Practice*, Cambridge University Press, (2005).

Jaggi et al. 2005 S. Jaggi, P. Sanders, P. A. Chou, M. Effros, S. Egner, K. Jain, and L. M. G. M. Tolhuizen, "Polynomial time algorithms for multicast network code construction", *IEEE Trans. Inf. Theory*, 51, 1973–1982 (2005).

Jain 1989 A. K. Jain, *Fundamentals of Digital Image Processing*, Prentice Hall, (1989).

Jajszczyk and Wagrowski 2005 A. Jajszczyk and M. Wagrowski, "OFDM for wireless communication systems", *IEEE Commun. Mag.*, 43(9), 18–20 (2005).

Jakes 1974 W. C. Jakes, *Microwave Mobile Communications*, IEEE Press (reprint), (1974).

Jamali and Le-Ngoc 1991 S. Jamali and T. Le-Ngoc, "A new 4-state 8PSK TCM scheme for fast fading, shadowed mobile radio channels", *IEEE Trans. Veh. Technol.*, 40, 216–222 (1991).

Jarvelainen and Haneda 2014 J. Jarvelainen and K. Haneda, "Sixty gigahertz indoor radio wave propagation prediction method based on full scattering model", *Radio Sci.*, 49(4), 293–305 (2014).

Jayant and Noll 1984 N. S. Jayant and P. Noll, *Digital Coding of Waveforms – Principles and Applications to Speech and Video*, Prentice Hall, Englewood Cliffs (1984).

Ji and Liu 2007 Z. Ji and K. J. R. Liu, "Dynamic spectrum sharing: A game theoretical overview", *IEEE Commun. Mag.*, 45(5), 88–95 (2007).

Ji et al. 2015 M. Ji, G. Caire, and A. F. Molisch, "Wireless device-to-device caching networks: Basic principles and system performance", *IEEE J. Sel. Areas Commun.*, 34(1), 176–189 (2015).

Jiang and Hanzo 2007 M. Jiang and L. Hanzo, "Multiuser MIMO-OFDM for next-generation wireless systems", *Proc. IEEE*, 95, 1430–1469 (2007).

Jiang and Wu 2008 T. Jiang and Y. Wu, "Peak-to-average power ratio reduction techniques for OFDM signals", *IEEE Trans. Broadcast.*, 54, 257–268 (2008).

Jiang et al. 2010 T. Jiang, L. Song, and Y. Zhang (eds), *Orthogonal Frequency Division Multiple Access Fundamentals and Applications*, CRC Press, (2010).

Jiang et al. 2021 W. Jiang, B. Han, M. A. Habibi, and H. D. Schotten, "The road towards 6G: A comprehensive survey", *IEEE Open J. Commun. Soc.*, 2, 334–366 (2021).

Jiang et al. 2022 S. Jiang, W. Wang, Y. Miao, W. Fan and A. F. Molisch, "A Survey of Dense Multipath and Its Impact on Wireless Systems," *IEEE Open Journal of Antennas and Propagation* 3, pp. 435–460 (2022)

Johannesson and Zigangirov 1999 R. Johannesson and K. S. Zigangirov, *Fundamentals of Convolutional Coding*, Wiley – IEEE Press (1999).

Johnson et al. 2001 D. B. Johnson, D. A. Maltz, and J. Broch, "DSR: The dynamic source routing protocol for multi-hop wireless ad hoc networks", in C. E. Perkins (ed.), *Ad Hoc Networking*, Chapter 5, pp. 139–172, Addison-Wesley, (2001).

Jovocic and Visvanath 2009 A. Jovocic and P. Viswanath, "Cognitive radio: An information-theoretic perspective", *IEEE Trans. Inf. Theory*, 55, 3945–3958 (2009).

Junhai et al. 2009 L. Junhai, Y. Danxia, C. Liu, and F. Mingyu, "A survey of multicast routing protocols for mobile ad-hoc networks", *IEEE Commun. Surv. Tutor.*, 11, 78–91 (2009).

Jurafsky and Martin 2000 D. Jurafsky and J. H. Martin, *Speech and Language Processing*, Prentice Hall, Upper Saddle River (2000).

Kahwaja et al. 2019 W. Khawaja, I. Guvenc, D. W. Matolak, U. -C. Fiebig, and N. Schneckenburger, "A survey of air-to-ground propagation channel modeling for unmanned aerial vehicles", *IEEE Commun. Surv. Tutor.*, 21, 2361–2391 (2019).

Kailath et al. 2001 T. Kailath, H. Vikalo, and B. Hassibi, "MIMO receive algorithms", in H. Boelcskei et al. (eds), *Space-Time Wireless Systems: From Array Processing to MIMO Communications*, Cambridge University Press, (2001).

Kalliola et al. 2002 K. Kalliola, K. Sulonen, H. Laitinen, O. Kivekas, J. Krogerus, and P. Vainikainen, "Angular power distribution and mean effective gain of mobile antenna in different propagation environments", *IEEE Trans. Veh. Technol.*, 51, 823–838 (2002).

Kaplan and Hegarty 2005 E. Kaplan and C. Hegarty, *Understanding GPS: Principles and Applications*, Artech House, (2005).

Karedal et al. 2009 J. Karedal, F. Tufvesson, N. Czink, N. Paier, C. Dumard, T. Zemen, C. F. Mecklenbrauker, and A. F. Molisch, "A geometry-based stochastic MIMO model for vehicle-to-vehicle communications", *IEEE Trans. Wirel. Commun.*, 8, 3646–3657 (2009).

Karttunen et al. 2016 A. Karttunen, C. Gustafson, A. F. Molisch, R. Wang, S. Hur, J. Zhang, and J. Park, "Path loss models with distance-dependent weighted fitting and estimation of censored path loss data", *IET Microw. Antennas Propag.*, 10, 1467–1474 (2016).

Karttunen et al. 2017 A. Karttunen, A. F. Molisch, S. Hur, J. Park, and C. J. Zhang, "Spatially consistent street-by-street path loss model for 28-GHz channels in micro cell urban environments", *IEEE Trans. Wirel. Commun.*, 16, 7538–7550 (2017).

Karunatilaka et al. 2015 C. Karunatilaka, F. Zafar, V. Kalavally, and R. Parthiban, "LED based indoor visible light communications: State of the art", *IEEE communications surveys & tutorials*, 17(3), 1649–1678 (2015).

Kattenbach 1997 R. Kattenbach, *Characterisierung zeitvarianter Indoor Mobilfunkkanäle mittels ihrer System- und Korrelationsfunktionen*, Dissertation an der Universität GhK Kassel, publiziert beim Shaker-Verlag, Aachen (1997).

Kattenbach 2002 R. Kattenbach, "Statistical modeling of small-scale fading in directional radio channels", *IEEE J. Sel. Areas Commun.*, 20, 584–592 (2002).

Kaur and Kumar 2020 A. Kaur and K. Kumar, "A comprehensive survey on machine learning approaches for dynamic spectrum access in cognitive radio networks", *J. Exp. Theor. Artif. Intell.*, 1–40 (2020) https://doi.org/10.1080/0952813X.2020.1818291.

Keller 1962 J. B. Keller, "Geometrical theory of diffraction", *J. Opt. Soc. Am.*, 2, 116–130 (1962).

Keller and Hanzo 2000 T. Keller and L. Hanzo, "Adaptive multicarrier modulation: A convenient framework for time-frequency processing in wireless communications", *Proc. IEEE*, 88, 611–640 (2000).

Kermoal et al. 2002 J. P. Kermoal, L. Schumacher, K. I. Pedersen, P. E. Mogensen, and F. Frederiksen, "A stochastic MIMO radio channel model with experimental validation", *IEEE J. Sel. Areas Commun.*, 20, 1211 (2002).

Keysight 2019 Keysight, *5G Boot Camp, Slideset*, Keysight Technologies, (2019).

Khandani et al. 2003 A. Khandani, J. Abounadi, E. Modiano, and L. Zhang, "Cooperative routing in wireless networks", *Allerton Conference on Communications, Control and Computing*, October (2003).

Khanzadi 2016 M. R. Khanzadi, *Phase noise in communication systems Modeling, Compensation, and Performance Analysis*, PhD thesis, Chalmers University (2016).

Khodjaev et al. 2010 J. Khodjaev, Y. Park, and A. S. Malik, "Survey of NLOS identification and error mitigation problems in UWB-based positioning algorithms for dense environments", *Ann. Telecommun.-annales des Télécommunications*, 65(5–6), 301–311 (2010).

Khorov et al. 2018 E. Khorov, A. Kiryanov, A. Lyakhov, and G. Bianchi, "A tutorial on IEEE 802.11 ax high efficiency WLANs", *IEEE Commun. Surv. Tutor.*, 21(1), 197–216 (2018).

Kim et al. 1999 Y. H. Kim, I. Song, H. G. Kim, T. Chang, and H. M. Kim, "Performance analysis of a coded OFDM system in time-varying multipath Rayleigh fading channels", *IEEE Trans. Veh. Technol.*, 48, 1610–1615 (1999).

Kim et al. 2014 J. Kim, D. Kim, Y. Cho, et al., "Analysis of envelope-tracking power amplifier using mathematical modeling", *IEEE Trans. Microw. Theory Techn.*, 62, 1352–1362 (2014).

Kivekäs et al. 2004 O. Kivekäs, J. Ollikainen, T. Lehtiniemi, and P. Vainikainen, "Bandwidth, SAR, and efficiency of internal mobile phone antennas", *IEEE Trans. Electromagn. Comp.*, 46, 71–76 (2004).

Kleijn 2005 W. B. Kleijn, *Information Theory and Source Coding*, Royal Institute of Technology (KTH), unpublished course notes, (2005).

Kleijn and Granzow 1991 W. B. Kleijn and W. Granzow, "Methods for waveform interpolation in speech coding", *Digit. Signal Process.*, 1, 215–230 (1991).

Kleijn and Paliwal 1995 W. B. Kleijn and K. K. Paliwal, *Speech Coding and Synthesis*, Elsevier, (1995).

Kleinrock and Tobagi 1975 L. Kleinrock and F. Tobagi, "Packet switching in radio channels: Part I – carrier sense multiple access modes and their throughput-delay characteristics", *IEEE Trans. Commun.*, 23, 1400–1416 (1975).

Klemenschits and Bonek 1994 T. Klemenschits and E. Bonek, "Radio coverage of road tunnels at 900 and 1800 MHz by discrete antennas", *Proc. PIMRC*, 1994, 411–415 (1994).

Knopp and Humblet 1995 R. Knopp and P. Humblet, "Information capacity and power control in single-cell multiuser communications", *IEEE Int. Conf. Commun.*, 1, 331–335 (1995).

Kobayashi and Caire 2006 M. Kobayashi and G. Caire, "An iterative water-filling algorithm for maximum weighted sum-rate of Gaussian MIMO-BC", *IEEE J. Sel. Areas Commun.*, 24, 1640–1646 (2006).

Koetter and Médard 2003 R. Koetter and M. Médard, "An algebraic approach to network coding", *IEEE/ACM Trans. Netw.*, 11, 782–795 (2003).

Kohno et al. 1995 R. Kohno, R. Meidan, and L. B. Milstein, "Spread spectrum access methods for wireless communications", *IEEE Commun. Mag.*, 33(1), 58–67 (1995).

Kondoz 2004 A. M. Kondoz, *Digital Speech: Coding for Low Bit Rate Communication Systems*, 2nd edition, Wiley, (2004).

Kosta et al. 2012 C. Kosta, B. Hunt, A. U. Quddus, and R. Tafazolli, "On interference avoidance through inter-cell interference coordination (ICIC) based on OFDMA mobile systems", *IEEE Commun. Surv. Tutor.*, 15(3), 973–995 (2012).

Kouyoumjian and Pathak 1974 R. G. Kouyoumjian and P. H. Pathak, "A uniform geometrical theory of diffraction for an edge in a perfectly conducting surface", *Proc. IEEE*, 62, 1448–1461 (1974).

Kozek 1997 W. Kozek, *Matched Weyl-Heisenberg Expansions of Nonstationary Environments*, Dissertation, Technical University Vienna, Austria (1997).

Kozek and Molisch 1998 W. Kozek and A. F. Molisch, "Nonorthogonal pulseshapes for multicarrier communications in doubly dispersive channels", *IEEE J. Sel. Areas Commun.*, 16, 1579–1589 (1998).

Kramer et al. 2005 G. Kramer, M. Gastpar, and P. Gupta, "Cooperative strategies and capacity theorems for relay networks", *IEEE Trans. Inf. Theory*, 51, 3037–3063 (2005).

Kramer et al. 2007 G. Kramer, I. Maric, and R. D. Yates, "Cooperative communications", *Found. Trends Netw.*, 1(3–4), 1–167 (2007).

Kraus and Marhefka 2002 J. D. Kraus and R. J. Marhefka, *Antennas: For All Applications*, 3rd edition, McGraw Hill, (2002).

Kreuzgruber et al. 1993 P. Kreuzgruber, P. Unterberger, and R. Gahleitner, "A ray splitting model for indoor propagation associated with complex geometries", *Proceedings of 43rd IEEE Vehicular Technology Conference*, Secaucus, NJ, pp. 227–230 (1993).

Krim and Viberg 1996 H. Krim and M. Viberg, "Two decades of array signal processing - the parametric approach", *IEEE Signal Proc. Mag.*, 13(4), 67–94 (1996).

Kristem et al. 2019 V. Kristem, A. F. Molisch, S. Niranjayan, and S. Sangodoyin, "Coherent UWB ranging in the presence of multiuser interference", *IEEE Trans. Wirel. Commun.*, 13, 4424–4439 (2019).

Ku and Walsh 2015 G. Ku and J. M. Walsh, "Resource allocation and link adaptation in LTE and LTE advanced: A tutorial", *IEEE Commun. Surv. Tutor.*, 17, 1605–1633 (2015).

Kubin 1995 G. Kubin, "Nonlinear processing of speech", in W. B. Kleijn, K. K. Paliwal (eds), *Speech Coding and Synthesis*, pp. 557–610, Elsevier, (1995).

Kuchar et al. 1997 A. Kuchar, J. Fuhl, and E. Bonek, "Spectral efficiency enhancement and power control of smart antenna system", *EPMCC 97*, Bonn, Germany, September 30–October 2 (1997).

Kumar et al. 2006 S. Kumar, V. S. Raghavan, and J. Deng, "Medium access control protocols for ad hoc wireless networks: A survey", *Ad hoc Netw.*, 4(3), 326–358 (2006).

Kunisch and Pamp 2003 J. Kunisch, and J. Pamp, "An ultra-wideband space-variant multipath indoor radio channel model", *IEEE Conference on Ultra Wideband Systems and Technologies*, pp. 290–294 (2003).

Kunz and Luebbers 1993 K. S. Kunz and R. J. Luebbers, *The Finite Difference Time Domain Method for Electromagnetics*, CRC Press, (1993).

Kurunathan et al. 2018 H. Kurunathan, R. Severino, A. Koubaa, and E. Tovar, "IEEE 802.15. 4e in a nutshell: Survey and performance evaluation", *IEEE Commun. Surv. Tutor.*, 20(3), 1989–2010 (2018).

Laneman and Wornell 2003 J. N. Laneman and G. W. Wornell, "Distributed space-time-coded protocols for exploiting cooperative diversity in wireless networks", *IEEE Trans. Inf. Theory*, 49, 2415–2425 (2003).

Laneman et al. 2004 J. N. Laneman, D. N. C. Tse, and G. W. Wornell, "Cooperative diversity in wireless networks: Efficient protocols and outage behavior", *IEEE Trans. Inf. Theory*, 50, 3062–3080 (2004).

Larmo et al. 2009 A. Larmo, M. Lindström, M. Meyer, G. Pelletier, J. Torsner, and H. Wiemann, "The LTE link-layer design", *IEEE Commun. Mag.*, 47(4), 52–59 (2009).

Larsen and Aarts 2004 E. R. Larsen and R. M. Aarts, *Audio Bandwidth Extension: Application of Psychoacoustics, Signal Processing and Loudspeaker Design*, Wiley, (2004).

Larsson and Stoica 2008 E. G. Larsson and P. Stoica, *Space-Time Block Coding for Wireless Communications*, Cambridge University Press, (2008).

Larsson et al. 2014 E. G. Larsson, O. Edfors, F. Tufvesson, and T. L. Marzetta, "Massive MIMO for next generation wireless systems", *IEEE Commun. Mag.*, 52(2), 186–195 (2014).

Laufer et al. 2009 R. Laufer, H. Dubois-Ferriere, and L. Kleinrock, "Multirate anypath routing in wireless mesh networks", *IEEE INFOCOM*, pp. 37–45 (2009).

Laurent 1986 P. A. Laurent, "Exact and approximate construction of digital phase modulations by superposition of amplitude modulated pulses", *IEEE Trans. Commun.*, 34, 150–160 (1986).

Laurila 2000 J. Laurila, *Semi-Blind Detection of Co-Channel Signals in Mobile Communications*, Dissertation TU Wien, Wien (2000).

Lawton and McGeehan 1994 M. C. Lawton and J. P. McGeehan, "The application of a deterministic ray launching algorithm for the prediction of radio channel characteristics in small-cell environments", *IEEE Trans. Veh. Technol.*, 43, 955–969 (1994).

Lee 1973 W. C. Y. Lee, "Effects on correlations between two mobile base-station antennas", *IEEE Trans. Commun.*, 21, 1214–1224 (1973).

Lee 1982 W. C. Y. Lee, *Mobile Communications Engineering*, McGraw Hill, New York (1982).

Lee 1995 W. C. Y. Lee, *Mobile Cellular Telecommunications: Analog and Digital Systems*, McGraw Hill, (1995).

Lee et al. 2005 J. W. Lee, R. R. Mazumdar, and N. B. Shroff, "Downlink power allocation for multi-class wireless systems", *IEEE/ACM Trans. Netw.*, 13, 854–867 (2005).

Lee et al. 2006 J. W. Lee, M. Chiang, and A. R. Calderbank, "Price-based distributed algorithms for rate-reliability tradeoff in network utility maximization", *IEEE J. Sel. Areas Commun.*, 24(5), 962–976 (2006).

Lee et al. 2022 M. C. Lee, M. Ji, and A. F. Molisch, "Scaling Laws for Cache-Aided Device-to-Device Networks for Wireless Video", in *Edge Caching for Mobile Networks* (W. Chen and H. V. Poor, eds.), Cambridge University Press, 2022.

Letaief and Zhang 2007 K. B. Letaief and W. Zhang, "Cooperative spectrum sensing", in E. Hossein, V. K. Barghava (eds), *Cognitive Wireless Communication Networks*, Springer, (2007).

Levie et al. 2021 R. Levie, Ç. Yapar, G. Kutyniok, and G. Caire, "RadioUNet: Fast radio map estimation with convolutional neural networks", *IEEE Trans. Wirel. Commun.*, 20, 4001–4015 (2021).

Li and Stuber 2006 Y. G. Li and G. L. Stuber, *Orthogonal Frequency Division Multiplexing for Wireless Communications*, Springer, (2006).

Li et al. 1997 J. Li, J. F. Wagen, and E. Lachat, "ITU model for multi-knife-edge diffraction", *Microwaves, Antennas Propag., IEEE Proc.*, 143, 539–541 (1997).

Li et al. 1998 Y. G. Li, L. C. Cimini, and N. R. Sollenberger, "Robust channel estimation for OFDM systems with rapid dispersive fading channels", *IEEE Trans. Commun.*, 46, 902–915 (1998).

Li et al. 1999 Y. G. Li, N. Seshadri, and S. Ariyavisitakul, "Channel estimation for OFDM systems with transmitter diversity in mobile wireless channels", *IEEE J. Sel. Areas Commun.*, 17(3), 461–471 (1999).

Li et al. 2013 J. Li, X. Wu, and R. Laroia, *OFDMA Mobile Broadband Communications: A Systems Approach*, Cambridge University Press, (2013).

Li et al. 2014 G. Y. Li, J. Niu, D. Lee, J. Fan, and Y. Fu, "Multi-cell coordinated scheduling and MIMO in LTE", *IEEE Commun. Surv. Tutor.*, 16(2), 761–775 (2014).

Li et al. 2017a Z. Li, S. Han, and A. F. Molisch, "Optimizing channel-statistics-based analog beamforming for millimeter-wave multi-user massive MIMO downlink", *IEEE Trans. Wirel. Commun.*, 16, 4288–4303 (2017).

Li et al. 2017b Y. Li, C. Tao, G. Seco-Granados, A. Mezghani, A. L. Swindlehurst, and L. Liu, "Channel estimation and performance analysis of one-bit massive MIMO systems", *IEEE Trans. Signal Process.*, 65(15), 4075–4089 (2017).

Liang et al. 2019 L. Liang, H. Ye, G. Yu, and G. Y. Li, "Deep-learning-based wireless resource allocation with application to vehicular networks", *Proc. IEEE*, 108(2), 341–356 (2019).

Liberti and Rappaport 1996 J. C. Liberti and T. S. Rappaport, "A geometrically based model for line of sight multipath radio channels", *Proceedings of IEEE Vehicular Technology Conference*, pp. 844–848 (1996).

Liberti and Rappaport 1999 J. C. Liberti and T. S. Rappaport, *Smart Antennas for Wireless Communications: IS-95 and Third Generation CDMA Applications*, Prentice Hall, (1999).

Liebenow and Kuhlmann 1993 U. Liebenow and P. Kuhlmann, "Determination of scattering surfaces in hilly terrain", *COST 231, TD (93)* pp. 119 (1993).

Lien et al. 2020 S. Y. Lien, D. J. Deng, C. C. Lin, H. L. Tsai, T. Chen, C. Guo, and S. M. Cheng, "3GPP NR sidelink transmissions toward 5G V2X", *IEEE Access*, 8, 35368–35382 (2020).

Liew et al. 2013 S. C. Liew, S Zhang, and L. Lu, "Physical-layer network coding: Tutorial, survey, and beyond," *Physical Communication*, 6, 4–42 (2013).

Lin and Costello 2004 S. Lin and D. J. Costello, *Error Control Coding*, 2nd edition, Prentice Hall, (2004).

Liu et al. 1996 H. Liu, G. Xu, L. Tong, and T. Kailath, "Recent developments in blind channel equalization: From cyclostationarity to subspaces", *Signal Process.*, 50, 83–99 (1996).

Liu et al. 2009 K. J. R. Liu, A. K. Sadek, W. Su, and A. Kwasinski, *Cooperative Communications and Networking*, Cambridge University Press, (2009).

Liu et al. 2012 L. Liu, C. Oestges, J. Poutanen, K. Haneda, P. Vainikainen, F. Quitin, F. Tufvesson, and P. De Doncker, "The COST 2100 MIMO channel model", *IEEE Wirel. Commun.*, 19(6), 92–99 (2012).

Liu et al. 2014a J. Liu, N. Kato, J. Ma, and N. Kadowaki, "Device-to-device communication in LTE-advanced networks: A survey", *IEEE Commun. Surv. Tutor.*, 17(4), 1923–1940 (2014).

Liu et al. 2014b Y. Liu, Z. Tan, H. Hu, L. J. Cimini, and G. Y. Li, "Channel estimation for OFDM", *IEEE Communications Surveys & Tutorials*, 16, 1891–1908 (2014).

Liu et al. 2019 J. Liu, Z. Luo, and X. Xiong, "Low-resolution ADCs for wireless communication: A comprehensive survey", *IEEE Access*, 7, 91291–91324 (2019).

Liu et al. 2021 Y. Liu, X. Liu, X. Mu, T. Hou, J. Xu, M. Di Renzo, and N. Al-Dhahir, "Reconfigurable intelligent surfaces: Principles and opportunities", *IEEE Commun. Surv. Tutor.*, 23, 1546–1577 (2021).

Lo 1999 T. K. Y. Lo, "Maximum ratio transmission", *IEEE Trans. Commun.*, 47, 1458–1461 (1999).

Loeliger 2004 H. A. Loeliger, "An introduction to factor graphs", *IEEE Signal Process. Mag.*, 21, 28–41 (2004).

Lohan et al. 2015 E. S. Lohan, H. Hurskainen, and J. Nurmi, "Galileo signals", in J. Nurmi, E. Lohan, S. Sand, H. Hurskainen (eds), *GALILEO Positioning Technology. Signals and Communication Technology*, Vol. 182, Springer, Dordrecht (2015).

Loncar et al. 2002 M. Loncar, R. Müller, T. Abe, J. Wehinger, and C. Mecklenbräuker, "Iterative equalizer using soft-decoder feedback for MIMO systems in frequency-selective fading", *Proceedings of URSI General Assembly 2002* (2002).

Lopez-Perez et al. 2011 D. Lopez-Perez, I. Guvenc, G. De la Roche, M. Kountouris, T. Q. Quek, and J. Zhang, "Enhanced intercell interference coordination challenges in heterogeneous networks", *IEEE Wirel. Commun.*, 18(3), 22–30 (2011).

Lott and Teneketzis 2006 C. Lott and D. Teneketzis, "Stochastic routing in ad-hoc networks", *IEEE Trans. Automat. Control*, 51, 52–70 (2006).

Love et al. 2003 D. J. Love, R. W. Heath, and S. Strohmer, "Grassmannian beamforming for multiple-input multiple-output wireless systems", *IEEE Trans. Inf. Theory*, 49, 2735–2747 (2003).

Love et al. 2008 D. J. Love, R. W. Heath, V. K. Lau, D. Gesbert, B. D. Rao, and M. Andrews, "An overview of limited feedback in wireless communication systems", *IEEE J. Sel. Areas Commun.*, 26(8), 1341–1365 (2008).

Lozano et al. 2008 A. Lozano, A. M. Tulino, and S. Verdu, "Multiantenna capacity myths and reality", in H. Boelcskei, D. Gesbert, C. B. Papadias, A.-J. van der Veen (eds), *Space-Time Wireless Systems*, Cambridge University Press, (2008).

Lu and Shen 2013 N. Lu and X. S. Shen, "Scaling laws for throughput capacity and delay in wireless networks: A survey", *IEEE Commun. Surv. Tutor.*, 16 (2), 642–657 (2013).

Lu et al. 2014 L. Lu, G. Y. Li, A. L. Swindlehurst, A. Ashikhmin, and R. Zhang, "An overview of massive MIMO: Benefits and challenges", *IEEE J. Sel. Top. Signal Process.*, 8(5), 742–758 (2014).

Lucky et al. 1968 R. W. Lucky, J. Salz, and E. J. Weldon Jr., *Principles of Data Communication*, McGraw Hill, (1968).

Luong et al. 2019 N. C. Luong, D. T. Hoang, S. Gong, D. Niyato, P. Wang, Y. C. Liang, and D. I. Kim, "Applications of deep reinforcement learning in communications and networking: A survey", *IEEE Commun. Surv. Tutor.*, 21(4), 3133–3174 (2019).

MacKay 2002 D. J. C. MacKay, *Information Theory, Inference and Learning Algorithms*, Cambridge University Press, (2002).

MacKay and Neal 1997 D. J. C. MacKay and R. M. Neal, "Near Shannon limit performance of low density parity check codes", *Electron. Lett.*, 33, 457–458 (1997).

Madan et al. 2009 R. Madan, N. B. Mehta, A. F. Molisch, and J. Zhang, "Energy-efficient decentralized control of cooperative wireless networks with fading", *IEEE Trans. Automat. Control*, 54, 512–527 (2009).

Mailloux 1994 R. J. Mailloux, *Phased Array Antenna Handbook*, Artech House, (1994).

Makki et al. 2020 B. Makki, K. Chitti, A. Behravan, and M. -S. Alouini, "A survey of NOMA: Current status and open research challenges", *IEEE Open J. Commun. Soc.*, 1, 179–189 (2020).

Malvar et al. 2003 H. Malvar, A. Hallapuro, M. Karczewicz, and L. Kerofsky, "Low-complexity transform and quantization in H.264/AVC", *IEEE Trans. Circ. Syst. Video Technol.*, 13, 598–603 (2003).

Mangel et al. 2011 T. Mangel, O. Klemp, and H. Hartenstein, "5.9 GHz inter-vehicle communication at intersections: A validated non-line-of-sight path-loss and fading model", *EURASIP J. Wirel. Commun. Netw.*, 2011, 1–11 (2011).

Manholm et al. 2003 L. Manholm, M. Johansson, and S. Petersson, "Antennas with Electrical Beamtilt for WCDMA: Simulations and Implementation", *Swedish National Conference on Antennas* (2003).

Mao et al. 2017 Y. Mao, C. You, J. Zhang, K. Huang, and K. B. Letaief, "A survey on mobile edge computing: The communication perspective," *IEEE communications surveys & tutorials*, 19(4), 2322–2358 (2017).

Mao et al. 2018 Y. Mao, B. Clerckx, and V. O. Li, "Rate-splitting multiple access for downlink communication systems: Bridging, generalizing, and out-performing SDMA and NOMA", *J. Wirel. Commun. Netw.*, 2018, 133 (2018).

Mao et al. 2022 Y. Mao, O. Dizdar, B. Clerckx, R. Schober, P. Popovski, and H. V. Poor, "Rate-splitting multiple access: Fundamentals, survey, and future research trends", *IEEE Communications Surveys & Tutorials* (2022).

Marcuse 1991 D. Marcuse, *Theory of Dielectric Optical Waveguides*, 2nd edition, Academic Press, Boston (1991).

Mardia et al. 1979 K. V. Mardia, J. T. Kent, and J. M. Bibby, *Multivariate Analysis*, Academic Press, London (1979).

Maric and Titlebaum 1992 S. V. Maric and E. L. Titlebaum, "A class of frequency hop codes with nearly ideal characteristics for use in multiple-access spread-spectrum communications and radar and sonar systems", *IEEE Trans. Commun.*, 40, 1442–1447 (1992).

Maric and Yates 2004 I. Maric and R. D. Yates, "Cooperative multihop broadcast for wireless networks", *IEEE J. Sel. Areas Commun.*, 22, 1080–1088 (2004).

Marsch and Fettweis 2011 P. Marsch and G. P. Fettweis (eds), *Coordinated Multi-Point in Mobile Communications: From Theory to Practice*, Cambridge University Press, (2011).

Martirosyan et al. 2008 A. Martirosyan, A. Boukerche, and R. W. N. Pazzi, "A taxonomy of cluster-based routing protocols for wireless sensor networks", *The International Symposium on Parallel Architectures, Algorithms, and Networks* (2008).

Marzetta 2010 T. L. Marzetta, "Noncooperative cellular wireless with unlimited numbers of base station antennas", *IEEE Trans. Wirel. Commun.*, 9(11), 3590–3600 (2010).

Marzetta et al. 2016 T. L. Marzetta, E. G. Larsson, H. Yang, and H. Q. Ngo, *Fundamentals of Massive MIMO*, Cambridge University Press, (2016).

Marzetta and Hochwald 1999 T. L. Marzetta and B. M. Hochwald, "Capacity of a mobile multiple-antenna communication link in Rayleigh at fading", *IEEE Trans. Inf. Theory*, 45, 139–157 (1999).

Matthaei et al. 1980 G. L. Matthaei, L. Young, and E. M. Jones, *Microwave Filters, Impedance-Matching Networks, and Coupling Structures*, McGraw-Hill (1964), reprint Artech House, (1980).

Matz 2003 G. Matz, "Characterization of non-WSSUS fading dispersive channels", *Proceedings of ICC'03*, pp. 2480–2484 (2003).

Matz and Hlawatsch 1998 G. Matz and F. Hlawatsch, "Time-frequency transfer function calculus (symbolic calculus) of linear time-varying systems (linear operators) based on a generalized underspread theory", *J. Math. Phys. (Special Issue on Wavelet and Time-Frequency Analysis)*, 39, 4041–4070 (1998).

Matz et al. 2002 G. Matz, A. F. Molisch, F. Hlawatsch, M. Steinbauer, and I. Gaspard, "On the systematic measurement errors of correlative mobile radio channel sounders", *IEEE Trans. Commun.*, 50, 808–821 (2002).

May and Rohling 2000 T. May and H. Rohling, "Orthogonal frequency division multiple access", in A. F. Molisch (ed.), *Wideband Wireless Digital Communications*, Prentice Hall, USA (2000).

Mayr 1996 B. Mayr, *Modulationsangepasste Codierung*, Lecture Notes, TU Vienna (1996).

McAulay and Quatieri 1986 R. J. McAulay and T. F. Quatieri, "Speech analysis/synthesis based on a sinusoidal representation", *IEEE Trans. Acoust., Speech, Signal Process.*, 34, 744–754 (1986).

McEliece 2004 R. McEliece, *The Theory of Information and Coding*, Student edition, Cambridge University Press, (2004).

McMillan 2005 R. W. McMillan, *Terahertz Imaging, Millimeter-Wave Radar*, NATO Advanced Study Institute: Advances in Sensing with Security Applications, (2005).

McNamara et al. 1990 D. A. McNamara, C. W. I. Pistorius, and J. A. G. Malherbe, *Introduction to the Uniform Geometrical Theory of Diffraction*, Artech House, Boston, MA (1990).

Mecklenbrauker et al. 2011 C. F. Mecklenbrauker, A. F. Molisch, J. Karedal, F. Tufvesson, A. Paier, L. Bernado, T. Zemen, O. Klemp, and N. Czink, "Vehicular channel characterization and its implications for wireless system design and performance", *Proc. IEEE*, 99, 1189–1212 (2011).

Médard and Sprintson 2011 M. Médard and A. Sprintson (eds), *Network Coding: Fundamentals and Applications*, Academic Press, (2011).

Mehlführer et al. 2009 C. Mehlführer, M. Wrulich, J. C. Ikuno, D. Bosanska, and M. Rupp, "Simulating the long term evolution physical layer", *17th European Signal Processing Conference*, pp. 1471–1478 (2009).

Mehta et al. 2007 N. B. Mehta, J. Wu, A. F. Molisch, and J. Zhang, "Approximating a sum of random variables with a lognormal distribution", *IEEE Trans. Wirel. Commun.*, 6, 2690–2699 (2007).

Mengali and D'Andrea 1997 U. Mengali and A. N. D'Andrea, *Synchronization Techniques for Digital Receivers*, Plenum, (1997).

van der Meulen 1971 E. van der Meulen, "Three-terminal communication channels", *Adv. Appl. Prob.*, 3, 120–154 (1971).

Meurling and Jeans 1994 J. Meurling and R. Jeans, *The Mobile Phone Book*, Communications Week International, London (1994).

Meyr and Ascheid 1990 H. Meyr and G. Ascheid, *Digital Communication Receivers, Phase-, Frequency-Locked Loops, and Amplitude Control*, Wiley, (1990).

Meyr et al. 1997 H. Meyr, M. Moeneclaeye, and S. A. Fechtel, *Digital Communication Receivers, Vol. 2: Synchronization, Channel Estimation, and Signal Processing*, Wiley, (1997).

Miller 2002 L. E. Miller, "Formulas for blocking probability", report of *Wireless Communications Technology Group, Advanced Network Technologies Division*, National Institute of Standards and Technology, (2002).

Milstein 1988 L. B. Milstein, "Interference rejection techniques in spread spectrum communications", *Proc. IEEE*, 76, 657–671 (1988).

Mirza et al. 2017 R. K. Mirza, Y. Zhang, D. Zrnic, and R. Doviak, "Investigating EM dipole radiating element for dual polarized phased array weather radars", in *Modern Antenna Systems*, M. A. Matin (ed.) IntechOpen, https://doi.org/10.5772/66502, (2017).

Mo and Heath 2015 J. Mo and R. W. Heath, "Capacity analysis of one-bit quantized MIMO systems with transmitter channel state information", *IEEE Trans. Signal Process.*, 63, 5498–5512 (2015).

Moeller et al. 2010 S. Moeller, A. Sridharan, B. Krishnamachari, and O. Gnawali, "Routing without routes: The backpressure collection protocol", *Proceedings of the 9th ACM/IEEE International Conference on Information Processing in Sensor Networks*, pp. 279–290 (2010).

Molisch 2000 A. F. Molisch (ed.), *Wideband Wireless Digital Communications*, Prentice Hall, (2000).

Molisch 2001 A. F. Molisch, *Base Station Cooperation*, AT&T Labs, internal report, (2001).

Molisch 2004 A. F. Molisch, "A generic model for the MIMO wireless propagation channel", *IEEE Proc. Signal Proc.*, 52, 61–71 (2004).

Molisch 2009 A. F. Molisch, "Ultrawideband propagation channels", *Proc. IEEE*, 97, 353–371 (2009).

Molisch and Hofstetter 2006 A. F. Molisch and H. Hofstetter, "The COST 273 MIMO channel model", in L. Correia (ed.), *Mobile Broadband Multimedia Networks*, Academic Press, New York (2006).

Molisch and Steinbauer 1999 A. F. Molisch and M. Steinbauer, "Condensed parameters for characterizing wideband mobile radio channels", *Int. J. Wirel. Inf. Netw.*, 6, 133–154 (1999).

Molisch and Tufvesson 2004 A. F. Molisch and F. Tufvesson, "Multipath propagation models for broadband wireless systems", in M. Ibnkahla (ed.), *Digital Signal Processing for Wireless Communications Handbook*, Chapter 2, pp. 2.1–2.43, CRC Press, (2004).

Molisch and Tufvesson 2005 A. F. Molisch and F. Tufvesson, "MIMO Channel capacity and measurements", in T. Kaiser (ed.), *Smart Antennas in Europe – State of the Art*, EURASIP Publishing, (2005).

Molisch and Tufvesson 2014 A. F. Molisch and F. Tufvesson, "Propagation channel models for next-generation wireless communications systems", *IEICE Trans. Commun.*, 97, 2022–2034 (2014).

Molisch and Win 2004 A. F. Molisch and M. Z. Win, "MIMO systems with antenna selection", *IEEE Microw. Mag.*, 5, 46–56 (2004).

Molisch and Zhang 2004 A. F. Molisch and X. Zhang, "FFT-based hybrid antenna selection schemes for spatially correlated MIMO channels", *IEEE Commun. Lett.*, 8(1), 36–38 (2004).

Molisch et al. 1995 A. F. Molisch, J. Fuhl, and E. Bonek, "Pattern distortion of mobile radio base station antennas by antenna masts and roofs", *Proceedings of the 25th European Microwave Conference*, Bologna, pp. 71–76 (1995).

Molisch et al. 2002a A. F. Molisch, M. Steinbauer, and H. Asplund, ""Virtual Cell Deployment Areas" and "Cluster Tracing"-new methods for directional channel modeling in microcells", *IEEE 55th Vehicular Technology Conference (VTC)*, Vol. 3, pp. 1279–1283 (2002).

Molisch et al. 2002b A. F. Molisch, M. Steinbauer, M. Toeltsch, E. Bonek, and R. Thoma, "Capacity of MIMO systems based on measured wireless channels", *IEEE JSAC*, 20, 561–569 (2002).

Molisch et al. 2003 A. F. Molisch, A. Kuchar, J. Laurila, K. Hugl, and R. Schmalenberger, "Geometry-based directional model for mobile radio channels–principles and implementation", *Eur. Trans. Telecommun.*, 14, 351–359 (2003).

Molisch et al. 2004 A. F. Molisch, K. Balakrishnan, C. C. Chong, et al, "IEEE 802.15.4a - final report", IEEE 802 standardization document (2004).

Molisch et al. 2005 A. F. Molisch, M. Z. Win, Yang-Seok Choi and J. H. Winters, "Capacity of MIMO systems with antenna selection," *IEEE Trans. Wireless Comm.* 4, 1759–1772 (2005).

Molisch et al. 2006a A. F. Molisch, H. Asplund, R. Heddergott, M. Steinbauer, and T. Zwick, "The COST259 directional channel model-part I: Overview and methodology", *IEEE Trans. Wirel. Commun.*, 5, 3421–3433 (2006).

Molisch et al. 2006b A. F. Molisch, D. Cassioli, C. C. Chong, et al., "A comprehensive model for ultrawideband propagation channels", *IEEE Trans. Antennas Propag.*, 54, special issue on wireless propagation, 3151–3166 (2006).

Molisch et al. 2007a A. F. Molisch, N. B. Mehta, J. Yedidia, and J. Zhang, "Cooperative relay networks with mutual-information accumulation", *IEEE Trans. Wirel. Commun.*, 6, 4108–4119 (2007).

Molisch et al. 2007b A. F. Molisch, M. Toeltsch, and S. Vermani, "Iterative methods for cancellation of intercarrier interference in OFDM systems", *IEEE Trans. Veh. Technol.*, 56, 2158–2167 (2007).

Molisch et al. 2017 A. F. Molisch, V. V. Ratnam, S. Han, Z. Li, S. L. H. Nguyen, L. Li, and K. Haneda, "Hybrid beamforming for massive MIMO: A survey", *IEEE Commun. Mag.*, 55(9), 134–141 (2017).

Molisch et al. 2021 A. F. Molisch, T. Choi, N. Abbasi, F. Rottenberg, and C. Zhang, "Millimeterwave channels", *Wiley 5G Encyclopedia* (2021).

Molnar et al. 1996 B. G. Molnar, I. Frigyes, Z. Bodnar, and Z. Herczku, "The WSSUS channel model: Comments and a generalisation", *Proc. Globecom*, 1996, 158–162 (1996).

Moon 2020 T. K. Moon, *Error Correction Coding: Mathematical Methods and Algorithms*, John Wiley & Sons, (2020).

Morelli et al. 2007 M. Morelli, C. C. J. Kuo, and M. O. Pun, "Synchronization techniques for orthogonal frequency division multiple access (OFDMA): A tutorial review", *Proc. IEEE*, 95, 1394–1427 (2007).

Moshavi 1996 S. Moshavi, "Multi-user detection for DS-CDMA Communications", *IEEE Commun. Mag.*, 34, 124–136 (1996).

Motley and Keenan 1988 A. J. Motley and J. P. Keenan, "Personal communication radio coverage in buildings at 900 MHz and 1700 MHz", *Electron. Lett.*, 24, 763–764 (1988).

Mouly and Pautet 1992 M. Mouly and M. B. Pautet, *The GSM System for Mobile Communications*, self-publishing, (1992).

MPEG 1 ISO/IEC 11172-2:1993, *Information Technology – Coding of Moving Pictures and Associated Audio for Digital Storage Media at up to about 1.5 Mbit/s – Part 2: Video.* (1993).

MPEG 2 International Telecommunications Union, *Information Technology - Generic Coding of Moving Pictures and Associated Audio Information - Part 2: Video*, 2nd edition, ITU-T Recommendation H.262 and ISO/IEC 13818-2:2000 (2000).

MPEG 4 ISO/IEC 14496-2:2004, *Information Technology – Coding of Audio-Visual Objects – Part 2: Visual*, 3rd edition (2004).

Mudumbai et al. 2009 R. Mudumbai, D. R. Brown, U. Madhow, and H. V. Poor, "Distributed transmit beamforming: Challenges and recent progress", *IEEE Commun. Mag.*, 47, 102–110 (2009).

Muirhead 1982 R. J. Muirhead, *Aspects of Multivariate Statistical Theory*, Wiley, (1982).

Mungara et al. 2015 R. K. Mungara, X. Zhang, A. Lozano, and R. W. Heath, "Performance evaluation of ITLinQ and FlashLinQ for overlaid device-to-device communication", *IEEE International Conference on Communication Workshop (ICCW)*, pp. 596–601 (2015).

Muquet et al. 2002 B. Muquet, Z. Wang, G. B. Giannakis, M. de Courville, and P. Duhamel, "Cyclic prefixing or zero padding for wireless multicarrier transmissions", *IEEE Trans. Commun.*, 50, 2136–2148 (2002).

Murota and Hirade 1981 K. Murota and K. Hirade, "GMSK modulation for digital mobile radio telephony", *IEEE Trans. Commun.*, 29, 1044–1050 (1981).

Nabar et al. 2004 R. U. Nabar, H. Bolcskei, and F. W. Kneubuhler, "Fading relay channels: Performance limits and space-time signal design", *IEEE J. Sel. Areas Commun.*, 22, 1099–1109 (2004).

Naderializadeh and Avestimehr 2014 N. Naderializadeh and A. S. Avestimehr, "ITLinQ: A new approach for spectrum sharing in device-to-device communication systems", *IEEE J. Sel. Areas Commun.*, 32(6), 1139–1151 (2014).

Nakagami 1960 M. Nakagami, "The M-distribution: A general of intensity distribution of rapid fading", in W. Hoffman (ed.), *Statistical Methods of Radio Wave Propagation*, Pergamon Press, (1960).

Namislo 1984 N. Namislo, "Analysis of mobile radio slotted ALOHA networks", *IEEE J. Sel. Areas Commun.*, 2, 583–588 (1984).

Narayanan et al. 2004 R. M. Narayanan, K. Atanassov, V. Stoiljkovic, and G. R. Kadambi, "Polarization diversity measurements and analysis for antenna configurations at 1800 MHz", *IEEE Trans. Antennas Propag.*, 52, 1795–1810 (2004).

Nazer and Gastpar 2008 B. Nazer and M. Gastpar, "Compute-and-forward: A novel strategy for cooperative networks", *42nd Asilomar Conference on Signals, Systems and Computers*, pp. 69–73 (2008).

Nazer et al. 2012 B. Nazer, G. Michael, S. A. Jafar, and S. Vishwanath, "Ergodic interference alignment", *IEEE Trans. Inf. Theory*, 58, 6355–6371 (2012).

Necker 2008 M. C. Necker, "Interference coordination in cellular OFDMA networks", *IEEE Netw.*, 22, 12–19 (2008).

Neely 2006 M. J. Neely, "Energy optimal control for time varying wireless networks", *IEEE Trans. Inf. Theory*, 52, 2915–2934 (2006).

Neely 2010 M. J. Neely, "Stochastic network optimization with application to communication and queueing systems", *Synth. Lect. Commun. Netw.*, 3(1), 1–211 (2010).

Neely and Urgaonkar 2009 M. J. Neely and R. Urgaonkar, "Optimal backpressure routing in wireless networks with multi-receiver diversity", *Ad Hoc Netw. (Elsevier)*, 7, 862–881 (2009).

Negenborn 2009 R. Negenborn, *Robot Localization and Kalman Filters*, Utrecht University, Utrecht, Netherlands, Master's thesis INF/SCR-0309 (2009).

Neubauer et al. 2001 Th. Neubauer, H. Jaeger, J. Fuhl, and E. Bonek, "Measurement of the background noise floor in the UMTS FDD uplink band", *European Personal Mobile Communications Conference* (2001).

Ngo et al. 2013a H. Q. Ngo, E. G. Larsson, and T. L. Marzetta, "Energy and spectral efficiency of very large multiuser MIMO systems", *IEEE Trans. Commun.*, 61, 1436–1449 (2013).

Ngo et al. 2013b H. Q. Ngo, M. Matthaiou, T. Q. Duong, and E. G. Larsson, "Uplink performance analysis of multicell MU-SIMO systems with ZF receivers", *IEEE Trans. Veh. Techn.*, 62, 4471–4483 (2013).

Ngo et al. 2014 H. Q. Ngo, E. G. Larsson, and T. L. Marzetta, "Aspects of favorable propagation in massive MIMO", *22nd European Signal Processing Conference (EUSIPCO)*, pp. 76–80 (2014).

Nilsson et al. 1997 R. Nilsson, O. Edfors, M. Sandell, and P. O. Börjesson, "An analysis of two-dimensional pilot-symbol assisted modulation for OFDM", *Proceedings IEEE International Conference on Personal Wireless Communications*, pp. 71–74, Bombay, India, December (1997).

Nitsche et al. 2014 T. Nitsche, C. Cordeiro, A. B. Flores, E. W. Knightly, E. Perahia, and J. C. Widmer, "IEEE 802.11 ad: Directional 60 GHz communication for multi-Gigabit-per-second Wi-Fi", *IEEE Commun. Mag.*, 52(12), 132–141 (2014).

Noor-A-Rahim et al. 2020, Md. Noor-A-Rahim, Z. Liu, H. Lee, M. O. Khyam, J. He, D. Pesch, K. Moessner, W. Saad, and H. V. Poor, "6g for vehicle-to-everything (v2x) communications: Enabling technologies, challenges, and opportunities", *Proc. IEEE* 110, 712–734 (2020).

Norklit and Andersen 1998 O. Norklit and J. B. Andersen, "Diffuse channel model and experimental results for array antennas in mobile environments", *IEEE Trans. Antennas Propag.*, 46, 834 (1998).

Nosratinia et al. 2004 A. Nosratinia, A. T. E. Hunter, and A. Hedayat, "Cooperative communication in wireless networks", *IEEE Commun. Mag.*, 42(10), 74–80 (2004).

O'Hara and Petrick 2005 B. O'Hara and A. Petrick, *The IEEE 802.11 Handbook: A Designer's Companion*, 2nd edition, IEEE Standards Publications, (2005).

O'Shaughnessy 2000 D. O'Shaughnessy, *Speech Communication: Human and Machine*, 2nd edition, IEEE Press, (2000).

Oehrvik 1994 S. O. Oehrvik, *Radio School*, Ericsson, Stockholm (1994).

Ogawa and Matsuyoshi 2001 K. Ogawa and T. Matsuyoshi, "An analysis of the performance of a handset diversity antenna influenced by head, hand, and shoulder effects at 900 MHz. I. Effective gain characteristics", *IEEE Trans. Veh. Technol.*, 50, 830–844 (2001).

Ogawa et al. 2001 K. Ogawa, T. Matsuyoshi, and K. Monma, "An analysis of the performance of a handset diversity antenna influenced by head, hand, and shoulder effects at 900 MHz. II. Correlation characteristics", *IEEE Trans. Veh. Technol.*, 50, 845–853 (2001).

Ogilvie 1991 J. A. Ogilvie, *Theory of Wave Scattering from Random Rough Surfaces*, IOP Publishing, (1991).

Okumura et al. 1968 Y. Okumura, E. Ohmori, T. Kawano, and K. Fukuda, "Field strength and its variability in VHF and UHF land mobile services", *Rev. Elec. Commun. Lab.*, 16, 825–873 (1968).

Oppenheim et al. 2015 A. V. Oppenheim, A. S. Willsky, and S. H. Nawab, *Signals and Systems*, 2nd edition, Pearson (2015).

Orfanidis 2016 S. J. Orfanidis, *Electromagnetic waves and antennas*, https://www.ece.rutgers.edu/~orfanidi/ewa/, 2016.

Ozgur et al. 2007 A. Ozgur, O. Leveque, and D. N. C. Tse, "Hierarchical cooperation achieves optimal capacity scaling in ad hoc networks", *IEEE Trans. Inf. Theory*, 53, 3549–3572 (2007).

Paetzold 2002 M. Paetzold, *Mobile Fading Channels: Modelling, Analysis, & Simulation*, Wiley (2002); 2nd edition to appear (2010).

Pajusco 1998 P. Pajusco, "Experimental characterization of D.O.A at the base station in rural and urban area", *Proceedings of the IEEE VTC'98*, pp. 993–998 (1998).

Papoulis 1985 A. Papoulis, "Predictable processes and Wold's decomposition: A review", *IEEE Trans. Acoust., Speech, Signal Process. ASSP*, 33, 933 (1985).

Papoulis 1991 A. Papoulis, *Probability, Random Variables, and Stochastic Processes*, 3rd edition, McGraw-Hill, New York (1991).

Parsons 1992 J. D. Parsons, *The Mobile Radio Channel*, Wiley, New York (1992).

Parsons et al. 1991 J. D. Parsons, D. A. Demery, and A. M. D. Turkmani, "Sounding techniques for wideband mobile radio channels: A review", *Proc. Inst. Elect. Eng. – I*, 138, 437–446 (1991).

Pathak and Dutta 2010 P. H. Pathak and R. Dutta, "A survey of network design problems and joint design approaches in wireless mesh networks", *IEEE Commun. Surv. Tutor.*, 13(3), 396–428 (2010).

Paulraj and Kailath 1994 A. J. Paulraj and T. Kailath, *"Increasing capacity in wireless broadcast systems using distributed transmission/directional reception (DTDR)"*, U.S. Patent 5,345,599 (1994).

Paulraj and Papadias 1997 A. J. Paulraj and C. B. Papadias, "Space-time processing for wireless transmissions", *IEEE Pers. Commun.*, 14(5), 49–83 (1997).

Paulraj et al. 2003 A. Paulraj, D. Gore, and R. Nabar, *Multiple Antenna Systems*, Cambridge University Press, (2003).

Paulraj et al. 2004 A. J. Paulraj, D. A. Gore, R. U. Nabar, and H. Bolcskei, "An overview of MIMO communications-a key to gigabit wireless", *Proc. IEEE*, 92(2), 198–218 (2004).

Pawula et al. 1982 R. F. Pawula, S. O. Rice, and J. H. Roberts, "Distribution of the phase angle between two vectors perturbed by Gaussian noise", *IEEE Trans. Commun.*, 30, 1828–1841 (1982).

Pedersen et al. 1997 K. Pedersen, P. E. Mogensen, and B. Fleury, "Power azimuth spectrum in outdoor environments", *IEEE Electron. Lett.*, 33, 1583–1584 (1997).

Pedersen et al. 1998 G. F. Pedersen, J. O. Nielsen, K. Olensen, and I. Z. Kovács, "Measured variation in performance of handheld antennas for a large number of test persons," COST259 TD(98)025, (1998).

Peel 2003 C. B. Peel, "On dirty-paper coding", *IEEE Signal Process. Mag.*, 20(3), 112–113 (2003).

Pelgrom 2017 M. Pelgrom, *Analog to Digital Conversion*, 3rd edition, Springer, (2017).

Peng et al. 2013 M. Peng, D. Liang, Y. Wei, J. Li, and H. H. Chen, "Self-configuration and self-optimization in LTE-advanced heterogeneous networks", *IEEE Commun. Mag.*, 51(5), 36–45 (2013).

Peng et al. 2016 W. Peng, F. G. Walls, R. A. Cohen, J. Xu, J. Ostermann, A. MacInnis, and T. Lin, "Overview of screen content video coding: Technologies, standards, and beyond", *IEEE J. Emerg. Sel. Topics Circuits Syst.*, 6(4), 393–408 (2016).

Perahia and Stacey 2015 E. Perahia and R. Stacey, *Next Generation Wireless LANs (802.11n and 802.11ac)*, 2nd edition, Cambridge University Press, (2015).

del Peral-Rosado et al. 2017 J. A. del Peral-Rosado, R. Raulefs, J. A. López-Salcedo, and S.- G. Granados, "Survey of cellular mobile radio localization methods: From 1G to 5G", *IEEE Commun. Surv. Tutor.*, 20(2), 1124–1148 (2017).

Perera et al. 2018 T. D. P. Perera, D. N. K. Jayakody, S. K. Sharma, S. Chatzinotas and J. Li, "Simultaneous Wireless Information and Power Transfer (SWIPT): Recent Advances and Future Challenges," *IEEE Communications Surveys & Tutorials*, 20(1), pp. 264–302 (2018).

Perkins and Royer 1999 C. E. Perkins and E. M. Royer, "Ad hoc on-demand distance vector routing", *Proceedings of 2nd IEEE Workshop on Mobile Computing Systems and Applications*, New Orleans, LA, February, pp. 90–100 (1999).

Peter 2017 M. Peter (ed.), "Measurement results and final mmmagic channel models: Specular wall reflections and diffused scattering measurements", *Millimetre-Wave Based Mobile RadioAccess Network for Fifth Generation Integrated Communications (mmMAGIC)*, report H2020-ICT-671650-mmMAGIC/D2.2 (2017).

Peterson and Davie 2003 L. L. Peterson and B. S. Davie, *Computer Networks: A Systems Approach*, 3rd edition, Academic Press, (2003).

Petrus et al. 2002 P. Petrus, J. H. Reed, and T. S. Rappaport, "Geometrical-based statistical macrocell channel model for mobile environments", *IEEE Trans. Commun.*, 50, 495 (2002).

van der Plassche 2003 R. van der Plassche, *CMOS Integrated Analog-To-Digital and Digital-To-Analog Converters*, 2nd edition, Kluwer, (2003).

Plenge 1997 C. Plenge, "Leistungsbewertung öffentlicher DECT-Systeme", Dissertation, RWTH Aachen (1997) [in German].

Polese et al. 2020 M. Polese, M. Giordani, T. Zugno, A. Roy, S. Goyal, D. Castor, and M. Zorzi, "Integrated access and backhaul in 5G mm wave networks: Potential and challenges", *IEEE Commun. Mag.*, 58(3), 62–68 (2020).

Polyanski et al. 2010 Y. Polyanskiy, H. V. Poor, and S. Verdú, "Channel coding rate in the finite blocklength regime", *IEEE Trans. Inf. Theory*, 56, 2307–2359 (2010).

Polydoros and Weber 1984 A. Polydoros and C. L. Weber, "A unified approach to serial search spread-spectrum code acquisition-part 1: General theory", *IEEE Trans. Commun.*, 32, 542–549; "Part II: Matched filter receiver", *IEEE Trans. Commun.*, 32, 550–560 (1984).

Poor 2001 H. V. Poor, "Turbo multiuser detection: A primer", *J. Commun. Netw.*, 3, 196–201 (2001).

Poor 2004 H. V. Poor, "Iterative multiuser detection", *IEEE Signal Process. Mag.*, 21, 81–88 (2004).

Pozar 2005 D. M. Pozar, *Microwave Engineering*, 3rd edition, Wiley, (2005).

Proakis 1968 J. G. Proakis, "On the probability of error for multichannel reception of binary signals", *IEEE Trans. Commun.*, 16, 68–71 (1968).

Proakis 1991 J. G. Proakis, "Adaptive equalization for TDMA digital mobile radio", *IEEE Trans. Veh. Technol.*, 40, 333–341 (1991).

Proakis and Salehi 2005 J. G. Proakis and M. Salehi, *Digital Communications*, 5th edition, McGraw Hill, New York (2005).

Qiu 2002 R. C. Qiu, "A study of the ultra-wideband wireless propagation channel and optimum UWB receiver design", *IEEE J. Sel. Areas Commun.*, 20, 1628–1637 (2002).

Qiu 2004 R. C. Qiu, "A generalized time domain multipath channel and its application in ultra-wideband (UWB) wireless optimal receiver design – Part II: Physics-based system analysis", *IEEE Trans. Wirel. Commun.*, 3, 2312–2324 (2004).

Qu et al 2019 Q. Qu, B. Li, M. Yang, Z. Yan, A. Yang, D. J. Deng, and K. C. Chen, "Survey and performance evaluation of the upcoming next generation WLANs standard-IEEE 802.11 ax", *Mob. Netw. Appl.*, 24(5), 1461–1474 (2019).

Quan et al. 2008 Z. Quan, S. Ciu, A. H. Sayed, and H. V. Poor, "Wideband spectrum sensing in cognitive radio networks", *Proceedings of the IEEE International Conference on Communication* (2008).

Quatieri 2002 T. F. Quatieri, *Discrete-Time Speech Signal Processing*, Prentice Hall, Upper Saddle River (2002).

Rabiner 1994 L. Rabiner, "Applications of voice processing to telecommunications", *Proc. IEEE*, 82, 199–228 (1994).

Rabiner and Schafer 1978 L. R. Rabiner and R. W. Schafer, *Digital Processing of Speech Signals*, Prentice Hall, Englewood Cliffs (1978).

Rajan and van der Veen 2011 R. T. Rajan, and A. J. van der Veen, "Joint ranging and clock synchronization for a wireless network", *IEEE International Workshop on Computational Advances in Multi-Sensor Adaptive Processing (CAMSAP)*, pp. 297–300 (2011).

Raleigh and Cioffi 1998 G. G. Raleigh and J. M. Cioffi, "Spatio-temporal coding for wireless communication", *IEEE Trans. Commun.*, 46, 357–366 (1998).

Ramachandran et al. 2004 I. Ramachandran, Y. P. Nakache, P. Orlik, J. Zhang, and A. F. Molisch, "Symbol spreading for ultrawideband systems based on multiband OFDM", *Proceedings of the Personal, Indoor, Mobile Radio Symposium*, pp. 1204–1209 (2004).

Ramo et al. 1967 S. Ramo, J. R. Whinnery, and T. van Duzer, *Fields and Waves in Communication Electronics*, Wiley, New York (1967).

Rao and Yip 1990 K. R. Rao and P. Yip, *Discrete Cosine Transform: Algorithms, Advantages, Applications*, Academic Press, (1990).

Rappaport 1996 T. S. Rappaport, *Wireless Communications – Principles and Practice*, 2nd edition 2001, Prentice Hall, (1996).

Rappaport 1998 T. S. Rappaport (ed.), *Smart Antennas: Adaptive Arrays, Algorithms, and Wireless Position Location*, IEEE Press, (1998).

Rappaport et al. 2013 T. S. Rappaport, S. Sun, R. Mayzus, H. Zhao, et al., "Millimeter Wave Mobile Communications for 5G Cellular: It Will Work!," in *IEEE Access* 1, 335–349 (2013).

Rappaport et al. 2015a T. S. Rappaport, G. R. MacCartney, M. K. Samimi and S. Sun, "Wideband Millimeter-Wave Propagation Measurements and Channel Models for Future Wireless Communication System Design," *IEEE Transactions on Communications*, 63, 3029–3056 (2015).

Rappaport et al. 2015b T. S. Rappaport, R. W. Heath, R. C. Daniels, J. N. Murdock, *Millimeter wave wireless communications*, Pearson (2015).

Rappaport et al. 2019 T. S. Rappaport, Y. Xing, O. Kanhere, et al., "Wireless Communications and Applications Above 100 GHz: Opportunities and Challenges for 6G and Beyond." *IEEE Access*, 7, pp. 78729–78757 (2019)

Rappaport et al. 2022 T. S. Rappaport, K. S. Remley, C. Gentile, A. F. Molisch, and A. Zajic, *Radio Propagation Measurements and Channel Modeling: Best Practices for Millimeter-Wave and Sub-Terahertz Frequencies*, Cambridge University Press, (2022).

Ratnam et al. 2020 V. V. Ratnam, H. Chen, S. Pawar, B. Zhang, C. J. Zhang, Y. J. Kim, S. Lee, M. Cho, and S. R. Yoon, "FadeNet: Deep learning-based mmwave large-scale channel fading prediction and its applications", *IEEE Access*, 9, 3278–3290 (2020).

Razawi 2011 B. Razawi, *RF Microelectronics*, 2nd edition, Prentice Hall, (2011).

Reed 2005 J. H. Reed, *An Introduction to Ultra Wideband Communication Systems*, Prentice Hall, (2005).

Ren et al. 2014 Y. Ren, L. Li, G. Xie, et al., "Experimental demonstration of 16 Gbit/s millimeter-wave communications using MIMO processing of 2 OAM modes on each of two transmitter/receiver antenna apertures," *2014 IEEE Global Communications Conference*, pp. 3821–3826 (2014).

Reznik et al. 2007 Y. A. Reznik, A. T. Hinds, C. Zhang, L. Yu, and Z. Ni, "Efficient fixed-point approximations of the 8x8 inverse discrete cosine transform", *Proc. SPIE*, 6696, 1–17 (2007).

Rice 1947 S. O. Rice, "Statistical properties of a sine wave plus random noise", *Bell Syst. Techn. J.*, 27, 109–157 (1947).

Rice 2008 M. D. Rice, *Digital Communications: A Discrete-Time Approach*, Prentice Hall, (2008).

Richards 2014 M. Richards, *Fundamentals of Radar Signal Processing*, 2nd edition, McGraw Hill, (2014).

Richardson 2005 A. Richardson, *WCDMA Design Handbook*, Cambridge University Press, (2005).

Richardson and Urbanke 2008 T. Richardson and R. Urbanke, *Modern Coding Theory*, Cambridge University Press, (2008).

Richardson et al. 2001 T. J. Richardson, M. A. Shokrollahi, and R. L. Urbanke, "Design of capacity-approaching irregular low-density parity-check codes", *IEEE Trans. Inf. Theory*, 47, 619–637 (2001).

Richter 2005 A. Richter, *"Estimation of Radio Channel Parameters: Models and Algorithms"*, PhD thesis, Technical University Ilmenau (2005).

Richter 2006 A. Richter, "The contribution of distributed diffuse scattering in radio channels to channel capacity: Estimation and modelling", *Proceedings of Asilomar Conference on Signals, Systems, and Computers*, Pacific Grove, CA, USA, October (2006).

Rohde and Rudolph 2013 U. L. Rohde and M. Rudolph, *RF/Microwave Circuit Design for Wireless Applications*, John Wiley & Sons, (2013).

Rogalin et al. 2014 R. Rogalin, O. Y. Bursalioglu, H. Papadopoulos, G. Caire, et al., "Scalable Synchronization and Reciprocity Calibration for Distributed Multiuser MIMO," *IEEE Trans. Wireless Comm.* 13, 1815–1831 (2014).

Röhrig and Müller 2009 C. Röhrig, and M. Müller, "Indoor location tracking in non-line-of-sight environments using a IEEE 802.15. 4a wireless network", *IEEE/RSJ International Conference on Intelligent Robots and Systems*, pp. 552–557 (2009).

Rommer et al. 2019 S. Rommer, P. Hedman, M. Olsson, L. Frid, S. Sultana, and C. Mulligan, *5G Core Networks*, Academic Press, (2019).

Roth et al. 2018 K. Roth, H. Pirzadeh, A. L. Swindlehurst, and J. A. Nossek, "A comparison of hybrid beamforming and digital beamforming with low-resolution ADCs for multiple users and imperfect CSI", *IEEE J. Sel. Topics Signal Process.*, 12, 484–498 (2018).

Roy and Fortier 2004 S. Roy and P. Fortier, "A closed-form analysis of fading envelope correlation across a wideband basestation array", *IEEE Trans. Wirel. Commun.*, 3, 1502–1507 (2004).

Roy et al. 1986 R. Roy, A. Paulraj, and T. Kailath, "ESPRIT – A subspace rotation approach to estimation of parameters of cisoids in noise", *IEEE Trans. Acoust., Speech, Signal Process.*, 34, 1340–1342 (1986).

Roy et al. 2004 S. Roy, J. R. Foerster, V. S. Somayazulu, and D. G. Leeper, "Ultrawideband radio design: The promise of high-speed, short-range wireless connectivity", *Proc. IEEE*, 92, 295–311 (2004).

Rubin 1979 I. Rubin, "Message delays in FDMA and TDMA communication channels", *IEEE Trans. Commun.*, 27, 769–777 (1979).

Rupp et al. 2016 M. Rupp, S. Schwartz, and M. Taranetz, *The Vienna LTE-Advanced Simulators: Up and Downlink, Link and System Level Simulation*, Springer, (2016).

Rusek et al. 2012 F. Rusek, D. Persson, B. K. Lau, E. G. Larsson, T. L. Marzetta, O. Edfors, and F. Tufvesson, "Scaling up MIMO: Opportunities and challenges with very large arrays", *IEEE Signal Process. Mag.*, 30(1), 40–60 (2012).

Sabharwal et al. 2014 A. Sabharwal, P. Schniter, D. Guo, D. W. Bliss, S. Rangarajan, and R. Wichman, "In-band full-duplex wireless: Challenges and opportunities", *IEEE J. Sel. Areas Commun.*, 32(9), 1637–1652 (2014).

Sadek et al. 2007 M. Sadek, A. Tarighat, and A. H. Sayed, "A leakage-based precoding scheme for downlink multi-user MIMO channels", *IEEE Trans. Wirel. Commun.*, 6, 1711–1721 (2007).

Sadr et al. 2009 S. Sadr, A. Anpalagan, and K. Raahemifar, "Radio resource allocation algorithms for the downlink of multiuser OFDM communication systems", *IEEE Commun. Surv. Tutor.*, 11(3), 92–106 (2009).

Sagias and Karagiannidis 2005 N. C. Sagias and G. K. Karagiannidis, "Gaussian class multivariate Weibull distributions: Theory and applications in fading channels", *IEEE Trans. Inf. Theory*, 51, 3608–3619 (2005).

Sahin et al. 2013 A. Sahin, I. Guvenc, and H. Arslan, "A survey on multicarrier communications: Prototype filters, lattice structures, and implementation aspects", *IEEE Commun. Surv. Tutor.*, 16, 1312–1338 (2013).

Sahinoglu et al. 2008 Z. Sahinoglu, G. Sinan, and G. Ismail, *Ultra-Wideband Positioning Systems*, Cambridge University Press, New York (2008).

Saito et al. 2013 Y. Saito, Y. Kishiyama, A. Benjebbour, T. Nakamura, A. Li, and K. Higuchi, "Non-orthogonal multiple access (NOMA) for cellular future radio access", *IEEE 77th Vehicular Technology Conference (VTC Spring)*, pp. 1–5 (2013).

Saleh and Valenzuela 1987 A. Saleh and R. A. Valenzuela, "A statistical model for indoor multipath propagation", *IEEE J. Sel. Areas Commun.*, 5, 128 (1987).

Salmi et al. 2009 J. Salmi, A. Richter, and V. Koivunen, "Detection and tracking of MIMO propagation path parameters using state-space approach", *IEEE Trans. Signal Process.*, 57, 1538–1550 (2009).

Salous 2013 S. Salous, *Radio Propagation Measurement and Channel Modelling*, John Wiley & Sons, (2013).

Sangodoyin and Molisch 2018 S. Sangodoyin and A. F. Molisch, "Impact of body mass index on ultrawideband MIMO BAN channels—measurements and statistical model", *IEEE Trans. Wirel. Commun.*, 17, 6067–6081 (2018).

de Santo and Brown 1986 J. A. de Santo and G. S. Brown, "Analytical Techniques for Multiple Scattering from Rough Surfaces" in E. Wolf (ed.), *Progress in Optics*, Vol. 23, Elsevier (1986).

Sari et al. 2000 H. Sari, F. Vanhaverbeke, and M. Moeneclaey, "Extending the capacity of multiple access channels", *IEEE Commun. Mag.*, 38(1), 74–82 (2000).

Sarieddeen et al. 2021 H. Sarieddeen, M. S. Alouini, and T. Y. Al-Naffouri, "An overview of signal processing techniques for terahertz communications", *Proceedings of the IEEE* 109, 10, 1628–1665 (2021).

Sarkar et al. 2013 S. K. Sarkar, T. G. Basavaraju, and C. Puttamadappa, *Ad hoc Mobile Wireless Networks: Principles, Protocols and Applications*, 2nd edition, CRC Press, (2013).

Sarwate and Pursely 1980 D. V. Sarwate and M. B. Pursley, "Crosscorrelation properties of pseudorandom and related sequences", *Proc. IEEE*, 68, 593–619 (1980).

Sato 1975 Y. Sato, "A method of self recovering equalization for multilevel amplitude modulation", *IEEE Trans. Commun.*, 23, 679–682 (1975).

Sayed and Kailath 2001 A. H. Sayed and T. Kailath, "A survey of spectral factorization methods", *J. Num. Linear Algebra Appl.*, 8, 467–496 (2001).

Sayed 2008 A. H. Sayed, "Adaptive Filters", John Wiley & Sons, 2008.

Sayeed 2002 A. M. Sayeed, "Deconstructing multiantenna fading channels", *IEEE Trans. Signal Process.*, 50, 2563 (2002).

Sayre 2008 C. W. Sayre, *Complete Wireless Design*, McGraw Hill, 2nd edition (2008).

Scaglione et al. 2006 A. Scaglione, D. L. Goeckel, and N. J. Laneman, "Cooperative communications in mobile ad-hoc networks", *IEEE Signal Process. Mag.*, 2006, 18–29 (2006).

Schaumann et al. 2008 R. Schaumann, H. Xiao, and M. van Valkenburg, *Design of Analog Filters*, 2nd edition, Oxford University Press, (2008).

Schiller 2003 J. Schiller, *Mobile Communications*, 2nd edition, Addison-Wesley, (2003).

Schlegel and Perez 2003 C. B. Schlegel and L. C. Perez, *Trellis and Turbo Coding*, Wiley, (2003).

Schmidl and Cox 1997 T. M. Schmidl and D. C. Cox, "Robust frequency and timing synchronization for OFDM", *IEEE Trans. Commun.*, 45, 1613–1621 (1997).

Schmidt 1986 R. Schmidt, "Multiple emitter location and signal parameters estimation", *IEEE Trans. Antennas Propag.*, 34, 276–280 (1986).

Schniter 2004 P. Schniter, "Low-complexity equalization of OFDM in doubly selective channels", *IEEE Trans. Signal Process.*, 52, 1002–1011 (2004).

Scholtz 1982 R. A. Scholtz, "The origins of spread spectrum communications", *IEEE Trans. Commun.*, 30, 822–854 (1982).

Schroeder and Atal 1985 M. Schroeder and B. Atal, "Code-excited linear prediction (CELP): High-quality speech at very low bit rates", *Proceedings of the IEEE International Conference Acoustics, Speech, Signal Process., ICASSP'85'*, pp. 937–940 (1985).

Schroeder et al. 1979 M. R. Schroeder, B. S. Atal, and J. L. Hall, "Optimizing digital speech coders by exploiting masking properties of the human ear", *J. Acoust. Soc. Am.*, 66, 1647–1652 (1979).

Schwarz et al. 2007 H. Schwarz, D. Marpe, and T. Wiegand, "Overview of the scalable video coding extension of the H.264/AVC standard", *IEEE Trans. Circ. Syst. Video Technol.*, 17, 1103–1120 (2007).

Sedaghat et al. 2014 M. A. Sedaghat, R. R. Mueller, and G. Fischer, "A novel single-RF transmitter for massive MIMO. In WSA 2014", *18th International ITG Workshop on Smart Antennas*, pp. 1–8 (2014).

Sendonaris et al. 2003 A. Sendonaris, E. Erkip, and B. Aazhang, "User cooperation diversity–Part I: System description and user cooperation diversity–Part II: Implementation aspects and performance analysis", *IEEE Trans. Commun.*, 51, 1927–1948 (2003).

Serbetli and Yener 2004 S. Serbetli, and A. Yener, "Transceiver optimization for multiuser MIMO systems", *IEEE Transactions on Signal Processing*, 52, 214–226 (2004).

Sesia et al. 2011 S. Sesia, I. Toufik, and M. Baker, *LTE-the UMTS Long Term Evolution: From Theory to Practice*, John Wiley & Sons, (2011).

Shafi et al. 2006 M. Shafi, M. Zhang, A. L. Moustakas, P. J. Smith, A. F. Molisch, F. Tufvesson, and S. H. Simon, "Polarized mimo channels in 3-d: Models, measurements and mutual information", *IEEE J. Sel. Areas Commun.*, 24(3), 514–527 (2006).

Shafi et al. 2017 M. Shafi, A. F. Molisch, P. J. Smith, T. Haustein, P. Zhu, P. De Silva, F. Tufvesson, A. Benjebbour, and G. Wunder, "5G: A tutorial overview of standards, trials, challenges, deployment, and practice", *IEEE J. Sel. Areas Commun.*, 35(6), 1201–1221 (2017).

Shafi et al. 2018 M. Shafi, J. Zhang, H. Tataria, A. F. Molisch, et al., "Microwave vs. Millimeter Wave Propagation Channels: Key Differences and Impact on 5G Cellular Systems," *IEEE Comm. Mag.* 56(12), 14–20 (2018).

Shannon 1948 C. E. Shannon, "A mathematical theory of communication", *Bell System Tech. J.*, 27, 379–423 and 623–656 (1948).

Shannon 1949 C. E. Shannon, "Communication in the presence of noise", *Proc. IRE*, 37, 10–21 (1949).

Shannon 1959 C. E. Shannon, "Coding theorems for a discrete source with a fidelity criterion", *IRE Natl. Convention Record*, 4, 142–163 (1959).

Shen et al. 2006 S. Shen, M. Guizani, R. C. Qiu, and T. L. Ngog, *Ultra-Wideband Wireless Communications and Networks*, Wiley, (2006).

Shen et al. 2010 Y. Shen, W. Henk, and M. Z. Win, "Fundamental limits of wideband localization---Part I: A general framework", *IEEE Trans. Inf. Theory*, 56, 4956–4980 (2010), "Fundamental limits of wideband localization – part II: Cooperative networks." *IEEE Trans. Inf. Theory*, 56, 4981–5000 (2010).

Shepard et al. 2012 C. Shepard, H. Yu, N. Anand, E. Li, T. Marzetta, R. Yang, and L. Zhong, "Argos: Practical many-antenna base stations", *18th Annual International Conference on Mobile Computing and Networking*, pp. 53–64 (2012).

Shi and Sun 2000 Y. Q. Shi and H. Sun, *Image and Video Compression for Multimedia Engineering: Fundamentals, Algorithms and Standards*, 2nd edition, CRC Press, (2000).

Shi et al. 2007 S. Shi, M. Schubert, and H. Boche, "Downlink MMSE transceiver optimization for multiuser MIMO systems: Duality and sum-MSE minimization", *IEEE Trans. Signal Process.*, 55, 5436–5446 (2007).

Shi et al. 2011 Q. Shi, M. Razaviyayn, Z. Q. Luo, and C. He, "An iteratively weighted MMSE approach to distributed sum-utility maximization for a MIMO interfering broadcast channel", *IEEE Trans. Signal Process.*, 59, 4331–4340 (2011).

Shiu et al. 2000 D. Shiu, G. J. Foschini, M. J. Gans, and J. M. Kahn, "Fading correlation and its effect on the capacity of multielement antenna systems", *IEEE Trans. Commun.*, 48, 502–513 (2000).

Shortle et al. 2018 J. F. Shortle, J.M. Thompson, D. Gross, and C. M. Harris, *Fundamentals of queueing theory* 5th ed., Wiley (2018).

Simeone 2018 O. Simeone, "A Brief Introduction to Machine Learning for Engineers", *Foundations and Trends in Signal Processing*, 12(3-4), 200–431 (2018).

Simon and Alouini 2004 M. K. Simon and M. S. Alouini, *Digital Communications Over Fading Channels*, 2nd edition, Wiley, (2004).

Simon et al. 1994 M. K. Simon, J. K. Omura, R. A. Scholtz, and B. K. Levitt, *Spread Spectrum Communications Handbook*, Revised edition, McGraw Hill, (1994).

Singh et al. 2010 S. Singh, N. B. Mehta, and A. F. Molisch, "Moment-matched lognormal modeling of uplink interference with power control and cell selection", *IEEE Trans. Wirel. Commun.*, 9, 932–938 (2010).

Sklar 1997 B. Sklar, "A primer on turbo code concepts", *IEEE Commun. Mag.*, 35, 94–102 (1997).

Sklar 2001 B. Sklar, *Digital Communications – Fundamentals and Applications*, 2nd edition, Prentice Hall, (2001).

Sklar and Harris 2004 B. Sklar and F. J. Harris, "The ABCs of linear block codes", *IEEE Signal Process. Mag.*, 21(4), 14–35 (2004).

Skolnik 2002 M. Skolnik, *Introduction to Radar Systems*, 3rd edition, McGraw Hill, (2002).

Skolnik 2008 M. Skolnik, *Radar Handbook*, 3rd edition, McGraw Hill, (2008).

Sodagar 2011 I. Sodagar, "The MPEG-DASH standard for multimedia streaming over the internet", *IEEE MultiMedia*, 18(4), 62–67 (2011).

Spencer et al. 2004a Q. H. Spencer, C. B. Peel, A. L. Swindlehurst, and M. Haardt, "An introduction to the multi-user MIMO downlink", *IEEE Commun. Mag.*, 42(10), 60–67 (2004).

Spencer et al. 2004b Q. H. Spencer, A. L. Swindlehurst, and M. Haardt, "Zero-forcing methods for downlink spatial multiplexing in multiuser MIMO channels", *IEEE Trans. Signal Process.*, 52, 461–471 (2004).

Spencer et al. 2006 Q. H. Spencer, J. W. Wallace, C. B. Peel, et al., "Performance of multi-user spatial multiplexing with measured channel data", in G. Tsoulos (ed.), *MIMO System Technology for Wireless Communications*, CRC Press, (2006).

Speth et al. 1999 M. Speth, S. A. Fechtel, G. Fock, and H. Meyr, "Optimum receiver design for wireless broadband systems using OFDM. I", *IEEE Trans. Commun.*, 47, 1668–1677 (1999); "II A case study", *IEEE Trans. Commun.*, 49, 571–578 (2001).

Spyropoulos et al. 2005 T. Spyropoulos, K. Psounis, and C. S. Raghavendra, "Spray and wait: An efficient routing scheme for intermittently connected mobile networks", *Proceedings of the 2005 ACM SIGCOMM Workshop on Delay-tolerant Networking*, pp. 252–259 (2005).

Stankovic et al. 2006 V. Stankovic, A. Host-Madsen, and Z. Xiong, "Cooperative diversity for wireless ad hoc networks", *IEEE Signal Process. Mag.*, 23(5), 37–49 (2006).

Steele and Hanzo 1999 R. Steele and L. Hanzo, *Mobile Communications*, 2nd edition, Wiley, (1999).

Steele et al. 2001 R. Steele, C. C. Lee, and P. Gould, *GSM, cdmaOne and 3G Systems*, John Wiley & Sons, (2001).

Steendam and Moenclaey 1999 H. Steendam and M. Moeneclaey, "Analysis and optimization of the performance of OFDM on frequency-selective time-selective fading channels", *IEEE Trans. Commun.*, 47, 1811–1819 (1999).

Stein 1964 S. Stein, "Unified analysis of certain coherent and noncoherent binary communications systems", *IEEE Trans. Inf. Theory*, 11, 239–246 (1964).

Steinbauer et al. 2001 M. Steinbauer, A. F. Molisch, and E. Bonek, "The double-directional radio channel", *IEEE Antenn. Propag. Mag.*, 43(4), 51–63 (2001).

Stojmenovic 2002 I. Stojmenovic, "Position-based routing in ad-hoc networks", *IEEE Commun. Mag.*, 40, 128–134 (2002).

Stojnic et al. 2006 M. Stojnic, H. Vikalo, and B. Hassibi, "Rate maximization in multi-antenna broadcast channels with linear preprocessing", *IEEE Trans. Wirel. Commun.*, 5, 2338–2342 (2006).

Studer and Bölcskei 2010 C. Studer and H. Bölcskei, "Soft–input soft–output single tree-search sphere decoding", *IEEE Trans. Inf. Theory*, 56, 4827–4842 (2010).

Stueber 2017 G. Stueber, *Principles of Mobile Communication*, 4th ed., Springer (2017).

Stueber et al. 2004 G. L. Stuber, J. R. Barry, S. W. Mclaughlin, Y. Li, M. A. Ingram, and T. G. Pratt, "Broadband MIMO-OFDM wireless communications", *Proc. IEEE*, 92(2), 271–294 (2004).

Stutzman and Thiele 1997 W. L. Stutzman and G. A. Thiele, *Antenna Theory and Design*, 2nd edition, Wiley, New York (1997).

Su and Pamarti 2009 P. Su and S. Pamarti, "Fractional-phase-locked-loop-based frequency synthesis: A tutorial", *IEEE Trans. Circuits Syst. II: Express Briefs*, 56(12), 8881–8885 (2009).

Sudarshan et al. 2006 P. Sudarshan, N. B. Mehta, A. F. Molisch, and J. Zhang, "Channel statistics-based RF pre-processing with antenna selection", *IEEE Trans. Wirel. Commun.*, 5, 3501–3511 (2006).

Sullivan et al. 2012 G. J. Sullivan, J. R. Ohm, W. J. Han, and T. Wiegand, "Overview of the high efficiency video coding (HEVC) standard", *IEEE Trans. Circuits Syst. Video Technol.*, 22, 1649–1668 (2012).

Sun and Reibman 2001 M. T. Sun and A. R. Reibman (eds), *Compressed Video Over Networks*, Marcel Dekker, (2001).

Sun et al. 2014 W. Sun, Z. Yang, X. Zhang, and Y. Liu, "Energy-efficient neighbor discovery in mobile ad hoc and wireless sensor networks: A survey", *IEEE Commun. Surv. Tutor.*, 16(3), 1448–1459 (2014).

Suzuki 1977 H. Suzuki, "A statistical model for urban radio propagation", *IEEE Trans. Commun.*, 25, 673–680 (1977).

Suzuki 1982 H. Suzuki, "Canonic receiver analysis for M-ary angle modulations in Rayleigh fading environment", *IEEE Trans. Veh. Technol.*, 31, 7–14 (1982).

Swarts et al. 1998 F. Swarts, P. van Rooyan, I. Oppermann, and M. P. Lötter, *CDMA Techniques for Third Generation Mobile Systems*, Kluwer, (1998).

Sweeney 2002 P. Sweeney, *Error Control Coding: From Theory to Practice*, Wiley, (2002).

Szyszkowicz et al. 2010 S. S. Szyszkowicz, H. Yanikomeroglu, and J. S. Thompson, "On the feasibility of wireless shadowing correlation models", *IEEE Trans. Veh. Technol.*, 59, 4222–4236 (2010).

Taga 1990 T. Taga, "Analysis for mean effective gain of mobile antennas in land mobile radio environments", *IEEE Trans. Veh. Technol.*, 39, 117–131 (1990).

Taga 1993 T. Taga, "Characteristics of space-diversity branch using parallel dipole antennas in mobile radio communications", *Electron. Commun. Jpn.*, 76 (Pt. 1), 55–65 (1993).

Tahir et al. 2017 B. Tahir, S. Schwarz, and M. Rupp, "BER comparison between Convolutional, Turbo, LDPC, and Polar codes", *2017 24th International Conference on Telecommunications (ICT)*, pp. 1–7 (2017).

Tamburini et al. 2012 F. Tamburini, E. Mari, A. Sponselli, B. Thidé, A. Bianchini, and F. Romanato, "Encoding many channels on the same frequency through radio vorticity: First experimental test", *New J. Phys.*, 14(3), 033001 (2012).

Tarokh et al. 1998 V. Tarokh, N. Seshadri, and A. R. Calderbank, "Space-time coding for high data rate wireless communication: Performance criterion and code construction", *IEEE Trans. Inf. Theory*, 44, 744–765 (1998).

Tarokh et al. 1999 V. Tarokh, H. Jafarkhani, and A. R. Calderbank, "Space–time block codes from orthogonal designs", *IEEE Trans. Inf. Theory*, 45, 1456–1467 (1999).

Tassiulas and Ephremides 1990 L. Tassiulas, and A. Ephremides, "Stability properties of constrained queueing systems and scheduling policies for maximum throughput in multihop radio networks", *IEEE Conference on Decision and Control*, pp. 2130–2132 (1990).

Tataria et al. 2021 H. Tataria, M. Shafi, A. F. Molisch, M. Dohler, H. Sjöland, and F. Tufvesson, "6G wireless systems: Vision, requirements, challenges, insights, and opportunities", *Proc. IEEE*, 109, 1166–1199 (2021).

Tech et al. 2016 G. Tech, Y. Chen, K. Müller, J. Ohm, A. Vetro, and Y. Wang, "Overview of the multiview and 3D extensions of high efficiency video coding", *IEEE Trans. Circuits Syst. Video Technol.*, 26, 35–49 (2016).

Tehrani and Caire 2014 A. S. Tehrani, and G. Caire, "Zigzag neighbor discovery in wireless networks", *IEEE International Symposium on Information Theory*, pp. 1386–1390 (2014).

Telatar 1999 I. E. Telatar, "Capacity of multi-antenna Gaussian channels", *Eur. Trans. Telecommun.*, 10, 585–595 (1999).

Thomae et al. 2000 R. S. Thomae, D. Hampicke, A. Richter, et al., "Identification of time-variant directional mobile radio channels", *IEEE Trans. Instrum. Meas.*, 49, 357–364 (2000).

Thomae et al. 2005 R. S. Thomae, M. Landmann, A. Richter, U. Trautwein, "Multidimensional high-resolution channel sounding", *Smart Antennas in Europe – State-of-the-Art*, EURASIP Book Series, p. 27, Hindawi Publishing Corporation (2005).

Tiedeman 2001 E. Tiedemann, "CDMA2000 1X: New capabilities for CDMA networks", *IEEE Veh. Technol. Soc. News*, 48(4), 4–12 (2001).

Thjung and Chai 1999 T. T. Tjhung and C. C. Chai, "Fade statistics in Nakagami-lognormal channels", *IEEE Trans. Commun.*, 47, 1769–1772 (1999).

Tobagi 1980 F. Tobagi, "Multiaccess protocols in packet communication systems", *IEEE Trans. Commun.*, 28, 468–488 (1980).

Toeltsch et al. 2002 M. Toeltsch, J. Laurila, K. Kalliola, A. F. Molisch, P. Vainikainen and E. Bonek, "Statistical characterization of urban spatial radio channels," *IEEE Journal on Selected Areas in Communications*, 20, 539–549 (2002).

Tomiuk et al. 1999 B. R. Tomiuk, N. C. Beaulieu, and A. A. Abu-Dayya, "General forms for maximal ratio diversitywith weighting errors", *IEEE Trans. Commun.*, 47(4), 488–492 (1999).

Tong et al. 1994 L. Tong, G. Xu, and T. Kailath, "Blind identification and equalization based on second-order statistics: A time domain approach", *IEEE Trans. Inf. Theory*, 40, 340–349 (1994).

Tong et al. 1995 L. Tong, G. Xu, B. Hassibi, and T. Kailath, "Blind identification and equalization based on second-order statistics: A frequency domain approach", *IEEE Trans. Inf. Theory*, 41, 329–334 (1995).

Tse and Visvanath 2005 D. Tse and P. Visvanath, *Fundamentals of Wireless Communications*, Cambridge University Press, (2005).

Tsoulos 2001 G. V. Tsoulos, *Adaptive Antennas for Wireless Communications*, IEEE Press, (2001).

Tsoulos 2006 G. Tsoulos (ed.), *MIMO Antenna Technology for Wireless Communications*, CRC Press, (2006).

Tu and Blum 2003 Z. Tu and R. S. Blum, "Multiuser diversity for a dirty paper approach", *IEEE Commun. Lett.*, 7(8), 370–372 (2003).

Turin et al. 1972 G. L. Turin, F. D. Clapp, T. L. Johnston, S. B. Fine, and D. Lavry, "A statistical model of urbran multipath propagation", *IEEE Trans. Veh. Technol.*, 21, 1–9 (1972).

Tuttlebee 1997 W. H. W. Tuttlebee (ed.), *Cordless Telecommunications Worldwide*, Springer, London (1997).

Ulukus et al. 2015 S. Ulukus, A. Yener, E. Erkip, O. Simeone, M. Zorzi, P. Grover, and K. Huang, "Energy harvesting wireless communications: A review of recent advances", *IEEE Journal on Selected Areas in Communications*, 33(3), 360–381 (2015).

Ungerboeck 1976 G. Ungerboeck, "Fractional tap-spacing equalizer and consequences for clock recovery in data modems", *IEEE Trans. Commun.*, 24, 856–864 (1976).

Ungerboeck 1982 G. Ungerboeck, "Channel coding with multilevel/phase signals", *IEEE Trans. Inf. Theory*, 28, 55–67 (1982).

Vaezi et al. 2019 M. Vaezi, Z. Ding, and H. V. Poor (eds), *Multiple Access Techniques for 5G Wireless Networks and Beyond*, Springer-Velag, Berlin (2019).

Vainikainen et al. 2002 P. Vainikainen, J. Ollikainen, O. Kivekas, and K. Kelander, "Resonator-based analysis of the combination of mobile handset antenna and chassis", *IEEE Trans. Antenn. Propag.*, 50, 1433–1444 (2002).

Valenti 1999 M. C. Valenti, *Iterative Detection and Decoding for Wireless Communications*, Ph.D. thesis, Virginia Tech University (1999).

Valenzuela 1993 R. A. Valenzuela, "A ray tracing approach to predicting indoor wireless transmission", *Proc. VTC*, 1993, 214–218 (1993).

Vanderveen and Paulraj 1996 A. J. Vanderveen and A. Paulraj, "An analytical constant modulus algorithm", *IEEE Trans. Signal Process.*, 44, 1136–1155 (1996).

Vanderveen et al. 1997 M. C. Vanderveen, C. B. Papadias, and A. Paulraj, "Joint angle and delay estimation (JADE) for multipath signals arriving at an antenna array", *IEEE Commun. Lett.*, 1, 12–14 (1997).

Vangala 2017 H. Vangala, *Efficient Algorithms for Polar Codes*, PhD thesis, Monash University (2017).

Vangala et al. 2015 H. Vangala, Y. Hong and E. Viterbo, "A comparative study of polar code constructions for the AWGN channel", arXiv preprint arXiv:1501.02473 (2015).

Vangala et al. 2016a H. Vangala, Y. Hong, and E. Viterbo, "Efficient algorithms for systematic polar encoding", *IEEE Commun. Lett.*, 20(1), 17–20 (2016a).

Vangala et al. 2016b H. Vangala, Y. Hong,and E. Viterbo, http://www.ecse.monash.edu.au/staff/eviterbo/polarcodes.html (2016b).

Vanghi et al. 2004 V. Vanghi, A. Damnjanovic, and B. Vojcic, *The CDMA2000 System for Mobile Communications*, Prentice Hall PTR, (2004).

Varshney and Kumar 1991 P. Varshney and S. Kumar, "Performance of GMSK in a land mobile radio channel", *IEEE Trans. Veh. Technol.*, 40, 607–614 (1991).

Vary and Martin 2006 P. Vary and R. Martin, *Digital speech transmission: Enhancement, coding and error concealment*, John Wiley & Sons (2006).

Vaseghi 2000 S. V. Vaseghi, *Advanced Digital Signal Processing and Noise Reduction*, 2nd edition, John Wiley & Sons, (2000).

Vaughan and Andersen 2003 R. Vaughan and J. B. Andersen, *Channels, Propagation and Antennas for Mobile Communications*, IEE Press, (2003).

Verdu 1998 S. Verdu, *Multiuser Detection*, Cambridge University Press, Cambridge (1998).

Vetro et al. 2011 A. Vetro, T. Wiegand, and G. J. Sullivan, "Overview of the stereo and multiview video coding extensions of the H.264/MPEG-4 AVC standard", *Proc. IEEE*, 99(4), 626–642 (2011).

Vieira et al. 2017 J. Vieira, F. Rusek, O. Edfors, S. Malkowsky, L. Liu, and F. Tufvesson, "Reciprocity calibration for massive MIMO: Proposal, modeling, and validation", *IEEE Trans. Wirel. Commun.*, 16, 3042–3056 (2017).

Vishwanath et al. 2003 S. Vishwanath, N. Jindal, and A. Goldsmith, "Duality, achievable rates, and sum-rate capacity of Gaussian MIMO broadcast channels", *IEEE Trans. Inf. Theory*, 49, 2658–2668 (2003).

Viswanath et al. 2002 P. Viswanath, D. Tse, and R. Laroia, "Opportunistic beamforming using dumb antennas", *IEEE Trans. Inf. Theory*, 48, 1277–1294 (2002).

Viterbi 1967 A. Viterbi, "Error bounds for convolutional codes and an asymptotically optimum decoding algorithm", *IEEE Trans. Inf. Theory*, 13, 260–269 (1967).

Viterbi 1995 A. J. Viterbi, *CDMA – Principles of Spread Spectrum Communication*, Addison-Wesley Wireless Communications Series, (1995).

Vitetta et al. 2000 G. Vitetta, B. Hart, A. Mammela, and D. Taylor, "Equalization techniques for single-carrier, unspread modulation", in A. F. Molisch (ed.), *Wideband Wireless Digital Communications*, Prentice Hall, (2000).

Vo and De 2015 Q. D. Vo and P. De, "A survey of fingerprint-based outdoor localization", *IEEE Commun. Surv. Tutor.*, 18(1), 491–506 (2015).

Waldschmidt et al. 2004 C. Waldschmidt, S. Schulteis, and W. Wiesbeck, "Complete RF system model for analysis of compact MIMO arrays", *IEEE Trans. Veh. Technol.*, 53, 579–586 (2004).

Walfish and Bertoni 1988 J. Walfish and H. L. Bertoni, "A theoretical model of UHF propagation in urban environments", *IEEE Trans. Antenn. Propag.*, 36, 1788–1796 (1988).

Wallace and Jensen 2004 J. W. Wallace and M. A. Jensen, "Mutual coupling in MIMO wireless systems: A rigorous network theory analysis", *IEEE Trans. Wirel. Commun.*, 3, 1317–1325 (2004).

Wang and Giannakis 2000 Z. Wang and G. B. Giannakis, "Wireless multicarrier communications", *IEEE Signal Process. Mag.*, 17(3), 29–48 (2000).

Wang and Poor 2003 X. Wang and H. V. Poor, *Wireless Communication Systems: Advanced Techniques for Signal Reception*, Prentice Hall, (2003).

Wang and Zhu 1998 Y. Wang and Q. F. Zhu, "Error control and concealment for video communication – A review", *Proc. IEEE*, 85, 974–997 (1998).

Wang et al. 2002 Y. Wang, J. Ostermann, and Y. Q. Zhang, *Video Processing and Communications*, Prentice Hall, (2002).

Wang et al. 2007 Y. K. Wang, I. Bouazizi, M. Hannuksela, and I. Curcio, "Mobile video applications and standards", *Proceedings of the ACM Multimedia, Workshop on Mobile Video*, Augsburg, Germany, September (2007).

Wang et al. 2015 C. Wang, A. Ghazal, B. Ai, Y. Liu and P. Fan, "Channel Measurements and Models for High-Speed Train Communication Systems: A Survey," *IEEE Comm. Surveys & Tutorials*, 18, 974–987 (2015).

Wang et al. 2019 R. Wang, O. Renaudin, O. C. U. Bas, S. Sangodoyin, and A. F. Molisch, "On channel sounding with switched arrays in fast time-varying channels", *IEEE Trans. Wirel. Commun.*, 18(8), 3843–3855 (2019).

de Weck 1992 J. P. de Weck, *Real-time Characterization of Wideband Mobile Radio Channels*, Dissertation an der TU Wien, Wien, Österreich (1992).

Weichselberger et al. 2006 W. Weichselberger, M. Herdin, H. Ö. Zcelik, and E. Bonek, "A stochastic MIMO channel model with joint correlation of both link ends", *IEEE Trans. Wirel. Commun.*, 5, 90–100 (2006).

Weiler et al. 2016 R. J. Weiler, M. Peter, W. Keusgen, A. Maltsev, I. Karls, A. Pudeyev, I. Bolotin, I. Siaud, and A. M. Ulmer-Moll, "Quasi deterministic millimeter-wave channel models in miweba", *EURASIP J. Wirel. Commun. Netw.*, 2016(1), 1 (2016).

Weinstein and Ebert 1971 S. Weinstein and P. Ebert, "Data transmission by frequency-division multiplexing using the discrete Fourier transform", *IEEE Trans. Commun.*, 19, 628–634 (1971).

Weisstein 2004 E. W. Weisstein. *Lindeberg-Feller Central Limit Theorem*. From MathWorld–A Wolfram Web Resource, http://mathworld.wolfram.com/Lindeberg-FellerCentralLimitTheorem.html (2004).

Wellens et al. 2007 M. Wellens, J. Wu, and P. Mahonen, "Evaluation of spectrum occupancy in indoor and outdoor scenario in the context of cognitive radio", *2nd International Conference in Cognitive Radio Oriented Wireless Networks and Communications*, pp. 420–427 (2007).

Wiegand et al. 2003 T. Wiegand, G. J. Sullivan, G. Bjontegaard, and A. Luthra, "Overview of the H.264/AVC video coding standard", *IEEE Trans. Circuits Syst. Video Technol.*, 13, 560–576 (2003).

Willner et al. 2012 A. E. Willner, J. Wang, and H. Huang, "A different angle on light communications", *Science*, 337(6095), 655–656 (2012).

Willner et al. 2015 A. E. Willner, H. Huang, Y. Yan, Y. Ren, N. Ahmed, G. Xie, C. Bao, L. Li, Y. Cao, Z. Zhao, J. Wang, M. P. J. Lavery, M. Tur, S. Ramachandran, A. F. Molisch, N. Ashrafi, and S. Ashrafi, 2015 "Optical communications using orbital angular momentum beams", *Adv. Opt. Photonics*, 7(1), 66–106 (2015).

Wilson 1996 S. G. Wilson, *Digital Modulation and Coding*, Prentice Hall, Upper Saddle River (1996).

Win and Chrisikos 2000 M. Z. Win and G. Chrisikos, "Impact of spreading bandwidth and selection diversity order on selective Rake reception", in A. F. Molisch (ed.), *Wideband Digital Communications*, pp. 424–454, Prentice Hall, (2000).

Win and Scholtz 1998 M. Z. Win and R. A. Scholtz, "Impulse radio: How it works", *IEEE Commun. Lett.*, 2, 36–38 (1998).

Win and Scholtz 2000 M. Z. Win and R. A. Scholtz, "Ultra -wide bandwidth time-hopping spread-spectrum impulse radio for wireless multiple-access communications", *IEEE Trans. Commun.*, 48, 679–691 (2000).

Win and Winters 1999 M. Z. Win and J. H. Winters, "Analysis of hybrid selection/maximal-ratio combining of diversity branches with unequal SNR in Rayleigh fading", *Proceedings of the VTC'99*, Spring, pp. 215–220 (1999).

Win et al. 2011 M. Z. Win, A. Conti, S. Mazuelas, Y. Shen, W. M. Gifford, D. Dardari, and M. Chiani, "Network localization and navigation via cooperation", *IEEE Commun. Mag.*, 49(5), 56–62 (2011).

Win et al. 2018 M. Z. Win, Y. Shen and W. Dai, "A Theoretical Foundation of Network Localization and Navigation," *Proc. IEEE*, 106, 1136–1165 (2018).

Winters 1984 J. H. Winters, "Optimum combining in digital mobile radio with cochannel interference", *IEEE JSAC*, 2, 528–539 (1984).

Winters 1987 J. H. Winters, "On the capacity of radio communications systems with diversity in Rayleigh fading environments", *IEEE J. Sel. Areas Commun.*, 5, 871–878 (1987).

Winters 1994 J. H. Winters, "The diversity gain of transmit diversity in wireless systems with Rayleigh fading", *Proceedings of the IEEE International Conference in Communications*, pp. 1121–1125 (1994).

Witrisal et al. 2009 K. Witrisal, G. Leus, G. Janssen, et al., "Noncoherent ultra-wideband systems", *IEEE Signal Process. Mag.*, 26(4), 48–66 (2009).

Witrisal et al. 2016 K. Witrisal, P. Meissner, E. Leitinger, et al., "High-accuracy localization for assisted living: 5G systems will turn multipath channels from foe to friend", *IEEE Signal Process. Mag.*, 33(2), 59–70 (2016).

Woerner et al. 1994 B. D. Woerner, J. H. Reed, and T. S. Rappaport, "Simulation issues for future wireless modems", *IEEE Commun. Mag.*, 32(7), 42–53 (1994).

Wolf 1978 J. K. Wolf, "Efficient maximum-likelihood decoding of linear block codes using a trellis", *IEEE Trans. Inf. Theory*, 24, 76–80 (1978).

Wong et al. 1999 C. Y. Wong, R. S. Cheng, K. B. Lataief, and R. D. Murch, "Multiuser OFDM with adaptive subcarrier, bit, and power allocation", *IEEE J. Sel. Areas Commun.*, 17, 1747–1758 (1999).

Wozencraft and Jacobs 1965 J. M. Wozencraft and I. M. Jacobs, *Principles of Communication Engineering*, Wiley, New York (1965).

Wu and Pan 2008 H. Wu and Y. Pan, *Medium Access Control in Wireless Networks*, Vol. 8, Nova Publishers, (2008).

Wu et al. 2011 J. Wu, N. B. Mehta, A. F. Molisch, and J. Zhang, "Unified spectral efficiency analysis of cellular systems with channel-aware schedulers", *IEEE Trans. Commun.*, 59, 3463–3474 (2011).

Wu et al. 2013 X. Wu, S. Tavildar, S. Shakkottai, T. Richardson, J. Li, R. Laroia, and A. Jovicic, "FlashLinQ: A synchronous distributed scheduler for peer-to-peer ad hoc networks", *IEEE/ACM Trans. Netw.*, 21, 1215–1228 (2013).

Wu et al. 2018 Y. Wu, A. Khisti, C. Xiao, G. Caire, K. K. Wong, and X. Gao, "A survey of physical layer security techniques for 5G wireless networks and challenges ahead", *IEEE J. Sel. Areas Commun.*, 36(4), 679–695 (2018).

Wymeersch 2007 H. Wymeersch, *Iterative Receiver Design*, Cambridge University Press, (2007).

Wymeersch et al. 2009 H. Wymeersch, L. Jaime, and M. Z. Win, "Cooperative localization in wireless networks", *Proc. IEEE*, 97(2), 427–450 (2009).

Xenakis et al. 2014 D. Xenakis, N. Passas, L. Merakos, and C. Verikoukis, "Mobility management for femtocells in LTE-advanced: Key aspects and survey of handover decision algorithms", *IEEE Commun. Surv. Tutor.*, 16(1), 64–91 (2014).

Xiao et al. 2006 C. Xiao, Y. R. Zheng, and N. C. Beaulieu, "Novel sum-of-sinusoids simulation models for Rayleigh and Rician fading channels", *IEEE Transactions on Wireless Communications*, 5, 3667–3679 (2006).

Xiao and Hu 2008 Y. Xiao and F. Hu (eds), *Cognitive Radio Networks*, CRC Press, (2008).

Xiong 2006 F. Xiong, *Digital Modulation Techniques*, 2nd edition, Artech, (2006).

Xu et al. 2000 Z. Xu, A. N. Akansu, and S. Tekinay, "Cochannel interference computation and asymptotic performance analysis in TDMA/FDMA systems with interference adaptive dynamic channel allocation", *IEEE Trans. Veh. Technol.*, 49, 711–723 (2000).

Xu et al. 2002 H. Xu, D. Chizhik, H. Huang, and R. A. Valenzuela, "A wave-based wideband MIMO channel modeling technique", *Proc. 13th IEEE Int. Symp. Pers., Indoor Mob. Radio Commun.*, 4, 1626 (2002).

Xu et al. 2018 C. Xu, L. Yang, and P. Zhang, "Practical backscatter communication systems for battery-free Internet of Things: A tutorial and survey of recent research", *IEEE Signal Process. Mag.*, 35(5), 16–27 (2018).

Yacoub 2007 M. D. Yacoub, "The kappa-mu distribution and the eta-mu distribution", *IEEE Antenn. Propag. Mag.*, 49(1), 68–81 (2007).

Yan et al. 2014 Y. Yan, G. Xie, M. P. J. Lavery, H. Huang, N. Ahmed, C. Bao, Y. Ren, Y. Cao, L. Li, Z. Zhao, A. F. Molisch, M. Tur, M. J. Padgett, and A. E. Willner, "High-capacity millimetre-wave communications with orbital angular momentum multiplexing", *Nat. Commun.*, 5(1), 1–9 (2014).

Yan et al. 2019 C. Yan, L. Fu, J. Zhang, and J. Wang, "A comprehensive survey on UAV communication channel modeling", *IEEE Access*, 7, 107769–107792 (2019).

Yang and Hanzo 2003 L. L. Yang and L. Hanzo, "Multicarrier DS-CDMA: A multiple access scheme for ubiquitous broadband wireless communications", *IEEE Commun. Mag.*, 41(10), 116–124 (2003).

Yang et al. 2009 Y. Yang, H. Hu, J. Xu, and G. Mao, "Relay technologies for WiMax and LTE-advanced mobile systems", *IEEE Commun. Mag.*, 47(10), 100–105 (2009).

Yang et al. 2013 C. Yang, S. Han, X. Hou, and A. F. Molisch, "How do we design CoMP to achieve its promised potential?", *IEEE Wirel. Commun.*, 20 (1), 67–74 (2013).

Yao and Padgett 2011 A. M. Yao and M. J. Padgett, "Orbital angular momentum: Origins, behavior and applications", *Adv. Opt. Photonics*, 3(2), 161–204 (2011).

Yates et al. 2021 R. D. Yates, Y. Sun, D. R. Brown, S. K. Kaul, E. Modiano, and S. Ulukus, "Age of information: An introduction and survey", *IEEE Journal on Selected Areas in Communications*, 39, 1183–1210 (2021).

Yeung 2006 R. W. Yeung, *A First Course in Information Theory*, Springer, (2006).

Ying and Anderson 2003 Z. Ying and J. Anderson, "Multi band, multi antenna system for advanced mobile phone", *Swedish National Conference on Antennas* (2003).

Yongacoglu et al. 1988 A. Yongacoglu, D. Makrakis, and K. Feher, "Differential detection of GMSK using decision feedback", *IEEE Trans. Commun.*, 36, 641–649 (1988).

Yoo and Goldsmith 2006 T. Yoo and A. Goldsmith, "On the optimality of multiantenna broadcast scheduling using zero-forcing beamforming", *IEEE J. Sel. Areas Commun.*, 24(3), 528–541 (2006).

Yu and Ottersten 2002 K. Yu and B. Ottersten, "Models for MIMO propagation channels: A review", *Wirel. Commun. Mob. Comput.*, 2, 653 (2002).

Yu et al. 2002 W. Yu, G. Ginis, and J. M. Cioffi, "Distributed multiuser power control for digital subscriber lines", *IEEE J. Sel. Areas Commun.*, 20, 1105–1115 (2002).

Yu et al. 2004 W. Yu, W. Rhee, S. Boyd, and J. M. Cioffi, "Iterative water-filling for Gaussian vector multipleaccess channels", *IEEE Trans. Inf. Theory*, 50, 145–152 (2004).

Yuan 2016 Y. Yuan, "Non-orthogonal multi-user superposition and shared access", in F. L. Luo, C. J. Zhang (eds), *Signal Processing for 5G: Algorithms and Implementations*, John Wiley & Sons, (2016).

Yucek and Arslan 2009 T. Yucek and H. Arslan, "A survey of spectrum sensing algorithms for cognitive radio applications", *IEEE Commun. Surv. Tutor.*, 11(1), 116–130 (2009).

Zafari et al. 2019 F. Zafari, A. Gkelias, and K. K. Leung, "A survey of indoor localization systems and technologies", *IEEE Commun. Surv. Tutor.*, 21(3), 2568–2599 (2019).

Zahir et al. 2013 T. Zahir, K. Arshad, A. Nakata, and K. Moessner, "Interference management in femtocells", *IEEE Commun. Surv. Tutor.*, 15(1), 293–311 (2013).

Zekavat and Buehrer 2019 R. Zekavat and R. M. Buehrer, *Handbook of Position Location: Theory, Practice and Advances*, 2nd edition, John Wiley & Sons, (2019).

Zhang and Dai 2004 H. Zhang and H. Dai, "Cochannel interference mitigation and cooperative processing in downlink multicell multiuser MIMO networks", *EURASIP J. Wirel. Commun. Netw.*, 2004(2), 202654 (2004).

Zhang et al. 2005 X. Zhang, A. F. Molisch, and S. Y. Kung, "Variable-phase-shift-based RF-baseband codesign for MIMO antenna selection", *IEEE Trans. Signal Process.*, 53, 4091–4103 (2005).

Zhang et al. 2008 H. Zhang, X. Zhou, and T. Chen, "Ultra-wideband cognitive radio for dynamic spectrum accessing networks", in Y. Xiao, F. Hu (eds), *Cognitive Radio Networks*, CRC Press, (2008).

Zhang et al. 2009 J. Zhang, P. V. Orlik, Z. Sahinoglu, A. F. Molisch, and P. Kinney, "UWB systems for wireless sensor networks", *Proc. IEEE*, 97(2), 313–331 (2009).

Zhang et al. 2013 Q. Zhang, C. Yang, and A. F. Molisch, "Downlink base station cooperative transmission under limited-capacity backhaul", *IEEE Trans. Wirel. Commun.*, 12, 3746–3759 (2013).

Zhang et al. 2016 Y. Zhang, J. Zheng, and H. H. Chen, eds., "Cognitive radio networks: architectures, protocols, and standards", CRC press (2016).

Zhao and Sadler 2007 Q. Zhao and B. M. Sadler, "A survey of dynamic spectrum access", *IEEE Signal Process. Mag.*, 24(3), 79–89 (2007).

Zhao et al. 2007 Q. Zhao, B. Krishnamachari, and K. Liu, "Low-complexity approaches to spectrum opportunity tracking", *2nd International Conference Cognitive Radio Oriented Wireless Networks and Communications*, pp. 27–35 (2007).

Zheng and Tse 2003 L. Zheng and D. L. C. Tse, "Diversity and multiplexing: A fundamental tradeoff in multipleantenna channels", *IEEE Trans. Inf. Theory*, 49, 1073–1096 (2003).

Zheng and Xiao 2003 Y. R. Zheng and C. Xiao, "Simulation models with correct statistical properties for Rayleigh fading channels", *IEEE Trans. Commun.*, 51, 920–928 (2003).

Zhou et al. 2018 P. Zhou, K. Cheng, X. Han, X. Fang, Y. Fang, R. He, Y. Long, and Y. Liu, "IEEE 802.11 ay-based mmWave WLANs: Design challenges and solutions", *IEEE Commun. Surv. Tutor.*, 20(3), 1654–1681 (2018).

Ziemer et al. 1995 R. E. Ziemer, R. L. Peterson, and D. E. Borth, *Introduction to Spread Spectrum Communications*, Prentice Hall, (1995).

Zienkiewicz and Taylor 2000 O. C. Zienkiewicz and R. L. Taylor, *Finite Element Method: Volume 1- The Basis*, Butterworth Heinemann, (2000).

Index

This index does not mention all occurrences of a word, but only the pages where it is material for the context. Also, words occurring frequently throughout the book, such as "bit error rate", are not included in the index.

Wireless Communications: From Fundamentals to Beyond 5G, Third Edition. Andreas F. Molisch.
© 2023 John Wiley & Sons Ltd. Published 2023 by John Wiley & Sons Ltd.
Companion website: www.wiley.com/go/molisch/wireless3e

About the Author

Andreas F. Molisch is the *Solomon Golomb – Andrew and Erna Viterbi* Chair Professor at the University of Southern California (USC). He joined USC in 2009 as professor and founding head of the Wireless Devices and Systems group. Before that, he spent 10 years in industry, at FTW, AT&T (Bell) Laboratories, and Mitsubishi Electric Research Labs, where he rose to Chief Wireless Standards Architect. He had received his PhD and habilitation degrees from TU Vienna (Austria) in 1994 and 1999, respectively.

Dr. Molisch has worked in the field of wireless communications since 1994, with topics ranging from channel measurement/ modeling, to MIMO transceivers, ultrawideband localization, to caching and mobile edge computing. He has made widely cited contributions to research (his h-index is >100), and standardization (including leading several international standardization activities). He is also the inventor of more than 70 granted patents, and many of his innovations have become part of international standards and widely used products.

Dr. Molisch's work has been recognized by multiple learned societies. He has been elected Fellow of the National Academy of Inventors, Fellow of the AAAS, Fellow of the IEEE, Fellow of the IET, and Member of the Austrian Academy of Sciences. He has received numerous awards, among them the IET Achievement Medal for Wireless Communications, the Technical Achievement Awards of the IEEE Vehicular Technology Society (Evans Avant-Garde Award) and the IEEE Communications Society (Armstrong Award), and the Technical Field Award of the IEEE for Communications (Eric Sumner Award).